Experimental Thermodynamics
Volume II

Experimental Thermodynamics of Non-reacting Fluids

Experimental Thermodynamics Volume II

Experimental Thermodynamics of Non-reacting Fluids

Prepared under the sponsorship of the
INTERNATIONAL UNION OF
PURE AND APPLIED CHEMISTRY
COMMISSION ON THERMODYNAMICS
AND THERMOCHEMISTRY

Editors
B. LE NEINDRE and B. VODAR
CNRS, Laboratoire des Interactions Moléculaires et des Hautes Pressions,
Meudon, France

LONDON
BUTTERWORTHS

ENGLAND:	BUTTERWORTH & CO. (PUBLISHERS) LTD. LONDON: 88 Kingsway, WC2B 6AB
AUSTRALIA:	BUTTERWORTH & CO. (AUSTRALIA) LTD. SYDNEY: 586 Pacific Highway, Chatswood, NSW 2067 MELBOURNE: 343 Little Collins Street, 3000 BRISBANE: 240 Queen Street, 4000
CANADA:	BUTTERWORTH & CO. (CANADA) LTD. TORONTO: 2265 Midland Avenue, Scarborough, M1P 4S1
NEW ZEALAND:	BUTTERWORTH & CO. (NEW ZEALAND) LTD. WELLINGTON: 26–28 Waring Taylor Street, 1
SOUTH AFRICA:	BUTTERWORTH & CO. (SOUTH AFRICA) (PTY) LTD. DURBAN: 152–154 Gale Street

A sequel to
Experimental Thermodynamics, Volume I:
Calorimetry of Non-reacting Systems
Edited by J. P. McCullough and D. W. Scott
Butterworths: London (1968)

IUPAC Publications
Chairman, Committee on Publications: G. OURISSON
Scientific Editor: B. C. L. WEEDON
Assistant Scientific Editor: C. F. CULLIS
Assistant Editor: E. G. F. BRIGGS

©

International Union of Pure and Applied Chemistry
1975

Suggested U.D.C. *number* 536·6/·7: 541·11

ISBN 0 408 70566 3

Printed in Great Britain by Page Bros (Norwich) Ltd, Norwich

Foreword

For several decades the Commission on Thermodynamics and Thermochemistry of the International Union of Pure and Applied Chemistry has had a concerned role in the definition and maintenance of standards as well as the enhancement of the quality of publication in the domain covered by its title. These concerns include—but are not limited to—the establishment and surveillance of international pressure and temperature scales, recommendations for calorimetric procedures, and the selection and evaluation of reference standards for calibration and testing of calorimeters of all types. In recent years the Commission has promoted three International Conferences on Chemical Thermodynamics on both sides of the Atlantic and the publication of seventeen annual issues of the *Bulletin of Thermodynamics and Thermochemistry*. Its dissemination of *A Guide to Procedures for the Publication of Thermodynamic Data* [*Pure and Applied Chemistry,* **29**, 397 (1972)] provided editors and referees with a set of carefully considered criteria and responded to the ultimate needs of the compiler, correlator, and critical evaluator of such data.

The Commission has also undertaken the creation of two books on *Experimental Thermochemistry*. The first of these appeared in 1956 (Interscience Publishers, Inc., New York) under the editorship of Frederick D. Rossini and the second in 1962 (Interscience–Wiley, New York) edited by Henry A. Skinner. Both have had a significant effect on the science and art of thermochemical calorimetry. In 1961, the Commission proposed that two additional volumes for important areas in *Experimental Thermodynamics* be prepared. John P. McCullough and Donald W. Scott edited Volume I (Butterworths, London) which appeared in 1968 and was concerned primarily with the experimental calorimetry of non-reacting systems. Now before you are the two parts of the second volume edited by Dr B. Le Neindre and Professor B. Vodar and concerned with a broad spectrum of thermophysical and calorimetric measurements under significant breadth of pressures. Since the primacy of pressure as a practical thermodynamic constraint is second only to that of temperature, the urgency of formulating and delineating experimental expertise is self-evident to accelerate and guide the orderly and purposeful exploitation of this domain in its manifold applications to scientific and technologically relevant problems.

Appreciation and thanks are due not only to the authors for enthusiastic cooperation but to the Editors as well for their synthesis of a coordinated volume of great breadth and depth from the many independent contributions.

We are confident that readers and researchers from many disciplines engaged in thermodynamic measurements at elevated pressures will be encouraged by what they find here, and that these volumes will enhance the achievement of more meaningful results over extended ranges of pressure. If it does so, its authors and editors will have been amply repaid for their yeoman endeavors.

        STIG SUNNER
        Past-Chairman
        EDGAR F. WESTRUM, JR
        Chairman
        Commission on Thermodynamics and Thermochemistry
        International Union of Pure and Applied Chemistry

December, 1973

Preface

This book was planned to contain an up to date description of experimental methods and procedures used for the study of the thermophysical properties of fluids, in a broad range of temperature from liquid helium to the highest temperatures at which experiments are feasible and in an extensive range of pressure. To achieve this aim, taking into account the variety of subjects, the matter was divided into chapters dealing with progressively more complex problems, i.e. starting from the measurement of fundamental parameters: temperature (a complement to Volume I), pressure (extensively developed to cover the range from ultra low to ultra high pressure) and volume; then single component and single phase systems, followed by two or many components and two or multiphase systems. Solids are not included specifically, but the border line phenomena of melting, liquid–solid phase equilibria, surface tension and adsorption are embraced. It had appeared useful to devote special chapters to experimental methods under extreme or particularly difficult conditions such as high temperatures (liquid metals or molten salt), high pressures combined with high temperatures (internal heating) and very low temperature. Methods using transient techniques are briefly described in two chapters.

To ensure the best expertise for each topic, the matter was divided into 26 chapters (including a total of 13 sub-chapters) and among about 50 authors or co-authors from 11 different countries who have contributed to this book.

In some cases it was thought necessary to provide sufficient theoretical background for understanding the basic implications of the experimental methods. On the other hand it appeared unavoidable, in certain cases at least, to go, without becoming too technical, into a sufficiently accurate description of experimental devices and procedures. It is hoped, that in general a proper balance between these two trends was achieved, so as to make the book useful to most of the experimental thermodynamicists. The editors would like to thank all the authors and contributors for their excellent cooperation. Special thanks are due to the past and present presidents of the Commission on Thermodynamics and Thermochemistry, Professor S. Sunner and Professor E. F. Westrum Jr for their helpful comments to improve the presentation of this book and their patient encouragement. Our thanks also to referenced authors for giving permission for the inclusion of curves, drawings or other information. In addition, the assistance of the publishers has been appreciated.

BERNARD LE NEINDRE
BORIS VODAR

Contributors to this Volume

M. Allemand (*France*)
D. Ambrose (*UK*)
Yu. A. Atanov (*USSR*)
C. W. Beckett (*USA*)
H. Braunstein (*USA*)
J. Braunstein (*USA*)
J. Brielles (*France*)
E. Brunner (*West Germany*)
A. Cezairliyan (*USA*)
A. H. Cook (*UK*)
D. L. Decker (*USA*)
A. Dédit (*France*)
Y. Doucet (*France*)
E. U. Franck (*West Germany*)
Y. Garrabos (*France*)
L. A. Guildner (*USA*)
G. Hammond (*USA*)
F. Hensel (*West Germany*)
P. L. M. Heydemann (*USA*)
W. E. Keller (*USA*)
A. S. Kertes (*Israel*)
H. Kienitz (*West Germany*)
W. Knoche (*West Germany*)
M. Lallemand
G. Lalos (*USA*)
A. Lascar (*France*)
R. Leclercq (*France*)

B. Le Neindre (*France*)
Y. Leroux (*France*)
J. M. H. Levelt Sengers (*USA*)
O. Levy (*Israel*)
R. D. McCarty (*USA*)
P. Malbrunot (*France*)
G. Y. Markovits (*Israel*)
C. W. F. T. Pistorius (*South Africa*)
P. P. Pugachevich (*USSR*)
A. Rauch (*France*)
P. A. Redhead (*Canada*)
S. Ruthberg (*USA*)
G. Saville (*UK*)
G. M. Schneider (*West Germany*)
E. E. Shpilrain (*USSR*)
B. L. Smith (*UK*)
H. Strehlow (*West Germany*)
T. Takaishi (*Japan*)
J. Terrien (*France*)
W. Van Dael (*Belgium*)
J. Vermesse (*France*)
D. Vidal (*France*)
B. Vodar (*France*)
B. E. Welch (*USA*)
E. Whalley (*Canada*)
A. G. Williamson (*New Zealand*)

Acknowledgements

The illustrations listed below have appeared elsewhere in the past. The present authors, editors and publishers are grateful to all those concerned in the original publications for permission to use their figures again. Sources of individual illustrations are indicated throughout the text by reference to the earlier relevant works.

Chapter 3
Figures 2, 4, 5, 7–11, 13, 14, 17–26
Chapter 4, Part 1
Figures 1–4, 7, 8
Chapter 4, Part 3
Figures 5, 17, 19
Chapter 4, Part 6
Figures 2–5, 7–11
Chapter 4, Part 7
Figures 1–3, 6–11
Chapter 7
Figures 1–4, 6–12, 14, 15
Chapter 8
Figures 3–14
Chapter 9
Figures 1, 5–11, 13, 14, 16, 20, 21–27, 29–34, 37–39
Chapter 11
Figures 3, 15–18
Chapter 12
Figures 5–14
Chapter 13
Figures 2–15

Chapter 15
Figures 1–12
Chapter 16, Part 1
Figures 6–14, 16–21, 23
Chapter 16, Part 2
Figures 2–9
Chapter 17, Part 1
Figures 2, 4, 5, 7, 10
Chapter 17, Part 2
Figures 9, 10, 14, 16–18, 21, 23
Chapter 19, Part 1
Figures 1–13
Chapter 19, Part 2
Figures 1–5, 7–14
Chapter 20
Figures 7, 8
Chapter 22
Figures 3, 7, 11, 18, 19
Chapter 23
Figure 31
Chapter 26
Figures 1, 2, 5, 8, 10–17, 19

Contents

Foreword	v
Preface	vii
Contributors to this Volume	viii
Acknowledgements	ix

1. General Introduction ... 1
B. Le Neindre and B. Vodar

    I. Introduction ... 3
    II. Recommended Mathematical Symbols ... 4
    III. Units and Symbols for Units ... 4
        1. The International System of Units ... 4
        2. Definition of SI Base Units ... 5
        3. Names and Symbols for SI Base Units ... 6
        4. Names and Symbols for SI Derived Units ... 6
        5. Supplementary Units ... 7
        6. Practical Realization of Some Important SI Units ... 8
        7. Decimal Multiples and Submultiples of SI Units ... 11
        8. Units Outside the International System ... 11
    IV. Physical Quantities and Symbols for Physical Quantities ... 13
        1. Definition ... 13
        2. Basic Physical Quantities ... 13
        3. Derived Physical Quantities ... 14
        4. Use of the Words 'Specific' and 'Molar' in the Names of Physical Quantities ... 23
        5. Symbols for Particular Cases of Physical Quantities ... 24
        6. Recommended Subscripts ... 24
        7. Recommended Superscripts ... 25
    V. Symbols for Chemical Elements, Nuclides and Particles ... 25
    VI. Values of the Fundamental Constants ... 25
    VII. Thermodynamic Definitions ... 28
        1. Thermodynamic System ... 28
        2. Thermodynamic Equilibrium ... 29
        3. Reversible and Irreversible Processes ... 29
        4. Zeroth Law of Thermodynamics ... 29
        5. Equation of State ... 29
        6. First Law of Thermodynamics ... 31
        7. Internal Energy, U ... 31
        8. Enthalpy ... 32
        9. Second Law of Thermodynamics—Entropy ... 32
        10. Third Law of Thermodynamics ... 33
        11. Free Energy ... 34
        12. Heat Capacity ... 34
        13. Joule–Thomson Coefficient ... 35
        14. Speed of Sound ... 35
    VIII. Thermodynamic Data Tables ... 36
        1. Table of Atomic Weights (1971) ... 36
        2. Critical Constants ... 38
        3. Pressure–Volume–Temperature Relationships of Gases—Virial Coefficients. Accuracy of these Tables ... 39
        4. Amagat Density ... 53
        5. Standard Enthalpies of Formation and Standard Entropies at 298.15 K ... 54

CONTENTS

	6. Density of Mercury		54
	7. Density of Water		55
IX.	Definition of Activities and Related Quantities		57
	1. Chemical Potential and Absolute Activity		57
	2. Pure Substances		57
	3. Mixtures		60
	4. Solutions		61
X.	Accuracy and Precision		63
	1. Definitions		63
	2. Expression of the Uncertainties of Final Results		64
XI.	Conversion Tables		66
XII.	References		69

2. Reference Materials for Thermometric Fixed Points .. 71
H. KIENITZ and E. BRUNNER

I.	Primary Fixed Points as Defined by the International Practical Temperature Scale of 1968	71
	1. Definition of the IPTS-68	71
	2. Range 13.81 K to 273.15 K	72
	3. Range 0°C to 630.74°C	75
	4. Range 630.74°C to 1064.43°C	76
	5. Range above 1064.43°C	76
II.	Secondary Reference Points as Recommended by the International Committee on Weights and Measures	76
III.	Supplementary Recommendations on Apparatus, Methods and Procedures	77
	1. Influence of Pressure on the Freezing Point Temperature	78
	2. Triple Point, 17.042 K Point and Boiling Point of Equilibrium Hydrogen	78
	3. Boiling Point of Neon	79
	4. Triple Point and Boiling Point of Oxygen	79
	5. Boiling Point of Water	79
	6. Freezing Points of Tin and Zinc	80
	7. Freezing Points of Silver and Gold	80
IV.	Practical Temperature Scales over the 0.2 K to 5.2 K Range	80
V.	Standard Reference Samples	81
	1. Gases	81
	2. Catalogue of Physicochemical Standard Substance	81
	3. Further Recommendations of Calibration Materials	81
VI.	References	84

3. Temperature Measurement under Pressure .. 87
B. LE NEINDRE and Y. GARRABOS

I.	Introduction	87
II.	Temperature Measurement at Atmospheric Pressure	88
	1. Temperature Measurement below 13.81 K	88
	2. Temperature Measurement from 13.81 K to 630.74°C	88
	3. Temperature Measurement from 630.74°C to 1064.43°C	89
	4. Temperature Measurement above 1064.43°C	89
III.	Temperature Measurement under Pressure	90
	1. Determination of Pressure Effects on Thermoelectricity	90
	2. Temperature Measurement in a Hydrostatically Pressurized Cell	91
	3. Temperature Measurement in Non-hydrostatic Systems	96
	4. Effect of Pressure on the Relative Difference between Various Thermocouples	106

CONTENTS

        5. Optical Pyrometry at High Pressure 108
        6. Computer Method 111
  IV. References 111

4. (Part 1). Pressure Measurements I—Mercury Absolute Manometers 115
L. A. Guildner and J. Terrien

    I. Introduction 115
   II. Optical Methods of Measuring the Distance between the Mercury Surfaces .. 117
        1. Cathetometer 117
        2. Lateral Shift 117
        3. Sensing by Interference Techniques 118
        4. Other Optical Devices 120
  III. Electrical Methods for Sensing the Position of the Mercury Surfaces .. 120
        1. Electrical Contacts 120
        2. Capacitance Techniques 121
   IV. Ultrasonic Sensing of the Meniscus Position 125
    V. Determination of the Height of the Mercury Column 126
   VI. General Considerations 127
  VII. Conclusion 130
 VIII. Abstract 130
   IX. References 130

4. (Part 2). Pressure Measurements II—Pressure Scale and Fixed Point .. 133
D. L. Decker

    I. Apparatus that Requires a Fixed Point Calibration 133
   II. Choice of Fixed Points 134
  III. Measurement of the Pressure at Fixed Points 135
   IV. How to Make Use of Fixed Points in High Pressure Calibration .. 138
    V. The Present Set of Fixed Points for Pressure Calibration 140
        1. Mercury Melting Curve 140
        2. Bismuth I–II 141
        3. Thallium II–III 142
        4. Barium I–II 142
        5. Bismuth III–V 143
        6. Fixed Points above 100 kbar 143
   VI. Conclusions 144
  VII. References 144

4. (Part 3). Pressure Measurements III—Piston Gages 147
P. L. M. Heydemann and B. E. Welch

       List of Symbols 147
    I. Introduction and Historical Review 149
   II. Basic Equations and Elastic Distortion 152
  III. Piston Gage Designs 155
        1. Simple Piston Gage 155
        2. Tilting Piston Gage 156
        3. Vacuum-backed Piston Gage 157
        4. Re-entrant Cylinder Design 159
        5. Differential Piston 159
        6. Controlled Clearance 161
        7. Ball Gages 162
        8. Grooved Pistons 163
        9. Very High Pressure Piston Gages 165
   IV. Calibration of Primary Standards 168
        1. Controlled Clearance Piston Gage 168

CONTENTS

        2. Similarity Method 184
    V. Calibration of Piston Gages 188
        1. Cross-float 188
        2. Reference Levels 191
        3. Evaluation 191
    VI. The Use of Piston Gages 193
        1. Measurement of Pressure 193
        2. Procedures and Methods 194
    VII. Abstract 200
    VIII. References 200

4. (Part 4). Pressure Measurements IV—Secondary Gage—Differential Manometers .. 203
    YURI A. ATANOV
    I. Introduction 203
    II. Mechanical Devices 204
    III. Electrical Resistance Gages 205
    IV. Differential Manometers 210
    V. References 211

4. (Part 5). Pressure Measurements V—Instruments for Relative Pressure Measurements 213
    R. LECLERCQ
    I. General Considerations 213
    II. Pressure Sensitive Elements; Sensors 214
        1. Aneroid Capsule 214
        2. Bellows and Springs 215
        3. Bourdon Tube 215
        4. Gimlet Shaped Tube 216
        5. Piezoelectric Quartz 216
    III. Methods of Measurement 217
        1. Force Measurements. The Pressure Effect 217
        2. Displacement Measurements. The Pressure Effect 218
    IV. Special Transducers 222
    V. Specific Designs 222
        1. General Considerations 222
        2. Flight-control Instruments 223

4. (Part 6). Pressure Measurements VI—Pressure Measurements for the Range 1kPa to 100μPa 229
    S. RUTHBERG
    I. Introduction 230
    II. Direct Measurement Procedures, Reference Standards 231
        1. Precision Liquid Columns 231
        2. Compression Manometer—McLeod Gage 245
        3. Systematic Errors Arising from the Use of a Cold Trap 254
        4. Viscosity Manometer 258
        5. Knudsen Radiometer Manometer 263
    III. Pressure Generators 264
        1. Volumetric Pressure Divider—Static Expansion 264
    IV. Transfer Gages. Precision 268
    V. References 270

4. (Part 7). Pressure Measurements VII—Very Low Pressures and Ultra Low Pressures (below 10^{-6} Torr) 273
    P. A. REDHEAD
    I. Introduction 273

CONTENTS

II.	General Problems of Low Pressure Measurements		274
	1. Gages as Sinks or Sources		274
	2. Measurement in Non-uniform Environments		277
	3. Residual Currents		278
	4. Relative Gage Sensitivities for Different Gases		284
	5. Calibration of Gages		286
III.	Pressure Measurements from 10^{-6} to 10^{-10} Torr		289
	1. Hot-cathode Ionization Gages		290
	2. Cold-cathode Gages		291
IV.	Pressure Measurements below 10^{-10} Torr		294
	1. Shielded-collector Gages		294
	2. Bent-beam Gage		294
	3. Hot-cathode Magnetron Gage		296
V.	Comparison of Gages		297
VI.	References		297

5. The Absolute Measurement of Volume .. 303
A. H. Cook

I.	Introduction		303
II.	Experimental Problems		305
III.	The Volume of a Cube of Tungsten Carbide		311
IV.	Volumes of Cubes of Fused Silica		315
V.	Single Crystals of Pure Silicon for the Measurement of Avogadro's Number		318
VI.	The Density of Water		318
VII.	Summary		319
VIII.	References		319

6. Measurement of p–V–T Properties of Gases and Gas Mixtures at Low Pressure .. 321
G. Saville

I.	Introduction		321
II.	General Principles of p–V–T Measurement		323
III.	Methods of Measuring Pressure		323
	1. Secondary Manometers		325
IV.	The Volume Problem		325
V.	The Experimental Volume/Manometer Interface		328
VI.	Temperature Measurement and Control		331
VII.	p–V–T Methods at Constant Volume		332
VIII.	p–V–T Methods Involving Expansion		334
IX.	Relative Methods		338
X.	Effect of Gas Adsorption on p–V–T Measurements		341
XI.	Gas Density Microbalance		342
XII.	p–V–T Properties of Mixtures		344
XIII.	References		345

7. Equation of State of Gases at High Pressures and Low or Moderate Temperatures .. 347
J. Brielles, A. Dédit, M. Lallemand, B. Le Neindre, Y. Leroux, J. Vermesse and D. Vidal

I.	Introduction		347
II.	Technical Features Common to Various Experimental Methods		349
	1. Cryostats and Thermostats		349
	2. Temperature Measurement		352
	3. High Pressure Vessels		354
	4. Pressure Measurements		354
	5. Determination of the Piezometer Volume		355

CONTENTS

III.	Various Methods of Measurement	356
	1. The Gas Expansion Method	356
	2. Isothermal Methods	357
	3. The Isochore Method	362
	4. The Weight Method	363
	5. The Burnett Method	365
IV.	List of the Most Important Reports	368
V.	Comparison between Theory and Experiment	375
	1. Virial Expansion of a Hard Sphere System	375
	2. Perturbation Method	376
VI.	Conclusion	377
VII.	References	377

8. p–V–T Relationships in Gases at High Pressures and High Temperatures ... 383
P. Malbrunot

I.	Introduction	383
II.	Techniques of Heating the Gas under Study	384
	1. External Heating	384
	2. Internal Heating	385
III.	Various Methods of Measurement	389
IV.	Constant Temperature Methods	389
	1. Variable Volume Techniques	389
	2. Methods with Weighing Technique and Constant Volume Piezometer	394
	3. Miscellaneous Techniques	397
V.	Constant Pressure Methods	399
VI.	Constant Volume Methods	400
VII.	Critical Comparison of the Various Measurement Methods	405
	1. Measurement Techniques	405
	2. Measured Quantities	405
VIII.	Results	407
IX.	Equations of State	409
	1. Law of Corresponding States	409
	2. Empirical Equations	409
	3. Equations Derived from Statistical Mechanics	412
	4. Tables of Thermodynamic Properties	416
X.	References	416

9. The Compression of Liquids ... 421
E. Whalley

I.	Introduction	422
II.	Thermodynamics of Volume Changes	425
	1. The more Important Thermodynamic Derivatives of Pressure, Volume, and Temperature	425
	2. Relations between the Derivatives	426
	3. Quantities that can be Determined by Measuring Volumes, Thermal Expansions and Compressions	428
III.	Historical Introduction	430
IV.	Some Experimental Considerations	431
	1. Relative Expansion and Compression	431
	2. Expansion, Compression, Expansivity and Compressibility	431
	3. Dilatation of the Vessel	432
	4. Seasoning of Pressure Vessels and Piezometers	434
	5. Heat of Compression	436
	6. Corrosive Liquids	436

CONTENTS

7. Gases at High Pressures		437
8. External and Internal Heating of Pressure Vessels		437
9. Summary of the more Important Sources of Error in the Measurement of the Compression of Liquids		439
10. Accurate Measurements		441
V. Piezometric Methods—Liquid Piston		441
1. Some General Considerations		441
2. Single-point Methods		445
3. Multiple-point Methods		450
VI. Piezometric Methods—Solid Piston		463
VII. Piezometric Methods—Bellows		465
VIII. Simple Piston–Cylinder Method		469
IX. Constant–Volume Vessel		474
1. Introduction		474
2. Displacement of a Solid Piston		476
3. Displacement of a Liquid Piston		480
4. Bellows Volumometer		482
5. Direct Weighing of Fluid Removed		482
6. Volumetric Measurement of the Gas at Low Pressure		483
7. Volumetric Measurement of Liquid at Low Pressure		483
X. Weight Methods		484
1. Direct Weighing Methods		484
2. Hydrostatic Weighing		485
XI. Ultracentrifuge Method		488
XII. Negative Pressures		488
XIII. Adiabatic Compression		489
XIV. Isochoric Thermal Pressure Coefficient		490
1. Piezometric Methods		490
2. Constant-volume Vessel with Direct Pressure Measurement		491
3. Constant-volume Vessel with Indirect Pressure Measurement		492
XV. Calorimetric Methods		493
XVI. Miscellaneous Methods		494
1. Methods Based on Radioactivity		494
2. Variable-volume Vessel		494
XVII. References		494

10. Determination of Thermodynamic Properties from the Experimental P–V–T Relationships .. 501

R. D. MCCARTY

I. Computational Methods—Introduction		501
II. The Equation of State		502
III. Estimation of the Parameters of an Equation of State		505
1. Linear Least Squares		506
2. The Round Off Problem		507
3. Least Squares with Constraints		508
4. Non-linear Parameter Estimation		509
5. The Simultaneous Use of Several Types of Property Data in Least Squares Parameter Estimation		510
IV. Statistical Aspects of Least Squares Estimation		511
1. Least Squares Conditions and Formation of Weights		511
2. Significance Tests		514
3. Confidence Intervals		515
V. Miscellaneous Techniques for Improving the Accuracy of Thermodynamic Properties Calculated from an Equation of State		516

CONTENTS

	1. Thermodynamic Equilibrium Conditions as Simultaneous Data	516
	2. Constrained Boundary Conditions	517
VI.	Thermodynamic Property Equations	518
VII.	Mathematical Formulas Useful in Thermodynamic Calculations	523
	1. Derivative Chain Rule	523
	2. Implicit Solutions of Equations of State	523
	3. Joining Independent p–V–T Surfaces	524
	4. A Solution of M Equations for M Unknowns	524
VIII.	Abstract	525
IX.	References	525

11. Thermodynamic Properties and the Velocity of Sound ... 527
W. Van Dael

	List of Symbols	528
	Introduction	529
I.	Thermodynamic Relations	530
	1. Adiabatic Properties	530
	2. Sound Velocity	532
II.	Absorption and Dispersion	533
	1. Translational Relaxation	534
	2. Rotational and Vibrational Relaxation	536
	3. Critical Dispersion	539
	4. Other Relaxation Phenomena	541
III.	Sound Velocity and the Equation of State	541
	1. Ideal Gas	541
	2. Virial Equation of State	541
	3. Van der Waals Equation of State	543
	4. BWR Equation of State	544
IV.	Sound Velocity in Mixtures	544
	1. Ideal Mixture	545
	2. Non-ideal Mixtures	548
V.	Experimental Methods	549
	1. Interferometer Methods	549
	2. Pulse Methods	552
VI.	Sound Velocity in Gases	555
	1. General Behavior	555
	2. The Absolute Value of W in the Low Pressure Limit	556
	3. The Initial Slope $(\delta W^2/\delta p)_T$	558
	4. High Density Data	559
	5. Generalized Behavior of Sound Velocity in Gases. Corresponding States Treatment	559
VII.	Sound Velocity in Pure Liquids	561
	1. General Behavior	561
	2. Liquids Coexisting with Their Saturated Vapor	562
	3. Single Phase High Density Fluids	563
	4. Critical Region	565
	5. Generalized Sound Velocity Behavior in Dense Fluids	568
VIII.	Sound Velocity in Liquid Mixtures	570
	1. Homogeneous Mixtures	570
	2. Mixtures Showing Phase Separation	572
IX.	Acknowledgements	574
X.	References	574

CONTENTS

12. Relation of the Dielectric Constant and the Refractive Index to Thermodynamic Properties .. 579
  B. L. SMITH
   I. Introduction .. 579
   II. Theoretical .. 580
       1. Lorentz Model .. 581
       2. Onsager–Böttcher Theory .. 582
       3. Statistical-mechanical Calculations .. 584
       4. Phenomenological Shell Model .. 586
       5. Variation of Polarizability with Density .. 587
       6. Generalization of Theory to Optical Frequencies .. 588
   III. Experimental Determinations of Dielectric Properties .. 588
       1. Methods for Determining Refractive Index .. 589
       2. Results of Refractive Index Measurements .. 592
       3. Methods for Determining Dielectric Constants .. 597
       4. Results of Dielectric Constant Measurements and Comparison with Refractive Index Data .. 599
   IV. Magneto-optical Properties .. 600
   V. Conclusions .. 604
   VI. References .. 605

13. Vapor Pressures .. 607
  D. AMBROSE
   I. Introduction .. 607
   II. Static Measurements .. 610
   III. The Isoteniscope and Related Methods .. 616
   IV. Static Measurements at Elevated Temperatures and Pressures .. 618
   V. The Critical Point .. 621
   VI. Effect of the Presence of Mercury .. 621
   VII. Vapor Pressures of Liquefied Gases .. 623
   VIII. Effect of Thermal Transpiration .. 624
   IX. Comparative Static Measurements .. 624
   X. Static Measurements at Very Low Pressures .. 625
   XI. Use of Radioactive Tracers .. 625
   XII. Ebulliometric Measurements .. 626
   XIII. Ebulliometric Measurements at Pressures below 2 kPa .. 633
   XIV. Method of Ramsay and Young .. 634
   XV. Dynamic Measurements without a Buffer Gas .. 635
   XVI. The Quasi-static Method .. 637
   XVII. Measurement of the Force Exerted by the Vapor .. 638
   XVIII. Evaporation Methods for Low Pressures .. 641
   XIX. Gas-saturation Method .. 645
   XX. Differential Thermal Analysis .. 648
   XXI. Gas Chromatography .. 649
   XXII. Mass Spectrometry .. 651
   XXIII. Vapor Pressures of Mixtures .. 651
   XXIV. References .. 652

14. Thermodynamic Properties near the Critical State .. 657
  J. M. H. LEVELT SENGERS
   I. Introduction .. 658
   II. Theoretical Background .. 660

CONTENTS

	1. The Origin of Critical Anomalies	660
	2. Power Laws	661
	3. Symmetry	664
	4. Homogeneity and Scaling	669
	5. Beyond Simple Scaling	675
III.	Special Experimental Difficulties	677
	1. Divergences and Their Consequences	677
	2. Gravity	678
	3. Equilibration	681
IV.	Refractive Index Measurements	684
	1. Principle	684
	2. The Use of Optical Techniques for Bulk Density Determination	685
	3. Local Density Determination Using Refractive Index	686
	4. Density Gradient Determination	687
V.	Dielectric Constant Measurements	689
	1. General	689
	2. $p\varepsilon T$ Measurements	690
	3. Density Profiles by Dielectric Constant Determination	690
VI.	Conventional pVT and Vapor Pressure Measurements	691
	1. pVT Measurements	691
	2. The Vapor Pressure	693
VII.	Calorimetry	693
	1. Experimental Problems in C_v Determination	693
	2. Reducing the Heat Capacity of the Container	695
	3. Long Relaxation Times	695
	4. Correcting for Gravity	696
	5. Increasing the Temperature Resolution	698
	6. Checking for Consistency	698
	7. Tests of Scaling	699
VIII.	Coexistence Curves	700
	1. General	700
	2. Gravity	700
	3. The Method of Meniscus Disappearance	702
	4. Young's Method of the Twin Cells	703
	5. Coexistence Curves from Isothermal and Isochoric Intercepts	703
	6. Coexistence Curves by Dielectric Constant and Refractive Index Techniques	705
	7. The Use of Floats	705
	8. Power Law Analysis of Coexistence Curves	706
	9. The Diameter of the Coexistence Curve	708
IX.	Scattering	709
	1. Introduction	709
	2. Intensity of Scattered Light	709
	3. Angular Dependence of the Intensity of Scattered Light	711
	4. Light Scattering and Small-angle X-Ray Scattering	714
	5. The Experimental Situation in Critical Opalescence	714
	6. The Spectrum of Scattered Light	715
X.	Sound	718
	1. Sound, Ultrasound and Hypersound	718
	2. Gravity Effects in Sound Velocity Measurements	718
	3. Sound Dispersion and Attenuation	719
XI.	Concluding Remarks	719
XII.	Acknowledgements	720
XIII.	References	721

CONTENTS

15. Solubility 725
A. S. KERTES, O. LEVY and G. Y. MARKOVITS
    I. Introduction 725
    II. General Considerations 726
    III. Concentration and Activity Coefficient Scales 727
    IV. Solubility of Gases in Liquids 729
        1. Manometric-volumetric Methods 730
        2. Chemical-analytical Methods 739
        3. Miscellaneous Methods 740
    V. Solubility of Liquids in Liquids 741
        1. Volume Reading 741
        2. Cloud Point 742
        3. Miscellaneous Methods 742
    VI. Solubility of Solids in Liquids 743
        1. Saturation Method 743
        2. Cloud Point 744
        3. Chemical and Instrumental Analysis 746
    VII. References 746

16. (Part 1). Phase Equilibria (General Procedure) I—Phase Equilibria of Two-component Systems and Multicomponent Systems 749
A. G. WILLIAMSON
    List of Symbols 749
    I. Introduction 749
    II. Thermodynamics of Mixtures 751
    III. Liquid Mixtures 753
    IV. Empirical Representation of Liquid–Vapor Equilibrium Data .. 755
    V. Tests for Thermodynamic Consistency of Liquid–Vapor Equilibrium Data 759
    VI. Experimental Methods of Determining Liquid–Vapor Equilibrium Data 761
        1. Dynamic Methods 763
        2. Static Methods 767
    VII. Static Measurements with Analyses of Both Phases 775
    VIII. The McBain Balance Method 778
    IX. Dewpoint and Bubble-point Measurements 780
    X. The Isopiestic Method 782
    XI. Differential Methods 783
    XII. Light-scattering Measurements 783
    XIII. References 784

16. (Part 2). Phase Equilibria (General Procedure) II—Phase Equilibria of Liquid and Gaseous Mixtures at High Pressures 787
G. M. SCHNEIDER
    I. Introduction 787
    II. Basic Phase–Theoretical Aspects 788
    III. Discussion of General Procedures 789
        1. The Analytical Method 789
        2. The Synthetic Method 791
        3. Miscellaneous Methods 794
    IV. Description of Special Equipments 795
        1. Mercury-operated Apparatus of Krichevskii and Tsiklis .. 796
        2. Apparatus of Tsiklis and Maslennikova 796
        3. Apparatus of Tödheide and Franck 797
        4. Optical Cell Used by de Swaan Arons and Diepen .. 797
        5. Optical Cell Developed by Oeder and Schneider .. 797

CONTENTS

		6. Optical Cell Developed by Alwani and Schneider	798
		7. Optical Cell Developed by Buback and Franck	798
		8. Apparatus Developed by Michels *et al.*	798
	V.	Conclusions	799
	VI.	References	800

17. (Part 1). Liquid–Solid Phase Equilibria I—Melting Points and Volume Changes upon Melting .. 803

C. W. F. T. PISTORIUS

	I.	Introduction	803
	II.	Methods used for Melting Point Determination at Normal Pressure	807
		1. Visual Methods	807
		2. Microscopical Methods	808
		3. Quenching Methods	809
		4. Pyrometric Methods for Use above 2 000°C	810
		5. Calorimetric Methods	811
		6. Thermal Arrest Methods	813
		7. Special Methods Used in Isolated Cases	816
	III.	Differential Thermal Analysis at Normal Pressure	816
		1. Historical	817
		2. Basic Method	818
		3. Critical Assessment of DTA	819
	IV.	Methods Used for Melting Point Determination at High Pressure	820
		1. Historical	821
		2. Methods of Measuring Melting Points at High Pressures in Apparatus Using Hydrostatic Pressure Media	821
		3. Methods of Measuring Melting Points at High Pressures in Apparatus Using Quasi-hydrostatic Pressure Media	824
		4. Methods of Measuring Melting Points at High Pressures in Opposed-anvil Apparatus	827
		5. Methods of Measuring Melting Points at High Pressures in Multiple-anvil Apparatus	828
		6. Conclusion	828
	V.	Methods for Determining Volume Changes on Melting	829
	VI.	Abstract	830
	VII.	References	830

17. (Part 2). Liquid–Solid Phase Equilibria II—Cryoscopy .. 835

Y. DOUCET

		List of Symbols	836
	I.	General Considerations	837
		1. Cryoscopic Law for Ideal, Non-ionic Solutions	837
		2. Cryoscopic Law for Non-ideal and Non-ionic Solvents	839
		3. Cryoscopic Law for Aqueous Electrolytes	839
		4. Cryoscopic Law for Electrolyte Solutions in Salt Media	841
		5. The Methods of Cryoscopy	846
	II.	Kinetic Cryoscopy Equipments	852
		1. Apparatus Used from − 100°C to 200°C	852
		2. Temperature Measurement	853
		3. Table of Organic and Inorganic Solvents	861
		4. Salt Hydrate and Aqueous Eutectic Cryoscopy	861
		5. Molten Salts Cryoscopy	871
	III.	Adiabatic Cryoscopy Equipments	878
		1. Aqueous Solution Equipment	878

CONTENTS

	2. Non-aqueous Solvents Equipment	881
	3. Concentration Measurements	882
	4. Pseudo-equilibrium Methods	886
IV.	A Few Applications of Cryoscopy	887
	1. Data Derived from Zero Concentration Extrapolation, $(\theta/m)_0$	887
	2. Data Derived from the (θ/m) versus m Cryoscopic Graph	889
	3. Data Derived from the Schröder Curve	895
V.	References	898

18. EMF Measurements in Molten Salts ... 901
J. Braunstein and H. Braunstein

I.	Introduction and Scope	902
II.	Cell EMF and Thermodynamic Properties	903
	1. Introduction	903
	2. Classification of EMF Cells	903
	3. Sign Convention for EMF and Cell Diagram	908
	4. Thermodynamic Expressions for Cell EMFs	909
	5. Principal Error Sources	920
III.	Experimental	924
	1. Reporting Data and Results	924
	2. Apparatus	924
	3. Electrodes	934
IV.	Special Applications	941
	1. Phase Diagrams	941
	2. Association Equilibria of Dilute Solutes	944
	3. Miscellaneous	949
V.	Acknowledgement	950
VI.	References	950

19. (Part 1). Thermodynamic Properties of Fluid Metals I—Medium and Low Pressures ... 953
E. E. Shpilrain

I.	Introduction	953
II.	Thermal Properties. Specific Heat	953
	1. The Direct Heating Method	953
	2. The Drop-calorimeter	954
	3. The Exploding Wire Method	956
III.	Density Measurements	957
	1. Direct Methods	957
	2. Indirect Methods	963
IV.	Vapor Pressures	966
V.	References	973

19. (Part 2). Thermodynamic Properties of Fluid Metals II—High Temperatures and High Pressures ... 975
F. Hensel and E. U. Franck

I.	Introduction	975
II.	Experimental Methods	976
	1. Density	976
	2. Vapor Pressures	980
III.	Results	983
	1. Density	983
	2. Vapor Pressure Curves and Critical Data	986
IV.	References	990

CONTENTS

20. Interphase Surface Tension .. 991
P. P. Pugachevich
    I. Introduction .. 991
    II. Static Methods of Determination of the Interphase Surface Tension .. 992
        1. Method of Capillary Rise or Depression .. 992
        2. Method of the Shape of a Sessile Drop or a Gas Bubble .. 995
    III. Semistatic Methods of Determination of the Interphase Surface Tension .. 1001
        1. Method of the Weight and Volume of a Bubble .. 1001
        2. Ring or Plate Rupture Method .. 1003
        3. Method of Maximal Pressure in a Gas Bubble or a Drop .. 1006
        4. Improved Gas Devices with One Calibrated Tube .. 1009
        5. Improved Gas Devices with Two Calibrated Tubes .. 1014
        6. Gravitational Devices .. 1015
    IV. References .. 1020

21. Adsorption .. 1023
T. Takaishi
    I. Introduction .. 1023
    II. Vacuum Systems .. 1024
    III. Pressure Measurements .. 1025
        1. Gages .. 1025
        2. Sources of Error and Their Correction .. 1028
    IV. Adsorption Isotherms .. 1033
        1. Cleaning the Adsorbent Surfaces .. 1033
        2. Equilibration .. 1034
        3. Volumetric Methods .. 1036
        4. Gravimetric Methods .. 1050
        5. Flow Methods .. 1060
    V. Adsorption Cryostats .. 1063
    VI. Acknowledgement .. 1065
    VII. References .. 1065

22. Chemical Relaxation in Liquid Systems .. 1069
H. Strehlow and W. Knoche
    I. Introduction .. 1069
    II. Chemical Relaxation Techniques .. 1071
        1. Jump Methods .. 1071
        2. Stationary Methods .. 1081
    III. The Information Obtainable from Chemical Relaxation Measurements .. 1088
    IV. Some Applications of Chemical Relaxation Techniques .. 1095
        1. The Neutralization Reaction $H^+ + OH^- \underset{k_2}{\overset{k_1}{\rightleftharpoons}} H_2O$ .. 1095
        2. The Formation of Metal Ion Complexes: $M^{m+} + L^{n-} \rightleftharpoons ML^{(m-n)+}$ .. 1096
        3. The Mechanism of Cooperative Ligand Binding on an Allosteric Enzyme .. 1099
        4. Ultrasonic Absorption in Water–Dioxane Mixtures .. 1103
    V. Summary .. 1107
    VI. References .. 1107

23. Thermodynamic Properties from Shock Waves .. 1109
A. Lascar and A. Rauch
    I. Introduction .. 1110
    II. Theoretical Aspects .. 1111
        1. The Basic Relations .. 1111
        2. Properties of the Hugoniot Curve; Stability of Shock Waves .. 1116

CONTENTS

        3. Reflection of Shock Waves and Rarefaction Waves 1125
   III. Experimental Techniques 1132
        1. Shock Wave Generators 1133
        2. Methods of Measurement 1138
   IV. Equation of State for Liquids at Very High Pressures 1146
        1. Experimental Results 1146
        2. Theoretical Models of the Equation of State 1152
   V. References 1157

24. Electrical Discharge Techniques for Measurements of Thermodynamic Properties of Fluids at High Temperatures 1161
A. CEZAIRLIYAN and C. W. BECKETT

      List of Symbols 1161
   I. Introduction 1162
   II. General Method 1162
   III. Capacitor Discharge Systems 1164
        1. Description of Systems 1164
        2. Design Considerations 1166
        3. Measurement of Experimental Quantities 1170
        4. Examples of Thermodynamic Measurements 1176
        5. Summary of Pertinent Literature since 1964 1177
   IV. A Millisecond-resolution System 1178
        1. Description of the System 1178
        2. An Example of Thermodynamic Measurements at and above Melting Points 1180
   V. Discussion 1182
   VI. Appendix 1183
   VII. References 1189

25. The Ballistic Compression and High Temperature Properties of Dense Gases 1193
G. T. LALOS and G. L. HAMMOND

   I. Introduction 1193
   II. The Ballistic Piston Compressor 1195
        1. General Description 1195
        2. Operation 1198
        3. Instrumentation 1202
   III. Physical Properties Studies 1210
        1. The Equation of State 1210
        2. Optical Studies 1212
        3. Other Studies 1215
   IV. Summary 1216
   V. References 1217

26. Thermodynamic Properties of Fluids below 20 K 1219
W. E. KELLER

   I. Introduction 1219
   II. Temperature Scales and Thermometry below 20 K 1220
        1. The Basis for Thermodynamic Thermometry below 20 K .. 1221
        2. Primary Thermometry from 1 to 20 K 1221
        3. Primary Thermometry below 1 K 1222
        4. Temperature Scales in Use below 20 K 1230
        5. Secondary Thermometry below 20 K 1234
   III. Refrigeration Techniques below 0.3 K 1240
        1. The ^{3}He–^{4}He Dilution Refrigerator 1241

CONTENTS

		2. Nuclear Cooling	1244
		3. Cooling by Adiabatic Freezing in Liquid ^{3}He	1246
		4. Heat Transfer and Isolation at ULT	1248
	IV.	Preparation of Helium and Hydrogen Samples	1251
	V.	Calorimetry of Fluids below 20 K	1253
		1. Liquid ^{4}He near the Melting Curve	1253
		2. Liquid ^{4}He near the λ-Point	1254
		3. Liquid ^{3}He at ULT and at Pressures up to the Melting Curve	1256
		4. Latent Heats of Vaporization of Hydrogen and ^{4}He	1256
		5. Liquid ^{3}He–^{4}He Mixtures	1257
	VI.	pVT Measurements below 20 K	1257
		1. Molar Volume	1258
		2. pVT Properties at Melting	1260
		3. Osmotic Pressure of ^{3}He–^{4}He Solutions	1262
	VIII.	References	1263
27.	Author Index		1267
28.	Subject Index		1299

CHAPTER 1

General Introduction

B. Le Neindre and B. Vodar
L.I.M.H.P., C.N.R.S.
1 Place A. Briand, Bellevue 92, France

Contents

I.	Introduction	3
II.	Recommended Mathematical Symbols	4
III.	Units and Symbols for Units	4
	1. The International System of Units	4
	2. Definition of SI Base Units	5
	A. Unit of Length—the Metre	5
	B. Unit of Mass—the Kilogram	5
	C. Unit of Time—the Second	5
	D. Unit of Electric Current—the Ampere	5
	E. Unit of Thermodynamic Temperature—the Kelvin	5
	F. Unit of Luminous Intensity—the Candela	6
	G. Unit of Amount of Substance—the Mole	6
	3. Names and Symbols for SI Base Units	6
	4. Names and Symbols for SI Derived Units	6
	A. Expression	6
	B. Recommendations	7
	5. Supplementary Units	7
	6. Practical Realization of Some Important SI Units	8
	A. Length	8
	B. Mass	8
	C. Time	8
	D. Electric Quantities	9
	E. Temperature	9
	F. Photometric Quantities	10
	G. Amount of Substance	10
	7. Decimal Multiples and Submultiples of SI Units	11
	A. Expression	11
	B. Recommendations	11
	8. Units Outside the International System	11
	A. Decimal Fractions and Multiples of SI Units having Special Names	11
	B. Other Units now Exactly Defined in Terms of the SI Units	12
	C. Units Defined in Terms of the Best Available Experimental Values of Certain Physical Constants	13
	D. 'International' Electrical Units	13
IV.	Physical Quantities and Symbols for Physical Quantities	13
	1. Definition	13
	2. Basic Physical Quantities	13
	3. Derived Physical Quantities	14
	4. Use of the Words 'Specific' and 'Molar' in the Names of Physical Quantities	23
	5. Symbols for Particular Cases of Physical Quantities	24
	6. Recommended Subscripts	24
	7. Recommended Superscripts	25

V.	Symbols for Chemical Elements, Nuclides and Particles	25
VI.	Values of the Fundamental Constants	25
VII.	Thermodynamic Definitions	28
	1. Thermodynamic System	28
	2. Thermodynamic Equilibrium	29
	3. Reversible and Irreversible Processes	29
	4. Zeroth Law of Thermodynamics	29
	5. Equation of State	29
	A. Definition	29
	B. Virial Coefficients	30
	C. The Equation of State of the Partition Function	30
	6. First Law of Thermodynamics	31
	7. Internal Energy, U	31
	8. Enthalpy	32
	9. Second Law of Thermodynamics—Entropy	32
	10. Third Law of Thermodynamics	33
	11. Free Energy	34
	A. Helmoltz Free Energy, A	34
	B. Gibbs Free Energy, G	34
	12. Heat Capacity	34
	13. Joule–Thomson Coefficient	35
	14. Speed of Sound	35
VIII.	Thermodynamic Data Tables	36
	1. Table of Atomic Weights (1971)	36
	2. Critical Constants	38
	3. Pressure–Volume–Temperature Relationships of Gases—Virial Coefficients—Accuracy of These Tables	39
	4. Amagat Density	53
	5. Standard Enthalpies of Formation and Standard Entropies at 298.15 K	54
	6. Density of Mercury	54
	7. Density of Water	55
IX.	Defininition of Activities and Related Quantities	57
	1. Chemical Potential and Absolute Activity	57
	2. Pure Substances	57
	A. Properties of Pure Substances	57
	B. Fugacity of a Pure Gaseous Substance	57
	3. Mixtures	60
	A. Definition of a Mixture	60
	B. Partial Pressure	60
	C. Fugacity of a Substance in a Gaseous Mixture	60
	4. Solutions	61
	A. Definition of a Solution	61
	B. Properties of Infinitely Dilute Solutions	61
	C. Activity Coefficient of a Solute Substance in a Solution	61
	D. Relative Activity of a Solute Substance in a Solution	61
	E. Osmotic Coefficient of the Solvent Substance in a Solution	62
	F. Relative Activity of the Solvent Substance in a Solution	62
X.	Accuracy and Precision	63
	1. Definitions	63
	2. Expression of the Uncertainties of Final Results	64
XI.	Conversion Factors	66
XII.	References	70

GENERAL INTRODUCTION

I. Introduction

In this book we describe several methods for measuring thermodynamic properties as well as tests for evaluating measurements which appear in the literature.

Because the proper understanding, evaluation and even interchangeability of tables of thermodynamic data require consistent and correct use of units by all investigators we have chosen to cover the recommended SI units in some detail. Since the same can be said about symbols, we shall spend some time in the discussion of various recommended symbols.

Many of the numerous methods which we shall cover have in recent years been described in the literature often without information essential to their evaluation. Because of this we present here some of the recommendations of the Thermodynamic Commission of the IUPAC (June 1971) on the description of such measurement systems. This should be considered to be a plea for such proper descriptions in future publications.

'A description of the apparatus including details of the reaction container or calorimeter vessel, the controlled environment, and measuring systems such as those for time, temperature, and pressure should be carefully described. The design of calibration heaters and heater lead placement and precautions as to shielding or isolation of electrical energy equivalent circuits should be presented including the method of calibration and the sensitivity of the instruments used in these measuring systems such as thermometers, bridges and potentiometers, flowmeters, weighing devices and methods of determining the energy equivalent of or otherwise calibrating the system, including certificate values for calibration substances or other information relating to establishment of the energy scale. The history of a particular apparatus which is used in an on-going series of researches should be maintained and documented as to modifications, improvements, etc., to the end that should corrections be made necessary by subsequent recalibrations or by revelation of systematic errors or bias, such corrections can be applied to all affected data either by the author or by compilers or reviewers. Information establishing the energy scale (preferably with traceability to a calibrating or standardizing laboratory), together with details of the observational procedures, the methods of evaluating the corrected temperature increment, methods of data reduction, and the precision of the measurements should be given. The reliability of the results should be established by the use of recognized reference substances such as the samples likely to be recommeneded by the IUPAC Sub-commission on Standard Calibration Materials, those provided by the US National Bureau of Standards[*], those authorized by the Calorimetry Conference for thermophysical or thermochemical measurements[†], or those systems generally recognized as standard for mixtures, e.g. the solubility of oxygen in water at 1 atm and 298.15 K for gas solubility[1], hexane plus cyclohexane for enthalpy increments on solutions, and benzene plus cyclohexane for volume increments on mixing[2].

'The source of and/or method of preparation for all materials used, including calibration, reference and auxiliary substances; experimental values for analysis and pertinent physical properties of materials, the criteria of characterization and purity, as well as the method, temperature, time interval, etc. of storing samples and preparing them for measurements should be stated whenever this is important. The density used in reduction of weighing to mass, and special procedures such as for dealing with partially filled ampoules, should be specified. For studies made on solutions, the source, preparation and quality of the solvent should always be included, and should include information as to dissolved gases (carbon dioxide, air, etc.) whenever these impurities may influence the results'.

---

[*] US National Bureau of Standards—Office of Standard Reference Materials, Washington, DC 20234.

[†] Calorimetry Conference standards such as *n*-heptane, benzoic acid, synthetic sapphire (aluminum oxide), and copper for heat capacity and for enthalpy measurements.

The first author wishes to thank the National Bureau of Standards for extending to him its hospitality in allowing him to be a guest worker during a twenty one month period. He wishes also to thank the members of the Equation of State Section as well as those in the Reactor Division for their collaboration and cooperation. He wishes particularly to thank M. Klein and J. M. H. Levelt Sengers for many useful conversations and for particular material relating to this chapter.

II. Recommended Mathematical Symbols†

equal to	$=$
not equal to	$\neq$
identically equal to	$\equiv$
corresponds to	$\triangleq$
approximately equal to	$\approx$
approaches	$\rightarrow$
asymptotically equal to	$\simeq$
proportional to	$\propto \quad \sim$
infinity	∞
less than	$<$
greater than	$>$
less than or equal to	$\leq \quad \leqslant \quad \leqq$
greater than or equal to	$\geq \quad \geqslant \quad \geqq$
much less than	$\ll$
much greater than	$\gg$
plus	$+$
minus	$-$
multiplied by	$\times$
a divided by b	$\dfrac{a}{b} \quad a/b \quad ab^{-1}$
magnitude of a	$\lvert a \rvert$
a raised to the power n	a^n
square root of a	$a^{1/2} \quad a^{\frac{1}{2}} \quad \sqrt{a} \quad \sqrt[n]{a}$
nth root of a	$a^{\frac{1}{n}} \quad a^{\frac{1}{n}} \quad \sqrt[n]{a} \quad \sqrt[n]{a}$
mean value of a	$\langle a \rangle \quad \bar{a}$
natural logarithm of a	$\ln a \quad \log_e a$
decadic logarithm of a	$\lg a \quad \log_{10} a \quad \log a$
binary logarithm of a	$\operatorname{lb} a \quad \log_2 a$
exponential of a	$\exp a \quad e^a$

III. Units and Symbols for Units

1. *The International System of Units*

In 1960, the Conférence Générale des Poids et Mesures (CGPM) adopted the International System of Units which is a comprehensive system of units based on a selected set of independent SI units. There are three classes of SI Units: base units, derived units, and supplementary units. The seven SI

---

† Taken from ISO Recommendation R 31, Part XI: Mathematical signs and symbols for use in physical sciences and technology (1961), where a more comprehensive list can be found.

base units are the meter, kilogram, second, ampere, kelvin, candela and mole. For each physical quantity in the International System of Units there is only one SI Unit. Derived units are contained in the second class of SI units; these units can be deduced from combinations of base units according to algebraic relations connecting the corresponding quantities. Several algebraic expressions in terms of base units have special names and symbols and they can be used to form other derived units. The third class of SI units is called supplementary units and comprises all the other units.

These three classes of SI units form a coherent system of units. The decimal submultiples and multiples of SI units formed by means of SI prefixes are distinct from the coherent set of SI units.

2. *Definition of SI Base Units*

A. UNIT OF LENGTH—THE METER

The meter is the length equal to 1 650 763.73 wavelengths in vacuum of the radiation corresponding to the transition between the levels $2p_{10} - 5d_5$ of the krypton-86 atom.

B. UNIT OF MASS—THE KILOGRAM

The international prototype of the kilogram was legalized by the first Conférence Générale des Poids et Mesures (1889) and was considered as the unit of mass (and not of weight or of force): it is equal to the mass of the international prototype of the kilogram.

The international prototype made of platinum–iridium is kept at the Bureau International des Poids et Mesures under specified conditions.

C. UNIT OF TIME—THE SECOND

The second is the duration of 9 192 631 770 periods of the radiation corresponding to the transition between the two hyperfine levels of the ground state of the cesium-133 atom.

D. UNIT OF ELECTRIC CURRENT—THE AMPERE

The ampere is that constant current which, if maintained in two straight parallel conductors of infinite length, of negligible circular cross section, and placed 1 meter apart in vacuum, would produce between these conductors a force equal to 2×10^{-7} newton per meter of length.

E. UNIT OF THERMODYNAMIC TEMPERATURE—THE KELVIN

The definition of the thermodynamic temperature is based on the triple point of water, which was selected by the tenth CGPM (1954) as the fundamental fixed point. The temperature 273.16 K was assigned to it by definition. The name kelvin (symbol K) was adopted by the thirteenth CGPM (1967) which defined the unit of thermodynamic temperature as follows:

The kelvin unit of thermodynamic temperature is the fraction 1/273.16 of the thermodynamic temperature of the triple point of water.

The unit kelvin and its symbol K must also be used to express an interval or a difference of temperature*.

F. Unit of Luminous Intensity—the Candela

The candela is the luminous intensity, in the perpendicular direction, of a surface of 1/600 000 square meter of a blackbody at the temperature of freezing platinum under a pressure of 101 325 newton per square meter.

G. Unit of Amount of Substance—the Mole

The mole† is the amount of substance of a system which contains as many elementary entities as there are atoms in 0.012 kilogram of carbon-12.

3. *Names and Symbols for SI Base Units*

Unit symbols are written in roman (upright) lower case type in general; or with a roman capital (for the first letter) if the symbol is derived from a proper name. These symbols are not followed by a full stop (period). Unit symbols remain unchanged in the plural.

Table 1. Names and symbols for SI Base Units

Symbol for SI Unit	Name of SI Unit	Physical quantity
m	meter	length
kg	kilogram	mass
s	second	time
A	ampere	electric current
K	kelvin	thermodynamic temperature
cd	candela	luminous intensity
mol	mole	amount of substance

4. *Names and Symbols for SI Derived Units*

A. Expression

Mathematical symbols of multiplication and division are used to express algebraically derived units in terms of base units. Several derived units have been given special names and symbols which may themselves be used to express other derived units in a simpler way than in terms of the base units.

---

* In addition to the thermodynamic temperature (symbol T), Celsius temperature (symbol t) defined by the equation $t = T - T_0$, where $T_0 = 273.15 \text{ K}$ by definition, is also used. The Celsius temperature is in general expressed in degrees Celsius (symbol °C). The unit 'degree Celsius' is thus equal to the unit 'kelvin' and an interval or a difference of Celsius temperature may also be expressed in degrees Celsius.

† When the mole is used, the elementary entities must be specified and may be atoms, molecules, ions, electrons, other particles, or specified groups of particles.

GENERAL INTRODUCTION

Table 2. Special names and symbols for certain SI Derived Units

Symbol for SI Unit	Name of SI Unit	Physical quantity	Definition of SI Unit
C	coulomb	electric charge	A s
F	farad	electric capacitance	$A^2 s^4 kg^{-1} m^{-2} (= A s V^{-1})$
H	henry	inductance	$kg\, m^2 s^{-2} A^{-2} (= V A^{-1} s)$
Hz	hertz	frequency	s^{-1}
J	joule	energy	$kg\, m^2 s^{-2}$
lm	lumen	luminous flux	cd sr
lx	lux	illumination	$cd\, sr\, m^{-2}$
N	newton	force	$kg\, m\, s^{-2}$
Pa	pascal	pressure	$kg\, m^{-1} s^{-2} (= N m^{-2})$
S	siemens	electric conductance	$kg^{-1} m^{-2} s^3 A^2 (= A V^{-1} = \Omega^{-1})$
T	tesla	magnetic flux density	$kg\, s^{-2} A^{-1} (= V s s\, m^{-2})$
V	volt	electrical potential difference	$kg\, m^2 s^{-3} A^{-1} (J A^{-1} s^{-1})$
W	watt	power	$kg\, m^2 s^{-3} (J s^{-1})$
Wb	weber	magnetic flux	$kg\, m^2 s^{-2} A^{-1} (= V s)$
Ω	ohm	electric resistance	$kg\, m^2 s^{-3} A^{-2} (= V A^{-1})$

NOTE: Dimensionless quantities are expressed by pure numbers. In this case the corresponding SI unit which is the ratio of the same two SI units may be expressed by the number 1.

B. RECOMMENDATIONS

To insure uniformity in the use of units, some recommendations have been issued.

(1) The product of two or more units is preferably indicated by a (raised) dot. When there is no risk of confusion with another unit symbol, the dot may be omitted. For example: N·M or N m but not mN.

(2) A derived unit formed from two others by division may be expressed by a solidus (oblique stroke), a horizontal line, or a negative index.

For example: m/s, $\frac{m}{s}$ or m s^{-1}.

(3) More than one solidus should never be used in the same expression; if necessary parentheses are used to eliminate ambiguity.

For example: m/s² or m · s^{-2} but not m/s/s

m · kg/(s³ · A) or m · kg · s^{-3} · A^{-1} but not m · kg/s³/A.

5. *Supplementary Units*

The third class of SI Units called 'supplementary units' may be regarded either as base units or as derived units.

Actually this class contains two purely geometrical units: the radian which is the SI Unit of plane angle, and the steradian which is the SI Unit of solid angle.

Table 3. SI Supplementary Units

Symbol for SI Unit	Name of SI Unit	Physical quantity
rad	radian	plane angle
sr	steradian	solid angle

6. Practical Realization of Some Important SI Units

A. Length

For the characteristics of the discharge lamp radiating the standard line of krypton-86, the following specification was adopted:

The CIPM recommends* that the line of krypton-86 adopted as primary standard of length be realized by means of a hot cathode discharge lamp containing krypton-86 of purity not less than 99 per cent in sufficient quantity to insure the presence of solid krypton at a temperature of 64 K. The lamp shall have a capillary of internal diameter 2 to 4 millimeters, and wall thickness approximately 1 millimeter.

It is considered that, provided the conditions listed below are satisfied, the wavelength of the radiation emitted by the positive column is equal to the wavelength corresponding to the transition between the unperturbed levels to within 1 part in 10^3:

(1) the capillary is observed end-on in a direction such that the light rays used travel from the cathode end to the anode end;
(2) the lower part of the lamp including the capillary is immersed in a bath maintained to within one degree of the temperature of the triple point of nitrogen;
(3) the current density in the capillary is 0.3 ± 0.1 ampere per square centimeter.

Other lines of krypton-86 and several lines of mercury-198 and of cadmium-114 are recommended as secondary standards†.

The wavelength of these lines varies with pressure, temperature and the composition of the air in which the light travels; the refractive index of the air must therefore be measured *in situ*.

B. Mass

The primary standard of the unit of mass is the international prototype of the kilogram kept at the BIPM. Balances whose precision can reach 1 part in 10^8 are used to compare the mass of 1kg secondary standards of platinum–iridium or of stainless steel with the mass of the prototype. Submultiples or multiples of the kilogram can be obtained by standardization of a series of masses.

C. Time

A very specific equipment is required to produce electric oscillations at the frequency of vibration of the atom of cesium-133 which defines the second. A quartz oscillator, frequency multipliers and synthesizers, a klystron, phase-sensitive detectors, an apparatus for producing an atomic beam of

---

* *Procès-Verbaux CIPM*, 28, 71 (1960) and *Comptes Rendus des Séances de la Onzième Conférence Générale des Poids et Mesures*, p85. Gauthier-Villars: Paris (1960).

† *Procès Verbaux CIPM*, 31 (1963); Recommendation 1,26 and *Comptes rendus XIIth CGPM*, p18 (1964).

cesium in a vacuum, cavity resonators, uniform and non-uniform magnetic fields, and an ion detector are necessary.

It is possible to obtain pulses at the desired frequencies, for instance 1 Hz, 1 kHz, etc. by division. The stability and the reproducibility can exceed 1 part in 10^{11}. Waves whose frequencies are known to about the same accuracy are broadcast by radio transmitters. Time signals also are broadcast by radio waves. These signals are given in a time scale called 'Coordinated Universal Time (UTC)'. In order that the day of 86 400 seconds should be approximately equal to the period of rotation of the Earth (this period is known to be irregular), the second thus defined is larger than that of the CGPM.

In addition to the cesium beam, there are other standards such as the hydrogen maser, rubidium clocks, quartz frequency standards and clocks, etc. Their frequency is controlled by comparison with a cesium standard.

D. Electric Quantities

The ampere is obtained from the force measurement between two coils, of measurable shape and size, that transport the current.

The ohm, the farad and the henry are accurately connected by impedance measurements at a known frequency, and their absolute value may be calculated (1) from the self-inductance of a coil, or the mutual inductance of two coils, in terms of their linear dimensions, or (2) from the change in capacitance of a capacitor in terms of the change in length of its electrodes (method of Thompson–Lampard).

The volt is deduced from the ampere and the ohm. The accuracy of these measurements lies between 1 and 3 parts in 10^6. Secondary standards are used to obtain absolute measurements such as:
(1) coils of manganin wire for resistance standards;
(2) galvanic cells with cadmium sulphate electrolyte for electromotive force standards;
(3) capacitors for capacitance standards (of 10 pF for example).

Recent techniques such as the measurement of the ratio h/e by the Josephson effect, provide means of checking the volt, with very good precision.

E. Temperature

Absolute temperature measurements are related to thermodynamics, for example by the gas thermometer. At 273.16 K the accuracy is of the order of 1 part in 10^6, but it is less at higher and at lower temperatures.

In 1968, the CIPM adopted a new International Practical Temperature Scale (IPTS-68) which agrees with the best thermodynamic results to date. This new scale (which replaces the 1918 scale, amended in 1960) is published in 'Comité Consultatif de Thermométrie', 8th session, 1967, Annexe 18; *Comptes rendus*, XIIIth CGPM, 1967–68, Annexe 2; and in *Metrologia*, **5**, 35 (1969).

The instruments employed to measure temperature on the IPTS-68 are the platinum resistance thermometer, the platinum–10% rhodium/platinum thermocouple and the monochromatic optical pyrometer. These instruments are calibrated at defined fixed points, the values of which are assigned by agreement.

Because temperature scales are of considerable importance for all thermodynamic measurements, the considerations and conversions cited by Rossini[34] and others should be observed.

F. Photometric Quantities

Absolute photometric measurements are realized by comparison with the luminance of a blackbody at the temperature of freezing platinum. The accuracy of these measurements is somewhat better than one per cent. The results of these measurements are maintained by means of incandescent lamps fed with d.c. in a specified manner. These lamps constitute standards of luminous intensity and of luminous flux.

Luminous sources, having a color other than that of the primary standard, utilize a procedure taking account of the spectral luminous efficiencies.

Photometric quantities are thereby defined as quantities proportional to the sum or integral of a spectral power distribution, weighted according to a specified function of wavelength.

G. Amount of Substance

All quantitative results of chemical analysis or of dosages can be expressed in moles, in other words in units of amount of substance of the constituent particles. Some definition principles of this unit are given below.

The simplest case is that of a pure substance composed of atoms having the chemical symbol X.

A mole of atoms X contains by definition as many atoms as there are ^{12}C atoms in 0.012 kilogram of carbon-12. As neither the mass $m(^{12}C)$ of an atom of carbon-12 nor the mass $m(X)$ of an atom X can be measured accurately, the ratio of these masses $m(X)/m(^{12}C)$, which can be accurately determined, is used. The mass corresponding to 1 mole of X is then $[m(X)/m(^{12}C)]$ (0.012 kg. The molar mass $M(X)$ of X (quotient of mass by amount of substance) is

$$M(X) = [m(X)/m(^{12}C)] (0.012) \text{ kg/mol}$$

In the case of a pure substance made up of molecules B which are combinations of atoms X, Y, ... according to the chemical formula $B = X_\alpha Y_\beta ...$ the mass of one molecule of B is $m(B) = \alpha m(X) + \beta m(Y) + ...$

This mass is not known with precision, but the ratio $m(B)/m(^{12}C)$ can be determined accurately. The molar mass of molecular substance B is then

$$M(B) = \frac{m(B)}{m(^{12}C)} (0.012) \text{ kg/mol} = \left\{ \alpha \frac{m(X)}{m(^{12}C)} + \beta \frac{m(Y)}{m(^{12}C)} + \cdots \right\} (0.012) \text{ kg/mol}$$

The same procedure is used in the more general case when the composition of the substance B is specified as $X_\alpha Y_\beta ...$ even if $\alpha, \beta, ...$ are not integers. If we denote the mass ratios $m(X)/m(^{12}C)$, $m(Y)/m(^{12}C)$, ... by $r(X)$, $r(Y)$, ... the molar mass of the substance B is given by the formula

$$M(B) = [\alpha r(X) + \beta r(Y) + \ldots](0.012) \text{ kg/mol}$$

There are other methods based on the laws of physics and physical chemistry for measuring amounts of substance. Since one mole of particles of any perfect gas occupies the same volume at a temperature T and a pressure p (0.0224136 m^3 at $T = 273.16$ K and $p = 101\ 325 \text{ N/m}^2$), measurements of volume ratios is a method of measuring the ratio of amounts of substance for any two gases (the corrections to apply if the gases are not perfect are well known).

7. Decimal Multiples and Submultiples of SI Units

A. Expression

Table 4. SI prefixes

Symbol	Factor	Prefix	Symbol	Factor	Prefix
T	10^{12}	tera	d	10^{-1}	deci
G	10^{9}	giga	c	10^{-2}	centi
M	10^{6}	mega	m	10^{-3}	milli
k	10^{3}	kilo	μ	10^{-6}	micro
h	10^{2}	hecto	n	10^{-9}	nano
da	10^{1}	deca	p	10^{-12}	pico
			f	10^{-15}	femto
			a	10^{-18}	atto

B. Recommendations

(1) Prefix symbols should be printed in roman (upright) type without spacing between the prefix symbol and the unit symbol.

(2) An exponent affixed to a symbol containing a prefix indicates that the multiple or submultiple of the symbol is raised to the power expressed by the exponent.

$$\text{For example: } 1 \text{ cm}^3 = 10^{-6} \text{m}^3$$
$$1 \text{ cm}^{-1} = 10^{2} \text{m}^{-1}$$

(3) Compound prefixes are to be avoided.

For example: 1 nm but not 1 mμm

(4) The names and symbols of decimal multiples and submultiples of the unit of mass, which already contains a prefix, are constructed by attaching the appropriate prefix to the word 'gram' and symbol g.

For example: mg not μkg, μg not nkg, Mg not kkg

8. Units Outside the International System

A. Decimal Fractions and Multiples of SI Units having Special Names

The units of *Tables 5* and *6* do not belong to the International System of Units. Their use is to be progressively discouraged.

Table 5. Units having special names exactly defined in terms of SI Units

Symbol for unit	Name of unit	Definition of unit	Physical quantity
Å	ångström	10^{-10} m	length
b	barn	10^{-28} m^2	area
bar	bar	10^5 N m^{-2}	pressure
dyn	dyne	10^{-5} N	force
erg	erg	10^{-7} J	energy
G	gauss	10^{-4} T	magnetic flux density (magnetic induction)
l	liter	10^{-3} m^3	volume
M	mole per liter	10^3 mol m^{-3}	concentration
Mx	maxwell	10^{-8} Wb	magnetic flux
P	poise	10^{-1} kg m^{-1} s^{-1}	dynamic viscosity
ph	phot	10^4 lx	illumination
rad	rad	10^{-2} J kg^{-1}	absorbed dose of ionizing radiation
sb	stilb	10^4 cd m^{-2}	luminance
St	stokes	10^{-4} m^2 s^{-1}	kinematic viscosity, diffusion coefficient
t	tonne	10^3 kg	mass
μ	micron	10^{-6} m = μm	length

B. Other Units now Exactly Defined in Terms of the SI Units

Table 6. Units having special names now exactly defined in terms of SI units

Symbol for unit	Name of unit	Definition of unit	Physical quantity
atm	atmosphere	101 325 N m^{-2}	pressure
cal$_{IT}$	IT calorie	4.1868 J	energy
cal$_{th}$	thermochemical calorie	4.184 J	energy
c	curie	3.7×10^{10} s^{-1}	radioactivity
in	inch	2.54×10^{-2} m	length
kgf	kilogramme-force	9.806 65 N	force
kW h	kilowatt-hour	3.6×10^6 J	energy
lb	pound (avoirdupois)	0.453 592 37 kg	mass
mmHg	conventional millimeter of mercury*	$13.5951 \times 980.665 \times 10^{-2}$ N m^{-2}	pressure
Oe	oersted	$1000/4\pi$ A m^{-1}	magnetic field strength
°R	degree Rankine	(5/9) K	thermodynamic temperature
Torr	torr	$(101\,325/760)$ N m^{-2}	pressure

* The conventional millimeter of mercury, symbol mmHg (not mm Hg), is the pressure exerted by a column exactly 1 mm high of a fluid of density exactly 13.5951 g cm^{-3} in a place where the gravitational acceleration is exactly 9.80665 m s^{-2}. The mmHg differs from the torr by less than 2×10^{-7} Torr.

C. UNITS DEFINED IN TERMS OF THE BEST AVAILABLE EXPERIMENTAL VALUES OF CERTAIN PHYSICAL CONSTANTS

The factors for the conversion of these units to SI Units are subject to change at any time based on new experimental measurements of the constants involved.

Table 7. Units defined in terms of experimental values of physical constants

Symbol	Name of unit	Conversion factor	Physical quantity	Definition
eV	electron volt	1.60219×10^{-19} J	energy	(1)
u	unified atomic mass unit	1.66041×10^{-27} kg	mass	(2)

Definitions: (1) one electron volt is the energy acquired by an electron in passing through a potential difference of one volt in a vacuum. (2) The unified atomic mass unit is equal to one twelfth part of the mass of an atom of the nuclide ^{12}C.

D. 'INTERNATIONAL' ELECTRICAL UNITS

These units are obsolete having been replaced by the 'absolute' (SI) units in 1948. The conversion factors which should be used with electrical measurements quoted in 'international' units depend on where and on when the instruments used to make the measurements were calibrated. The following two sets of conversion factors refer respectively to the 'mean international' units estimated by the ninth Conférence Générale des Poids et Mesures in 1948, and to the 'US international' units estimated by the National Bureau of Standards (USA) as applying to published measurements made with instruments calibrated by them prior to 1948.

$$1 \text{ 'mean international ohm'} = 1.00049 \, \Omega$$
$$1 \text{ 'mean international volt'} = 1.00034 \text{ V}$$
$$1 \text{ 'US' international ohm'} = 1.000495 \, \Omega$$
$$1 \text{ 'US' international volt'} = 1.000330 \text{ V}$$

IV. Physical Quantities and Symbols for Physical Quantities

1. Definition

A physical quantity is the product of a numerical value (a pure number) and a unit.

2. Basic Physical Quantities

Seven physical quantities are generally regarded as independent basic physical quantities. These seven quantities and the symbols used to denote them are presented in *Table 8*.

Table 8. Basic physical quantities

Symbol for quantity	Basic physical quantity
l	length
m	mass
t	time
I	electric current
T	thermodynamic temperature
I_v	luminous intensity
n	amount of substance

The definition of amount of substance, as of all other physical quantities, has nothing to do with any choice of unit, and in particular has nothing to do with the particular unit of amount of substance called the mole.

3. Derived Physical Quantities

All other physical quantities are regarded as being derived from, and as having dimensions derived from, the seven independent basic physical quantities by definitions involving only multiplication, division, differentiation, and/or integration.

Tables 9 and *10* contain the symbols for the most important quantities likely to be used by physicists and chemists. In a few cases where conflicts are foreseen, further flexibility can be obtained by the use of capital letters as variants for lower-case letters and vice versa in preference to subscripts. For example, the recommended symbol for power is P and for pressure p or P; but P and p may be used for two powers or for two pressures. If power and pressure appear together, however, P should be used only for power and p only for pressure, and necessary distinctions between different powers or between different pressures should be made by the use of subscripts or other modifying signs. Symbols within parentheses in *Tables 9* and *10* are reserved symbols.

We have chosen an alphabetical classification in Latin and Greek letters so that the inadequacies of such a classification can be better shown. To try to diminish conflicts, each quantity was characterized by a number in the column called: nature. Each number is related to the following sections, chosen to define a group of quantities, independently of the others.

(1) Space, time and related quantities
(2) Mechanical and related quantities
(3) Molecular and related quantities
(4) Thermodynamics and related quantities
(5) Chemical reactions
(6) Electricity and magnetism
(7) Electrochemistry
(8) Light and related electromagnetic radiation
(9) Transport properties

GENERAL INTRODUCTION

Table 9. Physical and chemical quantities—Latin symbols

Symbol	Name	SI Unit[a]	Definition or description	Alternative symbol	Dimension	Nature
A	area	m^2		S, A_s	L^2	(1)
A	Helmholtz energy		$U - TS$			(4)
A	affinity of a reaction		$-\Sigma_B \nu_B \mu_B$			(5)
A	internal transmission density, absorbance[b]		$\log(1/\tau_i)$	D_i		(8)
A_1	Alfvén number[c]		$vl(\rho\mu)^{\frac{1}{2}}/\mathbf{B}$			(9)
A_r	relative atomic mass of an element[d] (also called 'atomic weight')					(2)
A_s	area[e]	m^2		S, A		(1)
a	acceleration	$m\,s^{-2}$	v/t		LT^{-2}	(1)
a	thermal diffusivity	$m^2\,s^{-1}$	$\lambda/\rho c_p$		$L^2 T^{-1}$	(4)
a	(linear) absorption coefficient[f]		D_i/l			(8)
a, a_B	activity, relative activity of substance B					(4)
$\mathbf{B}$	magnetic flux density, magnetic induction					(6)
B	susceptance		$Y = G + iB$			(6)
B	concentration of solute substance B[f]			C_B		(3)
C	Heat capacity	$m^2\,kg\,s^{-2}\,K^{-1}$ ($J\,K^{-1}$)			$ML^2 T^{-2}\theta^{-1}$	(4)
C	capacitance					(6)
Co	Cowling number[c]		$\mathbf{B}^2/\mu\rho v^2$			(9)
c	velocity	$m\,s^{-1}$		$v, u, w,$		(1)
c	velocity of sound				LT^{-1}	(2)
c	specific heat capacity	$m^2\,s^{-2}\,K^{-1}$ ($J\,kg^{-1}\,K^{-1}$)	heat capacity divided by mass		$L^2 T^{-2}\theta^{-1}$	(4)
c	speed of light *in vacuo*					(8)
c_B	concentration of solute substance B[f]	$mol\,m^{-3}$	amount of B divided by the volume of the solution	[B]	L^{-3}	(3)
D	diffusion coefficient	$m^2\,s^{-1}$			$L^2 T^{-1}$	(2)
$\mathbf{D}$	electric displacement					(6)
D_i	internal transmission density, absorbance		$\log(1/\tau_I)$	A		(8)
d	diameter	m				(1)
d	relative density		ratio of the density to that of a reference substance			(2)
d	collision diameter of of a molecule			σ		(3)
E	energy	$N \cdot m$			$ML^2 T^{-2}$	(2)
E	modulus of elasticity		normal stress divided by linear strain, Young's modulus			(2)
(E)	internal energy			U		(4)
$\mathbf{E}$	electric field strength	$V\,m^{-1}$				(6)
E	electromotive force					(7)
E	irradiance[g]	$kg\,s^{-3}$ ($W\,m^{-2}$)	$d\Phi/dS$		MT^{-3}	(8)

Table 9—*continued*

Symbol	Name	SI Unit[a]	Definition or description	Alternative symbol	Dimension	Nature
E_k	kinetic energy			T, K		(2)
E_p	potential energy			V, Φ		(2)
Eu	Euler number[c]		$p/\rho v^2$			(9)
e	linear strain			ε		(2)
e	elementary charge (of a proton)					(6)
F	force		m a		MLT^{-2}	(2)
F	Faraday constant					(7)
Fo	Fourier number		$a\Delta t/l^2$			(9)
Fo^*	Fourier number for mass transfer[c]		Dt/l^2			(9)
Fr	Froude number[c]		$v/(lg)^{\frac{1}{2}}$			(9)
f	frequency	s^{-1}		ν		(1)
(f)	friction coefficient		frictional force divide by normal force	μ		(2)
f	fugacity			(p^*)		(4)
f_B	activity coefficient, mole fraction basis					(4)
G	weight			(W)		(2)
G	shear modulus		shear stress divided by shear angle			(2)
G	Gibbs energy		$H - TS$			(4)
G	conductance		$Y = G + iB$			(6)
Gr	Grashof number[c]		$l^3 g\alpha\Delta\theta\rho^2/\eta^2$			(9)
Gr^*	Grashof number for mass transfer[c]		$-l^3 g(\partial\rho/\partial x)_{T,p} \Delta x \rho/\eta^2$			(9)
g	statistical weight					(3)
H	Hamiltonian function					(2)
H	enthalpy		$U + pV$			(4)
H	magnetic field strength	$A m^{-1}$				(6)
H	exposure		$\int E\,dt$			(8)
Ha	Hartmann number[c]		$Bl(\kappa/\eta)^{\frac{1}{2}}$			(9)
h	height	m				(1)
h	coefficient of heat transfer	$W m^{-2} K^{-1}$	density of heat flowrate divided by temperature difference		$MT^{-3}\theta^{-1}$	(4)
h	Planck constant					(8)
$\hbar$	Planck constant divided by 2π					(8)
I	moment of inertia					(2)
I	ionic strength		$I_m = \frac{1}{2}\Sigma_i m_i z_i^2$ or $I_c = \frac{1}{2}\Sigma_i c_i z_i^2$			(4)
I	electric current					(6)
I	radiant intensity[g]	$W sr^{-1}$	$d\Phi/d\omega$		ML^2T^{-3}	(8)
J	Massieu function		$-A/T$			(4)
J	rate of reaction		$d\xi/dt$	$\dot{\xi}$		(5)
J_X, J	flux (of a quantity X)					(9)
j	electric current density					(6)
j_0	exchange current density					(7)
K	kinetic energy			E_k, T		(2)
K	compression (bulk) modulus		$p = -K\Delta V/V_0$			(2)
K	equilibrium constant					(5)

GENERAL INTRODUCTION

Table 9—continued

Symbol	Name	SI Unit[a]	Definition or description	Alternative symbol	Dimension	Nature
Kn	Knudsen number[c]		λ/l			(9)
k	mass transfer coefficient		mass divided by time and by cross-sectional area	k_m		(2)
k	Boltzmann constant			k_B		(4)
k	thermal conductivity	$\text{W m}^{-1}\text{K}^{-1}$		λ		(4)
k	rate constant					(5)
L	angular momentum					(2)
L	Lagrangian function					(2)
L	Avogadro constant			N_A		(3)
L	self inductance					(6)
L	radiance[g]	$\text{W m}^{-2}\text{sr}^{-1}$	$(dI/dS)\cos\theta$			(8)
L_{12}	mutual inductance			M		(6)
Le	Lewis number[c]		a/D			(9)
l	length	m			L	(1)
l	mean free path			λ	L	(3)
M	moment of force	$\text{m}^2\text{kg s}^{-2}$ (N m)				(2)
M	molar mass		mass divided by amount of substance			(3)
$\boldsymbol{M}$	magnetization		$(\boldsymbol{B}/\mu_0) - \boldsymbol{H}$			(6)
M	mutual inductance			L_{12}		(6)
M	radiant excitance[g]		$d\Phi/dS$			(8)
Ma	Mach number[c]		v/c			(2)
M_r	relative molecular mass of a substance[h] (also called 'molecular weight')					(3)
m	mass				M	(2)
$\boldsymbol{m}$	electromagnetic moment		$E_p = -\boldsymbol{m}\cdot\boldsymbol{B}$	μ		(6)
m_B	molality of solute substance B[i]		amount of B divided by mass of solvent			(3)
N	number of molecules					(3)
N_A	Avogadro constant			L		(3)
Nu	Nusselt number[c]		hl/k			(9)
Nu^*	Nusselt number for mass transfer[c]		$k_m l/\rho D$			(9)
n	amount of substance			(v)		(3)
n	refractive index					(8)
P	power	J/S	energy divided by time		ML^2T^{-3}	(2)
$\boldsymbol{P}$	dielectric polarization		$\boldsymbol{D} - \varepsilon_0\boldsymbol{E}$			(6)
P	pressure	N m^{-2}		p	$\text{ML}^{-1}\text{T}^{-2}$	(2)
Pe	Péclét number[c]		vl/a			(9)
Pe^*	Péclét number for mass transfer[c]		vl/D			(9)
Pr	Prandtl number[c]		$\eta/\rho a$			(9)
p	momentum					(2)
p	pressure	N m^{-2}		P	$\text{ML}^{-1}\text{T}^{-2}$	(2)
$\boldsymbol{p}$	electric dipole moment			$\boldsymbol{p}_e$		(6)

Table 9—continued

Symbol	Name	SI Unit[a]	Definition or description	Alternative symbol	Dimension	Nature
p	permanent dipole moment of a molecule			μ		(6)
p	induced dipole moment of a molecule			p_i		(6)
(p^*)	fugacity			f		(4)
Q	partition function (system)			Z		(3)
Q	quantity of electricity					(6)
Q	radiant energy[g]					(8)
q	partition function (particle)			z		(3)
q	heat[j]	N·m		Q	ML^2T^{-2}	(4)
R	resistance					(6)
Ra	Rayleigh number[c]		$l^3 g \alpha \Delta\theta \rho/\eta a$			(9)
Re	Reynolds number[c]		$\rho v l/\eta$			(9)
Re_m	magnetic Reynolds number[c]		$v\mu\kappa l$			(9)
R_m	molar refraction		$\dfrac{(n^2-1)}{(n^2+2)} V_m$			(8)
r	radius	m				(1)
r_B	rate of increase of concentration of substance B			v_B		(5)
S	area	m²		A, A_s		(1)
S	entropy	J K⁻¹				(4)
Sc	Schmidt number[c]		$\eta/\rho D$			(9)
Sr	Strouhal number[c]		lf/v			(9)
St	Stanton number[c]		$h/\rho v c_p$			(9)
St^*	Stanton number for mass transfer[c]		$k_m/\rho v$			(9)
s	path, length of arc					(1)
s	symmetry number			σ		(3)
T	period		$1/v$			(1)
T	kinetic energy			E_k, K		(2)
T	thermodynamic temperature, absolute temperature	K			θ	(4)
T	internal transmittance[b]		transmittance of the medium itself, disregarding boundary or container influence	τ_i		(8)
t	time	s			T	(1)
t	Celsius temperature	°C		θ	θ	(4)
t	transport number		transference number or migration number			(7)
U	internal energy			(E)		(4)
U	electric tension		iR			(6)
u	velocity	m s⁻¹		v, w, c		(1)
u	electric mobility		velocity divided by electric field strength	μ		(7)
V	volume	m³			L^3	(1)
V	potential energy			E_p, Φ		(2)

GENERAL INTRODUCTION

Table 9—continued

Symbol	Name	SI Unit[a]	Definition or description	Alternative symbol	Dimension	Nature
V	electric potential			Φ		(5)
v	velocity	$m\,s^{-1}$		u, w, c	LT^{-1}	(1)
v	specific volume	$m^3\,kg^{-1}$	volume divided by mass		$M^{-1}L^3$	(2)
v_B	rate of increase of concentration of substance B		dc_B/dt	r_B		(5)
(W)	weight			G		(2)
We	Weber number[c]		$\rho v^2 l/\gamma$			(9)
w	work[j]	$N \cdot m$	force times path	W	ML^2T^{-2}	(2), (4)
w	velocity	$m\,s^{-1}$		u, v, c	LT^{-1}	(1)
w	specific weight (volumic weight)					(2)
X	reactance					
x_B	mole fraction of substance B		$n_B/\Sigma_i n_i$	y_B		(3)
Y	Planck function		$-G/T$			(4)
Y	admittance (complex admittance)		$1/Z$			(6)
y_B	mole fraction of substance B			x_B		(3)
y_B	activity coefficient, concentration basis					(4)
Z	collision number		number of collisions divided by volume and by time			(3)
Z	partition function (system)			Q		(3)
Z	compression factor		pV_m/RT			(4)
Z	impedance (complex impedance)		$R + iX$			(6)
z	partition function (particle)			q		(3)
z	charge number of a cell reaction					(7)
z_B	charge number of an ion B		positive for cations, negative for anions			(7)

Table 10. Physical and chemical quantities—Greek symbols

Symbol	Name	SI Unit[a]	Definition or description	Alternative symbol	Dimension	Nature
α	plane angle			$\beta, \gamma, \theta, \phi$		(1)
α	cubic expansion coefficient	K^{-1}	$V^{-1}(\partial V)/(\partial T)_p$		θ^{-1}	(4)
α	degree of dissociation					(5)
α	electric polarizability of a molecule					(6)
α	electrochemical transfer coefficient					(7)

Table 10—continued

Symbol	Name	SI Unit[a]	Definition or description	Alternative symbol	Dimension	Nature
α	absorptance, absorption factor[b]		ratio of absorbed to incident radiant or luminous flux			(8)
α	angle of optical rotation					(8)
β	plane angle			$\alpha, \gamma, \theta, \phi$		(1)
β	pressure coefficient	K^{-1}	$p^{-1}(\partial p/\partial T)_V$		θ^{-1}	(4)
Γ	surface concentration, surface excess					(3)
γ	plane angle			$\alpha, \beta, \theta, \phi$		(1)
γ	surface tension	$kg\,s^{-2}$		σ	MT^{-2}	(2)
γ	ratio C_p/C_V	(Nm^{-1})		(κ)		(4)
γ_g	activity coefficient, molality basis					(4)
δ	loss angle					(6)
δ	thickness of diffusion layer					(7)
ε	linear strain (relative elongation)		$\Delta l/l_0$	e		(2)
ε	permittivity	$F\,m^{-1}$	$\boldsymbol{D} = \varepsilon \boldsymbol{E}$			(6)
ε	molar (linear) absorption coefficient[b,m]		D_1/lc			(8)
						(8)
ε_0	permittivity of vacuum					(6)
ε_r	relative permittivity[k]		$\varepsilon/\varepsilon_0$			(6)
ζ	electrokinetic potential (zeta potential)		τ/ε			(7)
η	viscosity (dynamic)	$m^{-1}kg\,s^{-1}$ or $Pa\,s$		(μ)	$ML^{-1}T^{-1}$	(2)
η	overpotential, overtension (also called 'overvoltage')					(7)
Θ	characteristic temperature					(3)
θ	plane angle			$\alpha, \beta, \gamma, \phi$		(1)
θ	volume strain (bulk strain)		$\Delta V/V_0$			(2)
θ	angle of contact					(2)
θ	Celsius temperature	°C		t		(4)
κ	**wavenumber in solid state studies**					(1)
κ	compressibility	$kg^{-1}m\,s^2$	$-V^{-1}(dV/dp)$		$M^{-1}LT^2$	(2)
κ	isothermal compressibility		$-V^{-1}(\partial V/\partial p)_T$			(4)
(κ)	ratio C_p/C_V			γ		(4)
κ	conductivity (formerly called specific conductance)		$\boldsymbol{j} = \kappa \boldsymbol{E}$	(σ)		(6)
κ	electrolytic conductivity (formerly called specific conductance)			(σ)		(7)
$\Lambda, \lambda^{(m)}$	molar conductivity of electrolyte or ion[n]		k/c			(7)

Table 10—continued

Symbol	Name	SI Unit[a]	Definition or description	Alternative symbol	Dimension	Nature
λ	wavelength					(1)
λ	mean free path			l		(3)
λ	thermal conductivity	$W\,m^{-1}\,K^{-1}$		k	$MLT^{-3}\theta^{-1}$	(4)
λ_B	absolute activity of substance B		$\exp(\mu_B/RT)$			(4)
(μ)	viscosity	$kg\,m^{-1}\,s^{-1}$				(2)
μ	friction coefficient		frictional force divided by normal force	(f)		(2)
μ	reduced mass				M	(2)
μ	Joule–Thomson coefficient		$(\partial T/\partial p)_H$			(4)
$\boldsymbol{\mu}$	electromagnetic moment		$E_p = -\boldsymbol{\mu}\cdot\boldsymbol{B}$	m		(6)
$\boldsymbol{\mu}$	permanent dipole moment of a molecule			p		(6)
μ	permeability	$H\,m^{-1}$	$\boldsymbol{B} = \mu\boldsymbol{H}$			(6)
μ	electric mobility		velocity divided by electric field strength	u		(7)
μ_B	chemical potential of substance B					(4)
μ_B	Bohr magneton					(6)
$\tilde{\mu}_B$	electrochemical potential of ionic component B		$\mu_B + Z_B F\phi$			(7)
μ_0	permeability of vacuum					(6)
μ_r	relative permeability		μ/μ_0			(6)
ν	frequency			f		(1)
$\bar{\nu}$	wavenumber	m^{-1}	$1/\lambda$	σ		(1)
ν	kinematic viscosity	$m^2\,s^{-1}$	η/ρ		$L^2 T^{-1}$	(2)
(ν)	amount of substance			n		(3)
ν_B	stoichiometric coefficient of substance B		negative for reactants, positive for products			(5)
Ξ	grand partition function (system)					(3)
ξ	extent of reaction		$d\xi = dn_B/\nu_B$			(5)
$\dot{\xi}$	rate of reaction		$d\xi/dt$	J		(5)
Π	osmotic pressure					(4)
ρ	density	$kg\,m^{-3}$			ML^{-3}	(2)
ρ	charge density	$m^{-3}\,s\,A$ $(c\,m^{-3})$				(6)
ρ	resistivity (formerly called specific resistance)		$\boldsymbol{E} = \rho\boldsymbol{j}$			(6)
ρ	reflectance, reflection factor[b]		ratio of reflected to incident radiant or luminous flux			(8)
ρ_B	mass concentration of substance B		mass of B divided by the volume of the solution			(3)
σ	wavenumber	m^{-1}	$1/\lambda$	$\bar{\nu}$		(1)

Table 10—continued

Symbol	Name	SI Unit[a]	Definition or description	Alternative symbol	Dimension	Nature
σ	normal stress					(2)
σ	surface tension	$kg\,s^{-2}$ ($N\,m^{-1}$)		γ	MT^{-2}	(2)
σ	collision diameter			d		(3)
σ	symmetry number			s		(3)
σ	surface charge density					(6)
(σ)	conductivity			κ		(6)
(σ)	electrolytic conductivity			κ		(7)
τ	characteristic time interval, relaxation time, time constant					(1)
τ	shear stress					(2)
τ	strength of double layer		electric moment divided by area			(7)
τ	transmittance, transmission factor[b]		ratio of transmitted to incident radiant or luminous flux			(8)
τ_i	internal transmittance[b]		transmittance of the medium itself, disregarding boundary or container influence	T		(8)
Φ	potential energy			E_p, V		(2)
Φ	magnetic flux					(6)
Φ	radiant flux, radiant power[g]	$kg\,m^2\,s^{-3}$	$\Phi = dq/dt$		ML^2T^{-3}	(8)
Φ	quantum yield					(8)
ϕ	plane angle			a, ξ, γ, θ		(1)
θ	fluidity		$1/\eta$			(2)
ϕ	osmotic coefficient					(4)
ϕ	electric potential			V		(6)
ϕ	inner electric potential					(7)
ϕ_B	volume fraction of substance B					(3)
χ	adiabatic compressibility		$-\dfrac{1}{V}\left(\dfrac{\partial V}{\partial p}\right)_s$			(4)
χ	magnetic susceptibility		$\mu_r - 1$	χ_m		(6)
χ	surface electric potential difference		$\phi - \psi$			(7)
χ_e	electric susceptibility		$\varepsilon_r - 1$			(6)
ψ	outer electric potential					(7)
ω	solid angle					(1)
ω	angular frequency, pulsatance		$2\pi\nu$			(1)
ω	angular velocity	$rad\,s^{-1}$				(1)

(a) The symbols in this column for some SI units of physical quantities are those currently used. In fact, neither the physical quantity, nor the symbol used to denote it, should imply a particular choice of unit. Operations on equations involving physical quantities, units and numerical values, should follow the ordinary rules of algebra. Thus the physical quantity called the critical pressure and denoted by p_c has the value for water:

$$p_c = 22.12 \text{ MPa}$$

This equation may equally well be written in the form

$$p_c/\text{MPa} = 22.12$$

GENERAL INTRODUCTION

These are especially useful for the headings in tables and as labels on the axes of graphs.

(b) These names and symbols are in agreement with those adopted jointly by the International Commission of Illumination (CIE) and the International Electrotechnical Commission (IEC). The term extinction [for internal transmission density] is unsuitable because extinction is now reserved for the diffusion of radiation rather than for absorption.

The term absorptivity [for (linear) absorption coefficient] should be avoided because the meaning absorptance per unit length has been accepted internationally for the term absorptivity.

(c) References to the nature of the symbols used in defining the transport property quantities (9) are as follows:
$a(4)$, $\boldsymbol{B}$ (6), c (2), C_p (4), D (2), f (1), g (1), h (4), k (4), k_m (2), l (1), p (2), t (1), v (1), x (3), α (4), γ (2), η (2), θ (4), κ (6), λ (3), μ (6), ρ (2).

(d) The ratio of the average mass per atom of the natural nuclidic composition of an element to $\frac{1}{12}$ of the mass of an atom of nuclide ^{12}C.

$$\text{Example: } A_r(\text{Cl}) = 35.453$$

The concept of relative atomic mass may be extended to other specified nuclidic compositions, but the natural nuclidic composition is assumed unless some other composition is specified.

(e) The symbol A_s may be used when necessary to avoid confusion with the symbol A for Helmholtz energy.

(f) Concentration is sometimes called 'molarity' but this name is both unnecessary and liable to cause confusion with molality and is therefore not recommended. A solution with a concentration of 0.1 mol dm^{-3} is often called a 0.1 molar solution or a 0.1 M solution.

(g) The same symbol is often used also for the corresponding luminous quantity. Subscripts e for energetic and v for visible may be added whenever confusion between these quantities might otherwise occur.

(h) The ratio of the average mass per formula unit of the natural nuclidic composition of a substance to $\frac{1}{12}$ of the mass of an atom of nuclide ^{12}C.

$$\text{Example: } M_r(\text{KCl}) = 74.555$$

The concept of relative molecular mass may be extended to other specified nuclidic compositions, but the natural nuclidic composition is assumed unless some other composition is specified.

(i) A solution having a molality equal to 0.1 mol kg^{-1} is sometimes called a 0.1 molal solution or a 0.1 m solution. If a symbol is needed for mol kg^{-1} then some symbol other than m, which is the SI unit-symbol for the metre, should be chosen.

(j) It is recommended that $q > 0$ and $w > 0$ both indicate increase of energy of the system under discussion. Thus $\Delta u = q + w$.

(k) Also called dielectric constant, and sometimes denoted by D when it is independent of $\boldsymbol{E}$.

(l) The formula unit whose concentration is c must be specified.

$$\text{Example: } \Lambda(\text{Mg}^{2+}) = 2\Lambda(\tfrac{1}{2}\text{Mg}^{2+})$$

(m) The word molar, contrary to the general rule given in section IV, 4, here means 'divided by concentration'.

4. Use of the Words 'Specific' and 'Molar' in the Names of Physical Quantities

The word 'specific' before the name of an extensive physical quantity is restricted to the meaning 'divided by mass'. For example, specific volume is the volume divided by the mass. When the extensive quantity is represented by a capital letter, the corresponding specific quantity may be represented by the corresponding lower case letter.

Examples: volume: V specific volume: $v = V/m$
heat capacity at constant pressure: C_p
specific heat capacity at constant pressure: $c_p = C_p/m$

The word 'molar' before the name of an extensive quantity is restricted to the meaning 'divided by amount of substance'. For example, molar volume is the volume divided by the amount of substance. The subscript m attached to the symbol for the extensive quantity denotes the corresponding molar quantity.

Examples: volume: V molar volume: $V_m = V/n$
Gibbs energy: G molar Gibbs energy: $G_m = G/n$

The subscript m may be omitted when there is no risk of ambiguity. Lower case letters may be used to denote molar quantities when there is no risk of misinterpretation.

The symbol X_B, where X denotes an extensive quantity and B is the chemical symbol for a substance, denotes the partial molar quantity of the substance B defined by the relation

$$X_B = (\partial X/\partial n_B)_{T, p, n_c} \ldots$$

where n_c indicates that the amount of all substances except B are held fixed. For a pure substance B the partial molar quantity X_B and the molar quantity

X_m are identical. The partial molar quantity X_B of pure substance B, which is identical with the molar quantity X_m of pure substance B, may be denoted by $X_B^\star$, where the superscript ★ denotes 'pure', so as to distinguish it from the partial molar quantity X_B of substance B in a mixture.

5. Symbols for Particular Cases of Physical Quantities

It is much more difficult to make detailed recommendations on symbols for physical quantities in particular cases than in general cases. This arises from an incompatibility between the need for specifying numerous details and the need for keeping the printing reasonably simple. Among the most awkward things to print are superscripts to subscripts and subscripts to subscripts. Examples of symbols to be avoided are:

$$\Lambda_{NO_3^-} \qquad \Delta H_{25°C} \qquad (pV)_{0°C}^{p=0}$$

The problem is vastly reduced if it is recognised that two different kinds of notation are required for two different purposes. In the formulation of general fundamental relations the most important requirement is for a notation which is easy to understand and easy to remember. In applications to particular cases, in quoting numerical values, and in tabulation, the most important requirement is the complete elimination of any possible ambiguity even at the cost of an elaborate notation.

The advantage of dual notation is already to some extent accepted in the case of concentration. The recommended notation for the formulation of the equilibrium constant K for the general reaction

$$0 = \Sigma_B \nu_B B$$

is

$$K = \Pi_B (c_B)^{\nu_B}$$

but when we turn to a particular example it is better to use a notation such as:

$$Br_2 + H_2O = HOBr + H^+ + Br^-$$

$$\frac{[HOBr][H^+][Br^-]}{[Br_2]} = K$$

$$K(25°C) = 6 \times 10^{-9} \text{ mol}^2 \text{ dm}^{-6}$$

Once the principle of dual notation is accepted, its adaptability and usefulness become manifest in all fields of physical chemistry.

6. Recommended Subscripts

The following subscripts are recommended:

ad adiabatic
c   critical quantity
i   isomer

k characterizes a polytropic evolution, i.e. when p and V are bound by $pV^k =$ constant, where k is a positive constant
m molar quantity
n normal
p pressure (characterizes an isobaric evolution)
q heat (characterizes an adiabatic evolution, i.e. when the system does not exchange heat with the surroundings)
r reduced quantity
S entropy (characterizes an isentropic evolution)
T temperature (characterizes an isothermal evolution)
v volume (characterizes an isochoric evolution)
∞ value at infinity.

7. *Recommended Superscripts*

The following superscripts are recommended:
° or ★   pure substance
∞ infinite dilution
id   ideal
° or ⊖   standard in general
+   activated complex, transition state
E excess properties.

V. Symbols for Chemical Elements, Nuclides and Particles

A nuclide is a species of atoms of which each atom has identical atomic number (proton number) and identical mass number (nucleon number). Different nuclides having the same value of the atomic number are named isotopes or isotopic nuclides. Different nuclides having the same mass number are named isobars or isobaric nuclides.

It has been recommended that symbols for chemical elements should be written in roman (upright) type. The symbol is not followed by a full stop except when it occurs at the end of a sentence in text.

Examples: Ca, C, H, He

The nuclide may be specified by attaching numbers. The mass number should be placed in the left superscript position; the atomic number, if desired, may be placed as a left subscript. The number of atoms per molecule is indicated as a right subscript. Ionic charge, or state of excitation, or oxidation number may be indicated in the right superscript space.

Examples: Mass number: $^{14}N_2$, $^{35}Cl^-$
Ionic charge: Cl^-, Ca^{2+}, PO_4^{3-} or $PO_4{}^{3-}$
Excited electronic state: He★, NO★
Oxidation number: $Pb_2^{II}Pb^{IV}O_4$, $K_6M^{IV}Mo_9O_{32}$ (where M denotes a metal).

VI. Values of the Fundamental Constants[28, 37]

A list giving a consistent set of fundamental constants has been published

Table 11. Fundamental constants

Symbol	Quantity	Definition	Value	Uncertainty ($\times 10^{-6}$)	SI Units
a_0	Bohr radius	$\left[\frac{\mu_0 c^2}{4\pi}\right]^{-1}\left(\frac{\hbar^2}{m_e e^2}\right) = \frac{\alpha}{4\pi R_\infty}$	5.291715	8.1	10^{-11} m
amu	Atomic mass unit (cf. u)		1.660531	11	10^{-27} kg
c	Velocity of light†		2.997924562	0.1	10^8 m sec^{-1}
c_1	First radiation constant	$8\pi hc$	4.992579	38	10^{-24} Jm
c_2	Second radiation constant	$\frac{hc}{k}$	1.438833	61	10^{-2} mK
e	Electron charge		1.6021917	7.0	10^{-19} C
e/m_e	Electron charge to mass ratio		1.7588028	5.4	10^{11} C kg^{-1}
F	Faraday constant	$N_A e$	9.648670	54	10^4 C mol^{-1}
G	Gravitational constant		6.6732	3100	10^{-11} N m^2 kg^{-2}
h	Planck's constant		6.626196	50	10^{-34} J sec
$\hbar$	$\hbar = h/2\pi$		1.0545919	8	10^{-34} J sec
$h/2m_e$	Quantum of circulation		3.636947	11	10^{-4} J sec kg^{-1}
k	Boltzmann's constant	R/N_A	1.380621	59	10^{-23} J K^{-1}
L, N_A	Avogadro's number		6.022169	40	10^{23} mol^{-1}
m_e	Electron rest mass		9.109558	54	10^{-31} kg
M_p	Proton rest mass		1.672614	11	10^{-27} kg
M_n	Neutron rest mass		1.674920	11	10^{-27} kg
M_p/m_e	Ratio of proton mass to electron mass		1836.109	11000	—
R	Gas constant		8.31434	350	J K^{-1} mol^{-1}
R_∞	Rydberg constant	$\left(\frac{\mu_0 c^2}{4\pi}\right)^2\left(\frac{m_e e^4}{4\pi\hbar^3 c}\right)$	1.09737312	0.11	10^7 m^{-1}
r_0	Classical electron radius	$\left(\frac{\mu_0 c^2}{4\pi}\right)\left(\frac{e^2}{m_e c^2}\right) = \frac{\alpha^3}{4\pi R_\infty}$	2.817939	16	10^{-15} m

GENERAL INTRODUCTION

Symbol	Name	Formula	Value	Uncertainty	Units
T_{ice}	'Ice point' temperature†		273.1500	100	K
RT_{ice}			2.271 06	120	10^3 J mol^{-1}
u	Atomic mass unit (cf. amu)		1.660 531	11	10^{-27} kg
Y	Einstein constant relating mass and energy	c^2	8.987 551 679	0.6	10^{16} J kg^{-1}
Z	Constant relating wavenumber and energy	$N_A hc$	1.196 255	38	10^{-1} J m mol^{-1}
α	Fine structure constant	$\left(\dfrac{\mu_0 c^2}{4\pi}\right)\left(\dfrac{e^2}{\hbar c}\right)$	7.297 351	11	10^{-3}
$\gamma_{p'}$	Gyromagnetic ratio of protons in H$_2$O		2.675 1270	8.2	10^8 rad sec^{-1} T^{-1}
γ_p	$\gamma_{p'}$ corrected for diamagnetism of H$_2$O		2.675 1965	8.2	10^8 rad sec^{-1} T^{-1}
ϵ_0	Permittivity of vacuum	$\mu_0^{-1} c^{-2}$	8.854 185	18	10^{-12} J^{-1} C^2 m^{-1}
λ_c	Compton wavelength of the electron	$\dfrac{h}{m_e c}$	2.426 3096	7.4	10^{-12} m
$\lambda_{c,p}$	Compton wavelength of the proton	$\dfrac{h}{M_p c}$	1.321 4409	9.0	10^{-15} m
$\lambda_{c,n}$	Compton wavelength of the neutron	$\dfrac{h}{M_n c}$	1.319 6217	9.0	10^{-15} m
μ_0	Permeability of vacuum		4π	—	10^{-7} J s^2 C^{-2} m^{-1}
μ_B	Bohr magneton	(c) $\left(\dfrac{e\hbar}{2m_e c}\right)$	9.274 096	65	10^{-24} J T^{-1}
μ_e	Electron magnetic moment		9.284 851	65	10^{-24} J T^{-1}
μ_n	Nuclear magneton	(c) $\left(\dfrac{e\hbar}{2M_p c}\right)$	5.050 951	50	10^{-27} J T^{-1}
μ_p	Proton magnetic moment		1.410 203	9.9	10^{-26} J T^{-1}
σ	Stefan–Boltzmann constant	$\dfrac{\pi^2 k^4}{60 \hbar^3 c^2}$	5.669 61	960	10^{-8} W m^{-2} K^4
Φ_0	Magnetic flux quantum	(c)$^{-1}\left(\dfrac{hc}{2e}\right)$	2.067 8538	69	10^{-15} T m^2

†Measured recently by K. H. Evenson (NBS, Boulder).

‡The 'ice point' temperature T_{ice} is the temperature of equilibrium of solid and liquid water saturated with air at a pressure of one atmosphere. The quantity RT_{ice} is identical to the quantity $\lim_{p \to 0}(pV_m)$ for a gas at that temperature.

by Taylor et al.[37]. For details concerning the development of this set of constants and their assigned uncertainties, the reader is referred to the original paper.

VII. Thermodynamic Definitions

Classical thermodynamics, as distinct from statistical thermodynamics, describes phenomena in which temperature plays an important part. The thermodynamic processes involve transformation of energy from one form to another, governed by the thermodynamic laws (i.e. the zeroth law, the first law, the second law, and the third law). Each of the laws constitutes an axiomatic generalization obtained on the basis of experience but these laws cannot be directly and exhaustively verified by suitable experiments. Their truth is ascertained from the fact that none of the consequences derived from them have failed to be verified experimentally.

Thermodynamics makes a clear and systematic distinction between real irreversible phenomena and idealized reversible processes.

1. Thermodynamic System

A thermodynamic system is any collection of matter, enclosed in a clearly defined boundary in terms of which a given analysis is performed. For the description of a particular thermodynamic process the choice of the system is arbitrary and, in fact, the boundary of the system does not need to be rigid and may be drawn in any manner. Neither does the system need to contain the same material object at all times. A system is homogeneous if it has the same properties throughout its extension, and heterogeneous if it is composed of several phases.

A system is called closed if its ideal or real walls are impermeable to matter. A close system contains the same collection of material objects all the time. A closed system whose boundary is also impermeable to the passage of heat and work is called an isolated system. Alternatively, the designation is applied also to closed systems whose boundaries are adiabatic and impermeable to the passage of heat only.

A system is called open if it is bounded by ideal or real walls some of whose parts are crossed by matter, and its boundary is called a control surface. The open system is said to be in a steady state if the state of any given point within its control surface is independent of time. In particular, the mass and energy must remain constant within the control surface of an open system in steady state.

The state variables associated with a system are the macroscopic variables that describe its behaviour. These may be either internal variables (pressure, volume, masses) or external variables (such as macroscopic velocities or position coordinates in an external field). An internal state variable that does not depend on the quantity of matter of the region considered is called an intensive property; such a property is constant throughout a homogeneous system (e.g. pressure, density, electric field strength). An internal state variable that depends on the quantity of matter in the region considered is called an extensive property; such a property is a first degree function of the masses (e.g. mass, volume, total electric moment).

If a system subject to external influences passes through a continuous series of equilibrium states and if all actions produced by the system are at all times exactly counterbalanced by external actions, the process is called quasistatic.

2. *Thermodynamic Equilibrium*

The general condition for thermodynamic equilibrium is that all rates of irreversible processes vanish, that is production of entropy vanishes.

The system is in internal equilibrium when all the intensive variables do not change in time by virtue of internal causes. A subsystem of a system can be in internal equilibrium without being in equilibrium within the system.

The steady state is a non-equilibrium state in which the intensive properties are maintained independent of time by external forces.

If a system is in internal equilibrium and is completely surrounded by a wall and if no change can be produced in the system except by work done on the system, the wall is called a thermally insulating wall and any process producing a change of state is called an adiabatic process.

3. *Reversible and Irreversible Processes*

A reversible thermodynamic process is an ideal transformation of state produced by the process such that the system can return from the final state to its initial state without any change in the surroundings.

If we consider a transformation along a path ABC, the change is reversible if there exists a path CBA such that the variables characterizing the state of the system return to the same values in the inverse order and exchanges of heat, work and matter with the surroundings are of the reverse sign and take place in the reverse order. All quasistatic processes are reversible.

If, after a change of state the initial state of the system cannot be restored without changes in the surroundings, the transformation is irreversible. All changes which take place in nature are irreversible.

4. *Zeroth Law of Thermodynamics*

If two systems are in thermal equilibrium with a third across a thermally conducting wall, then they are in thermal equilibrium with each other.

All systems in thermal equilibrium with one another have the same temperature.

5. *Equation of State*

A. Definition

An equation of state is any equation which connects a series of states of matter by a relationship between pressure, volume and temperature. Specifically:

$$p = T\left(\frac{\partial p}{\partial T}\right)_V - \left(\frac{\partial U}{\partial V}\right)_T \tag{1}$$

$$V = T\left(\frac{dV}{dT}\right)_p + \left(\frac{dH}{dp}\right)_T \tag{2}$$

For a perfect gas (any gas at a pressure sufficiently low to render its properties essentially indistinguishable from those which it would have at zero pressure) we have

$$pV = nRT \tag{3}$$

where n is the amount of substance.

B. Virial Coefficients

Virial coefficients are the coefficients in the expansion of the compressibility factor pV of an imperfect gas in powers of the density $1/V$

$$pV = RT[1 + B_{(V)}/V + C_{(V)}/V^2 \text{---}] \tag{4}$$

or in powers of the pressure p

$$pV = RT[1 + B_{(p)}p + C_{(p)}p^2 + \text{---}] \tag{5}$$

The density expansion is the more fundamental of the two. It can be proved that such an expansion exists for gases at moderate densities and that the consecutive coefficients can be in turn related to interactions between pairs, triplets, etc., of molecules.

The pressure virials are connected to the density virials through

$$B_{(p)} = B_{(V)}/RT \tag{6}$$

$$C_{(p)} = (C_{(V)} - B_{(V)}^2)/(RT)^2 \tag{7}$$

The Boyle temperature is the temperature at which the second virial coefficient, B, changes sign.

C. The Equation of State of the Partition Function

The partition function is an expression giving the distribution of molecules in different energy states in a system

$$Z = \Sigma q_r \exp(-\varepsilon_r/kT) \tag{8}$$

where q_r is the statistical weight of the rth state of energy level ε_r (the rth energy level is composed of q_r states), k is the Boltzmann constant, and T the absolute temperature.

The energy levels ε_r, may be attributed to rotation, translation, vibration and electronic energy.

In terms of the partition function the equation of state can be written

$$p = kT(\partial \ln Z/\partial V)_T \tag{9}$$

6. First Law of Thermodynamics

The first law can be interpreted as a law of energy conservation. The work performed adiabatically in any reversible or irreversible process between two states of equilibrium 1 and 2 of any closed system depends on the end states 1 and 2 and is independent of the particular details of the process $1 \rightarrow 2$ (Carathéodory's definition).

Mathematically this means that the integral $\int_1^2 dW_{ad}$ is that of a perfect differential and consequently

$$\int_1^2 dW_{ad} = U_1 - U_2 \tag{10}$$

We write $U_1 - U_2$ in order to follow the convention that work done by the system is positive and that in the process the energy of the system decreases. When the process is not adiabatic the previous equation can be generalized by introducing the heat Q

$$\int_1^2 dQ = U_2 - U_1 + \int_1^2 dW \tag{11}$$

7. Internal Energy, U

The total energy of a system may be considered as consisting of two parts:

(1) External energy: energy that may be associated with the motion and position of the center of gravity of the system with respect to the observer.

(2) Internal energy: energy that the particles composing the system have by virtue of their mutual attractions and motions relative to a set of axes at the center of gravity.

On a macroscopic scale the internal energy U is measured as the maximum amount of work in an adiabatic process required to transform the system from one state to a reference state at the absolute zero of temperature. In general the temperature of the reference state differs from the absolute zero, so we define only a variation of internal energy ΔU.

In non-adiabatic processes the difference between the work and the internal energy change is defined to be heat.

$$\Delta U = U_2 - U_1 = W + Q \text{ (closed system)} \tag{12}$$

For a given system the internal energy U is a function of the state of the system and not of the process, and is completely defined, apart from a constant generally taken as arbitrary,

$$U = \Delta U + \text{arbitrary constant} \tag{13}$$

For an isolated system, $U = $ constant (law of conservation of energy).
The internal energy can be calculated using

$$(\partial U/\partial V)_T = T(\partial p/\partial T)_V - p \tag{14}$$

or, for a multicomponent system, by writing

$$dU = T\,dS - p\,dV + \Sigma_i \mu_i\, dn_i \tag{15}$$

in which the quantity μ_i is the chemical potential of component i. Chemical potentials are defined by

$$\mu_i = (\partial U/\partial n_i)_{S,v,n_j} \tag{16}$$

and are intensive variables.

In terms of the virial coefficients:

$$U - U^\circ = -RT\left[\frac{T}{V}\frac{dB_{(V)}}{dT} + \frac{T}{2V^2}\frac{dC_{(V)}}{dT} + \cdots\right] \tag{17}$$

In terms of the partition function Z the internal energy is given by

$$U = kT^2(\partial \ln Z/\partial T)_V \tag{18}$$

8. Enthalpy

The enthalpy is a thermodynamic potential defined as

$$H = U + pV \tag{19}$$

which may be differentiated

$$(\partial H/\partial p)_T = V - T(\partial V/\partial T)_p \tag{20}$$

For a multicomponent system

$$dH = T\,dS + V\,dp + \Sigma_i \mu_i\,dn_i \tag{21}$$

In terms of the virial coefficients[15]

$$H - H^\circ = RT\left[\frac{1}{V}\left\{B_{(V)} - T\frac{dB_{(V)}}{dT}\right\}\right.$$
$$\left. + \frac{1}{V^2}\left\{C_{(V)} - \frac{1}{2}T\frac{dC_{(V)}}{dT}\right\} + \cdots\right] \tag{22}$$

In terms of the partition function

$$H = kT^2\left(\frac{\partial \ln Z}{\partial T}\right)_V + kTV\left(\frac{\partial \ln Z}{\partial V}\right)_T \tag{23}$$

9. Second Law of Thermodynamics—Entropy

The second law is contained in two statements:

A. The Carnot Theorem

The quantity of heat dQ exchanged by any (closed) thermodynamic system during an infinitesimal reversible process can be represented as a product $dQ = T\,dS$ of the thermodynamic temperature T and of the perfect differential dS of entropy. In other words the differential expression of heat dQ is integrable.

B. The Principle of Entropy Increase

In an isolated system, the entropy of all the bodies contained in it remains constant during a reversible process, increases during an irreversible (natural) process and can never decrease.

$$\Delta S \geqslant 0 \tag{24}$$

An alternative and equivalent formulation asserts that in any closed system (not necessarily isolated) the change in entropy dS is composed of two terms

$$dS = d_e S + d_i S \tag{25}$$

The term due to an exchange of heat with the surroundings

$$d_e S = (1/T)\,dQ \tag{26}$$

can be positive, negative or zero depending on dQ. The term $d_i S$ is due to processes taking place within the system, can never be negative and is positive when the process is irreversible and zero when it is reversible.

All formulations assert that natural processes are irreversible and occur spontaneously in one direction only. The Carathéodory formulation of the second law generalizes the common feature of all irreversible processes, i.e. the existence of inaccessible states in adiabatic processes. Thus, if an adiabatic process from state 1 to state 2 is irreversible, then the initial state 1 cannot be reached adiabatically from state 2, no matter how close state 2 is to state 1; state 1 is therefore inaccessible from state 2.

In one-component closed systems the total differential entropy is given by

$$dS = (1/T)dU + (p/T)\,dV \tag{27}$$

with

$$(\partial S/\partial p)_T = -(\partial V/\partial T)_p \tag{28}$$

and

$$(\partial S/\partial V)_T = (\partial p/\partial T)_V \tag{29}$$

Equation 27 can be generalized for multicomponent systems, resulting in the fundamental Gibbs formula

$$dS = (1/T)dU + (p/T)dV - \Sigma_i (1/T)\mu_i \,dn_i \tag{30}$$

In terms of the virial coefficients[15]

$$S - S^\circ = -R\left[\ln p + \frac{T}{V}\frac{dB_{(V)}}{dT} + \frac{B_{(V)}^2}{2V^2} - \frac{C_{(V)}}{2V^2} + \frac{T}{2V^2}\frac{dC_{(V)}}{dT} + \cdots\right] \tag{31}$$

In terms of the partition function

$$S = k \ln Z + (U/T) \tag{32}$$

10. Third Law of Thermodynamics

This principle, due to Nernst, can be enumerated as follows. It is impossible by any procedure, no matter how idealized, to reach absolute zero in a finite number of operations.

11. Free Energy

The free energy is the maximum amount of work that the system can do in coming to static equilibrium.

A. HELMOLTZ FREE ENERGY, A

The Helmholtz free energy is a thermodynamic potential defined by

$$A = U - TS \tag{33}$$

In differential form, this becomes

$$dA = -S\,dT - p\,dV$$

and so the decrease in the Helmholtz free energy is the maximum work done on a system in a constant temperature reversible process.
For a multicomponent system

$$dA = -S\,dT - p\,dV + \Sigma_i \mu_i\,dn_i \tag{34}$$

In terms of the partition function

$$A = -kT \ln Z \tag{35}$$

The Helmholtz free energy is of considerable importance in statistical mechanics because of the simple way in which it is related to the partition function.

B. GIBBS FREE ENERGY, G

By definition

$$G = U - TS + pV = H - TS \tag{36}$$

For a multicomponent system

$$dG = -S\,dT + V\,dp + \Sigma_i \mu_i\,dn_i \tag{37}$$

In terms of the partition function

$$G = -kT \ln Z + kTV\,(\partial \ln Z/\partial V)_T \tag{38}$$

12. Heat Capacity

If C_V is the heat capacity at constant volume and C_p the heat capacity at constant pressure:

$$C_V = T(\partial S/\partial T)_V = (\partial H/\partial T)_V \tag{39}$$

$$C_p = T(\partial S/\partial T)_p = (\partial H/\partial T)_p \tag{40}$$

$$C_p - C_V = T(\partial p/\partial T)_V (\partial V/\partial T)_p \tag{41}$$

or, for the coefficient of volume expansion,

$$\alpha = (1/V)(\partial V/\partial T)_p \tag{42}$$

and, for the coefficient of isothermal compressibility,

$$\kappa = -(1/V)(\partial V/\partial p)_T \tag{43}$$

so that

$$C_p - C_V = TV(\alpha^2/\kappa) \tag{44}$$

In an adiabatic change, however, the coefficient of adiabatic compressibility is given by

$$\chi = -(1/V)(\partial V/\partial p)_S \tag{45}$$

and

$$(\chi/\kappa) = (C_V/C_p) = (1/\gamma) \tag{46}$$

In terms of the virial coefficients

$$C_V - C_V^\circ = -R\left[\frac{2T}{V}\frac{dB_{(V)}}{dT} + \frac{T^2}{V}\frac{d^2B_{(V)}}{dT^2} + \frac{T}{V^2}\frac{dC_{(V)}}{dT} + \frac{T^2}{2V^2}\frac{d^2C_{(V)}}{dT^2} + \cdots\right] \tag{47}$$

$$C_p - C_p^\circ = -(R/V^2)\left[T^2V\frac{d^2B_{(V)}}{dT^2} - B_{(V)} - T\frac{dB_{(V)}}{dT} - C_{(V)} + \frac{T\,dC_{(V)}}{dT}\right.$$
$$\left. -\frac{T^2}{2}\frac{d^2C_{(V)}}{dT^2} + \cdots\right] \tag{48}$$

In terms of the partition function

$$C_V = (\partial/\partial T)(kT^2\{\partial \ln Z/\partial T\}_V) \tag{49}$$

13. Joule–Thomson Coefficient

From $\mu = (\partial T/\partial p)_H = -(\partial H/\partial p)_T/(\partial H/\partial T)_p$, we derive (50)

$$\mu = (\partial T/\partial p)_H = -V(1 - \alpha T)/C_p \tag{51}$$

In terms of virial coefficients

$$\mu = \frac{1}{C_p^\circ}\left\{T\frac{dB_{(V)}}{dT} - B_{(V)} + \frac{1}{V}\left[2B_{(V)}^2 - 2TB_{(V)}\frac{dB_{(V)}}{dT} - 2C_{(V)}\right.\right.$$
$$\left.\left. + T\frac{dC_{(V)}}{dT} + \frac{RT^2}{C_p^\circ}\frac{d^2B_{(V)}}{dT^2}\left(T\frac{dB_{(V)}}{dT} - B_{(V)}\right)\right] + \cdots\right\} \tag{52}$$

14. Speed of Sound

The speed of propagation of sound c_0 (at zero frequency) can be calculated by the following formula

$$c_0 = (V/M\chi)^{\frac{1}{2}} \tag{53}$$

VIII. Thermodynamic Data Tables

1. Table of atomic weights (1971)

Table 12. Atomic Weights 1971

Scaled to the relative atomic mass $A_r(^{12}C) = 12$

The values of $A_r(E)$ given here apply to elements as they exist in materials of terrestrial origin and to certain artificial elements. When used with due regard to the footnotes they are considered reliable to ± 1 in the last digit, or ± 3 if that digit is subscript.

Alphabetical Order in English[21]

Name	Symbol	Atomic number	Atomic weight	Name	Symbol	Atomic number	Atomic weight
Actinium	Ac	89	—	Mercury	Hg	80	200.5_9
Aluminum	Al	13	26.98154^a	Molybdenum	Mo	42	95.9_4
Americium	Am	95	—	Neodymium	Nd	60	144.2_4
Antimony	Sb	51	121.7_5	Neon	Ne	10	$20.17_9{}^c$
Argon	Ar	18	$39.94_8{}^{b,c,d,g}$	Neptunium	Np	93	237.0482^f
Arsenic	As	33	74.9216^a	Nickel	Ni	28	58.7_1
Astatine	At	85	—	Niobium	Nb	41	92.9064^a
Barium	Ba	56	137.3_4	Nitrogen	N	7	$14.0067^{b,c}$
Berkelium	Bk	97	—	Nobelium	No	102	—
Beryllium	Be	4	9.01218^a	Osmium	Os	76	190.2
Bismuth	Bi	83	208.9804^a	Oxygen	O	8	$15.999_4{}^{b,c,d}$
Boron	B	5	$10.81_6{}^{c,d,e}$	Palladium	Pd	46	106.4
Bromine	Br	35	79.904^c	Phosphorus	P	15	30.97376^a
Cadmium	Cd	48	112.40	Platinum	Pt	78	195.0_9
Calcium	Ca	20	40.08^g	Plutonium	Pu	94	—
Californium	Cf	98	—	Polonium	Po	84	—
Carbon	C	6	$12.011^{b,d}$	Potassium	K	19	39.09_8
Cerium	Ce	58	140.12	Praseodymium	Pr	59	140.9077^a
Cesium	Cs	55	132.9054^a	Promethium	Pm	61	—
Chlorine	Cl	17	35.453^c	Protactinium	Pa	91	231.0359^f
Chromium	Cr	24	51.996^c	Radium	Ra	88	$226.0254^{f,g}$
Cobalt	Co	27	58.9332^a	Radon	Rn	86	—
Copper	Cu	29	$63.54_6{}^{c,d}$	Rhenium	Re	75	186.2
Curium	Cm	96	—	Rhodium	Rh	45	102.9055^a
Dysprosium	Dy	66	162.5_0	Rubidium	Rb	37	$85.467_8{}^c$

GENERAL INTRODUCTION

Element	Symbol	Z	$A_r(E)$
Einsteinium	Es	99	—
Erbium	Er	68	167.2_6
Europium	Eu	63	151.96
Fermium	Fm	100	—
Fluorine	F	9	18.99840^a
Francium	Fr	87	—
Gadolinium	Gd	64	157.2_5
Gallium	Ga	31	69.72
Germanium	Ge	32	72.5_9
Gold	Au	79	196.9665^a
Hafnium	Hf	72	178.4_9
Helium	He	2	$4.00260^{b,c}$
Holmium	Ho	67	164.9304^a
Hydrogen	H	1	$1.0079_9^{b,d}$
Indium	In	49	114.82
Iodine	I	53	126.9045^a
Iridium	Ir	77	192.2_2
Iron	Fe	26	55.84_7
Krypton	Kr	36	83.80
Lanthanum	La	57	138.905_5^b
Lawrencium	Lr	103	—
Lead	Pb	82	$207.2^{d,g}$
Lithium	Li	3	$6.94_1^{c,d,e,g}$
Lutetium	Lu	71	174.97
Magnesium	Mg	12	$24.305^{c,g}$
Manganese	Mn	25	54.9380^a
Mendelevium	Md	101	—
Ruthenium	Ru	44	101.0_7
Samarium	Sm	62	150.4
Scandium	Sc	21	44.9559^a
Selenium	Se	34	78.9_6
Silicon	Si	14	28.08_6^d
Silver	Ag	47	107.868^c
Sodium	Na	11	22.98977^a
Strontium	Sr	38	87.62^g
Sulfur	S	16	32.06^d
Tantalum	Ta	73	180.947_9^b
Technetium	Tc	43	—
Tellurium	Te	52	127.6_0
Terbium	Tb	65	158.9254^a
Thallium	Tl	81	204.3_7
Thorium	Th	90	232.0381^f
Thulium	Tm	69	168.9342^a
Tin	Sn	50	118.6_9
Titanium	Ti	22	47.9_0
Tungsten	W	74	183.8_5
Uranium	U	92	$238.029^{b,c,e}$
Vanadium	V	23	$50.941_4^{b,c}$
Wolfram	W	74	183.8_5
Xenon	Xe	54	131.30
Ytterbium	Yb	70	173.0_4
Yttrium	Y	39	88.9059^a
Zinc	Zn	30	65.38
Zirconium	Zr	40	91.22

[a] Element with only one stable nuclide.
[b] Element with one predominant isotope (about 99 to 100 per cent abundance), errors in abundance determinations have a correspondingly small effect on the confidence in the value of $A_r(E)$.
[c] Element for which the value of $A_r(E)$ derives its reliability from calibrated measurements (i.e. from comparisons with synthetic mixtures of known isotopic composition).
[d] Element for which known variations in isotopic abundance in terrestrial material prevent a more precise atomic weight being given; $A_r(E)$ values should be applicable to any 'normal' material.
[e] Element for which values of A_r may be found in commercially available products that differ from the tabulated value of $A_r(E)$ because of inadvertent or undisclosed changes of isotopic composition.
[f] Element for which the value of A_r is that of the most commonly available long-lived nuclide.
[g] Element for which geological specimens are known in which the element has an anomalous isotopic composition.

where χ is the adiabatic compressibility and V the volume.

For a perfect gas

$$c_0 = (\gamma p/\rho)^{\frac{1}{2}} \tag{54}$$

where

$$\gamma = C_p/C_V.$$

In terms of virial coefficients

$$c^2 = \frac{\gamma^\circ RT}{M}\left\{1 + \frac{1}{V}\left[2B_{(V)} + 2(\gamma^\circ - 1)T\frac{dB_{(V)}}{dT} + \frac{(\gamma^\circ - 1)^2}{\gamma^\circ}\right.\right.$$
$$\left.\left. \times T^2\frac{d^2B_{(V)}}{dT^2}\right] + \cdots\right\} \tag{55}$$

2. Critical Constants

Table 13. Table of critical constants

Substance	Formula	T_c K	p_c 10^5 N m^{-2}	ρ_c kg m^{-3}	Ref.
Air		132.55	37.66	313	3
Ammonia	NH_3	405.55	113.5	234	10
Argon	A	150.86	48.97	535.9	27
Bromine	Br_2	575.1	133	1 180	35
Carbon dioxide	CO_2	304.14 (± 0.04)	73.81	468.4	25
Carbon monoxide	CO	132.92	34.98	301	24
Cesium[a]	Cs	2 050	117	407	30
Chlorine	Cl_2	417.1	77	573	35
Chlorotrifluoromethane	$CClF_3$	301.9	39.5	580	25
Deuterium (equilibrium)	D_2	38.262	16.50	66.86	20
Deuterium (para)	D_2	38.25	16.82		
Ethane	C_2H_6	305.4	48.8	203	24
Fluorine	F_2	144.31	52.2	574	33
Heavy water	D_2O	643.81 (± 0.03)	216.59 (± 0.03)	337	5
Helium	^{4}He	5.189 (± 0.002)	2.29	69.1	8, 35, 39
Helium-3	^{3}He	3.3094 (± 0.001)	1.147	41.38	40
Hydrogen (equilibrium)	H_2	32.994	12.94	30.77	20
Hydrogen (para)	H_2	32.934	12.96	31.42	16, 28, 35
Hydrogen deuteride	HD	35.908	14.84	48.15	20
Iodine	I_2	826.1			
Krypton	Kr	209.4	55.0	908.0	8
Lithium	Li	3 223 (± 600)	680	120 (± 33)	9
Methane	CH_4	190.53	45.947	162.7	17
Mercury	Hg	1 763 (± 15)	1 490	5 300	18
n-Butane	C_4H_{10}	425.2	38.0	228	24
Neon	Ne	44.40	26.9	483.0	8, 16
n-Hexane	C_6H_{14}	507.4	29.7	233	24

GENERAL INTRODUCTION

Table 13—continued

Substance	Formula	T_c K	p_c $10^5 \mathrm{Nm^{-2}}$	ρ_c $\mathrm{kg\,m^{-3}}$	Ref.
Nitrogen	N_2	126.0	33.9	311.0	35
Nitrous oxide	N_2O	309.57	72.7	452	25
n-Pentane	C_5H_{12}	469.6	33.7	237	24
Oxygen	O_2	154.576	50.35	436.2	41
Potassium[b]	K	2 198	155	194	9, 12
Propane	C_3H_8	369.8	42.5	217	24
Radon	R	378	62.8		35
Rubidium[c]	Rb	2 093	159	346	9
Sodium	Na	2 573	275	206	9
Sulfur hexafluoride	SF_6	318.707 (± 0.002)	37.6	725	11
Water	H_2O	647.06 (± 0.03)	220.45 (± 0.03)	315	5
Xenon	Xe	289.736	58.4	1112.4	39

(a) $T_c = 2047.7$; $p_c = 117.3$; $\rho_c = 420$
(b) $T_c = 2280.8$; $p_c = 164.0$
(c) $T_c = 2106$; $p_c = 134.0$
in *Proceedings of the Sixth Symposium on Thermophysical Properties*, pp 387, 358, 362. Americal Society of Mechanical Engineers: New York (1973).

3. Pressure–Volume–Temperature Relationships of Gases—Virial Coefficients

Tables of second virial coefficients as functions of temperature are presented for several gases[26]. These tables are based on a re-examination of the original pVT data. In the case of eight of these gases, the experimental second virial coefficients have been fitted to those predicted by the $(m, 6)$ potential function and potential parameters determined. In almost all cases, an optimum fit was obtained for $m = 18$. The optimum potential function was used for extrapolating the tables and for obtaining temperature derivatives of B. Of great interest is the fundamental relationship of virial coefficients to the molecular interaction. If the molecular field is represented by a function $\phi(r)$ where r specifies the relative coordinates of two molecules[19], then

$$B_V(T) = 2\pi N_A \int_0^\infty \{1 - \exp[-\phi(r)/kT]\} r^2 \, dr \qquad (56)$$

The virial $B_V(T)$ is uniquely determined through equation 56 if the molecular interaction $\phi(r)$ is known, but the reverse is not true. Higher virials can likewise be related to interaction between triplets etc. of interacting molecules. These expressions for the higher virials are less useful not only because the higher virials are less well known experimentally, but also because the influence of potential function non-additivity on these virials is less well known theoretically.

The average virials presented in these tables can be used for calculations of precise pV products at low pressures, and the initial density dependence of other thermodynamic properties. If precise pVT values are needed at higher densities, it is usually preferable to interpolate through the original data.

Table 14. Second virial coefficient of argon and its temperature derivatives

T K	B cm^3/mol	$T\,dB/dT$ cm^3/mol	$T^2\,d^2B/dT^2$ cm^3/mol	T K	B cm^3/mol	$T\,dB/dT$ cm^3/mol	$T^2\,d^2B/dT^2$ cm^3/mol
80.00	−288.0	577.	−1954.	172.00	−66.9	135.	−345.
82.00	−274.2	544.	−1820.	176.00	−63.8	130.	−331.
84.00	−261.4	514.	−1700.	180.00	−60.9	126.	−318.
86.00	−249.7	488.	−1592.	190.00	−54.4	116.	−290.
88.00	−238.7	463.	−1495.	200.00	−48.7	107.	−266.
90.00	−228.6	441.	−1408.	210.00	−43.7	100.	−245.
92.00	−219.1	420.	−1328.	220.00	−39.2	93.	−228.
94.00	−210.3	401.	−1256.	230.00	−35.2	88.	−212.
96.00	−202.0	384.	−1190.	240.00	−31.5	83.	−198.
98.00	−194.3	367.	−1129.	250.00	−28.2	78.	−186.
100.00	−187.0	352.	−1074.	260.00	−25.3	74.	−176.
102.00	−180.2	338.	−1025.	273.15	−21.7	69.	−163.
104.00	−173.8	325.	−976.	280.00	−20.1	67.	−157.
106.00	−167.7	313.	−932.	300.00	−15.7	61.	−142.
108.00	−161.9	302.	−892.	320.00	−11.9	56.	−130.
110.00	−156.5	292.	−855.	340.00	−8.7	51.	−119.
112.00	−151.3	282.	−820.	360.00	−5.8	48.	−110.
114.00	−146.4	272.	−788.	373.15	−4.2	46.	−105.
116.00	−141.7	264.	−757.	380.00	−3.4	44.	−102.
118.00	−137.3	255.	−729.	400.00	−1.1	42.	−96.
120.00	−133.1	247.	−702.	411.52	0.0	40.	−92.
124.00	−125.2	233.	−654.	450.00	3.4	36.	−82.
128.00	−118.0	220.	−612.	500.00	6.9	31.	−71.
132.00	−111.4	209.	−573.	550.00	9.7	28.	−63.
136.00	−105.4	198.	−539.	600.00	11.9	25.	−57.
140.00	−99.8	188.	−509.	700.00	15.4	20.	−47.
144.00	−94.6	180.	−481.	800.00	17.8	17.	−40.
148.00	−89.8	172.	−456.	900.00	19.7	14.	−34.
152.00	−85.3	164.	−433.	1000.00	21.1	12.	−30.
156.00	−81.1	157.	−413.	1100.00	22.2	11.	−27
160.00	−77.2	151.	−394.	1300.00	23.8	8.	−21
164.00	−73.5	145.	−376.	1500.00	24.8	7.	−18.
168.00	−70.1	140.	−360.				

Table 15. Second virial coefficient of carbon dioxide (CO_2)

T K	B cm^3/mol	T K	B cm^3/mol	T K	B cm^3/mol
250.00	−181.8	350.00	−84.7	540.00	−21.9
255.00	−174.1	360.00	−79.0	560.00	−18.4
260.00	−166.8	370.00	−73.8	580.00	−15.3
265.00	−160.0	373.15	−72.2	600.00	−12.4
270.00	−153.5	380.00	−68.9	620.00	−9.8
273.15	−149.7	390.00	−64.4	640.00	−7.4
275.00	−147.4	400.00	−60.2	660.00	−5.1
280.00	−141.7	410.00	−56.3	680.00	−3.1
285.00	−136.2	420.00	−52.6	700.00	−1.3
290.00	−131.1	430.00	−49.1	714.81	0.0
295.00	−126.2	440.00	−45.9	750.00	2.7
300.00	−121.5	450.00	−42.8	800.00	6.0
310.00	−112.8	460.00	−40.0	850.00	8.8
320.00	−104.8	480.00	−34.7	900.00	11.1
330.00	−97.5	500.00	−30.0	950.00	13.0
340.00	−90.8	520.00	−25.8	1000.00	14.6

Table 16. Second virial coefficient of dry CO_2-free air and its temperature derivatives

T K	B cm^3/mol	$T\,dB/dT$ cm^3/mol	$T^2\,d^2B/dT^2$ cm^3/mol	T K	B cm^3/mol	$T\,dB/dT$ cm^3/mol	$T^2\,d^2B/dT^2$ cm^3/mol
100.00	−167.3	318.	−935.	210.00	−34.5	95.	−230.
102.00	−161.2	307.	−893.	220.00	−30.2	89.	−214.
104.00	−155.3	295.	−854.	230.00	−26.4	84.	−200.
106.00	−149.8	285.	−818.	240.00	−22.9	79.	−187.
108.00	−144.6	275.	−785.	250.00	−19.8	75.	−176.
110.00	−139.6	266.	−754.	260.00	−16.9	71.	−166.
112.00	−134.9	258.	−725.	273.15	−13.5	66.	−155.
114.00	−130.4	249.	−698.	280.00	−11.9	64.	−150.
116.00	−126.1	242.	−673.	300.00	−7.7	58.	−136.
118.00	−122.0	235.	−649.	320.00	−4.1	54.	−124.
120.00	−118.2	228.	−627.	340.00	−1.0	50.	−114.
124.00	−110.9	215.	−586.	346.81	0.0	48.	−111.
128.00	−104.3	204.	−550.	360.00	1.7	46.	−106.
132.00	−98.1	193.	−517.	373.15	3.4	44.	−101.
136.00	−92.5	184.	−488.	380.00	4.2	43.	−98.
140.00	−87.3	176.	−462.	400.00	6.3	40.	−92.
144.00	−82.5	168.	−438.	450.00	10.7	35.	−79.
148.00	−78.0	161.	−416.	500.00	14.1	30.	−69.
152.00	−73.8	154.	−396.	550.00	16.8	27.	−61.
156.00	−69.9	148.	−378.	600.00	19.0	24.	−55.
160.00	−66.2	142.	−361.	650.00	20.8	21.	−50.
166.00	−61.1	134.	−339.	700.00	22.3	19.	−45.
172.00	−56.5	127.	−319.	800.00	24.7	16.	−38.
178.00	−52.2	121.	−301.	900.00	26.4	14.	−33.
184.00	−48.3	115.	−284.	1000.00	27.8	12.	−29.
190.00	−44.7	110.	−270.	1200.00	29.7	9.	−23.
200.00	−39.3	102.	−248.	1400.00	30.9	7.	−19.

GENERAL INTRODUCTION

Table 17. Second virial coefficient of deuterium

T K	B cm³/mol	T K	B cm³/mol	T K	B cm³/mol
84.00	−10.4	150.00	6.0	280.00	13.2
88.00	−8.7	160.00	7.1	300.00	13.5
92.00	−7.0	170.00	8.1	320.00	14.0
96.00	−5.6	180.00	8.9	340.00	14.4
100.00	−4.2	190.00	9.5	360.00	14.7
110.00	−1.3	200.00	10.2	373.15	14.9
115.00	0.0	220.00	11.3	380.00	15.0
120.00	1.0	240.00	12.2	400.00	15.2
130.00	3.0	260.00	12.8	420.00	15.5
140.00	4.6	273.15	13.1		

Table 18. Second virial coefficient of heavy water vapor (D_2O)

T K	B cm³/mol	T K	B cm³/mol	T K	B cm³/mol
432.00	−314.5	466.00	−227.6	570.00	−119.6
434.00	−307.8	468.00	−224.2	580.00	−113.6
436.00	−301.3	470.00	−220.5	590.00	−108.1
438.00	−295.0	475.00	−212.2	600.00	−103.0
440.00	−288.6	480.00	−204.4	610.00	−98.0
442.00	−282.6	485.00	−197.2	620.00	−93.3
444.00	−276.9	490.00	−190.3	630.00	−88.8
446.00	−271.3	495.00	−183.8	640.00	−84.6
448.00	−265.9	500.00	−177.6	650.00	−80.6
450.00	−260.8	505.00	−171.8	660.00	−76.7
452.00	−255.9	510.00	−166.3	670.00	−73.1
454.00	−251.4	515.00	−161.1	680.00	−69.5
456.00	−247.0	520.00	−156.4	690.00	−66.2
458.00	−242.8	530.00	−147.7	700.00	−63.0
460.00	−238.8	540.00	−140.0	710.00	−59.9
462.00	−234.8	550.00	−132.8	720.00	−57.0
464.00	−231.1	560.00	−126.1		

Table 19. Second virial coefficient of helium

T K	B cm³/mol	T K	B cm³/mol	T K	B cm³/mol
9.00	−26.0	22.00	−0.5	100.00	11.4
10.00	−21.7	22.64	0.0	120.00	11.8
11.00	−18.1	24.00	0.9	160.00	12.3
12.00	−15.2	26.00	2.0	200.00	12.3
13.00	−12.7	28.00	3.0	273.15	12.0
14.00	−10.5	30.00	3.8	373.15	11.3
15.00	−8.7	35.00	5.4	400.00	11.1
16.00	−7.1	40.00	6.6	600.00	10.4
17.00	−5.6	45.00	7.5	800.00	9.8
18.00	−4.3	50.00	8.2	1 000.00	9.3
19.00	−3.2	60.00	9.2	1 200.00	8.8
20.00	−2.2	80.00	10.6	1 400.00	8.4

Table 20. Second virial coefficient of krypton and its temperature derivatives

T K	B cm^3/mol	$T\,dB/dT$ cm^3/mol	$T^2\,d^2B/dT^2$ cm^3/mol	T K	B cm^3/mol	$T\,dB/dT$ cm^3/mol	$T^2\,d^2B/dT^2$ cm^3/mol
106.00	−394.3	807.	−2813.	215.00	−102.4	198.	−521.
108.00	−379.6	771.	−2659.	220.00	−97.9	191.	−499.
110.00	−365.8	737.	−2518.	230.00	−89.8	178.	−459.
112.00	−352.8	706.	−2389.	240.00	−82.4	166.	−425.
114.00	−340.5	677.	−2270.	250.00	−75.9	156.	−396.
116.00	−329.0	650.	−2160.	260.00	−69.9	147.	−369.
118.00	−318.1	625.	−2059.	270.00	−64.5	139.	−346.
120.00	−307.8	601.	−1965.	273.15	−62.9	136.	−340.
122.00	−298.0	579.	−1878.	280.00	−59.6	132.	−326.
124.00	−288.8	559.	−1797.	290.00	−55.1	125.	−308.
126.00	−280.0	539.	−1722.	300.00	−51.0	119.	−291.
128.00	−271.6	521.	−1651.	310.00	−47.2	113.	−276.
130.00	−263.7	504.	−1586.	320.00	−43.7	108.	−263.
132.00	−256.1	488.	−1524.	340.00	−37.4	100.	−239.
134.00	−248.9	473.	−1467.	360.00	−31.9	92.	−219.
136.00	−242.0	458.	−1413.	373.15	−28.7	87.	−207.
138.00	−235.4	444.	−1362.	380.00	−27.1	85.	−202.
140.00	−229.1	432.	−1314.	400.00	−22.9	80.	−187.
144.00	−217.3	408.	−1226.	420.00	−19.1	75.	−175.
148.00	−206.4	386.	−1148.	440.00	−15.8	70.	−163.
152.00	−196.4	366.	−1078.	460.00	−12.7	66.	−153.
156.00	−187.1	348.	−1015.	500.00	−7.5	59.	−137.
160.00	−178.5	332.	−959.	550.00	−2.2	52.	−120.
164.00	−170.5	317.	−907.	575.00	0.0	49.	−113.
168.00	−163.1	303.	−860.	600.00	2.0	47.	−107.
172.00	−156.1	290.	−817.	650.00	5.6	42.	−96.
176.00	−149.5	279.	−778.	700.00	8.5	38.	−88.
180.00	−143.4	268.	−742.	800.00	13.2	32.	−74.
186.00	−134.9	253.	−693.	900.00	16.7	28.	−64.
192.00	−127.1	239.	−650.	1000.00	19.5	24.	−56.
198.00	−119.9	227.	−611.	1100.00	21.6	21.	−50.
205.00	−112.2	214.	−570.	1300.00	24.8	17.	−40.
210.00	−107.2	206.	−545.	1500.00	27.0	14.	−34.

Table 21. Second virial coefficient of hydrogen

T K	B cm^3/mol	T K	B cm^3/mol	T K	B cm^3/mol
24.00	−112.8	46.00	−39.2	130.00	3.7
25.00	−106.2	48.00	−36.2	140.00	5.1
26.00	−100.3	50.00	−33.4	150.00	6.4
27.00	−94.8	54.00	−28.6	160.00	7.6
28.00	−89.6	58.00	−24.5	170.00	8.6
29.00	−85.0	62.00	−21.0	180.00	9.5
30.00	−80.7	66.00	−17.9	190.00	10.2
31.00	−76.7	70.00	−15.2	200.00	10.8
32.00	−73.0	74.00	−12.9	250.00	13.0
33.00	−69.5	78.00	−10.9	273.15	13.7
34.00	−66.2	82.00	−8.9	300.00	14.4
35.00	−63.2	86.00	−7.2	350.00	15.3
36.00	−60.2	90.00	−5.7	373.15	15.6
38.00	−55.0	100.00	−2.5	400.00	15.9
40.00	−50.3	110.04	0.0	420.00	16.1
42.00	−46.2	120.00	2.0		
44.00	−42.5				

Table 22. Second virial coefficient of methane and its temperature derivatives

T K	B cm³/mol	T dB/dT cm³/mol	T² d²B/dT² cm³/mol	T K	B cm³/mol	T dB/dT cm³/mol	T² d²B/dT² cm³/mol
110.00	−334.0	671.	−2244.	210.00	−95.3	193.	−505.
112.00	−322.2	643.	−2132.	220.00	−86.6	179.	−463.
114.00	−311.0	618.	−2029.	230.00	−78.9	167.	−428.
116.00	−300.5	594.	−1934.	240.00	−72.0	157.	−397.
118.00	−290.5	571.	−1846.	250.00	−65.8	147.	−369.
120.00	−281.1	550.	−1764.	260.00	−60.2	139.	−346.
122.00	−272.2	531.	−1688.	270.00	−55.2	131.	−324.
124.00	−263.7	512.	−1617.	273.15	−53.6	129.	−318.
126.00	−255.6	495.	−1551.	280.00	−50.5	124.	−306.
128.00	−248.0	479.	−1490.	290.00	−46.3	118.	−289.
130.00	−240.7	464.	−1432.	300.00	−42.3	113.	−273.
132.00	−233.7	449.	−1378.	320.00	−35.4	103.	−247.
134.00	−227.0	435.	−1327.	340.00	−29.4	94.	−225.
136.00	−220.7	422.	−1280.	360.00	−24.2	87.	−207.
140.00	−208.8	398.	−1193.	373.15	−21.2	83.	−196.
144.00	−197.9	377.	−1115.	380.00	−19.7	81.	−191.
148.00	−187.8	357.	−1046.	400.00	−15.7	76.	−177.
152.00	−178.5	340.	−984.	450.00	−7.4	65.	−150.
156.00	−169.9	323.	−928.	500.00	−1.1	56.	−130.
160.00	−161.9	309.	−877.	509.66	0.0	55.	−126.
164.00	−154.5	295.	−831.	550.00	4.0	50.	−114.
168.00	−147.5	282.	−789.	600.00	8.1	45.	−102.
172.00	−141.0	271.	−751.	650.00	11.5	40.	−92.
176.00	−134.9	260.	−716.	700.00	14.3	37.	−83.
180.00	−129.2	250.	−683.	800.00	18.8	31.	−70.
184.00	−123.8	241.	−653.	900.00	22.1	26.	−61.
188.00	−118.7	232.	−626.	1000.00	24.7	23.	−53.
192.00	−113.9	224.	−600.	1100.00	26.8	20.	−47.
196.00	−109.4	216.	−576.	1300.00	29.8	16.	−39.
200.00	−105.1	209.	−554.	1500.00	31.9	13.	−32.

Table 23. Second virial coefficient of neon and its temperature derivatives

T K	B cm^3/mol	$T\,\mathrm{d}B/\mathrm{d}T$ cm^3/mol	$T^2\,\mathrm{d}^2B/\mathrm{d}T^2$ cm^3/mol
80.00	−11.8	37.	−87.
90.00	−7.8	31.	−73.
100.00	−4.8	27.	−63.
110.00	−2.3	24.	−55.
120.00	−0.4	21.	−49.
122.11	0.0	21.	−48.
130.00	1.2	19.	−44.
140.00	2.6	18.	−40.
160.00	4.8	15.	−34.
200.00	7.6	11.	−26.
240.00	9.4	9.	−20.
273.15	10.4	7.	−17.
280.00	10.6	7.	−17.
320.00	11.5	6.	−14.
360.00	12.1	5.	−12.
373.15	12.3	5.	−12.
400.00	12.6	4.	−11.
500.00	13.3	3.	−8.
600.00	13.8	2.	−6.
700.00	14.0	1.	−5.
800.00	14.2	1.	−4.
900.00	14.3	1.	−3.
1000.00	14.3	0.	−3.

Table 24. Second virial coefficient of nitrogen and its temperature derivatives

T K	B cm³/mol	$T\,dB/dT$ cm³/mol	$T^2\,d^2B/dT^2$ cm³/mol	T K	B cm³/mol	$T\,dB/dT$ cm³/mol	$T^2\,d^2B/dT^2$ cm³/mol
100.00	−160.0	304.	−874.	210.00	−31.1	94.	−224.
102.00	−154.1	293.	−837.	220.00	−26.9	88.	−209.
104.00	−148.5	283.	−802.	230.00	−23.2	82.	−195.
106.00	−143.2	273.	−769.	240.00	−19.7	78.	−183.
108.00	−138.2	264.	−739.	660.00	−13.8	70.	−163.
110.00	−133.4	256.	−711.	273.15	−10.5	65.	−152.
112.00	−128.7	248.	−684.	280.00	−8.9	63.	−147.
116.00	−120.4	233.	−637.	200.00	−4.7	58.	−134.
120.00	−112.7	220.	−594.	320.00	−1.2	53.	−122.
124.00	−105.7	208.	−557.	327.22	0.0	52.	−119.
128.00	−99.3	197.	−524.	340.00	1.9	49.	−113.
132.00	−93.4	188.	−494.	360.00	4.6	46.	−105.
				373.15	6.2	44.	−100.
136.00	−87.9	179.	−467.	380.00	7.0	43.	−97.
140.00	−82.8	171.	−442.	400.00	9.1	40.	−91.
144.00	−78.1	163.	−420.				
148.00	−73.7	156.	−400.	450.00	13.5	34.	−78.
152.00	−69.7	150.	−381.	500.00	16.8	30.	−68.
156.00	−65.8	144.	−364.	550.00	19.5	26.	−61.
160.00	−62.3	139.	−349.	600.00	21.7	24.	−54.
166.00	−57.3	131.	−327.	700.00	25.0	19.	−45.
172.00	−52.7	125.	−308.	800.00	27.3	16.	−38.
178.00	−48.6	118.	−291.	900.00	29.1	14.	−33.
184.00	−44.7	113.	−276.	1000.00	30.4	12.	−29.
190.00	−41.2	108.	−262.	1200.00	32.3	9.	−23.
200.00	−35.9	100.	−242.	1400.00	33.5	7.	−19.

GENERAL INTRODUCTION

Table 25. Second virial coefficient of oxygen and its temperature derivatives

T K	B cm^3/mol	$T\,dB/dT$ cm^3/mol	$T^2\,d^2B/dT^2$ cm^3/mol	T K	B cm^3/mol	$T\,dB/dT$ cm^3/mol	$T^2\,d^2B/dT^2$ cm^3/mol
100.00	−197.5	383.	−1201.	210.00	−44.8	104.	−259.
102.00	−190.1	367.	−1141.	220.00	−40.1	97.	−240.
104.00	−183.1	352.	−1087.	230.00	−35.9	91.	−223.
106.00	−176.5	339.	−1036.	240.00	−32.1	86.	−208.
108.00	−170.3	326.	−989.	250.00	−28.7	81.	−195.
110.00	−164.4	314.	−946.	260.00	−25.6	77.	−184.
112.00	−158.9	303.	−906.	273.15	−22.0	72.	−171.
114.00	−153.6	293.	−868.	280.00	−20.2	69.	−164.
116.00	−148.6	283.	−833.	300.00	−15.7	63.	−148.
120.00	−139.3	265.	−770.	320.00	−11.8	58.	−135.
124.00	−130.9	249.	−715.	340.00	−8.4	53.	−124.
128.00	−123.2	235.	−667.	360.00	−5.5	49.	−114.
132.00	−116.2	222.	−624.	373.15	−3.7	47.	−109.
136.00	−109.7	210.	−585.	380.00	−2.9	46.	−106.
140.00	−103.8	200.	−551.	400.00	−0.6	43.	−99.
144.00	−98.3	190.	−520.	405.88	0.0	42.	−97.
148.00	−93.2	182.	−492.	450.00	4.1	37.	−85.
154.00	−86.2	170.	−454.	500.00	7.7	32.	−74.
160.00	−79.9	159.	−422.	550.00	10.6	29.	−65.
166.00	−74.2	150.	−393.	600.00	12.9	25.	−58.
172.00	−69.1	142.	−368.	700.00	16.5	21.	−48.
178.00	−64.3	134.	−346.	800.00	19.1	18.	−41.
184.00	−60.0	127.	−326.	1000.00	22.4	13.	−31.
190.00	−56.0	121.	−308.	1200.00	24.5	10.	−25.
200.00	−50.0	112.	−281.	1400.00	25.9	8.	−20.

Table 26. Second virial coefficient of water vapor (H_2O)

T K	B cm³/mol	T K	B cm³/mol	T K	B cm³/mol
432.00	−311.2	466.00	−225.4	570.00	−119.0
434.00	−304.5	468.00	−222.1	580.00	−113.1
436.00	−298.1	470.00	−218.5	590.00	−107.6
438.00	−291.9	475.00	−210.2	600.00	−102.5
440.00	−285.5	480.00	−202.5	610.00	−97.6
442.00	−279.7	485.00	−195.4	620.00	−93.0
444.00	−273.9	490.00	−188.6	630.00	−88.6
446.00	−268.5	495.00	−182.2	640.00	−84.4
448.00	−263.2	500.00	−176.2	650.00	−80.4
450.00	−258.2	505.00	−170.4	660.00	−76.6
452.00	−253.4	510.00	−165.0	670.00	−72.9
454.00	−248.9	515.00	−160.0	680.00	−69.4
456.00	−244.7	520.00	−155.3	690.00	−66.1
458.00	−240.5	530.00	−146.7	700.00	−62.9
460.00	−236.5	540.00	−139.1	710.00	−59.9
462.00	−232.6	550.00	−132.0	720.00	−57.0
464.00	−228.9	560.00	−125.3		

Table 27. Second virial coefficient of xenon and its temperature derivatives

T K	B cm³/mol	$T\,dB/dT$ cm³/mol	$T^2\,d^2B/dT^2$ cm³/mol
220.00	−230.7	429.	−1225.
225.00	−221.2	411.	−1166.
230.00	−212.4	395.	−1111.
235.00	−204.0	380.	−1061.
240.00	−196.2	366.	−1015.
245.00	−188.7	353.	−972.
250.00	−181.7	341.	−933.
255.00	−175.1	329.	−896.
260.00	−168.8	319.	−861.
265.00	−162.8	308.	−829.
270.00	−157.2	299.	−799.
273.15	−153.7	293.	−781.
280.00	−146.6	281.	−745.
290.00	−137.0	266.	−697.
300.00	−128.3	252.	−654.
310.00	−120.2	239.	−616.
320.00	−112.8	227.	−582.
330.00	−106.0	217.	−551.
340.00	−99.7	207.	−523.
350.00	−93.8	198.	−497.
360.00	−88.4	190.	−474.
370.00	−88.3	182.	−452.
373.15	−81.7	180.	−446.
380.00	−78.5	175.	−433.
390.00	−74.0	169.	−415.
400.00	−69.8	163.	−398.

Table 27.—continued

T K	B cm^3/mol	$T\,dB/dT$ cm^3/mol	$T^2 d^2 B/dT^2$ cm^3/mol
420.00	−62.2	152.	−368.
440.00	−55.4	142.	−342.
460.00	−49.2	133.	−319.
480.00	−43.7	126.	−299.
500.00	−38.8	119.	−282.
525.00	−33.1	111.	−262.
550.00	−28.1	104.	−245.
575.00	−23.6	98.	−230.
600.00	−19.6	93.	−215.
650.00	−12.5	84.	−193.
700.00	−6.6	76.	−175.
768.03	0.0	67.	−154.
800.00	2.7	64.	−146.
900.00	9.6	55.	−125.
1000.00	15.0	48.	−110.
1100.00	19.3	42.	−97.
1200.00	22.8	38.	−87.
1300.00	25.6	34.	−79.
1400.00	28.0	31.	−72.
1500.00	30.1	28.	−66.

Table 28. Third virial coefficients of various substances

T K						C in units of 10^2 cm^6/mol							
	He	Ne	Ar	Kr	Xe	N$_2$	O$_2$	Air	H$_2$	D$_2$	H$_2$O	D$_2$O	CO$_2$ CH$_4$
25									14.0				
30									16.0				
35									14.3				
40									12.1				
45									10.7				
50									9.6				
55									8.9				
60	2.7	4							8.4				
70	2.5	4							7.4				
80	2.4	4	7						6.9				
90	2.3	4	9						6.4				
100	2.2	4	12						6.1	6			
110	2.1	3	16						5.9	5			
120	2.0	3	20						5.7	5			
130	1.9	3	23						5.5	5			
140	1.8	3	25					28	5.4	5			
150	1.7	3	23					26	5.3	5			
160	1.6	3	22			26	23	24	5.2	5			
180	1.5	3	20			21	20	21	5.0	5			
200	1.3	3	18			19	17	19	4.8	5			

Table 28 —continued

T K	He	Ne	Ar	Kr	Xe	N$_2$	O$_2$	Air	H$_2$	D$_2$	H$_2$O	D$_2$O	CO$_2$	CH$_4$
						C in units of 10^2 cm^6/mol								
220	1.2	3	16	33		17	15	18	4.6	5				
240	1.1	3	15	30		16	13	17	4.5	5				
260	1.1	3	13	28		15	12	16	4.4	5				
273	1.1	3	12	27	62	15	11	15	4.2	5			57	29
280	1.0	3	12	26	59	15	11	15	4.1	5			56	28
300	1.0	2	11	24	54	14	10	15	3.9	5			52	26
320	1.0	2	11	23	50	14		14	3.6	5			49	24
340	0.9	2	10	21	46	14		14	3.4	5			45	22
360	0.8	2	9	20	41	13			3.2	5			42	21
380	0.8	2	9	19	36	13			3.0	4			38	19
400	0.7	2	9	18	34	13			2.9	4			36	18
420	0.7		9	18	32	12				3			32	17
440			8	17	30	12								16
460			8	16	28	12								16
480			8	16	26	12								15
500			7	15	24	12					−100	−150		15
525			7	15	22						−53	−64		14
550			7	14	20						−17	−20		14
575			7	14	18						+2	0		14
600			7	13							9	8		13
650				13							12	12		′
700				12							10	12		

Table 29. Boyle temperature and the inversion temperature of various substances

Substance	Boyle temperature	Inversion temperature	Substance	Boyle temperature	Inversion temperature
He	22.64		O$_2$	405.88	764.43
Ne	122.11	231.42	Air	346.81	658.79
Ar	411.52	779.91	H$_2$	110.04	
Kr	575.00	1089.72	D$_2$	115.30	
Xe	768.03	1455.79	CO$_2$	714.81	
N$_2$	327.22	620.63	CH$_4$	509.66	967.81

Table 30. Potential parameters for the m, 6 potential of selected substances

Substance	m	ε/k K	$b_0\ (=\tfrac{2}{3}\pi N_A \sigma^{-3})$ cm^3/mol
Neon	18	47.74	22.83
Argon	18	160.87	43.74
Krypton	18	224.78	53.78
Xenon	18	300.29	73.32
Nitrogen	21	139.41	54.41
Oxygen	21	172.93	44.49
Air	21	147.76	50.95
Methane	21	217.14	57.96

GENERAL INTRODUCTION
ACCURACY OF THESE TABLES

The main source of oxygen data (L. A. Weber)[41] is particularly precise, 0.1 cm^3/mol in B. For hydrogen and deuterium, problems with the temperature scale between 100 and 273 K may cause errors in B as large as 0.5 cm^3/mol. For H$_2$O and D$_2$O there is only one source for which the precision ranges from several cm^3/mol at the lower temperature to 0.2 cm^3/mol at the highest temperature. For CO$_2$, discrepancies of several cm^3/mol in B exist between data from different sources, and for CH$_4$ of 0.7 cm^3/mol in B and of ten per cent in C.

4. Amagat Density

A practical unit of volume frequently used is the Amagat unit.

The volume in Amagat units V_A is the ratio of the actual volume V of a gas over the normal volume V_N where the latter is that which the gas would occupy at 0°C and 1 atm (1.013250 × 10^5 N m^{-2}).

The normal volume V_N for a mole of a real gas differs slightly from the normal volume $V_0 = 22413.6$ cm^3/mol of a perfect gas due to deviations from ideality at 0°C and 1 atm.

The virial expansion used in conjunction with Amagat units of volume is

$$pV_A = A_A + B_A V_A + C_A V_A^2 + \ldots$$

The Amagat virials are:

$V_N = V_0/A_0,$ $\qquad B_A = B_V A_A/V_N,$

$A_0 = 1 - B_A(0°C) - C_A(0°C),$ $\qquad C_A = C_V A_A/V_N^2,$

$A_A = A_0 T/273.15,$ $\qquad D_A = D_V A_A/V_N^3.$

We calculate the normal density ρ_N at 0°C and 1 atm thus

$$\rho_N = \text{atomic weight}/V_N$$

where

$$V_N = RT_0 [(pV)_N/(pV)_0]$$

Table 31. Table of normal volumes and normal densities

Substance	V_N cm^3/mol	ρ_N kg m^{-3}	Substance	V_N kg m^{-3}	ρ_N kg m^{-3}
air	22 400.1	1.292 8	D$_2$	22 426.7	0.179 80
Ar	22 391.9	1.784 04	H$_2$	22 427.3	0.089 90
CH$_4$	22 360.0	0.717 00	He	22 425.6	0.175 00
CO	22 400.0	1.250 4	Kr	22 350.7	3.749 3
CO$_2$	22 263.9	1.976 24	N$_2$	22 403.1	1.250 42
C$_2$H$_2$	22 157.0	1.174 7	Ne	22 424.0	0.899 88
C$_2$H$_4$	22 245.9	1.260 99	NH$_3$	22 080.0	0.771 38
C$_2$H$_6$	22 178.6	1.355 7	O$_2$	22 391.6	1.429 04
C$_3$H$_8$	21 990.0	2.005	Xe	22 259.9	5.898 5

Note: To convert to kg m^{-3} if the density is given in Amagat units (ρ_A), multiply by ρ_N.

$$V_N = RT_0/\{1 - B_A(0°C) - C_A(0°C)\}$$
$$\approx RT_0[1 + B_A(0°C)]$$
$$\approx RT_0[1 + B_V(0°C)/V_0]$$

5. Standard Enthalpies of Formation and Standard Entropies at 298.15 K

The following table contains a set of proposed values by the ICSU–CODATA Task Group for the standard enthalpies of formation at 298.15 K, $\Delta H°_{f, 298.15}$, the standard entropies at 298.15 K, $S°_{298.15}$ and the standard enthalpy increment between 0 K and 298.15 K, $H°_{298.15} - H°_0$ of several species.

Table 32. Standard enthalpies of formation and standard entropies at 298.15°K

Substance	State	$\Delta H°_{f(298.15\,K)}$ kJ mol^{-1}	$S°_{(298.15\,K)}$ J K^{-1} mol^{-1}	$H°_{(298.15\,K)} - H°_0$ kJ mol^{-1}
O	g	249.17 ± 0.10	160.946 ± 0.020	6.728 ± 0.003
O$_2$	g	0	205.037 ± 0.033	8.682 ± 0.004
H	g	217.997 ± 0.006	114.604 ± 0.015	6.197 ± 0.002
H$^+$	aq	0	0	—
H$_2$	g	0	130.570 ± 0.033	8.468 ± 0.003
H$_2$O	l	−285.830 ± 0.042	69.950 ± 0.080	13.293 ± 0.021
H$_2$O	g	−241.814 ± 0.042	188.724 ± 0.040	9.908 ± 0.008
He	g	0	126.039 ± 0.012	6.197 ± 0.002
Ne	g	0	146.214 ± 0.016	6.197 ± 0.002
Ar	g	0	154.732 ± 0.020	6.197 ± 0.002
Kr	g	0	163.971 ± 0.020	6.197 ± 0.002
Xe	g	0	169.572 ± 0.020	6.197 ± 0.002
Cl	g	121.290 ± 0.008	165.076 ± 0.020	6.272 ± 0.003
Cl$^-$	aq	−167.080 ± 0.088	56.73 ± 0.16	—
Cl$_2$	g	0	222.965 ± 0.040	9.180 ± 0.008
HCl	g	−92.31 ± 0.13	186.786 ± 0.033	8.640 ± 0.004
Br	g	111.86 ± 0.12	174.904 ± 0.020	6.197 ± 0.002
Br$^-$	aq	−121.50 ± 0.15	82.84 ± 0.20	—
Br$_2$	l	0	152.210 ± 0.040	24.52 ± 0.13
Br$_2$	g	30.91 ± 0.11	245.350 ± 0.054	9.724 ± 0.012
HBr	g	−36.38 ± 0.17	198.585 ± 0.033	8.648 ± 0.004
I	g	106.762 ± 0.040	180.673 ± 0.020	6.197 ± 0.002
I$^-$	aq	−56.90 ± 0.84	106.70 ± 0.20	—
I$_2$	c	0	116.139 ± 0.080	13.196 ± 0.040
I$_2$	g	62.421 ± 0.080	260.567 ± 0.063	10.117 ± 0.012
HI	g	26.36 ± 0.80	206.480 ± 0.040	8.657 ± 0.006
N	g	472.68 ± 0.40	153.189 ± 0.020	6.197 ± 0.002
N$_2$	g	0	191.502 ± 0.025	8.669 ± 0.003
C	c	0	5.74 ± 0.12	1.050 ± 0.020
C	g	716.67 ± 0.44	157.988 ± 0.20	6.535 ± 0.006
CO	g	−110.53 ± 0.17	197.556 ± 0.032	8.673 ± 0.008
CO$_2$	g	−393.51 ± 0.13	213.677 ± 0.040	9.364 ± 0.008

6. Density of Mercury

At one standard atmosphere the density of mercury can be calculated using the following equation

$$\rho(t_{68}, p_0) = \rho(20°C, p_0)/\{1 + A(t_{68} - 20°C) + B(t_{68} - 20°C)^2\}$$

where $\rho(t_{68}, p_0)$ and $\rho(20°C, p_0)$ are the densities of mercury under a pressure of one standard atmosphere (101 325 N m^{-2}) at $t_{68} = t°C$ and at $t_{68} = 20°C$* respectively, and

$$A = 18\,115 \times 10^{-8} \,°C^{-1}$$
$$B = 0.8 \quad \times 10^{-8} \,°C^{-2}$$

The density of mercury at $t_{68} = 20°C$ and one standard atmosphere is

$$\rho(20°C, p_0) = 13\,545.86_7 \,\text{kg m}^{-3} \tag{ref. 7}$$

The complete relationship, which is intended for use in the determination of barometric pressure by means of a mercury column, is

$$\rho(t_{68}, \tfrac{1}{2}p) = \rho(t_{68}, p_0)/\{1 - \kappa(\tfrac{1}{2}p - p_0)\}$$

where $\rho(t_{68}, \tfrac{1}{2}p)$ is the mean density of mercury at $t_{68}°C$ in a barometric column supported by the pressure p being measured, and

$$\kappa = 3.8 \times 10^{-11} \,\text{N}^{-1}\text{m}^2 \tag{ref. 4}$$

In a column where p_1 and p_2 are the total pressures acting upon the lower and upper mercury surfaces respectively, the mean density of mercury is

$$\rho[t_{68}, \tfrac{1}{2}(p_1 + p_2)] = \rho(t_{68}, p_0)/\{1 - \kappa[\tfrac{1}{2}(p_1 + p_2) - p_0]\}$$

The error in the calculated density value is ± 0.02 kg m^{-3} in the range 0°C to 40°C.

7. Density of Water

At atmospheric pressure in the range 0°C to 42°C, Chappuis[6] was the first to determine accurately the density of water which was accepted as standard by the International Bureau of Weights and Measures. It is conveniently represented empirically by the Tilton–Taylor[38] equation

$$1 - d = \frac{(t - 3.9863)^2}{508\,929.2} \times \frac{(t + 288.941\,4)}{(t + 68.129\,63)}$$

where d is expressed in g/ml^{-3} to within about one part per million over the whole experimental range. Furthermore, these data have been extended to 80°C by Steckel and Szapiro[36], and to 85°C by Owen et al.[31]. Unfortunately these sets of data differ by as much as 27×10^{-6} g/ml at 80°C. The data of Owen et al.[31] were assumed to be correct until recently when Gildseth et al.[14] suggested that the densities of water above 40° should be redetermined, using mercury as a standard. A modified form of the Tilton–Taylor equation was proposed by these authors to fit the data over the entire 0°C to 80°C range

---

* t_{68} is referred to the IPTS-68.

Table 33. Density of water, g/ml, at atmospheric pressure

Degrees	Temperatures								
	0	10	20	30	40	50	60	70	80
0	0.9998396	0.9997001	0.9982056	0.9956504	0.9922205	0.9880420	0.9832054	0.9777762	0.918015
1	0.9998985	0.9996056	0.9979941	0.9953443	0.9918346	0.9875868	0.9826882	0.9772026	
2	0.9999398	0.9994981	0.9977723	0.9950297	0.9914414	0.9871250	0.9821653	0.9766336	
3	0.9999642	0.9993779	0.9975404	0.9947066	0.9910409	0.9866568	0.9816364	0.9760394	
4	0.9999720	0.9992452	0.9972986	0.9943753	0.9906332	0.9861822	0.9811019	0.9754496	
5	0.9999639	0.9991005	0.9970472	0.9940358	0.9902185	0.9857015	0.9805616	0.9748547	
6	0.9999403	0.9989441	0.9967861	0.9936883	0.9897968	0.9852144	0.9804035	0.9742545	
7	0.9999017	0.9987761	0.9965158	0.9933329	0.9893682	0.9847212	0.9794640	0.9736490	
8	0.9998485	0.9985968	0.9962363	0.9929697	0.9889329	0.9842204	0.9789069	0.9730384	
9	0.9997812	0.9984066	0.9959478	0.9925989	0.9884908	0.9837168	0.9783443	0.9724225	

(d is expressed in g/ml).

$$1 - d = \frac{(t - 3.9863)^2 (t + 288.9414)}{508929.2 \quad (t + 68.12963)} - 0.01144 \exp(-374.3/t)$$

The densities calculated from this equation are presented in *Table 33*. Although the densities in this table are quoted to the seventh decimal place, they are uncertain in the sixth decimal place.

Under pressure, measurements of the p–V–T of water with high accuracy up to 1 kbar and over a temperature range from 0°C to 150°C have been carried out by Kell and Whalley[22]. The compression of water relative to mercury has been measured at intervals of 10°C, to a reproducibility of 4 to 7 p.p.m. of the volume and a maximum uncertainty of about 40 p.p.m. at 1000 bar. Interpolated values of the specific volume are given in *Table 34*.

IX. Definition of Activities and Related Quantities

1. Chemical Potential and Absolute Activity

The chemical potential μ_B of a substance B in a mixture of substances B, C, ..., can be defined by

$$\mu_B = (\partial G/\partial n_B)_{T, p, n_c, ...}$$

where G is the Gibbs energy of the mixture, T is the thermodynamic temperature, p is the pressure, and n_B, n_c, ..., are the amounts of the substances B, C, ..., in the mixture. Compare, also, equation 16.

(In molecular theory the symbol μ_B is sometimes used for the quantity μ_B/L where L is the Avogadro constant.)

The absolute activity λ_B of the substance B in the mixture is a number defined by

$$\lambda_B = \exp(\mu_B/RT) \text{ or } \mu_B = RT \ln \lambda_B$$

where R is the gas constant.

2. Pure Substances

A. Properties of Pure Substances

The superscript $\star$ attached to the symbol for a property of a substance denotes the property of the pure substance. Sometimes, it is appropriate to treat a mixture of constant composition as a pure substance.

B. Fugacity of a Pure Gaseous Substance

The fugacity $f_B^\star$ of a pure gaseous substance B is a quantity which has the same dimensions as pressure. It is defined in terms of the absolute activity $\lambda_B^\star$ of the pure gaseous substance B by

$$f_B^\star = \lambda_B^\star \lim_{p \to 0} (p/\lambda_B^\star) \qquad (T \text{ const.})$$

Table 34. Specific volumes of water (cm³/g) interpolated using data by Kell and Whalley[22]

p(bar)	0.0	10.0	20.0	25.0	30.0	40.0	50.0	60.0	70.0
0.0	1.000212	1.000349	1.001844	1.003008	1.004415	1.007887	1.012153	1.017133	1.022789
1.0	1.000161	1.000301	1.001798	1.002962	1.004370	1.007843	1.012108	1.017093	1.022743
5.0	0.999957	1.000109	1.001614	1.002781	1.004190	1.007664	1.011930	1.016912	1.022559
10.0	0.999703	0.999871	1.001385	1.002555	1.003965	1.007441	1.011707	1.016686	1.022329
25.0	0.998942	0.999156	1.000698	1.001877	1.003293	1.006775	1.011041	1.016010	1.021641
50.0	0.997682	0.997973	0.999561	1.000755	1.002180	1.005670	1.009937	1.014891	1.020501
75.0	0.996432	0.996798	0.998432	0.999640	1.001075	1.004574	1.008840	1.013780	1.019369
100.0	0.995192	0.995632	0.997310	0.998534	0.999978	1.003486	1.007752	1.012677	1.018247
125.0	0.993963	0.994474	0.996197	0.997435	0.998888	1.002405	1.006671	1.011583	1.017695
150.0	0.992743	0.993325	0.995092	0.996344	0.997807	1.001332	1.005598	1.010497	1.016027
175.0	0.991533	0.992185	0.993995	0.995261	0.996733	1.000268	1.004532	1.009419	1.014930
200.0	0.990333	0.991053	0.992906	0.994185	0.995667	0.999210	1.003474	1.008349	1.013842
225.0	0.989142	0.989929	0.991824	0.993117	0.994608	0.998161	1.002424	1.007287	1.012761
250.0	0.987961	0.988814	0.990750	0.992056	0.993557	0.997119	1.001381	1.006233	1.011689
275.0	0.986790	0.987707	0.989684	0.991003	0.992513	0.996084	1.000345	1.005187	1.010625
300.0	0.985628	0.986608	0.988625	0.989957	0.991477	0.995057	0.999317	1.004148	1.009569
325.0	0.983332	0.984435	0.986531	0.987888	0.989426	0.993024	0.997282	1.002091	1.007480
350.0	—	—	—	—	—	—	—	—	—
400.0	0.981072	0.982293	0.984465	0.985847	0.987404	0.991019	0.995275	1.000069	1.005422
450.0	0.978848	0.980183	0.982429	0.983835	0.985409	0.989043	0.993297	1.998073	1.003395
500.0	0.976660	0.978104	0.980422	0.981850	0.983443	0.987094	0.991345	0.996105	1.001396
550.0	0.974506	0.976055	0.978442	0.979894	0.981503	0.985171	0.989420	0.994166	0.999427
600.0	0.972386	0.974037	0.976490	0.977964	0.979590	0.983275	0.987522	0.992253	0.997486
650.0	0.970299	0.972047	0.974565	0.976061	0.977702	0.981404	0.985649	0.990366	0.995572
700.0	0.968245	0.970087	0.972667	0.974184	0.975840	0.979558	0.983801	0.988506	0.993684
750.0	0.966222	0.968155	0.970794	0.972332	0.974003	0.977736	0.981978	0.986670	0.991823
800.0	0.964231	0.966250	0.968947	0.970506	0.972190	0.975938	0.980179	0.984858	0.989986
850.0	0.962269	0.964373	0.967125	0.968705	0.970401	0.974163	0.978404	0.983070	0.988174
900.0	0.960338	0.962523	0.965328	0.966927	0.968634	0.972410	0.976652	0.981305	0.986386
950.0	0.958435	0.960699	0.963554	0.965174	0.966891	0.970680	0.974922	0.979562	0.984621
1000.0	0.956561	0.958901	0.961804	0.963443	0.965169	0.968970	0.973214	0.977841	0.982879

T(°C)

GENERAL INTRODUCTION

p(bar)	80.0	90.0	100.0	110.0	120.0	130.0	140.0	150.0
0.0	1.029072	1.035978	1.043502	1.051647	1.060420	1.069850	1.079964	1.090802
1.0	1.029024	1.035929	1.043451	1.051594	1.060364	1.069791	1.079900	1.090735
5.0	1.028834	1.035733	1.043247	1.051381	1.060139	1.069553	1.079647	1.090464
10.0	1.028598	1.035488	1.042992	1.051114	1.059858	1.069256	1.079331	1.090126
25.0	1.027889	1.034756	1.042231	1.050318	1.059018	1.068369	1.078387	1.089118
50.0	1.026717	1.033545	1.040972	1.049001	1.057631	1.066904	1.076830	1.087455
75.0	1.025554	1.032344	1.039724	1.047697	1.056258	1.065454	1.075291	1.085813
100.0	1.024400	1.031153	1.038486	1.046405	1.054899	1.064021	1.073770	1.084192
125.0	1.023256	1.029972	1.037260	1.045124	1.053554	1.062602	1.072266	1.082591
150.0	1.022121	1.028801	1.036045	1.043855	1.052223	1.061199	1.070780	1.081010
175.0	1.020995	1.027640	1.034840	1.042598	1.050905	1.059810	1.069311	1.079447
200.0	1.019878	1.026488	1.033645	1.041353	1.049600	1.058436	1.067859	1.077904
225.0	1.018770	1.025346	1.032461	1.040119	1.048308	1.057076	1.066422	1.076379
250.0	1.017670	1.024213	1.031287	1.038896	1.047028	1.055731	1.065002	1.074872
275.0	1.016580	1.023089	1.030123	1.037684	1.045761	1.054398	1.063596	1.073383
300.0	1.015498	1.021974	1.028969	1.036483	1.044506	1.053080	1.062206	1.071910
350.0	1.013359	1.019772	1.026691	1.034113	1.042032	1.050482	1.059470	1.069015
400.0	1.011253	1.017606	1.024451	1.031784	1.039603	1.047935	1.056790	1.066184
450.0	1.009179	1.015473	1.022248	1.029496	1.037219	1.045437	1.054165	1.063414
500.0	1.007137	1.013375	1.020081	1.027248	1.034877	1.042987	1.051591	1.060703
550.0	1.005125	1.011308	1.017949	1.025037	1.032576	1.040582	1.049067	1.058047
600.0	1.003143	1.009274	1.015851	1.022863	1.030315	1.038221	1.046591	1.055445
650.0	1.001190	1.007270	1.013786	1.020725	1.028093	1.035903	1.044161	1.052893
700.0	0.999264	1.005296	1.011752	1.018621	1.025907	1.033625	1.041775	1.050391
750.0	0.997366	1.003352	1.009749	1.016551	1.023757	1.031386	1.039433	1.047936
800.0	0.995493	1.001435	1.007776	1.014512	1.021642	1.029186	1.037132	1.045527
850.0	0.993646	0.999546	1.005832	1.012505	1.019560	1.027021	1.034871	1.043161
900.0	0.991824	0.997683	1.003915	1.010526	1.017511	1.024892	1.032649	1.040838
950.0	0.990025	0.995845	1.002024	1.008576	1.015494	1.022797	1.030466	1.038555
1000.0	0.988249	0.994032	1.000159	1.006653	1.013507	1.020734	1.028320	1.036313

$T(°C)$

or in terms of the chemical potential μ_B by

$$RT \ln f_B^\star = \mu_B^\star + \lim_{p \to 0} (RT \ln p - \mu_B^\star) \qquad (T \text{ const.})$$

where p is the pressure of the gas and T is its thermodynamic temperature. It follows from this definition that

$$\lim_{p \to 0} (f_B^\star/p) = 1 \qquad (T \text{ const.})$$

and that

$$RT \ln (f_B^\star/p) = \int_0^p (V_B^\star - RT/p) \, dp \qquad (T \text{ const.})$$

where $V_B^\star$ is the molar volume of the pure gaseous substance B.

A pure gaseous substance B is treated as an ideal gas when the approximation $f_B^\star = p$ is used. The ratio $(f_B^\star/p)$ may be called the fugacity coefficient.

3. Mixtures

A. Definition of a Mixture

A mixture is a gaseous, liquid or solid phase containing more than one substance, when all substances are treated in the same way.

B. Partial Pressure

The partial pressure p_B of a substance B in a gaseous mixture is a quantity with the same dimensions as pressure defined by

$$p_B = y_B p$$

where y_B is the mole fraction of the substance B in the gaseous mixture and p is the pressure.

C. Fugacity of a Substance in a Gaseous Mixture

The fugacity f_B of the substance B in a gaseous mixture containing mole fractions $y_B, y_C, \ldots$, of the substances $B, C, \ldots$, is a quantity with the same dimensions as pressure, defined in terms of the absolute activity λ_B of the substance B in the gaseous mixture by

$$f_B = \lambda_B \lim_{p \to 0} (y_B p / \lambda_B) \qquad (T \text{ const.})$$

or in terms of the chemical potential μ_B by

$$RT \ln f_B = \mu_B + \lim_{p \to 0} \{RT \ln (y_B p) - \mu_B\} \qquad (T \text{ const.})$$

It follows from this definition that

$$\lim_{p \to 0} (f_B / y_B p) = 1 \qquad (T \text{ const.})$$

and that

$$RT \ln (f_B/y_B p) = \int_0^p (V_B - RT/p)\,dp \qquad (T \text{ const.})$$

where V_B is the partial molar volume of the substance B in the gaseous mixture.

A gaseous mixture of B, C, ..., is treated as an ideal gaseous mixture when the approximations $f_B = y_B p$, $f_C = y_C p$, ..., are used. It follows that $pV = (n_B + n_C + \ldots)RT$ for an ideal gaseous mixture of B, C, ...

The ratio $(f_B/y_B p)$ may be called the fugacity coefficient of the substance B.

When $y_B = 1$ the definitions given in this section for the fugacity of a substance in a gaseous mixture can be reduced to those given in section 2.B for the fugacity of a pure gaseous substance.

4. Solutions

A. Definition of a Solution

A solution is a liquid or solid phase containing more than one substance, when one of the substances (the solvent) is treated differently from the other substances (solutes). When the sum of the mole fractions of the solutes is small compared with unity, the solution is called a dilute solution. In what follows, the solvent substance will be denoted by A and the solute substances by B, C, ...

B. Properties of Infinitely Dilute Solutions

The superscript ∞ attached to the symbol for a property of a solution denotes the property of an infinitely dilute solution.

C. Activity Coefficient of a Solute Substance in a Solution

The activity coefficient γ_B of a solute substance B in a solution (especially in a dilute liquid solution) containing molalities m_B, m_C, ..., of solute substances B, C, ..., in a solvent substance A, is a number defined in terms of the absolute activity λ_B of the solute substance B in the solution by

$$\gamma_B = (\lambda_B/m_B)/(\lambda_B/m_B)^\infty \qquad (T, p \text{ const.})$$

or in terms of the chemical potential μ_B by

$$RT \ln (m_B \gamma_B) = \mu_B - (\mu_B - RT \ln m_B)^\infty \qquad (T, p \text{ const.})$$

It follows from this definition that

$$\gamma_B^\infty = 1 \qquad (T, p \text{ const.})$$

D. Relative Activity of a Solute Substance in a Solution

The relative activity a_B of a solute substance B in a solution containing molalities m_B, m_C, ..., of solute substances B, C, ..., in a solvent substance A, is a number defined in terms of the absolute activity λ_B by

$$a_B = (\lambda_B/m^\ominus)/(\lambda_B/m_B)^\infty = m_B\gamma_B/m^\ominus \qquad (T, p \text{ const.})$$

or in terms of the chemical potential μ_B by

$$RT \ln a_B = \mu_B - RT \ln m^\ominus - (\mu_B - RT \ln m_B)^\infty$$
$$= RT \ln (m_B\gamma_B/m^\ominus)$$

where $m^\ominus$ is a standard value of molality (usually chosen to be 1 mol kg^{-1}) and where the other symbols are as defined in section 4.C.

It follows from this definition of a_B that

$$(a_B m^\ominus/m_B)^\infty = 1 \qquad (T, p \text{ const.})$$

The name activity is often used instead of the name relative activity for this quantity.

E. Osmotic Coefficient of the Solvent Substance in a Solution

The osmotic coefficient ϕ of the solvent substance A in a solution containing molalities $m_B, m_C, \ldots,$ of solute substances B, C, ..., is a number defined in terms of the absolute activity λ_A of the solvent substance A in the solution by

$$\phi = (M_A \, \Sigma_i \, m_i)^{-1} \ln (\lambda_A^\star/\lambda_A)$$

where $\lambda_A^\star$ is the absolute activity of the pure solvent substance A at the same temperature and pressure and M_A is the molar mass of the solvent substance A or, in terms of the chemical potential $\mu_A^\star$, by

$$\phi = (\mu_A^\star - \mu_A)/RT M_A \, \Sigma_i m_i$$

where $\mu_A^\star$ is the chemical potential of the pure solvent substance A at the same temperature and pressure.

For an ideal dilute solution as defined in section 4.C or 4.D it can be shown that $\phi = 1$.

F. Relative Activity of the Solvent Substance in a Solution

The relative activity a_A of the solvent substance A in a solution containing molalities $m_B, m_C, \ldots,$ or mole fractions $x_B, x_C, \ldots,$ of solute substances B, C, ..., is a number defined in terms of the absolute activity λ_A of the solvent substance A in the solution by

$$a_A = \lambda_A/\lambda_A^\star = \exp(-\phi M_A \Sigma_i m_i)$$

or in terms of the chemical potential μ_A by

$$RT \ln a_A = \mu_A - \mu_A^\star = -RT\phi M_A \, \Sigma_i m_i$$

where the other quantities are as defined in section 4.E.

GENERAL INTRODUCTION

X. Accuracy and Precision [23]

1. Definitions

The thermodynamic data must be reported in a form as free from misinterpretation as possible. If a choice is to be made, rough data must be plotted or tabulated instead of smoothed final results. Small effects can be lost when data are smoothed or fitted to a predetermined function. In addition to the presentation of the data, estimates of the probable accuracy and precision indices must be given. A reported value whose accuracy is entirely unknown is worthless.

Accuracy refers to the success of estimating the true value of a quantity. It is often confused with precision which refers to the clustering of samples about their own average, which will be the true value if this average is not biased. If there is no bias in the experimental and estimation procedures, accuracy and precision are synonymous for all practical purposes.

The accuracy or uncertainty or actual error of a reported value (occurring in a practical measurement process) which is the magnitude and sign of its deviation from the true value defined conceptually by an ideal measurement process, is an unknowable limit to the error but can usually be determined from the precision of the measurement process by which the reported value was obtained and from reasonable limits to the possible bias of the measurement process.

The bias or systematic error of a measurement process is the magnitude and direction of its tendency to measure something other than what was intended. Among systematic errors[23] one can take into account:

(i) the sensitivity or resolution possible in the measurements,
(ii) the effect of assumptions made in processing the data,
(iii) the effect of possible systematic trends, both those for which corrections were made and those for which this could not be done,
(iv) uncertainties in auxiliary data taken from other work,
(v) bias in the method of computing.

The precision refers to the typical closeness of successive independent measurements of a single magnitude generated by repeated applications of the process under specified conditions. The precision can be expressed by the following values:

(a) the precision of a measurement process is measured by the standard deviation (σ) of a single determination,
(b) when σ is known, the arithmetic mean (x_n) of n determinations is measured by:
   (1) the standard error ($\sigma/n^{\frac{1}{2}}$) of the reported value,
   (2) two or three sigma limits (commonly used bounds of imprecision),
   (3) the confidence interval or half-width of the confidence interval (indicate one or two sides),
   (4) the probable error of the reported value ($=0.6745\sigma/n^{\frac{1}{2}}$) for normally distributed data points,
   (5) the mean deviation, or average deviation, of a measurement from the mean calculated from the sample. (Limiting mean of mean deviation $[=(2/\pi)^{\frac{1}{2}}\{(n-1)/n\}^{\frac{1}{2}}\sigma]$ for normally distributed data points).

(c) when σ is unknown, m means computed from n measurements, are represented by: (1), (2), (3) above. If the measurements are of equal precision, σ^2 can be estimated by $\rho_p^2 = (1/m) \sum_{i=1}^{m} s_i^2$ where s measures the statistical dispersion.

The terms standard deviation and standard error should be reserved to denote the canonical values for the measurement process, based on recent experience.

When there is insufficient recent experience an estimate of the standard error (standard deviation) must be computed. In such cases, to avoid possible misunderstanding, the term 'computed (or estimated) standard error' or 'computed standard deviation' could be used.

The estimate of the standard error employed should be that obtained from:

estimate of standard error = (sum of squared residuals/nv)$^{\frac{1}{2}}$

where n is the (effective) number of completely independent determinations of which the reported value a is the arithmetic mean (or other appropriate least-squares adjusted value) and v is the number of degrees of freedom involved in the sum of square residuals (that is, the number of residuals minus the number of fitted constants or other independent constraints on the residuals).

If the reported value a is the arithmetic mean then an estimate of the standard error is $(S^2/n)^{\frac{1}{2}}$ where $S^2 = \sum_{i=1}^{n} (x_i - a)^2/(n - 1)$, and n is the number of independent determinations of which a is the arithmetic mean.

2. Expression of the Uncertainties of Final Results

The uncertainty of a reported value is indicated by stating credible limits to its inaccuracy. No single form of expression for these limits is universally satisfactory. In fact, different forms of expression are recommended which will depend on the relative magnitude of the imprecision and likely bias, and their relative importance. Four distinct cases have to be examined:

(i) both systematic error and imprecision negligible,
(ii) systematic error not negligible, imprecision negligible,
(iii) neither systematic error nor imprecision negligible,
(iv) systematic error negligible, imprecision not negligible.

Final results and their respective uncertainties should be reported in sentence form whenever possible.

(i) If systematic error and imprecision are both negligible the reported result can be given after rounding, to the number of significant figures consistent with the accuracy requirements of the situation, together with an explicit statement of the accuracy. If no statement of accuracy or precision accompanies a reported number, then, in accordance with the usual conventions governing rounding, this number will ordinarily be interpreted as being accurate within plus or minus half a unit in the last significant figure given.

(ii) When the imprecision of a result is negligible but the inherent systematic error of the measurement process concerned is not negligible, qualification of a reported result should be limited to a single quasi-absolute type of statement that places bounds on its inaccuracy. These bounds should be stated to no more than two significant figures. The report result itself should be given (that is, rounded) to the last place affected by the stated bounds. Accuracy statements should be given in sentence form in all cases, except when several results of different accuracies are presented. The fact that the imprecision is negligible should be stated explicitly.

(iii) When neither the imprecision nor the systematic error of a result are negligible, then the following rules are recommended:

(a) A reported result should be qualified by a quasi-absolute type of statement that places bounds on its systematic error, and a separate statement of its standard error or its probable error or of an upper bound thereto, whenever a reliable determination of such value or bound is available. Otherwise a computed value of the standard error, or probable error, so designated should be given together with a statement of the number of degrees of freedom on which it is based.

(b) The bounds to its systematic error and the measure of its imprecision should be stated to no more than two significant figures.

(c) The reported result itself should be stated at most to the last place affected by the more precise of the two qualifying statements.

(d) The qualification of a reported result with respect to its imprecision and systematic error should be given in sentence form, except when results of different precisions or with different bounds to that systematic error are presented in tabular arrangement.

The term standard error is to be understood as signifying the standard deviation of the reported value itself, not as signifying the standard deviation of the single determination.

The above recommendations should not be construed to exclude the presentation of a quasi-absolute type of statement placing bounds on the inaccuracy, that is on the overall uncertainty of a reported value, provided that separate statements of its imprecision and its possible systematic error are included also. To be in good taste the bounds indicating the overall uncertainty should not be numerically less than the corresponding bounds placed on the systematic error outwardly increased by at least three times the standard error.

When a reliably established value for the relevant standard error is available and the dispersion of the present measurements is in keeping with this experience, then the canonical value of the standard error should be used. If such experience indicates that the standard error is subject to fluctuations greater than the intrinsic variation of such a measure, then an appropriate upper bound should be given.

(iv) When the systematic error of a result is negligible but its imprecision is not, the qualification of the reported value should be limited to a statement of its standard error or of an upper bound. The standard error or upper bound thereto should be stated to not more than two significant figures. The reported result itself should be stated at most to the last place affected by the stated value or bound to its imprecision. The qualification of a reported result with respect to its imprecision should be given in sentence form except when results of different precisions are presented in tabular arrangement. The fact that the systematic error is negligible should be stated explicitly. To be reliable the bounds of the inaccuracy should be numerically equal to at least three times the stated standard error. All the above statements have been widely developed in the following chapters.

XI. Conversion Factors

Tables 36 to *51* express the definitions of miscellaneous units of measure as exact numerical multiples of coherent SI units, and provide multiplying factors for converting miscellaneous units to corresponding SI units. Throughout the tables, the factor is written with one numeral preceding the decimal point while the 2-figure column on the right gives the appropriate power of ten by which the factor should be multiplied if this index is positive, or divided if the index is negative. Thus 1 angstrom = 10^{-10} m; cf. first entry in *Table 39*.

Table 36. Density

To convert to kg/m^3 from	multiply by	
g/cm^3	1.00	+03
lbm/in^3	2.7679905	+04
lbm/ft^3	1.6018463	+01
$slug/ft^3$	5.15379	+02

Table 37. Energy/area time

To convert to W/m^2 from	multiply by	
Btu†/ft^2 sec	1.1348931	+04
Btu†/ft^2 min	1.8914885	+02
Btu†/ft^2 h	3.1524808	00
Btu†/in^2 sec	1.6342462	+06
cal†/cm^2 min	6.9733333	+02
erg/cm^2 sec	1.00	−03
W/cm^2	1.00	+04

† thermochemical

Table 38. Force

To convert to N from	multiply by	
dyn	1.00	−05
kgf or kp	9.80665	+00
kip	4.4482216152605	+03
lbf	4.4482216152605	00
ozf	2.7801385	−01
pdl	1.38254954376	−01

Table 39. Length

To convert to m from	multiply by	
Å	1.00	−10
astronomical unit	1.4959789	+11
fermi	1.00	−15
ft	3.048	−01
ft (US survey)	3.048006096	−01
in	2.54	−02
micron	1.00	−06
mil	2.54	−05
mile (US statute)	1.609344	+03
mile (UK nautical)	1.853184	+03
mile (international nautical)	1.852	+03
parsec	3.08374	+16
yd	9.144	−01

Table 40. Mass

To convert to kg from	multiply by	
carat (metric)	2.00	−04
grain	6.479891	−03
gram	1.00	−03
$kgfsec^2$ meter (mass)	9.80665	00
kilogram mass	1.00	00
oz mass (avoirdupois)	2.8349523125	−02
oz mass (troy or apoth)	3.11034768	−02
dwt	1.55517384	−03
lbm (avoirdupois)	4.5359237	−01
lbm (troy or apoth)	3.732417216	−01
slug	1.45939029	+01
ton (assay)	2.9166666	−02
ton (long)	1.0160469088	+03
ton (metric)	1.00	+03
ton (short, 2000 lb)	9.0718474	+02
tonne	1.00	+03

GENERAL INTRODUCTION

Table 41. Power

To convert to W from	multiply by		To convert to W from	multiply by	
Btu (thermochemical)/second	1.054350264488	+03	lbf/second)	7.4569987	+02
Btu (thermochemical)/minute	1.7572504	+01	horsepower (boiler)	9.80950	+03
calorie (international)/second	4.1868	00	horsepower (electric)	7.46	+02
calorie (thermochemical)/second	4.184	00	horsepower (metric)	7.35499	+02
calorie (thermochemical)/minute	6.9733333	−02	horsepower (UK)	7.457	+02
foot lbf/hour	3.7661610	−04	horsepower (water)	7.46043	+02
foot lbf/minute	2.2596966	−02	kilocalorie (thermochemical)/minute	6.9733333	+01
foot lbf/second	1.3558179	00	kilocalorie (international)/second	4.1868	+03
horsepower (550 foot			kilocalorie (thermochemical)/second	4.184	+03
			watt (international of 1948)	1.000165	00

Table 42. Pressure

To convert to Newton/meter^2	multiply by		To convert to Newton/meter^2	multiply by	
atmosphere (normal = 760 torr)	1.01325	+05	inch of mercury (60°F)	3.37685	+03
atmosphere (technical = 1 kgf/cm^2)	9.80665	+04	inch of water (39.2°F)	2.49082	+02
bar	1.00	+05	inch of water (60°F)	2.4884	+02
barye	1.00	−01	kgf/centimeter^2	9.80665	+04
centimeter of mercury (0°C)	1.33322	+03	kgf/meter^2	9.80665	00
centimeter of water (4°C)	9.80638	+01	lbf/foot^2	4.7880258	+01
dyne/centimeter^2	1.00	−01	lbf/inch^2 (p.s.i.)	6.8947572	+03
foot of water (39.2°F)	2.98898	+03	millibar	1.00	+02
inch of mercury (32°F)	3.386389	+03	millimeter of mercury (0°C)	1.333224	+02
			pascal	1.00	00
			p.s.i. (lbf/inch^2)	6.8947572	+03
			torr (0°C)	1.33322	+02

Table 43. Specific enthalpy and specific energy

To convert to J/kg from	multiply by	
Btu (international)/lbm	2.326	+03
foot lbf/lbm	2.989066920	00
kcal (international)/g	4.1868	+06
kcal (international)/kg	4.1868	+03
kJ/kg	1.00	+03
kp m/g	9.80665	+03
J/g	1.00	+03
lbf ft^3/lbm in^2	4.30425636	+02

Table 44. Specific entropy or specific heat capacity

To convert to J/kg K from	multiply by	
bar cm^3/g K	1.00	+02
Btu (international)/lbm degF	4.1868	+03
Btu (international)/lbm°R	4.1868	+03
ft lbf/lbm deg F	5.38032046	00
kcal (international)/kg K	4.1868	+03
kJ/kg K	1.00	+03
kp m/g K	9.80665	+03
k W h/lbm deg F	1.42859546	+07
J/g K	1.00	+03

Table 45. Speed

To convert to m/s from	multiply by	
foot/hour	8.4666666	−05
foot/minute	5.08	−03
foot/second	3.048	−01
inch/second	2.54	−02
kilometer/hour	2.777778	−01
knot (international)	5.144444444	−01
mile/hour (US statute)	4.4704	−01
mile/minute (US statute)	2.68224	+01
mile/second (US statute)	1.609344	+03

Table 46. Temperature

To convert from	to	
Celsius	kelvin	$t_K = t_C + 273.15$
Fahrenheit	kelvin	$t_K = (5/9)(t_F + 459.67)$
Fahrenheit	Celsius	$t_C = (5/9)(t_F - 32)$
Rankine	kelvin	$t_K = (5/9) t_R$

Table 47. Thermal conductivity

To convert to W/m K from	multiply by	
Btu (international)/ft h deg F	1.73073467	+00
cal/sec cm °C	4.1868	+02
ft lbf/h ft deg F	2.22411081	−03
kcal (international)/m h deg F	1.163	00
kp/m h °C	2.72406944	−03
watt/ft deg F	5.90551181	00

Table 48. Time

To convert to second (mean solar) from	multiply by	
day (mean solar)	8.64	+04
day (sidereal)	8.6164090	+04
hour (mean solar)	3.60	+03
hour (sidereal)	3.5901704	+03
minute (mean solar)	6.00	+01
minute (sidereal)	5.9836174	+01
month (mean calendar)	2.628	+06
second (ephemeris)	1.000000000	00
second (sidereal)	9.9726957	−01
year (calendar)	3.1536	+07
year (sidereal)	3.1558150	+07
year (tropical)	3.1556926	+07

Table 49(a). Viscosity (kinematic)

To convert to m²/s from	multiply by	
centistoke	1.00	−06
stoke	1.00	−04
foot²/second	9.290304	−02

Table 49(b). Viscosity (dynamic)

To convert N s/m² or to kg/m s from	multiply by	
centipoise	1.00	−03
lbf second/foot²	4.7880258	+01
lbm/foot hour	4.13378873	+04
lbm/foot second	1.4881639	00
poise	1.00	−01
poundal second/foot²	1.4881639	00
slug/foot second	4.7880258	+01

Table 50. Volume

To convert to m³ from	multiply by		To convert to m³ from	multiply by	
acre foot	1.2334819	03	inch³	1.6387064	−05
barrel (petroleum, 42 gal)	1.589873	−01	liter	1.000028	−03
cup	2.365882365	−04	pint (US dry)	5.5061047135	−04
fluid ounce (US)	2.957352956	−05	pint (US liquid)	4.73176473	−04
foot³	2.8316846592	−02	quart (US dry)	1.1012209427	−03
gallon (UK liquid)	4.546087	−03	quart (US liquid)	9.4635295	−04
gallon (US dry)	4.4048837708	−03	stere	1.00	00
gallon (US liquid)	3.785411784	−03	ton (register)	2.8316846592	00
			yard³	7.6455485798	00

Table 51. Work, Energy

To convert to J from	Multiply by	
Btu (ISO/TC 12)	1 055 06	+03
Btu (International Steam Table)	1 055 04	+03
Btu (mean)	1 055 87	+03
Btu (thermochemical)	1 054 350 264 488	+03
Btu (39°F)	1 059 67	+03
Btu (60°F)	1 054 68	+03
calorie (International Steam Table)	4 1868	+00
calorie (mean)	4 190 02	+00
calorie (thermochemical)	4 184	+00
calorie (15°C)	4 185 80	+00
calorie (20°C)	4 181 90	+00
electron volt	1 602 10	−19
erg	1 00	−07
foot lbf	1 355 817 9	+00
foot poundal	4 214 011 0	−02
joule (international of 1948)	1 000 165	+00
kilocalorie (International Steam Table)	4 1868	+03
kilocalorie (mean)	4 190 02	+03
kilocalorie (thermochemical)	4 184	+03
kilowatt hour	3 60	+06
kilowatt hour (international of 1948)	3 600 59	+06
ton (nuclear equivalent of TNT)	4 20	+09
watt hour	3 60	+03

XII. References

1. Battino, R. and H. L. Clever. *Chem. Rev.* **66**, 395 (1966).
2. Battino, R. *Chem. Rev.* **71**, 5 (1971).
3. Beattie, J. A., B. E. Blaisdell, J. Kaye, H. T. Gerry and C. A. Johnson. *Proc. Amer. Acad. Arts Sci.* **74**, 371 (1941).
4. Bett, K. E., P. F. Hayes, and D. M. Newitt. *Phil. Trans. Roy. Soc.* **247**, 59 (1954).
5. Blank, G. *Wärme und Stoffübertragung*, **2**, 53 (1969).
6. Chappuis, P. *Trav. et Mem. Bur. Int. Poids Mes.* **13**, C10 (1907).
7. Cook, A. H. and N. W. B. Stone. *Phil. Trans. Roy. Soc. A*, **250**, 279 (1957).
8. Cook, G. A. *Argon, Helium and the Rare Gases*. Interscience: New York and London (1961).
9. Dillon, J. G., P. A. Nelson, and B. S. Swanson. *J. Chem. Phys.* **74**, 4229 (1966).
10. Din, F. *Thermodynamic Functions of Gases*. Butterworths: London (1961–62).
11. Feke, G. T., J. B. Lastovka, G. B. Benedek, K. H. Langley and P. B. Elterman. *Optics Commun.* **7**, 13 (1973).
12. Freyland, W. F. and F. Hensel. *Ber. Bunsenges. Phys. Chem.* **76**, 347 (1972).
13. Garvin, D. 'Guidelines for the reporting of numerical data and experimental procedures', submitted to the *J. Res. Nat. Bur. Stand.*
14. Gildseth, W., A. Habenschuss and F. H. Spedding. *J. Chem. Engng Data,* **17**, 402 (1972).
15. Ginnings, D. C. 'Heat', Vol. VI of *Precision Measurements and Calibration. Spec. Publ. US Nat. Bur. Stand. No. 300* (1970).
16. Goodwin, R. D. *J. Res. Nat. Bur. Stand.* **74A**, 221 (1970).
17. Goodwin, R. D. *J. Res. Nat. Bur. Stand.* **74A**, 655 (1970).
18. Hensel, F. and E. U. Franck. *Ber. Bunsenges. Phys. Chem.* **70**, 1154 (1966).
19. Hirschfelder, J. O., C. F. Curtiss and R. B. Bird. *Molecular Theory of Gases and Liquids*. Wiley: New York (1954).
20. Hoge, H. J. and J. W. Lassiter. *J. Res. Nat. Bur. Stand.* **47**, 75 (1951).
21. IUPAC. *Pure Appl. Chem.* **30**, 637 (1972).
22. Kell, G. S. and E. Whalley. *Phil. Trans. Roy. Soc. A*, **258**, 565 (1965).
23. Ku, H. H. 'Statistical concepts and procedures', Vol. I of *Precision Measurement and Calibration. Spec. Publ. US Nat. Bur. Stand. No. 300* (1969).
24. Kudchadker, A. P., G. N. Alani and B. Zwolinski. *Chem. Rev.* **68**, 659 (1968).
25. Levelt-Sengers, J. M. H., J. Straub and M. Vicentini-Missoni. *J. Chem. Phys.* **54**, 5034 (1971).
26. Levelt-Sengers, J. M. H., M. Klein and J. S. Gallagher. 'Pressure–volume–temperature relationships of gases—Virial coefficients', to be published in the next edition of the *American Institute of Physics Handbook*.
27. Michels, A., J. M. Levelt and W. De Graaff. *Physica,* **24**, 659 (1958).
28. McGlashan, M. L. *Pure Appl. Chem.* **21**, 1 (1970).
29. Natrella, M. G. 'Experimental statistics', *Nat. Bur. Stand. Handbook No. 91* (1963).
30. Oster, G. F. and C. F. Bonilla. *Fifth Symposium on Thermophysical Properties*, p 122. American Society of Mechanical Engineers, Purdue University: Lafayette (1960).
31. Owen, B. B., J. R. White and J. S. Smith. *J. Amer. Chem. Soc.* **78**, 3561 (1956).
32. Page, C. and P. Vigoreux. *Spec. Publ. US Nat. Bur. Stand. No. 330* (1971).
33. Prydz, R. and G. C. Straty. *J. Res. Nat. Bur. Stand.* **74A**, 747 (1970).
34. Rossini, F. D. *J. Chem. Thermodynamics,* **2**, 447 (1970).
35. Samsonov, G. V. *Handbook of the Physiochemical Properties of the Elements.* [Russian translation] Plenum: New York (1968).
36. Steckel, F. and S. Szapiro. *Trans. Faraday Soc.* **59**, 331 (1963).
37. Taylor, B. N., W. H. Parker and D. N. Langenberg. *Rev. Mod. Phys.* **41**, 375 (1969).
38. Tilton, L. W. and I. K. Taylor. *J. Res. Nat. Bur. Stand.* **18**, 205 (1907).
39. Vicentini-Missoni, M., J. M. H. Levelt-Sengers and M. S. Green. *Phys. Rev. Letters,* **22**, 389 (1969).
40. Wallace, B. and H. Meyer. *Phys. Rev.* **A2**, 1563 and 1610 (1970).
41. Weber, L. A. *J. Res. Nat. Bur. Stand.* **74A**, 93 (1970).

CHAPTER 2

Reference Materials for Thermometric Fixed Points

H. KIENITZ and E. BRUNNER

BASF Aktiengesellschaft 6700 Ludwigshafen, West Germany

Contents

I.	Primary Fixed Points as Defined by the International Practical Temperature Scale of 1968 (IPTS-68)	71
	1. Definition of the IPTS-68	71
	2. Range 13.81 K to 273.15 K	72
	3. Range 0°C to 630.74°C	75
	4. Range 630.74°C to 1064.43°C	76
	5. Range above 1064.43°C	76
II.	Secondary Reference Points as Recommended by the International Committee on Weights and Measures	76
III.	Supplementary Recommendations on Apparatus, Methods and Procedures	77
	1. Influence of Pressure on the Freezing Point Temperature	78
	2. Triple Point, 17.042 K Point and Boiling Point of Equilibrium Hydrogen	78
	3. Boiling Point of Neon	79
	4. Triple Point and Boiling Point of Oxygen	79
	5. Boiling Point of Water	79
	6. Freezing Points of Tin and Zinc	80
	7. Freezing Points of Silver and Gold	80
IV.	Practical Temperature Scales over the 0.2 K to 5.2 K Range	80
V.	Standard Reference Samples	81
	1. Gases	81
	2. Catalogue of Physicochemical Standard Substances	81
	3. Further Recommendations of Calibration Materials	81
VI.	References	84

I. Primary Fixed Points as Defined by the International Practical Temperature Scale of 1968 (IPTS-68)

1. *Definition of the IPTS-68*

The IPTS-68[17] is based on the assigned values of the temperature of a number of reproducible equilibrium states (fixed points) and on standard instruments calibrated at these temperatures. Interpolation between the temperatures of the primary fixed points is effected with the aid of formulae which establish the relationship between the indications of these standard instruments and the values of the IPTS.

The defining fixed points are established by producing specified equilibria between the phases of pure substances. The equilibria and the IPTS values assigned to them are listed in *Table 1*.

The standard interpolation instrument between the primary fixed points is, in the range 13.81 K to 903.89 K (630.74°C), the platinum resistance thermometer, whose resistance ratio $R\,(100°C)/R(0°C)$ must not be less than 1.3925. The standard instrument prescribed for the range 630.74°C to 1 064.43°C is the platinum–10 per cent rhodium/platinum thermocouple. Above 1 064.43°C radiation pyrometers are used.

2. Range 13.81 K to 273.15 K

From 13.81 K to 273.15 K the temperature T is defined by the equation

$$W(T) = W_{CCT}(T) + \Delta W(T) \tag{1}$$

where $W(T)$ is the resistance ratio $R(T)/R(273.15\text{ K})$ of the platinum resistance thermometer and represents the actual measured value. $W_{CCT}(T)$ is derived from a standard reference function. The resistance values W_{CCT} for whole-number values of T are given in *Table 2*. Tables including intermediate values may be obtained from the various metrological institutes, e.g. National Bureau of Standards, USA; Bureau International des Poids et Mesures, France; Physikalisch Technische Bundesanstalt, West Germany. For every platinum thermometer the W_{CCT} values of *Table 2* and the individual deviations $\Delta W(T)$ calculated from equations 2 to 5 below give a specific $W(T)$ table.

$\Delta W(T)$ is the deviation $W(T)$ from the reference function $W_{CCT}(T)$ and is represented in the various temperature ranges by the following polynomials:

(a) From 13.81 K to 20.28 K the deviation function is

$$\Delta W(T) = A_1 + B_1 T + C_1 T^2 + D_1 T^3 \tag{2}$$

The constants are determined from measured ΔW values at the triple point of equilibrium hydrogen, at 17.042 K and at the boiling point of equilibrium hydrogen, and from the derivation of the deviation function equation 3 at the boiling point of equilibrium hydrogen.

(b) From 20.28 K to 54.361 K the deviation function is

$$\Delta W(T) = A_2 + B_2 T + C_2 T^2 + D_2 T^3 \tag{3}$$

The constants are determined from the measured ΔW values at the boiling point of equilibrium hydrogen, at the boiling point of neon and at the triple point of oxygen, and from the derivation of the deviation function equation 4 at the triple point of oxygen.

(c) From 54.361 K to 90.188 K the deviation function is

$$\Delta W(T) = A_3 + B_3 T + C_3 T^2 \tag{4}$$

The constants are determined from measured ΔW values at the triple and boiling points of oxygen and from the derivation of the deviation function equation 5.

Table 1. Values of the temperatures of the primary fixed points on the IPTS-68, and their estimated uncertainties δT in terms of their thermodynamic temperature*[17]

Substance	Equilibrium	T, K	t, °C	δT, K	$W_{\mathrm{CCT}}(T)$
Hydrogen†	solid + liquid + vapor phase (triple point)	13.81	−259.34	0.01	0.00142206
Hydrogen†	liquid + vapor phase at a pressure of 33 330.6 Pa (= 25/76 atm)	17.042	−256.108	0.01	0.00253444
Hydrogen†	liquid + vapor phase (normal boiling point)	20.28	−252.87	0.01	0.00448517
Neon	liquid + vapor phase (normal boiling point)	27.102	−246.048	0.01	0.01221272
Oxygen	solid + liquid + vapor phase (triple point)	54.361	−218.789	0.01	0.09197252
Oxygen	liquid + vapor (normal boiling point)	90.188	−182.962	0.01	0.24379909
Water‡	solid + liquid + vapor (triple point)	273.16	0.01	exact by definition	1
Water‡§	liquid + vapor phase (normal boiling point)	373.15	100.00	0.005	1.39259668
Tin§	solid + liquid phase	505.118	231.968	0.015	1.89257086
Zinc	solid + liquid phase	692.73	419.58	0.03	
Silver	solid + liquid phase	1235.08	961.93	0.2	
Gold	solid + liquid phase	1337.58	1064.43	0.3	

* Except for the triple points and one fixed equilibrium hydrogen point, the assigned temperature values correspond to equilibrium states at a pressure of 1 atm. The abbreviation atm denotes standard atmosphere defined as 1 013.250 dyn cm^{-2} or 101 325 N m^{-2}, 1 Pa (Pascal) = 1 N m^{-2}.
† Hydrogen whose ortho and para forms are in equilibrium. At the normal boiling point the equilibrium composition of hydrogen is 0.21 per cent ortho and 99.79 per cent para; at room temperature it is 75 per cent ortho and 25 per cent para. Equilibrium between the ortho and para forms is achieved by employing ferric hydroxide as catalyst.
‡ Water whose isotopic composition is that of ocean water. The extreme temperature differences between the triple points of ocean and continental surface water have been found to be about 0.00025 K.
§ The equilibrium between the solid and liquid phases of tin may be used in place of the normal boiling point of water as one of the primary fixed points.

Table 2. Values of the resistance ratio $W_{CCT}(T)$ in accordance with equation 2 for whole numbers of T

T, K	$W_{CCT}(T)$	T, K	$W_{CCT}(T)$	T, K	$W_{CCT}(T)$	T, K	$W_{CCT}(T)$
13	0.001 230 61						
14	0.001 459 73	64	0.131 111 89	114	0.346 508 00	164	0.556 870 48
15	0.001 745 41	65	0.135 303 63	115	0.350 785 19	165	0.561 016 06
16	0.002 094 74	66	0.139 512 84	116	0.355 059 10	166	0.565 159 58
17	0.002 515 12	67	0.143 738 00	117	0.359 329 89	167	0.569 301 12
18	0.003 014 28	68	0.147 977 73	118	0.363 597 54	168	0.573 440 76
19	0.003 599 62	69	0.152 230 58	119	0.367 861 99	169	0.577 578 48
20	0.004 277 80	70	0.156 495 41	120	0.372 123 31	170	0.581 714 23
21	0.005 054 95	71	0.160 771 08	121	0.376 381 51	171	0.585 848 06
22	0.005 936 68	72	0.165 056 43	122	0.380 636 57	172	0.589 979 99
23	0.006 928 04	73	0.169 350 49	123	0.384 888 51	173	0.594 110 08
24	0.008 033 16	74	0.173 652 40	124	0.389 137 32	174	0.598 238 35
25	0.009 255 04	75	0.177 961 17	125	0.393 383 16	175	0.602 364 78
26	0.010 595 85	76	0.182 276 05	126	0.397 625 94	176	0.606 489 31
27	0.012 056 90	77	0.186 596 28	127	0.401 865 67	177	0.610 612 08
28	0.013 639 01	78	0.190 921 07	128	0.406 102 42	178	0.614 733 10
29	0.015 342 61	79	0.195 249 92	129	0.410 336 28	179	0.618 852 29
30	0.017 167 68	80	0.199 582 12	130	0.414 567 09	180	0.622 969 72
31	0.019 113 63	81	0.203 917 14	131	0.418 795 07	181	0.627 085 40
32	0.021 179 44	82	0.208 254 45	132	0.423 020 15	182	0.631 199 39
33	0.023 363 43	83	0.212 593 44	133	0.427 242 33	183	0.635 311 64
34	0.025 663 35	84	0.216 933 88	134	0.431 461 69	184	0.639 422 13
35	0.028 076 45	85	0.221 275 23	135	0.435 678 31	185	0.643 530 94
36	0.030 599 53	86	0.225 617 12	136	0.439 892 10	186	0.647 638 07
37	0.033 229 16	87	0.229 959 16	137	0.444 103 22	187	0.651 743 52
38	0.035 961 55	88	0.234 301 05	138	0.448 311 59	188	0.655 847 30
39	0.038 793 05	89	0.238 642 48	139	0.452 517 30	189	0.659 949 47
40	0.041 719 68	90	0.242 983 15	140	0.456 720 33	190	0.664 049 96
41	0.044 737 60	91	0.247 322 90	141	0.460 920 77	191	0.668 148 86
42	0.047 842 92	92	0.251 661 28	142	0.465 118 61	192	0.672 246 07
43	0.051 031 78	93	0.255 998 36	143	0.249 313 87	193	0.676 341 76
44	0.054 300 36	94	0.260 333 69	144	0.473 506 60	194	0.680 435 77
45	0.057 644 86	95	0.264 667 18	145	0.477 696 82	195	0.684 528 25
46	0.061 061 61	96	0.268 998 70	146	0.481 884 59	196	0.688 619 13
47	0.064 546 79	97	0.273 328 07	147	0.486 069 85	197	0.692 708 41
48	0.068 096 90	98	0.277 655 16	148	0.490 252 74	198	0.696 796 17
49	0.071 708 35	99	0.281 979 88	149	0.494 433 19	199	0.700 882 32
50	0.075 377 56	100	0.286 302 01	150	0.498 611 35	200	0.704 966 94
51	0.079 101 23	101	0.290 621 54	151	0.502 787 07	201	0.709 050 04
52	0.082 875 95	102	0.294 938 41	152	0.506 960 58	202	0.713 131 61
53	0.086 698 59	103	0.299 252 45	153	0.511 131 72	203	0.717 211 74
54	0.909 566 00	104	0.303 563 59	154	0.515 300 65	204	0.721 290 26
55	0.094 475 15	105	0.307 871 83	155	0.519 467 37	205	0.725 367 33
56	0.098 423 36	106	0.312 177 10	156	0.523 631 80	206	0.729 442 88
57	0.102 407 74	107	0.316 479 39	157	0.527 794 09	207	0.733 516 90
58	0.106 425 83	108	0.320 778 56	158	0.531 954 17	208	0.737 589 47
59	0.110 475 06	109	0.325 074 67	159	0.536 112 11	209	0.741 660 59
60	0.114 553 12	110	0.329 367 65	160	0.540 267 92	210	0.745 730 26
61	0.118 657 89	111	0.333 657 51	161	0.544 421 67	211	0.749 798 41
62	0.122 787 22	112	0.337 944 16	162	0.548 573 36	212	0.753 865 18
63	0.126 939 14	113	0.342 227 68	163	0.552 722 91	213	0.757 930 43

Table 2—continued

T, K	$W_{CCT}(T)$	T, K	$W_{CCT}(T)$	T, K	$W_{CCT}(T)$	T, K	$W_{CCT}(T)$
214	0.761 994 30						
215	0.766 056 72	230	0.826 825 31	245	0.887 292 00	260	0.947 471 52
216	0.770 117 70	231	0.830 865 61	246	0.891 312 69	261	0.951 473 51
217	0.774 177 30	232	0.834 904 61	247	0.895 332 24	262	0.955 474 30
218	0.778 235 45	233	0.838 942 24	248	0.899 350 49	263	0.959 473 85
219	0.782 292 23	234	0.842 978 57	249	0.903 367 44	264	0.963 472 19
220	0.786 347 56	235	0.847 013 53	250	0.907 383 09	265	0.967 469 31
221	0.790 401 51	236	0.851 047 26	251	0.911 397 53	266	0.971 465 13
222	0.794 454 09	237	0.855 079 63	252	0.915 410 74	267	0.975 459 80
233	0.798 505 23	238	0.859 110 69	253	0.919 422 74	268	0.979 453 25
224	0.802 555 06	239	0.863 140 46	254	0.923 433 43	269	0.983 445 41
225	0.806 603 52	240	0.867 168 94	255	0.927 442 83	270	0.987 436 42
226	0.810 650 54	241	0.871 196 11	256	0.931 451 01	271	0.991 426 14
227	0.814 696 25	242	0.875 221 99	257	0.935 458 05	272	0.995 414 71
228	0.818 740 59	243	0.879 246 57	258	0.939 463 71	273	0.999 401 99
229	0.822 783 64	244	0.883 269 94	259	0.943 468 22		

(d) From 90.188 K the deviation function is

$$\Delta W(T) = A_4 t + C_4 t^3 (t - 100°C) \tag{5}$$

$$t = T - 273.15 \text{ K}$$

The constants are determined from measured ΔW values at the boiling points of oxygen and water. If the freezing point of tin is used as a fixed point instead of the boiling point of water, $W(100°C)$ for the platinum resistance thermometer has to be calculated from equations 6 and 7.

3. Range 0°C to 630.74°C

From 0°C to 630.74°C, t is defined by

$$t = t' + 0.045 \left(\frac{t'}{100°C}\right)\left(\frac{t'}{100°C} - 1\right)\left(\frac{t'}{419.58°C} - 1\right)\left(\frac{t'}{630.74°C} - 1\right) °C \tag{6}$$

where

$$t' = \frac{1}{\alpha}[W(t') - 1] + \delta\left(\frac{t'}{100°C}\right)\left(\frac{t'}{100°C} - 1\right) \tag{7}$$

and

$$W(t') = R(t')/R(0°C) \tag{8}$$

The constants $R(0°C)$, α and δ are determined by measurement of the resistance at the triple point of water, the boiling point of water (or the freezing point of tin) and the freezing point of zinc.

Equation 7 corresponds with the following equation:

$$W(t') = 1 + At' + Bt'^2 \qquad (9)$$

where

$$A = \alpha(1 + \delta/100°C) \quad \text{and} \quad B = -10^{-4}\alpha\delta \,°C^{-2}$$

4. Range 630.74°C to 1064.43°C

From 630.74°C to 1 064.43°C, t is defined by

$$E(t) = a + bt + ct^2 \qquad (10)$$

where $E(t)$ is the electromotive force of a standard thermocouple of a platinum–rhodium (10 per cent Rh) alloy and platinum, when one junction is at 0°C and the other at t°C. The constants a, b and c are determined from the values of E at 630.74°C $\pm$ 0.2°C, as measured by a platinum resistance thermometer, and at the freezing points of silver and gold.

The thermocouple must be such that the electromotive forces $E(630.74°C)$, $E(Ag)$ and $E(Au)$ meet the following requirements:

$$E(Au) = (10\,300 \pm 50)\,\mu V \qquad (11)$$

$$E(Au) - E(Ag) = 1\,183\,\mu V + 0.158[E(Au) - 10\,300\,\mu V] \qquad (12)$$

$$E(Au) - E(630.74°C) = 4\,766\,\mu V + 0.631[E(Au) - 10\,300\,\mu V] \pm 8\,\mu V \qquad (13)$$

5. Range above 1 064.43°C

Above 1 064.43°C the temperature T is defined by

$$\frac{L_\lambda(T)}{L_\lambda[T(Au)]} = \frac{\exp[c_2/\lambda T(Au)] - 1}{\exp[c_2/\lambda T] - 1} \qquad (14)$$

In this equation $L_\lambda(T)$ and $L_\lambda[T(Au)]$ are the spectral concentrations of the radiance of a blackbody at an unknown temperature T and at temperature $T(Au)$, the freezing point of gold, at the wavelength λ; $c_2 = 0.014388$ m × K. It is not necessary to specify the wavelength to be employed. With an optical pyrometer there is measured the ratio of the intensity of monochromatic radiation in the visible wavelength range emitted by a blackbody at an unknown temperature to the intensity of the same radiation and wavelength emitted from a blackbody at the gold point.

II. Secondary Reference Points as Recommended by the International Committee on Weights and Measures

In addition to the defining fixed points given in *Table 1*, there are a number of secondary reference points available (*Table 3*). Except for the triple points and vapor pressure equations, the temperatures relate to an equilibrium pressure of 101 325 Pa ($= 1$ atm).

Table 3. Secondary reference points of the IPTS-68

Equilibrium	T, K	t, °C
Triple point of normal hydrogen	13.956	
Boiling point of normal hydrogen ($\log p/p_0 = 1.734\,791 - 44.623\,68/T$ $+ 0.023\,186\,9\,T - 0.000\,048\,017\,T^2$ within the range $13.956 \leqslant T \leqslant 30$ K)	20.397	
Triple point of neon	24.555	
Boiling point of neon ($\log p/p_0 = 4.611\,52 - 106.385\,1/T - 0.036\,833\,1\,T + 0.000\,042\,489\,2/T^2$ within the range $24.555 \leqslant T \leqslant 40$ K)	27.102	
Triple point of nitrogen	63.148	
Boiling point of nitrogen ($\log p/p_0 = 5.893\,139 - 404.131\,05/T - 2.3749 \log T/T_0 - 0.014\,250\,5\,T + 72.534\,2 \times 10^{-6}\,T^2$ within the range $63.148 \leqslant T \leqslant 84$ K)	77.348	
Boiling point of oxygen ($\log p/p_0 = 5.961\,546 - 467.455\,76/T - 1.664\,512 \log T/T_0 - 0.013\,213\,01\,T + 50.804\,1 \times 10^{-6}\,T^2$ within the range $54.361 \leqslant T \leqslant 94$ K)	90.188	
Sublimation point of carbon dioxide $T = [194.674 + 12.264\,(p/p_0 - 1) - 9.15\,(p/p_0 - 1)^2]$ K within the range $194 \leqslant T \leqslant 195$ K	194.674	
Freezing point of mercury	234.288	
Freezing point of water		0
Triple point of diphenyl ether		26.87
Triple point of benzoic acid		122.37
Freezing point of indium		156.634
Freezing point of bismuth		271.442
Freezing point of cadmium		321.108
Freezing point of lead		327.502
Boiling point of mercury ($t = [356.66 - 55.552\,(p/p_0 - 1) - 23.02\,(p/p_0 - 1)^2 + 14.0\,(p/p_0 - 1)^3]$°C within the pressure range $90 \times 10^3 < p < 104 \times 10^3$ Pa)		356.66
Freezing point of the copper–aluminum eutectic		548.23
Freezing point of antimony		630.74
Freezing point of aluminum		660.37
Freezing point of copper		1 084.5
Freezing point of nickel		1 455
Freezing point of cobalt		1 494
Freezing point of palladium		1 554
Freezing point of platinum		1 772
Freezing point of rhodium		1 963
Freezing point of iridium		2 447
Melting point of tungsten		3 387

III. Supplementary Recommendations on Apparatus, Methods and Procedures

The official publication on the IPTS-68[17] gives some recommendations regarding apparatus, methods and procedures covering the items:

Standard resistance thermometer;
Standard thermocouple;
Triple point and the normal boiling point of equilibrium hydrogen;
Normal boiling point of neon;
Triple point and normal boiling point of oxygen;
Normal boiling point of water;
Freezing point of tin;

Freezing point of zinc;
Freezing point of silver;
Freezing point of gold.

The papers in Part 1 of Volume III of *Temperature, Its Measurement and Control in Science and Industry*[2], are concerned especially with accurate measurements of the kelvin (thermodynamic) temperatures and standards, measuring instruments and apparatus for practical thermometry for the most precise and reliable temperature measurements.

Full details of precautions to be observed in freezing point determinations are given in various publications (for instance refs. 7–9).

1. Influence of Pressure on the Freezing Point Temperature

The hydrostatic pressure in the fixed point vessels causes small temperature effects which are summarized in *Table 4*.

Table 4. Effect of pressure on the freezing-point temperatures of metals

Metal	Freezing point at 101 325 Pa (= 1 atm) °C	Pressure coefficient	
		kelvin per 101 325 Pa (= 1 atm)	kelvin per 1 cm immersion in liquid
Mercury	−38.862	+0.0054	+0·000071
Indium	156.634	+0·0049	+0·000033
Tin	231.968	+0·0033	+0.000022
Bismuth	271.442	−0·0035	−0.000034
Cadmium	321.108	+0.0062	+0.000048
Lead	327.502	+0.0080	+0.000082
Zinc	419.58	+0.0043	+0.000027
Antimony	630.74	+0.00085	+0.000005

2. Triple Point, 17.042 K Point and Boiling Point of Equilibrium Hydrogen

The temperature of equilibrium between the solid, liquid and vapor phases of hydrogen can be determined by employing a sufficient quantity of liquid hydrogen with a catalyst (e.g. ferric hydroxide) in a cavity in a copper block in which platinum resistance thermometers are immersed and which is surrounded by an evacuated space. The temperature of the block is lowered until the hydrogen freezes; the temperature is then allowed to rise slowly and the transition at the triple point is observed. The flat portion of the time/temperature curve may be constant to approximately 0.0001 K for thirty minutes or longer.

The equilibrium temperature between liquid and gaseous hydrogen is normally measured by the static method. The vapor pressure is transmitted to a manometer located outside the cryostat. Comparative measurements are taken between the vapor-pressure thermometer thus formed and the immersed platinum resistance thermometers. The measurements may be considered valid when the values obtained are independent of the ratio of the volume of liquid hydrogen to the volume of the vapor in the cavity. This is also true of the boiling points of neon, oxygen and water.

The temperature as a function of the vapor pressure p of equilibrium hydrogen is given to an accuracy of a few millikelvins for the range 13.81 K to 23 K by the equation

$$\log p/p_0 = A + B/T + CT + DT^2 \qquad (15)$$

where

$A = 1.711\,466$ $\qquad\qquad B = -44.010\,46$ K

$C = 0.023\,590\,9$ K^{-1} $\qquad D = -0.000\,048\,017$ K^{-2}

$p_0 = 101\,325$ Pa

3. Boiling Point of Neon

The temperature T as a function of the vapor pressure p of neon is given to an accuracy of ± 0.0002 K for the range from 27 K to 27.2 K by the equation

$$T = [27.102 + 3.3144\,(p/p_0 - 1) - 1.24\,(p/p_0 - 1)^2 \\ + 0.74\,(p/p_0 - 1)^3]\ \text{K} \qquad (16)$$

4. Triple Point and Boiling Point of Oxygen

These two points may be determined in a manner similar to that described for hydrogen.

Particular attention should be paid to the purity of the oxygen used in the vapor-pressure thermometer. It is sufficiently pure when the normal boiling point remains constant with the stepwise removal of oxygen vapor fractions.

The temperature T as a function of the vapor pressure p of oxygen is given to an accuracy of ± 0.0001 K for the range 90.1 K to 90.3 K by the equation

$$T = [90.188 + 9.5648\,(p/p_0 - 1) - 3.69\,(p/p_0 - 1)^2 \\ + 2.22\,(p/p_0 - 1)^3]\ \text{K} \qquad (17)$$

5. Boiling Point of Water

The boiling point of water is usually determined by the dynamic method, the thermometer being located in the saturated water vapor. For exact measurements it is preferable to use closed systems connected to a manostat. The thermometer must be protected against radiation emitted by bodies at temperatures varying from the boiling-point temperature.

The observed temperature must, at constant pressure, be independent of variations in the rate of heat supply, measuring time and depth of immersion of the thermometer.

The temperature t as a function of the vapor pressure p of water is given to an accuracy of ± 0.0001 K for the range 99.9°C to 100.1°C by the equation

$$t = [100 + 28.0216\,(p/p_0 - 1) - 11.642\,(p/p_0 - 1)^2$$
$$+ 7.1\,(p/p_0 - 1)^3]°C \qquad (18)$$

6. *Freezing Points of Tin and Zinc*

Temperatures which are easily reproducible are obtained in the slow freezing of high purity metals. A criterion for checking the purity of a sample of zinc or tin is that its melting range should be less than 0.001 K. Similar requirements should be made on the other fixed points having melting-point temperatures below 630°C. Above 630°C the IPTS-68 scale may be attained by using as interpolation instrument the thermocouple which is much less accurate than to the extent of ±0.001 K.

A measure of the purity of the metal is the degree of supercooling in the freezing process. It is not possible to give a generally applicable value as it may be influenced by many factors. As the variation in freezing temperature depends to a considerable extent on the type of impurity, it is difficult to estimate the influence of small amounts of impurities on the freezing temperature.

For zinc and tin a purity of 99.9999 per cent is recommended, but not directly required.

The freezing points of zinc and tin may be determined in an inert atmosphere in a crucible of very pure graphite (99.999 per cent by weight) provided with an axial thermometer well.

7. *Freezing Points of Silver and Gold*

It is recommended that the freezing points of silver and gold should be measured in crucibles of very pure artifical graphite, ceramic material or vitreous silica.

The equilibrium temperature has been attained when the electromotive force of the thermocouple is independent of small variations in the depth of immersion into the metal, and when it remains constant for at least five minutes during one freeze. The crucible containing the gold must be such that it contains a cavity of uniform temperature immersed in the gold. This blackbody is required for measurements with radiation pyrometers. Because of its high emissivity, graphite is suitable for this purpose.

IV. Practical Temperature Scales over the 0.2 K to 5.2 K Range

The 1958 ^{4}He scale[4] and the 1962 ^{3}He scale[15] recommended by the International Committee of Weights and Measures are defined by the temperature dependence of the ^{4}He and ^{3}He vapor pressures.

The upper limits are set by the critical points of these gases and the lower limits by the vapor pressures becoming too low for practical measurement.

V. Standard Reference Samples

1. Gases

The fixed points of the temperature scale may be determined at low temperatures with gases which are supplied in extremely high purity by industry.

Table 5 lists various suppliers of high purity gases and the percentage purity specified by the supplier.

Table 5.

Supplier	H_2	O_2	Ne	N_2	CO_2
		purity in vol. % specified by the supplier			
L'air Liquide	99.9997	99.998	99.99	99.9992	99.9986
Messer Griesheim	99.9995	99.99	99.999	99.999	99.9985
Linde	99.999	99.95	99.993	99.999	
Baker	99.999	99.6	99.996	99.997	99.99
Lif-O-Gen	99.9995	99.997	99.998	99.999	99.995

It should be noted that the expression 'Research Grade', which is used by various suppliers, is not defined and is often employed to designate the purest product manufactured. Other suppliers who perhaps produce a purer gas do not use this term. Considerable differences have arisen between various state institutes regarding the determination of the boiling point of oxygen; these discrepancies are probably due first and foremost to impurities in the gas. Particularly in the case of oxygen, it is extremely difficult to ascertain a purity of 99.995 per cent.

The final criterion for establishing whether gases are sufficiently pure is the IPTS-68 definition that the fixed points should be independent of the ratio of the volume of the liquid phase to the volume of the gas phase. The changes in boiling point should not exceed 1 mK even if there is a considerable variation in the liquid portion. As the low-boiling gases hydrogen, neon and oxygen are isotope mixtures, the isotope composition varies in the isothermal expansion defined by IPTS-68, resulting in a change in boiling point of 0.5 mK. However, for practical realization of the temperature scale this effect should be of no significance.

2. Catalogue of Physicochemical Standard Substances

The IUPAC Commission on Physicochemical Measurements and Standards has published a *Catalogue of Physicochemical Standard Substances*[12] in which the samples obtainable from metrological state institutes for thermometric fixed points are listed. See *Table 6*.

3. Further Recommendations of Calibration Materials

The criterion for acceptability of the recommended materials is temperature reproducibility for the fixed points. Chemical purity of the substances is of the utmost importance. Methods of quality control vary from substance to substance. An important criterion for the purity of triple point or melting

Table 6

Purity moles %	Chemical name (Sample No.)	Value and accuracy	Source	Remarks
99.999	Aluminum metal	Freezing point (660°C)[b]	A	Thermodynamic temperatures
	Aluminum (44e)	Freezing point (660.3 ± 0.2°C)[b]	D	
	Aluminum oxide	Melting point (2053°C)	D	See refs. 13 and 14
99.99	Benzoic acid	Freezing point (122°C)[b]	C	Purity derived from temperature/enthalpy curves
99.99	Benzophenone	Freezing point (48°C)[b]	C	
99.99	Biphenyl	Freezing point (70°C)[b]	C	
99.999	Cadmium metal	Freezing point (321°C)[b]	A	Thermodynamic temperatures
99.999	Cadmium (SRM 749)	Freezing point (321.108°C)[b]	D	
	Copper (45d)	Freezing point (1084.5 ± 0.5°C)[b]	D	Standard for vapor pressure
99.99	Dimethyl terephthalate	Freezing point (142°C)[b]	C	Purity derived from temperature/enthalpy curves
99.998	Gold metal	Freezing point (1064.43°C)[a]	A	Thermodynamic temperatures
99.999	Gold (SRM 745)	Freezing point (1064.43°C)[a]	D	Standard for vapor pressure
	Lead (49e)	Freezing point (327.502 ± 0.005°C)[b]	D	
99.99	Naphthalene	Freezing point (80°C)[b]	C	Purity derived from temperature/enthalpy curves
99.996	Neopentane	Transition point (−132°C)[b]	B	Purified by using a spinning band type distillation tower of 3 m height and an adsorption column packed with molecular sieve
		Triple point (−16°C)[b]	B	

Table 6 —continued

Purity moles %	Chemical name (Sample No.)	Value and accuracy	Source	Remarks
99.99	Phenanthrene	Freezing point (100°C)[b]	C	Purity derived from temperature/enthalpy curves
99.999	Platinum	Freezing point (1772°C)[b]	D	In process as vapor pressure SRM
99.994	Silver metal	Freezing point (961.93°C)[a]	A	Thermodynamic temperatures
>99.9	Silver–copper eutectic	Freezing point (779°C)[b]	A	Thermodynamic temperatures
99.999	Silver (SRM 748)	Freezing point (961.93°C)[a]	D	Standard for vapor pressure
99.99	Sodium	Freezing point (97°C)[b]	C	Purity derived from temperature/enthalpy curves
99.999	Sulphur	Boiling point (444°C)[b]	A	Thermodynamic temperatures
99.99	1,2,4,5-Tetrachlorobenzene	Freezing point (140°C)[b]	C	Purity derived from temperature/enthalpy curves
99.999	Tin metal	Freezing point (231°C)[b]	A	Thermodynamic temperatures
	Tin (42f)	Freezing point (231.968 ± 0.005°C)[a]	D	
	Tungsten	Freezing point (3387°C)[b]	D	In process as vapor pressure SRM
99.999	Zinc metal	Freezing point (419.58°C)[a]	A	Thermodynamic temperatures
99.999	Zinc (SRM 740)	Freezing point (419.58°C)[a]	D	

NOTE. SRM = Standard Reference Material. Temperature of the primary fixed points for calibration on the IPTS-68 are indicated by the superscript a. Secondary reference points carry a nominal temperature value for general information only, and are indicated by the superscript b. The temperature certified by the standardizing laboratory appears only on the certificate provided with the sample.

Benzoic acid (triple point 122.37°C), naphthalene, phenol and phthalic anhydride were available as thermometric triple point cells from the US National Bureau of Standards. Stocks of these cells were not replenished, but special purity material for this use will still be made available.

A Physikalisch-Technische Bundesanstalt. 33 Braunschweig, Bundesallee 100, Federal Republic of Germany

B The Government Chemical Industrial Research Institute of Tokyo, 1-Chome, Honmachi, Shibuya-Ku, Tokyo, Japan

C Institute for Physical Chemistry TNO, Utrechtseweg 48, P.O. Box 108, Zeist, The Netherlands

D Office of Standard Reference Materials, US Department of Commerce, National Bureau of Standards, Washington, DC, 20234, USA

point substances is temperature constancy in the melting range, i.e. in the flat portion of the time/temperature curve. Below 630°C this melting range should generally be less than 0.005 K. A further purity criterion often employed is the degree of supercooling during the freezing of the substances. However, this varies considerably from substance to substance.

Table 7

Substance	Triple point temperature °C (m melting point temp.)	References
p-Nitrotoluene[a]	51.54 ± 0.02	3, 10
Naphthalene[b]	80.275 ± 0.010	3, 6, 10
Adipic acid[c]	151.42 ± 0.02	3
Anisic acid[d]	182.98 ± 0.02	3, 10
2-Chloranthraquinone[e]	209.035 ± 0.020	3, 10
Carbazole[f]	245.34 ± 0.02	3, 10
Anthraquinone[g]	284.59 ± 0.01	3, 10
Imidazole	89.50 (m)	10
Acetanilide	114.30 (m)	10
Phthalic anhydride	131.22 (m)	10
Dimethyl terephthalate	140.65 (m)	5, 10
Benzanilide	163.20 (m)	10
Triphenylene	198.40 (m)	10
Hexachlorobenzene	227.80 (m)	10
p-Iodobenzoic acid	271.60 (m)	10
1-Methylnaphthalene	−30.480 (m)	1
Chlorobenzene	−45.58 (m)	11
n-Heptane	−90.61 (m)	1
Carbon disulphide	−112.1 (m)	16

[a] Recommended methods for purifying technical grade material: recrystallization (from methanol), distillation and drying over phosphorus pentoxide; in each cell, it was possible to maintain the triple point temperature within ±0.002°C for at least three hours.

[b] The desired mush at the start of freezing is not readily obtained with this compound. Premature crystallization of the material must be avoided to obtain reproducible freeze-outs. A reduced cooling rate is necessary until freezing is initiated with the formation of a few crystals. At this point, the additional insulation must be removed and continued shaking of the cell produces a fine crystalline suspension. It was possible to maintain the triple point temperature within ±0.002°C for at least fifteen minutes. Purified grade naphthalene was used without further treatment.

[c] Using the same technique described for naphthalene, reproducible freeze-outs were obtained. It was possible to maintain the triple point temperature within ±0.002°C for at least twenty minutes. Certified grade material was employed without further treatment.

[d] The insulated cell technique as described for naphthalene is necessary to obtain consistent freeze-outs. It was possible to maintain the triple point temperature within ±0.002°C for at least ten minutes. Technical grade anisic acid was used, purified by acid hydrolysis of the recrystallized sodium salt.

[e] Colarusso and Semon[3] were able to maintain the triple point temperature within ±0.002°C for at least twenty minutes. They purified technical grade material by sublimation.

[f] It was possible to maintain the triple point temperature within ±0.002°C for at least ten minutes employing the insulated cell technique described for naphthalene. In spite of darkening of the material during prolonged heating periods, the triple point values were not appreciably affected. Certified grade carbazole, purified by sublimation, was used by the authors.

[g] It was possible to maintain the reproducible equilibrium temperature within ±0.002°C for at least ten minutes using the insulated cell procedure mentioned above. The authors used technical grade anthraquinone, purified by sublimation.

VI. References

[1] American Petroleum Institute *Research Project 44*.
[2] Brickwedde, F. G. (Ed.) *Temperature: Its Measurement and Control in Science and Industry*, Reinhold: New York (1962).
[3] Colarusso, V. G. and H. A. Semon. *Analyt. Chem.* **40**, 1521 (1968).
[4] van Dijk, H., M. Durieux, J. R. Clement and J. K. Logan. *J. Res. Nat. Bur. Stand.* **64A**, 1 (1960).

[5] Elliot, J. H. and M. D. Chris. *J. Chem. Engng Data*, **13**, 475 (1968).
[6] Enagonio, D. P., E. G. Pearson and C. P. Saylor. 'Thermometric cells for calibration of liquid-in-glass thermometers'. *Temperature: Its Measurement and Control*, Vol. III, Pt 1, 219. C. M. Herzfeld (Ed.) Reinhold: New York (1962).
[7] *J. Res. Nat. Bur. Stand.* **14**, 247 (1935).
McLaren, E. H. 'The freezing points of high-purity metals and precision temperature standards'. *Temperature: Its Measurement and Control in Science and Industry*, Vol. III, Pt 1, 185. Reinhold: New York (1962).
[9] 'Methods of testing thermocouples and thermocouple materials' (by W. F. Roesner and S. T. Lonberger), *US Nat. Bur. Stand. Circ. No. 590* (1958).
[10] Pella, E. and M. Nebuloni. *J. Thermal Anal.*, **3**, 229 (1971).
[11] *Physical Properties of Chemical Compounds*, R. R. Dreisbach Vols. I and II. American Chemical Society: Washington DC (1955–61).
[12] 'Catalogue of Physicochemical Standard Substances'. *Pure Appl. Chem.* **29**, 599 (1972).
[13] Schneider, S. J. *Pure Appl. Chem.* **21**, 115 (1970).
[14] Schneider, S. J. and C. L. McDaniel. *J. Res. Nat. Bur. Stand.* **71A**, 317 (1967).
[15] Sydoriak, S. G. and R. H. Sherman. *J. Res. Nat. Bur. Stand.* **68A**, 547 (1964);
Sydoriak, S. G., T. R. Roberts and R. H. Sherman. *J. Res. Nat. Bur. Stand.* **68A**, 559 (1964);
Sherman, R. H., S. G. Sydoriak and T. R. Roberts. *J. Res. Nat. Bur. Stand.* **68A**, 579 (1964).
[16] 'Selected values of chemical thermodynamic properties', *Circ. US Nat. Bur. Stand. No. 500* (1952).
[17] 'The International Practical Temperature Scale of 1968'. *Metrologia*, **5**, 35 (1969).

CHAPTER 3

Temperature Measurement under Pressure

B. LE NEINDRE and Y. GARRABOS
1 Place A. Briand, Bellevue 92, France

CONTENTS

I.	Introduction	87
II.	Temperature Measurement at Atmospheric Pressure	88
	1. Temperature Measurement below 13.81 K	88
	2. Temperature Measurement from 13.81 K to 630.74°C	88
	A. The Platinum Resistance Thermometer	88
	B. Thermocouples	89
	C. The Quartz Thermometer	89
	3. Temperature Measurement from 630.74 to 1 064.43°C	89
	4. Temperature Measurement above 1 064.43°C	89
III.	Temperature Measurement under Pressure	90
	1. Determination of Pressure Effects on Thermoelectricity	90
	2. Temperature Measurement in a Hydrostatically Pressurized Cell	91
	A. The Single Wire Method	91
	B. The Composite Thermocouple Method	92
	C. Results	93
	(1) Pt/Pt–10% Rh Thermocouples	93
	(2) Chromel–alumel Thermocouples	94
	(3) Copper–constantan Thermocouples	94
	(4) Electrical Feedthrough Techniques for High Pressure	95
	3. Temperature Measurement in Non-hydrostatic Systems	96
	A. Modified Bridgman Method	97
	B. Thermal Noise Method	100
	C. Indirect Self-consistent Methods	102
	D. Results	105
	(1) Pt/Pt–10% Rh Thermocouples	105
	(2) Chromel–alumel Thermocouples	105
	(3) Copper–constantan Thermocouples	105
	(4) Ni/Ni–18% Mo Thermocouples	105
	(5) Iron–vanadium, Iron–nickel and Iron–constantan Thermocouples	106
	(6) Au/Pt–40% Rh Thermocouples	106
	(7) Tungsten–iridium Thermocouples	106
	4. Effect of Pressure on the Relative Difference between Various Thermocouples	106
	5. Optical Pyrometry at High Pressure	108
	6. Computer Method	111
IV.	References	111

I. Introduction

One of the major problems in research in the area of high pressure has been the lack of an accurate and reliable method for continuous and simultaneous *in situ* evaluation of true pressure and temperature. Up to now, most of the temperature sensors have been developed for use at atmospheric pressure. On the other hand pressure calibration is difficult at temperatures far from room temperature, for instance close to absolute zero. The problem of temperature measurement in high pressure experiments has become more acute with the development of internally heated pressure cells. The great majority of sensors developed for use under pressure have been thermocouples, whose small size makes them suitable for high pressure experiments, even if such techniques are not free of problems as we will see. In the measurement of temperature at high pressure the biggest additional difficulty arises from the fact that it is difficult to get truly hydrostatic conditions; most of the high temperature, high pressure cells are composed of materials like levastone, talc or graphite which do not permit a close approximation to hydrostatic conditions.

II. Temperature Measurement at Atmospheric Pressure

A general survey of temperature measurement and control at atmospheric pressure has been edited by Herzfeld[44].

1. Temperature Measurement below 13.81 K

Germanium or carbon thermometers are specially useful in the temperature range below 20 K. In this temperature range the sensitivity of most thermocouples approaches zero. However, thermocouples using an alloy of gold with 2.11 atomic per cent of cobalt versus copper or 'normal' silver have a greater sensitivity than other types of thermocouples[54, 60, 61].

2. Temperature Measurement from 13.81 K to 630.74°C

A. THE PLATINUM RESISTANCE THERMOMETER

The standard instrument used from 13.81 K to 630.74°C is the platinum resistance thermometer. The standard platinum resistance thermometer has a four-terminal resistance element made of pure platinum wire, wound in such a way that it is free of strain initially and remains so in use. The four terminals allow the resistance of the resistor between the branch points to be measured independently of the resistance of the leads. When the thermometer has been calibrated, it is provided with a table which relates resistance ratios to temperature range expressed to a maximum precision equivalent to 0.0001 degree. It is usual to make calibrations by transfer from a standard thermometer to another thermometer. When the uncertainties of the primary calibration and the transfer are combined the total uncertainty of the calibration obtained by transfer is approximately ±0.002 per cent of the temperature in degrees C. The Mueller bridge is commonly used to measure carefully the resistance of the thermometer coil. To determine the temperature, one measures the ratio R_t/R_0 where R_t is the resistance of the resistor at the temperature to be determined and R_0 is the resistance of the resistor at the ice point. For accurate determinations R_0 should be determined with the same bridge that measures R_t, thus an ice bath or a triple point cell is necessary. The platinum resistance thermometer cannot be used under pressure.

B. Thermocouples

Iron–constantan thermocouples are widely used in the range $-190°C$ to $800°C$[54, 48, 7]. Copper–constantan thermocouples also have a large sensitivity in the range from $-200°C$ to $400°C$[54, 7]. Another thermocouple, called 'Platinel' [(3Au–83 Pd–14 Pt)/(65 Au–35 Pd)] has been developed recently. In the temperature range from $100°C$ to $900°C$ the chromel–constantan thermocouple has great sensitivity.

C. The Quartz Thermometer

In the temperature range $-80°$ to $250°C$, quartz thermometers, which provide a precise digital indication of temperature to a resolution of 0.1 millidegree, have been developed recently*.

3. Temperature Measurement from 630.74°C to 1064.43°C

The standard instrument for the range $630.74°C$ to $1064.43°C$ is the platinum/platinum–10% rhodium thermocouple[3, 55]. The reference junction is held at a fixed temperature, which is usually the ice point. For accurate measurements the wires must be homogeneous, unstressed and well insulated. After calibration at fixed points (silver, gold) the uncertainties of this thermocouple are about $\pm 0.3°C$ from 630.74 to $1064.43°C$ and do not increase more than $2°C$ up to $1450°C$.

4. Temperature Measurement above 1064.43°C

Above $1064.43°C$ the IPTS-68 is defined by Planck's radiation law as expressed by the relation

$$\frac{I_\lambda(T)}{I_\lambda(T_{Au})} = \frac{\exp[c_2/\lambda T_{Au}] - 1}{\exp[c_2/\lambda T] - 1} \tag{1}$$

where $I_\lambda(T)$ and $I_\lambda(T_{Au})$ are the spectral intensities at temperature T and at the freezing point of gold, (T_{Au}), of the radiance of a blackbody at the wavelength λ, and $c_2 = 0.014388$ meter kelvin. The optical pyrometer is the standard instrument in this temperature range. The uncertainty of the calibration of an optical pyrometer is about $\pm 6°C$. An automatic optical pyrometer which uses a photoelectric detector to balance the brightness of the pyrometer lamp filament with that of the target of unknown temperature has been developed recently (Leeds and Northrup Co.). The uncertainty of the photoelectric pyrometer is well below $\pm 1°C$. Corrections in optical pyrometry and photometry for the refractive index of air have been published by Blevin[11]. An extensive review of high temperature thermometry (above

---

* Hewlett–Packard, Palo Alto, California, USA.

1 000 K) has been prepared recently by Bedford[4] and by Zysk et al.[69, 71], which includes discussions of noble and refractory metal thermocouples (with emphasis on the platinum–rhodium and tungsten–rhenium alloys), recent developments in optical pyrometry and some other useful techniques. Several thermocouples have been developed for temperature measurements above 1000°C[4]. Platinum-based thermocouples such as the Pt/Pt–13% Rh[3, 5, 55], are stable up to 1 400°C, the Pt–20% Rh/Pt–40% Rh[2] or the Pt–30% Rh/Pt–6% Rh[19] have a practical upper limit of about 1 800°C. Iridium-based thermocouples are preferred at higher temperatures up to 2 100°C, and include the Ir/Ir–60% Rh[9] or Ir/Ir–50% Rh[10]. For temperatures up to 2 300°C and possibly as high as 3 000°C, tungsten-based thermocouples are appropriate, such as the W/Re[64], W/W–26% Re[47] or W–3% Re/W–25% Re[1, 71].

III. Temperature Measurement under Pressure

Since thermocouples provide the primary means for *in situ* temperature measurement in most high-pressure experiments carried out at elevated temperatures, a knowledge of the effect of pressure on the electromotive force of thermocouples is very important.

1. Determination of Pressure Effects on Thermoelectricity

Until now, the theoretical understanding of thermoelectricity and the electronic properties of metals and alloys have not been able to provide any reliable quantitative calculations of the effect of pressure on thermal e.m.f.s. The e.m.f. of a thermocouple is due to the temperature dependence of the Fermi energies of the metals which constitute the thermocouple. As the Fermi energy of a metal is pressure-sensitive, the calibration of a thermocouple changes with pressure. Some of the effects involved have been reviewed by Bourassa[14] *et al.* The voltage of a thermocouple generated over a given temperature interval is

$$E = \int_{T_0}^{T_j} S_{ab}(T,p) \, dT \tag{2}$$

where T_0 and T_j are reference temperature and temperature of the hot junction, respectively; $S_{ab}(T, p)$ is the relative Seebeck coefficient or the thermoelectric power[2]. The thermoelectric power of a given material is a function of both pressure and temperature and can be written as a sum of a pressure-independent term and a pressure-dependent term, $S_{ab}(T, p) = S_{ab}(T) + \Delta S_{ab}(p, T)$. Thus the e.m.f. of a pressurized thermocouple is given by

$$E = \int_{T_0}^{T_s} S_{ab}(T) \, dT + \int_{T_s}^{T_j} \Delta S_{ab}(p, T) \, dT \tag{3}$$

where $\Delta S_{ab}(p, T)$ is the pressure-modified Seebeck coefficient and T_s is the pressure seal temperature. The relative Seebeck coefficient applies to a pair of thermoelements a, and b is the difference of two absolute Seebeck coefficients, each of which is applied to one thermoelement.

$$S_{ab} = S_a - S_b$$

The voltage change introduced by pressurizing the thermocouple is the amount by which the e.m.f. changes over the same temperature range.

$$\Delta E = \int_{T_s}^{T_j} [S_a(T) - \Delta S_a(p, T)] dT - \int_{T_s}^{T_j} [(S_b(T) - \Delta S_b(p, T)] dT \qquad (4)$$

where $\Delta S_a(p, T)$ and $\Delta S_b(p, T)$ are the pressure-modified absolute Seebeck coefficients. The thermoelectric power S of a material can be evaluated at the Fermi energy ε_F by[53]

$$S = \frac{\pi^2}{3} \frac{k^2 T}{e\varepsilon_F} \left\{ \frac{\partial \ln A}{\partial \varepsilon} + \frac{\partial \ln \lambda}{\partial \varepsilon} \right\}_{\varepsilon_F} \qquad (5)$$

where k is the Boltzmann constant, T is the absolute temperature, e is the electron charge, A is the effective area of the Fermi surface, and λ is the electron mean free path. The pressure dependences of A and λ are not known sufficiently well in real metals and alloys to predict the effect of pressure on thermocouples. However, a rough estimation can be made by applying the free electron approximations to the terms in A and λ and differentiating with respect to pressure, so that

$$dS/dp \simeq \alpha(\pi^2 k^2 T/e)(\kappa/\varepsilon_F) \qquad (6)$$

where α is a factor of the order of unity and κ is the isothermal compressibility.

The simultaneous pressure gradient and temperature gradient, which generally exist in a pressure seal and throughout the high pressure cell, may modify this result. A solution to this problem has been given by Hanneman and Strong[41] when T and p are both functions of position within the cell and therefore can be written as functions of each other. In the case of a single wire, they approximate $\int_{T_s}^{T_j} \Delta S_a(p, T) dT$ by $A \int_{T_s}^{T_j} p(T) dT$, where A is a constant and $p(T)$ is the internal pressure written in terms of the temperature.

2. Temperature Measurement in a Hydrostatically Pressurized Cell

The hydrostatic conditions are in general limited to a pressure less than 15 kbar and a temperature no more than several hundred degrees. Essentially two types of methods have been used, namely single wire and composite thermocouple methods.

A. The Single Wire Method

A diagram of this method is shown in *Figure 1*. Here, a part of a continuous homogeneous wire is pressurized. One pressure seal is heated to T_j, while the other is cooled to T_0 by an external bath. As the internal pressure is increased from atmospheric pressure p_0 to a higher value p_j, the thermal e.m.f. output of the single wire will continuously deviate from its corresponding value at p_0 for fixed values of T_j and T_0. Thus, accurate, absolute corrections for pressure effects on the thermocouple can be obtained. In this method the pressure gradient in the seal regions from p_j to p_0 must be kept isothermal. Bridgman[16] was the first to measure the effect of pressure on thermal e.m.f.s of various materials using the single wire method. His measurements concerned some 20 metals and alloys from 0 to 12 kbar and in a limited temperature range (0° to 100°C). Unfortunately, these measurements did not include most of the thermocouple materials now in use, but the results are generally believed to have an accuracy of better than ten per cent. The single wire measurements of the pressure dependence of the thermoelectric power have been made hydrostatically by Freud and La Mori[32] for chromel,

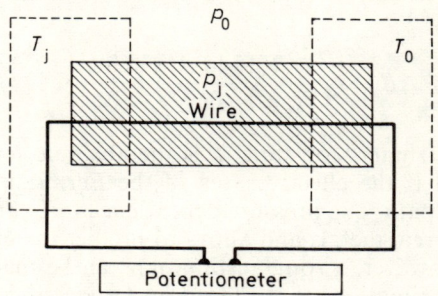

Figure 1. Schematic diagram of high-pressure single wire circuit.

alumel, copper and constantan to 8 kbar in the temperature range −195°C to 290°C. A schematic diagram of the hydrostatic apparatus is given in *Figure 2.* The wires were threaded through the 0.6 mm diameter bore of a stainless steel tube. One end was brought out to ambient pressure through a frozen gas **U**-tube seal at 78 K. At the other end the high temperature seal

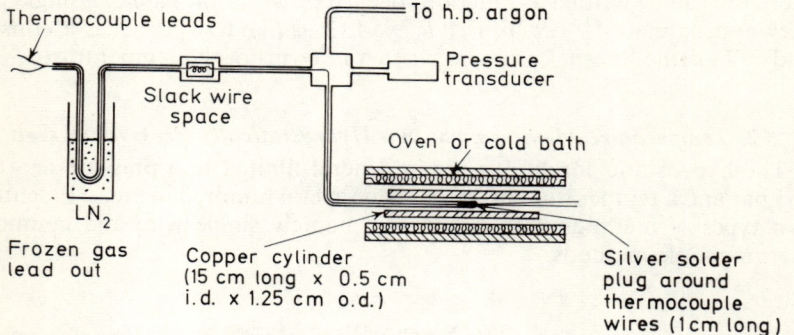

Figure 2. Hydrostatic pressure apparatus used by Freud *et al.* for measuring single wire high-pressure thermal e.m.f.

is made by silver soldering the thermocouple wire a small distance into the high pressure tubing.

B. The Composite Thermocouple Method

The related composite thermocouple method for measuring the pressure effects on thermocouples in a hydrostatic system is shown schematically in *Figure 3.* Birch[8] was the first to attempt to use this method from 0 to 4 kbar and up to 470°C, but he did not cool the pressure seal in his apparatus to 0°C and he did not observe an absolute effect as discussed by Bundy[18]. Accurate corrections on a Pt/Pt–10% Rh thermocouple up to 10 kbar and

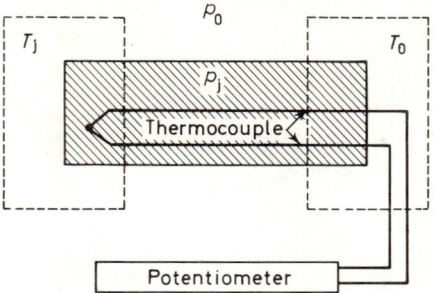

Figure 3. Schematic diagram of thermocouple circuit for high pressure apparatus.

500°C have been obtained by Bell, England and Boyd[6] using a composite thermocouple hydrostatic method. The thermocouple was subjected to nitrogen pressure in a tube (René 41), and heated by an electrical furnace constructed to have a thermal gradient less than 0.4°C/cm (*Figure 4*). The e.m.f. of the pressurized thermocouple was read as a function of pressure relative to a second Pt/Pt–10% Rh thermocouple heated to the same

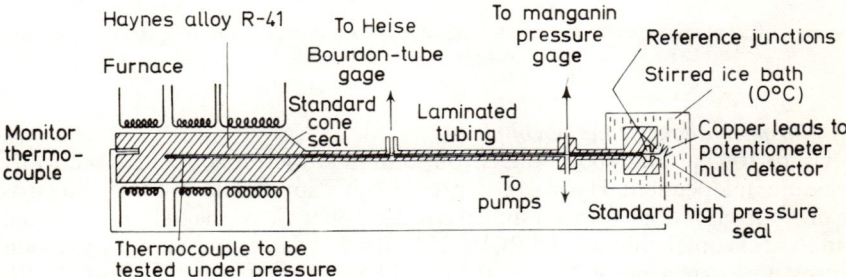

Figure 4. Gas apparatus used by Bell *et al.* for measurement of the effect of pressure on the thermal e.m.f. of the platinum/platinum–10 per cent rhodium thermocouple.

temperature in a small cavity at atmospheric pressure. Leads from both thermocouples were brought into a stirred ice bath in which the pressure seal was suspended. Two additional thermocouples were employed, one in the ice bath itself and another on the high-pressure side of the cold pressure seal.

C. Results

(1) Pt/Pt–10% Rh *Thermocouples*

Figure 5 shows the experimental data of Bell *et al.*[6] in a series of isobars. A tentative temperature correction of the Pt/Pt–10% Rh thermocouple was tried by Vidal et al.[66] using the indirect method of the 'melting wire' under hydrostatic pressure up to 8 kbar. Even if their results are not very accurate, the authors estimated the correction to be of the order of those of Getting and Kennedy[36] at 1 000°C.

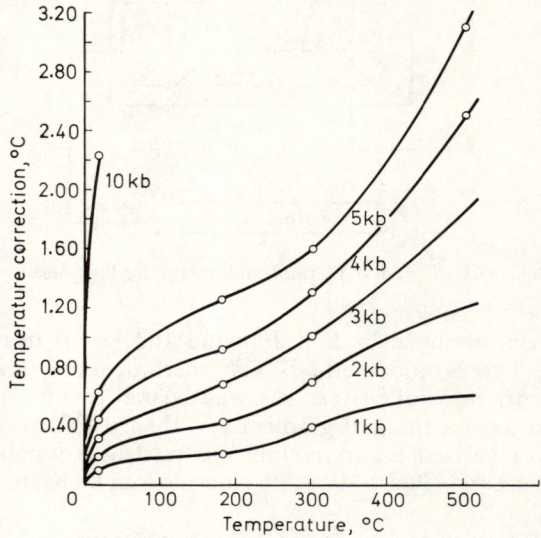

Figure 5. Effect of pressure on the thermal e.m.f. of the platinum/platinum–10% rhodium thermocouple (Bell *et al.*).

(2). Chromel–alumel Thermocouples

Up to 10 kbar and 600°C the chromel–alumel thermocouple is practically pressure independent. Hydrostatic-pressure measurements, made by Lazarus *et al.*[50] up to 7 kbar (argon) and from 500°–980°C, of the relative thermal e.m.f.s of chromel–alumel and Pt/Pt–10% Rh thermocouples have shown that a pressure correction of $(0.57 \pm 0.03°C)/kbar$ is required for Pt/Pt–10% Rh thermocouples from about 600° to 1 000°C. This value corresponds to slightly lower corrections than those given by Hanneman *et al.*[43].

Chromel–alumel thermocouples become pressure dependent at 720°C and degrade rapidly at 800°C. This behavior is borne out by the more recent results of Getting and Kennedy[36] who found major deviations from linearity in the output of their chromel–alumel thermocouples at 1 000°C. However, systematic differences of unknown origin were found between pressure corrections measured first with pressure increasing and then with decreasing pressure.

(3). Copper–constantan Thermocouples

Corrections deduced from the data of Freud and La Mori[32] in the temperature interval of 190 K to 560 K are presented *Figure 6*. Scatter in their data introduced an error of ± 0.25 μV/kbar. Bloch and Chaisse[12] have used the composite thermocouple method to determine the effect of pressure on copper–constantan thermocouples over the range 0 to 5 kbar and −196°C to 89°C. Their copper–constantan results are in excellent agreement with those of Bridgman and of Freud and La Mori over the common range of measurement.

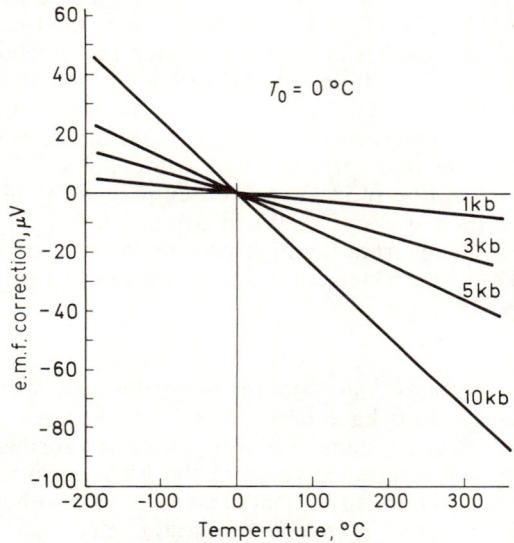

Figure 6. Effect of pressure on the thermal e.m.f. of the copper–constantan thermocouple.

(4). Electrical Feedthrough Techniques for High Pressure

Reviews of electrical feedthrough techniques for high pressure gas systems have been reported previously[25,51,62,65]. Pipestone cone seals, in which a central cone (the conductor) is insulated from the pressure vessel by a hollow cone of pipestone, have been traditionally used up to 10 kbar with hydrocarbons[17,59], and up to 5.5 kbar with hydrogen[24]. A variant is when the conical conductor is coated with a layer of epoxy[22] or some other insulating shell such as ivory, alumina, teflon or makrolon. Terry and Ruof[63] have described a convenient method for making plastic cone seals by heat shrinking Kynar* or other plastic tubing. Another variant is a technique in which insulated wire passes through a hole into a shell of epoxy which is sealed like the above[13,20]. Several conductors can be used for one epoxy shell. An unusual method, in which the conductor is a bolt screwed into the wall of the pressure vessel and insulated by a resin material with finely ground mica, has been used by Gibbs and Jarman[37].

At low temperatures, a different approach is to seal several leads in frozen oil (Silicone oil or Octoil S)[26,56]. However, it has the disadvantage that considerable lengths of electrical leads have to be used. The technique using

---

* Kynar is a trade name for a semi-rigid high temperature polyvinylidene fluoride tubing which is manufactured by Rayclad Tubes, a subsidiary of the Raychem Co.

epoxy has the disadvantage that the difference in thermal expansion coefficients between the epoxy and associated components of the seals causes the epoxy to crack open; this has been prevented by a simple technique developed by Hammons[40]. A method which does not require precise machining is the sealing of small copper wires into holes in the pressure vessel with some type of epoxy[39]. Insulated wires can be sealed in high pressure tubing[38]. A bundle of wires can also be cemented in a hole[48,49]. Stainless steel sheathed, magnesium oxide insulated thermocouple wire has been used up to 30 kbar. The stainless steel sheath is silver soldered or brazed into a hole in the wall of the vessel or into a high pressure fitting in a liquid medium[21,23,58]. If the pressure medium is a gas, leakage can be prevented by sealing the leads at the high pressure end with epoxy[27,46,57]. Another technique for the prevention of leakage is the impregnation of the ceramic insulation with pressurized epoxy[45]. When using thermocouple wires, the transition zone of the leads from the high pressure to atmospheric pressure must be kept at the reference temperature: a solution to this problem has been introduced by Brielles et al.[15].

3. Temperature Measurement in Non-hydrostatic Systems

Thermocouple corrections have been achieved with much experimental difficulty in non-hydrostatic systems. The sources of error are mainly pressure gradients and temperature gradients within the high pressure region, but errors may also be due to non-isothermal pressure seals or other effects such as chemical contamination, plastic deformation, electrical shunting or diffusion between thermoelements. The total pressure-induced contribution to the e.m.f. of a thermocouple may be expressed by[43]

$$\delta E_p = \int_{T_0}^{T_j} (\partial Q/\partial p)\, p(\tau)\, dT \qquad (7)$$

where $(\partial Q/\partial p)$ is the pressure dependence of the thermopower of the thermocouple. $p(\tau)$ is the pressure/temperature function along the thermocouple path into the cell. T_j is the internal couple junction temperature, and T_0 is the temperature where the pressure drops to atmospheric pressure p_0.

If $(\partial Q/\partial p)$ is assumed constant, δE_p is proportional to a temperature shift. A diagram of a typical temperature/pressure path of a thermocouple, i.e. (T_0, CD), is shown schematically in Figure 7[43]. The area ADT_0T_j corresponds to the maximum theoretical correction when seals are isothermal. The ratio CDT_jT_0/ADT_jT_0 is the fraction of the maximum correction applicable to a particular cell. Correction uncertainties due to superimposed pressure/temperature gradient effects, occasionally exceed ten per cent of the total temperature correction. The previous types of gradient modifications apply to both relative and absolute pressure effected thermocouple corrections in any kind of device. The measurement of the pressure/temperature gradient along a thermocouple path in a cell is a difficult problem. To determine approximate temperature gradients within a cell and its seal region, thermocouple junctions are placed sequentially in a series of positions from the center of the cell towards the seal. The e.m.f.s of all thermocouples along the path are measured when both the cell pressure and internal temperature are varied.

TEMPERATURE MEASUREMENT UNDER PRESSURE

Careful thermocouple arrangements must be made to avoid an alteration of the temperature gradients in the cell. To approximate the pressure gradients in the cell, a series of localized pressure calibrations can be used[43].

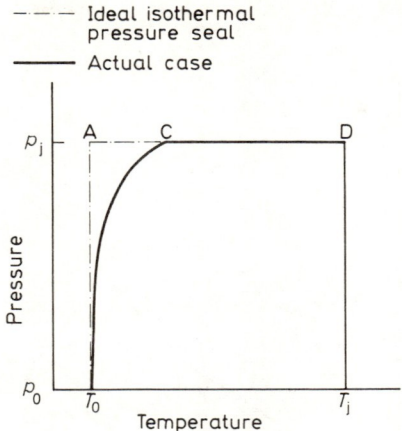

Figure 7. Diagram of the pressure/temperature profile along a thermocouple path in a typical, and ideal, high pressure cell.

However, as the pressure gradient calibration within the non-hydrostatic cell is also temperature dependent, the uncertainty increases. Three types of measurement methods have been used in non-hydrostatic systems to evaluate absolute e.m.f. corrections: A. a modified Bridgman method; B. direct thermal noise technique; C. an indirect calibration method.

A. Modified Bridgman Method

The known temperature gradient is set up within the high pressure cell using external heat sources, thus the corresponding range of experimental results is necessarily restricted to a limited range of temperature differences. Bundy[18] utilized this method to obtain thermocouple corrections up to 72 kbar in a maximum temperature gradient of 100°C for constantan, platinum, nickel, alumel, Pt–10% Rh, copper, chromel and Ni–18% Mo (Figure 8). Errors due to a small temperature gradient in the seal regions of the cell did not exceed 5 per cent of the thermocouple correction obtained. Similar measurements by Fujishiro et al.[33] in an opposed anvil apparatus have extended the single wire alumel data of Bundy up to 130 kbar for a 100°C gradient. The single wire technique has been also used by Getting and Kennedy[36] to determine the absolute effect of pressure on Pt/Pt–10% Rh and chromel–alumel thermocouples. Their experiments were conducted in an end-loaded piston–cylinder apparatus to a maximum pressure of 35 kbar and a temperature of 1000°C. One portion of the wire in the chamber is subjected to pressure whereas the other portion is at atmospheric pressure as it passes out of an axial hole through a tungsten carbide end-load plate.

The bushing supports the high pressure and provides inside it a 1 atm environment around the wires. A coaxial graphite resistance heating element was used to heat the inside end of the carbide bushing where the high temperature seal is located. The detailed diagram of the pressure cell is shown in

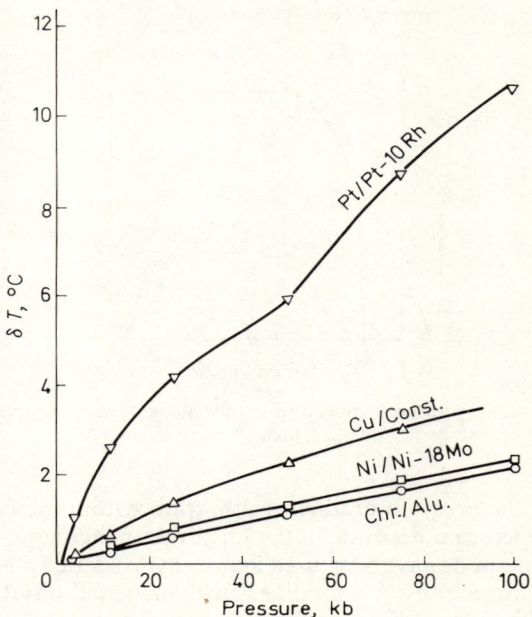

Figure 8. Pressure corrections to four common thermocouples for ΔT of 100°C.

Figure 9. Test wires entered an axial hole in the piston at one end and left at the other end through a hole in the carbide bushing. To permit uniform transmission of pressure to the wires, silver chloride was used up to 500°C and boron nitride above. To be sure that the pressure drop across the junction is isothermal, six thermocouples were located in two groups of three spanning the seal regions; no temperature variation across the seals larger than three per cent of the hot seal temperature was observed. The voltage correction is applied to the thermocouple path between the hot temperature junction and the point where the thermocouple wires emerge from the high pressure apparatus. The pressure uncertainty is estimated as ± 3 kbar over the entire pressure range.

Freud and La Mori have studied the e.m.f./pressure dependence using a single wire in a piston and cylinder apparatus to 40 kbar for chromel, alumel platinum and platinum–10 per cent rhodium. *Figure 10* shows a schematic cross section of the cell design for this experiment. Silver chloride (AgCl) is used as a pressure medium; the thermocouple wires were sheathed in boron nitride to be protected from the reactivity of AgCl.

TEMPERATURE MEASUREMENT UNDER PRESSURE

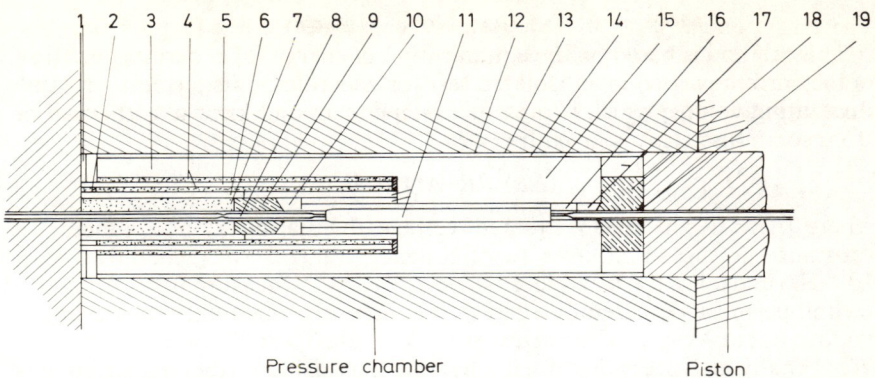

Figure 9. Detailed view of single-wire experiment pressure cell of Getting and Kennedy.

1. Hard fired pyrophyllite
2. Copper
3. Talc
4. Graphite
5. Soft fired pyrophyllite
6. Hard fired pyrophyllite
7. 3% Co tungsten carbide
8. 99% alumina
9. Molybdenum
10. Soft fired pyrophyllite
11. Dehydrated boron nitride
12. Molykote lubricant
13. 0.001 in lead foil
14. 0.020 in Teflon tubing
15. Talc
16. Dehydrated boron nitride
17. Pyrophyllite
18. 304 stainless steel
19. Hard fired pyrophyllite

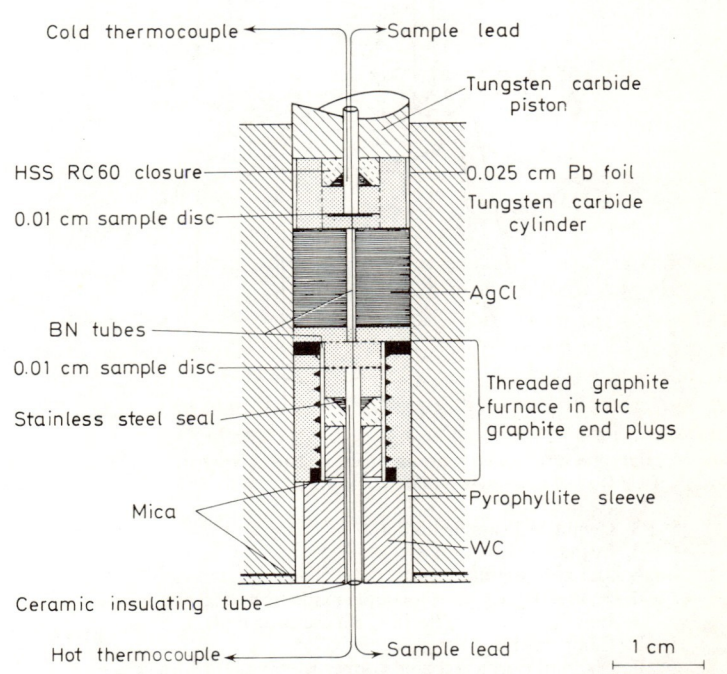

Figure 10. Piston and cylinder high pressure cell for measuring single-wire high pressure thermal e.m.f., of Freud and La Mori.

B. Thermal Noise Method

This method is based on the equipartition of energy. The thermal agitation of the current carried in an isolated resistor generates at its terminals a small fluctuating voltage which is pressure independent and known as thermal or 'Johnson' noise. This is given by

$$(\Delta E)^2/(R \cdot \Delta f) = 4kT \tag{8}$$

where $(\Delta E)^2$ is the mean squared voltage fluctuation in a conductor of resistance R measured over the frequency interval Δf (generally below 10^{14} Hz) and kT is the usual thermal energy term[35]. Measurements using this technique must be carefully carried out since the magnitude of the mean voltage fluctuations is of the order of a microvolt. To hold noise to a low level in internally high pressure cells, d.c. heating is used, but further improvements are still necessary to make the thermal noise method accurate. At high pressure and temperature the most important results are those of Wentorf[68], who determined corrections for Pt/Pt–10% Rh thermocouples up to 1 000°C and 40 kbar in a high pressure cell shown in *Figure 11*. The data are in

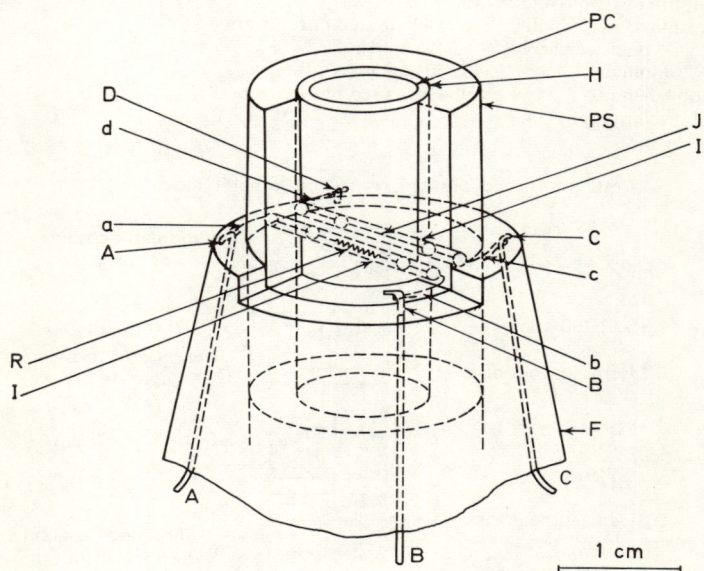

Figure 11. High pressure cell containing probe resistor (Wentorf).
  PC. Rod of fired pyrophyllite,
  H. Heater tube,
  PS. Cylindrical sample holder,
  I. Lucalox (Al_2O_3) tube,
  a, b. 0.25 mm platinum wire,
  c, d. Pt/Pt–10% Rh thermocouple made from 0.25 mm wires,
  J. Junction of the Pt/Pt–10% Rh thermocouple,
  F. Compressible stone gasket,
  A, B. 0.50 mm Formex-coated copper wires,
  C, D. 0.5 mm Pt and Pt–10% Rh wires,
  R. Probe resistor.

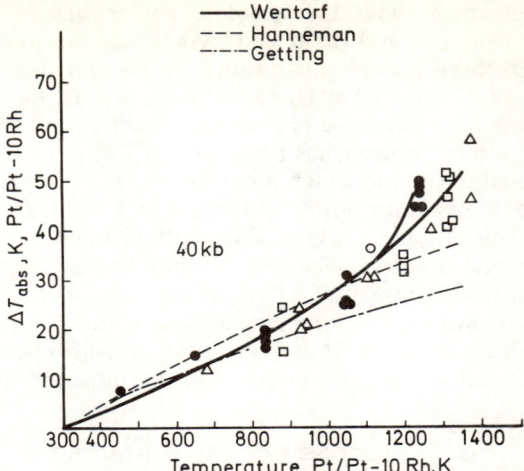

Figure 12. Thermal noise estimated temperature minus Pt/Pt–10% Rh indicated temperature, versus Pt/Pt–10% Rh indicated temperature at 40 kbar compared with the data of Hanneman and Getting.

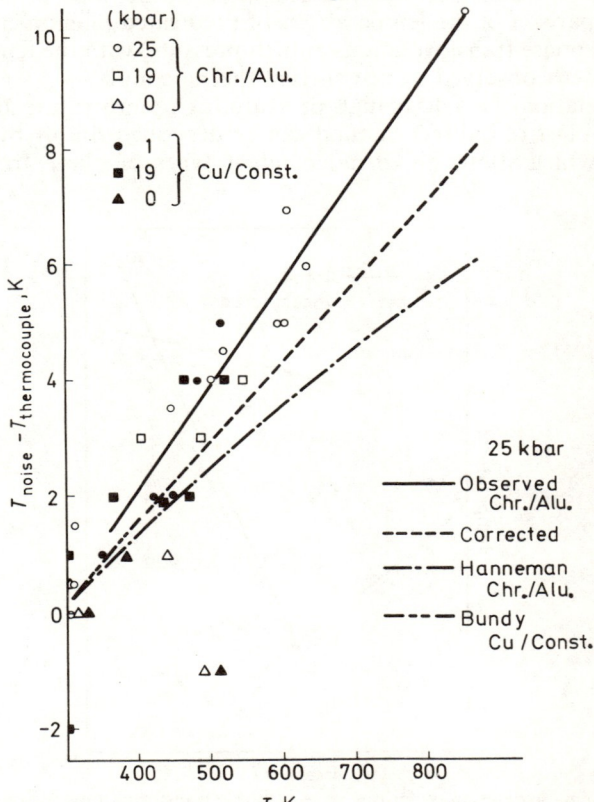

Figure 13. Pressure correction for thermocouples determined by Fujishiro *et al.*

reasonable agreement with the absolute corrections of Hanneman and Strong[43]. For example, at 40 kbar and 1 000°C the absolute correction to a Pt/Pt–10% Rh thermocouple determined by Wentorf is +40°C compared to a value of +36°C deduced by Hanneman and Strong (*Figure 12*). A method of balancing the Johnson noise of a sensing resistor with that of a reference resistor in noise thermometry has been used by Fujishiro et al.[34]. In place of the conventional technique which measures the mean square voltage of the thermal noise, the balance point is detected by counting the rate of pulses surpassing a constant gate voltage. This method was applied to correct the outputs of chromel–alumel and copper–constantan thermocouples embedded in a girdle-type high pressure cell. The accuracy of this thermometry is within 0.1 per cent at room temperature and within 0.3 per cent at 900 K under atmospheric pressure. At high pressure, pressure effects on the outputs of chromel–alumel and copper–constantan thermocouples are as shown in *Figure 13*.

C. Indirect Self-consistent Methods

Extensive approximate determinations of the absolute effect of pressure on the e.m.f. of thermocouples have been obtained mostly by Hanneman and Strong[41–43] who used the following indirect methods:

(1). Comparison of the temperatures of predicated phenomena (as single-component phase transformations at high pressure) with the temperatures of the phenomena observed by pressurized thermocouples.

(2). Correlation of results of high-pressure, high-temperature diffusion data.

The first class of indirect method can be described simply by referring to *Figure 14* which shows three independent types of phase transformation

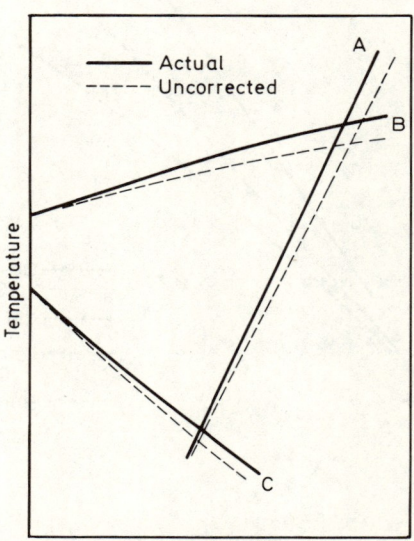

Figure 14. Schematic diagram of various types of phase transformations used in indirect determinations of thermocouple corrections.

common to various high pressure systems. The solid curve A shows a steep slope of the temperature/pressure transformation boundary. Phase changes such as graphite to diamond, with a large ratio of volume change to entropy change, are typical of this transformation. The solid curve B with a relatively low slope is typical of melting curves of metals with a low volume of fusion. The solid curve C which shows a transformation with a decreasing volume on going to the higher temperature phase is met with in the melting of many semiconductors and the $\alpha \rightarrow \gamma$ transformation in iron. The dashed lines represent uncorrected experimental curves corresponding to each type of transformation. From simultaneous measurements of two or more phase transformations (of different types) within the central isobaric pressure region, approximate thermocouple corrections to obtain a best fit of both

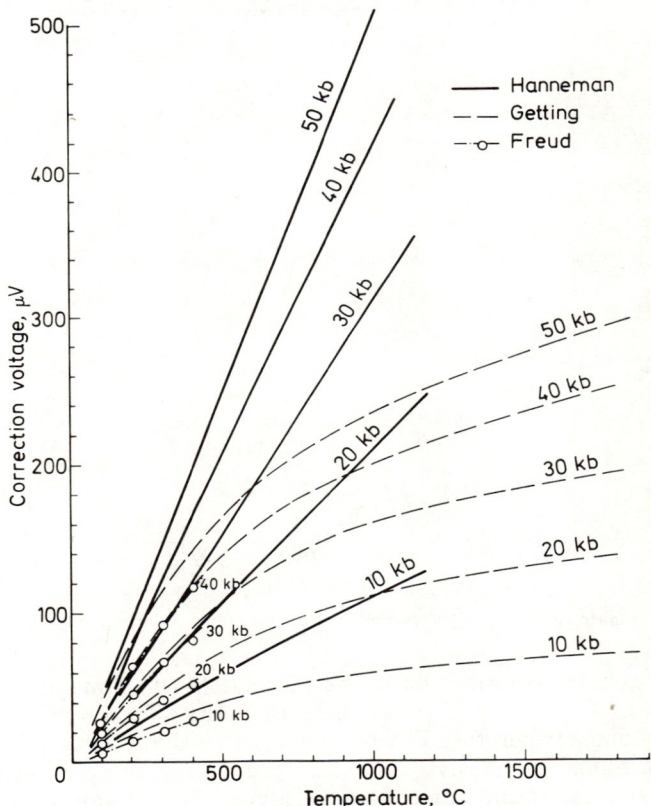

Figure 15. Pressure corrections for Pt/Pt–10% Rh thermocouples.

sets of data can be deduced. The most sensitive types of combinations of transformations are A versus C or B versus C since the C transitions have slopes opposite to those of the other two types. Such data provide both approximate temperature corrections at pressure, but also the temperature coefficient of the internal cell pressure at fixed load. A series of such experiments over various pressure/temperature ranges with a variety of different materials have been carried out by Hanneman and Strong who obtain a reasonably self-consistent approximation of absolute thermocouple corrections. The second independent method was based on a comparison between high pressure diffusion experiments and theory. Two approaches can be used to approximate the temperature correction. The first way is to find at a

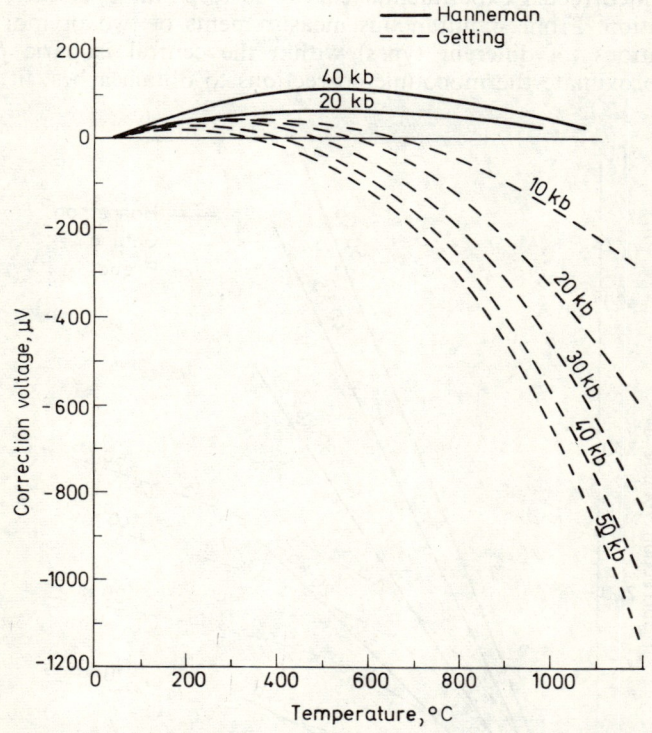

Figure 16. Pressure corrections for chromel–alumel thermocouples.

given pressure and composition the best fit of data which maintains a linear behavior log D_p versus $1/T$ plot (where D_p is the diffusion coefficient at a pressure p and temperature T) while giving a constant value of activation volume for diffusion, $V^\star$, over the temperature range of experimental data. The second temperature correction is deduced from a comparison of the value $V^\star$ calculated from the experimental atmospheric and high-pressure diffusion coefficient at fixed temperature with that obtained from experiments in the hydrostatic range and from theory. Getting and Kennedy suggest that there are sizable uncertainties associated with this procedure. As we shall see, substantial disagreement exists between the absolute corrections reported by Hanneman et al.[43] and by Getting and Kennedy[36] (*Figures 15* and *16*).

D. Results

(1). Pt/Pt–10% Rh *Thermocouples*

A complete plot of the approximate absolute corrections for Pt/Pt–10% Rh thermocouples deduced from the best fit of all the indirect data of Hanneman and Strong[43] and the thermal noise results of Wentorf[68] adjusted for a pressure gradient factor of unity are presented in *Figure 15*. The dashed lines represent the results of Getting and Kennedy[36]. The uncertainty claimed by these authors is $\pm 10\% \pm 10\,\mu V$. The corrections shown should be added to the uncorrected temperature. The values of Wentorf[68] are approximately 20 per cent greater than those of Hanneman at 50 kbar and 1 200°C. The results of Freud and La Mori[32] are also reported in this figure.

(2). *Chromel–alumel Thermocouples*

Approximate absolute corrections for chromel–alumel thermocouples deduced from the best fit of the indirect data of Hanneman and Strong[43] are shown in *Figure 16*. For comparison the Getting–Kennedy[36] corrections are shown as dashed lines. The uncertainty claimed by these authors is $\pm 20\% \pm 20\,\mu V$. The corrections shown should be added to the uncorrected temperature. As the absolute corrections for chromel–alumel thermocouples are small (less than 0.5°C up to 4 kb and 1 000°C), this thermocouple seems to be one of the most suitable for the measurement of temperature under pressure if care is taken to obtain good junctions.

(3). *Copper–constantan Thermocouples*

The results of Bundy[18] are shown in *Figure 8*.

(4). Ni/Ni–18% Mo *Thermocouples*

Results of Bundy[18] are shown in *Figure 8*.

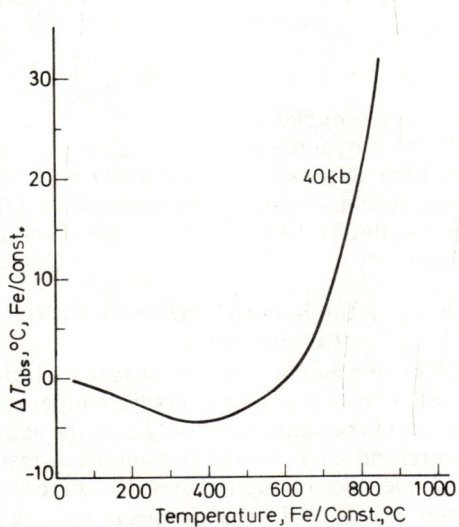

Figure 17. ΔT_{abs} versus observed T at various isobars for iron–constantan thermocouples.

(5). *Iron–vanadium, Iron–nickel and Iron–constantan Thermocouples*

The approximate absolute thermocouple corrections deduced from diffusion data in the iron–vanadium and iron–nickel systems are in reasonably good agreement with those obtained up to 1300°C by other methods (i.e. within 20 per cent)[41]. The effect of pressure for iron–constantan thermocouples is shown in *Figure 17*.

(6). Au/Pt–40% Rh *Thermocouples*

The results of Hanneman et al.[43] are reported in *Figure 18*.

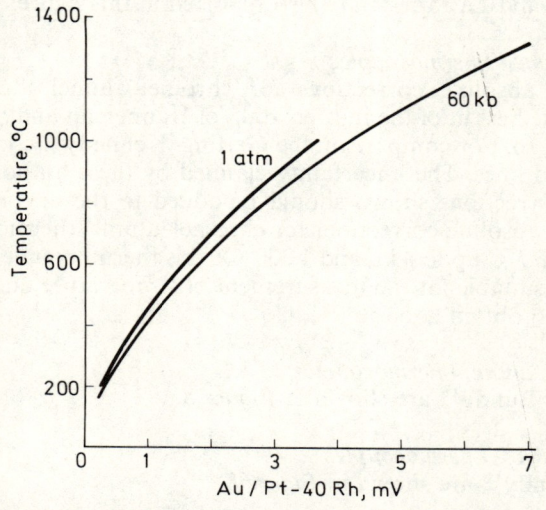

Figure 18. Pressure effect on calibration of Au/Pt–40 Rh thermocouples for temperature determination.

(7). *Tungsten–iridium Thermocouples*

A tungsten–iridium thermocouple has been used by Davies and Evans[22] up to 2200°C and 50 kbar to study the graphitization of diamond. These authors calibrated the tungsten–iridium thermocouple in terms of chromel–alumel and platinum/platinum–10% rhodium thermocouples using the data of Getting and Kennedy.

4. Effect of Pressure on the Relative Difference Between Various Thermocouples

Two different types of thermocouples are brought within an internally heated high-pressure cell along equivalent pressure/temperature paths. Their junctions are both placed in the same position, i.e. in the same conditions of pressure and temperature, and the wires are surrounded by insulating material leading out of the cell to the cold junctions. These leads are then connected to sensitive potentiometers or other devices for accurately reading the e.m.f. outputs of both thermocouples. The e.m.f. outputs of the two different thermocouples are compared to standard thermocouple tables and the apparent temperature deviation is noted.

Absolute corrections for one thermocouple can be deduced from the absolute data of the other one. Bundy[18] reported results of a relative pressure experiment intercomparing the readings of chromel–alumel thermocouples and Pt/Pt–10% Rh thermocouples over a pressure range of 50 kbar and a range of temperatures to 1 200°C. Getting and Kennedy[36] suggest that the relative pressure effect which Bundy attempted to determine was masked by chemical contamination of his thermocouples and other calibration effects at temperatures above several hundred degrees. An intercomparison of chromel–alumel versus Pt/Pt–10% Rh thermocouples has also been reported by Hanneman and Strong[43]. Their results are in reasonable agreement with those of Peters and Ryan[52]. Approximate relative corrections (for a pressure/temperature gradient correction factor of unity) for Pt/Pt–10% Rh versus chromel–alumel couples are shown in *Figure 19*. The estimated

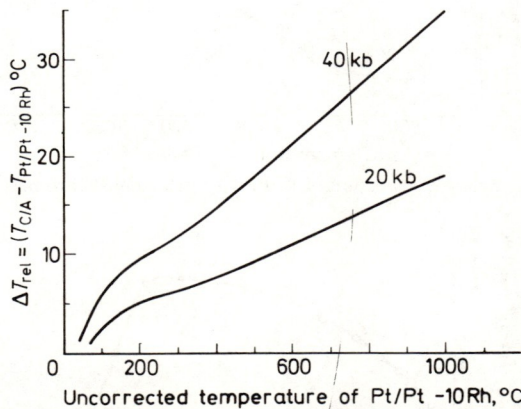

Figure 19. Approximate relative corrections between Pt/Pt–10% Rh and chromel–alumel thermocouples up to 1 000°C at 20 and 40 kbar intervals. Gradient factor equal to unity.

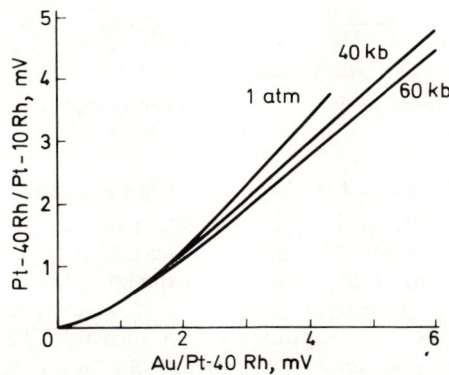

Figure 20. Direct Pt–40% Rh/Pt–10% Rh versus Au/Pt–40% Rh comparison.

accuracy of these relative measurements is approximately ±10 per cent, thus these data are more reliable than any absolute corrections at high temperatures.

Figure 20 shows a direct comparison of the Pt–40% Rh/Pt–10% Rh e.m.f. versus the Au/Pt–40% Rh e.m.f.[43]. Relative correction data have been determined for Pt/Pt–13% Rh against chromel–alumel (*Figure 21*) and iron–constantan against Pt/Pt–10% Rh at 40 kb (*Figure 22*)[42].

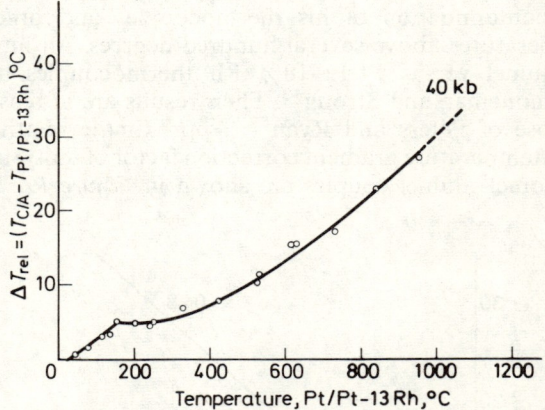

Figure 21. Relative correction of Pt/Pt–13% Rh against chromel–alumel.

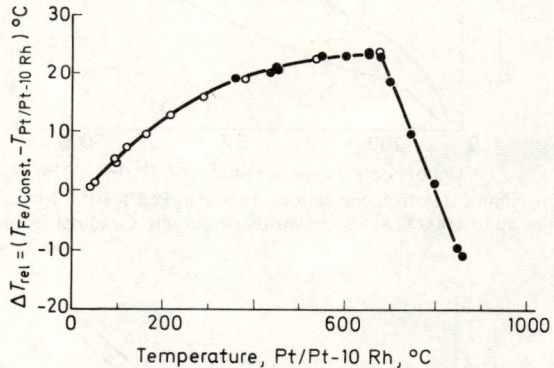

Figure 22. Relative correction of iron–constantan against Pt/Pt–10% Rh.

5. Optical Pyrometry at High Pressure

A cylindrical chamber provided with two plungers has been used by Fateeva and Vereshchagin[28] for optical research at high pressure (up to 100 kbar) and high temperature. A design of the chamber is shown in *Figure 23*. The chamber is equipped with two U-10 grade steel (tempered to $R_c = 60$ to 62) tapered plungers 2. Both plungers were movable. The insulation and sealing of the plungers was achieved by means of a thin layer of talc 1 which acts as an insulating material. The plungers were fitted, at their high-pressure end, with a cylindral termination 3 which protrudes into the cylindrical portion of the insert liner 4 of the chamber 6. A 0.5 mm clearance was left between the cylindrical portion of the plunger 3 and the cylindrical hole in the liner 4, with the diameter of the inner hole in the liner 5.0 mm. As the

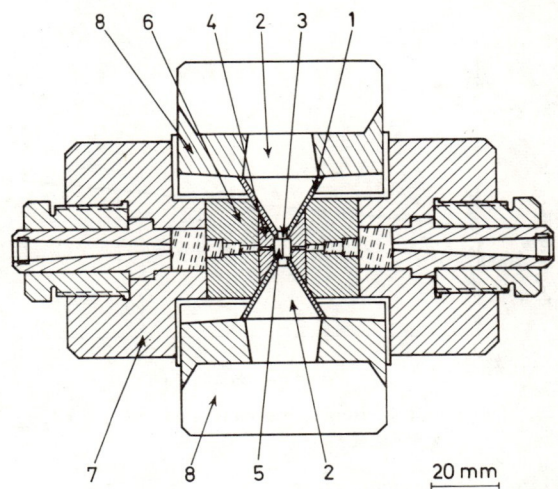

Figure 23. Chamber for optical studies, built to take pressures up to 100 kbar.

pressure rises, this clearance becomes filled with the pressure-transmitting medium, sodium chloride (NaCl) 5. Excess leakage loss of salt 5 is hindered by the talc seal 1. The chamber 6 was made of 4–8 grade steel, tempered to $R_c =$ 48 to 50. All chambers are intended for optical investigations and were provided with side cylindrical viewing ports, steps widening out from the center to the periphery. The ports filled with crystalline NaCl can be used to determine the temperature of the tested specimen from the radiation emitted by this specimen. With this type of cell, Fateeva and Vereshchagin have investigated the melting curve of tantalum up to 60 kbar[29], the melting curve of molybdenum up to 90 kbar[31], and the melting curve of graphite up to 90 kbar[30]. In the first experiment the melting temperature of tantalum[29] was determined in accordance with Planck's law, from a study of the test-piece on the basis of the intensity ratios of two narrow spectral regions corresponding to the wavelengths λ_1, and λ_2

$$I_1(\lambda_1)/I_2(\lambda_2) = f(T)$$

The starting point for the measurements was taken as the melting point of tantalum at atmospheric pressure. To take into account the selective absorption of radiation by the vapors of the material under study surrounding the test-piece, a correction was made during heating by comparing the I_1/I_2 and I_2/I_3 ratios obtained experimentally with those of a calculated calibration curve for this temperature. To reduce the errors associated with the variation in selective absorption in the vapor layer of the material being studied at different temperatures and pressures, each temperature on the melting curve was determined independently from the ratios of two intensities $I_1/I_2 = f_1(T)$ and $I_2/I_3 = f_2(T)$. The results of these measurements are given in *Figure 24*. The authors claimed that the probable error of the temperature does not exceed 6.0 per cent.

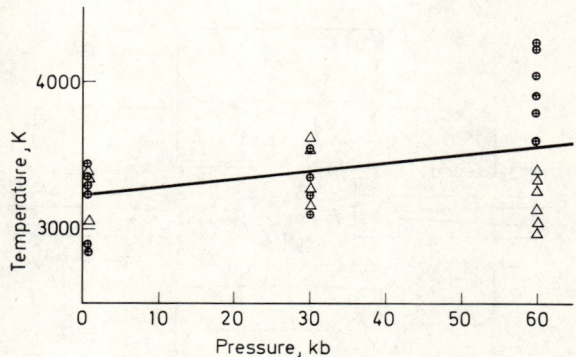

Figure 24. Melting curve of tantalum, $\triangle I_1/I_2$, $\oplus I_2/I_3$.

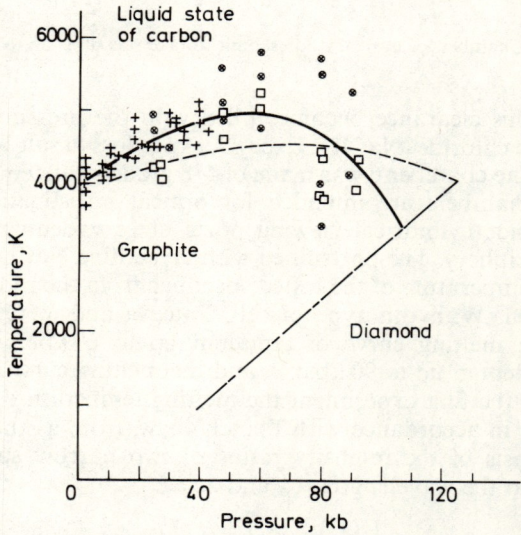

Figure 25. The melting temperature of graphite, solid line; Fateeva's data, dashed line; Bundy's data, chain dotted line: diamond/graphite equilibrium line, + measurements based on I_2/I_3 in the set up to 40 kbar, + measurements based on I_1/I_2 in sets up to 60 and 100 kbar, $\oplus$ measurements based on I_2/I_3 in sets up to 60 and 100 kbar.

Each melting temperature of graphite[30] was determined independently from two pairs of intensity ratios: $I_1/I_2 = f_1(T)$ and $I_2/I_3 = f_2(T)$ with $\lambda_1 = 420$ μ, $\lambda_2 = 622$ μ and $\lambda_3 = 825$ μ. The measurement results are shown in *Figure 25*. The authors claimed a precision of ±8 per cent in the temperature measurement. The same method was used to study the melting curve of molybdenum[31], and the results are shown in *Figure 26*. The authors claimed that the probable error in the measurement of the temperature is ±4 per cent.

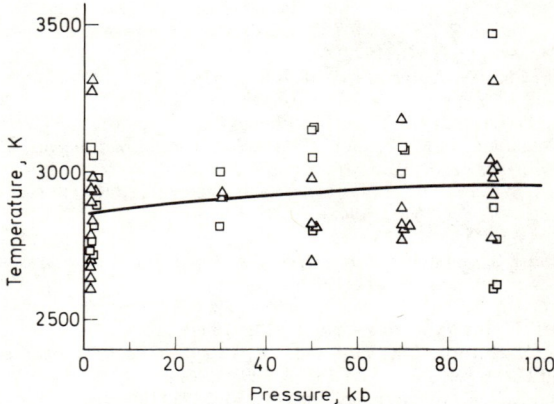

Figure 26. Melting curve of molybdenum up to 90 kbar, □ temperature determined from I_1/I_2, △ temperature determined from I_2/I_3.

6. Computer Method

Waxman and Hastings[67] have tested a new experimental procedure to determine the pressure effect on the behavior of the e.m.f. of thermocouples by simulation of the data on a computer. An important advantage of this procedure is that the hot junction temperature can be accurately evaluated without measuring it directly. Thus, the e.m.f. of the pressurized thermocouple is expressed parametrically as a function of several variables: (i) a measurable property related to the hot-junction temperature, such as the electric power to the pressurized furnace; (ii) the cold-junction temperature; and (iii) a reference temperature, such as 100°C at which the pressure effect is known. A set of redundant equations is obtained by variation of experimental conditions, and from the parameter evaluation the e.m.f. versus temperature relation of the pressurized thermocouple can be determined. The simulated data show that, at temperatures between 100° and 1 200°C and at pressures less than 70 kbar, the procedure should not introduce a temperature error greater than the error associated with the measurement of the e.m.f., the power and the cold-junction temperature.

IV. References

[1] Asamoto, R. R. and P. E. Novak. *Rev. Sci. Instrum.* **38**, 1047 (1967). Note errata, *Rev. Sci. Instrum.* **39**, 1233 (1968).
[2] Bedford, R. E. *Rev. Sci. Instrum.* **36**, 1571 (1965).
[3] Bedford, R. E. *I.S.A. Trans.* **9**, 248 (1970).
[4] Bedford, R. E. *High. Temp. High Press.* **4**, 241 (1972).
[5] Bedford, R. E., C. K. Ma, C. R. Barber, T. R. Chandler, T. R. Quinn, G. W. Burns and M. Scroger in *Plumb(Ed)*, Part 3 (1972).

[6] Bell, P. M., J. L. England and F. R. Boyd. 'Accurate characterization of the high-pressure environment'. *N.B.S. Spec. Publ. No. 326*, p 63 (1971).
[7] Benedict, R. P. and H. F. Ashby. *Temperature: Its Measurement and Control in Science and Industry*, Vol. III, Pt 2, p 51. Reinhold: New York (1952).
[8] Birch, F. *Rev. Sci. Instrum.* **10**, 137 (1939).
[9] Blackburn, G. F. and F. R. Caldwell. in *Temperature: Its Measurement and Control in Science and Industry*, Vol. III, Pt 2, p 161. Reinhold: New York (1962).
[10] Blackburn, G. F. and F. R. Caldwell. *J. Res. Nat. Bur. Stand.* **68c**, 41 (1964).
[11] Blevin, W. R. *Metrologia*, **8**, 146 (1972).
[12] Bloch, D. and F. Chaisse, *J. Appl. Phys.* **8**, 409 (1967).
[13] Blosser, L. G. and H. S. Young. *Rev. Sci. Instrum.* **33**, 1007 (1962).
[14] Bourassa, R. R., D. Lazarus and D. A. Blackburn. *Phys. Rev.* **165**, 853 (1968).
[15] Brielles, J., D. Vidal and P. Malbrunot. *J. Phys. E, Scientific Instruments*, **6**, 609 (1973).
[16] Bridgman, P. W. *Proc. Amer. Acad. Arts Sci.* **53**, 269 (1918).
[17] Bridgman, P. W. 'Studies in large plastic flow and fractures'. McGraw-Hill: London (1962); *Proc. Amer. Acad. Arts Sci.* **76**, 57 (1948).
[18] Bundy, F. P. *J. Appl. Phys.* **32**, 483 (1961).
[19] Burns, G. W. and J. S. Gallagher. *J. Res. Nat. Bur. Stand.* **70c**, 89 (1966).
[20] Corll, J. A. *Rev. Sci. Instrum.* **35**, 243 (1964).
[21] Cornish, R. H. and A. E. Ruoff. *Rev. Sci. Instrum.* **32**, 639 (1961).
[22] Davies, G. and T. Evans. *Proc. Roy. Soc. A*, **328**, 413 (1972).
[23] Davies, L. A., R. B. Gordon, J. K. Tien and J. R. Vaisnays. *Rev. Sci. Instrum.* **35**, 368 (1964).
[24] Dickson, A. and H. Meyer. *Phys. Rev.* **138**, A 1293 (1965).
[25] Downs, J. L. and R. T. Payne. *Rev. Sci. Instrum.* **40**, 1278 (1969).
[26] Dugdale, J. S. and J. A. Hulbert. *Canad. J. Phys.* **35**, 720 (1957).
[27] Duk, Y. *Rev. Sci. Instrum.* **37**, 1611 (1966).
[28] Fateeva, N. S. and L. F. Vereshchagin. *Pribori i Teknika Eksperimenta*, **3**, 222 (1970).
[29] Fateeva, N. S. and L. F. Vereshchagin, *Zh. Eksp. Teor. Fiz. SSSR*, **13**, 157 (1971).
[30] Fateeva, N. S. and L. F. Vereshchagin, *Zh. Eksp. Teor. Fiz. SSSR*, **14**, 233 (1971).
[31] Fateeva, N. S. and L. F. Vereshchagin, *Soviet Physics, Doklady*, **16**, 322 (1971).
[32] Freud, P. J. and P. N. La Mori. 'Accurate characterization of the high-pressure environment', *N.B.S. Spec. Publ. No. 326*, p 67 (1971).
[33] Fujishiro, I., H. Mii and S. Sakaida. *Bull. Jap. Soc. Appl. Phys.* **37**, 621 (1968).
[34] Fujishiro, I., H. Mii and M. Senoo, *Colloq. Int. Cent. Nat. Rech. Sci.* (1969), No. 188, p 457 (Publ. 1970).
[35] Garrison, J B. and A. W. Lawson. *Rev. Sci. Instrum.* **20**, 785 (1949).
[36] Getting, J. C. and G. C. Kennedy. *J. Appl. Phys.* **41**, 4552 (1970).
[37] Gibbs, D. F. and M. Jarman, *J. Sci. Instrum.* **35**, 472 (1958).
[38] Goree, W. S., B. McDowell and T. A. Scott. *Rev. Sci. Instrum.* **36**, 99 (1965).
[39] Gugan, D. *J. Sci. Instrum.* **33**, 160 (1956).
[40] Hammons, B. E. *Rev. Sci. Instrum.* **42**, 1889 (1971).
[41] Hanneman, R. E. and H. M. Strong, *J. Appl. Phys.* **36**, 523 (1965).
[42] Hanneman, R. E. and H. M. Strong, *J. Appl. Phys.* **37**, 612 (1966).
[43] Hanneman, R. E., H. M. Strong and F. P. Bundy. 'Accurate characterization of the high-pressure environment'. *N.B.S. Spec. Publ. No. 326*, p 53 (1971).
[44] Herzfeld, C. M. *Temperature: Its Measurement and Control in Science and Industry*, Vol. III, Pt 2. Reinhold: New York (1962).
[45] Heydemann, P. L. M. *Rev. Sci. Instrum.* **38**, 558 (1967).
[46] Lacam, A., D. Vidal and M. Lallemand. *Instruments et Laboratoires*, Dunod: Paris. **36**, 1 (1966).
[47] Lachman, J. C. and J. A. McGurty, in *Temperature: Its Measurements and Control in Science and Industry;* Pt 1, p 177 Reinhold: New York (1962).
[48] Lazarre, F. *J. Phys. Radium*, **14**, 213 (1953).
[49] Lazarre, F., J. R. Saurel and B. Vodar. *J. Rech. CNRS*, **5**, 324 (1954).
[50] Lazarus, D., R. N. Jeffrey and J. D. Weiss. *Appl. Phys. Letters*, **19**, 371 (1971).
[51] Magnien, C. and R. Bienaime. *Rev. Phys. Appl.* (Fr.), **3**, 283 (1968).
[52] Peters, E. T. and J. J. Ryan. *J. Appl. Phys.* **37**, 933 (1966).
[53] Powell, R. F. *Contemp. Phys.* **13**, 159 (1972).
[54] Powell, R. L., L. P. Caywood and M. D. Bunch. in *Temperature: Its Measurement and Control in Science and Industry*, Vol. III, Pt 2, p 65. Reinhold: New York (1962).

[55] Quinn, T. J. and T. R. D. Chandler. *Platinum Metals Rev.* **16**, 2 (1972).
[56] Schirber, J. E. *Phys. Rev.* **140 A**, 2065 (1965).
[57] Schirber, J. E. and D. W. Schanfoldt. *Rev. Sci. Instrum.* **9**, 270 (1968).
[58] Simon, I. *Rev. Sci. Instrum.* **28**, 963 (1957).
[59] Smith, A. H. in 'Second Annual Report to O.N.R. on High Pressure Research' by the Institute for the Study of Metals, University of Chicago, Ed. A. W. Lawson, p 3 (1949).
[60] Sparks, L. L. and J. G. Hust. *Spec. Publ. (US) Nat. Bur. Stand. No. 260*, 26 pp (1972).
[61] Sparks, L. L., R. L. Powell and W. J. Hall. *(US) Nat. Bur. Stand. Monogr. No. 124*, 56 pp (1972).
[62] Stepanov, V. A. *Pribory i Tekhnika Eksperimenta*, **3**, 179 (1961).
[63] Terry, R. E. and A. L. Ruof. *Rev. Sci. Instrum.* **43**, 1379 (1972).
[64] Thomas, D. B. *J. Res. Nat. Bur. Stand.* **67 C**, 337 (1963).
[65] Tsiklis, D. S. *Handbook of Techniques in High-pressure Research and Engineering*, p 258. Plenum: New York (1968).
[66] Vidal, D., J. Brielles, M. Lallemand and P. Malbrunot. *J. Phys. D, Applied Physics*, **6**, 1052 (1973).
[67] Waxman, M. and J. R. Hastings. *J. Appl. Phys.* **43**, 2629 (1972).
[68] Wentorf Jr, R. H. in 'Accurate characterization of the higher-pressure environment' *N.B.S. Spec. Publ. No. 326*, 81 (1971).
[69] Zysk, E. D. and A. R. Robertson. *Instrum. Technol.* **18**, 30 (1971).
[70] Zysk, E. D. and A. R. Robertson. *Instrum. Technol.* **19**, 42 (1972).
[71] Zysk, E. D. and D. A. Toenshoff. *Engelhard Ind. Tech. Bull.* **7**, 137 (1967).

CHAPTER 4

Part 1: Mercury Absolute Manometers

L. A. GUILDNER
National Bureau of Standards, Washington, DC, USA
and
J. TERRIEN
Bureau International des Poids et Mesures, Sèvres, France

Contents

I.	Introduction	115
II.	Optical Methods of Measuring the Distance between the Mercury Surfaces	117
	1. Cathetometer	117
	2. Lateral Shift	117
	3. Sensing by Interference Techniques	118
	4. Other Optical Devices	120
III.	Electrical Methods for Sensing the Position of the Mercury Surfaces	120
	1. Electrical Contacts	120
	2. Capacitance Techniques	121
IV.	Ultrasonic Sensing of the Meniscus Position	125
V.	Determination of the Height of the Mercury Column	126
VI.	General Considerations	127
VII.	Conclusion	130
VIII.	Abstract	130
IX.	References	130

I. Introduction

A manometer is an instrument for measuring pressures of gases and vapors. Pressure is force per unit area. The SI unit of pressure, with the symbol 'Pa', is the pascal, which is a newton per square meter. In this chapter, an 'absolute manometer' is one which operates as a 'primary' instrument; that is, it is constructed to operate so that the pressure can be evaluated directly in the base units of mass, length and time.

In different ranges of pressure, the highest accuracies of measurement are achieved by different kinds of manometers. From 5×10^2 Pa to 1.3×10^5 Pa, the most accurate device now available for the measurement of an absolute pressure is a mercury manometer. It is constructed so that a gas, exerting a pressure p at a lower surface of mercury, supports a column of mercury that is under a smaller pressure p' exerted by vapor and gas at an upper surface. The mercury column may be set up in a variety of ways, the most common being a U-tube. If the average density is designated as ρ, the gravitational attraction as g, and the vertical distance between the two mercury surfaces as h, the equation for the difference of pressures at the lower and upper

surfaces is
$$p - p' = \rho g h \tag{1}$$

The uncertainties involved in each quantity of equation 1 can be made small. The density of mercury depends upon its purity, isotopic composition, temperature and compression. The impurities in mercury can be reduced below 1 p.p.m. (part per million) by use of traditional cleaning techniques; with great care, the impurities can be reduced about a hundredfold more. Natural mercury has isotopes of mass 196, 198, 199, 200, 201, 202 and 204, with any variation of the abundance of 196 and 204 unlikely to be significant. The isotopic ratios of mercury may differ because of its origin or its subsequent treatment, and thereby leads to differences of density. The density of mercury has been measured with unprecedented accuracy at the National Physical Laboratory, Teddington, by Cook and Stone. This elegant work is reported in two papers[14,13]. Cook summed up the results of both by concluding that the density of all but two of their samples had a standard deviation of 0.2 p.p.m. from the average, and that '... there is a high probability that the density of any sample of pure mercury will be within 1 p.p.m. of this value'. The two samples excluded from the average came from NPL stock which had been repeatedly used and cleaned over many years[5]. They were higher in density than the average by 1 and 1.1 p.p.m., presumably because of a change in isotopic composition. The average value, recalculated to 20°C (IPTS-68) and normal atmospheric pressure, p_0, of 101 325 Pa[11] is $\rho(20°C, p_0) = 13\,545.87$ kg/m^3.

To calculate the density at temperatures other than 20°C, Cook[12] selected the equation of Beattie et al.[2]. The density of mercury at temperatures near 20°C, measured on the IPTS-68, can be expressed by the Beattie equation without additional uncertainty as

$$\rho(t_{68}, p_0) = \rho(20°C, p_0)/[1 + A(t_{68} - 20°C) + B(t_{68} - 20°C)^2]$$

with $A = 18115 \times 10^{-8}\,°C^{-1}$ and $B = 0.8 \times 10^{-8}\,°C^{-2}$. Regrettably, the thermal expansivity of mercury is so large that a change of 5.5 mK in the temperature changes its density 1 p.p.m.

The compressibility of mercury is $\kappa = 3.8 \times 10^{-11}\,Pa^{-1}$ [8,4]. When the measured absolute pressure is p, the mean pressure within the mercury column is $\frac{1}{2}p$, and the value of the density to be used, at the temperature t_{68}, is

$$\rho(t_{68}, \tfrac{1}{2}p) = \rho(t_{68}, p_0)[1 - \kappa(\tfrac{1}{2}p - p_0)]^{-1}$$

The gravitational attraction, g, may be found at the location of the manometer, by means of a comparison measurement with a gravimeter that has been calibrated at a station where g is already known within the Potsdam System. A negative correction amounting to $-14 \times 10^{-5}\,m\,s^{-2}$ (-14 mGal) was agreed upon internationally in 1968 and is to be applied in order to account for the difference between the Potsdam System and the latest results of absolute g-measurements. A higher accuracy is obtained by relative measurements between the location of the manometer and one or several of the sites included in the International Gravity Standardization Net (IGSN-71) recently adopted by the International Union of Geodesy and Geophysics. In terms of altitude h the value of g changes by about $(1/g)(\delta g/\delta h) = 3 \times 10^{-7}\,m^{-1}$.

PART 1: MERCURY ABSOLUTE MANOMETERS

The height of the mercury column can be measured by a variety of methods that are described in the following pages. A finite area in the central part of the meniscus exists where the capillary depression is less than 1 μm, if the diameters of the menisci are larger than $d = 3$ cm[30, 6]. This size probably suffices for optical techniques, but for capacitive and ultrasonic methods a larger size of meniscus is desirable in order that a central portion 3 cm across will be adequately planar.

The upper meniscus, when the containing cell is evacuated, is under pressure produced by the vapor of mercury plus pressures of any residual gases. In most pumping arrangements there will be concentration gradients of the gaseous components, so the total pressure at the meniscus itself will be difficult to determine accurately. A mercury vapor pump should be used in order that the upper cell system may be saturated with mercury vapor. The pressure of any other vapors or gases in the system (and ultimately at the meniscus) can be made insignificant by employing a proper pumping system. The actual pressure on the meniscus can then be evaluated as the vapor pressure of mercury at the temperature of the upper cell. Ernsberger and Pitman have made reliable measurements of the vapor pressure of mercury[17], which are satisfactorily confirmed by heat capacity measurements and thermodynamic calculations[15, 10]. From their work, the vapor pressure of mercury in pascals can be represented by

$$\log p' = 10.1621 - 3204/T \text{ for } T = 285 \text{ to } 326 \text{ K}$$

and this equation gives $p' = 0.171$ Pa, and

$$dp'/dT = 0.0147 \text{ Pa K}^{-1}, \text{ at } t_{68} = 20°C.$$

II. Optical Methods of Measuring the Distance between the Mercury Surfaces

1. Cathetometer

For the best of the older manometers which were in use at the Bureau International des Poids et Mesures (BIPM)[25, 29], the National Physical Laboratory (NPL)[27], Massachusetts Institute of Technology (MIT)[3], etc., the vertical distance between the two levels of mercury was obtained by means of a cathetometer. A cathetometer has two telescopes, each with an ocular micrometer. With the telescopes horizontal, the observer sets each micrometer to the mid-point between a point just above the center of the meniscus and its image by reflection on the mercury; this point is either a real one or an image projected by a collimator. As only the upper half of the telescope lens can receive light passing above the mercury, any error of focusing shifts the reading and causes systematic errors. The upper and the lower readings on the mercury surfaces are transferred to a vertical scale. When calibration and temperature corrections of the readings of the scale are made, the error is about 3–5 μm.

2. Lateral Shift

In this method, a collimator projects an image of a grid on to a surface of mercury; the reflected rays are focused by another lens to a real image of the grid. The angle of incidence is shifted by 5° from the vertical, so that any vertical displacement of the reflecting surface of mercury moves the second grid-image laterally. This displacement is detected by means of a sensitive photoelectric device[16], which helps to locate the level of mercury to within 1 µm. At the NPL two such assemblies, each mounted on a comparator, are used for absolute pressure measurements, in mercury tubes of large diameter. It is a precise but heavy and complicated instrument.

3. Sensing by Interference Techniques

The primary manometer now in use at the BIPM[7] is provided with an interferometric device by which the vertical distance between the mercury surfaces is transferred to the horizontal displacement of a carriage (*Figure 1*). Such a displacement is easily measured, e.g. by means of a graduated scale.

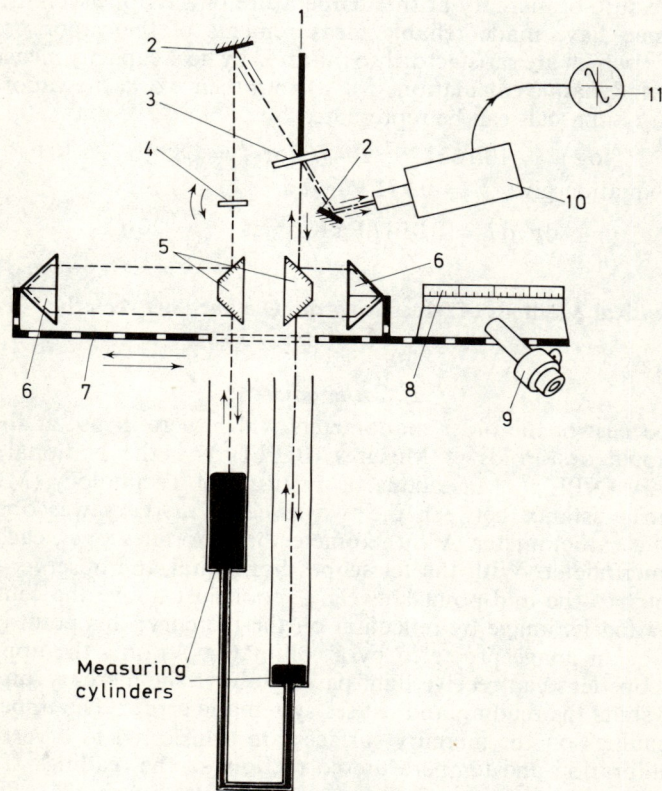

Figure 1. Principle of operation of BIPM manometer.
1. Source of white light.   2. Reflecting mirrors.   3. Separator.
4. Oscillating compensator. 5 and 6. Special prisms. 7. Carriage.
8. Graduated scale.   9. Reading microscope.   10. Photomultiplier.
11. Oscilloscope.

PART 1: MERCURY ABSOLUTE MANOMETERS

The two partial white light beams of a Michelson interferometer are reflected vertically by the surfaces of mercury, so that the path difference varies by twice the variation of the vertical distance between these surfaces. On the horizontal carriage are fastened two reflecting sets of three orthogonal mirrors by which the partial beams are lengthened or shortened until the path difference in the interferometer is zero. In this condition, a white fringe appears which is detected photoelectrically when an oscillating compensating plate sweeps across the zero path difference. The interferometer adjustment is not altered if the guiding ways of the carriage are not perfectly rectilinear because of the properties of the sets of three mirrors (cube corners). The reading is insensitive to the distance between the manometer and the carriage; it depends upon the difference in height of the surfaces, so that the part carrying the mercury can be suspended by springs in order to minimize the ripples on the mercury.

The zero reading is obtained on the graduated scale when the pressure is the same above both surfaces of mercury, or when a wide mercury disc is inserted in both vertical beams of light. When the pressure to be measured is applied, the displacement of the carriage from the zero position gives the height of the mercury column, divided by a factor four on account of the number of reflections of the measuring beams.

An additional correction is needed because the white fringe is obtained when differences in the refractive index dispersion make the paths differ in the two beams. This is the case when one beam travels in a vacuum, the other in a gas. If the length travelled in a medium i (refractive index n_i) is l_i, the white (achromatic) fringe is obtained when $\Sigma m_i l_i$ is the same in both beams, with $m = n + \sigma \, dn/d\sigma$ (σ is the mean wavenumber of the light). The corresponding correction with pure nitrogen in the pressure limb may vary typically between 0 and 30 µm; it is larger with helium.

This manometer is easy to handle. Its accuracy is of the order of 1 or 2 µm in the length of the mercury column but the refractive index of the gas must be known to calculate the above correction. It is probably the only commercially available optical mercury manometer of such an accuracy[18].

Other manometers using about the same principles of interferometric sensing have been made at the National Research Laboratory for Metrology (Tokyo)[24, 26]. One of these manometers, corresponding to *Figure 2*, demonstrates the design by which the positions of the menisci in U-tubes are located by a Michelson interferometer using white light. The change of position referenced to a pair of level menisci, located by the same interferometer, is measured by counting the interference fringes of 0.633 µm wavelength of the radiation produced by a He–Ne laser. This method is stated to define the positions of the menisci within ±0.3 µm on the average (±1 fringe). The wavelength of each light beam depends upon the index of refraction of its particular medium, the white light to be corrected as mentioned above, and the laser light according to the index of refraction of air. These corrections are discussed in detail in ref. 7.

In the foregoing interferometric manometers, the meniscus is located by a white light interferometer that enables the center of the fringe system to be determined in spite of the existence of substantial ripples. A new prototype standard manometer that has been built at the NBS uses long wavelength

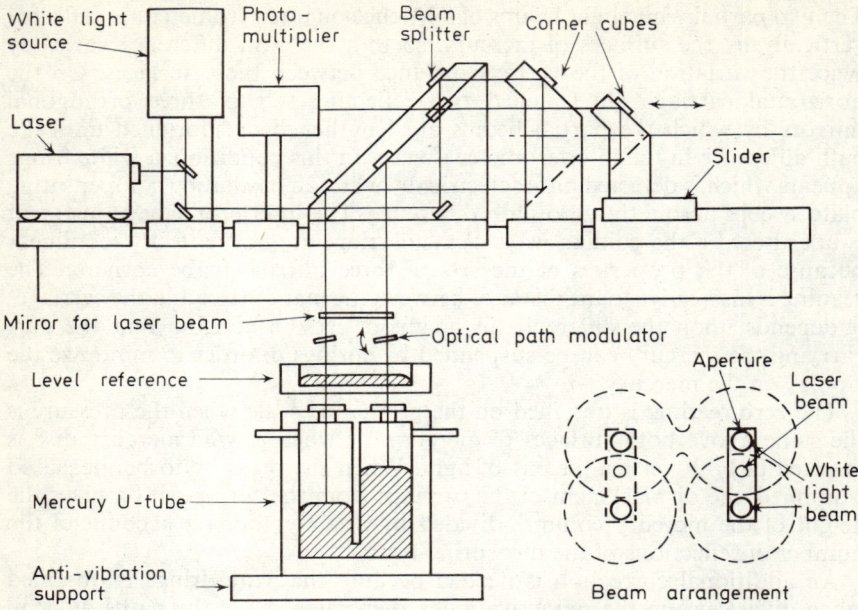

Figure 2. Principle of operation of NRLM manometer.

laser radiation so that the menisci can be located and their differential distance evaluated by a single interferometer. The detection system is devised to provide two signals 90° apart and thus to allow automatic up-down fringe counting. Careful vibration isolation is necessary. When ripples on the meniscus are of small enough amplitude relative to the wavelength of the laser radiation (10.6 μm), the mercury surfaces function nearly enough as mirrors. Some difficulty has been experienced with transient waves on the menisci as the result of irregular changes of meniscus contact angles. The vagaries of this effect have been reduced by evaporating nickel or chromium metal on to the surfaces of the 7.6 cm i.d. tubes used for the U-tube. These tubes permit fail-free fringe counting at speeds of 0.5 cm/min. Thus the manometer is capable of following fairly rapid changes of pressure automatically. Operation of the prototype indicates that a resolution of 0.1 μm or less is possible.

4. *Other Optical Devices*

Other optical devices, not all already realized, are described in ref. 31.

A general review of mercury barometers and manometers covering the commonly used, less precise, instruments is published in ref. 9.

III. Electrical Methods for Sensing the Position of the Mercury Surfaces

1. *Electrical Contacts*

The position of a mercury surface can be detected by electrical contact

PART 1: MERCURY ABSOLUTE MANOMETERS

with a metallic index. The mercury is raised to the contact, while being observed to assure a minimum penetration of the meniscus. The method at best is considered to have an uncertainty in locating the crown not less than 5 µm, which increases greatly if the surface of the mercury becomes contaminated.

2. *Capacitance Techniques*

An accurate mercury manometer, using capacitance sensing, is represented by an instrument built at the National Bureau of Standards[20]. Its present capacitance measuring system is capable of detecting changes of 1 nm in the average mercury height. The factors which contribute uncertainty to the determination of the meniscus position by capacitance have been reduced to 1 or 2 nm each. The main problems in locating the position of a meniscus by capacitance are considered in the following paragraphs.

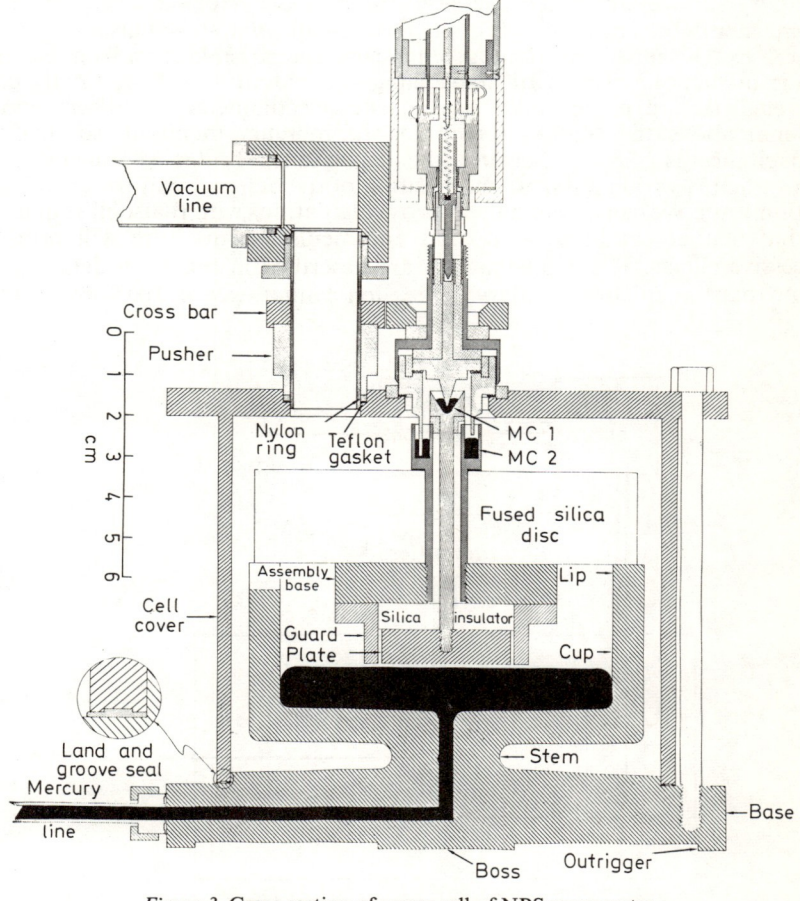

Figure 3. Cross section of upper cell of NBS manometer.

A cross section of an NBS manometer cell is shown in *Figure 3*. It has a 7.62 cm diameter mercury cup that is semi-isolated by an undercut to reduce transmission of stresses, and protected by a cover supported from the base. The electrical connections to the plate and guard assembly are made by contacts dipping into mercury. These minimize any variation of force on the fused silica disc from flexure of the lid of the cell case. All cell bases were made of gage block stock that was subsequently hardened. Both Inconel-X and 52-100 steel have been used; for any future cell fabrication, stainless steel types 17–4 PH or 440 C might be considered instead.

The position of a meniscus is determined by locating its crown. For the best measurements, a meniscus is chosen large enough that the maximum possible capillary depression is always insignificant. When the meniscus is located by capacitance, the curvature of that part of the meniscus under the plate must also be insignificant, but the choices of plate size and spacing are a compromise to minimize all uncertainties. A large plate size increases the sensitivity of the capacitance measurement to a displacement of the mercury surface, but it also increases the possible error which might occur from meniscus curvature. Similarly, a small plate-to-meniscus spacing increases the sensitivity, but it also increases the possible error from electrostatic attraction, ripples, tilt and change of mercury level. As finally constructed, the upper cell plate has an effective diameter of 3.08 cm and is 0.8 mm above the central portion of the mercury meniscus, so that the capacitance is ~ 8 pF. Because the capacitance bridge is sensitive and reproducible within 1 p.p.m., the position of the mercury can be reproduced within 1 nm. We can assure ourselves by calculations with Blaisdell's equation 6 that neither capillary depression nor meniscus curvature will produce excessive effects. These calculations are described in ref. 20 in detail. They show that the greatest capillary depression cannot exceed 0.035 nm, and the

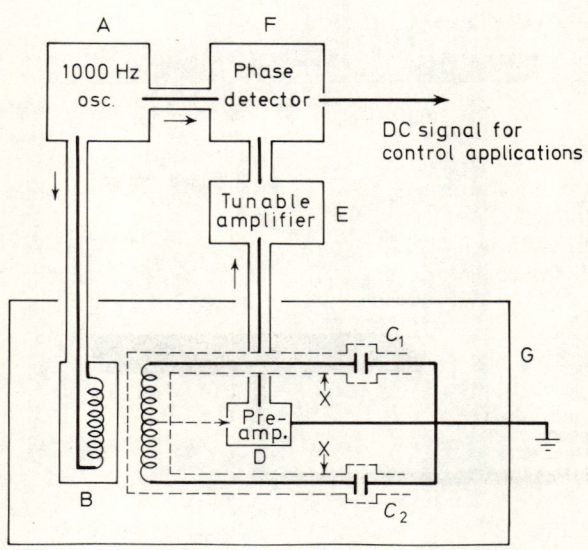

Figure 4. Schematic arrangement of capacitance bridge.

PART 1: MERCURY ABSOLUTE MANOMETERS

greatest effect from meniscus curvature cannot exceed 3 nm. Both quantities depend upon the angle of contact of the meniscus with the wall of the cup. Unless the level of the mercury in the manometer is distinctly changed, the meniscus contact angle remains nearly constant. Therefore, procedures were adopted to avoid large changes in mercury level, so that nearly the same meniscus shape exists both at zero level and at pressure measurements, and no significant net difference results.

The capacitance bridge used in the NBS manometer has been discussed in ref. 19. It is a transformer ratio-arm bridge with 3-lead capacitors. Because the mercury is at ground potential, the center tap has to be operated off the ground and an extra shield is required (see *Figure 4*). There are effects of capacitance coupling between the shields around the primary and secondary windings which must be accounted for in the operation of the bridge; details are given in the article cited. As the bridge is operated, the two lower cells are used in parallel in one arm versus the upper cell and a variable capacitor in the other arm.

This capacitance method of determining meniscus location reduces the effects of tilt, small misalignments and ripples to second order. We summarize these effects in terms of the relative change of capacitance, $\Delta C/C_0$, where $\Delta C/C_0 = 1.25 \times 10^{-6}$ is equivalent to an error of 1 nm when the spacing, s_0, is 0.08 cm:

(1) The effect of ripples of amplitude a irrespective of frequency is $\Delta C/C_0 = a^2/(2s_0^2) + \ldots$ Ripples with an amplitude of 1.26 μm give $\Delta C/C_0 = 1.25 \times 10^{-6}$.

(2) The effect of tilt of the capacitance plate of radius r_0, at an angle of θ radians with respect to level, is

$$\Delta C/C_0 = r_0^2 \theta^2/(4s_0^2) + \ldots$$

For $\theta \leqslant 135$ μrad, $\Delta C/C_0 \leqslant 1.25 \times 10^{-6}$.

Two other potential sources of error in the NBS manometer depend upon the arrangement of the manometer and will be discussed as items (3) and (4) subsequently.

The manner in which the mercury column is set up affects the accuracy of the pressure measurement. If the upper cell could be located directly above the lower cell, the column could tilt $\sqrt{2}$ mrad before an error of 1 p.p.m. in the vertical height is incurred. Practice has shown that it is better to have the lower cell displaced in order to measure the reference values of capacitance corresponding to the zero level. Much greater care must then be exercised to avoid significant errors from tilting of the manometer. In the NBS manometer, for example, the horizontal distance between centers of the cells is 15 cm. Consequently, a change of height of 1 p.p.m. in 10^5 Pa is produced by a tilt of only 5 μrad. For the same error at smaller pressures the tilt must be proportionately reduced. With careful techniques it would be possible to re-level the manometer within a tolerance of 1 or 2 μrad but it would be clumsy.

More accurate methods of accounting for changes of level have been developed for the NBS manometer. Stimson built a U-tube manometer and measured the change of level with a sensitive 'tiltmeter', supported on

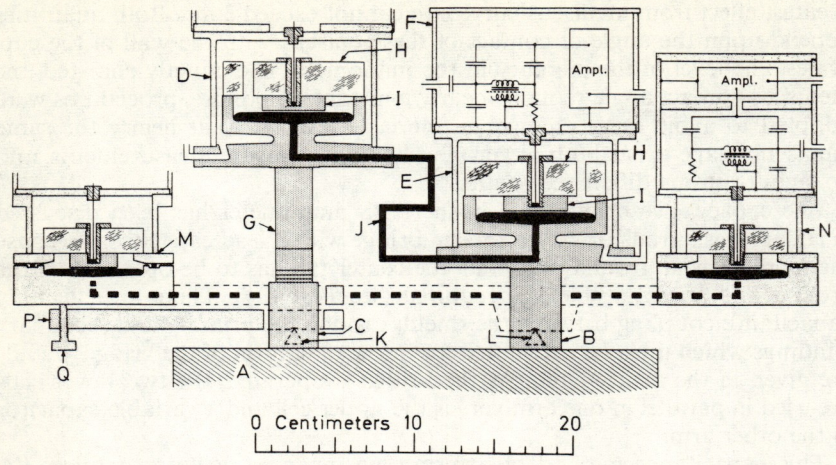

Figure 5. Earlier NBS capacitance manometer of Stimson.

knife edges which were part of the manometer cell pedestals (*Figure 5*). This device was reliable within several tenths of a μrad when carefully thermostatically controlled. This limit was inherent because the tiltmeter could not be positioned any more precisely on the knife edges. The tiltmeter was also susceptible to warping* from changes of temperature gradients. It was a relatively complicated device, and required separate readings from those for the manometer.

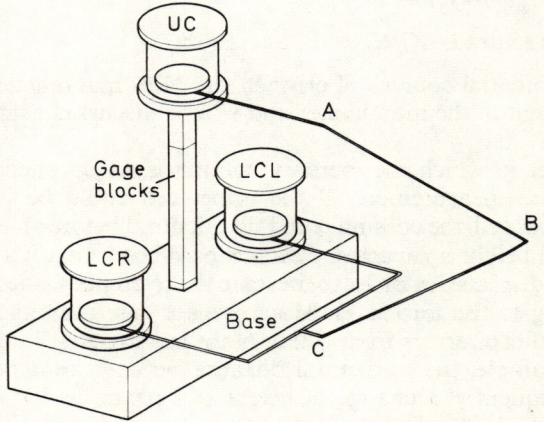

Figure 6. Principle of operation of NBS manometer.

---

* We define warping as the displacement of one part with respect to another as a result of a change of the temperature gradient in a material of relative linear thermal expansion α. If the positions of the two parts are a distance l_0 apart, the amount of warping is $\Delta x = \alpha l_0^2 \Delta t/h$, where $\Delta t/h$ is the thermal gradient normal to l_0.

PART 1: MERCURY ABSOLUTE MANOMETERS

The latest model of the NBS mercury manometer uses a W-tube configuration, with the long column in the middle (see *Figure 6*). At zero level all cells are wrung to pedestals, which are supported on a rigid base. The rigidity requirements were fulfilled by using a thick heavy slab of Invar, so installed as to minimize possible heat flow to the surroundings. For pressure measurement the upper cell is raised and then wrung to a stack of gage blocks on the center pedestal. These blocks serve to measure the displacement. This system provides precisely reproducible positioning of the cells and is relatively simpler to make than a U-tube plus a tiltmeter. When the sensing of the meniscus positions is carried out with the capacitors of both lower cells in parallel in one arm of a capacitance bridge versus the capacitor of the upper cell in the other arm, the combined capacitances of the lower cells compensate for tilts of the base. This arrangement is susceptible to two second order errors.

(3) The error caused by tilt of the base.
$\Delta C/C_0 = 2(\theta l/s_0)^2$, where $l = 15$ cm is the distance between cells and θ is the deviation in radians from level.
$\Delta C/C_0 \leq 1.25 \times 10^{-6}$ if $\theta \leq 7$ μrad.

(4) The error caused by a change in level of the mercury. Let the ratio of the sum of the effective areas of the two capacitance plates of the lower cells to the effective area of the capacitance plate of the upper cell be $1 + \alpha (\alpha \ll 1)$, and the average separation of the lower cell plates from the mercury surfaces be greater than that for the upper cell, s_0, by an increment s_1. If the mercury level changes by an amount m, then the relative uncertainty in the capacitance is

$$\Delta C/C_0 = m/s_0 (\alpha + 2s_1/s_0)$$

For plate areas equal within about 0.05 per cent and plate-to-meniscus spacing equal within $|s_1/s_0| < 1.3 \times 10^{-4}$, $\Delta C/C_0 \leq 1.25 \times 10^{-6}$ for $|m| \leq 1.3$ μm.

An accurate mercury absolute manometer using capacitance sensing was constructed at the Mendeleev Institute of Metrology (VNIIM, Leningrad) and reported by Izrailov and Kirenkov[23] (see *Figure 7*). Both it and the NBS mercury manometer have articulated arms to allow movement of the upper cell. The height of the upper cell above the lower cell of the VNIIM instrument is read from a scale.

The best commercial capacitance instruments available[22, 28] appear to lack the refinements that are required to obtain the highest accuracy, but are more reliable than most other commercial instruments.

IV. Ultrasonic Sensing of the Meniscus Position

The position of a mercury meniscus can be detected by an ultrasonic interferometer. This method is very sensitive to small changes; the positions of the menisci can be located with an imprecision less than 0.1 μm. The speed of sound in mercury, however, is uncertain by not less than two parts in 10^4 and it is affected by the dimensions and shape of the container. These limita-

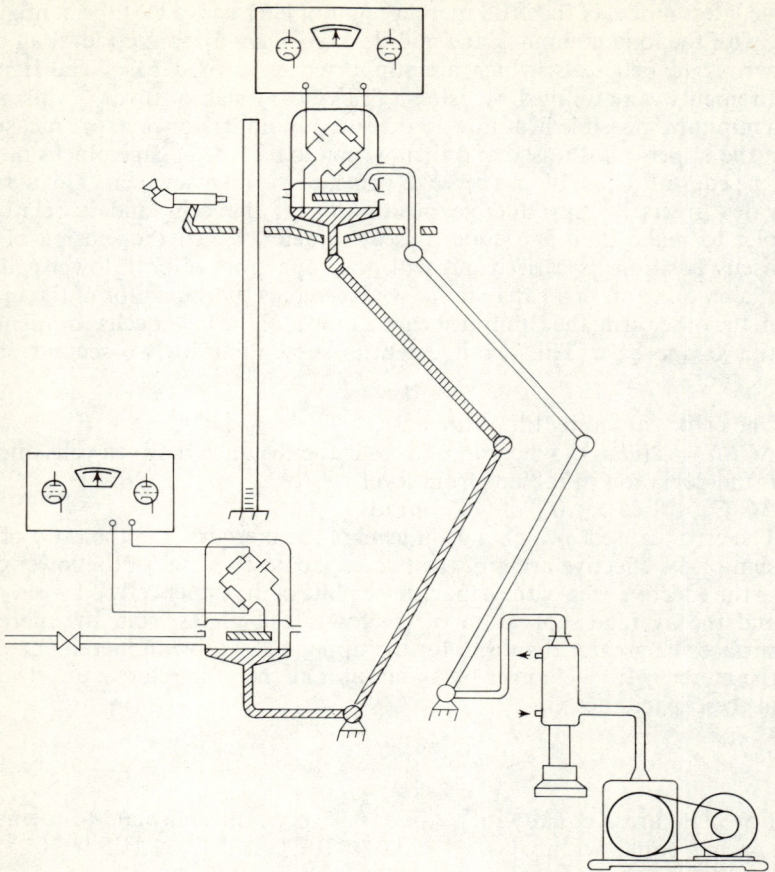

Figure 7. Principle of operation of VNIIM manometer.

tions make ultrasound useful, at present, only as a means of reproducibly locating the menisci.

The ultrasonic device reported by P. L. M. Heydemann operates with a resolution of 0.4 μm for mercury columns 4 to 20 cm long[21]. The signal is reflected from a central portion of the meniscus, 3 cm in diameter. Consequently, the mercury columns need to be at least 5 cm in diameter for the reflecting portion of the meniscus to be flat enough at all its possible contact angles.

A commercial mercury absolute manometer, for which ultrasonic techniques are employed[32] is offered with '0.2% precision'. It has a 15 mm i.d. U-tube, with a measuring range from 0 to 800 mm Hg.

V. Determination of the Height of the Mercury Column

Three techniques for determining the height of the mercury column have

been cited in the discussion:
1. The use of scales, employed for the BIPM and VNIIM manometers.
2. The use of gage blocks, as described for the NBS manometer.
3. The use of interference of laser radiation for manometers at NRLM and NBS. It should be recognized that, in principle, any of the techniques for locating the meniscus surfaces could be used with any of the means of measuring the height of the mercury column.

Technique 1 (scales) is the simplest; a very good quality divided scale, properly calibrated and read as is done at the BIPM, gives a value of any length with an accuracy of a few tenths of a µm.

Technique 2 requires the calibration of the gage blocks used to make up the length which is desired; the accuracy of these calibrations is of the order of a few hundredths of a µm.

Technique 3 is perhaps the easiest to operate, when a good stabilized laser and a reversible counting electronic device are available. The variation of the wavelength of the laser radiation as a function of pressure, temperature and composition of the air must be considered for the calculation of corrections.

VI. General Considerations

The accuracy of every mercury manometer is limited to a greater or lesser extent by errors resulting from the following:

1. Maintenance or measurement of level.
2. Vibrations.
3. Control and measurement of the mercury temperature.
4. Determination of the pressure above the upper cell meniscus.

The practices vary in each laboratory. At the BIPM, the mounting system was designed particularly to reduce vibrations (to which the optical method is sensitive). The U-tube was suspended by three spring and shock absorber arrangements, which are attached to a stiff upper plate. The long column of the BIPM manometer is in the center of the instrument, with three short columns symmetrically disposed around it. When the level changes, the center of gravity remains on the central axis, and no axial displacement results thereby.

The height of the VNIIM manometer is read by a microscope that is leveled by a carriage pivoted at the vertical axis of the upper cell (see *Figure 8*). This microscope is sighted on a scale aligned with the vertical axis of the lower cell. As we know, the ripples from vibration produce a second order error in the observed capacitance; no special precautions to avoid ripples are mentioned in Izrailov's article.

The two systems which have been used at the NBS for determining changes of level have already been discussed. The NBS manometer is installed at the Gaithersburg, Maryland site in a cellar two floors below the basement (see *Figure 9*). The cellar floor rests on original earth to provide vibration isolation from the main building. The manometer was set upon a concrete pier, resting on a pad to provide further vibration isolation from the cellar floor. In order to prevent the transmission of vibration from the laboratory, all

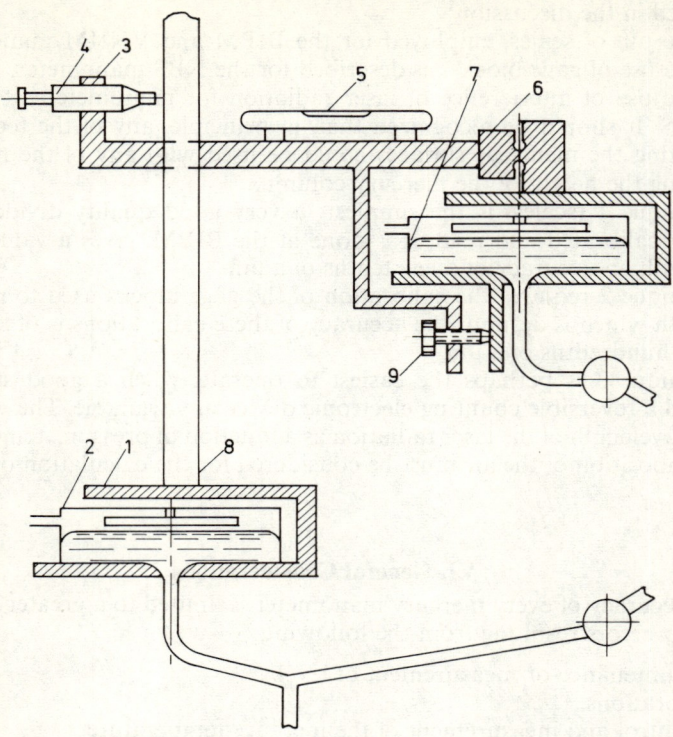

Figure 8. Schematic disposition of parts of the reading system (VNIIM). 1. Bracket, 2. Lower head, 3. Bracket, 4. Readout microscope, 5. Spirit level, 6. Spring hinge, 7. Upper head, 8. Readout scale, 9. Adjusting screw.

connections to the manometer from the floors above are flexible (including vacuum and gas lines).

For temperature control the cellar of the NBS manometer was designed to allow a circulation of air through ducts surrounding the manometer as shown by the arrows in the illustration. In the plenum, the air passes over a cooling coil and steady heaters, with a variable heat input added to control the temperature in the room below. The objective is that the temperature should be nearly uniform in the vicinity of the manometer, and should vary only slowly, in order that temperatures sensed on the surfaces of the mercury column will be the mean temperature of the mercury itself. Observations show that the temperature of the mercury is constant within about 2 mK, and the drift does not exceed 2 mK/day. The temperatures are measured by differential thermocouples with reference junctions attached to a copper block containing a calibrated capsule platinum resistance thermometer. Twenty-five copper–constantan thermocouples in series were distributed uniformly along the mercury column to sense an average temperature difference which is always within 0.025 kelvin of the temperature of the reference block. The sensitivity of the measurement is 1012 µV/K.

For temperature uniformity, the VNIIM manometer is encased to afford

PART 1: MERCURY ABSOLUTE MANOMETERS

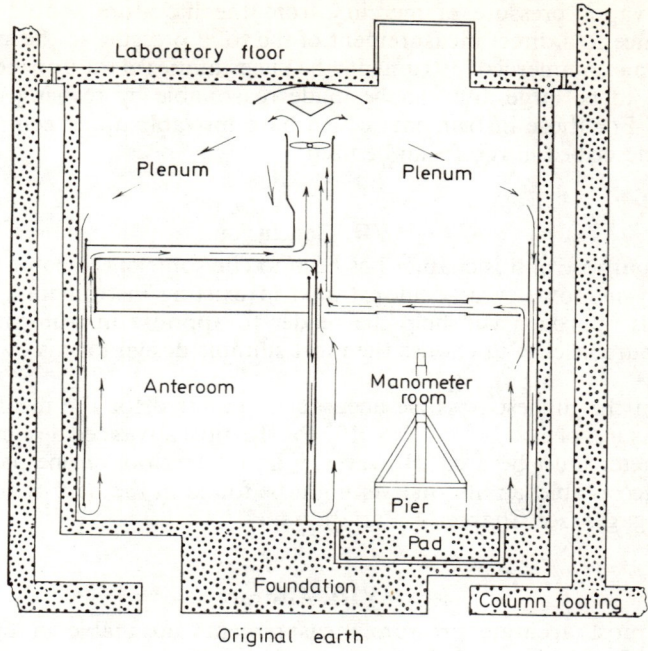

Figure 9. Schematic cross section of cellar (NBS).

air circulation. The mercury temperature is determined as the sum of a reference value and the temperature difference between it and the mercury arm as measured by twelve copper–constantan thermocouples. No further details are published; in their article Izrailov and Kirenkov state it was their objective to hold the temperature within 'about $\pm 0.01 - \pm 0.02$ deg', which is equivalent to ± 1.5 to ± 3 p.p.m. of the mercury density.

The BIPM manometer has five copper–constantan thermocouples with reference junctions in a copper block, the temperature of which is measured by a mercury-in-quartz thermometer. Three of the thermocouples are attached to the glass tube of the mercury column. The manometer is located in a thermostatically controlled room, and the mean temperature of the mercury column is stated to be known within a few thousandths of a kelvin.

The pressure in the upper cell is at least as great as the vapor pressure of mercury at the temperature of its meniscus, and may be more. The upper cells of all accurate manometers are dynamically vacuum-pumped for the determination of absolute pressures. The pump should be a mercury pump to saturate the space above the meniscus with mercury vapor. In order for the uncertainty in the pressure to be as small as possible, the components of the cell should be free of materials that are capable of evolving significant amounts of gases or vapors, and connections to the pump should be made with lines having high gas conductance and low real or virtual leakage. When the pressure as measured by a McLeod gage is sufficiently small, the principal pressure on the meniscus must come from mercury vapor. Values of the vapor pressure of mercury from the literature should be used in preference to a direct measurement of the total pressure in the system.

In some commercial instruments the uncertainties of the upper cell pressure are apt to be large, but can be made reasonable by revising the vacuum system. For those instruments which have movable upper cells, articulated joints, described in ref. 1, can be used.

VII. Conclusion

The purpose of this chapter has been to cite some of the better practices in mercury manometry as demonstrated in existing instruments. It is hoped that this approach will help the reader to appraise instruments which he might purchase, or to choose the most suitable design for his own construction.

When the highest possible accuracy is required for the measurement of gas pressures of 5×10^2 to 1.3×10^5 Pa, the most advanced design of mercury manometer must be used. However, if an uncertainty of one part in 10^4 or 10^5 is acceptable, an alternative might be found in the best quality of deadweight gages (see Chapter 4, Part 3).

VIII. Abstract

The most accurate pressure measurements attainable in the range of 5×10^2 Pa to 1.3×10^5 Pa can be made by using mercury manometers. In the design of a mercury manometer the principal aims are those of locating the crown of the meniscus, and of determining the height of the mercury column. Different techniques are cited by describing some accurate instruments in use at BIPM, NRLM, NBS and VNIIM.

IX. References

[1] Anderson, R. L., L. A. Guildner and R. E. Edsinger. *Rev. Sci. Instrum.* **41**, 1076 (1970).
[2] Beattie, J. A., B. E. Blaisdell, J. Kaye, H. T. Gerry and C. A. Johnson. *Proc. Amer. Acad. Arts Sci.* **74**, 371 (1941).
[3] Beattie, J. A., D. D. Jacobus, J. M. Gaines Jr, M. Benedict and B. E. Blaisdell. *Proc. Amer. Acad. Arts Sci.* **74**, 327 (1941).
[4] Bett, K. E., K. E. Weale and D. M. Newitt. *Brit. J. Appl. Phys.* **5**, 243 (1954).
[5] Bigg, P. H. *Brit. J. Appl. Phys.* **15**, 1111 (1964).
[6] Blaisdell, B. E. *J. Math. Phys.* **19**, 186 (1940).
[7] Bonhoure, J. and J. Terrien. *Metrologia*, **4**, 59 (1968).
[8] Bridgman, P. A. *Proc. Amer. Acad. Arts Sci.* **47**, 345 (1911).
[9] Brombacher, W. G., D. P. Johnson and J. L. Cross. *US Nat. Bur. Stand. Monogr.* 8 (1960).
[10] Busey, R. H. and W. F. Giauque. *J. Amer. Chem. Soc.* **75**, 806 (1953).
[11] Chattle, M. V., *NPL Report Qu9* (April 1970).
[12] Cook, A. H. *Brit. J. Appl. Phys.* **7**, 285 (1956).
[13] Cook, A. H. *Phil. Trans. Roy. Soc. London A*, **254**, 125 (1961).
[14] Cook, A. H. and N. W. B. Stone. *Phil. Trans. Roy. Soc. London A*, **250**, 279 (1959).
[15] Douglas, T. B., A. F. Ball and D. C. Ginnings. *J. Res. Nat. Bur. Stand.* **46**, 334 (1951).
[16] Elliott, K. W. T. and D. C. Wilson. *J. Sci. Instrum.* **34**, 349 (1957).
[17] Ernsberger, F. M. and H. W. Pitman. *Rev. Sci. Instrum.* **26**, 584 (1955).
[18] Etablissements Jaeger, 92-Levallois-Perret, France.
[19] Guildner, L. A. and R. E. Edsinger. *J. Res. Nat. Bur. Stand.* **69C**, 13 (1965).
[20] Guildner. L. A., H. F. Stimson, R. E. Edsinger and R. L. Anderson. *Metrologia*, **6**, 1 (1970).

PART 1: MERCURY ABSOLUTE MANOMETERS

[21] Heydemann, P. L. M. *Rev. Sci. Instrum.* **42**, 983 (1971).
[22] Ideal-Aerosmith Inc., 1505 East Fox Farm Road, Cheyenne, Wyo. 82001.
[23] Izrailov, K. S. and I. I. Kirenkov. *Trudy VNIIM*, **51** (**111**), 5 (1961).
[24] Kaneda, R., S. Sudo and K. Nishibata. *Bull. Nat. Res. Lab. Metr.* **9**, 24 (1964).
[25] Marek, W. J. *Trav. et Mem. Bur. Int. Poids Mes.* **3**, D22 (1884).
[26] Mitsui, K., H. Sakurai and T. Mochizuki. *Temperature: Its Measurement and Control in Science and Industry*, Vol. IV, Pt 1, Chap. 31. Instrument Society of America: Pittsburgh (1972).
[27] National Physical Laboratory, Teddington, *Notes on Applied Science No. 9* (1955).
[28] Schwien Engineering Inc., 2882 Metropolitan Place, Pomona, Calif. 91767.
[29] Sears, J. E. and J. S. Clark. *Proc. Roy. Soc. A*, **139**, 139 (1933).
[30] Stulla-Gotz, J. *Phys. Z.* **35**, 404 (1934).
[31] Terrien, J. *Rev. Opt.* **38**, 29 (1959).
[32] Wallace and Tiernan Division, 25 Main Street, Belleville, New Jersey 07109.

CHAPTER 4

Part 2. Pressure Scale and Fixed Point

D. L. DECKER

Brigham Young University, Provo, Utah 84602, USA

Contents

I.	Apparatus that Requires a Fixed Point Calibration	133
II.	Choice of Fixed Points	134
III.	Measurement of the Pressure at Fixed Points	135
IV.	How to Make Use of Fixed Points in High Pressure Calibration	138
V.	The Present Set of Fixed Points for Pressure Calibration	140
	1. Mercury Melting Curve	140
	2. Bismuth I–II	141
	3. Thallium II–III	142
	4. Barium I–II	142
	5. Bismuth III–V	143
	6. Fixed Points above 100 kbar	143
VI.	Conclusions	144
VII.	References	144

I. Apparatus that Requires a Fixed Point Calibration

A straightforward measurement of pressure involves measuring a force and an area. Several types of high pressure apparatus are very difficult to calibrate so as to know the pressure at the sample because a relatively unknown fraction of the force is dissipated in friction between moving parts of the pressure apparatus or lost to internal friction within the gasket material surrounding the sample. The area over which the pressure is applied is also a difficult problem at high pressures because of compression of the high pressure apparatus. These types of apparatus are most generally 'calibrated' by arriving at a correspondence between the load applied and the sample pressure. This is accomplished by replacing the sample with some material which experiences a phase change at some known pressure and noting the load at this change. This gives a point on the load pressure calibration curve. Several such points are measured and then connected with a smooth curve. Of course, the above presupposes that there has been a measurement of the pressure at which these phase changes take place.

The types of high pressure apparatus which are most dependent on this technique for pressure calibration are the flat anvil devices, the multianvil devices, and the supported anvil devices. The pressure in a piston and cylinder device, however, can be reasonably well calculated by force over area measurements, for the corrections for friction can be approximated. Thus this type of apparatus is often used without making a fixed point pressure calibration. Nonetheless the pressure in such a system should be checked for consistency against a fixed point calibration once such a calibration technique is well established.

In the following sections we will discuss criteria for selecting phase transitions which will be most suitable for a fixed-point pressure calibration. We will then turn to the problem of making the fundamental measurement of the pressure at the fixed points and suggest ways in which high pressure experimenters can best perform their calibration using the established fixed points. This section will also include common problems that are sometimes overlooked and give estimates of the best possible accuracy in the measurement of pressure with present capabilities. We finally conclude by discussing the best value to be assigned to the equilibrium pressure at the fixed points. The reader should also be informed of two other recent works containing reviews of fixed point pressure calibration[1, 28].

II. Choice of Fixed Points

There should be some criteria for choosing fixed points or at least selecting from the known phase transitions the ones that are most reliable and simple to use for pressure calibration. The items that should be considered are: (1) detectability of the transition, (2) rate of transformation from one phase to the second, (3) hysteresis and region of indifference between the phases, (4) effects of impurities upon the pressure at the phase change, (5) temperature dependence of the phase transition, and (6) availability of the material and ease of handling.

By detectability we mean the ease by which the phase change can be noted. Most useful are materials with a significant change in such properties as electrical resistance, volume, or with a large latent heat of transformation. There are other properties which might be easily detected by some apparatus or experimenters but these three mentioned are almost universally detectable in any laboratory. Of these, the electrical resistance is most easily measured so that the fixed points should be phase transitions exhibiting distinct changes in electrical resistance.

The transformation kinetics are important in detecting the phase change as well as the rapidity with which the measurement can be made. The accuracy of the measurement of a sluggish transition would be poor unless special care is taken to vary the pressure gradually so as not to overshoot the transition. The ideal situation is to have a rapidly progressing transition. There are few quantitative measurements on kinetics of phase transitions at high pressure, thus this is one area which should be investigated.

It often happens that nucleation and growth problems will cause a hysteresis in the initiation of the transition, i.e. the transition begins at different pressures depending upon whether the pressure is increasing or decreasing. This hysteresis has its origin in the fact that a certain amount of energy is required to form the strained surface boundary between the two phases. The work expended in pressurizing the sample beyond its phase equilibrium value must be sufficient to form the new boundary between the present phase and the nucleus of the new phase. If one establishes a situation in which a material is already a mixture of the two phases there is still a range of pressure over which the transformation progresses in neither direction for very small pressure changes. This pressure interval was called by Bridgman, the 'Region of Indifference'. This phenomenon must be related to the difficulty of moving a barrier between two regions of different phases and should be related to the amount of strain in the sample and to other physical properties of the material such as impurity content. For very precise work one would prefer transitions for which the region of indifference is small. It is also desirable to have the nucleation hysteresis as small as possible because it is not always possible to design the calibration experiment to detect the region of indifference properly.

An ideal calibration point would be one for which the pressure at the phase change is negligibly affected by small amounts of impurities in the sample material. Otherwise a maximum allowable impurity content would have to be specified and if this were very small it may make the material less readily accessible to all experimenters. Not much experimental work has been done in this area.

The phase transitions used for calibration should ideally be insensitive to temperature so that the pressure chamber need not be accurately monitored for temperature. In all cases the effects of temperature on the pressure of the transition should be carefully studied so that someone calibrating his apparatus is aware of the amount of care that he must use.

Finally we should mention the fact that some materials are not well suited for calibration purposes because they are not readily available to all researchers or they are chemically so active or decompose so easily that great care must be used in handling them.

To put the technique of fixed point pressure calibration on a firm basis, each of the above criteria should be carefully considered and thoroughly investigated in selecting the transitions to be used as primary fixed points. We note that for temperature calibration the measurement of the fixed points and the handling and selection of the materials used, including specification of required purity, have been carefully and completely delineated. This must also be done in pressure calibration.

III. Measurement of the Pressure at Fixed Points

Before one can use fixed points for calibration, some set of such points must first be established, and the transitions must be well characterized as described above. Thus the pressure at each chosen phase transition must be determined by direct measurement against a fundamental standard of pressure. Unfortunately such standards of accurate pressure measurement are not available at all high pressures and some other less direct means must be resorted to. The techniques which have been used in the past will be briefly discussed here. They include: (1) The mercury manometer, (2) the free piston gage, (3) the piston and cylinder, (4) flat anvil devices, (5) theoretical equations of state, (6) extrapolation from transition points at lower pressures, and (7) dynamic shock measurements.

The mercury manometer has been used for pressure measurement to 2.5 kbar[9,8] which is too low to be used for measurements of high pressure fixed points. It has been very useful, however, in the understanding of the corrections and errors inherent in the free piston gage and thus has been indirectly involved in the study of high pressure fixed points.

The free piston gage has been taken to pressures as high as 26 kbar[36] and is the most accurate device for establishing fixed points below that pressure. It is capable of accuracies of the order of 10 to 100 parts per million in pressure. The principal uncertainties come from estimating the change in dimensions of the piston with pressure. The value of the pressure along the melting curve of mercury and at the Bi I–II transition were established by measurements using a free piston gage.

The pressure in a piston and cylinder apparatus can be estimated by correcting the load for friction effects and by correcting the area of the piston for compression when under load. These corrections are far more difficult to make than those for the free piston gage and thus the pressure accuracy is at best only one or two per cent. If the sample is surrounded by some pressure-transmitting medium, a further uncertainty arises from the internal friction in this medium. Although the accuracy for measuring pressure is very poor, this is still the only direct static pressure-measuring instrument for pressures above 26 kbar that is presently being used. This type of apparatus has been extended to a pressure of 80 kbar by making the piston very short[35]. One of the measurements which help to establish the pressures at the Ba I–II and Bi III–V phase transitions involves this method[35,33]. One further drawback of this apparatus is that one does not easily find the region of indifference and thus does not unambiguously establish the equilibrium pressure of the transition. In fact the hysteresis observed in this apparatus for the measurement of the transition on increasing and decreasing pressure not only includes sample hysteresis but a hysteresis due to reversing the pressure gradients in the press and the pressure chamber. The midway point between the up and down pressure transitions has traditionally been chosen as the pressure at the transition. The dangers and possible errors in this assumption have been discussed by Pistorius et al.[51].

Bridgman[22] used force and area measurements to determine the pressure in an opposed anvil system. The friction corrections are enormous in this case so that it is amazing that Bridgman did as well as he did. One realizes the difficulty and uncertainty in this type of pressure measurement when we compare Bridgman's volume[21] and resistance pressure calibrations which differ by more than 30 kbar at the Bi III–V transition. It also appears that the strength of the material in Bridgman's volume apparatus altered his pressure calibration in such a way that his results with sodium chloride or cesium chloride in the pressure cell, for example, were very near the presently accepted value but his results with barium in the pressure cell gave a pressure that is too high*. Present high pressure researchers do not attempt to calculate the pressure in opposed anvil devices, but rather calibrate them from fixed points or use the extrapolation schemes discussed below.

Equations of state for alkali halide crystals[27] and alkali metals[50] can be calculated with fair accuracy in the region above 26 kbar. The alkali halide calculations are successful because the principal contribution to the interaction between the atoms is the well known coulomb interaction. The

---

* See p 267 of ref. 1.

PART 2. PRESSURE SCALE AND FIXED POINT

success of the calculations for alkali metals stems from the open type lattice with small overlap of the ions and thus simplicity of the electronic structure of these metals. The alkali halide calculations are semi-empirical with the empirical parameters being established from experimental measurements at atmospheric pressure. The principal uncertainty arises from anharmonic effects which are estimated using a volume-dependent Gruneisen parameter. The equation of state for sodium metal has been calculated on a more fundamental basis than for the alkali halides. The use of these equations in pressure calibration work involves measuring the volume of the alkali halide or the alkali metal at the instant of a phase transition for some other material. The material undergoing the transition must be embedded within the sodium chloride or sodium metal in such a manner that both experience the same pressure. Volume measurements at high pressures are most accurately accomplished using x-ray techniques. These at best yield volumes accurate to only 0.3 per cent. The uncertainty in the measured volume along with theoretical uncertainties in the equations of state make the pressure measurement about two per cent accurate. This technique has been valuable in helping to establish the value of the pressure at the Ba I–II and Bi III–V transitions and has been used in calibrating pressure up to 300 kbar[5].

Many experimenters have calibrated their pressure apparatus to pressures above those spanned by the now reasonably-well-established fixed points by using some kind of extrapolation procedure. Extrapolation is always potentially dangerous but the apparently inherent desire to extrapolate load versus pressure linearly has been the cause of many experimenters claiming pressures well beyond those actually attained. A graph of load versus sample pressure will normally curve toward the pressure axis as more of the load is taken by internal friction and compression of gasket material; thus a linear extrapolation always predicts too high a pressure. On the other hand it is difficult to predict the amount of curvature so that pressures at very high loads are extremely uncertain, at least until some fixed points are more accurately established in this range. Extrapolation is a first step, however, toward estimating pressures at the highest loadings attainable.

Another type of extrapolation procedure which should be mentioned involves the use of the manganin gage. The resistance of manganin versus pressure has been accurately characterized in the range available to free piston gages. It is found that one can represent the pressure accurately up to 25 kbar in terms of the resistance of manganin by the expression[12]

$$p = a(\Delta R/R_0) + b(\Delta R/R_0)^2 \qquad (1)$$

where R_0 is the manganin wire resistance at atmospheric pressure and ΔR is the change in resistance from this value. The parameters a and b can be determined by measuring the resistance at the mercury melting point and at the Bi I–II transition and one then uses the above equation to extrapolate the manganin gage readings to higher pressures, at least to the limit of pressures attainable in a liquid pressure medium, which is now about 60 kbar[4]. In this way the Tl I–II and Ba I–II transition pressures have been measured[63].

A final technique, that of dynamic shock measurements, will conclude this section. Phase transitions are detected as abrupt changes in slope of the Hugoniot curve. The pressure at these points can be calculated with a few per cent absolute accuracy by a well established technique[31]. Thus one would expect this to be a good procedure for establishing fixed points at very high pressures. There are two difficulties with this procedure. First the temperature increases with pressure along the Hugoniot causing the measured transition to be that at some elevated and not too accurately known temperature. Until one has established the effects of temperature on the phase transitions this adds considerable uncertainty to the value of the pressure at the transition as used by 'static' workers. Secondly the shock front moves through the crystal at high speed with times of the order of 10^{-6} to 10^{-9} sec in the pressure wave. Phase transitions normally involve a rearrangement of the atoms, which requires time. Thus the transition will not be likely to take place at the equilibrium value, but the sample will be greatly overpressurized in order for it to run in these short times.

IV. How to Make Use of Fixed Points in High Pressure Calibration

We now consider the practical experimental problems of what to assign as the pressure at a phase transition when one is not able to measure the equilibrium pressure, and how to interpolate the load versus pressure curve between the calibrated points.

In most high pressure apparatus one calibrates the load on the pressure system versus the pressure in the sample chamber. The sample may be encased in a solid pressure-transmitting medium and there may be gasketing material forming a pressure seal as well as considerable friction between the moving parts of the press. These systems have so much apparatus hysteresis that it is impossible to cycle the system to find the true nucleation hysteresis or the region of indifference in the sample's phase transition. Persons using the piston and cylinder apparatus often use the half way point between the initiation of the phase change detected on increasing the load and that on decreasing the load. This would be disastrous with opposed anvil systems, multianvil apparatus, or belt type apparatus. These systems are calibrated by noting the first instance of resistance change at the phase transitions as observed on the increasing pressure cycle. The actual pressure at the sample, however, is not uniquely given by a knowledge of the equilibrium pressure at the appropriate phase change. The difference between the pressure one should assign to the observed change and the equilibrium pressure depends upon the amount of inherent sample hysteresis, the pressure gradients at the sample, and the rate at which the pressure is increasing. In addition to these considerations one must also be aware of the fact that a phase change, which is accompanied by a sizeable volume change, as in the Bi I–II transition, will alter the local strain in the surrounding medium and thus will affect the nature and possibly even the pressure at the transition depending upon the mechanical properties of the surroundings. The latter effects are the probable reason why Jeffery et al.[41] found a difference in the phase transition pressures of flat strip versus round wire; their results are discussed in detail by Zeto and Vanfleet[64].

These problems must be worked out for each high pressure system and for each type of pressure cell individually and the experiments will probably involve measurements of consistency between different apparatus. As an

PART 2. PRESSURE SCALE AND FIXED POINT

example, with a tetrahedral anvil press, using a pyrophyllite pressure chamber, the pressures at the Bi I–II, Ba I–II, and the Bi III–V transitions are probably about 0.2, 1.2 and 2 kbar respectively above the equilibrium pressures when observed on the increasing-pressure cycle. These values are inferred from the x-ray measurements of Jeffery et al.[41] and an experiment by Zeto et al.[64, 65] wherein they compared the Bi I–II transition for a bismuth wire suspended freely in a liquid chamber and a similar wire embedded in epoxy and then the epoxy placed in the same liquid cell. This same article points out the importance of the rate of change of pressure on the pressure of the phase change. More of this type of study would be very valuable to the basic understanding of pressure calibration.

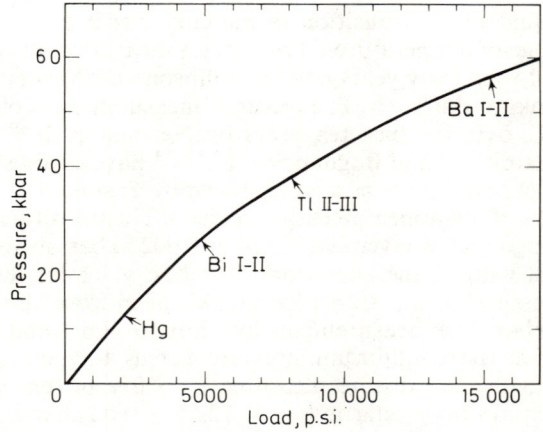

Figure 1. A typical calibration curve of sample pressure versus press load in a tetrahedral anvil pressure device.

Once one has established some calibration points of the load versus sample pressure for his apparatus he must next consider how to connect these points together, to form a calibration curve. Near the origin of zero pressure, where gasket material is moving and compression is large, the curve may well have a different character compared to the high pressure region where the gaskets are immobile and an increasing fraction of the load is being used to compress them. It is unlikely that there would be a discontinuity in the slope of the curve, and one should resist the temptation to assume that the calibration curve is made up of two linear regions for this will again lead to problems in extrapolating to very high pressures. A typical curve for a tetrahedral anvil device is shown in *Figure 1*. The data for this curve are taken from x-ray measurements*. A good method of extrapolation has been worked out by Contré[25] and was used by Drickamer[29] in his latest calibration.

---

* Results from several theses at Brigham Young University over the past few years (unpublished).

V. The Present Set of Fixed Points for Pressure Calibration

We will now consider the individual phase transitions that have been proposed and used in fixed point pressure calibrations. The compilations by Decker et al.[28] and by Contré[25] list all earlier measurements of pressure at these points and compare the work of various experimenters. The presentation here will be concerned with selecting the best value for the pressure to be assigned to these points, determining the precision with which these pressures are known, and discussing other considerations such as temperature effects, impurity effects, time effects, and hysteresis etc. Only individual measurements made since the above reviews will be discussed in any detail.

1. Mercury Melting Curve

The solid/liquid phase transition of mercury has been carefully studied over a wide range of temperatures. The earliest measurements were made by Bridgman[18] followed forty years later by Johnson and Newhall[42] and in the USSR by Zhokhovskii[66, 67]. The several measurements of the melting pressure at 0°C, over the past ten years by Newhall et al.[49], Dadson and Greig[26], Yasunami[60, 62] and Bogdanov et al.[10, 11] have established this point to an accuracy of about 3 bar as clearly shown by Yasunami[62]. The value by Bogdanov et al. did not appear in earlier reviews. The transition was detected by volume change and is given as 7.5724 ± 0.0025 kbar at 0°C which is in good agreement with all the later work. The best value for the equilibrium value of the pressure for this transition at 0°C taken from all recent work is 7.570 ± 0.003 kbar. The measurements by Zhokhovskii[66] and by Bogdanov et al.[11] also give the equilibrium pressure versus temperature along the melting curve, with the pressure measured by a free piston gage. The best value from their results would give $p = 12.55 \pm 0.02$ kbar at 25°C with a temperature gradient of $dp/dT = 0.201$ kbar/°C. This is also the value one

Table 1. Best averaged values for the equilibrium transition pressure, the up transition pressure, and the temperature derivative of the transition pressure. All values at 25°C

Transition	Equilibrium (kbar)	Equilibrium* (kbar)	Up pressure (kbar)	dp/dT (kbar/°C)
Hg (melting)	12.55 ± 0.02		12.6 ± 0.1	$+0.201 \pm 0.005$
Bi I–II	25.49 ± 0.07	25.50 ± 0.06	25.7 ± 0.3	-0.043 ± 0.003
Tl II–III	36.6 ± 0.3	36.7 ± 0.3	37.4 ± 1.2	$+0.007 \pm 0.009$
Ba I–II	55.3 ± 0.9	55 ± 2	56.5 ± 1.5	$+0.008 \pm 0.003$
Bi III–V	76.9 ± 1.5	77 ± 3	78.8 ± 2.2	-0.13 ± 0.01
Sn I–II			98 ± 4	

* As chosen by the Symposium on the Accurate Characterization of the High-Pressure Environment held at the National Bureau of Standards, Gaithersburg, Md, 1968. See ref. 1, pp 1–3.

would get from the equation proposed at the NBS symposium*. The room temperature point is a more useful value for the calibration of large high pressure apparatus using solid pressure-transmitting media. These results are summarized along with those of other transitions in *Table 1*. The melting

* See ref. 1, p 2.

transition in mercury is very sharp and can be repeatedly measured to a precision of 0.2 bar[62]; however, there is normally some supercooling in the fusion transition and thus it is not very reproducible. The melting point appears to be at the equilibrium phase boundary. Unfortunately, a system calibrated by noting phase transitions on increasing pressure will pass through this transition from the liquid to the solid phase and the indication of the phase change will take place above the equilibrium transition value. No measurements are available to indicate how large these effects might be in a solid pressure system, but when contained in a liquid the hysteresis is from 30 to a few hundred bars which probably depends, among other things, upon temperature, rate of pressure change, and purity of mercury used.

2. *Bismuth I–II*

The Bi I–II polymorphic phase transition was first observed by Bridgman in 1940[16, 19], who noted the volume change at the transition. He assumed the equilibrium pressure as that at the center of the region of indifference. The only other measurement that truly found the center of the region of indifference is that by Heydemann[36]. Several measurements have been made of the center of the I–II, II–I hysteresis notably by Boyd and England[13], Kennedy and La Mori[43, 44], Vereshchagin et al.[59], and La Mori[46]. The pressure at the discontinuity in the shock Hugoniot curve, corresponding to the Bi I–II transition was reported by Duff and Minshall[30] and by Larson[48]. An indirect measurement of the center of the hysteresis was made by Jeffery et al.[41] which, when using the most recent sodium chloride equation of state[27], gives the value 25.32 ± 0.77 kbar. All these values are tabulated by Decker et al.[28] and by Contré[25] except for the most recent result by La Mori[46] which gave 25.37 ± 0.20 kbar at 25°C using a piston gage. This measurement also yielded a temperature variation of -0.046 kbar/°C. The measurement by Heydemann[36] appears to be superior to the others in absolute accuracy so that the most probable value for the equilibrium pressure at this transition given in *Table 1* is almost exclusively determined from his measurement. Zeto and Vanfleet[64] have shown that the first indication of the Bi I–II transition with the bismuth in a solid medium is very close to the equilibrium pressure if the pressure is increased at a very slow rate. This indicates that the measurements of the midpoint of the I–II, II–I hysteresis are below the equilibrium value. This seems valid in that all these measurements are lower than those of Heydemann. It is difficult to control the pressure at the rates indicated by Zeto and Vanfleet[64] in a solid medium press and most measurements of the Bi I–II transition in these presses are under such circumstances that the I–II transition probably appears above the equilibrium value.

The Bi I–II transition is very sharp in a hydrostatic medium with a region of indifference of 20 bar[64]. The resistance and volume changes are both large and easily observed. The characteristics, such as time effects, growth rates, region of indifference, and hysteresis, have been carefully examined[64]. Impurity effects upon the transition are negligible[36, 14], also bismuth is readily available in high purity. Grain size does have a small effect on the transition[64] but this is probably not important in non-liquid systems.

Several measurements of the temperature variation of the Bi I–II phase line have been reported[13,16,19,24,37,38,45,46,53,57] with values ranging from -0.040 to -0.051 kbar/deg. The most recent by Houck[37] was carried out specifically to measure this effect with the result that $dp/dT = -40.6 \pm 1.2$ bar/°C. The value in *Table 1* is a most probable average over all measurements.

The Bi II–III transition is strongly affected by the strain in the surrounding environment of the bismuth created by the volume change at the Bi I–II transition in a non-liquid pressure chamber. For this reason, as well as the fact that it is so near the I–II transition, this point is of little use in pressure calibration and will not be considered here.

3. *Thallium II–III*

It is useful to have a fixed point for pressure calibration in the neighbourhood of 40 kbar to aid in determining the shape of the calibration curve. The thallium II–III transition has traditionally been used but it is a material that is not handled easily, the transition is very sluggish, and little experimental work has been done to analyse the nature of the transition. The results and techniques of the measurements of this point are discussed by Decker *et al.*[28]. The transition was first measured by Bridgman[15] in 1935 at which time he determined the equilibrium value for the pressure at the transition. Similar measurements of this point were made by Boyd and England[13], Kennedy and La Mori[43,44], Vereshchagin[59], and most recently by La Mori[46]. La Mori reported the result of 36.5 kbar at the transition, but felt his value was not as reliable as the earlier measurements[43] because not as much time was taken in passing through the transition. In addition to the piston and cylinder measurements there is the determination of the midpoint between the II–III, III–II hysteresis on the sodium chloride scale by Jeffery *et al.*[41], which gives the value 35.4 ± 2.1 kbar*. The hysteresis is relatively narrow, of the order of 2 kbar, but little is known about effects of impurities on the transition or of the position of the region of indifference relative to the hysteresis of the II–III, III–II transition in a solid medium. The region of indifference is only 6 bar. The variation with temperature of the pressure at this transition was given by La Mori[46] as 0.010 kbar/°C. Other measurements of the temperature effects are -0.016 kbar/°C by Bridgman[15], $+0.024$ kbar/°C by Jayaraman *et al.*[40], and $+0.004$ kbar/°C by Adler and Margolin[2].

This transition is not well suited to be a fixed point. Perhaps the cesium transition near 40 kbar would be better but it has not yet been studied with enough detail. In addition, cesium is not easily handled because of its chemical activity and one cannot make a definite choice at this time.

4. *Barium I–II*

The barium transition is important in that it is used as a stepping-off point for extrapolation to the higher points. Unfortunately, this transition cannot be determined accurately because it has a large region of indifference[63].

---

* Using the recent sodium chloride scale given in ref. 27.

PART 2. PRESSURE SCALE AND FIXED POINT

The value of the pressure at this point has undergone many revisions over the years. Zeto and Vanfleet alone determined the region of indifference[63]; all other measurements: Bridgman[17,20], La Mori[47], Haygarth et al.[35], Vereshchagin et al.[59], and Jeffery et al.[22] involved the midpoint of the hysteresis pattern. The earlier measurements probably estimated their pressures to be too high. Of the recent measurements those of Haygarth et al. and Jeffery et al. on the recent sodium chloride scale[27], and of Zeto and Vanfleet[63] are in good agreement but are definitely in opposition to the value reported by Vereshchagin et al. Measurements of the temperature derivative for this transition have yielded 0.005 kbar/°C by Jayaraman et al.[39], 0.010 kbar/deg by Bastide et al.[6,7] and 0 to 0.015 kbar/°C by Haygarth et al.[35]. Sample purity may have an effect on the transition pressure as indicated by Haygarth et al.; in their work a 99.5 per cent pure sample underwent the phase change about 0.4 kbar below a highly pure specimen.

5. Bismuth III–V

The pressure assigned to the upper bismuth, or Bi III–V, transition has been greatly reduced in recent years but much of the original problem came from adopting the resistance measurement by Bundy[23] in a belt apparatus which he reported as 122 kbar. The original volume measurement by Bridgman[17,20] of this transition in a two-stage piston apparatus was only about 10 kbar above the value accepted now. Of the recent measurements of this transition only those of Jeffery[41], which is 76.0 ± 1.8 kbar using the best sodium chloride scale[27], and of Stark and Jura[53] determined the center of the hysteresis. Klement et al.[45] and Haygarth et al.[33,34] measured the center of the hysteresis of the phase transition at lower pressures and then extrapolated the phase line to 25°C. All other measurements[32,59,25,29] were apparently of the up transition value, i.e. with increasing pressure. Vereshchagin et al.[59] did not give enough detail to enable others to assess what was measured and their results appear to be definitely in error. Drickamer has measured this point in opposed anvils by extrapolation from lower-pressure known points. In 1961 he reported 89–92 kbar[3] for this transition which in 1970 he revised to 74 ± 1 kbar[29]. The equilibrium value in *Table 1* was taken from the data of Jeffery et al. and of Haygarth et al. while the up transition value was estimated from these same measurements plus those of Giardini[32], Stark and Jura[53], and Contre[25].

The temperature derivative for this transition is more pronounced than most and is taken from the work of Haygarth et al.[34].

6. *Fixed Points above 100 kbar*

The Bi III–V transition is the highest-pressure phase transition that has been directly attained in a piston type apparatus. The pressure at all higher-pressure transitions has been determined either from the sodium chloride equation of state or from an extrapolation of a calibration passing through the lower-pressure fixed points. Of course, as the pressures at these lower pressure points are refined the extrapolations must be reworked. The points that have been considered in the 100 to 200 kbar range are Sn I–II, Fe α–ε, and the upper barium point. The iron transition is so sluggish and metastable that values all the way from 83 kbar[55] to 152.5 kbar[58] have been reported for this transition. It is of no use as a calibration point.

There are a number of measurements of the Sn I–II transition, which indicate that it is a sharp transition and could possibly make an acceptable fixed point, but the resistance change is not clear enough. Bundy[23] reported 114 kbar in a belt apparatus. Balchan and Drickamer[3] reported 113–115 kbar in an opposed anvil apparatus on the latter's early pressure scale, which is definitely too high[29]. Other results in opposed anvil apparatus are 99 ± 4 kbar by Stark and Jura[53] and 107 kbar by Stromberg et al.[54]. Contré[25] measured 104 ± 5 kbar using a multianvil device. The value on the sodium chloride scale[27,41] is 95 ± 3 kbar. All these measurements are up transition values and the variations may reflect the effects of the different strains from the different apparatus on the amount of super-compression.

The pressure at the upper barium transition, Ba II–III, has been measured only in Drickamer's anvils which on his old scale was 144 kbar[28] being revised to 120 ± 2 kbar on his new scale[29]. Another point that has been proposed is the Pb I–II transition. Balchan and Drickamer[3] reported 161 kbar, Takahashi[56] 130 kbar, and Drickamer[29] 130 ± 2 kbar. More work is needed to decide which of these two points Ba II–III or Pb I–II, might prove most useful for pressure calibration.

VI. Conclusions

Fixed point calibration is in a much better state today than it was five or six years ago and most of the leading high-pressure laboratories are beginning to agree on their pressures, at least below 100 kbar. The fixed points discussed in this chapter have been those traditionally and most commonly used. There may exist materials that would serve this purpose far better than those discussed. However, of the many other phase transitions which have been suggested as fixed points, none, as yet, have been widely adopted or for that matter have been demonstrated to be better than those discussed here.

VII. References

[1] 'Accurate characterization of the high-pressure environment', edited by E. C. Lloyd, *NBS Spec. Publ. No. 326* (US Government Printing Office: Washington, D.C., 1971) pp 1–3, 263–271, 313–340.
[2] Adler, P. N. and H. Margolin. *Trans. Met. Soc. AIME,* **230**, 1048 (1964).
[3] Balchan, A. S. and H. G. Drickamer. *Rev. Sci. Instrum.* **32**, 308 (1961).
[4] Barnett, J. D. and C. D. Bosco. *Rev. Sci. Instrum.* **38**, 957 (1967).
[5] Bassett, W. A., T. Takahashi, H. K. Mao and J. S. Weaver. *J. Appl. Phys.* **39**, 319 (1968).
[6] Bastide, J. P. and C. Susse. *High Temp. High Press.* **2**, 237 (1970).
[7] Bastide, J. P., C. Susse and R. Epain. *CR Acad. Sci., Paris,* **267**, 857 (1968).
[8] Bett, K. E. and D. M. Newitt. *The Physics and Chemistry of High Pressure,* p 99. (Symposium at Olympia, London, 1962.) Gordon and Breach: New York (1963).
[9] Bett, K. E., P. F. Hayes and D. M. Newitt. *Phil. Trans. Roy. Soc. London,* **254**, 59 (1954).
[10] Bogdanov, V. S. *Izmeritelnaya Tekhnika* No. **8**, 29 (1967).
[11] Bogdanov, V. S., Yu. L. Levin, S. S. Sekoyan and Yu. I. Shmin. 'Accurate characterization of the high-pressure environment', edited by E. C. Lloyd. *NBS Spec. Publ. No. 326,* p 297. US Government Printing Office: Washington, DC (1971).
[12] Boren, M. D., S. E. Babb and G. J. Scott. *Rev. Sci. Instrum.* **36**, 1456 (1965).

PART 2. PRESSURE SCALE AND FIXED POINT

[13] Boyd, F. R. and J. L. England. *J. Geophys. Res.* **65**, 741 (1960).
[14] Brandt, N. B. and N. I. Ginsburg. *Soviet Phys.–JETP,* **44**, 848 (1963).
[15] Bridgman, P. W. *Phys. Rev.* **48**, 893 (1935).
[16] Bridgman, P. W. *Phys. Rev.* **57**, 235 (1940).
[17] Bridgman, P. W. *Phys. Rev.* **60**, 351 (1941).
[18] Bridgman, P. W. *Proc. Amer. Acad. Arts Sci.* **47**, 321 (1911).
[19] Bridgman, P. W. *Proc. Amer. Acad. Arts Sci.* **74**, 1 (1940).
[20] Bridgman, P. W. *Proc. Amer. Acad. Arts Sci.* **74**, 399 (1942).
[21] Bridgman, P. W. *Proc. Amer. Acad. Arts Sci.* **74**, 425 (1942).
[22] Bridgman, P. W. *Proc. Amer. Acad. Arts Sci.* **81**, 167 (1952).
[23] Bundy, F. P. *Phys. Rev.* **110**, 314 (1964).
[24] Compy, E. M. *J. Appl. Phys.* **41**, 2014 (1970).
[25] Contré, M. 'Accurate characterization of the high-pressure environment', edited by E. C. Lloyd, *NBS Spec. Publ. No. 326*, p 291. US Government Printing Office: Washington, DC (1971).
[26] Dadson, R. S. and R. G. P. Greig. *Brit. J. Appl. Phys.* **16**, 1711 (1965).
[27] Decker, D. L. *J. Appl. Phys.* **42**, 3239 (1971).
[28] Decker, D. L., W. A. Bassett, L. Merrill, H. T. Hall and J. D. Barnett. *J. Phys. and Chem. Ref. Data,* **1**, 773 (1972).
[29] Drickamer, H. G. *Rev. Sci. Instrum.* **41**, 1667 (1970).
[30] Duff, R. E. and F. S. Minshall. *Phys. Rev.* **108**, 1207 (1957).
[31] Duval, G. E. and G. R. Fowles. *High Pressure Physics and Chemistry,* 2nd ed. (Bradley), Ch. 9. Academic Press: New York (1963).
[32] Giardini, A. A. and G. A. Samara. *J. Phys. Chem. Solids,* **26**, 1523 (1965).
[33] Haygarth, J. D., H. D. Luedemann, I. C. Getting and G. C. Kennedy. 'Accurate characterization of the high-pressure environment', edited by E. C. Lloyd, *NBS Spec. Publ. No. 326* p 35. US Government Printing Office: Washington, DC (1971).
[34] Haygarth, J. D., H. D. Luedemann, I. C. Getting and G. C. Kennedy. *J. Phys. Chem. Solids,* **30**, 1417 (1969).
[35] Haygarth, J. D., I. C. Getting and G. C. Kennedy. *J. Appl. Phys.* **8**, 4557 (1967).
[36] Heydemann, P. L. M. *J. Appl. Phys.* **38**, 2640, 3424 (1967).
[37] Houck, J. C. *J. Res. Nat. Bur. Stand.* **74** A, 51 (1970).
[38] Il'ina, M. A. and E. S. Itakevich. *Soviet Phys.-Solid State,* **8**, 1873 (1967).
[39] Jayaraman, A., W. Klement Jr. and G. C. Kennedy. *Phys. Rev. Letters,* **10**, 387 (1963).
[40] Jayaraman, A., W. Klement Jr., R. C. Newton and G. C. Kennedy. *J. Phys. Chem. Solids,* **24**, 7 (1963).
[41] Jeffery, R. N., J. D. Barnett, H. B. Vanfleet and H. T. Hall. *J. Appl. Phys.* **37**, 3172 (1966).
[42] Johnson, D. P. and D. H. Newhall. *Trans. Amer. Soc. Mech. Engrs.,* **75**, 301 (1953).
[43] Kennedy, G. C. and P. N. La Mori. *J. Geophys. Res.* **67**, 851 (1962).
[44] Kennedy, G. C. and P. N. La Mori. *Progress in Very High Pressure Research,* p 304. Wiley: New York (1961).
[45] Klement Jr, W., A. Jayaraman and G. C. Kennedy. *Phys. Rev.* **131**, 632 (1963).
[46] La Mori, P. N. 'Accurate characterization of the high-pressure environment', edited by E. C. Lloyd. *NBS Spec. Publ. No. 326,* p 279. US Government Printing Office: Washington, DC (1971).
[47] La Mori, P. N. *High Pressure Measurement,* edited by A. A. Giardini and E. C. Lloyd. Butterworths: Washington, DC (1963).
[48] Larson, D. B. *J. Appl. Phys.* **38**, 1541 (1967).
[49] Newhall, D. H., L. H. Abbot and R. A. Dunn. *High Pressure Measurements,* edited by A. A. Giardini and E. C. Lloyd, p 339. Butterworths: Washington, DC (1963).
[50] Pastine, D. J. *Phys. Rev.* **166**, 703 (1968); **175**, 905 (1968).
[51] Pistorius, C. W. F. T., E. Rapoport and J. B. Clark. *Rev. Sci. Instrum.* **38**, 1741 (1967).
[52] Ponyatovskii, E. G. *Soviet Phys.-Crystallogr.* (translation), 5, 147 (1960).
[53] Stark, W. and G. Jura. *Amer. Soc. Mech. Engrs Publication, 64-WA/Pt-28* (1964).
[54] Stromberg, H. D. and D. R. Stephens. *Amer. Soc. Mech. Engrs Paper 64-WA/Pt-13* (1964).
[55] Takahashi, T. and W. A. Bassett. *Science,* **145**, 3631 (1964).
[56] Takahashi, T., W. A. Bassett and H. K. Mao. *J. Geophys. Res.* **73**, 4714 (1969).
[57] Tikhomirova, N. A., E. Yu. Tonkov and S. M. Stishov. *Soviet Phys.–JETP Letters,* **3**, 60 (1966).

[58] Vereshchagin, L. F., A. A. Semerchan, N. N. Kuzin and Yu. A. Sadkov. *Soviet Phys.-Doklady,* **14**, 340 (1969).
[59] Vereshchagin, L. F., E. V. Zubova, I. P. Buimova and K. P. Burdina. *Soviet Phys.–Doklady*, **11**, 585 (1967) translation of *Dokl. Akad. Nauk SSSR*, **169**, 74 (1966).
[60] Yasunami, K. *Metrologia,* **4**, 168 (1968).
[61] Yasunami, K. *Proc. Imp. Acad. Japan,* **43**, 310 (1967).
[62] Yasunami, K. *Rev. Phys. Chem. Japan,* **37**, 1 (1967).
[63] Zeto, R. J. and H. B. Vanfleet. *J. Appl. Phys.* **40**, 2227 (1969).
[64] Zeto, R. J. and H. B. Vanfleet. *J. Appl. Phys.* **42**, 1001 (1971).
[65] Zeto, R. J., H. B. Vanfleet, E. Hryckowian and C. D. Bosco. 'Accurate characterization of the high-pressure environment', edited by E. C. Lloyd. *NBS Spec. Publ. No. 326*, p 25. US Government Printing Office: Washington, DC (1971).
[66] Zhokhovskii, M. K. *Izmeritelnaya Tekhnika,* No. 5, 3 (1955).
[67] Zhokhovskii, M. K. *Izmeritelnaya Tekhnika,* No. 7, 11 (1959); No. 8, 14 (1959).

CHAPTER 4

Part 3. Piston Gages

P. L. M. HEYDEMANN and B. E. WELCH
National Bureau of Standards, Washington, DC, USA

Contents

	List of Symbols	147
I.	Introduction and Historical Review	149
II.	Basic Equations and Elastic Distortion	152
III.	Piston Gage Designs	155
	1. Simple Piston Gage	155
	2. Tilting Piston Gage	156
	3. Vacuum-backed Piston Gage	157
	4. Re-entrant Cylinder Design	159
	5. Differential Piston	159
	6. Controlled Clearance	161
	7. Ball Gages	162
	8. Grooved Pistons	163
	9. Very High Pressure Piston Gages	165
IV.	Calibration of Primary Standards	168
	1. Controlled Clearance Piston Gage	168
	A. Load	168
	B. Area	171
	C. Temperature Correction	173
	D. Pressure Coefficient	173
	E. Change of Area with Jacket Pressure	175
	F. Zero Clearance Jacket Pressure	178
	G. Total Uncertainty of Pressure	183
	2. Similarity Method	184
V.	Calibration of Piston Gages	188
	1. Cross-float	188
	2. Reference Levels	191
	3. Evaluation	191
VI.	The Use of Piston Gages	193
	1. Measurement of Pressure	193
	2. Procedures and Methods	194
	A. Cleaning	194
	B. Temperature Control and Measurement	195
	C. Weights	196
	D. Corrections	197
	E. Fluids	198
	F. Rotation of Piston or Cylinder	199
	G. Mechanical Support and Covers	200
VII.	Abstract	200
VIII.	References	200

List of Symbols

a, a^* transducer coefficient

b_1, b^*	transducer coefficient
b, b_1, b_2	pressure coefficient of area
c_1, c	transducer coefficient
d_1, d^*	transducer coefficient
d	jacket pressure coefficient
g	acceleration due to gravity
g_ϕ	acceleration due to gravity, at sea-level
g_l	acceleration due to gravity, local
g	ratio of shear moduli
h	halfwidth of radial clearance
k	ratio of elastic moduli
l	length of clearance or engagement
p	pressure
p_{bar}	pressure, barometric
p_c	pressure, inside clearance
p_e	pressure, on end face of piston or cylinder
p_j	jacket pressure
p_{sat}	saturation pressure
p_{z0}	zero clearance jacket pressure
q_z	clearance versus jacket pressure coefficient
r	radius
r_c	radius, internal, cylinder
r_p	radius, piston
s	piston position
s_z	zero clearance jacket pressure coefficient
u	increase in piston diameter
u	voltage
x	reduction coefficient
y	reduction coefficient
z	coordinate along cylinder axis
A	area
A_c	area of cylinder
A_{eff}	area, effective
A_0	area, effective, at zero pressure
A_p	area of piston
B	temperature coefficient of saturation pressure
C	circumference of piston
C	constant
C_b	pressure correction
C_d	jacket pressure correction
C_t	temperature correction
C_z	clearance correction
D	jacket pressure coefficient
E	jacket pressure coefficient
E	Young's modulus
E_0, E_1, E_2	temperature coefficient of saturation pressure
F	force
F	jacket pressure coefficient
H	halfwidth of clearance

PART 3. PISTON GAGES

H	relative humidity
H	difference in reference levels
M	mass
Q	leak rate
R_c	outer radius of cylinder
T	tare
T	gage temperature
T_{ref}	reference temperature
T_w	tare weight
U	increase in internal cylinder diameter
V	volume
W	weight
α_p	thermal expansivity of piston
α_c	thermal expansivity of cylinder
γ	surface tension
η	viscosity
$\theta =$	$(3\mu - 1)/2E$
λ	pressure coefficient of area
ρ_{air}	density of air
ρ_{fl}	density of fluid
ρ_{M_i}	density of weight i
ϕ	latitude, geographical

I. Introduction and Historical Review

The early seventeenth century was a time of great progress in science and it is perhaps not surprising that the great contemporaries Galileo, Descartes, Boyle, and many others devoted some of their time to the study of the properties of the atmosphere. In 1644 Torricelli described his famous measurement of atmospheric pressure in a letter[50] to his friend Michelangelo Ricci. Within 25 years followed the famous experiments by Pascal and Perier (1648), Guericke (1672) and Boyle (1669). One hundred years later, in 1764, James Watt began to lay the foundation for the *Steam Tables* by making the first careful measurements of the properties of steam. During this time mercury manometers were used to determine pressure. About 1800 Richard Trevithik built the first high pressure steam engine which proved to be much more efficient than Watt's and Newcomen's machines; from then on new types of gages were required to cover the pressure range of interest for industrial applications. In 1846 the German railway engineer Schinz[34] had discovered that a curved tube of elliptical cross section would change its curvature when subjected to internal pressure and by 1848 steam pressure gages based on this principle were in use on locomotives in Germany[19]. In 1847 both Schinz[46] and Bourdon[6] patented devices which are now universally known as Bourdon gages. In 1846 Galy-Cazalat[20] described the first piston manometer, a combination of mercury manometer and hydraulic multiplier. Two pistons with a large area ratio were mechanically coupled. The high pressure acted on the small piston and the pressure under the large piston was measured with a mercury manometer. *Figure 1* shows a piston

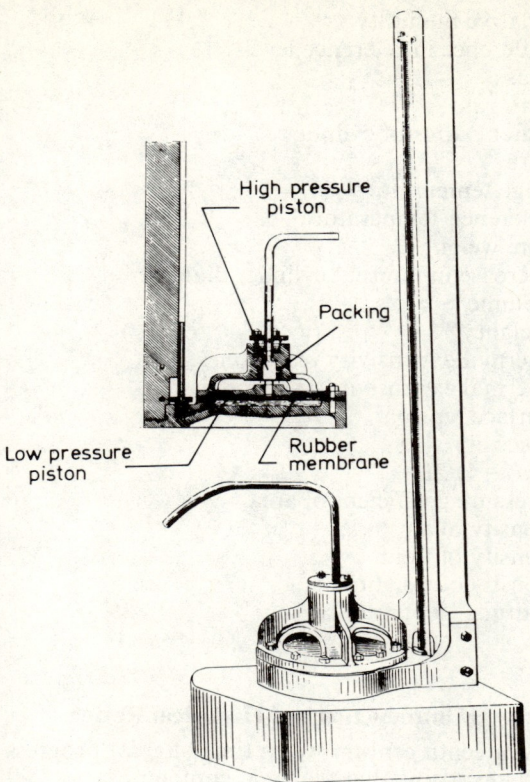

Figure 1. Piston manometer (Desgoffe, 1871).

manometer for pressures to 1 000 bar designed by Desgoffe[17] and manufactured by Pelletier in Paris. The high pressure piston was sealed with a packing; the low pressure piston was sealed with a rubber membrane. Amagat[2] used a similar instrument but he did away with the membrane and used rotation to relieve friction between the pistons and their cylinders.

At this time dead weight loaded piston gages had already reached a state of perfection that is best demonstrated by an instrument with automatic loading manufactured by L. Seyss[47]. *Figure 2* gives a schematic view of this piston gage. The cylinder (A) contained two concentric pistons (K). Force was applied to the pistons with the help of the rod (S) and a yoke. Unused weights were hanging from the supporting frame. As pressure was applied the pistons moved up and the weights were picked up one by one. If the retaining nut (D) was screwed into the cylinder (A), only the small piston moved. For lower pressures, nut (D) was fastened to part (C) and both pistons moved.

The simple piston gage patented by E. Ruchholz[45] shown in *Figure 3* has all the important features of gages still manufactured today: top-loading straight piston, rotation to relieve friction, integral valve and connector for

PART 3. PISTON GAGES

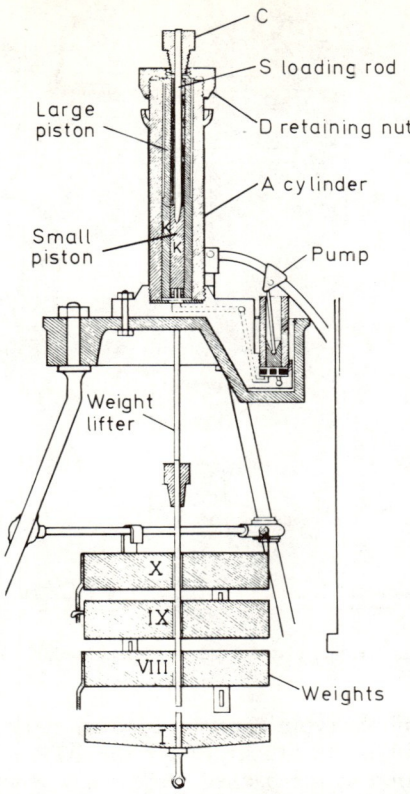

Figure 2. Piston gage with hydraulic weight loading and two concentric pistons (Seyss, 1869).

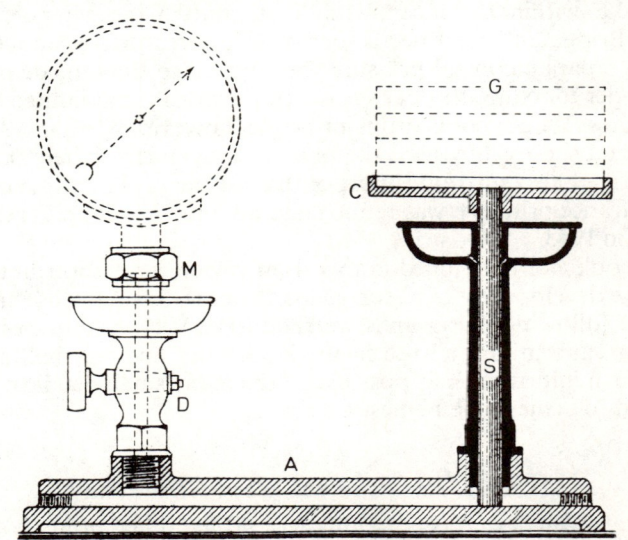

Figure 3. Manometer calibration stand with piston gage (Ruchholz, 1882).

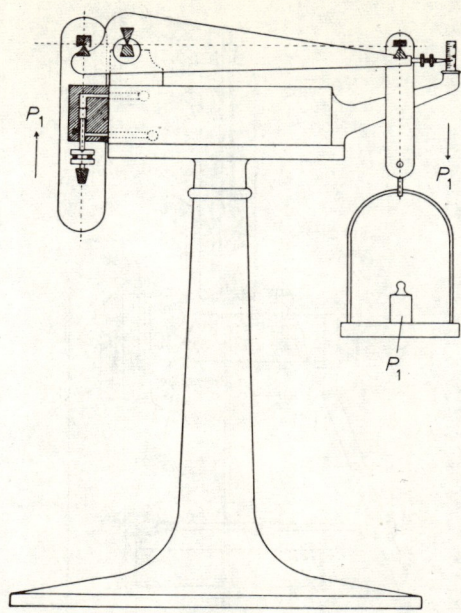

Figure 4. High pressure piston gage with beam loading (Stückrath, 1897).

test gages and an oil reservoir. A piston gage for 500 bar with beam loading (*Figure 4*) was built by P. Stückrath for the PTR at Berlin[18]. It had no provision for rotation of the packed piston; nevertheless its precision was estimated to be about 0.04 per cent at 250 bar[53]. At about 20 bar this piston gage agreed to within 0.2 per cent with the standard manometer of the PTR[44].

In 1908 Bridgman[8] described a generator for pressures to about 7000 bar in which a separate control pressure was applied to the outside of one part of the cylinder to reduce the leakage which previously had limited the attainable pressures. He abandoned this principle, however, when in 1911 he first suggested and successfully used the re-entrant cylinder[9], where the pressure to be generated also surrounds part of the outside of the cylinder. The controlled clearance principle was refined and adapted to piston gages by D. H. Newhall[41] in 1953.

Many people not mentioned in this short review have contributed significantly to the development of piston gages, their manufacture, use and evaluation. In the following paragraphs we shall discuss types of gages available today, their calibration and use in the laboratory and we shall attempt to relate as much information as possible on the many small details which must be observed to achieve the highest accuracy.

II. Basic Equations and Elastic Distortion

Pressure is defined as force per unit area. Many different instruments can be used for its indication but its absolute value can only be determined with a

PART 3. PISTON GAGES

direct force per unit area measurement or by involving basic laws of thermodynamics. Piston gages and manometers use the force per unit area principle. In piston gages the pressure is applied to the end face of a piston of known cross sectional area and the resulting force is measured by loading the piston with weights.

The pressure p generated by a piston gage at its reference level is given by equation 1

$$p = \frac{\sum_i M_i g(1 - \rho_{air}/\rho_{M_i}) + \gamma C + T_w}{A_0[1 + (\alpha_c + \alpha_p)(T - T_{ref})][1 + bp]} \tag{1}$$

where the symbols have the following meanings:

M_i, ρ_{M_i} mass and density of weight i,
ρ_{air} density of air at the temperature, barometric pressure and humidity prevailing in the laboratory,
γ surface tension of pressure-transmitting fluid,
C circumference of the piston where it emerges from the fluid,
T_w tare weight or error,
A_0 effective area of the assembly at zero pressure,
α_c, α_p thermal expansivities of cylinder and piston,
T temperature of the assembly,
T_{ref} temperature to which A_0 is referred,
b pressure coefficient of the effective area.

The term $(1 - \rho_{air}/\rho_{M_i})$ is the air buoyancy correction for weight i. γC is the force exerted on the piston by the surface tension of the fluid. T is a tare weight or error. The term $[1 + (\alpha_c + \alpha_p)(T - T_{ref})]$ corrects the area for thermal expansion. The term $[1 + bp]$ describes the change of the effective area with pressure. This is the most important correction term. It will be discussed here while the rest of the terms of equation 1 will be treated in greater detail in section IV which describes the calibration of primary gages.

Two effects contribute to the change of effective area with pressure: the distortion of the piston under the combination of the longitudinal stress due to the applied load and the hydrostatic pressure over part of its length, and the distortion of the cylinder due to the internal pressure over part of its length.

Several authors have contributed significantly to the calculation of these distortions[4, 14, 27, 31, 55]. The change in radius r_p of a solid piston subjected to end pressure p, surrounded by the pressure p_c in the clearance is given according to Johnson and Newhall[30] by

$$\frac{r_p(p) - r_p(0)}{r_p(0)} = \frac{\mu p}{E} + \frac{p_c}{E}(\mu - 1) \tag{2}$$

where μ is Poisson's ratio and E is the modulus of elasticity. The change in the radius r_c of a hollow cylinder of outside radius R_c is given by

$$\frac{r_c(p) - r_c(0)}{r_c(0)} = \frac{p_c}{E} \frac{(1 + \mu)R_c^2 + (1 - \mu)r_c^2}{R_c^2 - r_c^2} - \frac{p_0}{E} \frac{2R_c^2}{R_c^2 - r_c^2} + \frac{\mu p_e}{E} \tag{3}$$

where p_c is the pressure inside the cylinder, p_0 that on the outside and p_e is the pressure on the end faces.

p_c in equations 2 and 3 varies along the piston from a maximum at the lower end to zero in the area of contact or minimum clearance and the exact correction for piston and cylinder distortion depends on the position of the minimum clearance. In practice the position of minimum clearance may even shift with pressure.

Johnson et al.[31] have made a more detailed study of the distortion of a piston and arrive at the expression

$$\frac{r_p(p) - r_p(0)}{r_p(0)} = - p(z)\frac{(1 - 3\mu)}{2E} \qquad (4)$$

which is independent of the form of the function $p(z)$, where z is an axial coordinate, as long as the pressure gradient is in an area removed from the ends of the piston by one or more diameters. Equation 4 gives the same distortion as equation 2 for $p_c = 0.5\ p$.

The viscous drag of the pressure fluid streaming through the clearance causes an upward force on the piston. Dadson et al.[14] give this force as

$$F = 2\pi r_p \int_0^l \{- h\,(dp/dz)\}\,dz \qquad (5)$$

where $2h$ is the width and l the length of the clearance. Both p and h are unknown functions of the coordinate z. It is immediately obvious that for an undistorted system where h is constant this leads to

$$F = 2\pi r_p hp \qquad (6)$$

Except for terms quadratic in h this equals the force generated by the pressure p over a ring with the internal diameter $2r_p$ and the width h. To avoid separate corrections for the fluid friction it is therefore expedient to use as effective area the arithmetic mean between the areas of the piston and of the cylinder

$$A_{eff} = \tfrac{1}{2}(A_p + A_c) \qquad (7)$$

Equations 2 and 3 can now be introduced into equation 7 and, assuming for simplicity that $p_c = 0.5\ p$, we have for the pressure coefficient n of equation 1

$$b = -(1 - 3\mu)/2E_p \quad \text{(piston distortion)}$$

$$+ \{(1 + \mu_c)R_c^2 + (1 - \mu_c)r_c^2\}/\{2E_c(R_c^2 - r_c^2)\}$$

(cylinder distortion due to internal pressure)

$$- (p_0/pE_c)(2R_c^2/\{R_c^2 - r_c^2\})$$

(cylinder distortion due to external pressure)

$$+ (p_e/p)(\mu_c/E_c) \qquad (8)$$

(cylinder distortion due to end loading)

Let us get an estimate for the size of these corrections to the area. For steel let $E = 2 \times 10^{11}$ N/m^2 and $\mu = 0.28$; further let $R_c/r_c = 3$ and we have for the pressure coefficient

$$b_{\text{steel}} = -4 \times 10^{-13} + 38 \times 10^{-13} - 112 \times 10^{-13}(p_0/p)$$
$$+ 14 \times 10^{-13}(p_e/p) \quad (9)$$

all in m^2/N. For cemented tungsten carbide with $E = 6 \times 10^{11}$ N/m^2 and $\mu = 0.2$ the pressure coefficient is

$$b_{\text{TC}} = -3.4 \times 10^{-13} + 12 \times 10^{-13} - 38 \times 10^{-13}(p_0/p)$$
$$+ 3 \times 10^{-13}(p_e/p) \quad (10)$$

all in m^2/N. It is immediately obvious that depending on the design of the gage the pressure coefficient can be either positive or negative and that under certain conditions it could approach zero. Different ways of dealing with the pressure coefficient b have led to the design of various types of piston gages to be discussed in the following section.

III. Piston Gage Designs

1. *Simple Piston Gage*

The piston and cylinder combination used by Dadson and Greig[16] in their determination of the mercury freezing pressure is a good example of a well-designed, simple piston gage. A schematic diagram is shown in *Figure 5*. Force is applied to the head of the piston through an auxiliary piston and a flexible joint to prevent imparting a bending moment to the gage piston. The operating area of the assembly is removed by several diameters from the lower part of the cylinder, which is distorted by making a pressure seal. The cylinder is stressed only by the pressure inside the crevice and the

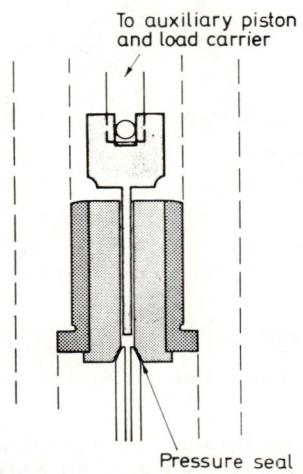

Figure 5. Simple piston gage, top loading (Dadson, 1965).

Table 1. Pressure coefficients b

Pressure coefficient (1/kPa)	Piston/cylinder material	Range (MPa)	Cylinder type
$+1.1 \times 10^{-8}$	steel/brass	17	simple
-1.3×10^{-8}	steel/brass	70	re-entrant
$+5.2 \times 10^{-9}$	steel/steel	34	simple
-6.4×10^{-9}	steel/steel	83	re-entrant
-8.1×10^{-9}	carbide/steel	17	re-entrant
-6.1×10^{-9}	carbide/steel	83	re-entrant
-2.9×10^{-9}	carbide/carbide	280	re-entrant
-5.1×10^{-10}	tungsten/carbide	—	controlled clearance

pressure coefficient of the assembly, described by the first two terms of equation 8, is positive. *Table 1* lists pressure coefficients for piston gages of this and other types.

2. *Tilting Piston Gage*

The extension of the operating pressure range towards lower pressures is limited by the weight of the piston itself. The weight can be reduced by using a hollow piston, by backing it with a vacuum, or by tilting the assembly. The tilting piston gage for very low pressures was developed by U. O. Hutton[28]. *Figure 6* demonstrates the principle. The effective weight of the

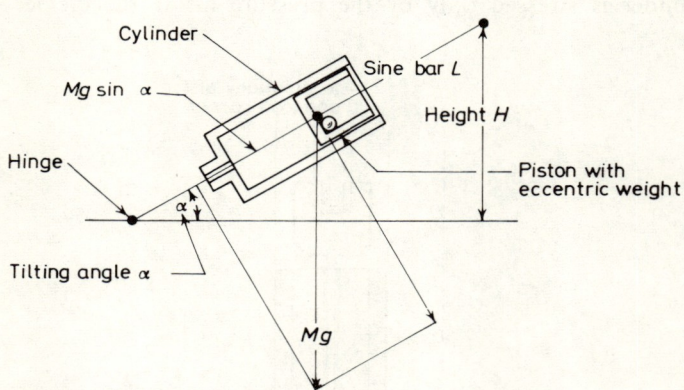

Figure 6. Tilting piston gage, schematic.

piston, Mg, is reduced to $Mg \sin \alpha$, by tilting the piston–cylinder assembly. α is the angle between the axis of the cylinder and the horizontal. To relieve friction the cylinder is rotated and the piston is held in position by a small

PART 3. PISTON GAGES

eccentric weight. The angle α on whose measurement the accuracy of the pressure measurement largely depends can be measured with a sine bar of known length L and a measurement with gage blocks or micrometer of the height H. It is less accurate but more convenient to use a carefully calibrated angle scale. Hutton[28] has provided a careful study of errors inherent in the tilting piston gage.

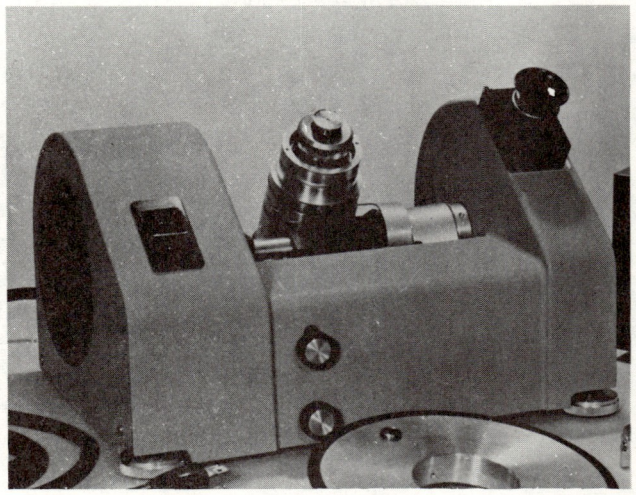

Figure 7. Tilting piston gage in tilted position, eyepiece for angle scale read-out on right, cylinder drive motor on left.

A photograph of a tilting piston gage is shown in *Figure 7*. This gage can be calibrated and operated as a normal piston gage with the piston–cylinder assembly locked into a vertical position and weights loaded on the piston in the usual manner. When operated in the tilting mode the tilt angle is measured with a calibrated angle scale. The angle scale can be read to 10 arcsec and calibrated to 30 arcsec. The pressure ranges in normal mode operation are $2 \, kN/m^2$ to $100 \, kN/m^2$ with a low range piston–cylinder assembly and $14 \, kN/m^2$ to $4 \, MN/m^2$ with a high range assembly. In the tilting mode 0 to $2 \, kN/m^2$ and 0 to $14 \, kN/m^2$ can be measured.

3. *Vacuum-backed Piston Gage*

Normally piston gages operate in the gage mode measuring pressures above atmospheric pressure. To measure absolute pressures, as required for the calibration of air data computers, the pressure on the free end of the piston must be reduced to a very small value. This is conveniently done with a bell jar placed over the gage as shown in *Figure 8*. The back pressure must be added to the pressure generated by the force per unit area to arrive at the total absolute pressure at the reference level of the instrument. The back pressure must therefore be known to better than the required accuracy for

Figure 8. Vacuum-backed piston gage, pumping port on left above pressure connection, weight stack under evacuated bell jar.

the pressure at the gage port. In many vacuum-backed piston gages this is difficult since no vacuum gage is provided inside the bell jar and the pumping ports are often too small in diameter. The back pressure should not be measured on the pumping line but inside the bell jar or at a separate port.

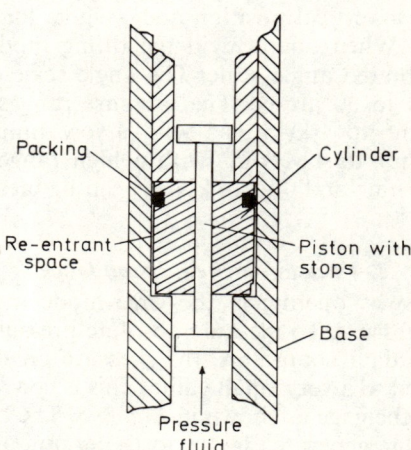

Figure 9. Re-entrant cylinder, pressure is supplied from below acting on the lower end face, most of the circumference and on the inside up to the line of contact.

PART 3. PISTON GAGES

4. Re-entrant Cylinder Design

Simple piston and cylinder assemblies are limited in their range of operating pressures by the excessive leakage of pressure fluid past the piston at high pressures. Inspection of equation 8 shows that, if pressure is applied to the outside and to the end of the cylinder, the inner diameter of the cylinder may actually decrease faster with pressure than the diameter of the piston thus reducing the clearance at high pressures to zero. This principle was first employed by Bridgman, who used it in this form in a generator for up to 1 320 MPa[9].

Figure 10. Re-entrant type piston gage with base, weight hanger and drive motor.

Figure 9 shows a cross section through a re-entrant cylinder where most of the outside of the cylinder is exposed to pressure. *Figure 10* is a photograph of a commercially available piston gage. The cylinder assembly is mounted on a triangular base with built-in spirit level and adjustment screws. Weights are placed on a weight hanger. A motor drive is used to rotate the piston.

5. Differential Piston

In the attempt to go to even higher pressures one encounters two limitations: the accuracy with which piston diameters are determined decreases inversely with the diameter, and the loads become more difficult to handle. The differential piston developed and used extensively by Michels[38, 39]

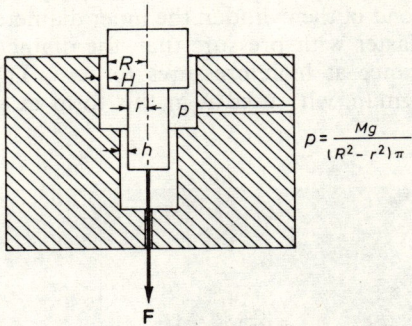

Figure 11. Differential piston, schematic.

allows one to use large diameter pistons and reasonably small loads. *Figure 11* shows a schematic of a differential piston gage. At low pressures the effective area of the device is the difference of the areas of the large and small piston–cylinder pairs

$$A = \pi[R^2 - r^2 + RH - rh] \tag{11}$$

For use at high pressures the piston–cylinder combinations are manufactured to have an interference fit at atmospheric pressure and can be used

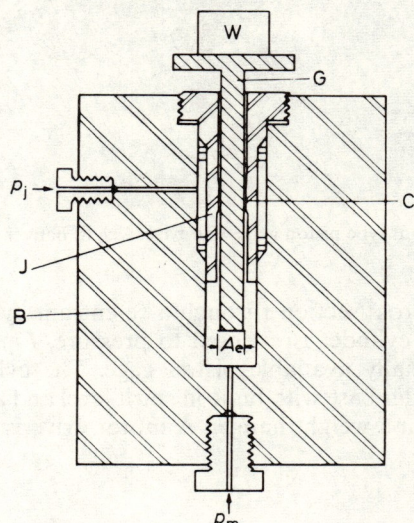

Figure 12. Controlled clearance piston gage, schematic. B gage body, G piston, J jacket, C line of contact, p_j jacket pressure, W load.

PART 3. PISTON GAGES

only above a certain pressure. Ebert[48] describes measurements with a differential piston which is freed at 140 MN/m² and can be used to 500 MN/m².

Presently the area A of a simple piston can be determined with greater accuracy than that of a differential piston of the same effective area and differential pistons have therefore lost much of their attractiveness.

Let dr be the uncertainty of the piston radius, R and r the radii of a differential piston and r_s the radius of a simple piston of the same area A, then we have for the difference of the uncertainties of the areas

$$dA_{\text{diff}} - dA_{\text{simple}} = 2\pi \, dr(R + r - R^2 - r^2) \tag{12}$$

which for positive $R \neq r$ is always greater than zero.

6. Controlled Clearance

Re-entrant cylinders were introduced by Bridgman in order to reduce the excessive leakage at high pressures. As discussed in section III.4 there is an upper limit of pressure at which the clearance in a piston and re-entrant cylinder assembly becomes zero. In the controlled clearance piston gage the elastic distortion of the cylinder is controlled by a separate pressure, the

Figure 13. Controlled clearance piston gage for 700 MPa with hydraulic weight loading.

jacket pressure, and the cylinder can thus be made to conform to the piston throughout the entire range of pressures for which the gage is equipped (equation 8, third term). *Figure 12* shows such a gage schematically.

Making the cylinder to conform to the piston brings about two important advantages. First, to determine the pressure coefficient we need to consider only the distortion of the piston and the change of clearance with jacket pressure. The former has been derived by Johnson[31] *et al.* as well as by other authors; the latter is determined empirically. The second advantage is that the clearance can be adjusted at will to obtain the best operating conditions for the particular fluid used over a wide range of operating pressures. In fact it appears that the more successful high pressure piston gages are all of the controlled clearance type[23, 32].

Figure 13 shows a commercially available controlled clearance piston gage. Weights are applied through a yoke. A hydraulic lift below the weight stack permits loading and unloading of the large weights. Small weights are moved by hand. A motor and pulley rotate the piston to relieve friction. Separate pressure connections are made to the jacket (upper left) and to the cylinder (upper center).

A rigorous analysis of the controlled clearance piston gage, that could be used to optimize the design, is not available. A reasonable assumption is that the length of the annulus to which the jacket pressure is applied should be at least twice as long as its diameter and that the line of contact (minimum clearance) should be about halfway between the jacket packings. Since the line of contact moves upwards as internal pressure rises it may be advantageous to relieve the cylinder very slightly to fix the line of contact at the halfway point.

The calibration of controlled clearance piston gages will be discussed in section IV.

7. *Ball Gages*

Balls made of a variety of materials, in many different sizes, are commercially available. These balls are manufactured to close diameter tolerances with very small deviations from true sphericity. A number of gages have been

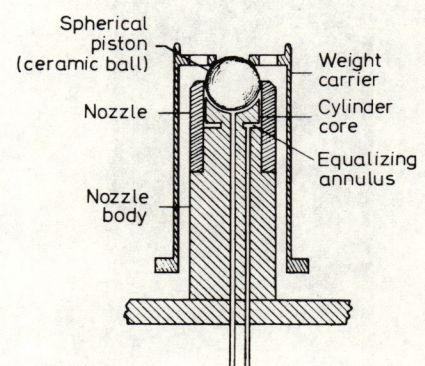

Figure 14. Ball type pressure regulator, schematic.

designed around such balls. While they may not truly be called piston gages, they measure pressure in terms of force per unit area and should at least be mentioned.

A gage which uses a ball instead of the piston was recently proposed by D. H. Newhall[42]. Its obvious advantage is that the line of contact is fixed near the equator of the ball and that it can therefore be held about halfway between the packings of the jacket. One would also expect that the viscous drag on the piston would be very small allowing operation at rather low pressures.

A schematic diagram of another ball-type piston gage is shown in *Figure 14*. Here air is supplied from a flow regulator to the equalizing annulus and from there, under the ball to the output port. In operation the ball, with weight hanger and weights, floats on a film of air with very little friction. Although the mathematical analysis may be difficult, this gage, once calibrated, is a very convenient source for stable, known pressures. Its reproducibility approaches a few parts per million and its sensitivity is even better.

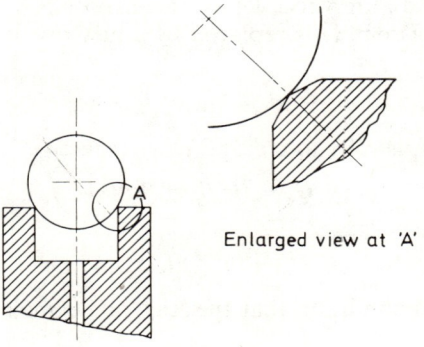

Enlarged view at 'A'

Figure 15. Ball gage after Wylie.

A ball gage suggested by P. Wylie[54] uses the principle shown in *Figure 15*. A ball covers the opening of a straight cylinder. The ball may be loaded by using a yoke. In operation ball and weights float on the compressed air in the cylinder. Excess pressure is released through the annulus between ball and cylinder. The edge of the cylinder is broken and ground to form an obtuse wedge to prevent permanent elastic deformation[11].

8. Grooved Pistons

For accurate measurements of pressure it is customary to use freely rotating or coasting pistons rather than to employ motor drives. At low pressures, when the inertia of the applied load is small, the coasting times are reduced. This very often sets the lower limit of pressure, especially with systems using oil as lubricant and pressure-transmitting medium. The deceleration of the piston and weights is caused by the viscous drag of the oil within the clearance between piston and cylinder. Obviously coasting times can be increased by using a less viscous oil, by increasing the clearance at the expense of sensitivity and accuracy, and by using special weights with high inertia. A substantial increase of the coasting times can be realized if the piston is modified according to a proposal by Dadson and Greig[15].

For any type of piston gage the leak rate Q is

$$Q = 4\pi r h^3 p / 3\eta l \qquad (13)$$

where r is the radius of the piston, $2h$ is the radial clearance, p is the pressure within the clearance, η is the viscosity of the fluid, and l is the length of the engagement of the working surfaces.

The friction force H is

$$H = \pi r l \eta v / h \qquad (14)$$

where v is the peripheral velocity. A reduction of the friction force at constant leak rate can be achieved in several obvious ways. However, variations of r are limited by considerations about the accuracy of the measurement of the area $A_o = \pi r^2$. The viscosity η is tied to the particular fluid in use. Consequently only h and l are variables at our disposal.

If the original assembly is replaced by a new one with the corresponding dimensions

$$h_o = x h_n \quad \text{and} \quad l_o = y l_n \qquad (15)$$

where the subscripts refer to old and new respectively, we have

$$Q_n / Q_o = (h_n^3 / l_n)(l_o / h_o^3)(y/x^3) \qquad (16)$$

and

$$H_n / H_o = (h_o / l_o)(l_n / h_n) = x/y$$

with our previous condition that the leak rate remain unchanged $Q_o = Q_n$

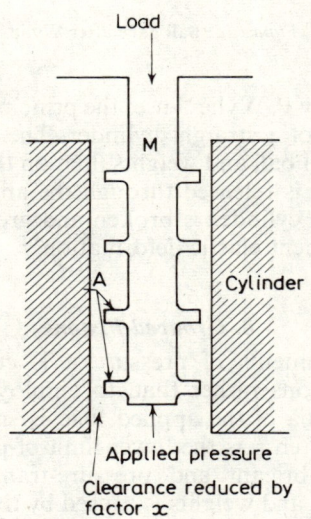

Figure 16. Reduction of viscous drag on piston.

and $y = x^3$. If the clearance h is reduced by a factor x and the length of engagement is reduced by x^3, then the rate of deceleration is reduced by x^2.

Figure 16 shows a schematic view of a practical way of reducing the length of engagement l without sacrificing the bearing length to diameter ratio. Several prototypes employing this principle have been used successfully[15]. Pistons of this type are also used in hydraulic load cells of the controlled clearance type, which show sensitivities of better than 1 part in 10^6 due in part to very low friction.

9. Very High Pressure Piston Gages

Several attempts have been made to operate piston gages at pressures above one GPa. We shall discuss here three typical devices.

Johnson and Heydemann[32] operated a controlled clearance piston gage at pressures up to 2.6 GPa. *Figure 17* shows the arrangement of this gage. The piston is loaded through a pivot with up to 8 200 N of dead weight (840 kg) balanced on the tip of the pivot and rotating with the piston to reduce friction. The clearance is adjusted by pressure, p_j, at the outside of the jacket. The end thrust ram loads the piston for longitudinal support and also generates radial support pressure by pressing the tapered cylinder into the tapered jacket. Pressure is generated by the integral hydraulic

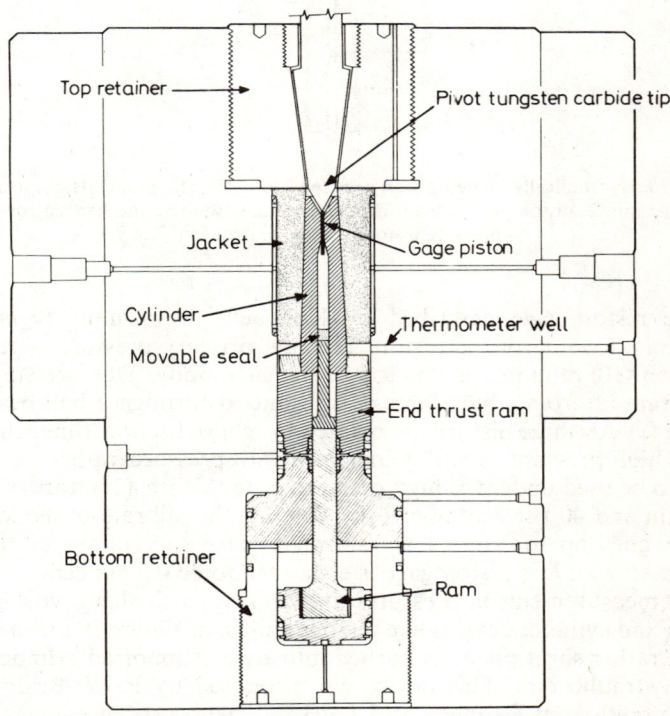

Figure 17. Controlled clearance piston gage for 2 500 MPa (Johnson and Heydemann).

intensifier in the lower part of the gage body. Electrical leads are brought out from the high pressure cavity through the intensifier rams. This gage was used to determine the Bi I–II transition pressure[23]. It is today the most accurate gage for the range to 2.6 GPa. The fluids used were i-pentane, mixtures of i-pentane and aviation instrument oil, and i-amyl alcohol.

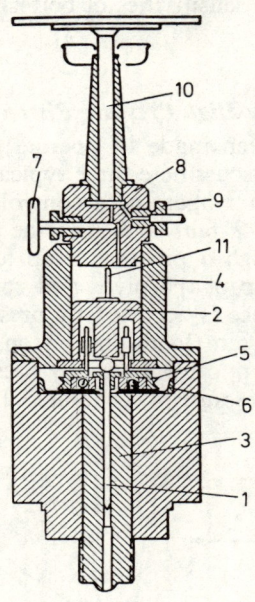

Figure 18. Hydraulically loaded piston gage for 2500 MPa (Konyaev). High pressure piston 1, hydraulic ram 2, piston gage generating ram pressure 10, drive mechanism for ram and high pressure piston rotation 5, manifold 7 to 9.

The piston gage described by Konyaev[35], shown in *Figure 18*, uses hydraulic loading of the piston. The low primary pressure is generated by a piston (10) mounted in the top of the gage body. This pressure acts on a large ram (2) from which force is transmitted through a ball bearing to the piston (1). All three pistons are rotated to relieve friction. Since the clearance of the high pressure assembly increases with pressure, rather viscous liquids have to be used under the high pressure piston. With a mixture of 60 per cent glycerin and 40 per cent glycol (by volume) the fall rate of the low pressure piston goes up to 5 mm/s at midrange. The uncertainty of the pressure generated with this piston gage is estimated to be 0.5 per cent.

For measurements at pressures to five GPa particularly with solid media, piston and cylinder devices like the one shown in *Figure 19* are in widespread use. A rather short piston is pushed into a well supported cylinder by means of a hydraulic ram. This device was proposed by P. W. Bridgman[10] and later modified by Kennedy and LaMori[33] who used it extensively for the measurement of properties of solids. Pressure is determined from the force exerted by the ram and the area of the high pressure piston. Pistons are rotated to relieve friction[8]. Measurements with fluids can be made by inserting containers into the cylinder[24, 29].

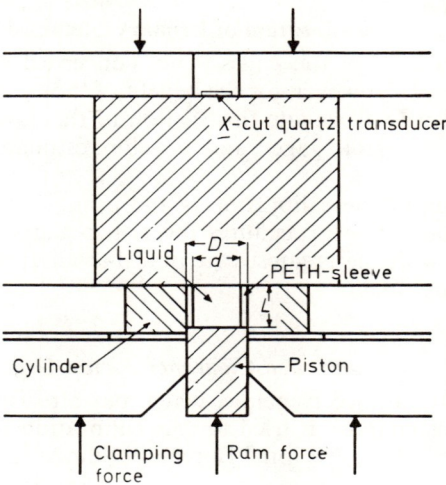

Figure 19. 5 000 MPa piston and die assembly (Heydemann and Houck) with polyethylene (PETH) fluid container.

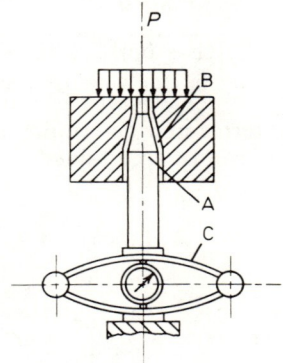

Figure 20. 10 GPa piston gage (Vereshchagin). High pressure piston A, clearance with special lubricant B, force gage C.

L. F. Vereshchagin and co-workers have proposed a piston–cylinder device for pressures as high as ten GPa[51]. *Figure 20* reproduces a sketch of Vereshchagin's apparatus. A tapered piston with two concentric cylindrical areas is fitted into the high pressure cylinder. The end face of the small cylindrical section is exposed to the high pressure. It is just long enough to prevent leakage from the high pressure chamber. Friction is reduced by rotation of the piston and by a special lubricant introduced into the gap B. Not enough experimental details are available to evaluate the performance of this gage.

IV. Calibration of Primary Standards

In general the user of piston gages is not concerned with the calibration of primary standards. However, for the benefit of those who need the highest possible accuracy, we will devote this section to the description of the calibration of primary piston gages and develop estimates of the accuracy attained.

In the discussion of the calibration of the primary piston gages we will assume that the operating procedures for piston gages outlined in section VI.2 are followed with the greatest care and that all applicable precautions mentioned there are taken.

1. Controlled Clearance Piston Gage

The controlled clearance principle, which has already been described in section III.6 of this chapter, is used for the calibration of primary pressure gages at the United States National Bureau of Standards and in many other standards laboratories.

The pressure generated by a controlled clearance piston gage at its reference level is given by

$$p = \frac{\sum_i M_i g(1 - \rho_{\text{air}}/\rho_{M_i}) + \gamma C + T_w}{A_0[1 + (\alpha_c + \alpha_p)(T - T_r)](1 + bp)[1 + (D + EW + FW^2) \times (p_{z0} + s_z W + q_z W^2 - p_j)]} \quad (17)$$

Let us inspect the various terms and determine how unknown coefficients and estimates of uncertainty can be ascertained.

A. Load

The term

$$W = \sum_i M_i g (1 - \rho_{\text{air}}/\rho_{M_i}) + \gamma C + T_w \quad (18)$$

describes the force applied to the piston to counterbalance the internal pressure p.

M_i are the individual weights having a density ρ_{M_i}. In the following text we use M as an abbreviation for $\sum_i M_i$. Methods used to calibrate weights for piston gages are described in detail by H. E. Almer[1]. True mass should always be used. The use of apparent mass values may lead to gross errors in gages where the weights are placed inside an evacuated belljar. As an indication of the accuracy routinely attained in the calibration of piston gage weights, *Table 2* lists mass and uncertainty values for a variety of sizes.

PART 3. PISTON GAGES

Table 2. Uncertainty of the mass of piston gage weights in parts per million (p.p.m.)

Mass [g]	Density [g/cm^3]	Uncertainty [p.p.m.]	Mass [g]	Density [g/cm^3]	Uncertainty [p.p.m.]
22 684.198	8.0	1	236.0034	7.8	20
4 536.8390	8.0	3	59.00085	7.8	10
590.0089	7.8	10	23.6007	7.8	20

FOR A GROUP OF FIVE WEIGHTS:

226 806.4	7.8	0.01			

The sea-level acceleration due to gravity g_ϕ can be computed from equation 19

$$g_\phi = 980.6160\,(1 - 0.0026373 \cos 2\phi + 0.0000059 \cos^2 2\phi) \qquad (19)$$

where ϕ is the latitude. For computation of the local acceleration due to gravity g_l a combination of free air and Bouguer reduction gives best results

$$g_l = g_\phi - 0.0003086\,h + 0.0001118\,(h - h^1) \qquad (20)$$

where h^1 is the elevation in meters of the general terrain for a radius of 70 km and h, in meters, is the elevation of the station above sea-level. This acceleration due to gravity is in the Meteorological Gravity System and all values are 0.013 cm s^{-2} lower than in the Potsdam System[48]. It may often be desirable to measure the local value of g in the laboratory where the primary standard is operated, since large buildings, subterranean rocks etc., tend to affect the value calculated from equation 20. In the USA such a local determination of g is available from the Coast and Geodetic Survey, US Department of Commerce. An absolute determination of the local gravity was made at the National Bureau of Standards site in Gaithersburg by D. R. Tate[49]. At the reference site NBS-2 in Room 129 of the Mechanical Engineering Building the absolute value of gravity is $g = (980.1018 \pm 0.0005)$ cm s^{-2}. The coordinates of NBS-2 are

    Lat. = 39° 07.85′ north
    Long. = 77° 13.20′ west
    Elev. = 131.042 m above mean sea-level

The NBS value for g is 0.0132 cm s^{-2} lower than the Potsdam value determined for this site.

The air buoyancy correction is quite large: 147.5 p.p.m. at 23°C, 30 per cent relative humidity, 1000 mb barometric pressure for a stainless steel weight of density 8 g/cm^3. For the determination of pressures it must therefore be carefully evaluated. The density of moist air has been tabulated[22, 40]. It can be computed from

$$\rho_{air} = 1.2929 \times 10^{-3}\,(273.13/T)\,p_{bar} - 0.3783\,H p_{sat} \qquad (21)$$

where p_{bar} and p_{sat} are the barometric and the saturation pressures respectively in N/m^2 and H is the relative humidity in per cent. The saturation pressure

p_{sat} according to A. Wexler and L. Greenspan[52] is given by

$$p_{sat} = \exp[E_0 T^{-1} + E_1 + E_2 T + B \ln T] \tag{22}$$

with

$$E_0 = -7.2465822 \times 10^3 \text{ K}$$
$$E_1 = 7.7641232 \times 10^1$$
$$E_2 = 5.7447142 \times 10^{-3} \text{ K}^{-1}$$
$$B = -8.2470402$$

H. A. Bowman and R. M. Schoonover[7] give a simpler expression for ρ_{air} which is of sufficient accuracy over the range 20 to 30°C. *Table 3* lists air density over a limited range of the parameters p_b, H and t. If ρ_{air} is computed from equation 21 we may neglect any uncertainty introduced by the air buoyancy correction.

Table 3. Air density in mg/cm³
Relative humidity 20 per cent

Barometric pressure mb	Temperature °C							
	22	22.5	23	23.5	24	24.5	25	25.5
1000.	1.178	1.176	1.174	1.172	1.170	1.168	1.166	1.164
1002.	1.181	1.178	1.176	1.174	1.172	1.170	1.168	1.166
1004.	1.183	1.181	1.179	1.177	1.175	1.173	1.170	1.168
1006.	1.185	1.183	1.181	1.179	1.177	1.175	1.173	1.171
1008.	1.188	1.186	1.183	1.181	1.179	1.177	1.175	1.173
1010.	1.190	1.188	1.186	1.184	1.182	1.180	1.177	1.175
1012.	1.192	1.190	1.188	1.185	1.184	1.182	1.180	1.178
1014.	1.195	1.193	1.190	1.188	1.186	1.184	1.182	1.180
1016.	1.197	1.195	1.193	1.191	1.189	1.187	1.185	1.182
1018.	1.199	1.197	1.195	1.193	1.191	1.189	1.187	1.185
1020.	1.202	1.200	1.198	1.195	1.193	1.191	1.189	1.187
1022.	1.204	1.202	1.200	1.198	1.196	1.194	1.192	1.189
1024.	1.206	1.204	1.202	1.200	1.198	1.196	1.194	1.192
1026.	1.209	1.207	1.205	1.202	1.200	1.198	1.196	1.194
Relative humidity 50 per cent								
1000.	1.175	1.172	1.170	1.168	1.166	1.164	1.162	1.164
1002.	1.177	1.175	1.173	1.170	1.168	1.166	1.164	1.166
1004.	1.179	1.177	1.175	1.173	1.171	1.168	1.166	1.168
1006.	1.182	1.179	1.177	1.175	1.173	1.171	1.169	1.171
1008.	1.184	1.182	1.180	1.177	1.175	1.173	1.171	1.173
1010.	1.186	1.184	1.182	1.180	1.178	1.175	1.173	1.175
1012.	1.189	1.187	1.184	1.182	1.180	1.178	1.176	1.178
1014.	1.191	1.189	1.187	1.184	1.182	1.180	1.178	1.180
1016.	1.194	1.191	1.189	1.187	1.185	1.182	1.180	1.182
1018.	1.196	1.194	1.191	1.189	1.187	1.185	1.183	1.185
1020.	1.198	1.196	1.194	1.192	1.189	1.187	1.185	1.187
1022.	1.201	1.198	1.196	1.194	1.192	1.189	1.187	1.189
1024.	1.203	1.201	1.198	1.196	1.194	1.192	1.190	1.192
1026.	1.205	1.203	1.201	1.199	1.196	1.194	1.192	1.194

PART 3. PISTON GAGES

Table 3—continued

Relative humidity 80 per cent

Barometric pressure mb	Temperature °C							
	22	22.5	23	23.5	24	24.5	25	25.5
1000.	1.171	1.169	1.166	1.164	1.162	1.160	1.157	1.164
1002.	1.173	1.171	1.169	1.166	1.164	1.162	1.160	1.166
1004.	1.176	1.173	1.171	1.169	1.167	1.164	1.162	1.168
1006.	1.178	1.176	1.174	1.171	1.169	1.167	1.164	1.171
1008.	1.181	1.178	1.176	1.174	1.171	1.169	1.167	1.173
1010.	1.183	1.181	1.178	1.176	1.174	1.171	1.169	1.175
1012.	1.185	1.183	1.181	1.178	1.176	1.174	1.171	1.178
1014.	1.188	1.185	1.183	1.181	1.178	1.176	1.174	1.180
1016.	1.190	1.188	1.185	1.183	1.181	1.178	1.176	1.182
1018.	1.192	1.190	1.188	1.185	1.183	1.181	1.178	1.185
1020.	1.195	1.192	1.190	1.188	1.185	1.183	1.181	1.187
1022.	1.197	1.195	1.192	1.190	1.188	1.185	1.183	1.189
1024.	1.199	1.197	1.195	1.192	1.190	1.188	1.185	1.192
1026.	1.202	1.199	1.197	1.195	1.192	1.190	1.188	1.194

Table 4. Surface tension γ of fluids at room temperature

Material	Spinesso 38 (mineral oil)	Glycerin	Petroleum	Water	Olive oil
Surface tension. γ, N/m	0.031	0.063	0.026	0.075	0.033

The term γC describes the force generated by the surface tension of the fluid acting on the piston where it emerges from the fluid. γ is the surface tension of the fluid and C is the circumference of the piston. *Table 4* lists values of surface tension for a variety of fluids. The value for Spinesso 38 is representative for many fine instrument oils used in piston gages. Over the temperature range from 20 to 30°C it is assumed to be accurate to ten per cent. For a piston of 32 mm² cross-sectional area this correction amounts to about 10^{-4} N. This constitutes a small correction for high pressure gages but for oil-operated gages at low pressures it may have to be taken into account. This correction does not apply to air-operated gages.

T_w is a tare force which may result from an error in the determination of the mass of one of the weights or in the computation of the head corrections; it might also be a characteristic of a gage.

B. Area A_0

A_0 is the cross-sectional area of the piston as determined from numerous measurements of the diameter along the length of the piston. No piston is perfectly round nor is any one perfectly straight. For a primary standard every effort is made to produce a straight and round cylinder. Polar plots of deviation of actual piston diameter from a nominal diameter are a facile way to check the roundness of the piston. An example of such a plot is reproduced in *Figure 21*. It was made for a piston of nominally 6.4 mm diameter approximately 20 mm from the upper end. The deviations from a perfectly round cylinder do not exceed 0.015 μm.

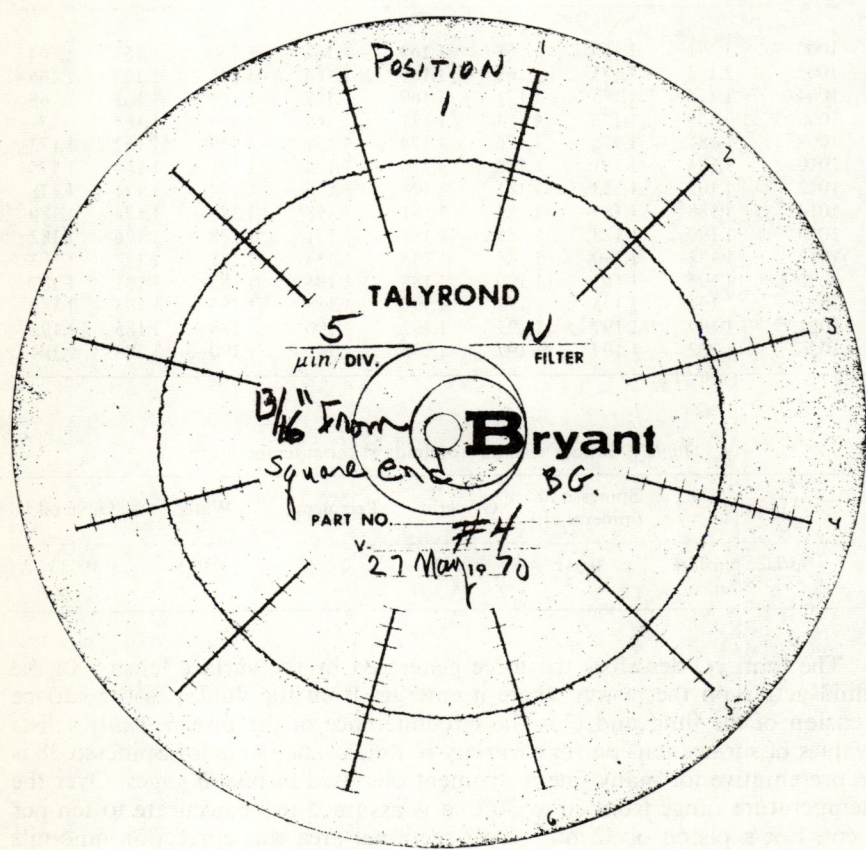

Figure 21. Talyrond of gage piston, 0.125 μm/div.

Cylindricity or straightness of the piston is ascertained from measurements of the diameter along the entire length of the piston. An upper limit can be set for the deviation from straightness of the cylinder: over the operating stroke—usually a few millimeters—the diameter must not vary by more than its own uncertainty. In view of the very high precision of which controlled clearance piston gages are capable, a better limit would be that the relative change in area over the operating stroke should not exceed the sensitivity. Let d be the radius of the piston and ε the deviation from d, then $\Delta A/A = 2\varepsilon/d$ and for a gage with a sensitivity of 10^{-6} and a diameter of 6.4 mm $\varepsilon = 6.4 \times 10^{-9}$ m. This change in radius occurs over the operating stroke of say 3 mm and a piston of 60 mm total length should not have more than 0.128 µm variation of its diameter over its entire length. Low pressure piston gages have diameters of say 25 mm and sensitivities of 10^{-7}. In this case the permissible variation of the diameter over the length of the piston (60 mm) would be 0.025 µm. This is difficult indeed to achieve.

The Dimensional Metrology Section of the National Bureau of Standards presently claims an uncertainty (one standard deviation) of 2.5×10^{-8} m in the determination of the diameter of pistons for primary gages irrespective of their diameter. This favors large diameters. The uncertainty on the area of the 25 mm diameter piston is 2 p.p.m., that for a 6 mm diameter piston is 8 p.p.m.

An alternative method to determine the area of a gage piston is by crossfloating against a mercury manometer of very high accuracy. We do not have any reports about recent cross-floats aimed at determining the area of gage pistons, but in the past one hundred years several attempts have been made not only to determine the area of a piston gage but also its change with pressure[5,37]. There are now one or two manometers available with a documented uncertainty of one or two parts per million[21] at about 100 kPa which undoubtedly will be used in the near future for the determination of the area of primary piston gages.

C. Temperature Correction

The temperature correction for the effective area is given by

$$C_T = 1 + (\alpha_c + \alpha_p)(T - T_r) \tag{23}$$

The reference temperature here is the temperature at which the piston area assumes the value A_0. For a tungsten carbide piston inside a steel cylinder the correction amounts to 17 p.p.m. with $T - T_r = 1°C$. The gage temperature can easily be measured to 0.05°C close to the cylinder. We have no evidence that the piston is not at this temperature except with very small or very large clearances where a rise in the temperature of the gage of the order of 0.1°C can be observed.

The thermal expansion coefficients can be determined to better than five per cent. The uncertainty for the temperature correction of a tungsten carbide piston inside a steel cylinder with $T - T_r = 1°C$ is

$$dC_T = (\partial C_T/\partial T\ dT) + (\partial C_T/\partial \alpha)\ d\alpha = 0.8\ \text{p.p.m.} \tag{24}$$

D. Pressure Coefficient

The pressure coefficient b is computed from the elastic constants of the piston material[31]

$$b = (3\mu - 1)/E$$

where μ is Poisson's constant and E is the modulus of elasticity. Both values can be obtained from ultrasonic measurements on the piston material with

Table 5. Elastic constants, density and thermal expansivities of piston and cylinder materials

Material	Shear modulus G (N/m^2)	Young's modulus E (N/m^2)	Poisson's ratio μ	Thermal expansivity α (K^{-1})	Density ρ_m (g/cm^3)	Pressure coefficient (piston) b (m^2/N)
K9-steel	7.92×10^{11}	2.05×10^{11}	0.295			
'Hydurax' aluminum-bronze	5.38×10^{11}	1.43×10^{11}	0.333			
GEC heavy metal, tungsten alloy	14.2×10^{11}	3.67×10^{11}	0.286	5.8×10^{-6}	17	
Tungsten carbide K68*		6.11×10^{11}	0.209	4.54×10^{-6}	14.9	-6.10×10^{-13}
Tungsten carbide C93*		6.31×10^{11}	0.210	4.5×10^{-6}	15.1	-5.86×10^{-13}
Tungsten carbide K96		6.32×10^{11}	0.211	4.5×10^{-6}	14.9	-5.81×10^{-13}
Stainless steel AM-350 (precip. hardening)		2.03×10^{11}	0.300	11.4×10^{-6}	7.81	-4.93×10^{-13}
Brass	4.2×10^{10}	1.1×10^{11}		18×10^{-6}	8.6	

* Data from ultrasonic measurements.

an uncertainty of less than one per cent. Both E and μ are functions of pressure and temperature and should be measured through the full operating range of p and T.

Pistons are frequently made from cemented tungsten carbide. Data for the elastic constants of this and other materials are listed in *Table 5*. With the data for tungsten carbide C93 the piston distortion, bp, amounts to 2.3 p.p.m. at 4 MN/m^2 and to 117 p.p.m. at 200 MN/m^2.

The uncertainty of the pressure coefficient $C_b = (1 + bp)$ is typically

$$dC_b = \frac{\partial C_b}{\partial \mu} d\mu + \frac{\partial C_b}{\partial E} dE + \frac{\partial C_b}{\partial p} dp \tag{25}$$

$dC_b = 3$ p.p.m. at $p = 200 \text{ MN/m}^2$.

No allowance has been made for the accuracy of the theory of elastic distortion. We may well want to multiply the value for dC_b with a factor of two or three to account for uncertainties due to an inadequate theory.

E. Change of Area with Jacket Pressure

For each load F or internal pressure p there is a jacket pressure p_z which reduces the clearance to zero. As the jacket pressure p_j is reduced below p_z the clearance, and thereby the effective area, increases. This functional dependence of the effective area on jacket pressure which we discussed in detail in section II of this chapter is described by

$$A_{\text{eff}} = A_0[1 + (\alpha_p + \alpha_c)(T - T_r)] [1 + bp] [1 + d(p_z - p_j)] \tag{26}$$

with $d = D + EW + FW^2$, where D, E, F and p_z are constants, W is the force acting on the piston and p_j is the jacket pressure. The change of the area with jacket pressure follows from a differentiation with respect to p_j

$$\partial A_{\text{eff}}/\partial p_j = - dA_0[1 + (\alpha_p + \alpha_c)(T - T_r)][1 + bp] \tag{27}$$

In principle d could be determined by measuring the change of the generated pressure with jacket pressure. In practice it is more convenient to keep the generated pressure constant and to adjust the load on the piston accordingly

$$\frac{\partial A_{\text{eff}}}{\partial p_j} \cdot \frac{\partial p_j}{\partial W} = \frac{1}{p} = \text{const.} \tag{28}$$

The pressure p can be monitored with any pressure transducer of sufficient sensitivity and short-term stability, in particular this can be done with a second piston gage cross-floated against the controlled clearance gage. The set-up used to cross-float a controlled clearance piston gage under calibration against a second very stable gage is the same as described in section V.1 for the cross-float calibration of piston gages using a differential pressure indicator of adequate sensitivity. In this case we have

$$-dp_j = \frac{W_S A_R (1 + bp)_R [1 + (\alpha_c + \alpha_p)(T - T_r)]_R}{W_R A_S (1 + bp)_S [1 + (\alpha_c + \alpha_p)(T - T_r)]_S} - dp_z - 1 = \frac{P_S}{P_R} - dp_z - 1 \tag{29}$$

where W are the applied forces and the subscripts R and S refer to the reference gages and the standard gage respectively. For a given load W_R the load W_S is measured as a function of the jacket pressure p_j. The resulting values for W_S, with the corrections for temperature and pressure, should ideally be a linear function of jacket pressure. The slope of the line is the jacket pressure coefficient d. *Figure 22* presents a plot of two sets of force

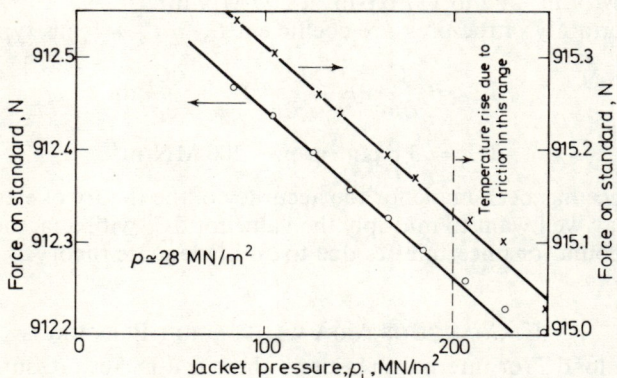

Figure 22. Controlled clearance piston gage: balancing force as a function of jacket pressure for constant output pressure.

versus jacket pressure data. A deviation from the straight line is clearly noticeable for very high jacket pressures or small clearances where also a sharp temperature rise was observed during the measurements. Experience shows that d is generally dependent on the applied load and is therefore expressed as a polynomial in W

$$d = D + EW + FW^2 \tag{30}$$

We rewrite equation 30 with the abbreviations given

$$(P_S/P_R) - 1 = dp_z - dp_j \tag{31}$$

and differentiate with respect to p_j

$$\partial\{(P_S/P_R) - 1\}/\partial p_j = -d = -(D + EW + FW^2) \tag{32}$$

For the evaluation of the data the differentiation is replaced by finite differences and a least squares fit using the OMNITAB[26] FIT instruction is made to determine the constants D, E and F.

A set of data which includes those shown in *Figure 22* was evaluated solving for three different combinations of coefficients. The results are shown in *Table* 6. The fit using a constant and a quadratic term yields the lowest standard deviation of the residuals using only two coefficients. The uncertainty of the correction term $C_d = 1 + (D + FW^2)(p_z - p_j)$ due to uncertainties in D and F may be estimated from the total differential. With the data in *Table* 6, we have $dC_d/C_d = 3.9 \times 10^{-6} + 7.5 \times 10^{-6} = 11.4$ p.p.m.

An alternative method to determine d which obviates the use of a second

Table 6. Jacket pressure coefficient d. Coefficients obtained from least squares fits to cross-float data

(a) $d = D_0$,
$D_0 = 1.696 \times 10^{-11} \text{ m}^2/\text{N}$, $\sigma D_0 = 5.70 \times 10^{-13} \text{ m}^2/\text{N}$,
SDR = $9.29 \times 10^{-12} \text{ m}^2/\text{N}$.

(b) $d = D_1 + E_1 W$,
$D_1 = 2.206 \times 10^{-11} \text{ m}^2/\text{N}$, $\sigma D = 4.51 \times 10^{-13} \text{ m}^2/\text{N}$,
$E_1 = -1.984 \times 10^{-15} \text{ m}^2/\text{N}^2$, $\sigma E = 1.59 \times 10^{-16} \text{ m}^2/\text{N}^2$,
SDR = $3.07 \times 10^{-12} \text{ m}^2/\text{N}$.

(c) $d = D_2 + F_2 W^2$,
$D_2 = 1.996 \times 10^{-11} \text{ m}^2/\text{N}$, $\sigma D = 3.00 \times 10^{-13} \text{ m}^2/\text{N}$,
$F_2 = 3.738 \times 10^{-19} \text{ m}^2/\text{N}^3$, $\sigma F = 2.94 \times 10^{-20} \text{ m}^2/\text{N}^3$,
SDR = $3.02 \times 10^{-12} \text{ m}^2/\text{N}$.

(d) $d = D_3 + E_3 W + F_3 W^2$,
$D_3 = 2.095 \times 10^{-11} \text{ m}^2/\text{N}$, $\sigma D = 9.02 \times 10^{-13} \text{ m}^2/\text{N}$,
$E_3 = -9.051 \times 10^{-16} \text{ m}^2/\text{N}^2$, $\sigma E = 7.79 \times 10^{-16} \text{ m}^2/\text{N}^2$,
$F_3 = -2.070 \times 10^{-19} \text{ m}^2/\text{N}^3$, $\sigma F = 1.46 \times 10^{-19} \text{ m}^2/\text{N}^3$,
SDR = $2.99 \times 10^{-12} \text{ m}^2/\text{N}$.

piston gage is used by the Harwood Engineering Company, the only manufacturer of controlled clearance piston gages. Here the coefficient d is computed from elastic theory[43]

$$d = 2w^2/E(w^2 - 1) \qquad (33)$$

with w the ratio of outer to inner radius of the cylinder and E the modulus of elasticity of the cylinder material.

For the gage, which we have used here repeatedly for illustration, d computed from equation 33 is

$$d = 1.19 \times 10^{-11} \text{m}^2/\text{N}$$

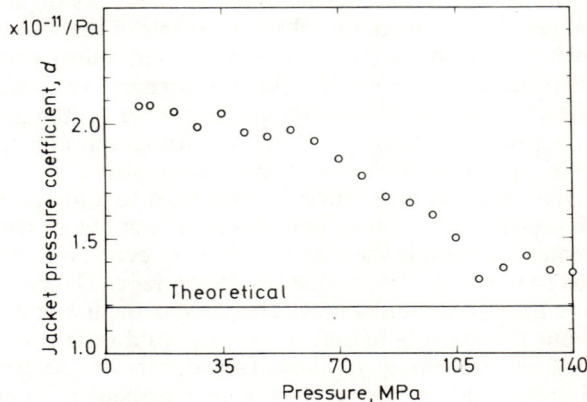

Figure 23. Controlled clearance piston gage: coefficient d-determined from across float and from elastic theory.

with $w = 2.5$ and $E = 2 \times 10^{11}$ N/m². In *Figure 23* we have plotted data obtained from individual cross-floats (equation 29) and the d computed from equation 33. For a gage of this particular design the agreement is quite unsatisfactory.

F. Zero Clearance Jacket Pressure p_z

To determine the jacket pressure for which the clearance is reduced to zero we make the assumption that the leak rate past the piston is proportional to the third power of the clearance or that

$$(\partial V/\partial t)^{\frac{1}{3}} = \text{const} \times C \qquad (34)$$

where V is the volume of oil inside the piston gage. The gage is valved off from the rest of the pressure system and is assumed to be leak-tight with the exception of the leakage past the piston through the clearance C. Equation 34 is, of course, the basic law for laminar flow between infinitely extended plates derived by Hagen in 1839 and by Poiseuille in 1840 for plane and parallel plates. Instead of observing the leak rate $\partial V/\partial t$ it is more convenient to measure the fall rate of the piston $\partial s/\partial t$, where s is the position of the piston. Now we can rewrite equation 34 as

$$(\partial s/\partial t)^{\frac{1}{3}} = \text{const} \times C = \text{const}\,(p_z - p_j) \qquad (35)$$

and determine p_z from a measurement of $\partial s/\partial t$ as a function of p_j. p_z depends on the internal pressure and this is reflected in equation 17 where we have expressed p_z as a function of the weight W

$$p_z = p_{z0} + s_z W + q_z W^2 \qquad (36)$$

The piston position of an operating piston gage can be conveniently measured with a linear voltage displacement transducer (LVDT) of the differential transformer type. In these transducers a magnetic core connected to the feeler rod moves freely up and down inside the transducer. The flat lower end of the feeler rod rests on a spherical button on the weight hanger with a radius equal to the distance between the top of the piston and the top of the weight hanger. This reduces the effect of wobble of the weights on the transducer. The feeler rod forms part of the load. Every effort was made to reduce friction between the feeler rod and the transformer to a minimum. It is important to know that the feeler exerts no other force on the piston than that due to its own weight. *Figure 24* shows a photograph of a transducer mounted on top of a piston gage with the feeler rod in place.

The output signal from the transducer is connected to a digital voltmeter and readings are taken in well known intervals for about five or ten minutes. We find it convenient to sample the digital voltmeter every 5 s with the help of a synchronous switch and to print the results on tape. The position data are then fitted as a linear function of time. The slope of the line is the fall rate.

Let us divert our attention here and discuss the calibration of the transducer. Because of the sensitivity of these transducers to the presence of large masses of metal and to temperature it is expedient to calibrate the transducer *in situ*. For this purpose gage blocks are inserted between a flat on the weight hanger and the spherical button and the output voltage

PART 3. PISTON GAGES

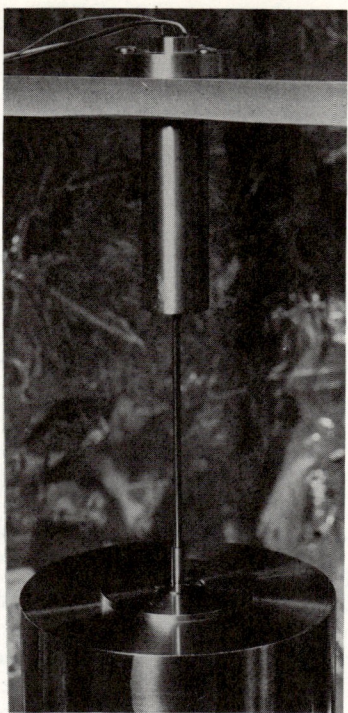

Figure 24. Position transducer mounted on controlled clearance piston gage.

u is noted. This should be done in sufficiently small steps over a range of about 1 mm preferably centered about the zero output voltage position. As we are interested in the change in position with time only any systematic error causing an offset in all points may be neglected. The accuracy of gage blocks wrung together far exceeds our requirements and does not contribute to the uncertainty. Temperature changes in the structure supporting the LVDT may affect the output reading by 0.1 µm. Wringing of gage blocks causes their temperature to rise. While gage blocks are changed the feeler rod is lifted up into the LVDT where it is heated up slightly, which leads to a noticeable time-dependent offset. Positioning of the feeler rod on the button is not completely reproducible. These and other random errors affect the calibration. Two third-order polynomials are fitted to the position versus voltage data.

$$u = a + bs + cs^2 + ds^3 \quad (37)$$

and

$$s = a^* + b^*u + c^*u^2 + d^*u^3 \quad (38)$$

where s and u are position and output voltage respectively. To illustrate typical performance we will present data obtained with an LVDT with integral oscillator and detector for which we have data from eight calibrations made over a span of more than one year.

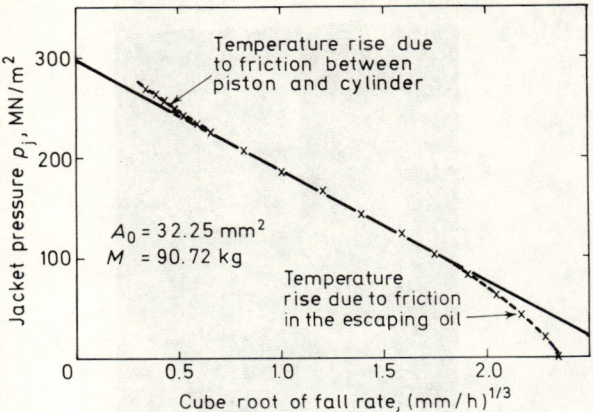

Figure 25. Controlled clearance piston gage: jacket pressure versus cube root of fall rate.

For the first fit (equation 37) the residual standard deviation was 4 mV. The voltmeter used had a resolution of 1 mV and was believed to be accurate to ± 2 mV and stable to ± 1 mV. The second fit (equation 38) rendered a standard deviation of the fit of 0.15 μm. The scatter of the data increases with the magnitude of the output voltage and may primarily be due to fluctuations in the performance of the electronics of the LVDT. The coefficient $b^* = 0.3645$ mm/volt with a standard deviation of $S_{b^*} = 0.0004$ mm/volt.

Fall rate measurements are made as a function of jacket pressure with load as a parameter. The clearance should at all times be large enough to maintain an oil or air film between piston and cylinder. For example, in an oil-operated piston gage with a piston of about 6 mm diameter we limit the jacket pressure to fall rates greater than about 0.025 mm/h. *Figure 25* shows data from a fall rate measurement taken with one particular load W on the piston. The cube root of the fall rate is in fact a linear function of the jacket pressure over part of its range. The data deviate from this linear relationship at very high and very low pressures. When either very high or

Table 7. Fall rate data

Run No.	$\partial s/\partial t$ mm/h	Std dev. mm/h	$(\partial s/\partial t)^{\frac{1}{3}}$ $(mm/h)^{\frac{1}{3}}$	Std dev. $(mm/h)^{\frac{1}{3}}$	Stroke mm	Time s
56	21.057	0.017	2.7614	0.0007	1.112	190
59	4.044	0.002	1.5933	0.00009	0.550	490
61	0.7403	0.0009	0.90464	0.00004	0.0750	365
62	0.7376	0.0013	0.90353	0.00005	0.0809	395
63	0.7200	0.0010	0.89626	0.00004	0.07110	355
64	0.7440	0.0009	0.90612	0.00004	0.07461	360
66	0.1192	0.0008	0.49210	0.00003	0.01128	340

PART 3. PISTON GAGES

very low jacket pressures are used the temperature of the piston and cylinder increases by several hundredths of one kelvin. At high values of p_j this is due to friction between piston and cylinder as is evident from the very short coasting times. At low jacket pressures it is due to friction in the oil passing through the clearance at high speed. Either range should be avoided. Friction between piston and cylinder may cause rapid deterioration of both.

Some typical fall rate data are shown in *Table 7*. The points 61 to 64 are repetitions of the same point under nominally the same conditions. They give an indication of the amount of random scatter. Note that the deviation from the mean of the points 61 to 64 constitutes only a small fraction of the total range of cube root of fall rate values used to extrapolate $p_z(W)$. We pointed out above that the change in position of the piston is known to about 0.15 μm. The time interval can be known to better than 1 s. The random error in $(\partial s/\partial t)^{\frac{1}{3}}$ is much larger than the uncertainty of either s or t. Random errors in setting and maintaining p_j are the largest single source of uncertainty in the fall rate measurement. Temperature fluctuations have a considerably smaller effect.

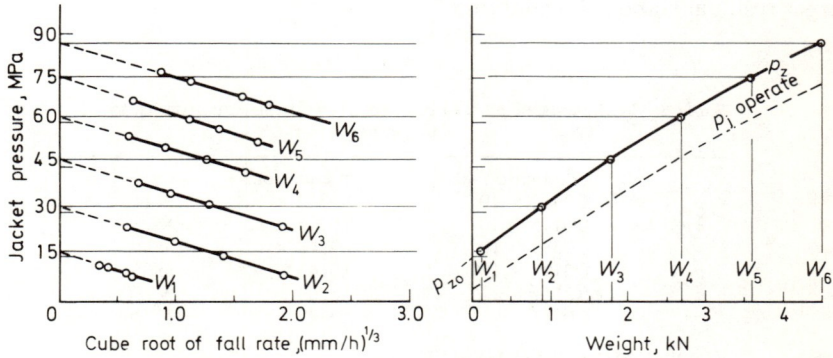

Figure 26. Controlled clearance piston gage: zero clearance jacket pressure versus load.

In *Figure 26* the cube root of the fall rate is plotted versus jacket pressure for several loads W. The individual fall rate lines are extrapolated to zero fall rate where they intersect the ordinate at $p_z(W)$. The RHS of *Figure 26* shows a plot of the p_z versus load. The extrapolation of this curve to zero load intersects the ordinate at p_{z0}. A dashed line in the graph indicates the operating jacket pressures. The operating jacket pressures are chosen to give fall rates of about 1 mm/h, consistent with small clearance and reasonably low friction.

In actual practice the graphical analysis is replaced by data analysis on a digital computer. To obtain p_z for a given pressure p or load M the function

$$p_j = p_z - (1/C)(\partial s/\partial t)^{\frac{1}{3}}, \quad C \text{ is constant} \tag{39}$$

is fitted to data from within the linear portions of the plot of $(\partial s/\partial t)^{\frac{1}{3}}$ versus p_j. As mentioned before, p_z is a function of W. It is convenient to fit a polynomial in W of the second degree (see equation 36) to the p_z data. A more accurate and instructive result is obtained when the functions

$$(\partial s/\partial t)^{\frac{1}{3}} = ap_j + b + cW + dW^2 \tag{40}$$

and

$$p_j = p_{z0} + s_z W + q_z W^2 + r(\partial s/\partial t)^{\frac{1}{3}} \tag{41}$$

are fitted to all of the available $p_z(p_j, W)$ data. The coefficients of equations 40 and 41 are related as

$$p_{z0} = -b/a, \quad s_z = -c/a, \quad q_z = -d/a \tag{42}$$

The random uncertainty of the data set to which a function is fitted will appear as a residual standard deviation of the dependent variable. This may be compared with the estimated uncertainty of the variable and will help to identify other possible sources of uncertainty. In both fits each point can be given equal weight or it may be weighted inversely proportional to the standard deviation of $(\partial s/\partial t)^{\frac{1}{3}}$. The latter method leads to much smaller standard deviations of the coefficients but necessarily is accompanied by larger residual standard deviations.

Table 8. Coefficients of $p_z(M)$ obtained from fitting equation 41 to 57 data points

	from equation 40	from equation 41	
p_{z0}	1.3412×10^7	1.3379×10^7	N/m^2
σp_{z0}	0.0175×10^7	0.0116×10^7	N/m^2
s_z	1.9583×10^4	1.9548×10^4	m^{-2}
σs_z	0.0194×10^4	0.0145×10^4	m^{-2}
q_z	-0.8131	-0.8069	N^{-1}m^{-2}
σq_z	0.0311	0.0329	N^{-1}m^{-2}

RESIDUAL STANDARD DEVIATIONS:
$\sigma(\partial s/\partial t)^{\frac{1}{3}}$ (equation 40) = 0.014 (mm/h)$^{\frac{1}{3}}$
σp_j (equation 41) = 4.9×10^5 N/m^2

To determine the estimated uncertainty of p_z, as computed from the fits to the data, we compare the standard deviations of the residuals of $(\partial s/\partial t)^{\frac{1}{3}}$ (equation 41) with our estimate of their uncertainty. *Table 7* indicates that the precision of the $(\partial s/\partial t)^{\frac{1}{3}}$ data used here for illustration is about 0.01 (mm/h)$^{\frac{1}{3}}$. From fitting equation 40 to the data follows 0.014 (mm/h)$^{\frac{1}{3}}$ in reasonable agreement with our estimate. For p_j we have estimated an uncertainty of 100 kPa and a precision of 30 kPa. The fit of equation 41 to the data renders a standard deviation of the residuals of p_j of 490 kPa. This large uncertainty which is due to the fall rate data should be reflected in the p_z data and we assume an uncertainty in p_z of about 500 kPa. In *Table 8* we list coefficients p_{z0}, s_z and q_z with their standard deviations as obtained by fitting equations 40 and 41 to 57 data points.

To obtain another estimate of the uncertainty of the jacket pressure term

PART 3. PISTON GAGES

$C_z = p_{z0} + s_z W + q_z W^2 - p_j$, we form the total differential

$$dC_z = \frac{\partial C_z}{\partial p_{z0}} dp_{z0} + \frac{\partial C_z}{\partial s_z} ds_z + \frac{\partial C_z}{\partial q_z} dq_z + \frac{\partial C_z}{\partial p_j} dp_j \tag{43}$$

where

$$C_z = p_{z0} + s_z W + q_z W^2 - p_j$$

For W we shall substitute the load applied for the highest pressure generated by this gage (140 MPa); for dp_{z0}, ds_z and dq_z we shall use the standard deviations from the fit (Table 8); and for dp_j we assume an uncertainty of 100 kPa. With these data the sum of the amounts of the individual contributions is $dC_z = 1.8 \times 10^6$ Pa. Under the same conditions $C_z = 13 \times 10^6$ Pa. This correction term is then known to about fourteen per cent.

G. Total Uncertainty of Pressure

In the course of the discussion of the calibration of a controlled clearance piston gage we have given estimates of uncertainty for a number of coefficients. We shall now combine all of these to obtain an estimate of the uncertainty of the highest pressure generated by the gage. For any other type of piston gage one would proceed in the same manner adding or dropping terms as appropriate.

The fractional uncertainty in p can be expressed either as the sum of all terms of the total differential

$$\frac{dp}{p} = \sum_1^i \left| \frac{1}{p} \frac{\partial p}{\partial x_i} \right| dx_i \tag{44}$$

Table 9. Terms of the total differential dp/p

x_i		$\frac{1}{p}\frac{\partial p}{\partial x_i}$	dx_i	$\frac{1}{p}\frac{\partial p}{\partial x_i} dx_i$
M	M^{-1}	$2.2 \times 10^{-6}\, g^{-1}$	$0.5\, g$	1.1×10^{-6}
g	g^{-1}	$1.02 \times 10^{-3}\, s^2/cm$	$5 \times 10^{-4}\, cm/s^2$	0.5×10^{-6}
ρ_M	$\rho_{air}\rho_M^{-2}$	$2 \times 10^{-5}\, cm^3/g$	$0.1\, g/cm^3$	2.0×10^{-6}
γ	C/gM	$4.5 \times 10^{-6}\, m^{-1}$	$3.1 \times 10^{-2}\, N/m$	0.1×10^{-6}
C	γ/gM	$7 \times 10^{-6}\, m^{-1}$	$5 \times 10^{-7}\, m$	0.0×10^{-6}
A_0	A_0^{-1}	$3.1 \times 10^4\, m^{-2}$	$2.5 \times 10^{-10}\, m^2$	7.8×10^{-6}
α_c	ΔT	$2°C$	$6 \times 10^{-7}\, °C^{-1}$	1.2×10^{-6}
α_p	ΔT	$2°C$	$2.2 \times 10^{-7}\, °C^{-1}$	0.4×10^{-6}
ΔT	$(\alpha_c + \alpha_p)$	$1.63 \times 10^{-4}\, °C^{-1}$	$0.05\, °C$	0.8×10^{-6}
b	p	$138\, MN/m^2$	$1.5 \times 10^{-14}\, m^2/N$	2.1×10^{-6}
p^*	b	$5.9 \times 10^{-13}\, m^2/N$	$1.4 \times 10^{-4}\, N/m^2$	0.0×10^{-6}
D		$1.3 \times 10^7\, N/m^2$	$3 \times 10^{-13}\, m^2/N$	3.9×10^{-6}
F	$(p_z - p_j)W$	$2.6 \times 10^{14}\, N^2/m^2$	$2.9 \times 10^{-20}\, m^2/N^3$	7.5×10^{-6}
p_z	d	$1.18 \times 10^{-11}\, m^2/N$	$5 \times 10^5\, N/m^2$	5.9×10^{-6}
p_j	d	$1.18 \times 10^{-11}\, m^2/N$	$10^5\, N/m^2$	1.2×10^{-6}

* In the term $(1 + bp)$.
Terms involving T and ρ_{air} have been neglected.

or as the root mean square of all terms of the total differential

$$\text{RMS}\left(\frac{dp}{p}\right) = \left(\sum_{1}^{i}\left[\frac{1}{p}\frac{\partial p}{\partial x_i}dx_i\right]^2\right)^{\frac{1}{2}} \qquad (45)$$

where x_i are the parameters and coefficients of equation 17. In *Table 9* we have listed the $(1/p)(\partial p/\partial x_i)\,dx_i$ using appropriate approximations. With the data of *Table 9* the sum of the terms of the total differential is

$$dp/p = 35 \text{ p.p.m.}$$

and the root mean square of all terms is

$$\text{RMS}\,(dp/p) = 13.5 \text{ p.p.m.}$$

The data used in these examples are taken from calibration records of a controlled clearance piston gage of vintage design. The data are not typical of the accuracy that can be achieved with more recent designs but they show clearly the difficulties that one may encounter in determining the zero clearance pressure p_z and the jacket pressure coefficient d.

2. Similarity Method

The similarity method, a method of calibration for primary standard piston gages based on a concept entirely different from the controlled clearance principle, was developed by R. S. Dadson[13]. A detailed account of it is given in ref. 14. This method is of great interest for two reasons. One is its ingenious simplicity; the other is the fact that it affords us an independent check on the calibration of primary gases. The similarity method together with the flow method, to be discussed below, are practised at the National Physical Laboratory in Teddington, UK.

A cross section through a simple piston and cylinder system under pressure is shown schematically in *Figure 27*. Clearance and distortion are grossly exaggerated. Following the notation used by Dadson we denote by r and R the radii of the undistorted piston and cylinder respectively and by $u(x)$ and

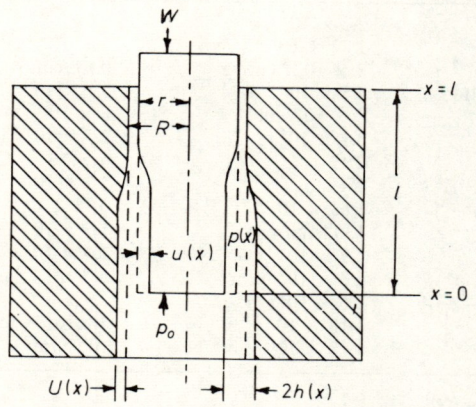

Figure 27. Distortion of a piston gage, schematic.

$U(x)$ the increases in these radii due to the pressure $p(x)$ in the space between piston and cylinder; $2h(x)$ is the radial separation.

The upward force due to the pressure acting on the base of the piston is

$$p_0 \pi r^2 [1 + 2u(0)/r] \qquad (46)$$

The upward force due to the vertical component of the pressure on the cylinder surface of the piston is

$$2\pi r \int_0^l \{p(x)\, \partial u(x)/\partial x\}\, dx \qquad (47)$$

and the upward force due to fluid friction is

$$2\pi r \int_0^l \{-h(x)\, \partial p(x)/\partial x\}\, dx \qquad (48)$$

With $h(x) = H + \frac{1}{2}[U(x) - u(x)]$ and some algebra, which may not be immediately obvious to the reader, we can combine and rewrite equations 46 to 48 and obtain for the effective area A_{eff}

$$A_{\text{eff}} = \pi r^2 \left[1 + \frac{2H}{r} + \frac{U(0) - u(0)}{r} + \frac{1}{rp_0} \int_0^l p\left(\frac{\partial U}{\partial x} + \frac{\partial u}{\partial x}\right) dx \right] \qquad (49)$$

As a limiting case for low pressures where the distortions $U = u = 0$ we have

$$A_{\text{eff}} = \pi r^2 \left[1 + 2H/r\right] \qquad (50)$$

which one recognizes as the first approximation of

$$A_{\text{eff}} = \pi (r + H)^2 \qquad (51)$$

where $r + H$ is the arithmetic mean of the diameters of the piston and cylinder.

The general theoretical solution of equation 49 is complicated by the fact that $U(x)$, $u(x)$ and $p(x)$ are interrelated and that to know $\partial p/\partial x$ one would also have to know η and $\partial \eta/\partial p$, where η is the viscosity of the fluid. To obtain a useful approximation Dadson assumes that the components of $U(x)$ and $u(x)$ due to the fluid pressure in the interspace between piston and cylinder are proportional to the pressure $p(x)$ at the same position. Without knowing the actual functional form of $U(x)$, $u(x)$ and $p(x)$ one may now proceed to integrate the last term on the RHS of equation 49. For a piston the diameter of which is small compared to its length and for a cylinder whose radius and wall thickness are small compared to its length one arrives at the following expression for the effective area

$$A_{\text{eff}} = A_0 \left[1 + \frac{p_0}{2E}(3\mu - 1) + \frac{p_0}{2E} \frac{(1+\mu)R_0^2 + (1-\mu)R^2}{R_0^2 - R^2} \right] \qquad (52)$$

R_0^2 is the outer radius of the cylinder, E is the Young's modulus, and μ is Poisson's ratio. The second term on the RHS of equation 52 describes the distortion of the piston, the last term describes that of the cylinder. As Dadson points out the computation of distortion of the cylinder is much less accurate and the idealized conditions are more difficult to meet than those for the piston.

In general the effective area of a piston gage could be described by

$$A_p = A_0[(1 + \lambda f(p)] \tag{53}$$

where λ is a constant, involving for example, the elastic constants and dimensions of piston and cylinder as given in equation 52. $f(p)$ is an unknown function of pressure assumed to be the same for gages made of different materials but with exactly the same dimensions. Experience shows that the distortion is commonly a linear function of pressure and $f(p)$ may then be replaced by p. Inspection of equation 52 suggests that the λ for different gages should be related as their inverse moduli of elasticity, E, provided that the materials of construction have the same Poisson's ratio.

If two gages of precisely the same dimensions are at hand with effective areas

$$A_p = A_0(1 + \lambda_A p) \tag{54}$$

and

$$B_p = B_0(1 + \lambda_B p) \tag{55}$$

one can determine the area ratio

$$\frac{A_p}{B_p} = \frac{A_0(1 + \lambda_A p)}{B_0(1 + \lambda_B p)} \approx \frac{A_0}{B_0}[(\lambda_A - \lambda_B)p] = \frac{A_0}{B_0}[1 + (1 - k)\lambda_A p] \tag{56}$$

from a cross-float of the two gages following the procedure outlined in section V.1. The ratio

$$\lambda_B/\lambda_A = E_A/E_B = k \tag{57}$$

of the elastic moduli can be determined from the results of measurements of the elastic constants using either quasi-static or ultrasonic measurements.

One of the conditions imposed above required that the Poisson's ratios of the two materials should be the same. *Table 5* includes data for the materials used by Dadson: K9 steel, aluminum alloy 'hydurax', tungsten alloy 'GEC heavy metal'. A correction for the small differences in the values of the Poisson's ratios can be made as follows. If the ratio R_0/R of the cylinder radii is large the last term in equation 52 is found to be proportional to

$$(1 + \mu)/2E = 1/G \tag{58}$$

where G is the shear (rigidity) modulus. We now write the distortion coefficient in the form

$$\lambda_A = \theta_A + \phi_A \text{ and } \lambda_B = \theta_B + \phi_B \tag{59}$$

where $\theta = (3\mu - 1)/2E$ and ϕ is that part of λ which is explicitly dependent upon the deformation of the cylinder previously found to be proportional to G^{-1}. Using g for the ratio of the two moduli of rigidity we now have as the result of the cross-float

$$\lambda_A - \lambda_B = \phi_A - \phi_B + \theta_A - \theta_B \tag{60}$$

PART 3. PISTON GAGES

We eliminate ϕ_B from equation 60 by introducing $\phi_B = g\phi_A$ and ϕ_A by introducing $\phi_A = \lambda_A - \theta_A$. Finally we have

$$\lambda_A = \{(\lambda_A - \lambda_B) + \theta_B - g\theta_A\}/(1 - g) \tag{61}$$

determining λ_A in terms of the measured quantity $(\lambda_A - \lambda_B)$, the ratio of the shear moduli $g = G_A/G_B = \phi_B/\phi_A$, and term $\theta_B - g\theta_A$ containing the moduli of elasticity and Poisson's ratio.

Dadson[14] has also described an extension of this method to three materials, which gives additional support to the validity of the method.

An estimate of the uncertainty in λ_A may be obtained from the total differential of equation 61 taking the absolute amount of each contribution

$$d\lambda_A = \frac{1}{1-g}[d(\lambda_A - \lambda_B) + d\theta_B - g\,d\theta_A]$$

$$+ \frac{1}{(1-g)^2}[\theta_A - \theta_B - (\lambda_A - \lambda_B)]\,dg \tag{62}$$

The error $d(\lambda_A - \lambda_B)$ follows from the total differential of equation 56, again disregarding the sign of the individual contributions,

$$d(\lambda_A - \lambda_B) = \frac{1}{p}\left[2\frac{A_p}{B_p}d\left(\frac{B_0}{A_0}\right) + \frac{B_0}{A_0}d\left(\frac{A_p}{B_p}\right)\right] = 1 \times 10^{-14}\,\text{m}^2/\text{N}$$

assuming that

$A_0 \simeq B_0 = 1.29 \times 10^{-5}\,\text{m}^2$
$dA_0 \simeq dB_0 = 1.5 \times 10^{-10}\,\text{m}^2$
$A_p/B_p = 1$
$d(A_p/B_p) = 10^{-6}$
$p = 2 \times 10^8\,\text{N/m}^2$

We shall further assume that assembly A is made of K9 steel and assembly B is made of hydurax with the following properties and respective uncertainties:

$E_A = 2 \times 10^{11}\,\text{N/m}^2$ $\qquad E_B = 1.4 \times 10^{11}\,\text{N/m}^2$
$\mu_A = 0.295$ $\qquad \mu_B = 0.333$
$dE_A = 1 \times 10^9\,\text{N/m}^2$ $\qquad dE_B = 7 \times 10^8\,\text{N/m}^2$
$d\mu_A = 0.001$ $\qquad d\mu_B = 0.001$
$\theta_A = -2.88 \times 10^{-13}\,\text{m}^2/\text{N}$ $\qquad \theta_B = -3.5 \times 10^{-15}\,\text{m}^2/\text{N}$
$d\theta_A = 8.9 \times 10^{-15}\,\text{m}^2/\text{N}$ $\qquad d\theta_B = 1.1 \times 10^{-14}\,\text{m}^2/\text{N}$
$g = 1.44$
$dg = 0.005$

For cylinders of steel and hydurax of infinite wall thickness $\lambda_A = 2.85 \times 10^{-12}$ m²/N and $\lambda_B = 4.62 \times 10^{-12}$ m²/N and the last term of equation 62 becomes

$$1/\{[\theta_A - \theta_B - (\lambda_A - \lambda_B)]/(1 - g^2)\,dg < 3.8 \times 10^{-14}\,\text{m}^2/\text{N}$$

The total uncertainty of λ_A then is (equation 62)

$$d\lambda_A \simeq 1.1 \times 10^{-13} \text{ m}^2/\text{N}$$

or if we use the r.m.s. of the individual contributions rather than the sum

$$\text{RMS } d\lambda_A \simeq 0.6 \times 10^{-13} \text{ m}^2/\text{N}$$

The precision of the determination of λ is of the same order. Dadson's measurements indicate a dispersion of the values of λ of less than four per cent using three different materials. The value of λ_A following from Dadson's measurements was about 4×10^{-12} m²/N.

The total uncertainty of the pressure generated with a piston gage calibrated with the similarity method can be estimated from the total differential of equation 17, replacing b with λ. Using the uncertainties discussed above, or the terms from *Table 9* as appropriate, we have

$$\frac{dp}{p} = \sum_1^i \frac{1}{p} \frac{\partial p}{\partial x_i} dx_i = 43 \times 10^{-6} \tag{63}$$

and

$$\text{RMS} \frac{dp}{p} = \left[\sum_1^i \left(\frac{1}{p} \frac{p}{x_i} dx_i \right)^2 \right]^{\frac{1}{2}} = 61 \times 10^{-6} \tag{64}$$

where x_i are the parameters and coefficients of equation 1.

V. Calibration of Piston Gages

1. Cross-float

In principle we can determine the effective area and the pressure coefficient of a piston gage from measurements of the internal diameter of the cylinder, the diameter of the piston and a computation of the pressure coefficient using the elastic distortion equations of section II. However, the resulting uncertainties are so large that in practice all piston gage calibrations are done by cross-floating against a primary standard. The object of this exercise is to determine the effective area and the pressure coefficients of the test gage in terms of those of the primary standards.

In equation 65 p is the pressure generated at the reference level of the test instrument by the standard gage.

$$p = \frac{F^T}{A_0^T(1 + b_1^T p + b_2^T p^2)} \tag{65}$$

where

$$F^T = \frac{Mg[1 - (\rho_{\text{air}}/\rho M)] + \gamma C + T^T}{1 + (\alpha_p + \alpha_c)(T - T_r)} \tag{66}$$

is the force exerted on the test gage piston, A_0^T is the effective area of the test gage, b_1^T is the fractional change of effective area with pressure of the test

PART 3. PISTON GAGES

gage, and b_2^T is the fractional change of effective area of the test gage with the square of the pressure.

Note that for simplicity the temperature correction of the area has been lumped with the force, F^T.

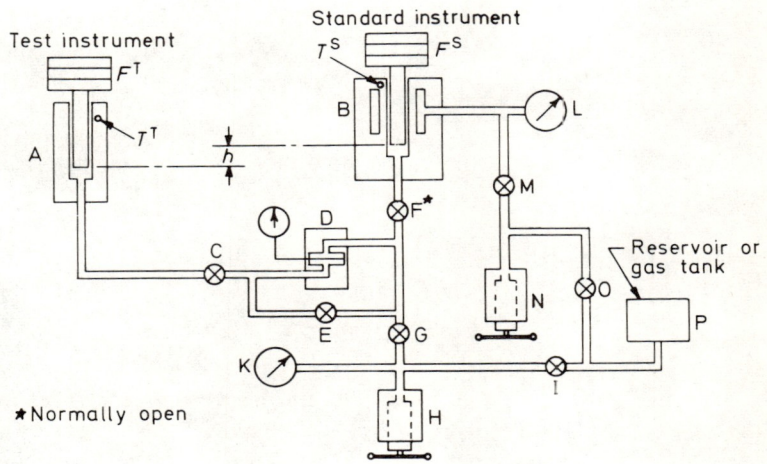

Figure 28. Cross-float set-up, schematic.

The RHS of equation 65 represents the pressure generated by the test gage at its reference level. By adjusting F^T this pressure is made equal to p. The effective area A^T and the coefficients b_1^T and b_2^T can then be obtained by fitting

$$F^T = A_0^T p(1 + b_1^T p + b_2^T p^2) - T^T \qquad (67)$$

to the (F^T, p) data obtained from the cross-float set-up. The test instrument (A in *Figure 28*) is connected to the standard (B) through a short length of large bore tubing. A valve (C) can interrupt this connection. The differential pressure indicator (D) can be short circuited by valve (E). Valve (G) connects to the fluid supply consisting of a screw pump (H), gage (K) and reservoir (P). If a controlled clearance piston gage serves as a standard, a separate supply of jacket pressure is required. Valves (C) and (E) must be constant volume valves. Markus[36] has described a pneumatically operated constant volume valve that gives very satisfactory service. The differential pressure indicator must have a sensitivity of at least one part in one million and should be capable of withstanding without damage the full line pressure when applied differentially.

The photograph, *Figure 29*, shows a view of a typical cross-float set-up with the controlled-clearance piston gage on the left and the test instrument on the right. Located in the line between the gages is the differential pressure transducer and indicator with two air-actuated valves.

For the set-up and preparation of the gages the reader is referred to the section on procedures. Needless to say that in order to attain a reproducibility

Figure 29. Cross float set-up; controlled clearance gage (left), differential pressure transducer and constant volume valves (center), test gage (right).

of one part per million or better in the data taken during a cross-float the greatest care must be taken and patience be exercised. A good cross-float may well take a week to complete.

Generally during a cross-float ten measurements are taken at seven pressures ranging from about ten per cent of the full range to the full range alternating the direction of rotation of the two instruments, as shown in *Table 10*. A list of weights to be used on either gage is prepared in advance and if a controlled clearance gage is used, the appropriate operating jacket pressures are chosen from the plot of jacket pressure versus fall rate with load as parameter (*Figure 26*).

With valves (C) and (E) open (see *Figure 28*) and the weight hangers down on the stop both gages are loaded and spun. Gages are always spun by hand and allowed to coast to avoid possible vertical components resulting from continuous motor drive. Both pistons are then raised above their operating level with the help of the screw pump H, and left coasting with valve (G) closed until both gages have reached temperature equilibrium. This may take 45 minutes or more. With valve (C) closed the differential pressure indicator is zeroed. Valve (E) is closed and (C) opened, and the weights on one of the gages are adjusted to zero differential pressure. When balance is attained the applied loads, the jacket pressure and the gage temperatures are recorded. Finally valves (E) and (G) are opened and the pressure is lowered to bring the weight hangers down to the stop before loading up for the next point.

PART 3. PISTON GAGES

Table 10. Schedule of points taken during a cross-float

Point	Pressure % of full range	Direction of rotation standard	test
1	10	CW	CW
2	10	CW	CCW
3	40	CW	CW
4	70	CW	CCW
5	100	CW	CW
6	100	CCW	CCW
7	85	CCW	CW
8	55	CCW	CCW
9	55	CCW	CW
10	25	CCW	CCW

2. Reference Levels

For piston gages with straight pistons (*Figure 3*) the reference level is placed at the lower end of the piston. For pistons of irregular shape of the submerged part an adjustment of the reference level is made as illustrated in *Figure 30*. In this example the piston has a flange at the lower end serving as a stop. The volume of the shaded, protruding part of the flange is

$$V = (h\pi/4)(D^2 - d^2)$$

If the piston were lengthened by this amount the length of the piston would increase by

$$l = \{(D^2 - d^2)/d^2\}h \tag{67}$$

The reference level for a piston of this shape would be defined as being l below the bottom of the piston.

In a piston gage cross-float, or whenever a piston gage is used to generate a known pressure at the reference level of another instrument, a correction has to be made for the fluid head in the connecting line. This correction has the form

$$gH\{\rho_{fl}[1 + p(1/\rho)(d\rho/dp)] - \rho_{air}]\} \tag{68}$$

It is subtracted from the pressure generated by the standard instrument at its reference level. The level difference, H, is counted positive upwards from the reference level of the standard instrument; it is given in meters. The term $[1 + p(1/\rho)(d\rho/dp)]$ adjusts for the change of density of the fluid with pressure; this is an approximation only, but sufficient for most applications. The term $-\rho_{air}gH$ is the buoyancy correction applied to the fluid head.

3. Evaluation

We can now return to equation 1 and 66 and determine the effective area and the coefficients of the test gage from a number of least squares fits[26] of equation 67 to the data (F^T, p).

If equations 69.1 to 69.8 are fitted to the data at very low pressures; the terms involving the pressure coefficients b_1 and b_2 are usually insignificant and either equation 69.1 or 69.2 is used to characterize the gage. Note that T may be either a tare error or a coefficient necessary to characterize the behavior of the gage properly. At higher pressures the coefficient b_1 and occasionally also b_2 become significant and must be included in the function fitted to the data. In a few types of gages the coefficient b_1 is insignificant while b_2 must certainly be considered.

$$F^T = pA_0^T$$
$$F^T = pA_0^T - T$$
$$F^T = pA_0^T(1 + b_1^T p)$$
$$F^T = pA_0^T(1 + b_1^T p) - T$$
$$F^T = pA_0^T(1 + b_1^T p + b_2^T p^2)$$
$$F^T = pA_0^T(1 + b_1^T p + b_2^T p^2) - T$$
$$F^T = pA_0^T(1 + b_2^T p^2)$$
$$F^T = pA_0^T(1 + b_2^T p^2) - T$$
(69.1–69.8)

A high speed digital computer will perform these computations in a few seconds and, apart from the coefficients, will determine the standard deviations of the coefficients, the residual standard deviations and the residuals. A plot of the residuals as functions of pressure will show at a glance whether any gross errors have been made in recording and entering the data. One may also separate the data according to the direction of rotation of either gage and check for effects depending on the direction of the rotation.

The proper fit is finally selected by comparing the residual standard

Table 11. Piston gage coefficients for a gage with tungsten carbide piston and cylinder; range to 1.4×10^8 N/m²

Fit	1	3	5	7	
Coefficients					Dimensions
A	8.40029	8.40379	8.40273	8.40239	$\times 10^{-6}$ m²
$3\sigma A$	138	71	109	26	p.p.m.
b_1	—	−3.619	−0.856	—	$\times 10^{-9}$ m²/N
$3\sigma b_1$	—	0.597	2.260	—	$\times 10^{-9}$ m²/N
b_2	—	—	−1.363	−1.777	$\times 10^{-14}$ m⁴/N²
$3\sigma b_2$	—	—	1.104	0.169	$\times 10^{-14}$ m⁴/N²
σFit	12.79	1.56	0.81	0.90	$\times 10^{-3}$ N/m²
$3\sigma p/p_{max}$	138	153	632	58	p.p.m.

deviations of the various fits. *Table 11* lists a typical set of results for a gage with tungsten carbide piston and cylinder. The standard deviation of the residuals, δ Fit, is reduced as more coefficients are used to characterize the gage. However, since the number of degrees of freedom is reduced simultaneously the uncertainty of the coefficients increases. In the last row the uncertainty in pressure, $3\sigma p/p_{max}$, due to the uncertainty of the coefficients is entered; it goes up rapidly as the number of coefficients used increases.

PART 3. PISTON GAGES

The appropriate fit for this particular gage is fit 7 with only a quadratic pressure coefficient. The quadratic term is quite typical for this type of gage but is rarely found in other designs.

The total uncertainty of each coefficient is the sum of the random uncertainty from the cross-float and the systematic uncertainty of the measurement (calibration of the standard). The total uncertainty of the pressure generated with a piston gage is easily obtained from the total differential of equation 1 together with that of equations 67 and 68. For tilting piston gages the uncertainty due to the imprecision of the angle scale calibration must be added. For near vertical operation this uncertainty amounts typically to a few p.p.m., at $\alpha = 60°$ it amounts to 80 p.p.m., at 30° to 240 p.p.m. and at 5° to about 1.6 per cent.

VI. The Use of Piston Gages

1. *Measurement of Pressure*

Piston gages may be used either as a generator of an accurately known pressure or as a device to measure a pressure generated by some other source. As long as the gage piston is floating the pressure at the reference port is essentially constant and may be computed from equation 1. Connected into a static or quasi-static system the piston gage will set the pressure prevalent in the system. If a variable pressure, as for example in a gas thermometer, is to be measured the system must be connected to the piston gage through a differential pressure indicator as shown in *Figure 28*. As in a cross-float the piston gage is balanced to match the system pressure. Differential pressure transducers and indicators for various ranges are available commercially. Most of these transducers employ a membrane separating the piston gage from the system. The deflection of the membrane is sensed by capacitative transducers, differential transformers, linear potentiometers, reluctance transducers, strain gages or other means. In each case the user must determine whether or not the differential pressure transducer is compatible with the

Table 12. Fluids compatible with piston gages for various pressures

Nominal pressure (MPa)	Fluid	Nominal pressure (MPa)	Fluid
15 (gas)	Breathing quality air Nitrogen	1500	Heptanes (mixtures) White gasoline Plexol 268
300	SAE 10 motor oil (Pennsylvania type) Instrument oil Spindle oil Turbine oil	2000	Isoamyl alcohol Pentanes (mixtures)
800	Varsol (pure white) Octoil-S Plexol 201	3000	Pentanes (mixtures)

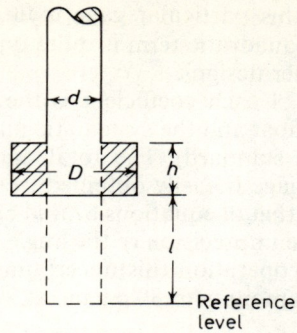

Figure 30. Adjustment of reference level for irregularly shaped pistons.

pressure fluids used. The gage constants of the transducer should be independent of the line pressure and the transducer should be able to withstand the full line pressure if applied to only one side of the diaphragm.

If only small variations of a pressure have to be measured it is advantageous to use a differential pressure transducer to measure the difference between the constant pressure generated by a piston gage and the variable systems pressure. In this case linearity is an additional requirement for the differential pressure transducer.

Differential pressure transducers are often used to separate a system from the piston gage if the pressure fluid is not compatible with the piston gage. For pressure fluids compatible with piston gages see *Table 12*.

Commercially available piston gages cover the range from about 1400 Pa to 1400 MPa. For pressures to about 15 MPa air or nitrogen are the most common pressure fluids. Above this pressure liquids are used almost exclusively. A listing of a number of commercially available piston gages can be found in ref. 3.

2. *Procedures and Methods*

The determination of pressure with piston gages requires knowledge of the gage parameters and consideration of a number of corrections which we have discussed in the section on calibrations. In addition, certain precautions must be taken to insure the most accurate results. Again some of these have been mentioned before, others will be discussed here.

A. CLEANING

In most high precision set-ups containing mechanical components with moving parts a clean system is desirable. The oil piston gage is more likely to function properly over a longer period of time if the instrument, lines, and fittings are carefully cleaned. The oil piston gage will usually operate satisfactorily under most conditions, but damage can result if particles of dirt become lodged between the piston and cylinder. Oil piston gages should be cleaned with solvents such as fluorinated carbons or trichloroethylene whenever a change is made from one oil to another.

Air piston gages on the other hand will not function properly when dirty and if forced to operate under such conditions damage to the piston and cylinder is likely.

To obtain the highest performance from an air piston gage a good cleaning technique is essential. One such method is simply to use mild soap and and warm water. The piston and cylinder should be scrubbed thoroughly, rinsed with warm water, air blasted to remove the water droplets and finally polished with lens tissue. Before assembling the piston and cylinder, dry, clean air (or nitrogen) is used to remove lint particles.

Another method that appears to give satisfactory results is to wash the piston and cylinder with mild soap and water, followed by a complete rinse with warm water. Next, in order, rinse with alcohol, trichloroethylene and freon (1 p.p.m. purity). The items are then polished with lens tissue and air blasted. Other cleaning substances have been used in the past and are quite acceptable if the desired results are obtained and no damage to the piston and cylinder results.

Some important points in a good cleaning technique are: (1) remove all foreign material, such as dirt and grease, (2) leave no residue, (3) polish, (4) remove all remaining lint before assembling.

It may be necessary to repeat the cleaning procedure to obtain satisfactory results. It is also important that the remaining portion of the system be clean.

Where the removal of a piston or cylinder is difficult or undesirable, and the parts must be cleaned in place, water should be used with extreme care; it is far safer to use a substance that will evaporate or boil off fairly rapidly.

The piston gage operator should be able to determine from the behavior of the instrument (spin time, sensitivity, etc.) whether the instrument is functioning properly.

B. Temperature Control and Measurement

The effect of temperature on a piston gage is given by the term $[1 + (\alpha_c + \alpha_p)(T - T_{ref})]$ of equation 1. Temperature coefficients for various materials are listed in *Table 5*. A precision of one part per million in the determination of the area requires an imprecision of the temperature measurement of less than 0.05°C for a steel piston and cylinder combination.

The quality of temperature control and measurement necessary in a piston gage set-up is determined by the level of precision expected. When operating in the field where temperature control is difficult and the required accuracy is of the order of a few parts in 10 000 or less, a mercury in glass thermometer is adequate for the measurement of temperature.

In the standards laboratory temperature control may be ± 0.5°C or better. Under these conditions and with good air circulation the temperature of the mass of a piston gage may remain constant within ± 0.05°C for several hours. It is also occasionally desirable to be able to detect temperature changes of ± 0.005°C or less. With this capability small changes in the temperature of the piston and cylinder can be observed. Observations of this type are quite valuable when calibrating controlled clearance piston gages; as an example, a sharp increase in temperature can indicate too close a working clearance between piston and cylinder. When measurements were made as a function of jacket pressure the gage temperature increased by several hundredths of a degree each time a jacket pressure close to the upper end of the operating range was chosen. This rise was subsequently used as an indication for the upper (and lower) limit of the jacket pressure range.

Platinum resistance thermometers, thermocouples, and thermistors are a few of the devices used to detect small changes in temperature.

To make accurate, absolute temperature measurements careful calibration of the sensing elements, including leads and read-out, must be made.

Although a temperature sensing element may be available, locating it on the piston gage is sometimes difficult. The optimum location for the sensor would be at the working area of the piston and cylinder but this is seldom possible. It is, on occasion, possible to place the thermometer in contact with the cylinder and through conduction small temperature changes can be detected.

The basic design of controlled clearance piston gages at the NBS has an oil relief hole of approximately 2 mm diameter drilled through the head of the gage. Small platinum resistance probes are inserted into this port and placed in contact with the cylinder; this arrangement has been found to be quite satisfactory.

C. Weights

High precision weights for the piston gage are usually made from a material that is hard, non-porous, non-magnetic and solid. The surface finish should be smooth. As a rule the diameter should be large, so that the height of the stack does not exceed the diameter. The center of mass of the weights should be located somewhat below the pivot point of the piston and weight table. Individual weights should be balanced and nest on the piston so that the load is balanced.

Depending on the basic design of the instrument, various methods of loading the weights have been devised. For most low pressure work weights are manually loaded for each pressure desired. When working at higher pressures and consequently larger masses, mechanical loading is more convenient. If weights are loaded through the use of an external mechanical means, it is important to see that no extraneous forces contribute to the load on the piston. Weights have been loaded above the piston, about the piston and below the piston; the weights may remain fixed while the piston oscillates or the piston and weights may rotate together. Regardless of the method chosen for loading the weights, the gage should be mechanically stable and well balanced at all pressures within the range.

Piston gage weights are usually calibrated by the manufacturer against standards calibrated by a national (or state) standards laboratory. In the past calibration results for piston gage weights were given in terms of apparent mass; that is, apparent mass versus 'normal' brass. In this sense normal conditions are an air density of 1.2 mg/cm^3, a temperature of 20°C, an ideal density for brass of 8.4 g/cm^3 at 0°C, and a coefficient of volume expansivity of 0.000054/°C. This leads to an ideal density for brass of 8.3909 g/cm^3 at

20°C. True mass values are now reported and it is therefore necessary to determine the true density of each item that is to be calibrated. If for a particular set of standard weights the corrections are given in terms of apparent mass versus brass, the true mass corrections may be computed as follows (Almer[1])

$$M_T = M_A + \rho_A(V_S - V_B)$$

where M_T is the true mass of the weight, M_A is the apparent mass versus brass of the weight, that is the nominal value plus the apparent mass correction, ρ_A is the density of normal air (1.2 mg/cm^3 at 20°C), V_S is the volume of the standard at 20°C, and V_B is the volume of an equivalent mass of normal brass at 20°C. The true mass correction is

$$\text{Correction} = M_T - \text{nominal value}$$

True volumes may be calculated by making hydrostatic weighting (Bowman[7]).

D. Connections

When connections are made in a pressure system extreme care should be taken to install the properly rated tubing, fittings, valves, etc. This is most important in a gas system where the fluid is highly compressible. Plumbing material should be chosen to be fully compatible with the pressure fluid. Stainless steels are widely used in liquid systems, whereas copper, plastics, rubber, etc., are commonly found in low pressure gas systems. When threading and coning high pressure tubing it is essential that the threads be carefully made and that cones have the correct angle and proper finish.

All plumbing should be carefully cleaned to remove particles of metal, lint, and dirt so as not to damage moving parts. If filters are used in a system, it is necessary to choose a type that will not break up or expel particles into the pressure fluid. Sometimes drying substances (silica gel, etc.) are used in a system to remove moisture and foreign materials. There again, particles of this material should not be freely introduced into the system.

To obtain the optimum in response time lines are usually kept short and internal diameters should be as large as is practicable. Damping of a system may be achieved through the use of needle valves, filters or other types of line restrictions.

Proper valve arrangement can make a system highly efficient and more than one type of valve may be found useful in such a system. The constant volume valve can be operated without disturbing the internal pressure of the system and is therefore quite convenient when comparing one piston gage against another. Damping valves may be employed in a piston gage intercomparison set-up where short or long term oscillations have been detected. It has been found that response time may suffer slightly as a result of using damping valves, but this situation may be acceptable if the system is made more stable by using the valves. Non-rotating stem valves appear to function very well within a given pressure range. The seat formed in this type of valve usually remains in good condition for a long time, if used carefully.

Manually operated valves are available which can be used in liquid systems to 1 400 MPa; however, some manual valves become very difficult to operate in the upper part of this range. At times it is almost impossible to determine when the valve is closed; damage to the stem and/or seat is not uncommon in such valves. Air actuated and electrically actuated hydraulic valves avoid some of these difficulties and have the added convenience of remote control. With valves of this type seals are of great concern; also, if movable stems are used precise adjustments are necessary to insure against leaks or seat damage.

Table 13. Examples of tubing and fitting sizes for various pressure ranges

Range (MPa)	Material	O.D. (mm)	O.D. (in.)	I.D. (mm)	I.D. (in.)	Fitting type
0.1	copper	9.5	3/8	0.81 (Wall)	0.032 (Wall)	Flare
	plastic	9.5	3/8	12.7	1/2	Swagelock
	rubber	19.0	3/4	6.4	1/4	
4	copper	6.4	1/4	0.76 (Wall)	0.030 (Wall)	Flare
100	stainless	4.8	3/16	0.89 (Wall)	0.035 (Wall)	Ruska
400	stainless	6.4	1/4	1.6	1/16	Aminco
700	composite	14.3	9/16	4.8	3/16	Harwood
1400	composite	19.0	3/4	1.6	1/16	Harwood

Occasionally, it is desirable to reduce the volume of fluid in a system to a minimum because of temperature effects. As an example, when taking rate of fall measurements on a controlled clearance gage the volume of the pressure fluid beneath the piston should be reduced to an absolute minimum. This can be accomplished by using short lengths of small bore tubing and valving off directly under the piston. In an existing system of large bore tubing it is sometimes possible to insert small rods into the tube bore where the small fluid volume is desired.

Table 13 lists common materials, sizes and fitting types for various pressure ranges.

E. Fluids

The pressure fluid used in a piston gage should be such that the optimum in fall rate and response time is achieved over the complete range of the instrument. The fluid should also be compatible with the components of the system.

Several factors must be considered for the selection of the proper pressure fluid. Viscosity is of primary importance when working with pressures above 300 MPa. It has been found from a study of fluid viscosity versus pressure, that in general the saturated hydrocarbons (e.g. paraffin series) and diesters appear to change more slowly with pressure than do the silicones and fluorocarbons. The use of the pure hydrocarbons is preferred over the use of mixtures. When using fluids of composite hydrocarbons made up of several constituents having a wide band of molecular weight, slow crystalline solidification can occur at high pressures. It is therefore important, when working with such fluids, that the pressure on the low side of all intensifiers be monitored.

Variations in fluid density at elevated pressures can lead to uncertainties in the buoyancy correction on some instruments. Usually it is possible to account for the buoyancy correction in terms of a shift in the reference level (Cross[12]), thus avoiding the change in density problem.

Surface tension needs to be known only at atmospheric pressure and room temperature. This would be the condition under which the piston emerges through the surface of the liquid to the operating position.

The conductivity of the pressure fluid is important in those cases where electrical measurements are to be made and the leads come into direct contact with the fluid.

The properties of several common fluids used with piston gages are listed in *Table 12*. Numerous other substances and mixtures have been used at various pressures with reasonable success. A 50/50 mixture of pentane and isopentane has been used[25] to 5 000 MPa. Mixtures of glucose, glycerin and water have been used to 1 500 MPa[30].

Filters may be used in the pressure fluid supply lines. Magnets are sometimes installed in the pressure ports of piston gages to catch magnetic particles. Used oil should be discarded rather than used again.

Once the proper pressure fluid has been chosen the entire system must be purged. This is particularly important when oil is used as a fluid at low pressures where large quantities of gas can give erratic results. As a test for gas in the system, the gage is unloaded, leaving only the piston assembly; the instrument is pumped up and valved off and the piston is depressed by hand. Due to the low compressibility of oil as opposed to air, the system will appear firm when little or no air is present. This method works best when the volume of oil is kept small and the piston area is relatively large.

At pressures above a few MPa the gas in a liquid system will go into solution. There are several methods of removing gas from a system. The simplest is the removal of the piston from the gage and pumping the liquid and gas out through the cylinder, when the gas no longer appears the piston may carefully be replaced. On some instruments a bleed hole is provided for this purpose. Occasionally, gas may be removed from a piston gage by leaving the system under pressure for several hours, this drives the gas into solution.

F. Rotation of Piston or Cylinder

In most modern piston gages either the piston or the cylinder is rotated or oscillated to relieve friction and to maintain a continuous film of air or oil between piston and cylinder. This is usually done with the help of electric motors and drive mechanisms.

There are several points concerning drive motors that one should be aware of; the heat given off by some drive motors may increase the temperature of the piston and cylinder by several degrees above the nominal room temperature over a period of several hours. Under these conditions it is difficult to measure the temperature of the piston and cylinder accurately. Shielding the motor or moving it to a different location may lessen the problem. Some commercial drive motors rotate the piston assembly many times faster than is necessary. A high speed of rotation on a gas gage can cause the piston and cylinder to become dirty more rapidly; also, there is the danger of galling at high speeds. Some manufacturers recommend the application of a thin film of oil to piston and cylinder to prevent galling and to reduce excessive fall rates. While this procedure may lead to an acceptable fall rate it will inevitably increase friction and generally reduce the sensitivity. Occasionally, piston assemblies are rotated by lines attached to a drive mechanism; it is necessary that these lines be kept horizontal when the gage is in operation. A vertical component of force applied to the load on the piston can occur when drive lines are short and not horizontal. Hand rotation of the piston and weight stack is quite satisfactory since the optimum speed of rotation can be achieved for each instrument. 15 to 30 revolutions per minute are adequate for most gages.

G. Mechanical Support and Covers

The piston gage should be placed on a table that is strong enough not to tilt under a varying load. It should not sway or rock. Table tops should be non-magnetic. Laminated wood or heavy aluminum plates serve well. Legs with adjustable length facilitate horizontal adjustment and add to the stability. Often auxiliary hydraulic or pneumatic equipment can be accommodated directly under the table very close to the piston gage.

Air currents in the laboratory may cause unbalances of up to 10 p.p.m. and piston gages should therefore be covered with appropriate shields during their use. Covers could be made of plastic, glass or metal. Hinged covers facilitate the operation. Windows should be provided to enable the operator to check the operating level. Plastic covers must be provided with metallic liners (Faraday cage) to avoid electrostatic forces on the weight stack.

VII. Abstract

Piston gages are instruments which measure pressure as force per unit area. There are a number of sources of error among which elastic distortion is the largest. Several types of piston gages have been developed to reduce the elastic distortion errors and to allow operation at a variety of pressures. A minimum of elastic distortion error is found with the controlled clearance piston gage. Other sources of error include temperature effects, air buoyancy, and head corrections. To reach the highest precision, operating procedures must be followed which insure adequate control of all important parameters.

VIII. References

[1] Almer, H. E., *Tech. Note US Nat Bur. Stand. No. 577* (1971), S. C. Cont. No. C13.46:577, Government Printing Office: Washington, DC 20402.
[2] Amagat, E. H., *Ann. Chim. (Phys.)*, **29**, 70 (1893).
[3] Aronson, M. H., Editor, 'Dead weight testers'. *Measurements and Data*, **30**, 72 (1971).

PART 3. PISTON GAGES

[4] Bennett, C. O. and B. Vodar. *High-Pressure Measurement Symposium,* American Society of Mechanical Engineers: New York (November 1962); Butterworths: Washington, DC (1963).
[5] Bett, K. E., P. F. Hayes and D. M. Newitt. *Phil. Trans. Roy. Soc. (London),* **247,** 59 (1954).
[6] Bourdon, E. *French Pat. No. 4408* (1849).
[7] Bowman, H. A. and R. M. Schoonover. *J. Res. Nat. Bur. Stand.* **71C,** 179 (1967).
[8] Bridgman, P. W., *Proc. Amer. Acad. Arts Sci.* **44,** 201 (1908).
[9] Bridgman, P. W., *Proc. Amer. Acad. Arts Sci.* **47,** 321 (1911).
[10] Bridgman, P. W., *Proc. Amer. Acad. Arts Sci.* **72,** 207 (1938).
[11] Caw, W. A., *J. Sci. Instrum.* Ser. 2, **2,** 73 (1969).
[12] Cross, J. L., 'Piston manometer', *Monogr. US Nat. Bur. Stand. No. 65*, Government Printing Office: Washington, DC (1964).
[13] Dadson, R. S., *Nature, London,* **176,** 188 (1955).
[14] Dadson, R. S., R. G. P. Greig and A. Horner. *Metrologia,* **1,** 55 (1965).
[15] Dadson, R. S. and R. C. P. Greig. *J. Sci. Instrum.* **42,** 331 (1965).
[16] Dadson, R. S. and R. C. P. Greig. *Brit. J. Appl. Phys.* **16,** 1711 (1965).
[17] Desgoffe, C., *Sucr. Indig. Colon.* **6,** 151 (1871).
[18] Ebert, H., *Z. Angew. Phys.* **1,** 331 (1949).
[19] *Eisenbahn Zeitung,* **7,** (March and April 1849).
[20] Galy-Cazalat, M., *Bull. Soc. Enc. Industr. Nat. Paris,* p 590 (November 1846).
[21] Guildner, L. A., H. F. Stimson, R. E. Edsinger and R. L. Anderson. *Metrologia,* **6,** 1 (1970).
[22] *Handbook of Chemistry and Physics,* F9 (52nd edition). Chemical Rubber Publishing Company: Cleveland, Ohio (1971).
[23] Heydemann, P. L. M., *J. Appl. Phys.* **38,** 2640 (1967).
[24] Heydemann, P. L. M. and J. C. Houck. *J. Appl. Phys.* **40,** 1609 (1969).
[25] Heydemann, P. L. M., *Procédés du Colloque International sur les Propriétés Physiques des Solides sous Pression,* Grenoble (1969).
[26] Hogben, D., S. T. Peavey and R. N. Varner. 'Omnitab II'. *Tech. Note US Nat. Bur. Stand. No. 552* Government Printing Office: Washington, DC 20402 (1971).
[27] Holborn, L. and H. Schultze. *Ann. Phys., Lpz.* **47,** 1089 (1915).
[28] Hutton, U. O., *J. Res. Nat. Bur. Stand.* **63C,** 47 (1959).
[29] Jayaraman, A., A. R. Hutson, J. H. McFee, A. S. Coriell and R. G. Maines. *Rev. Sci. Instrum.* **38,** 44 (1967).
[30] Johnson, D. P. and D. H. Newhall. *Trans. Amer. Soc. Mech. Engrs,* **75,** 301 (1953).
[31] Johnson, D. P., J. L. Cross, J. D. Hill and A. H. Bowman. *Industr. Engng Chem.* **49,** 2046 (1957).
[32] Johnson, D. P. and P. L. M. Heydemann. *Rev. Sci. Instrum.* **38,** 1294 (1967).
[33] Kennedy, G. C. and P. N. LaMori. In *Progress in Very High Pressure Research,* Wiley: New York (1961).
[34] Knowles Middleston, W. E. *The History of the Barometer,* John Hopkins Press: Baltimore (1964).
[35] Konyaev, Yu. S., *Prib. Tekh. Eksp.* **4,** 107 (1961).
[36] Markus, W., *Rev. Sci. Instrum.* **43,** 158 (1972).
[37] Meyers, C. H. and R. S. Jessup. *J. Res. Nat. Bur. Stand.* **6,** 1061 (1931).
[38] Michels, A., *Ann. Phys., Lpz.* **72,** 285 (1923).
[39] Michels, A., *Ann. Phys., Lpz.* **73,** 577 (1924).
[40] *Handbook US Nat. Bur. Stand. No. 77,* Vol. III, 'Precision measurement and calibration', Government Printing Office: Washington, DC (1961).
[41] Newhall, D. H., *US Pat. No. 2796 229* (18 June 1957).
[42] Newhall, D. H., *US Pat. Appl.* (1971).
[43] Newhall, D. H. and L. H. Abbot. *Measurements and Data,* **19,** 90 (1970).
[44] PTR, *Bericht über die Tätigkeit der, Z. Instrumentenkunde,* **14,** 301 (1894).
[45] Ruchholz, E., *German Pat.* K1.42 No. 18 626 (19 January 1882).
[46] Schinz, G., *Prussian Pat. No. 3* (1849).
[47] Seyss, L., *Z. Österr. Ing. Architon. Vereins,* **25,** (1869).
[48] *Smithsonian Meteorological Tables,* 6th ed. Smithsonian Institution: Washington, DC (1951).
[49] Tate, D. R., *J. Res. Nat. Bur. Stand.,* **72C,** 1 (1968).
[50] Torricelli, Evangelista. *Opere,* eds. G. Loria and G. Vassura. Faenza (1919).
[51] Vereshchagin, L. F., E. V. Zubova, I. P. Buimova and K. P. Burdina. *Dokl. Akad. Nauk SSSR,* **169,** 74 (1966).
[52] Wexler, A. and L. Greenspan. *J. Res. Nat. Bur. Stand.,* **75A,** 213 (1971).

[53] Wiebe, H. F., *Z. compr. flussige Gase*, **1**, 1 (1897).
[54] Wylie, P., private communication (1969).
[55] Zhokhovskii, M. K., *Theory and Design of Instruments with Packing-free Pistons* (in Russian). Moscow (1959).

CHAPTER 4

Part 4. Secondary Gage—Differential Manometers

Yuri A. Atanov
Physical and Radiotechnical Measurements Institute, Mendeleevo, Moscow Region, USSR

Contents

I.	Introduction	203
II.	Mechanical Devices	204
III.	Electrical Resistance Gages	205
IV.	Differential Manometers	210
V.	References	211

I. Introduction

All instruments and devices based on the functional or tabulated relationship of some measured quantity x and pressure p are known as secondary gages. Such a relationship or pressure scale is determined by calibration against a primary standard instrument such as a mercury column or free piston gage (see Chapter 4, Parts 1 and 3). Generally speaking, any physical, chemical or mechanical characteristic of a substance may be utilized in constructing such secondary gages. Practical convenience, however, rejects a considerable number of gages. For example, the variation of the chemical reaction rate cannot be utilized due to the difficulty of the direct measurement of the reaction rate. Hence, the obvious fundamental requirement for a secondary gage is the possibility of direct and convenient measurement of the chosen parameter.

The need for calibration stipulates the second requirement, which is the ability to obey the observed relationship during a sufficiently long period. The long-term stability of gage readings makes the interval between two consecutive calibrations longer and increases the reliability of pressure measurements. The short-term stability or reproducibility of readings and freedom from hysteresis are also essential. To some extent the gage stability may be judged by the zero stability, that is the reproducibility of gage readings at $p = 0$, p here being interpreted as the pressure above atmospheric pressure.

The next important property of an acceptable secondary gage is the relatively weak dependence of the measured parameter on other environmental variables, in particular, the temperature T. With all other characteristics equally demanding, there should be a preference for the gage with which the ratio of the temperature coefficient to the pressure coefficient $(\partial x/\partial T)/(\partial x/\partial p)$ is least.

The important characteristic of the secondary pressure scale or interpolation equation, as it is sometimes called, is its linearity. This facilitates pressure measurement and the application of automatic pressure control systems in which a secondary gage is used as a sensing element.

When pressures over 15 kb are to be measured, a small size of gage becomes desirable since the working volume of a high pressure cell is limited.

Apart from the general requirements mentioned above there are some special ones due to specific experimental conditions. For example, in studies of transient pressures, adequate high frequency response, minimum distortion of the pressure field and ruggedness become essential for the construction of a satisfactory gage. And for characterization of a high pressure–high temperature medium, thermally stable gages are needed.

The measurement of transient pressures lies mostly outside the scope of this book. Here we concentrate primarily on measurements of stationary and slowly changing pressures above 10 b. The numerous gages and methods used in industry for pressure measurements and controls are also omitted. Only accurate well-studied pressure gages used in a research laboratory are discussed.

II. Mechanical Devices

In all instruments of mechanical type the pressure is determined by measuring the deformation of a certain elastic element. This deformation is measured either directly as in a Bourdon gage or by means of various electromechanical transducers.

Bourdon gages have found a very extensive application in pressure measurements[3, 14, 25]. At present such gages are manufactured for all pressure ranges up to 12 kb and have accuracies from 0.2 to 5 per cent of the full scale reading.

The principal element of the gage is made of a steel tube with an elliptical or oval cross section, bent in the arc of a circle (Bourdon tube). One end of the tube is closed. Through a connector on the other end of the tube the pressure to be measured is applied. Under pressure the cross section of the tube distorts becoming more circular and the stresses generated force the tube partially to straighten out. The tube tip travel is converted by a gear-and-pinion arrangement to give a pointer rotation. At higher pressures thick-walled Bourdon tubes are used in which the tip movement is reduced. In such a case it is possible to increase the sensitivity of the gage without sacrificing the strength by bending the tube in the form of a spiral or helix. Due to inelastic effects and deviations from Hooke's law which occur in steels, Bourdon gages are liable to give rather inaccurate readings in the higher pressure ranges. Accordingly, high precision Bourdon gages giving accuracies of a few tenths of one per cent of the full scale reading need the utmost care in operation and frequent recalibration. It is not recommended that such gages be used at pressures exceeding half to three quarters of their maximum range. After accidental overload, the calibration of the gage should be repeated.

To reduce the hazards of tube rupture as a result of severe overloads the gage is usually provided with a rugged front cover and a thin back cover which can fly off readily under pressure.

A novel version of a Bourdon tube was suggested[2, 21], to increase the upper limit of the gage. In this new type the internal bore is shifted eccentrically relative to the neutral axis of the tube. When subjected to pressure, the tube tends to straighten out as a result of the bending moment developed in the plane of the cross section. Gages incorporating the new tube are now commercially available with a maximum range reaching 16 kb.

A straight tube with an eccentric bore may also serve as a secondary gage[13]. Under pressure the tube bends in the direction of the thicker wall, the tip movement being measured by a dial gage or any other displacement transducer.

There is a great variety of mechanical gages in which a membrane is used as an elastic element and the membrane deformation is converted into an electrical signal. Most of these transducers measure rapidly changing pressures. Their upper limit does not usually exceed a few hundred bars.

Strain gage transducers are applied in measurements of high stationary pressures. These gages are based on the variation in the electrical resistance of the wire with strain in the wire. The electrical resistance of the strain gage cemented to the membrane may be calibrated to indicate the pressure acting on the membrane. Depending on the conditions of measurement and the type of membrane, the size and number of strain gages may vary. Temperature compensation may be arranged easily. Gages of this type are comparatively simple, inexpensive and convenient to work with. Limitations on the range and accuracy of measurements are the same as in Bourdon gages, since the membrane behavior is not free from the inelastic effects mentioned above.

The simplest way to estimate pressure inside a pressure cylinder would be to measure the strain of the cylinder walls on the outside. To do this one might cement strain gages to the exposed surface or wind the wire around the cylinder itself. These measurements are far from being accurate, for the actual response of the gage depends markedly on the site at which it is fixed. Moreover, in compound or supported cylinders the strain may not vary directly with internal pressure. The method, therefore, has too many disadvantages and may be applied only for the rough estimations of pressure.

III. Electrical Resistance Gages

Very early on, it was suggested[16] that pressure should be measured in terms of the electrical resistance of a wire coil located inside a pressure vessel. The pressure coefficient of resistance $(1/R_0)(\partial R/\partial P)_T$ of most metals is of the order of 10^{-5} to 10^{-6} bar^{-1}. Unfortunately the ratio of the temperature coefficient to the pressure coefficient is equal to about 10^3 bar °C^{-1} for most metals which is far too high a value. This means that the gage in which pure metal wire is used should be very carefully thermostatically controlled, since temperature variations as small as ± 0.1°C shift gage readings by up to 100 bar. Besides, an excessive temperature coefficient defers pressure indication until the adiabatic heat of compression resulting from the application of pressure has been completely dissipated. This of course reduces the applicability of such gages for the measurement of non-stationary pressures. Another drawback of pure metals is their low electrical resistivity. Since for the sake of convenience a gage should have a resistance of at least a few ohms, the coils of pure metal wire become too large to be acceptable for pressure measurements. And gages made of a very fine wire are too fragile to withstand sudden pressure changes.

Manganin wire gages are practically free from the drawbacks mentioned above. Manganin is an alloy of copper (83 to 86.5 per cent), manganese (11 to 13.5 per cent) and nickel (2.5 to 3.5 per cent) which is commercially available and is widely used for the manufacture of stable high-precision resistors, standard coils etc. The pressure coefficient of a coil of manganin wire varies from 2.1 to 2.5×10^{-6} bar^{-1} for various brands. The resistance versus temperature curve of most brands of manganin reaches a maximum value near room temperature. Accordingly, the temperature coefficient of resistance is very small in this range ('the zero temperature coefficient range'). The ratio of the temperature coefficient to pressure coefficient is on the average about 4 bar °C^{-1} at room temperature. It is possible to displace the zero temperature coefficient range to higher temperatures through appropriate doping and adjustment of the manganin composition. Due to the high electrical resistivity of manganin (about 25 times that of copper) a manganin gage can be constructed which is compact enough to be accommodated in a very small volume.

The performance of manganin gages depends not only on the composition of the alloy but also on the ways in which the wire is manufactured and the coil is wound and seasoned, as well as on many other factors. There is unanimous agreement that gage coils must be wound freely on the holder. Probably the best configuration is the spiral-wound coil mounted in grooves on a ceramic holder which is fixed to the closure plug[20]. The free access of pressure-transmitting fluid to the wire results in better linearity and stability of the gage.

During the process of winding the coil the electrical resistance of the wire increases by up to a few per cent due to the generation of considerable mechanical stresses. The decay of these stresses may continue for years and is accompanied by the persistent decrease of the electrical resistance. To accelerate seasoning and to insure the stability of the zero-pressure resistance R_0 the coils must be annealed. The treatment used by Bridgman[9] required 48 hours at 140°C. This was the highest temperature which the silk insulation of the wire would stand. Further raising of the temperature as shown in later experiments may lead to some unfavorable effects, such as the oxidation of manganese resulting in higher copper concentration and a dramatic increase of the temperature coefficient $(1/R_0)(\partial R/\partial T)_{p=0}$. Besides it has been noted[22] that prolonged heating may considerably shift the zero temperature coefficient range toward higher temperatures. Satisfactory seasoning is accomplished also with high current techniques when the electrical energy stored in a capacitor bank is discharged through a resistance coil. By variation of the charging voltage, one may raise the temperature of the coil to 500°C for a few milliseconds.

In his work at 30 kbar Bridgman[10] modified the seasoning technique by introducing periodic cooling of the coils in dry ice. At the present time it is easy to increase further the temperature difference of seasoning by using liquid nitrogen (-196°C)[5].

It should be noted that the requirements for the free suspension of the wire and the temperature seasoning of the coils are similar to those encountered in the manufacture of platinum resistance thermometers.

When the seasoning is completed, that is the electrical resistance of the coil is stabilized, the ends of the wire are soldered to the insulated leads of the closure plug. The insulation resistance of the leads should be high enough to eliminate the appreciable shunting of the coil resistance. Usually, 10^8 to 10^9 ohms resistance is sufficient. Such insulation can be attained easily when materials like mica or ferrous oxide are used. Repeated applications of pressure weaken the soldered joint, making it brittle and unreliable. Better performance is obtained when short pieces of copper wire welded to the ends of manganin wire are used for making a soldered joint. Often only one lead is used, the other end being soldered to the body of the plug. In high precision experiments when coil resistance increments under pressure are measured by means of a potentiometer, four leads should be provided.

Excellent results are obtained when pressure seasoning of gages is carried out. In this treatment a preliminary application of pressure exceeding the maximum pressure of the working range of the gage is applied to the coil[4].

Careful calibration of manganin gages against a free piston gage has made it possible to establish that the pressure dependence of the electrical resistance is not as linear as was assumed earlier. *Figure 1* shows the typical

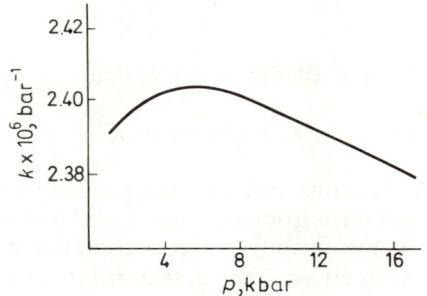

Figure 1. A typical k/p plot of a high-pressure manganin gage.

dependence on pressure of the pressure coefficient k derived from calibration data. We define the coefficient k here as $(1/R_0)(\Delta R/p)$, where ΔR is the resistance increment of the coil at pressure p and R_0 is the zero-pressure resistance. At low pressures the coefficient k increases with pressure. This portion of the curve probably characterizes how freely the wire is suspended from its holder and is not representative of the gage behavior at higher pressures. Indeed, when a wire is tightly wound on the holder or the gage is covered with some kind of lacquer, the maximum in the k/p curve is shifted to higher pressures.

Apparently the choice of the interpolation equation for a manganin gage depends on what pressure range it is supposed to work in. For lower pressures up to 5 or 6 kbar, one may fit calibration data[15] with a quadratic equation of the type $\Delta R = Ap + Bp^2$. On the other hand, when higher pressures are measured it is more satisfactory to use the interpolation formula $p = \alpha' \Delta R + \beta'(\Delta R)^2$. The difference between experimental values and those obtained from the equation does not exceed 0.1 per cent if the coefficients α' and β' are calculated by the least squares method[5]. The variation of these coefficients for gages manufactured from the same spool of manganin wire has been investigated[4].

Some 24 identical 40-ohm gages were manufactured from the same sample of 0.08 mm wire and seasoned by the modified Bridgman method and then calibration against a 15 kbar free piston gage. The interpolation equation normalized to the zero-pressure resistance was

$$p = \alpha(\Delta R/R_0) + \beta(\Delta R/R_0)^2$$

and the average values of α and β for the group were found to be:

$$\alpha = 406\,900 \pm 2\,240 \text{ bar}$$

and

$$\beta = 276\,000 \pm 83\,200 \text{ bar}$$

The values 2 240 and 83 200 bar are standard deviations of the coefficients given by the least squares treatment.

A further 22 identical 100-ohm gages manufactured from the same brand of 0.1 mm manganin wire were also studied. The scatter of α and β values turned out to be smaller:

$$\alpha = 407\,870 \pm 1\,540 \text{ bar}$$

and

$$\beta = 149\,500 \pm 40\,200 \text{ bar.}$$

The smaller β value testifies to the better linearity of 100-ohm gages compared with 40-ohm ones.

From these results conclusions useful in practice may be drawn. Indeed, one may calibrate just one gage from a group of identical gages and use its interpolation equation for any uncalibrated gage from that group to measure pressures up to 10 kbar with an accuracy of two or three per cent.

The calibrated manganin gage provides more accurate measurements of high pressures. Normally deviations from the average resistance increment observed in a series of calibrations of the gage including short-term zero wanderings amount to 0.003 ohm for a 100-ohm coil, which is equivalent to 12 bar uncertainty. Since our 15 kbar free piston gage has an accuracy better than 0·2 per cent, the absolute error of the calibrated manganin gage is estimated to be equal to about $(0.002p + 12)$ bar. To insure such high accuracy it is necessary to carry out periodic check calibrations of the gage.

One should bear in mind that the working temperature range of manganin gages does not usually exceed 70°C. When pressure is applied at higher temperatures, a gage undergoes non-reversible changes and its recalibration becomes necessary since the pressure coefficient k of the gage decreases in an unpredictable manner. The best results are obtained when the gage is kept inside a pressure vessel at the temperature of calibration.

When the expensive free piston gage is not available, fixed points (see Chapter 4, Part 2) on the pressure scale may be used for the calibration of secondary gages. The freezing of mercury at 0°C occurring at 7 569 bar[19]

PART 4. SECONDARY GAGE—DIFFERENTIAL MANOMETERS

provides a very convenient calibration point. First, let us estimate the accuracy of the manganin gage calibrated at this fixed point and linearly extrapolated to higher pressures. For the average 40-ohm gage cited earlier the relative resistance increment $(\Delta R/R_0)_M$ at the mercury freezing pressure at 0°C would be 18.37×10^{-3} and for the 100-ohm gage 18.43×10^{-3}[†]. The slopes κ of the linear interpolation equation would be 411 970 and 410 630 bar respectively. From the difference between quadratic and linear interpolation equations (see *Figure 2*) one may estimate errors involved in pressure measurements using the linear equation:

$$\Delta p = 2755 (\Delta R/R_0)^2 - 5040 (\Delta R/R_0) \text{ for 40-ohm gages}$$

$$\Delta p = 1495 (\Delta R/R_0)^2 - 2760 (\Delta R/R_0) \text{ for 100-ohm gages}$$

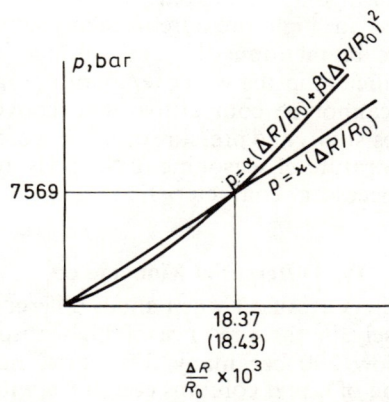

Figure 2. Comparison of linear and quadratic interpolation equations.

Table 1

p, kbar		2	4	6	8	10	15	20
$\Delta p/p$	40-ohm gage	−0.9	−0.6	−0.25	0.07	0.40	1.20	2.0
%	100-ohm gage	−0.5	−0.3	−0.15	0.04	0.20	0.65	1.1

Table 1 shows the probable errors of pressure measurements up to 20 kbar with manganin gages calibrated at one fixed point. When two fixed points are available, good results are obtained with the quadratic interpolation equation $p = \alpha(\Delta R/R_0) + \beta(\Delta R/R_0)^2$. The coefficients are easily found from the following expressions:

$$\alpha = \frac{p_1(\Delta R_2)^2 - p_2(\Delta R_1)^2}{\Delta R_1 \times \Delta R_2 (\Delta R_2 - \Delta R_1)} \times R_0$$

[†] The $(\Delta R/R_0)$ values are found from solution of the equation $\beta(\Delta R/R_0)^2 + \alpha(\Delta R/R_0) - 7569 = 0$ where α and β are average coefficients for 40-ohm and 100-ohm gages respectively.

$$\beta = \frac{p_2 \Delta R_1 - p_1 \Delta R_2}{\Delta R_1 \times \Delta R_2 (\Delta R_2 - \Delta R_1)} \times R_0^2$$

Here ΔR_1 and ΔR_2 are the resistance increments of the gage at the fixed point pressures p_1 and p_2.

At high pressures the most widely used fixed points are the freezing of mercury at 0°C and the Bi(I)–Bi(II) polymorphic transition occurring at at 25.50 kbar[19]. The Bi(I)–Bi(II) fixed point may be substituted with the pressure of mercury freezing at 75°C. According to careful studies described in ref. 8 this pressure is estimated to be 22.7 ± 0.1 kbar.

Since the second-degree term β is small the two-point calibration of the gage provides the measurement accuracy to better than one per cent throughout the whole pressure range up to 30 kbar.

Of other alloys used in high pressure measurements gold–chromium (2.1 per cent Cr) should be mentioned[11]. The only merit of this alloy is its lower temperature coefficient in the wider temperature range. However, the gold–chromium gage cannot be competitive with conventional manganin gages due to the lower resistivity and pressure coefficient of the gold–chromium wire. Besides, the temperature of seasoning of the coils must be strictly controlled to provide the necessary characteristics.

IV. Differential Manometers

Differential manometers are used when small differences of pressure are to be measured precisely in the presence of high absolute pressure. For absolute pressures below 200 bar many differential manometers involve direct visual observation of liquid columns contained within glass tubes[7,18]. At higher pressures various mechanical devices such as bellows-type gages or strain-indicating instruments are adapted to differential measurements. In fact, a great number of differential manometers incorporate some kind of diaphragm while displacements are measured by means of various transducers[12,17,23,24].

At still higher pressures a differential manganin manometer satisfies almost every possible requirement for precise measurements of small pressure differences. It consists of two identical manganin gages located in two different parts of a pressure-measuring apparatus[6]. The pressure difference between these two parts is determined by means of a four-arm resistance bridge, manganin gages being switched into adjacent arms. When all arm resistances and both gage initial resistances are equal ($R_{01} = R_{02} = R_0$), the measured pressure difference is $\Delta p = \kappa(\Delta R/R_0)$, where ΔR is the increment of the adjustable arm resistance needed to balance the bridge. In practice zero-pressure resistances are usually slightly different ($R_{01} \neq R_{02}$) and such differential manometers need a preliminary calibration in which the bridge indications are registered as a function of pressure applied to both gages. It has been reported[26] that a differential manganin manometer is capable of measuring a fraction of a bar pressure difference at an absolute pressure of kbar. The accuracy of measurement depends on the pressure difference and is estimated to be one or two per cent for pressure differences above 5 bar at absolute pressures over 2 kbar. No appreciable difficulties were encountered in our laboratory in extending differential manometers to the highest hydrostatic pressures now in use.

V. References

[1] Alexeyev, K. A., Y. A. Atanov and L. L. Burova. *Proceedings* of the Institutes of the USSR Standards Committee, **75**, 44 (1964).
[2] Andrews, L. E. *Elastic Elements of Instruments* (1966).
[3] ASME Publications: High Pressure Measurements. *Paper N 53-YKD-I* (1953).
[4] Atanov, Y. A. and E. M. Ivanova. *Measurement Technique*, **14**, 247 (1971).
[5] Atanov, Y. A. and E. M. Ivanova. *Spec. Publ. US Nat. Bur. Stand. No. 326*, 49 (1971).
[6] Bakhvalova, V. V. and M. K. Zhokhovsky. *Proceedings* of the Institute of the USSR Standards Committee, **75**, 55 (1964).
[7] Biles, M. B. *Instruments*, **24**, 159 (1952).
[8] Bogdanov, V. S., Y. L. Levin, S. S. Sekoyan and Y. I. Shimin. *Spec. Publ. US Nat. Bur. Stand. No. 326*, 297 (1971).
[9] Bridgman, P. W. *Proc. Amer. Acad. Arts. Sci.* **47**, 321 (1912).
[10] Bridgman, P. W. *Proc. Amer. Acad. Arts Sci.* **72**, 157 (1939).
[11] Darling, H. E. and D. H. Newhall. *Trans. Amer. Soc. Mech. Engrs*, **75**, 311 (1953).
[12] Ewing, C. T., J. R. Spann, J. P. Stone, E. W. Steinkuller and R. R. Miller. *J. Chem. Engng Data*, **15**, 508 (1970).
[13] Gielessen, J. *Z. Angew. Phys.* **8**, 193 (1956).
[14] Kardos, G. *J. Basic Engng*, **81**, 645 (1959).
[15] Lippmann, H. and M. Richard. *Feingerätechnik*, **19**, 368 (1970).
[16] Lisell, E. *Uppsala Univ. Årsskrift*, **1**, 1 (1903).
[17] Malbrunot, P. F., P. A. Meunier, G. M. Scatena, W. H. Mears, K. P. Murphy and J. V. Sinka. *J. Chem. Engng Data*, **13**, 16 (1968).
[18] Maslach, G. J. *Rev. Sci. Instrum.* **23**, 367 (1952).
[19] *Spec. Publ. US Nat. Bur. Stand. No. 326*, 314 (1971).
[20] *Tech. News. Bull., US Nat. Nat. Bur. Stand.* **40**, 96 (1956).
[21] Nogatkin, A. G. *Priborostroyenie (USSR)*, **5**, 13 (1956).
[22] Schulze, A. *Z. Metallkunde*, **32**, 317 (1940).
[23] Waxman, M. and T. Chen. *J. Res. Nat. Bur. Stand.* **69c**, 27 (1967).
[24] Waxman, M. and J. R. Hastings. *J. Res. Nat. Bur. Stand.* **75c**, 165 (1971).
[25] Zhokhovsky, M. K. *Pressure and Vacuum*. Mashgiz: Moscow (1952).
[26] Zhokhovsky, M. K. *Izmeritelnaya Tekhnika*, **7**, 11 (1959).

CHAPTER 4

Part 5. Instruments for Relative Pressure Measurements

R. LECLERCQ

Jaeger Co., 2 rue Baudin, 92-Levallois-Perret, France

Contents

I.	General Considerations	213
II.	Pressure Sensitive Elements; Sensors	214
	1. Aneroid Capsule	214
	A. The Restrained Capsule	214
	B. The Free Capsule	214
	2. Bellows and Springs	215
	3. Bourdon Tube	215
	4. Gimlet Shaped Tube	216
	5. Piezoelectric Quartz	216
III.	Methods of Measurement	217
	1. Force Measurements. The Pressure Effect	217
	2. Displacement Measurements. The Pressure Effect	218
	A. Mean Amplitude Displacements	218
	(1) The Potentiometer	218
	(2) Synchro-detection	218
	(3) The Variable Capacitor	219
	(4) The Differential Coil Transformer	220
	(5) Direct Reading	220
	B. Small Amplitude Displacements	220
	(1) The Differential Transformer	220
	(2) Variation of Self-inductance	221
	(3) Strain Gages	222
IV.	Special Transducers	222
V.	Specific Designs	222
	1. General Considerations	222
	2. Flight-control Instruments	223
	A. Altitude Measurement: the Altimeter	223
	B. Speed Measurement: the Airspeed Indicator and the Mach Meter	224
	Special Devices: Airspeed and Mach Number Indicators	225
	C. Examples of Applications	225
	The Altimeter	225
	The Airspeed and Mach number indicator	227

I. General Considerations

Secondary instruments used in pressure measurements are mainly transducers converting pressure into a suitable parameter such as the movement of a pointer over a dial, an electrical signal, etc. In common instruments, pressure measurements are reduced either to a measurement of forces acting on well-defined surfaces or to controlled device displacements which constitute a sensor.

These forces and displacements are converted into motions by mechanical or optical methods or into electrical signals by systems called transmitters. Both sensor and transmitter constitute a pressure transducer. The different ways of combining sensor and transmitter give rise to a very wide diversity of transducers. The instruments, which will be discussed later, are among the most commonly used. Compared with the number of transmitters, the number of sensors remains very small, since transmitter evolution is very closely connected with improvements in the electronic field.

In order to cover the greatest number of transducers, the following text is divided into three main parts, outlining the principal types of sensors and transmitters, and giving several examples of common practical applications.

II. Pressure Sensitive Elements; Sensors

1. *Aneroid Capsule* (*Figure 1*)

This system is used for either absolute or differential measurements, generally in the range between 0 and 10 bar. Depending on which type of transmitter is connected, the precision varies from 5×10^{-4} to 5×10^{-2} of full scale deflection. In absolute pressure determinations (input p_1), there is no tube p_2 and the pressure inside the capsule is zero.

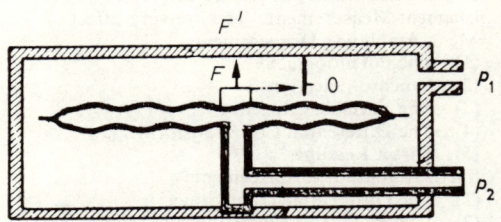

Figure 1. The aneroid box.

In differential pressure measurements, i.e. of p_1-p_2, the connection is made as shown in *Figure 1* and the force F is produced by the pressure difference $p_1 - p_2$.

There are two ways of using this force F.

A. THE RESTRAINED CAPSULE

The force F' created by an electrical system (coil and magnet for example) or by a mechanical system (a spring for example) balances F in such a way that no displacements occur on the sides of the capsule. In this case $F = F'$ and F' is measured indirectly in the transmitter part of the transducer.

B. THE FREE CAPSULE

Under the action of the pressure difference $p_1 - p_2$ the deformation of the capsule occurs and a balance position is reached due to the restoring elastic

PART 5. INSTRUMENTS FOR RELATIVE PRESSURE MEASUREMENTS

forces of the capsule. For a given pressure difference, $p_1 - p_2$ corresponds to a definite displacement of the sides of the capsule.

On each side a pin is attached. One of these is held to the frame, the other can move and control an element which measures the displacement, constituting the transmitter.

2. *Bellows and Springs* (*Figure 2*)

Generally these devices are used in a pressure range from 0 to 10 bars and constitute simpler instruments than those having capsules. Their accuracy ranges from one to five per cent of full scale deflection.

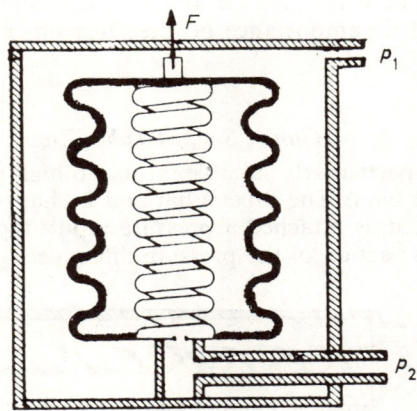

Figure 2. The bellows.

Here the principle is the same as in a capsule but the method of operation is different. The elastic force which balances the pressure force is produced essentially by the spring, the bellows providing the tightness and allowing change of pressure to be converted into a force.

In these devices either the force F or the bellows displacement is measured. The more usual method involves measuring the relative shift of the end faces of the bellows.

3. *Bourdon Tube* (*Figure 3*)

This device is very widely used. The system makes it possible to cover a large range of pressures up to several hundred bars depending on the nature of the working materials (metals, quartz) and on the dimensional characteristics of the tube.

The flattened tube is bent. One end is fixed and is fitted with a pressure input and the other (free) end is closed. Inside the tube the pressure is p_2, outside it is p_1. The action of the pressure difference $p_2 - p_1$ gives rise to a deformation. The displacement of the free end is converted, through a transmitter, into an electrical signal.

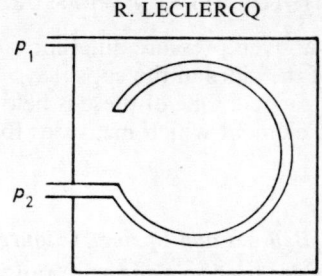

Figure 3. The curved tube.

With laboratory instruments the accuracy can reach 10^{-4} of the full scale deflection. With industrial devices the accuracy is generally of the order of 10^{-3}. In view of the importance of these instruments, they will be further described below.

4. Gimlet Shaped Tube (*Figure 4*)

This device is particularly suitable for use in high pressure measurements (several hundred bars). The tube is flat and is shaped like a gimlet. At one end, which is fixed, is attached a pressure input; the other end is free and closed. Under the action of the pressure difference, $p_2 - p_1$, the tube tends

Figure 4. The gimlet shaped tube.

to straighten out, producing a small rotation around its longitudinal axis. This rotation is measured with a transmitter sensitive to angular displacement.

5. Piezoelectric Quartz (*Figure 5*)

A quartz disc cut off along a direction perpendicular to its electrical axes is compressed under a pressure p. The resultant force can be written $Q = KF$ where K is the Curie constant of the quartz, and Q is the electric charge between the quartz plates.

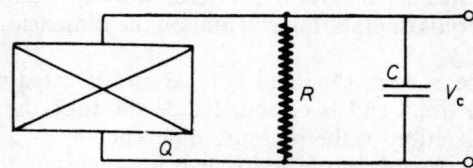

Figure 5. The piezoelectric quartz.

PART 5. INSTRUMENTS FOR RELATIVE PRESSURE MEASUREMENTS

The measurement of the charge is made between the terminals of a capacitor, C, connected in parallel to the two ends or directly between the terminals of the quartz plates, these latter being considered to be a capacitor. Then the voltage is given by $V = Q/C = (K/C)F$.

This voltage may be measured with an electrometer system having a very high input impedance.

The value of R, the internal resistance of the quartz, and the input impedance of the electrometer are not, however, infinite. Let us write

$$V_c = V_0 \exp(-t/RC)$$

or

$$V_c/V_0 = \exp(-t/RC)$$

III. Methods of Measurement

1. *Force Measurements. The Pressure Effect* (*Figure 6*)

These measurements are made with almost no displacement of the pressure sensor.

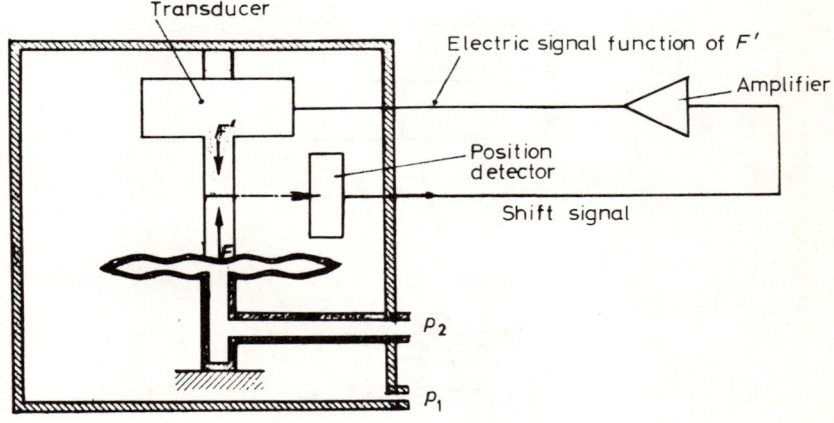

Figure 6. Force measurements.

The force F produced by the sensor under the influence of pressure is balanced by a force F'. A position-sensitive detector indicates each displacement of the movable part of the gage (pin attached to diaphragm of capsule, Bourdon tube end, etc...).

For any variation in pressure, and hence of the force F, there is a corresponding displacement. The signal measuring this displacement acts on F', tending to restore the equilibrium $F = F'$. Essentially two methods are used:

(a) The balancing force F' is produced by an accurate electromagnet. The force F' is proportional to the current I circulating through the winding, so that $F' = kI$, and I is then proportional to $p_1 - p_2$.

(b) The balancing force F' is produced by a compressed spring. The force produced by the spring is a function of the strain generated by a device with an endless screw. The strain in the spring depends upon the rotation of the endless screw. An electric transmitter measures this rotation, which thus depends on $p_1 - p_2$.

2. Displacement Measurements. The Pressure Effect

Among the very numerous ways known for measuring the displacements, the following systems are the most widely used:

A. Mean Amplitude Displacements

Examples: capsules, Bourdon tubes.

(1) The Potentiometer

The sliding contact is connected to the moving device (*Figure 7*), and the quantity measured is $\Delta R/R$, where R is the total resistance of the potentiometer. Generally, the potentiometer is connected across terminals 1 and 2 with a very stable supply voltage, and the voltage measured between 1 and 3 depends on the displacement of the sliding contact.

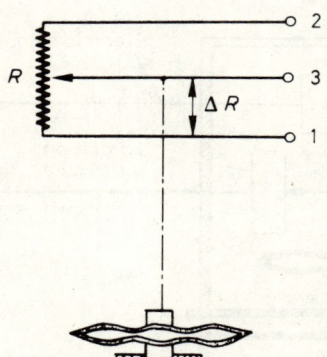

Figure 7. The potentiometer.

The disadvantage of using the potentiometer resides in the friction existing in the sliding motion of the contact, bringing a non-negligible disturbance to the mobility of the whole system.

(2) Synchro-detection (Figure 8)

Transmission systems of the 'synchro' type are based on the angular position measurements of a rotor attached to the moving device (capsule). Through the three windings of S_1 the rotor R_1 induces alternating magnetic fields whose components depend upon the position of R_1 with respect to the fixed windings of S_1. In S_2 the same field conditions are applied and their resultant with respect to S_2 has the same position as R_1 with respect to S_1. The induced

PART 5. INSTRUMENTS FOR RELATIVE PRESSURE MEASUREMENTS

voltages in R_2 reach a maximum when R_2 and the induced vector are parallel, and are zero when they are perpendicular.

The voltage provided by R_2 is applied to an amplifier A whose output current controls the motor M, driving R_2 in such a direction that its voltage becomes zero. The system then remains stable. The position of R_2 is firmly

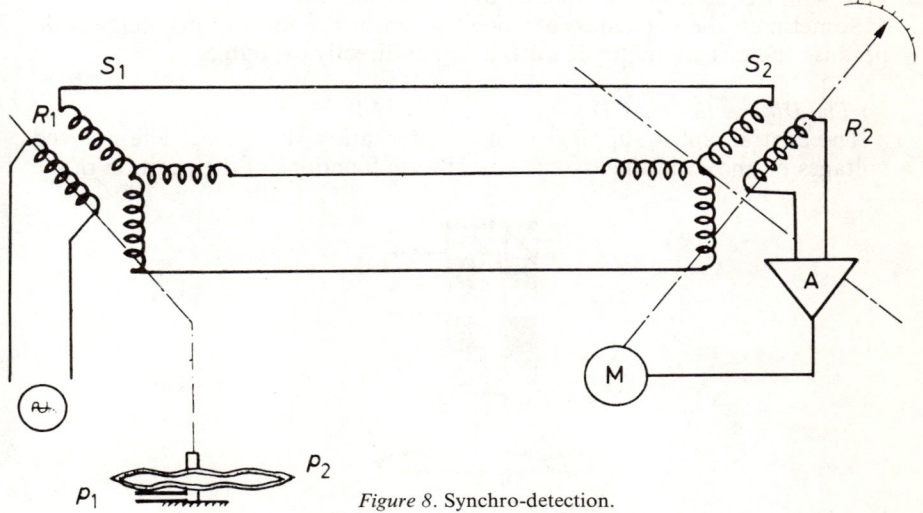

Figure 8. Synchro-detection.

related to that of R_1 and thus depends on the pressure difference $p_1 - p_2$.

This type of drive can be used to control complicated mechanisms. The disturbance in the motion due to the synchrotransmitter may be very small. The precision in angular displacement of such a system can approach 10^{-4} of one complete revolution of the transmitter. This method is frequently used in aircraft.

(3) *The Variable Capacitor (Figure 9)*

An adjustable capacitor can be connected to the moving-part of the sensor.

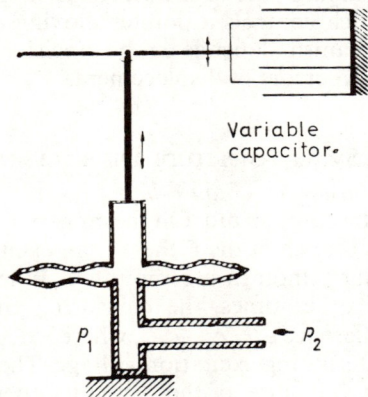

Figure 9. The variable capacitor.

219

Sometimes, the moving plates are attached directly to the sensor, and the variable capacitor is used to modify the frequency of an oscillator.

Pressure measurement is thus reduced to frequency measurement, and the stability of the oscillator is now a parameter which depends directly on measurement precision. When the precision has to be improved, systems are used with frequency ratios related to a reference capacitor.

Sometimes, the capacitance ratio is given in the form of a voltage ratio because there is an analogue output signal directly available.

(4) The Differential Coil Transformer (Figure 10)

The central coil is supplied with an alternating voltage e_0. The induced voltages e_1 and e_2 in the associated coils are functions of the position of the

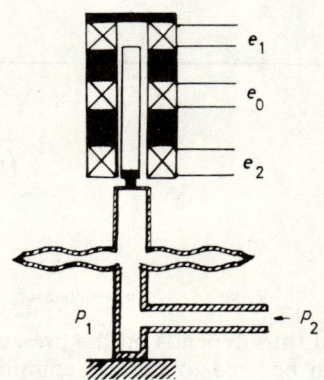

Figure 10. The differential coil transformer.

core and vary in opposite directions. The working signal is $e = e_1 - e_2$, and can be considered proportional to the displacement, in the normal range.

(5) Direct Reading (Figure 11)

In this system the moving part of the sensor is connected to a movement multiplying device which controls a pointer moving over a calibrated dial. The quality of the mechanism depends on the precision required. The driving torque can contribute to errors of displacement.

B. SMALL AMPLITUDE DISPLACEMENTS

(1) The Differential Transformer (Figure 12)

An I-shaped magnetic core can move in the air gap of an E-shaped magnetic circuit. On the central branch of the E there is an excitation coil. On each of two outer arms, similar windings are mounted. When the core is located centrally, $e_1 = e_2$. At other times, the measuring circuits detect a signal proportional to the difference $e = e_1 - e_2$, where e depends on the displacement and on e_0, the alternating excitation voltage. The common frequencies are 50 Hz and 400 Hz, but some of these circuits operate at a frequency as high as 10 kHz.

PART 5. INSTRUMENTS FOR RELATIVE PRESSURE MEASUREMENTS

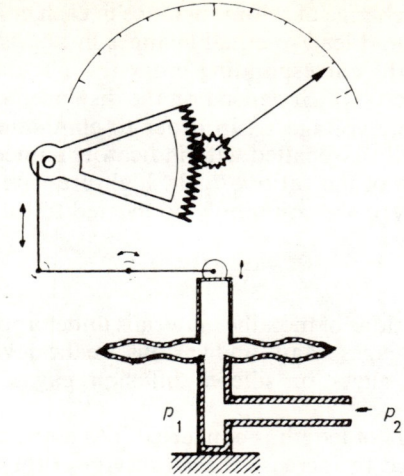

Figure 11. Direct reading.

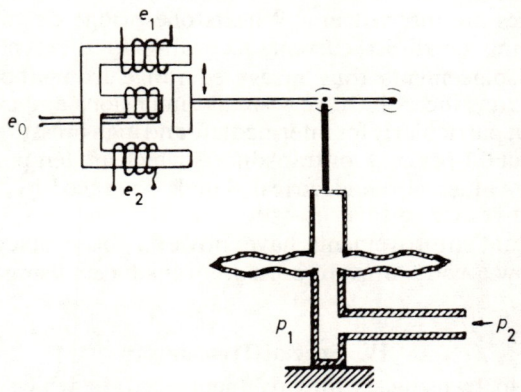

Figure 12. The differential transformer.

(2) *Variation of Self-inductance (Figure 13)*

The change of self-inductance is achieved with the variation of the air gap of two magnetic coils symmetrically connected. The pressure-sensitive device has a small displacement, so that with change of pressure there is an increase in the air gap of one coil and a reduction in that of the other. These

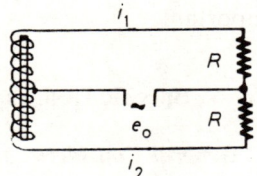

Figure 13. Variation of self-inductance.

variations produce a change of self-inductance in each coil, and therefore also of their impedances, and lead to equal changes in opposite directions of the currents i_1 and i_2 in the corresponding branches. The operating parameters are the currents i_1 and i_2 which depend on the displacement of the core in the gap and on the supply voltage e_0. In order to eliminate this last effect, the transmitters are usually associated with indicators named logometers whose position is a function of the ratio i_1/i_2. In high-pressure measurements, the transmitters of this type are commonly connected to tube gages of the kind shown in *Figure 4*.

(3) Strain Gages

The small deformations of metallic materials under applied pressure can be measured with strain gages. Many and various are the devices used, including: (i) resistance-related gages or silicon diffusion gages, (ii) strain threads, (iii) deposited films.

In each case, change of length is converted into a resistance change. Many combinations are used to increase the sensitivity of the transducers. One is obtained by combining several gages in such a way that according to the electrical connections used, the sensitivities will be additive and the secondary effects (temperature, for example) will cancel each other. Generally, strain threads or gages are inserted in a Wheatstone bridge circuit supplied with either alternating or direct current according to the type of measuring device. The displacements thus measured range from about one micron upwards. The transducers have very small dimensions and can be used over a large range but particularly for intermediate and high pressures. The accuracy attained is about 0.1 per cent for transducers which are temperature-compensated. The low values of the electrical signals produced by these indicators have somewhat restricted their growth.

Very important improvements have, however, been made to this aspect and there is now a wide range of pressure transducers using these methods.

IV. Special Transducers

In addition to transducers already mentioned, in which the pressure is measured through a sensitive metallic device, there are also other devices which respond in a complex manner to changes of pressure. For example, the 'piezo-transistors', made by Pitran, have electrical characteristics which depend on the pressure applied to a diaphragm connected to a suitably earthed emitter junction. Under certain conditions of supply and use, the output voltage is proportional to the pressure. This device is very sensitive to the differential pressure between the sides of the diaphragm. The sizes are small and the precision generally about one per cent. The sensitivity to temperature is relatively important.

V. Specific Designs

1. General Considerations

These specific devices are quoted as examples. They have many uses in

PART 5. INSTRUMENTS FOR RELATIVE PRESSURE MEASUREMENTS

aeronautics where the precision is very important. They can be classified into two distinct groups according to use:

(i) Flight control instruments whose main unit concerns altitude, speed, Mach number, climbing speed;

(ii) Instruments for engine control and for control of hydraulic and pneumatic accessories.

The technologies used are different for the two groups. The first set of instruments must above all be very accurate, that is to say, their precision must be better than 0.1 per cent of full scale deflection. Their structure is delicate but they must withstand all conditions likely to occur in an aircraft, where temperatures may vary from $-40°$ to $+70°C$ and there is much continuous vibration.

Primarily, the second set of instruments must be strongly built and very simple, only being used for general control purposes. Their precision varies from one to five per cent of full scale deflection. The environmental conditions of use are often very severe, from both temperature and vibration points of view because they may well be located in very critical places in the aircraft (for instance: engine bay, wings).

2. Flight-control Instruments

A. Altitude Measurement: the Altimeter

The altitude of an aircraft is the vertical height between its position and the position of a real or conventional reference level (ground, sea or arbitrary mark). Measurement of the static (atmospheric) pressure facilitates evaluation of the altitude if the relation between pressure and altitude is defined. This pressure/altitude relationship has been defined conventionally; it is called a 'standard atmosphere' and is used as an international reference. This conventional reference makes it possible to calibrate altimeters. The value of the altitude-pressure, as given by an altimeter, can change slightly from day to day, at any one point having a given height, according to atmospheric conditions. However, for this same point, the altimeters in an aircraft must indicate the same value. The following conventions are used.

In the standard atmosphere the values of a certain number of parameters depending on the altitude are defined (static pressure, temperature, air density, etc.)

The pressure parameter in which we are interested varies according to the law:

$$p_S = 1013.25\,(1 - 22.5569 \times 10^{-6}\,Z)^{5.25611} \text{ for } Z < 11\,000\,\text{m}$$

$$p_S = 226.32\,\exp(-157.6883 \times \Delta Z \times 10^{-6}) \text{ for } Z > 11\,000\,\text{m}$$

with $\Delta Z = Z - 11\,000$. p_S is measured in mbar and Z in meters. The following corresponding values are obtained:

$$Z = \quad 0\,\text{m for } p_S = 1013.25\,\text{mbar}$$
$$Z = 1000\,\text{m for } p_S = 898.75\,\text{mbar}$$

$$Z = 5000 \text{ m for } p_S = 540.21 \text{ mbar}$$
$$Z = 10\,000 \text{ m for } p_S = 264\cdot36 \text{ mbar}$$
$$Z = 15\,000 \text{ m for } p_S = 120.45 \text{ mbar}$$
$$Z = 20\,000 \text{ m for } p_S = 54.75 \text{ mbar}$$

Evaluation of this law is carried out in the mechanism of the altimeter, or with electronic computers when Z is given electronically; it can be fed into complex systems such as the air data computer. The precision ranges from ± 1 to ± 2 mbar for measurements between 2 mbar and 1050 mbar, account being taken of hysteresis and temperature conditions.

B. Speed Measurement: the Airspeed Indicator and the Mach Meter

In an indirect way the speed of an aircraft can be given by means of a differential pressure measurement: $p_t - p_S$, p_t being the total or impact pressure due to the speed and being measured at the level of a hole almost normal to the flow, and p_S being the static pressure existing at the altitude of the flight. The airspeed indicator is an instrument measuring a differential pressure $p_T - p_S$ and indicating a 'definite speed' corresponding to the speed of the aircraft, at ground level and in a standard atmosphere, that is to say: $p_S = 1013.25$ mbar, $\theta = +15°$C. Airspeed indicators are calibrated according to Saint Venant's law

$$p_T - p_S = \tfrac{1}{2}\rho v^2 \{1 + \tfrac{1}{4}M^2 + (2 - \gamma)M^4/24 + \ldots\}$$

where $\gamma = 1.4$, ρ is the density of air, and M is the Mach number corresponding to the speed of the aircraft. Note that $V = Mc$, where c is the velocity of sound at a given altitude depending on the temperature; at ground level: $c = 340.29$ m/s. The following relations are obtained at the indicated velocities (VI):

$$VI = 0 \text{ m/s for } p_T - p_S = 0 \text{ mbar,}$$
$$VI = 25 \text{ m/s } (90 \text{ km/h}) \text{ for } p_T - p_S = 3.8 \text{ mbar,}$$
$$VI = 50 \text{ m/s } (180 \text{ km/h}) \text{ for } p_T - p_S = 15.4 \text{ mbar,}$$
$$VI = 100 \text{ m/s } (360 \text{ km/h}) \text{ for } p_T - p_S = 62 \text{ mbar,}$$
$$VI = 200 \text{ m/s } (720 \text{ km/h}) \text{ for } p_T - p_S = 266.9 \text{ mbar,}$$
$$VI = 500 \text{ m/s } (1800 \text{ km/h}) \text{ for } p_T - p_S = 2329.1 \text{ mbar.}$$

The accuracy of the measurements ranges from ± 1 to ± 2 mbar, for a full scale deflection corresponding to 1500 mbar, account being taken of hysteresis and temperature conditions.

The Mach meter is an instrument which shows the value of the Mach number, that is to say the ratio V/c, where V is the speed of the aircraft, and c denotes velocity through the surrounding environment. For subsonic velocities, Saint Venant's relations can be written

PART 5. INSTRUMENTS FOR RELATIVE PRESSURE MEASUREMENTS

$$M^2 = \frac{2}{(\gamma - 1)} \left[\left(\frac{p_T}{p_S}\right)^{(\gamma-1)/\gamma} - 1 \right]$$

with $\gamma = 1.4$. For supersonic velocities Rayleigh's law is used

$$\frac{p_T}{p_S} = \frac{[\frac{1}{2}(\gamma + 1) M^2]^{\gamma/(\gamma-1)}}{[1 + \{2\gamma/(\gamma + 1)\} (M^2 - 1)]^{1/(\gamma-1)}}$$

where γ is the ratio of the specific heats of air at constant pressure and at constant volume. At $M = 1$, the representative curves of Rayleigh's and Saint Venant's laws are tangential and have the same curvature.

In a Mach meter the pressure $p_T - p_S$ is applied on the one hand and the pressure p_S on the other. Thus the above equations are generally written as the function $(p_T - p_S)/p_S = f(M)$, whose several values are given below:

For
$M = 0$, $(p_T - p_S)/p_S = 0$

$M = 0.5$, $(p_T - p_S)/p_S = 0.1863$

$M = 1$, $(p_T - p_S)/p_S = 0.8930$

$M = 1.5$, $(p_T - p_S)/p_S = 2.4133$

$M = 2$, $(p_T - p_S)/p_S = 4.6404$

In these instruments, one capsule measures $p_T - p_S$ which is a function of the airspeed while the other measures p_S which is a function of the altitude. A system of levers with adjustable arms converts the corresponding movements into a movement proportional to $(p_T - p_S)/p_S$. This latter movement acts on a needle which rotates in front of a dial calibrated in Mach numbers M.

(1) Special Devices: Airspeed and Mach Number Indicators

Airspeed and Mach number indicators include, in addition, a pointer-mask rotating in terms of p_S. The calibrations are on a logarithmic scale and set in such a way that in front of the speed pointer it is possible to read M on the dial of p_S, so that as with a slide rule, the relative rotations of the dials give directly the ratio $(p_T - p_S)/p_S$ and consequently the Mach number M.

C. Examples of Applications

(1) The Altimeter (Figure 14)

This instrument indicates height from − 1 000 to 80 000 ft in a range of temperature normally between −20°C and +50°C. The mechanism is complex.

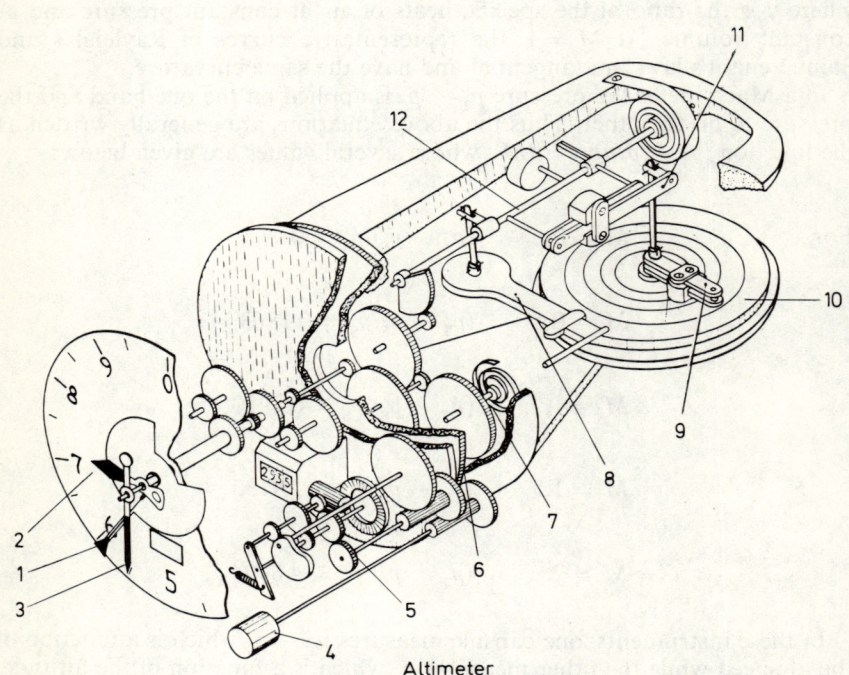

Figure 14. Jaeger type 81 altimeter calibrated in feet:
1 pointer-mask (10.000 ft); 2 Small pointer (1000 ft); 3 Large pointer (100 ft); 4 Barometric setting knob; 5 Ground pressure counter; 6 Mechanism frame rotating wheel; 7 Anti-backlash hairspring; 8 Balance assembly; 9 Bimetal strip; 10 Capsule; 11 Bimetal strip; 12 Rocking shaft.

PART 5. INSTRUMENTS FOR RELATIVE PRESSURE MEASUREMENTS

First the movement of the capsule is transformed to obtain the required relation $Z = f(p_S)$ and then largely magnified with a sector/gear train. The barometric setting mechanism, operated by a control knob, orientates the mechanism which is contained in a rotating frame.

(2) The Airspeed and Mach Number Indicator (Figure 15)

The principle of this instrument is based on that of a slide-rule; the indicated speed is given by a mechanism connected to an airspeed capsule (total

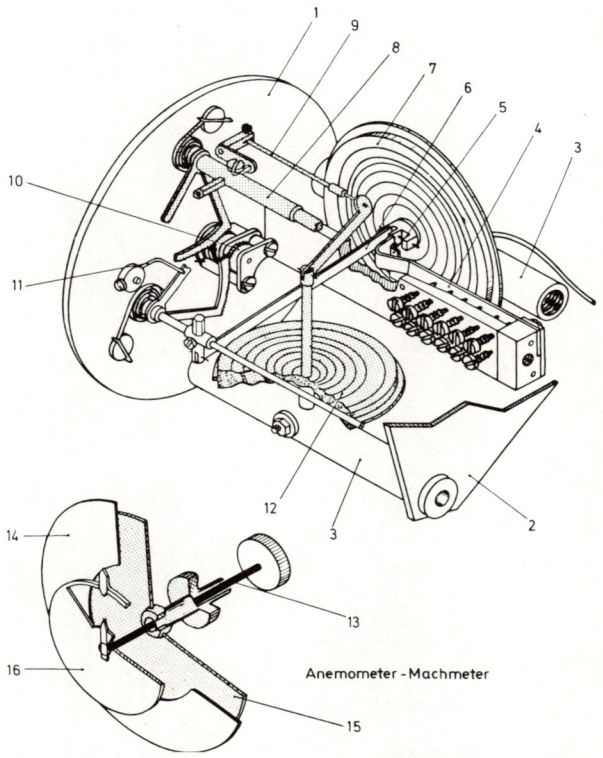

Figure 15. Jaeger type 120 anemometer-machmeter:
1 and 2 Mounting plates; 3 Pillar; 4 and 5 Restraining springs; 6 Link; 7 Airspeed capsule; 8 Altitude rocking shaft; 9 Bimetal rod; 10 Airspeed gear train; 11 Stop pin; 12 Altitude Capsule; 13 Shaft for pointer-mask (16); 14 Airspeed dial calibrated in knots; 15 Moving dial calibrated in Mach numbers; 16 Pointer-mask.

pressure: $p_T - p_S$). The Mach number is read on a dial driven by an altitude capsule (static pressure: p_S). The Mach number is a function of the ratio of the total pressure to the static pressure. The graduations are logarithmic in such a way that the relative rotations of dial, pointer-mask give the required ratio directly.

227

CHAPTER 4

Part 6. Pressure Measurements for the Range 1 kPa to 100 μPa*

STANLEY RUTHBERG
National Bureau of Standards, Washington, DC 20234

Contents

I. Introduction	230
II. Direct Measurement Procedures. Reference Standards	231
1. Precision Liquid Columns	231
A. General Concepts	231
(1) Pressure Balance	231
(2) Capillarity	232
(3) Vapor Pressure of Fluid	233
(4) Examples	233
B. U-Tube Micrometer Point Contact Manometer	233
(1) Description	233
(2) Density and Column Length	234
(3) Measurement of Column Heights	236
(a) Capillarity	236
(b) Temperature Effects on Micrometers	237
(c) Micrometer Setting and Reading	238
(d) Levelling	238
(4) Sorption and Leakage	239
(5) Vapor Pressure	239
(6) Uncertainty of Measurement	239
C. Interferometric Micromanometer	240
(1) Description	240
(2) Density and Column Length	243
(3) Measurement of Column Length	243
(a) Capillarity	243
(b) Fringe Displacement and Differential Height	243
(c) Temperature, Index of Refraction, and Zero Level	244
(d) Levelling	245
(4) Sorption	245
(5) Uncertainty of Measurement	245
2. Compression Manometer—McLeod Gage	245
A. Description	245
B. Density and Column Length	247
C. Measurement of Column Length	247
(1) Capillarity	247
(2) Charging Effects	248
(3) Determination of Differential Height	248
(4) Levelling	249
D. Vapor Pressure	249
E. Compression Ratio	249
(1) Capillary Cross Section	250
(2) Co and Distortion Volumes	251
F. Sorption	252

---

* Contribution of the National Bureau of Standards, not subject to Copyright.

G. Cold Trapping and Pressure Offset 252
H. Uncertainty of Measurement 252
3. Systematic Errors Arising from the Use of a Cold Trap 254
   A. Thermal Transpiration 254
   B. Mercury Vapor—Stream Pumping Effect 254
      (1) Description 254
      (2) Comparison of Theoretical Values 256
      (3) Compensation for the Pressure Offset 257
4. Viscosity Manometer 258
   A. Molecular Drag 258
   B. Oscillating Vane 260
      (1) Description 260
      (2) Accuracy 261
   C. Diamagnetic Levitation Manometer 262
      (1) Description 262
      (2) Accuracy 262
5. Knudsen Radiometer Manometer 263
III. Pressure Generators 264
1. Volumetric Pressure Divider—Static Expansion 264
   A. Description 264
   B. Operation 265
   C. Uncertainties 267
IV. Transfer Gages. Precision 268
V. References 270

I. Introduction

The measurement of pressure between 1 kPa and 100 µPa is complicated by the fact that the nature of the gas transport mechanism undergoes a distinct change within this range. The description of gas flow in rarefied gaseous systems is usually divided into three parts, with the division specified by the range in value of the ratio of the molecular mean free path to the characteristic dimension of the channel through which the gas passes. In the upper pressure range, for example at 1 kPa where the mean free path for all gases is less than 20 µm at 298 K, the characteristics of the flow are dominated by intermolecular collisions. Viscosity and thermal conductivity of the gases are independent of pressure; other properties such as temperature, density and flow velocity show small variation within a distance of one mean free path; therefore, the flow is hydrodynamical and viscous. In the low pressure range, for example at 100 µPa where the mean free path is of the order of 200 m at 298 K, the gas flow is characterized by molecular free flight, and transport is determined by gas–wall interactions. Discontinuities in temperature and variations in molecular flux may occur within the gas at a distance of one mean free path, and the flow is free-molecular flow. The transition from viscous to molecular flow at intermediate pressures is characterized by the influence of both types of collisions. No general derivations of flow equations are constructed from first principles for this transition range, and description is semi-empirical.

Direct measurement procedures within this range depend upon the extension into the transition region of schemes characteristic of the viscous flow domain for higher pressures or characteristic of the molecular flow domain for lower pressures, and no one single procedure is capable of covering the whole range. Numerous methods of measurement have been proposed for

PART 6. PRESSURE MEASUREMENTS FOR THE RANGE 1 kPa TO 100 μPa

use within the range of 1 kPa to 100 μPa. Some show adequate potential but need further development. Consideration is given below to those procedures which offer reasonable confidence in their use as based upon the existence of sufficient measurements data, uncertainty analyses and history of operation.

This manuscript comprises three parts: procedures for the direct measurement of pressure, procedures for the generation of pressure points, and transfer gages. Most attention is given to the first part where various methods are considered, including liquid column micromanometers, McLeod gages, viscosity manometers, and the radiometer manometer.

II. Direct Measurement Procedures, Reference Standards

1. Precision Liquid Columns

A. GENERAL CONCEPTS

Many forms of liquid column micromanometers of both **U**-tube and fixed cistern type have been developed for the pressure range 1 kPa to 100 μPa[10]. All balance the test pressure with a differential height of fluid of known density. The variation in form occurs with the method of height measurement, selection of manometer fluid, treatment of capillarity, and provision for vapor pressure effects.

(1) Pressure Balance

The pressure balance at the liquid-connecting port of two manometer tubes is indicated in *Figure 1* and represented by the equation

$$p_a + \rho_a g y_a + p_{ca} = p_b + \rho_b g y_b + p_{vpb} + p_{cb} \tag{1}$$

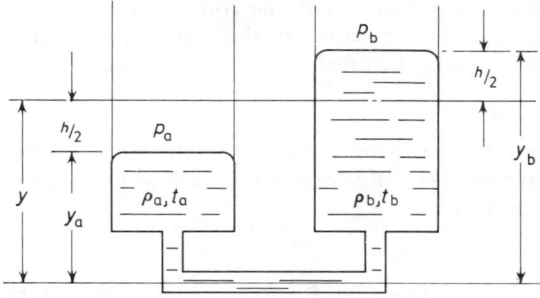

Figure 1. Pressure balance at the liquid-connecting port of a **U**-tube manometer.

where p, ρ, g and y are the pressure, density, acceleration due to gravity and column height for the particular column, respectively. p_{vp} is the vapor pressure of the manometer fluid, and p_c is the net pressure due to capillarity. The density varies with temperature as

$$\rho = \rho_r / \{1 + m(t - t_r)\} \tag{2}$$

231

where ρ_r, t_r are reference values and m is the coefficient of volume expansion. Thus, from equation 1,

$$\frac{p_a - p_b}{\rho_r g} = \frac{y_b}{1 + m(t_b - t_r)} - \frac{y_a}{1 + m(t_a - t_r)} + \frac{p_{vpb} - p_{vpa}}{\rho_r g} + \frac{p_{cb} - p_{ca}}{\rho_r g} \quad (3)$$

Column heights may be referenced to the average position, y, and temperature to a common manometer temperature, t_m, where

$$\begin{aligned} y_a &= y - fh & t_a &= t_m + \delta t_a \\ y_b &= y + (1 - f)h & t_b &= t_m + \delta t_b \end{aligned} \quad (4)$$

with h as the differential column height and f as the fractional drop of the a column from the average position. When $m(\delta t_a + \delta t_b)$ and $m(t_m - t_r) \ll 1$, to a first approximation, equation 3 takes the form

$$\begin{aligned}(p_a - p_b)/\rho_r g = {} & h[1 - m(t_m - t_r)] + hm[f \delta t_a + (1 - f) \delta_b] \\ & + ym[\delta t_a - \delta t_b] + (p_{vpb} - p_{vpa})/\rho_r g + (p_{cb} - p_{ca}/\rho_r g\end{aligned} \quad (5)$$

From this we see that a drift in the temperatures of the columns from the average manometer temperature t_m, or $[f \delta t_a + (1 - f) \delta t_b]$, introduces an uncertainty proportional to the reading. The drift in the temperatures of the columns from each other, $[\delta t_a - \delta t_b]$, introduces a minimum fixed uncertainty proportional to average column length. Uncertainties in the measurement of column lengths include those caused by cosine errors due to offset from the vertical, sine errors associated with changes in level, and gain or loss of fluid to the walls of the tubes. The capillarity is reflected into the height measurement where variation in capillary depression contributes an uncertainty to the height measurement. The fixed cistern type of instrument, i.e. one in which the height of only one liquid surface is measured, is subject to further uncertainties associated with the volume of the liquid, the volumes of the liquid menisci, the temperature of the manometer, and the change in volume with pressure of gas trapped beneath the surfaces.

(2) Capillarity

A capillary rise (or depression) will be present when a curved interface exists in a liquid at rest. The difference in pressure which occurs between the two sides of the interface is

$$\Delta p = \sigma(1/R_1 + 1/R_2) \quad (6)$$

where σ is the surface tension and R_1, R_2 are the radii of curvature at the point of examination[77,42]. If the surface has rotational symmetry, equation 6 has the form in Cartesian coordinates of

$$\ddot{y}/(1 + \dot{y}^2)^{3/2} + \dot{y}/x(1 + \dot{y}^2)^{1/2} = (2/\alpha^2)(C + y) \quad (7)$$

where the origin has been set at the top of the meniscus, the axis of rotation is the y axis, the $(+)$ y direction is downwards, C is the capillary depression or rise, and α is a constant of value

$$\alpha = \{2\sigma/(\rho_1 - \rho_2)g\}^{1/2} \quad (8)$$

PART 6. PRESSURE MEASUREMENTS FOR THE RANGE 1 kPa TO 100 μPa

for ρ_1 and ρ_2 the densities on the two sides of the interface[43]. Laplace's equation, equation 7, is not amenable to an exact solution; however, numerical solutions have been obtained by a number of investigators, and tables are available for mercury[1, 6, 8, 9, 11, 27, 38, 63].

These equations and their solutions relate the capillarity to the surface tension, the bore of the tubulation, and the meniscus height. Previous studies related to manometry have been primarily on mercury, which does not ordinarily wet the container walls. In practice, the capillarity of mercury tends to vary with the age of the manometer, with the direction of the change in pressure, and with local differences in the condition of the wall surface. The surface tension of freshly distilled mercury in a vacuum has been measured as $0.484 \, \text{N m}^{-1}$ at 298 K. It is less when in contact with air. Surface contamination, even apparently trivial, will affect the value. Kistemaker deduced the value of $0.430 \pm 0.005 \, \text{N m}^{-1}$ at 291 K for air contact in large bore, clean glass tubulations[38]. There is no certainty that this value holds in other tubulations. Thus the capillary depression cannot be given accurately for a particular bore of tube and meniscus height. The error is estimated to be about ± 20 per cent[9]. The procedure usually taken in precision manometry is to utilize a tubulation bore of such a diameter that the maximum capillarity is small compared to other uncertainties.

(3) Vapor Pressure of Manometer Fluid

Fluids are usually selected to have low vapor pressures. Mercury, the fluid predominantly used, has a vapor pressure in Pa according to Ernsberger and Pitman of

$$\log p_{vp} = 10.1621 - 3204/T, \qquad 285 \leqslant T \leqslant 326 \text{ K} \qquad (9)$$

for a value of 0.171 Pa (1.28 mTorr) and a dp_{vp}/dT of $0.0147 \, \text{Pa K}^{-1}$ (0.11 mTorr K^{-1}) at 293 K[23, *].

The vapor pressure of mercury may be of significant value compared to the pressure measured. Therefore, precautions must be taken to insure that $(p_{vpt} - p_{vpa})/\rho_r g$ of equation 5 is either negligible or known, both by suitable temperature control and by matched pumping speeds at the surfaces.

(4) Examples

Two types of **U**-tube micromanometers are considered here as examples of instruments that afford the precision and stability necessary for a reference standard. The **U**-tube Micrometer Point Contact Manometer can be used for high accuracy with convenience and simplicity of operation. The Interferometric Micromanometer obtains the smallest minimum detectable signal.

B. U-TUBE MICROMETER POINT CONTACT MANOMETER

(1) Description

Micrometer driven index points are employed for locating the liquid surfaces and for measuring the heights of the liquid columns[50, 61, 66, 69]. Appropriate manometer fluids are mercury and low vapor pressure diffusion

---

* Guildner and Terrien, this volume page 115.

pump oils. A schematic of the mercury manometer is given in *Figure 2*. Photographs of operable mercury and oil manometers are given in *Figures 3* and *4* respectively. In these versions, large diameter (38 mm i.d.) precision-bore glass tubes are clamped between rigid stainless steel end plates and sealed with flat Teflon gaskets. Liquid connection between the tubes is provided by holes drilled in the bottom plate, with diameters selected to give near critically damped motion of the fluid in the columns. Vacuum ports are

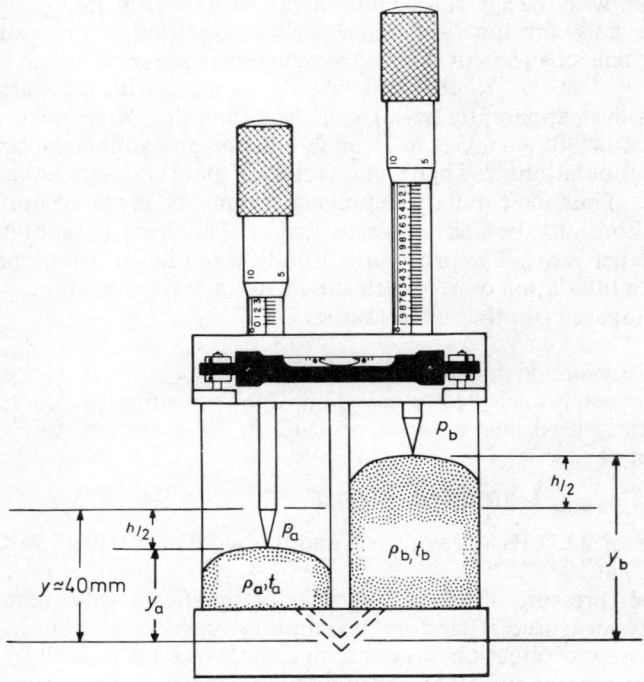

Figure 2. Schematic diagram of the mercury **U**-tube Micrometer Point Contact Manometer.

provided in the top plate. Sensitive levels are mounted on the platform to monitor sine and cosine errors. The manometer temperature is determined with a precision thermometer set into the lower end plate and closely coupled to the fluid. The micrometers are mounted in the top plate of the mercury manometer and in the bottom plate of the oil manometer. The index points, ground on the ends of the micrometer spindles, have a radius of $\leqslant 0.01$ mm. **O**-rings seal the micrometer spindles which are carefully polished to produce acceptable moving seals.

(2) Density and Column Length

The density of mercury and its variation with temperature are well documented. The density of DC704 oil, chosen as a discrete species[33], has been measured to be 1.0610 g/cm^3 at 298 K when saturated with air at one

PART 6. PRESSURE MEASUREMENTS FOR THE RANGE 1 kPa TO 100 μPa

Figure 3. Photograph of a mercury U-tube Micrometer Point Contact Manometer with valves, cold traps and optical background.

atmosphere pressure, with an expansion coefficient of 70×10^{-5}/K. The change in density between the saturated and unsaturated state was determined to be less than one part in 10^4 of the density[69]. Equation 5 indicates the magnitude of the effects due to the uncertainties in the temperatures of the liquid columns relative to each other and to the manometer temperature. If the average ambient temperature is held to $\pm 0.1°C$ under operational conditions, the deviations of the column temperatures from the average manometer temperature, δt_a and δt_b in equation 5, can be $\sim 0.05°C$. The deviation of the column temperatures from each other, $\delta t_a - \delta t_b$, can be $\sim 0.03°C$. These values introduce an uncertainty of about 8 mPa (6×10^{-5} Torr) plus four parts in 10^5 of the measured pressure for oil and about 26 mPa (2×10^{-4} Torr) plus one part in 10^5 of the measured pressure for

Figure 4. Photograph of an oil **U**-tube Micrometer Point Contact Manometer with valves and optical background.

mercury for 40 mm average length columns[61]. Compressibility effects for columns of this length are negligible.

(3) Measurement of Column Heights

(a) *Capillarity*—With the numerical solutions of Blaisdell[8] for equation 7 and appropriate values for fluid density, surface tension, and a contact angle of 0°, the resulting computed capillarity for 38 mm diameter tubes corresponds to 67 mPa (5×10^{-4} Torr) for mercury and 6.7 mPa (5×10^{-5} Torr) for DC 704 diffusion pump fluid. Since the oil wets clean glass, the difference in capillary rise in the two tubes has not appeared to exceed $\sim$20 per cent of the capillarity itself or $\sim$1.3 mPa. With change in pressure, individual oil columns can initially gain or lose fluid to oil films formed at the wall, particularly as oil droplets collect along the glass walls above the oil surface after extended operation. Subsequent restabilization in column heights is a

PART 6. PRESSURE MEASUREMENTS FOR THE RANGE 1 kPa TO 100 μPa

function of drainage and transfer of liquid between the columns. An interval of about one minute may be required for restabilization and about five minutes after large pressure changes. The capillarity of mercury is much more variable. Extensive lowering of a surface often produces a nearly flat meniscus, but by the avoidance of readings under such a condition or by the manipulation of the column level so that the meniscus is more normal, the difference in capillary depression between the two columns can apparently be kept to less than 40 mPa.

(b) *Temperature Effects on Micrometers*—The micrometers are initially set to the liquid surfaces with zero differential pressure on the columns. As indicated in *Figure 5*, the projection of an index point beyond the micrometer

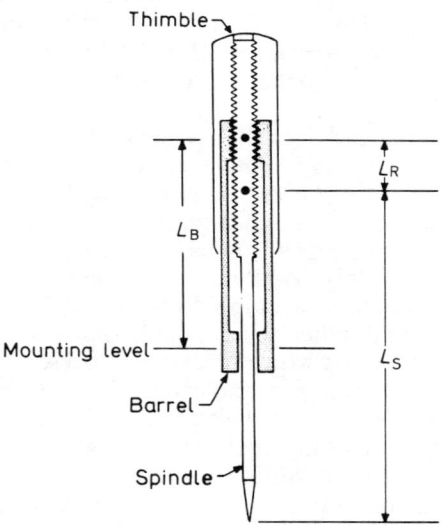

Figure 5. Schematic diagram of micrometer for the **U**-tube Micrometer Point Contact Manometer. L_B—length of barrel, L_S—length of spindle below top of barrel, L_R—change in spindle position with change in liquid level.

mounting level is $(L_S - L_B)$, where L_B is the length of the micrometer barrel and L_S is the length of micrometer spindle below the top of the barrel. After pressurization, the micrometers are reset to the new surface levels. The projection of an index point is then $(L_S + L_R - L_B)$, where L_R corresponds to the difference in the readings of that micrometer and is equal to half the differential column height, or $f = 0.5$ in equation 5. However, manipulation of the micrometers can cause a difference in temperature between the spindle and barrel and a drift in the average micrometer temperature during the time of measurements. When the temperature effects are applied, the differential column height becomes

$$h = [\Delta R'' - \Delta R'][1 + \alpha_S(t_m - t_0)] + [\Delta R'' - \Delta R']\alpha_S \delta t$$
$$+ 2[L_S^\circ \alpha_S \delta t_S - L_B^\circ \alpha_B \delta t_B] \quad (10)$$

where $\Delta R''$, $\Delta R'$ are the differences in readings of the two micrometers at pressurization and at the zero point, respectively; α_S, α_B are expansion coefficients for the micrometer spindles and barrels; t_0 is the reference temperature for micrometer calibration; L_B°, L_S° are the barrel length and the spindle length at the middle of the range of the micrometers at calibration temperature; δt_S, δt_B are uncertainties in the temperatures of the micrometer parts; and δt is the difference in temperature between the micrometer spindles and the manometer[69]. Such temperature variations should be measured, but can be reduced with thermal insulation on the thimble. Measurements have shown that a layer of cork 0.5 cm thick over the sides and ends of the thimble can reduce these drifts, δt, δt_S, and δt_B, to $<0.05°C$ Such insulation reduces the uncertainties for 50 mm range micrometers to ~ 2.7 mPa (2×10^{-5} Torr) for oil and 27 mPa (2×10^{-4} Torr) for mercury plus less than one part in 10^6 of the pressure indication.

(c) *Micrometer Setting and Reading*—The position of the liquid surface is detected by the formation of a dimple caused by the contact of the index point with the surface. The dimple is observed by the distortion of a geometrical pattern, reflected in the liquid surface as viewed with a magnifying lens from above with mercury and from below the surface level in the case of oil. The pattern is formed from an illuminated white card placed behind the columns, which card contains a number of sharply defined, dark, parallel bands, as seen in *Figure 4*. When the dimple is viewed at the boundary of a band, a phase contrast occurs which greatly enhances its observation. With experience and the aid of optical magnification, an observer can set and read large thimble micrometers with smallest scale division of 2 μm to $\frac{1}{4} \pm \frac{1}{8}$ scale division or to $\pm \frac{1}{4}$ μm. Many observations have shown that a maximum dispersion in a series of fixed surface readings can be obtained of 1.5 μm or ± 0.75 μm with the oil manometer and a slightly larger value with the mercury manometer. A probable uncertainty of setting and reading for both liquid surfaces is ~ 1.25 μm (13 mPa) for the oil manometer and somewhat larger or ~ 2 μm (0.27 Pa) for mercury. Determination of the length of the columns also requires a zero level measurement. The uncertainty of the zero level can be reduced by using the average value from many observations. Micrometer setting and reading introduces the largest uncertainty in the use of these instruments.

(d) *Levelling*—Deviation of the micrometer axes from the vertical introduces a second order (cosine) error in the measurement of column height. This error is negligible if moderate care is taken to insure that the micrometer axes are vertical. However, any change in the level of the instrument in the plane of the micrometer axes that occurs between measurements of the liquid surfaces introduces a sine error which is equal to the product of the distance between the axes and the sine of the angle of change. A stable mount and the use of a precision level can minimize this effect. In the par-

PART 6. PRESSURE MEASUREMENTS FOR THE RANGE 1 kPa TO 100 μPa

ticular instruments shown in *Figures 3* and *4* the distance between axes was 47.6 mm and the precision level had a sensitivity of 16 sec of arc/div (77.5 μrad/div). With optical magnification level sensing could be made to ± 0.1 division. This introduced an uncertainty of 49 mPa with the mercury manometer and 4 mPa with the oil manometer.

(4) Sorption and Leakage

In low pressure operation, leakage through the seals is a factor since this causes a pressure gradient in the connecting tubulation due to gas flow. This gradient can be reduced by the use of relatively high conductance lines. Measurements on the instruments described indicated typical leak rates of $\sim 5 \times 10^{-8}$ Pa m^3/s. A second potential source of gas flow exists in the oil manometer because of gas sorption by the fluid. When the DC 704 oil is kept at pressures <200 Pa, the maximum sorption rate (outgassing) was found to be $\sim 9 \times 10^{-8}$ Pa m^3/s with an average rate of $\sim 3 \times 10^{-8}$ Pa m^3/s. This connecting line conductances of 1 l./s would give a maximum pressure difference due to leakage and sorption of $\sim 1.3 \times 10^{-4}$ Pa.

(5) Vapor Pressure

For a $\delta t_a - \delta t_b$ as in equation 5 of 0.03°C in the columns, the difference in the mercury vapor pressures is insignificant, as calculated from equation 9.

It is usually desirable to exclude mercury vapor from the test chamber. This can be done by placing a barrier type differential gage between the test chamber and the manometer or by using a vapor trap. Both procedures have disadvantages. The first method requires an auxiliary pressure control on the manometer side for precise tracking of the test pressure. This can be done with a thermal element for pressures in the viscous flow range range as for the NBS gas thermometer*,[29], but this may lead to pressure gradients and/or discontinuities at pressures in the transition and molecular flow ranges. A direct pressure control may be necessary for these lower ranges[61]. The use of a trap sets up a mercury pressure gradient and diffusion current which can cause a pressure offset at the manometer for the test gas of as much as 50 per cent, see section II.3 below. This would affect the least count of the mercury point contact manometer.

(6) Uncertainty of Measurement

When ambient temperature is held to ± 0.1°C and levelling to ± 7.8 μrad, the magnitudes of the expected uncertainties are summarized in *Table 1* for both mercury and DC 704 as manometer fluids. Thus, the uncertainty (3 st. dev.) of measurement with these instruments can be 16 mPa (1.2 × 10^{-4} Torr) plus six parts in 10^5 of reading for the range 13 mPa (1 × 10^{-4} Torr) to about 500 Pa (~ 3.5 Torr) for the oil manometer and 0.26 Pa (2 × 10^{-3} Torr) plus three parts in 10^5 of the reading for the mercury version for the range to ~ 10 kPa (~ 70 Torr).

The above values exclude micrometer calibration uncertainty. As micrometers can be calibrated to better than 1 μm, operating uncertainties are as shown in *Figure 6*. The accuracy of measurement that can be obtained with the mercury manometer is equal to that obtained with the large standard instruments using white light fringe interferometry, while that obtained with

Table 1. Summary of analyses of errors for **U**-tube Micrometer Point Contact Manometers, oil and mercury, with 38 mm diameter columns. Ambient temperature constant to ± 0.1 K and levelling to ± 1.6 sec of arc.

	Mercury	Oil
I. Errors not prop. reading (Estimated bounds of errors)	Millitorr	
1. Level change	0.37	0.03
2. Capillarity	0.3	0.01
3. Setting and reading	2.0	0.10
4. Density (temp. variation)	0.2	0.06
5. Micrometer temperature	0.2	0.02
6. Pressure stability	<0.1	0.01
Total theoretical $(\Sigma E^2)^{\frac{1}{2}}$	2.1	0.12
Observed 3σ	1.9 to 2.2	0.07 to 0.13
II. Errors prop. reading	Parts per 10000 of reading	
1. Density (temp. variation)	0.1	0.4
2. Micrometer temperature	0	0
3. Manometer temperature	0.2	0.2
Total Σ	0.3	0.6
Total of I and II for 1 torr	2.13×10^{-3}	1.8×10^{-4}
Observed 3σ		1.8×10^{-4} torr

$y = 40$ mm.   Temp. measured to 0.02 K

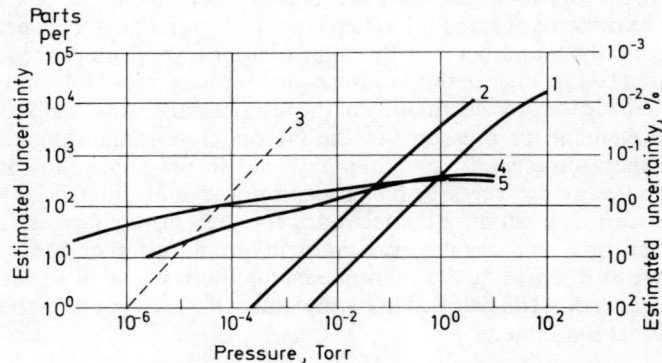

Figure 6. Measurement accuracy. Uncertainty versus pressure. 1. mercury **U**-tube Micrometer Point Contact Manometer. 2. Oil **U**-tube Micrometer Point Contact Manometer. 3. Interferometric micromanometer, oil. 4. McLeod gage. 5. Volumetric pressure divider—static expansion.

the oil manometer is not far from that of the 76 mm diameter mercury-capacitance manometer used for the NBS (US) Gas Thermometer [29, *].

C. Interferometric Micromanometer

(1) Description

Optical interferometric procedures have been used for increased sensitivity

---

* See this volume, pp 115, 121.

in the measurement of column length. Low density diffusion pump oils allow increased differential heights and in addition provide kinematic viscosities 100 times that of mercury so that surface ripples are heavily damped. A schematic diagram of a device employed by Thomas, Johnson and Little[70] is given in *Figure 7*. The retaining walls for a central reservoir were machined into the top surface of a brass plate. The plate was set into a large cylindrical cistern of Octoil S. Horizontal slits at the bottom of the retaining walls provided a liquid path between the inner and outer pools.

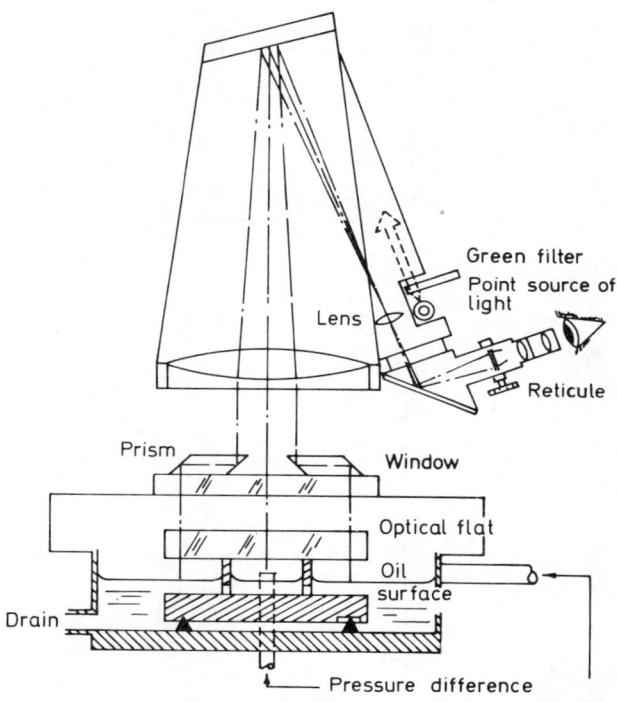

Figure 7. Schematic diagram, interferometric micromanometer—Thomas, Johnson and Little.

A quartz optical flat resting on the polished surface of the retaining walls isolated the inner pool from the outer pool. Sealing was accomplished with an oil film between the brass and quartz surfaces. The middle pool was 5 cm wide by 12 cm length with walls 3 mm thick, and the outer cistern was of 17 cm outer diameter. Collimated monochromatic light was projected into the manometer to form interference fringes between the reflections off the oil surfaces and the bottom face of the optical flat. Two symmetrically placed prisms served as periscopes to the outer pool. This formed a double **U**-tube or **W**-configuration to minimize levelling errors.

Aubry and Delbart[3] used a fixed cistern manometer with an optically flat mirror set below a DC 704 oil surface, as in *Figure 8*. The center pool was 4.29 cm in diameter. The outer pool was an annular ring of 6.24 cm inner

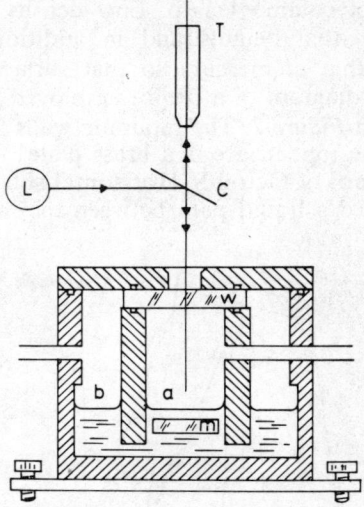

Figure 8. Schematic diagram, interferometric micromanometer—Aubry and Delbart.

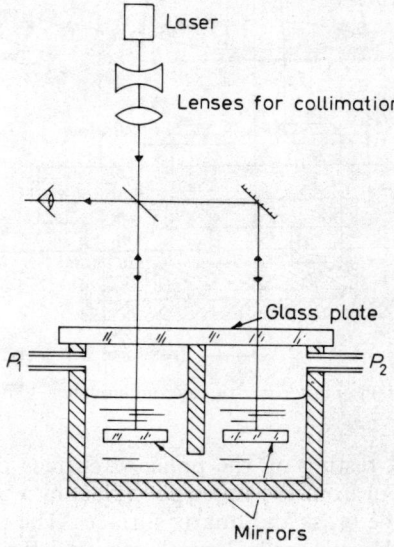

Figure 9. Schematic diagram, interferometric micromanometer—Stevenson and McFadden.

diameter with a 1.23 cm width to give a surface area twice that of the inner pool. The optical path difference for fringe formation was in the oil wedge between the oil surface and the mirror.

Stevenson and McFadden[65] utilized a laser source and a Michelson interferometer with two mirrors set below the liquid surfaces of a **U**-tube manometer which was filled with galvanometer damping oil, as sketched in

Figure 9. An optical path difference was generated as the levels changed. No dimensions were listed.

(2) Density and Column Length

Equation 5 indicates that a small temperature difference between columns could produce a relatively significant fixed uncertainty in these manometers. An average column height above the liquid connecting ports of 1 cm and a difference between column temperatures of 0.01 K would contribute an uncertainty for DC 704 and Octoil S of $\sim 7 \times 10^{-5}$ mm of oil or ~ 0.7 mPa (5×10^{-6} Torr). A deviation of 0.01 K in column temperature from the assigned manometer temperature, t_m, would contribute an uncertainty of 7 p.p.m. of the differential height, while an error of the same magnitude in the assigned manometer temperature would cause an uncertainty of 14 p.p.m.

In the **W**-tube, Thomas *et al.* kept the average column height, y, to 1 mm by utilizing the meniscus to help seal off the connecting ports. Their pools were shallow, in intimate contact with the brass block, and insulated by the surrounding oil; however, no temperature measurements were made in the columns. Data are not available for the average column heights in the fixed cistern of Aubry and Delbart or the **U**-tube of Stevenson and McFadden.

(3) Measurement of Column Heights

(a) Capillarity—From the numerical solutions of Blaisdell[8], the computed capillarity for a 42.9 mm diameter pool of DC 704 oil as in the Aubry–Delbart instrument corresponds to 1.3 mPa (1×10^{-5} Torr). An annular ring 62.4 mm i.d. by 12.3 mm width would have greater capillarity. Thus, a variation of 20 per cent would contribute an expected uncertainty of at least 0.3 mPa (2×10^{-6} Torr) to the fixed cistern instrument. The pools of the **W**-configuration of Thomas *et al.* were found to be flat to $\leqslant 0.1$ fringe per cm at the center. In a long pool, the surface is lifted by a distance

$$y = (\sigma/\rho g)(d^2 y/dx^2)$$

where $d^2 y/dx^2$ is the surface curvature. The surface tension, σ, is ~ 0.03 N m^{-1}, and the density is ~ 0.9 g/cm^3 for Octoil S so that the capillarity corresponded to < 10 µPa ($< 1 \times 10^{-7}$ Torr).

(b) Fringe Displacement and Differential Height—For the formation of the fringe systems, the stationary reflecting surface is oriented at a small angle to the liquid surface. Illumination and observation are along a normal to the surface. A change in level of the liquid surfaces causes a fringe to move to a new position that holds the optical path length constant. A horizontal displacement equal to an interfringe distance corresponds to a change in liquid level of $\lambda/2n$ where λ is the optical wavelength (free space) and n is the index of refraction of the medium between the reflecting surfaces.

The **W**-tube of *Figure 7* produced a fringe pattern as shown in *Figure 10*. When a pressure difference was applied, the center set of fringes moved in one direction and the outside sets moved in the opposite direction. For a relative shift of s fringes, the differential column height was

$$h = \tfrac{1}{2}\lambda s \tag{11}$$

with $n = 1$ for the vacuum interface. With such a fringe pattern reading was a fifth of a fringe at best, i.e. 0.2 ± 0.1. For the mercury green line (546.1 μm) a tenth of a fringe corresponded to ~ 0.24 mPa ($\sim 1.8 \times 10^{-6}$ Torr). For the fixed cistern of *Figure 8* the fringe shift was a measure of the change in level,

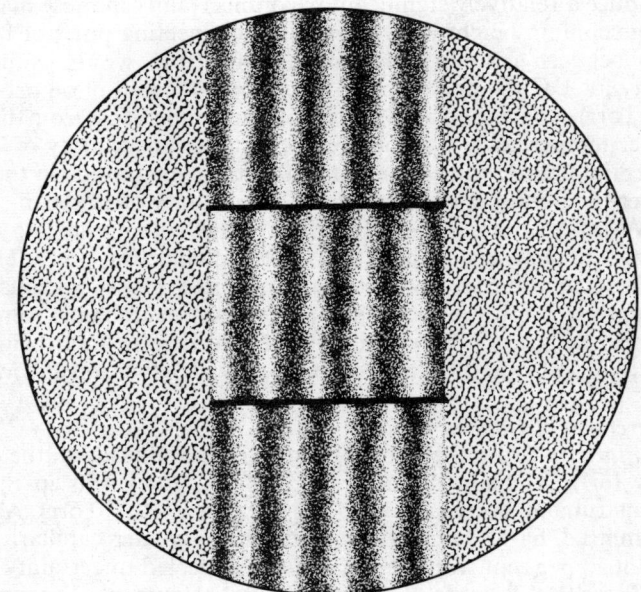

Figure 10. Interference fringe pattern—**W**-tube micromanometer of Thomas, Johnson and Little.

$(1 - f)h$, of the center pool only with f from equation 4 being $a_1/(a_1 + a_2)$ where a_1 and a_2 were the areas of the inner and outer pools; thus, the differential height measured was

$$h = (\lambda/2n)[1 + (a_1/a_2)]s \qquad (12)$$

With n approximately 1.5 and $a_1/a_2 = 0.5$ the least count of this instrument was about the same as that of the **W**-tube; however, an added uncertainty arises from the factor of a_1/a_2. An imprecision in the diameter of the pools of ± 0.02 mm contributes an error of ~ 0.26 per cent of the reading. For the laser instrument shown in *Figure 9* a differential height, h, in the **U**-tube produced a change in the path length from that at the zero position of $[h_{\text{liquid}} - h_{\text{gas}}]$ so that the height measured was

$$h = \{\lambda/2(n_1 - n_g)\}s \qquad (13)$$

At low pressures, with $n_g = 1$, $n_1 \sim 1.5$, and s determined to ± 0.1 fringe, sensitivity would have been half as good as that of the other two instruments.

(c) *Temperature, Index of Refraction, and Zero Level*—The index of refrac-

PART 6. PRESSURE MEASUREMENTS FOR THE RANGE 1 kPa TO 100 μPa

tion for these pump oils may be related to their density through the Lorentz–Lorenz formula so that

$$\{(n^2 - 1)/(n^2 + 1)\}(1/\rho) = k \qquad (14)$$

where k is the specific refraction. Therefore, for $\rho \sim 1$ and $n \sim 1.5$, the relative change in n with change in ρ is about 0.2. This contributes an uncertainty of about one part per 10^4 of the reading/K, i.e. 0.2 of $7 \times 10^{-4}\,\mathrm{K}^{-1}$.

The variation in manometric temperature, t_m, does not affect the zero differential reading in the **U**-tube instruments but does in the fixed cistern instrument. A change in liquid volume with change in temperature causes a change in the zero level of amount $ym\,\delta t_m$. A correction can be applied if the manometer temperature is monitored. An uncertainty remains of value $y(m/2)(\delta t_a + \delta t_b)$, which is of the same magnitude as the third term of equation 5. Frequent zero readings can reduce this uncertainty.

(d) *Levelling*—As the stationary reference surfaces were fixed within the instruments, these instruments were subject to a sine error caused by a change in level that occurs between measurements. The laser **U**-tube had a direct sine error equal to the product of the distance between axes and the sine of the angle of tilt just as is the case in the Point Contact manometer. The magnitude of this error is reduced in the fixed cistern manometer since both pools are coaxial; however, a rotation from the vertical axis does cause a change in level of the surfaces since the volume of fluid is constant. Hence, an error can occur in height measurement off the center pool, even though the differential height would remain constant. This is a cosine error $[y(\cos\alpha - 1)]/[\cos\alpha]$ which becomes negligible under modest levelling control. The **W**-tube essentially eliminates the error by averaging the displacements at the outer legs to give an equivalent axial value.

(4) Sorption

Diffusion pump oils may absorb gas to as much as about ten per cent of their volume. It is necessary to allow these oils to degas before operating at low pressure. The previous remarks on absorption in the micrometer Point Contact manometer are relevant, see Sorption and Leakage section, page 239.

(5) Uncertainty of Measurement

No detailed statistical data are available for these instruments, but an expected uncertainty curve is give as No. 3 in *Figure 6*.

2. Compression Manometer—McLeod Gage

A. Description

An alternative measurements approach which gives a wider pressure range of operation than do the micromanometers, but at a lower precision at the high pressures, is embodied in the compression manometer or McLeod gage. With this a gas is precompressed to a second pressure that can be measured by differential column heights according to standard manometric practice[47]. The structure of a McLeod gage is depicted in *Figure 11*. With

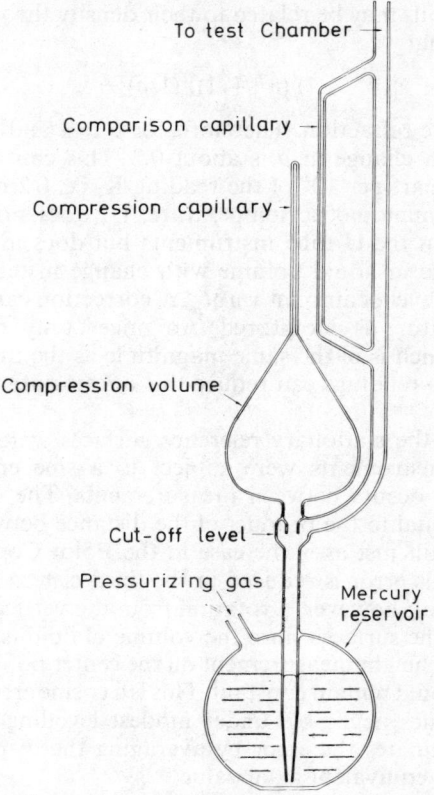

Figure 11. Schematic of precision McLeod gage.

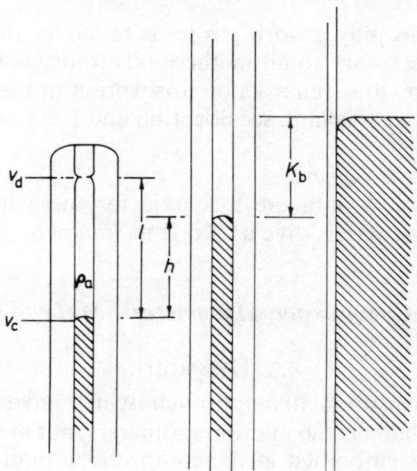

Figure 12. McLeod gage: Relative column heights in the compression and reference arms. v_d—distortion volume. v_c—covolume. l—length of compressed gas column. h—differential column height. K_b—capillarity of comparison capillary tube.

PART 6. PRESSURE MEASUREMENTS FOR THE RANGE 1 kPa TO 100 µPa

the test chamber at low pressure, gas admitted to the reservoir drives the mercury up the central tube. When the mercury reaches the cut-off level, a sample of test gas at an initial pressure p_b is captured within the compression volume V. With continued rise, the test gas is swept into the compression capillary so that a final pressure p_a is determined by the differential height h between the compression capillary and the comparison capillary, as shown in *Figure 12*. The initial pressure p_b is then computed from the final pressure p_a, the compression ratio, and the appropriate gas law. A typical precision McLeod gage intended for a least count of ~100 µPa would have a compression volume of two liters and capillary diameters of 1 mm.

B. Density and Column Length

If the variation in temperature over all parts of the gage were not greater than 1 K, the uncertainties due to the variation in the density of mercury would be relatively small. After compression the average column height y above the liquid connecting channel is about 40 cm in a two liter McLeod gage, which would contribute a fixed uncertainty in differential height of about 0.07 mm K^{-1}. A 1 K drift in column temperatures from the manometer temperature would lead to an uncertainty of 1.8 parts per 10^4 of the reading. Uncertainty due to measurement of actual manometer temperature t_m should be even less.

C. Measurement of Column Length

(1) Capillarity

Small diameter tubulations are required to obtain large compression ratios, but this causes capillarity to be a major source of uncertainty.

As the tubulation diameter becomes smaller, the meniscus shape approaches that of a spherical segment and $R_1 \to R_2$ in equation 6. If θ is the angle of contact of the liquid interface at the wall and r is the radius of the tube, equation 6 becomes

$$\Delta p = \rho g C = (2\sigma/r) \cos \theta \qquad (15)$$

or in terms of the meniscus height h_m

$$\rho g C = 4\sigma h_m/(h_m^2 + r^2) \qquad (16)$$

Calculations by Akiyama *et al.* have shown that the spherical segment approximation for meniscus shape is accurate to within one per cent for capillary tubes of ≤ 1 mm diameter[1].

In a study directed specifically to McLeod gage use, Akiyama *et al.* corroborated the 20 per cent variation in capillarity as a function of bore diameter, but found an agreement to within about ten per cent between experimental and computed values of capillarity as a function of meniscus height for untreated tubulations. Treatment of tubulations by the Rosenberg technique[58] increased capillarity but decreased relative variation to about five per cent. These data are in agreement with the behavior found in practice; namely, that the capillarity can vary with position along a given tube by ten per cent and can change in time as a function of exposure. However, it can also be demonstrated that a systematic variation reproducible to about one per cent can be obtained in carefully prepared, clean capillaries[35, 54].

Akiyama et al. found representative values for σ and θ to be 0.450 N m^{-1} and $140°$, respectively, for treated tubes; whence, a 1 mm diameter tubulation would evidence a capillarity computed from equation 15 to be 10.4 mm$\pm$ (0.5 to 1 mm), whereas an experimental repeatability could be ± 0.1 mm. For highest accuracy, then, the procedure is to chart actual capillarity as a function of position along the tube at a pressure below the response of the gage and to apply these data as a correction. That is, from equation 5 with $(p_a - p_b)/\rho_r g = 0$, $(p_{vpb} - p_{vpa})/\rho_r g$ kept small, and h_0 representing the difference in capillarity $(\kappa_a - \kappa_b)$ as a function of position along the capillary tubes,

$$-(p_{cb} - p_{ca})/\rho_r g = h_0[1 - m(t_m - t_r)] + h_0 m [f\delta t_a + (1 - f)dt_b] + ym(\delta t_a - \delta t_b)_0 \quad (17)$$

Then in measurement,

$$(p_a - p_b)/\rho_r g \approx (h - h_0)[1 - m(t_m - t_r)] + ym[\delta t_a - \delta t_b) - (\delta t_a - \delta t_b)_0] + (P_{vpb} - P_{vpa})/\rho_r g \quad (18)$$

where the relatively small temperature effects proportional to the reading have been omitted for $\delta ts \leqslant 1°C$. An uncertainty remains, however, due to the differences in column temperatures which exist at the time the capillarity is charted and at the time the measurement is made, which contributes an uncertainty of at least $(2^{\frac{1}{2}})(\delta t_a - \delta t_b)$ or a fixed uncertainty of ~ 0.1 mm per °C difference. It is therefore desirable to control the temperature variation across the capillaries to $<0.5°C$. Within this temperature restriction the effect of change in surface tension with temperature, and hence capillarity, is also negligible.

Where such precision is desired, it is advantageous to have the capillary diameter uniform along the length as evidenced by equation 15.

Since the effect of the variation in capillarity is compounded by the use of a comparison capillary, Keevil et al. eliminated the comparison capillary and used a large bore tubulation as a reference arm for which capillarity was then a relatively small error[36, 54, 22]. Then, $h_0 \equiv \dot{\kappa}_a$.

(2) Charging Effects

The movement of the mercury column in the capillary tube can cause an electrical charging of the wall which results in erratic behavior in the mercury level, particularly on downward transit. Rapid motion of the mercury column increases this effect which can be troublesome. Preferred operation allows the mercury level to approach the final rest position smoothly and slowly with tapping of the capillary tube until the mercury level has stabilized.

(3) Determination of Differential Height

The most accurate method makes use of a calibrated scale and a travelling telescope fitted with a reticule or filar eyepiece. The scale should be placed alongside the capillary and reference column so that the scale and viewpoint on the menisci fall in the plane of focus of the telescope. The telescope should be at such a distance that operator presence does not affect the temperature at the gage. The gage columns, scale axis and telescope ways should be aligned. Appropriate sources of illumination are needed, with the mercury surfaces backlighted through translucent screens. The telescope is first sighted on the meniscus and then rotated to the scale. The reticule or the filar eyepiece provides the vernier function needed to transfer to a scale graduation. The precision of sighting and reading for such apparatus can be ~0.01 mm (3 st. dev.)[10].

A sighting ring and vernier can also be used at somewhat reduced precision, ~0.05 mm.

(4) Levelling

Modest levelling of the gage columns is sufficient to reduce any cosine error to a negligible amount. As a zero reference level on the columns is not ordinarily available, any misalignment of the scale or telescope from the vertical causes a sine error which requires greater levelling control for minimization. As with the U-tube instruments, such error is a fixed value equal to the product of the distance between the columns and the sine of the angle of tilt from the vertical. For example, a 25 mm distance between column centers produces an error of ~0.4 mm per degreee of tilt.

D. Vapor Pressure

The difference in the mercury vapor pressures between the two columns of the gage is relatively small as compared to the differential column heights, within the restrictions on temperature gradients set in C.1 above. We may then drop this term from equation 18 with little consequence.

E. Compression Ratio

If the relationship between the *p-V-T* quantities before and after compression follows the ideal gas law, the initial test pressure has the value

$$p_b = (v/V)(T_b/T_a) p_a \tag{19}$$

where v is now the final compressed volume in the capillary. When combined with equation 18, we have

$$\frac{p_b}{\rho_r g} \approx \frac{v}{V} \frac{T_b}{T_a} \frac{[h - h_0][1 - m(t_m - t_r)]}{1 - (/V)(T_b/T_a)} \tag{20}$$

where the fixed uncertainty due to the uncertainty in the difference of column temperatures has been momentarily omitted. The factor V/v is the compression ratio. In a high vacuum McLeod gage, the V/v may range from 10^4 to 2×10^6; therefore, the second term of the denominator is neglected.

The ideal gas law is a reasonable approximation at temperatures above the critical temperature of the gas being measured. A better representation is the van der Waals equation from which the initial pressure would be, under isothermal compression,

$$p'_b = \frac{(p_a + a/v^2)(v - b)}{V - b} - \frac{a}{v^2} \qquad (21)$$

in place of that given by equation 19. The quantities a and b are the van der Waals constants for the particular gas. The relative error accrued by use of equation 19 as compared to equation 21 is

$$\frac{p'_b - p_b}{p_b} \approx \frac{(p_a + a/v^2)(v - b) - p_a v}{RT}$$

where we assume $b \ll V$, $a \ll V^2$, and $p_b V = p_a v = RT$, or to first order

$$(\Delta p_b/p_b) \approx (p_a/RT)(a/RT - b) \qquad (22)$$

Jansen and Venema[35] computed the relative errors with equation 22 for a number of gases as a function of the pressure in the compression capillary. For a final compression of up to 40 cm differential height h, the relative error for gases with a critical temperature below the operating temperature of the McLeod gage is $\leqslant \sim 0.1$ per cent. The pressure can be measured for critical temperatures above the operating temperature provided the saturation vapor pressure is greater than that in the compression capillary. For many such gases the use of the ideal gas law introduces an error of up to 0.5 per cent.

The compression volume V is a fixed quantity that is normally determined during the construction of the instrument through the measurement of the mass of that quantity of distilled water (or mercury) which fills it. A one or two liter compression volume can be determined readily to about 0.1 per cent.

The compressed volume v is nominally the product of the compression capillary cross section a and the length l of the compressed gas column, as indicated in *Figure 12*. Actual volume determination requires corrections for both a and l.

(1) Capillary Cross Section

Small bore capillary tubes are commercially available with diameters uniform to about 5 μm. Treatment by the Rosenberg technique can cause greater variation. It is therefore necessary, for precision use, to determine the average cross sectional area of the compression capillary as a function of the length of the compressed gas column. This is accomplished by placing a small pellet of mercury of measured mass into the capillary before it is sealed to the compression chamber and before the end plug is fused in. The pellet assumes a cylindrical shape with spherical segment end-caps. The cross sectional area is computed from

$$a = \{3m_{Hg} - \pi\rho h_m^3\}/3\rho(s - h_m) \qquad (23)$$

where m_{Hg} is the mass of the pellet, ρ is the density of mercury, h_m is the meniscus height, and s is the length of the cylinder between end points. Measurements on the length of the pellet as it is moved along the capillary lead to the value of a as a function of position. Cross sectional diameters can be determined in this way to within ~ 2 μm (3 st. dev.) for a 1 mm diameter tube. The average value of a is then determined for a given length l of gas column, or a longer pellet of mercury can be used to derive average values while the shorter pellet is used to derive variation in cross section.

(2) Co and Distortion Volumes

It is seen with reference to *Figure 13* that the shape of the compressed volume departs from that of a right circular cylinder due to the meniscus at the lower end and the distortion at the upper end caused by seal-off.

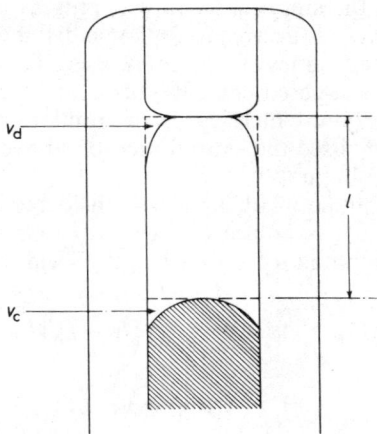

Figure 13. Top of compression capillary, high compression. v_d—distortion volume. v_c—co-volume. l—length of compressed gas column.

An additional volume or covolume v_c is to be added which is the difference between the volume of a right circular cylinder and a spherical segment both of a height equal to that of the meniscus, h_m, so

$$v_c = \tfrac{1}{2} a h_m - \tfrac{1}{6}\pi h_m^3 \tag{24}$$

and the effective increase in length is

$$l_c = v_c/a \tag{25}$$

A 1 mm diameter tube with a meniscus height of 0.25 mm contributes an l_c of 0.1 mm.

The distortion at the end of the tube may be handled in two ways. If the seal-off is smooth on the interior, a small amount of mercury can be introduced into the end of the capillary under vacuum and the volume to a given length from the end determined by weighing[18]. This obviously has to be done before the capillary is sealed to the compression volume. The deviation of this end volume from that of a right circular cylinder is the distortion volume v_d which has an effective length l_d.

An alternative procedure[54, 53, 15] is to determine the effective end-point of the compression capillary by gas measurement. With the zero graduation of the scale set to some convenient level such as at the top of the compression capillary, the length of the compressed gas column as measured to the top of the meniscus is $l_i - l_0$, where l_i is the scale reading at the meniscus top and l_0 corresponds to the effective end point. Then by equations 20 and 25 as a test sample of gas is compressed up the capillary,

$$l_i + l_c = (Vp_b/\rho_r g)(1/\{a(h - h_0)\}) + l_0 \qquad (26)$$

which is a linear relationship for $(l_i + l_c)$ and $1/a(h - h_0)$ from which l_0 is calculated.

A disadvantage of the smooth, fused end plug as indicated in *Figure 13* is that the treated surface is destroyed for some distance away from the fused portion so that the small values of l cannot be used. To remedy this, Podgurski and Davis employed a tapered end-plug fused at its upper end for minimal change to the capillary but which leaves an annular crevice at the end of the capillary volume. They used the second method above to locate the effective end of the capillary volume[54].

The compressed volume v can now be considered as the product of the effective length of the gas column, $(l + l_c - l_d)$ or $(l_i + l_c - l_0)$, and the average cross sectional area a for that length. Then in equation 20 to close approximation

$$p_b = \rho_r g(a/v)(T_b/T_a)\left[l + l_c - l_d\right]\left[h - h_0\right]\left[1 - m(t_m - t_r)\right] \qquad (27)$$

F. Sorption

Accurate use requires a minimum amount of outgassing or adsorption of gas within the compression volume. This effect can be monitored by holding the mercury surface just above the cut-off level and making periodic determinations of pressure. The rate of exchange may be computed and applied as a correction if small. Excessive outgassing should be eliminated. A further test is made by use of other gases for determination of effective end-point of the capillary.

G. Cold Trapping and Pressure Offset

The usual means for eliminating the mercury vapor from the test chamber is a cold trap between the McLeod gage and the test chamber. Again, this introduces a systematic offest in pressure due to both vapor stream pumping and thermal transpiration. Several studies have been made and many suggestions offered for the determination or the reduction of the amount of vapor stream pumping. Further discussion is given below (page 254). For the moment, the relevant fact is that reduction of this offset to about one per cent is possible. Thermal transpiration is reduced, presumably by use of a symmetrical cold trap and molecularly rough walls. It is an effect which should be tested for and represents additional uncertainty.

H. Uncertainty of Measurement

The observed pressure is derived from equation 27. Some of these variables are now of independent error and some dependent. Thus, a, V and g are

PART 6. PRESSURE MEASUREMENTS FOR THE RANGE 1 kPa TO 100 µPa

of independent error. As the temperature is to be controlled and if the gas is allowed to equilibrate before and after compression, we can relate T_b and T_a to the manometer temperature following the manner of equation 4 so that

$$(T_b/T_a) = (t_m + \delta t_a)/(t_m + \delta t_b)$$

or to first approximation

$$(T_b/T_a) = 1 + (\delta t_a - \delta t_b)/t_m \tag{28}$$

The variables h, l and l_c are measured in common during operation and are dependent in error. The quantity l_d may or may not be of independent error as determined by method of measurement. The capillarity h_0 is of independent error. The total imprecision in p_b would normally be treated as a r.m.s. of the dependent and independent parts. But as the relative magnitudes of these various imprecisions range widely in value, a more appropriate expression for relative uncertainty is

$$\frac{\delta p}{p} \approx \frac{\delta K}{K} + \frac{\delta l + \delta l_c + \delta l_d}{l + l_c - l_d} + \frac{\delta h + \delta h_0}{h - h_0} \tag{29}$$

where

$$K \equiv \rho_r g(a/v)(T_b/T_a)$$

By applying the guidelines and imprecisions given above to equation 29 for a high vacuum gage of two liter compression volume and 1 mm diameter capillary, we obtain estimated uncertainties such as those listed in *Table 2*. A bias of one per cent is to be added for the vapor stream pumping effect,

Table 2. Impression of model gage, high vacuum McLeod

p Torr	1×10^{-5}	1×10^{-4}	1×10^{-3}	1×10^{-2}
$\delta K/K \%$	0.5	0.5	0.5	0.5
$\dfrac{\delta h + \delta h_0}{h - h_0} \%$	~1	~0.3	0.1	<0.1
$\dfrac{\delta l + \delta l_c + \delta l_d}{l + l_0 - l_d} \%$	~2.5	~1	0.2	<0.1
Total $3\sigma \%$	~4	~1.8	0.8	~0.5

Compression volume 2*l*
Capillary diameter 1 mm, $\delta a/a = 0.4\%$
$\delta h_0 = 0.1$ mm

Reference arm 19 mm diameter
Heights measured to 0.01 mm

$\delta V = 0.5$ cm^3

$(\delta t_a - \delta t_b)/t_m = 0.5/300$

$l_c = 0.1$ mm

$\delta T = 0.5$ K

$l_d \sim l_c$

which bias is reduced with pressures in and beyond the transition region. Thus, the approximate three per cent uncertainty at 1×10^{-4} Torr for a single measurement is in agreement with values obtained experimentally. The value of five per cent at 1×10^{-5} Torr is somewhat optimistic. We have not included the erratic behavior which normally happens as the mercury level approaches the end of the compression capillary. Similar guidelines were given to a second model comprised of a 300 cm^3 compression volume and a capillary of 6 mm diameter. Results for the whole range are given as curve No. 4 in *Figure 6*.

Uncertainty can be reduced by using a multiplicity of data points for each test sample of gas as given in equation 26. The coefficient Vp_b/pg can then be more precisely determined and the value of p_b extracted from this[54, 22].

3. Systematic Errors arising from the Use of a Cold Trap

A. Thermal Transpiration

It was first discovered by Neumann[52] that a temperature gradient in a gaseous system could induced a pressure gradient. At high pressures, where the mean free path is small, the pressure in a closed system is the same everywhere and independent of such a temperature gradient. For low pressures, Reynolds[5F] and Maxwell[46] predicted that the gas pressures p_1 and p_2 in two separate chambers maintained at temperatures T_1 and T_2 connected by a small diameter tubulation would bear the ratio

$$p_1/p_2 = (T_1/T_2)^{\frac{1}{2}} \tag{31}$$

Subsequent theoretical and experimental studies by Knudsen[40] confirmed this value. This ratio p_1/p_2 is called the thermal transpiration ratio which, then, has the limiting values of unity and $(T_1/T_2)^{\frac{1}{2}}$. The dependence of the thermal transpiration ratio on pressure has the form of an **S**-curve with the major change in its value occurring in the transition flow region[21, 45, 74].

Since the thermal transpiration ratio is a function of the tubulation diameter, particularly in the transition flow region, the use of an asymmetrical cold trap can cause a systematic bias in pressure of several per cent of the measured pressure[60]. More recently, some experiments[31, 32, 55] have shown deviations from Knudsen's limiting value $(T_1/T_2)^{\frac{1}{2}}$ of the thermal transpiration ratio by ten per cent or more, which have been attributed to a departure from cosine scattering at the gas/wall interface[31, 21]. It is therefore necessary that the cold trap be both symmetrical in shape and molecularly rough.

B. Mercury Vapor Stream Pumping Effect

(1) Description

The consequent streaming of the mercury vapor from the manometer to the cold trap as depicted in *Figure 14* produces a pumping action which reduces the pressure of the test gas at the manometer relative to the test chamber. This effect was first analysed by Gaede[26] as a diffusion phenomenon; the magnitude of the effect was evaluated by Ishii and Nakayama[34] and by

PART 6. PRESSURE MEASUREMENTS FOR THE RANGE 1 kPa TO 100 µPa

Podgurski and Davis[54]; a formalism was derived by Takaishi[67] from kinetic theory; and a number of further studies were made both to test the nature of the effect and to derive corrections for it[12, 16, 17, 22, 48, 51, 56, 59, 68, 71, 73].

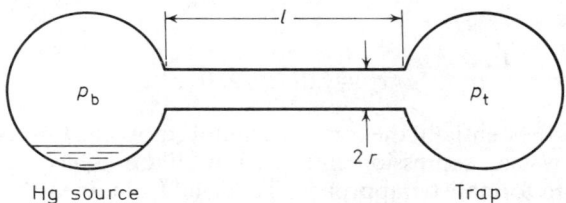

Figure 14. Mercury vapor—stream pumping effect—schematic. p_b—pressure measured at the manometer. p_t—pressure at the trap or test chamber. l—length of tabulation connecting mercury source to trap. r—radius of connecting tubulation.

The diffusion theory assumes that the molecules of the test gas become entrained in the mercury vapor stream and are swept toward the cold trap to create a gas concentration gradient. This gradient builds up until it is balanced by a counterflowing diffusion current. The entrainment velocity is assumed equal to the mass flow velocity of the mercury stream. Thus, at steady state

$$-D \, \text{grad} \, n = nu$$

where D is the interdiffusion coefficient, n is the gas concentration, and u is the mercury vapor stream velocity. Furthermore,

$$d(pV)/dt \equiv Q = P_m u a$$

$$p = nkT$$

$$D = D_0/p_m$$

$$Q = U(p_m + p_g - p_t)$$

where p_m, u and a represent values at any cross section of the stream of the mercury vapor pressure, the stream velocity, and the cross sectional area, while k is the Boltzmann constant, D_0 is the value per unit pressure, Q is the throughput, U is the conductance of the channel for mercury vapor, p_g is the pressure at the gage, and p_t is the test chamber pressure. Thus,

$$\ln(p_t/p_g) = (B/p_m)(p_m + p_g - p_t) \tag{32}$$

where

$$B = (U/D_0) p_m \int_0^l (1/a) dx \tag{33}$$

for a channel of length l. In practice the channel may be composed of a number of segments of various radii. If these are each long segments of uniform bore,

$$\frac{1}{U} = \Sigma_i \frac{1}{U_i} = \Sigma_i \frac{1}{(8/3)(r_i^3/l)(\pi/4)\langle c \rangle}$$

for $\langle c \rangle$ the average thermal velocity of the mercury molecules and r_i the radii. Thence, in equation 33

$$B = \frac{8}{3}\left(\frac{kT}{2\pi m_m}\right)^{\frac{1}{2}}\left(\frac{P_m}{D_0}\right)(r_e) \tag{34}$$

where

$$r_e = \Sigma_i(l_i/r_i^2)/\Sigma_j(l_j/r^3) \tag{35}$$

Equation 32 is essentially the form as employed by most investigators while equation 35 was an expression employed by Elliott et al.[22].

In his more formalistic approach, Takaishi[67] obtained the same form as equation 32 but with

$$B = \frac{3\alpha\pi^2 r\xi^2 P_m}{4kT}\left(\frac{m_g}{m_m + m_g}\right)^{\frac{1}{2}} \tag{36}$$

where ξ is the effective collision cross sectional diameter and α is a parameter to be obtained by experiment with the bounds $1 \leq \alpha \leq \frac{4}{3}$.

Both of these diffusion and kinetic theory expressions are fundamentally limited to the molecular flow range. However, the experimental studies of Nakayama[51], Rambeau[56], and Takaishi and Sensui[68] show that the form of equation 32 is applicable through the transition range, although Colgate and Genre[16] show a deviation.

(2) Comparison of Theoretical Values

Although a number of the investigators obtained agreement between their experimental values for pressure offset and the theoretical values to within a few per cent, examination of the data used in the calculations from theory indicates that differences of over 40 per cent should have been encountered, for calculation requires a selection of values for collision cross section and vapor pressure of mercury. It is also observed that the experimental data show a dispersion of the same order.

As examples, Ishii and Nakayama[34] evaluated equation 34 by using the Stefan–Maxwell form of the interdiffusion coefficient and the Ernsberger–Pitman[23] values for the mercury vapor pressure. In the first instance

$$D_0 = \frac{C_1^2 + C_2^2}{3\pi n \zeta_{12}^2} \rightarrow \frac{4kT}{3\pi^2 \zeta_{12}^2}\left\{\frac{\pi kT}{2}\left(\frac{m_m + m_g}{m_m \times m_g}\right)\right\}^{\frac{1}{2}} \tag{37}$$

Elliott et al.[22, 17] used a recipe for the diffusion coefficient of [30]

$$D_0 = \frac{3}{8}\frac{kT}{\pi\zeta^2}\left\{\frac{\pi kT}{2}\left(\frac{m_m + m_g}{m_m \times m_g}\right)\right\}^{\frac{1}{2}} \tag{38}$$

with values for ζ about two per cent larger than those of Ishii and Nakayama, and values of mercury vapor pressure as derived from the *International Critical Tables* which are about ten per cent lower than those of Ernsberger and Pitman, so that B as computed by them should be about ten per cent larger than those of Ishii and Nakayama. Nakayama[51] used the same recipe for D_0 as Elliott et al., the vapor pressure according to Ernsberger and Pitman,

PART 6. PRESSURE MEASUREMENTS FOR THE RANGE 1 kPa TO 100 μPa

and B from equation 36. When equation 38 is substituted into equation 36,

$$B \to \frac{9\pi^2}{32}\left(\frac{kT}{2\pi m_m}\right)^{\frac{1}{2}} \alpha p_m \frac{r}{D_0} \qquad (39)$$

Thus, Nakayama would have obtained values some 14 per cent larger than Elliott et al. where $\alpha = 1$, with an additional 30 per cent approximately were $\alpha = \frac{4}{3}$. Similar comparisons can be made in the computations of other workers.

(3) Compensation for the Pressure Offset

Since theory and experimental data cannot give us a sufficiently accurate calculation of the offset, the most reliable procedure is to reduce the offset to below some agreeable value by adjustment of the product $p_m r$, as in equations 34 and 39.

The vapor pressure of mercury is reduced by an order of magnitude when the surface temperature is reduced from 298 K to 273 K. Hence, the offset is reduced by about the same amount. For example a 1 cm radius tubulation at 298 K would produce an offset with nitrogen of about 29 per cent[34, 51]. But cooling to 273 K would reduce this offset to about 2.5 per cent. Ishii and Nakayama cooled the mercury surface alone; however, this can lead to error because of the unknown effect on gas temperature. To overcome this problem, Rothe[59] refrigerated the whole McLeod gage and applied a correction for thermal transpiration, while Colgate and Genre[16] used a heat exchanger to force the mercury surface back to room temperature before raising to the cut-off level of the McLeod gage, which procedure still leaves some doubt about the gas temperature.

In correcting for thermal transpiration at low pressure, unless one is assured that gas/wall scattering is cosine, equation 31 should be replaced by

$$p_1/p_2 = (T_1/T_2)^{b/2} \qquad (40)$$

where b can be different from unity by ten per cent or more[21]. Thus, because of this the uncertainty in the gage pressure is

$$(\delta p/p) = [\ln (T_1/T_2)^{\frac{1}{2}}](\delta b/b) \qquad (41)$$

Therefore, correction from 273 K to 298 K can contribute some 0.4 per cent uncertainty.

Alternatively, the value of the effective radius of the connecting tubulation needed to reduce the offset to one per cent for nitrogen gas at 298 K is computed as 0.4 mm. The disadvantage of using such a small radius is in the long response time created; that is, the conductance for a 5 cm length of capillary of this radius is about 0.001 l./s. With a volume of 2 l., the time constant for response (V/U) would be some 30 minutes. Any change in pressure in the test chamber can then require a long wait, e.g. about five times the time constant is necessary to stabilize the manometer pressure to within one per cent of the initial change; or, if a bypass line is used to shorten the response, the offset is large and one must still wait for some time after the capillary path is selected. Meinke and Reich[48] moved the capillary to a position between the mercury reservoir and the cut-off level of their McLeod gage and held the mercury within this capillary prior to measurement. At the time of measurement, they valved off the gage from the cold trap and then raised the mercury. For this procedure to be effective, all mercury must be eliminated above the capillary tube. Céspiro[12] reduced the constriction further by replacing the capillary tube with a shut-off valve and examined the time necessary for all mercury to evaporate from the system above this valve. Combining Meinke and Reich's procedure with that of Céspiro should reduce the offset to one per cent or less.

4. Viscosity Manometer

In the molecular flow range the viscosity of the gas, and hence the molecular drag on a moving element, is proportional to the pressure. Many forms of moving element instruments have been constructed. Designs include the vibrating cantilever beam, vibrating membrane, oscillating vane, torsion pendulum, rotating disc, and spinning sphere[20, 44]. Measurement of the damping force is through one of the following methods: rate of decay of oscillation, the driving force or power to retain constant amplitude of oscillation, the force imparted to a second fixed surface due to the momentum transfer from the moving element. Instruments typically have had an operating range of two to three decades in pressure for pressures between about 1 Pa and 10 µPa.

The advantages of the viscosity manometer are that interaction with the environment is minimal, materials of construction are relatively inert, and operation is compatible with a wide range of gases, vapors and corrosives. Although dependent upon gas composition, response at a fixed temperature is directly proportional to the square root of the molecular weight of the gas species, so that the relative sensitivity for any two pure gases can be calculated simply. On the other hand, most of the instruments have been delicate and somewhat tedious in operation so that application has been limited to the research laboratory. Most designs require calibration, although some forms such as the oscillating vane could serve as absolute reference standards.

A. Molecular Drag

Viscosity manometers may be categorized by the direction of motion of the moving element with respect to the active surface, i.e. normal or parallel.

Consider a small surface element of area $d\sigma$ to be in motion with a velocity u in the direction of its normal when immersed in a gas at low pressure. The drag acting on this element is then the net rate of momentum transfer due to the impinging molecules. The condition is imposed that the velocities of the incoming molecules obey a Maxwell–Boltzmann distribution with distribution function

$$f = f(v_x, v_y, v_z) = (m/2\pi kT)^{\frac{3}{2}} \exp(-mv^2/2kT) \tag{42}$$

for which

$$\int_{-\infty}^{\infty}\int\int f(v_x, v_y, v_z)\, dv_x\, dv_y\, dv_z = 1 \tag{43}$$

PART 6. PRESSURE MEASUREMENTS FOR THE RANGE 1 kPa TO 100 μPa

and the average speed $\langle v \rangle$ is

$$\int \int_{-\infty}^{\infty} \int v\, f(v)\, dv_x\, dv_y\, dv_z$$

or

$$\langle v \rangle = \int_0^\infty \int_0^\pi \int_0^{2\pi} v^3\, f(v)\, \sin\theta\, d\theta\, d\phi\, dv \tag{44}$$

where the v_i are velocity components, m is the molecular mass, k is the Boltzmann constant and T is the absolute temperature.

As indicated in the diagram of *Figure 15*, a molecule approaching the back

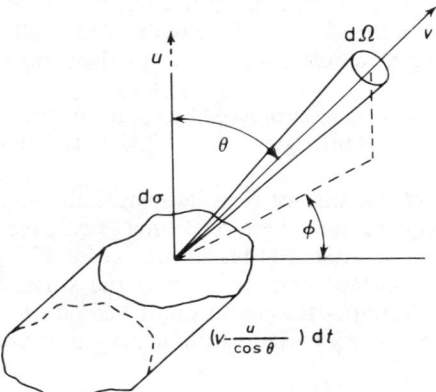

Figure 15. Momentum transfer to the moving element of a viscosity manometer—flux diagram.

of $d\sigma$ at an angle θ will only arrive if its speed v is greater than $u/\cos\theta$. Thus, the number of molecules which collide with the back face of $d\sigma$ in an interval of time dt from an angle of θ, ϕ with speed v must be that fraction of all those contained within the cylinder of slant height $(v - u/\cos\theta)$ and volume $(v - u/\cos\theta)\cos\theta\, d\sigma\, dt$ that have a probability $f(v)v^2 d\Omega\, dv$ of travelling in that direction, where $d\Omega$ is the elemental solid angle corresponding to a range of $d\theta$ at θ and $d\phi$ at ϕ. If n is the molecular concentration, the number of possible arrivals is then

$$nf(v)\, v^2\, (v - u/\cos\theta)\cos\theta\, d\sigma\, d\Omega\, dt\, dv \tag{45}$$

As each molecule transports a relative momentum normal to the surface of $m(v\cos\theta - u)$, the net flow of momentum per unit time, per unit area into the back surface is

$$F_{b\,in} = \int_{v=0}^{\infty} \int_{\theta=0}^{\pi/2} \int_{\phi=0}^{2\pi} nm(v\cos\theta - u)^2\, f(v)\, v^2\, \sin\theta\, d\theta\, d\phi\, dv \tag{46}$$

Anderson[2] and Christian[13] assume that the pressure exerted on the face is twice the momentum exchange of equation 46, or $2F_{bin}$.

The contribution at the top face of dσ is similarly

$$p_t = 2 \int_0^\infty \int_0^{\pi/2} \int_0^{2\pi} nm(v\cos\theta + u)^2\, f(v)\, v^2 \sin\theta\, d\theta\, d\phi\, dv \qquad (47)$$

so that the net force on dσ per unit area is the difference between equations 47 and 46 or

$$p_t - p_b = 8 \iiint nmuv^3\, f(v) \cos\theta \sin\theta\, d\theta\, d\phi\, dv \qquad (48)$$

which with the definition of $\langle v \rangle$ from equation 44 gives the damping force per unit area as

$$F_d = 2nmu\langle v \rangle \equiv 4(2m/\pi kT)^{\frac{1}{2}} up \qquad (49)$$

Evrard and Boutry[24] assume that the molecular impact is inelastic and that the molecules re-evaporate off the surface with a Maxwell–Boltzmann distribution for the surface as a reference plane. In their case the factor of 2 is replaced by 1.78.

The magnitude of the drag as expressed by equation 49 for a surface velocity in m/s and a gas pressure of nitrogen in Pa at 296 K is 0.269 µPaN per square meter of area.

When the motion of the surface is in its own plane, or parallel, as in the case of the suspended rotating sphere of Beams et al.[7], the drag is of smaller value. In this case, we may assume that all molecules with a normal velocity >0 can arrive at the surface and the flux to the surface per unit area is $n\langle v \rangle/4$. If cosine scattering is the situation, a velocity u is imparted to all evaporating molecules; whence, the resultant drag/unit area is

$$F_p = nmu\langle v \rangle / 4 \qquad (50)$$

or $\frac{1}{8}$ of that indicated by equation 49. If scattering were specular, no drag would occur.

B. Oscillating Vane

(1) Description

In this design a relatively large rectangular vane is hinged at the top and allowed to swing freely. As the damping torque on an elemental area dσ at a distance r from the pivot is $F_d r\, d\sigma$, the torque τ on the vane is by equation 49

$$\tau = 4(2m/\pi kT)^{\frac{1}{2}} \times p\,\dot\theta \int r^2\, d\sigma \qquad (51)$$

where $\dot\theta$ is the angular velocity. For a vane of width a and length b,

$$\int r^2\, d\sigma = \tfrac{1}{3}ab^3.$$

The equation of motion is then

$$I\ddot\theta + (R + R_0)\dot\theta + f(\theta)\theta + Mgl \sin\theta = 0 \qquad (52)$$

where I is the moment of inertia of the system, R is the resistive or damping factor from equation 51, $Mgl \sin\theta$ the restoring torque due to the weight of

PART 6. PRESSURE MEASUREMENTS FOR THE RANGE 1 kPa TO 100 μPa

the pendulum acting at the center of mass at a distance l from the pivot line, R_0 a resistive component due to the pivot, and $f(\theta)$ represents a mechanical stiffness in the suspension. If the vane swings through small angles with no flexure in the suspension, $f(\theta)$ may be neglected and $\sin \theta \sim \theta$. Since the damping force is small, the amplitude of oscillation is then

$$\theta = \theta_0 \exp[-\{(R + R_0)/2I\}] \cos(\omega t + \varepsilon) \tag{53}$$

with a frequency of

$$v = \frac{1}{2\pi}\left[\frac{Mgl}{I} - \left(\frac{R + R_0}{4I^2}\right)^2\right]^{\frac{1}{2}} \tag{54}$$

The logarithm of the ratio of two successive amplitudes is the decrement

$$\delta = (R + R_0)/2Iv \tag{55}$$

or alternatively the time necessary to fall to half the initial amplitude is

$$t_{0.5} = \{2I/(R + R_0)\} \ln 2 \tag{55^1}$$

so that from equation 51

$$Kp = (2I/t_{0.5}) \ln 2 - R_0 \tag{56}$$

where

$$K = (4/3)(2m/\pi kT)^{\frac{1}{2}} ab^3 \tag{57}$$

and R_0 may be evaluated by determining $t_{0.5}$ at a pressure $p_0 \ll p$.

Extension of the range to the lowest pressures requires a minimum moment of inertia with large vane area and length. Vane dimensions of about 10 cm on edge are required for a least count of about 10 μPa[75]. Such pressures require observation times of about 10^4 second.

To reduce observation time and increase convenience, Christian and Leck[14] used the vane as a moving capacitor plate in a frequency modulated circuit to apply a synchronous driving torque, while using the output voltage as a measure of amplitude of oscillation. Austin[4] modified the electronics so as to monitor power input. With an aluminum vane of 9 cm length by 5 cm width and 5 μm thickness in a pendulum of 10 cm length, they obtained a vane oscillation of about 2 Hz with a minimum detectable signal of about 70 μPa.

(2) Accuracy

Christian and Leck[14] intercompared their oscillating vane manometer with a conductance limited dynamic pressure generator for a number of gases. Although measurement reproducibility was quoted as ±70 μPa for pressures up to 13 mPa, adherence to the $M^{\frac{1}{2}}$ law was off by 15 per cent for xenon while measured values of K in equation 57 were different from computed values by factors of two to three. Their instrument, however, was housed in a relatively small enclosure. With increased enclosure volume and greater accuracy in the measurement of vane losses and pressure, Austin[4] obtained agreement between measured and computed values of K for a number of gases to be within five to twelve per cent, with the experimental values greater than the computed values.

C. DIAMAGNETIC LEVITATION MANOMETER

(1) Description

In order to eliminate the pivot problems of the oscillating vane Evrard and Boutry[24] supported a small rotating graphite disc with a non-uniform magnetic field with the axis of rotation being a diameter of the disc as in

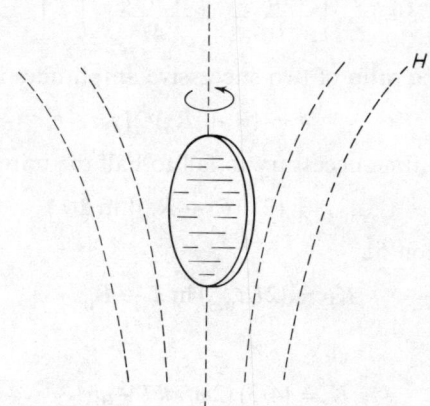

Figure 16. Diamagnetic suspension of the spinning disc viscosity manometer.

Figure 16. For this configuration the surface was in motion normal to its plane reducing equation 52 to

$$I\dot{\omega} + R\omega = 0$$

for $\dot{\theta} \equiv \omega$ such that

$$\dot{\omega} = -(R/I)\omega$$

with R from equation 51 as

$$R = 3.56(2m/\pi kT)^{\frac{1}{2}} p \int r^2 \, d\sigma$$

where the factor of 2 in equation 49 has been replaced by their factor of 1.78. Since the moment of inertia of a disc of thickness t is $\rho t \int r^2 \, d\sigma$ and the mass of a disc of density ρ is $\pi r^2 \rho t$, the solution for ω becomes

$$\ln(\omega/\omega_0) = -3.56(2m/\pi kT)^{\frac{1}{2}} \pi r^2 (pI/M)(t - t_0) \qquad (58)$$

The speed of rotation was determined photoelectrically.

Their disc was of 3 mm diameter by 0.05 mm thickness and was suspended within a glass tube of 1 cm diameter with an external magnet. This made for a relatively simple, rugged structure with an operating range to 10 µPa.

(2) Accuracy

In an intercomparison of this type of instrument with a liquid column

PART 6. PRESSURE MEASUREMENTS FOR THE RANGE 1 kPa TO 100 μPa

micromanometer and a conductance limited flow apparatus, Gourjault et al.[28] found agreement to within a few per cent. However, in their derivation of the damping torque, Evrard and Boutry neglected the contribution of the edge of the disc, where the direction of motion was parallel to the surface. Although equation 50 shows the drag to be $\frac{1}{8}$ of the amount on the faces of the disc, a comparison of $\int r^2 \, d\sigma$ for the faces and edge indicates that the edge value is 13 per cent of the face amount so that the edge could contribute some two per cent to the viscous drag.

5. Knudsen Radiometer Manometer

A gage of increased sensitivity independent of gas composition was originally formulated by Knudsen which preferentially directed momentum transfer between plates at different temperatures[20, 44]. In the diagram of *Figure 17* are two plates A, held transversely on a suspension S and close to

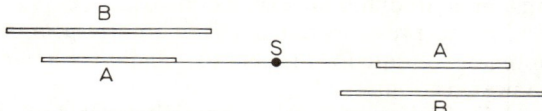

Figure 17. Schematic of the Knudsen gage vane structure. A—suspension and detector vanes. B—heated fixed vanes.

two fixed parallel plates B, surrounded by a gas at low pressure with the mean free path much larger than the gap between A and B. The plates A are presumably at gas temperature T while plates B are set to a known temperature $T_B > T$ with auxiliary heaters. Thus, molecules arriving at the side of vanes A and facing vanes B have greater velocities than those coming from the gas phase. This produces a torque on the suspension.

If one considers this a thermal transpiration effect[37] according to equation 31 with a temperature in the gap as an average value, then the pressure in the gap is

$$p_g = p\left[\frac{\frac{1}{2}(T + T_B)}{T}\right]^{\frac{1}{2}}$$

The net outward force per unit area on A is

$$P = p_g - p = p\left[\{(T + T_B)/2T\}^{\frac{1}{2}} - 1\right] \tag{59}$$

According to this the torque on the suspension depends only on the gas pressure and temperatures.

To eliminate the temperature of the movable plates A as a factor, one can place a third set of fixed plates on the opposite sides of A. When a detailed analysis is made including the energy transfer at the surfaces, equation 59 is replaced by[44]

$$P = \tfrac{1}{2}p[(T_1/T)^{\frac{1}{2}} - (T_3/T)^{\frac{1}{2}}][\alpha_1(2 - \alpha_2)/(\alpha_1 + \alpha_2 - \alpha_1\alpha_2)] \tag{60}$$

where α_1 and α_2 are the accommodation coefficients at the wall, i.e. a measure of the temperature equilibration between the impacting molecule and the wall. It is assumed that $\alpha_1 = \alpha_3$. The measured pressure is sensitive to the difference in the accommodation coefficients. The expression equation 60 is further dependent upon the assumption of a Maxwellian distribution between the plates. This is probably not valid[76].

The practical difficulties of construction have precluded this gage from extensive application. It is interesting to note that Evrard and Boutry[24] have combined the rotating disc with a vane geometry similar to that of Klumb and Schwarz[39] to obtain a response into the 0.1 µPa range, although theirs is not used as an absolute manometer.

III. Pressure Generators

Pressure point generators are employed to calibrate transfer gages or to establish a given environment within which the experiment may proceed. Two methods have been used which are basically gas expansion procedures, i.e. static expansion and dynamic expansion. Static expansion produces greater accuracy in the pressure range 1 kPa to 100 µPa, while dynamic expansion is more appropriate for the very low and ultra low pressures to be discussed in the next section.

A fundamental limitation with these procedures is that any output is always through a transfer gage with consequently reduced accuracy.

1. *Volumetric Pressure Divider—Static Expansion*

A. Description

Basically this procedure is the inverse of that of the McLeod gage. A small volume of permanent gas at a known low pressure, volume and temperature is expanded into a larger volume initially at low pressure. Large expansion

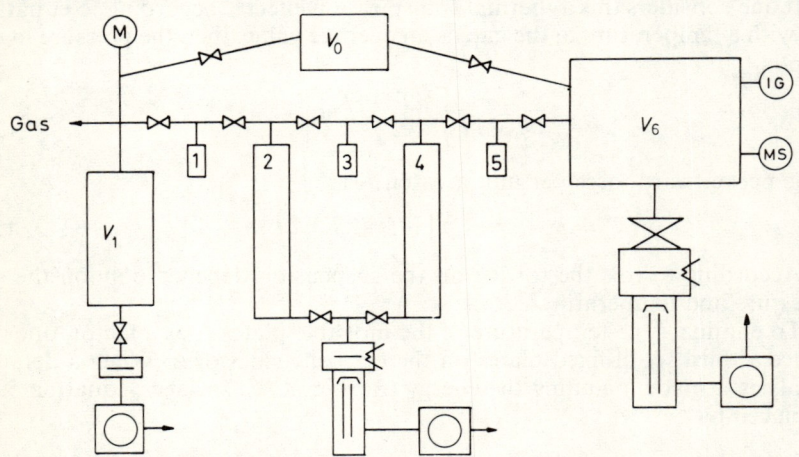

Figure 18. Schematic for a volumetric pressure divider—multiple expansion. Volumes 1, 3, 5 are sampling chambers. Volumes 2, 4, 6 are large expansion chambers. V_1—gas reservoir. V_0—secondary chamber of known volume used to calibrate large volume V_6.

PART 6. PRESSURE MEASUREMENTS FOR THE RANGE 1 kPa TO 100 μPa

ratios are possible. Initial pressure is measured independently of the expansion cycle, and cascading can be used for large expansion ratios. Operation is relatively rapid. It is, however, subject to a number of dynamic effects such as sorption, outgassing, and ancillary test gage pumping.

The attractiveness of the procedure is in the simplicity of the theory, without all the attendant effects that occur with the McLeod gage, and in the relative ease of obtaining adequate accuracy in the initial pressure measurement. Several variations on this theme have been practised in recent years, mostly concerning the method of obtaining large expansion ratios[19, 41]. Some have used multiple expansion[5, 22, 49, 62], while others employed single expansion into a very large chamber[25, 64, 72].

A multiple expansion apparatus for high precision has a structure as indicated schematically in *Figure 18*. Volumes 1, 3 and 5 are small sampling chambers, while volumes 2, 4 and 6 are large. The gas is first admitted into chambers V_1 and 1 and then sequentially expanded into V_6. Expansion ratios of 10^7 to 10^8 can be obtained from volume 1 to V_6.

Curve 5 of *Figure 6* is an uncertainty characteristic indicative of this type of procedure.

B. Operation

After the systems have been pumped and processed, the chambers are isolated from the pumps. Gas is admitted to volumes V_1 and 1, stabilized to temperature T_1, and pressure measured with the manometer. The small volume 1 is then isolated from V_1 and opened to volumes 2 and 3. Volume 3 is then isolated from 2 and opened to 4 and 5. Finally, 5 is isolated from 4 and opened to V_6. From the gas law the expansion ratios for each sequence are

$$R_1 = \frac{V_1}{T_1\left(\frac{V_1}{T_1} + \frac{V_2}{T_2} + \frac{V_3}{T_3} + \Delta\right)}$$

$$R_2 = \frac{V_3}{T_3\left(\frac{V_3}{T_3} + \frac{V_4}{T_4} + \frac{V_5}{T_5} + \Delta\right)} \quad (61)$$

$$R_3 = \frac{V_5 T_6}{V_6 T_5}$$

A volume, Δ, is included for the valves if these are not of constant volume and for correcting the volume of ionization gages due to operation at temperatures greater than ambient. Such effects are not negligible in small volumes. For example, valves are opened and closed twice for each cycle where each closure could introduce as much as 5 to 10 cm³ change. The ionization gage correction is due to the increased operating temperature so that the effective volume is $V_0 T_0/T$ for T_0 at room temperature; thus wall temperatures of 350 K introduce ten per cent correction or about 10 cm³.

In practice the pressure in each chamber is not constant but drifts due to outgassing from the walls and pumping by the attendant ionization gage as indicated in *Figure 19*. Expansion of the gas into that chamber then causes a jump in pressure followed by the same drift. *Figure 19* is the case for a well

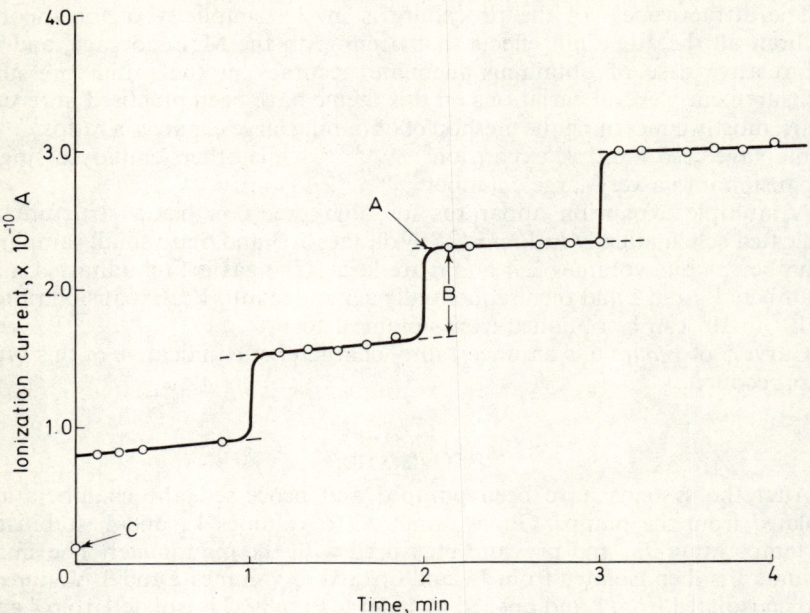

Figure 19. Pressure profile for volumetric divider with argon gas. A—expansion at time t'. B—pressure increment. C—residual gas 'pressure' in chamber before expansion.

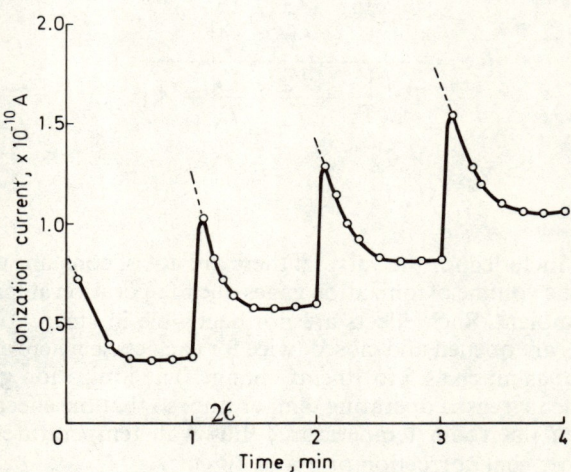

Figure 20. Pressure profile for volumetric divider with hydrogen gas.

PART 6. PRESSURE MEASUREMENTS FOR THE RANGE 1 kPa TO 100 μPa

behaved test gas[5]. Thus, for each expansion cycle such as from volume 1 into volumes 2 and 3.

$$p_3 = \frac{Rp_1}{1+R} + \frac{1}{1+R}\left[p_{30} + \frac{dp_{30}}{dt}(t' - t_0)\right] + \frac{dp_3}{dt}(t - t') - p_a \qquad (62)$$

where p_1 is the initial pressure in V_1, p_{30} that in V_3 before gas entry; dp_{30}/dt is the drift rate in $V_2 \mp V_3$ before, and dp_3/dt is that afterwards; $t' - t_0$ is the interval of drift before gas entry; $t - t'$ is the time of drift after entry until the time of observation at t; p_a is the equivalent pressure of adsorbed gas.

On the other hand a test gas such as hydrogen that is strongly adsorbed on the walls or pumped can have a behavior such as that indicated by *Figure 20*[5]. Extrapolation of the base time at each expansion is required to determine actual pressure change, which reduces accuracy accordingly.

C. Uncertainties

The characteristic curve No. 5 of *Figure 6* is derived as an exercise using a model representing this procedure by utilizing operationally found imprecisions. Let volumes 1, 3 and 5 be 0.1 l., volumes 0, 2 and 4 be cubical chambers 25 cm on edge, and V_6 be a cubical chamber 50 cm on edge.

All the components are constructed of stainless steel. All volumes and interconnecting lines except V_6 are determined by direct weighing with fill water. We assume careful procedures to achieve 0.1 per cent for the small volumes and 0.01 per cent for intermediate volumes. The large chamber V_6 is measured by gas expansion in turn from V_0. We control temperature to 300 ± 0.5 K. Chambers 2 and 4 along with associated lines are each pumped at 75 l./s. Chamber V_6 is pumped at 200 l./s. Baking at 400 K is used until the wall outgassing rate at room temperature is 10^{-12} Torr l./s cm^2 (10^{-9} Pa m^3/s m^2). After initial processing we would find the conditions in the chambers to be as in *Table 3*. Ionization gage pumping speed has been taken as 0.1 l./s.

Table 3. Volumetric pressure divider—initial conditions, multiple expansion

	V_2, V_4	V_6
Length on edge	25 cm	50 cm
Wall area	3 750 cm^2	15 000 cm^2
Volume	15.6 l.	125 l.
Outgassing rate	3.75×10^{-9} tl/s	1.5×10^{-8} tl/s
Pump speed	75 l./s	200 l./s
Blank-off	5×10^{-11} Torr	7×10^{-11} Torr
dp/dt after isolation, outgas	1.4×10^{-8} Torr/min	7.2×10^{-9} Torr/min
dp/dt from ion. gage pumping		
at 10^{-8} Torr	-3.8×10^{-9} Torr/min	-5×10^{-10} Torr/min
at 10^{-6} Torr	-3.8×10^{-7} Torr/min	-5×10^{-8} Torr/min

In the first expansion cycle, R and p_1 are independent variables, but this is not so for the second and third expansions. Normally, the first expansion would involve a r.m.s. of the imprecisions, but examination of data shows order of magnitude differences in some of the factors. Thus, for all expansions we choose the form

$$\delta p_3 = R_1 \delta p_1 + p_1 \delta R_1 + \delta p_{30} + (t' - t_0)\delta(dp_{30}/dt)$$
$$+ (t - t') \delta(dP_3/dt) \quad (63)$$

and

$$\frac{\delta R_1}{R_1} = \left[\left\{\frac{\delta V_1}{V_1}\right\}^2 + \left\{\frac{\delta(V_1 + V_2 + V_3)}{V_1 + V_2 + V_3}\right\}^2 + \left\{\frac{\delta T_2}{T_2}\right\}^2 + \left\{\frac{\delta T_1}{T_1}\right\}^2\right]^{\frac{1}{2}} \quad (64)$$

Valves are assumed to be of constant volume. During expansions the temperature for the sum of three volumes is taken as equivalent to the larger volumes V_2 and V_4. Error in $\delta(t' - t_0) \, dp_{30}/dt$ and $\delta(t - t') \, dp_3/dt$ is considered small. Initial pressures are to be in the 1 000 Torr range (1.3 × 10^5 Pa) where 3σ imprecision of measurements with a liquid manometer is 0.1 Torr or in the 1 Torr range where precision is easily one part in 10^3 as with a U-tube Micrometer Point Contact Manometer. We assume blank-off pressure in the chambers is measurable to 20 per cent (optimistic) and that the drift can be estimated to within five per cent. Sample results for curve 5 of *Figure 6* are given in *Table 4*.

As represented, curve 5 indicates a significant improvement over the MeLeod gage, curve 4. However, tests of precisions obtained have been made primarily with ionization gages. Comparisons have been made against a precision McLeod gage[22] and against a dynamic pressure divider through transfer gages[49]. In these comparisons via ionization gage, data showed no significant distinctions.

IV. Transfer Gages—Precision

Two types of instruments have shown sufficient stability and precision to serve this pressure range. These are the quartz Bourdon and capacitance–diaphragm manometers. The quartz Bourdon has a precision (3σ) with a low pressure cell of about 2 mTorr, a least count of <1 mTorr, may have small systematic deviations of 2–3 mTorr, and is stable for long periods. The capacitance–diaphragm gage extends to lower pressures. For a 1 Torr full scale range, some appear to have a capability of about two per cent precision at 1 mTorr increasing to some few parts in 10^4 at 1 Torr. The characteristic seems to be a quadratic fit of the form

$$p = A + Bx + Cx^2 \quad (65)$$

where x is the test gage reading and p is the measured pressure. An example has $A = 0.000005 \pm 0.000051$, $B = 1.02839 \pm 0.00044$, and $C = -0.00876 \pm 0.00049$ where the uncertainties given are the least square estimates of the standard deviation of the coefficients with an estimator of the standard deviation of the fit being 0.00020. These coefficients reproduced the test pressures to within a maximum error of ± 0.0006 Torr. Values vary

PART 6. PRESSURE MEASUREMENTS FOR THE RANGE 1 kPa TO 100 μPa

Table 4. Imprecisions after expansion—volumetric pressure divider

	1st expansion	A 2nd expn	3rd expn
p_{initial}	$p_{11} = 1580$ Torr	$p_{31} = 10$ Torr	$p_{51} = 6.3 \times 10^{-2}$ Torr
p_{final} R	$p_{31} = 10$ 6.33×10^{-3}	$p_{51} = 6.3 \times 10^{-2}$ 6.33×10^{-3}	$p_6 = 5 \times 10^{-5}$ 8×10^{-4}
$\delta R/R$ $R\,\delta p$ $p\,\delta R$	0.3% 6.33×10^{-4} 3.2×10^{-2}	0.3% 2.1×10^{-4} 2×10^{-4}	0.4% 3.3×10^{-7} 3.7×10^{-7}
δp_0	—	2×10^{-11}	2×10^{-11}
$\delta p'_0(t' - t_0)$		7×10^{-10}	3.6×10^{-10}
$\delta p'(t - t')$		7×10^{-10}	3.6×10^{-10}
$\delta p/p$	0.3%	0.65%	1.4%
		B	
p_i	$p_{11} = 1$	$p_{31} = 6.3 \times 10^{-3}$	$p_{51} = 3.97 \times 10^{-5}$
p_f	$p_{31} = 6.3 \times 10^{-3}$	$p_{51} = 3.97 \times 10^{-5}$	$p_6 = 3.18 \times 10^{-8}$
R	6.33×10^{-3}	6.33×10^{-3}	8×10^{-4}
$\delta R/R$ $R\,\delta p$	0.3% 6×10^{-6}	0.3% 1.6×10^{-6}	0.4% 1.37×10^{-9}
$p\,\delta R$	2×10^{-5}	1.3×10^{-7}	2.2×10^{-10}
δp_0		2×10^{-11}	2×10^{-11}
$\delta p'_0(t' - t_2)$	1.2×10^{-8}	7×10^{-10}	3.6×10^{-10}
$\delta p'(t - t')$	1.2×10^{-8}	7×10^{-10}	3.6×10^{-10}
$\delta p/p$	0.41%	4.3%	8%

with instruments. With reasonable treatment, this type of instrument can also hold its calibration for extended periods, but can evidence a sudden permanent shift of the order of a few parts in 10^3 at full scale of 1 Torr.

The ionization gage is the principal instrument of the low pressure region. Typical reported results with dynamic expanders indicate five per cent repeatability for a single run for an individual tube in the 1×10^{-6} Torr range (100 µPa), ten per cent agreement after lapsed time, and larger values if the tube were removed from the stand and then replaced at a later time.

References

[1] Akiyama, Y., H. Hashimoto and K. Nakayama. *J. Vac. Sci., Japan*, **7**, 12 (1964).
[2] Anderson, J. R. *Rev. Sci. Instrum.* **29**, 1073 (1958).
[3] Aubry, B. and R. Delbart. *Le Vide, Paris*, **117**, 194 (1965).
[4] Austin, W. E. *Vacuum*, **19**, 319 (1969).
[5] Barton, R. S. and J. N. Chubb. *Vacuum*, **15**, 113 (1965).
[6] Bashforth, F. and J. C. Adams. *An Attempt to Test the Theories of Capillary Action*, Cambridge University Press: London (1883).
[7] Beams, J. W., J. L. Young and J. W. Moore. *J. Appl. Phys.* **17**, 886 (1946).
[8] Blaisdell, B. E. *J. Math. Phys.* **19**, 217 (1940).
[9] Brombacher, W. G., D. P. Johnson and J. L. Cross. *US Nat. Bur. Stand. Monogr. 8* (1960).
[10] Brombacher, W. G. *US Nat. Bur. Stand. Monogr. 114* (1970).
[11] Burden, R. S. *Surface Tension and the Spreading of Liquids*, 2nd ed. Cambridge University Press: London (1949).
[12] Céspiro, Z. *Vakuum Tekhnik.* **17**, 68 (1968).
[13] Christian, R. G. *Vacuum*, **16**, 175 (1966).
[14] Christian, R. G. and J. H. Leck. *Vacuum*, **16**, 299 (1966).
[15] Clark, R. J. *J. Sci. Instrum.* **5**, 126 (1928).
[16] Colgate, S. O. and P. A. Genre. *Vacuum*, **18**, 553 (1968).
[17] Dadson, R. S., K. W. T. Elliott and D. M. Woodman. *Proceedings* of the Fourth International Vacuum Congress, p 1. Institute of Physics and the Physical Society: London (1968).
[18] Dunoyer, L. *Vacuum Practice*, p 68. Bell: London (1926).
[19] Dunoyer, L. *Vacuum Practice*, p 91. Bell: London (1926).
[20] Dushman, S. and J. M. Lafferty. *Scientific Foundations of Vacuum Technique*, 2nd ed., p 244. Wiley: New York (1962).
[21] Edwards, T. and J. P. Hobson. *J. Vac. Sci. Technol.* **2**, 182 (1965).
[22] Elliott, K. W. T., D. M. Woodman and R. S. Dadson. *Vacuum*, **17**, 439 (1967).
[23] Ernsberger, F. M. and H. W. Pitman. *Rev. Sci. Instrum.* **26**, 584 (1955).
[24] Evrard, R. and G. A. Boutry. *J. Vac. Sci. Technol.* **6**, 279 (1969).
[25] Flanick, A. P. and J. E. Ainsworth, *Rev. Sci. Instrum.* **32**, 408 (1961).
[26] Gaede, W. *Ann. Phys., Lpz.* **46**, 357 (1915).
[27] Gould, F. A. and T. Vickers. *J. Sci. Instrum.* **29**, 35 (1952).
[28] Gourjault, J. J., J. Lefévre and P. S. Choumoff. *J. Vac. Sci. Technol.* **9**, 206 (1972).
[29] Guildner, L. A., H. F. Stimson, R. E. Edsinger and R. L. Anderson. *Metrologia*, **6**, 1 (1970).
[30] Hirschfelder, J. O., C. F. Curtiss and R. B. Bird. *Molecular Theory of Gases and Liquids*. Chapman and Hall: London (1954).
[31] Hobson, J. P. *J. Vac. Sci. Technol.* **6**, 257 (1969).
[32] Hobson, J. P., T. Edwards and R. Verreault. *Canad. J. Phys.* **41**, 983 (1963).
[33] Huntress, A. H., A. L. Smith, B. D. Power and N. T. M. Dennis. *Transactions* of the Fourth National Vacuum Symposium, p 104. American Vacuum Society/Pergamon: New York (1958).
[34] Ishii, H. and K. Nakayama. *Transactions* of the Eighth Vacuum Symposium and Second International Congress of 1961, p 519. Pergamon: New York (1961).
[35] Jansen, C. G. and A. Venema. *Vacuum*, **9**, 219 (1959).
[36] Keevil, N. G., R. F. Errington and L. T. Newman. *Rev. Sci. Instrum.* **12**, 609 (1941).
[37] Kennard, E. H. *Kinetic Theory of Gases*, p 67. McGraw-Hill: New York (1938).
[38] Kistemaker, J. *Physica*, **11**, 270 (1945).
[39] Klumb, H. and H. Schwarz. *Z. Phys.* **122**, 418 (1944).
[40] Knudsen, M. *Ann. Phys., Lpz.* **33**, 1435 (1910); **83**, 797 (1927).
[41] Knudsen, M. *Ann. Phys., Lpz.* **44**, 525 (1914).
[42] Laplace, P. *Traité de Mécanique Céleste.* **10**, 64. Duprat: Paris (1798, 1823).

PART 6. PRESSURE MEASUREMENTS FOR THE RANGE 1 kPa TO 100 μPa

[43] Laplace, P. *Oeuvres Complètes*. **4**, 349. Gauthier-Villars: Paris (1880).
[44] Leck, J. M. *Pressure Measurement in Vacuum Systems*, 2nd ed., p 35. Chapman and Hall: London (1964).
[45] Liang, S. C. *J. Appl. Phys.* **22**, 148 (1952); *J. Phys. Chem.* **57**, 910 (1953).
[46] Maxwell, J. C. *Phil. Trans. Roy. Soc., London*, **170**, 231 (1879).
[47] McLeod, H. *Phil. Mag.* **48**, 110 (1874).
[48] Meinke, C. and G. Reich. *Vakuum-Technik*, **12**, 79 (1963); **11**, 86 (1962); *Vacuum*, **13**, 579 (1963).
[49] Meinke, C. and G. Reich. *J. Vac. Sci. Technol.* **4**, 356 (1967).
[50] Muendel, C. F. *Z. Phys. Chem.* **85**, 435 (1913).
[51] Nakayama, K. *Jap. J. Appl. Phys.* **7**, 1114 (1968).
[52] Neumann, C. *S.B. Akad. Wiss. Wien*, Abt. IIa, **24**, 49 (1872).
[53] Nottingham, W. B. and F. L. Torney Jr. *Transactions* of the Seventh National Vacuum Symposium, p 117. American Vacuum Society/Pergamon: New York (1961).
[54] Podgurski, H. H. and F. N. Davis. *Vacuum*, **10**, 377 (1960).
[55] Podgurski, H. H. and F. N. Davis. *J. Phys. Chem.* **65**, 1343 (1961).
[56] Rambeau, G. *Le Vide, Paris*, **24**, 219 (1969).
[57] Reynolds, O. *Phil. Trans. Roy. Soc. London*, **170**, 727 (1879).
[58] Rosenberg, P. *Rev. Sci. Instrum.* **9**, 258 (1938).
[59] Rothe, E. W. *J. Vac. Sci. Technol.* **1**, 66 (1964).
[60] Rusch, M. and O. Bunge. *Z. Tech. Phys.* **13**, 77 (1932).
[61] Ruthberg, S. *J. Vac. Sci. Technol.* **6**, 401 (1969).
[62] Schuhmann, S. *Transactions* of the Ninth National Vacuum Symposium, p 493. American Vacuum Society/Macmillan: New York (1962).
[63] Siu, M. C. I. *J. Res. Nat. Bur. Stand.* **73A**, 611 (1969).
[64] Smetana, F. O. and C. T. Carley Jr. *J. Vac. Sci. Technol.* **3**, 49 (1966).
[65] Stevenson, W. H. and P. W. McFadden. *Rev. Sci. Instrum.* **36**, 1272 (1965).
[66] Stillman, M. H. *Sci. Pap. US Nat. Bur. Stand.* **10**, 371 (1914).
[67] Takaishi, T. *Trans. Faraday Soc.* **61**, 840 (1965).
[68] Takaishi, T. and Y. Sensui. *Vacuum*, **20**, 495 (1970).
[69] Thomas, A. M. and J. L. Cross. *J. Vac. Sci. Technol.* **4**, 1 (1967).
[70] Thomas, A. M., D. P. Johnson and J. W. Little. *Transactions* of the Ninth National Vacuum Symposium, p 468. American Vacuum Society/Macmillan: New York (1962).
[71] Tunnicliffe, R. J. and J. A. Rees. *Vacuum*, **17**, 457 (1967).
[72] Vermandé, M. *J. Le Vide, Paris*, **7**, 1145 (1952).
[73] de Vries, A. E. and P. K. Rol. *Vacuum*, **15**, 135 (1965).
[74] Weber, S. and G. Schmidt. *Commun. Kamerlingh Onnes Lab., Leiden*, **246c**, 11 (1936).
[75] Wetterer, G. *Z. Tech. Phys.* **20**, 281 (1939).
[76] Wu, Y. *Ann. Phys., Lpz.* **18**, 321 (1966).
[77] Young, T. *Miscellaneous Works*, **1**, 418 (1805).

CHAPTER 4

Part 7. Very Low Pressures and Ultra Low Pressures (below 10^{-6} Torr)

P. A. REDHEAD

National Research Council, Ottawa, Canada

Contents

I.	Introduction	273
II.	General Problems of Low Pressure Measurements	274
	1. Gages as Sinks or Sources	274
	A. Pumping and Re-emmission in Gages	274
	B. Gas Interactions at Hot Surfaces	276
	C. Blears Effect	277
	2. Measurement in Non-uniform Environments	277
	3. Residual Currents	278
	A. Soft X-Ray Photoemission	279
	B. Electronic Desorption	280
	C. Measurement of Residual Current	283
	4. Relative Gage Sensitivities for Different Gases	284
	5. Calibration of Gages	286
	A. Absolute Calibrations	286
	B. Relative Calibrations	287
	(a) Static Expansion	287
	(b) Linear Rise Rate	287
	(c) Single-stage Pressure Division	287
	(d) Quadratic Rise Rate	287
	(e) Quadratic Rise Rate and Single-stage Division	287
	(f) Multiple-stage Division	289
III.	Pressure Measurements from 10^{-6} to 10^{-10} Torr	289
	1. Hot-cathode Ionization Gages	290
	2. Cold-cathode Gages	291
	A. Penning Gages	291
	B. Magnetron Gage	292
	C. Inverted-magnetron Gage	293
IV.	Pressure Measurements below 10^{-10} Torr	294
	1. Shielded-collector Gages	294
	2. Bent-beam Gage	294
	3. Hot-cathode Magnetron Gage	296
V.	Comparison of Gages	297
VI.	References	297

I. Introduction

Pressures below 10^{-6} Torr are measured almost exclusively with ionization gages. These gages have the advantages of high sensitivity, potentiality for measuring pressures as low as 10^{-18} Torr, and relative simplicity of the gage and its electronic controls. The attendant disadvantages are rather poor absolute accuracy, serious interactions between the gage and the system being measured, and the need for considerable skill and knowledge on the part of the operator to obtain reproducible, accurate results. Other types of gage have been developed which do not require ionization of the gas and which are potentially capable of measuring pressures in the ultrahigh vacuum (u.h.v.) range ($<10^{-9}$ Torr) which may not suffer from the disadvantages of ionization gages. None of these gages has been developed to the point where it is a serious competitor to the ionization gage at pressures below 10^{-6} Torr, nor have the operational problems of these gages been clearly established. This chapter will concentrate its attention on ionization gages with only very brief reference to other types.

Ionization gages measure molecular density rather than pressure, thus it is necessary to know the temperature and nature of the gas to convert the observed readings to measurements of total pressure. In many experiments, particularly at very low pressures, a measure of *total* pressure is inadequate and it becomes necessary to measure partial pressures with a suitable mass spectrometer. The general considerations of low pressure measurements outlined in section II below are equally applicable to measurements of partial pressure.

Serious difficulties still exist in making accurate low-pressure measurements, partly resulting from instrumental inadequacies, which are greatly increased with chemically active gases. As a result, the approach taken here is somewhat different to those chapters discussing more highly developed measurement methods. In this chapter more emphasis is placed on the methods of gage operation and the precautions required than on the detailed design of the instruments.

Several books and review articles on the subject of low pressure measurements are available to which the reader is referred for further details[59, 69, 98, 99, 100, 113, 114]. Brombacher[16, 17] has published useful bibliographies on low pressure measurements.

II. General Problems of Low Pressure Measurements

This section briefly discusses those problems which are general to the measurement of both total and partial pressures below 10^{-6} Torr.

1. Gages as Sinks or Sources

A. Pumping and Re-emission in Gages

All vacuum gages pump and/or evolve gas to some degree. To obtain accurate measurements of pressure in a vacuum system it is necessary to minimize these effects or to measure the pumping or emission rates so that the necessary corrections can be made. These effects are largest for chemically active gases and can be made relatively small for rare gases which are not pumped by chemisorption, nor are they usually physisorbed at room temperature.

Considering the simplest case for a single gas, the pressure in the gage is given by

$$V \, dp/dt = -p(S_E + S_C) + F_G + F_{EX} \qquad (1)$$

where V is the gage volume (liters), S_E the electronic pumping speed (i.e. the pumping speed resulting from the ionization of the gas), and S_C the pumping speed resulting from chemisorption or chemical interaction of the gas with the gage (liter sec^{-1}). F_G is the rate of evolution of gas from the gage itself and F_{EX} is the flow of gas into the gage from external sources (Torr. liter.sec^{-1}). The major part of the electronic pumping action results from the entrapment of positive ions at surfaces (ionic pumping). Also contributing to S_E is the increased rate of sorption of excited molecules or molecular fragments resulting from electron impact on the gas. Electronic pumping speeds for chemically active gases are similar in magnitude to those for rare gases.

The ionic pumping of gases into the metal and glass parts of ionization gages is followed by re-emission of some of the previously pumped gas. Pumping of rare gases in Bayard–Alpert gages has been studied by several authors[24, 56]. Following a sudden introduction of gas, the pressure is reduced by ionic pumping and when the ionizing electron current is turned off the pressure rises again as a result of re-emission of the pumped gas. During the pumpdown period, if E_G and F_{EX} are sufficiently small, the pumping speed can be measured by using the relation

$$p = p_0 \exp[-t(S_E + S_C)/V] \qquad (2)$$

where p_0 is the pressure at zero time, when gas is introduced. Similarly, the evolution rates can be determined from a measurement of the pressure increase after the electron current is turned off. This curve can be obtained by switching the electron current on intermittently for times sufficiently short that pumping is insignificant. In this case the electron current is turned off at $t = 0$ and $p = p_1$, then

$$V(p - p_1) = (F_G + F_{EX})t \qquad (3)$$

The re-emission process and experimentally determined magnitudes of the re-emission probabilities are discussed in ref. 99, section 7.2. The re-emission of ionically pumped gas is frequently the process limiting the lowest attainable pressure in a vacuum system, e.g. an ionization gage operated for a few minutes at 10^{-5} Torr cannot subsequently be used at pressures below 10^{-10} Torr without thermal treatment (degassing). The electronic pumping speed of a gage decreases with the amount of gas pumped, thus measured values of pumping speeds of gages reported in the literature must be used with caution unless the condition of the gage is clearly specified. A table of measured pumping speeds of various gages, in a so-called clean condition, for different gases can be found in ref. 99, p 268.

The electronic pumping speed of a hot-cathode gage is linearly related to the electron current and depends in a more complex way on the electron energy and the energy of the positive ions when they strike a surface. Measurements of gage pumping speed as a function of electron energy have been reported by several experimenters[56, 20]. To minimize the electronic pumping speed of an ionization gage the electron current, the electron accelerating voltage, and any ion accelerating voltages should be reduced to the minimum compatible with the required sensitivity. With hot-cathode gages it is often advantageous to lower the electron accelerating voltage rather than the electron current. Both electronic pumping speed and sensitivity decrease linearly with electron current, whereas pumping speed decreases much more rapidly than sensitivity when the electron accelerating voltage is reduced. With cold-cathode gages the pumping speeds are not easy to control and may be too high for many experiments.

Errors in pressure measurement arise when S_E, S_C or F_G in equation 1 become large in relation to similar terms for the rest of the vacuum system. These errors can be minimized by designing the connections to the gage to have a large conductance or by using a nude gage, i.e. a gage placed directly into the system to be measured without a separate envelope.

The processes discussed so far have only involved one gas. Another re-emission process becomes troublesome when more than one gas is present in significant quantities. The impact of one type of ion (A) on a surface may release a gas molecule (B) previously pumped. This effect is known as ionic replacement[19, 58] and can result in serious contamination of gas B if it is introduced after pumping gas A for some time. The magnitude of the effect can be estimated from the replacement coefficient which has been measured for various ions on both metal and glass surfaces (see ref. 99, p 272, for tabulated values).

When chemically active gases are present, the accurate measurement of pressure is much more complex. We must now take into account the chemical pumping speed S_C in equation 1 and include desorption of chemisorbed species in F_G. The maximum chemical pumping speed of a gage[2], if all surfaces had unit sticking probability, is several thousand liters.sec^{-1}; fortunately, the surfaces saturate in time (after the adsorption of about 10^{-2} Torr. liter of gas) and reach an equilibrium with the gas phase where the net chemical pumping speed decreases to zero. Sudden pressure changes may result in a disturbance of this equilibrium if the outer chemisorbed layers are reversibly adsorbed[50].

Other effects besides ionic pumping and chemisorption contribute to the pumping speed of gages. Interaction of molecular gases with a hot cathode can create appreciable pumping speeds. Hydrogen on a tungsten filament above 1 475 K yields a speed of 0.1 liter.sec^{-1} resulting from dissociation[46], oxygen[32] at filament temperatures of 1 700 K gives 0.04 liter.sec^{-1}, nitrogen[121] at a filament temperature of 2 100 K gives 10^{-4} liter.sec^{-1}. Pumping has also been observed to result from the formation of atomic nitrogen from molecular nitrogen[121] and the formation of metastables from nitrogen[57] by electron impact.

B. Gas Interactions at Hot Surfaces

Chemically active gases react with the hot cathodes of ionization gages resulting in a change in gas composition in the system which is often more of a problem than the pumping speed effects described in the previous section. Some possible reactions at a 2 200 K tungsten filament are: (a) $H_2 \rightarrow 2H \rightarrow H_2O + C_xH_y$, (b) $C_xH_y + X_aO_b \rightarrow CO + CO_2$, (c) $O_2 + W + C \rightarrow CO + CO_2 + WO_3$, (d) $H_2O + W + C \rightarrow CO + W_2 + WO_3 + H$; many others are possible[2, 92].

Atomic hydrogen is formed at a measurable rate when the filament temperature exceeds 1 000 K with an activation energy for desorption[47] of atomic hydrogen of 67 kcal mole^{-1}. The atomic hydrogen adsorbed readily on most surfaces giving an anomalously high pumping speed for hydrogen. Other molecules, in particular CO, are produced by hydrogen interaction with surfaces containing carbon[46, 110]. Tables showing the gas composition at various hydrogen pressures can be found in ref. 99, pp 276–277. The formation of carbon monoxide, water vapor, and methane, when hydrogen is introduced, has also been observed[77, 78].

Oxygen interacts with carbon impurities in hot filaments to form carbon oxides[105], yielding pumping speeds as high as 30 liters.sec^{-1} in a Bayard–Alpert gage at pressures of 10^{-7} Torr. The interaction of oxygen with hot tungsten has been studied in detail by Singleton[109].

Carbon dioxide reacts with hot tungsten filaments to yield oxygen and carbon monoxide[104], water vapor yields hydrogen, methane, carbon monoxide and carbon dioxide[35], and methane decomposes in a complex way[8]. The above brief list by no means exhausts the reactions studied; further bibliography may be found in ref. 99.

The effects resulting in contamination of chemically active gases by hot filaments can be reduced by:

(a) Reducing the carbon impurities in the hot filament. Heating to about 2000 K in 10^{-7} Torr of oxygen for more than 20 hours is effective for tungsten[12].
(b) Reducing the operating temperature of electron emitters by the use of low work function materials[120]. For a more detailed discussion and bibliography of different types of cathodes suitable for hot cathode gages, see ref. 99, p 299 et seq.
(c) Enclosing the hot cathode by a metal surface at which atomic species can readily recombine.

Reduction of the electron emission in hot-cathode gages to minimize pressure measurement errors is not always effective, particularly in hydrogen; for a discussion of this problem, see ref. 99, p 278.

C. Blears Effect

Blears[15] first observed that the reading on a nude ionization gage was higher by a factor of ten than that of an identical gage attached to the system through a tubulation. The discrepancies were only observed in the presence of diffusion pump oils. The effect appears to result from the adsorption of the oil molecules on the walls of the tubulation[43, 101]. Similar effects have been observed with water vapor[101] and with carbon monoxide, hydrogen and nitrogen[5], the effects for hydrogen and nitrogen being smaller.

The detailed mechanism of this effect is not clearly understood, at least for the smaller molecules. It is advisable to avoid this problem, when chemically active gases are present, by the use of a nude gage.

2. Measurement in Non-uniform Environments

In many experimental situations the flux density of gas per unit solid angle is *not* independent of position and direction, *nor* is the whole system isothermal.

Ionization gages measure molecular density and are not, in general, sensitive to direction or velocity of the molecules. Tubulated gages placed in an enclosure where a directed gas flow exists will give different readings depending on the orientation of the tubulation with respect to the gas flow[52, 115]. Another case often encountered is where a gage is immersed in a cryogenic fluid. Suppose a gage measures a static pressure p_1 at $T_1 = 300$ K and is then immersed in liquid nitrogen ($T = 77$ K) with a gas that is not physically adsorbed at that temperature. The pressure in the gage drops to p_2 where

$$p_2/p_1 = T_2/T_1 = 77/300 = 0.256 \tag{4}$$

The *density* in the ionization gage remains unchanged and hence the gage reading is unaltered.

Several examples of pressure measurements in systems where the gas flow is directed may be found in the literature, for a review of some of these examples see ref. 99, pp 283–287.

It is often necessary to measure pressure in systems where the gage is at one temperature, usually 300 K, while the rest of the system is at another temperature. If there is no net flow of gas across the aperture at the temperature transition, then

$$(p_1/p_2) = (T_1/T_2)^{\frac{1}{2}} \tag{5}$$

This law of thermal transpiration has been verified for an aperture[30] over the pressure range where free molecular flow occurs (the pressure range of concern of this chapter). In most experiments it is inconvenient to use an aperture and the temperature transition occurs at a tube. It has been shown experimentally[30] that equation 5 is not valid under these conditions, particularly when the transition tube is made of glass which is smooth on an atomic scale. In this case

$$(p_1/p_2) = b(T_1/T_2)^{\frac{1}{2}} \tag{6}$$

where $b < 1$ for $T_1 > T_2$ and the value of b depends on the geometry of the transition region and the smoothness of the surfaces. For atomically rough surfaces (metal or etched glass) b tends to unity.

Thermal effects in hot-cathode gages also result in small errors in pressure measurement caused by heating of the gas in the gage. The gas heating occurs mainly at the gage envelope rather than at the cathode. This effect can be made negligibly small by using a nude gage or reduced considerably by using a metal envelope rather than glass.

3. Residual Currents

A current to the ion collector exists in most ionization gages which is independent, to first order, of the pressure in the system; this is called the residual current and establishes a lower limit of pressure measurable with any particular gage design. The residual current is the collector current remaining if the pressure in the gage was instantaneously reduced to zero. Two processes contribute to the residual current; (a) photo-electron emission resulting from soft x-rays produced by the ionization electrons, (b) electronic desorption of positive ions from adsorbed layers on electron bombarded surfaces.

A. Soft X-Ray Photoemission

In hot-cathode gages, or any type of gage where the electron current is constant, a flux of soft x-rays is generated, which is independent of pressure, and which releases photo-electrons from the ion collector or other electrodes. The photocurrent is indistinguishable in the measuring circuit from the positive ion current used as an indication of pressure. Since the electron current to the anode of a cold-cathode gage is approximately proportional to pressure, there is no x-ray limit for this type of gage. We confine our discussion here to x-ray effects in hot-cathode gages of the Bayard–Alpert type; similar considerations apply to other hot-cathode gages. A general description of the behavior of soft x-rays and the photo-electric yields therefrom in the energy range of concern in gages can be found in ref. 99, pp 287–290.

Hot-cathode ionization gages were used for about 25 years before Nottingham[85] pointed out the limitation in the lowest pressure measurable caused by soft x-rays. Until Nottingham's comment, the lowest pressure measurable with conventional ionization gages was about 10^{-8} Torr. Bayard and Alpert[10] developed a new design of gage (the Bayard–Alpert gage, BAG hereafter) which reduced the x-ray limit by a large factor.

This gage with its modifications is now almost universally used in the pressure range 10^{-6} to 10^{-10} Torr. *Figure 1* shows schematically the original design of Bayard and Alpert. The principle of this design was to reduce the

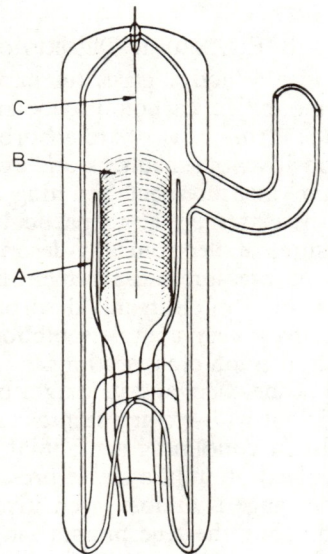

Figure 1. Schematic diagram of first Bayard–Alpert gage, showing filament (A), grid (B), and ion collector (C)[10].

solid angle subtended by the ion collector and thus minimize the x-ray flux at the collector surface. The ion collector was a wire of about 0.2 mm diameter, which only intercepted about 1/300 of the total x-ray flux. Various modifications to this design have been made (some are described in the following sections) which all involve the principle of reducing the x-ray flux at the collector surface or preventing the photo-electrons leaving the collector by suitable suppressor electrodes.

Standard BAGs have x-ray limits of about 10^{-10} Torr. By using the modulation technique (described later) the limit can be reduced to about 7×10^{-12} Torr. Other hot-cathode gage designs have reduced the limit to as low as 10^{-13} Torr.

The residual current resulting from x-rays is relatively constant for a given gage operated under fixed conditions and can be subtracted out, when operating near the x-ray limit, if it can be measured with reasonable accuracy.

So far, only the 'forward' x-ray effect has been discussed, i.e. the release of photo-electrons from the ion collector by soft x-rays resulting in a positive current at the collector. The 'reverse' x-ray effect may be observed in BAGs with envelopes operated at the same potential as the collector (usually earth potential). In this case, x-rays strike the envelope releasing photo-electrons, a small fraction of which are capable of reaching the collector resulting in a negative current[44]. Although the fraction of the photo-electrons that can reach the collector is small, almost all the x-ray flux is intercepted by the envelope. The net result is that the reverse x-ray photocurrent can be comparable, or larger than, the forward x-ray photocurrent. This effect can be completely prevented by applying a small bias between the collector and the envelope (about 10 V).

B. Electronic Desorption

When certain chemically active gases (in particular, oxygen, halogens, water vapor, carbon monoxide, carbon dioxide and hydrogen; in the order of the magnitude of the errors caused) are adsorbed on a surface and bombarded by electrons, positive ions, neutrals, and excited neutrals are released. The errors in pressure measurement resulting from these electronically desorbed species can be extremely large, particularly after a gage has been exposed to high pressures of electronically-desorbable gases. The effects of electronic desorption on pressure measurement have been reviewed in detail[97] and the whole subject of electronic desorption has been surveyed[70]; the latter paper contains a very complete bibliography and tabulation of measured electronic desorption cross sections.

The general nature of the effect and its magnitude may be exemplified by reference to an experiment with oxygen summarized in *Figure 2*[95]. Oxygen was admitted to a system containing a modulated BAG and an extractor gage with the pump valved-off until a stable pressure of 5×10^{-7} Torr was obtained. The extractor gage is almost insensitive to electronic desorption and indicates approximately the true pressure in the system. At time zero the oxygen supply was turned off and the pressure indications of the BAG and extractor gage agreed. At $t = 16$ minutes, the getter-ion pump was turned on and the pressure dropped rapidly. The pressure indicated by the

PART 7. VERY LOW PRESSURES AND ULTRA LOW PRESSURES

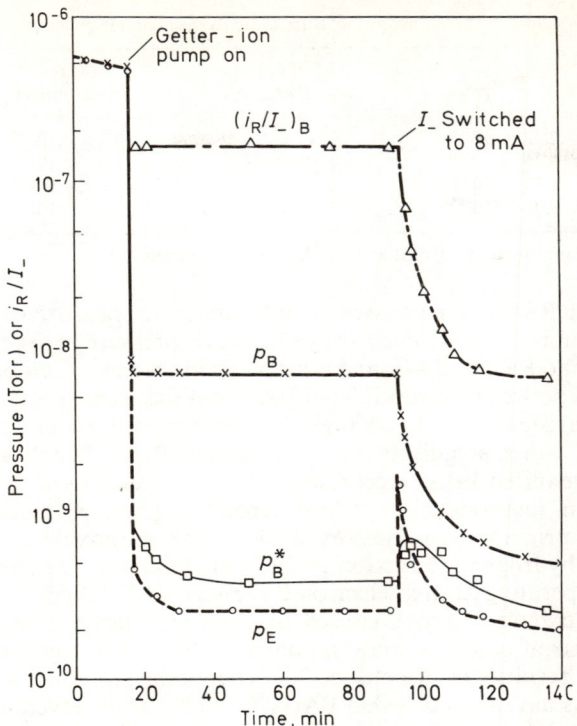

Figure 2. Present indications of a BAG (p_B), a modulated BAG (p_B^*), and an extractor gage (p_E) as a function of time after exposure to oxygen with initial electron currents of 8×10^{-5} A. $(i_R/I_-)_B$ is the residual current in the BAG divided by the electron current[95].

extractor gage had reached 2.5×10^{-10} Torr after ten minutes while the BAG was indicating a pressure of 7×10^{-9} Torr, a factor of 30 high. Pressure measured by modulating the BAG (p_B^*) was in reasonable agreement with the true pressure (p_E). At $t = 93$ minutes, the electron current in both gages was increased to 8×10^{-3} A. The true pressure p_E showed a burst of gas, released by electron bombardment, which was roughly followed by the modulated BAG (p_B^*). p_B dropped fairly rapidly as the oxygen was desorbed from the grid of the gate and finally approached p_E in 20 hours. Detailed measurements of the effects of oxygen on pressure measurements with a BAG[94] showed pressure readings too high by factors as large as 400.

The errors in pressure measurement with hot-cathode gages in the presence of electronically desorbable gases can be minimized by: (a) using a high electron current to keep the electron collector clean (10^{-2} A in a BAG), (b) use of the modulator technique, (c) use of a type of gage in which the electronic desorption error has been shown to be small (e.g. the extractor gage[95] or the bent-beam gage[45]), (d) use of a mass spectrometer (though care must be taken to recognize the peaks in the mass spectrometer resulting from electronic desorption in the ion source)[97], (e) proper choice of the material of the electron collector.

Table 1. Equilibrium pressure indications (Torr) in oxygen

BAG	Modulated BAG	Mass spectrometer
2.22×10^{-8}	7.6×10^{-9}	7.7×10^{-9}
3.25×10^{-8}	1.30×10^{-8}	1.28×10^{-8}
1.77×10^{-8}	6.0×10^{-9}	6.0×10^{-9}
2.04×10^{-8}	6.9×10^{-9}	7.0×10^{-9}

Gage electron current at 10^{-2} A (From Singleton[109]).

Modulated BAGs can be used to give accurate measurements of static oxygen pressures[109, 119]. Table 1 shows the static pressure of oxygen indicated by a BAG, a modulated BAG, and a mass spectrometer[109]. Good agreement was obtained between the modulated BAG and the mass spectrometer while the BAG consistently read too high. Stepwise pressure changes of oxygen are followed rather sluggishly by a modulated BAG[109] and some tens of minutes are required before accurate readings are once more obtained.

Modulation methods also reduce errors of pressure measurement in hydrogen[110] principally as the result of carbon monoxide contamination produced by hydrogen interaction at the heated filament. Errors can also be reduced by operating at high electron currents ($\sim 10^{-2}$ A)[48].

Attempts to reduce errors, caused by electronic desorption, by suitable choice of electron collector material have not been very successful and the experimental results are in conflict. Molybdenum, tantalum, tungsten and platinum grids have been tested in BAGs[68] and it was observed that tungsten and platinum did not cause errors from electronic desorption with carbon monoxide and water vapor. Measurements made by other experimenters[97] did not show any significant improvement of platinum over molybdenum in oxygen. Platinum, platinum–iridium, and molybdenum have been studied in hydrogen, water vapor, carbon monoxide and oxygen[53]; the evidence from these experiments does not permit a preferred choice of material to minimize electronic desorption errors. The bulk of the existing evidence suggests that platinum-clad molybdenum is the best candidate for general use but is no better than molybdenum or tungsten in oxygen.

Two types of hot-cathode gage have been proved to be insensitive to electronic desorption errors for most gases; these are the extractor gage[95] and the bent-beam gage[45, 117]. Other similar gage designs[40, 73] may be insensitive to these errors but this has not yet been demonstrated experimentally.

When adsorption and electronic desorption are non-dissociative and there is no significant thermal desorption of the gas in question, then it can be shown[97] that

$$I_G^+ / I_S^+ = c(KI^- / s) \qquad (7)$$

where I_G^+ is the ion current from the gas phase, I_S^+ is the electronically desorbed ion current, c is a constant for a particular gas–surface combination, K is the gage sensitivity factor $[K = (I_G^+ / I^-)(1/p)]$, I^- the electron current, and s the sticking probability of the gas on the bombarded surface. Note that equation 7 is independent of pressure. The conditions imposed to arrive at equation 7 are not generally applicable and thus this equation is only introduced to indicate general tendencies.

Equation 7 indicates that to minimize errors from electronic desorption (i.e. to maximize I_G^+/I_S^+) it is desirable to operate with a high electron current and as large a value of K as possible. Hot-cathode gages with high K are the Orbitron[74] ($K \approx 10^4$ Torr^{-1}) and the hot-cathode magnetron gage[61, 63] ($K \approx 10^6$ Torr^{-1}). The insensitivity of the Orbitron gage to electronic desorption errors has not been established experimentally; however, the hot-cathode magnetron is insensitive to these errors[28].

To minimize errors from electron desorption in cold-cathode gages (if any) it is advisable to use a gage whose anode can be degassed by electron bombardment[82, 83].

C. Measurement of Residual Current

Measurement of residual current is essential when working at pressures approaching the x-ray limit of the gage; it is useful at all times as an indication of the degree of contamination of the gage.

There are three basic methods of measuring the residual current (i_R):
(a) by rapidly reducing the pressure in the gage well below its x-ray limit,
(b) by measuring the collector current as a function of electron energy, and
(c) by modulation methods.

Method (a) is so obvious that it requires no comment other than that it is rarely possible. Method (b) was the original method used by Bayard and Alpert[10] to demonstrate the usefulness of their gage design. *Figure 3* shows

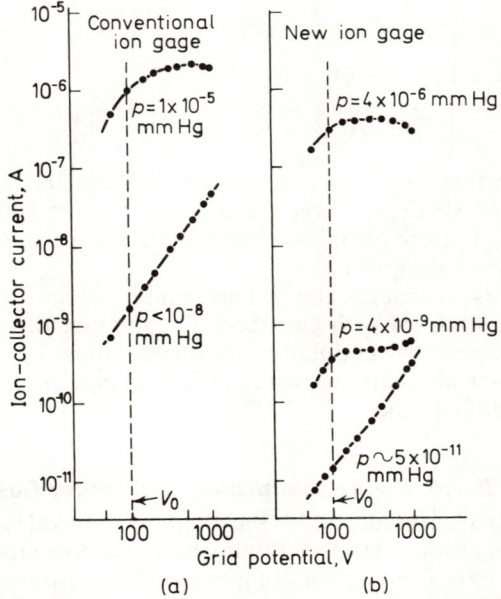

Figure 3. Ion-collector current versus grid voltage for (a) RCA type 1949 gage, and (b) Bayard–Alpert gage[10].

the original curves comparing a conventional (large collector) gage with a BAG. At low pressures, when the positive ion current is comparable to the x-ray photocurrent, the straight line portion of the curve at higher voltages can be extrapolated back to the operating voltage to estimate the x-ray photocurrent. Only the x-ray photocurrent is measured and, since the time required for the measurement is considerable, conditions may change during the measurement.

Method (c), the modulation method, is applicable to most types of hot-cathode gage and, in general, measures the sum of the x-ray photocurrent and the electronically desorbed ion current. The modulator electrode usually consists of a fine wire, whose potential can be switched, inserted into the ionization space. Modulation electrodes have been added to BAGs[6,93,100] and extractor gages[95].

The modulator is switched between two potentials, usually the upper is grid potential (V_g) and the lower is earth potential (Mode I). With the modulator at V_g, the ion-collector current is

$$I_1 = I_G^+ + i_R \tag{8}$$

where I_G^+ is the ion current from the gas phase. With the modulator earthed

$$I_2 = (1 - \beta)I_G^+ + (1 - \varepsilon)i_R \tag{9}$$

where $(1 - \beta)$ and $(1 - \varepsilon)$ are the modulation factors for I_G^+ and i_R respectively. β may be determined at high pressures when $I_G^+ \gg i_R$, then

$$\beta = (I_1 - I_2)/I_1 \equiv \Delta I/I_1 \tag{10}$$

When $\varepsilon = 0$, the usual case, then

$$I_G^+ = \Delta I/\beta \quad \text{and} \quad i_R = I_1 - I_G^+ \tag{11}$$

The assumption that $\varepsilon = 0$ is reasonable at pressures above 3×10^{-11} Torr with a modulated BAG. At lower pressures, or when $\varepsilon \neq 0$, the determination of I_G^+ and i_R is more complex; one method of measurement under these conditions has been described[51].

Modulation as described above may cause pressure bursts when the modulator is bombarded with electrons. To minimize this effect the potentials, between which the modulator is switched, must be carefully chosen. Various modes of modulation, to minimize possible pressure bursts, have been experimentally studied[41,65,88,96].

4. Relative Gage Sensitivities for Different Gases

Table 2 lists the measured, relative sensitivities of several types of ionization gages to various gases. The last column shows the ionization cross section for electrons of 100 eV energy for comparison[89]. In most cases the sensitivities are normalized to argon; where argon data are not available, the reference gas is nitrogen.

PART 7. VERY LOW PRESSURES AND ULTRA LOW PRESSURES

Table 2. Relative sensitivity of ionization gages to various gases, normalized to argon or nitrogen

Gas	1 Triode	2	3	4	5 Bayard–Alpert gage	6	7	8
Ar	1	1	1	1		1	1	1
He	0.1283	0.14	0.13	0.17			0.134	0.15
Ne	0.2407	0.22		0.26			0.258	
Kr	1.333			1.53			1.34	
Xe	2.190			2.22	2.5			
N_2		0.67	0.75		1	0.53		0.56
H_2	0.2808	0.28			0.43		0.30	
Cl_2								0.64
CO	0.823				1.1	0.48		
O_2			0.94			0.48		
NO						0.61		
Hg								
H_2O								
CO_2	1.163						0.90	
HCl							1.20	
NH_3							0.91	
SF_6		1.7					1.90	
Air			0.81					
CH_4	1.015							
C_2H_6	1.992							

Gas	9	10	11 Hot-cathode magnetron	12 Magnetron	13 Inverted magnetron	14 Trigger discharge	15 Total ionization cross section
Ar	1	1		1.76			1
He		0.127		0.24	0.15		0.128
Ne	0.22	0.220					0.233
Kr	1.43	1.39					1.47
Xe	2.0	2.02					1.92
N_2		0.705	1	1	1	1	0.885
H_2		0.298	0.43	0.52		0.32	0.324
Cl_2							
CO		0.780					0.93
O_2		0.616	0.73	0.99		0.66	0.940
NO							1.10
Hg							
H_2O							
CO_2		1.01		1.29			1.24
HCl							
NH_3							
SF_6							2.49
Air							
CH_4							1.28
C_2H_6							

Notes on Table 2

Column	Gage designation	V_{gf} (V)	I^- (A)	V_a (kV)	B (gauss)	p (Torr)	Refs
1	Leybold IM-1	180	10^{-3}				76
2	Westinghouse 5966	140	10^{-4}			10^{-5} up	107
3	Veeco RG-75	127.5	10^{-4}			10^{-5}–10^{-2}	80
4	Experimental	240	10^{-3}			10^{-9}–10^{-4}	25
5	Westinghouse 5966	145				$<5 \times 10^{-5}$	31
6	Veeco RG-75	150	10^{-2}			10^{-4}	4
7	Leybold IM-4	180	10^{-4}			10^{-5}–10^{-3}	13
8	Veeco RG-75	140	10^{-4}			10^{-6}–10^{-5}	108
9	Experimental	105	8×10^{-5}			10^{-8}–10^{-5}	99
10	NRC 553	145	9.3×10^{-5}			10^{-4}–3×10^{-3}	118
11	Experimental	300	5×10^{-8} ($B=0$)		250	10^{-10}–10^{-6}	27
12	NRC 552			5	1000	10^{-7}–10^{-5}	7
13	Experimental			6	2000	10^{-8}–10^{-5}	49
14	G.E., T.D.G.			2	1000	5×10^{-10}	27
15			Ratio of total ionization cross sections				89

5. *Calibration of Gages*

Ionization gages may be required to operate over a range of as much as 10^7. Absolute calibration of pressure gages over this wide range, particularly for chemically active gases, offers some formidable problems. The general principles only are discussed here and references to more detailed descriptions are given. In general, it is advisable to calibrate the gage *in situ*, since the gage–system interactions can affect the calibration.

A. ABSOLUTE CALIBRATIONS

Calibrations of ionization gages must be traceable to instruments capable of measuring absolute pressure. Some of these instruments have been described in earlier sections of this chapter. The manometer most commonly used to calibrate ionization gages absolutely is the McLeod gage. Also used less frequently are diaphragm gages and deadweight gages. Direct calibration of ionization gages against McLeod gages is only possible at pressures above 10^{-5} Torr with any reasonable accuracy (± 5 per cent or less). A modified form of McLeod gage has been developed[75,79] capable of measuring to 10^{-8} Torr. Detailed descriptions of the use of McLeod gages can be found in ref. 99, p 254, and ref. 69, p 3.

Two aspects of calibrations with McLeod gages are worthy of emphasis here and both result from the necessity to use a liquid nitrogen trap with a mercury-filled McLeod gage. First, only the pressure of non-condensable gases is measurable with a McLeod gage. Calibration with condensable gases can be done by using a diaphragm manometer which can be calibrated in turn against a McLeod gage. Second, the adsorption of mercury vapor by the cold trap produces a pumping action (the vapor stream effect)[26,55] which can cause substantial errors. These errors can be reduced to less than one per cent by cooling the McLeod gage[29, 103] or by reducing the conductance between the McLeod gage and the trap[71, 30].

B. RELATIVE CALIBRATIONS

Since most absolute pressure gages do not extend into the high vacuum or u.h.v. region, indirect methods of calibration are necessary. Two general methods can be used: (i) The variation of relative gage sensitivity with pressure is measured from the lower limit to about 10^{-3} Torr, the gage is then compared against an absolute gage above 10^{-4} Torr. (ii) Gas is allowed to flow from a vessel at high pressure through two or more apertures of accurately measured conductance (or conductance ratio) into a pumping port of known speed, thus creating a calculable pressure at the gage. Both methods rely on the simple gas laws being observed.

The various techniques used in method (ii) above are summarized in *Table 3* and the corresponding arrangements are shown schematically in *Figure 4*.

(a) *Static Expansion*

In mode I ($V_i \approx V_2$) an initial pressure p_a is established in V_1 and valve A opened, the pressure at the test gage (G_{II}) is given by the equation in *Table 3*. If the valve is closed and the cycle repeated n times, then

$$p_2 = p_a[V_1/(V_1 + V_2)]^n \tag{12}$$

In mode II ($V_2 \gg V_1$), the initial pressure p_1 is made sufficiently high to be measurable by an absolute gage[72, 106, 111].

(b) *Linear Rise Rate*

With p_1 constant and $p_1 \gg p_2$ the gas flowrate into V_2 is constant. With zero pumping speed in V_2, p_2 rises linearly with time.

(c) *Single-stage Pressure Division*[34]

A steady flow is established with p_1 measured directly and when the speed of the pump S is very large compared to the conductance C_2. C_2/C_1 must be very large. A thin-walled aperture is used for C_2. Various types of porous material[21, 86] have been used for C_1 which retain molecular flow behavior up to 10 Torr. Pumping effects in V_2 can be tolerated provided the pumping speeds are small compared to C_2.

(d) *Quadratic Rise Rate*

If $p_1 \gg p_2 \gg p_3$ then p_3 will increase quadratically with time. Again, $C_3 \ll S$. The advantage of this method is the reduction in time necessary to cover a wide pressure range compared with the linear rise method. As in (b) the pumping speed of the gage must be kept very small.

(e) *Quadratic Rise Rate and Single-stage Division*

This combines the advantages of (d) and (c). Pumping action at the gage can be tolerated provided it is small compared to C_4.

Table 3. Characteristics of gage calibration systems

Measurement method	Gage pressure	Valid time range	Measured pressure range (Torr)	Claimed accuracy %	Refs
(a) Static expansion Mode I	$p_a \dfrac{V_1}{V_1 + V_2}$	$t \gg \dfrac{V_2}{C_A^*}$	10^{-9} to 10^{-3}	± 2	30
Mode II	$p_1 \dfrac{V_1}{V_1 + V_2}$	$t \gg \dfrac{V_2}{C}$	10^{-6} to 10^{-4}	± 10	106
(b) Linear rise rate	$\dfrac{p_1 C_1 (t - t_L)}{V_2}$	$t_L < t \ll \dfrac{V_2}{C_1}$	10^{-8} to 10^{-2}	± 3	99
(c) Single-stage pressure division	$p_1 \dfrac{C_1}{C_1 + C_2}$	$t \gg \dfrac{V_2}{C_2}$	10^{-11} to 10^{-3}	$\sim \pm 5$ ($p > 10^{-8}$ Torr)	27
(d) Quadratic rise rate	$\dfrac{p_1 C_1 C_2}{V_2 V_3}(t - t_L)^2$	$t \ll \dfrac{V_2}{C_1}, \dfrac{V_3}{C_2}$	10^{-9} to 10^{-3}	$(\sim \pm 10)$	3
(e) Quadratic rise rate and single-stage division	$\dfrac{p_1 C_1 C_2}{V_2 V_3} \cdot \dfrac{C_3}{C_3 + C_4}(t - t_L)^2$	$\dfrac{V_3}{C_3} \ll t \ll \dfrac{V_2}{C_1}, \dfrac{V_3}{C_2}$	10^{-10} to 10^{-4}	$\sim \pm 10$	64
(f) Multistage pressure division	$\dfrac{p_1 C_1 C_2 C_3}{(C_2 + C_2')(C_3 + C_3')(C_4 + C_4')}$	$t \gg \dfrac{V_2}{C_2'}, \dfrac{V_3}{C_3'}, \dfrac{V_4}{C_4'}$	10^{-9} to 10^{-4}	± 10	102

* C_A is the conductance of valve A when open.

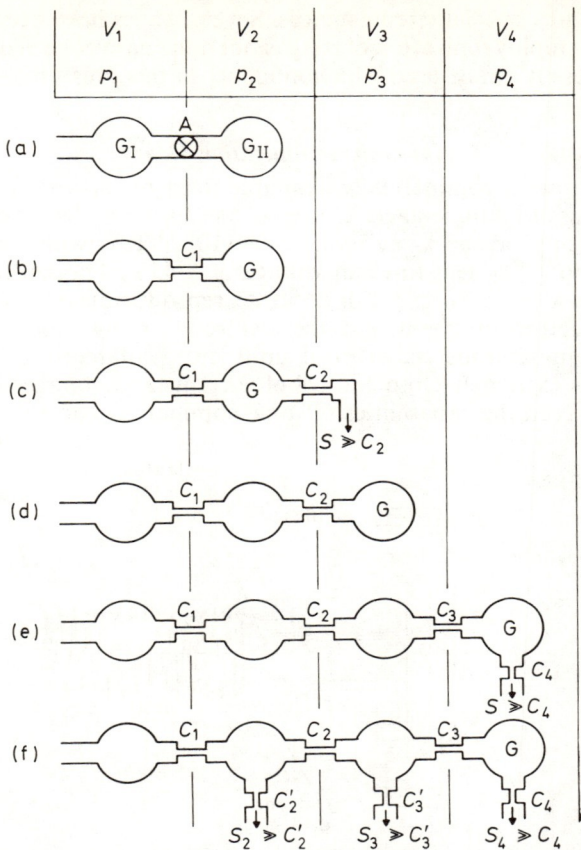

Figure 4. Schematic diagram illustrating various arrangements used for the calibration of ionization gages. The letters (a) to (f) correspond to those in *Table 3*.

(f) *Multiple-stage Division*

Provided p_1 is low ($<10^{-2}$ Torr) all the conductances can be thin-walled apertures whose conductance can be readily calculated[18].

Discussion of the limitations of these methods and detailed consideration can be found in ref. 99, pp 257–262.

When ionization gages are compared directly with absolute gages in the range 10^{-5} to 10^{-3} Torr it is unsafe to make the common assumption that the measured gage sensitivity will be the same at lower pressures, errors of 30 per cent or more may thus be introduced. Many types of hot-cathode ionization gages show sizeable changes of sensitivity above 10^{-6} Torr, see ref. 99, p 311.

III. Pressure Measurements from 10^{-6} to 10^{-10} Torr

In this pressure range the most widely used type of gage is the Bayard–Alpert gage in various modified forms. Cold-cathode gages may have specific advantages under special circumstances. Several gages have been developed which do not employ ionization of the gas but none has yet been developed to the point where it can be used with confidence in this pressure range.

1. Hot-cathode Ionization Gages

Figure 5 shows a commercially available form of Bayard–Alpert gage[10] designed for use as a nude gage. This gage has an ion-collector of 0.125 mm diameter which yields an x-ray limit of 2×10^{-11} Torr with a grid voltage of $+130$ V and a filament-to-earth voltage of $+45$ V. The sensitivity factor for nitrogen is $K = 25$ Torr^{-1}. For accurate, reproducible pressure measurements the electron emission and the electrode voltages must be closely regulated. Many commercial control units are available; for very precise measurements extremely close control of currents and voltages is necessary which may exceed the capabilities of some commercial control units[112].

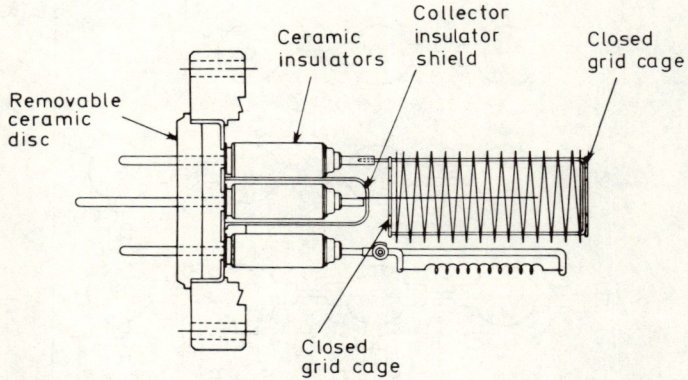

Figure 5. Design of commercially available form of Bayard–Alpert gage (courtesy of Varian Associates).

Rigorous degassing of the grid is essential for reproducible results. Some gage designs have grids in the form of an unsupported helix heated by passing current through the grid wire. A preferable arrangement is to heat the grid by electron bombardment, typically with 1 kV electrons at a power density of about 15 W cm^{-2} to yield a degassing temperature of about 1800 K.

The use of uncoated glass envelopes for the gage may result in several spurious effects (see ref. 99, p 308). It is preferable to enclose the gage in a metal envelope or to coat the inside of the glass envelope with a transparent conducting layer of tin oxide[36]. Nude gages should be surrounded by a metal mesh to isolate them electrically from other elements of the vacuum system.

It is most desirable to use gages with low work-function coatings on the filament which reduce the operating temperature. This minimizes chemical interactions at the hot filament and reduces heating of the gage electrodes. Suitable coating materials are lanthanum boride, thoria or yttria; detailed comparisons of these materials for cathodes of gages can be found in ref. 99, pp 299–393. Thorium or yttrium oxides appear to offer the most advantages.

PART 7. VERY LOW PRESSURES AND ULTRA LOW PRESSURES

A useful modification to the original Bayard–Alpert design is the addition of an electrode to modulate the ion current[93]. The modulator consists of a wire placed parallel to the ion-collector inside the grid of a BAG. By switching the potential of the modulator the collector current can be modulated by as much as 40 per cent while the residual current is unaltered. This technique was discussed in section II, 3C of this chapter. The modulator method permits the rapid determination of residual currents and is particularly useful in the presence of chemically active gases. Several methods of modulation have been examined experimentally[1, 88, 96, 116].

One other effect, specific to BAGs, should be considered when precision measurements are attempted. The current to the ion-collector (at pressures where the residual current is negligible) consists of the sum of the positive ion current to the collector and the secondary electron current from the collector. The secondary electrons result principally from the emission of Auger electrons caused by ion bombardment. Provided the secondary yield remains constant the effect is unimportant and is taken into account in the calibration process. However, the yield is strongly dependent on the presence of chemisorbed gas layers on the collector. At the high pressures used for most absolute pressure calibrations, the collector may be rapidly cleaned by sputtering, resulting in a sensitivity different from that in normal gage use. Errors as large as eight per cent have been observed to result from this process[37]. The error can be minimized by insuring that the total ion bombardment of the collector during calibration does not exceed about 10^{15} ion cm^{-2}.

The Orbitron gage[74] depends on the production of very long electron paths in the electric field by launching electrons from a hot filament into planetary orbits between two coaxial cylinders (inner cylinder positive). A typical gage has an anode diameter of 0.23 mm operated at $+500$ V, an anode current of 10^{-6} A, yielding a sensitivity factor K of 10^5 Torr^{-1} and an x-ray limit of 9×10^{-12} Torr. Several designs have been reported[33, 38]. Although this type of gage has some potential advantages it has not yet been tested sufficiently to prove its usefulness for precise pressure measurements.

2. Cold-cathode Gages

Cold-cathode, crossed-field (EXB) gages are not normally used for precision measurements because: (a) in general, the ion current varies nonlinearly with pressure, (b) the dense, trapped electron space-charge in these gages results in serious low-frequency instabilities and high-frequency oscillations, (c) their pumping speed is usually high and uncontrollable. In some special cases the advantages of these gages may be important; viz. (a) there is no x-ray limit, (b) electronic desorption effects are small and cause little error in pressure measurement.

A. PENNING GAGES

The Penning gage[97] consists of a tubular anode with two interconnected cathode plates spaced a short distance from the ends of the tubular anode; an axial magnetic field of 1–2 kgauss and an anode voltage of 1–6 kV is used.

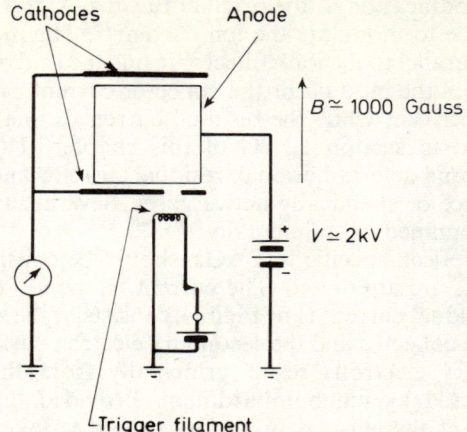

Figure 6. Schematic diagram of trigger discharge gage (Penning gage)[122].

Figure 6 shows schematically a variant of the Penning gage, called a trigger discharge gage[122], with an additional trigger filament that can be heated to start the discharge. The characteristics of the trigger discharge gage will be used as an example of the performance of Penning gages in the pressure range of interest here. This gage is operated with 2 kV anode voltage and a magnetic field of about 1 kgauss. The output current varies as $p^{1.2}$ so that the sensitivity at 10^{-6} Torr is 2.5 A Torr^{-1} decreasing to 0.1 A Torr^{-1} at 10^{-11} Torr[27]. The output current does not follow a smooth curve with pressure, but exhibits several sudden jumps and hysteresis in the current/pressure curves[27,66]. The position and magnitude of the jumps in output current appear to depend on the history of the gage's operation.

B. Magnetron Gage

The magnetron gage[91] is a geometric variant of the Penning gage; one version is shown schematically in *Figure 7*[84]. The anode is a cylinder (12.5 mm long and 30 mm diameter), the cathode is spool-shaped consisting of an axial cylinder (3 mm diameter by 12.5 mm long) joined to two circular end-discs (21 mm diameter). This gage is normally operated at 5.5 kV in an axial field of 1.16 kgauss. To reduce field emission from the cathode, two annular shield electrodes are used in some designs.

Several studies of the current/pressure characteristics of the magnetron gage have been published and are summarized in ref. 99, p331. Most experiments indicate that a linear current/pressure response is obtained over the range 10^{-6} to about 5×10^{-10} Torr with a sensitivity of 10 A Torr^{-1}, for an anode voltage of 6 kV, and 5 A Torr^{-1} at 4.8 kV. Below 5×10^{-10} Torr the output current/pressure obeys a power law ($i^+ = kp^n$) with n between 1.25 and 1.6 depending on conditions. For the gage shown in *Figure 7* $n = 1$ and $k = 2$ A Torr^{-1} in the pressure range 10^{-5} to 10^{-11} Torr[84].

Another version of the magnetron gage, similar to that shown in *Figure 7*, has the axial cylinder of the cathode replaced by a tungsten wire helix which

PART 7. VERY LOW PRESSURES AND ULTRA LOW PRESSURES

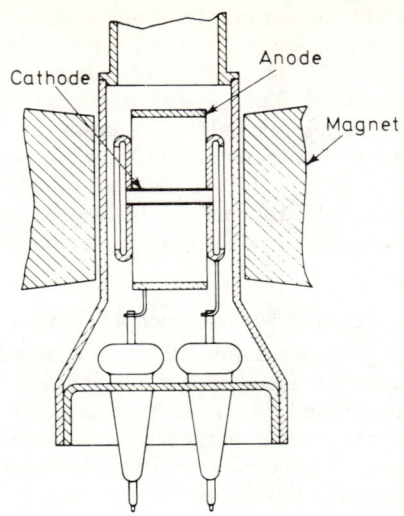

Figure 7. Schematic diagram of cold-cathode magnetron gage[83].

can be heated by passing current[83]. This allows outgassing of the gage by electron bombardment from the heated cathode and the gage can be started at very low pressures (down to 10^{-13} Torr) by flashing the wire helix. The calibration curve of this gage is linear from 10^{-5} to 5×10^{-13} Torr with an anode voltage of 4.5–5.5 kV and a magnetic field 1.12–1.18 kgauss.

C. Inverted-Magnetron Gage

This is another geometric variant of the Penning gage with the anode on the axis[11, 42, 49, 90]. *Figure 8* shows schematically the electrode arrangement of an inverted-magnetron gage (MM-14S) having a shield to prevent field emission from the cathode[82]. The cathode is a cylinder (8 mm diameter by

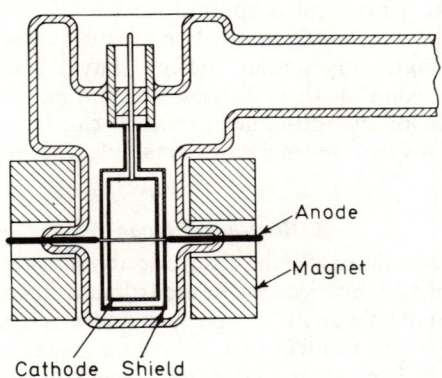

Figure 8. Schematic diagram of inverted magnetron gage[82].

293

3.5 mm long) partially closed at both ends and surrounded by an earthed shield electrode. The anode is an axial wire (0.25 mm diameter) passing through holes in the cathode end-plates, which can be heated by passing current through the wire. The gage is normally operated with an anode voltage of 6 kV and a magnetic field of 2 kgauss.

The current/pressure characteristic follows a power law ($i^+ = kp^n$) in the pressure range 10^{-4} to 10^{-12} Torr, with $n = 1.1$[49,39] and $n = 1.0$[82]. The inverted-magnetron gage appears to be less subject to mode jumping, hysteresis and noisy behavior than the magnetron gage.

IV. Pressure Measurements below 10^{-10} Torr

Measurement of pressure below 10^{-10} Torr is still something of an art and the instrumentation is not as well developed as for pressures above 10^{-10} Torr. The modulated Bayard–Alpert gage, the Orbitron gage, and the cold-cathode gages are capable of marking measurements below 10^{-10} Torr if very carefully used. This section will briefly describe three types of hot-cathode gages specifically developed for pressures below 10^{-10} Torr.

1. Shielded-collector Gages

This class of hot-cathode ionization gage uses an ionization region similar to a Bayard–Alpert gage but the ions are extracted from the ionization region and collected on an external collector. The earliest design of this type of gage, the extractor gage, is shown in *Figure 9*[95]. The grid and filament arrangement is similar to that of a BAG. Ions formed within the grid are attracted towards the shield, which is at filament potential. Most of the ions pass through the hole in the shield and are focused on to the earthed ion-collector by the hemispherical reflector which is at earth potential. The x-ray limit of this type of gage is reduced below that of a BAG because the collector wire subtends a very small solid angle at the source of x-rays (the grid). This gage is normally operated with filament, envelope and shield at +200 V, grid-filament voltage of 105 V, ion-collector at earth and an electron current of 2×10^{-3} A or less. The sensitivity factor K for nitrogen is 13 Torr^{-1} and the x-ray limit is about 3×10^{-13} Torr.

The extractor gage is relatively insensitive to errors caused by electronically desorbed ions. Gages of similar design have been developed and tested[9,14,22,23,73]. Some of these designs contain an insulating glass tube around the support for the ion-collector which can be bombarded by ions and/or electrons. This feature may cause unstable operation.

2. Bent-beam Gage

Figure 10 shows schematically the electrode arrangement of the bent-beam gage[45]. Ions are formed by electron bombardment inside the grid and the ions are then extracted through an aperture and electrostatically deflected through 90° by a pair of semicircular deflection plates on to the collector. A suppressor grid in front of the collector prevents secondary and photo-electrons from leaving the collector. The gage is operated with the potentials

PART 7. VERY LOW PRESSURES AND ULTRA LOW PRESSURES

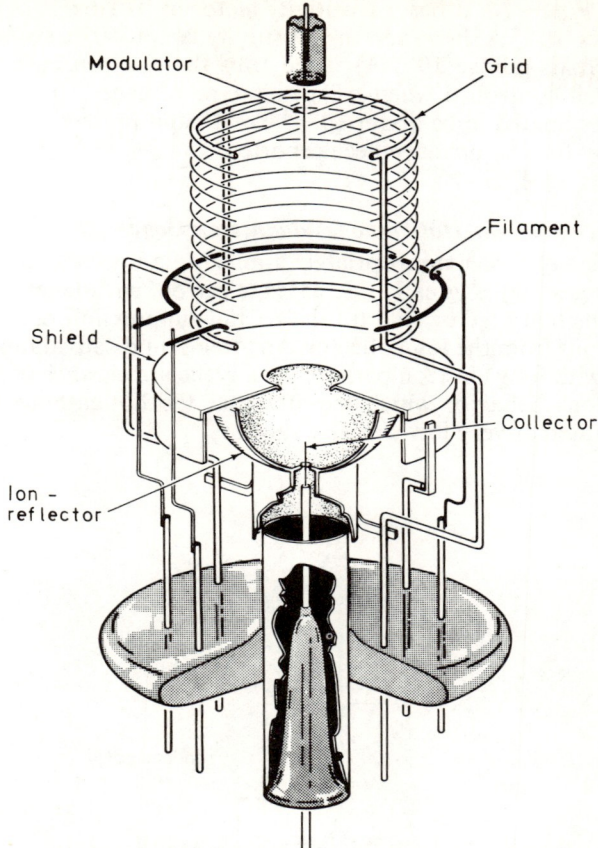

Figure 9. Schematic diagram of extractor gage[95].

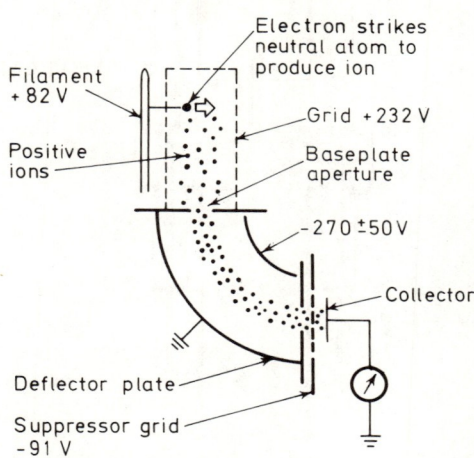

Figure 10. Schematic diagram of bent-beam gage[45].

indicated in *Figure 10*. It has a sensitivity factor of 10 Torr^{-1} at an electron current of 4×10^{-3} A; however, the sensitivity factor varies strongly at high electron currents (above 10^{-4} A)[73] and thus the electron current must be extremely well controlled when using this type of gage. The x-ray limit of this gage is estimated to be about 10^{-14} Torr[45] and pressure measurements as low as 3×10^{-12} Torr have been reported[67].

3. Hot-cathode Magnetron Gage

A magnetic field when appropriately applied to a hot-cathode gage yields two advantages: (a) a greatly increased electron path-length and hence increased sensitivity factor, and (b) the partial suppression of secondary or photo-electrons from the ion-collector. An attendant disadvantage, common to all gages with long electron path-lengths, is the appearance of instabilities and anomalous behavior which requires that the hot-cathode magnetron gage be operated at very low electron currents.

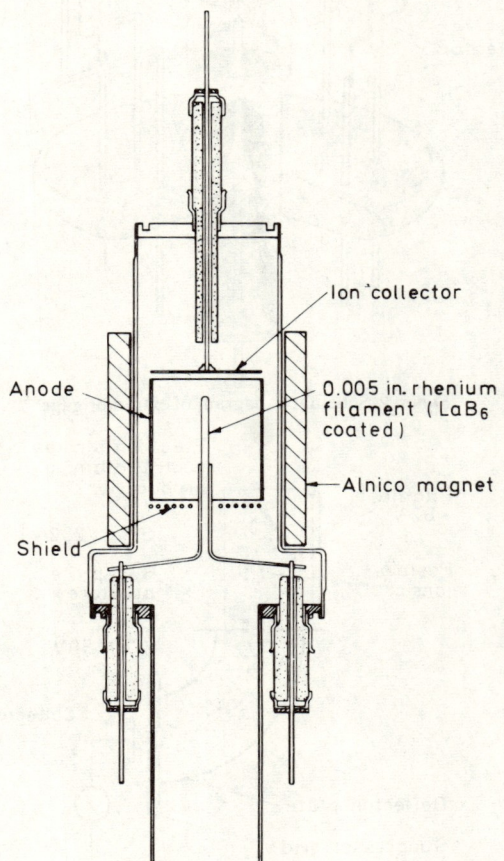

Figure 11. Cross-sectional view of hot-cathode magnetron gage[27].

PART 7. VERY LOW PRESSURES AND ULTRA LOW PRESSURES

The first successful design[61, 62] is shown schematically in *Figure 11*. It consists of a cylindrical magnetron with a hairpin filament on the axis operated in an axial magnetic field about 2.5 times the cut-off value (about 250 gauss). The electrons are prevented from escaping axially by end-plates negative with respect to the filament, and the ions are collected on one of these plates. The anode is 24 mm diameter by 32 mm long. Electron currents in the range 10^{-7} to 10^{-9} A are used to insure stable operation. Below 10^{-8} Torr the electron current to the anode is independent of pressure. The ion-collector current is linearly proportional to pressure over the range 10^{-6} to the lowest pressures measured. The preferred operating conditions are: filament $+90$ V, anode $+390$ V, shield $+35$ V, ion-collector earthed, electron emission $(B = 0)\ 1 \times 10^{-8}$. These conditions yield a sensitivity of 6.4×10^{-2} A Torr^{-1}. The x-ray limit under these conditions is about 3×10^{-15} Torr, and measurements have been made down to 10^{-13} Torr[27].

The sensitivity has been increased and the x-ray limit further reduced by replacing the ion-collector electrode with a lens system focusing the ions on to the first dynode of an electron multiplier[66]. The x-ray induced photo-emission is reduced to such a low level with this design that pressures as low as 3×10^{-18} Torr should be measurable with ion counting methods.

Various modifications to the basic design have been reported[54, 63, 81].

V. Comparison of Gages

The choice of an appropriate gage for a particular experiment depends on many factors, some of which have been briefly discussed above and more detailed considerations can be found in ref. 99, chapters 7 and 8. Some of the important characteristics and limitations of the gages discussed in this chapter are indicated in *Tables 4*(a) and (b).

VI. References

[1] Akaishi, K. *Jap. J. Appl. Phys.* **8**, 1061 (1969).
[2] Alpert, D. *Vide, Paris*, **17**, 19 (1962).
[3] Alpert, D. and R. S. Buritz, *J. Appl. Phys.* **25**, 202 (1954).
[4] Anderson, H. U. *Rev. Sci. Instrum.* **34**, 703 (1963).
[5] Apgar, E. *Proceedings* of the Second European Vacuum Symposium, p 223. R. A. Lang: Esch (Taunus) (1963).
[6] Appelt, G. *Vakuum-Technik*, **11**, 174 (1962).
[7] Barnes, G. J. Gaines and J. Kees, *Vacuum*, **12**, 141 (1962).
[8] Baronetsky, E. and A. Klopfer, *Advanc. Vac. Sci. Technol.* **1**, 401 (1960).
[9] Barz, A. and P. Kocian, *J. Vac. Sci. Technol.* **7**, 200 (1970).
[10] Bayard, R. T. and D. Alpert, *Rev. Sci. Instrum.* **21**, 571 (1950).
[11] Beck, A. H. and A. D. Brisbane, *Vacuum*, **2**, 137 (1952),
[12] Becker, J. A., E. J. Becker and R. G. Brandes, *J. Appl. Phys.* **32**, 411 (1961).
[13] Bennewitz, H. G. and H. D. Dohmann, *Vakuum-Technik*, **14**, 8 (1965).
[14] Beitel, G. A. and C. M. Gosselin, *J. Vac. Sci. Technol.* **7**, 580 (1970).
[15] Blears, J. *Proc. Roy. Soc. A*, **188**, 62 (1947).
[16] Brombacher, W. G. 'Bibliography and Index on Vacuum and Low Pressure Measurement', *NBS Monograph 35* (US Department of Commerce).
[17] Brombacher, W. G. *NBS Tech. Note 298* (1967).
[18] Bureau, A. J., L. S. Laslett and J. M. Keller, *Rev. Sci. Instrum.* **23**, 683 (1952).
[19] Carmichael, J. H. and E. A. Trendelenburg, *J. Appl. Phys.* **29**, 1570 (1958).
[20] Carter, G. *Nature, London*, **183**, 1619 (1959).

Table 4(a). Comparison of hot-cathode ionization gages, non-magnetic

Gage	(a) p_X (Torr) Measured	(a) p_X (Torr) Estimated	(b) p_{min} (Torr) Measured	(c) Electronic desorption	(d) I^- (A)	(e) V_g (V)	(f) K_{eff} (Torr^{-1})	(g) Sensitivity (A Torr^{-1})	Refs
Bayard–Alpert	3×10^{-11}			Yes	8×10^{-3}	105	25	0.2	98
Modulated BAG									
Mode I			3×10^{-12}	No	8×10^{-3}	105	8	6.4×10^{-2}	51
Mode II			$\sim 8 \times 10^{-12}$	No	10×10^{-3}	105	~ 4	4×10^{-2}	65
Mode III				No	8×10^{-3}	105	5	4×10^{-2}	96
Extractor (a)	2.6×10^{-12}	3×10^{-13}	$7 \times 10^{-13}*$	No	8×10^{-13}	105	13	0.1	95
(b)	$\sim 9 \times 10^{-12}$		4×10^{-12}	?	4×10^{-3}	150	20	8×10^{-2}	73
Orbitron			$\sim 1 \times 10^{-11}$	No	1.2×10^{-6}	540	9×10^4	0.1	74
Bent-beam		10^{-14}	$3 \times 10^{-12}*$	No	6×10^{-3}	150	10	4×10^{-2}	45

Table 4(b). Comparison of ionization gages; magnetic

Gage	(a) p_X (Torr)	(b) p_{min} (Torr) Measured	(d) I^- (A)	(h) V_a (kV)	(i) B (kgauss)	(g) Sensitivity (A Torr^{-1})	(j) Linearity	Refs
Hot-Cathode								
Hot-Cathode magnetron	4×10^{-14}	$4 \times 10^{-13}*$	3.5×10^{-9}	0.3	0.25	9×10^{-2}	1	61, 62
Hot-Cathode magnetron with multiplier	3×10^{-17}	3.5×10^{-9}		0.3	0.25	10^4	1	60
Cold-Cathode								
Trigger discharge	0	10^{-13}	—	2	1	$2.5 (P = 10^{-6})$ $0.1 (P = 10^{-11})$	1.13–1.2	27
Magnetron	0	3×10^{-13}	—	4.5–5.5	1.15	3.0	$(10^{-5}$–5.5×10^{-13} Torr$)$	83
Inverted magnetron (MM-14S)	0	3×10^{-13}	—	6	2	0.31	$(10^{-4}$–10^{-12} Torr$)$	82

PART 7. VERY LOW PRESSURES AND ULTRA LOW PRESSURES

Columns

(a) p_x — Pressure at which ion current equals x-ray photocurrent.
(b) p_{min} — Minimum pressure measured. Values marked (*) are lowest indicated pressures, not checked by other means.
(c) Electronic desorption — Indicates whether gage is subject to serious errors caused by electronic desorption of ions from the electron collector.
(d) I^- — Typical operating current to anode.
(e) V_{gf} — Grid-filament voltage.
(f) K_{eff} — Effective sensitivity factor for nitrogen. For unomdulated gages, $K_{eff} = K = (I^+/I^-)(1/p)$. For modulated gages $K_{eff} = K\beta$ where β is a modulation factor (see equation 9).
(g) Sensitivity — Measured at noted electron current.
(h) V — Anode voltage.
(i) B — Magnetic field.
(j) Linearity, n — Exponent in power law, $I^+ = kp^n$.

All pressures are equivalent nitrogen

[21] Christian, R. G. and J. H. Leck, *J. Sci. Instrum.* **43**, 229 (1966).
[22] Clay, F. P. and L. T. Melfi, *J. Vac. Sci. Technol.* **3**, 167 (1966).
[23] Cleaver, J. S. and W. H. Zakrzewski, *Vacuum*, **18**, 73 (1968).
[24] Cobic, B. G. Carter and J. H. Leck, *Brit. J. Appl. Phys.* **12**, 288 (1961).
[25] Cobic, B., G. Carter and J. H. Leck, *Vacuum*, **11**, 247 (1961).
[26] Dadson, R. S., K. W. Elliott and D. M. Woodman, *Proceedings* of the Fourth International Vacuum Congress, p 679. Institute of Physics and the Physical Society: London (1968).
[27] Davis, W. D. *J. Vac. Sci. Technol.* **5**, 23 (1968).
[28] Davis, W. D. *J. Vac. Sci. Technol.* **6**, 85 (1969).
[29] de Vries, A. E. and P. K. Rol, *Vacuum*, **15**, 135 (1965).
[30] Edmonds, T. and J. P. Hobson, *J. Vac. Sci. Technol.* **2**, 182 (1965).
[31] Ehrlich, G. *Advance. Catalysis*, **14**, 255 (1963).
[32] Eisinger, J. *J. Chem. Phys.* **30**, 412 (1959).
[33] Fitch, R. K. and W. J. Thatcher, *J. Phys. E.*, Ser. 2, **1**, 317 (1968).
[34] Fletcher, B. and J. F. Watts, *Vacuum*, **17**, 445 (1967).
[35] Garbe, S., K. Klopfer and W. Schmidt, *Vacuum*, **10**, 81 (1960).
[36] Gomer, R. *Rev. Sci. Instrum.* **24**, 993 (1953).
[37] Gopalaraman, C. P., R. A. Armstrong and P. A. Redhead, *J. Vac. Sci. Technol.* **7**, 195 (1970).
[38] Gosselin, C. M., G. A. Beitel and A. Smith, *J. Vac. Sci. Technol.* **7**, 233 (1970).
[39] Grigor'ev, A. M. *Instrums Exp. Tech.* 870 (1959).
[40] Groszkowski, J. *Bull. Acad. Polon. Sci; Ser. Sci. Techn.* **14**, 1023 (1966).
[41] Groszkowski, J. *Bull. Acad. Polon. Sci; Ser. Sci. Techn.* **18**, 931 (1970).
[42] Haefer, R. *Acta Phys. Austriaca*, **9**, 200 (1954).
[43] Haefer, R. and J. Hengevoss, *Trans. Amer. Vac. Soc. Nat. Vac. Symp.* **7**, 67 (1960).
[44] Hayward, W. H., R. L. Jepsen and P. A. Redhead, *Trans. Amer. Vac. Soc. Nat. Vac. Symp.* **10**, 228 (1963).
[45] Helmer, J. C. and W. H. Hayward, *Rev. Sci. Instrum.* **37**, 1652 (1966).
[46] Hickmott, T. W. *J. Appl. Phys.* **31**, 128 (1960).
[47] Hickmott, T. W. *J. Chem. Phys.* **32**, 810 (1960).
[48] Hobson, J. P. and J. Earnshaw, *Proceedings* of the Fourth International Vacuum Congress, p 619. Institute of Physics and the Physical Society: London (1968).
[49] Hobson, J. P. and P. A. Redhead, *Canad. J. Phys.* **36**, 271 (1958).
[50] Hobson, J. P. and J. Earnshaw, *J. Vac. Sci. Technol.* **4**, 257 (1967).
[51] Hobson, J. P. *J. Vac. Sci. Technol.* **1**, 1 (1964).
[52] Holkeboer, D. H. *Trans. Amer. Vac. Soc. Nat. Vac. Symp.* **10**, 292 (1963).
[53] Huber, W. K. and G. Rettinghaus, *J. Vac. Sci. Technol.* **6**, 89 (1969).
[54] Iapteff, B. and P. S. Choumoff, *Vide, Paris*, **146**, 103 (1970).
[55] Ishii, H. and K. Nakayama, *Trans. Amer. Vac. Soc. Nat. Vac. Symp.* **8**, 406 (1961).

[56] Ishikawa, K. *Jap. J. Appl. Phys.* **4**, 461 (1965).
[57] Jaeckel, R. and E. Teloy, *Trans. Amer. Vac. Soc. Nat. Vac. Symp.* **8**, 406 (1961).
[58] James, L. H. and G. Carter, *Trans. Amer. Vac. Soc. Nat. Vac. Symp.* **9**, 502 (1962).
[59] Kornelsen, E. V. *Trans.* of the Third International Vacuum Congress, Vol. I, p 65. Pergamon: London (1965).
[60] Lafferty, J. M. *Rev. Sci. Instrum.* **34**, 467 (1963).
[61] Lafferty, J. M. *Trans. Amer. Vac. Soc. Nat. Vac. Symp.* **7**, 97 (1960).
[62] Lafferty, J. M. *J. Appl. Phys.* **32**, 424 (1961).
[63] Lafferty, J. M. *Proceedings* of the Fourth International Vacuum Congress, p 647. Institute of Physics and the Physical Society: London (1968).
[64] Lange, W. J. and D. P. Eriksen, *J. Vac. Sci. Technol.* **3**, 303 (1966).
[65] Lange, W. J. and J. H. Singleton, *J. Vac. Sci. Technol.* **3**, 319 (1966).
[66] Lange, W. J. J. H. Singleton and D. P. Eriksen, *J. Vac. Sci. Technol.* **3**, 338 (1966).
[67] Lassiter, W. S. *J. Vac. Sci. Technol.* **6**, 418 (1969).
[68] Lawson, R. W. *Brit. J. Appl. Phys.* **18**, 1783 (1967).
[69] Leck, J. H. *Pressure Measurement in Vacuum Systems*, 2nd ed. Chapman and Hall: London (1964).
[70] Madey, T. E. and J. T. Yates, *J. Vac. Sci. Technol.* **8**, 525 (1971).
[71] Meinke, C. and G. Reich, *Vakuum-Technik*, **12**, 79 (1963).
[72] Meinke, C. and G. Reich, *J. Vac. Sci. Technol.* **4**, 356 (1967).
[73] Melfi, L. T. *J. Vac. Sci. Technol.* **6**, 322 (1969).
[74] Meyer, E. A. and R. G. Herb, *J. Vac. Sci. Technol.* **4**, 63 (1967).
[75] Miller, J. R. *Vacuum*, **17**, 387 (1967).
[76] Moesta, H. and R. Renn, *Vakuum-Technik*, **6**, 35 (1957).
[77] Moore, G. E. and F. C. Unterwald, *J. Chem. Phys.* **40**, 2626 (1964).
[78] Moore, G. E. and F. C. Unterwald, *J. Chem. Phys.* **40**, 2639 (1964).
[79] Moser, H. and H. Poltz, *Z. Instrumkde*, **65**, 43 (1957).
[80] McGowan, W. and L. Kerwin, *Canad. J. Phys.* **38**, 567 (1960).
[81] Nakeo, I. F. *J. Jap. Vac. Soc.* **9**, 49 (1966).
[82] Nichiporovich, G. A. *Instrums Exp. Tech.* No. 6, 1440 (1967).
[83] Nichiprovich, G. A. and I. F. Khanina, *Proceedings* of the Fourth International Vacuum Congress, p 666. Institute of Physics and the Physical Society: London (1968).
[84] Nichiporovich, G. and I. F. Khanina, *Instrums Exp. Tech.* No. 1, 159 (1968).
[85] Nottingham, W. B. *Proceedings* of the Seventh Conference on Physics and Electronics. Massachusetts Institute of Technology: Cambridge, Mass. (1947).
[86] Owens, C. L. *J. Vac. Sci. Technol.* **2**, 104 (1965).
[87] Penning, F. M. and K. Nienhuis, *Philips Tech. Rev.* **11**, 115 (1949).
[88] Poulter, K. F. *Vacuum*, **20**, 385 (1970).
[89] Rapp, D. and P. Englander-Golden, *J. Chem. Phys.* **43**, 1464 (1965).
[90] Redhead, P. A. *Canad. J. Phys.* **36**, 255 (1958).
[91] Redhead, P. A. *Canad. J. Phys.* **37**, 1360 (1959).
[92] Redhead, P. A. *Trans. Amer. Vac. Soc. Nat. Vac. Symp.* **7**, 108 (1960).
[93] Redhead, P. A. *Rev. Sci. Instrum.* **31**, 343 (1960).
[94] Redhead, P. A. *Vacuum*, **13**, 253 (1963).
[95] Redhead, P. A. *J. Vac. Sci. Technol.* **3**, 173 (1966).
[96] Redhead, P. A. *J. Vac. Sci. Technol.* **4**, 57 (1967).
[97] Redhead, P. A. *J. Vac. Sci. Technol.* **7**, 182 (1970).
[98] Redhead, P. A. and J. P. Hobson, *Brit. J. Appl. Phys.* **16**, 1555 (1965).
[99] Redhead, P. A., J. P. Hobson and E. V. Kornelsen, *The Physical Basis of Ultrahigh Vacuum*. Chapman and Hall: London (1968).
[100] Reich, G. *Z. InstrumKde*, **74**, 254 (1966).
[101] Reich, G. *Trans. Amer. Vac. Soc. Nat. Vac. Symp.* **7**, 112 (1960).
[102] Roehrig, J. R. and J. C. Simonds, *Trans. Amer. Vac. Soc. Nat. Vac. Symp.* **8**, 511 (1961).
[103] Rothe, E. W. *J. Vac. Sci. Technol.* **1**, 66 (1964).
[104] Schissel, P. O. *J. Appl. Phys.* **33**, 2659 (1962).
[105] Schuemann, W. C. *Rev. Sci. Instrum.* **34**, 700 (1963).
[106] Schuhmann, S. *Trans. Amer. Vac. Soc. Nat. Vac. Symp.* **9**, 463 (1962).
[107] Schulz, G. J. *J. Appl. Phys.* **28**, 1149 (1957).
[108] Shaw, M. L. *Rev. Sci. Instrum.* **37**, 113 (1966).
[109] Singleton, J. H. *J. Chem. Phys.* **45**, 2819 (1966).

PART 7. VERY LOW PRESSURES AND ULTRA LOW PRESSURES

[110] Singleton, J. H. *J. Vac. Sci. Technol.* **4**, 103 (1967).
[111] Smetana, F. O. and C. T. Carley, *J. Vac. Sci. Technol.* **3**, 47 (1966).
[112] Spencer, C. M. and D. Staheli, *J. Vac. Sci. Technol.* **5**, 105 (1968).
[113] Steckelmacher, W. *J. Sci. Instrum.* **42**, 63 (1965).
[114] Steckelmacher, W. *Vuoto Sci. & Tecnol.* **2**, 189 (1969).
[115] Stickney, W. W. and B. B. Dayton, *Trans. Amer. Vac. Soc. Nat. Vac. Symp.* **10**, 105 (1963).
[116] Tuzi, Y. S. Okada and M. Kim, *Jap. J. Appl. Phys.* **7**, 1415 (1968).
[117] Urbanek, K. *J. Vac. Sci. Technol.* **4**, 328 (1967).
[118] Utterback, N. G. and T. Griffiths, *Rev. Sci. Instrum.* **37**, 866 (1966).
[119] Van Oostrom, A. *J. Sci. Instrum.* **44**, 927 (1967).
[120] Werner, J. G. and J. H. Leck, *J. Phys. E.* **2**, 861 (1969).
[121] Winters, H. F., D. R. Denison, D. G. Bills and E. E. Donaldson, *J. Appl. Phys.* **34**, 1810 (1963).
[122] Young, J. R. and F. P. Hession, *Trans. Amer. Vac. Soc. Nat. Vac. Symp.* **10**, 234 (1963).

CHAPTER 5

The Absolute Measurement of Volume

A. H. COOK

Cavendish Laboratory, University of Cambridge, Cambridge, UK

Contents

I.	Introduction	303
II.	Experimental Problems	305
III.	The Volume of a Cube of Tungsten Carbide	311
IV.	Volumes of Cubes of Fused Silica	315
V.	Single Crystals of Pure Silicon for the Measurement of Avogadro's Number	318
VI.	The Density of Water	318
VII.	Summary	319
VIII.	References	319

I. Introduction

A particular volume is an arbitrary quantity of no general interest and the measurement of volume acquires importance only as a means to the determination of density in absolute terms. This chapter is concerned with the measurement of volume as a step in the determination of the density of a fluid or solid and thus towards the establishment of fundamental relations between physical constants. It is of interest, however, to look back over the history of measurements of volume for, in the past, units of volume had a more significant place in systems of measurement than they occupy today. Many substances of trade, such as corn, meal, milk or wine, were measured by volume rather than by mass and so reasonably accurate measures of capacity were provided in all pre-Metric systems of measurement from Babylonian and Egyptian times onwards. In the early scientific period it became possible to weigh liquids much more accurately than to measure volumes and so the measures of capacity came to be defined in terms of the mass of liquid they contained at a given temperature, as indeed the English gallon is legally defined today.

A new approach was introduced with the development of the Metric system. The original intention of the founders of that system was to base all units on natural standards and, as is well known, they took the length of the arc of a meridional quadrant of the Earth as the standard of length and then defined a unit of mass through the mass of water occupying a volume measured in terms of that unit. The procedure is exactly similar to that we could use to define quantities in atomic units. Instead of defining length in terms of the wavelength of a line in the krypton spectrum, it could be defined in terms of the cesium standard of frequency and the natural constant, the speed of light; in just the same way, the gram was defined in terms of the unit of length and a natural constant, the density of water. However, it turned out that neither the natural unit of length nor the density of water was well defined and so after a few years the meter came to be defined by a metal bar and the kilogram by a metal weight.

It is not easy to determine the precise volume of a liquid contained in a vessel even though the volume of that vessel might be quite accurately measured in terms of the unit of length and so as higher accuracy was increasingly demanded for volumetric measurements in chemistry, it was found desirable to establish a new unit of volume, the liter, which was almost a cubic decimeter but which was defined in terms of the mass of water contained. To establish the relation between the liter and the cubic decimeter, the density of water had to be measured with precision, work that was undertaken by Chappuis[4] and by de Lépinay, Buisson and Benoît[13] working at the International Bureau of Weights and Measures. The density of water was found from the mass displaced by a solid of known volume weighed in water; the solids used were glass cubes of which the dimensions were measured interferometrically in terms of wavelengths of light, the relations of which to the International Prototype Meter had been determined a few years earlier by A. A. Michelson and J. R. Benoît and by Benoît, Ch. Fabry and Pérot, also at the International Bureau of Weights and Measures.

Weighing the solid in water avoids the difficulty of knowing how much water is contained in a vessel and had been used in earlier work relating volume and mass in which the solid was a circular cylinder. A cylinder is more difficult to work accurately in glass by opticians' techniques and is more difficult to measure interferometrically than a cube, so that the cube was chosen for interferometric measurements, whereas when lengths were measured mechanically and the lathe was the most accurate machine tool, it was natural to choose a cylinder for the solid. These considerations are not of historical interest alone—if accurate measurements of volume are to be made, the form of the body to be measured, whether a solid body or a hollow container, must be chosen to exploit the most developed techniques of fabrication and measurement.

Following the work of Chappuis and of de Lépinay, Buisson and Benoît, no fundamental measurements of volume were made prior to the determinations of the density of mercury at the National Physical Laboratory[6,8]. The density of mercury was determined by a number of authors[2,3,10,14,16,17] by comparison with water; Scheel and Blankenstein weighed a sinker in mercury and water but every one else weighed a pyknometer filled in turn with water and mercury. The results cannot be relied on to much better than 5 p/M, to which the uncertainty in the density of water contributes about 2 p/M [p/M—parts per million].

The density of mercury is required for two purposes in thermodynamics —to enable the pressure in a barometer to be obtained in absolute units, and as entering determinations of the volumes of vessels (such as those used in thermodynamic experiments on gases) from the mass of mercury contained in them. Pressures of the order of 10^5 to 10^6 N/m^2 can still be obtained most accurately (to 1 p/M) with the mercury barometer but the mercury column has been superseded by the free piston gage at higher pressures. Mercury is a more satisfactory liquid to use than water in determining volumes of flasks or capillary tubes or similar gas handling apparatus to high accuracy, because its density is better defined (it contains no dissolved air) and because it does not wet glass, and it was for those reasons that when a new determination of the density of mercury was undertaken at the National Physical Laboratory it was decided to use absolute measurements of volume and not to work by comparison with the density of water. The National Physical Laboratory measurements have not been repeated, they are the basis of all accurate measurements of density and volume involving the use of mercury, and so it seems justifiable to devote much of this chapter to an account of the measurements of volume entailed in that work.

Methods of x-ray interference measurements developed in recent years by Hart[11] and Hart and Bonse[12], have led to the possibility of a new means of determining Avogadro's number which involves the measurement of the density of a single crystal of very pure silicon. Now, x-ray interference systems enable one slice of a single crystal of silicon to be displaced relative to other slices by a countable number of lattice spacings, the number being large enough for the displacement to be measured by an optical interferometer with light from a gas laser. Thus the lattice spacing of a single crystal of silicon should be measurable directly in terms of optical wavelength standards. If the atomic weight and the density are known, then Avogadro's number may be evaluated.

Because Avogadro's number is a key quantity in the system of fundamental atomic constants, and therefore in modern systems of measurements, a new determination along the lines just indicated is being prepared jointly by the National Physical Laboratory and the Department of Physics at Bristol University.

The bulk of this chapter comprises two sections (III and IV) describing the measurements of the volumes of a cubical sinker and a cubical box used at the National Physical Laboratory for the determination of the density of mercury. They are preceded by a discussion (Section II) of the experimental problems involved in the measurement of volume to high accuracy and are followed by an outline of the method to be used for the measurement of the density of silicon (section V) and by consideration of the lessons to be drawn from the work heretofore described.

II. Experimental Problems

The measurement of a volume through measurements of length requires that the technique for the measurement of length shall be of sufficient accuracy and, what is in practice of greater importance, it requires that the geometry of the space being measured shall be defined precisely. Measurements of length by themselves do not define a volume, they must be placed in known relations to each other. Three figures have sufficiently well defined geometries—the sphere, the circular cylinder with plane parallel ends, and the rectangular parallelepiped. Which of these figures is chosen will depend in part on the use to which the measurements of volume will be put but the decisive considerations will usually be the accuracy with which objects of a given geometry and size may be made, and their suitability for the most precise measurements of length. Solid spheres, cylinders and parallelepipeds can all be made with high precision by modern workshop techniques; if a typical dimension is 0.1 m (typical volume 10^{-3} m^3) it should be possible to construct such objects for which the actual surfaces will lie within 1 p/M of the ideal geometrical form. It should be possible to make hollow cubes and cylinders with like precision. While between these simple forms, workshop techniques impose no clear choice, methods of measurement of length do give a clear preference for parallelepipeds. Optical interference is the only method that need be seriously considered for measurement of length and by far the simplest, and hence the most accurate, arrangements are attained if the surfaces of the solid are themselves parallel reflecting surfaces incorporated in the interferometer. Rectangular parallelepipeds satisfy this condition alike for solid and hollow bodies and the most accurate measurements of volume so far achieved have been of objects that are nominally cubes. Work currently in progress also involves rectangular parallelepipeds.

The cube is the obvious shape to choose because for a given accuracy in the measurement of length, the best accuracy in the measurement of volume is achieved if all the sides have the same length. Cubes were used in the work described in sections III and IV but in work now in preparation (section V) they have been abandoned. To understand why, consider further the definition of the volume of a solid or hollow object and of the length measured by optical interference.

The problem is somewhat different for glasses and hard metals, the two materials that have been used in construction. A metal cube of tungsten carbide, having almost the same density as mercury, was used as a sinker in the measurement of the density of mercury by displacement (section III) while in a complementary measurement, a hollow cubical box to contain mercury was constructed out of fused silica (section IV). A solid cube of fused silica has also been used as a sinker in the measurement of the density of water by displacement[4,13]. When glass surfaces are finished by optical polishing, the glass flows over the surface in the final stages, as well as being ground off by the abrasive, and the surface shows no detectable small scale roughness. If one may imagine the surface to be defined locally by a plane laid across the highest points, the surface so defined will be indistinguishable, to within 0.01 μm, from the physical surface against which a liquid comes into contact. However, that is not necessarily the surface as seen by an interferometer. When light is reflected at the boundary between an insulator and a vacuum, and the incident and reflected light lie in the vacuum, then the effective plane of reflection, on which the phases of the incident and reflected light are equal and which is therefore the boundary of a length measured interferometrically, is the surface of the insulator, but if the incident and reflected light lie in the insulator, then the phases of the incident and reflected light differ by π at the boundary, corresponding to an effective plane of reflection lying at a distance of half a wavelength within the vacuum. Thus in some interferometric measurements, a half wavelength of the light used must be added to the apparent length.

With a metal surface, the situation is more complex. The surface of a hard metal as formed by lapping does not flow as do glasses and soft metals, but is irregular on a microscopic scale, with a characteristic range of depth of 0.02 μm and a characteristic horizontal dimension of 1–10 μm, depending on the metal and lapping abrasive. It is also covered with a network of fine scratches, perhaps up to 1 mm long and 0.02 μm deep and 0.1 μm wide. As a consequence the surface defined locally by a plane laid on the peaks of the actual surface lies above the mean position of the actual surface with which a liquid maintains contact, and the volume displaced by a solid sinker is somewhat less than that defined by planes placed across the peaks. Added to this geometrical complication is the optical problem of defining the phase shift on reflection at the surface of a material which, like tungsten carbide, may be a mixture of materials of different, relatively low, electrical conductivities. Two ways of dealing with the problem are described in sections III and V.

With modern workshop techniques, opposite faces of a parallel sided block can be worked to be parallel to better than 0·7 μm over an area 0.1 m square. Interferometric measurements made with light reflected normally to such faces define the perpendicular distance with high accuracy and the mean separation of a pair of such faces is well defined by measurements between a number of pairs of points (or small areas) evenly distributed over them.

It is evident that in order to obtain the volume of a solid from measurements of length, the accuracy with which the solid is constructed must satisfy certain geometrical constraints in order that measurements between any one pair of faces of a parallelepiped may be related in a sufficiently well determined manner to those between the other two pairs of faces. The necessary conditions may be examined by means of a formal transformation of the integral for the volume of a nearly cubical object to a set of line integrals round the edges and surface integrals over the faces. The working is set out in Cartesian coordinates for a nearly cubical object but it could readily be written down for a cylinder (in cylindrical polar coordinates) or a sphere (in spherical polar coordinates).

The volume in Cartesian coordinates is

$$\iiint dx\, dy\, dz$$

the integration being taken throughout the solid.

To introduce the form of the faces, use the divergence theorem:

$$\iiint \text{div}\, \boldsymbol{v}\, dx\, dy\, dz = \iint \boldsymbol{v}\, d\boldsymbol{S}$$

The left hand integral is the volume if the vector $\boldsymbol{v}$ is $\frac{1}{3}(x\boldsymbol{i}, y\boldsymbol{j}, z\boldsymbol{k})$ where $(\boldsymbol{i}, \boldsymbol{j}, \boldsymbol{k})$ are unit vectors parallel to the (x, y, z) coordinate directions.

Thus the volume is

$$\tfrac{1}{3} \iint x\, dy\, dz + \tfrac{1}{3} \iint y\, dx\, dz + \tfrac{1}{3} \iint z\, dx\, dy$$

where the first integral is taken over the faces perpendicular to the x direction, and so on.

Now let $x = x_0 + x'$ on one of the faces with its normal nearly parallel to the x direction. If the solid is nearly cubical, x' will be small. Then $\iint x\, dy\, dz = x_0 \int dy\, dz + \int x'\, dy\, dz$. x' is the amount by which the face departs from a plane perpendicular to the x direction. It includes both a tilt of the surface as a whole and departures of the surface from a plane. The tilt is taken into account if x_0 is chosen to be the mean value of x over the surface; the second integral then gives the effect of departures of the surface from a plane.

Now the surface integral

$$\iint dy\, dz$$

is equal to the line integral

$\frac{1}{2}\int y\, dz$ along the edges parallel to z

$-\frac{1}{2}\int z\, dy$ along the edges parallel to y.

The integrals may be written as

$$\tfrac{1}{2}\int \delta y\, dz + \tfrac{1}{2}\int \delta z\, dy + y_0 z_0$$

where δy is the difference between values y' at equal values of z on the opposite edges and δz is the corresponding difference between values of z'.

Thus, finally, the volume of the parallelepiped is

$$x_0 y_0 z_0 + \tfrac{1}{6} x_0 (\int \delta y\, dz + \int \delta z\, dy)$$
$$+ \tfrac{1}{6} y_0 (\int \delta x\, dz + \int dz\, dx)$$
$$+ \tfrac{1}{6} z_0 (\int \delta y\, dx + \int \delta x\, dy)$$
$$+ \tfrac{1}{3} \int x'\, dy\, dz + \tfrac{1}{3} \int y'\, dx\, dz$$
$$+ \tfrac{1}{3} \int z'\, dx\, dy$$

This explicit formula for the volume will be valid provided the quantities x_0, y_0, z_0, x', y', z', are measured in mutually perpendicular directions. The faces of the solid do not have to be perpendicular—the parallelepiped does not have to be a right parallelepiped and it is not even necessary for the faces to be parallel, but it is essential that the measurements should be made in three perpendicular directions. Otherwise, a relative error of order $(1 - \cos \theta)$ or $\tfrac{1}{2}\theta^2$ will be committed, where θ is the amount by which a direction of measurement departs from nominal. The condition is not a rigorous one—θ need not be less than a thousandth of a radian (3 min) for an error of 1 p/M —and it is most easily satisfied by making the solid with mutually perpendicular faces and measuring perpendicular to faces. Lapping techniques enable faces to be made parallel to high accuracy (within a few seconds of arc) and interferometers can be set up that automatically give measurements perpendicular to the appropriate surface.

The geometry of solids that are nominally cylindrical or spherical requires special attention; if we define the *diameter* of a circle or sphere as the separation of parallel tangent lines or planes and the *radius* as the radius of curvature, then it is possible for the diameter to be constant around the circle or sphere yet for the radius to differ from half the diameter. One possible condition for a circle is shown in *Figure 1*. Let P_1 be the common center of curvature of opposite segments, S_1 and S_4, having radii of curvature equal to r_1 and r_2. The diameter is then equal to $(r_1 + r_2)$. Now suppose segment S_2 has the radius r_2 and segment S_5 has radius r_1, with the common centre P_2, and segment S_3 has radius r_1 and S_6 has radius r_2 with the common centre P_3.

If then P_1, P_2 and P_3 are the vertices of an equilateral triangle, the various segments will have common tangents, for example, S_1 and S_2 will have a common tangent perpendicular to $P_1 P_2$, as will S_4 and S_5; then the diameter

THE ABSOLUTE MEASUREMENT OF VOLUME

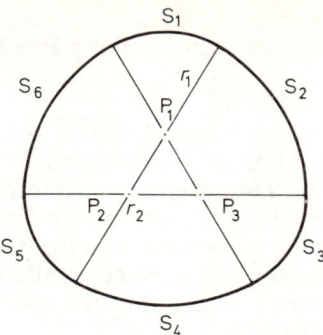

Figure 1. Lobed 'circle' of constant diameter but with two radii of curvature.

will be constant over the circle and there will be no discontinuity of tangent, yet the radii of curvature will differ. Such a figure is called lobed, and it is easy in certain methods of figuring cylinders and spheres to produce lobed cylinders or spheres.

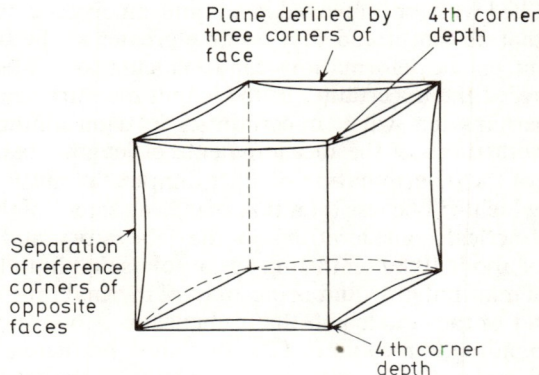

Figure 2. Geometrical framework for measurements of cube of tungsten carbide.

With the geometry shown in *Figure 2* the area of the circle is

$$(3/2)(\pi/3)(r_1^2 + r_2^2)$$

or

$$(\pi/2)(r_1^2 + r_2^2)$$

The diameter is $(r_1 + r_2)$ and so the apparent area that would be calculated from the measured separation of tangent planes is

$$(\pi/4)(r_1 + r_2)^2$$

Now let $r_1 = r - \rho$ and $r_2 = r + \rho$.
Then the apparent area is πr^2 while the true area is

$$\pi(r^2 + \rho^2)$$

Suppose then that the area is required correct to 1 p/M; ρ/r would have to be less than about 1/2000. Similar considerations apply to spherical surfaces. It is possible to measure the radii of curvature of surfaces to higher accuracy interferometrically, but the measurements on curved surfaces are more time-consuming than those on plane surfaces, and the geometry is not so readily checked as is that of solids bounded by plane parallel faces. For these reasons, the volumes of cylinders and spheres are less easy to measure with high accuracy than are those of parallelepipeds.

In summary, optical and metal lapping techniques enable solids and boxes to be made which are so close to ideal geometrical forms that the volumes can be defined by measurements of length to within 0.1 p/M and which are of such optical quality that they may be incorporated in interferometers.

The dimensions of a solid or a box are determined by three factors: deformation in use or in construction, problems of handling, and the best size for length measurements. Beyond a certain size, about 0.2 m characteristic dimension, it becomes difficult (though not impossible) to work metals and glasses to the highest accuracy because of deformation under their own weight. Again, when filled with, or immersed in a liquid, an object is subject to hydrostatic forces that deform it and that become greater as the size is increased. To some extent, elastic deformations can be calculated or measured but they should not exceed the uncertainty of the length measurement by more than a factor of perhaps ten, when uncertainties in determining them become comparable with those of the measurements of length (the relevant uncertainty is that of the determination of, for example, the mean separation of a pair of faces, which will be less than that of measurement of the separation at one point). Practical considerations in the laboratory and the workshop place a limit of about 10 kg on the masses of solids and liquids, both because of problems of manipulation and on account of thermal time constants. Such arguments lead to the conclusion that dimensions should not exceed about 0.2 m, corresponding to a volume of 8×10^{-3} m^3, and there is no difficulty at all in measuring the dimensions with uncertainties approaching 0.01 p/M, corresponding to uncertainties in the volume of between 0.03 and 0.1 p/M. These are ideal dimensions, making the optimum use of metal and glass working techniques and of interferometric methods; other conditions, such as the amounts or sizes in which materials are available, may restrict objects to smaller sizes (section V).

Unfortunately it is not possible to make metal or glass blocks that are perfect in form. With the greatest care in handling, some damage of the edges of blocks of brittle materials is bound to take place; the appearance of such defects is worse than their effect—for example, an obvious chip 1 mm × 0.1 mm × 0.1 mm has a volume of 10^{-11} m^3 or 10^{-8} of the volume of a cube of side 0.1 m, but even so, a careful examination of the edges and corners of objects must be made.

It is very improbable that the dimensions of an object can be measured at the same temperature and under the same pressure as it is immersed in or filled with liquid. Accordingly, the coefficient of thermal expansion and the compressibility must be determined, preferably over a range covering the conditions of measurement and use. It will be found that it is considerably easier to measure the thermal expansion than the compressibility (sections III, IV). In setting up equipment for measurements, the relatively long thermal time constant (about 10 min) of an object of size 0.1 m must be kept firmly in mind.

Lastly, the importance of strict attention to cleanliness cannot be too strongly emphasized. Films of dirt and grease can be difficult to detect but they systematically increase the volume of a sinker and decrease the volume of a box by amounts which may not be insignificant (section III).

III. The Volume of a Cube of Tungsten Carbide

In the first part of the program of determinations of the density of mercury carried out at the National Physical Laboratory, the density was found from the mass of mercury displaced by a 9 cm cube of tungsten carbide of which the volume was measured in terms of the wavelengths of radiations from mercury-198[6]. Tungsten carbide was selected because by a suitable choice of the amount of cobalt (15 per cent by weight) mixed with the tungsten carbide, the density was adjusted to be very slightly greater than that of mercury—in fact, the weight of a cube with a mass of nearly 10 kg only just exceeded 200 g in mercury. At the same time tungsten carbide sintered with cobalt is very hard and a cube made of it can be lapped to be very close to an ideal cubical form. Optical measurements with an autocollimator showed that all angles between faces lay within 1 sec of the nominal 180° or 90° while it was found in the course of the measurements of volume that all faces were plane to within 200 nm. All faces were slightly concave. The mass of 10 kg was about the most that could be handled conveniently and the length of the edges (9 cm) was about as large as could be conveniently measured interferometrically with the sharpest radiations (those from mercury-198) that were available at the time (1952) when the work was planned. Greater lengths could be measured with the krypton-86 sources that were developed later on, and of course with gas laser sources.

The cube was carefully examined on completion. The radii of curvature of the edges were about 1 μm and those of the corners about 20 μm. The largest chip was at one corner and amounted to 10^{-9} of the volume of the cube, while the total volume of chips and defects of edges and corners was less than 10^{-8} of the volume.

The forces on the cube were different when it was being measured and when it was being weighed in mercury. Being weighed, it was subject to nearly equal normal pressures on all the faces so that shear distortions were small and the hydrostatic compression from atmospheric pressure was about 0.03 p/M. Distortions during weighing in mercury could therefore be neglected. When being measured, the cube was placed upon a block rather smaller than a face. The size of the support was chosen by analogy with the bending of beams to minimize, approximately, the change of form due to shear, and in addition, experiments were made with different means of support to see if any change in the length of an edge could be detected. None was, and it was concluded that the change in volume due to shear distortion in measurement could be neglected.

The volume was derived in three stages. In the first place, the surface showed micro-roughness and the mean position had to be found. Secondly, the departures of the faces from the plane were measured and finally, the lengths of all edges were measured. The edges were measured at different temperatures to enable a coefficient of thermal expansion to be derived, for although all measurements and weighings were made as close to 20°C as possible, there were small differences of temperature for which corrections were required. The lengths of the edges were measured interferometrically in air under normal ambient conditions, so that no correction was needed for hydrostatic compression under normal atmospheric pressure (it would amount to about 0.4 p/M and would not be negligible).

A well defined geometrical framework is required into which to fit the measurements; it is shown in *Figure 2*. The shape of a face is described by its distance below a plane which passes through three of the corners, while the separation of a pair of faces is defined by the distances between opposite corners; of the four such distances, one is redundant. The distance between opposite corners was also not well defined because it is not susceptible of direct measurement, the distance is actually taken between points lying just inside each corner. Each face was divided into 100 squares and the form was defined by measurements at the center of each square; the edge lengths were the distances between the centers of those squares bounded by corners.

Not only the definition of the edge lengths but also the measurement of the roughness of the surfaces was determined by the method of measurement of edge lengths. In order to make use of an existing Kösters interferometer[7] engineers' block gages were wrung into direct contact with opposite faces (*Figure 3*) and the separation measured (*Figure 4*).

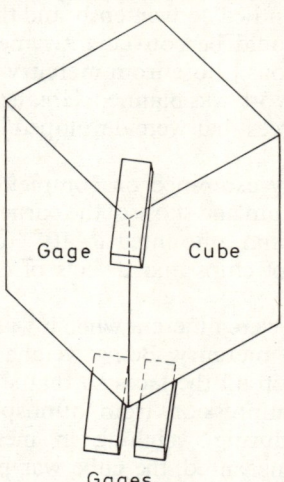

Figure 3. Arrangement of engineers' block gages wrung in optical contact with top and bottom faces of cube.

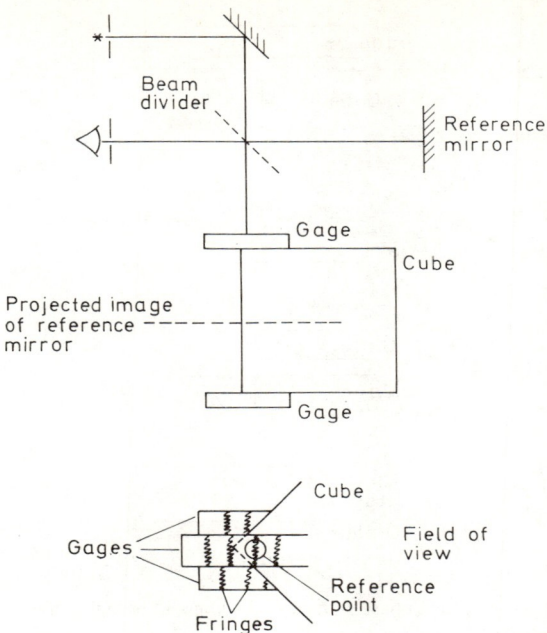

Figure 4. Arrangement of cube under Kösters interferometer. The projected image of the reference mirror of the interferometer falls midway between the upper surfaces of the upper and lower gages giving fringes of equal visibility upon them.

In order to define the edge length, a spot was marked on the upper gage which was wrung to the upper face of the cube so that the spot lay at the center of the corner square and the interference fringe value was evaluated for that position. The fringe displacements between the upper and lower gages were estimated by eye in four radiations from mercury-198.

Let L be the distance between opposite faces of the cube, measured between the peaks of the rough surface. Let l be the thickness of the upper gage and let w be the thickness of a film of oil between gage and cube or gage and gage. Then (*Figure 5*) the measured separation of the upper gage faces when wrung to the cube is

$$D = L + l + 2w$$

and when wrung together is

$$d = l + w$$

Hence
$$L = D - d - w$$

and the precise meaning of l is irrelevant.

Now if the volume of the cube is determined from L it will be overestimated by twice the volume defect corresponding to the surface irregularities. Let the mean depth of the irregularities be t'. Then the required edge length is

$$L - 2t'$$

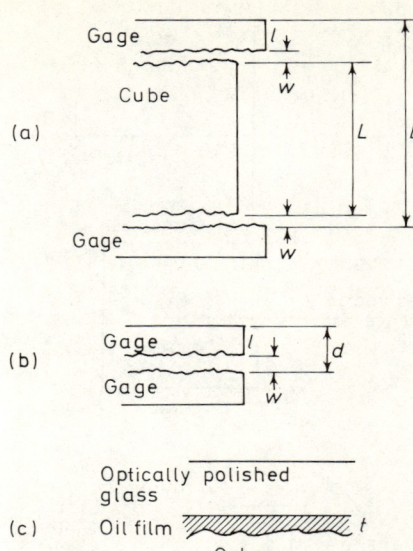

Figure 5. Elimination of optical phase shifts and thickness of wringing film in measurement of separation of faces of cube:
(a) Gages wrung to cube: (b) Gages wrung together: (c) Polished glass plane wrung to cube.

But t' is equal to the mean thickness t of a film of oil contained between the cube and a polished glass plane in wringing contact, less the wringing film thickness w.

Thus the required edge length is

$$D - d + w - 2t$$

The value of the wringing film thickness between polished surfaces was found by Rolt and Barrell[15] to be 5 nm. t was found by wringing an optically polished glass plate on to the cube and observing the area to which an oil drop of known volume spread out.

The mean value was 29.5 nm with a standard deviation of 0.5 nm. The form of each face was found by determining the distance of each of the 100 grid points from the horizontal surface of a shallow layer of liquid covering it. The liquid was a clear, viscous paraffin oil, and the separation was measured in a Fizeau interferometer, using the green radiation of mercury. The use of a liquid surface as a reference plane[1] requires considerable care to avoid disturbance from convection currents or leaks. Details, which may be found in the original papers, are not discussed here for the method would probably not now be used when interferometers using a gas laser source afford more flexible means of measurement (section VI). The results were expressed as the mean depth of the surface below the plane defined by three of the corner reference points, and the depth of the fourth corner reference point below the same plane.

The variances of all measurements were analysed in order to estimate the variances due to possible sources of error.

There was evidence that the apparent shape of a face depended on the way it was set up in the liquid-surface interferometer, while the measured edge lengths appears to depend on how well the gages were wrung to the cube. There was no evidence of any change in the dimensions of the cube over a period of some two years.

In the final combination of data, the separations of the three pairs of corner points defining the reference planes for opposite faces were calculated by least squares from the measured separations of the four pairs of corner points, together with the depths of the fourth corner point on each face below the plane defined by the other three. The volume of the material cube was equal to the volume enclosed by the reference planes less the volume enclosed between each and the corresponding face, less the volume corresponding to the surface roughness. The standard deviation of the estimated volume was 0.15 p/M.

Mention has already been made of the effects of surface films on the volume of mercury displaced by the cube. The problem of cleaning the cube was always a troublesome one, for oil films were deliberately placed on the cube, both to measure the roughness and for wringing gages to the surface. The most effective way of removing grease from a metal surface is to allow vapor of a substance such as isopropyl ether to condense upon it, but it was difficult to set up a still to contain the large and heavy cube of tungsten carbide. There was also the risk of changing the dimensions of the cube by heating it. It was therefore cleaned by swabbing with re-distilled isopropyl ether and fresh cotton wool; if no visible film was left on the surface, the cleaning was assumed to be satisfactory. Of all solvents tried isopropyl ether was the only one that appeared to leave no residue when it evaporated. There was no way of checking that all films had been removed, save visual inspection of the surface, and it was not until the measurements of the density of mercury described in the next section turned out to agree with those using the tungsten carbide cube, that it was possible to be sure that the cleaning had indeed been adequate.

IV. Volumes of Cubes of Fused Silica

In measurements complementary to those described in section III, the density of mercury was found by weighing the mass contained in a 7 cm cubical box of measured dimensions. Various possible systematic errors, in particular, any due to grease films upon the surfaces, are of opposite effect in the two determinations, so that the difference between the results obtained in the two ways should embrace the errors of both.

The cubes were assembled from rectangular blocks of fused silica worked to have optically flat polished surfaces so that they might adhere by molecular contact.

The top and bottom faces enclosed the other four (*Figure 6*) and were supported by them, so that on assembly they were closely parallel. The two longer side faces were also reasonably well supported by the smaller ones, and were also nearly parallel, but the two smallest faces were not well supported and were inclined at an appreciable angle ($4\frac{1}{2}$ sec). The construction of the cube required very great skill and patience and, as in all precise metrological work, the possibility of undertaking the measurements at all depended on having at the National Physical Laboratory technicians with superlative qualities.

Mercury was admitted to the cube under vacuum through a small capillary tube in the center of the upper face of the cube (*Figure 6*) and after weighing,

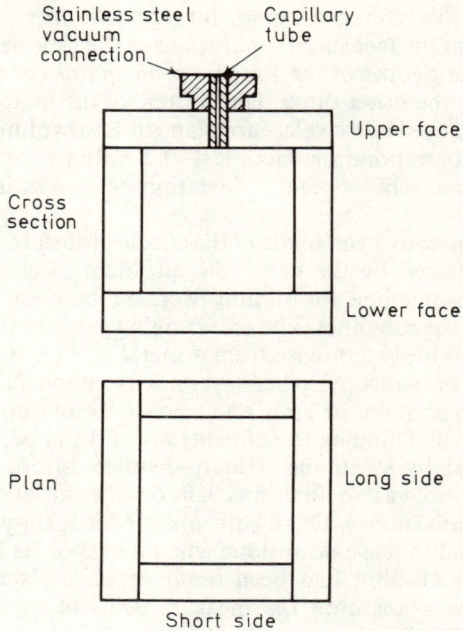

Figure 6. Construction of cube of fused silica. The capillary tube and stainless steel fitting are attached with epoxy resin.

it could be sucked out through a hypodermic tube inserted through the capillary.

The blocks forming the cube were assembled under isopropyl ether and a necessary condition for successful molecular adhesion was that they should be free of grease films. It could be confidently supposed, therefore, that on assembly there were no significant surface films on the inner faces that might reduce the effective volume of the box, leading to a low value of the measured density of mercury. Such films might accumulate during the course of filling and emptying the cube and a close watch was kept on the results to detect the effects of films. None was observed and the concordance (0.3 p/M) with the results obtained with the tungsten carbide cube (section III) indicated that the effects were not significantly greater than other errors of measurement, although the difference was consistent with the presence of films in one or both of the determinations but most likely on the cube of tungsten carbide.

THE ABSOLUTE MEASUREMENT OF VOLUME

It was inevitable that slight damage should be done to the edges of the blocks of fused silica during manipulation and a careful estimate of the total volume of chips was made. It was considerably greater than for the tungsten carbide cube, amounting to 3 p/M, with an uncertainty of about 0.1 p/M. The mercury was filled into the cube under vacuum and always appeared to fill it completely.

It had been intended to measure the separation of all pairs of faces by use of fringes of equal inclination (Haidinger rings[7]) formed by the interference of light reflected from opposite internal faces. The outer faces therefore had wedge plates cemented to them to eliminate confusing reflections.

The reflectivity of uncoated fused silica is about four per cent, so that the fringes have the two-beam $\cos^2$ profile as with the Michelson interferometer and not the sharp multiple beam profile of the Fabry–Perot interferometer. It was, however, possible to measure the diameters of the rings with an accuracy corresponding to about one per cent of the fringe spacing. Excellent fringes were obtained from the top and bottom faces and adequate ones from the long side faces (although they were in places difficult to measure in the violet radiations of mercury-198) but because of the relatively large angle between the small side faces, Haidinger rings, which require parallel reflecting faces, were not seen. A system of interference fringes formed between the three beams reflected from the opposite faces of the cube and the reference mirror of a Kösters interferometer[5,7] was therefore used. For all pairs of faces, measurements were made at the centers of 100 equal areas into which the faces were divided. Numerous observations were involved and the readings were recorded on punched cards and reduced with a computer program.

Because the interior of the cube communicated with the outside only through the small capillary tube, it was unlikely that the density of the air inside would be the same as that outside. The refractive index could not then be estimated from the pressure, temperature and humidity measured externally and so the cube was placed in an evacuated enclosure for the interferometric measurements. The hydrostatic pressure on the cube when it was filled with mercury was therefore more than one atmosphere greater than when it was measured and the change of volume had to be allowed for. It was thought that because of the composite construction of the cube, the change might not be given just by the simple hydrostatic compression of the component blocks, and so measurements of representative dimensions of the cube were made in air and *in vacuo*. The measurements were difficult: care had to be taken that the pressure and density of (dry) air inside and outside the cube were the same, while the major part of the observed change of order of interference was due to the difference of refractive index between air and vacuum. The volume coefficient that was found, 2.7×10^{-6} per atm, is in fact very close to that calculated from ultrasonic measurements on fused silica.

The temperature coefficient of linear expansion was also found from measurements of face separations at different temperatures.

Whereas there was no evidence that the cube of tungsten carbide changed dimensions, there is clear evidence of changes in separations of the faces of the silica cube, especially of the smaller side faces, and in estimating the volume to be used in any particular set of weighings, allowance had to be made for the changes. The major uncertainty in the volume of the hollow cube is in fact the effect of these changes, whereas for the solid cube, it comes from the method of measurement involving wringing gages to the surfaces of the cube.

The volume of mercury weighed in the cube included a small quantity in the capillary tube in the upper face. The volume of the tube was calibrated by measuring the length at various places within it of a small weighed quantity of mercury, and when the cube was filled with mercury, the height in the capillary was determined with a probe gage.

The volume of a solid object has no meaning by itself; the temperature must also be specified. Most observations described in sections III and IV were carried out at or near 20°C and although the coefficients of linear expansion are not great ($5.2 \times 10^{-6} \deg^{-1}$ for the cube of tungsten carbide and $0.52 \times 10^{-6} \deg^{-1}$ for fused silica), nonetheless they must be measured carefully. The important point to watch is that the range of temperature used in the measurements of the coefficients should be much larger, say five to ten times greater, than the range of temperature in any other measurements.

V. Single Crystals of Pure Silicon for the Measurement of Avogadro's Number

It was explained earlier that an x-ray interferometer can now be used to measure the lattice spacing of a single crystal of pure silicon in terms of optical wavelengths, leading to the possibility of determining Avogadro's number by a direct route, for if the lattice spacing, density and atomic weight of pure silicon are all known, the number of atoms in a kilogram-atom may be calculated. Preparations are therefore being made at the National Physical Laboratory to measure the density of a single crystal of pure silicon[9].

Silicon is a hard material that can be optically worked to an accurate right parallelepiped, but it is too brittle for engineers' block gages to be wrung to it as was done for the cube of tungsten carbide. So far as purely instrumental considerations go, there is no need nowadays for wringing auxiliary gages, since it is possible to devise schemes using gas laser sources that give directly the separation of the opposite plane parallel reflecting faces of a block. However, the optical phase shift on reflection from the block is not known, nor is the effect of surface roughness easy to determine. It is therefore intended to use a differential method; two blocks of silicon will be prepared, x-ray observations being used to check that the lattice spacings are effectively identical, and the differences of dimensions will be found interferometrically, using reflections from parallel faces. Thus the density will be found from the difference of volume and difference of mass of the two blocks. An uncertainty in the density not exceeding 0.1 p/M might be anticipated if the dimensions of the blocks were of the order of some centimeters.

VI. The Density of Water

The density of water has an uncertainty of about 2 p/M and a new determination may be needed in experimental thermodynamics, and even more for physical oceanography. Some remarks on the problems of measurement of volume likely to be encountered in the measurement of the density of water may therefore be of interest.

The critical problem in the measurement of the density of water seems to be the control of the quantity of air dissolved in the water. It is therefore important to allow the water to come into equilibrium at a controlled temperature with air at a controlled pressure, conditions that seem to rule out the measurement of density by weighing the water contained in a hollow box with a capillary inlet as was done for mercury.

A sinker of known volume could, however, be weighed in water in a thermostat at constant pressure. The mass could be adjusted to be very close to the mass of water displaced so that the sinker could be supported by an electromagnetic suspension, thus avoiding difficulties of getting a suspension wire into a closed vessel and of controlling the surface tension forces on the wire. The sinker would then be a hollow box fabricated from slabs of fused silica, containing an armature for the suspension. The problem of assembling and working the slabs would be far less than that for the hollow silica cube, and a box of outside dimensions of say 70–80 mm could be prepared with very high precision. The thermal expansion would be easy to measure and allow for.

It would probably be best, for cleanliness, to leave the surface of the silica in its natural state; there might be some optical advantage in giving it a metallic coating, but cleaning such a coating would be difficult, it might wear off, and the phase shift on reflection would have to be measured and allowed for in interferometric measurements. That from the clean silica surface could be taken to be zero.

It seems that the measurements of volume entailed in measurements of the density of water should present no problem and that accuracies of about 0.1 p/M should readily be obtained.

VII. Summary

A summary of the place of measurements of volume in fundamental metrology and thermodynamics is followed by a discussion of the main factors to be considered in designing accurate measurements of volume in terms of the fundamental standards of length, particular attention being paid to the relation between techniques of fabrication of solid objects and techniques of measurement. Two measurements of volume of high precision are described —the measurement of a dense cube of tungsten carbide and the measurement of a hollow cube of fused silica, both undertaken for the absolute measurement of the density of mercury. R ference is made to measurements of the density currently being prepared in connection with the determination of Avogadro's number and some comments are made on the problems of the absolute determination of the density of water.

VIII. References

[1] Barrell, H. and R. Marriner. *Brit. Sci. News*, **2**, 130 (1948).
[2] Batuecas, T. and J. I. F. Alonso. *An. Soc. Esp. Fis. Quim.* **44B**, 1101 (1948).
[3] Batuecas, T. and F. L. Casado, *J. Chem. Phys.* **33**, 41 (1936).
[4] Chappuis, P. *Trav. Bur. Int. Poids Mes.* **14**, B.1 (1910).
[5] Cook, A. H. *J. Sci. Instrum.* **34**, 455 (1957).

[6] Cook, A. H. *Phil. Trans. Roy. Soc. A*, **254**, 125 (1961).
[7] Cook, A. H. *Interference of Electromagnetic Waves*, ix + 253 pp. Clarendon Press: Oxford (1971).
[8] Cook, A. H. and N. W. B. Stone, *Phil. Trans. Roy. Soc. A*, **250**, 279 (1957).
[9] Curtis, I., I. Morgan, M. Hart and A. B. Milne. 'A new determination of Avogadro's number.' In *Precision Measurements and Fundamental Constants*, US National Bureau of Standards: *Sci. Publ. No. 343*, pp 285–290. Washington, DC (1971).
[10] Guye, P. A. and T. Batuecas. *J. Chem. Phys.* **20**, 308 (1923).
[11] Hart, M. *J. Phys. D*, **1**, 1405 (1968).
[12] Hart, M. and U. Bonse. *Acta Crystallogr. A.A*, **24**, 240 (1968).
[13] de Lépinay, J. M., J. M. Buisson and J. R. Benoît. *Trav. Bur. Int. Poids Mes.* **14**, C.1 (1910).
[14] Marek, W. J. *Trav. Bur. Int. Poids Mes.* **2**, D.54 (1883).
[15] Rolt, F. H. and H. Barrell. *Proc. Roy. Soc. A*, **116**, 401(1927).
[16] Scheel, K. and F. Blankenstein. *Z. Phys.* **31**, 202 (1925).
[17] Thiesen, M. and K. Schell. *Z. InstrumKde*, **18**, 138 (1898).

CHAPTER 6

Measurement of p–V–T Properties of Gases and Gas Mixtures at Low Pressure

G. SAVILLE

Department of Chemical Engineering and Chemical Technology, Imperial College, London S.W.7, UK

Contents

I.	Introduction	321
II.	General Principles of p–V–T Measurement	323
III.	Methods of Measuring Pressure	323
	1. Secondary Manometers	325
IV.	The Volume Problem	325
V.	The Experimental Volume/Manometer Interface	328
VI.	Temperature Measurement and Control	331
VII.	p–V–T Methods at Constant Volume	332
VIII.	p–V–T Methods Involving Expansion	334
IX.	Relative Methods	338
X.	Effect of Gas Adsorption on p–V–T Measurements	341
XI.	Gas Density Microbalance	342
XII.	p–V–T Properties of Mixtures	344
XIII.	References	345

I. Introduction

Low pressure p–V–T measurements on gases have a long history and date back to the work of Boyle and Charles in the seventeenth and eighteenth centuries when the foundations of what we now know as the perfect gas laws were laid down. It was not, however, until the nineteenth century that extensive use began to be made of such measurements when the need for the gas thermometer became apparent and likewise the need for determining atomic weights (or more correctly, molecular weights from which atomic weights could be derived). Both of these needs have continued through into the twentieth century and the gas thermometer remains to this day as the fundamental standard of temperature. The situation with regard to atomic weights has, however, changed considerably over the last thirty years. Mass spectrometry has improved to such an extent that it is unlikely that gas volumetric methods will be used again, in the foreseeable future, for the determination of atomic weights. Thus, for example, the ratio of the atomic masses ^{12}C to ^{16}O has been determined mass spectrometrically to better than one part in 10^7. This high precision has shown also that polynuclidic elements (elements with more than one naturally occurring isotope) have an overall atomic weight which varies with the source of the material. Thus in the 1969 IUPAC atomic weights table, oxygen is given as 15.9994 ± 0.0003 where the error bounds reflect variability of source and not the experimental uncertainty in determining the atomic weight of a particular sample of the gas. This variability of atomic weight clearly has implications in the determination of the gas constant R (see further below).

All of this does not mean that the experimental methods developed for determining molecular weights by gas volumetry are now redundant, merely that a change in emphasis has occurred. We can see this by considering the classic method of weighing the amount of gas enclosed in a known volume V at some measured pressure p and temperature T. If the mass so determined is m, and the measurement is repeated for various values of p, then it is found that provided the pressure is not too high the results follow the equation

$$pV/m = A(1 + \lambda p) \qquad (1)$$

where A and λ are constants, and both may be determined by plotting the appropriate graph. The constant A is immediately recognized as RT and is clearly given as the value of pV/m in the limit of zero pressure. Hence the name 'limiting density' which is usually associated with this method. If the measurements are performed at an accurately known temperature and the molecular weight of the gas is known the gas constant R may be calculated. It is only in this final step that the procedure differs from that used in the earlier part of this century for atomic weight determinations, when the limiting density for the unknown gas would be compared with that for oxygen to give the relative molecular weight. It must be pointed out that the method outlined above is one of the few methods available to us for determining the gas constant R.

As far as these older measurements were concerned, the non-ideality of the gas, expressed in the constant λ, was a nuisance and led only to lower accuracy in the molecular weight determinations. However, the non-ideality of a gas is now an important quantity in its own right since it leads to information on intermolecular forces. It is probably true to say that the majority of low pressure p–V–T measurements are carried out with this object in mind. Usually the non-ideality is expressed in the virial series

$$pV = nRT(1 + nB/V + n^2C/V^2 + \ldots) \qquad (2)$$

(where n is the number of moles, B, C etc. are the virial coefficients and are functions of temperature and of the nature of the gas) since statistical thermodynamics can relate the virial coefficients to the intermolecular forces in a particularly simple manner.

Provided n/V is sufficiently small, i.e. the pressure is low, it is justifiable to truncate the virial series at

$$pV = nRT(1 + nB/V) \qquad (3)$$

and in general, when low pressure p–V–T measurements are referred to, it is implied that this is so. This is the definition of low pressure which will apply in this review.

Modern usage of low pressure p–V–T measurements is therefore to determine: (a) the gas thermometric temperatures (by extrapolation to the perfect gas state), (b) the gas constant R (also by extrapolation), and (c) the second virial coefficient B; and any other $p-V-T$ property of the real gas can of course be derived from a combination of these measurements.

Examination of equation 3 above suggests that experimentally we must measure four quantities suitably precisely, namely pressure, volume, temperature and quantity of substance. Although this is certainly true if we wish to determine the gas constant R, it is an unnecessarily large number of properties to measure if all we require is the second virial coefficient B, or if we are trying to define the thermodynamic temperature scale. This is because we have the additional piece of information that as $p \to 0$, $pV \to nRT$. The practical result of this is that it is never necessary to measure accurately more than three properties. Consequently, several methods of performing $p-V-T$ measurements on gases have been developed depending on which property is chosen as not to be measured.

II. General Principles of $p-V-T$ Measurement

In virtually all methods, the gas under investigation is contained in a well defined volume and the pressure and temperature are measured and/or controlled. Measurements are then made by varying one quantity and determining the effect on the second. Thus one may vary the volume and examine the resulting pressure change, temperature remaining constant throughout. This is the method most frequently used for the determination of second virial coefficients. If volume is the quantity maintained constant and pressure is measured as a function of temperature one has the normal experimental arrangement for gas thermometry, but it has occasionally been used for determining second virial coefficients as well. Measurements at constant pressure can provide the same information as those at constant volume, but the method is infrequently used as it is experimentally rather more difficult to carry out. For obvious reasons, experiments involving simultaneous changes in p, V and T are rarely used.

The only method of $p-V-T$ measurement which does not follow the above principles is the gas density microbalance. This measures relative values of (m/V) at a given p and T.

III. Methods of Measuring Pressure

Since low pressure $p-V-T$ work rarely involves pressures of more than 2 to 3 bar, and frequently the maximum is only 1 bar, the most widely used manometer is the mercury column, the pressure being calculated as

$$p = \rho g h \qquad (4)$$

where ρ is the density of mercury, g is the acceleration due to gravity, and h is the height of the mercury column. The precision with which the absolute value of p may be determined depends therefore on the precision with which each of the three relevant quantities can be measured. Modern instrumentation can determine the absolute value of g at any particular location with an accuracy of rather better than one part in a million[43], but often g is measured relative to the National Standards Laboratory in a particular country. In doing this latter, one must be careful to use the correct base value. Prior to

1969, all g values were measured relative to the value at Potsdam and the latter was taken as 9.81274 m s^{-2}. This value was found to be in error and in 1969 the base value was changed to 9.81260 m s^{-2}. The situation with regard to the density of mercury has been reviewed by Bigg[6]. At 20°C it is known with an accuracy of about one part in 10^6 but at temperatures removed from this, the uncertainty increases substantially. Thus at 100°C it is nearer one in 10^5. The fact that the density of mercury is a function of temperature imposes a restriction on the temperature fluctuations which can be tolerated in a mercury column. Thus to measure pressures to one part in 10^6 requires temperature control to 0.005K.

The precision with which the height of a mercury column can be determined depends on (a) the precision of the measuring instrument: cathetometer, line standard, etc., and (b) the extent to which surface tension interferes with the measurement. The effect of (b) can be eliminated by use of wide bore columns. In practice this means 30 mm bore columns for pressure measurements to $\pm 1 \text{ N/m}^2$ and 40 mm bore for measurements to $\pm 0.1 \text{ N/m}^2$.

The reader should refer to articles on pressure measurement for full details[4, 5, 8, 25, 33, 47] but one can make the following general statements for manometers operated at about 20°C to measure pressures of up to one bar. Pressure measurement to $\pm 3 \text{ N/m}^2$ may be made with a high quality research cathetometer and manometer limbs of 20 mm precision bore glass tubing with relatively simple thermostatic control. Measurement to $\pm 0.3 \text{ N/m}^2$ requires rather more elaborate distance-measuring techniques and more careful thermostatic control but it is still possible in a normal research laboratory[47]. Measurement to $\pm 0.1 \text{ N/m}^2$ requires what is essentially a National Standards manometer[5, 8, 25].

At temperatures above 20°C, precision is usually substantially lowered, not only because of uncertainty in the density of mercury, but also because the vapor pressure of mercury increases from the value of about 0.1 N/m^2 at 20°C to 10 N/m^2 at 80°C and to 300 N/m^2 at 150°C. Some of the difficulties involved have been well described by Bottomley[12]. Mercury manometers are rarely used above 150°C.

In some p–V–T experiments, pressure ratios are required rather than absolute pressures. Uncertainties in g and ρ are then irrelevant although precise temperature control may still be necessary.

In recent years the gas-operated pressure balance has been used more and more frequently, although as yet, very few results of investigations using this instrument have been published. The principle of operation is essentially the same as that of the oil-operated balance but the unit is frequently arranged to measure absolute pressures when it is covered with a bell jar which is continuously evacuated. Very high pumping speeds are necessary to do this for there is, of course, a continual leak of gas past the piston. Most workers have found it necessary to attach a diffusion pump via a 5 cm diameter pipe in order to attain a sufficiently high vacuum.

The pressure as measured by a pressure balance with the piston and cylinder unit vertical may be calculated as

$$p = mg/A \tag{5}$$

where m is the mass of weights on the piston, and A is the effective area of the piston. However, various manufacturers have made balances in which the piston and cylinder can be tilted away from the vertical. With these the pressure is now calculated as

$$p = mg \cos \theta / A \qquad (6)$$

where θ denotes inclination to the vertical. This enables one particular loading of the balance to cover a range of pressures. This is particularly useful when measuring absolute pressures.

The precision with which p may be calculated is limited by the precision of A, which usually is limited to one or two parts in 10^5. In the case of the tilting balance the determination of θ may further reduce the accuracy attainable and, in general, the tilting balance does not have quite the same precision as the vertical balance. With the latter, the error in p is essentially a constant fraction of the measured pressure over the whole pressure range and use can often be made of this in the design of a given p–V–T experiment (see further below).

Commercial balances using the vertical arrangement of piston and cylinder are available for measurement of pressures above 2000 N/m². For pressures below this the tilting design must be used. One such balance (not commercial) has been described by Douslin and Osborn[22].

1. Secondary Manometers

The commercially available spiral Bourdon tube manometer, when calibrated against one of the primary manometers described above, is an exceedingly convenient and rapid means of measuring pressure. The unit consists of a multi-turn thin-wall quartz tube mounted on a torsional suspension so that the 'uncoiling' of the tube due to a pressure difference across it produces rotation of a galvanometer-type mirror. This rotation is detected by a photocell and a servomechanism follows the rotation round, the angle of rotation being measured by counting the number of revolutions executed by a worm driving the pinion on which the photocell is mounted. Although a sensitivity of about one in 10^5 of full scale is possible, slight hysteresis of the quartz spiral, imperfections in the worm and pinion and variations in supply voltage mean that only one in 10^4 can be relied upon unless very great care is taken in the calibration and use of the instrument. This would normally mean calibration at closely spaced intervals—some 500 over the whole range—and a mode of usage which reduced the effect of hysteresis[17].

IV. The Volume Problem

All methods of measuring p–V–T properties of gases have in common the necessity of confining the gas in one or more well defined vessels. It is always necessary also to be able to control them thermostatically and this requirement should therefore be borne in mind in making a design. Other requirements depend on the p–V–T method chosen. Thus sometimes it is sufficient to have a vessel whose volume remains constant, or which varies in a well characterized way, only over the duration of one single experiment. At the other extreme we have cases where it is necessary to calibrate the vessel during construction and to know that this calibration does not change over a long period of time.

p–V–T vessels are usually made from glass, quartz, copper and, more rarely, stainless steel. Glass and quartz are obvious choices for their ease of manufacture, but their low thermal conductivity invariably means that the associated thermostat must be of the liquid circulation type and this limits the temperature range over which the apparatus may be used. Copper has a high thermal conductivity and is usually chosen for the vacuum isolated thermostat (see more below). Stainless steel is a rarely used alternative.

To obtain stability in the experimental volume, some form of stress relief is frequently necessary. This presents no problem with glass and quartz but is more difficult with metals, particularly if soldered, rather than welded, joints are used and if the surface of the metal has been carefully polished to reduce adsorption (see below). An example of the lengthy procedure needed to do this can be seen in the paper by Aston et al.[38] on the construction of a gas thermometer for use at temperatures below room temperature. Design of the shape of the vessel is also important and if a cylindrical unit is required, domed ends are preferable to flat discs for the latter can show the 'oil drum effect' and this makes a nonsense of any precalibration. In some p–V–T methods, such as the Burnett (see below) stability is required only during the sequence of pressure measurements at one temperature and provided thermal stresses are avoided for this length of time, a flat end is satisfactory.

Except in such cases where measurements are being carried out at constant pressure, it is always necessary to consider the effect of elastic distortions of the vessel. If the pressure external to the vessel is maintained equal to the internal pressure then, provided the material used is homogeneous, the fractional change in volume is independent of the shape of the vessel and is equal to p/κ where κ is the bulk modulus of the material. *Table 1* gives the bulk moduli of commonly used materials. It is seen that the fractional change in volume when the pressure varies over the range 0 to 1 bar ($=10^5$ N/m^2) is a few parts in 10^6. In practice this means the volume change can either be ignored or, if necessary, allowed for with adequate accuracy.

Table 1. Elastic moduli of various materials

	Bulk modulus N/m^2	Young's modulus N/m^2	Poisson's ratio σ
Glass (Pyrex)	3.5×10^{10}	6.3×10^{10}	0.20
Quartz	3.69	7.31	0.170
Copper	13.78	12.98	0.343
Stainless steel	16.6	21.5	0.28

If there is a pressure difference across the wall of the vessel, the situation is by no means as satisfactory. Of the commonly used vessel shapes, an analytical solution for the change in volume exists only for a spherical vessel of uniform wall thickness. The popular cylinder, whether with flat or domed ends, does not have an analytical solution. Mechanical engineers have developed finite difference computer programs capable of solving the problem but the enormous amount of computing time required has precluded any attempt to produce tables of volume changes and so far each case has had to be treated on its merits. For this reason, it is tempting to obtain an approximate solution by treating the vessel as a closed cylinder with ends which do not constrain the diameter of the cylinder in any way because an analytical solution for this case does exist. It will of course give an upper bound to the volume change. It should be noted at this point that even a cylinder with hemispherical ends does not have an analytical solution, and it is not possible to represent it as the sum of a sphere and a cylinder with unconstrained ends. This is because the strains in the diameter of a cylinder do not follow the same variation with pressure as do the strains in a sphere. The relevant equations for the calculation of the change in volume, when the internal pressure in the vessel is p and the external pressure is zero, are:

Cylinder with flat ends which do not constrain the diameter

$$\delta V/V = [p/E(k^2 - 1)][3 - 6\sigma + 2k^2(1 + \sigma)] \qquad (7)$$

Sphere

$$\delta V/V = [3p/E(k^3 - 1)][\tfrac{1}{2}k^2(1 + \sigma) + 1 - 2\sigma] \qquad (8)$$

where E denotes Young's modulus, σ is Poisson's ratio, and k is the ratio of outer radius to inner radius. The magnitude of this volume change may best be put into perspective by considering an example. For a cylindrical copper vessel 10 cm in diameter with a wall thickness of 0.5 cm, the fractional change in volume is 1.5×10^{-5} per bar. This error is much too large to ignore in precise p–V–T work but furthermore, since the estimate may be too large by 10 to 20 per cent due to the approximations involved, it often happens that it cannot be applied as a correction since its precision is insufficiently known.

This uncertainty in the volume change when there is a pressure difference across the experimental vessel has led many workers to use pressure compensation[4, 47] in which the experimental vessel is surrounded by a second vessel whose pressure is maintained from a separate source. An alternative approach is to measure the change in volume in a calibration experiment. The experimental vessel is surrounded by a second vessel filled with fluid and closed with a capillary tube. The change in volume on pressurizing the first vessel is then reflected in a change in liquid level in the capillary. This method is usually satisfactory but one must be aware of its limitations: (a) the volume change in the external dimensions of a vessel are not exactly equal to the volume change in the internal dimensions, and (b) the elastic constants vary with temperature.

In some p–V–T experiments, it is necessary to precalibrate the volume of the experimental vessel during construction. This is usually done by weighing when filled with a fluid such as water or mercury, and again when empty. Provided reasonable precautions are taken, such as equilibrating the filled vessel while maintained under thermostatic control, and the shape of the vessel does not allow the formation of air pockets (alternatively the fluid can be distilled in), a precision of one in 10^4 is readily attainable. One in 10^5 requires much more effort and is the common practical limit, although slightly better accuracy than this has been achieved with a glass or quartz vessel calibrated by filling with mercury. With metal vessels, the residual errors probably arise from small cracks in the soldered joints which are large enough to contain gas but are too small to be filled with liquid at atmospheric pressure, the void being maintained by surface tension. Since the calibration of volume must take place at or near room temperature, use of the experimental volume at any other temperature requires correction for the coefficient of expansion. This is often the largest source of error in a volume calibration Thus for a copper vessel at 90 K, the uncertainty in volume on this count is about three parts in 10^5, although by measuring the coefficient of expansion of the actual sample of copper used, Preston-Thomas and Kirby[42] reduced the uncertainty to less than one part in 10^5. Of course, if the $p-V-T$ experiment is designed in such a way that volume ratios only are required and that all relevant vessels are at the same temperature, coefficient of expansion corrections nullify each other.

V. The Experimental Volume/Manometer Interface

Although most of the gas under investigation in a typical $p-V-T$ experiment is contained in the experimental volume described above, there is necessarily a small amount of gas in the lines connecting it to the manometer. In most papers in the literature these are described as 'dead spaces'. Sometimes the manometer is operated at the experimental temperature and the dead spaces will then usually also be at the experimental temperature. Otherwise, the manometer is operated at some convenient temperature close to room temperature and the interconnecting lines must then pass through a temperature gradient. By use of small bore capillary tubing and a number of calibrated thermocouples along any temperature gradient present there is usually no problem in calculating the amount of gas contained in these lines and hence computing the necessary corrections. There remains therefore the question of the rather larger dead space in the manometer unit itself. This dead space requires very careful consideration for often it contributes the largest error in a particular $p-V-T$ method.

If a mercury column is used as the manometer, it is tempting to use the mercury meniscus itself as the interface between the experimental gas and the manometer. However, before doing this it is imperative to consider what uncertainty in the dead space volume ensues. We may conveniently divide the dead space in the mercury manometer into two parts, (a) the volume above a horizontal plane which touches the top of the mercury meniscus and (b) the volume below this plane, the so-called 'meniscus co-volume'.

The error in a determination of (a) is directly proportional to the cross sectional area of the manometer tube and to the uncertainty in the location of the (hypothetical) plane. Sometimes it will be sufficient to use the same cathetometer for this as is used for making a pressure measurement, the same reading being used for both purposes. However, as the bore of the manometer tube is increased to make pressure measurements more precise the error in the determination of volume (a) increases rapidly and some more exact means of locating the top of the mercury meniscus is needed. A sharply pointed tungsten needle fused into the manometer tube is very convenient here as it has been observed that the mercury surface can be set against a tungsten point with a precision of 0.001 mm[11].

The error in a computation of the meniscus co-volume depends on how well we can define the shape of a mercury meniscus. In a tube of given bore, the shape is determined by: (i) the surface tension of the mercury, (ii) the contact angle between mercury and glass and (iii) deviations from ideal shape due to contaminated mercury or manometer tubing. To some extent (i) and (iii) are linked since, for example, oxygen contamination reduces the surface tension. Investigations of the shape of the mercury meniscus have been carried out both experimentally[30] and theoretically[7, 24, 45] and these show that the capillary depression is very sensitive to the magnitude of the surface tension although fortunately, the meniscus co-volume is less so. In order to apply the above results to a particular manometric problem it is usual to assume a value for the surface tension γ and by measuring the 'height' of the meniscus, to calculate the contact angle. In practice, of course, tables of capillary depression and meniscus co-volume are generated for a particular value of γ for various meniscus heights, the contact angle being effectively eliminated from the calculation. Highly pure mercury has a surface tension of about 0.50 N/m but when slightly contaminated, even though it is virtually indistinguishable from the highly pure, it falls as low as 0.40 N/m. This therefore is the range of surface tension which will be encountered in practice. *Table 2* gives values of capillary depression and menis-

Table 2. Effect of surface tension on the performance of a mercury manometer

Bore mm	$\gamma = 0.40$ N/m		$\gamma = 0.50$ N/m	
	Capillary depression mm	Meniscus co-volume cm^3	Capillary depression mm	Meniscus co-volume cm^3
10	0.39	0.046	0.54	0.048
15	0.10	0.091	0.16	0.096
20	0.027	0.134	0.048	0.144

cus co-volume for various sizes of tubing when the meniscus height is 1.5 mm. This latter value is approximately the one observed for well formed menisci in tubing of more than 10 mm bore. It is seen that for a 20 mm bore tube, the uncertainty in the meniscus co-volume is 0.01 cm^3 and, presumably, increases still further with larger bores (existing calculations for co-volume do not extend beyond 20 mm). In most p–V–T work this uncertainty is approaching the allowable limit, compatible with precision in pressure measurement, for a 20 mm bore tube. The table also shows that although the uncertainty in meniscus co-volume decreases substantially for a smaller bore tube, the capillary depression uncertainty is now too large to be neglected.

In view of the above difficulties with the mercury column, workers making very precise p–V–T measurements invariably use a null manometer designed specially to reduce dead space volume uncertainties but which is incapable of measuring any sizeable pressure difference. These manometers necessitate the use of a second gas-handling system to generate and maintain a pressure exactly or nearly exactly equal to that in the experimental volume, as detected by the null manometer, and the reference pressure can then be measured by any convenient full range manometer. Null manometers which have been used may be divided into three clases: (a) diaphragm, (b) glass or quartz Bourdon tube and (c) a mercury differential manometer of special design which reduces the effect of variations in the surface tension of mercury; (c) has been rarely used, for although it has a high performance—p to ± 0.003 mm mercury column and dead space volume constant to within about 0.0001 cm^3—it is rather difficult to use in practice. Reference is therefore made to the original papers[39, 40, 44].

Diaphragm null manometers are either of the tightly stretched variety or have a flexibly mounted diaphragm. For p–V–T work the latter are of limited use, in spite of their high sensitivity, for they have a relatively large, somewhat irreproducible dead space. The stretched diaphragm is very much to be preferred. Many of these have been described in the literature[2, 9, 18, 19, 36, 42, 48] and have usually been developed in connection with gas thermometers. A typical high performance unit is the one described by Preston–Thomas and Kirby[42] in which an 8 μm stainless steel diaphragm was stretched and clamped between two stainless steel plates both of which had a spherically ground area of 5 cm diameter and 50 μm deep to restrict its movement. Displacement of the diaphragm away from its null position was detected by the change in electrical capacitance between the diaphragm itself and one of the stainless steel plates. The null point of such a unit, defined as the reading on the capacitance detector when the diaphragm has an equal pressure on each side, is sensitive to many factors, including temperature, chemical nature and pressure of the gas on the capacitance side of the diaphragm (due to dielectric constant variations), pressure inside and outside the unit (due to elastic distortions) and also the magnitude of the last pressure difference to which the diaphragm had been subjected (due to hysteresis). It is therefore normal to control the unit thermostatically and to calibrate the null point as a function of pressure, using the gas to be investigated, and once calibrated, ensure that at no time is the diaphragm subjected to a sufficiently large pressure difference to produce any permanent deformation. After taking these precautions, the unit described above was capable of measuring pressures reliably to ± 0.1 N/m^2 (the sensitivity was actually ten times this) and the deadspace volume of 0.020 cm^3 was reproducible to better than 0.001 cm^3.

The performance described above probably approaches the limit of what has been achieved to date. The literature, of course, if full of descriptions of diaphragm type null-indicating manometers of similar or lower performance using either capacitance or other methods of detection. There are also many commercial instruments on the market although probably only a very few of them have been investigated from the point of view of low pressure p–V–T experiments.

The original Bourdon tube null-indicating manometer was the glass spoon gage in which the deflection of the index rod was observed with a microscope. Increased sensitivity was achieved by use of an optical lever[34] or by capacitance detection[1] but the limiting factor was sentivity to vibration. Replacement of the glass spoon by a helical glass coil of thin wall with a mirror attached for the optical lever was a considerable improvement but some sentivity to vibration still remained. There was also the problem of slight hysteresis. This latter can be reduced by the use of quartz rather than glass and sensitivity to vibration can be largely eliminated by use of a torsional suspension. These latter improvements have been incorporated in commercial instruments made by Texas Instruments and by Ruska Instrument Corp. A typical quartz Bourdon tube consists of a 20 to 30 turn helical coil, 7 cm long and 1 cm in diameter, made from a quartz tube of about 2 mm diameter and of the order of 0.1 mm wall thickness. One end of the coil had a rigid mounting and the other is suspended torsionally between two quartz fibres. A mirror is also attached at the free end. The sensitivity of such a unit is of the order of one degree of rotation per 300 N/m^2. By use of an optical lever and a photocell detector[47] a sensitivity of 0.1 N/m^2 is readily obtained. More recent investigations by these same workers have shown that the sensitivity can be increased to 0.03 N/m^2 and that the shift in the null point with pressure can be reduced to below 0.06 N/m^2 for total pressures in the range 1 N/m^2 to 10^5 N/m^2. Drift in the null point over a period of several days was also less than 0.06 N/m^2. Only very elementary precautions were needed to reduce the effect of vibration to below this sensitivity. The one remaining problem, that of hysteresis, was still present, but its effect was less than 0.03 N/m^2 provided the pressure difference did not deviate from the null point by more than 10^3 N/m^2.

The one major advantage of all of these Bourdon tube elements when operated as null-point instruments is the almost absolute constancy in the dead space, which certainly does not vary by more than 0.0001 cm^3.

If, however, the quartz spiral Bourdon tube is used not as a null-point instrument but as a direct means of measuring pressure (i.e. by using the mechanical device sold by the manufacturers for measuring the rotation of the coil) the constancy of the dead space volume is no longer assured and it may vary by up to 0.001 cm^3.

VI. Temperature Measurement and Control

In all $p-V-T$ experiments it is necessary to have available one or more reliable devices for thermostatic control. Whether it is also necessary to be able to measure the true thermodynamic temperature depends on the details of the experiment being performed. The International Practical Temperature Scale of 1968 is designed to reproduce the thermodynamic temperature scale to within the known accuracy of the latter. For temperatures between 13.81 K and 630.74 K the Scale is based on the platinum resistance thermometer. The uncertainty in the thermodynamic temperature of the various fixed points in this range is quoted as 0.01 K for those below 273.16 K, 0.005 K at the normal boiling point of water and rising to 0.03 K at the zinc point (692.73 K)[28]. For certain portions of the temperature range, these uncertainties may be a little pessimistic (see e.g. Preston-Thomas[42]) but nevertheless the error band for temperatures below 273.16 K cannot be reduced below ±0.002 K. These uncertainties must be borne in mind when designing a $p-V-T$ experiment in which a value for the thermodynamic temperature must be obtained from platinum resistance thermometer readings for incorporation in the calculation of the required $p-V-T$ properties.

In contradistinction to the uncertainties in the thermodynamic temperature scale, very high precision is possible in a single temperature measurement using a platinum resistance thermometer. Single reading a.c. bridges are now commercially available which have a reproducibility of 0.0001 K. Thus in other than gas thermometer p–V–T work the highest possible precision can only be achieved if the experiment is designed so as not to require accurate values of the thermodynamic temperature but to know that, for example, the temperature has remained constant to within 0.0001 K over the duration of the experiment is sufficient. Another technique which is often used is to build what is effectively a gas thermometer into the apparatus.

Numerous thermostatic devices suitable for p–V–T work have been described in the literature. The simplest is possibly the stirred ice bath or, more rarely, the ice triple point cell. These of course operate at only one temperature. The second most popular is the stirred liquid bath, the choice of liquid depending on the range of temperature of interest. Water is used at or around room temperature, oil for temperatures up to 200 or 250°C, molten salt baths up to 500°C[3] and low-boiling hydrocarbons below room temperature down to about 150 K. A variant on the stirred liquid bath is a liquid boiling under a controlled pressure. All of these baths when well designed have a precision of 0.001 K and by deliberately arranging for a thermal lag in the experimental vessel which is immersed in them, short term temperature stability of 0.0001 K is possible in the experimental volume.

In p–V–T work below room temperature, the metal block thermostat has been used successfully by many workers[2, 9, 15, 29, 38, 41, 42, 44, 47]. The basic principle here is to make the experimental volume from a massive copper block and to provide it with electrical heating windings and a resistance thermometer. This is surrounded by another copper vessel also provided with heater windings and a thermometer (thermocouples are often used here). This latter vessel serves a very similar role to the radiation shield used in low temperature heat capacity calorimetry. Both vessels are then contained in an outer jacket, evacuated for thermal isolation, and this in turn is surrounded by a Dewar vessel containing a suitable refrigerant. In more recent equipment, power to each set of heater windings comes from an electronic temperature controller which has both proportional and integral functions (the derivative function is not needed). The 'radiation shield' would normally be controlled about 0.5 K below the temperature of the experimental vessel with a precision of 0.01 K. The experimental vessel can then be controlled to 0.0001 K over periods of an hour or so, with a long term precision of nearer 0.0005 K. The method is not suitable for making measurements above about 300 K as radiation heat interchange then becomes too large, but at the other extreme it may be used at liquid helium temperatures.

VI. p–V–T Methods at Constant Volume

The basic arrangement is as shown in *Figure 1*, and uses any combination of experimental vessel, null-indicating manometer and absolute manometer described above, the final choice being made on the precision required. The experimental method is to fill the thermostatically controlled vessel with a quantity of gas, the amount of which may or may not be known depending

MEASUREMENT OF p–V–T PROPERTIES OF GASES

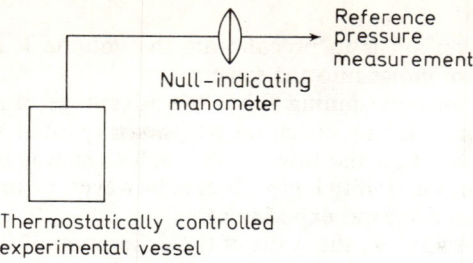

Figure 1. p–V–T methods at constant volume.

on the method chosen, and the pressure is measured as a function of temperature. Clearly, the volume will change with temperature due to thermal expansion of the vessel and to this extent the experiment is not truly at constant volume. However, corrections may be applied provided the coefficient of expansion is known. The method places a premium on the use of a stress-free experimental vessel as reproducibility of volume is of the utmost importance. As a rule, the null-indicating manometer is operated at room temperature and there will therefore be dead space corrections to be made for the interconnecting tubing. Occasionally, the null manometer will be at the same temperature as the experimental vessel; this can have the advantage of eliminating the uncertainty in the temperature gradient along the interconnecting tubing but in contrast one has the uncertainty in the position of the true null point of the manometer due to the more arduous conditions under which it operates. In general, provided the experimental vessel can be made sufficiently large, the null-indicating manometer at a constant (room) temperature is to be preferred.

The experimental configuration described above is the one normally employed in establishing the perfect gas, i.e. thermodynamic, temperature scale. It is not necessary to know the quantity of gas used, usually helium, and the volume of the experimental vessel is required only with low accuracy in order to evaluate the dead space corrections. However, values of the second virial coefficient are needed and it is usual to take these from the results of a separate experiment on equipment set up specially for the purpose. In principle this could be done on the gas thermometer itself by using different initial filling pressures. However, this is rarely done as the precision is usually insufficient.

If an already calibrated platinum resistance thermometer is used in the gas thermometer then the apparatus may be used to determine the variation of the second virial coefficient of the gas with temperature. The method requires a knowledge of the second virial coefficient at one temperature, say T_0 as this enables m/V to be obtained from

$$p_0/RT_0 = (m/V)[1 + B_0(m/V)] \qquad (9)$$

where V is the volume of the experimental vessel, m is the number of moles contained in it, and p_0 is the pressure at temperature T_0. Then, if the gas pressure is measured as a function of temperature the second virial coefficient B may be determined at the same experimental temperatures. An obvious, but little used, variant is to precalibrate the volume V and to introduce a known number of moles into the system.

This method for determining B has the advantage of rapidity but it does not, however, provide any check on whether the initial gas filling (i.e. m/V) was too large such that the third virial coefficient was having a significant effect on the compressibility factor. It can, however, be investigated by using a different m/V in a second experiment.

If adsorption of gas on the walls of the experimental vessel is anticipated, this may be investigated by the use of a second vessel with a greater surface area to volume ratio. This applies to both gas thermometer and second virial coefficient determinations.

VIII. p–V–T Methods Involving Expansion

This is by far the most popular way of measuring the second virial coefficient of a gas. It also embraces a rather less frequently used method of gas thermometry, which is particularly appropriate at high temperatures[39]. The apparatus consists of a main experimental vessel, V_0 in Figure 2, connected to a pressure measuring device and also, by means of stopcocks, to a series of one or more subsidiary vessels. The experimental procedure is to measure the pressure of the gas, initially contained in V_0 only, at each stage of expansion into the previously evacuated subsidiary vessels.

In the simplest possible example, $V_0, V_1, \ldots, V_n$ will all be operated at the same temperature and the volume of each vessel will have been predetermined (actually the ratios $V_1/V_0, \ldots, V_n/V_0$ will be sufficient). Since one has the series of equations:

$$p_0 V_0 = mRT(1 + Bm/V_0) \tag{10a}$$

$$p_1(V_0 + V_1) = mRT(1 + Bm/[V_0 + V_1]) \tag{10b}$$

$$p_n(V_0 + V_1 + \ldots + V_n) = mRT(1 + Bm/[V_0 + V_1 + \ldots + V_n]) \tag{10c}$$

they can be solved by some suitable least squares procedure for the second virial coefficient B. It can be seen that there is no need to determine the number of moles m if we reduce the $(n + 1)$ equations to n equations by dividing the first equation by the second equation, the first equation by the third equation etc., to obtain

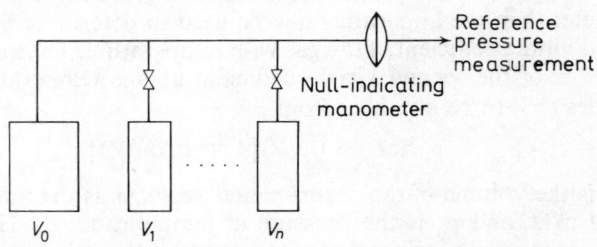

Figure 2. p–V–T methods involving expansion.

$$\frac{p_0 V_0}{p_1 (V_0 + V_1)} = \frac{1 + Bm/V_0}{1 + Bm(V_0 + V_1)} \tag{11}$$

and in as far as in low pressure $p-V-T$ work, Bm/V is always small compared with unity, m on the RHS of this equation can be obtained with sufficient accuracy from 10a by assuming some approximate value for B. The need for precision therefore arises only in the determination of pressure ratios and volume ratios. The precision on the temperature can be at least one and sometimes two orders of magnitude lower, provided that it remains constant over the duration of the experiment to within a precision comparable with that of the pressure measurement. This latter puts a premium on a good thermostatic device, which means a very sensitive thermometer to check that it is so, but there is no need for it to be accurately calibrated.

The first variant on the method is to remove the restriction of the vessels V_0 to V_n being at the same temperature. If V_0 is maintained at the temperature of interest, all of the other vessels could be at some convenient temperature close to room temperature. This will obviously simplify the design of that part of the apparatus which must be at the temperature of interest when the latter is well removed from room temperature. The penalty to be paid is rather more complex calculations to obtain B and also the need to know the second virial coefficient B_1 at the temperature T_1 of the vessels V_1 to V_n. Perhaps an even more serious requirement is the need to determine the ratio of the two temperatures being used to an accuracy which is comparable with the accuracy with which the volume calibrations have been carried out.

The final variant to be considered is the one in which only one subsidiary vessel is used, the apparatus consisting of V_0 and V_1 only. The method was first described by Burnett[14] and his name is usually associated with it. The procedure begins as before, the gas being allowed to expand from V_0 to V_1. The stopcock connecting V_0 to V_1 is then closed and V_1 evacuated. Reopening the stopcock allows a second expansion to take place into V_1 and the procedure can be repeated until the pressure in V_0 becomes too low to be measured with adequate precision. The method can be used with V_0 and V_1 either at the same temperature or at different ones.

The relative merits of the various expansion methods can be judged from the way the experimental errors in the various primary measurements propagate through to the final value of the second virial coefficient which is obtained. The type of pressure measuring equipment in use is also relevant here. In a given mercury column manometer, the absolute precision with which a pressure can be measured is essentially independent of the magnitude of the pressure itself. Thus pressures might be measured to (say) ± 1 N/m^2 for pressures in the range 10^4 to 10^5 N/m^2. With a pressure balance, however, the uncertainty is usually in the effective area of the piston rather than in the sensitivity with which balancing can be carried out. Thus in this case it is the relative accuracy which is more appropriate and, for example, a given unit might measure pressures in the range 10^4 to 10^5 N/m^2 to within one part in 10^5. This difference in behaviour of the two pressure measuring devices is relevant to the choice of null-indicating manometer at the experimental volume/pressure measurement interface, for if full advantage is to be taken of the precision of a given pressure balance, the sensitivity of the

Table 3. Comparison of expansion methods of p–V–T measurement at low pressures

		Mercury manometer		Pressure balance	
	p_1/p_n	Optimum expansion ratio	σ_p N/m^2	Optimum expansion ratio	σ_A
Burnett	10	1.3–1.6	0.25	1.4–1.75	0.6×10^{-5}
Burnett	50	1.4–2.0	0.25	1.5–2.15	1.0
Calibrated volumes		1.7–3.0	1.0	2.0–3.0	1.5

null-indicating manometer must be based on the lowest pressure which it is intended to measure with the balance.

Table 3 gives the results of an error analysis to determine the conditions necessary to determine the second virial coefficient B to ± 1 cm^3 at 0°C when the initial pressure in the experimental volume V_0 is 10^5 N/m^2. In this table, expansion ratio means the ratio of the pressure before, to the pressure after an expansion and applies irrespective of whether all experimental vessels are at the same temperature. σ_A is a relative accuracy criterion. It is to be noted that the precision of the Burnett method depends on the ratio of the first to the last pressure measured as well as on the individual expansion ratio. One might make the obvious statement that the penalty of avoiding volume measurements in the Burnett method is the need for more precise pressure measurements and this is particularly acute when a mercury column is used.

If the experimental temperature and initial pressure differ from 273 K and 10^5 N/m^2 respectively, the required precision in σ_p and σ_A for an accuracy of ± 1 cm^3 in B may be obtained from:

$$\sigma_p = \sigma_{p,\,273\,\text{K}}(273/T) \tag{12a}$$

$$\sigma_A = \sigma_{A,\,273\,\text{K}}(273/T) \tag{12b}$$

$$\sigma_p = \sigma_{p,\,10^5\,\text{N/m}^2}(p_1/10^5)^2 \tag{12c}$$

$$\sigma_A = \sigma_{A,\,10^5\,\text{N/m}^2}(p_1/10^5) \tag{12d}$$

The attraction of the pressure balance for very low pressure p–V–T measurements (i.e. $p \ll 10^5$) should be noted.

In the above discussion, no account has been taken of the effect of dead spaces. These do not in any way affect the precision provided that they can be calibrated with adequate accuracy and as a rule this presents no real problem. They do, however, complicate the calculations substantially, particularly when the two-temperature method is used. The reader is referred to the original papers for full details[20,41,47].

Experimental vessels V_0 to V_n may take any forms described in the section on the volume problem, choice being made on the basis of convenience and the temperature range under investigation. When one or more of the vessels is operated at or close to room temperature, mercury may be used as a confining fluid so as to provide a unit of variable volume. Figure 3 shows three glass bulbs joined together by short lengths of narrow bore tubing, the volumes between the four etch marks A, B, C and D having been pre-

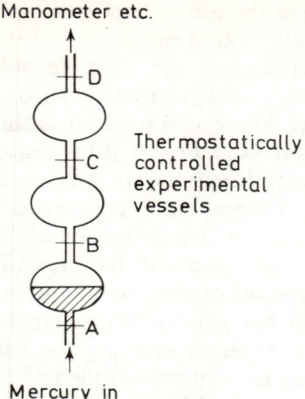

Figure 3. Precalibrated variable volume unit using mercury and glass.

calibrated. Injection of mercury from the bottom provides a unit of variable volume and the three bulbs could form vessels V_1, V_2 and V_3 in *Figure 2*. The bore of the short lengths of glass tubing interconnecting the bulbs is chosen so that errors in volume, either due to setting against the etch mark or due to meniscus co-volume variation, are negligibly small. On a few occasions researchers have found it worthwhile to improve the reproducibility here by use of tungsten index points rather than etch marks. Another variable volume vessel involving mercury utilizes the ease with which mercury can be weighed[11]. Here, a single vessel is used, initially full of mercury, and the expansions are carried out by running out mercury, the increase in gas volume being computed from the weight of mercury collected in a weighing bottle—see *Figure 4*.

One disadvantage of the use of mercury as a confining fluid in a variable volume unit is the unpredictability of the results which can occur when gas adsorption is significant. This is because the amount of glass surface exposed varies during an expansion and also because some stirring of the mercury necessarily takes place as it is removed from the variable volume unit. Both

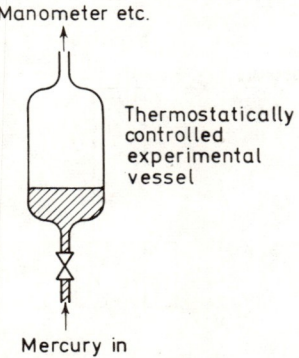

Figure 4. A single glass vessel used as a variable volume unit.

of these effects change the amount of gas adsorbed on the various surfaces involved. Bottomley[11] recommends that reproducibility may be assured only by prior evacuation of the variable volume unit before filling with mercury, thus ensuring completely desorbed surfaces. With mercury at its highest point, the gas being investigated is then admitted and the pressure recorded. Withdrawal of mercury thus exposes fresh glass surface and subsequent pressure measurements are more reproducible. Bottomley's recommendation clearly precludes the use of a compression procedure as the rising mercury meniscus will trap gas on the wall of the vessel. Of course, having obtained reproducibility in the pressure measurements one still has the problem of estimating or measuring the amount of adsorption. Usually this would have to be carried out in a separate experiment as the technique of packing the vessel so as to increase the surface area to volume ratio is usually impracticable in a mercury filled variable volume vessel. Suitable methods have been described by Bottomley[10].

A second disadvantage in the use of mercury arises at high temperatures where the vapor pressure is far from small. This means that in, for example, the three bulb unit shown in *Figure 3* the mercury vapor/gas composition may well be different in each bulb and since diffusion is slow, substantial errors can occur.

IX. Relative Methods

These are popular due to the speed with which measurements can often be made. The guiding principle is to compare the p–V–T properties of the unknown substance with those of a substance of known behavior. This always involves the simultaneous use of two sets of gas-containing or expansion vessels and usually the requirement is for essentially identical sets. Setting up the apparatus is therefore more difficult but the return for this effort comes in the simpler measurements which subsequently need to be made. The methods can be subdivided into (i) gas thermometer type, and (ii) isothermal measurements.

One might take as an example of (i) the differential constant volume gas thermometer (see *Figure 5*) as used successfully by Staveley[15] and by Beenakker[46]. Two nearly identical vessels V_1 and V_2 contained within a

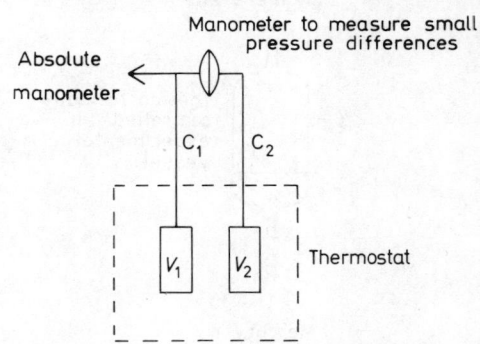

Figure 5. Differential constant volume p–V–T methods.

thermostat are connected via capillaries C_1 and C_2 to a manometer which is capable of measuring small pressure differences, and sometimes to an absolute manometer. If the differential manometer is operated within the thermostat the interconnecting tubing can be eliminated. In use, one volume (say V_1) is filled to some suitable pressure with the gas of known $p-V-T$ properties and simultaneously V_2 is filled to the same pressure with the gas under investigation. This operation would usually be carried out at some convenient temperature near to room temperature. The experimental procedure would then be to measure the pressure difference as a function of temperature. In the ideal case of identical vessels, no dead spaces and truly constant volume the pressure difference can be simply related to the change in $(B_{ref} - B_{unknown})$ from the filling temperature to the final temperature. In practice it is much more difficult to calculate this change in ΔB due to limitations of the apparatus. For example, choice of differential manometer presents a substantial problem. The diaphragm type, which is ideal from a constant volume point of view, rarely covers an adequate pressure range. The mercury manometer has exactly the opposite characteristics. The quartz spiral manometer, or any other glass or quartz Bourdon tube manometer, has the required constant volume and large range characteristics but the reference side of the gage necessarily has an enormous dead space. It is this absence of even an approximately ideal differential manometer that limits the applicability of the differential gas thermometer.

Examination of *Figure 5* shows that innumerable variants on the basic method are possible. Thus it can be converted into an 'equal pressure differential gas thermometer' by arranging for either one or both of the vessels to be variable in volume. This can be done most conveniently if the temperatures are around room temperature when mercury can be used as a confining fluid[12]. Choice of differential manometer is now simplified but in order to solve the necessary equations one does need to know rather more about the geometry of the apparatus than one does in the constant volume case.

Both of the methods give, essentially, changes in B with T. They therefore require a knowledge of B at one temperature if absolute values of B are to be obtained. As a rule this is obtained from other sources, or occasionally from a double expansion procedure of the type described below, built into the same apparatus as the differential thermometer.

The second type of differential measurement is performed isothermally and involves simultaneous expansion, or compression, of both the unknown and the reference gas. A schematic apparatus is shown in *Figure 6* in which a

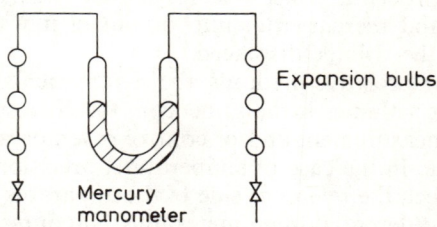

Figure 6. Differential $p-V-T$ methods involving expansion.

mercury U-tube represents the pressure measuring device. The method has been most frequently used for measurements at or about room temperature when mercury could be used as the confining fluid and the whole of the apparatus would be at the same temperature. Clearly, for measurements outside this temperature range it is possible to devise an experiment which is either completely mercury free, such as by use of valves for interconnecting the expansion bulbs and a diaphragm manometer, or a hybrid with (say) the mercury manometer at room temperature and expansion bulbs at the experimental temperature.

In the simplest variant on this general method a constant volume manometer would be used, such as a diaphragm, and the first bulb on each side of the apparatus filled to the same pressure with the unknown and the reference gas respectively. Pressure differences would be followed as a function of expansion into the other bulbs. Clearly, an enormous premium is placed on the accurate precalibration of the various bulbs used, although measurement would be fast provided a suitable constant volume manometer was available. The difficulties with the latter are exactly the same as for the differential gas thermometer.

If a mercury manometer must be used, then it is probably preferable to operate in a zero pressure difference mode, varying the volume of one or other of the halves of the apparatus by a small, known amount to maintain it so. Even simpler from an experimental point of view, but more complex from the computational one is to allow both volume and pressure to change. Of course, both uses of the mercury manometer involve the usual compromise between wide bore limbs for precision on pressure, and narrow bore limbs for precision on volume. It must be emphasized that the precision with which the virial coefficient is obtained is essentially the same whether the manometer is used in the constant volume, zero pressure difference or mixed modes.

In the literature will be found numerous descriptions of implementations of this basic method. McGlashan and Potter[37] used mercury exclusively as their confining fluid compressing the gas from etch mark to etch mark on the narrow bore tubes connecting the expansion bulbs, and also a mercury U-tube manometer to measure the pressure differences. Lichtenthaler, Schramm and Schäfer[35] used an essentially similar apparatus but operated in the zero pressure difference mode by the addition of a capillary to one side of the apparatus into which mercury was run to make the necessary small variations in volume. Bottomley, Reeves and Whytlaw-Gray[11] also operated in the zero pressure difference mode using a mercury manometer with tungsten reference points on each limb. Instead of using a series of bulbs for the expansion procedure, just one large bulb was used on each half of the apparatus and mercury run into or out of it was weighed as a means of calculating the volume displaced.

In all of the relative methods, the requirements on the precision of calibration of the volumes is the same as in the corresponding absolute method. It is in the measurement and/or control of temperature and pressure that the gain is made. In the case of temperature, precision may be relaxed substantially. In effect, the reference side of the apparatus is acting as a gas thermometer. The differential manometer must of course have the same sensitivity as the corresponding manometer in the absolute method but the required operating range is much smaller. Computation of the results from a differential experiment nearly always requires a knowledge of the initial filling pressure, but fortunately, not to very high accuracy.

X. Effect of Gas Adsorption on p–V–T Measurements

As a rule, gas adsorption becomes a significant error in p–V–T work when the experimental pressure reaches a significant fraction of the saturation pressure at experimental temperature. The magnitude of this error varies enormously and is of course at its worst with strongly adsorbing polar substances, although smaller errors have been observed with the rare gases[47]. The nature of the adsorbing surface is critical, and although there is little one can do with a glass surface except leave it in its 'as blown' state and not etch it with chromic acid[10], metal surfaces can be polished, ideally by electrochemical means, to make the actual adsorbing area correspond more closely to the geometrical surface area. In some recent work carried out by the author, substantial adsorption was detected with argon and krypton near their boiling points on 'as machined' copper but on electro-polished copper xenon did not adsorb to any detectable extent.

Adsorption affects p–V–T results not only by giving erroneous values but also by giving irreproducible ones. If consideration is given to the effect of adsorption at the design stage of p–V–T equipment the errors can be largely reduced although never totally eliminated. It must be remembered that p–V–T methods are not identical in the way that they are affected by adsorption. Consider the idealized case in which gas is being expanded from one vessel to a second identical one at the same temperature and that one is on that part of the adsorption isotherm where amount adsorbed is directly proportional to pressure. Clearly, the total amount of gas adsorbed on the walls of the vessel or vessels will be the same both before and after expansion and no errors in the experimental p–V–T results will be obtained. If, however, these same two vessels are used in a Burnett sequence, where, after the first expansion the gas in the expansion vessel is pumped away *together with the gas adsorbed on the wall*, cumulative errors will be seen to creep in at each stage of the expansion. But in the case of Burnett expansion from a lower temperature vessel where adsorption is taking place, into a higher temperature vessel where no adsorption occurs, these cumulative adsorption errors will be much reduced. Errors must also occur in the constant volume gas thermometer method as the temperature range covered usually extends from higher temperatures where there is no adsorption, down to temperatures near the boiling point where adsorption takes place.

Irreproducibility in p–V–T measurements due to adsorption is most frequently observed when mercury is used as the confining fluid such as in measurements on organic vapors. It occurs in two ways. First, mercury flowing through greased taps will carry with it a thin film of grease which is deposited on the glass surfaces with which it comes into contact so changing its adsorption characteristics. Grease-free taps are therefore vital. Secondly, as pointed out by Bottomley[10,11], if p–V–T measurements are made by compression rather than expansion, gas which is adsorbed on the glass walls of the vessel will be trapped there by the advancing mercury. On the other hand, if an expansion procedure is used it is possible for the retreating mercury surface to expose degassed glass surface. This leads to more reproducible results, although one still has to calculate or measure the amount of adsorption both on the glass and on the mercury surface.

Various ways have been employed to determine the magnitude of gas adsorption on $p-V-T$ measurements and so to correct for it. If only one well defined part of the apparatus is at the low temperature where adsorption takes place, it is possible in a separate experiment artificially to increase the surface area to volume ratio of the vessel in question and to repeat the whole series of $p-V-T$ measurements. Any change in the results will most likely be due to adsorption, and extrapolation to zero (measured) surface area can be made. This has been done by packing with tubing[29] or by making a double wall vessel for Burnett measurements[47]. Bottomley and co-workers[11] have preferred to determine the adsorption isotherms of the vapor on glass and on mercury in separate experiments and to correct their $p-V-T$ measurements in this way. Fender and Halsey[23] measured the complete adsorption isotherms for adsorption of krypton on copper at 77.5 K where the vapor pressure is low and used this to correct their $p-V-T$ measurements made on argon and krypton at higher temperatures.

XI. Gas Density Microbalance

The operating principles of a gas density balance can be seen from *Figure 7* which shows a modern microbalance built by Haworth and Sutton[26]. A buoyancy bulb of some 10 cm^3 capacity is suspended from a balance beam with a suitable counterpoise at the other end. The whole balance, which is some 10 to 15 cm in length, is operated inside a chamber into which the gas under investigation is admitted. The buoyancy forces exerted on the bulb are proportional to the density of the gas and may be measured or compensated in some suitable way. Haworth and Sutton used electromagnetic compensation. Other authors[16, 33] have used riders which could in some cases be moved along the balance beam with a magnet.

The microbalance was originally developed for the determination of

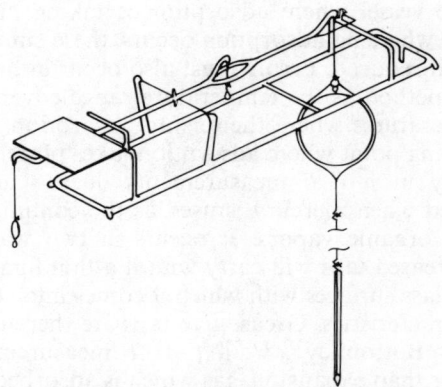

Figure 7. Gas density balance (after Haworth and Sutton[26]).

atomic weights and consequently, virtually all experimental measurements reported to date have been at or close to room temperature. Almost no experience has been gained of measurements well removed from 300 K. Balance sensitivities in the range 0.1 to 1 µg can be achieved by use of quartz fibre suspension and this is now more or less standard procedure. With gases of higher molecular weight, say about 100, with pressures of the order of 1 bar, this enables an accuracy of about 1 in 10^5 in the density to be achieved. Clearly, precision on pressure and temperature measurement must be comparable if the full accuracy is to be utilized. One problem associated with measurements on higher molecular weight gases, particularly if they are polar, is that of adsorption. The effect of adsorption can be lessened by (a) making the balance of quartz, since in general this material has the lowest adsorptive capacity for gases, (b) ensuring that the exposed quartz surface is as uniform as possible as far as adsorptive capacity is concerned, and (c) arranging that the moments of the gas adsorbed on the two arms of the balance are equal. It is generally considered that the effect of adsorption can be reduced in this way to between five and ten per cent of the amount which would be observed if adsorption took place on one side only of the balance.

Gas density measurements have in the past been carried out almost exclusively using relative methods. With earlier balances in which only a few balance points were possible the procedure was to determine the pressures at which the reference gas of known density and the unknown gas both balanced the microbalance. With continuously variable electromagnetic compensation for the buoyancy force, measurements can be made at any pressure and this is usually much more convenient. The method is still relative, of course, for the electromagnetic compensator would be calibrated by making measurements on a gas of known p–V–T properties.

There are three main sources of error in the gas density microbalance method: adsorption which has already been mentioned, change of volume of the buoyancy bulb with pressure, and the effect of impurities in the gas under investigation. The calculation of the change in volume with pressure of the silica bulb is particularly difficult as only rarely will the bulb have a uniform wall thickness. Whereas in most early work, adsorption and volume change of the buoyancy bulb were determined in separate experiments, Haworth and Sutton chose to carry out calibration experiments using two different gases of known equation of state and to obtain the correction coefficients from these measurements.

The need for a sample of high purity can be avoided if it is assumed that the gas under investigation obeys (say) the truncated virial equation

$$pV = mRT(1 + Bm/V) \tag{13}$$

and that an already calibrated quartz microbalance is used. Van Ness and co-workers[21] were the first to publish results using this method, although in place of a quartz microbalance they used an analytical semi-microbalance. This enabled them to add calibrated weights to the buoyancy bulb so as to achieve a balance. This is of course equivalent to the use of a calibrated quartz microbalance. They showed that ignoring adsorption etc. the variation of balancing mass (m) with pressure (p) should follow the equation

$$p/m = RT/MV + BmRT/(MV)^2 \tag{14}$$

where V is the volume of the buoyancy bulb, and M is the molar mass. Hence a plot of p/m against m gives B and so avoids the need for high purity gas.

Recently, Haworth and Sutton[27] have used their magnetically compensated microbalance in an analogous manner and by a suitable graphical analysis of their experimental measurements obtained values for the second virial coefficient B corrected automatically for adsorption, volume change of the buoyancy bulb and impurities in the gas sample.

Unfortunately, there is a penalty to be paid for the convenience in use of these graphical methods, and that is a certain loss of accuracy. If a gas of adequate purity is available and the adsorption and volume change corrections are known from independent sources, a direct determination of density will always give a more precise figure for both density and the second virial coefficient.

XII. p–V–T Properties of Mixtures

The experimental methods described so far in this review are specific neither to pure substances nor to mixtures, and may therefore be applied to both. Apart from the obvious engineering uses of a knowledge of the p–V–T properties of mixtures they are also a very useful starting point for studies on the intermolecular interaction between unlike molecules. A mixture obeys the virial equation just as does a pure substance but on statistical mechanical grounds it is convenient to represent the second virial coefficient of (say) a binary mixture as

$$B_{\text{mixture}} = x_1^2 B_{11} + 2x_1 x_2 B_{12} + x_2^2 B_{22} \tag{15}$$

where x_1 and x_2 are the mole fractions of species 1 and 2, B_{11} and B_{22} are the corresponding second virial coefficients for the pure components, and B_{12} is a pseudo virial coefficient for a gas having only unlike molecule interactions. The quantity which is most useful as far as the study of unlike molecule interactions is concerned is

$$\delta = 2B_{12} - B_{11} - B_{22} \tag{16}$$

For this reason, a number of experiments have been described which enable one to measure this quantity directly. Essentially, they involve measuring the volume change on mixing two gases, all initial and final pressures being identical, or they involve measuring the pressure change on mixing two gases under constant total volume conditions.

The most frequently used method is that of measurement of the pressure change at constant volume and has been described by Knobler[31,32] and by Brewer[13]. *Figure 8* shows the essentials of the apparatus used. In use, V_1 and V_2 would be filled with one gas and V_3 with the second gas to the same pressure. The gas in V_1 would be retained as a reference and the gases in V_2 and V_3 mixed by some suitable means (shown in *Figure 8* as a Toepler pump). The resulting pressure change would then be read off on the differential manometer. The required characteristics of the latter are not quite so stringent as were needed in the case of relative p–V–T methods because a large dead space can be tolerated on the reference side provided that it is adequately thermostatically controlled. Initially, Knobler used an oil filled U-tube

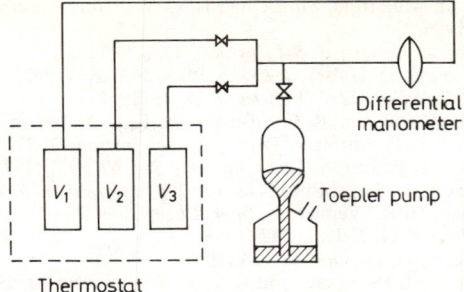

Figure 8. Apparatus for measurement of the pressure change on mixing of two gases.

manometer, which has the disadvantage of making the experiment not truly constant volume, but in his later work he used a diaphragm manometer, which overcomes this problem but introduces the disadvantage of limited pressure range. However, provided the molecular species do not have strong interactions between unlike pairs, the range is often adequate.

The method involving the measurement of the volume change on mixing two gases at constant pressure has not so far been applied to low pressure p–V–T work, although it has been used by Beenakker[49] at higher pressures. The basic apparatus is still as shown in *Figure 8* but with the addition of some means of changing the volume. In the case of the measurements made by Beenakker, the volume change on mixing was always positive and the excess gas was run off into a separate gas measuring system.

XIII. References

[1] Andon, R. J. L., J. D. Cox, E. F. G. Herington and J. F. Martin. *Trans. Faraday Soc.* **53**, 1074 (1957).
[2] Barber, C. R. *Temperature, Its Measurements and Control in Science and Industry*, Vol. III, Pt 1, p 103, Reinhold: New York (1962).
[3] Beattie, J. A., M. Benedict, B. E. Blaisdell and J. Kaye, *J. Chem. Phys.* **42**, 2274 (1965).
[4] Beattie, J. A., D. D. Jacobus, J. M. Gaines, M. Benedict and B. E. Blaisdell. *Proc. Amer. Acad. Arts Sci.* **74**, 327 (1940–42).
[5] Berry, R. J. *Canad. J. Phys.* **36**, 740 (1958).
[6] Bigg, P. H. *Brit. J. Appl. Phys.* **15**, 1111 (1964).
[7] Blaisdell, R. E. *J. Maths Phys.* **19**, 186, 217 and 228 (1940).
[8] Bonhoure, J. and J. Terrien. *Metrologia*, **4**, 59 (1968).
[9] Borovick-Romanov, A. C., P. G. Strelkov, M. P. Orlova and D. N. Astrov. *Temperature, Its Measurement and Control in Science and Industry*, Vol. III, Pt 1, p 113. Reinhold: New York (1962).
[10] Bottomley, G. A. and C. G. Reeves. *Trans. Faraday Soc.* **53**, 1455 (1957).
[11] Bottomley, G. A., C. G. Reeves and R. Whytlaw-Gray. *Proc. Roy. Soc. A.* **246**, 504 (1958).
[12] Bottomley, G. A. and T. H. Spurling, *Austral. J. Chem.* **17**, 501 (1964).
[13] Brewer, J. and G. W. Vaughn. *J. Chem. Phys.* **50**, 2960 (1969).
[14] Burnett, E. S. *J. Appl. Mech. Trans. Amer. Soc. Mech. Engrs*, A3, 136 (1936).
[15] Byrne, M. A., M. R. Jones and L. A. K. Staveley. *Trans. Faraday Soc.* **64**, 1747 (1968).
[16] Casado, F. L., D. S. Massie and R. Whytlaw-Gray. *Proc. Roy. Soc. A*, **207**, 483 (1951).
[17] Compton, J. P. *Metrologia*, **6**, 103 (1970).
[18] Cope, J. O. *Rev. Sci. Instrum.* **33**, 980 (1962).
[19] Cook, D. *Canad. J. Chem.* **35**, 268 (1957).
[20] Constabaris, G., J. H. Singleton and G. D. Halsey. *J. Phys. Chem.* **63**, 1350 (1959).

[21] Di Zio, S. F., M. M. Abbott, D. Zibello and H. C. Van Ness. *Industr. Engng Chem. (Fundamentals)*, **5**, 569 (1966).
[22] Douslin, D. R. and A. Osborn. *J. Sci. Instrum.* **42**, 369 (1965).
[23] Fender, B. E. F. and G. D. Halsey. *J. Chem. Phys.* **36**, 1881 (1962).
[24] Gould, F. A. and T. Vickers. *J. Sci. Instrum.* **29**, 85 (1952).
[25] Guildner, L. A., H. F. Stimson, R. E. Edsinger and R. L. Anderson. *Metrologia*, **6**, 1 (1970).
[26] Haworth, W. S. and L. E. Sutton. *J. Phys. (E), Sci. Instrum.* **3**, 271 (1970).
[27] Haworth, W. S. and L. E. Sutton. *Trans. Faraday Soc.* **67**, 2907 (1971).
[28] 'International Practical Temperature Scale of 1968'. *Metrologia*, **5**, 35 (1969).
[29] Johnston, H. L. and H. R. Weimer. *J. Amer. Chem. Soc.* **56**, 625 (1934).
[30] Kistemaker, J. *Physica*, **11**, 270 and 277 (1945).
[31] Knobler, C. M. *Rev. Sci. Instrum.* **38**, 184 (1967).
[32] Knobler, C. M., J. J. M. Beenakker and H. F. P. Knaap. *Physica*, **25**, 909 (1959).
[33] Lambert, B. and C. S. G. Phillips. *Phil. Trans. A*, **242**, 415 (1950).
[34] Lewis, A. and D. W. G. Style. *Nature, London*, **139**, 631 (1937).
[35] Lichtenthaler, R. N., B. Schramm and K. Schäfer. *Ber. Bunsenges. Phys. Chem.* **73**, 36 (1969).
[36] Lovejoy, D. R. *Rev. Sci. Instrum.* **32**, 41 (1961).
[37] McGlashan, M. L. and D. J. B. Potter. *Proc. Roy. Soc. A*, **267**, 478 (1962).
[38] Moessen, G. W., J. G. Aston and R. G. Ascah. *Temperature, Its Measurement and Control in Science and Industry*, Vol. III, Pt 1, p 91. Reinhold: New York (1962).
[39] Moser, H. *Temperature, Its Measurement and Control in Science and Industry*, Vol. II, p 103. Reinhold: New York (1955).
[40] Moser, H., J. Otto and W. Thomas. *Z. Phys.* **147**, 59 (1957).
[41] Pool, R. A. H., G. Saville, T. M. Herrington, B. D. C. Shields and L. A. K. Staveley. *Trans. Faraday Soc.* **58**, 1692 (1962).
[42] Preston-Thomas, H. and C. G. M. Kirby. *Metrologia*, **4**, 30 (1968).
[43] Taylor, B. N., W. H. Parker and D. N. Langenberg. *Rev. Mod. Phys.* **41**, 375 (1969).
[44] Thomaes, G. and R. van Steenwinkel. *Rev. Sci. Instrum.* **31**, 825 (1960).
[45] Ury, J. F. *Metrologia*, **5**, 11 (1969).
[46] Varekamp, F. H. and J. J. M. Beenakker. *Physica*, **25**, 889 (1959).
[47] Weir, R. D., I. Wynn Jones, J. S. Rowlinson and G. Saville. *Trans. Faraday Soc.* **63**, 1320 (1967).
[48] White, D. and J. Hilsenrath. *Rev. Sci. Instrum.* **29**, 648 (1958).
[49] Zandbergen, P. and J. J. M. Beenakker. *Physica*, **33**, 343 (1967).

CHAPTER 7

Equation of State of Gases at High Pressures and Low or Moderate Temperatures

J. Brielles, A. Dédit, M. Lallemand, B. Le Neindre, Y. Leroux,
J. Vermesse and D. Vidal

Group 1, Section of Thermodynamics, Molecular Interaction and High Pressure Laboratory, 92—Bellevue, France

Contents

I.	Introduction	347
II.	Technical Features Common to Various Experimental Methods	349
	(1) Cryostats and Thermostats	349
	A. Boiling Fluid Cryostat	349
	B. The Stirring Liquid Cryostat	349
	C. Thermostats	352
	(2) Temperature Measurement	352
	(3) High Pressure Vessels	354
	(4) Pressure Measurements	354
	(5) Determination of the Piezometer Volume	355
III.	Various Methods of Measurement	356
	(1) The Gas Expansion Method	356
	(2) Isothermal Methods	357
	A. Divided Piezometer Method	357
	B. Piezometer with Contacts	358
	C. Piston Displacement Method	359
	D. Tsiklis Piezometer Method	361
	(3) The Isochore Method	362
	(4) The Weight Method	363
	(5) The Burnett Method	365
IV.	List of the Most Important Reports	368
V.	Comparison between Theory and Experiment	375
	(1) Virial Expansion of a Hard Sphere System	375
	(2) Perturbation Method	376
VI.	Conclusion	377
VII.	References	377

Notice

This review paper does not cover experimental methods suitable for the determination of p–V–T values at very low (liquid helium) temperatures or close to the critical point. For these specialized subjects the reader can refer to the corresponding chapters of this volume.

Extensive surveys of p–V–T measurements of gases at high pressure have been reported previously by Saurel[165], Levelt Sengers[93] and Tsiklis[185].

I. Introduction

The accurate experimental determination of the specific volume of a fluid in equilibrium over a wide range of pressure and temperature, gives important information on the fluid state. This information is useful, both experimentally and on a theoretical level.

From a knowledge of the equation of state relating the pressure p, the absolute thermodynamic temperature T and the density ρ; $f(p, T, \rho) = 0$, one gets directly the state quantities of a fluid in equilibrium: the compressibility factor Z, the internal energy, the entropy, the enthalpy, the free energy, and the free enthalpy. The measurement of other equilibrium or non-equilibrium quantities needs a knowledge of the variation of ρ as a function of p and T with a precision of 0.5 per cent at least and often 0.1 per cent. To compensate for lack of data and to allow extrapolation, a great number of equations of state have been established, the most famous simple one is the Van der Waals equation; others like those of Benedict–Webb–Rubin, Beattie–Bridgeman and Martin–Hou need several constants to be determined from the experimental data available.

With different potentials of binary interaction types, statistical mechanics facilitates the computation of the compressibility factor following expansion in terms of the density. With the simplest potential the first terms of this expansion have been computed. With the hard sphere potential, the right value of the equation of state has been given. Two numerical methods, Monte Carlo and molecular dynamics, allow us to make an exact simulation of the properties of a fluid composed of some hundreds of molecules, perfectly defined by suitable laws of interaction.

Perturbation methods have been developed: in this case the compressibility factor has been approximated for a hard sphere system corrected by a contribution due to the action of attractive forces which is treated in successive interactions. Again with the help of computers the compressibility factor has been calculated using only a first order perturbation. A comparison between these various theoretical approaches and experiments can be made by numerically adjusting the parameters of the molecular interaction potential. As we will see for the simplest dense gases there is agreement between theoretical forecasts and experimental data at their ultimate precision. It seems that in practice workers must extend the experimental range and improve the precision, if they want to contribute to the refinement of the equation of state of real dense fluids.

In the temperature range covered by this survey, gases have different statistical behaviors. When the temperature is high enough and the density low enough the thermal wavelength is well below the mean distance between molecules, and the particles can be treated as classical objects. On the other hand, at temperatures and densities for which the wavelength is of the same order of magnitude as the intermolecular distance, quantum effects appear; that is to say when

$$\{\hbar/(2mT)^{\frac{1}{2}}\}\{N/V\}^{\frac{1}{3}} \ll 1 \tag{1}$$

In this case study of the equation of state is enhanced and may yield knowledge of density effects, or of the quantum effects at equilibrium.

EQUATION OF STATE OF GASES AT HIGH PRESSURES

II. Technical Features Common to Various Experimental Methods

1. Cryostats and Thermostats

The determination of the equation of state of a fluid is in general the result of accurate measurement of three parameters: pressure, density and thermodynamic temperature. The system must be set in well defined, controlled and measured conditions.

In the low temperature range, two types of cryogenic apparatus are in use, boiling fluid and stirring fluid cryostats. Above room temperature, conventional thermostats can be used.

A. Boiling Fluid Cryostat

The high pressure vessel which contains the sample gas is immersed in a cryogenic fluid which fills an insulated bath. The cryogenic fluid is boiling under its own vapor pressure. If the liquid is pure and if its pressure is fixed, the bath temperature is stable and can be checked by measuring the vapor pressure.

A liquid hydrogen bath can be used between 14 and 20 K, a liquid oxygen bath between 55 and 90 K.

B. The Stirring Liquid Cryostat

In this case the high pressure vessel is isolated from the cryogenic fluid and can be immersed in a gas or a liquid characterized by a large temperature interval between its fusion and its boiling point. A thermal element surrounding the vessel allows us to increase its temperature in relation to that of the cryogenic fluid. A uniform temperature is obtained on the one hand owing to the thermal mass of the high pressure vessel and on the other hand by stirring of the intermediate fluid. However, numerous precautions must be taken into account to avoid temperature gradients and loss of coldness by radiation and conduction of non-immersed parts. Sometimes the space between the cooler and the vessel is empty. This type of cryostat is especially suitable for working between liquid nitrogen temperature and room temperature. In *Figure 1* is shown the cryostat used by Michels et al.[126] between $-180°C$ and $0°C$, with a temperature control better than $0.001°C$ for several hours.

A double-walled metal vessel, containing a liquid which does not get too viscous at the operational temperature, is immersed in a cooling medium. The heat loss to the medium is balanced by the input of electrical energy. The nature of the gas between the walls and its pressure are the main factors which determine the thermal contact and their choice thus provides a rough temperature control. Fine adjustment is made by a heating element, whose current is regulated through a discontinuous temperature regulator. This regulator consists of a Wheatstone bridge, one of the arms of which is a platinum wire immersed in the cryostat liquid. The cryostat liquid is stirred mechanically to insure temperature equilibrium. The heating wire is wound over the full length of the outside of the inner wall of the vessel. The wall was made of heavy copper plate. To decrease the heat influx from the top of the

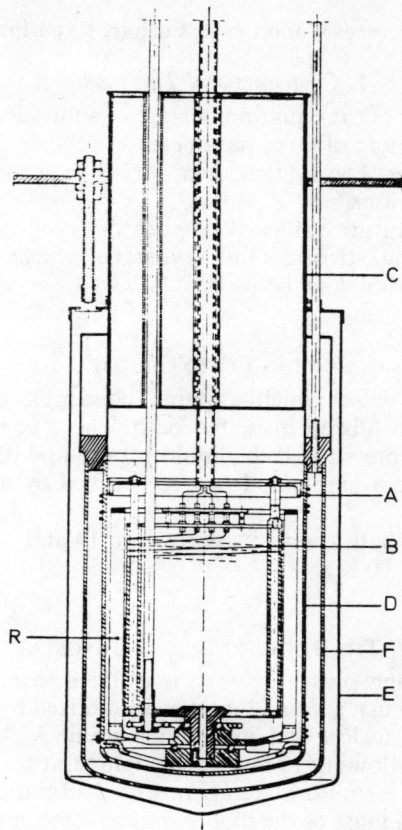

Figure 1. The cryostat of Michels *et al.*[126]. A Heater. B Regulator wire. C Evacuated cylinder. D Inner wall. E Outer wall. F Intermediate wall. R Stirrer.

cryostat an evacuated cylinder made of Berlin silver is fitted in the wide neck of the inner vessel. Through the top and bottom of this cylinder, tubes are soldered to provide access for the stirrer shaft, steel capillary, electrical leads and a platinum thermometer.

Figure 2 shows a cryostat used by Prydz and Straty[147] to measure the $p-V-T$ values of compressed gaseous and liquid fluorine between 53 and 300 K. The $p-V-T$ pipet is suspended in the lower portion of the cryostat by a thin-walled stainless steel reflux tube. Gaseous or liquid nitrogen or gaseous helium in this tube cools the copper pipet. A plug was silver soldered into the lower end of the pipet while the cavity was filled with helium to avoid oxidation of the copper. Helical grooves were machined on the outside of the cylinder to carry a 32 s.w.g., 500 ohm constantan wire heater. The platinum resistance thermometer was tinned and secured in the thermometer well of the cell with Rose's (Bi-Sn-Pb) alloy. A guard ring is supported about 5 cm above the pipet by the reflux tube. This ring is automatically controlled to match the cell temperature and thermally tempers the electrical leads which are wound around its copper cylinder before they are connected to the

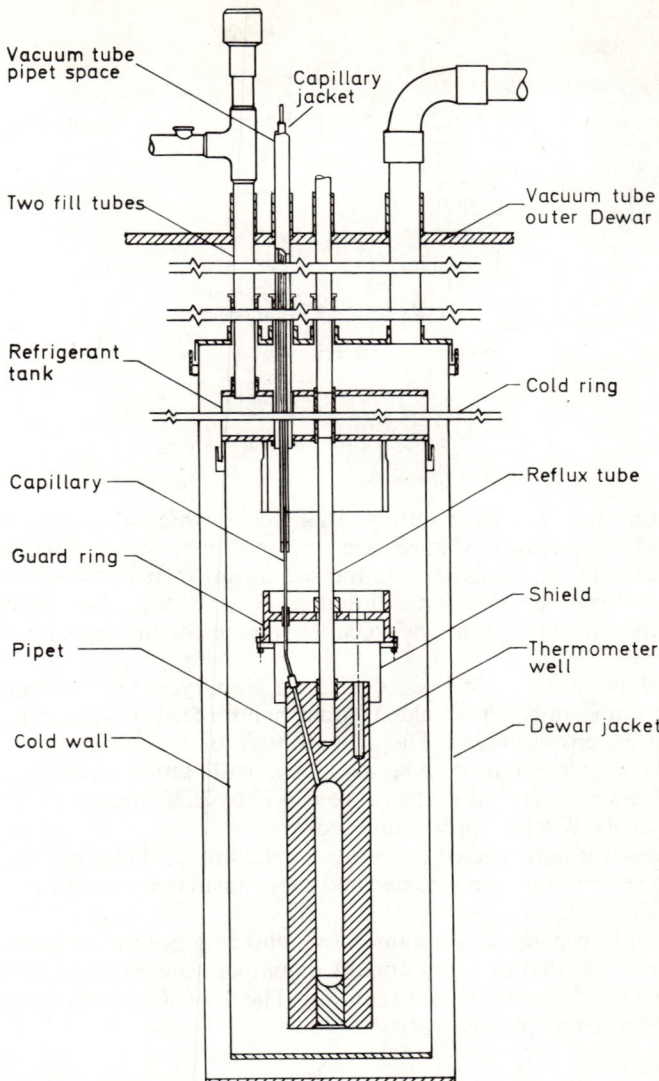

Figure 2. The cryostat of Prydz and Straty.

pipet. The temperature control of the guard ring is obtained by a gold–cobalt versus copper differential thermocouple between the pipet and the guard ring. Tempering of the electrical leads entering the cryostat at room temperature is accomplished by thermally anchoring them to a cold ring suspended from the liquid refrigerant tank. Good thermal contact with the ring is insured because all wires were assembled as one unit in a parallel plane and varnished together before they were mounted.

The pipet is surrounded by an evacuable copper jacket soldered to the

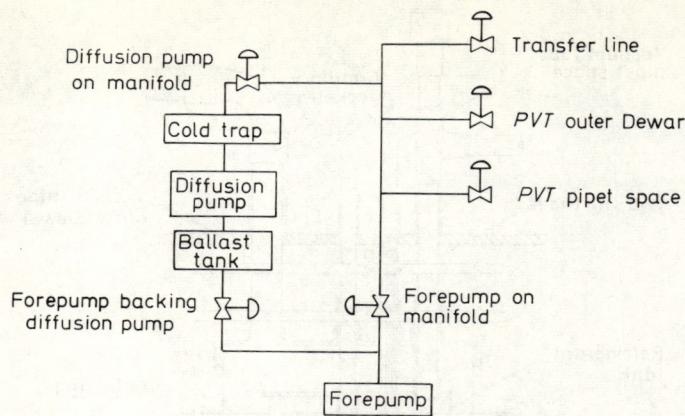

Figure 3. Cryostat vacuum valving.

refrigerant tank with Rose's alloy. This jacket, which also serves as the cold wall, and the refrigerant tank are enclosed in another vacuum-jacketed dewar. Finally the cryostat is immersed in an open liquid-nitrogen dewar having a three-step electronic liquid level indicator. The arrangement of forepump, diffusion pump and vacuum valving for the cryostat is indicated in *Figure 3*.

Several cryostats have been developed by Cryogenic Associates (USA); they are liquid nitrogen shielded liquid helium feed dewars in a controlled temperature environment. The temperature of the sample may be automatically controlled from 5 K to 300 K, with temperature regulation of ± 0.05 K at any set point in the range of 5 K to 26 K and 0.10 K in the range of 25 K to 300 K with appropriate sensors.

The dewar is constructed primarily of Heliarc welded aluminum with the helium reservoir nitrogen shielded and super-insulated to minimize cryogenic boil off.

Temperature control is obtained by balancing power dissipated through an electrical heater on the controlled temperature block against cooling from the dewar's liquid helium reservoir. The flow of gaseous helium coolant is controlled by a solenoid valve.

C. Thermostats

Cryostats can be used between 0°C and 200°C; however, classical oil bath thermostats having a temperature stabilization reaching 0.001°C are more suitable. The temperature uniformity is insured by mechanical stirring or forced circulation of the fluid.

2. *Temperature Measurement*

The equation of state must be determined with regard to the thermodynamic temperature scale. One must compare this scale with the International

Practical Temperature Scale (IPTS-68) to which each experimentor must refer. Since 1968 the deviation between these two scales has been reduced to the exactness of the thermodynamic temperature, and the practical temperature has been extended to lower temperatures. The IPTS-68 is such that temperatures between 13.81 K and 630.74°C are close approximations to the thermodynamic temperature. The 1948 IPTS is now of course obsolete; but the divergences between these two practical scales are known[33,178,179].

The fixed points defined in the temperature range suitable for calibration are: the triple point of equilibrium hydrogen: 13.81 K (ortho and para hydrogen in chemical equilibrium at the reference temperature); the liquid–vapor equilibrium of hydrogen at a pressure of 25/76 of a standard atmosphere: 17.042 K; the boiling point of equilibrium hydrogen: 20.28 K; the boiling point of neon: 27.102 K; the triple point of oxygen: 54.361 K; the boiling point of oxygen: 90.188 K; the triple point of water: 273.16 K; the boiling point of water: 373.15 K; the freezing point of zinc: 692.73 K.

Between 13.81 K and 630.74°C the platinum thermometer must be used as standard and interpolating instrument. When the platinum probe is pure, well annealed and free of all strain, the resistance measurement is well reproducible and sensitive over a wide range, the delay time is small and the temperature variation law is simple. The relationship between resistance and temperature is obtained above 0°C, with a standard function and known deviation functions. In the IPTS-68 between 90.188 K and 273.15 K; the Callendar–Van Dusen equation has been replaced by the W_{ccT} 68 (T68) standard function[178,179]. Between 0°C and 630.74°C the temperature is given by two polynomials and the Callendar equation is modified.

In p–V–T measurements at high pressure, it is not possible to know directly the temperature of the fluid under test; as a matter of fact the sensitive element cannot be put under pressure. The only possibilities available for its location are placement on the container wall or introduction inside a cavity drilled in the metallic mass; in both these cases, there is the problem of close contact with the vessel wall.

Several precautions must be taken to insure good accuracy; thus it is necessary to avoid the effect of thermal conduction along the thermometer threads on the sensing device and to avoid any condensation of steam between conductors and insulators. The reproducibility of the platinum resistance must be checked at the triple point of water and at the boiling point of helium (4.214 K) and in the intermediate range where it is easy to make comparison with the boiling point of a fluid (hydrogen in equilibrium, or oxygen for example).

In careful measurements with a platinum resistance thermometer, the fluid temperature can be measured with a precision of 0.001 K.

For the temperature control system and occasional temperature gradient detections in the high pressure vessel, cheaper devices than the platinum thermometer can be used. For instance, a carbon resistor or semi-conductor resistor such as doped germanium crystals confined in a helium atmosphere give good reproducibility. In general temperature differences between two points are indicated with differential thermocouples (copper–constantan for example).

3. High Pressure Vessels

The mechanical resistance of a high pressure vessel at low or very low temperature is a limiting factor of utilization. In fact, even if the tensile strength, the yield point and the hardness should increase, metals and alloys become brittle. This decrease of the resilience depends on the material and on the treatment it has received. Austenitic steels are frequently used at low temperature and high pressure, for instance carbon steel containing a high percentage of nickel has been used up to 15 kbar at liquid helium temperature. In recent work beryllium bronzes seem promising in this extremely low temperature range.

There is a further technical problem raised, namely the tightness of the sealing device and the output threads; but it can generally be resolved. Teflon is suitable for unsupported area high pressure seals between -100 and $+150°C$, up to several kbar; other seals can be made in treated leather. At very low temperature, crushed or unsupported seals in zinc or copper work correctly. For tight output threads, epoxy is suitable.

4. Pressure Measurements

Essentially three types of device have been used to determine the pressure in p–V–T measurements of fluids from atmospheric pressure up to 10 kbar: mercury gages, free piston gages and strain gages. Close to atmospheric pressure the most accurate measurements use a barometric mercury column, correctly thermostatically controlled and mounted on a rigid base. This barometer yields absolute measurements. The most elaborate devices use columns with large free surfaces of mercury; the level is indicated by a viewfinder microscope mounted on a micrometer; such an arrangement gives a precision of 10^{-6} bar, around atmospheric pressure. In the first experiments of Amagat and Cailletet, the pressure was measured with a mercury column 300 m high. As the most important reports on the equation of state refer to the determination of the specific volume in normal conditions, this measuring instrument is a basic device in p–V–T determinations.

In the low pressure range an instrument that combines extreme accuracy with versatility and unexcelled operating convenience has been developed by Texas Instruments. The gage consists basically of two parts, an interchangeable capsule containing the fused quartz pressure-sensitive element and the readout instrument. Bourdon tube deflection is measured directly by optical means so that no fragile force-balance linkages are required. Optical coupling between pressure sensor and readout eliminate reactive and frictional forces on the sensor, assuring optimum repeatability. Several tubes are available in the range 0 to 35 bar. Close to atmospheric pressure the minimum resolution in a standard gage is 0.008 mm Hg.

Free piston gages or deadweight gages are primary gages at high pressure. Commonly used devices are of two types, simple piston or differential piston.

In the first class, the pressure acts on the effective surface of the piston of the deadweight gage and this pressure is balanced by weights and the atmospheric pressure. In the second class of devices, the piston of the deadweight gage is built with two effective sections of different surface; a dead weight balances the differential pressure between the two sections. Most of the high pressure deadweight gages work with oil, thus the gas pressure must be transmitted by an oil/gas separator (tube filled with mercury, diaphragm). Pistons are built without rings, thus an oil film flows outside the cylinder when the instrument is operating; moreover, they rotate or oscillate to avoid friction. In the modern deadweight gage the clearance between the cylinder and the piston is controlled by a counterpressure acting on the external wall of the cylinder. When the measured pressure increases, the effective section of the piston can be modified, thus on the one hand some deadweight gages are equipped with several sets of piston-and-cylinder adapted to each range of pressure and on the other hand absolute calibrations of the apparatus are carried out when possible. One sees that pressure measurement with a deadweight gage is a fastidious operation and moreover the losses of the transmitting liquid, even under control, limit the operating time when one wants to work at constant volume. The precision of a good deadweight gage is about 10^{-4} in the range of measurements up to 14 kbar.

At very high pressures, pressure measurements are generally accomplished by means of secondary gages the most currently used of which is the strain gage. The effect of pressure on the electrical resistance of a manganin thread, wound on without strain and aged by the repeated application of pressure, is almost linear. These well thermostat gages have shown after several calibrations with deadweight gages a good reproducibility (of the order of 10^{-3}), their delay time is very short and their sensitivity of the order of 10^{-4}. They can only be used in p–V–T measurements if they can be calibrated frequently in the whole explored pressure range. As a matter of fact, some of them show, at intermediate pressure, a slight curvature of the electrical resistance variation which must be known. If a deadweight gage cannot be used for calibration, it is possible to calibrate these gages at fixed points: for instance: the freezing point of mercury at 273.15 K which is 7566.2 bar and the freezing point of argon at 113.648 K which is 1323.84 bar.

5. *Determination of the Piezometer* Volume*

From among the three fundamental parameters, the piezometer volume is perhaps the most often subject to errors of measurement. Several calibration methods can be used.

The volume can be calibrated by weighing the piezometer with a liquid (alcohol, water, mercury), the specific volume of which is well determined. These liquids must be distilled and boiled to remove any dissolved gas. Calculations to obtain the actual volumes involve corrections for thermal expansion and elastic stretching of the piezometer itself and for air buoyancy on the flasks and weights. This method is not always suitable because it can give rise to corrosive effects and moreover the filling of the whole cell is sometimes doubtful.

Measurements of the piezometer volume have also been obtained by expansion of a moderately compressed gas, whose equation of state is known, from a standard volume to the unknown volume.

---

* Called sometimes pyknometer or pipet.

The weight of a gas filling the piezometer whose specific volume is well known at the pressure and temperature of filling can also be used.

Finally the volume can be determined by means of a volumetric pump and a mercury gage.

When the utilization conditions are far from the calibration conditions, it is necessary to correct the piezometer volume. Corrections to piezometer deformations under high pressure are determined from the application of elasticity theory and knowledge of the elastic properties of the enclosing material. Now, Young's modulus of some steels and alloys is known under pressure, thus the correction of the volume of the enclosure which can reach 0.1 per cent per kbar, can be determined precisely.

In some experimental devices working at low or high temperature, one part of the tubulation filled with gas is at a temperature different from that of the piezometer. It is necessary to measure these dead spaces precisely to determine the amount of gas contained. However, the ratio of dead space to piezometer volume must be small.

III. Various Methods of Measurement

In general there are five different procedures that are used to obtain the specific volume, namely the gas expansion method, the isothermal method, the isochore method, the weight method and the Burnett method.

1. The Gas Expansion Method

The calibrated piezometer is set up in the temperature conditions chosen for the experiment and filled with the gas to be studied at approximately the desired density. The pressure and the temperature are noted (*Figure 4*). When either the desired maximum temperature or pressure is attained the gas is released into expansion vessels of calibrated volumes, combined in such a manner as to give a final pressure close to one atmosphere. These vessels are put in a thermostat whose temperature is close to room temperature. The pressure is generally measured with the aid of a manometer. The total amount of sample in the piezometer, the capillary and the pressure detector is deter-

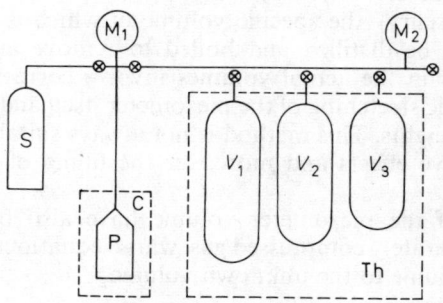

Figure 4. The gas expansion method. P Pyknometer. C Cryostat. S Gas supply. Th Thermostat. M_1 High pressure gage. M_2 Low pressure gage. V_1, V_2, V_3 expansion volumes.

mined by applying the gas law to the state of the sample gas in the expansion vessels. If the equation of state is not known it is possible to obtain the compressibility factor Z of the gas under these conditions from a Burnett expansion experiment. Another method is to reduce the gas to atmospheric pressure or below by expansion; data on the deviation from ideality or second virial coefficients can be used to calculate the whole mass of fluid enclosed in the piezometer. To obtain the actual amount of fluid in the piezometer alone, the portion of the sample in the capillary and the pressure detector, which is dependent upon temperature, is subtracted from the total amount of sample. The experimental density at each $p-T$ point is obtained by dividing the amount of sample in the piezometer by its calibrated volume corrected for temperature and pressure effects.

This method needs only one normal volume determination for every $p-T$ relationship measured. It has been used with various modifications by Holborn and Otto[58-61], Bartlett[6-9], Scott[168] and Goodwin[53].

2. Isothermal Methods

A. Divided Piezometer Method

A pressure-tested gasholder of known volume and calibrated stem is plunged into a thermostat whose temperature is close to room temperature (*Figure 5*). It is filled with the fluid under study at atmospheric pressure and controlled at a temperature where the isotherms of the fluid are known. Knowledge of its normal volume gives the amount of fluid introduced into the container. The stem is connected by a capillary to a piezometer of known volume, set up in a controlled temperature cryostat. Movement of a mercury column or a piston changes the pressure inside the container and the piezometer. The mass of the fluid enclosed in the piezometer is given by the difference between the known total amount of sample and the quantity remaining in the container. This last can be determined when previous $p-V-T$ measurements have been made at room temperature. Then setting the cryostat to a new temperature enables p and V for another isotherm to be measured.

As in the previous method, this one needs only one determination of the normal volume of the fluid to cover the whole range of temperature and pressure; moreover, it is possible to use it at low temperature. One disadvantage is due to the small amount of gas filling the container at the beginning, bringing some limitation to the pressure range, especially at low temperature. Another disadvantage is that known properties of the fluid at the container temperature are required. This method has been used by Kamerlingh Onnes *et al.*[67-74], to determine the $p-V-T$ of air, carbon dioxide, hydrogen, helium, neon, nitrogen, oxygen and agron.

In Beattie's apparatus, the volume of the piezometer is modified for mercury injection purposes by means of a calibrated volumetric pump. Beattie *et al.*[11-17] have studied ethane, isobutane, propane and xenon up to 500 bar and 300°C.

The work of Robertson and Babb on argon, nitrogen and methane, up to 10 kbar and from 35° to 400°C, has been undertaken with a method close to that of the divided piezometer. This method is described in detail in chapter 8.

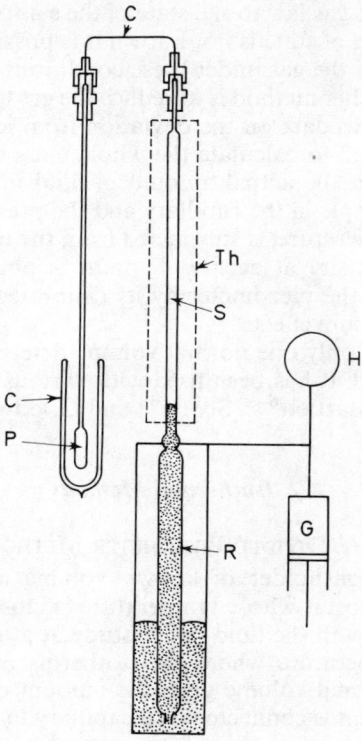

Figure 5. The divided piezometer of Kamerlingh Onnes. R Reservoir. S Calibrated capillary tubing. C Thin capillary. P Pyknometer. C Cryostat. G Pressure producer. Th Thermostat. H High pressure measurement.

B. Piezometer with Contacts

This method was first developed by Amagat[2] (1893) and perfected by Michels[104, 130] at the Van der Waals Institute. The gas is compressed with the help of a mercury piston in a glass piezometer which consists of several bulbs connected together by capillaries. A platinum wire is sealed through the wall of each capillary. Resistors are inserted between these wires. The bottom capillary of the piezometer dips into mercury. When the gas outside the piezometer is pressurized, the mercury is forced up into the piezometer. Whenever the mercury touches a platinum wire, its position can be determined by a suitable resistance measurement. These electrical contacts locate the mercury displacement and define precisely the volume occupied by the gas (*Figure 6*). The filling of the piezometer can be made at low pressure or with the help of a projecting attachment at higher pressure.

Isotherms of several gases have thus been determined in the Van der Waals Institute, with very good precision up to 3 kbar and 150°C. This device cannot be used at low temperature, the limitation being the freezing point of mercury at the operating pressure.

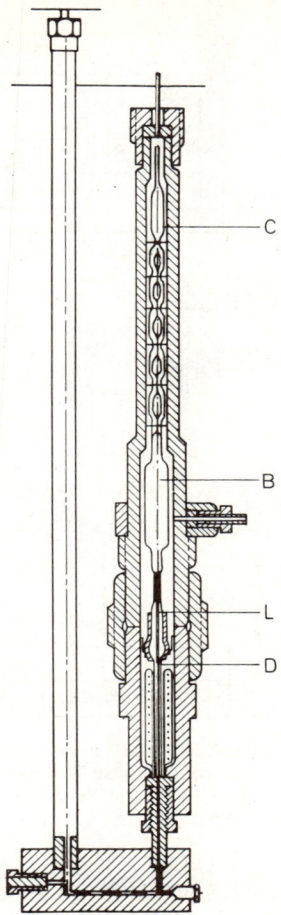

Figure 6. The Michels piezometer for pressures above 250 atm. B Bulb of precompression. C Calibrated bulbs. L Capillary. D Mercury level.

C. Piston Displacement Method

Figure 7 shows the experimental apparatus used by Lahr and Eversole[92]. The core of the pressure vessel was very nearly cylindrical; it was 100 mm high, 12.5 mm in i.d. and 75 mm in o.d. It was made of Bearcat tool steel (Rockwell C-55). A gas cor ection of 1.5 mm stainless steel tubing, secured to the protruding portion of the inner cylinder by a Midget high pressure fitting, emerged flush with the inner surface of the vessel at a point 24 mm from the top. Bridgman-type seals having a nominal 14 per cent unsupported area were used with Teflon gaskets and Bearcat steel (R-55) studs to close the vessel. A thermocouple was inserted into the lower seal in a hole drilled to within 3 mm of the gas space. A jacket around the assembly permitted the

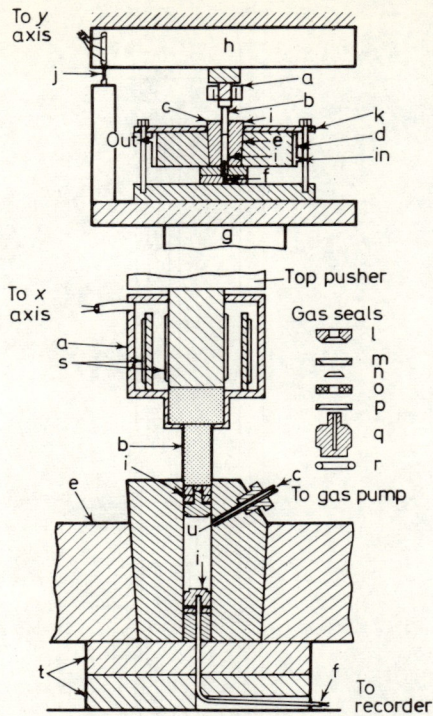

Figure 7. The apparatus of Lahr and Eversole. a Load cell. b Tungsten carbide piston. c Gas inlet. d Cooling jacket. e Vessel. f Thermocouple lead. g Moving platen. h Stationary platen. i Gas seals j Linear extensometer. k Flange. l Pusher. m Steel ring. n Steel inner ring. o Teflon spacer. p Brass ring. q Stud. r O-ring. s Strain gages. t Baseplates.

circulation of hot or cold fluid to maintain the desired temperature. The assembly was mounted between the platens of a 70ton press.

Pressure and press movement were recorded on an x/y recorder. Pressure was measured by a load cell mounted directly above the upper piston. The press movement was measured by a Bournes linear potentiometer rigidly mounted to the fixed upper platen of the press. Piston travel was taken as the difference between the press movement and extraneous compression of the piston, gaskets, and supporting members of the vessel.

To begin a run, the line and vessel were purged and the gas was compressed into the vessel at room temperature or below and at 1000 atm or less. The upper piston was advanced to seal off the inlet port in the vessel; then the piston was further advanced very slowly to reach the desired pressure. At the end of each test, the gas was removed and its quantity measured.

This very simple method may not be suitable at very high pressure, depending on the gas. Thus in a Bridgman experiment, hydrogen gave rise to a breaking of the cylinder at 9 kbar. Lahr and Eversole have used this method to study the compressibility of argon, krypton and xenon, near the freezing curve, between 134 K and 368 K and up to 22 kbar.

In this type of experiment, one of the most important difficulties is to connect piston displacements with volume variations of the gas. These variations can be masked by the compressibility of pistons, the extension of the press, and the deformation of the sealing ring.

After the first checks, Bridgman[23-25] further adapted the very high pressure divided piezometer, which allowed him to measure the p–V–T of hydrogen, helium, ammonia, nitrogen and argon up to 15 kbar.

A strong piezometer (*Figure 8*) holding a weighed amount of gas and equipped with a valve, is introduced inside a thermostatically controlled

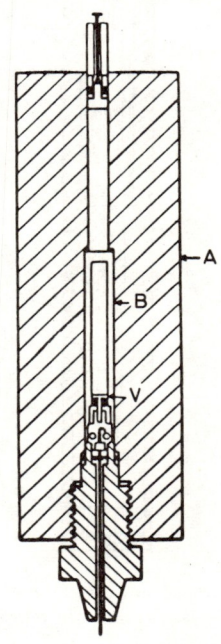

Figure 8. The Bridgman apparatus. A vessel. B Piezometer. V Inward opening valve.

high pressure vessel, connected to a moving piston. The high pressure vessel is filled with a compressing liquid (petroleum in the Bridgman's experiment) of known equation of state. The whole is compressed with the piston. When the liquid pressure equals the gas pressure inside the piezometer, the valve opens and one proceeds with the compression of the two fluids by pushing on the piston whose displacements are recorded. This method possesses several disadvantages, the distortion of the volume of the high pressure vessel must be calculated, some gas can be dissolved in the compressing liquid and inversely. Its utilization is limited to gases which do not react with the compressing fluids. Finally the precision is poor.

D. TSIKLIS PIEZOMETER METHOD

This is an isothermal experimental process combining the Beattie piezometer method and a partial expansion of the gas. It was developed by

Kritchevskii and Tsiklis[89]. A piezometer (*Figure 9*) contains n_1 moles of gas, at known temperature T and pressure p conditions. One injects a volume V_2 of mercury and one returns to the initial pressure p by expanding an amount

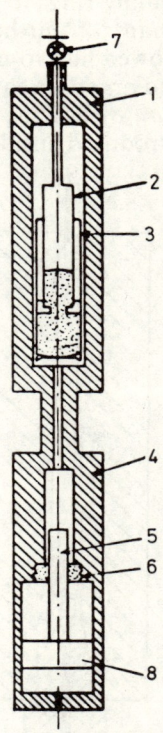

Figure 9. The Tsiklis apparatus. 1 High pressure vessel. 2 Piezometer. 3 Mercury tank. 4 Intensifier. 5 Piston. 6 Seals. 7 Valves. 8 Low pressure piston.

of gas in a calibrated vessel. Let V_3 be the volume of the gas realeased, at standard conditions. If V_0 is the normal molar volume, V the molar volume at T and P, it follows that

$$pV = ZRT$$
$$p(Vn_1 - V_2) = (n_1 - V_3/V_0) ZRT$$

from which we deduce

$$V = V_2 V_0 / V_3$$

3. *The Isochore Method*

An amount of gas whose mass is given by a previous measurement is confined at known temperature and pressure in a piezometer of constant but not necessarily known volume. The pressure is measured at a temperature where the p–V relationship of the fluid is known. This permits evaluation of the specific volume of the fluid in these conditions. p–T data for that specific isochore and for other isochores may be obtained in the same manner. Since pressures are usually measured at the same temperatures for each isochore the resulting p–V–T data are rearranged in isotherms. The fact that no volume calibration or measurement of normal volume is needed, makes this method quite attractive. However, as described, the method has the disadvantage that it gives results relative to the p–V–T of the gas at the filling conditions. When the temperature falls the maximum pressure to be reached decreases, thus the method is more suitable at high temperatures.

Michels et al.[126] used this method to study the p–V–T of gases, between 0°C and −180°C up to 1 200 bar (*Figure 10*).

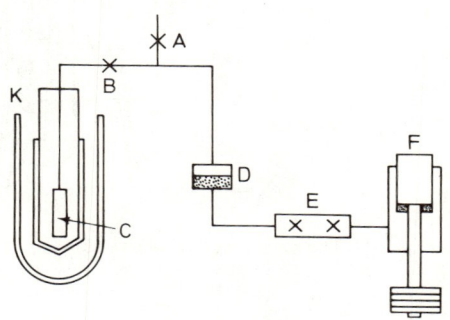

Figure 10. The isochore method. A Supply valve. B Valve. C Piezometer. D Differential manometer. E Intensifier. F Pressure balance. K Cryostat.

The gas/oil separator of the deadweight gage in this case was a diaphragm differential gage used as a null detector. It allowed them to fix a constant value of the piezometer volume, and its sensitivity was approximately 0.001 bar. The reproducibility of the data was always remarkably good, as shown by the studies of Michels *et al.*, on air, hydrogen, deuterium and argon.

If the amount of gas filling the calibrated piezometer can be weighed, measurements are absolute; this was done by Benedict[18, 19].

4. The Weight Method

A piezometer of known volume is suspended from a balance and weighed empty. It is filled with a sample of gas to a selected temperature and pressure, and the increase in weight is recorded. In these conditions the density of gas is determined directly without normal volume determination (*Figures 11* and *12*). When the piezometer is plunged into a cryostat, the friction between the piezometer and the cryogenic fluid can bring about weighing errors; to reduce these risks, Crawford and Daniels[35] used gaseous nitrogen to insure a good thermal contact in these experiments. On the other hand, the weight of the pressure producer and measuring mechanisms must be compensated. Depending on the type of balance used, it may be important that the ratio of gas weight to piezometer weight shall not be too small.

Crawford and Daniels have been able to determine the p–V–T of argon between 95 K and 210 K up to 6.3 kbar.

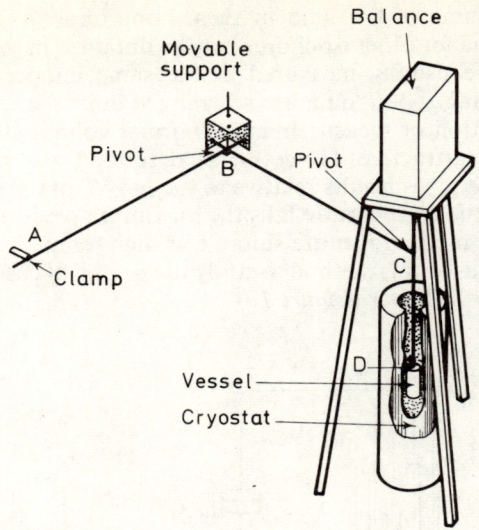

Figure 11. The weight method of Crawford and Daniels.

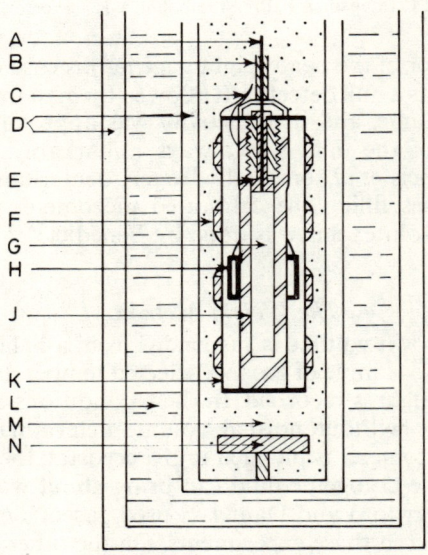

Figure 12. Details of the pressure vessel and the cryogenic apparatus of Crawford and Daniels. A Pressure tubing. B Vertical part of suspension. C Electrical leads. D Vacuum. E Teflon seal. F Heaters. G Piezometer. H Temperature sensor. J Pressure vessel. K Aluminum thermal enclosure for vessel. L Liquid nitrogen. M Dry nitrogen gas. N Heater for temperature gradient control.

5. *The Burnett Method*

The experimental procedure is begun by filling a sample volume V_1 to the initial pressure (*Figure 13*). A second vessel of volume V_2 which can be connected to the first vessel is evacuated. These two vessels are thermostatically controlled at the isotherm temperature T. When the volumes are isolated from each other the gas pressure p_0 is measured in the sample volume.

The corresponding compressibility factor Z_0 is

$$Z_0 = p_0 V_1(p_0)/n_0 RT \tag{2}$$

where n_0 represents the initial number of moles and $V_1(p_0)$ indicates that the sample volume is pressure dependent as a result of its distortion under pressure.

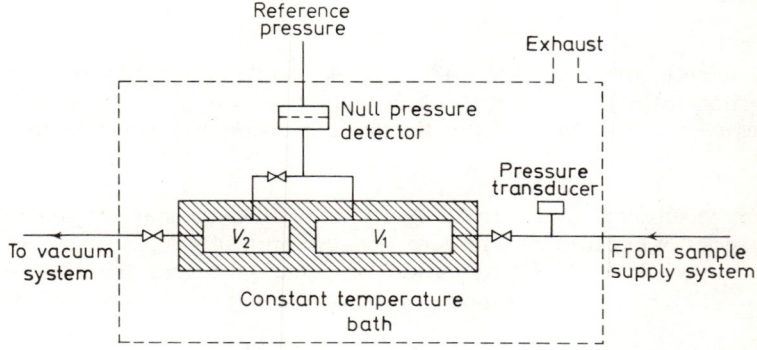

Figure 13. The Burnett apparatus.

In the next step, called the expansion step, the gas is allowed to expand to the expansion volume. After temperature and pressure equilibration has taken place the pressure p_1 of the gas that now occupies both volumes is measured. For this first expansion step the compressibility factor Z_1 is

$$Z_1 = p_1[V_1(p_1) + V_2(p_1)]/n_0 RT \tag{3}$$

If N_1 is the ratio of the combined sample and expansion volume to the sample volume i.e. $N_1 = [V_1(p_1) + V_2(p_1)]/V_1(p_0)$, one obtains

$$(p_0/p_1) = N_1(Z_0/Z_1) \tag{4}$$

The ratio of volumes N_1 depends slightly on both p_0 and p_1, and can be written as $N_1 = N[1 + \Delta N_1(p_0, p_1)]$, where N, known as the cell constant is the value of the volume ratio in the limit as the pressure goes to zero and $\Delta N_1(p_0, p_1)$ is a cell constant correction for the dependence of the cell volumes on pressure.

Generally the sample volume is isolated again and the expansion step is repeated for the reduced quantity of gas n_1, that is now in the sample volume. This leads to a new pressure ratio

$$(p_1/p_2) = N_2(Z_1/Z_2) \tag{5}$$

where $N_2 = N[1 + \Delta N_2(p_1, p_2)]$. This procedure is usually repeated to successively lower pressures with the successively smaller quantities of gas that remain in the sample volume after each expansion until the lowest pressure that can be measured precisely is reached. The generalized form of equation 5 for the rth expansion is

$$(p_{r-1}/p_r) = N_r(Z_{r-1}/Z_r) \tag{6}$$

By extrapolating (p_{r-1}/p_r) of equation 6 to zero pressure, N is obtained since $Z_{r-1} = Z_r = 1$ in the limit. The value at zero pressure can be determined by a least squares fit of a quadratic equation to these data. Thus, since N is known, the value of N_r at any other pressure may be obtained by using the elastic stretching formulas of Roark[155]. These equations relate the change of either volume from before to after the expansion as

$$(\Delta V/V) = k(p_{r-1} - p_r) \tag{7}$$

The pressure correction ΔN_1, ΔN_2 can be also determined from a separate distortion experiment. Substituting values of $(p/Z)_{r-1}$ for $r = 2, 3, \ldots$ successively into equation 6 (for the first expansion, $r = 1$) results in

$$(p_r N_1 N_2 \ldots N_r)/Z_r = (p/Z)_0 \tag{8}$$

where the subscript on the right-hand term denotes the starting pressure for each series of expansions. The zero pressure limit of $p_r N_1 N_2 \ldots N_r$ is $(p/z)_0$ since $\lim Z_r = 1.0$. Combination of equation 6 and the relationship $p = \rho ZRT$, leads to a corresponding Burnett equation in terms of the density

$$\rho_r = \rho_{r-1}/N_r \tag{9}$$

Equations 6 and 9 are basic equations in the reduction of Burnett data and may be transformed into different relationships in which the compressibility factor is represented by a finite virial expansion in either pressure or density. In some previous treatments authors using this method have dealt almost exclusively with the equation of state in the form of a pressure series which appears, at first sight, most suited to the experimental procedure depending as it does entirely on pressure measurements.

$$pV = A + Bp + Cp^2 + \ldots \tag{10}$$

However, as it is the virial coefficients in the density series which are needed to provide the link with the theory of intermolecular forces, a density expansion has been also derived

$$pV = a(1 + b/V + c/V^2 + \ldots) \tag{11}$$

in which it is known that

$$a = A = RT, \quad b = B, \quad c = B^2 + AC, \ldots$$

with increasingly complex relations for the higher virials. The deduction c from B and C clearly introduces an additional step in which accuracy may be lost.

Readers can refer to the circumstantial report of Hall and Canfield[55] of the mathematical method which points the way to obtain data of the compressibility factor and the virial coefficients from Burnett's method.

Non-linear statistics have been applied to Burnett's data to form a redundant set of equations for evaluating the virial coefficients and the cell constant by Waxman et al.[198].

This method was developed in 1936 by Burnett[27], who had worked on helium and air at ambient temperature and up to 150 bar. Schneider et al.[167] used it at high temperature. Its precision is based on the temperature stability of the thermostat, the precision of the pressure readings, and the number of expansions. With this method, Weier et al.[200] determined the second virial coefficients of argon and krypton between 80 K and 225 K with a precision of 10^{-3}. Care must be taken at low temperature to avoid adsorption of gas on the piezometer walls.

The apparatus shown schematically in *Figures 14* and *15* has been used by Waxman and Hastings[197] to measure the compressibility factors of argon and carbon dioxide. It consists of a sample volume V_I and an expansion volume V_{II} as well as components for (1), isolating, evacuating or filling both volumes, (2) measuring the pressure in the sample volume, and (3) thermostatically controlling and measuring the temperature of the apparatus.

The gas in the sample volume is isolated from the expansion volume by an expansion valve and from the external system by a metal diaphragm which is the pressure-responsive element of a null-type transducer. A counterbalancing gas acts on the other side of the diaphragm; the diaphragm is in its null position when the pressure on each side of the diaphragm is the same. The pressure of the sample gas is obtained indirectly from the measurement

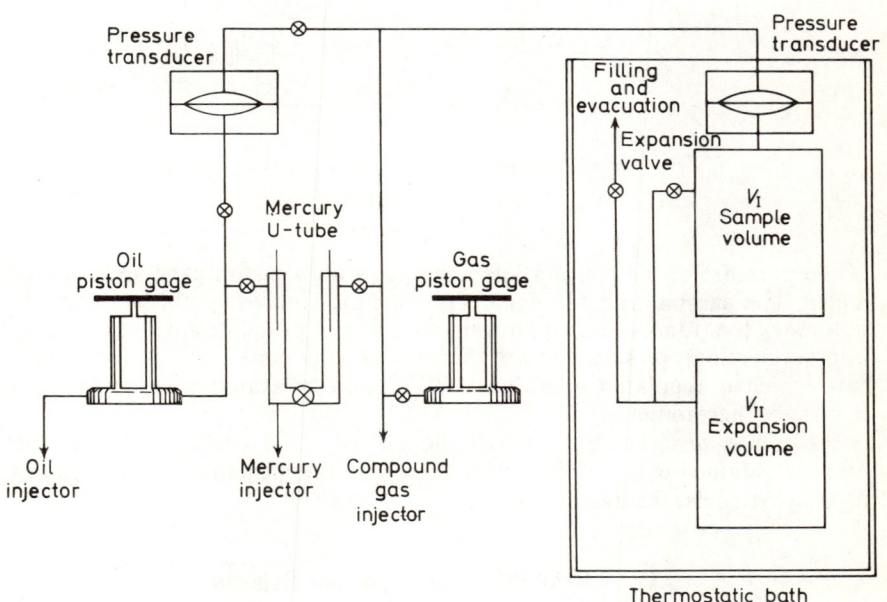

Figure 14. The Burnett apparatus of Waxman and Hastings.

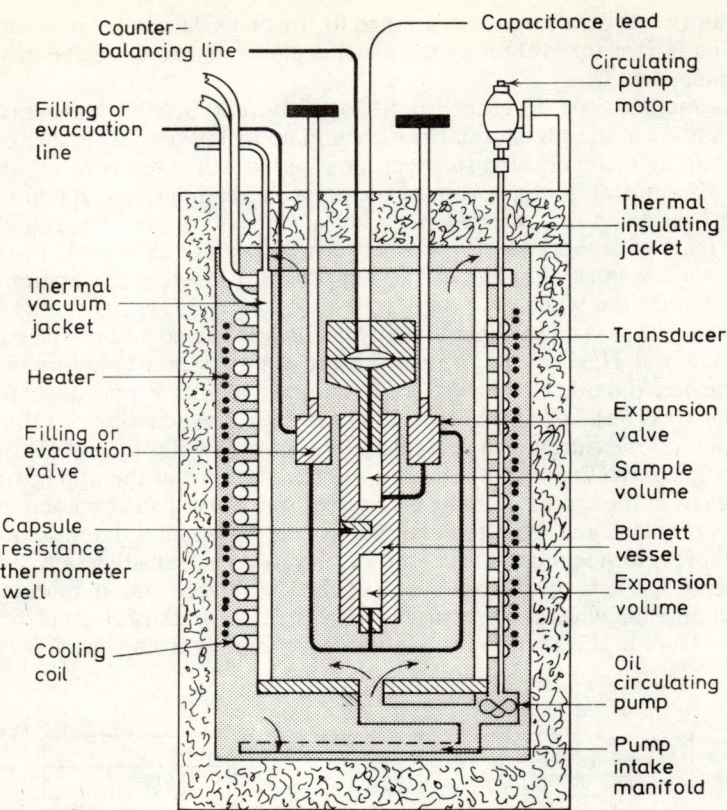

Figure 15. Oil bath used by Waxman and Hastings.

of the pressure of the counterbalancing gas with a piston gage and a barometer. The sample and the expansion volumes including the appropriate volume of the transducer, expansion and filling or evacuation valves, and the interconnecting pressure line are thermostatically controlled at a temperature which is regulated to about 0.001°C and measured with a platinum resistance thermometer.

Dadson *et al.*[39] have compared the second virial coefficient of carbon dioxide obtained with the Burnett method and the piezometer with contacts used by Michels; the agreement is 10^{-3}.

IV. List of the Most Important Reports

In the following table are listed the most important reports on p–V–T data. The temperature range is expressed in kelvin and the pressure maximum in bar.

EQUATION OF STATE OF GASES AT HIGH PRESSURES

Substance	Year	Author	Temp. range (K)	Press. max. (Bar)	Method	Precision	Ref.
Acetylene	1926	Sameshima	273–298	12	Piezometer		163
Air	1893	Amagat	273–473	3000			2
	1896	Witkowski	128–373	130			203
	1908	Koch	194–273	200			87
	1915	Holborn et al.	273–473	100			62
	1923	Penning	128–293	60			142
	1936	Burnett	303	60	Burnett		27
	1945	Kiyama	303	4400			86
	1953	Michels et al.	273–348	2200	Piezometer	2×10^{-4}	125
	1954	Michels et al.	118–248	1000	Isochore	2×10^{-4}	120
Ammonia	1923	Bridgman	373	12000	Piston	10^{-2}	23
	1925	Mayers et al.	238–673	28			101
	1930	Bartlett et al.	203–673	1000	Isochore		9
	1930	Beattie et al.	323–598	120			16
	1933	Keyes	303–473	1000	Isochore	25×10^{-4}	82
Argon	1901	Ramsay et al.	284–510	106			148, 149
	1910	Kamerlingh Onnes et al.	124–293	63	Piezometer	10^{-3}	70
	1923–30	Masson et al.	298–448	125			98, 99
	1924–25	Holborn et al.	173–673	100	Expansion	10^{-2}	59, 60
	1924–35	Bridgman	303–368	15000	Piston	10^{-2}	23, 24, 25
	1949	Oishi	273–373	Low			137
	1949	Michels et al.	273–423	2900	Piezometer	2×10^{-4}	128
	1958	Michels et al.	118–248	1050	Isochore	2×10^{-4}	108
	1961	Rogovaya et al.	90–248	200			160
	1962	Lahr et al.	137–368	18000	Piston	10^{-2}	92
	1967	Weier et al.	80–190		Burnett		200
	1967	Polyakov et al.	298–673	10000	Expansion	10^{-2}	145
	1969	Crawford et al.	95–210	6300	Weight	3×10^{-3}	35
	1969	Robertson et al.	308–673	10000	Piezometer	3×10^{-3}	159
	1971	Waxman et al.	273–598	250	Burnett	10^{-3}	197
Carbon dioxide	1893	Amagat	273–473	1000			2
	1935–37	Michels et al.	273–423	3000	Piezometer	2×10^{-3}	102, 110, 111, 112
	1944	Reamer et al.	346–546	680		2×10^{-4}	150

Substance	Year	Author	Temp. range (K)	Press. max. (Bar)	Method	Precision	Ref.
	1951	Kendall et al.	277–511	680	Piezometer		80
	1954	Kennedy	273–1273	1400	Burnett		81
	1964	Butcher et al.	263–473	Low	Burnett		28
	1971	Waxman et al.	273–598	250	Burnett	2×10^{-4}	197
Carbon monoxide	1929	Scott	298	170			168
	1929	Goig-Botella	273–298	130			52
	1930	Bartlett et al.	203–473	1000	Isochore		9
	1931	Townend et al.	273–298	600		2×10^{-3}	181
	1952	Michels et al.	273–423	3000	Piezometer	2×10^{-3}	109
	1970	Robertson et al.	308–573	10000	Piezometer	25×10^{-4}	158
Cyclopropane	1970	Lin	293–473	310	Piezometer		95
Deuterium	1941	Michels et al.	273–423	3000	Piezometer		105
	1959	Michels et al.	98–423	2800	Isochore		106
Ethane	1935–39	Beattie et al.	298–548	360	Piezometer		15, 17
	1937	Sage et al.	330–430	200			162
	1944	Reamer et al.	346–546	680		4×10^{-3}	151
	1954	Michels et al.	273–423	200	Piezometer		116
Ethylene	1893	Amagat	273–473	100			2
	1925	Masson et al.	298	125			98
	1927	Crommelin et al.	271–293				38
	1929	Dannel et al.	288–318	202		5×10^{-3}	40
	1936	Michels et al.	273–423	70		2×10^{-4}	107
	1942	Michels et al.	273–423	3000		2×10^{-4}	103
	1954	Walters et al.	266–308	40		5×10^{-3}	196
	1964	Butcher et al.	263–473		Burnett		28
	1966	Thomas et al.	273–323	21		10^{-2}	180
	1970	Babb et al.	308	8000	Piezometer	2×10^{-3}	4
Fluorine	1970	Prydz et al.	53–300	245	Expansion	2×10^{-4}	147, 173
Freon 12 and 13	1966	Michels et al.	273–423	400			119
Heavy water	1959	Rivkin	323–573	100	Piezometer	10^{-3}	154
Helium	1907–24	Kamerlingh Onnes et al.	14–373	100		10^{-3}	1, 22, 67, 68, 113

EQUATION OF STATE OF GASES AT HIGH PRESSURES

Substance	Year	Author	Temp. range (K)	Press. max. (Bar)	Method	Precision	Ref.
	1923	Bridgman	303–368	15000	Piston	10^{-2}	23
	1924–26	Holborn et al.	90–673	120			60, 61, 62
	1929	Gibby et al.	298–448	130			50
	1931	Wiebe et al.	203–473	1000			202
	1927–33	Keesom et al.	14–373	16			76, 78, 133, 135, 136
	1941	Michels et al.	273–423	300	Piezometer	2×10^{-4}	129
	1955	Keller	4.3–40	3000	Expansion	2×10^{-4}	79
	1965	Canfield et al.	133–273	520	Burnett	15×10^{-4}	29
	1966	Glassford et al.	4.2–20	1300			51
	1966	Sullivan	70–120	700			175
	1969	Briggs et al.	123–273	800	Burnett		26
	1972	Tsederberg et al.	123–293	400	Burnett		183
Hydrogen	1893	Amagat	273–473	3000			2
	1907–24	Kamerlingh Onnes et al.	56–393	100	Piezometer	10^{-3}	37, 69, 72, 74
	1923	Bridgman	303–368	13000	Piston	10^{-2}	23
	1925–26	Holborn et al.	173–373	100			60
	1926	Verschoyle	273–373	205	Piezometer	5×10^{-4}	192
	1927–30	Bartlett et al.	200–673	1000	Expansion	2×10^{-3}	6, 8
	1930	Basset et al.	273	5000			10
	1938	Wiebe et al.	273–573	1000			201
	1941	Michels et al.	273–423	3000			105
	1950	Johnston et al.	203–298	1000	Expansion		65
	1952	Bennett et al.	298–398	3000	Isochore		20
	1959	Michels et al.	98–423	2952			106
Hydrogen sulfide	1968	Lewis	53–373	1700	Isochore		94
Isopentane	1894	Young	283–553	60			204
	1954	Isaac et al.	408–608	200	Piezometer		64
	1959	Silberberg et al.	323–473	47	Burnett		169
	1959	Vohra et al.	448–473	182	Burnett		195
Krypton	1901	Ramsay et al.	284–510	100			149
	1952	Beattie et al.	273–573	415		2×10^{-3}	14
	1962	Lahr et al.	169–361	12000	Piston	10^{-2}	92
	1966	Trappeniers et al.	273–423	3000	Piezometer	2×10^{-4}	182
Methane	1922	Keyes et al.	273–473	300	Isochore		85

Substance	Year	Author	Temp. range (K)	Press. max. (Bar)	Method	Precision	Ref.
Methane cont.	1927	Keyes et al.	273–473	255			83
	1931	Freeth et al.	273–293	210			48
	1931	Kvalnes et al.	203–473	1000		10^{-3}	91
	1936	Michels et al.	273–423	385		10^{-4}	113
	1943	Olds et al.	294–511	680		10^{-3}	138
	1958	Schamp et al.	273–423	389			166
	1958	Pavlovich et al.					141
	1961	Mueller et al.	144–283	480	Burnett		132
	1964	Douslin et al.	273–623	400	Piezometer		45
	1965	Deffet et al.	323–423	3000	Piezometer		41
	1969	Robertson et al.	308–673	10000	Piezometer	3×10^{-3}	157
Methyl chloride	1964	Hsu	308–498	310	Piezometer		63
Methyl fluoride	1952	Michels et al.	273–423	150	Piezometer		117
Neon	1915–19	Kamerlingh Onnes et al.	143–673	100	Piezometer	10^{-3}	36, 71
	1925–26	Holborn et al.	90–673	100	Expansion	5×10^{-4}	60
	1960	Michels et al.	273–423	2900	Piezometer	2×10^{-4}	122
	1967	Sullivan et al.	70–120	700			175, 176
	1969	Gibbons	27–70	200	Piezometer		49
	1971	Street	80–130	2000	Piezometer	2×10^{-4}	174
Nitrogen	1893	Amagat	273–473	3000		2×10^{-3}	2
	1901	Ramsay et al.	284–510	100			149
	1922–24	Holborn et al.	90–673	100	Piezometer	2×10^{-4}	139
	1923	Bridgman	341	15000	Piston	10^{-2}	23
	1924	Kamerlingh Onnes et al.	124–293	60		10^{-3}	77
	1923–26	Smith et al.	273–473	320	Piezometer		171
	1926	Verschoyle	273–293	205		5×10^{-4}	192
	1927–30	Bartlett et al.	203–673	1000		2×10^{-3}	6, 8
	1930	Basset et al.	289	5000			10
	1934–36	Michels et al.	273–423	3000	Piezometer	2×10^{-4}	130, 140
	1935	Bridgman	133–297	6000	Weight		25
	1937	Benedict	90–273	1200			18
			98–473	6000	Piston		19
	1951	Tsiklis	323–423	10000			184

EQUATION OF STATE OF GASES AT HIGH PRESSURES

Substance	Year	Author	Temp. range (K)	Press. max. (Bar)	Method	Precision	Ref.
Oxygen	1965	Canfield et al.	133–273	541	Burnett	15×10^{-4}	29
	1969	Robertson et al.	308–673	10000	Piezometer	3×10^{-3}	156
	1893	Amagat	273–473	3000			2
	1915–24	Kamerlingh Onnes et al.	156–293	65	Piezometer	10^{-3}	73, 75
	1922–24	Holborn and Otto	273–373	100	Expansion	5×10^{-4}	58
	1923	Masson and Dolly	298				98
	1925	Nijhoff et al.	121–233	10			134
	1954	Michels et al.	273–323	135	Piezometer		115
	1965	Tsiklis et al.	298–673	1000	Expansion	10^{-2}	186
	1970	Weber	54.5–300	330			199
Para-hydrogen	1961–63	Goodwin et al.	15–100	350	Isochore	10^{-3}	53, 54
	1969	El Hadi et al.	12–23	Low	Expansion		46
Propane	1934	Sage et al.	294–378	204	Piezometer		161
	1935	Beattie et al.	370–548	100	Piezometer		15
	1940	Deschner et al.	303–609	680			42
	1949	Cherney et al.	323–398	50	Piezometer	25×10^{-4}	31
	1949	Reamer et al.	346–546	680	Piezometer	10^{-3}	152
	1962	Dittmar et al.	373–413	1000			43
	1970	Babb et al.	308–473	10000	Piezometer	5×10^{-3}	4
Propene	1940	Vaughan et al.	273–573	80	Piezometer	10^{-2}	189
	1949	Farrington et al.	277–555	690	Piezometer	3×10^{-3}	47
	1949	Marchman et al.	303–523	215	Piezometer	25×10^{-3}	96, 97
	1953	Michels et al.	276–423	2830	Piezometer	3×10^{-4}	123, 124
	1959	Pfennig et al.	305–343	2			144
	1962	Dittmar et al.	373–413	1000			43
	1969	Robertson et al.	308–473	10000	Piezometer		157
Propyne	1962	Vohra et al.	323–473	315	Piezometer	5×10^{-3}	194
Steam	1934–36	Smith et al.	400–750				84, 170
Tetrafluoromethane	1961	Douslin et al.	273–623	394	Piezometer	3×10^{-3}	44
Xenon	1901	Ramsay et al.	284–510	100			149
	1934–51	Beattie et al.	289–573	406		2×10^{-3}	12

Substance	Year	Author	Temp. range (K)	Press. max. (Bar)	Method	Precision	Ref.
Ar–He mixtures	1954	Michels et al.	273–423	3000	Piezometer	2×10^{-4}	121
	1962	Lahr et al.	224–362	6450	Piston	10^{-2}	92
	1973	Streett et al.	165–289	3815	Expansion	10^{-3}	174
	1930	Tanner et al.	298–447	125			177
	1967	Kalfoglou et al.	303–773	80	Burnett		66
	1970	Blancett et al.	223–323	700	Burnett		21
	1971	Provine et al.	143–183	706	Burnett		146
CH_4–C_2H_6 mixtures	1939	Michels et al.	273–323	60			114
CH_4–H_2 mixtures	1961	Mueller et al.	144–283	480	Burnett		132
CO_2–CH_4 mixtures	1944	Reamer et al.	346–546	680			150
CO_2–C_2H_4 mixtures	1964	Butcher et al.	173–313	500			28
	1967	Sass	311–373	300	Burnett		164
	1967	Ku et al.					90
D_2–He mixtures	1965	Cramer	2–298	1400	Burnett		34
N_2–H_2 mixtures	1949	Michels et al.	273–423	340			118
N_2–H_2–NH_3 mixtures	1951	Michels et al.	273–423	300			127
N_2–He mixtures	1957	Kramer et al.	65	125			88
	1960	Miller et al.	57	266	Burnett	10^{-3}	131
	1965	Canfield et al.	133–273	500	Burnett	15×10^{-4}	29
	1965	Cramer	298	1400	Burnett		34
	1967	Ku et al.	311–373	300	Burnett		90
	1970	Hall et al.	83–113	700	Burnett		55
	1971	Mathews and Morey	310–449	256	Burnett		100
Ne–He mixtures	1972	Vogl et al.	253–313	37	Burnett		193

V. Comparison between Theory and Experiment

With two types of comparison between theory and experiment, we will show the advantages of measuring precisely the specific volume of real gases at pressures from 1 to 11 kbar and temperatures below 200°C.

1. Virial Expansion of a Hard Sphere System

If $\beta(k)$ represents the cluster integrals of Mayer, the compressibility factor Z of a gas can be expressed by an expansion of the density number n ($n = N/V$), where

$$Z = 1 - \sum_{k \geq 1} \{k/(k+1)\} \beta(k) n^k \qquad (12)$$

and N is the number of molecules in the volume V.

In the case of a system of molecules similar to hard spheres, it is convenient to introduce the packing fraction y, defined by

$$y = \tfrac{1}{6}\pi d^3 n \qquad (13)$$

d being the diameter of the hard sphere.

Ree and Hoover[153] evaluated by computation the integrals up to the seventh coefficient of Z.

$$Z = 1 + 4y + 10y^2 + 18.36y^3 + 28.2y^4 + 39.5y^5 + 56.5y^6 + \ldots \qquad (14)$$

Carnaham and Starling[30] put forward a simple approximation of each coefficient of this expansion which showed a forming law of order r. They found

$$Z = 1 + 4y + 10y^2 + 18y^3 + 28y^4 + 40y^5 + \ldots + (r^3 + 3r)y^r \qquad (15)$$

This result is close to the previous one.

In this form, it is possible to find the sum of Z

$$Z = p/nkT = (1 + y + y^2 - y^3)/(1 - y)^3 \qquad (16)$$

which can be considered to be an exact equation of the compressibility coefficient of a hard sphere system.

This result is close to those given by the Percus–Yevick approximation, where the pressure is expressed in terms of the virial theorem,

$$Z_v = (1 + 2y + 3y^2)/(1 - y)^2 \qquad (17)$$

or the compressibility

$$Z_c = (1 + y + y^2)/(1 - y)^3 \qquad (18)$$

It is possible to compare the calculated values (equation 16) and the experimental data of argon at 35°C up to 10 kbar[159]. A numerical value of d can be calculated, for example, from the viscosity coefficient of argon at high temperature[188] ($d = 3.07 \times 10^{-10}$ m). *Figure 16* shows the relative deviations ($\rho_{cal} - \rho_{exp})/\rho_{exp}$ for each pressure.

At low pressure the deviation reaches 25 per cent, whereas from 3 800 to 10 000 bar, the deviation decreases and is less than three per cent. This example shows that equation 16 is a good approximation to computer calcula-

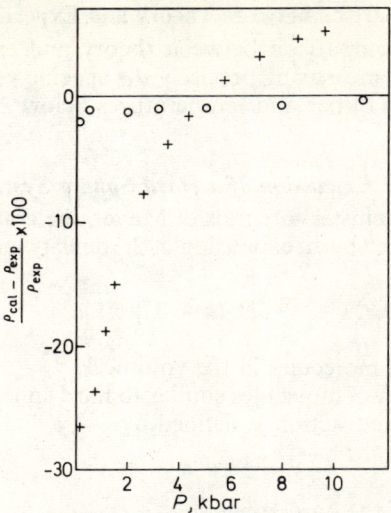

Figure 16. Deviation from theoretical and experimental density data for argon at 35°C. Crosses: hard spheres fluid ($d = 3.07 \times 10^{-10}$ m) (equation 16). Circles: perturbation method.

tions, and the effect of the attractive part of the potential vanishes at high density by screening.

Next, one is tempted to try to improve the hard sphere model by a perturbation method.

2. Perturbation Method

The thermodynamic properties of a classical simple fluid can be calculated from the partition function in terms of the potential energy of the molecules. The potential energy can be written as a sum of pair, triplet, quadruplet, etc., interactions, but it is usually assumed that only pair interactions are important. This is known to be false for real systems; however, in simple fluids the pair interactions provide most of the potential energy. Zwanzig[205] was the first to assume that the intermolecular potential can be written as a sum of the hard-sphere potential and a perturbation potential. Recently this theory has been developed largely by Henderson and Barker[57]. Verlet and Weiss[190, 191] gave a perturbation method suitable for calculating the compressibility factor of simple liquids. The compressibility factor is split into two parts

$$Z = Z_R + Z_A \qquad (19)$$

where Z_R is the compressibility factor associated with the repulsive part of a realistic potential of Lennard–Jones type and Z_A corresponds to the attractive part.

Z_R can be compared to Z^{HS} of a hard sphere system and Z_A to Z^1 which is the first order contribution to attractive forces. Then

$$Z = Z^{HS} + Z^1 \qquad (20)$$

The numerical values given by this method have been compared to the exact calculus given by a Monte Carlo method for a Lennard–Jones fluid of some hundreds of particles, and for various orders of density. The agreement is good at high density.

In *Figure 16* have been reported the values of ρ, deduced from Z as a function of p and compared with the previous experimental data[159]. In this case the diameter d varies, depending on the density from 3.30 to 3.34×10^{-10} m. The deviation is only two per cent at 100 bar and less than one per cent from 200 to 11 000 bar, it is 0.3 per cent for this last value.

VI. Conclusion

Consequently, the comparison between theory and experiment can be improved if the measurements of ρ as a function of p and T are given with a precision of:

— a few per cent, to show the contribution due to the repulsive part of the potential,
— some thousandths, to analyse the first order contribution to the attractive part of the potential,
— less than one thousandth, to test numerical values of characteristic parameters of a given potential of binary interaction.

A precision of one per cent on d produces an error of several per cent on ρ, if one wants to determine Z for a hard sphere system from an expansion in terms of ρ.

At the present time, pressure measurements at 10^{-4} in the intermediate pressure range, and at 10^{-3} in the high pressure range can be carried out easily. Careful measurements and control of the temperature can be close to one thousandth of one degree on the thermodynamic temperature scale. The limiting factor of precision of p–V–T measurements at high density is the piezometer volume determination in the extreme condition where it is working. In that case the precision will be improved by studies of elastic properties of piezometer materials at high temperatures and high pressures.

VII. References

[1] Van Agt, F. P. and H. Kamerlingh Onnes. *Commun. Phys. Lab. Univ. Leiden*, No. 176b (1925).
[2] Amagat, E. H., *Ann. Chim. (Phys.)*, **29**, 68 (1893).
[3] Babb, S. E., *J. Chem. Phys.* **52**, 5963 (1970); **54**, 4200 (1971).
[4] Babb, S. E. and S. L. Robertson. *J. Chem. Phys.* **53**, 1097 (1970).
[5] Babb, S. E., G. J. Scott, C. D. Epp and S. L. Robertson. *Rev. Sci. Instrum.* **40**, 670 (1969).
[6] Bartlett, E. P., *J. Amer. Chem. Soc.* **49**, 687 (1927).
[7] Bartlett, E. P. *J. Amer. Chem. Soc.* **49**, 1955 (1927).
[8] Bartlett, E. P., H. L. Cupples and T. H. Tremearne. *J. Amer. Chem. Soc.* **50**, 1275 (1928).
[9] Bartlett, E. P., H. C. Hetherington, H. M. Kvalnes and T. H. Tremearne. *J. Amer. Chem. Soc.* **52**, 1363 and 1374 (1930).
[10] Basset, J. and R. Dupinay. *CR Acad. Sci., Paris*, **191**, 1295 (1930).
[11] Beattie, J. A. *J. Amer. Chem. Soc.* **46**, 342 (1924), **49**, 1124 (1927).
[12] Beattie, J. A. *Proc. Amer. Acad. Arts Sci.* **69**, 389 (1934).
[13] Beattie, J. A., R. J. Barriault and J. S. Brierley. *J. Chem. Phys.* **19**, 1222 (1951).
[14] Beattie, J. A., J. S. Brierley and R. J. Barriault. *J. Chem. Phys.* **20**, 1615 (1952).
[15] Beattie, J. A., C. Hadlock and N. Poffenberger. *J. Chem. Phys.* **3**, 93 (1935).

[16] Beattie, J. A. and C. K. Lawrence. *J. Amer. Chem. Soc.* **52**, 6 (1930).
[17] Beattie, J. A., G. J. Su and G. L. Simard. *J. Amer. Chem. Soc.* **61**, 924 (1939).
[18] Benedict, M. *J. Amer. Chem. Soc.* **59**, 2224 (1937).
[19] Benedict, M. *J. Amer. Chem. Soc.* **59**, 2233 (1937).
[20] Bennett, C. O. and B. F. Dodge. *Industr. Engng Chem. (Industr.)*, **44**, 180 (1952).
[21] Blancett, A. L., K. R. Hall and F. B. Canfield. *Physica,* **47**, 75 (1970).
[22] Books, J. D. A. and H. Kamerlingh Onnes. *Commun. Phys. Lab. Univ. Leiden,* No. 170a (1924).
[23] Bridgman, P. W. *Proc. Amer. Acad. Arts Sci.* **59**, 173 (1923).
[24] Bridgman, P. W. *Phys. Rev.* **45**, 930 (1934).
[25] Bridgman, P. W. *Proc. Amer. Acad. Arts Sci.* **70**, 1 (1935).
[26] Briggs, T. C., B. J. Dalton and R. E. Barieau. *Rep. Invest. US Bur. Min. No.* 7287 (1969).
[27] Burnett, E. S. *J. Appl. Mech.* **3**, A 136 (1936).
[28] Butcher, E. G. and R. S. Dadson. *Proc. Roy. Soc. (London) A,* **277**, 448 (1964).
[29] Canfield, F. B., T. W. Leland and R. Kobayashi. *J. Chem. Engng Data,* **10**, 92 (1965).
[30] Carnaham, N. F. and M. E. Starling. *J. Chem. Phys.* **51**, 635 (1969).
[31] Cherney, B. J., H. Marchman and R. York. *Industr. Engng Chem. (Industr.)* **41**, 2653 (1949).
[32] Comings, E. W., in *High Pressure Technology* (1956).
[33] *Comptes Rendus* des Séances de la XIIIe Conférence Générale des Poids et Mesures, p A 1. Bureau International des Poids et Mesures: Sèvres (1968).
[34] Cramer, J. D. *Los Alamos, Lab. 3500 M S U C 34, Physics T I D 4500* (1965).
[35] Crawford, R. K. and W. B. Daniels. *J. Chem. Phys.* **50**, 3171 (1969).
[36] Crommelin, C. A., J. P. Martinez and H. Kamerlingh Onnes. *Commun. Phys. Lab. Univ. Leiden,* No. 154a (1919).
[37] Crommelin, C. A. and J. C. Swallow. *Commun. Phys. Lab. Univ. Leiden,* No. 172a (1924).
[38] Crommelin, C. A. and H. G. Watts. *Commun. Phys. Lab. Univ. Leiden,* No. 189c (1927); Crommelin, C. A. and H. G. Watts. *Z. Angew. Chem.* **42**, 1121 (1929).
[39] Dadson, R. S., E. J. Evans and J. H. King. *Proc. Phys. Soc.* **92**, 1115 (1967).
[40] Dannell, H., V. Munster and H. Stolzenberg. *Z. Angew. Chem.* **42**, 1121 (1929).
[41] Deffet, L. and F. Ficks. *Advances in Thermophysical Properties at Extreme Temperatures and Pressures: Purdue,* p 107. American Society of Mechanical Engineers (1965).
[42] Deschner, W. W. and G. G. Brown. *Industr. Engng Chem. (Industr.),* **32**, 836 (1940).
[43] Dittmar, P., F. Schulz and G. Strese. *Chem.-Ing.-Tech.* **34**, 437 (1962).
[44] Douslin, D. R., R. H. Harrison, R. T. Moore and J. P. McCullough. *J. Chem. Phys.* **35**, 1357 (1961).
[45] Douslin, D. R., R. H. Harrison, R. T. Moore and J. P. McCullough, *J. Chem. Engng Data,* **9**, 358 (1964).
[46] El Hadi, Z. E. M. A., J. A. Dorrepaal and M. Durieux. *Physica,* **41**, 320 (1969).
[47] Farrington, P. S. and B. H. Sage. *Industr. Engng Chem. (Industr.),* **41**, 1734 (1949).
[48] Freeth, F. and T. Verschoyle. *Proc. Roy. Soc. (London) A,* **130**, 453 (1931).
[49] Gibbons, R. M., *Cryogenics,* **9**, 251 (1969).
[50] Gibby, C. W., C. C. Tanner and I. Masson. *Proc. Roy. Soc. (London) A,* **122**, 283 (1929).
[51] Glassford, A. M. P. and A. J. Smith. *Cryogenics,* **6**, 193 (1966).
[52] Goig-Botella, S., *An. Soc. Esp. Fis. Quim.* **27**, 315 (1929).
[53] Goodwin, R. D. *J. Res. Nat. Bur. Stand.* **65C**, 231 (1961).
[54] Goodwin, R. D., D. E. Diller, H. M. Roder and L. A. Weber. *J. Res. Nat. Bur. Stand.* **67A**, 173 (1963).
[55] Hall, K. R. and F. B. Canfield. *Physica,* **47**, 99 (1970).
[56] Hansen, J. P. and L. Verlet. *Phys. Rev.* **184**, 151 (1969).
[57] Henderson, D. and J. A. Barker. *Physical Chemistry. An Advanced Treatise.* Volume VIII A: *Liquid State,* p 377. Edited by D. Henderson. Academic Press: New York (1971).
[58] Holborn, L. and J. Otto. *Z. Phys.* **10**, 367 (1922).
[59] Holborn, L. and J. Otto. *Z. Phys.* **30**, 320 (1924).
[60] Holborn, L. and J. Otto. *Z. Phys.* **33**, 1 (1925).
[61] Holborn, L. and J. Otto. *Z. Phys.* **38**, 359 (1926).
[62] Holborn, L. and H. Schultze. *Ann. Phys., Lpz.* **47**, 1089 (1915).
[63] Hsu, C. C. and J. J. McKetta. *J. Chem. Engng Data,* **9**, 45 (1964).
[64] Isaac, R., Kun Li and L. N. Canjar. *Industr. Engng Chem. (Industr.),* **46**, 199 (1954).
[65] Johnston, H. L. and A. D. White, *Trans. Amer. Soc. Mech. Engrs,* **72**, 785 (1950); Johnston, H. L., W. F. Keller and A. S. Friedman. *J. Amer. Chem. Soc.* **76**, 1482 (1954).

[66] Kalfoglou, N. K. and J. G. Miller. *J. Phys. Chem.* **71**, 1256 (1967).
[67] Kamerlingh Onnes, H. *Commun. Phys. Lab. Univ. Leiden*, No. 102a (1907); No. 102c (1908); No. 165c (1924).
[68] Kamerlingh Onnes, H. and J. D. A. Books. *Commun. Phys. Lab. Univ. Leiden*, No. 170b (1924).
[69] Kamerlingh Onnes, H. and C. Braak, *Commun. Phys. Lab. Univ. Leiden*, No. 99a, 2 (1907); No. 100a, 2 (1907); No. 100b, 13 (1907).
[70] Kamerlingh Onnes, H. and C. Crommelin. *Commun. Phys. Lab. Univ. Leiden*, No. 118b (1910).
[71] Kammerlingh Onnes, H. and C. A. Crommelin. *Commun. Phys. Lab. Univ. Leiden*, No. 147d (1915).
[72] Kamerlingh Onnes, H., C. A. Crommelin and E. I. Smod. *Commun. Phys. Lab. Univ. Leiden*, No. 146b (1915).
[73] Kamerlingh Onnes, H., C. Dorsman and G. Holst. *Commun. Phys. Lab. Univ. Leiden*, No. 146a (1915).
[74] Kamerlingh Onnes, H. and W. J. de Haas. *Commun. Phys. Lab. Univ. Leiden*, No. 127c (1912).
[75] Kamerlingh Onnes, H. and H. A. Kuypers. *Commun. Phys. Lab. Univ. Leiden*, No. 169a (1924).
[76] Kamerlingh Onnes, H. and F. M. Penning. *Commun. Phys. Lab. Univ. Leiden*, No. 165b (1923).
[77] Kamerlingh Onnes, H. and A. Th. van Urk. *Commun. Phys. Lab. Univ. Leiden*, No. 169d (1924).
[78] Keesom, W. H. and J. J. M. van Santen. *Commun. Phys. Lab. Univ. Leiden*, No. 227b (1933).
[79] Keller, W. E. *Phys. Rev.* **97**, 1 (1955); **98**, 1571 (1955); **100**, 1790 (1955).
[80] Kendall, B. J. and B. H. Sage. *Petroleum, London*, **14**, 184 (1951).
[81] Kennedy, G. C. *Amer. J. Sci.* **252**, 225 (1954).
[82] Keyes, F. *J. Amer. Chem. Soc.* **53**, 965 (1931).
[83] Keyes, F. and H. Burks. *J. Amer. Chem. Soc.* **49**, 1403 (1927).
[84] Keyes, F., L. B. Smith and H. T. Gerry, *Proc. Amer. Acad. Arts Sci.* **70**, 319 (1936).
[85] Keyes, F., L. Smith and D. Joubert. *J. Math. Phys. (Mass. Inst. Technol.)*, **1**, 191 (1922).
[86] Kiyama, R. *Rev. Phys. Chem. (Japan)*, **19**, 38 (1945).
[87] Koch, P. P. *Ann. Phys., Lpz.* **27**, 311 (1908).
[88] Kramer, G. M. and J. G. Miller. *J. Phys. Chem.* **61**, 785 (1957).
[89] Kritchevskii, I. R. and D. S. Tsiklis. *Dokl. Akad. Nauk SSSR*, **78**, 1169 (1951).
[90] Ku, P. S. and B. F. Dodge. *J. Chem. Engng Data*, **12**, 158 (1967).
[91] Kvalnes, H. and V. Gaddy. *J. Amer. Chem. Soc.* **53**, 397 (1931).
[92] Lahr, P. H. and W. G. Eversole. *J. Chem. Engng Data*, **7**, 42 (1962).
[93] Levelt Sengers, J. M. H. *Physics of High Pressure and the Condensed Phase*, p 60. Edited by A. Van Itterbeek. North-Holland: Amsterdam (1965).
[94] Lewis, L. C. *J. Chem. Engng Data*, **13**, 482 (1968).
[95] Lin, D. C. K. *J. Chem. Engng Data*, **15**, 483 (1970).
[96] Marchman, H., H. Prengle and R. L. Motard. *Industr. Engng Chem. (Industr.)* **41**, 11 (1949).
[97] Marchman, H., H. Prengle and R. L. Motard. *Industr. Engng Chem. (Industr.)* **41**, 2658 (1949).
[98] Masson, J. and L. G. F. Dolly. *Proc. Roy. Soc. (London) A*, **103**, 524 (1923).
[99] Masson, J. and C. Tanner. *Proc. Roy. Soc. (London) A*, **126**, 268 (1930).
[100] Mathews, J. F. and D. C. D. Morey. *Canad. J. Chem. Engng Data*, **49**, 282 (1971).
[101] Mayers, C. H. and R. S. Jessup. *Refrig. Engr*, **11**, 345 (1925).
[102] Michels, A., B. Blaisse and C. Michels. *Proc. Roy. Soc. (London), A*, **160**, 358 (1937).
[103] Michels, A. and M. Geldermans. *Physica*, **3**, 967 (1942).
[104] Michels, A. and R. Gibson. *Ann. Phys., Lpz.* **87**, 850 (1928).
[105] Michels, A. and M. Goudeket. *Physica*, **8**, 347 and 353 (1941).
[106] Michels, A., W. de Graaf, T. Wassenaar, J. M. H. Levelt and P. Louverse. *Physica*, **25**, 25 (1959).
[107] Michels, A., J. de Grueter and F. Niesen. *Physica*, **3**, 346 (1936).
[108] Michels, A., J. Levelt and W. de Graaf. *Physica*, **24**, 659 (1958).
[109] Michels, A., J. M. Lupton, T. Wassenaar and W. de Graaf, *Physica*, **18**, 121 (1952).
[110] Michels, A. and C. Michels. *Proc. Roy. Soc. (London) A*, **153**, 214 (1935).
[111] Michels, A. and C. Michels. *Proc. Roy. Soc. (London) A*, **160**, 348 (1937).
[112] Michels, A., C. Michels and H. Wouters. *Proc. Roy. Soc. (London) A*, **153**, 214 (1935).
[113] Michels, A. and G. Nederbragt. *Physica*, **3**, 569 (1936).
[114] Michels, A. and G. Nederbragt. *Physica*, **6**, 369 (1940).
[115] Michels, A., H. W. Schamp and W. de Graaf. *Physica*, **20**, 1209 (1954).
[116] Michels, A., W. van Staaten and J. Dawson. *Physica*, **20**, 17 (1954).
[117] Michels, A., A. Visser, R. J. Lunbeck and G. J. Wolkers. *Physica*, **18**, 114 (1952).

[118] Michels, A. and T. Wassenaar. *Appl. Sci. Res.* **A1**, 258 (1949).
[119] Michels, A., T. Wassenaar, G. J. Wolkers, C. Prins and L. Klundert. *J. Chem. Engng Data*, **11**, 449 (1966).
[120] Michels, A., T. Wassenaar, J. M. H. Levelt and W. de Graaf. *Appl. Sci. Res.* **A4**, 381 (1954).
[121] Michels, A., T. Wassenaar and P. Louverse. *Physica*, **20**, 99 (1954).
[122] Michels, A., T. Wassenaar and P. Louverse. *Physica*, **26**, 539 (1960).
[123] Michels, A., T. Wassenaar, P. Louverse, R. I. Lunbeck and G. J. Wolkers. *Physica*, **19**, 287 (1953).
[124] Michels, A., T. Wassenaar, P. Louverse, R. I. Lunbeck and G. J. Wolkers. *Appl. Sci. Res.* **A4**, 34 (1953).
[125] Michels, A., T. Wassenaar and W. van Seventer. *Appl. Sci. Res.* **A4**, 52 (1953).
[126] Michels, A., T. Wassenaar and Th. N. Zwietering. *Physica*, **18**, 67 (1952); **19**, 371 (1953).
[127] Michels, A., T. Wassenaar, G. J..Wolkers, W. de Graaf and P. Louverse. *Appl. Sci. Res.* **A3**, 1 (1951).
[128] Michels, A. and H. Wijker. *Physica*, **15**, 627 (1949).
[129] Michels, A. and H. Wouters. *Physica*, **8**, 923 (1941).
[130] Michels, A., H. Wouters and J. de Boer. *Physica*, **1**, 587 (1934); **3**, 585 (1936).
[131] Miller, J. E., L. Stroud and L. W. Brandt. *J. Chem. Engng Data*, **5**, 6 (1960).
[132] Mueller, W., T. Leland and R. Kobayashi. *Amer. Instr. Chem. Engrs Jnl*, **7**, 267 (1961).
[133] Nijhoff, G. P. *Commun. Phys. Lab. Univ. Leiden*, No. 64c and No. 64f (1928).
[134] Nijhoff, G. P. and W. H. Keesom. *Commun. Phys. Lab. Univ. Leiden*, No. 179b (1925).
[135] Nijhoff, G. P. and W. H. Keesom. *Commun. Phys. Lab. Univ. Leiden*, No. 227b (1933).
[136] Nijhoff, G. P., W. H. Keesom and B. Iliin. *Commun. Phys. Lab. Univ. Leiden*, No. 188c (1927).
[137] Oishi, J. *J. Sci. Res. Inst., Tokyo*, **43**, 220 (1949).
[138] Olds, R., H. Reamer, B. Sage and W. Lacey. *Industr. Engng Chem. (Industr.)*, **35**, 922 (1943).
[139] Otto, J. and L. Holborn. *Z. Phys.* **10**, 367 (1922).
[140] Otto, J., A. Michels and H. Wouters. *Z. Phys.* **35**, 97 (1934).
[141] Pavlovich, N. V. and D. L. Timrot. *Teploenergetika*, **4**, (1958).
[142] Penning, F. M. *Commun. Phys. Lab. Univ. Leiden*, No. 166, 3 (1923).
[143] Penning, F. M. and H. Kamerlingh Onnes. *Commun. Phys. Lab. Univ. Leiden*, No. 165c (1923).
[144] Pfennig, H. W. and J. McKetta. *Petrol. Refiner*, **36**, 309 (1957).
[145] Polyakov, E. V. and D. S. Tsiklis. *Russ. J. Phys. Chem.* **41**, 1278 (1967).
[146] Provine, J. A. and F. B. Canfield. *Physica*, **52**, 79 (1971).
[147] Prydz, R. and G. C. Straty. *J. Res. Nat. Bur. Stand.* **74**, 747 (1970).
[148] Ramsay, W. and M. Travers. *Trans. Roy. Soc. (London) A*, **197**, 47 (1901).
[149] Ramsay, W. and M. Travers. *Z. Phys. Chem.* **38**, 641 (1901).
[150] Reamer, H., R. H. Olds, B. H. Sage and W. N. Lacey. *Industr. Engng Chem. (Industr.)*, **36**, 88 (1944).
[151] Reamer, H., R. H. Olds, B. H. Sage and W. N. Lacey. *Industr. Engng (Industr.) Chem.* **36**, 956 (1944).
[152] Reamer, H., B. H. Sage and W. N. Lacey. *Industr. Engng Chem. (Industr.)*, **41**, 482 (1949).
[153] Ree, F. H. and W. G. Hoover. *J. Chem. Phys.* **40**, 939 (1964).
[154] Rivkin, S. L. *Atomnaya Energiya*, **7**, 457 (1959).
[155] Roark, R. J. *Formulas for Stress and Strain*, 4th ed., p 108. McGraw-Hill: New York (1965).
[156] Robertson, S. L. and S. E. Babb. *J. Chem. Phys.* **50**, 4560 (1969).
[157] Robertson, S. L. and S. E. Babb. *J. Chem. Phys.* **51**, 1957 (1969); **51**, 3152 (1969).
[158] Robertson, S. L. and S. E. Babb. *J. Chem. Phys.* **53**, 1094 (1970).
[159] Robertson, S. L., S. E. Babb and G. J. Scott. *J. Chem. Phys.* **50**, 2160 (1969).
[160] Rogovaya, J. A. and M. G. Kaganer. *Zh. Fiz. Khim.* **35**, 2135 (1961); English translation: *Russ. J. Phys. Chem.* **35**, 1049 (1961).
[161] Sage, B. H., J. G. Schaafsma and W. N. Lacey. *Industr. Engng Chem. (Industr.)*, **26**, 1218 (1934).
[162] Sage, B. H., D. C. Webster and W. N. Lacey. *Industr. Engng Chem. (Industr.)*. **29**, 658 (1937).
[163] Sameshima, J., *Bull. Chem. Soc. Japan*, **1**, 41 (1926).
[164] Sass, A. *J. Chem. Engng Data*, **12**, 168 (1967).
[165] Saurel, J. R. *Mémor. Artill. Fr.* **31**, 129 (1957).
[166] Schamp, H., E. Mason, A. Richardson and A. Altman. *Physics of Fluids*, **1**, 329 (1958).
[167] Schneider, W. G. *Canad. J. Res.* **B27**, 339 (1949); Nicholson, G. A. and W. G. Schneider. *Canad. J. Chem.* **33**, 589 (1955);

Whalley, E., Y. Lupien and W. G. Schneider. *Canad. J. Chem.* **31**, 722 (1953); **33**, 633 (1955); Whalley, E. and W. G. Schneider, *Trans. Amer. Soc. Mech. Engrs,* **76**, 1001 (1954).
168 Scott, G. A. *Proc. Roy. Soc. (London) A*, **125**, 330 (1929).
169 Silberberg, I. H., J. J. McKetta and K. A. Kobe. *J. Chem. Engng Data,* **4**, 323 (1959).
170 Smith, L. B. and F. G. Keyes. *Proc. Amer. Acad. Arts. Sci.* **69**, 285 (1934).
171 Smith, L. B. and R. S. Taylor. *J. Amer. Chem. Soc.* **45**, 2107 (1923); **48**, 3122 (1926).
172 Solbrig, C. W. and R. T. Ellington. *Chem. Engng Progr.* **59**, 127 (1963).
173 Straty, G. C. and R. Prydz. *J. Sci. Instrum.* **41**, 1223 (1970).
174 Streett, W. B. *J. Chem. Engng Data,* **16**, 289 (1971); *J. Chem. Thermodynamics,* **5**, 633 (1973).
175 Sullivan, J. A. 'p–V–T data for neon and helium at temperatures from 70 K and pressures to 690 atm.' Ph.D. Thesis. University of Michigan, 146 pp. (1966).
176 Sullivan, J. A. and R. E. Sonntag. *Advanc. Cryogenic Engng,* **12**, 706 (1967).
177 Tanner, C. C. and I. Masson. *Proc. Roy. Soc. (London) A*, **126**, 268 (1930).
178 *Temperature*, Edited by C. M. Herzfeld, Part 1, Sections VI and VII. Reinhold: New York (1962).
179 'The International Practical Temperature Scale of 1968'. *Metrology,* **5**, 35 (1969).
180 Thomas, W. and M. Zander. *Z. Angew. Phys.* **20**, 417 (1966).
181 Townend, D. T. A. and L. A. Bhatt. *Proc. Roy. Soc. (London) A*, **134**, 502 (1931).
182 Trappeniers, N. J., T. Wassenaar and G. J. Wolkers. *Physica,* **32**, 1503 (1966).
183 Tsederberg, H. B., V. H. Popov and V. R. Petrov. *Teploenergetika,* **6**, 87 (1972).
184 Tsiklis, D. S. *Dokl. Akad. Nauk. SSSR,* **79**, 289 (1951).
185 Tsiklis, D. S. *Handbook of Techniques in High Pressure Research and Engineering,* Plenum: New York (1968).
186 Tsiklis, D. S. and A. I. Kulikova. *Russ. J. Phys. Chem.* **39**, 928 (1965),
187 Tsiklis, D. S. and E. V. Polyakov. *Dokl. Akad. Nauk SSSR,* **17**, 62 (1967).
188 Vasilesco, V. *Ann. Phys., Paris,* **20**, 137 (1945).
189 Vaughan, W. E. and N. R. Graves. *Industr. Engng Chem.* **32**, 1252 (1940).
190 Verlet, L. *Cours Fac. Sci. Orsay, France* (November 1970).
191 Verlet, L. and J. J. Weiss. *Phys. Rev.* **A5**, 939 (1972).
192 Verschoyle, T. T. H. *Proc. Roy. Soc. (London) A*, **111**, 552 (1926).
193 Vogl, W. F. and K. R. Hall. *Physica,* **59**, 529 (1972).
194 Vohra, S. P., T. L. Kang, K. A. Kobe and J. J. McKetta. *J. Chem. Engng Data,* **7**, 150 (1962).
195 Vohra, S. P. and K. A. Kobe. *J. Chem. Engng Data,* **4**, 329 (1959).
196 Walters, R. I., J. H. Tracht, E. B. Weinberger and R. I. Rodgers. *Chem. Engng Progr.* **50**, 511 (1954).
197 Waxman, M. and J. R. Hastings. *J. Res. Nat. Bur. Stand.* **75c**, 165 (1971).
198 Waxman, M., J. R. Hastings and W. T. Chen. *Proceedings* of the Fifth Symposium on Thermophysical Properties, pp 248–261. Edited by C. F. Bonilla. American Society of Mechanical Engineers: New York (1970).
199 Weber, L. A. *J. Res. Nat. Bur. Stand.* **74A**, 93 (1970).
200 Weier, R. D., I. Wynn-Jones, J. S. Rowlinson and G. Saville. *Trans. Faraday Soc.* **63**, 1320 (1967).
201 Wiebe, R. and V. L. Gaddy. *J. Amer. Chem. Soc.* **60**, 2300 (1938).
202 Wiebe, R., V. L. Gaddy and C. Heins. *J. Amer. Chem. Soc.* **53**, 1721 (1931).
203 Witkowski, A. W. *Phil. Mag.* **41**, 288 (1896).
204 Young, S. *Proc. Phys. Soc., London,* **13**, 602 (1894).
205 Zwanzig, R. W. *J. Chem. Phys.* **22**, 1420 (1954).

CHAPTER 8

p–V–T Relationships in Gases at High Pressures and High Temperatures

P. MALBRUNOT

Laboratoire de Physique Appliquée*, Université des Sciences et Techniques du Languedoc, Monptellier, France

Contents

I.	Introduction	383
II.	Techniques of Heating the Gas under Study	384
	1. External Heating	384
	2. Internal Heating	385
III.	Various Methods of Measurement	389
IV.	Constant Temperature Methods	389
	1. Variable Volume Techniques	389
	A. Injection Method	389
	B. Metallic Bellows Method	391
	2. Methods with Weighing Technique and Constant Volume Piezometer	394
	3. Miscellaneous Techniques	397
V.	Constant Pressure Methods	399
VI.	Constant Volume Methods	400
VII.	Critical Comparison of the Various Measurement Methods	405
	1. Measurement Techniques	405
	2. Measured Quantities	405
VIII.	Results	407
IX.	Equations of State	409
	1. Law of Corresponding States	409
	2. Empirical Equations	409
	3. Equations Derived from Statistical Mechanics	412
	A. Virial Expansion	412
	B. Equations of the Perturbation Type	413
	C. Equations Deduced from Percus–Yevick Theory	414
	4. Tables of Thermodynamic Properties	416
X.	References	416

I. Introduction

A sufficient reason for measuring the equilibrium parameters p, V and T for gases at simultaneous high pressure and high temperature is simply to determine the phenomenological behavior of the gases under these unusual conditions. But in this range, as at lower pressure and temperature, the interest of pVT relationships is of primary importance on the fundamental level (for thermodynamics and for all physical or chemical properties that are

---

* Associé au Centre National de la Recherche Scientifique.

related to the gas density) and on the theoretical level (equilibrium statistical mechanics).

The application of these data on a practical basis has been well developed with the growth of the chemistry of synthesis and polymerization in which high pressure and temperature occur more and more often. These new applications have been added to the more traditional ones such as motors, heat exchangers, and ballistics.

It would be advantageous if the theory were sufficiently well developed to make the carrying out of pVT measurements under these very severe physical conditions unnecessary. Unfortunately, in spite of numerous recent developments[27, 32, 67, 85, 86], no general theory is capable of providing sufficiently precise and accurate data to be useful beyond the experimental range that has already been explored. The only recourse then is to make numerous and accurate experimental measurements.

The most satisfactory definitions of the 'high temperature' range for gases are evidently those which are based on physical criteria, for instance, the relative location of the critical point, the relative location of the minimum of the intermolecular potential, and the slight or even negligible importance of quantum effects. But such definitions correspond to temperature limits that are different for each gas so that they cannot be retained on the experimental level (at room temperature, helium is hot, argon moderately hot, xenon is near its critical temperature, water is liquid). Hence, from an experimental point of view, we will consider high temperatures to be those for which the method of heating can influence the choice of the method of measurement. Therefore, we will deal with the question of heating techniques before we examine the proper methods of measurement. Because thermostats become more difficult to build above 200°C, we will consider this temperature to be one lower limit of what we define as high temperature.

II. Techniques of Heating the Gas under Study

Heating of the gas sample may be accomplished by either of two possible processes: (1) external heating, the heating of the high pressure apparatus itself; or (2) internal heating, the heating of the gas by a heater located inside the high pressure chamber.

1. External Heating

This method does not involve any difficulty from a purely thermal point of view because it permits the use of a standard furnace. This method also presents important advantages in the determination of the temperature of the gas sample. Because of its large metallic mass, the high pressure enclosure constitutes both a thermal reservoir suitable for the stability of the temperature reached and a conducting mass suitable for the uniformity of this temperature. Thus very accurate measurement devices, such as platinum resistance thermometers or thermocouples, can be numerous (for temperature homogeneity checking) and can be used without any loss of accuracy due to the high pressure.

It is from a mechanical high pressure point of view that this method presents disadvantages. One difficulty is obtaining the necessary tightness of the high pressure enclosure since a non-metallic seal is destroyed at high temperatures. A. Leycuras and Y. Leycuras[63], in an extensive review of heating processes under pressure, have described the most important approaches used so far: metal–metal contact, squeezed metallic seal, tubular metallic O-ring, or standard seal put in a low temperature region. Another difficulty appears in the pVT measurements. When the wall supports a high internal pressure, the combined effect of temperature and pressure on the container volume is not accurately known. But it is mainly the limited strength at high temperature and pressure of the metals used for pressure vessels that is the chief disadvantage of this heating method. Because of this, the best alloys (based on molybdenum) can be used only up to 800° to 1 000°C and 4 000 bar for a limited time[24, 108], and the working life of the pressure vessel is restricted to a limited number of cycles.

2. Internal Heating

In this case the furnace is put directly in the compressed gas inside the high pressure vessel. With good thermal isolation and adequate cooling of the wall of the vessel, the wall can be maintained near room temperature. It is then possible to avoid the temperature and pressure limitations of external heating. Therefore, internal heating does not involve any additional difficulty for ordinary high pressure techniques. Moreover, the wall of the vessel that encloses the gas sample is subjected to both internal and external hydrostatic pressure. The corrections for the volume distortion due to the pressure are easy to calculate from the compressibility data of the piezometer wall material[22].

From a thermal point of view this method has two main disadvantages: (1) good thermal isolation of the heater is difficult to obtain, and (2) precise temperature measurements cannot be made because the effects of pressure on the temperature measuring devices (thermocouples or platinum resistance thermometers) are not well known.

Heat exchange in a compressed gas is very different from what it is at atmospheric pressure. Not only does the thermal conductivity of the gas increase with pressure but, more important, streams of natural convection appear. A. Leycuras and Y. Leycuras[63] have described all the approaches which have been applied to this problem. Their conclusions agree with those of Malbrunot, Meunier and Vidal[68] on this subject. The type of isolation suggested by Stepanov[92] has proved to be the most satisfactory as may be deduced from simple physical considerations. This isolation is formed by a series of cylindrical and coaxial metallic screens having a thin wall (0.2 mm). These screens are made from a metal with low thermal conductivity and are separated by a uniform thin space (0.1 mm). The gas layers so realized are thin enough to eliminate all convection between the screens and thereby constitute good thermal isolation since the thermal conductivity of most gases at high pressure and temperature is less than that of the best isolating ceramics*. The screens not only isolate the gas layers but play their usual role in the prevention of radiation losses. The only difficulty with this arrangement is the achievement of uniform spaces of the order of 0.1 mm between the metallic screens. The isolation suggested by Malbrunot et al.[68] gets around this difficult problem in a simple manner. It is achieved with a thin metallic sheet (0.05 mm), rolled in a spiral, with a very small space between the successive layers provided by small bossing obtained by stamping of the metallic sheet†. The difference in isolation between this spiral arrangement with contact points and Stepanov's device is small. This type of arrangement provides two or three times greater thermal efficiency than the best previous insulation.

The increase of the thermal conductivity of the gas with pressure has a smaller effect on the heat losses than the convection does. However, this increase must be taken into account in furnaces operating under pressure

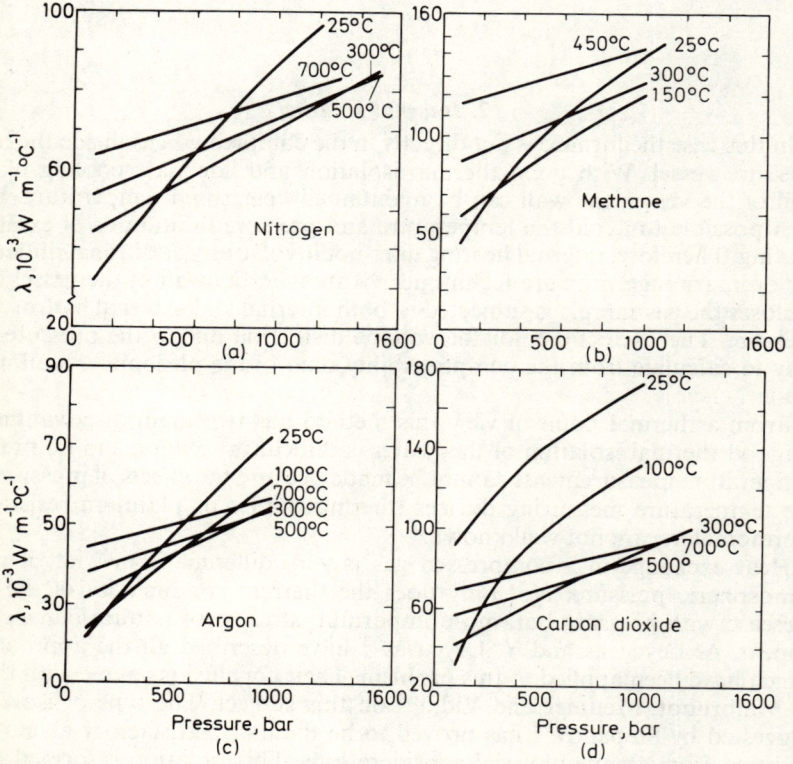

Figure 1. Variation with pressure of the thermal conductivity of some compressed gases.

---

* For instance, the thermal conductivity of nitrogen at 5000 bar and 1000°C is about 0.5 W/m . K^{-1}, while that of fused alumina at the same pressure and temperature is 5 W/m . K^{-1}.

† Satisfactory results have been obtained with a tungsten sheet which allows use of this arrangement at very high temperature.

when optimum conditions are required as in the case of pVT measurements. In fact, when almost all convective transfer has been eliminated by a system similar to that previously described, the thermal conductivity occurs in a way that affects the temperature uniformity of the heated volume. This cannot be neglected. When the gases are greatly compressed, their thermal conductivity increases with pressure more rapidly at moderate temperature than at high temperature, as the curves of *Figure 1*, taken from Johannin[49], Le Neindre et al.[59, 60], and Tufeu et al.[94] show. Therefore, at high pressure there is a large increase in thermal losses in the cooler regions, especially in the external part of the furnace and in the wall of the enclosure. Hence, the operation of a furnace under pressure is very different from the operation of the same furnace at atmospheric pressure. For instance, at 1 500°C the power is three times larger at 5 000 bar than at 1 bar where the insulation described previously minimizes the losses by convection. Consequently, notable temperature gradients exist inside the heated volume, i.e. along the measuring cell. In fact, in these furnaces, which generally have a cylindrical geometry, no heating is provided along the surfaces having a plane perpendicular to the cylindrical axis. This causes a temperature decrease at the ends of the heated zone. Careful insulation of the ends of the cells is necessary, but often it is not sufficient and guard arrangements are necessary. The best way to do this is to use closures with a winding which provide heat in the axial direction, but often they are difficult to use because of the numerous fittings that enter the furnace (thermocouples, capillary tubing, etc.). Then the approach of compensating windings coaxial to the main winding remains. This approach is always preferable to increasing the length of the main winding. This kind of compensation requires large amounts of input energy since the emitting directions are perpendicular to the direction of thermal losses.

To conclude, let us say that the size and the position of the resistance thermometers or the thermocouples in relation to the sample are very important when we want to make precise measurements. With a thermocouple, for instance, notable losses can occur through the thermocouple wires when they are too large in diameter or are mounted in their insulators with too large a clearance. A temperature difference can be introduced between the thermocouple or the resistance thermometer and the sample when they are not in close contact with each other, since even a very thin compressed gas layer is a good thermal insulator.

The use of a resistance thermometer under pressure is avoided as a consequence of certain difficulties that involve a loss of its accuracy. In the first place, its size, which must be large enough to keep the winding free, is in contradiction with the previous criteria. (A small coated probe cannot be used.) Furthermore, the insulation at high temperature and the necessity to provide four leads, i.e. four tight entering wires, for each probe is a major difficulty. Finally, this probe has the disadvantages of integrating the measured temperature over its volume and of being sensitive to the influence of the pressure on the gas inclusions and to the direct effect of the pressure on the resistivity of its material. This last effect, equivalent to about 0.5°C for 1 kbar, varies with the temperature and is not known with great accuracy.

The chromel–alumel and platinum/platinum–rhodium thermocouples are the most frequently used. The study of the effect of pressure has been mainly carried out for heating in solid media[25, 35, 40, 102] at pressures of several tens of kilobars. This effect is important and must be taken into account for experiments conducted under these conditions. In a fluid compressed at several kilobars the effect is so small that correction is justified only in very accurate measurements such as pVT measurements. The chromel–alumel thermocouple shows a small effect of pressure, Figure 2, but

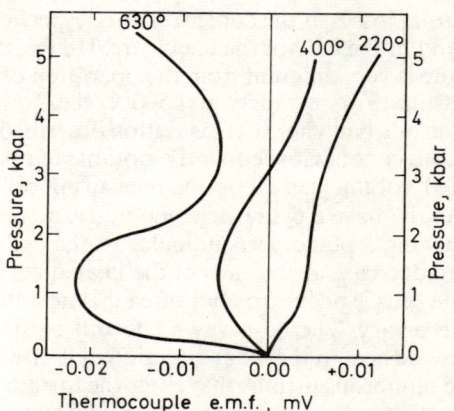

Figure 2. Effect of pressure on e.m.f. of the chromel–alumel thermocouple (from ref. 20).

the variations of this effect with temperature and pressure are relatively complex and not always well defined. Moreover, this thermocouple is not of the highest degree of reproducibility[57]. Thus, for accurate measurement it is preferable to use the noble metal thermocouples such as platinum/platinum–10 per cent rhodium. The effect at high temperature of a pressure of several kilobars on this thermocouple has been measured by Bell et al.[19] up to 400°C and in the neighborhood of 1 000°C by Lazarus et al.[57] and by Lallemand et al.[56]. Resulting from these measurements, it is found that the correction on the e.m.f. is a decrease which follows a slow rise from 0.2°C/kbar at 100°C to 0.6°C/kbar at 1 000°C. It must be said that the comparison of these last data with those obtained in solid media shows good agreement with the results of Getting et al.[35] but a notable discrepancy with those of Hanneman et al.[40] and of Wentorf[102] (their correction being of the order of 1°C/kbar at 1 000°C). So, other measurements using different methods will therefore be needed to decide between these discrepancies in order to confirm the final magnitude of this correction.

Concerning the other thermocouples, the reports are scarcer. Hanneman and Strong[40] have published some limited results on the platinum/platinum –13 per cent rhodium and iron–constantan thermocouples; we do not find anything about the most refractory thermocouples made of iridium, rhenium or tungsten.

The optical pyrometer, which is interesting at high temperature, has never been used in pVT measurements under pressure. This is because the high temperature ranges up to now did not justify it and also because it is difficult and expensive to use when great accuracy is needed.

III. Various Methods of Measurement

In general the procedures described in the previous section are found suitable for the determination of pVT parameters. In presenting particular devices applied to the measurement of high pressure and temperature, we will classify the methods of measurement in three principal sets:

(i) constant temperature measurements,
(ii) constant pressure measurements,
(iii) constant volume measurements.

The constant temperature methods consist of the compression of a known quantity of gas whose temperature is maintained at a constant value by a thermostat. Either the variation of the volume occupied by the gas, or the variation of the mass of the gas enclosed in a constant volume* is measured. In this way the pV isotherm of the gas is directly determinable. Amagat, Kamerlingh Onnes, Bridgman, and Michels have used this method.

In the case of measurements at constant pressure, the temperature variation of the gas occurs as either the volume or the mass is changed in order to keep the pressure constant. Thus, the TV isobars of the gas are determined.

Finally, the constant volume methods consist of temperature and pressure measurements of a fixed quantity of gas maintained in a constant volume. In this case the pT isochores are directly determined.

In describing in the following sections the apparatus corresponding to the different methods, we will limit ourselves to the more recent versions and to those having the best capabilities in both the extent and accuracy of the measurements. Saurel[88] published some years ago a complete review of all the arrangements used since the work of Boyle and Mariotte. Moreover, let us emphasize that several types of apparatus described in the following sections can be used without modification to make measurements according to more than one of the different procedures that have been classified previously.

A summary of experimental pVT data of gases at high pressure and temperature is presented in section VIII. Readers are referred to this section for the different results obtained with these devices. We will compare these different devices in section VII only after we have first described them in order to emphasize more easily their advantages or deficiencies.

IV. Constant Temperature Methods

1. Variable Volume Techniques

A. Injection Method

This method is so called because it consists of the measurement of the

---

* Since Amagat, it is customary to call the constant or variable volume container of the gas sample a piezometer. We will frequently use this term in the following text.

volume of a fluid that must be put into a previously heated enclosure to reach some desired pressure.

The arrangement, which has permitted Kennedy[53] to study steam and carbon dioxide up to 1000°C and 800 bar, is shown in *Figure 3**. It was

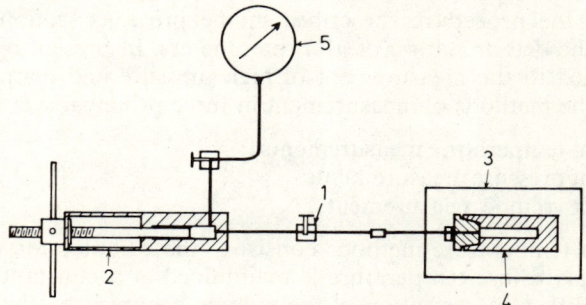

Figure 3. Apparatus of Kennedy: 1 valve, 2 screw-type injector, 3 furnace, 4 pressure vessel (capacity 20 or 100 cm^3), 5 pressure gage.

used in the following manner: The valve (1) was closed and the water was compressed to the required pressure with the screw-type injector (2); then the valve (1) was opened and the quantity of water which was just necessary to reach the specified pressure was injected into the enclosure. Only the enclosure was brought to high temperature by external heating. The maximum temperatures so reached were 600°C at 2500 bar, 900°C at 1400 bar, and 1000°C at 800 bar. The pressure was measured with the pressure gage (5) up to 1400 bar and with a manganin gage above that. The injector was

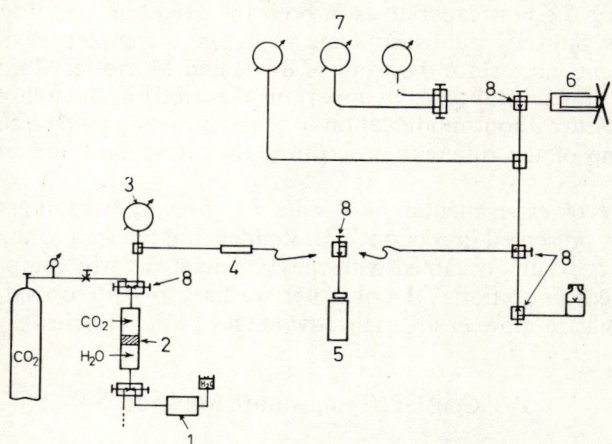

Figure 4. Apparatus of Greenwood: 1 pump, 2 separating cylinder, 3 pressure gage (carbon dioxide pressure measurements), 4 filter, 5 pressure vessel (capacity 56 cm^3), 6 screw-press, 7 pressure gages (water pressure measurements), 8 valves.

---

* To avoid loading the text we specify, if necessary, the size of each apparatus in the legend of the figure that describes it.

previously calibrated at different pressures to determine the mass of water expelled by each turn of the screw. Kennedy considered the precision of his results difficult to evaluate.

The apparatus used by Greenwood[36] to study the mixtures of carbon dioxide and steam involved three different parts (*Figure 4*). Carbon dioxide was injected into one part and water into another; the third was the measuring vessel proper. This vessel was heated externally by a cylindrical furnace that entirely surrounded it. In the first step, the vessel was filled with carbon dioxide by means of the pump and the separator; any trace of water was eliminated with a filter of magnesium perchlorate and glass wool. The temperature and the pressure were measured after the filling, but the mass of carbon dioxide was only determined by analysis at the end of the measurements. The carbon dioxide filling device was removed and was replaced by the water filling device. The water was introduced with the calibrated injector in several successive amounts; for each of them the pressure was measured. This allowed the isotherm of the mixture to be obtained as a function of several concentrations of water in the pure carbon dioxide.

Greenwood made measurements from 0 to 500 bar and from 450° to 800°C. Each part of his apparatus was calibrated separately. He estimated the error in the temperature to be 1°C (about 0.15 per cent), in the pressure to be 0.2 per cent, and in the quantities measured by the injector to be less than 0.1 per cent. The most important errors were those in the volume of the vessel (about 0.4 per cent) and in the mass of carbon dioxide (0.5 per cent). These numbers agree with the scatter of the experimental points with reference to the fitted values obtained by computer. They are also of the size of the deviation between Greenwood's data and the data of Kennedy and Holser[54] obtained with the previous arrangement. This fact verifies the consistency of the injection method and gives an idea of its precision.

B. METALLIC BELLOWS METHOD

Metallic bellows of high quality have allowed the production of volumetric devices that free one from the difficulties and limitations imposed by mercury in the former devices.

The apparatus used by Babb et al.[6] is shown in *Figure 5*. The volume of the

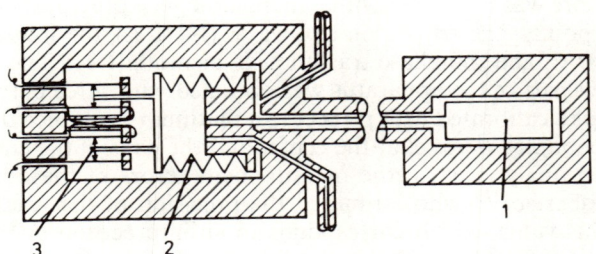

Figure 5. Apparatus of Babb: 1 constant volume container (10 cm³), 2 bellows container (1.9 cm o.d. with a maximum volume change of about 4.3 cm³) 3 device for detecting the bellows displacement.

gas in the experiment was distributed between a constant volume container brought to high temperature (up to 400°C) by external heating and a bellows container kept at constant temperature, which was taken as the reference temperature. The displacement of this bellows was detected electrically by a variable resistance circuit. A capillary tubing provided the connection between the two parts of the apparatus. By their arrangement the two containers were entirely subject to the same hydrostatic pressure; this simplified both the calibrations and the corrections.

At the working temperature chosen, it was possible, by changing the pressure of the confining gas, to change the pressure and thus the density of the gas studied and thereby establish the corresponding isotherm. For each pressure the density of the gas in the heated part, or rather its mass because the volume was constant, was obtained by subtracting from the entire mass the mass of the gas contained in the bellows, which was kept at the reference temperature. The filling of this piezometer was carried out when all the apparatus (bellows and constant volume) was at the reference temperature. Literature data gave the filling density, and a previous calibration of the bellows gave the corresponding volume.

In principle the calibration of the bellows was standard; its motions were transmitted by oil to a mercury column contained in a calibrated capillary, and the displacement of this column was measured with a cathetometer. We must emphasize that in the course of the calibration the bellows were not moved mechanically but by the effect of pressure or vacuum, in order to reproduce the exact conditions of its utilization. Thus, Babb observed a slight effect of non-linearity, of the order of 0.1 per cent, between the displacement and the corresponding volume variation. This effect was probably due to the influence on the lateral wall of the small pressure difference between the inside and outside of the bellows when it was displaced from its equilibrium position.

One of the main parts of this apparatus was the electrical system for detecting the bellows displacement. Metal contacts attached to the bellows moved along nickel–chromium wires placed in a Wheatstone bridge. Outside the high pressure container this circuit was completed by a potentiometric circuit. The determination of the position was very accurate (about 3 µ). It was not disturbed by the contact resistance between the metal contacts and the wires.

The pressure was measured with a manganin gage previously calibrated at three fixed points: the triple point of water (I–III–L), the freezing point of mercury at 0°C, and the phase transition of bismuth (I, II). The temperature of the heated part of the apparatus was obtained with three chromel–alumel thermocouples calibrated with respect to a platinum resistance thermometer. The thermocouples were set at the center and the ends of the enclosure.

A detailed analysis of all the measurement errors of the apparatus was made by Robertson[79] who estimated the precision of the results to be 0.2 per cent. This value, which corresponds to an imprecision in the measured temperatures of 0.1°C, might better be achieved with a platinum resistance thermometer which is more accurate and stable than the chromel–alumel thermocouples.

The device developed by Burnham et al.[26], shown in *Figure 6*, is identical

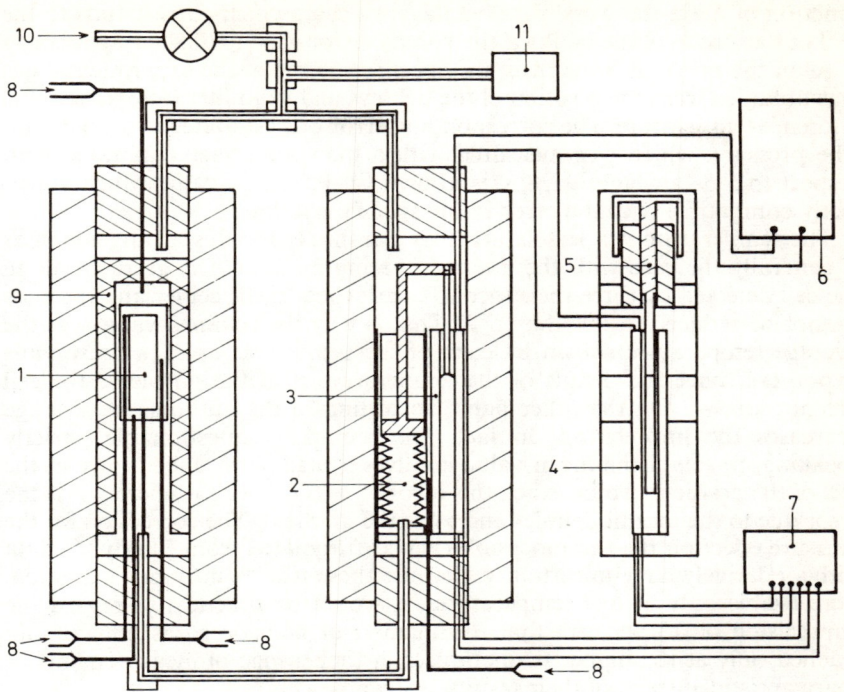

Figure 6. Apparatus of Burnham: 1 p–V–T cell (capacity 15 cm^3), 2 bellows (2.5 cm o.d. with a volume change of 11 cm^3), 3, 4 linear variable differential transformer, 5 micrometer stand, 6 audio oscillator, 7 tuned amplifier and null detector, 8 thermocouple, 9 furnace, 10 high pressure argon, 11 manganin cell.

to Babb's in principle. The most important difference is in the heating system. The system of Burnham *et al.* has an internal furnace that allows conditions to reach 900°C at 9 000 bar instead of 400°C at 10 000 bar. A second difference concerns the detection of the bellows displacement which was achieved in this system with a linear variable differential transformer. An iron core fixed at the free extremity of the bellows (*Figure 6*) was displaced inside two coaxial coils. One coil was supplied with a constant low frequency voltage (2 000 Hz); the voltage across the terminals of the second coil was therefore proportional to the bellows displacement. This voltage was not read directly but was compared outside the high pressure container with another produced by a similar arrangement in which the core was attached to a micrometer screw. Therefore, when the bellows was displaced, the balancing of the two voltages (which was detected with a null-detector), was restored by adjusting the micrometer screw. Burnham *et al.* claimed to detect a displacement of 1 μ. To determine exactly the relationship between the volume variations (ΔV) of the bellows and the position (ΔX) of the bellows, they carried out a series of calibrations in which they weighed the amount of water expelled in a given advance of the bellows. To these results, they fitted a polynomial that gave $\Delta V / \Delta X$ as a function of X. In the worst disadvantageous case, which corresponds to the end of the stroke of the bellows, the volume error was ± 0.004 cm^3 per cm.

As in the previous apparatus, the specific volume of the experimental gas was obtained from the position of the bellows and from literature data at the reference temperature. The corresponding error was estimated to 0.1 per cent. The pressure, which was measured with a manganin gage calibrated with respect to a deadweight gage, was known to ± 5 bar, a value which is perfectly compatible with the error in the specific volume.

The temperature is a less satisfactory parameter for this arrangement, as is generally the case with the internal heating method. The gradient along the cell detected by three thermocouples attached to the center and the ends cannot be reduced to less than 5°C. That is why the retained value was the average temperature shown by each of these thermocouples, a reasonable hypothesis since the details of the temperature distribution inside the cell are not known. On the other hand, according to the authors, the average decreases the imprecision. In fact, this remark is questionable; strictly speaking, to retain the mean value involves a maximum uncertainty in the size of the gradient. To be exact, this error is $\pm(\Delta T + \frac{1}{2}\delta T)$, where ΔT is the error due to the thermocouples and δT is the gradient. The correction for the pressure effect on the thermocouples was extrapolated from Bundy's[25] data with a relatively large uncertainty. Because these results, in Bundy's opinion, were usable only at low temperatures, it would be practical to regard the imprecision of 0.3 per cent that Burnham *et al.* considered as a maximum reached only at the higher temperatures in their range of measurements as characteristic, instead of their results as a whole.

2. Methods with Weighing Technique and Constant Volume Piezometer

Benedict[20] in his work on nitrogen from low temperatures to 200°C was the first to use this technique. Vilevich, Vereshchagin and Kalashnikov[97] improved this type of apparatus and reached 400°C and 3 500 bar. The piezometer was a small, high-pressure bulb provided with a miniature valve. This piezometer was hung from the plate of an accurate balance inside a furnace (*Figure* 7). At the required temperature it was filled to the chosen pressure; after the pressure and temperature were stabilized, their respective values were measured with the deadweight gage and chromel–alumel thermocouples. The volume of the piezometer had been previously determined at one bar and 25°C by weighing the bulb when it was filled with carbon tetrachloride. As stated in section II, one of the disadvantages of this method is the lack of data on the piezometer distortion that occurs when the piezometer is both heated and subjected to high internal pressure.

To resolve this difficulty, Vilevich *et al.* proceeded to a direct dilatometric measurement of the volume change of the outside of the piezometer under the corresponding experimental conditions by the use of a simple rod dilatometer. Moreover, they assumed that the change in internal volume was the same as the change in external volume. This approximation is reasonable when the wall of the piezometer is thin. Therefore, the use of their method is limited to pressures that can be sustained by the thin-walled container. They found this change of volume with pressure was larger then they expected, but, on the

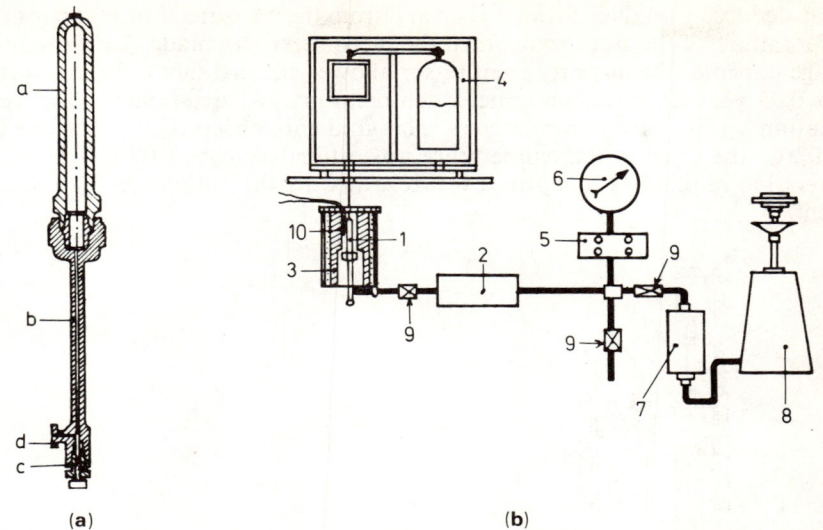

Figure 7. Apparatus of Vilevich: **(a)** piezometer (capacity 5 cm^3): a capsule, b capillary connection, c needle valve, d tube connection; **(b)** schematic diagram of the apparatus: 1 piezometer, 2 oil separator, 3 copper block for equalizing the temperature, 4 precision deadweight gage, 5 comb, 6 precision Bourdon gage, 7 separator with a floating piston, 8 piston manometer, 9 high pressure valve, 10 thermocouple.

other hand, it stayed at the same order of magnitude at each temperature in the experimental range. Using their own experimental corrections, they deduced their pVT values with a final precision of 0.1 per cent.

To measure the compressibility of air up to 600°C and 700 bar, Vukalovich et al.[101] built an apparatus of great applicability that allows absolute measurements. A mercury compressor (*Figure 8*) operating at room temperature was connected to a constant volume piezometer set in a saltpeter thermostatic bath which could reach 600°C. The connection between the compressor and the piezometer accomplished through an optical window cell in which the mercury level in the joining capillary could be observed. The gas could be forced back entirely into the piezometer. A set of three valves allowed one to (i) isolate the piezometer. (ii) empty the system, and (iii) join the piezometer to the gas bottle either for the filling or the recovery operation. Two types of gas cylinders were used. One of 500 cm^3 volume for the filling had a sufficient wall thickness to support 300 bar. This cylinder could contain up to 100 g of air, and its weight (900 g) allowed the use of a balance precise to 5 mg. The cylinder designed for the recovery was thinner, smaller (150 cm^3), and lighter (150 g); it could contain only 20 g of gas, but allowed the use of a very precise balance (0.1 mg). Then, by successive weighings of these cylinders, the mass of the gas sample was directly determinable when the apparatus was either filled or empty.

The piezometer volume measured at one bar and 20°C was known to 0.02 per cent. The correction as a function of the temperature and pressure was deduced from Nejmark's[73] data and from the measured linear expansion of a sample of the steel from which the piezometer was made. A deadweight gage joined to the mercury compressor allowed the pressure to be measured to 0.05 per cent. The piezometer temperature was determined with two platinum resistance thermometers. The gradients observed did not exceed 0.03°C; the error in the temperature was estimated to be 0.07°C. The final precision of the pVT quantities was regarded by the authors as 0.1–0.2 per cent.

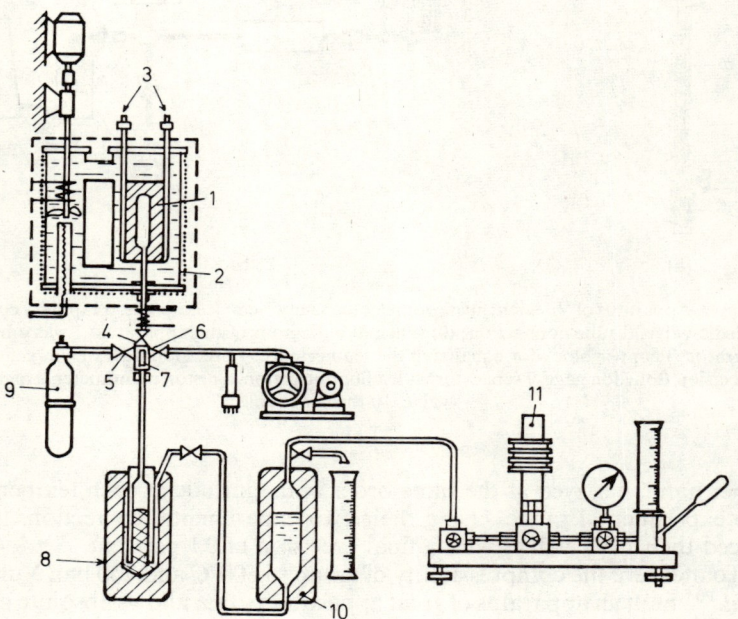

Figure 8. Apparatus of Vukalovich: 1 piezometer (capacity 334 cm^3), 2 thermostatic bath, 3 platinum resistance thermometers, 4, 5, 6 valves, 7 optical window cell, 8 mercury siphon water–gas separator, 9 bottle for weighing the gas under test, 10 water–oil separator, 11 deadweight pressure gage.

The measurement technique is deduced easily from the apparatus description. After the apparatus was initially emptied, it was filled while the mercury was kept in the low position. Then the isolating valve was closed, and the gas was compressed until the mercury reached the reference level seen through the optical window. Temperature and pressure were measured after their stability had been verified. Other points of the corresponding isotherm were obtained by emptying the piezometer into the recovery cylinder several times. At each step the mercury was readjusted to the reference level by changing the pressure. The cylinders were weighed after the filling and each of the successive recoveries.

This apparatus was also used for constant volume measurements. After the filling and the first measurement of pressure and temperature, the piezometer temperature was raised and the pressure was continuously adjusted so as to keep the mercury at the reference level. Setting the temperature at several values, a series of points on a pT isochore of the sample gas was obtained. Vukalovich et al. made such constant volume measurements to check the consistency of their constant temperature results.

3. Miscellaneous Techniques

Tsiklis and Polyakov[95] determined the pVT behavior of nitrogen up to 10 000 bar and 400°C. In their measurements they avoided the correction for the conjugate effect of temperature and pressure on the size of the high pressure container. They used a displacement method which was based on the presence or absence of a steel cylinder entirely subject to the pressure. Their measurements were performed in the following way (*Figure 9*). The amount of gas contained in the volume V of the enclosure at determined conditions of

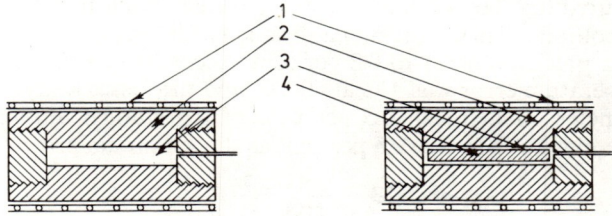

Figure 9. Apparatus of Tsiklis: 1 heating winding, 2 high pressure vessel, 3 volume occupied by gas under study, 4 metallic plug (volume 10 cm^3).

temperature and pressure was measured with an expansion device. Then a second measurement was carried out under the same conditions after the steel cylinder of calibrated volume V was set inside the enclosure. By difference the quantity of gas corresponding to the volume v of the steel cylinder, whose strain conditions were known, was deduced. In order to obtain several experimental points, the enclosure was emptied by steps, and for each of these steps, the corresponding pressure and amount of gas were measured; obviously, the same steps were repeated with the cylinder inside the enclosure.

The mass of the sample gas was measured by the standard technique of expansion (thermostatically controlled vessels of known volume and a mercury manometer). The pressure was measured with a manganin gage set in a separator filled with pentane to avoid contact with the gas (the adsorption of nitrogen modifies the properties of manganin). The gage had been calibrated only with reference to one point: the freezing of mercury at 0°C. That may be insufficient for accurate measurement. Boren et al.[23] and Atanov[1] found that such a linear approximation introduced fluctuating errors with a maximum of 18 bar at 3 500 bar and 70 bar at 10 000 bar. Thus the precision of ± 20 bar claimed by the authors appears over-valued. The volume of the steel cylinder was measured under normal conditions of temperature and pressure, and the corrections due to the effect of higher temperature and pressure were calculated from the literature data on the thermal expansion and compressibility of the steel used. The final precision of the measured molar volume was estimated to be 0.3 per cent. The enclosure was heated with an electric furnace whose temperature was measured to approximately 0.5°C by four chromel–alumel thermocouples distributed to measure the possible gradients.

The apparatus used by Presnall[77] to conduct experiments with hydrogen up to 1 800 bar and 600°C was very simple to operate quickly in order to leave the sample gas in the piezometer for as short a time as possible. The losses by diffusion, which are always very important with this gas at high temperature, were thus greatly reduced. The piezometer, which was a constant volume reservoir, was externally heated by an electric furnace. Each amount of gas introduced at given temperature and pressure was recovered by a system of vessels and a mercury manometer. The temperature was determined by a thermocouple, the container volume was corrected according to Kennedy's data[53], the pressure was measured by two Bourdon gages (manufactured by Heise) that were separated from the hydrogen by a mercury column. These gages, calibrated with respect to a deadweight gage, had a precision of 0.1 to 0.2 per cent. According to Presnall, the maximum temperature error was 4°C at 600°C, and the losses to hydrogen across the piezometer could reach 0.3 per cent of the studied amount; it seems reasonable to take the value 0.5 per cent as the maximum error of his results.

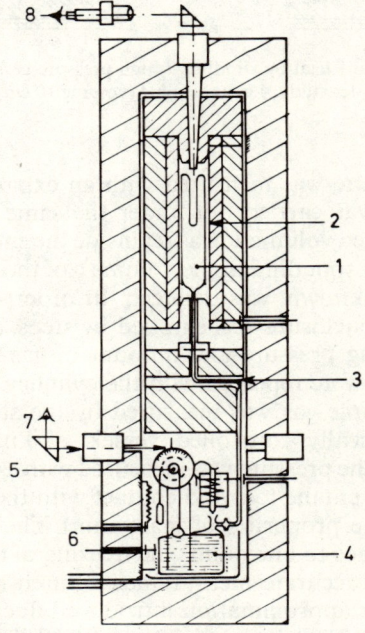

Figure 10. Apparatus of Basset: 1 piezometer (capacity 2.5 cm³), 2 furnace, 3 capillary tube, 4, 5 mercury volumetric device with a reading dial, 6 platinum resistance thermometers (to check the temperature of the mercury tank), 7 reading device with optical pressure windows, 8 optical pyrometer.

V. Constant Pressure Methods

The principle of the apparatus built by J. and J. Basset[12] consisted of the measurement of the expansion of a fluid sample enclosed in a container heated by an external furnace. The pressure was kept constant during the temperature variations. The method of adjusting the gas volume is understood easily by examination of *Figure 10*. The reservoir was filled when the valve was opened. The valve was then closed and the reservoir was heated. The amount of gas contained in the temperature controlled lower tank was read on the dial. The precision derived from an analysis of various causes of error was of the order of 0.5 per cent at 5 000 bar and 1 000°C.

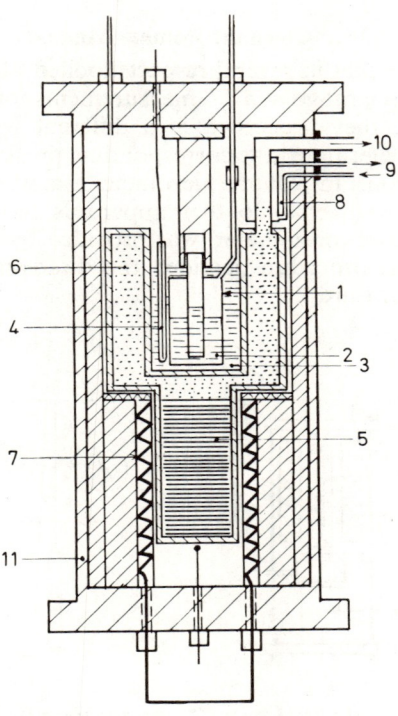

Figure 11. Apparatus of Luft: 1 measuring vessel (4 cm o.d., 12 cm length), 2 sealing liquid, 3 heat transfer liquid, 4 thermocouple, 5, 6 mercury boiler–condenser unit, 7 heating element, 8 auxiliary water condenser, 9 cooling water, 10 to vacuum, 11 high pressure vessel.

It should be noted that both Babb's and Burnham's apparatus could be used in the same way as Basset's. The bellows would take the place of the mercury siphon and its plunger in the determination of the amount of gas that passed from the piezometer to the cold zone.

Luft[64] recommended a very different procedure for making the constant pressure measurement. Into the piezometer (*Figure 11*) which had initially been filled and brought to the desired temperature and pressure, a supplementary amount of gas having a known mass was injected. The volume corresponding to this amount of gas was determined by the difference in the level of the piezometer liquid separator. This operation was repeated at different temperatures while the pressure was maintained constant. A differential pressure measurement between the piezometer and the enclosure was used to calculate the difference in the level of the separating liquid. This procedure eliminated the need for an internal level locating device. In spite of this advantage the method was not very precise since the author estimated that a five per cent error in the final value of the density could exist. However, Luft concluded that this constant pressure method should be developed and improved because of its simplicity and its ease of use.

VI. Constant Volume Methods

With an apparatus that in several respects looked like Kennedy's Franck et al.[33, 66] extended the range of pVT measurements for water to 800°C and 6000 bar. However, they proceeded in a different way. For a measured amount of gas they determined the corresponding pressure for each temperature reached by the constant volume piezometer that was heated in an electric furnace. In the improved version of their apparatus, the closure of the heated high pressure container (piezometer) was provided by a piston driven by a hydraulic press. This procedure avoided the disadvantage caused by the thread at high temperatures.

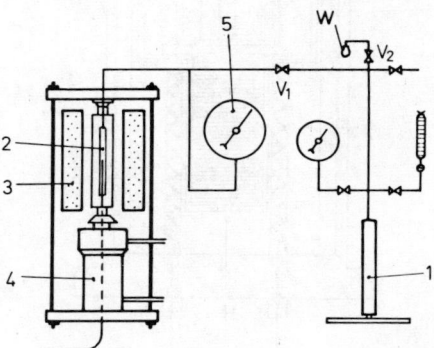

Figure 12. Apparatus of Franck: 1 screw-press, 2 autoclave (capacity 32 and 50 cm³), 3 furnace, 4 hydraulic press, 5 Bourdon pressure gage, V_1, V_2 valves, W recovery cylinder.

The filling was made with an injector (*Figure 12*) in the following way. After all the apparatus was evacuated and outgassed, the valve V_1 was closed. The injector was filled with water and its pressure was brought to the required value. The valve V_1 was opened slightly, and the water compressed by the injector was forced into the piezometer. When the injector was at the end of its stroke, V_1 was closed and the injector was drawn back to its initial position. The injector was then used to force water at the same pressure as before into the cylinder W through the valve V_2. This amount of water was determined by the weight of W. The latter processes were a calibration of the injector during its operation under pressure. The whole procedure was repeated as many times as necessary to fill the piezometer. As the pump had been calibrated during the filling, it was easy to deduce the amount of water enclosed in the piezometer.

The temperature was measured to within about 0.2°C at 200°C and 1°C at 800°C with nickel–chromium/nickel thermocouples previously calibrated with reference to the melting point of zinc, tin, antimony and silver. The authors specified that though the thermocouples were inside the high pressure enclosure, they did not need to correct for the effect of pressure; this effect was negligible, even zero, in their case because the part under pressure was not subject to a significant temperature gradient. A Bourdon gage (trademark 'Heise') which had been compared with a deadweight gage gave the pressure to about 0.1 per cent. After a detailed analysis of errors, the precision of the density was estimated to be 1 per cent.

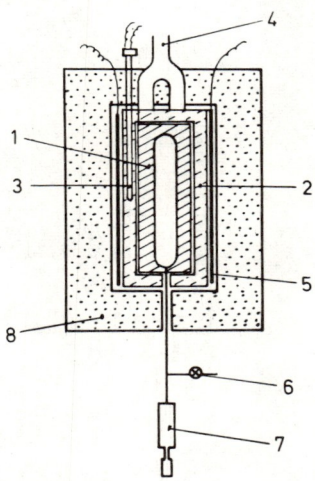

Figure 13. Apparatus of Juza: 1 pressure vessel or piezometer (capacity 100 cm^3), 2 copper block, 3 resistance thermometer, 4 cooling channel, 5 electrical resistance heater, 6 needle valve, 7 pressure gage, 8 thermal insulation.

Figure 13 shows schematically the apparatus of Juza *et al.*[50]. The gas contained in a constant volume piezometer was brought to the desired temperature by external heating, and its pressure was measured with a strain gage. The gage consisted of a tube. The variations in the external volume of this tube, due to the internal pressure that it was sustaining, were registered by the displacement of the mercury level in the tank in which the tube was placed. Electric contact allowed a micrometer with a needle tip to follow the mercury level with precision. Calibration with respect to a deadweight gage had shown that such a gage could be precise to ±20 bar at 4000 bar. This device was located outside the heated zone, of course; it was maintained at 25°C. The block surrounding the piezometer could be either cooled to liquid nitrogen temperature or heated to 600°C by an electrical resistance. Cooling during the filling permitted one to reach a high pressure by thermal compression when the piezometer was heated. Furthermore it was with another auxiliary thermal compressor that the filling was made at the desired pressure. The piezometer temperature was measured with a platinum resistance thermometer to ±0.2°C. The uniformity of the temperature was measured by several thermocouples; they showed no deviations greater than 0.1°C. Juza *et al.* also used the literature data to determine the filling mass and to calculate the correction for the effect of temperature and pressure on the piezometer. They found finally the maximum error of the specific volume to be 0.3 per cent.

For the determination of the pVT properties of sodium, potassium and cesium vapors. Stone *et al.*[93] have used a method whose principle is as simple as the Juza one. Although the pressure range of their measurements (30 bar) is not particularly high, their apparatus must be mentioned in this chapter for its simplicity, its suitability to the highest temperatures and the peculiarity of the substances studied.

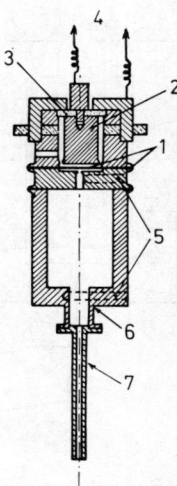

Figure 14. Apparatus of Stone (the volume of the chamber is 57 and 113 cm^3): 1 diaphragm, 2 electrical probe, 3 insulator, 4 to electrical circuit, 5 thermocouple wells, 6 filling port, 7 closure tube.

The apparatus shown in *Figure 14* was a small pot equipped with a pressure-sensing diaphragm and a small tube for filling, evacuating and closing. An electrically insulated probe was used to detect the diaphragm position. To meet the experimental conditions imposed by temperature and containment, the entire cell, except for the mullite insulator, was fabricated from a refractory alloy, columbium–1 per cent zirconium. To provide both protection and a means for external pressurization for diaphragm operation an electric furnace was built within a pressure shell. An Alundum core heater with three molybdenum windings was mounted at the center of the shell and insulated from it with a zirconia bubble.

The procedure for a typical pVT measurement was as follows. A capsule of predetermined volume was filled with vacuum-distilled alkali metal and placed in the apparatus through the filling port. The closure tube was welded on, the apparatus evacuated and heated, and the tube welded off. Then the loaded apparatus with thermocouples installed above and below the chamber was supported inside the thimble of the pressure furnace and equilibrium measurements of temperature and pressure were made over the required range. The furnace pressure was adjusted at each measurement temperature so that the diaphragm was alternately making and breaking electrical contact with the probe. Under these conditions the furnace pressure, observed externally with calibrated Bourdon tube gages, was equal to the pressure of the metal vapor.

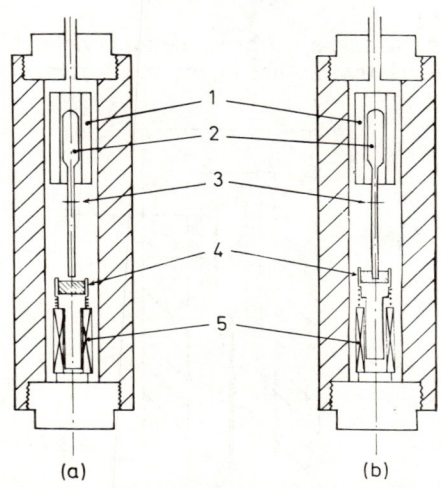

Figure 15. Apparatus of Saurel: 1 furnace, 2 piezometer (capacity 6 cm^3), 3 level mark, 4 mercury cup, 5 electromagnet; (a) filling position, (b) measurement position.

The authors have discussed the principal sources which contribute to the uncertainty of the measured quantities and have concluded that the final error on the compressibility factor is in the range from 0.25 to 0.44 per cent.

Table 1. Precision of the apparatus of ref. 70

Load	10^{-5}	piston	
Piston area	1.5×10^{-4}	gage	
Temperature effect	3×10^{-5}	2×10^{-4}	pressure 0.05%
Hydrostatic correction		10^{-4}	
Capillarity		10^{-4}	
Thermocouples		10^{-3}	temperature
Temperature gradient		1.5×10^{-3}	0.3%
Pressure effect on thermocouples		0.5×10^{-3}	
Corrections		negligible	density
Determination		3×10^{-3}	0.3%

The method used by Saurel and Lecocq has been described several times[58, 87, 98]. Its principle is shown in *Figure 15*. The piezometer, located in an internal furnace, was extended by a capillary tube whose end was immersed in a mobile mercury cup that was moved by an electromagnetic coil. In the filling position the interior of the piezometer was connected to the enclosure, but in the measurement position mercury separated the gas in the piezometer from that in the enclosure. An electric contact allowed detection of the mercury level, which was maintained constant during a cycle of experiments by the adjustment of the gas pressure in the enclosure.

Recently, a new apparatus was built on this principle[70] and gave results up to 5000 bar and 1000°C with a final precision of 0.3 per cent. This precision is analysed in *Table 1*.

More recently, on this last apparatus, the main defects of the method (the use of mercury, the relative nature of the measurements) were eliminated

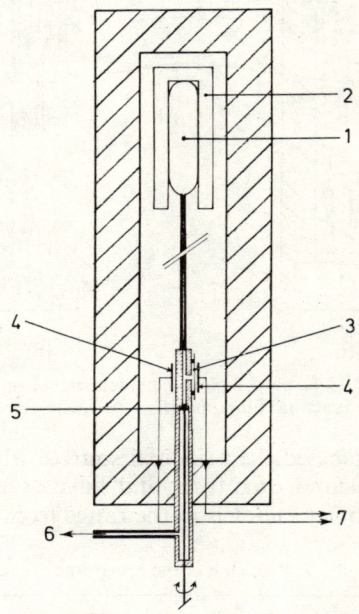

Figure 16. Apparatus of Malbrunot: 1 piezometer (capacity 6 and 10 cm^3), 2 heater, 3 flat metallic diaphragm, 4 measuring electrodes, 5 small needle valve, 6 filling or recovery circuit, 7 to capacitance bridge.

by the improvement shown in *Figure 16*[70]. The piezometer capillary was connected both to a plane metallic diaphragm separator of small volume and to a miniature valve inside the enclosure. The valve was connected to a gas supply that could be different from the one for the enclosure. This fact permitted the study of a gas different from the compressed gas in the enclosure and is an advantage in the case of expensive, dangerous, or very reactive gases. The valve and its electric driving mechanism were designed to permit the piezometer to be evacuated up to the valve stem, whatever the pressure in the enclosure might be. It was thus possible to recover the gas in the experiment and to determine its mass. The method is therefore absolute and independent of the literature data.

VII. Critical Comparison of the Various Measurement Methods

1. Measurement Techniques

The fact that most of the apparatus corresponds to the constant temperature method can be explained easily. In the first place the interest in obtaining isotherms directly is obvious because they give immediate access to the compressibility, a quantity particularly important for a compressed gas. Furthermore these measurements are easier and therefore faster to perform. Pressure changes in this type of measurement are obtained and stabilized more rapidly than temperature changes. This is especially true for the piezometer for which the thermal inertia is relatively important, particularly in the external heating method.

The constant pressure measurements, inasmuch as they require simultaneous measurement of the temperature and mass variations, are less used, especially as they do not offer the advantage of being more accurate.

The constant volume apparatus in which the pressure is measured by a manganin gage or a strain gage are simpler to use because they do not require extensive manipulations during the pressure measurement; at various temperatures the pressure is simply noted. This facility disappears when the pressure is measured by a deadweight gage because a separator between the gas sample and the fluid of the gage is necessary. Whatever its type, mercury or metallic diaphragm, this separator must have a small and almost constant volume; then the temperature of the piezometer is changed, a new balancing of the pressure is needed. The completion of a measurement cycle thus takes longer and is more demanding.

2. Measured Quantities

Among the corrections to be applied to the experimental volume to take into account the effects of temperature and pressure, those corresponding to hydrostatic corrections are the most precise, especially when the pressure exceeds several kilobars. From this point of view the displacement method of Tsiklis is interesting because his apparatus avoids the complications necessarily caused by the placement of a thin-walled piezometer within a confining gas whose pressure must be the same as that inside the piezometer. This displacement method has, however, a double disadvantage for those using piezometers in a hydrostatic medium: The measurements that require two operations to determine one point are more time-consuming and they are less accurate because the final value is a combination of the results of two successive temperature, pressure, and volume measurements, that is, the standard error is multiplied by two.

At high temperature the diffusion of the sample gas through the piezometer wall is an effect that cannot be avoided and therefore must be studied carefully. In the case where the piezometer is the pressure container, the thickness of the wall slows the process substantially, so that during a cycle of measurement the diffused amount is very small, even negligible. This can be verified easily by comparison of the mass of the sample gas before and after the experiment. When the piezometer is entirely under pressure and the sample gas is the same as the confining gas, the diffusion through the thin wall of the piezometer is extremely limited because of the slight pressure difference between the inside and the outside of the piezometer due only to the balancing error. On the other hand, the study of a gas different from the one in the high pressure vessel can cause more important effects, as the partial pressure is zero on one side and several thousand bars on the other side. In this case, care must be taken both in the choice of the piezometer material according to the existing works[8, 58] and in the conduct of the measurements, particularly with regard to the duration of the cycles which must be as short as possible.

When mercury was used as a separator liquid in the high pressure–high temperature devices, it was generally kept at room temperature (as was the case for the apparatus in references 101 and 87 so that the effect on the measured pressure of its vapor pressure and solubility in the sample gas would remain small. The effect on the measured pressure is also small when the warmer part of the gas is maintained at high temperature and a thermal diffusion effect occurs. For instance, the evaluation of these effects made by Malbrunot[67] from the data of Bergeon[21] for nitrogen at 5000 bar and 1000°C yielded a relative value of the order of 10^{-7} of the sample gas pressure. This result has been confirmed by Haar and Sengers[37] in a recent study of these phenomena in which they have evaluated these effects for mercury temperatures up to 400°C. Even in this extreme case, the influence on the pressure was still small, approximately 0.3 per cent. In conclusion, the use of mercury as a separator liquid does not involve a significant disturbance of the results. It is from a practical point of view that its use is more annoying. Its manipulation under pressure requires great care, and wrong operations generally have serious consequences.

Concerning the methods of determining the mass of the gas sample, the absolute character of the direct weighing method must be emphasized. The actual balances with their automatic handling of weights and their high sensitivity and accuracy, especially for heavy loads, make this measurement easy and quick. For this reason this technique is replacing more and more the expansion into calibrated volumes which is slower and more delicate, more difficult to use, and even more expensive for an order of accuracy comparable to that for the direct weighing method.

Pressure measurements by a deadweight gage are certainly the most desirable when gas pVT measurements are made by a static method. When a deadweight gage cannot be used, a manganin gage can give good results (0.2 per cent to 0.5 per cent) if it is calibrated carefully* and if its reproducibility is frequently verified[3].

---

* The calibration can be made with respect to a deadweight gage or to fixed points (melting point of mercury, phase transformation of ammonium fluoride or bismuth, triple point of ice) that are sufficiently numerous for the interpolation formula to be valid.

Finally, concerning the temperature measurement, one must insist in addition to the remarks of section II (probe setting, homogeneity control) on the importance of calibrating with respect to one or several fixed points under the conditions of the experimental measurements. This can be carried out easily in the case of thermocouples by the melting wire method (silver, gold, palladium) or the observation of the melting point* (zinc, aluminum). This calibration method is not used enough. It is the only one which makes obvious the validity of the thermocouple circuit and the measuring device. A calibration by comparison can mask a systematic error.

VIII. Results

The bibliography of experimental data on high temperature–high pressure gases is summarized in *Table 2*. In addition to the range covered by these data, the accuracy is given where possible. The lack of a complete analysis of the causes of the errors and even of a careful definition of these various errors leads in several cases to an overestimation of the accuracy, as the comparison between the results often shows. It would therefore be desirable if a minimum of common criteria were adopted in this field, for instance, of the type proposed by Kestin[55] at the Seventh Conference on Steam Properties.

According to Kestin, the accuracy of a physical quantity determination does not just result from an analysis of the method of measurement itself, but also from an analysis of various values of the quantity that arise from different and independent determinations. He defines the accuracy of such a quantity as being the residual and irreconcilable deviation between the best determinations. The precision, which is applied specifically to a method of measurement, can be determined in different ways: either it is the maximum possible error† obtained by an analysis of causes of errors for each reading, in which the physical law that the apparatus obeyed is taken into account, or it is the reproducibility corresponding to several independent measurements. If there is no anomaly, these two values must be of the same order.

To the precision so defined can be added systematic errors masked by good reproducibility if they are themselves reproducible. They often correspond to forgotten or poorly known corrections. Systematic errors, which are difficult to evaluate with certainty, could be one of the causes of the common situation in which the deviation between several independent measurements is larger than their respective precisions.

It is suitable to add that in the special case of the measurements of pVT parameters, the physical relation $pV = ZRT$, with Z equal to some units, requires that these three quantities have a similar precision. Otherwise the lack of precision of one of these quantities is repeated in the other two.

An examination of *Table 2* shows that there are few examples at high temperature in which corresponding data have been obtained by different methods. The comparisons from which the accuracy can be calculated follow-

---

* Here the thermocouple junction is surrounded by a small sample (wire or ribbon) of the metal to be melted.
† This error is the maximum deviation that the experimental value shows with regard to the absolute values (international standards). It will be the sum of the sensitivity, the calibration precision and the reproducibility between two successive calibrations of the measuring devices.

Table 2

Gas	Range of measurements Max. temp. °C	Max. pressure bar	Precision %	Reference
Air	600	700	0.1 to 0.2	101
Ammonia	300	1 600		51
	425	130	0.1	15
Argon	1 000	1 000	0.2	58
	600	80	0.05	103
	400	10 000	0.3	76
	400	10 000	0.2	83
Butane	300	350	0.1	18
Cesium	1 420	33	0.25	93
Carbon monoxide	300	B^*	2	28
	200	1 000	0.1	11
	300	10 000	0.2	82
Carbon dioxide	600	50	0.05	71
	1 000	1 400		53 and 54
	475	4 000	0.3	50
	500	200		99
Helium	400	125	0.05	45
	1 200	80	0.05	110
	500	100	0.05	109
Heptane	350	315	0.1	91
Hydrogen	400	1 000	0.1	9
	300	1 000		106
	600	1 800	0.5	77
Isobutane	300	300	0.1	17
	250	B^*	2	47
Krypton	600	80	0.05	105
	300	400	0.1	14
Methane	350	400	0.2	30
Neon	400	100	0.05	44 and 45
	700	80	0.05	75
Nitrogen	400	100	0.05	44 and 45
	400	1 000	0.1	9 and 10
	800	900	0.2	89
	400	10 000	0.3	95
	400	10 000	0.2	80
	1 000	5 000	0.3	67 and 69
Pentane	300	350	0.1	16
Potassium	1 420	27	0.3	93
Propane	200	10 000	0.2	81 and 5
	250	B^*	2	47
Propylene	200	10 000	0.2	81
Sodium	1 420	21	0.4	93
Water (steam and hypercritical gas)	1 000	1 400		46
	900	9 000	0.3	26
	850	6 000	1	66
	900	400	0.2	100
	450	100	0.2	52
Xenon	300	400	0.01	13
	700	100	0.05	104

* This reference gives the experimental second virial coefficient B.

ing the definition given previously, are limited. Let us mention the case of water for which the comparison made by Burnham et al.[26] revealed an accuracy of 0.3 per cent up to 900°C and 1 400 bar and one per cent up to 850°C and 4 500 bar; that of nitrogen in the range up to 400°C and 5 000 bar, which is covered by three different methods of measurement, was 0.3 per cent[67]. These numbers give an idea of what we can expect from experimental pVT results at high temperature and pressure in the present state of measuring techniques.

IX. Equations of State

Although the equation of state is a fundamental equation in physics (thermodynamics, statistical mechanics, molecular physics), it appears more often as a formal expression of correlation between experimental pVT data. It is for the most part the latter aspect that we will consider here, as is also the case in Chapter 10 of this book.

1. Law of Corresponding States

This simplest correlation was deduced from the similarities of the curves describing the experimental results. All gases obey the same equation of state expressed in the reduced coordinates p_r, V_r, T_r, which are related to the corresponding values at the critical point by $p_r = p/p_c$, $V_r = V/V_c$, $T_r = T/T_c$. In fact this law gives a better approximation if we add the Kamerlingh Onnes principle 'all substances having similarly shaped molecules have equivalent properties of state' and some peculiarities that take into account, for instance, the polarity of the molecules.

On this basis Nelson and Obert[74] presented generalized compressibility diagrams $Z = p_r V_r / RT_r$, so called because they are valid for all gases up to forty times the critical pressure and fifteen times the critical temperature. For many gases this covers a large part of the high temperature–high pressure range. The use of these diagrams leads to more precise and certain results for non-polar gases than for polar gases. Lydersen, Greenhorn and Hougen[65] presented similar diagrams up to $T_r = 15$ and $p_r = 30$, but with a third parameter, the critical compressibility, which made their diagrams a better approximation of the real behavior of each compound.

These diagrams are useful as a first approximation of gas behavior for which they usually describe the state properties with an error of a few per cent.

2. Empirical Equations

These are analytical formulations based on phenomenological, empirical, or sometimes theoretical considerations connecting the pVT parameters. Their great number, several tens, is proof of their insufficiency. They are only valid in limited ranges and often for a restricted number of substances. Among the most common, which are generally classified according to the number of constants[43, 78], we will mention three that yield reasonable results under pressure: the Redlich–Kwong, the Beattie–Bridgeman, and the Benedict–Webb–Rubin equations.

The Redlich–Kwong equation is a two-constant equation of the van der Waals type

$$[p + a/\{T^{\frac{1}{2}}V(V + b)\}](V - b) = RT$$

where a and b can be calculated from the critical point constants; one finds

$$T_c = \{(a/b)(0.0867/0.4278)(1/R)\}^{\frac{2}{3}}$$
$$p_c = \{(0.0867/b)^5(a/0.4278)^2 R\}^{\frac{1}{3}}$$
$$V_c = 3.847b$$

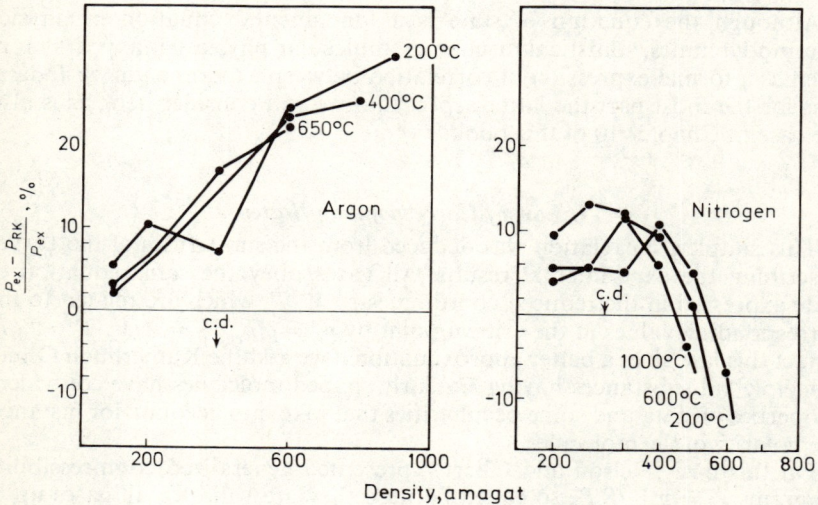

Figure 17. Comparison of the Redlich–Kwong equation of state with experimental results (from refs 45, 55 and 56): c.d. critical density.

Figure 17, which gives an idea of its validity in the case of nitrogen and argon, shows that it rapidly becomes useless at high pressure.

The Beattie–Bridgeman equation appears in the form

$$p = \frac{RT}{V^2}\left(1 - \frac{c}{VT^3}\right)\left(V + B_0 - \frac{bB_0}{V}\right) - \frac{A_0}{V^2}\left(1 - \frac{a}{V}\right)$$

The five adjustable constants, A_0, B_0, a, b and c, have been calculated for a large number of compouqds: noble gases, hydrocarbons and alcohols. In general, this equation does not lead to good results for densities which are of the order of the critical density, as is shown in *Figure 18* where the equation is compared with the experimental results for nitrogen. On the other hand, within the preceding limit the temperature behavior remains reasonable even at 800°C.

The Benedict–Webb–Rubin equation is an extension of the Beattie–Bridgeman equation to hydrocarbons. Relatively complex since it contains

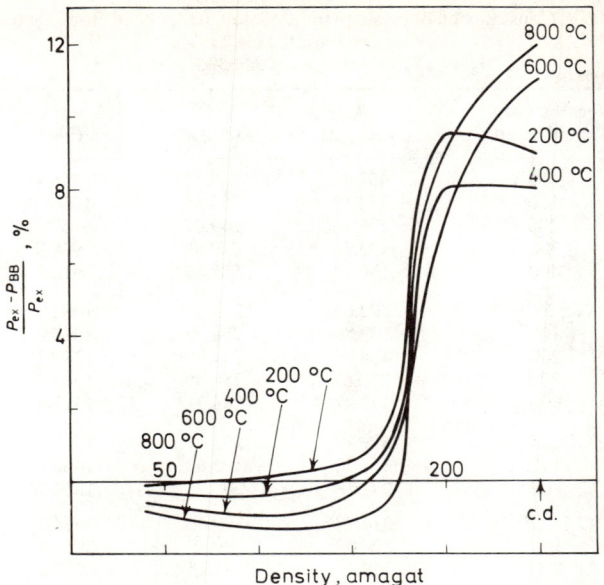

Figure 18. Comparison of the Beattie–Bridgeman equation of state with experimental results on nitrogen (from refs 67, 89 and 80): c.d. critical density.

eight constants, this equation describes pretty well the behavior of hydrocarbons and some gases such as carbon dioxide, nitrogen and sulfur dioxide up to about twice the critical density. Reid and Sherwood[78] have tabulated the constants for these different gases.

Recently Babb[4] suggested an equation for dense gases. This equation originated from the remark made by Robertson et al.[83] that the derivative $(\partial p/\partial \rho)$, where ρ is the density, becomes linear at high pressure. Then Babb[2] deduced from experimental considerations a reduced isotherm $\{C/(p + B)\}$ $\{dp/d\rho\}_T$ which is a function only of the pressure and is valid for all gases at all temperatures. C and B are the parameters of the Tait equation

$$\rho = \rho_0 + C \log[(p + B)/(p_0 + B)]$$

Babb's equation of state results from an integration by two successive approximations of this isotherm. Its form is

$$\rho = \rho_0 + C\{\log(p + B)/B + A_0 e^{B\alpha} \exp[-\alpha(p + B)] + 2A_0 e^{2B\alpha} \exp[-2\alpha(p + B)]\}$$

Three of the five constants depend on the particular gas and on the temperature while the others are independent. Babb has tabulated the values of these constants from 35° to 400°C. These values were obtained by fitting the above equation with his own experimental values by the method of least squares. In *Table 3* we have reproduced these data giving values of the constants for argon, nitrogen, methane, propane, ethylene, cyclopropane and carbon

Table 3. Constants of Babb's equation of state (calculated from his own experimental results, see Table 2)
$A_0 = 0.73100514 \quad \alpha = 0.65639644/\text{kbar}$

Substance and temp	d_0 (amagat)	C (amagat)	B (bar)
Ar, 35°C	457.43	246.35	1244.4
100	374.17	255.55	1083.3
200	272.46	269.07	967.3
300	193.31	279.53	908.3
400	136.39	289.46	913.2
N_2, 35	392.66	233.19	1482.9
100	332.89	238.65	1337.0
200	264.68	249.98	1287.0
300	210.67	258.49	1261.2
400	171.97	268.31	1313.4
CH_4, 35	429.14	193.11	1797.9
101.36	367.73	195.62	1462.4
200	291.88	202.13	1214.54
C_3H_6, 35	314.77	99.60	3664.4
100	290.03	90.89	2815.4
200*	259.31	92.18	2230.2
C_2H_4, 35	395.93	137.37	2738.8
CO, 35	391.63	229.77	1508.4
100	337.09	238.40	1442.7
200	270.05	250.17	1374.1
300	225.89	262.37	1439.9
C_3H_8, 35†	290.86	82.13	3534.5
100*	267.55	79.32	2618.1
200*	237.73	81.59	2091.6

* These three isotherms presented experimental difficulties and consequent high scatter. They were thus ignored in determining the best values of A_0 and α. The C_3H_6 isotherm at 200°C has been corrected for the polymerization encountered, and thus partially smoothed. See the original paper for details.

† This isotherm is at a temperature below the critical temperature and is thus for the liquid.

monoxide. The lower limit at which this equation can be used is 500 bar at room temperature and as much as two to three kbars at 1000°C. On the other hand, because its exponential terms decrease rapidly with pressure, it is well suited for the high pressure extrapolations. Furthermore, Babb verified that with parameters calculated at pressures below 5000 bar, his equation gave a maximum deviation at 10000 bar of 0.2 per cent.

3. Equations Derived from Statistical Mechanics

A. Virial Expansion

This is a density expansion of the compressibility factor Z

$$Z = p/\rho RT = 1 + B\rho + C\rho^2 + D\rho^3 + E\rho^4 + \ldots$$

Ursell-Mayer theory expresses formally the coefficients of this expansion in terms of the interacting molecules taken by two. However, the calculation of the coefficients from an interaction model becomes more and more difficult for the higher order coefficients. This is the reason why, in spite of computer help, only seven coefficients for the rigid-sphere model have been determined, four for the attractive–repulsive model of Lennard–Jones, and two for the more complex Stockmayer potential which takes into account the molecular orientation. Thus the virial expansion is only applicable at moderate density, which corresponds for most gases to a range of pressure of up to one or occasionally several kilobar[67].

For non-polar gases, the four coefficients corresponding to the interaction model of Lennard–Jones[7] used with the force constants calculated by Sherwood and Prausnitz[90] give reasonable results. For instance, with nitrogen the deviation from the experimental results is less than five per cent up to 1 000°C and 3 kbar[67].

In the more complex case of polar gases, the Stockmayer potential has been the most frequently used. The two calculated coefficients[43] were used with parameters tabulated by Monchick and Mason[72]. The range of application is in this case more limited as far as the pressure is concerned. The uncertainty of the results is generally larger and more difficult to estimate.

B. Equations of the Perturbation Type

These equations are obtained by a method which consists of using the well developed and exact theory of the rigid-sphere model as a basis and perturbing it by superposing a part borrowed from a more realistic potential.

The perturbation calculations can be relatively approximate because they have only a small effect on the final result. Zwanzig[111] was the first to propose this type of equation of state for high temperatures. He relied on the fact that in this range the repulsive part of the intermolecular potential was predominant because of the violent collisions. He therefore considered a hard-sphere gas perturbed by the attractive part of a Lennard–Jones potential. He obtained the following form of equation

$$\frac{pV}{RT} = \frac{p_0^* V^*}{T^*} + \frac{a_1^*(V^*)}{T^*} + O\left(\frac{1}{T^*}\right)^2$$

where

$$V^* = \frac{V}{Na^3 c^3}, \qquad T^* = \frac{k}{\varepsilon} T, \qquad p^* = p\left(\frac{a^3 c^3}{\varepsilon}\right)$$

a and ε are the force constants of the Lennard–Jones potential and c is the rigid-sphere diameter considered here as an adjustable parameter, $p_0^* V^*/T^* = p_0 V/RT$ is the equation of state of rigid spheres, and $a_1^*(V^*)$, which corresponds to the perturbation itself, was calculated from the Yvon–Born–Green–Kirkwood theory. Repeating this calculation using the more recent theory of Percus–Yevick, Frisch et al.[34] have extended the initial work of Zwanzig. Applied to argon[34] and nitrogen[67], the new equation gives good results. However, its use is not easy because an adjustment of the rigid-sphere diameter is required which depends on the experimental data. Recently, Levesque and Verlet[62] and Henderson and Barker[41] have extended this type of equation to extreme temperatures (100 times the reduced temperature $T^* = kT/\varepsilon$) and pressures (corresponding to liquid densities) inaccessible to experiments in the form of static measurements.

C. Equations Deduced from Percus–Yevick Theory

The equation suggested by Rowlinson[84] is an approximate solution of the Percus–Yevick equation for the Lennard-Jones potential, valid only at high temperature (reduced temperatures T^* equal to or greater than 12). Its form is simple

$$pV/RT = (1 + \xi + \xi^2)/(1 - \xi)^3$$

where

$$\xi = (b_m/4V)(x^{3/n})\{1 + (1/N_A)F(x)\}^3$$

and $x = \varepsilon/kT$. Here ε is the force constant of the Lennard-Jones potential (well depth) written in the form $4\varepsilon[(\sigma/r)^n - (\sigma/r)^{n/2}]$; and $b_m = 2^{6/n}\frac{2}{3}\pi N_A \sigma^3$. $F(x)$ is a function tabulated by the author. These values are reproduced in

Table 4. $F(x)$ function of Rowlinson's equation of state

$\tau = x^{-1}$	$F(x)$	$\tau = x^{-1}$	$F(x)$
2	−3.570959	50	+0.032279
3	−2.452316	60	0.083508
4	−1.890323	70	0.122792
5	−1.543227	80	0.154127
6	−1.303763	90	0.179861
7	−1.126681	100	0.201476
8	−0.989338	150	0.273769
9	−0.879049	200	0.316118
10	−0.788110	250	0.344706
12	−0.645997	300	0.365649
14	−0.539117	400	0.394818
16	−0.455144	500	0.414574
18	−0.387014	600	0.429080
20	−0.330360	700	0.440308
25	−0.222302	800	0.449331
30	−0.144601	900	0.456785
35	−0.085382	1000	0.463078
40	−0.038375	∞	0.577216

Table 4. The fact that the preceding equation includes only the force constant of the Lennard-Jones potential gives it a general character while making its application simple. The results that it gives for helium, hydrogen and nitrogen (*Figure 19*) show good agreement even at high density.

Haar and Shenker[38, 39] derived a simple equation of state by a rearrangement of the terms in the Ursell-Mayer density expansion of the configuration integral. This equation is of the van der Waals type with two temperature dependent parameters. It may be written:

$$Z = (1 + y + y^2)/(1 - y)^3 + (B - b)\rho$$

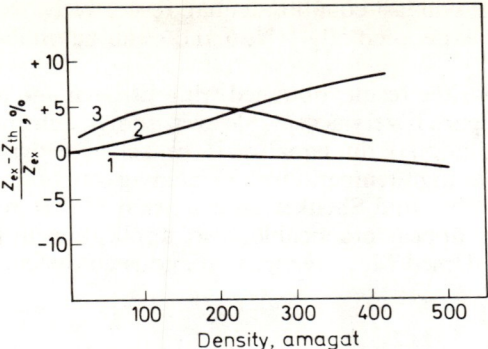

Figure 19. Comparison of the Rowlinson equation of state with experimental results (from refs 67, 77 and 106): 1 He at 200°C, 2 H_2 at 600°C, 3 N_2 at 1000°C.

where the first term is the Percus–Yevick equation of state, B is the second virial coefficient, ρ is the density and

$$4y = b\rho, \qquad b = \tfrac{2}{3}\pi N_A a^3$$

where b is the second virial coefficient for hard spheres and a, the sphere diameter.

Haar and Shenker calculated the sphere size as a boundary condition. In their analysis they formulated an expression for the second virial coefficient and its first temperature derivative for an effective interaction consisting of a hard-sphere repulsion plus a Lennard–Jones attractive bowl. These integrals were then set equal respectively, to values for B and dB/dT observed experimentally.

Thus, in the high temperature range they found:

$$B \simeq b\{1 - \tfrac{8}{3}(\beta\varepsilon) - 0.61(\beta\varepsilon)^2 - 0.15(\beta\varepsilon)^3 + \ldots\}$$

and

$$T dB/dT \simeq b\{\tfrac{8}{3}\beta\varepsilon + 1.22(\beta\varepsilon)^2 + 0.45(\beta\varepsilon)^3 + \ldots\}$$

with $\beta = 1/kT$ and ε the well depth of the Lennard–Jones type effective potential. In dB/dT, terms in the temperature derivative of $b(T)$ and $\varepsilon(T)$ were ignored because these quantities were slowly varying functions of the tempera-

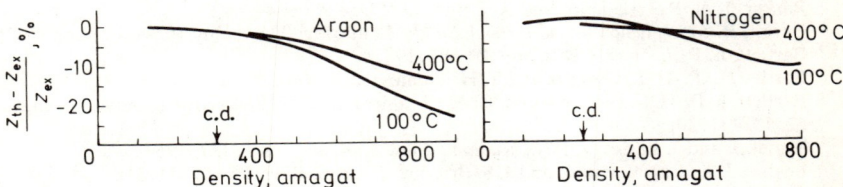

Figure 20. Comparison of the Haar–Shenker equation of state with experimental results (from refs 83 and 80): c.d. critical density.

ture. Setting these two last equations equal, respectively, to values for B and dB/dT observed experimentally[31, 61], b and ε can be obtained for each isotherm.

Figure 20 shows the results obtained with nitrogen and argon: the agreement with experience is very acceptable even at high densities.

The actual limitation of this equation is the lack of an experimental second virial coefficient at high temperatures for many gases.

According to Haar and Shenker an extension of this approach to polar substances would appear practicable; also, application to mixtures follows directly from the Ursell-Mayer expansion and is consistent with a one-fluid model.

4. Tables of Thermodynamic Properties

This outline would be incomplete if we did not mention the tables of thermodynamic properties. They are, in general, the result of correlations between the best experimental results extended when necessary by theoretical considerations. The best experimental pVT values provide the basic data for the correlations that lead to the establishment of the tables.

Let us mention for air and its components the Hilsenrath tables[42] (up to 100 bar and 3 000 K), those of Vassermann *et al.*[96] up to 1000 bar and 1300 K), and those published by the IUPAC[48] (up to 1 000 bar and 1 100 K) and for other gases, such as acetylene, ethylene, propane, methane, air, argon and nitrogen, the Din tables[29] (which have various limits in each case: in temperature from 500 to 700 K and in pressure from 1 000 to 10 000 bars).

In almost all these cases, the deviations between experimental and calculated values are of the order of 0.1 to 0.2 per cent and do not exceed 0.5 per cent. In the range where there has been extrapolation, no reliable estimate of the accuracy can be made. It is therefore advisable to use the tables of data with great care beyond the zone covered by experiments.

X. References

[1] Atanov, Yu. A. and E. M. Yvanova. 'Symposium on the accurate characterization of the high pressure environment', NBS, Gaithersburg, Maryland (1968). *Spec. Publ. Nat. Bur. Stand. No. 326* (1971).
[2] Babb, S. E. *J. Chem. Phys.* **52**, 5963 (1970).
[3] Babb, S. E. *Mechanical Behaviour of Materials Under Pressure*, Chap. 3. Edited by H. L. D. Pugh. Elsevier: London (1970).
[4] Babb, S. E. *J. Chem. Phys.* **54**, 4200 (1971).
[5] Babb, S. E. and S. L. Robertson. *J. Chem. Phys.* **53**, 1097 (1970).
[6] Babb, S. E., G. J. Scott, C. D. Epp and S. L. Robertson. *Rev. Sci. Instrum.* **40**, 670 (1969).
[7] Barker, J. A., P. J. Leonard and A. Pompe. *J. Chem. Phys.* **44**, 4026 (1966).
[8] Barrer, R. M. *Diffusion in and through Solids*. Cambridge University Press: London (1951).
[9] Bartlett, E. P. *J. Amer. Chem. Soc.* **49**, 687 (1927).
[10] Bartlett, E. P., H. L. Cupples and T. H. Tremearne. *J. Amer. Chem. Soc.* **50**, 1275 (1928).
[11] Bartlett, E. P., H. C. Hetherington, H. M. Kvalnes and T. H. Tremearne. *J. Amer. Chem. Soc.* **52**, 1374 (1930).
[12] Basset, J. and J. Basset. *J. Phys. Radium, Phys. Appliquée*, **15-A**, 147 (1954).
[13] Beattie, J. A., R. J. Barriault and J. S. Brierley. *J. Chem. Phys.* **19**, 1222 (1951).
[14] Beattie, J. A., J. S. Brierley and R. J. Barriault. *J. Chem. Phys.* **20**, 1613 (1952).
[15] Beattie, J. A. and C. K. Lawrence. *J. Amer. Chem. Soc.* **52**, 6 (1930).
[16] Beattie, J. A., S. W. Levine and D. R. Douslin. *J. Amer. Chem. Soc.* **74**, 4778 (1952).

[17] Beattie, J. A., S. Marple and D. G. Edwards. *J. Chem. Phys.* **18**, 127 (1950).
[18] Beattie, J. A. and W. H. Stockmayer. *J. Chem. Phys.* **10**, 473 (1942).
[19] Bell, P. M., R. J. Boyd and J. L. England. 'Symposium on the accurate characterization of the high pressure environment', NBS, Gaithersburg, Maryland (1968). *Spec. Publ. Nat. Bur. Stand. No. 326*, p 63 (1971).
[20] Benedict, M. *J. Amer. Chem. Soc.* **59**, 2224 and 2233 (1934).
[21] Bergeon, R. *J. Rech. CNRS*, **7**, 371 (1956).
[22] Birch, F. *Handbook of Physical Constants*, Section 7. Geological Society of America. Edited by S. P. Clarke. *Memoir 97*, New York (1966).
[23] Boren, M. D., S. E. Babb and G. J. Scott. *Rev. Sci. Instrum.* **36**, 1456 (1965).
[24] Bourassa, R. R., D. Lazarus and D. A. Blackburn. *Phys. Rev.* **165**, 853 (1968).
[25] Bundy, F. P. *J. Appl. Phys.* **32**, 483 (1961).
[26] Burnham, C. W., J. R. Holloway and N. F. Davis. *Amer. J. Sci.* **267 A**, 70 (1969).
[27] Cole, G. H. A. *An Introduction to the Statistical Theory of Classical Simple Dense Fluids*. Pergamon: Oxford (1967).
[28] Connoly, J. F. *Physics of Fluids*, **7**, 1023 (1964).
[29] Din, F. *Thermodynamic Functions of Gases*. Butterworths: London, Vols I and II (1956), Vol. III (1961).
[30] Douslin, D. R., R. H. Harrison, R. T. Moore and J. M. McCullough. *J. Chem. Engng Data*, **9**, 358 (1964).
[31] Dymond, J. H. and E. B. Smith. *The Virial Coefficient of Gases*. Clarendon Press: Oxford (1969).
[32] Eyring, H. *Statistical Mechanics* Vol. II of *Physical Chemistry: An Advanced Treatise*, chapters 6 and 7. Academic Press: New York (1967).
[33] Franck, E. U. and K. Todheide. *Z. Phys. Chem.*, N.F. **22**, 232 (1959).
[34] Frisch, H. L., J. L. Katz, E. Praesgaard and J. F. Lebowitz. *J. Phys. Chem.* **70**, 2016 (1966).
[35] Getting, I. C. and G. C. Kennedy. *J. Appl. Phys.* **41**, 4552 (1970).
[36] Greenwood, H. J. *Amer. J. Sci.* **267 A**, 191 (1969).
[37] Haar, L. and J. M. H. Levelt Sengers. *J. Chem. Phys.* **52**, 5069 (1970).
[38] Haar, L. and H. S. Shenker. 'Fifth symposium on thermophysical properties', Boston (1970), *Proceedings*, p 223. edited by C. F. Bonilla. American Society of Mechanical Engineers (United Engng Center), New York (1970).
[39] Haar, L. and S. H. Shenker. *J. Chem. Phys.* **55**, 4951 (1971).
[40] Hanneman, R. E. and H. M. Strong. *J. Appl. Phys.* **37**, 612 (1966).
[41] Henderson, D. and J. A. Barker. *Phys. Rev. A.*, **1**, 1266 (1970).
[42] Hilsenrath, J. *et al*. 'Tables of thermal properties of gases'. *Circ. US Nat. Bur. Stand. No. 564* (1955).
[43] Hirschfelder, J. O., C. F. Curtiss and R. B. Bird. *Molecular Theory of Gases and Liquids*. Wiley: New York (1954).
[44] Holborn, L. and J. Otto. *Z. Phys.* **33**, 1 (1925).
[45] Holborn, L. and J. Otto. *Z. Phys.* **38**, 359 (1926).
[46] Holser, W. T. and J. C. Kennedy. *Amer. J. Sci.* **257**, 71 (1959).
[47] Huff, J. A. and T. M. Reed. *J. Chem. Engng Data*, **8**, 306 (1963).
[48] *International Thermodynamic Tables of the Fluid State* (IUPAC). Edited by S. Angus and R. Armstrong. Butterworths: London (Argon: 1972).
[49] Johannin, P. *Thesis*, Paris (1958).
[50] Juza, J., V. Kmonicek and O. Sifner. *Physica*, **31**, 1735 (1965).
[51] Kazarnovskii, Ya. S. *Acta Phys.-chim. URSS*, **12**, 513 (1940).
[52] Kell, G. S., G. E. MacLaurin and E. Whalley. *J. Chem. Phys.* **48**, 3805 (1968).
[53] Kennedy, G. C. *Amer. J. Sci.* **252**, 225 (1954).
[54] Kennedy, G. C. and W. T. Holser. *Handbook of Physical Constants*, Geological Society of America, edited by S. P. Clarke, *Memoir 97*, p 371 (1966).
[55] Kestin, J. 'Seventh international conference on the properties of steam', Tokyo (1968), *Proceedings*. American Society of Mechanical Engineers (United Engng Center), New York (1970).
[56] Lallemand, M., J. Brielles, D. Vidal and P. Malbrunot. *J. Phys. D: Appl. Phys.* **6**, 1052 (1973).
[57] Lazarus, D., R. N. Jeffery and J. D. Weiss. *Appl. Phys. Letters*, **19**, 371 (1971).
[58] Lecocq, A. *Thesis*, Paris (1959).
[59] Le Neindre, B., R. Tufeu, P. Bury, P. Johannin and B. Vodar. *Proceedings* of the Eighth

International Conference on Thermal Conductivity, Purdue (1968), edited by C. Y. Ho and R. E. Taylor. Plenum: New York (1969).
60 Le Neindre, B., R. Tufeu, P. Bury, P. Johannin and B. Vodar. *Proceedings* of the Ninth International Conference on Thermal Conductivity, Iowa State University (1969), edited by H. R. Schanks. US Atomic Energy Commission, Division of Technical Information (1970).
61 Levelt Sengers, J. M. H., M. Klein and J. S. Gallagher. To appear in American Institute of Physics *Handbook*. McGraw-Hill: New York.
62 Levesque, D. and L. Verlet. *Phys. Rev.* **182**, 307 (1969).
63 Leycuras, A. and Y. Leycuras. 'Chauffage sous hautes pressions' from *Les Hautes Températures* edited by G. Chaudron, p 445. Masson: Paris (1973).
64 Luft, L. *Industr. Engng. Chem.* **49**, 2035 (1957).
65 Lydersen, A. L., R. A. Greenhorn and O. A. Hougen. 'Generalized thermodynamic properties of pure fluids'. *Coll. Eng., Univ. Wisconsin, Engng Exp. Sta. Rep. No. 4*: Madison (1955)
66 Maier, S. and E. U. Franck. *Ber. Bunsenges. Phys. Chem.* **70**, 639 (1966).
67 Malbrunot, P. *Thesis*, Paris (1970).
68 Malbrunot, P., P. Meunier and D. Vidal. *High Temp. High Press.* **1**, 93 (1969).
69 Malbrunot, P. and B. Vodar. *CR Acad. Sci., Paris*, **268**, 1337 (1969).
70 Malbrunot, P. and B. Vodar. *High Temp. High Press.* **3**, 225 (1971).
71 McCormack, K. E. and W. G. Schneider. *J. Chem. Phys.* **18**, 1269 (1950) and **19**, 849 (1951).
72 Monchick, L. and E. A. Mason. *J. Chem. Phys.* **35**, 1676 (1961).
73 Nejmark, B. E. *Mechanical Properties of Steel and Alloys used in Energetics*. Energia: Moscow (1967).
74 Nelson, L. C. and E. F. Obert. *Trans. Amer. Soc. Mech. Engrs*, **76**, 1057 (1954).
75 Nicholson, G. A. and W. G. Schneider. *Canad. J. Chem.* **35**, 589 (1955).
76 Polyakov, E. V. and D. S. Tsiklis. *Zh.Fiz. Khim.* **41**, 2370 (1967).
77 Presnall, D. C. *J. Geophys. Res.* **74**, 6026 (1969).
78 Reid, R. C. and T. K. Sherwood. *The Properties of Gases and Liquids*, 2nd ed. McGraw-Hill: New York (1966).
79 Robertson, S. L. *Thesis*, University of Oklahoma (1969).
80 Robertson, S. L. and S. E. Babb. *J. Chem. Phys.* **50**, 4560 (1969).
81 Robertson, S. L. and S. E. Babb. *J. Chem. Phys.* **51**, 1357 (1969).
82 Robertson, S. L. and S. E. Babb. *J. Chem. Phys.* **53**, 1094 (1970).
83 Robertson, S. L., S. E. Babb and G. J. Scott. *J. Chem. Phys.* **50**, 2160 (1969).
84 Rowlinson, J. S. *Molec. Phys.* **7**, 349 (1964).
85 Rowlinson, J. S. *Rep. Progr. Phys.* **28**, 169 (1965).
86 Rowlinson, J. S. *Liquids and Liquid Mixtures*, 2nd ed. Butterworths: London (1969).
87 Saurel, J. *J. Phys. Radium*, **14**, 215 (1953).
88 Saurel, J. *Memorial* de l'Artillerie Française, 1er fasc., p 129 (1957).
89 Saurel, J. *Thesis*, Paris (1958).
90 Sherwood, A. E. and J. M. Prausnitz. *J. Chem. Phys.* **41**, 429 (1964).
91 Smith, L. B., J. A. Beattie and W. C. Kay. *J. Amer. Chem. Soc.* **59**, 1587 (1937).
92 Stepanov, V. A. and L. F. Vereshchagin. *Pribory i Tekh. Eks.* **2**, 194 (1961).
93 Stone, J. P., C. T. Ewing, J. R. Spann, E. W. Steinkuller, D. D. Williams and R. R. Miller. *J. Chem. Engng Data*, **11**, 309 (1966).
94 Tufeu, R., B. Le Neindre and P. Bury. *Physica*, **44**, 81 (1969).
95 Tsiklis, D. S. and E. V. Polyakov. *Dokl. Akad. Nauk SSSR*, **176**, 308 (1967).
96 Vassermann, A. A., Y. A. Kazavchinskii and V. A. Rabinovich. *Thermophysical Properties of Air and its Components*. Nauka: Moscow (1966), (in Russian).
97 Vilevich, A. V., L. F. Vereshchagin and Ya. A. Kalashnikov. *Pribory i Tekh. Eks.* **3**, 146 (1961). Translated in *Instrums Exp. Tech. Pittsburgh*, **1**, 559 (1961).
98 Vodar, B. and J. Saurel. *High Pressure Physics and Chemistry*, edited by R. S. Bradley, Vol. I, Chap. 3. Academic Press: London (1963).
99 Vukalovich, M. P. and Ya. F. Masalov. *Teploenergetika*, **13**, 58 (1966), translated in *Heat Pwr Engng (Toronto) W*, **13**, 73 (1966).
100 Vukalovich, M. P., M. S. Trakhtengerts and G. A. Spiridonov. *Teploenergetika*, **14**, 65 (1970).
101 Vukalovich, M. P., V. N. Zubarev, A. A. Alexandrov and A. D. Kozlov. *Teploenergetika*, **1**, 70 (1970).

[102] Wentorf, R. H. 'Symposium on the accurate characterization of the high pressure environment' NBS, Gaithersburg, Maryland (1968). *Spec. Publ. Nat. Bur. Stand. No. 326*, p 81 (1971).
[103] Whalley, E., Y. Lupien and W. G. Schneider. *Canad. J. Chem.* **31**, 722 (1953).
[104] Whalley, E., Y. Lupien and W. G. Schneider. *Canad. J. Chem.* **33**, 633 (1955).
[105] Whalley, E. and W. G. Schneider. *Trans. Amer. Soc. Mech. Engrs*, **76**, 1001 (1954).
[106] Wiebe, R. and V. L. Gaddy. *J. Amer. Chem. Soc.* **60**, 2300 (1938).
[107] Wiebe, R., V. L. Gaddy and C. Heins. *J. Amer. Chem. Soc.* **53**, 1721 (1931).
[108] Williams, D. W. *Amer. Mineral.* **53**, 1765 (1968).
[109] Witonsky, R. J. and J. G. Miller *J. Amer. Chem. Soc.* **85**, 282 (1963).
[110] Yntema, J. L. and W. G. Schneider. *J. Chem. Phys.* **18**, 641 and 646 (1950).
[111] Zwanzig, R. W. *J. Chem. Phys.* **22**, 1420 (1954).

CHAPTER 9

THE COMPRESSION OF LIQUIDS*

E. Whalley
Division of Chemistry, National Research Council of Canada, Ottawa, Canada

Contents

I.	Introduction	422
II.	Thermodynamics of Volume Changes	425
	1. The more Important Thermodynamic Derivatives of Pressure, Volume and Temperature	425
	2. Relations between the Derivatives	426
	3. Quantities that can be Determined by Measuring Volumes, Thermal Expansions and Compressions	428
III.	Historical Introduction	430
IV.	Some Experimental Considerations	431
	1. Relative Expansion and Compression	431
	2. Expansion, Compression, Expansivity and Compressibility	431
	3. Dilatation of the Vessel	432
	4. Seasoning of Pressure Vessels and Piezometers	434
	5. Heat of Compression	436
	6. Corrosive Liquids	436
	7. Gases at High Pressures	437
	8. External and Internal Heating of Pressure Vessels	437
	9. Summary of the more Important Sources of Error in the Measurement of the Compression of Liquids	439
	10. Accurate Measurements	441
V.	Piezometric Methods—Liquid Piston	441
	1. Some General Considerations	441
	2. Single-point Methods	445
	A. Floating Index Methods	446
	B. Aimé's Method	446
	C. Electrical Contact Method	449
	3. Multiple-point Method	450
	A. Visual Method	450
	B. Electrical Contact Methods	456
	C. Differential Transformer Method	458
	D. Capacitance Method	460
	E. Magnetic Float Method	462
	F. Radiated Signal Method	462
VI.	Piezometric Methods—Solid Piston	463
VII.	Piezometric Methods—Bellows	465
VIII.	Simple Piston–Cylinder Method	469
IX.	Constant-volume Vessel	474
	1. Introduction	474
	2. Displacement of a Solid Piston	476
	3. Displacement of a Liquid Piston	480
	4. Bellows Volumometer	482

* N.R.C. No. 13660.

	5. Direct Weighing of Fluid Removed	482
	6. Volumetric Measurement of the Gas at Low Pressure	483
	7. Volumetric Measurement of Liquid at Low Pressure	483
X.	Weight Methods	484
	1. Direct Weighing Methods	484
	2. Hydrostatic Weighing	485
XI.	Ultracentrifuge Method	488
XII.	Negative Pressures	488
XIII.	Adiabatic Compression	489
XIV.	Isochoric Thermal Pressure Coefficient	490
	1. Piezometric Methods	490
	2. Constant-volume Vessel with Direct Pressure Measurement	491
	3. Constant-volume Vessel with Indirect Pressure Measurements	492
XV.	Calorimetric Methods	493
XVI.	Miscellaneous Methods	494
	1. Methods Based on Radioactivity	494
	2. Variable-volume Vessel	494
XVII.	References	494

I. Introduction

Methods for the determination of the density of liquids can be divided into three classes as follows.

1. Methods in which the density is measured in terms of the fundamental physical standards of measurement.

2. Methods in which it is measured relative to the density of a reference liquid or solid. Usually, although not always, the density of the reference liquid or solid will have been determined either directly or via intermediate liquids by a method in category 1.

3. Methods in which the change of density of a liquid, caused by an impressed change of the state of the liquid such as temperature, pressure, magnetic or electric field, etc., is measured. In these methods, either the change of volume is measured relative to the change of volume of the containing vessel or the change of density is measured relative to the change of density of a float or sinker.

4. Methods in which the volume, or more usually the change of volume, is deduced from measurements other than the direct measurement of change.

Methods in category (1) are reviewed in Chapter 5 of this volume, and will not concern us further. It is worth noting, however, that the difficulty of making accurate measurements by these methods leads to the liter being defined as the volume of 1 kg of water at its temperature of maximum density, and thus to be independent of the fundamental unit of length. It is only recently that this definition has been replaced by one (one cubic decimeter) relating it to the unit of length. Even now, the accepted relation between the old liter and the cubic decimeter is still based on the data considered by the Third General Conference on Weights and Measures in 1901[77].

Methods in category (2) that are suitable for use at ambient pressures, such as pyknometric and buoyancy methods, have been well reviewed recently in detail by Bauer and Lewin[23], and will not be systematically reviewed here. Methods that are particularly useful at high temperatures have been reviewed by White[255].

THE COMPRESSION OF LIQUIDS

This chapter will be concerned mainly with methods that fall into category (2) that are used under high pressure and all those that fall into category (3), that is, with the techniques for measuring the change of volume of liquids caused by change of temperature and pressure. The descriptions will stress particularly the measurement of the effects of pressure as this is usually more difficult than the measurement of the effects of temperature at ambient pressure. The changes of volume caused by changes of electric and magnetic fields are too small to measure accurately, and are more easily and more accurately obtained by measuring the effect of pressure on the electric and magnetic susceptibilities, and these measurements are outside the scope of this review. Gravitational fields do not in themselves change the volume of substances.

Liquids that are not in complete thermodynamic equilibrium can change in volume due to spontaneous processes occurring within them. Methods of measuring volume changes due to changes of pressure and temperature can be readily adapted to measuring these spontaneous volume changes. Many of the methods described can be adapted in reasonably obvious ways to measuring volume changes at phase transitions.

Several methods of investigating the p–V–T relations of liquids fall into category (4), that is methods in which changes of volume are deduced from other measurements. The most commonly used of these methods is the measurement of the speed u of small-amplitude sound, which is related to the adiabatic effect of pressure on the density by the thermodynamic relation

$$u^2 = (\partial \rho / \partial p)_S^{-1} \tag{I.1}$$

where ρ is the density, p the pressure, and S the entropy. By application of equation II.4 it can be shown that

$$\left(\frac{\partial \rho}{\partial p}\right)_T = \left(\frac{\partial \rho}{\partial p}\right)_S - \frac{T}{C_p \rho^2}\left(\frac{\partial \rho}{\partial T}\right)_p \tag{I.2}$$

where T is the temperature and C_p the heat capacity per unit mass at constant pressure. By integrating equation I.2 between pressures p_1 and p_2 it follows that

$$\rho(p_2) - \rho(p_1) = \int_{p_1}^{p_2} u^{-2}\, dp - T \int_{p_1}^{p_2} (1/C_p \rho^2)(\partial \rho / \partial T)_p^2\, dp \tag{I.3}$$

The first term on the right side can be obtained directly from the measured speeds of sound. The second term will rarely be more than about one third of the first term, and a first approximation can be obtained by using either the zero-pressure values or approximate high-pressure values if these are known or can be estimated. Thus an approximate set of densities as a function of temperature and pressure can be obtained which give a better approximation to the second term. By a continued iteration, the accuracy inherent in the speed of sound and the heat capacity at low pressure can be reached. Alternatively, the integration can be carried out over successive small pressure intervals, the pressure coefficients of the thermal expansivity and the heat capacity being obtained from equations II.1 and II.34. This procedure has been followed successfully by Davis and Gordon[83] for mercury, and is discussed in detail in Chapter 10. Kell[137] has discussed the isothermal compressibilities of liquid water as obtained from direct measurement of volume changes and from the speed of sound and has concluded that the speed of sound yields much more accurate results. It seems likely that in the foreseeable future, the most accurate pVT data on liquids at low and moderate temperatures, particularly on liquids like water and mercury for which the second term on the right side of equation I.3 is small compared with the first term, will be obtained from the speed of sound.

Isothermal volume changes can be obtained from the adiabatic pressure thermal coefficient

$$\beta_s = (\partial T/\partial p)_S \tag{I.4}$$

Since, according to equation II.15

$$\beta_s = T(\partial V/\partial T)_p/C_p \tag{I.5}$$

where V is the specific volume, and according to equation II.34

$$C_p(T, p) = C_p(T, p = 0) - T\int_0^p (\partial^2 V/\partial T^2)\,dp \tag{I.6}$$

it follows that

$$V(T, p) = V(T_0, p) + \int_{T_0}^T \{\beta_s(T, p)C_p(T, p = 0)/T\}\,dT \\ + \int_{T_0}^T \int_0^p (\partial^2 V/\partial T^2)_p\,dp\,dT \tag{I.7}$$

The first two terms on the right side of equation I.7 are directly measurable, and if the third term is small, the volume can be obtained as a function of temperature by successive approximations. However, this method does not appear to have been used in practice, partly because of the difficulty of measuring β_s sufficiently accurately.

The intensity of low-angle x-ray, electron, or neutron scattering by a molecular liquid when extrapolated to zero angle is, according to the classical fluctuation theorem, proportional to the quantity (see, for example, Sjölander[215])

$$RT\kappa_T/V - 1$$

where R is the gas constant, κ_T is the isothermal compressibility, and V is the molar volume. If $\kappa_T = 100 \times 10^{-6}$ bar^{-1}, $V = 100$ cm^3 mole^{-1}, the term $RT\kappa_T/V$ is about 0.024. The accuracy of measurement of the intensity of scattering at the present time is far from sufficient to give even approximate values of κ_T, although some attempts have been made (Harris and Clayton[111a]).

As far as I have been able to ascertain, no detailed review of the methods used in measuring the compression of liquids has ever been published, although there are accounts of more limited scope (Sengers[210], Verbeke and Van Itterbeek[240], Newitt[178], Tsiklis[236]). There are few properties of fluids that have been measured by such a wide variety of techniques as have been used for measuring volume changes, and because of the large number of measurements that have been done, and their importance in science and technology, a reasonably complete as well as a critical description of the various methods employed has been attempted. The ancillary techniques required for the measurements, such as the measurement and control of temperature and pressure, generation of pressure, design of pressure vessels, closures, electrical seals, and so on are not in general discussed in detail.

Not all measurements made in this field need have the highest attainable accuracy. There are, in fact, no reasonably standard techniques used for measurements pushed to the attainable limits of modern experimental accuracy, quite unlike, for example, the case in calorimetry. Consequently, methods that have yielded relatively inaccurate results are described as well as those capable of the highest accuracy. As so many papers have been published in the field, it is not easy to insure that all papers with significant descriptions of experimental techniques have been found. I apologize to any whose papers have been missed, and ask that significant omissions be brought to my attention.

Only relatively few of the methods described have been used widely. Some are particularly convenient for doing a few measurements of moderate accuracy using little specialized apparatus. Some are capable of high accuracy, although few of the techniques appear to have been pushed to their maximum accuracy.

Which of the many methods should one use for a particular measurement? The main considerations deciding this question are as follows.

(1) Temperature range
(2) Pressure range
(3) Chemical reactivity of the experimental liquid
(4) Accuracy required
(5) Apparatus and expertise already available in the laboratory.

Gases at high density are distinguished from liquids only in that a high pressure is required to maintain the high density. Techniques of measuring volume changes depend more on the magnitude of the volume changes to be measured than on the pressure, and so methods of measuring volume changes in dense gases are more akin to those used for liquids than to those used for gases at low pressures, and are included in this survey. The main experimental differences are in the methods of compressing the gas to the required density before the compression measurements are done.

II. Thermodynamics of Volume Changes

1. *The more Important Thermodynamic Derivatives of Pressure, Volume and Temperature*

To describe the thermodynamic state of homogeneous liquids not subjected to electric or magnetic fields, such as we are considering, requires two independent variables, such as the pressure and temperature. Consequently, two independent variables are required to fix the volume, and volume changes are determined only if the changes of two independent variables are specified. Pressure changes are most usually carried out either at constant temperature (isothermal) or constant entropy (adiabatic), and temperature changes at constant pressure (isobaric), constant volume (isochoric), or along the liquid/vapor equilibrium line (saturation). A vast number of other boundary conditions are of course possible.

The relative change in volume $[V(p_2) - V(p_1)]/V(p_1)$ caused by a change of pressure from p_1 to p_2 is called the compression, a name apparently introduced by Adams[2], and the relative change in volume caused by a change of temperature is called the expansion.

Although the finite quantities compressions and expansions for given changes of pressure or temperature are usually what is directly measured, it is the derivatives that are of most interest. The most important of these are as follows.

$-(\partial \ln V/\partial p)_T = \kappa_T$ is the isothermal compressibility,

$-(\partial \ln V/\partial p)_S = \kappa_S$ is the adiabatic compressibility,

$(\partial \ln V/\partial T)_p = \alpha_p$ is the isobaric thermal expansivity,

$(\partial \ln V/\partial T)_\sigma = \alpha_\sigma$ is the saturation thermal expansivity,

$(\partial T/\partial p)_S = \beta_S$ is the adiabatic pressure thermal coefficient,

$(\partial p/\partial T)_V = \gamma_V$ is the isochoric thermal pressure coefficient,

where p is the pressure, V the volume, T the temperature, S the entropy, and the subscript σ indicates that the derivative is to be taken along the saturation line. A saturation compressibility can in principle be defined; it is, however, not very useful since most liquids expand along their saturation line, and so the saturation compressibility is usually negative.

If a system is thermodynamically stable, that is, is not to separate spontaneously into parts having different properties, there are certain conditions on the variation of the volume with the pressure that must be obeyed. These are (Guggenheim[111], p 96)

$$(\partial V/\partial p)_T < (\partial V/\partial p)_S < 0$$

where T is the temperature and S the entropy. Thus the isothermal $-(\partial \ln V/\partial p)_T$ and the adiabatic $-(\partial \ln V/\partial p)_S$ compressibilities must be positive, and the isothermal compressibility must be the larger.

Because of the equality

$$\partial^2 \ln V/\partial p\, \partial T = \partial^2 \ln V/\partial T \partial p$$

$$(\partial \alpha_p/\partial p)_T = -(\partial \kappa_T/\partial T)_p \tag{II.1}$$

The thermal pressure coefficient at a fixed density is almost independent of the temperature, as noted by Ramsay and Young[191,192]. It is therefore particularly useful for extrapolating densities into regions where no measurements have been made.

2. Relations between the Derivatives

Relations between the various coefficients can be derived as follows. If X is the specification of the variation of one variable other than the pressure p during the volume change and Y is a variable, then the volume V is

THE COMPRESSION OF LIQUIDS

$$V = f(p, Y) \tag{II.2}$$

and so

$$dV = (\partial V/\partial p)_Y \, dp + (\partial V/\partial Y)_p \, dY \tag{II.3}$$

By dividing by dp and specifying the boundary condition X we have

$$(\partial V/\partial p)_X = (\partial V/\partial p)_Y + (\partial V/\partial Y)_p (\partial Y/\partial p)_X \tag{II.4}$$

There is of course a similar family of relations for temperature changes. Because of equation II.2 with $Y = T$ or S, it follows that

$$(\partial p/\partial T)_V (\partial T/\partial V)_p (\partial V/\partial p)_T = -1 \tag{II.5}$$

and

$$(\partial V/\partial p)_S (\partial p/\partial S)_V (\partial S/\partial V)_p = -1 \tag{II.6}$$

In addition

$$(\partial T/\partial p)_S (\partial p/\partial V)_S (\partial V/\partial T)_S = 1 \tag{II.7}$$

From equations II.4 and II.5 and the appropriate Maxwell equation, equation II.26,

$$(\partial S/\partial p)_T = -(\partial V/\partial T)_p = -\alpha V \tag{II.8}$$

it is readily shown that

$$\gamma_V = \alpha_p/\kappa_T \tag{II.9}$$

$$\beta_S = -\kappa_S/\alpha_S \tag{II.10}$$

where α_S is the little-used adiabatic thermal expansivity $(\partial \ln V/\partial T)_S$, and these are readily put in the more useful forms.

$$\beta_S = (\kappa_T - \kappa_S)/\alpha_p \tag{II.11}$$

and

$$\kappa_S/\kappa_T = C_V/C_p \tag{II.12}$$

where C_V and C_p are the heat capacities of a fixed amount of material at constant volume and pressure respectively. Since the heat capacities at constant pressure and constant volume are related by the equation (Guggenheim[111], p 105)

$$C_p - C_V = TV\alpha_p^2/\kappa_T \tag{II.13}$$

it follows that

$$\kappa_T - \kappa_S = T\alpha_p^2/C_p' \tag{II.14}$$

where C_p' is the heat capacity at constant pressure per unit volume. From equations II.11 and II.14

$$\beta_S = T\alpha_p/C_p' \tag{II.15}$$

By applying equation II.4 with T substituted for p and with X denoting saturation and $Y = p$, we find

$$\alpha_\sigma - \alpha_p = \kappa_T(\partial p/\partial T)_\sigma \tag{II.16}$$

If the Clausius–Clapeyron equation,

$$(\partial p/\partial T)_\sigma = \Delta S/\Delta V \tag{II.17}$$

where ΔS and ΔV are the entropy and volume of vaporization, is introduced into equation II.16 we obtain

$$\alpha_\sigma - \alpha_p = \kappa_T \Delta S/\Delta V \tag{II.18}$$

If the vapor is assumed to be ideal and the volume of the liquid negligible in comparison with it,

$$\Delta V = RT/p_\sigma \tag{II.19}$$

where p_σ is the saturation pressure, and equation II.18 becomes

$$\alpha_\sigma - \alpha_p = \kappa_T p_\sigma \Delta H/R \tag{II.20}$$

3. Quantities that can be Determined by Measuring Volumes, Thermal Expansions and Compressions

Thermodynamics yields many relations between volumes and their pressure and temperature derivatives and other thermodynamic quantities, and some of the more important are as follows.

The volume is of course the isothermal pressure derivative of the Gibbs energy G,

$$V = (\partial G/\partial p)_T \tag{II.21}$$

and is also the isentropic pressure derivative of the enthalpy

$$V = (\partial H/\partial p)_S \tag{II.22}$$

Furthermore, the pressure is the negative of the isothermal volume derivative of the Helmholtz energy A

$$p = -(\partial A/\partial V)_T \tag{II.23}$$

and the negative of the adiabatic volume derivative of the internal energy U

$$p = -(\partial U/\partial V)_S \tag{II.24}$$

From equation II.23 with the help of equations II.8 and II.27 below it follows that

$$(\partial U/\partial V)_T = T\gamma_V - p$$

so that the volume coefficient of the internal energy can be obtained from the thermal pressure coefficient.

According to equation II.15, a measurement of the adiabatic pressure thermal coefficient and the thermal expansivity yields the heat capacity at constant pressure, a possibility realized by Joule[133] who in fact measured all three quantities. Burlew[64,65,66] determined heat capacities in this way a few decades ago.

There are four Maxwell equations obtained by twice differentiating the equation

THE COMPRESSION OF LIQUIDS

$$dG = -S\,dT + V\,dp \tag{II.25}$$

and its analogues in two ways. They are

$$(\partial V/\partial T)_p = -(\partial S/\partial p)_T \tag{II.26}$$

$$(\partial p/\partial T)_V = (\partial S/\partial V)_T \tag{II.27}$$

$$(\partial T/\partial p)_S = (\partial S/\partial V)_p^{-1} \tag{II.28}$$

which is equivalent to equation II.15, and

$$(\partial T/\partial V)_S = -(\partial S/\partial p)_V^{-1} \tag{II.29}$$

$$= T\gamma_V/C'_V \tag{II.30}$$

where C'_V is the heat capacity per unit volume at constant volume. Thus, the pressure and temperature coefficients of the entropy can be determined by measuring volume changes. Equations II.15 and II.29 can be transformed to equation II.13 and

$$C'_V = T\alpha_p^2 \kappa_S/\kappa_T(\kappa_T - \kappa_S) \tag{II.31}$$

respectively. Thus, the heat capacity can be measured without measuring any temperature differences caused by heating the liquid. A brief review of the measurements of heat capacity in this way is given by Cruickshank, Ackermann and Giguère[81].

Equations for the pressure coefficient of the enthalpy H and the volume coefficient of the internal energy U are readily derived from the foregoing, and are

$$(\partial H/\partial p)_T = V - T(\partial V/\partial p)_T \tag{II.32}$$

$$(\partial U/\partial V)_p = p + T(\partial p/\partial T)_S \tag{II.33}$$

By differentiating equations II.26 and II.27 with respect to temperature, we have

$$T(\partial^2 V/\partial T^2)_p = -(\partial C_p/\partial p)_T \tag{II.34}$$

$$T(\partial \gamma_V/\partial T)_V = T(\partial^2 p/\partial T^2)_V = (\partial C_V/\partial V)_T \tag{II.35}$$

Thus, information about the variation of pressure, volume, and temperature can be obtained from the heat capacities. The heat capacity at nearly constant volume can be measured by enclosing the fluid in a pressure vessel which is used as the calorimeter vessel. The heat capacity of a liquid at constant high pressure can be measured by means of a flow calorimeter (see Sirota and Mal'tsev[214]).

The Joule–Thomson coefficient, which describes the ratio of the temperature change to the pressure change of a fluid when it is passed through an efficient throttle so that all its work of compression is converted into heat, is given by

$$(\partial T/\partial p)_H = -(1 - \alpha T)/C'_p$$

and so can in principle be used to determine properties of the volume. The method has been frequently applied to gases, but as far as I know has not been applied seriously to liquids. For a liquid having $\alpha = 10^{-3}\,\text{K}^{-1}$, $C_p = 2\,\text{J cm}^{-3}\,\text{deg}^{-1}$, the Joule–Thomson coefficient is $0.035\,\text{deg bar}^{-1}$, and so it should be readily measurable to a good accuracy.

III. Historical Introduction

It was first shown by Canton[71] that liquids are compressible and perfectly elastic. He placed water in a bulb to which a capillary tube was attached and placed it in the receiver of an air pump. He varied the pressure of air in the receiver and noted the movement of the meniscus in the capillary tube. To prevent a change of temperature of the bulb due to condensation or evaporation of moisture from its surface, the bulb was immersed in water. The compression measured was probably more nearly adiabatic than isothermal but the difference is insignificant for water. The experiments were carefully done, as shown by the agreement to two figures of his compressibility of water relative to glass, which he thought was incompressible, of $45 \times 10^{-6}\,\text{bar}^{-1}$ at 10°C with the best modern value[137]. He also noted that the compressibility of water decreased with increasing temperature, whereas he expected an increase as he found for spirit of wine and olive oil, and that sea-water was less compressible than pure water[72]. Whether water is compressible was not a solely academic question, but had great practical significance. From his results he calculated that an ocean two miles deep is 60 feet shallower than it would be if water was incompressible.

The first measurements at pressures appreciably greater than atmospheric were made by Perkins[183]. In place of Canton's receiver, he used either a cannon pressurized with water or, on the way by ship from America to England, the ocean to a depth of 500 fathoms. In place of Canton's glass bulb, he used a steel tube with a rod passing into it through a stuffing box. The tube was filled with water and pressurized either in the cannon or by dropping it into the ocean. The compression of the water caused the rod to move into the tube, and a collar that fitted tightly on the tube and could slide over it indicated the movement of the rod. He was concerned about the loss of pressure due to friction of the packing around the rod, and so replaced the tube by one which was flattened and had a non-return valve that allowed fluid to enter the tube but not to escape. The tube was weighed before and after compression to determine the compression of the water. In a second paper[184] he reported compressions to 2 kbar, which was a pressure record by a large margin at that time. In this way he found that the compressibility of water decreases with increasing pressure.

Perkins also coined the word 'piezometer' to describe the tube he used for measuring the compression of liquids. It is derived from the Greek πιέζειν 'to press' and apparently means an instrument to measure the press or the squeeze of the fluid. Both the *Shorter Oxford English Dictionary* and Webster's *New International Dictionary*, Second Edition, define it as an instrument for measuring the pressure. Apart from some specialized uses in hydraulics with this meaning, it is almost universally used to describe an instrument for measuring the compression of a fluid.

THE COMPRESSION OF LIQUIDS

IV. Some Experimental Considerations

1. Relative Expansion and Compression

In most methods of measuring the change of volume of a liquid when it is heated or compressed, it is the change of volume less the change of volume of the containing vessel or of a sinker that is measured. In some methods, the primary measurements are affected by the change of volume of another fluid, the change of dimension of some part, or perhaps the change of the conductance of an electrical conductor. All these must be allowed for when the experimental data are reduced. The compression of a confining fluid may need to be measured in a separate experiment. The change in dimension or conductance may be allowed for by known coefficients, or may need to be measured. These allowances are relatively straightforward and will not be discussed in detail; they do, however, need to be done carefully.

2. Expansion, Compression, Expansivity and Compressibility

The thermal expansion and the compression of liquids are measured for two main purposes, to determine the density under given conditions of temperature and pressure, or to determine the thermal expansivity or the compressibility. Since, as explained in section II, finite differences in the density are measured, the density relative to that at a suitable standard state can be determined directly. It is useful therefore to define finite-difference quantities like the compression, which is the relative change of volume $[V(p_2) - V(p_1)]/V(p_1)$ caused by a pressure change from p_1 to p_2, or the analogous thermal expansion $[V(T_2) - V(T_1)]/V(T_1)$. These quantities are particularly useful for technical purposes if, for example, the density at a certain pressure relative to zero pressure is required. Sometimes the compression is divided by the pressure or the pressure difference to give an average compressibility. In terms of the compressibility it is a complicated quantity, the relation being

$$[V(p_2) - V(p_1)]/V(p_1)(p_2 - p_1) = \overline{V\kappa}/V(p_1)$$

where

$$\overline{V\kappa} = (p_2 - p_1)^{-1} \int_{p_1}^{p_2} V\kappa \, dp$$

For scientific purposes the derivatives are more important. Unfortunately, there are few methods known of measuring the derivatives directly, the only method known at present being the measurement of the adiabatic pressure coefficient of the density by measurement of the speed of sound, as described in Chapter 11 of this volume. Compressibilities and thermal expansivities must therefore be determined by differentiating the measured finite differences of the density. This is done either by fitting an equation to some or all of the measurements or by graphical methods, and the techniques are described in Chapter 10 of this volume.

There is, however, a related experimental problem in determining the compressibility or thermal expansivity. It is to decide the finite intervals over which the compression is to be measured. If an isothermal volume change ΔV caused by a pressure change Δp is to be measured to an accuracy of a fraction a of the volume change, then the temperature of the sample must be controlled to better than

$$\Delta T = a \, \Delta p \kappa_T / \alpha_p$$

Typical values of the thermal expansivity and the compressibility of molecular liquids are 10^{-3} deg^{-1} and 100×10^{-6} bar^{-1}. If pressure intervals of one bar are to be used and an accuracy of 0.1 per cent in the compressibility is required, which is about the highest attainable at the present time, the volume change must be known to about 0.1 p.p.m. of the volume and so the temperature must be controlled to about 10^{-4} deg. It is very difficult to control temperatures so well over long periods. Since thermal equilibrium after a compression typically requires an hour, pressure changes much greater than one bar are required for isothermal measurements. It follows that for the highest accuracy, isothermal measurements with temperature control to 1 mdeg should be made at pressure intervals of about ten bars or more. Direct measurements of the adiabatic compressibility using small enough pressure intervals can, however, be made since short times are of the essence of the method.

An interval of ten bars or more is, however, a large interval from another point of view. The pressure coefficient of the compressibility of typical molecular liquids near room temperature is about 10^{-3} bar^{-2} so that a pressure change of ten bars corresponds to a change of about one per cent in the compressibility. Unless the temperature is controlled to much closer than 1 mdeg, it is clearly impossible to measure isothermal compressions in a range in which the volume is essentially linear in the pressure unless results to only one per cent accuracy or lower are required. For results of highest accuracy, the pressure range must be chosen so that reasonable values of the contribution to the volume of at least the first non-linear term in the pressure can be determined. The preferred pressure range for accurate compressibilities at zero pressure is therefore several tens of bars. The pressure coefficient of the compressibility usually decreases strongly with increasing pressure. The pressure interval required for accurate compressibility at high pressures can be higher because of this, and because the compressibility is required in the middle of the range of measurement, rather than at one extreme as it is at zero pressure. For accurate measurements, pressures should be measured with a good pressure balance calibrated to say 100 p.p.m.

Many measurements have been made with pressure intervals as high as 250, 500 or even 1000 bar. They clearly will not give accurate compressibilities at low pressure.

3. *Dilatation of the Vessel*

Since most compressions and expansions are measured relative to the compression and expansion of a vessel, the compression and expansion of the vessel must usually be determined. Sometimes, measurements are made relative to a solid body immersed in the fluid; then the compression and expansion of the solid body must be determined. This can be done from elasticity theory using the known elastic constants and thermal expansivity of the material of the vessel, or by measuring the change of volume relative to the expansion and compression of a reference liquid or solid.

According to elasticity theory, the fractional change $\delta V/V$ of the inside volume V of a long cylindrical vessel subjected to a hydrostatic pressure p_i inside and p_o outside, when the thrust of the pressure on the end is taken by the walls, assuming that the deformation is purely elastic, is

$$\delta V/V = \{1/E(k^2 - 1)\} \{p_i[2k^2(1 + v) + 3(1 - 2v)] - p_o k^2(5 - 4v)\} \quad \text{(IV.1)}$$

where k is the ratio of outside to inside diameter, E is Young's modulus, v is Poisson's ratio, and the effect of stress on these three quantities is ignored to first order. Because of equation IV.1 this and related formulae are only approximate at high pressure, and in fact are always approximate because the end effects in a vessel of finite length are ignored. The end effects of a simplified model were discussed by Verbeke and Van Itterbeek[240], and no doubt more elaborate calculations could be done if a good enough representation of the actual vessel could be prescribed. Most calculations assume that the thermal expansivity and the elastic constants are uniform and isotropic throughout the material, although an allowance for anisotropy could be made by assuming that the values parallel and perpendicular to the axis are different. In fact, as is discussed more fully in section IV.3, residual stresses often cause the constants to vary with position within the wall, so that the assumptions of equation IV.1 are not valid, and heating or applying pressure to the vessel causes it to distort. The methods of reducing this effect by seasoning are discussed in section IV.3. The change of volume of the vessel should be calculated from equation IV.1 only for work of low to moderate accuracy.

Wassenaar[245] has measured the external dilatation of a vessel pressurized internally and found it was only about 0.8 of the theoretical value neglecting end effects. He then assumed that the internal dilatation was smaller than theory by the same amount. Vilevich, Vereshchagin and Kalashnikov[242] have assumed that the internal dilatation of a thin-walled vessel was equal to the external dilatation, but did not comment on whether it agreed with simple theory. Crawford and Daniels[80] find, in fact, that it does not, but give no details.

If a vessel is subjected to pressure on the inside only, it may undergo permanent deformation by yielding under the non-hydrostatic stresses. This can be largely avoided by making the vessel from a material of high yield strength and keeping the stresses well below the yield stress. If the vessel is subjected to the same pressure inside and out, the stresses in the vessel wall are purely hydrostatic, and, except under some conditions discussed in section IV.4, will not deform permanently. Furthermore, the change of the inside volume can be accurately calculated from the compressibility of the material no matter what the shape, there being no end effects. It is therefore preferable to have equal pressures inside and outside the vessel rather than to have pressure inside only, and is necessary for accurate measurements unless the non-hydrostatic components of the stress in the wall of the vessel are much less than the yield or fracture stress.

For accurate measurements, the thermal expansion and the compression of the vessel must be measured. Direct methods of doing this will not be described in detail in this chapter; they are the standard methods of measuring the thermal expansion and compression of solids applied to the vessel or a part of it, or to material similar to the material of the vessel.

For the most accurate measurements, it is better to measure the expansion and compression of the vessel relative to that of a reference solid or liquid of accurately known density. Measurements are made with the vessel filled with the experimental fluid and with the fluid wholly or partly replaced by the reference substance. If the two sets of measurements are made at the same temperatures and pressures, the compression of the experimental substance relative to that of the reference substance can be obtained directly. If they are not so made, then the expansion and compression of the vessel can be obtained and interpolated by means of an equation of state or by graphs. In this method, the reference liquid or solid need be measured at only relatively few of the pressures and temperatures, as the volume of the vessel is usually nearly linear in the temperature and the pressure, and interpolation is straightforward. The most commonly used substances for this purpose are iron, whose compression has been measured accurately up to 30 kbar by Bridgman[57,59], mercury, the data for which have been reviewed by Bett, Weale and Newitt[29], and water, for which accurate compressions in the range 0° to 150°C, 1 to 1000 bar have been measured by Kell and Whalley[138]. Densities of mercury in this range based on Bett, Weale and Newitt's equation are listed by Kell and Whalley[138].

4. *Seasoning of Pressure Vessels and Piezometers*

It is standard practice to anneal carefully all piezometers (piezometers are vessels placed inside a pressure vessel and subjected to hydrostatic pressure inside and out that are used for measuring the compression of liquids) used for accurate measurements of volume changes in order to remove internal strains. This would not be necessary if the thermal expansivity and compressibility were independent of the local strain. Apparently, however, they are not independent, and changing the temperature or the pressure to which a strained vessel is subjected both inside and outside changes its shape, and because there is a change of shape there is a change of volume. This distortion can cause errors in the compression of the piezometer as calculated from the compressibility of a specimen of the same material. If the compression of the piezometer is measured relative to the known compression of a reference material, the distortion is not important if it is reproducible. The distortion is not, however, always reproducible, perhaps because the local stresses induced by the pressure and the non-uniform elastic constants may become high enough to cause local permanent deformation and so distort the vessel permanently. Careful annealing (annealing procedures should be obtained from the manufacturer of the material) is therefore necessary for all accurate work. In addition, the piezometer should be pressure seasoned by subjecting it several times to pressures at least as high as those to be used in the measurements before the measurements are begun.

It is well known that the effect of pressure on the electrical resistance of metals, and particularly of manganin, where it has been well documented by many workers since manganin is frequently used for measuring the pressure, is reproducible only when the metal has been pressure seasoned in this way[38].

The cause of the effect does not appear to be known. If the pressure is nominally hydrostatic, either different parts of the metal have different compressibilities or some parts have an anisotropic compressibility because of strain, and so shear stresses exceeding the elastic limit are induced which causes permanent deformation.

Irreproducible volumes due apparently to these causes have been reported for both metal[39] and glass[36, 98] vessels.

Even vessels that are annealed enough to remove all internal stresses can be distorted when subjected to hydrostatic pressure. This occurs if the grains of the metal are not randomly oriented, due for example to the operations used for forming it (see Barrett and Massalski[17]). In any forming by plastic deformation such as forging, drawing or rolling, all metals, even the cubic metals, show preferred orientation of their crystal grains. No crystals are mechanically isotropic, even cubic crystals requiring three elastic constants to describe their elastic properties, compared to two for isotropic materials. A piece of metal in which the crystals have a preferred orientation is therefore elastically anisotropic. A brief account of the influence of texture on the elasticity of metals has been given by Bradfield[33]. Plastic deformation can, of course, cause preferred orientation only if the crystals are plastically anisotropic.

Cubic metals have an isotropic compressibility, and consequently a polycrystalline piece that is well annealed will be compressed isotropically under uniform hydrostatic pressure even if the grains are preferentially oriented. Non-cubic materials are compressed non-isotropically, and a hydrostatic pressure applied to a polycrystalline piece of a non-cubic metal causes the individual crystal grains to deform if they are to maintain contact, and at high enough pressures they will yield plastically. Such materials should never, of course, be used for piezometers except perhaps for the work of lowest accuracy.

The dilatation of a vessel that is subject to internal pressure only is, of course, affected by all these processes. Frequently, the vessel cannot be annealed to remove internal stresses because it has been hardened by introducing some. A careful relief of stress at a temperature that will not cause an unwanted permanent softening is all that can be done. Hayward[114] has reported without giving details that the diameter of a particular vessel, presumably cylindrical, actually decreased at some locations when internally pressurized, and that its distortion caused errors of 'as much as 30 per cent in the compressibility of water' measured with the vessel.

In addition, the internal pressure causes microscopic non-hydrostatic stresses, and with the internal stresses they may cause permanent deformation even if they are well below the yield stress. Even in a polycrystalline cubic metal, the different crystal grains will have different elastic constants in a particular laboratory-fixed direction, and so a non-hydrostatic applied stress will cause different stresses in different grains depending on their orientation relative to the applied stress. Some grains may be stressed enough to be permanently deformed, and so cause hysteresis, particularly on the first few applications of stress. A pressure seasoning is therefore particularly important; its effects, sometimes at least, decay with time[42] and measurements have to be made immediately after the seasoning.

The change of the inside volume V of a closed-end heavy-walled vessel with inside pressure p is, according to the usual theory of elasticity and neglecting end effects,

$$\frac{\partial \ln V}{\partial p} = \frac{1}{E}\frac{3(1-2v) + 2(1+v)k^2}{k^2 - 1}$$

where E is Young's modulus, v is Poisson's ratio and k is the ratio of outside to inside diameters of the vessel. For typical steels, this is about 2×10^{-6} bar^{-1}. The change in volume of a piezometer is determined only by the compressibility if the compressibility is uniform, and is

$$\partial \ln V/\partial p = -3(1-2v)/E$$

which for typical steels is about -0.7×10^{-6} bar^{-1}. It is therefore more important, even for measurements of moderate accuracy, to measure the dilatation of an internally pressurized vessel, than it is of a piezometer.

5. Heat of Compression

When a liquid is compressed, it usually heats. If the compression is adiabatic, the increase in temperature with pressure is given by the adiabatic pressure thermal coefficient which is given by equation II.15. Since T and C'_p are necessarily positive, the sign of β_S is determined by the sign of α_p. A liquid that expands on heating, heats on adiabatic compression, and one that contracts, cools.

Since α is of the magnitude 10^{-3} deg^{-1} for many liquids, V about 100 cm^3 mole^{-1}, and C'_p about 1 J cm^{-3}, β_S is typically of the magnitude of 10 mdeg bar^{-1}. A pressure change of ten bars therefore causes a temperature increase of about 100 mdeg and about eight half-lives (if the decay is simple exponential) must be allowed for this to decay to less than 1 mdeg. This waiting time can be reduced by first compressing the fluid by twice the required pressure increment for one half-life and then reducing the pressure. For water, β_S is of course zero at 4°C, and it is about 1.8 mdeg bar^{-1} at 25°C, the low value being caused by both a low thermal expansivity and a high heat capacity per unit volume.

The decay of the temperature can be followed directly by means of a thermocouple or resistance thermometer immersed in the fluid or indirectly by following the decay of the pressure. The time required for temperature equilibrium within the required accuracy is readily calculated.

6. Corrosive Liquids

Liquids corrode in such a wide variety of ways that no general technique can be suggested for measuring their compression or thermal expansion. The thermal expansion of liquids that do not corrode glass can be measured by the usual pyknometric techniques provided no mercury or grease (if this is what they corrode) is used. The compression can be measured by piezometric methods using a mercury-filled piezometer provided that the liquid is enclosed in thin-walled flexible glass bulbs (Richards and Stull[199]) that are immersed in the mercury. A bellows piezometer can be used if material that is not corroded by the liquid is used for the bellows.

Water at high temperatures and pressures corrodes most metals that are strong enough for high-pressure vessels. The pressure rises if the hydrogen produced does not diffuse into the metal fast enough, and falls if it does[145]. A simple alternative to accepting the corrosion is to place some copper oxide in the vessel[141] for the hydrogen to reduce and reform water. It is probably not very successful and has been little used.

A platinum liner was tried by Coolidge and Mailey[78] many years ago, but was rejected as unsuitable for the accuracy required because the volume was not reproducible enough. Burnham and Wall[68] have swaged a platinum liner to the wall and have used a collapsible gold sack for measuring the pVT properties of water up to 900°C. The sack was used to separate water and carbon dioxide in the pressure vessel, and the volume of the sack was determined from the known pVT properties of carbon dioxide.

7. Gases at High Pressures

The main difference between liquids and gases at pressures high enough that their density is comparable to that of the liquid is that gases require a high pressure to maintain their density. Experimental measurements of the compressibility of dense gases can be made in the same way as for liquids, once the apparatus is filled with the dense gas at a suitable pressure. There are several ways of doing this, which can be summarized as follows.

(1) A heavy-walled metal piezometer fitted with an inward-opening valve is filled with gas at the required pressure[44,45]. The piezometer is placed in a pressure vessel and the pressure increased until the hydraulic fluid just enters the piezometer. The compression of the hydraulic fluid and gas together are measured as described in section IX. The compression of the hydraulic fluid, expansion of the vessel, etc., can be allowed for in separate measurements, and the compression of the gas obtained. If the gas is soluble in the hydraulic fluid, as it was in the kerosene that Bridgman[45] used, it is necessary to assume that no volume change occurs on solution.

(2) A glass or other piezometer consisting of a large bulb with a narrow tube attached to the upper part and a U-tube attached to the lower part is filled with gas at essentially atmospheric pressure which is confined by means of mercury in the U-tube. The piezometer is then placed in a pressure vessel containing mercury and the pressure increased until the gas is compressed into the upper tube. The position of the mercury meniscus in the upper tube can be measured in a suitable way. The method appears to have originated with Cailletet[70] and has been used by Michels and his collaborators for many accurate measurements up to 3 kbar[166,167].

(3) The piezometer is connected to a capillary tube through which gas can be introduced from outside the pressure vessel.

8. External and Internal Heating of Pressure Vessels

At temperatures up to a few hundred degrees Centigrade it is convenient to heat the pressure vessel and its contents by placing it in a thermostatically.

controlled container. But hot vessels will not withstand high internal pressure, and so at high temperatures and pressures it is necessary to heat the sample and its container but not the vessel. A vessel with a furnace inside it and a cooling coil to keep the vessel cool appears to have been first used by Watson[246] for measuring the density of water up to 1000°C and 1300 bar. His apparatus is shown schematically in *Figure 1*. Smyth and Adams[219] used the

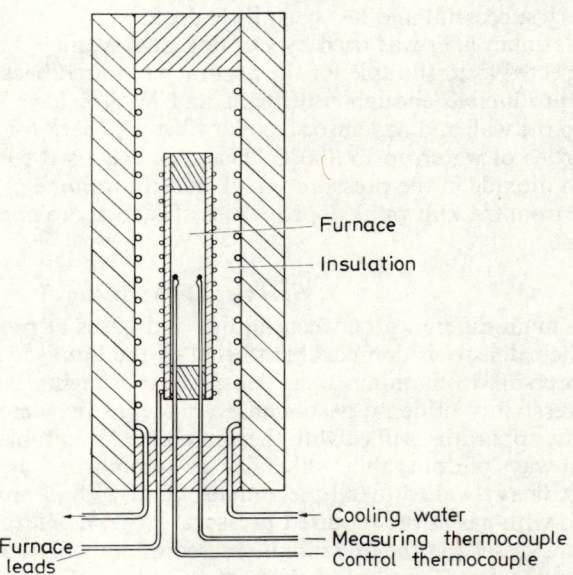

Figure 1. Schematic diagram of Watson's[246] internally heated pressure vessel.

technique for measuring the equilibrium pressure of the system $CaCO_3$–CaO–CO_2. Yntema and Schneider[261] increased the range to 1200°C at 100 bar, and Saurel and Vodar[205] to the design conditions of 1500°C and 10 kbar. The technique has been used in hydrostatic apparatuses in recent years in several other laboratories, for example, by Baker[14], Mangold and Franck[161], Bannard and Hills[16], Barton *et al.*[19], Postill, Ross and Cusack[187], Burnham, Holloway and Davis[67], and these references should be consulted for details. A review has recently been published by Holloway[122]. Externally heated vessels have been reviewed by Presnall[189].

The main experimental problem is, of course, the high loss of heat by conduction and convection through the pressurized fluid. Watson's apparatus, sketched in *Figure 1*, is still reasonably typical. Leads to the internal vessel and for the pressurizing gas are not shown. The pressurized furnace is insulated from the wall of the vessel and the wall is cooled by a cooling coil. The tubing of the coil must, of course, be strong enough to withstand the pressure inside the vessel. For the lower temperatures or the lower pressures when a relatively thin-walled vessel is used, the thermal resistance of the wall may be low enough to permit external cooling to be used to keep the inside cool. The variation of temperature in the furnace caused by convection is frequently reduced by winding the furnace in several parts and controlling and measuring each part independently.

The heating of the inside wall of the vessel has two opposing effects, namely to reduce the strength of the material and to induce thermal stresses that oppose the pressure stresses (Whalley[252, 253, 254]). Only the reduction in the strength of the material is important in most apparatuses.

9. Summary of the more Important Sources of Error in the Measurement of the Compression of Liquids

As well as the usual sources of error in physicochemical measurement, such as the measurement of temperature and pressure, purity of materials, etc., there are a number that are more or less peculiar to the measurement of compression. The more important are as follows.

(1) Inaccurate knowledge of the compression of the reference material, such as another liquid or solid, a sinker, etc., or of the dilatation of the pressure vessel or of the piezometer.

The dilatation of the vessel or piezometer is discussed in section IV.3.

(2) Incomplete filling of the vessel or piezometer with liquid, so that air or other gas is trapped.

If there is not too much for the amount of liquid near it, the gas can usually be dissolved in most liquids by pressurizing for a time. If gas is trapped in narrow re-entrants, enough time may need to be allowed for diffusion of the dissolved gas into the bulk of the liquid before reducing the pressure; otherwise, bubbles may be reformed. Mercury will not, of course, dissolve gases appreciably, and all air must be removed by evacuating the vessel before filling or the mercury must be boiled.

The error is easily tested for since if gas bubbles are present, the rate of compression at low pressures will be very high. If a fraction α of the vessel is occupied by gas at a pressure p, the apparent compressibility κ_{app} of the liquid, assuming that the gas does not dissolve in the liquid, is

$$\kappa_{app} = \kappa_l + \alpha(\kappa_g - \kappa_l)$$

where κ_l and κ_g are the compressibilities of the liquid and gas respectively. For a perfect gas $\kappa_g = p^{-1}$, and at one bar, the apparent compressibility of a liquid whose compressibility is 100×10^{-6} bar^{-1} would be too high by one per cent if α were 1 p.p.m. For mercury, 1 p.p.m. of air by volume would cause an error in the compressibility at one bar of about 40 per cent and at ten bars about four per cent. If a vessel is to be filled with mercury in a single operation, it is clear that it must be evacuated to about 10^{-8} bar or less for accurate work if the remaining air cannot be removed after the filling. Hayward[114] gives some examples of apparent errors that might be due to air in the vessel.

(3) Leaks in the vessel or piezometer.
(4) Improper design of seals.

Seals for piezometers and for pressure vessels whose volume is important in the measurements should be carefully designed. It is advisable, except perhaps for less accurate work, and necessary for the most accurate work,

Table 1. List of some of the most precise measurements of the density of liquids under pressure

Authors	Method	Substance	Temperature range, °C	Pressure range, bar	Relative to	Reproducibility p.p.m.
Holder and Whalley[120,121]	liquid piston, electrical contact	benzene, etc.	25–75	0–100	glass	2–3
Kell and Whalley[138]	constant-volume vessel	water	0–150	1–1000	mercury	6
Wilson and Bradley[257]	bellows piezometer	sea-water	0–40	0–1000		20
Fahey, Kupke and Beams[95]	magnetic float	aqueous solutions	20	0–400	water at 20°C	3
Millero, Curry and Drost-Hansen[169], Millero and Lepple[171]	liquid piston, visual	water, D_2O	2–65	2–34	glass or water	5
Hayward[114]	liquid piston, visual	mercury	20	30–162	iron	6
Millero, Knox and Emmet[170]	magnetic float	aqueous solutions	25	1–1200	water	30

that seals do not introduce stresses that might slowly relieve themselves and so cause a volume change. High sealing forces at a metal-to-metal contact should therefore be avoided if possible, and self-sealing seals should be used whenever possible. The volume enclosed by the seal must be reproducible. One of the most convenient seals for temperatures not far from room temperature is a rubber O-ring, and it has few rivals for sealing moving shafts. The groove in which the O-ring fits must be carefully designed so that the ring occupies essentially the whole volume of the groove when the apparatus is assembled, but is not so tight, particularly when mounted on a moving shaft, that the friction between the shaft and the ring causes heavy wear. The usual manufacturer's design recommendations allow for appreciable movement of the ring parallel to its axis, and this causes an indeterminacy in the volume at low pressures. In addition, triangular-shaped backing rings should be used to prevent extrusion of the O-ring.

Welded seals are satisfactory if they are strong enough to stand the stresses, if they are well annealed, and if the oxidation that occurs when they are made (unless they are in noble metals) can be reduced by circulating argon or is unimportant.

(5) Lack of thermal equilibrium after pressurizing, as discussed in section IV.5.

10. Accurate Measurements

A summary of the measurements of density under pressure that have an accuracy greater than 50 p.p.m. and are known to the author is given in Table 1.

V. Piezometric Methods—Liquid Piston

1. Some General Considerations

A major class of methods of measuring the p–V–T properties of liquids is that in which a piezometer is used, that is a vessel with the experimental fluid confined within it in a suitable way that is placed inside a pressure vessel and that allows pressure to be transmitted to the fluid. The contraction of the fluid in the piezometer when the pressure vessel is pressurized is measured in one of several ways, the most important of which are the movement of a liquid or solid piston that confines the liquid in the piezometer, or the collapse of the piezometer itself, as with the bellows piezometer. The solid-piston and the bellows piezometer are described in sections V.6 and V.7 respectively. In the methods described in this section, the fluid is confined by a liquid piston, which is usually mercury, and the compression is measured by the movement of the meniscus in the piezometer. The method is of course limited to pressures below the freezing pressure of the confining liquid. The freezing pressure of mercury as a function of temperature is shown in Figure 2 (Bridgman[39, 50], Zhokovsky[262]). The lowest temperature at which mercury is liquid is −38.9°C, and the maximum pressure at 0°C is about 7.5 kbar and at 25°C is about 13 kbar. This limitation can sometimes be overcome at the expense of an increased complexity of the equipment by having the main body of the piezometer at the experimental temperature and the confining liquid in its capillary tube at a different temperature, as was described, for example, by Saurel and Vodar[205].

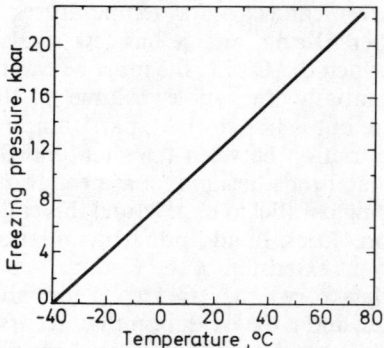

Figure 2. Freezing pressures of mercury according to Bridgman[39, 50] and Zhokovksy[262].

Many versions of liquid-piston piezometers have been made. In some of them, only one position of the meniscus can be detected—the so-called single-point methods. In others, measurements can be made at many or even a continuous range of points—the so-called multiple-point methods. Each sub-class of methods can be further sub-divided according to the method used for detecting the position of the meniscus.

The position of the mercury surface can be detected by looking at it, by noting when it touches an electrical contact, by locating a piece of iron floating on it by means of a differential transformer, by measuring the capacitance between the mercury in a capillary tube and a metallic coating outside it, by locating a magnet floating on it by means of a compass needle or a coil, or by applying a high-frequency voltage to the mercury and detecting the signal radiated through a gap in an electrical screen surrounding the capillary tube. All except the last two methods appear to be capable of high precision, and results of the highest accuracy have been obtained by the visual and electrical contact methods.

Electrical contacts that can be moved to make contact with a mercury surface can, of course, be used in principle, but appear to have been used only rarely. In the simplest designs, the volume of the system changes when the contact moves and so the pressure changes. This can be avoided by several designs (Sage and Lacey[204], Kell and Whalley[138], Barton et al.[19]). In Sage and Lacey's[204] (see section V.9) and Kell and Whalley's[138] designs, the electrical contact was moved through a packing and the consequent volume change compensated by a similar but opposite volume change elsewhere. In Barton et al.'s[18, 19] design the electrical contact was moved by means of a nut rotated by a magnet external to the pressure vessel.

A possible source of error in all liquid-piston methods is a film of liquid between the mercury (say) and the wall of the piezometer. This film can form in two quite distinct ways, as an equilibrium film or as a non-equilibrium film.

Jessup[130] has shown both experimentally and theoretically that under certain conditions a film of liquid can creep against the forces of gravity between a mercury and a solid surface. *Figure 3*(a) represents the situation before creep has occurred and *Figure 3*(b) after it has occurred. If the film forms spontaneously there must be an increase in the Gibbs energy. Consider

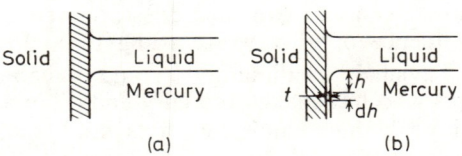

Figure 3. The formation of an equilibrium film of liquid between a mercury and a solid surface.

an elementary horizontal strip of unit length, height dh, and depth h below the mercury surface. The thickness of the film at this depth is t. The increase in the Gibbs energy due to the formation of this elementary strip of film is (Adams[2])

$$d\Delta G = d(G_2 - G_1)$$
$$= [(\rho_m - \rho_l)ght + (\gamma_{ml} + \sigma_{sm} - \sigma_{ls})] \, dh \quad (V.1)$$

where ρ_m and ρ_l are the densities of mercury and liquid, g is the acceleration due to gravity, γ_{ml} is the Gibbs energy of the mercury/liquid surface, and σ_{sm} and σ_{ls} are the works of adhesion of the solid/mercury and liquid/solid surfaces. The increase in the Gibbs energy due to the formation of the whole film is therefore

$$\Delta G = \int (\Delta \rho g h t - \Gamma) \, dh \quad (V.2)$$

where

$$\Delta \rho = \rho_m - \rho_l \quad (V.3)$$

and

$$\Gamma = -(\gamma_{ml} + \sigma_{sm} - \sigma_{ls}) \quad (V.4)$$

Since g and t are inherently positive and $\Delta\rho$ and h have the same sign, a necessary but not sufficient condition for the formation of a stable film is that

$$\Gamma > 0 \quad (V.5)$$

If the film is stable, the Gibbs energy is stable with respect to variations in the thickness at constant temperature and pressure. This leads to the condition that the equilibrium thickness t_e of the film is such that at this thickness

$$\Delta \rho g h = \partial \Gamma / \partial t \quad (V.6)$$

if the density of the liquid film is independent of its thickness. Unfortunately, $\partial \Gamma / \partial t$ is not known for any liquid, and so predictions cannot be made at present. Nevertheless, it is clear that films can form spontaneously in some circumstances.

Jessup[130] observed that toluene or diphenyl ether did not creep past mercury in a glass apparatus, but that it did so in a stainless-steel tube at the rate of about 10 mm³ per day. The creeping could be prevented by inserting in the mercury line a length of platinum capillary with a thin coating of gold on the inside.

Whether or not a film is thermodynamically stable, one will always form when mercury is raised in a tube which it does not wet. The film will then drain under the influence of gravity according to hydrodynamic theory (see, for example, Levich[153], Tallmadge[231], and references cited there). But as anyone knows who has shaken mercury and oil in a glass vessel and left it to stand, a uniformly thick film is not stable, but breaks up into isolated drops. The reason for this is as follows.

The quantity Γ defined by equation V.4 is the Gibbs energy of formation of unit area of solid/mercury surface by removing solid/liquid and liquid/mercury surface. If it is negative, a uniform film will tend to break into drops in order to form more solid/mercury surface. The condition therefore for the break-up of a uniform film is

$$\Gamma < 0 \tag{V.7}$$

if the formation of the drops does not change the gravitational potential energy.

To see how a uniform film might be unstable to small fluctuations of its thickness, we consider two equal elementary areas 1 and 2 of film, both at the same height and of thickness t and density ρ. A small mass of liquid is transferred from area 1 to area 2, causing the thickness of 1 to decrease by dt_1 and the thickness of 2 to increase by dt_2. The change of the Gibbs energy per unit area due to this transfer is

$$dG = -(\partial \Gamma/\partial t)(dt_2 - dt_1) \tag{V.8}$$

To terms of the first order,

$$dt_2 = -dt_1 = dt \tag{V.9}$$

so that

$$(\partial G/\partial t)_{T,p} = 0 \tag{V.10}$$

where dt and $-dt$ are the change of thickness of the elementary areas, as it must be by symmetry. The surface Gibbs energy of a uniform film is therefore at either a maximum or a minimum.

If G is expanded in a Taylor series in the thickness, we find for the second derivative

$$\partial^2 G/\partial t^2 = -\partial^2 \Gamma/\partial t^2 \tag{V.11}$$

If the quantity on the right side of equation 5.11 is positive, G is a minimum when the film has uniform thickness and the film is thermodynamically stable. On the other hand, if the right side is negative, then G is a maximum for uniform thickness, and the film is thermodynamically unstable. The likely variations of Γ with t for the cases in which Γ at height t is positive and negative respectively are sketched in *Figure 4*. If the limiting value of Γ is negative, $-\partial^2 \Gamma/\partial t^2$ is negative, and the uniform film is unstable, as predicted above.

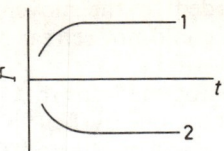

Figure 4. Schematic diagram of the likely variation of Γ with t if Γ is positive (1) and negative (2) at high thicknesses.

The unstable film will tend to drain according to hydrodynamic laws, and at the same time to break up into droplets confined by the mercury which contacts the solid everywhere except at the droplets. The film will therefore be pinched off and will eventually cease draining.

A few preliminary tests of this effect have been done by Kell and McLaurin (unpublished). Mercury was injected into a long capillary tube with two electrical contacts sealed across it. The volume defined by the contacts was measured by injecting mercury into the tube and was reproducible to ~ 100 p.p.m. It was apparently smaller by ~ 500 p.p.m. when dioctyl sebacate was floating on the mercury than when the mercury was dry, and the trapped oil film broke into drops. When bulbs were blown in the tube, as in the apparatus of Holder and Whalley[120], illustrated in *Figure 14*, about 4 500 p.p.m. of oil was trapped, presumably because more oil was trapped by the shallow slope of the upper part of the bulbs than by the vertical walls. With water or methanol, less than one half of this fraction was trapped. It is recommended therefore that for the most accurate work, and if the liquid is compressed by mercury, liquid-piston piezometers should be calibrated with a liquid of the same viscosity as that of the experimental liquid, and bulbs should not be blown in the capillary tube. Nothing is known at present about the rate of draining or pinching-off of the film.

It seems likely that the more slowly the mercury is raised the less fluid will be trapped if the film is not thermodynamically stable. This point was commented upon by the Earl of Berkeley and C. V. Burton[28] who showed that a slow and uniform increase in the pressure could be obtained by means of a pressure balance. The weight to be added to the rotating balance was suspended by three long extensible springs from a cage that was mounted on a ball bearing. The cage and associated parts were lowered slowly so that the weight was only slowly taken by the rotating balance.

2. *Single-point Methods*

In single-point methods, one compression is measured for each set-up of the apparatus. Perkins[183, 184] was the first to use them and had three versions, one using a piezometer with a solid piston, which has already been mentioned in section III[183]. The second[183] used a cylindrical steel piezometer whose wall was partly collapsed and which was fitted with an inward-opening valve. On compression, fluid entered the piezometer, but could not escape on decompression, and so forced out the walls. In the third method[184], the liquid was confined by a short piston that fitted closely in a tube. The maximum movement of the piston was recorded by the movement of a light spring that advanced with the piston but could not retreat because of friction with the tube wall.

Other single-point liquid-piston methods that have been used but are not now of much practical interest are one by Tait[229] who used an index similar to Perkins's floating on mercury, and one by both Cailletet[69] and Tait[227] who determined the maximum movement of mercury by the length of gilding or silvering dissolved from the inside of a tube. Only two methods in this class seem to have practical interest at the present time. These are Aimé's method and the electrical contact method. The floating-index method has been revived recently (Kumagai and Toriumi[148]) and is mentioned for completeness.

A. Floating Index Method

The method (Perkins[184], Tait[229]) has been rarely used in recent times because it is inconvenient and the only use I know of is that of Kumagai and Toriumi[148] who measured the density of liquid ammonia to 1.8 kbar between $-20°$ and $40°C$ to an accuracy of about 200 to 700 p.p.m. The piezometer was a bulb of about 4 cm^3 volume with a capillary tube attached that was dipped in mercury. The index spring was moved by a steel float on the mercury.

B. Aimé's Method

In this method (Aimé[6]), the piezometer opening is closed by mercury in such a way that when the liquid inside it is compressed, the contraction carries mercury into the piezometer. The mercury then falls away from the inlet tube so that only the experimental fluid flows out when the piezometer is decompressed. The piezometer is then removed from the pressure vessel and weighed. An important source of error is clearly the finite size that the mercury drop

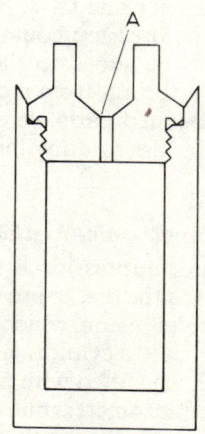

Figure 5. Bridgman's[39] version of Aimé's piezometer.

attains before falling away from the inlet. According to Bridgman[39] it was used in many modified forms by other experimenters between 1843 and 1911.

Bridgman's[39] piezometer, which was used to measure the compressions of mercury (Bridgman[39]) and water (Bridgman[40]) up to about 10 kbar, is shown in *Figure 5*. It consisted of a steel vessel with a very fine capillary tube A for the entry of mercury. The capillary was made by drilling a hole in the head and plugging it with a pin along which a fine scratch had been made. The mercury drop that fell from the hole was said to be barely perceptible to the eye. The piezometer was used to measure the compression of mercury by filling it with mercury, placing it upside down in the pressure vessel, and allowing water to enter on compression and mercury to leave on decompression. The compressibility of the steel was directly measured in another experiment, so that runs with water and mercury as the piezometer fluids, which are necessary to determine the compression of one, yield the compression of both. The method will work only if during a run the compression increases monotonically with the time, except for a change of volume equal to the volume of the capillary tube. The accuracy of the compressions appears to be about 100 p.p.m. It was during these measurements of the compression of water that ice VI (Bridgman[40]) was discovered.

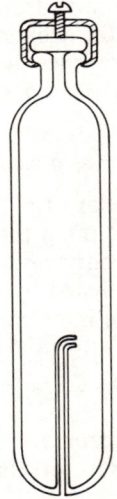

Figure 6. Adams's[2] version of Aimé's piezometer, as modified by Gibson[99].

Adams's[2] version of Aimés piezometer is illustrated in *Figure 6*, as modified by Gibson[99] by closing the upper end with a ground disc held by a clamp instead of by a conical stopper sealed with mercury. The piezometer was filled with the experimental fluid, the lower end was dipped in mercury, and the pressure was increased. Mercury entered the piezometer through the re-entrant capillary and was trapped. The volume of the piezometer used by Adams[2] was 8.5 cm^3 and the diameter of the capillary at the tip was 0.5 mm. The volume of each drop was about 0.1 mm^3 or about 120 p.p.m. of the volume of the piezometer. Pressures were measured to about 1000 p.p.m., which is just about accurate enough for the precision of the compressions. According to Gibson[99], the diameter of the capillary at the lower end should also be about 0.5 mm, presumably to define well the amount of liquid in the piezometer. The pressure range was 0 to 1 kbar, but there seems to be no reason why pressures up to the freezing point of mercury should not be used. Glass piezometers should not, of course, be used at pressures where there is appreciable hysteresis in the compression of the glass. Although the pressure vessel must be opened and the piezometer weighed after every compression, the method has been widely used because of its simplicity, particularly if only a few measurements are required.

Gibson[100] later placed his piezometer inside a pressure vessel fitted with a window so that he could see the mercury drop. The pressure was adjusted until the mercury was just flush with the tip of the capillary tube. This resulted in an increase in the accuracy to about 40 or 50 p.p.m. which corresponded to the accuracy of his pressure measurement at 1 kbar. No doubt the accuracy could be improved by measuring the pressure more accurately, for example by using a pressure balance.

There is a fundamental limitation to the accuracy attainable by this method unless the meniscus can be seen as in Gibson's[100] apparatus. The smaller the capillary tip the smaller the mercury drop and so the smaller the error in the volume. But at the same time, the drop in pressure across the surface of the drop is greater and so the pressure is less accurate. The volume v of the drop just before it breaks away from the tip of the capillary tube is given by (Adam[1], p 368)

$$v = r\gamma/g\,\Delta\rho\,F \tag{V.12}$$

where r is the radius of the capillary tip, γ the interfacial tension between mercury and the experimental liquid, g the acceleration due to gravity, $\Delta\rho$ the difference in density of mercury and the liquid, and F an empirical function of v/r^3 whose values are listed by Adam.

The pressure drop δp across the interface just as the drop falls is approximately

$$\delta p = 2\gamma/r \tag{V.13}$$

The pressure of the liquid in the piezometer is less than the measured pressure by this amount. Unless it is allowed for, this corresponds to an error in the volume of

$$\delta v = 2\gamma\kappa V/r \tag{V.14}$$

where V is the internal volume of the piezometer. The volume of the drop and the error in the pressure have equal effect when

$$r^2 = 2g\,\Delta\rho\,F\kappa V \tag{V.15}$$

For a piezometer volume of 10 cm^3 with $\Delta\rho = 12$ g cm^{-3},

$$r = 50F^{\frac{1}{2}}\,\mu\text{m} \tag{V.16}$$

Unfortunately, values of F have not been tabulated for such small tubes, but by extrapolation from known values (Adam[1], p 378) they are probably much less than 0.1. The optimum value of r is therefore of the order of magnitude of one micron. Only Bridgman[39] appears to have approached this optimum.

For a capillary of radius 10 μm and an interfacial tension of 500 dyn cm^{-1}, the pressure difference across the interface is about 1 bar. If V is 10 cm^3, then the volume of the drop relative to that of the piezometer is

$$v/V = 2 \times 10^{-6}/F \tag{V.17}$$

It would appear therefore that the maximum accuracy in the compression that can be expected by this method, unless the mercury meniscus is observed, is of the magnitude of 50 to 100 p.p.m. as attained by Bridgman[39].

C. Electrical Contact Method

Many measurements were made by Richards during the first quarter of this century using the piezometer shown in *Figure 7* (Richards and Jones[196]).

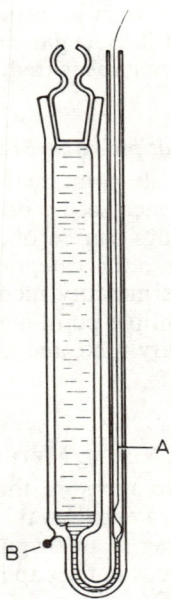

Figure 7. Richards and Jones's[196] electrical contact piezometer.

It consisted of a U-tube with a wide and a narrow leg. The wide leg was stoppered and held the liquid being investigated. The bottom of the U contained a small quantity of mercury that made contact with the platinum rod A and the platinum seal B. The piezometer was placed in a pressure vessel and pressurized until the mercury just touched the rod A. The pressure was then released, the vessel opened, and successive accurately known amounts of mercury added to the U. The pressure at which the mercury just touched the rod A was measured after each addition. The compression relative to mercury can be obtained by repeating the measurements with the piezometer full of mercury. Several variants of the piezometer were described (Richards and Shipley[197], Richards and Stull[198,199]).

The maximum pressure is limited by the metal–glass seal. The glass can be compressed more than the metal and so as the pressure is increased, the glass surrounding the metal is put into tension and it eventually cracks. Richards's maximum pressure was 500 bar, but no doubt the method could be used to at least 3 kbar if platinum tubing rather than rod or wire were sealed through the glass (Baliga and Whalley[15]), and measurements could probably be made near 5 kbar if a short life were acceptable. The glass–metal seal can be eliminated completely if a.c. techniques are used and a capacitive admittance is provided through the wall of the vessel via a metal coating on the outside or by immersing it in mercury (see section V.3.D.). No report of this having been done appears to have been published, but it would extend the usefulness of the method to the freezing pressure of mercury. A version of this method in which a metallic piezometer was used was described by Talbott[230].

Liquids that attack mercury, such as bromine, were enclosed (Richards and Stull[199]) in thin-walled flat flexible glass bulbs which were placed in the piezometer, and their compression measured.

3. Multiple-point Methods

A serious disadvantage of single-point methods is, of course, that only one reading can be obtained with each set-up of the apparatus. Multiple-point methods in which many readings can be obtained have many advantages. Canton's[71,72] apparatus falls into this category (see section III); the pressure vessel was transparent and the mercury meniscus could be seen. The first attempt to use a non-visual multiple-point method was Tait's[228] who sealed electrical contacts into a capillary tube, and many other methods have been developed since that time.

A. Visual Methods

Canton[71,72] was the first to measure the compressibilities of liquids. He placed a glass bulb, to which a capillary tube was attached and which was filled with liquid, into the receiver of an air pump and visually observing the change of position of the meniscus in the capillary tube with pressure. Many have used this method since. Downer and Gardiner[87], for example, compressed liquids using nitrogen pressure in a glass pipette not pressurized externally. Somewhat surprisingly, a patent on the cell was applied for. Oersted[180] extended the technique to 50 bar by replacing the air receiver by a heavy-walled glass tube filled with water and using a pellet of mercury in the capillary tube to separate the experimental fluid from the water. A more recent version of this apparatus was used by Suchodski[226] and is illustrated in *Figure 8*. The piezometer C contains the experimental liquid confined by the mercury into which the capillary neck at the lower end dips. It was suspended by wires inside the glass tube B which was cemented to the steel block A, through which pressure was introduced by a suitable connection.

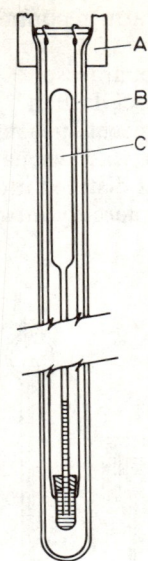

Figure 8. Suchodski's[226] piezometer for visual detection of the movement of the mercury meniscus.

The change in the position of the mercury with pressure was observed. Pressures up to 19 bar were used and the accuracy of the compressibility as estimated from the accuracy of the various measurements was 1.7 to 3.5 × 10^{-6} bar^{-1}.

Amagat[8] in some of his experiments used a piezometer that was not pressurized externally, and visually observed the compression of the fluid contained in it by the movement of mercury in a capillary tube. The maximum pressure was limited to 400 bar by the bursting of the glass capillary tubes, conditions also reached by Barus[20]. Later, Amagat[10] placed the piezometer in a pressure vessel and the mercury surface was observed through windows fitted to plugs screwed into the pressure vessel. The method could be used to 360°C and 1 kbar, and it was perhaps the earliest use of such windows at high pressure. A similar apparatus was described by Smith[217] and used with the unusually low pressure increments of 0.2 bar and a total range of about 1.4 bar.

A more elaborate and accurate version for pressures up to 80 bar was described by Jessup[129]. A long narrow bulb that was pressurized internally only was filled with the experimental liquid and confined by mercury in a U-tube. The mercury was pressurized by gas whose pressure was measured by a pressure balance. The change of volume of the fluid caused by a change of either temperature or pressure was determined by measuring the movement of the mercury meniscus in the U-tube. The usual corrections were made. The dilatation of the glass bulb caused by changes of pressure and temperature were measured relative to that of mercury. The accuracy of the final results was not stated but a comparison of the mean compressibility over the range 0 to 50 bar with the measurements of Kell and Whalley[138] and the compressibilities at 1 atm recommended by Kell[137] suggests that they are accurate to about 0.5×10^{-6} bar^{-1}.

Several relatively accurate apparatuses of this kind have been described in recent years. Diaz Peña and McGlashan[86] have used the glass piezometer shown in *Figure 9*. As in all recent measurements, the capillary tube was precision-bore. The piezometer was evacuated through the ground joint A and the experimental liquid was distilled into it. through a tube sealed to D. The tube D was sealed off and mercury introduced through A to seal off the

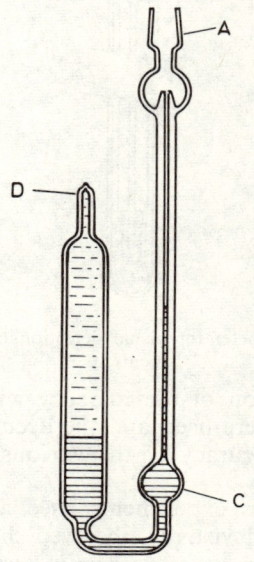

Figure 9. Diaz Peña and McGlashan's[86] visual detection piezometer.

liquid. The bulb C acted to prevent air or water from being drawn into the piezometer when it was cooled. The piezometer was placed in a pressure vessel made from a piece of Perspex [poly(methyl methacrylate)] tubing, and the position of the mercury meniscus at 5-bar intervals up to 30 bar was measured by a cathetometer to ± 10 μm. The reproducibility and accuracy were not carefully assessed but can perhaps be judged from the statement that the displacement of the mercury meniscus, when the bulb contained carbon tetrachloride, was linear in the pressure up to 30 bar. At this pressure, the contribution of the square term in the compression as a function of pressure is 63 p.p.m.[120] or about 1/50 of the compression. An accuracy of about one to two per cent in the compressibilities can therefore be expected. Diaz Peña and Cavero[85] modified this apparatus by making the pressure vessel from a block of Perspex with two holes drilled in it, connected along their length, to take the bulb and the capillary tube. The range was 0–20 bar.

Millero, Curry and Drost-Hansen[169] in the apparatus illustrated schematically in *Figure 10* placed the body D of their piezometer, which had a volume of about 450 cm³, in a brass cylinder C and the capillary tube B in a much narrower glass boiler tube A attached to the end closure of the piezometer. Although their maximum pressure was 15 bar, it should be possible to go to much higher pressures by this technique. They greatly increased the

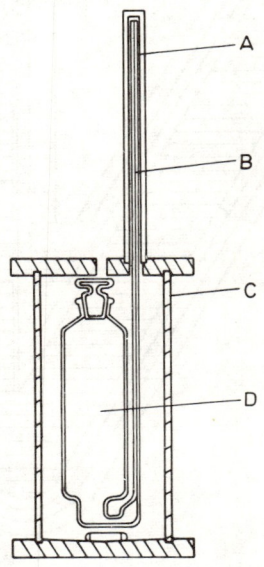

Figure 10. Millero, Curry and Drost-Hansen's[169] piezometer shown schematically. Bolts were used to hold the heads on the pressure vessel and to hold the sight glass A to the vessel.

accuracy by controlling the temperature inside the pressure vessel to ±300 μdeg over the 45 to 60 min required for a measurement, by measuring the pressure with a Texas Instruments fused-quartz pressure gage to an accuracy of 200 p.p.m., and by locating the mercury with an Ole–Dick cathetometer to ±3 μm. Measurements were made over the range 2 to 15 bar at about 5-bar intervals and the compressibility of water was obtained with an average deviation and an estimated experimental error of 0.1 to 0.2 × 10^{-6} bar^{-1}. The compressibility of water was assumed to be independent of pressure in the range 2 to 15 bar, which is reasonable since it changes by only 0.0128 × 10^{-6} bar^{-2} at 25° [138]. In later measurements[171] on both H_2O and D_2O, the pressure intervals were increased to about 17 bar and the maximum pressure to 34 bar. The accuracy was increased slightly to about 0.1 × 10^{-6} bar^{-1} in the compressibility of both H_2O and D_2O, and to about ±0.06 × 10^{-6} bar^{-1} in the difference of the compressibility of H_2O and D_2O. The earlier compressions were measured relative to the compression of glass, the values of which were attributed to Jolley[132]. The later measurements were made relative to the compression of H_2O[138].

A similar apparatus illustrated in *Figure 11* was used by Hayward[115] to measure the compression of mercury at 20°C and 192 bar relative to that of

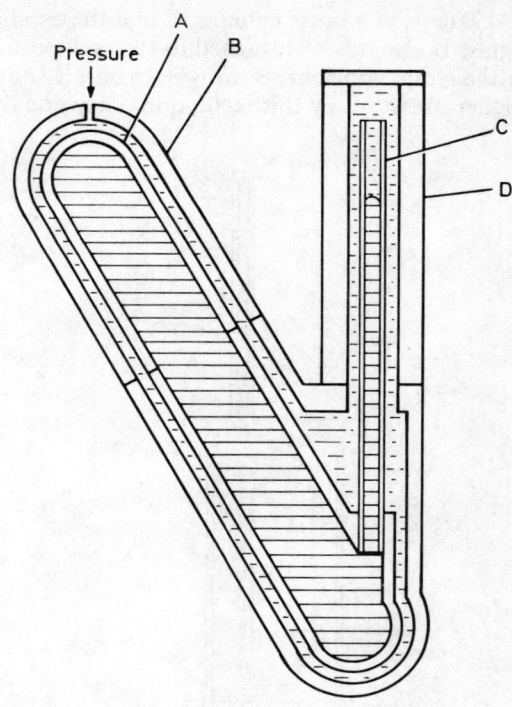

Figure 11. Hayward's[115] visual detection piezometer.

iron to a high accuracy. The piezometer A was made of so-called isotropic titanium and was ~1 l. in volume. It was jointed in the middle so that the mercury could be replaced by a cylinder of iron and the compression of mercury relative to iron measured, and was placed in the outer jacket B. The mercury meniscus was observed in the precision-bore capillary tube C, which was of about 2 mm inside diameter and 50 cm long, through the Pyrex cylinder D. Neither the joint in the piezometer, that between the piezometer and the capillary tube, nor that between the pressure vessel B and the sight glass D were described. The temperature was controlled to $\pm 0.01°C$ and the pressure measured with a pressure balance. The compression was measured about thirty times between 30 and 162 bar when the vessel contained mercury, but only four times when it was nearly filled with a large cylinder of iron. The difference between the two sets of measurements, if the initial and final pressures are the same, is the compression relative to iron of a volume of mercury equal to that of the iron. Both should, of course, be measured with the same care. The results were extrapolated to yield the compression at 192 bar to an accuracy of 0.8 per cent or 6 p.p.m. in the density. The claims made by Hayward for the unique method of analysing the results that he used may not be universally accepted. Ewing, Marsh and Stokes[94] have described another version of this design.

Yet another version, which is made completely of glass and is illustrated in *Figure 12*, has been used by Coleman and Cruickshank[81]. The vessel B, which was contained in a pressure jacket C, was filled with the experimental liquid which was confined by means of a mercury pellet in the precision-bore

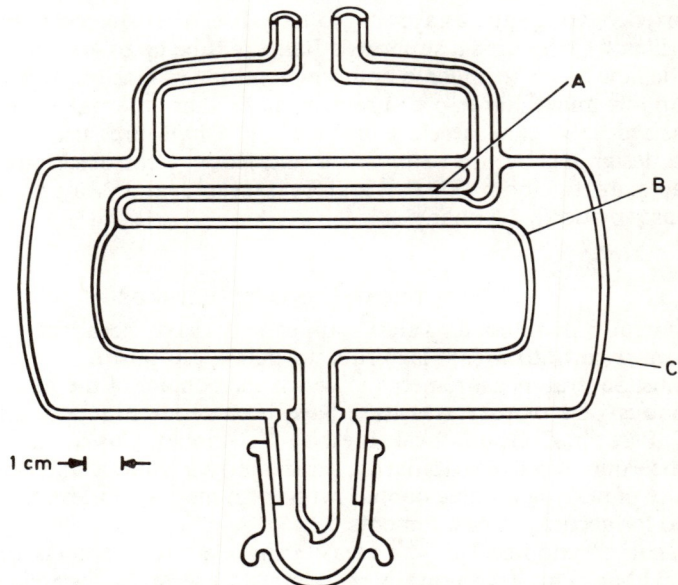

Figure 12. Coleman and Cruickshank's (Cruickshank, Ackermann and Giguère[81]) visual detection piezometer for measuring both isothermal and adiabatic compressibilities up to 5 bar.

capillary tube A. The space between B and C and on the left side of the mercury pellet in A was filled with a pressurizing fluid whose pressure was increased quickly by 1 or 2 bar and the movement of the mercury pellet noted. The apparatus was used for both adiabatic and isothermal compression measurements by noting the movement of the mercury within a few seconds of the rapid release of pressure, and again after thermal equilibrium had been attained. The liquid in the pressure jacket was chosen so that for it $(\partial T/\partial p)_S$ was slightly greater than that for the experimental liquid to compensate for the low adiabatic heating of the glass. Then the mercury meniscus would stay in the same position for several minutes after release of the pressure. The apparatus was designed primarily for measuring the heat capacity of liquids by means of equation II.14, and a precision of one to two per cent in the heat capacity was claimed. It seems clear that with appropriate modifications, including using a metal pressure vessel and detecting the mercury meniscus by non-visual methods, the method could be used under high pressure.

So far as I know, no one has used the visual method at pressures higher than Amagat's[8] and Barus's[20] 400 bar, and this was with a capillary tube that was not pressurized externally, except for Amagat's later[10] use of conical windows to 1 kbar. The use of small windows could undoubtedly greatly extend the pressure range but at the expense of accuracy because of the distortion of the windows. The maximum pressure for a purely piezometric viewing method appears to be Hayward's[115] 191 bar. Greater pressures could be reached using a commercial armoured sight gage or an externally supported viewing tube as appears to have been first suggested by Roebuck and Miller[201], who used a supported Bakelite tube up to 400 bar. Since glass is brittle, it is much stronger in compression than in tension, and a glass tube will support much higher pressures externally than internally. An externally pressurized tube can therefore be used as a high-pressure viewing tube. Such a design has been used by Wells and Roof[250] for observing liquid/gas interfaces up to about 500 bar and by Stryland and May[225] for Raman spectroscopy up to 2.5 kbar.

B. Electrical Contact Methods

As already mentioned, Tait[228] appears to have been the first to seal platinum wires into a capillary tube to detect the position of the mercury meniscus. Several measurements in which the contact of the mercury with a continuous wire or tube was used have been reported (Carnazzi[73], Bridgman[45], Essex[93], Cohen and van Lieshout[76], Skinner, Cussler and Fuoss[216]). The uncertain effects of capillarity, particularly around a fine wire, and the difficulty of making reliable contact with most metals, makes this not a good method for accurate measurements.

Amagat[9, 10] adopted Tait's[228] idea of using electrical contacts for pressures up to 3 kbar. This technique is in common use at the present time. His piezometer was a glass bulb with a capillary tube attached, into which were sealed several electrical contacts as shown in *Figure 13*. The capillary tube dipped into mercury. For measurements on gases at densities of about those of liquids, a large bulb was attached to the capillary tube as in *Figure 13*(b), to accommodate the higher initial compression of the gas; measurements were made only when the gas had been compressed into the upper bulb and capillary tube. The highest pressure reached by this technique appears to be 5 kbar by Eduljee, Newitt and Weale[92]. This limit is imposed by the failure of the platinum–glass seals.

Accurate measurements on several organic liquids and liquid mixtures were made by Holder and Whalley[120, 121] by this technique. The main aim of the experiments was to determine the isothermal compressibilities at zero pressure, and the pressure range was chosen to give accurate values of the curvature of the p–V isotherms at low pressures without making measurements at very high pressures. The range of the measurements was therefore about 100 bar and the pressure interval 5–10 bar (see section IV.1). The apparatus is shown in *Figure 14*. The bulb A was about 0.3 l. in volume and was joined to a capillary tube D of 0.8-mm internal diameter on which were blown ten bulbs, each of internal volume 0.33 cm³. Eleven electrical contacts made of 0.005-in.-diameter platinum wire were sealed into the capillary and were joined electrically by platinum strip painted on the outside. The short-circuiting of successive lengths of strip by the rising mercury column was

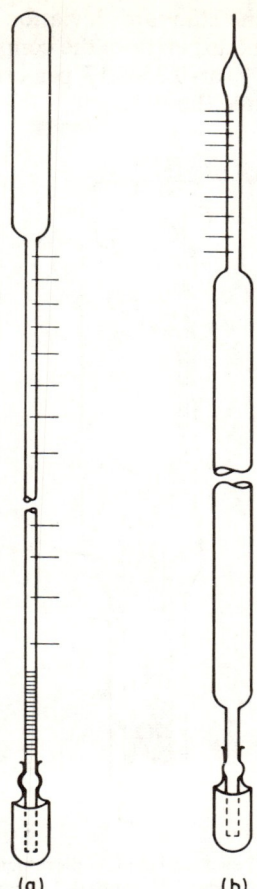

Figure 13. Amagat's[9] two versions of Tait's[228] electrical-contact piezometer for liquids (a) and gases (b).

measured with a Wheatstone bridge arranged so that the potential difference before contact was <10 mV and the current carried by a contact <15 μA. The volume of each of the bulbs was measured to ~1500 p.p.m. (or about 1.5 p.p.m. of the volume of the main bulb A) by injecting mercury from a calibrated screw injector. The bulb A was supported on a copper support block C, and an aluminum spacer B and aluminum chips were inserted to help thermal contact. The half-life for the approach to thermal equilibrium was about 10 min, and about 1.5 h was required for equilibrium. The temperature was controlled to 2 mK and the pressure was measured to 300 p.p.m. with a pressure balance, the reproducibility of the contact pressure was 10 mbar and the difference in pressure between making and breaking contact was about 5 mbar. When the volumes were fitted to a power series in the pressure, usually a cubic, the standard deviation was about 2 p.p.m. In the best measurements, the standard error of the compressibility at zero pressure was 0.05–0.17×10^{-6} bar^{-1} or 0.05–0.17 per cent. The actual uncertainty is thought to be about double these values.

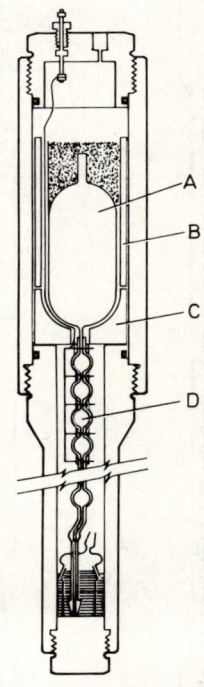

Figure 14. Holder and Whalley's[120] electrical contact piezometer for measuring the compression of liquids in the range 0–100 bars.

Extensive and accurate measurements on gases at pressures up to 3 kbar have been made by Michels and his collaborators[166, 167] in the range 0 to 150°C using Amagat's method. Two apparatuses were used, one for the range 16 to 250 bar[166] and one for the range 70 to 3000 bar[167]. Although these techniques are discussed in more detail in Chapters 6 and 8, they are mentioned here because the techniques used after the gas has been compressed to liquid densities are directly applicable to liquids.

Several measurements of the compression of molten salts have been reported using this technique (Pollard, Crowe and Strauss[186]).

C. Differential Transformer Method

The position of a mercury surface can be measured by locating a piece of iron floating on it by means of a differential transformer, whose principle is illustrated in Figure 15. The secondary of the transformer consists of two coils connected back to back, so that the output is zero when the iron core is exactly in the middle. If the transformer is movable, it can therefore be used to locate the float.

A piezometer based on this principle was first described by Doolittle, Simon and Cornish[89] for use to 4 kbar and their apparatus is illustrated in

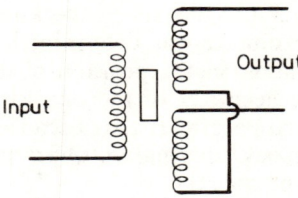

Figure 15. Principle of the differential transformer.

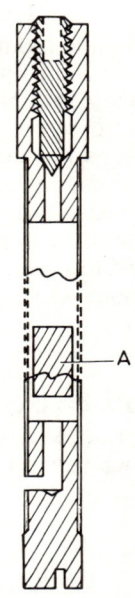

Figure 16. Doolittle, Simon and Cornish's[89] piezometer using a differential transformer to detect an iron float on the mercury.

Figure 16. It was made of precision-drawn thin-walled stainless steel, type 321, about 475 mm long, 2.4 mm o.d., 0.75 mm wall terminated by a screw closure at the top and a plug valve at the bottom for retaining mercury. The float A was of steel, 1.75 mm diameter and 2.5 mm long. A little mercury was added and the remaining volume was filled with liquid via the upper opening. The loaded piezometer was then placed in a vessel made of a length of $\frac{3}{8}$-in.-o.d., $\frac{1}{8}$-in.-i.d. stainless-steel tubing so that the lower end dipped in mercury, and the plug valve at the lower end was opened. When the vessel was pressurized, the movement of the steel float was followed by two differential transformers a fixed distance apart, one of which was used to locate the upper closure of the capillary tube, which was used as a reference point, and one to locate the float. The accuracy of location was about 10 μm. The mass of liquid in the piezometer was determined by extrapolating the length of the column to low pressure and using the known density[88]. The cross-sectional area of the piezometer was determined by weighing the mercury released from it for a change in the position of the meniscus as measured by the transformer. The accuracy of the density was about 300 p.p.m. A major advantage of this apparatus is that because the pressure vessel is so small, it being only $\frac{3}{8}$ in. o.d., thermal equilibrium is reached rapidly and measurements can be done quickly.

A similar apparatus was used by Grindley and Lind[110] to pressures of 8 kbar. Only one transformer was used and it was moved along the pressure tubing by the lead screw of a tool-room lathe with an accuracy of about 25 μm. The maximum error in the density is thought to be about 200 p.p.m.

There is no doubt that this method has not yet been taken to its maximum accuracy nor has it been fully developed. Greatly increased accuracy would be obtained if the capillary tube containing the float were connected to a larger piezometer and if the temperature were controlled and the pressure measured to a suitable accuracy, as was done by Bradshaw and Schleicher[34] in measurements of the thermal expansion of sea-water relative to that of mercury up to 1 kbar using a fused quartz dilatometer. The large piezometer could be kept at a different temperature from the tube containing the mercury, and so measurements at high and low temperatures could be made, and if the mercury was maintained at about 75°C, pressures up to 20 kbar could be obtained.

D. Capacitance Method

The position of a mercury surface in a non-conducting tube can be measured by means of the capacitance between the mercury and an electrode on

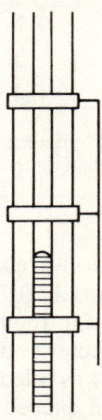

Figure 17. Discontinuous capacitance method of detecting the position of a mercury surface.

the outside of the tube. The electrode can be several narrow bands of metal, and the mercury detected at fixed positions, or it can be continuous and the position of the mercury detected continuously.

If narrow bands of metal are used as in *Figure 17*, the mercury is at the midpoint of a band when the capacitance changes most rapidly with pressure, and the magnitude of the capacitance tells which band the mercury is near (Whalley, unpublished work). The accuracy can probably be made comparable to that of the electrical contact method used by Holder and Whalley (see section V.3.C) and should be usable at all pressures below the freezing pressure of mercury since there are no glass–metal seals to fail. Two advantages of this, as against the continuous method described below, are that the location of the mercury surface does not depend on knowing the effect of pressure on the permittivity of the glass, and the calibration of the capillary tube can be done between fixed points.

In the continuous method, the capillary tube is coated with a platinum film and the capacitance between the mercury and the platinum is measured. The method is in principle extraordinarily sensitive. The capacitance per unit length of the capillary tube is $2\pi\varepsilon\varepsilon_0/\ln k$ where ε is the permittivity of the glass, ε_0 the permittivity of free space, and k the ratio of the outer to the inner diameter of the capillary tube. If $\varepsilon = 5$, a typical value for a glass at audio frequencies, and $k = 5$, the capacitance of the capillary tube is 0.12 pF mm^{-1}. Capacitance bridges having a sensitivity of 10 aF can now be purchased, and so a nominal sensitivity of 0.1 µm or 1000 Å is readily obtained. If the bore of the capillary tube is 1 mm in diameter, this corresponds to a nominal volume sensitivity of about 80×10^{-6} mm^3. The curvature of the mercury

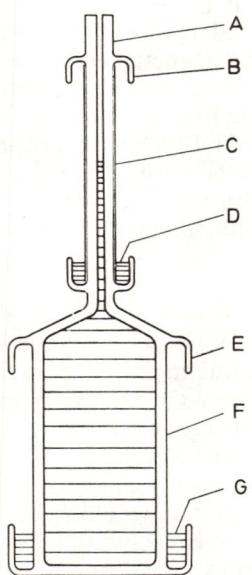

Figure 18. Adams and Whalley's[5] piezometer using capacitance detection of the mercury meniscus.

meniscus and the possibility of fluid being trapped between the mercury and the glass introduce greater errors.

The method has been used by Adams and Whalley[5] to measure the compressibility of fused quartz relative to mercury and of water relative to fused quartz to pressures of 10 kbar. When the compression of mercury was measured, electrical connection to the mercury was made by a capacitance method as illustrated in *Figure 18*. The precision-bore capillary tube A was coated with a fused platinum coating C supplied by Engelhard Industries, and contact to it was made by mercury in the annular cup D. Contact to the mercury in the bulb F was made by a capacitance connection via the mercury in the annular cup G.

The piezometer was placed in the pressure vessel and the change of capacitance between the cups D and G was measured when the pressure was changed. When other liquids such as water were being measured, the piezometer was turned upside down and the end of the capillary tube was dipped into mercury in a cup. Electrical contact was made to the mercury in the capillary tube directly via the mercury in the cup and to the platinum coating via mercury in the annular cup B.

A more complicated method using several ring electrodes has been used by Pribadi[190] for detecting the passage of the boundary between two ionic solutions of nearly the same conductivity. The method was applied to the measurement of ionic transference numbers, but may have application in compression measurements.

E. Magnetic Float Method

A magnet that floats at the interface of two liquids in a non-magnetic piezometer can be detected by a coil or a compass needle. Sherman and Edeskuty[212] have used the magnetic reluctance of a coil to detect the position of a magnet floating on a mercury surface for approximate purposes to about 0.5 mm. A compass needle has been used by Rugta, Stager and Mathur[203]. The sample was confined in a tube with a rubber piston to which an Alnico magnet was attached, and the position of the magnet was detected to 0.3 mm. Many varieties of this technique are, of course, possible, but it seems unlikely to compete in accuracy with, for example, the related differential transformer method.

F. Radiated Signal Method

If a mercury column to which a high-frequency voltage is applied is advanced along a non-conducting capillary tube which is electrically screened for part of its length, an electrode outside the capillary tube picks up a signal as soon as the mercury emerges above the screen[205]. Its sensitivity is about 200 μm as used by Saurel and Vodar[205] for measuring $(\partial p/\partial T)_V$ for gases at high temperature and pressure. The method has not apparently been used for isothermal measurements. It would need modification to be used as a multi-point method, but it is included here for completeness. In a later version of the same apparatus (Malbrunot and Vodar[159]) the mercury was eliminated and pressure balance was detected by a capacitance method using a thin membrane.

VI. Piezometric Methods—Solid Piston

In this method (Perkins[183]) the piezometer is fitted with a piston which transmits the pressure of the hydraulic fluid to the contents of the piezometer. The movement of the piston is measured in some way.

Single-point methods were used by Perkins[183, 184] (see section III). Bridgman[37] used Perkins's design in nearly the original form to measure the compression of mercury up to about 6 kbar. The piezometer is illustrated in *Figure 19*. The cylinder D contained the fluid under investigation and was

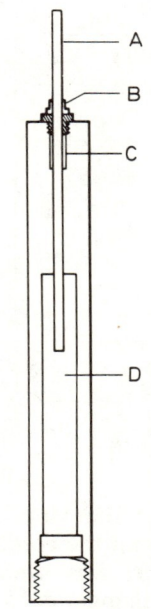

Figure 19. Perkins's[183] solid-piston piezometer as modified by Bridgman.

closed at the lower end by a steel plug and a nut. The piston A slid through the packing C and transmitted the external pressure to the fluid. A collar B slid along the rod A as it moved into the piezometer, but stayed in place as it moved out. The movement of the collar therefore measured the compression of the liquid. The method has all the disadvantages of single-point methods, plus friction in the packing and, as used by Bridgman, a leak at the packing. The leak could undoubtedly be avoided, and the friction could probably be reduced by using O-rings.

Hyde[126] has described a solid-piston piezometer in which the piston is a well fitting steel rod without packing. In test measurements on water, the compressibility was about ten per cent too small, and the apparatus cannot be considered satisfactory. The method does not appear to have been used in recent years, presumably because it is too inconvenient.

The method was converted into a continuous one by Bridgman[46, 47] who used it to measure the change of volume at a solid–solid transition in a solid immersed in a liquid in the piezometer. The piezometer is illustrated in *Figure 20*. The piston D slid in the cylinder E which was closed at the lower end and contained the experimental fluid. The movement of the piston was

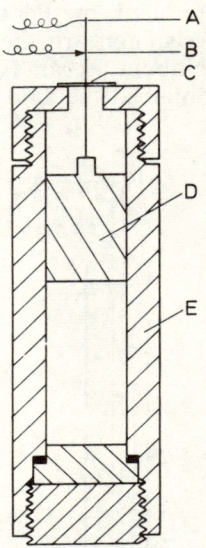

Figure 20. Bridgman's[37] piston piezometer with potentiometric detection of the movement of the piston.

determined[43] by the resistance of a slide wire A, which was measured potentiometrically to eliminate errors from the contact resistance between the wire B and the sliding contact C which was attached to, but insulated from, the cylinder. The wire was made of manganin to reduce thermal e.m.f.s with the copper lead wires. The sensitivity was about 0.5 μm (Bridgman[57], p 125). The details of construction of the slide wire are described in an earlier paper when the method was used to measure the linear compression of solids[43]. Corrections must be made for the change in the resistivity and length of the slide wire with pressure, and in the compression of the various parts of the piezometer. To prevent leaks, the piston and cylinder must fit very well; the diametrical clearance was smaller than 0.25 μm. The seal of the lower plug was a ring of gold, chosen because it is soft and so forms a good seal and because it is slightly less compressible than steel and so the joint becomes tighter the higher the pressure. Later, Bridgman[48] rejected the method as being insufficiently sensitive for the measurement of the thermal expansion and replaced it by the bellows piezometer method described in section VII. A piston piezometer and associated equipment for 30 kbar based on Bridgman's design can be purchased (Newhall and Abbot[177]). In this apparatus, leaks past the piston are prevented by an O-ring seal rather than by a close clearance. Vereshchagin and Galaktionov[241] have used a similar apparatus with a slightly different arrangement of the potentiometer for pressures up to 3 kbar.

The movement of the piston can also be measured by capacitance methods, one of which is described by Skinner, Cussler and Fuoss[216]. The accuracy of the density in this work is, however, only about 0.5 per cent.

VII. Piezometric Methods—Bellows

Instead of measuring the compression of a liquid in a piezometer by noting the movement of a liquid or solid piston, it can be measured by enclosing it in a collapsible piezometer, subjecting the vessel to pressure, and measuring the extent of the collapse. Richards has used this technique to isolate reactive fluids from mercury by enclosing them in a thin-walled glass capsule (see section V.2.C), and the simple piston–cylinder technique with the liquid in a collapsible capsule (see section VIII) is a variation of this method.

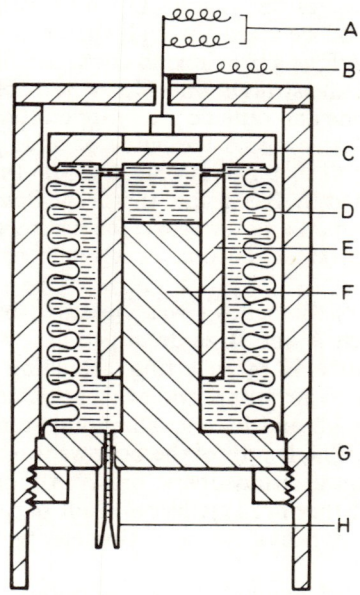

Figure 21. Bridgman's[48] bellows piezometer with potentiometric measurement of the displacement.

Bridgman[48] showed that a flexible metal bellows could be used as a collapsible vessel. The bellows shortens until the internal pressure is equal to the external pressure, apart from the relatively small pressure required to deform the bellows, and the change of volume can be determined from the shortening. A major advantage of the method is that the liquid is completely isolated from the pressure-transmitting fluid, and the corrections to be applied to the measurements are small.

Three methods have been used for measuring the change of length of the bellows, a potentiometric method[48], one using a linear differential transformer (Boelhouwer[31], Shakhovskoi et al.[211]), and a direct visual method (Blagoi and Sorokin[30]). In addition, the bellows can be used essentially as a null device to separate the pressure-transmitting and experimental fluids (Gilchrist, Earley and Cole[101]). Many other methods of detecting the motion of the bellows could be used, such as modifications of those used for liquid-piston piezometers. However, the potentiometric and linear-differential-transformer methods appear to be the most accurate and convenient.

Bridgman's original apparatus is illustrated in *Figure 21*. It was used to 12 kbar, and so the bellows had to accommodate a compression of about one third of their length. Commercial bellows that would do this were not available in 1931 (they are now, from several companies) and so Bridgman made his own. The unextended length was 1 inch and the allowable movement was $\frac{3}{16}$ inch in both directions or about 32 per cent of the unextended volume. Its capacity was about 5 cm^3.

The apparatus consisted of the bellows D which was soldered to the end pieces C and G and could be filled through the entrance hole H. A guide piston and cylinder E and F was provided to constrain the bellows to shorten and lengthen along the same straight line. Holes were, of course, provided to equalize the pressure inside and outside the cylinder. Later workers using the same technique have either omitted the guide or have placed it outside the bellows (Shakhovskoi et al.[211]). If it is placed inside the bellows, it has the disadvantage, as was noted by Bridgman, that if the fluid was compressed to the freezing point and a small amount froze, the piston and cylinder jammed and subsequent contraction of the fluid was taken up by the diametrical collapse of the bellows.

The change in length of the bellows was measured by the potentiometric method using the contacts A and B, as described in section VI. The bellows was filled through the tube H by inverting, attaching a reservoir of fluid, placing in the receiver of an air pump, and evacuating and filling the receiver several times slowly enough and by a small enough amount to prevent the bellows from collapsing. The length of the bellows was linear in the volume to better than 0.1 per cent at atmospheric pressure. Corrections were made for the compression and thermal expansion of all parts of the bellows and for the change in the resistance of the potentiometer wires with temperature and pressure.

The accuracy is not stated explicitly but the smoothness of the compression as a function of pressure appears to be about 1 to 2.5 parts per thousand in the compression, or about 300 to 800 p.p.m. in the volume, an accuracy which has only recently (Wilson and Bradley[257]) been surpassed by this technique. A similar apparatus was used to 10 kbar by Cutler et al.[82], and slightly modified by Nelson et al.[176], in which the compression is accurate to at worst 1.6 per cent at 345 bar and 0.4 per cent at 10 kbar, corresponding to an error in the density of about 1 p.p.t. at 345 bar and 1.4 p.p.t. at 10 kbar. Snyder and Winnick[220] have described an apparatus for 13 kbar which yields an accuracy of about 600 p.p.m. in the density. Mopsik[175], instead of using a wire potentiometer, used a strip of platinum on a fused silica rod and obtained a reproducibility of 5 µm. An unusually detailed description of equipment was given by the American Society of Mechanical Engineers[11].

A bellows piezometer and the necessary equipment for pressures up to 30 kbar can be purchased (Newhall and Abbot[177]).

Boelhouwer[31] modified apparatus due to Seeder[209] and attached the bellows to a piezometer, as shown in *Figure 22*, to increase the sensitivity at

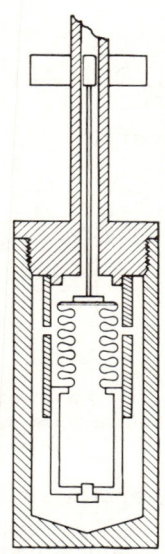

Figure 22. Boelhouwer's[31] modification of the bellows piezometer for lower pressures and with the movement measured by a linear differential transformer.

low pressures, the maximum pressure being 1200 bar. He attached a soft-iron armature to the bellows by means of a stiff brass wire. The armature moved inside a stainless-steel capillary tube and its position was located by means of a differential transformer mounted outside the capillary tube on a micrometer screw. An accuracy of about 400 p.p.m. in the relative volume was claimed. Madigosky[157] has described a similar apparatus except that a fused quartz rod was used to connect the bellows to the magnetic core. The position of the core could be measured to about 2.5 μm. The pressure was measured by three Heise Bourdon gages up to 7 kbar to a reproducibility of about 0.1 per cent, and the accuracy in the compression was about 250 p.p.m.

The differential transformer can also be mounted inside the pressure vessel, and this arrangement was first described by Shakhovskoi *et al.*[211]. Their piezometer is illustrated in *Figure 23*. The transformer is not used as a null instrument, as is usual when it is placed outside the pressure vessel, but the output potential was proportional to, and was used to measure, the displacement of the core from the null point. An accuracy of about 0.1 per cent in the density was claimed. Millet and Jenner[173] also mounted their differential transformer inside the pressure vessel in an apparatus designed for the range 200 to 500 K and up to 15 kbar to an accuracy of about 0.2 per cent in the volume, but it was not tested at either high pressures

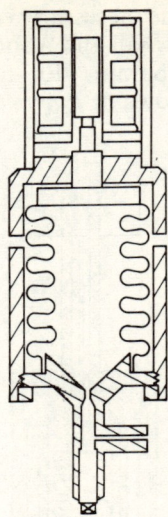

Figure 23. Shakhovskoi *et al.*'s[211] bellows piezometer with measurement of the movement by a linear differential transformer inside the vessel.

or temperatures. The ancillary equipment for this work is described in another paper[172].

The movement of the bellows was measured visually by means of a window in the side of the pressure vessel to a claimed accuracy of 10 μm in an appara-

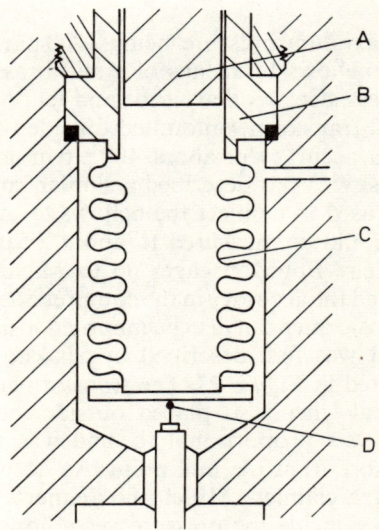

Figure 24. Gilchrist, Earley and Cole's[101] apparatus for using a bellows as a null device separating the experimental and the pressure-transmitting fluids.

tus described by Blagoi and Sorokin[30]. The distortion of the windows adds an uncertain error, and the method is not to be recommended for accurate work.

Gilchrist, Earley and Cole[101] used the apparatus illustrated in *Figure 24*, in which the volume enclosed by the bellows C could be varied by means of a piston A that passed through a packing in the head B of the pressure vessel, and whose location could be measured by a micrometer. A compression of the bellows by an excess external pressure was compensated by a displacement of the piston A, the end of the bellows being located by an electrical contact pin D attached to the base of the pressure vessel. The accuracy of the measurements was not stated, but does not appear to have been high.

At least two cells have been described (Mopsik[175], Schornack and Eckert[208]) for measuring simultaneously the change of volume and the change of dielectric constant using a bellows piezometer.

The most accurate measurements by this technique appear to be those of Wilson and Bradley[257]. The bellows was attached to a pressure vessel to increase the sensitivity, and a differential transformer mounted inside the pressurizing vessel was used to measure the displacement of the end of the bellows. The secondary of the transformer was attached to the bellows, and the primary was fixed in position. The movement of the secondary caused by the movement of the bellows generated a signal that was compensated by the movement of the secondary in a closely similar reference transformer. The secondary of the reference transformer was moved by a micrometer graduated at 100 microinch intervals and readable to 20 microinches, which corresponds to about 10 p.p.m. of the volume. The standard deviation of the densities obtained with the instrument was 20 p.p.m. although the deviations from the measurements of Kell and Whalley[138] are frequently several times this value. Nevertheless it is clear that the method is a promising one and could undoubtedly yield results accurate to 20 p.p.m. if the apparatus were calibrated carefully relative to a standard liquid and if the piezometer were pressurized over the whole exterior, as appears not to have been done in this apparatus.

A related technique has been used by Duedall and Paulowich[90] for measuring the difference in the compression of two liquids. The two liquids were placed in two bellows mounted back to back and the contraction of each was measured by means of a linear differential transformer. The difference in the contractions was determined by electrical methods and was recorded directly. A maximum pressure of 900 bar was used and the precision was between 0.03 and 0.09 × 10^{-6} bar^{-1} or about 80 p.p.m. in the density at 900 bar.

The use of bellows to measure the amount of fluid injected into a constant-volume vessel, and hence not as a simple piezometer, is described in section IX.

VIII. Simple Piston–Cylinder Method

The simplest method in principle of measuring the compressibility of liquids is to confine the liquid in a cylinder plugged at one end and fitted with a packed piston at the other. The pressure of the liquid is increased by pushing in the piston, and the volume change measured by measuring the displacement of the piston. Correction must be made for the dilatation of the cylinder and the compression of the piston. This is usually most conveniently done by replacing some of the liquid by a solid of known compression, such as iron, and repeating the measurements. The compression relative to the solid can then be obtained. Alternatively, measurements can be made on a liquid of known compression. In the cataloguing of the various kinds of apparatus, a distinction has been made between apparatus in which only one cylinder is used and its volume is varied by means of a piston, and apparatus in which two cylinders are used, one of fixed volume and one whose volume can be varied by means of a piston. The one-cylinder method is treated in this section, and the two-cylinder method is described under the heading 'Constant-volume vessel' in section IX.

The simple piston–cylinder method appears to have been first used by Parsons and Cook[181] at pressures up to 4.5 kbar. They used a cylinder, 4-in. inside diameter and 18-in. outside diameter with a capacity of about 2 l., mounted in a 2000-ton press. The dilatation of the cylinder was eliminated by making measurements with half the liquid replaced by steel so that the compression of the liquid relative to steel could be obtained. In the simple form a major source of error is the friction in the packing, which causes the pressure calculated from the force applied to the piston and its area to be in error. Parsons and Cook[181] and others since have taken measurements with both ascending and descending pressures and taken the average, assuming that the friction is independent of the direction of motion of the piston, an assumption that may not always be valid. The effects of friction can often be greatly reduced by slowly oscillating or rotating the piston relative to the cylinder in both the experimental piston–cylinder and the jack (Bridgman[58]). The relative motion of the piston and cylinder about their common axis establishes kinetic friction in the vertical direction.

Because of the friction and because the compression of the liquid is measured without the magnification of the motion afforded by a capillary tube attached to a bulb, the method in its simplest form is not suitable for low pressures. If the effect of pressure on the cross-sectional area of the vessel is ignored for approximate purposes, the fractional change in the length of the column of liquid is the fractional change in its volume. The compressibility of molecular liquids even at low pressures is about 100×10^{-6} bar^{-1} and usually decreases rapidly with increasing pressure. The relative change of length of the liquid column is therefore less than about 100×10^{-6} bar^{-1} or about $1\,\mu\text{m cm}^{-1}$ bar^{-1}. Compressions accurate to one per cent can be measured therefore only for pressure changes of the magnitude of some hundreds of bars. The usual pressure range of the experiments will therefore be a kilobar or more. This disadvantage can be partly overcome by pushing a small-diameter piston into a large-diameter vessel[113] (see below).

Bridgman[41] improved the method in the following ways. He replaced the piston seal by his unsupported area seal (Bridgman[35]), and eliminated the effect of the friction of the seal by measuring the pressure of the fluid directly by means of a manganin resistance gage (Bridgman[38]) which was inserted in the lower end of the cylinder. In order to avoid short-circuiting the manganin coil when electrically conducting fluids were used, it was immersed in kerosene, and the experimental fluid was placed in a steel piezometer to which pressure could be transmitted through an opening in one end.

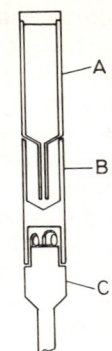

Figure 25. Bridgman's[41] piezometer for a simple piston–cylinder apparatus.

In the measurements (Bridgman[41]) that were made on water up to 80°C and 12 kbar, the water was placed in a steel tube open at the top which was then immersed in the kerosene. In somewhat later work on other liquids, some of which were miscible with kerosene, the piezometer shown in *Figure 25* was used. The steel tube A terminated in a capillary tube that dipped under mercury in the cup B. The manganin gage was attached to the plug C, which was inserted in the lower end of the cylinder, and a hole bored in the end of the bulb B served to protect the manganin wire. Glass bulbs were tried but always broke, presumably because the kerosene became so viscous at the higher pressures that it would temporarily support appreciable non-hydrostatic stresses. The use of a less viscous liquid such as isopentane would no doubt cure this problem. The compression of the kerosene and the mercury if it was present, and the dilatation of the cylinder, were allowed for by replacing the steel tube by a solid steel rod of the same size. The amount of kerosene in the cylinder was then the same since care was taken to insure that all dimensions were reproducible from run to run. The difference in the motion of the piston then gives directly the compression of the experimental liquid relative to steel.

The pressure range used was zero to 12 kbar, and it was necessary to season the vessel. For isothermal compression measurements, the pressure was raised to the maximum and lowered several times. The effect of the seasoning decayed with time, and measurements had to be made immediately after the seasoning. Two to three hours were required to measure the compression at about 20 points with both increasing and decreasing pressure largely because 5–7 min was allowed for the heat of compression to dissipate after each compression. When thermal expansions at constant pressure were measured, by moving the piston out so as to keep the pressure constant when the temperature was increased, both a pressure and a temperature seasoning were required, the pressure seasoning being on account of the small change in pressure that inevitably occurred when the temperature was changed.

The manganin gage was calibrated at various pressures and numerous other corrections were made as described in the original papers. No detailed estimate of the accuracy of the compressions or expansions was given but it appears to be a few parts per thousand.

A similar apparatus has been used by Adams, Williamson and Johnston[3] for the compression of solids, but could equally well be used for liquids. This was done by Adams[2] to 12 kbar in which n-butyl ether was used as the hydraulic fluid and the piezometers were of glass. A more recent version has been described by Weir[248]. A simple version in which the pressure was measured from the force on the piston and the hysteresis between runs with increasing and decreasing pressure, after the manner of Parsons and Cook[181], has been described by Bridgman[58] for pressures up to 5 kbar and by Schamp et al.[207] for 900 bars. Bridgman's[58] measurements required about one hour for a complete run and the hysteresis was reduced by oscillating the piston. Montgomery[174] has extended the range to 100 kbar using a tapered piston and cylinder, primarily for solid samples although the technique should be adaptable to liquid samples. The displacements of the pistons were measured by linear differential transformers.

Bridgman's early[41] apparatus was later[45] adapted for measuring the compression of dense gases. The main new experimental problem was to fill the cylinder with dense gas at the required high initial pressure. Three methods were tried. In the first, the gas at a pressure of about 2 kbar was introduced into the cylinder through a hole in the wall near the end of the piston. The method worked mechanically, but failed in that both hydrogen and air reacted chemically with the cylinder and its contents. No doubt it could be used for other gases and has in fact been used for many experiments (see Babb[12] for an account of the techniques used). The method has been adapted to the measurement of the compression of low-temperature gases by Stewart[222]. The piston and cylinder are shown in *Figure 26*. The cylinder 4 was sealed by a sheet of copper foil 2 held in place by a cap. The vessel was filled with liquid by condensing it through the inlet hole and capillary tube 3. The piston 1, which had an unsupported-area seal, was pushed into the cylinder and sealed the hole, and compressions were measured in a straightforward way.

The second method also failed and has already been mentioned in section V.3.B. In the third and successful method, the gas was compressed to 2 kbar into a cylinder capable of withstanding the pressure and fitted with an inward-opening valve. This cylinder was placed in the compression cylinder

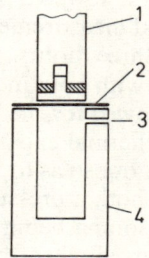

Figure 26. Stewart's[222] simple piston–cylinder apparatus for measuring the compression of low-temperature liquids and gases.

and compressed in the usual way. When the external pressure reached the internal pressure, fluid could enter the cylinder and compress the gas. In these experiments, pressures up to 16 kbar were used.

Basset[22] used a cylinder open at both ends. One end was closed by a movable piston and the other by a valve through which gas was introduced by a compressor. More details of the apparatus are given in Basset[21], but there is little in either paper, and few measurements seem to have been done.

In order to increase the accuracy of the measurements at relatively low pressures, Hayward[113] has made an apparatus in which a small-diameter piston is pushed into a vessel of much larger diameter, and has rotated the piston to reduce the friction at the packing. If the load is measured by a load cell, the friction of the jack that produces the load does not affect the results. The apparatus is illustrated in *Figure 27*. Hayward recommended that it be

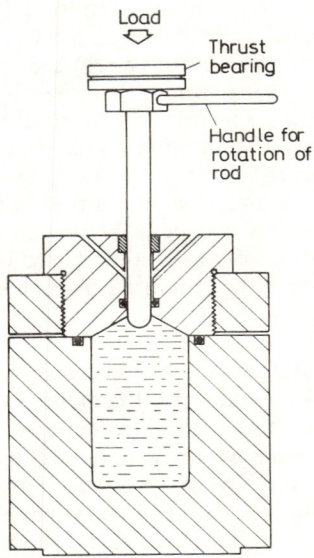

Figure 27. Hayward's[113] simple piston–cylinder apparatus with a small piston.

used with an automatically recording compressive testing machine. Slow compression results in isothermal measurements, particularly if the vessel is immersed in a liquid bath, and rapid compression gives at least an approximation to adiabatic measurements. A better approximation to an adiabatic compression curve can be obtained from a rapid-loading measurement followed immediately by an unloading at the same rate and averaging the loading and unloading measurements. An internally heated vessel based on the same principle and usable to 200°C at 5 kbar has been described by Marenko et al.[162].

A particularly simple form of piston–cylinder compression apparatus was described by Bridgman[52]. The liquid was contained in a lead capsule of the form shown in *Figure 28*(a). The base was fused to the cylinder in an atmosphere of hydrogen, and after filling the capsule with liquid, the opening at the top was sealed by a lead plug fused by means of a condenser discharge. Capsules used later (Bridgman[58]) were cylindrical on the inside and were joined at the base by cold soldering amalgamated surfaces. The amount

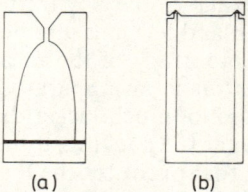

Figure 28. Capsules for use with a piston–cylinder apparatus.

of liquid was determined by weighing. Indium (Bridgman[51]) gives lower friction but it can be used only below 156.6°C, its melting point at zero pressure. The friction could no doubt be reduced and the accuracy increased by oscillating the piston (Bridgman[57]), either by hand (Bridgman[58]) or mechanically (Kennedy and LaMori[143]). No doubt the pressure range could be increased by adapting the method to Bridgman's 100-kbar apparatus (Bridgman[53, 56]); a commercial version is available (Newhall and Abbot[177]). An adaptation of this method that was suitable for solids and could undoubtedly be used for liquids was described by Bridgman[54] for rapid measurements on small quantities of material and used later for liquids to 40 kbar (Bridgman[58]), and 6 kbar (Bridgman[60]). About 0.1 cm^3 of material was required and the capsule was only about $\frac{1}{8}$ in. thick. A complete run, including assembly, dismantling and 30 readings in the range up to 25 kbar required only two hours. The dilatation of the vessel was reduced to about one third of that of a steel vessel by making it of cemented tungsten carbide. Heydemann and Houck[117] and Houck and Heydemann[125] used a similar method but replaced the lead capsule with a Teflon cylinder open at both ends. The fluid was sealed successfully by the pressure between the pistons and the Teflon, but the possibility of a slight leak when the seal is initially made may reduce the accuracy, and it seems likely that the fluid will leak on reducing the pressure. A Teflon cup and cap as illustrated in *Figure 28*(b) has been used by several groups for various measurements (see for example Kuriakose and Whalley[149]) and can undoubtedly be used for the compression of liquids. The phase changes in Teflon (Rigby and Bunn[200], Bridgman[55], Weir[249], Pistorius[185]) need to be taken account of.

Lentz[151], Lentz and Franck[152], and Alwani and Schneider[7] have fitted a window to the end of their cylinder and determined the position of the piston optically. The uncertainty of the variation of the refractive index of the fluid with pressure and of the distortion of the windows limits the accuracy.

IX. Constant-volume Vessel

1. Introduction

In this method, which appears to have been invented by Cowper and

Tammann[79] in connection with their measurements of the compressibility of amorphous materials like amorphous dextrose, a vessel of essentially constant volume, which is kept at the temperature of the experiment, is filled with fluid and known further amounts of fluid are injected into it or removed from it from or to a reservoir at some convenient temperature, and the resulting pressure measured. It is essentially an extension of the simple piston–cylinder apparatus in which the vessel is separated from the compressor so that the accuracy of measurement and the temperature range can be increased. It is distinguished for purposes of classification from some of the piezometric methods which it resembles in that in the piezometric methods the volume measurements are made on the piezometer itself, whereas in the constant-volume vessel method, the volume measurements are made in another part of the apparatus. The distinction is sometimes arbitrary. For example, a liquid-piston method is classified as piezometric if the mercury (say) surface in the piezometer is directly measured, and as a constant-volume vessel method if the other mercury surface is in another vessel and is directly measured. Nevertheless the classifications are useful to keep some order in the descriptions. Thermal expansion can be measured in this way by measuring the amount of fluid that must be injected or ejected to keep the pressure constant when the temperature is changed.

The method is in principle usable over a wide temperature range, the lower limit being set by the freezing of the fluid and the higher limit by the chemical instability of the fluid either to decomposition or to reaction with the container wall. It is undoubtedly the most versatile of all the methods, and can be made very precise.

There are many variants. Sometimes the vessel and the reservoir are filled with the experimental fluid. Sometimes, only the vessel contains the experimental fluid, which is confined either in a piezometer or directly in the vessel itself by another fluid such as kerosene (Bridgman[40]) or mercury, which is injected in known amounts into the vessel. One advantage of this procedure is that the density of the experimental liquid at the temperature and pressure of the reservoir need not be known since only the confining liquid is under these conditions. A given apparatus can then be used for many fluids without calibrations being necessary for each fluid. Other advantages are that the pressurizing fluid can be chosen to be compatible with any pressure seals that might be required in the injector vessel and with the pressure measuring device, which might be, for example, a manganin resistance gage which can be used only in non-conducting fluids.

A serious disadvantage is that the experimental and confining fluids might be mutually soluble to a significant extent. Kerosene, for example, can be used to confine water for measurements of moderate accuracy but can be used with few other molecular liquids. Even mercury can dissolve in the experimental fluid in significant amounts, particularly at high temperatures and in fluids of moderate density. This can significantly affect the measurements as pointed out by Keyes[145] and discussed in more detail by Jepson and Rowlinson[128], and there is no way of accurately allowing for the effect. Another disadvantage is that the measured volume change is only partly due to the compression of the experimental fluid and this reduces the accuracy.

The amount of fluid added to or removed from the vessel can be measured in many ways. It can be measured by the movement of a piston that completely confines the fluid (Cowper and Tammann[79]) or of a metallic bellows (see section VII) or by the volume or weight of the fluid released from the vessel. The vessel and its contents can be weighed: simple weighing methods are described in section X.1. The piston can be moved hydraulically or by a screw thread and its movement can be measured by a dial gage, a micrometer, a linear differential transformer, or, if a screw thread is used, by counting turns or by a micrometer reading device. The movement of a bellows can be measured by any of the methods described in section VII for bellows piezometers.

Mixtures of two or more components can be measured either by injecting or removing the mixture or by using two injectors, one for one component and one for the other. Mixtures of carbon dioxide and water to 500 bar and 800°C have been studied by Greenwood[107] using two injectors.

The pressure can be measured directly in the injector vessel, for example, by a manganin gage, or by any suitable pressure gage attached to it.

The variants of the method have been classified according to the ways of measuring the amount injected into or ejected from the vessel.

2. *Displacement of a Solid Piston*

In this method the change of volume of the fluid is measured by the movement of a solid piston. In the main variants of the method, the piston is moved hydraulically or by a screw thread, and displaces either the experimental fluid or an intermediate fluid.

This is the method originally used by Cowper and Tammann[79] who measured the pressure by means of a Bourdon tube connected to the vessel. The method was used by Bridgman[40] to measure the compression of water below 0°C to about 6.5 kbar, and these are still the standard data in this region. The water was placed in a steel tube which was itself placed in the pressure vessel and pressure was transmitted by kerosene. The kerosene was compressed by a hydraulically operated piston in a separate intensifier and the displacement of the piston of the intensifier was measured to 100 microinches in 3 inches. The pressure of the kerosene was measured by means of a coil of manganin wire, which was in the pressure vessel rather than in the intensifier, to an accuracy of about 300 p.p.m., which limited the accuracy of the experiment. A disadvantage of the technique is that water was in direct contact with kerosene. In measurements on organic liquids (Bridgman[42]) which are miscible with kerosene, the kerosene and the liquid were separated by a mercury piston. In measurements on liquids at low temperatures (Bridgman[49]) the vessel and injector were filled with gas.

The apparatus used by Keyes[145] for measuring the density of water up to 460°C and 350 bar has been described in great detail. Known amounts of water were injected into a vessel from a screw injector and the pressure was measured by a pressure balance, the oil of which was separated from the experimental water by a mercury U-tube. The screw injector was called a volumnometer, but more recently the 'n' has been dropped and the spelling 'volumometer' (Kell and Whalley[138]) adopted. The pressure vessels were made of nickel, austenitic (18 per cent chromium, 8 per cent nickel) stainless steel, or for the lower temperatures a chrome–vanadium steel. Coolidge and Mailey[78] had earlier attempted to avoid corrosion of the vessel by the water by lining it with a spun platinum liner. It was unsuccessful because the volume of the liner was not reproducible to the accuracy required. Consequently, the corrosion, which replaces water and iron by hydrogen and iron oxide, was allowed for as well as possible. Recently, Burnham and Wall[68] have swaged a platinum liner to an Inconel outer vessel at 900°C and 7 kbar and found it satisfactory. The equilibrium ratio of the pressure of hydrogen to that of water vapor at low pressures in equilibrium with the system Fe_3O_4–Fe was stated by Keyes[145] to be 3.4 at 500°C, but the reaction was too slow to yield so much hydrogen. Both a cylindrical and a spherical vessel were used, the spherical vessel being used in an attempt to increase the accuracy with which its dilatation under pressure could be calculated from elastic data. The dilatation of the vessel was measured by filling the system completely with mercury, measuring the compression of mercury relative to the dilatation of the apparatus, and so calculating the dilatation of the apparatus from the known compression of mercury.

The construction and pressure seasoning of the volumometers was described in considerable detail. They varied in size from 50-cm^3 displacement for pressures to 3 500 bar to 250-cm^3 for 1 500 bar and 1 000-cm^3 for 100 bar. They were calibrated at atmospheric pressure to an accuracy of ~ 200 p.p.m. in ten turns by expelling mercury and weighing it, and at high pressures by connecting the volumometer only to the U-tube and pressure balance and noting the variation of pressure with piston displacement. The measurements made with this apparatus were reported by Smith and Keyes[218] to a claimed accuracy of 100 p.p.m. in the density. Their measurements, at least below 150°C, are at least as accurate as this and usually much better (Kell and Whalley[138]).

A similar technique with somewhat less acuracy was used by Kennedy[140], Kennedy et al.[142], and Holser and Kennedy[123, 124] for measurements on water to a maximum pressure of 2 500 bar and a maximum temperature of 400°C. An Inconel vessel was used to reduce the effects of corrosion, and an attempt was made to further reduce the effects by adding powdered cupric oxide to the pressure vessel. It was hoped that hydrogen formed from the reaction of water with the vessel would reduce the cupric oxide so that the net result would be, for example,

$$CuO + Fe \rightarrow FeO + Cu$$

which would occur with little volume change.

Measurements on water by this technique were made by Maier and Franck[158] to 800°C at 6 kbar using a vessel made of Udimet 700, a proprietary alloy of Special Metals Inc., New Hartford, New York, USA. The end closures were held in place by a hydraulic press, rather than by the more usual screw threads, and the temperature of the water was measured by a thermocouple inside the pressure vessel. The accuracy of the density under these extreme conditions was about one per cent.

Keyes's apparatus has been adapted by Beattie[25] for measurements on gases. A weighed amount of gas was placed in the pressure vessel and compressed by injecting or removing known amounts of mercury from a screw

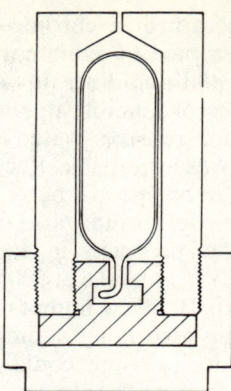

Figure 29. Beattie's[25] method of introducing into a pressure vessel a substance whose vapor pressure is less than 15 bar.

injector or volumometer. An accuracy of 500–1 000 p.p.m. in the density was claimed. An ingenious method of introducing into the pressure vessel substances whose vapor pressure is less than about 15 bar at 30°C was used. It is illustrated in *Figure 29*. The substance was introduced into the glass bulb under a vacuum and frozen. The capillary tip was bent, sealed off, and scratched with a file near the tip. The bulb was placed in the pressure vessel as shown in *Figure 29*. The vessel was placed on its side, evacuated, filled with mercury and turned upright. The liner floated up and the tip broke at the file mark. A similar apparatus was used by Benedict[27] to temperatures of 600°C to investigate the properties of saturated aqueous potassium chloride. The solubility of mercury in the fluid, particularly at high temperatures, limits the accuracy of these measurements.

The most accurate apparatus of this kind that has been described is that of Kell and Whalley[138]. It was used to measure the density of water in the range 0–1 000 bar, 0–150°C to a reproducibility (systematic errors, mainly the error of the compressibility of mercury that was used to measure the compressibility of the pressure vessel, were greater at the higher temperature and pressures) of 6 p.p.m. It is shown diagrammatically in *Figure 30* and followed the general principle of Keyes's apparatus with many refinements to increase the accuracy.

The pressure vessel was jacketed by a second pressure vessel so that it could be subjected to the same pressure inside and out, and became nearly a piezometer. The dilatation under pressure is then smaller and more reproducible than when the vessel is pressurized on the inside only.

The volumometer, which was used for injecting known volumes of water into the pressure vessel, is illustrated in *Figure 31*. Its cylinder was also pressurized externally to increase the reproducibility. The piston of the volumometer was advanced into the cylinder by means of a screw thread. To reduce backlash and insure that the piston advanced nearly linearly in the rotation, the two threads were carefully lapped together with a small clearance, and the female thread was 5 in. long. The readability and reproducibility were 0.1 mm^3 and 0.5 mm^3 respectively, there was no significant backlash, and the deviation from linearity of the amount delivered was less than 3 mg. To reduce wear on the screw thread, the piston was balanced by a piston of the same diameter subjected to the same pressure, as shown in *Figure 30*. The piston seals in the volumeter were O-rings, and to insure they were properly lubricated and to eliminate changes of volume due to extrusion they were balanced by oil pressure generated by the pressure balance.

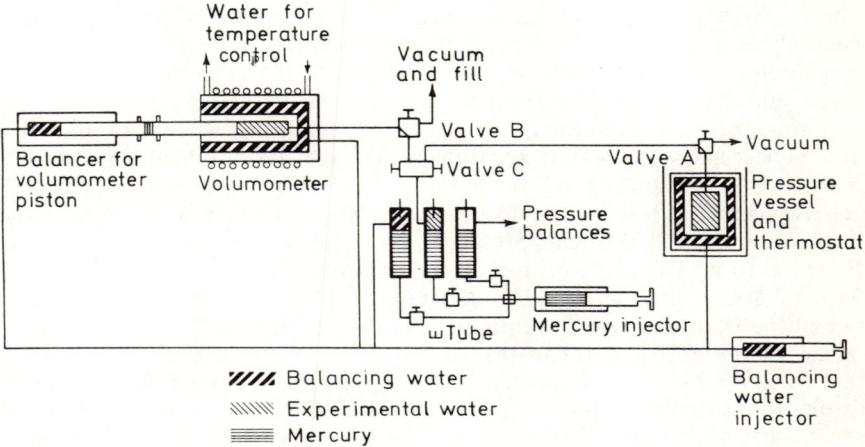

Figure 30. Schematic diagram of Kell and Whalley's[138] apparatus.

The pressure of the experimental water was measured by a pressure balance, and the oil of the balance was separated from the water by mercury in two legs of a three-legged Ш-tube fitted with electrical contacts for detecting the mercury surface. The third leg was used for equilibrating the pressure of the experimental and the balancing-water systems, the pressure of the balancing water being generated by the balancing-water injector. Later the balancer for the volumometer piston and the volumometer itself were pressurized by oil from the pressure balance, and only the pressure vessel was balanced by the balancing water system.

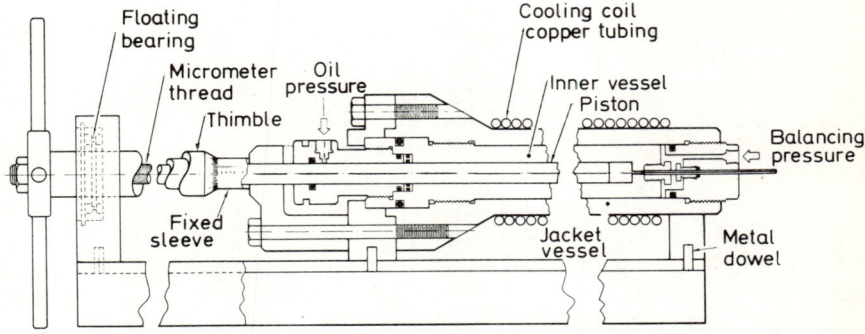

Figure 31. Kell and Whalley's[138] volumometer.

The compression of the vessel was measured relative to that of mercury, and the uncertainty of ~40 p.p.m. in the density of water at 1 000 bar in the range 0–150°C is due largely to the uncertainty in the compression of mercury. Accurate measurements of the compression of mercury would be very valuable.

The same apparatus fitted with a larger pressure vessel has been used by Kell, McLaurin and Whalley[139] for measuring the density of gases up to about 100 bar. The volumometer readings as a function of pressure could be analysed to yield the coefficients of an equation of state without separate measurements of the volume of the vessel, provided that both the measurements and the equation of state approach the ideal gas values at low pressure.

Simple versions of this apparatus are readily made and when compressions are made relative to water, accurate results can be obtained. Heath and Whalley[116] have measured the compression of glycerol–water mixtures relative to that of water to an accuracy of a few tens p.p.m. in the density by dispensing with the balancing pressures, and measuring pressure with a Heise Bourdon tube. The displacement of the volumometer piston was measured with a dial gage graduated at 100-μinch intervals. The whole pressure system except the Bourdon tube was immersed in the thermostat. The displacement of the piston when the apparatus was filled with water was compared with that when it was filled with glycerol–water mixtures, and the difference in the displacement analysed directly. Adams and Laidler[4] used a similar technique except that a commercial screw injector with an attached micrometer scale was used as the volumometer.

3. *Displacement of a Liquid Piston*

Sage and Lacey[204] have made extensive measurements of the pressure–volume–temperature of hydrocarbons in the range 20–370°C to about 700 bar in a very elaborate and highly successful apparatus. The experimental

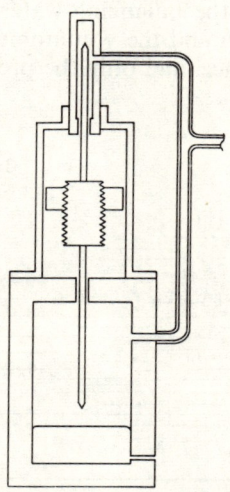

Figure 32. Sage and Lacey's[204] constant-volume movable electrical contact.

vessel was one leg of a mercury U-tube. The other leg, which is illustrated schematically in *Figure 32*, was a similar vessel at room temperature to which it was connected by a capillary tube, and which was fitted with a movable electrical contact. The contact was driven by a worm, not shown in *Figure 32*, which drove the driving nut and screw. Gas pressure was applied to the mercury surface to compress the experimental fluid, and its position was determined to about 6 μm. To insure that the pressure did not change when the needle was moved, a similar compensating needle was moved in a small vessel that was also connected to the gas supply as shown in *Figure 32*. Another way of accomplishing the same objective was described by Kell and Whalley[138] and is illustrated in *Figure 33*. As the shaft 2 which carries the needle is screwed in, a space A, which is connected to the interior of the vessel, opens and the dimensions are chosen to keep the volume constant. The space C is connected directly to the atmosphere.

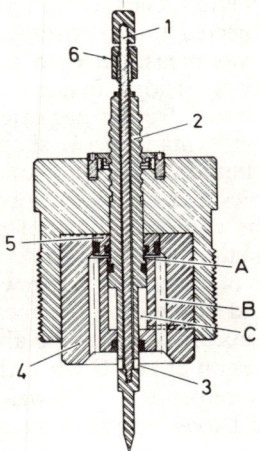

Figure 33. The working part of Kell and Whalley's[138] constant-volume movable electrical contact.

The amount of mercury in the vessel at low pressure was determined to about 10 p.p.m. of the capacity of the vessel by weighing the ejected mercury as a function of the position of the surface. The calibration was done under pressure by freezing the mercury in a part of the connecting tubing that was bent into a U-shape, and measuring the position of the mercury meniscus as a function of pressure. As far as I know, this was the first use of the frozen-capillary valve. The amount of fluid introduced into the experimental vessel was determined by weighing. Since the apparatus was designed for the study of multiphase as well as single-phase systems, the experimental vessel was provided with a stirrer operated magnetically from outside the bath. The accuracy was claimed to be between 1 and 2.5 p.p.t. in the density in the temperature range 20–240°C.

Köster and Franck[147] have used this technique to 600°C and 10 kbar.

The mercury level was detected by a float that was attached to a magnet that could move freely in the capillary tube attached to the U-tube. The position of the magnet was detected by a coil.

Michels, Blaisse and Michels[165] also confined the gas in a piezometer with mercury and weighed the mercury displaced from the vessel when the pressure was reduced.

4. Bellows Volumometer

The motion of a bellows (see section VII) can be used to measure the volume of sample injected into or removed from the pressure vessel. In the apparatus described by Babb et al.[13] for measurements on gases to 10 kbar and 400°C, the bellows was enclosed in one pressure vessel and the constant-volume vessel enclosed in a second. The capillary tube that connected the bellows and the constant-volume vessel was enclosed in the tubing that connected the two pressure vessels. The measuring system was thus completely enclosed and could be subjected to hydrostatic pressure throughout. The movement of the bellows was measured by a potentiometric arrangement using slide wires as described by Bridgman (see section VII). The bellows was filled at an appropriate pressure from a gas intensifier and sealed off by a freeze valve (Sage and Lacey[204]) made of a U-shaped piece of pressure tubing that could be immersed in liquid nitrogen to solidify the fluid. The pressure inside and outside the bellows had of course to be kept the same within one bar during the loading. The compressibility of nickel, which was used for the bellows and the experimental vessel, was determined in a separate experiment. The volume and mass of gas in the system were determined by measuring a reference isotherm and comparing the measurements with measurements made by others. The accuracy should be about 2 p.p.t., but agreement with the data of others on argon at 100°C when the initial mass and volume were determined from measurements at 35°C was about 15 p.p.t.

Burnham, Holloway and Davis[67] have described a similar apparatus for use with both liquids and vapors, particularly water, up to 900°C. The motion of the bellows was determined by a differential transformer, and the connecting capillary between the two pressure vessels was not externally pressurized. The bellows was calibrated by weighing the water discharged, and the compressibility of the pressure vessel was calculated from its elastic constants. To allow high temperatures and pressures to be reached, the furnace was placed inside the jacket (Watson[246]) so that the hot experimental vessel was subjected to hydrostatic pressure only. The uncertainty of the density was about 3 p.p.t. over the whole range. More recently, Burnham and Wall[68] have put their experimental vessel and bellows in the same long pressure vessel and have measured the bellows displacement by the slide wire method. The experimental vessel was lined with platinum, and it contained a collapsible gold bag into which water–carbon dioxide mixtures were injected.

5. Direct Weighing of Fluid Removed

This method has been used by Tanishita[233] for measuring the compression of steam. The vessel was made of platinum and was externally pressurized. After filling and heating, successive small amounts of water were removed and weighed. It seems likely that increased accuracy could be obtained and the evaporation of water avoided if mercury was displaced by the water and was itself weighed.

6. *Volumetric Measurement of the Gas at Low Pressure*

The compression of low-temperature liquids cannot easily be measured by the constant-volume vessel method using piston displacement because if the piston is at room temperature, the fluid is gaseous and the compression of the gas is too large for accurate measurements, and if it is cooled, it is difficult to maintain the piston packing. Consequently, the compression of low-temperature liquids is frequently measured by compressing the liquid into a vessel and letting off small amounts to low pressure. The liquid let off is evaporated and warmed to room temperature and the volume and pressure of the gas measured.

The method was first used by Holborn and Otto[119] primarily for gases up to 200 bar. A review of the experimental methods used by various laboratories has been given by Sengers[210], and need not be repeated here. Buchmann[63] made measurements on helium to 2 kbar by this method, and Wiebe and Gaddy[256] described an apparatus for 1 000 bar.

Johnston, Keller and Friedman[131] have described an apparatus for measuring the compression of hydrogen at low temperatures by this method to an apparent accuracy of a few hundred p.p.m. in the density, and the method has also been used by Sherman and Edeskuty[212] to 1 K, van Itterbeek and Verbeke[127, 240], and Rogovaya and Kaganer[202].

In the apparatus used by Streett and Staveley[224] in several investigations up to 680 bar a small cell of volume about 3.7 cm^3 was filled with the liquid at various pressures and expanded to the vapor at low pressure and room temperature. The accuracy appears to be of the magnitude of 1 p.p.t. in the density. It can, of course, also be used for gases at high temperatures, and Presnall[189] has measured hydrogen to 600°C and 1 800 bar by this method. Higher temperatures could not be used because of diffusion of hydrogen into the vessel. Usually, it appears that more information could be obtained with less effort by making multiple serial expansions.

7. *Volumetric Measurement of Liquid at Low Pressure*

The compression of a substance that is liquid can be measured by compressing it into a vessel, measuring the pressure, and letting out small amounts into a burette and measuring the pressure each time. This method has been used by Kerr, Kessler and Gamet[144] to a claimed accuracy of about 250 p.p.m. in the density at 140 bar. The method does not appear to promise high precision because of the uncertain meniscus errors and the difficulty of visually estimating volumes. But it might be developed into a useful method if enough effort is made.

X. Weight Methods

There are two main weight methods of determining the density of a liquid, the determination of the mass of a known volume of liquid by weighing it directly or by determining it indirectly from the buoyant force on a sinker of known volume that is immersed in the liquid. Both methods have been used under pressure.

1. Direct Weighing Methods

In principle, all that is necessary is to weigh a vessel of known volume, usually called a pyknometer, both empty and filled with a fluid at a known temperature and pressure. It is one of the commonest techniques for liquids at low pressures (Bauer and Lewin[23], White[255]), is capable of high accuracy, and can be used for absolute measurements (see Chapter 5). Its accuracy under pressure has not so far been made very high, partly because the vessel may be an order of magnitude or more heavier than the contents. Although the method has been used for many years for gases at low densities, it appears to have been first used for gases at nearly liquid densities and for liquids under pressure by Benedict[26].

In Benedict's[26] apparatus a nickel vessel of internal volume about 1 cm^3 was filled with fluid at a known temperature and pressure, which was measured by a pressure balance. The capillary tube that connected the vessel to the rest of the apparatus was then closed permanently by a screw clamp at a point that had previously been annealed, was cut on the side of the clamp away from the weighing vessel, and the cut end was sealed with silver solder. The vessel was allowed to warm to room temperature and was weighed. The gas was released by cutting the capillary tube, and the empty vessel weighed. When measurements were being made at low temperatures and high pressures, the pressures would become too great for the weighing vessel when it warmed to room temperature. To avoid this, the capillary tube was not soldered closed, but was soldered to a second weighing vessel. When the clamp on the capillary tube was released, some gas escaped from the main vessel into the second vessel and so the pressure did not become too great. Temperatures down to $-150°C$ and pressures up to 1.5 kbar were used. Allowance was made for the volume of the capillary tube and the change of volume of the vessel due to changes of temperature and pressure. The accuracy of the density was 1–4 p.p.t. Vilevich, Vereshchagin and Kalashnikov[242] described the use of an apparatus based on Benedict's but using a valve to confine the gas. Its range was 3.5 kbar and 400°C. The external dilatation of the pyknometer under pressure was measured by immersing the vessel in a dilatometer and pressurizing it. The movement of the meniscus of the fluid in the dilatometer capillary gave the external expansion of the vessel and this was assumed to be equal to the internal expansion because the wall of the vessel was thin. To be accurate, the compression of the material of the vessel should have been subtracted. An accuracy of the specific volume of about 1 p.p.t. was claimed.

Vukalovich and Altunin[244] have increased the accuracy of the weighing in this technique by adsorbing the gas on to activated carbon in another vessel at ~ 170 K. When this vessel was heated to room temperature for weighing the pressure was not high because the gas was adsorbed on the carbon, and a thin-walled and so light weighing vessel could be used.

Van Witzenburg and Stryland[259] have adapted Michels, Blaisse and Michels's[165] apparatus to the determination of the density of liquid and solid argon at pressures up to 2 kbar by direct weighing. The weighing vessel was suspended in a cryostat from a balance and filled with argon by means of a flexible capillary tube so arranged that it did not greatly affect the sensitivity of the balance. The vessel weighed about 3 kg and the contents 46–87 g. The sensitivity of the balance was about 5 mg corresponding to about 100 p.p.m. of the mass of the contents. The effect of the capillary tube on the apparent weight was determined by weighing a dummy vessel with the capillary tube attached. The volume of the vessel as a function of pressure was determined by weighing the vessel filled with mercury. The maximum volume change due to internal pressure was 0.3 per cent. The accuracy of the final results was thought to be about 1 p.p.t.

A similar method was used by Crawford and Daniels[80], also for argon. The compression was measured relative to iron by making measurements with the vessel entirely filled with argon and with most of the volume occupied by an iron rod. It was stated without further details that the internal volume did not behave as simple theory indicates. The accuracy appears to be about 1 p.p.t.

2. Hydrostatic Weighing

Several methods based on the principle of hydrostatic weighing have been described, most having been used only once so far. They use beam, tension, torsion or magnetic balances, magnetic lifting of a sinker that is then weighed directly, a series of floats used essentially as Cartesian divers, and hydrometers. Few of the methods have been well developed, and the only one so far shown to be capable of high accuracy is the magnetic weighing method of Fahey, Kupke and Beams[95] and Millero, Knox and Emmet[170]. The methods used have not been classified because of the few descriptions so far published.

The first instrument of this type was apparently due to Golubev[103] and is illustrated in *Figure 34*. It consisted of quartz sinker C in a pressure vessel D connected to an iron solenoid core A in a pressure vessel B which was connected to vessel D by a capillary tubing. A solenoid coil surrounded vessel B and was suspended from a balance. The apparent weights of the solenoid with and without enough current passing through it to raise the sinker gave the apparent weight of the sinker.

Pavlovich and Timrot[182] suspended their sinker on a thin constantan wire and used the change of resistance of the wire with tension to determine the apparent weight of the sinker. Their apparatus was designed primarily for gases, although it could easily be adapted for liquids. Boiko and Voityuk[32] have used a similar apparatus for liquids up to 60 bar. The accuracy is, however, not very high, and for ethylene, for example, in the range 0° to −90°C is only about 5 p.p.t.

Razumikhin[193] employed a more direct method of weighing up to 5 kbar in the apparatus illustrated in *Figure 35*. An arm balance 3 fitted with weights 2 and 4 of greatly different density was placed inside a pressure vessel that was

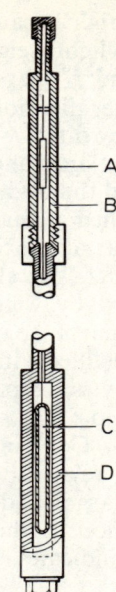

Figure 34. Schematic diagram of Golubev's[103] hydrostatic weighing apparatus.

fitted with observing windows perpendicular to the section shown. Additional weights could be placed in a depression on weight 2 by means of the loading device 1 even when the vessel was under pressure. The balance was equilibrated at atmospheric pressure by adding weights to one arm. Additional weights were then added, each time increasing the pressure until balance was obtained. Yekhlakov and Rodionov[260] used an elastic beam balance inside a pressure vessel. The vessel was tipped until a weight at the end of the beam touched an electrical contact. The angle of tilt was related to the density of the liquid. An accuracy of about 5 p.p.t. in the density was achieved.

A somewhat different but related technique has been used by Goldman[102] to measure the density of liquids saturated with gases. He floated a hollow

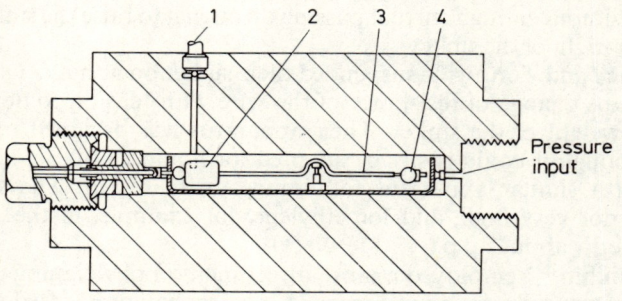

Figure 35. Diagram of Razumikhin's[193] pressurized balance.

486

hydrometer which was weighted by various weights as appropriate in the liquid which was pressurized by the gas. The stem of the hydrometer was open so that the gas could enter and equalize the internal and external pressures. The hydrometer was observed through windows in the pressure vessel. An accuracy of 1 p.p.t. was claimed.

A series of sinkers has been used by Postill, Ross and Cusack[187] to measure the density of liquid mercury in the range 20–1 000°C, 0–1 kbar to an accuracy of about 3 p.p.t. The sinkers were made of tantalum–molybdenum alloys of different densities to cover the range 11 to 13.5 g cm^{-3}, and were made γ-radioactive so that their positions could be detected with a Geiger counter and slit system outside the pressure vessel. The temperature and pressure were adjusted until the density of the mercury was nearly equal to that of the float. This was detected by slightly changing the temperature (presumably the pressure could be changed instead) and causing the float to move between the bottom and top of the vessel.

Fahey, Kupke and Beams[95] have adapted to high pressures the magnetic float method (Lamb and Lee[150], Beams and Clarke[24], Ulrich, Kupke and Beams[239]) of measuring the density of liquids. The principle of the method is

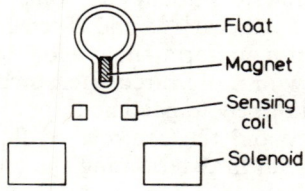

Figure 36. Principle of the magnetic float method.

illustrated in *Figure 36*. A float that contains a magnet is immersed in the fluid and acted upon by a solenoid. The inductance of a sensing coil depends upon the position of the magnet, and a servo circuit adjusts the current in the solenoid so that the Q of the sensing coil acquires a fixed value which corresponds to the float being held in a fixed position. The current through the solenoid then depends upon the density of the solution. The current must be calibrated with liquids of known density. Water (Kell and Whalley[138]) at 20°C was used as the standard, and the average deviation was 3 p.p.m. in the density.

Millero, Knox and Emmet[170] have used a related method, in which the current of the solenoid was adjusted manually to keep the float just touching the bottom of the vessel as determined by observation through two Plexiglas windows. Their solenoid was not subjected to the hydrostatic pressure as was that of Fahey, Kupke and Beams. The pressure was measured with a Heise Bourdon gage. Platinum weights could be added to the float to vary its density, and three were used to cover the range 1 to 1 200 bar at 25°C. It was calibrated with reference to water (Kell and Whalley[138]) to a precision of about 15 p.p.m., although the Heise Bourdon gage used to measure the pressure was accurate at best to 0.1 per cent which corresponds to 30 p.p.m.

The float was about five times as compressible as solid glass, presumably because it was hollow.

XI. Ultracentrifuge Method

In almost all the methods of compressing liquids described in this chapter, the liquid is compressed hydrostatically. Pressure can also be generated by subjecting the liquid to more general forces, for example, by centrifuging it, and the resulting compression can in principle be measured. Cheng and Schachman[74] pointed out that this was a practical possibility as the pressures generated were 200 bar and more. Since the ultracentrifuge generates a pressure gradient in the sample, it generates a density gradient. The sedimentation equilibrium of small solid particles could therefore be used to determine the relative compressions of the particles and the liquid. The measurements made by Cheng and Schachman[74] were at least of the right magnitude. Schachman[206] pointed out that the compressibilities of liquids could be measured directly in a cell that consisted of a large flat reservoir connected to a thin vertical slit in which the meniscus was observed. This method of augmenting the movement of the meniscus over that observed in a tube of uniform cross section does not appear to have been tried, but Dayantis[84], Richard, Glick and Burkat[194], and Richard and Rogers[195] have reported measurements in a uniform tube, and obtained isothermal compressibilities that agreed with more direct measurements to a few per cent. The method does not seem likely to become important because it is difficult to obtain a high accuracy and the compressed fluid is not at a constant density. It is, however, useful in determining the compressibility of macromolecules in solutions.

An adaptation of the method to the measurement of compressibilities at negative pressure has been described by Winnick and Cho[258] (see section XII).

XII. Negative Pressures

Since zero pressure is not a special pressure for a liquid, the density is expected to be continuous from positive to negative pressures. There appear to be only three measurements of densities at negative pressures.

Meyer[163, 164] confined the liquids in an evacuated heavy-walled glass capillary tube that was sealed off. The liquid was heated until it filled the tube. When it was cooled, it did not break away from the tube walls, and a negative pressure was generated. The negative pressure was measured by a two-turn glass spiral which was connected to the capillary tube. Pressures to -72 bar were reached. The same principle was used by Hayward[113]. A thin-walled metal vessel was filled with fluid and the change of pressure with temperature was measured by strain gages attached to the outside (see section XIV.2).

Winnick and Cho[258] adopted an ingenious apparatus due to Briggs[61, 62]. It consisted of a C-shaped glass capillary tube filled symmetrically with liquid extending into the short arms. The capillary tube was spun about its axis in an ultracentrifuge so that a negative pressure was generated in the central part of the tube and a positive pressure in the ends. The position of the meniscus at various speeds of rotation was measured by a Schlieren technique. The measurements, which were reported in more detail by Cho[75] in his thesis, have been reanalysed in detail by MacDonald[156].

XIII. Adiabatic Compression

It was pointed out in section IV.2 that accurate compression measurements over pressure intervals of the magnitude of one bar cannot be made isothermally unless the temperature is controlled to 100 μdeg or so, but that accurate adiabatic measurements can be made because the temperature needs to be controlled only to a range such that the compressibility (not the volume) changes insignificantly over it.

This advantage of adiabatic compression measurements appears to have been first exploited by Tyrer[237,238] who used a piezometric method with visual detection of a mercury piston. Two designs were used that differed only in detail. The later one is illustrated in *Figure 37*. Two piezometers were used, one made of copper and one of glass, and the glass one is illustrated in

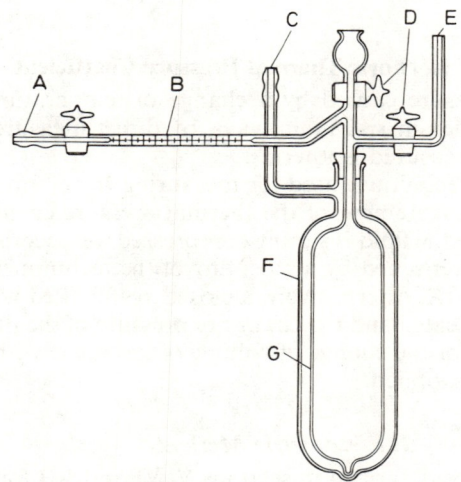

Figure 37. Tyrer's[238] glass apparatus for measurement of the adiabatic compression.

Figure 37. It consisted of an inner vessel G of about $\frac{1}{2}$ l. capacity enclosed in an outer vessel F. The vessel G was filled completely to the taps D and a mercury thread in the horizontal graduated capillary B. The two tubes A and C were attached by rubber tubing to an air compressor and the pressure increased by about one bar. When thermal equilibrium was reached the pressure was released quickly and the change in position of the mercury meniscus quickly observed. The copper apparatus was similar in design.

If the air that was compressed into the vessel F was not dried, a film of liquid sometimes condensed on the outside of the vessel G which caused noticeable temperature changes when it evaporated. The space between vessels F and G was therefore filled with oil to eliminate this effect. The accuracy of the compressibilities was said to be about 0.1 per cent, but the values for water appear to be lower by about one per cent than the best values from the speed of sound (Kell[137]).

A similar method in which a piston was driven rapidly into a cylinder by a ram was discussed by Villey[243] based on some unpublished measurements by R. Kling, but no data were presented.

A glass piezometer about one third the size of Tyrer's was used by Staveley and Parham[221] and one about one sixth the size was used by Harrison and Moelwyn-Hughes[112]. Coleman and Cruickshank[81] (see section V.3.A) have described a related piezometer in which both adiabatic and isothermal compressibilities could be determined.

The adiabatic compressibility does not appear to have been measured at high pressures, although there appears to be no fundamental reason why it should not be. The piezometer should be a thin-walled vessel pressurized with a liquid having the same or slightly higher value of $(\partial T/\partial p)_S$ as the experimental fluid, as used by Coleman and Cruickshank[81] at low pressures, and the change of volume should be measured rapidly.

XIV. Isochoric Thermal Pressure Coefficient

The change of pressure caused by a change of temperature at constant volume can be obtained in two main ways, by direct measurement and by the interpolation of measured isotherms.

Almost any of the techniques used for measuring the compression can be used for the direct measurement of the thermal pressure coefficient. After a change of temperature the fluid is simply compressed (or decompressed) to its original volume as determined by almost any of the techniques described in sections V, VI, VII or IX. Alternatively, a closed vessel filled with the experimental liquid can be heated and the change of pressure of the fluid measured. The usual correction for the change of volume of the vessel with temperature and pressure must be applied.

1. Piezometric Methods

Most of the methods described in sections V, VI and VII for the measurement of the compression can in principle be used for measuring the thermal pressure coefficient by reasonably obvious adaptations. Only the electrical contact piezometer appears in fact to have been used for liquids and it is probably the method most commonly used. Visual detection of a mercury meniscus has been used for vapors by Keys and Felsing[145a].

The electrical contact piezometer was introduced by Barus[20] and was used by Westwater, Frantz and Hildebrand[251]. Some recent users of the method are Malcolm and Ritchie[160] and MacDonald and Hyne[155]. The measured pressure change is influenced by the thermal expansion and the compression of the piezometer. The apparent thermal pressure coefficient γ' is

$$\gamma' = (\partial p/\partial T)_{V=v} \qquad (XIV.1)$$

where the subscript $V = v$ indicates that the volume V of the fluid is equal to the volume v of the vessel. The true thermal pressure coefficient can be calculated as follows. Since

$$p = f(V, T)$$

$$dp = (\partial p/\partial T)_V \, dT + (\partial p/\partial V)_T \, dV$$

It follows that

$$\gamma' = (\partial p/\partial T)_{V=v} = \gamma + (\partial p/\partial V)_T \, dv/dT \qquad \text{(XIV.2)}$$

Now

$$dv/dT = (\partial v/\partial T)_p + (\partial v/\partial p)_T \, dp/dT \qquad \text{(XIV.3)}$$

By substituting equation XIV.3 into XIV.2 we find

$$\gamma = \gamma'(1 - \kappa_T^v/\kappa_T) + \alpha^v/\kappa_T \qquad \text{(XIV.4)}$$

where α^v and κ_T^v are the thermal expansivity and compressibility of the vessel. Sometimes values of κ_T are not known. Equation XIV.4 can then be transformed (Barton et al.[18, 19], MacDonald and Hyne[155]) to

$$\gamma = \gamma'\alpha/(\gamma'\kappa_T^v + \alpha - \alpha^v)$$

The accuracy of the measurements (Westwater et al.[251]) is about 0.03 bar deg^{-1} or about 0.3 per cent and has not significantly improved in 45 years.

Most usually only one electrical contact is provided so that each density requires a different set-up of the apparatus. There is no reason in principle why several contacts should not be used so that several densities could be measured with the same set-up. This has been done by Barton et al.[19] for some measurements of the compressibility of molten salts up to 1.5 kbar and 520°C, using a quartz piezometer consisting of a bulb attached to a capillary tube with sealed-in platinum contacts. The piezometer was placed in a pressure vessel and pressurized with gas that was directly in contact with the molten salt in the capillary tube, and the pressure was varied at each temperature to cause the salt to just touch the contacts in turn. The high temperatures were attained using an internally heated pressure vessel.

2. Constant-volume Vessel with Direct Pressure Measurement

In this method the fluid is confined in a closed vessel of essentially constant volume and the change of pressure caused by a change of temperature is measured. The dilatation of the vessel caused by the change of pressure must, of course, be allowed for. The various methods differ mainly in the manner of measuring the pressure.

Simon and Kippert[213] used the apparatus shown in *Figure 38* primarily for low temperatures and 3 kbar. It consisted of a pressure vessel C of volume about 9 cm^3 closed by the tip B of a needle valve whose packing A was well away from the cold region. The pressure of the liquid was measured by a manganin resistance wire D that was calibrated at each temperature. The maximum error was about six per cent.

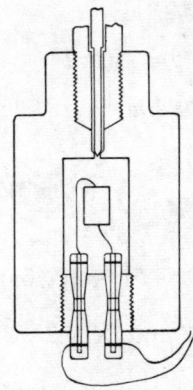

Figure 38. Diagram of Simon and Kippert's[213] apparatus for measuring the isochoric thermal pressure coefficient.

A very simple apparatus was used by von Nieuwenburg and Blumendal[179] to measure isochores of water to a relatively low accuracy ($\pm 1°C$, ± 2 bar). It consisted simply of an autoclave connected to a Bourdon gage which was filled with various amounts of water, and the pressure was noted as a function of temperature in the range 350 to 480°C. Tammann and Rühenbeck[232] kept the Bourdon tube free of water by using a mercury U-tube. A similar apparatus was used by Franck and Tödheide[97] for mixtures of carbon dioxide and water to 750°C and 2 kbar. For measurements on hydrogen chloride (Franck, Brose and Mangold[96]) the Bourdon gage contained glycerol and was separated from the hydrogen chloride by silver-plated bellows. A similar apparatus was used by Keesom and Keesom[135, 136] to measure the isochores of liquid helium to a pressure of 25 bar. Various densities were obtained by letting off small amounts of helium into an evacuated bulb of known volume and measuring the pressure with a mercury column.

Hayward[113] confined the fluid in a relatively thin-walled vessel which was $\frac{1}{4}$ in. o.d., 0.025 in. wall thickness and had strain gages attached to the outside. The experimental vessel and a similar unpressurized one fitted with similar strain gages were placed in a bath which was heated by increments. The ratio of the resistances of the strain gages as measured by a bridge gave directly the increase in pressure. As Hayward pointed out, one of the merits of this method is that it can be used to measure the expansions under negative pressure by first heating the vessel to generate a positive pressure and dissolve air bubbles, and then cooling to below the temperature at which it was sealed. Measurements to at least -14 bars were reported (Hayward[113]). No doubt Simon and Kippert's[213] apparatus could be used for the same purpose.

3. Constant-volume Vessel with Indirect Pressure Measurement

In this method a vessel of essentially constant volume is connected to a pressure null detector such as an elastic diaphragm or a mercury U-tube kept at laboratory temperature. It is related to the methods described in section IV for the measurement of compressions.

Michels, Wassenaar and Zweitering[168] have described a very accurate apparatus based on this method for measuring the compression primarily of gases, although it is often used for liquids, mainly at low temperature (Verbeke and van Itterbeek[240]). An elastic diaphragm was used as a null detector. For gases, the experimental values of pV were accurate to about 100 p.p.m. Goodwin[104] has described a similar apparatus and Straty and Prydz[223] have adapted it for measurements of fluorine. Malbrunot and Vodar[159] have described the use of an elastic diaphragm with a quartz piezometer for conditions up to 1 000°C and 5 kbar. This replaced an earlier version (Saurel and Vodar[205]) in which a mercury piston was used whose position was determined by the radiated-signal method described in section V.3.F. A commercial diaphragm of unstated origin and suitable for use to 350°C was used by Lo and Stiel[154]. Thomas[234] has described an apparatus for measurements on hydrogen chloride using a mercury U-tube.

Hill and Lounasmaa[118] made measurements on liquid helium with the helium confined in a copper vessel by means of mercury in a glass U-tube that was attached to it by a steel capillary. The pressure required to position the mercury at a fixed point was measured by a pressure balance or a Bourdon tube. The amount of helium in the vessel was determined by allowing it to expand into a known volume at room temperature and measuring the pressure. The accuracy claimed for $(\partial p/\partial T)_V$ is about 0.7 per cent or about 40 mbar deg^{-1}. Values of the pressure, volume and temperature were, of course, obtained at the same time. The heat capacity at constant volume was also measured in the same vessel. A similar apparatus was described by Timrot and Borisoglebski[235] for measurements on low-temperature gases.

XV. Calorimetric Methods

According to equations II.35 and II.36

$$T(\partial^2 p/\partial T^2)_V = (\partial C_V/\partial V)_T \tag{XV.1}$$

$$T(\partial^2 V/\partial T^2)_p = -(\partial C_p/\partial p)_T \tag{XV.2}$$

so that measurement of the heat capacity at constant volume as a function of temperature at constant volume yields the curvature of the isochore for that volume. The slope of the isochore can be obtained by a single integration, and the isochore itself by a double integration, provided two points on the isochore or a point and the slope at a point are known. There appears to have been no application of this method to liquids, but Dugdale and Simon[91] have applied it to solid helium. The principles are of course exactly the same for solids and liquids, and indeed, since helium is a soft solid, Dugdale and Simon's apparatus could be used for liquids. The techniques used in measuring the heat capacity under pressure will not concern us here.

In a similar way, isotherms can in principle be obtained from the pressure coefficient of C_p.

Goodwin and Weber[105,106] have measured the heat capacities of liquid oxygen both at saturation and at constant volume and from these have obtained values of the thermal expansion $(\partial \rho/\partial T)_\sigma$ along the saturation curve and of the constant-volume thermal pressure coefficient $(\partial p/\partial T)_V$.

XVI. Miscellaneous Methods

1. Methods Based on Radioactivity

The intensity of the radioactivity from a radioactive source is proportional to the density of the source if there is no self-absorption. Consequently, relative densities can be measured by the relative intensities of the radiation. The method was used by Weinberger and Schneider[247] to measure the effect of the gravitationally-induced pressure on the density of xenon in the critical region, and by Korshunov et al.[146] to measure cesium in the range 100 to 2200°C and 20 to 130 bar.

2. Variable-volume Vessel

An ingenious cell used by Grilly[108] for measuring the compression of liquid helium is shown in *Figure 39*. It consists of three beryllium–copper

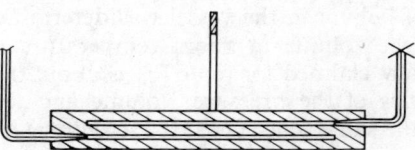

Figure 39. Diagram of Grilly's[108] variable volume pressure vessel.

discs welded around the edges to form two flat cavities. The upper cavity was the pressure vessel and the helium was confined to it by the valve. The lower cavity was connected to a helium supply at room temperature whose pressure could be varied. Variation of its pressure changed the volume of the sample cell by deflection of the central diaphragm. The pressure in the upper cell was measured by the deflection of the upper diaphragm as determined by a differential transformer. When both diaphragms were calibrated the volume of the upper cell as a function of the pressure of the fluid inside it could be determined. The standard deviation of the compressibility from a fitted line was about 1.5 per cent or about $5 \times 10^{-6}\,\mathrm{bar}^{-1}$. Later (Grilly[109]), the deflection of the upper diaphragm was measured by a capacitance method instead of by the differential transformer.

XVII. References

[1] Adam, N. K. *The Physics and Chemistry of Surfaces*, third edition. Oxford University Press: London (1941).
[2] Adams, L. H. *J. Amer. Chem. Soc.* **53**, 3769 (1931).
[3] Adams, L. H., E. D. Williamson and J. Johnston. *J. Amer. Chem. Soc.* **41**, 12 (1919).
[4] Adams, W. A. and K. J. Laidler. *Canad. J. Chem.* **45**, 123 (1967).

[5] Adams, W. A. and E. Whalley. Unpublished work.
[6] Aimé, G. *Ann. Chim. (Phys.)*, **8**, 257 (1843).
[7] Alwani, Z. and G. M. Schneider. *Ber. Bunsenges. Phys. Chem.* **73**, 294 (1969).
[8] Amagat, E. H. *Ann. Chim. (Phys.)*, **11**, 520 (1877).
[9] Amagat, E. H. *C.R. Acad. Sci., Paris,* **103**, 429 (1886).
[10] Amagat, E. H. *Ann. Chim. (Phys.)*, **29**, 68 (1893).
[11] American Society of Mechanical Engineers. *Pressure–viscosity report, I: Viscosity and density of over 40 lubricating fluids of known composition at pressures to 150 000 psi and temperatures to 425°F*, New York (1953).
[12] Babb, S. E. 'Techniques of high pressure experimentation', in *Technique of Inorganic Chemistry*, Vol. IV, Ed. H. B. Jonassen and A. Weissberger, p 83. Interscience: New York (1966).
[13] Babb, S. E., G. J. Scott, C. D. Epp and S. L. Robertson. *Rev. Sci. Instrum.*, **40**, 670 (1969).
[14] Baker, E. H. *J. Chem. Soc.* 464 (1962).
[15] Baliga, B. T. and E. Whalley. *J. Phys. Chem.* **73**, 654 (1969).
[16] Bannard, J. E. and G. J. Hills. *High Temp. High Press.* **1**, 571 (1969).
[17] Barrett, C. S. and T. B. Massalski. *Structures of Metals*, third edition, Chapter 20. McGraw-Hill: New York (1966).
[18] Barton, A. F. M. *J. Chem. Educ.* **48**, 161 (1971).
[19] Barton, A. F. M., G. J. Hills, D. J. Fray and J. W. Tomlinson. *High Temp. High Press.* to be published (1972). Referred to by Barton[18].
[20] Barus, C. *Amer. J. Sci.* **39**, 478 (1890).
[21] Basset, J. *C. R. Acad. Sci., Paris*, **185**, 343 (1927).
[22] Basset, J. *C.R. Acad. Sci., Paris,* **191**, 928 (1930).
[23] Bauer, N. and S. Z. Lewin. 'Determination of density', Chapter IV of *Technique of Organic Chemistry*, Vol. I: third edition, *Physical Methods Part I*, Ed. A. Weissberger, Interscience: New York (1959).
[24] Beams, J. W. and A. M. Clarke. *Rev. Sci. Instrum.* **33**, 750 (1962).
[25] Beattie, J. A. *Proc. Amer. Acad. Arts Sci.* **69**, 389 (1934).
[26] Benedict, M. *J. Amer. Chem. Soc.* **59**, 2224 (1937).
[27] Benedict, M. *J. Geol.* **47**, 252 (1939).
[28] Berkeley, Earl of, and C. V. Burton. *Phil. Mag.*, 6th Ser.; **32**, 153 (1916).
[29] Bett, K. E., K. E. Weale and D. M. Newitt. *Brit. J. Appl. Phys.* **5**, 243 (1954).
[30] Blagoi, Yu. P. and V. A. Sorokin. *Zh. Fiz. Khim.* **42**, 546 (1968). English translation *Russ. J. Phys. Chem.* **42**, 290 (1968).
[31] Boelhouwer, J. W. M. *Physica,* **26**, 1021 (1960).
[32] Boiko, N. V. and B. V. Voityuk. 'Experimental investigation of the density of liquid hydrocarbons by hydrostatic weighing on a tensometric balance' in *Thermophysical Properties of Gases and Liquids*, No. 1, Ed. V. A. Rabinovich (1968). English translation published by Israel Program for Scientific Translations: Jerusalem (1970).
[33] Bradfield, G. 'Use in industry of elasticity measurements in metals with the help of mechanical vibrations'. *National Physical Laboratory Notes on Applied Science No. 30*, p 134. Her Majesty's Stationery Office: London (1966).
[34] Bradshaw, A. and K. E. Schleicher. *Deep-Sea Research,* **17**, 691 (1970).
[35] Bridgman, P. W. *Proc. Amer. Acad. Arts Sci.* **44**, 201 (1909).
[36] Bridgman, P. W. *Proc. Amer. Acad. Arts Sci.* **44**, 221 (1909).
[37] Bridgman, P. W. *Proc. Amer. Acad. Arts Sci.* **44**, 255 (1909).
[38] Bridgman, P. W. *Proc. Amer. Acad. Arts Sci.* **47**, 321 (1911).
[39] Bridgman, P. W. *Proc. Amer. Acad. Arts Sci.* **47**, 347 (1911).
[40] Bridgman, P. W. *Proc. Amer. Acad. Arts Sci.* **47**, 441 (1911).
[41] Bridgman, P. W. *Proc. Amer. Acad. Arts Sci.* **48**, 309 (1912).
[42] Bridgman, P. W. *Proc. Amer. Acad. Arts Sci.* **49**, 3 (1913).
[43] Bridgman, P. W. *Proc. Amer. Acad. Arts Sci.* **58**, 165 (1923).
[44] Bridgman, P. W. *Rec. Trav. Chim. Pays-Bas,* **42**, 568 (1923).
[45] Bridgman, P. W. *Proc. Amer. Acad. Arts Sci.* **59**, 173 (1924).
[46] Bridgman, P. W. *Proc. Amer. Acad. Arts Sci.* **61**, 57 (1926).
[47] Bridgman, P. W. *Z. Kristallogr.* **67**, 363 (1928).
[48] Bridgman, P. W. *Proc. Amer. Acad. Arts Sci.* **66**, 185 (1931).
[49] Bridgman, P. W. *Proc. Amer. Acad. Arts Sci.* **70**, 1 (1935).
[50] Bridgman, P. W. *Proc. Amer. Acad. Arts Sci.* **74**, 1 (1940).

[51] Bridgman, P. W. *Proc. Amer. Acad. Arts Sci.* **74**, 21 (1940).
[52] Bridgman, P. W. *Proc. Amer. Acad. Arts Sci.* **74**, 399 (1942).
[54] Bridgman, P. W. *Proc. Amer. Acad. Arts Sci.* **74**, 425 (1942).
[54] Bridgman, P. W. *Proc. Amer. Acad. Arts Sci.* **76**, 9 (1945).
[55] Bridgman, P. W. *Proc. Amer. Acad. Arts Sci.* **76**, 71 (1948).
[56] Bridgman, P. W. *Proc. Amer. Acad. Arts Sci.* **76**, 55 (1948).
[57] Bridgman, P. W. *The Physics of High Pressure*, Bell: London (1949).
[58] Bridgman, P. W. *Proc. Amer. Acad. Arts Sci.* **77**, 129 (1949).
[59] Bridgman, P. W. *Proc. Amer. Acad. Arts Sci.* **77**, 189 (1949).
[60] Bridgman, P. W. *J. Chem. Phys.* **19**, 203 (1951).
[61] Briggs, L. J. *J. Appl. Phys.* **21**, 721 (1950).
[62] Briggs, L. J. *J. Chem. Phys.* **19**, 970 (1951).
[63] Buchmann, E. *Z. Phys. Chem. A*, **163**, 461 (1933).
[64] Burlew, J. S. *J. Amer. Chem. Soc.* **62**, 681 (1940).
[65] Burlew, J. S. *J. Amer. Chem. Soc.* **62**, 690 (1940).
[66] Burlew, J. S. *J. Amer. Chem. Soc.* **62**, 696 (1940).
[67] Burnham, C. W., J. R. Holloway and N. F. Davis. *Amer. J. Sci.* **267-A**, 70 (1969).
[68] Burnham, C. W. and V. J. Wall. Private communication (1972).
[69] Cailletet, L. *C.R. Acad. Sci., Paris,* **75**, 77 (1872).
[70] Cailletet, L. *Ann. Chim. (Phys.)*, **15**, 132 (1878).
[71] Canton, J. *Phil. Trans.* **52**, 640 (1762).
[72] Canton, J. *Phil. Trans.* **54**, 261 (1764).
[73] Carnazzi, P. *Nuovo Cim.* **5**, 180 (1903).
[74] Cheng, P. Y. and H. K. Schachman. *J. Amer. Chem. Soc.* **77**, 498 (1955).
[75] Cho, S. J. 'The pressure, volume, and temperature behavior of water at negative pressures', *Ph.D. Thesis*, University of Missouri, University Microfilms, Ann Arbor, Michigan, USA (1970).
[76] Cohen, E. and A. K. W. A. van Lieshout. *Proc. Acad. Sci. Amst.* **39**, 38 (1936).
[77] Conférence Générale des Poids et Mesures (Troisième), *Comptes Rendus des Sciences* de la Conférence Générale des Poids et Mesures, 38 (1901).
[78] Coolidge, W. D. and R. D. Mailey. (1910). Unpublished work quoted by Keyes (1933).
[79] Cowper, A.D. and G. Tammann. *Z. Phys. Chem.* **68**, 281 (1910).
[80] Crawford, R. K. and W. B. Daniels. *J. Chem. Phys.* **50**, 3171 (1969).
[81] Cruickshank, A. J. B., Th. Ackermann and P. A. Giguère. 'Heat capacity of liquids and solutions near room temperature', Chapter 12 of *Experimental Thermodynamics*, Vol. I: *Calorimetry of Non-Reacting Systems*, Ed. J. P. McCullough and D. W. Scott, Butterworths: London (1968).
[82] Cutler, W. G., R. H. McMickle, W. Webb and R. W. Schiessler. *J. Chem. Phys.* **29**, 727 (1958).
[83] Davis, L. A. and R. B. Gordon. *J. Chem. Phys.* **46**, 2650 (1967).
[84] Dayantis, J. *C.R. Acad. Sci., Paris,* **267C**, 223 (1968).
[85] Diaz Peña, M. and B. Cavero. *Anal. Fis. Quim.* **60-B**, 357 (1964).
[86] Diaz Peña, M. and M. L. McGlashan. *Trans. Faraday Soc.* **55**, 2018 (1959).
[87] Downer, L. and K. E. S. Gardiner. *J. Inst. Petrol.* **58**, 1 (1972).
[88] Doolittle, A. K. and D. B. Doolittle. *Amer. Inst. Chem. Engrs Jnl*, **6**, 153 (1960).
[89] Doolittle, A. K., I. Simon and R. M. Cornish. *Amer. Inst. Chem. Engrs Jnl*, **6**, 150 (1960).
[90] Duedall, I. W. and S. Paulowich. *Rev. Sci. Instrum.* **44**, 120 (1973).
[91] Dugdale, J. S. and F. E. Simon. *Proc. Roy. Soc. A*, **218**, 291 (1953).
[92] Eduljee, H. E., D. M. Newitt and K. E. Weale. *J. Chem. Soc.* 3086 (1951).
[93] Essex, H. *Z. Anorg. Chem.* **88**, 189 (1914).
[94] Ewing, M. B., K. N. Marsh and R. H. Stokes. *J. Chem. Thermodynamics*, **4**, 637 (1972).
[95] Fahey, P. F., D. W. Kupke and J. W. Beams. *Proc. Nat. Acad. Sci. Wash.* **63**, 548 (1969).
[96] Franck, E. U., M. Brose and K. Mangold, *Prog. Int. Res. Thermodyn. Transport Prop.* (January 1962), p 159. Ed. J. F. Masi and D. H. Tsai, American Society of Mechanical Engineers: New York (1962).
[97] Franck, E. U. and K. Tödheide. *Z. Phys. Chem.*, N. F. **22**, 232 (1959).
[98] Gibson, R. E. *J. Amer. Chem. Soc.* **56**, 4 (1934).
[99] Gibson, R. E. *J. Amer. Chem. Soc.* **57**, 284 (1935).
[100] Gibson, R. E. *J. Amer. Chem. Soc.* **59**, 1521 (1937).
[101] Gilchrist, A., J. E. Earley and R. H. Cole. *J. Chem. Phys.* **26**, 196 (1957).

[102] Goldman, K. *Brit. J. Appl. Phys.* **9**, 40 (1958).
[103] Golubev, I. F. *Tr. Gos. Inst. Azo. Prom. No. 7*, p 47. Goskhimizdat: Moscow (1957). A short description in English is given in Tsiklis[236], p 409
[104] Goodwin, R. D. *J. Res. Nat. Bur. Stand.* **65C**, 231 (1961)
[105] Goodwin, R. D. and L. A. Weber. *J. Res. Nat. Bur. Stand.* **73A**, 1 (1969).
[106] Goodwin, R. D. and L. A. Weber. *J. Res. Nat. Bur. Stand.* **73A**, 15 (1969).
[107] Greenwood, H. J. *Amer. J. Sci., Schairer Vol.*, **267A**, 191 (1969).
[108] Grilly, E. R. *Phys. Rev.* **149**, 97 (1966).
[109] Grilly, E. R. *J. Low Temp. Phys.* **4**, 615 (1971).
[110] Grindley, T. and J. E. Lind. *J. Chem. Phys.* **54**, 3983 (1971).
[111] Guggenheim, E. A. *Thermodynamics*, fourth edition. North-Holland: Amsterdam (1959).
[111a] Harvis, R. W. and G. T. Clayton, *Phys. Rev.* **153**, 229 (1967).
[112] Harrison, D. and E. A. Moelwyn-Hughes. *Proc. Roy. Soc. A*, **239**, 230 (1957).
[113] Hayward, A. T. J. *Acta Imeko*, 249 (1964).
[114] Hayward, A. T. J. *J. Phys. D: Appl. Phys.* **4**, 938 (1971).
[115] Hayward, A. T. J. *J. Phys. D: Appl. Phys.* **4**, 951 (1971).
[116] Heath, J. B. R. and E. Whalley. Unpublished work (1967).
[117] Heydemann, P. L. M. and J. C. Houck. 'Ultrasonic and dilatometric measurements at very high pressures' in *Accurate Characterization of the High-pressure Environment*. Ed. E. C. Lloyd, *US Nat. Bur. Stand. Spec. Publ. No. 326*, p 11 (1971).
[118] Hill, R. W. and O. V. Lounasmaa. *A*, **252**, 44 (1960).
[119] Holborn, L. and J. Otto. *Z. Phys.* **30**, 320 (1924).
[120] Holder, G. A. and E. Whalley. *Trans. Faraday Soc.* **58**, 2095 (1962).
[121] Holder, G. A. and E. Whalley. *Trans. Faraday Soc.* **58**, 2108 (1962).
[122] Holloway, J. R. 'Internally heated pressure vessels' in *Research Techniques for High Pressure and High Temperature*, Ed. G. C. Ulmer, p 217. Springer: Berlin (1971).
[123] Holser, W. T. and G. C. Kennedy. Part III, *Amer. J. Sci.* **256**, 744 (1958).
[124] Holser, W. T. and G. C. Kennedy. Part IV, *Amer. J. Sci.* **256**, 744 (1958).
[125] Houck, J. C. and P. L. M. Heydemann. *J. Res. Nat. Bur. Stand.* **75A**, 121 (1971).
[126] Hyde, J. H. *Proc. Roy. Soc. A*, **97**, 240 (1920).
[127] Van Itterbeek, A. and O. Verbeke, *Cryogenics*, **1**, 77 (1960).
[128] Jepson, W. B. and J. S. Rowlinson. *J. Chem. Phys.* **23**, 1599 (1955).
[129] Jessup, R. S. *J. Res. Nat. Bur. Stand.* **5**, 985 (1930).
[130] Jessup, R. S. *J. Appl. Phys.* **23**, 543 (1952).
[131] Johnston, H. L., W. E. Keller and A. S. Friedman. *J. Amer. Chem. Soc.* **76**, 1482 (1954).
[132] Jolley, E. L. (1968). Personal communication to Millero, Curry and Drost-Hansen[169].
[133] Joule, J. P. *Phil. Mag.* **17**, 364 (1858).
[134] Kay, R. L. 'Transference number measurements' in *Techniques in Electrochemistry*, Vol. II, Ed. A. Salkind, and E. Yeager, Interscience: New York, in press (1972).
[135] Keesom, W. H. and A. P. Keesom. *Proc. Acad. Sci. Amst.* **36**, 612 (1933). Also issued as *Commun. Phys. Lab. Univ. Leiden No. 224d*.
[136] Keesom, W. H. and A. P. Keesom, *Proc. Acad. Sci. Amst.* **36**, 612 (1933). Also issued as *Commun. Phys. Lab. Univ. Leiden No. 224e*.
[137] Kell, G. S. *J. Chem. Engng Data*, **15**, 119 (1970).
[138] Kell, G. S. and E. Whalley. *Phil. Trans. A*, **258**, 565 (1965).
[139] Kell, G. S., G. E. McLaurin and E. Whalley. *J. Chem. Phys.* **48**, 3805 (1968).
[140] Kennedy, G. C. *Amer. J. Sci.* **248**, 540 (1950).
[141] Kennedy, G. C. *Amer. J. Sci.* **255**, 724 (1957).
[142] Kennedy, G. C., W. L. Knight and W. T. Holser. *Amer. J. Sci.* **256**, 590 (1958).
[143] Kennedy, G. C. and LaMori, P. N. *Progress in Very High Pressure Research*, Ed. F. P. Bundy, W. R. Hibbard and H. M. Strong, p 304. Wiley: New York (1961).
[144] Kerr, S. L., L. H. Kessler and M. B. Gamet. *Trans. Amer. Soc. Mech. Engrs,* **72**, 1143 (1950).
[145] Keyes, F. G. *Proc. Amer. Acad. Arts Sci.* **68**, 505 (1933).
[145a] Keyes, F. G. and W. A. Felsing, *J. Amer. Chem. Soc.* **41**, 589 (1919).
[146] Korshunov, Yu. S., A. P. Scnchenkov, E. I. Asinovskii and A. T. Kunavin, *Teplofiz. Vysok. Temp.* **8**, 1288 (1970). English translation in *High Temperature (Moscow)*, **8**, 1207 (1970).
[147] Köster, H. von and E. U. Franck. *Ber. Bunsenges. Phys. Chem.* **73**, 716 (1969).
[148] Kumagai, A. and T. Toriumi. *J. Chem. Engng Data*, **16**, 293 (1971).
[149] Kuriakose, A. K. and E. Whalley. *J. Chem. Phys.* **48**, 2025 (1968).
[150] Lamb, A. B. and R. E. Lee, *J. Amer. Chem. Soc.* **35**, 1668 (1913).

[151] Lentz, H. *Rev. Sci. Instrum.* **40**, 371 (1969).
[152] Lentz, H. and E. U. Franck. *Ber. Bunsenges. Phys. Chem.* **73**, 28 (1969).
[153] Levich, V. G. *Physicochemical Hydrodynamics,* p 674. Translated by Scripta Technica, Inc. Prentice-Hall Inc.: New Jersey (1962).
[154] Lo, H. Y. and L. I. Stiel. *Industr. Engng Chem. Fundamentals,* **8**, 713 (1969).
[155] MacDonald, D. B. and J. B. Hyne. *Canad. J. Chem.* **49**, 611 (1971).
[156] MacDonald, J. R. *J. Chem. Phys.* (in press) (1972).
[157] Madigosky, W. M. *Rev. Sci. Instrum.* **37**, 227 (1966).
[158] Maier, S. and E. U. Franck. *Ber. Bunsenges. Phys. Chem.* **70**, 639 (1966).
[159] Malbrunot, P. and B. Vodar. *High Temp. High Pres.* **3**, 225 (1971).
[160] Malcolm, G. N. and G. L. D. Ritchie. *J. Phys. Chem.* **66**, 852 (1962).
[161] Mangold, K. and E. U. Franck. *Ber. Bunsenges. Phys. Chem.* **73**, 21 (1969).
[162] Marenko, I. N., V. A. Ivanov and S. M. Stishov. *Dokl. Akad. Nauk SSSR*, **188**, 564 (1969). English translation in *Soviet Phys. Doklady*, **14**, 924 (1970).
[163] Meyer, J. *Z. Elektrochem.* **17**, 743 (1911).
[164] Meyer, J. *Abh. Bunsenges. Phys. Chem.* **3**, Nr 1 (1911).
[165] Michels, A., B. Blaisse and C. Michels. *Proc. Roy. Soc. A*, **160**, 358 (1937).
[166] Michels, A. and C. Michels. *Proc. Roy. Soc. A*, **153**, 201 (1935).
[167] Michels, A., C. Michels and H. Wouters. *Proc. Roy. Soc. A*, **153**, 214 (1936).
[168] Michels, A., T. Wassenaar and Th. N. Zwietering. *Physica*, **28**, 67 (1952).
[169] Millero, F. J., R. W. Curry and W. Drost-Hansen. *J. Chem. Engng Data*, **14**, 422 (1969).
[170] Millero, F. J., J. H. Knox and R. T. Emmet. *J. Solution Chem.* In press (1972).
[171] Millero, F. J. and F. K. Lepple. *J. Chem. Phys.* **54**, 946 (1971).
[172] Millet, M. and G. Jenner. *J. Chim. Phys.* **67**, 1667 (1970).
[173] Millet, M. and G. Jenner. *J. Chim. Phys.* **67**, 1766 (1970).
[174] Montgomery, P. W. *Rev. Sci. Instrum.* **37**, 1526 (1966).
[175] Mopsik, F. I. *J. Res. Nat. Bur. Stand.* **71A**, 287 (1967).
[176] Nelson, R. R., W. Webb, and J. A. Dixon. *J. Chem. Phys.* **33**, 1756 (1960).
[177] Newhall, D. H. and L. H. Abbot. *Proc. Inst. Mech. Engrs*, **182**, Pt 3c, 288 (1968).
[178] Newitt, D. M. *The Design of High Pressure Plant and the Properties of Fluids at High Pressure,* Clarendon Press: Oxford, Chapters VII and XVIII (1940).
[179] von Nieuwenburg, C. J. and H. B. Blumendal. *Rec. Trav. Chim. Pays-Bas,* **51**, 707 (1932).
[180] Oersted, M. *Ann. Chim. (Phys.),* **22**, 192 (1823).
[181] Parsons, C. A. and S. S. Cook. *Proc. Roy. Soc.* **85**, 332 (1911).
[182] Pavlovich, N. V. and D. L. Timrot. *Teploenergetika*, No. 4, 69 (1958).
[183] Perkins, J. *Phil. Trans.* **110**, 324 (1820).
[184] Perkins, J. *Phil. Trans.* **116**, 541 (1826).
[185] Pistorius, C. W. F. T. *Polymer*, **5**, 315 (1964).
[186] Pollard, L. J., M. L. Crowe and W. Strauss. *J. Chem. Engng Data*, **16**, 134 (1971).
[187] Postill, D. R., R. G. Ross and N. E. Cusack, *Advanc. Phys.* **16**, 493 (1967).
[188] Presnall, P. C. *J. Geophys. Res.* **74**, 6026 (1969).
[189] Presnall, P. C. 'Compressibility measurements of gases using externally heated pressure vessels' in *Research Techniques for High Temperature*, Ed. G. C. Ulmer, p 259. Springer: Berlin (1971).
[190] Pribadi, K. S. 'A radio frequency moving boundary detector', *J. Solution Chem.* (in press) (1972). Kay[134] also describes the method.
[191] Ramsay, W. and S. Young. *Proc. Roy. Soc.* **42**, 3 (1887).
[192] Ramsay, W. and S. Young. *Phil. Mag.* **23**, 435 (1887).
[193] Razumikhin, V. N. *Trudy Inst. Kom. Standentov. Mer i Izmerit. Prib.* **46**, 96 (1960). An account is given in Tsiklis[236].
[194] Richard, A. J., J. Glick and R. Burkat. *Anal. Biochem.* **37**, 378 (1970).
[195] Richard, A. J. and K. S. Rogers. *Canad. J. Chem.* **49**, 3956 (1972).
[196] Richards, T. W. and G. Jones. *J. Amer. Chem. Soc.* **31**, 158 (1909).
[197] Richards, T. W. and R. Shipley. *J. Amer. Chem. Soc.* **38**, 989 (1916).
[198] Richards, T. W. and W. N. Stull. 'New method of determining compressibility', *Publ. Carnegie Instn,* **7** (1903). This contains a more detailed account of Richards and Stull[199]
[199] Richards, T. W. and W. N. Stull. *J. Amer. Chem. Soc.* **26**, 399 (1904).
[200] Rigby, H. A. and C. W. Bunn. *Nature, Lond.* **164**, 583 (1949).
[201] Roebuck, J. R. and E. E. Miller, *Rev. Sci. Instrum.* **10**, 179 (1939).

[202] Rogovaya, I. A. and M. G. Kaganer. *Zh. Fiz. Khim. SSSR*, **34**, 917 (1960). English translation, *Russ, J. Phys. Chem.* **34**, 917 (1960).
[203] Rugta, K. K., R. A. Stager and G. P. Mathur. *Canad. J. Chem. Engng*, **49**, 886 (1971).
[204] Sage, B. H. and W. N. Lacey. *Trans. Amer. Instr. Min. Metall. Engrs*, **136**, 136 (1940).
[205] Saurel, J. and B. Vodar. *J. Rech. CNRS*, **33**, 386 (1955).
[206] Schachman, H. K. *Ultracentrifugation in Biochemistry*, p 32. Academic Press: New York (1959).
[207] Schamp, H. W., J. R. Hastings and S. Weissman. *Physics of Fluids*, **8**, 8 (1965).
[208] Schornack, L. G. and C. A. Eckert. *J. Phys. Chem.* **74**, 3014 (1970).
[209] Seeder, W. A. *Thesis*, Utrecht (1943), referred to by Boelhouwer[31].
[210] Sengers, J. H. M. Levelt, Chapter 3 of '*Physics of High Pressure and the Condensed State*', Ed. A. von Itterbeek, p. 60 North-Holland: Amsterdam (1965).
[211] Shakhovskoi, G. P., I. A. Lavrov, M. D. Pushkinskii and M. G. Gonikberg. *Pribory i Tekh. Eks.* **1**, 181 (1962). English translation *Instr. Exp. Techn.* **1**, 184 (1962).
[212] Sherman, R. H. and F. J. Edeskuty. *Annals of Physics*, **9**, 522 (1960).
[213] Simon, F. and F. Kippert. *Z. Phys. Chem.* **135**, 113 (1928).
[214] Sirota, A. M. and B. K. Mal'tsev. *Teploenergetika*, **6**, 7 (1959). English translation *S-137*, Morris D. Friedman, Inc., 1383A Washington St., West Newton 65, Mass, USA.
[215] Sjölander, A. 'Theory of neutron scattering by liquids', Chapter 7 of *Thermal Neutron Scattering*, Ed. P. A. Egelstaff. Academic Press: London (1965).
[216] Skinner, J. F., E. L. Cussler and R. M. Fuoss. *J. Phys. Chem.* **72**, 1057 (1968).
[217] Smith, A. W. *Proc. Amer. Acad. Arts Sci.* **42**, 421 (1906).
[218] Smith, L. B. and F. G. Keyes. *Proc. Amer. Acad. Arts Sci.* **69**, 286 (1934).
[219] Smyth, F. H. and L. H. Adams. *J. Amer. Chem. Soc.* **45**, 1167 (1923).
[220] Snyder, P. S. and J. Winnick. *Proceedings of the Fifth Symposium on Thermophysical Properties*, p 115. American Society of Mechanical Engineers: New York (1970).
[221] Staveley, L. A. K. and D. N. Parham. 'La compressibilité de la glace, du benzène solide et du cyclohexane solide à leur point de fusion', in *Changements de Phases*, p 366. Société de Chimie Physique: Paris (1952).
[222] Stewart, J. W. 'High pressures at low temperatures', Chapter 10 of *Modern Very High Pressure Techniques*, Ed. R. H. Wentorf, Butterworths: London (1962).
[223] Straty, G. C. and R. Prydz. *Rev. Sci. Instrum.* **41**, 1223 (1970).
[224] Streett, W. B. and L. A. K. Staveley. *Advanc. Cryogenic Engng*, **13**, 362 (1968).
[225] Stryland, J. C. and A. D. May. *Rev. Sci. Instrum.* **31**, 414 (1960).
[226] Suchodski, W. A. *Z. Phys. Chem.* **74**, 257 (1910).
[227] Tait, P. G. *Proc. Roy. Soc. Edinburgh*, **12**, 223, 224 (1883).
[228] Tait, P. G. *Proc. Roy. Soc. Edinburgh*, **13**, 2 (1884).
[229] Tait, P. G. 'Compressibility of water, glass and mercury,. *Report* on the scientific results of the voyage of HMS Challenger 1873–76. Ed. C. W. Thomson and J. Murray, *Physics and Chemistry*, Vol. II, Part IV, p 3. HMSO: London (1889).
[230] Talbott, A. C. *Phil. Mag.* **19**, 1126 (1935).
[231] Tallmadge, J. A. *J. Phys. Chem.* **75**, 583 (1971).
[232] Tammann, G. and A. Rühenbeck. *Ann. Phys., Lpz.* **13**, 63 (1932).
[233] Tanishita, I. 'Experimental studies on the specific volume of steam at high temperature and pressure (first report)'. *Jap. Soc. Mech. Engrs. Rep. No. 4.* (undated).
[234] Thomas, W. *Progress of International Research on Thermodynamic Transport Properties*, (January 1962), Ed. J. F. Masi and D. H. Tsai, p 166. American Society of Mechanical Engineers: New York (1962).
[235] Timrot, D. L. and V. P. Borisoglebski. *J. Exp. Theor. Phys. USSR*, **38**, 1729 (1960). English translation *Soviet Phys. JETP*, **11**, 1248 (1960).
[236] Tsiklis, D. S. *Handbook of Techniques in High-pressure Research and Engineering.* Translated by A. Bobrowsky, Plenum: New York (1968).
[237] Tyrer, D. *J. Chem. Soc.* **103**, 1675 (1913).
[238] Tyrer, D. *J. Chem. Soc.* **105**, 2534 (1914).
[239] Ulrich, D. V., D. W. Kupke and J. W. Beams, *Proc. Nat. Acad. Sci., Wash.* **52**, 349 (1964).
[240] Verbeke, O. and A. van Itterbeek, Chap. 4 of *Physics of High Pressures and the Condensed State*, p 99. Ed. A. van Itterbeek, North-Holland: Amsterdam (1965).
[241] Vereshchagin, L. F. and V. A. Galaktionov, *Pribory i Teckn. Eksperim.* **1**, 98 (1957). A brief description in English is given in Tsiklis[236], p 398.

[242] Vilevich, A. V., L. F. Vereshchagin and Ya. A. Kalashnikov. *Pribory Tekh. Eksp.* **3**, 146 (1961). English translation in *Instrum. Exptl. Techn.* **1**, 559 (1961).
[243] Villey, J. *C.R. Acad. Sci., Paris*, **206**, 655 (1938).
[244] Vukalovich, M. P. and V. V. Altunin, *Teploenergetika*, **6**, No. 11, 58 (1959). A long abstract is published in *Chem. Abstr.* **54**, 21905 (1960).
[245] Wassenaar, T. *Thesis*, p 70. University of Amsterdam (1952). Quoted by Sengers[210], p 85.
[246] Watson, W. *Proc. Roy. Soc. Edinburgh*, **31**, 456 (1911).
[247] Weinberger, M. A. and W. G. Schneider, *Canad. J. Chem.* **30**, 847 (1952).
[248] Weir, C. E. *J. Res. Nat. Bur. Stand.* **45**, 468 (1950).
[249] Weir, C. E. *J. Res. Nat. Bur. Stand.* **50**, 95 (1953).
[250] Wells, F. W. and J. G. Roof. *Rev. Sci. Instrum.* **26**, 403 (1955).
[251] Westwater, W., H. W. Frantz and J. H. Hildebrand. *Phys. Rev.* **31**, 135 (1928).
[252] Whalley, E. *Canad. J. Technol.* **34**, 368 (1956).
[253] Whalley, E. *Canad. Technol.* **34**, 268 (1956).
[254] Whalley, E. *Internat. J. Mech. Sci.* **1**, 379 (1960).
[255] White, J. L. Chapter 8 of *Physicochemical Measurements at High Temperatures*. Ed. J. O'M. Bockris, J. L. White and J. D. Mackenzie. Butterworths: London (1959).
[256] Wiebe, R. and V. L. Gaddy. *J. Amer. Chem. Soc.* **60**, 2300 (1938).
[257] Wilson, W. and D. Bradley. *Deep-sea Research*, **15**, 355 (1968).
[258] Winnick, J. and S. J. Cho, *J. Chem. Phys.* **55**, 2092 (1971). Full details and results are given by Cho[75].
[259] van Witzenburg, W. and J. C. Stryland. *Canad. J. Phys.* **46**, 811 (1968).
[260] Yekhlakov, A. D. and K. P. Rodionov, *Fiz. Metal. i Metalloved.* **9**, 982 (1960). English translation in *Phys. Met. Metallog.* **9**, 932 (1960).
[261] Yntema, J. L. and W. G. Schneider, *J. Chem. Phys.* **18**, 641 (1950).
[262] Zhokovsky, M. K. *Izmeritel'-naya Tekhnika*, 3 (1955).

CHAPTER 10

Determination of Thermodynamic Properties from the Experimental $p-V-T$ Relationships*

R. D. MCCARTY

Cryogenics Division, National Bureau of Standards, Boulder, Colorado 80302, USA

Contents

I.	Computational Methods—Introduction	501
II.	The Equation of State	502
III.	Estimation of the Parameters of an Equation of State	505
	1. Linear Least Squares	506
	2. The Round Off Problem	507
	3. Least Squares with Constraints	508
	4. Non-linear Parameter Estimation	509
	5. The Simultaneous Use of Several Types of Property Data in Least Squares Parameter Estimation	510
IV.	Statistical Aspects of Least Squares Estimation	511
	1. Least Squares Conditions and Formation of Weights	511
	2. Significance Tests	514
	3. Confidence Intervals	515
V.	Miscellaneous Techniques for Improving the Accuracy of Thermodynamic Properties Calculated from an Equation of State	516
	1. Thermodynamic Equilibrium Conditions as Simultaneous Data	516
	2. Constrained Boundary Conditions	517
VI.	Thermodynamic Property Equations	518
VII.	Mathematical Formulas Useful in Thermodynamic Calculations	523
	1. Derivative Chain Rule	523
	2. Implicit Solutions of Equations of State	523
	3. Joining Independent $p-V-T$ Surfaces	524
	4. A Solution of M Equations for M Unknowns	524
VIII.	Abstract	525
IX.	References	525

I. Computational Methods—Introduction

The engineering applications of classical thermodynamics have grown steadily over the years. Today the need for thermodynamic properties in science and industry is greater than ever before in history. Further, their needs are continually changing to include properties of a greater variety, at pressures and temperatures of an increasing range and with accuracies that tax the state of the art. In recent years these needs have been accelerated by the tremendous worldwide effort in the field of space exploration.

Traditionally, the approach to obtaining thermodynamic properties for a

---

* Contribution of the National Bureau of Standards, not subject to copyright.

fluid has been via the equation of state for that fluid. The equation of state is usually defined by the experimental determination of the p–V–T surface of the fluid. Other thermodynamic properties which may be measured are specific heat capacity, Joule–Thomson coefficient, enthalpy, speed of sound, and internal energies. Any of these combined with two of the three p–V–T variables are sufficient to define the equation of state. The phrase 'equation of state' is often applied to a mathematical model of the thermodynamic equation of state. From such a mathematical model, the necessary thermodynamic relationships may easily be formed to obtain other needed thermodynamic properties. The procedures and techniques for translating one type of thermodynamic data to another by means of a mathematical equation of state is the sole purpose and substance of this chapter.

II. The Equation of State

The equation of state of a substance is usually described in terms of a set of p–V–T values, for that substance. Textbooks on thermodynamics usually lead the student into the concept of the equation of state by describing a p–V–T system and pointing out that once two of the three variables are chosen, one may not choose the third, rather it is a fixed value determined by the choice of the other two. This implies an equation of state which connects the thermodynamic variables. Scientists and engineers have been trying to describe the equation of state for various substances with mathematical models for many years. Almost all of these mathematical equations of state are empirical and are at best only as good as the experimental data to which they were fit.

Before the appearance of the high speed digital computers, it was extremely time-consuming to determine large numbers of adjustable parameters from experimental data. As a consequence, most of the early equations of state were of a simple nature. The virial equation of state was one of the earliest forms and is still widely used today because of its relationship to molecular theory.

$$pV = RT(1 + B\rho + C\rho^2 + \ldots) \qquad (1)$$

Equation 1 is the most commonly used form; however, the expansion in pressure is also used.

$$pV = RT(1 + BP + CP^2 + \ldots) \qquad (2)$$

Both of the expansions are infinite in length, and are theoretically only valid for the gas phase of a fluid.

The virial expansion represents the changes in volume or pressure along a line of constant temperature, and the number of terms needed to represent a given isotherm depends on the range of the p–V variation, and on the precision of the experimental measurements.

The virial coefficients may be calculated from the molecular potential function. For this reason the virial coefficients from experimental data continue to be of interest to the theoretician. Unfortunately, the accuracy of the virial coefficients obtained from experimental data is not good. In the case of the second virial coefficient, B, the uncertainty is probably no better than two to three per cent in the best of cases. For C the uncertainty is probably no better than ten per cent, and for the higher order terms, i.e. $D, E, \ldots$, the imprecision in most data is sufficiently high that no physical significance can be inferred.

The van der Waals equation of state[18] is probably one of the earliest attempts to represent both the liquid and vapor phases with a single mathematical function. The van der Waals equation is

$$p = RT/(V - b) - (a/V^2) \quad (3)$$

More recently the equation of state

$$p = \{RT(1 - \varepsilon)/V^2\}[V + B] - A/V^2 \quad (4)$$

was developed by Beattie and Bridgeman[1]. This equation of state works quite well up to twice the critical density which in its time was quite an accomplishment.

In 1940, Benedict et al.[2] published an empirical equation of state which they used to represent the p–V–T surface of light hydrocarbons. Their equation is

$$p = RT/V + (B_0 RT - A_0 - C_0/T^2)/V^2 + (bRT - a)/V^3$$
$$+ a\alpha/V^6 + [C(1 + \gamma/V^2)e^{-\gamma/V^2}]/(V^3 T^2) \quad (5)$$

The exponential term $e^{-\gamma/V^2}$ is the term which makes this equation of state different from the others listed here. As one follows the development of equations of state which have been proposed they become more and more complicated (more terms) as time goes on. These five equations are only a sample of the historical development of the mathematical form of the equation of state, literally hundreds of others have been proposed; however, most of them were modifications of those listed here.

With the advent of high speed digital computers, it became possible to estimate parameters for more complicated mathematical functions. As a result, some empirical equations of state have 35 or more adjustable parameters, some of which appear non-linearly in the equation. For the purposes of this work, four general types of equations of state which are now being used to represent calculated thermodynamic properties will be considered.

Many investigators perform isothermal p–V–T experiments, and represent their results in mathematical form by the virial expansion of equation 1. There are a number of reasons why the virial expansion is still used. Probably the most important reason is that it works well for experimental results taken at a constant temperature when there are no phase changes present. The biggest handicap of the virial representation is finding a suitable temperature dependence for the various coefficients.

Another problem with the virial expansion is the proper truncation of the expansion. The tendency to use too many terms is prevalent today because it is so easy to add another term when a computer is available to perform the necessary matrix inversions. The problem of estimating the virial coefficients and the associated problem of when to truncate the expansion will be discussed later.

One advantage of using the virial expansion is that theoretical values for the coefficients may be calculated. When thermodynamic data are needed for a fluid in a pressure and temperature range where no experimental data exist, it is possible to generate these data via the virial expansion using coefficients calculated from a molecular potential function. This is a valuable procedure when *no experimental data exist*; however, the virial coefficients estimated from experimental data are usually to be preferred to those calculated from theory. The virial coefficients calculated from a molecular potential function are also useful in detecting gross systematic errors in experimental data.

A second model of the p–V–T equation of state in widespread use today is the modified Benedict–Webb–Rubin[2] one. There are many modifications of the BWR equation of state but they may all be characterized by three common features: they all have one or more exponential terms which contain a non-linear undetermined parameter; they all have forms which are designed to allow explicit isothermal calculation of the derived thermodynamic properties; and they all are analytical at the critical point. When the parameters in these equations are determined properly they work well over wide ranges of pressure and temperature. They give good representation of both the liquid and vapor phases with a single function and they have the definite advantage of being adaptable to using many different kinds of experimental data in the determination of the parameters. This technique, of determining the parameters from several types of data, will be discussed in more detail later on.

The primary disadvantage of the modified BWR equation of state is incorrect representation of the critical point. If an equation of state is to give at least a qualitatively correct description of the region near the critical point

$$(\partial^2 p/\partial T^2)_{\rho_c} \to \infty \text{ (or becomes very large)} \qquad (6)$$

as P and $T \to p_c, T_c$.

This is not so for the modified BWR equation. The first modification of the latter was probably due to Strobridge[17] and is

$$\begin{aligned}
p = {} & RT\rho(Ra_1 T + a_2 + a_3/T + a_4/T^2 + a_5/T^4)\rho^2 \\
& + (Ra_6 T + a_7)\rho^3 + a_8 T\rho^4 \\
& + \rho^3(a_9/T^2 + a_{10}/T^3 + a_{11}/T^4)\,e^{-a_{16}\rho^2} \\
& + \rho^5(a_{12}/T^2 + a_{13}/T^3 + a_{14}/T^4)\,e^{-a_{16}\rho^2} \\
& + a_{15}\rho^6
\end{aligned} \qquad (7)$$

where the a_is are all adjustable parameters. The Strobridge fit of equation 7 to a selected set of p–V–T data for nitrogen obtained good results. Since then many modifications of equation 7 have been used to represent various p–V–T surfaces (see, for example, refs 6, 10, 16). The modified BWR equation of state is very useful in correlating data from many sources and of different kinds.

A third type of equation of state takes into account the non-analytical behavior of fluids shown by recent experimental work in the vicinity of the critical point. These equations of state are non-analytical and do give at least qualitatively correct behavior in the critical region, i.e. they satisfy equation 6. Their primary advantage is more realistic predictions in the region of the critical point. Their primary disadvantage is a direct result of their non-analyticity, that is, the integrations necessary for the calculation of the derived thermodynamic properties are not possible in closed form. This means that numerical methods must be employed to calculate properties such as entropy, enthalpy and specific heats which are time-consuming even on a fast computer.

It also means that the newer techniques of simultaneous use of different kinds of data such as specific heats and p–V–T cannot be used with these equations of state. Even when using p–V–T data alone these equations are hard to fit because they are very sensitive to systematic deviations in the data. For an example of such an equation of state, see Goodwin[7].

Another method of calculating thermodynamic data from p–V–T data was used quite successfully by Roder et al.[15] and later by Weber[11], and by Prydz and Straty[14]. This particular method works well on precise, internally consistent data but is not useful for correlating data from many different sources. The method essentially consists of curve fitting a power series in temperature to lines of constant density and a power series in density to lines of constant temperature. Interpolations are then made between the polynomials. Roder et al.[15] used 39 isotherms and 90 isochores together with a virial surface for low densities to represent the equation of state for hydrogen to 100 K with pressures to 340 atm. This method is probably the most accurate in representing precise data. It is not flexible and is difficult to program on a computer because of the large amount of input information necessary.

The matter of choosing a mathematical form for representation of a p–V–T surface is highly dependent upon the application and the extent of the p–V–T surface available. The purpose of the previous discussion of four types of mathematical forms was to help the reader to choose a particular form for his situation.

In summary, the power series method offers the greatest accuracy, the modified BWR offers an excellent compromise of speed, versatility and accuracy. The non-analytical equation of state offers more realistic behavior in the critical region at the sacrifice of speed and versatility and finally, the virial representation offers ease and a link with molecular theory.

III. Estimation of the Parameters of an Equation of State

After a mathematical equation of state has been chosen, the next step is to estimate the values of the adjustable parameters which appear in that equation of state. The number of adjustable parameters may be fixed or variable, depending upon the equation of state chosen. Suppose, for example, there is an equation of state which may be symbolized by

$$y = \sum_{m=1}^{M} a_m f_m(x_1, x_2, \ldots, x_k) \qquad (8)$$

where $y, x_1, x_2, \ldots, x_k$ are all physical variables but for convenience the y is chosen as the dependent variable and the x_k are chosen as the independent variables. In the case of the p–V–T surface, p may be chosen as y and V and T are then x_1 and x_2. The a_m represent the adjustable parameters which are to be estimated.

1. Linear Least Squares

Further suppose N discrete experimental values

$$(Y_n, X_{n_1}, X_{n_2}, \ldots, X_{n_k}), \quad n = 1, N \tag{9}$$

(Note: capital letters will be used to indicate experimental values of the physical variables.) The problem is to estimate the a_m from the experimental data set. The BLUE (best linear unbiased estimator) estimators are obtained by minimizing the sum of the squares, s', of the residuals r_n, $n = 1, N$ with respect to the a_m

$$s' = \sum_{n=1}^{N} r_n^2 = \sum_{n=1}^{N} (y_n - Y_n)^2 \tag{10}$$

The conditions under which the resulting estimators are BLUE will be discussed in more detail later; however, for purposes of developing the 'least squares' equations it is sufficient here to mention that the theoretical conditions for BLUE usually are most nearly satisfied by introducing a weighting factor, W_n, and minimizing the weighted sum of squares, S,

$$S = \sum_{n=1}^{N} W_n r_n^2 = \sum_{n=1}^{N} W_n (y_n - Y)^2 \tag{11}$$

W_n is the reciprocal of the effective variance of the dependent variable

$$W_n = 1/\sigma_{r_n}^2 \tag{12}$$

and

$$\sigma_{r_n}^2 = \sigma_{Y_n}^2 + \sum_{k=1}^{K} [(\partial y/\partial x_{nk}) \sigma_{x_{nk}}]^2 \tag{13}$$

where the partial derivatives $\partial y/\partial x_{nk}$ are evaluated at each data point. To minimize S with respect to the a_m ($m = 1, 2, \ldots, M$), the usual condition is imposed on the partial derivatives, i.e.

$$\partial S/\partial a_m = 0 \quad (m = 1, 2, \ldots, M) \tag{14}$$

Replacing the y_n in equation 10 by its functional form S becomes

$$S = \sum_{n=1}^{N} W_n (\sum_{m=1}^{M} a_m f_{nm} - Y_n)^2 \tag{15}$$

then forming the partial derivatives in equation 14 gives

$$\sum_{n=1}^{N} W_n (\sum_{m=1}^{M} a_m f_{nm} - Y_n) f_{nm'} = 0 \quad (m' = 1, 2, \ldots, M) \tag{16}$$

Equation 16 yields M equations in M unknowns and may be written as

$$\sum_{m=1}^{M} a_m \sum_{n=1}^{N} W_n f_{nm} f_{nm'} = \sum_{n=1}^{N} W_n f_{nm'} Y_n \quad (m' = 1, 2, \ldots, M) \tag{17}$$

The equations from 17 are commonly called the 'normal equations' and may be written in matrix notation as

$$[F]\{a\} = \{B\} \tag{18}$$

where $[F]$ is a M by M matrix containing the $W_n f_{nm} f_{nm'}$ terms, the $\{a\}$ is a column matrix containing the parameters to be estimated, and $\{B\}$ is a column matrix containing the $W_n f_{nm'} Y_n$ terms*.

The solution of equation 18 in matrix notation is ...

$$\{a\} = [F]^{-1}\{B\} \tag{19}$$

where $[F]^{-1}$ denotes the inverse of $[F]$.

2. The Round Off Problem

The actual method of forming the inverse of $[F]$ is an important step in the estimation procedure. Many methods of inverting a matrix are described in the literature. Most of these methods are satisfactory for certain problems but one also must be aware that none of them is satisfactory for certain other problems. The problem arises from arithmetic round off. In general, the more significant figures carried, the less significant is the round off problem. Usually, the larger the matrix, the more significant is the round off problem. However, this is not always true, because the 'ill conditioned' matrix will result in round off errors even when small matrices are being inverted. The 'condition' of a matrix is determined by the relative size and position of the elements, but a detailed treatment of causes and prediction of 'ill conditioning' will not be given here. For a discussion of the problem and a set of test matrices, the reader is referred to Newman and Todd[13]. It is best to consider every least squares matrix potentially 'ill conditioned'.

How does one detect round off effects in the results? Round off effects are not always obvious. For a detailed treatment of calculating error bounds due to round off effects in the matrix inversion problem see Newman and Todd[13] and Newman[12]. If round off errors are suspected they often may be detected by repeating the problem using more significant figures in the inversion calculations.

What are the best methods of matrix inversion? Probably the least likely to produce round off problems are the orthogonalization methods, such as the Gram–Schmidt method[4]. These methods, however, have other drawbacks; they require the formation of the overdetermined set of equations. When working with large data sets and equations of state with many parameters, this can be a serious problem. For example, if one has 1 000 data points and 30 parameters to estimate, the overdetermined matrix alone requires 30 000 locations in a single precision configuration and 60 000 locations in double precision. Double precision would probably be necessary for such a large number of parameters.

Any of the more traditional non-orthogonal methods having the pivotal

---

* The normal equations are often formed from the so-called 'overdetermined set' in matrix notation $[F] = [f]^T[W][f]$ where $[f]$ is an N by M matrix and $[f]^T$ is the transpose, $\{B\} = [f]^T[W]\{Y\}$. $[W]$ is a diagonal matrix of weights of order N. The indicated matrix multiplications will result in the exact same 'normal equations' as 17 and 18.

feature usually gives good results when an adequate number of significant figures is carried through the calculations. The pivotal feature, to minimize the possibility of round off, rearranges the normal matrix, placing the larger elements where they will be used early in the calculations. A few good methods of inverting matrices are the Gauss–Jordon, the Crout and the Newton–Raphson. Most textbooks on numerical analysis will describe one or more of these methods. See for example, Hildebrand[8]. See section VII.4 for a simple method of solving M equations for M unknowns.

The primary advantage of these later methods over an orthogonalization method is the small amount of computer storage needed to solve the normal equations. By comparison, with the example given in the orthogonalization case, the entire 60 000 matrix required there is not necessary when using a non-orthogonal method as the normal equations may be formed data point by data point, eliminating the need for permanent storage of the data in the computer memory. Regardless of what method is used, values calculated with the resulting parameters and the input data should always be compared.

3. Least Squares with Constraints

Very often in estimating parameters for an equation of state, it is desirable for the resulting mathematical model to pass exactly through some point on the p–V–T surface or to have a specific slope or curvature at a point on the surface. This may be accomplished by placing appropriate constraints on the estimating procedure; but of course the number of constraints on an equation of state may not exceed the number of parameters to be estimated.

Such constraints are most easily accomplished by using the methods of Lagrangian multipliers in which

$$Q = S + \sum_{l=1}^{L} \lambda_l \left(\sum_{m=1}^{M} a_{mn} g_{lm} - C_l \right) \tag{20}$$

are minimized, where the

$$\sum_{m=1}^{M} a_m g_{lm} = C_l \quad (l = 1, 2, \ldots, L) \tag{21}$$

represent constraints which are to be satisfied exactly by the equation of state and S is taken from equation 15. The formulation of the simultaneous equations is much the same as previously described. The Q differentiated with respect to the a_m gives

$$\sum_{m=1}^{M} a_m \sum_{n=1}^{N} W_n f_{nm} f_{nm'} + \sum_{l=1}^{L} \lambda'_l g_{lm'}$$

$$= \sum_{n=1}^{N} W_n f_{nm'} Y_n \quad (m' = 1, 2, \ldots, M) \tag{22}$$

where $\lambda'_l = \tfrac{1}{2}\lambda_l$.

The constraint equations are not differentiated with respect to the a_m because the S will be minimized by satisfying the constraining equations exactly. Equation 22 gives M equations in $M + L$ unknowns; therefore the L equations from equation 21 must be used to make the solution possible.

The inclusion of equation 21 gives $M + L$ equations in $M + L$ unknowns. In a partitioned matrix notation

$$\begin{bmatrix} [F] & [g]^T \\ [g] & [0] \end{bmatrix} \begin{Bmatrix} a \\ \lambda \end{Bmatrix} = \begin{Bmatrix} B \\ C \end{Bmatrix} \tag{23}$$

where $[F]$, $\{B\}$ and $\{a\}$ are the same as in equation 18, and $[g]$ and $\{C\}$ are defined by the constraining equations (see Hust and McCarty[9]).

After the normal equations with constraints are formed the solution for the a_ms is the same as before.

A common use of the constraining technique in thermodynamics is to fix the critical point of an equation of state in the following manner:

$$p_c = p(V_c, T_c) \tag{24}$$

$$(\partial p/\partial \rho)_T = 0 \tag{25}$$

$$(\partial^2 p/\partial \rho^2)_T = 0 \tag{26}$$

$$(\partial p/\partial T)_\rho = (\mathrm{d}p/\mathrm{d}T) \text{ (vapor pressure)} \tag{27}$$

4. Non-linear Parameter Estimation

Some of the more recent equations of state have one or more parameters which are non-linear. For example, the BWR equation of state has a term $e^{\gamma \rho^2}$. The γ is a non-linear adjustable parameter which must be estimated in some manner. If some of the parameters to be estimated appear non-linearly in the equation of state, the following iterative procedure may be used. If

$$y = y(x_1, x_2, \ldots x_k, a_1, a_2, \ldots a_m) \tag{28}$$

then initial estimates of the parameters, a_m, are determined and denoted by $a_{10}, a_{20}, \ldots, a_{mo}$. If the difference between the calculated value of $y = y_0$ and the experimental Y is small, then

$$Y - y_0 = \left(\frac{\partial y}{\partial a_1}\right) \Delta a_1 + \left(\frac{\partial y}{\partial a_2}\right) \Delta a_2 + \ldots + \left(\frac{\partial y}{\partial a_m}\right) \Delta a_m \tag{29}$$

where the Δa_m is the correction to the initial estimates a_{mo}, i.e.

$$a_{m_1} = a_{mo} + \Delta a_{mo} \tag{30}$$

Equation 29 is obtained from the Taylor expansion of equation 28, neglecting the higher order terms. The normal equations are formed from equation 29 in exactly the same manner as before since the corrections Δa_m are now linear. The procedure is repeated until the Δa_m become small relative to the a_m. However, the procedure does not always converge on a set of a_m and if it does there is no assurance that the convergence is on an absolute minimum. Similarly the constraining equations for the non-linear case

$$g_l(a_1, a_2, \ldots, a_m) = C_l \quad (l = 1, 2, \ldots, L)$$

may be written as

$$C_l - g_{l_0} = \sum (\partial g_l / \partial a_m) \Delta a_m \quad (l = 1, 2, \ldots, L) \tag{31}$$

where g_{l_0} is g_l evaluated with the $a_{10}, a_{20}, \ldots, a_{m0}$.

The resulting $M + L$ normal equations for the constrained non-linear case are therefore given by:

$$\sum_{m=1}^{M} \Delta a_m \sum_{n=1}^{N} W_n \left(\frac{\partial y_n}{\partial a_m}\right)\left(\frac{\partial y_n}{\partial a_{m'}}\right) + \sum_{l=1}^{L} \lambda'_l \left(\frac{\partial g_l}{\partial a_{m'}}\right)$$
$$= \sum_{n=1}^{N} W_n \left(\frac{\partial y_n}{\partial a_{m'}}\right)(Y - y_n) \qquad (m' = 1, 2, \ldots, M) \qquad (32)$$

and

$$\sum_{m=1}^{M} \Delta a_m \left(\frac{\partial g_l}{\partial a_m}\right) = C_l - g_l \qquad (l = 1, 2, \ldots, L) \qquad (33)$$

5. The Simultaneous Use of Several Types of Property Data in Least Squares Parameter Estimation

The determination of a mathematical representation of an equation of state is greatly aided by the simultaneous inclusion of several different kinds of property data. Higher order thermodynamic data such as heat capacities, enthalpies, velocity of sound, and heats of vaporization, will greatly improve the resulting equation of state when they are used simultaneously with p–V–T data in the least squares determination of the parameters. The theoretical conditions for the least squares estimators to be BLUE (see the next section) do not restrict the use of different kinds of data. The only requirement is the existence of a mathematical relationship between the data and that the resulting relationship is of closed form and still linear in the parameters being estimated.

For example, consider the virial equation of state

$$pV/RT = 1 + B/V + C/V^2 + \ldots \qquad (34)$$

The usual procedure is to estimate the Bs, Cs, etc., called virial coefficients, by a least squares procedure from the p–V–T data for each isotherm or, perhaps, by assuming some temperature dependence for the virials and using the data from several isotherms all at once. For the purposes of an example of simultaneous estimation, the B of equation 34 will be taken as

$$B = \sum_{i=1}^{I} a_i T^{(1.5 - i/2)} \qquad (35)$$

Now if the temperature dependence of B is to be estimated over a wide range of temperature using p–V–T data, second virial data, and speed of sound data simultaneously, the following procedure applies. For the second virial data, equation 35 is used and the B_n corresponds to the y in equation 8 and the a_i and $T^{(1.5-i/2)}$ correspond to the a_m and f_m, respectively.

To include the p–V–T data, the equation

$$pV/RT - 1 = \sum_{i=1}^{I} a_i \rho T^{(1.5 - i/2)} \qquad (36)$$

is used where the y in equation 8 is now $pV/RT - 1.0$ and the a_i and $\rho T^{(1.5 - i/2)}$ are the a_m and f_m, respectively. The inclusion of the speed of sound data is a little more complicated; however, if no higher order virial contribution is present it may be shown that

$$C^2 = C_0^2(1 + \alpha p) \tag{37}$$

where C^2 is the speed of sound squared, $C_0^2 = (C_p/C_v) RT/M$ is the ideal gas speed of sound, and

$$\alpha = \frac{1}{RT}\left(2B + \frac{4}{3}T\frac{dB}{dT} + \frac{4}{15}T^2\frac{d^2B}{dT^2}\right) \tag{38}$$

Substituting equations 35 and 38 into equation 37 results in:

$$\left(\frac{C^2}{C_0^2} - 1\right)\frac{RT}{p} = \sum_{i=1}^{I} a_i\left[2 + \frac{4}{3}\left(1.5 - \frac{i}{2}\right) + \frac{4}{15}\left(1.5 - \frac{i}{2}\right)\left(0.5 - \frac{i}{2}\right)\right]$$
$$\times T^{(1.5 - i/2)} \tag{39}$$

Now the y of equation 8 becomes $(C^2/C_0^2 - 1.0)RT/p$, the a_i again corresponds to the a_m and the coefficient of the a_i corresponds to the f_m. Since the three different ys described above all have different units and may vary in numerical value over a wide range, the weighting of these data becomes of much greater importance than in the case of a single type of property data.

More examples and greater detail in weighting procedures will be given in a later section. In matrix notation the general problem is formulated by:

$$[F_1]\{a\} = \{B_1\}, [F_2]\{a\} = \{B_2\} \ldots [F_i]\{a\} = \{B_i\} \tag{40}$$

where each matrix equation represents a set of normal equations for a different kind of data. The simultaneous minimization of these sets of data is then accomplished by forming a single set of normal equations by

$$[[F_1] + [F_2] + \ldots [F_i]]\{a\} = \{B_1\} + \{B_2\} + \ldots + \{B_i\} \tag{41}$$

where the $[F_j]$ and $\{B_j\}$ have been weighted appropriately*.

IV. Statistical Aspects of Least Squares Estimation

1. Least Squares Conditions and Formation of Weights

The method of least squares estimation is widely used in all kinds of curve fitting. Although the methods and techniques given here are perfectly general, the problems of estimating parameters for a mathematical model of an equation of state will be emphasized. Why is the least squares technique so widely used? There are two basic reasons. First, the mathematical condition of minimizing the sum of the squares of deviations leads to a convenient method of handling large amounts of experimental data and, second, under certain conditions the method of least squares estimation will yield BLUE (best linear

---

* The weighting procedures are described briefly in this section and in more detail in the next section.

unbiased estimators, best in the sense of minimum variance) results. The conditions for BLUE estimators are:

$$E(Y_i) = f(a_1, a_2, \ldots, a_m, x_{1i}, x_{2i}, \ldots, x_{ki}) \tag{42}$$

$$\text{Cov}(Y_i, Y_j) = 0 \text{ for all } i \neq j \tag{43}$$

$$\text{Var}(Y_i) = \sigma^2 \text{ for all } i \tag{44}$$

$(x_1, x_2, \ldots, x_k)$ are not subject to random error.

The first condition, equation 42, states that the expected value of an observed dependent variable is equal to some function of the independent variables and adjustable parameters. Basically, this condition requires knowledge of the true mathematical form of the equation of state, a condition which is usually only approximated. The second condition, equation 43, requires zero covariance between the random parts of the observations. In the case of p–V–T measurements this is usually satisfied. The third condition, equation 44, requires a constant variance, σ^2, for all observed Y_i. This may or may not be satisfied. If the p–V–T experimental data are from a single source and if the same apparatus is used over the entire range of measurement, then it is probably satisfied. If, however, the apparatus or instrumentation is changed during the experiment or if multiple sources of data are used then the condition can be more nearly satisfied by a transformation to a probability distribution which is of constant variance. From elementary statistics it is known that if a random variable, say Y_i, having a mean μ_y and a variance σ_y^2, is transformed to the random variable Z by the relationship

$$Z = (Y_i - \mu_y)/\sigma_y \tag{45}$$

The random variable Z will have a mean of 0 and a variance of 1. Since

$$E(Y_i) = \mu_y = f(a_1, a_2, \ldots, a_m, x_{1i}, x_{2i}, \ldots, x_{ki}) \tag{46}$$

then minimizing

$$\sum_{i=1}^{I} [Y_i - f(a_1, a_2, \ldots, a_m, x_{1i}, x_{2i}, \ldots, x_{ki})]^2 / \sigma_{y_i}^2 \tag{47}$$

will satisfy the least squares conditions when a non-constant variance over the data set applies. The weight then is

$$W_i = 1/\sigma_{y_i}^2 \tag{48}$$

The fourth condition for least squares is never satisfied for the p–V–T case, as all three variables are measured and therefore subject to random error, i.e. one of the three, p–V–T, must be chosen as the dependent variable and the other two become the x or independent variables. However, these random errors in the independent variables may be taken into account by including their effect on Y_i by the error propagation formula, again from elementary statistics. The variance in Y_i now becomes:

DETERMINATION OF THERMODYNAMIC PROPERTIES

$$\sigma^{2*}_{y_i \text{(effective)}} = \sigma^2_{y_i \text{(experimental)}} + \sum_{k=1}^{K} \sigma^2_{x_{ki}} (\partial f/\partial x_k)^2 \qquad (49)$$

The weight for an experimental point then becomes

$$W_i = 1/\sigma^2_{y_i \text{(effective)}} = 1/(\sigma^2_{y_i \text{(experimental)}} + \sum_{k=1}^{K} \sigma^2_{x_{ki}} (\partial f/\partial x_k)^2) \qquad (50)$$

If p is chosen as Y then

$$W_i = 1/\sigma^2_{p_i \text{(effective)}} = 1/[\sigma^2_{p_i \text{(experimental)}} + (\partial f/\partial V)^2_T \sigma^2_{V_i} + (\partial f/\partial T)^2_V \sigma^2_{T_i}]$$

In general, the W_i must be estimated for each data point. To accomplish this, the $\sigma^2_{y_i}$ and $\sigma^2_{x_{ki}}$ are estimates of the precision of the experimentally measured variables. For example, if the experimental uncertainty of the temperature measurement for a set of p–V–T data is assessed at 0.01 K, the σ^2_T would then be $(0.01)^2$. Experimenters' estimates of the uncertainty in their measurements are usually low. An experimental procedure which is rarely done in practice but which would be extremely valuable in subsequent analysis is to measure repeatedly an experimental point. If this were done for several points distributed uniformly over the ranges of X and Y, the standard formula for estimating variance could be applied at each point.

$$\sigma^2_{y_i} = \left(\sum_{i=1}^{I} y_i^2 - \left(\sum_{i=1}^{I} y_i \right)^2 \bigg/ I \right) \bigg/ (I-1) \qquad (51)$$

where I is the number of repetitions of a single data point.

The partial derivatives are easily estimated by first obtaining estimates of the a_m from a non-weighted fit and then iterating using weighted fits. The procedures to follow in determining the weighting for the simultaneous use of different kinds of data are now obvious. The estimates of the variance of the experimental variables are applied to equation 50 as before and the partial derivatives are formed from the explicit equations for the particular property being fit. In equation 50 the weight for the p–V–T surface is given when p is the dependent variable. If, for example, C_v were the dependent variable then the equations would be:

$$C_{vi} = g(a_1, a_2, \ldots, a_m, \rho_i, T_i) \qquad (52)$$

$$W_{cV_i} = 1/\sigma^2_{cV_i \text{(effective)}} = 1/(\sigma^2_{cV_i \text{(experimental)}} + (\partial g/\partial \rho)^2_T \sigma^2_{\rho_i} + (\partial g/\partial T)_\rho \sigma^2_{T_i}) \qquad (53)$$

Often the weights may be estimated more accurately by an application of the least squares procedure to smaller segments of the data set such as isotherms or isochores and then using the variance of the fit (see equation 55) as an estimate of the weight for that segment of the data.

An indication of how well the weighting has been estimated may be obtained from the variance of the fit (see equation 55) since it should be in the neighborhood of one if the weighting is correct.

---

* To be perfectly general equation 49 should have a term $\sum_{k=1}^{K} \sum_{j=1}^{J} (\partial f/\partial x_k)(\partial f/\partial x_j)\sigma_{x_{kji}} \quad k \neq j$ where $\sigma_{x_{kji}}$ is the covariance between x_k and x_j. However, for the p–V–T case, $\sigma_{x_{kji}}$ is assumed to be 0.

All of the preceding is valid for random errors only. No statistical techniques are applicable to systematic errors in the data. The first condition, equation 42, assumes all of the systematic errors have been removed. In practice there is almost always some systematic error present in the data and one can only hope that it is small enough to be ignored. There are experimental procedures which will detect, and in some cases randomize, the systematic error; however, these procedures will not be discussed here.

2. Significance Tests

The least squares estimation procedure and the related weighting procedures as previously outlined here are independent of probability distribution. Therefore, regardless of how the random variate is distributed, in a probability sense, the procedures are valid under the conditions set forth. One can never hope to satisfy these conditions exactly; the best that can be done is to approximate them as nearly as possible. The statistical procedures which will now be described are distribution dependent and the assumption is made that the random variate is normally distributed. Whether this assumption is correct or not is not critical as the central limit theorem tells us all probability distributions may be approximated quite well by the normal distribution if a reasonable number of independent points is taken and if one does not try to work at very high or very low probabilities.

The question of whether a particular term is justified or not in an equation should always be answered when performing least squares estimation of parameters to an equation of state. The question may be posed as the hypothesis of $a_m = 0$. To test this hypothesis the statistic

$$t^2 = \hat{a}_m^2 / d_{mm} \tag{54}$$

may be formed, where $\hat{a}_m^2$ is the square of the estimated value of the parameter a_m and the d_{mm} is the mth diagonal element of the dispersion matrix D. In this case:

$$[D] = S^2[F]^{-1}$$

$$S^2 = \left(\sum_{i=1}^{I} [Y_i - f(\hat{a}_1, \hat{a}_2, \ldots, \hat{a}_m, x_{1i}, x_{2i}, \ldots)]^2 / \sigma_{Y_i}^2 \right) / \text{d.f.} \tag{55}$$

The d.f. (degrees of freedom) is the number of points plus the number of constraints minus the number of parameters being estimated and the F has the same meaning as in section III. The $[F]^{-1}$ or inverse of the normal equations is formed in the solution of the normal equation. Therefore merely multiplying by the variance of the fit, S^2, one obtains with very little effort the dispersion matrix. The statistic t^2 is equally valid for the weighted or unweighted case, provided the least squares conditions are satisfied.

To test an original hypothesis of $a_m = 0$: if

$$t^2 > F(\alpha)_{1, I-M} \tag{56}$$

the hypothesis is rejected and the mth term would be retained. The $F(\alpha)_{1, I-M}$ is obtained from tabulations of the F distribution, the indices $1, I - M$ are the degrees of freedom of the numerator and denominator of t^2 respectively, and α is the confidence level. If one has 60 points or more in the fit and α is chosen as 95 per cent, then the number four may be used as an approximate value for $F(\alpha)_{1, I-M}$.

When doing least squares estimation of parameters for an equation of state, the significance testing of parameters is an important procedure for arriving at the form of an equation of state. This procedure should always be used in truncating the virial expansion. Many virial expansions are not truncated in this way and usually contain too many terms. When statistically non-significant terms are included in an equation of state, their function is to try to fit the random scatter in the data. The extrapolation of any P–V–T surface usually leads to large errors and the presence of non-significant terms greatly enhances the possibility of these errors.

3. Confidence Intervals

The dispersion matrix may also be used to form confidence intervals on the estimated parameters and on the calculations using the parameters in the equation of state.

$$(\hat{a}_m - t(\alpha)\hat{\sigma}_{a_m}, \hat{a}_m + t(\alpha)\hat{\sigma}_{a_m}) \tag{57}$$

where the $\hat{a}_m$ is the estimated value of the true value a_m; the $t(\alpha)$ is the Student t statistic for $I - M$ degrees of freedom exceeded in absolute value with probability α and $\hat{\sigma}_{a_m}$ is the square root of the ith diagonal element of the D matrix*, gives a $(1 - \alpha)$ confidence interval for a_m.

To form a confidence interval on a calculation using the estimated parameters, the dispersion matrix D is used. Suppose there is some property Z such that

$$\hat{Z} = f(\hat{a}_1, \hat{a}_2, \ldots, \hat{a}_m, x_1, x_2, \ldots, x_k) \tag{58}$$

then a confidence interval is formed on Z by

$$(\hat{Z} - Q, \hat{Z} + Q) \tag{59}$$

where

$$Q = t_{I-M}(\alpha)\hat{\sigma}_Z \tag{60}$$

and

$$\hat{\sigma}_Z^2 = \sum_{m=1}^{M} \sum_{r=1}^{M} (\partial Z/\partial a_m)(\partial Z/\partial a_r) \hat{\sigma}_{m,r}^2 \tag{61}$$

The partial derivatives are evaluated at the same set of x_k used for Z and the $\hat{\sigma}_{m,r}^2$ is the m, rth element of the dispersion matrix D. The $t_{I-M}[\alpha]$ is the Student t statistic for $I - M$ degrees of freedom exceeded in absolute value with probability α, and $\hat{Z}$ is the calculated value of Z resulting from the evaluation of $Z(\hat{a}_1, \hat{a}_2, \hat{a}_m, x_1, x_2, x_k)$ at some set of x and $\hat{a}$.

---

* This may be shown to be equivalent to the F test for significance, i.e. if the confidence interval includes 0 then the hypothesis of $a_m = 0$ is accepted.

The technique of weighting to more nearly approximate the least squares conditions for BLUE estimation is a very useful one. In some cases, such as using different kinds of data simultaneously and in the liquid region where $(\partial P/\partial \rho)_T$ is very large, it is absolutely necessary to weight the input data when using least squares estimation. The significance testing of parameters in an equation of state is also an essential procedure to follow, and valid significance tests may be carried out only if proper weighting has been performed.

V. Miscellaneous Techniques for Improving the Accuracy of Thermodynamic Properties Calculated from an Equation of State

There are many techniques to apply which will ultimately result in a better mathematical representation of an equation of state. Some of the more useful of these will be discussed here. The technique of simultaneous use of different kinds of property data[9] may be applied to parameter estimation in many ways. The use of higher order experimental thermodynamic data has already been discussed. The technique may also be applied in other ways.

1. Thermodynamic Equilibrium Conditions as Simultaneous Data

If the parameters for an equation of state have been estimated using only p–V–T data, the isothermal calculation of the derived thermodynamic

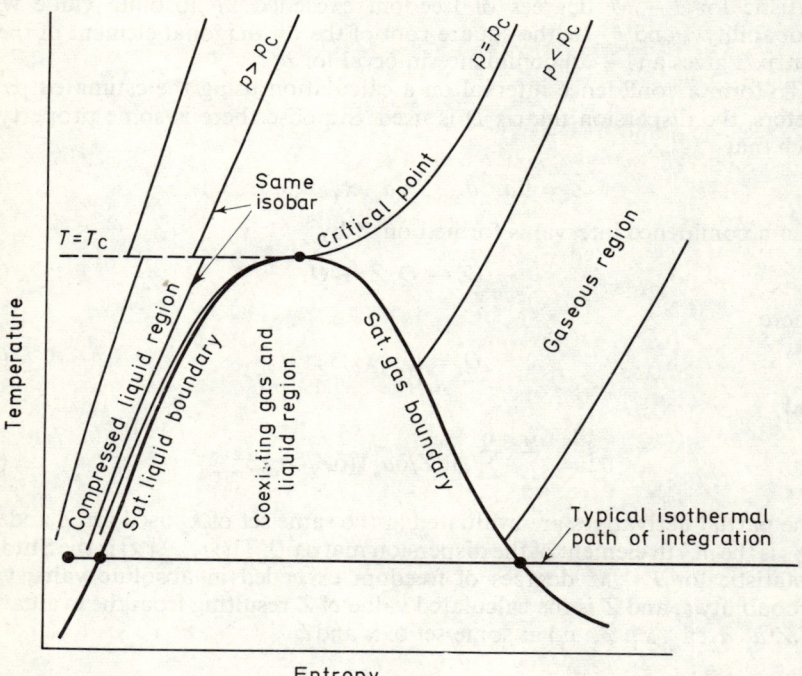

Figure 1. A schematic temperature/entropy diagram depicting discontinuities at the critical temperature.

properties from the gaseous phase through the region of the coexisting gaseous and liquid phases into the compressed liquid region is usually not possible using only the equation of state. For example, the entropy calculation indicated by equation 96 would have to be stopped at the saturated gaseous phase boundary and some other method of crossing the two-phase region used. This is usually accomplished by using the Clausius–Clapeyron relationship of equation 103. This requires the introduction of new information in the form of a vapor pressure equation and its derivatives, the usual result being discontinuities at or near the critical temperature for all densities greater than the critical density, see *Figure 1*.

However, if the derived thermodynamic equilibrium conditions are inserted into the parameter estimation along with the p–V–T and other data, the equation of state becomes a valid means of crossing the coesistence region. This means the Clausius–Clapeyron relationship does not have to be introduced and all of the derived thermodynamic property equations of section VI become valid for any point where the fluid is in a single phase. This application of simultaneous fitting was made by Bender[3]. He used the conventional least squares techniques and p–V–T data. In addition, he inserted simultaneously with the p–V–T data, the Gibbs energy relationship of

$$G_l - G_g = 0 = \int_{T^*\rho_g}^{\rho_l} \left[\frac{1}{\rho}\left(\frac{\partial p}{\partial \rho}\right)_T\right] d\rho \qquad (62)$$

where the G_l and G_g are the Gibbs energies of the coexisting liquid and vapor phases. The resulting mathematical model allows the continuous integration of the equations for entropy, enthalpy, etc., through the two-phase region. The author applied this technique in estimating parameters for helium, again for a modified BWR equation of state. It was found that in addition to equation 62 it was helpful also to use

$$S_g - S_l = \left(\frac{dP}{dT}\right)_{sat} (V_g - V_l) = -\int_{T^*\rho_l}^{\rho_g}\left[\frac{1}{\rho^2}\left(\frac{\partial p}{\partial T}\right)_\rho\right] d\rho \qquad (63)$$

which injects the entropy of vaporization into the parameter estimation.

2. Constrained Boundary Conditions

A second advantage of the use of a single function to calculate derived thermodynamic properties in both gas and liquid is the speed at which the calculations may be performed. In many cases, speed is a very important factor.

Occasionally the equation of state may be improved if the p–V–T surface is divided into regions. This is not often done because of the problem of continuity at the boundaries. However, it is possible to constrain one surface to another at a given line at the time the estimation of the parameters is made. For example, in *Figure 2*, if one wishes a continuous transition from region I to region II, where the boundary is the critical isochore (or the saturated liquid line for $T < T_c$), the p–V–T surface of region I is determined first and the following equations used in estimating parameters for region II.

---

* Denotes a constant temperature path for the integration.

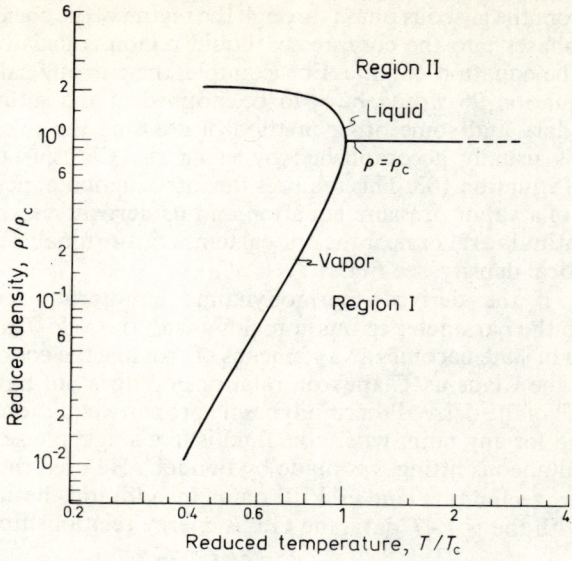

Figure 2. A generalized phase diagram arbitrarily divided into two regions.

$$\text{Prop}\,(\rho, T)_{\exp} = \text{Prop}\,(\rho_L, T)_I + [\text{Prop}\,(\rho, T)_{II} - \text{Prop}\,(\rho_L, T)_{II}] \quad (64)$$

where the property being used in the fit is denoted by Prop, the regions are denoted by the subscripts I and II, and L denotes the boundary line. The actual equation being fit in region II is the Δ Prop from the boundary line. Thus, the transition is constrained to be continuous across the boundary. See equation 113 for an alternative method of joining two surfaces.

VI. Thermodynamic Property Equations

After the mathematical model of the equation of state is complete the next step is to calculate related thermodynamic properties using the model. From basic thermodynamic definitions some of the more useful relationships are:

$$\left(\frac{\partial S}{\partial p}\right)_T = -\left(\frac{\partial V}{\partial T}\right)_P \quad (65)$$

$$\left(\frac{\partial S}{\partial V}\right)_T = \left(\frac{\partial p}{\partial T}\right)_V \quad (66)$$

$$C_V = T\left(\frac{\partial S}{\partial T}\right)_V = \left(\frac{\partial U}{\partial T}\right)_V \quad (67)$$

$$C_p = T\left(\frac{\partial S}{\partial T}\right)_P = \left(\frac{\partial H}{\partial T}\right)_P \quad (68)$$

DETERMINATION OF THERMODYNAMIC PROPERTIES

$$C_\sigma = T\left(\frac{\partial S}{\partial T}\right)_\sigma = \left(\frac{\partial U}{\partial T}\right)_\sigma + p\left(\frac{\partial V}{\partial T}\right)_\sigma \tag{69}$$

$$\left(\frac{\partial U}{\partial V}\right)_T = T\left(\frac{\partial p}{\partial T}\right)_V - p \tag{70}$$

$$\left(\frac{\partial H}{\partial p}\right)_T = V - T\left(\frac{\partial V}{\partial T}\right)_p \tag{71}$$

$$G = H - TS \tag{72}$$

$$H = U + PV \tag{73}$$

$$A = U - TS = G - pV \tag{74}$$

where S, H, U, G, A have the usual meanings and C_p is the specific heat capacity at constant pressure, C_V, the specific heat capacity at constant volume, and C_σ, the specific heat capacity at constant saturation.

From these equations it is seen that p–V–T relations are sufficient to determine the changes in enthalpy, entropy, Gibbs energy, internal energy and the Helmholtz energy, along isothermal paths. From equations 66 and 70 one obtains:

$$\Delta S_T = {}_{T^*}\!\!\int_{V_1}^{V_2} \left(\frac{\partial p}{\partial T}\right)_V dV \tag{75}$$

and

$$\Delta U_T = {}_{T}\!\!\int_{V_1}^{V_2} \left[T\left(\frac{\partial p}{\partial T}\right)_V - p\right] dV \tag{76}$$

From equations 65 and 71 one obtains:

$$\Delta S_T = {}_{T}\!\!\int_{p_1}^{p_2} -\left(\frac{\partial V}{\partial T}\right)_p dp \tag{77}$$

$$\Delta H_T = {}_{T}\!\!\int_{p_1}^{p_2} \left[V - T\left(\frac{\partial V}{\partial T}\right)_p\right] dP \tag{78}$$

The changes in S and H along an isobar are given by:

$$\Delta S_p = {}_{p}\!\!\int_{T_1}^{T_2} \frac{C_p}{T} dT \tag{79}$$

$$\Delta H_p = {}_{p}\!\!\int_{T_1}^{T_2} C_p \, dT \tag{80}$$

The changes in S and U along an isometric are given by:

$$\Delta S_V = {}_{V}\!\!\int_{T_1}^{T_2} \frac{C_V}{T} dT \tag{81}$$

$$\Delta U_V = {}_{V}\!\!\int_{T_1}^{T_2} C_V \, dT \tag{82}$$

* Variable in front of integral sign denotes path of integration.

The changes in S and U along the saturation boundary with respect to temperature are given by:

$$\Delta S_\sigma = \int_{\sigma \, T_1}^{T_2} \frac{C_\sigma}{T} \, dT \tag{83}$$

$$\Delta U_\sigma = \int_{\sigma \, T_1}^{T_2} \left[C_\sigma - p \left(\frac{dV}{dT} \right)_\sigma \right] dT \tag{84}$$

where σ denotes constant saturation and the p and V are saturation properties.

The changes in C_V and C_p along an isothermal path are given by:

$$\Delta C_V = \int_{T \, V_1}^{V_2} T \left(\frac{\partial^2 p}{\partial T^2} \right)_V dV \tag{85}$$

$$\Delta C_p = \int_{T \, p_1}^{p_2} T \left(\frac{\partial^2 V}{\partial T^2} \right)_p dp \tag{86}$$

It is important to note from the preceding equations that the only means of calculating changes in S, H and U with respect to temperature is by integration of the heat capacities. Therefore a typical equation fo the calculation of entropy from some point V_0, T_0 to some other point V, T, is:

$$S = S_0 + \int_{T \, V_0}^{V} \left(\frac{\partial p}{\partial T} \right)_V dV + \int_{p \, T_0}^{T} \frac{C_p}{T} dT \tag{87}$$

and similarly for enthalpy

$$H = H_0 + \int_{T \, V_0}^{V} \left[T - \left(\frac{\partial p}{\partial T} \right)_V - p \right] dV + \int_{p \, T_0}^{T} C_p \, dT \tag{88}$$

Ideal gas properties are generally about an order of magnitude more accurate than the best real gas property data. Because of the greater accuracy, especially for heat capacity data, and because ideal gas heat capacities are more generally available, it is common to use ideal gas heat capacity data in equations 87 and 88.

To make use of the ideal gas heat capacities for entropy and enthalpy calculations, a specific path of integration is adopted. To avoid some of the confusion to follow, the superscripts (*) and (°) refer to the ideal gas at 1 atm and 0 atm. respectively. The subscript 0 refers to a reference or datum point. Now hypothesize a point $p_0 = 1$ atm and T_0 on the ideal gas surface. The isothermal change in entropy from $p = 1$ atm to $p = 0$ atm is given by applying equation 75.

$$S^\circ_{T_0} = S^*_{T_0} + \int_{T \, RT/p}^{V = \infty} \left(\frac{\partial p}{\partial T} \right)_V dV \tag{89}$$

since at $p = 0$, $V = \infty$ and at $p = 1$ atmosphere, $V_0 = RT/p$, and for the ideal gas $(\partial p/\partial T)_V = R/V$. Thus,

$$S^\circ_{T_0} = S^*_{T_0} + R \ln (V_\infty / RT) \tag{90}$$

DETERMINATION OF THERMODYNAMIC PROPERTIES

Similarly for the real gas, the isothermal change in entropy is given by again applying equation 75.

$$S_{V, T_0} - S_{V_\infty, T_0} = \int_{V=\infty}^{V} \left(\frac{\partial p}{\partial T}\right)_V dV \bigg|_T \quad (91)$$

In the limit as $p \to 0$ the $S_{V_\infty, T_0} \to S_{T_0}^\circ$ and substituting $S_{T_0}^\circ$ for S_{V_∞, T_0} in equation 91 gives

$$S_{V, T_0} = S_{T_0}^* + R \ln(V_\infty/RT) + \int_{V_\infty}^{V} \left(\frac{\partial p}{\partial T}\right)_V dV \quad (92)$$

but noticing that

$$R \ln(V_\infty/RT) = R \ln(V/RT) + R \int_{V}^{V_\infty} \frac{dv}{V} \bigg|_T \quad (93)$$

and substituting in equation 92 gives

$$S_{V, T_0} = S_{T_0}^* + R \ln(V/RT) + \int_{V_\infty}^{V} \left[\left(\frac{\partial p}{\partial T}\right)_V - \frac{R}{V}\right] dV \bigg|_{T_0} \quad (94)$$

Equation 94 allows the isothermal calculation of the entropy from $p = 1$ atm and $T = T_0$ on the ideal gas surface to some arbitrary point of p and T_0 on the real gas surface. A means of moving in the temperature direction must now be found. Using equation 79 and the ideal gas heat capacities, the entropy change for the ideal gas from T_0 to T at 1 atm is

$$\Delta S_T^* = \int_{T_0}^{T} C_p^\circ \frac{dT}{T} \bigg|_p \quad (95)$$

Since the ideal gas heat capacity is a function of temperature only, adding equation 95 to equation 94 and substituting $\rho = 1/v$ gives

$$S_{\rho, T} = S_{T_0}^* + \int_{T_0}^{T} \frac{C_p^\circ}{T} dT \bigg|_p - R \ln(\rho RT/p_0) + \int_{0}^{\rho} \left[\frac{R}{\rho} - \frac{1}{\rho^2}\left(\frac{\partial p}{\partial T}\right)_\rho\right] d\rho \bigg|_T \quad (96)$$

Equation 96 allows the calculation of the entropy at an arbitrary point on the p–V–T surface if one has the ideal gas specific heat, a mathematical model of the p–V–T surface which will allow the calculation of the integral over density, and a value for the ideal gas entropy at some reference temperature T_0 and 1 atm. The integral over density in equation 96 requires the mathematical model of the equation of state to be reduced to the ideal gas equation of state as $\rho \to 0$.

Following the same derivation, one finds the equation for enthalpy to be

$$H = H_{T_0}^* + \int_{0}^{\rho} \left[\frac{p}{\rho^2} - \frac{T}{\rho^2}\left(\frac{\partial p}{\partial T}\right)_\rho\right] d\rho \bigg|_T + \frac{(p - \rho RT)}{\rho} + \int_{T_0}^{T} C_p^\circ dT \bigg|_p \quad (97)$$

Other useful thermodynamic formula include:
heat capacity at constant volume

$$C_V = C_V^\circ - \int_{0}^{\rho} \frac{T}{\rho^2}\left(\frac{\partial^2 p}{\partial T^2}\right)_\rho d\rho \bigg|_T \quad (98)$$

heat capacity at constant pressure

$$C_p = C_V + \left(\frac{\partial p}{\partial T}\right)_V^2 \bigg/ \left[\rho^2 \left(\frac{\partial p}{\partial \rho}\right)_T\right] \qquad (99)$$

speed of sound

$$C = \left[\left(\frac{C_p}{C_v}\right)\left(\frac{\partial p}{\partial \rho}\right)_T\right]^{1/2} \qquad (100)$$

or

$$C = \left[-V\left(\frac{\partial p}{\partial V}\right)_s\right]^{1/2} \qquad (101)$$

Joule–Thomson coefficient

$$\mu = \frac{1}{C_p}\left[T\left(\frac{\partial V}{\partial T}\right)_p - V\right] \qquad (102)$$

If the mathematical model of the equation of state being used will predict the thermodynamic equilibrium conditions of equation 62 adequately, and if the model is well behaved in the region of the coexisting gaseous and liquid phases, equations requiring isothermal integrations through the coexisting gaseous and liquid phases may be used directly in the compressed liquid region, i.e. $p > p_\sigma$, $T < T_c$†. If this is not so, other means of reaching the compressed liquid region are required. One of the most commonly used methods is to cross the two-phase coexistence region isothermally using the Clausius–Clapeyron relationships

$$S_{\sigma_g} - S_{\sigma_1} = \left(\frac{dp}{dT}\right)_\sigma (V_{\sigma_g} - V_{\sigma_1}) \qquad (103)$$

$$H_{\sigma_g} - H_{\sigma_1} = T(S_{\sigma_g} - S_{\sigma_1}) \qquad (104)$$

where σ_g denotes constant saturation of the gaseous phase and σ_1 denotes constant saturation of the liquid phase.

The use of equation 103 to establish the entropy and enthalpy along the saturated liquid line very often introduces other problems. In addition to those mentioned in section V.1, the heat capacities calculated for the region of $T < T_c$ and $p > p_c$ are usually in substantial error. The erroneous heat capacity calculation results from the dp/dT of the vapor pressure curve being in error at the lower temperatures. Since dp/dT is small at the lower temperatures, large percentage errors are common which in turn causes equally large percentage errors in the entropy and enthalpy along the saturated liquid line, and all of the derived thermodynamic property calculations for the compressed liquid are based on the saturated liquid line when this method is used. The heat capacities are especially sensitive to this type of error because they are functions of the slopes of the entropies and enthalpies.

Thermodynamic properties of the liquid and vapor phases in coexistence

---

† See section V.1.

are usually defined as the fraction of the total mass existing in the gaseous phase which is called the quality.

$$Q = (V - V_{\sigma_1})/(V_{\sigma_g} - V_{\sigma_1}) \tag{105}$$

The quality may be also defined by substituting for V in equation 105 any of the other thermodynamic properties except density, temperature, pressure or Gibbs function. The temperature, pressure and Gibbs function are equal for the saturated liquid and gas in coexistence and definition by the density is not by mass.

VII. Mathematical Formulas Useful in Thermodynamic Calculations

1. Derivative Chain Rule

Certain mathematical manipulations of an equation of state or the partial derivatives are sometimes useful in thermodynamic calculations. For obtaining derivatives, the two formulas:

$$\left(\frac{\partial X}{\partial Y}\right)_f \left(\frac{\partial Y}{\partial Z}\right)_f \left(\frac{\partial Z}{\partial X}\right)_f = 1 \tag{106}$$

and

$$\left(\frac{\partial X}{\partial Y}\right)_z \left(\frac{\partial Y}{\partial Z}\right)_x \left(\frac{\partial Z}{\partial X}\right)_y = -1 \tag{107}$$

are useful where X, Y, and Z represent the state variables and f may be any other thermodynamic variable. Bridgman[5] gives a very good extensive treatment of thermodynamic formulas.

2. Implicit Solutions of Equations of State

Mathematical equations of state are usually of the form

$$p = f(\rho, T)$$

where p is the dependent variable and ρ and T are independent variables. Solution for one of the independent variables is usually not possible in an explicit form. When no explicit solution exists, numerical techniques are employed to solve for either of the two independent variables. One of the most popular for thermodynamic properties is the Newton iteration. For example, if density were needed for a particular p and T, an initial guess for density, ρ_0, is made, and

$$\rho_1 = \rho_0 - \frac{f(\rho_0, T) - p}{\partial f(\rho_0, T)/\partial \rho} \tag{108}$$

is repeated until the required accuracy is achieved. The extension for a temperature solution with a p, ρ input is straightforward. If multiple solutions exist, the initial guess will determine which solution is obtained. For example, with an equation of state which is valid for both gaseous and liquid phases at temperatures below critical, the temperature obviously has multiple solutions for $p < p_c$ and $T < T_c$. In general, if the solution in the liquid region is desired, the initial guess on density should be larger than the solution density, and if the gaseous solution is desired, the initial density should be smaller than the solution density.

Where solutions of the equation of state for both independent variables are required, a simultaneous iterative technique will work. Take, for example, the solution of ρ and T for a particular p and H, an initial guess of ρ_0 and T_0 is made.

$$\Delta p = p - p(\rho_0, T_0) = \left(\frac{\partial p(\rho_0, T_0)}{\partial \rho}\right)_T \Delta \rho + \left(\frac{\partial p(\rho_0, T_0)}{\partial T}\right)_\rho \Delta T \quad (109)$$

$$\Delta H = H - H(\rho_0, T_0) = \left(\frac{\partial H(\rho_0, T_0)}{\partial \rho}\right) \Delta \rho + \left(\frac{\partial H(\rho_0, T_0)}{\partial T}\right) \Delta T \quad (110)$$

All the information in equations 109 and 110 is known, except the $\Delta \rho$ and ΔT, which may be calculated, and the new ρ and T become

$$\rho_1 = \rho_0 + \Delta \rho \quad (111)$$

$$T_1 = T_0 + \Delta T \quad (112)$$

The procedure is repeated until the $\Delta \rho$ and ΔT become small relative to ρ and T. The extension of this procedure to any input pair of thermodynamic variables is straightforward.

3. Joining Independent p–V–T Surfaces

In some instances it is desirable to join p–V–T surfaces at a common boundary.

The method outlined in section V is applicable if the joining of the two surfaces is planned before the parameter estimation takes place. Frequently, the two surfaces are prepared independently or the decision to join is made after the parameter estimation is accomplished. A smooth transition from one surface to another may be effected by using a weighted average of the two surfaces in a transition region. Take, for example, regions I and II in *Figure 2*. If a second line of constant density were drawn at, say, $\rho = 1.5 \rho_c$, then the following equation would effect a smooth transition from region I to region II for all temperatures and densities in the range of $\rho_c < \rho < 1.5 \rho_c$

$$p(\rho, T) = p(\rho, T)_I (1.5\rho_c - \rho)/\Delta\rho + p(\rho_c, T)_{II}[1 + (\rho - 1.5\rho_c)/\Delta\rho] \quad (113)$$

where $\Delta \rho$ is the difference in density between the two boundary lines. At $\rho = \rho_c$ the second term is 0 and at $\rho = 1.5\rho_c$ the first term is 0, thereby effecting the continuous smooth transition. If the two regions being joined are similar in the transition region, this technique works well.

4. A Solution of M Equations for M Unknowns

Let us write the equations 17 in another way:

$$a_1 x_{11} + a_2 x_{12} + a_3 x_{13} + \ldots a_m x_{1m} = C_1$$

DETERMINATION OF THERMODYNAMIC PROPERTIES

$$a_1 x_{21} + a_2 x_{22} + \ldots\ldots a_m x_{2m} = C_2$$
$$\vdots$$
$$a_1 x_{m1} + a_2 x_{m2} + \ldots\ldots a_m x_{mm} = C_m \quad (114)$$

where the $C_m = \sum_{n=1}^{N} W_n f_{nm'} Y_n$ and the $x_{nm} = \sum_{n=1}^{N} W_n f_{nm} f_{nm'}$.

The simplest method of solving these M equations for the a_m unknowns is one proposed by Gauss. It consists of first dividing the first equation by x_{11} and using the result to eliminate a_1 from the succeeding equations. Next, the modified second equation is divided by the x_{22} and the result is again used to eliminate the a_2 from the succeeding equations. After the elimination has been effected M times, equation 114 results in

$$a_1 + a_2 x'_{12} + a_3 x'_{13} + \ldots a_m x'_{1m} = C'_1$$
$$a_2 + a_3 x'_{23} + \ldots a_m x'_{2m} = C'_2$$
$$\ldots\ldots \quad (115)$$
$$a_{m-1} + a_m x'_{m-1, m} = C'_{m-1}$$
$$a_m = C'_m$$

The solution is completed by working backward from the last equation to obtain the other $a_{m-1}, a_{m-2}, \ldots, a_1$ successively. To minimize the effect of round off, it is desirable that the $x_{mm'}$ element of an equation be small (but not zero) with respect to the other x_{mm} in that equation. This may be optimized by re-arranging the remaining equations after each elimination.

VIII. Abstract

Methods and techniques for calculating the thermodynamic properties of a fluid from a mathematical model of the p–V–T equation of state are given. Criteria for choosing a mathematical form for the equation of state are discussed. Numerical and statistical parameter estimation techniques presented include, linear and non-linear least squares; constraints; simultaneous use of different kinds of data; significance tests and statistical weighting. Formulas for many of the derived thermodynamic properties are also presented.

IX. References

[1] Beattie, J. A. and O. C. Bridgeman. *Proc. Amer. Acad. Arts Sci.* **63**, 229 (1929).
[2] Benedict, M., G. B. Webb and L. C. Rubin. *J. Chem. Phys.* **8**, 334 (1940).
[3] Bender, E., *Proceedings* of the Fifth Symposium on Thermophysical Properties, p 227. American Society of Mechanical Engineers: New York (1970).
[4] Björck, Ake. *Nordisk Tidskrift for Informationsbehandling (Copenhagen)*, **7**, 1 (1967).
[5] Bridgman, P. W. *The Thermodynamics of Electrical Phenomena in Metals and a Condensed Collection of Thermodynamic Formulas*. Dover Publications: New York (1961).
[6] Coleman, T. C. 'The thermodynamic properties of nitrogen'. *Res. Rep. No. 11* Idaho University: Moscow (1970).
[7] Goodwin, R. D. and R. Prydz. 'Densities of compressed liquid methane, and the equation of state'. *Publ. US Nat. Bur. Stand.* (in press) (1971).

[8] Hildebrand, F. B. *Introduction to Numerical Analysis*. McGraw-Hill: New York (1956).
[9] Hust, J. G. and R. D. McCarty. *Cryogenics*, **7**, 200 (1967).
[10] McCarty, R. D. and R. B. Stewart. *Advances in Thermophysical Properties at Extreme Temperatures and Pressures*, p 84. American Society of Mechanical Engineers: New York (1965).
[11] McCarty, R. D. and L. A. Weber. 'Thermophysical properties of oxygen from the freezing liquid line to 600 R for pressures to 5000 psia'. *Tech. Note US Bur. Stand. No. 384* (1971).
[12] Newman, M. *Matrix Computations, Survey of Numerical Analysis* Vol. VI, p 222. Edited by John Todd. McGraw-Hill: New York (1962).
[13] Newman, M. and J. Todd. *J. Soc. Industr. Appl. Math.* **6**, 466 (1958).
[14] Prydz, R. and G. R. Straty. 'The thermodynamic properties of compressed gaseous and liquid fluorine; *Tech. Note US Nat. Bur. Stand. No. 392* (1970).
[15] Roder, H. M., L. A. Weber and R. D. Goodwin. 'Thermodynamic and related properties of parahydrogen from the triple point to 100 K at pressures to 340 atmospheres'. *US Nat. Bur. Stand. Monogr. No. 94* (1965).
[16] Stewart, R. B. 'The thermodynamic properties of oxygen'. *Ph.D. Thesis* Iowa University: Ames (1966).
[17] Strobridge, T. R. 'The thermodynamic properties of nitrogen, from 64 to 300 K between 0.1 and 200 atmospheres;. *Tech. Note US Nat. Bur. Stand. No. 129* (1962).
[18] Van der Waals, J. D. *Over de Continuiteit van den Gasen*. Vloeistofloestand: Leyden (1873).

CHAPTER 11

Thermodynamic Properties and the Velocity of Sound

W. Van Dael

Laboratorium voor Molekuulfysika, Universiteit te Leuven, Heverlee, Belgium

Contents

List of Symbols		528
Introduction		529
I. Thermodynamic Relations		530
1. Adiabatic Properties		530
2. Sound Velocity		532
II. Absorption and Dispersion		533
1. Translational Relaxation		534
2. Rotational and Vibrational Relaxation		536
3. Critical Dispersion		539
4. Other Relaxation Phenomena		541
III. Sound Velocity and the Equation of State		541
1. Ideal Gas		541
2. Virial Equation of State		541
3. Van der Waals Equation of State		543
4. BWR Equation of State		544
IV. Sound Velocity in Mixtures		544
1. Ideal Mixture		545
2. Non-ideal Mixtures		548
V. Experimental Methods		549
1. Interferometer Methods		549
A. Fixed Path Interferometer		550
B. Variable Path Interferometer		552
2. Pulse Methods		552
VI. Sound Velocity in Gases		555
1. General Behavior		555
2. The Absolute Value of W in the Low Pressure Limit		556
3. The Initial Slope $(\delta W^2/\delta p)_T$		558
4. High Density Data		559
5. Generalized Behavior of Sound Velocity in Gases. Corresponding States Treatment		559
VII. Sound Velocity in Pure Liquids		561
1. General Behavior		561
2. Liquids Coexisting with Their Saturated Vapor		562
3. Single Phase High Density Fluids		563
4. Critical Region		565
5. Generalized Sound Velocity Behavior in Dense Fluids		568
VIII. Sound Velocity in Liquid Mixtures		570
1. Homogeneous Mixtures		570
2. Mixtures Showing Phase Separation		572
IX. Acknowledgements		574
X. References		574

List of Symbols

a, b	constants in the Van der Waals equation
B, B'	second virial coefficient
B	molecular compressibility [equation 7.18].
C, C'	third virial coefficient
C_i	internal energy part of the molar specific heat
C_p	molar specific heat at constant pressure
C_p°	the low pressure limit of C_p
C_p^E	excess molar specific heat at constant pressure [equation 4.26]
C_v	molar specific heat at constant volume
C_v°	the low pressure limit of C_v
C_v^E	excess molar specific heat at constant volume
C_σ	molar specific heat along the saturated vapor pressure curve
D, D'	fourth virial coefficient
F	molar Helmholtz free energy
f	frequency
H^E	excess enthalpy of a mixture
k	wavenumber $2\pi/\lambda$
k, k'	constants in equations 7.15 and 7.19
k_B	Boltzmann constant
K, L, M	coefficients in the series expansion of $W(1/V)$
K', L', M'	coefficients in the series expansion of $W(p)$
$\overline{L}$	mean free path
m	molecular mass
M	molecular weight
n	number density
p	pressure
p_0	$= \varepsilon/\sigma^3$ [equation 6.6]
r	intermolecular distance
R	gas constant
R	molecular sound velocity [equation 7.16]
S	entropy
T	temperature in Kelvin
T_0	$= \varepsilon/k_B$ [equation 6.4]
T_c	critical temperature
U	internal energy per mole
$\overline{V}$	average molecular thermal velocity
V	molar volume
V^E	excess molar volume [equation 4.25]
V_0	$= N_A\sigma^3$ [equation 6.5]
W	sound velocity
W_{mig}	sound velocity in a mixture of ideal gases
$W(\omega)$	sound velocity at the angular frequency ω
$W(0)$	sound velocity in the low frequency limit
$W(\infty)$	sound velocity in the high frequency limit
W_0	$= (N_A\varepsilon/M)^{\frac{1}{2}}$ [equation 6.7]
W°	sound velocity in a gas in the low pressure limit
x	dimensionless quantity $x = k\xi$
x_i	molar fractions

THERMODYNAMIC PROPERTIES AND THE VELOCITY OF SOUND

X	dimensionless quantity defined in equation 2.3
Y	dimensionless quantity defined in equation 2.4
Z	collision number
Z_W	dimensionless quantity defined in equation 6.3
α_p	expansion coefficient at constant pressure
α_S	adiabatic expansion coefficient
α_σ	expansion coefficient along the saturated vapor pressure line
α'	'anomalous' absorption $\alpha' = \alpha - \alpha_{cl}$
α	amplitude absorption coefficient
α_{cl}	classical amplitude absorption coefficient [equation 2.1]
β_T	isothermal compressibility
β_S	adiabatic compressibility
γ	ratio of the specific heats C_p/C_v
γ°	low pressure limit of γ
γ_S	adiabatic thermal pressure coefficient
γ_v	isochoric thermal pressure coefficient
γ_σ	thermal pressure coefficient along the saturated vapor pressure line
ε	$\lvert T - T_c\rvert/T_c$
ε	energy parameter in the intermolecular potential
η_s	coefficient of shear viscosity
η_v	coefficient of volume viscosity
$\varkappa$	coefficient of thermal conductivity
λ	acoustical wavelength
ν	frequency
ξ	correlation length
ρ	density
σ	molecular diameter
σ	index characterizing derivatives taken along the saturated vapor pressure curve
τ	relaxation time [equation 2.11]
τ'	modified relaxation time [equation 2.14]
$\bar{\tau}_c$	mean time between molecular collisions [equation 2.8]
φ_i	volume fraction [equation 4.13]
ϕ	intermolecular potential
ψ_i	weight fraction [equation 4.24]
ω	angular frequency
ω_D	inverse relaxation time for critical entropy fluctuations [equation 2.23].

Introduction

The measurement of the velocity of sound in fluids provides a means for obtaining some equilibrium thermodynamic data which are not readily accessible by other experimental methods. The unique feature of sound propagation is the fact that, in the large majority of the experimental situations, it is an adiabatic process: sound velocity data give direct and precise information on the adiabatic properties of the fluid.

There is another important aspect of sound velocity, namely that it is very closely related to derivatives of the equation of state. Consequently, the precision on these derivatives is often substantially better when they are deduced from the velocity rather than when they are obtained from the analysis of classical pVT data.

The relations between sound velocity and the other thermodynamic properties are treated in section I. This purely thermodynamic approach can only be applied if the frequency of the sound wave is so low that the system can be considered at each moment as being in local equilibrium. This means that the period of the acoustic oscillation should be long compared to a characteristic relaxation time. When both these times are comparable in magnitude the sound velocity becomes a function of the frequency. This point has been worked out in section II with a general discussion about these dispersion effects for different relaxation processes. The purpose was to give an idea of the limitations on the applicability of the formulas of section I and to sketch the procedure for an eventual transformation of high frequency sound velocity data to their quasi-static values.

In section III are the general thermodynamic relations applied on some widely used equations of state with special emphasis on virial expansion. A generalization for mixtures is given in section IV: some attention has been paid to the behavior of sound velocity in the ideal mixture model and to the deviations from it in real systems.

A very schematic survey of the relevant experimental techniques is given in section V. No attempt at all is made to be complete: there are several textbooks treating this topic and it is outside the scope of this contribution to duplicate this largely technical information.

Sections VI, VII and VIII discuss experimental facts about sound velocity respectively in gases, liquids and liquid mixtures. Emphasis is laid on the thermodynamic aspects: interpretations based on models for the dense fluid state, although they provide some valuable qualitative insight, are not discussed because they will in general not allow one to derive reliable numerical data. The aim of these last three sections is to illustrate, by a selection of examples, how sound velocity data have been of some use in the study of the thermodynamics of fluids.

I. Thermodynamic Relations

1. Adiabatic Properties

Adiabatic properties of a fluid may be important from a thermodynamic point of view. Adiabatic (or isentropic, i.e. with constant entropy S) changes on the pVT surface are described by three coefficients:
the expansion coefficient

$$\alpha_S \equiv \frac{1}{V}\left(\frac{\delta V}{\delta T}\right)_S$$

the compressibility

$$\beta_S \equiv -\frac{1}{V}\left(\frac{\delta V}{\delta p}\right)_S$$

and the thermal pressure coefficient

$$\gamma_S \equiv \left(\frac{\delta p}{\delta T}\right)_S$$

The three adiabatic coefficients are interrelated by

$$\alpha_S = -\beta_S \gamma_S$$

The most interesting from these three is β_S due to its relation to other properties:

$$\beta_S = \frac{C_v \alpha_p}{C_p \gamma_v} = \frac{\beta_T}{\gamma} \qquad (1.1)$$

$$\beta_S = \beta_T - \frac{TV\alpha_p^2}{C_p} \qquad (1.2)$$

$$\frac{1}{\beta_S} = \frac{1}{\beta_T} + \frac{TV\gamma_v^2}{C_v} \qquad (1.3)$$

The other adiabatic coefficients give rise to the following equations:

$$\gamma_S = \frac{C_p}{TV\alpha_p} = \gamma_v + \frac{C_v}{TV\alpha_p} \qquad (1.4)$$

$$\alpha_S = -\frac{C_v}{TV\gamma_v} = \alpha_p - \frac{C_p}{TV\gamma_v} \qquad (1.5)$$

The adiabatic derivatives at a given point on the PVT surface can of course be derived from the equation of state making use of the formulas 1.1 to 1.5 and involving knowledge of the specific heat. The latter can in principle be calculated by an integration of the equations:

$$\left(\frac{\delta C_p}{\delta p}\right)_T = -T\left(\frac{\delta^2 V}{\delta T^2}\right)_p \qquad (1.6)$$

$$\left(\frac{\delta C_v}{\delta V}\right)_T = T\left(\frac{\delta^2 p}{\delta T^2}\right)_V \qquad (1.7)$$

It is quite clear that in order to proceed in this way the available pVT data have to be extremely precise and that some integration constants must be known. Unless the equation of state is extremely accurate the equations 1.6 and 1.7 may lead to very unreliable specific heat values.

The quantity β_S can in principle also be measured in a direct way. Some experiments have been carried out, for example, by Staveley et al.[113]. The measurement starts with a compression of the fluid: when equilibrium is reached, the pressure is suddenly released and an almost adiabatic change in volume occurs which is small but measurable. Various corrections have to be made due principally to the compressibility of the vessel and to the heat flow to the surroundings. The precision on these data is of the order of a few per cent and they are only available for a very limited number of substances in temperature and pressure ranges that do not present too much experimental difficulty.

2. Sound Velocity

Some of the experimental problems in a direct β_S measurement may be circumvented if the changes in p, V and T can be made oscillating around an equilibrium state with a period that is on the one hand small enough to prevent a net heat flow to the surroundings but on the other hand large enough to allow the system to reach at each moment thermodynamic equilibrium. A sound wave can fulfil these requirements.

The commonly used relation between the isentropic compressibility and the sound velocity W is given by

$$W^2 = \left(\frac{\delta p}{\delta \rho}\right)_S = \frac{1}{\rho \beta_S} \tag{1.8}$$

or with the pressure written as a partial derivative of the internal energy per mole U

$$p = -\left(\frac{\delta U}{\delta V}\right)_S$$

one obtains also

$$W^2 = \frac{V^2}{M}\left(\frac{\delta^2 U}{\delta V^2}\right)_S \tag{1.9}$$

We may remark here that all the symbols for extensive variables we use in this chapter have the meaning of molar properties. This simplifies the expressions for W which itself is not an extensive property.

Other equivalent expressions may be found by straightforward transformations with equations 1.1 to 1.3

$$W^2 = \frac{\gamma}{\rho \beta_T} \tag{1.10}$$

$$W^2 = \frac{1}{\rho(\beta_T - TV\alpha_p^2/C_p)} \tag{1.11}$$

$$W^2 = \frac{1}{\rho}\left(\frac{1}{\beta_T} + \frac{TV\gamma_v^2}{C_v}\right) \tag{1.12}$$

Making use of $p = -(\delta F/\delta V)_T$ with F the molar Helmholtz free energy one gets also

$$W^2 = \frac{\gamma V^2}{M}\left(\frac{\delta^2 F}{\delta V^2}\right)_T \tag{1.13}$$

It should be emphasized, however, that in the derivation of equation 1.8 two important assumptions have been made:
(a) the deviations of the density and the temperature from their mean values

are small so that the hydrodynamic equations can be linearized. This implies that the amplitude of the sound wave should be small, a condition that is readily realized.

(b) the absorption coefficient is small, which means, in general, that the frequency has to be sufficiently low. This criterion is more ambiguous than the foregoing; we will discuss it in some detail in section II in order to be aware of the fact that not all sound velocity data may be analysed with the thermodynamic relations of section I.

II. Absorption and Dispersion

For small amplitude sound waves the acoustic energy is absorbed by the classical mechanism of viscosity and thermal conductivity. In a first approximation both absorption effects are additive, giving what is usually called the classical amplitude absorption coefficient[57]

$$\alpha_{cl} = \frac{\omega^2}{2\rho W(\omega)W^2(0)}\left[\frac{4}{3}\eta_s + (\gamma - 1)\frac{M\varkappa}{C_p}\right] \quad (2.1)$$

$W(\omega)$ is the sound velocity at the angular frequency ω while $W(0)$ is the corresponding value in the low frequency limit where α_{cl} vanishes. As will be shown further, the dispersion due to classical absorption is almost always negligible and one replaces $W(\omega)W^2(0)$ in equation 2.1 often by $W^3(\omega)$ or by $W^3(0)$.

The first term in equation 2.1 with η_s, the coefficient of shear viscosity, has been calculated by Stokes under the assumption that the coefficient of bulk viscosity is zero. The second term with $\varkappa$, the coefficient of thermal conductivity, is named after Kirkwood.

Equation 2.1 gives a rather good description of the experimental absorption coefficient for low density monatomic gases. For a more general description, however, one needs the higher order approximations of α which according to Truesdell[122] can be written in the following form

$$\alpha = \frac{\omega}{2W(\omega)} X[1 + (\gamma - 1)Y + O(X^2)] \quad (2.2)$$

with X and Y dimensionless quantities defined by

$$X \equiv \frac{\omega(\frac{4}{3}\eta_s + \eta_v)}{\rho W^2(0)} \quad (2.3)$$

$$Y \equiv \frac{\varkappa M}{(\frac{4}{3}\eta_s + \eta_v)C_p} \quad (2.4)$$

with η_v the coefficient of volume viscosity. In the expressions 2.1 and 2.2 for $\alpha(\omega)$ the factor $W(\omega)$ occurs and it is already clear that both quantities are strongly interrelated. The same feature appears in the explicit expression of W as a function of the frequency[122]

$$\frac{W^2(\omega)}{W^2(0)} = \{1 + \tfrac{1}{4}X^2[3 + 10(\gamma - 1)Y - (\gamma - 1)(7 - 3\gamma)Y^2] + O(X^4)\} \quad (2.5)$$

The numerical value of Y is of the order of unity. So if $X^2 \ll 1$, all the terms

$O(X^2)$ may be neglected in equations 2.2 and 2.5 and one obtains again the starting formulas for the sound velocity (equation 1.8) and for the classical absorption coefficient (equation 2.1) (assuming the Stokes condition $\eta_v = 0$).

There may be several factors that make X^2 not negligibly small: high frequencies, low densities or high values of the coefficient of volume viscosity.

The sign of the dispersion effect is determined by the square bracket term in equation 2.5. For low density gases the value of Y can easily be estimated by using the Eucken relation

$$\varkappa M/\eta C_v = \tfrac{1}{4}(9\gamma - 5) \tag{2.6}$$

This indicates that the dispersion is positive.

1. Translational Relaxation

Until now the fluid has been considered as a continuum: a kinetic treatment will introduce collisions and relaxation effects. For low density gases this transition can be made by expressing the quantities in equation 2.5 in terms of the mean free path $\bar{L}$ and the mean time between collisions $\bar{\tau}_c$.

For rigid spherical molecules the coefficient of shear viscosity is given to a first approximation by

$$\eta_s = (5\pi/32)mn\bar{L}\bar{v} \tag{2.7}$$

with the mean free path $\bar{L} = (\sqrt{2}\pi n\sigma^2)^{-1}$ and the average thermal velocity

$$\bar{v} = (8kT/\pi m)^{\tfrac{1}{2}}$$

Higher order approximations will multiply the numerical factor in equation 2.7 by 1.016 which gives 0.499 for the prefactor[64, 27]. With $\bar{\tau}_c = \bar{L}\bar{v}^{-1}$ equation 2.7 can be written as

$$\eta_s = (5/4)p\bar{\tau}_c \tag{2.8}$$

The time 1.25 $\bar{\tau}_c$ can be interpreted as a translational relaxation time τ_{tr}: it means that on the average there is slightly more than one collision needed to decrease an excess in kinetic energy to $1/e$ of its value. Following this same idea the expression 2.3 can be transformed into

$$X = (10\pi/3\gamma)(\bar{\tau}_c/\tau_s) \tag{2.9}$$

where τ is the period of the sound wave.

It follows that the parameter X is proportional to the product $\omega\bar{\tau}_c$. According to kinetic theory $\bar{\tau}_c$ is inversely proportional to the density or to the pressure. Consequently the characteristic parameter in the dispersion relation 2.5 is ω/p.

For calculation purposes it is easy to express $\bar{\tau}_c$ directly in terms of η/p as indicated by equation 2.8 and to substitute the Eucken factor in Y.

The dispersion formula 2.5 can then be simplified to the form

$$\frac{W^2(\omega)}{W^2(0)} = 1 + a\left(\frac{\eta\omega}{p}\right)^2 + \ldots \tag{2.10}$$

with a equal to 1.41 for monatomic and 1.48 for diatomic gases. The higher order approximations for η_s in equation 2.7 will multiply the value of X in equation 2.9 by a factor 1.016 and that of a in equation 2.10 by a factor 1.032. Equation 2.10 shows clearly that some dispersion will be present in every experimentally determined sound velocity. The amount is, however, in many situations several orders of magnitude smaller than the experimental accuracy. This has been illustrated in *Figure 1* showing the dispersion data obtained by Greenspan[48]. The simplified equation 2.10 describes the experimental data very well up to $\omega\eta/p = 1$; this corresponds for all gases at atmospheric pressure with a frequency of about 1 GHz. The ratio $W^2(\omega)/W^2(0)$ has increased at that frequency up to a value of two.

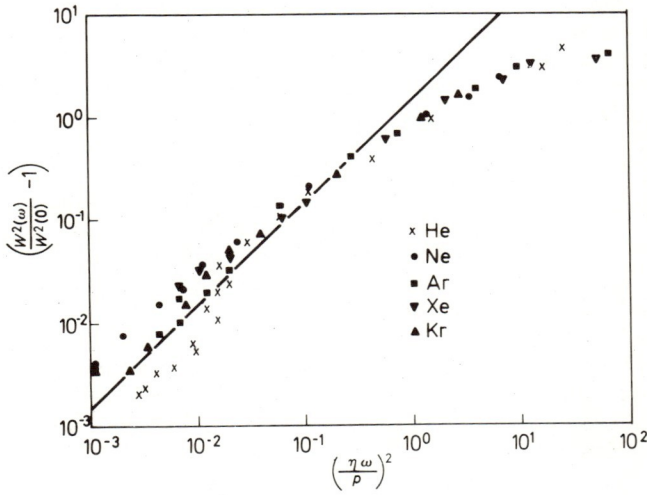

Figure 1. Dispersion due to translational relaxation. The points are experimental data by Greenspan[48]. The straight line represents equation 2.10.

The conclusion is that at ultrasonic frequencies translational dispersion can only be detected in very dilute gases and that it will be completely negligible in dense fluids. This may not be so if the frequency increases in the hypersonic range and caution is needed in the thermodynamic interpretation of hypersonic sound velocity data in dense fluids. In this situation the meaning of $\bar{\tau}_c$ becomes somewhat doubtful and one can expect at least that one should prefer to put X proportional to ω/ρ rather than to ω/p.

Experimental hypersonic velocities in simple liquids do not confirm unambiguously the existence of dispersion effects. For liquid nitrogen no systematic difference has been found between the ultrasonic data at 1 MHz[125, 114] and the hypersonic results at 3–5 GHz[100]. In liquid ammonia and in liquid sulfur dioxide a small positive dispersion effect seems to be present at 5 GHz[135]. On the other hand the hypersonic velocities obtained at 3 GHz by Fleury and Boon[42] are for liquid argon slightly and for liquid neon definitely lower than the corresponding ultrasonic data[125, 77]. It may be added that in some of these fluids a coefficient of volume viscosity has been found comparable in magnitude with the coefficient of shear viscosity.

For a more thorough discussion of translational relaxation in dilute gases we refer to review articles by Greenspan[49] and by Bauer[7,8].

2. *Rotational and Vibrational Relaxation*

Experimental absorption coefficients in polyatomic fluids are often much larger than the values predicted by equations 2.1 and 2.2. A formal solution can be found by assuming that this excess absorption is simply due to a non-zero volume viscosity, whatever its physical meaning may be. Although this formalism has the advantage of unification of the formulas it is subject to some criticism on its internal consistency[122] and it hides the real nature of the excess absorption, which is microscopic.

Let us consider a polyatomic fluid: a sound wave will force the local temperature to oscillate around an equilibrium value. The internal energy of the molecules is quantized. For simplicity we will assume that only two energy levels are occupied. The population of these levels has to adapt itself continuously to the temperature fluctuation by transitions between the two states. These transitions occur when in a molecular collision a sufficient amount of energy can be transferred from the external to the internal degrees of freedom or vice versa. Apparently not every collision will be a suitable one for this transfer: one defines a collision number Z as the average number of collisions occurring before the energy to be transferred, ΔE, has decreased to the e^{-1} part of its initial value. The quantity ΔE is the difference between the momentary value of the internal energy E and the value E_0 it would have if there was equilibrium with the energy of the external degrees of freedom.

This collision number can also be related to a time scale with the help of a relaxation time τ defined by

$$- dE/dt = (1/\tau)(E - E_0) \tag{2.11}$$

giving

$$Z = \tau/\bar{\tau}_c \tag{2.12}$$

The derivation of the equations for the absorption coefficient and the dispersion effect may be found elsewhere[57,11]. The results are

$$\alpha'\lambda = \pi \frac{W^2(\omega)}{W^2(0)} \frac{(C_p - C_v)C_i}{C_v(C_p - C_i)} \frac{\omega\tau'}{1 + \omega^2\tau'^2} \tag{2.13}$$

where α' is the 'anomalous' absorption equal to $\alpha_{\text{total}} - \alpha_{\text{cl}}$, C_i is the internal energy part in the specific heats and

$$\tau' = \{(C_p - C_i)/C_p\}\tau \tag{2.14}$$

The dispersion formula is

$$\frac{W^2(0)}{W^2(\omega)} - \frac{\alpha'^2 W^2(0)}{\omega^2} = 1 - \frac{(C_p - C_v)C_i}{C_v(C_p - C_i)} \frac{\omega^2\tau'^2}{1 + \omega^2\tau'^2} \tag{2.15}$$

In almost every experimental situation the second term on the left is small and can be neglected.

A further approximation valid for low values of $\omega\tau'$ is

$$\frac{W^2(0)}{W^2(\omega)} = 1 - \frac{(C_p - C_v)C_i}{C_v(C_p - C_i)}\omega^2\tau'^2 \qquad (2.16)$$

The factor containing the specific heats in equations 2.13 and 2.15 gives an idea of the maximum dispersion effect that can occur. At the highest frequencies the internal specific heat C_i is completely blocked and one gets

$$W^2(\infty) = \frac{1}{\rho\beta_T}\frac{C_p - C_i}{C_v - C_i} \qquad (2.17)$$

and consequently

$$\frac{W^2(\infty) - W^2(0)}{W^2(0)} = \frac{(C_p - C_v)C_i}{C_p(C_v - C_i)} \qquad (2.18)$$

The symbols C_p and C_v in these formulas retain their usual meaning of thermodynamic variables corresponding with low frequency processes.

Choosing a special value for the angular frequency $\omega_m = \tau'^{-1}$, the function $W^2(0)\alpha'\lambda/W^2(\omega)$ reaches a maximum value

$$\frac{W^2(0)}{W^2(\omega_m)}\alpha'(\omega_m)\lambda(\omega_m) = \frac{\pi}{2}\frac{W^2(\infty) - W^2(0)}{W^2(0)} \qquad (2.19)$$

At this same frequency the curve

$$\frac{W^2(0)}{W^2(\omega)} - \frac{\alpha'^2 W^2(0)}{\omega^2} - 1$$

goes through its halfpoint

$$\frac{W^2(0)}{W^2(\omega_m)} - \frac{\alpha'^2(\omega_m)W^2(0)}{\omega_m^2} - 1 = \frac{1}{2}\frac{W^2(\infty) - W^2(0)}{W^2(0)} \qquad (2.20)$$

It is common practice to simplify equation 2.19 by putting $W^2(0)/W^2(\omega_m) \sim 1$ and equation 2.20 by assuming $\alpha'^2 W^2(0)\omega_m^{-2} \sim 0$. This can be allowed when the dispersion and the anomalous absorption effects are very small so that the errors introduced by those assumptions are not larger than the experimental uncertainties in α' and W.

The dispersion formulas contain, exactly as for translation relaxation, the frequency of the sound wave only in the combination $\omega\tau$ with $\tau = Z\bar{\tau}_c$. So ω and p have again a complementary role and only the ratio ω/p determines the location on the dispersion/absorption curves. It may be emphasized here that on dispersion curves measured as a function of pressure the trivial pressure effect on W, as discussed in section VI, has first to be removed.

An illustration of the properties of vibrational relaxation phenomena is given in *Figure 2* with experimental data in gaseous sulfur hexafluoride by Haebel[56].

The relaxation times, even for simple molecules, may vary over as much as ten decades. A survey of some experimental data is given in *Table 1* both for rotational and vibrational relaxation. More detailed compilations have been made by Matheson[82], Cottrell and McCoubrey[31], Stevens[114] and Gordon, Klemperer and Steinfeld[47].

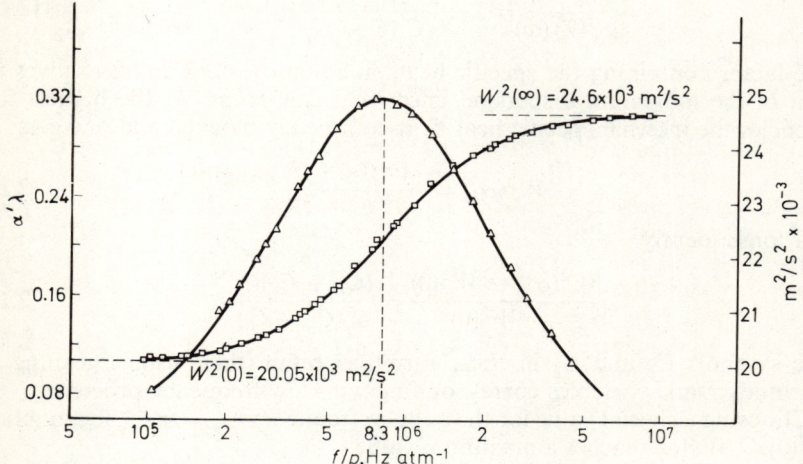

Figure 2. Anomalous absorption and dispersion due to vibrational relaxation. Data points were obtained by Haebel in gaseous sulfur hexafluoride at 52°C[56]. Curves are drawn according to equations 2.13 and 2.15 with $(f/p)_m = 8.3 \times 10^5$ Hz/atm.

Those data are important even in the context of a thermodynamic interpretation of the sound velocity: they make it possible to evaluate how far experimental velocity data in polyatomic fluids deviate from the thermodynamic limit.

Two quite different pieces of information are needed for this purpose: (1) a characteristic f/p (or τp or Z) indicating where the center of the

Table 1. Relaxation times and collision numbers for vibrational and rotational energy transfer in polyatomic molecules at atmospheric pressure [from A. J. Matheson[82], p 60ff]

Substance	T K	τ_{vib} sec	Z_{vib}	T K	$\tau_{rot} \times 10^9$ sec	Z_{rot}
CH_4	298	2.0×10^{-6}	1.8×10^4	303	1.1	12
CO	290	6.0	5×10^{10}	298	0.28	2.1
CO_2	293	7.0×10^{-6}	5.4×10^4	298	—	1.6
C_2H_6	303	7.1×10^{-10}	10	286	—	4.0
D_2	1100	1.2×10^{-5}	—	773	45	225
HCl	290	$> 1 \times 10^{-2}$	—	300	—	6.2
H_2	300	1.1×10^{-3}	1.6×10^7	873	43	303
H_2O	410	8.0×10^{-9}	76	323	0.32	4.0
NH_3	298	$\leqslant 4 \times 10^{-10}$	$\leqslant 5$	300	—	2.3
N_2	476	1.6×10^{-1}	8.2×10^8	295	0.76	5.5
NO	296	4×10^{-7}	2.7×10^3	296	0.12	1.0
O_2	303	1.6×10^{-2}	9.8×10^7	295	0.51	3.3
SF_6	284	7.6×10^{-7}	2.1×10^3	284	0.6	5

dispersion curve is located on the ω/p axis. These data are normally given by the ω/p values where the maximum in the anomalous absorption occurs (*Figure 2*). It appears that the low density data also can be used for the dense fluid without introducing large errors: $(\rho\tau)_{\text{gas}} \simeq (\rho\tau)_{\text{liquid}}$.

(2) the magnitude of the dispersion effect depending on the ratio of the internal specific heat C_i to the total specific heat. C_i can be calculated if the internal energy levels between which the transitions occur are identified. The dispersion magnitude can, however, also be derived, as has been shown in equations 2.19 and 2.20, from the height of the maximum in the experimental absorption coefficient per wavelength (*Figure 2*). Such a procedure may be very helpful in cases where it is not so evident which molecular mechanism gives rise to the relaxation. This may happen when several relaxation processes overlap: although the formulas given in this chapter are strictly only valid for a system with transitions between only two energy levels it appears that they can sometimes also be used for more complicated situations with the introduction of an effective relaxation time.

A more complete discussion of internal relaxation phenomena can be found in the textbooks of Herzfeld and Litovitz[57] and of Bhatia[11] and in review articles by Kneser[70, 71] and by Sette[110].

3. Critical Dispersion

A remarkable example of a relaxation phenomenon with a great number of relaxation times can be found in critical fluid systems (see also Chapter 14, section X.3).

Chynoweth and Schneider[29] observed in xenon in the critical state an anomalous absorption and dispersion. They gave a qualitative explanation based on the existence in the critical state of clusters or molecular aggregates whose dynamic equilibrium is disturbed by the sound wave.

Later on a more complete theoretical study of this effect was made by Fixman[39, 40, 41] and by Kawasaki[66, 67, 68]. An essential point in Kawasaki's derivation is the mode–mode coupling idea leading to the result that a sound wave of wavenumber k mainly decays by breaking up into two heat-conduction modes with wavenumbers $k - k'$ and k'. We will not enter into details of this problem and rather refer to some recent survey articles on this subject by Sette[11], Mistura[93] and Garland[45].

According to Garland et al.[46, 36], the dispersion is described by

$$W^2(0)/W^2(\omega) = 1 - B(\varepsilon)J(\omega^*) \tag{2.21}$$

in which B is a function of thermodynamic quantities varying with

$$\varepsilon \equiv (T - T_c)/T_c$$

J is an integral over an infinite set of relaxation processes each with a frequency dependence of the same form as that for the rotational dispersion described in II.2

$$J = \int_0^\infty \frac{x^2}{(1 + x^2)^2} \frac{\omega^2\tau^2}{1 + \omega^2\tau^2} dx \tag{2.22}$$

The integration goes over all values of x, being the product of the correlation length ξ and the wavenumber k which, according to the mode–mode coupling theory, may have all possible values. The time τ in equation 2.22 equals $2/K(x)\omega_D$ where ω_D^{-1} is the relaxation time for the diffusion of entropy fluctuations over one correlation length with

$$\omega_D = (\varkappa/\rho C_p) \cdot (1/\xi^2) \tag{2.23}$$

$K(x)$ is the Kawasaki function

$$K(x) = \tfrac{3}{4}[1 + x^2 + (x^3 - 1/x)\tan^{-1} x] \tag{2.24}$$

Using $\omega^* \equiv \omega/\omega_D$ one can then write

$$J(\omega^*) = \int_0^\infty \frac{x^2}{(1+x^2)^2} \left[\frac{\{2\omega^*/K(x)\}^2}{1 + \{2\omega^*/K(x)\}^2} \right] dx \tag{2.25}$$

J is a universal function of ω^* which in turn is a function of the frequency and of $T - T_c$. This is similar to the situation for thermal relaxation where the variable was f/p.

In *Figure 3* a comparison is made between experimental results in xenon

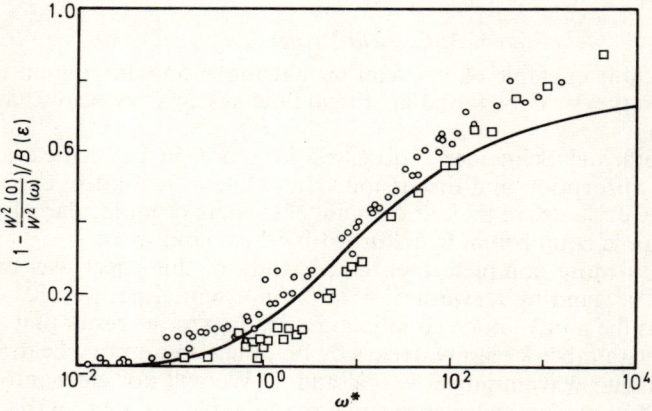

Figure 3. Reduced sound velocity dispersion in xenon as a function of the reduced frequency ω^*: ○ ultrasonic data (0.4–5 MHz), □ hypersonic data (170–500 MHz). The curve represents $J(\omega^*)$ of equation (2.25) [After Eden et al.[36]].

along the critical isochore and the calculated function $J(\omega^*)$[36]. There is very good agreement for data points taken over a large range of frequencies and over different temperature intervals. It follows from the graph that large dispersion effects are present in the critical region: even at low values of ω the value of ω^* may increase enormously as the critical point is closely approached and it seems impossible to measure the thermodynamic sound velocity right at the critical point.

For a critical binary mixture similar formulas apply except that the heat diffusivity has to be replaced by mass diffusivity. The characteristic frequency is now defined by

$$\omega'_D \equiv 2D/\xi^2 \tag{2.26}$$

where D is the binary diffusion coefficient.

4. Other Relaxation Phenomena

There exist many other situations where anomalous absorption and sound velocity dispersion occur: they all have in common that some equilibrium depends on pressure and on temperature. A sound wave tries to shift this equilibrium periodically: relaxation effects are present each time this equilibrium lags behind the varying p or T value. Some examples are: chemically reacting fluids, liquids showing rotational isomerism or associations and highly viscous liquids.

All these effects are related to rather complex molecules and fall outside the scope of this paper. An extensive discussion of these phenomena can be found in textbooks by Herzfeld and Litovitz[57], Bhatia[11], Matheson[82] and in review articles by Lamb[75, 76], Sette[110] and Litovitz[79].

III. Sound Velocity and the Equation of State

The thermodynamic relations in section I can be applied to equations of state. A variety of pVT relations exist with different degrees of exactness and complexity. For some of the most used ones the expressions for the sound velocity will be given now.

1. Ideal Gas

For the ideal gas, i.e. a hypothetical fluid consisting of point particles with no mutual interaction, the equation of state is

$$pV = RT \tag{3.1}$$

and the sound velocity

$$W^{\circ 2} = \gamma^{\circ} RT/M \tag{3.2}$$

2. Virial Equation of State

For real gases the virial expansion is an exact equation of state provided that an infinite number of terms is taken into account. The practical use is limited to low density gases where the series converges rapidly.

$$pV = RT(1 + B/V + C/V^2 + \ldots) \tag{3.3}$$

The sound velocity is given by

$$W^2 = \frac{\gamma RT}{M}\left(1 + \frac{2B}{V} + \frac{3C}{V^2} + \ldots\right) \tag{3.4}$$

The ratio of the specific heats γ can be written according to equation 1.3 as

$$\gamma = 1 - \frac{T}{C_v}\left(\frac{\delta V}{\delta p}\right)_T \left(\frac{\delta p}{\delta T}\right)_V^2 \tag{3.5}$$

and the specific heat C_v, by integration of equation 1.7, as

$$C_v = C_v^\circ - T\int_V^\infty \left(\frac{\delta^2 p}{\delta T^2}\right)_V dV \tag{3.6}$$

Applying these expressions to equation 3.4 one obtains for the sound velocity

$$W^2 = \frac{\gamma^\circ RT}{M}\left(1 + \frac{K}{V} + \frac{L}{V^2} + \frac{M}{V^3} + \ldots\right) \tag{3.7}$$

with

$$K = 2B + 2(\gamma^\circ - 1)T\frac{dB}{dT} + \frac{(\gamma^\circ - 1)^2}{\gamma^\circ}T^2\frac{d^2B}{dT^2} \tag{3.8}$$

$$L = \left[B + (2\gamma^\circ - 1)T\frac{dB}{dT} + (\gamma^\circ - 1)T^2\frac{d^2B}{dT^2}\right]^2\frac{\gamma^\circ - 1}{\gamma^\circ}$$

$$+ \frac{1 + 2\gamma^\circ}{\gamma^\circ}C + \frac{\gamma^{\circ 2} - 1}{\gamma^\circ}T\frac{dC}{dT} + \frac{(\gamma^\circ - 1)^2}{2\gamma^\circ}T^2\frac{d^2C}{dT^2} \tag{3.9}$$

$$M = \left(2T\frac{dB}{dT} + T^2\frac{d^2B}{dT^2}\right)\left[B + (2\gamma^\circ - 1)T\frac{dB}{dT} + T^2\frac{d^2B}{dT^2}(\gamma^\circ - 1)\right]^2 \times$$

$$\frac{(\gamma^\circ - 1)^2}{\gamma^\circ} + \frac{1}{\gamma^\circ}\left[B + 2(\gamma^\circ - 1)T\frac{dB}{dT} + (\gamma^\circ - 1)T^2\frac{d^2B}{dT^2}\right] \times$$

$$\left[2C + 2\gamma^\circ T\frac{dC}{dT} + (\gamma^\circ - 1)T^2\frac{d^2C}{dT^2}\right] + \frac{2(\gamma^\circ - 1)}{\gamma^\circ}D$$

$$+ \frac{2(\gamma^\circ + 2)(\gamma^\circ - 1)}{3\gamma^\circ}T\frac{dD}{dT} + \frac{(\gamma^\circ - 1)^2}{3\gamma^\circ}T^2\frac{d^2D}{dT^2} \tag{3.10}$$

Although equation 3.3 is the most logical form of the virial expansion it has the disadvantage that the molar volume is not a suitable experimental parameter: one generally prefers the pressure. The corresponding expansion in terms of p gives

$$pV = RT(1 + B'p + C'p^2 + \ldots) \tag{3.11}$$

The sound velocity becomes

$$W^2 = (\gamma RT/M)\left[1 + 2B'p + (3C' + B'^2)p^2 + (4D' + 4B'C')p^3 + \ldots\right] \tag{3.12}$$

and

$$W^2 = (\gamma^\circ RT/M)(1 + K'p + L'p^2 + M'p^3 + \ldots) \tag{3.13}$$

THERMODYNAMIC PROPERTIES AND THE VELOCITY OF SOUND

The coefficients of the density and pressure expansion are interrelated by

$$B' = (1/RT)B \tag{3.14}$$

$$C' = \{1/(RT)^2\}(C - B^2) \tag{3.15}$$

$$D' = \{1/(RT)^3\}(D - 3BC + 2B^3) \tag{3.16}$$

$$K' = (1/RT)K \tag{3.17}$$

$$L' = \{1/(RT)^2\}(L - BK) \tag{3.18}$$

$$M' = \{1/(RT)^3\}[M - 2BL + (2B^2 - C)K] \tag{3.19}$$

3. Van der Waals Equation of State

For dense fluids the Van der Waals equation gives a qualitative description of the pVT surface.

$$(p + a/V^2)(V - b) = RT \tag{3.20}$$

The corresponding sound velocity is given by

$$W^2 = \frac{\gamma RT}{M}\left[\frac{1}{(1 - b/V)^2} - \frac{2a}{RTV}\right] \tag{3.21}$$

or by

$$W^2 = \frac{\gamma RT}{M}\left[\frac{1}{(1 - b/V)^2} - \frac{2}{1 - b/V} + \frac{2pV}{RT}\right] \tag{3.22}$$

with

$$\gamma = 1 - \frac{(\gamma_0 - 1)(1 - b/V)^{-2}}{2a/VRT - 1/(1 - b/V)^2} \tag{3.23}$$

leading to

$$W^2 = \frac{RT}{M}\left[\frac{\gamma_0}{(1 - b/V)^2} - \frac{2a}{RTV}\right] \tag{3.24}$$

and

$$W^2 = \frac{RT}{M}\left[\frac{\gamma_0}{(1 - b/V)^2} - \frac{2}{1 - b/V} + \frac{2pV}{RT}\right] \tag{3.25}$$

Other equations of state may follow the pVT surface much closer at the cost of an increased number of constants and much more complicated expressions for W. Thus, in the equation of Berthelot a temperature dependence is introduced for the Van der Waals constant a

$$(p + a'/TV^2)(V - b) = RT \tag{3.26}$$

and the corresponding sound velocity is

$$W^2 = \frac{RT}{M}\left\{\frac{1}{(1 - b/V)^2} - \frac{2a'}{RT^2V} + \frac{[(1 - b/V)^{-1} - (a'/RT^2V)]^2}{(\gamma_0 - 1)^{-1} - (2a'/T^2V)}\right\} \tag{3.27}$$

4. BWR Equation of State

Another approach is given by the Benedict–Webb–Rubin equation

$$p = RT\rho + (B_0 RT - A_0 - C_0/T^2)\rho^2 + (bRT - a)\rho^3$$
$$+ a\alpha\rho^6 + (C/T^2)\rho^3 (1 + \beta\rho^2) e^{-\beta\rho^2} \qquad (3.28)$$

with a sound velocity[74]

$$W^2 = \frac{\beta T^4 [R + (B_0 R + 2C_0/T^3)\rho + bR\rho^2 - (2C\rho^2/T^3)(1 + \beta\rho^2)e^{-\beta\rho^2}]^2}{\{R\beta T^3/(\gamma° - 1)\} - 6C + 6C_0\beta\rho + 3Ce^{-\beta\rho^2}(2 + \beta\rho^2)}$$
$$+ RT + 2\rho(B_0 RT - A_0 - C_0/T) + 3\rho^2(bRT - a)$$
$$+ 6a\alpha\rho^5 + (Ce^{-\beta\rho^2}/T^2)(3\rho^2 + 3\beta\rho^4 - 2\beta^2\rho^6) \qquad (3.29)$$

Another series of analytical expressions for the sound velocity is based on models for the liquid state. Some of them have been treated by Van Dael and Van Itterbeek[123]; they will not be discussed here because their predictive qualities are rather poor.

The general conclusions of this section are:
(a) The expressions for the sound velocity derived from empirical equations of state are rather complicated and one cannot hope that they are more reliable than the starting formulas themselves.
(b) Experimental W data form a severe test for the validity of an equation of state. It appears, however, to be a frustrating task to improve the coefficients in a pVT equation, or even the equation itself, by making use of W data.
(c) These two remarks do not hold when the equation of state is exact as for the low density virial expansion. This topic will be fully discussed in section VI.

IV. Sound Velocity in Mixtures

In mixtures of gases or liquids an additional variable, the composition, appears besides the temperature and the density. In principle this should not give rise to serious problems. The thermodynamic relations of section I remain valid and it is sufficient to add to all symbols the suffix 'mix'.

Difficulties may start, however, as soon as one tries to express the properties of the mixture as a sum of the properties of the pure components weighted with some concentration-dependent factor. This can only be done in an almost exact way for the low density gas with the aid of the virial expansion as in equation 3.3 with

$$B_{\text{mix}} = x_1^2 B_{11} + 2x_1 x_2 B_{12} + x_2^2 B_{22} \qquad (4.1)$$

where x_1 and x_2 are the molar fractions. Similar relations apply for the higher order virial coefficients. The values of B_{11} and B_{22} are identical to those of the pure gases at the mixture temperature and B_{12} refers to a hypothetical gas in which only encounters between molecules of different species occur. The numerical values of B_{12} can be determined by experiment or may be calculated by applying the corresponding states arguments treated in section

THERMODYNAMIC PROPERTIES AND THE VELOCITY OF SOUND

VI.5 provided that some assumptions are made on the interaction potential $\phi_{12}(r_{12})$.

The sound velocity in a gaseous mixture is then given by a formula similar to equation 3.4

$$W^2_{mix} = \frac{\gamma_{mix} RT}{x_1 M_1 + x_2 M_2} \left(1 + \frac{B_{mix}}{V_{mix}} + \frac{C_{mix}}{V^2_{mix}} + \cdots \right) \qquad (4.2)$$

with

$$\gamma_{mix} = \gamma^\circ_{mix} \left[1 + (\gamma^\circ_{mix} - 1) \frac{2T}{V_{mix}} \frac{dB_{mix}}{dT} + \frac{\gamma^\circ_{mix} - 1}{\gamma^\circ_{mix}} \frac{T^2}{V_{mix}} \frac{d^2 B_{mix}}{dT^2} \cdots \right] \qquad (4.3)$$

and

$$\gamma^\circ_{mix} = \frac{x_1 (C^\circ_p)_1 + x_2 (C^\circ_p)_2}{x_1 (C^\circ_v)_1 + x_2 (C^\circ_v)_2} \qquad (4.4)$$

or in terms of the coefficients introduced in equations 3.8 to 3.10

$$W^2_{mix} = \frac{\gamma^\circ_{mix} RT}{x_1 M_1 + x_2 M_2} \left(1 + \frac{K_{mix}}{V_{mix}} + \frac{L_{mix}}{V^2_{mix}} + \cdots \right) \qquad (4.5)$$

with

$$K_{mix} = 2B_{mix} + 2(\gamma^\circ_{mix} - 1) T \frac{dB_{mix}}{dT} + \frac{(\gamma^\circ_{mix} - 1)^2}{\gamma^\circ_{mix}} T^2 \frac{d^2 B_{mix}}{dT^2} \qquad (4.6)$$

If both components have the same values of C°_p and C°_v some simplification can be made by the fact that $\gamma^\circ_1 = \gamma^\circ_2 = \gamma^\circ_{mix}$. In this special case

$$K_{mix} = x^2_1 K_{11} + 2x_1 x_2 K_{12} + x^2_2 K_{22} \qquad (4.7)$$

with the same meaning of the indices as for the volumetric virial coefficients.

For higher densities the problem is in general attacked in two successive steps. One introduces in a first stage a simplified mixture model, the ideal mixture, which is not too unrealistic provided that the two components have rather similar properties. In a second step one considers property deviations from this simplified model called 'excess' properties. This procedure is analogous to the common practice of considering the properties of a gas to be those of an ideal gas plus a correction term which can be calculated numerically if some assumptions are made.

In this presentation we will concentrate only on the purely thermodynamic aspects of the problem and we will not insist on the model calculations.

1. Ideal Mixture

The ideal mixture is a hypothetical one whose properties are a straightforward combination of those of the pure components. On the other hand, the ideal mixture may be considered as a standard of normal behavior even if there is probably no real mixture which will be normal in this sense over a large temperature and density range.

The definition of an ideal mixture is discussed elsewhere[106, 52, 129, 58]. For our purpose the important properties for a binary mixture are

$$V^{im} = x_1 V_1 + x_2 V_2 \tag{4.8}$$

$$C_p^{im} = x_1 (C_p)_1 + x_2 (C_p)_2 \tag{4.9}$$

$$C_v^{im} = x_1 (C_v)_1 + x_2 (C_v)_2 \tag{4.10}$$

V_i, $(C_p)_i$ and $(C_v)_i$ are the properties of the pure components at the same pressure, temperature and phase as the mixture. This may of course lead to difficulties in the coexisting liquid phase, because the saturated vapor pressure of one of the components will be higher than that of the mixture. Consequently one set of liquid data in equations 4.8 to 4.10 has to be extrapolated below the lowest pressure where the equilibrium liquid exists. Also, T_{mix} may be greater than T_c of the lighter component, in which case it is common to extrapolate the vapor pressure curve and use hypothetical properties. This situation is physically rather unsatisfying and also causes large uncertainties at the higher temperatures where the extrapolations may cover rather large Δp or ΔT ranges.

By differentiating equation 4.8 one obtains

$$\left(\frac{\delta V^{im}}{\delta p}\right)_{T,x} = x_1 \left(\frac{\delta V_1}{\delta p}\right)_{T,x} + x_2 \left(\frac{\delta V_2}{\delta p}\right)_{T,x} \tag{4.11}$$

and

$$\beta_T^{im} \equiv -\frac{1}{V^{im}} \left(\frac{\delta V^{im}}{\delta p}\right)_{T,x} = \varphi_1 (\beta_T)_1 + \varphi_2 (\beta_T)_2 \tag{4.12}$$

with φ_i the volume fraction

$$\varphi_i = x_i V_i / (x_i V_i + x_j V_j) \tag{4.13}$$

The adiabatic compressibility of the ideal mixture is then given by

$$\beta_S^{im} = \varphi_1 \frac{\gamma_1}{\gamma^{im}} (\beta_S)_1 + \varphi_2 \frac{\gamma_2}{\gamma^{im}} (\beta_S)_2 \tag{4.14}$$

and the sound velocity is consequently

$$W^{im^2} = \frac{\gamma^{im} V^{im^2}}{(x_1 M_1 + x_2 M_2)[x_1 V_1 \gamma_1 (\beta_S)_1 + x_2 V_2 \gamma_2 (\beta_S)_2]} \tag{4.15}$$

or

$$\frac{\gamma^{im} V^{im}}{(x_1 M_1 + x_2 M_2) W^{im^2}} = \varphi_1 \frac{\gamma_1 V_1}{M_1 W_1^2} + \varphi_2 \frac{\gamma_2 V_2}{M_2 W_2^2} \tag{4.16}$$

The ratio of the specific heats can be avoided in the following way

$$W^{im^2} = \frac{1}{\rho^{im}} \left[\frac{1}{\beta_T^{im}} + \frac{T V^{im} \gamma_v^{im^2}}{C_v^{im}} \right] \tag{4.17}$$

$$\frac{1}{W^{im^2}} = \rho^{im} \left[\beta_T^{im} - \frac{T V^{im} \alpha_p^{im^2}}{C_p^{im}} \right] \tag{4.18}$$

THERMODYNAMIC PROPERTIES AND THE VELOCITY OF SOUND

These expressions for the sound velocity in an ideal mixture contain no other properties than those of the pure fluids: unfortunately real mixtures are not ideal. This, however, does not exclude the possibility that some valuable qualitative insight about the properties of real mixtures may be gained by studying the ideal case.

A mixture of ideal gases shows all the properties of an ideal mixture. In this case $V_1 = V_2 = V^{im}$ and also, provided that the two components have the same number of internal degrees of freedom, $\gamma_1 = \gamma_2 = \gamma^{im}$. The sound velocity in this mixture of ideal gases is given by

$$\frac{1}{W^2_{mig}} = x_1 \frac{x_1 M_1 + x_2 M_2}{M_1} \frac{1}{W_1^2} + x_2 \frac{x_1 M_1 + x_2 M_2}{M_2} \frac{1}{W_2^2} \quad (4.19)$$

or

$$W^2_{mig} = \frac{\gamma° RT}{x_1 M_1 + x_2 M_2} \quad (4.20)$$

It may be emphasized that the ideal gas mixture is a special case of an ideal mixture: however, it is not true that in every ideal mixture the sound velocity is given by equation 4.19 or 4.20.

A further point is that although the low pressure limit of the sound velocity in a gaseous mixture is correctly given by the ideal gas mixture formula 4.20 it would be an erroneous generalization to say that every gaseous mixture becomes an ideal mixture at low density.

The condition for being ideal at low density is that $B_{12} = \frac{1}{2}(B_{11} + B_{22})$ which is equivalent to $K_{12} = \frac{1}{2}(K_{11} + K_{22})$ (provided that $\gamma_1° = \gamma_2°$): in that case

$$B_{mix} = x_1 B_{11} + x_2 B_{22} \quad (4.21)$$

and also

$$K_{mix} = x_1 K_{11} + x_2 K_{22} \quad (4.22)$$

The sound velocity for this ideal mixture of non-ideal gases is then

$$W^{im2} = x_1 \frac{\gamma° RT}{x_1 M_1 + x_2 M_2} \left(1 + \frac{K_{11}}{V^{im}} + \ldots\right) + x_2 \frac{\gamma° RT}{x_1 M_1 + x_2 M_2} \times \left(1 + \frac{K_{22}}{V^{im}} + \ldots\right) \quad (4.23)$$

and if $V_1 \simeq V_2$

$$W^{im2} = \psi_1 W_1^2 + \psi_2 W_2^2 \quad (4.24)$$

where

$$\psi_i = x_i M_i/(x_i M_i + x_j M_j)$$

is the weight fraction.

2. Non-ideal Mixtures

Real mixtures have properties that scatter around those of an ideal mixture. One defines excess functions giving the difference between the numerical value of some property in the real mixture and the hypothetical value it would have if the mixture were ideal. The excess properties we need are:

$$V^E = V - V^{im} \tag{4.25}$$

$$C_p^E = C_p - C_p^{im} \tag{4.26}$$

with

$$C_p^E = (\delta H^E/\delta T)_p \tag{4.27}$$

The compressibility β_T is given by

$$(\beta_T)_{mix} = \frac{1}{V^{im} + V^E}\left[\beta_T^{im} V^{im} - \left(\frac{\delta V^E}{\delta p}\right)_{T,x}\right] \tag{4.28}$$

and the expansion coefficient α_p by

$$(\alpha_p)_{mix} = \frac{1}{V^{im} + V^E}\left[\alpha_p^{im} V^{im} + \left(\frac{\delta V^E}{\delta T}\right)_{p,x}\right] \tag{4.29}$$

These derivatives, with the intensive variables p or T kept constant, are equal to the sum of two terms: a first one representing the ideal mixture contribution and a second one giving the way in which the deviation from ideality changes with p or T.

Other derivatives, where an extensive variable is kept constant, will not obey such simple additivity rules. This applies for the adiabatic compressibility and consequently also for the sound velocity.

Indeed the fact that the mixture carries out a change along a curve of constant total entropy does not imply that each subsystem in the mixture changes adiabatically. For this reason β_S is transformed with the equations 1.1 to 1.3 giving:

$$\beta_S - \beta_S^{im} = \gamma\beta_T - \gamma^{im}\beta_T^{im} \tag{4.30}$$

$$\beta_S - \beta_S^{im} = (\beta_T - \beta_T^{im}) - T\left[\frac{V\alpha_p^2}{C_p} - \left(\frac{V\alpha_p^2}{C_p}\right)^{im}\right] \tag{4.31}$$

$$\frac{1}{\beta_S} - \frac{1}{\beta_S^{im}} = \left(\frac{1}{\beta_T} - \frac{1}{\beta_T^{im}}\right) + T\left[\frac{V\gamma_v^2}{C_v} - \left(\frac{V\gamma_v^2}{C_v}\right)^{im}\right] \tag{4.32}$$

with

$$\beta_T - \beta_T^{im} = -\frac{V^E}{V}\beta_T^{im} - \frac{1}{V}\left(\frac{\delta V^E}{\delta p}\right)_T \tag{4.33}$$

and

$$\alpha_p - \alpha_p^{im} = -\frac{V^E}{V}\alpha_p^{im} + \frac{1}{V}\left(\frac{\delta V^E}{\delta T}\right)_p \tag{4.34}$$

In order to make sure that the normal thermodynamic relations can be

THERMODYNAMIC PROPERTIES AND THE VELOCITY OF SOUND

applied on all the excess functions we prefer not to use the name excess compressibility for $\beta_T - \beta_T^{im}$, nor excess expansion coefficient for $\alpha_p - \alpha_p^{im}$.

The sound velocity in the mixture can then be compared with the ideal mixture value by

$$\frac{1}{W^2} = \frac{1}{\gamma V^2}\left[\frac{\gamma^{im} V^{im2}}{W^{im2}} - M\left(\frac{\delta V^E}{\delta p}\right)_T\right] \tag{4.35}$$

$$\frac{1}{W^2} - \frac{1}{W^{im2}} = \left\{\rho\beta_T - \rho^{im}\beta_T^{im} - TM\left[\frac{\alpha_p^2}{C_p} - \frac{\alpha_p^{im2}}{C_p^{im}}\right]\right\} \tag{4.36}$$

$$W^2 - W^{im2} = \frac{1}{\rho\beta_T} - \frac{1}{\rho^{im}\beta_T^{im}} + \frac{T}{M}\left[\frac{V^2\gamma_v^2}{C_v} - \left(\frac{V^2\gamma_v^2}{C_v}\right)^{im}\right] \tag{4.37}$$

Equations 4.31 and 4.36 offer the best perspective for interpretation because all the variables they contain can be expressed in a relatively simple form in terms of V^E, $(\delta V^E/\delta p)_T$, $(\delta V^E/\delta T)_p$ and $(\delta H^E/\delta T)_p$.

V. Experimental Methods

In this section we will discuss some experimental aspects of sound velocity measurements. Let us start with the remark that in the framework of this chapter, which emphasizes the relation between W and other thermodynamic properties, the problem 'how to measure' is in general a minor one compared with 'how to interpret the results'. Often the currently obtained accuracies on W cannot even be completely exploited by lack of other thermodynamic data with a comparable accuracy.

A great number of textbooks and review articles treat the physical and technical aspects of acoustic and ultrasonic experiments at length. We mention the work of Mason[81], Bergmann[9], Kudrjawzew[72], Richardson[105], Blitz[18], Nozdrev[96], Schaaffs[107], Beyer and Letcher[10], Truell et al.[121] and Matheson[82]. Further general reference works are the *Physical Acoustics* series[85] and the *Encyclopedia of Physics*, Vol. XI, *Acoustics*[110].

We will only give a qualitative description and discussion of the most suitable actual methods for the measurement of the thermodynamic sound velocity in fluids. Hypersonic techniques, although extremely interesting from other points of view, will not be included.

There are two quite different ways to measure the velocity. The first one determines the time a signal needs to cover some accurately known distance. This method is in principle straightforward but in order to make measurement on a laboratory scale, in samples with well-defined properties, the acoustic path will be rather small and the transit times to be measured, will be in the microsecond range.

The second way makes use of the wave character of the sound and it obtains W as the product of a frequency and the corresponding wavelength.

1. Interferometer Methods

The classical acoustical interferometer contains essentially a transducer that produces a plane sound wave and a reflector parallel to the source at a distance d. The incident and the reflected wave interfere to form standing waves. Provided that the absorption coefficient of the fluid is not too high and that the loss by reflection is small a great number of waves will add in the vessel and resonance will occur if the length of the resonator equals an integral number of half wavelengths

$$d = n\lambda/2 \tag{5.1}$$

There are different ways of observing that the interferometer is in resonance. One can use as reflector a second acoustic transducer: its signal will reach maximum amplitude at resonance. This is the solution one prefers generally for a fixed path interferometer. Another possibility is to look at the acoustical load the fluid exerts on the source, which for this application will normally be a quartz crystal. It has been shown by Hubbard[62, 63] that a freely vibrating quartz has an electric equivalent a series resonance circuit containing a self-inductance, a capacity C and a resistance R. If the crystal radiates into the resonator volume, the acoustic load will alter the electric properties of the source into R' and C' given by

$$R' = R + AP \tag{5.2}$$

$$1/C' = 1/C + A\omega Q \tag{5.3}$$

where A is a constant determined completely by the properties of the transducer and the load. P and Q are the real and imaginary part of the acoustic impedance of the resonator: they are both periodic functions of $2d/\lambda$. Each time that $d = n\lambda/2$ the value of Q is zero and P reaches a maximum. The variations in the resistive part of the quartz impedance can easily be detected by monitoring the current through the quartz. This is the technique usually applied in the variable path interferometer. A thorough discussion of the equivalent circuits for an operating interferometer is given by Mason[81].

There are of course a lot of other techniques that can be used to detect resonances[9] but, except in very special situations, they will not give a better precision than the two methods we described.

Let us now take a resonator filled with a fluid from which we want to measure its sound velocity. The procedure will be to bring the interferometer into different situations of resonance and to use the parameter values for these resonances in order to deduce λ from them. Remembering the condition of equation 5.1 there will be two ways to make the interferometer resonant, either by changing d at constant λ, or by changing λ at constant d. Both methods can be applied and are referred to as the variable path and the fixed path interferometer.

A. Fixed Path Interferometer

From the basic relation $W = \lambda v$ it follows that the wavelength can be changed either by a variation of v at constant W or by changing W at constant v.

We will start with the first case: it corresponds with the normal experimental situation where one wants to know the velocity of sound in a fluid at a given value of the state parameters. The resonant frequencies are interrelated with the sound velocity and the resonator length d by

$$v_n = nW/2d \qquad (5.4)$$

Theoretically it is sufficient to know $\Delta v = v_{n+1} - v_n$ in order to derive directly

$$W = 2d\,\Delta v \qquad (5.5)$$

Practice is, however, more complicated. In this method the transducers have to show a nearly constant electroacoustic conversion efficiency over a certain frequency range. This excludes the use of quartz crystals at their resonant frequency. Some other transducers may, however, adequately fulfil this requirement. First, ceramic electrostrictive materials, such as barium titanate and lead zirconate titanate[81], and condenser microphones or Sell transducers[109, 73, 86]. They will then be used in a double transducer interferometer.

A further requirement is that the resonance pattern as a function of the frequency must not be disturbed by modes other than the longitudinal one in the fluid. This condition is rather hard to realize except at the very lowest resonant frequencies, where the different modes can easily be identified. A systematic study of such fixed path resonators for audio frequency applications has been made by Fritsche[44] and by Shields et al.[112]. A serious drawback of this method is the divergence of the acoustic wave, which starts when the path length goes out of the Fresnel region limited by r^2/λ with r the radius of the transducer. This results in extra absorption at the walls of the resonator and an important decrease in the apparent sound velocity. Although the sound velocity readings can be corrected for this effect[44, 112] some precision is lost and this method can only be recommended for measurements in gases with relaxation times so long that ultrasonic results are subject to dispersion.

The other solution for avoiding spurious resonances is to use sufficiently high frequencies so that the divergence remains negligible and the unwanted resonance modes are not activated. Such an apparatus has been used by Eggers[37] for frequencies up to 15 MHz.

The two crystals are separated by a metal tube which acts as a spacer. A difficulty may be the direct acoustical coupling between the two crystals by other means than the fluid. This can of course be avoided but at the cost of precision on the distance d and consequently on the absolute value of W.

The alternative possibility to make measurements in a fixed path interferometer is to keep the frequency constant and slowly to change W by varying the temperature, the pressure or the composition of the fluid. This method is very useful for continuous recording of changes in the sound velocity: it is, however, not well suited for absolute measurements.

Although most fixed path interferometers have the form of a tube with two transducers at the ends there are other very convenient arrangements. Ceramic electrostrictive materials can be molded in almost any shape so that in addition resonators of another geometry can be constructed. A hollow barium titanate cylinder vibrating radially has been used by Guptill et al.[53] and by Aziz et al.[4]. The high order resonances are separated in frequency by an amount $\Delta v = W/2r$ with r the inner radius of the cylinder. The resonances are detected on a part of the transducer tube that is used as a receiver. The reproducibility and the precision of such a system are very good. The accuracy on the final data depends strongly on the exact knowledge of the inner diameter of the cylinder which has to be machined very carefully in order to avoid ellipticity.

B. Variable Path Interferometer

The variable path interferometer has certainly been the most widely used apparatus for sound velocity measurements in gases. The reflector can in this arrangement be moved but must remain, however, parallel to the source. For a sound velocity of $1\,000$ m s^{-1} and a frequency of 1 MHz the value of $\lambda/2$ is only 0.5 mm so that a displacement Δd can be made over quite a large number of half wavelengths. Further it is not necessary to know the total value of d itself. In most interferometers the displacement is effectuated and monitored with a micrometer screw which is usually at room temperature and atmospheric pressure. Frequently the resonator itself is in quite a different situation. This may cause experimental difficulties, and the displacement of the reflector may be different from that at the micrometer screw if the temperature distribution along the driving rod changes with the displacement. High pressure in the resonator may also deform the rod. It should be emphasized, however, that these possible errors are second-order effects as they are proportional only to the changes in the temperature and pressure effects during the displacement. The main problems, however, reside in the detection of the resonances: a lack of parallelism between the reflector and source, or diffraction effects, make it difficult to determine exactly the resonance conditions. A lot of experimental work has been done in an effort to minimize these problems. An extensive bibliography may be found in papers by Del Grosso et al.[35] and by Stewart[115,116]. A description of a typical high precision interferometer for gases has been given by Plumb and Cataland[102]. The accuracy they get is of the order 10^{-5}.

2. Pulse Methods

The method most widely used in practice for sound velocity measurements in dense fluids is the pulse technique. The acoustic part of the experimental set-up is very similar to the single or double crystal interferometer. The continuous wave oscillator is replaced by a 'burst' generator delivering a pulse of sinusoidal voltage to the crystal. If the duration of the pulse is not much longer than the period of the sound wave the Fourier transform in the frequency domain will have large sidebands and eventually dispersion phenomena will occur. In the single crystal arrangement the acoustic pulse travels over a distance d, hits a reflector and reaches the transducer again after a time $t = 2d/W$. In this version of the apparatus there is always the problem that the applied signal and the received one, which is several orders of magnitude smaller, appear at the same electrodes and that precautions have to be taken to avoid overload of the receiver. This difficulty does not arise in the two-crystal arrangement where both functions are well separated. The transit time can be measured roughly on the screen of an oscilloscope; it can be done with better precision by putting on the same screen calibrated time marks. With modern electronics (e.g. Hewlett-Packard computing counter 5360A with time interval plug-in) it is also possible to measure directly the time interval between the emitted and the received pulse with an accuracy of the order of 1 nanosecond and with a resolution that is still substantially better. This would give for a transit time of 10 μsec, which for a velocity of 1 000 m sec^{-1} corresponds with a resonator length of 1 cm, a sound velocity accurate to 10^{-4}. The possibilities of this apparatus have been described by Lacy and Daniel[75]: its accuracy can be further considerably increased by the computing capabilities of the instrument. It can digitally average the transit time for up to 10^4 measurements and in that way eliminate the noise and jitter of the electronics. The resolution of the system is then of the order of 20 psec. Moreover the transit time can be enlarged by increasing the distance d: one can obtain the same effect by programming the counter in a way that the first few echoes are overlooked and that the nth echo can trigger the end of the time measurement. The total time can then be made more that 100 μsec. The time measurement will then have an estimated accuracy better than 10^{-5} and resolution of the order of 10^{-7}. The limitations are clearly no longer due to the time measurement but rather to the knowledge of the effective distance. This technique has further all the possibilities of digital readout and automation.

There are similar methods which are in principle capable of reaching about the same accuracy but at the cost of more labour and more time for each measuring point. They all make use of the fact that, the transit time being very small, a much higher precision can be obtained by measuring the inverse of this time.

This can be achieved in its simplest form by making the ultrasonic pulse repetitive with a period that is equal to the transit time or to a multiple of it. This can in a two-crystal arrangement be done by manual adjustment of the repetition frequency v_r of the pulses until complete superposition is detected on an oscilloscope screen by constructive interference[84]. This method only works well if all the subsequent pulses are identical, i.e. they start with the same phase of the r.f. signal. An automated version of this apparatus is the 'sing around system' which is usually applied with a single transducer[61, 26]; the first returned echo of a train is used to trigger the pulse generator and to refire the transmitter. The repetition rate is again in direct relation to the transit time. As no manual adjustments are needed in this apparatus it is very well suited for continuously monitoring W values. There are other versions of this arrangement[43] specially adapted for observing changes in velocity with a sensitivity of a few parts in 10^7.

Other pulse-technique applications make a more explicit use of the periodic nature of the r.f. signal. Williams and Lamb[132] use in a two-crystal arrangement a continuous wave generator which is gated so that two pulses which are coherent in phase go through the sample. By choosing the time interval between these pulses approximately equal to the transit time, the echoes are superimposed again but this time the phase relation between the two r.f. signals at the front edge of the pulses will depend on the ratio of the transit time to the period of the r.f. wave. By slowly changing the frequency of the latter this ratio changes and the amplitude of the received signal will show a characteristic interference pattern with sharp peaks. The frequencies where destructive interference occurs are recorded and they are related to the transit time t by

$$v_n t = (2n + 1)/4$$

with n an integer that can be determined from comparison between v_n, v_{n+1}, v_{n+2}. A variant of this method has been used by McSkimin[83] where one pulse goes through a variable delay line. Again a phase comparison can be made of the two signals arriving at the same amplifier.

A further alternative is the echo-overlap method used by Papadakis[97] where the repetition frequency of the r.f. generator is very low so that all echoes from one burst die out before the next starts. The echo-overlap is effected optically on an oscilloscope by driving the timebase at a frequency equal to the reciprocal of the time between two echoes.

Other methods of observing interferences compare a received pulse with a c.w. generator. The acoustical path acts simply as a variable delay line. The transit times can be found in absolute values but the method is especially useful for high resolution measurements of changes in velocity with a precision of 5×10^{-6} [133, 1].

A final possibility is to make the pulse so long that echoes overlap again showing interference with maxima and minima for values of the r.f. frequency from which W can be deduced[83].

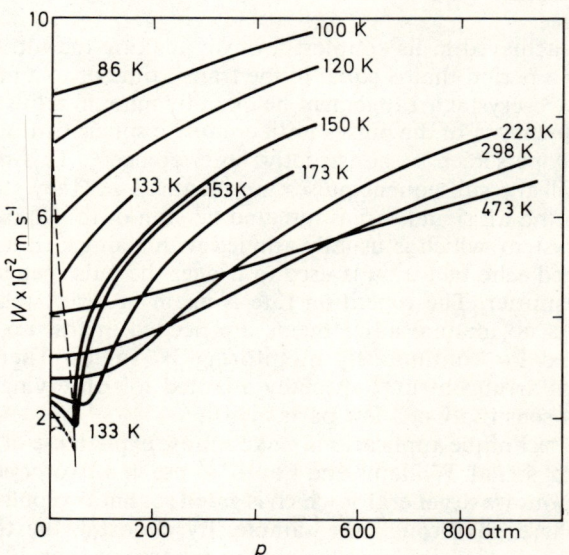

Figure 4. Sound velocity data in argon as a function of pressure for different temperatures[125, 118, 79, 89]: – – coexisting liquid, ···· coexisting gas, —— gas or liquid (T_c = 150.7 K).

VI. Sound Velocity in Gases

1. General Behavior

A great number of experimental data on the temperature and pressure dependence of sound velocity in gases exists. They have been summarized up to the year 1963 by Schaaffs[108].

For the cryogenic gases (helium, hydrogen, neon, oxygen, carbon monoxide, argon, methane, air) a very complete compendium of the data available up to 1961 has been published[117].

The general behavior may be seen in *Figures 4* to *6* where argon data are

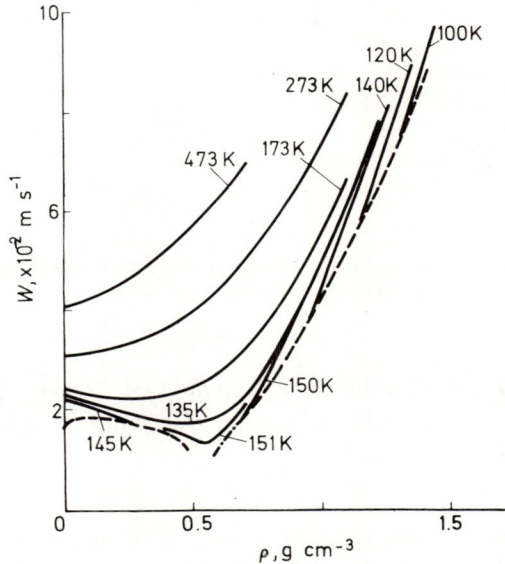

Figure 5. Sound velocity data in argon as a function of the density for different temperatures (same data and same symbols as in *Figure 4*).

collected from different experiments. For other gases similar more or less complete pictures can be drawn.

There are, however, also other sources of information available on sound velocity. Several recent tables of thermodynamic properties (among others, those published by the Van der Waals Laboratory in Amsterdam and by the National Bureau of Standards in Washington) also contain a value of W derived from analytical expressions fitting a large number of experimental pVT data[123, 59].

In the following paragraphs we will discuss the relevant information that can be gained from experimental W data in different regions of the diagrams in *Figures 4* to *6*.

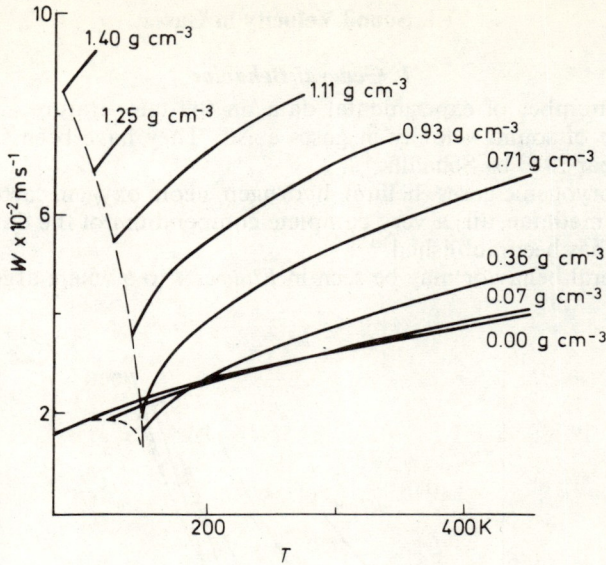

Figure 6. Sound velocity data in argon as a function of the temperature for constant densities (same data and same symbols as in *Figure 4*).

2. The Absolute Value of W in the Low Pressure Limit

According to equation 3.13 the low pressure limiting value of the sound velocity is given by

$$W^{\circ 2} = \gamma^\circ RT/M \tag{6.1}$$

W° depends on three parameters γ°, T and M: if one of them is unknown a careful determination of W° and of the two other variables allows us to deduce the unknown. W° can be found by extrapolation of a $W(p)$ curve approaching as closely as possible to the zero pressure limit. Let us first consider the factor $\gamma^\circ = C_p^\circ/C_v^\circ$. The specific heats in the dilute gas are completely determined by the external and internal degrees of freedom of the molecules. Due to the quantization of the rotational and vibrational energy the values of C_p° and C_v° may change with temperature. These specific heat values are needed in thermodynamics as integration constants: low pressure sound velocities can yield these data.

A second application is the determination of M. This is especially attractive for composition analysis in a low density gas mixture. The main advantage is that the composition can be monitored continuously without disturbing the system by sampling. Disadvantages are that the acoustical resonator is quite large compared to, say, a cell for thermal analysis and also that the sensitivity may be small if both components do not differ much in sound velocity. This technique has been used by Van Itterbeek and Nihoul for measurement of the shift in concentration in a thermal diffusion experiment[128].

A third possibility exists in the use of $W°$ in order to determine T. The important feature is that T in equation 5.1 is expressed according to the thermodynamic temperature scale: a sound velocity measuring system for low density gases may be used as a primary thermometer. It has several properties in common with that other primary temperature standard, the gas thermometer. They both give thermodynamic temperatures and the measurements have to be made in both cases as a function of pressure in order to extrapolate the data into the region where ideal gas laws apply. The acoustical thermometer is, however, completely insensitive to some of the major experimental difficulties that limit the accuracy of the gas thermometer, among others the dead volume at room temperature, the thermomolecular pressure difference occurring in the narrow connecting tubes and finally, the adsorption effects.

The idea for this application was already present in the work of Keesom and Van Itterbeek[69, 127]. It has been taken up again by Van Itterbeek and De Laet[126] with special attention to the liquid helium temperature range. The precision they obtained with their low frequency fixed path interferometer was not high enough to allow meaningful conclusions about the temperature scale they used.

Plumb and Cataland[102, 103, 25] using a very sophisticated variable path

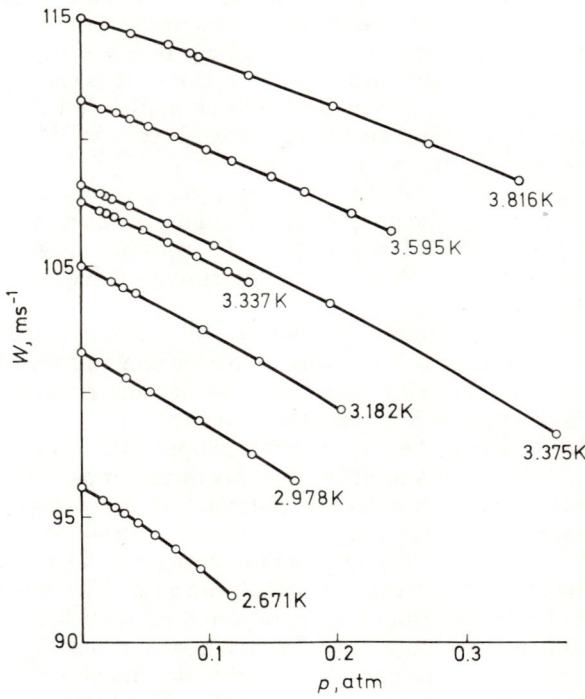

Figure 7. Sound velocity data in ^{4}He as a function of pressure at temperatures below the normal boiling point[51].

interferometer (1 MHz) were able to measure W with a precision better than 10^{-5} in the range 4 to 20 K: temperatures could be derived with an accuracy of 0.002 K.

Grimsrud and Werntz[51] made similar measurements both in ^{3}He and ^{4}He with a variable path interferometer working at audio frequencies (*Figure 7*). They confirm the Cataland and Plumb conclusion that the 1958 temperature scale, based on helium vapor pressure, is too low by 5 to 10 mK in the range 2 to 5 K.

This thermometer is of course quite a cumbersome apparatus: the temperature sensing volume is relatively large and it has a large heat capacity. Moreover each T measurement takes a long time during which T has to be constant in spite of the mechanical motion in the interferometer and the changes in pressure.

3. The Initial Slope $(\delta W^2/\delta p)_T$

According to equations 3.13 and 3.17 the initial slope of the quantity W^2 versus p is given by $\gamma° K/M$ with K defined by equation 3.8.

It has for a long time been realized[69] that an experimental determination of K over a certain temperature range can in principle provide data on the virial coefficient $B(T)$. Unfortunately the functional form of $B(T)$ is not known exactly. The usual procedure has been to assume a function $B(T)$ of a general form that was successful in describing the temperature dependence of virial coefficients in other cases, e.g. $B = a + bT^{-1}$: the coefficients a and b were then obtained by least squares methods applied on the experimental K data. This procedure has been applied on the data of the acoustical thermometer experiments discussed in section VI.2 and also on other related work[20, 78] (*Figure 7*).

Some criticism of this method has been formulated by Bruch[22, 23]: he proposed a way to obtain the unique solution of the problem by avoiding the imposition of a functional form of $B(T)$. It appeared, however, that his method is rather impracticable regarding the unrealistic requirements on the accuracy of the data[21].

This method of determining B from sound velocity measurements is especially suitable for very low temperatures where the saturated vapor pressure is low. Normal pVT measurements in this situation are often hampered by large systematic errors due to adsorption.

Another way of using the experimental slopes $(\delta W^2/\delta p)_T$ is to derive directly from a series of data at different temperatures most probable values of the constants in the intermolecular potential. It is not required for this purpose to make the detour over the volumetric virial coefficients. This procedure will be discussed in some detail in paragraph VI.5.

Gaseous mixtures can be treated much as pure gases. The function $B_{mix}(T)$ can in principle be derived from a series of sound velocity data as a function of pressure at different temperatures: the fact, however, that B_{mix} is a sum of three terms, each with its different temperature dependence, does not make it easy to guess the right functional form for $B_{mix}(T)$. Moreover the quantity of interest in the mixture is B_{12} rather than B_{mix} [see equation 4.1]. Indeed the function $B_{12}(T)$ allows us to establish the intermolecular potential function ϕ_{12}. It will be shown in section VI.5 that this information on Φ_{12} can also be drawn directly from the K_{12} data.

4. High Density Data

The curvature in the $W^2(p)$ isotherms of *Figure 4* corresponds with the $L', M' \ldots$ terms in the virial expansion [equation 3.13]. Similar technique as described in section VI.3 can be used to obtain the form of $C(T), D(T) \ldots$. It is clear that the complexity of the analysis and the uncertainty of the results will increase drastically for the higher order terms. For this reason one prefers the following ways of interpretation:

(a) a direct comparison of experimental $L(T)M(T)$ or $L'(T)M'(T)$ values with the theoretical numbers for some reasonable potential function. From this comparison numerical values of the potential parameters may follow (see also section VI.5).

(b) a purely thermodynamic treatment identical with that applied for liquids (see also section VII.3).

5. Generalized Behavior of Sound Velocity in Gases. Corresponding States Treatment

The virial expansions of W^2 in equations 3.7 and 3.13 show that the sound velocity always equals the ideal gas value multiplied by a dimensionless correction factor

$$W^2 = W^{\circ 2} Z_W \tag{6.2}$$

with

$$Z_W = [1 + (K/V) + (L/V^2) + \ldots] \tag{6.3}$$

This Z_W is mainly a function of the density. To a smaller extent it depends also on the temperature in its coefficients $K, L, \ldots$ which are completely determined by γ° and by the volumetric virial coefficients and their first and second temperature derivatives.

From theoretical arguments and confirmed by experimental evidence it is known that all gases with a similar intermolecular potential have $B(T)$, $C(T) \ldots$ functions which can be reduced to a universal set $B^*(T^*), C^*(T^*) \ldots$ provided that all these functions are scaled by means of molecular units. For a two-parameter potential of the form $\Phi = \varepsilon f(r/\sigma)$ with ε an energy parameter and r/σ the intermolecular distance divided by a characteristic length parameter, the reduced properties are defined by [34]

$$T^* \equiv T/T_0 \qquad T_0 = \varepsilon/k_B \tag{6.4}$$

$$V^* \equiv V/V_0 \qquad V_0 \equiv N_A \sigma^3 \tag{6.5}$$

$$p^* \equiv p/p_0 \qquad p_0 = \varepsilon/\sigma^3 \tag{6.6}$$

By dimensional analysis the reduction constant for the sound velocity is found to be

$$W_0^2 = N_A \varepsilon/M \quad \text{and} \quad W^{*2} \equiv W^2/W_0^2 \tag{6.7}$$

and equations 3.7 and 3.13 may be transformed into

$$W^{*2} = \gamma^\circ T^*[1 + (K^*/V^*) + (L^*/V^{*2}) + \ldots] \tag{6.8}$$

$$W^{*2} = \gamma^\circ T^*(1 + K'^* p^* + L'^* p^{*2} + \ldots) \tag{6.9}$$

with

$$K^* \equiv K/V_0, \qquad L^* \equiv L/V_0^2 \tag{6.10}$$

$$K'^* \equiv K'/p_0, \qquad L'^* \equiv L'/p_0^2 \tag{6.11}$$

Frequently it is especially important to consider how far W^2 deviates from $W^{\circ 2}$, i.e. to evaluate the non-ideality contribution. For this purpose the trivial γ° and T^* dependence in equations 6.8 and 6.9 can be eliminated from W^{*2} by introducing a new reduced sound velocity W^+ defined by

$$W^{+2} \equiv W^{*2}/\gamma^\circ T^* \tag{6.12}$$

For all gases and for all temperatures W^+ is equal to one in the low density limit. It can easily be seen that

$$W^{+2} = W^2/W^{\circ 2} \tag{6.13}$$

The coefficients $K^*, L^* \ldots$ and $K'^*, L'^* \ldots$ can be written as functions of the reduced virial coefficients by combining equation 3.8 with equation 6.10 so that one obtains

$$K^* = 2B^* + 2(\gamma^\circ - 1)T^* \frac{dB^*}{dT^*} + \frac{(\gamma^\circ - 1)^2}{\gamma^\circ} T^{*2} \frac{d^2 B^*}{dT^{*2}} \tag{6.14}$$

with $B^* \equiv B/V_0$ and also from equations 3.17 and 6.11

$$K'^* = K^*/T^*$$

The expressions for L^*, M^* and L'^*, M'^* can be obtained in a similar way. Numerical values for these coefficients have been calculated[123] for a Lennard–Jones 12–6 potential making use of the volumetric virial coefficients calculated by Bird and Spotz[12,13] and tabulated by Hirschfelder[60]. Some of the results are shown graphically in *Figure 8*. Another universal set of reduced virial coefficients based on direct experimental data has been tabulated by Gyorog and Obert[55]. The derived values of K^*, L^*, M^* do not show systematic deviations from the theoretical values drawn in *Figure 8*.

In this way universal charts of the reduced sound velocity can be made for each group of gases having the same γ°[123]. The general appearance of such a diagram is the same as that of *Figures 4* and *7*. The accuracy of the sound velocities calculated according to this procedure depends strongly of course on the convergence of the series expansion and on the suitablity of the potential parameters used for a given gas.

For relatively low densities the agreement between experimental and calculated data is in general extremely good[123].

Attention may be drawn to the fact that experimental $K(T), L(T)\ldots$ functions offer the same possibilities as the $B(T), C(T)\ldots$ curves with respect to the determination of intermolecular potential constants. The methods of analysing $B(T)$ for this purpose have been reviewed by Mason and Spurling[80]:

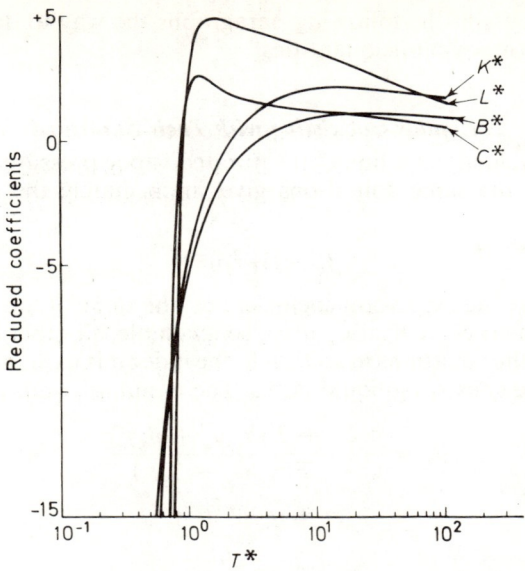

Figure 8. The change with T^* of: (1) B^* and C^*: the reduced volumetric virial coefficients in the density expansion of pV/RT; (2) K^* and L^*: the reduced coefficients in the density expansion of $W^2/W^{\circ 2}$.

they can be applied on $K(T)$ in exactly the same way. The most common approach brings an experimental plot of $\log |K|$ versus $\log T$ into coincidence with a calculated $\log |K^*|$ versus $\log T^*$ graph: the relative displacement of both pairs of axes gives the numerical value for $\log \varepsilon/k$ and for $\log N_A \sigma^3$ respectively. This method has not yet found many applications but it can be expected that sound velocity measurements, obtained with modern high precision techniques, offer in this respect at least equivalent and probably better perspectives than the classical pVT measurements. This conclusion is also valid for the study of gaseous mixtures and more specifically for the investigation of the interaction potential $\phi_{12}(r_{12})$.

VII. Sound Velocity in Pure Liquids

1. General Behavior

A systematic survey of experimental data in liquids may be found in the work of Schaaffs[108]. For cryogenic liquids the reader can be referred to a very complete NBS compendium[117].

The general behavior of sound velocity in a liquid has already been illustrated over a broad temperature and pressure range in *Figures 4* to *6*. The remarkable features are that in the liquid $(\delta W/\delta p)_T$ is normally positive and $(\delta W/\delta T)_p$ negative: the low density gas can have slopes of the opposite sign.

We will discuss in the following paragraphs the way to derive from such data other thermodynamic quantities.

2. Liquids Coexisting with Their Saturated Vapor

The sound velocity in a liquid at saturated vapor pressure combined with the density at the same conditions gives immediately the adiabatic compressibility by

$$\beta_S = (W^2 \rho)^{-1} \tag{7.1}$$

Provided that the expansion coefficient α_σ, the slope of the vapor pressure curve γ_σ and the specific heat C_σ are also available, all other derivatives can be found from the equations in section I. The index σ is used for all derivatives taken along the coexisting liquid curve. The useful relations are[106]:

$$\beta_T = \frac{\beta_S C_\sigma + TV\alpha_\sigma(\alpha_\sigma + \beta_S \gamma_\sigma)}{C_\sigma - TV\gamma_\sigma(\alpha_\sigma + \beta_S \gamma_\sigma)} \tag{7.2}$$

$$\alpha_p = \alpha_\sigma + \beta_T \gamma_\sigma \tag{7.3}$$

$$\gamma_v = \alpha_p/\beta_T \tag{7.4}$$

$$C_p = C_\sigma + TV\alpha_p \gamma_\sigma \tag{7.5}$$

$$C_v = C_\sigma - TV\alpha_\sigma \gamma_v \tag{7.6}$$

These calculations have been carried out for a great number of simple fluids in the complete range between the triple point and the critical point[106]. The procedure may easily be adapted when the data to start with are different[118,95].

The general appearance of the derived data is illustrated in *Figures 9, 10* and *11* for liquid carbon dioxide[98]. A comparison with the direct data avail-

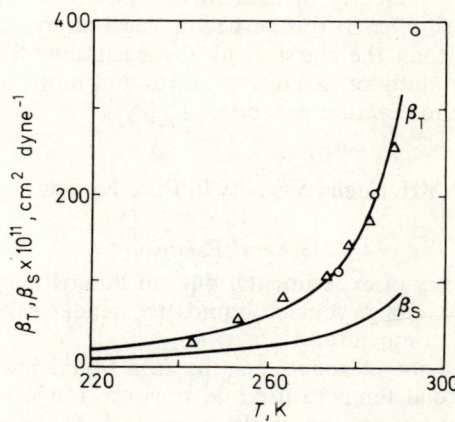

Figure 9. Compressibilities of liquid carbon dioxide derived from sound velocity measurements by Pecceu and Van Dael[98].
△ ○ data obtained respectively by Jenkin[65] and by Michels et al.[87].

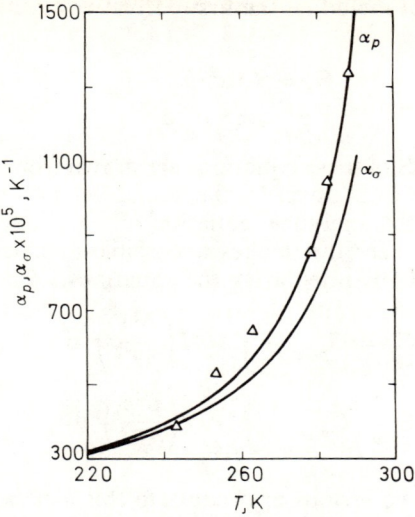

Figure 10. Thermal expansion coefficients of liquid carbon dioxide derived from sound velocity measurements[98]. △ data from Jenkin[65].

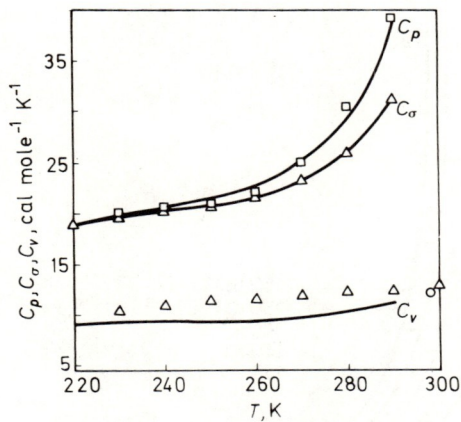

Figure 11. Specific heats in liquid carbon dioxide derived from sound velocity measurements[98]. △ □ ○ data respectively by Eucken and Hauck[38], TPRC[120] and Michels et al.[88].

able gives a good idea of the merits of sound velocity data in the study of liquids.

3. Single Phase High Density Fluids

The way of approach is somewhat different in this case: one needs a complete set of $\rho(pT)$ data including the first derivatives α_p and β_T. The specific heats are then given by:

$$C_p = TV\alpha_p^2/\beta_T - \beta_S \qquad (7.7)$$
$$C_v = C_p\beta_S/\beta_T \qquad (7.8)$$

All data on the RHS of these equations are drawn from a known equation of state except β_S which is given by the sound velocity measurement. It is clear also that from the analytical equation of state function second order derivatives may be taken: this makes it possible to calculate specific heat values without W data by integrating the equations 1.6 and 1.7:

$$C_p = C_{p_{ref}} - \int_{p_{ref}}^{p} T\left(\frac{\delta^2 V}{\delta T^2}\right)_p dp \qquad (7.9)$$

$$C_v = C_{v_{ref}} + \int_{v_{ref}}^{v} T\left(\frac{\delta^2 p}{\delta T^2}\right)_v dV \qquad (7.10)$$

There are, however, two serious drawbacks to this method:

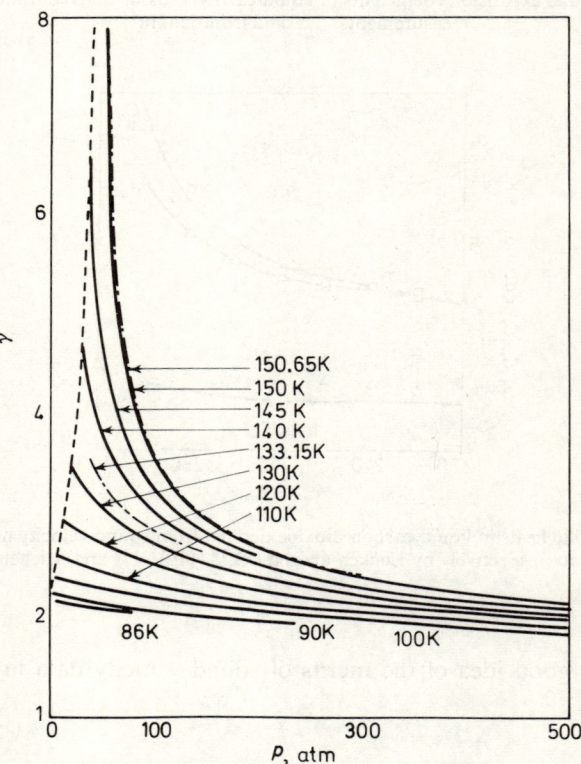

Figure 12. The specific heat ratio γ in liquid argon derived from sound velocity measurements by Thoen et al.[118]. —·— data obtained by Michels et al.[89].

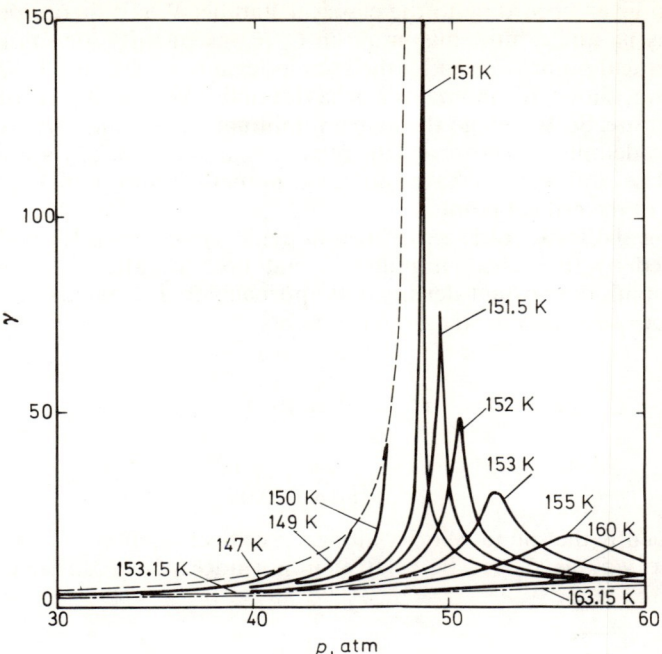

Figure 13. The specific heat ratio in gaseous argon derived from sound velocity measurements[119].
— · — data obtained by Michels et al.[89].

(i) an integration constant at some reference point has to be known: this may be the data point for the liquid coexisting with its vapor or in the more general case simply the low density limit C_p° and C_v°.

(ii) the precision on the data may decrease by an order of magnitude for each differentiation. Therefore equations 7.7 and 7.8 will give specific heat data with a higher reliability than equations 7.9 and 7.10.

These comments are of course not only valid for the liquid under overpressure but also for the high density gas state. The general behavior of the ratio of the specific heats can well be illustrated by *Figure 12* giving data for liquid argon over its whole temperature range up to pressures of 500 atm[118]. This picture is completed by *Figure 13* for gaseous argon[119].

4. Critical Region

At the gas–liquid critical point both γ and β_T diverge (see also Chapter 15). So it becomes difficult to evaluate W^2 from equation 1.10. This problem can be avoided by using equation 1.12

$$W^2 = \frac{V}{M\beta_T} + \frac{TV^2\gamma_v}{MC_v} \tag{7.11}$$

For a long time the interpretation of sound velocity in the critical region has been given according to the classical Van der Waals theory predicting a diverging β_T and a finite maximum in C_v. Consequently the sound velocity at the critical point is given by the second term in equation 7.11. The factor γ_v does not show any anomalous behavior and is very nearly constant along any isochore. So W will go through a minimum value which in this theory is, however, definitely non-zero. This picture was until recently applied generally and several authors have tabulated numerical values of W_c, the sound velocity at the critical point.

During the last decade experimental evidence has been formed, strongly supported by theoretical arguments, that both β_T and C_v are divergent. Along a path of constant density and approaching T_c from above, the divergence goes according to the power laws as

$$\beta_T \sim |T - T_c|^{-\gamma} \tag{7.12}$$

$$C_v \sim |T - T_c|^{-\alpha} \tag{7.13}$$

with

$$\gamma \simeq 1.2 \text{ and } \alpha \simeq 0.05. \tag{7.14}$$

Consequently the sound velocity along the critical isochore will go to zero in the same way as $C_v^{-\frac{1}{2}}$. The relative importance of the different terms in

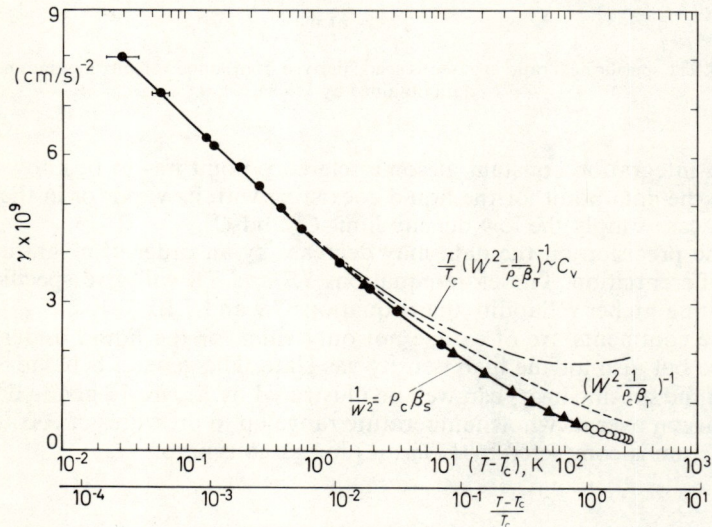

Figure 14. The contribution of the different terms in equation 7.11 to the temperature dependence of W^2 at the critical density. ●▲○ data respectively from refs 119, 89 and 74.

equation 7.11 can be compared in *Figure 14* where an analysis is given of the experimental sound velocity data obtained in argon by Theon et al.[119]. The result is that α is small, but that a logarithmic singularity, which means $\alpha = 0$, may not be excluded. Measurements in helium by Williamson and Chase[133] and by Barmatz[6] led to the same conclusion.

A general picture of the sound velocity in the critical region of argon is given in *Figure 15*. Similar data were obtained also in many other substances: they have been reviewed by Garland[45].

Until now data have been considered along the critical isochore. A similar analysis can be made following other lines on the pVT surface, e.g. along

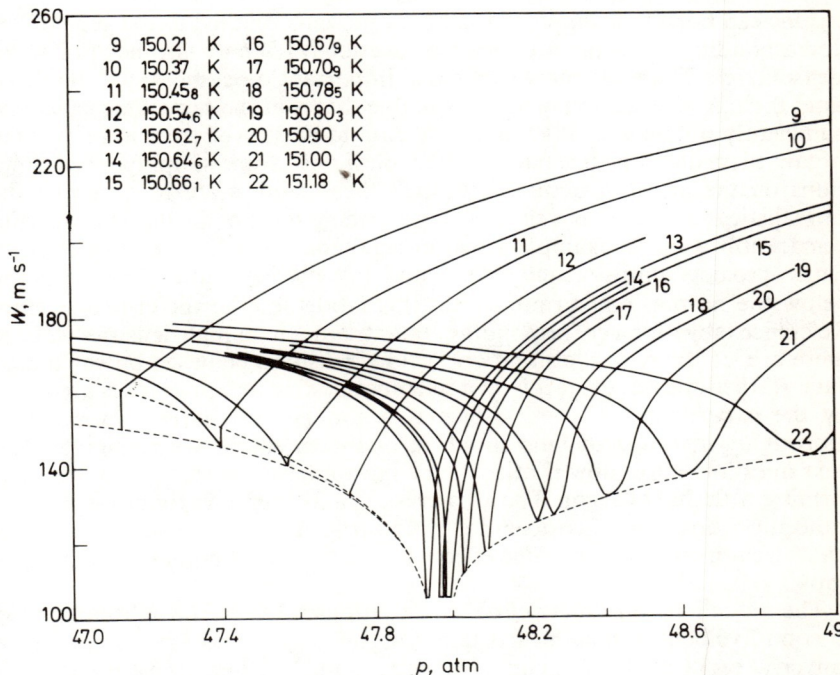

Figure 15. Sound velocity isotherms in the critical region of argon by Thoen *et al.*[119].

the gas–liquid coexistence line. For the same arguments as given before only the second term in equation 7.11 will determine the value of W^2 close to T_c: ρ and γ_v are not constant along this curve but in the limit is will again be the factor C_v^{-1} that determines completely the slope and the value of the exponent which is along this path called α'.

This method of obtaining the exponents compares favorably with direct specific heat measurements. In the latter case the apparent specific heat is a mean value over a temperature interval: in this way some details of the divergent behavior may be masked. A further problem is that the time for reaching thermal equilibrium increases sharply close to T_c.

A feature that may disturb the analysis of sound velocity data is critical dispersion (see also section II, 3). This may be a very suitable tool for studying the dynamic aspects of critical phenomena but from the thermodynamic point of view it is just a source of errors. This effect has been avoided as much as possible in some of the experiments we discussed[119, 6] by using sound waves in the audio frequency range but it is clear from the treatment in section II.3 and from *Figure 3* that no frequency is low enough to avoid completely all dispersion effects up to the critical point itself.

A thorough discussion of sound velocity and critical phenomena may be found in review articles by Garland[45], Sette[111] and Mistura[93].

5. Generalized Sound Velocity Behavior in Dense Fluids

One can hope that the same kind of arguments that led to the generalized corresponding states picture for gases (section VI.5) may also be applied for dense fluids. There are, however, some important differences: for the dense fluid there is no equation of state available that is at the same time exact and sufficiently simple to enable numerical data to be derived from it. In order to obtain these data one has had to make, until now, some simplifying assumptions on the liquid structure. Although these models are in general quite unrealistic they sometimes succeed in describing some of the liquid properties remarkably well. An example is the tunnel theory for a 12–6 fluid by Barker[5] which predicts almost exactly the sound velocity in saturated liquid argon below the normal boiling point. But this model fails in the critical region. Reference tables for a corresponding states treatment at high densities will not therefore, contrary to the situation for low densities, be based on calculated data. Rather one has to start with a reference fluid whose properties are given by the experiment; a good choice is argon because it is a monatomic thoroughly investigated fluid in which quantum effects are negligible. The next question is to know if other fluids have intermolecular potentials conforming with that of argon. The answer is certainly negative for non-spherical molecules: however, this does not necessarily mean that a corresponding states treatment based on some effective intermolecular potential should not apply.

The reduction constants defined in equations 6.4 to 6.6 have been applied to sound velocities in saturated liquid argon, xenon and krypton: a single universal reduced $W^*(T^*)$ curve has been found[3]. Other simple fluids, such as oxygen, nitrogen and methane, give similar but definitely different curves[123,125]. This is not surprising considering the fact that the critical temperature of these fluids reduced by ε/k_B is not a constant: so complete coincidence of the $W^*(T^*)$ curves is *a priori* excluded.

Some improvement may be found by the empirical rule of using the critical data as reduction constants. This has been tried by Nozdrev[96] with W_c, the sound velocity at the critical point, as scale factor for W. Of course this choice no longer makes sense regarding the recent evidence that W_c goes to zero. This is, however, a minor problem since one can choose W at any given T/T_c as reference; this has been done by Nozdrev[96] for $T/T_c = 0.6$ and by Mikhailenko and Blagoy[90] for $T/T_c = 0.777$.

A further remark can be made on the role of the ratio of specific heats. It has been shown in section III.2 that the density dependence of W in gases is different for monatomic and for diatomic molecules. A similar distinction should be made for liquids: γ depends on the internal degrees of freedom and this is an accidental property which to a first approximation does not have much to do with the other thermodynamic properties. So the quantity W^2/γ can be expected to follow more closely a corresponding states principle than

W^2. This statement is confirmed by Van Dael et al.[125] for argon, methane, nitrogen and oxygen data.

It will, however, be clear that in order to obtain a reasonable agreement for a large class of fluids at least a third molecular parameter should be added to the set of reduction constants. This extra parameter can be the empirical Pitzer factor[101], or it can be a more fundamental property, e.g. the factor describing how deviations from a spherically symmetrical intermolecular potential can be converted into a temperature and density dependent effective Lennard–Jones 12–6 potential[30]. An analysis of a large number of experimental data is now in progress and preliminary results indicate that W and its derived data might be suitable properties for obtaining the characteristic values in a three-parameter corresponding states treatment[99].

For the sake of completeness we will also mention some empirical rules giving a direct relation between the sound velocity and some other thermodynamic properties. The first one was proposed by Rao[104] as an empirical relation between the temperature derivative of the sound velocity and the expansion coefficient

$$\frac{1}{W}\left(\frac{\delta W}{\delta T}\right)_p = -k\frac{1}{V}\left(\frac{\delta V}{\delta T}\right)_p \tag{7.15}$$

with k a constant commonly equal to three. An integration gives

$$W^{1/k}V = R \tag{7.16}$$

with R the so-called 'molecular sound velocity' which should be independent of temperature and pressure. For complex liquids, with a great number of internal degrees of freedom, k has been found to be approximately three in the temperature range below the normal boiling point. It has, however, repeatedly been shown that R is not a constant for all liquids and not even for one liquid over its whole temperature range[123, 19, 4]. It is indeed quite evident that approaching the critical point R will go to zero. From equation 7.15 a relation between the adiabatic compressibility and the expansion coefficient has been derived by Wada[131]

$$\frac{1}{\beta_S}\left(\frac{\delta \beta_S}{\delta T}\right)_p = -\frac{1}{7}\frac{1}{V}\left(\frac{\delta V}{\delta T}\right)_p \tag{7.17}$$

or

$$B = \beta_S^{1/7} V \tag{7.18}$$

where B is the 'molecular compressibility'. This equation contains strictly the same information as equation 7.15 and its validity is consequently also restricted.

A similar type of empirical rule has been proposed by Carnavale and Litovitz[24] for the pressure derivative of W

$$\frac{1}{W}\cdot\left(\frac{\delta W}{\delta p}\right)_T = -k'\frac{1}{V}\left(\frac{\delta V}{\delta p}\right)_T \tag{7.19}$$

with $k' \simeq 3$.

Other empirical findings are that the sound velocity in the saturated liquid changes almost linearly with temperature or that it depends linearly on the density over quite a large range[125, 2]. None of these statements is valid over the whole fluid range and they fail badly in the critical region. The conclusion is that all these empirical 'rules' apply only in special cases where the general thermodynamic relations obtained by differentiating the expressions of section I.2, can be simplified by omitting or approximating some terms. They lack generality and should not be used as thermodynamic relations.

VIII. Sound Velocity in Liquid Mixtures

1. Homogeneous Mixtures

There is a very large number of sound velocity data available for binary liquid mixtures[108, 96]. The interpretation of the composition dependence of W seems, however, not to be very simple. Schaaffs[108] classifies the mixtures in different groups on the basis of the graphical representation of the $W(x)$ curves. Every combination seems to be possible: straight lines, convex or concave curves with maxima or minima, etc. Part of this complexity is due to the fact that in many of the mixtures investigated association effects occur which are strongly composition dependent. We prefer to limit this discussion to some mixtures of simple fluids. Experimental data are available from Blagoy et al. for argon–oxygen[16] and argon–methane[15], by Mikhailenko et al. for argon–krypton[91] and krypton–methane[92], by Güsewell et al. for neon–hydrogen and neon–deuterium[54] and by Van Dael and Vangeel for nitrogen–oxygen[124].

The exact formulas for W_{mix} are complex (see also section IV.2). This has led several authors to make simplifying assumptions. The most common is that the adiabatic compressibility of the mixture should be given[17, 105] by

$$\beta_S = \varphi_1(\beta_S)_1 + \varphi_2(\beta_S)_2 \qquad (8.1)$$

According to equation 4.14 this is true if the mixture is an ideal one and if moreover $\gamma_1 = \gamma_2 = \gamma^{im}$: only exceptionally and fortuitously will these conditions be met for liquids. Equation 8.1 can be transformed into a linear combination of the mole fractions if the additional assumption $V_1 = V_2$ is made

$$\beta_S^{im} = x_1(\beta_S)_1 + x_2(\beta_S)_2 \qquad (8.2)$$

The sound velocities with the equations 8.1 and 8.2 are respectively

$$\frac{x_1 V_1 + x_2 V_2}{x_1 M_1 + x_2 M_2} \frac{1}{W^{im2}} = \varphi_1 \frac{V_1}{M_1 W_1^2} + \varphi_2 \frac{V_2}{M_2 W_2^2} \qquad (8.3)$$

and

$$\frac{1}{x_1 M_1 + x_2 M_2} \frac{1}{W^{im2}} = \frac{x_1}{M_1 W_1^2} + \frac{x_2}{M_2 W_2^2} \qquad (8.4)$$

THERMODYNAMIC PROPERTIES AND THE VELOCITY OF SOUND

Still further simplifications may lead to empirical relations that appear to be valid in some cases[16, 110]:

$$W^2 = \psi_1 W_1^2 + \psi_2 W_2^2 \qquad (8.5)$$

$$W = \psi_1 W_1 + \psi_2 W_2 \qquad (8.6)$$

$$W = x_1 W_1 + x_2 W_2 \qquad (8.7)$$

$$W = \frac{1}{\varphi_1/W_1 + \varphi_2/W_2} \qquad (8.8)$$

It is clear that these formulas have no general predictive power: the number of assumptions to be made in order to derive them is so large that even an eventual agreement with experiment does not allow us to make any physical conclusion.

A more fundamental, but more tedious, approach starts from the equations in section IV.2. In many respects the situation for a mixture is not different from that of a pure fluid: the density has to be known in order to derive β_S and some information on the specific heats is needed for deducing β_T. There are only a very limited number of mixtures where all these data are available. For this reason it seems desirable to design combined experiments where in addition properties other than the velocity of sound are measured in the same mixture. This idea has been realized to some extent in the experiments of Blagoy et al.[15, 16] and of Mikhailenko et al.[91, 92]. The most important results emerging from their data are the derivatives $(\delta V^E/\delta p)_T$ and $(\delta V^E/\delta T)_p$ in equations 4.33 and 4.34: they appear to be respectively positive and negative in agreement with mixing theories. When the other properties of the mixture are not measured the sound velocity alone may give some qualitative but valuable information by considering the ratio W^2/W^{im2}.

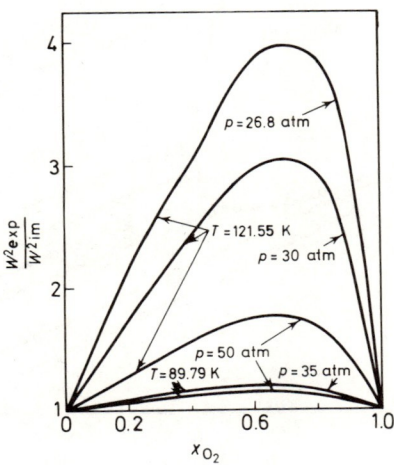

Figure 16. The ratio W^2_{exp}/W^{im2} as a function of T, p and x in liquid nitrogen–oxygen mixtures as obtained by Van Dael and Vangeel[124].

The deviation of W^2/W^{im2} from unity is a direct measure of the product $(V/V^{im})(\gamma/\gamma^{im})(\beta_T^{im}/\beta_T)$.

The available experimental data suggest that in general $W \geqslant W^{im}$ and that the deviations from ideal mixture behavior decrease with increasing density due either to an increase of the pressure at constant temperature, or to a decreasing temperature at constant pressure. This behavior is illustrated in *Figure 16* for experimental sound velocities in nitrogen–oxygen mixtures at different temperatures and pressures[124].

2. Mixtures Showing Phase Separation

We will treat briefly the sound velocity in binary mixtures which show

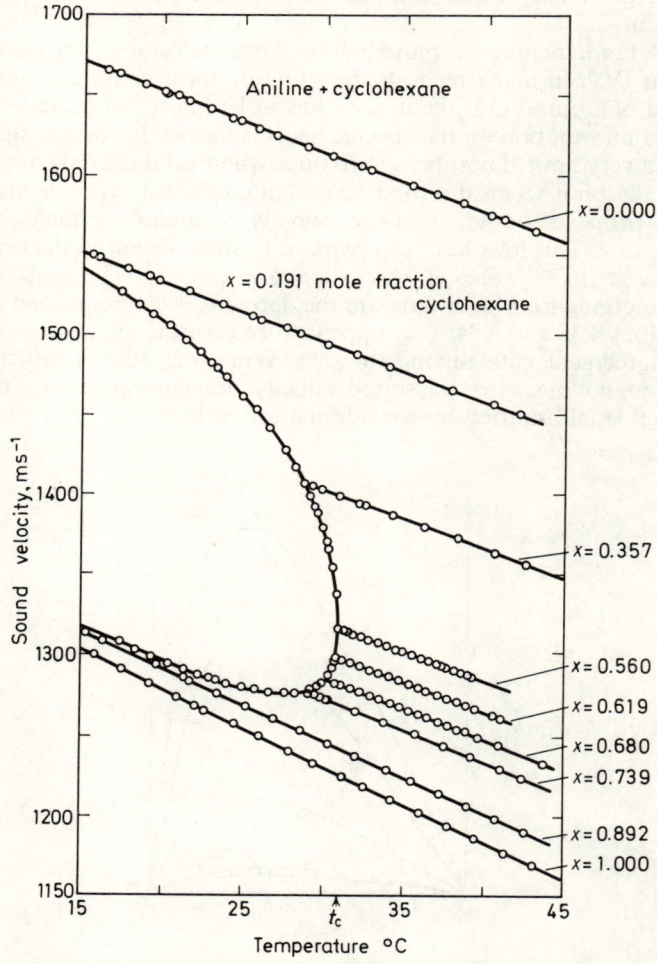

Figure 17. The velocity of sound as a function of T and x in the liquid mixture aniline–cyclohexane with a lower critical solution temperature at 31°C. Data by Vangeel and Van Dael[130].

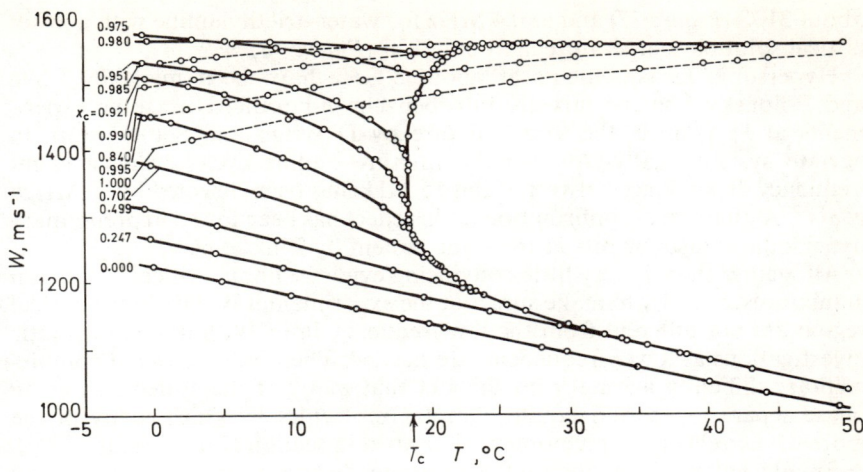

Figure 18. The velocity of sound as a function of T and x in the liquid mixture water–triethylamine with an upper critical solution temperature at 18°C. Data by Vangeel and Van Dael[129].

phase separation above or below some critical temperature. As shown also in Chapter 15 this phase transition can be treated in the general framework of critical phenomena. The power laws describing the singular behavior of various properties at a gas–liquid critical point can be translated for the binary mixture critical point by changing the variables. According to Griffiths and Wheeler[50] β_T and C_p are in general weakly divergent when the critical point is approached along a constant concentration path while the values of β_S and C_v remain finite. As for the gas–liquid critical system the sound velocity can be split into two parts using equation 1.12

$$W^2 = \frac{V}{M\beta_{T,x}} + \frac{TV^2}{MC_{v,x}}\left(\frac{\delta p}{\delta T}\right)^2_{v,x} \qquad (8.9)$$

At the mixture critical solution point one expects the first term on the RHS to vanish so that a finite critical velocity will be found[94], given by

$$W_c^2 = \frac{TV^2}{MC_{v,x}}\left(\frac{\delta p}{\delta T}\right)^2_{v,x} \qquad (8.10)$$

Again it has to be emphasized that strong dispersion effects may be present (see also section II.3). There is only a limited amount of experimental W data in the critical region of binary mixtures: part of these results is quite old and was not obtained with actual knowledge on critical phenomena in mind. A critical survey of the available data is given by Garland[45], by Sette[111] and by Mistura[93]. All these results agree with the prediction that no strong anomaly is present and that W remains finite.

A recent systematic study of the sound velocity over the whole concentration range has been carried out by Vangeel and Van Dael[130] at 1 MHz for the mixture cyclohexane–aniline with an upper critical solution temperature at about 31°C (*Figure 17*) and at 0.4 MHz for water–triethylamine with a lower critical solution temperature at about 18°C (*Figure 18*).

Hypersonic measurements at about 5 GHz have been made by Chen and Polonsky[28] in the mixture nitrobenzene–*n*-hexane. Ultrasonic experiments at 15 MHz in the same mixture by D'Arrigo and Sette[33] seem to deviate systematically. Also in the mixture aniline–cyclohexane a slight frequency dependence between 2 and 15 MHz has been detected by D'Arrigo *et al*.[32]. A qualitative confirmation of this effect has been found in preliminary hypersonic measurements in the same system by Zink *et al*.[136].

Altogether there is very little convincing evidence about critical dispersion in mixtures: in order to make sure that the experimental W data in the critical region are not different from the low frequency limit $W(0)$ more systematic investigations at lower frequencies are needed. There is, however, a complementary difficulty arising from the fact that many of the systems showing phase separation contain quite complex molecules in which many of the more trivial relaxation phenomena described in section II may occur.

Finally let us point out that if accurate dispersion-free sound velocity data are available along characteristic paths one can derive directly the power law exponents. This can be illustrated by studying the coexistence lines in *Figures 17* and *18*. Close to T_c the variation of W is completely determined by $(\delta W/\delta x)_T (dx/dT)$ which indicates that in the limit for $T \to T_c$ and with $(\delta W/\delta x)_T$ a slowly varying function of x and T,

$$|W_{\text{phase 1}} - W_{\text{phase 2}}| \sim |T - T_c|^\beta \tag{8.11}$$

IX. Acknowledgements

The author wishes to thank everybody in the Laboratorium voor Molekuulfysika at Leuven University who contributed in one way or another to the achievement of this chapter.

Special thanks go to R. Hellemans, K. E. Starling, J. Thoen, E. Vangeel and H. Zink for reviewing and criticizing parts or the whole of the manuscript.

X. References

[1] Abraham, B. M., Y. Eckstein, J. B. Ketterson, M. Kuchnir and P. Roach. *Phys. Rev.* **A1**, 250 (1970).
[2] Aziz, R. A., D. H. Bowman and C. C. Lim. *Canad. J. Phys.* **50**, 646 (1972).
[3] Aziz, R. A., C. C. Lim and D. H. Bowman. *Canad. J. Chem.* **45**, 1037 (1967).
[4] Aziz, R. A., D. H. Bowman and C. C. Lim. *Canad. J. Chem.* **45**, 2079 (1967).
[5] Barker, J. A. *Proc. Roy. Soc.* **259**, 442 (1961).
[6] Barmatz, M. *Phys. Rev. Letters*, **24**, 651 (1970).
[7] Bauer, H. J. 'Phenomenological theory of the relaxation phenomena in gases'. *Physical Acoustics*, Vol. II, part A, pp 48–131. Edited by W. R. Mason. Academic Press: New York (1965).
[8] Bauer, H. J. *Advanc. Molec. Relaxation Processes*, **2**, 319–376 (1972).
[9] Bergmann, L. *Der Ultraschall*, S. Hirzel: Stuttgart (1954).
[10] Beyer, R. T. and S. V. Letcher, *Physical Ultrasonics*, Academic Press: New York (1969).
[11] Bhatia, A. B. *Ultrasonic Absorption (An Introduction to the Theory of Sound Absorption and Dispersion in Gases, Liquids and Solids)*. Clarendon Press: Oxford (1967).
[12] Bird, R. B. and E. L. Spotz, University of Wisconsin, CM599 (1950).
[13] Bird, R. B., E. L. Spotz and J. O. Hirschfelder. *J. Chem. Phys.* **18**, 1395 (1950).

14. Blagoy, Y. P., A. E. Butko, S. A. Mikhailenko and V. V. Yakuba. *Akust. Zh.* **12**, 405 (1966); *Soviet. Phys. Acoust.* **12**, 355 (1967).
15. Blagoy, Y. P., A. E. Butko, S. A. Mikhailenko and V. V. Yakuba. *Ukr. Fiz. Zh.* **13**, 1605 (1968).
16. Blagoy, Y. P., A. E. Butko, S. A. Mikhailenko and V. V. Yakuba. *Ukr. Fiz. Zh.* **13**, 1805 (1968).
17. Blandamer, M. J., and D. Waddington, *J. Phys. Chem.* **74**, 2569 (1970).
18. Blitz, J. *Fundamentals of Ultrasonics*, Butterworths; London (1963).
19. Bowman, D. H., C. C. Lim and R. A. Aziz, *Canad. J. Chem.* **46**, 1175 (1968).
20. Boyd, M. E., S. Y. Larsen and H. Plumb, *J. Res. Nat. Bur. Stand.* **72A**, 155 (1968).
21. Boyd, M. E. and R. D. Mountain, *Phys. Rev.* **A2**, 2164 (1970).
22. Bruch, L. W. *Phys. Rev.* **178**, 303 (1969).
23. Bruch, L. W. *Phys. Rev.* **A2**, 2164 (1970).
24. Carnavale, E. H. and T. A. Litovitz. *J. Acoust. Soc. Amer.* **27**, 547 (1955).
25. Cataland, G. and H. Plumb, *J. Res. Nat. Bur. Stand.* **69A**, 531 (1965).
26. Cedrone, N. P. and D. R. Curran. *J. Acoust. Soc. Amer.* **23**, 627 (1951).
27. Chapman, S. and T. G. Cowling. *The Mathematical Theory of Non-uniform Gases*. Cambridge University Press: London (1960).
28. Chen, S. H. and N. Polonsky, *Phys. Rev. Letters*, **20**, 909 (1968).
29. Chynoweth, A. G. and W. G. Schneider. *J. Chem. Phys.* **20**, 1777 (1952).
30. Cook, D. and J. S. Rowlinson. *Proc. Roy. Soc. (London) A*, **219**, 405 (1953).
31. Cottrell, T. L. and J. C. McCoubrey, *Molecular Energy Transfer in Gases*. Butterworths: London (1961).
32. D'Arrigo, G., L. Mistura and P. Tartaglia, *Phys. Rev.* **A1**, 286 (1970).
33. D'Arrigo, G. and D. Sette. *J. Chem. Phys.* **48**, 691 (1968).
34. De Boer, J. *Physica*, **14**, 139 (1948).
35. Del Grosso, U. A., E. J. Smura and P. F. Fougere. US Naval Research Laboratory *Rep. No. NRL-4439* (1954).
36. Eden, D., C. W. Garland and J. Thoen. *Phys. Rev. Letters*, **28**, 726 (1972).
37. Eggers, F. *Acustica*, **19**, 323 (1967).
38. Eucken, A. and F. Hauck, *Z. Phys. Chem.* **134**, 161 (1928).
39. Fixman, M. *J. Chem. Phys.* **32**, 1363 (1960).
40. Fixman, M. *J. Chem. Phys.* **36**, 1957 (1962).
41. Fixman, M. *J. Chem. Phys.* **36**, 1961 (1962).
42. Fleury, P. A. and J. P. Boon. *Phys. Rev.* **186**, 244 (1969).
43. Forgacs, R. L. *J. Acoust. Soc. Amer.* **32**, 1697 (1960).
44. Fritsche, L. *Acustica*, **10**, 189 (1960).
45. Garland, C. W. 'Ultrasonic investigation of phase transitions and critical points'. *Physical Acoustics*, edited by W. P. Mason. Vol. VII, Chap. 2. Academic Press: New York (1970).
46. Garland, C. W., D. Eden and L. Mistura. *Phys. Rev. Letters*, **25**, 1161 (1970).
47. Gordon, R. G., W. Klemperer and J. I. Steinfeld. *Annu. Rev. Phys. Chem.* **19**, 215 (1968).
48. Greenspan, M. *J. Acoust. Soc. Amer.* **28**, 644 (1956).
49. Greenspan, M. 'Transmission of sound waves in gases at very low pressure'. *Physical Acoustics*, edited by W. P. Mason, Vol. II. part A, pp 1–43. Academic Press: New York (1965).
50. Griffiths, R. B. and J. C. Wheeler. *Phys. Rev.* **A2**, 1047 (1970).
51. Grimsrud, D. T. and J. H. Werntz. *Phys. Rev.* **157**, 181 (1967).
52. Guggenheim, E. A. *Mixtures*. Clarendon Press: Oxford (1952).
53. Guptill, E. W., C. K. Hoyt and D. K. Robinson, *Canad. J. Phys.* **33**, 397 (1955).
54. Güsewell, D., F. Schmeissner and J. Schmid. *Cryogenics*, **10**, 150 (1970).
55. Gyorog, D. A. and E. F. Obert. *Amer. Instn Chem. Engrs Jnl*, **10**, 625 (1964).
56. Haebel, E. U. *Acustica*, **20**, 65 (1968).
57. Herzfeld, K. F. and T. A. Litovitz. *Absorption and Dispersion of Ultrasonic Waves*. Academic Press: New York (1959).
58. Hildebrand, J. H. and R. L. Scott. *Regular Solutions*, Prentice-Hall: Englewood Cliffs (1962).
59. Hilsenrath, J. et al. *Circ. US Nat. Bur. Stand. No. 564* (1955).
60. Hirschfelder, J. O., C. F. Curtiss and R. B. Bird. *Molecular Theory of Gases and Liquids*. Wiley: New York (1954).
61. Holbrook, R. D. *J. Acoust. Soc. Amer.* **20**, 590 (1948).
62. Hubbard, J. C. *Phys. Rev.* **38**, 1011 (1931).
63. Hubbard, J. C. *Phys. Rev.* **41**, 523 (1932).
64. Jeans, J. J. *Kinetic Theory of Gases*. Cambridge University Press: London (1960).
65. Jenkin, C. F. *Proc. Roy. Soc. (London) A*, **98**, 170 (1921).

[66] Kawasaki, K. *Progr. Theor. Phys.* **40**, 930 (1968).
[67] Kawasaki, K. *Ann. Phys. (NY)*, **61**, 1 (1970).
[68] Kawasaki, K. *Phys. Rev.* **A1**, 1750 (1970).
[69] Keesom, W. H. and A. Van Itterbeck, *Commun. Kamerlingh Onnes Lab. Univ. Leiden*, No. 213b (1931).
[70] Kneser, H. O. 'Schallabsorption und -dispersion in Gasen', *Encyclopedia of Physics*, Vol. XI, 1 Acoustics I, pp 129–195. Edited by S. Flügge. Springer: Berlin (1961).
[71] Kneser, H. O. 'Relaxation processes in gases'. *Physical Acoustics*, Vol. III, part A, pp 133–199. Edited by W. P. Mason. Academic Press: New York (1965).
[72] Kudrjawzew, B. B. *Anwendung von Ultraschallverfahren bei Physikalisch-Chemischen Untersuchungen.* VEB Deutscher Verlag der Wissenschaften: Berlin (1955).
[73] Kuhl, W., G. R. Schodder and F. K. Schröder. *Acustica*, **4**, 519 (1954).
[74] Lacam, A. *J. Rech. CNRS*, **34**, 25 (1965).
[75] Lacy, L. L. and A. C. Daniel. *J. Acoust. Soc. Amer.* **52**, 189 (1972).
[76] Lamb, J., 'Thermal relaxation in liquids', *Physical Acoustics*, Vol. II. part A, pp 203–280. Edited by W. P. Mason. Academic Press: New York (1965).
[77] Larson, E. U., D. G. Naugle and T. W. Adair. *J. Chem. Phys.* **54**, 2429 (1971).
[78] Lestz, S. S. and R. N. Grove. *J. Chem. Phys.* **43**, 883 (1965).
[79] Litovitz, T. A. and C. M. Davis. 'Structural and shear relaxation in liquids'. *Physical Acoustics*, Vol. II. part A, pp 282–349. Edited by W. P. Mason. Academic Press: New York (1965).
[80] Mason, E. A. and T. H. Spurling. *The Virial Equation of State*. Pergamon: Oxford (1969).
[81] Mason, W. P. *Piezoelectric Crystals and their Application to Ultrasonics*. Van Nostrand: New York (1950).
[82] Matheson, A. J. *Molecular Acoustics*. Wiley–Interscience: London (1971).
[83] McSkimin, H. J. *J. Acoust. Soc. Amer.* **22**, 413 (1950).
[84] McSkimin, H. J. *J. Acoust. Soc. Amer.* **33**, 12 (1961).
[85] McSkimin, H. J. 'Ultrasonic methods for measuring the mechanical properties of liquids and solids'. *Physical Acoustics*, edited by W. P. Mason, Vol. I, pp 271–334. Academic Press: New York (1964).
[86] Meyer, E. and G. Sessler. *Z. Phys.* **149**, 15 (1957).
[87] Michels, A., B. Blaisse and C. Michels. *Proc. Roy. Soc. (London) A*, **160**, 358 (1937).
[88] Michels, A., A. Byl and C. Michels. *Proc. Roy. Soc. (London) A*, **160**, 376 (1937).
[89] Michels, A., J. M. Levelt and G. J. Wolkers. *Physica*, **24**, 769 (1958).
[90] Mikhailenko, S. A. and Y. P. Blagoy. *Russ. J. Phys. Chem.* **42**, 566 (1968).
[91] Mikhailenko, S. A., Y. P. Blagoy, A. E. Butko and V. A. Sorokin. *Ukr. Fiz. Zh.* **15**, 563 (1970).
[92] Mikhailenko, S. A., Y. P. Blagoy, A. E. Butko and V. A. Sorokin. *Ukr. Fiz. Zh.* **15**, 571 (1970).
[93] Mistura, L. *Proceedings of the Internation School of Physics Enrico Fermi*, Course L.I, edited by M. S. Green, pp 563–577, Academic Press: New York (1971).
[94] Mistura, L. *J. Chem. Phys.* **57**, 2311 (1972).
[95] Naugle, D. G. *J. Chem. Phys.* **56**, 5730 (1972).
[96] Nozdrev, V. F. *Application of Ultrasonics in Molecular Physics*. Gordon and Breach: New York (1963).
[97] Papadakis, E. P. *J. Acoust. Soc. Amer.* **42**, 1045 (1967).
[98] Pecceu, W. and W. Van Dael. *Physica*, **63**, 154 (1973).
[99] Pecceu, W. *Doctoral Thesis*, University of Leuven (1973).
[100] Pine, A. S. *J. Chem. Phys.* **51**, 5171 (1969).
[101] Pitzer, K. S., D. Z. Lipmann, R. F. Curl, C. M. Huggins and D. E. Petersen. *J. Amer. Chem. Soc.* **77**, 3433 (1955).
[102] Plumb, H. H. and G. Cataland. *Science*, **150**, 155 (1965).
[103] Plumb, H. H. *J. Res. Nat. Bur. Stand.* **69A**, 375 (1965).
[104] Rao, M. R., *Indian J. Phys.* **14**, 109 (1940).
[105] Richardson, E. G. *Ultrasonic Physics*, Elsevier: Amsterdam (1962).
[106] Rowlinson, J. S. *Liquids and Liquid Mixtures*, 2nd ed. Butterworths: London (1969).
[107] Schaaffs, W. *Molekularakustik*, Springer: Berlin (1963).
[108] Schaaffs, W. *Landolt-Börnstein Tables*, Group II, Vol. V, 'Molecular acoustics', Springer: Berlin (1967).
[109] Sell, H. *Z. Tech. Phys.* **18**, 3 (1937).
[110] Sette, D. 'Dispersion and absorption of sound waves in liquids and mixtures of liquids', *Encyclopedia of Physics*, Vol. XI, 1 Acoustics I, pp 275–359. Edited by S. Flügge. Springer: Berlin (1961).

[111] Sette D. *Proceedings* of the International School of Physics Enrico Fermi, Course LI, pp 508–562. edited by M. S. Green. Academic Press: New York (1971).
[112] Shields, F. D., K. P. Lee and W. J. Wiley. *J. Acoust. Soc. Amer.* **37**, 724 (1965).
[113] Staveley, L. A. K., W. I. Tupman and K. R. Hart. *Trans. Faraday Soc.* **51**, 323 (1955).
[114] Stevens, B. *Collisional Activation in Gases*, Pergamon: Oxford (1967).
[115] Stewart, J. L. *Rev. Sci. Instrum.* **17**, 59 (1946).
[116] Stewart, J. L. and E. S. Stewart. *J. Acoust. Soc. Amer.* **24**, 22 (1952).
[117] Stewart, R. B. and V. J. Johnson. 'A compendium of the properties of materials at low temperature (Phase II)', *Wadd. Tech. Rep. No. 60–56*, Part IV.
[118] Thoen, J., E. Vangeel and W. Van Dael. *Physica*, **45**, 339 (1969).
[119] Thoen, J., E. Vangeel and W. Van Dael. *Physica*, **52**, 205 (1971).
[120] Touloukian, Y. S. (Ed).. *Thermophysical Properties of Matter*, Vol. VI. IFI/Plenum: New York (1970).
[121] Truell, R., C. Elbaum and B. Chick. *Ultrasonic Methods in Solid State Physics*. Academic Press: New York (1969).
[122] Truesdell, C. *J. Rat. Mech. Anal.* **2**, 643 (1953).
[123] Van Dael, W. and A. Van Itterbeek. Chap. 7 in *Physics of High Pressures and the Condensed Phase*, Edited by A. Van Itterbeek. North Holland: Amsterdam (1965).
[124] Van Dael, W. and E. Vangeel. *Proceedings* of the First International Conference on Calorimetry and Thermodynamics, Warsaw, 1969, p. 555.
[125] Van Dael, W., A. Van Itterbeek. A. Cops and J. Thoen. *Physica*, **32**, 611 (1968).
[126] Van Itterbeck, A. and W. De Laet *Physica*, **24**, 59 (1958).
[127] Van Itterbeck, A. and W. H. Keesom. *Commun. Kakerlingh Onnes Lab. Univ. Leiden*, No. 209c (1931).
[128] Van Itterbeck, A. and J. Nihoul. *Acustica*, **5**, 142 (1955).
[129] Van Ness, H. C. *Classical Thermodynamics of Non-electrolyte Solutions*. Pergamon: Oxford (1964).
[130] Vangeel, E. and W. Van Dael. (To be published).
[131] Wada, Y. *J. Phys. Soc., Japan.* **4**, 280 (1949).
[132] Williams, J. and J. Lamb. *J. Acoust. Soc. Amer.* **30**, 308 (1958).
[133] Williamson, R. C. and C. E. Chase. *Phys. Rev.* **176**, 285 (1968).
[134] Whitney, W. M. and C. E. Chase. *Phys. Rev.* **158**, 200 (1967).
[135] Wong, L. Y. and A. Anderson. *J. Opt. Soc. Amer.* **62**, 1112 (1972).
[136] Zink, H., F. Nys and W. Van Dael. (To be published).

CHAPTER 12

Relation of the Dielectric Constant and the Refractive Index to Thermodynamic Properties

B. L. SMITH

School of Mathematical and Physical Sciences, University of Sussex, Falmer, Sussex, UK

Contents

I.	Introduction	579
II.	Theoretical	580
	1. Lorentz Model	581
	2. Onsager–Böttcher Theory	582
	3. Statistical-mechanical Calculations	584
	4. Phenomenological Shell Model	586
	5. Variation of Polarizability with Density	587
	6. Generalization of Theory to Optical Frequencies	588
III.	Experimental Determinations of Dielectric Properties	588
	1. Methods for Determining Refractive Index	589
	2. Results of Refractive Index Measurements	592
	3. Methods for Determining Dielectric Constants	597
	4. Results of Dielectric Constant Measurements and Comparison with Refractive Index Data	599
IV.	Magneto-optical Properties	600
V.	Conclusions	604
VI.	References	605

I. Introduction

The refractive index (n) or dielectric constant (ε) of a dielectric material is a function of the thermodynamic state. For a non-polar substance, in which the constituent molecules do not possess a permanent dipole moment, $\varepsilon = n^2$ (at the same frequency), and the dielectric constant is related to the molecular polarizability α, Avogadro's number N_A, and the density ρ, by the Clausius–Mossotti relation (CM).

$$\{(\varepsilon - 1)/(\varepsilon + 2)\}\{1/\rho\} = \tfrac{1}{3} N_A \alpha \tag{1}$$

The optical analogue is the Lorentz–Lorenz relation (LL)

$$\{(n^2 - 1)/(n^2 + 2)\}\{1/\rho\} = \tfrac{1}{3} N_A \alpha \tag{2}$$

Dielectric constant and refractive index data are frequently used to establish p–V–T relationships for non-polar fluids, for example the slope of the critical isotherm and the shape of the liquid–vapor coexistence curve, using equations 1 and 2. These techniques are very attractive, particularly at extreme pressures or in the neighborhood of phase transitions where conventional density methods are difficult.

Such procedures are valid only if the Lorentz–Lorenz function or the Clausius–Mossotti function remain constant throughout the region under study. In particular it is generally assumed that the Lorentz–Lorenz function, and hence the molecular polarizability $\alpha = 3LL/N_A$, are independent of both density and temperature. Experimental investigations[16,33,35] have shown that, although LL is nearly constant over a limited density range, small but significant changes (say, one to five per cent) do occur as the density changes radically, for example, from a low-pressure gas to a liquid. It is thus important to establish the magnitude of any deviations from equations 1 and 2 from a practical viewpoint. But before discussing the main experimental techniques for determining thermodynamic properties from n and ε however, it is necessary to examine the theoretical basis for equations 1 and 2.

II. Theoretical

In this section the theory of non-polar dielectrics, in which the static dielectric constant is related to the density and polarizability, leading to relations of the form given in equation 1, is reviewed. Since this book is concerned with measurements on non-reacting fluids, the discussion will be mainly limited to the simple case where the polarization is isotropic and due to elastic displacement of the electrons only; in other words to optical or electronic polarization.

A molecule subject to an external field, but not isolated from other molecules will have an induced dipole moment **m** given by

$$\boldsymbol{m} = \alpha \boldsymbol{E}'$$

where $\boldsymbol{E}'$, the local field, is the field intensity in the neighborhood of the molecule due to the applied field and also to the presence of the other molecules.

For gases at low densities, the molecules are sufficiently far apart for the local field intensity to be equated with the applied field. However, when intermolecular distances become comparable with molecular dimensions, an appreciable fraction of the local field intensity is contributed by the electric fields of other molecules. This is true of gases when they are sufficiently dense to depart significantly from the ideal gas law, and of solids and liquids in general.

Provided that the local field may be regarded as being uniform over the small region occupied by the molecule, i.e. the molecules are not too close, the concepts of local field and molecular polarizability are useful and valid. Under conditions of strong interaction, however, these quantities become more questionable, except as average properties of the system. Much effort has gone into calculating the local field and relating it to the observed refractive index n and the dielectric constant ε of dielectrics. The best known method of calculating the local field intensity in a dense medium was originated by H. A. Lorentz[30]. His derivation has set the pattern for most theoretical treatments of the dielectric constant ever since. For this reason it is worth considering the argument in some detail.

RELATION OF THE DIELECTRIC CONSTANT AND THE REFRACTIVE INDEX

1. Lorentz Model

Consider a piece of dielectric subject to an external applied field E, the shape being such that the field strength in the material is uniform (*Figure 1*). Select a particular molecule and about it draw a sphere of radius R, large in

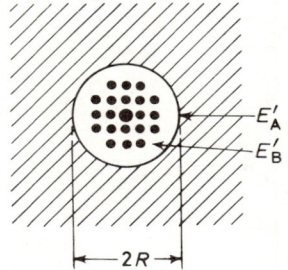

Figure 1. Lorentz model for a dielectric.

comparison with intermolecular distances but small compared with the macroscopic dimensions of the specimen. The local field intensity E' in the neighborhood of the molecule may be considered in two parts:

$$E' = E'_A + E'_B$$

where E'_A is the field due to the material outside the sphere, treated as a continuous medium, and E'_B is the field due to the molecules inside the sphere.

From a macroscopic argument it may be shown that

$$E'_A = E + P/3$$

where P is the polarization, or dipole moment per unit volume and

$$P = nm = (\varepsilon - \varepsilon_0)E \qquad (3)$$

To compute E'_B, some definite distribution of the molecules must be assumed. The case treated by Lorentz was a cubic lattice of molecules, each electrostatically equivalent to a dipole, and each having the same dipole moment m. He showed that for this model $E'_B = 0$ and therefore

$$E' = E'_A = E + P/3 \qquad (4)$$

He applied this formula, as an approximation, under somewhat more general conditions. The resulting expression for the dielectric constant is very simple and has been widely used to describe the properties of gases, liquids and solids. It is interesting to note that although the limitations on its validity have been largely ignored, Lorentz himself recognized them.

Assuming then that $E'_B = 0$, on the basis that it vanishes for a cubic lattice of dipoles, and is presumably small for any isotropic material, from equations 3 and 4 we obtain

$$(\varepsilon - 1)/(\varepsilon + 2)\rho = N_A \alpha/3 \qquad (5)$$

which predicts that for a given substance the function $(\varepsilon - 1)/(\varepsilon + 2)\rho$ is a constant, independent of density and temperature.

A. Michels and collaborators were the first to have established experimental deviations from equation 5 for the inert gases (*Figure 2*). Both dielectric constant and refractive index measurements suggest that $(\varepsilon - 1)/(\varepsilon + 2)\rho$ or $(n^2 - 1)/(n^2 + 2)\rho$ increase slowly with increasing

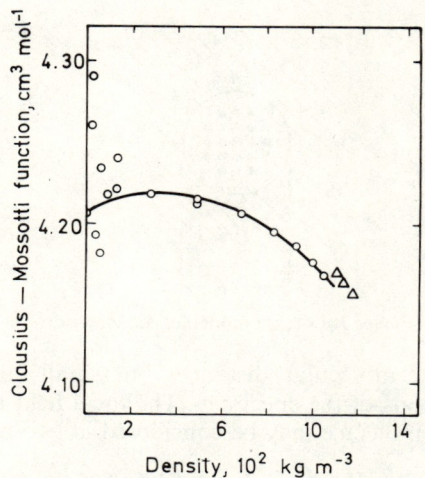

Figure 2. Variation in the Clausius–Mossotti function $(\varepsilon - 1)/(\varepsilon + 2)\rho$ for compressed argon gas with density. (Δ—measurements on the liquid under saturated vapor pressure near the triple point) (Michels and Botzen[33]).

density up to roughly critical density and then decrease at higher densities. More recent measurements on the gas at relatively low pressures and on the liquid confirm this trend. It is interesting to speculate why equation 5 is not valid. The most obvious defect in using the Lorentz model for fluids is the assumption concerning cubic symmetry and in the two next sections more acceptable methods of calculating the influence of molecular neighbors on the local field are given, one based on an alternative empirical model and the other based on a rigorous statistical mechanical approach.

2. Onsager–Böttcher Theory

Böttcher[5] has made use of another model to calculate the local field intensity. One particular molecule is considered which, it is assumed, inhabits a spherical cavity in the material. The field surrounding the cavity is treated on a strictly macroscopic basis, the calculation being an adapted form of Onsager's theory of polar fluids[37]. A brief derivation of Böttcher's formula is as follows.

Consider a selected molecule in a dielectric. Assume that it may be represented by an isotropically polarizable sphere of polarizability α and radius a. The local field intensity $\boldsymbol{E}'$ that acts on the molecule can be analysed into two terms

$$\boldsymbol{E}' = g\boldsymbol{E} + f\boldsymbol{m} \tag{6}$$

RELATION OF THE DIELECTRIC CONSTANT AND THE REFRACTIVE INDEX

The term gE represents the effect of the applied field and the moments induced by it in the surrounding molecules and the term fm represents the effect of the moments induced in the surrounding molecules by the field of the selected molecule.

The cavity field, gE, is the field acting in an empty spherical cavity in a piece of dielectric of such a shape that at points remote from the cavity the polarization is uniform and parallel. If E is the field strength in the dielectric at points far removed from the cavity, then

$$gE = [3\varepsilon/(2\varepsilon + 1)] \times E \tag{7}$$

The reaction field, fm, is related to the induced dipole moment by the expression

$$fm = (m/a^3)\{2(\varepsilon - 1)/(2\varepsilon + 1)\} = \alpha E'\mathcal{R} \tag{8}$$

From equations 6, 7 and 8

$$E' = [3\varepsilon/(2\varepsilon + 1)] \times E + E'\mathcal{R}\alpha$$
$$E' = [3\varepsilon/(2\varepsilon + 1)] \times E/(1 - \alpha\mathcal{R}) \tag{9}$$

Combining equations 3 and 9

$$n\alpha/(1 - \alpha\mathcal{R}) = (\varepsilon - 1)(2\varepsilon + 1)/3\varepsilon \tag{10}$$

At this stage Onsager eliminated a from the equivalent relation for polar fluids, by using the approximation $4\pi na^3/3 = 1$. This is equivalent to the assumption that the molecules fill the whole available space. Böttcher, however, retained a in the formula and assumed that it was a molecular constant independent of density.

Equation 10 may be conveniently written in two forms

$$\frac{1}{\alpha} - \frac{2}{a^3}\frac{(\varepsilon - 1)}{(2\varepsilon + 1)} = \frac{3n\varepsilon}{(\varepsilon - 1)(2\varepsilon + 1)} \tag{11}$$

which implies a linear relationship between $2(\varepsilon - 1)/a^3(2\varepsilon + 1)$ and $3n\varepsilon/(\varepsilon - 1)(2\varepsilon + 1)$, and

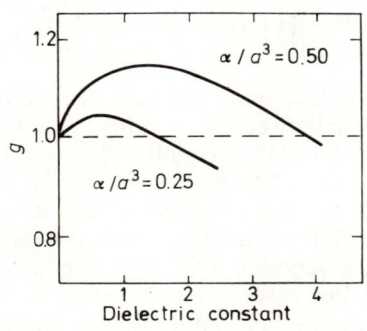

Figure 3. The factor g for $\alpha/a^3 = 0.25$ and $\alpha/a^3 = 0.50$ (equation 12) (Onsager–Böttcher theory[5]).

$$\frac{(\varepsilon - 1)}{(\varepsilon + 2)} \times \frac{1}{\rho} = \frac{\alpha}{3}\left[\frac{9\varepsilon}{(\varepsilon + 2)(2\varepsilon + 1) - \alpha/a^3(2\varepsilon - 2)}\right] = \frac{g\alpha}{3} \quad (12)$$

which is similar to the Clausius–Mossotti formula. g may thus be regarded as a correction factor to the Clausius–Mossotti formula. For $\varepsilon = 1$ the factor is equal to one. For larger values of ε the correction depends on the ratio α/a^3. In *Figure 3*, g is plotted as a function of ε for two values of α/a^3. It can be seen that with increasing ε (increasing density) g will increase, pass through a maximum, and then decrease again. Böttcher's formula thus differs from the Clausius–Mossotti formula in so far as it predicts that the function $(\varepsilon - 1)/(\varepsilon + 2)\rho$ does not remain constant, but will vary in the manner indicated.

3. Statistical-mechanical Calculations

To develop a rigorous theory the statistical-mechanical system of all the molecules must be considered. It should still be possible, however, to evaluate the mean moment of a single representative molecule; for statistically they are all equivalent. The method outlined below was originated by Kirkwood[29] and Yvon[47].

The molecules in the system are considered to behave in an electric field as point dipoles of polarizability α, and to be spherically symmetric in their other interactions. It is assumed that the dipoles are in instantaneous equilibrium with the field; that is that the instantaneous moment m_i of molecule i is $\alpha E'_i$, where E'_i is the local field intensity acting on the molecule. If, in the statistical-mechanical averaging over configurations of the system, the molecules are treated as point centres of force subject to classical laws, then for this simple model a rigorous calculation of the dielectric constant is possible.

The result may be expressed in the form

$$\frac{(\varepsilon + 2)}{3(\varepsilon - 1)} = \frac{1}{v^{(1)}\alpha}(1 - v^{(1)}\alpha^2 G + \ldots) \quad (13)$$

where

$$G = 8\pi \int_0^\infty \frac{\rho(r)\,dr}{r^4} + 2\int\left(\frac{v_{123}^{(3)}}{[v^{(1)}]^2} - \frac{v_{12}^{(2)}v_{23}^{(2)}}{[v^{(1)}]^3}\right)\frac{P_2\cos\theta}{r_{12}^3 r_{23}^3}\,dv_2\,dv_3 \quad (14)$$

$\rho(r)$, $v^{(1)}$, $v^{(2)}$, $v^{(3)}$ are the radial, one, two and three molecule distribution functions, and θ is the angle between the vectors r_{12} and r_{23}.

If only the first term on the RHS of equation 13 is retained, the expression can be reduced to the Clausius–Mossotti formula. The second term gives a first order correction. To calculate this correction it is necessary to have values not only of the radial distribution function $\rho(r)$ (or equivalently, of the two molecule distribution function $v^{(2)}$), but also of the three molecule distribution function $v^{(3)}$. For gases at low pressure, it is possible to express these distribution functions as series; but for liquids this method fails, and the calculation of even $\rho(r)$ is, in general, very difficult.

For $v^{(3)}$, the following approximation has sometimes been used

$$v^{(3)}_{123} = v^{(2)}_{13}v^{(2)}_{23}v^{(2)}_{31}/[v^{(1)}]^3 \qquad (15)$$

Brown[8] has used equation 15 together with the following very crude approximation for the radial distribution function $\rho(r)$,

$$\begin{aligned}\rho(r) &= 0 \quad r < a \\ \rho(r) &= 1 \quad r > a\end{aligned} \qquad (16)$$

Physically, this corresponds to the assumption that the molecules are hard spheres subject to no attractive forces, and that only binary encounters are important. The resultant formula is

$$\frac{(\varepsilon + 2)}{(\varepsilon - 1)}\frac{1}{\rho} = \frac{1}{Z} - \frac{2\alpha}{a^3} + \frac{15}{16}Z \qquad (17)$$

where $Z = \alpha/3$.

De Boer, van der Maesen and ten Seldam[4] have evaluated the integrals in equation 14 by a series method for monatomic molecules subject to a Lennard–Jones interatomic potential. As can be seen in *Figure 4*, comparison

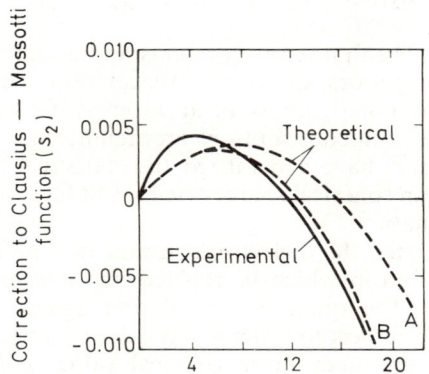

Figure 4. Comparison of the statistical-mechanical correction to the Clausius–Mossotti function (s_2) for argon (de Boer, van der Maesen and ten Seldam[4]) with experimental data (Michels and Botzen[33]). The two curves A and B are obtained by using alternative values for the parameters in the Lennard–Jones equation.

of the calculated curves with experimental data on argon shows significant deviations, although the general course of the curves is predicted correctly.

Buckingham and Pople[10] have extended the statistical theory by introducing a Lennard–Jones potential function and also by allowing for the contribution from dipole moments induced by the molecular quadrupole of neighboring molecules. For gases which have large quadrupole moments, such as carbon dioxide, the quadrupole-induced term becomes very important. Their expression for the 'second dielectric virial' coefficient B defined by

$$\{(\varepsilon - 1)/(\varepsilon + 2)\}(1/\rho) = A + B/p + \ldots \qquad (18)$$

is

$$B = \frac{N_A^2 \alpha^2}{18} \left\{ \frac{\alpha}{r_0^3} y^{-4} H^6(y) + \frac{1}{8} \frac{H^2}{\varepsilon_0 r_0^5} y^{-2} H^8(y) \right\} \qquad (19)$$

where H is the molecular quadrupole moment, ε_0 and r_0 the interaction parameters in the Lennard–Jones potential, and $y = 2(\varepsilon_0/kT)^{\frac{1}{2}}$.

An expansion of the Clausius–Mossotti function in a form similar to that given by Buckingham and Pople has been developed independently by Zwanzig[49] in a theoretical discussion of the effect of quadrupole moments on the dielectric constants of gases. The deviation from the function is found to be proportional to the square of the quadrupole moment, and is of the same order of magnitude as the deviation due to the variation of the polarizability with density (see section II.5).

Later papers by Buckingham[9] expand the dipole moment of a molecule in terms of the polarizability and hyperpolarizability c_μ. For large distances apart, an expression for the polarizability of a point dipole in the presence of its neighbors gives as the mean polarizability of two atoms

$$\alpha_1 + \alpha_2 = 2 \left\{ \alpha_0^3 + \left(2\alpha_0^3 + \frac{10 C_\mu U r_0^6}{9 x_0} \right) \right\} r^{-6} \qquad (20)$$

where $y = 2(U/kT)^{\frac{1}{2}}$. The first term represents additional moments induced in a molecule by its neighbors, and is the classical term: the other term arose from the fact that the polarizability of an atom is not independent of its neighbors. For argon, both contributions are significant.

Mandel and Mazur[32] have generalized the statistical theory to include specimens of other than spherical shape. Their result for non-polar molecules can be reduced to equation 13.

Van Vleck[46] calculated the dielectric constant of a non-polar fluid by a series expansion method in which he replaces the internal coordinates by harmonic oscillators. The quantum calculation agrees with Kirkwood's result for rigid spheres subject to a van der Waals attractive potential $-C/r^6$ provided C is given the quantum-mechanical value $+3\hbar\omega_0\alpha^2/4$. A more recent quantum-mechanical calculation has been carried out by Jansen[23].

4. Phenomenological Shell Model

The absorption spectra of condensed argon, krypton and xenon have recently been studied[2]. In each case, doublets have been observed close to the atomic resonance doublets. In the case of the solids, these may be interpreted in terms of Wannier excitons[18] with $r = 1$, built from a spin–orbit split valence band and the lowest conduction band. Calculation of the radius[42] suggests that they are only slightly greater than the interatomic spacings, hence the excitons may be considered to be tightly bound. Data for the polarizability and u.v. dispersion of argon, for example, indicate that the number of electrons that contribute to the dielectric properties is $N \approx 8.7$. Since the ground-state configuration of argon is $KL3s^2 3p^6$, this indicates that only the outer electrons contribute significantly, and thus a tight-binding model based on lowest-energy transitions only might be expected to describe the dielectric properties to a reasonable approximation. Since doublets close to the atomic resonance doublets are observed in absorption spectra for the liquid also, it is interesting to determine to what extent such an approach might be applied to fluids.

The most appropriate method is to adapt the semiphenomenological shell model used by Doniach and Huggins[14] for the solidified inert gases. In this method the atoms are represented by localized electron oscillators and are coupled via dipole–dipole and nearest-neighbor forces. The dielectric properties are assumed to result mainly from the lowest level u.v. band ($^3P_1 \to {}^1S_0$ transition), which is represented by a single density-dependent Frenkel exciton line. The Lorentz–Lorenz function is given by

$$\frac{3(n^2-1)}{(n^2+2)}\frac{1}{\rho} = \frac{\omega_p^2}{\omega_0^2 + V_0^{111}(T) - \omega^2} + \chi_1 = \chi_0 + \chi_1 \qquad (21)$$

where ω_p^2 represents the dipole–dipole interaction of the lowest atomic states, ω_0 is the natural frequency of the non-interacting atomic oscillators ($\omega_0 \gg \omega$), and $V_0^{111}(T)$ corresponds to the shift in ω_0^2 observed with increasing density and produced by the shell-core forces. The term χ_1, assumed to be frequency-independent, is introduced to take account of the core polarizability and possible contributions to the exchange integrals from higher atomic states. Comparing equation 21 with equation 1, if χ_1 is ignored, ω_p^2 is equivalent to $N_A \alpha$ and $(\omega_0^2 - \omega^2)$, modified slightly by $V_0^{111}(T)$, describes the dispersion.

Equation 21 has been found to describe the properties of the solidified inert gases within experimental error, and fitting procedures have been used to extract the most appropriate values for the parameters from the existing experimental data. Since the solids have the cubic symmetry required by the Lorentz model, comparison of these values with data for the fluids could allow the configurational contribution to the dielectric properties to be estimated.

5. *Variation of Polarizability with Density*

In the previous sections it has been assumed that the polarizability of a molecule is independent of density. Brown[7] has shown that use of the equilibrium values of the dipole moment is legitimate for the harmonic oscillator model of the internal degrees of freedom. A rigorous theory of molecular polarizability should, however, take into consideration the variation of α with density.

Strictly speaking, the deviation of α from the equilibrium value should be calculated using quantum-mechanical methods. For example, Jansen and Mazur[24] have carried out a third order perturbation calculation in order to evaluate the lowest order deviation. It is possible to express the result by giving a formula for the effective polarizability of an individual molecule as a function of the molecular positions; this is the quantity which, if substituted for α, gives the correct mean moments. Jansen and Mazur point out that from this point of view, the effect can be described as a variation of the polarizability of a molecule with distance from its neighbors, or with density; but from a more fundamental point of view, the polarizability of an individual molecule ceases to be a useful concept when the molecules are tightly coupled. Jansen and Mazur restricted their calculation to hydrogen and helium molecules. Jansen and Solem[25] have extended the work to include the heavy rare gases.

At high densities perturbation methods fail and approximate methods have to be used. Ten Seldam and de Groot[41] have calculated the variation of polarizability with density for the argon atom. They used as their model Jensen and Gombás's modification of the statistical Thomas–Fermi atom; and treated the atom as a quantum-mechanical system in which the wave function, instead of satisfying the usual conditions at $r = \infty$, is required to vanish on the surface of the finite sphere $r = a$.

An alternative approach is due to Keil[28], who has calculated the static polarizability α of an argon atom by estimating the contribution to α of each electron shell by means of a linear variational technique. The calculation confirms that about 95 per cent of the total polarizability is due to the $3p^6$ electrons, and most of the density dependence is due to variations in this contribution, as implied by the tight-binding model. Values for the change in polarizability calculated by this method are larger than those observed experimentally, but this is mainly due to the difficulty in estimating an accurate absolute value for the free polarizability.

6. Generalization of Theory to Optical Frequencies

For non-polar substances, the lowest frequency at which appreciable absorption occurs is usually in the visible or ultra-violet. For all frequencies ω which are less than ω_0 by a sufficient amount, the dielectric constant should be equal to the static dielectric constant ε_0 and should satisfy the Maxwell relation $\varepsilon = n^2$.

If, as is the case for argon, the electronic vibration frequencies are in the far ultra-violet (~ 1100 Å); then the refractive index measured in the visible region should not vary much with wavelength. A simple dispersion formula should be adequate to describe the variation and to extrapolate the refractive index to zero frequency.

Böttcher[5,6] has extended Lorentz and Onsager–Böttcher theory to the case where the applied field is the field of a light wave. Yvon[47] has likewise carried out a classical statistical-mechanical calculation. The quantum-mechanical calculation of Jansen and Mazur has been generalized to optical frequencies by Mazur and Mandel[32].

In each case the conclusion reached was that the dielectric constant formula is valid at optical frequencies, if ε is replaced by n^2 and α is expressed as a function of frequency.

III. Experimental Determinations of Dielectric Properties

Both refractive index and dielectric constant methods have been used to establish the dielectric properties of non-polar fluids. In general it has been found easier to obtain accurate absolute values using optical methods, particularly at higher densities, but there have also been some extremely accurate measurements of the dielectric constant of gases at low pressure using a capacitance bridge.

1. Methods for Determining Refractive Index

Several workers have used the spectrometric method of minimum deviation[3]. In this method, the sample is contained in a hollow cell with windows set at an angle to each other. The refractive index is deduced from

$$n = \sin\tfrac{1}{2}(A + D)/\sin\tfrac{1}{2}A$$

where A is the effective prism angle and D is the angle of minimum deviation.

Smith and co-workers[11, 20, 38, 42] have used a stainless-steel prism with sapphire windows to contain inert gas samples. A cell suitable for such a

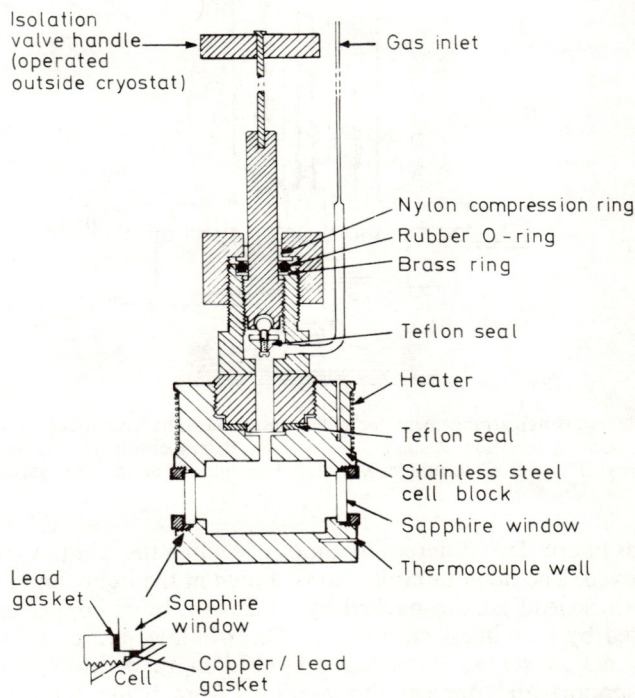

Figure 5. Optical cell for refractive index measurements; the cell is mounted inside a cryostat located at the axis of rotation of an optical spectrometer; the two windows are located at 45° to each other in the horizontal plane.

purpose[12] is shown in *Figure 5*. It consists of a stainless steel block in which two sapphire windows are set at an angle of about 45° to each other in the horizontal plane. The sample is thus confined in the shape of a prism. The cell is mounted in a cryostat, located at the axis of rotation of an optical spectrometer (*Figure 6*). The refractive index is found from measurements of the prism angle A and the angle of minimum deviation D.

The sapphire windows are 12 mm in diameter and 5 mm thick, cut with the c axis perpendicular to the face. They are polished to a specified flatness of 25 μm or better, the two faces of each window being parallel to within

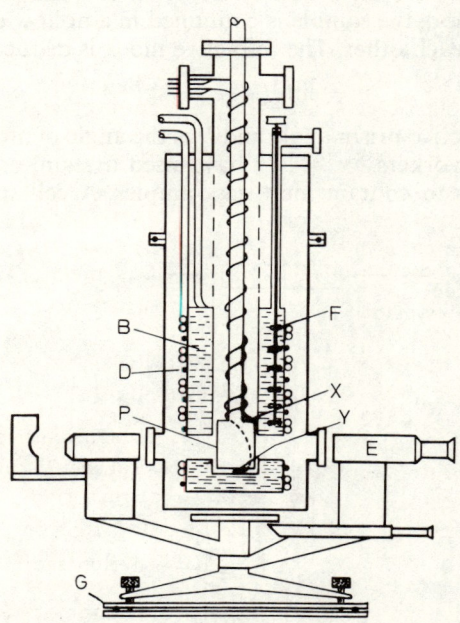

Figure 6. Experimental arrangement for determining the refractive index of a fluid by the method of minimum deviation: P, stainless-steel prism; B, temperature control bath; D, heaters; E, spectrometer; F, stirrer; G, turntable; X, Y, thermocouples. (Garside, Mølgaard and Smith[20]).

30 seconds of arc. Two different methods of sealing the windows to the block are employed. The demountable seal is shown in the figure and consists of a 0.25 mm thick lead gasket, backed by a 0.25 mm thick copper annulus and compressed by a stainless-steel screw. The windows are cushioned against the compression screws by means of 0.12 mm thick lead gaskets. These provide support and prevent the windows from being subjected to local stresses due to possible non-uniformity of the contact surface of the keeper.

This method of sealing windows was found to be completely satisfactory at pressures of up to 200 bar provided that the temperature range did not exceed about 40 K. However, thermal cycles of greater amplitude caused creep of the lead, and necessitated tightening the keeper at frequent intervals. For experiments covering a larger temperature range a different cell was used in which the windows were permanently sealed to the block with ATI Araldite epoxy resin. This enabled pressures of up to 150 bar to be studied over a temperature range of about 100 K. Teague and Pings[44] have used a similar cell in which the sapphire windows are brazed to the stainless steel.

The isolation valve is a modified form of a Hoke type 3000 valve with a Teflon needle. It is sealed to the cell block by means of a 0.5 mm thick Teflon washer. In general Teflon and other plastic materials are considered unsuitable for use as seals in systems to be used in critical work (Weinberger and Schneider[48]). However, Smith and Pings (1967 private communication) have found that provided that the Teflon is baked under vacuum at 373 K for 12 h prior to use and then handled with grease free implements it is unlikely to contaminate the system appreciably.

The experimental procedure is as follows. The cryostat containing the cell is maintained at a temperature corresponding to a known isotherm ($T \gg T_c$) of the fluid to be studied, and the cell filled with fluid, the pressure is varied until the density corresponds to a predetermined value. The cell is then isolated by slowly closing the needle valve. It is important to avoid perturbation of the sample as it is isolated. Density changes may be detected by making visual observations of the angle of minimum deviation during this procedure.

Once the sample has been isolated, measurements of D may be made at convenient temperature intervals. If the temperature approaches that of the coexistence curve and the sample density is not too close to the critical density then the two-phase region may be detected by a discontinuity in the slope of a graph in which minimum deviation is plotted against temperature. Close to the critical point ($T - T_c \lesssim 0.2$ K) gravitational effects become important and periods of up to several hours may be necessary for thermal equilibrium to be reached. By masking the cell with a narrow horizontal slit the gradient in the cell may be determined, allowing refractive indices to be measured to within about 1 mm of the two-phase interface.

The prism angle may be determined prior to the experiment by filling the prism with water and measuring the angle of minimum deviation as a function of pressure up to 100 bar. Chapman and Smith used experimental data for the pressure dependence of the refractive index by Rosen[40] to deduce the effective prism angle and also to investigate the possibility of prism distortion under pressure. The angle A may be determined to ± 30 seconds by this method. No significant (above 20 seconds of arc) deviation of A with pressure was found.

By use of a Bellingham and Stanley Ltd optical spectrometer the angle of minimum deviation may be measured with a relative accuracy to ± 10 seconds or an absolute accuracy to ± 30 seconds, thus determining the refractive index to ± 0.012 per cent. Relative measurements, for example the specific temperature dependence of the refractive index, may be determined under favorable conditions to ± 0.004 per cent. The effect of the temperature coefficient of expansion on the cell is to change its volume by 0.003 per cent/K. In practice this necessitates a small correction to the Lorentz–Lorenz function.

The cell described here has been successfully used to investigate the temperature dependence of xenon isochores over a temperature range of 70 K to within 0.1 K of the critical point (57.6 bar, 289.7 K). The Lorentz–Lorenz function is determined to ± 0.3 per cent, a factor of three better than has been found possible by other methods. The results are discussed in the next section.

Another important technique that has been used to establish dielectric properties via the refractive index is the interference method. For example, Edwards and co-workers[17] have used a Jamin interferometer in a series of experiments with helium. A copper cell is used, fitted with glass end windows and in contact with a liquid helium bath. The cell itself is filled with helium. Light is split by an interferometer mirror into two beams, one travelling directly through the cell, the other through a vacuum space. Interference fringes are seen as the two beams are recombined. The absolute value of the refractive index for helium vapor can be determined by counting the number of fringes that pass a reference point as the cell is evacuated. Absolute values of n for liquid helium can be made by placing an optical compensator in the comparator beam and locating the zero order of interference by observing white light fringes. Absolute values of n are obtained by these methods to a precision of about 0.1 per cent. In later modifications, the Jamin telescope was adapted by the addition of a beam splitter and photomultiplier tube so that the fringes could be both observed visually and also recorded on a chart-recorder.

One further optical technique that ought to be mentioned here is the use of refractive index observations to determine density gradients in systems. This method has been used mainly to study carbon dioxide[31] and assumes the validity of the Lorentz–Lorenz relationship. The sample is contained in a vertical cell and the refractive index is investigated as a function of height. In experiments close to the critical point, periods of several days are required in order to insure that the sample reaches equilibrium.

2. Results of Refractive Index Measurements

Extensive results exist for the refractive indices of fluid argon and xenon over a wide range of temperature and pressure using the spectrometric method of minimum deviation. For measurements involving the refractive

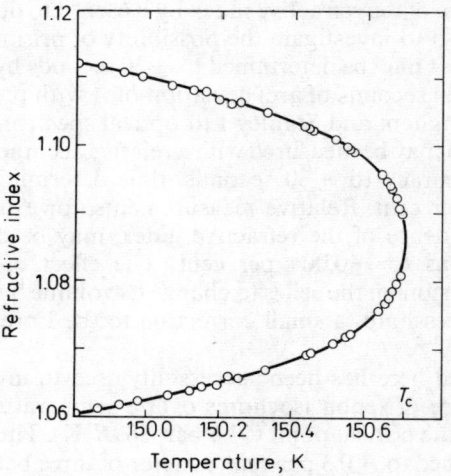

Figure 7. Refractive index of argon liquid and vapor in coexistence near the critical point (Teague and Pings[44]).

indices of gas and liquid states in coexistence, it is possible to load a specimen and bring the whole cryostat to a stable control point such that the gas/liquid interface is near the centre of the cell. When analysing the coexistence curve, differences in the values of $(n^2 - 1)/(n^2 + 2)$ for the gas and liquid phases are required. Since it is possible to measure the angles of minimum deviation of both phases at the same time, the error in the difference is only about one-half the normal error associated with two independent measurements.

Teague and Pings[44] have reported measurements for the coexistence curve of fluid argon from 85 K to 150.697 K and for isotherms between 133 K to 173 K, chosen to correspond with isotherms for which p–V–T data are available[34]. The results will be discussed in some detail since they provide a good illustration of many of the problems of interpretation of dielectric data.

The results for the coexistence curve are shown in *Figure 7* for values of temperature near the critical region. By graphical means the critical temperature was estimated to be $T_c = 150.709$ K and using the law of rectilinear diameters the critical refractive index is $n_c = 1.08587$. The shape of the coexistence curve in the neighborhood of the critical region is of considerable theoretical interest. If one assumes that the Lorentz–Lorenz theory is correct, values of the refractive index may be used to ascertain coexisting densities. Since the shape of the coexistence curve in the critical region[19] is given by a critical exponent β, defined by

$$\rho_1 - \rho_g = A(T_c - T)^\beta \tag{22}$$

then

$$\ln\left[\left(\frac{n^2-1}{n^2+2}\right)_1 - \left(\frac{n^2-1}{n^2+2}\right)_g\right] = A' + \beta \ln(T_c - T) \tag{23}$$

and β may be obtained from the slope of a ln–ln logarithmic plot. In the present case $\beta \approx 0.375$. However, as Teague and Pings point out, this analysis requires refinement in two respects. First, there is no obvious reason why a simple expression of the form given in equation 22 need hold. For example, higher order terms in $(T_c - T)$ could be required to fit the data. Secondly, the effective polarizabilities of the gas and liquid might vary significantly. Indeed, earlier work[1] on argon suggests that this might be so. However, both these possibilities can be taken into account by modifying equation 23 by including higher-order terms, with coefficients B and C which may be determined from the data, i.e.

$$\ln\left[\left(\frac{n^2-1}{n^2+2}\right)_1 - \left(\frac{n^2-1}{n^2+2}\right)_g\right] = A' + \beta \ln(T_c - T) + B(T_c - T) + C(T_c - T)^2 \tag{24}$$

Coefficients in equation 24 were evaluated by Teague and Pings using linear regression analysis for values of T_c in the range ± 0.01 K, in steps of 0.001 K. The striking result of this procedure was that the standard deviation of the fitted equation from the experimental points passed through a minimum when $T_c = 150.704$ K, a temperature which also corresponded to the minimum standard deviation in β and negligible values for B and C. Teague and Pings thus conclude that the best critical values are $T_c = 150.704$ K, $\beta = 0.364 \pm 0.007$ and $n_c = 1.08587 \pm 0.0001$.

The refractive index results for argon may be combined with density data for argon[34] to evaluate the Lorentz–Lorenz function. The major problem is that of matching the data. Because Levelt's critical temperature is higher than that found by Teague and Pings, the Lorentz–Lorenz function for the gas is found to increase rapidly as T_c is approached. Values of the Lorentz–Lorenz function for the liquid become smaller. However, if the two sets of measurements are adjusted to the same critical temperature, the values for the gas and liquid meet continuously at the critical point. Hence what might appear at first sight to be an anomaly is in fact probably a data-matching problem.

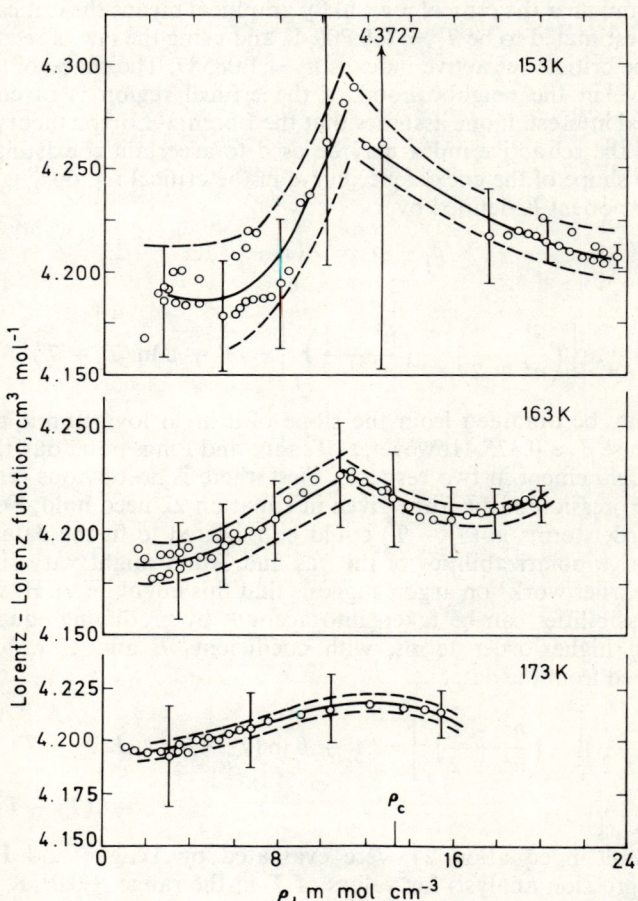

Figure 8. The variation of Lorentz–Lorenz function of argon for three gas-phase isotherms (Teague and Pings[44]).

RELATION OF THE DIELECTRIC CONSTANT AND THE REFRACTIVE INDEX

The same problem is encountered when interpreting refractive index results for argon isotherms. *Figure 8* shows three gas-phase isotherms. The large error bars are due to the data-matching problems mentioned above. Teague and Pings conclude that any anomaly in Lorentz–Lorenz function near the critical point is small[43], that the temperature dependence of the Lorentz–Lorenz function at constant density is also small and that there might be a small increase ($\lesssim$ 1.5 per cent) in the value of the Lorentz–Lorenz function of the liquid compared with that of the gas.

Probably the best test of the constancy of the Lorentz–Lorenz function to date has been the measurements on xenon by Parpia and Smith[38]. In this investigation an isochoric method was used, in which the sample cell was filled with xenon and isolated at a temperature $(T - T_c) \approx 10$ K. at which the density could be determined with great accuracy. Measurements of refractive index were then carried out on the isolated sample as a function of temperature using the spectrometric method of minimum deviation. This method has two advantages over conventional isothermal techniques, namely (a) the density and refractive index are determined on the same sample concurrently, thus greatly reducing data-matching problems, and (b) the specific temperature dependence of the Lorentz–Lorenz function is established, a property that may be more significant and useful from a theoretical viewpoint than isothermal data.

Measurements were made on 20 samples corresponding to densities in the range 1 to 15 k mol m^{-3} (*Figure 9*). Following isolation of the sample, the angle of minimum deviation was measured as a function of temperature from 299 K to slightly below the point at which the isochore reached the coexistence curve (*Figure 10*). As the coexistence curve was reached, the

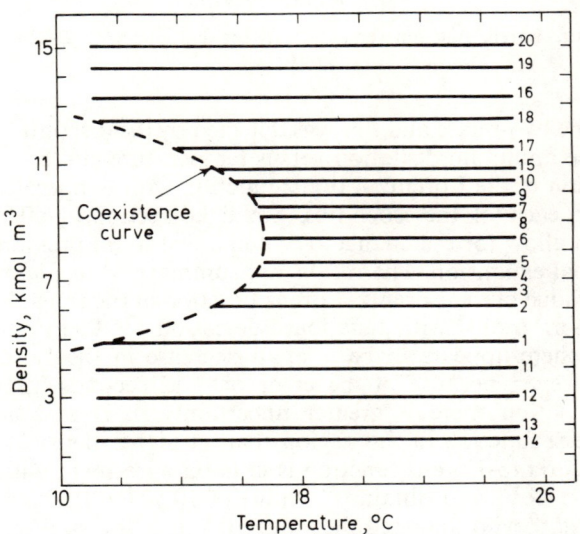

Figure 9. Isochores of xenon showing the temperature range on which refractive index measurements have been made (Parpia and Smith[38]).

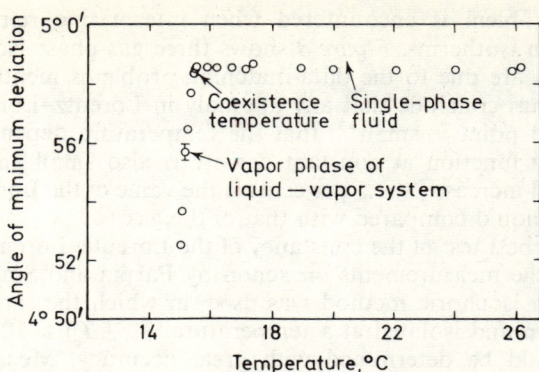

Figure 10. Angle of minimum deviation versus temperature for a typical isochore of fluid xenon (2 in *Figure 9*—density = 6.1 k mol m^{-3}) (Parpia and Smith[38]).

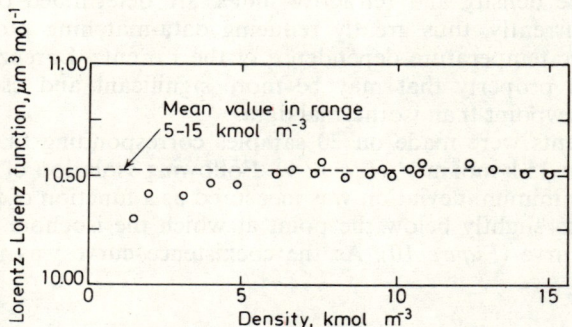

Figure 11. Lorentz–Lorenz function versus density for fluid xenon at 298.15 K. (Parpia and Smith[38]).

sample became two-phase and this was detected by the discontinuous change in slope of the minimum deviation versus temperature curve.

The variation of the Lorentz–Lorenz function with temperature was then evaluated for each of the isochores. Small corrections (<0.01 per cent) were made to allow for the change of volume due to (a) pressure distortion and (b) thermal expansion. The results are summarized in *Figure 11*.

The mean value of the Lorentz–Lorenz function in the range 5 to 15 k mol m^{-3} is 10.524 m^3 mol^{-1} with a standard deviation of 0.02 m^3 mol^{-1}. Below 5 k mol m^{-3} there appears to be a small decrease in the Lorentz–Lorenz function. However, because of the error of ±12 seconds in the angle of minimum deviation there is greater uncertainty in D and hence in the Lorentz–Lorenz function in this region than at higher densities. The mean value of the Lorentz–Lorenz function is in good agreement with the results of Chapman *et al.*[11] who obtained a value of 10.527 ± 0.02 m^3 mol^{-1} and of Garside *et al.*[20] who obtained the value 10.5 ± 0.1 m^3 mol^{-1}.

The average slopes of the Lorentz–Lorenz functions are plotted as a function of density in *Figure 12*. There is some indication that the slope

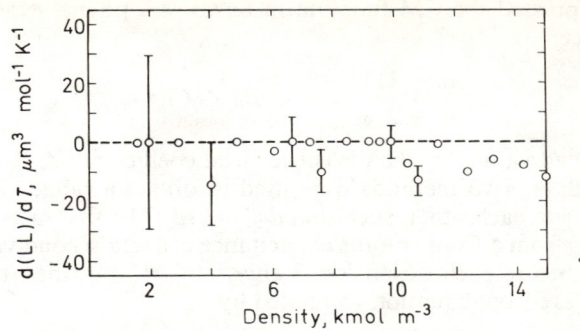

Figure 12. Average slope of the Lorentz–Lorenz function (d LL/dT) versus density for fluid xenon. (Parpia and Smith[38]).

becomes negative at higher densities, in accordance with expectations, but the change is not much greater than the estimated experimental error, and hence may not be significant.

Parpia and Smith therefore conclude that (a) it is unlikely that the Lorentz–Lorenz function for xenon varies by more than ± 0.5 per cent throughout the entire fluid range, (b) there is no evidence for the existence of a large anomaly near the critical point, and (c) the temperature dependence of the Lorentz–Lorenz function is small, except possibly at higher densities.

3. Methods for Determining Dielectric Constants

The problems and techniques associated with accurate measurement of the dielectric constant are well illustrated by considering the work of Cole and co-workers[26]. In this method the dielectric constant is measured by a transformer bridge method, in which the capacitance difference between two identically constructed capacitors, one containing the gas and the other evacuated, is balanced by calibrated variable capacitances or capacitance networks. By the use of closely coupled transformer windings as ratio arms, only direct capacitances between ungrounded electrodes of the cells are measured, as effects from shielded lead or other capacitances to earth are eliminated to a very high degree of approximation by the guard circuit action such ratio arms provide. Cole estimates that variations in residual capacitances with pressure have largely been responsible for much of the random and systematic deviations in resonance substitution methods used in several of the earlier investigations.

Both sample and reference cells contained parallel plate capacitors of similar design to a standard air capacitor. These were mounted in the pressure vessels in a way which insured that the parts determining the measured direct capacitances were subjected to hydrostatic pressure with a minimum of distortion from pressure-induced stresses. The cells were of stainless steel construction with quartz insulation. Teflon and soapstone washers were used as pressure seals for the connecting leads.

Cole expressed the CM function in terms of a power series of molecular density d, i.e.

$$\frac{(\varepsilon - 1)}{(\varepsilon + 2)}\frac{1}{d} = A + Bd + Cd^2 + \ldots \qquad (25)$$

and extracted values for the dielectric virial coefficients A, B, C etc. from the dielectric data. Two methods were used to obtain a range of densities d_i in fixed ratio r to each other, such that $d_{i+1} = rd_i$. The first involved successive expansions from a fixed volume capacitance cell to a second volume which is evacuated before each expansion. Values of CM after the ith and $(i + k)$th expansion are from equation 25 related by

$$\frac{CM_i}{CM_{i+k}} = r^{-k} + (r^{-k} - 1)\frac{B}{A^2}CM_i + (r^k - r^{-k})\left[\frac{C}{A^3} - \frac{B^2}{A^4}\right]CM_i^2 + \ldots \quad (26)$$

Using this procedure, a series of $2k$ densities is generated to give k independent ratios CM_i/CM_{i+k} and these are analysed in terms of equation 26 to determine the expansion ratio r and from it the dimensionless ratio B/A^2. Although this method has been employed by Cole quite successfully, there are difficulties associated with determining r to a few parts per million via a sequence of ten to fourteen measurements, because of the possibility of systematic errors or drift in the components.

In a second method Cole used three capacitance cells of identical construction, connected so that any two could be expanded to fill all three, and a cycle of expansions into each cell in turn is carried out. If the cell volumes are $V_A = V(1 + \delta_A)$, $V_B = V(1 + \delta_B)$ and $V_C = V(1 + \delta_C)$ respectively, where V is the average volume, then after three expansions the density ratio is

$$d_{i+3}/d_i = (\tfrac{2}{3})^3 \left[1 - \tfrac{1}{4}(\delta_A\delta_B + \delta_B\delta_C + \delta_C\delta_A) + \tfrac{1}{8}\delta_A\delta_B\delta_C\right] \qquad (27)$$

Since the linear term in the deltas vanishes by the condition $\delta_A + \delta_B + \delta_C = 0$, the expansion ratio is known to second order in advance, and analysis of the intermediate data gives the two independent values δ_A and δ_B via equation 26. The major advantage of this method is that the effects of systematic errors are to a great extent eliminated.

During the past decade, capacitance measuring techniques have increased the absolute accuracy of measurements on gases at low pressure by about an order of magnitude[15], and the precision of measuring small capacitance changes by several orders of magnitude. Another important development is the computable cross capacitor[45]. The advantage of this device is that the capacitance per unit length of a cross capacitor in the form of a cylindrical bore is unaffected by a symmetrical change in radial dimensions of the bars. Thus it is necessary only to consider the variation in length of a uniform bar with change in ambient pressure in order to determine the value of $\partial c/\partial p$ for such a configuration. This calculation is straightforward and leads to

$$\frac{1}{c}\left(\frac{\partial c}{\partial p}\right)\Delta p = \frac{\Delta l}{l} = \frac{\Delta p}{E}(1 - 2\sigma)$$

where E is Young's modulus, σ is Poisson's ratio and Δp is the pressure change. Because the variation $(1/c)(\partial c/\partial p)$ may be reliably computed, the increased accuracy and precision of capacitance measuring equipment may be fully exploited. Two additional advantages of the cross-capacitance configuration are that there is no unwanted supporting solid dielectric in the active field of the capacitor and possible contamination of the electrode surfaces by thin films of unknown permittivity has a negligible effect on the measured capacitance.

Many of the experimental determinations of the dielectric constant of fluids at higher densities have used relatively simple capacitors. In the experiments by Roach and Douglas[39] on helium for example, two flat horizontally mounted copper plates spaced 0.025 mm apart were used. The capacitor was mounted in a heavy copper cell filled with helium. The cell was submerged in a liquid helium bath, with accurate temperature regulation. Measurements were made on the liquid and vapor in coexistence by filling the cell so that the meniscus was either just above or just below the capacity gap.

4. Results of Dielectric Constant Measurements and Comparison with Refractive Index Data

Cole and co-workers determined the dielectric constants of helium, neon, argon and krypton as a function of relative density with a precision of about 10^{-6} at 49°C and at pressures up to 150 atmospheres. The first and second dielectric virial coefficients were estimated from data taken at pressures below seven atmospheres and are given in *Table 1*. The results of the low pressure runs for helium and krypton are shown in *Figure 13*, representing the gases with the smallest and largest polarizabilities respectively. The results for helium and neon show a small but consistent decrease

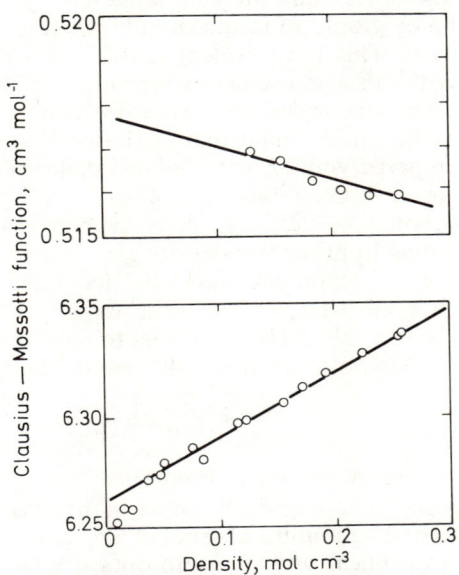

Figure 13. The Clausius–Mossotti function versus density for helium (above) and krypton (below) at 50°C. (Orcutt and Cole[26]).

of the Clausius–Mossotti function with increasing density whereas the Clausius–Mossotti functions for argon and krypton increase with density in this region.

Table 1. First and second dielectric virial coefficients as estimated from the dielectric constant measurements of Orcutt and Cole[26].

Gas	A (cm^3 mol^{-1})	B (cm^3 mol^{-1})
He	0.519 ± 0.001	−0.06 ± 0.04
Ne	0.998 ± 0.001	−0.30 ± 0.10
Ar	4.142 ± 0.002	−0.39 ± 0.20
Kr	6.267 ± 0.003	5.6 0.30

Significant decreases in both the Clausius–Mossotti[33] and the Lorentz–Lorenz[35] functions with increasing density at higher densities have been reported for argon. It is interesting to compare the value of the functions at the triple point with those of a low density gas. The value of n obtained for the triple point of argon by Sinnock and Smith is $n = 1.2334$ at 546.1 nm. This value is in agreement with an earlier measurement by Jones and Smith[27] to within 0.0002 (±0.02 per cent). Refractive index data due to Michels and Botzen[33] and Teague and Pings[44] may be extrapolated to the triple point from low density data extending to 0.90 ρ_{TP} and 0.96 ρ_{TP}, respectively. These, when adjusted to 546.1 nm, give values of 1.2316 and 1.2334. However, Michels and Botzen carried out their measurements at 298 K and thus their low value might indicate a slight temperature dependence of the refractive index. The dielectric constant determination by Amey and Cole, converted to 546.1 nm by graphical means, is equivalent to 1.2336, and thus also in good agreement. This is equivalent to $LL_{TP} = 4.1707$ cm^3 mole^{-1} and may be compared with a molecular polarizability of the free atom of 4.222 cm^3 mole^{-1} at the same wavelength, i.e. a decrease of 1.2 per cent.

Similarly, Sinnock and Smith found that for liquid krypton at the triple point $n = 1.3032$, compared with an extrapolated dielectric value of 1.298, equivalent to a Lorentz–Lorenz function of 6.45 cm^3 mol^{-1}. The low pressure gas value is about 6.46 cm^3 mole^{-1}, hence there is a decrease of less than 0.1 per cent on going from the low density gas value to the triple point. The equivalent values for xenon are $n = 1.392$ (cf. 1.385 from dielectric measurements) which gives $LL_{TP} = 10.475$ cm^3 mole^{-1} and is about 0.1 per cent lower than the gas value. There appears to be about a two per cent change in the Clausius–Mossotti function in helium in going from a gas to a liquid.

Magneto-optical Properties

The inert gas molecules are not optically active in the sense that they cause rotation of plane polarized light in the absence of a magnetic or electric field. However, small electro-optical and magneto-optical effects are observed and the study of these can provide useful additional information about molecular polarizabilities and in the condensed state, intermolecular forces.

The only property that has been systematically studied is the Faraday effect, i.e. the rotation of the plane of polarization of plane polarized light when it passes through a substance in the direction of an applied magnetic field. The degree of rotation is described by the Verdet constant V, which is the rotation per unit path length per unit field strength for a substance and is defined by

$$V = 2.98 \times 10^{-4} \, E \, dn/dE \text{ radians}$$

where E is the energy of the incident radiation.

The dependence of V on density can be calculated using a model very similar to the Lorentz local field model discussed in section II.1. The model leads to a quantity Ω which should be independent of state and hence is analogous to the Lorentz–Lorenz function:

$$\Omega = \{9n/(n^2 + 2)^2\} \{V/\rho\}$$

where n is the refractive index and ρ is the density. V/ρ is sometimes referred to as the specific rotation. Changes in Ω result from changes in the energies and strengths of the electronic transitions and also deviations from the Lorentz local field.

A careful investigation of the Verdet constant of the condensed inert gases has recently been carried out by H. V. Mølgaard[36], using a cryostat with the sample placed between the poles of a powerful electromagnet

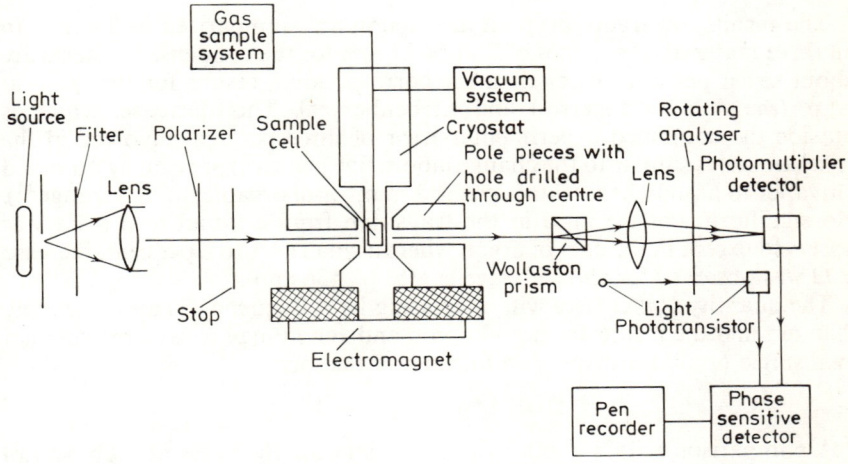

Figure 14. Apparatus used to determine the Faraday rotation in condensed inert gases. (Mølgaard[36]).

(*Figure 14*). Monochromatic light from a spectral lamp is made parallel by means of a lens and the beam is then carefully collimated. The light passes through a polarizing Nicol prism and is made plane polarized. It then passes through holes drilled in the electromagnet pole pieces and through the sample between them. Before reaching the analyser, the beam is split into two plane polarized components by a Wollaston prism. The analyser is a second Nicol prism, kept rotating at about 20 revolutions per second. The light signal is detected by a photomultiplier tube and the output fed to an electronic phase-sensitive detector. As the analyser rotates it passes alternately the two components from the Wollaston prism. A switching device, synchronized to the analyser, allows the phase-sensitive detector to receive and compare the magnitudes of the two signals.

The Wollaston prism is first set with the magnetic field at zero so that the two components are exactly equal and the phase-sensitive detector gives a small reading. When the magnetic field is applied, the small rotation of the plane of polarization causes the two components to differ by an amount indicated on the detector. The angle by which the Wollaston prism has to be rotated in order to restore the null value, recorded from an accurate circular scale in which it is mounted, is a direct measure of the Faraday rotation.

Table 2. Values of Ω for argon, krypton and xenon at 546.1 nm (10^{-1} min cm^2 gauss^{-1} mol^{-1}).

	Ω (gas) n.t.p.	Ω (liquid)	Ω (solid)
Argon	2.295	2.141 ± 0.04 (87 K)	2.113 ± 0.04 (81.8 K)
Krypton	4.70	4.50 ± 0.09 (119.2 K)	4.44 ± 0.09 (112 K)
Xenon	10.95	10.2 ± 0.2 (165 K)	10.14 ± 0.2 (159 K)

The results for argon, krypton and xenon are summarized in *Table 2*. In all three materials the values of Ω at 546.1 mm for the condensed systems are about seven per cent lower than the corresponding results for the gases at n.t.p. (the data of Ingersoll and Liebenberg[22]). This decrease, which is outside the combined experimental error of three per cent involved in the comparison, is similar to the change (about five per cent) previously observed for vapor to liquid phase transitions in a number of organic fluids (Grange[21]). No significant change in Ω in the transition from a liquid to a solid was observed except in the case of argon where a small (~ 1 to 2 per cent) increase in Ω was observed for shorter wavelengths (<436 mm).

The quantity Ω increases with increasing light frequency, ω more strongly than ω^2, and the results for liquid argon and xenon may be well represented by a single oscillator dispersion formula of the form

$$\Omega = Af\omega^2/(\omega_0^2 - \omega^2)^2$$

A comparison with a similar fit to the data for the gases at n.t.p. shows that the change in Ω between the gas and the condensed state appears in this type of fit as an increase in the single oscillator frequency and a decrease in the single oscillator strength, f. Though several independent sets of measurements were made on liquid krypton and the results have a form similar to those for the other materials, none of the sets of values of Ω are sufficiently self-consistent to allow accurate single oscillator parameters to be deduced.

The values of Ω appear to be nearly independent of temperature. In the liquid range, lying between the triple point and normal boiling point for each substance (~ 4 K), Ω changed by one to two per cent only. In the solid state Ω remained constant within experimental error between the triple points and the lowest temperatures reached (40 K below the triple point in the case of krypton and xenon, and 11 K below the triple point for argon). This temperature independence is in contrast to the quite strong temperature dependence shown by the values of Ω derived from the refractive index values of Sinnock and Smith using the Becquerel equation

$$V = (e/2mc^2)\omega \, dn/d\omega$$

The corresponding changes in the derived values of Ω at 546.1 mm for the liquids are three per cent for argon and 1.5 per cent for krypton and xenon. While the measured values of Ω in all three materials change by $\leqslant 1$ per cent in the transition from liquid to solid, values derived using the Becquerel equation show an increase of ten per cent for argon, a decrease of ten per cent for krypton and an increase of 1.5 per cent for xenon on solidification. Although strictly speaking the Becquerel equation applies only to a collection of free atoms and may not be expected to apply exactly to the condensed inert gases, in which the spin–orbit coupling is strong and transitions to the continuum are important, it seems more likely that the inconsistent differences in the derived values are due to uncertainties in the refractive index data. The uncertainty in refractive index is probably least in xenon, which of the three materials has the strongest dispersion. The derived increase of 1.5 per cent is compatible with the measured lack of change within the likely error (± 3 per cent). This very small change for the liquid to solid transition suggests that there are no large changes in the electronic states. This is supported by studies of the reflection spectrum of liquid and solid xenon near the triple point by Beaglehole, which showed that most of the important features of the solid spectrum persisted in the spectrum for the liquid.

The differences observed between the Lorentz–Lorenz functions for the condensed inert gases and the corresponding values for the gases at low densities, and also the changes observed with changing density may be described in terms of changes in the lowest excited states, the exciton states below the band gap (see section II.4). Although this may not be a totally valid approximation at optical frequencies, nevertheless the small changes in the Lorentz–Lorenz function and the relatively larger changes in Ω can be accounted for in terms of the observed or expected changes in the lowest states.

Although it is not possible to deduce the oscillator strengths from present absorption measurements on the condensed inert gases, one would expect a reduction of oscillator strengths from the free atom values of about five to ten per cent on theoretical grounds (Dexter and Knox[13]). Those of lower states which can be related to atomic states show small shifts to higher energy in the solids, e.g. about four per cent in argon and two per cent in krypton at 20 K. The effect of these changes on the polarizability and the Faraday effect can be estimated from using the known oscillator strengths for the free atoms by assuming that the distribution of oscillator strength over the various transitions is not very different in the condensed state from that in the gas state.

In argon, for example, a ten per cent drop in f values and a four per cent increase in energy of the oscillators leads to a drop of 1.8 per cent in the polarizability and 5.2 per cent in Ω, in good agreement with the observed decreases in the Lorentz–Lorenz function and in Ω from the condensed state to the gas values at n.t.p. This argument appears to be less successful in the case of xenon where the shift in energy of the lowest states is ~ 0.1 per cent. However, the observed changes in the Lorentz–Lorenz function and Ω can be well accounted for by a change in the continuum or band states amounting to a shift in effective energy of less than one per cent.

The Faraday effect is clearly more sensitive than the refractive index to the effect of interatomic interactions on the atomic properties, and the larger changes observed in Ω should help to establish the size and sign as well as the origins of the small changes in polarizability. The simple models which are currently used to describe the optical properties can correctly predict or be made to fit the sign and order of magnitude of the observed changes, but the approximations involved limit their physical meaning and likelihood of quantitative agreement between theory and experiment. Optical measurements made outside the region of absorption are difficult to interpret since a large number of electronic transitions are involved. However, as these measurements are extended into the ultra-violet region closer to the lowest transition energies, these transitions begin to dominate and the information becomes more specific. An extension of both the refractive index and Faraday rotation measurements further into the ultra-violet region should give more information on these lowest transitions and their changes with density and state. However, a more realistic theoretical treatment of the Faraday effect, which takes into account all the types of electronic transitions in the inert gases, will be required before these results can be interpreted in greater detail.

V. Conclusions

Measurements of the dielectric constant and refractive index are often used to deduce the variation of the density of a dielectric with temperature or pressure, particularly at extreme pressures or in the neighborhood of phase transitions where conventional density methods are difficult. Such techniques appear to be valid for non-polar materials, provided that appropriate corrections are made for the change in polarizability at higher densities. The specific influence of temperature appears to be very small.

The Lorentz–Lorenz and Clausius–Mossotti relations (equations 1 and 2) may be assumed to hold to within about ± 1 per cent throughout most of the fluid range, even to within ~ 1 K of the critical point. There is a small rise (e.g. krypton) or fall (e.g. helium) in the CM function at low density (~ 1 atm) with increasing density, presumably due to statistical fluctuations on the induced dipole moment, but at higher densities the CM and LL functions for all fluids show a decrease with increasing density. This is due to a reduction in the effective polarizability. Corrections for both effects may be made either empirically, using, for example, the Onsager–Böttcher relationships or by using the appropriate theoretical estimates. By combining dielectric data for the solids, in which the statistical effects may be assumed to be negligible, with results for the liquids, it should prove possible to distinguish between the statistical and polarizability contributions to the variation in the CM function.

Both dielectric constant and refractive index methods may be used to obtain results of high precision, although optical techniques are easier to apply to systems under pressure, and yield accurate absolute measurements with fewer corrections. In both optical and dielectric methods care must be taken to avoid distortion of components under pressure.

Magneto-optical measurements are still in their infancy. These, together with electro-optical measurements, should yield a great deal of information concerning the effects of local field on molecular properties.

VI. References

[1] Abbiss, C. P., C. M. Knobler, R. K. Teague and C. J. Pings. *J. Chem. Phys.* **42**, 4145 (1965).
[2] Baldini, G. *Phys. Rev.* **128**, 1562 (1962);
Schnepp, O. and K. Dressler. *J. Chem. Phys.* **33**, 49 (1960);
Bostanjoglo, O. and L. Schmidt. *Phys. Letters*, **22**, 130 (1966);
Steinberger, I. T. and O. Schnepp *Solid State Commun.* **5**, 417 (1967);
Beaglehole, D. *Phys. Rev. Letters*, **15**, 551 (1965).
[3] Baly, E. C. C. *Spectroscopy*. Longmans Green: New York (1929).
[4] De Boer, J., F. Van der Maesen and C. A. ten Seldam. *Physica*, **19**, 265 (1953).
[5] Böttcher, C. J. F. *Physica*, **9**, 937 (1942).
[6] Böttcher, C. J. F. *Theory of Electric Polarization*. Elsevier: Amsterdam (1952).
[7] Brown, W. F. *Handbuch der Physik*. **17**, 69 (1956).
[8] Brown, W. F. *Handbuch der Physik*. **17**, 74 (1956).
[9] Buckingham, A. D. *Trans. Faraday Soc.* **52**, 747 (1956).
[10] Buckingham, A. D. and J. A. Pople. *Trans. Faraday Soc.* **51**, 1029 (1955): *J. Chem. Phys.* **27**, 820 (1957).
[11] Chapman, J. A., P. C. Finnemore and B. L. Smith. *Phys. Rev. Letters*, **21**, 1306 (1968).
[12] Chapman, J. A. and B. L. Smith. *J. Sci. Instrum.* **2**, 100 (1969).
[13] Dexter, D. L. and R. S. Knox. *Excitons*. Interscience: Paris (1965).
[14] Doniach, S. and R. Huggins. *Phil. Mag.* **12**, 393 (1965).
[15] Dunn, A. F. *Canad. J. Phys.* **42**, 53 (1964).
[16] Eatwell, A. J. and G. O. Jones. *Phil. Mag.* **10**, 1059 (1964).
[17] Edwards, M. H. and W. C. Woodbury. *Phys. Rev.* **129**, 1911 (1963).
[18] Elliott, R. J. *Phys. Rev.* **108**, 1384 (1957).
[19] Fisher, M. E. *US Nat. Bur. Stand. Misc. Publ. No. 273*, 108 (1966).
[20] Garside, D. H., H. V. Mølgaard and B. L. Smith. *J. Phys. B*, **1**, 449 (1968).
[21] Grange, J. *Ann. Phys., Paris*, **3**, 48 (1959).
[22] Ingersoll, L. R. and D. H. Liebenberg. *J. Opt. Soc. Amer.* **46**, 538 (1956).
[23] Jansen, L. *Phys. Rev.* **112**, 434 (1958).
[24] Jansen, L. and P. Mazur. *Physica*, **21**, 193 (1955).
[25] Jansen, L. and J. Solem. *Phys. Rev.* **104**, 1291 (1956).
[26] Johnston, D. R., G. J. Oudemans and R. H. Cole. *J. Chem. Phys.* **33**, 1310 (1960);
Amey, R. L. and R. H. Cole. *J. Chem. Phys.* **40**, 146 (1964);
Orcutt R. H. and R. H. Cole. *Physica*, **31**, 1779 (1965).
Orcutt, R. H. and R. H. Cole. *J. Chem. Phys.* **46**, 697 (1967).
[27] Jones, G. O. and B. L. Smith. *Phil. Mag.* **5**, 355 (1960).
[28] Keil, T. H. *J. Chem. Phys.* **46**, 4404 (1967).
[29] Kirkwood, J. G. *J. Chem. Phys.* **4**, 592 (1936).
[30] Lorentz, H. A. *The Theory of Electrons*. Teubner: Liepzig (1909).
[31] Lorentzen, H. L. *Acta Chem. Scand.* **7**, 1336 (1953).
Straub, J. *Z. Tech. Chem.* **5**, 291 (1956).
[32] Mandel, M. and P. Mazur. *Physica*, **22**, 289 (1956).
[33] Michels, A. and A. Botzen. *Physica*, **15**, 769 (1949).
[34] Michels, A., J. M. Levelt and G. J. Wolkers. *Physica*, **24**, 769 (1958).
[35] Michels, A., C. A. ten Seldam and S. D. J. Overdijk. *Physica*, **17**, 781 (1951).

[36] Mølgaard, H. V. *D. Phil. Thesis.* University of Sussex (1971).
[37] Onsager, L. *J. Amer. Chem. Soc.* **58**, 1486 (1936).
[38] Parpia, D. Y. and B. L. Smith. *J. Phys. C.* **4**, 2254 (1971).
[39] Roach, P. R. and D. H. Douglas Jr. *Phys. Rev. Letters.* **19**, 287 (1967).
[40] Rosen, J. S. *J. Opt. Soc. Amer.* **37**, 832 (1947).
[41] ten Seldam, C. A. and S. R. de Groot. *Physica.* **18**, 905 (1952).
[42] Sinnock, A. C. and B. L. Smith. *Phys. Rev.* **181**, 1297 (1969).
[43] Taylor, L. S. *J. Math. Phys.* **4**, 824 (1963);
Larsen, S. Y., R. D. Mountain and R. Zwanzig. *J. Chem. Phys.* **42**, 2187 (1965).
[44] Teague, R. K. and C. J. Pings. *J. Chem. Phys.* **48**, 4973 (1968).
[45] Thompson, A. M. and D. G. Lampard. *Nature, London,* **177**, 888 (1956).
[46] van Vleck, J. H. *J. Chem. Phys.* **5**, 991 (1937).
[47] Yvon, J. *Recherche sur la Théorie Cinétique des Liquides.* Herman: Paris (1937).
[48] Weinberger, M. A. and W. G. Schneider. *Canad. J. Chem.* **30**, 422 (1952).
[49] Zwanzig, R. W. *J. Chem. Phys.* **25**, 211 (1956).

CHAPTER 13

VAPOR PRESSURES

D. Ambrose

Division of Chemical Standards, National Physical Laboratory, Teddington, Middlesex, UK

Contents

I.	Introduction	607
II.	Static Measurements	610
III.	The Isoteniscope and Related Methods	616
IV.	Static Measurements at Elevated Temperatures and Pressures	618
V.	The Critical Point	621
VI.	Effect of the Presence of Mercury	621
VII.	Vapor Pressures of Liquefied Gases	623
VIII.	Effect of Thermal Transpiration	624
IX.	Comparative Static Measurements	624
X.	Static Measurements at Very Low Pressures	625
XI.	Use of Radioactive Tracers	625
XII.	Ebulliometric Measurements	626
XIII.	Ebulliometric Measurements at Pressures below 2 kPa	633
XIV.	Method of Ramsay and Young	634
XV.	Dynamic Measurements without a Buffer Gas	635
XVI.	The Quasi-static Method	637
XVII.	Measurement of the Force Exerted by the Vapor	638
XVIII.	Evaporation Methods for Low Pressures	641
XIX.	Gas-saturation Method	645
XX.	Differential Thermal Analysis	648
XXI.	Gas Chromatography	649
XXII.	Mass Spectrometry	651
XXIII.	Vapor Pressures of Mixtures	651
XXIV.	References	652

I. Introduction

The pressure exerted by an unsaturated vapor is dependent upon the volume of the vapor and its temperature. If such a vapor, containing a single component only, is confined at constant temperature in a vessel of variable volume and the volume is decreased, the pressure will rise as the isotherm is traced until the vapor becomes saturated and there is a discontinuity of slope at the *dew-point*, where condensation begins. Further decrease in volume leads to further condensation at constant pressure, the *vapor-pressure*, until the last bubble of vapor is condensed at the *bubble-point* and the sample is entirely liquid. (Condensation must take place slowly so that the heat of condensation may be dissipated in the surrounding enclosure, as otherwise the temperature, and therefore the pressure, will rise). An alternative description of the behavior when two phases are present is that, in equilibrium conditions, the pressure exerted by the vapor above a pure liquid is dependent upon temperature only and is independent of the fraction of the liquid which has vaporized. If, however, the sample consists of more than one component, the pressure is not independent of the fraction vaporized since the more volatile components evaporate more readily than the less volatile components and the compositions of the vapor and liquid differ: in these circumstances the bubble-point pressure will be higher than the dew-point pressure (except in the circumstances that the overall composition is that of an *azeotrope*; the complications arising in azeotropic systems have been ignored in this discussion)[103]. Purity of the sample is therefore of great importance if the vapor pressure of a single substance is being measured while if a solution is being studied the compositions of the liquid and vapor phases enter as additional variables.

Change in composition may arise as a result of the vaporization process. For example, when ethanol evaporates at ambient temperature the gaseous molecules are of the same chemical species as those in the liquid whereas when bismuth telluride (Bi_2Te_3, m.pt 850 K) is heated to 900 K dissociation into the elements takes place and the gaseous species differ from those present in the coexisting liquid. In the second instance there is a chemical equilibrium to be considered, and the total pressure exerted is the sum of the partial pressures of all gaseous species. Inorganic substances vaporizing at high temperatures frequently behave in this way—see, for example, the discussion by Gilles[72]—and the specialist in these studies is always concerned with what molecular species are present in the gas phase not only because of the occurrence of chemical reactions but also because many of his measurements are made by effusion methods, and for evaluation of these the molecular weights of the effusing species, i.e. the chemical types and their polymeric state, must be known. Organic molecules when heated may break down irreversibly so that the vapor pressure is indeterminate but otherwise, in general, they vaporize at relatively low temperatures without chemical change, and it is not necessary for the exact molecular species to be known unless this is required by the experimental technique in use—in the example quoted above of ethanol, the question whether the molecules are monomers or polymers is of no importance in the methods which would normally be used for determination of the vapor pressure at ambient temperatures, but would become so if lower vapor pressures at subambient temperatures were being determined by Knudsen's method or by the transpiration method.

In this chapter only measurement of the vapor pressures of substances which vaporize without chemical change will be considered, and little will be said about measurements at temperatures greater than 800 K. Above this temperature, not only does the chemical equilibrium become paramount but also the factors affecting the experimental methods differ from those obtaining at lower temperatures—above 800 K thermal equilibrium is reached by radiation rather than by conduction and convection, special methods for the measurement of temperature are required, and a major part of any research may be the correct choice for materials of construction of the apparatus; whereas below 800 K temperatures may be routinely measured by platinum resistance-thermometers as well as by thermocouples and mercury-in-glass thermometers, and glass is almost always an acceptable material of construction unless metal apparatus is required for strength in experiments at elevated pressures, or for its superior thermal properties.

Another limit to the subject matter of this chapter is that it is mainly concerned with the determination of the vapor pressures of liquids; and the study of solids, which in general have lower vapor pressures, is not covered so completely. Much of what is said applies to solids, but solid + vapor systems do differ significantly from liquid + vapor systems in respect of the attainment of equilibrium within the condensed phase. A liquid may normally be assumed to be approximately uniform in composition if it is boiling or is stirred mechanically whereas no mixing can occur in a solid except by diffusion—a relatively slow process, the effect of which is negligible in a vapor-pressure experiment; if therefore a solid comprises more than one component and the components differ in volatility, evaporation will lead to a change in composition at the surface, which will differ from that obtaining in the body of the phase. Evaporation or condensation may also change the temperature of the surface of a solid sample to a greater extent than it will a liquid sample because the process of heat transfer within a solid, and to and from its surroundings, is slow.

Methods of vapor-pressure measurement can be broadly classified under two headings—static and dynamic—and dynamic methods may be further subdivided into those in which a steady state is set up and those in which there is transfer of the sample through the vapor phase from one part of the apparatus to another. In a dynamic method the conditions must approach as nearly as possible to those of a static equilibrium if we are to obtain a correct result. On the other hand, a perfect static equilibrium (which, on a molecular scale, is itself a dynamic process) cannot be achieved in reality. All static methods inherently have some dynamic features which are eliminated to a greater or lesser degree in different apparatuses, whereas in a method such as that in which the pressure in the equilibrium chamber is measured by means of a piston there is of necessity movement of the sample, but it is

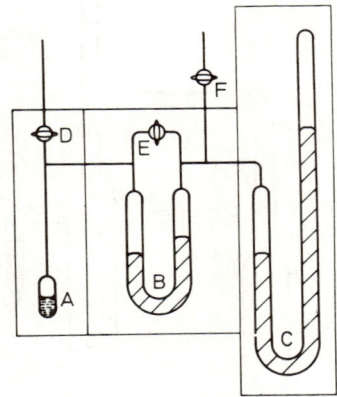

Figure 1. Schematic diagram of components of apparatus for static measurements. A: equilibrium chamber; B: null-manometer; C: measuring manometer; D, E, F: valves or stopcocks.

made as small as possible so that the determination is nominally static. In this chapter the broad differentiation between static and dynamic methods is maintained, but the detailed classification adopted is in accordance with the types of apparatus used (and their range of use) rather than strictly in accordance with their underlying principles.

The account given here is not encyclopaedic: for detailed discussion of measurements at high temperatures (and at low pressures) other sources should be consulted[102, 104, 113, 114, 126, 141], and, within its own limits, the present account is supplemented in some respects by several others[6, 45, 173, 174]. In particular, the present author has discussed elsewhere[6] the correlation of vapor pressure by equations, and the corrections arising from hydrostatic heads of gas or vapor that are applicable in vapor-pressure measurements. The discussion of the interrelation between vapor pressure and purity in another book published under the auspices of IUPAC, "*The Characterization of Chemical Purity*", is applicable, *mutatis mutandis*, in the present context[103].

II. Static Measurements

In the static method for vapor-pressure measurement a sample of the substance is confined in a vessel attached to a manometer and the pressure exerted by the vapor, when the vessel is maintained at a fixed temperature with both liquid and vapor phases present, is measured. The original method in which the sample is introduced into the evacuated space above the mercury in a barometer tube is now only suitable as a pedagogic demonstration and will not be discussed. The principle underlying modern equipment is illustrated in *Figure 1* where the apparatus is shown divided into three parts, an equilibrium chamber A, a null-manometer B, and a measuring manometer C: each of these is in an enclosure maintained at its appropriate temperature

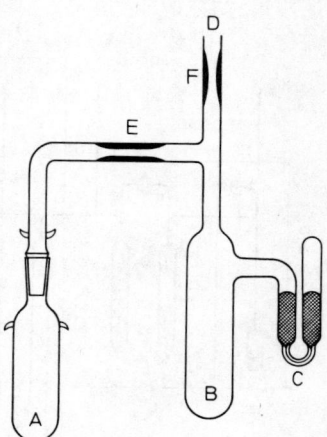

Figure 2. Apparatus for static measurements in the range 0.1 to 3kPa[145]. A: vessel in which sample is degassed; B: equilibrium vessel; C: mercury manometer; D: connection to vacuum; E, F: constrictions for sealing after sample has been distilled into B.

T_A, T_B, T_C. When the apparatus is set up the sample is charged into A, and all the air in the system must then be removed through the stopcock D before measurements can be made. Inclusion of the bypass stopcock E is desirable for convenience in manipulation of the apparatus.

The null-manometer B is not essential and the equilibrium vessel may be attached directly to the measuring manometer, but this apparently simpler arrangement has the disadvantage that T_C must then be at least as high as T_A. The inconvenience of building a thermostat containing a full-sized manometer for operation at temperatures much above ambient makes this method generally unsuitable for use except with a comparatively short manometer when equilibrium vessel and manometer may be immersed in the same thermostat bath. A simple apparatus of this type suitable for pressures between 0.1 and 3 kPa is shown in *Figure 2*[145]. If the null-manometer is used the pressure across it is balanced and transmitted to the measuring manometer by gas introduced through F (*Figure 1*), and either F or some other valve in the gas manifold must allow fine adjustment of the gas flow in and out so that the adjustment may be made slowly and delicately. With this arrangement the measuring manometer may be kept at room temperature and may be a standard instrument, e.g. a Kew-type (fixed cistern) barometer. The manometer in *Figure 2* may be used as a null-instrument if a connection is provided to the sealed limb. Apparatus incorporating mercury manometers requires thermostat baths provided with high-quality glass windows so that the mercury levels may be observed undisplaced by optical distortion.

A gas-operated pressure balance, in which the force exerted by the gas on a piston of known area is balanced by the addition of weights to the piston, may be substituted for the manometer C in *Figure 1*. The disadvantages of this type of instrument are that pressures below about 1.5 kPa cannot be measured directly, and that for measurements below atmospheric pressure the weights must be placed in an evacuated chamber. This means that they cannot be adjusted easily, and once the balance has been set up to measure a particular pressure the other variables in the experiment (temperature of the equilibrium vessel, condition of balance of the null-manometer) must be adjusted to fit that pressure. In respect of accuracy the pressure balance is probably superior to all but the most refined mercury manometers, and it is not difficult to envisage modification of some models so that at least minor adjustment to the weights applied could be made during the course of the experiment.

Frequently an elastic gage will be chosen in place of a mercury-filled U-tube as the null-manometer B. Many types of gage have been described which are based on elastic deformation—of a diaphragm[27, 166, 182], or of a curved tube[29, 111], or of a metal bellows[45, 77]—for use at pressures ranging from 10^{-3} Pa to beyond 10 MPa. These instruments can be made sensitive, but all need calibration, and their accuracy depends upon the absence of effects due to hysteresis. Elastic gages will not be discussed in detail here because instruments are now available commercially which will in many instances be superior in performance to any that can be made in the laboratory. Most of the commercial gages incorporate a metal diaphragm, but one of the most successful types is that in which the sensitive element is a helical quartz tube: the elastic properties of quartz are very good, and the hysteresis effects are small, so that this type of gage may be used for measurement as well as for null-detection, i.e. it can take the place of B and C in *Figure 1*. For example, Clarke and Glew have used this type of gage for measurement of the vapor pressures of deuterium sulfide and hydrogen sulfide through the range from about 30 kPa to 2.2 MPa (with two quartz tubes covering different ranges of pressure)[39].

Direct measurement of pressure below 100 Pa by means of a mercury manometer is difficult, and for low pressures the measuring manometer may be a McLeod gage—the type used, for example, by Martin and Biddiscombe for measurements on phenols in the range 1 to 100 Pa[115]. Advances in the manufacture of metal diaphragm gages suggest that these also may be useful for measurement of low pressures, i.e. to take the place of B and C, as well as for null-detection[176].

If the temperature of a mercury null-manometer changes, the zero positions of the menisci will rise or fall, but if the temperature is uniform their relative positions will be unchanged and the difference in pressure is obtained from the difference in height of the two menisci, corrected appropriately according to the density of the mercury. The zero of an elastic gage, however, will normally change with change in temperature, and the change can be determined by use of the valve E (*Figure 1*) which allows the two sides of the gage to be interconnected at the same pressure (as shown in *Figure 1*, this check would be done under vacuum with the temperature T_A sufficiently low for the vapor pressure of the sample to be negligible). Furthermore, the sensitivity of the gage will be affected by temperature changes and this will be important if the gage is being used for the actual measurement rather than as a null-instrument. Because of these effects it will be desirable to keep an elastic gage at a constant temperature greater than that at which any measurements are made, i.e. there will be two thermostats for A and B, and it will be necessary to insure that the tubing connecting the two parts of the apparatus is also kept at a temperature greater than T_A. An example of the use of a quartz-helix in this way above ambient temperature is provided by the work of Singh and Benson on the vapor pressures of mixtures of alcohols at temperatures up to 330 K[161]. In order to reach higher temperatures (450 K) for measurement of the vapor pressure of naphthalene, Fowler, Trump and Vogler used two quartz-helix gages, one at a high temperature as the null-instrument B and the other at ambient temperature as the measuring instrument C[64].

The key points about the static method are: (1) The measured pressure will be too high if air or other low-boiling component is present and too low if any high-boiling component is present. (2) In work above room temperature condensation will take place in the coolest part of the apparatus occupied by the vapor and it is therefore essential that all connecting tubing and the valves D and E are heated at least to the temperature T_A. (3) The method is called static, and the endeavor is made to determine equilibrium conditions. These will only be attained to the extent that the system is isolated from external disturbances. It may be possible to maintain a thermostat sufficiently constant for changes in its temperature to be insignificant, but volume changes are unavoidable when the pressure is balanced because manometers operate by movement, either of the mercury surface or of the diaphragm. Compression of the vapor leads to condensation and a rise in the temperature of the sample, expansion to evaporation and a fall in temperature. These excursions must be reduced to a minimum by careful balancing with progressively smaller adjustments over a period of time if equilibrium is to be approached. The difficulty of reaching and preserving equilibrium will be increased if a mercury manometer of very large diameter is used (as is desirable for exact measurement of the pressure); the volume change of a diaphragm gage may also be appreciable, and in this respect the quartz-helix gage may be advantageous because the volume change required to operate it is very small.

Some authors[2, 161] have recommended use of a stirrer in the sample vessel A as an aid to reaching equilibrium, and this will be particularly desirable when mixtures rather than single substances are being studied. The attainment of reproducible equilibrium conditions is especially difficult when two condensed phases are present, for example, if the vapor pressure of a saturated solution or the triple-point pressure of a pure substance is being measured, and stirring may then be essential[3]. (The remark about the triple point applies to the apparatus typified by *Figure 1* where the sample is small, not to a triple-point cell containing water such as is used for realization of the temperature 273.16°K[18] where the sample is large and the external temperature is easily controlled by surrounding the cell with melting ice.) For a solid condensed phase, approach to equilibrium via condensation of the sample rather than via evaporation has been recommended[3, 120a].

Degassing of the sample is frequently carried out by freezing it, pumping away the non-condensable gases, allowing it to melt, and then repeating the sequence of operations, the number of repetitions being determined by the operator's experience or exhaustion—if vapor pressure measurements are made at one temperature after each sequence the value obtained after an initial fall should eventually become constant at the true value. Provision must be made for this operation either by making the thermostat round A removable or by making the equilibrium vessel with its attached manometer B detachable from the remainder of the apparatus. It is difficult to insure that all the air is removed, and since careful observations must take time the possibility of leakage of air into the apparatus must be considered; it is therefore advantageous if this can be checked during the course of the experiments by repetition of the initial pumping.

It is the writer's belief that procedures for degassing which include distillation or sublimation are always more effective than that just described. For example, in one apparatus[25, 82, 148a] the frozen sample is allowed to sublime on to a cold finger above the sample bulb while the system is pumped; the authors recommend that conditions (i.e. the temperatures of the sample and of the cold finger) should be adjusted so that a sample of 40 cm^3 takes from 1 h to 2 h to sublime, and the sublimate remains solid at all times— because if the sublimate melted they found some of the gas redissolved. One sublimation is normally adequate but, if it is not, the cold finger is allowed to warm so that the sample runs back into the sample bulb and is then ready for repetition of the process. Loss from the sample may be minimized if the pumping line passes through a conventional cold trap from which any material carried over and condensed may be distilled back when the pumping has been discontinued. In another apparatus[50] the sample bulb incorporates a condenser, and immediately above this a tap, so that the sample may be degassed by boiling under reflux at atmospheric pressure; if the condenser is then allowed to warm, the rising vapor expels the remaining air from the sample container, and the tap is shut (with simultaneous discontinuance of the heating) at the moment the condensing ring reaches it. However, it does not seem that this method will lead to elimination of the last trace of air as easily as the other method just described. Degassing of liquids of high viscosity and low volatility has been discussed by Battino and Evans[19a].

Absence of other impurities to the degree required will be demonstrated if the pressure can be measured (and is found to be unchanged) with different vapor volumes so that the conditions of the sample range from as near to the dewpoint to as near to the bubble-point as possible, i.e. from it being nearly all in the vapor phase to being nearly all in the liquid phase. This is standard practice in the apparatus to be described in section IV for measurements at pressures above atmospheric but is not practicable in the apparatus shown in *Figure 1* (although it is possible to arrange for some variation in the volume available to the sample by displacing mercury into the equilibrium vessel). An equivalent method is to distill the sample and to verify that the vapor pressures of the first and last fractions are the same—for example, Shepherd distilled gases isothermally for this purpose in the vapor pressure apparatus itself, a method which avoids the possibility of contamination of the fractions during transfer from one apparatus to another[159].

The last paragraph has been written on the assumption that no azeotrope is formed with impurities present—if one is formed and the sample being examined is of the azeotropic composition (as it could be if purification has been carried out by fractional distillation) the presence of impurity will not be revealed by this test. Although azeotropic composition changes with temperature—and consequently the sample might not pass the test if it were repeated at another temperature—the change in composition is small, and the sensitivity of detection of an azeotrope-forming impurity will usually be low.

Absence of volatile impurities is particularly important for measurements at very low pressures. In many instances, the relative volatility of two volatile substances increases as the temperature is lowered, i.e. their vapor-pressure curves, obtained when $\log p$ is plotted against $1/T$, are more widely separated at low temperatures than they are at higher temperatures; consequently, any impurities present which are more volatile than the substance under study may have a much larger effect at low pressures than they do at higher pressures. For example, Hopke and Sears made measurements with a Rodebush manometer (to be described in section XVII) on trimethylbenzenes in the range 1.3 to 133 Pa, and found it necessary to pump the samples for 36 h while they were maintained at $-78°C$ before steady readings could be obtained; they made calculations showing that an observable effect would be caused by the presence of only 10 p.p.m. of a volatile impurity[88]. For such measurements greater care will be needed in insuring that air is removed from the apparatus and preliminary baking out may be desirable. On the other hand, Balson reported no special difficulties of this nature in his work on bromobenzyl cyanide at considerably lower pressures[15].

VAPOR PRESSURES

For measurements on aniline in the pressure range 0.2 to 3 kPa Röck has described a simple apparatus (*Figure 3*)[145]. The degassed sample was sealed in a U-tube with two vertical limbs 12 cm high joined by a horizontal tube of 2 mm bore. The two limbs were surrounded by jackets through which liquids were circulated, one at the temperature at which the vapor pressure was required and one at a temperature at which the vapor pressure was very

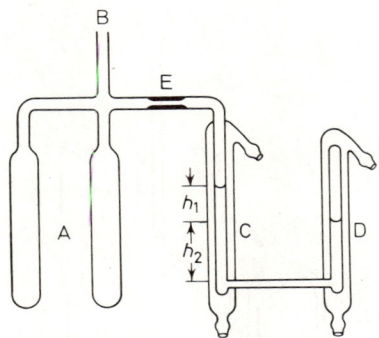

Figure 3. Apparatus for static measurements in the range 0.05 to 1 kPa[145]. A: vessels for introduction and degassing of sample; B: connection to vacuum; C: manometer arm kept at low temperature t_1; D: manometer arm kept at temperature of measurement t_2; E: constriction for sealing after sample has been distilled into manometer. If the vapor pressure at t_1 is negligible that at t_2 is given by the expression $p = \rho_1 g h_1 + (\rho_1 - \rho_2) g h_2$ where ρ_1, ρ_2 are the liquid densities at t_1, t_2.

low. The difference between the vapor pressures at the two temperatures was, therefore, obtained in terms of the height of a column of the aniline itself. The vapor pressure at the lower temperature, which was a small correction to be added to the difference, had in this investigation been measured by another method. Hackspill first used this type of apparatus for measurement of the vapor pressures of the alkali metals[78]; in this instance the two limbs were identical so that the one first heated could be interchanged with the other which had been kept relatively cool, i.e. at just above the melting point of the metal, and check made whether there had been any outgassing of the glass at the higher temperature. Conduction of heat between the two limbs will be one of the factors which limit the accuracy attainable.

Röck also described the application in the range 7 to 64 Pa—again to aniline—of a different type of static measurement[145]. If a bulb containing the sample is connected to a relatively large evacuated vessel of known volume and the system is allowed to come to equilibrium, assay of the amount vaporized will allow calculation of the vapor pressure from the gas laws provided the effect of adsorption can be ignored. The method is simple in principle but it has been little used, and only in exceptional instances would it seem to offer advantages in comparison with the gas-saturation method to be described in section XIX.

III. The Isoteniscope and Related Methods

One of the simplest ways of applying the static method is by the use of the *isoteniscope* in which, as in the method just described, the liquid under study itself acts as a manometric fluid. The original design of Smith and Menzies[163] and a modification used by the writer are shown in *Figure 4*. The isoteniscope

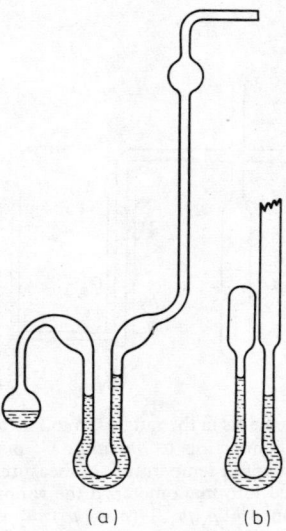

Figure 4. Two designs of isoteniscope. (a) Smith and Menzies[163]; (b) Ambrose.

is charged with the sample, and is then attached to a manifold with a gas reservoir in which the pressure of an inert gas can be adjusted and maintained constant. The isoteniscope is placed in a thermostat, and the conditions of temperature and pressure are adjusted so that the sample boils: the air is gradually carried out of the closed side of the isoteniscope by the vapor, and when it is judged by the operator that all the air has been removed the pressure is raised slowly or the temperature is lowered, whichever is the more convenient, until boiling ceases and the levels of the liquid in the two limbs are approximately equal—at this point the vapor pressure of the same exerted in the sealed limb is equal to the gas pressure applied to the system plus any hydrostatic head due to the difference in the two levels of liquid. (The two limbs of the apparatus must be of the same diameter so that the effects of capillarity are equal and cancel one another.) The boiling-out process and observation are then repeated until constancy of the result obtained confirms that in fact all the air has been expelled. A new temperature is then selected and the process of boiling and observation is repeated—it is simpler to start afresh each time, unless successive temperatures are to be very close together, than to attempt to adjust the rates of change of temperature and pressure so that no gas is forced round into the closed limb. Smith and Menzies found it essential to stir the thermostat bath very vigorously, but it is not clear whether the object of this vigor was achievement of uniformity in the temperature or the resultant vibration of the isoteniscope and agitation of the sample. Good results can be obtained by the method but it is laborious if a precision comparable to that easily obtained by ebulliometric methods is to be achieved.

A modified form of isoteniscope has also been used in which the manometric fluid is not the liquid itself but mercury[30]. Since use of mercury entails about a 15-fold loss of sensitivity in measurement of the hydrostatic head in comparison with most organic liquids, it does not seem advantageous unless there is a special reason for its adoption. For example, if mercury is used as the confining fluid in the manometric limbs the method may be adapted for measurement of the vapor pressures of solids.

If two isoteniscopes, one of which contains water and each of which is in a separately controlled bath, are both balanced against the same pressure by adjustment of the temperatures of the baths, measurements may be made in the comparative manner recommended below in section XII and exact determination of the pressure applied is then not necessary (the two isoteniscopes should be guarded by cold traps in the connecting lines to the pressure main so that there is no cross-contamination of the samples). However, a simpler way of making comparative measurements is to follow the example of Goncharov and Karapet'yants who used an isoteniscope containing the substance under study and, in place of a second isoteniscope, an ebulliometer containing a reference substance for determination of the pressure[73]. This mode of operation makes use of one of the advantages of the isoteniscope—that no more than 2 ml of the substance is required—but it lacks the symmetry between the two halves of the apparatus which is a valuable feature of the comparative method as normally applied in ebulliometry.

Of the methods available for determination of the vapor pressures of very small samples, refinement of the one first proposed by Siwoloboff[162] for atmospheric boiling points—which has a close resemblance in principle to the isoteniscope—seems to offer good possibility of achieving fair accuracy. A sealed capillary tube is inverted in a small quantity of the liquid in a test tube to form a focus for the formation of bubbles. The test tube is heated in a stirred bath to the boiling temperature of the sample, and after a period of boiling during which the air trapped in the capillary tube is displaced by vapor, the temperature at which boiling just ceases, i.e. the temperature at which the liquid begins to be drawn into the capillary tube, is determined. Sunner and Magnusson[169] described apparatus based on this method with which they were able to detect differences in condensation temperatures of 0.001 K between closely related samples. e.g. between successive fractions obtained by distillation. For the actual temperature, as distinct from temperature differences between samples, the method suffers from the disadvantage that bubble formation is affected to an unknown extent by surface tension, but determinations can be made with samples as small as 0.1 ml and it is unlikely that any other method which has been used at that scale is more accurate. However, there seems no reason why an isoteniscope should not be constructed on a micro-scale, and this would have the advantage over Siwoloboff's method that no problem arises for operation at a series of pressures since boiling will always begin in the isoteniscope, if it is suitably designed, as soon as the temperature is high enough. With the inverted sealed capillary tube used at atmospheric pressure or below, on the other hand, once the tube has filled with degassed liquid and the final bubble has collapsed, it is virtually impossible to proceed with further experiments without dismantling the apparatus so that the liquid can be shaken out of the tube.

The range of application of the inverted sealed capillary apparatus has been extended by Silva, Johnson and Cubicciotti who described apparatus intended for use at temperatures up to 1 250 K, which they tested by making measurements on carbon tetrachloride at pressures above atmospheric extending up to the critical pressure (corresponding to a temperature of 556 K)[160]. At these elevated pressures no problem arises in restarting the boiling, as a bubble will form in the tube as soon as the pressure is reduced a little below the equilibrium value.

Other micro-methods (some of them dynamic, i.e. boiling-point measurements) are reviewed by Thomson and Douslin[173, 174].

IV. Static Measurements at Elevated Temperatures and Pressures

A special form of apparatus suitable for use at pressures above atmospheric and temperatures above ambient is that originally used by Andrews for his work on carbon dioxide[13]. It was subsequently developed by Hannay and Hogarth[80, 81] and by Ramsay and Young[139, 191] (interesting sidelights on the history of this apparatus are given by Flint[62]), and in principle is unchanged in the modern apparatus used by Kay[93, 94] and by Connolly and Kandalic[44]. It allows determination both of the vapor pressure of the liquid and of the pressure–volume–temperature relationships of the liquid and the vapor, including the values of the three variables at the critical point.

The sample is confined over mercury in a thick-walled glass tube sealed at its upper end, and the tube is attached to a steel U-tube containing mercury; the pressure is transmitted through the mercury to the pressure gage and a piston on a screw allows the level of the mercury, and hence the volume occupied by the sample, to be adjusted. The experimental tube is heated by means of a vapor jacket. (Alternatively, the glass tube itself may be bent into the form of a U, often then known as a Cailletet tube, and the apparatus is heated by immersion in a normal thermostat bath[31].) The differences between the original designs and that used today, for example, by Kay lie in the details of the methods used for sealing the glass tube to the steel U-tube, in the methods of measuring pressure (by Bourdon or piston gage in place of the tube containing air on which Boyle's law measurements were made by the early experimenters), in the availability of glass tubing of uniform bore which simplifies and improves the volumetric measurements often made at the time, and, of course, in greatly improved methods of temperature measurement. Kay's apparatus has also always included in the glass tube a steel ball which may be moved up and down by a magnet in order to stir the sample. The glass tube is sealed to the metal U-tube by means of a union incorporating an unsupported-area seal of the type shown in *Figure 5*[7]. An O-ring seal was used by Connolly and Kandalic whose apparatus is illustrated in *Figure 6*[44]. Air was removed from this apparatus by freezing the sample in the tip of the glass tube with liquid nitrogen so that the tube could be evacuated, and the

VAPOR PRESSURES

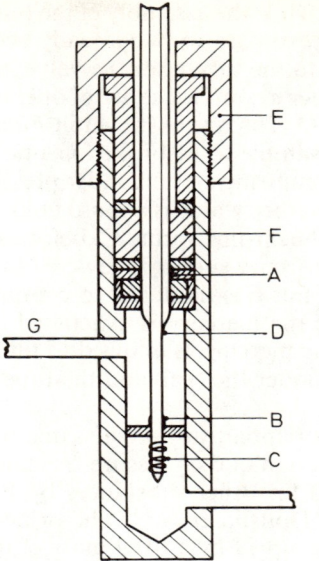

Figure 5. Union for joining a sealed tube to metal apparatus and breaking *in situ*[7]. A, B: swellings on glass tube; C: Nichrome spring; D: cut in glass; E: nut; F: elastomer sealing ring; G: connection to valve and vacuum system.

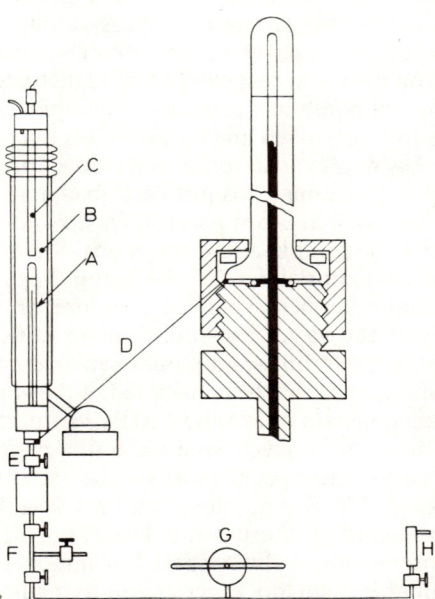

Figure 6. Apparatus for measurement of vapor pressures greater than atmospheric pressure by the static method[44]. A: experimental tube; B: vapor jacket; C: thermometer; D: O-ring sealed glass-to-metal union; E: charging valve; F: valves; G: mercury injector; H: mercury/oil interface.

sample was kept frozen while the assembly of the tube and valve was mounted as shown. The rest of the apparatus had already been charged with mercury, which was propelled into the tube after the valve had been opened. Air may also be removed by rather a complex series of operations in which the sample is boiled in the tube[93, 94]. The union shown in *Figure 5* allows a sealed tube containing a degassed sample to be broken *in situ*. The swelling of the glass tube at A prevents it from being ejected under pressure, and another swelling at B serves to locate a washer which is a loose fit in the outer steel tube and is held in position by the Nichrome spring C. Before assembly, a cut is made in the glass at D and the mercury surface is lowered to the level of C; the tube is put in position and the nut E is tightened to compress the elastomer sealing ring F. The space over the mercury is evacuated through valve G and the valve is then closed. The mercury is allowed to fill the evacuated space (any residual air is trapped under the seal) and the tube is broken at D by gentle rocking in the seal.

An alternative form of apparatus for the same purposes is that originally developed for determination of the pressure–volume–temperature properties of water, but later used for other substances by Beattie and Edwards[21, 23] and, more recently, by Douslin *et al.*[54]. The principle of the method is the same, i.e. the volume occupied by a fixed mass of the substance is varied by means of a screw pressing a plunger into a reservoir containing mercury so that the variation of pressure with volume at a series of fixed temperatures may be found, but the apparatus is made wholly of metal, and the volumetric measurements depend upon calibration of the screw and not upon visual observation of the position of the mercury. The size of the sample container is no longer limited by the strength of glass tubing, and a larger sample may be used—for this reason, the results obtained with this apparatus may probably (but by no means necessarily) be of higher accuracy. The sample is contained in the metal bomb in a glass liner, and the experimental procedure allows the sample to be degassed and sealed in this liner, which is then broken *in situ* after assembly and evacuation of the whole apparatus.

Since the two pieces of apparatus just described may be used for the determination of pressure–volume–temperature properties as well as of vapor pressure, the size of the sample and dimensions of the sample container are chosen so that total vaporization of the sample can be effected and the properties of the vapor studied. There is, therefore, normally no difficulty in determining whether the vapor pressure remains constant on compression from the dewpoint to the bubble-point, but in an assessment of the results the temperature requirements must be considered: if the vapor pressure of a pure sample is to remain constant at around 1 MPa to 1 part in 5 000, this implies that the temperatures at the levels where the first condensation occurs and where the final bubble disappears must be the same within about 0.01 K and, for example, at 500 K and upwards, this is a high performance to demand of the surrounding thermostat. The vapor jacket used in the first apparatus is advantageous in this respect but imposes a lower limit on the temperature attainable than do other methods of heating the apparatus.

Disadvantages of this method are that the initial degassing is final, and that any products of the decomposition which may occur at the higher temperatures cannot be removed.

V. The Critical Point

The vapor-pressure curve terminates at the critical point. The definition and identification of this point are of interest both practically and theoretically, and the critical region has been the subject of much study because at no time in the past hundred years has the behavior of fluids in the critical region been free of anomalies according to the then current theories. No attempt is made to survey the subject here and for further study the reader is directed to accounts of the early difficulties[37,110], of the practical investigations up to the present time[105], of modern theoretical ideas about the critical state[157], and to the book by Rowlinson, *Liquids and Liquid Mixtures*[153]. It is clear now that the grosser anomalies once thought to occur in the critical region were only anomalies because there was a lack of understanding of the effect of gravity—which causes a measurable change in density over the height of a column of fluid at temperatures near the critical because the fluid is at this point highly compressible—and of the long time required for equilibrium to be reached when the two co-existing phases have almost the same properties.

The determination of the critical temperature is most simply carried out by observation of the disappearance and reappearance of the meniscus between the liquid and vapor phases when a suitable amount of the substance is confined in a strong sealed glass tube[101]. If the measurement is made with apparatus of the type shown in *Figure 6* then the critical pressure may be determined in the same experiment. When the investigation is carried out in metal apparatus, on the other hand, the disappearance of the meniscus cannot be observed and the critical temperature must be found by analysis of the inflections of the isotherms—it is the temperature at which $(\partial p/\partial V)_T = 0$, and $(\partial^2 p/\partial V^2)_T = 0$, where p is the pressure, V the volume and T the temperature, and the corresponding pressure is the critical pressure. If, as would be usual, determinations of the critical values are combined with measurements of the vapor pressure at lower temperatures then the observed critical pressure should fall at the end of a smooth curve obtained when $\log p$ is plotted against $1/T$.

VI. Effect of the Presence of Mercury

When mercury is present in the apparatus consideration must be given to the effect its vapor may have on the apparent vapor pressure. In *Figure 1*, mercury is present in all parts of the apparatus and its vapor does not affect the balance of the null-manometer, and unless the temperature T_B very much exceeds T_C any correction for the hydrostatic heads of mercury vapor in the right-hand limb of B and the left-hand limb of C will be negligible. On the other hand, an apparatus used by Cruickshank and Cutler incorporated a mercury-sealed cut-off on the sample side of an elastic gage[48]. There was, therefore, mercury on one side of the null-manometer at the temperature of measurement but not on the other; the mercury vapor pressure, therefore, contributed to the pressure acting on the gage and had to be subtracted from the measured value. This procedure is satisfactory provided the mercury vapor pressure is small in comparison with that being measured, but it seems undesirable when low vapor pressures are being measured at high temperatures because the correction may then be an appreciable fraction of the quantity being measured.

At the higher temperatures and pressures prevailing in Kay's apparatus[93,94] and in Beattie's apparatus[21] an additional factor enters into the account because vapor pressure is increased by external pressure. Thermodynamic argument leads to the well known Poynting equation[137,138] which relates the increase in vapor pressure to the applied pressure as follows.

For equilibrium between the liquid (L) and gaseous (G) phases the chemical potential μ of the mercury must be the same in both phases. For mercury under its own vapor pressure,

$$\mu(L, p^*) = \mu(G, p^*, 1) \tag{1}$$

where p^* is the vapor pressure and the number 1 on the RHS is the mole fraction of mercury in the vapor. If now a second component is introduced so that the pressure rises to p and the mole fraction of mercury in the gas phase is y,

$$\mu(L, p) = \mu(G, p, y) \tag{2}$$

Subtraction of equation 1 from equation 2 gives

$$\mu(L, p) - \mu(L, p^*) = \mu(G, p, y) - \mu(G, p^*, 1) \tag{3}$$

The LHS of equation 3 is given by $\int_{p^*}^{p} V_m \, dp$ where V_m is the molar volume of liquid mercury, and since the liquid is virtually incompressible

$$\mu(L, p) - \mu(L, p^*) = V_m(p - p^*) \tag{4}$$

If the mixed vapor behaves as a perfect gas, the RHS of equation 3 is given by

$$\mu(G, p, y) - \mu(G, p^*, 1) = RT \ln(yp/p^*) \tag{5}$$

$$\therefore RT \ln(yp/p^*) = V_m(p - p^*) \tag{6}$$

Now $yp = p'$, the enhanced partial pressure of mercury in the vapor,

$$\therefore \ln(p'/p^*) = V_m(p - p^*)/RT = V_m \Delta P/RT \tag{7}$$

where ΔP is the excess applied pressure. In fact the enhancement is small and $p'/p^* \approx 1$ so that

$$\ln(p'/p^*) \approx (p'/p^*) - 1 = (p' - p^*)/p^* = \Delta p/p^* \tag{8}$$

where Δp is the enhancement in the vapor pressure due to ΔP. Then

$$\Delta p/p^* \approx V_m \Delta P/RT \tag{9}$$

No account is taken in this derivation of molecular interaction, i.e. of the fact that the mercury (or any other condensed component present) may be soluble in the gas. It has been found[143,183], for example, that the concentration of water in a compressed gas saturated with water vapor is not only greater than the concentration corresponding to the normal equilibrium vapor pressure of water on ice at the prevailing temperature, but is also greater than the concentration calculated according to equation 7 and, as far as has been investigated, this behavior is general—another example is provided by an investigation of the vaporization of naphthalene in compressed ethylene[52].

Rowlinson and his co-workers have determined the equilibrium concentrations and rates of diffusion of mercury in several compressed gases, but such experiments have not yet shown how a more accurate correction in the general case is to be determined[90, 91, 142, 168]. A theoretical discussion of the enhancement of the mercury vapor pressure in dense gases, in which the experimental evidence is reviewed, has been published by Haar and Levelt Sengers[76]. The practical situation may also be complicated by doubt whether the vapor is saturated with mercury or not. Young (who had no stirrer in his apparatus) pointed out that the rate of diffusion of mercury is very slow through the layer of liquid covering it when the sample is in two phases, and that when observations were made in a series in which the volume was successively allowed to expand the concentration of mercury could be expected to be low, whereas in a series in which the volume was made successively to contract the concentration could approach saturation[190]. In Kay's apparatus where the sample is stirred, or in Beattie's apparatus where the ratio of the diameter to the height of the volume occupied by the sample is much greater, saturation must be reached more quickly. Until a better founded method is proposed the best course in making these measurements seems to be to aim at saturation and to correct by subtraction of the mercury vapor pressure adjusted in accordance with equations 7 or 9—extension of the method to the correction of volumetric measurements will be found in Beattie's original description of his apparatus[21].

Ambrose, Broderick and Townsend avoided the problems associated with the presence of mercury in the heated zone by heating only the upper end of the tube of apparatus otherwise generally similar (but without the stirrer) to that used by Kay. This is a satisfactory procedure for vapor-pressure measurement but precludes the possibility of making volumetric measurements at the same time[7].

It is possible that the presence of mercury will affect the values found for the critical temperature and pressure. Some measurements on decane made by Kay with mercury present in the system and also with it absent are reported by Kudchadker, Alani and Zwolinski[105]; the results suggest that the effect will be measurable but not practically important, as the differences found would be within the experimental errors of many determinations.

VII. Vapor Pressures of Liquefied Gases

The vapor pressures of gases have most frequently been determined by the static method either with[121] or without[9, 89] use of a stirrer. In such investigations T_A (*Figure 1*) is automatically less than T_B or T_C and no problem can arise from condensation of the substance in manometers or connecting lines outside the cryostat. It is, however, necessary to insure that the equilibrium chamber is in fact the coldest part of the apparatus inside the cryostat, because cryostats are often constructed of nested Dewar vessels and have solid-carbon-dioxide slush or liquid nitrogen in the outer vessel so that large temperature differences may exist at closely neighboring points. In such conditions, through faulty design or operation, the lead-out tubes may be cooled below the intended, measured, operating temperature and if this happens condensation will occur at the cold point with the result that the apparent vapor pressure is lower than its true value.

It is possible to apply the gas directly to the measuring manometer C without interposition of the null-manometer B but it will usually be found more convenient for it to be retained; whatever arrangement is adopted, for high precision the volume of the system external to the cryostat should be no larger than necessary and must be protected from rapid changes in temperature, because otherwise the expansion and contraction of the gas will cause continual disturbance of conditions in the vapor-pressure vessel so that equilibrium will never be reached.

Many of the most accurate vapor-pressure measurements on gases have been made in the course of calorimetric experiments or in apparatus whose design is based on calorimetric practice (i.e. the thermal isolation of the equilibrium vessel by means of vacuum jackets and the elimination by careful design of unwanted flows of heat through connecting tubes). The apparatus is still essentially the same as that shown in *Figure 1* except that the calorimeter and its shields take the place of the equilibrium vessel A and its thermostat[74, 86, 167].

Spauschus has described an automatic apparatus for measurements of vapor pressures greater than atmospheric in the temperature range 233 to 373 K[165].

VIII. Effect of Thermal Transpiration

When wide differences in temperature exist between different parts of the apparatus account may have to be taken at low pressures of the effect of thermal transpiration. If an equilibrium vessel at 70 K is connected to a manometer at ambient temperature by a narrow tube, because of the difference in the two temperatures, the pressure will not in fact be uniform throughout the system but will be higher in the hotter part. The difference in pressure between the hot and cold parts is determined by the geometry of the apparatus, i.e. the diameter of the connecting tubing, and by the properties of the gas[1, 26, 150]. The effect becomes significant at pressures below 400 Pa and Liang has shown, for example, that the apparent vapor pressure of xenon at 77.3 K may be nearly twice the true value of about 250 Pa if the diameter of the connecting tubing is as small as 1.6 mm[109].

IX. Comparative Static Measurements

The vapor pressures of two substances may be compared if two equilibrium cells are set up side by side in the same thermostat and communicate with each other through a differential mercury manometer or diaphragm gage. This comparative method has the inherent advantage that, provided the temperature of the thermostat is uniform, its value need not be determined so exactly as is necessary when an absolute measurement of the vapor pressure is made because, for small temperature changes, both vapor pressures will to a first approximation change by the same amount, so that the difference between the two will remain nearly constant. But measurement of the temperature is not difficult, and in making the comparison in this way there are the disadvantages that two systems must be degassed instead of one, and that it is difficult to measure the difference in pressure exactly if this is large. Comparative static measurements are, therefore, only attractive if the pressures in the two sides of the apparatus do not differ greatly over the whole range of temperature studied; for example, if the vapor pressure of a solution of an involatile solute is being compared with that of the pure solvent. Thomson and Douslin[174] describe the setting up of a simple apparatus for comparative measurements and quote as an example of its use the comparison of the vapor pressure of 2,2,3,3-tetramethylbutane (triple-point temperature 374·0 K, normal boiling temperature 379.6 K) with that of water[34]. In a different temperature range, Clarke, Din and Robb[38] measured the vapor pressure of argon by comparison with that of oxygen (which was well established). Today, direct static measurement of the vapor pressure of the argon using a modern design of pressure balance and a platinum resistance-thermometer (at that time, along with the associated bridge, unobtainable) is certainly to be preferred.

X. Static Measurements at Very Low Pressures

Methods of pressure measurement well known in vacuum technology, viz. those depending on ionization and on the rate of heat loss from a hot wire (Pirani gage), have also been used for the determination of very low vapor pressures. There are also several other methods for the measurement of very low pressures which have been used for this purpose; for example, Langmuir suggested measurement of the viscosity of the vapor by observation of the rate of damping of a horizontal oscillating disc suspended on a quartz fiber[106], and Knudsen developed a gage, based on Crookes's radiometer, in which the difference between the momentum of molecules reflected from the bright polished side of a vane and that of those reflected from the opposite blackened side is measured[98]. Klumb and his co-workers designed a cylindrical form of this gage and used it for the determination of the vapor pressures of several alcohols in the range 10^{-4} to 10^{-1} Pa[96,97], and Miller used a similar gage for measurements on naphthalene[122].

The range of use of these gages is beyond the limits set by the writer for detailed discussion in this chapter and for more information about them the reader is recommended to the accounts by Dushman of the gages themselves[56], and by Clopper, Altman and Margrave and by Cooper and Stranks of their application to vapor-pressure measurement[41,45].

XI. Use of Radioactive Tracers

If a radioactive tracer can be introduced into the substance the amount present in the vapor phase may be determined from the radioactivity of the vapor. Ideally, the isotope used should be capable of preparation in high concentration free from other isotopes, and should emit only one type of radiation, which should be as hard as possible, and should possess a suitable halflife. The concentration of molecules in the vapor will then be directly proportional to the counting rate after the necessary corrections have been made to the latter. The proportionality constant may be calculated if all the relevant parameters are known but is more simply determined from measurement of the counting rate when the equilibrium vessel is maintained at a temperature at which the vapor pressure is already known. The method has been used by Dainton and Kimberley[49] for measurements on white phosphorus in the range 0.05 to 0.15 Pa, and by Rosen and Wallace[149] for measurements on phosphorus pentafluoride in the range 3 to 45 kPa.

Radioactive tracers may also be used for assay of the amounts of substance transported in the gas-saturation and Knudsen effusion methods (Sections XVIII and XIX). Where it is applicable, radioactive assay can be very much more sensitive than any other method, and by its use the lower limits for measurement of vapor pressure have been much extended. A full account of the use of radioactive tracers is given by Cooper and Stranks[45].

XII. Ebulliometric Measurements

To a first approximation the temperature at which a liquid boils in an open vessel is the temperature at which its vapor pressure is equal to the atmospheric pressure. If the boiling is carried out under reflux, and the pressure in the apparatus can be varied, measurement of the boiling temperatures at differing pressures is a convenient way of determining the vapor pressure of the liquid. However, many liquids do not boil smoothly, particularly at reduced pressures, and bumping takes place, i.e. there is superheating of the liquid above its equilibrium temperature and this is followed by a violent evolution of vapor, which discharges the superheat so that the temperature falls. Even if boiling is smooth, some superheating must take place because boiling is a dynamic process not an equilibrium state[170]. Of importance also is the fact that the pressure at any point in the liquid is the sum of that imposed by the buffer gas plus that due to the hydrostatic head of the liquid itself, so that the temperature registered is affected by the depth of immersion of the thermometer. Direct immersion of the thermometer in the boiling liquid is not therefore a satisfactory way of making exact measurements, and this presumably is the reason why it has sometimes been stated that ebulliometric methods of vapor-pressure measurement are not as good as static methods.

A steady temperature reading may be obtained in a boiling-point apparatus (an *ebulliometer*) either by arranging for the liquid to approach as nearly as possible to the equilibrium state before its temperature is measured, or by measuring the temperature not of the liquid but of the condensing vapor. If the liquid being boiled is pure these two temperatures will be very close together, but if it is a (non-azeotropic) mixture they will differ because the composition of the vapor will differ from that of the liquid. Although therefore measurement of the vapor condensation-temperature, as will be described below, is to be preferred for vapor-pressure determinations on pure substances, liquid temperatures must be measured whenever mixtures are being studied. Measurements of both boiling and condensation temperatures are often also made with single-component samples so that the identity of the two temperatures may be demonstrated, and some indication thereby obtained of the purity of the sample.

Cottrell designed the first satisfactory apparatus for the measurement of the boiling temperature of a liquid[46, 181]. In a *Cottrell pump* a mixture of liquid and vapor is pumped up a tube and on to the thermometer by the boiling action itself, in a way now widely familiar because of its use in the domestic coffee percolator. If the pumping rate is correctly adjusted, by suitable design of the percolator tube and adjustment of the boil-up rate, there is time for the liquid and vapor to come to equilibrium so that the desired temperature is measured. The method has been used in most of the more recent designs of ebulliometer, although it is now more usual for the tube up which the liquid–vapor mixture is pumped to be placed outside the boiler rather than inside it, as in Cottrell's original apparatus, because this arrangement is more effective in removing the superheat (and even then, according to Swietoslawski, the equilibrium temperature is not finally reached until after the mixture has expanded into the thermometer chamber)[170].

After boiling has started in an ebulliometer the vapor rises in the apparatus and condenses on the upper colder parts, thereby raising their temperature to its condensation temperature, and when the vapor has reached the condenser a steady condition of reflux is set up. If the thermometer is located in the vapor the temperature recorded will be the equilibrium temperature provided the thermometer does not lose heat by radiation and is adequately protected from the boiler heater. The boiler heater may affect the thermometer directly by radiation, but if the ebulliometer is in an enclosure the boiler heater may also affect the thermometer indirectly through convection of heated air or by conduction along the wall of the enclosure: both these processes may raise the temperature of the outer wall of the ebulliometer and, if their effects are detectable, suitable screening must be provided for the thermometer, and redesign of the apparatus may be necessary. The descending reflux, on the other hand, may have been cooled below its equilibrium temperature in the condenser: the length of the path for the reflux below the condenser should therefore be long enough for thermal equilibrium between the rising vapor and descending liquid to be reached before the latter arrives in the region where the temperature is measured, or else the reflux should descend to the boiler by a separate channel, and it may sometimes be necessary for the condenser temperature to be increased above its usual ambient value.

Ebulliometry was long used for the establishment of the temperature scale, by measurement of the condensation temperatures of water and sulfur at atmospheric pressure, and was developed by Swietoslawski as a general method of precision suitable for measurement of vapor pressures and the study of solutions over a wide pressure range. Swietoslawski designed an ebulliometer in which the temperatures both of the boiling liquid (reduced as far as possible from its superheated state by the pumping action just described) and of the condensing vapor were measured[170], and the value of these ebulliometers for the determination of the vapor pressures of organic compounds and of water was demonstrated by Zmaczynski[192]. These two workers were pioneers in the application of the comparative method, i.e. the relating of the unknown vapor pressure to the known vapor pressure of a reference substance boiling at the same pressure. If two ebulliometers connected to the same pressure main are used—one containing the substance under investigation, and the other a reference substance (usually water) for which the relationship between vapor pressure and temperature is accurately known—two condensation temperatures may be determined, and from that of the reference substance the pressure may be calculated. In this way necessity for determination of the pressure may be avoided—a considerable advantage because temperatures can more easily be measured accurately than can pressures, and the same measuring equipment is used for both variables. This comparative method also helps to reduce any systematic errors caused by imperfections in the design of the apparatus or in the measuring equipment—because they tend to cancel out—and there is the advantage that, if measurements have been made on a series of substances with a single reference substance, the values can all easily be recalculated and brought up to date if later more accurate work shows that the values for the reference substance need amendment. Zmaczynski reported in his paper the condensation temperatures both for the substance investigated and for water: this method of presentation makes later checking and recalculation particularly easy (and it is to be regretted that the example Zmaczynski set has not been universally followed). The method is applicable with either the simplest or the most refined temperature-measuring equipment, and if a compound boils smoothly in the apparatus, no other method will yield results of comparable precision so easily, the precision being determined by the quality of the thermometry. As examples of modern ebulliometric equipment, that in use at the US Bureau of Mines, Bartlesville, and that in use at the National

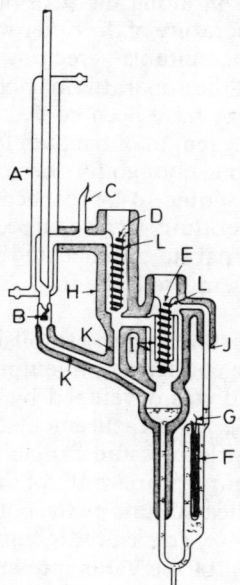

Figure 7. Swietoslawski-type ebulliometer[131]. A: condenser; B: drop counter; C: filling tube; D: thermometer pocket (condensation temperature); E: thermometer pocket (boiling-liquid temperature); F: heater pocket (wrapped with glass thread); G: heater; H: lagging; I: glass shields; J: percolator tube; K: condensate return tubes; L: glass helices.

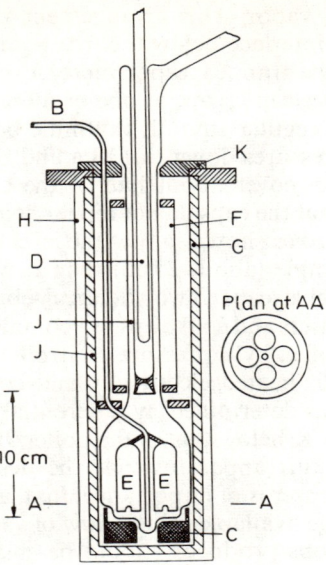

Figure 8. Boiler for measurement of condensation temperature[4]. AA: liquid level; B: filling tube; C: heater; D: thermometer pocket; E: bubble caps; F: radiation shield; G: heated jacket; H: outer canister; J, J: differential thermocouples; K: Sindanyo plate and lid.

Physical Laboratory, Teddington will be described; the first follows Swietoslawski and Zmaczynski in providing for measurement of both liquid and vapor temperatures, the second follows the later practice of Rossini and his co-workers in providing for measurement of the vapor temperature only[151,186].

In the Bartlesville apparatus[131], two identical ebulliometers (*Figure 7*), of which one contains the sample and the other water, are each connected through their condensers to a buffer volume of about 16.5 liters, and through a diffusion barrier and cold trap to a common line through which the pressure is adjusted. The boiler is arranged so that vapor rising from the boiling liquid pumps a mixture of liquid and vapor over the thermometer well E; the boiler and percolator tube J are not lagged, and the surroundings are maintained at least 10 K below the boiling temperature so that any superheating caused by the boiling can be dissipated during the passage of the mixture to the thermometer well. The boiler-heater well is wrapped with glass thread to promote steady boiling, and the heater itself is immersed in the well in a silicone oil. The baffles I assist disengagement of the vapor from the liquid, and the vapor passes over the thermometer well D, where the condensation temperature is obtained, before it is condensed and returned to the boiler. The purity and stability of the sample are considered to be adequate if the difference between the two temperatures does not exceed 0.005 K.

The special feature of the NPL apparatus[4] for measurement of condensation temperatures is the boiler (*Figure 8*) in which four inverted cups (about 18 mm in diameter and 40 mm high) are held so that their open ends just dip below the surface of the liquid and entrap, first gas and later, after the gas has diffused out, vapor. This arrangement insures that there is at all times a vapor/liquid interface below the free liquid surface in the boiler. As heat is applied and the trapped vapor increases in volume it pushes down the liquid until bubbles can escape at one or more of the indentations in the rim of each cup, and regular smooth rhythmic boiling takes place from all the bubble-caps at pressures down to 1 kPa and below. The diameter of the caps is such that they cover about half of the cross-sectional area of the boiler. It is essential that the caps do not dip far below the surface of the liquid and the boiler is shaped as shown so that the level of the liquid is less affected by the size of the sample (about 25 ml) than it would be if the outer tube closely fitted the bubble caps. Two identical ebulliometers are connected through reflux condensers and cold traps to a common main in which currently the pressure is controlled by a pressure controller (Texas Instrument Company). The jacket G is heated electrically and controlled automatically so that its temperature, as determined by differential thermocouples located at J, J is between 1 and 5 K below that of the boiler. It has been found in the use and development of this apparatus that the design of any drop counter fitted below the condenser is of importance; first, and obviously, that it must not restrict the passage available for the flow of vapor to the condenser and, secondly, that the drops produced should be small, and should run down the wall of the boiler rather than fall straight back into the boiling liquid if they are not to cause irregularity in the temperature record. It is also doubtful whether the radiation shield F as shown in *Figure 8* is useful: at low temperatures (for example, those measured in the water ebulliometer at pressures up to 200 kPa) its presence or absence appears to make no difference to the results, whereas at high temperatures (for example those measured in an ebulliometer containing mercury[10]) the apparatus did not give satisfactory results, and it was necessary to place radiation shields (modelled on the design given by Beattie, Benedict and Blaisdell[22]) inside the ebulliometer.

Mueller and Burgess[125] demonstrated that temperatures measured at the normal boiling point of sulfur—717.824 K (IPTS-68)—might be as much as 0.2 K below the true value if the radiation shields for the thermometer were incorrectly designed, and it is now standard practice[17] for ebulliometric equipment used for establishing the temperature scale to have internal radiation shields—these are essential at high temperatures. It is not clear what is the upper limit of temperature for the application of the ebulliometric method; Ambrose and Sprake[10] used a metal ebulliometer satisfactorily for measurement of the vapor pressure of mercury up to 770 K (800 kPa), and Rodebush and his co-workers, Fogler[63] and Dixon[147], reported satisfactory measurements (with a refractory apparatus) on cadmium up to 867 K (10 kPa) and on zinc up to 960 K (6.5 kPa); but the latter authors concluded that their measurements on lead in the range 1430 to 1600 K (1.3 to 6.5 kPa) were not satisfactory, and they expressed doubts about the method for the work because of errors due to radiation losses from the thermocouple used for determination of the temperature, and to the tendency of the vapor to superheat. However, the radiation shield they used was probably not adequate for its purpose, and the design of their apparatus seems susceptible of improvement. Barton and Bloom[19], on the other hand, in a study of the vapor pressures of several metal chlorides at pressures up to 65 kPa, claimed that bubbling nitrogen slowly through the liquid prevented it from superheating, and that measurement of the temperature of the boiling liquid in the range 1 000 to 1 500 K gave reliable results (i.e. to within ± 1 K).

Temperatures in both the apparatuses described, i.e. *Figures 7* and *8*, are measured by means of platinum resistance-thermometers used with high quality bridges, and a sensitivity of 0.001 K or better is attained. The results obtained are in close agreement; for example, the boiling temperature of hexafluorobenzene at 101.325 kPa calculated from the results reported from Bartlesville is 353.398 K[55] while the most recent NPL value is 353.401 K; another value obtained in the NPL with a different sample of hexafluorobenzene in an earlier design of apparatus is 353.406 K[47] (these temperatures are expressed on IPTS-68 and the numbers given for the two values already published have been obtained by re-computation in NPL on the new scale). At the precision attained in these measurements, differences of more than 0.01 K between the values of the normal boiling points for one substance obtained by different workers are likely to be due to differences in the purities of the samples used—it is purity which is probably the limiting factor in the accuracy attainable in ebulliometric measurements.

Apparatus of the type shown in *Figure 8* has been modified for use at temperatures below ambient by surrounding it with a cooled jacket and using solid carbon dioxide as the condensing agent. With the modified apparatus the vapor pressure of diethyl ether was measured in the range 7 to 200 kPa (250 to 328 K)[12].

In an ebulliometer of the Swietoslawski-type the difference between the temperatures of the boiling liquid and of the condensing vapor may indicate whether impurity is present, or whether the substance is decomposing in the conditions of the experiment. It does not give positive evidence of purity because an azeotrope may not be detected, and—to the extent that the determination of purity is now firmly based on other methods, in particular on chromatography and on the study of the solid–liquid phase change—the indications to be obtained in this way from the ebulliometer, although still useful, are less necessary. Furthermore, if the temperature-measuring equipment is sufficiently sensitive, the behavior of the condensation temperature alone may reveal inadequacy of the sample. It is standard practice in the determinations made at the NPL for a small quantity of the sample to be distilled off before measurements are made: with a pure sample, the change arising in the condensation temperature (which alone is measured) is barely detectable. If water is present in the sample the temperature record is usually unsteady (as, for example, it was during many measurements made on methanol, a substance from which it proved difficult to eliminate traces of water[8]). A stringent test of the experiment (and therefore of the sample) is the fitting of an equation—if measurements have been made over the pressure range 10–200 kPa then 95 per cent of the points will usually lie within 0.005 K of the best curve passing through all the points, whereas the scatter if the sample is wet or otherwise impure may exceed 0.01 K. With high-boiling compounds the onset of decomposition at the upper end of the range will be apparent from unsteadiness in the temperature record during the experiment, and later study during the curve-fitting procedure may suggest that the scatter of some of the topmost points is unduly large and that they should be rejected, even though they were apparently acceptable when the experiment was in progress. Experience at NPL therefore suggests that in most instances there is little advantage to be gained by measuring the boiling temperature in addition to the condensation temperature provided the ebulliometer (and its associated equipment) is properly designed.

In ebulliometric methods a second component is always present—the gas by which the pressure in the system is controlled. It is assumed, and the assumption is borne out in practice, that at low pressure this gas does not affect the behavior of the substance under study, but at higher pressures the basis for the assumption is less secure. If the gas dissolves in the liquid the boiling and condensation temperatures of the latter could be affected—and for this reason helium is in principle to be preferred as the gas because its solubility in liquids is smaller than that of other gases[68a]. Equally it can be demonstrated, both theoretically and practically as has already been discussed, that pressure applied to a liquid or solid will raise its vapor pressure. However, at pressures up to 200 kPa no differences have been found between values obtained by static measurements and values obtained by ebulliometry that are to be attributed to these effects. Ambrose, Sprake and Townsend used the comparative ebulliometric method for determination of the vapor pressure of benzene in the range 100–500 kPa and found no difference between the values obtained when helium was used as the buffer gas and those obtained when nitrogen was used, and all the values were in satisfactory agreement with an interpolation between values at lower pressures and values obtained by static measurements at higher pressures[11]. For a higher range, a metal ebulliometer has been described[129, 130] for use in the comparative manner (with water as reference substance) at pressures up to critical: at these pressures, however, there are few sets of results more precise than 1 part in 10^3, and it is not possible to say from any comparison whether the presence of helium or any other gas at a pressure of 1 MPa or more will have a significant effect on vapor pressures calculated from the measured temperatures. Since in the comparative method both liquids are subjected to the same pressure by the buffer gas, errors due to the effects now under consideration, like errors due to other causes, will tend to cancel rather than reinforce each other.

The upper limit of pressure at which a glass apparatus may be used is fixed by considerations of safety, but the lower limit is fixed by one of several factors: (1) The tendency of the boiling liquids, particularly the water, to bump. (2) If water (as is to be preferred) is used as the reference substance, it cannot be refluxed below a pressure of around 2–3 kPa because the condenser temperature cannot be taken below 0°C. (3) When the pressure of the gas is reduced below 10 kPa diffusion of the sample is little hindered and a volatile sample is liable to move from the boiler to the cold trap. (4) There must be a pressure drop in any dynamic system, and at some pressure this will become significant in relation to the equilibrium vapor pressure—if the boil-up rate is reduced to reduce the pressure drop the rising vapor may not carry enough heat to control the temperature of the thermometer and its reading will not then be meaningful. Of these factors (1) the onset of bumping has usually occurred at pressures of 15 kPa or more: this problem has been eliminated in the NPL apparatus, smooth boiling has been obtained with all substances tried, and (2) and (3) have become the limiting factors. It is not yet clear at what pressure the final limit of (4) is reached, but Ambrose and Sprake have made satisfactory ebulliometric determinations of the vapor pressure of mercury by this method down to a pressure of 200 Pa (the pressure being measured not comparatively by means of a reference substance but directly with the aid of a mercury manometer in which the positions of the menisci were determined by means of probes attached to micrometers[10]).

XIII. Ebulliometric Measurements at Pressures below 2 kPa

The ebulliometer for the measurements on mercury referred to in the preceding paragraph was generally similar to that shown in *Figure 8*, with the same arrangement of four inverted bubble-caps in the boiler, but was of somewhat larger diameter (85 mm compared with 65 mm for the original design, and with bubble-caps of 25 mm diameter in place of 18 mm). There was, however, one important difference—the diameter of the ebulliometer was maintained at the full 85 mm over nearly its whole height. About six turns of lead tubing 6 mm in diameter were wrapped round the ebulliometer immediately above the heated jacket to act as a condenser and to prevent undue heating of the greased joint used to join the thermometer pocket to the body of the ebulliometer. Since the temperature of the thermometer never exceeded 505 K, a single radiation shield made of thin stainless steel sheet wrapped around the glass pocket and wired on tightly was adequate. Observations made with this apparatus at pressures down to 200 Pa showed that the method gives good results at low pressures if the pressure drop is kept as low as possible between the boiling and condensing zones by use of the wide tube. Measurements could only be made with difficulty, however, at pressures of 100 Pa or less, and this appeared to be the lower limit for the meaningful use of the apparatus. Two ebulliometers of the type just described may be used in the comparative manner, with mercury as reference substance, for determination of vapor pressures in the range 200 Pa to 2 kPa[10].

Hickman, many years previously, had made ebulliometric measurements at low pressures, and his apparatus is shown in *Figure 9*[84]. The boiler-heater is a bare resistance wire immersed in the liquid, and there is a very short wide path for the vapor between the boiling liquid and the thermometer. With this apparatus the temperature recorded was plotted against the heat input to the boiler: after an initial rise the temperature was found to be independent of boil-up rate over a fair range and then rose again with further increase in the heat supplied. The plateau temperature was taken as the correct value and measurements were made at pressures ranging from 30 to 500 Pa, the upper limit being set by the onset of splashing of the superheated liquid on to the thermometer (a fault which did not affect the experiments on mercury described above because of the larger size of the apparatus). Hickman studied viscous high-boiling liquids such as dibutyl phthalate, and his original papers should be consulted if work is planned on substances of this type; because of their peculiar behavior when boiling Hickman called them *phlegmatic*. It is not yet known whether the bubble-cap ebulliometer is satisfactory for phlegmatic liquids.

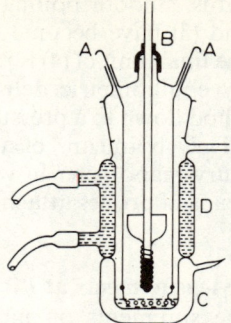

Figure 9. Ebulliometer for measurements at low pressures[84]. A, A: leads to heater; B: thermometer pocket; C: vacuum-jacketed boiler; D: condenser.

Ebulliometric measurements can evidently be made at pressures as low as 30–100 Pa if the apparatus is suitably designed, but the design may have to be a compromise between conflicting requirements: it is difficult to establish satisfactory conditions at low pressures and the accuracy falls markedly. In any event, accuracy in the measurements begins to fall as the pressure is reduced below 1 000 Pa—whereas at atmospheric pressure a precision of 1 part in 10^5 or better can be looked for (and much greater errors arise from lack of purity in the sample), at 1 000 Pa it will be difficult to maintain a precision of 1 in 10^4. In terms of temperature, Ambrose and Sprake estimated the uncertainty of their work on mercury to be ± 0.05 K at 3 kPa increasing to ± 0.2 K at 100 Pa[10].

XIV. Method of Ramsay and Young

In the well-known method of Ramsay and Young[140], the temperature of evaporation of a small sample held on a suitable absorbent material tied round the bulb of the thermometer is measured. Although convenient, the method as originally described is not a precise one, because the liquid flowing in to replace that evaporated will tend to make the temperature of the thermometer fall unless the apparatus is elaborated so that the liquid is preheated to the correct temperature. This is most easily achieved if the mode of operation is changed from one of intermittent evaporation to one of steady state reflux as in the modification described by Thomas, Smith and Meatyard (*Figure 10*)[172]. A glass-fiber skirt, out of which an inverted V is cut to allow free escape of the vapor, is tied round the bulb of a mercury thermometer. The skirt dips into the pool of liquid at the bottom of the wide boiling-tube so that the liquid rises to the thermometer bulb by capillary action. The apparatus is immersed in a bath heated to the approximate temperature at which it is desired to make the measurement, and when the pressure is reduced to the appropriate value evaporation occurs from the wick, the thermometer takes up the corresponding steady temperature, and reflux drips from the copper cooling coil. In these conditions, therefore, the point at which evaporation occurs is continuously fed with liquid preheated to the correct temperature—in fact slightly superheated since no visible ebullition

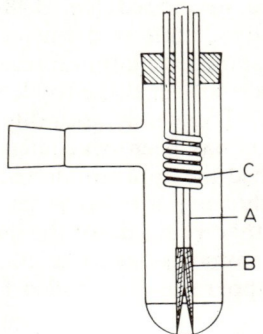

Figure 10. Apparatus for measurement of evaporation temperature[172]. A: mercury-in-glass thermometer; B: glass-fiber skirt dipping into sample; C: condensing coil.

should take place in the bulk of the liquid, and, as in an ebulliometer provided with a Cottrell pump, the superheat is discharged by the evaporation. In the procedure adopted by Thomas *et al.*, after reflux has started the bath temperature is raised slowly until it is 30 to 40 K above the temperature recorded by the thermometer, and it is claimed that over this range the latter temperature is independent of that of the surroundings. The results reported for several compounds in the range from 5 Pa up to normal atmospheric pressure agree with the values obtained by other methods with errors estimated to be no more than ± 0.2 K, and the design is one which should be suitable for operation at the lowest pressures at which a dynamic method is applicable, since smooth evaporation is obtained and the dimensions of the apparatus are such that pressure-drop effects are minimized.

XV. Dynamic Measurements without a Buffer Gas

An advantage of the ebulliometric methods already described is that, within limits, the conditions are self-regulating: excess cold is supplied to the condenser, and change in the boil-up rate is balanced by a change in the upper level at which the vapor condenses, i.e. in the area of the condenser utilized to effect condensation, with the result that the heat input to the boiler may be varied widely (perhaps by a factor of two) before any change is detectable in the measured temperature. Two pieces of apparatus have been described for measurements at pressures above atmospheric in which boiling takes place in the absence of any buffer gas, and the heat supplied to the boiler is exactly balanced by that removed in the condenser; in this mode of operation the pressure developed in the system, and consequently its temperature, is determined by the amount of heat supplied. Change in the boil-up rate is therefore rapidly followed by change in the measured temperature.

The first apparatus of this type is that described by Edwards and Johnson for the special purpose of determining the vapor pressure of carbon dioxide in the neighborhood of 273 K (as a means of calibrating a pressure balance)[58]. The apparatus, which is simply constructed from metal tubing capable of withstanding the pressure developed (*ca.* 3.48 MPa) and from standard couplings, consists of a tube 25 mm in diameter containing a thermometer pocket long enough for a standard platinum resistance-thermometer, with a heater at the bottom and a side arm at the top leading to a condenser cooled with solid carbon dioxide. The whole assembly is lagged and surrounded by ice. After the apparatus has been evacuated it is charged with carbon dioxide. When it has reached the appropriate temperature, the boiler-heater is switched on and a steady state of reflux is set up, the reflux liquid in this instance being directed so that it runs down the thermometer pocket, because for some settings the thermometer has to be cooled below the temperature of its surroundings. The apparatus is intended for use with a gas-operated

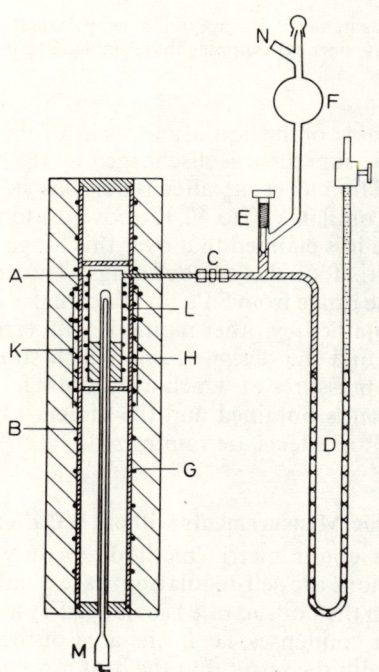

Figure 11. Apparatus for determination of vapor pressure by dynamic method[9]. A: Monel boiler provided with central thermometer pocket; B: aluminum tube-oven; C: coupling linking metal connecting tube to glass; D: mercury-filled manometer connected to oil-filled line leading to piston gage; E: glass valve with PTFE stem (Fisher and Porter Company, USA); F: reservoir for introduction of the sample; G, H: electrical heater windings; K, L: control resistor windings, insulated by suitable wrappings from G and H; M: platinum resistance-thermometer; N: connection to vacuum system.

pressure balance, i.e. operated with the carbon dioxide itself, and the balance is connected to the apparatus at a point near the lower end of the thermometer pocket. Results of very high precision have been obtained, and there seems no doubt that the apparatus could be adapted for measurements on other easily condensable but less innocuous gases from, say, 220 K upwards if a diaphragm-gage or mercury-filled U-tube were interposed between it and the pressure balance in the manner employed in the apparatus about to be described.

Ambrose and Sprake used the second apparatus (*Figure 11*) for the measurement of the vapor pressures of organic liquids from about 500 kPa nearly to the critical at temperatures of 400 K and upwards[9]. The metal vessel A, suitable for operation at pressures up to 5 MPa, contains the sample and is provided with a pocket for the thermometer which reaches up into the vapor space; the vessel A is surrounded by the jacket B and the two are provided with heater windings G, H and resistance temperature-sensors K, L. Heat applied to the liquid in A causes it to evaporate, and the pressure and temperature to rise; the vapor condenses on the upper wall and the thermometer pocket, raising their temperatures, and a steady state is reached when the temperature of A exceeds that of B by an amount sufficient for transfer to B of the heat supplied to A. The tube, C, which passes outside the heated enclosure, fills with condensed liquid and this transmits the pressure to the mercury-filled U-tube D (from where the line leads to a pressure balance). The size of sample used is sufficient to fill the connecting tubing as shown and to fill about one third of A at room temperature. After the start of the experiment, when the temperature has been raised sufficiently for the pressure in the system to exceed the atmospheric pressure, air initially present may easily be removed by operation of the connected hydraulic system to the space below the valve E and bled off; in a similar manner volatile decomposition products may be removed—a valuable advantage (which is also possessed by the ebulliometric method) over most of the static methods, from which satisfactory results cannot be obtained if there is any decomposition of the sample.

A much more complex apparatus from which air (or other volatile impurities) may be withdrawn after the start of the experiment is the calorimeter with which Osborne, Stimson and Ginnings determined the heat capacity and vapor pressure of water from 373 K up to its critical temperature[132]. It was used both as a constant-volume calorimeter (static measurements) and as a flow calorimeter; in the latter mode its operation was similar to that just described, with the exception that there was no reflux because the vapor was drawn off at exactly the same rate as it was formed by vaporization of the liquid.

XVI. The Quasi-static Method

Rodebush and Dixon devised a method which they called *quasi-static*[147]. Although boiling takes place, the temperature measured is that of the thermostat containing the sample, as in a static determination, and not that of the liquid or vapor, as in an ebulliometric determination. The principle of the method is illustrated by *Figure 12*. The sample is contained in the chamber A inside a furnace, and the two tubes B and C are connected through the differential manometer D; the pressure is adjusted through the stopcock E and is measured by a manometer attached at F. At the start of the experiment the apparatus contains an inert gas at a pressure greater than the vapor pressure of the sample, and this gas is gradually withdrawn through E until the

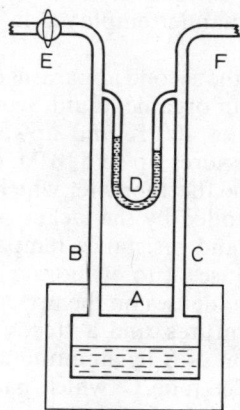

Figure 12. Apparatus for the quasi-static method[147]. A: sample boiler in furnace; B, C: tubes in which condensation takes place; D: differential manometer; E: connection to vacuum; F: connection to manometer for measurement of the pressure in the system.

liquid begins to boil; once the level of the condensing vapor has reached the tubes B and C it acts as a barrier to further loss of the gas contained in C, which can only take place by diffusion through the vapor and whereas, after the initial withdrawals of gas, brief fluctuations in pressure are apparent from observation of the manometer D, after the vapor seal has been established across tubes B and C, an appreciable time is required for pressure equilibrium to be reached throughout the system. At this point the vapor pressure of the sample is equal to the pressure of the gas in the system.

Although the authors reported a test of the method with mercury in the range 0.8 to 2.5 kPa, it has been used primarily at higher temperatures for measurement of the vapor pressures of less volatile metals, e.g. of lead and sodium[61,147], and was in fact developed because of the difficulties Rodebush had found in using an ebulliometric method at 800 K and upwards. The method (to which more references will be found in a chapter by Clopper, Altman and Margrave[41]) does not appear to offer any advantages at lower temperatures where normal ebulliometric methods are satisfactory.

XVII. Measurement of the Force Exerted by the Vapor

The use of the pressure balance for measurement of pressure has already been mentioned in section II. Except in the apparatus for the determination of the vapor pressure of carbon dioxide described in section XV, however, it is not suitable in its usual form for direct application of the vapor to the piston; but several specially designed apparatuses for the determination of low vapor pressures have been described which employ the same principle. This can be applied in two ways: either with a piston moving in a closely fitting cylinder, as is the normal manner of operation of a pressure balance, or (historically, the earlier) with a flat lid fitting exactly against an orifice of known area. In both methods the balance is an integral part of the vapor-pressure apparatus.

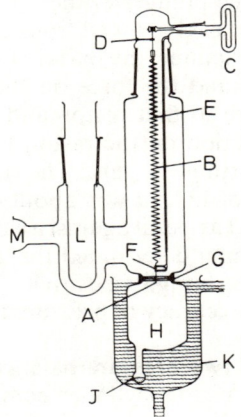

Figure 13. Apparatus for measurement of force exerted by the vapor[60]. A: piston made from mica discs; B: quartz spring supported from windlass, C; D: guide for spring; E: upper reference point; F: lower reference point; G: heater winding; H: vapor source-chamber with sample bulb, J; K: thermostat; L: condenser; M: connection to vacuum.

Ernsberger and Pitman used the apparatus shown in *Figure 13* for measurement of the vapour pressure of mercury in the range 0.5 to 2.5 Pa (285 to 299 K)[60]. A piston made of mica discs suspended on a helical quartz spring fits in an accurately ground glass cylinder that is part of the equilibrium chamber. The space above the piston is evacuated, and the force exerted by the vapor against the piston is balanced by unwinding the supporting thread for the quartz spring from the windlass. If it can be assumed that all the vapor escaping past the piston is pumped away and exerts no back pressure, the upward force on the piston is due to the vapor and the suspension while the downward force is due to its own weight. The method is nearly, but not quite, static because there must be some loss of vapor; this is minimized as far as possible by the good fit of the piston in its cylinder and by the use of multiple discs to form the piston, and since there is loss from the equilibrium vessel it will be clear when degassing is complete if observations are continued until a steady value of the pressure is obtained. The authors discussed the sources of error in the method in detail and estimated that the results obtained were good to ±1 per cent. One important requirement is that the sample must have an adequate surface to evaporate at the rate required—this is arranged during the preliminary manipulation of the apparatus because before measurements are made the temperatures are adjusted so that the sample distils out of the lower bulb J and condenses to form a layer covering all the walls of the vapor source-chamber H. Kaufman and Whittaker used the same apparatus for measurement of the vapor pressure of sebaconitrile in the range 0.03 Pa to 1 Pa (*ca.* 300 to 340 K)[92].

Back and Betts used an apparatus of similar type, in which the piston–cylinder combination was constructed from a glass syringe of 5 cm^3 capacity, for measurement of the vapor pressure of hydroxylamine between 10 and 65 Pa (*ca.* 295 to 340 K)[14]. The plunger of the syringe was cut to a length of about 20 mm (mass, 3 g) and was suspended from a torsion balance, modified to allow its operation under vacuum; the barrel of the syringe was connected to the evaporation chamber, and the force on the piston was balanced by adjustment of the torsion wire so that piston and balance returned to their null position, the effect of friction on the piston being minimized by gentle thumping of the bench supporting the gage. The rate of leakage of the vapor past the piston at a pressure of 133 Pa was about 4 µmol/h. The piston and torsion balance were calibrated as a unit against a sloping oil-filled manometer (whereas Ernsberger and Pitman determined the area of the piston from its geometry and determined the weight-sensitivity of the quartz helix in a separate experiment) and the accuracy of pressure measurement claimed was about ±0.15 Pa.

Successful operation of the two pressure balances just described depends upon making—or buying—a piston–cylinder combination with sufficiently small tolerances that the piston does not stick and that the rate of escape of the vapor is low. The techniques necessary for doing this were not well developed when Rodebush and his co-workers, Coons[146] and Henry[148], first used the lid-and-orifice design. This was then a more practical approach because the two mating surfaces have only to be made flat, and subsequently different investigators[15, 53, 84, 156] have used several variants of this type of manometer, i.e. with the disc in a horizontal plane either above or below the orifice, or with the disc in a vertical plane. With the first arrangement the disc is suspended and weighed, for example, magnetically or by means of a quartz spring, while with the second arrangement[84] the disc is attached to a pendulum which is balanced by tilting the whole apparatus. Whichever arrangement is used, the gage has the disadvantage, in comparison with the piston–cylinder gage, that its effective area is less certain since the wall of the orifice has a finite thickness, and it is not clear whether the inner or outer diameter is to be taken for calculation of the area. Furthermore, the evaporation is intermittent, i.e. the lid is either shut (provided the apparatus has been correctly assembled) or open, and instability can easily arise which at best makes the point of balance difficult to determine, and at worst may cause damage to the apparatus.

Douslin and Osborn have described a precisely engineered inclined piston gage for use in vapor-pressure measurement[55]. In this instrument the force exerted by the piston is adjusted by tilting the assembly of piston and cylinder. The angle of tilt may be measured very accurately, and the pressure balance is calculated from the accurately measured diameter of the piston by application of the appropriate principle of mechanics. The cylinder is rotated at constant angular velocity, but because the piston is weighted eccentrically it is always moving about its axis relative to the cylinder so that it does not stick. The piston is lightly lubricated, and the film of oil provides a seal between the lower end of the cylinder, which communicates with the vessel containing the sample (at temperatures below ambient), and the upper end which is under vacuum; there is therefore no loss of vapor, and measurements made with this instrument can be truly static. Pressures up to 5 kPa may be measured, and the instrument may be used like any other pressure gage for measurement of vapor pressures at temperatures above ambient if a suitable null-manometer is used to confine the sample.

XVIII. Evaporation Methods for Low Pressures

Vapor pressures below about 10 Pa may be determined from the rate at which the substance evaporates or from the rate at which it effuses through an orifice. So many investigators have used these methods that it is impossible to survey their work comprehensively here, and only an elementary account of the principles and different types of apparatus is given. *The Characterization of High Temperature Vapors* should be consulted for a more detailed account of, in particular, the underlying theory, whatever the range of temperature in which the investigator is interested[114]. Other useful sources of information are the reviews by Cater[36], by Cooper and Stranks[45], by Komarek[102], by Margrave[113], and by Winterbottom[188].

In normal conditions molecules evaporate from the surface of a condensed phase but do not move very far because they collide with other molecules in the gas phase, while molecules moving in the gas phase strike the surface and are re-absorbed: at equilibrium the rates at which the two processes occur are equal. If, however, the total pressure is sufficiently low—and the container is sufficiently large—the molecules evaporating from the surface will move sufficiently far before collision for it to be unlikely that they will be directed back towards the surface, and there will then be a measurable transfer of the substance from the condensed to the vapor phase.

The mean free path λ of a molecule in the gas phase at low pressures is given by

$$\lambda = 1/\sqrt{2\pi n\sigma^2} \qquad (10)$$

where n is the number of molecules in unit volume and σ is the molecular diameter. From this equation it may be calculated that at room temperature and pressures around 1 Pa the mean free path is greater than 1 cm. Under these conditions therefore (and at higher temperatures and lower pressures) the distances molecules move before collision are comparable with the dimensions of normal apparatus. If it can be insured that molecules which have evaporated are removed from the space around their source so that they never return, it will be possible to measure the rate at which evaporation is taking place by determining the loss in mass of the source sample, or by determining the amount of substance collected.

The vapor pressures of high-melting metals may be found by direct application of this method (which was originally used by Langmuir[107, 108]). The metal in the form of a filament of known area is heated electrically in high-vacuum conditions to the desired temperature, and the rate of loss is determined by collection of the evaporated metal on a cold surface. Then the pressure p is given by

$$(p/\text{Pa}) = \frac{m/\text{kg}}{\alpha(A/\text{m}^2)(t/\text{s})} \left\{ \frac{2\pi(R/\text{J mol}^{-1}\text{K}^{-1})(T/\text{K})}{M/\text{kg mol}^{-1}} \right\}^{\frac{1}{2}} \quad (11)$$

where m is the mass of metal evaporated in time t from a wire of area A, M is the molar mass of the evaporating species, T is the temperature, and, as an illustration, the equation has been written in terms of SI units. The evaporation coefficient α (which was originally assumed by Langmuir to be unity) must be found by other studies if the results obtained are to be accurate. Langmuir's method is one of the most sensitive and has been used for measurement of vapor pressures down to 10^{-5} Pa, but it is only suitable for use at high temperatures for samples which can be heated resistively and can be prepared with known surface areas.

Of much wider application is the method, first used by Knudsen, in which the sample is contained in a vessel that is sealed except for a small hole of known area through which the vapor escapes into the surrounding evacuated space[99, 100]. The determining area is then the area of the hole and not of the sample, and the determining rate is not the rate of evaporation but the rate of effusion from the vessel. The vapor pressure is given by

$$p = \frac{m}{KAt}\left(\frac{2\pi RT}{M}\right)^{\frac{1}{2}} \quad (12)$$

where the symbols have the same meaning as before except that A is now the area of the hole. The place of the evaporation coefficient is taken by the Clausing factor K (see *Table 1*), which is introduced in order to take account

Table 1. Factors for use in equations 12 and 13 for cylindrical orifices of various sizes (specified in the first column by the ratio of length to radius).

Ratio	0.1	0.2	0.6	1.0	2.0	4.0	10.0
K	0.952	0.909	0.771	0.672	0.514	0.357	0.191
f	0.968	0.937	0.825	0.733	0.572	0.402	0.218

These values are abstracted from a more comprehensive table given by Freeman and Edwards[67].

of the fact that the hole is made in a sheet of finite thickness; the hole is therefore cylindrical in form, and some of the molecules may strike its wall instead of escaping to the surroundings[40]. Although the evaporation coefficient no longer enters directly into the expression for the vapor pressure, the rate of evaporation is still important: since vapor is being lost from inside the cell it must be replaced, and the sample must have sufficient surface exposed for it to evaporate sufficiently fast to make good the loss if the apparent vapor pressure is not to fall below the true value. For example, the ratio of the surface area of the sample to the area of the orifice must not be less than 100 if this effect is to be less than one per cent[126].

There are two direct ways in which Knudsen's method may be applied: first, by collection of the substance after it has effused from the cell and determination of its amount (for example, by condensation on a cold finger and subsequent chemical assay); and, second, by determination of the loss in mass of the cell and its contents. For the first way, the cell may be in contact with the wall of the apparatus and no problem arises in the achievement of thermal equilibrium, but there may be uncertainty in the time t because of the need to heat up and cool down the sample at the beginning and end of the experiment (a shutter is sometimes incorporated in the apparatus to minimize this effect, or else the size of the orifice may be chosen so that the duration of the experiment is long compared with the time required for the apparatus to come to thermal equilibrium or two experiments may be conducted, one of long duration and one of short duration, from which the effect due to the time taken for equilibrium to be reached may be determined). For the second way, the cell itself is weighed, for example, by suspension from a quartz spring; a continuous record may be obtained and observations are only made when the temperature is steady. The time may therefore be fixed exactly, but because the cell is suspended in an evacuated space the achievement of thermal equilibrium will be difficult; and at relatively high rates of effusion the sample is likely to fall below that of the surroundings, because the heat of evaporation or of sublimation cannot be replaced unless the temperature for the experiment is sufficiently high for thermal equilibrium to be reached primarily by radiation. Whichever way is adopted, it is important that the orifice itself should be maintained neither above nor below the operating temperature. Both ways are applicable at pressures down to about 10^{-3} Pa, a lower limit set by the time required for the observations.

The evaporation methods just discussed require knowledge of the molar masses of the evaporating species—at high temperatures these may not be the species present in the condensed phase, while at lower temperatures they may not be the monomers of the simplest assumption, but in fact dimers or higher polymers.

Many measurements have been made in which the calculation of the pressures was based on the simple theory that has been outlined, but in fact the exact theory of the kinetic processes that take place is complicated[35, 123, 134, 152, 179, 184], and Komarek suggests that many of the measurements made prior to 1958 are unreliable because factors now better understood had been neglected[102]. At the simplest level, one question at issue is whether the Clausing factor is a sufficient correction for the orifice geometry, and an alternative way of determining this correction is to use the apparatus for measurements on a substance whose vapor pressure is known. For this purpose, benzophenone, which was studied by Neumann and Völker by both the Knudsen method and that about to be described[127], has been suggested as a reference substance. Thomson and Douslin have given equations to represent the vapor pressure of benzophenone together with a table of numerical values[174].

A third effusion method, of somewhat lower sensitivity than those already described, is that in which the recoil force due to the beam of molecules issuing from an orifice is measured[66, 178]. The effusion cell, suspended on a torsion wire, is provided with two horizontal orifices arranged symmetrically so that the two molecular beams exert a couple, and the torque can be determined by measurement of the angular displacement of the cell. The pressure is then given by the expression

$$p = 2W\phi/(A_1 f_1 d_1 + A_2 f_2 d_2) \tag{13}$$

where ϕ is the angular displacement, W is the torsion constant of the suspension, $A_1 A_2$ are the areas of the orifices which lie at distances $d_1 d_2$ from the point of suspension and $f_1 f_2$ are correction factors for the orifice geometry (see *Table 1*)[67]. As with the suspended Knudsen cell, this method may present difficulties for the achievement of thermal equilibrium, but it has the advantage that the molar masses of the vapor species need not be known because the force of the beam depends on the mass effusing, and not on the masses of the individual molecules. If the loss in mass of the cell can be determined under the same conditions, the two measurements provide a way of determining the molar mass of the vapor species, and this value can then be used for measurements at lower pressures by the Knudsen method.

Two examples of the type of apparatus used for effusion measurements are

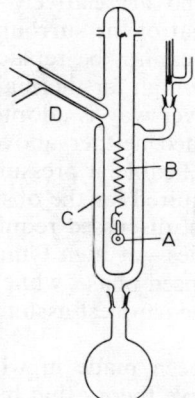

Figure 14. Knudsen apparatus[75]. A: glass effusion cell; B: quartz spring; C: vapor jacket; D: condenser.

illustrated in *Figures 14* and *15*. *Figure 14*[75] is typical of the relatively simple approach adopted for many studies of organic compounds. The glass effusion cell, provided with an orifice about 1 mm in diameter, is suspended from the quartz spring and its movement as the sample effuses is observed; from the previously determined sensitivity of the spring the change in mass may be calculated. It is important that no condensation should take place on the spring, which must therefore be kept at the temperature of the measurement—in this instance by immersion in the vapor jacket. *Figure 15*[187] shows a more complex device for combined Knudsen and torque-effusion studies. The graphite cell is attached to a quartz rod, to which is fixed a mirror for observation of the torsional movement, and the assembly is suspended by the tungsten torsion fiber (20 cm long) to an automatic recording microbalance. The lower part of the apparatus is heated in a tubular furnace; during experiments on lead in the range 875 to 975 K vapor condensed on the walls of

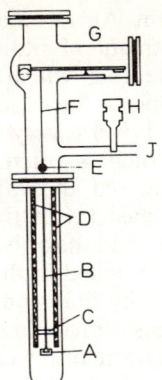

Figure 15. Apparatus for Knudsen-effusion and torque-effusion measurements[187]. A: sample cell; B: quartz rod; C: radiation shield; D: thermocouple pockets; E: mirror; F: torsion fiber; G: microbalance; H: vacuum gage; J: connection to vacuum.

the upper part of the apparatus, but no condensation on the suspension was observed at pressures below 1 Pa. The areas of the orifices were determined from photographs taken at a magnification of 75 (other workers have reported projecting a magnified image of the orifice on to a screen for this purpose)[177]. The whole apparatus was calibrated by using it for measurements on mercury at room temperature. Wiedemann has described an automatic apparatus (for Knudsen-type measurements only) suitable for use in the range 200 K to 420 K in which the cell is provided internally with a thermocouple so that it is possible to determine the temperature of the sample itself[185].

The validity of these methods as described so far depends upon the existence of high-vacuum conditions in the apparatus outside the cell. In contrast, Motzfeldt proposed a variation of the torque-effusion method in which a null-position is found by varying the pressure of the gas surrounding the cell—if this pressure is greater than the vapor pressure, the vapor escapes only by diffusion and the recoil force is negligible, but as soon as the pressure falls below the vapor pressure, the vapor forms a directed molecular beam and movement of the cell takes place[124]. This method was suggested for use at high temperatures, and at higher pressures (13 kPa) than have been considered so far in this section as measurable by evaporation methods.

The force of a molecular beam need not be measured only by its recoil—Klumb and Lückert have described a method in which the molecular beam is generated in the body of the apparatus; after passing through a collimating diaphragm the beam strikes a vane supported on a torsion wire, and the force is calculated from the movement of the vane[96]. The glass effusion chamber has a number of vertical metal rods sealed through the base to assist attainment of thermal equilibrium. The authors report use of this apparatus for measurements on several alcohols in the range 0.01 to 1 Pa.

XIX. Gas-saturation Method

In the methods just considered, the vapor emanating from a condensed phase is removed under vacuum. An alternative method for removal is to sweep the vapor away in a stream of indifferent gas. This *transpiration, transport,* or *gas-saturation* method, which is applicable at pressures of around 3 kPa downwards, is one of the oldest methods for vapor-pressure measurement and has been used over a very wide range of temperature.

To a first approximation, which is often adequate, the equilibrium at modest pressures between condensed and vapor phases is unaffected by the presence of an uncondensable gas; i.e. in time, gas over a liquid or solid becomes saturated with its vapor, and the only effect of the gas is to slow down the rate at which equilibrium is reached in the vessel because transfer to the vapor phase can only take place by diffusion through the gas. If, instead of being stationary, the saturated gas is made to move over the condensed phase it will carry away the vapor and more of the condensed phase will evaporate, the process continuing until all of the condensed phase has been removed (a fact of common observation in relation to wind and water or ice). If the amounts of carrier gas saturated with vapor and of the substance transported are determined and their molar masses are known, the corresponding number of moles n_c, n_v may be calculated, and the vapor pressure is given by

$$p = [n_v/(n_v + n_c)] P \qquad (15)$$

where P is the total pressure.

Apparatus for determination of vapor pressure in this way comprises: (1) a source of gas (which may need purification, in particular, by removal of water vapor often present in laboratory compressed air and in commercial cylinders of gases); (2) means for controlling the flow of the gas; (3) a saturator maintained at constant temperature; (4) collection traps in which the transported material is removed from the carrier gas for subsequent assay; and (5) a means of measuring the volume of gas passing during an experiment. The accuracy achieved is directly related to the accuracy of the assay and of the measurement of the gas volume—the operative volume is that of the dry the gas is saturated with water, appropriate adjustment of the measured value the volume is measured at a different temperature and pressure, often when the gas is saturated with water, appropriate adjustment of the measured value is required (the procedure is given in detail by Thomson and Douslin[174]). It is also essential to insure that the gas is in fact saturated at the desired temperature, that the substance is quantitatively recovered, and that there is no entrainment of the condensed phase nor premature condensation between the saturator and collection traps. Various aspects of these requirements and of the apparatus will now be discussed in greater detail.

If there is no objection to the presence of oxygen in the carrier gas, dried air may be aspirated through the train of saturators and absorbers, and its volume can then be obtained from measurement of the volume of water run out of the aspirator. Air has often been used as the carrier gas in studies on the vapor pressures of solutions of salts, and for this purpose it has been found convenient to pass the same stream of air through saturators containing, first, water, and then the solution, with an absorber for the water vapor in between the two[180]. In such an experiment the vapor pressures of the solution and pure water are in effect compared, and it is not necessary to determine the exact volume of air that has passed because the term disappears from the equations. If a positive pressure of the carrier gas is applied, i.e. the gas is drawn from a cylinder or the laboratory compressed air supply, the volume may again be determined by displacement of water from an aspirator, or by means of a rotary wet gas meter. Whichever method is used, the temperature of the measuring device must be known and kept constant because the volume measured will be affected by the amount of water vapor in the gas. The gas meter must be calibrated[33] but its use may be advantageous if it is necessary for very large volumes of carrier to be used; though normally filled with water, a gas meter may be used with an involatile liquid such as butyl phthalate[133] and one of the corrections required in evaluation of the results is then eliminated. Pearce and Snow suggested electrolytic generation of the carrier gas and were able to obtain an exact value for the amount by means of a silver coulometer[135], but one may perhaps interpret some later investigators' description of their method as the use of 'detonating gas' as sensible advice[70].

Alternatively, the volume of the carrier gas may be obtained as a product of the flowrate and the duration of the experiment. It is now necessary for the flow to be kept constant by use of a good quality pressure regulator and possibly a constant mass-flow regulator, the flowrate being observed continuously by means of a capillary flowmeter or intermittently by means of a soap-film meter. Control and measurement of constant gas-flows is discussed in texts on gas chromatography[5], but one point may be emphasized here—the methods normally depend upon flow of the gas through restrictors, either capillary tubes or needle valves, and constancy depends upon constancy in the viscosity of the gas. Since the viscosity is temperature-dependent, restrictors must be kept at constant temperature for the most constant flow to be obtained.

At moderate temperatures the saturator may be immersed in a thermostat bath; no problem arises in determining its temperature, but it will be necessary to insure that the gas reaches the correct temperature. Saturation of the gas must be checked by finding whether the value obtained for the vapor pressure varies with flowrate or, alternatively, by finding whether a different value is obtained when the gas is presaturated at a higher temperature. Although increase in flowrate should ultimately lead to incomplete saturation of the gas it is possible that in fact the apparent vapor pressure will rise because, due to unsatisfactory design of saturator, droplets of liquid or fine particles of solid are entrained in the gas. It is advantageous if the pressure drop required to force the gas through the saturator is low, and a simple example of a saturator for liquids meeting this requirement is that described by Shaw[87, 158]. A much more complicated example is that described by Smith, Combs and Googin for use with aqueous solutions in which a mechanically driven rotor exposes a film of liquid to the gas[164]. Solids may be coated on a support such as glass helices[115], ground up with sand[51], or crushed and packed in beds without any intimately mixed support[133]. Baxter and Lansing stated that crushing of hydrated salts is essential if correct values for their aqueous vapor pressures are to be obtained, and that it is desirable for some of the salt to be present in a lower state of hydration so that any free water from the mother liquor of the crystallization is taken up[20]. A gas chromatography column is an efficient saturator, but it should be noted that here the particles are so small that curvature of the exposed surface of a liquid affects its vapor pressure and the effective value may not be the normal equilibrium value[83,136]. For inorganic substances at high temperatures thermal equilibrium is reached rapidly by radiation, and in many experiments the pre-heated gas is passed over the substance in a boat heated in a furnace[45].

After saturation the gas passes to the absorber—premature condensation must be avoided by keeping connecting tubing at a sufficiently high temperature. Alternatively, the gas stream may be diluted immediately after it leaves the saturator by addition of a subsidiary stream of gas[16], or else the contents of the tubing must be added for analysis to that collected in a trap[51]. Water is generally collected on a drying agent and assayed by weighing; other types of substance may be collected by refrigeration or by scrubbing with a suitable solution, and assayed by many different methods of analysis —since their sensitivities vary very greatly, the lower limits of applicability of the method and the amounts to be collected for a satisfactory determination vary correspondingly. Consideration must be given in setting up the apparatus to the possibility of losses of material due to adsorption or through connections between different parts—Menzies showed, for example[119], that glass-wool or asbestos-wool interfered with the accuracy of water determinations, and before the general availability of ground glass joints and couplings for metal tubing, it was recommended that joints made with elastomer tubing should have the glass tubes butted closely together so that as little as possible of the permeable material was exposed to the gas and vapor[24].

An alternative to collection of the substance is the continuous analysis of the gas stream. Use of the flame-ionization detector for this purpose has been proposed by Eggertsen, Seibert and Stross and by Franck[59,65].

Some complications which may sometimes need consideration are that calculation of the results on the basis of the ideal gas laws and the absence of molecular interaction between the carrier and the vapor is an oversimplification[69]; and that at relatively high vapor pressures, at low flowrates of the carrier gas, or where large temperature gradients exist the effect of diffusion may be significant[45,120,128].

XX. Differential Thermal Analysis

When heat is applied continuously to a body which is isolated from its surroundings, its temperature rises steadily unless a transition involving an enthalpy change takes place: if such a transition occurs the temperature/time record will exhibit a marked change in curvature, and this is most clearly revealed if the temperature of the sample is compared, by means of a differential thermocouple, with that of a reference sample, subjected to the same conditions, which does not undergo any transition. Commercial equipment is available for application of the method, *differential thermal analysis*, and Kemme and Kreps demonstrated its suitability for the measurement of vapor pressures[95]. They mixed samples of from four to twelve microliters of various organic compounds (aliphatic alcohols and alkyl chlorides) with about 40 mg of glass beads (size 100 μm) and used an equal mass of beads as the reference material. After the sample block had been placed in the analyser, the pressure in the chamber was adjusted to the required value and the temperature of the oven was raised at about 10 K min^{-1}. As soon as the instrument trace showed that evaporation was taking place the temperature of the sample was determined. Measurements were made in the pressure range 0.5 to 103 kPa, and tests with water at a pressure of about 31 kPa showed that the accuracy to be expected was ± 0.1 K. Advantages of the method are that it is fairly rapid, that it requires only a small sample, and that the trace from the instrument may give indication of the presence of impurity, or of decomposition of the sample at elevated temperatures.

XXI. Gas Chromatography

Gas chromatography may be used as an analytical method in vapor-pressure studies. For example, Friedrich and Stammbach measured the vapor pressures between 10^{-4} and 10 Pa (320 to 400 K) of some triazine herbicides by the gas-saturation method, and collected the transpired material in a cooled chromatographic column[68]. After a suitable volume of gas had passed through the saturator the column was disconnected and heated to a temperature suitable for chromatographic separation. At the same time as the chromatogram was started a measured amount of a reference material of similar character, e.g. another triazine, was injected into the column so that the detector response, and hence the amount of transpired material, could be determined. The advantages of the method are, first, that the analysis depends on a separation process, and small quantities of volatile impurities do not affect the result—as they would if the assay were done by a non-specific method such as weighing—and, secondly, that, as was demonstrated in the investigation, the vapor pressures of more than one substance could be determined in one run.

In a study of some thia-alkanes by static measurements, Mackle and McClean attached a small sampling chamber to the equilibrium vessel; when equilibrium had been reached the contents of the sampling chamber (which could be isolated from the rest of the system by means of a stopcock) were flushed on to a gas-chromatographic column for analysis[112]. These investigators did not calibrate the detector so that the actual vapor pressure could be found (although, in principle, this could have been done), but used the changes in peak size found for different temperatures of the equilibrium vessel, which are proportional to the changes in vapor pressure if the sample size is exactly reproducible, for calculation from the Clapeyron equation of the enthalpy of vaporization. The enthalpy of vaporization obtained in this way could be used for the construction of an equation which would allow extrapolation for the vapor pressure from values already known at higher temperatures.

The chromatographic process in gas–liquid chromatography is, however, itself a manifestation of vapor-pressure effects, and the technique is therefore applicable to vapor-pressure measurements in a more fundamental way. The basic parameter in gas–liquid chromatography is the *specific retention volume* V_g, the volume of carrier gas at the column pressure (less the volume of gas in the column) required to elute a substance divided by the mass of the stationary liquid phase in the column. The quantity is related to the vapor pressure p of the substance by the equation

$$\log V_g = \log(R/M) - \log p - \log \gamma \qquad (16)$$

where M is the molar mass of the substance and γ is its activity coefficient in solution in the stationary phase[5]. Since the determination of specific retention volume is difficult it is common practice for the retention to be expressed as a relative value $r_{1/2}$, the ratio of the retention of a substance 1 to that of a reference substance 2. The relative retention is related to the respective vapor pressures and activity coefficients by

$$\log r_{1/2} = \log p_2/p_1 + \log \gamma_2/\gamma_1 \qquad (17)$$

Closely related compounds have similar activity coefficients and for such pairs the second term of the RHS of equation 17 is nearly zero or, if not zero, varies in a regular manner with change of temperature or with change in substance, for example, as we proceed to successive members of a homologous series. Two practical consequences follow from these relationships; first, a straight line of slope approximately -1 is obtained if, over a range of temperature, the logarithms of the vapor pressure and of any retention parameter at the same temperature are plotted against each other, and furthermore the lines for different members of one homologous series coincide; and, second, a straight line is obtained if the logarithm of a retention parameter is plotted against boiling point. The chromatographer's normal interest is to relate unknown retentions to known vapor pressures (or their ratios) or boiling points, but there has now been enough study for the reverse process to be useful. Matukuma measured the retentions of 324 alkanes and alkenes and related their retentions to their boiling points sufficiently exactly to warrant his suggesting that some of the values given for the latter in the API tables might be in error[117]. Since much study has been devoted to relating retentions to molecular structure[155, 171] the method has potentialities as a basis for estimating unknown boiling points, or for identifying measured values which should be suspect either because they were obtained with samples which were impure or, no doubt in some instances, to which an incorrect structure had been assigned.

If the activity coefficient is known the vapor pressure may be calculated from equation 16, but the equation has been more frequently used for the calculation of activity coefficients from retention measurements and known vapor pressures. However, Duty and Mayberry determined the vapor pressures of solutions of hydrocarbons on the basis of equation 16 and suggested that the method might be suitable for industrial quality control[57].

Gas–liquid chromatography has considerable potentialities in the study of vapor–liquid equilibria but its use for straightforward vapor-pressure measurement has not been extensive. The complications of an exact treatment will therefore only be indicated here—equation 16 has been derived on the basis of ideality of the gas phase and this is not adequate if accurate results are sought, and the value of the specific retention volume is affected by the experimental conditions and its exact determination is difficult (see the reviews by Conder, by Giddings and Malik, by Martire and Pollara and by Young[43, 71, 116, 189]). When they are applicable, methods requiring measurement of relative retentions only are to be preferred.

XXII. Mass Spectrometry

This section is no more than a bare outline of the application of mass spectrometry to the determination of low vapor pressures. Suitable texts for further study are those by Beynon[28] and, for high temperatures, by Büchler and Berkowitz-Mattuck, and by Margrave[32,114].

Over a wide range of pressures up to a maximum of about 0.5 Pa the height of a peak in the mass spectrum of a compound is proportional to its partial pressure in the ionization chamber of the spectrometer. If a suitable equilibrium chamber and inlet system are attached to the mass spectrometer, measurements made on the spectra obtained for different temperatures of the equilibrium chamber allow calculation of heats of evaporation or sublimation. One advantage of the method is that measurements can be made despite the presence of impurities, including air if the sample has been incompletely degassed, provided the impurities do not contribute to the peaks chosen for study—comparison of several peaks in different parts of the spectrum will assist in checking this because, if they are due to a single component only, their ratios should remain constant as the temperature of the equilibrium chamber is changed.

The further step of converting the readings obtained to actual vapor pressures requires knowledge of the proportionality constant relating ion-current to the partial pressure of a species in the ionization chamber, which depends upon the instrumental parameters in a manner too complex to be dealt with in a brief discussion. Tickner and Lossing avoided the necessity of knowing this constant by using a calibrated leak between the equilibrium chamber and the ion chamber, i.e. they determined the response of the apparatus as a whole by independent experiments; they measured the vapor pressures of several hydrocarbons at low temperatures and, by inclusion of secondary reservoirs into which the vapor could be expanded before its admission to the mass spectrometer, were able to extend their measurements up to a pressure of 1.3 kPa[175].

A more widespread use of the mass spectrometer is for identification of the vapor species effusing from a Knudsen cell, and this has been particularly valuable in high-temperature studies of inorganic compounds[114].

XXIII. Vapor Pressures of Mixtures

The vapor–liquid equilibrium of multicomponent mixtures in which all the components are volatile is a subject warranting extended treatment such as is given by Hála, Pick, Fried and Vilím[79]. Static and ebulliometric methods are both applicable, but composition is now one of the variables and, for example, the design of an ebulliometer must allow sampling for analysis of both vapor and liquid phases (such an apparatus is generally called an *equilibrium still*). However, for technological purposes, all that may be needed is the boiling temperature of the mixture, and this may be determined in an ebulliometer, in which the temperature of the liquid is measured, similar to that shown in *Figure 7* but slightly redesigned so that there is a relatively large liquid sample, and therefore little change in its composition as a result of the formation of vapor when boiling takes place. Another technological need is for the vapor pressures of relatively involatile oils and here the gas-saturation method may be useful[42].

Another type of multicomponent mixture, in which an involatile component is dissolved in a volatile solvent—as in an aqueous salt solution—has been given passing mention earlier in the chapter. Static, ebulliometric and gas-saturation methods are applicable to measurement of the vapor pressure. Ebulliometry has been extensively used for determination of the increase in boiling point of a solvent due to solution of an involatile solute— and hence of the molecular weight of the solute. Swietoslawski has discussed the method in detail, and there is much of interest in his book relating both to this and to other applications of ebulliometry—for example, to the measurement of vapor pressures of mixtures of immiscible components[170]. Vapor pressures of solutions may be compared by the *isopiestic* or *isotonic* method[154]. If solutions (in the same solvent) of different salts are exposed in a vessel held at constant temperature, there will be a transfer of the solvent through the vapor phase until the resultant solutions all exert the same vapor pressure; the *isotonic* compositions may then be obtained by analysis of each solution. In a related method devised by Hill for determination of molecular weights, samples of a solution and the solvent are placed in the vessel and the temperatures of the two samples are measured by means of thermocouples or thermistors[85,118]. Since the vapor pressure of the pure solvent is higher than its vapor pressure over the solution evaporation from the former and condensation on the latter takes place, and the difference developed in the temperatures of the samples allows calculation of the difference in the two vapor pressures.

XXIV. References

[1] Adamek, J. and V. Ponec. *Coll. Czech. Chem. Commun.* **35**, 2477 (1970).
[2] Allen, P. W., D. H. Everett and M. F. Penney. *Proc. Roy. Soc. (London), Ser. A*, **212**, 149 (1952).
[3] Ambrose, D. *Trans. Faraday Soc.* **52**, 772 (1956).
[4] Ambrose, D. *J. Phys. E*, **1**, 41 (1968).
[5] Ambrose, D. *Gas Chromatography*, Butterworths: London (1971).
[6] Ambrose, D., 'Vapour pressures', in M. L. McGlashan, senior reporter, *Specialist Periodical Report: Chemical Thermodynamics*, Vol. I, p 218. The Chemical Society: London (1972).
[7] Ambrose, D., E. Broderick and R. Townsend. *J. Chem. Soc. A*, 633 (1967).
[8] Ambrose, D. and C. H. S. Sprake, *J. Chem. Thermodyn.* **2**, 631 (1970).
[9] Ambrose, D. and C. H. S. Sprake, *J. Chem. Soc. A*, 1263 (1971).
[10] Ambrose, D. and C. H. S. Sprake, *J. Chem. Thermodyn.* **4**, 603 (1972).
[11] Ambrose, D., C. H. S. Sprake and R. Townsend. *J. Chem. Thermodyn.* **1**, 499 (1969).
[12] Ambrose, D., C. H. S. Sprake and R. Townsend, *J. Chem. Thermodyn.* **4**, 247 (1972).
[13] Andrews, T., *Phil. Trans. Roy. Soc. London*, **159**, 575 (1869).
[14] Back, R. A. and J. Betts, *Canad. J. Chem.* **43**, 2157 (1965).
[15] Balson, E. W., *Trans. Faraday Soc.* **43**, 48 (1947).
[16] Balson, E. W., K. G. Denbigh and N. K. Adam, *Trans. Faraday Soc.* **43**, 42 (1947).
[17] Barber, C. R., *The Calibration of Thermometers*, HMSO: London (1971).
[18] Barber, C. R., R. Handley and E. F. G. Herington, *Brit. J. Appl. Phys.* **5**, 41 (1954).
[19] Barton, J. L. and H. Bloom, *J. Phys. Chem.* **60**, 1413 (1956).
[19a] Battino, R. and F. D. Evans, *Analyt. Chem.* **38**, 1627 (1966).
[20] Baxter, G. P. and J. E. Lansing, *J. Amer. Chem. Soc.* **42**, 419 (1920).
[21] Beattie, J. A., *Proc. Amer. Acad. Arts. Sci.* **69**, 389 (1934).
[22] Beattie, J. A., M. Benedict and B. E. Blaisdell, *Proc. Amer. Acad. Arts. Sci.* **71**, 327 (1936).
[23] Beattie, J. A. and D. G. Edwards, *J. Amer. Chem. Soc.* **70**, 382 (1948).
[24] Bechtold, M. F. and R. F. Newton, *J. Amer. Chem. Soc.* **62**, 1390 (1940).

[25] Bell, T. N., E. L. Cussler, K. R. Harris, C. N. Pepela and P. J. Dunlop, *J. Phys. Chem.* **72**, 4693 (1968).
[26] Bennett, M. J. and F. C. Tompkins, *Trans. Faraday Soc.* **53**, 185 (1957).
[27] Beynon, J. D. E. and R. B. Cairns, *J. Sci. Instrum.* **41**, 111 (1964).
[28] Beynon, J. H., *Mass Spectroscopy and its Application to Organic Chemistry*, Elsevier: Amsterdam (1960).
[29] Blend, H., *Rev. Sci. Instrum.*, **38**, 1527 (1967).
[30] Booth, H. S. and H. S. Halbedel, *J. Amer. Chem. Soc.* **68**, 2652 (1946).
[31] Booth, H. S. and C. F. Swinehart, *J. Amer. Chem. Soc.* **57**, 1337 (1935).
[32] Buchler, A. and Joan B. Berkowitz-Mattuck, 'Mass spectrometry in the measurement of vapor pressure', in R. A. Rapp, ed., *Techniques in Metals Research*, Vol. IV, Part 1, p 161. Interscience: New York (1970).
[33] Burwell, R. L., M. Metlay and F. W. Pfohl, *Rev. Sci. Instrum.* **21**, 681 (1950).
[34] Calingaert, G., H. Soroos, V. Hnizda and H. Shapiro, *J. Amer. Chem. Soc.* **66**, 1389 (1944).
[35] Carlson, K. D., 'The Knudsen effusion method' in J. L. Margrave, ed., *The Characterization of High-Temperature Vapors*, p 115. Wiley: New York (1967).
[36] Cater, E. D., 'Measurement of gross equilibrium vaporization rate (Knudsen methods)', in R. A. Rapp, ed., *Techniques of Metals Research*, Vol. IV, Part 1, p 21. Interscience: New York (1970).
[37] Clark, A. L., *Chem. Rev.* **23**, 1 (1938).
[38] Clarke, A. M., F. Din and J. Robb, *Physica*, **17**, 876 (1951).
[39] Clarke, E. C. W. and D. N. Glew, *Canad. J. Chem.* **48**, 964 (1970).
[40] Clausing, P., *Ann. Phys., Lpz.*, **12**, 961 (1932).
[41] Clopper, P. R., R. L. Altman and J. L. Margrave, 'Static techniques, dew points, vapor densities, and boiling point measurements', in J. L. Margrave, ed., *The Characterization of High-Temperature Vapors*, p 48. Wiley: New York (1967).
[42] Coburn, J. F., *ASLE (Amer. Soc. Lubric. Engrs), Trans.* **12**, 129 (1969).
[43] Conder, J. R., 'Physical measurements by gas chromatography', in J. H. Purnell, ed., *Progress in Gas Chromatography (Advanc. Analyt. Chem. Instrum.*, Vol. VI), p 209. Interscience: New York (1968).
[44] Connolly, J. F. and G. A. Kandalic, *Physics of Fluids*, **3**, 463 (1960).
[45] Cooper, R. and D. R. Stranks, 'Vapor pressure measurements', in H. B. Jonassen and A. Weissberger, eds., *Technique of Inorganic Chemistry*, Vol. VI, p 1. Interscience: New York (1966).
[46] Cottrell, F. G., *J. Amer. Chem. Soc.* **41**, 721 (1919).
[47] Counsell, J. F., J. H. S. Green, J. L. Hales and J. F. Martin, *Trans. Faraday Soc.* **61**, 212 (1965).
[48] Cruickshank, A. J. B. and A. J. B. Cutler, *J. Chem. Engng Data*, **12**, 326 (1967).
[49] Dainton, F. S. and H. M. Kimberley, *Trans. Faraday Soc.* **46**, 912 (1950).
[50] Davison, R. R., W. H. Smith Jr and K. W. Chun, *Amer. Inst. Chem. Engrs Jnl*, **13**, 590 (1967).
[51] Dickinson, W., *Trans. Faraday Soc.* **52**, 31 (1956).
[52] Diepen, G. A. M. and F. E. C. Scheffer, *J. Amer. Chem. Soc.* **70**, 4085 (1948).
[53] Dietz, V., *J. Chem. Phys.* **4**, 575 (1936).
[54] Douslin, D. R., R. T. Moore, J. P. Dawson and G. Waddington, *J. Amer. Chem. Soc.* **80**, 2031 (1958).
[55] Douslin, D. R. and A. Osborn, *J. Sci. Instrum.* **42**, 369 (1965).
[56] Dushman, S., *Scientific Foundations of Vacuum Technique*, 2nd ed. Wiley: New York (1962).
[57] Duty, R. C. and W. R. Mayberry, *J. Gas Chromatogr.* **4**, 115 (1966).
[58] Edwards, J. L. and D. P. Johnson, *J. Res. Nat. Bur. Stand.* **72C**, 27 (1968).
[59] Eggertsen, F. T., E. E. Seibert and F. H. Stross, *Analyt. Chem.* **41**, 1175 (1969).
[60] Ernsberger, F. M. and H. W. Pitman, *Rev. Sci. Instrum.* **26**, 584 (1955).
[61] Fischer, A. K., *Rev. Sci. Instrum.* **37**, 717 (1966).
[62] Flint, E. P. *Chem. & Ind. (London)*, 1618 (1968).
[63] Fogler, M. F. and W. H. Rodebush, *J. Amer. Chem. Soc.* **45**, 2080 (1923).
[64] Fowler, L., W. N. Trump and C. E. Vogler, *J. Chem. Engng Data*, **13**, 209 (1968).
[65] Franck, A. *Chemiker-Ztg*, **93**, 668 (1969).
[66] Freeman, R. D., 'Momentum sensors', in J. L. Margrave, ed., *The Characterization of High-Temperature Vapors*, p 152. Wiley: New York (1967).
[67] Freeman, R. D. and J. G. Edwards, 'Transmission probabilities and recoil force correction factors for conical orifices', in J. L. Margrave, ed., *The Characterization of High-Temperature Vapors*, p 508. Wiley: New York (1967).

68 Friedrich, K. and K. Stammbach, *J. Chromatogr.* **16**, 22 (1964).
68a Gerrard, W., *J. Appl. Chem. Biotechnol.* **22**, 623 (1972).
69 Gerry, H. T. and L. J. Gillespie, *Phys. Rev.* **40**, 269 (1932).
70 Gibson, R. E. and L. H. Adams, *J. Amer. Chem. Soc.* **55**, 2679 (1933).
71 Giddings, J. C. and K. L. Malik, *Industr. Engng Chem.* **59**, No. 4, 18 (1967).
72 Gilles, P. W., 'Vaporization processes', in J. L. Margrave, ed., *The Characterization of High-Temperature Vapors*, p 19. Wiley: New York (1967).
73 Goncharov, A. K. and M. Kh. Karapet'yants, *Tr. Mosk. Khim.-Tekhnol. Inst.* No. 62, 17 (1969).
74 Goodwin, R. D., *J. Res. Nat. Bur. Stand.* **65C**, 231 (1961).
75 Greenwood, N. N., P. G. Perkins and M. E. Twentyman, *J. Chem. Soc. A.* 2109 (1967).
76 Haar, L. and J. M. H. Levelt Sengers, *J. Chem. Phys.* **52**, 5069 (1970).
77 Hackham, R., W. E. Austin and R. D. Thomas, *J. Sci. Instrum.* **42**, 344 (1965).
78 Hackspill, M. L., *Ann. Chim. (Phys.)*, **28**, 613 (1913).
79 Hála, E., J. Pick, V. Fried and O. Vilím. *Vapour–Liquid Equilibrium* (trans. G. Standart), 2nd ed. Pergamon: Oxford (1968).
80 Hannay, J. B., *Proc. Roy. Soc (London)*, Ser. *A*, **30**, 484 (1880).
81 Hannay, J. B. and J. Hogarth, *Proc. Roy. Soc. (London)*, Ser. *A*, **30**, 178 (1880).
82 Harris, K. R. and P. J. Dunlop, *J. Chem. Thermodyn.* **2**, 805 (1970).
83 Hawkes, S. J. and J. C. Giddings, *Analyt. Chem.* **36**, 2229 (1965).
84 Hickman, K. C. D., J. C. Hecker and N. D. Embree, *Analyt. Chem.* **9**, 264 (1937).
85 Hill, A. V., *Proc. Roy. Soc., (London)*, Ser. *A*, **127**, 9 (1930).
86 Hoge, H. J., *J. Res. Nat. Bur. Stand.* **44**, 321 (1950).
87 Holm-Jensen, I., *Analyt. Chim. Acta*, **23**, 13 (1960).
88 Hopke, E. R. and G. W. Sears, *J. Amer. Chem. Soc.* **70**, 3801 (1948).
89 Jenkins, A. C. and C. M. Birdsall, *J. Chem. Phys.* **20**, 1158 (1952).
90 Jepson, W. B., M. J. Richardson and J. S. Rowlinson, *Trans. Faraday Soc.* **53**, 1586 (1957).
91 Jepson, W. B. and J. S. Rowlinson, *J. Chem. Phys.* **23**, 1599 (1955).
92 Kaufman, M. H. and A. G. Whittaker, *J. Chem. Phys.* **24**, 1104 (1956).
93 Kay, W. B., *J. Amer. Chem. Soc.* **68**, 1336 (1946); **69**, 1273 (1947).
94 Kay, W. B. and G. M. Ranbosch, *Industr. Engng Chem.* **45**, 221 (1953).
95 Kemme, H. R. and S. I. Kreps, *J. Chem. Engng Data*, **14**, 98 (1969).
96 Klumb, H. and J. Lückert, *Vakuum-Tech.* **8**, 62 (1959).
97 Klumb, H. and H. Schwarz. *Z. Phys.* **122**, 418 (1944).
98 Knudsen, M., *Ann. Phys., Lpz.*, **32**, 809 (1910).
99 Knudsen, M., *The Kinetic Theory of Gases.* Methuen: London (1950).
100 Knudsen, M., *Ann. Phys., Lpz.*, **29**, 179 (1909); **34**, 593 (1911).
101 Kobe, K. A. and R. E. Lynn. *Chem. Rev.* **52**, 117 (1953).
102 Komarek, K. L., in O. Kubaschewski, ed., *Metallurgical Chemistry*, p. 75. H.M.S.O.: London (1972).
103 Kreglewski, A., 'Vapour pressure and boiling temperature measurements', in L. A. K. Staveley, ed., *The Characterization of Chemical Purity: Organic Compounds*, p 51. Butterworths: London (1971).
104 Kubaschewski, O., E. Ll. Evans and C. B. Alcock, *Metallurgical Thermochemistry*, 4th ed. Pergamon: Oxford (1967).
105 Kudchadker, A. P., G. H. Alani and B. J. Zwolinski, *Chem. Rev.* **68**, 659 (1968).
106 Langmuir, I, *J. Amer. Chem. Soc.* **35**, 105 (1913).
107 Langmuir, I, *Phys. Rev.* **2**, 239 (1913).
108 Langmuir, I. and C. M. Mackay, *Phys. Rev.* **4**, 377 (1914).
109 Liang, S. Chu, *J. Appl. Phys.* **22**, 148 (1951).
110 Maass, O. *Chem. Rev.* **23**, 17 (1938).
111 Machin, W. D., *Canad. J. Chem.* **45**, 1904 (1967).
112 Mackle, H. and R. T. B. McClean, *Trans. Faraday Soc.* **60**, 817 (1964).
113 Margrave, J. L., 'Vapour pressure', in J. O'M Bockris, J. L. White and J. D. Mackenzie, eds., *Physicochemical Measurements at High Temperatures*, p. 225. Butterworths: London (1959).
114 Margrave, J. L., ed., *The Characterization of High-Temperature Vapors*, Wiley: New York (1967).
115 Martin, J. F. and D. P. Biddiscombe, *Trans. Faraday Soc.* **54**, 1316 (1958).
116 Martire, D. E. and L. Z. Pollara. 'Interactions of the solute with the liquid phase', in J. C.

Giddings and R. A. Keller, eds., *Advances in Chromatography*, Vol. I, p 335. Dekker: New York (1965).
117 Matukuma, A., *Gas Chromatography 1968* (ed. C. L. A. Harbourn), p 55. Institute of Petroleum: London (1969).
118 Meeks, A. C. and I. J. Goldfarb, *Analyt. Chem.* **39**, 908 (1967).
119 Menzies, A. W. C., *J. Amer. Chem. Soc.* **42**, 978 (1920).
120 Merten, U. and W. E. Bell, 'The transpiration method', in J. L. Margrave, ed., *The Characterization of High-Temperature Vapors*, p 91. Wiley: New York (1967).
120a Meves, C. H. and M. S. Van Dusen, *J. Res. Nat. Bur. Stand.* **10**, 381 (1933).
121 Michels, A. and T. Wassenaar, *Physica*, **14**, 104 (1948).
122 Miller, G. A., *J. Chem. Engng Data*, **8**, 69 (1963).
123 Motzfeldt, K., *J. Phys. Chem.* **59**, 139 (1955).
124 Motzfeldt, K., *Acta Chem. Scand.* **18**, 1795 (1964).
125 Mueller, E. P. and H. A. Burgess, *J. Amer. Chem. Soc.* **41**, 745 (1919).
126 Nesmeyanov, A. N., *Vapor Pressures of the Elements* (trans. J. I. Carasso). Inforsearch: London (1963).
127 Neumann, K. and E. Völker, *Z. Phys. Chem. A*, **161**, 33 (1932).
128 Norman, J. H. and P. Winchell, 'Measurement of vapor pressures by transpiration, isopiestic, and other techniques', in R. A. Rapp, ed., *Techniques of Metals Research*, Vol. IV, Part 1, p 131. Interscience: New York (1970).
129 Oliver, G. D. and J. W. Grisard, *Rev. Sci. Instrum.* **24**, 204 (1953).
130 Oliver, G. D. and J. W. Grisard, *J. Amer. Chem. Soc.* **78**, 561 (1956).
131 Osborn, A. G. and D. R. Douslin, *J. Chem. Engng Data*, **11**, 502 (1966).
132 Osborne, N. S., H. F. Stimson and D. C. Ginnings, *J. Res. Nat. Bur. Stand.* **23**, 197 (1939).
133 Overberger, J. E., W. A. Steele and J. G. Aston, *J. Chem. Thermodyn.* **1**, 535 (1969).
134 Paule, R. C. and J. L. Margrave, 'Free-evaporation and effusion techniques', in J. L. Margrave, ed., *The Characterization of High-Temperature Vapors*, p 130. Wiley: New York (1967).
135 Pearce, J. N. and R. D. Snow, *J. Phys. Chem.* **31**, 231 (1927).
136 Petsev, N. and C. Dimitrov. *J. Chromatogr.* **34**, 310 (1968).
137 Porter, A. W., *Proc. Roy. Soc. (London), Ser. A*, **79**, 519 (1907); **80**, 457 (1908).
138 Poynting, J. H., *Phil. Mag.* **12**, 32 (1881).
139 Ramsay, W., *Proc. Roy. Soc. (London), Ser. A*, **30**, 328 (1880).
140 Ramsay, W. and S. Young, *Phil. Trans. Roy. Soc. London*, **175**, 37 (1884).
141 Rapp, R. A., ed., *Techniques in Metals Research*, Vol. IV, Part 1. Interscience: New York (1970).
142 Richardson, M. J. and J. S. Rowlinson, *Trans. Faraday Soc.* **55**, 1333 (1959).
143 Robin, S. and B. Vodar, *J. Phys. Radium*, **13**, 264 (1952).
144 Robinson, R. A. and D. A. Sinclair. *J. Amer. Chem. Soc.* **56**, 1830 (1934).
145 Röck, H., *Z. Phys. Chem.* **4**, 242 (1955).
146 Rodebush, W. H. and C. E. Coons, *J. Amer. Chem. Soc.* **49**, 1953 (1927).
147 Rodebush, W. H. and A. L. Dixon, *Phys. Rev.* **26**, 851 (1925).
148 Rodebush, W. H. and W. F. Henry, *J. Amer. Chem. Soc.* **52**, 3159 (1930).
148a Rondeau, R. E., *J. Chem. Educ.* **44**, 530 (1967).
149 Rosen, F. D. and D. Wallace, *Rev. Sci. Instrum.* **24**, 349 (1953).
150 Rosenberg, A. J. and C. S. Martel, *J. Phys. Chem.* **62**, 457 (1958).
151 Rossini, F. D., B. J. Mair and A. J. Streiff, *Hydrocarbons from Petroleum*. Reinhold: New York (1953).
152 Rossmann, M. G. and J. Yarwood, *J. Chem. Phys.* **21**, 1406 (1953).
153 Rowlinson, J. S., *Liquids and Liquid Mixtures*, 2nd ed. Butterworths: London (1969).
154 Scatchard, O., W. J. Hamer and S. E. Wood, *J. Amer. Chem. Soc.* **60**, 3061 (1938).
155 Schomburg, G., *Analyt. Chim. Acta*, **38**, 45 (1967).
156 Sears, G. W. and E. R. Hopke, *J. Phys. Chem.* **52**, 1137 (1948).
157 Sengers, J. V. and A. L. Sengers. *Chem. Engng News*, **46**, 104 (10 June 1968).
158 Shaw, J. A., *Industr. Engng Chem. (Analyt. Ed.)*, **6**, 479 (1934).
159 Shepherd, M., *J. Res. Nat. Bur. Stand.* **12**, 185 (1934).
160 Silva, W. J., J. W. Johnson and D. Cubicciotti, *Rev. Sci. Instrum.* **36**, 1505 (1965).
161 Singh, Jaswant and G. C. Benson, *Canad. J. Chem.* **46**, 1249 (1968).
162 Siwoloboff, A., *Ber. Dtsch. Chem. Ges.* **19**, 795 (1886).
163 Smith, A. and A. W. C. Menzies, *J. Amer. Chem. Soc.* **32**, 1412 (1910).
164 Smith, H. A., R. L. Combs and J. M. Googin, *J. Phys. Chem.* **58**, 997 (1954).

[165] Spauschus, H. O., *Rev. Sci. Instrum.* **32**, 1279 (1961).
[166] Spedding, F. H. and J. L. Dye, *J. Phys. Chem.* **59**, 581 (1955).
[167] Straty, G. C. and R. Prydz, *Rev. Sci. Instrum.* **41**, 1223 (1970).
[168] Stubley, D. and J. S. Rowlinson, *Trans. Faraday Soc.* **57**, 1275 (1961).
[169] Sunner, S. and N. Magnusson, *Acta Chem. Scand.* **4**, 1464 (1950).
[170] Swietoslawski, W., *Ebulliometric Measurements*, Reinhold: New York (1945).
[171] Takács, J., C. Szita and G. Tarján, *J. Chromatogr.* **56**, 1 (1971).
[172] Thomas, L. H., H. Smith and R. Meatyard, *J. Phys. E*, **1**, 1119 (1968).
[173] Thomson, G. W., 'Determination of vapor pressure', in A. Weissberger, ed., *Physical Methods of Organic Chemistry*, Vol. I, Part 1, p 357. 3rd ed. Interscience: New York (1959).
[174] Thomson, G. W. and D. R. Douslin, 'Determination of pressure and volume', in A. Weissberger and B. W. Rossiter, eds., *Physical Methods of Chemistry*, Vol. I, Part 1, p 23. Wiley: New York (1971).
[175] Tickner, A. W. and F. P. Lossing, *J. Phys. Chem.* **55**, 733 (1951).
[176] Turcotte, R. P., T. D. Chikalla and L. Eyring, *Analyt. Chem.* **43**, 958 (1971).
[177] Verhoek, F. H. and A. L. Marshall, *J. Amer. Chem. Soc.* **61**, 2737 (1939).
[178] Volmer, M. Z., *Phys. Chem., Bodenstein-Festbound*, 863 (1931).
[179] Ward, J. W. and M. V. Fraser, *J. Chem. Phys.* **50**, 1877 (1969).
[180] Washburn, E. W. and E. O. Heuse, *J. Amer. Chem. Soc.* **37** 309 (1915).
[181] Washburn, E. W. and J. W. Read, *J. Amer. Chem. Soc.* **41**, 729 (1919).
[182] Waxman, M. and W. T. Chen, *J. Res. Nat. Bur. Stand.* **69C**, 27 (1965).
[183] Webster, T. J., *J. Soc. Chem. Ind. (London)*, **69**, 343 (1950).
[184] Whitman, C. I., *J. Chem. Phys.* **20**, 161 (1952); **21**, 1407 (1953).
[185] Wiedemann, H. G., *Thermochim. Acta*, **3**, 355 (1972).
[186] Willingham, C. B., W. J. Taylor, J. M. Pignocco and F. D. Rossini, *J. Res. Nat. Bur. Stand.* **35**, 219 (1945).
[187] Wilson, A. E., J. H. Kim and A. Cosgarea, *Rev. Sci. Instrum.* **36**, 1428 (1965).
[188] Winterbottom, W. L., 'Free-vaporization measurements (Langmuir technique)', in R. A. Rapp, ed. *Techniques of Metals Research*, Vol. IV, Part 1, p 95. Interscience: New York (1970).
[189] Young, C. L., *Chromatogr. Rev.* **10**, 129 (1968).
[190] Young, S., *J. Chim. Phys.* **4**, 425 (1904).
[191] Young, S., *Sci. Proc. Roy. Dublin Soc.* **12**, 374 (1910).
[192] Zmaczynski, A., *J. Chem. Phys.* **27**, 503 (1930).

CHAPTER 14

Thermodynamic Properties near the Critical State

J. M. H. LEVELT SENGERS

Institute for Basic Standards, National Bureau of Standards

Contents

I.	Introduction	658
II.	Theoretical Background	660
	1. The Origin of Critical Anomalies	660
	2. Power Laws	661
	3. Symmetry	664
	4. Homogeneity and Scaling	669
	5. Beyond Simple Scaling	675
III.	Special Experimental Difficulties	677
	1. Divergences and Their Consequences	677
	2. Gravity	678
	3. Equilibration	681
IV.	Refractive Index Measurements	684
	1. Principle	684
	2. The Use of Optical Techniques for Bulk Density Determination	685
	3. Local Density Determination Using Refractive Index	686
	4. Density Gradient Determination	687
V.	Dielectric Constant Measurements	689
	1. General	689
	2. $p\varepsilon T$ Measurements	690
	3. Density Profiles by Dielectric Constant Determination	690
VI.	Conventional pVT and Vapor Pressure Measurements	691
	1. pVT Measurements	691
	2. The Vapor Pressure	693
VII.	Calorimetry	693
	1. Experimental Problems in C_v Determination	693
	2. Reducing the Heat Capacity of the Container	695
	3. Long Relaxation Times	695
	4. Correcting for Gravity	696
	5. Increasing the Temperature Resolution	698
	6. Checking for Consistency	698
	7. Tests of Scaling	699
VIII.	Coexistence Curves	700
	1. General	700
	2. Gravity	700
	3. The Method of Meniscus Disappearance	702
	4. Young's Method of the Twin Cells	703
	5. Coexistence Curves from Isothermal and Isochoric Intercepts	703
	6. Coexistence Curves by Dielectric Constant and Refractive Index Techniques	705
	7. The Use of Floats	705
	8. Power Law Analysis of Coexistence Curves	706
	9. The Diameter of the Coexistence Curve	708
IX.	Scattering	709
	1. Introduction	709

	2.	Intensity of Scattered Light	709
	3.	Angular Dependence of the Intensity of Scattered Light	711
	4.	Light Scattering and Small-angle X-Ray Scattering	714
	5.	The Experimental Situation in Critical Opalescence	714
	6.	The Spectrum of Scattered Light	715
X.	Sound	718
	1.	Sound, Ultrasound and Hypersound	718
	2.	Gravity Effects in Sound Velocity Measurements	718
	3.	Sound Dispersion and Attenuation	719
XI.	Concluding Remarks	719
XII.	Acknowledgements	720
XIII.	References	721

I. Introduction

The study of thermodynamic properties in the vicinity of the critical point has a venerable history of about a century. The *Ph.D. thesis* of van der Waals[148], treating the continuity of the gaseous and liquid states, has provided the theoretical guidelines for the early experimental studies. They were carried out principally at the University of Leiden by Kamerlingh Onnes, Keesom, Verschaffelt, Crommelin and others, and with less intensity, at other institutions, such as the Physikalisch-Technische Reichsanstalt in Germany. After 1920, the interest in the subject slackened; there were some occasional flashes of very good experimental work, such as that of Michels and co-workers in Holland and of Maass and co-workers in Canada in the 'thirties; of Murray and Mason, and of Schneider and co-workers in Canada in the 'fifties; however, the true importance of this work became apparent only recently, after a considerable growth in theoretical understanding.

A revival of interest had to wait for two other strands of developments. One was the finding of very high λ-shaped specific heat curves at certain solid state transitions and in liquid ^{4}He at the superfluid transition. The second was the refinement of calculations on the Ising model for the magnetic transition, finally resulting in Onsager's famous exact solution of the two-dimensional model in zero field (1944)[105]. The surprising aspect of this solution was a logarithmic divergence of the specific heat C_v in the lattice gas. This had profound consequences because it implies that the free energy does not possess a Taylor expansion in volume and temperature at the critical point. This result is in complete contradiction to classical theories like that of van der Waals. Although this possibility of a non-analyticity of the free energy at the critical point caused a considerable amount of discomfort to experimentalists, it could really have been raised half a century earlier. Shortly after 1900 it was already abundantly clear that the coexistence curve of gases is approximately cubic, in contrast to the prediction of a quadratic curve by van der Waals; (the assumption of an analytic free energy leads to even-powered coexistence curves). Historically, however, the idea that the gas–liquid transition might have a non-analytic character required the insight of Onsager[105], Tisza[138] and Lee and Yang[83] that there is an analogy between the gas–liquid critical point and the order–disorder transitions in binary alloys, magnets, liquid ^{4}He and the Ising model. A crucial step in the application of Ising model results to the gas–liquid transition was the discovery of a weakly diverging specific heat C_v in the critical regions of argon, nitrogen and oxygen by Voronel and co-workers[5, 142–145]. The situation is now such that most experimental work on thermodynamic properties in the critical region is carried out in anticipation of non-analytic behavior. In such research, one can distinguish two stages of analysis. First, experimental properties are analysed for deviations from analytic behavior on approaching the critical point along certain preferred paths, such as the coexistence curve, the critical isotherm, isochore or isobar. The mathematical description used is that of the power laws[45] here described in section II. The first power-law analysis in the critical region of gases was made by Verschaffelt around 1900[139]. Next, an attempt is made at describing the thermodynamic behavior throughout the critical region. This is done by means of the so-called scaling laws, introduced by Widom in 1965[153]. For gases, such a scaled analysis is possible only after the symmetry character of the critical region is assessed. Section II provides the necessary background for present-day experimentation in the critical region, by summarizing the ideas of power laws, symmetry and scaling. We finish that section with a discussion of some new ideas about thermodynamic behavior beyond the range of validity of the present scaling theory.

The discussion of the experimental situation starts in section III with a treatment of the basic experimental difficulties. These difficulties are principally due to the divergence of several properties at the critical point. The divergence of the *expansion coefficient* makes the pressure a very awkward parameter for controlling the system near the critical point. The divergence of the *compressibility* results in density gradients in the field of gravity, due to the weight of the gas. This effect, which reduces the precision of classical thermodynamic measurements near the critical point, can be turned to advantage by studying these properties as functions of height in the cell; this way it is possible to boost pressure resolution by two orders of magnitude. Finally, the divergence of the *thermal diffusivity* causes the long equilibration times which have, over the years, tried the patience of many experimenters studying critical phenomena.

Sections IV and V are devoted to two powerful and rather new techniques used in critical region studies, refractive index and dielectric constant methods. Both have been used to study properties as functions of height. Section VI discusses pVT measurements by conventional methods. Although Volume I in this series treats calorimetry in general, calorimetry in the critical region presents such peculiar challenges and offers such great rewards, that section VII has been devoted to this subject. Methods of coexistence curve determination are discussed in section VIII. This gives us a chance to review a variety of classical and modern methods of densimetry on gases and liquids.

A chapter on critical phenomena could hardly be called complete if the most powerful new tool, laser light scattering, were not discussed. A summary of the method, its potential, and some of the most important results, is given in section IX. A brief summary of x-ray scattering in the critical region is found in the same section. A short discussion of sound propagation in the critical region is given in section X. Concluding remarks, in section XI, complete the Chapter.

II. Theoretical Background

1. *The Origin of Critical Anomalies*

If in a system in equilibrium small changes are made in some of the intensive variables or fields, the system reacts by changes in the extensive variable. Thus, if pressure is exerted on a gas its volume will decrease. The more stable the system, the smaller its response to a disturbance. The characteristic feature of a critical point is that at this point the system reaches the limit of its stability[138]. This is evident in the fact that responses to disturbances become very large. Thus, the compressibility, the volume response to a pressure change at constant temperature, diverges strongly at the critical point. Likewise, the specific heat C_p, which is an entropy response to a temperature change at constant pressure, diverges strongly; so does the expansion coefficient, the volume response to a temperature change at constant pressure. In the ferromagnet, the susceptibility, which is the response of the magnetization to the applied field, is a strongly diverging quantity. There are also properties which exhibit more subtle divergences at the critical point. Thus the specific heat C_v in gases and C_M in magnets, which are energy responses to temperature changes, diverge slowly.

The strongly diverging responses are associated with large fluctuations in the corresponding extensive variables. In ferromagnets for example, there is partial local alignment of the spins above the Curie temperature, even though there is no spontaneous magnetization. The average diameter of these regions of local ordering is called the *correlation length* ξ. In gases there are strong fluctuations of the local density and entropy. Near the critical point, the correlation length becomes so large that visible light is strongly scattered; this is the phenomenon of critical opalescence. The presence of these large fluctuations, which may extend over many interparticle distances, is the origin of the critical phenomena, with the behavior of the system determined by the size and nature of the fluctuations, rather than by the particular features of the molecular interaction. This is why critical phenomena have a character of *universality*[73]. Universality means that in systems in which the fluctuations have the same nature, the critical phenomena have the same character. The nature of the fluctuation is determined by a few very general features, such as the dimensionality of the system, the range of the interparticle forces, the symmetry character of the Hamiltonian and, of course, the distance from the critical point. Universality implies that identical critical behavior is expected within the class of three-dimensional Ising systems; for all three-dimensional Heisenberg magnets; and within the class of gases and binary fluids, with a possible exception for quantum fluids. The so-called *power laws* are used for the description of this behavior (section II.2). The distance from the critical point is characterized by two variables which are, for the two independent thermodynamic fields, their differences from the critical values. One variable describes a displacement of the system away from the critical point along the coexistence curve. Its conjugate is an energy, a weakly fluctuating quantity. The other variable, H in the magnet, $p - p_c$, or, more precisely, $\mu - \mu_c$ in the gas, drives the system away from the critical point in a direction not parallel to the coexistence curve. Its conjugate is, respectively, the magnetization M or the density $\rho - \rho_c$, a strongly fluctuating quantity. The *scaling laws* (section II.4) reduce the dependence of the fluctuations and the thermal behavior on these two variables to a dependence on a single variable, which is a combination of the original two.

After this general introduction, the rest of this chapter will be devoted to more specialized topics of direct interest to the experimentalist working in the critical region of gases. We will discuss the power laws (section II.2) and the scaling laws (section II.4). Before introducing the latter, a study of the special symmetry of the gas–liquid case must be made (section II.3). The chapter ends with some speculation about possible extensions of scaling farther away from critical points.

2. Power Laws

Power laws are the means by which the asymptotic thermodynamic behavior near a critical point is described. As an example, consider the van der Waals equation[148]. At the critical point, the pressure can be expanded in a Taylor series in density and temperature. Since $(\partial p/\partial \rho)_c$ and $(\partial^2 p/\partial \rho^2)_c$ equal zero, one finds, retaining only low-order terms:

$$p - p_c = p_{30}(\rho - \rho_c)^3 + p_{10}(T - T_c) + p_{11}(\rho - \rho_c)(T - T_c) + \ldots \quad (1)$$

where p_{30}, p_{10} and p_{11} are constants related to pressure derivatives with respect to ρ and T taken at the critical point[84]. The asymptotic form of the critical isotherm, $T = T_c$, is then given by $p - p_c = p_{30}(\rho - \rho_c)^3$, or along the *critical isotherm*,

$$|\Delta p| \sim |\Delta \rho|^\delta \quad (2)$$

with $\delta = 3$, $\Delta p = (p - p_c)/p_c$, and $\Delta \rho = (\rho - \rho_c)/\rho_c$. Likewise, we derive from equation 1 that on the critical isochore ($\rho = \rho_c$) the compressibility $K_T = [\rho(\delta p/\delta \rho)]^{-1}$ behaves as

$K_T = [\rho_c p_{11}(T - T_c)]^{-1}$, or along the *critical isochore*,

$$K_T \sim |\Delta T|^{-\gamma} \quad (3)$$

with $\gamma = 1$ and $\Delta T = (T - T_c)/T_c$. With some effort[4, 79, 84] the shape of the coexistence curve is found to be $\Delta \rho = \rho_c - \rho_G = \rho_L - \rho_c \sim |T_c - T|^{\frac{1}{2}}$ or along the *coexistence curve*;

$$\Delta \rho \sim |\Delta T|^\beta \quad (4)$$

with $\beta = 0.5$. The coexisting densities can be expanded in powers of $|\Delta T|^{\frac{1}{2}}$ so that the diameter of the coexistence curve, $\bar{\rho} = (\rho_L + \rho_G)/2 = \rho_c + \Delta \rho_L - \Delta \rho_G$ will start off as

$$\bar{\rho} = \rho_c + a|\Delta T|^{2\beta} \quad (5)$$

or $\bar{\rho} = \rho_c + a|\Delta T|$, the 'law of the rectilinear diameter'. According to the van der Waals equation, the constant-volume properties C_v and $(\partial^2 p/\partial T^2)_v$ are finite at the critical point; however, these properties exhibit a sudden change at T_c along the critical isochore[4, 7].

The exponent values obtained above follow from the fact that a Taylor expansion was used and that $(\partial p/\partial \rho)_c$ and $(\partial^2 p/\partial \rho^2)_c$ are zero. They have nothing to do with the special form of the van der Waals equation and are therefore a universal property of all equations that possess a Taylor expansion and in which only two derivatives are zero. These equations are called classical. Most equations used by engineers and chemists are classical in this sense.

However, this classical thermodynamic behavior is in conflict with experiment. The best-known example of such a conflict is that of the gas–liquid coexistence curve, equation 4; for 70 years it has been known to be approximately cubic rather than quadratic[52,139]. The second example is the specific heat C_v which has been found to diverge weakly[5,142–145,102], instead of remaining finite. For describing thermodynamic anomalies in non-classical systems one assumes that power laws still hold, but that the

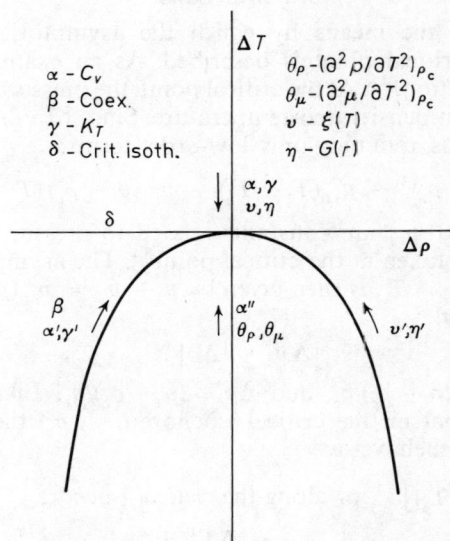

Figure 1. The definition of the most common critical exponents is indicated in a $T-T_c$, $\rho-\rho_c$ plane. The exponents describe the power-law behavior of the respective properties along the preferred curves indicated: critical isochore, critical isotherm and coexistence curve.

critical exponents may have non-classical values[154,45]. The definitions of these exponents (*Figure 1*) and their approximate values found by analysis of experimental data for real systems are listed in *Table 1*. For gases, these values turn out to be virtually independent of the substance chosen, a first example of the universality of critical behavior (section II.1). The definition of the specific heat exponent α in *Table 1* deserves a few comments. In the classical case, the specific heat remains finite; its critical anomaly is a jump when the two-phase boundary is crossed. The specific heat C_v of real gases, however, has a weak divergence at the critical point. This divergence could be asymptotically described by

$$C_v \simeq A^\circ |\Delta T|^{-\alpha} \tag{6}$$

THERMODYNAMIC PROPERTIES NEAR THE CRITICAL STATE

Table 1. Critical exponents.

Exponent	Definition		classical	3-dim. lattice gas	real fluid
α	$T > T_c$	$C_v \simeq A^+(\|\Delta T\|^{-\alpha} - 1)/\alpha$	0	0.125 ± 0.003	
α'	$T < T_c$	$C_v \simeq A_I^-(\|\Delta T\|^{-\alpha'} - 1)/\alpha'$	0	$\left.\begin{array}{c}\\ 0.15 \pm 0.04\end{array}\right\}$	0.07 ± 0.05
α''	$T < T_c$	$C_v \simeq A_{II}^-(\|\Delta T\|^{-\alpha''} - 1)/\alpha'$	0		
β	$T = T_c$	$\Delta\rho \simeq B\|\Delta T\|^\beta$	0.5	$0.312 {+0.002 \atop -0.005}$	0.35 ± 0.01
γ	$T > T_c$	$\rho_c^2 K_T \simeq \Gamma \|\Delta T\|^{-\gamma}$	1	1.250 ± 0.003	
γ'	$T < T_c$	$\rho^2 K_T \simeq \Gamma' \|\Delta T\|^{-\gamma'}$	1	$1.251 {+0.008 \atop -0.005}$	1.22 ± 0.04
δ	$T = T_c$	$\Delta p \sim \Delta\mu \simeq D(\Delta\rho)\|\Delta\rho\|^{\delta-1}$	3	5.0 ± 0.1	4.4 ± 0.2
θ_p	$T < T_c$	$(\partial^2 p/\partial T^2)_{\text{vap}} \sim \|\Delta T\|^{-\theta_p}$	0	0.125	0.1 ± 0.1
θ_μ	$T < T_c$	$(\partial^2 \mu/\partial T^2)_{\text{vap}} \sim \|\Delta T\|^{-\theta_\mu}$	0	0	0
ν	$T > T_c$	$(\xi)_{\rho_c} \sim \|\Delta T\|^{-\nu}$	0.5	0.644 ± 0.003	0.61 ± 0.03
η	$\rho = \rho_c$, $r \to \infty$ $T = T_c$	$G(r) \sim 1/r^{d-2+\eta}$	0	0.060 ± 0.007	0.1 ± 0.05

rigorous exponent inequalities[62,115,116]   conjectured exponent equalities.

$\alpha + 2\beta + \gamma' \geq 2$ $\alpha = \alpha' = \alpha''$ $\alpha + 2\beta + \gamma = 2$

$\alpha + \beta + \beta\delta \geq 2$ $\gamma = \gamma'$ $\gamma = \beta(\delta - 1)$

$\theta_p, \theta_\mu \leq \alpha'' + \beta$ $\theta_p = \alpha''$ $\gamma = (2 - \eta)\nu$

---

with α near zero. This practice has two drawbacks. In fitting equation 6 to experimental data, the value of $A°$ varies strongly with the choice of α. Furthermore, equation 6 cannot describe the special case $C_v \simeq \ln|\Delta T|$, or $\alpha = 0$. It is preferable to define the asymptotic behavior of C_v as

$$C_v \simeq A[|\Delta T|^{-\alpha} - 1]/\alpha \qquad (7)$$

The coefficient A, equal to $\alpha A°$, is found to vary slowly, while equation 7 goes to $\ln \Delta T$ in the limit $\alpha \to 0$. Even with the expression 7, a reliable fit to experimental data can rarely be made. The divergence of C_v is so slow that non-diverging background terms remain non-negligible even for close approaches to the critical point. Ideally, these background terms should be included in the fit. The mathematical structure of these terms, however, is not fully known. For a discussion, see section II.5.

The power-law exponents defined in *Table 1* are not completely independent. Thermodynamics imposes certain inequalities between them; some of these are also listed in *Table 1*. As we will see later, the scaling laws assume equalities in exponent relations such that only two exponents are independent. We give these equalities in *Table 1*. In *Table 1*, primes refer to exponents for $T < T_c$ as distinguished from those for $T > T_c$; indices I and II refer to one- and two-phase regions, respectively. In defining exponents, we have assumed that they are the same irrespective of whether the critical point is approached from the low or the high density side; thus we have assumed a symmetry between gas and liquid phase. This point will be discussed more fully in section II.3. The actual determination of a power-law exponent is by no means trivial. The most frequent pitfalls are, in order of descending occurrence, the following. First of all, it is not always realized that the experimental accuracy diminishes in approaching the critical point. The accuracy is determined not only by the error made in measuring the experimental quantity, but also by the propagated error arising from the error in the independent variables, $T - T_c$, $\rho - \rho_c$, or others. The latter error is often preponderant in close approaches to the critical point, and it is magnified when logarithms are taken in order to fit data with straight lines. No least-squares results for critical exponents can be trusted when a proper error analysis has not been made. A second source of error arises from *a priori* ignorance of the range of asymptotic behavior. This range can only be found by trial and error. This is done by reducing the interval around the critical point while testing whether the data are fitted within their error and whether the exponent and coefficient remain stationary within their errors. The apparent asymptotic interval will clearly depend on the quality of the data. If the precision of the data is improved, deviations from power-law behavior on extending the range will show up sooner, and the exponents obtained will have less bias. The third source of error is ignorance of the critical values of the thermodynamic parameters of the system when power-law fits are made. Ideally, critical conditions should be directly observed in the course of the experiment. If this is not feasible, the critical parameters have to be varied within reason in the analysis. In practice this implies that non-linear rather than linear least-squares methods must be used, which is an essential complication[34]. Finally, substantial errors in power-law exponents and coefficients are made in cases in which there is an unknown but non-negligible background contribution, that is, a contribution which varies more slowly than the leading anomalous term. A striking example is the specific heat as already mentioned. Until theory provides the expressions for the background terms in the specific heat, there is no hope for a satisfactory precision in the exponent α (see section II.5).

3. Symmetry

In this section, we wish to clarify the concept of gas–liquid symmetry at the critical point. Before defining this term, let the reader be reminded that there are two equivalent ways in which a thermodynamic description of one-component fluid systems can be given. Starting with the fundamental relation between variations of intensive variables

$$V\,dp - S\,dT - N\,d\mu = 0 \qquad (8)$$

we obtain, on dividing by N, an equation in which the extensive properties are molar ones

$$d\mu = V_m\,dp - S_m\,dT \qquad (9)$$

Molar volume and entropy are obtained by partially differentiating the chemical potential $\mu(p, T)$ with respect to pressure and temperature. The most widely used free energy derived from $\mu(p, T)$ is the Helmholtz free energy $A(V_m, T)$ for which

$$dA = -p\,dV_m - S_m\,dT \qquad (10)$$

Here pressure and molar volume are conjugate variables. The corresponding equation of state is $p(V_m, T)$.

However, an alternative system is obtained if equation 8 is divided by V to yield

$$dp = \rho\, d\mu + s\, dT \tag{11}$$

in which $\rho = N/V$, $s = S/V$ are number and entropy densities, respectively; they are derived from the potential $p(\mu, T)$ by differentiation with respect to μ and T, respectively. The free energy corresponding to equation 11 is the Helmholtz free energy density $a(\rho, T) = A/V$ for which

$$da = \mu\, d\rho - s\, dT \tag{12}$$

here chemical potential and density are conjugate variables. The corresponding equation of state is $\mu(\rho, T)$. The two descriptions, one in terms of molar properties and the other in terms of densities, are thermodynamically equivalent.

Before the new theories of critical phenomena were developed, the term gas–liquid symmetry would have had the following meaning. At the critical point thermodynamic functions such as the Helmholtz free energy, chemical potential and pressure can all be developed in Taylor expansions in density and temperature, or in volume and temperature. It then follows that the coefficients and exponents of the coexistence curve must be the same whether approaching the critical point from low or from high density, and in density as well as in volume variables. Likewise, on the critical isotherm the coefficient D and exponent δ are the same on the low- and high-density side, and for all choices of variables, p–V, p–ρ, μ–V and μ–ρ. Thus the analyticity of the free energy entails a symmetry property between gas and liquid *on the coexistence curve* and *along the critical isotherm*. However, since the critical point in a real system has been found to have a non-analytic character, one can no longer take this symmetry for granted. Before reviewing the experimental situation in relation to this symmetry property, we shall first discuss a different and more stringent concept of symmetry which has arisen from the study of the Ising model. As is well known, the ferromagnetic and the Ising-model phase transitions have complete symmetry in the sense that, on reversing the field H, the magnetization M is reversed. The field is an antisymmetric function of the magnetization while the magnetic free energy A and entropy S are symmetric in M; as a consequence, there is no difference in entropy between opposite points on the phase boundary. The Ising model is translated into the lattice gas model by replacing the field by the chemical potential difference $\mu(\rho, T) - \mu(\rho_c, T)$ and the magnetization by the density difference $\rho - \rho_c$; it follows that the lattice gas has a well-defined but curious symmetry structure, namely antisymmetry of the μ–ρ relation at all temperatures, not only at T_c. The p–V and p–ρ relations do not have any special symmetry character (except asymptotically near the critical point on the critical isotherm); thus the lattice gas model has complete symmetry, but only in the second choice of thermodynamic variables, $\mu(\rho, T)$.

The coexisting *densities* of the lattice gas are symmetric with respect to ρ_c everywhere, and the density diameter $\bar{\rho} = (\rho_L + \rho_G)/2$ equals ρ_c at all temperatures. The coexisting *volumes* lack special symmetry except asymptotically near T_c. The antisymmetry of the $\mu-\rho$ relation implies analyticity at ρ_c, because $\mu(\rho_c, T)$ can always be written as the average of $\mu(\rho_c + \Delta\rho, T)$ and $\mu(\rho_c - \Delta\rho, T)$, with the latter points chosen in the one-phase region where the $\mu(\rho,T)$ relation is well behaved. Thus $\mu(\rho_c, T)$ and its derivatives all exist at T_c. Consequently, using the Yang–Yang relation[160],

$$\frac{C_v}{T} = V\left(\frac{\partial^2 p}{\partial T^2}\right)_v - N\left(\frac{\partial^2 \mu}{\partial T^2}\right)_v \tag{13}$$

the anomaly in C_v on the critical isochore is related to that in $(\partial^2 p/\partial T^2)_v$ alone, $(\partial^2\mu/\partial T^2)$ being regular[160]. This is so both above and below T_c. However, just as we cannot take the classical symmetry for granted for the critical region of real systems, neither can we expect the lattice–gas symmetry to hold. In fact, the lattice–gas model seems highly artificial and its analogy to the real gas cannot be expected to go beyond yielding non-classical critical exponents.

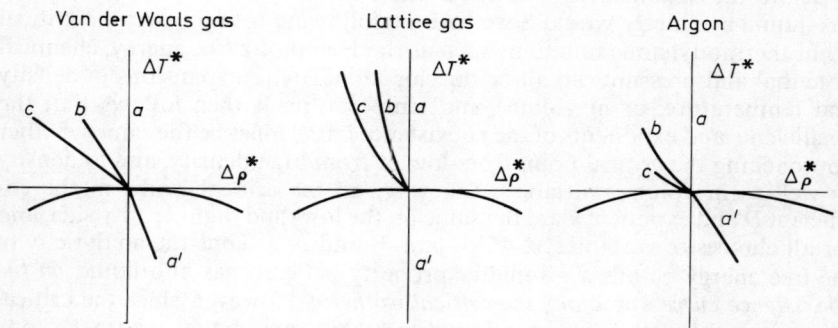

Loci: a $\mu-\rho$ inflection
     b $p-\rho$ inflection
     c   Max. compressibility

                                                 d' Diameter of coex. curve

Figure 2. The symmetry character of the van der Waals equation, the lattice gas and the real gas (argon) is indicated by showing, for isotherms, the loci of the $\mu-\rho$ and $p-\rho$ inflection points and the maxima of the compressibility. That the locus of the isothermal $\mu-\rho$ inflection points coincides with the critical isochore in the van der Waals equation is accidental; it is not generally so in classical equations.

Let us now review the experimental situation with these two notions of symmetry in mind (*Figure 2*). A careful analysis of the asymptotic form of the top of the coexistence curve of several gases[86] has shown that this top is indeed symmetric; coefficients as well as exponents are asymptotically the same on the two branches; the top of the coexistence curve has, asymptotically, the symmetry of the classical case and of the Ising model (*Figure 3*). In addition, however, analysis of the $\mu-\rho$ isotherms of a number of gases has shown that these curves have asymptotic antisymmetry with respect to the critical isochore over a large temperature range above T_c[140, 149] (*Figures 4, 5*). This means that the stringent lattice-gas symmetry, with its particular choice

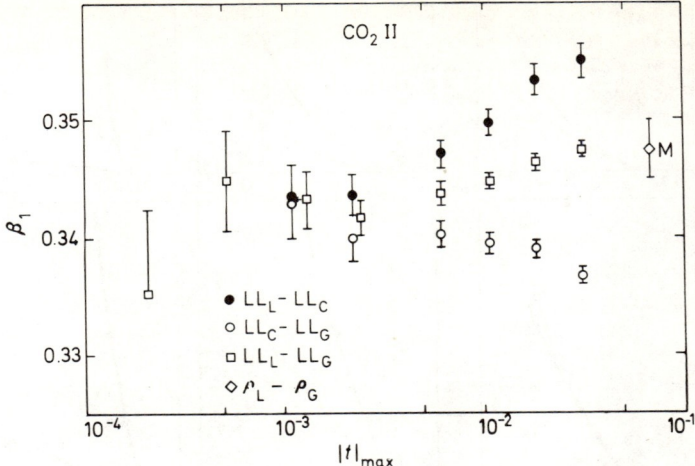

Figure 3. The apparent exponent β, in $\rho - \rho_c = B|\Delta T|^\beta$, as a function of the range $t = (T_c + T)/T_c$ of the data, for coexistence curve data for carbon dioxide obtained by refractive index measurements[117, 131]. The exponents on the liquid side, indicated by full circles, become asymptotically equal to those on the gas side, indicated by open circles. A much larger asymptotic range is obtained if the density difference between gas and liquid is analysed (open squares). After ref. 86.

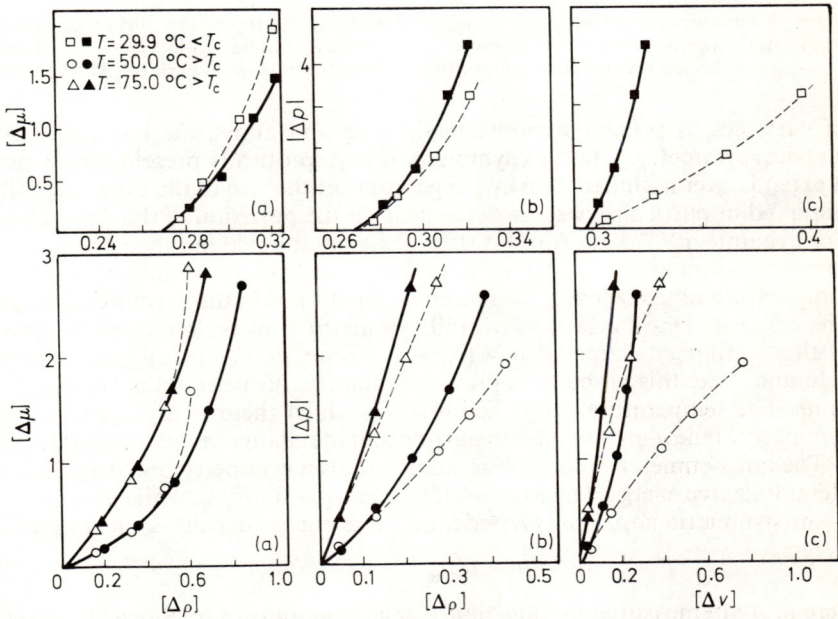

Figure 4. The symmetry of the $\mu - \rho$, $p - \rho$ and $p - V$ relations for carbon dioxide along isotherms near the critical point[96]. The low-density side (open symbols, dashed curve) has been rotated over 180° to compare it with the high-density side (solid symbols, full curve). It is seen that the $\mu - \rho$ antisymmetry is the most pronounced of the three. The three plots in the top row are for a slightly subcritical isotherm which terminates at the coexistence curve. After ref. 140.

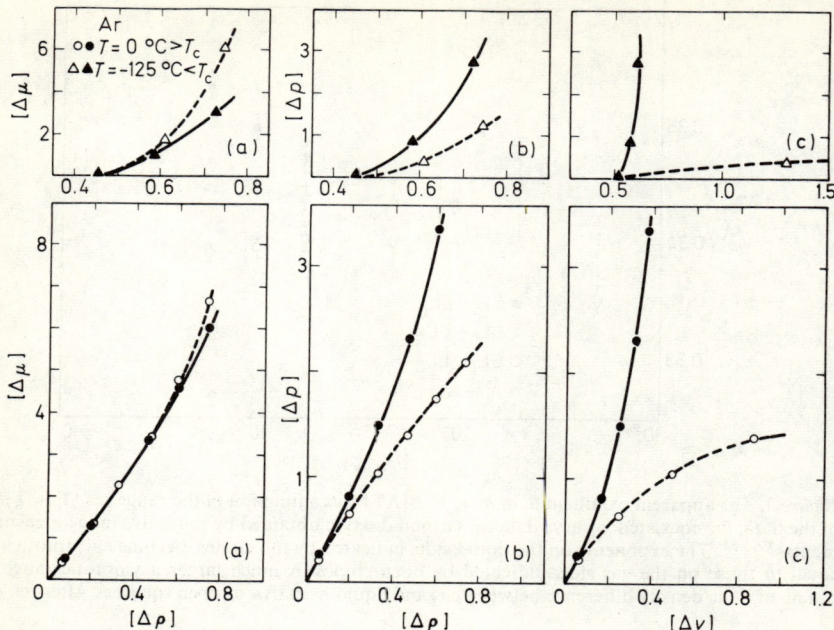

Figure 5. The symmetry of the μ–ρ, p–ρ and p–V relations for argon[98] at a slightly subcritical temperature (top row) and at a temperature of roughly $2T_c$. Symbols as in *Figure 4*. The antisymmetry of the μ–ρ relation seems to persist to temperatures far above critical. After ref. 140.

of variables, is present asymptotically over and above the less restrictive classical symmetry. That this symmetry is asymptotically present means that it extends over a limited density range from ρ_c; the size of the range actually observed in particular cases is dependent on the precision of the data taken. For accurate pVT data, $\Delta\rho = \pm 0.25\rho_c$ is a good guide number. In practice this means that in view of the width of the coexistence dome only a very small temperature range below T_c can be considered to be in the asymptotic range. The clearest sign of a lack of overall symmetry is to be found in the slope of the 'rectilinear diameter' $\bar{\rho} = (\rho_L + \rho_G)/2$ versus T. For all gases except ^{3}He and ^{4}He, this slope is appreciable, that is, of order unity in reduced units. The temperature range below T_c in which there is asymptotic antisymmetry of the μ–ρ relation is so narrow that the change in $\bar{\rho}$ can be neglected.

The antisymmetry of the μ–ρ relation entails a symmetry property for the Helmholtz free energy density $a = A/V$. Since $\mu = (\partial a/\partial \rho)_T$, while $\mu - \mu(\rho_c, T)$ is antisymmetric in $\rho - \rho_c$, we redefine a free energy density a_s such that

$$\Delta\mu = \mu - \mu(\rho_c, T) = (\partial a_s/\partial \rho)_T; \qquad (14)$$

here μ, A are measured in units of $p_c V_c$, and ρ in units of ρ_c. Since $(\partial a_s/\partial \Delta\rho)_T$ is antisymmetric in $\Delta\rho$, a_s must be symmetric. It is related to a by

$$a = a_s + \mu(\rho_c, T)\Delta\rho \qquad (15)$$

The symmetric part of the free energy density is closely related to that of the magnetic free energy A, for which $dA = H\, dM - S\, dT$. However, in contrast to the full symmetry of the magnetic free energy, that of a_s is limited to densities near critical.

4. Homogeneity and Scaling

As Widom[153] first showed, the scaling laws, which formulate thermodynamic behavior in the critical region, follow from the hypothesis of homogeneity. Thermodynamic properties, apart from some regular background terms, are assumed to be generalized homogeneous functions of their arguments. The magnetic free energy $A(T, M)$ introduced in the previous section is assumed to be of the asymptotic form $A(T, M) = A_0(T) + A'$, with

$$A'(\lambda^{c_T} \Delta T, \lambda^{c_M} M) = \lambda A'(\Delta T, M) \qquad (16)$$

in which c_T, c_M are fixed exponents and λ an arbitrary constant. In contrast to ordinary homogeneous functions, the two exponents c_T, c_M are not equal. For a discussion of the properties of generalized homogeneous functions, see ref. 130. Homogeneity implies that if A' is known on one contour in the T,M plane it is known everywhere. All thermodynamic properties are assumed to be free of non-analyticities in the entire one-phase region. As was shown by Kadanoff[72], the homogeneity assumption is connected with the existence of a correlation length which becomes large near the critical point.

In the gas–liquid case, the homogeneity assumption is made for the symmetric part of the free energy density, a_s, defined in equation 15, after subtracting a regular background term $A_0(T)$. Thus, for $a'_s = a_s - A_0(T)$ we have

$$a'_s(\lambda^{c_T} \Delta T, \lambda^{c_\rho} \Delta \rho) = \lambda a'_s(\Delta T, \Delta \rho) \qquad (17)$$

The power laws can be readily derived from the homogeneity assumption 17. There are only two exponents c_T and c_ρ in terms of which all power law exponents can be expressed. Of the exponents α, β, γ, δ and θ only two are independent, with the others related to them by the equalities listed in Table 1. As an example, consider the form of the critical isotherm $\Delta\mu(\Delta\rho, \Delta T = 0)$. Since $\Delta\mu = (\partial a_s/\partial \rho)_T$, we find from expression 17

$$\lambda^{c_\rho}[\partial a'_s/\partial(\lambda^{c_\rho} \Delta\rho)] = \lambda[\partial a'_s/\partial \rho]$$

or

$$\lambda^{c_\rho - 1} \Delta\mu(\lambda^{c_T} \Delta T, \lambda^{c_\rho} \Delta\rho) = \Delta\mu(\Delta T, \Delta\rho) \qquad (18)$$

Note that the chemical potential difference $\mu - \mu(\rho_c T)$ is also a homogeneous function of its arguments. Along the critical isotherm $\Delta T = 0$, so that

$$\lambda^{c_\rho - 1} \Delta\mu(0, \lambda^{c_\rho} \Delta\rho) = \Delta\mu(0, \Delta\rho) \qquad (19)$$

Choosing $\lambda^{c_\rho} = 1/|\Delta\rho|$, we find

$$\left(\frac{1}{|\Delta\rho|}\right)^{(c_\rho - 1)/c_\rho} \Delta\mu(0, \pm 1) = \Delta\mu(0, \Delta\rho) \qquad (20)$$

However, because of antisymmetry $\Delta\mu(+1, 0) = -\Delta\mu(-1, 0)$ and thus

$$(\text{sign } \Delta\rho)|\Delta\rho|^{(1-c_\rho)/c_\rho} \Delta\mu(0,1) = \Delta\mu(0,\Delta\rho) \tag{21}$$

in which $\Delta\mu(0,1)$ is a constant. By comparing with the power law $\Delta\mu = D\,\Delta\rho|\Delta\rho|^{\delta-1}$ we see immediately that

$$\begin{aligned} D &= \Delta\mu(0,1) \\ \delta &= (1-c_\rho)/c_\rho \end{aligned} \tag{22}$$

Reasoning along similar lines for the inverse compressibility

$$\chi^{-1} = [\rho^2 K_T]^{-1} = (\partial^2 a'_s/\partial\rho^2)_T$$

we find

$$\chi^{-1}(\Delta T, 0) = |\Delta T|^{(1-2c_\rho)/c_T} \chi^{-1}(1,0) \tag{23}$$

so that, with $\chi = \Gamma |\Delta T|^{-\gamma}$,

$$\begin{aligned} \Gamma &= \chi(1,0) \\ \gamma &= (1-2c_\rho)/c_T \end{aligned} \tag{24}$$

The anomalous behavior of the specific heat above T_c on the critical isochore is found by observing that $C'_{vs}/VT = -(\partial^2 a'_s/\partial T^2)_\rho$. Using the homogeneity relation one derives

$$\frac{C'_{vs}}{VT} = (\Delta T)^{(1-2c_T)/c_T} \frac{C'_{vs}(1,0)}{VT} \tag{25}$$

and recalling the definition

$$\frac{C'_{vs}}{VT} \sim \frac{A^+}{\alpha}[(\Delta T)^{-\alpha} - 1]$$

we have

$$\begin{aligned} \frac{A^+}{\alpha} &= \frac{C'_{vs}(1,0)}{VT} \\ \alpha &= (2c_T - 1)/c_T \end{aligned} \tag{26}$$

The coexistence curve is defined by $\Delta\mu = 0$ while $\Delta\rho, \Delta T \neq 0$. Using the homogeneity assumption one finds

$$\Delta\mu(-1, |\Delta T|^{-c_\rho/c_T}\Delta\rho) = 0 \tag{27}$$

so that

$$|\Delta T|^{-c_\rho/c_T}\Delta\rho = B \tag{28}$$

in which B is a constant. Thus we retrieve the coexistence-curve power law $\Delta\rho = B|\Delta T|^\beta$ with

$$\beta = c_\rho/c_T \tag{29}$$

The power law exponents γ', α' in the one-phase region along the coexistence curve are obtained by combining the homogeneity assumption 17 with the equation of the coexistence curve 28. For the compressibility $\chi = \rho^2 K_T$ one obtains

$$|\Delta T|^{(1-2c_\rho)/c_T}\chi^{-1}(-1, B) = \chi^{-1}(\Delta T, \Delta\rho) \tag{30}$$

so that, recalling $\chi = \Gamma'|\Delta T|^{-\gamma'}$,

$$\Gamma' = \chi(-1, B)$$
$$\gamma' = (1 - 2c_\rho)/c_T \tag{31}$$

For the anomalous part of the specific heat one finds

$$|\Delta T|^{(1-2c_T)/c_T}\frac{C'_{vs}(-1, B)}{VT} = \frac{C'_{vs}(\Delta T, \Delta\rho)}{VT} \tag{32}$$

which, with

$$\frac{C'_{vs}}{VT} \sim \frac{A^-}{\alpha'}[|\Delta T|^{-\alpha'} - 1]$$

yields

$$\frac{A^-}{\alpha'} = \frac{C'_{vs}(-1, B)}{VT}$$
$$\alpha' = (2c_T - 1)/c_T \tag{33}$$

The specific heat anomaly in the two-phase region is obtained by observing that the symmetric part of the free energy on the phase boundary is given by

$$a'_s(\Delta T, \Delta\rho) = |\Delta T|^{1/c_T} a'_s(-1, B) \tag{34}$$

However, this is also the value of the symmetric free energy contribution on the critical isochore. We obtain the anomalous part of the specific heat by differentiating, thus,

$$\left(\frac{C'_{vs}}{VT}\right)_{II} = -a'_s(-1, B)(1/c_T)(1/c_T - 1)|\Delta T|^{(1/c_T)-2} \tag{35}$$

so that, in

$$\left(\frac{C'_{vs}}{VT}\right)_{II} \sim \frac{A^-_{II}}{\alpha''}[|\Delta T|^{-\alpha''} - 1]$$

$$\frac{A^-_{II}}{\alpha''} = -(1/c_T)(1/c_T - 1)a'_s(-1, B)$$
$$\alpha'' = (2c_T - 1)/c_T \tag{36}$$

All critical exponents have now to be expressed in terms of the pair c_T, c_ρ. An inspection of the relations 24, 31, and 26, 33, 36 reveals immediately that

$$\gamma = \gamma'$$
$$\alpha = \alpha' = \alpha'' \tag{37}$$

Moreover, one readily derives the equalities of *Table 1*,

$$\beta(\delta + 1) = 2 - \alpha$$

and

$$\beta(\delta - 1) = \gamma \qquad (38)$$

Finally, because of built-in μ–ρ antisymmetry (section II.3), the vapor pressure exponent θ_p equals α while $\theta_\mu = 0$.

Thus the homogeneity assumption implies the power laws with certain relations between exponents, leaving only two of them independent. We now proceed to show that homogeneity implies scaling.

Using the homogeneous equation 17 for the free energy a'_s and setting $\lambda^{c_\rho} = 1/|\Delta\rho|$, we find

$$a'_s(\Delta T, \Delta \rho) = |\Delta \rho|^{1/c_\rho} a'_s(\Delta T/|\Delta \rho|^{c_\rho/c_T}, 1) \qquad (39)$$

Replacing the exponents c_ρ and c_T by more familiar ones, we obtain the scaling law for the free energy[63]

$$a'_s(\Delta T, \Delta \rho)/|\Delta \rho|^{\delta+1} = a(\Delta T/|\Delta \rho|^{1/\beta}) \qquad (40)$$

Thus the free energy, a'_s, scaled with $|\Delta\rho|^{\delta+1}$, is a function of only one variable, the scaled temperature

$$x = \Delta T/|\Delta \rho|^{1/\beta} \qquad (41)$$

Similarly, the scaling law for the chemical potential follows from the homogeneity property 17. It is

$$\Delta\mu(\Delta T, \Delta \rho)/(\Delta \rho)|\Delta \rho|^{\delta-1} = h(x) \qquad (42)$$

The chemical potential, scaled by $(\Delta\rho)|\Delta\rho|^{\delta-1}$, is again a function of only one scaled variable x.

The homogeneity assumption is, at this stage, no more than a plausible hypothesis. A key function of experimental thermodynamic studies in the critical region is to provide tests of the homogeneity assumption and the range of thermodynamic variables over which it is valid. As a first step, knowledge of the individual exponents is obtained by taking data along the preferred curves, after which the validity of the equalities in *Table 1* is tested. So far, in gas–liquid critical regions, no evidence has been found for violation of these equalities. Next, one may try to scale the data in the *entire* density/temperature plane. In principle, if the quantity $h(x) = \Delta\mu/|\Delta\rho|^\delta$ is plotted as a function of $x = \Delta T/|\Delta\rho|^{1/\beta}$, the data should all fall on a single curve. There are, however, a number of reasons why this is not practical. First of all, the precisions of $h(x)$ and x vary considerably from point to point. The errors have to be estimated before the quality of scaling can be judged. Furthermore, this quality will also depend on the choices made for a number of the parameters, such as T_c, ρ_c, δ and β. Finally both $h(x)$ and x range up to infinity so that graphical methods are hard to apply. We have found it advantageous to devise a functional form for the $h(x)$ versus x relation in order to fit the data by least-squares techniques, linear or non-linear, depending on the circumstances. A goodness-of-fit criterion is then readily established in the form of the standard deviation; varying precision is taken care of by weighting; the deviations of the individual points are compared with their estimated error to check whether systematic errors are present; and a number of parameters are readily varied in order to optimize the fit. The MSG (or NBS) equation[140] proposed for this purpose is

$$h(x) = E_1 \left(\frac{x + x_0}{x_0}\right) \left[1 + E_2 \left(\frac{x + x_0}{x_0}\right)^{2\beta}\right]^{(\gamma - 1)/2\beta} \tag{43}$$

in which E_1 and E_2 are adjustable parameters and $-x_0$ is the value of x on the coexistence boundary, $x_0 = B^{-1/\beta}$.

Although this equation is not analytic across the critical isochore, the fact that the lowest two derivatives exist implies that equation 43 is adequate for most practical purposes. Equation 43 has been shown to fit critical region data both for gases and for magnetic substances within their estimated

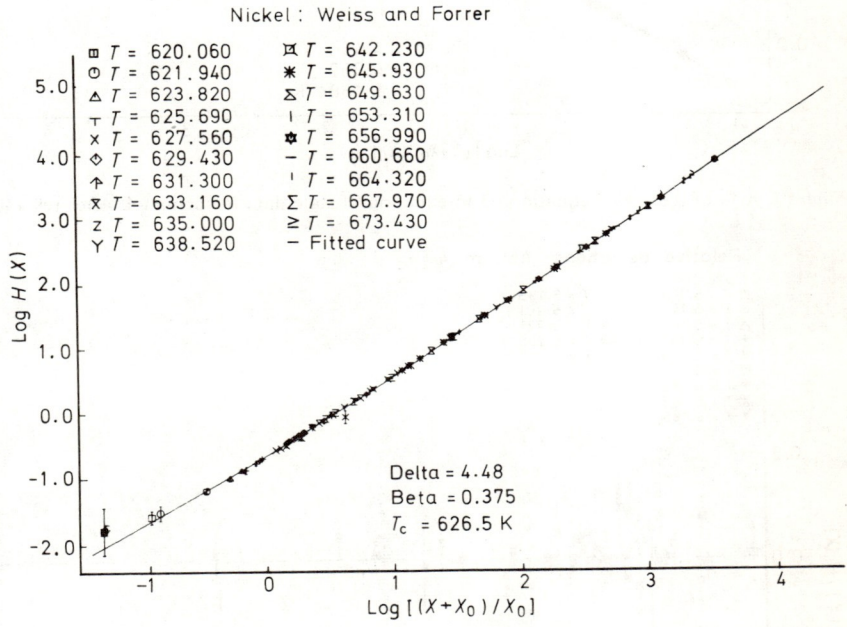

Figure 6. A fit of the scaled equation 43 to the magnetic equation-of-state data for nickel. After ref. 141.

errors[140, 141] (Figures 6–8). The use of the NBS equation is the most direct way available for fitting critical-region equation-of-state data, with proper error analysis. The equation is, however, difficult to integrate for obtaining free energies and specific heats. An alternative description, the parametric representation, introduced by Schofield[119], has the advantage of being entirely free of non-analyticities in the one-phase region while a special case,

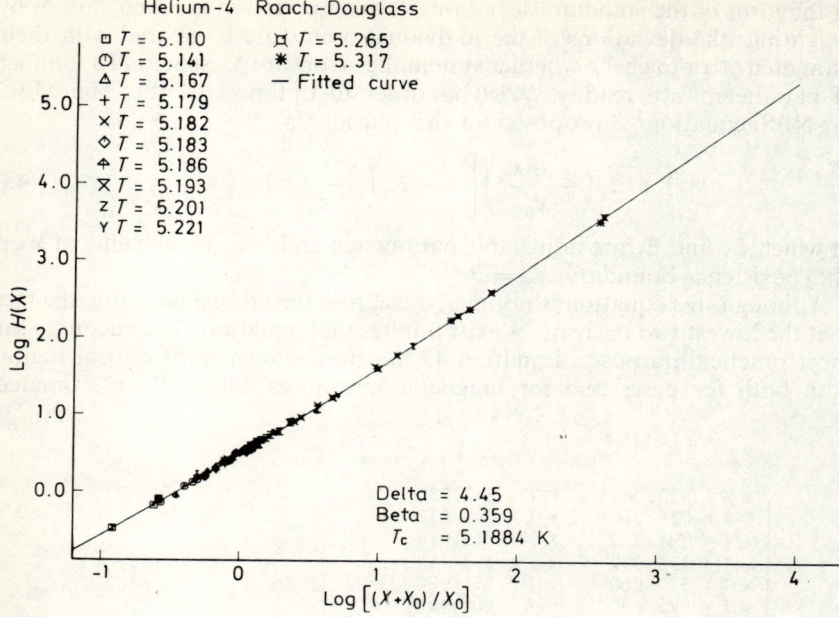

Figure 7. A fit of the scaled equation 43 to equation-of-state data for helium-4. After ref. 140.

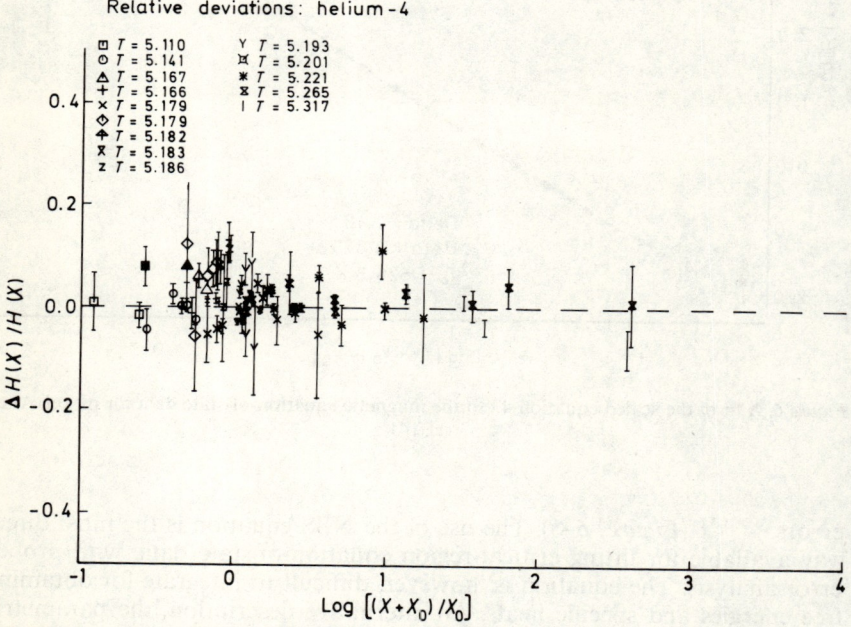

Figure 8. Deviations of the experimental data for helium-4 from the fitted equation 43. The error bars are based on estimates of the experimental errors in all variables.

the so-called linear model, introduced by Schofield, Litster and Ho[120], has the advantage of being readily integrated.

In the parametric form, the thermodynamic variables expressed in terms of two new variables r and θ, r denoting a distance from the critical point and θ a location on a contour of constant r. θ runs from -1 to $+1$; it equals ± 1 on the coexistence curve and 0 on the critical isochore. The transformations proposed by Schofield are:

$$\Delta \mu = a\theta(1 - \theta^2) r^{\beta\delta}$$
$$\Delta T = (1 - b^2\theta^2) r \qquad (44)$$

in which a and b are adjustable constants. The density is transformed as follows

$$\Delta \rho = m(\theta) r^\beta \qquad (45)$$

The critical anomalies are all described by means of the r-dependence; the θ-dependence is assumed to be regular. It is easily shown that the proposed transformations imply both the power and the scaling laws. The latter are valid because the combinations $\Delta\mu/|\Delta\rho|^\delta$ and $x = \Delta T/|\Delta\rho|^{1/\beta}$ are both functions of θ only.

The 'linear model' is obtained by assuming that $m(\theta)$ is linear in θ. In addition, a special choice is usually made for the parameter b namely $b^2 = (\delta - 3)/(\delta - 1)(1 - 2\beta)$. Apart from β, δ, T_c and ρ_c, this leaves only two adjustable constants, one less than in the NBS equation.

The linear model is readily integrable and yields fairly simple closed expressions for the free energy and specific heat. Nevertheless, the linear model has rarely been tested by direct fit to experimental data. The constants are usually obtained from data along special curves or from the NBS equation. An $m(\theta)$ plot is then made which exhibits approximately linear behavior. In view of the considerable advantage and rather widespread use of the linear model a critical analysis of its power to represent critical region data seems imperative. Such an analysis is now in progress[61].

At the present time, the scaling hypothesis seems to hold quite satisfactorily for fluid systems. The only difficulties encountered are related to the specific heat. Ahlers's[1] very precise data on C_p near the λ-line of ^{4}He seem to indicate a slightly different coefficient of the logarithmic anomaly above and below T_λ, while scaling predicts these coefficients to be equal. In magnetic systems, the exponents α and α'' are usually very different. In our opinion, the weakness of the specific heat anomaly implies that the apparent value of the exponent depends strongly on the choice of background terms and on the presence in the fluid of disturbing effects such as gravity. Careful and well-controlled experiments in which close attention is paid to background and side-effects are sorely needed.

5. Beyond Simple Scaling

The assumptions of simple scaling are only asymptotically valid in model and in real systems. These assumptions break down when higher order terms become important. For fluids, the onset of higher order contributions is demonstrated dramatically by departures from symmetry. Thus, for all substances except helium, the diameter of the coexistence curve, $\bar{\rho}=(\rho_L+\rho_G)/2$, shows considerable temperature dependence. According to experiment, the diameter appears to be straight for temperatures which are within ten per cent of the critical. The 'law of the rectilinear diameter' was established by Cailletet and Mathias[16] in 1886 as an experimental fact; it also follows from classical theory in a straightforward manner (section II.1); however, it is unintelligible if non-classical critical points are considered. Model calculations of Widom and Rowlinson[155] and of Mermin[93] on lattice models lacking particle–hole symmetry result in anomalous behavior of the coexistence curve diameter. There are also thermodynamic arguments which suggest that the diameter should be singular. Thus, as Griffiths and Wheeler[64] pointed out, the coexistence curve is the only direction in the space of intensive variables p, μ, T which is singled out for special behavior. In all directions not parallel to this curve, a thermodynamic property should behave in the same way when approaching the critical point. Since the entropy density is given by $(\partial p/\partial T)_\mu$ while the number density is $(\partial p/\partial \mu)_T$, both derivatives are taken in directions not parallel to the coexistence curve and these two densities should behave equivalently[94, 95]. Now the temperature derivative of the sum of coexisting entropy densities is a C_v-like quantity, diverging as $|\Delta T|^{-\alpha}$ when approaching the critical point. Likewise one expects the temperature derivative of the sum of the coexisting number densities to diverge as $|\Delta T|^{-\alpha}$ or

$$\bar{\rho} = \rho_c + |\Delta T|^{1-\alpha} \tag{46}$$

Green and co-workers[60, 26] have used a generalization of the parametric representation, combined with an invariance principle conserving the 'special direction' in the space of intensive variables, to extend scaling for thermodynamic properties around the critical point, with the power-laws as the lowest-order term. In the interest of the experimentalist we cite, from the new results of Cook and Green[26], the leading terms in the expansions of thermodynamic properties along preferred curves.

Coexistence curve:

$$\Delta\rho^\pm = \pm B_1|\Delta T|^\beta + B_2^\pm|\Delta T|^{1-\alpha} + B_3\Delta T + \ldots$$
$$p = p_c + c_1\Delta T + p_1^-|\Delta T|^{2-\alpha} + p_2(\Delta T)^2 + p_3^-|\Delta T|^{2\beta\delta-1} + \ldots$$
$$\mu = \mu_c + c_2\Delta T + H_1(\Delta T)^2 + H_2^-|\Delta T|^{2\beta\delta-1} + \ldots$$
$$\frac{C_v}{VT} = A_1^-|\Delta T|^{-\alpha} + A_2 + A_3^-|\Delta T|^{1-\beta-2\alpha} + \ldots \tag{47}$$
$$\rho^2 K_T = G_1^-|\Delta T|^{-\gamma} + G_2^\pm|\Delta T|^{\beta-1} + \ldots$$

Superscript $\pm$ refers to sign of $\Delta\rho$
Superscript $-$ refers to $T < T_c$

Critical isotherm:

$$\mu = \mu_c + D_1(\Delta\rho)|\Delta\rho|^{\delta-1} + D_2^\pm|\Delta\rho|^{2\delta-1/\beta} + \ldots$$

$$p = p_c^+ + D_1\rho_c(\Delta\rho)|\Delta\rho|^{\delta-1} + \frac{D_1\delta}{\delta+1}|\Delta\rho|^{\delta+1} + D_2^\pm \rho_c |\Delta\rho|^{2\delta-1/\beta} + \cdots$$

Superscript ± refers to sign of $\Delta\rho$ (48)

Critical isochore:

$$\mu = \mu_c + c_2\Delta T + H_1(\Delta T)^2 + H_2^+|\Delta T|^{2\beta\delta-1} + \cdots$$
$$p = p_c + c_1\Delta T + p_1^+|\Delta T|^{2-\alpha} + p_2(\Delta T)^2 + p_3^+|\Delta T|^{2\beta\delta-1} + \cdots$$
$$\frac{C_v}{VT} = A_1^+|\Delta T|^{-\alpha} + A_2 + A_3^+|\Delta T|^{1-\beta-2\alpha} + \cdots \tag{49}$$

Superscript + refers to $T > T_c$

In this approach, the chemical potential is no longer analytic along the critical isochore, but the second derivative $(\partial^2\mu/\partial T^2)_{\rho_c}$ does not diverge. Thus the anomaly in C_v is the same as that in $(\partial^2 p/\partial T^2)_{\rho_c}$, cf. equation 13. Note that the specific heat expansion, after the diverging term and the constant, has a term which contains a power of ΔT much lower than unity. The inclusion of this term has a powerful effect on the fit of specific heat data[15].

III. Special Experimental Difficulties

1. Divergences and Their Consequences

The principal experimental problems in critical region studies are directly related to the fact discussed in section II, that two thermodynamic derivatives, the isothermal compressibility $K_T = (1/\rho)(\partial\rho/\partial p)_T = (1/\rho^2)(\partial^2\mu/\partial p^2)_T$, and the isobaric expansion coefficient, $\alpha_p = (1/\rho)(\partial\rho/\partial T)_p = (1/\rho)(\partial^2\mu/\partial p \, \partial T)$ diverge strongly at the critical point; while a dynamic coefficient, the thermal diffusivity $\lambda_T' = \lambda/\rho C_p$, goes to zero.

The divergence of the isobaric expansion coefficient implies that a system maintained at constant pressure near the critical point will undergo large changes in density when the temperature changes. Even the slight temperature variations under conditions of temperature control can produce excessive changes in the thermodynamic state. Experiments near critical points should therefore never be run at constant pressure, since this can leave the thermodynamic state of the system ill-defined. Rather, the experimental volume should be sealed off so that the density cannot vary.

The large values attained by the isothermal compressibility offer both problems and challenges. The problems come about when one tries to deduce the experimental density from the observed pressure using known pVT data. Because of the large compressibility, small pressure errors will result in very wrong densities. In fact, there have been many apparent critical anomalies in physical properties reported that were due solely to erroneous density assessments. If the density has to be obtained from known pVT data, this should be done at a point remote from the critical, where the compressibility is not excessive and the error in density due to pressure uncertainty slight. The cell should then be sealed and brought to critical conditions.

The second consequence of the large compressibility is that the small pressure gradient caused by the weight of the gas column in the field of gravity is enough to create substantial density gradients in a fluid in near-critical conditions. As a consequence, the average or bulk density may differ substantially from the density at which the pressure or some other physical property is measured. It is sometimes possible to correct for this effect (sections III.2 and VII.2). However, when noxious volumes are present in the form of long gas lines leading upwards to pressure registration equipment, the gas leads in these lines contribute to the compression in a way for which it is very hard to correct. Such experiments cannot yield reliable data in the close vicinity of the critical point.

The gravity effect thus can give rise to serious error in critical region measurements; it can, however, also be used to great advantage. By studying properties as a function of height and hence of the small pressure gradient, a very fine resolution of this pressure dependence is obtained. Some of the most detailed critical region experiments have been designed in this way. We shall discuss these in section III.2.

A third consequence of the large compressibility is the presence of large density fluctuations. These fluctuations will scatter electromagnetic radiation. With the development of the laser with its ability to produce beams of narrow bandwidth, light scattering has rapidly become one of the most informative tools for studying the critical region. We devote section IX to this subject.

The last major experimental difficulty in critical region studies has to do with the slowness with which fluctuations decay; the entropy fluctuations die out with the thermal diffusivity $\lambda'_T = \lambda/\rho C_p$ where λ is the thermal conductivity. The thermal diffusivity goes to zero at the critical point. This is at least part of the reason for the long waiting times that have to be employed near the critical point. We shall discuss equilibration times in section III.3.

2. Gravity

In the vicinity of the critical point the response of the system to a change in the chemical potential becomes very large. In a one-component system this response is given by the compressibility $(\partial \mu/\partial \rho)^{-1} = \rho^2 K_T$, in a binary fluid by the osmotic susceptibility $(\partial \mu/\partial c)^{-1}$. The variation of the chemical potential in the field of gravity equals mgh, where m is the molecular weight, g is the gravity constant and h the height of the cell. In general, the response of the system to this small variation is negligible. However, under critical conditions the response diverges. As a consequence, density gradients are to be expected in gases in near-critical conditions. This was already predicted by Gouy in 1892[58]. (The same author made some observations of this effect in glass cells filled with carbon dioxide and controlled in temperature to a fraction of a milli Kelvin[58].) As in gases, the response of a binary fluid to a chemical potential difference diverges strongly. Consequently, concentration gradients are expected near the consolute point. This was pointed out by Yvon in 1937[162].

The actual size of the gravity effect in one-component fluids depends on the distance from the critical point, the height of the cell, the molecular weight m and the critical parameters of the fluid. The chemical potential difference across the cell is given by $\Delta\mu = mgh$, which, in reduced units, becomes

$$\Delta\mu^* = N\,\Delta\mu/p_c V_c = h^* = h/h_c \tag{50}$$

with

$$h_c = p_c V_c / Nmg \tag{51}$$

For a given cell height, and a given reduced distance from the critical point, the system with the largest value of $1/h_c$ will have the largest gravity effects.

Values of $1/h_c$ are listed in *Table 2* for a number of substances. It is obvious that molecular weight is only one of the factors determining the size of $1/h_c$. Relatively speaking, gravity effects are most serious in the light gases ^{3}He

Table 2. Characteristic height $h_c = p_c V_c / Nmg$

Substance	p_c, bar	ρ_c, g/cm^3	$10^5\,(1/h_c)$ in cm^{-1}
^{3}He	1.14	0.0414	3.54
^{4}He	2.29	0.0698	2.99
Ne	27.3	0.484	1.75
Ar	48.5	0.530	1.07
Kr	55.0	0.91	1.61
Xe	58.6	1.10	1.84
H_2	13.0	0.031	0.23
N_2	33.9	0.31	0.89
O_2	50.7	0.43	0.83
CO	35.0	0.30	0.84
CH_4	46.4	0.162	0.34
CO_2	73.9	0.468	0.63
N_2O	72.7	0.452	0.61
$CClF_3$	39.5	0.58	1.44
H_2O	221.2	0.315	0.14

and ^{4}He because their critical pressures are so low. Typical values of actual density differences observed in cells a few centimeters high may amount to ten per cent of the average density if $|T - T_c|/T_c < 10^{-4}$.

Experiments in the critical region may be conveniently divided into three classes depending on the way in which the gravity effect enters. First, consider those experiments that, because of design problems or incomplete settling, cannot be properly corrected for gravity. In this category, which we shall call Class C, belong most (but not all) of the pVT and specific heat experiments carried out before 1960, along with some carried out after that time. Typical examples are pVT experiments in which the experimental vessel is of appreciable height, and is connected to the pressure device by a long vertical capillary of uncertain temperature distribution. In analysing these data, a region has to be excluded in which pressure differences along isotherms are not appreciably larger than the hydrostatic head. Other cases are those in which the cell, large or small, is stirred shortly before data are taken. This has been rather common practice in specific heat experiments. If during the acquisition of the data appreciable density gradients were destroyed by stirring, those data should be excluded in further analysis.

In the second category, which we call Class B, we place those experiments that can be corrected for gravity because the cell is of limited height and because sufficient time has been allowed for the density profile to develop. The best known pVT experiment in this class is that of Habgood and Schneider[65] on xenon. In this experiment the temperature was controlled to ± 1 millidegree, the height was limited to 1 cm and there were no noxious volumes; in principle, these data could be corrected for gravity. In general, correction for gravity effects is rather straightforward. Data not beset with gravity effects are first fitted to an appropriate scaled equation (section II). The equation is then used to calculate the density profile in the region where density gradients do exist, with a guess being made about the position of the maximum gradient in the cell. Integration over the height of the cell then yields the average density. The estimated position of the point where the density gradient is largest is then moved up and down until the calculated average density equals the filling density. Next, the correction to the observed pressure is obtained by computing $\Delta p = \int \rho g \, \mathrm{d}h$ between the level at which the pressure was measured and the level at which the average density was realized. After thus correcting all experimental points affected by gravity, the fit of the scaled equation is repeated, this time to all data points.

When correcting specific heat experiments, it must be realized that not only are different thermodynamic states simultaneously present in the cell, but also that matter is redistributed during heating, giving rise to a kind of latent-heat effect. Thus, the specific heat of a vessel of non-zero height is not a height-averaged specific heat but, rather, is proportional to the second derivative of the height-averaged free energy or the first derivative of the height-averaged entropy[118]. The corrections to specific heat are much more appreciable than are those to pressure, even in cells of quite small height. In several cases, they have been calculated in detail. This will be discussed in section VII.

Thus far, we have discussed experiments in which gravity is a disturbing effect; in the next category we classify those experiments in which gravity is an asset. In these experiments, which we will call Class A, thermodynamic properties such as the density or the compressibility are measured as a function of height. If this height is resolved to a fraction of a millimeter, the equivalent pressure resolution obtained is of the order of 10^{-5} bar, which is one to two orders of magnitude beyond what can be obtained with conventional pressure devices. Optical experiments are carried out in this way in a straightforward manner, with the detecting device moved along the height of the cell (see section IV). Also, a stack of capacitors has been used in a dielectric constant experiment[150] (section V). Alternatively, local densities have been measured using floats[159], and local density gradients using Schlieren techniques[107,126,127] and an interference method[157]. The Schlieren method has also been used for the study of concentration gradients in binary mixtures[14] (section IV). These Class A experiments are potentially the most powerful high-resolution tools for close approach to the critical point. In practice, they have not as yet yielded much quantitative information because very long waiting times and extraordinarily good temperature control are needed in order to obtain stable and reproducible density profiles (see section III.3).

3. Equilibration

The critical point is a limiting point at which the stability of the system breaks down. On approaching this point, the forces driving the system to equilibrium diminish and the times needed for reaching equilibrium increase. The slowness of equilibration was immediately noticed in the early experiments in the critical region; descriptions abound of the changes in visual appearance of fluids maintained in near-critical conditions for many hours, and of the hysteresis effects observed when systems are cycled through the critical point. These descriptions do not contribute much to the understanding of the processes governing equilibration because the conditions under which the early experiments were carried out were not always well enough defined. With the development of methods for more refined temperature control with better long-term stability, reproducible behavior is generally obtained even in near-critical conditions. The time needed for equilibration, however, is found to vary widely. In *Table 3*, information pertaining to equilibration is collected for some of the most carefully performed critical region experiments. Two groups of experiments are considered, namely local density determinations and calorimetric measurements. In the first case, we list the statement of the author regarding the time he had to wait before stable readings of the local density could be made. In the second case, we list the relaxation time obtained from the changes in drift rate.

Table 3 shows a tremendous range in equilibration times, from seconds in helium calorimetry to many hours in local density experiments. Several authors have observed that equilibration in the two-phase region takes much longer than it does in the single phase[15, 32, 87, 163].

The equilibration will probably take place in a number of stages, the slowest one determining the relaxation time. From a non-equilibrium initial state, the pressure will be the first to relax, through shear- and damped longitudinal modes or sound waves. Since the damping of sound waves increases near the critical point (section X), this relaxation process should be even faster here than away from critical. Once the pressure is equalized, the entropy variations have to damp out through thermal diffusivity (section IV.6). This is a slow process because the thermal diffusivity coefficient

$$\lambda'_T = \lambda/\rho c_p \tag{52}$$

goes to zero at the critical point (section X).

It seems of interest to determine whether the observed relaxation times could be explained in terms of the mechanism of thermal diffusivity. For the process of relaxation by thermal diffusion, a characteristic time will be taken as that for a relaxing sphere of radius R,

$$\tau_{\lambda'_T} = R^2/\pi^2 \lambda'_T \tag{53}$$

For R, we take the largest thermal path, that is, the largest value of the distance from any point in the fluid to the nearest wall.

In *Table 3* we list, for each experimental situation, typical values of the thermal path and the height. An estimate of λ'_T, the thermal diffusivity is available for carbon dioxide and xenon from Rayleigh linewidth experiments

Table 3(a). Equilibration times: local density experiments

Ref.	Gas	$(T - T_c)/T_c$	Observed eq. time τ_{exp}	Height	Typical thermal path	Estimated $\lambda'_T = \dfrac{\lambda}{\rho C_p}$	Equilibration times in seconds $\tau_{\lambda r}$	τ_{exp}
{117, 131}	CO_2, N_2O, $CClF_3$	10^{-5}	10–12 h	2 cm	1.5 cm	4×10^{-7}	5.6×10^6	4×10^4
149	^{3}He	3×10^{-4}	15 min	1 mm	0.25 mm	1×10^{-6}	63	900
10	^{3}He	10^{-3}	1–2 h	1 cm	0.05 mm	4×10^{-6}	63	4×10^3
150	O_2	6×10^{-5}	4 h	15 cm	0.5 cm	3×10^{-6}	8×10^3	1.4×10^4
157	CO_2	10^{-4}	several hours	2 cm	3 mm	1.5×10^{-6}	6×10^3	$\sim 10^4$
163	^{3}He	-10^{-3}	few min	1 cm?	1 mm?	8×10^{-6}	130	200
		-10^{-4}	1–2 h			1.4×10^{-6}	710	4×10^3

Table 3(b). Relaxation times; calorimetric experiments

Ref.	Gas	$(T - T_c)T_c$	Observed relaxation time	Height	Typical thermal path	Estimated $\lambda'_T = \dfrac{\lambda}{\rho C_p}$	Relaxation times in seconds $\tau_{\lambda r}$	τ_{exp}
102	^{4}He	2×10^{-4}	minutes	3 mm	0.05 mm	1.5×10^{-6}	2	200
36	Xe	10^{-4}	500 s	1 cm	0.4 mm	5×10^{-6}	30	500
		3×10^{-5}	1500 s			2×10^{-6}	80	1.5×10^3
87	CO_2	-6×10^{-6}	1000 s	1 mm	0.5 mm	2×10^{-7}	0.7×10^3	10^3
15	^{3}He	-4×10^{-5}	100 s	0.5 mm	0.25 mm	6×10^{-7}	100	100
		-4×10^{-4}	20 s			1.5×10^{-6}	40	20
		+, any value	negligible				same order as above	0
32	^{3}He	-10^{-2}	10 s	0.087 cm	0.044 cm	5×10^{-5}	4	10
		-10^{-3}	100 s			10^{-5}	20	100
		-10^{-4}	1000 s			1.4×10^{-6}	140	1000
		$+10^{-4}$	<35 s			7×10^{-7}	280	<3

(section IX). If not, Kawasaki's equation[74]

$$\lambda'_T = KT/6\pi\eta\xi \qquad (54)$$

is used. A known or estimated value of the (at most weakly anomalous) viscosity is substituted; the correlation length ξ for the heavier molecules has been assumed to be the same as that of carbon dioxide at the same distance $(T - T_c)/T_c$ from critical. For helium, it has been taken a factor two smaller. Below T_c, ξ is assumed to be $(\Gamma/\Gamma')^{\frac{1}{2}}$, or about two times smaller than above at the same $|T - T_c|$ value. Comparing the observed equilibration times with the estimated time $\tau_{\lambda'_T}$ for *local density* experiments, we observe that, in general, the experimental values are of the order of the estimated time characteristic of thermal diffusion. Exceptions are the surprisingly long relaxation time in ^{3}He, which may be related to the fact that the authors did the experiment in isobaric conditions (cf. section III); and the 'short' 10 h relaxation time reported for the optical experiments in carbon dioxide, nitrous oxide and chlorotrifluoromethane, in which perhaps convection may have speeded up the equilibration in the early stages.

In the *calorimetric* experiments, the following regularities can be detected. In the carbon dioxide and ^{3}He experiments, the relaxation times in the one-phase region are much shorter than in the two-phase region, and also shorter than estimated. As pointed out by Dahl and Moldover[32], this may be due to convective processes speeding up equilibration by effectively reducing the thermal path. The relaxation times in the two-phase region are of the same order as those estimated for these three experiments. Although the ^{4}He and xenon experiments do not seem to follow the same pattern as the other three experiments, the difference may well be apparent only. In the ^{4}He experiment, only two thirds of the gas was in the 0.1 mm slots for which the estimate was made. The other third had a ten times larger thermal path and its estimated relaxation time would therefore be several minutes. In the xenon experiment, the temperature differences used in the estimate of λ'_T were taken with respect to the reported value of T_c, which lies near the maximum of C_v but considerably below the maximum in the relaxation time[87]. Taking T_c to coincide with the maximum in the relaxation time would drastically increase our estimated $\tau_{\lambda'_T}$ above T_c.

Concluding, the local density experiments seem to equilibrate as fast as dictated by thermal diffusivity or perhaps a little faster. The one-phase calorimetric experiments relax much faster than dictated by thermal diffusivity, and the two-phase calorimetric experiments relax perhaps a little more slowly. In calorimetric experiments, thermal gradients occur naturally, so that convection should speed up the equilibration; in the two-phase region, the presence of a barrier, the interface, might counteract the convective process[32].

Further investigation of equilibration under well-controlled conditions seems necessary. Elucidation of the equilibration processes would help in optimizing experimental design. The study of relaxation times by Buckingham and co-workers[36, 87] is an excellent example of a systematic study. It is hoped that more work of this caliber will be undertaken.

IV. Refractive Index Measurements

1. Principle

In most refractive index studies in the critical region, the purpose is to obtain information about the *density*. This can be a global density determination (Class B or Class C experiment) as an alternative to conventional methods; it can be a local density determination (Class A experiment) in which the density is studied as a function of height; or in which, in the Schlieren experiments, density gradients as functions of height are measured directly. In all cases, use is made of the Lorentz–Lorenz relation between density and refractive index n

$$(n^2 - 1)/(n^2 + 2) = 4\pi\alpha\rho/3 \tag{55}$$

The molar polarizability α is usually assumed to be constant. Two questions can be raised regarding the validity of equation 55. First, to what precision is this relation valid, and second, is a critical anomaly to be expected in $(n^2 - 1)/(n^2 + 2)\rho$? An affirmative answer to the second question would preclude the use of optical methods for critical region density determinations whereas modest variations in the quantity $(n^2 - 1)/(n^2 + 2)\rho \equiv$ LL could well be tolerated in such studies. The global variation of LL has been studied over wide density intervals in several cases, namely in carbon dioxide by Michels et al.[97], in argon by Knobler et al.[78], and in hydrogen by Diller[35]. For these substances at temperatures from below the critical to several times the critical, the variation of LL is limited to one or two per cent.

The other question, that of a possible anomaly in the Lorentz–Lorenz function, has been answered by Larsen, Mountain and Zwanzig[82]. They observed that the occurrence of critical opalescence indicates an anomaly in the imaginary part of the refractive index; which raises a question as to what happens to the real part. Larsen et al. calculated the first-order correction term to the Lorentz–Lorenz relation for values of the correlation length up to that of the wavelength of light ($|\Delta T|/T_c < 10^{-5}$) and found that the imaginary part of the refractive index, and the correction to the real part, are of the order of a part in 10^4 of n. (Wilcox and co-workers[68] recently reported the detection of an anomaly of this size in xenon.) Thus, the use of expression 55 for deriving densities from measured refractive indices appears to be justified throughout the critical region. Nevertheless, in an ideal experiment the weak dependence on density and temperature of the Lorentz–Lorenz function $(n^2 - 1)/(n^2 + 2)\rho$ should first be measured, while the density is measured independently. A word of caution is in order here. In several experiments, the Lorentz–Lorenz function was measured as a function of temperature and pressure, while known pVT data were used to calculate the density from observed pressures. This method fails in the critical region where the compressibility is large so that small pressure errors cause large density errors. Outside the critical region, however, it can be used with caution. Once the weak temperature and density dependence of LL are established by measurements outside the critical region, the values of $(n^2 - 1)/(n^2 + 2)$ determined near the critical point can be used to calculate the density.

In practice, the LL function is usually assumed to be a constant over the limited range of temperature and density used in critical region studies. Although asymptotic behavior is not affected by this practice, some caution is needed in comparison with results derived from more direct density determinations over larger density intervals[86].

There are three experimental problems which complicate near-critical local refractive index studies, namely critical opalescence, curvature of the beam and stringent requirements of temperature stability and uniformity. Critical opalescence may reduce the intensity of the transmitted beam to the point where measurements are made impossible. Its effect can be reduced by shortening the length of the light path through the vessel. Downward curvature of the light beam is caused by the refractive index gradient. For light entering in the horizontal plane, the angle of deflection of the beam from the horizontal, $d\phi$, after it has passed a distance ds in the medium, is given by

$$d\phi = (1/n)(dn/dz)ds \qquad (56)$$

The relation 55 implies that dn/dz varies proportionally to the gradient $d\rho/dz$; since the latter diverges at the critical point, beam curvature becomes pronounced. Its effect can be reduced by keeping the optical path short. On the other hand, beam curvature is used to advantage in Schlieren experiments for direct determination of the density gradient; even in these experiments, the cells are designed as thin slabs in order to keep the curvature limited to measurable values and to minimize the complications that arise in the analysis when the light beam is no longer perpendicular to the density gradient.

Temperature stability and homogeneity present more formidable problems in optical experiments than in most other critical region experiments. The cell and the surrounding thermostat have to be provided with windows. Not only will heat be radiated into the cell by the light source, but also, the experimental configuration is of low symmetry so that temperature gradients are hard to avoid. On the other hand, no critical region experiments are more easily spoiled by slight temperature disturbances than density profile determinations. In these experiments, reliance on stable and sensitive temperature control systems is not enough; the use of a thermocouple or pile to indicate temperature differences across the cell is mandatory.

2. *The Use of Optical Techniques for Bulk Density Determination*

In this type of experiment, the refractive index determination, combined with relation 55, replaces more orthodox density determinations. We refer to Chapter 12 of this volume for experimental details. The coexisting densities of ^{4}He were determined in this way by Edwards and co-workers[37, 38] using a Jamin interferometer. The advantage of an interferometer is its high precision, but a drawback in critical region work is the difficulty of keeping track of the fringe count. For absolute measurements these fringes have to be counted starting with the evacuated cell and continuing up to the final filling density, which requires delicate filling procedures; especially in the critical region the system is very easily disturbed to the point where the fringes temporarily disappear. Pings and co-workers[108] have used refraction of light by a prism-shaped cell to obtain *PVT* data in the critical region of argon and krypton.

3. Local Density Determination Using Refractive Index

Experiments of this nature have been carried out by Lorentzen[88] and by Schmidt and Straub[117, 131]. In both experiments, a prism was immersed in a fluid at near-critical conditions. When the density is no longer uniform the deflection of the beam varies as a function of height. Lorentzen used a vertical slit as a source and recorded the S-shaped image obtained. Schmidt and Straub moved a horizontal slit in the outcoming rays and measured the angle of deflection as a function of height. The data reported by Schmidt and Straub cover several substances at various temperature intervals from critical.

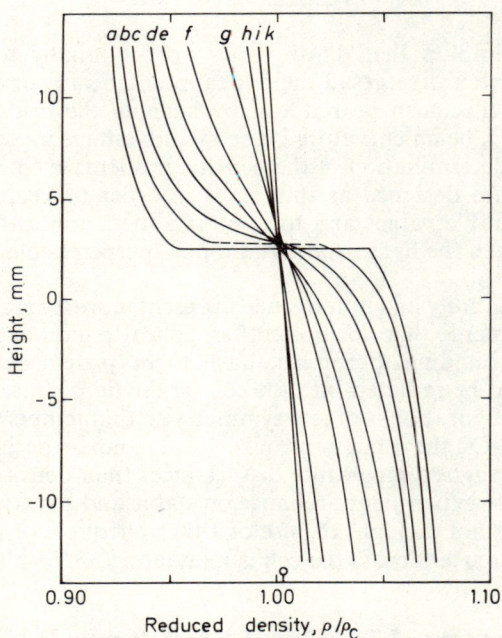

Figure 9. Density profiles in nitrous oxide obtained by the refractive index technique of refs. 117, 131. The isotherms have been obtained for the following temperature differences, in mK, with the critical temperature: a, -2; b, $+3$; c, $+10$; d, $+20$; e, $+30$; f, $+38$; g, $+63$; h, $+119$; i, $+167$; k, $+218$.

A typical example is given in *Figure 9*. Since data of this kind offer the most direct and detailed test of scaling symmetries (directly portraying the μ–ρ relation and the top of the coexistence curve), it would be welcome indeed if these measurements were repeated and extended using the best temperature control equipment now available.

4. Density Gradient Determination

When a light beam passes through an inhomogeneous medium, it undergoes a deflection which depends on the magnitude of the refractive index gradient and on the angle between the light beam and this gradient. Thus, if light enters horizontally into a cell in which a vertical density gradient exists, it will be deflected downwards, and for small curvature the angle of deflection will be given by equation 56. This curvature, a hindrance in ordinary refractive index determinations, is the basis for sophisticated optical methods for determining refractive index gradients. These methods have been developed in the framework of diffusion studies in liquids[71]. They can be classified according to whether or not interferometry was used. First application of either class dates back to the nineteenth century.

Starting with non-interferometric methods, many ingenious experiments have been designed through the years in which the deflections of the light beam are used for quantitative measurement of the refractive index gradient throughout the cell. Thus Wiener[156], in 1893, made the vertical concentration gradient in a diffusion cell visible by forming an image of a 45° slit with a small-aperture lens, while passing the light through the diffusion cell. The vertical displacements of the points on the actual distorted image from the undistorted one was a measure of the gradient all along the cell. Lamm, in the 1940s (see ref. 71) devised a variety of methods for quantitative measurement of refractive index gradients. One of them, the slit method, was a variant of the much older Schlieren method. The Schlieren method originated as a qualitative means for the display of inhomogeneities in a medium. In this method, an image of the medium is formed on a screen. The undistorted light is intercepted by an appropriate screen or knife edge. The inhomogeneities are directly visible as light streaks on a black background. Toepler, in the nineteenth century, and several other investigators (see ref. 71) perfected the method into a quantitative one. The basic arrangement of *Figure 10* is used. A parallel beam of light is passed through the cell. The light is deflected downwards in regions where a refractive index gradient (here resulting from a density gradient) exists. The lens L_1 forms images of the source at the position G. These images will be displaced from the center by an amount

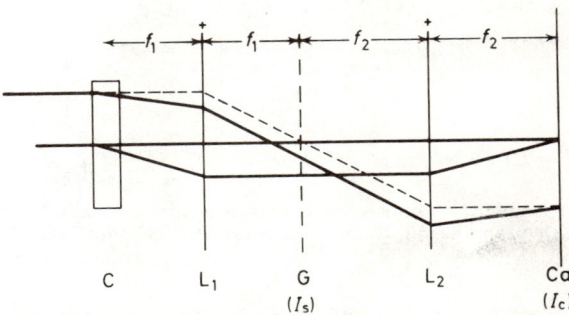

Figure 10. Schlieren arrangement of studying refractive index gradients. A point source and condenser are used to illuminate the cell C with parallel light. The grid G alternately transmits and intercepts light deflected downwards in the cell due to the presence of density gradients. A system of light and dark bands is seen in the cell image at Ca. After ref. 157.

depending on the refractive index gradient that the image-forming beam has experienced. An objective lens L_2 then forms an image of the cell on the camera back Ca. This image is not distorted by the presence of density gradients. If, however, a grid is placed at the location of the source image G, then light from regions of increasing gradients is intermittently passed and intercepted, and thus a system of black bands appears in the cell image at C. If a grid with equidistant lines is used, each band, in lowest order, corresponds to a fixed increment in gradient.

The Schlieren method has occasionally been used in critical region studies. Palmer[107] was the first to use it in the USA, in carbon dioxide and ethylene. Since then, it has been rather widely used in the USSR, principally by Shimanskaya and co-workers[126, 127, 104, 3] in one-component systems and recently, by Blagoi et al.[14] near a consolute point in a two-component fluid. The analysis of the density gradient data is still at an early stage. Only recently[3, 17] have attempts been made at a description in terms of scaling laws. The Schlieren method is the only one of the class of one-interferometric refractive index gradient determinations that has been used for critical region studies.

This limited usage of well-established experimental techniques is even more evident when we turn to the interferometric methods. In diffusion studies, it was realized quite early that these methods should be capable of increased precision. The first one to observe an interference pattern in light that had passed through a region of refractive index gradients was Gouy in 1880[57]. Gouy's arrangement is shown in *Figure 11*. Interference occurs between

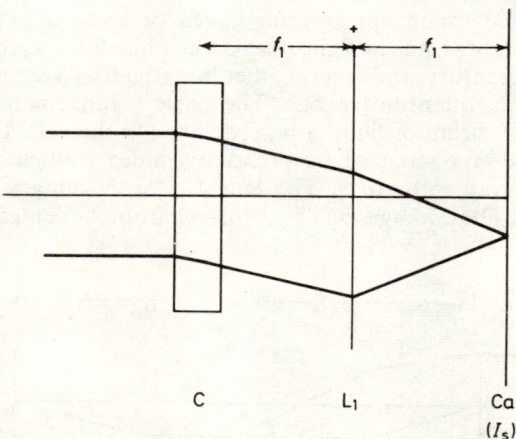

Figure 11. Gouy's method[57, 157] for the study of refractive index gradients. Interference is produced between light beams passing through regions of identical refractive index gradients.

beams that have been deflected through equal angles. This interference is alternately constructive and destructive, thus giving rise to a rather complex fringe pattern. The theory of this phenomenon was worked out by Kegeles, Gosting and Onsager[75, 56]. Apart from Gouy's method, there are several other interesting interferometric methods for diffusion studies, such as the Rayleigh double slit method which portrays directly the refractive index profile and is capable of high precision (see ref. 71).

To date, Wilcox, Balzarini[157] and co-workers are the only experimenters who have used an interferometric method (Gouy's method) for density gradient studies in the critical region. This way, they have obtained the most precise and detailed density profile data presently available. Wilcox[158] developed a scaled equation closely related to the linear model for describing interference patterns he obtained for xenon. Wilcox's scaled equation is in good agreement[41] with the NBS equation[140] for xenon, although the latter was obtained for pVT data which were less precise and farther away from the critical point than Wilcox's data. This confirms the notion that the asymptotic range of scaling is rather large so that information about the region very close to the critical point can be safely extrapolated from a knowledge of the behavior farther away.

It is hoped that the potential of interferometric methods for critical region studies will be more extensively explored in the near future.

V. Dielectric Constant Measurements

1. General

Dielectric constant methods for density determination, in which use is made of the Clausius–Mosotti (CM) relation

$$(\varepsilon - 1)/(\varepsilon + 2) = \tfrac{4}{3}\pi\alpha\rho \tag{57}$$

have some definite advantages over refractive index methods. In fact, an important share of our quantitative knowledge about thermodynamic behavior near the critical point has come from recent dielectric constant experiments. First of all, in contrast to refractive index measurements, the dielectric constant experiments are carried out in windowless vessels so that temperature gradients are more readily avoided. Secondly, capacitances can be measured with much higher accuracy than that required for these experiments. Finally, the special problems affecting optical experiments, light beam curvature and critical opalescence (section IV), have no counterpart in dielectric constant determinations. On the other hand, refractive index measurements are somewhat more informative because, as we saw in section IV, they can be made to yield bulk densities, as well as density profiles and density gradients, as continuous functions of height. Dielectric constant measurements give only local densities between the capacitor plates; however, by stacking capacitors[150,133] one can also obtain stepwise information about the density profile.

Dielectric constant measurements have been one of our principal sources of information about the critical regions of ^{3}He[163,149] and ^{4}He[113,114]. In these gases the critical pressures are low and problems caused by the pressure deformation of the capacitors, and by the need for bringing out the leads, are minor. The use of the Clausius–Mosotti relation 57 for calculating the density from the dielectric constant introduces questions which are similar to those raised in connection with the Lorentz–Lorenz relation (section IV).

The quantity $(\varepsilon - 1)/(\varepsilon + 2)\rho$ shows a dependence on density and temperature which, in an ideal experiment, should first be studied outside the critical region[76], and then used to correct the relation 57 in the critical region. The density dependence of $(\varepsilon - 1)/(\varepsilon + 2)\rho$ is expected to be weaker than that of $(n^2 - 1)/(n^2 + 2)\rho$ because the first order correction term cancels for the former quantity but not for the latter[162].

Just as discussed in Section IV for the Lorentz–Lorenz relation, we warn against tests of the Clausius–Mosotti relation in which ε is measured as a function of pressure and temperature after which known pVT data are used to calculate the density; this practice may lead to severe errors in regions of large compressibility.

In less-than-optimum experiments, the ratio $(\varepsilon - 1)/(\varepsilon + 2)\rho$ is simply assumed to be constant. The constant can be eliminated by using reduced units. For applications in which deviations from the critical density are small this practice is eminently sound; much hard and monotonous work needed for obtaining pVT data is thereby avoided.

Because capacitance measurements always yield a local density, the fact has sometimes been overlooked that gravity effects in these measurements are just as severe as in more orthodox pVT measurements, unless care is taken that the pressure is measured at the level of the capacitor.

2. $p\varepsilon T$ Measurements

The early studies of the equation of state in the critical region by means of $P\varepsilon T$ measurements are those of Chase and Zimmerman on ^{3}He[20,163] and those of Roach and Douglass on ^{4}He[113,114]. In these experiments, gravity effects were not always small because of the presence of sizable gas heads above the capacitor. After some initial difficulties with interpretation, the ^{4}He data were shown to conform nicely to the scaling laws and scaling symmetries for densities within 35 per cent from critical[140], with exponents close to those of heavier gases. An extensive set of $p\varepsilon T$ data for ^{3}He has been obtained recently by H. Meyer's group at Duke[10,149]. Here, the pressure head was virtually eliminated by passing the fill capillary horizontally through the range of the temperature gradient between the ^{3}He cell and a ^{4}He bath at 4.2 K. These data were shown[149] to obey the symmetry and scaling properties for densities within 25 per cent from critical with exponents close to those of heavier gases. Moreover, data along the critical isotherm out to 60 per cent from the critical density were used to obtain an estimate of the first correction to simple scaling[149].

3. Density Profiles by Dielectric Constant Determination

There is only one example in the literature of a density profile study by means of a stack of capacitors. This Class A experiment was carried out by L. Weber[150] near the critical point of oxygen. Because of the higher critical pressure of oxygen, this experiment was therefore performed at much higher pressures than the helium experiments discussed earlier. This required a detailed study of the pressure effect on the five capacitors. Notwithstanding these additional complications, the experiment was carried out in optimized form, i.e. the full density and temperature dependence of the CM function $(\varepsilon - 1)/(\varepsilon + 2)\rho$ was obtained in an independent experiment and taken into account in the interpretation of the data. The capacitors were spaced about 2.5 cm apart. For measurements below T_c, two capacitors were located in one phase and three in the other. The location of the sensors could be estimated with good precision from the known pVT data and the geometry of the cell. Whenever a density gradient was present, the average compressibility could be obtained by means of the relation $\rho^2 K_T \sim \Delta\rho/\Delta h$ between consecutive capacitors in the same phase. Since three capacitors were present in one of the phases, deviation of $\rho^2 K_T$ from constancy very near the critical point could be detected and $\rho^2 K_T$ extrapolated to the location of the meniscus. Above T_c, the compressibility was obtained from capacity differences between two adjacent capacitors on either side of the location of the critical density.

Weber's measurements are, to date, the most direct and reliable confirmation of the scaling ideas obtained by density profile measurements. The symmetry of $\rho^2 K_T$ was confirmed, and the values obtained for the critical exponents by these detailed measurements very near the critical point agree with those obtained from scaled analyses of gross pVT data. It is hoped that variants of this method will soon be in use in other laboratories[133].

VI. Conventional pVT and Vapor Pressure Measurements

1. pVT Measurements

If some precautions are taken in order to overcome the special difficulties presented by critical region studies (section III), conventional techniques of pVT determination can be quite successful. Precise temperature control and the elimination of gradients present demanding challenges in these studies. With new developments in the construction of strain-free high-resistance, small-size temperature sensors, semiconductor devices and a variety of a.c. and d.c. control equipment, and the possibilities offered by the new a.c. thermometry bridges[31], these challenges can be met.

Since the pV isotherms are very flat, better than average reproducibility and sensitivity is certainly needed in the pressure measurements. In the classical work on carbon dioxide by Michels, Blaisse and Michels[96], and that on xenon by Habgood and Schneider[65], a pressure reproducibility of about 10^{-3} bar was obtained using piston gages and mercury separators. With the best modern piston gages, combined with very good barometry (see Chapter 4, Part 3) this precision can perhaps be boosted to a few times 10^{-4} bar, a rather impressive precision for critical pressures in the neighborhood of 50 bar. In the helium range, mercury manometers rather than piston gages have been used[20,149] because of the low critical pressure. A quartz Bourdon gage with resolution of 3×10^{-5} bar has been used satisfactorily by several experimenters[149,77,124], although the long-term stability of this gage seems to leave something to be desired. This problem was circumvented elegantly by Kierstead who reported $(dp/dT)_v$ values for ^{4}He along isochores, by measuring pressure increments rather than absolute pressures[77].

Gravity effects present the largest problem for measurements near the critical state. There are two aspects to the solution of this problem, namely limiting the height of the vessel and eliminating large uncontrolled gas heads in the connecting capillaries. Since gravity effects can be corrected for in principle (section VIII), the first problem is minor with respect to the second. The limitation on the height of the vessel can be achieved by placing a long narrow container with its axis in a horizontal plane. In this way, the gas head was limited to slightly over 1 cm in the work of Habgood and Schneider[65], and in the recent work on carbon dioxide carried out at the National Bureau of Standards[85]. In most conventional pVT work, a capillary connects the vessel with a pressure measuring device outside the thermostat. This capillary runs through regions of uncontrolled temperature gradients. It may contain a rather large and fluctuating gas head which can be a major source of error

Figure 12. A pVT apparatus of the Burnett type, with limited height. The transducer is placed in the bath, thus eliminating noxious volumes and spurious gas heads. After ref. 85.

in critical region studies. In the work of Habgood and Schneider[65], and in that of Michels et al.[96], no such problem existed because a mercury separator was used which was placed right in the same thermostat. In the recent NBS work, a pressure transducer was used as a separator; this transducer was placed in the cryostat with its membrane coplanar with the axes of the horizontal vessels (*Figure 12*). In the work of Wallace and Meyer on ^{3}He[149], a capillary connected the helium vessel with the pressure device; however, it was brought out of the controlled section horizontally, and its vertical section was heated well above the critical temperature. Volumometry presents no special problems in the critical region. Experiments are most profitably run along isochores because pressure and density equilibration after expansion along isotherms may require a long time.

Some recent good pVT work in the critical region includes work on cryogenic fluids at the National Bureau of Standards in Boulder, Colorado[54, 151, 110], that on steam performed by Rivkin in the 'sixties'[112] and that of the Louvain group of Verbeke and co-workers on argon and methane[70, 146]. In all of these experiments, however, noxious volumes were present and gas heights were not minimized.

2. The Vapor Pressure

The form of the vapor pressure curve is expected to be peculiar near the critical point. Thermodynamic and power-law considerations lead to the expectation of a weak anomaly in d^2p/dT^2 (Table 1), namely $d^2p/dT^2 \sim |T - T_c|^{-\theta_p}$ with $\theta_p \leqslant \alpha'' + \beta$, or about 0.4. If the symmetry conditions of lowest-order scaling prevail (section II.3), then the anomaly in d^2p/dT^2 is the same as that of C_v, namely $\sim |T - T_c|^{-\alpha}$, with α about 0.1. Proposed generalizations of scaling[60, 26] give for the vapor pressure, in addition to the terms arising from lowest-order symmetry scaling, non-analytic terms of powers higher than two in $|T - T_c|$, cf. equation 47.

It is clear that precise vapor pressure measurements are required if a weak anomaly in a second derivative is to be verified experimentally. The usual problems with vapor pressure determination, cold spots in the fill capillary and slow redistribution of matter between the two phases, are compounded with those of slow equilibration and gravity effects due to the weight of the dense vapor phase. As discussed in the previous subsection, the best results are again expected when no noxious volumes or fill capillaries are part of the experimental system, and when the height of the cell is small.

The improvements in fitting near-critical vapor pressure data when, in addition to an expression analytic in temperature, a term of the form $(T - T_c)^{2-\theta_p}$ is used, were noted by several experimenters and for a number of substances[140, 55, 147, 84, 85]. Regarding the value of θ_p, Kierstead[77] showed that his results for dp_{vap}/dT of ^{3}He were compatible with a $\ln|T - T_c|$ anomaly in the second derivative $d^2 p_{vap}/dT^2$. The 'best' value of the exponent θ_p to be obtained from a fit to experimental data depends to some extent on the assumptions made about the 'background'. For the recent NBS data on carbon dioxide[85], however, the values of θ_p minimizing the standard deviation of the fit, for different assessments of the background, were always in the range 0 to 0.12; this suggests the equality of θ_p and α as required by lowest-order gas–liquid symmetry.

VII. Calorimetry

1. Experimental Problems in C_v Determination

In thermodynamic studies of the critical region, the subtle divergence of the constant volume specific heat C_v has turned out to be a fundamental and revealing property. This is in contrast with C_p, the strong divergence of which was predicted by classical and non-classical theories alike, since it is linked to the divergence of K_T by the relation $C_p = C_v + TVK_T(\partial p/\partial T)_v^2$. Historically, weakly anomalous specific heats were first discovered in so-called λ transitions in solids, and in liquid ^{4}He. In his basic paper on the two-dimensional Ising model and its logarithmic specific heat[105] Onsager, in 1944, mentioned the analogy with the λ transition in liquid helium. In 1948 Tisza[138] then presented a generalized treatment of both λ transitions and critical points. Nevertheless, it took till 1962 before the existence of a logarithmic anomaly in the specific heat C_v near the critical point of fluids was first experimentally established by Voronel and co-workers[5, 142-145]. The difficulty of C_v measurements in pressurized gases is the main cause for the delay. As long as strength requirements are met by the use of heavy walls, the heat capacity of the vessel will be much larger than that of the gas, thereby greatly reducing the accuracy of the experiment. In C_p measurements, this problem is solved using flow methods in which large quantities of gas are circulated through the calorimeter. Precise measurements of C_p in pressurized gases became possible however, as soon as strength requirements were met by means other than thick walls.

The second difficulty affecting C_v measurements near a critical point is that of long relaxation times. The relaxation times increase because the thermal diffusivity $\lambda'_T = \lambda/\rho C_p$ goes to zero when the critical point is approached. This problem was discussed in section III. Several experimental solutions will be discussed in section VII.3. The third problem arising in close approaches to the fluid critical point is, as always, due to the effect of gravity. When the sample temperature is changed, matter has to be redistributed over the height of the cell, giving rise to the latent-heat type contributions to C_v. This leads to major distortion of the specific heat versus temperature relation at all temperatures at which density gradients are not negligible. It has been suggested that stirring the contents of the vessel will obliterate the gravity effect. However, it is by no means clear that an actual system being stirred in the presence of gravity corresponds to a hypothetical system in true equilibrium in the absence of gravity. In fact, stirring might well affect the long-range correlations responsible for the specific heat divergence. If a C_v experiment is run under properly controlled thermodynamic conditions without stirring, it is possible to correct for gravity accurately[9, 69, 118] while data taken in stirred containers are beyond correcting. Nevertheless, it is imperative that a specific heat experiment be made with and without stirring, in order to settle this question.

A further challenge in C_v determinations is the high resolution needed in the steep parts of the C_v versus T curve. A number of experimental solutions to this problem will be discussed below.

A factor which is often overlooked in the excitement of determining the 'spike' in C_v is the so-called background contribution, i.e. the sum of all constant and slowly-varying contributions to C_v. In hardly any other thermodynamic property does the background play such an important role. The mathematical character ascribed to the leading anomalous term depends on the assessment of this background in a sensitive way. It is therefore necessary to measure this background in addition to the anomalous part. This requires that attention be paid to such routine matters as the heat capacity of the empty container, assessment of the temperature scale used, corrections for noxious spaces etc. Proper determination of the total specific heat, in addition to simply obtaining the anomalous contributions, is imperative for another reason, namely that of insuring thermodynamic consistency. Ultimately, the C_v data will have to be imbedded in the entire body of thermodynamic data for the same substance. This is near to impossible if temperatures were measured on an unknown scale or if not all of the necessary corrections to the specific heat data were determined and applied.

The ideal specific heat experiment, moreover, is not limited merely to taking some data on the critical isochore very near T_c. A number of considerations make it imperative that data be taken at other densities as well, and over a considerable temperature range. Data taken at different densities permit, in addition to power law analysis, tests of scaling and of some of its basic underlying assumptions (section VII.7). They also permit certain consistency checks to be made on the method used (section VII.6). By extension of the temperature and density ranges, moreover, imbedding of the critical region data in the existing body of thermodynamic data can more readily be achieved (section VII.6).

2. Reducing the Heat Capacity of the Container

The C_v data in the critical regions of argon, nitrogen and oxygen reported by Voronel and co-workers[5, 142–145] were the first C_v data precise enough to suggest the existence of a divergence. This was at least partly due to the design of the calorimeter. A spherical stainless steel vessel with a wall thickness of only 0.18 mm was used by them in place of the bulky high-pressure vessels previously used. Amirkhanov[2] used a sphere with 1 mm wall in the even more demanding C_v measurements on steam ($P_c \sim 220$ bar), but he supported the wall by a jacket, mechanically connected to (but thermally insulated from) the calorimeter sphere by powdered cuprous oxide. Edwards, Lipa and Buckingham[36] in their C_v measurements on carbon dioxide used a thin-walled pillbox reinforced with steel ribs. In all these cases the heat capacity of the calorimeter was less than that of the gas. In none of these cases were corrections made for departure from isochoric conditions due to the varying pressure of the fluid.

The heat capacity of the container was also less than that of the fluid in the C_v measurements in helium. Here one profits from the combined effect of the vanishing heat capacity of the vessel and the low critical pressure of the fluid. In this category belong the experiments near the ^{4}He λ line of Fairbank et al.[42, 43] and of Ahlers[1]; those of Moldover et al. in the critical region of ^{4}He and ^{3}He and their mixtures[50, 102] and those in ^{3}He by Brown and Meyer[15].

3. Long Relaxation Times

The only sure way of 'beating' the long relaxation times present near the critical point is to use a static method such as drop calorimetry. In this approach, a sample is thermostatically controlled for as long as needed, in a near-critical state, then dropped into an ice bath, and the amount of heat released is calculated from the amount of ice melted. By taking different starting points, the specific heat can be obtained. This method is being perfected by Schmidt[118]. It is obvious that the measurement of even one

C_v point by this method is a lengthy process, demanding high reproducibility of the experimental conditions. Therefore, most calorimetric experiments in the critical region are performed in adiabatic calorimeters with stepwise or continuous heating techniques.

The effect of the long relaxation time can be reduced by making thermal paths in the system short. The theoretical lower limit to the permissible length of the thermal path is the correlation length. However, this length does not exceed the wavelength of light[30,91,132] even within a few millidegrees from the critical point. In practice, calorimeters with thermal paths of 0.5 mm have shown relaxation times of up to 20 minutes; these are manageable times from the point of view of adiabatic calorimetry. In Moldover's experiments on ^{4}He and ^{3}He, the calorimeter volume consisted of a set of 0.1 mm wide slots milled in a copper block[102]. His relaxation times were a few minutes. Brown and Meyer used a flat disc 0.5 mm high for their C_v measurements in ^{3}He[15] and mention equilibration times of up to 30 minutes in the two-phase region. In their xenon experiment Edwards, Lipa and Buckingham[36] used 0.8 mm holes in a copper block while their carbon dioxide experiment was done in a 1 mm high pillbox-shaped calorimeter[77]. In this last experiment, the relaxation time was measured separately using ordinary adiabatic calorimetry. It was assessed to behave roughly as $\tau \sim |T - T_c|^{-\frac{2}{3}}$. Specific heats were then measured continuously by 'ramping', sample cell and shield being heated at a uniform rate and the heat supply to the cell being monitored so that the cell temperature follows that of the shield. The heat supplied to the cell is then proportional to its heat capacity. Ramp rates were adjusted to the thermal relaxation time using the condition

$$\tau \dot{T}(dC_v/dT) \ll C_v \tag{58}$$

which guarantees that an interval of temperature over which the change in C_v is still small compared to C_v itself is crossed in a time longer than the relaxation time. In view of the power laws $C_v \sim |\Delta T|^{-\alpha}$ and $\tau \sim |\Delta T|^{-\frac{2}{3}}$ it follows that the ramp rate $\dot{T}$ has to be reduced as fast as $|\Delta T|^{\frac{5}{3}}$. In practice, 10^{-6} K/s was a limit on $\dot{T}$, precluding approaches to critical closer than 0.01 K. Significant differences between cooling and heating curves were observed at temperatures closer to critical.

A third method has been proposed for rapidly reaching equilibrium[142-145], namely by continuous stirring. This method, however, is likely to prevent rather than to help the reaching of a true equilibrium state and should be avoided unless thoroughly tested.

4. *Correcting for Gravity*

The problem of gravity effects in specific heat measurements has been a fascinating one and the last decade has witnessed a sequence of papers on the subject. In most of them, classical or semi-classical equations of state were used; in several the thermodynamics employed was erroneous. Only recently, two independent satisfactory treatments of this matter have appeared. Barmatz and Hohenberg[9] used the linear model of the parametric representation (section II), with parameters obtained from a fit to the NBS equation of state, to calculate the specific heat of xenon for the experimental configuration of Buckingham et al. In a recent paper[69] they did extensive calculations for several gases and different sample heights. Schmidt[118] integrated the NBS equation and calculated the gravity effect in xenon for his own drop calorimeter, for Buckingham's cell and for several other hypothetical sample heights. In calculating the gravity effect, the arguments used are basically the following. When heating a sample of finite height in near-critical conditions two effects are present. First, the local specific heat C_H varies from point to point along the height of the cell because the sample represents a series of thermodynamic states (section III). These variations of C_H can be obtained by integration of the scaled equation of state; this integration then yields the free energy and its second temperature derivative at each height in the cell. There is, however, a second quite important effect, and that is the redistribution of matter during heating. Thus, the true average specific heat is obtained not by averaging the local specific heat C_H over the height of the cell, but rather by taking a height average of the temperature derivative of the local entropy in the cell. Thus, the heat capacity density C/V at the level H (with respect to the level of the critical density in the cell), following ref. 118, is given by

$$(C/V)_H = T(\partial s/\partial T)_H = T(\partial s/\partial T)_\rho + T(\partial s/\partial \rho)_T(\partial \rho/\partial T)_H \quad (59)$$

where s is the entropy density. The first term on the right, $T(\partial s/\partial T)_\rho$, represents the local heat capacity density, while the second represents the mass redistribution effect. However, since $s = -(\partial a/\partial T)_\rho$ and $(\partial \rho/\partial T)_H = (\partial \rho/\partial T)_\mu$, we have

$$T^{-1}(C/V)_H = T^{-1}(C_v/V)_\rho - (\partial^2 a/\partial T \partial \rho)(\partial \rho/\partial T)_\mu \quad (60)$$

$(\partial \rho/\partial T)_\mu$ is readily obtained from a scaled equation such as 43

$$(\partial \rho/\partial T)_\mu = -\Delta\rho|\Delta\rho|^{-1/\beta} h'(x)/\{\delta h(x) - xh'(x)\beta^{-1}\} \quad (61)$$

and likewise,

$$\partial^2 a/\partial \rho \partial T = (\partial \mu/\partial T)_\rho = -(\partial \mu_c/\partial T)_\rho + |\Delta\rho|^{\delta-1/\beta} h'(x) \quad (62)$$

while $(C_v/V)_p = -(\partial^2 a/\partial T^2)_p$ is calculated from the free energy density a which can be obtained by integrating the scaled equation.

Finally, the heat capacity of the cell is obtained by integrating $(C/V)_H$ over the height of the cell while properly taking care of the varying diameter.

It is found[118] that if the average filling density equals ρ_c, the average specific heat is finite at T_c, and so is its slope $(\partial \bar{C}/\partial T)_{\rho_c}$. The curvature $(\partial^2 \bar{C}/\partial T^2)_{\rho_c}$ diverges at T_c. The maximum in $\bar{C}$ occurs below T_c. Typical numbers are for xenon, a shift of the maximum of 0.01 K in a cell 1 cm high and 0.05 K in a cell 10 cm high[118]. In a cell 1 cm high, gravity corrections in xenon are important within a range of 0.06 K from T_c, or for $|\Delta T|/T_c < 2 \times 10^{-4}$.

If C_v data have not been corrected for gravity, those within this range should be excluded from power law analysis. Since, empirically, a simple power law for C_v holds for ranges not much larger than 5×10^{-3} in $\Delta T/T_c$, it follows that only a decade and a half is available for power law analysis of C_v. This is one of the reasons for the wide spread in reported α values. The apparent specific heat in a field of gravity, calculated according to the scaling laws, is in quite good agreement with the experimental values measured by Edwards et al.[36]. It seems urgent to apply gravity corrections to all C_v data measured in the critical region in unstirred cells and see if a more uniform value of α emerges.

5. Increasing the Temperature Resolution

Since a specific heat experiment is usually a differencing experiment, it suffers from the same problem as do other determinations of a slope: if the slope varies strongly, the interval over which the difference is taken has to be narrowed; however, if the interval is narrowed precision is lost. In view of the rapid increase of C_v near the critical point the temperature resolution has to be improved. For critical points near room temperature, the best resolution obtained in C_v measurements is about $1\,\text{mK}$ ($\Delta T/T_c = 3 \times 10^{-6}$) using regular 25Ω platinum thermometers. Using thermistors, the sensitivity of the thermometry can be increased by an order of magnitude; however, thermistors do not have the stability of a platinum resistance thermometer. There have been static experiments in binary liquids near their consolute points in which the temperature has been stabilized and measured to $0.1\,\text{mK}$[137]. A resolution of $0.1\,\text{mK}$ is therefore not an unreasonable goal for C_v experiments. The development of high-resistance strain-free miniature platinum thermometers and of a.c. bridges for detection should prove helpful in attaining this goal. With $0.1\,\text{mK}$ resolution, temperature increments as small as $0.01\,\text{K}$ would still yield C_v values precise to one per cent. For a general discussion of thermometry in the context of calorimetry we refer to Chapter 2 in the first volume of this series.

In the liquid helium range, the use of carbon and germanium resistors has improved the resolution by perhaps three orders of magnitude, $10^{-6}\,\text{K}$. Thus, Moldover[102] reports resolutions of $2 \times 10^{-6}\,\text{K}$ at the critical point of ^4He, using a carbon resistor, while Brown and Meyer[15] report resolutions of $5 \times 10^{-7}\,\text{K}$ at the critical point of ^3He. On a reduced scale, however, this resolution is about one order of magnitude better than that for substances with critical points near room temperature. The finest relative resolution in temperature has been reached in C_p measurements near the λ-line. Fairbank et al.[42,43] reached a resolution of $2 \times 10^{-7}\,\text{K}$ using carbon resistors. Ahlers[1] pushed the resolution to $10^{-7}\,\text{K}$ by using the amount of superfluid in a miniature thermal conductivity cell located below the experimental chamber as an interpolating device. This technique is of course applicable only for measurements near the λ-line.

6. Checking for Consistency

A direct test of the reliability of the experimental method used is made by by comparing C_v data obtained in the two-phase region at various filling densities. According to the Yang–Yang relation[160]

$$\frac{C_v}{T} = V\left(\frac{\partial^2 p}{\partial T^2}\right)_v - N\left(\frac{\partial^2 \mu}{\partial T^2}\right)_v \tag{63}$$

the C_v values along isotherms should vary linearly with the molar volume.

This test is so simple that it is surprising to see how many published data fail it. From the slope of the straight line $d^2 p/dT^2$ is obtained, permitting another consistency test to be made, if precise vapor pressure data for the same substance are available. Since there is also a theoretical reason for obtaining C_v data in the two-phase region (section VII.7) this practice is strongly recommended.

Specific heat values not too close to the critical point are reliably calculated from precise pVT or speed-of-sound data (cf. Chapters 10 and 11). It is therefore desirable that C_v measurements along at least a few of the isochores be extended into temperature ranges where such C_v values have been calculated. This is a good check on consistency and facilitates the imbedding of the data in the larger body of thermodynamic values.

7. Tests of Scaling

The scaling property of the specific heat can be readily derived from that of the symmetrized free energy a'_s in equation 17. This equation implies that the scaled symmetrized specific heat after subtraction of the background, $(C'_{vs}/VT)/|\Delta T|^{-\alpha}$, should be a universal function of $x = \Delta T/|\Delta \rho|^{1/\beta}$. In order to test the scaling law, C_v data are needed over a range of densities and temperatures. Moreover, these data have to be symmetrized. Since $a'_s = a - A_0(T) + \mu(\rho_c T)\Delta\rho$, we have

$$\frac{C'_{vs}}{VT} = \frac{C_v}{VT} + \frac{d^2 A_0}{dT^2} + \Delta\rho \frac{d^2 \mu(\rho_c T)}{dT^2} \tag{64}$$

and before scaling can proceed, an assessment of the analytic density-independent background term $d^2 A_0/dT^2$ and of the quantity $d^2 \mu_c/dT^2$ will have to made. This latter quantity, as explained in section II, plays a fundamental role in the scaling of fluid properties. It is assumed to be regular at the critical point. Specific heat measurements in the two-phase region permit an estimate of this quantity, again by using the Yang–Yang relation 63. The slope of C_v/VT versus ρ data in the two-phase region equals $d^2\mu_c/dT^2$. This is the most direct way this quantity has been obtained hitherto[15, 102, 140] and there has thus far been no reason to believe it to be divergent. The other unknown in equation 64, $d^2 A_0/dT^2$, has usually been obtained by fitting C_v data along the critical isochore using

$$C_v = A^{\pm} |\Delta T|^{-\alpha} + B \tag{65}$$

in which, using reduced units, B is to be identified with $d^2 A_0/dT^2$, the latter assumed to be constant. Note that in the context of scaling it is imperative that B be the same above and below T_c. Fits of the form $C_v = A^{\pm}|\Delta T|^{-\alpha} + B^{\pm}$ with $B^+ \neq B^-$ are frequently used without realizing that they violate scaling. Unfortunately, the values obtained for the 'background' $d^2 A_0/dT^2$ vary considerably with the choice of α; this background, therefore, has to be considered the most elusive quantity of all. Probably the best practice is to consider it to be an adjustable parameter in the process of scaling C_v. In fitting the analytic background, an expression linear in temperature, rather than the constant term B, is sometimes used. However, according to extended scaling (section II.4), a term $|\Delta T|^{1-\beta-2\alpha}$ should be inserted between B and the term proportional to ΔT. This term, roughly $|\Delta T|^{0.6}$, dominates the ΔT term near T_c.

Specific heat data have been scaled successfully in only a few instances, namely for the heliums[15, 102]. It is desirable that this be tried for the heavier gases as well.

VIII. Coexistence Curves

1. General

Along the coexistence curve of a fluid, two phases of different number and entropy density coexist. The curve of coexisting densities versus temperature shows a considerable amount of symmetry. The curve of average densities, $\bar{\rho} = (\rho_L + \rho_G)/2$, is experimentally found to be linear in temperature (law of the rectilinear diameter[16]). The slope of the diameter varies considerably with the complexity of the substance. It is near unity, in reduced units, for carbon dioxide and nitrous oxide[86] but larger for water vapor and near zero for the heliums. In the volume representation, the symmetry of the coexistence curve is less pronounced and the diameter is curved. The shape of the top of the coexistence curve has been known for 70 years to be almost cubic[52, 139]. Thus

$$\frac{\rho_L - \rho_G}{2\rho_c} = B \left| \frac{T - T_c}{T_c} \right|^\beta \tag{66}$$

with $\beta \sim 0.35$.

Except very near the critical point, densities along the coexistence curve are readily measured, so that data have been obtained by a variety of methods during the past century. Some of these methods are reviewed in this section. The principal difficulties occur only in the near-vicinity of T_c and are twofold: the effect of gravity and the need for excessively good temperature control. The gravity effect and correction for it are discussed in section VII.2. The need for good temperature control is evident by considering the density changes caused by temperature fluctuations:

$$\delta\left(\frac{\rho - \rho_c}{\rho_c}\right) = \beta B \left|\frac{T - T_c}{T_c}\right|^{\beta-1} \frac{\delta T}{T_c} \tag{67}$$

It is seen that this propagated density error diverges at T_c and thus, near T_c, becomes more important than the direct error made in the measurement of the density. We shall discuss this effect more fully when treating the coexistence curve data analysis in section VIII.7.

2. Gravity

In near critical states, the density gradients introduced by gravity in the highly compressible coexisting phases cause the average densities in the two phases to be farther apart from each other than are the true coexisting densities at the interface. Due to the gradient, there is a range of average filling densities for which the critical state will be reached somewhere in the cell. Thus, the curve of bulk coexisting densities will be wider than the true coexistence curve, and will have a flat top. The temperature region over which the top of the coexistence curve is distorted will depend on the nature of the substance, specifically, its value of h_c (Table 2), and on the height of the experimental vessel.

This flattening was demonstrated convincingly in the experiments of Weinberger and Schneider[152] in xenon. They measured the coexistence curve in a 19 cm high cell and also in a cell of 1.2 cm diameter in the horizontal position. With a cell height of 19 cm, for temperatures within a few tenths °C from critical, the bulk coexisting densities differed by more than five per cent from those obtained in a cell 1 cm high. In the tall cell, the dome had a flat portion over a density range more than ten per cent from the critical density.

Table 4(a). Density profile in xenon

$\Delta\rho = (\rho - \rho_c)/\rho_c$	$h(x)$	$10^6 \, h/h_c$	h cm
0.042	0.1339	0.0622	0.003
0.043	0.3246	0.1680	0.009
0.044	0.5020	0.2888	0.016
0.051	1.4349	1.6280	0.088
0.052	1.5336	1.9025	0.103
0.053	1.6256	2.2013	0.120
0.054	1.7114	2.5256	0.137
0.075	2.6844	17.953	0.976
0.076	2.7080	19.249	1.04
0.077	2.7305	20.611	1.12

$\Delta T = 2.07 \times 10^{-5}$
$T - T_c = -0.006$ K

Table 4(b). Density profile in xenon, continued

h cm	$\Delta\rho = (\rho - \rho_c)/\rho_c$	weight $w \sim$ area	$w \Delta\rho$
0	0.0416	0	0
0.1	0.0518	0.0601	0.00311
0.2	0.0570	0.0811	0.00464
0.3	0.0608	0.0943	0.00573
0.4	0.0638	0.1028	0.00655
0.5	0.0663	0.1070	0.00710
0.6	0.0687	0.1088	0.00745
0.7	0.0705	0.1070	0.00755
0.8	0.0723	0.1028	0.00745
0.9	0.0740	0.0943	0.00699
1.0	0.0755	0.0811	0.00614
1.1	0.0767	0.0601	0.00461
1.2	0.0781	0	0

$\Delta T = 2.07 \times 10^{-5}$

$T - T_c = -0.006$ K

$(\bar{\rho} - \rho_c)/\rho_c$ 0.0673
$(\rho - \rho_c)/\rho_c$ 0.0416
difference 0.0257

Table 5. Density error in coexisting densities of xenon due to gravity, in a cell 1.2 cm high

$T_c - T$, K	$(T_c - T)/T_c$	$(\rho - \rho_c)/\rho_c$	$(\bar{\rho} - \rho_c)/\rho_c$	$(\bar{\rho} - \rho)/\rho_c$
0.001	3.45×10^{-6}	0.0240	0.0637	0.0397
0.006	2.07×10^{-5}	0.0416	0.0673	0.0257
0.020	6.9×10^{-5}	0.0630	0.0768	0.0138
0.045	1.55×10^{-4}	0.0836	0.0909	0.0073
0.095	3.28×10^{-4}	0.1087	0.1120	0.0033
0.254	0.877×10^{-3}	0.1534	0.1545	0.0011
0.496	1.712×10^{-3}	0.1938	0.1943	0.0005

In a well-determined experimental configuration and in the absence of noxious volumes, it is possible to correct for these gravity effects. As an example, the gravity effect in the xenon experiment is worked out for the horizontal cell in *Tables 4* and *5*. Basically, for any temperature below T_c, the density profile is calculated using the scaled equation of state 43, with known or estimated values of the exponents and coefficients. In the xenon experiment, points on the coexistence curve were obtained by observing the temperature of meniscus disappearance at the top or bottom of the cell for the particular filling density. Thus in calculating the density profile the meniscus is located either at the top or at the bottom. Integrating the density profile over the height of the cell, while taking account of the varying cross section then yields the average or bulk density, to be compared with the density at the location of the meniscus. The difference is the correction to the observed bulk density at this particular temperature. These corrections are listed in *Table 5*. Realizing that the density was measured to 0.1 per cent in this experiment, one notices that at all temperatures within 0.25 K from critical, $(\Delta T/T_c < 8 \times 10^{-4})$, even in a cell of such a low height, density corrections due to gravity exceed the experimental error. In all those cases in which such corrections have not been made, data points affected by gravity should be excluded from power law analysis in order to avoid bias of the value of β[69].

3. *The Method of Meniscus Disappearance*

This is one of the most venerable and direct ways of determining the coexistence curve. A known amount of fluid is confined to a transparent vessel of known volume. The vessel is maintained at a temperature at which two phases are present. The temperature is then raised gradually and, depending on the average density, the meniscus will then fall or rise. The temperature at which the meniscus is seen to disappear out of the cell (at bottom or top) is noted. The coexisting density of the one phase remaining at this temperature is then just equal to the known filling density. The experiment is repeated for various other filling densities. The method obviously has several drawbacks. As many different samples are required as there are data points needed, since each filling density corresponds to only one point on the coexistence curve. Each sample requires careful volume calibration of the cell, and determination of an amount of gas. Also, in this method, coexistence curve densities are not obtained in pairs coexisting at the same temperature, which complicates the data analysis. Nevertheless, in the hands of careful workers, this method has yielded some of the most reliable results. Its principal advantage is that the meniscus is directly observed. In this respect it is clearly superior to several of the more indirect methods that claim higher precision but are often beset with unchecked systematic errors (section VIII.4). There are several variants of the method in which the number of volume calibrations and measurements is reduced. Thus, a single cell can be used in conjunction with a gas injector system attached to the cell. The meniscus disappearance temperature is again established for different quantities of gas in the cell. The amounts of gas are obtained from volume calibration of the gas injector and the known pVT properties of the gas in the injector. Another alternative is to fill the cell at a high initial density and bleed off gas in steps, measuring the volume of the released gas at one temperature.

The method of multiple sealed cells was used by Cook for nitrous oxide and carbon dioxide[25]. The method of the gas injector was the one used in Leyden in all the early measurements of coexistence curves: the gas injector was a piezometer of the type used in pVT experiments. The method of successive expansion was used in the experiment of Weinberger and Schneider, an outstanding example of coexistence curve determination. Most standard density determinations are good to at least 0.1 per cent. Since the factor limiting the accuracy of coexistence curve determination near T_c is not density but temperature (section VIII.2), it would seem that the precision of the classical method of meniscus disappearance could be increased by improving temperature control in the region near critical.

4. Young's Method of the Twin Cells

In the method introduced by Young[161] in 1891, coexisting densities are obtained over wide temperature ranges by measuring the meniscus height in two samples of different average densities and confined to two graduated tubes of uniform bore. From the location of the meniscus and the total volumes of the tubes, the volumes V_L, V_G of the liquid and gas phases in the two tubes can be calculated. If, in addition, the total mass of fluid in each tube is known, then two simultaneous equations of the form $\rho_L V_L + \rho_G V_G = m$ can be solved for the two unknowns ρ_L, ρ_V. Young's method has obvious advantages over the meniscus disappearance method in that only two samples are used and true coexisting densities are obtained. However, the method increases in precision when the two filling densities are farther apart and is therefore not suitable for measurements very close to the critical point. Cornfeld and Carr[27] have recently used Young's method for accurate determination of the coexistence curve of xenon over a large temperature range.

5. Coexistence Curves from Isothermal and Isochoric Intercepts

Coexistence curve data are often obtained as fringe benefits of pVT experiments. Experimental $p-V$ isotherms are extrapolated to the vapor pressure at the same temperature, yielding a gas or liquid volume on the coexistence curve; alternatively, $p-T$ isochores are extrapolated and intersected with the vapor pressure curve, yielding a T, ρ point on the coexistence curve. Both methods are precise and reliable as long as the temperature is far from critical. However, they have frequently led to severe error for coexistence curve determination near the critical point.

Table 6. Direct and propagated temperature errors in mK, along the coexistence curve of argon, obtained from isochoric intercepts

$T_c - T$, K	(a) σ_T, mK	(b) σ_{TP}, mK isochore	(c) σ_T, mK isochore	(d) σ_T(int), mK isochore	(e) σ_{TP}, mK Vap.	(f) σ_T, mK Vap.	(g) σ_T(int)
Gas							
0.275	2	6	6	30	5	5	30
1.090	2	7	7	19	5	6	20
2.574	2	8	8	17	5	6	18
4.625	2	9	10	11	5	5	12
7.957	2	11	11	9	5	5	10
Liquid							
0.346	2	4	5	23	5	6	24
1.090	2	3	4	11	5	6	12
1.512	2	3	4	9	5	5	11
2.736	2	3	3	7	5	5	9
4.444	2	2	3	5	5	5	7
6.315	2	2	3	4	5	6	7

(a) estimated error in temperature measurement.
(b) propagated temperature error along isochore, due to pressure error of two parts in 10^4.
(c) total temperature error on isochore.
(d) temperature error in intercept due to error (c).
(e) propagated temperature error along vapor pressure curve, due to pressure error of two parts in 10^4.
(f) total temperature error on vapor pressure curve.
(g) total temperature error in intercept, due to temperature errors along isochore and vapor pressure curve.

In this region, the isotherms become very flat and the slightest error in pressure measurement or extrapolation causes large errors in the coexisting volume. The isochores become almost confluent with the vapor pressure curve and, moreover, exhibit some curvature very near the phase boundary. It is readily shown (Table 6) that data precise to a few millikelvins can give errors of several hundredths of a kelvin in the intercept because of the near-equality of the slopes; in addition, there are the uncertainties introduced by extrapolation.

If this method is to be used at all in the critical region, the following points should be kept in mind. First of all, an independent determination of T_c by direct observation is an invaluable asset. Secondly, finely spaced points should be used along the isochore near the transition. Thirdly, scaling laws should be used to describe the form of the critical isotherm, critical isochore and vapor pressure curve in determining the intercepts. Finally, a careful error assessment should be made, and the data points weighted accordingly in any further analysis. An example of such an error assessment is given in Table 6 for the isochoric argon data of ref. 98. It was assumed that the pressure error was two parts in 10^4 and the temperature error 0.002 K. The error in the temperature of the intercept is compounded of temperature and pressure errors along the isochore and along the vapor pressure curve, neglecting density errors along the isochore. The observed difference in slope is used to estimate the error in the intercept. Note that the estimated total error $\sigma_T(\text{int})$ does not contain the effects of the extrapolation, which are much harder to assess.

6. Coexistence Curves by Dielectric Constant and Refractive Index Techniques

The methods of critical region studies discussed in sections III and IV are applicable and have been applied to coexistence curve studies. Thus, recent results on the coexistence curves of ^{3}He[20] and ^{4}He[113,114] were obtained by capacitive methods. The coexistence curves of carbon dioxide, nitrous oxide and chlorotrifluoromethane were obtained by measuring the refractive index[117,131]; likewise that of argon[109]. Substantial improvement over more traditional methods is obtained near the critical point, if the density is measured locally, near the interface, removing the need for gravity corrections. This was done in the refractive index measurements of Schmidt and Straub[117,131] mentioned above and the capacitor experiment of Weber[150] on oxygen. The limitations of these methods have been discussed in sections III and IV to which we refer the reader for details.

7. The Use of Floats

The most precise determination of coexisting densities was done in binary liquid mixtures near the consolute point, using a magnetic densitometer[159]. The current needed to maintain the float at a given level is a measure of the density of the medium. The float is calibrated using a number of liquids of known density. The support current is readily measured to one part in 10^5. The method is particularly well-suited to binary liquids, because their density differences are modest; it has not yet been successfully used in gases. Coexisting densities in the gas–liquid system are quite far apart except near critical and vary rapidly with temperature. Large variations in the support current would therefore be needed, reducing the precision of the measurement. The height of these magnetic floats is usually several millimeters, so that bulk rather than local densities are measured.

An alternative method, that of suspending a number of small-size glass spheres of varying density, has been proposed and used in gases. When density gradients are present, each float will find its own level where the local density equals that of the float. The floats can be calibrated by successively suspending them in a gas of known pVT properties far from critical conditions. This method of microfloats has been used successfully by Shimanskaya and co-workers, in benzene[104]. A float-and-spring method was used extensively by Maass and co-workers in the 1930s for measuring coexistence curves and pV isotherms in one-component systems. A float of calibrated weight and volume is suspended by a quartz spring. The extension of the spring is measured; it is related to the buoyancy force exerted on the float by the medium. The same method was perfected by Thompson and Rice[137] in their classical study of the coexistence curve of the perfluoromethane–carbon tetrachloride system. Recently, densities of the liquid phase along the coexisting curves of oxygen, nitrogen and argon obtained by the float-and-spring method were reported by Goldman and Scrase[53].

Float methods in densimetry have definite advantages over optical and dielectric constant measurements because of the direct relationship between the buoyant force and the fluid density. The use of microfloats for local density determination seems especially attractive.

8. Power Law Analysis of Coexistence Curves

In the power law analysis of coexistence curves the aim is to obtain values of $B^{\pm}$, $\beta^{\pm}$ in the asymptotic expression

$$\left|\frac{\rho^{\pm} - \rho_c}{\rho_c}\right| \sim B^{\pm} \left|\frac{T - T_c}{T_c}\right|^{\beta^{\pm}} \tag{68}$$

Here ρ^+ is the density of the liquid, ρ^- that of the gas at coexistence, and ρ_c the critical density. There are two kinds of problems to be faced in this analysis. The first class of problems is of a mathematical nature: how large is the asymptotic range; are the coefficients B^+ and B^- and the exponents β^+ and β^- identical; what would be the character of higher order terms in the expansion 68? The second class of problems has to do with statistical analysis. How are errors in density and temperature assessed in the particular experiments; how do these errors propagate; and how are data points weighted in fitting equation 68?

As to the functional form of equation 68 and its range of validity, the following facts have emerged in recent years. First of all, in the cases studied, the top of the coexistence curve has, within experimental error, shown the same symmetry as that of the classical case and the Ising model, namely $B^+ = B^-$ and $\beta^+ = \beta^-$ [86]. The range of asymptotic validity of equation 68, however, is quite small, typically about 3×10^{-3} in $\Delta T/T_c$, or about one degree for a gas like carbon dioxide (*Figure 3*). The next higher order terms in expression 68 are likely to be two terms of the form $|\Delta T|^{1-\alpha}$ [155, 60, 93–95] and ΔT^{2} [6], respectively (see section VIII.8). It is an empirical fact that these higher order terms cancel to a large extent when the difference $\rho^+ - \rho^-$ is formed. For this difference, a power law with the same exponent $B = B^+ = B^-$ and $\beta = \beta^+ = \beta^-$ is found to hold, but the asymptotic range of validity is of order 0.1 in $\Delta T/T_c$[86]. The latter fact is of great help in the analysis of coexistence curves. The data within the asymptotic range of equation 68, 3×10^{-3}, are often beset with errors due to gravity and they are quite sensitive to temperature errors. The data for $3 \times 10^{-3} < (T_c - T)/T_c < 0.1$ are usually more plentiful and more reliable, thus permitting much better estimates of B and β.

It is not always possible to form the difference $\rho^+ - \rho^-$, simply because in many of the techniques discussed the two coexisting densities are not measured at the same temperature. In that case, the apparent straightness of the coexistence curve diameter $\bar{\rho} = (\rho_L + \rho_G)/2$ (see, however, section VIII.8) can be used to extend the asymptotic range. The following expression is then used:

$$\frac{\bar{\rho} - \rho^-}{\rho_c} = \frac{\rho^+ - \bar{\rho}}{\rho_c} = B\left|\frac{T_c - T}{T_c}\right|^{\beta} \tag{69}$$

with

$$\bar{\rho} = \rho_c \left[1 + a \left(\frac{T_c - T}{T_c} \right) \right]$$

This expression has the same asymptotic range as that for $\rho^+ - \rho^-$, namely $(T_c - T)/T_c < 0.1$. The constant a can be considered as an additional adjustable parameter in the fit.

The statistical aspects of the fit are twofold. In an analysis of the experimental procedure, an assessment must be made of the errors in the measurement of the density and the temperature. In a least squares fit, these errors have to be used to weight the data. If non-linear fitting procedures are used the weights for the two variables can be assigned[34]. Since linear least-squares techniques are usually preferred, B and β are generally obtained from an expression of the form

$$\ln \frac{(\rho^+ - \rho^-)}{2\rho_c} = \ln B + \beta \ln \frac{T_c - T}{T_c} \qquad (70)$$

with weights assigned to the densities only. In this case, the temperature error δT has to be propagated into the density using the relation:

$$\delta(\rho^+/\rho_c) = \delta(\rho^-/\rho_c) = \beta B \left| \frac{T_c - T}{T} \right|^{\beta - 1} \frac{\delta T}{T_c} \qquad (71)$$

The variance of the density difference $\rho^+ - \rho^-$ is a sum of variances due to direct density errors and to propagated temperature errors with the propagated error diverging at T_c. An example of the density error arising from a temperature change of 1 mK, for various distances from T_c in carbon dioxide, is shown in *Table 7*. It is seen that within a few hundredths of a kelvin from critical, the temperature error propagated into density, surpasses the direct density error (which is typically less than 1×10^{-3}) even

Table 7. Coexistence curve of carbon dioxide. Propagated temperature error

$|T - T_c|$, K	Change in $(\rho - \rho_c)/\rho_c$ due to 1 mK error in T	Value of $(\rho - \rho_c)/\rho_c$
0.3	1.9×10^{-4}	1.8×10^{-1}
0.03	8.4×10^{-4}	8.0×10^{-2}
0.003	3.7×10^{-3}	3.6×10^{-2}

for such a slight temperature variation. It is clear that for close approach to critical, a standard density determination will be sufficient while the temperature determination needs extraordinary care.

When the logarithm is taken in equation 68 for fitting purposes, another error enlargement results

$$\delta \ln \frac{(\rho^+ - \rho^-)}{2\rho_c} = \frac{\delta(\rho^+ - \rho^-)}{\rho^+ - \rho^-} \qquad (72)$$

where $\delta(\rho^+ - \rho^-)$ is the total estimated error in $\rho^+ - \rho^-$. Since $\rho^+ - \rho^-$ goes to zero at T_c, the error in the logarithm diverges faster than the error in $\rho^+ - \rho^-$, and the weight of the data points near T_c has to be reduced accordingly.

If only a few data points are taken near T_c, and some of the factors reducing their weight are overlooked, quite erroneous values of β may result, as has been witnessed repeatedly in the recent literature. Carefully analysed data for coexistence curves in gases result in β values in the range 0.345 to 0.365 and with typical errors of about 0.005. Thus, the observed variation of β from substance to substance is barely outside the experimental error.

9. The Diameter of the Coexistence Curve

The most intriguing question to be studied in relation to coexistence curves is the shape of the diameter $\bar{\rho}$. The law of the rectilinear diameter[16] is about a century old and is in agreement, within error, with the experimental

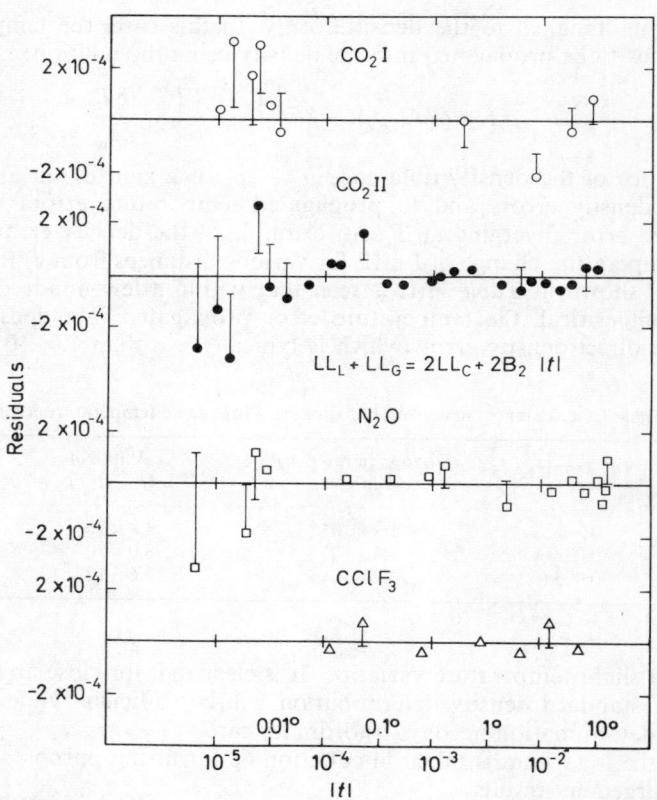

Figure 13. Deviations of the coexistence curve diameter from a straight line, for the data of refs 117, 131. The bars denote experimental error. There is no indication of a violation of the law of the rectilinear diameter. After ref. 86.

facts[86,150] even for the most precise data (*Figure 13*). Nevertheless, theoretically it is very unlikely that the diameter is straight (section II.5). Recent predictions give $|\Delta T|^{1-\alpha}$ for the leading term in the sum of the densities[26,60,93–95,155]. An increase in precision is necessary if this prediction is to be tested. We feel that such an increase is possible given the present state-of-the-art in temperature control. Ideally, the two coexisting densities should be measured simultaneously at each temperature, as close to the interface as possible, and over a sizeable range of reduced temperatures, e.g. from 3×10^{-6} to 10^{-1}. Such an experiment seems to deserve urgent attention.

IX. Scattering

1. Introduction

The theory of light scattering from density fluctuations in fluids was developed at the beginning of this century. The names of Einstein and Smoluchowski[39,129] are attached to the equation which relates the intensity of scattered light to the compressibility of the medium. Ornstein and Zernike[106] modified the theory by introducing correlations between density fluctuations in the medium, thereby explaining the anisotropy of the scattered intensity in the vicinity of a critical point. Although an assumption of analyticity is made for the angle and wavelength dependence of the scattering function, the Ornstein–Zernike theory has not been definitely disproved by experiment and describes the scattered intensity of light in the critical region with surprising accuracy. Until a decade ago, quantitative observations of scattered light were very difficult to make and there were only a few qualitative results available in the literature. The development of laser techniques, however, has given rise to a sudden increase in experimental possibilities. In the past few years, laser light scattering has been one of the fastest-growing and most highly productive sources of information about the critical region of fluids. The reason is that the laser beam is so narrow in frequency that in addition to the intensity, the spectrum of scattered light can now be measured with good accuracy. This spectrum, as was pointed out by Landau and Lifshitz[80], and worked out in detail by Mountain[103], contains a wealth of information about thermodynamic and transport properties of the fluid. The emphasis in laser scattering is heavily concentrated on transport properties. The subject has been treated recently in several review articles[21,22,92]. It is our feeling, however, that a chapter on critical region thermodynamics would be incomplete without at least a survey of the potential of light scattering experiments. In the treatment of the intensity of scattered light, we shall also discuss x-ray scattering from density fluctuations.

2. Intensity of Scattered Light

When a polarized electromagnetic wave of intensity I_0 falls on a scattering medium, the intensity scattered into unit solid angle, per unit of scattering volume, is expressed as a fraction of the incident intensity by means of the Rayleigh ratio $R_{\theta,\phi} = I_{\theta,\phi}R^2/I_0 V$. Here ϕ is the angle between the direction

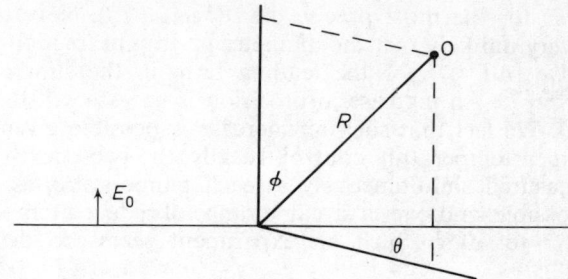

Figure 14. A light scattering experiment. The incident electric vector is indicated by E_0. The scattered intensity is observed at O, at an angle θ with the forward direction and an angle ϕ with the direction of polarization.

of polarization of the incident light and the direction in which the scattered light is observed (Figure 14). If there are N independent scatterers, of polarizability α, then the Rayleigh ratio is given by

$$R_{\theta,\phi} = (N/V)\alpha^2 k_0^4 \sin^2 \phi \tag{75}$$

The scattering is isotropic, i.e. it does not depend on θ.

In a dense medium, light is scattered from the fluctuations in the dielectric constant. The mean square of these fluctuations, $\langle(\delta\varepsilon)^2\rangle$, is related to the fluctuations in density and temperature by

$$\langle(\delta\varepsilon)^2\rangle = (\partial\varepsilon/\partial\rho)_T^2 \langle(\delta\rho)^2\rangle + (\partial\varepsilon/\partial T)_\rho^2 \langle(\delta T)^2\rangle \tag{76}$$

Near the critical point, the density fluctuations are large while those in temperature are small. Moreover, for non-polar fluids $(\partial\varepsilon/\partial T)_\rho$ is much smaller than $(\partial\varepsilon/\partial\rho)_T$. As a result, the second term in equation 76 can be neglected with respect to the first. The density fluctuations in a unit volume are related to the compressibility by the fluctuation theorem

$$\langle(\Delta\rho)^2\rangle = \rho^2 k_B T K_T \tag{77}$$

If correlations between density fluctuations in different volume elements are ignored, one obtains the Einstein–Smoluchowski scattering equation

$$R_{\theta,\phi}^{ES} = (k_0^4/16\pi^2) \sin^2 \phi \rho^2 (\partial\varepsilon/\partial\rho)^2 k_B T K_T \tag{78}$$

Einstein scattering is still isotropic, as is seen from equation 78. In a dilute gas, $(\partial\varepsilon/\partial\rho)_T$ approaches $4\pi\alpha$, according to the Clausius–Mosotti relation, while K_T approaches $(\rho k_B T)^{-1}$; thus, equation 78 goes over into equation 75 for independent scatterers. The scattered intensity in equation 78 is proportional to the compressibility. Although the derivation of equation 78 is no longer valid when correlations exist between density fluctuations in different volume elements, the Einstein–Smoluchowski relation will still hold in the limit of $\theta = 0$ (forward scattering). Thus, the measurement of the intensity of scattered light in the low angle limit yields the compressibility of the fluid.

In binary mixtures, the dielectric constant fluctuates because of concentration fluctuations. The quantity $\rho^2(\partial\varepsilon/\partial\rho)^2 K_T$ in equation 78 is here replaced by $(\partial\varepsilon/\partial c)^2(\partial c/\partial\mu)_T$.

3. Angular Dependence of the Intensity of Scattered Light

The angular dependence of the scattering intensity arises from the correlations in the fluid. Density correlations are conveniently described by means of the radial distribution function g(r), defined such that the particle density of a point at distance r from a given molecule is equal to ρ g(r). In a fluid, g(r) is different from unity only for r no larger than a few molecular diameters; however, near the critical point g(r) becomes long-ranged, which follows immediately from the fluctuation theorem

$$k_B T \rho K_T = 1 + \rho \int [g(r) - 1] \, d\vec{r} \tag{79}$$

when one realizes that K_T diverges. Light scattered from density fluctuations at r_i and r_j has a phase difference described by a phase factor $\exp(i\vec{q}\cdot\vec{r}_{ij})$ where, cf. Figure 15,

$$\begin{aligned} \vec{q} &= \vec{k}_o - \vec{k}_s \\ |\vec{k}_s| &\simeq |\vec{k}_o| \\ |\vec{q}| &= 2k_o \sin\tfrac{1}{2}\theta \end{aligned} \tag{80}$$

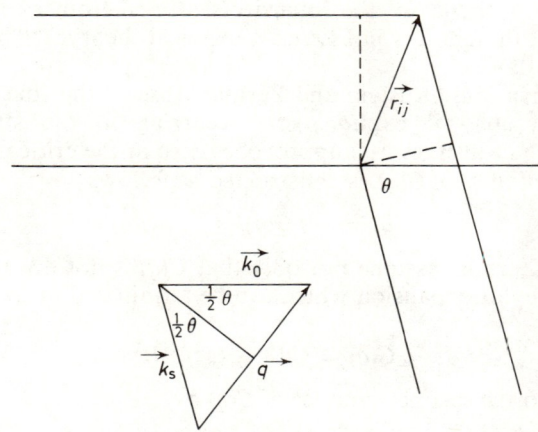

Figure 15. The wave vector $\vec{q} = \vec{k}_o - \vec{k}_s$ determines the phase difference $\vec{q}\cdot\vec{r}_{ij}$ of light scattered through an angle θ from two points at $\vec{r}_i$ and $\vec{r}_j$; $\vec{r}_{ij} = \vec{r}_j - \vec{r}_i$.

Then, the Rayleigh ratio for the scattered light is given by

$$R_{\theta,\phi} = (k_0^4 \sin^2\phi/16\pi^2)\rho(\partial\varepsilon/\partial\rho)^2 [1 + \rho\int\{g(r) - 1\}\exp(i\vec{q}\cdot\vec{r})\,d\vec{r}] \tag{81}$$

The first term in square brackets is the isotropic contribution from independent scatterers. The second term depends on the angle θ through q. In the limit of zero angle, $\exp(i\vec{q}\cdot\vec{r})$ equals unity and the Einstein result 78 is recovered, relating the intensity scattered in the forward direction to the compressibility.

The structure factor $S(q)$, defined as

$$S(q) = 1 + \rho \int \{g(r) - 1\} \exp(i\vec{q}\cdot\vec{r})\,d\vec{r} \tag{82}$$

can be considered as a generalized compressibility $k_B T\rho_- K_T(q)$ (cf. equation 79). In the case of visible light, with $k_0 \sim 10^5/\text{cm}$, at $\theta = 60°$ or $|\vec{q}|=|\vec{k_0}|$, r has to be of the order of 10^{-5} cm or 1000 Å before $\exp(i\vec{q}\cdot\vec{r})$ becomes substantially different from unity. Ordinarily, $g(r) - 1$ will have reached the value 0 long before r becomes so large, and we retrieve the isotropic Einstein scattering. Only near the critical point is $g(r)$ sufficiently long-ranged for $g(r) - 1$ not to be negligible when $\exp(i\vec{q}\cdot\vec{r})$ begins to deviate from unity, and scattering is then no longer isotropic. The Einstein formula will fail first for large angles; it remains strictly valid in the limit of zero angle.

For x-rays, wavenumbers are typically a factor 10^3 higher than for visible light. Therefore, $\exp(i\vec{q}\cdot\vec{r})$ will fall away from unity, at $\theta = 60°$, for r only a few Ångström units; this is a region of interest for intramolecular structure, and for the behavior of $g(r)$ at small distance. The long-wavelength critical density fluctuations will, however, be probed at small angles, typically a few degrees or less, and will manifest themselves in a strong angular dependence of the scattered intensity.

The theory of Ornstein and Zernike[106] developed in 1914–1916, describes the angular dependence of the intensity scattered from correlated density fluctuations. Although it is in essence a classical theory[59, 46] it has survived to the present day.

In their derivation, Ornstein and Zernike observe that the Fourier transform $G(q) = \int \{g(r) - 1\} \exp(i\vec{q}\cdot\vec{r})\,d\vec{r}$ occurring in the structure factor $S(q)$, equation 82, diverges as q approaches zero at the critical point. A non-diverging function of q, $C(q)$, is constructed as follows

$$\rho C(q) = -\rho G(q)/\{1 + \rho G(q)\} \tag{83}$$

Ornstein and Zernike assume not only that $C(q)$ is not diverging, but also, that it has a Taylor expansion which can be terminated at the q^2 term

$$C(q) = C(0) - (q^2/\rho)R^2 \tag{84}$$

R^2 is related to the second moment of $C(q)$ by

$$R^2 = (\rho/6) \int C(r)r^2\,dr \tag{85}$$

It is hoped that this moment exists and that it is substantially constant in the critical region. We have called the quantity $S(q) = 1 + \rho G(q)$ the generalized compressibility $k_B T K_T(q)$. The Rayleigh ratio of the scattered light may then be written as

$$R_{\theta,\phi} = A^2 k_B T\rho^2 K_T(q) = A^2\rho[1 + \rho G(q)] \tag{86}$$

with

$$A^2 = (k_0^4/16\pi^2)(\sin^2\phi)(\partial\varepsilon/\partial\rho)^2 \tag{87}$$

However, it follows from equation 83 that
$$1 + \rho G(q) = 1/\{1 - \rho C(q)\} \qquad (88)$$
so that
$$1 - \rho C(q) = [k_B T \rho K_T(q)]^{-1} \qquad (89)$$
and
$$1 - \rho C(0) = [k_B T \rho K_T]^{-1} \qquad (90)$$

By using equation 84 for $C(q)$, we find for the generalized compressibility
$$k_B T \rho K_T(q) = 1/\{1 - \rho C(0) + q^2 R^2\} = 1/R^2[\{1 - \rho C(0)\}/R^2 + q^2] \qquad (91)$$

Defining a length ξ by
$$\xi^{-2} = \{1 - \rho C(0)\}/R^2 = 1/(k_B T \rho K_T R^2) \qquad (92)$$

we obtain the famous result
$$(A^2 \rho) R_{\theta, \phi}^{-1} = R^2 \xi^{-2}[1 + q^2 \xi^2] \qquad (93)$$

The inverse intensity as a function of q^2 is a straight line, generally referred to as an Ornstein–Zernike or O–Z plot. A typical O–Z plot is shown in Figure 16. The slope equals R^2, the quotient of slope and intercept gives

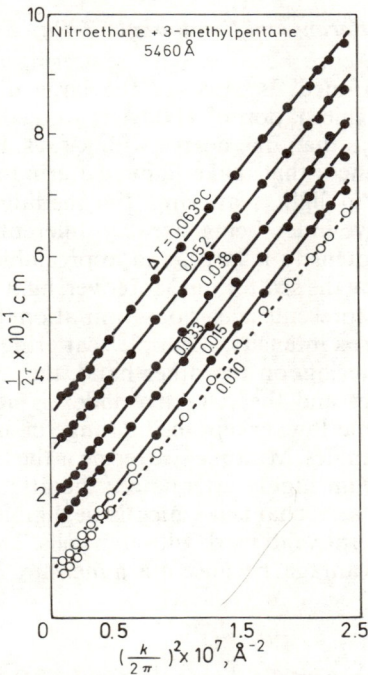

Figure 16. O–Z plot of the inverse scattering intensity as a function of $(k/2\pi)^2$ for a critical mixture of 3-methylpentane and nitroethane. The various isotherms are labelled with the respective values of $\Delta T = T - T_c$. After ref. 92.

ξ^2 [and, with equation 92, K_T]. Thus the study of O–Z plots yields the correlation length, which can be thought of as the average size of the cluster; this is a physical concept playing a central role in scattering near the critical point.

If the Rayleigh ratio is written as

$$R_{\theta,\phi} = (k_0^4 \sin^2 \phi/16\pi^2)\rho^2(\partial\varepsilon/\partial\rho)^2 k_B T K_T (1 + q^2\xi^2)^{-1} \tag{94}$$

it is seen that the $q^2\xi^2$ term in the denominator will dominate for large ξ. Since q is proportional to k_0, the Rayleigh ratio will have a k_0^2 rather than k_0^4 dependence near the critical point.

Equation 92 shows that the correlation length behaves as $(K_T)^{\frac{1}{2}}$. Therefore, along the critical isochore, $\xi \sim |\Delta T|^{-\nu}$, with $\nu = \frac{1}{2}\gamma$. However, since the theory of Ornstein and Zernike is classical, this result has been questioned[59,46] and

$$\xi \sim K_T^{\frac{1}{2}(1-\eta)} \tag{95}$$

has been proposed instead, a non-zero value of η reflecting deviations from the O–Z theory. In fluids, the experimental information from light scattering does not yet allow the firm conclusion that η is non-zero. If curvature of the O–Z plots is observed, it is usually in regions where sizeable multiple scattering is expected to be present.

4. Light Scattering and Small-angle X-Ray Scattering

A characteristic difference between light scattering and x-ray scattering has already been mentioned. Because of the large difference in incident wavenumber, a typical fluctuation of 1 000 Å is studied at a 60° angle with light, and at an angle less than one degree with x-rays. However, there are a number of other differences that make light scattering and x-ray scattering complementary. Thus, in light scattering, the medium is 'soft' or highly polarizable, the refractive index being sizeably different from unity. Consequently, the scattered intensity is large, and appreciable attenuation of the incident light is caused by the scattering. Moreover, light is multiply scattered in the medium, which presents a serious limitation to interpretation of measurements of scattered intensity. Multiple scattering is reduced by using thin cells. In x-ray scattering, on the other hand, the medium is very 'stiff' at these high frequencies and the refractive index is near unity. Thus, the scattered intensity is quite low, except in the range of interest namely near the critical point at low angles. Multiple scattering is the least of the problems. The incident wave is again highly attenuated; the attenuation, however, is now due to absorption, a mechanism which is negligible in light scattering. This makes it again imperative to work with thin cells. The attenuation of the beam can be used to advantage, because it is a measure of the density of the fluid.

5. The Experimental Situation in Critical Opalescence

Critical opalescence studies with visible light have, until recently, been confined to binary liquids near their consolute points. The reasons are clear.

The experiments are carried out at pressures near atmospheric, so cells are relatively easy to build. Temperature control is simplified because those systems are studied that have consolute points near room temperature. Multiple scattering can be reduced by using components with closely matching refractive indices. The molecules are usually larger than those of simple gases; thus, the scale of the system (that is, the typical values of R in equation 93), is such that the O–Z plots have considerable slope. The literature on light scattering in binary fluids is so rich that we must confine ourselves merely to the mention of some of the pioneers, e.g. Debye, Chu and McIntyre and refer to the literature[21,22,92] for further information. In all of the more recent work, the O–Z plots are straight lines for temperatures to within 0.01 K from critical (*Figure 16*); closer in, multiple scattering is usually present. A consensus is forming as to the value of the compressibility exponent γ, placing it near 1.22 to within a few hundredths.

Intensity studies in gases are much more scarce than in binary liquids because of the much greater experimental difficulties. In fact, the first quantitative studies of critical opalescence were not performed with light but with x-rays, in the pioneering studies of Eisenstein and Gingrich[40] on argon. This work has been carried out and perfected by P. Schmidt and co-workers[135,136]; in their recent work on argon[6], the O–Z theory was confirmed and values for the compressibility and for the correlation length were obtained. Chu and co-workers[23] have reported results of x-ray scattering in carbon dioxide.

Quantitative determinations of the intensity of scattered visible light in gases have become feasible after the development of the laser. The angular dependence of this scattered intensity is usually slight. Studies in xenon have been made by Smith, Giglio and Benedek[128]; recent measurements in carbon dioxide are those by Maccabee and White[90,91], and by Lunacek and Cannell[89]. Puglielli and Ford[111] measured the turbidity of sulfur hexafluoride with ordinary light; using the O–Z theory, this turbidity can be related to the compressibility. In such an experiment, multiple scattering is of no concern. Chalyi and Alekhin[17] measured the intensity of scattered light as a function of height in *n*-pentane and deduced values for both γ and δ from their data. The results of the measurements of gases can be summarized as follows. A consensus to within ten per cent has been reached as to the absolute value of the scattered intensity at given conditions in the same gas. The γ values obtained tend to cluster around 1.22 with the exception of the results of refs. 90 and 91. The data close to critical (except for turbidity data), are usually somewhat distorted by multiple scattering. Thus, the γ values obtained may depend on the lower cut-off of the range $|T - T_c|$. These values may also depend on the higher cut-off of this range, since the extent of the true asymptotic range for the validity of the simple power law is not well known. The range dependence has not yet been critically studied. The ratio of the compressibility coefficients Γ, Γ' above and below T_c is usually reasonably close to that predicted by scaled equations of state.

6. The Spectrum of Scattered Light

Since several excellent treatises on the subject exist[22,28,92] we shall only give a concise summary, with emphasis on the importance for thermodynamic studies near the critical point.

The optical mixing techniques developed for detecting laser light have facilitated study not only of the intensity but also of the spectrum of scattered light[11, 29]. This spectrum contains a wealth of information about thermodynamic and transport properties. In one-component fluids the spectrum, at any given $\vec{q}$ value (angle with the incident), contains three components: a central unshifted line, the Rayleigh line, and two shifted lines, the Brillouin lines, at frequencies $\omega_0 \pm \Delta\omega$. The shifted lines are caused by reflection from sound waves (adiabatic pressure fluctuation) of the appropriate wavelength $\Lambda = 2\pi/q$; the frequency shift $\Delta\omega$ equals $c_q q$ where c_q is the speed of sound at frequency q. Thus, the shift of the Brillouin lines yields the speed of hypersound. At infinite wavelength the speed of sound is expected to go to zero slowly, as $|\Delta T|^{\alpha/2}$. Consequently, one expects the Brillouin line to move in to the central line when the critical point is approached, up to the point where the correlation length becomes larger than the wavelength of the sound wave. The width of the Brillouin lines, $\Gamma_B q^2$, gives the damping of hypersound. In one-component fluids the dominant term in the damping is $(\lambda/\rho C_p)(C_p - 1)$ where λ is the thermal conductivity. This term goes to infinity roughly as $|\Delta T|^{-\nu}$. Consequently, the Brillouin lines become very wide. The central line, the Rayleigh line, was diagnosed by Landau and Placzek to be caused by isobaric entropy fluctuations[81]. These do not propagate and are damped by thermal diffusion. In a one-component system, the thermal diffusivity $\lambda/\rho C_p$ goes to zero roughly as $|\Delta T|^{\nu}$ at the critical point. Consequently the central line becomes very sharp. The intensity ratio of Rayleigh and Brillouin lines was given by Landau and Placzek to be

$$I_c/2I_B = (C_p - C_v)/C_v \qquad (96)$$

Measurement of the intensity ratio therefore provides information about yet another important thermodynamic quantity. Since C_p diverges so much faster than C_v, the entire spectral intensity is concentrated in the central line close to the critical point.

The full expression for the spectrum in the hydrodynamic limit in a one-component fluid was first given by Mountain[103] and it is

$$R(q, \omega) = (k_0^4/16\pi^2)(\sin^2\phi)(\rho\partial\varepsilon/\partial\rho)^2 k_B T K_T$$
$$\times \left[\left(1 - \frac{C_v}{C_p}\right)\frac{2\lambda' q^2}{(\lambda' q^2)^2 + \omega^2} + \frac{C_v}{C_p}\left\{\frac{\Gamma_B q^2}{(\Gamma_B q^2)^2 + (\omega + c_0 q)^2}\right.\right.$$
$$\left.\left. + \frac{\Gamma_B q^2}{(\Gamma_B q^2)^2 + (\omega - c_0 q)^2}\right\}\right] \qquad (97)$$

The lines are Lorentzians and λ' is the thermal diffusivity $\lambda/\rho C_p$. For binary mixtures, the shape of the central line is more complicated because there are several damping mechanisms. However, very near the critical point, the diffusion process is the principal damping mechanism and the line shape is the same as that of equation 97, with λ' replaced by D, the diffusion constant. The diffusion constant goes to zero roughly as $|\Delta T|^{\nu}$ just like λ'. A full discussion of the experimental features of the spectrum, its q dependence and the theories developed for explaining the experimental results is outside the scope of this paper. For the interested reader, we shall give some of the highlights, and references for further study.

The spectrum of scattered light has been intensively studied in recent years, both in binary fluids and in simple gases. Pioneering work in gases was done by Benedek and co-workers[51, 128], and by Cummins and co-workers[30, 67, 132]; in binary fluids, the first studies were those of Bergé, Volochine and co-workers[12, 13], and of Chu[21]. The principal results obtained were those for the width of the Rayleigh lines $\lambda' q^2$ in equation 97, where λ' is the diffusivity. This width is usually measured along the coexistence curve and on the critical isochore above T_c. In the one case, carbon dioxide, where all the parameters entering into the thermal diffusivity $\lambda'_T = \lambda/\rho C_p$ are known from independent thermodynamic data[121, 99, 100] it was soon established that the diffusivity obtained from light scattering agrees with that from thermal conductivity measurements in the range of overlap (*Figure 17*). The

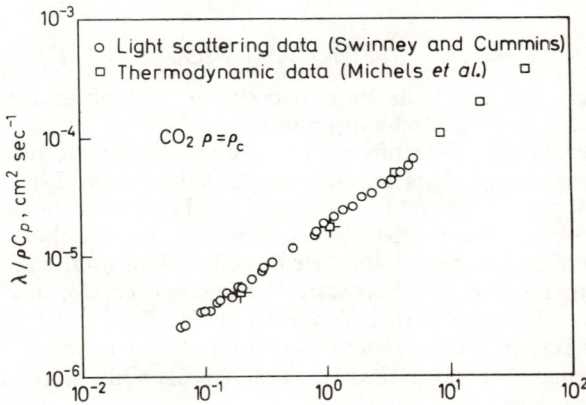

Figure 17. The thermal diffusivity of carbon dioxide on the critical isochore as a function of temperature. The circles represent data obtained by light scattering techniques[132], while the squares are results from thermodynamic measurements[121].

exponent ψ, describing the width of the Rayleigh line along the critical isochore as a function of $|T - T_c|$, showed sizeable variations from system to system, until it was shown[122] how to correct for background contributions which are quite appreciable in this case. A consensus about the behavior of the anomalous part of the Rayleigh linewidth has since formed: this width behaves as $|T - T_c|^{\gamma - \nu}$ with $\gamma - \nu = 0.62 \pm 0.02$ for binary fluids as well as for gases[122, 123].

The Rayleigh linewidth shows this behavior only in the hydrodynamic limit, $q\xi \ll 1$, when the probing wavelength is much longer than the correlation length. In the regions where the correlation length exceeds the probing wavelength (close to the critical temperature, large angles) new phenomena arise which have been qualitatively described by the theory of dynamic scaling[66], and virtually quantitatively by the theory of mode–mode coupling[74].

We refer the interested reader to these original papers, and also to refs. 18 and 122 for experimental details.

X. Sound

1. Sound, Ultrasound and Hypersound

The general subject of sound measurements is treated in Chapter 11. We shall confine ourselves to a summary of the methods used and results obtained very near the critical point. There are three different methods in which values of the speed and attenuation of sound can be obtained. They can be characterized by the frequency employed. Using thermodynamic and transport properties, c_0, the *zero-frequency* speed and α_0, the attenuation of sound, can be calculated, using

$$c_0^2 = (\rho K_s)^{-1} = (\rho K_T)^{-1} + (\partial P/\partial T)_v^2 (T/\rho^2 C_v) \tag{98}$$

and

$$\alpha_0 = (2\pi^2/c_0^3 \rho)\left[(\tfrac{4}{3}\eta + \zeta) + \lambda'(C_v^{-1} - C_p^{-1})\right] \tag{99}$$

Here η is the viscosity, ζ the bulk viscosity, λ' the diffusivity; $\lambda' = \lambda/\rho C_p$ in the gas and $\lambda' = D$ in the binary liquid.

Experimental determinations of the speed of sound are usually done at ultrasonic frequencies, *from the kHz to the MHz range*. For experimental details, we refer to Chapter 11 of this volume. These experiments have been carried out over a long period of time. The data most important from the point of view of critical phenomena are those by Chynoweth and Schneider[24] in xenon from 200 kHz to 2 MHz; by Carome and co-workers[44] in carbon dioxide at a few kHz; by Thoen, Vangeel and Van Dael[134] in argon between 20 and 100 kHz; by Garland and co-workers[48,49] in xenon in the MHz range; by Barmatz[8] in ^{4}He in the kHz range. In this range of frequencies some dispersion is usually observed.

The third method is that of light scattering. As discussed in section IX, the shift of the Brillouin lines is proportional to the speed of sound while their width reflects the attenuation. Typical frequencies used in Brillouin scattering experiments are in the *100 MHz range*. In this range of frequencies, dispersion is considerable. Measurements on Brillouin shifts and lineshapes have been performed by Cummins, Swinney and Henry in xenon and carbon dioxide[30,67]; by Cannell and Benedek in xenon[19].

2. Gravity Effects in Sound Velocity Measurements

Sound velocity measurements by ultrasonic techniques have to be corrected for gravity effects in that range around the critical point in which the density is not uniform. In order to find the average velocity in this medium, the wave equation has to be solved for the inhomogeneous fluid. The local thermodynamic state can be described by the scaling laws. Only in one case has this program been carried through, namely, for the speed-of-sound measurements in ^{4}He by Hohenberg and Barmatz[69]. Significant distortion of the velocity versus temperature curve, in a cell only 0.5 cm high, occurs at the critical density in the region of $\Delta T \leqslant 3 \times 10^{-4}$; at $\Delta T = 10^{-5}$, the correction to c_0 is about twenty per cent. These corrections depend on the mode present in the cell; thus, part of the dispersion present in the uncorrected data is only apparent.

The Brillouin line measurements are essentially local. If they are carried out at the level in the cell where the intensity is a maximum, the local density is the critical and no gravity corrections are needed.

3. Sound Dispersion and Attenuation

It has been known for a long time[24] that the zero-frequency speed of sound along the critical isochore passes through a deep minimum at the critical temperature. The equation 98 leads us to expect that c_0 will go to zero slowly, as $|\Delta T|^{\alpha/2}$. This seems to be in agreement with the experimental data, obtained at low frequencies ($\leqslant 100\,\text{kHz}$)[8, 24, 44, 134]. At the somewhat higher ultrasonic frequencies, dispersion is found, not only in cases where internal degrees of freedom are present, such as carbon dioxide, but also in the monatomic substances xenon[24] and helium[8]. This dispersion is accompanied by an anomalous absorption, much larger than that of equation 99. These effects are much more pronounced in the high frequency range of the Brillouin spectra. This dispersion can be qualitatively explained by the presence of 'structural relaxation' (Fixman[47]). We follow a description of this process by Mistura[101]. Local thermodynamic equilibrium in the passing sound wave has to be achieved in regions of the size of the correlation length by diffusion of the entropy fluctuations. The characteristic time for such a diffusion process is $\tau \approx \xi^2/\lambda'$, where λ' is the diffusivity (thermal, $\lambda' = \lambda/\rho C_p$, in gases; diffusion constant D in binary liquids). For frequencies of the sound near the inverse of this characteristic time, the sound wave loses energy because it is not in local equilibrium with the critical fluctuations. In gases, this characteristic frequency is

$$\omega_D = \lambda' \xi^{-2} \qquad (100)$$

which goes to zero, roughly as $|\Delta T|^{3\nu}$ or $|\Delta T|^2$. That this picture makes sense was shown convincingly by Garland, Eden and Mistura who plotted the observed anomalous attenuation in xenon, from ultrasound and from Brillouin linewidth measurements, as a function of ω/ω_D and showed that these two sets of data coalesced on a single curve[48]. Kawasaki has predicted the shape of this curve from his mode–mode coupling theory[74]. At high frequencies systematic deviations have been reported[101, 48, 29]; by carrying the calculations to higher order, these discrepancies disappear and satisfactory agreement with Kawasaki's theory is obtained for the attenuation of sound[49].

XI. Concluding Remarks

This survey of thermodynamic studies in the critical region of gases has, we hope, given the impression of a field of vivid interest and widespread activity. The interest is caused by a rapidly growing understanding of critical phenomena and by the awareness of their universality. The activity is at least partly due to several 'breakthroughs' in the experimental possibilities. The principal one is undoubtedly the development of the laser; this has permitted the study of the spectrum of scattered light, which contains a wealth of information about thermodynamic and transport properties. Other factors contributing to the refinement of experimental possibilities are the improved means of temperature stabilization, the development of thin-walled calorimeters, and the application of the potential of capacitance measurements to density determination.

As was already sensed by Gouy in 1890, the study of physical properties as functions of height in the field of gravity yields the most detailed information about experimental behavior very near the critical point. However, really reliable quantitative information has only begun to appear in the literature in recent years. It would seem that there is still a large need and a wide potential for this type of critical region study. Scattering measurements as a function of height, although quite feasible, have not made their appearance in the literature yet. Interference methods for measurement of refractive index gradients were developed many years ago; however, their application to critical region studies to date is limited to two experiments.

There is also need for further study of the equilibration process. An approximate experimental assessment of equilibration times has been achieved in a number of calorimetric experiments, however, with somewhat contradictory results. A knowledge of these times is essential to proper design of the calorimeter.

Concerning the description and interpretation of experimental results, there is need for work in two directions. Thus far, the experimental data that have been shown to scale have been mainly data in an intermediate range, $10^{-4} < \Delta T/T_c < 3 \times 10^{-2}$, $5 \times 10^{-2} < \Delta \rho/\rho_c < 3 \times 10^{-1}$. It is necessary that data much closer to critical be obtained from measurements of properties as functions of height, and be examined for their scaling properties. There is an urgent need for assurance that scaling in the intermediate range is indeed truly asymptotic. Besides study of experimental behavior closer to the critical point, there is also need for consideration of the behavior farther away. The vicinity of the critical point, presently described by the thermodynamic scaling laws, will have to be imbedded in the larger equation-of-state, presently described by a variety of more or less empirical equations-of-state. 'Extending scaling' is one way of enlarging the range of applicability of the thermodynamic description of the critical region and of accounting for the lack of symmetry of this region in gases. Equivalent and alternative approaches certainly deserve attention.

XII. Acknowledgements

The writing of the section on x-ray scattering was facilitated by enlightening discussion with Professor P. Schmidt. Professor G. Allen pointed out to me how refractive index gradient methods developed for diffusion measurements could be applied to critical region studies. F. J. Cook and Professor M. S. Green provided the expansion of extended scaling prior to publication. Professor J. V. Sengers, Dr M. Moldover and Dr R. D. Mountain contributed to the section on equilibration times and made useful suggestions about other

XIII. References

[1] Ahlers, G. *Phys. Rev. Letters*, **21**, 1159 (1968); **23**, 464 (1969); Ahlers, G. *Phys. Rev.* **3**, 696 (1971).
[2] Amirkhanov, Kh. I. and A. M. Kerimov. *Teploenergetika*, **8**, 64 (1963); **9**, 61 (1963).
[3] Artyukhovskaya, L. M., E. T. Shimanskaya and Yu. I. Shimanskii. *Soviet Phys. (JETP)*, **32**, 375 (1971).
[4] Baehr, H. D. *Forsch. Ingenieurw.* **29**, 143 (1963); *Brennstoff-Wärme-Kraft*, **15**, 514 (1963).
[5] Bagatskii, M. I., A. V. Voronel and V. G. Gusak. *Zh. Eksp. Teor. Fiz. SSSR*, **43**, 728 (1962).
[6] Bale, H. D., B. C. Dobbs, J. S. Lin and P. W. Schmidt. *Phys. Rev. Letters*, **25**, 1556 (1970).
[7] Barieau, R. E. *J. Chem. Phys.* **45**, 3175 (1966); **49**, 2279 (1968); Barieau, R. E. *Phys. Rev. Letters*, **16**, 297 (1966).
[8] Barmatz, M. *Phys. Rev. Letters*, **24**, 651 (1970).
[9] Barmatz, M. and P. C. Hohenberg. *Phys. Rev. Letters*, **24**, 1225 (1970).
[10] Bendiner, W., D. Elwell and H. Meyer. *Phys. Letters*, **26A**, 421 (1968).
[11] Benedek, G. B. *Polarisation, Matière et Rayonnement*, p 49. Livre de Jubilé en l'honneur du Professeur A. Kastler. Presses Universitaires de France: Paris (1969).
[12] Bergé, P., P. Calmettes, C. Laj and B. Volochine. *Phys. Rev. Letters*, **23**, 693 (1969).
[13] Bergé, P., P. Calmettes, M. Tournarie and B. Volochine. *Phys. Rev. Letters*, **24**, 1223 (1970).
[14] Blagoi, Yu. P., V. I. Sokhan and L. A. Pavlichenko. *JETP Letters*, **11**, 190 (1970).
[15] Brown, G. R. and H. Meyer. *Phys. Rev.* **A6**, 364 (1972).
[16] Cailletet, L. and E. C. Mathias. *CR Acad. Sci., Paris*, **102**, 1202 (1886); **106**, 1563 (1887).
[17] Chalyi, A. V. and A. D. Alekhin. *Soviet Phys. (JETP)*, **32**, 181 (1971).
[18] Chang, R. F., P. H. Keyes, J. V. Sengers and C. O. Alley. *Phys. Rev. Letters*, **27**, 1706 (1971).
[19] Cannell, D. S. and G. B. Benedek. *Phys. Rev. Letters*, **25**, 1157 (1970).
[20] Chase, C. E. and G. O. Zimmerman. *Phys. Rev. Letters*, **15**, 483 (1965).
[21] Chu, B. *Misc. Publ. US Nat. Bur. Stand. No. 273**, 123 (1966).
[22] Chu, B. *Annu. Rev. Phys. Chem.* **21**, 145 (1970).
[23] Chu, B. and J. S. Lin. *J. Chem. Phys.* **53**, 4454 (1970).
[24] Chynoweth, A. G. and W. G. Schneider. *J. Chem. Phys.* **20**, 177 (1952).
[25] Cook, D. *Trans. Faraday Soc.* **49**, 716 (1953).
[26] Cook, F. and M. S. Green. *Bull. Amer. Phys. Soc.*, Ser. II, **17**, 276 (1972), Abstract CC 1; *Phys. Rev.* in preparation.
[27] Cornfeld, A. B. and H. Y. Carr. *Bull. Amer. Phys. Soc.*, Ser. II, **17**, 277 (1972), Abstract CC 11; *Phys. Rev. Letters*, **29**, 28 (1972).
[28] Cummins, H. Z., in *Proceedings of the Varenna Summer School on Critical Phenomena*, edited by M. S. Green. Academic Press: New York (1972).
[29] Cummins, H. Z. and H. L. Swinney, in *Progress in Optics*, edited Emil Wolf, Vol. VIII. North Holland: Amsterdam (1970).
[30] Cummins, H. Z. and H. L. Swinney. *Phys. Rev. Letters*, **25**, 1165 (1970).
[31] Cutkosky, R. D. *J. Res. Nat. Bur. Stand.* **74C**, 15 (1970).
[32] Dahl, D. and M. R. Moldover. *Phys. Rev. Letters*, **27**, 1421 (1971); *Phys. Rev.* **A6**, 1915 (1972).
[33] Degiorgio, V. and J. B. Lastovka. *Phys. Rev.* **A4**, 2033 (1971).
[34] Deming, W. E. *Statistical Adjustment of Data*. Wiley: New York (1943).
[35] Diller, D. E. *J. Chem. Phys.* **49**, 3096 (1968).
[36] Edwards, C., J. A. Lipa and M. J. Buckingham. *Phys. Rev. Letters*, **20**, 496 (1968).
[37] Edwards, M. H. *Phys. Rev.* **108**, 1243 (1957).
[38] Edwards, M. H. and W. C. Woodbury. *Canad. J. Phys.* **39**, 1833 (1961); *Phys. Rev.* **129**, 1911 (1963).
[39] Einstein, A. *Ann. Phys., Lpz.* **33**, 1275 (1910).

---

* *Misc. Publ. US Nat. Bur. Stand. No. 273.* "Critical phenomena", Proceedings of a Conference at Washington, DC (1965). Edited by M. S. Green and J. V. Sengers.

[40] Eisenstein, A. S. and N. S. Gingrich. *Phys. Rev.* **62**, 261 (1942).
[41] Estler, W. T., L. R. Wilcox, R. Hocken and T. Charlton. *Bull. Amer. Phys. Soc.*, Ser. II, **17**, 614 (1972), Abstract KJ2.
Estler, W. T. *Ph.D. Thesis*, Stoneybrook (1972).
[42] Fairbank, W. M., M. J. Buckingham and C. F. Kellers. *Proceedings of the Fifth International Conference on Low Temperature Physics, Madison, Wis.* p 50 (1957).
[43] Fairbank, W. M. and C. F. Kellers. *Misc. Publ. US Nat. Stand. No. 273**, 71 (1966).
[44] Feke, G. T., K. Fritsch and E. F. Carome. *Phys. Rev. Letters*, **23**, 1287 (1969).
[45] Fisher, M. E. *Misc. Publ. US Nat. Bur. Stand. No. 273**, 21 (1966).
[46] Fisher, M. E. *Misc. Publ. US Nat. Bur. Stand. No. 273**, 108 (1966).
[47] Fixman, M. *Advanc. Chem. Phys.* **6**, 175 (1964).
[48] Garland, C. W., D. Eden and L. Mistura. *Phys. Rev. Letters*, **25**, 1161 (1970).
[49] Garland, C. W., D. Eden and J. Thoen. *Phys. Rev. Letters*, **28**, 726 (1972).
[50] Gasparini, F. and M. R. Moldover. *Phys. Rev. Letters*, **23**, 749 (1969).
[51] Giglio, M. and G. B. Benedek. *Phys. Rev. Letters*, **23**, 1145 (1969).
[52] Goldhammer, D. A. *Z. Phys. Chem.* **71**, 577 (1910).
[53] Goldman, K. and N. G. Scrase. *Physics*, **44**, 555 (1969); **45**, 1 (1969).
[54] Goodwin, R. D. *J. Res. Nat. Bur. Stand.* **65C**, 231 (1961);
Goodwin, R. D., D. E. Diller, H. M. Roder and L. A. Weber. *J. Res. Nat. Bur. Stand.* **67A**, 173 (1963).
[55] Goodwin, R. D. *J. Res. Nat. Bur. Stand.* **73A**, 487 (1969).
[56] Gosting, L. J. and L. Onsager. *J. Amer. Chem. Soc.* **74**, 6066 (1952).
[57] Gouy, M. *CR Acad. Sci., Paris*, **90**, 307 (1880).
[58] Gouy, M. *CR Acad. Sci., Paris*, **115**, 720 (1892); **116**, 1289 (1893); **121**, 201 (1895).
[59] Green, M. S. *J. Chem. Phys.* **33**, 1403 (1960).
[60] Green, M. S., M. J. Cooper and J. M. H. Levelt Sengers. *Phys. Rev. Letters*, **26**, 492 (1971).
[61] Greer, W. L., J. M. H. Levelt Sengers and J. V. Sengers, *Bull. Amer. Phys. Soc.*, Ser. II, **17**, 277 (1972). Abstract CC8. *J. Phys. Chem. Ref. Data* to be published.
[62] Griffiths, R. B. *J. Chem. Phys.* **43**, 1958 (1965).
[63] Griffiths, R. B. *Phys. Rev.* **158**, 176 (1967).
[64] Griffiths, R. B. and J. C. Wheeler. *Phys. Rev.* **A2**, 1047 (1970).
[65] Habgood, H. W. and W. G. Schneider. *Canad. J. Chem.* **32**, 98 and 164 (1954).
[66] Halperin, B. I. and P. C. Hohenberg. *Phys. Rev.* **177**, 952 (1969).
[67] Henry, D. L., H. L. Swinney and H. Z. Cummins. *Phys. Rev. Letters*, **25**, 1170 (1970).
[68] Hocken, R. and L. R. Wilcox. *Bull. Amer. Phys. Soc.*, Ser. II, **17**, 614 (1972). Abstract KJ1.
[69] Hohenberg, P. C. and M. Barmatz. *Phys. Rev.* **A6**, 289 (1972).
Hocken, R. *Ph.D. Thesis*, Stoneybrook (1973).
[70] Jansoone, V., H. Gielen, J. De Boelpaep and O. B. Verbeke. *Physica*, **46**, 213 (1970).
[71] Jost, W. *Diffusion in Solids, Liquids and Gases*, pp 444 and 458. Academic Press: New York (1960).
[72] Kadanoff, L. P. *Physics Letters*, **2**, 263 (1966).
[73] Kadanoff, L. in *Proceedings of the Varenna Summer School on Critical Phenomena*, edited by M. S. Green, Academic Press: New York (1972).
[74] Kawasaki, K. *Ann. Phys. (NY)*, **61**, 1 (1970).
[75] Kegeles, G. and L. J. Gosting. *J. Amer. Chem. Soc.* **69**, 2516 (1947).
[76] Kerr, E. C. and R. H. Sherman. *Proceedings of the Eleventh International Conference on Low Temperature Physics,* St. Andrews, Scotland, Vol. I, p 236 (1968).
[77] Kierstead, H. A. *Phys. Rev.* **3**, 329 (1971); **A7**, 242 (1973).
[78] Knobler, C. M., C. P. Abbiss and C. J. Pings. *J. Chem. Phys.* **41**, 2200 (1964).
[79] van Laar, J. J. *Proc. Acad. Sci. Amst.* **14^I**, 428 (1911/12); **14^{II}**, 1091 (1911/12).
[80] Landau, L. D. and E. M. Lifshitz. *Electrodynamics of Continuous Media*, Addison-Wesley: Reading, Mass. (1960).
[81] Landau, L. D. and G. Placzek. *Phys. Z. Sowjet.* **5**, 172 (1934).
[82] Larsen, S. Y., R. D. Mountain and R. Zwanzig. *J. Chem. Phys.* **42**, 2187 (1965).
[83] Lee, T. D. and C. N. Yang. *Phys. Rev.* **87**, 410 (1952).
[84] Levelt Sengers, J. M. H. *Industr. Engng Chem. (Fundamentals)*, **9**, 470 (1970).
[85] Levelt Sengers, J. M. H. and W. T. Chen. *J. Chem. Phys.* **56**, 595 (1972).

* *Misc. Publ. US Nat. Bur. Stand. No. 273.* "Critical phenomena", Proceedings of a Conference at Washington, DC (1965). Edited by M. S. Green and J. V. Sengers.

[86] Levelt Sengers, J. M. H., J. Straub and M. Vicentini-Missoni. *J. Chem. Phys.* **54**, 5034 (1971).
[87] Lipa, J. A., C. Edwards and M. J. Buckingham. *Phys. Rev. Letters*, **25**, 1086 (1970).
[88] Lorentzen, H. L. *Acta Chem. Scand.* **7**, 1335 (1953); **9**, 1724 (1953).
[89] Lunacek, J. H. and D. S. Cannell. *Phys. Rev. Letters*, **27**, 841 (1971).
[90] Maccabee, B. S. and J. A. White. *Phys. Rev. Letters*, **26**, 1468 (1971); **27**, 495 (1971).
[91] Maccabee, B. S. and J. A. White. *Physics Letters*, **35A**, 187 (1971).
[92] McIntyre, D. and J. V. Sengers in *Physics of Simple Liquids*, p 447. Edited by H. N. V. Temperley, J. S. Rowlinson and G. S. Rushbrooke. North Holland: Amsterdam (1968).
[93] Mermin, N. D. *Phys. Rev. Letters*, **26**, 169, 957 (1971).
[94] Mermin, N. D. and J. J. Rehr. *Phys. Rev. Letters*, **26**, 1155 (1971).
[95] Mermin, N. D. and J. J. Rehr. *Phys. Rev.* **A4**, 2408 (1971).
[96] Michels, A., B. Blaisse and C. Michels. *Proc. Roy. Soc. A*, **160**, 358 (1937).
[97] Michels, A. and J. Hamers. *Physica*, **4**, 995 (1937).
[98] Michels, A., J. M. H. Levelt and W. de Graaff. *Physica*, **24**, 659 (1958).
[99] Michels, A., J. V. Sengers and P. S. van der Gulik. *Physica*, **28**, 1201 and 1216 (1962).
[100] Michels, A and J. V. Sengers. *Physica*, **28**, 1238 (1962).
[101] Mistura, L., in *Proceedings of the Varenna Summer School on Critical Phenomena*, edited by M. S. Green. Academic Press: New York (1972).
[102] Moldover, M. R. *Ph.D. Thesis*, Stanford (1966);
Moldover, M. R. and W. A. Little. *Misc. Publ. US Nat. Bur. Stand. No. 273**, 79 (1966);
Phys. Rev. **182**, 342 (1969).
[103] Mountain, R. D. *J. Res. Nat. Bur. Stand.* **70A**, 207 (1966); *Rev. Mod. Phys.* **38**, 205 (1966).
[104] Naumenko, Zh. P., E. T. Shimanskaya and Yu. I. Shimanskii. *Ukr. Fiz. Zh.* **12**, 143 (1967).
[105] Onsager, L. *Phys. Rev.* **65**, 117 (1944).
[106] Ornstein, L. S. and F. Zernike. *Proc. Acad. Sci. Amst.* **17**, 793 (1914).
[107] Palmer, H. B. *J. Chem. Phys.* **22**, 625 (1954).
[108] Pings, C. J. Private communication (August 1971).
[109] Pings, C. J. and R. K. Teague. *Phys. Letters*, **26A**, 496 (1968).
[110] Prydz, R. and G. C. Straty. *J. Res. Nat. Bur. Stand.* **74A**, 747 (1970).
[111] Puglielli, V. G. and N. C. Ford. *Phys. Rev. Letters*, **25**, 143 (1970).
[112] Rivkin, S. L. and T. S. Achundov, *Teploenergetika*, **9**, 57 (1962); **10**, 66 (1963);
Rivkin, S. L. and G. V. Troianovskaya, *Teploenergetika*, **11**, 72 (1964);
Rivkin, S. L., T. S. Achundov, E. A. Kremnevskaia and N. N. Assadulaieva, *Teploenergetika*, **13**, 59 (1966).
[113] Roach, P. R. *Phys. Rev.* **170**, 213 (1968).
[114] Roach, P. R. and D. H. Douglass Jr. *Phys. Rev. Letters*, **17**, 1083 (1966).
[115] Rushbrooke, G. S. *J. Chem. Phys.* **39**, 842 (1963).
[116] Rushbrooke, G. S. *J. Chem. Phys.* **43**, 3439 (1965).
[117] Schmidt, E. H. W. *Misc. Publ. US Nat. Bur. Stand. No. 273**, 13 (1966).
[118] Schmidt, H. H. *J. Chem. Phys.* **54**, 3610 (1971).
[119] Schofield, P. *Phys. Rev. Letters*, **22**, 606 (1969).
[120] Schofield, P., J. D. Litster and J. T. Ho. *Phys. Rev. Letters*, **23**, 1098 (1969).
[121] Sengers, J. V. *Ph.D. Thesis*, Amsterdam (1962).
[122] Sengers, J. V. and P. H. Keyes. *Phys. Rev. Letters*, **26**, 70 (1971).
[123] Sengers, J. V. *Ber. Bunsenges. Phys. Chem.* **76**, 260 (1972).
[124] Sherman, R. H. *Phys. Rev. Letters*, **15**, 141 (1965).
[125] Sherman, R. H. and F. J. Edeskuty. *Ann. Phys. (NY)*, **9**, 522 (1960).
[126] Shimanskaya, E. T. *Ukr. Fiz. Zh.* **3**, 542 (1958).
[127] Shimanskii, Yu. I. and E. T. Shimanskaya. *Ukr. Fiz. Zh.* **7**, 861 (1962).
[128] Smith, I. W., M. Giglio and G. B. Benedek. *Phys. Rev. Letters*, **27**, 1556 (1971).
[129] Smoluchowski, M. V. *Ann. Phys., Lpz.* **25**, 205 (1908).
[130] Stanley, H. E. *Introduction to Phase Transitions and Critical Phenomena*, Chap. 9. Oxford University Press: New York (1971).
[131] Straub, J. *Chem.-Ing.-Tech.* **5/6**, 291 (1967).
[132] Swinney, H. L. and H. Z. Cummins. *Phys. Rev.* **171**, 152 (1968).
[133] Thoen, J., W. Van Dael and M. Caubergh. Private communication (1971).
[134] Thoen, J., E. Vangeel and W. Van Dael. *Physica*, **52**, 205 (1971).

---

* *Misc. Publ. US Nat. Bur. Stand. No. 273.* "Critical phenomena", Proceedings of a Conference at Washington, DC (1965). Edited by M. S. Green and J. V. Sengers.

[135] Thomas, J. E. and P. W. Schmidt. *J. Chem. Phys.* **39**, 2506 (1963).
[136] Thomas, J. E. and P. W. Schmidt. *J. Amer. Chem. Soc.* **86**, 3554 (1964).
[137] Thompson, D. R. and O. K. Rice. *J. Amer. Chem. Soc.* **36**, 3547 (1964).
[138] Tisza, L. Cornell (1948). In: *Phase Transformation in Solids*, edited by R. Smoluchowsky, J. E. Mayer and W. A. Weyl. Wiley: New York (1951).
[139] Verschaffelt, J. E. *Commun. Phys. Lab. Univ. Leiden*, **28**, (1896); **55**, (1900).
[140] Vicentini-Missoni, M., J. M. H. Levelt Sengers and M. S. Green. *J. Res. Nat. Bur. Stand* **73A**, 563 (1969).
[141] Vicentini-Missoni, M., R. I. Joseph, M. S. Green and J. M. H. Levelt Sengers. *Phys. Rev.* **B1**, 2312 (1970).
[142] Voronel', A. V. and Yu. R. Chashkin. *Soviet Phys.* (*JETP*), **24**, 263 (1967).
[143] Voronel', A. V., Yu. R. Chashkin, V. A. Popov and V. G. Simkin. *Soviet Phys.* (*JETP*), **18**, 568 (1964).
[144] Voronel', A. V., V. G. Gorbunova, Yu. R. Chashkin and V. V. Shchekochikhina. *Soviet Phys.* (*JETP*), **23**, 597 (1966).
[145] Voronel', A. V., V. G. Snigirev and Yu. R. Chashkin. *Soviet Phys.* (*JETP*), **21**, 653 (1965).
[146] Verbeke, O. B., V. Jansoone, H. Gielen and J. De Boelpaep. *J. Phys. Chem.* **73**, 4076 (1969).
[147] Verbeke, O. B. *Cryogenics*, **10**, 486 (1970).
[148] van der Waals, J. D. *Ph.D. Thesis*, Amsterdam (1873).
[149] Wallace, B. and H. Meyer. *Phys. Rev.* **A2**, 1563, 1610 (1970).
[150] Weber, L. A. *Phys. Rev.* **A2**, 2379 (1970).
[151] Weber, L. A. *J. Res. Nat. Bur. Stand.* **74A**, 93 (1970).
[152] Weinberger, M. A. and W. G. Schneider. *Canad. J. Chem.* **30**, 422 (1952).
[153] Widom, B. *J. Chem. Phys.* **43**, 3898 (1965).
[154] Widom, B. and O. K. Rice. *J. Chem. Phys.* **23**, 1250 (1955).
[155] Widom, B. and J. S. Rowlinson. *J. Chem. Phys.* **52**, 1670 (1970).
[156] Wiener, O. *Ann. Phys. Chem.*, N.F., **49**, 105 (1893).
[157] Wilcox, L. R. and D. Balzarini. *J. Chem. Phys.* **48**, 753 (1968).
[158] Wilcox, L. R. and W. T. Estler. *J. Phys., France*, **32**, C5A-175 (1971).
[159] Wims, A. M., D. McIntyre and F. Hynne. *J. Chem. Phys.* **50**, 616 (1969).
[160] Yang, C. N. and C. P. Yang. *Phys. Rev. Letters*, **13**, 303 (1964).
[161] Young, S. *Trans. Chem. Soc.* **59**, 37 (1891).
[162] Yvon, J. 'Actualités scientifiques et industrielles': *No. 542*, 'Fluctuations en densité'; *No. 543*, 'La propagation et la diffusion de la lumière'. Hermann: Paris (1937).
[163] Zimmerman, G. O. and C. E. Chase. *Phys. Rev. Letters*, **19**, 151 (1967).

CHAPTER 15

SOLUBILITY

A. S. KERTES[1], O. LEVY[2] and G. Y. MARKOVITS[2]

(1) Institute of Chemistry, The Hebrew University, Jerusalem, Israel

and

(2) Department of Chemistry, The University of the Negev, Beer Sheba, Israel

Contents

I.	Introduction	725
II.	General Considerations	726
III.	Concentration and Activity Coefficient Scales	727
IV.	Solubility of Gases in Liquids	729
	1. Manometric-volumetric Methods	730
	A. Degassing	730
	B. Calibration and Filling	731
	C. Mixing and Equilibration	734
	D. Manometric Reading	735
	E. Solubility at High Pressure	737
	F. Solubility at High Temperature	738
	2. Chemical-analytical Methods	739
	A. Specific Chemical Reactions	739
	B. Gas Chromatography	739
	C. Mass Spectrometry	740
	3. Miscellaneous Methods	740
V.	Solubility of Liquids in Liquids	741
	1. Volume Reading	741
	2. Cloud Point	742
	3. Miscellaneous Methods	742
VI.	Solubility of Solids in Liquids	743
	1. Saturation Method	743
	2. Cloud Point	744
	3. Chemical and Instrumental Analysis	746
VII.	References	746

I. Introduction

The solubility of a substance is the maximum amount of it which will dissolve in a given amount of solvent, at given pressure and temperature, yielding a homogeneous molecular or ionic dispersion. If solubility must be known accurately, it must be determined experimentally. There is no reliable way to calculate solubility from the physical and chemical properties of the pure compound. Quite frequently, however, there are criteria for the prediction of solubilities, based on creditable equations meeting the rigorous requirements of sound thermodynamic principles. While an ever-increasing reliance can be placed on theoretical calculations—as molecular data become available— it will not be possible to replace experiment with theory in the determination of solubilities. It is only fortunate that it is relatively easy to measure solubilities with adequate precision.

When considering solubility, nine possible combinations of binary systems with the three states of matter can occur. In the present review, we will be concerned with the experimental methods, techniques and procedures used in binary systems where the solubility of the solute is relatively low, yielding true solutions as defined above. Such are the systems of solubility of gases, liquids and solids in liquids.

The first part of this chapter surveys briefly the basic principles and general considerations which govern the heterogeneous equilibria in two phase binary systems. The second part describes the methods which have been used to measure equilibrium solubilities in systems where the resulting solution is a dilute one. It is divided according to the three main types of heterogeneous equilibria: gas in liquid, liquid in liquid and solid in liquid. For each type a few of the best and most popular techniques are described and their limitations discussed.

In addition to the proper choice of experimental device and the careful execution of the measurement, the accuracy and precision of a solubility determination will depend on the purity of the materials employed, successful temperature control and true attainment of equilibrium. These topics are of course common to many, if not all, types of measurements reviewed in this volume, and will thus not be treated here specifically.

The reader is advised to consult reviews on errors, accuracy and precision of chemical and instrumental analysis[65], on purification of solvents[92,96], on chemical equilibria[65,97], and control and measurement of temperature[78,108] prior to carrying out a solubility determination.

II. General Considerations

For our purpose, the most important thermodynamic quantity associated with a solution is the Gibbs free energy.

$$G = G^\circ + RT \ln a \tag{1}$$

At a given temperature and pressure it depends only on the composition of the system. At unit solute activity, which is the usual hypothetical standard state, $G^\circ = G$. For electrolytes, the standard state is usually defined as a solution of unit concentration, where the mean ionic activity coefficient becomes unity. With non-electrolytes, the pure liquid is chosen as the standard state, and the activity defined as unity. For solid non-electrolytes, the activity of the 'supercooled liquid' is calculated from the heat of fusion ($\ln a_2 = -(\Delta H_f/R)(1/T - 1/T_m)$). For a gaseous solute, the standard state is defined as that at unit fugacity ($a = f/f^\circ$).

A binary mixture is defined ideal if the heat of mixing and volume change of mixing both equal zero. Thermodynamically this means that the free energy of mixing contains only an entropy term, the enthalpy of mixing being zero.

$$G_{id}^M = -TS^M = RT(x_1 \ln x_1 + x_2 \ln x_2) \tag{2}$$

where x_1 and x_2 stand for the mole fraction of solvent and solute respectively.

Real solutions deviate from the requirements of equation 2, and the free energy of mixing becomes

$$G^M = G^M_{id} + G^E \tag{3}$$

where the excess free energy, in terms of activity coefficients, is

$$G^E = RT(x_1 \ln f_1 + x_2 \ln f_2) = Bx_1 x_2 \tag{4}$$

where B is a constant independent of the composition[50] and temperature[33]. In practice B is found constant only for small temperature intervals and a limited number of mixtures. In the case of mixtures of non-electrolytes, the regular solutions concept[49] requires that the excess free energy, equation 4, be equal to the heat of mixing, e.g. the excess entropy of mixing be zero.

The solubility of a non-electrolyte in a non-polar medium can be approximated reasonably well from the solubility parameter relationship[49]

$$\ln a_2 = \ln \phi_2 + \phi_1(1 - V_2/V_1) + V_2\phi_1^2(\delta_1 - \delta_2)^2/RT \tag{5}$$

where ϕ, V and δ are the volume fraction, molar volume and the solubility parameter respectively. Values for δ are available for a large number of non-polar compounds[49, 50] or can be evaluated from direct solubility determinations or from heat of vaporization data. The solubility parameter of gases is usually low, their solubility will thus be relatively high in solvents of low δ, such as the aliphatic and aromatic hydrocarbons[107].

No such relatively simple relationship exists for the estimation of solubilities of electrolytes. The process consists of an endothermic step of dissolution, overcoming the lattice energy of the solid, and of an exothermic process of solvation (usually hydration) of the ion(s) formed. If the lattice and hydration energies are known, or can be evaluated[9, 57], at most a trend in solubility can be predicted.

III. Concentration and Activity Coefficient Scales

There are several possibilities of expressing concentrations in binary systems relevant to this review:

Mole fraction, x, is defined as the number of moles of solute divided by the sum of the number of moles of the solvent (subscript 1) and solute (subscript 2).

Mole percentage, is the mole fraction multiplied by 100.

Weight fraction, w, is defined as the weight (grams) of solute divided by the sum of the weights of the solvent and the solute.

Weight percentage, is the weight fraction multiplied by 100.

Volume fraction, ϕ, is defined as the volume (ml) of solute divided by the sum of the volumes of the solvent and the solute.

Volume percentage, is the volume fraction multiplied by 100.

Molality, m, is defined as the number of moles of the solute dissolved in 1 000g of solvent.

Molarity, c, is defined as the number of moles of the solute dissolved in one liter of solution.

The conversion relationships for several scales are listed in *Table 1*.

In the dilute solutions we are concerned with, the standard state can be chosen on the mole fraction, molality or molarity scale. Accordingly, the

Table 1. Conversion relationships for various scales of expressing concentration of solutes

Concentration scale given	Concentration scale sought				
	x_2	w_2	ϕ_2	m_2	c_2
x_2	—	$\dfrac{x_2 M_2}{x_1 M_1 + x_2 M_2}$	$\dfrac{x_2 \bar{V}_2}{x_1 \bar{V}_1 + x_2 \bar{V}_2}$	$\dfrac{1000 x_2}{x_1 M_1}$	$\dfrac{1000 d_{12} x_2}{x_1 M_1 + x_2 M_2}$
w_2	$\dfrac{w_2 M_1}{w_1 M_2 + w_2 M_1}$	—	$\dfrac{w_2 \bar{V}_2 M_1}{w_1 \bar{V}_1 M_2 + w_2 \bar{V}_2 M_1}$	$\dfrac{1000 w_2}{w_1 M_2}$	$\dfrac{1000 w_2 d_{12}}{(1 + w_2) M_2}$
ϕ_2	$\dfrac{\phi_2 \bar{V}_1}{\phi_1 \bar{V}_2 + \phi_2 \bar{V}_1}$	$\dfrac{\phi_2 \bar{V}_1 M_2}{\phi_1 \bar{V}_2 M_1 + \phi_2 \bar{V}_1 M_2}$	—	$\dfrac{1000 \phi_2 \bar{V}_1}{\phi_1 \bar{V}_2 M_1}$	$\dfrac{1000 \phi_2 \bar{V}_1 d_{12}}{\phi_1 \bar{V}_2 M_1 + \phi_2 \bar{V}_1 M_2}$
m_2	$\dfrac{m_2 M_1}{m_2 M_1 + 1000}$	$\dfrac{m_2 M_2}{1000 + m_2 M_2}$	$\dfrac{m_2 \bar{V}_2 M_1}{1000 \bar{V}_1 + m_2 \bar{V}_2 M_1}$	—	$\dfrac{m_2 d_{12}}{1 + 0.001 m_2 M_2}$
c_2	$\dfrac{c_2 M_1}{1000 d_{12} + c_2 M_1 - c_2 M_2}$	$\dfrac{c_2 M_2}{1000 d_{12} - c_2 M_2}$	$\dfrac{c_2 \bar{V}_2 M_1}{(1000 d_{12} - c_2 M_2) \bar{V}_1 + c_2 \bar{V}_2 M_1}$	$\dfrac{c_2}{d_{12} - 0.001 c_2 M_2}$	—

M—molecular weight
d—density
$\bar{V}_2 = \dfrac{1}{m_2}\left(\dfrac{1000 + m_2 M_2}{d_{12}} - \dfrac{1000}{d_1}\right)$ for dilute solutions

subscript 1—solvent
subscript 2—solute
subscript 12—solution

activity coefficient of the solute is equally expressed in these scales. The following interconversion relationships exist for f, γ and y, the activity coefficients on the rational (mole fraction), molal and molar concentration scales, using the notation in *Table 1*:

$$f_2 = \gamma_2(1 + 0.001\, M_1 m_2)$$
$$\quad = y_2/d_1[d_{12} + 0.001\, c_2(M_1 - M_2)] \quad (6)$$
$$\gamma_2 = y_2/d_1(d_{12} - 0.001\, c_2 M_2)$$
$$\quad = y_2 c_2/m_2 d_1 \quad (7)$$
$$y_2 = \gamma_2 d_1/d_{12}(1 + 0.001\, M_2 m_2)$$
$$\quad = \gamma_2 d_1 m_2/c_2 \quad (8)$$

The rational scale is usually considered as a thermodynamic one, emphasizing the ratio between the numbers of molecules in the system. f, the activity coefficient on that scale is a dimensionless quantity, as is the activity on the scale. If γ and y are regarded as dimensionless quantities, the corresponding activities $a(m)$ and $a(c)$ have the dimensions of the corresponding concentrations. In this case, the standard chemical potentials contain concealed terms in $RT \ln (\text{mol kg}^{-1})$ and $RT \ln (\text{mol liter}^{-1})$. Alternatively, if γ and y are taken to have the dimensions of reciprocal concentration, the corresponding activities are dimensionless.

IV. Solubility of Gases in Liquids

The comprehensive evaluations[7, 8, 79] of solubility data of gases in liquids reveal that the early instrumentation used in solubility determinations was not really reliable. In the last decade and a half several important modifications of the classical manometric method have been reported which makes it possible to achieve precisions as high as ±0.05 per cent.

Methods of determination can be divided into two broad groups, bearing no similarity of one to another. One, the physical manometric-volumetric procedures, are based on p–V determinations, while the other, the chemical-analytical methods measure specific chemical properties of the gas investigated. It is our belief that the physical methods are more precise, and of course, have a more general applicability. Instances are known, however, where the determination of quantities by chemical analysis using mass-spectrometry or gas chromatography will yield equally reliable results. Both groups of methods will be discussed.

In addition to the concentration scales of expressing solubilities discussed under the previous heading, gas solubilities are frequently expressed in three additional terms[79]:

(i) The Ostwald coefficient, $l = V_g/V_1$, representing the ratio of the volume of gas absorbed to the volume of the absorbing liquid at constant temperature;

(ii) The Bunsen coefficient, $\alpha = V_g/V_1(273/T)$, being defined as the ratio between the volume of gas at standard temperature and pressure and the unit volume of the absorbing liquid; and

(iii) The Henry law states $P_g = K_1 X_1$, or in a dilute solution, $P_g = K_2 C_1$ and thus $C_g = K_2 C_1$, where K_1 is Henry's constant. The interconversion formulas are $l = \alpha(T/273)$ and $K_2 = 1/l$.

1. Manometric-volumetric Methods

Most manometric methods are based on saturation of the liquid, and the measurement is that of the volume of the gas necessary to saturate a volume of initially gas-free solvent, at a precisely known pressure and temperature. Assuming a high purity of the components, the gas and the liquid, and an adequate temperature control, the following experimental factors will govern the accuracy and precision of the results obtained.

A. Degassing

An incomplete degassing of the liquid is probably the most serious source of error[49, 63]. The simplest and most frequently used method of degassing the solvent is by boiling it, preferentially under a vacuum, for various lengths of time[1, 17, 80]. An adequate check for a complete degassing is made by trapping the distilled fraction, usually ten to twenty per cent of the total volume, and introducing a gage between the trap and the pump. If the pressure drops to 5–10 μ, as measured by the gage, the degassing is satisfactory[4, 20]. Alternative methods of degassing are based on shaking[69, 93] or stirring[36] under suction[112], or freezing the solvent[74]. The latter procedure is followed by evacuation and allowing the solvent to thaw.

For low-vapor pressure solvents (oils) different procedures can be recom-

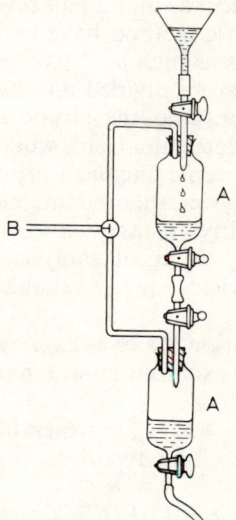

Figure 1. Apparatus[6] for degassing liquids of low vapor pressure. A—evacuation vessel; B—to vacuum pump.

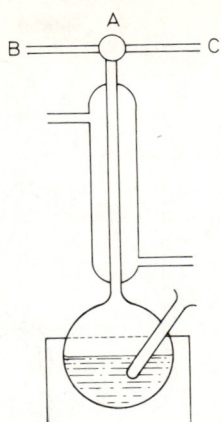

Figure 2. Degassing assembly[16] directly connected to the line. A—rotatable conical joint; B—to vacuum pump; C—to absorption vessel.

mended. As shown in *Figure 1*, the liquid is slowly dripped into a two-stage evacuation vessel (less than 1 mm Hg). A 97 per cent degassing can be achieved in the first stage[6]. In the alternative design[16], the degassing flask is fitted with a water-cooled reflux condenser directly to the vacuum line by a rotatable conical joint, as shown in *Figure 2*. This arrangement allows for a safe transfer of the degassed liquid into the absorption vessel. A similar way of degassing has been employed for hydrogen fluoride solutions[21].

In none of these methods of degassing can the volume of the liberated gas be measured. If needed, the volume can be measured by a procedure similar to one used for removing the gas after equilibration, described below.

B. Calibration and Filling

A careful calibration of all the components of a manometric device is an integral part of gas solubility measurements. Conventionally, the calibration is carried out by filling the various parts of the equipment with mercury[17,69], water[1] or other suitable liquid[55], at a constant temperature, and by weighing the withdrawn liquid.

The filling of the absorption vessel, A in *Figures 3–5*, with the degassed solvent can be done either by dislocating the mercury[69] and weighing it, or by a complete filling of the vessel by the solvent[81]. In the latter case, a certain amount of the solvent must be withdrawn and its volume precisely known, in order to allow the penetration of the gas into the absorption vessel. It is perhaps more convenient to leave some mercury in the bottom of the absorption vessel and let it be displaced by the gas transferred into the vessel[55,69,110].

The incoming gas may be presaturated with the solvent vapors—the wet method, or kept unsaturated in the buret, B in *Figures 4–5*—the dry method. There are merits and disadvantages to both methods[55]. In the wet method, the whole apparatus must be kept at the same temperature in

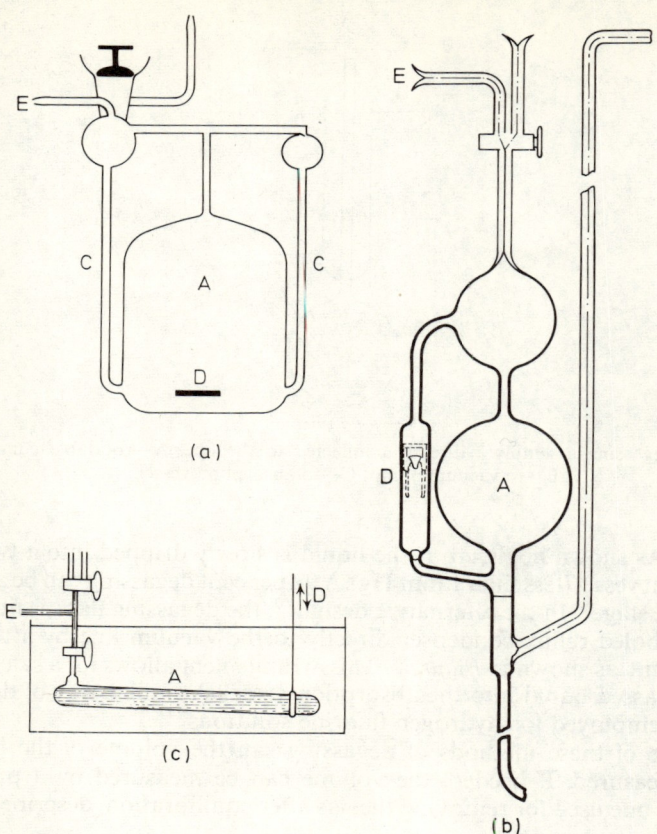

Figure 3. Mixing devices.
    (a) (from ref. 11): A—absorption vessel; C—capillaries; D—magnetic stirrer; E—to gas system
    (b) (from ref. 27): A—absorption vessel; D—magnetic plunger; E—to gas system.
    (c) (from ref. 80): A—absorption vessel; D—mechanical shaking device; E—to gas system.

order to avoid vapor condensation in the cold parts of the buret and connecting capillaries. A clear advantage of the wet method is that it does not require knowledge of the vapor pressure of the solvent at the given temperature, in order to calculate the amount of the dissolved gas. The method is more suitable for low-vapor pressure solvents, since an increased amount of solvent vapors in the gas phase in the case of high-vapor pressure liquids, may in reality, cause a marked deviation from its ideality.

The dry method, of course, has not the disadvantage of the possible non-ideality of the gas phase[4, 55]. Since the vapor penetrates into the gas phase during equilibration, the dry method requires a precise knowledge of the vapor pressure in order to calculate the amounts of the dissolved gas.

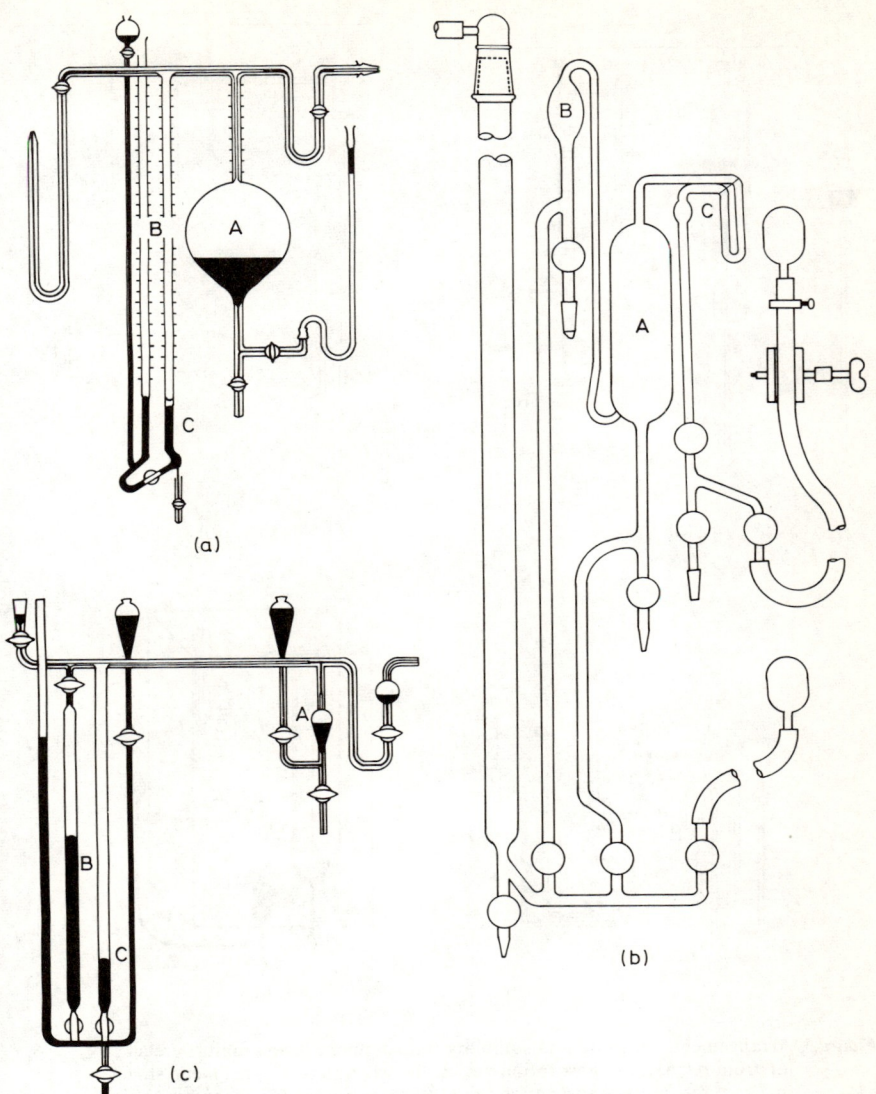

Figure 4. Arrangements for gas-solubility measurements
(a) (from ref. 69): A—absorption vessel; B—gas buret; C—liquid drop to presaturate the gas.
(b) (from ref. 20): A—absorption vessel; B—primary gas buret; C—secondary gas buret (measures the amount of undissolved gas—about five per cent of the total amount—and its size is designed accordingly).
(c) (from ref. 112): A—absorption vessel; B—gas buret; C—liquid drop to presaturate the gas.

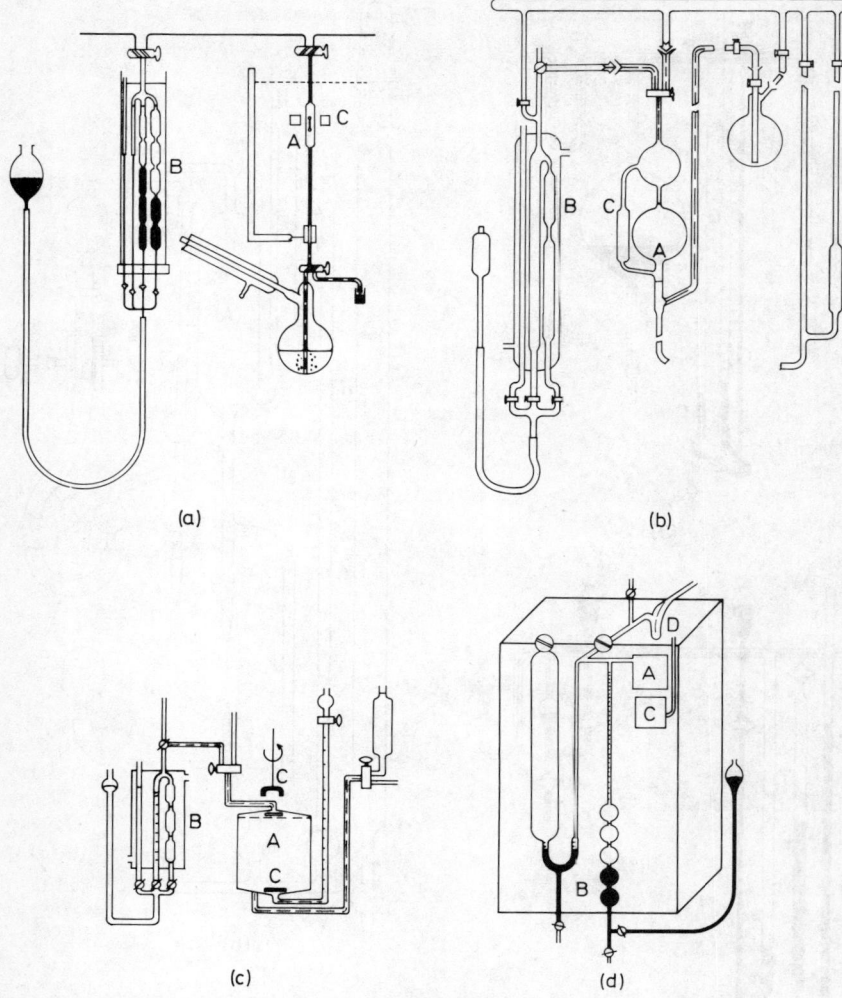

Figure 5. Arrangements for gas-liquid solubility measurements using multi-burets
(a) (from ref. 55): A—absorption vessel; B—gas burets; C—magnetic stirrer
(b) (from ref. 27): A—absorption vessel; B—gas burets; C—magnetic plunger
(c) (from ref. 4): A—absorption vessel; B—gas burets; C—magnetic stirrer.
(d) (from ref. 11): A—absorption vessel; B—gas burets; C—magnetic stirrer; D—presaturator.

C. Mixing and Equilibration

Having the gas in the buret, B in *Figures 4–5*, and the degassed solvent in the absorption vessel, A in *Figures 3–5*, the following step is to bring them into contact and to reach a true equilibrium. An equilibrium is supposed to be reached when no further change in the gas pressure occurs during an observation period. The rate of attaining equilibrium depends on shaking efficiency and intensity, and the interfacial area of contact[20], in addition to the nature of the system investigated and the experimental device employed. In the argon–water system[3], for example, a 97 per cent saturation is attained in a few minutes by a rapid saturation method, or 30 min by a slow saturation method, while a complete equilibrium is reached after an additional 4 h of stirring. The choice of method of mixing, whether shaking, stirring or bubbling, will be dictated essentially by the apparatus employed.

The frequency of shaking is that of the natural swinging period of mercury[69], and a horizontal motion is preferred in order to avoid a vertical acceleration of mercury and disturbance in the pressure measurements[20]. Stirring, mechanical[1] or magnetic[17,55], offers certain advantages. In one device[11], shown in *Figure 3*(a), three capillaries are sealed to the absorption vessel. A vigorous magnetic stirring will cause the liquid to circulate through two side capillaries and return through the middle one. In a similar device[27,73,85] shown in *Figure 3*(b), the solvent circulates between two interconnected absorption bulbs by means of a magnetically operated plunger enclosed in the side arm. Both procedures insure a continuous contact between the incoming gas and a new liquid surface. In order to avoid or minimize the possibility of a local overheating due to the magnetic stirring, Maxted and Moon[80] use a specially-shaped absorption vessel, shown in *Figure 3*(c), with a small quantity of mercury, and leave a small free space over the solvent. By rocking the vessel, the mercury and the gas bubble move in opposite directions, increasing considerably the rate of saturation.

Mixing through bubbling the gas into the solvent gives satisfactory results in a reasonable equilibration time, provided the liquid is viscous enough at room temperature[6], or made such by lowering the temperature[56]. If the gas is introduced under pressure into the solvent[116,117], a high viscosity of the liquid is not a pre-requisite for efficient equilibration. Morrison and Billett[88] equilibrate the liquid with the presaturated gas by a continuous flow of a liquid film through a spiral, thus increasing the contact surface significantly. The authors also claim that the procedure eliminates some of the disadvantages of using presaturated gas (wet method).

D. Manometric Reading

As the gas dissolves in the solvent, an underpressure results in the manometric buret. Assuming that the gas phase obeys the ideal gas law and that the vapor pressure of the pure solvent equals its pressure in the gas-saturated solution (a fair assumption for dilute solutions), the volume of the absorbed gas can be calculated from the difference of the mercury columns in the buret and in the open arm of the manometer. If the open arm is flexible, the pressure in the buret can be adjusted to the external barometric pressure, and the reading yields directly the volume of the absorbed gas. Several of the various possible arrangements are illustrated in *Figure 4*. The size, but especially the diameter, of the manometer affect, of course, the precision of the volume readings. A reduced diameter will increase the precision of reading but will limit the general applicability of the apparatus[55].

Accuracy of the equipment can be increased also by selecting absorption vessels according to the extent of gas solubility[4, 112].

In order to operate a general-purpose apparatus, but without loss in the precision of reading, additional burets, connected in series, can be used[4, 55]. As shown in *Figure 5*, the burets are connected to the mercury container by separate stopcocks, but interconnected by capillary tubes. When equilibrium is reached, the mercury level is brought to one of the marks on the capillary by means of the the stopcock and then by connection to the open end, the pressure in the gas buret is readjusted to the barometric pressure.

It has been shown[55] that the error in manometric readings is proportional to the dead-space volume in the absorption vessel. There is, however, a limit to the reduction of the volume of the dead space, while still insuring an efficient mixing and a sufficient area of interface[11, 80].

The liquid film-saturation method of Morrison and Billett[88], shown in

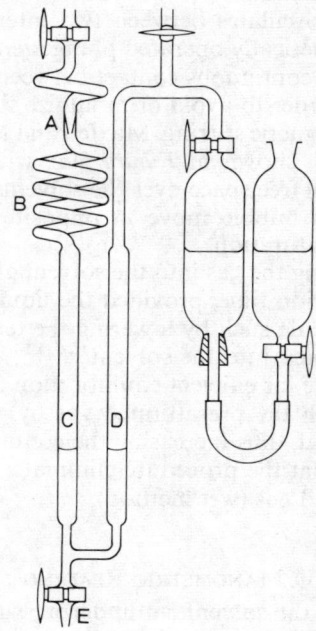

Figure 6. Apparatus[88] for measuring gas-liquid solubility by liquid film and saturation method. A—capillary; B—spiral; C, D—gas burets; E—exit for saturated liquid.

Figure 6, applies a modified method of measurement, and has been offered as an alternative to the commonly used Ostwald method[81]. The solvent, introduced dropwise through capillary A, forms a thin film in its passage through the spiral, which also serves as the saturation vessel. At the same rate the saturated solution is withdrawn at E. The liquid levels in the two arms C and D of the gas buret are thus kept equal. The reading in buret C

Table 2. Accuracies in gas–liquid solubility determinations (at 1 atm pressure)

Gas	Solvent	Accuracy, %	Temp. (°C)	Ref.
O_2	water	1	25	88
N_2	water	1	25	88
Ar	water	0.2	3–28	11
CH_4	water	1	25	88
C_2H_4	water	1	25	88
CF_4	water	0.5	0–50	4
SF_4	water	0.5	0–50	4
NF_3	water	0.5	0–50	4
CO_2	hydrogen fluoride	1	20–35	21
H_2	heptane	0.1	−35−+35	20
H_2	heptane	1	35	74
CO	heptane	1	25	38
H_2	perfluoroheptane	1	25–50	38
CO	perfluoroheptane	1	25	38
H_2	carbon disulfide	1	25	38
O_2	carbon disulfide	1	25	38
H_2	cyclohexane	1	21–36	27
N_2	cyclohexane	1	18–33	27
CO_2	cyclohexane	1	20–37	27
Ar	cyclohexane	1	17–36	27
C_2H_6	cyclohexane	1	19–35	27

gives the volume of the gas dissolved, while from the extracted amount of the saturated solution and the increase of the liquid level in the buret, the volume of the liquid is calculated.

In conclusion, the accuracy which can be achieved in gas–liquid solubility determinations using the manometric method varies both with the system investigated and the apparatus used. Accuracy claims for some specific systems are listed in *Table 2*. Comprehensive tabulations of solubility data have been published previously[7, 79].

E. SOLUBILITY AT HIGH PRESSURE

The essential difference between the technique and apparatus employed for gas solubility determinations at high pressure, and that at atmospheric pressure, is in the sampling method. The apparatus needs a suitably designed solubility bomb[68, 93, 116, 117, 120] which contains the liquid under pressure. The gas, under pressure, is usually introduced through bubbling. Samples are withdrawn through a high-pressure valve and a solubility pipet. *Figure 7* shows a typical high-pressure apparatus, consisting of the high-pressure line, solubility bomb and sampling valve, through which solution samples are withdrawn. The liquid sample enters the manometric buret system shown in *Figure 8*, where it is reduced to atmospheric pressure. The volume of the liquid and that of the evolved gas are measured simultaneously.

In calculating the solubility, correction should be made to the gas solubility at atmospheric pressure and liquid volume change due to the gas solubility. The volume change of the liquid as a function of pressure is usually insignificant[116].

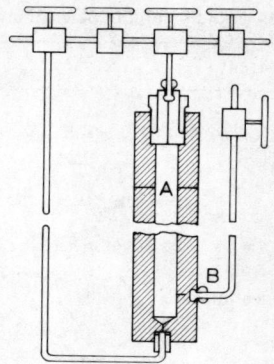

Figure 7. High-pressure apparatus[116, 117] for solubility of gases in liquids; A—absorption vessel; B—sampling device.

A modification of this principle consists of separating the components in the measuring device, and determining the gas by volume and the liquid by weight[23, 120].

F. Solubility at High Temperature

Most frequently high pressures are required for solubility determinations at increased temperatures in order to keep the solvent in the liquid state.

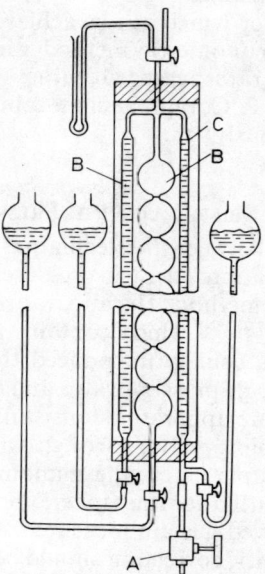

Figure 8. Buret system[116, 117] for measuring solubility of gases in liquids from samples reduced to normal pressure. A—connectors to the solubility bomb; B—gas buret; C—buret for measuring the volume of liquid.

SOLUBILITY

The design of the apparatus employed[93,116] is thus similar to that shown in *Figure 7* and *8*, where the solubility of hydrogen, oxygen, nitrogen and helium in water has been measured. The apparatus of Dymond and Hildebrand[27], shown in *Figure 5*, has been used for measuring the solubilities of gases in liquids at pressures in the vicinity of 1 bar over the temperature range 25° to 200°C[23].

2. Chemical-analytical Methods

In using analytical methods in the determination of gas solubilities, a great variety of chemical reactions are utilized. The choice of the method is dictated by the chemical nature of the components, and no fixed set of rules can be applied to facilitate the choice. Nevertheless, there are several commercially available instruments, such as the gas-chromatograph or the mass-spectrograph, which are frequently used as analytical tools for the determination of the dissolved gas.

A. Specific Chemical Reactions

A straightforward chemical analysis employs the reactivity of the gaseous component. One widely investigated system is the solubility of oxygen in water. The dissolved oxygen oxidizes manganese(II) in basic media to manganese hydroxide $[Mn(OH)_3]$, which is then determined iodometrically. The results obtained compare favorably with those of the manometric method, provided all necessary analytical steps are executed carefully[87].

The oxidizing nature of chlorine has been used in its determination in various solvents[39,111]. Hydrogen cyanide can be determined gravimetrically as its silver salt[72]. Ammonia is determined by acid titration[10].

It should be noted that these and similar direct chemical determinations are used only exceptionally, and their popularity is sharply decreasing.

Several possible methods of determination are based on sweeping the soluble gas from solution. Hydrocarbons are displaced from solution by oxygen, and their amount determined by combustion[18,99].

B. Gas Chromatography

The general applicability of gas-chromatographic analysis, the commercial availability of the equipment and the simplicity of its operation, make this method attractive. The accuracy of the method is generally lower than that attained by the manometric or even the direct chemical methods. It has been successfully applied for the determination of oxygen, nitrogen, methane, carbon dioxide in water[110], oxygen in lubricating oil[28], hydrocarbons in water[5], helium in liquid argon, carbon monoxide and oxygen[103], and in several other instances[7].

The usual procedure is to inject the saturated solvent into the gas chromatograph fitted with a suitable fractionator, which contains also a dehydrating agent, and the amount of the gas is calculated from the peak areas[5]. The lack of a complete removal of the gas from the liquid is the main source of error. The device shown in *Figure 9* has successfully been used to strip the gas

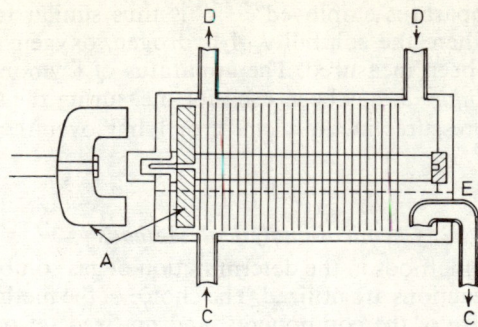

Figure 9. Stripper[118] for gas-chromatographic analysis. A—motor-operated magnets; C,C,—circulation of liquid; D,D—circulation of gas (countercurrent); E—optimum water level.

completely from the solution in less than 30 seconds[118]. The carrier flows counter-current to the sample at a rate of 500 ml per minute. The device can be used for continuous operation of the chromatographic column.

The gas-chromatographic technique can be used directly for determination of the solubility of gases in high-boiling point liquids. The solubilities of gaseous hydrocarbons in various solvents have been determined[122]. In this procedure the solvent is supported in a column on an inert stationary phase. First a carrier gas is passed through the column until saturation is reached, and then the gas under study is injected into the column. Though the method has several limitations[7], the results obtained agree within five per cent with literature data[67].

C. Mass Spectrometry

The method is frequently employed when the relative solubilities of two or more gases in a liquid are measured[13, 62]. The solubility ratio of the gases in the mixture, stripped from the solution and trapped, is determined, and the absolute solubility values evaluated when the solubility of one of the gases is independently available. The technique has been used for the determination of solubilities of isotopes of oxygen and nitrogen in water[62].

The procedure has also been employed in measuring solubilities in volatile liquids, determining the solute/solvent ratio[41].

3. *Miscellaneous Methods*

Several highly specialized methods were used for gas solubility measurements, many of them discussed by Battino and Clever[7]. In addition, mention should be made of methods utilizing the radiation properties of radioactive gases. Two equilibration techniques are mentioned, the static[94] and the dynamic[45, 107]. In the improved version of the latter procedure, the radioactive gas, or inactive gas containing the tracer, is bubbled through the solvent, and saturation checked by continuous counting at the inlet and outlet of the gas.

SOLUBILITY

A microgasometric method is used for the determination of respiratory gases dissolved in water[24, 98].

V. Solubility of Liquids in Liquids

Solubility measurements in liquid–liquid systems do not require experimentation of any complexity. This is probably the reason why the variety of experimental designs appearing in the literature is not great. Apart from some specific procedures, discussed under the heading of miscellaneous methods, only two methods have been suggested for general applicability. The procedures which have been most often followed are based on measuring volume changes upon equilibration, or on the determination of the 'cloud point', though other criteria for classification have also been discussed[76, 121].

1. Volume Reading

The method initially employed for the measurement of mutual solubilities of liquids[52] consists of mixing known volumes of liquids and measuring the volume of the two phases after equilibrium has been reached. In order to calculate the mutual solubility, several measurements have to be carried out with different initial volume ratios. The solubility coefficients are calculated on the assumption that no change in the combined volume occurs on mixing. The assumption, of course, may or may not be proved correct.

Density measurements, in addition to volume readings, insure higher precision[84]. From two parallel experiments, with m and m' weights of one component and volumes a, a' and b, b' of the phases, the saturation concentrations of the component in the upper, x, and the lower, y, can be calculated[53, 60] using the mass balance $ax + by = m$ and $a'x + b'y = m'$. In a similar way, the concentration of the second component can be calculated.

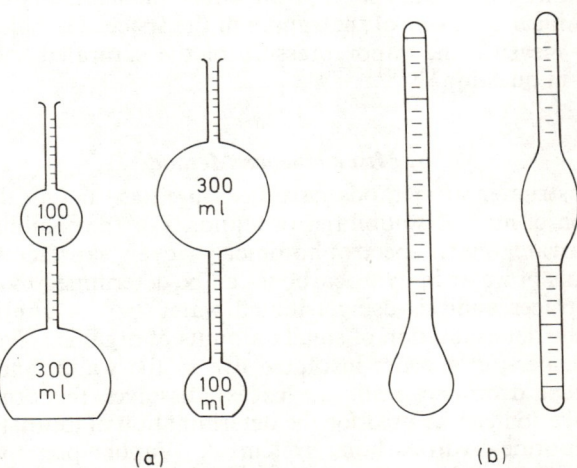

Figure 10. Apparatus for measuring mutual solubilities by volume method (a) from ref. 52; (b) from ref. 12.

A typical design of the volume determination device, shown in *Figure 10*(a), consists of two precisely calibrated bulbs connected by a graduated buret[52,60]. The optimum volumes of the bulbs can be calculated[53]. For small volumes of one of the components, the device shown in *Figure 10*(b) has been recommended[12]. Centrifugation of the whole flask improves phase separation and eliminates possible errors caused by retention of liquid on the vessel's walls or by emulsification[30]. If the two liquids have similar densities, centrifugation becomes mandatory.

2. Cloud Point

The method consists of observing the temperature at which a measured weight of one liquid is visually soluble in a measured weight of another liquid. This is done by superheating the mixture sufficiently to bring all of the first liquid into solution and then cooling until a fine mist is formed. From the temperature of appearance and disappearance of this fine cloud, solubility data at the equilibrium temperature are obtained. Since sealed tubes, bulbs or capillaries, depending on the volumes of the liquids[32,54,104], are used in these determinations, there is the advantage of eliminating loss by evaporation[42,47] and contamination of the system in the case of hygroscopic liquids[31]. The most serious disadvantage of the method is that it requires several determinations with varying amounts of the solute liquid, before solubility data at a desired temperature can accurately be obtained[2,105]. It is equally important that a true equilibrium be established at any temperature in adjusting the heating and cooling rates.

The visual observation of the equilibrium point of mist appearance is facilitated by employing a gas flame[100], or by using the Tyndall effect[114], or by microscopic observation[119].

The vapor pressure in the dead space above the liquid in the sealed tube may affect the equilibrium solubility, though not necessarily in all systems[86]. One can minimize the dead space, or introduce the necessary corrections[61], assuming an ideal behavior of the vapors in the space. The safest method is, however, to measure the vapor pressure of the saturated solution at the temperature in question[83,101].

3. Miscellaneous Methods

Various instrumental methods have been used as analytical tools in the determination of mutual solubilities of liquids. For the solubility of hydrocarbons in water, u.v. spectrophotometry gives satisfactory results[15]. Small amounts of water in hydrocarbons can be determined by Karl Fischer titration or tracer analysis using tritiated water[14,59]. A non-instrumental method for the determination of small amounts of organic solvents in water consists of dispersing a water-insoluble dye in the water, and adding the organic solvent dropwise, until the excess dissolves the organic dye[106]. Acid–base titration can be used for the determination of mutual solubility in hydrogen fluoride–hydrocarbon systems[77]. Vapor pressure measurements[82,102] and gas chromatography[22,51] have been used for measuring activities in liquid–liquid systems.

VI. Solubility of Solids in Liquids

The most particular feature of solid–liquid systems which strongly affects the choice of the solubility method is the degree of ionicity of the solid solute. As a result, surveys and reviews[71, 76] are divided accordingly into solubilities of electrolytes and non-electrolytes. As mentioned earlier, reasonable predictions on the solubility of non-electrolytes can be made on the basis of the solubility parameter concept[49].

The dependence of solubility on the particle size of the solute is an additional important feature of the systems[26, 58]. It is thus generally safer to reach true equilibrium by overheating the system, though in some systems supersaturation may be quite stable. Another possibility is to have the solid in small particle size. This is easily obtained by a rapid cooling of a supersaturated solution, which usually yields a precipitate in a finely divided state.

The time needed to reach equilibrium in solid–liquid systems is usually much longer than in either gas–liquid or liquid–liquid systems. For example, equilibrium is attained only after 20 h at room temperature in the iodine–perfluoroheptane system[40], and as much as ten days may be needed for dimethylglyoxime–alcohol systems[109]. It is thus most important that tests be made to insure that a true equilibrium has been reached and that it can be established from both sides.

1. Saturation Method

Probably the simplest saturation device, based on a percolation principle[34], is shown in *Figure 11*. A slightly more complex device, operating on the same principle[37], but applicable at increased external pressure, is shown in *Figure 12*.

In systems which do not reach equilibrium rapidly, solubility measurements are best carried out in circulating saturators, such as the one[35] shown in *Figure 13*, or using a saturation device with continuous stirring[40], as shown in *Figure 14*. The latter has a siphon for sampling.

Figure 11. Percolation apparatus[34] for solubility studies.

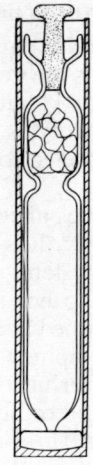

Figure 12. Device for percolation[37] at increased pressure.

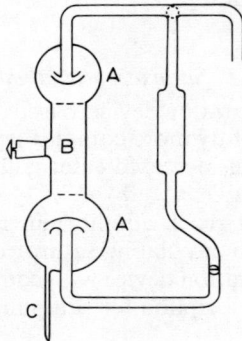

Figure 13. Circulating percolator[35]. A—saturation bulbs; B—solid container provided with supporting perforated glass plates; C—sampling stopcock.

A special device to reach saturation at constant pressure has been described[115]. It is useful for systems where the solute decomposes in the solute state to yield gaseous products, as in the case of solubilities of carbonates and bicarbonates in water.

2. *Cloud Point*

The procedure is similar to that used for liquid–liquid solubility systems. It gives reliable results in systems where the solid has medium solubility and requires a positive temperature coefficient of solubility. It is desirable to run two parallel experiments with a different solute-to-solvent ratio, one slightly oversaturated and the other slightly undersaturated at a given

SOLUBILITY

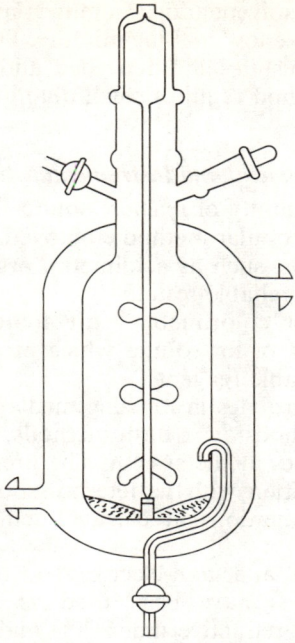

Figure 14. Saturator device[40].

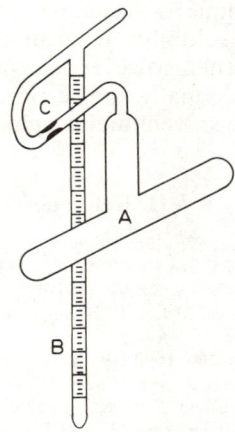

Figure 15. Solubility apparatus[83] designed for cloud method. A—solute container; B—pipet for solvent; C—sealing.

temperature. Both samples are kept at a constant temperature for several hours. If phase changes occur; the temperature must be readjusted[95].

A method combining cloud detection and volume measurements[83] uses the device shown in *Figure 15.* A known weight of the solute is placed in the solubility cell A, and the solvent added from the pipet D. The connecting tube is sealed at C by a suitable low-melting mixture. The solubility is determined by the appearance of a mist in cell A. If needed, additional solvent is added by unsealing at C. The method requires small quantities of the components.

3. Chemical and Instrumental Analysis

Determination of solubility of reactive solutes by specific chemical reactions is by far the most popular method employed. Conventional volumetric and gravimetric methods, such as acidimetry, argentometry or iodometry, yield probably the most reliable results.

Spectrophotometry or colorimetry[40] are frequently used as analytical tools for colored solutes, or for solutes which produce colored compounds by the addition of a suitable reagent.

The solubility of electrolytes in ionizing media can be measured by conductivity and e.m.f. methods[64, 91]. Both methods, but especially the former, are sensitive to impurities in the system. The presence of solid particles at equilibrium does not interfere with the determination. Both electrical methods are suitable for determinations at elevated temperatures[90] or increased pressure[19].

For both electrolytic and non-electrolytic solutes, several additional physicochemical methods have been used as analytical tools. Surface tension[89], density[25, 113], refractive index[43, 44] and polarimetric[75] measurements have been employed with varying success. The methods required calibration curves, and inherent in the procedure is the uncertainty of extrapolation to saturated solution.

The use of a tracer technique as the analytical tool is of course the simplest and a quicker method of solubility determination whenever feasible. The precision of the measurements, however, is usually lower than that of direct chemical analysis[46, 66]. An equally simple procedure, of medium accuracy, involves evaporation of the solvent and weighing the residue[29, 48, 70].

VII. References

[1] Akerlof, G. *J. Amer. Chem. Soc.* **57**, 1196 (1935).
[2] Alexejev, W. *J. Prakt. Chem.* **25**, 518 (1882).
[3] von Antropoff, A. *Proc. Roy. Soc. A*, **83**, 474 (1910).
[4] Ashton, J. T., R. A. Dave, K. W. Miller, E. B. Smith and B. J. Stickings, *J. Chem. Soc. (A)*, 1793 (1968).
[5] McAuliffe, C. *Nature, London*, **200**, 1092 (1963).
[6] Baldwin, R. R. and S. G. Daniel, *J. Appl. Chem.* **2**, 161 (1951).
[7] Battino, R. and H. L. Clever, *Chem. Rev.* **66**, 395 (1966).
[8] Battino, R. and E. Wilhelm, *Chem. Rev.* **73**, 1 (1973).
[9] Benjamin, L. and V. Gold, *Trans. Faraday Soc.* **50**, 797 (1954).
[10] Bell, J. M. and A. L. Field, *J. Amer. Chem. Soc.* **33**, 1940 (1911).
[11] Ben-Naim, A. and S. Baer, *Trans. Faraday Soc.* **59**, 2735 (1963).
[12] Bennett, G. M. and W. G. Philip, *J. Chem. Soc.* 1928 (1930).
[13] Benson, B. B. and P. D. M. Parker, *J. Phys. Chem.* **65**, 1489 (1961).
[14] Black, C., G. G. Joris and H. S. Taylor, *J. Chem. Phys.* **16**, 537 (1948).
[15] Bohon, R. L. and W. F. Claussen, *J. Amer. Chem. Soc.* **73**, 1571 (1951).
[16] Burrows, G. and F. H. Preece, *J. Appl. Chem.* **3**, 451 (1953).
[17] Cady, H. P., H. M. Elsey and E. V. Berger, *J. Amer. Chem. Soc.* **44**, 1456 (1922).

SOLUBILITY

[18] Claussen, W. F. and M. F. Polglase, *J. Amer. Chem. Soc.* **74**, 4817 (1952).
[19] Cohen, E. and L. R. Sinnige, *Z. Phys. Chem.* **69**, 102 (1909).
[20] Cook, M. W. and D. N. Hanson, *Rev. Sci. Instrum.* **28**, 370 (1957);
Cook, M. W., D. N. Hanson and B. J. Alder. *J. Chem. Phys.* **2**, 748 (1957).
[21] Cox, J. D. and A. J. Head. *Trans. Faraday Soc.* **58**, 1839 (1962).
[22] Cruickshank, A. J. B., B. W. Gainey and C. L. Young, *Trans. Faraday Soc.* **64**, 337 (1968).
[23] Cukor, P. M. and J. M. Prausnitz, *Industr. Engng Chem. (Fundamentals)*, **10**, 638 (1971).
[24] Douglas, E. *J. Phys. Chem.* **68**, 169 (1964); **69**, 2608 (1965).
[25] Dundon, M. L. and W. E. Henderson, *J. Amer. Chem. Soc.* **44**, 1196 (1922).
[26] Dundon, M. L. and E. Mack Jr. *J. Amer. Chem. Soc.* **45**, 2479 (1923).
[27] Dymond, J. H. and J. H. Hildebrand. *Industr. Engng Chem. (Fundamentals)*, **6**, 130 (1967).
[28] Elsey, P. G. *Analyt. Chem.* **31**, 869 (1959).
[29] Emerson, W. H. *J. Amer. Chem. Soc.* **29**, 1750 (1907).
[30] Evans, T.W. *Industr. Engng Chem. (Anal. Ed.)*, **8**, 206 (1936).
[31] Ewins, A. G. *J. Chem. Soc.* **105**, 350 (1914).
[32] Flaschner, C. and B. McEven, *J. Chem. Soc.* **93**, 1000 (1908).
[33] Fowler, R. H. and E. A. Guggenheim, *Statistical Thermodynamics*, pp 355–358. Cambridge University Press: London (1960).
[34] Fox, J. J. *J. Chem. Soc.* **95**, 878 (1909).
[35] Frear, G. L. and J. Johnston. *J. Amer. Chem. Soc.* **51**, 2082 (1929).
[36] Friedman, H. L. *J. Amer. Chem. Soc.* **76**, 3294 (1954).
[37] Gibson, R. E., *J. Amer. Chem. Soc.* **56**, 864 (1934).
[38] Gjaldbaek, J. C. *Acta Chem. Scand.* **6**, 623 (1952).
[39] Gjaldbaek, J. C. and J. H. Hildebrand. *J. Amer. Chem. Soc.* **72**, 609 (1950).
[40] Glew, D. H. and J. H. Hildebrand, *J. Phys. Chem.* **60**, 616 (1956).
[41] Gorodetskii, J. G., V. E. Skurat and V. L. Tal'rose, *Russ. J. Phys. Chem.* **44**, 1049 (1970).
[42] Groot, C. and J. H. Hildebrand. *J. Amer. Chem. Soc.* **70**, 3815 (1948).
[43] Heric, H. L. and C. D. Posey, *J. Chem. Engng Data*, **9**, 35 (1964).
[44] Heric, H. L. and C. D. Posey. *J. Chem. Engng Data*. **10**, 25 (1965).
[45] Hevesy, G. *J. Phys. Chem.* **16**, 429 (1912).
[46] Hevesy, G. and F. A. Paneth, *Z. Anorg. Chem.* **82**, 322 (1913).
[47] Hildebrand, J. H. and D. R. F. Cochran. *J. Amer. Chem. Soc.* **71**, 22 (1949).
[48] Hildebrand, J. H., E. T. Eleffson and C. W. Beebe. *J. Amer. Chem. Soc.* **39**, 2297 and 2301 (1971).
[49] Hildebrand, J. H., J. M. Prausnitz and R. L. Scott. *Regular and Related Solutions*, Van Nostrand: New York (1971).
[50] Hildebrand, J. H. and R. L. Scott, *Regular Solutions*, Prentice Hall: Englewood Cliffs, N.J. (1962).
[51] Hicks, C. P. and C. L. Young. *Trans. Faraday Soc.* **64**, 2675 (1968).
[52] Hill, A. E. *J. Amer. Chem. Soc.* **45**, 1143 (1923).
[53] Hill, A. E. and W. M. Malisoff. *J. Amer. Chem. Soc.* **48**, 918 (1926).
[54] Hoerr, C. W. and A. W. Ralston. *J. Amer. Chem. Soc.* **64**, 2824 (1942).
[55] Horiuti, J. *Sci. Pap. Inst. Phys. Chem. Res., Tokyo*, **17**, 125 (1931).
[56] Hu, J. H. and G. E. MacWood, *J. Phys. Chem.* **60**, 1483 (1956).
[57] Huggins, M. L. *J. Chem. Phys.* **5**, 143 (1937).
[58] Hulett, A. *Z. Phys. Chem.* **37**, 385 (1901).
[59] Joris, G. G. and H. S. Taylor, *J. Chem. Phys.* **16**, 45 (1948).
[60] Kablunov, I. A. and V. T. Malischeva. *J. Amer. Chem. Soc.* **47**, 1553 (1925).
[61] Keyes, D. B. and J. H. Hildebrand. *J. Amer. Chem. Soc.* **39**, 2126 (1917).
[62] Klots, C. E. and B. B. Benson. *J. Chem. Phys.* **38**, 890 (1963);
Klots, C. E. and B. B. Benson. *J. Marine Res.* **21**, 1 (1963).
[63] Kobotake, Y. and J. H. Hildebrand. *J. Phys. Chem.* **65**, 331 (1961).
[64] Kolthoff, I. M. *J. Phys. Chem.* **35**, 2711 (1931).
[65] Kolthoff, I. M. and P. J. Elving. *Treatise on Analytical Chemistry*, Part I, Volume I. Wiley-Interscience: New York (1959).
[66] Kolthoff, I. M. and C. Rosenblum. *J. Amer. Chem. Soc.* **56**, 1264 (1934).
[67] Kurkchi, G. A. and A. V. Jogansen. *Dokl. Akad. Nauk SSSR*, **145**, 1085 (1962).
[68] Lachowicz, S. K. *J. Imp. Coll. Chem. Engng Soc.* **8**, 51 (1954).
[69] Lannung, A. *J. Amer. Chem. Soc.* **52**, 68 (1930).
[70] Leighton, D. A. and J. B. Wilkins. *J. Amer. Chem. Soc.* **70**, 2600 (1948).

71. Leussing, D. L. in ref. 65, p 675.
72. Lewis, G. N. and D. B. Keyes. *J. Amer. Chem. Soc.* **40**, 472 (1918).
73. Linford, R. G. and J. H. Hildebrand. *J. Phys. Chem.* **73**, 4410 (1969).
74. Loprest, F. J. *J. Phys. Chem.* **61**, 1128 (1957).
75. Lowry, T. M. *J. Chem. Soc.* **85**, 1555 (1904).
76. Mader, W. J., R. D. Vold and M. J. Vold. In *Physical Methods in Organic Chemistry*, 3rd ed., Part I, p. 655. Ed. A. Weissberger. Wiley-Interscience: New York (1959).
77. Marcus, Y., J. Shamir and J. Soriano. *J. Phys. Chem.* **74**, 133 (1970).
78. Margrave, J. L. In *Physicochemical Measurements at High Temperatures*, p. 6. Edited by J. O'M. Bockris, J. L. White and J. D. Mackenzie. Butterworths: London (1959).
79. Markham, A. E. and K. A. Kobe. *Chem. Rev.* **28**, 519 (1941).
80. Maxted, E. M. and C. H. Moon. *Trans. Faraday Soc.* **32**, 769 (1936).
81. McDaniel, A. S. *J. Phys. Chem.* **15**, 587 (1911).
82. McGlashan, M. L. and K. W. Morcom. *Trans. Faraday Soc.* **57**, 581, 588 (1961).
83. Menzies, A. W. C. *J. Amer. Chem. Soc.* **58**, 934 (1936).
84. Merriman, R. W. *J. Chem. Soc.* **103**, 1774 (1913).
85. Miller, K. W. *J. Phys. Chem.* **72**, 2248 (1968).
86. Moles, E. and E. Jimeno. *Anal. Soc. Esp. Fis. Quim.* **11**, 393 (1913); *Chem. Abstr.* **9**, 743 (1915).
87. Montgomery, H. A. C., N. S. Thom and A. Cockburn. *J. Appl. Chem.* **14**, 28 (1964).
88. Morrison, J. T. and F. Billett. *J. Chem. Soc.* 2033 (1948).
89. Motylewski, S. *Z. Anorg. Chem.* **38**, 410 (1904).
90. Noyes, A. A. and W. D. Coolidge. *Z. Phys. Chem.* **46**, 321 (1903).
91. Owen, B. B. and S. R. Brindley. *J. Amer. Chem. Soc.* **60**, 2229 (1938).
92. Perrin, D. D., W. L. F. Armarego and D. R. Perrin. *Purification of Laboratory Chemicals.* Pergamon: Oxford (1966).
93. Pray, H. A., C. E. Schweickert and B. H. Minnich. *Industr. Engng Chem.* **44**, 1146 (1952); Goddman, J. B. and N. W. Krase. *Industr. Engng Chem.* **23**, 401 (1931).
94. Ramsteadt, F. E. *Le Radium*, **8**, 253 (1911).
95. Rhodes, F. H. and F. S. Eisenhauer. *Industr. Engng Chem.* **19**, 414 (1929).
96. Riddick, J. A. and W. B. Bunger. *Organic Solvents*, 3rd ed. Wiley-Interscience: New York (1970).
97. Richardson, F. D. and C. B. Alcock. In ref. 78, p 135.
98. Scholander, P. F. *J. Biol. Chem.* **167**, 235 (1947).
99. Shepherd, M. *J. Res. Nat. Bur. Stand.* **6**, 121 (1931); **26**, 351 (1941).
100. Sidgwick, W. V., P. Pickford and B. H. Wilsdon, *J. Chem. Soc.* **99**, 112 (1911).
101. Sidgwick, M. V. and E. K. Ewbank. *J. Chem. Soc.* **119**, 979 (1921).
102. Simons, J. H. and R. D. Dunlop. *J. Chem. Phys.* **18**, 335 (1950).
103. Sinor, J. E. and F. Kurata. *J. Chem. Engng Data*, **11**, 537 (1966).
104. Smith, R. A. *Mikrochemie*, **11**, 227 (1932).
105. Smith, A. and A. W. C. Menzies. *J. Amer. Chem. Soc.* **31**, 1183 (1909).
106. Sobotka, H. and J. Kahn. *J. Amer. Chem. Soc.* **53**, 2935 (1931).
107. Steinberg, M. and B. Manowitz. *Industr. Engng Chem.* **51**, 49 (1959).
108. Stimson, H. F., D. R. Lovejoy and J. R. Clement. In *Experimental Thermodynamics*, Vol. I, p 15. Edited by J. P. McCullough and D. W. Scott. Butterworths: London (1968).
109. Stromberg, A. G. and A. I. Zelyanskaya. *Analyt. Chem. USSR*, **4**, 286 (1949).
110. Swinnerton, J. W., V. J. Linnenbom and C. H. Check. *Analyt. Chem.* **34**, 483, 1509 (1962).
111. Taylor, N. W. and J. H. Hildebrand. *J. Amer. Chem. Soc.* **45**, 682 (1923).
112. Thomsen, E. S. and J. Gjaldbaek. *Acta Chem. Scand.* **17**, 127 (1963).
113. Vilbrandt, F. E. and J. A. Bender. *Industr. Engng Chem.* **15**, 967 (1923).
114. Vold, R. D., *J. Phys. Chem.* **43**, 1213 (1939).
115. Walker, A. C., U. B. Bray and J. Johnston. *J. Amer. Chem. Soc.* **49**, 1235 (1927).
116. Wiebe, R., V. L. Gady and C. Heins Jr. *Industr. Engng Chem.* **24**, 823 and 927 (1932).
117. Wiebe, R., V. L. Gady and C. Heins Jr. *J. Amer. Chem. Soc.* **55**, 947 (1933).
118. Williams, D. D. and R. R. Miller. *Analyt. Chem.* **34**, 657 (1962).
119. Wright, R. *J. Chem. Soc.* 1203 (1926).
120. Zelhoffer, G. F. *Industr. Engng Chem.* **29**, 548 (1937).
121. Zimmerman, H. K., Jr. *Chem. Rev.* **51**, 25 (1952).
122. Zorin, A. D., A. E. Ezheleva and G. G. Devyatykh. *Zavod. Lab.* **29**, 659 (1963). *Chem. Abstr.* **59**, 8181 (1963).

CHAPTER 16

Part 1. Phase Equilibria of Two-component Systems and Multicomponent Systems

A. G. WILLIAMSON

Department of Chemical Engineering, University of Canterbury, Christchurch 1, New Zealand

Contents

	List of Symbols	749
I.	Introduction	750
II.	Thermodynamics of Mixtures	751
III.	Liquid Mixtures	753
IV.	Empirical Representation of Liquid–Vapor Equilibrium Data	755
V.	Tests for Thermodynamic Consistency of Liquid–Vapor Equilibrium Data	759
VI.	Experimental Methods of Determining Liquid–Vapor Equilibrium Data	761
	1. Dynamic Methods	763
	2. Static Methods	767
VII.	Static Measurements with Analyses of Both Phases	775
VIII.	The McBain Balance Method	778
IX.	Dewpoint and Bubble-point Measurements	780
X.	The Isopiestic Method	782
XI.	Differential Methods	783
XII.	Light-scattering Measurements	783
XIII.	References	784

List of Symbols

The following symbols have the meanings listed:

p — pressure
T — temperature
μ_i — chemical potential of component i
V — volume
n_i — number of moles of component i
G — Gibbs free energy
s_i — partial molal entropy of component i
h_i — partial molal enthalpy of component i
y_i, x_i — mole fraction of component i
p^* — fugacity
V_i° — molar volume of pure component i
B_{ii} — second virial coefficient of component i
B_{ij} — mixture second virial for interaction between components i and j
γ_i — activity coefficient of component i
G^E — excess Gibbs free energy of mixing

H^E enthalpy of mixing
μ_i^E excess chemical potential of component i

I. Introduction

The main concern in the study of phase equilibria in multicomponent fluid systems is in determining the compositions of the phases when values have been assigned to sufficient variables to establish a thermodynamically well-defined state. The number of variables to which values must be assigned can be described in terms of the phase rule[28]. If we ignore, as we usually may, gravity, electric and magnetic fields and surface forces, the phase rule takes the form

$$P + F = C + 2$$

where P is the number of phases, F is the number of degrees of freedom, and C is the number of components. For systems of the type to be discussed here, in which no chemical reactions take place, the number of components is simply the number of chemically distinguishable species. The number of degrees of freedom is the number of intensive variables required to fix the thermodynamic state of the system. The variables of interest are usually the temperature, pressure and the compositions of the phases.

Even with the constraints of the phase rule, there is a wide variety of phase relations of interest. Two-phase systems may be classified according to the physical states of the phases as follows:

(1) solid–solid
(2) solid–liquid
(3) solid–gas
(4) liquid–liquid
(5) liquid–gas
(6) gas–gas

Of these, classes 1, 2, 3 and 5 are commonly found in both single and multi-component systems. Class 4 is commonly found only in multicomponent systems†. Class 6 systems are found only in multicomponent systems at high densities[64].

The discussion in this section will be confined to systems of classes 4 and 5, that is to fluid systems at low and medium pressures. The special behavior of class 6 systems will be discussed elsewhere[71]. Most of the discussion in this section will be confined to two-component systems. The extension to multi-component systems is usually conceptually simple though it may become algebraically complicated.

The thermodynamic behavior of a multicomponent system can usually be described most conveniently in terms of the changes in properties which occur on mixing rather than in terms of the total properties of the mixtures. For binary mixtures of classes 4 and 5, the components of the mixture and one of the mixed phases are all in the same physical state at the temperature and pressure of mixing and it is usually possible to examine the phase behavior over the whole composition range from pure component 1 to

---

† The exception being liquid ^{4}He [54].

PART 1. PHASE EQUILIBRIA OF TWO-COMPONENT SYSTEMS

pure component 2. It is thus convenient in describing the thermodynamics of these systems to use the same reference state for both components. This leads to a simplifying symmetry in the equations describing the behavior of the individual components. In systems of classes 2 and 3, on the other hand, because the pure components are in different physical states, it is usually more convenient to take different reference states. This leads to unsymmetrical formulations. Experimentally, such systems are usually limited to a region of low concentration of one of the components and it becomes convenient to distinguish between a solvent (usually the major component) which does not change state on forming the solution and the solute (usually the minor component) which does change state on forming the solution. It then becomes convenient to use pure solvent as the reference state for solvent behavior and the infinitely dilute solution (pure solvent) as the reference state for the solute. Furthermore, because of the limited composition range to be studied, the experimental methods for the study of such systems are usually rather different from those used with systems of classes 4 and 5.

Despite these differences which lead to the treatment of different classes of systems in different parts of this volume, it should be remembered that the separation is one of convenience rather than one of principle.

II. Thermodynamics of Mixtures

Because the experimental approach to the study of phase equilibria is very much dependent on the kind of thermodynamic information required, we shall begin by outlining the basic thermodynamic relations.

The starting points for a discussion of phase equilibrium are the conditions for hydrostatic, thermal, and chemical equilibrium respectively between the phases

$$p^\alpha = p^\beta = p^\gamma = \qquad (1)$$

$$T^\alpha = T^\beta = T^\gamma = \qquad (2)$$

and

$$\mu_i^\alpha = \mu_i^\beta = \mu_i^\gamma = \qquad (3)$$

where the superscripts refer to phases and the subscripts refer to components. The general presentation of the thermodynamics of fluid phase equilibria has been dominated by the perfect gas equation of state

$$pV = nRT \qquad (4)$$

This domination arises because the perfect gas behavior represents an accessible limiting behavior common to all substances and because the relations based on this equation are usually simple. The remaining relation of key importance is that for the pressure dependence of the Gibbs free energy

$$(\partial G/\partial p)_T = V \qquad (5)$$

This relation, in its integrated form, becomes

$$G(T, p) = G(T, p^\dagger) + \int_{p\dagger}^{p} V \, dp \qquad (6)$$

in which the conditions are indicated in parentheses and $p^\dagger$ refers to the pressure in some reference state. For the perfect gas, equation 6 takes the form

$$G(T, p) = G(T, p^\dagger) + RT \ln(p/p^\dagger) \qquad (7)$$

For the single-component perfect gas system in which the distinction between the molar Gibbs free energy and the chemical potential vanishes, we may write the expressions

$$\mu_i^{pg}(T, p) = \mu_i^{pg}(T, p^\dagger) + RT \ln (p/p^\dagger) \qquad (8)$$

$$s_i^{pg}(T, p) = s_i^{pg}(T, p^\dagger) - R \ln (p/p^\dagger) \qquad (9)$$

and
$$h_i^{pg}(T, p) = h_i^{pg}(T, p^\dagger) \qquad (10)$$

in which the superscript pg refers to perfect gas behavior. In dealing with multicomponent systems it is necessary to introduce one further simplifying concept, that of ideal mixing. This has some experimental justification as an empirical low-pressure limit to real behavior, but no rigorous proof in classical thermodynamics. The best justification for the ideal mixing concept lies in statistical mechanics[34]. For the purposes of the present discussion, ideal mixing will be introduced as an arbitrary reference of behavior for mixtures. It will be defined by the set of relations

$$\mu_i(T, p, x_i) = \mu_i^\circ(T, p, 1) + RT \ln x_i \qquad (11)$$

where the superscript $^\circ$ refers to pure component i in the same physical state (real or hypothetical) as the mixture and x_i is the mole fraction of component i in the mixture. The chemical potential of a component of an ideal mixture of perfect gases then becomes

$$\mu_i^{pg}(T, p, x_i) = \mu_i^\circ(T, p^\dagger) + RT \ln (p/p^\dagger) + RT \ln x_i \qquad (12)$$

It should be re-emphasized at this point that equation 12 contains two distinct approximations, viz:
 (1) the perfect gas assumption,
 (2) the ideal mixing assumption.

These two approximations may be relaxed separately to give equations relating to real behavior. First, we may envisage ideal mixing of components which are not themselves perfect gases. This can be done formally by replacing the pressures in equation 7 by fugacities denoted by p^* and defined by the relation

$$(\mu_i^{\circ\,rg} - \mu_i^{\circ\,pg})(T, p) = RT \ln p^*/p \qquad (13)$$

$$\lim_{p \to 0} p^*/p = 1 \qquad (14)$$

which in turn lead to

$$\mu_i^{\circ\,rg}(T, p) - \mu_i^{\circ\,pg}(T, p) = RT \ln p^*/p + \int_0^p (V_i^\circ - \{RT/p\})\,dp \qquad (15)$$

and
$$\mu_i^{\circ\,rg}(T, p) - \mu_i^{\circ\,pg}(T, p = 1) = RT \ln p + \int_0^p (V_i^\circ - \{RT/p\})\,dp \qquad (16)$$

PART 1. PHASE EQUILIBRIA OF TWO-COMPONENT SYSTEMS

The corresponding relation for gas mixtures which takes account of both the trends to ideal mixing and to perfect gas behavior as p tends to zero is

$$\mu_i^{rg}(T, p, y_i) - \mu_i^{\circ\, pg}(T, p = 1) = RT \ln py_i + \int_0^p (V_i - \{RT/p\}) \, dp \quad (17)$$

where y_i is the mole fraction of component i in the mixture. In the special case of a binary gas mixture whose behavior may be represented by a truncated virial equation in the form

$$pV_m = \Sigma n_i RT + P(n_1^2 B_{11} + 2n_1 n_2 B_{12} + n_2^2 B_{22})/(n_1 + n_2) \quad (18)$$

we get

$$(V_i - RT/p) = (\partial/\partial n_i) \Sigma n_i n_j B_{ij}/\Sigma n_i \quad (19)$$

and by substitution in 17

$$\mu_1^{rg}(T, p, y_i) - \mu_1^{\circ\, pg}(T, p=1) = RT \ln py_1 + B_{11}p + y_2^2(2B_{12} - B_{11} - B_{22})p \quad (20)$$

These relations are particularly useful in discussing liquid–vapor equilibrium in the low and medium pressure range.

III. Liquid Mixtures

In dealing with liquid mixtures, the convenient reference of behavior is again ideal mixing as defined by equation 11. However, in this case, because there is no simplifying general equation of state corresponding to the perfect gas equation, it is more convenient to express the real mixture behavior by an activity coefficient so that

$$\mu_i^l(T, p, x_i) = \mu_i^{\circ l}(T, p, 1) + RT \ln x_i \gamma_i \quad (21)$$

The condition for liquid–vapor equilibrium for each component of a mixture is

$$\mu_i^l(T, p, x_i) = \mu_i^g(T, p, y_i) \quad (22)$$

where x_i is the mole fraction of component i in the liquid and y_i is its mole fraction in the vapor phase. Equations 17, 21 and 22 then allow us to express liquid–vapor equilibrium in terms of a knowledge of the activity coefficients in the liquid phase, the equation of state for the gaseous phase and the vapor pressures of the pure components

$$RT \ln x_i \gamma_i = RT \ln py_i/p_i^\circ + \int_0^p (V_i^g - \{RT/p\}) dp$$
$$- \int_0^{p_i}(V_i^{\circ g} - \{RT/p\}) dp - \int_{p_i^\circ}^p V_i^{\circ l} dp \quad (23)$$

where p is the vapor pressure of the mixture, p_i° is the vapor pressure of pure component i, V_i^g, $V_i^{\circ g}$ are the partial molar volumes of component i in the vapor mixture and in the pure vapor respectively, and $V_i^{\circ l}$ is the molar volume of component i in the liquid, all at temperature T. For a binary mixture there will be two equations of type 23 which will allow p and y to be determined in terms of values of p_i°, x_i, γ_i and the equations of state for the vapors. The greatest hindrance to the exact solution of these equations lies in our inability to predict the γ_i as functions of x_i. On the other hand, experimental observations of liquid–vapor equilibrium can be most compactly expressed in terms of empirical functions relating x_i and γ_i. Again, in the special case of the vapor-phase behavior being expressed in terms of the truncated virial equation, then equation 23 becomes

$$RT \ln \gamma_i(T, p, x_i) = RT \ln (py_i/p_i^\circ x_i) + \int_0^p (V_i^g - V_i^{\circ g}) \, dp$$
$$+ \int_{p_i^\circ}^p (V_i^{\circ g} - \{RT/p\} - V_i^{\circ l}) \, dp \quad (24)$$

and, if the range $p_i^\circ - p$ is small,

$$RT \ln \gamma_i(T, p, x_i) = RT \ln (py_i/p_i^\circ x_i) + (B_{ii} - V_i^{\circ l})(p - p_i^\circ)$$
$$- p(1 - y_i)^2 \Delta \quad (25)$$

where $\Delta = (2B_{12} - B_{11} - B_{22})$.

The activity coefficients are frequently expressed in terms of an excess chemical potential μ_i^E defined by

$$\mu_i^E = RT \ln \gamma_i \quad (26)$$

As determined by equation 25, the activity coefficients and consequently the excess chemical potentials for a series of experimental measurements would each refer to the temperature T and the vapor pressure of the particular mixture. It is often more useful to reduce each result to a common reference pressure p^r via the relation

$$\mu_i^E(T, p^r, x_i) = \mu_i^E(T, p, x_i) + \int_p^{p^r} V_i^E dp \quad (27)$$

and to consolidate the excess chemical potentials into an overall excess Gibbs free energy of mixing

$$G^E(T, p^r, x_i) = \Sigma x_i \mu_i^E(T, p^r, x_i) = \Sigma x_i \mu_i^E(T, p, x_i) + \int_p^{p^r} V^E dp \quad (28)$$

where V^E and V_i^E are the excess volume and the partial molar excess volume of mixing respectively. Generally, when $p^r - p$ is small, the integral in equation 28 can be replaced by

$$\int_p^{p^r} V^E \, dp \approx V^E(p^r - p) \quad (29)$$

So far, the discussion has been related to the experimental observations of vapor pressures as functions of composition at a given temperature, the isothermal liquid–vapor equilibrium studies. An alternative and widely used method of experimental study is to examine the variation of boiling points with composition at constant pressure, the *isobaric* liquid–vapor equilibrium studies. In the isobaric measurements, each determination of G^E, or μ^E for a given component, is at a different temperature from the others in a given set. Such measurements are of considerable practical importance, e.g. in the design of fractionating columns which usually operate substantially at constant pressure. However, the results of these studies are not so convenient for thermodynamic analysis. The reduction of isothermal data to constant pressure involves only the small correction given in equation 29. The reduction of isobaric data to constant temperature involves the use of a relation such as

$$G^E/T(T^r, p, x_i) = G^E/T(T, p, x_i) - \int_T^{T^r} (H^E/T^2) \, dT \quad (30)$$

or

$$(\mu_i^E/T)(T^r, p, x_i) = \mu_i^E/T(Tpx_i) - \int_T^{T^r}(H_i^E/T^2)\,dT \qquad (31)$$

in which the term in the integral is frequently large compared with G^E or μ_i^E. A detailed knowledge of the enthalpy of mixing is thus an important adjunct to the thermodynamic analysis of isobaric liquid–vapor equilibrium data and, indeed, to the extrapolation of isothermal data from one temperature to another. Generally, because the experimental determination of liquid–vapor equilibrium is much more tedious than that of enthalpies of mixing, it is better to measure one set of liquid–vapor equilibrium data carefully either at constant T or at constant p and to extrapolate these results to other conditions using experimentally measured V^E and H^E data. Recent extensive reviews of the methods for determining V^E and H^E have been published by Battino[4] and by McGlashan[51].

IV. Empirical Representation of Liquid–Vapor Equilibrium Data

The phase boundaries for binary liquid-vapor equilibrium can be presented graphically in the conventional phase diagrams as shown in *Figures 1* and *2*, in which the values of p, x, y at given T or T, x, y at given p are shown. Such representation is sufficient for measurements of low accuracy. For precise data, tabular presentation of the same information is more appropriate and is necessary when precise data on the gas-phase imperfections are not available. When the gas-phase imperfections are known and/or when the data over the whole composition range are to be smoothed, presentation in the form of excess chemical potentials or excess Gibbs free energies of mixing is the most compact method. Presentation of the data as equations relating chemical potentials to composition is subject to the important, and necessary, constraint that the algebraic forms for the chemical potential versus composition relations shall satisfy the Gibbs–Duhem relation[27].

$$\sum n_i\,d\mu_i + V\,dp - S\,dT = 0 \qquad (32)$$

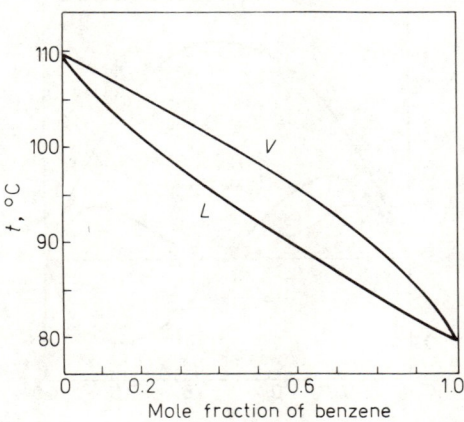

Figure 1. Boiling points of benzene and toluene mixtures at 750 mm Hg.

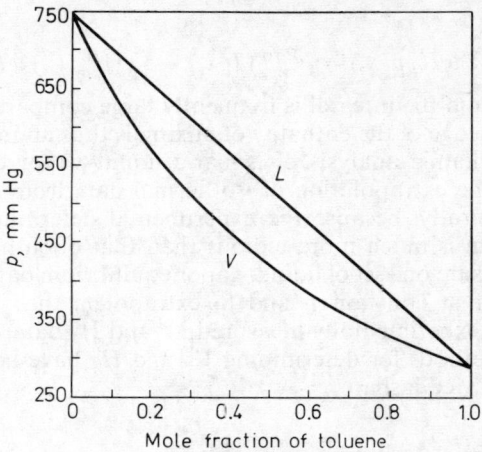

Figure 2. Vapor pressures of benzene and toluene mixtures at 79°C.

or, in terms of excess functions[32],

$$\sum x_i \, d\mu_i^E + V^E \, dp - (H^E/T) \, dT = 0 \tag{33}$$

which at constant T, p can be reduced to

$$\sum x_i \, d\mu_i^E = 0 \tag{34}$$

The empirical presentation of data as excess Gibbs free energies has the advantages that it automatically satisfies equation 34 and that the excess Gibbs free energy can usually be presented within experimental accuracy by a fairly simple function of composition. The disadvantage of the use of excess Gibbs free energies is that differentiation to give chemical potentials

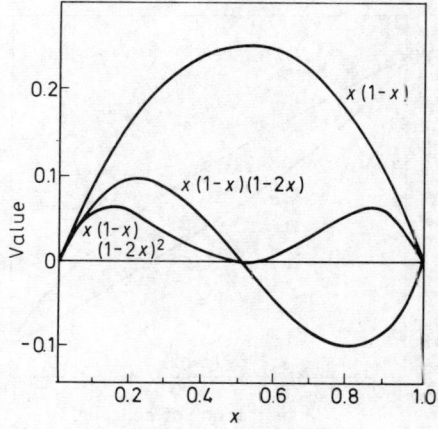

Figure 3. Form of the first three terms of equation 35.

has to be regarded with some suspicion, particularly near the ends of the composition range (x_i near 0, 1).

One of the more commonly used expressions for the excess Gibbs free energy of mixing which was popularized for a number of mixing functions by Redlich and Kister[60], Scatchard[66] and Guggenheim[26] is

$$G^E = x(1-x) \sum_{i=0}^{n} A_i(1-2x)^i \tag{35}$$

This expression has successive terms of alternating symmetry, the first three of which are shown in *Figure 3*. Since the excess Gibbs free energy of many binary mixtures is fairly symmetrical in composition, a reasonably good fit to the experimental data can usually be obtained with not more than three or four terms of equation 35. The expressions for μ_i^E or $RT \ln \gamma_i$ corresponding to equation 35 can be written in generalized forms as

$$\mu_1^E = RT \ln \gamma_1 = (1-x_1)^2 \sum_i A_i(1-2x_2)^{i-2}\{2i(1-x_2)-1\} \tag{36}$$

and

$$\mu_2^E = RT \ln \gamma_2 = x_1^2 \sum_i A_i(1-2x_1)^{i-2}\{2i(1-x_1)-1\} \tag{37}$$

In the simple case when $A_i = 0$ (all $i \geq 1$), these equations become:

$$G^E = A x_1 x_2 \tag{38}$$

$$\mu_1^E = A x_2^2 \tag{39}$$

and

$$\mu_2^E = A x_1^2 \tag{40}$$

which are the well-known expressions for the so-called strictly regular behavior of a simple mixture[29]. The two parameter form of equation 35 ($A_i = 0$ all $i \geq 2$) are

$$\mu_1^E = RT \ln \gamma_1 = x_2^2\{A_1 - A_2(3-4x_2)\} \tag{41}$$

and

$$\mu_1^E = RT \ln \gamma_2 = x_1^2\{A_1 + A_2(3-4x_1)\} \tag{42}$$

equivalent to the Margules 3 suffix equations[44] which are probably more familiar in the forms

$$\mu_1^E = (A-3B)x_2^2 + 4Bx_2^3 \tag{43}$$

and

$$\mu_2^E = (A+3B)x_1^2 - 4Bx_1^3 \tag{44}$$

Myers and Scott[55] have suggested the modified version of equation 35

$$G^E = x(1-x) \sum_{i=0}^{n} A_i(1-2x)^i/\{1-B(1-2x)\} \tag{45}$$

for fitting very skewed data. However, this expression has been relatively little used.

The other popular set of equations for representing activity coefficients may be derived from the regular solution equation of Scatchard and Hildebrand[33]

$$G^E = AV\phi_1\phi_2 \tag{46}$$

where V is the molar volume of the mixture and ϕ_1, ϕ_2 are the volume fractions of the components. This gives (neglecting volume changes on mixing) for the excess chemical potentials

$$\mu_1^E = RT \ln \gamma_1 = AV_1^\circ/\{(x_1/x_2)(V_1^\circ/V_2^\circ) + 1\}^2 \tag{47}$$

and

$$\mu_2^E = RT \ln \gamma_2 = AV_2^\circ/\{1 + (x_2/x_1)(V_2^\circ/V_1^\circ)\}^2 \tag{48}$$

where V_1°, V_2° are the molar volumes of the components. If we relax the requirement that V_1°, V_2° shall be the molar volumes equations 47 and 48 become:

$$\mu_1^E = RT \ln \gamma_1 = a/(1 + (ax_1/bx_2))^2 \tag{49}$$

$$\mu_2^E = RT \ln \gamma_2 = b/(1 + (bx_2/ax_1))^2 \tag{50}$$

These are the van Laar equations[40] for which the corresponding expression for the excess Gibbs free energy of mixing is

$$G^E = abx_1x_2/(ax_1 + bx_2) \tag{51}$$

The van Laar equations have been widely used and, considering their limitation to two adjustable parameters, have been very successful over a wide range of data. However, a common misuse of the van Laar equations has been in the representation of isobaric liquid–vapor equilibria with temperature-independent values of a and b. This implies that the enthalpy of mixing is zero. While this might not introduce a serious error in the representation of a single set of data covering a narrow range of boiling points, it could lead to serious errors in extrapolation from results so fitted.

Recently, an equation introduced by Wilson[80] has become popular and appears to be very successful in fitting the data for binary mixtures. Wilson's equation

$$G^E = -RT \left[\sum_{i=1}^{n} x_i \ln \left(1 - \sum_{j=1}^{n} x_j A_{ji}\right) \right] \tag{52}$$

which is general for any number of components and gives for a binary mixture

$$G^E = -RT[x_1 \ln(1 - A_{21}x_2) + x_2 \ln(1 - A_{12}x_1)] \tag{53}$$

Wilson and others (e.g. Prausnitz[57]) have shown that for a range of binary mixtures, equation 53 with but two parameters gives as good or a better fit to the data as does equation 35 with up to five parameters. A comparison of the two equations is contained in *Table 1* in which σ is the r.m.s. deviation of the data from the smooth curve.

PART 1. PHASE EQUILIBRIA OF TWO-COMPONENT SYSTEMS

Table 1

System		Number of experimental points	Equation 35. No. parameters	σ	Equation 53. No. parameters	σ
MeOH/CCl$_4$	35°C	9	3	4.7	2	1.9
MeOH/C$_6$H$_6$	35°C	9	5	3.4	2	3.0
CCl$_4$/CH$_3$CN	45°C	12	4	1.3	2	1.4
CCl$_4$/CH$_3$NO$_2$	45°C	11	3	1.7	2	1.1

A defect of equation 53 is that it is incapable of producing values of G^E sufficiently large to lead to liquid–liquid phase separation and is therefore not very suitable for handling activity coefficients near critical solution temperatures. Wilson has pointed out that this defect can be overcome by adding a third parameter to give the expression

$$G^E/RT = -C[x_1 \ln(1 - A_{21}x_2) + x_2 \ln(1 - A_{12}x_1)] \quad (54)$$

The equation suggested by Renon[61]

$$G^E/RT = x_1 x_2 A_{21} g_{21}/(x_1 + x_2 g_{21}) + A_{12} g_{12}/(x_2 + x_1 g_{12}) \quad (55)$$

with

$$A_{12} = (\beta_{12} - \beta_{22})/RT \qquad A_{21} = (\beta_{12} - \beta_{11})/RT$$

and

$$g_{12} = \exp(-\alpha_{12} A_{12}) \qquad g_{21} = \exp(-\alpha_{12} A_{21})$$

contains essentially three parameters, α_{12}, A_{12} and A_{21}, offers no advantages over the simpler equations 35 and 51 for systems with only small deviations from ideality, but like Wilson's equation it is more successful than the simpler expressions for systems in which G^E is strongly asymmetrical. Renon's equation has the advantage over the two-parameter Wilson expression that it is still valid in the two-liquid regions. Both Wilson's and Renon's equations are readily extended to multicomponent mixtures.

V. Tests for Thermodynamic Consistency of Liquid-Vapor Equilibrium Data

When independent values are available for the activity coefficients of the individual components of a mixture, it is possible to test the internal consistency of the experimental data with the aid of the Gibbs–Duhem relation. For binary mixtures at constant T, p, equation 34 takes the form

$$x_1(\partial \mu_1^E/\partial x_1)_{T,p} + x_2(\partial \mu_2^E/\partial x_1)_{T,p} = 0 \quad (56)$$

or

$$x_1(\partial \ln \gamma_1/\partial x_1)_{T,p} + x_2(\partial \ln \gamma_2/\partial x_1) = 0 \quad (56a)$$

Usually, because of the spacing of the experimental points, it is difficult to determine gradients with sufficient accuracy to use equations 56 or 56a directly, except to detect gross inconsistencies in the data.

More commonly, the Gibbs–Duhem relation is used in an integrated form. At the crudest level, the equality of the areas A_1 and A_2 given by

$$A_1 = \int_{x_1=0}^{1} x_1 \, d\mu_1^E = \int_{x_1=0}^{1} \mu_1^E \, dx_1 \tag{57}$$

and

$$A_2 = -\int_{x_1=0}^{1} x_2 \, d\mu_2^E = -\int_{x_1=0}^{1} \mu_2^E \, dx_1 \tag{57a}$$

can be taken as a rough measure of the internal consistency of the data. While equality of these areas is not a guarantee of consistency, inequality of the areas is certainly an indication of inconsistency. Some inequality is, of course, to be expected and differences of up to five per cent between A_1 and A_2 are observed even in good quality measurements.

A typical set of results at constant T is shown in *Figure 4*.

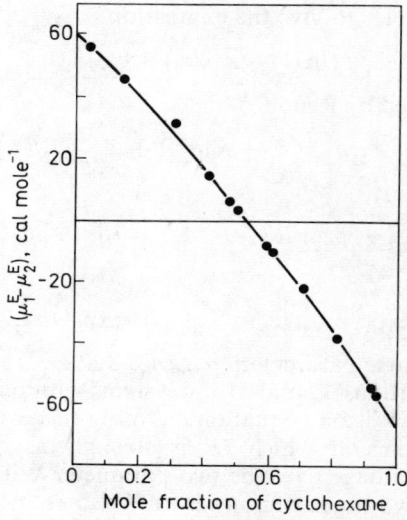

Figure 4. Gibbs–Duhem relation test for the liquid–vapor equilibrium data of mixtures of carbon tetrachloride and cyclohexane at 40°C.

Recent work by Herrington[31, 32] extends the concept of symmetrical area tests to a much more detailed analysis of thermodynamic consistency which may be used over parts of the composition range for both isothermal and isobaric data.

For isothermal data which have been corrected to constant pressure, Herrington integrates the expression

$$\Sigma x_i \, d \ln \gamma_i = 0 \tag{58}$$

to get

$$I_1 = \int_{x_1=0}^{x_1} x_1 \, d \ln \gamma_1 = \int_{\ln \gamma_{1,0}}^{\ln \gamma_{1, x_1'}} x_1 \, d \ln \gamma_1 \tag{59}$$

PART 1. PHASE EQUILIBRIA OF TWO-COMPONENT SYSTEMS

and

$$I_2 = \int_{x_1=1}^{x_1} x_2 \, d\ln \gamma_2 = \int_{\ln \gamma_2, 1}^{\ln \gamma_2, x_1'} x_2 \, d\ln \gamma_2 \tag{60}$$

and notes three properties of the integrals of use in testing the thermodynamic consistency of experimental data.
(1) $I_1 + I_2 = $ constant.
(2) The curves of I_1 and I_2 versus x are mirror images about $(I_1 + I_2)/2$.
(3) Properties 1 and 2 apply to any restricted range of composition.
The extension of this approach to isobaric data uses the equation

$$RT \, \Sigma x_1 \, d\ln \gamma_i - (H^E/T) \, dT = 0 \tag{61}$$

which gives the integrals

$$I_{1P} = \int_{\ln \gamma_1, 0}^{\ln \gamma_1, x_1'} x_1 \, d\ln \gamma_1 \tag{62}$$

$$I_{2P} = \int_{\ln \gamma_2, 1}^{\ln \gamma_2, x_1'} x_2 \, d\ln \gamma_2 \tag{63}$$

and

$$I_H = \int_{T_1}^{T'} (H^E/RT^2) \, dT \tag{64}$$

where T_1 is the boiling point of component 1 and T' is the boiling point of the mixture at x_1. The property

$$I_{1P} + I_{2P} + I_H = \text{constant} \tag{65}$$

is then used to test the consistency of the data.

Frequently data are not available for H^E. Herrington points out that if the data are examined in terms of I_{1P} and I_{2P} the magnitude of the value H^{E1} which must be assumed to satisfy equation 65 can itself be used as a guide to the thermodynamic consistency of the data. Known values of H^E are usually of the order of a few hundred J mol^{-1} (at $x = 0.5$) except for mixtures in which strong hydrogen bonding or other specific interactions occur. Unreasonably large values of H^{E1} can therefore often be taken as an indication of inconsistency in the data.

VI. Experimental Methods of Determining Liquid-Vapor Equilibrium Data

A two-phase, two-component system has two degrees of freedom. The state of the system can therefore be characterized by the determination of three variables. The accessible variables of interest are pressure, temperature and the phase compositions. Measurement of any three of these four quantities is sufficient to characterize the system. The decision about which variables to examine experimentally will depend on the physical and chemical properties of the components, on the accuracy with which the data are required and on the ease with which the various measurements can be made in particular cases. There are some advantages in attempting to measure all four variables —pressure p, temperature T, liquid composition x and vapor composition y. In particular, this leads to the independent evaluation of the individual activity coefficients and permits the use of the Gibbs–Duhem relation as a check on the consistency of the experimental results.

The accuracy with which the measurements need to be made can be discussed in terms of equation 25.

$$RT \ln \gamma_i(T, p, x_i) = RT \ln (py_i/p_i^\circ x_i) + (B_{ii} - V_i^\circ)(p - p_i^\circ) + pA(1 - y_i)^2 \tag{66}$$

from which the individual contributions to the error in $\ln \gamma_i$ may be calculated. The various contributions will, of course, depend on the methods used for pressure, temperature and composition measurements and on the actual regions of temperature and pressure being investigated. The various methods of measuring pressure and temperature are discussed elsewhere in this volume[58, 76]. Methods of measuring composition range from refractive index through density to gas–liquid chromatography and, for some sub-

Table 2. Errors in the measurement of activity coefficients from liquid–vapor equilibrium data
$p \approx 400$ torr $p_1^\circ \approx 500$ torr $T \approx 300$ K $y \approx 0.5$

Source of error	Accuracy of observation	Contribution to error (eq. 66)	Corresponding error in $\ln \gamma_i$
Pressure	$\delta p = \pm 0.05$ torr	$(\partial \ln \gamma_1/\partial p) = 1/p$	± 0.0001
Pressure	$\delta p = \pm 0.05$ torr	$(\partial \ln \gamma_1/\partial p_1^\circ) = 1/p_1^\circ$	± 0.0001
Composition	$\delta y = \pm 0.0001$	$(\partial \ln \gamma_1/\partial y_1) = 1/y_1$	± 0.0002
Composition	$\delta = \pm 0.0001$	$(\partial \ln \gamma_1/\partial x_1) = 1/x_1$	± 0.0002
Gas imperfection I	$\delta(B_1 - V_1^\circ) = \pm 50$ cm^3 mol^{-1}	$(\partial \ln \gamma_1/\partial (B-V)) \approx (p - p_1^\circ)/RT$	± 0.0002
Gas imperfection II†	$\delta(A) \approx \pm 100$ cm^2 mol^{-1}	$(\partial \ln \gamma/\partial A) \approx py_2^2$	± 0.0002
Temperature‡	$\delta T \approx 0.002$ K	$(\partial \ln p_1^\circ/\partial T) \approx 20$ torr K^{-1}	± 0.0001

† This term is frequently omitted entirely, usually for lack of knowledge of A.
‡ The temperature coefficient of γ_1 itself is usually small and the chief effect of temperature is related to uncertainties about the constancy of the temperature between the time when p_1° is measured and when p is measured.

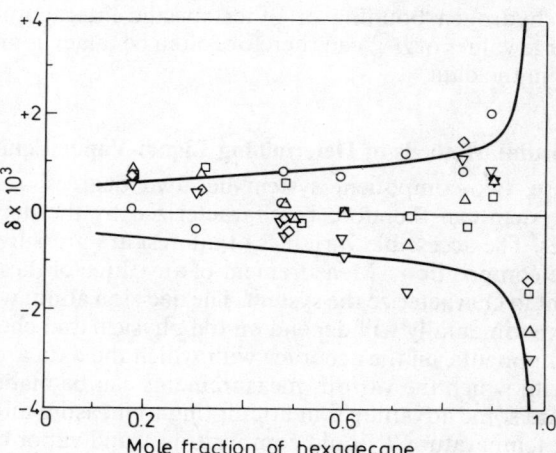

Figure 5. Deviations of $\ln \gamma$(hexane) from smooth curve plotted versus mole fraction of hexadexane: ○ 20°C; ▽ 30°C; ◇ 40°C; △ 50°C; □ 60°C. The solid lines are the limits of error calculated by the method illustrated in Table 2.

PART 1. PHASE EQUILIBRIA OF TWO-COMPONENT SYSTEMS

stances, to chemical methods. These all differ in their accuracy depending on the nature of the system and it is not possible to discuss the accuracy of composition measurements in general terms. As an example of the various contributions, *Table 2* gives typical contributions for reasonably accurate work in the region of atmospheric temperature and pressure.

The error in even a given set of measurements of γ varies with the composition. An example of the variation is shown in *Figure 5* for measurements on the system n-hexane and n-hexadecane. The wide range of methods used to obtain liquid–vapor equilibrium data defies classification. However, the most widely used experimental methods fall into two main groups which can be loosely called the dynamic (or ebulliometric) methods and the static methods, and we shall discuss these two approaches in turn with examples of the various designs of equipment used. In addition, there are a number of individual, less direct techniques which will be discussed separately at the end of this section.

1. Dynamic Methods

The ebulliometric method is used chiefly at low and moderate pressures and when it is proposed to determine the compositions of both liquid and vapor phases. Under these conditions, the vapor density is usually sufficiently

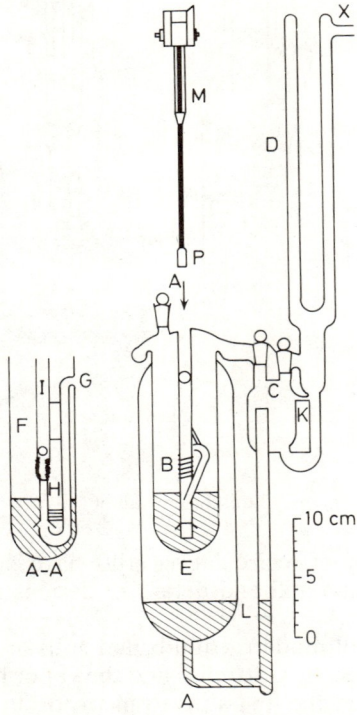

Figure 6. Equilibrium still of Scatchard, Raymond and Gilman[67].

low that a very large sample volume would be required to provide sufficient material (*ca.* 10 cm^3 liquid) for analysis by one of the conventional methods such as density measurements. This difficulty is circumvented by the use of a recirculating still. In the liquid recirculating still a mixture is boiled to produce a flow of liquid and vapor in intimate contact. The liquid and vapor flows are separated and the liquid flow is returned via a hold-up trap directly to the boiler. The vapor is totally condensed and the condensate is also returned to the boiler via a hold-up trap. The still is operated until a steady state is reached at which the pressure and boiling temperature are constant. In this state the condensate and liquid composition represent equilibrium vapor and liquid respectively.

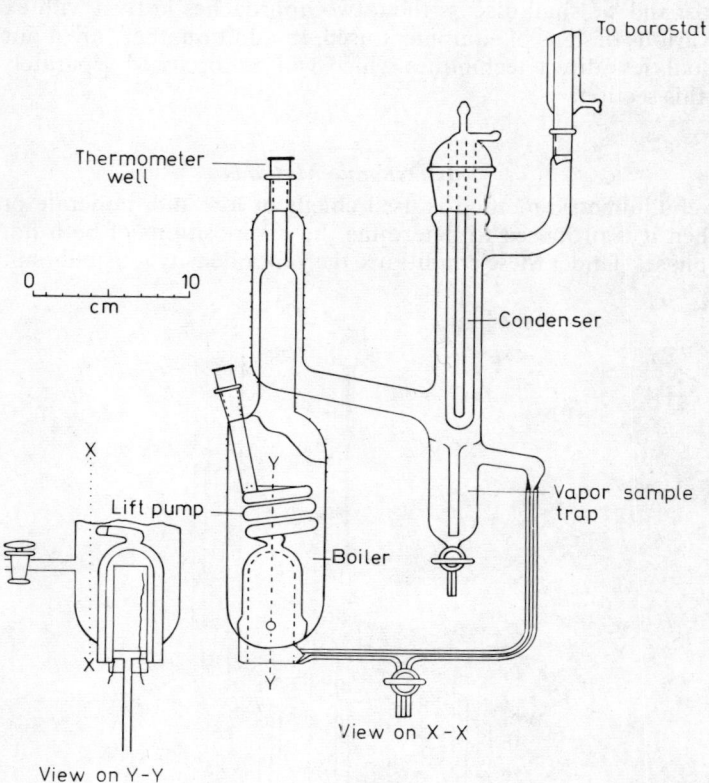

Figure 7. Equilibrium still of Ellis[20].

A very large number of recirculating stills have been designed and the history of the equilibrium still and details of designs up to 1948 have been reviewed by Fowler[23].

The key features of all modern equilibrium stills are the use of the vapor lift pump, first introduced by Cottrell[14] and the vapor hold-up trap first used by Sameshima[65]. One of the first stills to incorporate both of these devices was that of Scatchard, Raymond and Gilman[67] which is shown in *Figure 6*.

PART 1. PHASE EQUILIBRIA OF TWO-COMPONENT SYSTEMS

In this apparatus, the liquid boiling in the outer vessel E produces vapor which passes through the connecting tube into the inner vessel B where the vapor lift produces a stream of vapor and liquid which flows over the thermometer well. The liquid falls back into B and the vapor passes on to the total condenser D, the condensate from which passes through the hold-up trap C, and returns to E. At steady state, the liquid compositions of both E and B should be the same. Samples of the equilibrium compositions can be obtained from C and B via the stoppered ports shown in the diagram. The use of the double-boiler provides thermal insulation for the inner vessel and helps to avoid deviations from equilibrium due to superheating in E. The whole apparatus may be connected via X to a pressure control and manometer so that the apparatus can be operated at any desired temperature (or pressure). For isothermal studies, the pressure can thus be controlled to fix the boiling point of the mixture.

A design which has become popular for chemical engineering studies at constant pressure is that of Ellis[20] shown in *Figure 7*. Perhaps the most successful equilibrium still of the liquid recirculating type is that of Brown[10] which traces its origins back to the design of Gillespie[25]. Brown's still shown in *Figure 8* incorporates many refinements designed to insure that the samples of condensate and liquid accurately reflect the equilibrium compositions at a well-defined temperature and pressure. The liquid is boiled in the vessel A for which the main heat source is an external heater wound on to the outer

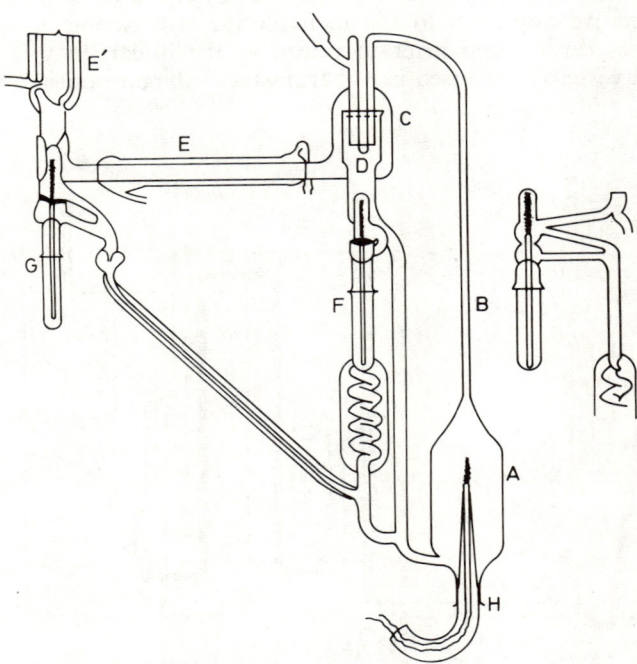

Figure 8. Brown's still[10].

surface. The interior surface of A is coated with a layer of sintered glass chips to promote nucleation. Even boiling is further promoted by a small internal heater of platinum wire inserted through the ground joint H. The liquid–vapor mixture pumps up the Cottrell pump B and flows over the thermometer well D. The liquid and vapor are separated at this point and the system of baffles insures that no liquid droplets are carried over into the condenser E. Liquid returns to the boiler via the trap F and the subsidiary cooling coil shown in the diagram. The vapor is condensed in E and E', the coolant flows in which are adjusted so that a sharp condensation line is achieved in E'. Condensate returns via the trap G and is mixed with the returning liquid flow just prior to re-entering the boiler. The slight cooling of the returning liquid flow insures that no flash vaporization occurs when the liquid and condensate flows mix. The traps F and G are fitted with magnetically operated valves and bypasses so that, when the still has reached a steady state, the samples of condensate and liquid may be isolated before the still is turned off. In this way there is no contamination of the samples with non-equilibrium material. This feature is particularly valuable when the still is operated at sub-ambient pressures and the boiling must be stopped and the still returned to atmospheric pressure before the sample traps are removed. The connection to pressure control and measurement apparatus is via the top of condenser E'.

Brown's still operates well with boiler sizes down to 150 cm^3 and on binary mixtures with relative volatilities up to about seven. With small boiler size or high relative volatility, it is difficult to achieve stable operation.

An alternative approach to the recirculating still is one in which the condensate is revaporized prior to return to the boiler. In this case the equilibrium vessel is immersed in a thermostat at the appropriate tempera-

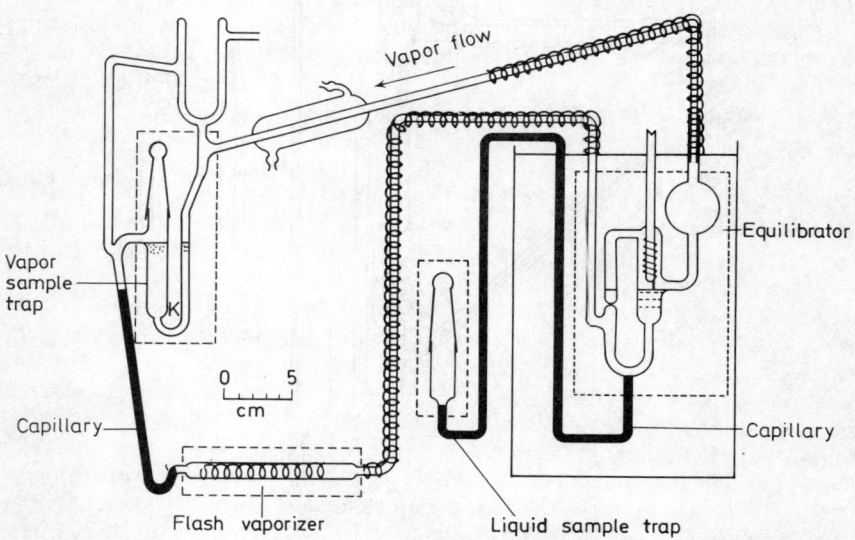

Figure 9. Vapor recirculation apparatus of McGlashan, Prue and Sainsbury[52].

ture, usually very slightly above the boiling temperature. The vapor is separated from the liquid and passes through heated lines to a condenser and hold-up trap. The condensate is returned from the hold-up trap through a vaporizer and thence to the Cottrell pump where liquid is pumped over the thermometer well. An example of this type of apparatus used by McGlashan, Prue and Sainsbury[52] and based on the designs of Prigogine and Desmyter[59] and of Kretschmer, Nakowska and Weibe[39] is shown in *Figure 9*. A variation on this type of apparatus in which the analysis of the mixtures was carried out by gas density measurements and therefore does not require a condensed sample, is described by Dunlap et al.[19]. In this apparatus, vapor was pumped through a one-liter vapor space which could be isolated from the rest of the equilibrium apparatus and connected to the gas density apparatus, and was then recirculated through the liquid. The whole apparatus was immersed in the thermostat. An arrangement of valves allowed a small quantity of the liquid to be isolated and vaporized into the gas density balance for analysis.

Dynamic methods are usually successful when the relative volatility of the mixture is not too high. At high relative volatilities, however, the recirculating stills tend to become unstable in operation. The other defect of most of the recirculating devices is that they are tedious to operate and take some hours to achieve equilibrium.

2. Static Methods

When one component is very involatile, as is the case, for example, in polymer solutions, then the relative volatilities become extremely high and the dynamic methods are of little use. On the other hand, when one component of a binary mixture is virtually involatile, the vapor phase consists entirely of the volatile component and there is no need to obtain a sample for analysis of the equilibrium composition. In this case, the composition of the liquid phase can readily be deduced from a knowledge of the total amounts of the two components in the system, the vapor pressure and the vapor volume.

In a simple static measurement, the mixture is merely confined to a vessel attached to a manometer and immersed in the thermostat. The mixtures are generally prepared synthetically by weighing the components into the vapor pressure cell. This eliminates the need for analysis of the mixtures. The chief difficulty with static methods is that it is necessary completely to remove all air from the system. This degassing of the system and of the materials is the major experimental difficulty. With volatile compounds, degassing can be achieved by multiple distillations in a vacuum with the condensate cooled by dry ice or liquid nitrogen. The air is removed by pumping after each distillation step. Usually six to twelve such distillations are required. Recently a good deal of attention has been paid to rapid efficient degassing methods and the reader is directed to the publications of Battino et al.[5] and of Bell et al[7]. Bell et al. report very efficient degassing by a sublimation method in which losses of material can be as low as five per cent.

Involatile compounds must be degassed by the less efficient method of successive freezing, pumping and melting.

The essential features of a static vapor pressure apparatus are the vapor pressure cell, thermostat, a manometer and a means of introducing the components into the vapor pressure cell. Typical methods of arranging the apparatus are shown in *Figure 10*. If measurements are to be made only at temperatures below ambient, then no distillation will occur from the cell and the system within the thermostat can be connected directly to an external manometer as in *Figure 10*(a). If measurements are to be made at temperatures higher than ambient, then the whole of the vapor phase must be confined to the thermostat. This can be achieved by putting the whole of the manometer in the thermostat as in *Figure 10*(b) or by introducing a low-range manometer in the thermostat and transferring the pressure to an external manometer by means of a gas column as in *Figure 10*(c).

Generally, for systems with vapor pressures below atmospheric, the mercury manometer is the most convenient and certainly the most widely used method of measuring the vapor pressure. In methods of type *10*(c), a narrow-range mercury manometer is often used in the thermostat. However, with the advent of precise mechanical differential pressure gages with sensitivities of ±0.02 torr or better, and the ability to withstand large over-range pressures, these devices are becoming more popular.

Volatile compounds are usually introduced by arranging for the manometer to be drained to provide a path to a vacuum line from which the material may be distilled into the cell. Typical arrangements of the manometer draining system are shown in *Figures 14* and *16*. More recent designs have used greaseless, all-metal or Teflon vacuum valves to obtain direct connection from the cell to the loading line.

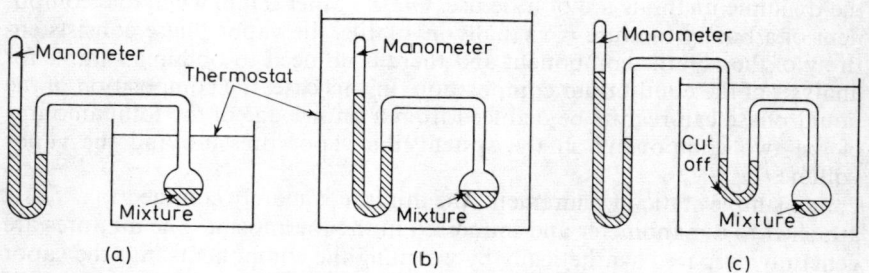

Figure 10. Types of static vapor pressure apparatus for mixtures.

Baxendale, Enüstün and Stern[6] used an apparatus of the type illustrated in *Figure 10*(c) for their studies of mixtures of benzene (volatile) and biphenyl (involatile). The actual apparatus is shown in *Figure 11*. The involatile component is weighed into the cell B_3 which is attached to the apparatus at the mercury sealed ground-glass joint. The mercury in the cut-off G is lowered and the involatile component is degassed by vacuum. The volatile component is degassed by multiple distillations between B_1 and B_2 and finally into B_2 where it is weighed. It is then distilled into B_3. With the cell cooled in dry ice, the mercury is raised in the cut-off G until it just touches the pointer. The thermostat is then filled and the temperature adjusted to the required value and the pressure adjusted in the connecting link between the manometer M

PART 1. PHASE EQUILIBRIA OF TWO-COMPONENT SYSTEMS

and the cut-off G until the mercury in G again just touches the pointer. A small correction for the effect of temperature on the apparent zero of G was necessary.

The apparatus of Allen, Everett and Penney[1] shown in *Figure 12*, illustrates

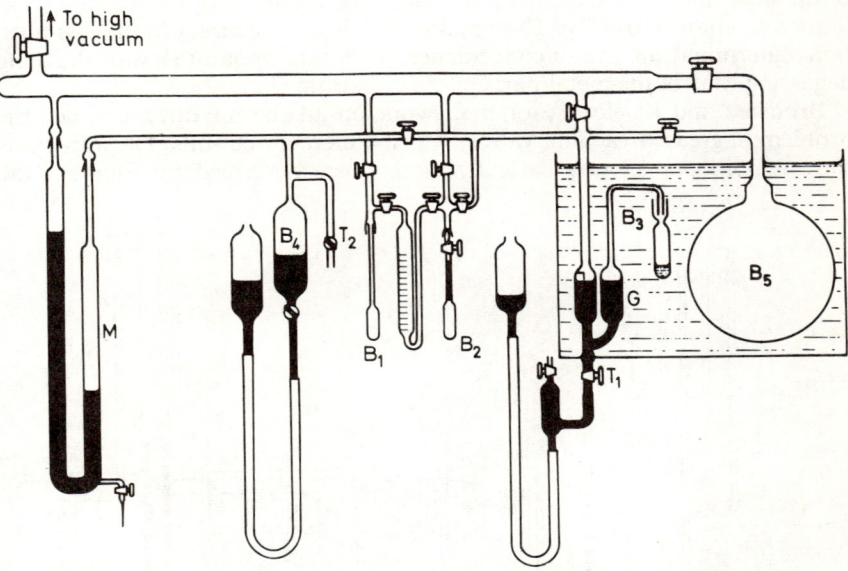

Figure 11. The static vapor pressure apparatus of Baxendale, Enüstün and Stern[6].

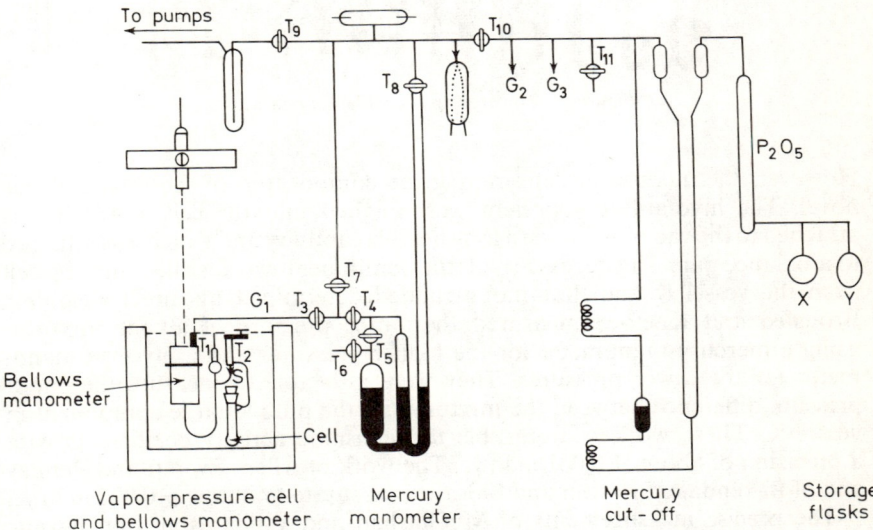

Figure 12. General view of the apparatus of Allen, Everett and Penney[1].

an early use of the all-metal vacuum tap and a mechanical null gage using a bellows as the sensitive element and an optical magnification of the movement of the bellows. The two components were introduced to the cell by first weighing the involatile component and degassing, followed by distillation into the cell of the volatile component and reweighing the cell. Because of the high mass of the cell and null gage, the accuracy of the weighing was limited to about ± 0.003 g. This resulted in a loss of accuracy in the composition determination. An inconvenience with this apparatus was the long degassing time of the metal parts of the apparatus.

Brønsted and Koefoed[9], for their work on alkane mixtures, avoided the problem of greased vacuum valves by using mercury cut-offs. The apparatus shown in *Figure 13* was basically of the type described by *Figure 10*(a).

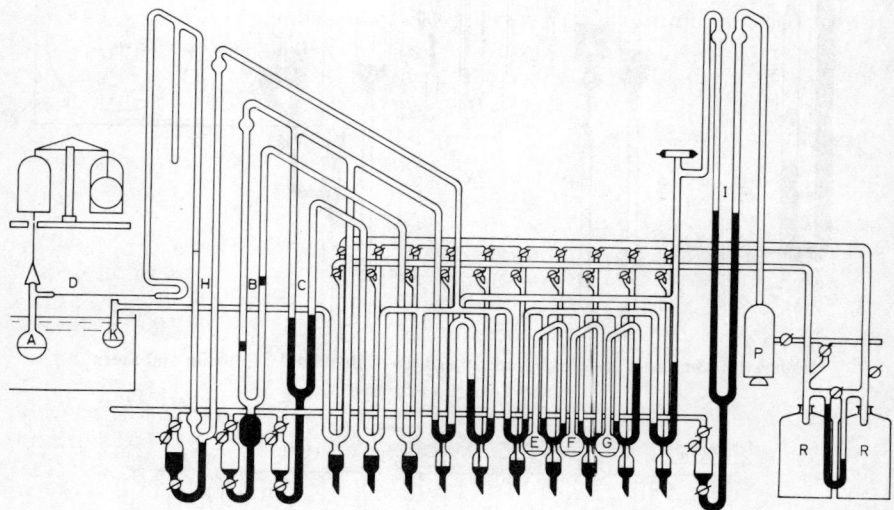

Figure 13. Brønsted and Koefoed's apparatus[9].

However, the method of determining the composition of the mixtures was novel. The involatile component was weighed into the cell A which was attached to the vacuum line via a long flexible capillary and which was attached to a balance pan. The degassed volatile component was distilled into the cell from the vessel K and the amount added determined by direct weighing. Brønsted and Koefoed measured the vapor pressure of alkane mixtures using a mercury manometer for the higher pressures and a glycerin manometer for the lower pressures. They were also able to measure the vapor pressure difference between the mixture and the pure volatile component in vessel K. These workers were able to determine activity coefficients with a precision of about ± 0.001 in $\ln \gamma_1$. The work of Allen, Everett and Penney and of Baxendale, Enüstün and Stern was accurate to about ± 0.002 in $\ln \gamma_1$.

The precise measurements of McGlashan and Williamson[53] on hexane and hexadecane were carried out in an apparatus of the type illustrated in

PART 1. PHASE EQUILIBRIA OF TWO-COMPONENT SYSTEMS

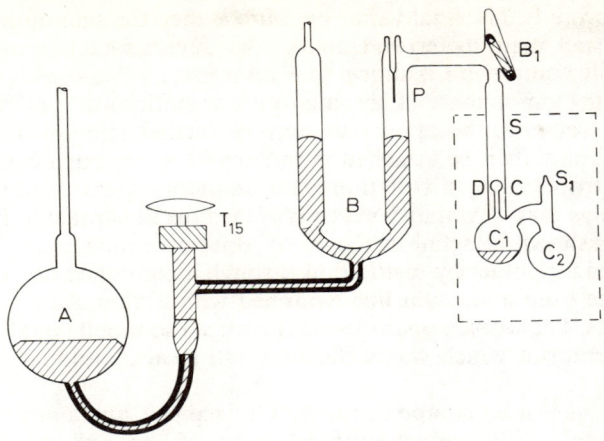

Figure 14. The cell and mercury cut-off of McGlashan and Williamson's apparatus[53].

Figure 10(c). These workers used a mercury cut-off and both compounds were introduced to the apparatus from pre-weighed ampoules equipped with break seals. The apparatus is shown in *Figures 14* and *15*.

The 'involatile' compound (hexadecane) was placed in the weighed double bulb C and attached to the vacuum line via a ground-glass joint at S_1. After degassing by distillation between C_1 and C_2, the cell was sealed at S_1 and the mass of degassed hexadecane determined by weighing. The cell was then attached to the apparatus at S and evacuated via B. The break seal D was then broken with the aid of the magnetically operated breaker B_1, and a pre-weighed amount of the volatile component (hexane) distilled into the cell. The mercury was then raised in B via tap T_{15}. The movable thermostat was raised around the apparatus and the pressure on the connecting line between B and an external manometer was adjusted to give only a small

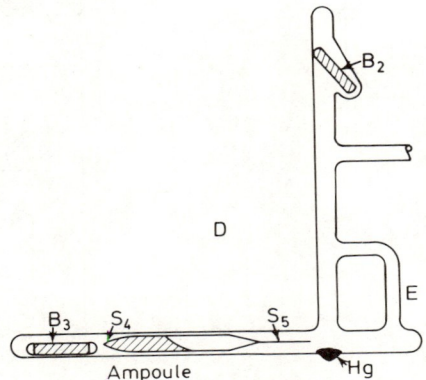

Figure 15. Device for introducing weighed quantities of volatile liquids[53].

difference across B. The final vapor pressure is then the sum of the pressures on the external manometer and on B. The device for opening ampoules of the volatile component is shown in *Figure 15*.

When both components of the mixture are sufficiently volatile to distil readily in a vacuum, the apparatus may be further simplified. The vapor pressure cell may then be attached permanently to the apparatus and both materials introduced by distillation. The simplest apparatus of this type of recent design is that of Marsh[45] which is of the type illustrated in *Figure 10*(a) and which is shown in detail in *Figure 16*. Both components are introduced from weighed ampoules by distillation through a combined manometer and cut-off device from a vacuum line equipped with Teflon/glass valves at the critical points. The whole apparatus was arranged so that it could be lowered into the thermostat which was sufficiently tall to accommodate the manometer.

Watson *et al.*[78] used an apparatus with a fixed cell, and a multi-turn glass Bourdon gage[82] with capacitance detection of the null position. Their

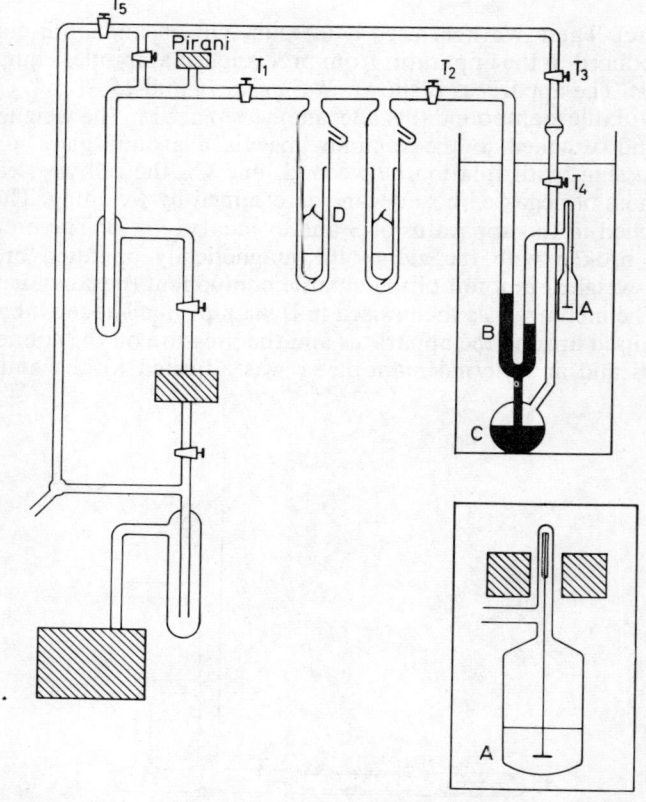

Figure 16. Marsh's apparatus for static vapor pressure measurements on mixtures of two volatile components[45].

PART 1. PHASE EQUILIBRIA OF TWO-COMPONENT SYSTEMS

apparatus is shown in *Figure 17*. One of the difficulties encountered with this apparatus arose from capillary condensation in the spiral. This was eliminated by maintaining the spiral at a temperature approximately 1°C above that

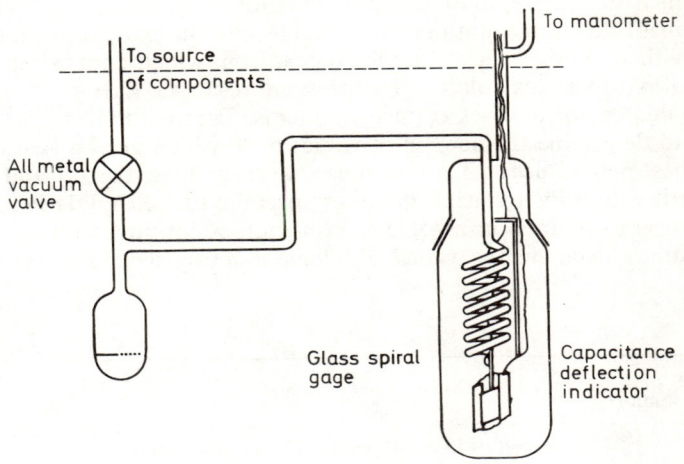

Figure 17. The cell and glass spiral null gage of Watson et al.[78].

of the rest of the apparatus. An advantage of using a mechanical cut-off is that it eliminates the difficulties associated with the reaction of some materials with the mercury in the manometer. Problems of this kind have frequently been encountered; for example, in work involving the chlorinated hydrocarbons[8,42]. Apparatus using similar techniques to those described here have been used by Scatchard, Wilson and Satkiewicz[69] who used a bellows null gage between the vapor pressure cell and the external manometer, and enclosed their cell and manometer in a very well-designed air thermostat.

When both components are volatile, the measurement of total vapor pressure, liquid composition and temperature are sufficient to characterize the system. However, the calculation of the individual activity coefficients is somewhat more complicated than when the vapor compositions are known. Barker[3] has described a method of successive approximations which can be used to establish the parameters in an empirical equation for the excess Gibbs free energy of mixing as follows. The partial pressures can be written in the form (see equations 24 and 25)

$$p_1 = \gamma_1 x_1 p_1^\circ \exp(V_1^\circ - \beta_{11})(p - p_1^\circ)/RT - p\Delta y_2^2/RT \qquad (67)$$

and

$$p_2 = \gamma_2 x_2 p_2^\circ \exp(V_2^\circ - \beta_{11})(p - p_2^\circ)/RT - p\Delta y_1^2/RT \qquad (68)$$

and the total pressure is then

$$p = p_1 + p_2 \qquad (69)$$

Barker chooses to express the excess Gibbs free energy by equation 35

$$G^E/RT = x_1 x_2 \sum_{i=1}^{m} A_i (1 - 2x_1)^{i-1} \qquad (70)$$

from which the corresponding expressions for γ_1, γ_2 can be derived. These when substituted into equations 67, 68 and 69 give the expression for $(\partial p/\partial A_i)$. We may then choose trial values for the A_i from which approximate values of γ_i, p, $(\partial p/\partial A_i)$ are calculated. The pressure residuals $R = p_{\text{expt}} - p_{\text{calc}}$ are then evaluated for a series of measurements. The method of least squares is used to determine the new set of values of A_i which gives a best fit to the data. These new values of A_i are then used to re-evaluate the γ_i, p and $(\partial p/\partial A_i)$ for a further iteration (which is necessary because the second derivatives of p with respect to A_i are non-zero). The procedure is continued until no significant changes occur in the values A_i. Table 3 shows the results of Barker's

Table 3

x_1	$p_{\text{expt}}/$ mmHg	$p_{\text{expt}} - p_{\text{calc}}$/mmHg			y_1 (calc)	y_1 (calc)
		$A_0 = 0.3859$ $A_1 = A_2 = 0$	$A_0 = 0.3592$ $A_1 = 0.0845$ $A_2 = 0.0366$	$A_0 = 0.3620$ $A_1 = 0.0891$ $A_2 = 0.0261$		
0.0464	454.62	−3.35	−0.48	−0.11	0.0993	0.0988
0.0861	476.25	−5.73	−0.60	−0.15	0.1734	0.1729
0.2004	534.38	−7.30	+0.25	+0.31	0.3465	0.3473
0.2792	569.49	−7.04	−0.11	−0.42	0.4404	0.4412
0.3842	613.53	−3.27	+1.08	+0.44	0.5458	0.5464
0.4857	650.16	−2.72	+1.07	−0.31	0.6310	0.6304
0.5824	679.74	+1.51	−0.33	−0.87	0.7016	0.7009
0.6904	708.78	+4.37	+0.15	−0.14	0.7749	0.7759
0.7842	729.77	+5.29	+0.32	+0.21	0.8363	0.8384
0.8972	748.46	+3.95	+0.41	−0.05	0.9148	0.9144

calculations for three iterations on the data of Brown[10] for the system benzene and n-heptane at 80°C.

In their static measurements on benzene and cyclopentane, Hermsen and Prausnitz[30] arranged the apparatus so that a sample of the liquid could be withdrawn from the cell and analysed. They then used the method just described to establish the equation describing G^E. However, if the vapor phase is kept small and if the total volume of the cell is known, then only the total composition of the system (liquid and vapor) need be known. One may then assume as a first approximation that the liquid composition is the same as the overall composition, and calculate an expression for G^E as described. This is then used to estimate the vapor compositions and, knowing the cell and liquid volumes, the amounts of each component in the vapor phase. The corrected liquid compositions thus determined are used in a further iteration of Barker's procedure and new values of A_i determined. Repetition of this procedure leads to the accurate values of A_i. Tedious as this double iteration procedure may seem, it is quickly and easily carried out on a computer and has been used successfully both by Marsh[45] and by Watson et al.[78]. McDermott and Ellis[50] have described in detail the extension of Barker's method to ternary liquid–vapor equilibrium measurements.

VII. Static Measurements with Analyses of Both Phases

With the advent of gas–liquid chromatography, it became possible to carry out reasonably accurate (better than one per cent and approaching 0.1 per cent) analyses of mixtures of volatile compounds using very small samples (down to 0.001 cm³ of liquid or 0.1 cm³ vapor). It is thus possible to use a static method without a large vapor phase to obtain T, p, x, y data. The method is also admirably suited to the determination of multicomponent liquid–vapor equilibrium data where previously physical methods of analysing mixtures were both tedious and not very accurate. Wichterle and Hála[79] used a low-pressure cell (shown in *Figure 18*) with a small vapor space and a

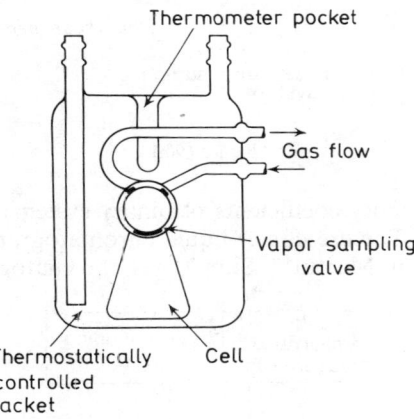

Figure 18. The miniature static vapor pressure apparatus of Wichterle and Hála designed to permit gas chromatographic analysis of the vapor[79].

specially designed sample valve to direct a small quantity of vapor into a gas–liquid chromatograph. The amount of material in the vapor phase is sufficiently small that the cell can be loaded with a mixture of known composition and it can be assumed that the liquid composition is the same as the overall composition. The apparatus used by Dastur and Thodos[17] in their study of the ethane–*n*-pentane–*n*-heptane system up to 174°C and up to 75 bar is an example of the use of gas–liquid chromatographic analysis for multicomponent high-pressure studies with direct analyses of both phases.

A very elegant apparatus for high-pressure measurements has recently been described by Rogers and Prausnitz[63] and is shown diagrammatically in *Figure 19*(a, b, c).

A more interesting use of gas–liquid chromatography is in the direct

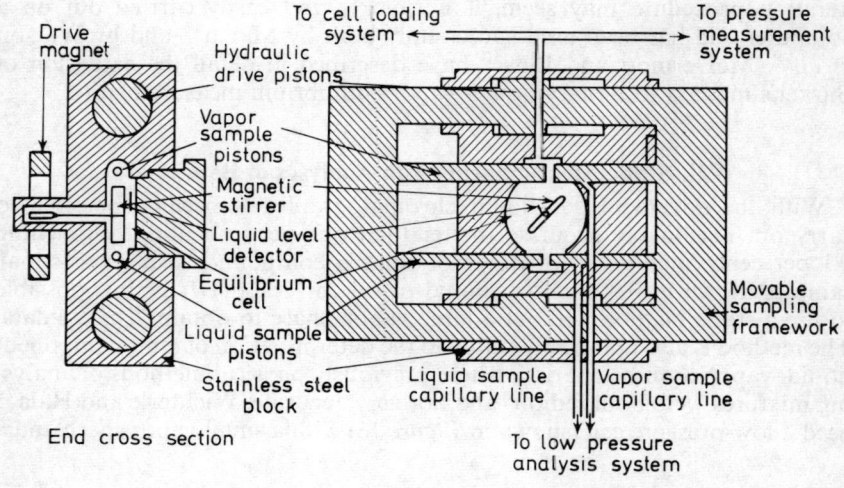

Figure 19(a)

determination of activity coefficients of binary systems in which one component is involatile. The use of gas–liquid chromatography for this purpose was first suggested by Martin[46]. Simple gas chromatographic theory gives

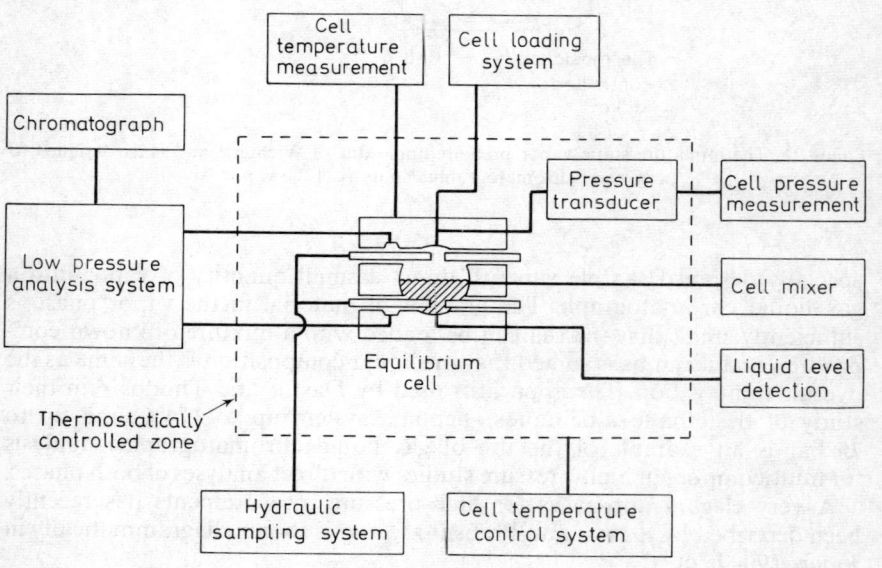

Figure 19(b)

PART 1. PHASE EQUILIBRIA OF TWO-COMPONENT SYSTEMS

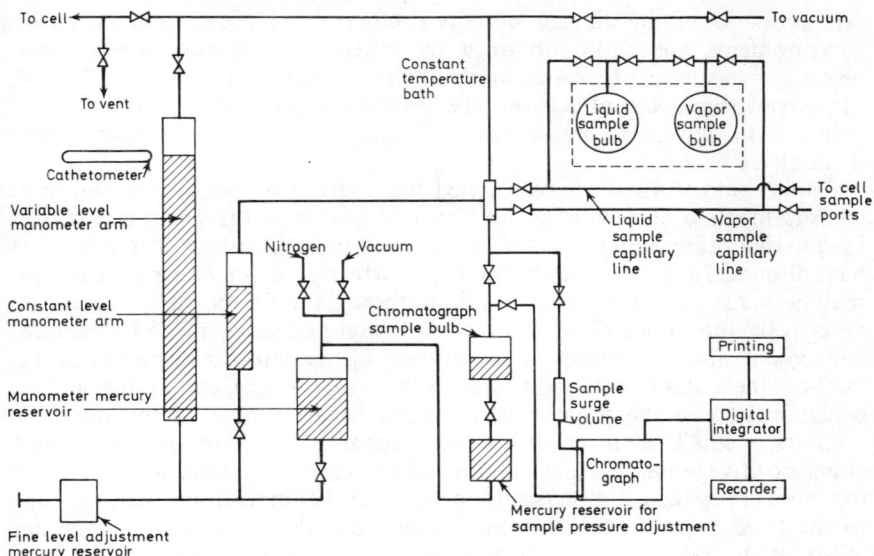

Figure 19(c)

Figure 19. Schematic layout of (a) high pressure cell, (b) auxiliary equipment, and (c) low-pressure analysis, apparatus of Rogers and Prausnitz[63].

the expression for the retention volume V_g of a volatile component at negligible concentration in the stationary phase

$$V_g = RT/Mp_1^\circ \gamma_1^\circ \tag{71}$$

where M is the molecular weight of the stationary liquid, p_1° is the vapor pressure of the volatile solute, and γ_1° is the activity coefficient of the volatile component in the stationary phase at infinite dilution. V_g is the retention volume of the sample corrected for the dead space in the injection and detection region and for the pressure variation in the column. Following earlier studies[21, 24, 41] in which the simple expression was used to determine activity coefficients with fair success, Cruickshank, Windsor and Young[16] carried out a detailed analysis of the column behavior and derived equations which allow for the pressure drop in the column and for the non-ideality of the gas phase. These workers obtained excellent agreement between the extrapolated

Table 4

System	$T/°C$	$\ln \gamma^\infty$ g.l.c.	$\ln \gamma^\infty$ v.p.
n-hexane + n-hexadecane	20	$\bar{1}.956$	$\bar{1}.958$
	30	$\bar{1}.956$	$\bar{1}.954$
	60	$\bar{1}.949$	$\bar{1}.952$
n-hexane + dinonyl phthalate	37.4	0.076	0.070
	50	0.063	0.060
	65.4	0.056	0.050

zero pressure infinite dilution activity coefficients obtained from their g.l.c. measurements and those obtained by other workers from direct vapor pressure measurements. An example of this agreement is shown in *Table 4*. Moreover, they were able, from the pressure coefficients of the retention volumes, to determine reasonably accurate data on the gas imperfections themselves.

Conder and Purnell[12] have carried the analysis of gas chromatographic measurements a stage further and have shown how the technique can be used to determine activity coefficients at finite concentrations. These workers have discussed in detail a number of ways in which experimental measurements may be made and interpreted to obtain these data. Of the various methods described, elution of a peak on a concentration plateau seems to be the most convenient since it requires no calibration of the detector response. In this method the column is equilibrated with a carrier gas containing a fixed concentration of the volatile component. Measurement of the retention volumes of small perturbations of the concentration of volatile component, either positive (volatile injection) or negative (carrier gas injection), then lead to values of γ in the solution (stationary phase) of composition corresponding to the fixed vapor concentration. Conder and Purnell claim an accuracy of about ± 1 per cent in γ which is comparable with or better than many of the static vapor pressure measurements with which they compare their results. The chief advantages of the gas chromatographic method are its speed and the fact that the accuracy of the results varies little with concentration, whereas in static methods the accuracy of the data (in terms of γ) falls off badly at low solute concentrations. On the other hand, gas chromatographic methods cannot be used successfully at solute concentrations corresponding to p_2/p_2° much greater than about 0.7. The g.l.c. methods thus form a valuable complement to the more conventional methods for binary systems with one volatile component. Stalkup and Deans[73] have examined the high-pressure case where solubility of the carrier gas in the stationary phase must be allowed for and Stalkup and Kobayashi[74] and Yadorich, Robinson and Chao[81] have used their treatment in high-pressure studies of the systems, ethane, propane and n-butane at infinite dilution in the methane–n-decane system using methane carrier gas at high pressure.

VIII. The McBain Balance Method

For binary mixtures with only one volatile component, it is possible to use the technique originally developed by McBain[49] for gas–solid adsorption measurements. In this method a weighed quantity of the non-volatile component is placed on the pan of a spring balance (usually a quartz spiral) and thoroughly degassed. The volatile component is then introduced into the vessel containing the spring balance and when the system has reached equilibrium, the amount of volatile component in the liquid phase and hence the liquid composition can be determined from the extension of the spring. The vapor pressure is measured on a manometer attached to the equilibrium vessel. This technique has been used by a number of workers, including Ashworth and Everett[2] whose apparatus[22] is shown in *Figure 20*. This technique has been particularly popular with workers interested in liquid–

PART 1. PHASE EQUILIBRIA OF TWO-COMPONENT SYSTEMS

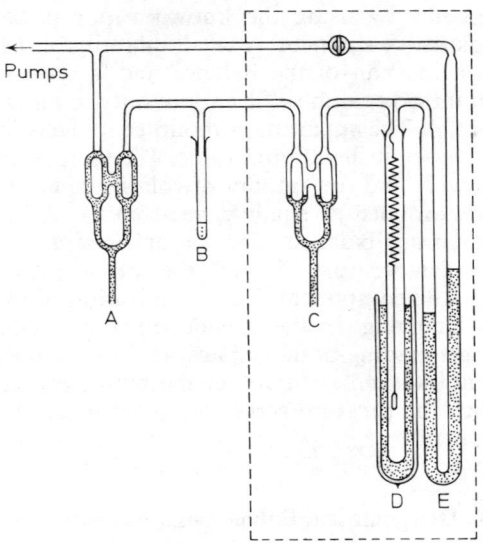

Figure 20. The McBain balance apparatus of Ashworth and Everett[2]; A and C are mercury float valves, B is the volatile component, D is the quartz spiral and E is the manometer.

vapor equilibrium data related to gas chromatographic work[47] and is routinely capable of an accuracy in γ of about ± 1 per cent.

An interesting variation on this approach was used by van der Waals[77] in his studies of the system n-heptane and polyethylene at 109°C. The van der Waals method which is shown *Figure 21* avoids the need for precise

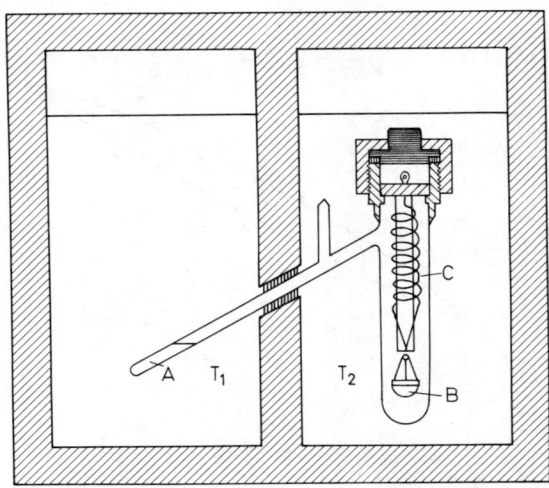

Figure 21. The van der Waals apparatus: A, volatile component; B, mixture; C, quartz spring balance.

pressure measurements by using the known vapor pressure/temperature relation for the volatile component. A small quantity of the involatile component is placed in the pan of the balance and is degassed. The volatile component is then introduced in sufficient quantity to allow excess liquid to be present in the arm A. The apparatus is mounted in the twin thermostat and the temperature T_2 is set at the desired value. The temperature T_1 is then set to some value below T_2 and the amount of volatile component in the liquid mixture phase is determined at equilibrium from the extension of the previously calibrated spring balance. The vapor pressure of the mixture is deduced from the temperature T_1 and the vapor pressure/temperature relation for the volatile component. The composition of the liquid mixture can be altered by changing T_1 and a wide range of compositions can be studied using a single loading of the apparatus. The accuracy of the method depends on the stability and accuracy of the temperatures T_1, T_2 and the accuracy with which the pressure versus temperature relation is known for the volatile component.

IX. Dewpoint and Bubble-point Measurements

The direct methods described so far have been essentially ones in which conditions are arranged so that both phases are present at equilibrium and the compositions of one or both phases are determined in some way. This is, in effect, establishing the phase boundary conditions for a number of horizontal sections through *Figure 22* of which the line AB is an example. It is, however, equally possible to establish a composition in the one-phase region and, by slowly altering the pressure, to traverse through the diagram in the direction CD. On such a traverse, the phase boundaries would appear as

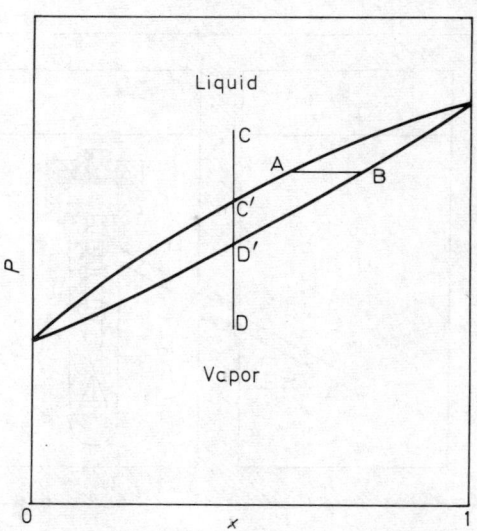

Figure 22. Comparison of direct vapor pressure and dewpoint methods.

PART 1. PHASE EQUILIBRIA OF TWO-COMPONENT SYSTEMS

discontinuities in the pV isotherm at the points C′, D′. An experiment beginning in the one-liquid region, say at C, with slowly reducing pressure, would provide a discontinuity at C′, the bubble point, and another at D′, the dewpoint. The experiment could, of course, equally well be performed in

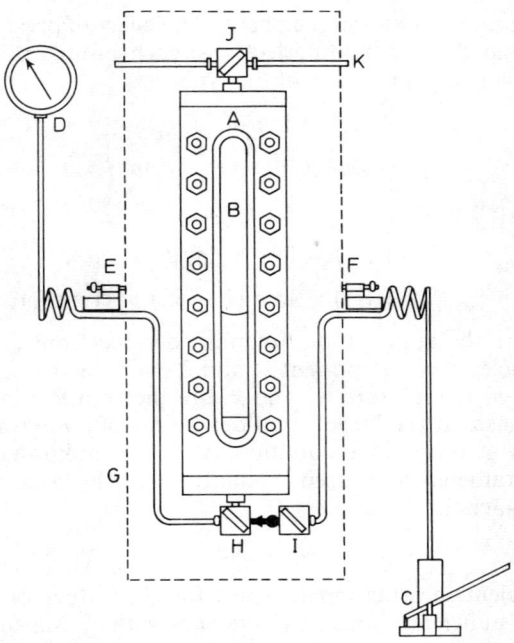

Figure 23. High pressure apparatus of Katz and Kurata[36].

the direction DC to provide first the dewpoint and then the bubble point. A series of such experiments carried out at different overall compositions would define the whole liquid–vapor phase diagram.

This method has been widely used in chemical engineering studies, particularly those at high temperature and pressure. A typical apparatus for such studies is that of Katz and Kurata[36], shown in *Figure 23*. The mixture is loaded into the pressure tube through the valve J; the pressure is varied with the mercury pump C and is measured on the Bourdon gage D. A similar procedure, but with a rather more sophisticated loading technique and with pressures measured on a deadweight gage was used by Kay and Ranbosch[37] in their studies of the propane–hydrogen sulfide system.

Dixon and McGlashan[18] have proposed an interesting method based on the observation of dewpoints and bubble points. These workers point out that it is possible to use a set of dewpoints and bubble points to determine the excess Gibbs free energy of mixing and thus the liquid–vapor diagram without a knowledge of the composition of the system. If both the dewpoint and the bubble point are observed for each of a series of mixtures and if one assumes that the excess Gibbs free energy can be adequately represented by an empirical expression, say of the type

$$G^E = x(1-x) \sum_{i=1}^{n} A_i(1-2x)^{(i-1)} \qquad (72)$$

then to a first approximation (neglecting the effect of pressure on the activity coefficients and the vapor imperfections) each bubble point may be represented by a pair of expressions of the type:

$$f_\alpha(x_1) = \ln p_b y_1 / p_1^\circ x_1 \qquad (73)$$

$$f_\beta(x_1) = \ln p_b(1-y_1) p_2^\circ \qquad (74)$$

and each dewpoint by:

$$f_\alpha(z_1) = \ln p_d x_1 / p_1^\circ z_1 \qquad (75)$$

$$f_\beta(z_1) = \ln p_d(1-x_1)/p_2^\circ (1-z_1) \qquad (76)$$

where f_α, f_β are the appropriate functions derived from equation 72 for the activity coefficients of components 1 and 2, x_1 is the overall mole fraction of component 1 in the mixture, y_1 and z_1 are the compositions of the gaseous and liquid phases at the bubble and dewpoints of a mixture of composition x_1. One thus acquires four equations in $3+n$ unknowns where n is the number of parameters to be used in equation 72. Any m pairs of dewpoint and bubble point such that

$$4m \geqslant 3m + n$$

provide sufficient information to solve for the values of n. Where $m = n$, there are just sufficient data and where $m > n$, there is a surplus of data and the n parameters may be chosen by the method of least squares. Dixon and McGlashan claim that the method is more convenient than, and at least as accurate as, the conventional methods. However, the experimental work which substantiates this claim has not yet been published[35].

X. The Isopiestic Method

In the isopiestic method a solution of known vapor pressure versus composition characteristics and containing only one volatile component is confined in a sealed vessel with a solution of unknown properties (but again with involatile solutes). The solvent will distil between the solutions until their compositions are such that they have the same vapor pressure. When equilibrium has been reached, samples of the two solutions are analysed. The analysis of the reference solution along with its known properties provides the vapor pressure and the vapor pressure versus composition data are thus established for the experimental solution.

This technique has been used extensively in the study of electrolyte solutions[62]. However, it has been relatively little used for non-electrolyte solutions. Corneliussen, Rice and Yamakawa[13] have used the technique in an interesting form for their studies of solutions of toluene with poly-p-methylstyrene and a number of other substituted styrenes and it has also been used by Christian, Affsprung and Johnson[11] for benzene plus water mixtures at high benzene concentrations.

XI. Differential Methods

Where the vapor pressure of the solvent is known accurately, the differential method may be used. This technique is particularly useful for dilute solutions of an involatile component in a volatile solvent. The basic apparatus is shown in *Figure 24* and, for dilute solutions in which the vapor pressure of the solu-

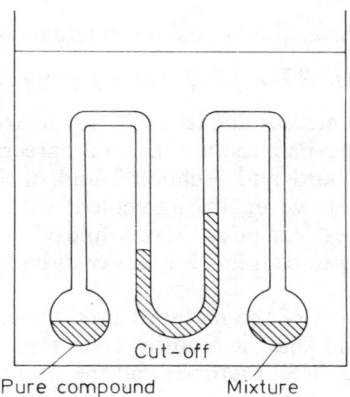

Figure 24. Schematic arrangement for differential vapor pressure method.

tion is not very different from that of pure solvent, the differential manometer is used, but a calibrated mechanical differential pressure gage of small range and high precision may also be used. The advantage of this type of equipment is that the expression (neglecting gas imperfections) for the activity coefficient in terms of the observations is for a single volatile.

$$\ln \gamma_1 = \ln \left[(p_1^\circ - \Delta P)/p_1^\circ x_1\right]$$

Since, usually, $\ln \gamma_1$ is much less temperature dependent than the vapor pressures themselves, the error due to temperature differences from one measurement to another among a set of supposedly isothermal measurements is much reduced compared with a set of direct total vapor pressure measurements. Recent uses of the differential method are illustrated by the work of Taylor and Rowlinson[75] on glucose–water solutions and of Malcolm and Rowlinson[43] on polyethylene–glycol–water mixtures. Kershaw and Malcolm[38] used an interesting modifiction of this method, incorporating a magnifying manometer of the type described by Sirianni and Puddington[72].

XII. Light-scattering Measurements

Light-scattering measurements have been widely used in the study of dilute polymer molecules for examining the microscopic state of the solution.

However, very little use has been made of this method for mixtures of small molecules.

Coumou and Mackor[15] have analysed the Rayleigh scattering of binary mixtures to obtain expressions for the activity coefficients:

$$\ln \gamma_1 = \int_0^x (1/x_1) \{1 - (R_{id}/R_c)\} \, dx_2 \qquad (77)$$

$$\ln \gamma_2 = \int_0^x (1/x_2) \{1 - (R_{id}/R_c)\} \, dx_1 \qquad (78)$$

where R_{id} is the ideal isotropic scattering,

$$R_{id} = (2\pi^2/N\lambda^4) \, Vx_1 x_2 (n \, \partial n/\partial x_2)^2 \qquad (79)$$

and R_c is the scattering due to composition fluctuations

$$R_c = (2\pi^2 RT/\lambda^4) \, Vx_1 \, (n \, \partial n/\partial x_2)^2/(\partial \mu_2/\partial x_2) \qquad (80)$$

Coumou and Mackor's measurements of G^E for mixtures of methanol and benzene and cyclohexane–benzene are in good agreement with the directly measured data of Scatchard and Tichnor[68] and of Scatchard, Wood and Mochel[70] respectively. However, the agreement with the data on benzene and neopentane mixtures[48] is poor. Myers and Clever[56] have used light-scattering measurements to examine a number of hydrocarbon and alcohol mixtures.

This is the most indirect of the methods used to determine excess Gibbs free energy of mixing and since it requires equipment and skills at least as complicated as the more direct methods and the results are, at least in some cases, doubtful, it is unlikely to achieve great popularity.

XIII. References

[1] Allen, G., D. H. Everett and F. M. Penney. *Proc. Roy. Soc.* A, **212**, 149 (1952).
[2] Ashworth, A. J. and D. H. Everett. *Trans. Faraday Soc.* **56**, 1609 (1960).
[3] Barker, J. A. *Austral. J. Chem.* **6**, 207 (1953).
[4] Battino, R. *Chem. Rev.* **71**, 5 (1971).
[5] Battino, R. and F. D. Evans. *Analyt. Chem.* **38**, 1627 (1966);
Battino, R., F. D. Evans and M. Bogan. *Analyt. Chim. Acta*, **43**, 518 (1968);
Battino, R., M. Banzhof, M. Bogan and E. Wilhelm. *Analyt. Chem.* **43**, 806 (1971).
[6] Baxendale, J. H., B. V. Enüstün and J. Stern. *Phil. Trans. Roy. Soc.* A, **243**, 169 (1951).
[7] Bell, T. N., E. L. Cussler, K. R. Harris, C. N. Pepela and P. J. Dunlop. *J. Phys. Chem.* **72**, 4693 (1968).
[8] Bottomley, G. A. and C. G. Reeves. *Trans. Faraday Soc.* **53**, 1455 (1957).
[9] Brønsted, I. J. N. and J. Koefoed. *Mat.-Fys. Medd., Danske Vid. Selsk*, **2**, No. 17 (1946).
[10] Brown, I. *Austral. J. Sci. Res.* **A5**, 530 (1952).
[11] Christian, S. D., H. E. Affsprung and J. R. Johnson. *J. Chem. Soc. London*, 1896 (1963).
[12] Conder, J. R. and J. H. Purnell. *Trans. Faraday Soc.* **64**, 1505 (1968); **64**, 3100 (1968); **65**, 825 (1969); **65**, 839 (1969).
[13] Corneliussen, R., S. A. Rice and H. Yamakawa. *J. Chem. Phys.* **38**, 1768 (1963).
[14] Cottrell, F. G. *J. Amer. Chem. Soc.* **41**, 721 (1919).
[15] Coumou, D. S. and E. L. Mackor. *Trans. Faraday. Soc.* **60**, 1539 (1964); **60**, 1726 (1964).
[16] Cruickshank, A. J. B., M. L. Windsor and C. L. Young. *Proc. Roy. Soc. London*, A, **295**, 259 and 271 (1966);
[17] Dastur, S. P. and G. Thodos. *Chem. Engng Sci.* **19**, 935 (1964).
[18] Dixon, D. T. and M. L. McGlashan. *Nature, London*, **206**, 710 (1965).
[19] Dunlap, R. D., R. G. Bedford, J. C. Woodbrey and S. D. Furrow. *J. Amer. Chem. Soc.* **81**, 2927 (1959).

PART 1. PHASE EQUILIBRIA OF TWO-COMPONENT SYSTEMS

[20] Ellis, S. R. M. *Trans. Inst. Chem. Engrs, London*, **30**, 58 (1952).
[21] Everett, D. H. and C. T. H. Stoddart. *Trans. Faraday, Soc.* **57**, 746 (1961).
[22] Everett, D. H. and W. I. Whitton. *Proc. Roy. Soc. A*, **230**, 91 (1955).
[23] Fowler, R. J. *The Industrial Chemist.* 717 (November 1948) and 824 (December 1948).
[24] Freeguard, G. F. and R. Stock. *Gas Chromatography*, p 103. Butterworths: London (1962).
[25] Gillespie, D. T. C. *Industr. Engng Chem. (Analyt.)* **18**, 575 (1946).
[26] Guggenheim, E. A. *Trans. Faraday Soc.* **33**, 151 (1937).
[27] Guggenheim, E. A. *Thermodynamics*, 3rd ed. p 215–7. North Holland: Amsterdam (1957).
[28] Guggenheim, E. A. *Thermodynamics*, 3rd ed., p 230–1. North Holland: Amsterdam (1957).
[29] Guggenheim, E. A. *Thermodynamics*, 3rd ed., p 250 et seq. North Holland: Amsterdam (1957).
[30] Hermsen, R. W. and J. M. Prausnitz. *Chem. Engng Sci.* **18**, 485 (1963).
[31] Herrington, E. F. G. *J. Appl. Chem.* **18**, 285 (1968).
[32] Herrington, E. F. G. *Instn Chem. Engrs Symposium Series*, **32**, 3 and 17 (1969).
[33] Hildebrand, J. H. *Solubility of Non-electrolytes*, 3rd ed., p 149. Reinhold: New York (1950).
[34] Hill, T. L. *Introduction to Statistical Thermodynamics*, p 81. Addison–Wesley: London (1960).
[35] Judd, N. F. and M. L. McGlashan. Private communication.
[36] Katz, D. L. and F. Kurata. *Industr. Engng Chem.* **32**, 817 (1940).
[37] Kay, N. B. and G. M. Ranbosch. *Industr. Engng Chem.* **45**, 221 (1953).
[38] Kershaw, R. W. and G. N. Malcolm. *Trans. Faraday Soc.* **64**, 323 (1968).
[39] Kretschmer, C. B., J. Nakowska and R. Weibe. *J. Amer. Chem. Soc.* **70**, 1785 (1948).
[40] van Laar, J. J. *Z. Phys. Chem.* **72**, 723 (1910).
[41] Langer, S. H. and J. H. Purnell. *J. Phys. Chem.* **67**, 263 (1963).
[42] Malcolm, G. N. and R. W. Kershaw. *Trans. Faraday Soc.* **64**, 323 (1968).
[43] Malcolm, G. N. and J. S. Rowlinson. *Trans. Faraday Soc.* **53**, 921 (1957).
[44] Margules, M. *SB Akad. Wiss. Wien. Math.-Nat.* Kl. [2], **104**, 1243 (1895).
[45] Marsh, K. N., *Trans. Faraday Soc.* **64**, 883 (1968).
[46] Martin, A. J. P. *Analyst*, **81**, 52 (1956).
[47] Martire, D. E., R. L. Pecsok and J. H. Purnell. *Trans. Faraday Soc.* **61**, 2496 (1965).
[48] Mathot, V. and A. Desmyter. *J. Chem. Phys.* **21**, 782 (1953).
[49] McBain, J. W. and A. M. Bakr. *J. Amer. Chem. Soc.* **48**, 690 (1926).
[50] McDermott, C. and S. R. M. Ellis. *Chem. Engng. Sci.* **20**, 545 (1965).
[51] McGlashan, M. L. *Experimental Thermochemistry*, Vol. II, Chap. 15, Interscience: New York (1962).
[52] McGlashan, M. L., J. E. Prue and J. E. J. Sainsbury. *Trans. Faraday Soc.* **50**, 1284 (1954).
[53] McGlashan, M. L. and A. G. Williamson. *Trans. Faraday Soc.* **57**, 588 (1961).
[54] Mendelssohn, K. *The Quest for Absolute Zero.* Weidenfeld and Nicolson: London (1966).
[55] Myers, D. B. and R. L. Scott. *Industr. Engng Chem.* **55**, 43 (1963).
[56] Myers, R. S. and H. L. Clever. *J. Chem. Thermodyn.* **2**, 53 (1970).
[57] Prausnitz, J. M. *Molecular Thermodynamics of Fluid Phase Equilibria*, p 229. Prentice Hall: New Jersey (1969);
Orye, R. V. and J. M. Prausnitz. *Industr. Engng Chem.* **57**, 19 (1965).
[58] Pressure measurement, see chapter 4 in this volume, p 115.
[59] Prigogine, I. and A. Desmyter. *Trans. Faraday Soc.* **47**, 1137 (1951).
[60] Redlich, O. and A. T. Kister. *Industr. Engng Chem.* **40**, 341–5 (1948).
[61] Renon, H. and J. M. Prausnitz. *A.I.Ch.E. Journal*, **14**, 135 (1968).
[62] Robinson, R. H. and D. A. Sinclair. *J. Amer. Chem. Soc.* **56**, 561 and 830 (1934);
Stokes, R. H. *J. Amer. Chem. Soc.* **69**, 1291 (1947).
[63] Rogers, B. L. and J. M. Prausnitz. *Industr. Engng Chem. (Fundamentals)*, **9**, 174 (1970).
[64] Rowlinson, J. S. *Liquids and Liquid Mixtures*, 2nd ed., p 219. Butterworths: London (1969).
[65] Sameshima, J. *J. Amer. Chem. Soc.* **40**, 1483 (1918).
[66] Scatchard, G. *Chem. Rev.* **44**, 7 (1949).
[67] Scatchard, G., C. L. Raymond and H. H. Gilman. *J. Amer. Chem. Soc.* **60**, 1275 (1938).
[68] Scatchard, G. and L. B. Tichnor. *J. Amer. Chem. Soc.* **74**, 3724 (1952).
[69] Scatchard, G., G. M. Wilson and F. G. Satkiewicz. *J. Amer. Chem. Soc.* **86**, 125 (1964).
[70] Scatchard, G., S. E. Wood and J. M. Mochel. *J. Amer. Chem. Soc.* **62**, 712 (1940).
[71] Schneider, G., see Chapter 16 (Part 2) of this volume, p 787.
[72] Sirianni, A. F. and I. E. Puddington. *Canad. J. Chem.* **33**, 755 (1955).
[73] Stalkup, F. I. and H. A. Deans. *A.I.Ch.E. Journal*, **9**, 108 (1963).
[74] Stalkup, F. I. and R. Kobayashi. *A.I.Ch.E. Journal*, **9**, 121 (1963).
[75] Taylor, J. B. and J. S. Rowlinson. *Trans. Faraday Soc.* **51**, 1183 (1955).

[76] Temperature measurement, see Chapter 3 in this volume, p 87.
[77] van der Waals, J. H., *Thesis*, University of Groningen (1950).
[78] Watson, I. D., R. J. Knight, I. R. McKinnon and A. G. Williamson. *Trans. Faraday Soc.* **64**, 1763 (1968).
[79] Wichterle, I. and E. Hála. *Industr. Engng. Chem. (Fundamentals)*, **2**, 155 (1963).
[80] Wilson, G. M., *J. Amer. Chem. Soc.* **86**, 127 (1964).
[81] Yadorich, A., R. L. Robinson and K. C. Chao. *A.I.Ch.E. Journal*, **17**, 1152 (1971).
[82] York, S. G., *J. Sci. Instrum.* **22**, 196 (1954).

CHAPTER 16

Part 2. Phase Equilibria of Liquid and Gaseous Mixtures at High Pressures

G. M. SCHNEIDER
University of Bochum, Bochum, West Germany

Contents

I.	Introduction	787
II.	Basic Phase–Theoretical Aspects	788
III.	Discussion of General Procedures	789
	1. The Analytical Method	789
	2. The Synthetic Method	791
	3. Miscellaneous Methods	794
	A. Phase Behavior from pVT Data	794
	B. The Dynamic Method	795
	C. The Glass Bulb Technique	795
IV.	Description of Special Equipments	795
	1. Mercury-operated Apparatus of Krichevskii and Tsiklis	796
	2. Apparatus of Tsiklis and Maslennikova	796
	3. Apparatus of Tödheide and Franck	797
	4. Optical Cell Used by de Swaan Arons and Diepen	797
	5. Optical Cell Developed by Oeder and Schneider	797
	6. Optical Cell Developed by Alwani and Schneider	798
	7. Optical Cell Developed by Buback and Franck	798
	8. Apparatus Developed by Michels *et al.*	798
V.	Conclusions	799
VI.	References	800

I. Introduction

One of the most important properties of a mixture is its phase behavior. It may be of considerable practical interest for example, for separation methods and much information about the thermodynamics of a mixture can be deduced from its phase behavior even in ranges of temperature and pressure where direct calorimetric and volumetric measurements are difficult.

Especially for the investigation of gas–liquid equilibria at pressures below atmospheric pressure and up to several bars a large variety of apparatus has been described in literature that has been reviewed elsewhere (see Hála[5], Williamson[40]). In the same way for the pressure range up to some ten or hundred bars much apparatus has been developed especially in order to study the gas–liquid critical range (see Hála[5]).

For the investigation of liquid and gaseous mixtures at still higher pressures, that is to say up to several thousand bars, however, relatively little apparatus has been described up to now because of many additional experimental and constructional difficulties concerning materials, seals, stirring, optical windows, electrical connections, sampling, analysis etc. They are greatest for gas–gas equilibria where often highly corrosive substances have to be investigated under elevated temperatures and pressures (e.g. aqueous solutions above the critical pressure and temperature of pure water).

It is the aim of this review to discuss the experimental techniques for the investigation of the phase behavior of fluid mixtures at pressures above 300 bar. In part II some basic phase–theoretical aspects will be reviewed. In part III general experimental procedures will be discussed whereas some special equipments will be described in part IV.

II. Basic Phase–Theoretical Aspects

Whereas in one-component fluid systems only gas–liquid equilibria exist, three different types of two-phase equilibria have to be considered in fluid mixtures: gas–liquid, liquid–liquid, and the so-called 'gas–gas' equilibria. Afterwards only some phase–theoretical aspects that are necessary for the understanding of the experimental techniques and of some constructional details of the measuring devices will be briefly discussed, the discussion being limited to binary systems.

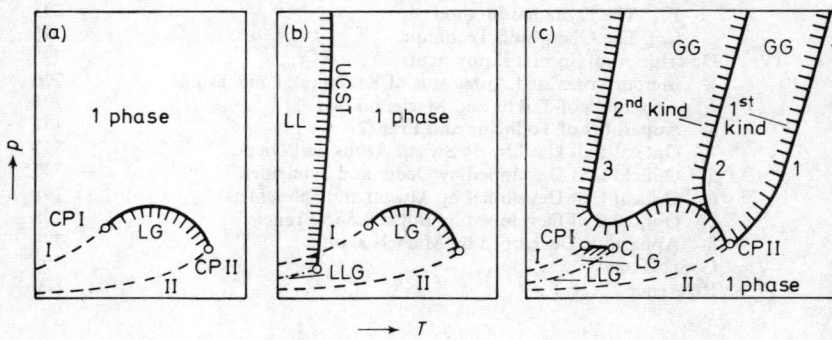

Figure 1. $p(T)$ projections of the phase diagrams of fluid binary systems (schematically, see text; CP = critical point): (a) System showing critical phenomena liquid–gas (LG), (b) System showing critical phenomena liquid–gas (LG) and liquid-liquid (LL) (UCST = upper critical solution temperature), (c) System showing gas–gas equilibria (GG) of the first kind (curve 1) and second kind (curve 2) (curve 3 corresponds to a transition type; C = critical end point).

The phase behavior and critical phenomena of fluid binary mixtures are most easily understood with the aid of pressure(p)/temperature(T) projections of the phase diagrams. Some particularly important types are represented schematically in *Figures 1*(a–c). The dashed curves are the vapor pressure curves of the pure components I and II respectively. The thick full lines are the projections of the critical curves; at temperatures and pressures beyond these curves, components I and II are miscible in all proportions.

For type 1a the *gas–liquid* critical curve is not interrupted and runs through a pressure maximum. Curves with a temperature minimum or maximum as well as with a monotonous or nearly linear shape have also been found. Since the critical pressure of pure components is

PART 2. PHASE EQUILIBRIA OF LIQUID AND GASEOUS MIXTURES

rarely higher than 100 bar, the pressure range necessary for these investigations is rather limited.

In *Figure 1*(b) a system is shown with a *gas–liquid* critical curve that is not interrupted and runs through the usual pressure maximum. At much lower temperatures additionally separation into two liquid phases (LL) is assumed, the upper critical solution temperature (UCST) rising steeply with increasing pressure. No superposition of the critical phenomena gas–liquid and *liquid–liquid* is found for systems of this type. More complicated types have also been found.

In *Figure 1*(c) $p(T)$-diagrams are schematically represented for systems showing *gas–gas* equilibria. Here the critical curves are interrupted and consist of two branches. The branch starting from the critical point CP I of the more volatile component I corresponds to gas–liquid equilibria (LG) and ends at a so-called critical end-point C on the three-phase line LLG where two liquid and one gaseous phases are in equilibrium whereas the branch beginning at the critical point CP II of the less volatile component either immediately tends to higher temperatures and pressures (curve 1; so-called 'gas–gas equilibria of the first kind') or goes through a temperature minimum first and then runs steeply to increasing temperatures and pressures (curve 2; so-called 'gas–gas equilibria of the second kind'). Curve 3 corresponds to a transition type between types 1 and 2 in *Figure 1*(c) and the type of *Figure 1*(b).

Gas–gas equilibria were predicted by van der Waals from the fold theory and were discussed in detail by Kamerlingh Onnes and Keesom in 1907. Type 2 was found for the first time in 1940 by Krichevskii and co-workers for the nitrogen–ammonia system. Type 1 was reported for the first time in 1952 by Tsiklis for the system helium–ammonia. In the meantime similar phase separation effects have been found in some forty other systems. The most elevated pressure being more than 14 kbar up to now obtained in the measurements for ammonia–nitrogen by Tsiklis. For a detailed discussion and examples see[18, 19, 22–25, 35–37].

It has been shown elsewhere[23–25] that there are continuous transitions between all types of two-phase equilibria in fluid mixtures at high temperatures and pressures. Thus gas–gas equilibria are equally as important as gas–liquid and liquid–liquid equilibria, gas–liquid equilibria being the type of two-phase equilibrium in fluid systems that is found at rather low pressures over a great range of temperatures near the vapor pressure curves of the pure components up to their critical points whereas liquid–liquid equilibrium is the normal type at low temperatures and relatively high pressures and gas–gas equilibrium the normal type at relatively high temperatures and high pressures.

Since this article is limited to high pressure work it will deal mainly with the experimental techniques that have been developed for the investigation of gas–gas and liquid–liquid equilibria.

III. Discussion of General Procedures

For the investigation of phase equilibria in fluid mixtures at high pressures essentially two different methods are used.

1. The Analytical Method

The autoclave is charged with the substances under test and values of temperature and pressure in the *heterogeneous* region are adjusted. In most cases effective stirring is necessary for equilibrating the two phases. Then the concentrations of the two phases in equilibrium are determined:

(a) by taking samples from each phase and analysing them outside the autoclave at normal or moderate pressure by an appropriate method.

(b) by determining the concentrations of the coexisting phases in the autoclave under pressure by an appropriate physicochemical method of analysis.

Advantages: Several isotherms (or isobars) and in special cases the whole system can be measured with one filling only. Systems with more than two components can be studied. The concentrations of the coexisting phases are determined directly.

Disadvantages: The method is not suitable near critical states or for barotropic systems where the equilibrating phases do not separate well. pVT data cannot be obtained simultaneously. Reliable sampling can be difficult especially if the substances under test show very different volatility, e.g. for gas–gas equilibria. Additionally the samples have to be small and the internal volume of the autoclave great in order to prevent the equilibrium from being too much disturbed by sampling (pressure drop, equilibrium shift). A great internal volume, however, requires heavy and expensive autoclaves. Usually effective stirring is indispensable; the increase of viscosity with increasing pressures has to be taken into account. On the other hand determination of concentration *in situ* according to method (b) requires an appropriate method and tedious calibration work.

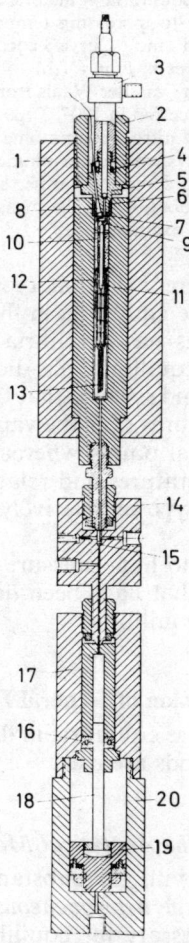

Figure 2. Apparatus for investigating gas–gas equilibria up to 20 kbar and 200°C (reprinted from Krichevskii and Tsiklis[9]): (1) double-walled high pressure vessel, (2) screw, (3) valve, (4) isolating ring, (5) seal, (6) packing, (7) exhaust ring, (8) electrical entry, (9) piezometer, (10) solenoid for magnetic stirrer, (11) stirrer, (12) mercury reservoir, (13) spring, (14) valve, (15) non-return valve, (16) packing, (17) cylinder of high-pressure multiplier, (18) high-pressure piston, (19) low-pressure piston, (20) low-pressure cylinder (for details see ref. 35).

PART 2. PHASE EQUILIBRIA OF LIQUID AND GASEOUS MIXTURES

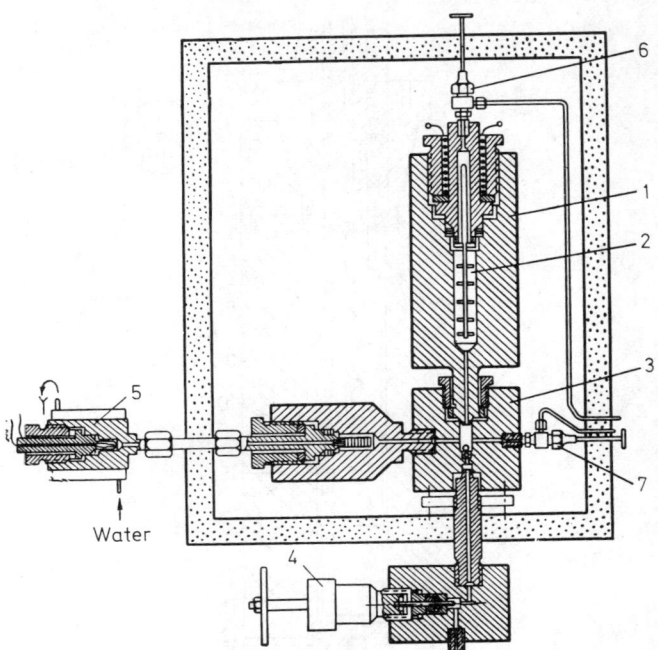

Figure 3. Apparatus for investigating gas–gas equilibria at high pressures and temperatures up to 500°C (analytical method, reprinted from Tsiklis and Maslennikova[38]): (1) high-pressure vessel, (2) magnetically driven stirrer, (3) cross, (4) valve to pressurizer, (5) thermostated manganin manometer, (6) valve for removing upper phase, (7) valve for removing lower phase (for details see ref. 35).

The analytical method has been used by several authors, e.g. Krichevskii and Tsiklis[9] (see *Figure 2*), Tsiklis and Maslennikova[38] (see *Figure 3*), Lindroos and Dodge[11], Takenouchi and Kennedy[32], Tödheide and Franck[33] (see *Figure 4*), Danneil, Tödheide and Franck[33], Peter and Reinhartz[16], Streett and co-workers[30], Schouten[26].

2. The Synthetic Method

The autoclave is charged with a mixture of known concentration and values of temperature and pressure in the *homogeneous* region are adjusted. Temperature is varied at constant pressure until a second phase appears (method of temperature variation). More points of the $p(T)$-isopleth are found by repeating the experiment under different (constant) pressures; the method fails where the absolute value of $(\partial p/\partial T)_c$ is very small or zero (c = concentration). It is equally possible to vary pressure at constant temperature (method of pressure variation); this method fails where the absolute value of $(\partial p/\partial T)_c$ is very great or infinite. Immediately after the appearance of the second phase, temperature or pressure should quickly be readjusted into the homogeneous region in order as much as possible to avoid complete demixing and layering of phases.

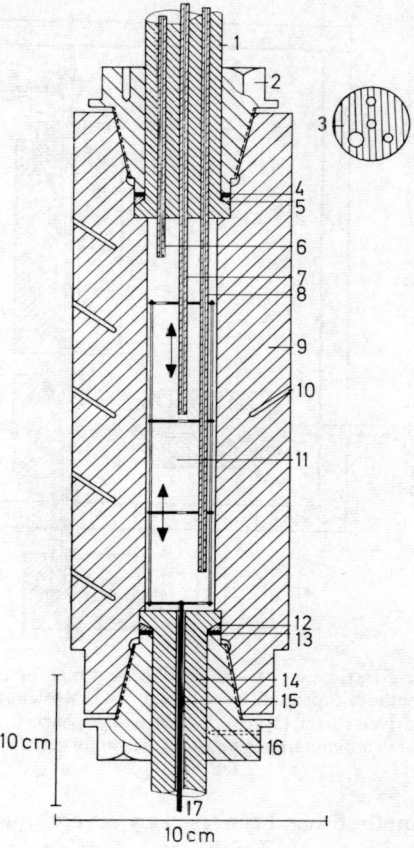

Figure 4. Apparatus for investigating fluid phase equilibria at high pressures and temperatures (analytical method, reprinted from Tödheide and Franck[33]): (1, 14) Bridgman piston, (2, 16) conical screw, (3) horizontal section through piston (1), (4, 5, 12, 13) packings, (6, 7, 8) sampling capillary tubes, (9) high-pressure vessel, (10) bore hole for thermocouple, (11) internal volume, (15, 17) stirrer.

Usually the appearance of the second phase is determined by direct visual observation of the resulting turbidity or meniscus in an optical cell. Visual observation is in general better than photoelectric registration since contours and details can be discriminated. In rare cases other methods are used (e.g. determination of dielectric constant[29], electrical conductivity[3, 6] etc).

Advantages: The method is particularly suitable for measurements near critical states. Barotropic systems can also be studied. Simultaneous determination of pVT data is possible. The high pressure cell can be small and inexpensive since no sampling is necessary. Cells for other optical (e.g. spectroscopic) high pressure measurements can often be used with only slight modifications. A whole $p(T)$-isopleth can be obtained with one filling. Normally the experimental procedure is easy and quick.

Disadvantages: The method is not suitable for measurements far away from critical states, e.g. where the absolute value of $(\partial T/\partial c)_p$ or of $(\partial p/\partial c)_T$ is great or infinite. Visual observation is impossible for iso-optic systems where the coexisting phases have approximately the same refractive index. For systems with more than two components the information obtained from

PART 2. PHASE EQUILIBRIA OF LIQUID AND GASEOUS MIXTURES

the experiments is smaller than with the analytical method since the connodes and by this the concentrations of the coexisting phases cannot be determined. As a rule, effective stirring is indispensable.

This method has been used by numerous authors for the investigation of all kinds of fluid two-phase equilibria under high pressure up to approximately 8 kbar, e.g. de Swaan Arons and Diepen[31] (see *Figure 5*), Alwani and Schneider[1] (see *Figure 7*), Schneider[21], Oeder and Schneider[14] (see *Figure 6*), Lentz[10], Steiner and Schadow[29], Ehrlich and Kurpen[4], Buback and Franck[3]

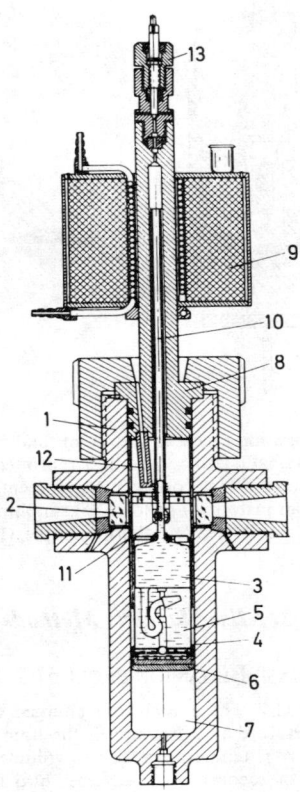

Figure 5. Apparatus for investigating fluid phase equilibria at high pressures designed for mixtures with small concentrations of the light component (synthetic method, reprinted from de Swaan Arons and Diepen[31]) (1) high-pressure vessel, (2) window, (3) pyknometer with mixture, (4) inner part of telescope tube, (5) mercury, (6) piston, (7) water, (8) seal, (9) solenoid, (10) iron bar, (11) permanent magnet, (12) bore hole for thermocouple, (13) valve.

(see *Figure 8*). A simple and inexpensive cell has been described by Schneider[20]. Much of the classical work has been effectuated in the apparatus of Kohnstamm and Timmermans[8, 17] (pressures up to 3 kbar being reached in this optical cell as long ago as 1912!). Even the Cailletet method using glass capillary tubes is still being used, e.g. by Pak and Kay[15].

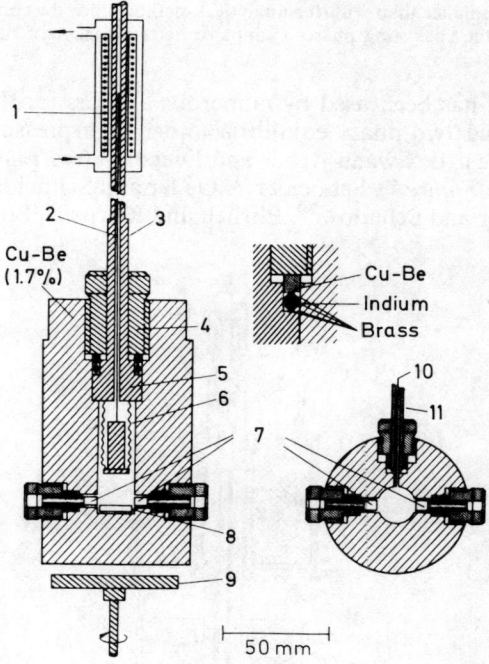

Figure 6. Optical cell for investigating phase equilibria in fluid mixtures at temperatures down to $-200°C$ and pressures up to 4 kbar (synthetic method, reprinted from Oeder and Schneider[14]): (1) thermostatically controlled solenoid for volume measurement, (2) iron wire, (3) high pressure inlet tube, (4) screw, (5) Bridgman piston, (6) bellows, (7) sapphire window, (8) magnetic stirrer, (9) rotating magnet, (10) steel-sheathed thermocouple, (11) capillary inlet tube

3. Miscellaneous Methods

A. PHASE BEHAVIOR FROM pVT DATA

This method resembles method 2. The autoclave is charged with a mixture of known concentration, and values of temperature and pressure in the homogeneous region are adjusted. From the three variables pressure p, temperature T, and volume V one is varied, the resulting value of a second is measured or recorded whereas the third is maintained constant. Three modifications can be used:

(α) $V(p)$-plot for $T =$ const. or $p(V)$-plot for $T =$ const.
(β) $p(T)$-plot for $V =$ const.
(γ) $V(T)$-plot for $p =$ const.

At a phase transition the plots show discontinuities of their slopes.

Advantages: Cells for the determination of pVT data under high pressure can often be used with slight modifications only. Of course simultaneous pVT measurements are possible.

Disadvantages: Often this method does not give clear results especially for gas–gas and liquid–liquid equilibria. It cannot be used near critical states.

The variation of T, p or V can be controlled by an electronic programmer (e.g. a temperature or pressure programmer). X/Y recorders or normal X-time recorders (for variations with uniform speed) are appropriate. The volume can be uniformly varied by a motor-driven rotating pump. Continuous determination of the volume is possible by the displacements of a bellows

PART 2. PHASE EQUILIBRIA OF LIQUID AND GASEOUS MIXTURES

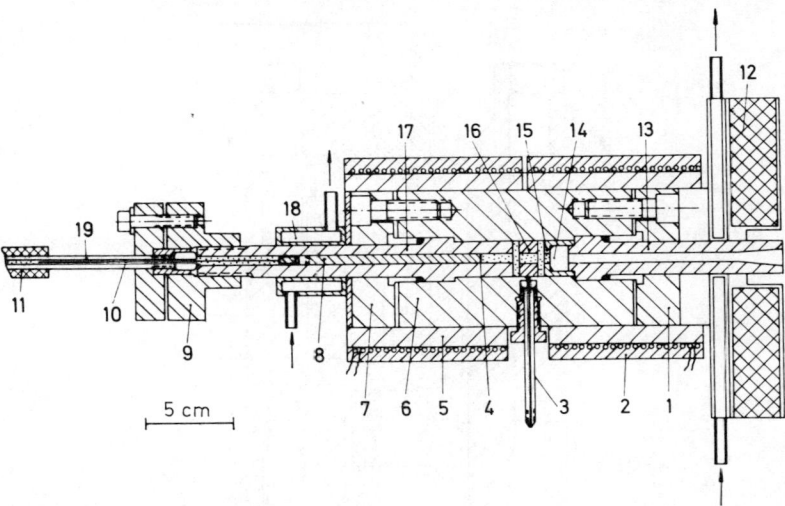

Figure 7. Optical cell for investigating phase equilibria in fluid mixtures up to 400°C and 4 kbar (synthetic method, reprinted from Alwani and Schneider[1]): (1, 7) flange, (2) electrical heating and thermal insulation, (3) capillary inlet tube, (4) platinum mirror, (5) aluminum block, (6) pressure vessel, (8) piston, (9) connection, (10) iron wire, (11, 12) solenoid, (13) window plug, (14) sapphire window, (15) steel cap, (16) ring magnet, (17) high-pressure tube, (18) cooler, (19) capillary tube.

measured inductively or from the height of a mercury meniscus in a container or a capillary tube.

The method has so far been used for the investigation of fluid phase equilibria by only a few authors, e.g. Lentz[10], Alwani and Schneider[1], whereas it is a common experimental procedure where solid phases occur.

B. THE DYNAMIC METHOD

This method has been widely used for the investigation of gas–liquid equilibria at low or moderate pressures. The pure gas is passed slowly under pressure through the liquid and the concentrations of both liquid and gaseous phases are determined. Since it is not easy to keep the pressure constant and to measure it exactly and since it is difficult to obtain true equilibrium this method is not very suitable for high pressure investigations and has been used only in rare cases, e.g. by Michels, Skelton and Dumoulin[13] (see *Figure 19*; for a detailed discussion see[13]).

C. THE GLASS BULB TECHNIQUE

The exhibition of gas–gas equilibria of the first kind can be checked with a simple technique developed by Khodeeva[7]. The less volatile component I is introduced into a glass bulb of known capacity in such a way that it is at its critical volume. The critical temperature $T_c(I)$ of the pure substance I is measured and then a small quantity of the second component is added. If the meniscus between the two phases disappears at a higher temperature than $T_c(I)$ then a region of gas–gas equilibrium of the first kind must exist in the system.

IV. Description of Special Equipments

A compilation of the experimental techniques in the field of high-pressure

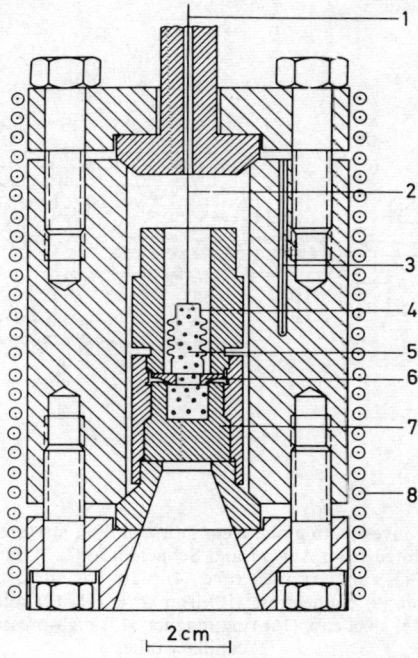

Figure 8. Optical cell for the determination of high-pressure high-temperature phase equilibria in molten salts (synthetic method, see text, reprinted from Buback and Franck[3]): (1) connection to pressurizing system, (2) inductive detector, (3) thermocouple, (4) gold–platinum bellows, (5) internal volume, (6) gold–platinum disc, (7) sapphire window, (8) heating.

phase equilibria is presented in two important books by Tsiklis[35, 37] which are highly recommended to the reader. Besides these books only a few review articles exist where experimental procedures are treated in more detail, e.g. by Tsiklis[36], Schneider[24]. Readers are also referred to the books of Prausnitz[18], Weale[39] and others.

In the following some special measuring devices for the determination of phase equilibria in fluid mixtures under high pressure will be described.

(1) Mercury-operated apparatus of Krichevskii and Tsiklis[9] for investigating gas–gas equilibria up to 20 kbar (analytical method, see *Figure 2*):

The high-pressure piezometer is charged with a mixture of known composition. Temperature and pressure are adjusted in the heterogeneous region and layering takes place. The upper phase is forced out through the upper valve (3) and is analysed. After all the upper phase has left, the lower phase is equally forced out and analysed. The phases are forced out by pumping mercury from beneath into the piezometer, the pressure being kept constant. The apparatus cannot be used for temperatures higher than 200°C because of safety risks and of the increasing solubility of mercury in the compressed gas.

(2) Apparatus of Tsiklis and Maslennikova[38] for investigating gas–gas equilibria at high pressures and temperatures up to 500°C (analytical method, see *Figure 3*):

PART 2. PHASE EQUILIBRIA OF LIQUID AND GASEOUS MIXTURES

The measuring autoclave consists of a thick-walled vessel (1) made from temperature and corrosion resistant stainless steel. The vessel is mounted with a magnetic stirrer (2) and connected to a cross (3) to which a pressure intensifier (not shown on the figure), a manganin manometer (5) and a sampling valve (7) are also connected. The manganin manometer is separated from the fluid under test by a bellows filled with dry kerosene. Samples of the upper and lower phase are withdrawn from the vessel through the valves (6) and (7) respectively and are analysed.

The equipment has been used for many measurements by the Russian team, e.g. in the system n-butane–hydrogen up to 6 kbar and 430°C.

(3) Apparatus of Tödheide and Franck[33] for investigating fluid phase equilibria at high temperatures and pressures (analytical method, see *Figure 4*):

The equilibrium vessel (9) is constructed from a temperature and corrosion resistant stainless steel. It is sealed by two Bridgman pistons (1, 14) and conical screws (2, 16). The mixture in the cell (11) can be efficiently agitated by the strokes of a stirrer that is driven magnetically in a special small autoclave not shown in the figure. The vessel is electrically heated; temperature is measured and controlled with thermocouples that are mounted in bore holes (10). With rotating pumps the pure components are filled into the vessel through one of the four holes in the upper piston, (1) at the normal temperature. Pressure is created during the heating process. After equilibrating the phases samples are drawn at different heights of the vessel through three capillary tubes (6, 7, 8) and analysed outside the autoclave. The apparatus was used for investigating the phase behavior of aqueous mixtures of carbon dioxide, ethane and n-butane up to 270°C and 3.5 kbar[33].

(4) Optical cell used by de Swaan Arons and Diepen[31] for investigating high-pressure phase equilibria in fluid mixtures (synthetic method, see *Figure 5*):

The fluid mixture under test (3) is enclosed in a glass pyknometer that is mounted in an autoclave (1). It is separated from the pressure transmitting medium water (7) by mercury (5). The phase equilibria in the upper capillary tube of the pyknometer can be observed through two windows of safety glass (2), the apparatus being designed especially for measurements at small concentrations of the specifically lighter component. The mixture is stirred by a small permanent magnet (11). This stirring magnet is pushed up and down from outside the pyknometer by a second permanent magnet that is fastened at the lower end of an iron tube (10), the latter being operated by a magnetic solenoid (9).

The cell was used for investigating gas–gas equilibria in the system helium–xenon up to 2 kbar[31].

(5) Optical cell developed by Oeder and Schneider[14, 21] for investigating fluid phase equilibria at low temperatures and pressures up to 4 kbar (synthetic method, see *Figure 6*):

The cell is constructed from copper–beryllium (1.7 per cent). All phase transitions within the cell can be observed through two sapphire windows (7). The liquid mixture in the cell is agitated by a small magnetic stirrer (8) driven from outside the cell by a rotating magnet (9). The pressure is initially created using isopentane or compressed nitrogen and transmitted to the fluid mixture under test by a bellows (6). An iron wire (2) is attached to this bellows, the displacement of this wire in the non-magnetic capillary inlet tube (3) can be determined by measurements of the inductance of a thermostatically controlled solenoid (1). Thus the determination of pVT data is equally possible. Indium is used for seals. The cell is thermostated in a brass block not shown in *Figure 6*; this block is cooled by pumping liquid nitrogen through holes in the block. The temperature is measured directly in the fluid mixture under test with a steel-sheathed thermocouple (10) introduced into the cell through a capillary inlet tube (11) through which the cell can also be filled.

The cell was designed for measurements down to -200°C and up to 4 kbar. It was used for the determination of the pressure dependence of upper critical temperature solutions in the

n-octane–carbon dioxide system[21] and for intensive investigations of the phase relationships and the pVT behavior of hydrocarbon–methane systems[14].

(6) Optical cell developed by Alwani and Schneider[1] for investigating fluid phase equilibria at high temperatures and high pressures (synthetic method, see *Figure 7*):

The optical cell is constructed from a corrosion and temperature resistant stainless steel (Nimonic 90). The phase transitions in the cell are observed through a single window of synthetic sapphire (14); the light for observation is focused into the cell by a medical mirror. During the measurements, the fluid mixture in the cell is efficiently agitated by a magnetic stirring device: a direct current of slowly alternating polarity is attached to a large solenoid (12) making the top of the window plug (13) by turns a magnetic north or south pole. By this the permanent ring magnet (16) in the cell is pushed back and forth, and the mixture under test is stirred efficiently. The pressure is created by a rotating pump with a pressure transmitting medium (mostly water) and measured with Heise gages. The pressure is transmitted to the fluid mixture under test by a piston (8) fitted in the bore of a thick high pressure tube (17), the former being additionally tightened with perbunan and viton O-rings. To this piston, an iron wire (10) is fastened whose position in the capillary inlet tube (19) can be determined by measurements of the inductance of a magnetic solenoid (11). Thus the determination of pVT data is equally possible. The cell is mounted in a large aluminum block (5) fitted with electrical heating and thermal insulation (2). The temperature of the mixture is measured directly with a steel-sheathed thermocouple introduced into the cell by the capillary tube (3) through which the cell can also be filled.

The apparatus was designed for pressures up to 4 kbar and temperatures up to 400°C and was used for the investigation of the phase behavior and critical phenomena of hydrocarbon–water mixtures[1].

(7) Optical cell developed by Buback and Franck[3] for the investigation of fluid phase equilibria at high temperatures and pressures (synthetic method, see *Figure 8*):

The autoclave for operation to 750°C and 2 kbar is shown in *Figure 8*. The essential part of the internal cell is a cylinder of synthetic sapphire (7), pressed on to a steel plug, thus providing a seal for the high pressure space using Poulter's principle. The inner flat surface of the sapphire boring is polished. A thin-walled (0.1 mm) gold–platinum bellows (4) is soldered to a gold disc (6), which is pressed against the sharp edge of the sapphire bore by means of a hollow steel screw, thus sealing the internal space (5) of the sapphire bore from the remaining part of the autoclave.

The interior of the cell is filled with a desired quantity of the substance under test at room temperature. On heating the cell, thermal expansion causes an elongation of the bellows. This elongation can be compensated by raising the pressure of carbon dioxide, which was used as pressurizing gas in most of these experiments. Thus the internal cell volume can be kept constant. The position of the bellows can be determined by an external inductive detector (2) to ± 0.03 mm (or 0.5 percent of the internal volume). Through the sapphire window (7) the phase equilibria can be observed. With slight modifications the cell can also be used for pVT, conductivity and spectroscopic measurements.

The cell was used for the determination of the vapor pressure curves of ammonium halides. It is also suitable for the investigation of mixture properties.

(8) Apparatus developed by Michels, Skelton and Dumoulin[13] for investigating gas–liquid equilibria up to 800 bar (dynamic method, see *Figure 9*):

The pure gaseous component enters through valve (16), passes through the liquid and with special precautions (e.g. the gas bubbles are first distributed by grooves on the circumference of a conical plug (13) and then by a fiber glass filter (10)) saturates the liquid while taking up molecules from it. Fine throttling valves (1, 2) allow samples of the gas (1) and liquid (2) phase respectively to be withdrawn continuously at the rate of about 1 000 cm³ n.t.p. per hour for the gas and 1 or 2 cm³ for the liquid phase which are analysed separately.

PART 2. PHASE EQUILIBRIA OF LIQUID AND GASEOUS MIXTURES

The apparatus was used for the investigation of the systems ammonia–krypton, ammonia–argon, and nitrogen–hydrogen–ammonia between $-30°$ and $120°C$ and up to 800 bar.

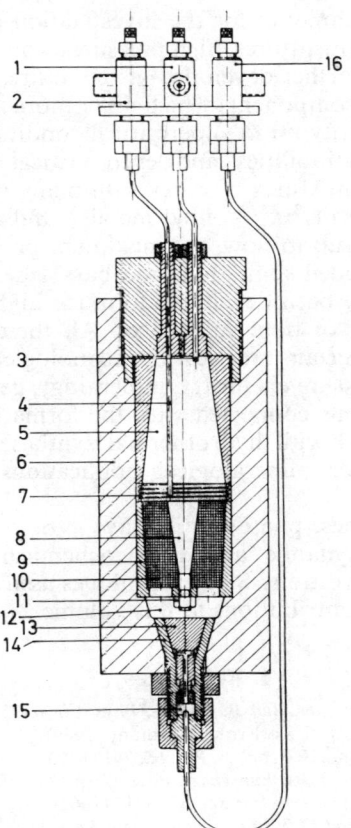

Figure 9. Apparatus for investigating gas–liquid phase equilibria up to 800 bar (dynamic method, reprinted from Michels, Skelton and Dumoulin[13]): (1,2) valves for outlet of gas (1) and liquid (2), (3, 4) capillary tubes for outlet of gas (3) and liquid (4), (5) reservoir, (6) high-pressure vessel, (7) spring, (8) conical steel plug, (9) hollow steel cylinder, (10) porous plug (Fiberglas), (11) nut, (12) reservoir, (13) steel cone with grooves on circumference (each less that 0.05 mm deep), (14) conical sleeve, (15) small reservoir, (16) throttling valve for gas inlet.

V. Conclusions

In the future the number of investigations on fluid phase equilibria under high pressure will certainly increase considerably, the experiments being extended to higher pressures and both to lower and to higher temperatures.

Investigations of highly compressed or condensed gas mixtures at *low* temperatures will be interesting from a theoretical point of view, since mixtures of small and highly symmetrical molecules can be studied that are suitable for a theoretical treatment. Measurements at low temperatures may also have a considerable practical and technical interest, for the petroleum and gas industry, for gas solubility, cryogenic applications, separation problems, space research etc.

The experimental techniques for the investigation of phase equilibria in fluid systems at *high* temperatures, high pressures, and possibly under strong corrosion will also be further developed in the future, in order to allow the study of mixtures with components which differ more and more in volatility, structure, size and polarity up to supercritical conditions. Phase equilibria, miscibility relations, solubilities and even critical phenomena will be increasingly studied in mixtures of water with many inorganic and organic compounds, in molten salts[3, 28, 34], fluid metals[6], and other substances. It is an interesting fact that, up to now, the maximum pressures in such experiments have rarely exceeded some thousand bars, whereas the experimental techniques have already been developed to much higher pressures in some cases, e.g. in the study of pure substances. All these measurements may additionally present a certain practical or technological interest for separation processes, high pressure chemistry, mineralogy, geochemistry and earth sciences, e.g. for problems connected with the formation and migration of oil and natural gas and with hydrothermal syntheses; indeed some high pressure equipments for mineralogical applications have already been described[2, 12, 27, 41].

It is evident that all these phase studies have to be completed by measurements of the thermodynamic and physicochemical properties of fluid mixtures under these conditions. Some techniques used for these experiments are described in other contributions to this volume.

VI. References

[1] Alwani, Z. and G. M. Schneider. *Ber. Bunsenges. Phys. Chem.* **73**, 294 (1969);
Alwani, Z. *Thesis*, University of Karlsruhe, Germany (1969).
[2] Boyd, F. R. and J. L. England. *J. Geophys. Res.* **65**, 741 (1960).
[3] Buback, M. and E. U. Franck. *Ber. Bunsenges. Phys. Chem.* **76**, 350 (1972).
[4] Ehrlich, P. and J. J. Kurpen. *J. Polymer Sci.* **A1**, 3217 (1963).
[5] Hála, E., J. Pick, V. Fried and O. Vilim. *Gleichgewicht Flüssigkeit-Dampf.* Akademie Verlag: Berlin (1960).
[6] Hensel, F. and E. U. Franck. *Ber. Bunsenges. Phys. Chem.* **70**, 1154 (1966);
Freyland, W. F. and F. Hensel. *Ber. Bunsenges. Phys. Chem.* **76**, 16 and 347 (1972).
[7] Khodeeva, S. M. *Zh. Fiz. Khim.* **40**, 1973 (1966): *Russ. J. Phys. Chem.* 1061 (1966).
[8] Kohnstamm, Ph. and J. Timmermans. *Proc. Acad. Sci. Amst.* **B15**, 1021 (1912).
[9] Krichevskii, I. R. and D. S. Tsiklis. *Zh. Fiz. Khim.* **17**, 115 (1943).
[10] Lentz, H. *Rev. Sci. Instrum.* **40**, 371 (1969).
[11] Lindroos, A. E. and B. F. Dodge. *Chem. Engng Progr. Symp. Ser. No. 3*, **48**, 10 (1953).
[12] Luth, W. C. and O. F. Tuttle. *Amer. Mineral.* **48**, 1401 (1963).
[13] Michels, A., G. F. Skelton and E. Dumoulin. *Physica*, **16**, 831 (1950);
Michels, A., E. Dumoulin and J. J. Th. van Dijk. *Physica*, **25**, 840 (1959); **27**, 886 (1961).
[14] Oeder, D. and G. M. Schneider. *Ber. Bunsenges. Phys. Chem.* **73**, 229 (1969);
Oeder, D. *Thesis*, University of Karlsruhe, Germany (1969).
[15] Pak, J. C. and W. B. Kay. *Industr. Engng Chem. Fundamentals*, **11**, 255 (1972).
[16] Peter, S. and K. Reinhartz. *Z. Phys. Chem. (Frankfurt)*, **24**, 103 (1960).
[17] Poppe, G. *Bull. Soc. Chim. Belg.* **44**, 640 (1935).
[18] Prausnitz, J. M. *Molecular Thermodynamics of Fluid Phase Equilibria*. Prentice Hall: Englewood Cliffs, NY (1969);
Prausnitz, J. M. *Advanc. Chem. Engng*, **7**, 139 (1968).
[19] Rowlinson, J. S. *Liquids and Liquid Mixtures*. 2nd ed., Butterworths: London (1969).

PART 2. PHASE EQUILIBRIA OF LIQUID AND GASEOUS MIXTURES

[20] Schneider, G. M. *Z. Phys. Chem. (Frankfurt)*, **37**, 333 (1963).
[21] Schneider, G. M. *Ber. Bunsenges. Phys. Chem.* **70**, 10 (1966).
[22] Schneider, G. M. *Ber. Bunsenges. Phys. Chem.* **70**, 497 (1966).
[23] Schneider, G. M. *Advanc. Chem. Phys.* **17**, 1 (1970).
[24] Schneider, G. M. *Fortschr. Chem. Forsch.* **13**, 559 (1970).
[25] Schneider, G. M. *Ber. Bunsenges. Phys. Chem.* **76**, 325 (1972).
[26] Schouten, J. A. *Thesis*, University of Amsterdam (1969).
[27] Seifert, F. *Thesis (Habil.)*, University of Bochum (1971).
[28] Silva, J., J. W. Johnson and D. Cubicciotti. *Rev. Sci. Instrum.* **36**, 1505 (1965);
Johnson, J. W., W. J. Silva and D. Cubicciotti. *J. Phys. Chem.* **69**, 3916 (1965).
[29] Steiner, R. and E. Schadow. *Z. Phys. Chem. (Frankfurt)*, **63**, 297 (1969).
[30] Streett, W. B. and C. H. Jones. *Advanc. Cryog. Engng*, **2**, 356 (1966);
Streett, W. B. *Cryogenics* **5**, 27 (1965);
Streett, W. B. and J. L. E. Hill. *J. Chem.* **54**, 5088 (1971).
[31] de Swaan Arons, J. and G. A. M. Diepen. *J. Chem. Phys.* **44**, 2322 (1966).
[32] Takenouchi, S. and G. C. Kennedy. *Amer. J. Sci.* **262**, 1055 (1964);
Sourirajan, S. and G. C. Kennedy. *Amer. J. Sci.* **260**, 115 (1962).
[33] Tödheide, K. and E. U. Franck. *Z. Phys. Chem. (Frankfurt)*, **37**, 387 (1963); *Ber. Dtsch. Keram. Ges.* **41**, 60 (1964);
Danneil, A., K. Tödheide and E. U. Franck. *Chem.-Ing.-Tech.* **39**, 816 (1967).
[34] Treiber, G. and K. Tödheide. *Ber. Bunsenges. Phys. Chem.* **77**, 540 (1973).
[35] Tsiklis, D. S. *Techniques for Physicochemical Studies under High and Super-high Pressure*, 3rd ed., Khimiya: Moscow (1965);
Tsiklis, D. S. *Handbook of Techniques in High-Pressure Research and Engineering*. Plenum: New York (1968).
[36] Tsiklis, D. S. and L. A. Rott. *Russ. Chem. Rev.* **36**, 351 (1967); *Uspekhi Khim.* **36**, 869 (1967).
[37] Tsiklis, D. S. *Phase Separations in Gas Mixtures*, Khimiya: Moscow (1969).
Tsiklis, D. S. *Phasentrennung in Gasgemischen*. Deutscher Verlag für Grundstoffindustrie: Leipzig (1972).
[38] Tsiklis, D. S. and V. Ya. Maslennikova. *Dokl. Akad. Nauk SSSR*, **157**, 426 (1964).
[39] Weale, K. E. *Chemical Reactions at High Pressures*. Spon: London (1967).
[40] Williamson, A. G. Chapter 16 (Part 1) of this book, p 749.
[41] Yoder, H. S., Jr. *Trans. Amer. Geophys. Un.* **31**, 827 (1950).

CHAPTER 17

Part 1. Melting Points and Volume Changes upon Melting

C. W. F. T. Pistorius

*Chemical Physics Group of the National Physical and National Chemical Research Laboratories,
South African Council for Scientific and Industrial Research,
Pretoria, South Africa*

Contents

I.	Introduction	803
II.	Methods used for Melting Point Determination at Normal Pressure	807
	1. Visual Methods	807
	2. Microscopical Methods	808
	3. Quenching Methods	809
	4. Pyrometric Methods for Use above 2000°C	810
	5. Calorimetric Methods	811
	A. Drop Calorimeters	811
	B. Adiabatic Calorimeters	812
	6. Thermal Arrest Methods	813
	7. Special Methods Used in Isolated Cases	816
III.	Differential Thermal Analysis at Normal Pressure	816
	1. Historical	817
	2. Basic Method	818
	3. Critical Assessment of DTA	819
IV.	Methods Used for Melting Point Determination at High Pressure	820
	1. Historical	821
	2. Methods of Measuring Melting Points at High Pressures in Apparatus Using Hydrostatic Pressure Media	821
	3. Methods of Measuring Melting Points at High Pressures in Apparatus Using Quasi-hydrostatic Pressure Media	824
	4. Methods of Measuring Melting Points at High Pressures in Opposed-anvil Apparatus	827
	5. Methods of Measuring Melting Points at High Pressures in Multiple-anvil Apparatus	828
	6. Conclusion	828
V.	Methods for Determining Volume Changes on Melting	829
VI.	Abstract	830
VII.	References	830

I. Introduction

The melting point of any pure substance, at any given pressure, can be simply defined as that temperature at which the Gibbs free energy of the solid and the liquid phases are equal. However, it is here necessary to make a distinction between a glass, a liquid, and a crystalline solid. Consider the relations between volume and temperature for a typical glass-forming substance shown in *Figure 1*.

If one starts with the crystalline solid at a low temperature and heats it, thermal expansion occurs along line AB. At T_B (the melting point) the crystal melts sharply with, usually, an increase in volume to reach point C in *Figure 1*. Further heating merely causes thermal expansion of the liquid along line CD. If the liquid is now cooled along line DC, freezing need not necessarily occur

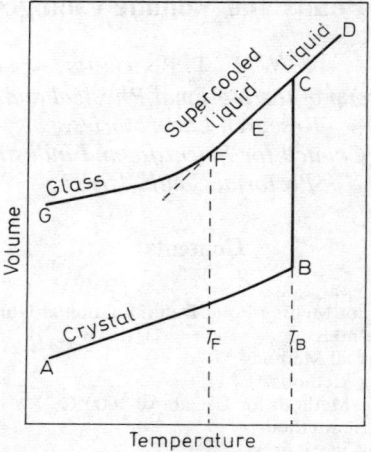

Figure 1. Typical relations between volume and temperature for the liquid, crystalline and glassy phases. Essentially after Jones[87].

at T_B (i.e. at point C), since supercooling is possible and, in fact, common. For most substances the liquid will be found to supercool to some extent until freezing to the crystalline state occurs at point E. T_E depends quite strongly on factors such as the rate of cooling, the environment of the sample and the previous thermal history. If, however, no nucleation of crystallites occurs, it is occasionally possible that the undercooled liquid may become steadily more viscous with falling temperature, and the line DCEFG is traversed. In such a case a glass is formed. The liquid/glass transition is not thermodynamically first-order, while the liquid/crystal transition is always first-order. Furthermore, the position of T_F depends on the experimental conditions, such as cooling rate and previous thermal history, while the melting point T_B of the crystal does not. In addition the final volume of the glass depends on the experimental conditions. The glass can therefore neither be considered as a stable solid nor as a metastable supercooled liquid, but rather as a non-equilibrium state. A comprehensive discussion of the structure and physical properties of glasses can be found elsewhere[163].

Polymers should ordinarily be regarded as mixtures of a large number of substances with different molecular weights, and therefore do not, strictly speaking, qualify to be considered pure substances. Nevertheless, a definite crystalline nature is common in high polymers, and so is a more or less well-defined melting point where the crystallinity disappears. On the other hand, very few polymers can be prepared in a state of complete crystallinity, and it is probably more realistic to regard the 'melting point' of a polymer as being the equivalent of T_F rather than of T_B.

It follows from the above that thermodynamically our definition of melting point applies only to the first-order transition between the crystalline and liquid states. It further follows that, in practice, observed freezing points may be subject to supercooling, while melting points of pure crystalline substances are not subject to superheating.

Some well-defined pure crystalline substances melt to liquids of a different composition, with a complementary solid phase remaining. This is termed incongruent melting. In general the methods given below for determining congruent melting points apply equally well to incongruent melting points, since incongruent melting is also a well-defined first-order transition. However, the freezing of an incongruently melted assemblage may be a more complex process altogether, since it is usually a chemical reaction between a liquid and a solid phase, and therefore dependent on rate of cooling, previous thermal history, particle size of the solid, and so on. This is outside the scope of this chapter.

Liquid crystals or mesophases in some pure high molecular weight organic compounds exist between first-order transitions to the stable crystalline solid and to the normal isotropic liquid. These mesophases have liquid-like properties, but possess long-range molecular order, and can therefore not be considered as liquids[11].

This brings us to the definition of a liquid. The essential features of the liquid state can be described best by comparison with those of the solid and gaseous states. A liquid is usually conceived as a substance with no rigidity but possessing a free surface. However, the example above of glasses, which, although almost perfectly rigid at low temperatures, may have negligible rigidity at higher temperatures, and yet be very different from the metastable supercooled liquid at the same temperature, shows that rigidity is not a universal criterion. Similarly a highly compressed substance just above its critical temperature has no free surface but is indistinguishable from corresponding states below the critical temperature which may or may not possess a free surface, so that the second criterion is also inadequate. Indeed, it seems that there is no satisfactory way of defining a liquid except by reference to its molecular structure. Comparison with solids and gases leads one to conclude that liquids have a molecular structure devoid of long-range order, but sufficiently closely packed to insure that any molecule is in continual kinetic interaction with its neighbors. A glass can be said to differ only in the qualitative sense that the molecular neighbors in a glass remain the same over a more or less long period of time, while in a true liquid they are continually changing. Nevertheless, there is considerable evidence that local order is not severely changed by melting, and that, on a short-range order basis, the instantaneous molecular configuration in a liquid is markedly similar to that in the solid[69], especially near the melting point.

For practical purposes the above suffices to distinguish sufficiently clearly between crystalline solids (including mesophases or liquid crystals), glasses and true liquids. In principle, therefore, the determination of a melting point is relatively simple.

There are nevertheless some aspects of the problem which may give rise to difficulties. The importance of purity cannot be sufficiently emphasized. The behavior of binary systems is treated in Chapter 16 of this book. It suffices here to mention that the extent of the melting point depression by impurity can be approximated by the Raoult– van't Hoff equation for dilute ideal solutions[44]

$$\Delta T = (RT^2/\Delta H_m)(N_L - N_S)$$

where ΔT is the change in melting point, T is the melting point of pure substance, R is the gas constant, N_L is the mole fraction of impurity in liquid solution, N_S is the mole fraction of impurity in solid solution, and ΔH_m denotes molar latent heat of melting.

The simplest case is when $N_L \gg N_S$, i.e. when the impurity is essentially insoluble in the solid. A simple melting point depression then occurs, amounting in the case of alumina (Al_2O_3) to $\sim 3°C$ for every mole per cent contaminant, and in the case of silica (SiO_2) to $\sim 80°C$ for every mole per cent contaminant. However, if $N_S > N_L$, an elevation of the melting point will result. It is clear that the effect of even small amounts of impurities may be appreciable. For the same reason it is important to contain the substance to be melted in a container with which the substance will not react before or after melting. This can, in some cases, be a very difficult requirement to meet.

A less obvious source of error can be found in a breakdown of unary phase behavior. Many ceramic materials seemingly behave as single components in that they exhibit all the obvious characteristics of one-component systems, and melt at defined temperatures. The behaviour may nevertheless be complex. Some compounds, for example, will melt at a given temperature in one atmosphere oxygen, but simply by changing the environment to one atmosphere of air, a different melting point will be observed. The x-ray analysis of the samples will show that compositional changes have taken place. The resulting equilibrium can therefore not be described in terms of the starting material. Titania (TiO_2), for instance, is known to lose oxygen upon heating in a supposedly oxidizing environment such as air. The determined melting point will not be representative of stoichiometric TiO_2 and will lie elsewhere in the Ti–O binary system. In order to obtain the melting point of the stoichiometric substance, the oxygen partial pressure must continually change during the course of the heating or cooling[147]. Failure to pay due attention to phenomena of this kind is a common cause for widely diverging oxide or chalcogenide melting points reported in the literature.

The most obvious problem inherent in the determination of a melting point is the accurate measurement of temperature. The thermodynamic temperature scale is the fundamental scale to which all temperatures should be referred. Measurements are usually carried out by platinum resistance thermometry, thermocouples or optical pyrometry (at very high temperatures), but there is no reason, except convenience, why less sophisticated instruments, such as ordinary mercury-in-glass thermometers, cannot give reasonably accurate results in suitable temperature ranges. Temperature measurement in general has been fully discussed in the previous volume in this series[111].

A measured melting point may be stable or metastable. Consider *Figure 2*, where hypothetical phase diagrams, illustrating typical metastable behavior, are shown. Case A contains a phase II with no stability field whatsoever. The vapor pressure curve ef of II intersects the metastable extension fb of

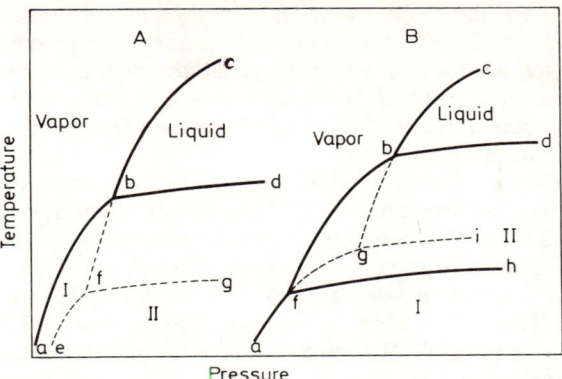

Figure 2. Typical phase diagrams illustrating metastable melting behavior. Solid lines indicate stable and dashed lines metastable boundaries. Derived from Schneider[147].

the stable liquid–vapor curve bc at f, a metastable triple point. The melting curve fg of II indicates metastable melting at temperatures lower than those for the stable melting curve bd of I. Case B illustrates metastable melting (curve gi) as influenced by a presence of two stable polymorphs I and II. Here solid I melts instead of transforming to solid II as required by the stable relations. Metastable melting cannot occur at a higher temperature than stable melting without postulating two immiscible liquids in a one-component system. For the same reason it appears to be impossible to superheat a solid. Numerous examples of metastable melting points are known, but *Figure 2* above shows that these points are as well-defined as stable melting points and, provided they are correctly identified, their occurrence can cause no ambiguity in melting point measurements.

II. Methods used for Melting Point Determination at Normal Pressure

In essence a melting point can be determined by observing the change of almost any property of the substance as a function of temperature. Since melting is a first-order phase change, the property being observed will change discontinuously at the melting point. It may often be more convenient to measure the freezing point than the melting point. In such a case it is always necessary to be certain that supercooling is avoided by maintaining equilibrium between solid and liquid phases.

1. Visual Methods

Historically the first method used by man to observe melting was visual in nature. In most cases the appearance of a crystalline solid and of a liquid is visually different, and can be easily distinguished.

When a substance melts at a temperature in the range 200–600 K and is available only in small quantities, its melting point may be determined by putting a small quantity of it in a thin-walled capillary tube which is then sealed at one end and attached to the bulb of a thermometer. The thermometer and tube are placed in an air or preferably a liquid bath which is stirred and very gradually heated while the appearance of the sample is closely watched. The temperature at which the substance begins to melt is read, and so is the temperature at which solidification begins when the temperature of the bath is gradually reduced. The latter temperature will usually be low due to supercooling.

An elaboration of this method is extensively used in organic chemistry, and can yield results correct to 0.01°C if due care is taken, and a thermometer (usually a Beckmann thermometer) capable of reading to this accuracy is used. The method consists of surrounding the thermometer bulb with a relatively large amount of sample (~ 20 g)[101], and roughly determining the melting point. The bath is then cooled to not more than ~ 0.2°C below the melting point of the sample. Freezing is induced by gently tapping the outside wall of the sample tube. As soon as crystals of the sample begin to separate, the thermometer reading will increase. The maximum temperature reached is the melting point.

Both methods described above are common procedures in organic syntheses, and the simple apparatus needed is commercially available. The ease and simplicity of the methods are attractive, but unless the proper precautions are taken, results may be in error by several degrees.

An extensive list of references to methods of this type is given elsewhere[126], and will not be repeated here.

2. *Microscopical Methods*

These methods are mainly used qualitatively when the amount of sample is extremely small, or when it is necessary to observe the behavior of, say, a single crystal grain in a mixed assemblage. As such, there is some application in mineralogy.

The simplest method under this classification is the observation of a crystal in a hot-stage microscope as the temperature is raised. Melting can be observed visually, and the temperature at that point is recorded. Alternatively, in the case of non-cubic transparent crystals, melting can be recognized by the onset of total extinction under crossed Nicols. However, when using this procedure care should be taken to insure that actual melting is observed, and not a polymorphic transition to a cubic high-temperature phase.

A slightly more sophisticated method is to mount the crystal in the V of a thermocouple junction, preferably of very fine film or wire (~ 0.01 cm diameter). The sample can be made to adhere to the thermocouple by means of a non-reacting glue, or by dipping the thermocouple into a molten mass of the sample, or, if the crystals are sufficiently small, merely by dipping the thermocouple into a collection of crystals until one crystallite adheres to the thermocouple. The thermocouple is slowly heated in the microscope, and a reading is taken when melting occurs.

The high-temperature microscope developed by Welch[181,182] used a Pt–5% Rh/Pt–20% Rh electrically heated thermocouple as a microfurnace, specimen holder and temperature gage simultaneously. The upper temperature limit was 1800°C. Welch and Gutt[183] raised this limit to 2400°C by using a hot stage of pure iridium, but the temperature of the specimen had to be measured by optical pyrometry. The use of Ir/Ir–40% Rh thermocouples, electrically heated, as simultaneous microfurnace, specimen holder and temperature gage allows temperatures up to 2150°C to be reached[72]. Van Tets and Wiedemann[173] describe an apparatus which allows simultaneous microscopic and DTA (see section III) investigations.

The success of these methods depends on sufficiently low thermal gradients near the sample and the temperature-measuring device, and sufficiently low heating rates. In practice it is generally found, however, that it is inconvenient to meet these conditions satisfactorily and results are generally reliable only to within a few degrees even at low temperatures. A very small sample often tends to supercool considerably more than a larger sample. The measurement of freezing points of such samples is therefore unreliable. In general, microscopic methods can only be recommended for exploratory work, or when accuracy is not of prime importance, or in some cases where the amount of sample available is simply insufficient for any other method to be used.

3. *Quenching Methods*

The quenching method, first employed by Shepherd, Rankin and Wright[152], and elaborated later[104, 119, 137, 149], has been specifically developed for the study of ceramic materials. The method consists of placing a small crystalline sample in a suitable container, usually a small platinum envelope which may be sealed. This is held at a selected precise temperature in a furnace for a desired period of time, and the sample is then quenched by dropping the container into cold mercury. In some cases rapid cooling of the furnace suffices for quenching. Examination of the cooled sample reveals whether or not melting has occurred. If there are no signs of melting, heating is repeated at successively higher temperatures. The melting point is bounded between the highest temperature with no evidence of melting and the lowest temperature with complete melting. The success of the method clearly depends on being able to distinguish between samples which had melted, and those that had not. The problem is relatively simple in the case of glass-forming substances. Evidence of melting can in such cases be found by microscopic examination of the quenched specimens. The presence of liquid at high temperatures will be revealed by the presence of optically isotropic material (glass) unless the substance crystallizes with a cubic structure. The problem of liquid detection is more difficult for substances which do not form glasses. The physical appearance of the quenched specimens is a valuable guide to the experienced observer. A sample which had lost its original shape and flowed to surrounding areas can be considered to have melted. Many ceramic substances tend to form viscous liquids upon melting, and consequently little flow occurs. In such cases the problem can be solved if single crystals are used as test specimens[148]. The single crystal will remain single unless melted. If, however, melting occurs, the cooled specimen will appear as a polycrystalline mass which can easily be distinguished from the original single crystal. This, of course, presupposes that there are no polymorphic transitions below melting which could also cause the disintegration of the single crystal. However, even in such cases the textural features of the daughter crystals can be useful guides as to whether or not melting has occurred.

Slight modifications of the methods referred to above are necessary when substances are to be studied that require inert or special atmospheres. This is a common problem when iron, chromium, etc. is present in the substance, and the heating must be done at controlled oxygen potentials[18, 124].

The accuracy obtainable by quenching methods is limited only by the number of bracketing runs the experimenter is willing to carry out, the Pt/Pt–10% Rh thermocouple calibration, and the degree of temperature regulation of the oven. Temperature regulators in use at the Geophysical Laboratory of the Carnegie Institute of Washington are capable of regulation to $\pm 0.3°C$ over periods of many months[16, 145]. Instead of calibrating the thermocouples separately, and then insuring that thermal gradients are kept negligible, calibration is usually done *in situ* against convenient reference points. Recalibration is necessary from time to time, especially at temperatures above $\sim 1300°C$, or when the thermocouple may have been contaminated. Details of the procedure are given by Schairer[145]. In general one can conclude that the method is capable of yielding results correct to $\sim \pm 1°C$ if the purity and homogeneity of the sample justify the effort involved. Above $\sim 1600°C$ the drift of Pt/Pt–10% Rh thermocouples becomes prohibitive, and other thermocouple materials, such a W–Re or Ir–Rh alloy combinations, or pyrometric means of temperature measurement become necessary. This usually entails a considerable lowering in the attainable accuracy, but does not seriously modify the basic methods described above.

4. *Pyrometric Methods for Use above 2000°C*

For the accurate determination of the melting point of a solid, using a pyrometric technique, it is essential to use furnaces in which the specimen can be melted under blackbody conditions. All these methods involve visual observation of the specimen during heating. The temperature at which a specimen is seen to melt is taken to be the melting point. It is therefore also necessary to determine the precise moment of collapse of the specimen without altering the blackbody environment. The measurement of temperature thus becomes the major obstacle in obtaining an accurate value. The problem is that these methods depend on the ability of an observer to distinguish a specimen from its surroundings in order to detect melting. This, however, is not compatible with the achievement of blackbody conditions, since a specimen located in a true blackbody enclosure becomes essentially invisible. The usual recourse[32, 161] is to allow the necessary deviation from perfect blackness to see the specimen, and to incorporate a correction factor obtained from emissivity data or from *in situ* calibration. Published emissivity data as well as many calibration procedures, however, are not always reliable and are often not applicable to specific experiments.

The quenching method described under the previous heading is suitable for determinations above 2000°C, and it is not necessary here to observe the melting process directly, so that the problem is reduced to finding a suitable blackbody furnace. However, as mentioned before, furnace instability and temperature control tend to become troublesome at these temperatures.

Perhaps the most suitable method has been described by Riley[139, 140]. Instead of observing directly the collapse of the sample upon melting, this effect is observed by having the solid sample between the optical pyrometer and a pinhole in the opposite side of the furnace. This pinhole appears black because it leads to a cold area. Before melting, the sample is seen as a bright disc in the pyrometer view, but upon melting, when the sample collapses, the pinhole is uncovered and appears as a black spot in the field of view. The taking of accurate optical pyrometer readings requires considerable time. In order to allow for this, simultaneous readings are made at a series of temperatures below the melting point by means of optical and continuous recording radiation pyrometers, so that the radiation pyrometer can be calibrated against the optical pyrometer. At the moment of melting no optical reading is taken, but the radiation pyrometer output is marked, and can later be related to the true temperature. In principle the accuracy of this method is limited mainly by the approach to blackbody conditions and to the reliability of the pyrometer used[108, 175].

The usual problems of purity of the sample, reaction of the sample with its container or with its gaseous environment, deviations from stoichiometry of the sample due to incongruent evaporation of one constituent, etc. all become major problems at very high temperatures, and no melting point should be accepted unless due precautions have been taken in its determination. These problems must be considered and solved afresh for each substance to be studied[140].

Finally, reference should be made here to the work of Foex and co-workers[54-57] which will be discussed in more detail in the section on thermal arrest. They obtain pyrometric cooling curves from ~3000°C down, using a solar furnace, and detect freezing points by means of thermal arrest.

5. *Calorimetric Methods*

The latent heat of melting affords an excellent and convenient indication of melting. It follows, therefore, that the melting point is determined more or less incidentally during any series of calorimetric heat-capacity measurements that includes the melting temperature. Details of calorimetric methods have been given in Volume I of this series, and need not be repeated here. It suffices at present merely to refer to the two main groups of calorimeters used in practice.

A. Drop Calorimeters

Drop calorimeters are usually used for specific heat determinations above room temperature. The sample is sealed in a suitable capsule, heated in a furnace to a precisely known temperature, and rapidly dropped into, e.g., a Bunsen ice calorimeter, where the heat evolved by the sample plus capsule in cooling to 0°C is measured. The corrections necessary and experimental details can be found elsewhere[49, 59, 62, 63].

The accuracy of melting point determination by means of a drop calorimeter depends mainly on the magnitude of the temperature intervals used near the melting point. While there is no reason why melting points cannot be determined, by this means, to essentially the same accuracy as the temperature of the furnace, the time involved for the numerous repeated heatings of the sample capsule becomes prohibitive if results to $\pm 0.01°C$ are desired. This is, in essence, merely a quenching method with the heat content of the sample used as an indicator of whether or not melting has occurred.

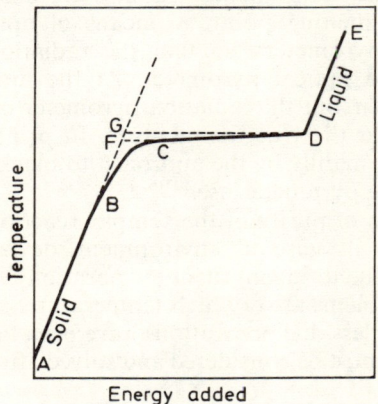

Figure 3. Typical melting curve obtained in an adiabatic calorimeter.

B. Adiabatic Calorimeters

Adiabatic calorimeters are usually used for low-temperature specific heat determinations, but are capable of results up to $500°C^{185}$. A variety of designs are in use. In essence they consist of calorimeters which are heated over a small temperature interval keeping the temperature of the environment, usually a jacket, as near as possible to that of the calorimeter[31, 63, 85, 159, 184, 185].

Accurate melting points can, however, be rather simply obtained by using an adiabatic calorimeter containing the solid sample at a temperature T_A slightly below the expected melting point (see Figure 3). If the calorimeter is now heated by adding measured amounts of electrical energy, and the temperature of the calorimeter is observed after thermal equilibrium had been established following each successive heating, a curve such as ABCDE in Figure 3 is obtained. This gives the temperature of the calorimeter and contents as a function of the energy added. If the sample had been perfectly pure, BCD would have been a horizontal straight line. In order to correct for the effect of impurity, the following procedure is followed. The line AFG, being simply an extrapolation of the part of the curve relating to solid material, is drawn. The horizontal distance FC from any observed point to the line AFG represents the energy that has been used to melt part of the sample, since F represents the temperature–energy condition which would have existed if no material had melted. The distance GD represents the total heat of melting. If the heat of melting is assumed to be constant, the fraction melted at point C is FC/GD. If it is further assumed that the impurity is soluble in the liquid and insoluble in the solid, the curve BCD will have approximately the form

$$T_m - T = a/F \qquad \text{or} \qquad T_m - T_D = [F/(1 - F)](T_D - T)$$

where T is the temperature at a point on the curve, T_m is the melting point of pure material, F is the fraction melted at T, and $a = T_m - T_D$ denotes a constant for the curve. The above relation follows directly from the Raoult–Van't Hoff equation discussed in section I. $T_m - T_D$, and therefore T_m, can now be obtained as the mean of the values for the different points[150]. This treatment can be easily modified if small amounts are present of an impurity soluble in the solid but not in the liquid.

Results correct to $\pm 0.010°C$ or better can be obtained by the above method if the accuracy of temperature measurement is sufficiently good.

6. Thermal Arrest Methods

These methods make use of the change in heat content of the solid/liquid transition. The experimental arrangement, in its simplest form, consists of a crucible containing the sample, and a temperature measuring device centrally located in the sample. A continuous record of the temperature during controlled heating or cooling of the sample will result in a temperature versus time plot referred to as a heating or cooling curve. Typical cooling curves are shown in *Figure 4*. At the freezing point heat will be evolved which causes

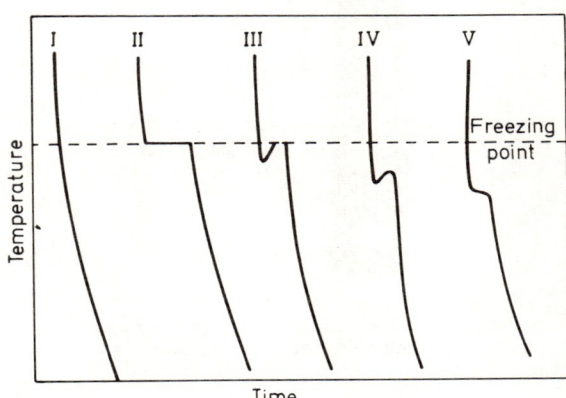

Figure 4. Typical cooling curves derived from Weber[180]. I—no freezing; II—ideal cooling curve with thermal arrest upon freezing; III—cooling curve with supercooling; IV, V—cooling curves with excessive rate of cooling, showing freezing points which are too low.

an arrest in the cooling process. Ideally, the sample temperature will remain constant until freezing is completed (curve II in *Figure 4*), but in practice supercooling is encountered (curve III in *Figure 4*). If the rate of cooling exceeds the rate of heat evolution, the freezing point observed will be too low (curves IV and V in *Figure 4*). In general superheating does not occur when pure substances are melted. This makes heating curves more desirable than cooling curves in this respect. However, heating curves are usually less well defined than cooling curves, since a cooling liquid is more likely to possess a uniform temperature because of the larger thermal conductivity and homogeneity of the melt, and furthermore small variations in the power input to the furnace may cause spurious abnormalities in the heating curve. The final choice between the use of heating and cooling curves is material dependent. Ceramics, in general, are easier to study by means of heating curves, but cooling curves should be preferred for most other substances.

In the case of pure metals this technique is capable of yielding results to about 0.001°C by means of heating curves[116] and to about 0.0002°C by

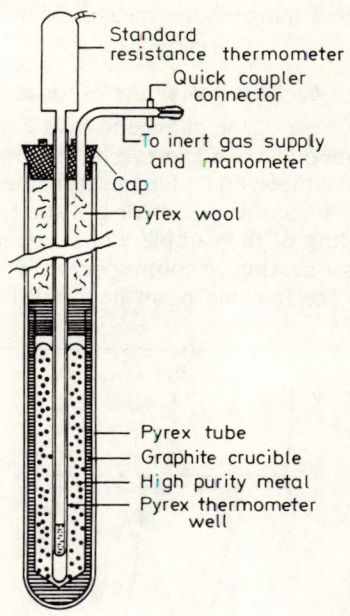

Figure 5. Typical sample holder for ingot of high-purity metal for accurate freezing-point determination. After McLaren[116].

means of cooling curves[2, 9, 113–115, 117, 118]. The experimental procedure is relatively simple in the platinum resistance thermometry range (up to 630°C, although work to 1 000°C is possible[53]). A typical sample set-up is shown in *Figure 5*. A graphite crucible is contained in a Pyrex glass tube which provides a means for removing the crucible from the furnace and also allows control of the inert atmosphere surrounding the melt. The resistance thermometer fits into a Pyrex well that extends to the bottom of the crucible. The type of furnace used is not important provided that the heat loss from the crucible is nearly radial when the axial length of the ingot is great enough to insure adequate immersion of the resistance thermometer.

Liquidus breaks on heating curves are, in practice, rounded because in a partly melted but unstirred ingot one finds convective thermal gradients in the melt which raise the temperature of the melt above the equilibrium liquidus temperature before all the solid has disappeared. On cooling, radial heat flow through a thickening solid shell and an absence of any spurious temperature gradients in the remaining liquid is easily assured by slowly solidifying the melt in a suitable furnace. The primary arrest on the cooling curve is always masked by supercooling and slow recalescence. During the recalescence, growth of the solid occurs at such a low rate that the latent heat released can only with difficulty raise the temperature of the liquid metal towards its liquidus temperature (*Figure 4*, III). For high-purity samples the melt temperature eventually becomes constant to better than 0.0001°C for periods of time that depend on the rate of freezing. This temperature can be identified with the liquidus point.

Two methods are used to shorten the duration of recalescence. Metals that nucleate their solid phase after a supercooling of less than about 1°C are solidified by means of induced freezing. The melt is allowed to cool normally until nucleation begins. The thermometer is then withdrawn and allowed to cool to room temperature, after which it is replaced. This results in rapid nucleation of a mantle of solid metal on the well which releases sufficient heat to raise the melt rapidly to its liquidus temperature. The recalescence of metals that supercool by large amounts is mainly retarded by the furnace itself. In such cases the furnace is allowed to cool slowly as usual, but when the temperature of the melt approaches the liquidus the sample holder is extracted into a cooler zone of the furnace. Upon nucleation the melt is immediately lowered into the hot zone of the furnace which is still close to the freezing temperature. After a few minutes of temperature equalization the familiar plateau results[113]. More complete experimental details can be found elsewhere[116]. The procedure in the case of molecular substances, such as organic materials, is not very different[65]. At higher temperatures use is made of thermocouples instead of platinum resistance thermometers, but once again with little modification in procedure.

It is possible to obtain a value of the freezing point of the pure material even if no pure material is available, using a correction procedure somewhat similar to that described in section 5 above. Complete details can be found elsewhere[110, 172].

Foex and his co-workers[54–57] have developed a sample cell for use at temperatures up to 3000°C. The sample, usually an oxide, is contained in a cooled capsule which is made to rotate rapidly. The central portion of the sample is then melted by means of a solar furnace. Centrifugal force creates a layer of molten material which only comes into contact with the unmelted powder itself, and so is not contaminated by the container. The shape of the molten layer is such as to approach a blackbody cavity. Upon masking off the solar radiation, rapid (20–100°C/second) cooling occurs, and a thermal arrest due to freezing is visible on a recorder plot of pyrometric output versus time. The method is convenient, and a large amount of useful data has been obtained, but it cannot be considered a precision method by any means. Supercooling probably always occurs, and the shape of the arrest is similar to those shown in *Figure 4* (V) and (IV). The results obtained can usually be expected to be too low. Compare, for instance, the almost 200°C supercooling observed at much lower cooling rates in the case of alumina[64].

7. Special Methods Used in Isolated Cases

Almost any property of a material will undergo a change upon melting. It follows that the melting point can be measured by measuring any such property as a function of temperature.

Predel and Arpshofen[136] measured the melting points of some metals by observing the rapid change of viscosity upon melting or freezing. Melting points of most substances can be measured with comparative ease by observing the discontinuous change of volume at the melting point (see section V for further details). In the case of metals, semi-conductors or ionic substances it may sometimes be convenient to measure melting points by observing the change of electrical resistivity with temperature. In general these methods are not seriously used at atmospheric pressure. At high pressures, however, several of these methods are routinely used for the measurement of melting points (see section IV below for further details).

III. Differential Thermal Analysis at Normal Pressure

Thermal arrest methods have been described in subsection II (6) above. These methods, although extremely accurate, are time-consuming. Furthermore, it is often found that the heat of melting or freezing is small in comparison with the total heat content of the sample and container, with the result that the thermal arrest is not very pronounced. There are three alternative ways of recording a heating or cooling curve in such a way that

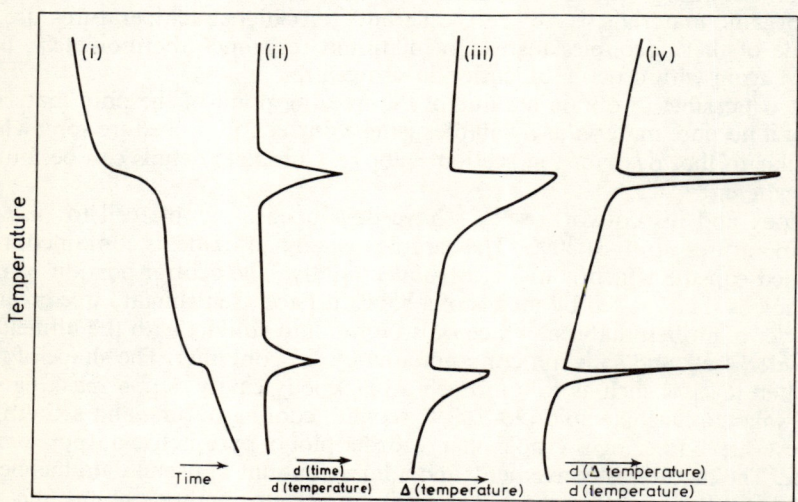

Figure 6. Typical examples of various methods of thermal analysis. (i) Thermal arrest; (ii) Inverse rate; (iii) DTA; (iv) Derived differential.

the arrest is made more obvious[58]: (a) inverse rate, (b) differential (DTA), (c) derived differential.

Figure 6 shows typical examples of the relative sensitivities encountered in these cases for a freezing point followed by a solid–solid phase transition. In the regular thermal arrest method, as we have seen, time versus sample temperature is recorded. The experimental assembly for the inverse rate method is similar, but here the quantity d(time)/d(temperature) is recorded versus temperature. A considerable improvement in the signal results. The experimental assembly for the DTA method is slightly different—two different thermocouples measure, in essence, the sample temperature and the oven temperature, and the difference ΔT between those two readings is recorded versus the reading of the sample thermocouple. Finally, the derived differential method, which uses the same experimental assembly as the DTA method, plots the quantity $d(\Delta T)/dT$ versus temperature. The method to be used depends on the requirements of the specific experiment, but methods (a) and (c) are very rarely used—perhaps because the simpler DTA method (b) is usually sufficiently sensitive, easy to record electronically, and commercially freely available. Only the DTA method will be described here in detail, but, if extra sensitivity is needed, it should be borne in mind that any commercial DTA arrangement can be easily converted to a derived differential method by suitable changes in the electronics of the system.

1. Historical

Roberts-Austen[142] was the first to modify the thermal arrest method in such a way that the difference of temperature ΔT between the sample and a non-transforming body at closely the same temperature in the furnace is measured directly. A typical early arrangement[36, 37] is diagrammatically shown in *Figure 7*. The use of a neutral body as reference material is import-

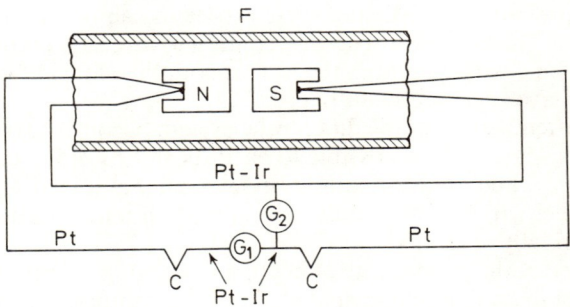

Figure 7. Typical early DTA arrangement in diagrammatical representation. After Carpenter[36]. F Furnace, N Neutral reference, S Sample, C Cold junctions, G_1 Reference galvanometer, G_2 Temperature potentiometer.

ant, since the effect of small variations in the furnace temperature is largely eliminated thereby. The reference body should have no phase changes in the temperature range of interest, and should have heat characteristics similar to those of the sample. Calcined alumina is a popular reference material for use with ceramics, but a non-reactive non-transforming metal is often used in metallurgy. Early workers used mechanical or photographically recording galvanometers to record results[78–80, 143], but with the gradual introduction of improved electronic X/Y and multipen recorders, together with d.c. microvolt amplifiers for the differential, older methods have become obsolete.

2. *Basic Method*

The basic method of making DTA measurements has not changed much since the time of the early workers mentioned above. A modern DTA set-up typically consists of a furnace which contains a block of metal with two usually axial cavities, one for the sample container and one for an identical container with the reference material. One thermocouple measures the temperature of the sample and another measures the temperature of the reference material. The sample and reference containers are usually suspended on the thermocouples in order to avoid direct contact with the metal block. The thermocouples are usually connected in such a way (see *Figure* 7) that the difference between the output voltages, i.e. the differential voltage, is directly amplified by means of a d.c. amplifier. The amplified signal is fed to, typically, one pen of a two-pen potentiometric recorder, while the output voltage from the sample thermocouple is fed to the other pen of the recorder. The furnace temperature can be changed at a smooth, and preferably constant, rate.

Simple as this is, there are nevertheless precautions to be taken. In the study of melting points it is often necessary to contain the sample in order to prevent reaction with the environment, decomposition or leakage. It is also desirable not to have direct contact between the thermocouple and the sample in order to avoid chemical attack on or deterioration of the thermocouple alloys, and to eliminate the possibility of contamination of a fresh sample by traces of a previous one. Most of these problems can be solved by sealing the sample in a suitable non-reacting container, usually of metal, which is equipped with a well for taking the thermocouple. However, glass or ceramic containers have been successfully used[95].

At higher temperatures the differential system becomes extremely sensitive to stray hum pickup. It is possible to earth both legs of the couples through condensors[127], and all leads should be shielded, but it is also possible to use a d.c. amplifier which is essentially insensitive to hum. Several such amplifiers are commercially available.

The choice of the proper rate of heating or cooling is important. The most accurate results can be obtained at slow rates of temperature change, but the peak heights decrease with decreasing rate, and this may make it difficult to observe the DTA signal at all. Commercial DTA set-ups usually are capable of variation in at least the range 1–20°C/minute.

The reference substance should be inert, and have thermal characteristics as similar as possible to those of the sample. Calcined alumina[71], calcined magnesia[123], sodium chloride[83], potassium chloride[77], or suitable metals are popular choices.

Before melting occurs, the temperatures of the sample and reference

thermocouples are essentially identical and no differential e.m.f. is generated. As soon as the phase change starts, a certain amount of heat must be exchanged before the sensitivity of the apparatus allows it to be detected. Therefore, the first deviation of the DTA curve from its baseline is more representative of the melting temperature than the peak maximum. It is also found that a change in rate of heating shifts the position of a peak maximum. From a theoretical analysis of the heat-flow situation Smyth[158] found that:
(a) If the differential temperature is plotted against the surface temperature of the sample, the point of initial departure from the baseline corresponds to the temperature of the phase change;
(b) If the differential temperature is plotted against the temperature of the centre of the sample while the outside is being heated uniformly, the peak maximum corresponds to the temperature of the phase change.

In addition, the indicated transition temperature is less than the true transition temperature by an amount called the thermal gradient constant[128]. This constant depends on the heating/cooling rate and on the construction of the sample container and block. In practice, it is simplest to calibrate the equipment by means of melting points which have been previously measured to high precision, since in general neither of Smyth's cases[158] (a) or (b) above applies exactly. It is notable, however, that, given low thermal gradients around the sample and reference material, the above uncertainties will be diminished at low heating/cooling rates. It is almost always desirable to use thermocouples which are small in comparison to the sample, so that loss of heat through the thermocouples is negligible.

Heating cycles are more desirable than cooling cycles for the measurement of melting points by means of DTA techniques. Supercooling is often difficult to recognize in the differential recorder trace upon cooling, while no such difficulty attends the measurement of a melting point upon heating.

In recent years it has often become necessary to use only a small amount of sample for DTA studies. For sample amounts of the order of 50 mg or less, heating rates are increased in order to obtain sharper peaks, but with the disadvantages mentioned above. Alternatively, the sample can be diluted with a large amount of an inert material[12], with attendant decrease in sensitivity as well as contamination problems. A more convenient method is to use microcells, often of platinum, with a hole on top for the sample (1–2 mg) and a hole at the bottom for the thermocouple. Heating rates of 0.2–1.0°C/minute are conveniently used[171].

Assessments of the effects of most variables mentioned here, as well as descriptions of detailed arrangements for specific cases, can be found elsewhere[4, 157].

3. Critical Assessment of DTA

DTA is perhaps the most generally convenient method of determining melting points. It is usable, without serious modification, up to temperatures of about 1 600°C, and can probably be used at even higher temperatures if more exotic thermocouple materials are utilized. It is relatively rapid, yields relatively accurate results, and requires relatively simple equipment. It can be used for materials available only in 1–2 mg quantities, and can easily be modified to study materials which require special atmospheres. Why, then, are other methods used at all? The answer is to be found in the word 'relative' above. DTA is rapid, but not as rapid as, e.g., an approximate melting-point determination by means of visible observation. DTA is accurate, but several orders of magnitude less so than a precision determination by means of thermal arrest methods. DTA does yield, with suitable calibration techniques, a simultaneous value for the latent heat of melting, but this value is considerably less accurate than a similar value determined by means of adiabatic calorimetry.

We can summarize by saying that DTA is a sound general-purpose method of investigating thermal phenomena, but in no case is it superior to a method designed specifically with a single aim in mind. However, if a temperature measurement accuracy of 0.2–1°C is sufficient, which it usually is in exploratory work, it is a method which can be strongly recommended, especially since it is easily adaptable to work that needs to be carried out under exotic conditions. In actual practice it is probably the most common method of determining melting points.

It should be pointed out that solid–solid phase changes and even some chemical reactions can be observed by means of DTA without any modification of the method. In fact, it is then often unnecessary to seal the sample in a closed container, because the likelihood of leakage is largely eliminated if no melting occurs.

IV. Methods Used for Melting Point Determination at High Pressure

The scope of this volume does not justify a detailed description of the many types of high-pressure high-temperature apparatus in use, and this can be found elsewhere[26,176]. Since rather different methods of melting point measurement are used in different types of apparatus, it is, however, necessary to classify briefly the main types of apparatus in general use:

(a) Apparatus using hydrostatic pressure media (gas or liquid), e.g. simple piston–cylinder types with seals[26,86,155,189], or more complex types working from compressors[15,88].

(b) Apparatus using quasi-hydrostatic pressure media (soft solids) but with pressure simply related to the piston force, e.g. internally heated piston–cylinder type[19,90,92].

(c) Opposed-anvil apparatus[7,14,25,70,129,141,156,177].

(d) Multi-anvil apparatus, including the tetrahedral anvil[10,66,74,107], the cubic anvil[144], the stepped-piston type devices[21,60], the belt[75,179], the girdle[186] and the multi-stage apparatus[178].

Little attention will be paid in what follows to the problem of pressure measurement—it will be assumed that that particular problem is satisfactorily solved, although this is not always so[134]. It will be convenient to discuss the methods of melting-point measurement with reference to the type of high-pressure apparatus used rather than with reference to the actual method.

In general, however, one finds that melting point determinations at high pressures do not differ substantially from the methods used at atmospheric pressure. Containment of the molten samples can be a problem, since most melting points rise rapidly with pressure. Containers which are ordinarily considered to be non-reactive often tend to react with the sample at the very much higher melting temperatures reached at even nominal pressures. Contamination is therefore a more serious problem than usual, especially in view of the fact that quite small samples are usually studied. Furthermore, the fragmentary data available indicate[26] that the entropy of melting either remains roughly constant or decreases along a high-pressure melting curve. This implies, using the Raoult–Van't Hoff relation, that for a normally rising melting curve the melting-point depression due to a given amount of impurity will increase considerably with pressure. However, decomposition of the samples is usually opposed by pressure, so that a sample which, for instance, loses a gaseous component upon melting under normal pressure, will usually behave perfectly at high pressures.

Melting normally occurs with an increase in specific volume, i.e. the liquid is less dense than the solid at the melting point. The influence of pressure on a first-order transition such as melting is given by the Clapeyron–Clausius relation

$$dT_m/dp = \Delta V/\Delta S$$

Since the entropy change ΔS is always positive, the sign of the slope of the melting curve is given by the sign of ΔV, which is usually positive. The majority of melting points therefore rise with pressure. Experimentally it is found that ΔV decreases with pressure considerably more rapidly than ΔS, and the slope therefore also decreases with pressure. In recent years a number of melting curves have been discovered which first rise with pressure, pass through a maximum and then drop with pressure. Examples of such substances are: cesium[89], europium[84], potassium nitrite[138], lithium chromate[131] and a number of semi-conducting materials, among others.

1. Historical

Probably the earliest measurements of melting points under pressure were made by Bunsen[30] and Thomson[174] to pressures of only a few bars. Dewar[48] studied the melting point of water to 0.7 kbar, but work in the 2–3 kbar pressure range was first done by Amagat[3], Damien[42,43], Barus[13] and Mack[109], followed by extensive work by Tammann[170] and Bridgman[26] to higher pressures. This work was carried out in hydrostatic apparatus, mostly by means of volume displacement, as will be described below.

2. Methods of Measuring Melting Points at High Pressures in Apparatus Using Hydrostatic Pressure Media

Since these types of apparatus are the only ones in which the pressure is highly accurately known, it is clear that the various precision methods of melting-point measurement are specially suitable in these cases. Perhaps the most common method used at temperatures below $\sim 250°C$ is the observation of the volume discontinuity upon melting[26,109,170].

The measurements are very simple in the piston–cylinder type of apparatus. The sample, suitably separated from the pressure-transmitting liquid if it is miscible with it, is immersed in this liquid, the whole apparatus, or at least the sample-containing part, is thermostatically controlled at the desired temperature, and a series of readings taken of piston displacement versus pressure. If supercooling is sufficiently severe that difficulty is encountered in obtaining a well-defined two-phase region, it is usually more convenient first to freeze the sample completely, and to make the measurements with decreasing pressure. A typical curve obtained on freezing a pure liquid is shown in *Figure 8*, as well as a typical curve obtained when impurities are present. A represents the freezing points and B the melting points (if the

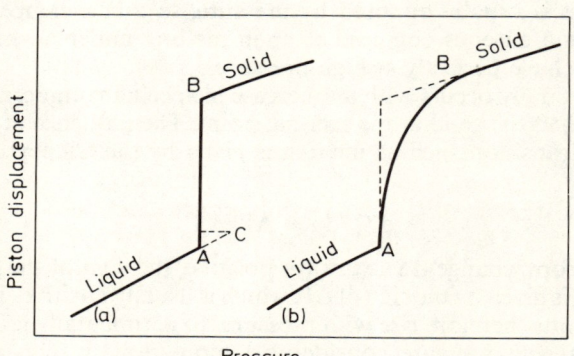

Figure 8. Typical curves of piston displacement versus pressure upon freezing (a) a pure liquid, and (b) an impure liquid (impurities insoluble in the solid phase), for the case where the liquid is less dense than the solid.

liquid is less dense than the solid). C indicates supercooling. It is clear from these curves that the volume change upon melting is directly obtained from measurements of this type. There is no reason why melting points obtained by this method should not be as accurate as the accuracy of pressure and temperature measurement allows. One of the best values for the freezing pressure of mercury—a fixed point on the pressure scale—was determined essentially in this way[125], yielding a total uncertainty of better than 0.05 per cent at 0°C.

A slight modification of this method consists of maintaining the pressure vessel at a constant temperature and releasing pressure from a value so high that the material is completely solidified. On crossing the melting curve melting, with automatic recovery of pressure, occurs. This is really a pressure-arrest method, as compared with the thermal-arrest method commonly used at atmospheric pressure. Deffet[46, 47] used this method extensively in studies of organic substances.

If the sample is electrically conducting, e.g. such as a metal, the melting point can conveniently be measured by measuring the electrical resistance of the sample as a function of either pressure or temperature, and observing the resistance discontinuity at the phase change. If the apparatus is externally heated, pressure is by far the more convenient parameter to change, and Bridgman[22, 23] used this technique, with the metallic sample contained in a non-conducting tube, to determine the melting curves of several low-melting metals.

The plugged-capillary method[38, 106, 121, 122, 154] is peculiar to high pressure, and enjoyed some vogue during the 1920–40 period. The sample is contained in two vessels connected by a capillary, the temperature of which is lowered until a pressure increment in one vessel fails to be transmitted to the other because of solidification in the capillary. The method can be criticized[24] in view of the fact that the solid plug in the capillary must be subjected to shearing stress, but in some cases it may be a convenient method if the highest accuracy is not desired.

The quantitative thermal arrest method is not very suitable for high-pressure studies in view of the virtual impossibility of obtaining good thermal insulation between the sample and the usually massive pressure cell. The apparatus acts as a heat sink which masks the desired effects. Dow and Hibsham[50] maintained their pressure vessel at constant temperature in a bath, lowered the pressure across the melting curve and detected the beginning of melting by the temperature reaction of a thermocouple embedded in the sample, with good agreement with volumetric methods. However, this is not an equilibrium method and requires rather awkward manipulation. More recently, however, this method has once again come into its own in conjunction with internal heating in more elaborate set-ups with inert gas pressure transmission. A relatively large sample is used (usually several grams) with a thermocouple embedded in it, but separated from it by means of, e.g., a thin ceramic tube. With a non-metal, the sample can be contained in a sealed thin-walled platinum tube equipped with a well for the thermocouple. The sample cell is surrounded by a furnace. Due to the relatively high thermal losses between the sample and the pressure-transmitting gas no quantitative latent heat values can be obtained, but the precision of melting-point determination is excellent. The limiting factor at present is the uncertainties in the effect of pressure on the thermocouples[76], but apart from this correction the results obtained can be within $\pm 0.01°C$. Unfortunately the *uncertainty* in the thermocouple pressure correction is as much as several degrees in the region 1 000°C, 30 kbar. Clark[40], Babb and his co-workers[5, 6, 112] and a number of Russian workers[17, 33–35] use the above technique. Particularly in the case of Babb and his co-workers the results can be considered precise. A thorough description of a suitable experimental procedure can be found elsewhere[112].

Differential thermal analysis is as simple and convenient a method as at atmospheric pressure. It is desirable to use an internally heated assembly, involving therefore a furnace which is also at high pressure. If an inert gas is used as pressure-transmitting fluid, this is relatively easy to obtain. Gibson[61] used this method as early as in 1928, and high-pressure DTA has since become one of the standard and reliable methods of studying melting phenomena at high pressures[67, 68, 96, 187, 188]. A popular method is to seal the sample into several thin-walled platinum tubes which surround, and are in thermal contact with, one thermocouple junction, while the reference junction measures the furnace temperature. The high thermal conductivity of the pressure-transmitting fluid usually necessitates somewhat higher heating/cooling rates than can be used at atmospheric pressure, but this drawback can be minimized by using larger samples. It is relatively simple to obtain a melting-point precision of $\sim 0.1°C$ by means of high-pressure DTA, but the accuracy is limited by the uncertainty in the effect of pressure on the thermocouples.

3. Methods of Measuring Melting Points at High Pressures in Apparatus Using Quasi-hydrostatic Pressure Media

The type example of this apparatus is the internally-heated piston–cylinder arrangement[19, 90, 92]. The most accurate pressure measurements in this type of apparatus are carried out when internal heating is eschewed and the apparatus considered solely as a compressor. The bore of the pressure vessel immediately above the piston is filled with a closed thin-walled cup of indium or Teflon containing the sample, and the remainder of the bore with talc or pyrophyllite cylinders containing the thermocouple in a steel well. Temperatures differing from ambient can be obtained by external heating or cooling of the bore but are usually limited to $\sim 200°C$. However, recently a slight modification of the arrangement was described[99, 100] which allows temperatures up to $\sim 450°C$ to be obtained. When the desired temperature is obtained, pressure is raised until the sample begins to freeze, as indicated by a discontinuity in the curve of piston displacement versus piston load. The friction can be largely relieved by rotating the piston. When a steady pressure reading is obtained, the sample is completely frozen and pressure is slowly

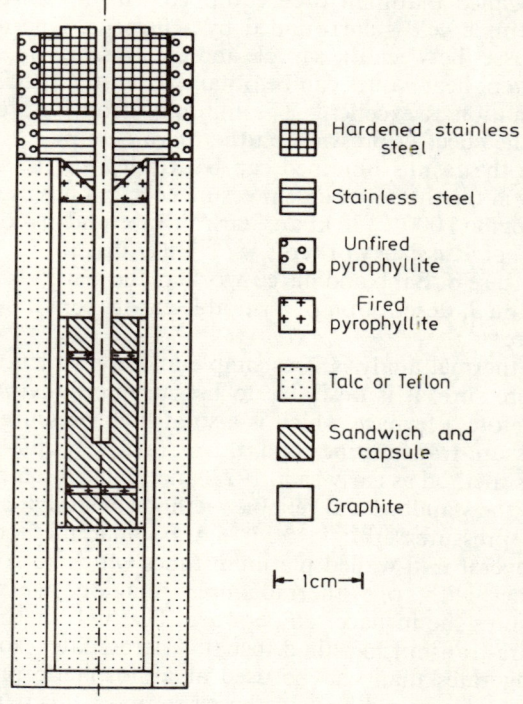

Figure 9(i)

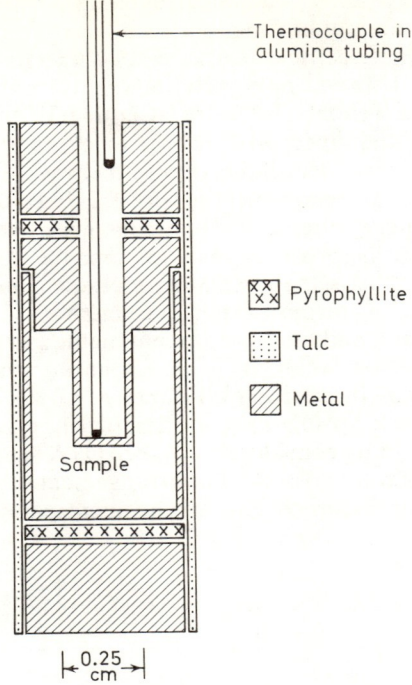

Figure 9(ii)

Figure 9. Arrangement for DTA in a piston–cylinder apparatus: (i) Complete furnace assembly; (ii) Detail of sample capsule.

released until melting starts, when the piston is once again rotated to relieve friction. Complete descriptions of the method can be found elsewhere[91, 132]. The method is accurate, with typical uncertainties of between 0.025 and 0.1 kbar, but it is also laborious, and is not often used for melting points, although it has the advantage of yielding the volume change upon melting directly. The melting curve of ammonium fluoride was recently determined in essentially this way[97].

Usually, however, the apparatus is used in the internally-heated mode. Under these conditions pressure measurement is reliable to ~0.5 kbar if suitable corrections are made[134]. Several methods of melting-point determination now become possible. The simplest one is probably the quenching technique described before, usually applied mainly to ceramic systems. The sample is sealed in a thin-walled platinum tube and brought to the desired pressure. Temperature is raised to the desired value, and then rapidly lowered while maintaining pressure. After release of pressure the sample is examined to determine whether or not melting has occurred[20, 45, 82, 105, 153]. In most studies the workers contented themselves with bracketing the melting point within ~10°C. The quenching method is, of course, also used in hydrostatic type apparatus with internal heating.

Measurement of the electrical resistance of a metallic sample is often a convenient method. If the sample melts at relatively low temperatures, it can be contained in a Teflon tube sealed at both ends by the electrical contacts. The resistance jump upon melting or freezing, changing either pressure or temperature, is a convenient indication of melting[94]. High melting point metals can be studied in the form of a wire[162]. Since the pressure is only quasi-hydrostatic, such a wire generally breaks upon melting, thus yielding a catastrophic increase in apparent resistance upon melting. This is, of course, a one-shot procedure, as is the quenching method. Some ceramic systems have also been studied by means of resistance methods[93].

By far the commonest and most convenient method in this type of apparatus is differential thermal analysis. DTA in the internally-heated piston–cylinder apparatus was pioneered by Kennedy and his co-workers[41, 89, 92] and especially the sample capsule was considerably improved by other workers afterwards[130, 138], yielding considerably higher DTA sensitivity and making lower heating/cooling rates possible. The arrangement presently in use in this laboratory is shown in *Figure 9*. This arrangement permits the detection of

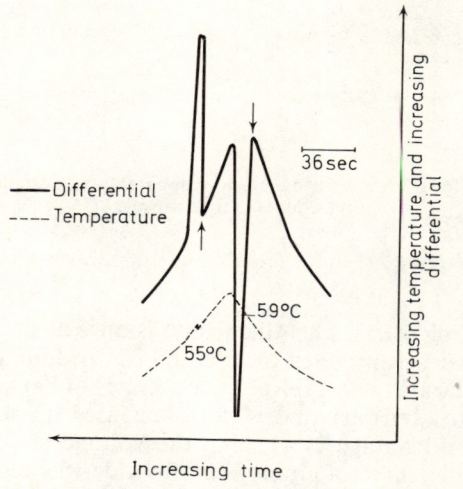

Figure 10. Typical DTA signals obtained upon melting and freezing ice VI at 16.0 kbar in a cell similar to that shown in *Figure 9*. After Pistorius, Rapoport and Clark[135].

a sharp phase change such as melting at heating/cooling rates of as low as 0.05°C/second if the latent heat is of the order of 2 kJoules/mole or more. Higher heating/cooling rates are, of course, necessary for smaller or more sluggish changes. A typical heating/cooling cycle, representing the melting and freezing (with super-cooling) of ice VI at 16.0 kbar, is shown in *Figure 10*. The filled sample capsules are sealed by metal-to-metal cold welding *in situ* at ~15 kbar. Pressure is then released to ~2 kbar and measurements are started. Particularly at low temperatures the friction-corrected pressure values obtained do not yield true sample pressures, and further corrections must be made to obtain agreement with hydrostatic results[134]. An important precaution is to use a d.c. amplifier for the differential signal that is completely insensitive to a.c. hum pickup, since the graphite furnace, heated by up to 500 A, is sufficiently close to the thermocouples to allow several volts of a.c. hum pickup. Perhaps the biggest problem in this type of work is to find a capsule metal which does not react with the sample at elevated pressures and temperatures.

DTA at atmospheric pressure is not quantitative, i.e. the area under the peak is not an exact indication of the latent heat. Nevertheless it is possible to obtain incidental latent heat values which are within five to ten per cent of the true values. This is not so with high-pressure DTA in a piston–cylinder apparatus. The peak area appears to depend especially strongly on the capsule metal, and on slight geometrical differences in arrangement. Calibration of peak areas has not yet been successfully done.

The precision of the method is ~0.5 kbar and ~0.2°C, but the temperature errors are further subject to the uncertainty in the effect of pressure on the thermocouples[76]. It is not possible here to refer to all papers on melting point determination at high pressures by means of DTA. Suffice it to say that Kennedy and his co-workers at the UCLA have studied most of the elements, Jayaraman and his co-workers at the Bell Telephone Laboratories have studied many rare earth elements and other magnetic or semi-conducting materials, and Pistorius and his co-workers in Pretoria have studied many simple inorganic substances. There are also a number of other groups, marginally less active, in the field.

4. *Methods of Measuring Melting Points at High Pressures in Opposed-anvil Apparatus*

The opposed-anvil type of apparatus can be designed to yield pressures of up to 500 kbar. However, pressure is not well known. Only extremely small samples (1–10 mg) can usually be accommodated. Contamination of samples is therefore a serious problem, especially if they are uncontained. These factors unite to make melting-point determination in this type of apparatus quite difficult.

The quenching technique is the most obvious one to use. However, the large pressure gradients present induce recrystallization, fracture, etc. and it is usually very nearly impossible to distinguish between a sample which has melted and one which has not. Melting can be visually detected in the heated optical device[7] and also in the externally heated diamond cell[177], but this can hardly be considered *measurements* of melting points. The melting point of ice VII has been measured to ~170 kbar in a simple squeezer[133,135] by dissolving a small amount of an ionic salt in the water and measuring the resistance jump upon melting. While temperatures were probably within ~5°C, it is doubtful whether the pressure was known within ten per cent. Perhaps it is simplest to conclude that reliable melting-point determinations have not, as yet, been made in apparatus of this type. If this should ever be done, some modification of either the resistance jump method or DTA would be the most promising of success.

5. *Methods of Measuring Melting Points at High Pressures in Multiple-anvil Apparatus*

The quenching method is a reliable, if laborious, method also for this type of apparatus. The technique needed is not essentially different from that used in the piston–cylinder, except where minor modifications are caused by apparatus design[1, 98, 151].

Melting points of metals and semi-conductors can be studied by means of resistance methods. The parting of the metal wire upon melting is usually taken as the indication of melting. If parting does not occur, as for instance when the sample is encapsulated, the resistance jump upon melting can often be used as an indication[27, 51, 73, 164–166].

Lees and Williamson[102] used simple thermal arrest to measure the freezing points of a number of metals and semi-conductors. The experimental arrangement is described in their paper. However, detectable arrests necessitated cooling rates of $\sim 400°C$/second, and the likelihood of gross supercooling frequently rules out this method, although it can occasionally be useful[167].

Perhaps the most sophisticated, as well as the most accurate, method for the family of multiple-anvil apparatuses is differential thermal analysis. The detailed arrangement is considerably more complex than with a piston–cylinder, but does not differ in essence[39, 168, 169], and very satisfactory DTA signals can be obtained at acceptable heating/cooling rates.

Finally, the tetrahedral anvil x-ray diffraction apparatus[10] allows melting point measurement simply by noting the disappearance of x-ray powder diffraction peaks upon melting. This is a particularly useful method in cases where a thermal arrest or a resistance jump had been found by other methods, but where doubts remain as to whether this represents melting or a solid–solid transition, as in the case of the event near 190 kbar at room temperature in rubidium[8, 28].

In general one can conclude that the resistance, quenching and DTA methods are best suited for melting point measurement in this family of apparatuses. The largest remaining problem is pressure measurement, since the room temperature calibrations are not valid at elevated temperatures[29, 103] and pressures at these temperatures may be in error by up to ~ 10 kbar.

6. *Conclusion*

In addition to the methods described above, the measurement of melting points at extemely high pressures by means of shock waves must be mentioned[81, 120]. However, the whole question of shock waves will be discussed in more detail elsewhere in this volume.

No claim is made that all methods that have been, or can be, used to measure melting points at high pressures are mentioned above. The field offers a wide scope for the ingenuity of the experimenter, and the measurement of almost any property of the sample as a function of pressure or temperature may reveal the melting point. In fact, measurements of the dielectric properties, magnetic properties and thermal conductivity, to mention but a few, have been used to indicate the occurrence of melting. The apparatus at the experimenter's disposal and his particular interests may largely dictate the method most suitable for his case.

V. Methods for Determining Volume Changes on Melting

The most obvious method for determining the volume change upon melting is to measure the density of the liquid as a function of temperature by any of a number of established methods discussed in Chapter 9 of this volume, and to measure the density of the solid, preferably by means of x-ray techniques, as a function of temperature, and to extrapolate these two curves to the known melting point. Unfortunately errors can often be introduced by anomalously high thermal expansion of the solid just below the melting point. Furthermore, one is in such a case measuring two large quantities, with concomitant errors, when only the difference between them is required.

Once again referring to Chapter 9 of this volume in order to save space, it is clear that the immersion of the solid in a non-reacting liquid with known p–V–T properties will yield a reliable value for the volume change upon melting if the thermal expansion of the mixture is then determined. In essence, this is the most generally used group of methods[126]. The pitfalls are found in the properties of the sample. If, for instance, it slowly decomposes near its melting point, a value for the volume change upon melting may be obtained that is wildly in error. It is probable that the poor agreement between different determinations for, for instance, the alkali halides[52, 146, 160] is due to this, or to related phenomena such as the difficulty of completely outgassing the sample.

A number of methods of determining the melting point at atmospheric and elevated pressures described above depend on the volume change, and these methods will incidentally yield good values for the volume change upon melting. It is hardly necessary to describe these once again.

Especially at atmospheric pressure it appears that, while not difficult in principle, the accurate measurement of volume changes upon melting is attended by a number of sample-dependent experimental problems. On the other hand, the latent heat of melting is easily and accurately measurable by calorimetric methods. Furthermore, the melting curve, i.e. the melting point as a function of pressure, is easily and relatively accurately measurable especially in the hydrostatic pressure range, that is, below ~ 30 kbar. In such cases the most accurate and certainly the most reliable values for the volume change will usually be yielded by the Clapeyron–Clausius relation

$$dT_m/dp = T\,\Delta V/\Delta H$$

using the initial slope of the melting curve together with the melting point and latent heat of melting at atmospheric pressure. At higher pressures, of course, the latent heat of melting is not known. However, many of the sources of error involved in a direct atmospheric pressure determination of ΔV become of negligible importance at elevated pressures. Complete outgassing of the specimen is not vital, and decomposition is very often prevented by pressure. In such a case, therefore, the Clapeyron–Clausius relation will yield good values for the latent heat of melting as a function of pressure, when T_m and ΔV are measured as functions of pressure.

VI. Abstract

The various methods used to determine melting points and volume changes upon melting of pure substances at atmospheric as well as elevated pressures are reviewed. Particular attention is paid to precision methods, with brief assessments of the possible sources of error.

VII. References

[1] Akimoto, S., E. Komada and I. Kushiro. *J. Geophys. Res.* **72**, 679 (1967).
[2] Alieva, P. Z. *Trudy Inst. Metrol. D. I. Mendeleev No. 36*, (96), 9 (1958).
[3] Amagat, E. H. *CR Acad. Sci., Paris*, **105**, 165 (1887).
[4] Arens, P. L., *A Study of the Differential Thermal Analysis of Clays and Clay Minerals*, Excelsiors Foto-offset: S'Gravenhage, Holland (1951).
[5] Babb, S. E., Jr. *J. Chem. Phys.* **37**, 922 (1962).
[6] Babb, S. E. Jr., P. E. Chaney and B. B. Owens. *J. Chem. Phys.* **41**, 2210 (1964).
[7] Balchan, A. S. and H. G. Drickamer. *Rev. Sci. Instrum.* **31**, 511 (1960).
[8] Balchan, A. S. and H. G. Drickamer. *Rev. Sci. Instrum.* **32**, 308 (1961).
[9] Barber, C. R. and A. Horsford. *Proc. Roy. Soc. (London) A*, **247**, 214 (1958).
[10] Barnett, J. and H. T. Hall. *Rev. Sci. Instrum.* **35**, 175 (1964).
[11] Barrall, E. M., R. S. Porter and J. F. Johnson. *Molec. Cryst.* **3**, 299 (1968).
[12] Barrall, E. M. and L. B. Rogers. *Analyt. Chem.* **34**, 1106 (1962).
[13] Barus, C. *Bull. US Geol. Surv. No. 92* (1892).
[14] Bassett, W., T. Takahashi and P. Stook. *Rev. Sci. Instrum.* **38**, 37 (1967).
[15] Birch, F., E. C. Robertson and S. P. Clark. *Industr. Engng Chem.* **49**, 1965 (1957).
[16] Bogue, R. H. *The Chemistry of Portland Cement*, p 237. Reinhold: New York (1947).
[17] Boksha, S. S. and G. P. Shakhovskoi. *Pribory i Tekh. Eksp.* **3**, 86 (1958).
[18] Bowen, N. L. and J. F. Schairer. *Amer. J. Sci.* **24**, 177 (1932).
[19] Boyd, F. R. and J. L. England. *J. Geophys. Res.* **65**, 741 (1960).
[20] Boyd, F. R. and J. L. England. *Yearb. Carnegie Instn*, **62**, 134 (1963).
[21] Bradley, R., D. Munro and M. Whitfield. *J. Sci. Instrum.* **42**, 714 (1965).
[22] Bridgman, P. W. *Proc. Amer. Acad. Arts Sci.* **56**, 61 (1921).
[23] Bridgman, P. W. *Proc. Amer. Acad. Arts Sci.* **60**, 385 (1925).
[24] Bridgman, P. W. *Phys. Rev.* **46**, 930 (1934).
[25] Bridgman, P. W. *Proc. Amer. Acad. Arts Sci.* **71**, 387 (1937).
[26] Bridgman, P. W. *The Physics of High Pressure*, Bell: London (1949).
[27] Bundy, F. P. *Phys. Rev.* **110**, 314 (1958).
[28] Bundy, F. P. *Phys. Rev.* **115**, 274 (1959).
[29] Bundy, F. P. *Spec. Publ. US Nat. Bur. Stand. No. 326*, 263 (1971).
[30] Bunsen, R. *Ann. Phys., Lpz.* **81**, 153 (1850).
[31] Bunsen, R. *Ann. Phys., Lpz.* **141**, 1 (1870).
[32] Burgess, G. K. *Sci. Pap. US Nat. Bur. Stand. No. 198*, 205 (1920).
[33] Butuzov, V. P. and M. G. Gonikberg. *Dokl. Akad. Nauk SSSR*, **91**, 1083 (1953).
[34] Butuzov, V. P., M. G. Gonikberg and S. P. Smirnov. *Dokl. Akad. Nauk SSSR*, **89**, 651 (1953).
[35] Butuzov, V. P., E. G. Ponyatovskii and G. P. Shakhovskoi. *Dokl. Akad. Nauk SSSR*, **109**, 519 (1956).
[36] Carpenter, H. C. H. *Coll. Res. Nat. Phys. Lab., London*, **2**, 52 (1907).
[37] Carpenter, H. C. H. and B. F. E. Keeling. *J. Iron Steel Inst., London*, **65**, 224 (1904).
[38] Chester, P. F. and J. S. Dugdale. *Phys. Rev.* **95**, 278 (1954).
[39] Churagulov, B. R., E. M. Feklichev, Ya. A. Kalashnikov and L. F. Vereshchagin. *Dokl. Akad. Nauk. SSSR*, **163**, 629 (1965).
[40] Clark, S. P., Jr, *J. Chem. Phys.* **31**, 1526 (1959).

41. Cohen, L. H., W. Klement Jr and G. C. Kennedy. *J. Phys. Chem. Solids*, **27**, 179 (1966).
42. Damien, B. *C. R Acad. Sci., Paris*, **108**, 1159 (1889).
43. Damien, B. *C. R Acad. Sci., Paris*, **112**, 785 (1891).
44. Darken, L. S. and R. W. Gurry. *Physical Chemistry of Metals*, 535 pp. McGraw-Hill: New York (1953).
45. Davis, B. and J. England. *J. Geophys. Res.* **69**, 1113 (1964).
46. Deffet, L. *Bull. Soc. Chim. Belg.* **44**, 41 (1935).
47. Deffet, L. *Bull. Soc. Chim. Belg.* **49**, 223 (1940).
48. Dewar, J. *Proc. Roy. Soc. (London)*, **30**, 533 (1880).
49. Douglas, T. B. and W. H. Payne. 'Physical properties of high-temperature materials. I. New apparatus for the precise measurement of heat content and heat capacity from 0° to 1500°C', *WADC Tech. Rep. No. 57-374*, Part I; *ASTIA Document No. AD 142119*, 36 pp. (1957).
50. Dow, R. B. and H. B. Hibsham. *J. Chem. Phys.* **5**, 960 (1937).
51. Dudley, J. D. and H. T. Hall. *Phys. Rev.* **118**, 1211 (1960).
52. Dworkin, A. S. and M. A. Bredig. *J. Phys. Chem.* **64**, 269 (1960).
53. Evans, J. P. and G. W. Burns. in *Temperature*, Vol. III, p 313. Edited by F. G. Brickwedde. Reinhold: New York (1962).
54. Foex, M. *Bull. Soc. Chim. France*, 137 (1962).
55. Foex, M. *Solar Energy*, **9**, 61 (1965).
56. Foex, M. and J. Traverse. *Bull. Soc. Franç. Minér. Cristallogr.* **89**, 184 (1966).
57. Foex, M. and J. Traverse. *Rev. Hautes Temp. Réfract.* **3**, 429 (1966).
58. Foote, F. D., C. O. Fairchild and T. R. Harrison. *Technol. Pap. US Nat. Bur. Stand.* **14**, 1–326 (1920).
59. Furukawa, G. J., T. B. Douglas, R. E. McCoskey and D. C. Ginnings. *J. Res. Nat. Bur. Stand.* **57**, 67 (1956).
60. Giardini, A., J. E. Tydings and S. B. Levin. *Amer. Mineral.* **45**, 217 (1960).
61. Gibson, R. E. *J. Phys. Chem.* **32**, 1197 and 1206 (1928).
62. Ginnings, D. C. and R. J. Corruccini. *J. Res. Nat. Bur. Stand.* **38**, 583 and 593 (1947); **39**, 309 (1947).
63. Ginnings, D. C., T. B. Douglas and A. F. Ball. *J. Res. Nat. Bur. Stand.* **45**, 23 (1950).
64. Gitlesen, G. and K. Motzfeldt. *Rev. Hautes Temp. Réfract.* **3**, 343 (1966).
65. Glasgow, A. R. Jr, A. J. Streiff and F. D. Rossini. *J. Res. Nat. Bur. Stand.* **35**, 355 (1945).
66. Gonikberg, M., D. Tsiklis and A. Opekunov. *Dokl. Akad. Nauk SSSR*, **129**, 88 (1959).
67. Göranson, R. W. and F. C. Kraček. *J. Phys. Chem.* **36**, 913 (1932).
68. Göranson, R. W. and F. C. Kraček. *J. Chem. Phys.* **3**, 87 (1935).
69. Green, H. S., 'The structure of liquids', in S. Flügge (Ed.), *Encyclopedia of Physics*, Vol. X, pp 1–133. Springer: Berlin (1960).
70. Griggs, D. and G. C. Kennedy. *Amer. J. Sci.* **254**, 722 (1956).
71. Grimshaw, R. W., E. Heaton and A. L. Roberts. *Trans. Brit. Ceram. Soc.* **44**, 76 (1945).
72. Gutt, W. *J. Sci. Instrum.* **41**, 393 (1964).
73. Hall, H. T. *J. Phys. Chem.* **59**, 1144 (1955).
74. Hall, H. T. *Rev. Sci. Instrum.* **29**, 267 (1958).
75. Hall, H. T. *Rev. Sci. Instrum.* **31**, 125 (1960).
76. Hannemann, R. E., H. M. Strong and F. P. Bundy. *Spec. Publ. US Nat. Bur. Stand. No 326*, 53 (1971).
77. Harvey, A. E., Jr, M. T. Edmison, E. D. Jones, R. A. Seybert and K. A. Catto. *J. Amer. Chem. Soc.* **76**, 3270 (1954).
78. Hoffmann, F. and R. Rothe. *Z. InstrumKde*, **25**, 273 (1905).
79. Hoffmann, F. and R. Rothe. *Z. Phys. Chem.* **55**, 113 (1906).
80. Hoffmann, F. and R. Rothe. *Z. Phys. Chem.* **59**, 448 (1907).
81. Horie, Y. *J. Phys. Chem. Solids*, **28**, 1569 (1967).
82. Hsu, L. *J. Geophys. Res.* **72**, 4235 (1967).
83. Jaffray, J. *J. Rech. CNRS*, 153 (1947).
84. Jayaraman, A. *Phys. Rev.* **135A**, 1056 (1964).
85. Jessup, R. S. *J. Res. Nat. Bur. Stand.* **55**, 317 (1955).
86. Johnson, D. and P. Heydemann. *Rev. Sci. Instrum.* **38**, 1294 (1967).
87. Jones, G. *Glass*, Methuen: London (1956).
88. Kats, G. and I. Ryabchikov. *Izvest. Akad. Nauk SSSR, Ser. Geol.* **5**, 121 (1968).
89. Kennedy, G. C., A. Jayaraman and R. C. Newton. *Phys. Rev.* **126**, 1363 (1962).

90 Kennedy, G. C. and P. N. LaMori in *Progress in Very High Pressure Research*, edited by F. P. Bundy, W. R. Hibbard and H. M. Strong. Wiley: New York (1961).
91 Kennedy, G. C. and P. N. LaMori. *J. Geophys. Res.* **67**, 851 (1962).
92 Kennedy, G. C. and R. C. Newton in *Solids under Pressure*, edited by W. Paul and D. M. Warschauer. McGraw-Hill: New York (1963).
93 Khitarov, N. T. and A. B. Slutskii. *J. Chim. Phys.* **64**, 1085 (1967).
94 Klement, W., A. Jayaraman and G. C. Kennedy. *Phys. Rev.* **131**, 1 (1963).
95 Kraček, F. C. *J. Phys. Chem.* **33**, 1281 (1929).
96 Kraček, F. C. *Trans. Amer. Geophys. Union*, **27**, 364 (1946).
97 Kuriakose, A. K. and E. Whalley. *J. Chem. Phys.* **48**, 2025 (1968).
98 Kushiro, I., Y. Syono and S. Akimoto. *J. Geophys. Res.* **73**, 6023 (1968).
99 LaMori, P. N., 'Compressibility of rocks and minerals to 450°C and 36 kb and their application to the upper mantle', *Ph.D. Thesis*, Northwestern University, 205 pp (1967).
100 LaMori, P. N. *Spec. Publ. US Nat. Bur. Stand. No. 326*, 279 (1971).
101 Landolt, H. *Z. Phys. Chem.* **11**, 349 (1889).
102 Lees, J. and B. H. J. Williamson. *Nature, London*, **208**, 278 (1965).
103 Leger, J. M., C. Susse and B. Vodar. *Spec. Publ. US Nat. Bur. Stand. No. 326*, 251 (1971).
104 Levin, E. M., C. R. Robbins and H. F. McMurdie. *Phase Diagrams for Ceramists*, 601 pp. American Ceramic Society: Columbus, Ohio (1964).
105 Lindsley, D. *Amer. Mineral.* **51**, 1793 (1966).
106 Lisman, J. H. C. *Smeltlijnen van Gecondenseerde Gassen*. Eduard Ijdo: Leiden (1934).
107 Lloyd, E., U. O. Hutton and D. Johnson. *J. Res. Nat. Bur. Stand.* **63C**, 59 (1959).
108 Lovejoy, D. R. in *Temperature*, Vol. III, p 487. Edited by F. G. Brickwedde. Reinhold: New York (1962).
109 Mack, E., *CR Acad. Sci., Paris*, **127**, 361 (1898).
110 Mastrangelo, S. V. R. and R. W. Dornte. *J. Amer. Chem. Soc.* **77**, 6200 (1955).
111 McCullough, J. P. and D. W. Scott (Eds.), *Experimental Thermodynamics*, Vol. I: *Calorimetry of Non-reacting Systems*, 606 pp. Butterworths: London (1968).
112 McDaniel, M. L., S. E. Babb Jr and G. J. Scott. *J. Chem. Phys.* **37**, 833 (1962).
113 McLaren, E. H., *Canad. J. Phys.* **35**, 1086 (1957).
114 McLaren, E. H., *Canad. J. Phys.* **36**, 585 (1958).
115 McLaren, E. H., *Canad. J. Phys.* **36**, 1131 (1958).
116 McLaren, E. H. in *Temperature*, Vol. III, p 185, edited by F. G. Brickwedde. Reinhold: New York (1962).
117 McLaren, E. H. and E. G. Murdock, *Canad. J. Phys.* **38**, 100 (1960).
118 McLaren, E. H. and E. G. Murdock, *Canad. J. Phys.* **38**, 577 (1960).
119 McNamara, E. P., *Ceramics*, 350 pp. Pennsylvania State University Press: State College, Pennsylvania, (1949).
120 McQueen, R. G., W. J. Carter, J. N. Fritz and S. P. Marsh. *Spec. Publ. US Nat. Bur. Stand. No. 326*, 219 (1971).
121 Michels, A., B. Blaisse and J. Hoohschagen. *Physica*, **9**, 565 (1942).
122 Mills, R. L. and E. R. Grilly. *Phys. Rev.* **99**, 480 (1955).
123 Mishin, V. P. and A. I. Garbuzov. *Biokhimiya*, **16**, 416 (1951).
124 Muan, A., *J. Metals*, **7**, 965 (1955).
125 Newhall, D. H., L. H. Abbot and R. A. Dunn, in *High-pressure Measurement*, Edited by A. A. Giardini and E. C. Lloyd. Butterworths: Washington (1962).
126 Partington, J. R., *An Advanced Treatise on Physical Chemistry*, Vol. III, pp 461 et seq. Longmans, Green: London (1952).
127 Pask, J. A. and M. F. Warner. *Bull. Amer. Ceram. Soc.* **33**, 168 (1954).
128 Penther, C. J., S. T. Abrams and F. H. Stross. *Analyt. Chem.* **23**, 1459 (1951).
129 Perez-Albuerne, E., K. Forsgren and H. Drickamer. *Rev. Sci. Instrum.* **35**, 29 (1964).
130 Pistorius, C. W. F. T., *J. Phys. Chem. Solids*, **26**, 1543 (1965).
131 Pistorius, C. W. F. T., *J. Phys. Chem. Solids*, **28**, 1811 (1967).
132 Pistorius, C. W. F. T. and J. B. Clark. *High Temp. High Press.* **1**, 561 (1969).
133 Pistorius, C. W. F. T., M. C. Pistorius, J. P. Blakey and L. J. Admiraal. *J. Chem. Phys.* **38**, 600 (1963).
134 Pistorius, C. W. F. T., E. Rapoport and J. B. Clark. *Rev. Sci. Instrum.* **38**, 1741 (1967).
135 Pistorius, C. W. F. T., E. Rapoport and J. B. Clark. *J. Chem. Phys.* **48**, 5509 (1968).
136 Predel, B. and I. Arpshofen. *Z. Naturf.* **23a**, 2052 (1968).

[137] Rankin, G. A. and F. E. Wright. *Amer. J. Sci.* **39**, 1 (1915).
[138] Rapoport, E. *J. Chem. Phys.* **45**, 2721 (1966).
[139] Riley, B. *J. Sci. Instrum.* **41**, 504 (1964).
[140] Riley, B. *Rev. Hautes Temp. Réfract.* **3**, 327 (1966).
[141] Ringwood, A. and A. Major. *Earth Planet. Sci. Lett.* **1**, 241 (1966).
[142] Roberts-Austen, W. C. *Proc. Inst. Mech. Engrs, London*, **1**, 35 (1899).
[143] Saladin, E. *Iron Steel Metall. Metallogr.* **7**, 237 (1904).
[144] Samara, G., A. Henius and A. Giardini. *J. Bas. Engng*, 729 (1964).
[145] Schairer, J. F. in *Physicochemical Measurements at High Temperatures*, p 117. Edited by J. Bockris, J. L. White and J. D. MacKenzie. Butterworths: London (1959).
[146] Schinke, H. and F. Sauerwald. *Z. Anorg. Allg. Chem.* **287**, 313 (1956).
[147] Schneider, S. J. *Spec. Publ. US Nat. Bur. Stand. No. 303*, 19 (1969).
[148] Schneider, S. J. and C. L. McDaniel. *J. Res. Nat. Bur. Stand.* **71A**, 317 (1967).
[149] Schneider, S. J. and J. L. Waring. *J. Res. Nat. Bur. Stand.* **67A**, 19 (1963).
[150] Scott, R. B., C. H. Meyers, R. D. Rands Jr, F. G. Brickwedde and N. Bekkedahl. *J. Res. Nat. Bur. Stand.* **35**, 39 (1945).
[151] Sharp, W. E. *J. Geophys. Res.* **74**, 1645 (1969).
[152] Shepherd, E. S., G. A. Rankin and F. E. Wright. *Amer. J. Sci.* **28**, 293 (1909).
[153] Shimada, M. *Spec. Contrib. Geophys. Inst. Kyoto Univ.* **6**, 303 (1966).
[154] Simon, F., M. Ruhemann and W. Edwards. *Z. Phys. Chem.* **B6**, 331 (1930).
[155] Skinner, J., E. Cussler and R. Fuoss. *J. Chem. Phys.* **72**, 1057 (1968).
[156] Slutskii, A. B. *Eksperimen. Issled. v Oblasti Glubinnykh Protsessov*, Inst. Geokhim. i Analit. Khim., Akad. Nauk SSSR, Materialy Simpoziuma, 212 (1960).
[157] Smothers, W. J. and Y. Chiang. *Differential Thermal Analysis: Theory and Practice*. Chemical Publishing Co.: New York (1958).
[158] Smyth, H. T. *J. Amer. Ceram. Soc.* **34**, 221 (1951).
[159] Southard, J. C. and F. G. Brickwedde. *J. Amer. Chem. Soc.* **55**, 4378 (1933).
[160] Spindler, H. and F. Sauerwald. *Z. Anorg. Allg. Chem.* **335**, 267 (1965).
[161] Statton, W. O. *J. Chem. Phys.* **19**, 33 (1951).
[162] Sterrett, K. F., W. Klement Jr and G. C. Kennedy. *J. Geophys. Res.* **70**, 1979 (1965).
[163] Stevels, J. M. 'The structure and the physical properties of glass', in S. Flügge (Ed.), *Encyclopedia of Physics*, Vol. XIII, pp 510–645. Springer: Berlin (1962).
[164] Strong, H. M. *J. Geophys. Res.* **64**, 653 (1959).
[165] Strong, H. M. in *Progress in Very High Pressure Research*, p 182. Edited by F. P. Bundy, W. R. Hibbard and H. M. Strong. Wiley: New York (1961).
[166] Strong, H. M. and F. P. Bundy. *Phys. Rev.* **115**, 278 (1959).
[167] Strong, H. M. and F. P. Bundy. *Spec. Publ. US Nat. Bur. Stand. No. 326*, 283 (1971).
[168] Susse, C. and R. Epain. *CR Acad. Sci., Paris*, **259**, 3756 (1964).
[169] Susse, C., R. Epain and B. Vodar. *CR Acad. Sci., Paris*, **258**, 4513 (1964).
[170] Tammann, G. *Aggregatszustände*. Voss: Leipzig (1922).
[171] Tanaka, S. *Bull. Chem. Soc. Japan*, **38**, 795 (1965).
[172] Taylor, W. J. and F. D. Rossini. *J. Res. Nat. Bur. Stand.* **32**, 197 (1944).
[173] van Tets, A. and H. G. Wiedemann. in *Thermal Analysis*, Vol. I, p 121, edited by R. Schwenker and P. Garn. Academic Press: New York (1969).
[174] Thomson, W. *Ann. Phys., Lpz.* **81**, 163 (1850).
[175] Treiman, L. H. in *Temperature*, Vol III, p 523, edited by F. G. Brickwedde. Reinhold: New York (1962).
[176] Tsiklis, D. S. *Handbook of Techniques in High-pressure Research and Engineering*. Plenum: New York (1968).
[177] van Valkenburg, A. *Rev. Sci. Instrum.* **33**, 1462 (1962).
[178] Vereshchagin, L., A. Semerchan, N. Kuzin and Y. Sadkov. *Dokl. Akad. Nauk SSSR*, **183**, 565 (1968).
[179] Vereshchagin. L., V. Galaktionov, A. Semerchan and V. Slesarev. *Dokl. Akad. Nauk SSSR*, **132**, 1059 (1960).
[180] Weber, R. L. *Heat and Temperature Measurement*, 422 pp. Prentice-Hall: New York (1950).
[181] Welch, J. H. *J. Sci. Instrum.* **31**, 458 (1954).
[182] Welch, J. H. *J. Sci. Instrum.* **38**, 402 (1961).
[183] Welch, J. H. and W. Gutt. *J. Amer. Ceram. Soc.* **42**, 11 (1959).
[184] West, E. D. and D. C. Ginnings. *Rev. Sci. Instrum.* **28**, 1070 (1957).

[185] West, E. D. and D. C. Ginnings. *J. Res. Nat. Bur. Stand.* **60**, 309 (1958).
[186] Wilson, W. *Rev. Sci. Instrum.* **31**, 331 (1960).
[187] Yoder, H. S. *Trans. Amer. Geophys. Union*, **31**, 827 (1950).
[188] Yoder, H. S. *J. Geol.* **60**, 367 (1952).
[189] Zhokhovskii, M. K., Y. S. Konyaev and V. G. Levchenko. *Pribory i Tekh. Eksp.* 118–22 (1959).

CHAPTER 17

Part 2. Cryoscopy

Y. DOUCET

Laboratoire de Thermodynamique, Université de Provence, Centre de St Jerome, 13013 Marseille, France

Contents

	List of Symbols	836
I.	General Considerations	837
	1. Cryoscopic Law for Ideal, Non-ionic Solutions	837
	2. Cryoscopic Law for Non-ideal and Non-ionic Solvents	839
	3. Cryoscopic Law for Aqueous Electrolytes	839
	4. Cryoscopic Law for Electrolyte Solutions in Salt Media	841
	A. Temkin Theory	842
	B. Förland Theory	843
	C. Haase Theory	845
	5. The Methods of Cryoscopy	846
	A. Kinetic Methods	847
	(1) Rüddorf Method	847
	(2) Other Related Methods	848
	(3) End Melting Point Method	850
	B. Equilibrium Method	851
II.	Kinetic Cryoscopy Equipments	852
	1. Apparatus used from $-100°C$ to $200°C$	852
	2. Temperature Measurement	853
	A. Thermocouples	854
	B. Fixed Point Obtainment	856
	C. Quartz Thermometer	858
	D. Platinum Resistor Thermometer	858
	E. Thermistor Resistance Thermometer	859
	F. Microcryoscopy	860
	G. Thermometric Probe Calibration	860
	3. Table of Organic and Inorganic Solvents	861
	4. Salt Hydrate and Aqueous Eutectic Cryoscopy	861
	A. Table of Salt Hydrate Solvents	868
	B. Tables of Aqueous Eutectic Solvents	869
	5. Molten Salts Cryoscopy	871
	A. Equipment	871
	B. Temperature Measurement	873
	C. Procedure	874
	D. Table of Molten Salt Cryoscopic Constants	878
III.	Adiabatic Cryoscopy Equipments	878
	1. Aqueous Solution Equipment	879
	2. Non-aqueous Solvents Equipment	881
	3. Concentration Measurements	882
	A. Conductivity Method	882
	B. Potentiometric Methods	882
	C. Interferometric Method	883
	D. Polarimetric Method	884
	E. Photometric Method	885
	F. Precongelation Method	886

		G. Chemical Analysis Methods	886
		4. Pseudo-equilibrium Methods	886
IV.	A Few Applications of Cryoscopy		887
	1. Data Derived from Zero Concentration Extrapolation, $(\theta/m)_0$		887
		A. Molar Masses	887
		B. Entropy and Enthalpy of Fusion	888
		C. Mixed Crystals Study	888
		D. Particle Numeration	888
	2. Data Derived from the (θ/m) versus m Cryoscopic Graph		889
		A. Osmotic Coefficient Calculation	889
		B. Solute Activity Coefficient Calculations	890
		C. Solvation Calculations	891
		D. Electrolyte Ionic Dissociation	891
		E. Hydrolysis Constant Calculations	893
		F. Cryoscopic Titrations	894
		(1) Acidimetry and Alkalimetry	894
		(2) Complex Formation	894
	3. Data Derived from the Schröder Curve		895
		A. Activity Coefficients	895
		B. Excess Chemical Potential Calculations	896
		C. Other Thermodynamic Calculations	897
V.	References		898

List of Symbols

A	Coefficient of Debye–Hückel theory
a	Ionic parameter of Debye–Hückel theory
a_i	Activity of particle i
a_+	Ideal activity of ionic substance
B	Coefficient of Debye–Hückel theory
c	Molarity (number of moles per liter of solvent)
d	Density
E	e.m.f. of one couple
f_i	Rational symmetric activity coefficient of component i
G	Molar Gibbs function (free enthalpy)
g	Gibbs function for n molecules or moles
I	Ionic strength
i	van't Hoff coefficient
j	Lewis cryoscopic variable
L_i	Heat of fusion, component i
l	Length of wire
ln	Natural or Napierian logarithm
log	Decimal or common logarithm
m	Molality (number of moles per kilogram of solvent)
N	Number of junctions
R	Universal ideal gas constant
R_g	Charging resistance
R_j	Resistance of wires
S or S'	Cross-sectional area of a wire
S_f	Molar melting entropy
T	Kelvin temperature
T_{0i}	Melting point of pure component i

PART 2. CRYOSCOPY

W	Interaction parameter for regular solutions
X_i	Mole fraction, component i
Z_i	Ionic charge of ith ion
α	Molecular dissociation coefficient
γ	Practical dissymmetrical activity coefficient of solute
$\gamma^{\pm}$	Mean activity coefficient of ions
δ	Hydrolysis ratio
Δc	Difference of the molar specific heats of the liquid and the solid component
ΔS	Entropy of mixing
θ	Cryoscopic depression $(T_{01} - T)$
λ	Cryoscopic constant in molal scale
ν	Total number of ions in the molecule
ν_i	Number of ions of type i
ν'_i	Number of ions of the solute common to the solvent of type i
π	Osmotic pressure
π'	Osmotic pressure in milliosmols
ϕ	Practical osmotic coefficient

I. General Considerations

As a professor at the University of Grenoble, François-Marie Raoult formulated many of the basic laws of physical chemistry. One of these laws provides a definition for so-called 'ideal solutions'; another one has a much broader content than was at first thought. It is the 'cryoscopic' law, which was so named by Raoult himself, from the Greek words χρνος (ice) and σκοπεω (I observe)[64]. Chemists, however, just saw in it one of the methods of determination of molar masses. It was only when van't Hoff[84] gave an expression for the solvent cryoscopic constant that Raoult's experimental work gained universal acceptance.

If van't Hoff's conclusion is correct, his argument has long been recognized as somewhat dubious. The osmotic pressure, which was supposed to follow ideal gas laws in dilute solutions, is no longer in use. The starting point, which is of great importance, is a purely phenomenological one. One considers the conditions for equilibrium between the pure, one-component solid phase and the solution, the concentration of which is constant and well-defined. If the solid is made of pure solvent crystals, one deals with a cryoscopic experiment. If the solid is one of the solutes, one deals with a solubility experiment. The van Laar–Schröder equation for solubility is therefore identical to the cryoscopic law.

1. Cryoscopic Law for Ideal, Non-ionic Solutions

The van Laar equation is given in all physical chemistry textbooks. The starting point is the adiabatic equilibrium condition at constant concentration, i.e. the equality of the chemical potentials of the common component in both phases. This need not be recalled here. Let X_1 be the solvent mole fraction, L_1 its heat of fusion at equilibrium temperature T kelvin. We have

$$\ln X_1 = \int_{T_{01}}^{T} (L_1/RT^2)\, dT \tag{1}$$

For integration purposes, one can use the Kirchhoff relationship, which amounts to assuming that the difference of the molar specific heats of the liquid and the solid solvents (Δc) is a constant in the T_{01}–T temperature interval. This leads to

$$\ln X_1 = \frac{L}{R}\left(\frac{1}{T_{01}} - \frac{1}{T}\right) + \frac{\Delta c}{R}\left(\frac{T_0}{T} - 1 - \ln \frac{T_0}{T}\right) \tag{2}$$

Δc is actually a function of temperature

$$\Delta c = \Delta a + \Delta bT + \Delta cT^{-2} \tag{3}$$

Unfortunately, the coefficients Δb and Δc are very often unknown. If one takes takes into account the term Δb, the above equation becomes

$$\ln X_1 = \frac{L}{R}\left(\frac{1}{T_{01}} - \frac{1}{T}\right) + \frac{\Delta a}{R}\left(\frac{T_0 - T}{T} - \ln \frac{T_0}{T}\right) + \frac{\Delta b}{2R}\left(\frac{T_0^2}{T} - 2T_0 + T\right) \tag{4}$$

Considerable temperature depressions are required in order that the second term (with Δc) be appreciable. It then represents no more than two or three per cent of the first term, and in such a case the second formula with both Δa and Δb is to be preferred. It gives the cryometric equation for an ideal solution from the dilute up to the eutectic concentration.

When the second term can be neglected, the equation then becomes

$$X_1 = K \exp(-L/RT) \tag{5}$$

where

$$\ln K = L/RT^2 \tag{6}$$

This latter formula lends itself to a simple graphic interpretation: it is a straight line in the log X_1 versus $1/T$ coordinates. Its slope is $L/4.58$ if L is expressed in cal mol^{-1} and is *independent of the nature of the solute*. If we now consider binary dilute solutions only, Raoult's law is experimentally observed. On the one hand, $X_1 = 1 - X_2$ is close to 1 as X_2 approaches zero. On the other hand, $\theta = T_{01} - T$ also approaches zero, whence, expressed as a power series

$$\theta = AX_2 + BX_2^2 + \ldots \tag{7}$$

In dilute solutions, concentrations are usually measured in terms of molality (number of moles per kilogram of solvent) rather than in mole fractions. One therefore obtains

$$(\theta/m) = \lambda(1 + am + bm^2 + \ldots) \tag{8}$$

where

$$\lambda = RT_0^2 M_1/L \tag{9}$$

and:

$$a = -\frac{M_1}{2} + \lambda\left(\frac{1}{T_0} - \frac{\Delta c}{2L}\right)$$

PART 2. CRYOSCOPY

$$b = M_1\left[\frac{M_1}{3} + \lambda\left(\frac{1}{T_0} - \frac{\Delta c}{2L}\right)\right]$$

For instance, this equation gives, for an ideal solution in benzene

$$(\theta/m) = 5.12 - 0.255m + 0.017m^2$$

and in acetic acid

$$(\theta/m) = 3.72 - 0.160m + 0.0073m^2$$

2. Cryoscopic Law for Non-ideal and Non-ionic Solvents

Since the time of Lewis (1913) the activity concept has been in use. This concept was introduced so as to keep the same formal expression as in the case of ideal solutions *for the molar Gibbs function*;

$$G = G_0 + RT \ln a$$

Raoult's law concerning vapor pressures thus becomes

$$p_1 = p_{01} a_1$$

(the subscript 1 refers to the solvent). The cryoscopy formula then becomes

$$\ln a_1 = \frac{L}{R}\left(\frac{1}{T_{01}} - \frac{1}{T}\right) + \frac{\Delta c}{R}\left(\frac{T_{01} - T}{T} - \ln \frac{T_{01}}{T}\right) \qquad (10)$$

In very dilute solutions X_1 approaches unity, and the same is true of a_1. The above series is therefore valid as a limiting law for $m = 0$. Therefore, in the molality scale

$$\lim (\theta/m) = \lambda \qquad (11)$$

whatever the solute.

Such is Raoult's cryoscopic formula. It shows that it is necessary to extrapolate towards the origin ($m = 0$) in order to obtain the cryoscopic constant experimentally. Raoult often stressed that point in his work[65].

3. Cryoscopic Law for Aqueous Electrolytes

In this paragraph a non-dissociated solvent such as water is considered (as far as water is concerned, the dissociation into H^+ and OH^- ions is absolutely negligible). These considerations can apply to other solvents as well, such as liquid ammonia, cyclohexanol, dimethylsulfoxide, etc. Aqueous electrolyte solutions are never ideal. The above formulae are, however, applicable, but one has to take the ionic dissociation into account. In an infinitely dilute solution, the electrolyte, whether strong or weak, is completely dissociated. Let v be the total number of ions in the molecule. The limiting freezing point depression is therefore $v\lambda$ instead of λ. This fact was seen (though not clearly) by Raoult[67] and was explained satisfactorily by Arrhenius[3] in about 1900. A statistical proof was given by Prigogine and Defay[61]. The evaluation of the number of configurations leads to the statistical entropy of mixing. Let B_i be the number of configurations of one particular ion i within all the solvent molecules. One has

$$\Omega = \Pi(B_i^{n_i}/n_i!) \qquad (12)$$

and

$$\mu_1 - \mu_{01} = -kT\partial(\ln \Omega)/\partial n_1 \qquad (13)$$

so that

$$\mu_1 - \mu_{01} = -RTX_2 v \qquad (14)$$

The equality of the chemical potentials of component 1 in the solid and liquid phases leads to

$$(\partial \mu_1/\partial X_1)_{T,p}\, dX_1 = (L/T)\, dT \qquad (15)$$

whence

$$dT/dX_1 = vRT^2/L \qquad (16)$$

Integration in the vicinity of T_{01} (dilute solutions) yields

$$\lim_{X_2 \to 0}(\theta/X_2) = v\lambda_1 \qquad (17)$$

This can equally be expressed in terms of molalities

$$\lim_{m \to 0}(\theta/m) = v\lambda \qquad (18)$$

This relationship has been verified experimentally to a high precision (within $10^{-5}\,°C$) in the cryoscopic experiments of Bedfort[4], at the beginning of this century, thus conforming to the predictions of Arrhenius (*Figure 1*).

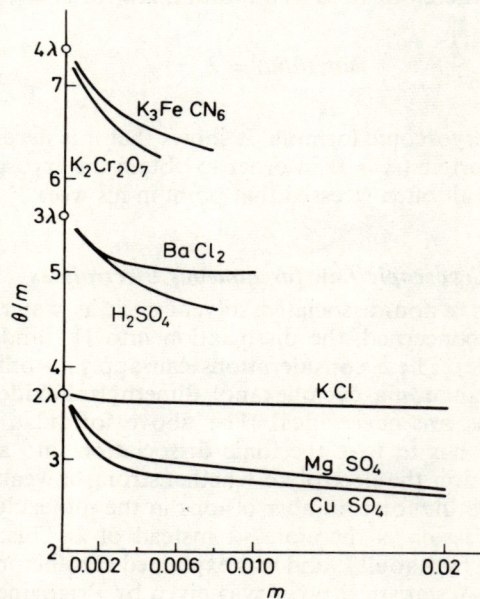

Figure 1. Cryoscopic data from J. G. Bedfort and verification of the Arrhenius limiting law.

PART 2. CRYOSCOPY

At finite concentrations, one has to know the solvent activity coefficient in order to determine the equation of the cryoscopic curve; this coefficient is, in turn, dependent upon the ionic activity coefficients. The latter are calculable within the limits of application of the Debye–Hückel theory. On the other hand, the aspect of the curve also depends upon possible ionic associations. Let α be the molecular dissociation coefficient. Let us define, following van't Hoff,

$$i = 1 + (\nu - 1)\alpha \tag{19}$$

One then obtains, from the Debye equations, with concentration c expressed on the molarity scale

$$\theta/c = \lambda[i - (\tfrac{1}{2}M_1 + \lambda/T_{01})i^2 c - \tfrac{1}{3}A\nu\alpha\sigma' z_1 z_2 I^{\frac{1}{2}}] \tag{20}$$

where

$$\sigma' = \{3/(BaI^{\frac{1}{2}})^3\}(1 + BaI^{\frac{1}{2}} - \ln(1 + BaI^{\frac{1}{2}})^2 - 1/\{1 + BaI^{\frac{1}{2}}\}) \tag{20a}$$

and the A and B coefficients appear in the classical Debye–Hückel expression

$$\ln f_{\pm} = -Az_1 z_2 I^{\frac{1}{2}}/\{1 + BaI^{\frac{1}{2}}\} \tag{21}$$

where z_1 and z_2 are the ionic charges, I the ionic strength and a an ionic parameter.

4. Cryoscopic Law for Electrolyte Solutions in Salt Media

The expression 'salt media' refers to cryoscopic solvents such as salt hydrates at an indifferent point, with congruent melting or at a transition point, binary or ternary aqueous eutectics and, finally, molten salts at either their melting point or at indifferent points or eutectics. In all these cases, one has to consider a multicomponent mixture where particles are essentially ions.

Two experimental facts are paramount as one uses those solvents:
(1) The ions common to the solvent and to the solute have no cryoscopic effect.
(2) The activity coefficient of the solvent is always close to unity. Ideal solutions may often be encountered.

The 'common ion' problem was not well understood by early investigators[50,7]; it was better seen by Livingston and his co-workers[46] but it was mostly under the supervision of Darmois[14] that a huge number of such cryoscopic experiments were carried out. Making use of the Stortenbecker theorem[79] concerning indifferent points, Darmois has shown that the chemical potential of the solvent remains unchanged as common ions are added. In 1952 Haase[32] gave a cryoscopic formula which showed both that there is a horizontal tangent at an indifferent point and that the common ions have no effect. There is also a statistical proof of the above propositions, similar to that given above. Let us suppose that the ions of type 1 are the common ions and let n'_1 be their number in the solvent and n''_1 their number in the solute. The ν other ions (a, b, ... k), are different in the solvent and the solute. With the above notations the number of complex ions (from the point of view of indiscernible particle statistics) is

$$\Omega = \frac{n_1''! \, \Pi_a^l (\beta_i)^{n_i}}{(n_1' + n_1'')! \, \Pi_a^k (n_i!)} \tag{22}$$

As in the previous case one obtains the following *limiting law*

$$\mu_1 - \mu_{01} = -RTX_2 v$$

But v is now no longer the total number of ions, as the common ions are excluded. *Figure 2* shows one of the verifications of this limiting law

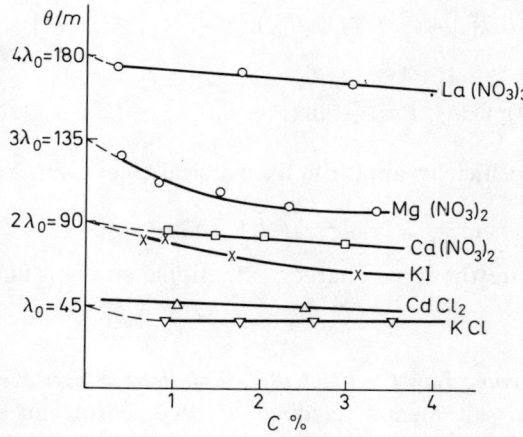

Figure 2. Verification of the limiting law in $CaCl_2 \cdot 6H_2O$ solvent, for common and non-common ions.

$\lim_{m \to 0} (\theta/m) = v\lambda$ for cryoscopic experiments in hydrated calcium chloride at 29.8°C.

At finite concentrations one must estimate the activity a_1 of a mixture of ions, and define what is meant by ideality.

A. Temkin Theory

In 1945 Temkin[81] formulated the following proposition: salts are completely dissociated into ions and the electric charges of both cations and anions are separated. Because of the signs of electrostatic forces, an anion is surrounded by more cations than anions. Hence the liquid (at least in the vicinity of the melting point) remains in a quasi-crystalline state. Since 1956 this is not only an assumption, but an experimental fact, because Zarzycki[92] obtained x-ray diffraction spectra for liquid drops and was able to compute coordination numbers of 3.7 for lithium fluoride, 4.1 for sodium fluoride, 5 for potassium fluoride; 4.7 for sodium chloride, and 5.5 for barium chloride.

Let us consider a monovalent alkali halide A^+X^- in a mixture with a second one, say B^+Y^-. This mixture is said to be ideal when it is both: (1) athermal, and (2) regular.

(1) These salts having the same type of structure, the anions may interchange with each other: the same is true of the cations. The interaction forces being similar in both the pure salt and the mixture: the enthalpy of formation of the solution is zero.

(2) The entropy of mixing may easily be derived from the number of complex ions in the two sublattices (a cationic one and an anionic one), as the anions on the one hand, and the cations on the other hand, are interchanged. One then obtains

$$\Delta S = - R(n_A \ln X_A + n_B \ln X_B + n_X \ln X_X + n_Y \ln X_Y) \quad (23)$$

where n_j is the number of equivalents of species j and X_j is the 'ionic fraction' of the j ion, defined, for the anions, with respect to the total number of anions and, for the cations, with respect to the total number of cations:

$$X_A = n_A/(n_A + n_B) \qquad X_X = n_X/(n_X + n_Y) \quad (24)$$

The above expression is identical to the entropy of an ideal solution. Let us calculate now the partial molal entropy of the solvent (AX)

$$\partial(\Delta S)/\partial n_A + \partial(\Delta S)/\partial n_X = - R \ln X_A X_X \quad (25)$$

As $\Delta H = 0$, one has, for the solvent

$$\mu_1 - \mu_{01} = RT \ln X_A X_X \quad (26)$$

Thus the activity of an ideal solution is

$$a_1^\star = X_A X_X \quad (27)$$

The ionic fraction of a common ion is equal to one. This can be generalized for two different salts with similarly charged ions $A_r X_s$ and $B_r Y_s$, and because

$$\frac{\partial(\Delta S)}{\partial n_1} = r \frac{\partial(\Delta S)}{\partial n_A} + s \frac{\partial(\Delta S)}{\partial n_X} \quad (28)$$

one finds

$$a_1^\star = X_A^r X_X^s \quad (29)$$

B. Förland Theory

If now the ions in the solvent and the solute differ in charge (for example, A_2X and BX), there arises a difficulty in interchanging ions A^+ and B^{2+}. (For simplification purposes, only one common ion X has been assumed to exist.) This case has been studied by Förland[29]. The first assumption is that in the A^+, B^{2+}, X^{2-} system there is still one cationic sublattice and one only. With each divalent ion B^{2+} a monovalent hole (site vacancy) would be associated. As the number of these cationic holes in the liquid is extremely large, their addition does not change the thermodynamic properties of the mixture. The total entropy is

$$\Delta S = - R(n_A \ln X_A + n_B \ln X_B) \quad (30)$$

whence
$$\overline{\Delta S_1} = -R \ln X_A^2 \tag{31}$$
The activity of an ideal solution is, therefore:
$$a_1^\star = X_A^2 \tag{32}$$
which is equivalent to the Temkin relationship.

A second assumption is that the structure of the liquid mixture is very close to that of a solid solution where a divalent salt BX is scattered among the molecules A_2X of the solvent, leading to the occupation of a B^{2+} site, thus creating a hole which remains associated with B^{2+}. Förland then expresses the ionic fractions:
$$X'_A = \frac{n_A}{n_A + 2n_B} \qquad X'_B = \frac{2n_B}{n_A + 2n_B} \tag{33}$$
The partial differentiation of the total entropy of mixing leads to
$$\overline{\Delta S_1} = -R(2 \ln X'_A + X'_B) \tag{34}$$
One can, however, consider, with Förland, the case of a dissociated cationic vacancy from a B^{2+} ion; then
$$\Delta S = -R(n_A \ln X'_A + 2n_B \ln X'_B) \tag{35}$$
This gives
$$\overline{\Delta S_1} = -2R \ln X'_A \tag{36}$$
The ideal activity is therefore
$$a_1^\star = X'^2_A \tag{37}$$
This is identical to the above formula, the X' replacing the X. We can therefore write that the activity of an ideal solution in a solvent $A_r X_s$ is, whatever the charge structure of the solute,
$$a_1^\star = X_A^r X_X^s \tag{38}$$
in the Temkin–Förland formulation, the ionic fraction of the common ions being equal to one.

The solvent activity coefficient f_{1T}, in the Temkin–Förland formulation is, by definition
$$f_{1T} = a_1/a_1^\star \tag{39}$$
The cryoscopic formula can therefore be written as
$$\ln a_1^\star f_{1T} = (L/R)(1/T_{01} - 1/T) + \varepsilon(1/T) \tag{40}$$
Let B_r, Y_s, be the solute; as a function of the stoichiometric mole fractions X_1 and X_2, $a_1^\star$ can be expressed in the following form
$$a_1^\star = \left(\frac{rX_1}{rX_1 + r'X_2}\right)^r \left(\frac{sX_1}{sX_2 + s'X_2}\right)^s \tag{41}$$
as A and B on the one hand, and X and Y on the other hand are distinct.

PART 2. CRYOSCOPY

If v_i denotes r and s successively; v_j, r' and s' successively, and v'_i the number of ions of the solute common to the solvent, one can write

$$\ln a_1^\star = \Sigma\, v'_i \ln \frac{X_1 + (v'_i/v_i)X_2}{X_1 + (v_j/v_i)X_2} \qquad (42)$$

A simpler formula is obtained for a dilute solution. If X_2 is small enough, one has

$$\ln a_1^\star = \Sigma(v'_i - v_j)X_2 \qquad (43)$$

$\Sigma v_j - \Sigma v'_i$ is the total number of ions in the solute, distinct from the solvent ions. Let v be that number. One obtains

$$\ln a_1^\star = \ln X_1^v \qquad (44)$$

This relationship will be used later.

C. Haase Theory

This theory was first formulated at a Phase Change Conference in Paris and was published in 1953[32]. It was reformulated in 1969[33]. No assumptions whatsoever are made regarding the structure of the molten salt and the latter is not necessarily completely dissociated into ions. The theory can be applied whatever the nature of the particles being studied, whether ions, ion pairs, simple or complex molecules. The 'particle concentration' is then related to the total number of particles in solution. If ions are of interest, one does not have to consider their sign.

Let us first consider the simple case of a total dissociation of the solvent A_rX_s, whose 'macroscopic mole fraction is X_1 and of the solute $B_{r'}Y_{s'}$, whose mole fraction is X_2. The activity of an ion A is defined as

$$\ln a_A = (\mu_A - \mu_{0A})/RT \qquad (45)$$

Its 'ionic' fraction is defined as

$$X_A = rX_1/(v_1 X_1 + v_2 X_2) \qquad (46)$$

if $v_1 = r + s$ and $v_2 = r' + s'$. If the solute is $A_{r'}Y_{s'}$ (A is a common ion) one has, for example,

$$X_A = (rX_1 + r'X_2)/(v_1 X_1 + v_2 X_2)$$

In the same manner, one has

$$\ln a_X = (\mu_X - \mu_{0X})/RT \qquad (47)$$

As usual, the stoichiometric activity of the solvent is such that

$$\ln a_1 = (\mu_1 - \mu_{01})/RT \qquad (48)$$

On the other hand

$$\mu_1 = r\mu_A + s\mu_X \qquad (49)$$

Accordingly

$$RT \ln a_1 = RT \ln (a_A^r a_X^s) + r\mu_{0A} + s\mu_{0X} - \mu_{01} \qquad (50)$$

The chemical potential μ_{01} is the value taken by μ_1 for the pure solvent. Let X^r_{0A} and X^s_{0X} be the ionic concentrations as N_1 approaches unity. One has

$$\mu_{01} = RT \ln X^r_{0A} X^s_{0X} + r\mu_{0A} + s\mu_{0X} \tag{51}$$

whence

$$a_1 = (a_A/X_{0A})^r \times (a_X/X_{0X})^s \tag{52}$$

The ideal activity is therefore

$$a^\star_{1H} = (X_A/X_{0A})^r \times (X_X/X_{0X})^s \tag{53}$$

the solvent activity coefficient being defined, according to Haase, as

$$f_{1H} = a_1/a^\star_{1H} \tag{54}$$

The ideal activity is easily expressed as a function of X_1 and X_2. As has been done previously, let us denote r and s successively by v_i, and the number of ions of the solute common to the solvent by v'_i. One obtains

$$\ln a^\star_{1H} = \sum v_i \ln \frac{X_1 + (v'_i/v_i) X_2}{X_1 + (v_2/v_1) X_2} \tag{55}$$

This formula becomes simpler if X_2 is small enough to enable one to take only the first two terms of the logarithmic power series. This gives

$$\ln a^\star_{1H} = - X_2(v_2 - \Sigma v'_i) \tag{56}$$

The bracketed terms represent the number of ions of the solute, distinct from solvent ions, say v. Hence

$$a^\star_{1H} = X^v_1 \tag{57}$$

Thus, as far as the limiting law is concerned, the ideal activity according to Haase is similar to the ideal activity according to Temkin. For finite concentrations, second-order differences only are observed. These differences become zero when the solvent and the solute have similar charge structures, (I–I) or (II–II).

The Haase particle fraction notation is particularly adaptable to partial dissociation studies for either the solvent or the solute. The number of ions of species i becomes $\alpha_1 v_i$ or $\alpha_2 v_i$ where α_1 and α_2 are the dissociation coefficients of the solvent and the solute, respectively.

5. The Methods of Cryoscopy

All cryoscopic formulae have been derived for states of equilibrium between pure solvent crystals and a solution of fixed composition. It would therefore seem that an adequate technique would consist of the association of a temperature and a concentration measurement. A very precise method does use this technique, it is the 'equilibrium method'. There are, however, other methods which are kinetic techniques: the starting freezing point and the end melting point.

PART 2. CRYOSCOPY

A. Kinetic Methods

(1) Rüddorf Method[71]

A tube, containing a solution of known concentration, is placed within a thermostat, the temperature of which is lower than the expected freezing point. If the solution is in a supercooled state by a few tenths of a degree, this can be stopped immediately by introducing a pure solvent crystal in the solution. The solid phase appears in the form of spangles, the number of which increases rapidly, and which are maintained amidst the liquid by thermal agitation, in spite of the density difference. At the same time, the temperature increases up to a certain value T_f where it is at a maximum. This maximum gives, as a first approximation, the starting temperature of freezing (*Figure 3*).

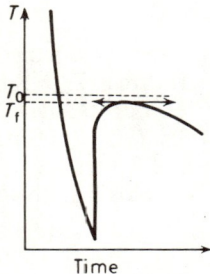

Figure 3. The starting-freezing point is not the equilibrium temperature between solid and liquid phases.

The concentration, however, is systematically greater than initially, as the pure solvent crystals come out of solution. This error can be reduced to a negligible term if one greatly increases the ratio of the volume of the solution to the volume of the crystals. This was done by Loomis[49]. The latter author has been criticized by Raoult, because as there is a decrease of the error in concentration, at the same time the error in temperature increases by a non-calculable quantity. The temperature of a mixture of a liquid with ice is subject to fluctuations due to heat transfers; the smaller the quantity of ice present the greater the fluctuations. It is therefore desirable to have a large enough quantity of ice present and make an adequate correction of the concentration.

Another correction is needed for the temperature T_f (*Figure 3*). The above maximum is not, actually, the freezing point of the solution. This maximum is systematically too low, as it corresponds to the calorimetric equilibrium between the heat evolved as the supercooled state is made to cease and the heat absorbed by the refrigerating bath. The stronger the refrigerating effect, the more depressed the temperature maximum. Nernst[57] and Abegg[1] have been able to calculate the difference $T_0 - T'$ between the true equilibrium temperature T_0 of both phases and the temperature T' of the maximum observed. Raoult prefers an experimental approach towards evaluating the above correction. Let m be the concentration of a given solution (*Figure 4*).

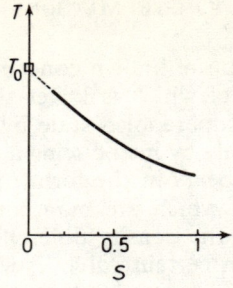

Figure 4. Raoult's graph to correct for the supercooling error.

The cryoscopic experiment is then performed for a supercooling S_1 of, say, 1.5°C; the maximum temperature is T_1, it is too low for the reasons stated above. The experiment is carried out again with a supercooling S_2 of 1°C and the maximum temperature is $T_2 > T_1$. With some other similar experiments, one can obtain the graph of $\bar{T}$ versus S which is approximately linear. Its extrapolation towards $S = 0$ yields T_0, which is the true freezing point of the solution of molality m.

(2) *Other Related Methods*

Let us first examine the method of Moulin[55]. This author plots temperature T versus time z. If the refrigerating bath is at a temperature of at least 10°C

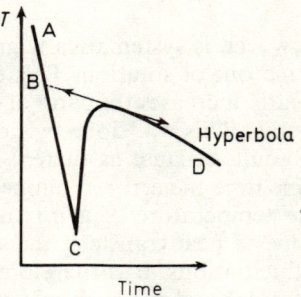

Figure 5. Moulin's graph to correct for the supercooling error.

below the solvent freezing point, a sharp linear decrease is observed, whereby the line AC is described (*Figure 5*). Its equation is

$$T = -az + K \qquad (58)$$

If no supercooling occurred, there would be an angular point B, and a subsequent hyperbolic decrease. For this effect, let us assume that the law $\theta = \lambda m$ is valid as a first approximation; the concentration at a given instant is

$$m' = m/(1 - r) \qquad (59)$$

the initial concentration being m, and r the relevant quantity of ice formed, with respect to the total weight of solvent. Let r be a linear function of time z:

$$r = k(z - z_0) \qquad (60)$$

the refrigerating bath temperature being constant and much lower than that of the experimental solution. One has $r = 0$ at the origin of time $(z = z_0)$. Thus, the law of variation of θ with time is

$$\theta = \lambda m/\{1 - k(z - z_0)\} \qquad (61)$$

Or, as $\theta = T' - T$

$$T = T' - \lambda m/\{1 - k(z - z_0)\} \qquad (62)$$

This is a hyperbola, with asymptotes the equations of which are $z = z_0 + 1/k$ for the vertical one, and $T = T'$ for the horizontal one. This curve intercepts the straight line AC at the point B. Moulin assumes that this point B may be taken as the intercept of a tangent to the hyperbola which is actually not distinguished from the hyperbola itself in that small domain. Moreover, the latter author takes into account a discrepancy between thermometer readings and the solution temperature, with which the thermometer is never at equilibrium.

Following White[88], Mair, Glasgow and Rossini[52] do not draw the tangent and keep the actual hyperbola (*Figure 6*). If no supercooling were to occur, the cooling curve would be A'B'EF, and B' the point sought. The part EF of the curve is to be extrapolated. For that purpose, one can write that the enthalpy change is the same between T_B and T_E whatever the way chosen, that is

$$H_B - H_E = H_{B'} - H_E$$
(through point C)   (direct)

If T_j designates the outer bath temperature and A a transfer coefficient, one has, at any given instant z:

$$dQ = A(T - T_j)\,dz$$

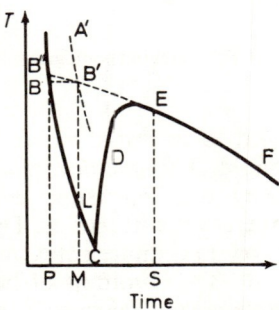

Figure 6. The NBS graph to correct for the supercooling error.

Therefore:

$$H_B - H_E = A_X \text{ area BCDESP}$$
$$H_B - H_E = A_X \text{ area B'ESM}$$

from which result the identity of the areas BB'MP and B'LCDE, and a graphical method of extrapolation. According to the above authors, the difference between B" (graphical extrapolation) and B' is 0.001 for $\theta = 0.06$ and 0.27°C for $\Delta t = 3$°C when the supercooling is 5°C and the cooling rate is 0.02°/min for 40 minutes.

Taylor and Rossini[80] have shown how, by a graphical construction, the point in question could be obtained, with the help of three points chosen on the experimental hyperbola. This will not be discussed further as, actually, these supercooling methods are extremely critical; many systematic errors occur, so that experiments are scarcely reproducible within less than a hundredth of a degree. In particular, in aqueous solutions, there is a large difference in temperature between the solution and the refrigerating bath, and this leads to the formation of an ice layer, which screens the heat transfers and renders the above corrections somewhat deceptive. Very careful experimentation is required for acceptable results.

(3) End Melting Point Method

Instead of using a thermostat at a temperature lower than that of the cryometric vessel, one can tackle the problem the other way round. The heating hyperbola is symmetrical with the cooling hyperbola with respect to a vertical axis (*Figure 7*). In B', symmetrical with B, the fusion is terminated and

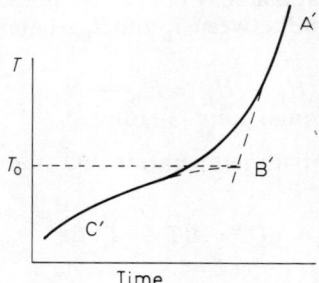

Figure 7. Final melting point graph.

the concentration has returned to its initial value. The temperature at this point is the temperature sought. This method has been used by Müller[56] for eutectic solution cryoscopy. It is interesting in that it does not require either concentration for temperature corrections. The slope of C'B', which is zero for the pure solvent, is low in a dilute solution, which is an advantage for obtaining the intercept B'. The duration of heating C'B' depends, of course, upon the mass of solvent and its specific heat, and also upon the warm bath temperature. It is easy to carry out the experiment in such a manner that the portion C'B' lasts about ten minutes. But it is more difficult to avoid a curved segment. This depends upon the agitation and mainly upon the rate of solvent crystallization.

Although it is to be preferred to the various starting-freezing methods, the final fusion technique leads to a systematic error as the crystals are in small quantity in a large mass of liquid. The solution temperature, even at B', is not that of the equilibrium considered, as this equilibrium temperature corresponds to equal thermal agitation of both solid and liquid molecules. Such is actually the case only in a thin liquid film near the crystals. In the bulk of the solution, the temperature is greater because of the heat transfer, whatever the amount of agitation. Thus, the C'B' curve does not correspond to a succession of instantaneous states of equilibrium. The above gap increases and the crystals disappear and finally, when it is large enough, the curve departs from its hyperbolic shape. The extrapolation, up to B', of the hyperbola, gives therefore too large a temperature and thus also too small a cryometric depression. Systematic comparisons with the equilibrium method, however, showed us that, for example, for aqueous potassium chloride, the discrepancy is no greater than one per cent for $m = 0.05$ and negligible at higher concentrations. In conclusion, this method is certainly the best within kinetic methods, as it avoids the complications due to supercooling.

B. Equilibrium Method

In 1895 Ponsot[59] gave a precise definition of this method and actually used it. This procedure may be described as follows: one takes a constant temperature medium and a test-tube containing a partly frozen solution, at a temperature close to that of this medium. After stirring, one waits till a stationary state is obtained, takes the solution temperature and decants the liquid part of it, which is titrated. Roloff[70] a student of Nernst, in Germany, applied the same method to hydrochloric acid solutions; the hydrochloric acid was titrated against barium hydroxide. The first advantage of having achieved an adiabatic equilibrium lies in the fact that one has a fixed temperature because there is a fusion plateau. Another advantage is that the mass of the solid phase is of the same order as the mass of the liquid phase, which avoids a systematic error in temperature. The procedure runs as follows.

The fusion or freezing temperature of the pure solvent is first obtained. For that purpose, to a certain volume of solvent, a corresponding quantity of thinly ground pure crystals is added. This quantity should not be too great, for the agitation would be inadequate, and this would result in too great a temperature reading in the liquid. It should, however, be large enough, according to what has been said earlier.

The temperature plateau being reached, a small quantity of solute is added. The solution is formed, the temperature goes down and if concentration is constant, another plateau is observed. The invariable concentration is linked to the adiabatic character of the vessel. A Dewar vessel is not always enough, because of its orifice. One must therefore put it in an adequate thermostat. The agitation brings about some heat; however, temperatures constant within 10^{-5} degree can be obtained in aqueous solutions, if the heat capacity is large enough and if the agitation does not correspond to a turbulent state, but rather to forced convection. Whereas the temperature is known with a high accuracy, the concentration on the plateau is not known. It is a definite drawback of this method to be obliged to take a sample of the solution and titrate it. The various methods used and their applications in particular cases will be discussed later.

II. Kinetic Cryoscopy Equipments

1. *Apparatus used from* $-100°C$ *to* $200°C$

The following apparatus is particularly adapted to organic and inorganic solvents, and, of course, to aqueous solutions. The final melting point method is used in preference. It is essentially composed (*Figure 8*) of a test tube A in which the solvent may be weighed. This tube is closed by means of a cap through which a rotating stirrer, and the thermometric probe, may be inserted. Solute known quantities may also be introduced through a tube.

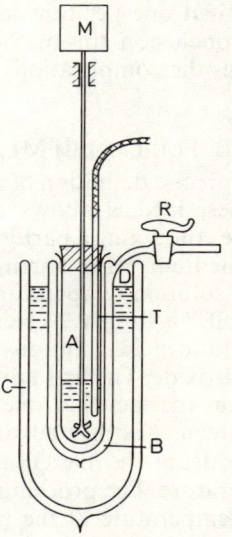

Figure 8. Universal cryoscopic apparatus: A Inner test tube, B Non-silvered dewar vessel, C Silvered dewar, D To vacuum, R Vacuum tap, T Thermometric probe, M Micromotor.

The liquid quantity introduced should not be too great, as otherwise the agitation would be faulty. It should not, on the other hand, be too small, for the fusion plateau might be too short and so, inaccurate. The tube A is first dipped into a refrigerating mixture, the level of which, however, is below the solution level. The agitator is switched on. In this manner, the formation of a solid layer that would screen thermal transfers is avoided. Then to the refrigerating bath, the non-metallized dewar vessel B is substituted; it may itself be contained within another dewar vessel C. It contains a liquid at a temperature a few degrees above the expected fusion temperature. The length

PART 2. CRYOSCOPY

of the fusion plateau may be adjusted by either the bath temperature C, and the vacuum in dewar B, to which tube D supplied with a vacuum stopcock S is connected.

The bath C is not necessary for cryoscopic experiments close to room temperature (water, benzene, ethyldibromide, dioxane, acetic acid), if the dewar B is well evacuated. On the other hand, if an all-purpose apparatus is desired, the vessel C must be supplied with liquid circulating from a thermostat, or the whole apparatus must be thermostatically controlled. Commercial equipments are available, which include a refrigerating unit and an electrical (resistor) heating unit. These equipments are also supplied with a thermostatic system operating over a broad temperature interval. If the start-of-freezing method is being used, one should start with a solution at a slightly higher temperature than that of the expected melting point; the thermostat should be set one or two degrees below that point. It often happens that the solvent is hygroscopic, it is then necessary to operate under inert atmosphere in a closed vessel. The cryoscopic vessel must then be entirely closed. See, for example, the apparatus of Finel and co-workers[26]. Instead of constructing a leakproof joint for the introduction of the rotating agitator and the thermometric probe, it is easier to set the latter in a thermometric well and to use

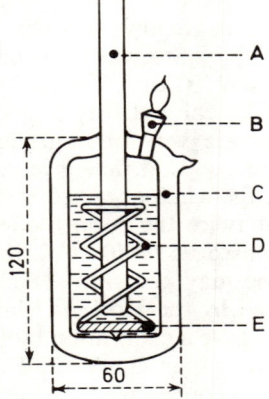

Figure 9. Apparatus for volatile or hygroscopic solvent: A Thermometric probe well, B Sample addition, C Vacuum, D Rotatable spiral stirrer, E Magnet.

magnetic stirring (*Figure 9*). Commercially available magnetic stirrers are generally not powerful enough to provide adequate agitation at a distance of 4 to 5 cm (double wall and thermostatically controlled liquid). It is possible to built magnetic stirrers with two crossed magnets.

2. *Temperature Measurement*

In supercooling cryoscopic methods, the measurement of concentration is no problem as it is achieved before the experiment itself, either by weighing or by volumetric determination. It is, however, still necessary to determine the temperature. For that purpose, the Roberteau differential thermometer is used, together with thermocouples, platinum resistor or thermistor thermometers, or quartz thermometers.

The differential thermometer will be mentioned only in terms of an historical survey, as it is no longer used in laboratories nowadays. It is, however, very sensitive; the usual equipments are graduated to a hundredth of a degree, and they enable one to assess a thousandth of a degree. The precision Beckmann thermometer has graduations of a thousandth of a degree. Its measurements are highly stable and reproducible with just a few precautions. The emergent column, calibration, inner pressure and 0–100 interval corrections are practically zero as they are identical in both the pure solvent and the solution. It has, however, a prohibitive drawback as it does not lend itself to graphical recording: photocell methods might be applied, but they have no interest compared to other electrical methods such as thermocouples, resistor or quartz thermometers.

A. Thermocouples

First there is a problem as far as the metals being used are concerned. Earlier investigations used to take iron–constantan and copper–constantan couples at room temperature; their homogeneity had to be tested. Very satisfactory commercial alloys are now available, giving highly reproducible e.m.f.s. In France, the Imphy Steelworks produce, under the names of BTE–CTE, nickel–chromium (BTE) and nickel–copper (CTE) wires. Their melting points are quite different, but they may easily be welded with silver or otherwise. Those wires are very homogeneous and their resistance is constant at any temperature, so that they give very stable couples for cryoscopic uses. Furthermore, the e.m.f. obtained is high, as it approaches 70 to 80 µV near 0°C, i.e. about twice the copper–constantan couple value. The BTE–CTE couple may be used up to 700°C. If the CTE alloy is replaced by the ATE alloy alumel, one may go above 1000°C. In the USA chromel–alumel couples are supplied by the Hoskins Manufacturing Co., for which the NBS (*Circular No. 561*) gave a calibration table from −200°C to 1371°C. Around 0°C, the sensitivity is about 40 µV/°C. The same firm also supplies a chromel–constantan (type E) couple, which is similar to the BTE–CTE.

As one couple does not give, generally speaking, a large enough e.m.f., several couples may be mounted in series. The whole circuit is U-shaped; in one branch are the cold connections, and in the other one the hot connections. The copper wires of the circuit form, together with the BTE and the CTE, two connections at 0°C (*Figure 10*). The wires are insulated with glyptal lacquer which is polymerized by evaporating at 140° for two hours, with all the wires covered in this way. All the similar connections are then put together, relacquered and introduced into a thin glass tube with some transformer oil which has two apparently contradictory functions: electrical insulation and thermal conductivity. As the system must be symmetrical, equal volumes must be put in both branches.

Both glass tubes are linked by a glass tubing bridge and the wires are enclosed in paraffin. It is important to avoid temperature variations along the wires. Lange[42] in his high precision apparatus, even uses an evacuated space,

PART 2. CRYOSCOPY

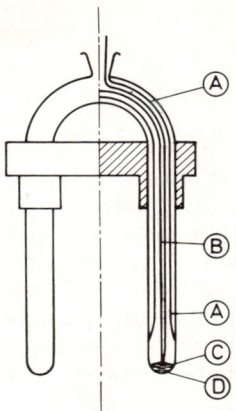

Figure 10. Thermopile: A Outer vacuum jacket, B Thermocouples, C Oil drop, D Thin glass.

made of two concentric glass tubes. It is desirable to divide the thermocouple cell into two equal parts. After the tenth welding, for example, two output copper wires are connected and a second cell with ten connections is made. If both these cells are opposed, the resulting e.m.f. must be zero. Series and parallel connections enable one to verify that the cell works satisfactorily, and resistance measurements enable one to localize insulation defects. Should both thermocouple wires be chosen of the same diameter? Let ρ_B and ρ_c be the resistivities, Sx and $S(1-x)$ the cross-sectional areas, and let each wire be of length l. The resistance is then given by

$$R_j = (l/S)\{\rho_B/x + \rho_c/(1-x)\} \tag{63}$$

This resistance is a minimum for

$$(\rho_c - \rho_B)x^2 + 2\rho_B x - \rho_B = 0 \tag{64}$$

For the BTE wire, we have: $\rho_B = 65.6 \times 10^{-6}$ ohm/cm, and for the CTE wire, $\rho_c = 49.5 \times 10^{-6}$; the term in x^2 being small, one obtains, quite approximately $x = \frac{1}{2}$. Wires of the same diameter are therefore chosen.

What is the optimum number of junctions N? Let E be the e.m.f. of one couple, and R_g the charging resistance of the d.c. amplifier used. The current is

$$i = NE/(R_g + NR_j) \tag{65}$$

It is at a maximum for

$$R_g = NR_j \tag{66}$$

As the resistance of a couple is low, a low-impedance charge should be chosen.

On the other hand, the cross section S is limited by the dimensions of the tubing. The total section S' at one's disposal ($S' = NS$ approximately), and the junction length l are imposed Then

$$R_g = N^2(l/S')\{\rho_B/x + \rho_c/(1-x)\} \tag{67}$$

if
$$\rho_B = \rho_c$$
so that
$$R_g = 4N^2 \rho l/S' \tag{68}$$

There exists along the thermocouple wires a temperature gradient and thereby a heat flux proportional to that gradient, to the cross sectional area of the wires, and to their heat conductivity. This heat flux should be minimized as much as possible. It is not a drawback in kinetic cryoscopic methods, where the cryoscopic vessel receives or loses heat, according to the technique in use. The problem is different with the equilibrium methods which require an almost perfect adiabatic situation. One might think that a thermocouple, giving the temperature difference between cold and hot connections, gives in one measurement only the cryoscopic depression, i.e. the difference between the pure solvent melting point and the solution melting point. This is not so, however, as the hot connection must be maintained at a very stable fixed temperature, generally 0°C.

(1) The pure solvent (except if aqueous solutions are considered) does not have, at its melting point, the stability of an international temperature standard.

(2) A thermocouple with many connections always has a residual e.m.f. when the hot and cold connections are dipped in the same bath and this residual e.m.f. is not stable enough to be considered as a constant for an individual thermocouple. In conclusion, measurements of cryoscopic depressions with a thermocouple are the same as with other thermometric probes. One of the connections is put in a fixed point thermostat (triple point cell, for example); the starting-freezing point or the end melting point should be taken on (a) the pure solvent and (b) a solution of known concentration, under the same conditions.

The e.m.f. to be measured is always small, even with many thermocouples in series, as the cryoscopic depression is rarely more than 1°C. An elongation method must therefore be used with a volt-sensitive recording galvanometer. Excellent equipments are available as far as the recorder is concerned, but in elongation measurements there is an essential quality: the precision of the return to zero. No parasitic e.m.f. should be introduced by the equipment; this requires copper connections with copper wiring and gold input wires to the amagnetic moving coil.

Another method consists of connecting the thermocouple to a d.c. amplifier, the input impedance being adequately adapted. Then a sturdier recording galvanometer may be used. A better solution is a recording potentiometer, as the potentiometric measurement of a thermocouple e.m.f. is always to be preferred because of the Thomson effect which always exists in a multi-connection thermocouple which is supplying some current.

B. Fixed Point Obtainment

Below 0°C an easily obtainable fixed point is the mercury freezing point at $-38.86°C$; above $0°C$ there is phenyl oxide at $26.877°C$. Those temperatures are, in fact, triple point temperatures which may be obtained as in the case of water. Still using the same technique, the standard temperature of $122.362°C$ may easily be obtained with benzoic acid, available in a state of very high purity. Further details are given by Schwab and Wichers[75].

Let us give a few details about the obtainment of the water triple point.

The method which consists of dipping the couple into thinly ground ice is faulty. The connections may be in contact with melted water which is above $0°C$ as it causes the fusion. A better method consists of using a water–ice mixture, adequately stirred. Fusion is avoided if this mixture is placed in a dewar vessel. The difficulty lies in stirring. In order that the water shall not increase its temperature above the equilibrium temperature, it should be submitted to forced convection movements so that the small ice fragments are under continuous flow of water. This problem will be discussed with equilibrium methods.

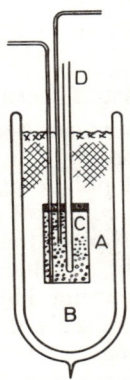

Figure 11. Water freezing point under atmospheric pressure: A Aluminum or silica container, B Ice, C Washed ground ice from twice distilled water, D Thermocouple well.

It is possible to dispense with the agitation, if melting is avoided. For this purpose, an aluminum or quartz container is completely filled with thinly ground ice, carefully washed (*Figure 11*). A paraffined cork stoppers the vessel; two tubes and the couple are inserted through the cork. The tubes are used to fill the interstitial gaps with water. After a reasonably long time, the temperature is stabilized. To insure that there is no cryoscopic effect due to impurities in ice, or coming from the cork on the glass, the liquid is pumped away and the vessel is filled again. The twice-distilled water is saturated with air drop-by-drop distillation, and ice is made from this water, in aluminum or Pyrex vessels. If care is taken to change the outer ice every evening, this zero point equipment is able to yield constant readings to better than $10^{-4}°C$ for several consecutive days. Another method consists of preparing in advance several triple point cells which are frozen just before use (*Figure 12*). The fabrication has been very well described by Moser[54] and by

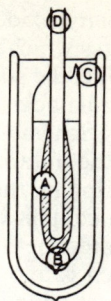

Figure 12. Triple point water cell: A Air-free ice, B To washing tube before sealing, C To vacuum before sealing, D Thermocouple well.

White[89]. The triple point method is thus easy to use; it can easily and safely give a $10^{-4}\,°C$ precision.

C. Quartz Thermometer

It is well known that the resonance frequency of a quartz crystal is a function of its temperature, according to a law which is represented over a broad interval, by a third degree polynomial. The coefficients of this polynomial depend upon the angle of cutting of the plate by the crystallographic axes of quartz. A detailed study showed that by careful cutting of the plate, the terms in T^2 and T^3 could be simultaneously brought to zero; this is called the LC (linear coefficient) cutting. The equipment includes a standard oscillator at 28 MHz, perfectly stabilized ($10^{-7}\,°C$/month drift) and a measuring oscillator of resonance frequency 1 000 Hz/°C around 28 MHz. Both frequencies are compared in a mixer which detects the differential frequency. The current is converted into impulses and displayed on 'Nixie' tubes. The reading frequency is controlled by the reference oscillator and goes from 10^{-2} to 1 second. Four digits may be read.

A second probe may be used. The beat frequencies of the two probe oscillators are then measured, which gives a differential temperature measurement just as in the case of a thermocouple. Six digits may then be read, with a $10^{-4}\,°C$ precision. There also is an output on a decoder, followed by an amplifier, which enables one to record the data with a recording galvanometer. The quartz thermometer thus yields a very high precision on instantaneous direct reading and graphic recording. The measurable temperature range is $-40°C$ to $230°C$.

D. Platinum Resistor Thermometer

The platinum resistance is the method that gives the best time stability and the best calibration reproducibility. It is made of a non-inductive winding surrounded by glass of the same expansion coefficient. The resistance at 0°C is of about 25 ohms, for some types and 100 ohms for other types. The tem-

PART 2. CRYOSCOPY

perature coefficient $(1/R_0)(dR/dt)$ is 3.96×10^{-3}. Therefore the dR variation for a dT variation of 10^{-3} is only 4×10^{-4} ohm. One must therefore have precision apparatus, such as modified Wheatstone bridges, or a comparison potentiometer with a well determined constant resistance. Many investigators used Wheatstone bridges in the early stages of cryoscopy, such as Griffiths and Chroustuff; Bedfort[4] used this method with a high precision, followed by

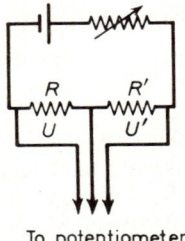

Figure 13. Potentiometric measurement of thermometric resistances.

Elliot, Chadwell, Getmann and many others. Lange and Berga[43] also used a platinum resistor. The best available equipment today is the Mueller temperature bridge.

In another method (*Figure 13*), the platinum resistance R is mounted in series with another known one, made of manganin, thermostatically controlled and absolutely constant R'. Let U and U' be the potential difference between the ends of R and R'. One has

$$R = R'(U/U') \qquad (69)$$

The system comprises a big dry cell, which is able to supply a constant current in a total resistance of about 1 000 ohms, including the platinum resistance R, an R' resistance equal to the latter at 0°C, made of a resistance box and a variable manganese resistance. Commercial potentiometers have two voltage ranges with a switch connecting the circuit to either. These outputs are connected to the ends of R and R'. The power dissipated as heat in the resistance is negligible and therefore the adiabatic character of the cryoscopic system is maintained.

E. THERMISTOR RESISTANCE THERMOMETER

Thermistors are semiconductors; their resistance decreases exponentially as temperature increases. Let R_T be the resistance at T and R_∞ the limiting value as $1/T$ approaches zero. One has

$$\ln(R_T/R_\infty) = B/T \qquad (70)$$

B being a constant. The temperature coefficient $-B/T^2$ is negative and about ten times greater than the platinum one. The measuring equipments are thus easy to build, but the time stability of the resistance is poor, compared to that of platinum. Calibration should therefore be repeated frequently. Another disadvantage is the non-linearity of R and T variations, an electronic or mechanical device may solve this problem. It should also be noted that, for small temperature variations, as in the case of cryoscopy, the variation is roughly linear. For example, the following variation has been observed around 0°C

$$R = 189 \times 10^{-4} \exp 3943/(331 + t)$$

Let us expand this into a power series around $t = 0$

$$R = 2816.4 - 102t + 2.13t^2$$

If $t = 1°C$, the error made by using the linear formula is only 0.07 per cent.

F. Microcryoscopy

As the thermistors are very small, microcryoscopic systems may be constructed by using only 1 cm³ of solution, which is quite interesting for some organic compounds. The apparatus of Meyer and Metzger[53] is made of a small double wall tube, which is evacuated. The (a) type (*Figure 14*) is stirred

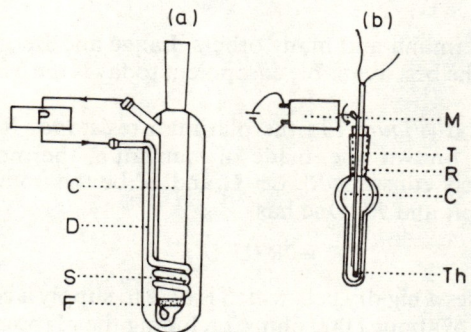

Figure 14. Microcryoscopes, from R. Meyer and J. Metzger [*Bull. Soc. Chim.* 1, 66 (1967)]: (a) P Peristaltic pump, C Cryoscopic cell, D Double envelope, S Coil, F Fritted glass. (b) M Micromotor, T Metal stirrer, R Standard taper joint, C Cryoscopic cell, Th Thermistor.

by nitrogen bubbling through a fritted glass disc. The (b) type is stirred mechanically. A micromotor provokes the vibration of a stainless steel rod. As is the case in the previously described equipments, these systems are enclosed in a thermostat set at either a temperature above the expected temperature (starting-freezing point) or below (final melting point). The thermistor is placed in one branch of a Wheatstone bridge, one diagonal of which contains a 1.5 V cell and a resistance which is able to change, by its variation, the bridge sensitivity. In the other diagonal is a recording millivoltmeter which gives, for a temperature variation of $10^{-2}°C$, a 10 cm deviation.

G. Thermometric Probe Calibration

Apart from international fixed points, calibration by comparison is feasible if a NBS-calibrated platinum resistance is available, together with a Mueller bridge. It is convenient to have the thermostat slowly cooled, or warmed, for example at a 10^{-2} degree per minute rate. The platinum resistance and the probe to be calibrated are then put in the same place and their signals are *simultaneously* recorded. The calibration curve results directly from the comparison of these two curves.

Another method consists of doing cryoscopic measurements in an aqueous solution of a strong electrolyte that has been very well studied, with a precision higher or equal to the precision sought. For example, there is potassium chloride, which is available in the crystalline form at a high degree of purity. Lange[44], in dilute solutions, and Cavallaro[11], in concentrated solutions, carried out very precise cryoscopic tests by the equilibrium method. Temperature depressions are given to $10^{-5}\,°C$ and concentrations to ± 0.2 per cent, even for 10^{-3}M concentrations.

If cryometry is carried out by using a supercooling method, this procedure is interesting in that it avoids systematic errors due to the method. The temperature corresponding to the e.m.f. e is, indeed, not that of the experimental solution, at the final melting point, but rather the temperature the solution would have in the Lange measurements, i.e. at thermodynamic equilibrium.

3. *Table of Organic and Inorganic Solvents*

This *Table 1* gives a list of about 170 solvents given in the order of increasing melting points, from $-116°C$ to $+246°C$. All of them have actually been used in cryoscopic experiments. The molal limiting depression, measured directly with various solutes, is given with the precision supplied by the authors. References are given in the last column. In earlier investigations, the cryoscopic constant was not always obtained by extrapolation towards zero concentration, so that it seemed desirable to give calculated cryometric constants every time basic data (such as heats of fusion) were available. When the dielectric constant is given, it is with two subscripted figures. On the left is the temperature of measurement which is not, quite often, the temperature of fusion. On the right is the frequency at which it has been determined, unless a static method has been used.

4. *Salt Hydrate and Aqueous Eutectic Cryoscopy*

Instead of using a compound, one may well use an indifferent point or a transition point in salt hydrate phase diagrams, or a eutectic point. Let us recall the essential characteristics of this type of cryoscopy:

(1) There are only small variations of the activity coefficient; this leads to almost straight cryoscopic curves, the extrapolation of which is easy.

(2) There is the common ion theorem; this explains why the solute ions, in common with the solvent ions, do not give a temperature depression.

The use of a eutectic as a cryoscopic solvent is an original idea of Müller[56] which is very well developed in his thesis work. In many experimental examples, the cryoscopic depressions are seen to have the same properties for eutectic solvents as for salt hydrates, the experimental technique being much

Table 1

Compound	Formula	M	T_f	λ meas.	λ calc.	ΔH_f	S_f	D	Ref.
Ethyl ether	$(C_2H_5)_2O$	74.12	−116.3	1.79			12.9	−116 9.79 [4]	9
Hydrogen chloride	ClH	36.47	−112	4.98	3.71	13.9	3.14		9
Acetone	$(CH_3)_2CO$	58.08	−94.82	2.40	2.7	23.4	7.63		65
Hydrogen bromide	BrH	80.92	−88.5	9.4			3.36	−80 31.5	9
Hydrogen sulfide	SH_2	34.08	−82.9	3.83	8.9	7.67	3.36		9
Sulfur chloride	S_2Cl_2	135.05	−80	5.36			9.7	5.75	10
Ammonia	NH_3	17.03	−77.79	1.32		81.2	7.06		83
Sulfur dioxide	SO_2	64.07	−72.7	3.01	0.53		8.45		10
Chloroform	$CHCl_3$	119.39	−63.49	4.90			1.01		11
Hydrogen iodide	IH	127.93	−50.8	20.3	17.5	5.68	3.26	−63 6.62 [4]	9
Dimethyl 2,4-thiazole	$SNC_3H(CH_3)_2$	113.8	−50.31	16.1			3.1		62
2,4,6-Trimethyl-pyridine	$(CH_3)_3C_5H_2N$	121.18	−44.19		5.53	18.8	10		60
Methyl 4-thiazole	$SNC_3H_2CH_3$	99.15	−44.13	4.85			9.3		62
Pyridine	$N(CH)_5$	79.10	−42	4.97			9.26	20 12.5	9
Methyl 5-thiazole	$SNC_3H_2CH_3$	99.15	−40.41	5.82			7.9		62
Thiazole	SCHNCHCH	85.12	−33.57	5.7			3.3		62
Trimethyl thiazole	$SNC_3(CH_3)_3$	127.21	−32.46	6.7			8.9		62
Benzotrifluoride	$C_6H_5CF_3$	146.11	−28.16	4.90			14.5		48
Methyl 2-thiazole	$SNC_3H_2CH_3$	99.15	−24.72	4.7			10.3		62
Carbon tetrachloride	CCl_4	153.84	−22.96	29.8	28.2	4.16	2.54	−23 2.29	9
Quinoline	$C_6H_4N(CH)_3$	129.15	−15.22		6.66	19.85	9.93		60
Hydrocyanic acid	CNH	27.03	−13.317 (triple pt)	1.845		73.6	7.65	20 95	53
o-Nitrotoluene β	$NO_2C_6H_4CH_3$	137.13	−10.56	5.08			14.0	25 26.07 [3]	66
Nitrogen dioxide	NO_2	46.01	−9.3	4.1	4.3	32.3	5.63		75
Bromine	Br_2	159.83	−7.3	9.71	8.67	16.18	9.74		12
Aniline	$C_6H_5NH_2$	93.12	−6.2	5.87	6.75	21	7.28	0 7.78 [8]	2
o-Nitrotoluene α	$NO_2C_6H_4CH_3$	137.13	−4.14	7.18			1.02	25 26.07 [3]	66
Hydrogen peroxide	H_2O_2	34.02	−1.70	2	1.97	74	9.26		84
Water	H_2O	18	0	1.858	1.858	79.5	5.24	0 87.6	14
Tetrabromoethane	$C_2H_2Br_4$	345.70	0.13	21.7			8.67	20 7.1	54
Phosphorus oxychloride	$POCl_3$	153.39	2	7.68	7.60	19.8	11.1	22 12.4	86

PART 2. CRYOSCOPY

Compound	Formula								
Formamide	HCONH$_2$	2.45	45.04	3.37		5.51			52
Dimethylaniline	C$_6$H$_5$N(CH$_3$)$_2$	2.5	121.18	5.85		11.3	20	4.48	2
Dicyclohexyl	(C$_6$H$_{11}$)$_2$	2.75	166.30	14.5		6.29	20	4.48	59
Benzyl ether	(C$_6$H$_5$CH$_2$)$_2$O	3.6	198.23	6.27		17.4			15
Benzene	C$_6$H$_6$	5.533	78.11	5.088	5.102	8.46	−5.5	2.31	29
Nitrobenzene	C$_6$H$_5$NO$_2$	5.72	123.11	6.89	6.87	9.95	25	34.09	78
d-Fenchone	C$_{10}$H$_{16}$O	6	152.23	6.8		12.4			52
Cyclohexane	C$_6$H$_{12}$	6.554	84.16	20.2	21	2.33	20	2.05	28
Bromoform	CHBr$_3$	7	252.77	14.43		9.84	17	4.43	1
o-Chlorophenol α	ClC$_6$H$_4$OH	7	128.56	7.72		9.29	58	5.41	51
p-Chlorotoluene	ClC$_6$H$_4$CH$_3$	7.5	126.58	5.6		12.6	58	5.55	7
Formic acid	HCO$_2$H	8.4	46.03	2.77	2.68	9.63	16	58.5	36
1,2-Dibromoethylene	(CH$_2$Br)$_2$	9.97	187.88	11.9	11.78	8.95	18	8.9	36
Sulfuric acid	SO$_4$H$_2$	10.25	98.08	6.154	6.62	8.31		large	63
Dichloroacetic acid	CHCl$_2$CO$_2$H	10.8	128.95		11.2	6.45	20	8.22	30
Ethylene diamine	(NH$_2$CH$_2$)$_2$·H$_2$O	11.3	78.12	2.43	2.08	21.2			72
p-Dioxane	(CH$_2$)$_4$O$_2$	11.65	88.10	4.63	4.70	77			79
Paraldehyde	[OCH(CH$_3$)]$_3$	12.6	132.16	7.05	6.55	34.25	20	2.23	55
p-Xylene	C$_6$H$_4$(CH$_3$)$_2$	13.2	106.16	4.33	4.15	10.6	20	14.5	69
m-Nitrotoluene	CH$_3$C$_6$H$_4$NO$_2$	16.1	137.13	6.78		11.5	20	2.42	81
Acetic acid	CH$_3$CO$_2$H	16.63	60.05	3.62	3.57	14.5	20	21.86	35
Octamethylcyclotetrasiloxane	(CH$_3$)$_3$Si$_4$O$_4$	17.4	296.64	11.2		11.6	58	6.12	76
Glycerol	(CH$_2$OH)$_2$CHOH	18.18	92.09	3.29	2.82	8.95	17		74
Dimethylsulfoxide	(CH$_3$)$_2$SO	18.54	78.13	4.4	3.85	15.3	21	15.3	19
Methyl succinate	(CH$_2$CO$_2$CH$_3$)$_2$	19.5	146.14	5.55	4.8	15.0	25	47	24
Phenyl hydrazine	C$_6$H$_5$NHNH$_2$	19.6	108.14	5.86	4.7	11.7			64
Methanesulfonic acid	CH$_3$SO$_3$H	19.66	96.10	5.69		17.8			31
Acetophenone	CH$_3$COC$_6$H$_5$	19.7	120.14	5.65		18.9	20	7.15	41
Anethole	CH$_3$(CH$_2$)C$_6$H$_4$OCH$_3$	22.5	148.2	6.2	6.73	9.84		large	37
1,2-Veratrole	C$_6$H$_4$(OCH$_3$)$_2$	22.5	138.16	6.4	6.3	12.3	21	15.6	37
Cyclohexanol	C$_6$H$_{11}$OH	22.6	100.16	38.2	43	12.9			71
Triphosphorus oxide	P$_2$O$_3$	23.8	109.94	11.45		12.8			84
Tetrachlorodifluoroethane	CCl$_2$FCCl$_2$F	24.65	203.85	37.7		1.42	25	15	63
tert-Butyl alcohol	(CH$_3$)$_3$COH	25.5	74.12	8.37	8.45	5.69			16
Diphenyl methane	(C$_6$H$_5$)$_2$CH$_2$	27	168.23	6.8	7.1	5.21	14	19.3	46
Diphenyl ether	(C$_6$H$_5$)$_2$O	27.5	170.20	8		14.1	18	2.47	37
p-Bromotoluene	BrC$_6$H$_4$CH$_3$	27.6	171.04	8.21	8.6	12.7	20	2.68	34
m-Chlorophenol	ClC$_6$H$_4$OH	28.5	128.56	8.30		11.9	58	5.49	71
					20.9	9.34			51

Table 1—continued

Compound	Formula	T_f	M	λ meas.	λ calc.	ΔH_f	S_f	D	Ref.
Heptachloropropane	$(CCl_3)_2CHCl$	29.5	285.24	12			10.7		17
o-Cresol	$CH_3C_6H_4OH$	30	108.13	5.6	6.06	30.1	15.8	17 10.3 [8]	77
o-Nitrobenzoic ethyl ester	$NO_2C_6H_4CO_2CH_5$	30	195.17	7.4			6.6	30 1.88	20
N-Methylacetamide	$CH_3CONHCH_3$	30.45	73.08	5.6	5.77	31.78	8.83		18
Quinoxaline	$C_6H_4N(CH)_2N$	30.5	130.14	8.9			9.47		67
Tin(IV) bromide	$SnBr_4$	31	438.36	28			22.0		43
Capric acid	$C_9H_{19}CO_2H$	31.5	172.26	4.7		38.9	8.03		37
Propane dinitrile	$CH_2(CN)_2$	32	66.06	4.89			12.7		82
o-Chloronitrobenzene	$ClC_6H_4NO_2$	32.5	157.56	7.5			13.3		51
p-Iodo-toluene	$CH_3C_6H_4I$	35	218.05	10.3	10.05	18.8	10.35		7
Pyrosulfuric acid	$H_2S_2O_7$	35.07	178.15	10.5	10.56	17.9	11.8	40 89.6	88
Ethylene carbonate	$C_3H_4O_3$	35.69	88.06	4.55			22.4		33
Methyl cinnamate	$C_6H_5(CH)_2CO_2CH_3$	36	162.18	7.1	7.17	42.7	9.20	24 5.6 [8]	23
p-Cresol	$CH_3C_6H_4OH$	36	108.13	7.4	7.25	26.3	13.8		77
Azoxybenzide	$(C_6H_5N)_2O$	36	198.22	8.5	8.78	21.6	9.24		23
p-Chlorophenol	ClC_6H_4OH	37	128.56	8.58			11.8		51
o-Nitrobenzaldehyde β	$NO_2C_6H_4CHO$	37.9	151.12	7.9			12.9		20
Phenylbenzylamine	$(C_6H_5)_2CH_2NH$	38	181.24	8.7			13.0		44
o-Nitrobenzaldehyde α	$NO_2C_6H_4CHO$	40	151.12	7.2			8.69	48 9 [8]	20
Phenol	C_6H_5OH	40.95	94.11	6.9	6.6	29.7	13.9		60
o-Bromonitrobenzene	$C_6H_4BrNO_2$	42	202.02	9.1			9.23		51
L-Menthol	$C_{10}H_{19}OH$	42.5	156.26	12.4	10.65	18.6	10.9		44
Salicylic acid	$OHC_6H_4CO_2C_6H_5$	43	214.21	12.3			27.5		45
Lauric acid	$CH_3(CH_2)_{10}CO_2H$	43.85	200.31	4.4	4.57	43.7	1.94		36
Phosphorus	P_4	44.1	124.08	32.2	40.2	4.97	14.6		80
m-Chloronitrobenzene	$ClC_6H_4NO_2$	44.4	157.56	6.07	6.81	29.4	13.4	45 5.48 [8]	51
p-Toluidine	$CH_3C_6H_4NH_2$	45	107.15	5.2	5.05	39.9	6.61		7
Cyanamide	$CNNH_2$	45	42.04	2.76			11.7		47
o-Nitrophenol	$NO_2C_6H_4OH$	45	139.11	7.45	7.51	26.8	14.5		3
Trichloroethylacetal	$CCl_3CH(OH)OC_2H_5$	46.2	193.47	8.0	8.45	24	13.3		37
Benzophenone α	$(C_6H_5)_2CO$	48.25	182.21	8.86	8.74	23.5	11.0		37
Phenylpropanoic acid	$C_6H_5(CH_2)_2CO_2H$	48.6	150.17	8.7					24

PART 2. CRYOSCOPY

Compound	Formula									
Urethane	$NH_2CO_2C_2H_5$	48.7	89.09	5.14	5.04	40.9	11.3			37
Cetyl alcohol	$C_{16}H_{33}OH$	49.3	242.44	6.1	6.1	33.8	25.4			37
α-Naphthylamine	$C_{10}H_7NH_2$	50.1	143.18	7.9	8.32	24.9	11.0			36
9-Octadecenoic acid trans	$C_{17}H_{33}CO_2H$	51	282.46	3.9	4.01	52.1	45.4			24
p-Nitrotoluene	$NO_2C_6H_4CH_3$	51.3	137.13	7.8			11.4	58	21.22	4
Thymol	$CH_3(C_3H_7)C_6H_3OH$	51.5	150.21	8.0	7.6	27.5	12.7		6	37
1,2-Diphenyl ethane	$(C_6H_5CH_2)_2$	52.5	182.25	7.28	6.8	31	17.3			57
Maleic anhydride	OCOCHCHCO	52.85	98.06		6.35	33.25	10			60
4-Aminobiphenyl	$(C_6H_5)_2NH$	53.4	169.22	8.67	8.41	25.2	13.0			36
Oxalic dimethyl ester	$(CO_2CH_3)_2$	54	118.09	5.2	5.00	42.7	15.4			5
Ethylene dicyanide	$(CNCH_2)_2$	54.5	80.09	18.3	18.2	11.7	2.85			27
2-Iodonaphthalene	$C_{10}H_7I$	54.5	254.08	15			1.10			25
Isoapiol	$(C_3H_5)C_6H_2H(OCH_3)_2(CH_2O_2)$	55	222.23	8			18.1			41
m-Bromonitrobenzene	$C_6H_4BrNO_2$	56	202.02	8.75			15.0			51
p-Dichlorobenzene	$C_6H_4Cl_2$	56	147.01	7.7	7.05	30.5	13.6	56	2.86	24
2-Chloronaphthalene	$C_{10}H_7Cl$	56	162.61	9.76			10.6		4	25
Trichloroacetic acid	CCl_3CO_2H	57.5	163.40	12.2	25.1	8.65	4.25	61	4.55	87
Nitronaphthalene α	$C_{10}H_7NO_2$	58.8	173.16	9.1	8.63	25.4	13.2			25
2-Bromonaphthalene	$C_{10}H_7Br$	59	207.07	12.4			11.0	19	5.17	25
Palmitic acid	$CH_3(CH_2)_{14}CO_2H$	59.92	256.42	4.3	4.41	50.3	38.7			49
Acetoxime	$(CH_3)_2CNOH$	61	73.09	5.6			8.68			37
Chloroacetic acid	CH_2ClCO_2H	62	94.5	5.24	5.42	41.22	11.5	62	20	56
1,3,5-Trichlorobenzene	$C_6H_3Cl_3$	63	181.46	8.7			13.9			22
3,5-Xylenol	$(CH_3)_2C_6H_3OH$	63.44	122.16	10.7	6.60	34.15	12.4			60
p-Bromophenol	BrC_6H_4OH	63.5	173.02	13.32	11	20.5	10.5			37
Tartaric diethyl ester diacetate	$[CH(CO_2CH_3)CO_2C_2H_5]_2$	67	290.27	9.2	9.85	23.4	14.7			70
p-Bromochlorobenzene	BrC_6H_4Cl	67.4	191.47	8.05	8.07	28.6	13.1			4
Azobenzene	$(C_6H_5N_2)_2$	68	182.22	4.5	4.73	49.6	15.2			37
Stearic acid	$CH_3(CH_2)_{16}CO_2H$	69.4	284.47	8			41.2			37
Benzoic phenyl ester	$C_6H_5CO_2C_6H_5$	70	198.21	8.9	8.93	26.4	16.8			45
2,4-Dinitrotoluene	$(NO_2)_2C_6H_3CH_3$	70.5	182.13	5.0	5.14	45.6	14.0			4
Stearin	$(C_{17}H_{35}CO_2)C_3H_5$	70.8	891.46	8.2	9.02	26.1	11.8			37
Diphenyl	$(C_6H_5)_2$	71	154.20	11.2			11.6	17	2.57	37
Diethyl-diphenylurea	$OC[N(C_2H_5)(C_6H_5)]_2$	71.5	268.35	6.2			16.3			85
Crotonic acid	$CH_3(CH)_2CO_2H$	72	86.09	15.6	9.3	25.3	6.30			41
Antimony trichloride	$SbCl_3$	73.2	228.13	9.0		15.3	10.1	75	30	73
Phenylacetic acid	$C_6H_5CH_2CO_2H$	76.7	136.14	11.87	8.1	30	11.6			21
α-Bromo.d-camphor	$C_{10}H_{15}OBr$	78	231.14				13.6			68

Table 1—continued

Compound	Formula	T_f	M	λ meas.	λ calc.	ΔH_f	S_f	D		Ref.
Naphthalene	$C_{10}H_8$	80.22	128.16	6.9	6.96	35.6	12.9			61
2,4,6-Trinitrotoluene	$(NO_2)_3C_3H_2CH_3$	80.7	227.13	11.5	11.2	22.3	14.3	2.30	100	4
Acetamide	CH_3CONH_2	81	59.07	3.63			11.4			26
p-Bromobenzoic methyl ester	$BrC_6H_4CO_2CH_3$	81	215.06	8.4			18.0	4.0 [8]	22	25
1-Chloro-4-nitrobenzene	$ClC_6H_4NO_2$	83.5	157.56	10.9	11.8	21.4	9.47			4
1,4-Dibromobenzene	$C_6H_4Br_2$	86.9	235.92	12.4	12.5	20.5	13.4	4.57	88	6
Phthalyl chloride	$C_6H_4(COCl)_2$	89	203.03	9.85			14.8			23
m-Dinitrobenzene	$C_6H_4(NO_2)_2$	89.57	168.11	10.6	10.6	24.7	11.4			4
DL-Tartaric dimethyl ester	$(CO_2CH_3)_2(CHOH)_2$	90	178.14	4.95	7.5	35.1	17.2			39
Triphenyl-methane	$(C_6H_5)_3CH$	92.5	244.32	12.45	14.84	17.8	11.9			44
Diphenylglyoxal	$(C_6H_5)_2(CO)_2$	95	210.22	10.5	12.1	22.2	12.6			4
2,4,6-Tribromophenol	$Br_3C_6H_2OH$	96	330.83	20.4	20.2	13.4	12.0			22
Pentachloro-fluoroethane	CCl_3CCl_2F	99.9	220.31	42						16
Phenanthrene	$C_{14}H_{10}$	100	178.22	12	11.37	24.3	11.6	2.80 [8]	20	40
p-Nitrobenzaldehyde	$NO_2C_6H_4CHO$	106.5	151.12	7.0			16.2			21
Resorcinol	$C_6H_4(OH)_2$	110	110.11	6.0	6.4	46.2	13.2	3.2 [8]	20	42
N-Phenylacetamide	$C_6H_5NHCOCH_3$	113.94	135.16	6.93			15.0			61
Iodine	I_2	114	253.84	20.4	20	14.87	9.75			13
Succinic anhydride	$(CH_2CO)_2O$	119.6	100.07	6	6.31	48.7	12.3			41
β-Naphthol	$C_{10}H_7OH$	122	144.16	11.25	10	31.3	11.4			20
Benzoic acid	$C_6H_5CO_2H$	122.36	122.12	8.79	9.12	33.9	10.4			61
Stilbene	$(C_6H_5CH)_2$	124	180.24	8.38	7.86	39.9	18.1			24
p-Bromonitrobenzene	$C_6H_4BrNO_2$	127	202.02	11.53			13.9			51
Urea	$CO(NH_2)_2$	132.7	60.06	21.5			2.25			50
Cinnamic acid-trans	$C_6H_5(CH)_2CO_2H$	133	148.15	10	8.97	36.5	13.3	3.5 [8]	22	38
Hexachlorocyclohexane	$C_6H_6Cl_6$	157	290.85	16.5			15.0			58
Benzanilide	$C_6H_5NHCOC_6H_5$	161	197.23	9.65			17.6			58
d-Camphor	$C_{10}H_{16}O$	177.7	152.23	39.7	48.2	8.39	2.83			8
Anthracene	$(C_6H_4CH)_2$	216.55	178.22	11.65	12.3	38.7	14.1			41
Hexachlorobenzene	C_6Cl_6	227	284.20	20.75			13.57			58
Dibenzopyrrole	$(C_6H_4)_2NH$	246	167.20	12.3	12.7	42.1	13.5			41

PART 2. CRYOSCOPY

Units: T_f, normal melting point, °C.
M, molecular weight, 10^{-3} kg.
λ, cryoscopic constant, deg (mol kg)$^{-1}$
ΔH_f, melting enthalpy, cal g^{-1} (cal = 4.184 J).
S_f, molar melting entropy, cal mol^{-1} K^{-1}.

References

1. Ampola, G. and C. Manuelli. *Gazz. Chim. Ital.* **25b**, 91 (1895).
2. Ampola, G. and C. Rimatori. *Gazz. Chim. Ital.* **27a**, 35–51 (1897).
3. Ampola, G. and C. Rimatori. *Gazz. Chim. Ital.* **27b**, 31 (1897).
4. Auwers, K., H. M. Smith and W. Bartsch. *Z. Phys. Chem.* **30**, 300 (1899).
5. Auwers, K. and H. M. von Dorhn. *Z. Phys. Chem.* **32**, 55 (1900).
6. Auwers, K. *Ber. Dtsch. Chem. Ges.* **33**, 1302 (1900).
7. Auwers, K. *Z. Phys. Chem.* **42**, 513 (1902).
8. Becker, J. *J. Chemist Analyst* **40**, 80 (1951).
9. Beckmann, E. and P. Wrentig. *Z. Anorg. Chem.* **67**, 17 (1910).
10. Beckmann, E. *Z. Phys. Chem.* **65**, 289 (1909).
11. Beckmann, E. and O. Faust. *Z. Phys. Chem.* **89**, 246 (1915).
12. Beckmann, E. *Z. Phys. Chem.* **51**, 96 (1906).
13. Beckmann, E. *Z. Phys. Chem.* **77**, 200 (1912).
14. Bedfort, J. G. *Proc. Roy. Soc. (London)* **83A**, 459 (1910).
15. Bennett, G. and G. H. Willis. *J. Chem. Soc. (London)*, 2309 (1918).
16. Berstein, J. and W. T. Miller *J. Amer. Chem. Soc.* **62**, 948 (1940).
17. Boeseken, J. and J. Benedictus. *Rec. Tran. Chim. Pays-Bas*, **37**, 121 (1918).
18. Bonner, O. D. and G. B. Woolsey, *J. Phys. Chem.* **75**, 2879 (1971).
19. Bonner, O. D., C. F. Jordan and K. W. Bunzl. *J. Phys. Chem.* **65**, 1443 (1961).
20. Bruni, G. *Gazz. Chim. Ital.* **28b**, 322 (1898).
21. Bruni, G. and F. Gorni. *Gazz. Chim. Ital.* **31a**, 49 (1901).
22. Bruni, G. and M. Padoa. *Gazz. Chim. Ital.* **33a**, 78 (1903).
23. Bruni, G. and C. Mascarelli. *Gazz. Chim. Ital.* **33a**, 89 (1903).
24. Bruni, G. and F. Gorni. *Gazz. Chim. Ital.* **30b**, 55 (1900).
25. Bruni, G. and M. Padoa. *Gazz. Chim. Ital.* **34a**, 133 (1904).
26. Bruni, G. and A. Trovanelli. *Gazz. Chim. Ital.* **34b**, 350 (1904).
27. Bruni, G. and C. Manuelli. *Z. Elektrochem.* **11**, 860 (1905).
28. Bruni, G. and M. Amadori. *Atti Ist. Veneto*. **70**, 1113 (1910).
29. Copenhafer, D. T. and C. A. Kraus. *J. Amer. Chem. Soc.* **73**, 4557, (1951).
30. Craig, R. A. *Thesis*, Ohio State University (1948).
31. Craig, R. A., A. B. Garrett and M. S. Newman. *J. Amer. Chem. Soc.* **72**, 163 (1950).
32. Dawson, L. R. and E. J. Griffiths. *J. Phys. Chem.* **56**, 281 (1952).
33. Doucet, J., H. Porta and G. Finiels. *CR Acad. Sci, Paris*, **270c**, 1208 (1970).
34. Durand, J. and E. Rouge. *Bull. Soc. Chim. France*, **37**, 697 (1925).
35. Eichelberger, W. C. *J. Amer. Chem. Soc.* **56**, 799 (1934).
36. Eykman, J. F. *Z. Phys. Chem.* **3**, 203 (1889).
37. Eykman, J. F. *Z. Phys. Chem.* **4**, 497 (1889).
38. Falciola, P. *Gazz. Chim. Ital.* **52**, 175 (1922).
39. Findlay, A. and A. N. Campbell. *J. Chem. Soc. (London)*, 1770 (1928).
40. Garelli, F. and A. Ferratini. *Gazz. Chim. Ital.* **23a**, 442 (1893).
41. Garelli, F. *Gazz. Chim. Ital.* **24b**, 229 and 263 (1894).
42. Garelli, F. *Gazz. Chim. Ital.* **25b**, 179 (1895).
43. Garelli, F. *Gazz. Chim. Ital.* **28b**, 253 (1898).
44. Garelli, F. and F. Calzolari. *Gazz. Chim. Ital.* **29b**, 258 and 357 (1899).
45. Garelli, F. and F. Gorni. *Gazz. Chim. Ital.* **34b**, 101 (1904).
46. Getman, F. H. *J. Amer. Chem. Soc.* **62**, 2179 (1940).
47. Gondberg, H. and B. Gobv. *Zh. Vseignz. Kim. Obshelet* **6**, 467 (1961).
48. Hals, L. J. and H. G. Bryce. *Analyt. Chem.* **23**, 1695 (1951).
49. Hock, L. and C. I. Nottebohm. *Kolloidchem. Beih.* **31**, 205 (1930).
50. Howello, W. *J. Chem. Soc. (London)*, 915 (1929).
51. Jona, T. *Gazz. Chim. Ital.* **39b**, 289 (1909).
52. Jona, T. *Gazz. Chim. Ital.* **47b**, 87 (1917).
53. Lange, J. and J. Berga. *Mh. Chem.* **81**, 921 (1950).
54. Lespiau, R. *CR Acad. Sci., Paris*, **169**, 31 (1919).
55. Louginine, W. and G. Dupont. *Bull. Soc. Chim. France*, **4**, 904 (1912).
56. Mameli, E. *Gazz. Chim. Ital.* **39**, 579 (1909).
57. Mascarelli, L. and I. Musatty. *Gazz. Chim. Ital.* **41a**, 107 (1911).
58. Mascarelli, L. and V. Babini. *Gazz. Chim. Ital.* **41a**, 89 (1911).
59. Mascarelli, L. and L. Vecchiotti. *Gazz. Chim. Ital.* **42a**, 110 (1912).
60. Mastrangello, S. V. *Analyt. Chem.* **29**, 841 (1957).
61. Mathews, J. *J. Amer. Chem. Soc.* **39**, 1125 (1917).
62. Meyer, R. and J. Metzger. *CR Acad. Sci., Paris*, **263c**, 1333 (1966).
63. Newman, M. S., H. G. Kuivila and A. B. Garrett. *J. Amer. Chem. Soc.* **67**, 704 (1945).
64. Oddo, G. *Gazz. Chim. Ital.* **43b**, 263 (1913).
65. Osaka, H. *Bull. Inst. Phys. Chem. Res. Japan*, **80** (1938).
66. Ostromisslensky, P. *Z. Phys. Chem.* **57**, 341 (1906).
67. Padoa, M. *Gazz. Chim. Ital.* **34a**, 146 (1904).
68. Padoa, M. and G. Rotondi. *RC Accad. Lincei*, **21**, II, 629 (1912).
69. Paterno, E. and C. Montemartini. *Gazz. Chim. Ital.* **24b**, 197 (1895).
70. Paterno, E. and C. Manuelli. *RC Accad. Lincei*, **6a**, 401 (1894).
71. Paterno, E. *Gazz. Chim. Ital.* **26b**, 1 and 9 (1896).
72. Pettit, L. D. *J. Chem. Phys.* **24**, 1301 (1962).
73. Porter, G. B. and E. C. Baughan. *J. Chem. Soc. (London)*, 744 (1958).
74. Pushin, N. A. and A. A. Giagoleva. *J. Chem. Soc. (London)*, **121**, 2820 (1922).
75. Ramsay, W. *Z. Phys. Chem.* **5**, 221 (1890).
76. Reutler, M. and E. Rosenbaum *Chem.-Ing.-Tech. Berlin*, **8**, 359 (1956).
77. Richardson, G. M. and P. W. Robertson. *J. Chem. Soc. (London)*, 1777 (1928).
78. Roberts, H. M. and G. R. Bury. *J. Chem. Soc. (London)*, **125**, 2219 (1924).
79. Roth, W. and L. Mayer. *Z. Elektrochem.* **41**, 229 (1935).
80. Schenck, R. *Ber. Dtsch. Chem. Ges.* **37**, 917 (1904).
81. Scheuer, O. *Z. Phys. Chem.* **72**, 523 (1910).
82. Scherick, R., H. Finken and F. Pleuger. *Liebigs Ann.* **462**, 278 (1928).
83. Schmid, L. and B. Becker. *Ber. Dtsch. Chem. Ges.* **58**, 1968 (1925).
84. Schreiner, E. and O. E. Frivoli. *Z. Phys. Chem.* **124**, 1 (1926).
85. Tettamanzi, A. and F. Arnaldi. *Atti Accad. Torino*, **77**, 268 (1941).
86. Walden, P. *Z. Anorg. Chem.* **68**, 307 (1910).
87. Walden, P. *Rec. Trav. Chim. Pays-Bas*, **48**, 880 (1929).
88. Wyatt, P. A. and J. R. Brayford. *Trans. Faraday Soc.* **52**, 642 (1956).

easier. The formation of an aqueous eutectic is very easy if one uses a salt with a large heat of dissolution. For example, in 100 g of water at 28°C contained in a dewar vessel, let us add in one operation 25 g of potassium nitrate, finely ground. A rotating stirrer makes the dissolution faster. The solution temperature decreases up to 13°C, which is the saturation point temperature. With 19°C water, the temperature decreases to 7°C, and the salt is in excess. Finally, if the water and the salt are at an initial temperature of 4°C, the final temperature is -4°C, much lower than the eutectic temperature (-2.84°C). The supercooling generally stops spontaneously; ice crystals go up to the surface of the solution, while excess salt falls to the bottom; the temperature reaches a constant -2.84°C value. The length of the plateau obtained depends upon the adiabatic character of the vessel and upon the quantity of ice formed.

If, now, the heat of dissolution of the salt being used is not sufficient, the water-containing test tube is dipped into a refrigerating bath at a temperature lower than the eutectic temperature. If care is taken to stir the liquid and to cool it by means of a bath with a level below the solution level in the test tube, no ice spangles are formed too early. The refrigerating bath may then be replaced by a dewar vessel of a diameter slightly greater than that of the test tube, whereon supercooling stops. The salt weight required is, anyway, about twice the weight corresponding to the eutectic composition. The latter is given in *Table 2*, for 100 g of water.

A. Table of Salt Hydrate Solvents

Only one study of these solvents has been made, by Livingston, so that the number of salt hydrates with a known cryoscopic constant is small. The constants given by Livingston and Boutaric are the result of averaging at finite concentrations. As Livingston et al.[47] published all their experimental

Table 2

Compound	Formula	T_f	M	λ meas.	λ calc.	L for 1g	Ref.
Sodium chromate	$Na_2CrO_4 \cdot 10H_2O$	19.92	342.16	3.81		39.2	3
Calcium chloride	$CaCl_2 \cdot 6H_2O$	29.48	219.09	4.25	4.50	40.7	3
Lithium nitrate	$LiNO_3 \cdot 3H_2O$	29.88	123	2.72			3
Sodium sulfate	$Na_2SO_4 \cdot 10H_2O$	32.384	322.22	3.27	3.20	58	2/5/6
Manganese nitrate	$Mn(NO_3)_2 \cdot 3H_2O$	34.81	232.99	6.74			4
Calcium nitrate	$Ca(NO_3)_2 \cdot 4H_2O$	42.17	236.16		5.94	33.9	4
Zinc nitrate	$Zn(NO_3)_2 \cdot 3H_2O$	45.5	243.44	5			4
Sodium thiosulfate	$Na_2S_2O_3 \cdot 5H_2O$	48.5	248.21	4.26	4.28	47.9	1

[1] Boutaric, A. *Bull. Soc. Chim. France*, 13, 651 (1913).
[2] Jahr, K. F. and R. Kubens. *Z. Elektrochem. Angew. Phys. Chem.* 56, 65 (1952).
[3] Livingston, J., R. Morgan and H. K. Benson. *Z. Anorg. Chem.* 55, 261 (1907); *J. Amer. Chem. Soc.* 29, 1168 (1907).
[4] Livingston, J., R. Morgan and P. T. Owen. *Z. Anorg. Chem.* 56, 168 (1908); *J. Amer. Chem. Soc.* 29, 1439 (1907).
[5] Pierret, E. *Bull. Soc. Chim. France.* 39, 590 (1926).
[6] Rolla, M. and M. Cola. *Ann. Chim. Ital.* 42, 651 (1952).

PART 2. CRYOSCOPY

results, we have been able to interpret them with an extrapolation towards the origin. The corrected values thus obtained are given here. The molar depression corresponds to 1000 g of salt. Lemaire[45] prefers to calculate them for 1000 g of the liquid phase, which actually forms the solvent. The enthalpy variation being unknown, a theoretical formula which will be given below for eutectics is used by this author.

B. Tables of Aqueous Eutectic Solvents

The eutectics for which the limiting molal depression is known from cryoscopic measurements, are listed in *Tables 3* and *4*. They are given in order of decreasing temperature. The value of the cryoscopic constant is given by the van't Hoff formula. The heat of fusion is here the isothermal enthalpy variation of 1 g of solvent as it goes from the solid to the liquid state, at the equilibrium temperature T. This variation is composed of a heat of dilution term which is the heat of dilution of the solvent from the mole fraction 1 to the mole fraction x_1 (this quantity approaches zero in an infinitely dilute solution), and of the heat of solidification of the pure solvent at temperature T.

In his thesis work, Lemaire[45] gave a value for λ, calculated from calorimetric and densimetric data, which are better known than the heat of fusion of a eutectic. He shows that

$$1/\lambda = 1/K_1 + 1/K_2 \tag{71}$$

with

$$K_1 = (RT_e^2/1000q_0)d$$

where q_0 is the heat of fusion of 1 g of ice at the eutectic temperature T_e, i.e. in calories

$$q_0 = 79.7 + 0.5t_e \tag{72}$$

and d is the eutectic liquid density.

The other constant, K_2, which contributes little to λ, is

$$K_2 = \{(c_0 - c)/c\}d\,(RT_e^2/1000q) \tag{73}$$

where q is the heat of fusion of the salt component of the eutectic, c is the ratio of the mass of anhydrous salt to that of the eutectic solution, and c_0 is the ratio of the anhydrous salt mass to the hydrated salt mass if it forms crystals with water, otherwise $c_0 = 1$.

Table 3 sets out the densities of the eutectics, together with the K_1 values as given by Lemaire. K is the experimental cryoscopic constant for 1000 g of water, whereas λ is the value for 1000 g of eutectic solvent.

The composition of the eutectic is given in terms of the mass of salt in 100 g of water, column P per cent. The heat of dissolution Q in calories is that of one mole of salt in n moles of water at 18°C. The number n is given in parentheses.

Those values enable one to predict the temperature depression occurring during the dissolution, if it is adiabatic. They also enable one to calculate the eutectic heat of fusion, if the specific heats are known. For example, Cohen–Adad[12] gives $\Delta H = 76.6$ cal/g, for urea at -11.4°C, which gives quite

Y. DOUCET

Table 3. Binary aqueous eutectics

Name	Formula	M	T_e	K	λ	K_1	$P\%$	d	Q	Ref.
Potassium perchlorate	$KClO_4$	138.56	−0.163	1.811			6.73			4
Glauber salt	$Na_2SO_4 \cdot 10H_2O$	322.22	−1.135	1.72	1.79	1.91	3.94	1.034	−18900 (400)	2
Potassium sulfate	K_2SO_4	174.26	−1.51	1.81	1.93	1.95	7.24	1.055	−6547 (400)	6
Potassium nitrate	KNO_3	101.10	−2.88	1.68	1.86	1.98	10.8	1.070	−8459 (200)	3
Potassium ferricyanide	$Fe(CN)_6K_3$	329.25	−3.50	1.63	2.04	2.06	25.2	1.116	−12400 (1000)	6
Seignette salt	$KNaC_4H_4O_6 \cdot 4H_2O$	282.23	−4.55		2.08	2.12	41	1.152	−3580 (∞)	5
Urea	$CO(NH_2)_2$	60.06	−11.60	1.20	1.78		48.4			7
Ammonium chloride	$(NH_4)Cl$	53.50	−15.44	1.43	1.78	1.93	24.5	1.058	−3895 (200)	2
Ammonium nitrate	$(NH_4)NO_3$	80.05	−17.20	1.29	2.30	2.18	77.7	1.198	−6332 (200)	2
Sodium nitrate	$NaNO_3$	85.01	−18.50	1.55	2.46		58.5		−5018 (200)	3
Perchloric acid	$HClO_4$	100.47	−59.7		4.43		106			1

[1] Ardon, M. and A. Linensberg. *J. Phys. Chem.* **65**, 1443 (1961).
[2] Doucet, Y. *J. Phys. Radium.* **4**, 51 (1943).
[3] Jahr, K. F., A. Brechlin, M. Blanke and R. Kubens. *Z. Anorg. Allgem. Chem.* **270**, 240 (1952).
[4] Kentamaa, J. *Suomen Kemistil.* **29**, 59 (1956).
[5] Lemaire, M. Thesis, Lyon (1956).
[6] Müller, H. J. Thesis, Paris (1937).
[7] Rollet, A. P. and R. Cohen-Adad, *CR Acad. Sci., Paris*, **227**, 554 (1948) and *Publ. Univ. Alger.* **1B**, 85 (1955).

PART 2. CRYOSCOPY

Table 4. Ternary aqueous eutectics

Formula	%	T_e	λ	d	Ref.
$Na_2B_4O_7$	2.5	−2.30	1.95	1.082	1
K_2SO_4	8.2				
Na_2HPO_4	3.3	−2.61	1.98	1.094	1
K_2SO_4	9.2				
$Na_2P_2O_7$	6.7	−3.27	2.25	1.148	1
K_2SO_4	12.8				
KNO_3	10.3	−3.32	1.90 ($K = 1.65$)	1.091	2
K_2SO_4	4.5				
Na_2HPO_4	2.86	−3.72	1.90	1.096	1
KNO_3	12.5				
Na_2SO_4	6.85	−4.82	2.02	1.129	1
KNO_3	14				
$Na_2C_2O_4$	2.10	−6.15	2.22	1.157	1
$K_2C_2O_4$	23.2				
NH_4Cl	25.1	−18.09	1.8 ($K = 1.31$)	1.113	2
KNO_3	11.9				

[1] Lemaire, M. *Thesis*, Lyon (1956).
[2] Müller, H. J. *Thesis*, Paris (1937).

exactly the cryometric $\lambda = 1.775$. The numbers following the salt formulas are the masses of these salts in the eutectic, for 100 g of water.

5. *Molten Salts Cryoscopy*

A. Equipment

As measurements are made in the 200 to 1 200°C range, the previous equipment is modified. The cryostat–thermostat is replaced by a vertical tube-furnace (*Figure 15*). The nature of the resistance and the heating power (1 to 5 kW) depend on the required temperature. Special alloy wires are commercially available for that purpose, or tungsten bars. The thermal insulation should not be over-efficient; since one works during a cooling cycle; a programming device, controlled by an additional resistance, must be able to slow up that cooling to a predetermined rate of the order of one degree per minute.

A crucible, containing a known weight of salt, is put in the center of the tube-furnace, in the region of the smallest temperature gradient. It must be deep enough for the immersion of the thermometric probe to be sufficient. Its essential quality is its chemical inertia towards the salt. Below 500°C, borosilicate glasses may be used. Above that temperature, pure nickel, platinum with an alumina crucible, or some superior brands of stainless alloys may be used. High purity graphite is also convenient.

A metallic crucible insures a good temperature equalization within the solution because of the high thermal conductivity of the metal. Furthermore, its low specific heat is an advantage as the heat transfers between the furnace and the crucible are quick.

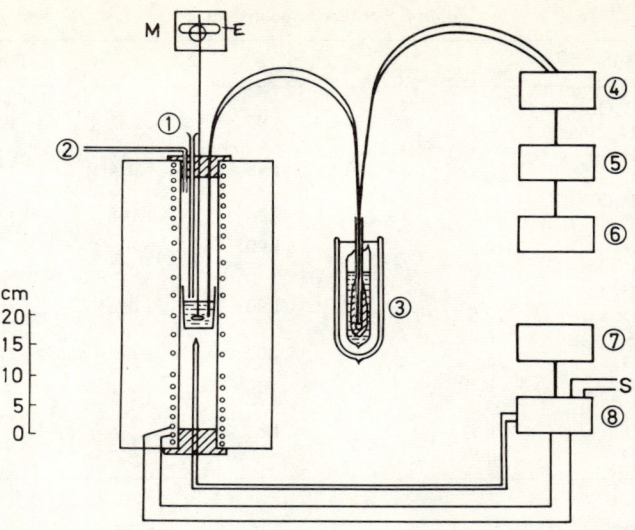

Figure 15. Cryoscope for molten salt solvents: 1 Sample addition tube, 2 Argon tube, 3 Triple point cell, 4 Manual potentiometer, 5 Preamplifier, 6 Recording potentiometer, 7 Regulator thermocouple, 8 Programmer, E Eccentric-wheel, M Micromotor, S Alternating current.

One may have to operate under an inert atmosphere. The central tube of the furnace is then flushed out by a nitrogen or argon flow. This makes the use of graphite crucibles easier at elevated temperatures.

If one has to operate under a vacuum, the easiest thing is to use sealed glass cells (*Figure 16*), as in the case of mercuric bromide, with a 'well' where the thermometric probe is introduced.

Solomons and Janz[76] gave details about the making of glass cells enabling one to work under the atmosphere desired and with a detachable test tube. The crucible is placed on a suitable stainless frame which may be lifted or lowered by means of a geared rod. If a sealed cell is being used, the latter rod undergoes an up-and-down periodic movement which insures agitation of the liquid. Such agitation is quite important. It has a twofold purpose, first to reduce the temperature gradient and then to insure a good contact between the crystals and the liquid, as the temperature sought is that of the solid–liquid interphase and the probe is dipped into the liquid. Forced convection movements play an essential part. An efficient rotating agitation is very difficult to obtain. Grjotheim[31], however, uses a small broad-threaded square screw made of pure graphite. This screw can be rotated and can start an upward convective movement of the liquid. The crucible A is also made of graphite (*Figure 17*).

Vertical alternating stirring may be achieved by either a magnetic induction coil with a soft iron core or a small motor supplied with an eccentric in control of a ringed rod in a stainless material. The heat quantity brought about by this rather strong stirring is of no importance for kinetic methods, which precisely require thermal exchanges with the outside of the system.

PART 2. CRYOSCOPY

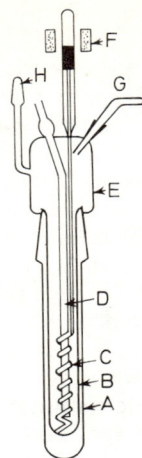

Figure 16. Cryoscopic tube for controlled atmosphere, from C. Solomons and G. J. Janz [*US Dept Comm., Wash., Publ. No. 131499* (1957)]: A Outer jacket, B Inner test tube, C Spiral stirrer, D Thermocouple well, E Standard taper joint, F Solenoid, G Sample addition tube, H To vacuum.

Some authors also agitate their melts by bubbling an inert gas through them.

Another point, which is also a difficult one because of the furnace, is the introduction, without loss, of a known weight of solute in the molten salt. Special devices should be used, which vary according to the type of cell (glass or metal). Very often, the salt added is put in a small glass ampoule which is broken in the crucible. One must, of course, insure that the quantity of salt added is completely soluble.

B. TEMPERATURE MEASUREMENT

Temperatures are taken by means of a thermocouple. The Le Chatelier

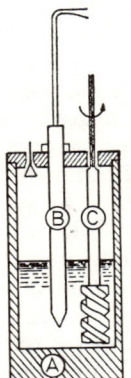

Figure 17. Graphite cryoscope with rotatable stirrer: A graphite test tube, B Thermocouple, C Helicoidal stirrer.

873

couple (Pt/Pt–10% Rh), the e.m.f. of which is only 5.65 µV. deg^{-1} at 0°C, reaches more than 10 µV. deg^{-1} at 600°C. Its use is therefore justified, the more so since it gives a much more reproducible e.m.f. than the chromel–alumel thermocouple. The cold junction is put in a water triple-point cell, while the hot one is put at a constant depth within the melt; it is coated with nickel or platinum if glass is not convenient. The e.m.f. at 600°C is then 5.200 mV; the potentiometer gives its full deflection for 1 mV. A scale change of the order of 5 mV is thus necessary; it may be obtained by either a built-in potentiometer or an external potentiometer which is set manually once and for all, near the equilibrium. If a 0.01°C sensitivity is required one must either mount 10 or 15 couples in series or use a preamplifier. The latter solution is to be preferred as it avoids multiplying by a factor of 10 or 15 the parasitic e.m.f.s (Thomson effect) and because then the recording potentiometer works in the best condition for reproducibility.

The calibration is done for international fixed points in the range investigated, and by comparison with the NBS table. The temperature of fusion of gold is one of those points. On the other hand, the manual potentiometer, and the recording potentiometer should be calibrated with respect to a standard potentiometer, such as the Mueller potentiometer for example, in order to insure that their readings are reproducible.

C. Procedure

The final melting point technique is difficult to use because of the agitation in the presence of a large quantity of solid. The starting-freezing method is, within the kinetic method, universally used among investigators. One must start from the pure solvent freezing point. It might be desirable, in order to find out any random error, to use two furnaces, two crucibles, and two thermocouples, the cold junctions of which are dipped in the same standard bath (water triple-point cell).

A recording potentiometer with several channels gives values alternately from one thermocouple and from the other. The freezing curve shows a slight supercooling which may reach 0.1°C but which, quite often, is extremely small. The important point is the intercept of the two cooling lines, the one from the pure liquid, and the other one from the liquid in the presence of crystals.

Then some solute is added in a pure solvent crucible, this solvent being entirely liquid, but at one or two degrees above its melting point only. The freezing curve is recorded on the second channel, at the same time as the pure solvent curve. Permanent comparison with the pure solvent channel enables one to find out systematic errors and to check on the cryoscopic depressions obtained. Any alteration or thermal decomposition of the solvent is seen at once. With a six-channel recording potentiometer, one can measure the temperatures of four crucibles of similar or different concentrations, the fifth channel being used for the pure solvent and the sixth one, short-circuited, giving at all times a check on the zero of the potentiometer.

After a first addition of solute salt, one comes back to a wholly liquid mixture and solute is added for the second time. One can go on in this way, with increasing concentrations of the salt B in solvent A, up to the eutectic

PART 2. CRYOSCOPY

Table 5

Salt name	M†	T_f	λ	ΔH_f	S_f	$C_L - C_S$	Ref.
Ammonium nitrate NH_4NO_3	80.05	442.6	20.5	1530	3.5	1.6	21,
Barium bromide $BaBr_2$	297.19	1130	98.8	7630	6.8		6
Barium chloride $BaCl_2$	208.27	1235	161	3900	4.7	3*	6
Barium fluoride BaF_2	175.36	1619	163	5100	3.5		2
Barium iodide BaI_2	391.19	984	118.6	6340	6.4		6
Barium nitrate $Ba(NO_3)_2$	261.35	865	39.3	9947	11.5	9.76	4
Barium orthophosphate $Ba_3(PO_4)_2$	602.04	2000	257	18600	9.3		38
Barium sulfate $BaSO_4$	233.43	1853	125.9	9700	4.7		38
Bismuth trichloride $BiCl_3$	315.37	506.6	28.3	5680	11.2		41
Calcium bromide $CaBr_2$	109.91	1038	58.8	6950	6.7		6
Calcium chloride $CaCl_2$	110.99	1045	35.5	6780	6.5	4	6
Calcium fluoride CaF_2	78.08	1683	61	7100	4.2		37
Calcium iodide CaI_2	293.92	848	64.6	10000	11.8		6
Calcium oxide CaO	56.08	2853	76.6	12000	4.2		38
Calcium pyrophosphate $Ca_2P_2O_7$	254.12	1626	55.4	24100	14.7		38
Calcium sulfate $CaSO_4$	136.15	1573	99.8	6700	4.3		38
Cesium bromide $CsBr$	212.83	909	61.9	5640	6.2		5
Cesium chloride $CsCl$	168.37	919	58.2	4840	5.3	1.1	5
Cesium fluoride CsF	151.91	987	55.4	5190	5.3		5
Cesium iodide CsI	259.83	899	73.9	5640	6.3	1.6	5
Cesium nitrate $CsNO_3$	194.92	687	53	3370	5		24
Cesium sulfate Cs_2SO_4	361.89	1277	122.6	9577	7.5		27
Lithium bromide $LiBr$	86.86	819	27.7	4220	5.1	1.5*	5
Lithium carbonate Li_2CO_3	73.89	891	13.6	10700	12		17
Lithium chloride $LiCl$	42.40	881	13.8	4760	5.4	0.3	5
Lithium fluoride LiF	25.94	1119	10	6460	5.8	0.8	7
Lithium hydride LiH	7.95	962	2.77	5280	5.5		20
Lithium iodide LiI	133.86	742	41.8	3500	4.7		5
Lithium metaborate $LiBO_2$	49.76	1113	15	8260	7.4		43
Lithium nitrate $LiNO_3$	68.95	525	6.17	6120	12	0.6	3
Lithium sulfate Li_2SO_4	133	1133	142	1975	1.75		29
Lithium eutectic $LiCl-KCl$	109.95	627	13.6	3200	5.1		40

875

Table 5—continued

Salt name	M†	T_f	λ	ΔH_f	S_f	$C_L - C_S$	Ref.
Lithium eutectic LiF–KF		557	16.4	2930	5.2		13
Lithium eutectic (Li$_2$/K$_2$)CO$_3$		778	14.7	8700	11		19
Lead chloride PbCl$_2$	278.12	768	57	4400	5.7	7	11
Magnesium fluoride MgF$_2$	62.32	1530	22	13900	8.6		2
Magnesium oxide MgO	40.32	3173	43.6	18500	5.9		38
Magnesium silicate MgSiO$_3$	100.40	1830	44	15000	8.2		33
Magnesium silicate MgSiO$_4$	140.71	2183	41	32300	14.8		33
Magnesium sulfate MgSO$_4$	120.39	1397	134	3500	2.5		38
Mercury bromide HgBr$_2$	360.44	509	43.7	4280	8.4	1.6	16
Mercury chloride HgCl$_2$	271.52	549	39.3	4150	7.6	8	14
Mercury iodide HgI$_2$	454.45	532	56.7	4500	8.4		14
Potassium bromide KBr	119.01	1005	39.3	6100	6	1.6*	5
Potassium carbonate K$_2$CO$_3$	138.20	1163	56.8	6600	5.7		17
Potassium chloride KCl	74.55	1049	25.4	6340	6	0.6	5
Potassium fluoride KF	58.10	1132	21.9	6750	4.6	1.6	5
Potassium fluoride acid KHF$_2$	78.10	512	19.6	1570	3		38
Potassium iodide KI	166.02	954	52.4	5740	6		5
Potassium nitrate KNO$_3$	101.1	611	31.5	2800	4.6	0.7	8
Potassium sulfate K$_2$SO$_4$	174.26	1341	68.8	9060	6.7		38
Potassium thiocyanate KSCN	97.18	451	12.7	3070	5		32
Potassium eutectic KF–NaF	51.5	993	20	5040	5.1		35
Silver nitrate AgNO$_3$	169.89	482.6	26.5	2985	6.2	5.1	15
Sodium bromide NaBr	102.91	1020	34.1	6240	6.1	1.2*	5
Sodium carbonate Na$_2$CO$_3$	106	973	30.7	6700	6.7		36
Sodium cyanide NaCN	49.02	836.8	16.3	4190	5		38
Sodium chlorate NaClO$_3$	106.45	528	10.9	5250	9.9	0.5	21a
Sodium chloride NaCl	58.45	1074	20.4	6690	6.2	0.8	28
Sodium chromate Na$_2$CrO$_4$	161.97	1065	62.5	5840	5.8		31
Sodium dichromate Na$_2$Cr$_2$O$_7$	262	593	25.1	8170	13.8		13
Sodium fluoaluminate Na$_3$F$_6$Al	209.96	1284	41	27640	21	−12.8	34
Sodium fluoride NaF	42	1268	16.7	8030	6.2	1.1	1
Sodium hydroxide NaOH	40.01	592	28.3	1520	2.6		39
Sodium iodide NaI	149.92	923	45.9	5600	6		5

PART 2. CRYOSCOPY

Compound	M, 10^{-3} kg	T_f, K	ΔH_f	S_f	$C_L - C_s$	λ	Ref.
Sodium nitrate NaNO$_3$	85	579	16.1	3520	6.1	1.3		18
Sodium nitrite NaNO$_2$	69.01	554.6	17	2480	4.5		23
Sodium sulfate Na$_2$SO$_4$	142.06	1158	63.1	5670	4.9	5.3*	12
Sodium thiocyanate NaSCN	81.07	580.7	9.4	5770	10		9
Sodium eutectic Na$_3$F$_6$Al–NaF	14%AlF$_3$	1158	27	4730	4.1		26
Sodium eutectic Na$_4$P$_2$O$_7$–NaCl	77.8%NaCl	998	30.7	6750	6.8		30
Rubidium bromide RbBr	165.40	965	35.2	5570	5.8		5
Rubidium chloride RbCl	120.94	998	41.9	5670	5.7		5
Rubidium fluoride RbF	104.48	1068	38.5	6150	5.8		5
Rubidium iodide RbI	212.40	915	67.7	5270	5.7		5
Rubidium nitrate RbNO$_3$	147.49	583	89	1110	2		25
Strontium bromide SrBr$_2$	247.46	916	170	2500	2.7		6
Strontium chloride SrCl$_2$	158.54	1146	106	3880	3.4		6
Strontium fluoride SrF$_2$	125.63	1736	106	7100	4.1		2
Strontium iodide SrI$_2$	341.47	811	94.9	4700	5.8		6
Strontium nitrate Sr(NO$_3$)$_2$	211.65	918	169	9900	10.8		42
Thalium nitrate TlNO$_3$	266.37	482	58.7	1960	4		10
Vanadium oxide V$_2$O$_5$	181.90	937.7	22	14500	15.4		22

Units: †M, Molecular weight, 10^{-3} kg.
T_f, Melting point, K.
λ, cryoscopic constant, deg (mol kg)$^{-1}$
ΔH_f molar enthalpy of fusion, cal mol^{-1}
S_f molar entropy of fusion, cal mol^{-1} K^{-1}.
$C_L - C_s$, molar heat capacity, cal mol^{-1}.

1 Cantor, S. and T. S. Carlton. *J. Phys. Chem.* **66**, 2711 (1962).
2 Delbove, F. *Silic. Ind.* **7**, 257 (1967).
3 Doucet, Y. and C. Vallet. *CR Acad. Sci., Paris*, **259**, 1517 (1964).
4 Doucet, Y. and C. Vallet. *CR Acad. Sci., Paris*, **261**, 2884 (1965).
5 Dworkin, A. S. and M. A. Bredig. *J. Phys. Chem.* **64**, 269 (1960).
6 Dworkin, A. S. and M. A. Bredig. *J. Phys. Chem.* **67**, 697 (1963).
7 Flood, H., O. Fykse and S. Urnes. *Z. Elektrochem.* **59**, 364 (1955).
8 Franzosini, P. and C. Sinistri. *Ric. Sci.* **33**, 411 (1963).
9 Franzosini, P. and R. Riccardi. *Ann. Chim., Roma*, **53**, 558 (1963).
10 Franzosini, P. and C. Sinistri. *Ric. Sci.* **IIA**, 439 (1963).
11 Goodwin, H. M. and H. T. Kalmus. *Phys. Rev.* **28**, 1 (1909).
12 Grjotheim, K., T. Malvorsen. and S. Urnes. *Canad. J. Chem.* **37**, 7, 1170 (1959).
13 Isbell, J. *Dissert. Abstr.* **20**, 3529 (1960).
14 Janz, G. J. and J. D. E. Macintyre. *Ann. NY Acad. Sci.* **79**, 790 (1959).
15 Janz, G. J., D. W. James and J. Goodkin. *J. Phys. Chem.* **64**, 937 (1960).
16 Janz, G. J. and J. Goodkin. *J. Phys. Chem.* **64**, 808 (1960).
17 Janz, G. J., E. Neuenschwander and F. J. Kelly. *Trans. Faraday Soc.* **59**, 841 (1963).
18 Janz, G. J., F. J. Kelly and J. L. Perano. *J. Chem. Engng Data*, **9**, 134 (1964).
19 Janz, G. J. and J. L. Perano. *Trans. Faraday Soc.* **60**, 1742 (1964).
20 Johnson, C. E., S. E. Wood and C. E. Crouthamel. *Inorg. Chem.* **3**, 1487 (1964).
21 Keenan, A. G. *J. Phys. Chem.* **60**, 1356 (1956).
(a) Kelley, K. K. *Bull. US Bur. Min. No. 601* (1962).
22 Kohlmuller, R. and N. Leitner. *Rev. Chim. Gen.* **1**, 275 (1964).
23 Kozlowski, T. R. and N. Bartholomew. *J. Electrochem. Soc.* **114**, 937 (1967).
24 Mustajoki, A. *Ann. Acad. Sci. Fenn.* A.vi, No. 7 (1957).
25 Mustajoki, A. *Ann. Acad. Sci. Fenn.* A.vi, No. 8 (1958).
26 Petit, G. *CR Acad. Sci., Paris*, **234**, 1281 (1957).
27 Petit, G. and C. Bourlange. *CR Acad. Sci., Paris*, **245**, 1788 (1957).
28 Riccardi, R. and C. Benaglia. *Gazz. Chim. Ital.* **91**, 315 (1961).
29 Riccardi, R. *Gazz. Chim. Ital.* **91**, 1479 (1961).
30 Riccardi, R. *Gazz. Chim. Ital.* **92**, 34 (1962).
31 Riccardi, R. *Bull Sci. Indust. Chim. Bologna*, **20**, 92 (1962).
32 Riccardi, R. and P. Franzosini. *Gazz. Chim. Ital.* **92**, 386 (1962).
33 Richardson, F. D. *Trans. Faraday Soc.* **52**, 1312 (1956).
34 Rolin, M. *Thesis*, Masson: Paris (1950).
35 Rolin, M. *Colloque de l'aluminium, Paris* (Juin, 1954).
36 Rolin, M. and J. M. Recapet. *Bull. Soc. Chim. France*, 2504 (1964).
37 Rolla, M., P. Franzosini and R. Riccardi. *Disc. Faraday Soc.* **32**, 84 (1961).
38 Rossini, F. D., D. D. Wagman, W. H. Evans, S. Levine and I. Jaffe. *Circ. US Nat. Bur. Stand No. 500* (1952).
39 Seward, R. P. *J. Amer. Chem. Soc.* **72**, 5507 (1955).
40 Solomons, C., J. Goodkin, H. J. Gardner and G. Janz. *J. Phys. Chem.* **62**, 248 (1958).
41 Topol, L. E., S. W. Meyer and L. D. Ransom. *J. Phys. Chem.* **64**, 862 (1960).
42 Vallet, C. and M. Gaune. *CR Acad. Sci., Paris*, **261**, 5455 (1965).
43 Zarzycki, J. *Thesis*, Masson: Paris (1953).

concentration. If the phase diagram is of the simple eutectic type, the second branch may be obtained with B as a solvent and A as a solute.

D. TABLE OF MOLTEN SALT CRYOSCOPIC CONSTANTS

The salts used as solvents are given in alphabetical order. Their melting points are given in kelvin and the experimental cryoscopic constant in deg. mol^{-1}. The molal heat of fusion is sometimes calculated from the cryoscopic constant, but when calorimetric measurements have been made they are given preferentially.

The column giving the difference in the molal specific heats of the liquid and the solid, in the vicinity of the melting point, is far from being complete. Data are often taken from the old work of Goodwin and Kalmus[30] or from the more recent work of Dworkin and Bredig[23]. When C_L is not known, it may be calculated from the rough empirical relationship $C_L/C_S = 1.1$ resulting from averaging the known data for 17 halides and 3 nitrates. This value, which is questionable, is denoted by a star,*. Kelley[39] and Douglas[22] give liquid-state specific heats. Unfortunately the latter are scarcely interpretable because C_S (solid state value) is not known in the vicinity of the melting point and depends strongly upon temperature.

III. Adiabatic Cryoscopy Equipments

The principle of achieving a state of equilibrium between the solution and the pure solvent crystals was understood by the pioneers of cryoscopy, Roloff[70] and Ponsot[59], Raoult, who criticized[66] the latter investigators, nevertheless considered their method 'theoretically faultless though very difficult to achieve in practice'. As a matter of fact it requires perfect adiabatic conditions, which was not easy to realize at that time.

1. Aqueous Solution Equipment

The first precision apparatus, which served as a model for further investigations, was built by Adams[2]. Essentially it is made of two dewar vessels, one of which contains the ice–water mixture (50% ice), while the other contains the solution–ice mixture (also 50 per cent ice). A thermocouple, composed of

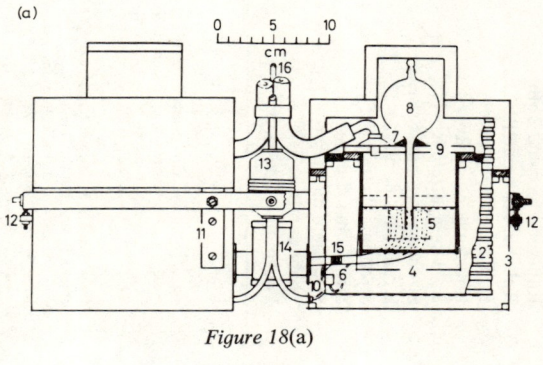

Figure 18(a)

PART 2. CRYOSCOPY

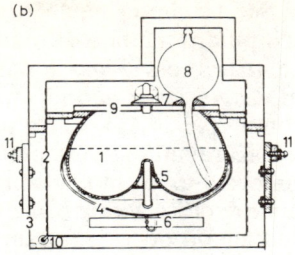

Figure 18(b)

Figure 18. Adiabatic cryoscope: (a) front view; (b) end view. From G. Scatchard, B. Vonnegut and D. W. Beaumont [*J. Chem. Phys.* **33**, 1292 (1960)]: 1 Freezing-point vessel, 2 Adiabatic shield, 3 Housing, 4 Thermal baffle, 5 Thermopiles mounted in Wood's metal. 6 Control thermopile mounted on shield, 7 Vacuum cups, 8 Sampling pipet, 9 Glass cover, 10 Leads to adiabatic-shield heater and control thermopile, 11 Swivel supports of unit in yoke, 12 Adjustable point of suspension of unit, 13 Shielded plug for adiabatic-shield heater and control thermopile leads, 14 Conduit for main thermopile leads, 15 Thermal ground of main thermopile wire to adiabatic shield, 16 Flexible electrical leads and evacuation tubes.

fifty U-shaped copper–constantan junctions, is dipped in both vessels. The agitation is performed by means of two little gilded brass pumps, which pour on the ice the liquid taken from the bottom of the vessel. The whole apparatus is surrounded by a bath of melting ice. Improvements concern the agitation, which is quite important because of the accuracy required. If the dewar vessels contain about a 10 cm depth of liquid, the hydrostatic pressure leads to a depression of the equilibrium temperature of $10^{-4}\,°C$, i.e. 10 or 100 times the thermocouple sensitivity. An adequate agitation must therefore bring the liquid into close contact with the ice crystals, avoid 'dead volumes' and obtain a stable temperature distribution. Furthermore, such an agitation should not destroy the adiabatic character of the apparatus by bringing about too much heat. Turbulent motion should be avoided and a stable convection state must

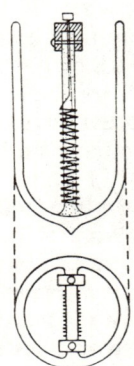

Figure 19. Cryoscope with entirely submerged thermocouples.

be achieved. Randall and Scott[63] are using an Archimedian screw. Lange[42] uses a small turbine which pumps the liquid from the bottom of the vessel. Scatchard et al.[74] built cylindrical cells (*Figure 18*) the vertical cross-section of which is of the shape of the Greek letter ω, and which undergoes oscillations of $\pm 30°$ about the horizontal axis in a 6–12 Hz rocking movement. An air agitation may also be used. Karagunis et al.[38] prefer the solution vapor and air pressures to be equal in order to avoid any condensation or evaporation. The air goes through a mixture of ice with a brine solution at the concentration wanted. The dissolved air produces a freezing point depression which would be the same in pure water and in the solution. This is not, however, the case, so that Robertson and La Mer[69] take this effect into account and in the same manner, Brown and Prue[9], and Scatchard and Prentiss[73] have shown that this solubility effect is lower than expected. The solution, however, is certainly saturated with air by this means of agitation, whereas mechanical agitation does not lead to such a certainty. Lange[42] has been able to avoid this problem; his apparatus is kept under vacuum and is operated at the triple-point pressure. This procedure was also that of Scatchard, Vonnegut and Beaumont[74]. With the U-shaped thermocouple, one must wait for a long time (more than one hour) for the residual e.m.f. to become stable within a few tenths of one microvolt, when both branches are dipped in two water–ice mixtures with the same proportion of ice and identical agitation. We prefer the following apparatus: only one dewar vessel is used; it is divided into two parts by an axial asbestos wall (*Figure 19*). This is constructed in such a manner that a hundred thermocouple junctions (made of BTE–CTE alloy). 0.1 mm in diameter, go through that wall, their hot and cold junctions being alternately on one side or the other of it. The system is made leak-proof with epoxy resin, and is covered with a bakelite layer polymerized in an oven at 130°C. The junctions are evenly scattered along the wall surface and should extend beyond the wall only a very little, in order to avoid impeding small ice crystals.

This arrangement has many advantages. The difference in melting points due to the hydrostatic pressure is averaged, the thermocouples being evenly scattered from the bottom to the top. Therefore a more stable temperature reading is observed. Another advantage lies in the quickness with which the zero is reached. The e.m.f. is stable when all the thermocouple wires are at the same temperature. A 'zero' plateau may be reached in half an hour or even less, within 0.3 µV. The short length of wire (10 to 12 mm only between hot and cold junctions) reduces the errors due to heterogeneity and brings about a remarkable stability. This is a great advantage.

There is, however, a lack of adiabaticism between the two compartments. An asbestos wall with a thickness of 1 cm is by no means a complete obstacle to heat transfer; the more so as two hundred metal wires penetrate through that wall, their total section being almost 2 mm². It should be noticed, however, that the difference in temperature is always less than 0.1°C; moreover, it is useless to go on with measurements on a plateau for more than a quarter of an hour. Finally, this wall is at 0°C, and therefore at a higher temperature than the solution, and so does not disappear with the use of a U-shaped thermocouple, as not all the inner surface of a dewar vessel dipped in ice is at a temperature below 0°C.

PART 2. CRYOSCOPY

The dewar vessel with its two compartments (*Figure 20*) adequately stoppered and supplied with a solution inlet and a sampling tube together with stirring equipment, is immersed completely in a water–ice thermostat. Each agitator is supplied with its own motor and adjustable rheostat; a single motor would not be convenient for the reasons stated above. The calibration

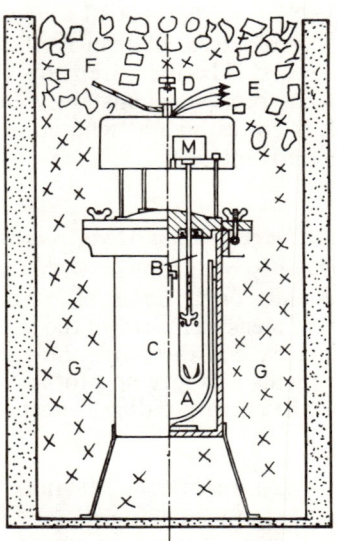

Figure 20. Adiabatic apparatus for aqueous cryoscopy: A Pure water compartment, B Turbine stirrer, C Thermopile, D Copper short circuit of thermopile, E to motor rheostat, F To potentiometer, G Ice, M Micromotor.

of the thermocouple may be performed by doing cryoscopic measurements in a potassium chloride solution or by comparison with a NBS platinum resistor thermometer and a Mueller bridge. For the above equipment, the calibration relationship, which is linear over the 0 to 0.1°C interval, is $\theta = 216e$ in units of °C and volts. With an easily accessible e.m.f. difference of 0.1 microvolt one associates a cryoscopic depression of 21.6 microdegree.

2. Non-aqueous Solvents Equipment

The equipments described above may be adapted to non-aqueous solvents with just slight modifications. Kraus and Vingee[41] have been using the equilibrium method with solvents such as benzene, dioxane and triphenyl methane. Eichelberger[24] used acetic acid.

At high temperatures in molten salt media, as was discussed above, Flood, Förland and Nesland[27] achieve an adiabatic equilibrium at a given temperature. The concentration is determined by means of a chemical analysis on the decanted liquid obtained by turning the crucibles upside down. In salt media, this method may also be of interest, though difficulties frequently arise as far as concentration measurements are concerned. We have tested this method, however, with the water–ammonium nitrate eutectic[17]. Kentamaa has used the water–potassium perchlorate eutectic[40]. At the Glauber salt transition point, and at the melting point of the aqueous eutectic of the same salt, Sabbah[72] built an adiabatic apparatus with two dewar vessels and a U-shaped thermocouple. The inside of this apparatus is maintained thermostatically at either $-1.13°C$ (eutectic point) or $32.38°C$ (transition point). The 'pseudo-zero' plateau is obtained after a somewhat long time (such as two hours). The measurements (by successive additions of solute) begin when it has been at a constant value for 20 to 30 minutes. The precision, of course, is not as high as with aqueous solutions, but these measurements have other advantages.

3. Concentration Measurements

In kinetic methods there is no problem with the determination of the concentrations as it depends on a series of weighings; difficulties arise, however, with temperature measurements (of the starting-freezing point and of the final melting point). By contrast, with the equilibrium methods, it is the temperature measurements that give no problem, whereas the difficulties are found in determining the corresponding concentration.

A. Conductivity Method

Randall and Vanselow[62], and later Lange[42], put the measuring cell inside the pumping tube of the small turbine, at a place where the liquid is not in the presence of ice crystals. Calibration of this apparatus with potassium chloride solutions gives from the work of Lange a constant of 0.2 cm^{-1}. The resistance versus concentration plot for the salt under investigation is made with stirring in the very conditions which prevail during a cryoscopic measurement. The concentration scale runs from $m = 10^{-3}$ to $m = 10^{-1}$ with a 0.1 per cent accuracy given.

Brown and Prue[9], who use an air agitation, and Scatchard and his co-workers[74], who use a rocking agitation, must do their conductivity testing wholly outside the cryoscopic vessel, in the water–ice thermostat, for example. A first sampling must be performed during finding of the 'pseudo-zero', so that the conductivity of the water under investigation shall be known. The interpolation on the resistance versus molality graph gives m within $3 \times 10^{-5} \text{ mol kg}^{-1}$.

B. Potentiometric Methods

These methods apply when the solutes are electrolytes and, generally, in salt hydrates or aqueous eutectics where the salt would considerably hamper a conductivity measurement. Sabbah[72], for example, in cryoscopic investigations at the $Na_2SO_4 \cdot 10H_2O$ transition point, uses the following equipment. In a thermostat at $32.38°C$, he puts a hydrogen electrode together with an electrometric vessel containing the solution under investigation and a series of junction bridges containing a saturated brine solution at $32.38°C$, and a potassium chloride solution at the same temperature and at room temperature (25°C) and a saturated calomel electrode. In a medium containing many Na^+ ions, the hydrogen electrode is to be preferred to the glass electrode. It takes 60 to 80 minutes for the solution to reach saturation and for the electrode potential to become stabilized. The e.m.f. is read on a millivoltmeter within 0.1 mV; this may give the molality of, say, boric acid and borates within 0.5 per cent.

This method is also very interesting in molten salts. A concentration cell is then built within the cryoscopic cell, thus avoiding any sampling that might disturb the state of equilibrium. A typical example is the study of solutions of beryllium fluoride in lithium fluoride, and inversely, with the cell

$$\text{Be} \left| \begin{array}{c} \text{LiF} \\ \text{BeF}_2 \\ \text{(I)} \end{array} \right| \left. \begin{array}{c} \text{LiF} \\ \text{BeF}_2 \\ \text{(II)} \end{array} \right| \text{Be}$$

The half-cell (I) is the constant-concentration reference cell, the other half (II) being at a variable concentration, the temperature of the whole cell being constant and determined by means of a thermocouple. J. Braunstein and K. A. Romberger[8] show that the above e.m.f. is a linear function of the lithium fluoride added, inasmuch as the solution is not saturated. If saturation takes place, the liquid phase composition remains unchanged in spite of further lithium fluoride additions, and the e.m.f. is constant. The e.m.f. versus concentration graph therefore shows, at a given temperature, a sharp break for the equilibrium concentration.

If, on the other hand, a constant concentration is maintained in (II) and the temperature is varied, a break is also obtained in crossing the liquidus line. It should be emphasized that this is by no means a starting-freezing kinetic method, but rather a series of equilibria at several fixed temperatures. This method is very sensitive and does not depend upon the liquidus slope, whether it is steep or gradual.

C. Interferometric Method

It can be applied to any type of solution, but on the condition that this solution be transparent enough. The concentration is linked to the difference in refractive index between the solution and the solvent. The Zeiss apparatus improvement of the gas apparatus of Lord Rayleigh, where Fraunhofer infinite fringes are observed. Haber and Lowe[34] have modified this equipment

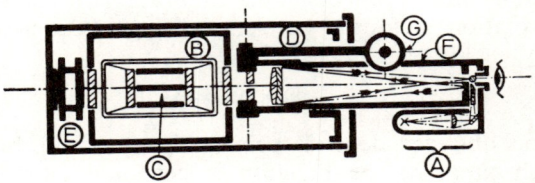

Figure 21. Rayleigh–Zeiss interferometer: A Electric bulb, B Water thermostat, C Plane and parallel faced cell, D Compensation plates, G Graduation drum, F Telescope.

and adapted it for solution analysis (*Figure 21*). The refractive index difference is compensated by the variable slope of a glass plate D which is moved by a micrometric screw G, the angle of rotation of which may be read on a graduated drum. The central fringe is replaced by the achromatic fringe given by the white light of a small electric bulb A. Finally, in order to obtain stable fringes, these authors put both cells in a water container B.

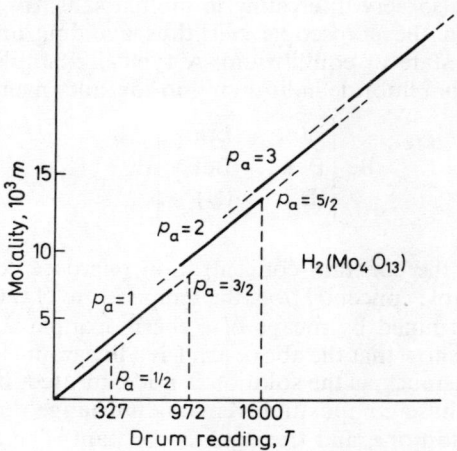

Figure 22. Interferometer standardization for aqueous molybdic acid solutions.

If an experimental plot of the molalities of the solution in the container versus the compensating drum readings T is made, one gets several short straight-line portions, forming a steplike graph, the step height being constant (*Figure 22*). This phenomenon was investigated by us[18], after comments by several authors, and the ratio β/μ between the dispersion difference and the solution–solvent refractive index appeared to be the key parameter. We have tested this theory experimentally, and obtained a continuous graph over the whole experimental interval[19].

D. Polarimetric Method

If Biot's law may be applied to the dilute solution, its molarity c (number of moles of mass M_2 per liter of inactive solvent) may be obtained by measuring the rotation under l decimeters. The calibration then consists of measuring the specific rotation $[\alpha]$ for the wavelength being used at a given temperature. One has

$$c = \{1000/[\alpha]lM_2\}\alpha \tag{74}$$

The precision obtained depends principally upon $[\alpha]$, but also upon the solution transparency. The titration is generally more precise than by chemical methods, but the sample taken must be of a large enough volume so that a polarimetric tube of at least 2 dm could be filled with it.

PART 2. CRYOSCOPY

This method is of particular interest for cryoscopy in a eutectic. The interferometer is useless because a large quantity of salt is present anyway. It is, however, good practice not to use values taken from tables, but to calibrate the apparatus with salt solutions at the eutectic concentration.

The substance studied is sometimes inactive; however, if the addition of a constant quantity of a given salt gives an active complex, the method may be used again. For example, the sodium and ammonium paramolybdates are not optically active; but Darmois showed that with malic and tartaric acids they form an optically active complex of high specific rotation, such as ammonium dimolybdotartrate $C_4O_6H_4,2MoO_3(NH_4)_2$, which is dextrorotatory. Its concentration is proportional to the molybdate concentration if a constant excess of tartaric acid is added. It is by this method that we have measured[21] the ammonium paramolybdate concentrations in the aqueous ammonium nitrate eutectic (44 g of salt in 100 g of solution). With a polarimetric tube of 50 cm, the precision of 0.1 degree on α gives the molality within 0.5×10^{-4} [21].

E. Photometric Method

The Lambert–Beer law gives a very simple relationship between the optical density of a totally dissociated solution (when a salt is considered) and its molarity in moles per liter

$$d = \varepsilon c l \tag{75}$$

If the molar extinction coefficient ε is given in tables, one just has to know the optical density of a tray of known thickness l. If Beer's law is not obeyed or if the salt under investigation is partially dissociated, a calibration curve $d = f(c)$ for a given tray and a given wavelength must be made. The precision depends on the apparatus. Between the visual photometer, using colored glasses and the electron multiplier spectrophotometer, there is a variety of equipments. The investigator has to choose the equipment giving acceptable cryoscopic precision.

These methods yield additional information in ionic solutions, namely,

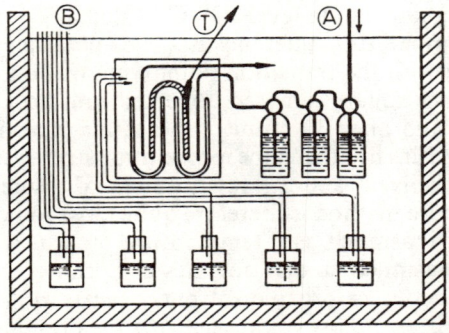

Figure 23. Cryoscopic apparatus for precongelation method of Robertson and La Mer: A. Gas-bubbling agitation, B. Stoppered flasks, T. Thermocouple.

the degree of association. Some basic references may be found in a work by van Halban and Kortum[86] about α-dinitrophenol dissociation.

F. PRECONGELATION METHOD

It is necessary to measure concentration after the adiabatic equilibrium temperature is obtained, because the salt or the small quantity of concentrated solution added to the solvent is not at the equilibrium temperature and brings about either fusion or freezing.

Hovorka and Rodebush[36] and later Robertson and La Mer[69] endeavoured to make this phenomenon small enough so as to be a calculable (or even negligible) correction. In the water–ice thermostat (*Figure 23*) ten carefully stoppered flasks were placed, containing solutions of known concentrations (by weighing). After a long enough time, at least four hours, these solutions reach a temperature equilibrium. Robertson and La Mer estimate that they are within no more than $+0.003°C$. This precongelation method is very interesting for dilute solutions. It avoids using the costly apparatus necessary for very low concentrations, but may lead to serious systematic errors in beginners' hands.

G. CHEMICAL ANALYSIS METHODS

These methods rarely give as good a precision as the temperature-taking method. However, a precision of 0.1 to 1 per cent may be attained in some particular cases. Let us give a recent example: cryoscopic measurements with halides and nitrates in N-methylacetamide[90]. Alkaline halides are titrated against standard silver nitrate in a micropipet and the nitrate solutions are weighed by evaporating at 140°C.

4. *Pseudo-equilibrium Methods*

In a preliminary experiment Lemaire[45], in the course of cryoscopic experiments at the sodium sulfate decahydrate transition point (32.38°C), starts by putting 35 g of water in the cryoscopic vessel and tries to determine the temperature at which the surroundings must be so that the water shall remain in a stationary state at 32.38°C. Adiabatic conditions are thus achieved with a temperature difference Δt. The operation is performed again with 35 g of liquid with the transition point composition, to which a certain weight of anhydrous sulfate is added, together with decahydrate. The same Δt is maintained, and after 30 minutes the equilibrium is reached and the temperature read. Finally, the same procedure is repeated with a solution containing 35 g of solvent and a known weight of solute salt, with still the same Δt. The Lemaire method is therefore quite close to both the equilibrium adiabatic method, inasmuch as a temperature plateau is obtained, and the kinetic methods inasmuch as one operates at a known concentration. One can also, in an aqueous or organic solvent, operate with a solution of predetermined concentration, provoking freezing by stopping the supercooling and then obtain a temperature plateau by adequate thermostatic regulation. During the temperature plateau, a solute addition can be made, another

PART 2. CRYOSCOPY

plateau then being reached. The quantity of solid formed out of the solution may be calculated if one knows by how much the solution is supercooled and its thermal data. The calculation may be verified, at the end of the experiment, by rapidly decanting at the solution temperature.

These methods, which associate kinetic and equilibrium techniques, are sometimes advantageous. We have been using them, for example, in ethylene carbonate which gives tenacious supercooling[16].

IV. A Few Applications of Cryoscopy

The applications of cryoscopy are extremely numerous. The list given here is by no means complete but underlines the importance of this technique within the fields of physical chemistry and thermodynamics of solutions.

1. Data Derived from Zero Concentration Extrapolation, $(\theta/m)_0$

A. Molar Masses

This is the oldest and the most universally known application of cryoscopy, i.e. its application to the determination of solute molar masses. If the solution is not ideal (which is never known, *a priori*) the extrapolation towards the origin is absolutely indispensable. As early as 1886, i.e. at the beginning of cryoscopic studies, Raoult[65] perceived quite well that this was necessary. Some of his followers, however, did not conform to this rule; this is why one can find frequently in earlier literature molar mass calculations which rest on three measurements at different finite concentrations. This leads to gross inaccuracies, especially if the solution departs strongly from ideal behavior. It was shown[20] that the calculation of M at molality of 0.5 in aqueous solution leads to an error of one per cent only in the case of glycerol; the error is five per cent for saccharose and 21 per cent with aniline.

Difficulties may occur with electrolytes on account of their ionic dissociation. Let us take, for example, the case of molybdic acid[21]. One form of this acid may be prepared by sulfuric acid attack on barium molybdate. This acid is known as Graham acid. Yellow molybdic dihydrate solutions are also known. Both acids give the same cryoscopic curve, which by extrapolation at $m' = 0$ gives $(\theta/m')_0 = 1.4 \pm 0.15$, the molalities m' being expressed in terms of Mo atoms. The acid formula being $(MoO_3)_x(H_2O)_y$ it should give, at infinite dilution, $(1 + 2y)$ ions. Therefore, with x Mo atoms, the limiting freezing point depression is

$$(\theta/m')_0 = 1.86 (1 + 2y)/x$$

One therefore obtains from cryoscopic data

$$(1 + 2y)/x \simeq \tfrac{3}{4}$$

With the hypothesis that $y = 1$, the so-called 'Graham acid' and the dihydrate solution are identical forms of a tetramolybdic acid $H_2Mo_4O_{13}$.

The dissociation problem may sometimes be disposed of by choosing a 'common ion' solvent. Let us take, for example, sodium paramolybdate, for which two formulae were proposed: $Na_6(Mo_7O_{24})$ and $Na_{10}(Mo_{12}O_{41})$. In aqueous cryometry the molar freezing point depressions would be $7\lambda_0$ or $11\lambda_0$ but, as the molecular masses would be in the ratio 12/7, the cryoscopic curves would be almost identical. Alternatively, if sodium sulfate at its transition point is taken as a solvent[21], the molar freezing point depressions would be approximately in the ratio 7/12. Experimental data yield unambiguously a figure of seven.

B. Entropy and Enthalpy of Fusion

The van't Hoff formula gives the molar enthalpy of fusion and thus also the molar entropy of fusion. This method is interesting, especially when calorimetry is not applicable. For example, before the cryoscopic measurements of Zarzycki[91] the enthalpy of lithium borate at 840°C was not known. Petit[58] performed systematic measurements at high temperatures. He thus determined enthalpies of fusion of many sulfates, molybdates and tungstates of alkali metals, alkaline earth halides and heavy metal salts. Grjotheim[31] gives the enthalpy of fusion of cryolite as 20250 cal mol^{-1} (at its indifferent point); Harrison[35] gives a value of 6970 cal mol^{-1} for sodium chloride.

C. Mixed Crystals Study

The cryoscopic equation relating to mixed crystals may be obtained by writing the solvent equilibrium condition, i.e. chemical potential identity for the component (1) in both phases. If a_1 and a'_1 are the solvent activities in the liquid and solid phases, one obtains

$$\ln(a_1/a'_1) = \int_{T_{01}}^{T} (\Delta H_1/RT^2) \, dT \tag{76}$$

Let X_2 and X'_2 be the mole fractions of the solute in, respectively, the liquid and the solid phase; by definition (Nernst) the partition coefficient is $\rho = X'_2/X_2$. The power series, after integration, and introduction of the molality m instead of X_2, leads to

$$(\theta/m) = \lambda(1 - \rho) + \rho M_1 \theta \tag{77}$$

where M_1 is the molar mass (in kg) of the solvent. At $m = 0$, ρ takes the value ρ_0, and it can be seen that

$$1 - \rho_0 = \lim_{m \to 0} \{(\theta/m)/\lambda\} \tag{78}$$

Petit[58] studied particularly such solvents as sodium sulfate, and barium, calcium and strontium sulfates. Cryoscopy enables one to study mixed crystals. Although it does not replace x-ray analysis, it may yield valuable information.

D. Particle Numeration

Particle numeration depends simply on the limiting law

$$\lim_{m \to 0} (\theta/m) = \nu\lambda \tag{79}$$

PART 2. CRYOSCOPY

when the cryoscopic constant λ is known. It is worthwhile recalling that v is the number of particles (ions or ion groups) as distinct from solvent ions. This number, of course, gives no information about the nature of these particles. Cryoscopy has been criticized on this ground; which is not always justified, as one may take into account ion solvation in the solute and the possible existence of several dissociation processes leading to the same particle number. Let us consider, for example, sodium pyrophosphate in solution in sodium fluoride, for which $\lambda = 17.5$. One finds $(\theta/m)_0 = 35$. Then, $(P_2O_7)^{IV-}$ gives two ions. Both dissociation schemes shown below are compatible with the cryoscopic data:

$$(P_2O_7)^{4-} \rightarrow (PO_4)^{3-} + (PO_3)^{-}$$
$$(P_2O_7)^{4-} \rightarrow (P_2O_6)^{2-} + (O)^{2-}$$

However, if an $(O)^{2-}$ ion is added to the solution, we do not find three ions but always only two. Such an addition of the $(O)^{2-}$ ion is clearly incompatible with the second scheme which gives and does not incorporate $(O)^{2-}$. The only possible process is the upper one.

When some solvation occurs, a second cryoscopy in a non-solvating solvent will often enable one to find out the nature of the ions. Further details are given in the work of Petit[58].

Cryoscopy may not only give the nature of the solute ions, but also the nature of the solvent ones, by appropriate application of the 'common ions' rule. Let us mention the characteristic results of Zarzycki[15]. His solvent is lithium metaborate ($LiBO_2$) at its indifferent point, 840°C. Its cryoscopic constant is $\lambda = 15$. The solute nickel oxide solute (NiO) also leads to $\lambda = 15$, whereby two assumptions may account for the data: (a) NiO is not dissociated; (b) It is dissociated into Ni^{2+} and O^{2-} ions, but there are already some O^{2-} ions in the molten metaborate. The results are similar with titanium oxide (TiO_2) and magnesia (MgO); cryoscopy of alumina (Al_2O_3) gives a limiting freezing point depression $(\theta/m)_0 = 30$. There are, in the latter case, two distinct particles. If (a) were true, these would be AlO_2^- and AlO^+, whereas if (b) were true, they would be $2Al^{3+}$. In order to solve this problem, Zarzycki made cryoscopic experiments with lithium oxide (Li_2O), which gave him a zero molal limiting freezing point depression. The conclusion is that molten lithium metaborate dissociates into O^{2-} ions.

2. *Data Derived from the* (θ/m) *versus* m *Cryoscopic Graph*

A. OSMOTIC COEFFICIENT CALCULATION

Let ϕ be the practical osmotic coefficient, as defined by the van't Hoff relationship

$$\pi = cRT\phi \tag{80}$$

where c is the molarity of the solute. It can be proved (this is done in most physical chemistry textbooks) that ϕ is linked to the solvent activity a_1, of density d, by

$$\phi = -(\ln a_1)/M_1 cd \tag{81}$$

On the other hand, the Schröder–van Laar cryoscopic law may be written as

$$\ln a_1 = -A\theta(1 + B\theta) \tag{82}$$

where

$$A = L/RT_0^2$$

and

$$B = (1/T_0) - (\Delta c/2L_1)$$

Hence

$$\phi = \{(\theta/m)/\lambda\}(1 + B\theta) \tag{83}$$

In aqueous solution, for example,

$$\phi = \{(\theta/m)/1.86\}(1 + 0.54 \times 10^{-3}\theta)$$

As the second term is generally negligible, the osmotic coefficient is merely equal to the ratio between the molar freezing-point depression observed, at molality m, and the limiting freezing-point depression. Osmotic pressure is therefore calculable from the cryoscopic depression. One has

$$\pi = (RTd/\lambda)\theta(1 + B\theta) \tag{84}$$

For medical purposes, osmotic pressures are usually measured in terms of milliosmoles. This unit represents the osmotic pressure in a solution which is ideal at the concentration of one millimole per liter

$$\pi_0 \times 1000 = 0.001\, RT \tag{85}$$

Let $\pi' = \pi/\pi_0$, one has

$$\pi' = 1000\,(d/\lambda)\theta(1 + B\theta) \tag{86}$$

whence, in aqueous solutions,

$$\pi' = 538\,\theta + 0.29\,\theta^2$$

A cryometric depression of 0.1°C corresponds to an osmotic pressure of 53.8 milliosmoles. It is, or course, much easier, for a serum or biological liquid, to measure its freezing point than its actual osmotic pressure.

B. Solute Activity Coefficient Calculations

From the experimental (θ/m) versus m graph, one can calculate the solute activity coefficient. The derivation by Lewis is given in most physical chemistry textbooks; in this derivation, the following variable is introduced

$$j = 1 - (\theta/m)/\lambda \tag{87}$$

for a non-dissociated solute, and

$$j = 1 - (\theta/m)/\nu\lambda \tag{87 bis}$$

for a ν-ion solute. One finally obtains

$$\ln \gamma = -j - \int_0^m j\,d(\ln m) + (2B/\lambda)\int_0^\theta (\theta/m)\,d\theta \tag{88}$$

where

$$B = (1/T_0 - \Delta c/2L_1)$$

For a salt, $\gamma_\pm$ is actually obtained, and the third term must be multiplied by v. It is to be noticed that γ and $\gamma_\pm$ are dissymmetrical practical coefficients, in the molality scale. If f_2 is the symmetrical rational coefficient for the solute, defined by $f_2 = a_2/X_2$ in the mole fraction scale, it is well known that the rational coefficient γ_2, which approaches unity when X_2 approaches zero, may be expressed as

$$\gamma_2 = f_2/f_{\infty 2} \tag{89}$$

where $f_{\infty 2}$ is the value of f_2 for $N_2 = 0$. The practical cryoscopic coefficient, defined in the molality scale $\gamma = a_2/m$, is linked to γ_2 by the relationship

$$\gamma = \gamma_2/(1 + M_1 m) \tag{90}$$

C. Solvation Calculations

The activity coefficients of a non-electrolyte (γ) and of an electrolyte ($\gamma_\pm$) give the magnitude of the departure from ideality, but no information about the causes for this departure. One of these is ion solvation; ion hydration, in aqueous solutions, is a particular case in point.

A great many theoreticians tried to approach the problem of the number of water molecules attached to an ion. Let us give here the point of view of Stokes and Robinson[78] as an example of how the $\gamma_\pm$ coefficient may be used. These authors assume that the lowering of the cryoscopic curve is due only to water molecule–ion interactions. The basic idea is taken from an earlier theory of Bjerrum: water of hydration is considered as taken away from the solvent, thereby modifying the molality of the solute. Let h be the number of water molecules linked to one cation, j the number of water molecules linked to one anion and $n = h + j$. The stoichiometric molality is m, but there are nm water molecules less in 1 000 g of water (or 55.51 moles). The true molality is, therefore

$$m = 55.51\, m/(55.51 - nm) \tag{91}$$

The activity of water is given directly by cryoscopy. The Gibbs–Duhem relationship gives the activity of the solute m' and the practical mean activity coefficient $\gamma_\pm$ which is given, too, by the Debye–Hückel relationship. Finally

$$\log \gamma_\pm = \frac{-0.488 \times Z_1 Z_2 I^{\frac{1}{2}}}{1 + 0.324 a I^{\frac{1}{2}}} - 7.824 \times 10^{-3} \frac{mn}{v} \frac{\theta/m}{1.86}$$
$$- \log[1 - 0.018(n-v)m] \tag{92}$$

The authors were able to find theoretical activity coefficients close to experimental ones for many strong electrolytes, up to concentrations corresponding to ionic strengths I of four.

D. Electrolyte Ionic Dissociation

Let us consider a weak electrolyte in aqueous solution, and its cryoscopic

(θ/m) versus m graph. The decreasing curve observed depends upon at least three factors:
  (a) Partial dissociation, which lowers the number of particles;
  (b) Interionic forces, which bring about a lowering of the curve;
  (c) Solute–solvent interactions, solvation, complex formation, etc.

It is well known that if α is the dissociation coefficient of an electrolyte composed of ν ions, it gives i particles, where

$$i = 1 + (\nu - 1)\alpha \tag{19}$$

the cryoscopic molality thereby being im. If such is the case, it is easy to see that

$$\frac{\theta}{m} = i\lambda \left[1 - \left(\frac{M_1}{2} + \frac{\lambda}{T_0}\right) im\right] - \frac{\lambda \ln f_1}{M_1} \tag{93}$$

The symmetrical rational solvent activity coefficient f_1 is related to the dissymmetrical ionic activity coefficients by the Gibbs–Duhem relationship. Only interionic forces are of interest here, the interactions discussed in C being excluded. Therefore the undissociated molecules activity coefficient is taken equal to one. Finally

$$\mathrm{d}(\ln f_1) = -M_1 \nu \alpha\, m\, \mathrm{d}(\ln f_\pm) \tag{94}$$

Let us take the value of $f_\pm$ resulting from the Debye–Hückel law

$$\ln f_\pm = \frac{-Az_1 z_2 I^{\frac{1}{2}}}{1 + BaI^{\frac{1}{2}}} \tag{21}$$

where the symbols retain their usual meaning, and the ionic strength is

$$I = \tfrac{1}{2}\nu\alpha\, m d z_1 z_2 \tag{21 bis}$$

One obtains

$$\ln f_1 = \tfrac{2}{3} M_1 A I^{\frac{3}{2}} \sigma \tag{95}$$

where

$$\sigma = \{3/(BaI^{\frac{1}{2}})^3\}\left[1 + BaI^{\frac{1}{2}} - \ln(1 + BaI^{\frac{1}{2}})^2 - 1/(1 + BaI^{\frac{1}{2}})\right] \tag{95 bis}$$

From this the cryoscopic curve equation is

$$(\theta/m) = i\lambda\left[1 - (\tfrac{1}{2}M_1 + \lambda/T_0)\, im\right] - \tfrac{1}{3}A\lambda\, \nu\alpha\, z_1 z_2\, dI^{\frac{1}{2}}\sigma \tag{96}$$

In the particular case of aqueous solutions at 273 K, this becomes

$$(\theta/m) = 1.86\,(i - 0.016\, i^2 m - 0.376\, \nu\alpha\, z_1 z_2 I^{\frac{1}{2}}\sigma)$$

where

$$\sigma = 1 - 0.846\, aI^{\frac{1}{2}} + 0.189\, a^2 I - 0.068\, a^3 I^{\frac{3}{2}}$$

and a is expressed in 10^{-10} m. If the salt is reasonably strong the Debye–Hückel correction term is not negligible as it measures the error of van't Hoff who ignored interionic forces.

PART 2. CRYOSCOPY

E. Hydrolysis Constant Calculations

The hydrolysis of a salt introduces some new ions into the solution. The cryoscopic depressions are greater, and, in some cases, equilibrium constant calculations are possible, at least in dilute solutions. Let us consider a salt composed of v_1 cations and v_2 anions. The hydrolysis reaction may be written as the reciprocal reaction of the reaction of an acid with a base. If the hydrolysis ratio, which is the number of moles of salt hydrolysed at equilibrium for one mole of salt present, is denoted as δ, there remains $m(1 - \delta)$ moles. These are partially dissociated. Let i_1 be the corresponding van't Hoff factor. The number of particles supplied by the salt itself is $m(1 - \delta)i_1$, at molality m. There are also $mv_1\delta$ moles of base. If this base is dissociated with factor i_2, it gives $mv_1\delta i_2$ particles. Without any assumptions regarding the nature of the electrolyte (acidic or basic), or, in other words, regarding the relative strengths of the acid and the base, one can give the acid a van't Hoff factor i_3. Its particle concentration is $mv_2\delta i_3$. The solvent (water) mole fraction is therefore

$$N_1 = 55.51/\{55.51 + m[(1 - \delta)i_1 + v_1\delta i_2 + v_2\delta i_3]\} \tag{97}$$

or

$$N_1 = 1/\{1 + 0.018\,m\,[i_1 + \delta(v_1 i_2 + v_2 i_3 - i_1)]\}$$

Let us define Σi as

$$\Sigma i = i_1 + \delta(v_1 i_2 + v_2 i_3 - i_1) \tag{98}$$

so that

$$N_2 \simeq 0.018\,m\,\Sigma i\,(1 - 0.018\,m\,\Sigma i) \tag{99}$$

and finally

$$(\theta/m) = 1.86\,\Sigma i\,(1 - 0.018\,m\,\Sigma i) \tag{100}$$

It is, of course, assumed that the functional form of the graph is due to hydrolysis only; in other words f_1 is equated to unity; in this calculation, the van't Hoff coefficients of the salt, the acid and the base are supposed to be given for each m molality. It is therefore possible to derive (θ/m) from experimental values. At infinite dilution, for a I–I salt, one has $i_1 = i_2 = i_3 = 2$ and

$$\lim_{m \to 0} (\theta/m) = 3.72\,(1 + \delta)$$

If $\delta = 1$ the limiting molal depression is doubled. For a II–I salt, such as copper acetate

$$\lim_{m \to 0} (\theta/m) = 4.58\,(1 + \tfrac{4}{3}\delta)$$

Similar formulae were used by Walden[87] when he treated the hydrolysis of 2,4-dimethyl-α-pyrone chlorhydrate, and also by Zawidski[93] for urea chlorhydrate. The latter author, however, used them at finite concentrations; the constant thus obtained (1.3×10^{-14}) is approximately correct.

F. Cryoscopic Titrations

When two distinct substances, not reacting chemically with each other are dissolved in a given solvent, forming an ideal ternary mixture, it may be assumed that the molal temperature depressions are additive in dilute solution. This is true inasmuch as only the first term in the Schröder formula power series is appreciable. A first difficulty appears with non-ideal mixtures; a second when the solutes react with each other. The problem is complicated when the electrolytes present are ionized, and when the activity coefficients vary. Cornec[13], however, has been able to treat such problems with success.

(1) Acidimetry and Alkalimetry

Cornec investigates the aqueous solution cryoscopy of an acid–base mixture. He takes a cm^3 of a standard sodium hydroxide solution (0.5 N) and $(100 - a)$ cm^3 of an acidic solution at the same concentration. The freezing point depression of this mixture is taken within 10^{-3} °C with a Beckmann thermometer, and plotted against a. There is always a sharp break in the curves when neutralization is obtained. The freezing point depression, for example, is 1.865°C for 0.5N hydrochloric acid. If 0.5N sodium hydroxide is added, it decreases up to 0.890°C at neutralization, increasing again up to 1.705°C beyond neutralization. Cryoscopy may thus serve as a neutralization indicator. Cornec showed that this method was more precise than the use of colored indicators, as it gives the second carbonic acidity, the three phosphoric and arsenic acidities, and the phenol and resorcine acidities, etc.

Later on, Cornec investigated salt medium cryoscopy, when with Müller[56], he used freezing point depression in a eutectic. Such media are very convenient for titration purposes, as they are almost ideal. The interpretation is eased by common ion effects. If, for example, an acid is to be titrated against sodium hydroxide. In order to avoid the cryoscopic effect due to Na$^+$, one may use as a solvent Glauber salt at its transition point. A known quantity of acid may be added, which leads to a freezing point depression. The depression is measured again with increasing quantities of sodium hydroxide added. The various cases that may be encountered have been discussed by Jahr and Kubens[37].

(2) Complex Formation

An earlier investigator, Byé[10], had worked out a similar cryometric titration method in a more complicated case. If a neutral salt is acidified, an acid salt is not always obtained. A progressive anion condensation may lead to complex formation; thus, paramolybdates Na$_6$(Mo$_7$O$_{24}$) result from the union of several molybdic anions; similarly, the molybdomalate anion Na$_2$(4 MoO$_3$, 2 C$_4$H$_4$O$_5$) is the result of the condensation of molybdic and malic anions.

The method consists of dissolving in Glauber salt a constant weight of sodium salt, to which increasing sulfuric acid quantities x are added. The molal depression decreases because of the condensation phenomenon; then, the new salt being formed, this depression increases, or remains constant, whether dissociation occurs or not. The abscissa of the angular point on the λ versus x curve gives the acid/base ratio, thereby fixing the salt composition. As for the degree of condensation, it may be obtained

PART 2. CRYOSCOPY

by taking the ratio of the limiting molal depressions of the initial salt ($x = 0$) and of the condensed salt. As Glauber salt is a nearly ideal medium, Byé takes the ratio of temperature depressions at finite concentrations, without any error on the integral value of the coefficient. The paramolybdate formula is thus in Mo_7 instead of Mo_6 (the latter formula having been proposed by Rosenheim).

Souchay[77] has successfully applied the above method to silicates and stannates in Glauber salt. He also uses the ice–potassium nitrate eutectic for nitrate titrations against potassium hydroxide. The method is then generalized to study some 'heterocomplexes' such as: selenito-molybdates, methylarsinomolybdates, molybdomalates, molybdotartrates. This showed that the cryoscopic method is faster and easier, and leads to more reliable interpretations than the diffusion and dialysis techniques used formerly. Cryoscopic titration curves are related to potentiometric ones. The attention of the reader is directed to an important theoretical work by Byé[10] and to measurements by Souchay[77].

3. Data Derived from the Schröder Curve

A. Activity Coefficients

As has been mentioned above, the use of the Schröder–van Laar formula at finite concentrations dates back from the first molten salt thermodynamic studies. The coordinates used are the same as those previously used by Schröder in solubility studies, with $\ln a$, plotted versus $1/T$. The ideal activity a_1 of an ionic mixture has been defined above. The activity coefficient of the solvent in the mole fraction scale is f_1, such that

$$a_1 = a_1^* f_1 \tag{101}$$

The definition in non-ionic media where $a_1^* = X_1$ is thereby extended. As a first approximation, one finds that $a_1^* = X_1^v$ where v denotes the number of solute ions as distinct from solvent ions; some authors have sometimes used a f_1' coefficient, such that $a_1 = (X_1 f_1')^v = a_1^* f_1'$. We finally obtain

$$\ln f_1 = (L/R)(1/T_{01} - 1/T) + \Sigma (1/T) - \ln a_1^* \tag{102}$$

The experimental cryoscopic graph where a_1^*, which is a known function of X_1, is plotted versus $1/T$, gives immediately $\ln f_1$ by comparison with the ideal curve which is, approximately, a straight line of slope L/R. When the axes are oriented as usual, if the experimental curve is above the ideal curve, the f_1 coefficient is smaller than unity. This applies, for example, with lithium nitrate as a solvent and sodium or potassium nitrates as solutes. The same solvent lithium nitrate has an activity coefficient above unity (graph below ideal graph) with silver nitrate as a solute[82]. Several mixtures have been found such that f_1 is very close to unity along the corresponding liquidus branch. If f_1 is known, one may theoretically calculate f_2, as the Gibbs–Duhem relationship gives

$$\ln f_2 = - \int_1^{X_1} (X_1/X_2)\,d(\ln f_1) \tag{103}$$

Difficulties, however, arise, when X_1/X_2 is plotted versus $\ln f_1$, because the above function asymptotically approaches infinity for $\ln f_1 = 0$. Lewis proposed to use the following variable

$$h = 1 + (\ln a_1)/(X_2/X_1) \qquad (104)$$

which leads to

$$-\ln f_2 = h + \int_0^{X_2}(h/X_2)\,dX_2 \qquad (105)$$

This avoids infinite values of the function being integrated. The difficulty lies elsewhere, as it is necessary to extrapolate the results from one eutectic branch to the other. In silver nitrate, for example, if lithium nitrate is dissolved, the eutectic point is observed at $X_{Li} = 0.25$. The experimental

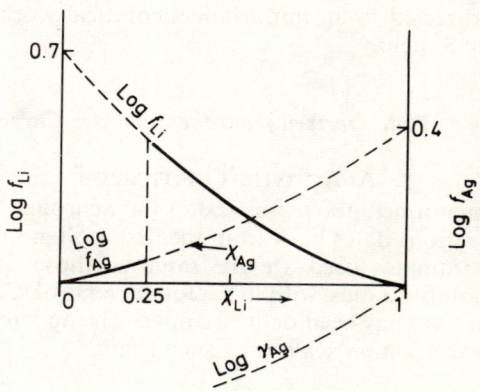

Figure 24. Graphical calculation by the Lewis method for lithium nitrate in silver nitrate solution.

values for f_{Ag} are not obtained beyond $X_{Li} = 0.25$. One must therefore extrapolate up to $X_{Li} = 1$, which requires a very reliable experimental equation.

The graphical method of Potier[60] on large-scale graph paper seems more accurate. It consists of drawing the tangents in order to have a graphical interpretation of the equation

$$d(\ln f_2)/dX_2 = (-X_1/X_2)d(\ln f_1)/dX_1 \qquad (106)$$

One can also use the values of $\ln \gamma$ calculated from θ/m by the Lewis method. *Figure 24* is an example of such a calculation, due to C. Vallet[82]. The dashed lines represent calculated points of the curves.

B. Excess Chemical Potential Calculations

The excess chemical potential of component i

$$\mu_i^E = RT \ln f_i \qquad (107)$$

is an isothermal quantity, defined at temperature T. When the excess chemical potential of a salt along the liquidus line is mentioned, this refers actually to another quantity, as T varies from T_f (melting point) to T_E (eutectic point). Let us consider a mixture of two salts with common cations (or anions). Let $n_1 = v_A X_1$ be the number of cations (or anions) of the solvent and $n_2 = v_B X_2$ be that of the solute. A classical derivation shows that the excess free energy of a regular solution ($S^E = 0$) is

$$g^E = W n_1 n_2 / (n_1 + n_2) \tag{108}$$

where W is the interaction energy of the two species A and B. The chemical potential of ions A is then

$$\mu_A^E = \left(\frac{\partial g}{\partial n_1}\right)_{n_2} = W \frac{n_2^2}{(n_1 + n_2)^2} \tag{109}$$

Let us calculate μ_A with the mole fractions of the salts X_1 and X_2. One may use the Temkin formulation which requires the existence of quasi-lattices, as the same assumption is made in the evaluation of g^E. The Haase formulation, which is valid whatever structural assumptions are made, may also be used. In both cases, however, one finds that

$$\mu_A^E = W \left(\frac{v_B X_2}{v_A X_1 + v_B X_2}\right)^2 \tag{110}$$

In the case of ions with the same valency ($v_A = v_B$), there remains

$$\mu_A^E = W X_2^2 \tag{111}$$

In the original Hildebrand theory, W does not depend upon temperature. This property will enable one to tell from cryoscopic measurements the regular character of a molten salt mixture.

At each temperature T of the liquidus line, the cryoscopic excess chemical potential is calculated. If its variation against X_2^2 is linear, the slope gives the interaction energy between ions A and B. Examples of such calculations are extremely numerous. Let us recall the already mentioned example of silver nitrate solutions in lithium nitrate, where

$$\mu_{Li}^E = 610 \, X_{Ag}^2$$

The other liquidus branch gives approximately the same value for W (here in cal. mol^{-1}).

C. Other Thermodynamic Calculations

The knowledge of W in regular solutions may lead to further developments. For example C. Vallet studied alkali–alkaline earth nitrate mixtures and tried to interpret W in terms of a model suggested by Förland[29] which relates coulombian potential energy of neighboring cations to a multiplicative factor (function of the coordination number upon melting). Results are satisfactory for a number of pairs of cations. In other cases, a polarization term must be introduced in order to take into account the NO_3^- polarization in the local electric field, which was calculated by Lumsden[51]. Finally, another additive term takes into account the London dispersion effect for low ionic radius cations. Such calculations have been carried out by Blander[5].

Let us mention also the 'significant structure theory' of Eyring[25], which, as modified by Vilcu and Misdolea[85] and by van Artsdalen[83] was applied to alkali halides.

Cryoscopies have also been performed on salts without a common ion. Their interpretation is somewhat difficult; for example, silver sulfate in a mixture with potassium nitrate, which yields four ions: K^+, Ag^+, NO_3^- and SO_4^{2-}. The interpretation requires silver nitrate–potassium sulfate cryoscopy, as well as nitrate–nitrate and sulfate–sulfate data. The 'reciprocal mixture' theory of Flood, Förland and Grjotheim[28] is quite useful in this respect, and is an important contribution to the theory of liquids.

For multicomponent systems Longuet-Higgins[48] describes a new statistical approach to the theory and defines a 'conformal solution' as one satisfying certain conditions. Reiss, Katz and Kleppa[68] successfully applied the latter theory to binary molten salt mixtures. Blander and Yosim[6] generalized this 'Conformal Ionic Solution' (CIS) to reciprocal mixtures.

Recently, M. Gaune-Escard developed a statistical theory which, leaving aside pair-interactions, introduces a 'surrounded-ion' concept[29a]. This encompasses as a particular case, and with a smaller number of parameters, the numerical results of the above theories[28, 68, 6].

In conclusion, we have seen how powerful a method cryoscopy is, in thermodynamic investigations.

V. References

[1] Abegg, R. *Z. Phys. Chem.* **20**, 207 (1896).
[2] Adams, L. H. *J. Amer. Chem. Soc.* **37**, 481 (1915).
[3] Arrhenius, S. *Z. Phys. Chem.* **1**, 631 (1887) and **2**, 491 (1888).
[4] Bedfort, J. G. *Proc. Roy. Soc. (London) A*, **83**, 459 (1910).
[5] Blander, M. *J. Chem. Phys.* **36**, 1092 (1962).
Blander, M. and J. Braunstein, *Ann. NY Acad. Sci.* **79**, 838 (1960).
[6] Blander, M. and I. J. Yosim. *J. Chem. Phys.* **39**, 2610 (1963).
[7] Boutaric, A. *Bull. Soc. Chim. France*, **13**, 651 (1913).
[8] Braunstein, J. and K. A. Romberger, *Inorg. Chem.* **9**, 1273 (1970).
[9] Brown, P. G. and J. E. Prue. *Proc. Roy. Soc. (London)*, **232**, 320 (1955).
[10] Byé, J. *Bull. Soc. Chim. France* **9**, 517 (1942).
[11] Cavallaro, L. *RC Accad. Ital.* **11**, 697 (1940).
[12] Cohen-Adad, R. and A. P. Rollet, *CR Acad. Sci., Paris*, **227**, 554 (1948) and *Publ. Univ. Alger*, **1B**, 85 (1955).
[13] Cornec, E. *Ann. Chim. Phys.* **29**, 490 (1913).
[14] Darmois, E. 'L'activité des solutions électrolytiques'. *Actualités. Sci., Paris (Herman) No. 945* (1943);
Darmois, E. *Bull. Un. Phys., Paris*, **260–261**, 193 (1933);
Darmois, E. and J.-J. Cessac. *CR Acad. Sci., Paris*, **191**, 1091 (1930);
Darmois, E. and R. Chalin. *CR Acad. Sci., Paris*, **195**, 786 (1932);
[15] Darmois, E. and J. Zarzycki. *CR Acad. Sci., Paris*, **234**, 95 (1952).
[16] Doucet, J., G. Porta and G. Finiels. *CR Acad. Sci., Paris*, **270C**, 1208 (1970).
[17] Doucet, Y. *J. Phys. Radium*, **4**, 53 (1943).
[18] Doucet, Y. *J. Phys. Radium*, **4**, 204 (1943).
[19] Doucet, Y. and Mlle J. Defretière, *CR Acad. Sci., Paris*, **224**, 337 (1947).
[20] Doucet, Y. *Bull. Un. Phys., Paris*, **361**, 76 (1947).
[21] Doucet, Y. *J. Phys. Radium*, **4**, 41 (1943).

PART 2. CRYOSCOPY

[22] Douglas, T. B. *Trans. Amer. Soc. Mech. Engrs*, **79**, 23 (1957).
[23] Dworkin, A. S. and M. A. Bredig, *J. Phys. Chem.* **64**, 269 (1960).
[24] Eichelberger, W. C. *J. Amer. Chem. Soc.* **56**, 799 (1934).
[25] Eyring, H., T. Ree and N. Hirai. *Proc. Nat. Acad. Sci., Wash.* **44**, 683 (1958).
[26] Finel, A., P. J. Gardner, R. D. G. Lane and B. Smethurst. *Lab. Pract.* **14**, 446 (1965).
[27] Flood, H., T. Förland and A. Nesland. *Acta Chem. Scand.* **5**, 1193 (1951).
[28] Flood, H., T. Förland and K. Grjotheim. *Z. Anorg. Allgem. Chem.* **276**, 289 (1954).
[29] Förland, T. *J. Phys. Chem.* **59**, 152 (1955).
[a] Gaune-Escard, M. J. C. Mathieu and R. Desre. *J. Chim. Phys.* **9**, 1390 and 1397 (1972) and **7–8**, 1033 (1973).
[30] Goodwin, H. M. and H. T. Kalmus. *Phys. Rev.* **28**, 1 (1909).
[31] Grjotheim, K. *Thesis*, Norges Teckniske Hogskole Trondheim (1956).
[32] Haase, R. *Z. Naturf.* **8a**, 380 (1953).
[33] Haase, R. *J. Phys. Chem.* **73**, 1160 (1969).
[34] Haber, F. and F. Lowe. *Z. Angew. Chem.* **23**, 1393 (1910).
[35] Harrison, J. P. *Thesis*, Paris (1956) and *CR Acad. Sci., Paris*, **241**, 298 (1955).
[36] Hovorka, F. and H. Rodebush, *J. Amer. Chem. Soc.* **47**, 1614 (1925).
[37] Jahr, K. F. and R. Kubens. *Z. Elektrochem.* **56**, 65 (1952).
[38] Karagunis, G., A. Hawkinson and G. Damköhler. *Z. Phys. Chem.* **151A**, 433 (1930).
[39] Kelley, K. K. *Bull. US Bur. Min. No. 476* (1949).
[40] Kentamaa, J. *Suomen Kemistilehti*, **29**, 59 (1956).
[41] Kraus, C. A. and R. A. Vingee. *J. Amer. Chem. Soc.* **56**, 511 (1934).
[42] Lange, J. *Z. Phys. Chem.* **177A**, 193 (1936).
[43] Lange, J. and J. Berga. *Mh. Chem.* **81**, 921 (1950).
[44] Lange, J. *Z. Phys. Chem.* **168A**, 147 (1934).
[45] Lemaire, M. *Thesis No. 225* (1956), Faculté des Sciences Lyon, France.
[46] Livingston, J., R. Morgan and H. K. Benson. *Z. Anorg. Chem.* **55**, 261 (1908) and *J. Amer. Chem. Soc.* **29**, 1168 (1907).
[47] Livingston, J., R. Morgan and P. T. Owen. *Z. Anorg. Chem.* **56**, 168 (1908) and *J. Amer. Chem. Soc.* **29**, 1439 (1907).
[48] Longuet-Higgins, H. C. *Proc. Roy. Soc. (London) A*, **205**, 247 (1951)
[49] Loomis, E. H. *Phys. Rev.* **1**, 199 (1893).
[50] Löwenhertz, R. *Z. Phys. Chem.* **18**, 71 (1895).
[51] Lumsden, J. *Disc. Faraday Soc.* **32**, 138 (1961).
[52] Mair, B. J., A. R. Glasgow and F. D. Rossini. *J. Res. Nat. Bur. Stand.* **26**, 591 (1941).
[53] Meyer, R. and J. Metzger. *Bull. Soc. Chim. France*, **1**, 66 (1967).
[54] Moser, H. *Ann. Phys., Lpz.* **5**, 343 (1929).
[55] Moulin, M. *J. Chim. Phys.* **8**, 321 (1910).
[56] Müller, H. *Ann. Chim.* **8**, 143 (1937).
[57] Nernst, W. and R. Abegg. *Z. Phys. Chem.* **15**, 681 (1894).
[58] Petit, G. *La Cryoscopie à Haute Température.* Masson: Paris (1965).
[59] Ponsot, A. *CR Acad. Sci., Paris*, **118**, 977 (1895).
[60] Potier, A. *CR Acad. Sci., Paris*, **240**, 1080 (1955).
[61] Prigogine, I. and R. Defay, *Chemical Thermodynamics*. Longmans Green: New York (1954).
[62] Randall, M. and A. P. Vanselow. *J. Amer. Chem. Soc.* **46**, 2418 (1924).
[63] Randall, M. and G. N. Scott. *J. Amer. Chem. Soc.* **49**, 647 (1927).
[64] Raoult, F. M. *CR Acad. Sci., Paris*, **101**, 1056 (1885).
[65] Raoult, F. M. *Ann. Chim. (Phys.)*, **6**, 289 (1886).
[66] Raoult, F. M. *Bull. Soc. Chim. France*, **3**, 21 (1899).
[67] Raoult, F. M. *Ann. Chim. (Phys.)*. **2**, 99 and 115 (1884).
[68] Reiss, H., J. L. Katz and O. J. Kleppa. *J. Chem. Phys.* **36**, 144 (1962).
[69] Robertson, C. and V. K. La Mer. *J. Phys. Chem.* **35**, 1953 (1931).
[70] Roloff, M. *Z. Phys. Chem.* **18**, 572 (1895).
[71] Rüdorff, F. *Ann. Phys., Lpz.* **114**, 63 (1861); **116**, 55 (1862) and **122**, 337 (1864).
[72] Sabbah, R. *Thesis.* University of Marseille (France) (1965).
[73] Scatchard, G. and S. Prentiss. *J. Amer. Chem. Soc.* **54**, 2696 (1932).
[74] Scatchard, G., B. Vonnegut and D. W. Beaumont. *J. Chem. Phys.* **33**, 1292 (1960).
[75] Schwab, F. W. and E. Wichers. *J. Res. Nat. Bur. Stand.* **34**, 33 (1945).
[76] Solomons, C. and G. J. Janz, *US Dept. Comm., Wash., Publ. No. 131 499* (1957) and *Rev. Sci. Instrum.* **29**, 302 (1958).

[77] Souchay, P. *Bull Soc. Chim. France*, **15**, 143 (1948).
[78] Stokes, R. H. and R. A. Robinson. *J. Amer. Chem. Soc.* **70**, 1870 (1948).
[79] Stortenbecker, W. *Z. Phys. Chem.* **10**, 183 (1892).
[80] Taylor, W. J. and F. D. Rossini. *J. Res. Nat. Bur. Stand.* **32**, 197 (1944).
[81] Temkin, M. *Acta Phys.-Chim. URSS.* **20**, 411 (1945).
[82] Vallet, C. *Thesis.* University of Marseille (France) (1970).
[83] van Artsdalen, E. R. *J. Phys. Chem.* **72**, 4155 (1968).
[84] van't Hoff, J. H. *Z. Phys. Chem.* **1**, 481 (1887).
[85] Vilcu, R. and C. Misdolea. *J. Chem. Phys.* **46**, 906 (1967).
[86] von Halban, H. and G. Kortüm. *Z. Phys. Chem.* **170A**, 351 (1934).
[87] Walden, P. *Ber. Dtsch. Chem. Ges.* **34**, 4192 (1901).
[88] White, W. P. *J. Phys. Chem.* **24**, 393 (1920).
[89] White, W. P. *J. Amer. Chem. Soc.* **56**, 20 (1934).
[90] Wood, R. H., R. K. Vicker and R. W. Kreis. *J. Phys. Chem.* **75**, 2313 (1971).
[91] Zarzycki, J. *Thesis.* University of Paris (1953).
[92] Zarzycki, J. *J. Phys. Radium*, **19**, 13A (1958).
[93] Zawidski, J. *Ber. Dtsch. Chem. Ges.* **37**, 2294 (1904).

CHAPTER 18

EMF Measurements in Molten Salts*

JERRY BRAUNSTEIN and HELEN BRAUNSTEIN

Chemistry Division, Oak Ridge National Laboratory, Oak Ridge, Tennessee, 37830, USA

Contents

I.	Introduction and Scope		902
II.	Cell EMF and Thermodynamic Properties		903
	1.	Introduction	903
	2.	Classification of EMF Cells	903
		A. One Component (Pure Molten Salt)	904
		B. Binary Systems	905
		C. Multicomponent Systems	908
		D. Jacobi–Daniell Cells	908
	3.	Sign Convention for EMF and Cell Diagram	908
	4.	Thermodynamic Expressions for Cell EMFs	909
		A. Electrode and Transport Processes	909
		B. Binary Systems with Liquid Junctions	910
		C. Multicomponent Systems	916
		D. Membrane Cells	918
	5.	Principal Error Sources	920
		A. Introduction	920
		B. Junction EMF	921
		C. Mixed Electrode Processes	922
		D. Thermal EMF	922
		E. Metal Solubility and Electronic Conduction	923
III.	Experimental		924
	1.	Reporting Data and Results	924
	2.	Apparatus	924
		A. Measuring Instruments	924
		B. Cells and Furnaces	926
		C. Materials	928
		D. Mixing	931
		E. Purification of Chemicals	932
	3.	Electrodes	934
		A. Gas Electrodes	934
		(a) The Halogen Electrode	934
		B. Metal Electrodes	937
		(a) The Silver Electrode	937
		(b) The Silver–Silver Halide Electrode	938
		C. Alloy Electrodes	939
IV.	Special Applications		941
	1.	Phase Diagrams	941
	2.	Association Equilibria of Dilute Solutes	944
		A. Measurements with Silver or with Silver–Silver Halide Electrodes	944

---

* Research sponsored by the US Atomic Energy Commission under contract with the Union Carbide Corporation.

B. Electrodes of the Third Kind 948
3. Miscellaneous 949
V. Acknowledgement 950
VI. References 950

I. Introduction and Scope

An e.m.f. measurement is capable, in principle, of providing directly the free energy of the cell process, and therefore all the thermodynamic information obtainable from the free energy and its derivatives. E.m.f. measurements may therefore provide free energies of formation of pure molten salts, free energies of mixing in binary molten salt mixtures, excess chemical potentials of selected components in multicomponent molten salt mixtures, and equilibrium constants for homogeneous and heterogeneous reactions of dissolved species. Because of the inherent accuracy of electrical measurements, e.m.f. measurements are capable of yielding accurate thermodynamic quantities. The essential requirements for accuracy include the establishment of local equilibrium within each phase in the cell and between adjacent phases; and, simultaneously, the maintenance of constraints against irreversible processes such as direct chemical reaction of the electrode materials and electronic conduction through the electrolyte. Such irreversible processes constitute chemical or electrical partial short circuits of the cell e.m.f., and invalidate the basic equilibrium thermodynamic equation relating the electromotive force of a cell, E (volts), to the electrical work obtainable from the cell reaction isothermally and isobarically, by reversible transfer of nF coulombs of charge, and hence to the Gibbs free energy change of the process occurring in the cell.

$$nFE = (W_{\text{elec, rev}})_{T, p} = -\Delta G \tag{1}$$

In principle, temperature coefficients of e.m.f. (and free energy) can lead to the corresponding entropies and enthalpies, but it is not unusual for the uncertainty in a derivative to be an order of magnitude greater than the uncertainty in the source data. This may occasionally be masked in least squares analysis of the data. Furthermore, systematic errors may well predominate over random errors in e.m.f. measurements. For example, dissolution of a metallic electrode in a molten salt may occur, followed by diffusion to the other electrode and direct chemical reaction. This effect may vary systematically with temperature and composition. Trace impurities or unsuspected reactions may lead to mixed electrode reactions and an e.m.f. which is the resultant of more than one (the desired) process. Therefore, enthalpies, and especially heat capacities, may be expected to be obtained more accurately from more direct calorimetric measurements rather than from e.m.f. measurements.

It is among the purposes of this article to: summarize the manner in which the measured electromotive forces of molten salt cells may be related to thermodynamic properties of the molten salts; discuss criteria for experimental conditions conducive to reproducible e.m.f. measurements; discuss sources of error in the measurements and interpretation; and present typical examples of the application of e.m.f. measurements to the determination of thermodynamic properties of molten salts.

The scope of the subject is too broad to permit a completely general or completely exhaustive treatment. Rather, it is hoped that the discussion and examples will serve to indicate the kinds of considerations needed in the experimental design and theoretical analysis of e.m.f. measurements whose purpose is to provide accurate thermodynamic data for molten salts.

Various aspects of e.m.f. measurements and their relation to the thermodynamics of molten salts have been discussed by Laity[59], Wagner[90], Alabyshev et al.[1], Rossotti[77], and Delimarskii and Markov[25]; the first two of the above references give particularly cogent treatments of the thermodynamic theory.

II. Cell EMF and Thermodynamic Properties

1. Introduction

A principal application of e.m.f. measurements is in the evaluation of thermodynamic properties of bulk phases. Interfaces between metallic phases and between metallic and electrolytic phases are unavoidable experimental concomitants, but the net effect of the detailed processes occurring within a few molecular diameters of such an interface, e.g. in a double layer, must vanish if the measurements are to have thermodynamic significance for the bulk phases. The measurements and equations are therefore best discussed in terms of thermodynamic quantities for neutral components, rather than single electrode potentials and single ion activities whose definition and interpretation would require non-thermodynamic assumptions. Even if unambiguous measurements were possible for quantities such as single electrode potentials, they probably would not provide a convenient reference for thermodynamic properties. The latter would require the differences between quantities containing uncertainties in the interface properties—the least accurate part of the measurement. It is of course essential to consider the changes of macroscopic properties and the transport of matter across each of the interfaces, and to establish that the interfacial properties do not affect the measured electromotive force. However, one may make the analogy to, say, vapor pressure measurements. The details of vaporization and condensation kinetics and of surface layer structure need not (and should not) be incorporated into the reported vapor pressures and activities; but it must be ascertained that these interface properties do not interfere with measurements of the bulk properties. Formal tables of single electrode potentials are useful as a shorthand notation to the extent that they are defined and measured in such a manner that thermodynamic quantities result from the combination of pairs of entries.

2. Classification of EMF Cells

E.m.f. cells may be classified in terms of the type of process that would occur on short circuiting the cell, as formation cells, mixing cells, and concentration cells with or without transference.

Thus, formation cells provide the free energy of formation of a pure molten salt, or of a component in a molten salt mixture, from the elements in their standard states. Mixing cells provide free energies of mixing, from which excess chemical potentials may be determined. Concentration cells with or without transference may also be employed to determine excess chemical potentials, although the latter require a knowledge of the transport numbers of the constituents.

For many purposes it is convenient also to classify molten salt cells in terms of the number of components in the molten salt system under investigation.

A. ONE COMPONENT (PURE MOLTEN SALT)

In a one component system, an appropriate e.m.f. cell may provide the free energy of formation of the salt*. This requires two different reversible electrodes, one for the cation and one for the anion of the salt

$$M\,|\,MX_z\,|\,X_{2(g)},\ C$$

M represents a metal such as silver and X_2 a halogen such as chlorine. The electrode reactions, overall cell reaction, Gibbs free energy change and electromotive force of such a cell may be written

$$M \to M^{z+} + Ze$$

$$\underline{Ze + \tfrac{1}{2}ZX_2 \to ZX^-}$$

$$M + \tfrac{1}{2}ZX_2 \to MX_z$$

$$-ZFE = \Delta G = \mu_{MX_z} - \mu_M - \tfrac{1}{2}Z\mu_{X_2} \tag{2}$$

The complete cell reaction always involves differences of chemical potential of neutral components even though the half reactions may be written in terms of charged species. If MX_z is a pure molten salt, M is pure metal (solid or liquid) and X_2 is a gas, equation 2 becomes

$$\Delta G = -ZFE = \mu^\circ_{MX_z} - \mu^\square_{M(solid)} - \tfrac{1}{2}Z(\mu^\circ_{X_2} + RT\ln a_{X_2})$$

or, for a liquid metal

$$\mu^\circ_{MX_z} - \mu^\circ_{M(liquid)} - \tfrac{1}{2}Z(\mu^\circ_{X_2} + RT\ln a_{X_2})$$

If all substances are in their standard states, the e.m.f. gives directly the standard free energy of formation. Generally the gas is not in the standard state and the expression may be written

$$\Delta G f^\circ[MX_z(l)] = -ZFE + \tfrac{1}{2}Z\,RT\ln a_{X_2} \simeq -ZFE + \tfrac{1}{2}Z\,RT\ln P_{X_2} \tag{3}$$

At the elevated temperatures of most molten salts, non-ideality of gases at

---

* A class of molten salt cells that will not be considered here is that in which the molten salt only facilitates investigation of the electrode materials as in an electrode concentration cell. A pure molten salt may serve as the electrolyte in a cell with two alloy electrodes or with a pure metal electrode and an electrode consisting of an alloy of the metal with a more noble metal $M\,|\,MX_z\,|\,M(M')$. Such a cell may be employed to determine activity coefficients of the metal M in the alloy, or the free energy of transfer of metal M between alloys of differing composition. It provides no thermodynamic information on the molten salt, however. Furthermore, to avoid the generally greater solubility of a metal in its own pure salt than in a mixture, a molten salt mixture generally would be preferable to a pure salt for application to alloy concentration cells; and a solid electrolyte would reduce convection[90].

pressures of one atmosphere and below is generally small enough to be neglected without contributing significant error. Frequently, an inert gas carrier is used with corrosive gases, and the activity calculated from the Dalton's law partial pressure. In gases with significant deviations from ideality, the fugacity corrections must be applied, but may be insufficiently well known. Experimental conditions of pressure, gas composition and flowrate should be reported explicitly so that such corrections may be made subsequently, and results from different sources compared.

B. Binary Systems

(a) Free energy of formation of one of the components of a mixture

$$\text{Be} \left| \begin{matrix} \text{LiF,} & \text{BeF}_2 \\ (1-x) & x \end{matrix} \right| \text{HF, H}_2 \left| \text{Pt} \right.$$

$$\text{Be} + 2\text{HF} \rightarrow \text{BeF}_2(x) + \text{H}_2$$

$$-2FE = \Delta G = \mu_{\text{BeF}_2}(x) - \mu_{\text{Be}} - 2\mu_{\text{HF}} + \mu_{\text{H}_2}$$
$$= \Delta G_f[\text{BeF}_2(x)] - 2\Delta G_f(\text{HF}) \qquad (4)$$

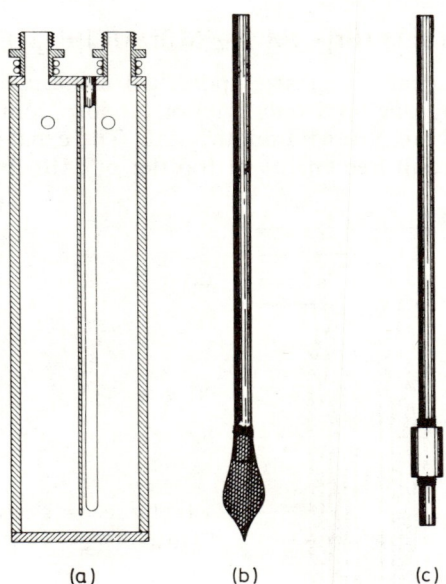

Figure 1. (a) All-nickel compartmented container for BeF_2–LiF mixtures used for e.m.f. studies of the formation cell[46].

$$\text{Be} \left| \begin{matrix} \text{BeF}_2 \\ \text{LiF} \end{matrix} \right| \text{HF, H}_2 \left| \text{Pt} \right.$$

(b) HF–H_2 electrode. The HF–H_2 gas mixture passes through the $\frac{1}{8}$ in. nickel tube and is forced through the platinum gauze bag dipping into the melt.
(c) Beryllium electrode. A $\frac{1}{2}$ in. cylinder of beryllium dipping into the melt is threaded on to a $\frac{1}{8}$ in. diameter nickel tube through which argon gas is passed to mix the melt.

Since the free energy of formation of hydrogen fluoride (HF) is known, it can be added to the free energy of the cell reaction to obtain the free energy of formation of dissolved BeF_2[46]. If the free energy of formation of pure BeF_2 is known, the chemical potential of BeF_2, and the activity coefficients are calculable. By application of the Gibbs–Duhem equation, the activity coefficient of LiF in solution may be determined, relative to an arbitrary assignment of the activity coefficient of LiF at one concentration[46]. The experimental arrangement of such a cell is indicated in *Figure 1*. The free energy of formation of LiF in solution is not accessible from measurements with this cell. In general, the free energy of formation of the second component would not be directly accessible by e.m.f. methods because the more active metal would reduce the ions of the more noble metal irreversibly, unless the activity of the more active metal can be reduced sufficiently by alloying. In the system considered here, it would be possible to accomplish this with a dilute alloy of lithium in bismuth. Such an electrode has been employed in a concentration cell but has not yet been employed in a formation cell in the above system[84]. The free energy of formation of the alloy would need to be known to then calculate the free energy of formation of LiF.

(b) *Direct measurement of chemical potential change on mixing; membrane cells.*

$$\text{C, } Br_2(g) | MBr | \xleftarrow{\text{glass}}(M^+)\rightarrow | MBr, MCl | Br_2(g), \text{C}$$

The cell above contains a glass or porcelain membrane, through which current is carried by one (and only one) of the ions common to the pure molten salt and the mixture with another salt. The e.m.f. provides a direct measure of the (partial) free energy of transfer of MBr from the pure salt

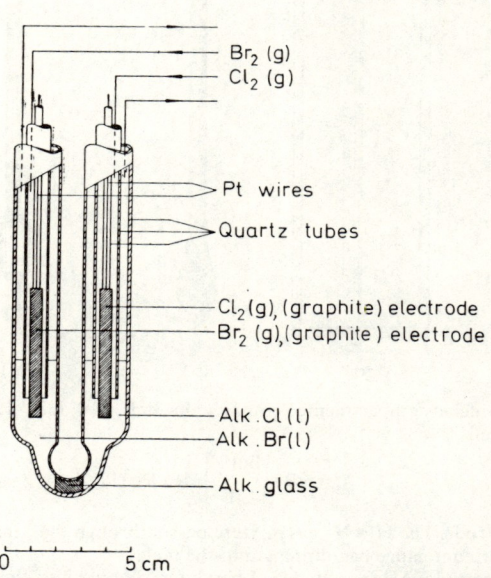

Figure 2. Quartz cell[70(b)] containing a membrane of alkali ion conductive glass.

phase to the mixture, or the change of chemical potential on mixing. This thermodynamic quantity can be calculated also from the difference between the e.m.f.s of two formation cells, one with pure salt and one with a mixture. The membrane cell, however, employed extensively by Egan[28] and by Østvold[70] eliminates the errors of subtraction of two e.m.f. readings. Furthermore, elimination of a metallic electrode eliminates the possibility of metallic dissolution and electronic conductivity. A cell of this type with a Cl_2 and a Br_2 electrode is indicated in *Figure 2*.

(c) *Concentration cells with transference*. If the two liquids in the cell above are separated by a porous frit, permitting constrained interdiffusion of the two liquids rather than selective ionic transport through a membrane, the cell is designated a concentration cell with transference, and its e.m.f. depends on the ionic mobilities in the liquid as well as on the chemical

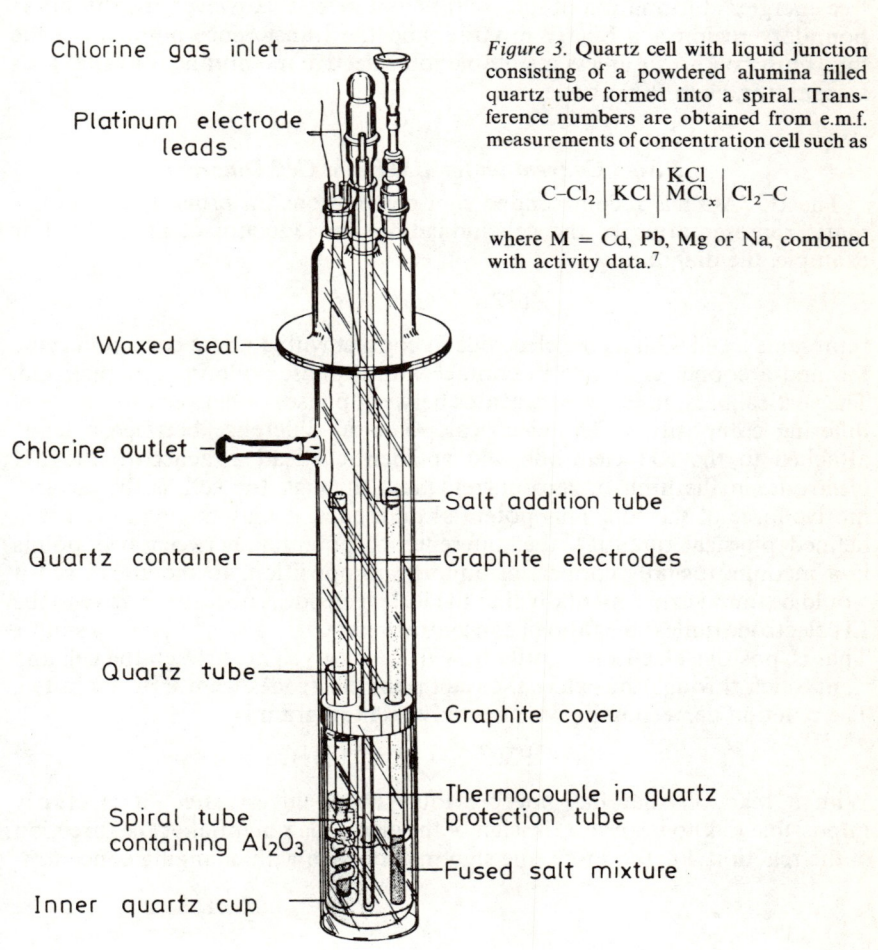

Figure 3. Quartz cell with liquid junction consisting of a powdered alumina filled quartz tube formed into a spiral. Transference numbers are obtained from e.m.f. measurements of concentration cell such as

$$C-Cl_2 \left| KCl \left| \begin{array}{c} KCl \\ MCl_x \end{array} \right| Cl_2-C \right.$$

where M = Cd, Pb, Mg or Na, combined with activity data.[7]

potentials of the components. *Figure 3* illustrates such a cell in which a spiral tube filled with Al_2O_3 powder, rather than a frit, provides the diffusion barrier[7].

C. Multicomponent Systems

The same kinds of cells are possible for multicomponent as for binary systems, but the effects of liquid junctions may be more complicated.

D. Jacobi–Daniell Cells

Cells of the type

$$M|MY|NY|N$$

where M and N are two different metals are employed in the comparison of free energies of formation of salts of different metals. However, the salt in the boundary region is a binary mixture, and the transference numbers of the ions must be known unless a suitable ion-selective membrane is available as a separator, as in B(b) above.

3. Sign Convention for EMF and Cell Diagram

The IUPAC has recommended sign conventions in terms of a diagrammatic representation of the cell indicating the sequence of phases[63]. For example, the diagram

$$Zn|Zn^{2+}|Cu^{2+}|Cu$$

represents a cell with a zinc electrode in contact with a phase containing zinc ion and a copper electrode in contact with a phase containing copper ion. The vertical bars indicate separations between phases or between solutions of differing composition. The electrical potential difference between a lead* attached to the RH electrode and an identical lead attached to the LH electrode, in the limit of zero current flow through the cell, is the electromotive force of the cell. This potential difference, it may be noted, is a well defined physical quantity, the difference of potential between two points in a medium (usually copper) of uniform composition. In the above cell it would be found experimentally that the RH electrode is positive relative to the LH electrode (unless the ratio of concentrations $[Cu^{2+}]/[Zn^{2+}]$ is very small). That is, positive electricity would flow from left to right through the cell and right to left through the external circuit if the electrodes were short circuited. The reaction corresponding to the above cell diagram is

$$\tfrac{1}{2}Zn + \tfrac{1}{2}Cu^{2+} \rightarrow \tfrac{1}{2}Zn^{2+} + \tfrac{1}{2}Cu$$

With a reaction that has not previously been investigated, it is clearly impossible to know *a priori* the sign of the electrode potential or the direction of the reaction. Furthermore, the sign may change with changing concentra-

---

* Usually copper.

tions in the two half-cells. Thus, if $[Cu^{2+}]/[Zn^{2+}] \ll 1$, the sign of the e.m.f. of the above cell diagram could become negative and should be so recorded.

If, on the other hand, the reaction had been written initially

$$\tfrac{1}{2}Zn^{2+} + Cu \rightarrow \tfrac{1}{2}Zn + \tfrac{1}{2}Cu^{2+}$$

the cell diagram should be written in a manner such that the RH electrode would be made positive by the occurrence of the above reaction.

$$Cu|Cu^{2+}|Zn^{2+}|Zn$$

$$\tfrac{1}{2}Zn^{2+} + e \rightarrow \tfrac{1}{2}Zn$$

$$\tfrac{1}{2}Cu \rightarrow \tfrac{1}{2}Cu^{2+} + e$$

Experimentally, the e.m.f. of the cell diagram written in this manner would be found negative (unless $[Cu^{2+}]/[Zn^{2+}] \ll 1$), and the free energy change of the corresponding reaction positive.

The 'electrode potential' of an electrode or half-cell has been defined[63] as the electromotive force of a cell in which a standard hydrogen electrode is the reference half-cell,

$$Pt, H_2|H^+|Zn^{2+}|Zn$$

The above convention is clearly derived from dilute aqueous electrolytes, and is of little utility for the greatly disparate media encountered in the investigation of molten salts.

Cell diagrams of the type indicated above do not uniquely describe the cell process (even if neutral components rather than ionic constituents are specified), but serve only to designate the sign convention. The magnitude (and possibly the sign) of the e.m.f. will be affected by the concentrations of the electroactive ions, the nature of the media, and (with certain exceptions noted below) the nature of the diffusional gradient resulting from the manner of establishment of the junctions between molten salts of differing composition. Some authors employ additional notation to indicate the nature of a junction between phases. A double vertical bar is frequently employed to indicate an arrangement designed to eliminate the effect of a junction between two liquids of differing composition, although the absence of such an effect is not always demonstrable. Here we shall employ a dashed single vertical line to indicate a liquid junction, requiring consideration of ionic transport across that junction.

4. Thermodynamic Expressions for Cell EMFs

A. Electrode and Transport Processes

The most general expressions for the e.m.f. of galvanic cells in terms of thermodynamic quantities and the transference numbers of neutral and charged species, including electrons, have been given by Wagner[90]. However, the large number of possible kinds of electrodes, phase boundaries, membranes and varieties of materials makes a single general expression cumbersome. Examples of the analysis for several types of cells will be given to illustrate the principles.

The expression for a cell e.m.f. in terms of thermodynamic quantities requires a knowledge of the cell process. This requires consideration both of the electrochemical reaction occurring at the electrode–melt boundary and of transport by ions, electrons and neutral species (which do not transfer charge but may result in free energy changes) across boundaries between phases of differing composition.

We consider first cells with liquid junctions since in fact most molten salt e.m.f. cells will, by either design or circumstance, include a liquid junction. These cells are considered in some detail since they illustrate the principles that must be extended to multicomponent systems and membrane systems, and the results may be then reduced to those for apparently simpler formation cells.

Although the e.m.f. is measured as the electric potential difference between two phases of identical composition (the copper leads attached to the electrodes) and the only observable changes are those of electrically neutral components, it has become customary, and may be convenient, to write the e.m.f. expression in terms of ionic constituents.

The e.m.f. is often formally written as the sum of hypothetical electric potentials attributed to the electrode reactions, each involving charged species:

$$L \rightarrow L^+ + e; E_l$$

$$e + R^+ \rightarrow R; E_r$$

and any liquid junction between the liquids in contact with each of the electrodes.

B. Binary Systems with Liquid Junctions

Considering a cell

$$M \left| \begin{array}{c|c} MX & MX \\ MY & MY \end{array} \right| X_2 \qquad (A)$$
$$\ \text{I} \quad\ \ \text{II}$$

one may write formally, for the contributions to the e.m.f. in terms of hypothetical free energy changes,

$$E = E_m + E_x + E_j$$

$$M \rightarrow M_I^+ + e \qquad \Delta G_m = \mu_{M_I^+} - \mu_M + \mu_e$$

$$e + \tfrac{1}{2}X_2 \rightarrow X_{II}^- \qquad \Delta G_x = \mu_{X_{II}^-} - \tfrac{1}{2}\mu_{X_2} - \mu_e$$

$$t_X X_{II}^- \rightarrow t_X X_I^- \qquad \Delta G'_j = t_X(\mu_{X_I^-} - \mu_{X_{II}^-}) = -\int_I^{II} t_X \, d\mu_X$$

$$(1-t_X)Y_{II}^- \rightarrow (1-t_X)Y_I^- \quad \Delta G''_j = (1-t_X)(\mu_{Y_I^-} - \mu_{Y_{II}^-}) = -\int_I^{II} (1-t_X) \, d\mu_Y$$

The chemical potentials of the charged species have no thermodynamic significance. They should be replaced by electrochemical potentials[40],

$\tilde{\mu}_i = \mu_i + Z_i\phi$. However, the correct thermodynamic result will be obtained in terms of chemical potentials of neutral components if electrode processes and transport processes of all charged species involved in the cell reaction are accounted for and summed.

The net change of free energy and the cell e.m.f. neglecting composition dependence of the transference numbers, becomes

$$-FE = \Delta G = \mu_{M_I^+} + \mu_{X_{II}^-} + \mu_{Y_I^-} - \mu_{Y_{II}^-} + t_X[(\mu_{X_I^-} - \mu_{Y_I^-})$$
$$-(\mu_{X_{II}^-} - \mu_{Y_{II}^-})] - \mu_M - \tfrac{1}{2}\mu_{X_2}$$
$$= \mu_{MY_I} + (1 - t_X)(\mu_{MX_{II}} - \mu_{MY_{II}}) + t_X(\mu_{MX_I} - \mu_{MY_I}) - \mu_M - \tfrac{1}{2}\mu_{X_2}$$

Since the transference numbers are not in general independent of composition,

$$-FE = \Delta G = \mu_{MX_{II}} - \mu_M - \tfrac{1}{2}\mu_{X_2} + \mu_{MY_I} - \mu_{MY_{II}} + \int_I^{II} t_X (d\mu_{MY} - d\mu_{MX}) \quad (5)$$

If x and $(1 - x)$ are the mole fractions of MX and MY, the Gibbs–Duhem relation leads to the expression

$$-FE_A = \Delta Gf_{MX_{II}} + \int_I^{II} \{(x - t_X)/(1 - x)\} d\mu_{MX} \quad (6)$$

The three hypothetical 'potentials' E_m, E_x and E_j are not measurable individually, and only their sum for an electrically neutral combination is measurable and relatable to thermodynamic quantities. The liquid junction potential is given by

$$FE_j = -\int_I^{II} t_X d\mu_{X^-} - \int_I^{II} (1 - t_X) d\mu_{Y^-} \quad (7)$$

The integrals on the RHS of equations 5 or 6 are sometimes erroneously called the liquid junction potential. It should be noted also that the correct thermodynamic equations can be arrived at either with the simpler quasi-thermodynamic (thermostatic) formalism or with the more rigorous formalism of the thermodynamics of irreversible process[59, 90, 35, 65].

Førland et al.[35] point out that cell e.m.f.s can be written entirely in terms of thermodynamic and transport properties of neutral components only.

The same formalism may be applied to the cell (B) with the salts of two metals with a common anion.

$$M|MX, NX \vdots MX, NX|X_2|C \quad (B)$$
$$\quad\quad I \quad\quad\quad II$$

Writing the electrode reactions and transport processes,

$$M \rightarrow M^+(I) + e$$
$$e + \tfrac{1}{2}X_2 \rightarrow X^-(II)$$
$$\bar{t}_M M^+(I) \rightarrow \bar{t}_M M^+(II)$$
$$(1 - \bar{t}_M)N^+(I) \rightarrow (1 - \bar{t}_M)M^+(II)$$

$$M + \tfrac{1}{2}X_2 + \bar{t}_M M^+(I) + (1 - \bar{t}_M)N^+(I) \rightarrow M^+(I) + X^-(II) + \bar{t}_M M^+(II)$$
$$+ (1 - \bar{t}_M)N^+(II)$$

The overall process involves only neutral components, as can be seen by adding equal numbers of $X^-(I)$ to both sides of the equation.

$$M + \tfrac{1}{2}X_2 + \bar{t}_M MX(I) + (1 - \bar{t}_M)NX(I)$$
$$\to MX(I) + \bar{t}_M MX(II) + (1 - \bar{t}_M)NX(II)$$

The $\bar{t}$ designate average transference numbers (since they may be concentration dependent) unless the concentration differences between I and II are infinitesimal. The transference numbers indicated are internal transference numbers and correspond here to mobilities measured relative to X^- ions as the reference, hence no transference number of X^- appears. Any ionic constituent or neutral component can be taken as the reference, but where X^- is common to both half-cells, it provides a convenient reference constituent. There is one less transference number than there are components in the mixture. The reaction could be written also in terms of *external* transference numbers corresponding to mobilities measured relative to a frame of reference external to the melt, e.g. the electrodes or a porous plug. The transport equations would then include one for transference of X^- relative to the external reference*.

$$\bar{t}_M^E M^+(I) \to \bar{t}_M^E M^+(II)$$
$$\bar{t}_N^E N^+(I) \to \bar{t}_N^E N^+(II)$$
$$(1 - \bar{t}_M^E - \bar{t}_N^E)X^-(II) \to (1 - \bar{t}_M^E - \bar{t}_N^E)X^-(I)$$

Since the external transference numbers are not properties of the melt alone[83], it is preferable when possible to employ the internal transference numbers, which are the Hittorf transference numbers.

If the compositions in the half cells are identical, the work (per mole of electrons) performed by the cell in discharging isothermally, isobarically and reversibly (e.g. through a very high resistance) is the negative of the free energy change of the process of formation of MX in the melt.

$$FE_{B_1} = -\Delta G = -[\mu_{MX(I)} - \mu_M - \tfrac{1}{2}\mu_{X_2}] = -\Delta Gf[MX(I)] \qquad (8)$$

If the composition in compartment II is changed infinitesimally, the free

---

* The internal (t_M^X) and external (t_M^E) transference numbers are related through the velocity of the reference constituent for internal transference numbers relative to the external frame of reference, $u_X - u_E$, and the equivalent concentrations, c.

External:

$$t_M^E = \frac{c_M(u_M - u_E)}{c_M(u_M - u_E) + c_N(u_N - u_E) + c_X(u_X - u_E)}$$

$$= \frac{c_M(u_M - u_X) + c_M(u_X - u_E)}{c_M(u_M - u_X) + c_N(u_N - u_X) + (c_M + c_N + c_X)(u_X - u_E)}$$

Internal, relative to X^-:

$$t_M^X = \frac{c_M(u_M - u_X)}{c_M(u_M - u_X) + c_N(u_N - u_X)}$$
$$1 = t_M^X + t_N^X = t_M^E + t_N^E + t_X^E$$
$$t_X^E \neq t_X^X = 0$$

energy change must include the work of transference through the concentration gradient; and the e.m.f. becomes

$$-F(E + dE) = \Delta G + dG = \Delta Gf[MX(I)] + t_M d\mu_{MX} + (1 - t_M) d\mu_{NX}$$

The chemical potential variations are not independent, but related by the Gibbs–Duhem equation

$$y\, d\mu_{MX} + (1 - y)\, d\mu_{NX} = 0$$

where y is the mole fraction of MX

$$-F \frac{dE}{dy} = \frac{dG}{dy} = \frac{t_M - y}{1 - y} \left(\frac{d\mu_{MX}}{dy} \right)$$

For a finite composition difference, the e.m.f. becomes

$$-FE_B = \Delta G = \Delta Gf[MX(I)] + \int_{y_I}^{y_{II}} \frac{t_M - y}{1 - y} \left(\frac{\partial \mu_{MX}}{\partial y} \right) dy \qquad (9)$$

The integral,

$$\int_{I}^{II} \left(\frac{t_M - y}{1 - y} \right) \frac{\partial \mu_{MX}}{\partial y} dy = -F(LJE) \qquad (10)$$

represents the contribution to the e.m.f. of the concentration gradient between solutions I and II. As such it defines a kind of operational liquid junction e.m.f., as discussed by Førland et al.[35]. It must be distinguished, however, from the immeasurable 'liquid junction potential' (LJP) which is defined as $-F(LJP) = \int t_M d\mu_M + t_N d\mu_N$.

The liquid junction e.m.f. (LJE) for a binary mixture has the useful property that it is independent of the form or diffuseness of the concentration gradient or of the mode of establishment, so long as the electrodes see only solutions I and II, respectively, and the intervening solution contains only MX and NX*. The concentration gradient, however, must be such that the

---

* This may be illustrated with a case typical of many binary molten salt mixtures. For many molten salt mixtures, the cation transference numbers nearly proportional to the mole fractions of the two salts[7]. Suppose $t_M \simeq ay + by^2$, $a + b = 1$. If the excess chemical potential is approximated as a symmetrical function of concentration

$$\mu_{MX} = \mu^\circ_{MX} + RT \ln y + \alpha(1 - y)^2$$

The integral $-F(LJE)$ becomes

$$\int_{I}^{II} \frac{t_M - y}{1 - y} d\mu_{MX} \simeq \int \frac{ay + by^2 - y}{1 - y} \left[\frac{RT}{y} - 2\alpha(1 - y) \right] dy$$

$$RT \int \{(a - 1) + by\}/\{1 - y\}\, dy - 2\alpha \int [a - 1)y + by^2]\, dy$$

$$F(LJE) \simeq bRT(y_{II} - y_I) + \alpha[a - 1](y_{II}^2 - y_I^2) + \tfrac{2}{3}\alpha b(y_{II}^3 - y_I^3) \qquad (10')$$

The integral LJE is exact for all values of y, including the case of pure NX in the half cell with the M electrode and pure MX in the other. (It must be noted, however, that if the M electrode

chemical potentials are definable throughout the region between the electrodes.

Equation 9 for the cell e.m.f. may be rearranged to give the free energy of formation of MX(II), with a different apparent LJE.

$$-FE_B = \Delta Gf[\text{MX(II)}] + \int_I^{II} \left[\frac{t_M - y}{1 - y} - 1\right] d\mu_{MX}$$

$$= \Delta Gf[\text{MX(II)}] - \int_I^{II} \frac{1 - t_M}{1 - y} d\mu_{MX} \qquad (11)$$

For the concentration cell with two X_2 electrodes,

$$X_2 | \text{MX, NX} \;\vdots\; \text{MX, NX} | X_2 \qquad (C)$$
$$\quad\;\; I \qquad\qquad II$$

the cell e.m.f. is the LJE of equation 9, for cell B.

$$FE_C = -\Delta G = -\int_I^{II} \frac{t_M - y}{1 - y} d\mu_{MX} \qquad (12)$$

This equation may be rearranged to one involving the partial molar free energy of transfer of MX or NX from I to II:

$$FE_C = -(\mu_{\text{MX(II)}} - \mu_{\text{MX(I)}}) + \int_I^{II} \frac{1 - t_M}{1 - y} d\mu_{MX} \qquad (13)$$

---

were in pure $NX(y_I = 0)$, the cell e.m.f. E_B (equations 8, 9) would not be meaningful; the calculated 'theoretical' e.m.f. corresponding to formation of MX in pure NX would be infinite, but in practice the e.m.f. would reach a finite value corresponding to some possibly indeterminate process involving impurities.) The integral LJE would be exact, i.e. independent of the form of the concentration gradient, for any physically meaningful concentration dependence of the transference numbers and activity coefficients.

For the Jacobi–Daniell cell,

$$M | \text{MX} \;\vdots\; \text{NX} | N$$
$$\;\; I \qquad II$$

the electrode and transport processes are:

$$M \rightarrow M_I^+ + e$$
$$e + N_{II}^+ \rightarrow N$$
$$\bar{t}_M M_I^+ \rightarrow \bar{t}_M M_{II}^+$$
$$\bar{t}_M N_I^+ \rightarrow \bar{t}_N N_{II}^+$$
$$M + (1 - \bar{t}_N) N_I^+ \rightarrow N + (1 - \bar{t}_M) M^+$$
$$M + N_{II}^+ + t_{M_I} M_I^+ + t_{N_I} N_I^+ \rightarrow N + M_I^+ + t_{M_{II}} M_{II}^+ + t_{N_{II}} N_{II}^+$$
$$X_{II}^- + (t_{M_I} + t_{N_I}) X_I^- \rightarrow X_I^- + (t_{M_{II}} + t_{N_{II}}) X_{II}^-$$

$$-FE = \Delta G = \Delta Gf(\text{MX})_I - \Delta Gf(\text{NX})_{II} + \int_I^{II} t_M d\mu_{MX} + \int_I^{II} t_N d\mu_{NX}$$

$$= \int_I^{II} \frac{t_M - y}{y} d\mu_{NX} = \int_I^{II} \frac{1 - t_N - y}{y} d\mu_{NX}$$

$$= -(\mu_{NX(II)} - \mu_{NX(I)}) + \int_I^{II} \frac{1 - t_N}{y} d\mu_{NX} \qquad (14)$$

For the cell with two metal electrodes

$$M|MX, NX \;\vdots\; MX, NX|M \qquad\qquad (D)$$
$$\quad\; I \qquad\qquad II$$

the analogous electrode and transport processes lead to the result

$$FE_D = -\Delta G = \mu_{MX(II)} - \mu_{MX(I)} - \int_I^{II} \frac{t_M - y}{1 - y} d\mu_{MX}$$

$$= \int_I^{II} \frac{1 - t_M}{1 - y} d\mu_{MX} \qquad (15)$$

We note that

$$F(E_D - E_C) = \mu_{MX(II)} - \mu_{MX(I)} \qquad (16)$$

Thus the combined cell $[(D) - (C)]$

$$X_2|MX, NX \;\vdots\; MX, NX|X_2^- M|MX, NX \;\vdots\; MX, NX \;|\; M \qquad (E)$$
$$\quad II \qquad\quad I \qquad\qquad\qquad I \qquad\qquad II$$

would provide the composition dependence of the chemical potentials of MX and (via the Gibbs–Duhem equation) NX. The same information could be obtained from the formation cell B, but concentration cells avoid possible direct reaction of the two different electrode materials.

Some special cases of the e.m.f. expressions are of interest. First, either pure MX or pure NX (with an X_2 but not an M electrode) can be used as the reference salt if its physical and chemical properties are convenient—but dissolution of the electrode in its own salt must be guarded against.

The e.m.f. of the concentration cell C, with electrodes reversible to the common ion, is zero if the transference numbers (relative to the common ion) of the non-common ions are equal to their mole fractions: $t_M = y, t_N = 1 - y$, as seen from equation 13.

The e.m.f. of cell D, with electrodes reversible to one of the non-common ions, is zero if the transference number of the ion to which the electrodes are reversible is unity. If the transference number of that ion is zero, the e.m.f. will lead to the chemical potential difference of one of the components.

In very dilute solutions of MX in NX, in most cases t_M approaches zero and equation 15 becomes

$$FE_D \simeq \mu_{MX(II)} - \mu_{MX(I)} \tag{17}$$

Such a cell can be useful, therefore, for determining activity coefficients in dilute binary mixtures.

The e.m.f. of cell C, however, is relatively insensitive to changes of chemical potential of dilute MX in the solvent NX (or of NX as a dilute solution in MX) unless the transference numbers of M and N differ significantly from y and $1 - y$.

In the above cells, it has been implied that M and N are cations, X an anion, but the relations apply equally well to common cation binary mixtures with X representing the metal.

The expressions have been derived for mixtures of uni–univalent salts. They are easily transformed to apply to charge unsymmetric mixtures by simply replacing the mole fraction, y, by the equivalent fraction and the molar chemical potentials by chemical potential per equivalent. For example, for cell D (equation 15)

$$FE_D = \int_I^{II} \frac{1 - t_M}{1 - y} d\mu_{MX}$$

If MX is BeF_2 and NX is LiF, with the mole fractions x and $1 - x$, respectively[75],

$$FE_D = \int_I^{II} \frac{1 - t_{Be}}{1 - \{2x/(1 + x)\}} d(\mu_{BeF_2}/2)$$

$$2FE_D = \int_I^{II} t_{Li} \left(\frac{1 + x}{1 - x}\right) d\mu_{BeF_2} \tag{18}$$

From the form of the above equations, it may be seen that decomposition of a cell e.m.f. into electrode potentials and liquid junction potential may not always be convenient. Furthermore, the term 'diffusion potential', carried over from the terminology of dilute aqueous electrolyte solutions with the tacit assumption of the Nernst–Einstein relation, may be misleading. The term 'mobility potential' may be more accurate, since the contribution to the cell e.m.f. depends on the transference numbers, determined by the relative mobilities[7]. Thus a value unity for the transference number of the cation to which the electrodes are reversible in a common anion binary molten salt mixture corresponds to zero e.m.f. for the concentration cell. There may be an appreciable diffusion gradient, but it is not coupled to the ionic mobilities.

C. Multicomponent Systems

The e.m.f. expression for cells with multicomponent systems can be written in a manner analogous to that for binary systems, but the junction e.m.f. is no longer independent of the form of the junction. Consider the cell E with a ternary molten salt system.

$$
M \left|\begin{array}{c} MX(y) \\ NX(x) \\ PX(1-y-x) \end{array}\right| \left.\begin{array}{c} NX \\ NX \\ PX \end{array}\right| M
$$
$$\qquad\qquad\text{I}\qquad\quad\text{II}\qquad\qquad\qquad\text{(E)}$$

With X^- as the reference constituent, the electrode and transport processes are:

$$M \to M^+(I) + e$$
$$e + M^+(II) \to M$$
$$\bar{t}_M M^+(I) \to \bar{t}_M M^+(II)$$
$$\bar{t}_N N^+(I) \to \bar{t}_N N^+(II)$$
$$(1 - \bar{t}_M - \bar{t}_N)P^+(I) \to (1 - \bar{t}_M - \bar{t}_N)P^+(II)$$

$$-(1 - \bar{t}_M)M^+(I) + \bar{t}_N N^+(I) + (1 - \bar{t}_M - \bar{t}_N)P^+(I) \to \{\ldots \bar{t}_N N^+(II)\ldots\}$$

and the cell e.m.f. becomes

$$-FE_E = \Delta G = \int_I^{II} (1 - t_M)[d\mu_{PX} - d\mu_{MX}] + t_N[d\mu_{NX} - d\mu_{PX}]$$

$$= [\mu_{PX}^{II} - \mu_{PX}^{I}] - [\mu_{MX}^{II} - \mu_{MX}^{I}] - \int_I^{II} (t_N + t_M)\,d\mu_{PX}$$
$$+ \int_I^{II} t_M\,d\mu_{MX} + \int_I^{II} t_N\,d\mu_{NX} \qquad (19)$$

One of the three chemical potentials may be eliminated by means of the Gibbs–Duhem equation, but there remain two independent composition variables and two transference numbers dependent upon composition, so that the integrals will depend on the way in which the gradient is formed. Special cases may arise that lead to simpler expressions corresponding to binary mixtures and to independence of the form of the diffusion gradient.

Such simplification will arise if all but one independent transference number are zero, either by chance, by virtue of the imposition of a membrane permeable to only one kind of ion, or by virtue of the concentration of one of the components being vanishingly small. Similarly solubility or other equilibria that maintain all but one of the chemical potential variations vanishingly small would eliminate all but one of the integrals.

Thus, if MX and NX are very dilute solutes (say $<10^{-3}$ mole fraction) in the solvent PX, $t_M \simeq t_N \simeq 0$ and the activity of the solvent PX is essentially that of pure solvent, whence

$$FE_E \simeq \mu_{MX}^{II} - \mu_{MX}^{I} \qquad (20)$$

A particularly important case here is in dilute reciprocal molten salt mixtures where association constants of dilute oppositely charged species may be evaluated.

$$M \left| \begin{array}{c|c} MX & MX \\ MZ & MZ \\ NX & NX \end{array} \right| M$$
$$\quad\quad\quad I \quad\quad II \quad\quad\quad\quad\quad\quad\quad\quad (F)$$

Taking X^- as the reference constituent, which would be convenient if it were the major anionic constituent, the electrode and transport processes give

$$-FE_F = \Delta G = \int_I^{II} (1 - t_M)[d\mu_{NX} - d\mu_{MX}] - \int_I^{II} t_Z \, d\mu_{NZ} \quad (21)$$

With the solute ions M^+ and Z^- sufficiently dilute ($<10^{-3}$) that μ_{NX} becomes that of pure solvent NX, $d\mu_{NX}$, t_Z and t_M become negligible, and

$$FE_F \simeq \mu_{MX}^{II} - \mu_{MX}^{I} \quad (22)$$

D. Membrane Cells

In the above cells the liquid junctions have been such as to permit diffusion, albeit constrained, of all constituents of the electrolyte. Membrane cells are those in which one or more of the ionic constituents is or are prevented from diffusing or migrating by means of a blocking component. In a generalized sense, any e.m.f. cell can be considered a 'membrane cell' permeable to the ions to which the electrodes are reversible. The electrodes block migration of other ions, and the electrolyte blocks (or should block) migration of electrons.

Equations for the e.m.f. of membrane cells and the relationship to thermodynamic properties have been developed by extension of the general equations for cells with transference[80, 81].

The membrane cells providing the most direct thermodynamic information are those in which the membrane permits passage of only a single kind of ionic constituent, but not electrons.

$$M \left| \begin{array}{c|c|c} MX & & \\ MZ & Z^- & MZ \\ NX & \text{(or } M^+) & \end{array} \right| Z_2 \left| C \right. \quad (G)$$
$$\quad\quad\quad I \quad\quad\quad\quad II$$

The cell reactions for the case of a membrane permeable to one of the ions present on both sides of the membrane are shown below.

(a) Membrane permeable to Z^- only ($t_Z = 1$; $t_M = t_X = 0$)

$$M \to M^+(I) + e$$

$$e + \tfrac{1}{2}Z_2 \to Z^-(II)$$

$$Z^-(II) \to Z^-(I)$$

$$\overline{M + \tfrac{1}{2}Z_2 \to MZ(I)}$$

$$-FE_{aG} = \Delta G = \Delta Gf[MZ(I)]$$

(b) Membrane permeable to M^+ only ($t_M = 1$; $t_Z = t_X = 0$)

$$M \to M^+(I) + e$$

$$e + \tfrac{1}{2}Z_2 \to Z^-(II)$$

$$M^+(I) \to M^+(II)$$

$$\overline{M + \tfrac{1}{2}Z_2 \to MZ(II)}$$

$$-FE_{bG} = \Delta G = \Delta Gf[MZ(II)]$$

A system corresponding closely to case (a) above is based on the use of a lanthanum fluoride (LaF_3) crystal, a fluoride conductive crystal with very low electronic conductivity, as a membrane in molten fluorides[21]. In the cell

$$\text{Be}\,|\,\text{LiF, BeF}_2\,|\,\text{LaF}_3\,|\,\text{LiF, BeF}_2\text{, NiF}_2\,|\,\text{Ni} \qquad\qquad \text{(H)}$$
$$\phantom{\text{Be}\,|\,}\text{I} \phantom{\text{LiF, BeF}_2\,|\,\text{LaF}_3\,|\,}\text{II}$$

the halogen electrode of cell G was replaced by a nickel–nickel fluoride electrode in cell H, leading to the following electrode and transport processes.

$$\text{Be} \to \text{Be}^{2+}(I) + 2e$$

$$2e + \text{NiF}_2 \to \text{Ni} + 2F^-(II)$$

$$2F^-(II) \to 2F^-(I)$$

$$\overline{\text{Be} + \text{NiF}_2 \to \text{Ni} + \text{BeF}_2(I)}$$

$$-2FE_4 = \Delta G = \Delta Gf[\text{BeF}_2(I)] - \Delta Gf^\circ(\text{NiF}_2) \qquad\qquad (23)$$

The construction of the LaF_3 reference half cell is illustrated in *Figure 4*.

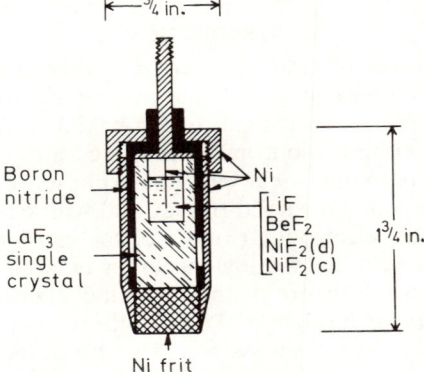

Figure 4. Reference half-cell for use in fluoride melts[21(a)]. A LaF_3 single crystal membrane, which is also the cup containing the reference electrolyte, separates the fluoride melt from the reference electrolyte, and supports conduction by fluoride ion transport exclusively.

Frequently a membrane may not be truly impermeable to all ions except one, and transport through the membrane must be considered, as through a diffusion gradient. Since the membrane effectively constitutes an additional component, one additional transference number is added. It then becomes convenient to choose the membrane as the reference for mobilities of the salt ions rather than to employ the Hittorf internal frame of reference of the molten salt. Ionic transport in glass membranes has been investigated in connection with the application of membrane cells to free energies of formation and partial molar free energies of mixing in binary molten salt mixtures[28, 70, 34, 69] primarily with glasses or porcelains conductive to a single ion. The partial molar free energy of mixing is obtainable also from membrane cells with glasses conductive to more than one ion, as with the cell[89]

$$Cl_2 | NaCl | Glass | NaCl, KCl | Cl_2 \qquad (I)$$
$$(Na^+, K^+)$$
$$\text{I} \qquad \text{II}$$

the cell equation is (see equation 13)

$$FE_I = \int_I^{II} \frac{t_{K^+} - x_{KCl}}{x_{KCl}} d\mu_{NaCl} \qquad (24)$$

For purposes of computing the chemical potential

$$\mu_{NaCl} - \mu^\circ_{NaCl} = F \int_I^{II} \frac{x_{KCl}}{t_{K^+} - x_{KCl}} dE_I \qquad (25)$$

The transference number t_{K^+} is that of potassium ion in the glass membrane ($t_K + t_{Na} = 1$).

5. *Principal Error Sources*

A. INTRODUCTION

Two principal sources of systematic error in thermodynamic quantities determined from e.m.f. measurements that are specific to the e.m.f. method (as distinct from more general sources of error related to the measurement of temperature and composition, impurities, etc.) are associated with ion transport across the boundaries between liquid phases of differing composition (or between a liquid and a solid phase), and with electron transfer processes occurring at the electrodes. (Errors of the e.m.f. measurement itself are generally minor because of the high accuracy of electrical measurements and electrical standards.) Errors of the first kind are attributed to uncertainties in the *liquid junction potential*. Those of the second kind are attributed to a *mixed potential*. Errors in composition can arise either from uncertainties in overall composition or from incomplete mixing, so that an electrode does not see the intended composition. A temperature gradient between the electrodes, difficult to eliminate in high temperature molten salt studies, may give rise to a thermoelectric e.m.f. beyond the uncertainty in the measurement of temperature. Solubility of a metallic electrode in a molten salt may give rise to electronic conduction and a partial short circuit of the cell.

B. Junction EMF

The effect of liquid junctions has been described above in terms of binary mixtures and multicomponent systems. It may be necessary to consider a junction e.m.f. also in formation cells involving a single electrolyte, since in practice it is rarely possible to measure the e.m.f. of a cell without separating the melts that are in equilibrium with the two different electrodes. The composition difference may often be negligible, but should be considered in estimating errors in the desired free energy changes.

Cells such as I have been employed for mixtures of alkali chloride with alkaline earth chloride, using glass membranes conductive only to alkali ions. If the membrane is not perfect, but should conduct some Ca^{2+} ion as well, the term

$$F\,\delta E = \int_I^{II} \frac{t_{Ca^{2+}}}{x_{CaCl_2}}\,d\mu_{NaCl} \tag{26}$$

includes an estimate of the error. If the transference number and concentration of Ca^{2+} in the membrane are constant, and the activities of Ca^{2+} silicate and Na^+ silicate in the membrane nearly ideal, the correction to the partial molar free energy of NaCl (calculated on the initial assumption that the glass is only Na^+-conductive) is [34, 69, 90]

$$\delta\mu_{NaCl} \sim -\frac{t_{Ca^{2+}}}{x_{Ca\text{-sil}}} RT \ln x_{Ca\text{-sil}} \tag{27}$$

With a glass containing 57 per cent Ca-silicate even one per cent transport by Ca^{2+} ($t_{Ca^{2+}} = 0.01$) would lead to an error of only $10^{-2}\,RT$ in the chemical potential.

Glass membranes without ions common to the molten salts and without a clearly specified conductive mechanism are often interposed between two molten salts in an e.m.f. cell in order to prevent reactive materials from contacting each other (e.g. to prevent Cl_2 of a gas electrode from reaching a metallic electrode in a formation cell). If the same salt is on either side of the membrane the presumption is that the junction e.m.f. will cancel. Asymmetry e.m.f.s, however, are the rule rather than the exception for glass membranes, and the results may be expected to be uncertain by the order of the difference of e.m.f. resulting from exchange of electrolytes on the two sides of the membrane.

If a cell such as B is employed to determine the free energy of formation of a salt MX, or the composition dependence of its chemical potential, equation 10′ provides an estimate of the junction e.m.f.

Note that ideality of mixing ($\alpha = 0$) and simple equality of the transference number to mole fraction ($a = 1$, $b = 0$) leads to $LJE = 0$.

C. Mixed Electrode Processes

The e.m.f. of a cell is equal to the potential difference at the terminals of a potentiometer connected to the electrodes of the cell under conditions (extrapolated) such that no current flows. Zero net current through an electrode, however, need not signify zero current of all electrode processes. With a metal electrode dipping into a melt containing ions of that metal, there may be additional possible processes than the metal–metal ion charge transfer process. Such additional processes may be redox reactions of impurities that are dissolved in the melt, or sorbed on either the electrode or on an 'inert' metal in contact with both active electrode material and the melt[39]. If such processes occur, zero net current at the electrode is the algebraic sum of the currents corresponding to these processes. If only one process occurs in addition to the 'main' electrode reaction, there are equal and opposite currents corresponding to the oxidation process in one reaction and the reduction process in the other. The resultant e.m.f. will not then correspond to thermodynamic equilibrium with respect to the presumed electrode reaction and will not therefore yield a thermodynamic property of the system. The contribution of each process to the resultant e.m.f. will depend on the activities of the reactants and products, and on the exchange current density of each process[39, 58]. The exchange current density is the specific rate of the forward and reverse reactions of an electrode process. It determines the rate of establishment of the e.m.f. of a cell (in which only a single well defined electrode process occurs at the other electrode), and the insensitivity to interference by spurious electrode reactions.

A high exchange current density is therefore desirable for an electrode. The smaller the exchange current density, the greater the care that must be exerted to remove impurities from the melt, to maintain a suitable gas atmosphere, and to minimize current flow during measurement in order to avoid a mixed potential.

Some typical values of exchange current[58] densities (i_0 A/cm^2) in molten salts are, in a solvent LiCl–KCl: Bi^{3+}/Bi, 8; V^{3+}/V^{2+}, 30; Pt^{2+}/Pt, 40; Ni^{2+}/Ni, 110; Ag^+/Ag, 190; Cd^{2+}/Cd, 210. For silver in $LiNO_3$–KNO_3–$NaNO_3$[44]: $i_0 = 5$ A/cm^2; for silver in aqueous perchlorate solution[91] at room temperature: $i_0 = 1$ A/cm^2. Other aqueous solution values[91] are: Zn^{2+}/Zn, 3×10^{-8}; Pb^{2+}/Pb, 8×10^{-4} A/cm^2.

The possibility of mixed electrode reactions can give rise to corrosion[91] of an electrode in the absence of current flow. Mixed electrode reactions may be manifested by failure to follow the Nernst equation for dilute solute ions, hence this criterion provides a valuable test. Further tests include varying the ratio of surface area of active and noble metal, the surface treatment of electrodes (to remove sorbed gases), and the atmosphere.

D. Thermal EMF

A thermal e.m.f. may arise from either of two sources: the thermoelectric power of the molten salt with a temperature gradient between the electrodes in the cell; or, in an isothermal cell, from dissimilar metal leads (e.g. a metal and graphite) passing through a temperature gradient to the measuring instrument. The former effect may be of the order of hundreds of microvolts per degree[85, 51], so that temperature uniformity within tenths of a degree is desirable, and this can be best achieved with stirring. For example, the magnitude of the thermally produced potential difference between identical electrodes at differing temperatures in the cell $Ag|AgNO_3|Ag$ is 0.34 mV per degree at 305°C[85].

The thermal e.m.f. arising from dissimilar metal electrodes can be eliminated by making the connection between each of the electrodes and its lead (both leads being of the same material) in the uniform high temperature region of the cell. There may be situations where this presents extreme design difficulties. In such cases the thermal e.m.f. may be measured directly and subtracted from the measured e.m.f. to obtain the cell e.m.f. The measurement should be preferably made *in situ* with the electrodes shorted together in the hot region. The direct reaction may make it inconvenient to carry out the measurement in the presence of salt. In the event that a measurement of the thermal e.m.f. under experimental conditions is not practical, the thermal e.m.f. should be estimated from available tables of thermoelectric power of metals[56(b)(c)(d)], but it may vary with relatively small changes of impurity or trace constituents.

E. Metal Solubility and Electronic Conduction

The properties of mixtures of metals with molten salts have been extensively reviewed[18, 24]. Here we are concerned only with those aspects relevant to errors in e.m.f. measurements in molten salts. In general two different types of solubility of a metal in its molten salt are observed. Alkali metals and alkaline earth metals tend to go into solution as the ions, the electrons being associated with a cavity or moving via a hopping mechanism. Metals such as cadmium or bismuth tend to go into solution as subvalent species, possibly in clusters of normal ions (e.g. Cd_2^{2+}). Either type species may constitute a partial short circuit if the reduced species ($Na°$, e^-, Cd_2^{2+}) can diffuse into an oxidizing environment. The solubilities of the metals generally are significantly smaller in salts of foreign ions or in mixtures than in salts of the metal. The solubilities generally are higher in chlorides and bromides than in fluorides. The transference number of electrons adds a junction e.m.f. to the e.m.f. of a cell such as

$$Ag|AgBr|Br_2(g), C \qquad (J)$$
$$III$$

Hence (see ref. 90, p 34)

$$FE_J = -\Delta Gf(AgBr) - \int_I^{II} t_e \, d\mu_{Ag} \qquad (28)$$

where

$$t_e = \frac{\kappa_e}{\kappa_{ionic} + \kappa_e} \simeq \frac{\kappa_e}{\kappa_{ionic}} = 10^{-3} \qquad (29)$$

leading to a correction of about 4×10^{-3} volt to the measured e.m.f. of about 0.9 volt. κ_e and κ_{ionic} are the electronic and ionic conductivities respectively.

III. Experimental

1. Reporting Data and Results

Recommended symbols, terminology and units have been adopted by the IUPAC Council[63]. The Commission on Thermodynamics and Thermochemistry has issued a *Guide to Procedures for the Publication of Thermodynamic Data*[54]. In addition to general careful attention to the description of apparatus and procedures, particular attention is required for the sources of error listed in Section II.5. Detailed description of cell components and configuration is needed to evaluate and compare results of different authors with regard to liquid junctions, thermal e.m.f., and mixed potentials.

2. Apparatus

A. Measuring Instruments

The assembly and utilization of apparatus for electromotive force measurements in a molten salt system should be designed to insure results characteristic of the system and independent of the apparatus. In addition to the molten salt system whose thermodynamic properties are desired, the required assembly includes a cell for containment of the melt, a furnace, electrodes and measuring instrumentation. The most severe limitation on experimental design is the restricted availability of reversible, non-polarizable electrodes. Reliable instrumentation for accurate e.m.f. measurement is readily available. Complexity of containment and furnace construction may arise from problems of compatibility of the molten medium with container materials, as with fluoride melts, or from the study of systems with exceptionally high melting points, as with silicates.

The most direct and accurate method of measuring e.m.f. still is the Poggendorf potentiometer[6], such as the Leeds and Northrup, Types K3 or K4, but the convenience and availability of high impedance (10^6 ohm and higher) electronic digital and differential voltmeters make them attractive competitors, even with cell resistances low enough to make a high input impedance unnecessary. Application of either type instrument requires a standard e.m.f. source against which the instrument is routinely calibrated[43]. The saturated Weston cell is the most commonly used working standard, and although it should be calibrated frequently (at least yearly) it requires little care other than protection from sudden temperature changes by enclosing it in an insulated vessel such as a polyfoam container. Under standard laboratory conditions, this cell may provide a voltage reference accurate within $1-2 \times 10^{-6}$, but the uncertainty may be an order of magnitude greater under ordinary laboratory conditions. The unsaturated Weston cell sometimes is used because of its lower temperature coefficient, although it is less stable and requires more frequent checking. This is rarely a limiting factor in the determination of thermodynamic properties.

Bates[6] suggests a potentiometer as suitable for e.m.f. measurement in cells of 10^5 ohms or less. Inman[47] has used a potentiometer to measure the e.m.f. of cells containing a glass electrode of 16×10^3 ohms resistance in nitrate melts at 250°C, although the reduced overall sensitivity may then correspond to a galvanometer deflection of only 1 cm per mV. The authors find a high impedance electronic differential voltmeter desirable when the cell resistance exceeds 10^4 ohms. Many molten salt cells containing glass membrane electrodes or diffusion barriers of low porosity operating between 400 and 700°C have typical resistances less than 5000 ohms[11, 23, 26], so that the ordinary potentiometer is adequate. Cells containing certain separators or membranes other than glass (e.g. boron nitride[52] or lanthanum trifluoride used in fluoride melts[21(a)]) do have sufficiently high resistances to require a high impedance instrument for accurate e.m.f. measurements. In the examples cited, the high resistance may result either from very low porosity of a salt bridge, as with the boron nitride barrier, or from inherent high resistivity of the electrolytic conductor, as with the lanthanum trifluoride crystal, a fluoride ion conductor. For convenience, the authors have routinely used a high impedance potentiometric voltmeter such as the Keithley 662 for e.m.f. measurement but e.m.f. readings in each experiment are checked on a Leeds and Northrup type K3 potentiometer to insure that the cell resistance is not unexpectedly high due to a gas bubble or a non-conducting solid lodged at the junction. The cause of a spurious e.m.f. resulting from a change in the nature of the junction might go unnoticed on the Keithley, but the sudden occurrence of a very high resistance becomes immediately apparent with the loss of sensitivity of the type K3 potentiometer.

A comparison of some of the typical characteristics of several types of e.m.f. measuring instruments is given in *Table 1*. The values, for particular

Table 1. Characteristics* of e.m.f. measuring instruments

	Poggendorf type potentiometer	Electronic potentiometric voltmeter	Electronic electrometer	Digital voltmeter
Detector	Galvanometer or electronic null detector	Electronic null indicator	Vibrating-reed null detector	Direct reading
Internal resistance	~150 ohms	10^6 ohms or greater	10^{13} ohms or greater	10^{10} ohms
Resolution	10 μV	1 μV	3 μV	1 μV
Limit of error	±15 μV	±10 μV	±30 μV	±7 μV
Source of standard	Weston cell	Zener diode reference	Zener diode reference	Zener reference and oscillator
Mode of reading	Intermittent	Continuous	Intermittent or continuous	Continuous
Type of circuit	Resistance only	Vacuum tube or solid state	Vacuum tube or solid state	Solid state

* When reading an e.m.f. of about 100 mV.

commercial instruments, are intended primarily for comparison purposes, and any given instrument would differ slightly from others of its kind. The stability and convenience of the differential voltmeter and electrometer developed rapidly when the art of solid state electronics provided the extremely stable Zener diode-voltage reference source. The development of solid state integrating circuits gave rise to the digital voltmeter. Although the passive circuit potentiometer[42] remains the 'standard' for e.m.f. measurements, the electronic instrument provides a rapid, convenient, stable measuring instrument in a small, often portable, package. For cells with electrodes subject to polarization, the high impedance of an electrometer is indispensable for measurements without current drain from the source. Continuous monitoring of the e.m.f. may be provided by connecting a high impedance recorder to the output terminals of the measuring instrument, but it is important to determine just how much current can be drawn from the source without polarizing the electrodes, since the impedance of many instruments decreases rapidly as the imbalance departs from the null position. With a standard potentiometer of low input resistance the e.m.f. is balanced with intermittent rapid tapping of the key to minimize current flow from the source. Any e.m.f. measuring instrument in constant use should be calibrated every three months. Sharp temperature changes should be prevented. With these considerations the measuring portion of the e.m.f. experiment should present the minimum experimental problems.

B. Cells and Furnaces

The design and construction of cells for e.m.f. measurements are tied closely to the furnace design. At low temperatures (below 400°C), where a circulating molten salt bath may be feasible, more freedom of design is possible[38]. Laboratory furnaces have been discussed in several reviews[66, 53], the problem of temperature measurement and control being an integral part of furnace design. Design of an electric resistance furnace, the most common type used in the laboratory, consists largely of selection of materials: the

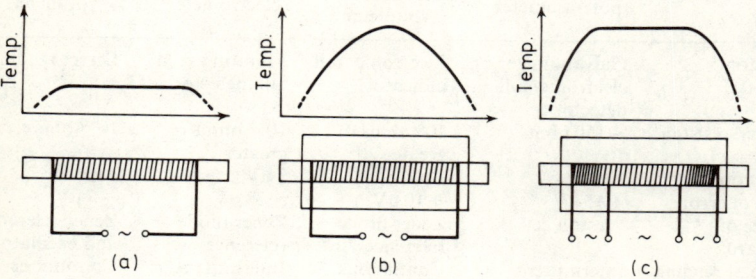

Figure 5. Temperature distribution in a cylindrical furnace[66]: (a) un-insulated; (b) insulated; (c) insulated furnace with middle windings shunted.

resistor material (usually, but not always, a metallic winding), an electrically insulating support for the winding (usually ceramic or other refractory material) and the thermal insulation. As described by Motzfeld[66], an ideal furnace is a closed envelope with a uniform temperature at all points in the interior; this ideal is not achievable in practice. Many furnaces are in the shape of an open cylinder, a convenient configuration for insertion of the cell. Even spacing of the windings along the length of the cylinder, with no thermal insulation, results in a temperature distribution such as that shown in *Figure 5*(a). Insulation of the walls of the cylinder worsens the distribution, to that shown in *Figure 5*(b), where the energy dissipation from the middle of the winding is less than at the ends. A compensatory increase of the power input to the ends of the cylinder results in a distribution like that in *Figure 5*(c), with uniformity comparable to that of *Figure 2*(a) but less susceptible to external temperature perturbations. In many commercially available tube furnaces* the winding is fitted with external terminals at intervals of two inches or so, to permit distribution of the power input by the insertion of shunts. The adjustment procedure is empirical, and should be made with a completely assembled cell in place in the furnace. Insertion of internal insulation or baffles at the ends of the furnaces to hinder heat flow along the cylinder walls is sometimes useful in reducing the temperature gradient. The authors have used a 4 in. diameter tube furnace into which an all-metal 3 in. diameter cell was inserted and the annular region backfilled with loosely packed vermiculite. In spite of a somewhat longer time constant (for establishment of equilibrium between the furnace wall and the molten salt—about 6–8 h for a 500°C change in temperature) the temperature oscillations due to an on–off type controller were virtually eliminated. The relative power input immediately above the level of molten salt in the cell was increased by shunting the windings below the cell in order to compensate heat losses via conduction along metal risers that communicate with the melt and extend from the hot region to ambient temperatures. A less important source of heat loss, convection through the cover gas, can be reduced by judicious choice of gas with low thermal conductivity, that of helium being 7.5 times that of argon and 5.5 times that of nitrogen.

Control of temperature can be accomplished manually[9] or automatically[48]. In either case control to ± 0.1 degree is attainable, but resettability of temperature is facilitated by an automatic controller and is preferred. The degree of temperature control required varies with the temperature coefficient of the process under investigation. The simultaneous heterogeneous equilibria required for reversibility of an electrode of the third kind, Pd–PdO–CdO–Cd^{2+}[48], resulted in a temperature coefficient of 1 mV per degree. This requires control to $\pm 0.1°C$ for resolution of ± 0.1 mV in the cell e.m.f.[16]. We have occasionally found that a slow steady drift of temperature (within the permissible limits of temperature variation) leads to less noise in the measured e.m.f. than a smaller oscillatory variation. Temperature control may be facilitated by a constant voltage power source and two sets of windings, one of which carries a fixed current providing most of the power while current through the second winding is controlled manually or automatically. All heater elements should be non-inductively wound to prevent interference with the e.m.f. readings. Problems associated with pickup of stray electrical signals into the e.m.f. measuring circuit from the furnace windings or control circuit have rarely been discussed in detail, but an earthed metal envelope or a Faraday screen around the cell serves to isolate the electrodes from external interference[49]. Temperature in the cell is generally monitored with thermocouples, although mercury thermometers, thermistors or platinum resistance thermometers are adequate and useful for certain applications. Temperature

---

* Marshall Products Company, Columbus, Ohio.

scales and measurement have been extensively discussed elsewhere [ref. 43, chapter 2; ref. 56(a), section V].

Design and construction of the cell proper and the cell envelope must take cognizance of the corrosiveness of the molten salt, the temperature of measurement, and the possible requirements of an inert or reactive gas atmosphere over the system. Most designs[78, 19] incorporate a closed cell envelope whose length is several times as great as its diameter for better temperature uniformity in a relatively short melt region; however, e.m.f. measurements in nitrate melts at 350°C have been successful in open Pyrex beakers[17, 13].

C. MATERIALS

The high temperatures and corrosivities associated with most molten salts may impose stringent requirements on the selection of materials employed in e.m.f. cells.

The materials problems associated with molten salt e.m.f. cells may be classified in terms of (a) inherent stability of the molten salt, (b) reactivity of the molten salt with its container, (c) reactivity of the molten salt with its atmosphere, (e) reactivity of molten salt with electrode materials, (e) electrode-atmosphere interactions, and (f) insulating materials for electrode leads.

Thermal instability of molten salts to be investigated may be repressed by equilibration with a decomposition product. Nitrates decompose reversibly to nitrites, evolving oxygen[5]; an oxygen atmosphere may be employed to decrease the nitrite concentration. An NO_2–O_2 gas mixture bubbled over platinum has been employed as a reversible nitrate ion electrode in molten nitrates[87]. Impurities may catalyse decomposition of a molten salt, as do chloride and metal ion impurities in molten ammonium nitrate. A high vapor pressure may be considered a special case of thermal instability. If the electrolyte is a pure molten salt, this poses little problem, as it results in no composition change, provided the cell is designed to prevent condensation of salt in regions where it may short circuit the electrodes, react with insulating materials, or interfere with gas flow. For mixtures with volatile components, the vapor space should be maintained small. If it can be maintained at a temperature above the melt temperature, there will be no transport of salt from the liquid to cool regions. It may, however, be difficult to heat the entire vapor space if electrode leads pass through fittings with materials of limited thermal stability such as Teflon. Application of an inert gas pressure, while it does not lower the vapor pressure, can slow the transport of material sufficiently to permit a series of measurements without appreciable composition change.

Container materials must be selected on the basis of their inherent thermal stability and their inertness to the molten salt and to gases that may be employed as an atmosphere, e.g. vacuum (the container should have a low vapor pressure); 'inert' atmosphere such as nitrogen; reducing or oxidizing gases needed to repress salt decomposition; or a gas employed in an electrode reversible to an ion of the molten salt.

Pyrex cells have been employed extensively for thermodynamic studies of molten alkali nitrates at temperatures up to 500°C. At higher temperatures or for melts containing lithium ion, fused silica is preferable; Li^+ ions will exchange with the cations in Pyrex.

Fused silica is a satisfactory material for many molten salts, such as nitrates, chlorides, bromides, chloroaluminates or sulfates. It has limited applicability for short periods of time to some acidic mixtures of beryllium fluoride with alkali fluorides (i.e. the Be-rich mixtures), but not for the *basic* (alkali-rich) fluoride[4] mixtures.

Boron nitride has been employed[52], but it contains oxide, which is employed as a binder and is leached out of the boron nitride. *Figure 6* shows a boron

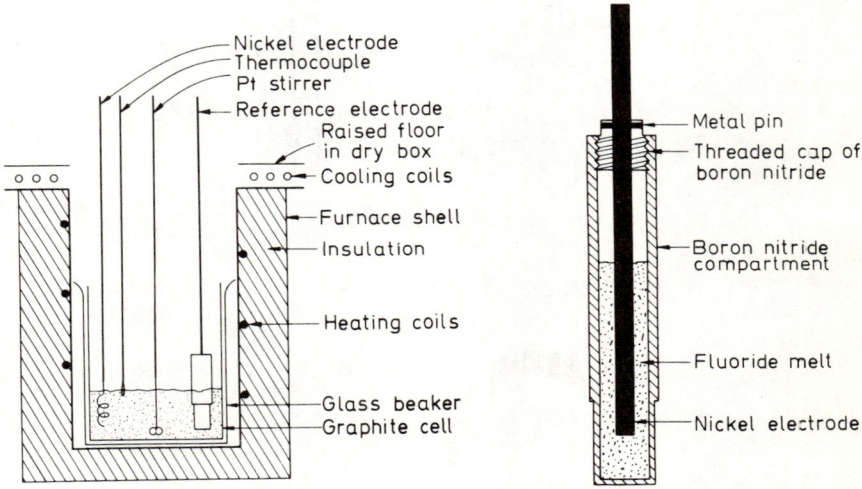

Figure 6. (a) Furnace and cell assembly for e.m.f. measurements of the concentration cell[52]

$$Ni \left| \begin{array}{c} Ni(II) \\ in\ (Li, Na, K)F \\ reference \end{array} \right| BN\ sheath \left| \begin{array}{c} Ni(II) \\ (Li, Na, K)F \end{array} \right| Ni$$

The furnace is mounted in the floor of a dry box to provide an inert atmosphere and to allow manipulation of electrodes and addition of reagents. (b) Detail of the Ni–Ni(II) reference electrode in a boron nitride compartment.

nitride sheathed nickel–nickel fluoride (unsaturated) half cell for use in fluorides[52].

Graphite is a satisfactory container material for crucibles for containment of molten fluorides, and its use also is indicated in *Figure 6*.

Inert metal containers probably are the most satisfactory. The metal chosen must be one which does not dissolve in the molten salt nor react with it nor with the atmosphere or other electrode materials. All-metal cells (i.e. except for insulators around electrode leads) are convenient for cleaning, outgassing, loading and assembly. The metal will not act as a short circuit between electrodes provided there is no reaction it can undergo with an e.m.f. smaller in magnitude than the e.m.f. of the cell reaction under investigation. The container material may be one of the electrodes, or in electrical contact with one of the electrodes; or both electrodes or half-cell compartments may be immersed in the salt without electrical contact with the container. An all-metal cell employed for the investigation of molten rare earth halide–rare earth metal systems[19] is shown in *Figure 7*.

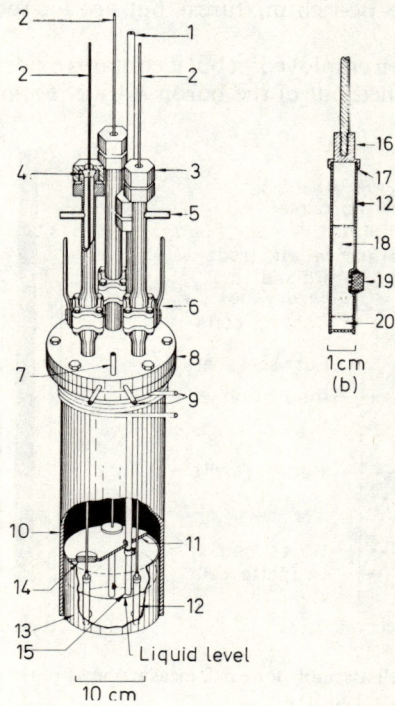

Figure 7. All metal cell for e.m.f. measurements of molten metal–metal halogenide solutions corrosive to ceramics[19]. 1, stainless steel thermocouple tube; 2, tantalum electrodes; 3, gasket tightener; 4, Teflon seal; 5, vacuum and pressure line to entry ports; 6, worchester valve; 7, vacuum and pressure line to tank; 8, Teflon O-ring seal; 9, water coolant tubes; 10, stainless steel tank; 11, stainless steel spacer plate attached to thermocouple tube; 12, electrode tube, see (b); 13, tantalum vessel; 14, lavite insulators; 15, tantalum jacketed thermocouple tube; 16, tantalum threaded cap; 17, weld; 18, solution molten $Cd-CdCl_2$; 19, porous tantalum plug; 20, excess cadmium in reference cell.

Reactivity of the molten salt with its atmosphere includes reactions with the gas itself, and with impurities in the gas. Generally an inert gas, such as argon, can be employed as a cover gas. With helium, removal of water can be accomplished with a liquid nitrogen–charcoal trap. Alternatively, passage through titanium sponge heated to 850°C will remove water and oxygen from either helium or argon. The latter, being denser, is the better blanket. A flowing gas blanket is frequently employed in a system difficult to maintain free of leaks. However, even with very low impurity contents, a flowing system will expose the cell to large total quantities of impurity. Maintaining a static excess pressure of inert cover gas in a leak-free system is therefore preferable when feasible. Where the gas is required for the electrode reaction, a flowing system may be necessary, but the flowrate may be made small. For chlorine electrodes, it has been found that saturation of a graphite electrode with chlorine gas will permit a steady e.m.f. without additional gas flow [57].

Reactivity of a molten salt with electrode materials may occur via dissolution of the metal in the molten salt, reduction of metal ions in the salt to a lower oxidation state, or direct chemical reaction between the salt and an electrode reversible to a solute in the melt. Electrode–atmosphere interactions may be more severe than container–atmosphere reactions, since the electrodes will be more reactive.

Insulating materials for the electrodes include Pyrex, silica, alumina and boron nitride, the compatibility considerations being analogous to those for container and melt. In all-metal cells, the stringency of the compatibility requirements for insulators may be greatly reduced. An inert porous metal frit may separate melts of two different compositions, and behave as a 'quasi-insulator' so long as the cell reaction is insufficiently energetic to electrolyse salt (whether solvent, solute or impurity) on either side of the metal 'quasi-insulator'. In a closed container it is still necessary to insulate the electrode leads from the metal container, and this is best done with Teflon-lined Swage-Lok fittings in tubes in a cooled region above the furnace.

D. Mixing

A mechanical stirrer is a valuable, often essential, adjunct of a molten salt e.m.f. cell, particularly for studies of mixtures. The stirrer serves the dual purpose of hastening mixing equilibrium, and minimizing temperature gradients. The elevated temperatures required for molten salts generally dictate an elongated configuration for furnace and cell, with cooling provided for gaskets or electrode insulators. Relatively small melt depths help to minimize temperature gradients.

Stirring greatly reduces the temperature gradient in the melt. Dissolution and convective mixing can be surprisingly slow, even at elevated temperatures, as indicated in *Figure 8*. This may be an extreme case. With crystals having lower melting temperatures than the bulk melt, dissolution will be

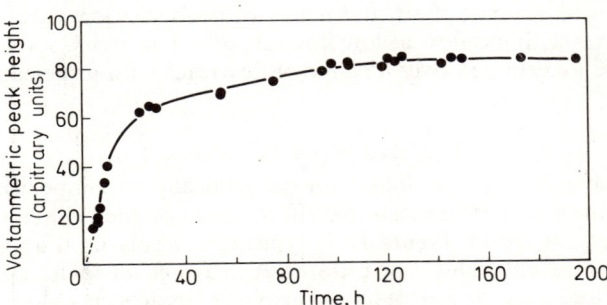

Figure 8. Time dependence of unstirred dissolution and mixing of solid LiF–UF$_4$ in molten LiF–BeF$_2$ at 550°C to form 0.008 mole fraction solution of UF$_4$.

much more rapid, but still hastened by stirring. In the electrolysis of lithium into molten bismuth to form a dilute Li–Bi alloy ($\sim 10^{-3}$ mole fraction Li), a constant e.m.f. results within minutes of the electrolysis, even in the absence of stirring[84]. Nevertheless, composition uniformity needs to be ascertained, and incorporation of a stirrer in the cell design is worthwhile.

Stirring may be continuous or intermittent. Gas bubbling is less effective than a mechanical stirrer because of channelling at low flowrates, splatter of material at high flowrates and increased transpiration of volatile materials (unless a presaturator is employed). For melts not sensitive to the atmosphere, a simple motor-driven glass paddle, as has been employed in molten nitrates, is effective. In enclosed systems the stirrer shaft may be inserted through a Teflon gland fitted to a cooled riser. It is generally unnecessary to stir small electrode compartments dipping into a bulk melt.

Related to the problem of mixing equilibrium is the kinetic problem of the three phase equilibrium at a gas electrode, involving equilibrium among the gas phase, the melt and the metallic electrode. In some systems, such as

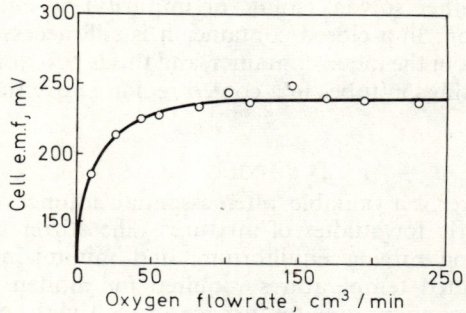

Figure 9. Dependence of cell e.m.f. on oxygen flowrate at an Au/O_2, CO_2 electrode in molten alkali carbonates at 660°C[60].

the halogen electrode in halide melts, equilibrium is observed with very low flowrates—and even in static systems, after simply treating a graphite electrode with the halogen gas and immersing it in the melt with no additional gas flow. In a carbonate melt, the e.m.f. with an oxygen–carbon dioxide electrode varies with gas flow at low flowrates[60], as in *Figure 9*. Constancy of e.m.f. must be ascertained over a range of flowrates with gas electrodes.

E. Purification of Chemicals

Purity of chemicals is required both for accuracy of composition and in order to eliminate specific undesirable effects on electrode behavior such as a mixed electrode reaction. *Figure 10* is typical of widely used apparatus for the purification of chlorides[45] or nitrates. It provides for melting the salt or salt mixture under an appropriate atmosphere (hydrogen chloride gas for chlorides; nitrates might be initially treated with nitric acid, and then melted under vacuum), followed by filtration under vacuum or under an inert gas

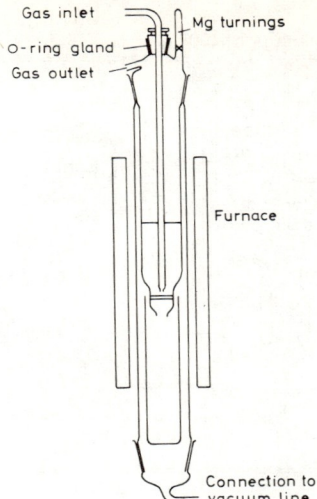

Figure 10. Pyrex cell for purification of salts by fusion under controlled atmosphere filtration and solidification under vacuum[45].

atmosphere. Fused silica rather than Pyrex is often employed, and a pair of inert electrodes may be inserted for 'pre-electrolysis' to remove electroactive impurities, including water, and to monitor the diffusion current.

Figure 11 shows a gas purification and handling system for gaseous sulfur–argon mixtures employed in a silver sulfide formation cell for silver sulfide–silver chloride mixtures[88]. The cell is shown in *Figure 12*. It may be noted that this formation cell includes an asbestos diaphragm as a diffusion barrier to prevent contact between dissolved sulfur in the sulfur–graphite half cell with the silver electrode.

Alkali fluorides are available as chips of single crystal material of high purity[76]. Beryllium fluoride, which is hygroscopic, is extremely viscous and difficult to crystallize from the liquid state. It has been purified by repeated sublimation[21(b), 46, 75] in a nickel apparatus, in which it can also be treated with hydrogen fluoride. *Figure 13* is a schematic of the nickel sublimation apparatus. Additional details on the handling of molten fluorides have been reviewed recently[3].

3. Electrodes

A. Gas Electrodes

(a) *The Halogen Electrode*

A crucial part of any gas electrode is the inert conductor at which the requisite three phase equilibrium must be established. Graphite is used almost exclusively as the electrode for halogen half-cells in molten salts, although

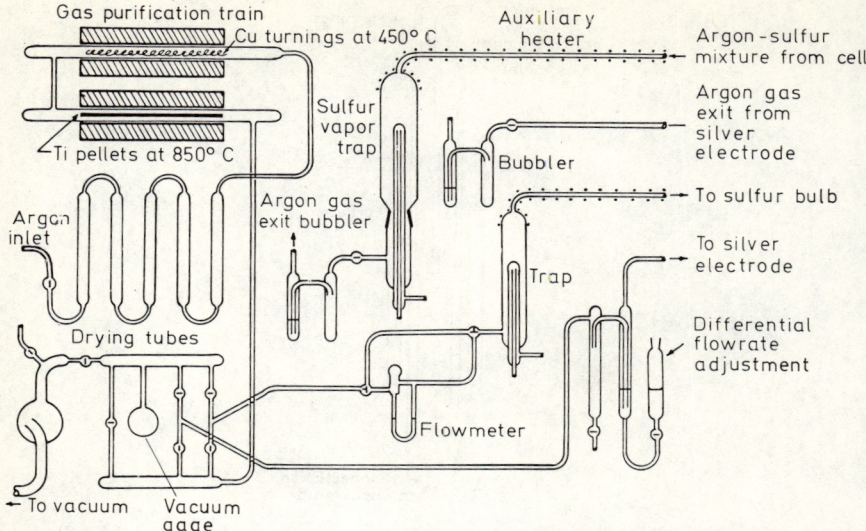

Figure 11. Apparatus for argon gas purification and argon–sulfur gas mixture control [88].

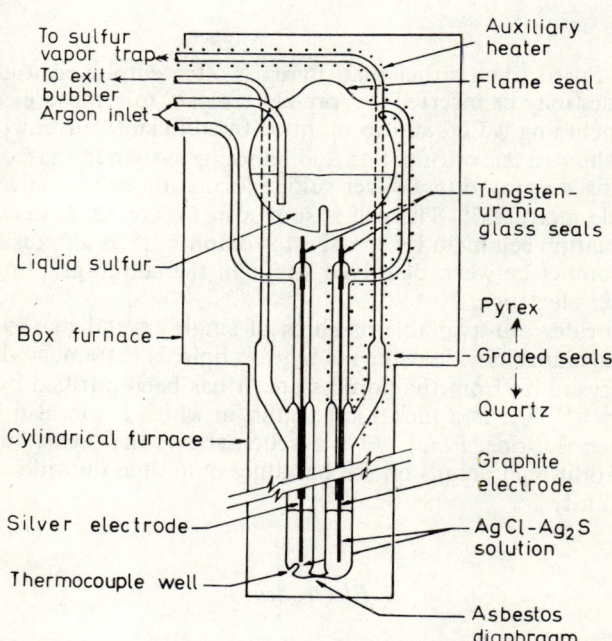

Figure 12. Detail of the sulfur electrode and the formation cell [88].

$$Ag \left| \begin{array}{c} Ag_2S \\ AgCl \end{array} \right| S \left| graphite \right.$$

Attack on the silver electrode by dissolved sulfur is prevented by the asbestos diaphragm.

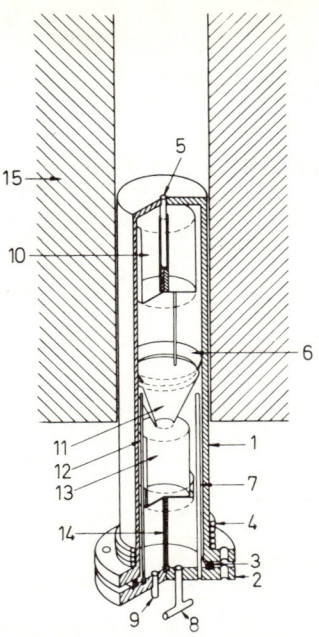

Figure 13. Nickel apparatus for purification of BeF$_2$ by hydrofluorination, sublimation and melting under vacuum[21(b)]. 1 Nickel container vessel. 2 Bottom flange. 3 Teflon O-ring. 4 Cooling coils for water circulation. 5 Thermocouple well which fits into the upper cup (10) sleeve. 6 Permanent guide welded to container wall to prevent material from dropping elsewhere than into the funnel (11) and thence into the lower cup (13). 7 Thermocouple well for measuring temperature near funnel. 8 Gas or vacuum entry. 9 Exit for gases. 10 Upper (loading) cup. 11 Funnel. 12 Support members for funnel and upper cup. 13 Lower (receiving) cup. 14 Support and holder for lower cup (13). 15 Furnace, into which vessel (1) is raised by steps to heat first the cup (10) and then the funnel (11).

the electrode configuration[72, 82, 20, 57], pretreatment of the graphite, and the degree of porosity vary widely. Spectroscopic or other high purity graphite rod has been treated[57] by first boiling in concentrated hydrogen chloride and rod deionized water, drying at 140°C, vacuum desiccation at 500°C, followed by alternate flushing with chlorine and evacuation at 500°C; the rod is then allowed to cool in a chlorine atmosphere. Anodization of graphite for several minutes in a chloride melt in the presence of chlorine gas often improves the stability of the electrode and hastens the approach to equilibrium. During treatment, a large volume of halogen is sorbed by the graphite; this is released when the rod is inserted into a melt. If the electrode compartment is small (e.g. contains no more than about 6 cm^3 of melt), merely dipping the chlorine-treated graphite into the melt is sufficient to insure saturation. Halogen reference electrodes were illustrated in *Figures 2* and *3*. A third, for use in metal–molten salt mixtures, is illustrated in *Figure 14*. The essential features of this electrode are: (1) the outer compartment of Vycor to provide ionic rather than electronic conduction between the melt inside and outside the compartment; (2) the inner quartz tube, surrounding the graphite, into which the chlorine is admitted, and (3) the graphite rod, which is frequently connected

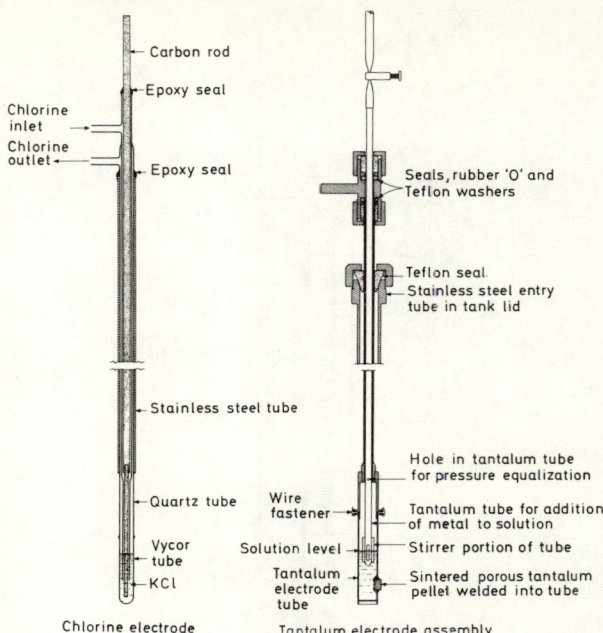

Figure 14. Chlorine electrode half cell and tantalum electrode half cell with Vycor barrier to electronic conduction for use with rare earth metal–rare earth halide mixtures[20].

to a platinum wire for electrical connection, but here extends up through the compartment to provide electrical connection externally. In the latter case, a thermal e.m.f. will be generated if the opposing electrode is not graphite. When a platinum lead is used, the junction should be at the same temperature as the cell to avoid a thermoelectric effect. Although some investigators[70] enclose the platinum wire in a quartz sheath to guard against corrosion by the chlorine gas, the absence of corrosion of unshielded platinum at 450°C has been reported[57]. The electrode compartment, like the cell proper, is preferably long (~ 60 cm) relative to its diameter (~ 25 mm) to minimize heat conduction between the melt and the external electrode leads. For the same reason, the dried halogen is sometimes preheated before admission into the compartment, although the long pathway in the compartment is usually sufficient to allow the gas to reach the temperature of the melt.

Various separators have been used between the chlorine–chloride electrode and the melt when necessary to prevent the halogen from reaching the anode. The junction has been formed by a glass frit[57], two fritted quartz discs[67], ion-selective glasses[69], as in *Figure 2*, ion-selective porcelains[55] a Vycor membrane[20] (*Figure 14*), and a spiral tube of silica containing alumina powder (*Figure 3*)[7]. A sodium ion-selective porcelain half-cell is illustrated in *Figure 15*[55].

The halogen gas is usually allowed to flow slowly through the melt, at a rate of about one bubble per minute, although, as pointed out above, a

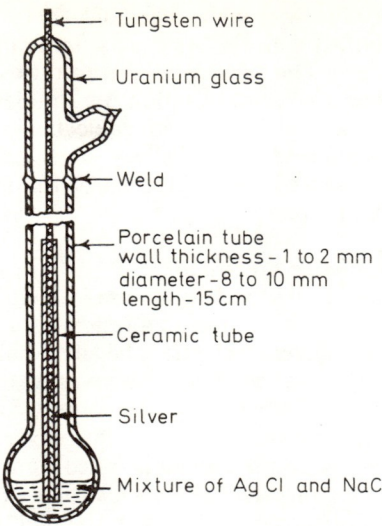

Figure 15. Sodium ion selective porcelain reference half cell with silver–silver chloride electrode[55].

stationary atmosphere of halogen over a saturated melt is often sufficient to produce a stable potential at the graphite electrode. The observation of an overvoltage for chlorine at carbon electrodes in the electrolysis of fused salts, as discussed by Drossbach[27], demonstrates the need for thorough pretreatment of the graphite, possibly including anodization, to insure reversible behavior.

B. Metal Electrodes

(a) *The Silver Electrode*

The silver electrode is widely used in molten salt studies, both directly as a silver ion indicator, Ag/Ag^+[9,32], and as a halide ion indicator, $Ag-AgX-X^-$[17,30,31]. Enclosed in an envelope containing a melt with a fixed concentration of indicator ion, the silver electrode may provide an excellent reference for use in many molten salts with melting points below that of silver metal. At high temperatures, the silver may be utilized as a liquid metal electrode.

Little effort is required for preparation of the silver electrode. High purity wire or foil is available, and, although the surface should be cleaned of sulfide, by abrasion if necessary, the oxide is unstable above 300°C. In low temperature aqueous melts, such as $Ca(NO_3)_2 \cdot 4H_2O$ which melts below 50°C, the oxide is removed by light flaming of the silver wire in a cool blue flame followed by quenching in the melt; this exposes a fresh silver surface to the melt. Either flaming or immersion in a high temperature melt (above 250°C) serves also to anneal the wire or foil. For this purpose small diameter wire (about 20–25 gage) and thin foils are preferred over silver rod. The heat leak likewise is diminished by use of a fine wire with a helix of 8–10 turns at the end to provide sufficient metal in contact with the melt to assure uniform temperature at the melt/electrode interface. The silver ion concentration is generally kept low, almost always less than 10^{-2} mole fraction Ag^+ (moles Ag^+ per mole solvent) for use as a reference half cell. When a silver electrode is employed in a halide melt it is sometimes referred to as a silver–silver halide electrode, but this terminology is best reserved for electrodes of the second kind, where the silver halide is a separate phase.

(b) *The Silver–Silver Halide Electrode*

The silver–silver halide electrode is for use in melts in which silver halides are only slightly soluble, such as the alkali nitrates. Preparation of the silver is similar to that for the silver electrode. The silver halide can be precipitated directly on the silver wire, in the melt to be investigated, by immersion of the metal in a dilute solution of silver ion in the solvent, followed by addition of a halide in sufficient concentration to precipitate silver halide. If the electrode is to be used in a reference compartment isolated from the indicator compartment by an asbestos plug or glass frit, the electrode is best prepared in a separate vessel, and then transferred along with the melt after ageing of the (coagulated) precipitate, to avoid plugging the pores of the asbestos or frit on coagulation of the colloid. Precipitation of silver halide in the asbestos or frit may interfere with ionic conduction between compartments through the salt phase and lead to e.m.f. contributions from exchange reactions involving the crystalline precipitate or the glass envelope. The silver halide electrode is limited in its temperature range by the increasing solubility of the silver halides with temperature, the solubility product increases about four orders of magnitude [from 2×10^{-10} to 5×10^{-6} $(mol/kg)^2$ for AgCl] from an aqueous solution at 25°C to equimolar (Na, K)NO$_3$ at 250°C. The solubility of silver halides in (Na, K)NO$_3$ has been investigated to 350°C[32].

The silver–silver halide electrode, like any saturated electrode, has a large temperature coefficient; but the concentration of silver ion, buffered by the solid halide, is constant at a constant temperature, promoting a stable reproducible e.m.f. The unsaturated silver–silver ion electrode, on the other hand, has a low temperature coefficient but may suffer variations in composition due to slow but possibly continuing impurity reactions at the electrodes. Enclosed in a glass membrane, the silver has a tendency to exchange with the sodium ions in the glass, but this is a very slow process and would only be detectable after prolonged use (months to years). This problem does not arise with a quartz envelope, but the temperature must be high enough to assure conduction through the quartz. In either case, prolonged use of a silver electrode is accompanied by recrystallization of the silver until the wire becomes a mass of crystals. The recrystallization is undoubtedly related to the high exchange current for the silver electrode, but a temperature gradient at the electrode probably enhances the process. Silver is deposited in the cold regions and dissolved from regions of the wire where its activity is increased by a higher temperature or mechanical strain. To combat the possible non-uniformity of the wire surface, silver is sometimes plated from solution on to the surface of a platinum foil. A liquid silver electrode likewise presents a uniform, reproducible strain-free surface.

C. Alloy Electrodes

In a number of cases pure metal electrodes may be unsuited to accurate e.m.f. measurements. Pure metal electrodes may dissolve in their own molten salts, leading to electronic conduction or a chemical short circuit of the cell. The more reductive metals may be too active for convenient containment and use as an electrode. A pure solid metal electrode may have too low an exchange current density for convenient e.m.f. measurements, or may be strained irreproducibly. Lowering the activity of the metal by solution and dilution in an inert liquid metal may be sufficient to provide a reversible stable metal electrode. An alloy electrode may be unsaturated with respect to the electrode metal, in which case a knowledge of the activity may be needed, or it may be saturated with respect to the pure solid or an intermetallic compound, providing a constant activity. Advantages of a saturated alloy over a solid metal electrode include a generally higher exchange current and elimination of the free energy of strain. An unsaturated alloy electrode generally exhibits a smaller temperature coefficient of e.m.f., but is more susceptible to changes in composition (and hence e.m.f.) as a result of minor oxidation–reduction surface reactions due to oxygen, impurities, polarization, etc. The saturated electrode, with a stable composition, however, suffers a larger temperature coefficient as a result of the additional contribution of the heat of fusion to the total free energy change with temperature, only the smaller heat of mixing contributing to the temperature coefficient in the unsaturated alloy. Bismuth has been widely used as a diluent, as have noble metals in alloys with lead, cerium, magnesium, uranium, lithium and sodium in alkali halide melts. Activity coefficients of these metals at infinite dilution in bismuth are reported[29] from measurements of molten chloride galvanic cells.

A dilute solution of lithium in molten bismuth has been used as an alloy electrode[84] for studies of the cell.

$$\text{Be} \left| \begin{array}{c} \text{LiF} \\ \text{BeF}_2 \end{array} \right| \text{Li(Bi)} \tag{K}$$

The cell reaction per Faraday,

$$\tfrac{1}{2}\text{Be} + \text{LiF} \rightarrow \tfrac{1}{2}\text{BeF}_2 + \text{Li(Bi)}$$

corresponds to an e.m.f. of ~ 300 mV at 550°C in a melt containing 67 mole % LiF and with 3×10^{-3} mole fraction of lithium in bismuth (activity of lithium $\simeq 10^{-8}$). Establishment of a stable e.m.f. in a cell containing two different metal electrodes contacting a solution of both metal ions appears to violate a fundamental principle of electrochemistry that the more active metal will reduce the ion of the more noble metal. In this case, however, the activity of the more active metal, lithium, has been reduced below that of beryllium by solution in bismuth. The more active beryllium, then, should reduce lithium into bismuth. That this does not occur is attested by the stable reproducible e.m.f. of cell K, and is considered to be due to the very low solubility of either lithium metal or beryllium metal in the LiF–BeF$_2$ melt[84]. *Figure 16* shows a LiF–BeF$_2$ concentration cell without transference, one-half of which is cell K; the crucible and reference compartment, in contact with both the fluoride melt and the bismuth alloy, are constructed of molybdenum although tantalum would be equally compatible. The beryllium electrodes are $\frac{1}{16}$ in. diameter rods that extend the entire length of the containment vessel and pass through Teflon seals, with electrical leads connected to them in air at ambient temperature. Brazed or welded joints of the beryllium to an inert conductor failed, presumably from attack at the metal/metal interface due to creep of the electrolyte, although a gold–nickel braze to nickel rod had been employed successfully in the absence of bismuth[75]. Purity of the melt is essential to avoid corrosion of the beryllium electrodes.

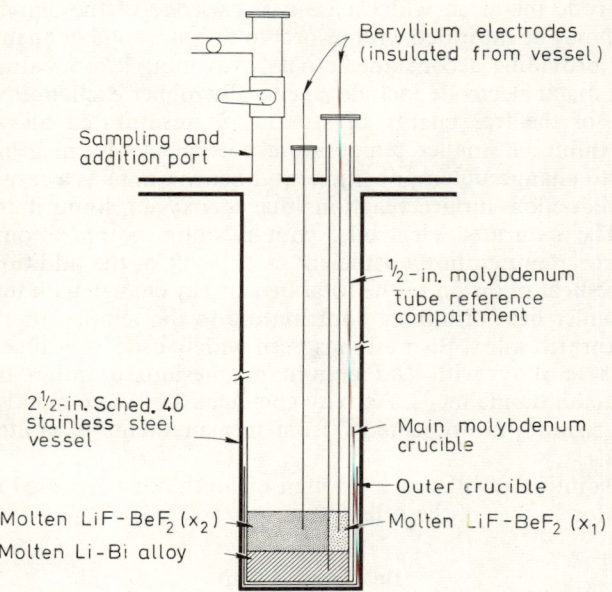

Figure 16. All metal concentration cell with Li–Bi alloy electrode[84].

Preparation of the lithium electrode *in situ* by electrolysis of the melt in contact with molten bismuth was the most expedient way of obtaining an accurate evaluation of the lithium concentration in the alloy.

The potential drop across a calibrated standard series resistor was recorded as a function of time and integrated to determine the quantity of electricity passed and of lithium deposited in the bismuth. The measured cell e.m.f., stable and constant within a minute after electrolysis, was corrected to account for the change in melt composition due to the electrolysis, the maximum correction amounting to 4.5 mV. The magnitude of this final correction was confirmed by addition of sufficient LiF to the melt to reproduce the initial composition. The e.m.f. of the cell K as a function of the concentration of lithium demonstrated Nernst behavior of this alloy electrode, which at higher concentrations has been investigated with an electrolyte mixture LiCl–LiF[36].

IV. Special Applications

1. Phase Diagrams

E.m.f. measurements have been used frequently for studies of phase diagrams of alloys. They have been less extensively used for molten salt systems, but are extremely useful when adaptable to binary or multicomponent systems. When a suitable cell is available, the e.m.f. method may be employed either as an adjunct to thermal analysis, or as an analytical method with isothermal composition variations. It probably is the most accurate method available for locating the relative positions of closely spaced phase field boundaries; it may be used to resolve and characterize invariant points in close proximity to one another, as well as to provide unusually accurate liquidus data for the calculation of chemical potentials[76(c)].

Two binary molten salt systems in which e.m.f. measurements have been employed to characterize complex parts of the phase diagrams are Li_2SO_4–Ag_2SO_4[71] and LiF–BeF_2[76(a)]. In both cases a particularly simple cell was used—a concentration cell with transference employing silver electrodes in the first case, and beryllium electrodes in the second. The latter case illustrates the principles and application of the method.

The e.m.f. of the concentration cell with transference

$$\text{Be} \left| \begin{array}{c|c} \text{LiF, BeF}_2 & \text{LiF, BeF}_2 \\ & 1-x, \ x \\ \text{Reference, I} & \text{II} \end{array} \right| \text{Be} \tag{L}$$

is

$$2FE_L = \int_I^{II} t_{Li}\left(\frac{1+x}{1-x}\right)\left(\frac{\partial \mu_{BeF_2}}{\partial x}\right) dx \tag{30}$$

Changing the composition of the melt on the right isothermally, while the melt composition on the left is held constant, leads to a change of cell e.m.f. Separation of a pure phase of either BeF_2 or LiF leads to a constant composition and constant chemical potentials of the components in the melt, and hence a constant e.m.f. In this case it is not essential to know the activity coefficients and transference numbers, since sharp breaks will result, and a constant e.m.f. The accuracy of phase data obtained in this manner is limited by the accuracy of the temperature and composition measurements, and of the phase equilibration. Stirring is desirable to accelerate equilibrium. Separation of solid solutions, as in the Li_2SO_4–Ag_2SO_4 system, is observed as a discontinuity of slope rather than as a constant e.m.f.

Alternatively, the e.m.f. of the cell may be monitored as a function of temperature. A discontinuity of slope of the e.m.f./temperature relation is observed on separation of a pure phase, unless that phase happens to be a compound having the composition of the melt.

The cell e.m.f. may be written

$$\frac{dE}{dT} = \left(\frac{\partial E}{\partial T}\right)_x + \left(\frac{\partial E}{\partial x}\right)_T \frac{dx}{dT} \tag{31}$$

The last term vanishes for a homogeneous melt, but results in an abrupt change of slope dependent on the slope of the phase boundary when appearance of the phase changes the composition of the remaining melt. This method is capable of much higher accuracy than kinetic methods such as thermal analysis by means of cooling curves. The e.m.f. measurements are virtually equilibrium measurements, taken at a series of temperatures. Phase equilibrium must be ascertained by changing the temperature cyclically and reproducing the measured e.m.f. Supercooling does not interfere with the accuracy of the measurements provided the equilibrium precipitating phase does nucleate, since the phase boundary may be determined from the e.m.f. change on disappearance of the solid phase with increasing temperature. The sharpness of the slope change depends in part on the composition difference between the reference compartment and the bulk compartment. The sensitivity can be shown to increase as the composition difference decreases. At a temperature $T + \delta$ just above the liquidus temperature, the slope of the e.m.f./temperature plot is

$$\left[\frac{dE}{dT}\right]_{\substack{\text{homogeneous} \\ (T+\delta)}} = \left(\frac{\partial E}{\partial T}\right)_x \bigg|_{(T+\delta)} \tag{32a}$$

At a temperature $T - \delta$ just below the liquidus,

$$\left[\frac{dE}{dT}\right]_{\substack{\text{2-phase} \\ (T-\delta)}} = \left(\frac{\partial E}{\partial T}\right)_x \bigg|_{(T-\delta)} + \frac{RT}{2F} t_{\text{Li}} \left(\frac{1+x}{1-x}\right) \left(\frac{\partial \mu_{\text{BeF}_2}}{\partial x}\right)_T \left[\frac{dx}{dT}\right]_{\text{phase boundary}} \tag{32b}$$

The sensitivity depends on the relative magnitudes of the two terms on the RHS of equation 32b. The first term on the RHS can be made almost arbitrarily small by maintaining the reference composition close to that of the composition in the bulk compartment, but the reference composition must of course remain liquid while solid phase precipitates in the bulk compartment.

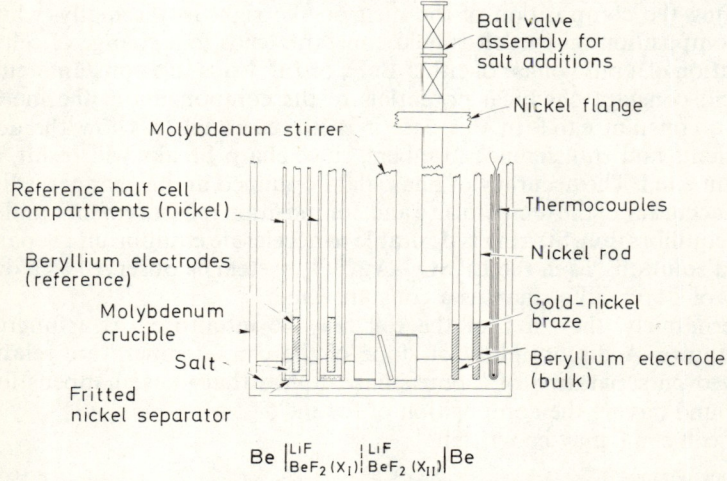

Figure 17. All metal concentration cell for e.m.f. measurement with transference with Be electrodes, in fritted nickel compartments and in bulk melt (LiF–BeF$_2$ mixtures)[75].

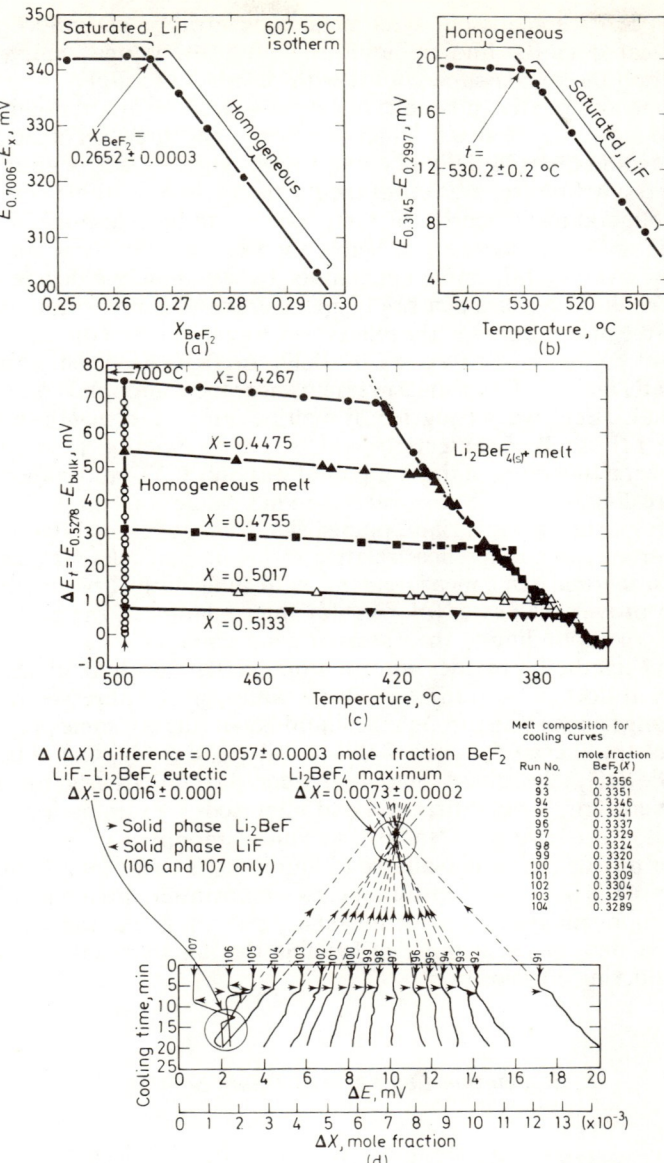

Figure 18. Phase data from e.m.f. measurements of concentration cell with transference[75, 76(a)]. Determination of liquidus composition and temperature from slope discontinuities: (a) isotherm at 607.5 ± 0.1 °C, e.m.f. versus total salt composition; (b) e.m.f. versus temperature, constant total salt composition 0.2997 mole fraction BeF_2; (c) E_t versus temperature for LiF–BeF_2 system near the Li_2BeF_4 liquidus. Reference compositions: □ 0.05278 mole fraction BeF_2. Total bulk compositions: ▲ 0.5133, △ 0.5017, ■ 0.4755 and ● 0.4267 mole fraction BeF_2. The points ○ are LiF additions at 498 °C; (d) cooling curves in region between composition Li_2BeF_4 and the LiF primary phase field. (Total temperature change ∼0.1°C).

943

Figure 17 is a schematic of the all metal concentration cell, and *Figure 18* shows typical data for finding liquidus compositions and temperatures, extending into the region supercooled with respect to Li_2BeF_4.

The nature and relative positions of closely spaced invariant points can be determined accurately from e.m.f. measurements. The direction of the composition changes in melts initially of composition on either side of an invariant point that result from separation of a eutectic mixture, a stable compound, or a compound above its peritectic temperature can be determined from the direction of the e.m.f. change just below the liquidus. For invariant points differing by several tenths mole per cent, or by less than one degree in temperature, absolute measurements of temperatures and composition may lead to appreciable uncertainties in the relative positions and even on the nature of the invariant points. In the system $LiF-BeF_2$, Li_2BeF_4 has been reported as incongruently melting. E.m.f. measurements near the compound composition showed that Li_2BeF_4 was a congruently melting compound, that the composition of the $LiF-Li_2BeF_4$ eutectic was 0.0053 mole fraction lower in BeF_2 and that the melting point of the compound was only 0.2°C above the eutectic temperature. *Figure 18*(d) demonstrates the e.m.f. changes at compositions and temperatures near the invariant points. The composition differences are calculated from e.m.f. differences relative to a common reference electrode, and from isothermal e.m.f. measurements with sufficiently large composition changes to provide values of $(\partial E/\partial x)_T$ for interpolation of the composition differences corresponding to the observed e.m.f. changes.

The e.m.f. method provides information on the location of the phase boundaries. It does *not* identify the phases, although the observed direction of e.m.f. change on passing through the liquidus can rule out some possibilities by virtue of the thermodynamic stability criteria, e.g. $(\partial \mu_i/\partial x_i) > 0$. In the system $LiF-BeF_2$, the phases had been identified previously by thermal gradient quenching, x-ray diffraction and microscopic examination.

The method is of course extendable to multicomponent systems if formation cells or mixing cells are available to monitor the chemical potential of a component of the melt. Concentration cells with transference are less reliable for multicomponent systems. Only for binary systems is the transference cell e.m.f. independent of the manner of forming the junction between the two liquids of differing composition.

2. *Association Equilibria of Dilute Solutes*

A. Measurements with Silver or with Silver–Silver Halide Electrodes

The solvent properties of molten salts vary widely and differ from those of other solvents, providing impetus to the study of solution chemistry of dilute solutes in molten salt solvents. Concentration cells with electrodes that are reversible to dilute solute ions[13,14] constitute an important source of information, both in anhydrous molten salts and in those containing water (hydrous melts). In these cells, the effect of the liquid junction is minimized by maintaining the solute ion to solvent ion ratio at about 10^{-3}, so that the current is carried virtually entirely by solvent ions. For the concentration cell

$$\text{Ag} \left| \begin{array}{c} \text{KNO}_3 \\ (\text{AgNO}_3) \end{array} \right| \left| \begin{array}{c} \text{KNO}_3 \\ (\text{AgNO}_3) \\ (\text{KCl}) \end{array} \right| \text{Ag} \qquad \qquad (\text{M})$$

$$ \text{I} \qquad \text{II}$$

where the dilute components are shown in parenthesis, the free energy change per Faraday is given by

$$FE_M = -\Delta G_M = \int_I^{II} (1 - t_{Ag^+}) d\mu_{AgNO_3} + (t_{Cl^-}) d\mu_{KCl} - (t_{K^+} + t_{Cl^-}) d\mu_{KNO_3} \quad (33)$$

The t are Hittorf transference numbers of the ions relative to nitrate ion, and the μ are chemical potentials of the components. (Equation 33 would include a term in the transference number of nitrate ion if the ion mobilities were relative to a stationary reference such as the cell wall, but the conclusions are unchanged.) At the solute concentrations considered, $t_{Ag^+} \simeq t_{Cl^-} \simeq 10^{-3}$, and therefore, $t_{K^+} \simeq 1$; however, the activity of KNO_3 is virtually unity in both half cells, hence $d\mu_{KNO_3} \simeq 0$; with neglect of the terms containing the multiplicative factors t_{Ag^+}, t_{Cl^-} and $d\mu_{KNO_3}$, the resulting equation

$$-FE_M = \Delta G \simeq \mu^I_{AgNO_3} - \mu^{II}_{AgNO_3} = -RT \ln \{a^{II}_{AgNO_3}/a^I_{AgNO_3}\} \quad (34)$$

expresses the e.m.f. expected for this concentration cell, in dilute solutions, with an *LJE* error less than 0.5 mV[9]. Activities of dilute components such as $AgNO_3$ in the absence of Cl^- (or of KCl in the absence of Ag^+) have been shown to obey the Nernst law[9,13], so that equation 34 becomes, in the presence of Cl^-,

$$E_M = \frac{RT}{F} \ln \frac{R^{II}_{AgNO_3}}{R^I_{AgNO_3}} + \frac{RT}{F} \ln \gamma_{AgNO_3^{II}} \quad (35)$$

where the activity coefficient may be defined by [9,13,14]

$$a_{AgNO_3} = R_{AgNO_3} \gamma_{AgNO_3} \quad (36)$$

and is unity for dilute solutions of $AgNO_3$ in the absence of Cl^-; R_{AgNO_3} is the mole ratio of $AgNO_3$ in the melt, moles of $AgNO_3$ per mole of solvent salt†.

The composition dependence of the activity coefficients has been interpreted in terms of the formation of associated species of the dilute solutes

---

† Activities and concentration scales for molten salt solutions are discussed in detail in refs. 8, 13 and 14 (footnote continued overleaf).

in the molten salt solvents[13]. The association equilibria have then been interpreted in terms of a quasi-lattice model of molten salt mixtures[14]. *Figure 19* illustrates a cell employed for such studies. A Pyrex electrolytic

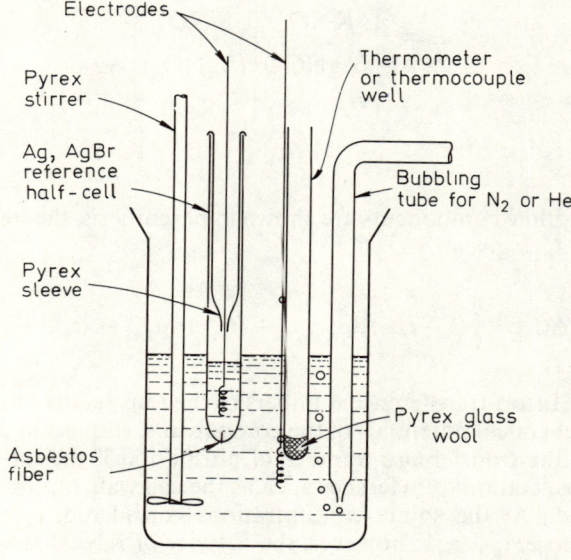

Figure 19. Pyrex e.m.f. cell for association equilibria of dilute solutes in molten nitrates[13].

For the cells with electrodes of the second kind, e.g.

$$\text{Ag, AgBr} \left| \begin{array}{c} \text{LiNO}_3 \\ \text{LiBr} \\ \\ \text{I} \end{array} \right| \begin{array}{c} \text{LiNO}_3 \\ \text{LiBr} \\ \text{Cd(NO}_3)_2 \\ \text{II} \end{array} \left| \text{AgBr, Ag} \right. \quad (M')$$

the electrode and transport reactions are:

$$\text{Ag} + \text{Br}_\text{I}^- \rightarrow \text{AgBr} + e$$
$$e + \text{AgBr} \rightarrow \text{Ag} + \text{Br}_\text{II}^-$$
$$t_{\text{Br}}\text{Br}_\text{II}^- \rightarrow t_{\text{Br}}\text{Br}_\text{I}^-$$
$$\tfrac{1}{2}t_{\text{Cd}}\text{Cd}_\text{I}^{2+} \rightarrow \tfrac{1}{2}t_{\text{Cd}}\text{Cd}_\text{II}^{2+}$$
$$t_{\text{Li}}\text{Li}_\text{I}^+ \rightarrow t_{\text{Li}}\text{Li}_\text{II}^+$$

The cell e.m.f. is given by

$$-FE_{M'} = \Delta G_{M'} = \int_\text{I}^\text{II} (1 - t_{\text{Br}})\, d\mu_{\text{LiBr}} - \tfrac{1}{2}t_{\text{Cd}}\, d\mu_{\text{Cd(NO}_3)_2} + t_{\text{Cd}}\, d\mu_{\text{LiNO}_3} \quad (33')$$

$$-E_{M'} \simeq \frac{RT}{F} \ln \frac{R^\text{II}_{\text{LiBr}}}{R^\text{I}_{\text{LiBr}}} + \frac{RT}{F} \ln \gamma^\text{II}_{\text{LiBr}} \quad (35')$$

In the absence of $\text{Cd(NO}_3)_2$, $\gamma_{\text{LiBr}} = 1$, and the Nernst equation is followed for LiBr.

or tall form beaker (250 ml to 1000 ml) is an adequate cell for the alkali nitrates and for mixtures of these with alkaline earth nitrates except (perhaps) lithium nitrate and mixtures containing high concentrations of lithium nitrate. Lithium ion exchanges with sodium ion in the glass, causing etching of the glass surface with eventual collapse of the glass structure. Any commercial pot furnace that is rated to 500°C (500–700 watts) and will accommodate the cell is suitable. An aluminum block, machined to accept the 250 ml beaker and inserted between the furnace walls and the cell, serves the dual purpose of reducing temperature variations and providing a uniformly heated surface for the cell. It is often necessary to earth the aluminum since small commercial furnace elements may not be non-inductively wound. Alternatively, furnaces have been fabricated[13] by winding (non-inductively) Nichrome wire (about 4 meters of approximately 0.5 mm diameter Nichrome wire) around an asbestos core, the wire held in place with moistened asbestos paper, water glass or refractory cement. Here too, an earthed aluminium insert is of considerable help in maintaining a constant temperature. Temperature may be controlled manually with a Variac or with a commercial temperature controller and a control thermocouple.

Electrodes for cell (M) consist of silver wires (approximately 0.5 mm diameter) dipping directly into the melt. Half cell I, the reference half cell, is separated from the main compartment, II, by an asbestos fiber sealed into the tip of a 9–11 mm diameter, 20–25 cm long Pyrex tube drawn to a point. An asbestos fiber, of about 0.3 mm diameter and 0.5–1.0 cm length, is inserted into the capillary tip, allowing half the fiber to protrude, then heated in the gas–air flame of a Meker burner until the molten glass seals around the glowing fiber. If a gas–oxygen flame is used, it is difficult to prevent fusion of the fiber. The portion of the silver wire that dips into the melt, a 1–2 mm diameter and 0.5 cm long helix, is prevented from contacting the compartment walls by threading the wire through a Pyrex capillary tube that fits closely into the compartment and extends to about 0.5 cm above the melt. The capillary tube also reduces convection above the melt and in the cooler regions of the reference compartment.

The temperature reversibility of the cell is ascertained by monitoring the return of the e.m.f. to its original value after a temperature cycle, usually, 10–20°C. Irreversibility may be caused by impurities trapped in the silver halogenide precipitate at a low temperature, or by removal of strain in the metal electrode after annealing at the higher temperature of a cycle.

Electrodes have been prepared by deposition of a silver halide on to a platinum substrate by electrolysis in aqueous solution. These were preconditioned before use for one hour in a molten nitrate bath containing the dilute ions involved in the cell reaction[7,8]. The silver halide coated electrode often provides more stable e.m.f. readings and Nernst response over a wider range of concentration than a bare silver wire.

The use of a silver halide electrode as both a silver ion and halogenide ion indicator depends upon a low solubility of the silver halogenide if competitive interference with dilute ion association is to be avoided. Solubilities of some silver halides in nitrate melts are given in *Table 2*.

The solubility product of the silver halides increases more than an order of magnitude for a 40°C rise in temperature in $LiNO_3$–KNO_3 eutectic and for a 100°C rise in equimolar $NaNO_3$–KNO_3. This increase in solubility restricts the temperature range in which the electrode can be used for investigation of association equilibria. A halide concentration of 10^{-3} moles/kg solvent is usual for a dilute solute. If the solubility product is, e.g. 10^{-7}, the silver ion concentration, $\sim 10^{-4}$ moles/kg solvent, will interfere with associating metal ions at a concentration less than 10^{-3} mole/kg

Table 2. Estimates of solubility product $(mol/kg)^2$ for the reaction $AgX(s) = Ag^+ + X^-$.

Solvent/Halogenide	AgCl	AgBr	AgI
$LiNO_3$–KNO_3 (a) 0.43–0.57 (150°C)	4.03×10^{-8}	1.90×10^{-10}	1.02×10^{-13}
$LiNO_3$–KNO_3 (b) 0.43–0.57 (190°C)	4.65×10^{-7}	4.40×10^{-9}	
$NaNO_3$–KNO_3 (c) equimolar (250°C)	5.25×10^{-6}	7.61×10^{-8}	1.86×10^{-10}
$NaNO_3$–KNO_3 (c) equimolar (350°C)	3.44×10^{-5}	6.79×10^{-7}	4.0×10^{-9}
KNO_3 (d) (350°C)	8.75×10^{-7}		
KNO_3 (d) (436°C)	5.46×10^{-6}		

(a) Ref. 12, 79.
(b) Ref. 64.
(c) Ref. 32.
(d) Ref. 10. Activity product calculated using the activities of the solute components obtained by e.m.f. measurements.

solvent unless the studied equilibrium is stronger at least by an order of magnitude than the silver halide association. Since extraction of the thermodynamic association constants for dilute solute species depends upon extrapolations to infinite dilution of these species[14, 15], it is important to obtain reliable data at low solute concentrations. It is helpful, then, to have information about the solubility of the intermediate solid phase of an electrode of the second kind, especially, the increase in solubility with increasing temperature.

B. Electrodes of the Third Kind

The electrode Pd, PdO, CdO, involving equilibrium among melt and three solid phases has been used to study Cd^{2+}–Br^- association in alkali nitrate melts at temperatures where the solubility of AgBr precludes use of an Ag, AgBr electrode[16, 48].

In the absence of a significant junction potential, minimized by maintaining solute concentrations below $\sim 10^{-3}$ mol/kg solvent, the e.m.f. of the cell

$$\text{Ag} \left| \begin{array}{c} MNO_3 \\ AgNO_3 \\ AgBr \end{array} \right| \left. \begin{array}{c} MNO_3, MBr \\ Cd(NO_3)_2 \\ PdO, CdO \end{array} \right| \text{Pd} \qquad (N)$$

I   II
Reference

where M is K or Na or a mixture of these two, is given by

$$E_N = \text{constant} + (RT/2F) \ln a^{II}_{Cd(NO_3)_2} \tag{37}$$

The palladium ion concentration, which controls the potential at the palladium electrode, is related to the cadmium ion concentration through the two oxide equilibria:

$$PdO \rightleftharpoons Pd^{2+} + O^{2-} \qquad K_1 = R_{Pd^{2+}} R_{O^{2-}}$$
$$CdO \rightleftharpoons Cd^{2+} + O^{2-} \qquad K_2 = R_{Cd^{2+}} R_{O^{2-}}$$

so that

$$R_{Pd^{2+}} = (K_1/K_2) R_{Cd^{2+}}$$

where the R_i are the concentrations (in dilute solutions, often conveniently expressed as mole ratios, moles solute per mole solvent) and the K_i are solubility product constants. In the absence of bromide ion, the e.m.f. of cell (M) was found[16] to follow the Nernst limiting law

$$E_N(R_{Br} = 0) = \text{constant} + (\mathbf{R}T/2F) \ln \mathbf{R}^{II}_{Cd(NO_3)_2}$$

in the solvents $NaNO_3$ at 331°C and KNO_3 at 358°C. In equimolar $NaNO_3$–KNO_3 at 258°C, however, although plots of e.m.f. versus the logarithm of cadmium nitrate concentration were straight lines, the slopes were in excess of the expected value by 10–20 per cent[16]. This behavior has been attributed to a mixed potential resulting from the closeness of the oxygen–oxide ion and palladium–palladium ion electrode potentials[16,48]. Exclusion of oxygen was considered undesirable as the oxygen activity was thought to separate the two potentials. Although the presence of even a small irreversible e.m.f. at an electrode is sufficient to question the validity of the measured e.m.f., the activity coefficients and association constants calculated using the 'calibrated' or experimentally determined slope in place of the 'Nernst' slope were in agreement with previous measurements in the mixed solvent using an indicator electrode of the second kind, a silver–silver halide electrode. This suggested that the oxygen potential remained reasonably constant over a wide range of composition. Proper response of the electrode can be expected only if PdO and CdO are both sparingly soluble, with PdO ten times less soluble than CdO (estimated at $\sim 10^{-5}$ moles/kg)[48] to prevent interference by palladium complex formation in the presence of halide ion. In fact, the PdO is so insoluble that it was necessary to equilibrate the solid with a melt for several days before use[48] or, alternatively, to pulverize the oxide by grinding to a fine powder on a Wig-L-Bug* to increase the oxide surface–melt contact[16]. The use of vigorous mechanical stirring is necessary here to facilitate attainment of equilibrium involving two conjugate solid phases.

3. Miscellaneous

One of the sources of interest in molten salt thermodynamics, to which e.m.f. measurements have been most directly applicable, is the development

---

* Crescent Dental Manufacturing Company, Chicago, Illinois.

of molten salt batteries and fuel cells[86]. E.m.f. measurements at temperatures above 1000°C, particularly in silicates and slags, are of considerable technological interest. Although the experimental difficulties are greatly increased, activities of components in molten slags reported within 20 per cent are valuable thermodynamic data[74]. As discussed by Laity[59], each molten salt has its own electromotive force series for dilute solute ions. E.m.f. series have been reported for single salt systems of halides[41] and in a number of solvents[73] including acetates[62], chlorides[37] and lithium fluoroberyllate[2]. Additional sources of information on molten salt systems to which e.m.f. measurements have contributed include the compilations of Lumsden[61], the *JANAF Tables*[50], the NSRDS–NBS[68] and of Bruneaux et al.[22].

Molten salt e.m.f. studies have contributed significantly to theoretical investigations of electrolytes. Molten salt theories have been reviewed extensively by Blander[8], Førland[33], Lumsden[61] and Braunstein[14]. In this chapter we have therefore avoided extensive theoretical discussion that might associate the measurements with particular theoretical formulations, but have attempted, rather, to present the means for relating the measured electrical quantities to chemical potentials.

V. Acknowledgement

We are pleased to acknowledge valuable comments and suggestions of Drs C. E. Vallet, C. E. Bamberger and A. L. Bacarella.

VI. References

[1] Alabyshev, A. F., M. F. Lantratov and A. G. Morachevskii. *Reference Electrodes for Fused Salts*, transl., The Sigma Press: Washington, DC (1965).
[2] Baes, C. F. 'The chemistry and thermodynamics of molten salt reactor fuels' in *Nuclear Metallurgy*. Vol. XV, *Symposium on Reprocessing of Nuclear Fuels* edited by P. Chiotti, USAEC Division of Technical Information, Conf. 690801, p 617 (1969).
[3] Bamberger, C. E., in *Advances in Molten Salt Chemistry*, Vol III, edited by J. Braunstein, G. Mamantov and G. P. Smith. Plenum: New York (to be published).
[4] (a) Bamberger, C. E. and C. F. Baes. *J. Amer. Ceram. Soc.* **55**, 564 (1972).
(b) Bamberger, C. E., C. F. Baes and J. P. Young. *J. Inorg. Nucl. Chem.* **30**, 1979 (1968).
[5] Bartholomew, R. F. *J. Phys. Chem.* **70**, 3442 (1966).
[6] Bates, R. G. *Electrometric pH Determinations*, Chap. 9. Wiley: New York (1954).
[7] Behl, W. K. and J. J. Egan. *J. Phys. Chem.* **71**, 1764 (1967).
[8] (a) Blander, M. 'Thermodynamic properties of molten salt solutions', in *Molten Salt Chemistry*, p. 127, edited by M. Blander. Interscience: New York (1964).
(b) Blander, M. 'Some fundamental concepts in the chemistry of molten salts', in *Molten Salts*, p 1, edited by G. Mamantov. Dekker: New York (1969).
[9] Blander, M., F. F. Blankenship and R. F. Newton. *J. Phys. Chem.* **63**, 1259 (1959).
[10] Blander, M., J. Braunstein and M. D. Silverman. *J. Amer. Chem. Soc.* **85**, 895 (1963).
[11] Bockris, J. O'M., G. J. Hills, D. Inman and L. Young. *J. Sci. Instrum.* **33**, 438 (1956).
[12] Bombi, G. G., M. Fiorani and G. A. Mazzocchin. *J. Electroanal. Chem.* **9**, 457 (1965).
[13] Braunstein, J. *J. Chem. Educ.* **44**, 223 (1967).
[14] Braunstein, J. 'Thermodynamics of molten salts and concentrated aqueous electrolytes' in *Ionic Interactions*, Vol. I, p 179. Edited by S. Petrucci. Academic Press: New York (1971).
[15] Braunstein, J., M. Blander and R. M. Lindgren. *J. Amer. Chem. Soc.* **84**, 1529 (1962).
[16] Braunstein, H., J. Braunstein and D. Inman. *J. Phys. Chem.* **70**, 2726 (1966).
[17] Braunstein, J. and R. M. Lindgren. *J. Amer. Chem. Soc.* **84**, 1534 (1962).
[18] Bredig, M. A. Ref. 8(a), p 367.
[19] Bronstein, H. R. *J. Electrochem. Soc.* **112**, 1032 (1965).
[20] Bronstein, H. R. *J. Phys. Chem.* **73**, 1320 (1969).

[21] (a) Bronstein, H. R. and D. L. Manning. *J. Electrochem. Soc.* **119**, 125 (1972).
[21] (b) Vallet, C. E., H. R. Bronstein and J. Braunstein, to be published.
[22] Bruneaux, M., J. Hladik, Y. Pointud and G. Morand. *Electrochim. Acta*, **13**, 1591 (1968).
[23] Caton, R. D., Jr and C. R. Wolfe. *Analyt. Chem.* **43**, 660 (1971).
[24] Corbett, J. 'The solution of metals in their molten salts', in *Fused Salts*, p 341. Edited by B. Sundheim. McGraw-Hill: New York (1964).
[25] Delimarski, Y. K. and B. F. Markov. *Electrochemistry of Fused Salts*, transl., The Sigma Press: Washington, DC (1961).
[26] Dijkhuis, C. G. M. and J. A. A. Ketelaar. *Electrochim. Acta*, **11**, 1607 (1966).
[27] Drossbach, P. *J. Electrochem. Soc.* **103**, 700 (1956).
[28] Egan, J. J. *J. Electrochem. Soc.* **112**, 79 C (1965);
Egan, J. J. Brookhaven National Laboratory, report *BNL 6589*.
[29] Egan, J. J. and R. H. Wiswall. *Nucleonics*, **15**, 104 (1957).
[30] Flengas, S. N. and T. R. Ingraham. *Canad. J. Chem.* **35**, 1139 (1957).
[31] Flengas, S. N. and T. R. Ingraham. *J. Electrochem. Soc.* **106**, 714 (1959).
[32] Flengas, S. N. and E. Rideal. *Proc. Roy. Soc. (London) A*, **233**, 443 (1956).
[33] Førland, T. Ref. 24, p 63.
[34] Førland, T. and T. Østvold. *Acta Chem. Scand.* **20**, 2086 (1966).
[35] Førland, T., L. U. Thulin and T. Østvold. *J. Chem. Educ.* **48**, 741 (1971).
[36] Foster, M. S., S. E. Wood and C. E. Crouthamel. *Inorg. Chem.* **3**, 1428 (1964).
[37] Gaur, H. C. and H. L. Jindal. *Electrochim. Acta*, **13**, 835 (1968).
[38] Graves, A. D., G. J. Hills and D. Inman. 'Electrode processes in molten salts' in *Advances in Electrochemistry and Electrochemical Engineering*, Vol IV, pp 117–183. Edited by P. Delahay and C. W. Tobias, Interscience: New York (1966).
[39] Gray, D. and A. Cahill. *J. Electrochem. Soc.* **116**, 443 (1969).
[40] (a) Guggenheim, E. A. *J. Phys. Chem.* **33**, 822 (1929).
(b) Guggenheim, E. A. *J. Phys. Chem.* **34**, 1540 (1930).
[41] (a) Hamer, W. J., M. S. Malmberg and B. Rubin. *J. Electrochem. Soc.* **103**, 8 (1956).
(b) Hamer, W. J., M. S. Malmberg and B. Rubin. *J. Electrochem. Soc.* **112**, 750 (1965).
[42] Harris, F. K. *Electrical Measurements*, Chap. 6. Wiley: New York (1952).
[43] Hartshorne, L. and A. G. McNish. 'Energy measurement and standardization' in *Experimental Thermodynamics*, Vol. I, p 59. Edited by J. P. McCullough and D. W. Scott, Butterworths: London (1968).
[44] Hill, D. L., G. J. Hills, L. Young and J. O'M. Bockris. *J. Electroanal. Chem.* **1**, 83 (1959).
[45] Hill, D. L., J. Perano and R. A. Osteryoung. *J. Electrochem. Soc.* **107**, 698 (1960).
[46] Hitch, B. F. and C. F. Baes. *Inorg. Chem.* **8**, 201 (1969).
[47] Inman, D. *J. Sci. Instrum.* **39**, 391 (1962).
[48] Inman, D. *Electrochim. Acta*, **10**, 11 (1965).
[49] Ives, D. J. G. and Janz, G. J. *Reference Electrodes*, Chap. 1, p 57. Academic Press: New York (1961).
[50] *JANAF Thermochemical Tables*, edited by D. R. Stull, PB-168370, Clearing House for Federal Scientific and Technical Information: Springfield, Va (August 1965).
[51] Janz, G. J. *Molten Salts Handbook*, p 357. Academic Press: New York (1960).
[52] Jenkins, H. W., G. Mamantov and D. L. Manning. *J. Electroanal. Chem.* **19**, 385 (1968).
[53] Kingery, W. D. *Property Measurements at High Temperatures*, Chap. 3, pp 59–87. Wiley: New York (1959).
[54] Kolesov, V. P., M. L. McGlashan, J. Rouquerol, S. Seki, C. E. Vanderzee and E. F. Westrum, *J. Chem. Engng Data*, **18**, 3 (1973).
[55] Labrie, R. J. and V. A. Lamb. *J. Electrochem. Soc.* **106**, 895 (1959).
[56] (a) Lachman, J. C. and J. A. McGurty. In *Temperature, Its Measurements and Control*, Vol. III, Pt 2, p 177. Edited by C. M. Herzfeld. Reinhold: New York (1962).
(b) Landolt–Bornstein, *Zahlenwerte und Funktione*, Band. II, Teil 6, p 984; Springer: Berlin (1959).
(c) *Handbook of Chemistry and Physics*, 36th ed., p 2380. Chemical Rubber Publishing Co.: Cleveland (1954).
(d) Wagner, H. J. and J. C. Stewart, Ref. 56(a), Vol. III, Pt 1, p 245.
[57] Laitinen, H. A. and J. W. Pankey. *J. Amer. Chem. Soc.* **81**, 1053 (1959).
[58] Laitinen, H. A., R. P. Tischer and D. K. Roe. *J. Electrochem. Soc.* **107**, 546 (1960).
[59] Laity, R. W. 'Electrodes in fused salt systems' in *Reference Electrodes*, edited by D. J. G. Ives and G. J. Janz. Academic Press: New York (1961).

[60] Lorenz, P. K. and G. J. Janz. *Electrochim. Acta*, **15**, 2001 (1970).
[61] Lumsden, J. *Thermodynamics of Molten Salt Mixtures*. Academic Press: New York (1966).
[62] Marassi, R., V. Bartocci, P. Cescon and M. Fiorani. *J. Electroanal. Chem.* **22**, 215 (1969).
[63] McGlashan, M. L. *Pure Appl. Chem.* **21**, 1 (1970).
[64] Mendez, J., I. J. Gal and J. W. Irvine Jr. *Inorg. Chem.* **7**, 1329 (1968).
[65] (a) Miller, D. G. *Chem. Rev.* **60**, 15 (1960).
(b) Pikal, M. J. and D. G. Miller. *J. Phys. Chem.* **74**, 1337 (1970).
[66] Motzfeld, K. 'Means of attaining and controlling temperature' in *Physicochemical Measurements at High Temperatures*, edited by J. O'M. Bockris, J. L. White and J. D. Mackenzie. Chap. 3, pp 47–86. Butterworths: London (1959).
[67] Neil, D. E., J. M. Clark and R. H. Wiswall, *J. Chem. Engng Data*, **10**, 21 (1965).
[68] NSRDS-NBS 28, *Molten Salts* Vol. II, Section 1, 'Electrochemistry of molten salts'. Superintendent of Documents, US Government Printing Office: Washington, DC 20402 (1969).
[69] Østvold, T. *Acta Chem. Scand.* **20**, 2187 (1966).
[70] (a) Østvold, T. 'On the application of glass membranes as alkali electrodes at elevated temperatures', *Thesis*, Institutt for Fysikalsk Kjemi, Norges Tekniske Høgskole: Trondheim (1966).
(b) Frøyland, K., T. Førland, N. H. Lundberg and T. Østvold. 'Free enthalpy measurements ments of fused alkali halides', in *Selected Topics in High Temperature Chemistry*, edited by T. Førland, K. Grjotheim, K. Motzfeld and S. Urnes. Universitetsforlaget: Oslo (1966).
[71] (a) Øye, H. *Acta Chem. Scand.* **18**, 361 (1964).
(b) Øye, H. *Acta Chem. Scand.* **21**, 111 (1967).
[72] Panish, M. B., F. F. Blankenship, W. R. Grimes and R. F. Newton. *J. Phys. Chem.* **62**, 1325 (1958).
[73] Plambeck, J. A. *J. Chem. Engng Data*, **12**, 77 (1967).
[74] (a) Richardson, F. D. 'Thermodynamic aspects of molten slags' in *The Physical Chemistry of Melts*, pp 75, 106. Nuffield Research Group in Extraction Metallurgy, Institute of Mining and Metallurgy: London (1953).
(b) Jeffes, J. H. E. and R. Sridhar. 'Activities in PbO-containing melts' in *Electromotive Force Measurements in High Temperature Systems*, p 199. American Elsevier: New York (1968).
[75] Romberger, K. A. and J. Braunstein. *Inorg. Chem.* **9**, 1273 (1970).
[76] (a) Romberger, K. A., J. Braunstein and R. E. Thoma. *J. Phys. Chem.* **76**, 1154 (1972).
(b) Romberger, K. A. and J. Braunstein. Molten Salt Reactor Program. *Semiannu. Prog. Rep. No. ORNL-4449*, p 138 (February 1970).
(c) Braunstein, J., K. A. Romberger and R. Ezell. *J. Phys. Chem.* **74**, 4383 (1970).
[77] Rossotti, H. *Chemical Applications of Potentiometry*, Chap. 14, p 176, Van Nostrand: London (1969).
[78] Sacchetto, G. A., C. G. Bombi and C. Macca. *J. Electroanal. Chem.* **36**, 47 (1972).
[79] Sacchetto, G. A., G. A. Mazzocchin and G. G. Bombi, *J. Electroanal. Chem.* **20**, 435 (1969).
[80] Scatchard, G. *J. Amer. Chem. Soc.* **75**, 2883 (1953).
[81] Scatchard, G. *Ion Transport Across Membranes*, pp 128–143, edited by H. T. Clarke, Academic Press: New York (1954).
[82] Senderoff, S. and G. W. Mellors. *Rev. Sci. Instrum.* **29**, 151 (1958).
[83] Sinistri, C. *J. Phys. Chem.* **66**, 1600 (1962).
[84] Sood, D. D. and J. Braunstein, *J. Electrochem. Soc.* In press.
[85] Sundheim, A. R. and J. Rosenstreich. *J. Phys. Chem.* **63**, 419 (1959); Pezzati, E. *Z. Naturforsch.* **A25**, 898 (1970).
[86] Swinkels, D. 'Molten salt batteries and fuel cells' in *Advances in Molten Salt Chemistry*, Vol. I, p 165, edited by J. Braunstein, G. Mamantov and G. P. Smith. Plenum: New York (1971).
[87] (a) Swofford, H. S. and H. A. Laitinen. *J. Electrochem. Soc.* **110**, 814 (1963).
(b) Ketelaar, J. A. A. and A. Dammers-DeKlerk. *Rec. Trav. Chim. Pays-Bas*, **83**, 322 (1964).
[88] Thompson, W. T. and S. N. Flengas. *Canad. J. Chem.* **46**, 1611 (1968).
[89] Thulin, L. *Acta Chem. Scand.* **26**, 225 (1972).
[90] Wagner, C. 'The electromotive force of galvanic cells involving phases of locally variable composition' in *Advances in Electrochemistry and Electrochemical Engineering*, edited by Paul Delahay, Vol. IV, pp 1–46. Interscience: New York (1966).
[91] West, J. M. *Electrodeposition and Corrosion Processes*, p 17. Van Nostrand; Reinhold: London (1970).

CHAPTER 19

Part 1. Thermodynamic Properties of Fluid Metals at Medium and Low Pressures

E. E. SHPILRAIN

Institute of High Temperature, Academy of Sciences of the USSR, Moscow, USSR

Contents

I.	Introduction	953
II.	Thermal Properties. Specific Heat	953
	1. The Direct Heating Method	953
	2. The Drop-calorimeter	954
	3. The Exploding Wire Method	956
III.	Density Measurements	957
	1. Direct Methods	957
	2. Indirect Methods	963
IV.	Vapor Pressures	966
V.	References	973

I. Introduction

The reason why liquid metals are treated separately in a book devoted to liquids in general lies in their rather special features. First, we should consider those features likely to make the experimental procedures more complicated and lead to the use of original experimental techniques.

In general, when metals are in a liquid state they are also usually at a high temperature. On the other hand, one must remember that metals are often chemically very active. Both these features lead to the development of special sample containers, require a controlled very pure atmosphere in the equipment, and need some specific measuring equipment.

Secondly, liquid metals which have a high electrical conductivity invite the development of new experimental methods which are characteristic only for metals.

It is obvious that in preparing this chapter and dealing with methods used to determine the thermodynamic properties of liquid metals it is difficult to avoid some repetition with other chapters; the author apologizes in advance for any such overlap.

II. Thermal Properties. Specific Heat

1. *The Direct Heating Method*

In the making of thermal measurements, the specific properties of metals make it possible to use direct sample heating by the conduction of electric current through the sample. Such methods were mostly developed for solid metals but recently they have also been used for liquid metals. As an example, the work described in ref. 1 may be mentioned, dealing with the measurement of the heat capacity of liquid metals by use of a kind of well-known temperature modulation technique which was previously developed for the study of the heat capacity of solids at elevated temperatures[22, 26]. The main point of this method can be explained as follows. The liquid metal under investigation was filled into a thin-walled niobium capillary tube with i.d. either 0.155 or 0.285 mm and o.d. 0.28 or 0.485 mm, and 260 to 280 mm in length. A second tube with the same dimensions was empty. Both tubes were placed in a glass vessel, where a vacuum of 3×10^{-6} torr was maintained. They were connected in series and an 18 Hz a.c. current was passed through them. The frequency was chosen to provide equal temperature of the capillary wall and the metal under investigation. The amplitude of the oscillating temperature was measured by means of a brightness record of the capillary tube using an electronic photomultiplier. From the measurement one can obtain the specific heat of a sample metal if the specific heat of the wall material is known. The precision of this method according to the measurement of the heat capacity of tin in the temperature range from 900° to 1 700°C is quoted as being between five and seven per cent.

2. *The Drop-calorimeter*

To measure the heat capacity of liquid metals, the well known drop-calorimeter technique can be also used. But as mentioned above, in putting this method into practice, one must take account of the high temperature conditions which are specific for metals. A study of the enthalpy of silver and tin in the temperature range from 1 243 to 1 588 K and from 1 222 to 1 745 K respectively, can be cited as an example[15]. Experimental details can be found in refs. 23, 24.

The metal under investigation was placed in a crucible and held in a graphite furnace for half an hour to reach steady-state conditions. Then the crucible was dropped into a copper calorimeter block. The furnace volume was first evacuated and a pressure of 23 torr of argon was maintained in the furnace during the experiment. The temperature of the liquid metal was measured by an optical pyrometer using a blackbody hole in the crucible wall. The length to diameter ratio of the hole was ten; this was sufficient to provide blackbody conditions. Corrections were made to allow for radiation absorption in a quartz window and a prism. The crucible in the runs with silver was machined from a graphite block. To protect it against failure when it is dropped into the calorimeter, the bottom of the calorimeter sample receiver was covered with graphite felt. As a consequence the time needed to reach temperature equilibrium in the calorimeter amounted to between 60 and 80 min for silver in the graphite crucible, whereas 30 min was sufficient for tin in a tantalum crucible.

With a graphite crucible the authors could not use an ordinary metal wire for its suspension. Therefore the crucible was fastened to a graphite rod which had a thin slot cut with a jewel saw in a cross section adjacent to

PART 1. THERMODYNAMIC PROPERTIES OF FLUID METALS

the crucible. A dropping device controlled by a solenoid shook the system; the graphite rod was broken at the slot and the crucible fell into the calorimeter. Another peculiarity of this high-temperature apparatus arose because of the special arrangements used to reduce radiation losses from the sample being dropped into the calorimeter. The precision of the results obtained can be expressed in terms of the standard deviation from the least squares equation. With tin, it was ± 28.8 cal/mol, whereas with silver it was ± 94.6 cal/mol, due to the long main calorimeter period.

References 34 and 35 give another example of exploring the drop-method to measure the enthalpy of liquid metals. To measure the amount of heat carried by the sample, a calorimeter with an evaporating liquid was used. The idea of this calorimeter was similar to that of the well known ice-calorimeter. The amount of heat dissipated by the sample was determined by measuring the mass of vapor which was produced by this heat from the calorimeter liquid maintained under saturation conditions. Although this calorimeter was of an isothermal type it could easily be made adiabatic by fitting the calorimeter into another vessel with the same liquid boiling in it. While some flow of heat might occur, due to the slight pressure differences in the vessels and some different superheatings of the boiling liquids, it was considerably reduced by fitting some radiation shield screens in the air gap between the vessels.

Using this method for the study of liquid metals, one must take into account the high thermal diffusivity of metals. In conjunction with high temperatures, it can lead to very large heat fluxes to the liquid in the calorimeter at the beginning of the main calorimeter period. This is undesirable, because high heat fluxes lead to tremendous boiling, and this increases the superheating and the splashing of the liquid and causes a transfer of liquid droplets to the vapor stream. All this reduces the accuracy of the measurement of the amount of heat. To avoid these shortcomings in the work of ref. 34, special devices were developed. The receiver tube of the calorimeter into which the crucible with liquid metal was dropped was provided with a heavy enclosure; its heat capacity was large enough to make the heat flux during the main period more uniform (see *Figure 1*). In this same figure, a large set of reflecting screens driven by an electromagnet can be seen; they diminish radiation heat losses from the container.

An interesting variation of the drop method is reported in refs 4, 9 and 10. As stated above, the container problem is very serious in the study of liquid metals, and becomes worse still with such liquid refractory metals as molybdenum, niobium and tungsten. One solution of the problem is to use no container at all but rather to apply electromagnetic levitation, when the metal under investigation is melted in a high frequency coil, the liquid produced being held in an appropriate position by the electromagnetic forces created by a properly shaped electromagnetic field. In particular, in ref. 4, this idea is used to measure the enthalpy of liquid molybdenum at temperatures around 2890 to 2925 K. It should be mentioned that when using the levitation method to study the enthalpy of liquid metals, one must overcome some difficulties in measuring the temperature of the liquid metal cluster when it is suspended in the furnace by electromagnetic forces. This is connected with the fact that it is impossible to produce a blackbody hole

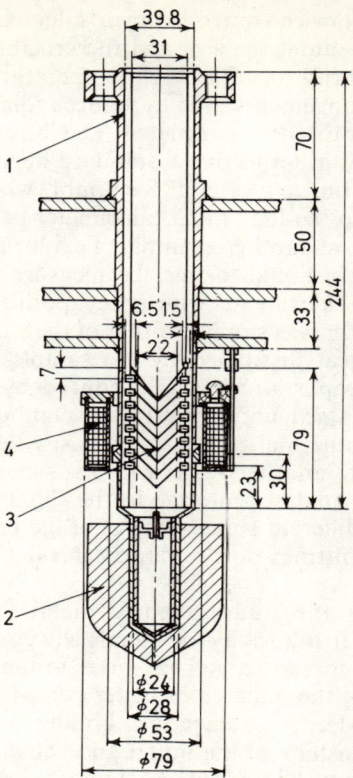

Figure 1. Receiver tube with screens: 1 receiver tube; 2 heavy enclosure; 3 reflecting screens; 4 electromagnet

in a liquid cluster. At the same time the emissivity of the liquid metal is normally unknown. Having this in mind, those authors determined the luminance temperature of the liquid by means of an optical pyrometer, EOP-51.

To recalculate this temperature in real terms at the melting point of the metal, its spectral emissivity was determined, which was then regarded as being independent of the temperature.

3. *The Exploding Wire Method*

It is of great interest to determine the heat capacity of liquid metals using the method of direct heating by an electrical current, in its modification known as the exploding wire method. In refs 11 and 12, the utilization of this method to determine thermal and other properties of molybdenum and tungsten up to temperatures of 3 700 and 4 500 K respectively is considered. The main features of the method lie in leading through a wire made of the metal to be investigated, a current pulse of high density, namely 4×10^6 A/cm^2. The discharge of a 5.7 μF capacitor through a ballast resistor of 50 Ω was used to generate this impulse. The wire was heated by the high density current; it melted and rose to the predicted temperature during 50 μsec.

It is obvious that such a process can be considered as adiabatic and all heat losses from the wire can be neglected. During the discharge, oscillograms from the voltage drop across the wire as well as across a standard resistor connected in series with the wire were recorded. Using these measurements it was possible to calculate at each moment the electrical power liberated in the wire. The wire temperature was determined from the measurement of its luminosity. The light flux was transformed in voltage by means of a photomultiplier, FEU-29, and recorded on the same oscillogram. The luminosity of the wire at the beginning and at the end of the melting process served as reference points to convert the voltage readings into temperatures.

The authors claim to have measured heat capacity data with an error that does not exceed ten per cent, in the region where the specific heat dependence on temperature is weak. In the vicinity of the melting point, the error can be greater.

III. Density Measurements

1. Direct Methods

Among methods which are used for density measurement, the principal one which is specific for liquid metals, depends on a considerable difference between the electrical conductivities of the liquid metal and its vapor. This method is especially convenient when it is necessary to obtain the density above the normal boiling point of the metal. Such a method was used in the work of ref. 18 to measure the density of liquid cesium and its vapor up to 1800 K. The experimental tube, made of a 90% Ta–10% W alloy, had a wall thickness of 1.5 mm, an external diameter of 25 mm and a length of 300 mm. It was filled with a measured amount of metal to be investigated and arc-welded. At a given temperature the liquid level in the tube was determined by measuring the voltage distribution along the tube when direct current was conducted through it. For this purpose the tube was provided with six spot-welded voltage terminals made from the same alloy. During the experiment, the different sections of the tube between two adjacent terminals could be filled either with liquid or with vapor. Only one section was partly filled with both liquid and vapor. When measuring the electrical resistance of this section and of sections filled only with one phase, one can calculate the level position inside the tube. A correction was made to allow for the fact that the liquid surface is not an equipotential one because between the wall and the tube axis there is a certain resistance. The magnitude of this correction was determined by means of a two-dimensional large scale electrical equivalent circuit which consisted of resistances to simulate the tube wall and liquid metal resistances having a flat surface. The possible influence of the liquid meniscus was considered. Using a simplified form of Rayleigh equation, the meniscus shape for a zero wetting angle was calculated. This shape was used in the above-mentioned electrical model and it was shown that the error in determining the liquid level due to the existence of a meniscus does not exceed 0.3 mm, at the lowest temperatures. At higher temperatures and in real wetting conditions, this error appears to be negligible. Thus the method described allows for a given temperature and a given amount of liquid metal filling the tube to obtain volumes of liquid and vapor. Following the same process, with two different amounts of liquid at the same temperature, it is possible to calculate saturated liquid and vapor densities. Data were taken for three runs when the tube was charged with 48.404, 49.154 and 73.517 g respectively. Calculations were carried out using charges 1–3 and 2–3. The average error of the results is claimed to be 0.02 g/cm^3 for the liquid metal and 0.014 g/cm^3 for the vapor.

To measure the density of liquid metals, classical methods such as the well known pyknometer method are also used. But the equipment used in those investigations is in general rather specific and it is worth mentioning it here. In ref. 21, the concept of an overflowing pyknometer is used to study the density, the compressibility and the thermal expansion of gallium at temperatures up to 600°C and pressures up to 2500 bar. The pyknometer having a well determined volume V_0 was filled with the metal to be investigated and its mass M_0 was determined by weighing. Thereupon the pyknometer was placed inside a furnace which was pressurized to maintain the required pressure. Then the temperature was raised by steps and the excess metal, which flowed out of the pyknometer due to thermal expansion during one step of heating, was collected in a special vessel. For the following step the collecting vessel was replaced by an empty one by means of a rotatable device. Thus without opening the furnace, it was possible to accomplish five steps of heating. After opening the apparatus, all the collecting vessels were weighed and the amount ΔM_i of metal in each of them was determined. The density corresponding to a given temperature was calculated using the formula

$$\rho_n(p, T) = (M_0 - \sum_{}^{n} \Delta M_i)/(V_0 + \Delta V(T))$$

where $\Delta V(T)$ is the correction taking account of the thermal expansion of the tungsten pyknometer. It was calculated using the thermal expansion data available for tungsten. The initial pyknometer volume V_0 was determined by calibration using distilled mercury. The maximum error in obtaining the density data of gallium is estimated to be ± 0.03 per cent.

Almost the same method was used to study the density of liquid alkali metals[40]. The pyknometer (see *Figure 2*) had a volume of 30 cm^3 and was machined from a Nb–1% Zr alloy. Its capillary had a 2.5 mm internal diameter. In each experiment, only one density value could be obtained. As the predicted temperature in the furnace was reached, the excess of metal due to thermal expansion overflowed from the pyknometer and was collected in a top chamber. After that the furnace was cooled, the amount of metal which remained in the pyknometer was determined using two different techniques. To suppress evaporation of metal from the pyknometer inside the furnace, an argon pressure was maintained which was about 1 atm higher than the saturation pressure of the metal under investigation. Temperature measurements were taken with a Pt–30% Rh/Pt–6% Rh thermocouple. Taking into account various sources of errors and corrections, the probable

PART 1. THERMODYNAMIC PROPERTIES OF FLUID METALS

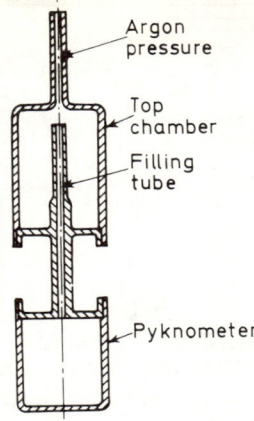

Figure 2. Specific volume apparatus.

error was estimated to range from ±0.25 to 0.30 per cent, in the temperature interval 850 to 1 260°C.

The same idea of an overflowing pyknometer was also used in refs 37 and 38, along with a continuous weighing of the pyknometer. The pyknometer was an argon arc-welded stainless steel vessel and was provided with a long capillary tube. The initial volume of the pyknometer was determined by several different methods, and its thermal expansion was calculated using thermal expansion data obtained experimentally for the pyknometer material. The pyknometer was carefully completely filled with the liquid metal to be investigated and suspended in a high temperature furnace with the capillary tube directed downwards. The inner furnace space was filled with an inert gas at a pressure exceeding the saturation pressure of the metal to be tested. The pyknometer suspension wire was attached to a balance beam. The balance was mounted on a platform above the furnace and enclosed in a thick-walled vessel at the same internal pressure as the furnace (*Figure 3*). The balance was remotely controlled by means of a selsyn set. The pyknometer capillary tube protruded out from the isothermal space, where the pyknometer was placed and reached a region where the temperature was slightly above the melting point of the metal. As the furnace temperature was increased the excess metal fell in droplets from the open end of the capillary and was collected in a vessel lying on the furnace bottom. The forming and falling of droplets could be watched through a window. When steady state conditions were reached, the pyknometer with the remaining amount of metal was weighed. If a droplet was hanging at the capillary end, it was blown away with a gas jet. Weighing corrections were introduced, allowing for buoyancy forces of the pyknometer, weights etc. The maximum relative error on the density of several alkali metals, up to 1 500°C, was estimated to be 0.2 per cent. At higher temperatures, the precision decreases slightly probably on account of temperature measurement errors.

The pyknometer method turns out to be very suitable, even with such a

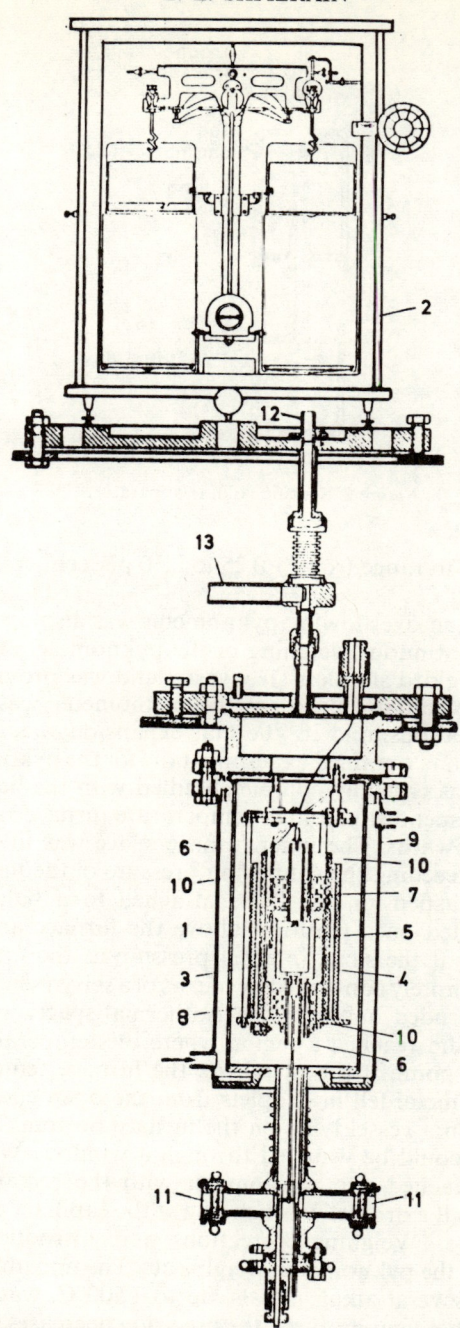

Figure 3. Apparatus for liquid metal density measurements: 1 pyknometer; 2 balance; 3 furnace; 4 alumina tube; 5 molybdenum cylinder; 6 connecting tube; 7, 8 top and bottom heaters; 9 water jacket; 10 thermocouple; 11 window; 12, 13 gas supply.

PART 1. THERMODYNAMIC PROPERTIES OF FLUID METALS

troublesome substance as uranium. In this case the main difficulty arises from the proper choice of the material of the pyknometer. In ref. 27, for the study of the density of the liquid uranium, in the temperature interval from 1137 to 1245°C, a substoichiometric zirconia was used as pyknometer material. The details of the method are given in ref. 32. The authors do not give any information about the precision but the error in this case is probably larger than in the case of a more common metallic pyknometer.

A lot of experiments have been carried out to determine liquid metal densities, using methods based upon x-radiation or neutron irradiation of the specimen.

In ref. 13, the density of liquid alkali metals as well as of their vapors was determined using a radioactive source. A measured amount of metal under investigation was placed in a capsule prepared from a molybdenum or tungsten alloy. Then it was placed in a neutron flux which transformed a small

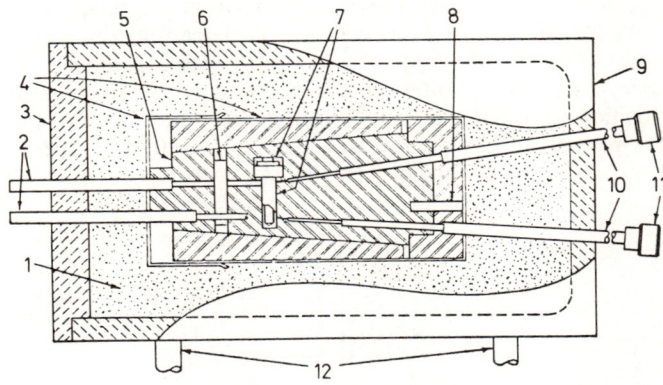

Figure 4. Induction furnace assembly for vapor–liquid density measurements: 1 insulation; 2 tantalum tubes ($\frac{1}{4}$ in. o.d.) connecting to temperature-measurement holes; 3 fiberfrax lid; 4 tantalum can and lid; 5 tapered Mo–30 wt% W cylinder in 3 in. o.d. Mo–30 wt% W sleeve; 6 locating pin; 7 capsule containing test material; 8 locating pin; 9 silicon carbide crucible (6.5 in. o.d. by 11 in. high by $\frac{1}{2}$ in. wall); 10 tantalum tubes ($\frac{1}{4}$ in. o.d.) connecting to collimation holes; 11 fitting mounted in vacuum tank wall and sealed with 0.010 mil diaphragm; 12 refractory supports.

fraction of atoms into radioactive isotopes. The intensity and the time of exposure to neutrons was chosen to obtain 5000 to 50000 count/min from the capsule. The irradiated capsule was then placed into a slightly tapered Mo–30% W cylinder which was inductively heated. An x-ray beam passed through the thin capsule wall (0.75 mm thick) and a collimating hole in the cylinder, and impinged on a counting device comprising a NaI crystal. The recorded number of counts was supposed to be proportional to the density of the substance in the capsule. Temperature measurements were made with W–Re (5–26) thermocouples. A sectioned assembly diagram of the apparatus is shown *Figure 4*. Vapor and liquid density data were obtained from room temperature up to or near the critical temperature, for three of the four metals investigated. For potassium having a short half-life isotope, data were obtained only up to 1 200°C. Strictly speaking this method is a relative one, because to establish the relation between density and measured count/min, density data of metals investigated below 1 200°C were used. In the quoted work, no correction allowing for capsule thermal expansion was introduced. For the entire liquid range measured, the error was two per cent; within a 95 per cent confidence level for vapor it ranged from ± 2 to ± 20 per cent. The critical density was obtained by means of the rectilinear diameter rule using corrections suggested by Rowlinson[28].

A similar idea was used in work reported in ref. 19, to obtain mercury density in the vicinity of the critical point. This investigation as a whole was a study of some physical problems connected with the transition from metallic to non-metallic states. But density measurements conducted in the course of this investigation were of independent interest. Measurements were carried out in a temperature range from 0 to 2 000°C and pressure range from 200 to 5 000 bar. The temperature was measured by means of W–Re thermocouples and the pressure with manganin gages calibrated against the mercury melting curve. The accuracy of the calibration was no less than 0.2 per cent. The metal under investigation was contained in a specially shaped beryllia capillary tube on which the same pressure of argon gas is applied inside and outside. Density readings were deduced from the measurement of x-rays from the ^{203}Hg isotope incorporated in the metal under investigation. The x-ray flux passed through two tungsten collimators and

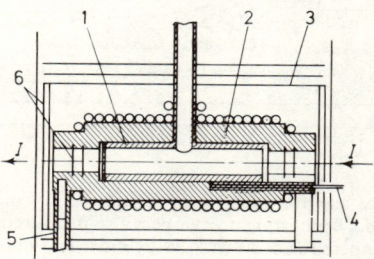

Figure 5. Apparatus for liquid metal density x-ray investigations: 1 crucible; 2 copper block; 3 heater; 4 thermocouple; 5 connection tube; 6 shields. *I* denotes path of x-rays.

a sintered magnesia window. To measure the x-ray intensity, a crystal of CsI was used as detector, connected with a photomultiplier, FEU-42. The transducer and counting device provided a counting stability better than 0.1 per cent, within twelve hours. To convert the relative x-ray intensity measurements into density values, the saturated mercury density and its compressibility at 400°C were used. The precision of density values is claimed to be two per cent and even better at temperatures above 1 600°C, 1.5 per cent in the range 1 400° to 1 600°C and less than one per cent at lower temperatures.

Sometimes the x-ray method is used with an external source of radiation, for example in the work reported in refs 2 and 3 an isotope ^{60}Co of 1 curie was used. The x-ray beam was produced by means of two collimators, 5 mm in diameter, one adjacent to the source and the other to the detector of radiation. A crystal detector with a photomultiplier were used to detect the x-ray flux. The metal under investigation was placed in a metallic vessel with parallel front surfaces, and introduced in a massive copper block (*Figure 5*) provided with holes forming a passage for radiation and with Pt/Pt–Rh thermocouples. All this assembly was mounted in a high temperature, high pressure furnace. To measure the initial x-ray intensity the whole apparatus was placed within the furnace, the metallic vessel being empty. The gas pressure in the furnace was changed to allow for radiation absorption in the gas. The assembly was checked by measuring the density of liquid lead of 99.96 per cent purity.

This method also is a relative one and is valid only when the density is well known at least at one temperature. In this work, the authors studied the densities of liquid gallium, cesium and rubidium at temperatures up to 1 400 K. They claimed that the maximum relative error is 0.25 per cent, without taking into account the accuracy of density values of the calibrating metal.

Of some interest in the study of liquid metal densities is the dilatometer method described in ref. 25. To measure potassium and some Na–K alloy densities the dilatometer was made of molybdenum. It was immersed in a thermostatically controlled enclosure filled either with a cooling mixture or with water. The liquid metal level in the dilatometer was measured through a window mounted in the enclosure, by means of a cathetometer. To measure the level, the whole dilatometer was moved up or down using a screw with captive nut, until it came into a reference position in front of the cathetometer. The level position was read by means of ten slits drawn on the dilatometer tube. The volume of the dilatometer tube was only five per cent of the overall volume, thus no precautions were taken to hold it at uniform temperature. The volume and the thermal expansion coefficient of the dilatometer were calibrated using distilled mercury and checked using distilled water. The maximum relative error of the data ranged from ± 0.15 per cent at 200°C to ± 0.25 per cent at elevated temperatures.

A classical method based upon the measurement of the pressure maximum in a gas bubble was also developed for liquid metals. In ref. 29, it is reported that this method was used to study the density of liquid Ag–Pb and Ag–Bi alloys. The gas used was hydrogen. At each temperature, each tube was immersed in the liquid metal to be investigated at different depths (10, 40, 50, 30 and 10 mm). From three to five different tubes calibrated with mercury were used. In the main runs, only those tubes showing a value of 13.54 ± 0.5 per cent for the density of mercury were used.

2. Indirect Methods

Side by side with direct density measurement methods, indirect ones may also be used, for example based upon sound velocity measurements. The well known Laplace formula gives the relation between sound velocity and adiabatic compressibility of a substance. The latter can be used to calculate the isothermal compressibility when the specific heats C_p and C_v are available. As regards liquid metals, the most frequently used techniques for measuring the sound velocity are different types of pulse methods.

In ref. 31, the sound velocities of liquid mercury and aluminum are measured in the temperature range from -35 to $48.3°C$ and from 660 to $1\,000°C$ respectively, using a pulse method. A sound wave was simultaneously conducted through two tubes, one of which was filled with the liquid metal to be investigated and the other with a standard liquid having a known sound velocity. The pulses were recorded by detectors set at distances S_x and S_v from the sound source end; the distances were chosen to insure simultaneous arrival of the sound waves at the detectors. In this case the sound velocity in the liquid to be investigated is calculated as

$$U_x = U_v S_x / S_v$$

When dealing with liquid aluminum, this method raises some additional difficulties because the temperatures encountered are too high to permit utilization of common quartz components for sound velocity measurements. Therefore, in ref. 31, the sound wave was conducted to the liquid under investigation through a sound-conducting rod made of sintered alumina. The end of the rod was used as the base of the liquid alumina container (*Figure 6*). The sound receiver also was built from an alumina rod which

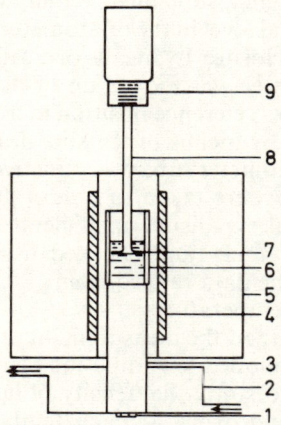

Figure 6. Schematic diagram of apparatus for liquid mercury and aluminum compressibility measurements: 1 quartz; 2 water cooler; 3, 8 sound-conducting rods; 4, 5 heaters; 6 vessel with liquid metal; 7 sound receiver; 9 micrometer screw.

could be shifted by means of a micrometer screw. The vessel containing the liquid metal under investigation was placed in the furnace. Temperature readings were taken with Pt/Pt–Rh thermocouples; the temperature non-uniformity along the working space did not exceed 1–2 K. As a standard liquid, water was used, its temperature being held constant within 0.05 K. Some precautions were taken to eliminate the influence of pulses which could penetrate to the receiver directly from the sound-conducting rod, as well as harmful resistances at quartz/sound-conducting rod/liquid metal interfaces. Nevertheless with the liquid aluminum the errors in sound velocity values were estimated to be ± 0.03 per cent, while for liquid mercury they were less than 0.02 per cent at the melting point.

PART 1. THERMODYNAMIC PROPERTIES OF FLUID METALS

The same method was used in work reported in ref. 17, to measure sound velocities in liquid indium, bismuth, mercury, lead, tin and cadmium at temperatures up to 350 to 450°C.

Another type of pulse method, including an interferometer, was used,

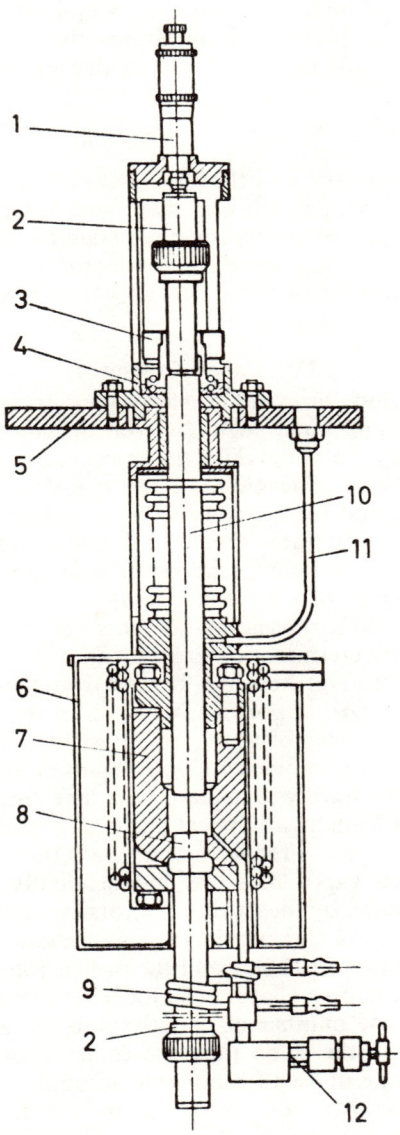

Figure 7. High-temperature pulse interferometer: 1, 3 micrometer screw; 2 quartz holder; 4, 5 supports; 6 furnace; 7 sample vessel; 8, 10 sound-conducting rods; 9 water cooler; 11 gas supply; 12 manifold.

as reported in ref. 41, to measure sound velocities in liquid sodium, potassium and some Na–K alloys. This method is based on direct use of the formula

$$u = f\lambda$$

where f is the sound frequency produced by a generator, ranging about $(2\,000 \pm 3) \times 10^3$ Hz. The wavelength λ was determined by moving the sound receiver until a node or maximum amplitude of the sound wave occurred at its face. To obtain these conditions, the receiver was shifted by a micrometer screw and the distance between the nth and mth maxima was measured. Then

$$\lambda = 2\,(l_m - l_n)/(m - n)$$

Due to the high temperature and the high chemical activity of metals under investigation, direct contact between them and quartz components was impossible; in consequence an intermediate stainless steel rod was used as a sound conductor. The tightness of the conductor was provided by bellows (*Figure 7*). The estimated error of the results was ± 0.6 per cent.

IV. Vapor Pressures

All the methods which are used to study the vapor pressure of various substances can be used for liquid metals as well. Specific properties of metals, however, have an effect upon the different components used.

At low pressures the most developed is the well known Knudsen method. According to this method the vapor pressure is obtained by measuring the amount of substance evaporated from the Knudsen cell per unit time. In such work the molecular weight of the vapor is assumed to be known. But often, one must recognize that metallic vapors are partly dimerized or even polymerized. Thus in the Knudsen method, it is necessary to make independent molecular weight determinations.

An interesting example of how the Knudsen method is used to obtain the pressure of a dimerized vapor can be found in ref. 20, this work being devoted to the measurement of the vapor pressures of lead and bismuth. The authors have used a combined torsion Knudsen method. According to this method, the capsule filled with metal under investigation was suspended in a vacuum by a fiber from an analytical balance. The capsule was provided with two holes eccentrically bored in the side walls through which the vapor could effuse. The total vapor pressure, independently from its molecular weight, could be obtained by measuring the torsion angle of the suspension fiber due to the impetus of effusing vapor jets. It should be mentioned that to take account of the non-ideal shape of the orifice holes, correction factors different from the well known Clausing factor had to be used, since the force is a function of both the number of particles and the angular distribution of effusing particles. At the same time the total vapor pressure could be calculated by the usual Knudsen formula including the unknown molecular weight. So, from these two equations, the molecular weight could also be derived.

The apparatus used is described in detail in ref. 44. The Knudsen cell was made of graphite stock 17 mm in length and 12 mm in width and height.

PART 1. THERMODYNAMIC PROPERTIES OF FLUID METALS

This stock was drilled along its length to provide the capsule for the metal to be investigated. The hole was closed with a small graphite plug. The dimensions of the cell cavity and cell orifices were chosen to conform to Knudsen effusion conditions. The orifice sizes were determined by taking photographs with a metallograph at a calibrated magnification factor of × 70. The orifice length was measured by means of a metallographic microscope which was focussed on the external edge of the orifice and on a brass plate pressed to the inner edge. The weight loss due to evaporation was measured by an automatically controlled microbalance. The microbalance arm constructed of quartz rod was leant against a quartz fiber and was provided at one end with a permanent magnet and at the other with a silver vane. When the capsule weight diminished, the balance position remained unchanged due to current change from a control coil coupled with the permanent magnet attached to the balance arm. The control coil current was adjusted by means of a high frequency induction coil having the silver vane as a core.

The Knudsen method was used to measure the vapor pressure of holmium from 10^{-8} to 10^{-1} torr[43]. The specific feature of this work was the use of a tube-type orifice at the higher pressures (*Figure 8*).

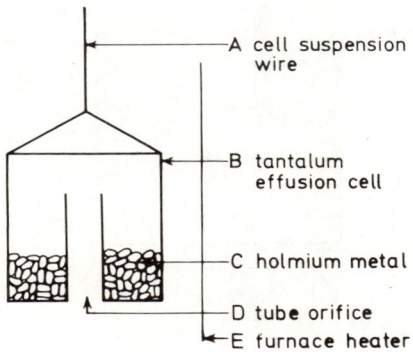

Figure 8. Effusion cell.

An interesting peculiarity of alkali metals having low ionization potentials was used in work devoted to the vapor pressure measurements of sodium, potassium and rubidium[8]. This method allows direct measurement of atom concentration in a saturated vapor. Proceeding from the Langmuir–Taylor method, a new relative one using cesium as reference substance was developed by the authors. According to this new method an atomic cesium beam effusing from a Knudsen cell passed through the space containing the saturated metallic vapor under investigation. Due to scattering by the vapor atoms, the initial cesium beam was attenuated; the intensity of the beam was determined by measuring the ionic current produced by the ionization of cesium on an incandescent tungsten filament.

A lot of work devoted to vapor pressure measurements of metals has been carried out using various types of static methods.

For example in ref. 14, a classical static method was used to obtain cesium vapor pressures up to 1 660°C. According to this method the liquid metal under investigation and its vapor are maintained in equilibrium in a closed piezometer which is carefully thermostatically controlled. The vapor pressure was determined by measuring simultaneously temperature and pressure in equilibrium conditions. The temperature was measured with tungsten–3 per cent rhenium and tungsten–25 per cent rhenium thermocouples; the vapor pressure was balanced with an inert gas by means of a zero-position diaphragm (*Figure 9*). The inert gas pressure was observed externally at

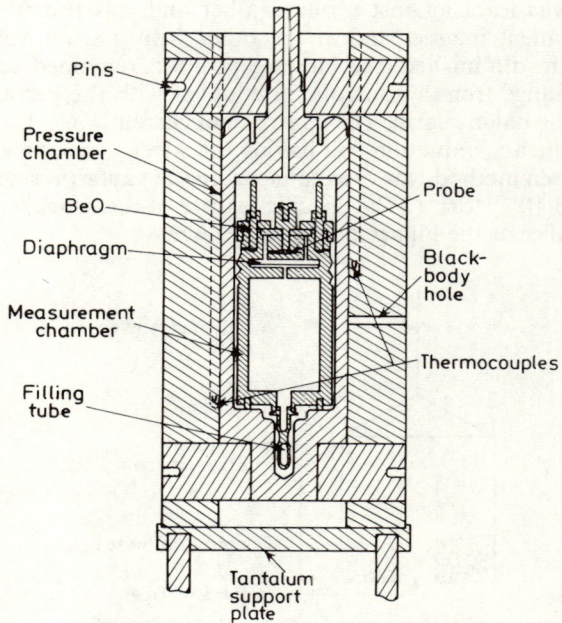

Figure 9. Vapor pressure measurement apparatus.

room temperature with a Bourdon-tube gage; the accuracy of this work was estimated to be $\pm$ 0.8 per cent.

The same method was utilized in work described in ref. 36 to obtain the vapor pressures of potassium and sodium near their boiling points. In contrast to ref. 35 where the zero position diaphragm was adjusted by making or breaking an electrical contact between the diaphragm and an insulated probe, in ref. 36 an inductive sensor was used (*Figure 10*).

Another static method was used in refs 37 and 38, to measure the vapor pressure of alkali metals. The metal under investigation filled a U-tube gage (*Figure 11*), the left hand limb of which is closed and the right hand one connected to vacuum and inert gas supply lines. A part of the U-tube adjacent to the top of the left hand limb was made of a refractory metal and was heated by an electrical resistance heater. The heater was made by depositing a thin layer of tungsten upon an alumina tube. The right hand glass limb

PART 1. THERMODYNAMIC PROPERTIES OF FLUID METALS

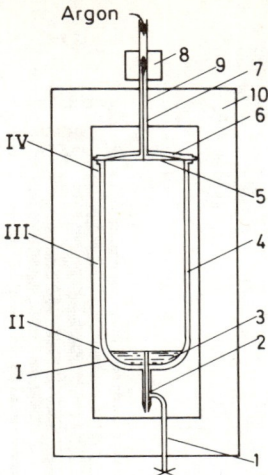

Figure 10. Schematic diagram of apparatus: I–IV thermocouples; 1 vacuum connection; 2 thermocouple well; 3 liquid metal; 4 equilibrium vessel; 5 membrane; 6 top cover; 7, 8 inductive sensors; 9 gas inlet; 10 thermostatically controlled enclosure.

of the U-tube was maintained at a temperature slightly higher than the melting point of the metal under test. Initially the inert gas pressure was adjusted so as to exceed the metal saturation pressure at an arbitrary temperature. At this high pressure there was no free surface of metal in the left hand limb of the U-tube. The whole tube was full to the top of liquid.

When the chosen temperature had been reached at the top of the hot left hand limb and had been held constant for a while, the gas pressure was diminished so that finally the metal in the left hand limb began to evaporate forming a free surface. The position of the surface in the left hand limb could be adjusted by watching the displacement in the right hand glass limb and changing the gas pressure. Usually the level in the left hand limb was held 5 to 10 mm away from the top of the tube. To make this position stable a slight temperature drop was established from the top of the left hand limb, not exceeding 0.15 K/cm. The temperature in the hot region was measured by three Pt/Pt–Rh thermocouples so that the temperature of the metal surface could be obtained by simple interpolation. The accuracy of the data for alkali metals up to 1 500°C was estimated to be 1.5 per cent.

In work reported in ref. 39 essentially the same method as above was used, but to record the level position a different method was developed. The evaporation and condensation of metal in the left hand limb was fixed here by a sharp change of the electrical conductivity and the thermoelectric power of the metal under test. The diagram of the set-up used is shown as *Figure 12*.

The metal under investigation was placed in a thin-walled (0.12 mm) tungsten (75 per cent) rhenium capsule, closed at the top. At the bottom, a capillary tube was attached to the capsule connecting it to the pressure-balancing system. The capsule was placed in a cylinder pressurized by argon

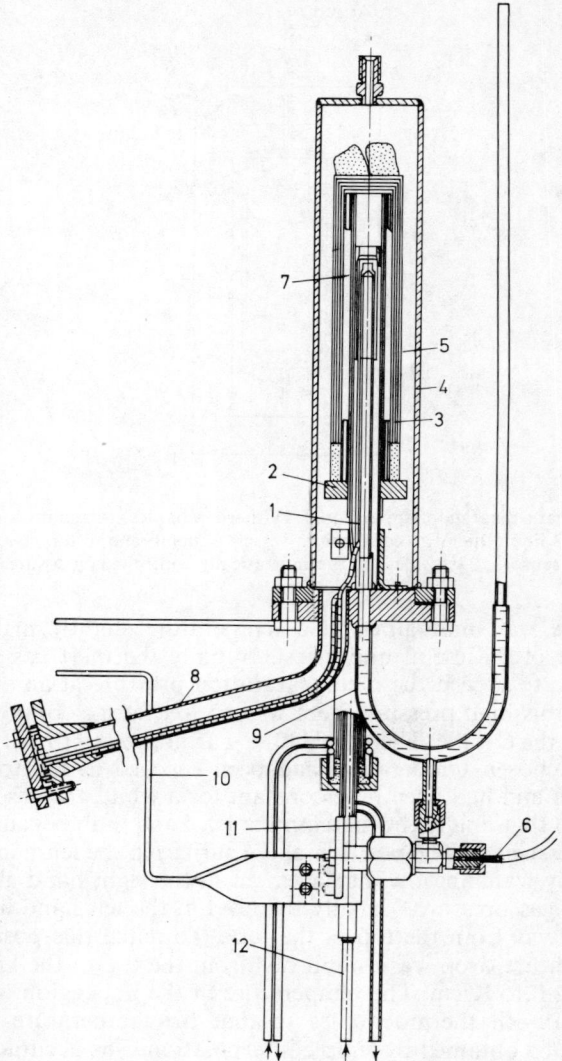

Figure 11. Apparatus for saturation pressure measurements: 1 casing; 2 heater; 3 shield; 4 safety vessel; 5 thick-walled tube; 6 to the filling line; 7 thermocouple; 8 tube for thermocouple electrodes; 9 U-tube; 10 a.c. current supply; 11 tube to water cooling jacket; 12 water supply.

and heated internally. Four 0.25 mm thick tungsten–rhenium wires labelled 2, 3, 4, 5 in *Figure 12* were fixed to the sidewall and the top of the capsule. The wires 2 and 3 consisted of 97% W, and 4 and 5 of 75% W. Thus different combinations of these wires formed several thermocouples to measure the temperature of the capsule as well as the temperature gradient along it.

PART 1. THERMODYNAMIC PROPERTIES OF FLUID METALS

Simultaneously those four wires made it possible to measure the electrical resistance of the metal specimen using conventional potentiometers.

The operating procedure to obtain vapor pressure data included several steps. A given argon pressure was established in the cylinder and a slow temperature rise of the capsule by means of the main heater initiated. This heater and the second one (see *Figure 12*) also maintained a slight temperature gradient from the capsule top to the bottom. During capsule heating,

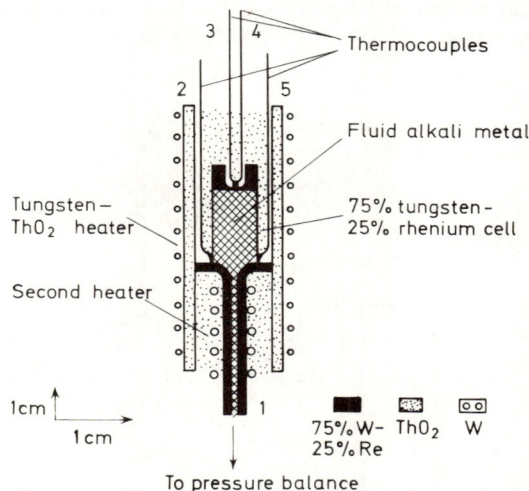

Figure 12. Cell for conductivity, thermoelectric power and vapor pressure measurements.

readings of electrical resistance of the metal specimen were taken. As soon as the temperature of the capsule reached the saturation temperature, a portion of metal was evaporated and the measured electrical resistance sharply increased. By contrast, when the temperature was decreased, complete condensation could be detected by a slight but sensitive increase of thermoelectric power in the WRe–K–WRe circuit. It should be mentioned that condensation could be detected by a slight but sensitive increase of thermothose of evaporation. This temperature difference decreased towards the critical point. The authors estimated the accuracy to be ± 0.5 per cent in temperature measurements and ± 1 per cent in pressure measurements.

A large amount of experimental work in metallic vapor pressure measurements was carried out by the boiling point method. This method is one of the most accurate but when it is used at high temperatures for metals, some difficulties arise. The point is that when a metal is boiling at elevated temperatures, there exists a pronounced temperature difference between the boiling liquid and its vapor. Having this in mind and to obtain reproducible and reliable vapor pressure data, one has to measure not the boiling liquid temperature but the vapor temperature by slight condensation on the temperature sensor. In the measurement of vapor temperature another obstacle arises due to radiant heat fluxes between the container walls and the sensor through the transparent vapor. As an example, some work where these peculiarities have been taken into consideration can be mentioned, see ref. 36. This work was devoted to the measurement of the vapor pressures of mercury and some alkali. At the heart of the experimental assembly, the boiler was made of a cylinder 42 mm in diameter and 88 mm in height. To the top of the boiler a condenser tube was welded, being 18 mm in o.d. and 335 mm in length. In the lower section of the boiler a horizontal tube of 11 mm i.d. and 36 mm in length, closed with a plug, was used as a heater shell. The outer surface of the tube was provided with notches to insure

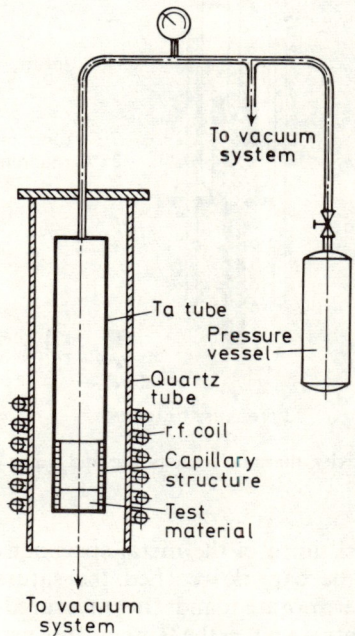

Figure 13. Experimental arrangement for measuring vapor pressures of different metals.

nuclear boiling. Along the condenser axis a tube 4 mm in i.d. having a tapered plug at the bottom, was inserted. It was used as a shell for a Pt/Pt–Rh thermocouple. To this tube, six conical caps were welded at 50 mm intervals, to hinder formation of too thick a condensate film on the thermocouple shell. Close to the liquid surface in the boiler, a screen was set up which prevented the direct penetration of the radiant heat flux from the superheated liquid surface to the thermocouple shell. A screen also protected the shell from radiant heat exchange to the boiler walls which may also have a temperature deviating from the real saturation temperature. The reliability of data which had been obtained with this technique was checked by changing over a wide range the boiling conditions in the boiler and the temperature distribution along the boiler walls and a part of the condenser tube adjacent to the boiler. The accuracy of data obtained was estimated to be 0.5 to 1.0 per cent, depending on the temperature range.

Recently in connection with high temperature heat pipe problems, vapor pressure data for a large number of metals became necessary. It was logical therefore that researchers dealing with heat pipes attempted to develop methods for measuring vapor pressure based on heat pipe designs[5, 6, 7, 30]. In particular, in ref. 30, with a technique of this kind, the vapor pressure of twelve metals in a pressure range 10 to 10^4 torr was measured.

The idea of this method is based upon the fact that in a gas which filled the heat pipe there exists a sharp threshold between condensing vapor and gas. This threshold can be identified because of a steep change of the tube wall temperature from the saturation temperature of the condensing vapor to the temperature of the gas above.

The apparatus used to carry through this method is described in ref. 5. The experimental tantalum heat pipe (*Figure 13*) was connected to a vessel containing pressurized argon. The tube worked in a vertical position; its lower section was filled with the metal under investigation and could be heated by a radio-frequency coil. At a sufficiently high heat input into the tube, a zone of constant temperature was established with a transition region to the cold upper part. It was ascertained that the temperature in the constant temperature zone was a function of the measured gas pressure only and was independent of the heat input. Thus this temperature could be correlated with the saturation temperature. The authors did not claim any accuracy values for the data obtained, but it is obvious that the method described can be used only for approximate estimations.

V. References

[1] Akhmatova, I. A. *Dokl. Akad. Nauk SSSR*, **162**, 127 (1965).
[2] Basin, A.C. *Sbornik 'Issledovanie teplofizicheskikh svoistv vestchestv'*, p. 81. Nauka: Novosibirsk (1970).
[3] Basin, A. S. and A. N. Solov'ev. *Sbornik 'Issledovanie teplofizicheskikh svoistv vestchestv'*, p 56. Nauka: Novosibirsk (1967).
[4] Berezin, B. Ya., V. Ya. Chekhovskoi and A. E. Sheindlin. *High Temp. High Press.* **3**, 287 (1971).
[5] Bohdansky, J. and H. E. J. Schins. *J. Appl. Phys.* **36**, 3683 (1965).
[6] Bohdansky, J. and H. E. J. Schins. *J. Less-Common Metals*, **13**, 248 (1967).
[7] Bohdansky, J. and H. E. J. Schins. *J. Phys. Chem.* **71**, 215 (1967).
[8] Buck, U. and H. Pauly. *Z. Phys. Chem.* **44**, 345 (1965).
[9] Chekhovskoi, V. Ya. and B. Ya. Berezin. *High Temperature (Moscow)*, **8**, 1244 (1970).
[10] Chekhovskoi, V. Ya., A. E. Sheindlin and B. Ya. Berezin. *High Temp. High Press.* **2**, 301 (1970).
[11] Dikhter, I. Y. and S. V. Lebedev. *High Temperature (Moscow)*, **8**, 51 (1970).
[12] Dikhter, I. Y. and S. V. Lebedev, *High Temperature (Moscow)*, **9**, 929 (1971).
[13] Dillon, I. G., P. A. Nelson and B. S. Swanson. *J. Chem. Phys.* **44**, 4229 (1966).
[14] Ewing, C. T., J. R. Spann, J. P. Stone, E. W. Steinküller and R. R. Miller. *J. Chem. Engng Data*, **15**, 508 (1970).
[15] Feber, R., C. C. Herrick and L. S. Levinson. *J. Chem. Thermodynamics*, **1**, 169 (1969).
[16] Freyland, W. F. and F. Hensel. *Ber. Bunsenges. Phys. Chem.* **76**, 17 (1972).
[17] Hill, J. E. and A. L. Ruoff. *J. Chem. Phys.* **43**, 2150 (1965).
[18] Hochman, J. M. and C. F. Bonilla. *Advanced Thermophysical Properties at Extreme Temperatures and Pressures*, p 122. American Society of Mechanical Engineers: New York (1965).
[19] Kikoin, I. K. and A. R. Sechenkov. *Fiz. Metal. i Metalloved.* **24**, 843 (1967).
[20] Kim, J. H. and A. Cosgarea. *J. Chem. Phys.* **44**, 806 (1966).
[21] von Köster, H., F. Hensel and E. U. Franck. *Ber. Bunsenges. Phys. Chem.* **74**, 43 (1970).
[22] Kraftmakher, Y. A. *Zh. Prikl. Mekh. Tekh. Fiz.* **5**, 176 (1962).

[23] Levinson, L. *Rev. Sci. Instrum.* **33**, 639 (1962).
[24] Levinson, L. *J. Chem. Phys.* **40**, 3584 (1964).
[25] Liu, C. S. *Acta Phys. Sin.* **22**, 758 (1966).
[26] Lowenthal, G. C. *Austral. J. Phys.* **16**, 47 (1963).
[27] Rohr, W. G. and L. J. Wittenberg. *J. Phys. Chem.* **74**, 1151 (1970).
[28] Rowlinson, J. S. *Liquids and Liquid Mixtures.* Butterworths: London (1959).
[29] Sauerwald, F. *Advanc. Physics*, **16**, 545 (1967).
[30] Schins, H. E. J., R. W. M. van Wijk and B. Dorpema. *Z. Metallk.* **62**, 330 (1971).
[31] Seemann, H. J. and F. K. Klein. *Z. Angew. Phys.* **19**, 368 (1965).
[32] Serpan, C. Z. and L. J. Wittenberg. *Trans. Metall. Soc. AIME*, **221**, 1017 (1961).
[33] Shpilrain, E. E. and A. M. Belova. *Sbornik 'Teplofizicheskie svoistva gazov'*, p 145. Nauka: Moscow (1970).
[34] Shpilrain, E. E. and D. N. Kagan. *Vypusk GSSSD, No. 3, 'Teplofizicheskie Svoistva vestchestv i materialov'*, p 144.
[35] Shpilrain, E. E. and D. N. Kagan, *High Temperature (Moscow)*, **7**, 328 (1969).
[36] Shpilrain, E. E. and E. V. Nikanorov. *Sbornik 'Teplofizicheskie svoistva gazov'*, p 141. Nauka: Moscow (1970).
[37] Shpilrain, E. E. and K. A. Yakimovich. *High Temperature (Moscow)*, **1**, 173 (1963).
[38] Shpilrain, E. E. and K. A. Yakimovich. *High Temperature (Moscow)*, **5**, 239 (1967).
[39] Shpilrain, E. E. and A. M. Zvereva (Belova), *Inkh.-fiz. Zh.* **6** (1963).
[40] Stone, J. P., C. T. Ewing, J. R. Spann, E. W. Steinküller, D. D. Williams and R. R. Miller. *J. Chem. Engng Data*, **11**, 320 (1966).
[41] Trelin, J. S., I. N. Vasilyev, V. B. Proskurin and T. A. Tsyganova. *High Temperature (Moscow)*, **4**, 364 (1966).
[42] Vinogradov, Y. K. and L. D. Volyak. *High Temperature (Moscow)*, **4**, 50 (1966).
[43] G. F. Wakefield, A. H. Daane and F. H. Spedding, *J. Chem. Phys.* **47**, 4994 (1967).
[44] Wilson, A. E., J. H. Kim and A. Cosgarea. *Rev. Sci. Instrum.* **36**, 1428 (1965).

CHAPTER 19

Part 2. Thermodynamic Properties of Fluid Metals at High Temperatures and High Pressures

F. HENSEL and E. U. FRANCK

Institut für Physikalische Chemie und Elektrochemie, Universität Karlsruhe, Germany

Contents

I. Introduction	975
II. Experimental Methods	976
1. Density	976
2. Vapor Pressures	980
III. Results	983
1. Density	983
2. Vapor Pressure Curves and Critical Data	986
IV. References	990

I. Introduction

Knowledge of thermodynamic properties of fluid metals at elevated temperatures and pressures has for a long time remained very limited. Although nearly four fifths of the elements are metallic and although the solid phases of these metals and alloys have been thoroughly investigated, comparatively very little experimental work has been done to determine thermodynamic data of the fluid phases over a wide range of temperatures. This was probably mainly due to technical difficulties.

New construction materials and developments of high pressure and high temperature techniques, however, have permitted a variety of new experiments in this field in recent years. These efforts were stimulated by an increased interest in fluid metals at high temperatures. The study of their properties provides important contributions to the understanding of metallic behavior in general of disordered systems. The metal–insulator transition can be studied with supercritical metals. Close relations exist with disordered solid semiconductors and metal–ammonia solutions. The study of gas–liquid critical behavior is of importance for theories on critical phenomena. pVT-data of the dense gas state should provide valuable information on interaction of particles with far reaching intermolecular forces.

Technological considerations have also created increased interest in experimental research with fluid metals. Alkali metals serve as heat-transfer media in nuclear reactors. These and other metals are considered as working fluids for advanced power plants and certain applications for energy storage may be envisaged.

Most non-metallic elements and compounds have liquid–gas critical temperatures which are approximately by a factor of 1.5 higher than the absolute normal boiling temperatures. This is not so for metals. It appears as if the factor is between 2.5 and 3 for many metals which can accordingly exist as liquids within a very wide temperature range. Experimental determinations of vapor densities and pressures at saturation conditions extending to the critical region have first been made with mercury. Pioneering work was done by Bender (1915)[2] who determined vapor densities to about 1400°C by enclosing mercury in strong quartz capillaries. Birch (1932)[3] was the first to measure the vapor pressure and give realistic values of critical temperature and pressure, namely 1460°C and 1610 bar. Predictions of critical data of higher boiling metals are difficult and depend to some extent on estimation and conjecture. The rectilinear diameter concept for the densities of coexisting states is of considerable importance. A number of quantitative predictions and discussions of critical data of metals have been given by Grosse[14].

A rapid increase of efforts to determine thermodynamic and other properties of fluid metals at high temperatures and pressures has occurred during the last decade in several countries. The state of knowledge in 1968 has been reviewed by Ross and Greenwood[24]. Selected methods and results with a certain emphasis on work from the Institute of Physical Chemistry of the University of Karlsruhe will be discussed below.

II. Experimental Methods

1. Density

Density measurements of liquid metals under pressure below 600°C have been made using externally heated autoclaves with oil or rare gases as pressurizing media. Different methods have been applied for these measurements.

Bridgman[4] determined the isothermal compressibility of liquid rubidium and cesium near the melting point by observing the displacement of a piston correcting the results for dilatation of the container.

Endo[9] studied the temperature dependence of the compressibility of sodium, potassium, rubidium and cesium near the normal melting point. He detected the volume changes by direct observation of the position of the meniscus of the liquid metal using an autoclave with high pressure windows.

Certainly one of the most reliable methods is the pyknometer technique which involves weighing an accurately known volume of the liquid metal. A special version of this method was used by Köster et al.[20, 27] for the determination of the equation of state of gallium, sodium and potassium up to 600°C and in part to 3000 bar. The apparatus is shown in *Figure 1*. The pyknometer, a tungsten cylinder with a volume of 10 cm^3, was located within an externally heated autoclave and the free surface of the liquid metal was in contact with pure argon at variable pressures. The capillary through which the overflow of the liquid metal occurred was located directly above a pivoted circular disc with five small beakers. The disc could be rotated from outside the autoclave. Thus it was possible to measure five successive density values with one pyknometer filling by collecting the liquid metal overflow in the different beakers. Temperatures were measured with thermocouples

PART 2. THERMODYNAMIC PROPERTIES OF FLUID METALS

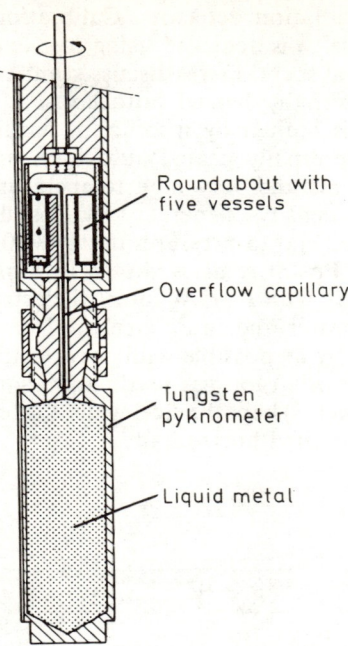

Figure 1. Pyknometer for density measurements up to 600°C at high pressures (Köster et al.[20]).

attached to the top and bottom of the pyknometer. This method, although time-consuming, is probably the most accurate one. There are only three quantities to be measured: volume, weight, and temperature and each quantity could be determined with high accuracy. However, this technique is applicable only at temperatures well below the boiling point. As the boiling point is approached, evaporation begins to interfere with the measurements.

A different technique was used by Dillon et al.[7] for the measurement of the densities of the alkali metals sodium, potassium, rubidium and cesium in the liquid and gaseous state at temperatures approaching the critical points. Externally heated vessels were employed at higher temperatures up to 2 200 K and at pressures up to 500 bar. In this method the alkali metal was usually confined within a thin-walled molybdenum (30 wt%)–tungsten capsule. Other capsule materials such as molybdenum, tungsten or molybdenum alloys were also used. The capsule was in a cylindrical block of the same material which provided external support against the internal metal vapor pressure. For thermal insulation the block with the capsule was surrounded by an alumina–silica crucible. The assembly was mounted in a vacuum tank and heated inductively.

The density measurement applied in this study involved activating the metal by irradiation with thermal neutrons to produce isotopes which emit gamma radiation suitable for counting. The supporting cylinder (block) which provided also shielding from radiation had tapered holes leading from the vapor and liquid phase regions of the capsule wall to the outside of the vacuum tank with radiation detectors. Calibration of fluid metal density versus irradiation level was achieved using low temperature liquid density data. The effect of self-absorption was discussed and was found to be negligible.

Application of externally heated autoclaves at high temperatures and very high pressures is limited by insufficient tensile strength of autoclave material. Therefore internally heated autoclaves are preferable for experiments at high pressures and very high temperatures. An example are the measurements of the density of mercury by Postill et al.[22] to a maximum temperature of 1 100°C and to pressures up to 1 000 bar.

The equipment of Postill et al. is shown in Figure 2(a). It consists of a stainless steel autoclave with 1 500 W internal electrical resistance heater (1). The space between heater and inner steel wall of the high pressure vessel was filled as completely as possible with pyrophyllite to reduce convection of the highly compressed argon gas used as pressure transmitting medium. An external water jacket (3) kept the steel walls at a temperature below 150°C. The autoclave was mounted horizontally.

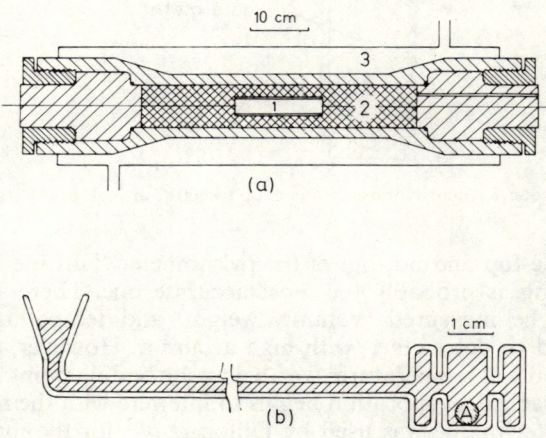

Figure 2. High pressure–high temperature apparatus for density measurements up to 1 100°C and 1 000 bar. (a) internally heated pressure vessel, (b) silica cell (Postill et al.[22]).

The mercury sample was inside a transparent silica cell shown in Figure 2(b) which had wells in the walls into which thermocouples were inserted. It has a capillary extending from the hot into the cold region of the autoclave. The silica cell was located so that the free surface of mercury being in contact with the argon gas was at about room temperature. For the measurement of the density buoyancy method was employed. A small spherical sinker (A) of polished tantal–molybdenum alloy was immersed in the mercury. When the density of the liquid was decreased by heating or decreasing the pressure the sphere (A) sank to the bottom, whilst an increase of the density of the liquid mercury made it rise. With a given sinker many combinations of pressure and temperature could be determined for which the sinker hovered between bottom and top. For these pressures and temperatures the mercury density was just equal to the density of the sinker, i.e. these values of pressure and temperature traced out an isochore. The position of the sinker which was made radioactive by irradiation with neutrons was detected by counting the emitted γ-rays from the ^{182}Ta isotope using a geiger counter and slit system outside the pressure vessel. This method of Postill *et al.* had the advantage of being simple and very accurate. It is time-consuming, however, and does not allow continuous measurements. It is not applicable in the low density region of mercury, in the superheated vapor or in the supercritical region.

An extension of the density measurements of mercury to higher temperatures, especially to the critical and supercritical region, was possible with the experimental method developed by the present authors[16]. Their results were obtained at temperatures and pressures up to 1700°C and up to 2100 bar.

The construction of the high pressure, high temperature equipment is shown schematically in *Figure 3*. The mercury sample was inside a cylinder of 5 cm^3 internal volume machined from solid molybdenum which was found to be compatible with mercury up to 1700°C. Since the molybdenum cell cannot withstand the high pressure of more than 2000 bar at temperatures above 1200°C, it was placed inside an internally heated high pressure vessel[16] which was pressurized with argon. The pressures of argon and liquid mercury were balanced outside the high pressure vessel at low temperature within an auxiliary vessel, a stainless steel 'buret'[16] which was connected by a narrow mercury-filled capillary tube with the molybdenum cell. To

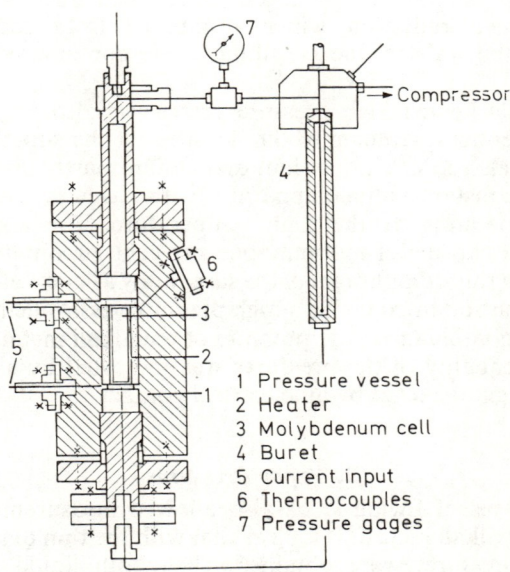

Figure 3. High pressure–high temperature apparatus for density measurement up to 2200 bar and 1700°C. (Hensel and Franck[16]).

obtain the density, the molybdenum cell, which is closed at one end, the connecting capillary and in part the buret were filled with mercury. Since the buret contained an axially mounted platinum–iridium wire, the mercury level in the buret could be determined by a simple measurement of the resistance of the wire. This permitted determination of the amount of mercury which remained inside the heated molybdenum cell as a function of pressure and temperature. Temperatures were measured by four platinum–rhodium thermocouples inserted into holes at different positions in the walls of the molybdenum cell. Pressures were determined by calibrated precision Bourdon gages connected to the argon-filled branch of the assembly. A small correction has to be made for the volume of mercury which was in the capillary in the region of the gradient. This correction was very small, however, because the volume of this region was less than one per cent of that of the molybdenum cell.

Equation of state data of mercury to supercritical temperatures up to 2000°C and pressures in part up to 5000 bar were also investigated by Kikoin and Sechenkov[19]. In these experiments again an internally heated autoclave was used much in the same way as in the experiment of Postill et al. The cell containing the mercury was an open beryllium oxide container under hydrostatic pressure with a free surface of liquid mercury in contact with argon. For the measurement of the density an indirect method similar, in principle, to that used by Dillon et al. was employed. The intensity of gamma-ray emission from the mercury specimen enriched with the radioactive isotope ^{203}Hg was measured through a magnesium oxide window in the wall of the autoclave. Many corrections had to be made to the results obtained in this experiment since self absorption by the mercury amounted to 20 per cent, absorption by the pressurized argon to between three and five per cent. An uncontrolled radiation which originated from radiactive mercury diffused into the walls of the beryllium oxide container caused important errors.

An internally heated high pressure vessel was also employed by Ewing et al.[30] for the measurements of the densities of the superheated vapors of the alkali metals sodium, potassium and cesium partly up to 1400°C. They used a single constant-volume apparatus with a thin pressure-sensing metal diaphragm operating at the high temperature. The alkali metals were contained in a capsule of niobium–one per cent zirconium alloy equipped with a 0.12 mm thick diaphragm of the same alloy as one wall of the container. The capsule was mounted inside a high pressure vessel filled with pressurized argon. The argon balanced the pressure of the alkali metal vapor inside the capsule. The equality of the pressures was indicated by a defined position of the diaphragm detected by an electrical contact.

2. *Vapor Pressures*

This apparatus of Ewing et al. also allowed measurement of the vapor pressure of the alkali metals. It is clear that with the thin diaphragm arrangement vapor pressures were obtained when both liquid and vapor were present in the capsule.

For a determination of the vapor pressure curve of liquid metals up to

PART 2. THERMODYNAMIC PROPERTIES OF FLUID METALS

very high temperatures, i.e. up to the critical point, one of the most reliable methods is the observation of the drastic change in the electrical conductivity connected with phase change. This technique was used to determine the vapor pressure curve of mercury, cesium and potassium up to the critical point[12, 17, 23]. *Figure 4* shows the general arrangement and details of an electrical resistance cell used for vapor pressure measurements with mercury[17]. The mercury sample was inside a molybdenum cell of the design shown in the right hand part of *Figure 4*. Thin tubes of sintered, non-porous alumina

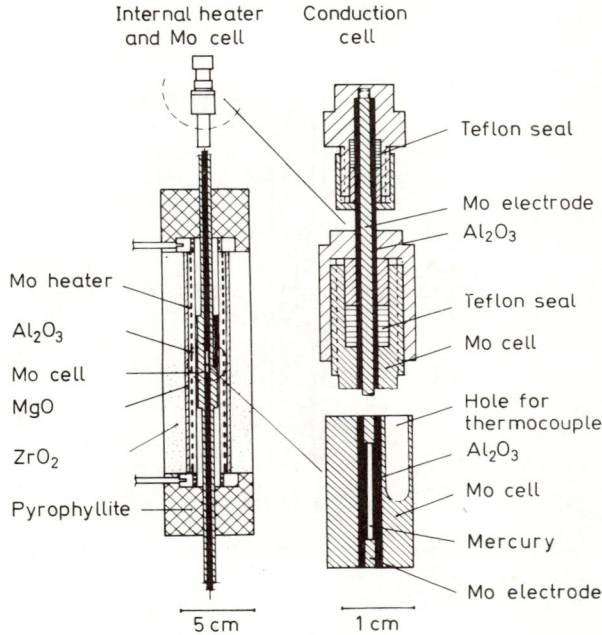

Figure 4. Conductivity cell used for measurement of the vapor pressure curve of mercury. (Hensel and Franck[16]).

were used as electrical insulators. The molybdenum cell with the internal heater was placed inside an autoclave, using argon as the pressurizing medium. The pressures of argon and liquid mercury were balanced outside the internally heated high pressure vessel at low temperature within a stainless steel auxiliary vessel. To obtain the conductivity, the potential differences at various currents between the two molybdenum electrodes in the mercury were determined. For the determination of the vapor pressure curve of mercury the following procedure was applied. At fixed pressures the temperature of the liquid metal in the cell was raised and the slowly increasing resistance observed. With subcritical pressures a discontinuous change to very large resistances was recorded at certain temperatures which was taken as an indication for the phase change from liquid to vapor. The corresponding pair of values for temperature and pressure belonged to the vapor pressure curve.

At high temperature, however, the application of the method described to the alkali metals involved difficulties. No electrically insulating container material compatible with potassium or sodium[12] at temperatures higher than 1000°C is available. Therefore a new method was developed by Freyland et al.[12]. Metallic tungsten–rhenium containers with thin walls were used as conductor in parallel with the fluid alkali metal samples. Again the cells were located inside an internally heated vessel pressurized with argon gas.

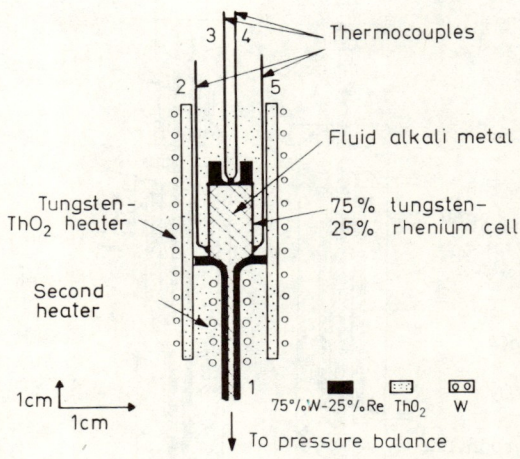

Figure 5. Conductivity—and thermoelectric power—cell used for the measurement of the vapor pressure curve of potassium. (Freyland and Hensel[12]).

The main part of the cell shown in Figure 5 is the thin walled tungsten–rhenium tube (75% W–25% Re). It is closed at the top. At the bottom a capillary ① which has a very small borehole connects the cell with the pressure-balancing system. With four tungsten–rhenium wires labelled ②, ③, ④, ⑤ the electrical resistance of the specimen is measured in the form of a four-terminal resistance, the tube is parallel to the fluid metal sample. The wires ② and ③ consist of 97% W–3% Re and the wires ④ and ⑤ of 75% W–25% Re. Thus the combinations of wires ② and ⑤ and of wires ③ and ④ form thermocouples for the measurement of the temperature directly at the bottom and at the top of the cell. In addition the two wires ④ and ⑤ and the fluid metal sample form a thermocouple consisting of tungsten–rhenium 75%–25%:fluid alkali metal:tungsten–rhenium 75%–25%. The thermoelectric voltage of this couple can be measured between ④ and ⑤ for a given temperature gradient along the fluid metal sample.

The vapor pressure curve of potassium was obtained[12] by heating the specimen inside the vessel at constant pressure and measuring its electrical resistivity and thermoelectric voltage. Evaporation was detected by a dis-

III. Results

1. Density

The intention of this section is to summarize some new density results of fluid metals under pressure at very high temperatures, especially at supercritical temperatures. This range is of special interest. The reason is that at supercritical temperatures a simple continuous decrease of the density to sufficiently low values causes the fluid to change from metallic to non-metallic behavior, and the type of chemical bonding changes. Density results in the relatively well explored range near the melting points of simple metals are not included. A complete summary of such results is given in the comprehensive review article by Ross and Greenwood[24].

Mercury and cesium are the only metals for which density data at high pressures and high temperatures have been measured for the liquid, the vapor and the supercritical phase. *Figure 6* shows isotherms of the density of liquid, gaseous and supercritical mercury as a function of pressure. The dotted line gives the extrapolated values for the subcritical liquid at equilibrium pressure. The critical density of mercury was estimated[22] to 5.3 g/cm^3 using density data of liquid and vapor along the saturation curve and assuming the validity of the rectilinear diameter law. The diagram *Figure 6* presents

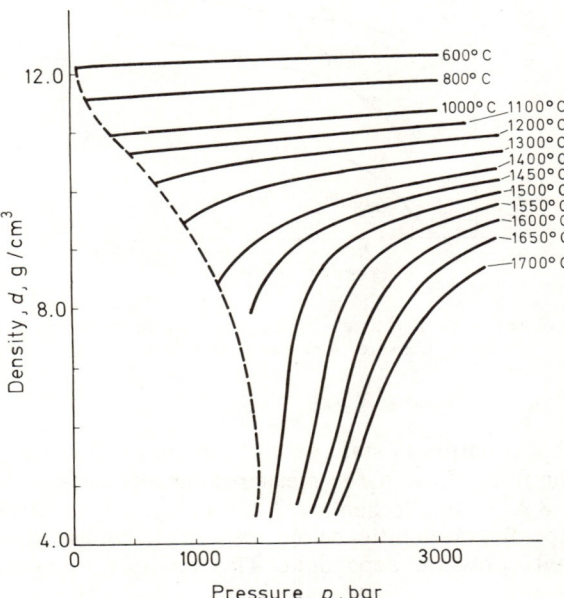

Figure 6. Isotherms of the mercury density at sub- and supercritical temperatures as a function of pressure. (Hensel and Franck[16], Kikoin and Sechenkov[19]).

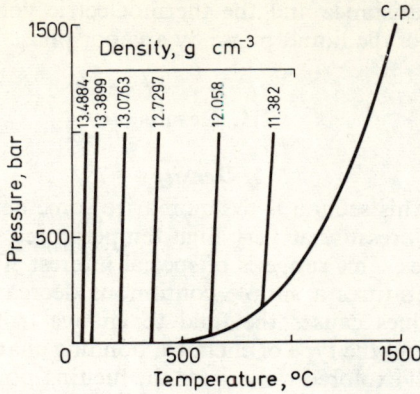

Figure 7. Mercury isochores up to pressures of 1 000 bar and subcritical temperatures of 1 100°C. (Postill et al.[22]).

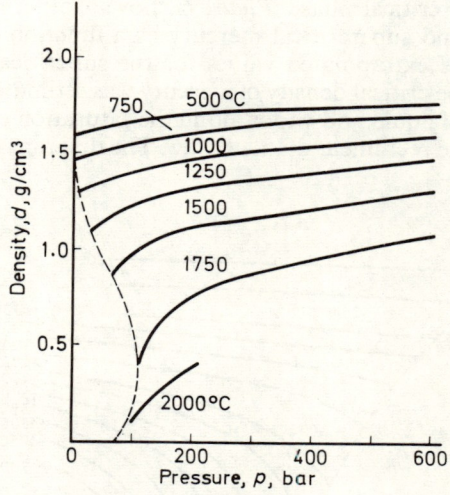

Figure 8. Isotherms of cesium density at sub- and supercritical temperatures as a function of pressure. (Alekseev et al.[1]).

the most probable equation of state data of mercury selected and critically analysed by Schmutzler[25] from the measured density data of Hensel and Franck[16] and of Kikoin and Sechenkov[19]. It is only at supercritical temperatures or at temperatures slightly below the critical temperature that the density is strongly pressure dependent. This is the normal behavior of most substances.

At lower subcritical temperatures, the compressibility of liquid mercury is relatively small. This is demonstrated by the results of Postill et al.[22]

represented in *Figure 7*. This is a pressure/temperature diagram for mercury giving the lines of constant density, and the vapor pressure curve up to the critical point as measured by Hensel and Franck[16].

Very recently measurements of the density of cesium at high pressures and temperatures were performed by Alekseev et al.[1] using the technique of counting the γ-rays of the ^{134}Cs isotope. Their apparatus and measuring method was similar to that employed by Kikoin et al. for the determination of the mercury densities. The results are represented in *Figure 8* at constant temperature as a function of pressure. They cover the subcritical, critical and supercritical regions up to 2 000°C and pressures up to 600 bar. A comparison of these results with the mercury data of *Figure 6* shows that in the liquid

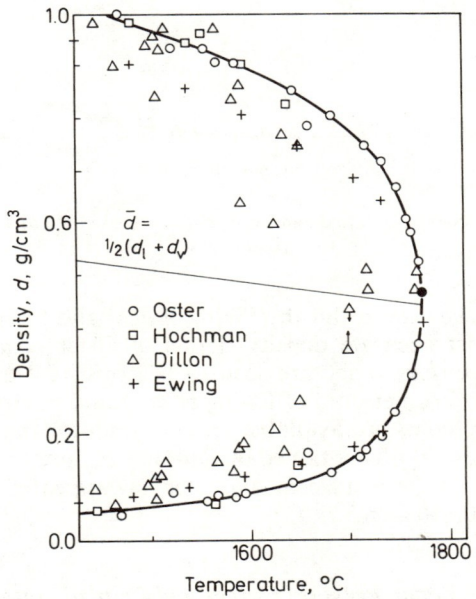

Figure 9. Densities of coexisting liquid and vapor phases of cesium as a function of temperature. (Oster and Bonilla[21]).

metallic phase at comparable reduced temperatures and pressures the compressibility of cesium considerably exceeds that of mercury. On the other side in the non-metallic, insulating low density vapor phase the compressibilities of cesium and of mercury are comparable. The dotted line in *Figure 8* gives the values for the subcritical liquid and vapor phase at equilibrium vapor pressure. The critical density of cesium, determined by the linear diameter method from these results, is 0.4 ± 0.02 g/cm³ in very good agreement with results obtained by other authors[6, 18, 21].

The linear diameter law referred to above has until now been tested only for a few metals[6, 22]. Using it, Oster et al.[21] and Dillon et al.[6] obtained the critical density of cesium as 0.407 g/cm³. Their results are shown in *Figure 9*

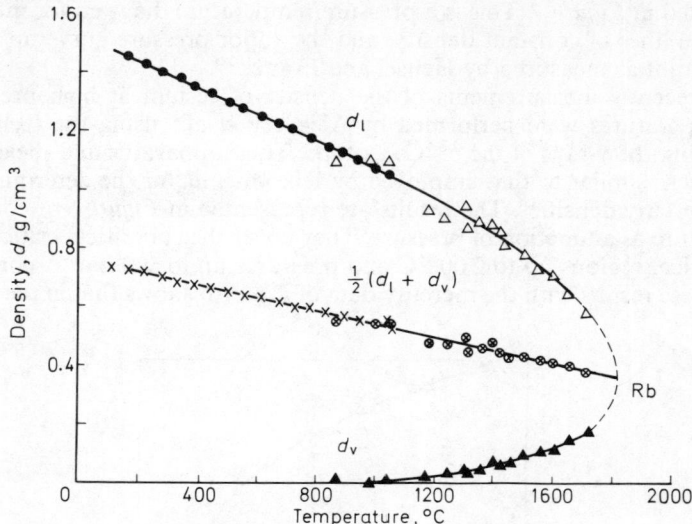

Figure 10. Densities of coexisting liquid and vapor phases of rubidium as a function of temperature. (Dillon et al.[6]).

together with some earlier results by Ewing et al.[11] and by Hochman et al.[18]. The probably most accurate density points of Oster et al. are shown by *Figure 9* as open circles. They are seen to be consistent and in very good agreement with the former data of Ewing et al.[11] and of Hochman et al.[18].

Similar measurements are available for rubidium for the saturated liquid and the vapor. These results obtained by Dillon et al.[6] are shown in *Figure 10*. Again the validity of the rectilinear diameter law appeared to be valid. The critical density is 0.346 g/cm^3.

2. Vapor Pressure Curves and Critical Data

A knowledge of the vapor pressure is of fundamental importance for any liquid, but for metallic liquids it is of special interest. This is because the thermodynamic transition from liquid to vapor across the saturation curve and the transition in the electronic structure from a highly conducting, possibly metallic liquid to a low density non-conducting vapor must be considered simultaneously. This is demonstrated by *Figure 11*, which shows on the left side the vapor pressure curve of liquid cesium up to the critical point, which was found at $T_c = 1750°C$ and $P_c = 110$ bar. The right side of *Figure 11* shows the electrical conductivity of cesium at sub- and supercritical temperatures as a function of pressure. The dotted line gives the conductivities of the coexisting liquid and vapor phases. At 1200°C and saturation pressure, for example, the conductivity of liquid cesium is 5900 ohm^{-1} cm^{-1}, which is quite high and surely metallic. At vaporization of the liquid the conductivity decreases discontinuously to a value of about 10 ohm^{-1} cm^{-1}, which is very small and caused by thermal ionization of the cesium vapor. The critical point must

PART 2. THERMODYNAMIC PROPERTIES OF FLUID METALS

be the limit for observing the sharp transition between a metallic and a non-metallic fluid, since at this point any distinction between liquid and vapor disappears. This is demonstrated by the experimental results. At supercritical temperatures a very steep but continuous increase of the conductivity with increasing pressure is observed. This behavior of the electrical conductivity across the saturation curve and in the critical region is typical for

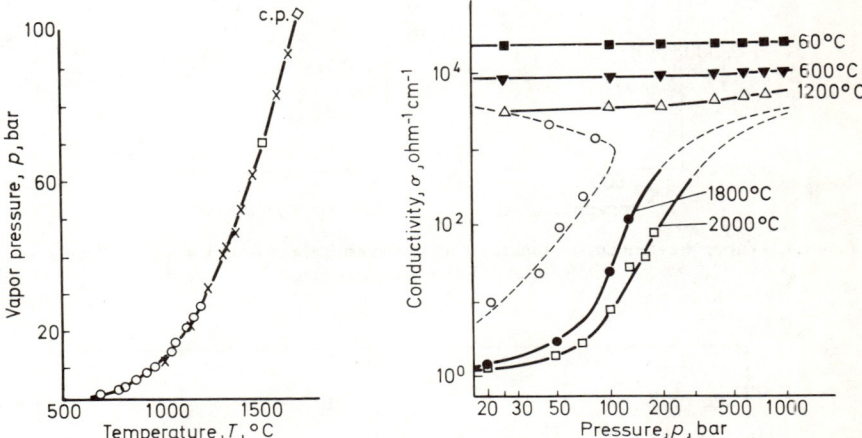

Figure 11. The vapor pressure curve of cesium up to the critical point (c.p.) and the electrical conductivity of cesium at sub- and supercritical temperatures as a function of the pressure. (Renkert et al.[23]).

liquid metals and has also been observed for mercury[16], cesium[23] and potassium[13]. The vapor pressure curves of mercury[16] and potassium[12] shown in *Figure 12* and *Figure 13* were also determined by studying the electrical conductivity as a function of pressure and temperature. These results are found to be in good agreement with earlier measurements by Cailletet et al.[5] and by Douglas et al.[8] for mercury and of Shpilrain and Nikanorov[28] for potassium at temperatures far below the critical as indicated in *Figures 12 and 13*. Further it is demonstrated that for both the metals mercury and cesium, over the entire range of temperature, the logarithms of the saturation vapor pressures are linear functions of the reciprocal temperatures.

As described above, the data of the critical point, which terminates the vapor pressure curve, were determined using conductivity measurements. In order to test the anticipation that the critical point detected by conductivity measurements coincides with the normal thermodynamic critical point the present authors also examined the form of their density isotherms of mercury and confirmed their data from the conductivity results. *Table 1* shows critical parameters of eleven metals. The critical temperatures range from about 1 500°C to 20 000°C. The measured data were either obtained by analysing conductivity data[17, 16, 19, 23, 12] or density data of the coexisting liquid and vapor phases[16, 19, 5, 21]. The critical data of the other metals have been

987

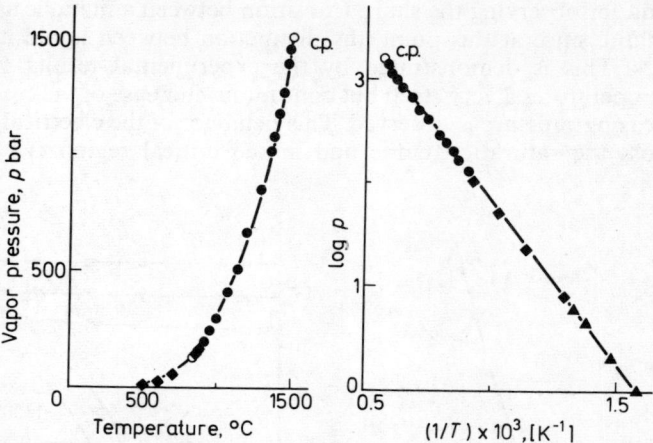

Figure 12. Vapor pressure curve of mercury up to the critical point: this work, Cailletet et al.[5], Douglas et al.[8]. (Hensel and Franck[16]).

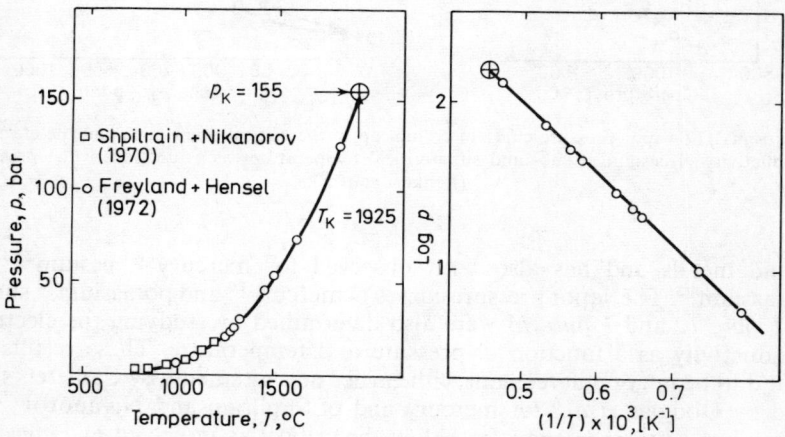

Figure 13. Vapor pressure curve of potassium up to the critical point: this work, Shpilrain et al.[28]. (Freyland and Hensel[12]).

estimated by Grosse[14] using the existing experimental data on liquid densities of these metals at lower temperatures and the linear diameter law.

It is possible to derive from the measured data some interesting thermodynamic properties. *Figure 14* shows as an example the vaporization entropy of potassium as a function of the reduced temperature up to the critical point in comparison with the non-metal argon. The ΔS-values are calculated from the slopes of the vapor pressure curve and the density data of the vapor and liquid at saturation pressure determined by Ewing et al.[29] and employing

PART 2. THERMODYNAMIC PROPERTIES OF FLUID METALS

the Clausius–Clapeyron equation. The curve for the non-metal in *Figure 14* can be clearly distinguished from that of the metal. This demonstrates that

Table 1. Critical data of some selected metals

Metal	T_c, K	p_c, bar	d_c, g/cm^3	Investigators	Remark
Mercury	1763	1490	5.3	Hensel, Franck[17]	measured
Mercury	1753	1480	5.7	Kikoin, Sechenkov[19]	measured
Cesium	2033	110	—	Renkert et al.[23]	measured
Cesium	2050	117	0.407	Oster, Bonilla[21]	measured
Potassium	2198	155	—	Freyland, Hensel[12]	measured
Rubidium	2093	159	0.346	Dillon et al.[6]	estimated
Sodium	2573	275	0.206	Dillon et al.[6]	estimated
Cadmium	2700	—	—	Grosse[14]	estimated
Tin	8720	2100	—	Grosse[14]	estimated
Copper	8500	—	—	Grosse[14]	estimated
Lead	5400	850	—	Grosse[14]	estimated
Iron	10000	—	—	Grosse[14]	estimated
Tungsten	23000	>10000	—	Grosse[14]	estimated

metals and non-metals cannot be included together in a group obeying a principle of corresponding states, which was extensively discussed by Ross and Greenwood[24]. This is not surprising. The reason is that for metallic liquids the situation is less simple than for insulating liquids. The interacting forces between atoms or molecules in the gaseous state of metallic elements

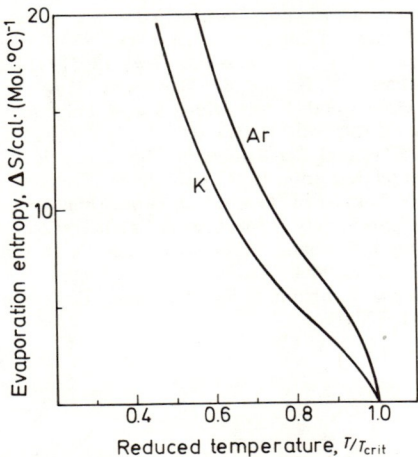

Figure 14. Evaporation entropy of potassium and argon as a function of the reduced temperature. (Freyland and Hensel[12]).

are very different from the forces between the ions in the liquid state. A sufficiently large increase in the density by condensation at subcritical temperatures or by a continuous compression at supercritical temperatures can change the fluid from non-metallic to metallic behavior. All properties of the fluid are affected by this transition and some in a drastic manner. This was recently demonstrated for cesium, potassium, and mercury by measurements of the electrical conductivity, optical absorption, the thermoelectric power and Hall coefficient[17, 13, 15, 26, 10].

IV. References

[1] Alekseev, V. A., V. G. Ovcharenko, Yu. F. Ryzhkov and A. R. Sechenkov. *Zh. Eksp. Teor. Fiz., Dis. Red.* **12**, 306–309 (1970).
[2] Bender, J. *Phys. Z.* **16**, 246 (1915); **19**, 410 (1918).
[3] Birch, F. *Phys. Rev.* **41**, 641 (1932).
[4] Bridgman, P. W. *Proc. Amer. Acad. Arts. Sci.* **60**, 385 (1925).
[5] Cailletet, L., E. Collardeau and C. A. Rivière. *CR Acad. Sci., Paris*, **130**, 1585 (1900).
[6] Dillon, I. G., P. A. Nelson and B. S. Swanson. *J. Chem. Phys.* **44**, 4229 (1966).
[7] Dillon, I. G., P. A. Nelson and B. S. Swanson. *Rev. Sci. Instrum.* **37**, 614 (1966).
[8] Douglas, T. B., A. F. Ball and J. Ginnings. *J. Res. Nat. Bur. Stand.* **46**, 334 (1951).
[9] Endo, H. *Phil. Mag.* **8**, 1403 (1963).
[10] Even, U. and J. Jortner. *Physics Letters A*, (1972) (in press).
[11] Ewing, C. T., J. R. Spann, J. P. Stone and R. R. Miller. *J. Chem. Engng Data*, **16**, 27 (1971).
[12] Freyland, W. F. and F. Hensel. *Ber. Bunsenges. Phys. Chem.* **76**, 17 (1972).
[13] Freyland, W. F. and F. Hensel. *Ber. Bunsenges. Phys. Chem.* **76**, 347 (1972).
[14] Grosse, A. V. *J. Inorg. Nucl. Chem.* **22**, 23 (1961);
Grosse, A. V. *Rev. Hautes Temp. Refractaires*, **3**, 115 (1966).
[15] Hensel, F. *Ber. Bunsenges. Phys. Chem.* **75**, 847 (1971).
[16] Hensel, F. and E. U. Franck. *Ber. Bunsenges. Phys. Chem.* **70**, 1154 (1966).
[17] Hensel, F. and E. U. Franck. *Rev. Mod. Phys.* **40**, 697 (1968).
[18] Hochman, J. M. and C. F. Bonilla. *Third Symposium on Thermophysical Properties*, pp 122–130. American Society of Mechanical Engineers: Purdue University (1960).
[19] Kikoin, I. K. and A. R. Sechenkov. *The Physics of Metals and Metallography*, **24**, 74 (1967).
[20] Köster, H. von, F. Hensel and E. U. Franck. *Ber. Bunsenges. Phys. Chem.* **74**, 43 (1970).
[21] Oster, G. F. and C. F. Bonilla. *Fifth Symposium on Thermophysical Properties*, pp 468–474. American Society of Mechanical Engineers: New York (1970).
[22] Postill, D. R., R. G. Ross and N. E. Cusack. *Phil. Mag.* **18**, 519 (1968).
[23] Renkert, H., F. Hensel and E. U. Franck. *Ber. Bunsenges. Phys. Chem.* **75**, 507 (1971).
[24] Ross, R. G. and D. A. Greenwood. *Progr. Mater. Sci.* **14**, 173 (1969).
[25] Schmutzler, R. W., *Thesis*, Karlsruhe (1971).
[26] Schmutzler, R. W. and F. Hensel. *Ber Bunsenges. Phys. Chem.* **76**, 531 (1972).
[27] Scholten, A. and F. Hensel. Karlsruhe (to be published).
[28] Shpilrain, E. E. and E. V. Nikanarov. *Proceedings of the Fifth Symposium on Thermophysical Properties*, p 450. American Society of Mechanical Engineers: New York (1970).
[29] Stone, J. P., C. T. Ewing, J. R. Spann, E. W. Steinküller, D. D. Williams and R. R. Miller. *J. Chem. Engng Data*, **11**, 306 (1966).
[30] Stone, J. P., C. T. Ewing, J. R. Spann, E. W. Steinküller, D. D. Williams and R. R. Miller. *J. Chem. Engng Data*, **11**, 320 (1966).

CHAPTER 20

Interphase Surface Tension

P. P. PUGACHEVICH
Institute for High Temperatures, Krasnokazarmennaya 17a, Moscow E250, USSR

Contents

I.	Introduction	991
II.	Static Methods of Determination of the Interphase Surface Tension	992
	1. Method of Capillary Rise or Depression	992
	2. Method of the Shape of a Sessile Drop or a Gas Bubble	995
III.	Semistatic Methods of Determination of the Interphase Surface Tension	1001
	1. Method of the Weight or Volume of a Drop	1001
	2. Ring or Plate Rupture Method	1003
	3. Method of Maximal Pressure in a Gas Bubble or a Drop	1006
	4. Improved Gas Devices with One Calibrated Tube (IGASDICT)	1009
	5. Improved Gas Devices with Two Calibrated Tubes (IGASD2CT)	1014
	6. Gravitational Devices	1015
IV.	References	1020

I. Introduction

Surface phenomena have been known by humanity from the remote past. However, the first qualitative observations of capillarity phenomena were made by the famous Arab scientist Algacini, in the thirteenth century and described in his book 'On the Balance of Wisdom'.

Later on, capillarity problems were studied by Leonardo da Vinci, Newton, Thomas Young, Laplace, Gauss, Poisson, Mendeleev, Gibbs, Rayleigh, Van der Vaals, Nils Bohr, Einstein, Schrödinger and other scientists.

Now, the field of research on surface tension (σ) of different liquids is very large and concerns problems related to: the production of rocket fuels, radiation protection, the use of surface phenomena in atomic energy, the creation of metallic heat transfers, working in weightless cosmic conditions, the production of ultra-pure metals, semiconductor materials and heat-resisting alloys, the study of fluorescence properties of amalgams which are replacing mercury in fluorescent tubes and the development of an elaborate technique for using and treating polymer materials, etc.

Finally, thermodynamics always deals with interphase surface tension (IST) and especially with surface tension between liquid/vapor or gas (σ_{10}), between liquid/liquid (σ_{11}) and liquid/solid state (σ_{12}). Taking this into account, we will examine some of the most common methods of measuring IST, especially in these cases and we will not consider problems related to the measurement and estimation of the surface tension of solid substances.

It is interesting to remark that the first measurements of surface tension of metals (mercury) were conducted by Mervo, ten years before Watt discovered the vapor engine. However, the real expansion of science in the field of surface phenomena coincides with the beginning of the nineteenth century. At this time, Young and Laplace developed one of the first theories of the surface tension of liquids. The famous Laplace equation[25,26], elaborated about 170 years ago, established a relation between the difference of pressures Δp on the outer and inner sides of a curved surface of a liquid, its IST σ_{10} or σ_{11} and the main radii of curvature R_1 and R_2 at a given point on the interface.

$$\Delta p = \sigma_{11}(1/R_1 + 1/R_2) \tag{1}$$

This equation was the starting point for elaborating numerous measurement methods of IST.

All these methods can be divided conditionally into static, semistatic and dynamic methods.

The first group includes the method of capillary rise or depression, the method of the shape of a sessile drop or bubble, the method of the shape of a hanging drop or bubble, and the method of the floating plate.

The second group includes the methods of ring or disc rupture, the methods of the weight or the volume of a drop, and the method of maximal pressure in a drop or a bubble.

Finally, the third group includes the method of capillary waves, the method of oscillating jets, and the method of oscillating drops.

There are other methods also which might be related to some other group of methods (for instance, method of thread constriction, of two sphere rupture, or two liquid streams opposed to each other, etc.). However, we will not describe these methods inasmuch as they are not sufficiently widely used.

However, the methods we mentioned above are not equally useful for measuring IST. The most widespread are: the method of capillary rise or depression, of the maximal pressure in a drop or bubble, and the method of the sessile drop, whereas, for instance, dynamic methods are seldom used for measuring IST.

II. Static Methods of Determination of the Interphase Surface Tension

1. Method of Capillary Rise or Depression

As it results from the Laplace equation 1, if solved to a first approximation, the height of rise h of a heavy liquid of density ρ_2 in a very narrow capillary tube, with inner radius r (*Figure 1*) will be given by

$$h = 2\sigma_{11} \cos \Theta / gr(\rho_2 - \rho_1) \tag{2}$$

in which Θ is the contact angle of wetting, g is the acceleration due to gravity, and ρ_1 denotes density of a lighter liquid. The solution of the Laplace equation, to a second approximation, leads to a more complex equation for h

$$h = \frac{2\sigma_{11} \cos \Theta}{gr(\rho_2 - \rho_1)} - \frac{r}{\cos \Theta}\left[1 - \frac{2}{3} \cdot \frac{1 - \sin^2 \Theta}{\cos^3 \Theta}\right] \tag{3}$$

and if we consider tubes with a sufficiently large diameter, as was done by Rayleigh[58] and Verschaffelt[63], we obtain even more complicated expressions.

Knowing h, r, ρ_2, ρ_1, g and Θ from equations 2 and 3, we are able to determine σ_{11}. However, we have to note that these formulas are valid only when

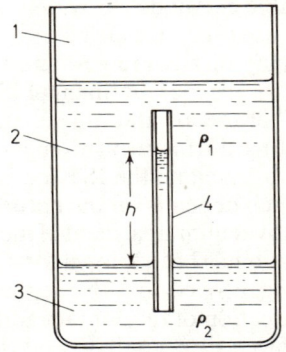

Figure 1. IST determination on the interface liquid/liquid by the capillary rise method. 1. Liquid container; 2. Light liquid of ρ_1 density; 3. Heavy liquid of ρ_2 density; 4. Capillary tube.

the surface of the studied liquid outside the capillary has infinite extent. In the opposite case, the capillary rise of liquid between tube 4 and the wall of the container 1 (*Figure 1*), will greatly distort IST. Thus, to measure IST and calculate it according to the above-mentioned formulas, we need relatively large quantities of liquid and this involves experimental difficulties.

To reduce the quantity of liquid under investigation as much as possible, we sometimes use two capillary tubes (*Figure 2*) of different radii r_1 and r_2 and the following formula to determine the difference of level of liquid in these capillaries

$$\Delta h = \frac{2\sigma_{11} \cos \Theta}{g(\rho_2 - \rho_1)} \left[\frac{1}{r_1} - \frac{1}{r_2} \right] \tag{4}$$

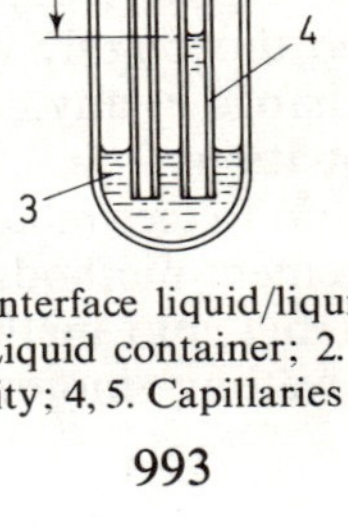

Figure 2. IST determination on the interface liquid/liquid by the capillary rise method using two capillaries of different radii. 1. Liquid container; 2. Light liquid of ρ_1 density; 3. Heavy liquid of ρ_2 density; 4, 5. Capillaries of different radii.

i.e. we get a very simple equation to calculate IST. The theory of the determination of IST with two communicating capillaries was first elaborated by N. D. Pilchikov[36], as early as 1888. Later on, this method was used by many workers to measure the IST of various liquids.

Thus, the method of capillary rise, which, according to Bakker[1] 'is one of the earliest and most reliable methods', is widely applied. However, being one of the earlier methods used for measuring IST and though it looks very simple, this method has many disadvantages, and these may make it sometimes impossible to apply it for the measurement of IST. It cannot, for instance, be used for measuring IST of metals and alloys, and of liquids that change their contact angles of wetting during measurement processes. In fact, as we see from equations 2 to 4, to calculate the IST for a given capillary, we must know the wetting angle, which depends on the nature and composition of the wetting liquid, on the thermovacuum treatment of the measuring device and on temperature, etc. This relation has been known for a long time, as when Metz, according to Thomas Young, proved that the depression of mercury in glass tubes depends on the imperfection of the contact and that after a long boiling of mercury in tubes, its surface grows concave and that depression turns into rise. Schumacher[60] after accurate experiments, found that in Pyrex tubes, the mercury meniscus turns completely flat, after a long thermovacuum treatment, and in quartz tubes—slightly concave. He has also observed a rise of mercury in capillaries of different kinds of Pyrex glass and has determined the qualitative bond between wetting and chemical composition of glass. The present author[42] has also observed a concave meniscus for pure mercury, in glass devices for IST measurement. However, we can see a concave meniscus growing convex as soon as the device is opened and air enters. We must suppose that the phenomenon between contact wetting angles and different factors will take place not only for glass and touching liquids, but also for other solid substances in contact with liquids. This has been proved by numerous experiments. In particular, we have found[55] that after a severe thermovacuum treatment, at 1 200°, of graphite and metals such as tantalum, tungsten, molybdenum, etc., they are thoroughly wetted by tin though, in ordinary conditions, molten tin does not spread on their surface. So it is necessary, in the capillary rise method, to take into account the contact angles of wetting, but it is impossible to determine the contact angle of wetting in the capillary tube simultaneously with the measurement of the height of the liquid rise, except in special cases.

On the other hand, it is not worth measuring contact angles of wetting, separately, on a flat support of the same material as that of the capillary and then reporting them in an IST calculation according to equations 2 to 4, because of the inadequacy of the measurement of Θ and h.

Thus, to avoid taking contact angles of wetting into account, we very often suppose that the studied liquid thoroughly wets the capillary material and that this condition does not change in any circumstances. This assumption leads to a simplification of equations 2 to 4, but, unfortunately, results obtained in this way, for IST, are no longer reliable. We can agree with Fergusson's assumption that measurement methods of IST should not be divided into static and dynamic ones, but into methods which do or do not take account of contact angles of wetting; moreover, all methods leading to a reckoning formula and including contact angles of wetting should be set aside and never used for the study of IST. Thus, we shall not describe the devices employed for determining IST according to the method of capillary rise.

2. *Method of the Shape of a Sessile Drop or a Gas Bubble*

The method of the shape of a sessile drop has been very widespread for measuring the IST of different liquids, from melted metals to polymer solutions. Numerous works have been specially devoted to the theory and the performance of this method[16, 17]. We know that the shape of a heavy liquid drop on a solid support immersed in a liquid [*Figure 3*(a)] or the shape of a light liquid drop on a solid support immersed in a light liquid [*Figure 3*(a)] or the shape of a light liquid drop on a solid support immersed in a heavy liquid [*Figure 3*(b) depends on the IST, the difference of density, the mass of

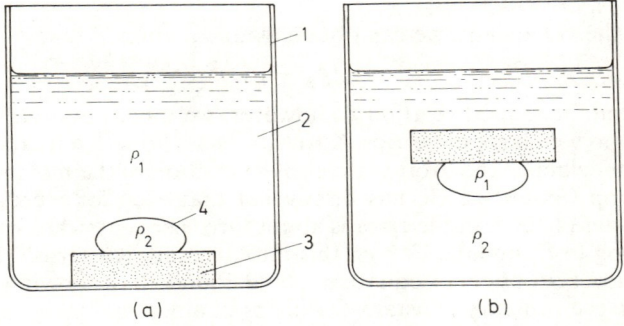

Figure 3. IST determination on the interface liquid/liquid, by the sessile drop shape method. 1. Liquid container; 2. Light liquid; 3. Solid support; 4. Heavy liquid.

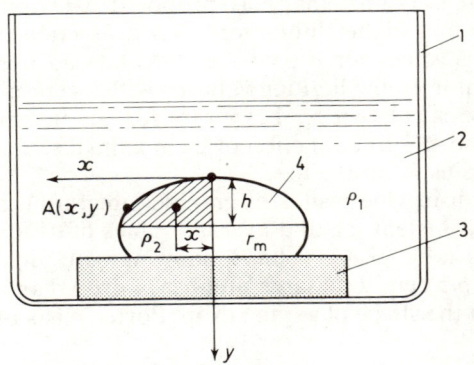

Figure 4. IST determination by the sessile drop shape method. 1. Liquid container; 2. Light liquid; 3. Solid substrate; 4. Heavy liquid.

the drop and the acceleration due to gravity. From equation 1 we can prove that for any point (x, y) on the surface of revolution of the drop (*Figure 4*)

$$\frac{y}{a^2} = \frac{1}{R_1} + \frac{1}{R_2} - \frac{2}{R_0} \tag{5}$$

where $a^2 = \sigma/(\rho_2 - \rho_1)g$ is a capillary constant; and R_0 is the radius of curvature at the top of the drop. Quincke[57] applied equation 5 for measuring IST according to the shape of a flat drop. In this case, $R_2 = \infty$, $R_0 = \infty$, so that equation 5 may be written

$$y/a^2 = 1/R_1 \tag{6}$$

If we express R_1 in equation 6 through a differential form and if we solve the resulting equation, we find that

$$2a^2 = h^2 \tag{7}$$

where h is the distance between the top of the drop and the equatorial surface (see *Figure 4*).

If we use the value of a^2, we can obtain from equation 7, that

$$\sigma_{11} = \tfrac{1}{2}g(\rho_2 - \rho_1)h^2 \tag{8}$$

Many workers have used relation 8; however, we must note that it can be applied only to very large flat drops. Kasterin[19] has shown from experimental data that the height of the drop is a complex function of the maximal radius r_m of the drop (*Figure 4*). He has discovered that even for mercury drops with a diameter of 100 mm the error is about three per cent, when calculating IST according to formula 8. Taking this into account, Kasterin[19] has given a more rigorous solution of equation 6 and found the following formula, to determine the capillary constant, for drops of any size

$$a^2 = \{2\bar{x}S - hr_m^2\}/\{2r_m(r_m/R_0 - 1)\} \tag{9}$$

where S is the crosshatched surface on *Figure 4*, and $\bar{x}$ is the abscissa of the center of gravity of this surface. He has also given simpler ways of calculating $\bar{x}$, S, R_0 values. We know of many articles related to the precision of the Quincke's formula. Lohnstein[27] has proposed a system of equations to determine lower and higher limits for a^2 as a function of h and r_m. Then, knowing limiting values for a^2, we can calculate an average value of the capillary constant from the limiting value a^2 with a certain error probability. Siedentopf[61], to calculate σ_{11}, has proposed an interpolation equation, allowing us to find IST as a function of the maximal radius r_m of a drop and the curvature radius R_0 at the top.

Heydweller[14], using the results of graphic integration of Laplace's equation obtained by Siedentopf, and average values of a^2 found by Lohnstein, has established a table of $a/h = f(h/r)$, allowing us to determine σ_{11} with a precision of one per cent for a large number of drop diameters. To calculate IST according to the shape of a lying drop, Porter[41] has used the formula

$$\frac{a^2}{r^2} = \frac{1}{2} \frac{h^2/r^2}{(1 - h^2/r^2) + \Delta} \tag{9}$$

where Δ is a correction depending on h/r. This correction has been calculated for small drops from Bashfort's and Adams's tables[2], presenting a numerical integration of the Laplace equation. For large drops, the Δ correction has been calculated by the approximate formula.

Tawde and Parvatikar[65], applying Bashfort's and Adams's tables, have established a relation between a^2/r^2 and h/r

$$a^2/r^2 = f(h/r) \tag{10}$$

An analogous table

$$a^2/4r^2 = f(r/h) \tag{11}$$

estimated from Bashfort's and Adams's tables has been proposed by Koshevnik, Kusakov and Lubman[22]. Popel and his colleagues have integrated Laplace's equation 1 graphically and published the results through diagrams[39, 40] allowing us to determine IST according to h and r_m, with a precision of 1.5 to 2.5 per cent.

As has been shown by experiment, the problem of the most precise calculation of IST through the parameters of a sessile drop is related to deductions from drop outline measurements on photographs using tables[2, 22] or diagrams[40].

We can easily see from *Figure 4* that the precision of IST determination depends on the precision of measurement of the distance h between the pole of the drop and the equatorial surface. Small deviations of this surface from the real position will lead to large errors in IST determinations.

Dorsey[11] has proposed an interesting and precise way of measuring photographs of drops. Instead of the height h, Dorsey has proposed measuring the segment H (*Figure 5*) equal to the distance between the top of the drop and the top of the 2α angle whose bisector coincides with the axis of symmetry of the drop and whose sides touch its outline. The value of the 2α angle between tangents may be arbitrary and greater than $\frac{1}{2}\pi$. In this case we

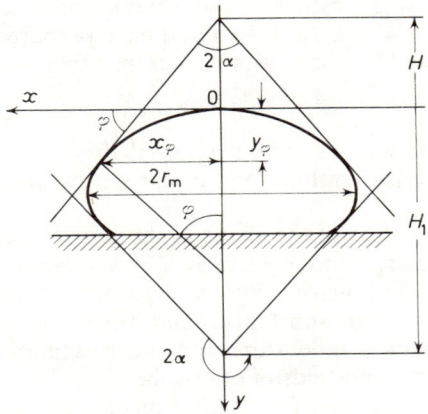

Figure 5. IST determination according to Dorsey.

must measure the segment H_1 (*Figure 5*). We can easily see from the figure, that

$$H = x_\varphi \tan \varphi - y_\varphi \tag{12}$$

This relationship is valid not only for tangential points but for any point of x and y coordinates located on the tangent; thus it is useless to look for a tangential point on the photograph of the drop but to determine H according to Ivashchenko's method[16, 17]. By this method and with the help of a microscope, for instance UIM-23, we determine coordinates x_3, x_4, z_4, z_5

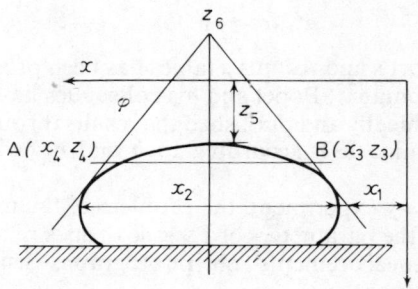

Figure 6. IST determination according to Ivashchenko *et al.*

and the angle φ (*Figure 6*) and then we calculate H according to the formula

$$H = \tfrac{1}{2}(x_4 - x_3) \tan \varphi - (z_4 - z_5) \tag{13}$$

From H and r_m values, we calculate the capillary constant; then, knowing the density, we derive the IST.

Dorsey has used an empirical formula elaborated from Bashfort's and Adams's table for $\varphi = \pi/4$

$$\frac{a^2}{r^2} = \frac{0.05200}{[(H/r) - 0.41421]} - 0.12268 + 0.0481\,[(H/r) - 0.4121] \tag{14}$$

Kozakevitch *et al.*[23], using Bashfort's and Adams's tables, have estimated $H\varphi/r$ values for angles $\pi/4$, $\pi/3$, $2\pi/3$, $3\pi/4$ and have prepared corresponding tables. Ivashchenko *et al.*[16, 17] have established the table for the relationship

$$a^2/r^2 = \mathrm{f}(H/r) \tag{15}$$

for angles $\phi = \pi/4$ and $\phi = \pi/3$ and have used this table to calculate a^2, and then IST, from considerations on similar drops and Bashfort's and Adams's tables.

In many works the method of the shape of the sessile drop has been used for IST measurements (σ_{10} and σ_{11}) with a great variety of substances: liquid metals and semiconductors, alloys, organic and inorganic liquids and their mixtures, polymer substances and their solutions, over wide ranges of concentration and temperature, and under vacuum or high pressure conditions reaching some hundreds of atmospheres.

In particular, to measure the IST of liquid metals, the sessile drop shape method has been used by Poindexter and Kernaghan[37] who have studied the IST of alkali metals. Burdon[6,7] has used this method to study σ_{10} of mercury; White[70], to measure σ_{10} of zinc; Kingery and Humenik[20], and later Popel et al.[38], to measure the IST of liquid alloys at temperatures exceeding 1400°C. Kozakevitch and Urbain[24] have applied this method to study different liquid ferrous alloys at high temperature, etc.

The popularity of this method is not accidental and is related to the fact that in the best version, the equipment developed by Ivashchenko and co-authors[16,17] for instance, it allows IST determination with a precision up to 0.5 per cent and measurements of temperature up to 2000°C.

Figure 7 is a diagram of the apparatus[17] developed by Ivashchenko et al. for the σ_{10} measurement of liquid metals and *Figure 8* part of this installation: the vacuum chamber and the furnace.

As shown in *Figure 7*, on a steel beam 10 are fixed a photocamera 1,

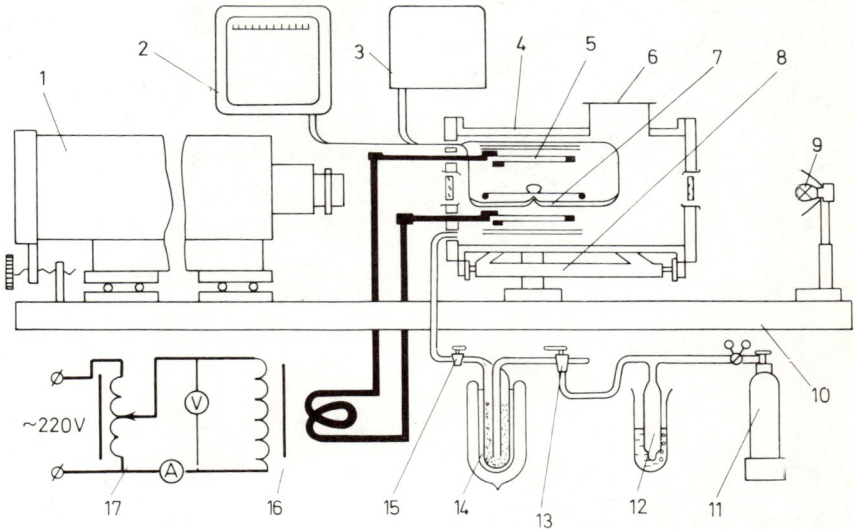

Figure 7. Diagram of the Ivashchenko apparatus for IST measurement of liquid metals. 1. Photocamera; 2. Potentiometer EPP.09; 3. Potentiometer KP.59; 4. Vacuum chamber; 5. Furnace of molybdenum foil with a melting drop below; 6. Connections. See text for items 7 to 17 inclusive.

an illuminator 9, a vacuum chamber 4 connected to a regulating system 8, allowing us to place the chamber horizontally. In the middle part of the vacuum chamber, there is a horizontal furnace 5, with the metal drop for study on a support, an equipment (not illustrated), its mounting and also thermocouples 7. The furnace with coaxial heater of molybdenum foil is fed by a step-down transformer 16 through a variable voltage transformer 17. The chamber 4 is connected through a pipe 6 to a vacuum system and can be filled up with purified gas, hydrogen or helium, through the valve 15 which connects the chamber to the purification gas system. The photo-camera 1, similar to the FK 13 × 18 camera, is equipped with an objective

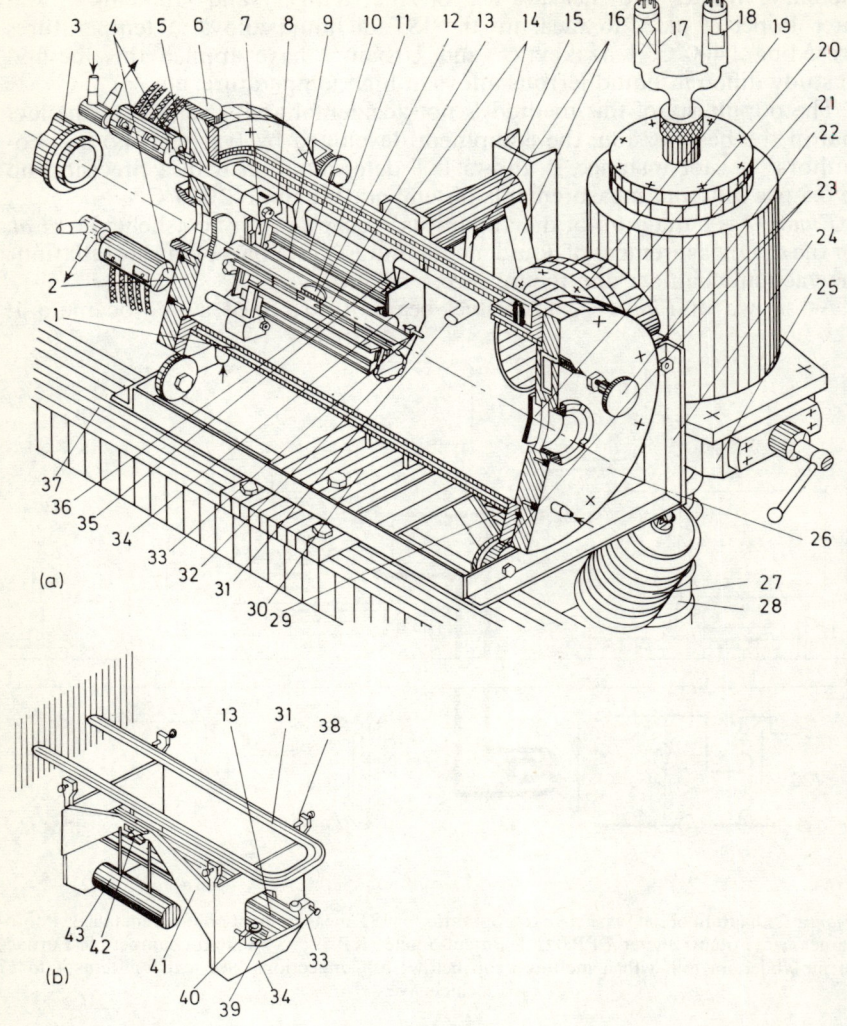

Figure 8. The Ivashchenko apparatus to measure IST of liquid metals. 1. Cylindrical chamber with flange packing of Teflon interlayers; 2, 26. Lateral caps; 3. Photocamera; 4. Water-cooling system of Teflon layers; 5. Optical quartz window; 6. Level; 7. Outlet; 8. Contact copper shoe with notches; 9. Coaxial heater of molybdenum foil; 10. Cylindrical screens; 11. Drop of liquid metal; 12. Support; 13. Layer of molybdenum foil with bent back edges; 14. Columns; 15. Revolving socket; 16. Rods; 17. Vacuum generator LT2; 18. Vacuum lamp LM2; 19. Connecting pipe; 20. T-shape head; 21. Revolving lock; 22. Handle for regulating the revolving lock; 24. Framework; 25. Bolt; 27, 29. Bellows; 28. Diffusion pump CVL-100; 30. Steel plate; 31. Watercooling U-shaped pipe; 32. Thermocouples; 33. Shoe; 34. Rocking jaw; 35. Suspension arms; 36. Aluminum oxide pipe; 37. Steel beam; 38, 39. Screws with conical tips; 40. Screws tightening the molybdenum plate 13; 41. Crank for the jaws; 42. Link hinge; 43. Load.

Industar 11-M = (f 300 mm), so the distance between the objective and the plate may be 1 800 mm.

The temperature under the support on which the drop is lying is measured by thermocouples 7, connected to the potentiometers 2 and 3. An automatic recording potentiometer 2 allows us to survey the temperature régime as the drop is being heated to a given temperature and another more precise potentiometer 3 is used to check the temperature measurements.

As shown in *Figure 7*, inert gas from the cylinder 11 is supplied through a reducer into a manostat 12, filled with D-IA oil. From the manostat, gas proceeds through a three-way valve 13 into an activated charcoal trap 14, cooled by liquid nitrogen. After a thermovacuum treatment, all systems including the furnace 4 are filled with inert gas (if IST measurement is done in an inert atmosphere), up to 1 atm. During the thermovacuum treatment, the furnace is heated to a temperature slightly lower than the melting point of the metal under test. Then the temperature of the system is raised so that the metal in the furnace 8 melts into a drop, which is photographed, then measured (parameters) as previously described.

To give a revolving shape to the drop, we use a refractory material crucible

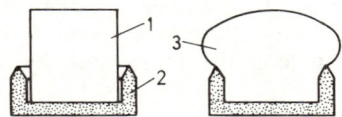

Figure 9. Constraining formation of a liquid metal drop. 1. Metal before melting; 2. Crucible of high-melting oxides; 3. Drop of melted metal.

with a cylindrical section and sharpened edges (*Figure 9*). This way of constraining formation of the drop has been reported[61, 33, 35].

III. Semistatic Methods of Determination of Interphase Surface Tension

1. *Method of the Weight or Volume of a Drop*

The method of the weight of a drop was the first one, as a result of which in 1868 Quincke[56] began to study the IST of liquid metals in an inert gas medium (σ_{10}). To measure IST, Quincke heated the tips of wires of known diameter and considered that the weight P of the fallen drop of the melted metal was

$$P = 2\pi r \sigma_{10} \quad (16)$$

where r is the radius of the wire.

Knowing the number of fallen drops and their total weight, we can find the weight of one drop and thus calculate

$$\sigma_{10} = P/2\pi r \quad (17)$$

the IST of the metal under investigation.

By this method, Quincke measured the surface tension of metals such as platinum, gold, zinc, lead, tin, sodium, potassium, etc.

Later on, this method was used to measure the IST of organic and inorganic substances and their mixtures.

But Quincke's method, which at first sight looks very simple and fault-free, in fact has important defects. As we can see from the relationship 16, expressing the so-called Tate law, the drop of liquid will fall when its weight is equal to the resultant of forces of surface tension along the drop ring outline. However, in fact, the weight of the fallen drop p_f is less than the weight of the hanging drop P, and so to calculate the surface tension as a function of the weight of the drop, we must know the correction $K = p_f/P$ and then

$$\sigma_{10} = p_f/2\pi r K \tag{18}$$

Whereas Antonov, Duclos and Morgan (see ref. 10) consider in every case that $K = 0.61$, others, such as Rayleigh, Lohnstein and Harkins (see ref. 10) suppose that this correction has different values as a function of the liquid studied and the diameter of the pipes. No other method has received so much attention as the method of the drop weight. Lots of articles have been devoted to it, but nevertheless it still has no rigorous theoretical basis.

Lohnstein[28–31] has calculated a table of corrections $K = f(r/a)$ where r is the radius of the calibrated tube and a^2 is the capillary constant, from the differential equation of the drop surface meridian curve and the unchecked assumption that the remainder of the drop forms with the surface to which it is attached, the same angle as the angle made by the whole drop before detachment (*Figure 10*).

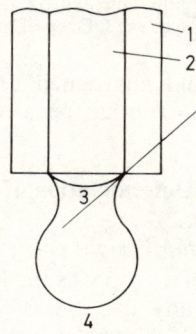

Figure 10. Surface tension determination by drop weight method. 1. Calibrated pipe; 2. Test liquid; 3. Surface of the drop excess; 4. Surface of the drop at the moment of detachment.

Harkins and Humphery[12, 13] have experimentally determined correction coefficients, using the method of capillary rise. However, the experimental curve of corrections due to these authors does not coincide with Lohnstein's theoretical curve so, using different tables of correction to reckon IST, it is possible to obtain different values of IST for the same liquid under the same conditions.

This ambiguity, added to the lack of theoretical basis of the method, has often been observed by searchers who have tried to improve the method of IST measurement as a function of the weight of the drop. Molody and Pavlov[32] have set up special experiments to check Tate's law, determining the drop ring diameter at the moment of breaking on photographs. They have found that neither Tate's law, nor the assumption on the proportionality of drop weight to the ring diameter, at the moment of detachment, is satisfied. The authors have given recommendations on the determination of the IST of liquids, using the method of the weight of the drop, but unfortunately, these recommendations are not universal.

Brown and McCormick[5] have developed a theory and described a special conical capillary to determine IST, according to the weight of the drop, which does not need correction tables. However, Brown and McCormick's device may be applied only to liquids which wholly wet the capillary, i.e. its applications are limited.

The lack of theoretical basis of the method of the drop weight is not the only reason against its application to the measurement of the IST of different liquids. In fact, when dealing with wetting liquids flowing out from narrow calibrated pipes, it must be noted that they produce drops which are too large, and if liquids do not wet the pipes, the drops are too small. Even Quincke noticed this, more than 100 years ago, and showed some cases of the formation of such drops.

To eliminate these difficulties, certain workers propose, when calculating IST, to take the outer or inner diameters of pipes into account whether the test liquid does or does not wet the material of the calibrated pipes. However, these suggestions cannot really be applied when the liquid, which at the beginning does not wet the pipe, progressively acquires this property, for instance at the expense of increasing the concentration of the additions under examination. So the method of the weight or volume of drops cannot be considered a sufficiently reliable one for IST measurement and cannot, in general, be recommended.

2. Ring or Plate Detachment Method

This method is based on IST determination from the strain necessary for the detachment of a ring or plate of known radius from the surface of the liquid, in which the previously balanced ring or plate was partially immersed. Numerous workers have developed the theory of ring rupture[68, 8, 66, 67, 34, 21] and by now there are a lot of tensometers, the most famous of which is du Noüy's.

Figure 11 shows the theoretical diagram of a tensometer based on Hooke's law. As the test liquid flows out from glass 5, ring 4 goes down and the spring stretches. Finally, there is a moment when the elastic tension of the spring has decreased to equal the tension (IST) of the liquid, wetting ring 4. Due to the IST applied to the inner and outer perimeters of the ring, this latter will cause the liquid layer to rise above the free surface of the liquid (*Figure 13*). It is evident that in the equilibrium state, this liquid layer will be equal to global sum F, conditioned by IST. Further very small lowering of the liquid in glass 5 involves detaching the ring from the surface of the liquid and this moment, corresponding to the lowest position of mark 3, is determined by a cathetometer.

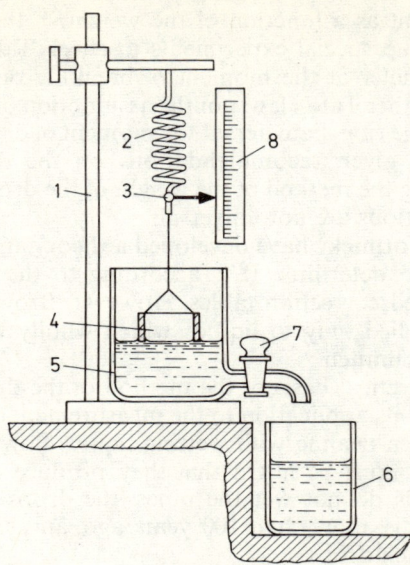

Figure 11. Theoretical diagram of the spring tensometer for IST determination of liquids. 1. Support; 2. Spring; 3. Mark; 4. Ring; 5. Glass reservoir containing the test liquid; 6. Glass reservoir to contain excess liquid; 7. Valve; 8. Scale.

Knowing the value of the extension of the spring 2, its coefficient of elasticity found by earlier calibration, and also the inner radius of the ring and its thickness, we can determine the IST according to one of the many formulas put forward in the reports cited.

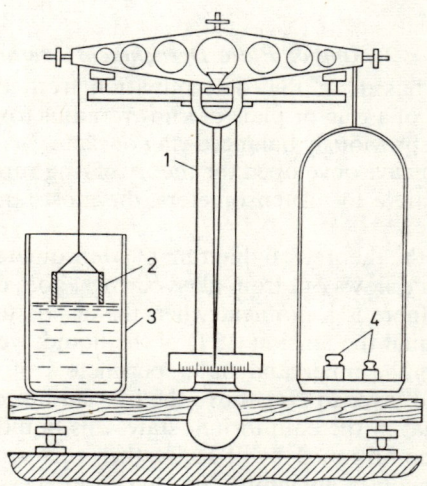

Figure 12. Theoretical diagram of the beam balance device to measure IST. 1. Balance; 2. Ring; 3. Glass reservoir with test liquid; 4. Equilibrating weights.

One of these formulas was proposed by Tichanovsky who found that, for a thick ring for which $a/n \ll 1$,

$$F = (8/\sqrt{2})\pi g\rho(R_1 + R_2)a^2[1 + \tfrac{1}{8}(a/n) - \tfrac{1}{32}(a/n)^2 + \ldots] \qquad (19)$$

where R_1 and R_2 are the respective outer and inner radii of the ring; a^2 is the capillary constant, $4n$ is the thickness of the ring; and ρ denotes density of the liquid under test.

There are also tensometers without springs, which use a beam balance (*Figure 12*). To one end of the beam is hung the ring, balanced, then put into contact with the surface of the sample liquid. After that, the right hand

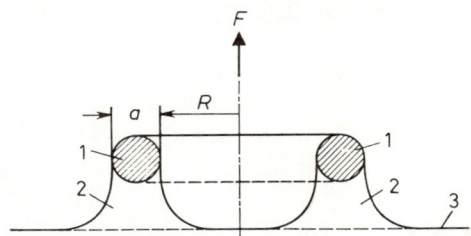

Figure 13. IST determination according to the ring detachment method. 1. Ring; 2. Raised liquid layer; 3. Liquid free surface.

pan is loaded complimentarily and then the strain F, necessary to detach the ring from the surface of the liquid, is reckoned. Knowing the parameters of the ring, the strain F and density of test liquid, we can determine its IST.

The tensometers, represented on *Figures 11* and *13*, are very complex and particularly so when the IST is studied over a large temperature interval, the upper limit of which may reach or exceed 100°C, while in addition some liquids react with air[63]. Unfortunately, even in simpler cases, this method cannot be considered to be perfect because it is necessary to take wetting contact angles into account, which theoretically cannot be determined simultaneously with IST measurement of the test liquid and depend on additions, experimental temperature, etc. Therefore, as for the method of capillary rise, we suppose that the test liquid wholly wets the material of the ring, so to determine the IST, we apply a formula which does not include cos Θ. But this is not the only problem encountered when using the method of ring or plate detachment. As a rule, the IST measurement of liquids by tensometer is done in air, assuming that the presence of water vapor in the air, of different organic mixtures, or various gases, acid vapors and other impurities, does not affect the IST determination. This point of view cannot be connected in any way to the requirements for precise IST measurement—an over-small physicochemical value depending sometimes on the smallest impurities. Moreover, if the test liquid has an appreciable vapor pressure, in closed devices when measuring the IST (this often applies for tensometers), there is a continuous evaporation of liquid. Consequently, the superficial layer composition is continuously renewed and thus the measured IST value does not correspond with an equilibrium state of the liquid–vapor system.

3. *Method of Maximal Pressure in a Gas Bubble or a Drop*

This method is one of the best and most precise for IST determination. It was used first by Simon in 1851[62], although the theoretical basis of the method was given by Cantor[9] in 1892.

This method consists of determining the maximal pressure, p_m, necessary to push a drop of a lighter liquid through a calibrated tube of radius r, which is plunged into a heavier liquid [*Figure 14*(a)] or to push a heavier liquid into a lighter one, through a vertical pipe [*Figure 14*(b)].

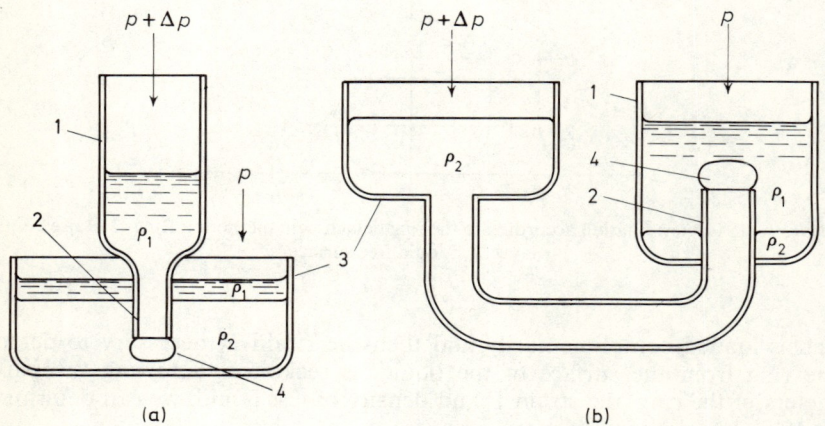

Figure 14. IST determination according to the method of maximal pressure in a drop. (a) Expression of a drop of light liquid into a heavy one; (b) Expression of a drop of heavy liquid into a light one. 1. Reservoir for light liquid; 2. Calibrated tube; 3. Reservoir for heavy liquid; 4. Drop of test liquid.

Cantor[9] found that for sufficiently narrow tubes, when the drop forms in the section of the calibrated pipe, it could be considered to be spherical.

$$\sigma_{11} = \tfrac{1}{2} r p_m \left[1 - \frac{2}{3} \frac{g(\rho_2 - \rho_1)}{p_m} - \frac{1}{3} \frac{g^2(\rho_2 - \rho_1)^2 r^2}{p_m^2} \right] \quad (20)$$

If we take a sufficiently wide calibrated tube, the drop in its section will be deformed under the action of the force of gravity. In this case, instead of the radius r, we take an effective radius X, which we find by the method of approximations in series with the help of Sugden's table[64]; then we deduce IST from the formula

$$\sigma_{11} = \tfrac{1}{2} p_m X \quad (21)$$

The most important advantages of this method reside in the fact that the formula to determine the IST does not include the value of the wetting contact angle; the number of parameters experimentally measured and necessary to determine the IST is not high and all these parameters can be determined with sufficient precision.

By now, in the literature, many different devices have been described for measuring the maximal pressure in a drop or a gas bubble. However, all these devices may be divided into four main types (*Figure 15*), the most common and simplest device for IST measurements by the method of maximal pressure in a gas bubble being illustrated in *Figure 15*(a).

We can show that when using this device, the IST of the test liquid can be

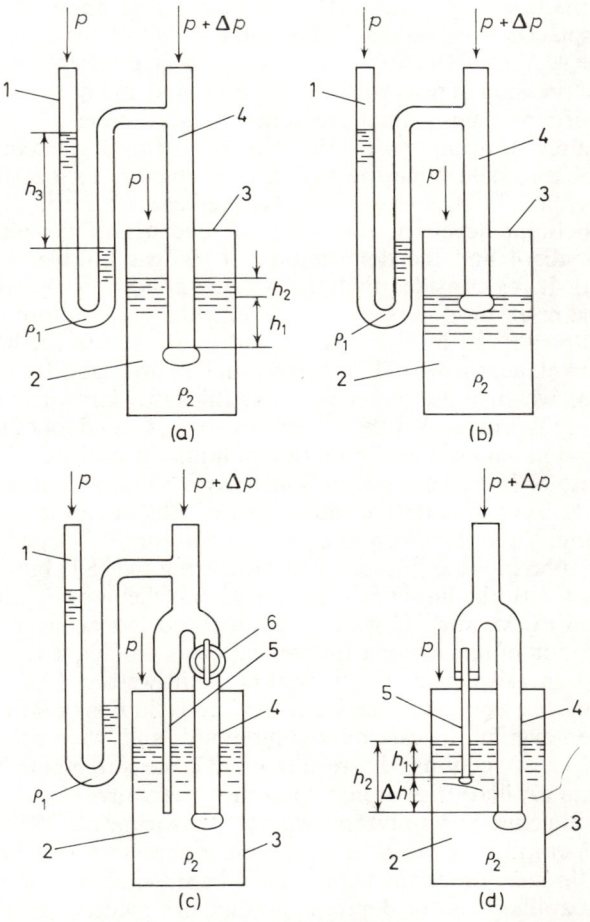

Figure 15. Diagrams of devices for IST measurement by the method of maximal pressure in a gas bubble: (a) Most frequent type of device; (b) Rehbinder's device; (c) Sugden's device; (d) Jaeger's device; 1. Liquid manometer; 2. Gas bubble; 3. Reservoir for test liquid; 4, 5. Calibrated tubes; 6. Valve. h_1 Initial immersion depth of the tube (4) h_2 Increase of immersion depth of the tube (4) after formation of a bubble on the tip of the tube; ρ_2—density of test liquid, ρ_1–density of manometric liquid; h_3 Height of manometric liquid.

deduced from

$$\sigma_{10} = \tfrac{1}{2}[g\rho_1 h_3 - g\rho_2(h_1 + h_2) \pm p_{kr}]X \qquad (22)$$

where p_{kr} is a quantity determining the capillary rise of the test liquid in the container 3; the sign ($+$) before p_{kr} applies to capillary rise; the sign ($-$), to depression and the meaning of all the other quantities may be understood from *Figure 15*(a) and reference to the diagrams.

As we can see from formula 22, to determine σ_{10} by the method of maximal pressure in a gas bubble, when using a calibrated tube, one must know the depth of immersion of this tube into the test liquid, the variation of the position of the level of liquid in the reservoir 3, depending on the amount of liquid displaced from the calibrated tube 4 and of the formation of the bubble in the section of this tube; one must also take into account the capillary rise or depression in reservoir 3 of the test liquid and have a special manometer to determine the maximal pressure in a gas bubble.

Unfortunately, in ordinary devices, using the method of maximal pressure in a gas bubble to determine the IST, it is impossible to measure capillary rise or depression in the reservoir 3. Neither can we normally determine variations in liquid levels in this reservoir because of the elimination of liquid from tube 4 and the deformation of the gas bubble at its tip [see *Figure 15*(a)]. If we consider that the wetting condition and the variation of level in the reservoir 3 depend on temperature, on conditions of thermo-vacuum treatment, and on the effect of additions, we can see how complex is the problem of measuring IST by this method in ordinary devices. To avoid these troubles, we must use reservoirs 3 of sufficiently large diameter, so that the variation of liquid level in these reservoirs is independent of the capillary rise or depression, and of the elimination of liquid from the calibrated tubes. In fact the use of large volumes of liquid leads to important experimental complications. To eliminate the immersion depths of calibrated tubes into the test liquid, from the formula, certain authors[59, 3, 4] propose not to immerse the tube into the liquid, when measuring the IST, but just to bring it into contact with the liquid [*Figure 15*(b)] and they give methods for IST determination in this way. However, this method has no general meaning, without speaking of its lack of a theoretical basis. In fact, it is impossible to use it for IST measurement of non-wetting or ill-wetting liquids. As stated above, it cannot be applied to the study of viscous liquids or liquid polymers, etc. Moreover, even in the case of wetting liquids and when using reservoirs 3 with small cross sections, the results of IST measurements are strongly affected by the capillary rise of liquids in this reservoir.

The most effective and universal way of measuring the IST seems to be the Sugden method[64], in which it is no longer necessary to take account of the depth of immersion of the calibrated tube into the test liquid or of the value of the capillary rise or depression. Sugden's method consists of using two calibrated tubes of different diameters 4 and 5 [*Figure 15*(c)], which plunged into the test liquid to an arbitrary depth, the same for both tubes. Knowing the maximal pressures p_1 and p_2, necessary to push gas bubbles throughout these tubes and their effective radii X_1 and X_2, we can easily determine the superficial tension of the liquid according to the formula

$$\sigma_{10} = \tfrac{1}{2}(p_1 - p_2)/(1/X_1 - 1/X_2) \tag{23}$$

However, Sudgen's method is not fault-free either. In fact, if the calibrated tubes 4 and 5 are sufficiently wide, and the reservoir 3 has small transverse dimensions, the liquid level position in the reservoir 3 will depend on which tube appears to hold the gas bubble at a particular moment. Consequently, in that case, we must take variations of liquid level in reservoir 3 into account or take a reservoir with a larger cross section.

Finally, there is one more way to measure IST according to the method of maximal pressure in a gas bubble, proposed by Jaeger[18] in 1892.

This method uses two calibrated tubes of different radii, which dip into the test liquid, to different depths [*Figure 15*(d)]. The distance between the tips of the pipes is selected so that gas bubbles formed on the tips of these tubes, are simultaneously sprayed out of the tubes, when reaching maximum pressure. Then, knowing the distance Δh between the sections of calibrated tubes, their effective radii X_1 and X_2 and the density of the test liquid ρ_2, we can calculate the IST, according to the formula

$$\sigma_{10} = \tfrac{1}{2}g\rho_2\Delta h/(1/X_1 - 1/X_2) \tag{24}$$

Jaeger's method, nearly forgotten, is the only way of determining the surface tension of the liquid when there is no effect on the σ_{10} value of the capillary rise or liquid depression in reservoir 3 and when it is not useful to know the depth of immersion of the calibrated tubes into the liquid under investigation. Moreover, as we see from equation 24, to determine the IST, it is not necessary to know maximal pressures in gas bubbles, i.e. to have a separate manometer for these measurements. But, this is really the weak point of Jaeger's method. In fact, it is nearly impossible to choose a height Δh, so that the propagation of gas bubbles through both calibrated tubes shall be simultaneous. However, the alteration of this condition leads to important errors, so it is doubtful if this method will ever be used for precise IST measurements. The common defect of these devices, described above and represented in *Figure 15*, is that they cannot be used for precise IST measurements at the liquid/liquid interface.

Taking this into account[45, 47], we have developed new devices, eliminating all the defects of previously described equipments (*Figure 15*) and using the method of maximal pressure in a gas bubble or drop. These new devices, arbitrarily called improved gas devices with one or two calibrated tubes (IGASD1CT and IGASD2CT) are based on the idea of using one branch of a U-shaped manometer filled with the test liquid as a reservoir, to study the liquid.

4. Improved Gas Devices with One Calibrated Tube (*IGASD1CT*)

Let us examine the general case of measuring IST in IGASD1CT (*Figure 16*) at the interface of two liquids, with restricted ability of dissolving each in the other, wetting the apparatus and having densities equal to ρ_1 and ρ_2, and assuming that $\rho_2 > \rho_1$. Let us suppose that initially the pressure will be equal to p in every part of the device, whereas liquid levels are located as shown in *Figure 16*(a). When we create an excess pressure Δp on the surface of the liquid of ρ_1 density in reservoir 1 [*Figure 16*(c)] after elimination of liquid of ρ_2 density, from the section of the calibrated tube 3, we start to form a drop of liquid of ρ_1 density. For an arbitrary pressure $p + \Delta p$ [*Figure 16*(c)], we can write the following equation

$$p + \Delta p + g\rho_1(H + h_1) = p + g\rho_1 h_3 + g\rho_2(h_1 + h_2) - p_{1kr} - p_{2kr} \quad (25)$$

where p_{1kr} and p_{2kr} are quantities taking capillary rise into account in relation to light or heavy liquid in reservoir 4. The manometric reservoir 2 is chosen to be large enough so that the value of the liquid capillary rise may be neglected.

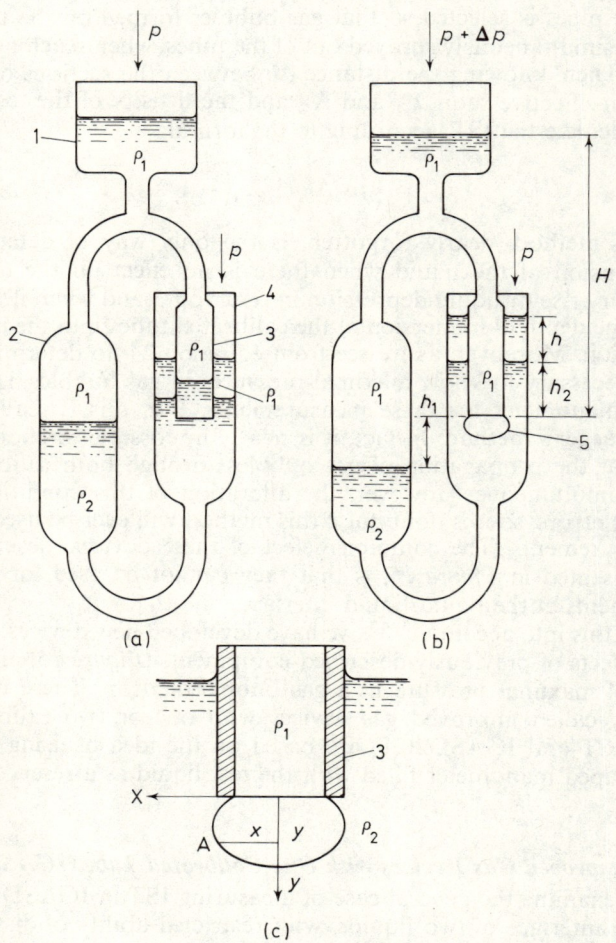

Figure 16. IGASDICT diagram. (a) Initial position; (b) Position corresponding to an arbitrary pressure in a drop of light liquid; (c) Drop of light liquid on the tip of the calibrated tube. 1. Reservoir for light liquid; 2. Manometric reservoir; 3. Calibrated tube; 4. Reservoir for calibrated tube; 5. Drop of light liquid.

INTERPHASE SURFACE TENSION

On the other hand, for any point A of the surface of the drop [*Figure 16(b)*], forming on the section of tube 3, we shall have

$$p + \Delta p + g\rho_1(H + y) = p + g\rho_1 h_3 + g\rho_2(h_2 + y) \\ + \sigma_{11}(1/R_1 + 1/R_2) - p_{1kr} - p_{2kr} \quad (26)$$

where R_1, R_2 are the main radii of curvature at the point A.

In solving equations 25 and 26 simultaneously, we find the following expression

$$g(\rho_2 - \rho_1)h_1 = g(\rho_2 - \rho_1)y + \sigma_{11}(1/R_1 + 1/R_2) \quad (27)$$

If we integrate this equation according to Cantor's method[9] for the case of a spherical drop of a light liquid, formed on the section of the calibrated tube 3 which dipped into a heavy liquid, we can show that IST is given by

$$\sigma_{11} = \tfrac{1}{2}g(\rho_2 - \rho_1)rh_1[1 - \tfrac{2}{3}(r/h_1) - \tfrac{1}{3}(r^2/h_1^2)] \quad (28)$$

or

$$\sigma_{11} = \tfrac{1}{2}g(\rho_2 - \rho_1)h_1 X \quad (29)$$

where h_1 is the height of the column of heavy liquid, corresponding to the maximal pressure in the drop, and X is the effective radius of the calibrated tube.

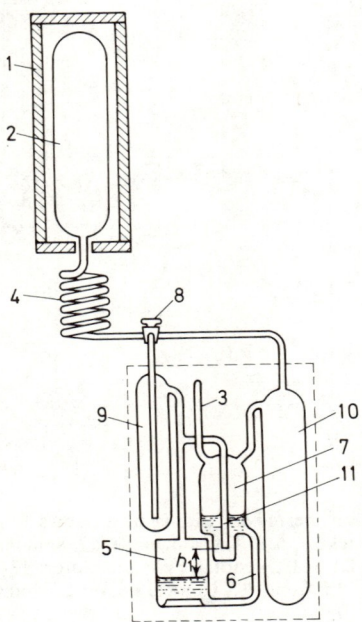

Figure 17. Improved gas device with one calibrated tube (IGASDICT). 1. Electrical furnace; 2. Reservoir with inert gas; 3. Tube to fill the device with test liquid; 4. Spiral tube; 5. Manometric reservoir; 6. Connecting pipe; 7. Reservoir for calibrated tube; 8. Threeway valve; 9. Intermediate reservoir; 10. Buffer reservoir; 11. Calibrated tube.

As we can see, equation 27 does not include the hydrostatic pressure associated with the immersion of the calibrated tube 3 into the test liquid; it also ignores values of the capillary rise of this liquid in the reservoir 4. To measure the maximum pressure in the drop, when using IGASD1CT, we don't need manometers to measure the pressure $p + \Delta p$ on the surface of the liquid in reservoir 1 (see *Figure 16*).

We have used IGASD1CT for various measurements of the IST of double

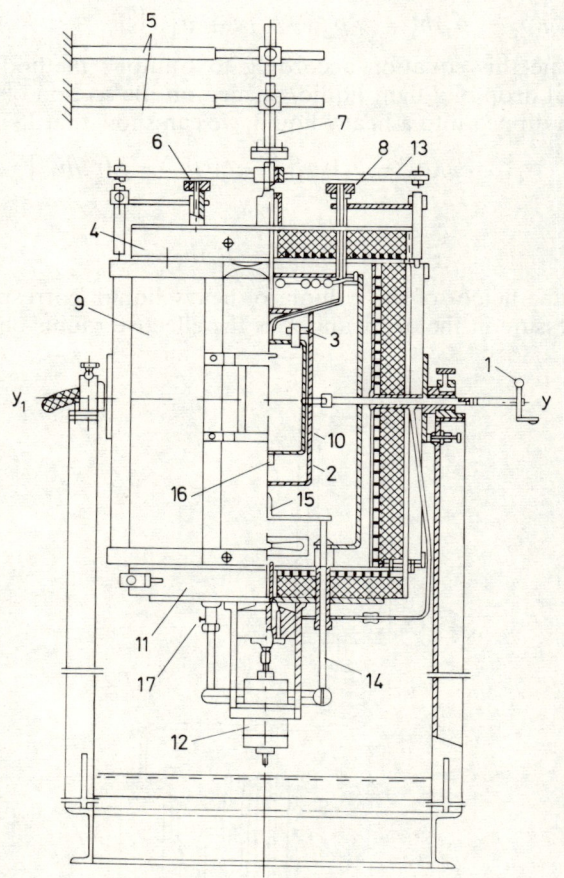

Figure 18. Diagram of an air thermostat with sight windows. 1. Handle; 2. Rotating frame; 3. Ratchet; 4. Split lid; 5. Brackets; 6, 8. Thermocouples; 7. Spindle of the frame; 9. Framework; 10. Immovable frame; 11. Base; 12. Ventilator electric motor; 13. Transverse rack; 14. Spindle; 15. Cone; 16. Conical sleeve; 17. Screw.

and triple solutions of oligomer and polymer substances in organic solvents, at the interface of inert gases and liquids, and to study the IST of polymer liquids, mixtures of organic substances with many components, aqueous solutions of inorganic salts, etc. For instance, to study the IST of solutions of rubber of low molecular weight in organic solvents, we have used a device represented in *Figure 17*. In this device, the maximal pressure in the gas bubble formed on the section of the calibrated tube 11, is created by the inert gas in bottle 2 and heated by electric furnace 1.

Before measurement of IST, the previously carefully washed and dried device is fitted to a metallic motionless frame 10 (*Figure 18*) and to tube 3 which supplies the necessary quantity of liquid; then it is soldered. After filling up the device, we place its measuring part into an air thermostat[52] with sight windows (*Figure 18*), the inside part of which is shown in *Figure 17* by a dotted line. In the thermostat, we have provided for compulsory mixing of the air and installed an electronic temperature controller.

After having controlled the liquid thermostatically in the device for two to three hours, we start measuring its IST. We then turn on furnace *1* (*Figure 17*) and slowly raise the temperature of the gas in bottle 2. The valve 8 is set so that bottle 2 is linked to the reservoir 9 (*Figure 17*). This induces gas to move from bottle 2 into the reservoir 9, acquire the thermostat temperature and then go to the manometer reservoir 5. The excess of pressure pushes the liquid out from the reservoir 5 along the tube 6 into the reservoir 7, but, simultaneously, the liquid is eliminated from the calibrated tube 11 and a gas bubble forms at its tip. When maximum pressure is reached, the bubble starts growing quickly, breaks away from tube 11 and flows up slowly. The started expulsion of drops goes on, without stopping, in a viscous liquid. To avoid it, excess pressure in the reservoir 5 is eliminated by turning the valve 8 so that the reservoir 9 is linked to the reservoir 10, when the first gas bubble is pushed out through tube 11. Thus after having pushed out the first two or three bubbles through the tube, we start measuring h_1 heights, corresponding to the maximal pressure in the gas bubble by means of a cathetometer. Knowing the average value of these heights, the density of liquid at a given temperature and the radius of the calibrated tube 11, we determine IST according to formula 29. After a series of measurements of heights h_1, the valve 8 is adjusted so that reservoirs 9 and 10 are connected to bottle 2, the furnace 4 is turned off and its lid is removed. The result is that pressure in the system levels off, and then IST measurement can continue, at a given temperature or after change of temperature and thermostatic control to permit measurement at another temperature (for instance, a higher one). IGASD1CT devices may be applied to IST measurement of corrosive, toxic and fairly hydrolysable liquids, but if they are, the system must be seamless soldered without valves or joints. To measure the IST of viscous liquids, we must use large calibrated tubes, the inner diameter of which may reach 5–8 mm, depending on the IST, density and viscosity of the test liquid. This is related to the fact that, using narrow tubes, it is impossible to get reproducible results because of the presence of a viscous liquid layer inside the tube ebbing away gradually during the process of measurement of IST.

Thus, in the case when the diameter of the calibrated tube has been chosen large enough so that, in comparison, the layer thickness of the viscous liquid on the inner wall of the tube is small, the height h_1 practically does not change.

5. Improved Gas Devices with Two Calibrated Tubes (IGASD2CT)

We have to note that the diameter of the manometric reservoir 5 in IGASD1CT must be large enough to enable us to neglect the capillary rise of liquid with density in this reservoir. This requires relatively large volumes of liquid, reaching 15–20 ml, to measure the IST. To reduce these quantities to 3–4 ml, we must employ improved gas devices with two calibrated tubes (IGASD2CT) and a narrow manometric reservoir 3 (*Figure 19*). Then, the

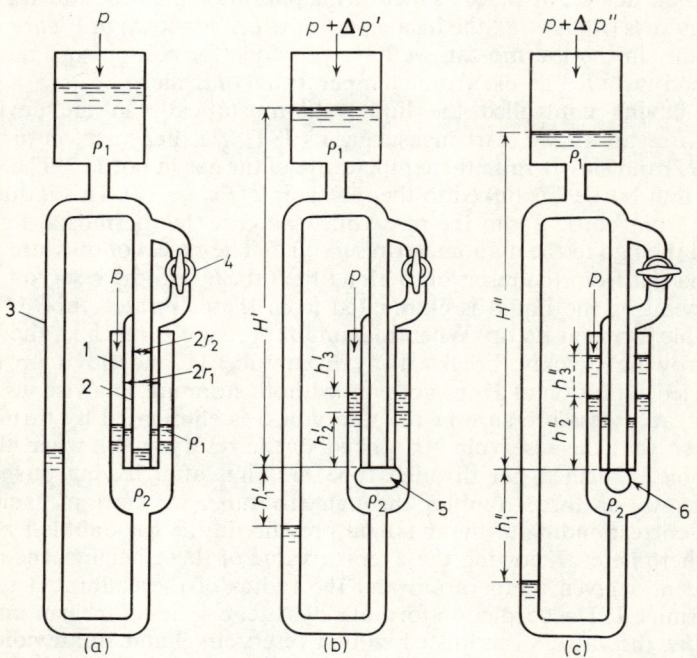

Figure 19. Diagram of improved gas device with two calibrated tubes (IGASD2CT). (a) Initial position; (b) Position of light liquid drops; (c) Position of heavy liquid drop. 1, 2. Calibrated tubes of different diameters; 3. Narrow manometric tube; 4. Valve; 5. Drop, forming on the tip of the wide calibrated tube; 6. Drop, forming on the tip of the narrow calibrated tube.

forming alternately of light liquid drops on sections from wide 1 and narrow 2 manometric tubes, we find heights h'_1 and h''_1, corresponding to maximal pressure in corresponding drops and we calculate the IST according to the formula

$$\sigma_{11} = \tfrac{1}{2}g(\rho_2 - \rho_1)(h''_1 - h'_1)/(1/r_2 - 1/r_1) \tag{30}$$

where r_1 and r_2 are respectively the radii of the wide and narrow calibrated tubes. One of these devices has been used to measure the IST of melted liquid crystals[48]. This soldered device (*Figure 20*) requires 3–4 ml of liquid, for measurements. As we can see from the figure, the forming of gas bubbles on sections of calibrated tubes is forced by gas heating in bottle

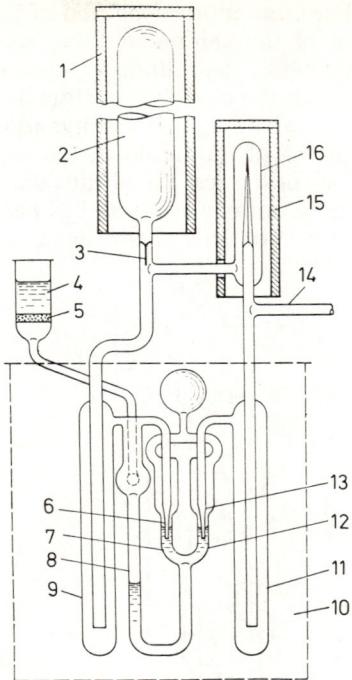

Figure 20. Improved gas device with two calibrated tubes (IGASD2CT). 1, 15. Electrical furnace; 2. Bottle with inert gas; 3. Capillary tube; 4. Liquid crystal melt; 5. Glass filter; 6, 13. Calibrated tubes; 7, 12. Reservoir for calibrated tubes; 8. Manometric tube; 9, 11. Intermediate reservoir; 10. Thermostat; 14. Tube; 16. Thermal valve.

2. To allow bubbles to form by turns, on the large and narrow tubes, we use a thermal valve 16. When the valve is heated up to a high temperature, the flowing out of gas through the very narrow capillary of the valve is considerably slowed down and gas bubbles form on the narrow tube 6; with a non-heated valve, gas bubbles form on the wide tube 13.

6. *Gravitational Devices*

Up to now, we have examined devices in which the pressure of the drop, forming on the section of the calibrated tube, rises with unchanged pressure above the surface of the liquid into which the calibrated tube has been immersed. This peculiarity requires special equipment to create a pressure within the drop, for instance bottles filled up with inert gas, heated and cooled from outside (*Figure 17*), or electromagnetic devices using an iron float by the aid of which pressure drops may be created in measuring systems[44]. In this case, measuring devices are tightly fixed to principal walls or shock-absorbers and thus must be considered as the assembling deficiency of these devices.

However, we sometimes use another method of formation of gas bubble or drop on the section of the calibrated tube, and especially above the surface of the liquid, into which the calibrated tube dips, so that rarefactions are made while maintaining the equilibrium state unchanged inside the gas bubble or drop. The main advantages of this method are obvious, when the discharge above the surface of the liquid in the calibrated tube reservoir is due to the spraying out of the test liquid into the intermediate reservoir. This method, called a gravitational method, has been used by the authors to build up a series of very compact closed devices, rotating around horizontal or vertical axes[15, 43, 46, 50, 51, 54].

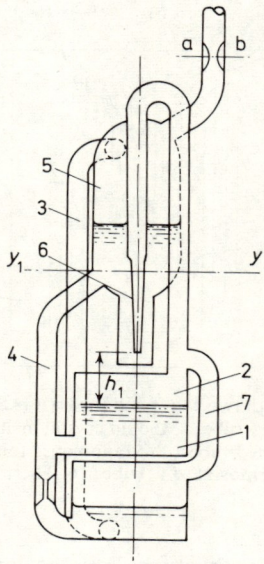

Figure 21. Improved gravitational gas device with one calibrated tube (GIGASD1CT). 1. Intermediate reservoir; 2. Manometric reservoir; 3. Hydraulic lock; 4. Connecting pipe; 5. Reservoir for the calibrated tube; 6. Calibrated tube; 7. Capillary.

One of these devices[46, 50], a gravitational improved gas device with one calibrated tube (GIGASD1CT), has been used to measure the IST of easily hydrolysable and toxic chlorides, aqueous salt solutions of inorganic substances, organic mixtures and other liquids. These devices (*Figure 21*) are useful for measuring the IST of non-viscous liquids. In that case, a certain quantity of liquid is transferred into the device, which is filled with inert gas up to a low pressure and unsoldered along a line ab (*Figure 21*) from the system, which is used for pumping and filling up the device with liquid and gas. Then the device with the test liquid is fixed to a rotating frame 2 (*Figure 18*), the frame is placed in an air thermostat with sight windows and after thermostatic control for two to three hours, the IST is measured. For this, handle 1 is pushed inside the thermostat to a stop position, the frame 2, with

GIGASD1CT fixed on it, is leant and rotated around axes yy_1, clockwise for 180°. The result is that liquid flows from reservoirs 1 and 2 (*Figure 21*) along tubes 3 and 4 into reservoir 5, then the device comes back to its initial position and handle 1 is turned right, to a stop position. The counterbalance is fixed to the lower part of the rotating frame 2 (not shown in *Figure 21*) and ratchet 3 maintains the frame in a strict vertical position and does not allow it to revolve counterclockwise around axis yy_1. Thanks to a small gap in lid 4 and the shift of handle 1 to the right after that the device returned to the initial position, the frame ceases to touch the thermostat and this is why the vibrations of the thermostat supplied with alternating current do not influence the process of measurement of IST.

When the device was returning to its initial position, the liquid slowly started to transfer from reservoir 2 (*Figure 21*) via capillary 7) into the intermediate reservoir 1. As soon as the level of the liquid in reservoir 2 falls to height h_1, corresponding to the maximal pressure in the gas bubble forming on the section of calibrated tube 6, gas bubbles begin to go through the tube. It follows that the pressure above the surface of the liquid in reservoir 5 will increase, and a part of this liquid will move along tube 4 to reservoir 2; the level of the liquid in this reservoir increases and eventually gas can no longer move through tube 6. As liquid from reservoir 2 flows continuously into reservoir 1, its level in reservoir 2 drops, and its distance to the section of the tube 6 reaches height h_1 for the second time, which corresponds to the maximal pressure in the gas bubble so that gas bubbles will proceed through tube 6 once more. We go on determining heights h_1, for which gas bubbles escape from the calibrated tube until all the liquid has flowed out from reservoir 5. Then, the device is tilted by a rotation around the axis yy_1 of 180°. The result is that liquid from reservoirs 1 and 2 is poured into reservoir 5, and when the device comes back to its initial position, it is possible to measure the IST, at a given temperature or after a further period under thermostatic control and then to make measurements at another temperature.

It is obvious that knowing average values of h_1, the density of the liquid and the radius of the calibrated tube 6, we can easily calculate the IST from formula 29.

The gravitational principle may be used not only to determine the IST at the interface liquid/inert gas, according to the method of maximal pressure in a gas bubble, but also at the interface liquid/saturated vapor, according to the method of maximal pressure in the drop. The different constructions of these devices (gravitational vacuum devices with one or two calibrated tubes (GIVD1CT and GIVD2CT) have been described[54, 52]. These devices have been used by numerous authors to measure the IST of mercury and many amalgams, low-melting metals and their alloys. One of these, a so-called combined gravitational vacuum device with one calibrated tube (CGVD1CT) has been constructed by the author[43] to measure the IST of alloys of low-melting metals, in a liquid state or forming a solid phase, at ambient temperature. In this device, it is possible not only to measure the IST, but also to proceed with hundreds of alloys, different from each other, by millionth or ten-millionth fractions in concentration of addition weight percentage, without cooling or opening the device. We must observe that prepared alloys do not need chemical, spectral or chemico-spectral analysis.

Finally, the capacity of work with CGVD1CT is ten times higher than for other devices described in literature. In this case the relative error does not exceed 0.2 per cent.

The accurately washed and dried device is fixed to a rotating frame 2 (*Figure 18*), soldered to the vacuum system and treated under vacuum in a special furnace, for a long period. Then, a determined quantity of pure metal according to its weight, is melted again into glass 1, through tube 2 (*Figure 22*), and into glass 3, while through tube 4 is added a certain quantity of sufficiently concentrated alloy (for instance, alloy of the examined metal with an addition of another metal), the concentration of which has been previously determined. Then, in hot condition, the device is unsoldered from the vacuum system and is placed into a previously heated rotating thermostat with sight windows (*Figure 18*). By use of handle 1, as described above, frame 2 with CGVD1CT is turned about the axis yy_1, through an angle

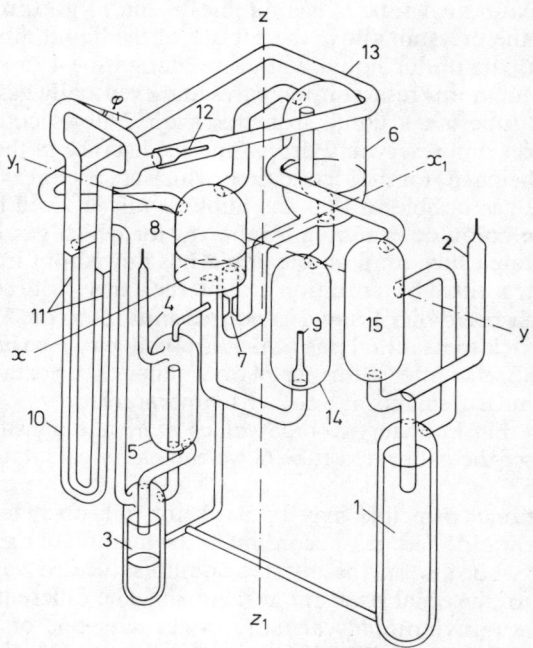

Figure 22. Combined gravitational vacuum device with one calibrated tube (CGVD1CT). 1. Glass for metallic solution; 2, 4, 13. Connecting pipes; 3. Glass for concentrated alloy; 5, 15. Intermediate reservoir; 6. Reservoir for alloy preparation; 7. Capillary; 8. Manometric reservoir; 9. Calibrated tube; 10. Buret; 11. Marks; 12. Dosing capillary; 14. Reservoir for calibrated tube.

equal to about 180°. Thus, metal from glass 1 (*Figure 22*) flows down into the reservoir 6, and alloy from glass 3 flows into the intermediate reservoir 5. When the device has returned to its initial position (*Figure 22*), handle 2 is turned right to the stop position, and the alloy from reservoir 5 flows into the glass 3, while metal flowing through the narrow capillary 7, starts slowly to flow from the reservoir 6 into the manometric reservoir 8 and the calibrated tube 9. When the maximum pressure in the drop of melted metal formed over the section of the tube 9, is reached, metal drops start falling down from the tip of the tube. Knowing the height of the metallic column h_1 (which has been measured with a cathetometer), corresponding to a given pressure and equal to the distance from the section of tube 9 and the surface of the melted metal in manometric reservoir 8, the density of the melted metal at the temperature of the experiment, and also the radius of the calibrated tube 9, it is possible to calculate the IST of the melted metal according to formula 29.

After numerous IST measurements on pure metals, at a given temperature, measurements have been made at other temperatures and after the first, dilute solutions were prepared. In this case spindles 14 (*Figure 18*) were moved up so that cone 15 came into the gap of the rotating frame sleeve 2, when spindles 14 were fixed with the screws 17. Later, the upper part of spindle 7 is disconnected from the lower one and this latter is fixed, by a moving cone (not shown in the figure) to a transverse rack 13 having a conical sleeve (not shown in figure). The thermostat with frames 2 and 10 attached to it, and also CGVD1CT, is turned about axis yy_1, counterclockwise, so that a part of the initial alloy from glass 3 (*Figure 22*) can fall into the buret 10. Then, the thermostat returns to its original position, the levels of alloy are measured with a cathetometer in the branches of buret 10, by comparison with the marks 11, and the frame with device is rotated about the axis zz_1, clockwise, by about 90°; the thermostat is revolved again about axis yy_1, counterclockwise so that the alloy can now fall from buret 10 into the design capillary 12.

Gently tapping the lower part of spindle 7 (*Figure 18*) by hand, we introduce a certain quantity of alloy into tubes 13 (*Figure 22*) and then we return the device to the initial position, represented in *Figure 22*.

The alloy poured into tube 13 then flows into the manometric reservoir 8, because separate parts of tube 13 relatively the the surface xy are placed under certain sharp angles, whereas the alloy remaining in the dosing capillary 12 is poured into the buret 10 where its levels are once again measured in terms of the marks 11. We find the volume of poured alloy according to the difference of levels before and after pouring. Knowing this volume, the concentration of alloy and its density at a given temperature and also knowing the quantity of pure metal remelted into the device, we determine the concentration of prepared solution.

Before measuring the IST of the prepared solution, we fix the lower part of spindle 7 with frames and CGVD1CT to the upper part of spindle 7, spindles 14 are lowered down into the position shown in *Figure 18*. After that, we start measuring the IST of the solution as has been described for the measurement of the IST of the pure metal.

As the surface of buret 10 forms a sharp angle φ with the plane zy when measuring the IST of the solution, i.e. when the device revolves about the axis yy_1, clockwise, through an angle exceeding 90°, the alloy remaining in the buret is poured into glass 3.

The preparation of the second and further alloys and the measurements of their IST are made in the same way as described above. At the end of the IST measurement of all the prepared solutions we fixed the frame with the device to the thermostat as has been described for preparation of the solutions and make the thermostat revolve about the axis yy_1, clockwise, by 360°. This causes metal solution to flow from reservoirs 6, 8 and 14 into, first (rotation to 270°), the intermediate reservoir 15, then, revolving further into glass 1, whereas the alloys remaining from experiments in glass 3 first flow into reservoir 5 (rotation to 180°), then, revolving further, into glass 3. After cooling the device down to ambient temperature, the alloy in glass 1 and unused alloy in glass 3, the glasses were destroyed, but as indicated in *Figure 22* they could easily be replaced by new ones and the device prepared for the study of the IST of other alloys.

IV. References

[1] Bakker, G., 'Kapillarität und Oberflächenspannung' *Handbuch der Experimentalphysik (Leipzig)*, **6**, (1928).
[2] Bashfort, F. and J. C. Adams, 'An attempt to test the theories of capillary action by comparing the theoretical and measured forms of drops of fluid'. Cambridge University Press (1883).
[3] Belton, J. W., *Trans. Faraday Soc.* **31**, 1413 (1935).
[4] Belton, J. W. and M. G. Ewans, *Trans. Faraday Soc.* **41**, 1 (1945).
[5] Brown, R. C. and H. McCormick, *Phil. Mag.* **39**, 420 (1948).
[6] Burdon, R. S., *Trans. Faraday Soc.* **28**, 866 (1932).
[7] Burdon, R. S., *Surface Tension and the Spreading of Liquids*. Cambridge University Press: London (1949).
[8] Cantor, M., *Ann. Phys., Lpz.* **47**, 380 (1892).
[9] Cantor, M., *Ann. Phys. Lpz.* **47**, 399 (1892).
[10] Convers, L., *Recherches sur l'Adsorption en Solution Métallique*. Nancy (1938).
[11] Dorsey, N. E., *J. Wash. Acad. Sci.* **18**, 505 (1928).
[12] Harkins, W. D. and E. C. Humphery, *J. Amer. Chem. Soc.* **38**, 228 (1926).
[13] Harkins, W. D. and E. C. Humphery, *J. Amer. Chem. Soc.* **38**, 236 (1926).
[14] Heydweller, A., *Ann. Phys., Lpz.* **65**, 311 (1898).
[15] Ibraguimov, Kh. I., N. L. Pokrovskij and P. P. Pugachevich, *Zh. Fiz. Khim., Mosk.* **40**, 266 (1966).
[16] Ivashchenko, Ju. P., B. B. Bogatyrenko and V. N. Eremenko, *Surface Phenomena in Melted Substances and Powder Metallurgy Processes*. Academy of Sciences of the USSR: Kiev (1963).
[17] Ivashchenko, Ju. P. and V. N. Eremenko, *Accurate Measurements of Surface Tension of Melted Substances by the Sessile Drop Method*. Naukova dumka: Kiev (1972).
[18] Jaeger, F. M. *SB Akad., Wiss. Wien*, Abt. IIa, **101**, 954 (1892).
[19] Kasterin, N. P., *Zh. Russk. Fiz.-Khim. Obshch. A*, **25**, 203 (1893).
[20] Kingery, W. D. and M. Humenik, *J. Phys. Chem.* **57**, 359 (1953).
[21] Kljachko, Ju. A. *Zavod. Laborat.* **6**, 1376 (1937).
[22] Koshevnik, Ju. U., M. M. Kusakov and N. M. Lubman, *Zh. Fiz. Khim., Mosk.* **27**, 1887 (1953).
[23] Kozakevitch, P., S. Chatel, G. Urbain and M. Sage, *Rev. Métall.* **52**, 139 (1955).
[24] Kozakevitch, P. and G. Urbain, *Rev. Métall.* **58**, 401 (1961).
[25] Laplace, P. S. 'La Mécanique Celeste' supplément au livre X. *Theorie de l'Action Capillaire*. (1806).
[26] Laplace, P. S. *Ann. Phys., Lpz.* **33**, 1 (1809).
[27] Lohnstein, Th., *Ann. Phys., Lpz.* **53**, 1062 (1894).
[28] Lohnstein, Th., *Ann. Phys., Lpz.* **20**, 237 (1906).
[29] Lohnstein, Th., *Ann. Phys., Lpz.* **20**, 606 (1906).
[30] Lohnstein, Th., *Ann. Phys., Lpz.* **21**, 1030 (1906).
[31] Lohnstein, Th., *Ann. Phys., Lpz.* **22**, 767 (1907).
[32] Molody, T. K. and P. P. Pavlov, *Izvestja Fiz. Inst. Pri Moskovskom Nauchnom Institute*, **1**, 21 (1919).
[33] Najdich, Ju. V. and V. N. Eremenko, *Fiz. Metal. i Metalloved.* **11**, 883 (1961).
[34] Nietz, A. H. and R. H. Lambert, *J. Phys. Chem.* **33**, 1460 (1929).
[35] Nijenko, V. I., V. N. Eremenko and L. I. Skljarenko, 'Surface phenomena in melted substances and derived solid phase'. *Kabardino Balkarskoe*. Nalchik: KBGU (1965).
[36] Pilchikov, N. D. *Zh. Russk. Fiz.-Khim. Obshch. A*, **20**, 83 (1888).

[37] Poindexter, F. E. and M. Kernaghan, *Phys. Rev.* **33**, 837 (1929).
[38] Popel, S. I., O. A. Esin and P. V. Geld, *Dokl. Akad. Nauk SSSR* **74**, 1097 (1950).
[39] Popel, S. I., O. A. Esin and Ju. P. Nikitine, *Dokl. Akad. Nauk SSSR*, **83**, 258 (1952).
[40] Popel, S. I., Ju. P. Nikitine and S. M. Ivanov, *Graphs for Calculating Surface Tension from Drop Dimensions*. Inst. Polytech: Oural Sverd Porsk (1961).
[41] Porter, A. W., *Phil. Mag.* **21**, 703 (1936).
[42] Pugachevich, P. P., *Zh. Fiz. Khim., Mosk.* **25**, 1365 (1951).
[43] Pugachevich, P. P., *Zh. Fiz. Khim., Mosk.* **33**, 1880 (1959).
[44] Pugachevich, P. P., *Analysis for Physics and Physical Chemistry*. MIS Moscow (1957).
[45] Pugachevich, P. P., *Aut. svid. SSSR No. 147023, Bull. Isobr. No. 9* (1962).
[46] Pugachevich, P. R. *Aut. svid. SSSR No. 149 943. Bull. Isobr. No. 17* (1962).
[47] Pugachevich, P. P., *Zh. Fiz. Khim., Mosk.* **36**, 1107 (1962).
[48] Pugachevich, P. P., *Aut. svid. SSSR, No. 158 144. Bull Isobr. No. 20* (1963).
[49] Pugachevich, P. P., *Surface Phenomena in Melted Substances and Powder Metallurgy Processes*. Academy of Sciences of the USSR: Kiev (1964).
[50] Pugachevich, P. P., *Zh. Fiz. Khim., Mosk.* **38**, 1377 (1964).
[51] Pugachevich, P. P., *Aut. svid. SSSR No. 232 597, Bull. Isobr. No. 34* (1968).
[52] Pugachevich, P. P., *Aut. svid. SSSR No. 231 212, Bull. Isobr. No. 35* (1968).
[53] Pugachevich, P. P. and V. P. Bitchkou, *Aut. svid. SSSR No. 147 344, Bull. Isobr. No. 10* (1962).
[54] Pugachevich, P. P. and V. A. Constantinov, *Dokl. Nauk SSSR*, **57**, 797 (1947).
[55] Pugachevich, P. P. and V. V. Lazarev, *Aut. svid. SSSR No. 137 003, Bull. Isobr. No. 6* (1961).
[56] Quincke, G., *Ann. Phys., Lpz.* **135**, 621 (1868).
[57] Quincke, G., *Ann. Phys., Lpz.*, **139**, 1 (1870).
[58] Rayleigh (Lord), *Proc. Roy. Soc. (London) A*, **93**, 184 (1916).
[59] Rehbinder, P. A., *J. Exp. Biol. Med., Moscow*, **4**, 939 (1927).
[60] Schumacher, E. E., *J. Amer. Chem. Soc.* **45**, 2255 (1923).
[61] Siedentopf, H., *Ann. Phys. Chem.* **61**, 235 (1897).
[62] Simon, N., *Ann. Chim. (Phys.)*, **33**, 5 (1851).
[63] Soloviev, A. N. and O. P. Makarova, *Teplofiz. Vysok. Temp.* **4**, 189 (1966).
[64] Sugden, S., *J. Chem. Soc.* **121**, 858 (1922).
[65] Tawde, N. R. and K. G. Parvatikar, *Indian J. Phys.* **25**, 473 (1951).
[66] Tichanovsky, J. J., *Phys. Z.* **25**, 299 (1924).
[67] Tichanovsky, J. J., *Phys. Z.* **26**, 522 (1925).
[68] Timberg, G., *Ann. Phys., Lpz.* **30**, 545 (1887).
[69] Verschaffelt, J. E., *Proc. Acad. Sci. Amst.* **21**, 366 (1919).
[70] White, D. W., *Trans. Amer. Soc. Metals*, **55**, 253 (1962).

CHAPTER 21

Adsorption

T. TAKAISHI

Institute for Atomic Energy, Rikkyo (St Paul's) University, Yokosuka, Japan

Contents

I.	Introduction	1023
II.	Vacuum Systems	1024
III.	Pressure Measurements	1025
	1. Gages	1025
	A. Pirani Gages	1026
	B. Mechanical Manometer	1027
	2. Sources of Error and Their Correction	1028
	A. Thermal Transpiration	1028
	B. Mercury Vapor Drag Effect	1031
IV.	Adsorption Isotherms	1033
	1. Cleaning the Adsorbent Surfaces	1033
	2. Equilibration	1034
	3. Volumetric Methods	1036
	A. General	1036
	B. Very Low Pressure Region (below 10^{-6} Torr)	1038
	C. Low Pressure Region	1038
	D. Intermediate Pressure Region ($10^{-3} \sim 10$ Torr)	1040
	E. Subatmospheric Pressure Region	1042
	F. High Pressure Region	1044
	G. Determination of Dead-space Volume	1046
	H. Adsorption Cells	1048
	I. Gas Dosing Systems	1049
	4. Gravimetric Methods	1050
	A. General	1050
	B. Helical Spring Balance	1051
	C. Beam-type Balances	1054
	D. Resonating Quartz Crystals	1057
	E. Sources of Error	1057
	(1) Buoyancy	1058
	(2) Convection and TMF	1058
	(3) Temperature Differences	1060
	(4) Electrostatic Charge	1060
	5. Flow Methods	1060
	A. Very Low Pressure Region	1060
	B. Gas Chromatographic Method	1062
V.	Adsorption Cryostats	1063
VI.	Acknowledgement	1065
VII.	References	1065

I. Introduction

Recently, various aspects of the gas/solid surface interaction have been studied with great success with the aid of modern experimental techniques.

However, no marked progress has occurred in the study of the thermodynamic properties of adsorption. Only a limited number of cases[19, 29, 60, 75, 94] are known in which the molecular image of adsorption has been elucidated from analysis of the thermodynamic properties. The main difficulties arise from surface contamination and surface heterogeneity. Two sources of surface contamination are considered. One is due to gaseous impurities, which are preferentially adsorbed, or which reduce or oxidize the surface of the adsorbent. The trouble caused by this type of contamination has been greatly reduced by modern vacuum techniques. Another source of contamination is the surface accumulation of impurities originally contained in the bulk of the solid[31, 53, 102]. At the present time, this problem is difficult to solve, and efforts are required to investigate the surface composition of adsorbents in their working state. Surface heterogeneity originates not only from impurities but also from structural defects, i.e. point defects, dislocations, steps, kinks and edges. Apart from point defects, the concentration of structural defects cannot be quantitatively or thermodynamically controlled. The divergence of experimental data from different laboratories may be ascribed mainly to such surface heterogeneity. However, if the preparation and treatment of an adsorbent are strictly controlled, not only in one's own laboratory but also internationally, reproducible data can be obtained.

This chapter describes experimental methods of measuring adsorption isotherms which are accurate, reproducible and give values suitable for theoretical analysis. Orthodox calorimetric methods of measuring adsorption heats and heat capacities of adsorbed films were described in Volume I of this series by Chihara and Morrison[8], or in Holme's review[39]. As shown by Kington and Aston[48], adsorption heats obtained from adsorption isotherms, using the Clapeyron–Clausius equation, agree with the calorimetric results. To obtain heats of adsorption, the method most suitable to the system under investigation can be chosen, but usually adsorption isotherms are preferable because they give much more information. For the theoretical aspects of the problem, refer to the monograph of Young and Crowell[104], Hill's review[36] and the original paper by Everett[26], in which thermodynamic and statistical mechanical methods of analysing the experimental data are thoroughly discussed.

Various techniques, such as e.s.r., n.m.r., spectroscopy, l.e.e.d., f.e.m. and f.i.m. are used in studies on adsorbed films, and some properties have been elucidated. However, these techniques will not be discussed here, since by themselves they do not determine the thermodynamic properties of adsorbed films, but are used as auxiliary methods.

II. Vacuum Systems

Most adsorption experiments have used glass vacuum systems, but recently metal systems have been becoming popular. Both systems have their respective merits; glass is convenient for the construction of sophisticated systems, and metal is suitable for obtaining an ultrahigh vacuum. Important points in constructing glass systems will be described.

The main component of the residual gas in a glass system at pressures below 10^{-4} Torr, is water vapor. To obtain good adsorption data, the residual water vapor pressure must be minimized, since most solid surfaces are affected by residual water. Before constructing a vacuum system, all parts such as McLeod gages, stopcocks, traps and connecting pipes, must be baked in a furance at 500°C for about ten hours with intermittent lowering of the temperature. This releases water and carbon monoxide occluded near the surface of the glass, which would otherwise be gradually released into the vacuum system. Water subsequently adsorbed on the surface of the treated glass during storage can easily be baked out after construction of the vacuum system. When the vacuum system has been constructed, connecting tubes are heated by winding Nichrome ribbon directly on to them, and large parts such as a McLeod gage by wrapping them with wide ribbon heaters. Except for ultrahigh vacuum systems, conventional adsorption apparatuses contain greased stopcocks which cannot be baked and, during electrical heating of the tube, must be protected with wet asbestos. The use of Apizon-T is advisable as a stopcock grease, since it is suitable for temperatures up to 70°C. If the vacuum system is properly constructed according to the above notes, a vacuum of 10^{-6} Torr is reached approximately four hours after the start of the first evacuation.

Not only residual water vapor but also hydrocarbon vapor from the stopcock grease or pump-oil must be eliminated as far as possible. Hydrocarbon vapor is effectively removed by cold traps containing adsorbents such as Molecular Sieve 13X. This method, however, cannot be applied to stopcocks between the main adsorption system and the pumping system or the gas-doser. Therefore, such stopcocks must be replaced by mercury cut-offs or greaseless valves. Recently, miniature greaseless metal valves have become available, which use gaskets of Viton, polyimide or copper, are bakable up to 200°C and are easy to operate. These valves facilitate the construction of glass–metal hybrid systems, which permit adsorption measurements in the 10^{-7} Torr region or lower. Physisorption studies at lower pressures, and chemisorption measurements must be carried out in vacuum systems capable of attaining this degree of evacuation. In chemisorption, only a limited part of the surface participates in adsorption, and these sites are easily contaminated by small amounts of residual gas. Furthermore, powder samples are often reduced by hydrocarbon vapor. For example, zinc oxide when baked-out in a conventional vacuum system becomes gray and partly reduced, but if a greaseless system is used this does not occur.

Ultrahigh vacuum systems are frequently used in studies on the dynamic characteristics of adsorption[80], but are rarely used in studies on equilibrium properties. The reasons for this are as follows. With powder samples, an ultrahigh vacuum cannot be attained because of the evolution of gases from the sample itself. With wire or film adsorbents, on the other hand, accurate determinations of the equilibrium amount of adsorption is very difficult on account of the self-pumping action of ionization gages.

III. Pressure Measurements

1. Gages

Pressure gages commonly used in adsorption experiments are the McLeod gage ($5 \times 10^{-4} \sim 10$ Torr), the Pirani gage and its modifications ($10^{-8} \sim 10^{-1}$ Torr), gages of the Bourdon type ($10^{-2} \sim 760$ Torr and $1 \sim 200$ atm), and U-tube manometers ($10^{-1} \sim 760$ Torr). Ionization gages have a self-pumping action and can be used only for limited kinds of gases at very low pressures where other gages cannot be used. Since primary gages are fully described elsewhere in this volume, only secondary gages which are important in low-pressure adsorption experiments will be discussed here.

A. Pirani Gages

Pirani gages of the conventional type, with a working range of $10^{-5} \sim 10^{-1}$ Torr, are fully described in various monographs[57, 66]. The lower pressure limit measurable with these gages is determined by the noise in their electronic circuits. Oguri et al.[71], succeeded in reducing the noise level to obtain an outstandingly stable Pirani gage giving a linear response in the pressure range $3 \times 10^{-8} \sim 10^{-2}$ Torr. This gage may be useful in adsorption experiments, inasmuch as its self-pumping action, if any, is negligibly small, and a good vacuum can easily be obtained and maintained. According to Oguri, most of the electrical noise originates from the spot-weld between the gage's filament and lead wire. Usually, the filament is held between thin platinum foil, fastened with thin platinum wire to the lead (made of tungsten or molybdenum), and then spot-welded as shown in *Figure 1*(a). If the geometry of the welded part is as shown in *Figure 1*(b), vibration of the filament may cause irregular changes in its electrical resistance by forming a short circuit. A desirable geometry of the welded part is shown in *Figure 1*(c). Care must also be taken in connecting the lead to the external circuit. If the vacuum system is to be baked-out at 400°C in order to attain an ultrahigh vacuum, silver or gold wire must be used to connect the gage to the bridge circuit to avoid oxidation of the leads. Furthermore, the usual care must be taken to minimize thermal noise generated at junctions, and the use of a low noise battery such as a mercury battery is required.

Another problem with Pirani gages is a tendency of the surface properties of the filament to change with time, which may cause a drift in the calibration curve. In order to solve this difficulty, use is made of a gold-plated[71] or

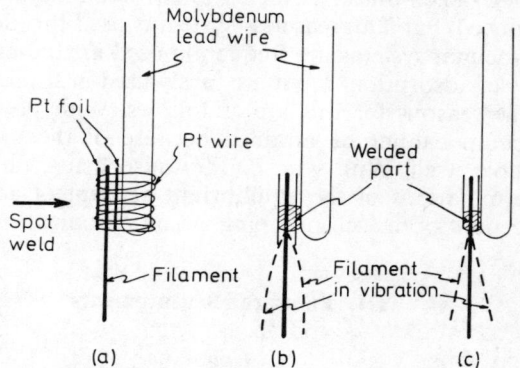

Figure 1. Method of spot-welding the filament of the Pirani gage.

platinum-clad[72] tungsten (or molybdenum) filament, which is relatively unreactive with respect to the ambient gases. Gold-plated filament, however, can be used only in vacuum systems free from mercury vapor. Oguri used a tungsten filament 100 mm in length, 10 μm in diameter plated with a 0.3 μm thick layer of gold.

B. Mechanical Manometer

Since the introduction of a glass Bourdon gage in 1906, various types of mechanical manometers have been developed[66], and have been used in adsorption measurements of condensable gases or organic vapors which would react with stopcock grease. These gages are also indispensable for clean adsorption apparatus because they can be baked-out at 400°C, while McLeod gages or U-tube manometers cannot be baked-out and become a source not only of mercury vapor but also of impurity gases such as water. Almost all classical gages, described in Melville and Gowenlock's monograph[66], are manually operated, while modern ones are of the recording type[5, 21, 40, 61], though some manual adjustments are required.

A sensitive recording spoon gage has been designed by Machin[61], and is shown in *Figure 2*. This gage is fairly robust and has a sensitivity of 0.004 Torr

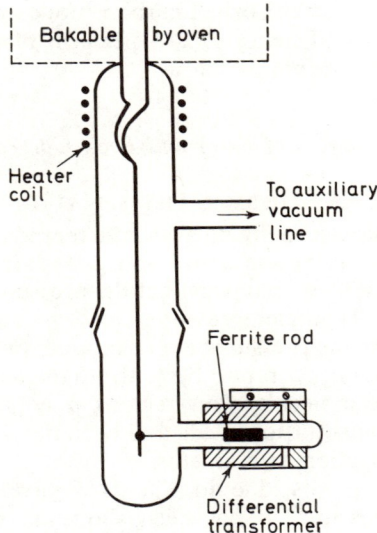

Figure 2. Recording spoon gage due to W. D. Machin[61].

in favorable cases, and the principle of operation is as follows. The movement of the spoon is converted to a horizontal displacement of a ferrite rod cemented to the top of the spoon through a 1 mm silica rod, and the location of the ferrite rod is detected by the differential transformer and recorded. According to Machin, the output was linear with pressure up to 4 Torr, and room temperature changes of 2 ~ 3°C did not affect the calibration of the gage.

A commercially available capacitance manometer has a sensitivity of 10^{-5} Torr in favorable cases and may be used in clean vacuum systems. This gage is actually a difference manometer and can fill the gap between the pressure ranges of the spoon and Pirani gages. Lamers and Rony gave a comprehensive report on them[55], including specific details on instruments commercially available (cf. Chapter 4 where further details are given).

The use of radioactive isotopes is occasionally very effective in adsorption measurements. If the specific activity of the gas concerned and the counting efficiency of the radiation detector are known, the counter can be used as a secondary manometer. The well type scintillation counter is commonly used to detect γ-rays, and a finger tube connected to the vacuum system is inserted into the well. In a conventional measuring system, ^{133}Xe at 10^{-8} Torr may produce a counting rate of the order of 1000 counts per minute and pressure measurements are made very easily in ultra-low pressure regions. On the other hand, detection of β-rays, especially of lower energies, is subject to some difficulties, since β-rays are easily absorbed by the walls of the vacuum system. Recently, Klier constructed a G–M counter with a very thin (2 mg/cm^2) mica window[50], which could be baked-out at 400°C and permitted an ultimate vacuum better than 1×10^{-10} Torr. This is designed for detection of ^{14}C in chemisorbed carbon monoxide on clean metal surfaces, and its counting efficiency is 24.6 per cent of the β-rays striking the external surface of the counter window.

2. *Sources of Error and Their Correction*

A. Thermal Transpiration

Adsorption of gases is usually measured at temperatures other than room temperature; so the adsorption vessel and gages are kept at different temperatures, say T_1 and T_2, respectively. Let the pressures in the corresponding parts be p_1 and p_2. At higher pressures, $p_1 = p_2$ according to the hydrostatic equilibrium. As the pressures are decreased and the mean free path of the gas molecules becomes comparable to the diameter of the tube connecting the two parts (in the molecular flow region), $p_1 \neq p_2$. This phenomenon is called the thermal transpiration effect, and becomes a source of serious error in low-pressure adsorption experiments.

A simple kinetic theory due to Knudsen gave the relation, $p_1/p_2 = (T_1/T_2)^{\frac{1}{2}}$ at lower pressures. It has been shown in recent years, however, that this relation does not hold[22, 77, 85], and at present there is no theoretical equation which perfectly describes the experimental data on thermal transpiration. Hence, corrections must be made by using experimental curves or empirical equations. One of the most reliable empirical equations, obtained by Takaishi and Sensui[91], is a modification of that due to Liang[58], and is expressed as

$$\frac{1-(p_1/p_2)}{1-(T_1/T_2)^{\frac{1}{2}}} = \frac{1}{\alpha\xi^2 + \beta\xi + \gamma\sqrt{\xi} + 1} \quad \text{for } T_1 < T_2 \qquad (1)$$

and

$$\frac{1-(p_2/p_1)}{1-(T_2/T_1)^{\frac{1}{2}}} = \frac{1}{\alpha\xi^2 + \beta\xi + \gamma\sqrt{\xi} + 1} \quad \text{for } T_1 > T_2 \quad (1')$$

with

$$\xi = 2p_2 d/(T_1 + T_2) = P_2 d/\overline{T}$$

where d denotes the diameter of the connecting tube along which the temperature gradient exists, α, β and γ are specific constants for the appropriate gas as given in *Table 1*.

Table 1. Values of constants contained in equation 1 and 1' from Takaishi and Sensui[91].

Gas	α 10^5 . K^2 . Rorr^2 . mm^{-2}	β 10^2 . K . Torr . mm^{-1}	γ [K . Torr . mm^{-1}]$^{\frac{1}{2}}$
H$_2$	1.24	8.00	10.6
Ne	2.65	1.88	30.0
Ar	10.8	8.08	15.6
Kr	14.5	15.0	13.7
CH$_4$	14.5	15.0	13
Xe	35	41.4	10
He	1.4–1.6	1.2–1.1	18–20
N$_2$	12	10	10–18
O$_2$	9–7	16–19	—

These values for α, β and γ are considered reliable, being deduced from the best fit of many workers' data. However, the values for helium, xenon, nitrogen and oxygen are obtained from rather few data, and their reliability is inferior compared to the others, but may be adequate to give a first-order approximation. For gases not given in *Table 1*, one is obliged to use the approximate estimated values for α, β and γ, using the relations

$$\alpha = 1.4 \times 10^4 \exp(0.507D) \qquad \text{K}^2 . \text{Torr}^{-2} . \text{mm}^{-2}$$

$$\beta = 5.66 \exp(0.607D) \qquad \text{K} . \text{Torr}^{-1} . \text{mm}^{-1}$$

and

$$\gamma = (110/D) - 14 \qquad \text{K}^{\frac{1}{2}} . \text{Torr}^{-\frac{1}{2}} . \text{mm}^{-\frac{1}{2}}$$

where D denotes the collision diameter (in Å, 1Å = 10^{-10} m) of the gaseous molecule as calculated from the viscosity η, by the relation[32]

$$\eta = \frac{5}{16}\left(\frac{mkT}{\pi}\right)^{\frac{1}{2}} \frac{1}{D^2}$$

As an example, the plots of equations 1 and 1' for hydrogen are shown in *Figure 3*, which also contains experimental points from various sources covering a temperature range from 14 to 673 K. One can see from this figure, the validity of the proposed empirical equation, and the pressure range in which thermal transpiration is appreciable.

Edmond and Hobson[22] studied thermal transpiration effects at pressures as low as 10^{-8} Torr, and their results at very low pressures are not in harmony with the extrapolation of the above. The validity of equation 1 or 1' may be limited to a region where $p_2 d/\overline{T} > 10^{-5}$ (Torr·mm/degree). Refer to Chapter 4 of this work for details of methods of correction of these effects at pressures lower than this.

A cold trap of the usual type, inserted between a gage and an adsorption

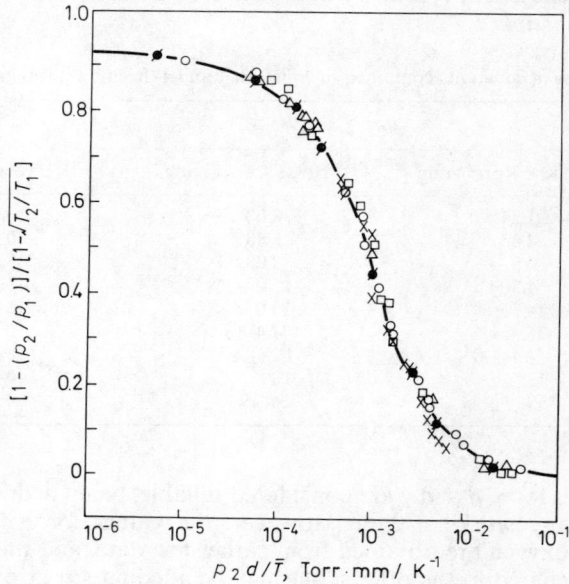

Figure 3. Thermal transpiration effect of hydrogen[91]. ○, $T_1 = 673.3$ K; □, $T_1 = 577.3$ K; △, $T_1 = 473.3$ K; ×, $T_1 = 90.2, 20.4, 14.15$ K. For the data at $T_1 < T_2$, the ordinate expresses the value of $[1 - (p_1/p_2)]/[1 - \sqrt{T_1/T_2}]$.

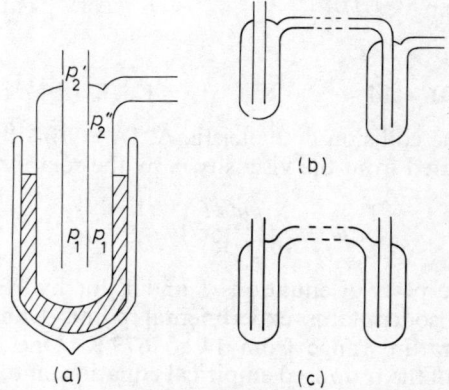

Figure 4. Pressure gradients in a conventional trap and two methods of connection.

vessel, introduces further errors in the pressure measurements. If pressure values in various parts of the trap are denoted as shown in *Figure 4*(a), then we have the relations $p'_2 > p_1, p''_2 > p_1$ and $p''_2 > p'_2$ owing to the differences in temperature and tube diameters. If two traps are connected in series, errors are accumulated in the arrangement in *Figure 4*(b), but compensated in that of *Figure 4*(c). However, even the latter arrangement gives rise to errors in volumetric adsorption measurements, since the amount of gas existing in the dead-space cannot be calculated accurately owing to the unfavorable pressure gradients. Consequently, it is advisable to use U-tube traps, even though they are inferior to the conventional type of traps with respect to trapping efficiency.

B. Mercury Vapor Drag Effect

In conventional apparatus for adsorption measurements, the McLeod gage is used as a standard. Mercury vapor from the gage is usually removed by cold or gold foil traps to protect the surface of the adsorbents from contamination by mercury. In this situation, mercury vapor flows continuously from the gage to the trap, dragging gas molecules in the direction of flow; that is to say, a weak pumping action operates. Even if other kinds of gages are used, the same situation arises when the gage is calibrated against a McLeod gage. This effect was first studied by Gaede as early as 1915. It is surprising that since then this effect was ignored by many workers, until about ten years ago Podgurski and Davis[77], and Ishii and Nakayama[41] stressed its importance in low pressure measurements. Takaishi[90] studied the effect on the basis of the kinetic theory of gases, correcting a numerical factor involved in Gaede's phenomenological equation.

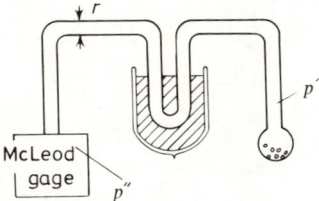

Figure 5. Pressure gradient in an adsorption apparatus due to the mercury vapor drag effect.

Consider a vacuum system as shown in *Figure 5* in which the connecting tube has a uniform radius r. There is a relation between p' and p'' as

$$\frac{\Delta p}{p''} = \frac{p' - p''}{p''} = \frac{p_{Hg}}{p''}\left\{1 - \frac{1}{B}\ln\left(1 + \frac{\Delta p}{p''}\right)\right\} \quad (2)$$

with

$$B = \frac{3\pi^2 d^2 \cdot \alpha r p_{Hg}}{4kT}\sqrt{\frac{m_2}{m_1 + m_2}} \quad (3)$$

where πd^2 is the collision cross section in the mutual diffusion between mercury vapor and the gas concerned, p_{Hg} is the mercury vapor pressure over the McLeod gage, m_1 and m_2 are the mass of a mercury atom and of the gas molecule respectively, and α is a numerical parameter with a value of $1 \leqslant \alpha \leqslant 4/3$. In this decade, many workers have investigated the applicability of these equations, arriving at several different conclusions, but the general validity of the equations is now accepted[23]. There remain, however, some ambiguities as to values for α and d. There is little data about the mutual diffusion between mercury vapor and other gases, and the required cross section must be estimated from other sources such as viscosity data. The use of the following equation is recommended

$$d = (0.9/2)\{d_{vis}(gas) + d_{vis}(Hg)\}$$

with $\alpha = 1$, where d_{vis} designates the cross-sectional diameter of the gas derived from viscosity measurements. Detailed studies[92] have shown that these empirical values hold for xenon, but their applicability for other gases has not been strictly tested so far. The recommended values of B for various gases have been given by Sensui[87]. The vapor pressure of mercury (in Torr) is calculated using the equation due to Ernsberger and Pitman[25]

$$\log p_{Hg} = 11.0372 - 3204/T$$

Experimental and calculated curves of $\Delta p/p''$ against log P'', for some representative cases, are given in *Figure 6*, which may serve to show the validity of equation 2 and the magnitude of the errors.

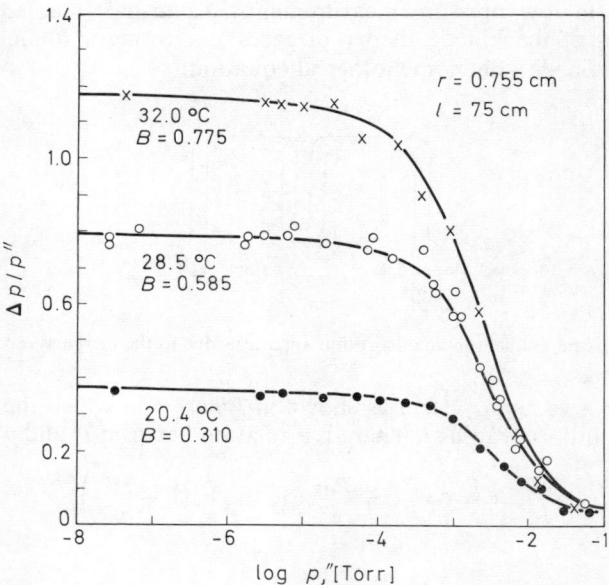

Figure 6. Mercury vapor drag effects in xenon at various ambient temperatures after T. Takaishi and Y. Sensui[92]. Connecting tube, radius, 0.755 cm; length, 75 cm; p'', the pressure indicated by the McLeod gage; Δp, the pressure difference between the adsorbent cell and the McLeod gage. Solid curves are theoretical values.

In the next place, consider the case where the connecting tube is not uniform but consists of n kinds of pipes of radius r_i and length l_i ($i = 1 \cdots n$). Then, r in equation 3 must be replaced by r_{eff}, where

$$r_{\text{eff}} = \left(\sum_{i=1}^{n} \frac{l_i}{r_i^2}\right) \bigg/ \left(\sum_{i=1}^{n} \frac{l_i}{r_i^3}\right)$$

This relation is valid only when there is no additional and unexpected source of mercury vapor, e.g. a mercury droplet on the wall, in the connecting tube.

V. Adsorption Isotherms

1. Cleaning the Adsorbent Surfaces

All solid surfaces, on exposure to air, become covered with chemisorbed or physisorbed layers of oxygen, carbon monoxide, water etc., and the first step in measuring the adsorption of a gas is to clean the surface of the adsorbent. To achieve this, adsorbents are baked-out at higher temperatures in a good vacuum. The choice of baking conditions, however, is a very difficult problem, inasmuch as the adsorbent may undergo irreversible changes in the course of baking-out. Finely divided metals, for example, sinter at relatively low temperatures. Materials such as silica and alumina contain a lot of water, partly as adsorbed species, and partly as a constituent of the solid, there being no strict distinction between the two types. Hence, vigorous out-gassing destroys the surface structure. Most metallic oxides are stable even at higher temperatures, but they can become oxygen-deficient on baking-out if a lower valency state is possible for the cation. Consequently, the outgassing conditions must be determined empirically for each adsorbent according to the desired state of the surface, and the conditions must be clearly described.

In physical adsorption studies, it is very useful to wash the adsorbent with the adsorbate gas[73]. After moderate baking-out, the adsorbate gas is introduced into the adsorption vessel, cooled to liquid nitrogen temperature, and then heated and evacuated. By repeating this cycle several times, the surface of the adsorbent is efficiently cleaned, and reproducible data are obtained. For chemisorption studies, the applicability of washing depends on the system concerned. In hydrogen–metal powder systems, this method is highly recommended, since the surface oxide on the metal is efficiently removed. On the other hand, in oxygen–metal systems, washing with oxygen would be ridiculous. In some cases, heating under a flow of helium or hydrogen is more effective for cleaning the adsorbent. Cylinder hydrogen or helium must be purified by diffusing it through a heated palladium or quartz tube, respectively.

An adsorption vessel made of quartz is often used, when outgassing at very high temperatures is required (above 500°C), and when there is no danger of sintering. Care must be taken on introducing hydrogen into such a quartz vessel kept at higher temperatures, as hydrogen will react with the quartz walls at temperatures above 600°C to produce silicon monoxide. This evaporates, and if the adsorbent has a high affinity for silicon, the silicon monoxide can be dissociatively adsorbed. For example, most metals may be contaminated by such treatment. With Pyrex glass, on the other hand, boron is transferred from the wall to the adsorbent by hydrogen reduction even at 400°C. Some materials, such as metals, contain carbon which aggregates on the surface[31]. Furthermore, prolonged outgassing, in a conventional vacuum system with greased stopcock, produces carbon specks on the surface[93]. These carbon contaminations may be removed by baking-out after the introduction of a small amount of oxygen or hydrogen.

Infra-red lamps are often used in place of conventional outgassing furnaces. This is of particular value for the efficient baking-out of a sample when the gravimetric method of adsorption measurements is to be used, where the sample is suspended in a high vacuum and radiation plays the leading role in heating. A radiation reflector of aluminum foil around the adsorption vessel insures efficient and homogeneous heating of the sample.

Special attention must be paid to the start of the outgassing, when fine powders or porous materials are used as adsorbents. Sudden evacuation or heating can cause the sample to distribute itself throughout the vacuum system, and a mishap such as this can be so fatal that the whole system must be reconstructed. To avoid this trouble, it is advisable to adopt the following procedure. The sample is lightly pressed into the form of a disc or pellet, with a non-reactive and removable binding agent such as water or alcohol. After inserting the sample into the adsorption vessel, the air is replaced by an inert gas, and the vessel is gradually heated up to 200°C (provided the sample is not affected at that temperature) under a gentle flow of the gas. After several hours of baking, the temperature is gradually raised to the required level and ambient gas is slowly evacuated.

It is traditionally considered that clean and ideal surfaces can be obtained by cleaving a single crystal in a high vacuum. This method certainly gives clean surfaces, but not ideally homogeneous ones. The fresh surfaces thus obtained contain a considerable number of steps and other structural imperfections, and furthermore the stoichiometry of the surface layer becomes non-ideal in materials such as alkali halides[28].

In conclusion, it is almost impossible to obtain a clean and perfect solid surface, except in special cases such as l.e.e.d. and f.i.m. experiments. We are obliged to be contented with *cleaned* surfaces, whose cleanliness depends on the properties and form of the adsorbents. Experimental data must be reproducible in the sense that the surface states are well controlled, even though the surface might not be ideally clean and perfect. Here the question may arise, what do the experimental data mean? Indeed, molecular interpretations cannot be derived straightforwardly from adsorption isotherms or heats by themselves, without additional knowledge about the nature of the surface. In this connection, information about pretreatment conditions of the adsorbent is most important, and must be definitely established for individual adsorbents.

2. *Equilibration*

Equilibration times in adsorption depend on the individual system, and vary considerably. In some cases, equilibrium is reached almost instantaneously, while in others it takes as long as twenty hours to attain a reproducible equilibrium. Generally speaking, the equilibration time is longer at lower temperatures, on porous adsorbents and in some kinds of chemisorption. At lower pressures, the heat liberated on adsorption dissipates mainly through radiation, and equilibration takes a much longer time. A typical example was described by Corrin and Rutkowski[12], who measured the adsorption of krypton on calcium halophosphate at 77.3 K. With the adsorbate contained in a cylindrical bulb, the first increment of gas came to *apparent* equilibrium within two hours, and subsequent increments within thirty

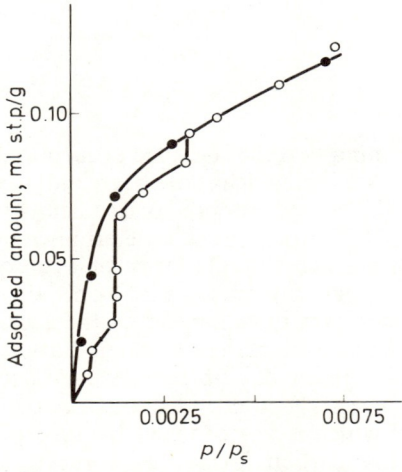

Figure 7. Adsorption isotherms of krypton on calcium halophosphate at 77.3 K from M. L. Corrin and C. P. Rutkowski[12]. ○, conventional adsorbent cell of glass; ●, adsorbent cell in the form of a metallic tray.

minutes. The resulting isotherm is shown by the open circles in *Figure 7*. When the adsorbent was spread on a tray to facilitate heat dissipation, the initial equilibration time appeared longer, sixteen hours for the first increment of gas. The subsequent increments attained equilibrium in five minutes. The resulting isotherm shown in the figure by solid points is reasonably smooth. These two separate isotherms show the importance of insuring that the system has truly attained equilibrium. Special care must be taken when the adsorption isotherm shows a two-dimensional phase change or hysteresis.

If possible, it is advisable to check the equilibration by measuring adsorption and desorption points alternatively. It may be concluded that equilibrium has been established if two sets of points lie in the same isotherm. This method, however, cannot be applied when the isotherm exhibits hysteresis. The pressure of adsorbate, after admission, decreases exponentially with time, and sometimes confirmation of equilibrium becomes difficult and time-consuming. One method of confirmation is as follows. When the pressure shows an apparently steady value, the temperature of the adsorbent is lowered from T_1 to T_2, and after some time the temperature is raised to its original level T_1, the pressure changes being as shown in *Figure 8*. If the decreasing and increasing curves approach the same value, this is probably the equilibrium pressure.

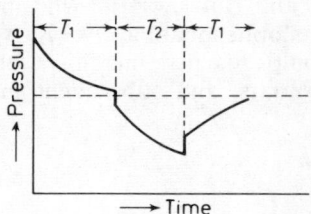

Figure 8. Trends of pressure change with time and attainment of equilibrium.

It must be borne in mind that the so-called adsorption equilibrium is not a true equilibrium, but a state in which some degrees of freedom are quenched. This situation may easily be understood if one remembers the following facts. The most stable external form of a crystal is determined by Wulff's theorem[35], but no real crystals have such a stable form; that is, crystals are not in the true equilibrium state as far as their external forms are concerned. Furthermore, the stable external form must depend on the atmosphere surrounding the crystal, and the external form of the adsorbent must be drastically changed to attain the true adsorption equilibrium. Most phenomena in physical adsorption can be successfuly interpreted on the assumption of an *inert adsorbent* the form of which is not changed by adsorption. Exceptions are porous glass and active carbon formed by compression and sintering[88]. In chemisorption, some deformations of adsorbents are expected to take place with a fairly long relaxation time. Hence, it sometimes becomes difficult to judge when the equilibrium (even apparent) of adsorption has been attained.

3. *Volumetric Methods*

A. General

Adsorption isotherms are conventionally measured by volumetric or gravimetric methods, and the respective merits of these two methods will be discussed. One microgram of nitrogen has a volume of 8×10^{-4} ml at s.t.p., or 600 ml at 10^{-3} Torr. This may be easily and accurately measured at an equilibrium pressure of 10^{-2} Torr, but not at 10 Torr. On the other hand, it is not particularly easy to weigh 1 µg. Hence, as far as non-condensable gases are concerned, the volumetric method is superior to the gravimetric one in adsorption measurements at the lower equilibrium pressures. The reverse applies with adsorption measurements at high equilibrium pressures and for condensable gases.

In the volumetric method, the number of moles of gas or vapor adsorbed is calculated from measurements of pressure, volume and temperature.

ADSORPTION

The procedure is as follows. A measured amount of gas is admitted into the adsorption vessel. After the attainment of equilibrium, the number of moles present in the gas phase is calculated from the equilibrium pressure. The difference between the number of moles admitted and the number of moles remaining in the gas phase gives the amount adsorbed; namely,

$$\text{(amount adsorbed)} = \text{(amount admitted)} - \text{(amount unadsorbed)} \tag{4}$$

The LHS can only be evaluated accurately if the third term of equation 4 is far smaller than the second. This condition limits, on the one hand, the applicability of the volumetric method, and on the other hand affords a guide for the construction of vacuum systems for adsorption measurements. If the adsorbent has a large surface area, and if the relative pressure of adsorbate is not too high, the ratio of (amount unadsorbed)/(amount adsorbed) becomes small, and the third term of equation 4 may be a minor correction. There is, in this situation, no need to diminish the dead-space volume, and in fact the use of wide connecting tubes is desirable for rapid evacuation. In the reverse situation, i.e. at higher relative pressures with adsorbents of lower surface areas, the third term correction becomes important. In order to overcome this unfavorable trend, it is desirable to design the apparatus so as to minimize the dead-space volume. Furthermore, each range of relative pressures poses inherent technical problems, such as sources of error, choice of valves and so on. Hence, it is convenient to classify adsorption apparatuses and to describe them individually according to the pressure range for which they are intended.

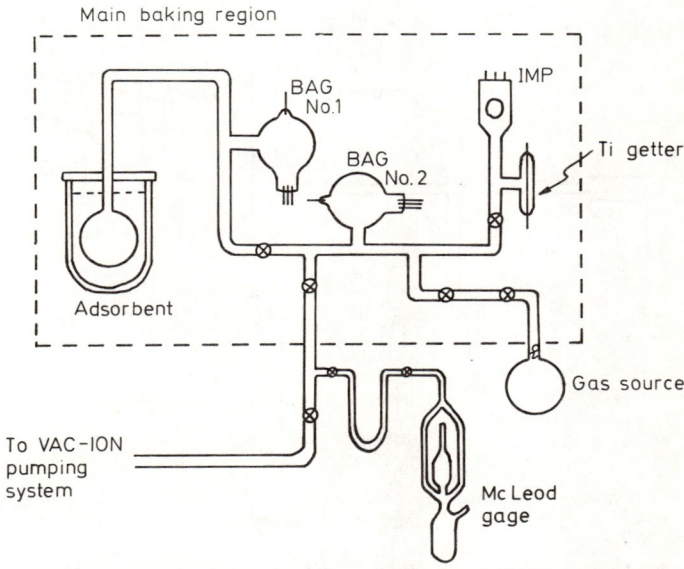

Figure 9. Ultrahigh vacuum system for adsorption measurements at very low pressures from J. P. Hobson and R. A. Armstrong[38]. BAG, Bayard–Alpart gage; IMP, inverted magnetron pump.

B. Very Low Pressure Region (below 10^{-6} Torr)

Hobson and Armstrong[38] measured adsorption isotherms of helium, argon and nitrogen on Pyrex glass, a part of the vacuum system, down to 10^{-10} Torr, by the standard volumetric method. Their apparatus is illustrated in *Figure 9*. In this pressure region, careful corrections must be made for the self-pumping action of the B–A gage and the thermal transpiration effect. Hydrogen and oxygen readily react with hot filaments and hence adsorption measurements of these gases in this pressure region may be impossible.

C. Low Pressure Region

In this section, techniques used in the pressure range of 10^{-7} up to 10^{-1} Torr will be discussed. *Figure 10* shows a versatile apparatus for use in this pressure range, which was constructed in the author's laboratory for the study of physical adsorption but it may also be used for chemisorption studies. All of the valves are made of metal; in the large one Viton gaskets are used, and copper gaskets in the small ones. Metal valves with Viton gaskets

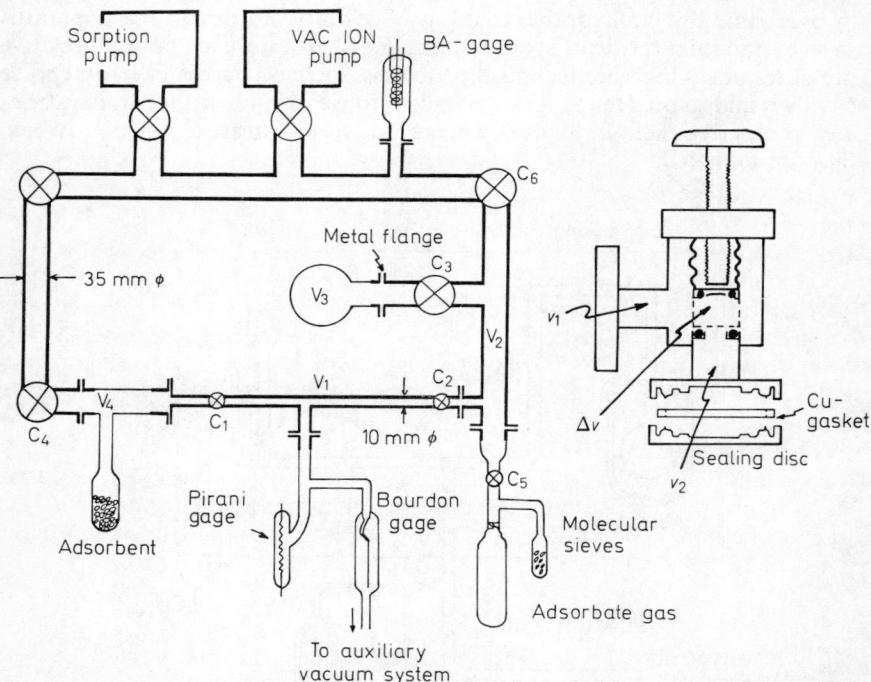

Figure 10. Apparatus for adsorption measurements at low pressures. Thin lines, glass; thick lines, stainless steel.

cannot be based-out above 200°C and are not appropriate for ultrahigh vacuum systems. However, they have the advantage of easy operation. This apparatus is a metal–glass hybrid system and may be modified easily according to the requirements of the experiment. A vacuum of 10^{-8} Torr is easily attained after moderate baking-out, and adsorption isotherms above 10^{-6} Torr can be measured.

In the metal valve shown in *Figure 10*, the dead-space volume is different in the open and closed states. This shortcoming, however, may be overcome by care in the design and operation of the apparatus. The valve pushrod must be located at a fixed position when the valve is open, so that the volume change, Δv, has a constant value. The values for v_1, v_2 and Δv in the figure can be measured by the water-filling method, taking advantage of the sealing disc of the flange.

Various parts of the system are designated as shown in *Figure 10*, and appropriate volumes of each parts may be $V_1 \simeq 20$–30 ml, $V_2 \simeq 200$ ml, $V_3 \simeq 2000$ ml and $V_4 \simeq 100$ ml. V_1 was chosen as the standard reference volume for succeeding calibrations, and its determination was carried out with a device and procedure to be explained in a later section. The value of V_3 was easily determined, with an accuracy of ± 0.1 per cent, by measuring the volume of the bulb (about 2000 ml) before connecting it to the valve C_3. The values for V_2 and V_4 were determined by the gas expansion method using the Bourdon gage, which had been calibrated against a standard manometer and McLeod gage before the apparatus was constructed.

The Pirani gage was calibrated against the Bourdon gage using the gas expansion method. The procedure was as follows. Adsorbate gas was introduced into the space between C_1 and C_2 up to a pressure of 10 to 30 Torr, and its pressure p_1 was measured accurately with the Bourdon gage. The gas was then expanded by successively opening the valves C_2 and C_3, the pressure being reduced to $p_1 V_1/(V_1 + V_2 + V_3)$. After closing C_2, the space between C_2 and C_6 was thoroughly evacuated and the valves C_3 and C_6 were closed. Repeating this cycle two or four times, a pressure between 10^{-7} and 10^{-2} Torr was attained and the calibration of the Pirani gage was carried out with confidence. In the second step of this cycle, an error is caused by closing the valve C_2 because of the change in the dead-space volume Δv_{2c}. However, the magnitude of the relative error is of the order of $\Delta v_{2c}/(V_1 + V_2 + V_3)$ or 0.1 per cent per one expansion cycle and could be negligible.

This apparatus does not contain any sources of mercury or organic vapors apart from the Viton gaskets, and the adsorbent surface and Pirani gage filament were almost free from contamination. When an inert gas such as helium, neon or argon is used as the adsorbate, the sorption pump must be replaced by a high efficiency diffusion pump.

Wooten and Brown[101], Lauterbach, Laskin and Leach[56] and Ross and Oliver[86] have designed adsorption systems intended for the study of physisorption of ethylene or ethane vapor at the temperature of liquid nitrogen on adsorbents with small surface areas. These systems have the advantage that the adsorbate does not come into contact with stopcock grease which dissolves organic vapors, but has the disadvantage that the measured pressure values at lower pressures must be corrected for the mercury vapor drag effect. The apparatus may also be used in the study of chemisorption.

There are special techniques for adsorption measurements using radioactive ^{133}Xe or ^{85}Kr, though the range of application is quite limited[1,9,10,11].

Chnèbault and Schürenkämper[9] designed an apparatus, shown in *Figure 11*, to measure a surface area as small as 1 cm^2 with an accuracy of ten per cent. ^{133}Xe decays with a half-life of 5.27 days emitting 0.347 Mev β-rays and 81 keV γ-rays. Almost completely pure ^{133}Xe can be extracted from neutron-

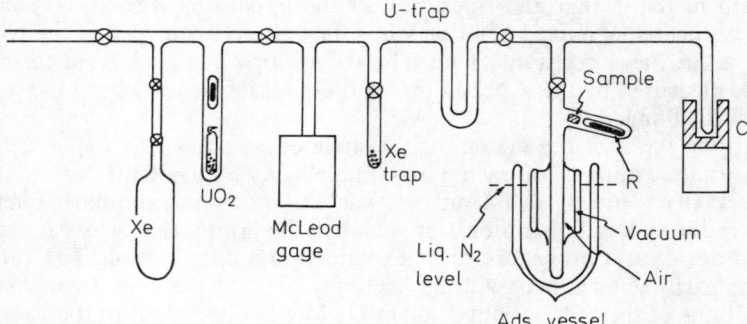

Figure 11. Apparatus for adsorption measurements of ^{133}Xe on solids with very low surface areas from P. Chnèbault and A. Schürenkämper[9]. C, counting device composed of a well-type scintillator and a photomultiplier; R, iron rod sealed in a glass tube.

irradiated uranium compounds such as uranium dioxide (UO_2), by partially reducing it with hydrogen prior to the irradiation[92]. With this pretreatment, oxygen evolved during the irradiation is re-absorbed into the UO_2 on heating it to extract the ^{133}Xe. The extracted ^{133}Xe is further purified by bringing it into contact with a titanium getter film. This careful treatment might not be necessary for surface area measurements, but is essential in detailed studies of adsorption phenomena.

The measuring procedure with this apparatus is as follows. Isotopically diluted ^{133}Xe, held in the trap, is supplied to the measuring system as required. Since the counting rate of γ-rays measured at C is proportional to the pressure measured with the McLeod gage, the calibration curve of counting rate versus pressure can be extrapolated with confidence. If the specific activity of xenon is suitably high, pressure measurements with a precision of one per cent or better are quite feasible at about 10^{-4} Torr. By comparing two calibration curves obtained with and without refrigerant in the U-trap, the mercury drag effect can be estimated. When the sample is bulky but has a very small surface area, competitive adsorption on the walls of the vessel cannot be neglected and must be carefully allowed for. The adsorption isotherm is first measured by keeping the sample in the limb as shown in the figure. The sample is then moved to the bottom of the adsorption cell with the aid of a magnet, the adsorption isotherm is repeated, and the required isotherm is obtained as the difference of these two. In this procedure, errors in the effective volume of the dead-space due to thermal transpiration are implicitly compensated.

D. Intermediate Pressure Region (10^{-3}–10 Torr)

Rosenberg[84] has designed an apparatus shown in *Figure 12*, that is

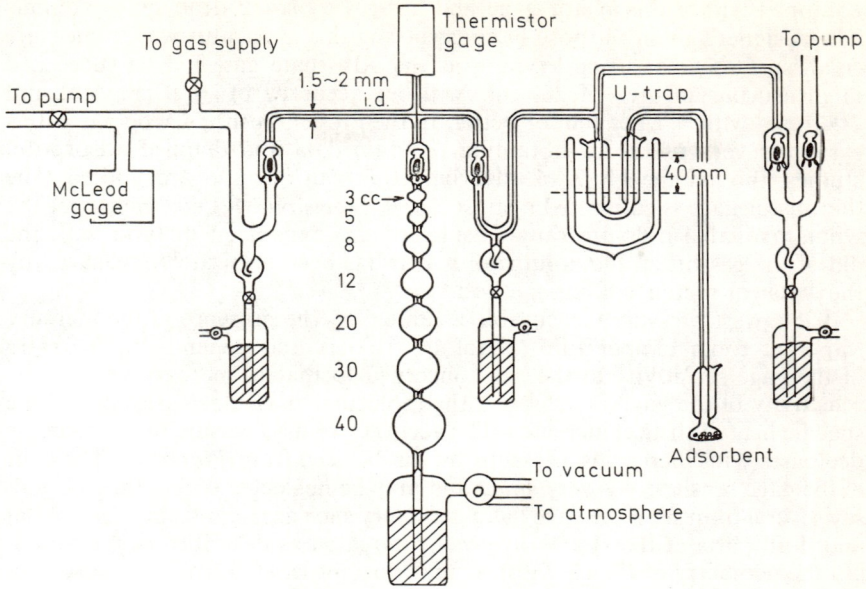

Figure 12. Rosenberg's type adsorption apparatus.

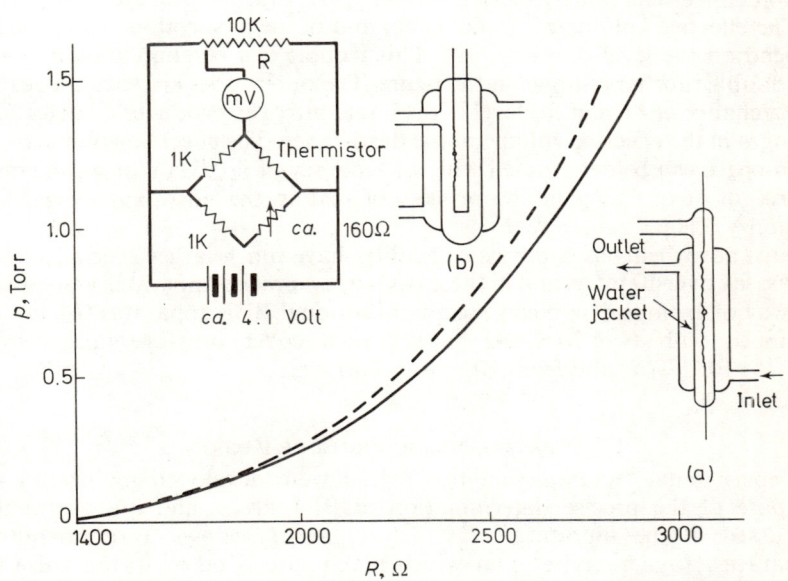

Figure 13. Electric circuit for a thermistor gage, a calibration curve, and geometries of gage construction.

equipped with a thermistor gage and has a very small dead-space volume. The designer's main purpose in constructing this apparatus was to measure small surface areas using krypton as the adsorbate gas, and he succeeded in measuring an area of 200 cm^2 with an accuracy of ± 10 per cent and 2000 cm^2 with ± 1 per cent or better. In the present author's experience, this system is very useful for obtaining accurate data and detailed adsorption studies, though the kinds of adsorbate that can be used are limited. The thermistor gage is calibrated against a high precision McLeod gage over the whole available pressure range, using the gas expansion method with the aid of the gas buret, the volume of which has been accurately measured by the water or mercury filling method.

The lowest pressure measurable is limited by the presence of the mercury vapor at room temperature (about 10^{-3} Torr), and depends on the ratio of the gage sensitivity to the gas concerned to that of mercury vapor. The sensitivity of the gage increases as the molecular mass decreases and as the specific heat of the gas increases. The accuracy of the pressure measurements decreases with increasing pressure, as can be seen from *Figure 13*. The drift of the gage sensitivity is very small and may be neglected over a long period, say several months, if the gage has a geometry such as is shown in *Figure 13*(a) and if all parts of the system are well pretreated as described in section II. [In the geometry of *Figure 13*(b), a small current leaks through an adsorbed film of mercury on the glass wall of the gage, and the calibration curve drifts toward the dotted curve in the figure.] The sensitivity, however, depends slightly on the room temperature which controls the vapor pressure in the gage. Since the bore diameter of the connecting tube is small, say 2 mm i.d., the magnitude of the mercury drag effect is negligibly small in the pressure region concerned, but corrections for thermal transpiration are essential.

The effective volume of the cold trap and of the adsorption vessel usually depend on the level of refrigerant. This trouble can be eliminated by using the double tube type shown in the figure. The outer tubes are vacuum-sealed, and a change of 4 cm in the level of the liquid nitrogen produces no detectable changes in the effective volume of the dead-space. The effective volume of the dead-space can be determined with an accuracy of ± 0.03 ml or so, the main source of error being in the process of sealing the adsorbent vessel [see section V.3.H].

Such adsorbents as catalysts or zeolites have much larger specific surface areas, say over 10 m^2/g, and in these cases a less precise apparatus is adequate. By way of example[94], we can use a conventional BET apparatus (*Figure 14*) equipped with two McLeod gages which cover pressure ranges from 10^{-3} to 10^{-1} Torr and from 10^{-1} to 5 Torr.

E. Subatmospheric Pressure Region

Experimental techniques in this region were developed mainly for the purpose of the precise determination of BET areas, and are thoroughly discussed in the literature[27, 46, 47, 73]. *Figure 14* shows a representative apparatus (Joyner's type[47]), in which much care is taken to minimize the dead-space volume and keep it constant. The zero level of the U-tube manometer can be easily detected if use is made of a device similar to that shown in the figure. The temperature of the gas buret must be kept constant and this is checked by inserting a thermometer into the water jacket. The sizes of the various bulbs of the buret are not critical, and a convenient range of volumes for most applications is 5, 15, 25, 50 and 150 ml. With such a buret,

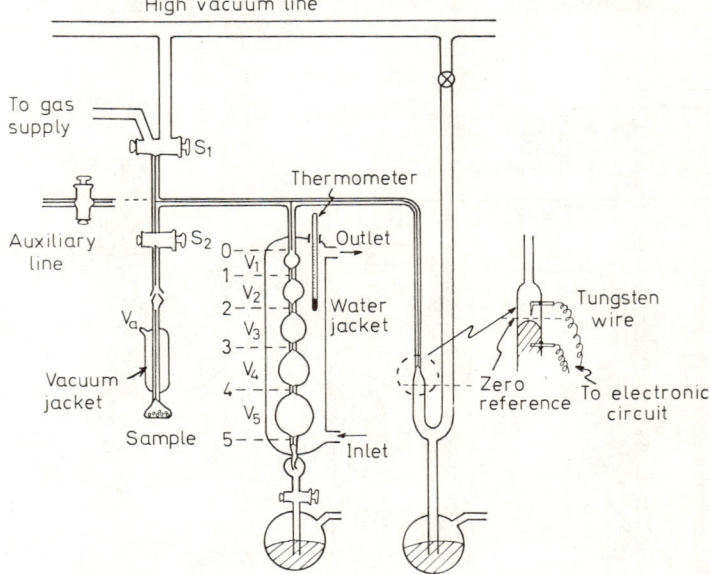

Figure 14. Joyner's type adsorption apparatus which is mainly used for determination of BET surface area[47].

the smallest surface area that can be measured with confidence is about 2 m² or so.

Halsey and his co-workers[13] constructed a highly precise and sophisticated system, in which the errors are an order of magnitude smaller than normal. Their system was designed for adsorption measurements of inert gases on adsorbents with low specific areas, at low coverages, and at temperatures above the critical temperature of the adsorbate. Such measurements require outstandingly accurate determinations of the dead-space volume and amount of gas admitted. They minimized errors arising from the manometer, the fluctuation of the water jacket temperature of the gas buret and the accumulation of errors due to successive gas dosage.

The adsorption of vapors on porous materials exhibits hysteresis in the region near the saturation pressure. In order to determine the distribution of pore sizes by analysing the hysteresis curves, accurate data are necessary. One of the most elaborate pieces of apparatus appropriate for this purpose was designed by de Boer and his co-workers[59], and is shown in *Figure 15*. Their design points were that: (a) The total amount of gas adsorbed must be able to be recovered quantitatively on desorption, (b) the volume of the dead-space is minimized, and (c) the temperature of the liquid nitrogen bath is checked continuously.

Condition (a) is fulfilled by use of a modified Töpler pump B. For condition (b), use is made of narrow connecting tubes and the specially designed manometer E, which has a main capillary tube and a reference one as well. The level of the reference tube can be independently adjusted to coincide with

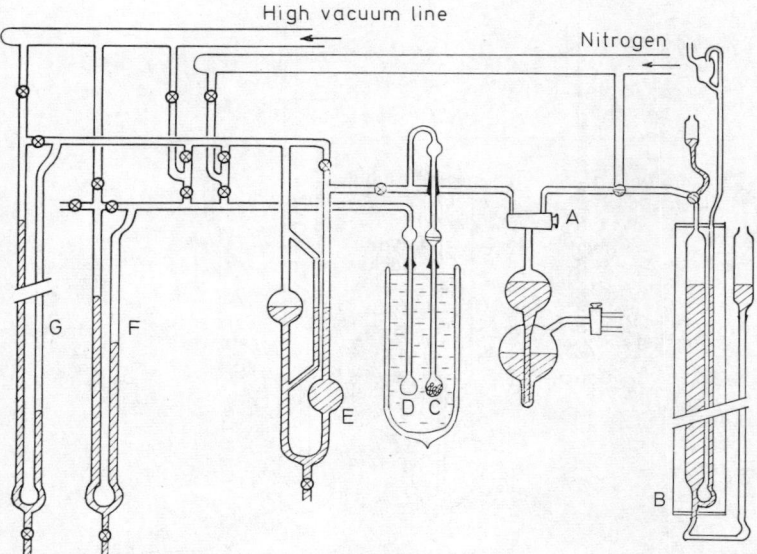

Figure 15. Adsorption apparatus appropriate for determination of distribution of pore sizes from B. C. Lippens, B. G. Linsen and J. H. de Boer[59]. C. sample cell; D, vapor pressure thermometer.

the main tube, and the pressure in the reference arm is measured by the auxiliary manometer G. The temperature of the bath can be accurately measured with the vapor pressure manometer F.

F. HIGH PRESSURE REGION

The higher the equilibrium pressure, the more the amount of unadsorbed gas remaining in the dead-space of an adsorption apparatus. Hence, adsorption measurements at higher pressures require very precise determination of the dead-space volume and also use of adsorbents with fairly large specific surface areas. Such adsorbents may be porous materials of which the pore volume v_p may not be small enough to be neglected in comparison with the volume of the adsorbent solid, v_{solid}. Here the question arises as to whether the pore space is to belong to the dead-space of the apparatus or not. In connection with this question, Coolidge[14] has given two kinds of definitions for the amount of adsorbed gas, m'_{ads} and m''_{ads},

(i) the difference between an amount of admitted gas, m_{admit}, and that remaining in the free space of the apparatus; namely,

$$m'_{ads} = v_s \rho_{ads} = m_{admit} - v_f \rho_{gas} \quad (5)$$

where v_s and v_f denote the volume of the adsorbed phase and that of the free space in the adsorption cell, respectively, and ρ_{ads} and ρ_{gas} the density of adsorbates in the adsorbed and gaseous phases, respectively, and

(ii) $m''_{ads} = v_s(\rho_{ads} - \rho_{gas})$

By virtue of the relations

$$v_t = v_f + v_s + v_{solid} \text{ and } m_{admit} = v_f \rho_{gas} + v_s \rho_{ads},$$

we have

$$m''_{ads} = m'_{ads} - \rho_{gas} v_s = m_{admit} - \rho_{gas}(v_t - v_{solid}), \quad (6)$$

where v_t denotes the total volume of the whole space in the adsorption cell, v_t and v_{solid} being determined unambiguously. Curves of m''_{ads} against pressure have a maximum point in the higher pressure regions, since ρ_{ads} and ρ_{gas} increase with increasing pressure, and converge to the same value, say, the density of the solid, that is, $\rho_{ads} - \rho_{gas}$ diminishes.

From a theoretical point of view, values for m'_{ads} are required, inasmuch as statistical thermodynamic properties of the adsorbed phase are expressed

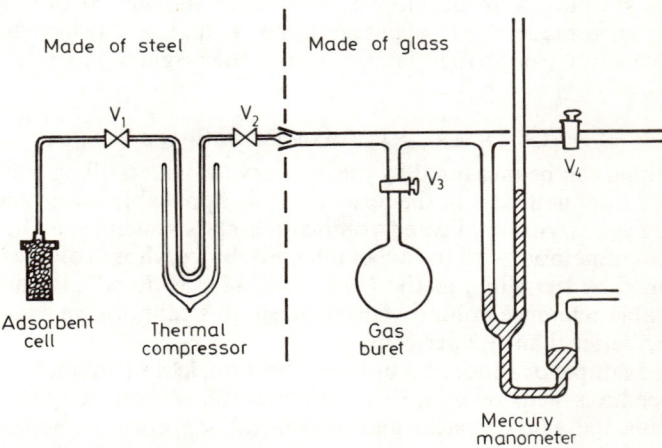

Figure 16. Apparatus for measurement of high pressure adsorption from Vasil'ev et al.[20].

as functions of a concentration of adsorbate. Thus a proper choice of the value for v_s becomes important in order to deduce values for m'_{ads} from those for m''_{ads}. In adsorbents such as zeolites having crystallographically regular and small pores, v_s may be equated to v_p with a good approximation. The value for v_p can be calculated from the total amount of adsorbent and crystallographic data. In amorphous adsorbents such as active charcoal and silica gel, on the other hand, the value for v_s may be considerably different from that of v_p. Pores in aggregates of amorphous fine particles have widely varied radii, and some space in pores may not be subject to the influence of the adsorbent. In such cases, the determination of v_s is open to some ambiguities and various empirical methods have been devised[67].

Recently, Menon[67] reviewed experimental techniques for adsorption measurements at high pressures, and here only representative apparatus is briefly discussed. Vasil'ev[20] designed a simple apparatus which has neither a compressor to raise pressure, nor a gage for measuring high pressures as shown in *Figure 16*. After baking-out adsorbent, a known amount of gas is transferred into a calibrated U-tube thermal compressor of steel by cooling with an appropriate refrigerant. Removing refrigerant and opening a valve V_1 between the U-tube and an adsorption cell, gas is brought into contact with adsorbents. When the adsorption equilibrium is reached, V_1 is closed and gas contained in the U-tube is expanded, by opening V_2, into a pressure (subatmospheric) measuring system made of glass. Using calibrated volumes of each part of the system, the equilibrium pressure and the amount adsorbed can easily be calculated from the pressure of the expanded gas.

Various kinds of apparatus have been designed which make use of a gas pipet and a manometer to measure the expanded gas[17,49,74]. With this equipment, adsorption isotherms can be measured up to 200 bar with an accuracy of ± 5 per cent or so depending on their design. Isotherms at higher pressures (up to 3000 bar) can be measured by using the glass piezometer techniques which were developed by Michel and his co-workers. In this method, use is made of adsorbent contained in a glass capsule which insures a clean vacuum. For further details, refer to the original paper[68].

G. Determination of Dead-Space Volume

A volume can be measured by the mercury (or water) filling method or the gas expansion method, but the former is only applicable to certain parts of a vacuum system. A complicated volume in a glass system must be measured by the latter method, with the aid of reference bulbs whose volumes have been determined by the filling method. The bulb of a McLeod gage may be used as a reliable reference volume, but, by itself, this cannot give results with an accuracy better than ± 2 per cent.

As an example of a more accurate calibration, let us explain some measuring procedures with reference to the apparatus shown in *Figure 14*. After evacuating the whole system and closing the stopcock S_2, helium is introduced into the gas buret and S_1 is closed. The mercury level in the buret is then adjusted to the ith mark, and the pressure, p_i, is measured with the U-manometer. Since the relation: $pV = $ constant, can be applied in common circumstances, plots of $1/p_i$ against $\sum_{j=1}^{i} V_j$ lie on a straight line as shown in *Figure 17*. The extrapolated line intersects the abscissa at V_0, the absolute value of which gives the volume of the space partitioned by the zero-mark of the buret, the two stopcocks, and the zero-level of the U-manometer. After repeating the procedure several times with different amounts of gas, the mean value for V_0 can be calculated. If the system is designed so that the ratios of $V_0 : V_1 : \ldots : V_5$ have the proper magnitudes, i.e. $V_1, V_2 > V_3, V_4$, the volume of V_0 can be determined with an accuracy of ± 0.1 per cent or better.

The dead-space volume of the cooled adsorption vessel containing the adsorbent may be measured in a similar manner by opening the stopcock S_2, but a correction must be made in order to take into account the non-ideality

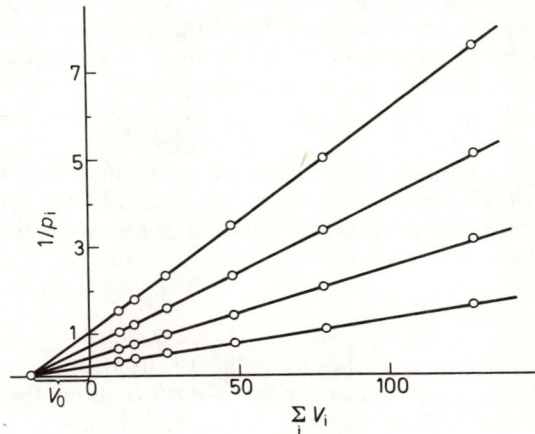

Figure 17. Diagram for determination of a dead-space volume.

of the gas at the bath temperature. Let the dead-space volume of the adsorption vessel measured by helium with and without refrigerant be V'_a and V''_a, respectively, and the effective volume for the adsorbate with refrigerant be V_a. Then, we have the relation,

$$V'_a = V_r + V_c$$

and

$$V''_a = V_r + V_c T_c/T_r$$

where T_r and T_c are temperatures of the room and the cold bath, respectively, and V_r and V_c denote the *true volumes* of the parts kept at T_r and T_c, respectively. Then, V_a becomes

$$V_a = V_r + (T_c/T_r)(1 + \alpha p)V_c$$

where α is the correction factor for non-ideality of the adsorbate gas. Some values of α for various gases are given in *Table 2*[104]. Solving these equations, we get

$$V_a = V''_a + \{T_c/(T_c - T_r)\} \alpha p(V''_a - V'_a) \tag{7}$$

With this effective volume V_a, we can calculate the amount of unadsorbed gas in the adsorption vessel as pV_a.

In adsorption measurements below 1 Torr, the effective dead-space volume determined by the helium expansion method must be corrected in a complicated manner, because both the thermal transpiration and mercury drag effects operate to different extents with helium and the adsorbate gas. Theoretical corrections for these factors are probably inaccurate though not impossible, and it is preferable to determine the effective dead-space volume using the adsorbate gas. The effective volume of the adsorption cell without adsorbent, V_a (blank), is measured over the pressure and temperature ranges concerned using the adsorbate gas. Then, the adsorbate is inserted into the cell, the volume of adsorbent, V_s, being calculated from its weight and density. The effective volume required, V_a, is given by

$$V_a = V_a(\text{blank}) - V_s(T_c/T_r)(1 + \alpha p) \tag{8}$$

If the adsorption cell is directly connected to the main line without using a normal joint, the change in the volume introduced by the process of re-connecting the tube is estimated to be ± 0.03 ml or less, if the bore radius of the connecting tube is as little as 2 mm.

H. Adsorption Cells

A good adsorption cell should fulfil the following conditions: small dead-space volume, easy evacuation and the prevention of powdery samples leaving the cell on sudden evolution of occluded gases. These conditions, however, cannot be satisfied simultaneously, and which should be stressed depends on the aim of the experiment. Several examples of cells are shown in *Figure 18*, all of which are designed to minimize the dead-space volume. Cell (a) is a common type, the sample inlet of which is made of a thin narrow tube which is easily sealed. The degassing of the adsorbent is accelerated by

Table 2. Values for the correction factor α of the gas imperfection from Young and Crowell[104].

Gas	Temp. (°C)	After Emmett and Brunauer atm^{-1}	From second virial coeff. atm^{-1}	From viscosity atm^{-1}
N_2	-195.8	6.6×10^{-2}	5.0×10^{-2}	4.5×10^{-2}
	-183	3.78	3.3	2.9
O_2	-195.8	—	6.3	5.3
	-183	4.17	4.2	3.4
Ar	-195.8	11.4		
	-183	3.9		
CO	-183	3.53		
CH_4	-183.1	—	7.1	6.6
	-140	7.79		
NO	-140	5.26		
N_2O	-78	7.68		
CO_2	-78	2.75		
	$+25$	0.76		
NH_3	-36	3.47		
	$+25$	1.56		
$n\text{-}C_4H_{10}$	0	14.2		
	25	4.2		

passing a gentle flow of helium in through the sample inlet during the baking-out. Cell (b) is designed to reduce the possibility of splashing of the adsorbent, and to trap mercury vapor from other parts of the system[18]. Cell (c) is used for samples in the form of thin plates with very small surface areas[65].

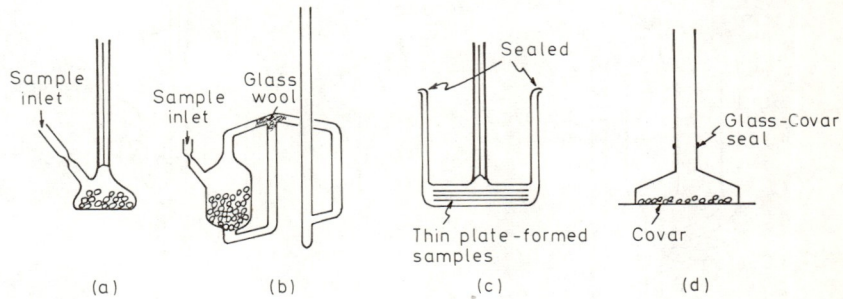

Figure 18. Various kinds of adsorption cells: (a) common type; (b) from V. R. Deitz and L. F. Gleysteen[18], (c) from D. E. Meyer and J. E. Wells[65]; (d) from J. A. Morrison and D. M. Young[70].

Cell (d) is a tray-shaped adsorption cell made of Covar, and designed to facilitate thermal conduction. This type of cell was used in Morrison's adsorption cryostat[70] and its thermal contact with a metal heat reservoir was facilitated by silicone grease. Such a contact can be improved by the use of indium or indium–gallium solders.

I. Gas Dosing Systems

The gas dosing systems shown in *Figures 10, 12* and *14* are of the internal dosing type. This, though the simplest, has the disadvantage that the volume of the doser is included in the volume of the adsorption system and hence successive dosing causes an accumulation of errors. This disadvantage becomes especially serious in the flat region of an adsorption isotherm. *Figure 19*(a) shows an external dosing system of Young's type[103]. The method of operation is as follows: mercury in A is raised up to the mercury float valve, and then adsorbate gas is introduced by lowering the mercury level in B to just below the cut-off. After measuring the pressure, and raising the mercury in B above the cut-off, the mercury in A is lowered slowly down to the level marked in the figure. Finally, the mercury in B is raised up to the top of the bulb, pushing the adsorbate through the fritted glass into the adsorption system. Care must be taken in the design so that the mercury level in the bulb can be lowered easily. For this, the radius of the opening of the cut-off must be wider than 6 mm i.d., and the smaller the volume between S_1 and F_1, the better. A doser of about 20 ml capacity can accurately admit known quantities of gas ranging from 1 to 10^{-4} ml s.t.p.

For adsorption measurements at lower pressures much smaller capacity dosing systems are required. Various types have been designed, and one is shown in *Figure 19*(b)[99]. The tubing between the cut-offs B and C in the figure consists of Pyrex capillary 1 mm i.d., and tube 6 mm i.d. The volumes

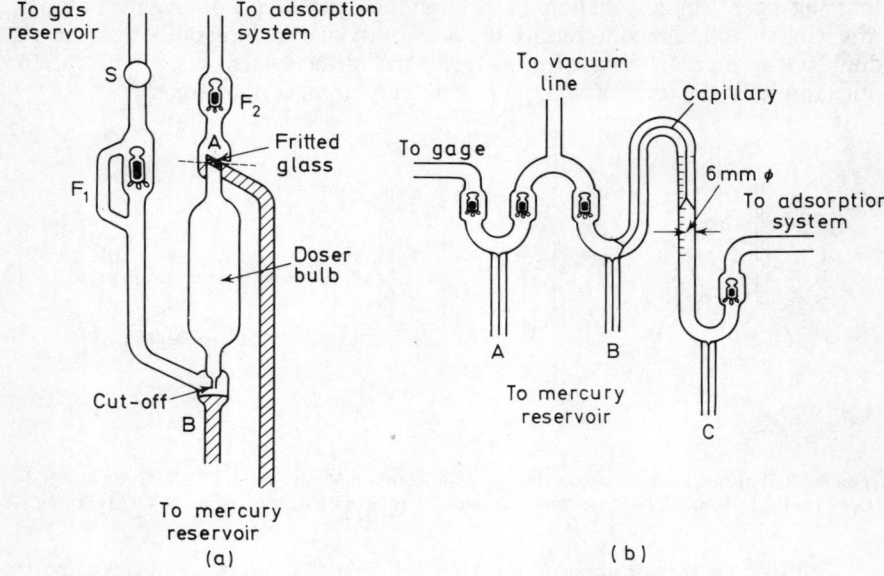

Figure 19. Gas dosing systems: (a) Young's type one[103]; (b) Micro-doser (from A. Weinstein and H. C. Friedman[99]).

between the etched marks are calibrated by the mercury filling method prior to the construction of the doser. With a precision McLeod gage and a standard mercury manometer, an amount of gas ranging from 10^{-7} to 1 ml s.t.p. can be dosed.

4. Gravimetric Methods

A. General

The gravimetric method of adsorption measurement has been used successfully not only in experiments with condensable vapors but also in the ultrahigh vacuum region. Its merit lies in the fact that the amount of adsorption is measured directly, in contrast to the volumetric method in which the amount adsorbed is calculated from the dead-space volume, and various pressures and temperatures. Thus, adsorption isotherms can be measured without accumulating errors that were mentioned so frequently in the preceding sections. Apart from the common sources inherent in weighing, the main errors arise from the effects of buoyancy which is complicated by the temperature inhomogeneity. The apparatus must be designed to minimize such effects.

Another advantage of the method is that the effects of the cleaning process of the sample's surface can easily be observed by following the decrease in weight of the sample. In the volumetric method, the adequacy of the outgassing may be checked by isolating the adsorbent and by measuring the pressure that builds up.

ADSORPTION

Important factors in the design of a balance are the sensitivity, sensibility, range and capacity. The sensitivity is the ratio of the reversible response to the change of mass which produces that response, e.g. the extension of a spring due to an increase in the load [the units are mm/g]. The term sensibility stands for the minimum variation in mass which can be reproducibly detected with a given accuracy. The range designates the maximum variation in mass that the balance can take for a given load, and the capacity means the maximum permissible load of the balance. In detailed studies of adsorption, isotherms at lower coverages have fundamental significance, and to have accurate results in that region, requires not a high sensibility but a high relative sensibility [(relative sensibility) = (sensibility)/(capacity)], inasmuch as coverage is proportional to the relative weight change.

Generally, two types of balances are used in adsorption experiments; namely, the helical spring balance and the beam type balance. The former has a wider range but a smaller capacity and a lower relative sensibility, and is used in studies of adsorbents with specific surface areas larger than 100 m^2/g. On the other hand, the latter has a higher relative sensibility but a narrower range, and is appropriate for use with adsorbents with small surface areas.

B. Helical Spring Balance

A helical quartz spring sorption balance of the type used by McBain and Bakr, is simple, reliable and convenient, although its sensitivity is rather low. Using a sealed-off apparatus, McBain and Sessions[64] carried out observations for twenty years, and concluded that no appreciable stretching of the quartz spirals occurred even when they were exposed to a variety of organic vapors, but exposure to water vapor over a long period did give rise to some slight stretching. Such high stability as this cannot be expected in springs made of other materials, e.g. Pyrex glass, beryllium–copper alloy

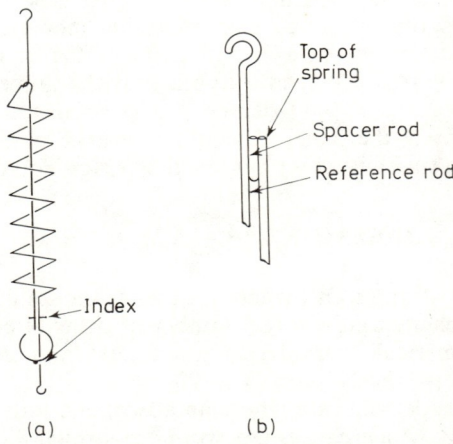

Figure 20. Reference rod for helical spring balance. (a) Figure of a reference rod, from M. J. Rand[79]; (b) Method of fixing the spring to the reference rod.

or molybdenum, and this is the reason why quartz springs are preferable in spite of their fragile nature.

When higher precision is required in measuring the stretching of a spring, it is customary to use a straight silica reference fiber suspended from the spring support. This cancels out variations in the height of the spring support with respect to the cathetometer that may be introduced by various causes. Sometimes, however, vibrations of the spring may cause serious damage. In order to avoid this trouble, Rand[79] designed a reference fiber the lower end of which forms a large circle, as shown in *Figure 20*(a). Since the diameter of the large circle is larger than that of the helix, they cannot interpenetrate each other. However, the method of suspension of the reference fiber shown in *Figure 20*(a) is not recommended, since the relative position with respect to the spring may be changed by shocks and vibrations reaching the apparatus. In the author's laboratory, the spring is fixed to the reference fiber with silver chloride solder as shown in *Figure 20*(b). The quartz rod spacer prevents the reference fiber from coming into contact with the spring at the lower end. According to our experience, the rigidly combined system is easy to operate and the chance of destruction is greatly reduced.

The sensitivity of the spring balance, K, is related to the helix parameters by the relation,

$$K = (8ND^3/Gd^4) \tag{9}$$

where N is the number of turns of the coil, D is the helix diameter, d is the fiber diameter, and G is the shear modulus for the material of which the spring is made. The value of G depends on the ambient temperature and load. The effect of the load is discussed first, since it limits the capacity of the spring. An increase in the load of course stretches the spring, causing an increase in the helix angle of the spring. [The helix angle is defined as the angle between a line tangent to a turn of the helix, and a plane perpendicular to the axis of the helix.] As the helix angle increases, the effect of the load changes from a torsional deformation to a higher modulus tensional deformation, and thus the effective shear modulus increases. Ernsberger and Drew[24] measured the decrease in sensitivity of the quartz spring balance as a function of helix angle θ, and found that the amount of the decrease reaches one per cent at $\theta = 11$ degree and at room temperature, independent of the design of the helix. Giving a tolerance of one per cent to the constancy of the sensitivity, they obtained a design equation for the quartz spring balance,

$$0.0783\,(L^5 K/n^2)^{\frac{1}{4}} + KM - L = 0 \tag{9'}$$

with $G = 3.11 \times 10^9$ mg/mm^2, where L denotes the total length of the helix, M is the load capacity and n is the number of turns of coil per unit length. By way of a numerical example, let $n = 5$ turn/mm, $K = 5$ mm/mg and $L = 500$ mm, and we then obtain $M \simeq 90$ mg.

Let us now consider the case where an adsorbent with a mass M adsorbs vapor of mass ΔM which causes the spring to stretch a distance ΔL. Then, we have

$$\Delta L = K \cdot \Delta M$$

Rearranging equation 9', we get

$$KM = L\{1 - 0.078(KL/n^2)^{\frac{1}{4}}\} \simeq L$$

since the second term inside the brackets is considerably smaller than unity. From these, we have the simple relation

$$\frac{\Delta M}{M} \simeq \frac{\Delta L}{L} \text{ or } \frac{\Delta M_{min}}{M} \simeq \frac{\Delta L_{min}}{L}$$

where ΔM_{min} is the sensibility or the minimum detectable variation in the mass, ΔL_{min} the corresponding variation in the length of the spring, and $\Delta M_{min}/M$ the relative sensibility. Thus the relative sensibility is wholly determined by the value of L, because ΔL_{min} is limited to about 0.01 mm. On the other hand, it is inconvenient, from the standpoint of easy manipulation, for the length L to exceed 500 mm. Consequently, relative sensibilities much higher than 2×10^{-5} cannot be expected in the spring balance. A commercially available quartz spring balance of the highest quality (manufactured by Houston Tech. Lab.) has the following specifications: Capacity 100 mg, Sensitivity 7 mm/mg. Its relative sensibility becomes 1.4×10^{-5} if the length of the spring can be measured with an accuracy of ± 0.01 mm.

As a numerical example, let us consider the adsorption of nitrogen on an adsorbent whose specific surface area is 100 m²/g. A monolayer film of nitrogen on 100 mg of this adsorbent amounts to 2.9 mg and film at a coverage of 0.01 causes a relative weight change of 2.9×10^{-4}. Therefore, detailed adsorption studies with a spring balance cannot be carried out on an adsorbent with specific surface areas smaller than 100 m²/g.

The shear modulus of quartz, G, increases with increasing temperature and the temperature coefficient of $\Delta G/G$ has a value of $1.359 \times 10^{-4} \sim 1.573 \times 10^{-4}$ deg^{-1} near room temperature. If the change in the sensitivity of the balance with temperature arises only through G, we have from equation 9

$$\Delta K/K = -\Delta G/G = -\alpha \Delta T$$

with $\alpha = 1.359 \times 10^{-4} \sim 1.573 \times 10^{-4}$ deg^{-1}. Ishimura[42] thoroughly investigated the temperature dependence of the sensitivity of a quartz spring balance, and in the region near room temperature obtained the relation

$$\Delta K/K = -1.48 \times 10^{-4} \Delta T$$

which proves the validity of the above assumption. At first sight, it might be considered that such a temperature effect as this is of little practical significance, since fluctuation in room temperature may not exceed 5°C and a sensitivity of four significant figures is neither required nor expected. But this is not so, as can be seen from the following consideration. In the gravimetric method of adsorption experiments, we measure a small deviation of the balance from its rest point which is determined by the large masses of the adsorbent, bucket and the spring itself. Let the sum of these effective masses be M_{syst}, and the mass of adsorbate be m_{ads}. Then a change in the sensitivity of the balance, ΔK, causes a drift of $M_{syst} \Delta K$ in the rest point, and errors in m_{ads} are not given by $m_{ads} \Delta K$ but by $(m_{ads} + M_{syst}) \Delta K$; that is to say, the error is magnified by a factor of M_{syst}/m_{ads}. Therefore, the higher the relative sensibility, the more severely the ambient temperature of the balance must be controlled. According to Ishimura, an error of several per cent can sometimes be introduced by fluctuations in the ambient temperature.

The quartz spring balance can be highly effective in adsorption studies of adsorbates which are relatively involatile and/or are reactive with respect to mercury or stopcock grease, such as Br_2, I_2, S, $AlCl_3$, $FeCl_3$ and $HgCl_2$. Barrer and Wasilewski[2] designed an apparatus suitable for work on these materials. The pressure of adsorbate is measured with a spiral Bourdon gage made of Pyrex glass. The temperature of adsorbent and that of the adsorbate source are independently controlled with an accuracy of $\pm 0.01°C$. The upper limit of temperature, 350°C, is set by the thermal properties of the helical spring and the spiral pressure gage, because higher temperatures may introduce some hysteresis effects. The sensitivity at high temperatures is considerably different from that at room temperature, and careful calibrations are required.

Moreau[69] constructed a recording spring balance which has a ferrite rod attached to the fiber and a differential transformer outside the sample tube. If the armature connected to the spring moves from a position of balance, an alternating voltage appears at the linear differential transformer. This signal is fed back to control the current in the differential transformer, restoring the armature to its original position. A signal proportional to the variation of the sample weight is taken from the feedback loop and recorded. Using a helical spring with a sensitivity of 1 mm/mg and a capacity of 100 mg, he succeeded in obtaining a sensitivity of several micrograms, with a response time of 0.01 mg/sec.

Bushuk and Winkler[7] designed a quartz spring torsion microbalance, in which a torsion fiber supports the load and the increase in weight is determined with a high-sensitivity quartz spiral. The torsion fiber insures a high capacity and the quartz spiral acts as a damper to increase the stability of the balance. This apparatus, however, is not suitable for obtaining a good vacuum.

Jones, Isaac and Phillips[45] have developed a quartz spring balance installed in a metal housing, which may be used up to 80 bar pressure. The use of this balance at higher pressures has two disadvantages arising from the buoyancy correction: (i) the boyuancy depends on the quantity being measured, that is the amount of adsorbate in the pores of the adsorbent, and (ii) near the critical point, density gradients exist in the gas phase and give rise to uncertainties in the density value in the vicinity of the adsorbent.

C. Beam-type Balances

Beam-type vacuum balances have symmetrical geometry, and the effect of buoyancy can be diminished. This advantage is very significant in detecting very small changes caused by adsorption. One of the most commonly employed vacuum balances is the Gulbransen type torsion balance[33], refinements of which have been made by Rhodin[81], and others[51,63]. A representative structure is shown in *Figure 21*(a), in which the tungsten torsion and suspension wires are cemented with fused silver chloride at right angles to the quartz beam.

Gulbransen obtained a sensibility of 3×10^{-7} g with a tungsten torsion

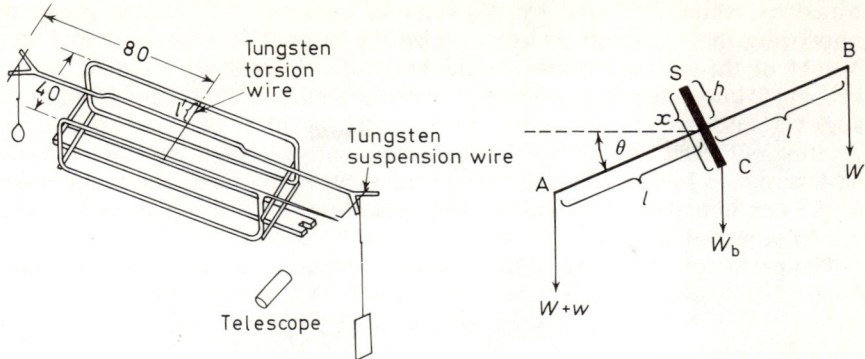

Figure 21. Gulbransen's type torsion balance: (a) Structure; (b) Condition of equilibrium.

wire of 25 μϕ, while Kolenkow and Zitzewitz[51] obtained 5 × 10^{-9} g with a quartz fiber of 75 μϕ. In soldering with silver chloride, use is made of a reducing flame in order to prevent oxidation of the tungsten wire and the silver chloride. In spite of this precaution, some silver oxide may be produced; this acts as a source of residual gas, and a vacuum better than 10^{-9} Torr may be difficult to obtain.

Let us now discuss the sensitivity of this type of balance with reference to *Figure 21*(b). Point S denotes the primary fulcrum (i.e. the axis of the torsion wire supporting the beam system), and the point C is the center of gravity of the balance itself, the weight of which is W_b. From simple geometrical considerations, we have at equilibrium the relation

$$(W + w)(l \cos \theta - h \sin \theta) = W(l \cos \theta + h \sin \theta) + W_b x \sin \theta + \sigma(\theta - \theta_0)$$

with

$$\sigma = 2Z\pi r^4/(2l')$$

where σ is the torsion constant, Z the rigidity modulus of the suspension fiber, l' and r its length and radius, respectively, and θ_0 is a constant to allow for the fact that the fiber is not free from torsion at $\theta = 0$ owing to deviation from an ideal construction. Neglecting higher order terms of θ, the sensitivity of the balance S is given by

$$S = l(d\theta/dw) = l^2/(W_b x + 2Wh + \sigma) \tag{10}$$

if the pointer is located at the end of the arm as shown in the figure. In order that the sensitivity shall be independent of the load, h must therefore be zero; namely, the three fulcrums A, B and S must lie in a straight line. In practice, however, it is not easy to make $|h|$ smaller than 20 μm. Furthermore, as shown by Ishimura[43], in a Gulbransen type balance the arm of which is a simple quartz beam, the load causes distortions in the beam. Thus, a change in sensitivity with load is inevitable and the balance must be calibrated prior to its use. Several examples of the calibration have been illustrated by Ishimura[43], and by Wolsky and Zdanuk[100]. The next point in improving the sensitivity is to minimize the term $W_b x$; namely, x and the weight of the balance beam should be made as small as possible, while keeping C lower than S in order to maintain stability. Lambert and Phillips[54] added a silica weight above the center of the beam in order to adjust the position of C, and this method has been frequently adopted. Several workers have designed highly sensitive torsion balances by reducing the magnitude of the denominator of equation 10 by making point S lower than point C (i.e. by adjusting to $x < 0$).[4]

The period of oscillation of the balance, τ, is given as a crude approximation by[6]

$$\tau = 2\pi S[2W + w + W_b \sigma^2]^{\frac{1}{2}}.$$

This equation neglects the distortion of the beam under load, the anharmonicity of the oscillation, and the damping. τ values for commonly used Gulbransen balances are generally about 5 sec or so in the absence of a load, and increase to 20–40 sec under a load of several grams. These long oscillation periods become a serious problem in high-sensitivity balances, since there is negligible damping due to the friction of the gas in a vacuum better than 10^{-4} Torr.

A beam balance can be operated as a null-point instrument, with a force of some kind operating to restore the beam to its original horizontal position when the beam is displaced. The restoring force also effectively damps the oscillations, partly solving the foregoing problem. Several kinds of automatic recording torsion balance have been developed along these lines, and some of them are commercially available, e.g. Chan's RG and RH types and the Sartorius Model 4142. Catalogues for the commercial instruments claim that a vacuum of 10^{-6} to 10^{-7} Torr can be attained; however, this vacuum is reached only by evacuating the system with a highly efficient pump. When the system is isolated, the vacuum deteriorates to 10^{-4} Torr or worse owing to water vapor evolved from parts of the balance. In order to use such a balance in accurate adsorption experiments, the system must be drastically improved to allow it to be baked-out at least at 180°C or so, for long periods. The author's experiences with the Cahn RG-type balance will be described. The anodized aluminum chassis is the worst source of water vapor, and its anodized layer was removed by sand-blasting it with alumina. As many electronic parts as possible were moved out of the housing of the balance. The coil, which acts as an electromagnet to restore the beam to its original position, cannot be baked-out above 120°C, and it was replaced by a new one made of polyimide-coated wire, which could be baked at 200°C. Furthermore, by using a housing made of stainless steel with metal valves, a vacuum of 10^{-7} Torr was attained after prolonged baking-out at 180°C, and a vacuum of 10^{-6} Torr was maintained overnight. It is considered that a better vacuum than this cannot be expected in such a complicated system. From the point of view of obtaining a good vacuum, simple handmade systems are superior to the commerical instruments, and it is recommended that the latter be used only in the intermediate pressure range for adsorbents with large surface areas.

Pivotal microbalances for use in a high vacuum have been developed by Czanderna and Honig[16]. The instrument pivots about two electropolished tungsten points supported in parabolic quartz cups. Since the tungsten tips rest precisely at the minima of the quartz cups, the balance is insensitive to vibrations. Care must be taken to keep the surface of the tungsten tips as clean as possible, insuring that there are no obvious points of friction. A sensibility of 10^{-7} g can rather easily be obtained, and a recording null instrument can be constructed by using the electromagnetic method for automatic compensation of mass changes. The pivotal balance is less sensitive than the torsion type, but can be set up and used without requiring special experimental techniques.

D. Resonating Quartz Crystals

In this decade resonating quartz crystals have been used as sensors for mass measurements in a vacuum. The principle is based on the fact that the frequency of a piezo-electric crystal is dependent on the total vibrating mass. Commercial instruments are now available which are commonly used to measure the thickness of evaporated films, or evaporation rates of solids. However, these commercial models are not sufficiently sensitive to be used in studies of gas adsorption on a small surface area. Currently, a change in the resonant frequency of the order of one part in 10^{10} or a change in the mass of 10^{-11} g can be detected with best quality custom-built instruments[98]. This means that oxygen adsorbed on a 0.1 cm^2 surface can be measured at coverages as low as 0.001. In order to attain such precision, Warner and Stockbridge thoroughly investigated the sources of errors which limit the sensibility of the instruments[89,98]. Even with the most favorable crystal geometry, the temperature of the resonator must be controlled within $\pm 0.01°C$ near 27°C. Baking-out at 400°C introduces stress between the quartz surface and the overlying electrode, and relaxation proceeds very slowly at room temperature. Hence, such precise adsorption measurements are not only painstaking but also restricted to well equipped laboratories.

Physical adsorption at large coverages can be measured rather easily with very simple electronic circuits, as demonstrated by Wade and Slutsky[97]. They installed the resonator chamber and electronic circuit in the same bath, the temperature of which was controlled within $\pm 0.001°C$. In their case, 1 p.p.m. shift in frequency corresponds to 2.65×10^{-8} g of adsorbent on a 1 cm^2 surface, and adsorption isotherms for hexane and water were obtained. Recently, Hillecke and Mayer carried out measurements at lower temperatures[37].

E. Sources of Error

Among the various sources of error in adsorption measurements with a vacuum microbalance, the serious ones are mechanical vibrations of the apparatus, the convection of gas in the system, the buoyancy of the adsorbent, thermal transpiration effects (in this case, usually called the thermomolecular force—abbreviated as TMF), forces due to electrostatic charges, temperature differences between the adsorbent and the tube wall, and the temperature gradients along the beam of the balance. At present, a balance sensibility of 0.1 μg may be attained without much difficulty. However, it is extremely difficult, if not impossible, to measure confidently 0.1 μg of adsorbate, owing to the errors arising from the sources mentioned above. When there is a temperature difference larger than 100 K between the adsorbent and the balance itself, the detection of even 1 μg of adsorbate may be subject to serious errors, in spite of careful efforts to reduce them. Therefore, in order to obtain precise data, reducing and correcting the errors are more important than improving the sensibility of the balance.

(1) Buoyancy

Buoyancy corrections in adsorption measurements become very important at higher pressures. According to the Archimedes principle, the correction term ΔW for a beam-type balance is given by

$$\Delta W = [(\text{volume of adsorbent}) - (\text{volume of counterweight})] \times (\text{density of ambient gas}).$$

In a spring balance, the equation needs several modifications, as there is no counterweight, and a correction must be applied to the spring itself. There is no ambiguity in the correction method for adsorbents with low surface areas, but some arise with high specific surface area adsorbents. The situation may be easily understood by referring to section IV.3.F. In the weighing method, we measure m''_{ads} defined by equation 6, and we must derive m'_{ads} from m''_{ads} by the same procedure mentioned in that section.

Recently, Pierotti[76] discussed this problem from a statistical mechanical viewpoint but there are as yet no experimental studies to test this theory.

(2) Convection and TMF

If one limb of a balance is heated or cooled, apparent mass changes are observed, and their dependence upon the ambient pressure is shown sche-

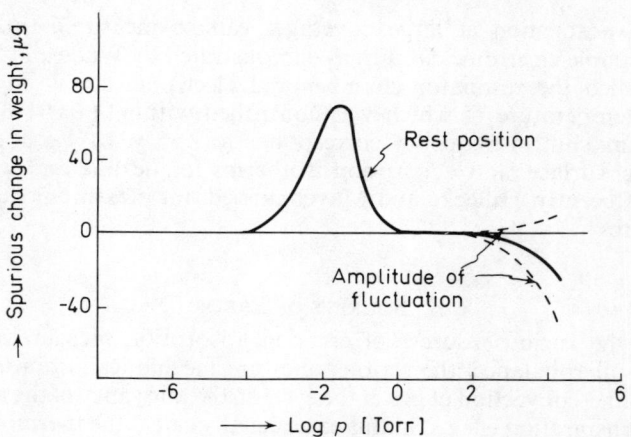

Figure 22. Spurious weight changes due to thermomolecular force and convection of gases. Solid curve shows rest position or midpoints; broken curves show the amplitude of fluctuation.

ADSORPTION

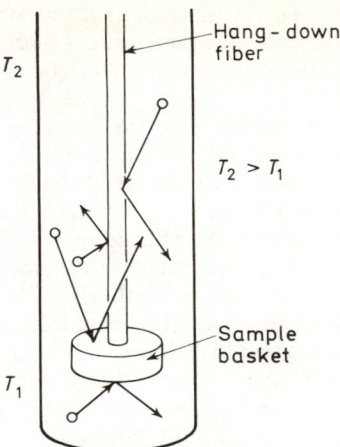

Figure 23. Momentum transfer of gaseous molecules to the balance at low pressures. The magnitude of the momentum of a molecule is proportional to the length of the arrow.

matically in *Figure 22*. These changes sometimes exceed 50 μg depending on the geometry of the balance system, and their correction is a decisive factor in obtaining precise data in adsorption measurements. By constructing and temperature-controlling the balance as symmetrically as possible, this spurious weight change may be substantially reduced, but not completely eliminated. Above 10 Torr the convection effect is predominant, while TMF

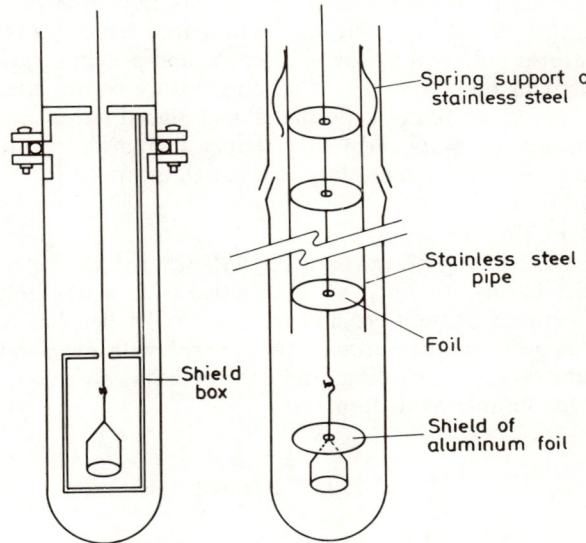

Figure 24. Shield for convection, thermomolecular force and radiation. The one on the left is from L. Cahn and H. Schultz[8a].

1059

is more important below 1 Torr with its maximum effect occurring between 10^{-1} and 10^{-2} Torr. The former also introduces fluctuations in the balance, the amplitude of which increases steeply above 200 Torr[52], as shown by the broken lines in *Figure 22*.

When gaseous molecules collide with the hanging balance, two kinds of forces exist which cause apparent weight changes; namely, longitudinal and transverse forces, as shown in *Figure 23*. If the fiber is placed in a region of inhomogeneous temperature, the longitudinal forces exerted on the fiber result in an apparent weight change. This spurious effect can be eliminated only by keeping all parts of the apparatus at the same temperature, but such a situation occurs very rarely. Some methods of reducing this effect are to keep the radius of the fiber as small as possible, and to use the balance in as symmetrical a state as possible. The transverse forces, on the other hand, operate on the horizontal faces of the bucket containing the adsorbent, or on the counterweight. Their effect can be greatly reduced by using a device such as that shown in *Figure 24*, where the bucket is shielded from the action of the transverse forces. In *Figure 24*(b), the aluminum foil shield serves not only to reduce fluctuations due to convection as pointed out by Robens *et al.*[83], but also as a shield for thermal radiation.

In spite of the effort of various authors[62, 78, 95], theoretical studies on TMF are still not able to give results that can be used for practical correction purposes. At present, errors caused by TMF or convection must be corrected empirically, and the situation may not improve in the future.

(3) *Temperature Differences*

Thermal equilibration between the adsorbent and the tube walls is not easily attained, owing to lack of direct contact. The temperature of the adsorbents cannot be measured directly and must be estimated from measurements in similar situations. Cutting[15] compared adsorption isotherms at low temperatures obtained by the volumetric and gravimetric methods, and concluded that in the worst case, the temperature of the adsorbent is 1 K higher than that of the bath. A thermal shield placed immediately above the bucket, as shown in *Figure 24*(b), may reduce such a temperature difference to a few millidegrees only according to a rough estimate[83].

(4) *Electrostatic Charge*

Electrostatic charging on parts of the balance and the tube walls can be avoided either by inserting a piece of β-active solid in the apparatus or by making the surfaces of the fiber and the tube walls electrically conducting. For the latter purpose, the surfaces are covered with evaporated gold film or transparent stannic oxide film, and Gomer[30] has described a method of preparation for stannic oxide film.

5. *Flow Methods*

A. Very Low Pressure Region

Chemisorption measurements at very low pressures on clean surfaces are most commonly made by the flow method[80]. Chemisorption studies using this method have been restricted to kinetic processes, and no isotherms have been obtained. On the other hand, physisorption isotherms have been obtained at very low pressures by the flow method[34, 82, 96]. Tuzi and Saito's apparatus[96], which can be baked-out at 450°C to 2×10^{-10} Torr, is shown in *Figure 25*. The operation procedure is as follows. The pumping speed at

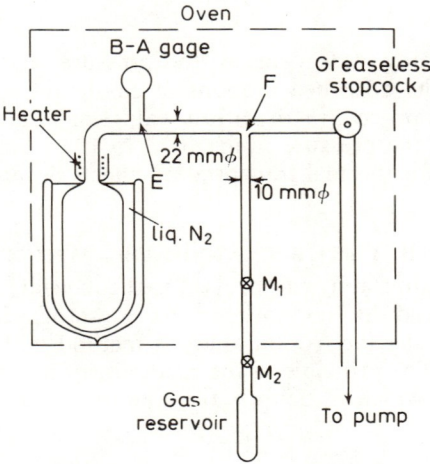

Figure 25. Apparatus for measurement of adsorption at very low pressure by the flow method (from Y. Tuzi and T. Saito[96]). M_1, variable leak made of metal; collar and heater coil at the top of the bulb are equipped to maintain a sharp temperature jump.

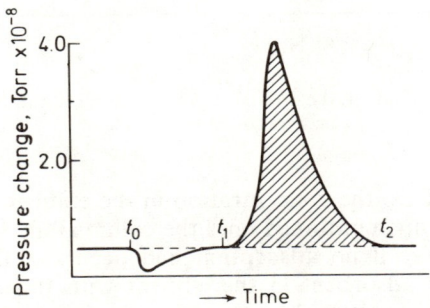

Figure 26. Pressure change in the apparatus shown in *Figure 25*. At t_0 the bulb is cooled; at t_1 the refrigerant is removed.

point E, S_E, is measured by the constant volume method, prior to the adsorption measurements. By adjusting the variable leak valve M_1 which is fitted to the gas reservoir, a steady flow of the adsorbate gas is obtained by the operation of the pumps. If the bulb B is cooled, say, by liquid nitrogen at time $t = t_0$, the pressure momentarily decreases and then gradually returns to its original value because of the occurrence of adsorption on the cold surfaces of the bulb. After attaining a steady state, the refrigerant is removed at $t = t_1$ and the adsorbed molecules are thermally desorbed. A representation of the pressure changes is shown in *Figure 26*. The amount of gas adsorbed, v_{ads}, is obtained by graphical integration as

$$v_{ads} = cS_E \int_{t_1}^{t_2} [p(t) - p_0] \, dt$$

where p_0 is the steady state value of the pressure, and c is a conversion factor to express the adsorbed amount of gas in the units of ml at s.t.p. Repeating the same procedure for various values of p_0, adsorption isotherms can be obtained in the pressure range 10^{-9} to 10^{-7} Torr. In these experiments, corrections for thermal transpiration effects become important.

B. Gas Chromatographic Method

Gas/solid chromatography can be used in the measurement of adsorption isotherms. Beebe and his co-workers[3] have established a procedure to obtain isotherms using the frontal analysis method without assuming any adsorption model. The principle of the method can be understood with the aid of *Figure 27*, in which C_i denotes the concentration of absorbate in the

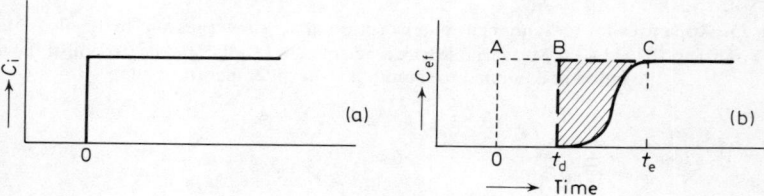

Figure 27. Gas chromatogram of the frontal analysis: (a) Concentration at the inlet. (b) Concentration at the outlet.

inlet gas stream, and C_{ef} the concentration in the effluent. The adsorbate is started at time $t = 0$ and continued until the column (and hence the effluent) reaches a steady state. If no adsorption had occurred, the change in the concentration of the adsorbent in the effluent with time would be in the form of a step, shifted to the right by the dead time t_d which is required for the carrier gas to reach the detector. This is shown by the heavy broken line in the figure. The dead time t_d can be measured by injecting a non-adsorbed gas into the carrier gas. Now, the amount of gas adsorbed by the column adsorbent is calculated from the [area of hatched portion in *Figure 27*(b)] × [total flowrate], and the pressure of adsorbate in equilibrium with the adsorbed phase, p, is given approximately by[44]

$$p = \tfrac{2}{3} p_{ef} \frac{(p_i/p_{ef})^3 - 1}{(p_i/p_{ef})^2 - 1} \simeq \frac{p_i + p_{ef}}{2}$$

where p_i and p_{ef} are the partial pressures of the adsorbate at the entrance and exit of the column, respectively. With this method one can, in principle, measure adsorption isotherms without any restrictions concerning the nature of adsorbents or adsorbates, or equilibration mechanisms.

In *Figure 27*(b), the area of the hatched region t_d CB is obtained as the difference between the areas of the regions $0t_d$CBA and $0t_d$ BA. The desired quantity can be accurately measured only when $t_e \gg t_d$, i.e. when the retention time is large compared to the dead time. This condition can be most easily satisfied by selecting a low column temperature, since, as the column temperature decreases, the retention time increases markedly while the dead time hardly alters. However, the following point must be taken into consideration. At lower column temperatures the equilibration time for adsorption becomes longer, and hence the adsorbate concentration in the effluent approaches the steady state value very slowly. The detectors used for measuring the concentration of adsorbate cannot be completely free from drift, so there is effectively a lower limit for the equilibration speed and hence for the column temperature.

In measuring adsorption isotherms an accuracy of two per cent or better can easily be attained, if the usual care is taken in the gas chromatographic techniques. According to Beebe *et al.*, the heats of adsorption, obtained by this method from the Clapeyron–Clausius relationship, agree quite well with those obtained by the calorimetric method. An important advantage of the gas chromatographic method is that adsorption isotherms and heats at extremely low coverages, say $\theta = 0.005$ or so, can be obtained rather easily and quickly.

V. Adsorption Cryostats

The construction of adsorption cryostats poses many difficult problems. It must be possible to bake-out the cell to at least 400°C, and then to cool it to low temperatures, and this cycle must be repeatable. The choice of appropriate construction material is severely restricted as compared with cryostats for other purposes. Further requirements for adsorption cryostats are the minimizing of the heat transfer along the connecting tube and the rapid attainment of a good vacuum, but these requirements conflict with each other. Thus, in the practical design of adsorption cryostats, some requirements are sacrificed for the sake of others. Especially with cryostats used for vacuum microbalances, fine control of the temperature is very difficult owing to their open structure. The cryostat designed by Morrison and Young[70] gives the highest degree of performance with regard to temperature control (an accuracy of 0.001 K in the range 78 K to 273 K), but is not so appropriate for baking-out and for cleaning the adsorbents. Here, simple and easily constructed cryostats are described, which may be recommended for common use. The cryostat shown in *Figure 28* is an improvement of the Fisher–McMillan type and may be used in the temperature range 65 to 90 K with a temperature constancy of 0.005 K. The temperature in the cryostat is lowered by reducing the vapor pressure of the refrigerant, the vapor pressure being finely controlled by the Cartesian manostat. All of the valves shown in the evacuation line are necessary for easy adjustment. The connecting tube of the adsorption cell is an evacuated dual tube, and

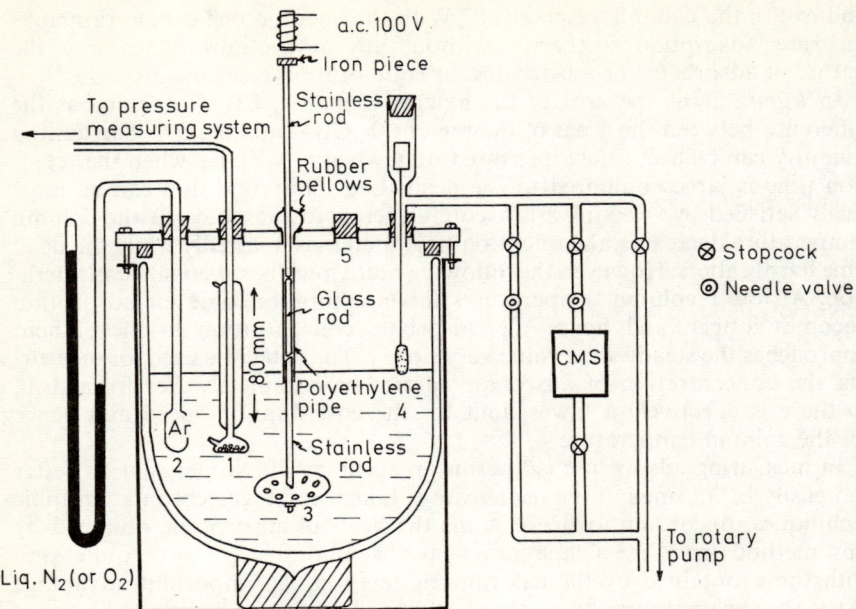

Figure 28. Simple and precise cryostat used in a temperature range 65 ~ 90 K after T. Takaishi and M. Mohri[90]. 1, adsorbent cell; 2, argon vapor pressure thermometer; 3, vibrating stirrer; 4, level indicator; 5, inlet of refrigerant; CMS, cartesian manostat.

the changes in the level of refrigerant (up to 40 mm) induce no detectable change in the effective dead-space volume of the adsorbent cell. Thus, there is no need to supply refrigerant frequently in order to keep its level constant, and the cryostat can be operated for about five hours without adding refrigerant. On account of this favorable situation, the temperature of the cryostat is held constant for long periods. Refrigerant is circulated efficiently by a vibrating stirrer which has a disc with many tapered holes at its lower end.

As an illustration of cryostats which are used for temperatures above the liquid oxygen temperature, *Figure 29* shows an adsorption cryostat for a vacuum microbalance. The temperature of the cryostat is kept constant within ± 0.02 K or so, by p.i.d. controlled electric current in the heater coil wound non-magnetically around the aluminum block. Better control can be achieved by use of a platinum resistor as the temperature sensor. However, in a vacuum microbalance, the temperature difference between the adsorbent and the tube walls is a more serious problem than the fine control of the wall temperature. At the start of the operation, the cryostat is cooled near to the required temperature by pouring liquid nitrogen into the inner and outer dewar vessel through inlets 3 and 4, respectively. Inlet 1 is used to blow cooled nitrogen into the inner dewar for fine temperature adjustment. The aluminum foil, which is laid over the adsorbent bucket, shields it from thermal radiation from above and prevents the adsorbent from being warmed.

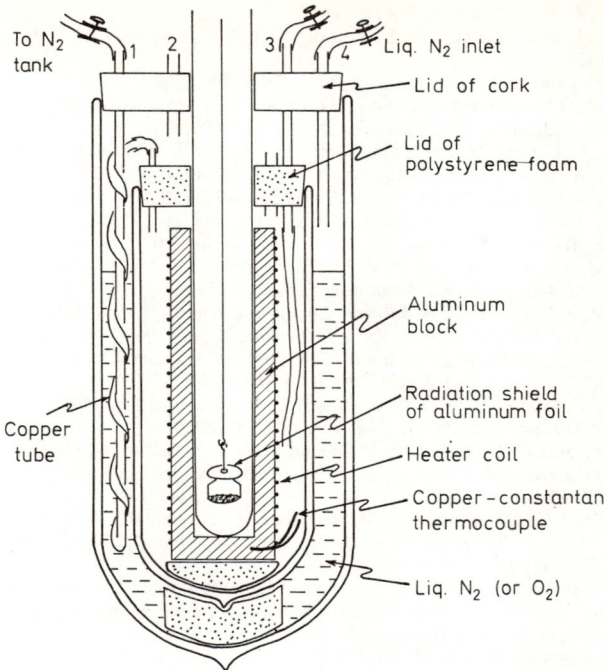

Figure 29. Adsorption cryostat used for vacuum microbalance in a temperature range 78 ~ 273K

For details on sophisticated cryostats, refer to the relevant original papers[13, 70].

VI. Acknowledgement

The author expresses his sincere thanks to Dr A. C. Herd of the Tokyo University who carefully read the manuscripts and corrected the language.

VII. References

[1] Aylmore, D. W. and W. B. Jepson. *J. Sci. Instrum.* **38**, 156 (1961).
[2] Barrer, R. M. and S. Wasilewski. *J. Sci. Instrum.* **37**, 432 (1960).
[3] Beebe, R. A., P. L. Evans, T. C. Wleisteuber and L. W. Richards, *J. Phys. Chem.* **70**, 1009 (1966).
[4] Bosch, A. van den, *Vacuum Microbalance Techniques*, Vol. V, p. 77. Plenum: New York (1966).
[5] Boschi, L. A. and E. A. Garcia. *Rev. Sci. Instrum.* **38**, 1610 (1967).
[6] Bradley, R. S. *J. Sci. Instrum.* **29**, 84 (1952).
[7] Bushuk, W. and C. A. Winkler. *Canad. J. Chem.* **33**, 1729 (1955).
[8] Chihara, H. and J. A. Morrison. *Experimental Thermodynamics*, Vol. I, Chap. 13, IUPAC, Butterworths: London (1968).
[8a] Cahn, L. and H. Schultz, *Analyt. Chem.* **35**, 1729 (1963).
[9] Chnèbault, P. and A. Schürenkämper, *J. Phys. Chem.* **69**, 2300 (1965).

[10] Clark, J. T. *J. Phys. Chem.* **68**, 884 (1964).
[11] Cochrane, H., P. L. Walker, W. S. Piethorn and H. C. Friedman. *J. Colloid Interface Sci.* **24**, 405 (1967).
[12] Corrin, M. L. and C. P. Rutkowski, *J. Phys. Chem.* **58**, 1089 (1954).
[13] Constabaris, G., J. H. Singleton and G. D. Halsey Jr. *J. Phys. Chem.* **63**, 1350 (1959).
[14] Coolidge, A. S. *J. Amer. Chem. Soc.* **56**, 554 (1934).
[15] Cutting, P. A., *Vacuum Microbalance Techniques*, Vol. VII, p. 71. Plenum: New York (1970).
[16] Czanderna, A. W. and J. M. Honig, *Analyt. Chem.* **29**, 1206 (1957).
[17] Czaplinski, A. and E. Zielinski, *Przem. Chem.* **37**, 640 (1958).
[18] Deitz, V. R. and L. F. Gleysteen. *J. Res. Nat. Bur. Stand.* **29**, 191 (1942).
[19] Drain, L. E. and J. A. Morrison. *Trans. Faraday Soc.* **48**, 316 (1952).
[20] Dubinin, M. M., B. P. Bering, V. V. Serpinsky and B. N. Vasil'ev, *Surface Phenomena in Chemistry and Biology*, p 172. Pergamon: Oxford (1958).
[21] Edgecombe, F. H. C. and D. A. Jardine. *Canad. J. Chem.* **39**, 1728 (1961).
[22] Edmond, T. and J. P. Hobson, *J. Vac. Sci. Technol.* **2**, 182 (1965).
[23] Elliott, K. W. T., D. M. Woodman and R. S. Dadson. *Vacuum*, **17**, 439 (1967).
[24] Ernsberger, F. M. and C. M. Drew *Rev. Sci. Instrum.* **24**, 117 (1953).
[25] Ernsberger, F. M. and H. W. Pitman. *Rev. Sci. Instrum.* **26**, 584 (1955).
[26] Everett, D. H. *Trans. Faraday Soc.* **46**, 453, 942, 957 (1950).
[27] Faeth, P. A. and C. B. Willingham. *Technical Bulletin on the Assembly, Calibration and Operation of a Gas Adsorption Apparatus*, etc., Mellon Institute for Industrial Research (September 1955).
[28] Gallon, T. E., I. G. Higginbotham, M. Prutton and H. Tokutaka, *Surf. Sci.* **21**, 224 (1970).
[29] Garden, L. A., G. L. Kington and W. Laing, *Proc. Roy. Soc. A*, **234**, 35 (1956).
[30] Gomer, R. *Rev. Sci. Instrum.* **24**, 993 (1953).
[31] Gomer, R. *Advanc. Catalysis*, Vol. VII, p 93. Academic Press: New York (1955).
[32] Guggenheim, E., *Elements of the Kinetic Theory of Gases*, Pergamon: Oxford (1960).
[33] Gulbransen, E. A. *Rev. Sci. Instrum.* **15**, 201 (1944).
[34] Hansen, N. *Vakuum-Technik*. **11**, 70 (1961).
[35] Herring, C. *Structure and Properties of Solid Surfaces*. Chap. 1, p 5. ed. R. Gomer and C. S. Smith, University of Chicago Press (1953).
[36] Hill, T. L. *Advanc. Catalysis*, Vol. IV, p 212. Academic Press: New York (1952).
[37] Hillecke, D. and H. Mayer, *Vacuum Microbalance Techniques*, Vol. VII, p 135. Plenum: New York (1970).
[38] Hobson, J. P. *Canad. J. Phys.* **37**, 300 (1959); *J. Chem. Phys.* **34**, 1850 (1961); J. P. Hobson and R. A. Armstrong, *J. Phys. Chem.* **67**, 2000 (1963).
[39] Holmes, J. M. *The Solid–Gas Interface*, Vol. I, Chap. 5, p 127. Ed. E. A. Flood, Marcel Dekker: New York (1967).
[40] Hooley, J. G. *Canad. J. Chem.* **35**, 1414 (1957).
[41] Ishii, H. and K. Nakayama, *J. Vacuum Soc. Japan*, **4**, 414 (1961); *Proc. 2nd Int. Congr. Vacuum Sci. Tech.*, p. 519. Pergamon: Oxford (1961).
[42] Ishimura, H. *Japanese J. Appl. Phys.* **4**, 934 (1965).
[43] Ishimura, H. *J. Vac. Soc. Japan*, **6**, 268 (1963) [in Japanese].
[44] James, A. T. and A. J. P. Martin. *Biochem. J.* **50**, 679 (1952).
[45] Jones, W. M., P. J. Isaac and D. Phillips. *Trans. Faraday Soc.* **55**, 1953 (1959).
[46] Joy, A. S. *Vacuum*, **3**, 254 (1953).
[47] Joyner, L. G. *Scientific and Industrial Glass Blowing*, p 257. Ed. by W. E. Barr and V. J. Anhorn, Instruments Publ. Co.: Pittsburgh (1949).
[48] Kington, G. L. and J. G. Aston. *J. Amer. Chem. Soc.* **75**, 1929 (1951).
[49] Kini, K. A. *Fuel*, London, **43**, 173 (1964).
[50] Klier, K. *Rev. Sci. Instrum.* **40**, 372 (1969).
[51] Kolenkow, R. J. and P. W. Zitzewitz. *Vacuum Microbalance Techniques*, Vol. IV, p 195. Plenum: New York (1965).
[52] Kuhn, W., E. Robens, G. Sandstede and G. Walter. *Vacuum Microbalance Techniques*, Vol. VII, p 161. Plenum: New York (1970).
[53] Kummers, J. T. and J. D. Young. *J. Phys. Chem.* **67**, 107 (1963).
[54] Lambert, B. and C. S. G. Phillips. *Phil. Trans. A*, **242**, 415 (1950).
[55] Lamers, K. W. and P. R. Rony. Lawrence Radiation Laboratory Report, *UCRL–11218*, Parts I and II (1964–65).
[56] Lauterbach, K. E., S. Laskin and L. Leach. *J. Franklin Inst.* **250**, 13 (1950).

[57] Leck, J. H. *Pressure Measurements in Vacuum Systems*, Chapman and Hall, London (1964).
[58] Liang, S. Chu. *J. Appl. Phys.* **22**, 148 (1951); *J. Phys. Chem.* **56**, 660 (1952); **57**, 910 (1953).
[59] Lippens, B. C., B. G. Linsen and J. H. de Boer. *J. Catalysis*, **3**, 32 (1964).
[60] Machin, W. D. and S. Ross. *Proc. Roy. Soc. A*, **265**, 455 (1962).
[61] Machin, W. D. *Canad. J. Chem.* **45**, 1904 (1967).
[62] Massen, C. H., J. A. Poulis and J. M. Thomas. *J. Sci. Instrum.* **41**, 302 (1964).
[63] Mayer, H., R. Niedermayer, W. Schroen, D. Stünkel and H. Göhre, *Vacuum Microbalance Techniques*, Vol. III, p 75. Plenum: New York (1963).
[64] McBain, J. W. and R. F. Sessions. *J. Colloid Sci.* **3**, 213 (1948).
[65] Meyer, D. E. and J. E. Wells. *J. Colloid Interface Sci.* **22**, 503 (1966).
[66] Melville, S. H. and B. G. Gowenlock. *Experimental Methods in Gas Reactions*, MacMillan: London (1964).
[67] Menon, P. G. *Chem. Rev.* **68**, 277 (1968); *Advances in High Pressure Research*, Vol. III, Chap. 5, Academic Press: Oxford (1969).
[68] Michel, A., P. G. Menon and C. A. Ten Seldan. *Rec. Trav. Chim. Pays-Bas*, **80**, 483 (1961).
[69] Moreau, C. *Vacuum Microbalance Techniques*, Vol. IV, p 21. Plenum: New York (1965).
[70] Morrison, J. A. and D. M. Young. *Rev. Sci. Instrum.* **25**, 518 (1954).
[71] Oguri, T. and I. Kanomata. Read before the Ninth Meeting of the Vacuum Society of Japan (November 1968).
[72] Ohtsuki, T. Private communication.
[73] Orr, C. and J. M. Dalla Valle. *Fine Particle Measurements*, p 175. MacMillan: London (1959).
[74] Ozawa, S. *Thesis*, Tohoku University, Japan (1971).
[75] Pace, E. L., W. T. Berg and A. R. Siebert. *J. Amer. Chem. Soc.* **78**, 153 (1956).
[76] Pierotti, R. A. *Vacuum Microbalance Techniques*, Vol. VI, p 1. Plenum: New York (1967).
[77] Podgurski, H. H. and F. W. Davis. *J. Phys. Chem.* **65**, 1343 (1961).
[78] Poulis, J. A., B. Pelupessey, C. H. Massen and J. M. Thomas. *J. Sci. Instrum.* **41**, 295 (1964).
[79] Rand, M. J. *Rev. Sci. Instrum.* **32**, 991 (1961).
[80] Redhead, P. A., J. P. Hobson and E. V. Kornelson. *The Physical Basis of Ultrahigh Vacuum*, Chapman and Hall: London (1968).
[81] Rhodin, T. N. *J. Amer. Chem. Soc.* **72**, 4343 (1950).
[82] Ricca, F. and R. Medana. *Ric. Sci.* **4**, 617 (1964).
[83] Robens, E., G. Sandstede, G. Walter and G. Wurzbacher. *Vacuum Microbalance Techniques*, Vol. VII, p 195. Plenum: New York (1970).
[84] Rosenberg, A. J. *J. Amer. Chem. Soc.* **78**, 2929 (1956).
[85] Rosenberg, A. J. and C. S. Martel Jr. *J. Phys. Chem.* **62**, 457 (1958).
[86] Ross, S. and J. P. Oliver. *On Physical Adsorption*, Interscience: New York (1964).
[87] Sensui, Y. *Vacuum*, **20**, 539 (1970).
[88] Sereda, P. J. and R. F. Feldman. *The Solid–Gas Interface*. Vol. II, Chap. 24, ed. E. A. Flood, Marcel Dekker: New York (1967).
[89] Stockbridge, C. D. *Vacuum Microbalance Techniques*, Vol. V, pp 147, 179, 193. Plenum: New York (1966).
[90] Takaishi, T. *Trans. Faraday Soc.* **61**, 840 (1965).
[90a] Takaishi, T. and M. Mohri, *JCS Faraday Trans. I*, **68**, 1921 (1972).
[91] Takaishi, T. and Y. Sensui. *Trans. Faraday Soc.* **59**, 2503 (1963).
[92] Takaishi, T. and Y. Sensui. *Vacuum*, **20**, 495 (1970).
[93] Takaishi, T. and Y. Sensui. *Surface Sci.* **19**, 339 (1970).
[94] Takaishi, T., A. Yusa and F. Amakasu. *Trans. Faraday Soc.* **67**, 3565 (1971).
[95] Thomas, J. M. and J. A. Poulis. *Vacuum Microbalance Techniques*, Vol. III, p 15. Plenum: New York (1963).
[96] Tuzi, Y. and T. Saito. *J. Vac. Sci. Technol.* **6**, 238 (1969); *J. Vac. Soc. Japan*, **14**, 83 (1971) [in Japanese].
[97] Wade, W. H. and L. T. Slutsky, *Vacuum Microbalance Techniques*, Vol. II, p 115. Plenum: New York (1962).
[98] Warner, A. W. and C. D. Stockbridge, *Vacuum Microbalance Techniques*, Plenum: New York. Vol. II (1962), pp 71, 93; Vol. III (1963), p 55.
[99] Weinstein, A. and H. C. Friedman. *Rev. Sci. Instrum.* **35**, 1083 (1964).
[100] Wolsky, S. P. and E. J. Zdanuk. *Vacuum Microbalance Techniques*, Vol. II, p 37. Plenum: New York (1962).
[101] Wooten, L. A. and J. R. C. Brown. *J. Amer. Chem. Soc.* **65**, 113 (1943).

[102] Yao, Y. F. Y. and J. T. Kummers. *J. Phys. Chem.* **73**, 2262 (1969).
[103] Young, D. M. *Rev. Sci. Instrum.* **24**, 77 (1953).
[104] Young, D. M. and A. D. Crowell. *The Physical Adsorption of Gases*, Butterworths: London (1962).

CHAPTER 22

Chemical Relaxation in Liquid Systems

H. STREHLOW and W. KNOCHE

Max-Planck Institut für Biophysikalische Chemie, Göttingen, Germany

Contents

I.	Introduction	1069
II.	Chemical Relaxation Techniques	1071
	1. Jump Methods	1071
	A. Temperature—Jump Methods	1075
	B. Pressure—Jump Methods	1078
	C. Field—Jump Methods	1080
	2. Stationary Methods	1081
	A. Acoustical Methods	1081
	B. Dielectric Methods	1086
III.	The Information Obtainable from Chemical Relaxation Measurements	1088
IV.	Some Applications of Chemical Relaxation Techniques	1095
	1. The Neutralization Reaction $H^+ + OH^- \underset{k_2}{\overset{k_1}{\rightleftarrows}} H_2O$	1095
	2. The Formation of Metal Ion Complexes: $M^{m+} + L^{n-} \rightleftarrows ML^{(m-n)+}$	1096
	3. The Mechanism of Cooperative Ligand Binding on an Allosteric Enzyme	1099
	4. Ultrasonic Absorption in Water–Dioxane Mixtures	1103
V.	Summary	1107
VI.	References	1107

I. Introduction

The rates of chemical reactions vary over many orders of magnitude. Some reactions are so slow at a given temperature that they practically do not occur, though the free enthalpy would decrease considerably. On the other hand, many chemical reactions occur very rapidly so that in some textbooks they are still called unmeasurably fast. This, however, is no longer true. Especially during the past twenty years a wealth of techniques has been developed to study reaction times much shorter than a second. Though the development and improvement of new techniques is still under way, the reaction rate of practically all types of chemical reactions can be measured by one or more methods.

On a logarithmic scale *Figure 1* shows the reaction times for some typical reactions at 25°C. Fast reactions, those with reaction times less than a second, are very common in inorganic chemistry (ionic reactions), in some elementary steps of organic reactions, and especially in enzymatically catalysed biochemical reactions. *Figure 1* also shows the range of times for which the experimental techniques so far developed can be applied. Here we are concerned only with chemical relaxation techniques[15]. For reviews on the other methods such as electrochemical techniques[52], flash photolysis[46], fluorescence

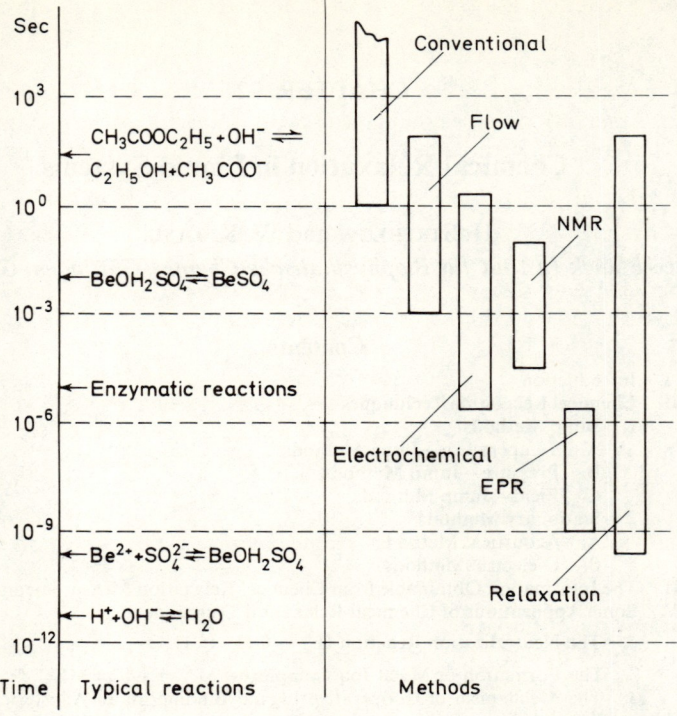

Figure 1. Typical reaction times for some reactions (25°C, all concentrations equal to 1M) and the approximate range of applicability of different techniques.

decay[43, 57], flow techniques[49, 23, 25], and magnetic resonance techniques[33], the quoted literature should be consulted.

In chemical relaxation techniques a chemical equilibrium is perturbed rapidly by changing the temperature for example. Immediately after the perturbation the system is not in equilibrium. For small perturbations the rate of reestablishment of an equilibrium state is proportional to the instantaneous deviation from equilibrium as in any relaxation process. That is, the change in concentration c of a reactant is proportional to its deviation from the corresponding equilibrium value $\bar{c}$

$$-dc/dt = (c - \bar{c})/\tau \qquad (1)$$

The time constant τ of the relaxation depends on the mechanism of the reaction, on the rate constants k_i and, in general, also on involved equilibrium constants K_j, and on concentrations c_k of the reaction partners

$$\tau = \tau(k_i, K_j, c_k, \text{mechanism}) \qquad (2)$$

Besides the relaxation time, the amplitude x_0 of the relaxation effect provides valuable information on equilibrium constants and on thermodynamic data of the reaction, as the reaction enthalpies ΔH or the reaction volumes ΔV

$$x_0 = x_0(K_j, \Delta H_j, c_k, \Delta V_j, \text{mechanism}) \qquad (3)$$

It will be shown that a complete analysis of the experimental results of chemical relaxation experiments supplies not only kinetic data but also thermodynamic information which is often difficult to obtain by any other method.

In section II the various relaxation techniques which so far have been proved useful in practice will be explained. Section III deals with the problem of evaluating relaxation experiments. In section IV four very different types of chemical reactions which have been investigated will be discussed to demonstrate the potential of chemical relaxation techniques.

II. Chemical Relaxation Techniques

The perturbation of a chemical equilibrium is performed by rapidly changing an external parameter—such as temperature T, pressure p, or electric field strength E—on which the chemical equilibrium constants depend. Thus, the kind of external parameter used to shift the equilibrium is the first general feature for a relaxation technique. The detection of the concentration changes involved may be performed by various means, e.g. by measuring the optical density or the electrical conductivity or—more indirectly—the dissipation of acoustical energy. Other physical quantities have also been applied for the detection of chemical relaxation (e.g. fluorescence[47], optical rotation[22], phase angle measurements[58], temperature[27]). The detection method is therefore the second general feature of chemical relaxation techniques. The third feature is the time function of the external parameter $[T(t), p(t) \text{ or } E(t)]$. Though many such time functions are conceivable, only two have been used extensively thus far, the step function (jump methods) and the sinusoidal perturbations (stationary methods).

1. Jump Methods

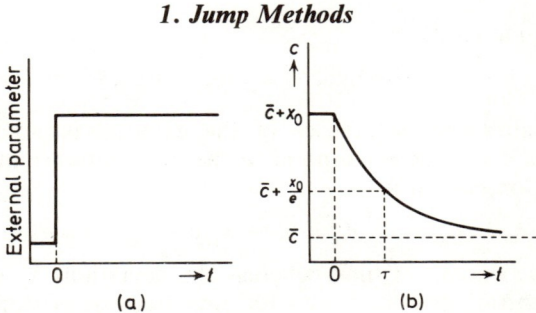

Figure 2. Step function chemical relaxation: (a) The external parameter (T, p or E) as a function of time; (b) The response of the concentration as a function of time.

As schematically indicated in *Figure* 2(a), an external parameter is changed rapidly thereby shifting the equilibrium constant according to

$$(\partial \ln K / \partial T)_p = \Delta H / RT^2 \qquad (4a)$$

$$(\partial \ln K/\partial p)_T = -\Delta V/RT \tag{4b}$$

$$(\partial \ln K/\partial E)_{p,T} = \Delta M/RT \tag{4c}$$

where ΔH is the enthalpy of reaction, ΔV the volume of reaction, and ΔM the 'polarization' of reaction. (This 'polarization' of reaction is the conjugate variable of the electric field strength E). *Figure 2*(b) shows how the concentration of the reactants changes after the sudden perturbation. The deviation of the actual concentrations from their equilibrium values are called x_i. x_i^0 is the difference between the equilibrium concentrations before and after the perturbation.

In general, the equilibrium constant does not change under isobaric (isothermal or isobaric-isothermal) conditions. This point will be clarified in a discussion of the pressure step function.

Since the pressure step is fast and, therefore, adiabatic, the proper equation to be used for the pressure dependence of K reads

$$\left(\frac{\partial \ln K}{\partial p}\right)_S = -\frac{\Delta V}{RT} + \frac{\alpha}{C_p \rho} \frac{\Delta H}{RT} \tag{5}$$

The second term takes into account the temperature jump due to the adiabatic pressure change. α is the coefficient of thermal expansion, C_p the specific heat, and ρ the density of the solution. Of course, for reaction times long compared with the thermal equilibration time of the solution in the measuring device, isothermal conditions apply and the orginal equation 4b should be used.

In a temperature-jump experiment a similar situation arises. However, the time constant of pressure equilibration after a fast temperature jump is of the order of l/v, where l is the length of the measuring cell and v is the velocity of sound. This time constant l/v is usually shorter than the relaxation times measured in an actual experiment so that equation 4a is applicable. Corresponding considerations apply to a field-jump experiment.

Temperature jumps of a few degrees centigrade, pressure jumps of about 100 bar, and electric field jumps of several times 10^4 V/cm are used to produce changes in K of a few per cent which result in concentration changes that can still be conveniently determined as a function of time.

To be specific we shall discuss a reaction of the type

$$A + B \underset{k_2}{\overset{k_1}{\rightleftharpoons}} C \tag{6}$$

The rate equation reads

$$-dc_A/dt = k_1 c_A c_B - k_2 c_C \tag{7}$$

Immediately after the occurrence of the external parameter jump, the concentrations c_i do not correspond to the new equilibrium values $\bar{c}_i$ [see *Figure 2*(b)]. Consequently,

$$c_A = \bar{c}_A + x_A,\ c_B = \bar{c}_B + x_B \text{ and } c_C = \bar{c}_C + x_C \tag{8}$$

The $\bar{c}_i$ are independent of time, whereas the deviations x_i and the instantaneous concentrations c_i do vary with time. Because of stoichiometry and conservation of mass we have the equations

$$x_A = x_B = -x_C \equiv x \tag{9}$$

Inserting equations 8 and 9 into 7 we obtain the following relation

$$-dx/dt = \{k_1(\bar{c}_A + \bar{c}_B) + k_2\}x + k_1 x^2 + [k_1 \bar{c}_A \bar{c}_B - k_2 \bar{c}_C] \tag{10}$$

The last term in brackets vanishes, since the $\bar{c}_i$ refer to equilibrium. In chemical relaxation experiments the relative change in concentration is very small ($x \ll c_i$). Therefore, $k_1 x^2$ may be neglected in equation 10. Thus, we are left with the linear differential equation

$$-dx/dt = x/\tau \tag{11}$$

with the relaxation time τ being given by

$$\tau = \{k_1(\bar{c}_A + \bar{c}_B) + k_2\}^{-1} \tag{12}$$

It should be emphasized that the quadratic differential equation 10 has been linearized thereby considerably reducing the mathematical complexity. Therefore, it may also be advantageous to apply relaxation techniques to slow reactions, although only a few applications of this kind have been reported[45, 56]. Furthermore, the relaxation rate $1/\tau$ in equation 12 is an additive combination of the rate constants (whereas equilibrium constants are ratios of rate constants).

The solution of equation 11 is

$$x(t) = x_0 e^{-t/\tau} \tag{13}$$

By plotting $1/\tau$ against the total concentration $c_0 = c_A + c_C$ we may obtain the two rate constants from the intercept and from the initial slope. Thus, the ratio of the rate constants, i.e. the equilibrium constant, is determined from kinetic measurements alone. However, it is usually more precise to calculate $c_A + c_B$ from conventionally determined equilibrium constants and to evaluate k_1 and k_2 from the intercept and the slope of the straight line $1/\tau$ against $c_A + c_B$ (eventually corrected for activity coefficients) (see section III). The comparison of the ratio k_1/k_2 with the equilibrium constant checks the consistency of the measurements.

The amplitude x_0 will now be calculated for reaction 6. By the shift of the external parameter, e.g. by the temperature jump δT, $\ln K$ is shifted by

$$\delta \ln K = (\partial \ln K/\partial T)_p \delta T = (\Delta H/RT^2) \delta T \tag{14}$$

and the equilibrium concentration $\bar{c}_A$ by x_0

$$x_0 = \Gamma(\Delta H/RT^2) \delta T \tag{15}$$

with

$$\Gamma = d\bar{c}_A/d \ln K \tag{16}$$

For reaction 6 K is given by $(\bar{c}_A \bar{c}_B/\bar{c}_C)(f_A f_B/f_C) \equiv (\bar{c}_A \bar{c}_B/\bar{c}_C)\Pi$, where Π is the activity coefficient product. Therefore,

$$d \ln K = d \ln \bar{c}_A + d \ln \bar{c}_B - d \ln \bar{c}_C + d \ln \Pi \tag{17}$$

and since (compare equation 9) $d\bar{c}_A = d\bar{c}_B = -d\bar{c}_C$

$$d \ln K = \left\{ \frac{1}{\bar{c}_A} + \frac{1}{\bar{c}_B} + \frac{1}{\bar{c}_C} + \frac{d \ln \Pi}{d\bar{c}_A} \right\} d\bar{c}_A \tag{18}$$

i.e. for reaction 6

$$\Gamma = \left\{ \frac{1}{\bar{c}_A} + \frac{1}{\bar{c}_B} + \frac{1}{\bar{c}_C} + \frac{d \ln \Pi}{d\bar{c}_A} \right\}^{-1} \tag{19}$$

If a is the physical quantity used to detect the change in concentration, x, and if the dependence of a on x is known, the enthalpy of reaction can be obtained according to

$$\Delta H = (RT^2/\Gamma)(\partial a/\partial x)^{-1}(\delta a/\delta T) \qquad (20)$$

For example, a may be the electric conductivity. The specific conductivity $\varkappa$ of a solution is given by

$$\varkappa = 10^{-3} \Sigma c_i \lambda_i^0 f_{\lambda i} \qquad (21)$$

where λ_i^0 are the equivalent conductivities at infinite dilution and $f_{\lambda i}$ are the conductivity coefficients (which are functions of the concentrations c_i). The measurable quantity is the relative change in resistance R[37]

$$-\frac{\delta R}{R} = \frac{\delta \varkappa}{\varkappa} = \frac{\Sigma \lambda_i^0 f_{\lambda i}\left(1 + \dfrac{\partial \ln f_{\lambda i}}{\partial \ln c_i}\right) \delta \bar{c}_i}{\Sigma c_i \lambda_i^0 f_{\lambda i}}$$

$$+ \frac{\Sigma c_i f_{\lambda i} \delta \lambda_i^0 + \Sigma c_i \lambda_i^0 (\delta f_{\lambda i})_{\delta \bar{c}_i = 0}}{\Sigma c_i \lambda_i^0 f_{\lambda i}} \qquad (22)$$

The second term on the RHS of equation 22 decays to zero very rapidly and does not supply kinetic information. The effect of the reaction is contained in the first term which, as a slower process, can be observed separately:

$$\frac{\delta a}{a} = \frac{\Sigma \lambda_i^0 f_{\lambda i}\left(1 + \dfrac{\partial \ln f_{\lambda i}}{\partial \ln c_i}\right) \delta \bar{c}_i}{\Sigma c_i \lambda_i^0 f_{\lambda i}} \qquad (23)$$

If non-reacting electrolytes are present in the solution (the $\delta \bar{c}_i$ of which vanish), the sensitivity of the conductivity detection method decreases, since the summation in the denominator of equation 23 has to be extended over all ions in the solution. From equation 23 we can calculate $\partial a/\partial x$ provided the λ_i^0, $f_{\lambda i}$ and $\partial \ln f_{\lambda i}/\partial \ln c_i$ are known or can be estimated from theory.

If optical absorption is used to monitor the change in concentration, the temperature coefficient of the extinction coefficient (and of the volume) has to be considered in calculating $\delta \bar{c}_i$ from observed changes in extinction, δE. If the reaction is observed by coupling it to a pH indicator, the dependence of the optical density of the indicator on x_0 has to be known as well.

As will be shown below, we may also obtain equilibrium constants from amplitude measurements. The precision of this technique is usually not very good. Sometimes, however, the data thus obtained are not easily arrived at by any other technique[41, 40]. Of course, amplitude measurements performed with the pressure- or field-jump techniques and with stationary methods may also be evaluated in a similar manner.

For the determination of τ and x_0 an apparatus is required which allows the application of a sudden temperature-, pressure- or electric field jump, and the measurement of the fast change in concentrations. The relative change in the concentrations amounts to a few per cent at best.

A. Temperature—Jump Methods

The technique used most often is the temperature-jump method with spectrophotometric detection as developed by Eigen and co-workers[15, 5]. The principle of the apparatus is outlined in *Figure 3*. The condenser C is charged to a voltage of 20 to 100 kV. Then, the spark gap G is fired so that C discharges through the cell S containing the solution to be investigated.

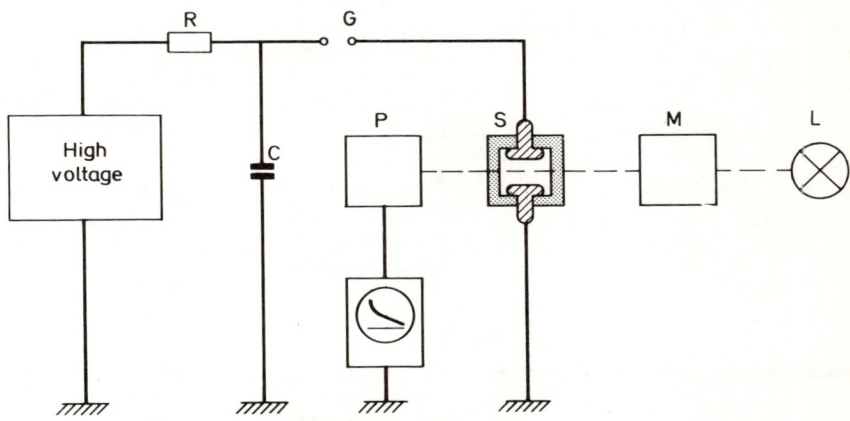

Figure 3. The temperature-jump relaxation apparatus (schematic).

The temperature of the solution is raised by Joule heating. The temperature increases exponentially with a time constant $R_x C/2$, where R_x is the resistance of the cell. Therefore, the electrical conductivity of the solution has to be high in order to get a fast discharge of C. A change in temperature of four to eight degrees can be achieved within about 10^{-6} sec. The heating time has been reduced to about 5×10^{-8} sec by using a high voltage cable instead of the condenser C^{28}. With a matched impedance of cell S a steplike discharge of the cable is achieved.

The detection of the change in concentration is performed photometrically. L is an intense lamp with constant luminosity, M is a monochromator and P is a photomultiplier, whose output is recorded by an oscilloscope. The wavelength is chosen so that the optical density changes as much as possible as a result of the reaction. In complicated reactions the change in concentration of a particular species can be observed at the appropriate wavelength. This selectivity is an important advantage of the optical absorption compared with other detection parameters such as conductivity or temperature. If the progress of the chemical reaction investigated does not cause a change in the optical density it is often possible to couple another chemical reaction to the system which changes the light absorption and which is fast enough not to be rate determining. Since proton transfer occurs in many reactions, pH indicators are often used as coupled systems. Also the ΔH of the overall reaction may be increased by coupling it to a suitable reaction with a large enthalpy of reaction, thereby increasing the sensitivity of this method.

Table 1. Survey of Chemical

Technique	$\Delta T, \Delta p, \Delta E$ produced by	Detection of Δc by	Time range Lower limit (sec)	Time range Upper limit (sec)
Temperature jump	Discharge of capacitor	Optical absorption (rotation, fluorescence)	5×10^{-6}	0.2
	Cable discharge	Optical absorption	5×10^{-8}	10^{-2}
	Flash light	Conductivity	5×10^{-6}	2×10^{-2}
	Laser	Conductivity	2×10^{-6}	2×10^{-2}
		Optical absorption	3×10^{-8}	
	Microwave	Conductivity, optical absorption	2×10^{-6}	2×10^{-2}
Pressure-jump	Autoclave with rupture disc	Conductivity	10^{-4}	100
		Thermometry	10^{-2}	100
		Optical absorption	10^{-4}	100
	Shock wave in liquids	Conductivity	2×10^{-6}	2×10^{-3}
		Optical absorption	2×10^{-6}	2×10^{-3}
Field-jump	Capacitor switch-on switch-off	Conductivity	10^{-7}	10^{-3}
	Cable discharge	Optical absorption	5×10^{-8}	5×10^{-5}
Stationary, ultrasonic	Ultrasonic wave	Dissipation of acoustical energy	10^{-10}	10^{-5}
		Phase angle in conductivity bridge	10^{-7}	10^{-3}
Stationary, dielectric	High electric field with super imposed h.f. field	Dielectric loss	5×10^{-9}	10^{-6}

CHEMICAL RELAXATION IN LIQUID SYSTEMS

Relaxation Techniques

Volume of solution needed (cm^3)	Type of reaction	Advantages	Disadvantages	Ref.
3–10	Biochemical reactions, complex formation, proton transfer	Very widely applicable	High electrolyte concentration necessary	5
0.1				28
0.05	Ionic reactions	Fast, simple, small volumes	Possibility of flash photolysis	55
0.01	Ionic reactions	Very fast, extremely small volumes	Not simple in application	30, 38 2
0.1	Ionic reactions	Conductivity and optical absorption possible, gentle disturbance of equilibrium	Expensive equipment	20 4
0.3	Ionic reactions, biochemical reactions	Simple, small volume, sensitive	Not very fast	41, 53, 37
50			Rather slow	27
5			Not very fast	24
0.1	Ionic reaction, proton transfer	Fast, small volume,	Time per experiment	29
5		Fast, small volume	Rates are measured at high pressures	34
10	Dissociation of weak electrolytes, proton transfer	High dilution possible, very fast	High dilution necessary. Sophisticated equipment	16, 14
2	Dissociation of weak electrolytes, proton transfer	High dilution possible, very fast. Measurements in non-aqueous media	High dilution necessary Sophisticated equipment	32
2–1 000	Ionic reactions, association of uncharged molecules, biochemical reactions	Very fast	Not sensitive, large volumes. A series for apparatus for a time range 10^{-9} to 10^{-5} sec	15, 16
0.1	Ionic reactions	Fast, small volume	Cumbersome to use	58
1	Biochemical reactions, hydrogen bonding reactions	Very fast	Experimentally difficult	3

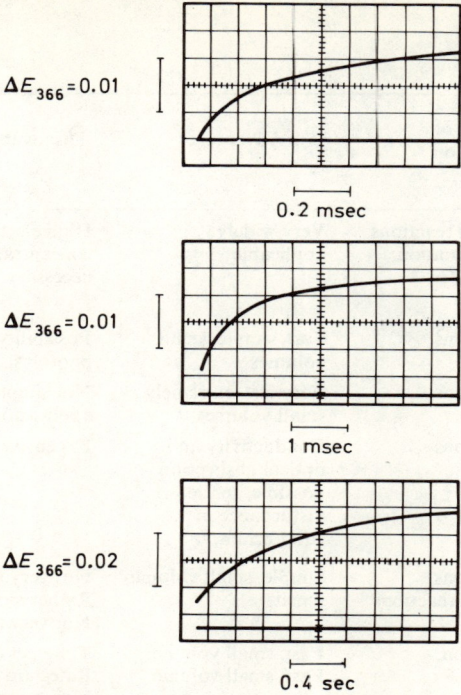

Figure 4. Temperature-jump relaxation oscillograms. The solution investigated: 6×10^{-4} M NAD plus 20 mg/ml GAPD, pH 8.5, 40°C (cf. section IV.3).

Figure 4 shows some typical temperature-jump oscillograms for reactions to be discussed in the last section of this chapter. For a more detailed discussion of this technique and also of the other versions of relaxation methods, the quoted literature should be consulted. The range of applicability of different relaxation techniques is summarized in *Table 1*.

In other types of temperature-jump techniques, pulses of electromagnetic waves: flash light[55], giant pulse laser[30, 38, 2], or microwave pulse[20, 4] have been occasionally used for rapid heating of the solution.

B. Pressure—Jump Methods

Another relaxation technique often used is the pressure-jump method[53], the principle of which is shown in *Figure 5*. In an autoclave A, which is closed by a thin metal membrane M, the pressure may be increased by pumping a liquid (e.g. kerosene) into the autoclave. At about 100 bar the metal membrane bursts spontaneously and the pressure decreases rapidly (in about 5×10^{-5} sec) to 1 bar. The solution is enclosed in a small conductivity cell C. This cell and a similar one containing a non-relaxing solution with about the same conductivity are subjected to the pressure jump inside the

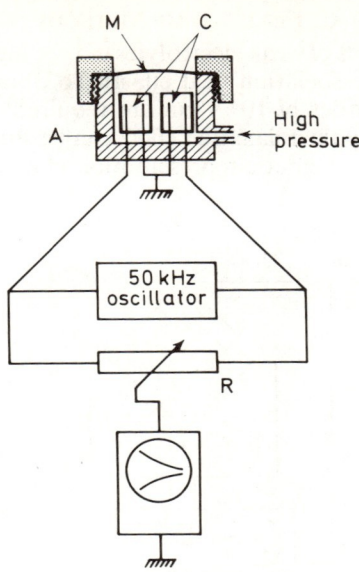

Figure 5. The pressure-jump relaxation apparatus (schematic).

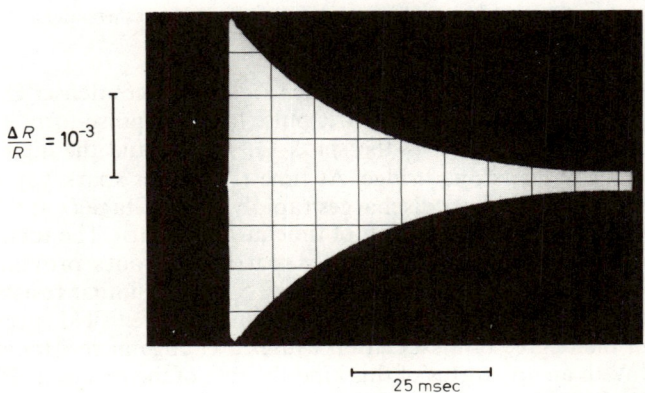

Figure 6. A pressure-jump relaxation oscillogram. The solution investigated: 0.1M BeSO$_4$, pH 2.84, 1.8°C (cf. section IV.2).

autoclave. The two cells are closed with soft polyethylene membranes. Together with the potentiometer R they form a Wheatstone bridge operated at about 50 kHz. A typical pressure-jump relaxation oscillogram is shown in *Figure 6*.

Another type of pressure-jump relaxation apparatus is a shock wave tube filled with liquid (useful in the time range from 10^{-6} to 10^{-3} sec). The detection is performed either optically[34] or conductometrically[29].

C. Field—Jump Methods

Dissociation constants of weak electrolytes may be increased by the application of electric field (dissociation field effect). To obtain measurable effects field strengths of the order of 10^5 V/cm are required*. Because of the high field strengths necessary the relaxation effect can be studied only in solutions of low conductivity. A schematic representation of a field-jump apparatus[14] is shown in *Figure 7*.

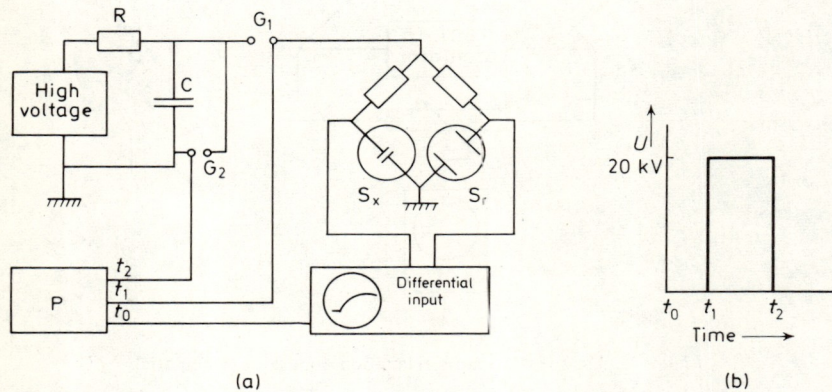

Figure 7. The field-jump relaxation apparatus (schematic).

As in the temperature-jump technique (*Figure 5*), a condenser C is charged to a high voltage. At time t_0 an electric pulse from the pulse-forming device P triggers the oscilloscope. At t_1 the spark G_1 is fired and the high voltage is applied to the Wheatstone bridge. At time t_2 another spark gap switch G_2 is closed and the condenser discharges rapidly. The voltage U at the bridge is schematically shown as a function of time in *Figure 7*(b). The relaxing liquid is contained in the cell S_x with closely spaced electrodes providing a high electric field, whereas the compensating cell S_r with a similar resistance has a large electrode spacing and a correspondingly lower field strength. The increase of the degree of dissociation causes a change in resistance which is observed. With an apparatus of this kind the rate of the reaction $H^+ + OH^- \rightleftarrows H_2O$ has been determined (see section IV).

Other types of field jump apparatus have used either a field pulse in the form of a critically damped single oscillation[16] or—for very fast measurements—the discharge of a coaxial cable instead of a condenser[32]. In the latter case the detection of the changing concentrations is performed spectrophotometrically.

---

* The dissociation field effect is a non-equilibrium process. Therefore equation 4c is not applicable. The theory of this effect has been developed by Onsager[44] yielding

$$\frac{d \ln K}{d|E|} = \frac{|z_1\lambda_1 - z_2\lambda_2|}{\lambda_1 + \lambda_2} \frac{|z_1 z_2| e_0^3}{8\pi\varepsilon k^2 T^2} \qquad (24)$$

2. Stationary Methods

A. Acoustical Methods

The transient methods discussed above exhibit favorable features for the determination of relaxation times. In a single experiment the relaxation time τ is obtained from an oscillogram. However, for reactions with relaxation times shorter than a few microseconds difficulties arise with these techniques. Especially, a sufficiently undistorted recording of the signals requires a bandwidth Δv of the electronic system which is at least equal to $1/\tau$. On the other hand, the signal to noise ratio S/N depends on Δv according to $S/N \sim \Delta v^{-\frac{1}{2}} \sim \tau^{\frac{1}{2}}$. That is, for short τ the signal to noise ratio will be unfavorably small. For very fast reactions, therefore, stationary methods, especially sound absorption techniques, are preferable. In these techniques a stationary perturbation with periodically changing external parameter is applied. The response of the system to chemical relaxation can now be studied with small bandwidth equipment. The principle of the stationary methods will be made clear qualitatively by considering a square wave pressure with frequency v applied to a relaxing solution [*Figure* 8(a)].

For the sake of simplification, in this section we call x the deviation of the actual concentration, c, from the equilibrium concentration at 1 atm, and $\bar{x}$ the deviation of the equilibrium concentration, $\bar{c}$, from $\bar{c}$ (1 atm). Thus, $x = c - \bar{c}$ (1 atm) and $\bar{x} = \bar{c} - \bar{c}$ (1 atm). The rate of change of x is now proportional to the difference $(x - \bar{x})$, i.e.

$$-dx/dt = (x - \bar{x})/\tau \qquad (25)$$

Curve 1 in *Figure* 8(b) shows how $\bar{x}$ follows the change in pressure caused by the square wave. For a very fast reaction ($v\tau \ll 1$), x follows the forcing function p without measurable delay, i.e. $x = \bar{x}$. A very slow reaction ($v\tau \gg 1$)

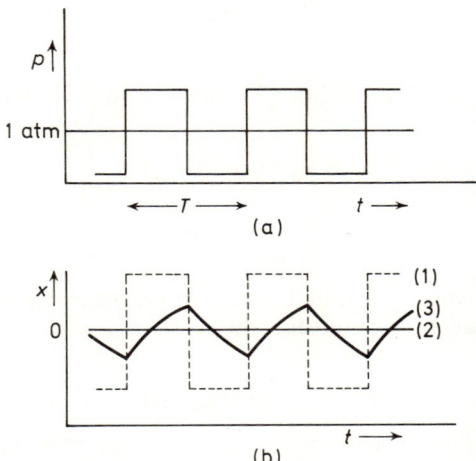

Figure 8. (a) A square wave pressure with frequency $v = 1/T$; (b) The change in concentration caused by the square wave pressure for a reaction with $v\tau \ll 1$ (curve 1), with $v\tau \gg 1$ (curve 2), and with $v\tau = 1$ (curve 3).

cannot follow the fast perturbation and the concentration does not change measurably: $x = 0$ [curve 2, *Figure 8*(b)]. In the most interesting case with $v\tau \approx 1$, an exponential relaxation is observed in each half period [curve 3, *Figure 8*(b)]. Two facts should be emphasized which also appear in the case of a sinusoidal perturbation. First, x does not attain its equilibrium value; thus, we have amplitude damping. Secondly, there is a phase shift between the forcing function, p, and the response, x, which is zero for $v\tau \ll 1$ and increases with increasing $v\tau$. [The phase shift is visualized as the time difference between the centers of gravity of the pulse, p, (at $t = \frac{1}{4}T$) and the corresponding point for the response function x (at $t > \frac{1}{4}T$)].

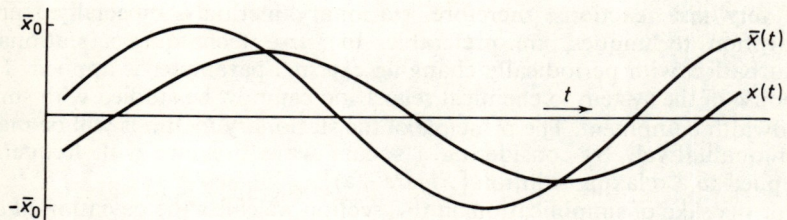

Figure 9. The change in equilibrium concentration ($\bar{x}(t)$) and actual concentration ($x(t)$) caused by a sinusoidal pressure perturbation for a reaction with $2\pi v\tau = 1$.

In *Figure 9* the corresponding difference between the equilibrium function, $\bar{x}$, and x for a reaction with $\omega\tau = 1$ is shown for a sinusoidal pressure perturbation. Again, we have an amplitude depression and a phase shift between x and $\bar{x}$, where the latter is in phase with p. Thus, we have:

$$p = p_1 + p_0 \sin \omega t \qquad (26)$$

$$\bar{x} = \bar{x}_0 \sin \omega t \qquad (27)$$

$$x = x_0 \sin (\omega t - \varphi) \qquad (28)$$

Introducing equations 27 and 28 into 25 and solving the differential equation we obtain

$$x_0 = \bar{x}_0 / \{1 + \omega^2 \tau^2\}^{\frac{1}{2}} \qquad (29)$$

and

$$\varphi = \arctan \omega\tau \qquad (30)$$

Both the damping factor $x_0/\bar{x}_0$ and the phase angle φ depend on the relaxation time. Due to the shift of the concentrations, there is a periodical change in volume δV:

$$\delta V = (\Delta V)Vx = (\Delta V)Vx_0 \sin (\omega t - \varphi) \qquad (31)$$

where V is the volume of the solution and ΔV the reaction volume. The phase shift φ between pressure and volume causes an energy dissipation. Using equations 26 and 31, the energy dissipated during one period can be calculated

$$-\int_{t=0}^{t=2\pi/\omega} p\, dV = \pi p_0 x_0 V(\Delta V) \sin \varphi \qquad (32)$$

Inserting equations 29 and 30 into equation 32 and setting

$$\bar{x}_0 = -(\Delta V/RT)\Gamma p_0 \qquad (33)$$

(cf. equations 4b and 16), we obtain

$$-\int_{t=0}^{t=2\pi/\omega} p\, dV = \frac{\pi p_0^2 (\Delta V)^2 \Gamma \omega \tau V}{RT(1 + \omega^2 \tau^2)} \qquad (34)$$

Dividing equation 34 by the volume and multiplying it by the frequency $\omega/2\pi$ yields the energy dissipated per unit volume per unit time

$$\frac{dE}{dt} = -\frac{\omega}{2\pi V}\int_{t=0}^{t=2\pi/\omega} p\, dV = \frac{p_0^2 (\Delta V)^2 \Gamma \omega^2 \tau}{2RT(1 + \omega^2 \tau^2)} \qquad (35)$$

dE/dt can be measured by the attenuation of a plane sound wave.

In a plane sound wave (in direction z) the displacement, a, of a particle is determined by the wave equation

$$\partial^2 a/\partial t^2 = (\partial^2 a/\partial z^2) v^2 \qquad (36)$$

where v is the velocity of sound. The (undamped) stationary solution of equation 36 is

$$a = a_0 \sin(\omega t - 2\pi z/\lambda) \qquad (37)$$

The energy per volume, E, is

$$E = E_{kin} + E_{pot} = 2\bar{E}_{kin} = \rho \overline{\left(\frac{\partial a}{\partial t}\right)^2} = \tfrac{1}{2}\rho \omega^2 a_0^2 \qquad (38)$$

For the gradient of the pressure we have, from Newton's law and the definition of pressure as force per area F,

$$-\frac{dp}{dz} = \frac{dm}{F\, dz}\frac{\partial^2 a}{\partial t^2} = \rho \frac{\partial^2 a}{\partial t^2} \qquad (39)$$

By integrating we obtain for the pressure

$$p = p_1 + p_0 \cos(\omega t - 2\pi z/\lambda) \qquad (40)$$

with

$$p_0 = \rho v \omega a_0 \qquad (41)$$

The energy flux, I, of the sound wave is given by

$$I = vE \qquad (42)$$

(I is called the sound intensity). Combining equations 38, 41, and 42 yields

$$p_0^2 = 2\rho v I \qquad (43)$$

The condition of continuity requires that the loss in energy density equals the gradient of the energy flux

$$-dE/dt = dI/dz \qquad (44)$$

Combining equations 35 with 44 and 43 we obtain

$$-dI = \frac{(\Delta V)^2 \Gamma \rho v \omega^2 \tau}{RT(1+\omega^2\tau^2)} I \, dz \qquad (45)$$

and by integrating

$$I = I_0 e^{-2\alpha z} \qquad (46)$$

with α, the damping coefficient, given by

$$\alpha = \rho v (\Delta V)^2 \Gamma \omega^2 \tau / \{2RT(1+\omega^2\tau^2)\} = B[\omega^2/(1+\omega^2\tau^2)] \qquad (47)$$

If the pressure is measured instead of the sound intensity we have according to equation 43

$$p_0 = p_{0,z=0} \exp(-\alpha z) \qquad (48)$$

So far it has been assumed that a single chemical relaxation causes the sound attenuation. In general, the expression for α is

$$\alpha = A\omega^2 + \sum_i [B_i \omega^2/(1+\omega^2\tau_i^2)] \qquad (49)$$

The term $A\omega^2$ is due to the dissipation of energy caused by the viscosity and by thermal equilibration processes in the sound wave. The summation in equation 49 is extended over all chemical relaxation processes. The τ_i can be obtained by measuring α as a function of the frequency. In *Figure 10* α/v^2 is plotted against frequency for some 2–2 valent electrolytes[15].

Two related damping coefficients often used should be mentioned. The absorption per wavelength μ

$$\mu = 2\alpha\lambda = 4\pi\alpha v/\omega \qquad (50)$$

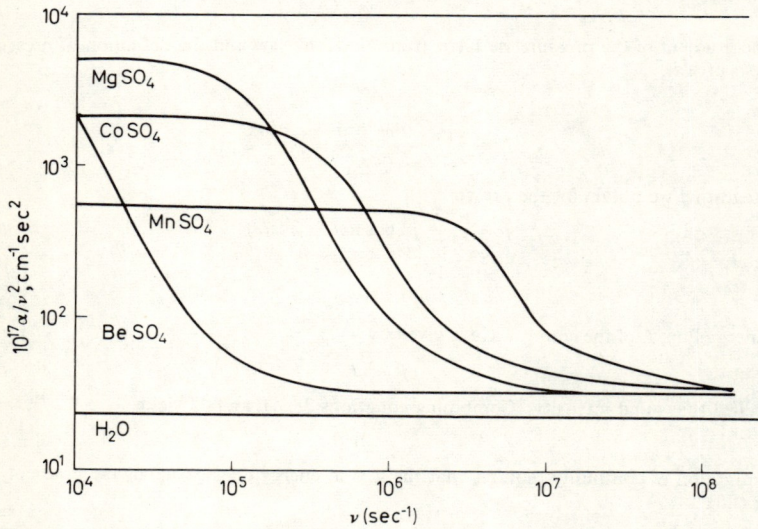

Figure 10. The sound absorption of 0.1M aqueous solutions of some 2–2 valent electrolytes.

and the chemical absorption cross section

$$Q = 2000\alpha/N_A c_0 \qquad (51)$$

where N_A is Avogadro's number and c_0 is the total number of moles of solutes (unreacted) per volume (in mol liter). Often $Q\lambda$ is plotted against frequency. $Q\lambda$ has the dimensions of a volume. μ and $Q\lambda$ plotted against frequency exhibit a maximum at $\omega\tau = 1$. (By inserting Q as defined by equations 51 and 47 into 46, we obtain an equation analogous to Lambert–Beer's law for light absorption). τ is measured at $\omega\tau \approx 1$. The corresponding attenuation coefficient α at $\omega = 1/\tau$ is still proportional to ω (see equation 47). Therefore, with slow reactions and correspondingly low frequencies α becomes exceedingly small and the distance $z = 1/2\alpha$ at which the sound intensity is damped to $1/e$ of its initial value becomes inconveniently large. Therefore, with relatively slow reaction ($\tau > 1\ \mu\text{sec}$) large volumes of solution are required for the measurement. (With special equipment[9], however, the measurement of relaxation times as long as 1 μsec has been achieved with samples of only 40 cm³; see below.) The measurement of φ and $\bar{x}/\bar{x}_0$ with detection methods other than dissipation of acoustical energy has been proposed[58].

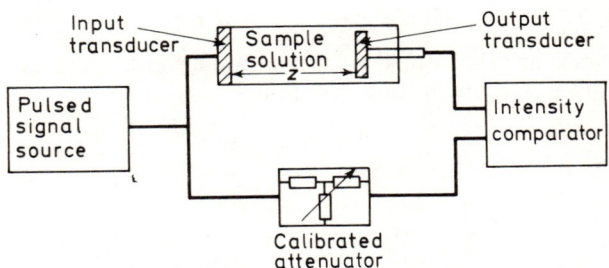

Figure 11. Apparatus used to measure ultrasonic absorption.

Figure 11 demonstrates the principle for measuring ultrasonic absorption. The input transmitter is energized by a pulsed oscillator and creates a sound wave in the solution investigated. (The oscillator is pulsed so as not to cause excessive heating of the solution which is especially serious at high frequencies; see equation 38). The intensity of the sound wave is measured as a function of distance z by comparing it with the signal passed through a calibrated attenuator. Thus, α is obtained by equation 46.

Another method applicable in the frequency range 1 to 100 MHz is based on the Debye–Sears effect[8]. In a travelling sound wave the density alternates at distances of half a wavelength. Since the refractive index varies with density, the solution in an ultrasonic field acts as a diffraction grating on a perpendicular light beam. The intensity of the first order diffraction turns out to be proportional to the square of the pressure p_0 which thus can be determined as a function of z. α is then obtained from equation 48.

With another technique[9] already mentioned above, in a resonating system transmitter quartz/solution/receiver quartz, the halfwidth of the intensity/

frequency curve Δv at resonance is determined by the absorption coefficient α according to

$$\Delta v = 2v\alpha/\pi \qquad (52)$$

Since there are (up to some hundred) multiple reflections in the resonator, comparatively small volumes of solution are required for the measurement of τ even at relatively low frequencies (~ 200 kHz to 100 MHz).

Unfortunately, a given ultrasonic absorption measuring device only covers a comparatively small frequency range. Therefore, quite an arsenal of different apparatus has been developed for the details of which the quoted literature must be consulted[10].

B. Dielectric Methods

A related stationary method takes advantage of the influence of chemical relaxation on dielectric energy losses in a condenser filled with the solution under investigation[3].

With a finite change in macroscopic electric moment ΔM due to the reaction the equilibrium constant of the reaction will be shifted by the application of an electric field according to

$$(\partial \ln K/\partial E)_{p,T} = \Delta M/RT \qquad (4c)$$

If the species in an electric field are in rotational equilibrium, as they usually are, $\Delta M = \Delta \varepsilon V E$ and integrating equation 4c leads to

$$\ln K/K_0 \sim E^2 \qquad (53)$$

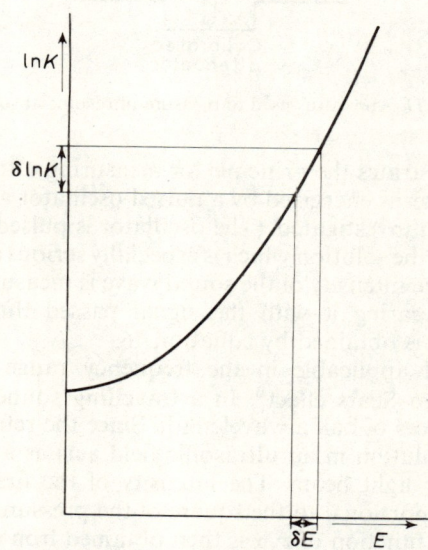

Figure 12. The equilibrium constant against the electric field strength for a reaction causing a change in macroscopic electric moment.

Therefore, in small alternating fields the change in ln K vanishes. To attain a measurable shift an intense static field of the order of 10^7 V/m with a superimposed alternating field of small amplitude has to be applied (see *Figure 12*). Therefore, stationary perturbations of this kind can only be performed in solutions of very low conductivity to avoid excessive heating in the stationary field. For a reaction of the type A $\rightleftarrows$ B the molar change in electric moment ΔM is proportional to $E \Delta(\mu^2)$

$$\Delta M = M_B - M_A \sim E(\mu_B^2 - \mu_A^2)/kT \tag{54}$$

where μ_A and μ_B are the dipole moments of species A and B respectively.

The proportionality in equation 54 is obtained by integration over a Boltzmann distribution of N dipole moments μ_i per unit volume which make an angle ϑ with the external field E

$$M_i \sim \int_0^\pi \exp(-\mu_i E \cos\vartheta/kT) \times (\mu_i \cos\vartheta) \sin\vartheta \, d\vartheta \tag{55}$$

and since $\mu E/kT \ll 1$ under all experimentally attainable conditions

$$M_i \sim \int_0^\pi (1 - \mu_i E \cos\vartheta/kT)\mu_i \cos\vartheta \sin\vartheta \, d\vartheta \sim E\mu_i^2/kT \tag{56}$$

In an electric field of frequency $\omega \approx 1/\tau$, as in ultrasonics, there will be a phase shift φ between the field and corresponding current due to the relaxing chemical reaction. The phase angle φ or rather its complement, the loss angle $\delta = \frac{1}{2}\pi - \varphi$ can be measured with a suitable impedance bridge.

Assuming a negligible conductivity $\varkappa$, the loss angle δ is given by

$$\tan \delta = \Delta\varepsilon^{chem} \omega\tau/(1 + \omega^2\tau^2) \tag{57}$$

where $\Delta\varepsilon^{chem}$ is the change in dielectric constant due to the chemical reaction

$$\Delta\varepsilon^{chem} = \frac{\Gamma(\Delta M)^2}{RT} \sim \frac{\Gamma E^2(\mu_B^2 - \mu_A^2)^2}{T^3} \tag{58}$$

Even at the highest static fields $\tan \delta$ is only of the order 10^{-6} to 10^{-4}, but sufficiently precise measurements are possible. According to equation 58 $\Delta\varepsilon^{chem}$ is proportional to the fourth power of the dipole moment μ_B (if $\mu_A \ll \mu_B$) so that a large change of dipole moment due to the reaction is essential for a good sensitivity of the technique. The influence of chemical reaction rates on dielectric relaxation has been used especially for dimerization reactions of the type $2A \rightleftarrows A_2$ in media of low dielectric constant[31].

Under normal conditions, the assumption of rotational equilibrium is a very good approximation, since for small molecules in a medium of low viscosity rotational relaxation times are of the order 10^{-12} to 10^{-9} sec. With large macromolecules and in more viscous media, however, the rotational reorientation takes a considerably longer time (10^{-6} to 10^{-3} sec). In such a situation the orientation is practically constant for times much longer than relaxation times due to fast chemical reactions. Schwarz[50] theoretically and experimentally investigated the behavior of such systems. In this case, we have to define an equilibrium constant $K(\vartheta)$ for a reaction of species the molecular axis of which makes an angle ϑ with the external field. That equilibrium constant depends on the electric field as

$$\left(\frac{\partial \ln K(\vartheta)}{\partial E}\right)_{p,T} = \frac{\Delta M(\vartheta)}{RT} = \frac{(\mu_A - \mu_B)\cos\vartheta}{kT} \tag{59}$$

In this rotationally frozen situation $\ln K(\vartheta)$ is proportional to E and not to E^2. Therefore, this kind of chemical relaxation can be investigated without a high static field using an ordinary bridge to measure dielectric constants and dielectric loss angles. Taking into account the relaxational contribution of the slow reorientation process, two relaxation times are observed:

$$\frac{1}{\tau_I} = \frac{1}{\tau_{chem}} + \frac{1}{\tau_{or}} \quad \text{and} \quad \frac{1}{\tau_{II}} = \frac{1}{\tau_{or}} \tag{60}$$

(If the chemical relaxation time τ_{chem} is much longer than the reorientation time τ_{or}, only the normal dielectric dispersion is observed.) With this technique the determination of τ_{chem} has been performed for the evaluation of the rate of helix/coil transformation in macromolecular biopolymers[51].

III. The Information Obtainable from Chemical Relaxation Measurements

As implied in equations 2 and 3 thermodynamic and kinetic information can be obtained by measuring relaxation rates and relaxation amplitudes with the techniques outlined above. Using the rate equation and the stoichiometric and mass balance conditions expressions for τ may be derived. This has been done in equations 7 to 12 for a simple reaction of the type

$$A + B \underset{k_2}{\overset{k_1}{\rightleftharpoons}} C \tag{6}$$

A similar derivation for the reaction

$$A \underset{k_2}{\overset{k_1}{\rightleftharpoons}} B \tag{61}$$

leads to

$$1/\tau = k_1 + k_2 \tag{62}$$

Another important reaction is the dimerization

$$2A \underset{k_2}{\overset{k_1}{\rightleftharpoons}} A_2 \tag{63}$$

With $x_A = -2x_{A_2}$ the expression for τ is

$$1/\tau = 4k_1 \bar{c}_A + k_2 \tag{64}$$

Introducing the total concentration $c_0 = c_A + 2c_{A_2}$, a more useful equation for τ is obtained

$$1/\tau^2 = k_2^2 + 8k_2 k_1 c_0 \tag{65}$$

Thus far, in the derivations for $1/\tau$, activity coefficients have been neglected. In practice, this will often be permissible. Even for ionic reactions the activity coefficients can be made constant during the course of the reaction by adding an excess of inert electrolyte. However, with conductometric detection the addition of an inert electrolyte leads to a corresponding decrease in the sensitivity of the method (see equation 23). Therefore, these measurements are performed with no or only a small addition of non-reacting electrolytes so that activity corrections have to be performed. This procedure will be outlined again for reaction 6. Considering activity coefficients, we have

$$-dc_A/dt = k_1 c_A c_B \Pi_f - k_2 c_C \Pi_b \tag{66}$$

where $\Pi_f = f_A f_B / f^+$ and $\Pi_b = f_C / f^+$ with f^+ the activity coefficient of the activated complex. (If the activity coefficients depend only on the charge and the ionic strength $f_C = f^*$ in the example given: $\Pi_b = 1$ and $\Pi_f = \Pi = f_A f_B/f_C$. This simplification is not always possible as for example for a reaction of the type $A^{3+} + B^+ \rightleftarrows 2C^{2+}$).

After the perturbation not only the c_i change to $\bar{c}_i + x_i$ but also the activity coefficient ratios change from Π_i to $\bar{\Pi}_i + \Delta\Pi_i$. Introducing these expressions into equation 66 and again neglecting all quadratic terms in the deviations, we arrive at (see equations 7 to 12)

$$\frac{1}{\tau} = k_1 \Pi_f \left(\bar{c}_A + \bar{c}_B + \bar{c}_B \frac{\partial \ln \Pi_f}{\partial \ln c_A}\right) + k_2 \Pi_b \left(1 + \frac{\partial \ln \Pi_b}{\partial \ln c_C}\right) \quad (67)$$

The dependence $\partial \ln \Pi_i / \partial \ln c_K$ is usually not available from experimental data and has to be estimated from semi-empirical formulas for the activity coefficients as for example[6]

$$-\log f_i = 0.5 z_i^2 (\{I^{\frac{1}{2}}/(1 + I^{\frac{1}{2}})\} - 0.3I) \quad (68)$$

where z_i are the valencies of the ions and I is the ionic strength. With equation 68 the Π_i and $\partial \ln \Pi_i / \partial \ln c_K$ may be evaluated.

The next important type of reaction to be discussed is a reaction coupled to a much faster one, e.g.

$$\begin{array}{c} A + B \rightleftarrows C \rightleftarrows D \\ 1 \quad\; 2 \quad\; 3 \end{array} \quad (69)$$

The fast equilibrium $1 \rightleftarrows 2$ will be established before the concentration of species D has changed to a measurable extent. Therefore, for the fast reaction step, the system may be treated as above (equation 12) and the relaxation time τ_I is given by

$$1/\tau_I = k_{12}(\bar{c}_A + \bar{c}_B) + k_{21} \quad (70)$$

(again neglecting activity corrections). On the other hand, for the calculation of the slow relaxation, the equilibrium $1 \rightleftarrows 2$ may be assumed to be established, i.e.

$$\delta(c_C/c_A c_B) = 0 \quad (71)$$

or because $x_A = x_B$

$$x_A/\bar{c}_A + x_A/\bar{c}_B = x_C/\bar{c}_C \quad (72)$$

In the slow relaxation we observe the formation of D from C, where C is in equilibrium with A and B. The rate equation can be written as

$$d(x_A + x_C)/dt = -k_{23} x_C + k_{32} x_D \quad (73)$$

With

$$x_A + x_C + x_D = 0 \quad (74)$$

and equation 72 the relaxation time τ_{II} is obtained:

$$\frac{1}{\tau_{II}} = \frac{-1}{x_A + x_C} \frac{d(x_A + x_C)}{dt} = k_{32} + k_{23} \frac{K_{12}(\bar{c}_A + \bar{c}_B)}{1 + K_{12}(\bar{c}_A + \bar{c}_B)} \quad (75)$$

with the equilibrium constant

$$K_{12} = \bar{c}_C/\bar{c}_A \times \bar{c}_B \tag{76}$$

The above treatment is permissible, if $\tau_{II} \gtrsim 20\tau_I$. If the two relaxation times are comparable, the two consecutive reactions cannot be treated separately. Furthermore, the experimental discrimination is difficult and must be handled with care. (A useful procedure for the evaluation of two comparable relaxation times has been proposed[54].) The following general treatment of reaction 69 is intended to outline a procedure also applicable to more complicated cases. A more sophisticated method is indicated in ref. 15.

For small deviations from equilibrium the rate equations for reaction 69 are given by

$$\begin{aligned} dx_A/dt &= -k_{12}(\bar{c}_A + \bar{c}_B)x_A + k_{21}x_C \equiv -k'_{12}x_A + k_{21}x_C \\ dx_C/dt &= k'_{12}x_A - k_{21}x_C + k_{32}x_D - k_{23}x_C \end{aligned} \tag{77}$$

x_D can be eliminated by the mass balance equation 74 with the result that

$$dx_A/dt = \alpha_{11}x_A + \alpha_{12}x_C$$

and
$$\tag{78}$$

$$dx_C/dt = \alpha_{21}x_A + \alpha_{22}x_C$$

where

$$\alpha_{11} = -k'_{12}; \quad \alpha_{12} = k_{21}; \quad \alpha_{21} = k'_{12} - k_{32}$$

and

$$\alpha_{22} = -k_{21} - k_{23} - k_{32} \tag{79}$$

(For other reactions with two consecutive steps α_{ik} must be changed correspondingly.)

To solve the simultaneous differential equations 78 they must be transformed to equations of the form

$$(dy_i/dt) = -(1/\tau_i)y_i \tag{80}$$

where the y_i are linear combinations of the x_i

$$y_i = \beta_{iA}x_A + \beta_{iC}x_C \tag{81}$$

These new variables y_i are called normal concentrations (in formal analogy to, say, the normal vibrations in spectroscopy). It is the change of these normal concentrations to which the observed relaxation times refer.

For the sake of simplification we set $\beta_{iC} = 1$ without loss in generality

$$y_i = \beta_i x_A + x_C \tag{82}$$

Introducing 82 and 78 into equation 80, we obtain

$$(1/\tau_i)(\beta_i x_A + x_C) + \beta_i(\alpha_{11}x_A + \alpha_{12}x_C) + \alpha_{21}x_A + \alpha_{22}x_C = 0 \tag{83}$$

Since the coefficients of the independent variables x_A and x_C must vanish separately, two equations for $1/\tau_i$ and β_i are obtained the solutions of which are (after resubstituting the α_{ik} pertaining to reaction 69 and rearranging)

$$1/\tau_{I, II} = \tfrac{1}{2}(\Sigma \pm [\Sigma^2 - 4\Pi]^{\tfrac{1}{2}}) \tag{84}$$

with

$$\Sigma = k'_{12} + k_{21} + k_{23} + k_{32} = 1/\tau_I + 1/\tau_{II} \tag{85}$$

and

$$\Pi = k_{21}k_{32} + k'_{12}(k_{23} + k_{32}) = (1/\tau_I) \times (1/\tau_{II}) \tag{86}$$

It is important to note that the two relaxation times observed are not due to the two steps separately, but both depend on all four rate constants.

If both τ_I and τ_{II} depend measurably on concentration, all four rate constants of the consecutive reactions 69 can be obtained from equations 85 and 86. The overall stability constant

$$K_a = \frac{\bar{c}_C + \bar{c}_D}{\bar{c}_A \times \bar{c}_B} = \frac{k_{12}}{k_{21}}\left(1 + \frac{k_{23}}{k_{32}}\right) \tag{87}$$

may be determined by any suitable conventional technique and compared with the corresponding value calculated from the rate constants. Since the accuracy attainable for one or more of the four rate constants may be rather poor, usually a combination of rate measurements and equilibrium data leads to the best set of rate constants.

Another way to get information on thermodynamic data is the quantitative evaluation of the amplitudes of the relaxation effects. From equation 82 (i = I, II) and the integrated equation 80 we arrive at

$$x_A = \{-y_I^0 \exp(-t/\tau_I) + y_{II}^0 \exp(-t/\tau_{II})\}/(\beta_{II} - \beta_I) \tag{88}$$

and

$$x_C = \{\beta_{II} y_I^0 \exp(-t/\tau_I) - \beta_I y_{II}^0 \exp(-t/\tau_{II})\}/(\beta_{II} - \beta_I) \tag{89}$$

The y_I^0 are the values of the normal variables for time $t = 0$.

As in the derivation of equation 84, for the β_i we get

$$\beta_i = (k_{32} - k'_{12})(\tau_i^{-1} - k'_{12})^{-1} \tag{90}$$

To measure the progress of the reaction experimentally, a physical quantity, a (e.g. the optical density or the electrical conductivity), is observed. In a relaxation experiment the change in a due to the changes in concentrations is

$$\delta a = \frac{\partial a}{\partial c_A} x_A + \frac{\partial a}{\partial c_B} x_B + \frac{\partial a}{\partial c_C} x_C + \frac{\partial a}{\partial c_D} x_D \tag{91}$$

If the concentration changes are measured conductometrically, the derivates $\partial a/\partial c_i$ are essentially the mobilities and in optical absorption they correspond to the extinction coefficients.

For the reaction 69 two relaxation times are observed; i.e. the observable a changes accordingly to

$$\delta a = \delta a_I \exp(-t/\tau_I) + \delta a_{II} \exp(-t/\tau_{II}) \tag{92}$$

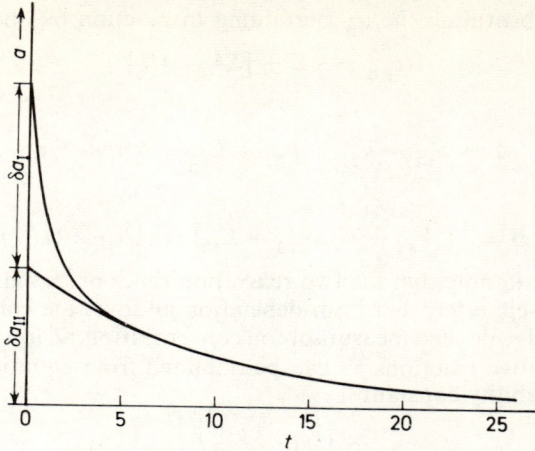

Figure 13. The change of the observable a with time according to equation 92.

(see *Figure 13*). Combining equations 74, 82, 88, 91, and 92, the total change x_A^0 and x_C^0 in concentration can be calculated

$$x_A^0 = \delta a_I \left[\frac{\partial a}{\partial c_A} + \frac{\partial a}{\partial c_B} - \frac{\partial a}{\partial c_D} + \beta_{II}\left(\frac{\partial a}{\partial c_D} - \frac{\partial a}{\partial c_C}\right) \right]^{-1}$$

$$+ \delta a_{II} \left[\frac{\partial a}{\partial c_A} + \frac{\partial a}{\partial c_B} - \frac{\partial a}{\partial c_D} + \beta_I\left(\frac{\partial a}{\partial c_D} - \frac{\partial a}{\partial c_C}\right) \right]^{-1} \quad (93)$$

and

$$x_C^0 = \delta a_I \beta_{II} \left[-\frac{\partial a}{\partial c_A} - \frac{\partial a}{\partial c_B} + \frac{\partial a}{\partial c_D} + \beta_{II}\left(\frac{\partial a}{\partial c_C} - \frac{\partial a}{\partial c_D}\right) \right]^{-1}$$

$$+ \delta a_{II} \beta_I \left[-\frac{\partial a}{\partial c_A} - \frac{\partial a}{\partial c_B} + \frac{\partial a}{\partial c_D} + \beta_I\left(\frac{\partial a}{\partial c_C} - \frac{\partial a}{\partial c_D}\right) \right]^{-1} \quad (94)$$

With x_A^0 and x_C^0 and the equilibrium constants k_{12}/k_{21} and k_{23}/k_{32} known, some thermodynamic functions can be calculated; e.g. for a temperature-jump experiment we get the enthalpies of reaction by

$$\delta \ln K_{12} = -\frac{x_A^0}{\bar{c}_A} - \frac{x_A^0}{\bar{c}_B} + \frac{x_C^0}{\bar{c}_C} = \frac{\Delta H_{12}}{RT^2} \delta T \quad (95)$$

and

$$\delta \ln K_{23} = -\frac{x_C^0}{\bar{c}_C} + \frac{x_D^0}{\bar{c}_D} = -\frac{x_C^0}{\bar{c}_C} - \frac{x_A^0 + x_C^0}{\bar{c}_D} = \frac{\Delta H_{23}}{RT^2} \delta T \quad (96)$$

If reaction $1 \rightleftarrows 2$ is fast compared to $2 \rightleftarrows 3$, i.e. $k_{12} \gg k_{23}$ and $k_{21} \gg k_{32}$, equations 70 and 75 are immediately obtained from equations 84 to 86, and from equation 90 we get $\beta_I = -K_{12}(\bar{c}_A + \bar{c}_B)$ and $\beta_{II} = 1$. Under these conditions the relaxation amplitudes are given by

$$\delta a_\mathrm{I} = \frac{K_{12}(\bar{c}_\mathrm{A} + \bar{c}_\mathrm{B})x_\mathrm{A}^0 - x_\mathrm{C}^0}{1 + K_{12}(\bar{c}_\mathrm{A} + \bar{c}_\mathrm{B})}\left[\frac{\partial a}{\partial c_\mathrm{A}} + \frac{\partial a}{\partial c_\mathrm{B}} - \frac{\partial a}{\partial c_\mathrm{C}}\right] \quad (97)$$

and

$$\delta a_\mathrm{II} = \frac{x_\mathrm{A}^0 + x_\mathrm{C}^0}{1 + K_{12}(\bar{c}_\mathrm{A} + \bar{c}_\mathrm{B})}\left[\frac{\partial a}{\partial c_\mathrm{A}} + \frac{\partial a}{\partial c_\mathrm{B}} - \frac{\partial a}{\partial c_\mathrm{D}} + K_{12}(\bar{c}_\mathrm{A} + \bar{c}_\mathrm{B})\right.$$
$$\left. \times \left(\frac{\partial a}{\partial c_\mathrm{C}} - \frac{\partial a}{\partial c_\mathrm{D}}\right)\right] \quad (98)$$

The procedure outlined in some detail for the treatment of reaction 69 may be applied to other types of reaction. In ultrasonic relaxation measurements

Table 2. For some important types of reaction the expressions for $1/\tau$ are summarized

Reaction	Reciprocal relaxation times
$A \underset{k_{21}}{\overset{k_{12}}{\rightleftharpoons}} B$	$\frac{1}{\tau} = k_{12} + k_{21}$
$A + B \underset{k_{21}}{\overset{k_{12}}{\rightleftharpoons}} C$	$\frac{1}{\tau} = k_{12}(\bar{c}_\mathrm{A} + \bar{c}_\mathrm{B}) + k_{21}$
$2A \underset{k_{21}}{\overset{k_{12}}{\rightleftharpoons}} A_2$	$\frac{1}{\tau} = \sqrt{k_{12}^2 + 8k_{12}k_{21}c_0}$ (with $c_0 = c_\mathrm{A} + 2c_{\mathrm{A}_2}$)
$A + C \underset{k_{21}}{\overset{k_{12}}{\rightleftharpoons}} B + C$ (C = Catalyst)	$\frac{1}{\tau} = k_{12}\bar{c}_\mathrm{C} + k_{21}\bar{c}_\mathrm{C}$
$A + B \underset{k_{21}}{\overset{k_{12}}{\rightleftharpoons}} C$ (B = buffered)	$\frac{1}{\tau} = k_{12}\bar{c}_\mathrm{B} + k_{21}$
$A + B \underset{k_{21}}{\overset{k_{12}}{\rightleftharpoons}} C + D$	$\frac{1}{\tau} = k_{12}(\bar{c}_\mathrm{A} + \bar{c}_\mathrm{B}) + k_{21}(\bar{c}_\mathrm{C} + \bar{c}_\mathrm{D})$
$A + B + C \underset{k_{21}}{\overset{k_{12}}{\rightleftharpoons}} D$	$\frac{1}{\tau} = k_{12}(\bar{c}_\mathrm{A}\bar{c}_\mathrm{B} + \bar{c}_\mathrm{A}\bar{c}_\mathrm{C} + \bar{c}_\mathrm{B}\bar{c}_\mathrm{C}) + k_{21}$
$A + B \underset{k_{21}}{\overset{k_{12}}{\rightleftharpoons}} AB \underset{k_{32}}{\overset{k_{23}}{\rightleftharpoons}} C$	$\frac{1}{\tau} = \tfrac{1}{2}(\Sigma \pm \sqrt{\Sigma^2 - 4\Pi})$ with $\Sigma = k_{12}(\bar{c}_\mathrm{A} + \bar{c}_\mathrm{B}) + k_{21} + k_{23} + k_{32}$ and $\Pi = k_{21}k_{32} + k_{12}(\bar{c}_\mathrm{A} + \bar{c}_\mathrm{B})(k_{23} + k_{32})$. If $k_{12}(\bar{c}_\mathrm{A} + \bar{c}_\mathrm{B}) \gg k_{23}$ and $k_{21} \gg k_{32}$ $\frac{1}{\tau} = \frac{(k_{23} \times k_{12}/k_{21})(\bar{c}_\mathrm{A} + \bar{c}_\mathrm{B})}{1 + (k_{12}/k_{21})(\bar{c}_\mathrm{A} + \bar{c}_\mathrm{B})} + k_{32}$
$AB + C \underset{k_{21}}{\overset{k_{12}}{\rightleftharpoons}} AC + B$ $\underset{k_{31}}{\overset{k_{13}}{\diagdown}} \quad \underset{k_{32}}{\overset{k_{23}}{\diagup}}$ $A + B + C$	With $\bar{c}_\mathrm{A}$ small compared with other concentrations $\frac{1}{\tau} = [k_{12} + k_{13}k_{32}/(k_{31}\bar{c}_\mathrm{B} + k_{32}\bar{c}_\mathrm{C})]$ $\times (\bar{c}_{AB} + \bar{c}_\mathrm{C}) + [k_{21} + k_{23}k_{31}/(k_{31}\bar{c}_\mathrm{B} + k_{32}\bar{c}_\mathrm{C})](\bar{c}_{AC} + \bar{c}_\mathrm{B})$

thermodynamic information is available from the value of $\Gamma(\Delta V)^2$ as obtained from the damping coefficient, α (see equation 48). A more detailed discussion of this problem has been given by Eigen and Tamm[17].

In *Table 2* a compilation of the expressions for relaxation times for different types of reactions is given.

An interesting application of chemical relaxation, proposed by Eigen and Winkler[18, 59], is a titration technique with relaxation times and/or relaxation amplitudes as an end-point indicator. Let a substance, A, be titrated with B forming a strong complex

$$A + B \underset{k_2}{\overset{k_1}{\rightleftharpoons}} AB \tag{99}$$

with a (small) dissociation constant

$$K = c_A c_B / c_{AB} \tag{100}$$

The relaxation time is given by

$$1/\tau = k_1(\bar{c}_A + \bar{c}_B) + k_2 \tag{12}$$

If the total concentration of A is $c_0 = c_A + c_{AB}$ and the total concentration of B added is $c_t = c_B + c_{AB}$, we have for the relaxation time

$$1/\tau = k_1 \sqrt{(c_0 - c_t + K)^2 + 4Kc_t} \tag{101}$$

τ has a maximum at the point of equivalence $c_t = c_0$. Similarly, the amplitude x_0 of the relaxation effect is proportional to the function Γ (see equation

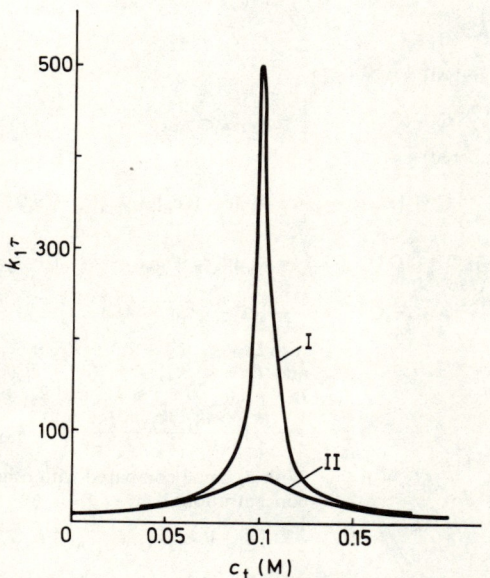

Figure 14. $k_1 \times \tau$ against c_t with $c_0 = 0.1$M according to equation 101. I: $K = 10^{-5}$M; II: $K = 10^{-3}$M.

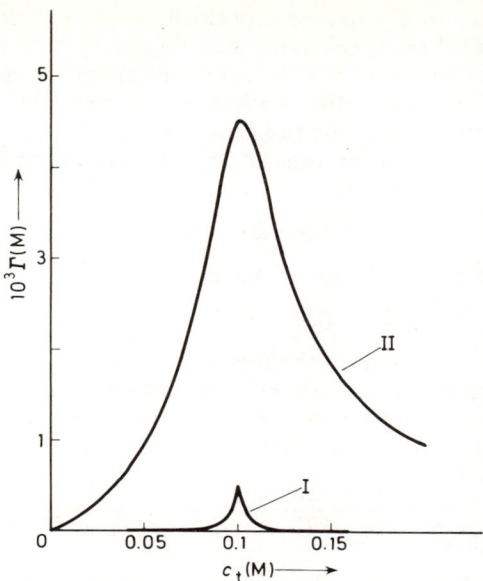

Figure 15. Γ against c_t with $c_0 = 0.1\,\text{M}$ according to equation 102. I: $K = 10^{-5}\,\text{M}$; II: $K = 10^{-3}\,\text{M}$.

16) which in terms of c_0, c_t and K reads

$$\Gamma = \frac{K(c_0 + c_t + K - \{(c_0 - c_t + K)^2 + 4Kc_t\}^{\frac{1}{2}})}{2\{(c_0 - c_t + K^2) + 4Kc_t\}^{\frac{1}{2}}} \quad (102)$$

In *Figures 14* and *15* calculated relaxation titration curves are shown with different parameters. This technique has not yet been extensively used (however, see ref. 18), but it certainly has interesting possibilities.

IV. Some Applications of Chemical Relaxation Techniques

In this section a few typical examples of chemical relaxation investigations are reported to give an impression of the kind of information which can be obtained by applying the methods outlined in the previous chapters.

1. The Neutralization Reaction $H^+ + OH^- \underset{k_2}{\overset{k_1}{\rightleftharpoons}} H_2O$

One of the most important reactions in chemistry is the recombination of hydrogen ions and hydroxyl ions in aqueous solution and the dissociation of water molecules into these two ions. The equilibrium constant is well known

$$\frac{c_{H^+} c_{OH^-}}{c_{H_2O}} = 1.8 \times 10^{-16}\,\text{M} \quad (103)$$

the enthalpy of the reaction is $-13.7\,\text{kcal/mol}$ and $\Delta V = 23.5\,\text{cm}^3/\text{mol}$. The reaction was first studied kinetically by Eigen and De Maeyer[14] using the field-jump method (see above, section II, *Figure 7*). The conductivity cells were filled with extremely purified water of specific conductivity $\varkappa \geq 6 \times 10^{-6} \Omega^{-1} m^{-1}$, i.e. more than 90 per cent of the conductivity is due to H^+ and OH^- ions. Under these conditions, the reaction of H^+ with other bases or OH^- with other acids does not contribute measurably to the observed relaxation time. The resistance of the cell amounted to more than $10^6 \Omega$. At 22°C a relaxation time of 35 ± 5 μsec was observed leading to a recombination rate constant

$$k_1 = (1.3 \pm 0.2) \times 10^{11} \, M^{-1} \, sec^{-1}$$

and for the rate of dissociation of water

$$k_2 = (2.4 \pm 0.4) \times 10^{-5} \, sec^{-1}$$

The rate determining step in the recombination reaction is the diffusion of the ions to a distance, a, at which in a very fast step the proper recombination occurs. For such a diffusion controlled reaction, between ions of charge z_+ and z_-, the rate constant has been calculated by Debye[7]

$$k_1 = \frac{4\pi N_A z_+ z_- e_0^2}{\varepsilon k T} \frac{(D_+ - D_-)}{\exp(z_+ z_- e_0^2/\varepsilon a k T) - 1} \tag{104}$$

where D_+ and D_- are the diffusion coefficients of the two ions.

Introducing $k_1 = 1.3 \times 10^{11} \, M^{-1} \, sec^{-1} = 1.3 \times 10^{14} \, cm^3 \, mol^{-1} \, sec^{-1}$ and the other known constants into equation 104, the distance of approach a turns out to be about 7 Å. We thus have to conclude that the hydration water of the ions has not to be removed before the recombination step proper, but that these molecules take part in the reaction as indicated in *Figure 16*.

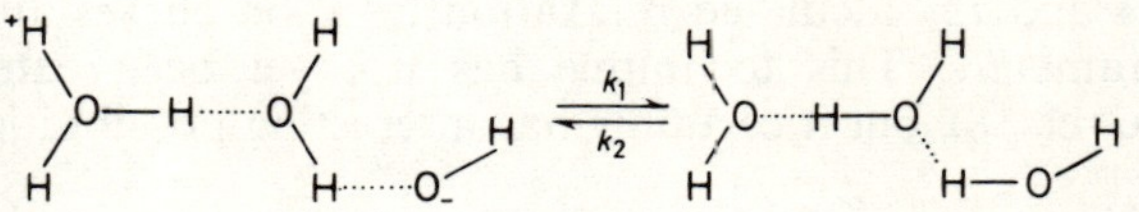

Figure 16. The mechanism of the reaction $H^+ + OH^- \rightleftarrows H_2O$.

This mechanism is also the reason for the high mobility of H^+ and OH^- in water. Very similar results also at other temperatures and with D_2O have been obtained by Ertl and Gerischer[20] using a temperature-jump apparatus with microwave heating.

Most reactions of the kind $H_3O^+ + A^- \rightleftarrows HA + H_2O$ and $OH^- + HA \rightleftarrows A^- + H_2O$ are also diffusion controlled. A summary on the kinetics of proton transfer reactions has been given by Eigen[11, 12, 19]

2. The Formation of Metal Ion Complexes: $M^{m+} + L^{n-} \rightleftarrows ML^{(m-n)+}$

The results of many investigations of complex formation reactions with chemical relaxation techniques are summarized in equation 105

$$\underset{1}{Me^{m+}} + L^{n-} \underset{k_{2'1}}{\overset{k_{12'}}{\rightleftarrows}} \underset{2'}{MeOH_2OH_2L^{(m-n)+}} \underset{k_{22'}}{\overset{k_{2'2}}{\rightleftarrows}} \underset{2}{MeOH_2L^{(m-n)+}} \underset{k_{32}}{\overset{k_{23}}{\rightleftarrows}} \underset{3}{MeL^{(m-n)+}} \qquad (105)$$

In a very fast reaction, the solvated ions Me^{m+} and L^{n-} form a complex **2'** in which the two ions are still separated by two layers of solvent. Also in a very fast, though slower reaction, the outer-sphere complex **2** is formed with only one layer of solvent separating the two ions. In a last step, which may be slower by orders of magnitude, the inner-sphere complex **3** is formed. Since step **1** ⇌ **2'** is only with difficulty experimentally discernible from **1** ⇌ **2** (see, however, ref. 1), we shall abbreviate scheme 105 to

$$\underset{1}{Me^{m+}} + L^{n-} \underset{k_{21}}{\overset{k_{12}}{\rightleftarrows}} \underset{2}{MeOH_2L^{(m-n)+}} \underset{k_{32}}{\overset{k_{23}}{\rightleftarrows}} \underset{3}{MeL^{(m-n)+}} \qquad (106)$$

which is of the type of reaction 69.

There are two experimental arguments which require the assumption of the more complicated mechanisms 105 or 106, respectively, instead of

$$Me^{m+} + L^{n-} \underset{k_{21}}{\overset{k_{12}}{\rightleftarrows}} MeL^{(m-n)+} \qquad (107)$$

which is of the type of reaction 6. The first argument is that two relaxation times (or even three with mechanism 105) are observed so that at least two consecutive steps for the reaction must be involved. The second piece of evidence supporting 106 rather than 107 is the dependence of the slower relaxation time on concentration. $1/\tau_{II}$ is not given by equation 12, which applies to a reaction of the type of equation 107, but rather by equation 75, thus indicating the mechanism 106.

How to evaluate the rate and equilibrium constants from the experiments will be shown for the formation of the complex $BeSO_4$ in aqueous solution. The relaxation time for the slow step, considering the activity coefficients, is given by

$$\frac{1}{\tau_{II}} = k_{32} + k_{23} \frac{2K_{12}\alpha c_0 f_\pm^2 \varkappa}{1 + 2K_{12}\alpha c_0 f_\pm^2 \varkappa} \qquad (108)$$

where α is the degree of dissociation and $\varkappa = 1 + (\partial \ln f_\pm)/\partial \ln \alpha$. The total association constant has been determined[21] at ionic strength $I = 0$ and 25°C.

$$K_a = \frac{[BeOH_2SO_4] + [BeSO_4]}{[Be^{2+}][SO_4^{2-}]} = K_{12}(1 + K_{23}) = 145 \, M^{-1}$$

Using Davies's equation[6] for the activity coefficients, α has been calculated for different total concentrations c_0 of $BeSO_4$. The relaxation times τ_{II} have been determined with the pressure-jump technique (see *Figure 6*). In *Figure 17* the relaxation rate is plotted against c_0^7.

From the intercept in *Figure 17* we have k_{32}, from the initial slope $2k_{23}K_{12}$, and from the asymptotic value of $1/\tau$ at high concentration $k_{23} + k_{32}$ is obtained.

Thus, all constants are known, leading to

$$k_{23} = 185 \text{ sec}^{-1}, \quad k_{32} = 285 \text{ sec}^{-1},$$
$$K_{23} = k_{23}/k_{32} = 0.65 \quad \text{and} \quad K_{12} = 100 \text{ M}^{-1}$$

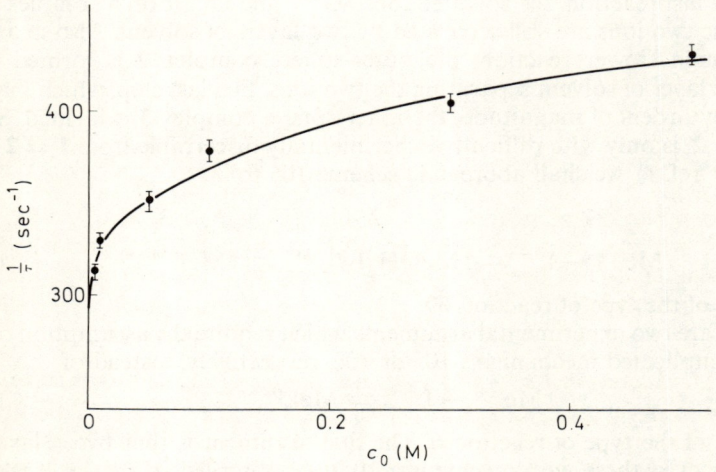

Figure 17. τ^{-1} against c_0 for aqueous solutions of BeSO$_4$ at 25°C.

The total association constant turns out to be

$$K_a = K_{12}(1 + K_{23}) = 165 \text{ M}^{-1}$$

The reasonable agreement with the thermodynamically determined value $K_a = 145 \text{ M}^{-1}$ indicates the consistency of the treatment[21]. Though the experimental precision of such a kinetic determination of equilibrium data is rather poor, it is clear that this technique is an independent, new method and, especially, also leads to an estimate of the inner-sphere/outer-sphere equilibrium constant K_{23}.

Since for BeSO$_4$ only the slow relaxation has been investigated in detail, equations 94 and 95 cannot be used to determine ΔV_{12} and ΔV_{23}. However, if we rearrange those equations the relative change in resistance ($\delta R/R$) due to the slow reaction can be calculated[40]

$$\frac{\delta R}{R} = \left(1 + \frac{\partial \ln f_\lambda}{\partial \ln \alpha}\right) \frac{\delta p}{RT} K_{23} \left[\frac{\alpha(1-\alpha)\Delta V_{12}}{[\alpha + 2\varkappa(1-\alpha)][\alpha(1+K_{23}) + 2\varkappa(1-\alpha)]} \right.$$
$$\left. + \frac{(1-\alpha)\Delta V_{23}}{(1+K_{23})\alpha + 2\varkappa(1+K_{23})(1-\alpha)} \right] \quad (109)$$

where f_λ is the coefficient of conductivity. In principle, the concentration dependence of $\delta R/R$ supplies ΔV_{12} and ΔV_{23}. f_λ may be calculated, for example, by the equation of Robinson and Stokes[48]. However, due to the experimental error, it was not possible to split $\delta R/R$ into the two terms indicated in equation 109.

A precision in the relaxation rate measurements of at least ± 1 per cent at low concentration of electrolyte is required to obtain reasonably accurate data of ΔV_{12} and ΔV_{23} from equation 109. This task, of improving the experimental precision of chemical relaxation measurements, is still a challenge for the experimentalist.

It should be mentioned that stability constants for complexes as weak as $BeSO_4$ obtained from conventional techniques also have poor accuracy. Improved chemical relaxation methods, therefore, may well compete with purely equilibrium methods in providing thermodynamic data of acceptable accuracy.

For complex-forming reactions involving Al^{3+} and Ga^{3+} and weak ligands[35] where τ_{II} does not vary measurably with concentration, an approximate value of K_i has been obtained by inserting estimated ΔV_{12} and ΔV_{23} data into equation 109.

3. The Mechanism of Cooperative Ligand Binding on an Allosteric Enzyme[36]

Certain enzymes are composed of a small number of identical subunits each of which may react with one molecule of a suitable ligand. Furthermore, the enzyme subunits undergo a conformational transformation. To be specific, we shall discuss as an example the enzyme yeast glyceraldehyde-3-phosphate dehydrogenase (GAPD) which is available in pure form. GAPD is known to consist of four subunits which exist in two conformationally different forms, the R-form and the T-form, which will be indicated by circles and squares, respectively, in the diagrams below. The enzyme forms rather stable complexes with nicotinamide-adenine-dinucleotide (NAD). The equilibrium constants of the complex formation have been studied spectrophotometrically. The results of these studies can be described by either of two reaction schemes the first of which is indicated in *Figure 18* where D is a ligand molecule of NAD. This scheme proposed by Koshland et al.[39] assumes that the binding of a ligand induces the conformational change of the occupied subunit of the enzyme ('induced fit') so that a sequence of enzymes from pure T-form over mixed conformations to pure R-form occurs.

Another interpretation of the reaction has been provided by Monod et al.[42] as shown in *Figure 19*. In this mechanism it is assumed that there is an equal affinity between D and each site and that the conformational transformation process is an all or none process, i.e. this process is an extremely cooperative one. Unfortunately, both mechanisms are consistent with equilibrium measurements, since there are enough adjustable parameters to fit the predictions of either reaction scheme.

To discriminate between the two possibilities, the elementary steps have to be investigated kinetically. Since the detection of intermediates requires the use of relatively high concentrations of the enzyme, the reactions studied are rather fast. Using the temperature-jump and stopped-flow technique,

Figure 18. The 'induced fit' mechanism I[39] for the reaction GAPD + NAD (see text).

Kirschner et al.[36] have investigated the GADP–NAD system. If scheme I were correct, four relaxation times would result, some of which, of course, might be undetectable. However, each observed relaxation rate $1/\tau_i$ should increase with increasing concentrations of the ligand. Experimentally, this turns out not to be so (see below $1/\tau_{III}$ in Figure 20) so that model I has to be discarded and model II will be used to interpret the measurements. In the actual experiments three relaxation times have been observed (see Figure 4) in the time ranges (depending on ligand concentration D): $\tau_I = (1\ \text{to}\ 5) \times 10^{-4}$ sec; $\tau_{II} = (5\ \text{to}\ 20) \times 10^{-4}$ sec and $\tau_{III} = (0.06\ \text{to}\ 5)$ sec. The two fast relaxation rates have been assigned to the bimolecular steps involved in forming either R- or T-complexes. Since the reaction with R is faster than that with T and a common ligand D is used up in both reactions, we have for the shortest relaxation time

$$1/\tau_I = k_D + k_A(D + \Phi R) \qquad (110)$$

where the rate constants are explained in Figure 19, D is the concentration of the ligand, and

$$\Phi R = 4R_0 + 3R_1 + 2R_2 + R_3 \qquad (111)$$

is the sum of the concentrations of the free binding sites of the R-state. For the sake of brevity, in this section the symbols D, R_i, T_i are also used to represent the concentration. The slower relaxation rate is described by

$$\frac{1}{\tau_{II}} = k'_D + k'_A \left(D + \Phi T \frac{K+D}{K+D+\Phi R}\right)^* \qquad (112)$$

where

$$\Phi T = 4T_0 + 3T_1 + 2T_2 + T_3 \qquad (114)$$

is the sum of the concentrations of free binding sites of the T-state. The equilibrium constant $K = k_D/k_A$ is given by:

---

* Equation 112 is valid, if $D + R \underset{k_D}{\overset{k_A}{\rightleftharpoons}} C$ is much faster than $D + T \underset{k'_D}{\overset{k'_A}{\rightleftharpoons}} E$. Under these conditions we have

$$dx_E/dt = k'_A(Dx_T + Tx_D) - k'_D x_E \qquad (113)$$

With the mass balance equations and the pre-equilibrium condition

$$x_R + x_T = x_D;\ x_R = -x_C;\ x_T = -x_E \quad \text{and} \quad \frac{x_D}{D} + \frac{x_R}{R} - \frac{x_C}{C} = 0$$

x_D may be expressed in terms of x_E and equation 112 is obtained. For a more detailed analysis of τ_I and τ_{II} see Eigen[13].

$$K = 4R_0 \times D/R_1 = 3R_1 \times D/2R_2 = 2R_2 \times D/3R_3 = R_3 \times D/4R_4$$
$$= k_D/k_A \quad (115)$$

since four subunits can associate with D but only one can dissociate in the equilibrium $R_0 + D \rightleftarrows R_1$ and correspondingly for the other equilibria $R_n + D \rightleftarrows R_{n+1}$.

For T the analogous equilibrium constant K' is:

$$K' = 4T_0 \times D/T_1 = 3T_1 \times D/2T_2 = 2T_2 \times D/3T_3 = T_3 \times D/4T_4$$
$$= k'_D/k'_A \quad (116)$$

From the measurement of τ_I and τ_{II} the rate constants (at 40°C and pH = 8.5) k_D, k_A, k'_D and k'_A and thereby the equilibrium constants were evaluated to be $K = 1 \times 10^{-4}$ M and $K' = 2.5 \times 10^{-3}$ M.

The relaxation time τ_{III} has been found to be independent of the concentration of the enzyme at constant concentration D of NAD. At constant enzyme concentration, however, $1/\tau_{III}$ varies with D as shown in *Figure 20*. This fact is the reason that mechanism I (*Figure 18*) had to be rejected but is consistent with the allosteric transformation mechanism (*Figure 19*), if we identify τ_{III} with the relaxation time due to the conformational transformation $R \rightleftarrows T$. During the measurement of τ_{III} the equilibrium between the

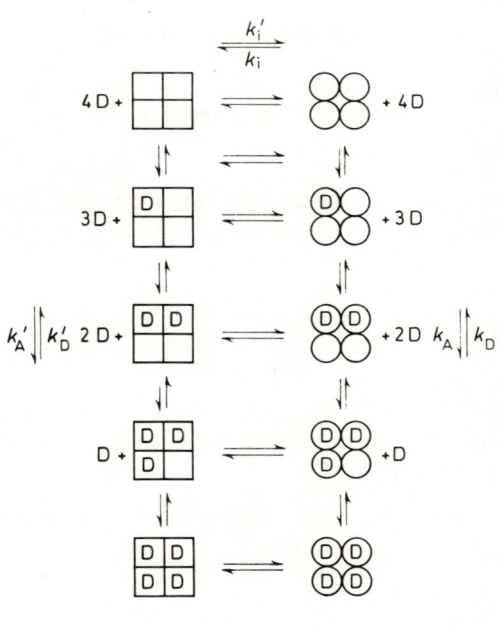

T-state   R-state

Figure 19. The 'all or none' mechanism II[42] for the reaction GAPD + NAD (see text).

different species R_i is established. The same is true for the species T_i. Thus, the rate equation is

$$\frac{d\Sigma R_i}{dt} = -k_0 R_0 + k'_0 T_0 - k_1 R_1 + k'_1 T_1 - \ldots - k_4 R_4 + k'_4 T_4 \quad (117)$$

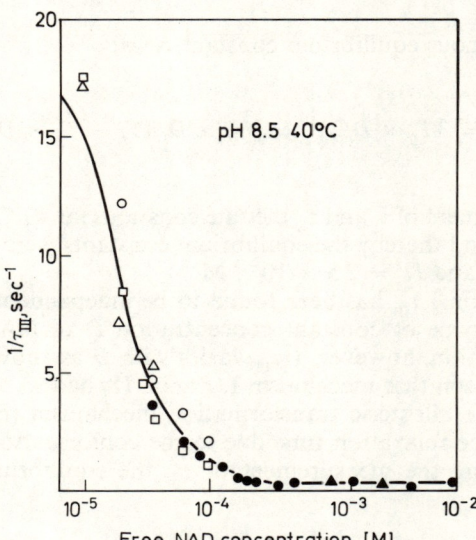

Figure 20. The dependence of $1/\tau_{III}$ on NAD concentration. $T = 40°C$, pH $= 8.5$. The full symbols represent data from stopped-flow experiments.

δR_i and δT_i, the changes in concentrations of the R_i and T_i after the temperature jump, are substituted in terms of $\delta \Sigma R_i$ and $\delta \Sigma T_i$, respectively. For example, with equation 116 δR_0 is given by

$$\delta R_0 = \frac{\delta \Sigma R_i}{1 + \frac{\delta R_1}{\delta R_0} + \frac{\delta R_2}{\delta R_1}\frac{\delta R_1}{\delta R_0} + \ldots} \quad (118)$$

From equations 117 and 118 and with $\delta \Sigma R_i = -\delta \Sigma T_i$ we obtain for $1/\tau_{III}$ (with the justifiable assumption that the concentration changes of D are negligible during the slow step)

$$\frac{1}{\tau_{III}} = \frac{k_0 + 4k_1(D/K) + 6k_2(D/K)^2 + 4k_3(D/K)^3 + k_4(D/K)^4}{(1 + D/K)^4}$$

$$+ \frac{k'_0 + 4k'_1(D/K') + 6k'_2(D/K')^2 + 4k'_3(D/K')^3 + k'_4(D/K')^4}{(1 + D/K')^4} \quad (119)$$

The sharp initial decrease of $1/\tau_{III}$ with D and its constant value at $D \geqslant 2 \times 10^{-4}$ M leads to the conclusion that all k'_i must be approximately equal so that the last term in equation 119 degenerates into $k'_0 \, (= k'_1 = k'_2 = k'_3 = k'_4)$. Correspondingly, the k_i values decrease with i as

$$k_i = k_0 (K/K')^i \tag{120}$$

or

$$\frac{1}{\tau_{\text{III}}} = k_0 \left(\frac{1 + D/K'}{1 + D/K}\right)^4 + k'_0 \tag{121}$$

By plotting $\log (1/\tau_{\text{III}} - k'_0)$ against $\log ((1 + D/K')/(1 + D/K))$, a straight line with slope four was obtained confirming the argument and the fact that *four* subunits of the enzyme are involved in the reaction. (The exponent four in equation 121 stems from the four subunits in the enzyme.) The data obtained are summarized in *Table 3*. The data in *Table 3* are obtained ex-

Table 3. Kinetic and thermodynamic data for the interaction of GAPD and NAD at pH = 8.5 and at 40°C.

$k_D = 1.1 \times 10^3 \text{ sec}^{-1}$	$K = 10^{-4} \text{ M}$
$k_A = 1.1 \times 10^7 \text{ M}^{-1} \text{ sec}^{-1}$	$K' = 2.5 \times 10^{-3} \text{ M}$
$k'_D = 8 \times 10^2 \text{ sec}^{-1}$	
$k'_A = 3.2 \times 10^5 \text{ M}^{-1} \text{ sec}^{-1}$	$L_0 = \dfrac{k_0}{k'_0} = 60$
$k_0 = 16.8 \text{ sec}^{-1}$	
$k'_0 = 0.28 \text{ sec}^{-1}$	$L_4 = \dfrac{k_4}{k'_4} = L_0 \left(\dfrac{K}{K'}\right)^4 = 1.5 \times 10^{-4}$
$= k'_1 = k'_2 = k'_3 = k'_4$	

clusively from kinetic measurements. They are consistent with conventionally determined equilibrium values.

Without NAD in the solution, the conformational equilibrium constant L_0 favors the T-form and the relaxation time $1/\tau_{\text{III}} \approx k_0$, whereas with an excess of NAD the R-form is predominant and the relaxation time $1/\tau_{\text{III}} \approx k'_0$. By adding NAD the fast reaction steps $R_0 \to T_0$ are blocked, thus reducing the relaxation rate until it is finally determined by the reaction $T_i \to R_i$ (with a rate 0.28 sec^{-1}). This example clearly shows the potential of fast reaction techniques in elucidating complicated biochemical reaction systems.

4. Ultrasonic Absorption in Water–Dioxane Mixtures

The measurement of ultrasonic absorption in mixed solvents can give information about specific intermolecular interactions. As an example such measurements in the system water–dioxane will be discussed[26].

The quantity α/v^2 (see equation 47) was found to be frequency dependent in mixtures containing 5 to 50 wt% water. The data at a given mixture were not consistent with the assumption of a single relaxation process; however, all experimental results could fit two relaxation processes. Accordingly, the data were analysed by the equation

$$\alpha/v^2 = \{C_{\text{I}}/[1 + (\omega\tau_{\text{I}})^2]\} + \{C_{\text{II}}/[1 + (\omega\tau_{\text{II}})^2]\} \tag{122}$$

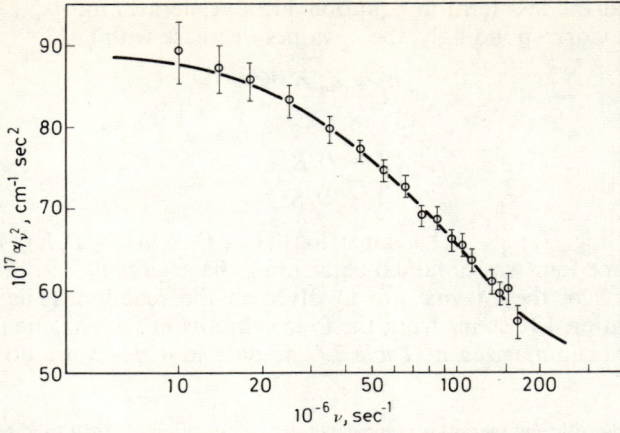

Figure 21. α/ν^2 against frequency for a dioxane:water mixture, 60 wt% water, 25°C. The curve is calculated using equation 122 with $C_I = 30 \times 10^{-17}$ cm^{-1} sec^2, $C_{II} = 12 \times 10^{-17}$ cm^{-1} sec^2, $B = 47 \times 10^{-17}$ cm^{-1} sec^2. $\tau_I = 1.4 \times 10^{-9}$ sec; $\tau_{II} = 5.3 \times 10^{-9}$ sec.

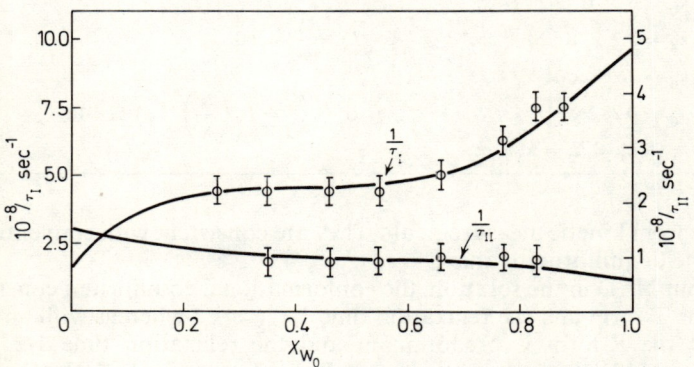

Figure 22. The reciprocal relaxation times $1/\tau_I$ (left scale) and $1/\tau_{II}$ (right scale) against the analytical mole fraction of water at 10°C. The curves are calculated for the mechanism described in the text.

(cf. equation 49). *Figure 21* shows a typical set of data together with the theoretical curve obtained by use of equation 122. In *Figure 22* the reciprocal relaxation times $1/\tau_I$ and $1/\tau_{II}$ are plotted against the mole fraction of water, and *Figure 23* shows the corresponding plot of the amplitude factors C_I and C_{II}. As will be shown below, all of the data can be explained by the mechanism

$$\underset{1}{2W + 2D} \underset{k_{21}}{\overset{k_{12}}{\rightleftarrows}} \underset{2}{Q + D} \underset{k_{32}}{\overset{k_{23}}{\rightleftarrows}} \underset{3}{Z} \tag{123}$$

where D represents dioxane, W water, Q a 2W:1D complex, and Z a 2W:2D complex.

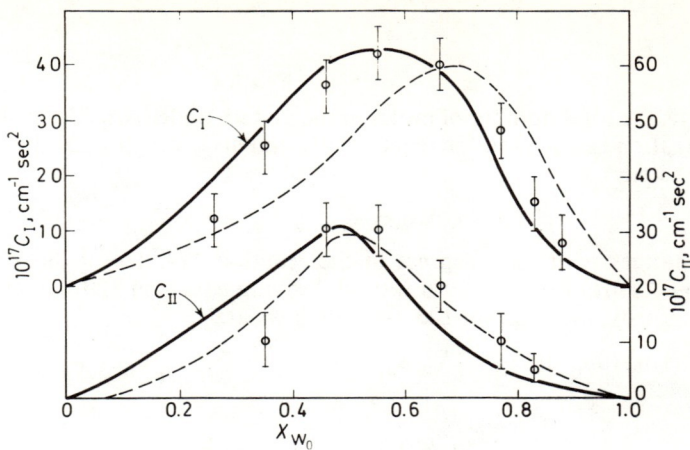

Figure 23. The amplitude factors C_I (left scale) and C_{II} (right scale) against the analytical mole fraction of water at 10°C. The curves are calculated for the mechanism described in the text: ——— with the assumption $\Delta V_I = \Delta V_{II} = 0$; ------ with the assumption $\Delta H_I = \Delta H_{II} = 0$.

In the following, the assumption is made that the mixture of the substances W and D and the complexes Q and Z is an ideal solution, i.e. the activity of each component is taken to be equal to its mole fraction. That means, the deviations from the usual definition of an ideal solution are due only to the complexes formed by the pure substances.

The equilibrium of reaction 123 is determined by the two stability constants

$$K_1 = X_Q/X_W^2 X_D$$

and

$$K_2 = X_Z/X_Q X_D$$

where X_i is the mole fraction of species i.

The rate equations can be obtained from transition state theory. In this theory, the reactants form an activated complex which decays according to a first order rate law. For example, step **1 ⇌ 2** of reaction 123 proceeds as

$$2W + D \underset{k_{\mp 1}}{\overset{k_{\pm 1}}{\rightleftarrows}} W_2D^{\ddagger} \underset{k_{\mp 2}}{\overset{k_{\pm 2}}{\rightleftarrows}} Q \qquad (124)$$

and $W_2D^{\ddagger}$ decays according to

$$-dn^{\ddagger}/dt = k_{\mp 1}n^{\ddagger} + k_{\mp 2}n^{\ddagger} \qquad (125)$$

$n^{\ddagger}$ is the number of moles of the activated complex. The rate of formation of the activated complex is

$$(1/n_t)(dn^{\ddagger}/dt) = k_{\pm 1}X_W^2 X_D + k_{\pm 2}X_Q \qquad (126)$$

obtained from the equilibrium conditions

$$K_1^{\ddagger} = X^{\ddagger}/X_W^2 X_D = k_{\pm 1}/k_{\mp 1} \qquad (127)$$

and

$$K_2^\ddagger = X_Q/X^\ddagger = k_{\ddagger 2}/k_{\ddagger 2} \tag{128}$$

where n_t is the total number of moles in the system under consideration. Only one relaxation is observed for reaction 124 and according to 126 the rate law can be written as

$$(1/n_t)(dn_Q/dt) = k_{12} X_W^2 X_D - k_{21} X_Q \tag{129}$$

(As n_t changes with the progress of the reaction 124, $1/n_t \times dn_Q/dt$ is not equal to dX_Q/dt.) For a general way of deriving equation 129 see ref. 26.

Analogously, for reaction 123, the rate laws are

$$(1/n_t)(dn_Q/dt) = k_{12} X_W^2 X_D - k_{21} X_Q - k_{23} X_Q X_D + k_{32} X_Z \tag{130}$$

and

$$(1/n_t)(dn_Z/dt) = k_{23} X_Q X_D - k_{32} X_Z \tag{131}$$

If reaction $1 \rightleftarrows 2$ is fast compared to reaction $2 \rightleftarrows 3$, the relaxation times are

$$1/\tau_I = k_{12} X_W^2 + 4 k_{12} X_W X_D - 4 k_{21} X_W^2 X_D + k_{21} \tag{132}$$

and

$$1/\tau_{II} = k_{32} + k_{23} X_D X_Q$$

$$\times \frac{4X_W + 4X_D + 4X_Q - 4X_D X_Q - 9X_W X_D - X_W X_Q}{X_W X_Q + 4X_Q X_D - 4X_W X_D X_Q + X_W X_D} \tag{133}$$

As *Figure 22* shows, τ_I is only about five times smaller than τ_{II}, i.e. the assumption made to derive equations 132 and 133 is not entirely valid. However, exact calculation indicates that the error involved is only ten per cent, which is about the experimental error.

The four rate constants and, thereby, the equilibrium constants obtained from the best fit of the measurements to equations 132 and 133 are $k_{12} = 8 \times 10^6$ sec^{-1}, $k_{21} = 1.6 \times 10^6$ sec^{-1}, $k_{23} = 0.9 \times 10^6$ sec^{-1}, $k_{32} = 0.6 \times 10^6$ sec^{-1}, $K_1 = 5$, and $K_2 = 1.5$. (Since the rate equations are written in mole fractions, the rate constants of the bi- and trimolecular reaction also have the dimension sec^{-1}.) From the temperature dependence of the rate constants the activation energies have been determined to be about 6 kcal mol^{-1} for all four steps.

Since ultrasound is an adiabatic pressure wave, it not only causes a periodic change in pressure but also in temperature. Taking that into consideration, in equation 48 the expression $(\Delta V)^2$ has to be replaced by $[\Delta V - (\alpha/\rho c_p)\Delta H]^2$ (cf. equation 5). $\alpha/\rho c_p$ depends strongly on the composition of the solution and so does $[\Delta V - (\alpha/\rho c_p)\Delta H]$. Therefore, it was not possible to calculate the volumes and enthalpies of reaction from the amplitude factors C_i. Assuming $\Delta H_i = 0$, the best fit of the C_i to the theoretical curve is obtained with $(\Delta V_{12})^2 = 1.5$ cm^6 mol^{-2} and $(\Delta V_{23})^2 = 0.25$ cm^6 mol^{-2}, and assuming $\Delta V_i = 0$ the best fit gives $(\Delta H_{12})^2 = 0.6$ kcal^2 mol^{-2} and $(\Delta H_{23})^2 = 0.1$ kcal^2 mol^{-2}. In *Figure 23* the curves calculated with these parameters are shown and the experimentally determined values of C_i indicate that both ΔV_i and ΔH_i contribute to sound absorption.

To summarize, in dioxane–water mixtures 2:1 and 2:2 water:dioxane complexes are formed. The reaction rates are too small to be diffusion controlled and the activation energies indicate that the complexes are relatively stable. Molecular models show that a possible structure of the 2:2 complex is one in which the planes of the dioxane molecules are parallel with a single water molecule attached to each end to form a total of four hydrogen bonds. In the 2:1 complex, a single water molecule is probably hydrogen bonded to an oxygen of each dioxane molecule.

V. Summary

An introduction is given to chemical relaxation spectrometry. The experimental techniques are described briefly and the methods of evaluating relaxation experiments are discussed. The kinetic and thermodynamic information which may be obtained is often not easily arrived at by other techniques. At the end of this chapter, four experimental investigations are reported which demonstrate the scope of the different techniques and the kinds of chemical reactions which can be studied conveniently by chemical relaxation methods.

VI. References

[1] Bechtler, A., K. G. Breitschwerdt and K. Tamm. *J. Chem. Phys.* **52**, 2975 (1970).
[2] Beitz, J. V., G. W. Flynn, D. H. Turner and N. Sutin. *J. Amer. Chem. Soc.* **92**, 4130 (1970).
[3] Bergmann, K., M. Eigen and L. De Maeyer. *Ber. Bunsenges. Phys. Chem.* **67**, 819 (1963).
[4] Caldin, E. F. and J. E. Crooks. *J. Sci. Instrum.* **44**, 449 (1967).
[5] Czerlinsky, G. and M. Eigen. *Z. Elektrochem.* **63**, 652 (1959).
[6] Davies, C. W. *Ion Association.* Butterworths: London (1962).
[7] Debye, P. *Trans. Electrochem. Soc.* **82**, 265 (1942).
[8] Debye, P. and F. W. Sears. *Proc. Nat. Acad. Sci. Wash.* **18**, 410 (1932).
[9] Eggers, F. *Acustica*, **19**, 323 (1967/68).
[10] Eggers, F. and K. Kustin. *Methods in Enzymology*, Vol. 16, Chap. 3. (Ed.: K. Kustin). Academic Press: New York (1969).
[11] Eigen, M. *Angew. Chemie.* **75**, 489 (1963).
[12] Eigen, M. *Nobel Symposium*, **5**, 245. (Ed.: S. Claesson). Interscience: New York (1968).
[13] Eigen, M. *Nobel Symposium*, **5**, 333. (Ed.: S. Claesson). Interscience: New York (1968).
[14] Eigen, M. and L. De Maeyer. *Z. Elektrochem.* **59**, 986 (1955).
[15] Eigen, M. and L. De Maeyer. *Technique of Organic Chemistry*, Vol. 8, Pt 2, Chap. 18. (Eds.: S. L. Friess, E. S. Lewis and A. Weissberger). Interscience: New York (1963).
[16] Eigen, M. and J. Schoen. *Z. Elektrochem.* **59**, 483 (1955).
[17] Eigen, M. and K. Tamm. *Z. Elektrochem.* **66**, 93 (1962).
[18] Eigen, M. and R. Winkler. To be published.
[19] Eigen, M., W. Kruse, G. Maass and L. De Maeyer. In *Progress in Reaction Kinetics*, Vol. II, p 285. (Ed.: G. Porter). Pergamon: New York (1964).
[20] Ertl, G. and H. Gerischer. *Z. Electrochem.* **65**, 629 (1961); **66**, 560 (1962).
[21] Firth, C. A., D. Hess and W. Knoche, submitted for publication in *Adv. Mol. Relax. Proc.*
[22] Forster, H. J. *Dissertation.* Braunschweig (1971).
[23] Gibson, Q. H. *Methods in Enzymology.* Vol. 16, Chap. 6. (Ed.: K. Kustin). Academic Press: New York (1969).
[24] Goldsack, D. E., R. E. Hurst and J. Love. *Analyt. Biochem.* **28**, 273 (1969).
[25] Gutfreund, H. *Methods in Enzymology.* Vol. 16, Chap. 7. (Ed.: K. Kustin). Academic Press: New York (1969).
[26] Hammes, G. G. and W. Knoche. *J. Chem. Phys.* **45**, 4041 (1966).

[27] Helisch, J. and W. Knoche. *Ber. Bunsenges. Phys. Chem.* **75**, 951 (1971).
[28] Hoffmann, G. W. *Rev. Sci. Instrum.* **42**, 1643 (1971).
[29] Hoffmann, H. and E. Yeager. *Rev. Sci. Instrum.* **39**, 1151 (1968).
[30] Hoffmann, H., E. Yeager and J. Stuehr. *Rev. Sci. Instrum.* **39**, 649 (1968).
[31] Hopmann, R. *Ber. Bunsenges. Phys. Chem.* **74**, 935 (1970).
[32] Ilgenfritz, G. *Dissertation.* Göttingen (1966).
[33] Johnson, C. S. *Advances in Magnetic Resonance.* Vol. 1, Chap. 2. (Ed.: J. Waugh). Academic Press: New York (1965).
[34] Jost, A. *Ber. Bunsenges. Phys. Chem.* **70**, 1057 (1966).
[35] Kalidas, C., W. Knoche and D. Papadopoulos. *Ber. Bunsenges. Phys. Chem.* **75**, 106 (1971).
[36] Kirschner, K., E. Gallego, I. Schuster and D. Goodall. *J. Molec. Biol.* **58**, 29 (1971).
[37] Knoche, W. 'Technique of Chemistry', in *Investigations of Rates and Mechanisms*, Part II, Chap. V. (Eds.: A. Weissberger and G. G. Hammes), Interscience: New York (1973).
[38] Koffer, H. *Ber. Bunsenges. Phys. Chem.* **75**, 1245 (1971).
[39] Koshland, D. E., G. Nemethy and D. Filmer. *Biochemistry*, **5**, 365 (1966).
[40] Kuehn, C. and W. Knoche. *Trans. Faraday Soc.* **67**, 2101 (1971).
[41] Ljunggren, S. and O. Lamm. *Acta Chem. Scand.* **12**, 1834 (1958).
[42] Monod, J., J. Wyman and P. Changeux. *J. Molec. Biol.* **12**, 88 (1965).
[43] Noyes, M. and A. Weller. *Technique of Organic Chemistry*, Vol. 8, Pt 2, Chap. 16. (Eds.: S. L. Friess, E. S. Lewis and A. Weissberger). Interscience: New York (1963).
[44] Onsager, L. *J. Chem. Phys.* **2**, 599 (1934).
[45] Pohl, F. M. *Europ. J. Biochem.* **4**, 373 (1968).
[46] Porter, G. *Technique of Organic Chemistry*, Vol. 8, Pt 2, Chap. 19. (Eds.: S. L. Friess, E. S. Lewis and A. Weissberger). Interscience: New York (1963).
[47] Rigler, R., C.-R. Rabl and T. M. Jovin, submitted for publication in *Rev. Sci. Instrum.*
[48] Robinson, R. A. and R. H. Stokes. *J. Amer. Chem. Soc.* **76**, 1991 (1954).
[49] Roughton, G. W. and B. Chance. *Technique of Organic Chemistry*, Vol. 8, Pt 2, Chap. 14. (Eds.: S. L. Friess, E. S. Lewis and A. Weissberger). Interscience: New York (1963).
[50] Schwarz, G. *J. Phys. Chem.* **71**, 4021 (1967).
[51] Schwarz, G. and J. Seeling. *Biopolymers*, **6**, 1263 (1968).
[52] Strehlow, H. *Technique of Organic Chemistry*, 'Investigations of Rates and Mechanisms', Vol. 8, Pt 2, Chap. V (Eds.: S. L. Friess, E. S. Lewis and A. Weissberger). Interscience: New York (1963).
[53] Strehlow, H. and M. Becker. *Z. Elektrochem.* **63**, 457 (1959); ref. 41.
[54] Strehlow, H. and J. Jen. *Chem. Instrum.* **3**, 47 (1971).
[55] Strehlow, H. and S. Kalarickal. *Ber. Bunsenges. Phys. Chem.* **70**, 139 (1966).
[56] Swinehart, J. H. and G. W. Castellan. *Org. Chemistry*, **3**, 278 (1964).
[57] Weller, A. In *Progress in Reaction Kinetics*, Vol. I, Chap. 7. (Ed.: G. Porter). Pergamon: New York (1961).
[58] Wendt, H. *Ber. Bunsenges. Phys. Chem.* **70**, 556 (1966).
[59] Winkler, R. *Dissertation.* Göttingen-Wien (1969).

CHAPTER 23

Thermodynamic Properties from Shock Waves

A. LASCAR and A. RAUCH

*Commissariat à l'Energie Atomique,
Centre d'Etudes de Vaujours, B.P.7, 93—Sevran, France*

Contents

I.	Introduction	1110
II.	Theoretical Aspects	1111
	1. The Basic Relations	1111
	(A) Acoustic Waves	1111
	(B) Shock Waves	1113
	(a) Conservation of Mass	1114
	(b) Conservation of Momentum	1114
	(c) Conservation of Energy	1114
	2. Properties of the Hugoniot Curve—Stability of Shock Waves	1116
	(A) Definition of the Hugoniot Curve	1116
	(B) Weak Shock Waves	1116
	(C) Shocks of Any Magnitude—General Properties	1119
	(a) Behavior of the Entropy	1120
	(b) Behavior of the Specific Volume	1121
	(c) Behavior of the Energy	1121
	(d) Behavior of the Pressure	1122
	(e) Behavior of the Velocities	1122
	(D) Phase Changes	1123
	3. Reflection of Shock Waves and Rarefaction Waves	1125
	(A) Evolution of the State of the Medium after a First Shock	1138
	Compression	1126
	(a) Representation in the (p, v) Plane. Rarefaction Waves	1126
	Second Shock Wave	1127
	(b) Representation in the (p, u_p) Plane. Rarefaction Waves	1128
	Reflected Shock Waves	1130
	(B) Transmission of Shock Waves at the Interface of Two Media	1130
	(C) Rarefaction Waves	1132
III.	Experimental Techniques	1132
	1. Shock Wave Generators	1133
	(A) Explosive Charge Generators	1133
	The Plane Wave Generator	1133
	Shock Accumulation	1135
	(B) Guns	1136
	Compressed Gas Guns	1137
	Powder Guns	1137
	Guns Driven by Light Gases	1137
	2. Methods of Measurement	1138
	(A) Measuring Velocities	1138
	Optical Method	
	Electronic Methods	1140
	(B) Direct Measurements	1142
	Pressure Measurements: Manganin Wire Transducer	1142
	Quartz Transducer	1143

	Direct Measurement of the Material Velocity Behind a Shock Wave	1144
	Direct Measurement of Density Behind a Shock Wave	1144
	Direct Measurement of Temperature	1145
IV.	Equation of State for Liquids at Very High Pressures	1146
	1. Experimental Results	1146
	(A) The Hugoniot Curve	1146
	Organic Liquids	1149
	Liquefied Gases	1150
	(B) Other Shock Wave Measurements	1150
	2. Theoretical Models of the Equation of State	1152
	(A) The Thomas–Fermi or Thomas–Fermi–Dirac Model	1152
	(B) Cell Models	1153
	(C) The Mie–Grüneisen Equation of State	1155
	(D) The Equation of State of Water by Rice and Walsh	1156
V.	References	1157

I. Introduction

Shock wave techniques allow the investigation of properties of matter at very high pressures, which cannot be reached by any other experimental means. While static means permit pressures up to a few hundred kilobars†, one can produce shock waves behind which pressures rise to several tens of megabars in the case of solids; this corresponds to energies of the order of one megajoule per cm^3, and this for durations of the order of a microsecond.

The results obtained by this technique pertain to several disciplines. In solid state physics they contribute information on electrical, optical, magnetic or mechanical properties at very high pressures. More specifically, investigation of the electronic structure of crystalline lattices become possible; thus one can observe the transition of insulators or semiconductors to the metallic state. Pressures like those existing inside the earth (1.4 Mb at the interface core/mantle, 3.5 Mb in the center) can experimentally only be attained by shock waves. Therefore the importance of this technique for geophysics can be clearly seen, especially for the identification of the state of those materials presumed to exist in the interior of our globe, and to determine their mechanical and magnetic properties and the speed of acoustic waves. For all these studies we refer to review articles on shock waves in solids[1, 17, 18, 27, 28, 34, 45, 52, 64]. The history of the physics of shock waves is outlined in the introduction of Vodar and Kieffer[60].

The main area of application of shock waves, however, is thermodynamics, especially the study of the equation of state of substances under very high pressure. The method is particularly suitable for these studies, while by using velocity measurements and conservation equations only, one can obtain information on the thermodynamic state of the material behind the shock wave.

More precisely, one obtains thus the specific energy e as a function of the pressure p and specific volume v, along a curve called the Hugoniot curve. These results are not sufficient to determine the equation of state but they allow us to verify theories or models and fix their parameters. Related to the

---

† 1 bar = 10^6 baryes (c.g.s.) $\simeq$ 1 atm; 1 kilobar = 1 kb = 10^3 bar; 1 megabar = 1 Mb = 10^6 bar.

equations of state, one can also determine the phase diagram at very high pressures in the region of (p, v) values near the Hugoniot. One can furthermore complete these results experimentally by measuring certain thermodynamic quantities as temperature or speed of sound.

Most of the work has been done with solids, especially metals. However, the techniques developed have also been applied to fluids and it is on those that we will focus our attention in this article, leaving aside what pertains specificallly to solids, especially phenomena connected with rigidity.

In the next section we will recall the theoretical foundations of the method. We will then review the principal experimental methods that have been applied. Finally, in the last section, we give a bibliography of studies about thermodynamic properties of fluids at very high pressures, obtained by way of shock waves.

II. Theoretical Aspects

1. The Basic Relations

It is known that in a fluid there can exist surfaces of discontinuity of the pressure p, the specific volume $v = 1/\rho$ (ρ denotes mass per unit volume) and of the material velocity u_p. The study of these discontinuity surfaces, called shock waves, is the subject of this chapter. In order to understand the problems with respect to their stability it is useful to recall first some results about the propagation of infinitely small perturbations, acoustic waves, which form a limiting case. In what follows we consider only plane waves; the medium in which they travel is thus supposed to be infinite, and the thermodynamic or kinematic quantities describing it depend only on one single space coordinate x and on the time t.

(A) Acoustic Waves

Let us consider a wave (*Figure 1*) behind which the pressure p, the specific volume v and the material velocity u_p increase by infinitely small amounts

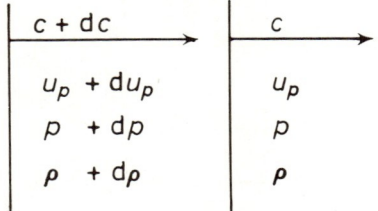

Figure 1. Infinitely small waves.

dp, dv and du_p. This wave has the following properties:

(i) it is adiabatic, for a perfect fluid at least, i.e. it does not change the entropy of the fluid;

(ii) it travels through the fluid with the speed of sound c, given by

$$c = (\partial p/\partial \rho)_s^{\frac{1}{2}} = v(-\partial p/\partial v)_s^{\frac{1}{2}} \qquad (1)$$

(iii) the changes in p, ρ and u_p are related by

$$dp = (c/v)\,du_p = \rho c\,du_p \tag{2}$$

or the equivalent relation

$$du_p = cd\rho/\rho = -cd\,v/v$$

Relation 2 shows immediately that for a compression wave, $dp > 0$, and the medium is accelerated in the direction of travel of the wave. For a rarefaction wave it is accelerated in the opposite direction.

Let us now consider a second wave following the first one. It travels with respect to the medium at a velocity $c + dc$. The velocity increase, dc, is related to the change of specific volume of the fluid. Taking the logarithmic derivative of equation 1 we get

$$\frac{dc}{c} = \frac{dv}{v} + \frac{1}{2}\left[\frac{-(\partial^2 p/\partial v^2)_s\,dv}{-(\partial p/\partial v)_s}\right]$$

or, by using equations 1 and 2

$$dc = -du - \frac{1}{2}\frac{v^2}{c}\left(\frac{\partial^2 p}{\partial v^2}\right)_s dv \tag{3}$$

Note the appearance here of the quantity $(\partial^2 p/\partial v^2)_s$ which plays a very important role in shock wave theory, as we will see later. This quantity is positive† for all substances, except in the case of a phase change, which we shall not discuss for the time being. If the first wave is a compression wave, $dv < 0$, and equation 3 tells us that this wave will be overtaken by another acoustic

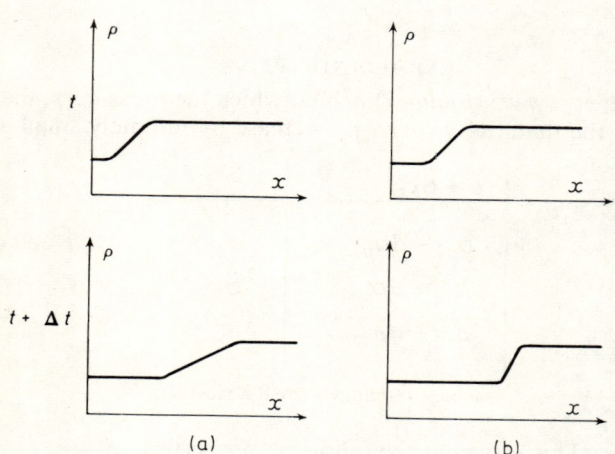

Figure 2. Evolution of density in a wave train: (a) Rarefaction waves; (b) Compression waves.

---

† It should be noted, however, that no principle of thermodynamics requires $(\partial^2 p/\partial v^2)_s$ to be positive.

wave coming up from behind. If the first wave is a rarefaction wave, the distance separating it from a following one will increase with time.

Now consider the case of a succession of waves, a wave train. The previous reasoning shows that a train of rarefaction waves will spread out in the course of time [*Figure 2*(a)] because their respective distances increase. But for a train of compression waves the reverse phenomenon takes place [*Figure 2*(b)]. Each wave gains on the previous one and the density gradient increases. When this gradient takes on too high a value, the fluid can no longer be considered to be perfect. Even though the coefficients of viscosity μ and thermal conductivity k may be very small, their effect cannot be neglected when the gradients of velocity and temperature become too large. Irreversible phenomena will begin to occur and the processes involved will stop being adiabatic. With increasing gradients the distance between the first and the last wave will tend to decrease towards a limiting value. This results in a permanent flow and the formation of a shock wave. The thickness of the shock wave depends on the values of μ and k. It will be such that the entropy increase generated by the irreversible phenomena is compatible with the conservation equations to be established in the next section. One has not been able to determine this thickness for the case of solids and of liquids. Different hypotheses have been suggested but the thickness appears to remain very small†, less than one micron[22, 5]. Moreover its magnitude does not enter into the basic formulas, as long as it is constant.

(B) Shock Waves

Suppose a shock wave travels through a fluid of specified volume v_1 and pressure p_1. Behind this wave the material is in the state p_2, v_2. For a compression wave one has $p_2 > p_1$ and $v_2 < v_1$. We will show later that in a medium not undergoing any phase changes, only shock waves of this type can exist. In this section, however, we do not make any restrictions in order to study the stability conditions of shock waves in general.

For an observer travelling with the wave the motion of the medium is stationary. In front of the wave the fluid moves with a speed u_1, behind it with

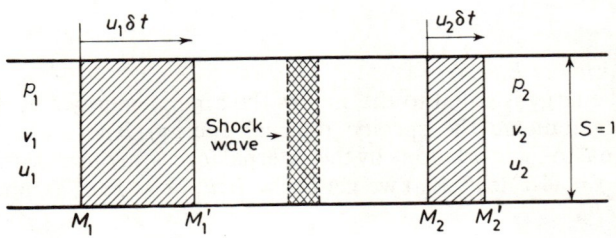

Figure 3. The fluid flows from left to right (u_1 and $u_2 > 0$). The initial state is therefore the state (p_1, v_1).

---

† Provided no phase change takes place on either side of the shock front. We return to them later.

a speed u_2. No supposition is made with respect to thickness or structure of the shock wave except for it to remain identical in the course of time. Let us consider a cylinder whose axis is perpendicular to the wave plane and whose cross section is of unit area (*Figure 3*). Two sections connected to the medium are situated at M_1 and M_2 at the time t, and at M'_1 and M'_2 at the time $t + \delta t$.

The fundamental equations of the shock wave are obtained by applying the conservation laws of mass, of momentum and of energy to the slice of material which evolves from $M_1 M_2$ (time t) to $M'_1 M'_2$ (time $t + \delta t$). Let G be the value of some extensive quantity (here either mass, momentum or energy) for the slice of material in question and let g be its value per unit volume. Then $G = \iiint_{slice} g \, d\tau$, where $d\tau$ is a volume element. Or since the cross section $S = 1$, $G = \int g \, dx$. The increase δG of G from t to $t + \delta t$ is

$$\delta G = \int_{M'_1}^{M'_2} g \, dx - \int_{M_1}^{M_2} g \, dx = -\int_{M_1}^{M'_1} g \, dx + \int_{M_2}^{M'_2} g \, dx$$

Thus

$$\delta G = g_2 u_2 \, \delta t - g_1 u_1 \, \delta t$$
$$\delta G / \delta t = g_2 u_2 - g_1 u_1 \qquad (4)$$

(a) *Conservation of Mass*

Take $G = M$ and $g = \rho = 1/v$. $\delta M = 0$, and therefore $u_1/v_1 = u_2/v_2$ (5)

(b) *Conservation of Momentum*

Take $G = Mu$, then $g = \rho u = u/v$. The increase of momentum per unit time is equal to the resultant of all external forces and thus $\delta G/\delta t = p_1 - p_2$. From equation 4

$$p_1 - p_2 = u_2^2/v_2 - u_1^2/v_1$$
$$p_2 + u_2^2/v_2 = p_1 + u_1^2/v_1 \qquad (6)$$

and, using equations 5, we can rewrite this as

$$-(p_2 - p_1)/(v_2 - v_1) = u_1^2/v_1^2 = u_2^2/v_2^2 \qquad (7)$$

(c) *Conservation of Energy*

The total energy is equal to the sum of the kinetic energy and the internal energy (let e indicate the specific internal energy). The increase of total energy is equal to the work done by the external forces, in our case the pressure. Thus, with $g = (1/v)(\tfrac{1}{2}u^2 + e)$ we get $\delta G = P_1 u_1 \, \delta t - P_2 u_2 \, \delta t$, and equation 4 gives

$$p_1 u_1 - p_2 u_2 = \frac{1}{2}\frac{u_2^3}{v_2} + \frac{e_2 u_2}{v_2} - \frac{1}{2}\frac{u_1^3}{v_1} - \frac{e_1 u_1}{v_1}$$

or, by using equation 5,

$$p_2 v_2 + e_2 + \tfrac{1}{2}u_2^2 = p_1 v_1 + e_1 + \tfrac{1}{2}u_1^2 \qquad (8)$$

THERMODYNAMIC PROPERTIES FROM SHOCK WAVES

Equations 7 and 8 lead to

$$e_2 - e_1 = \tfrac{1}{2}(p_1 + p_2)(v_1 - v_2) \quad (9)$$

This is the equation of Hugoniot. It is a very important one because it relates the thermodynamic states of the material in front of and behind the shock wave. Equations 5, 6 and 9 constitute together the fundamental shock wave equations.

In practice it is often more convenient to use a frame of reference fixed with respect to the laboratory. Then let u_s be the wave velocity in such a reference frame and u_p the velocity of the medium behind the wave. In front of the wave the velocity of the medium is of course zero (*Figure 4*).

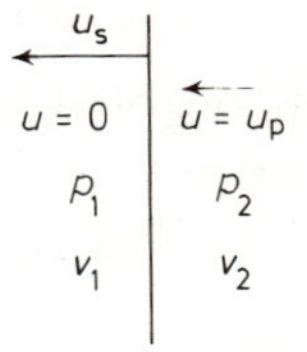

Figure 4. Shock wave in the laboratory coordinate system.

The shock wave equations are obtained by subjecting equations 5, 6 and 9 to a change of variables:

$$u_1 = -u_s \qquad u_2 = u_p - u_s$$

resulting in equations 10, 11 and 9 respectively:

$$(v_2/v_1) = (u_s - u_p)/u_s \quad (10)$$

$$p_2 - p_1 = (1/v_1)u_s u_p \quad (11)$$

$$e_2 - e_1 = \tfrac{1}{2}(p_2 + p_1)(v_1 - v_2) \quad (9)$$

Knowing the initial state p_1, v_1, e_1 and the velocities u_s and u_p, equations 9, 10 and 11 allow us to determine the final state p_2, v_2, e_2. The velocities are the most readily measurable ones of these quantities (cf. section III, 2A).

By inversion one obtains from equations 10 and 11:

$$u_p = [(p_2 - p_1)(v_1 - v_2)]^{\frac{1}{2}} \quad (12)$$

$$u_s = v_1[(p_2 - p_1)/(v_1 - v_2)]^{\frac{1}{2}} \quad (13)$$

A pressure p_2 of 100 kbar (about 10^5 atm) behind a shock wave can be considered as relatively small. If p_2 is the atmospheric pressure it is totally negligible compared with p_2. Equations 9 and 12 can then be combined to give

$$e_2 - e_1 = \tfrac{1}{2}u_p^2 = \tfrac{1}{2}p_2(v_1 - v_2)$$

In that case the work done by the wave, $p_2(v_1 - v_2)$, is equally divided between kinetic energy and internal energy.

2. Properties of the Hugoniot Curve—Stability of Shock Waves

(A) Definition of the Hugoniot Curve

Consider again equation 9

$$e_2 - e_1 = \tfrac{1}{2}(p_2 + p_1)(v_1 - v_2) \qquad (9)$$

This equation is symmetrical with respect to the states 1 (in front of the wave) and 2 (behind the wave) of the fluid. If one of these states is given, equation 9 provides a relation between the thermodynamic quantities e, p and v of the other state. For the sake of convenience we will from now on use the index 0 for the state which we consider to be given, and we will not 'index' quantities pertaining to the other state about which we seek information. Equation 9 then reads

$$e - e_0 = \tfrac{1}{2}(p + p_0)(v_0 - v) \qquad (14)$$

The equation of state of the fluid provides another relation

$$e = e(p, v) \qquad (15)$$

Combining equations 14 and 15 the implicit equation of the Hugoniot curve in the (p, v) plane is obtained

$$e(p, v) - e_0 = \tfrac{1}{2}(p + p_0)(v_0 - v)$$

This curve represents all states that can exist behind or in front of a shock wave, the other state being characterized by e_0, p_0, v_0 with, of course, $e_0 = e(p_0, v_0)$. We will first study the Hugoniot curve in the neighborhood of the point (p_0, v_0) and will then develop its general properties.

(B) Weak Shock Waves

For an infinitely small shock wave, first order development of equations 9, 11 and 13 gives:

$$de = -p\,dv \qquad (16)$$

$$u_s = v(-dp/dv)^{\frac{1}{2}} \qquad (17)$$

$$dp = +(1/v)u_s\,du_p \qquad (18)$$

Equation 16 expresses the notion that the process is adiabatic, because for an arbitrary process

$$de = -p\,dv + T\,ds$$

The speed of the wave according to equation 17 is the speed of sound

$$c = v(-\partial p/\partial v)_s^{\frac{1}{2}}$$

In equation 18 one can recognize equation 2, derived for acoustic waves. Thus an infinitely small shock wave is an acoustic wave.

The study of weak but finite shocks gives information on the Hugoniot curve in the neighborhood of the point (p_0, v_0). If we substitute $e - e_0 = \Delta e$, $v - v_0 = \Delta v$ and $p - p_0 = \Delta p$, equation 9 becomes

$$\Delta e = -(p_0 + \tfrac{1}{2}\Delta p)\Delta v \qquad (19)$$

THERMODYNAMIC PROPERTIES FROM SHOCK WAVES

We will develop the two sides of equation 19 in Taylor series. For that purpose we consider e and p as functions of specific volume v and specific entropy s; we will break off the development at these terms of the lowest degree in Δv and $\Delta s = s - s_0$ which are not identical for the two sides of equation 19. Keeping in mind the identities $(\partial e/\partial v)_s = -p$ and $(\partial e/\partial s)_v = T$ we get for the LHS of equation 19:

$$\Delta e = \left(\frac{\partial e}{\partial v}\right)_s \Delta v + \frac{1}{2}\left(\frac{\partial^2 e}{\partial v^2}\right)_s \Delta v^2 + \frac{1}{6}\left(\frac{\partial^3 e}{\partial v^3}\right)_s \Delta v^3 + \left(\frac{\partial e}{\partial s}\right)_v \Delta s + \ldots$$

$$\Delta e = -p_0 \Delta v - \frac{1}{2}\left(\frac{\partial p}{\partial v}\right)_s \Delta v^2 - \frac{1}{6}\left(\frac{\partial^2 p}{\partial v^2}\right)_s \Delta v^3 + T_0 \Delta s + \ldots$$

(the derivatives being taken at v_0, s_0) and for the RHS:

$$\Delta p = \left(\frac{\partial p}{\partial v}\right)_s \Delta v + \frac{1}{2}\left(\frac{\partial^2 p}{\partial v^2}\right)_s \Delta v^2 + \left(\frac{\partial p}{\partial s}\right)_v \Delta s + \ldots$$

$$-\left(p_0 + \frac{\Delta p}{2}\right)\Delta v = -p_0 \Delta v - \frac{1}{2}\left(\frac{\partial p}{\partial v}\right)_s \Delta v^2 - \frac{1}{4}\left(\frac{\partial^2 p}{\partial v^2}\right)_s \Delta v^3$$
$$- \frac{1}{2}\left(\frac{\partial p}{\partial s}\right)_v \Delta s \Delta v + \ldots$$

Equating, and retaining only non-cancelling terms of lowest degree, one gets

$$-\frac{1}{6}\left(\frac{\partial^2 p}{\partial v^2}\right)_s \Delta v^3 + T_0 \Delta s = -\frac{1}{4}\left(\frac{\partial^2 p}{\partial v^2}\right)_s \Delta v^3$$

$$\Delta s = -\frac{1}{12 T_0}\left(\frac{\partial^2 p}{\partial v^2}\right)_s \Delta v^3 \qquad (20)$$

For a compression wave $\Delta v < 0$. As the process takes place without heat exchange with the surroundings, the entropy change is due to internal irreversibility and must thus be positive, so $\Delta s > 0$. The necessary condition for the stability of compression waves is therefore

$$(\partial^2 p / \partial v^2)_s > 0 \qquad (I)$$

We encountered this inequality previously in Section II.1.A. It expressed the condition under which infinitely small compression waves will have the tendency to group together. It expressed therefore a condition for mechanical stability of compressional shock waves. Just now we found it back again as a thermodynamic stability condition of weak shock waves. We will see that for arbitrary shock this condition is not sufficient. Other conditions have to be added.

Condition I has been found to be true for all known substances, certain phase changes excluded. For rarefaction shock waves one would have $\Delta s < 0$, so these cannot exist†.

---

† We will later come across an exception, in the neighbourhood of certain phase changes.

Equation 20 shows in addition that the entropy changes with the third power of Δv, from which it follows that at the point (p_0, v_0) the Hugoniot curve and the isentrope through this point osculate. If p_H and p_s represent respectively the pressure along the Hugoniot curve and the isentrope through (p_0, v_0), we have close to this point

$$p_H(v_0 + \Delta v) - p_0 = \left(\frac{\partial p}{\partial v}\right)_s \Delta v + \frac{1}{2}\left(\frac{\partial^2 p}{\partial v^2}\right)_s \Delta v^2 + \frac{1}{6}\left(\frac{\partial^3 p}{\partial v^3}\right)_s \Delta v^3$$
$$+ \left(\frac{\partial p}{\partial v}\right)_v \Delta s + \ldots$$

and

$$p_s(v_0 + \Delta v) - p_0 = \left(\frac{\partial p}{\partial v}\right)_s \Delta v + \frac{1}{2}\left(\frac{\partial^2 p}{\partial v^2}\right)_s \Delta v^2 + \frac{1}{6}\left(\frac{\partial^3 p}{\partial v^3}\right)_s \Delta v^3 + \ldots$$

Replacing Δs by its value from equation 20 we get

$$p_H(v_0 + \Delta v) - p_s(v_0 + \Delta v) = -\frac{1}{12 T_0}\left(\frac{\partial p}{\partial s}\right)_v \left(\frac{\partial^2 p}{\partial v^2}\right)_s \Delta v^3 \qquad (21)$$

So, $p_H - p_s$ does indeed vary with the third power of Δv. The relative position of the two curves depends on the sign of $(\partial p/\partial s)_v$ which is the same as that of the thermal expansion coefficient at constant pressure $\beta = (1/v)(\partial v/\partial T)_p$. We can write

$$\left(\frac{\partial p}{\partial s}\right)_v = \left(\frac{\partial p}{\partial T}\right)_v \left(\frac{\partial T}{\partial s}\right)_v$$

and

$$\left(\frac{\partial p}{\partial T}\right)_v = -\left(\frac{\partial v}{\partial T}\right)_p \left(\frac{\partial p}{\partial v}\right)_T$$

from which

$$\left(\frac{\partial p}{\partial s}\right)_v = -\left(\frac{\partial T}{\partial s}\right)_v \left(\frac{\partial p}{\partial v}\right)_T \left(\frac{\partial v}{\partial T}\right)_p$$

Now

$$\left(\frac{\partial T}{\partial s}\right)_v = \frac{cv}{T} > 0 \quad \text{and} \quad \left(\frac{\partial p}{\partial v}\right)_T < 0$$

both being established thermodynamic inequalities. Therefore $(\partial p/\partial s)_v$ and $\beta = (1/v)(\partial v/\partial T)_p$ have the same sign. In general, β is positive, $p_H - p_s$ has the opposite sign of Δv and the Hugoniot curve lies above the isentrope for $\Delta v < 0$, below it for $\Delta v > 0$ (*Figure 5*). Still, one knows that β can be negative, e.g. for water at atmospheric pressure between 0 and 4°C. In that case the respective positions of the Hugoniot curve and the isentrope are reversed. There are thus shock waves possible that cool the fluid or where the temperature remains constant. This case has been studied in a paper by Thomsen[56].

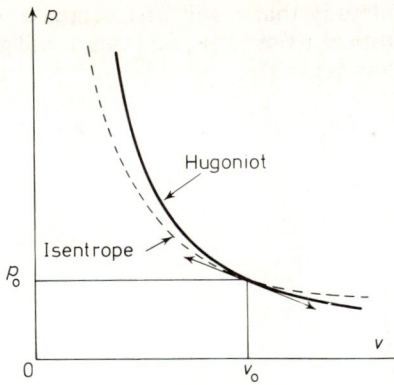

Figure 5. Relative positions of the Hugoniot curve and the isentrope, for a normal fluid ($\beta > 0$).

(C) Shocks of Any Magnitude—General Properties

In the preceding section we have studied the behavior of the Hugoniot curve near the point (p_0, v_0). Let us now study its asymptotic behavior for very intense shock waves. Behind such shocks the specific energy of the material is very high. The molecules are dissociated and the atoms will be at least partially ionized. The fluid is then composed of particles in Coulomb interaction, and one can apply the virial theorem.

$$3pv = 2e_{kin} + e_{pot} \tag{22}$$

or

$$3pv = 2e - e_{pot} \tag{23}$$

where e_{kin} is the kinetic and e_{pot} the potential energy.

Let us substitute into equation 22 the value of the energy taken from equation 14 in which the known state (p_0, v_0) is considered to be the initial state. Because we discuss very intense shock waves, we may neglect p_0 with respect to p. One obtains then

$$e_{pot}/e = 2(v_0 - 4v)/(v_0 - v) \tag{24}$$

Because the material is compressed $e_{pot} > 0$, and therefore $v < \tfrac{1}{4}v_0$.

At very high energies, the totally ionized material tends to behave like an ideal gas[7] and thus $e_{pot}/e_{kin} \to 0$ and by equation 24 $v \to \tfrac{1}{4}v_0$. We have neglected energy in the form of radiation, which becomes dominant at very high temperatures, because it varies as T^4, whereas the energy of the material varies in proportion with T. There exists, however, a level at which the temperature is high enough for the atoms to be completely ionized, and still low enough for the radiative energy component to be small compared with the total energy of the material†.

When we speak of high specific energy, we refer to such a temperature level.

---

† For example in the case of water at $T = 3.4 \times 10^6 °C$, the atoms will be completely ionized, but the radiation energy is only about one thousandth of the kinetic energy.

We have shown already that v will first decrease to values smaller than $\tfrac{1}{4}v_0$ and then asymptotically rise to $\tfrac{1}{4}v_0$ and thus v will go through at least one minimum value, below $\tfrac{1}{4}v_0$.

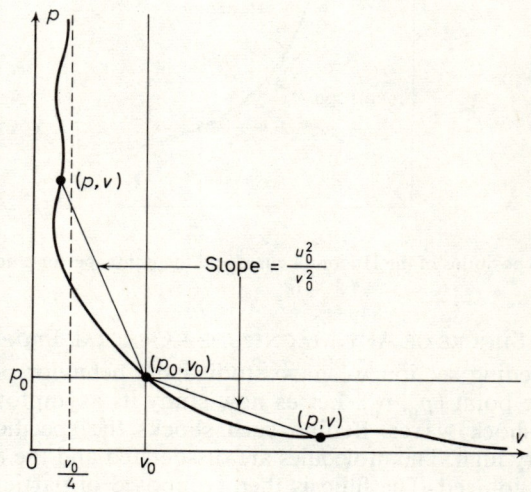

Figure 6. Hugoniot curve in the (p,v) plane.

The general appearance of the Hugoniot curve is shown in *Figure 6*. However, the first minimum of v already corresponds to shock waves too intense to be within experimental reach.

It would take too long here to establish in detail all the other properties of the Hugoniot curve and thus we will just indicate and briefly discuss them. The more demanding reader is referred to the article by Bethe[7] for complete derivations.

(a) Behavior of the Entropy

It can be shown that for $v < v_0$ one has $s > s_0$, and similarly for $v > v_0$ one has $s < s_0$, provided the conditions I and II are met (conditions which are sufficient but not necessary)

$$(\partial^2 p/\partial v^2)_s > 0 \qquad \text{(I)}$$

and

$$v(\partial p/\partial e)_v > -2 \qquad \text{(II)}†$$

---

† Landau and Lifshitz[15] merely demand that condition I must hold: meanwhile they admit that a certain quantity does not go to zero. For that reason Bethe's reasoning seems more complete.

We have already discussed condition I. Condition II has been verified for most materials. We can write, as we did at the end of B:

$$\left(\frac{\partial p}{\partial e}\right)_v = \left(\frac{\partial p}{\partial T}\right)_v \left(\frac{\partial e}{\partial T}\right)_v = c_v \left(\frac{\partial p}{\partial T}\right)_v$$

$$\left(\frac{\partial p}{\partial T}\right)_v = -\left(\frac{\partial v}{\partial T}\right)_P \left(\frac{\partial p}{\partial v}\right)_T$$

$(\partial p/\partial v)_T$ is negative and therefore $(\partial p/\partial e)_v$ has the same sign as $\beta = (1/v)(\partial v/\partial T)_P$, the thermal expansion coefficient at constant pressure. This coefficient is generally positive, a notable exception being water from 0° to 4°C at atmospheric pressure. But in that case $v(\partial p/\partial e)_v = -0.015$, a value much higher than that allowed by condition II.

In addition, and with the same hypotheses, one can demonstrate that there cannot be two different solutions for the shock equations with the same entropy s. This indicates that entropy is a 'suitable parameter' to characterize in an unequivocal way the intensity of the shock wave, which is true neither, as we have seen, for the specific volume nor, as we will see later, for the pressure. For states behind a shock wave, $v < v_0$, and the entropy may reach very high values. For states in front of a shock wave, the lower limit of s is either zero or a value set by the intersection of the Hugoniot curve and a phase change boundary.

(b) *Behavior of the Specific Volume*

For states of compression, $v < v_0$, we have seen that v goes through at least one minimum, below $\frac{1}{4}v_0$ and then approaches $\frac{1}{4}v_0$ from the lower side. For states in front of the wave, $v > v_0$, there is a continuous increase of v from v_0 to a finite value v_B, while the entropy decreases. The Hugoniot curve is shown in the (v, s) plot in *Figure 7*.

(c) *Behavior of the Energy*

With the help of the supplementary condition

$$(\partial p/\partial v)_e < 0 \tag{III}$$

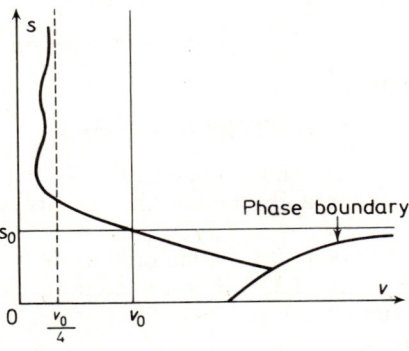

Figure 7. Hugoniot curve in the (s,v) plane.

one can show that the energy is a monotonical increasing function of the entropy s, for both states of compression, $v < v_0$, as well as for states in front of the shock, $v > v_0$. Condition III, in addition to the first two, and although it does not follow from any principle of thermodynamics, seems to hold for all homogeneous systems (no phase changes). An elaborate discussion on the subject can be found in Bethe's article.

(d) Behavior of the Pressure

For all compressed states, $v < v_0$, is the pressure of a monotonically increasing function of entropy s. On the other hand, for states of expansion, $v > v_0$, the behavior of p is not necessarily monotonic with respect to s or v. However, one always finds $p < p_0$. In *Figure 6* the Hugoniot has been drawn in the (p, v) plane.

(e) Behavior of the Velocities

Let us take equation 7 again, which applied to a reference frame moving with the shock wave and let us rewrite that expression in the notation specified at the beginning of the present chapter, namely

$$-(p - p_0)/(v - v_0) = u_0^2/v_0^2 \qquad (25)$$

Consider two nearly identical shock waves, such that the state (p_0, v_0) is the same for both, whereas the other state is either (p, v) or $(p + dp, v + dv)$. Let u_0 be the velocity of the substance in the state 'zero' with respect to the shock waves, which equals, apart from the sign, the velocity of the wave with respect to the medium. Differentiating equation 25 we get

$$d(u_0^2)((v - v_0)^2/v_0^2) = (v_0 - v)\,dp + (p - p_0)\,dv \qquad (26)$$

On the other hand, differentiation of equation 13, the equation for conservation of energy, gives

$$2\,de = -(p + p_0)\,dv + (v_0 - v)\,dp$$

For any process, we can write

$$T\,ds = de - p\,dv$$

This, combined with the previous relation, leads to

$$2T\,ds = (p - p_0)\,dv + (v_0 - v)\,dp$$

Then, according to equation 26,

$$d(u_0^2) = 2T\{v_0^2/(v - v_0)^2\}\,ds \qquad (27)$$

One sees that $|u_0|$ is an increasing function of s. For the compressed states s and p vary in the same way, and therefore the velocity is an increasing function of the pressure.

Let c_0 be the sound velocity for the material in the state (p_0, v_0), then for $s = s_0$ we have $u_0 = c_0$ (cf. II.2.A). For the compressed states $s > s_0$, and then, according to equation 27, $|u_0| > c_0$. For states in front of the shock wave $s < s_0$, and thus $|u_0| < c_0$.

The following important statement can therefore be made: a shock wave is supersonic with respect to the material in front of it and subsonic with respect to the material behind it.

A shock wave, then, will be overtaken by any wave following it; a shock wave is in effect subsonic with respect to the material it leaves behind while in the same material each wave moves with a velocity equal to or higher than the speed of sound.

A geometrical interpretation can be given of the results obtained here. Consider, in *Figure 6*, the lines connecting the states (p_0, v_0) and (p, v) (Rayleigh lines). According to equation 25 the slope of these lines represents, apart from a factor $(-1/v_0^2)$, the velocity of the shock wave with respect to the material in the state 0. Multiplied by the same factor, the speed of sound at (p_0, v_0) is equal to the slope of the tangent to the Hugoniot curve at this point, which is also the tangent to the isentrope. Thus one finds graphically the rules stated above. We may note that the Rayleigh lines can intersect the Hugoniot curve at one point only, because we have seen that their slope, in absolute value, is a continuously increasing function of s.

(D) Phase Changes

When phase changes take place the inequalities I, II and III, on which all previously established properties were based, cease to be valid. Consider the phase diagram represented in *Figure 8*, which holds for a first order phase change like melting or liquefaction. One can show that at the boundary in the (p, v) diagram between a two-phase and a single-phase region, the quantity $-(\partial p/\partial v)_s$ is always larger on the single-phase side. At point A (*Figure 8*) one then has $(\partial^2 p/\partial v^2)_s = -\infty$; a shock wave is not stable anymore in that region (II.2.B). Experience shows that shock wave splitting will then take place according to a process that we will describe later.

In that case one means by 'Hugoniot' usually the family of points representing the states that can exist behind the last shock wave. As we have seen, the entropy increase Δs from a shock wave is proportional to Δv^3. For not too strong shock waves the entropy varies little and the Hugoniot is close to the

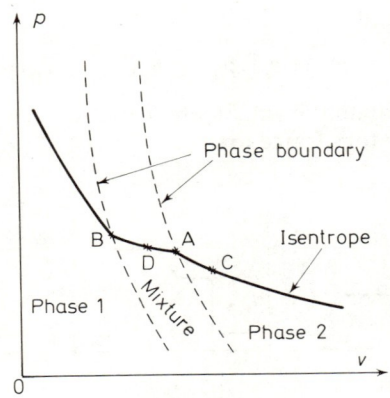

Figure 8. Phase diagram.

isentrope or at least the two curves look much alike. Such a curve is shown in *Figure 9*. Consider the state D, above the point C that indicates the beginning of the phase change; if that state could be reached by a single shock, starting from the state A, (p_A, v_A), the velocity of this shock would be $v_A[(p_D - p_A)/(v_A - v_D)]^{\frac{1}{2}}$, whereas the shock wave bringing the substance in the state C corresponding to a lower pressure, would have a higher

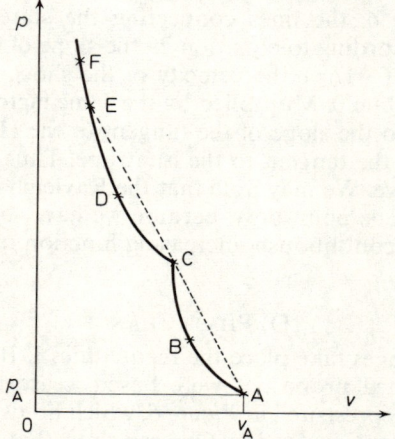

Figure 9. Hugoniot curve, with a phase change.

velocity, $v_A[(p_C - p_A)/(v_A - v_C)]^{\frac{1}{2}}$. One sees readily that this situation is impossible. Experience shows that a two-wave structure will arise. A first shock wave brings the material in the state C, without phase change; then follows another shock wave leading from state C to state D. This second shock brings about the phase change.

With respect to the material in state C, i.e. between the two waves, the velocity of the first shock wave is, according to equation 7

$$u'_{s_1} = v_C[(p_C - p_A)/(v_A - v_C)]^{\frac{1}{2}}$$

and that of the second

$$u'_{s_2} = v_C[(p_D - p_C)/(v_C - v_D)]^{\frac{1}{2}}$$

and one sees immediately from *Figure 9* that $u'_{s_2} < u'_{s_1}$. The velocities with respect to the laboratory frame are

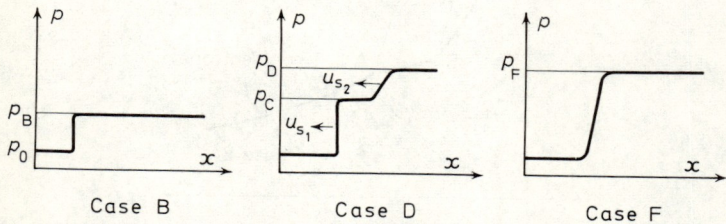

Figure 10. Profile of a shock wave for the three final states B, D and F of *Figure 9*.

$$u_{s_1} = v_A \left[\frac{p_C - p_A}{v_A - v_C}\right]^{\frac{1}{2}} = v_C \left[\frac{p_C - p_A}{v_A - v_C}\right]^{\frac{1}{2}} + u_{p_C}$$

and

$$u_{s_2} = v_C \left[\frac{p_D - p_C}{v_C - v_D}\right]^{\frac{1}{2}} + u_{p_C}$$

where u_{p_C} is the material velocity between the two waves.

For state D the specific energy is

$$e_D - e_0 = (e_D - e_C) + (e_C - e_0)$$
$$= \tfrac{1}{2}(p_D + p_C)(v_C - v_D) + \tfrac{1}{2}(p_C + p_0)(v_0 - v_C)$$

For states F, for which $(p_F - p_A)/(v_A - v_F) > (p_C - p_A)/(v_A - v_C)$, it is shown by experience that there is only a single shock wave with a phase change.

In solids, at pressures low enough for the rigidity of the material not to be negligible, one also observes a splitting of the shock wave. An elastic shock wave which brings the material into a state of anisotropic strain precedes a plastic shock wave. A study of these phenomena can be found elsewhere[45, 65].

For the three final states B, D and F of *Figure 9* the character of the shock wave is shown in *Figure 10* by plotting the pressure versus x. A shock wave accompanied by a phase change has probably a greater thickness than a common shock wave†; this thickness is related to the relaxation time of the transition.

The preceding considerations are in fact based on the assumption that the material immediately behind the shock wave is in thermodynamic equilibrium, which is not always so; metastable phases can exist and the transition of these into stable phases can only take place with a certain delay, depending on the pressure rise. The angular shape of the curve at C (*Figure 9*) may then be rounded off, changing it to concave downwards with $d^2p/dv^2 < 0$. Duvall et al.[21] have studied such phenomena.

Let us finally note the existence of rarefaction shock waves; in a medium for which the isentropes would be concave downwards, i.e. $(\partial^2 p/\partial v^2)_s < 0$, rarefaction shock waves would be stable because they would correspond to $\Delta s > 0$ (cf. II.2.B). For a phase change, say near state A of *Figure 8*, the situation is somewhat analogous, because there are segments (CD) joining two points of the curve and completely below it; such a configuration makes the appearance of a rarefaction shock wave possible. The formation mechanism of such waves is rather complicated, and the reader is referred to the book by Zel'dovitch and Raizer[65] for a complete discussion. The existence of rarefaction shock waves has been experimentally established for solid–solid transitions[30].

3. Reflection of Shock Waves and Rarefraction Waves

The pressures generated behind shock waves will not remain constant for more than a few microseconds at most. After that the material will be affected

---

† This has in reality been observed in the case of solid–solid transitions.

by reflected waves or rarefaction waves, due to the finite dimensions of the samples (of the order of 1 mm to 10 cm). The study of these waves, following a first shock wave, and of the changes which they induce in the state of the material, will be considered now.

(A) Evolution of the State of the Medium after a First Shock Compression

(a) Representation in the (p, v) Plane

Consider a material, brought into state 2 (*Figure 11*) by shock compression. The state 2 lies on the Hugoniot H_1 corresponding to the initial state 1. From this state 2 the material can evolve towards lower pressures if it is subjected to rarefaction waves, or towards higher pressures if subjected to a second shock wave.

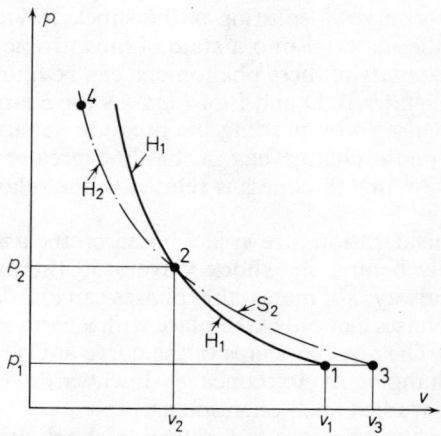

Figure 11. Locus of the points which can be reached after a first shock compression (point 2).

Rarefaction waves. Such waves are known to be isentropic and thus the state of the material will evolve downwards along the isentrope S_2 through 2, *Figure 11*. The position of S_2 with respect to H_1 depends on the sign of

$$\left(\frac{\partial v}{\partial s}\right)_P = \left(\frac{\partial v}{\partial T}\right)_P \left(\frac{\partial T}{\partial s}\right)_P$$

$$\left(\frac{\partial T}{\partial s}\right)_P = \frac{C_p}{T} > 0$$

and thus $(\partial v/\partial s)_P$ and $(\partial v/\partial T)_P$ have the same sign. For most fluids $\beta = (1/v)(\partial v/\partial T)_P$ is positive (cf. II.2.B). Then the rarefaction isentrope S_2 lies above the Hugoniot H_1.

According to the basic relation:

$$e_2 - e_1 = \tfrac{1}{2}(p_2 + p_1)(v_1 - v_2) \qquad (9)$$

the internal energy change during the first shock wave is given by the area of the trapezium 1 2 2' 1' in *Figure 12*. During the expansion from state 2 to state 3, $ds = 0$ and thus $de = -p\,dv$ which means that the energy change in absolute value is equal to the area of the curved trapezium 2 3 3' 2'.

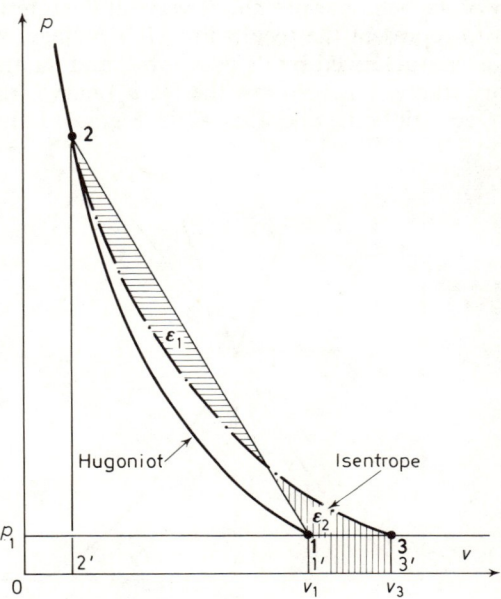

Figure 12. Determination of the energy trapped in the material after the passage of a shock wave followed by an isentropic release down to initial pressure.

From state 1 to state 3 the energy of the material has increased by an amount represented by the difference between the horizontally barred area ε_1, and the vertically barred ε_2. This energy, $\varepsilon_1 - \varepsilon_2$ has served to heat the material of which the final volume v_3 is larger than the initial volume.

In the case of vaporization during the expansion, when the volume of the gas formed is much greater than the initial volume of the liquid, the isentrope 2–3 (*Figure 12*) runs very differently from the Hugoniot curve. The value of ε_2 increases, as it corresponds to the work done against the exterior pressure: $\varepsilon_2 - \int_{v_1}^{v_3} p\,dv$. The energy $\varepsilon_1 - \varepsilon_2$ not only heats but also evaporates the fluid. One must therefore have $\varepsilon_1 - \varepsilon_2 > Q$, with Q being the heat of vaporization. In the textbook of Zel'dovitch and Raizer[65] a more precise discussion of a vaporization criterion can be found.

Second Shock Wave. Once in state 2, the material can be subjected to a second shock wave. The locus of its possible states is then the Hugoniot H_2, corresponding to the initial state 2. We have seen that H_2 and S_2 osculate in 2. For normal fluids $(\partial v/\partial T)_p > 0$ and thus H_2 (*Figure 11*) has a lesser slope than H_1. The complete locus of possible states when starting with state 2 consists of the two curves H_2 ($p > p_2$) and S_2 ($p < p_2$). Taken as one curve it is continuous in point 2, and so are its first two derivatives; the third derivative has a discontinuity (cf. II.2.B).

(b) Representation in the (p, u_p) Plane

For the study of the behavior of a shock wave at the interface of two media, it is convenient to represent the results in a (P, u_p) diagram. where the state of the material is characterized by its pressure P and its material velocity u_p in the laboratory frame. One obtains the (P, u_p) curve from the Hugoniot with the help of equations 12 and 13; 1–2 in *Figure 13* shows such a curve

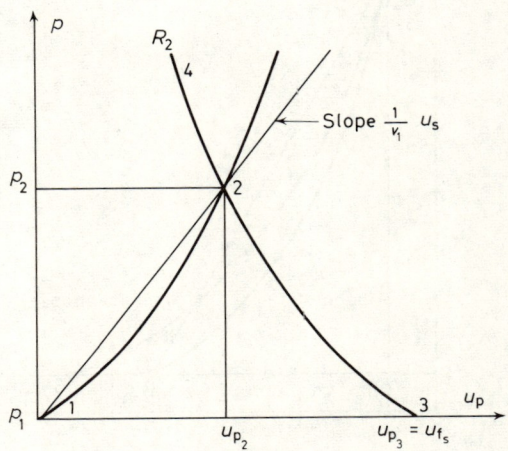

Figure 13. Pressure/velocity diagram.

schematically. The (p, u_p) curves of some liquids have been drawn in *Figure 14* together with, for comparison, the curves for some metals. Equation 11 reads $p_2 - p_1 = (1/v_1) u_s u_p$, and thus the slope of the line joining state 1 and state 2 is $(1/v_1) u_s$. This quantity is sometimes called 'shock impedance'; its value depends on the intensity of the shock wave.

Let us consider in the (p, u_p) diagram the locus of possible states after a first shock compression, i.e. starting with state 2.

Rarefaction Waves. Consider rarefactions coming from ahead, that is to say in the direction opposite to the acquired material velocity. Such rarefactions will originate when the shock wave reaches a free surface; the reasoning of this section applies uniquely to such waves. As we have seen above in II.1.A, these waves will accelerate the material in the opposite direction from that in which the wave is propagated, thus in the direction of the already acquired velocity u_n. Equation 2 must then be written as $du_p = c\, dv/v = -(v/c)\, dp$ with the understanding that $c > 0$ and $u_p > 0$. Along the rarefaction isentrope (2–3 in *Figure 13*) we can replace c by v $(-\partial P/\partial v)_s^{\frac{1}{2}}$ and we get

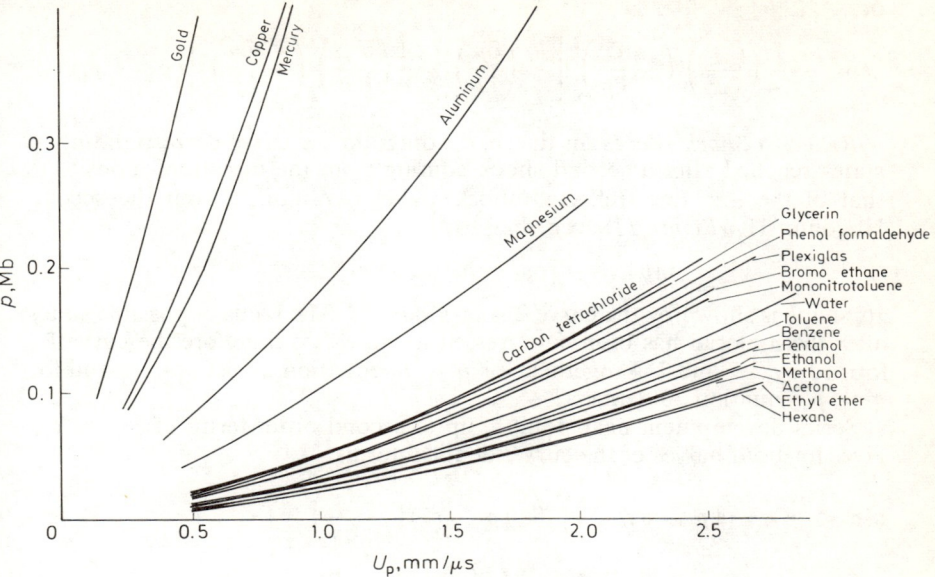

Figure 14. Pressure/velocity diagram for some liquids and metals.

$$du_p = -(-\partial v/\partial p)_s^{\frac{1}{2}} dp \tag{28}$$

The equation for the rarefaction curve is then given by

$$u(p) = u_{P_2} - \int_{P_2}^{p} \left(-\frac{\partial v}{\partial p}\right)_s^{\frac{1}{2}} dp = u_{P_2} + \int_{p}^{P_2} \left(-\frac{\partial v}{\partial p}\right)_s^{\frac{1}{2}} dp \tag{29}$$

For expansion down to the initial pressure p_1 the material velocity will be equal to the velocity u_{fs} of the free surface, or

$$u_{fs} = u_{P_2} + \int_{P_1}^{P_2} \left(-\frac{\partial v}{\partial p}\right)_s^{\frac{1}{2}} dp = u_{P_2} + u_R$$

The integral

$$u_R = \int_{P_1}^{P_2} \left(-\frac{\partial v}{\partial p}\right)_s^{\frac{1}{2}} dp$$

is the Riemann integral.

For shocks of medium intensity the quantity $(v_1 - v_2)$ is small: one can show that the shock curve 1–2 (*Figure 13*) and the expansion isentrope 2–3 are symmetrical with respect to the vertical through 2, up to and including quadratic terms in $(v_1 - v_2)$. This means that $u_{P_2} \simeq u_R$ and thus $u_{fs} \simeq 2u_{P_2}$, still up to quadratic terms in $(v_1 - v_2)$. If we write $u_{fs} = 2u_{P_2}(1 + \varepsilon)$ then calculation shows

$$\varepsilon = -\frac{1}{24}\left(\frac{\partial^2 p}{\partial v^2}\right)_s \left(-\frac{\partial v}{\partial p}\right)_s \left[-\left(\frac{\partial p}{\partial e}\right)_s + \frac{1}{4}\left(\frac{\partial^2 p}{\partial v^2}\right)_s \left(-\frac{\partial v}{\partial p}\right)_s\right](v_1 - v_2)^2$$

or

$$\varepsilon = -\frac{1}{24}\left(\frac{\partial^2 p}{\partial v^2}\right)_s \left(-\frac{\partial v}{\partial p}\right)_s^3 \left[-\left(\frac{\partial p}{\partial e}\right)_s + \frac{1}{4}\left(\frac{\partial^2 p}{\partial v^2}\right)_s \left(-\frac{\partial v}{\partial p}\right)_s\right](p_2 - p_1)^2$$

Reflected Shock Waves. In the (p, u_p) diagram the curve corresponding to states reached after a second shock coming from the direction opposite to that of the first one (reflected shock wave) is obtained from the second Hugoniot H_2 (*Figure 11*) according to

$$u(p) = -[(p - p_2)(v_2 - v)]^{\frac{1}{2}} + u_{p_2} \tag{30}$$

Its shape is shown by the curve 2–4 in *Figure 13*. The locus of possible states after the material has been compressed to state 2, is therefore the curve R_2 formed by 2–3 and 2–4 together; for $p < p_2$ equation 29 defines 2–3 and for $p > p_2$ equation 30 defines 2–4.

Series development near point 2, up to second order terms in $(P_1 - P_2)$, gives for both halves of the curve identical terms

$$u(p < p_2) = u(p > p_2) = u_{p_2} - (p - p_2)\left(-\frac{\partial v}{\partial p}\right)_{s,2}^{\frac{1}{2}}$$
$$+ \frac{1}{4}(p - p_2)^2 \left(\frac{\partial^2 v}{\partial p^2}\right)_{s,2} \left(-\frac{\partial v}{\partial p}\right)_{s,2}^{-\frac{1}{2}}$$

This means that R_2 and its first two derivatives are continuous at point 2. The third derivative has a discontinuity.

(B) Transmission of Shock Waves at the Interface of Two Media

Here we want to discuss what happens when a shock wave arrives at the interface between two media A and B (*Figure 15*); the shock front and the interface are supposed to be parallel. After the shock wave has passed, the two media A and B must, on both sides of the interface, have the same pressure and the same material velocity; it is convenient therefore to make use of the (p, u_p) diagram.

Let us consider a shock wave travelling through medium A, with the material behind the wave in the state 2 (*Figures 15* and *16*). When the wave

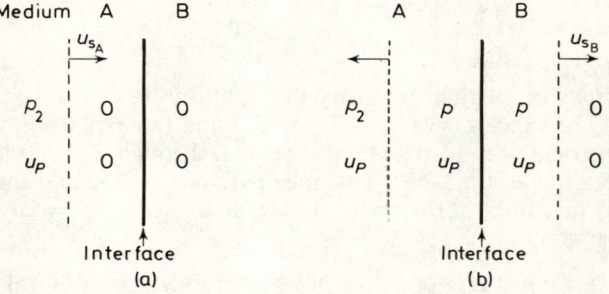

Figure 15. Configuration (a) before, and (b) after the wave has reached the interface.

reaches the interface a shock is transmitted into medium B. The locus of possible states for that medium is its shock curve P_B, whereas the locus of states in A is the curve R_2 (cf. II, 3, A). The pressure and the material velocity u_p at the interface are thus given by the intersection of P_B and R_2 (point 3 in *Figure 16*). There are now two possibilities. If 3 lies above 2 (e.g. 3' on P'_B) then a shock wave will be reflected into medium A. In that case material B is

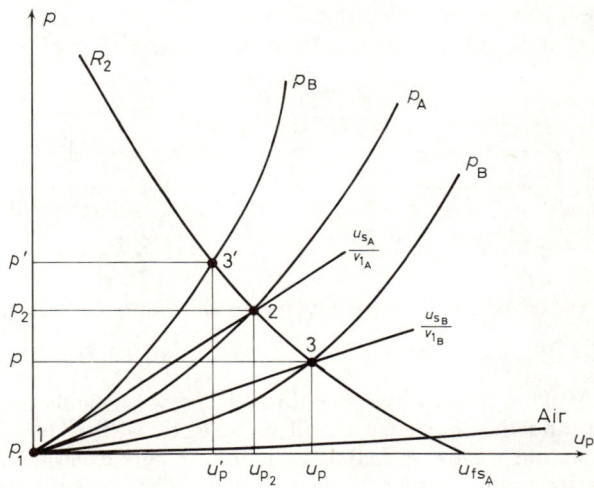

Figure 16. Transmission of a shock wave at the interface of two materials A and B.

said to have a higher shock impedance than material A. In the other case (3 below 2) rarefaction waves will travel back through medium A. If medium B is replaced by a vacuum, the curve p_B has to be replaced by $p_1 = 0$, i.e. the u_p axis, and the velocity acquired by the surface of A is the free surface velocity u_{fsA}. If the medium B is air, one usually treats this as if it was a vacuum, thereby making a small error only because the shock curve for air is so close to the u_p axis (*Figure 16*).

Returning to the general case, we have seen that the slope of the line 1–2, joining states 1 and 2, equals $(1/v_{1A})u_{sA}$ and the slope of 1–3 equals $(1/v_{1B})u_{sB}$, where v_{1A} and v_{1B} are the specific volumes of A and B in state 1 and u_{sA} and u_{sB} are the shock velocities. If one knows the equation of state of material A, and thus its shock curve and all curves R_2, then it is sufficient to measure u_{sA} and u_{sB} to determine states 2 and 3. First, point 2 is determined by the intersection of p_A and the line of slope u_{sA}/v_{1A} through point 1; subsequently point 3 is found as the intersection of R_2 and the line of slope u_{sB}/v_{1B} going through 1. Instead of u_{sA} one can measure the free surface velocity u_{fsA}. This leads to results of higher precision for the case of weak shock waves. This method of using a reference substance with a known equation of state, is called the 'impedance match method'.

(C) Rarefaction Waves

Thus far we have only considered those rarefaction waves which are created when a shock wave comes to the end of a sample, in contact with the atmosphere or with another material of lower shock impedance. But rarefaction waves can be of different origin. They can come from the other end of the sample, in which case they travel in the same direction as the shock wave and will overtake and attenuate it. They can also originate at the lateral faces of the sample; in that case they cause perturbations or 'lateral unloading' which we will consider in more detail.

Suppose that at the time 0 face AMB of a sample (*Figure 17*) is subjected

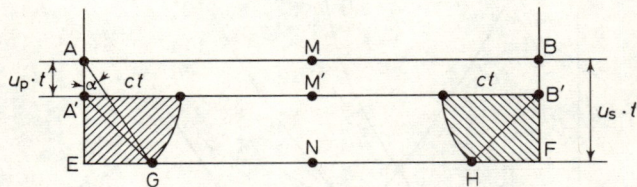

Figure 17. The shaded area corresponds to the zone reached by the lateral release waves.

to a perfectly plane shock wave. At a time t, plane AMB will be at A'M'B' with $MM' = u_p t$, and the shock wave will be at ENF with $MN = u_s t$. The first lateral rarefaction waves, started at A', will have perturbed behind the shock wave a circular zone of center A' and radius ct, where c is the speed of sound in the compressed material (the shaded area in *Figure 17*). Applying Pythagoras's theorem to the triangles AEG and A'EG we find

$$\tan^2\alpha = c^2/u_s^2 - (u_s^2 - u_p^2)/u_s^2$$

The rest of the material (the unshaded area) knows nothing of the perturbations coming from the edges. For that reason it is advisable to carry out experiments with samples of small thickness as compared to their lateral dimensions; in that way one has a large zone in which the phenomena are the same as for an infinite medium.

III. Experimental Techniques

The study of shock waves requires the use of very special and rather unusual experimental techniques. This is because shock wave phenomena are characterized by very high, quickly changing and destructive pressures. The physical quantities concerned must therefore be measured in very short times, of the order of one tenth of a microsecond, sometimes even one hundredth or one thousandth of a microsecond. For that reason the experimentalist, as his field of investigation broadened, was forced to develop new equipments to allow him to obtain ever more and also more precise measurements.

Mechanical or electromechanical instruments have too long a reaction time to tackle this domain of high speed, so that one has to have recourse to optics or electronics. With the present state of the art of electronics it becomes actually possible to take precise measurements within one thousandth of a microsecond. The gradual perfection of semiconductors and their considerably decreasing prices even allow the use of certain electronic devices in contact with explosives, without raising too much the total cost of the experiment. There are many quantities that can be directly or indirectly measured:

velocity of shock waves and rarefaction waves,

velocity of a projectile or of the material behind a shock wave,

pressures: either directly with pressure-sensitive detectors or indirectly by velocity measurements,

densities: directly by measuring electromagnetic absorption, indirectly behind a shock wave by velocity measurements,

temperatures: this measurement is difficult but in certain cases it can be performed by recording an emission spectrum.

We will begin our review with a description of the different types of shock wave generators, after which we will discuss the variety of experimental equipment being used in practice.

1. Shock Wave Generators

Of the many existing shock wave generators those employing high explosives are the most frequently used. With these generators one can reach pressures of several hundreds of kilobars. For lower pressures guns driven by gunpowder or by compressed gas can be used; higher pressures are produced in guns driven by light gases.

(A) Explosive Charge Generators

When an explosive in contact with some medium is detonated, the explosion will induce in that medium a shock wave, the amplitude and duration of which will depend on the nature, the geometrical form and the dimensions of the explosive charge. Since the measurements will be much easier when the shock wave is a plane one, generators capable of bringing about such plane waves are mostly used.

The Plane Wave Generator

This generator consists of a conical nucleus of slow-burning explosive, covered with a cone of rapidly burning explosive (*Figure 18*). If D_1 is the detonation speed of the fast-burning explosive, and D_2 the speed of the slow-burning one, the semi-angle of the cone must be such that $\cos \alpha = D_2/D_1$. These generators have generally a diameter $\emptyset$ of the order of 150 mm. By careful machining and the use of an explosive of homogeneous characteristics one can obtain a flatness or uniformity of firing of about 20 to 30 nanoseconds for a diameter of 150 mm. Usually such a generator serves to create a plane detonation wave in a cylinder of explosive (*Figure 19*) and this second detonation wave imposes its characteristics on the shock wave in the adjoining medium. Such a cylinder, often called the booster also serves to 'smooth out' the local perturbations created by the generator.

The intensity of the shock wave induced in the medium is a function of the

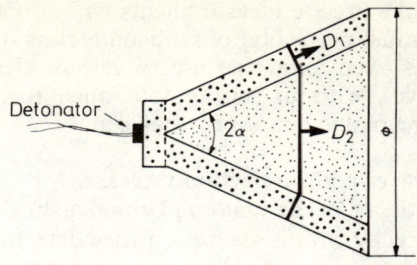

Figure 18. Plane wave lens.

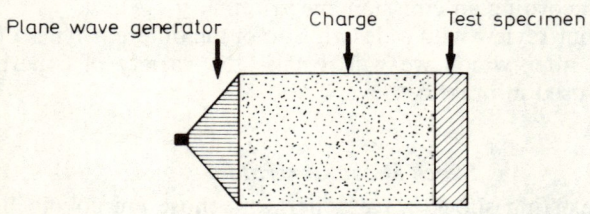

Figure 19. Schematic representation of a high explosive plane wave generator.

pressure behind the detonation wave in the booster, called the Chapman–Jouguet pressure p_{CJ}. *Table 1* lists for some simple explosives the speed of the detonation wave, together with the Chapman–Jouguet pressure.

Table 1

Explosive	Density	D, mm/μsec	p_{CJ}, kb
Trinitrotoluene (TNT)	1.64	5.01	190
Tetryl	1.7	7.7	253
RDX (Hexogene)	1.8	8.8	347

Because of technological or safety reasons one rarely uses pure explosives but rather a mixture of explosives and inert substances, which permits more precise machining and higher mechanical stability.

In order to find the pressure created by a given explosive charge in a certain adjacent medium, one must on the one hand consider the (p, u_p) curve of the medium and, on the other hand, the expansion curve of the explosive's combustion gases. Knowing that at each instant there must be equality of pressure and of material velocity at the boundary of medium and explosive, it is easy to find the equilibrium state between the two substances (see *Figure 20*). Equality of pressures and velocities at the interface is reached by either a reflected shock wave back into the burnt explosive (for medium 1 in *Figure 20*) or by a reflected expansion wave (medium 2). For fluid media the reflected wave is generally an expansion.

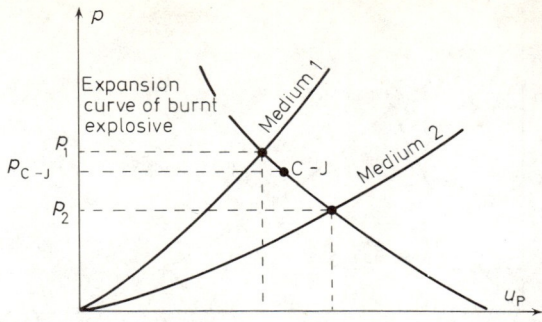

Figure 20. Determination of the pressure induced by an explosive in contact with a medium 1 or 2.

Shock Accumulation

We have just seen that for a given combination of explosive and adjoining medium, the magnitude of the pressure reached is fixed; there exists, however, a simple means of increasing the intensity of the shock wave, called 'shock accumulation'. The method consists of accelerating a metal plate with the help of an explosive and this plate then hits the sample in which the shock wave is created. For a better understanding of the mechanism of acceleration, let us consider jointly the space-time diagram and the (p, u_p) diagram of the metal plate and the burnt explosive (*Figures 21* and *22*). The shock wave induced in the metal plate emerges from it at point B. Here the shock encounters air of which the shock curve may be assumed to coincide with the velocity axis in the (p, u_p) diagram. A rarefaction wave BC coming from B travels back through the plate towards the combustion gases, bringing the metal into state R, given by the intersection of the metal's isentrope through Q and the shock curve of air (*Figure 22*). When this expansion wave passes

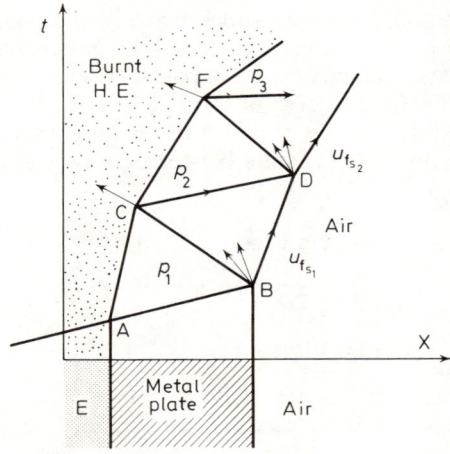

Figure 21. Shock accumulation: space/time diagram.

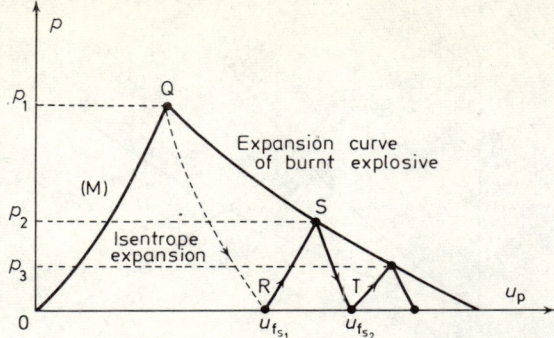

Figure 22. Shock accumulation: pressure/velocity diagram.

through the interface metal/gas at C, a compression wave CD must be reflected to the right in order to establish equality of pressure and velocity. Behind the compression wave CD the metal is in state S. This series of alternating compression and expansion waves will go on, accelerating the metal plate all the time until the gas pressure of the burnt explosive is reduced to ambient pressure. We will see next how such a plate, after acquiring a high velocity, can be used to create a shock wave.

(B) Guns

This second group of generators is used less frequently than explosives, because they require a rather heavy apparatus and infrastructure, that only few laboratories can afford. Before describing the different guns, we will examine the generation of a shock wave by the impact of a projectile hitting a target. A projectile moving with velocity V generates after impact a shock wave Ω_1 in the target and another shock wave in the projectile itself, such that at the interface pressure and velocity will be equalized (*Figure 23*). The values of pressure and velocity, after impact can easily be found by considering the respective (p, u_p) curve of target and projectile. Curve I (*Figure 24*) is the locus of states for the target and curve II that of the projectile. Their intersection represents the state between the two shocks Ω_1 and Ω_2.

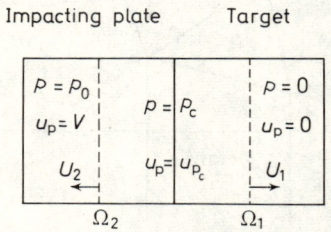

Figure 23. Generation of a shock by the impact of a projectile hitting a target.

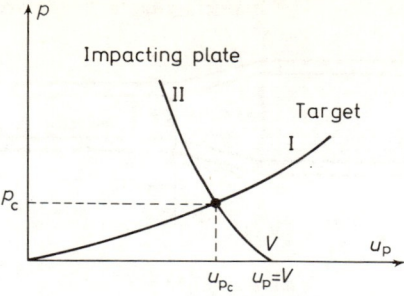

Figure 24. Shock (p_c, u_{p_c}), induced in the target (I) by an impacting plate (II) of velocity V.

Compressed Gas Guns

The operation of these is illustrated in *Figure 25*. At the start the projectile P blocks the entry hole of the high pressure gas and on the two sides A and B of the projectile the pressure is the same. By turning valve V in such a way that trigger gas is admitted to chamber A, the projectile will move to the

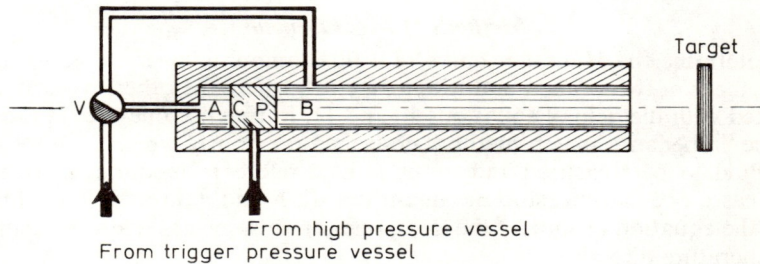

Figure 25. Compressed gas gun.

right, port C opens and the propulsion gas enters chamber A. Usually air or nitrogen is employed for propulsion. Such guns are used to obtain moderate projectile velocities (a few hundred m/s at most).

Powder Guns

The projectile is driven by combustion of gunpowder. These guns must have a smooth bore. The speed of the projectile can be in excess of 1 500 m/s. These guns not only allow higher velocities to be reached than with gas-driven guns, but they are also easier to operate.

Guns Driven by Light Gases

This type of gun has been developed in recent years to obtain plane shock waves of equally high or even higher intensity than those created in systems with explosives. Such guns generally work in two stages (*Figure 26*): in the first one a gunpowder propels a heavy piston which compresses a light gas until a diaphragm ruptures; in the second stage the expansion of the light

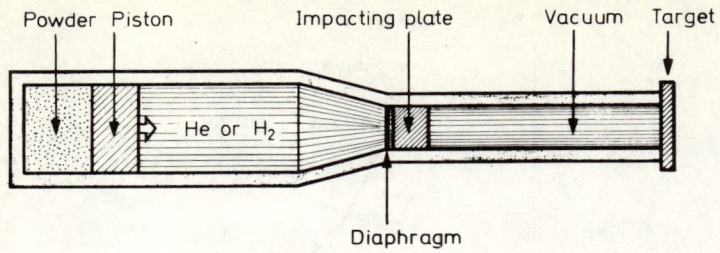

Figure 26. Light gas gun.

gas accelerates the projectile. The overall dimensions of these guns are impressive: up to several tens of meters in length. Their construction is expensive, their maintenance is delicate and the cost per shot is very high; as a result only a limited number exist across the world. For the past few years such equipment has been in operation in the USA[31].

2. *Methods of Measurement*

To determine the Hugoniot curve of a fluid amounts in principle to the measurement of two of the quantities u_s, u_p, p, v where the other two can be calculated with the help of equations 10 and 11. In practice one uses in most cases the 'impedance match method' described previously which requires u_s of the fluid to be measured and u_{fs} or u_s of a reference medium. In some special cases one can measure the quantities u_p, p and v directly. To define exactly the equation of state of the fluid, one can in some cases also measure the temperature directly.

(A) Measuring Velocities

Velocity measurement consists essentially of measuring time.

To measure the speed of a shock wave or that of a free surface the times have to be recorded at which the wave or moving matter passes two check points. Depending on the specific case, an optical or an electronic method may be used.

Optical Method

Passing by of the shock wave is recorded by using argon-filled cavities, lighting up when the shock goes by. The argon chamber is formed by turning a little cavity in a piece of Plexiglas, closed off by a steel cover of about 0.1 mm thickness (*Figure 27*). The cavity has a depth of something like 0.8 mm and contains argon at atmospheric pressure. When a shock wave hits the cavity, the argon is strongly compressed and heated. The ionization of the gas produces an intense light pulse the duration of which is the shorter the shallower the cavity. *Figure 28* shows an example of the use of argon chambers in a fluid. The device includes a metal base that transmits the shock to the

THERMODYNAMIC PROPERTIES FROM SHOCK WAVES

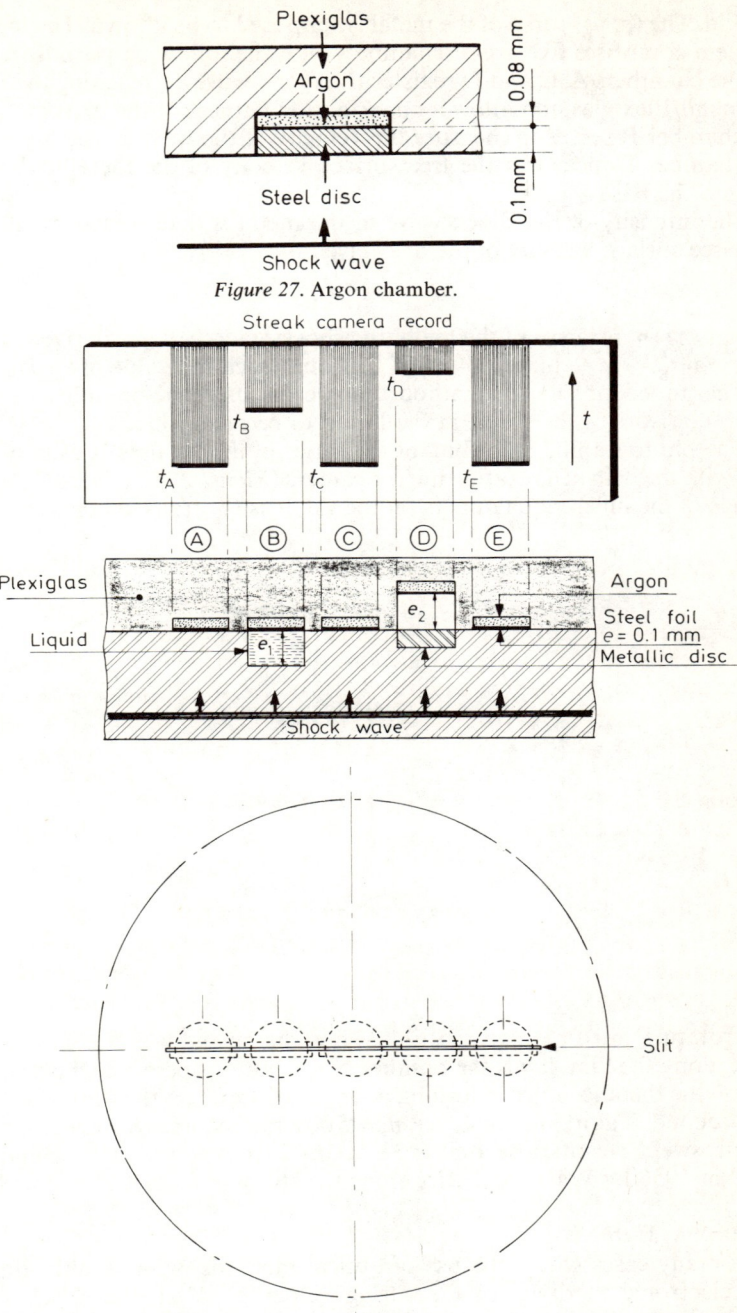

Figure 27. Argon chamber.

Figure 28. Device for measuring velocities with oxygen chambers.

liquid. The (p, u_p) curve of the metal is supposed to be known. The measuring system comprises five argon chambers with the following purposes:

the chambers A, C and D register the shock wave on reaching the interface metal/Plexiglas and allow a check on the flatness of the wave,

chamber B serves to measure the velocity of the wave in the liquid,

chamber E measures the free surface velocity of the metallic disc fitting into the base.

The intensity of the shock wave in the metal is determined by measuring the free surface velocity of the disc, given by

$$u_{fs_M} = e_2/(t_D - t_C)$$

From the (p, u) curve of the metal the shock velocity u_{sM} is derived. The time interval $t_B - t_C = (e_1/u_{sL}) - (e_1/u_{sM})$ from which the value of u_{sL} is found.

The impedance match method that we discussed before allows the other characteristics of the shock in the liquid to be determined.

The photographic recording of the series of light pulses is usually carried out with the help of a rotating mirror camera (*Figure 29*). The lens O_1 forms an image of the different chambers on the slit, lens O_2 transfers these images via

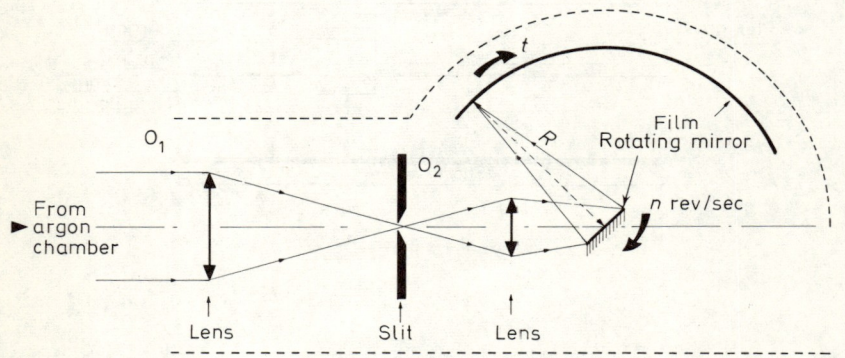

Figure 29. Streak camera.

the rotating mirror on to the film. Thus the slit images are spread out in time along the film. If n is the number of revolutions per second of the mirror and R the distance mirror-to-film, a length of $4\pi nR$ on the film is scanned in one second. Taking for instance $n = 5000$ rev/sec and $R = 50$ cm, the film length swept per microsecond is 31.4 mm. This experimental technique was used in 1956 by Walsh and Rice[63] for the study of fourteen organic liquids.

Electronic Methods

In many cases where the use of optical methods is impossible, one must use electronic methods of measuring velocities. Here the shock wave is detected by coaxial electrical contacts, as is schematically indicated in *Figure 30*. The contact consists of a coaxial conductor with a copper core and an outer tube of nickel or stainless steel. The conductor is closed off at one

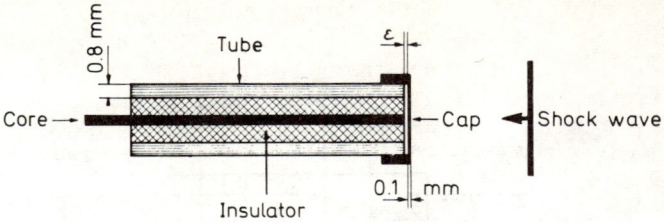

Figure 30. Coaxial electrical contact.

end by a brass cap in such a way that a gap ε remains between the central conductor and the cap. Upon arrival of a shock wave the cap will collapse and make electric contact with the core. A series of these contacts can be placed along the route of the shock wave (*Figure 31*) with their successive closure generating a series of electrical impulses that can be registered on an oscilloscope. Measurement of the time intervals between the pulses allows calculation of the shock wave velocity.

There exist variants on this contact where the air interval ε is replaced by a disc of piezo-electric or ferro-electric material.

The electronic method is especially useful with liquefied gases, where the optical method is difficult to apply.

We will briefly describe the device of Van Thiel and Alder[54] used in 1965 for the study of liquid argon. The argon is kept in the liquid state by cooling of the metallic container, which serves at the same time as reference element. The device, cooled by liquid nitrogen, is shown in *Figure 32*. The quantities defining the shock wave in argon are determined in the same way as was described for the optical method. The sequence of electric pulses is registered on a cathode ray oscillograph, an instrument very well suited for these fast phenomena. To obtain a longer timebase the electron beam is made to describe a sawtooth pattern on which the contact pulses show up as little 'pips'. A photograph of the series looks as is shown schematically by *Figure 33*.

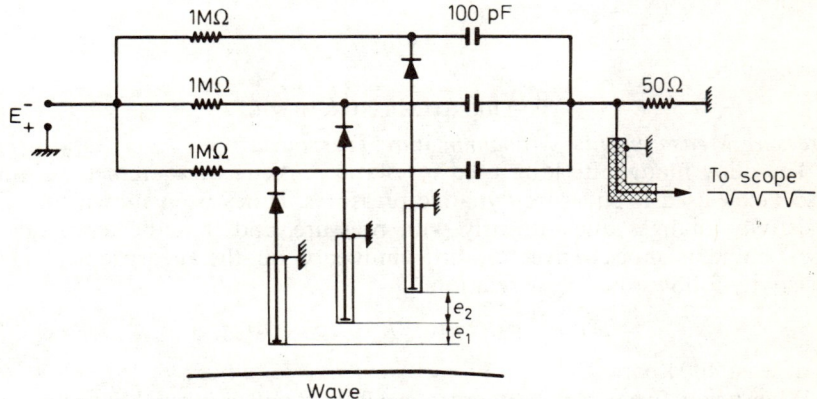

Figure 31. Electronics associated with electrical contacts.

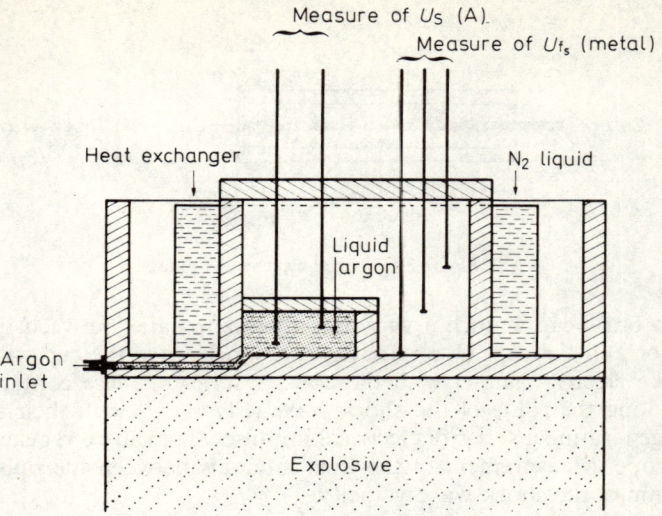

Figure 32. Device for measuring velocities with electrical contacts.

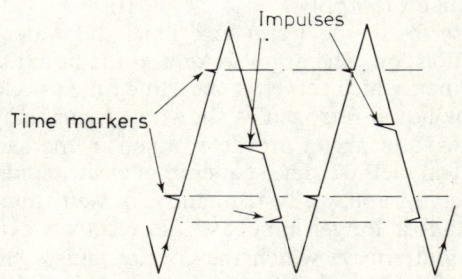

Figure 33. Sawtooth pattern.

(B) Direct Measurements

Pressure Measurements: Manganin Wire Transducer

The alloy manganin, long used for static pressure measurement, is now also being used to measure dynamic pressures. It has been shown that the resistivity changes quasi-linearly with pressure, and that the temperature rise behind a shock wave has little influence on the measurement. The resistivity follows the linear relation

$$\rho = \rho(1 + kp) \quad \text{with} \quad k = 2.5 \times 10^{-4} \, \text{kb}^{-1}$$

to at least 400 kbar.

When a very fine wire is submerged in a fluid, it will very quickly assume the pressure of that fluid. Pressure variations in the fluid at the location of the wire are detected by continuous resistance measurement. This is done by measuring the potential across the wire, which carries a constant current.

Quartz Transducer

When a shock wave passes through a quartz plate with metallized end faces (*Figure 34*), an electric current is generated in the circuit; this current is proportional to the difference in stress at the two faces.

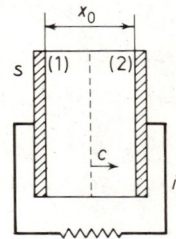

Figure 34. Quartz transducer.

When a quartz plate, cut perpendicular to the X axis is used, the charge created is directly proportional to the force exerted. Graham[25] has shown that up to 21 kbar one may assume: that the elastic coefficients are constant, that the conductivity is zero, and that the piezo-electric polarization is directly proportional to the stress. The proportionality factor is:

$$k = 2.04 \times 10^{-8} \text{ coulomb cm}^{-2} \text{ kbar}^{-1} \text{ for } p < 7 \text{ kbar}$$

and $\quad k = 2.16 \times 10^{-8}$ coulomb cm^{-2} kbar^{-1} for $p > 9$ kbar.

If one makes the following hypotheses: that the wave is parallel to the quartz plate, that the stress is one dimensional, that the wave velocity in the quartz is a constant c, that the electric constant does not change, one has in the region compressed by the shock wave (*Figure 34*).

$$\vec{D}_1 = \varepsilon \vec{E}_1 + \vec{P}$$

where $\vec{P}$ is the polarization vector. In the non-compressed region 2 one has

$$\vec{D}_2 = \varepsilon \vec{E}_2$$

With the displacement vectors supposed to be perpendicular to the wave, one has

$$D_1 = D_2 = D$$

Thus the electric fields are:

$$\vec{E}_1 = \vec{D}_1/\varepsilon - \vec{P}/\varepsilon$$

$$\vec{E}_2 = \vec{D}_2/\varepsilon$$

The potential differences between the end faces of the quartz are:

$$V(r) = E_1 ct + E_2(x_0 - ct)$$
$$= (D/\varepsilon)x_0 - (1/\varepsilon)\int_0^x \vec{P}\,dx$$

If the time constant RC is very small with respect to the travel time of the wave through the quartz, one may consider the circuit to be short-circuited, from which it follows that

$$D = (1/x_0)\int_0^x \vec{P}(x, t)\,dx$$

If the electric polarization varies linearly with the stress σ, $\vec{P} = k\sigma$, we get

$$D = (k/x_0)\int_0^x \sigma(x, t)\,dx$$

The current is given by

$$i = S(dD/dt) = S(kc/x_0)\int_0^x (\partial\sigma/\partial x)\,dx$$

For $t < x_0/c$ and $\sigma(x_0, t) = 0$.

$$i = (kSc/x_0)[\sigma(0, t) - \sigma(x_0, t)]$$

where S is the area of the sample/quartz interface.

We see that the current i is directly proportional to the stress at the sample/quartz interface.

Such a device, in a fluid, can measure the shock wave intensity in that fluid.

Finally, Kedrinskii et al.[32] have recently described a semiconductor pressure transducer for measurement of strong shock waves in liquids.

Direct Measurement of the Material Velocity behind a Shock Wave

We will describe an electromagnetic method developed in the USSR by Dremin and Shredov[19]. This method consists of registering the instantaneous velocity of a thin metallic foil suspended in the fluid. A U-shaped foil of 10^{-3} mm is placed in the fluid as indicated in *Figure 35*. An exterior magnetic field B is applied at right angles to the plane of the U. When the shock wave passes by, the horizontal part of length l will be quickly carried along with the speed of the surrounding material. The potential difference, v, between the terminals will be

$$v = u_p \times B \times l$$

This method enables measurement of the material velocity right behind the shock wave. It is equally used in the study of explosive liquids and solids.

Direct Measurement of the Density behind a Shock Wave

Some investigators have used x-rays to measure density variations behind shock waves. The x-ray tube is of the field emission type working at about 1000 or even 2000 kV. Such high voltages are obtained with a triggered generator of the Marx type, where condensers are charged in parallel across resistors and discharged in series across spark gaps. The duration of a discharge can be as short as 0.1 μsec or less (*Figure 36*). The density distribution of the photographic film is measured on a micro-densitometer and compared

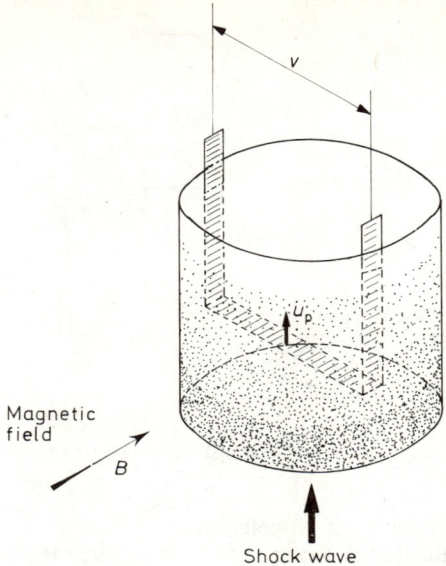

Figure 35. Device for measuring particle velocity with the electromagnetic method.

with that obtained under the same conditions with metallic references. The method is difficult because a certain number of errors can intervene (diffusion, diffraction). It has been applied in the study of water[13] and of liquid argon[14].

Direct Measurement of Temperature

The only direct methods of measuring temperature behind a shock wave are optical methods. These can only be used for transparent media. One calculates the temperature from the brightness of the material behind the

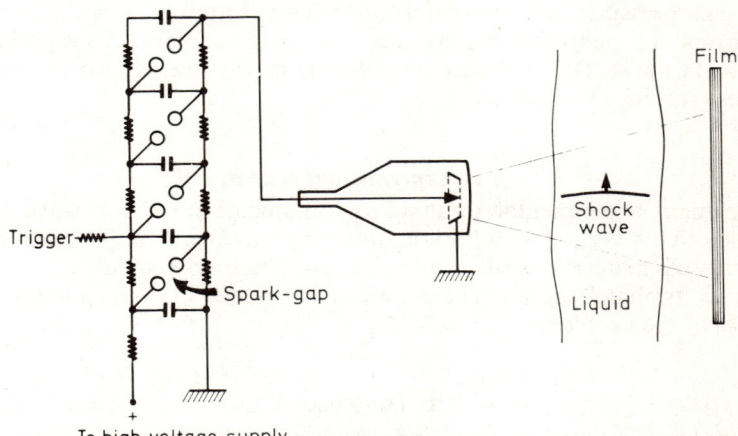

Figure 36. Device for measuring density with x-rays.

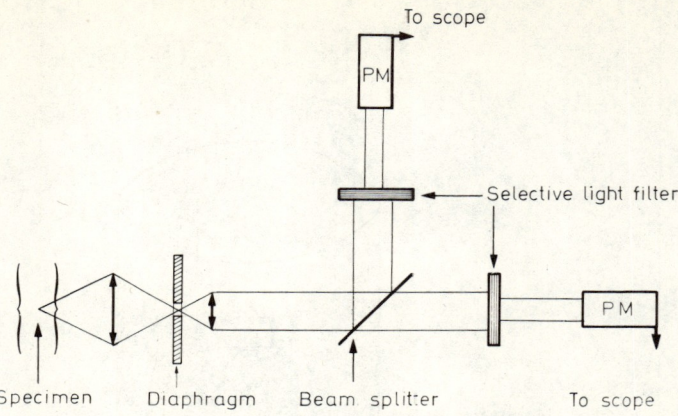

Figure 37. Experimental set-up for direct measurement of temperature.

shock wave at two different wavelengths, by comparing the luminous flux of the material with that of a standard source. The method has been used by Russian workers[35] for the study of the melting curve of sodium chloride and potassium chloride behind a shock wave. These authors have measured the brightness temperature at 4180 and 6250 Å. The experimental set-up is shown in *Figure 37*. With suitable optics the emitted light is divided into two beams. Each beam passes through an interference filter to select the proper wavelength which is then detected via a photomultiplier or an oscilloscope. The same principle has been applied by other authors to measure the temperature in liquids[36, 61].

IV. Equation of State for Liquids at Very High Pressures

With shock wave techniques the equation of state of some twenty liquids at very high pressures has been determined. This section deals with the bibliography of the published results. The first part concerns the experimental results. In the second part equation of state models, developed on the basis of these results, are discussed.

1. *Experimental Results*

The main experimental result to be obtained with shock waves is the Hugoniot curve, $p_H(v)$. In addition, other measurements can be made on the compressed material, to determine certain thermodynamic quantities, or to clarify its phase diagram; in this part we will review in succession those two aspects of the problem.

(A) The Hugoniot Curve

This curve is generally obtained from the measurement of the shock velocity u_s and the material velocity u_p. Experience shows that for most

Table 2

Liquid	$v_0/\mathrm{cm^3/g}$	T_0/K	p_0/bar	Hugoniot u_s and u_p in mm/μs	Validity	Refs.	Other refs.
O_2	0.832	76.9	1	$u_s = 1.282 + 1.723\, u_p - 0.05724\, u_p^2$	$0 < u_p < 4.72$	62	15
N_2	1.22	75	1	$u_s = 1.49 \pm 0.06 + 1.49 \pm 0.02\, u_p$	$2.5 < u_s < 7.4$	16	
				$u_s = -1.0 \pm 2.0 + 2.0 \pm 0.4\, u_p$	$7.4 < u_s < 8.4$		
				$u_s = 4.06 \pm 0.05 + 0.92 \pm 0.01\, u_p$	$8.4 < u_s < 9$		
N_2	1.24	77.4	1.9	$u_s = 1.588 + 1.360\, u_p$	$3.14 < u_s < 9.05$	66	
Ar	0.7117	86		$u_s = 0.84 + 2.07\, u_p$	$0 < u_p < 1.0$	54	
				$u_s = 1.19 + 1.627\, u_p$	$1.0 < u_p < 3.65$		
Ar	1.088	148.2	70	$u_s = 1.036 + 1.368\, u_p$	$1.6 < u_p < 4.1$	54	14
Xe	0.3231	165.2	2.24	$u_s = 0.480 + 2.238\, u_p - 0.225\, u_p^2$	$1.1 < u_p < 3.14$	35	
CO_2	0.649	196		$u_s = 1.65 + 1.93\, u_p - 0.080\, u_p^2$	$1.03 < u_p < 4.79$	66	
CCl_4	0.63			$u_s = 1.45 \pm 0.06 + 1.65 \pm 0.05\, u_p - 0.04 \pm 0.01\, u_p^2$	$2.2 < u_p < 8.3$	16	10, 63
C_6H_6	1.14			$u_s = 1.88 \pm 0.05 + 1.58 \pm 0.03\, u_p$	$2.7 < u_p < 5.7$	16	10, 63
				$u_s = 4.77 \pm 0.29 + 0.43 \pm 0.09\, u_p$	$5.7 < u_s < 6$		
				$u_s = 1.68 \pm 0.12 + 1.34 \pm 0.03\, u_p$	$6.2 < u_s < 9$		
CS_2	0.792			$u_s = 1.64 \pm 0.08 + 1.46 \pm 0.08\, u_p$	$2.4 < u_p < 3.5$	16	10, 63
				$u_s = 3.5$	$1.3 < u_s < 1.8$		
H_2O	1.00			$u_s = 1.25 \pm 0.06 + 1.32 \pm 0.02\, u_p$	$3.5 < u_s < 8.1$	53	2, 9, 13 37, 44, 63
				$u_s = 3.09 + 1.164\, u_p - 0.1138\,(u_p - 3.76)^2$	$0 < u_p < 3.76$		
				$u_s = 3.09 + 1.164$	$3.76 < u_p < 8.71$		
Ethyl ether	1.41			$u_s = 1.69 + 1.42$	$3.50 < u_p < 8.52$	53	63

homogeneous substances the relation $u_s(u_p)$ can be approximated well by linear dependence

$$u_s = A + Bu_p \qquad (31)$$

Attempts have been made to justify the validity of this relationship theoretically, in particular by Berger and Joigneau[6]; also, when experimentally a

Table 3

Liquid	V_0 cm³/g	u_s mm/μs	u_p mm/μs	Refs.
Mercury	0.0739	2.752	0.608	63
		3.101	0.772	
		3.504	0.978	
Acetone	1.274	5.37	2.510	63
	1.279	3.97	1.495	
Bromoethane	0.685	4.68	2.300	63
	0.682	3.40	1.363	
Ethyl alcohol	1.275	5.63	2.500	63
	1.267	4.03	1.487	
Glycerin	0.798	6.07	2.240	63
	0.794	4.58	1.328	
Hexane	1.499	5.54	2.590	63
	1.471	4.02	1.517	
Methanol	1.217	5.51	2.525	63
	1.255	3.95	1.483	
N-Amyl alcohol	1.236	5.81	2.465	63
	1.227	4.26	1.466	
Toluene	1.138	5.73	2.412	63
	1.141	4.12	1.443	
Mononitrotoluene	0.856	5.64	2.300	63
		4.20	1.340	
Methyl alcohol	1.27	5.50	2.46	10
		5.30	2.30	
		5.34	2.42	
Hydrogen $T_0 = 20.45$ $P_0 = 1.86$	0.555	0.940	0.592	55

sudden change in the slope of the curve $u_s(u_p)$ is observed, the existence of a phase change is deduced from it.

In each of the areas where u_p changes in accordance with the above relation, the explicit equation for the Hugoniot curve, obtained by using equations 10 and 11, is

$$p - p_0 = A^2(v_0 - v)/[v_0 - B(v_0 - v)]^2 \qquad (32)$$

For weak shocks u_s tends towards the speed of sound, c_0 (cf. II.2.B, equation 17). So when equation 31 remains valid, it must be that $A = c_0$. If this is not so, certain authors conclude that a phase change occurs between $P = 0$ and the lowest measured pressure.

There are, however, substances for which $u_s(u_p)$ cannot be approximated by a linear function; recourse is then generally made to a quadratic relationship $u_s = A + Bu_p + Cu_p^2$. The best function $u_s(u_p)$, either linear or quadratic, is determined by a least squares method. Whatever the result, the function $u_s(u_p)$ completely determines the Hugoniot curve. This function has been chosen for a recapitulation of results in *Table 2*. We have listed the equation given by the authors, whenever they give one, or the one calculated from their data in a *Compendium of Shock Data*[8], or as a last resort, the one calculated by ourselves for the case of xenon. When a liquid has been the subject of several studies we have not always quoted all results, but instead chosen either the most recent or the most complete results. The references to other papers are then listed in the column 'other references'. In the column 'region of validity', the limits are indicated between which the given $u_s(u_p)$ functions are valid. The upper limit corresponds to the most intense shock wave reported by the authors; of course the equation might hold beyond that. The lower limit corresponds to the least intense shock wave reported; however, if for $u_p = 0$ the equation results in the speed of sound, $u_s(0) = c_0$, we have assumed that the relation remains valid to zero pressure. For a few cases, finally, there are not enough data to derive a valid function $u_s(u_p)$; for those cases we have just listed the untreated results in *Table 3*.

The liquids that have been studied can be divided into two main categories, organic liquids and liquefied gases.

Organic Liquids

As early as 1957, Walsh and Rice[63] published results for fourteen organic liquids, obtained according to an experimental method described in the previous chapter, a method which later inspired several other investigators. The authors reported numerous data for water, to which we will return later, and only two measurements for each of the other liquids; their results have been included in *Table 3*, with the exception of those for ethyl ether, benzene, carbon tetrachloride, and carbon disulfide, all of which have been the object of later experiments.

The Hugoniot curve of ethyl ether has been established by Skidmore and Morris[53], while Cook and Rogers[10] published results for the other three liquids. In *Table 2* we have preferred, however, to list the results of Dick[16] for those same liquids, because they are based on very many experiments and are in agreement with the results of Walsh and Rice. According to Dick, benzene undergoes a first order phase change between 125 and 180 kbar accompanied by a change in specific volume of about ten per cent. Extrapolation of the $u_s(u_p)$ curve to $u_p = 0$ leads to a value different from the speed of sound at zero pressure. The author concludes from this, that another phase change takes place, probably solidification below 15 kbar.

Carbon disulfide undergoes a phase change at about 64 kbar, corresponding to $\Delta v/v_0 \simeq 15$ per cent. As with benzene, extrapolation of $u_s(u_p)$ to $u_p = 0$ results in $u_s(0) \neq c_0$ (40 per cent too high) which would indicate a first phase change below 20 kbar. As regards carbon tetrachloride, the experimental $u_s(u_p)$ results can be described either by a continuous quadratic relation, or by two straight sections of slightly different slope intersecting at $u_s = 4.7$ mm/μsec, corresponding to a pressure of 150 kbar. The existence of a phase change cannot therefore be ascertained, but the optical experiments of Walsh and

Rice show a loss of transparency of the substance between 70 and 130 kbar. Also, there is a strong rise of the electric conductivity between 100 and 150 kbar. It is thus probable that there is a phase change in this region. Furthermore extrapolation leads to a value of $u_s(0)$ which is 60 per cent above the speed of sound and this would indicate solidification below 20 kbar.

Water has been the subject of many publications; references concerning its Hugoniot curve are given in *Table 2*; the different results are all very close. We choose the relation $u_s(u_p)$ of Skidmore and Morris[53], who used their own results and those of Rice and Walsh; Skidmore obtained points up to a pressure of 1.14 Mbar. The point at 800 kbar measured by Al'tshuler et al.[2] agrees with those results. We did not take into account the quite recent results published by Podurets et al.[44], that we received after our work was written. These Russian authors determined experimentally a point of the Hugoniot curve for water at 14 megabar. The other published data are at lower pressures and are in good agreement with those of Rice and Walsh; in addition these authors have, by the reflected shock method, determined a secondary Hugoniot curve H_2, the initial state of which is a state reached by the first shock (cf. II.3); knowing the two curves H_1 and H_2 they were able to establish the equation of state of water (cf. IV.2.D).

Liquefied Gases

The experimental methods used are the same as those used for other liquids, with the addition of cooling devices; the results are grouped together in *Table 2*. Dick[16] noted anomalies in the Hugoniot curve of liquid nitrogen which could result from this curve passing close by the triple point; those anomalies have not been seen by Zubarev and Telegin[66]. In addition to the main oxygen Hugoniot curve, Wackerle et al.[62] have measured, using a reflected shock wave method, a series of secondary Hugoniot curves.

We will not emphasize the experimental results for other liquefied gases; these substances are amenable to a much less empirical theoretical approach than the organic liquids, and it is this theoretical aspect which in their case is the far more interesting one.

(B) Other Shock Wave Measurements

When the Hugoniot curve of a material has been obtained, additional measurements can be carried out in order to determine certain thermodynamic quantities or to complete the phase diagram.

Al'tshuler et al.[3] have measured the velocity of sound propagation, c, in water by observing lateral rarefaction waves: the equation given in II.3.C allows one to calculate c from the measured value of α. Keeler et al.[33] measured the sound velocity in acetone by an optical method, based on a stimulated Brillouin effect.

The temperature T can also be measured by an optical method for transparent media (cf. III.2.B). Kormer et al.[36] have thus determined the temperatures of sodium chloride and potassium chloride along the Hugoniot curve up to 800 kbar. They observe that the temperature remains nearly constant between 540 and 700 kbar for sodium chloride and between 330 and 480

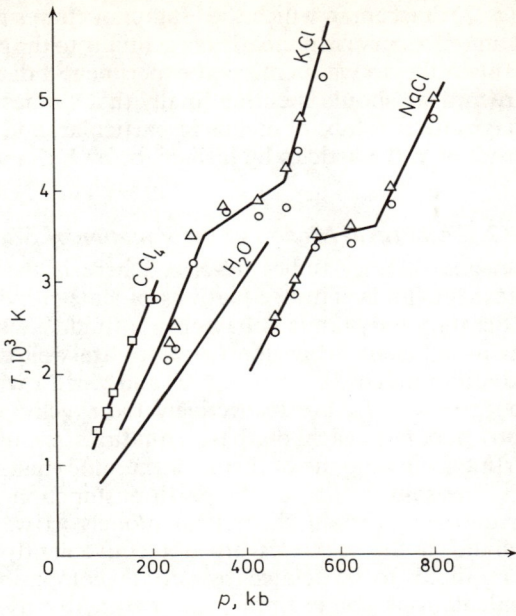

Figure 38. Temperature behind the shock wave versus pressure. The plateau in the curves for potassium chloride and sodium chloride corresponds to melting.

kbar for potassium chloride which enables them to pinpoint the intersections of the Hugoniot curve with the solidus and liquidus curves.

Besides Voskoboinikov and Bogomolov[61], Ramsay has measured the temperature of carbon tetrachloride. These results are all reported in an article by Cowperthwaite and Shaw[12]. Kormer et al.[37] published data for water; their results agree with the theoretical predictions of Rice and Walsh[46]. All these results are reproduced together on *Figure 38*.

Solidifying under shock of initially transparent fluids can be detected by measuring the light absorption coefficient. We saw the example of carbon tetrachloride in the preceding paragraph. Benzene, glycerin and alcohol[63, 37] remain transparent under shock and thus the authors conclude that no solidification takes place. The case of water is more complex. According to Walsh and Rice[63] the absorption coefficient does not increase between 30 and 100 kbar. Al'tshuler et al.[2] have continued to higher pressures and observed a decrease of transparency above 115 kbar, coinciding with a plateau in the $u_s(u_p)$ curve; they conclude that a transition from water to ice VII occurs at this pressure. Kormer et al.[36] have performed a greater number of experiments. They determined between 40 and 300 kbar the principal Hugoniot curve, several secondary ones (reflected shock method), and loss of transparency for the end states. According to these authors water remains transparent along the main Hugoniot curve, so there is no solidification. On the other hand, water submitted to two successive shocks will solidify if the first shock has an intensity between 20 and 40 kbar. The authors present in the

1151

same article a (p, T) diagram in which the Hugoniot curves have been drawn. The principal Hugoniot curve runs nearly tangentially to the phase change line, which could explain the above-mentioned experimental disagreement.

Just for the record we should mention finally that studies have been made of the phase diagrams of solids, of metals in particular, and of their melting. One could consult, say, the articles by Urlin[58] or by Urlin and Ivanov[59].

2. Theoretical Models of the Equation of State

The common goal of the studies described here is the investigation of equations of state for fluids. These equations of state provide the complete description of the thermodynamic behavior of a fluid. We may recall that a single equation is sufficient, when the variables are well chosen. Consider for example the equations $e(v, s)$, $h(p, s)$, $a(v, T)$ and $g(p, T)$, where $h = e + pv$, $a = e - Ts$ and $g = h - Ts$, are respectively the specific enthalpy, Helmholtz and Gibbs energies; each of these equations is sufficient by itself, because by starting from any one of them, all thermodynamic quantities can be calculated[40]. The same holds for the partition function Z, which can be calculated starting from certain theoretical models as we will see below. On the other hand, if the variables are not conveniently chosen, several equations are required, to be related by certain compatibility conditions. In computer calculations, for example, one frequently uses the equations $e = e(v,T)$ and $p = p(v,T)$. The compatibility relation is then $(\partial e/\partial v)_T = T(\partial p/\partial T)_v - p$. For all liquids appearing in *Table 2*, at least one equation of state has been proposed. In many cases more than one model has been suggested, often even by the same author. These models can be of very different origin, from theoretical physics to purely empirical models, with in between, those models of which the form is inspired by physical considerations but containing parameters to be adjusted experimentally. Liquefied gases are in general more suitable for theoretical studies than organic liquids, which are accessible only by more or less empirical models. The present section will deal with the bibliography of results published in this field. As it is impossible to study the theory of fluids here in detail, we will restrict ourselves to some brief remarks regarding the principal models that have been used and to some books or papers for appropriate reference.

(A) THE THOMAS–FERMI OR THOMAS–FERMI–DIRAC (TFD) MODEL[39,23]

This model is based on the calculation of the electron distribution around the nucleus in the semi-classical approximation. Its validity increases with the atomic number. The Thomas–Fermi–Dirac model takes into account the exchange interaction between electrons; it is especially valid at very high pressures and temperatures, generally much higher than those obtainable by shock waves. Nevertheless it can be used, in the case of rare gases, from a few kilobars upwards, as was shown by Ross and Alder[50]. These authors studied the equation of state of argon and xenon by using the TFD model. Nuclear motion is taken into account by an (exp-6) type cell model (cf. following paragraph) of which the parameters are listed in *Table 4*. The calculation was carried out with and without electronic excitation. For

Table 4

Substance	n	σ Å	ε/k K	Refs.
Ar	12	3.833	119.3	24
N_2	9	3.73	91.5	66
CO_2	9	3.78	290	66
Ar	8.5	3.405	124.0	51
Hg	8.5	2.898	851.0	51
N_2	7.0	3.71	95.9	51
H_2	6.9	2.87	29.2	51
C_6H_6	6.9	5.270	440.0	51
CCl_4	6.4	5.881	327.0	51

argon the calculation with excitation is in excellent agreement with the experimental results. To obtain agreement for xenon, the calculation must be done without electronic excitation for the lower pressures ($P < 300$ kbar) and with excitation for the higher pressures. This results from the fact that the TFD model permits continuous electronic excitation, and therefore gives too high energy values for those temperatures where kT is much below the first electronic energy gap.

(B) Cell Models[29,4]

These models are built on the interaction potential between two molecules. At large distances the interaction is due to induced dipoles and the interaction varies with R^{-6}. At short distances the molecules overlap and the potential is strongly repulsive. The Lennard–Jones potential (LJ) takes both phenomena into account and is written

$$u(R) = \frac{6}{n-6}\left(\frac{n}{6}\right)^{n/(n-6)} \varepsilon\left[\left(\frac{\sigma}{R}\right)^n - \left(\frac{\sigma}{R}\right)^6\right] \quad (33)$$

where ε (dimension of energy), σ (length dimension) and n are adjustable parameters.

This potential is indicated in short form by LJ ($n - 6$). Originally Lennard–Jones proposed a potential with $n = 12$, but it appeared that this potential was often too strongly repulsive, so that it was necessary to change the first exponent. One can also replace the R^{-n} part by an exponential term and thus obtain the modified Buckingham potential referred to as (exp $- 6$) which is written

$$u(R) = \varepsilon[\{6/(\alpha - 6)\} \exp\{\alpha(1 - R/\sigma)\} - \{\alpha/(\alpha - 6)\}(\sigma/R)^6] \quad (34)$$

In that case the three adjustable parameters are ε, α and σ.

The binary interaction potential has been derived theoretically for hydrogen by a quantum mechanical calculation. For more complex molecules it can be determined experimentally, e.g. by molecular beam diffraction. It should be noted that neither of the equations 33 or 34 fully represents the experimental results for all values of R. These expressions are thus only valid for limited regions.

Starting with the potential $u(R)$ and introducing certain hypotheses, the equation of state of a fluid can be established. In the cell models one assumes that the volume of the fluid can be divided into identical cells with their centers forming a lattice, and that each molecule occupies a cell within which it can move about independently of the others. The energy of each molecule is calculated from the binary interaction potential, assuming that all other molecules are at the centers of their cells and that the potential energies are additive. The energy thus calculated is a rather complicated function of the coordinates which is replaced by a spherically symmetrical function. In that way one can find the partition function Z and thus the equation of state of the fluid. This model is called (LJD) after its inventors Lennard–Jones and Devonshire.

The capacity of computing machines has made it possible to apply random methods. In molecular dynamics one starts with an arbitrary configuration of the position and velocity of each molecule; the evolution of the system is calculated by solving the equations of motion and the thermodynamic properties are obtained by averaging over the time. In the Monte Carlo method, averaging is carried out over all configurations. These are determined by a statistical procedure such that the occurrence of a configuration of total energy U is proportional to $\exp(-U/RT)$.

In *Tables 4* and *5* we have brought together some of the published results. The simplest theoretical case is that of hydrogen, which has been studied by Ross[48]. The potential $u(R)$ has been calculated by quantum mechanics and has been approximated in the region of interest by an $(\exp - 6)$ equation. The equation of state was obtained by the Monte Carlo method and gives a Hugoniot curve in very good agreement with experimental results (cf. preceding paragraph). It should be stressed that there is no adjustable parameter, as the theory for the equation of state is complete. If variation of the parameter α is allowed, perfect agreement between calculated and experimental values is obtained. The values of the parameters, with and without adjustment, are given in *Table 5*.

Table 5

Substance	α	σ Å	ε/k K	Remarks	Refs.
Ar	14	3.866	123.2	Calculated from second virial coefficient	24, 57
Ar	12	3.87	116	Molecular beam diffraction	24
Ar	14	4.18	123.2	Adjusted on Hugoniot curve	24
Ar	13.5	3.85	122	Used with TFD	50
Ar	13.5	3.80	123	⎧ Valid in the range of corresponding	35
Xe	13.5	4.45	236	⎩ states	35
H_2	11.5	3.407	34.22	Quantum mechanical calculation	48
H_2	11	3.407	34.22	Slightly adjusted	48
O_2	12.8	3.94	132		62

Argon has been frequently investigated. In a series of publications first Van Thiel and Alder, and later Ross and Alder[47, 49, 50, 54] have reported experimental results and the implications of these results obtained by a variety of theoretical approaches, one of which is the TFD model, have already been mentioned. The Monte Carlo method has been used to study the validity of the additivity hypothesis for potentials. The binary potential is known from molecular beam diffraction work. The lower part of the Hugoniot curve is well reproduced with this potential, but for very high pressures there is disagreement. This might prove that for small interatomic distances the potentials are no longer additive. Thus for a compression factor two the error committed is of the order of 30 per cent.

Keeler et al.[35] came to the same conclusions by a totally different method. If the potentials are additive, argon and xenon must have the same equation of state, expressed in reduced quantities (quantities divided by their value at the critical point); when starting at corresponding initial states the Hugoniot curves, drawn in a reduced pressure versus reduced specific volume diagram, must coincide. This has indeed been observed for sufficiently low reduced pressures ($p_r < 3000$); at higher pressures there is divergence which indicates that the hypothesis of additivity of potentials ceases to be valid.

According to Tung et al.[57] the experimental values for argon would be better reproduced when not considering the atomic motions to be independent, but rather coupled. They used an (exp − 6) model with the parameters given in *Table 4*. Fickett and Wood[24], also for argon, have applied the cell model with a potential deduced from molecular beam experiments. The agreement with the experimental results of Dapoigny et al.[14] is poor.

Saltzman et al.[51] have suggested a model based on the interaction of binary molecules that differs from the models explained above and which is specifically appropriate to account for shock compression. The potential used is of the LJ ($n - 6$) type, n being adjustable. The results concerning argon, benzene, hydrogen, mercury and nitrogen, agree well with experimental data.

We should finally mention the potential proposed by Van Thiel and Alder[55] to reproduce their results for hydrogen. For interatomic distances between 1.8 and 2.9 Å these authors give $u(R) = 18(2.90/R)^{8.54} \times 10^{-15}$ erg with R in Å.

All these different results are summarized in *Tables 4* and *5*. *Table 5* applies to the (exp − 6) potential; the parameters listed are those from equation 34 of the text. *Table 4* is in relation to the LJ ($n - 6$) potential. The listed parameters are those appearing in equation 33 of the text, except for the first line where the expression used is slightly different

$$u(R) = \varepsilon[(\sigma/R)^{12} - 2(\sigma/R)^6]$$

(C) The Mie–Grüneisen Equation of State[45,65]

This equation of state, although conceived for solids, has also been used for fluids; it is based on the hypothesis that the Grüneisen constant $\gamma = v(\partial p/\partial e)_v$ depends only on the specific volume v. That hypothesis is verified for crystalline lattices at sufficiently high temperatures. The equation of state can now be written in one of the following forms:

$$p = p_K + (\gamma/v)(e - e_K) \quad \text{or} \quad p = p_H + (\gamma/v)(e - e_H)$$

where p_K and e_K refer to the isothermal at 0 K and p_H and e_H refer to the Hugoniot curve.

One must then know the function $\gamma(v)$, one of the functions $e_H(p_H)$ or $e_K(p_K)$ and one complementary relation because the equation of state $P(v, e)$, as we have seen, is not sufficient. This complementary relation may take the form of a simple hypothesis such as $C_v = (\partial e/\partial T)_v = $ constant. The function $e_H(p_H)$ is derived from the Hugoniot curve, whereas the function $\gamma(v)$ is determined through physical considerations or by a supplementary hypothesis.

Thus, with the hypotheses $\gamma/v = \gamma_0 v_0$ and $C_v = $ constant, Dick[16] formulated equations of state for benzene, carbon disulfide, nitrogen and carbon tetrachloride. Duvall and Fowles[20] gave, with the same hypotheses, the equation of the isotherm passing through the initial state in the form of a limited series expansion in $x = v_0/v - 1$, together with the equation for the Hugoniot curve; they did this for acetone, ethyl alcohol, mercury, glycerin, benzene, ether, methanol, carbon tetrachloride and water.

For oxygen, Wackerle et al.[62] could calculate the same secondary Hugoniot curves as were found experimentally, by assuming that $\gamma/v = \gamma_0/v_0$.

Cowperthwaite and Shaw[14] have calculated the equation of state of carbon tetrachloride, nitromethane and water, by assuming $(\partial p/\partial T)_v$ to be constant, but with C_v as a function of T. The Grüneisen constant is given by $\gamma = v(\partial p/\partial T)_v \times (1/C_v)$. The temperature dependence of C_v is determined by theoretical considerations.

Gurtman et al.[26], starting from a more general point of view, established an equation of state for water valid up to 300 kbar. They assumed that C_v is constant and that $p(v,T) = h_1(v) + T h_2(v)$; from these hypotheses it follows that $\gamma = h_2(v) \cdot v/C_v$. The functions $h_1(v)$ and $h_2(v)$ are calculated from experimental data especially those of Walsh and Rice.

Cowperthwaite and Blackburn[11] have formulated an equation of state for liquid silicone. These authors intended to base this equation only on experimental results starting from a series of Hugoniot curves with different initial states. However, when the precision of the measurements turned out to be insufficient they had to use an equation of the Mie–Grüneisen type, with C_v and $(\partial P/\partial T)_v$ variables.

(D) The Equation of State of Water by Rice and Walsh[46]

This equation is based on the hypothesis that above 25 kbar the quantities $(\partial v/\partial T)_p$ and $(\partial h/\partial T)_p = C_p$ only depend on the pressure. This hypothesis is justified in the original paper. It is also justified, above 20 Mb, by the calculations of Latter and Latter[42]. It is shown in that case that C_p is constant and that $(\partial h/\partial v)_p = \xi(P)$ depends on the pressure only. The value of C_p has been estimated by Walsh and Rice from experimental data at static pressures. They take $C_p = 0.86$ cal/g^{-1} K^{-1}. The function $\xi(P)$ has been determined experimentally by Rice and Walsh up to 250 kbar. The principle consists of reaching the same pressure either by one single shock taking water to the state (p, v_2) or by two consecutive shocks taking the water first to the state (p_1, v_1) and then to (p, v'_2) (see Figure 39). The enthalpy, the specific volume and the pressure can be determined at the points 1, 2 and 2' with the help of

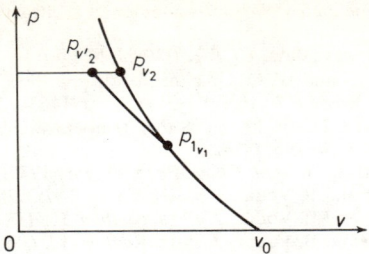

Figure 39. The state (p,v'_2) is reached by two consecutive shocks, while the state (p,v_2) is reached by one single shock.

the conservation equations. One gets then

$$\xi = (h'_2 - h_2)/(v'_2 - v_2)$$

The equations of state are determined fully from $v_H(p)$ and $\xi(p)$. Walsh and Rice were thus able to draw isentropes and isothermals and to calculate sound velocity, the entropy s_H and the temperature T_H along the Hugoniot curve. The resulting temperatures could later be verified experimentally (cf. IV.1.B); this is shown in *Figure 38*.

Papetti and Fujisaki[43], starting with the same hypotheses, gave an explicit form to the equations of state and also to the functions T_H and v_H. They also extended the validity of these equations by interpolating the functions $v_H(p)$ and $\xi(p)$ to 20 Mb, at which pressure, judged on the theoretical results of Latter and Latter who used the TF model, they are valid. As we mentioned in the theoretical part this model predicts a minimum of $v_H(p)$ around 20 Mb.

Lascar and Baconin[41] did measurements of $\xi(p)$ up to a pressure of 1.5 Mb by a method identical to that of Walsh and Rice; their results are very close to the interpolated data of Papetti and Fujisaki. Taking account of the equation for $\xi(p)$ of Walsh and Rice, valid to 250 kbar, these results can be put in the following forms:

$\text{Log } \xi = 2.17943 + 0.0030338p$, for $25 < p < 250$,

$\text{Log } \xi = -1.640463 (\log p/250)^2 + 1.746399 \times \log p/250 + 2.93788$,

$$\text{for } 250 < p < 438.19,$$

$\text{Log } \xi = 0.946765 \log p + 0.765039$, for $438.19 < p < 100000$,

where ξ and p are in kilobars.

V. References

[1] Al'tshuler, L. V. *Soviet Phys. Uspekhi*, **8**, 52 (1965).
[2] Al'tshuler, L. V., A. A. Bakanova and R. F. Trunin. *Soviet Phys. Doklady*, **3**, 761 (1958).
[3] Al'tshuler, L. V., S. B. Kormer, M. J. Brazhnik, L. A. Vladimirov, M. P. Speranskaya and A. I. Funtikov. *Soviet Phys. JETP*, **11**, 766 (1960).
[4] Barker, J. A. *Lattice Theories of the Liquid State*. Pergamon: Oxford (1963).
[5] Becker, R. *Z. Phys.* **8**, 321 (1921).
[6] Berger, J. and S. Joigneau. *CR Acad. Sci., Paris*, **259**, 2506 (1959).

[7] Bethe, H. A. *Report:* 'The theory of shock waves for an arbitrary equation of state'. *OSRD 545* (1942).
[8] Compendium of shock wave data. *UCRL 50108*, I, 524 pp; II, 231 pp (1967).
[9] Cook, M. A., R. T. Keyes and W. O. Ursenbach. *J. Appl. Phys.* **33**, 3413 (1962).
[10] Cook, M. A. and L. A. Rogers. *J. Appl. Phys.* **34**, 2330 (1963).
[11] Cowperthwaite, M. and J. H. Blackburn. 'Shock temperature calculations for silicone fluid'. *SRI Spec. Tech. Rep. No. 14* (1967).
[12] Cowperthwaite, M. and R. Shaw. *J. Chem. Phys.* **53**, 555 (1970).
[13] Dapoigny, J., J. Kieffer and B. Vodar. *CR. Acad. Sci., Paris*, **238**, 215 (1954).
[14] Dapoigny, J., J. Kieffer and B. Vodar. *J. Phys. Radium*, **16**, 733 (1955).
[15] Dapoigny, J., J. Kieffer and B. Vodar. *J. Phys. Radium*, **17**, 606 (1956).
[16] Dick, R. D. *LA 3915*, Los Alamos Scientific Laboratory of the University of California, Los Alamos, New Mexico (1968).
[17] Doran, D. G. and R. K. Linde. *Solid State Physics*, F. Seitz and D. Turnbull, Eds., New York, Vol, 18, p 85 (1966).
[18] Dremin, A. N. and O. N. Breusov. *Russ. Chem. Rev.* **37**, 392 (1968).
[19] Dremin, A. N. and K. K. Shredov. *Zh. Prikl. Mekh. Tekh. Fiz.* **2**, 154 (1964).
[20] Duvall, G. E., G. R. Fowles and Y. Horie. 'Equation of state in solids'. *WSU-SDI-67-01*, Vol. I (February 1967).
[21] Duvall, G. E., G. R. Fowles, M. H. Miles and C. T. Tung. 'Equations of state in solids'. *WSU-SDI-68-01* (February 1968).
[22] Eyring, H., R. E. Powell, A. H. Duffey and R. B. Parlin. *Chem. Rev.* **46**, 69 (1949).
[23] Feynman, R. P., N. Metropolis and E. Teller. *Phys. Rev.* **74**, 1561 (1949).
[24] Fickett, W. and W. W. Wood. *Physics of Fluids*, **3**, 204 (1959).
[25] Graham, R. A. *Bull. Amer. Phys. Soc.* **7**, 123 (1962).
[26] Gurtman, G. A., J. W. Kvish and C. R. Martings. *J. Appl. Phys.* **42**, 851 (1971).
[27] Hamann, S. D. *Rev. Pure Appl. Chem. (Austral.)*, **10**, 139 (1960).
[28] Hamann, S. D., *Advances in High Pressure Research*. R. S. Bradley, Ed., New York, Vol. I, p 85 (1966).
[29] Hirschfelder, J. O., C. F. Curtiss and R. B. Bird. *Molecular Theory of Gases and Liquids*. Wiley: New York (1954).
[30] Ivanov, A. G. and S. A. Novikov. *Soviet Phys. JETP*, **13**, 1321 (1961).
[31] Jones, A. H., W. M. Isbell and C. J. Maiden. *J. Appl. Phys.* **37**, 3493 (1966).
[32] Kedrinskii, V. K., R. J. Soloukhin and S. V. Stebnovskii. *J. Appl. Mech. Tech. Phys.* **10**, 607 (1969).
[33] Keeler, R. N., G. H. Bloom and A. C. Mitchell. *Phys. Rev. Letters*, **17**, 852 (1966).
[34] Keeler, R. N. and E. B. Royce. *UCRL 7/846*. University of California, Livermore (1970).
[35] Keeler, R. N., M. van Thiel and B. J. Alder. *Physica*, **31**, 1437 (1965).
[36] Kormer, S. B., M. V. Sinitsyn, G. A. Kirillov and V. D. Urlin. *Soviet Phys. JETP*, **21**, 689 (1965).
[37] Kormer, S. B., K. B. Yushko and G. V. Krishkevich. *Soviet Phys. JETP*, **27**, 879 (1968).
[38] Landau, L. D. and E. M. Lifshitz. *Fluid Mechanics*, p 326. Pergamon: London (1959).
[39] Landau, L. D. and E. M. Liftshitz. *Quantum Mechanics*. Pergamon: Oxford (1965).
[40] Landau, L. D. and E. M. Liftshitz. *Statistical Physics*, p 78. Pergamon London (1969).
[41] Lascar, A. and J. Baconin. To be published.
[42] Latter, A. and R. Latter. 'Equation of state of water'. *RH 1492 AEC*, Rand Corporation: Santa Monica, Calif. (1955).
[43] Papetti, R. A. and M. Fujisaki. *J. Appl. Phys.* **39**, 5412 (1968).
[44] Podurets, M. A., G. V. Simakov, R. F. Trunin, L. V. Popov and B. N. Moiseev. *Soviet Phys. JETP*, **35**, 375 (1972).
[45] Rice, M. H., R. G. McQueen and J. M. Walsh. *Solid State Physics*, F. Seitz and D. Turnbull, Eds., New York, Vol. VI, p 1 (1958).
[46] Rice, M. H. and J. M. Walsh. *J. Chem. Phys.* **26**, 824 (1957).
[47] Ross, M. *Phys. Rev.* **171**, 777 (1968).
[48] Ross, M. *UCRL 50911*. University of California, Livermore (1970).
[49] Ross, M. and B. J. Alder. *J. Chem. Phys.* **46**, 4203 (1967).
[50] Ross, M. and B. J. Alder. *J. Chem. Phys.* **47**, 4129 (1967).
[51] Saltzman, P. K., A. F. Collings and C. J. Pings. *J. Chem. Phys.* **50**, 935 (1969).
[52] Skidmore, I. C. *Appl. Mater. Res.* **4**, 131 (1965).
[53] Skidmore, I. C. and E. Morris. *Thermodynamics of Nuclear Materials*, p 173. IAEA: Vienna (1962).

[54] Van Thiel, M. and B. J. Alder. *J. Chem. Phys.* **44**, 1056 (1966).
[55] Van Thiel, M. and B. J. Alder. *Molec. Phys.* **10**, 427 (1966).
[56] Thomsen, J. S. *Physics of Fluids*, **11**, 1338 (1968).
[57] Tung, C. T., D. McLachlan and G. E. Duvall. 'Equations of state of liquids'. *WSU-SDL-69-01* (April 1970).
[58] Urlin, Y. D., *Soviet Phys. JETP*, **22**, 341 (1966).
[59] Urlin, Y. D. and A. A. Ivanov. *Soviet Phys. JETP*, **8**, 380 (1963).
[60] Vodar, B. and J. Kieffer. *Mechanical Behavior of Materials under Pressure*, p 1. H. L. Pugh, ed. Elsevier: Amsterdam, London, New York (1970).
[61] Voskoboinikov, I. M. and V. M. Bogomolov. *JETP Letters*, **27**, 879 (1968).
[62] Wackerle, J., Q. W. L. Seitz and J. C. Jamieson. 'Behavior of dense media under high pressure'. *Symposium on Hydrodynamic Phenomena*, p 85. IUTAM: Paris (1965).
[63] Walsh, J. M. and M. M. Rice. *J. Chem. Phys.* **26**, 815 (1957).
[64] Wentorf, R. H. *Brit. J. Appl. Phys.* **18**, 865 (1967).
[65] Zel'dovich, Y. B. and Y. P. Raizer. *Physics of Shock Waves and High-temperature Hydrodynamic Phenomena.* Academic Press: New York (1967).
[66] Zubarev, V. N. and G. S. Telegin. *Soviet Phys. Doklady*, **7**, 34 (1962).

CHAPTER 24

Electrical Discharge Techniques for Measurements of Thermodynamic Properties of Fluids at High Temperatures

ARED CEZAIRLIYAN and CHARLES W. BECKETT

National Bureau of Standards, Washington, DC 20234, USA

Contents

	List of Symbols	1161
I.	Introduction	1162
II.	General Method	1162
III.	Capacitor Discharge Systems	1164
	1. Description of Systems	1164
	2. Design Considerations	1166
	A. Electrical Circuit Equations	1166
	B. Consideration of Various Phenomena	1168
	(a) Skin Effect	1168
	(b) Magnetic Forces	1169
	(c) Temperature Distribution in the Specimen	1169
	(d) Other Phenomena	1169
	3. Measurement of Experimental Quantities	1170
	A. Power Measurements	1170
	B. Temperature Measurements	1172
	C. Other High-speed Measurements and Techniques	1174
	D. Recording of Experimental Quantities	1175
	4. Examples of Thermodynamic Measurements	1176
	5. Summary of Pertinent Literature since 1964	1177
IV.	A Millisecond-resolution System	1178
	1. Description of the System	1178
	2. An Example of Thermodynamic Measurements at and above Melting Points	1180
V.	Discussion	1182
VI.	Appendix	1183
VII.	References	1189

List of Symbols

Symbol	Definition
C	capacitance
E, e	voltage
F	force
f	frequency
i	current
L	inductance
p	pressure
R	resistance
T	temperature

t	time
U	energy
μ	permeability
ρ	resistivity

I. Introduction

Increasing interest in understanding the behavior of matter at high temperatures and demand for properties in applications related to aerospace, nuclear, and other high temperature fields necessitate the development of new techniques that can extend the measurements to temperatures above 2 500 K, the limit of accurate conventional (steady-state and quasi steady-state) methods.

In particular, transient techniques (shock, capacitor discharge, or similar rapid heating techniques) have shown considerable promise for measurements above 2 000 K. Since in a transient experiment the specimen is exposed to high temperatures for a very short time (usually lesss than 1 s), the undesirable contributions of most phenomena, such as chemical reactions, heat transfer, evaporation, diffusion, etc., are reduced to a negligible level.

In an earlier publication[8] the needs for the high-speed measurement methods were discussed in detail and various transient techniques for the measurement of experimental quantities, such as power and temperature that are required for the determination of selected properties, were reviewed. The above publication covered the pertinent literature prior to 1964.

In recent years, only a very limited amount of research has been performed in the application of high-speed techniques to the measurement of thermodynamic properties of fluids. However, advances have taken place in areas related to equipment and instrumentation. These have included development of diagnostic, detection, measurement, and recording systems, which are essential for the successful development of thermodynamic measurement systems.

The objective of this chapter is to review some of the recent advances in areas relevant to short-time thermodynamic studies at high temperatures, and to present, in an integrated form, various techniques that may be used in developing high-speed thermodynamic measurement systems. The presentations are generally limited to the following techniques: (1) those utilizing rapid electric discharges, and (2) those applicable to measurements of properties of fluids.

II. General Method

The method generally used for conducting high-speed thermodynamic studies is based on resistive self-heating of the specimen by the passage of high currents through it and measuring the pertinent quantities with appropriate time resolution. In general, the required quantities are: current through the specimen, potential difference across the specimen, and specimen temperature. Depending on the nature of the experiment, additional quantities (electrical and thermal) may be required.

Basically the measurement systems consist of an electrical power-pulsing circuit and associated high-speed measuring circuits. The pulsing circuit includes the specimen in series with an electric pulse power source, an adjustable resistance, a standard resistance, and a fast acting switch. The specimen is contained in a controlled-environment chamber. The high-speed measuring circuits include detectors and recording systems. A simplified block diagram of a generalized high-speed measurement system is shown in *Figure 1*.

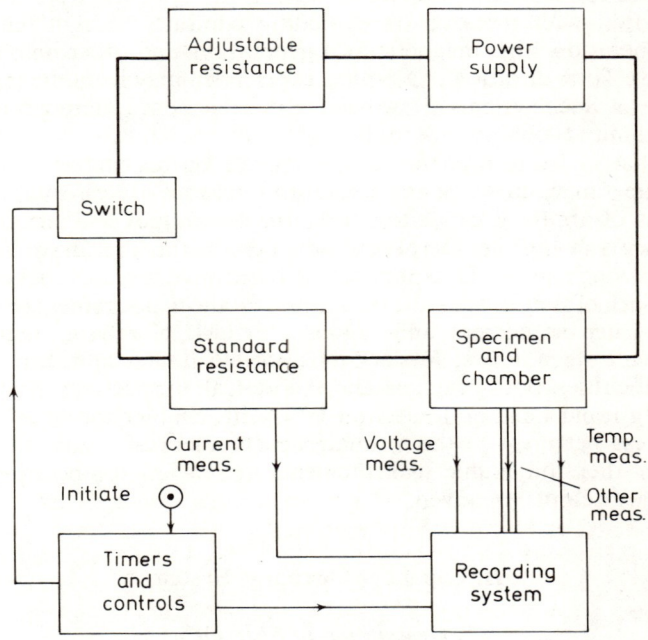

Figure 1. Block diagram of a generalized high-speed measurement system.

The measurement systems may be classified according to the type of equipment and instrumentation which, in turn, depends on the desired speed and time resolution of the experiments. A convenient dividing line may be placed at the millisecond-resolution level. Based on this, the measurement systems may be classified in two categories, namely millisecond- and sub-millisecond-resolution systems. Each category may be further subdivided depending on the type of equipment and the specific measurement techniques employed.

In this chapter, detailed presentations are limited to: (1) battery power-pulse systems for the millisecond-resolution case; and (2) microsecond-resolution, capacitor-discharge systems for the submillisecond resolution case. Other systems are briefly discussed for the sake of completeness and continuity.

Battery power-pulse systems have been developed primarily in connection with measurements of thermodynamic and transport properties of solid electrical conductors up their melting points (up to approximately 4000 K). Total duration of a typical experiment may be in the range from 0.1 to 1 s. Experimental quantities are measured and recorded at one millisecond intervals (approximately). Recent preliminary work indicates the possibility of extending the capabilities of such systems to measurements of liquid metals near their melting points. The details of an accurate millisecond-resolution battery-pulse system are presented in section IV of this chapter.

Microsecond-resolution capacitor discharge systems were developed in connection with research on exploding conductors, high temperature plasma generation, high magnetic field generation, and other high temperature studies. Total duration of a typical experiment may be in the range from 10 to 1000 μs. Such systems allow heating of specimens to temperatures well above the limit attainable with millisecond systems. Thus, such systems have the potential of being used for measurements on specimens over a wide temperature range, covering the liquid and gaseous states. In spite of the advantages of applying capacitor discharge techniques to thermodynamic measurements at high temperatures, advances in this area have been slow for the following reasons: (1) complexity and unconventionality of instrumentation; (2) lack of simple, accurate techniques for the measurement of electrical quantities, such as current and voltage; (3) lack of reliable temperature measurements techniques; (4) lack of accurate data acquisition systems; and (5) difficulties, both practical and theoretical, in specimen characterization during rapid heating. In section III of this chapter recent advances in various areas pertinent to the ultimate development of accurate discharge systems for thermodynamic measurements are reviewed, and examples of preliminary attempts are given.

III. Capacitor Discharge Systems

1. Description of Systems

Electrical power pulsing circuits and some of the instrumentation required in microsecond-resolution thermodynamic measurement systems may be similar to those used in exploding conductor or related research. A variety of such capacitive energy storage and discharge systems were developed in the 1950s and early 1960s for various applications. Their capacity ranged from joules to megajoules, and their discharge duration ranged from nanoseconds to several hundred microseconds. In recent years, only minor changes have taken place in such systems.

As of the present time, capacitive storage and discharge systems have been developed for reliable and almost routine operation at voltages as high as 30 kV. In spite of the difficulties that are encountered above this voltage, a few systems have been developed and have been operated at several hundred kilovolts.

The functional diagram of a typical capacitor discharge system that may be used for thermodynamic measurements is shown in *Figure 2*. The capacitors are connected in parallel to reduce the overall resistance of the main circuit. The charging circuit includes a conventional high voltage d.c. supply (usually with 1 to 100 mA current rating), an electrostatic voltmeter, and a high-

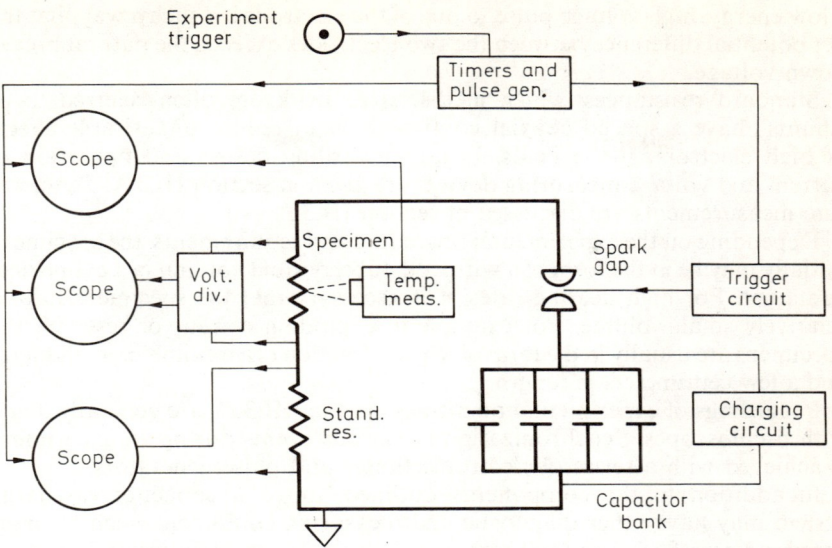

Figure 2. Functional diagram of a capacitor discharge system for thermodynamic measurements.

voltage switch for disconnecting the supply from the main circuit prior to the discharge. The capacitors used in such systems are designed to have very low inductance (1 to 100 nanohenries). The majority of capacitors used and reported in the literature fall in the 1 to 15 µF range with voltage ratings from 5 to 30 kV. Energy stored in a capacitor is expressed as $U = CE^2/2$, where C is capacitance, and E is the charging voltage. Since energy depends on the square of voltage, higher energy discharges are achieved more efficiently by increasing the charging voltage.

Because of their high current capacity and simplicity, spark gaps are frequently used as switching devices in capacitor discharge systems. A variety of types are reported in the literature. The type most commonly used for voltages up to 50 kV consists of two high current electrodes separated by a gap (air or other gas). Closing of the switch is achieved by a triggering circuit which ionizes the gas in the gap. As a result of ionization, resistance of the gap decreases to a negligible level, thereby causing initiation of the discharge. Several techniques may be used to ionize the gas. The most commonly used method is to supply a relatively low energy but high voltage pulse to a trigger electrode located near or coaxial with one of the main electrodes. The resultant spark between the trigger and one of the main electrodes is sufficient to ionize the gas. Another method is to expose the gas to the radiation from a pulsed laser. Above 50 kV, the switching of high currents becomes more difficult. It has been found that gas-type spark gaps are not reliable due to their sensitivity to gas composition (primarily to moisture content). Vacuum type switches, composed of two electrodes placed in a vacuum container, have shown higher reliability. In this case, switching is achieved by supplying a low energy, high voltage pulse to one of the electrodes in such a way that the net potential difference between the two electrodes exceeds the natural breakdown voltage.

Standard resistances, which in discharge work are often referred to as 'shunts', have a special coaxial configuration to reduce undesirable effects of high electromagnetic fields. Details regarding design and operation of current and voltage measuring devices are given in section III.3.A. Temperature measurements are discussed in section III.3.B.

Depending on the experimental procedure and requirements, the specimens initially may be in the form of a wire, rod, tube, or fluid column in a cylindrical container. For high heating rates, it is necessary that the specimens have a relatively small volume. For example in exploding conductor research the specimens are usually in the form of wires, a fraction of a millimeter in diameter and a few centimeters in length.

Recordings of experimental quantities (section III.3.D) are generally made with oscilloscopes. Synchronization of various events during an experiment is achieved with a series of electronic timers and pulse generators.

In addition to the components outlined above, a capacitor discharge system may have other diagnostic and measuring equipment, such as high-speed cameras, flash x-ray and high-speed spectroscopic facilities, and pressure transducers (section III.3.C.).

2. Design Considerations

A. Electrical Circuit Equations

The capacitor discharge system described in the previous section and shown in *Figure 2* can be represented electrically as a series *RLC* circuit (*Figure 3*). Dynamic characteristics of such a circuit after closing of the switch may be represented by the following equation

$$L \, d^2i/dt^2 + R \, di/dt + i/C = 0 \tag{1}$$

where, i is current, t is time, and R, L, C are resistance, inductance, and capacitance of the entire circuit, respectively. The solution of equation 1 is given in *Table 1* for three different cases. The results are also represented graphically in *Figure 4*.

It should be noted that in the solution of equation 1, the quantities R, L, and C are considered to be constant. In an actual capacitor discharge experiment this assumption is reasonably satisfied for L and C. However,

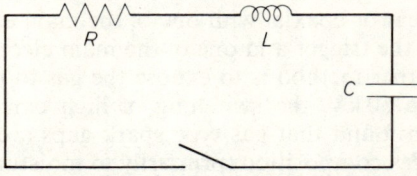

Figure 3. Simplified *RLC* (resistance–inductance–capacitance) circuit representing capacitor discharge systems.

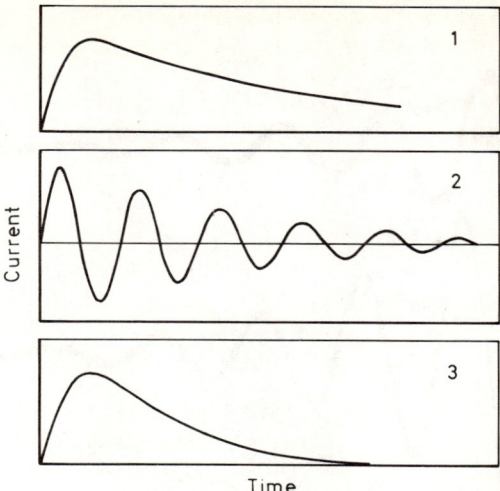

Figure 4. Typical discharge characteristics of a capacitor system. (1) overdamped, (2) underdamped, (3) critically damped.

Table 1. Summary of characteristics of a series RLC circuit (solution of equation 1) for three cases*.

Case	Designation	Criteria	Equation for current	Parameter in equation for current
1	Overdamped	$R^2 > \dfrac{4L}{C}$	$i = \dfrac{E_0}{2L\eta} e^{(-Rt/2L)}(e^{\eta t} - e^{-\eta t})$	$\eta = \left[\left(\dfrac{R}{2L}\right)^2 - \dfrac{1}{LC}\right]^{\frac{1}{2}}$
2	Underdamped	$R^2 < \dfrac{4L}{C}$	$i = \dfrac{E_0}{L\omega} e^{(-Rt/2L)} \sin \omega t$	$\omega = \left[\dfrac{1}{LC} - \left(\dfrac{R}{2L}\right)^2\right]^{\frac{1}{2}}$
3	Critically Damped	$R^2 = \dfrac{4L}{C}$	$i = \dfrac{E_0 t}{L} e^{(-Rt/2L)}$	

* The symbols are defined in the text or in the table. The quantity E_0 designates the capacitor voltage before the discharge.

R varies as a function of time because of the change in specimen resistance during heating. The degree of departure of actual results from those given in *Table 1* depends on the relative magnitude of the change in specimen resistance with respect to the resistance of the entire circuit.

Critical damping is the most desirable case to have in a discharge experiment. However, because of various conflicting design and operational parameters in an actual system, this is not always possible to achieve. In general, a typical performance corresponds to the 'underdamped' case, with the circuit oscillating at a natural angular frequency ω (defined in *Table 1*).

The results of an actual capacitor discharge experiment showing current through the specimen and voltage across the specimen are presented in *Figure 5*.

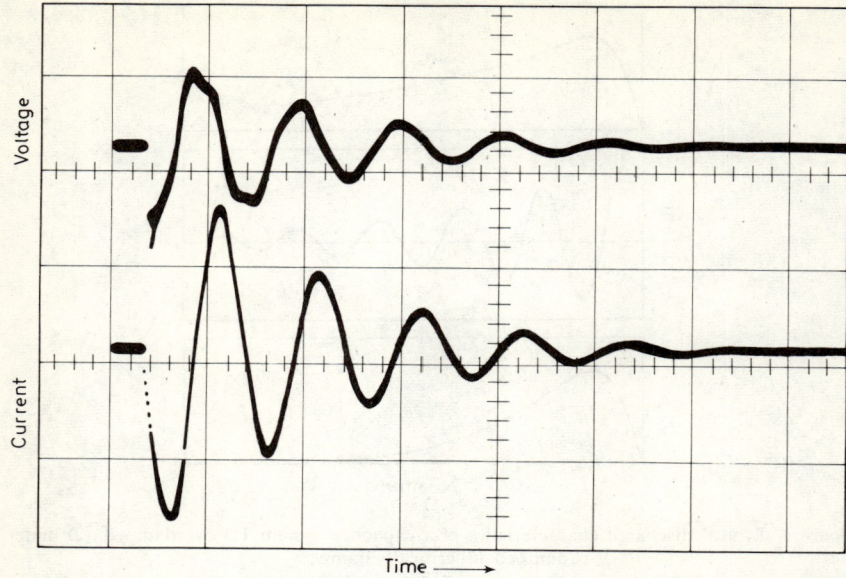

Figure 5. Oscilloscope trace record of voltage and current in a capacitor discharge experiment. Equivalence of each major division is: time = 50 μs, voltage = 2 kV, current = 50 kA.

B. Consideration of Various Phenomena

The successful design and operation of high-speed measurement systems and the meaningful interpretation of the data depend on detailed consideration of various thermal, electrical, and magnetic phenomena. In the following paragraphs some of the important items are discussed.

(*a*) *Skin Effect.* The skin effect, which is caused by changing currents in a circuit, alters the resistance of the specimen or other elements in the circuit and creates temperature non-uniformities in the specimen.

For a cylindrical conductor the skin effect is a function of the following parameter

$$x = 2\pi a \sqrt{(2\mu f/10^9 \rho)} \qquad (2)$$

where, a is the radius of the conductor, f is the frequency of current, ρ is the electrical resistivity of the conductor, and μ is the permeability of the conductor (for non-magnetic materials $\mu = 1$).

In the case of tubular (thin wall) conductors, the skin effect is a function of the following parameter

$$\beta = x\tau/a\sqrt{2} \qquad (3)$$

where, x is the quantity defined by equation 2, a is the outer radius of the conductor, and τ is the thickness of the conductor.

In general, the contribution of the skin effect in a specimen is expressed in terms of the ratio of its electrical resistance at the frequency in question to that under d.c. conditions. Results on the resistance ratio in terms of the parameter x and β are given in the literature[65].

(b) Magnetic Forces. High currents through a conductor create magnetic forces which act toward the center of the conductor. The effect of these forces is especially critical for tubular conductors.

Magnetic force (per unit length) on a tubular specimen resulting from the passage of currents through it is expressed as

$$F = (4i^2/3b)[(1 + 2m)/(1 + m)^2] \qquad (4)$$

where, i is the current, $m = a/b$, b is the outside radius, and a, the inside radius. The corresponding magnetic pressure is given as

$$p = (2i^2/3\pi b^2)[(1 + 2m)/(1 + m)^2] \qquad (5)$$

(c) Temperature Distribution in the Specimen. Heat transfer by conduction and convection are relatively slow processes in comparison to submillisecond duration heating. Thus, they do not affect temperature distribution in the specimen. Heat transfer by radiation may be considerable at very high temperatures, in which case radial temperature gradients may be established in the specimen. However, the major source of temperature non-uniformities in the specimen during heating may be the hot spots or zones that result from cross-sectional (geometrical) non-uniformities in the specimen. Thus, uniformity of the specimen's cross-sectional area along its entire length is a very important consideration in conducting accurate high-speed experiments. If the specimen geometry is not properly selected, skin effect may also be a serious source of temperature gradients.

(d) Other Phenomena. Other phenomena that should be considered in the design, operation, and interpretation of results are:

(1) *Thermionic Emission.* Excessive emission may create a current path parallel to that in the specimen, or it may alter the true voltage across the potential probes, thereby causing measurement errors. In addition, energy loss by thermionic emission may become appreciable at high temperatures.

(2) *Thermoelectric Effects.* The Thomson effect is probably the most significant thermoelectric effect in discharge experiments. Its contribution (absorption or evolution of heat) occurs only at the extreme ends of the specimen (near the clamps) where sharp temperature gradients exist.

(3) *Thermal Expansion.* A material expands (due to thermal effects) approximately with the speed of sound in that material. In order to avoid complications (practical and theoretical), it is advisable to limit the heating rate of the specimen to a value that allows the free expansion of the specimen at any time during the discharge.

(4) *Electrochemical Processes.* Chemical processes may take place in a fluid (solution) as a result of current flowing through it. Their contribution depends on the discharge conditions and the composition of the solution.

3. Measurement of Experimental Quantities

A. Power Measurements

The most frequently needed quantities in thermodynamic experiments are power absorbed by the specimen and specimen temperature. Absorbed power is determined from measurements of imparted power and power losses. In experiments in which the specimen is heated resistively, imparted power may be obtained from measurements of current through the specimen and voltage across the specimen. In capacitor discharge experiments, the only significant power loss from the specimen is that due to thermal radiation. In most cases, this loss comprises less than one per cent of imparted power. Therefore, a correction based on an estimated emittance value is generally considered to be satisfactory.

Accurate measurement of current and voltage in capacitor discharge experiments is a difficult task. This difficulty stems primarily from a high degree of interaction (due to electromagnetic fields) between signal carrying circuits and the main discharge circuit. Since the effect of this interaction is proportional to the quantity di/dt (where i is current and t is time), increasing difficulties with increasing discharge speeds are encountered.

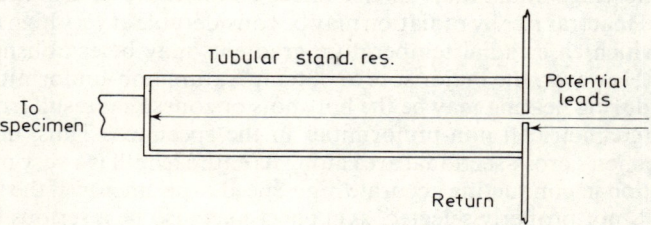

Figure 6. Schematic diagram of a tubular standard resistance used in capacitor discharge experiments.

The state of the art of current measurement techniques is more advanced than that for voltage. The coaxial shunt developed by Park[112], shown schematically in *Figure 6*, has the feature of eliminating, to a large extent, the effect of induced voltages in the signal carrying circuit. This is accomplished by having the resistance element in tubular form and by placing one of the voltage probes along the axis of the tube. Because of radial symmetry, the net electromagnetic force and the net induced voltage in the probe are therefore zero. Inaccuracy in current measurements with this scheme is estimated to be approximately one per cent. Due to the skin effect in the resistance element, this inaccuracy may increase at very high oscillatory discharge frequencies (above 1 MHz).

Several methods have been used to measure voltage across the specimen during a discharge experiment. However, as of now, no practical and accurate technique has been developed. The most frequently used technique is based on measurements with a voltage divider. The divider ratio is dependent upon the voltage across the specimen and the input voltage rating of the recording instrument. Such a measurement scheme is subject to errors due to induced voltages in the measuring circuit. The attempts that have been made to eliminate this error or to reduce it to a negligible level may be classified in two categories: (1) by extensive shielding of all the measuring circuits, and (2) by compensating for all the induced voltages. The first method is not satisfactory because there are always portions of the circuit (those near the specimen) that cannot be properly shielded. The second method, in principle, is more promising. Several different types of compensation (mathematical and physical) may be used. Mathematical compensation is based on recording the gross voltage signal (sum of true and induced components) in addition to current through the specimen, and adjusting the voltage data (with the use of a computer) after the experiment to bring it in phase with the current. The final voltage obtained by this approach represents the true resistive component across the specimen. The same result may be accomplished by physical compensation using an adjustable coil placed in series with the voltage divider circuit[22]. The function of the coil is to cancel erroneous voltage during the discharge by generating an induced voltage equal in magnitude but opposite in sign to that induced in the measurement circuit and the specimen. Equality between the erroneous induced voltage and the generated compensating volt-

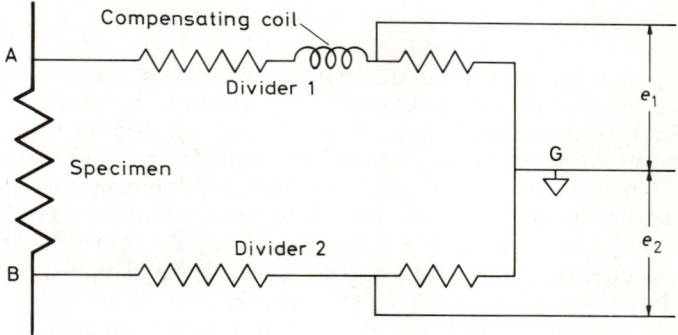

Figure 7. Electrical circuit of a voltage measuring technique utilizing two dividers and a compensating coil.

age may be established by rotating the coil (trial and error) until voltage and current traces are in phase with each other. A schematic diagram of a voltage measuring circuit based on a physical compensation technique is shown in *Figure 7*. An additional improvement incorporated in the circuit is the use of two voltage dividers. The signals corresponding to voltages e_1 and e_2 are applied differentially to an oscilloscope. The true voltage across the specimen is proportional to the quantity $(e_1 - e_2)$. The advantage of the two-voltage-divider system is that it allows measurement of specimen voltage even though both terminals of the specimen may be at a high potential with respect to earth. This provides flexibility (in the earthing procedure) in the design of the main discharge circuit. Inaccuracy in voltage measurements with proper compensation techniques is estimated to be approximately three per cent.

In recent years, techniques based on electro-optical and magneto-optical effects for the measurement of pulse voltages and currents have been explored.

B. Temperature Measurements

The ultimate development of capacitor discharge techniques for thermodynamic measurements depends on the success in developing accurate microsecond-resolution temperature measurement techniques. Pyrometric (for solids and liquids) and spectroscopic (for gases and plasmas) methods may be utilized to achieve this.

Although at the present time there is no commercially available instrument to qualify for high-speed temperature measurements, several prototype instruments have been developed in various laboratories for research purposes. However, almost all of these are relatively slow (time resolution in the range 0.5 to 50 ms) for measurements in capacitor discharge experiments. They were developed primarily in connection with transient temperature measurements in: (1) rocket motors, jet engine exhausts, ablation; and (2) millisecond resolution thermodynamic experiments. They are reviewed in a recent publication[32].

Most of the attempts to measure temperature of gases and plasmas at high temperatures with microsecond resolution were in connection with research on shock compression of gases, exploding wires, and pulsed plasmas. They are generally based on photoelectric or photographic detection of radiance. Conversion of radiance to temperature is accomplished by calibrating the detector with reference radiances before or after the transient experiments.

The fact that calibration and the transient experiment are not made simultaneously may introduce uncertainties. For example, in the case of photomultiplier tube detectors, fatigue, instability, or the effect of high magnetic fields during a discharge may invalidate the calibration. In the case of photographic films, variations in emulsion and development may introduce similar uncertainties. To avoid these difficulties it is necessary to develop systems in which the detector is calibrated during the transient experiment. In photoelectric systems this may be accomplished using a high-speed chopper which allows the alternate exposure of the detector to radiation from the unknown and from a reference. In photographic systems, no chopping is necessary and it suffices to photograph simultaneously one or more reference radiances in addition to the unknown on the same film during the transient experiment.

Another alternative that avoids the use of chopping schemes, is the development of photoelectric or similar detectors which are very stable and whose behavior is invariant under transient electromagnetic fields. Recent developments in the area of solid-state detectors show considerable promise. Their submicrosecond response characteristics are attractive; however, no careful study has yet been made regarding their stability and behavior under strong electromagnetic fields. Most solid-state detectors developed so far have a maximum response in the near infra-red and infra-red regions, which are not very practical for very high temperature studies. Discussions on high-speed solid-state detectors may be found in the literature[2, 3, 47, 95].

In steady-state temperature measurements, detectors are used generally to match the radiance of the unknown with that of the reference. However, in high-speed measurements this approach cannot be used because of time limitations. In the latter case, the unknown temperature has to be determined by interpolation of the unknown radiance from bracketing reference radiances. Accuracy of this approach depends upon, among other factors, the linearity of the detectors. Photoelectric detectors are generally linear to better than one per cent over a radiance range of three decades, which is sufficient for most measurements. If properly exposed, processed, and interpreted, linearity in film response may be maintained within one per cent over a radiance range of two decades.

In the design of high-speed pyrometers, it is important to optimize the speed with respect to temperature. If approximately 0.2 s of exposure time is required to make accurate temperature measurements (by photoelectric methods) at the gold point (1 337.58 K) corresponding to an effective wavelength of 650 nm, the following approximate exposure times are computed to be needed to perform the measurements at higher temperatures (at the same wavelength): 1000 μs at 2 000 K, 1 μs at 5 000 K, 0.1 μs at 10 000 K, and 0.01 μs at 50 000 K. Of course, by optimizing the effective wavelength as temperature is increased, the exposure times can be further reduced.

Discussions of various criteria important in developing high-speed pyrometers are given in earlier publications[8, 32].

In relatively slow experiments (upper microsecond and millisecond resolution), the accuracy of photographic methods is, in general, lower than that of photoelectric methods. However, in faster experiments, photographic methods show great promise. During the last decade, a considerable amount of research and developmental work on high-speed photography and related techniques, with resolution from the millisecond to the nanosecond range, was performed. Application of available technology on high-speed photography to temperature measurements may result in the successful development of photographic pyrometers with microsecond resolution. Even after the development of accurate photoelectric temperature measuring equipment with microsecond resolution, photographic methods will continue to be exploited and will be useful in detecting temperature non-uniformities and measuring temperature gradients in the specimen, photoelectric methods are likely to be more desirable in most applications.

The discussions in the above paragraphs have pertained to methods of detecting radiation rather than specific temperature measurement methods (such as pyrometric, spectroscopic), the reason being that detectors present most of the difficulties in high-speed measurements. Other components and measurement considerations may not be too different from their more conventional (slow-speed or steady-state) counterparts. An important item, however, is the consideration of the possible departure of the specimen from thermodynamic equilibrium, which may have particularly significant effects on spectroscopic measurements of gas and plasma temperatures. In a publication on a study of equilibrium in argon arcs (at pressures of 0.2 to 5 bar) Shumaker and Popenoe[127] have reported possible departures from equilibrium at electron densities below 5×10^{16} cm^3. Discussions of various subjects related to plasma spectroscopy may be found in a book by Griem[63]. The physics of the electric arc is discussed by Hoyaux[74]. A review of spectroscopic measurements of high temperatures is given by Drawin[51].

An important item that needs to be considered in high-speed temperature measurements is the availability of reference radiance sources. At moderately high temperatures (up to 1 500 K) vacuum tungsten filament lamps are generally used; in the range from 1 500 to 2 500 K gas-filled lamps can be used. The gas-filled lamps are convenient to use, and their calibration usually stays within one degree at 2 000 K after operating for ten hours. For temperatures above 2 500 K, there are no reliable and well characterized radiance sources. The best available source is the low-current graphite arc which yields a brightness temperature of approximately 3 800 K. The temperature level is dependent on the composition of the graphite rods and the operating conditions, such as current, electrode separation, and atmospheric pressure.

Investigations on the radiance temperature of the graphite arc and its relatively long-time (seconds to minutes) variations are reported in the literature[40, 92, 108]. The most recent measurements[92], made with a photoelectric pyrometer having $\frac{1}{6}$s time constant, gave 3 K for the standard deviation of temperature fluctuations from the mean, and up to 10 K for temperature variations over a period of 15 minutes. A study[29] on the short-time behavior of a graphite arc with a millisecond resolution photoelectric pyrometer showed that the standard deviation of radiance temperature fluctuations was 3 to 4 K; maximum deviations as large as 10 K from the mean were also observed.

Above 3 800 K, lasers and high pressure arcs may become useful as reference radiance sources. However, it is difficult to control their radiance stability.

C. Other High-speed Measurements and Techniques

In addition to the measurements of imparted power to the specimen and specimen temperature, time-resolved measurements of other quantities may be required during a discharge experiment. For example, the shape, expansion, density, and composition of the specimen, and in some cases, pressure in the experiment chamber may be needed.

For the measurement of quantities related to specimen geometry, high-speed photography and flash x-ray techniques may be employed. However, in some cases considerable discrepancies have been observed between the results obtained simultaneously by the two methods. In general, results from x-ray data are considered to be more reliable in measurements related to the diameter of an expanding specimen. Both x-ray and interferometric techniques may be used for rapid measurement of the density of gases. The nature of the products resulting from the electrical explosion of a specimen in a gas may be determined spectroscopically. Time resolved spectra may be obtained utilizing a high-speed camera in conjunction with a spectrograph. Recent advances in pulsed lasers have made it possible to conduct spectroscopic studies with a time resolution of approximately 1 ps[121, 137]. Transient pressure measurement techniques have been developed primarily in connection with shock tube research. In general they utilize transducers whose operation is based on piezoelectric or piezoresistive effects. A recent publication[66] describes a pressure gage (0.1 μs risetime) that can be used in shock tubes with ionized gases.

High-speed photography is used for both diagnostic and measurement purposes. Improvements in films and optical components, including utilization of electro-optical devices and pulsed lasers, have enhanced the contribution of high-speed photography to capacitor discharge experiments. Recent advances in this field may be found in the proceedings of a conference on high-speed photography[107].

Techniques for modulating optical radiation play an important role in high-speed experiments not only in connection with temperature measurements but also in other optical studies as in high-speed photography, spectroscopy, and interferometry. They may be divided into two general classes: (1) mechanical modulation, and (2) electro-optical or magneto-optical modulation. Most commonly used mechanical modulators are disc-shaped choppers, and turbine mounted mirrors; they may be used for modulating light to 10^4 and 10^7 Hz, respectively. For higher frequencies (up to 10^9 Hz) electro-optical modulators (Kerr cell, etc.) may be used. However, as of the present time, very little accurate work has been done to assess the operational characteristics of electro-optical modulators, such as transmission stability, and behavior under varying internal and external parameters.

Generation of short-duration light pulses, which may be needed in certain high-speed experiments, may be accomplished either by modulating a steady light source or by having a pulsed light technique. The latter has the advantage of delivering higher intensities. Flash lamps were used in the past for this purpose. However, advances in pulsed lasers have overshadowed all other pulsed light sources. They are capable of providing intense monochromatic light, and their pulse duration may be made as short as a few picoseconds.

D. Recording of Experimental Quantities

The ultimate utility of a measurement system lies in the proper recording of the experimental quantities. One of the major reasons for the delay in the development of high-speed thermodynamic and related measurement techniques has been the difficulties encountered in accurate recording of the signals.

The first truly high-speed recording of quantities was realized with the development of modern oscilloscopic techniques. Because of the very fast response characteristics (as low as nanoseconds) oscilloscopes were used for a wide range of experimental conditions, requiring from millisecond to nanosecond resolution. However, they have limitations which stem primarily from their relatively poor recording accuracy. In general, under optimum conditions oscilloscopic recording inaccuracy can hardly be less than one to two per cent of full scale. In special cases where pulses to be recorded are of rectangular or trapezoidal shape, accuracy may be improved by suppressing the major portion of the pulse and recording only the remainder of the signal. Since oscilloscopic recording (on film) is in analog form, additional effort is required to reduce the information to numeric (digital) form using micrometer microscopes or similar reading instruments. A device for reading (with digital output) oscilloscope trace photographs is described in the literature[78].

Advances in the electronics and computer fields during the last decade have made the development of accurate high-speed digital recording systems a possibility. Such systems generally consist of a multiplexer, an analog-to-digital converter, a core memory, pertinent interfacing equipment, and an output device (such as a magnetic tape deck, a teletypewriter, or a computer).

At the present time, several commercial and custom built digital data acquisition systems (with data collection at millisecond rates) are available. Such a system, having a full-scale signal resolution of approximately 0.01 per cent (13 bits) at a data recording rate of a group of three samples every 0.4 ms, is described in section IV.1. The speed of digital recording systems is limited primarily by the operation of analog-to-digital converters. Thus, in order to increase speed, accuracy requirements have to be relaxed. A high-speed digital data acquisition system is described in the literature[13] which has a data collection rate of one sample every 1.25 µs. The system is composed of a 16-channel multiplexer, an 8-bit analog-to-digital converter, a core memory (4096-word capacity), and a computer compatible tape transport. The full scale signal resolution of the system is approximately 0.4 per cent. The overall recording inaccuracy is reported to be one per cent.

It is likely that within a decade advances in the electronics field will permit development of systems for digital recording of quantities with microsecond resolution and with full-scale signal resolution of approximately 0.01 per cent. Such capabilities will undoubtedly make capacitor discharge techniques attractive research tools.

4. Examples of Thermodynamic Measurements

In spite of the extensive work performed in connection with capacitor discharge techniques, efforts so far in utilizing these techniques for thermodynamic measurements have been minimal, and in all cases they have not extended beyond the preliminary stage. Investigations in this direction are summarized in the following paragraphs.

The design of a capacitor discharge system for performing thermodynamic measurements in liquids at high pressures is described by Kroepelin and Neumann[84]. The technique is based on the generation of high dynamic pressures (through shock waves) in the liquid by exploding a wire immersed in the liquid. Electrical energy to the wire is supplied by a capacitor bank (four capacitors with 15 µF each, rated 20 kV). Only preliminary experiments on the performance of the discharge system were reported. In subsequent publications[105, 106, 123] several changes and refinements in the capacitor discharge system were described. However, no attempt was made in any of the publications to compute thermodynamic quantities.

A capacitor discharge system is described by Dikhter and Lebedev[49] for high temperature measurements of specific heat of metals (both in solid and liquid phases) as well as their heats of fusion. The system is similar to that used in exploding wire experiments. The technique is based on measuring, during the discharge, power imparted to the specimen (by measuring current through the specimen and voltage across the specimen), and specimen temperature (by measuring specimen radiance at two wavelengths). Oscilloscopes are used for recording the data. Experimental results are reported for the specific heat of solid tungsten (above 2600 K) and of liquid tungsten (up to 4500 K), as well as heat of fusion of tungsten. The specimen was a wire (diameter, 0.08 mm, length, 20 mm) and was pulse heated from room temperature to its liquid phase in short times (of the order of 20 μs) by the passage of high-density currents (approximately $4 \times 10^4 \, \text{A mm}^{-2}$). The estimated error in measurements of the specific heat of solid tungsten is reported to be ten per cent. Among a number of sources of experimental uncertainties, the assumption (as reported by the investigators) of constant normal spectral emittance over the entire range of temperature measurements probably contributed significantly to the overall uncertainty of the results. However, the experimental work was an important pioneering step toward the utilization of capacitor discharge techniques for thermodynamic measurements at high temperatures.

Application of a capacitor discharge technique for the measurement of volume, enthalpy, and resistance of liquid metals at high temperatures and high pressures is described by Henry[70]. The specimen, which is in the form of a wire (diameter, 1 mm; length, 25 mm), is mounted between two current leads such that it is axially concentric with a high-pressure cell. The cell is filled with helium to the desired working pressure. A current pulse of controlled magnitude and duration, from an overdamped discharge of a high-voltage capacitor bank (20 kV, 17 kJ), heats the wire to the desired state at a rate such that its expansion is nearly isobaric. The energy deposited in a segment of the sample is calculated from data obtained by measuring the current through the sample and the potential difference between two probes (spaced about 5 mm apart) near the center of the specimen. These data also allow calculation of the sample resistance during a major portion of the time that current flows. Diameter of the expanding specimen is measured using a flash x-radiographic technique. Preliminary measurements were performed on lead at temperatures up to 5000 K and at pressures up to 2000 bar. Duration of heating pulses were in the range 15 to 40 μs. Recommendations were made for additional improvements and developments, which included techniques of optical volumetric measurements and spectrophotometric temperature measurements.

Because of their pertinence to thermodynamic and related studies, other investigations, although not on fluids, may be appropriate to reference here. Parker[113] used a capacitor discharge technique to measure the energy of solid–solid phase transformation (α to β) and the specific heat of solid titanium. Ayers and Barnes[7] described a capacitor discharge system for the study of rapid phase transformation kinetics in metals. More recently, Forgacs[58] described a thermal pulser for similar studies. A general discussion on rapid heating techniques by capacitor discharge methods and their application to studies of substances in their solid phase (measurements of specific heat, energy of solid–solid phase transformations, and determination of Young's modulus) was given by Parker and Austin[114].

5. Summary of Pertinent Literature since 1964

A summary of the literature since 1964 pertinent to capacitor discharge and related high-speed experiments is given in the Appendix. These are in addition to the following two major publications: (1) by Früngel[60] on high-speed pulse technology in general, and (2) by Chace and Moore[39] on exploding wires. The two-volume publication by Früngel may be considered the most complete work on the design and operational characteristics of capacitor discharge systems. The publication edited by Chace and Moore is a set of symposium volumes that have resulted from conferences on exploding wires (two such conferences since 1964). The volumes contain a variety of papers on design and operational characteristics of discharge systems and their application to exploding wire research.

IV. A Millisecond-resolution System

1. *Description of the System*

In this section, a relatively slow system (millisecond resolution) for conducting thermodynamic studies at high temperatures is described. This system was developed originally for the accurate measurement of selected thermophysical properties of solid electrical conductors in the temperature range 1500 K to the melting point of the specimen. Recent preliminary investigations have shown the potential of this technique in extending the measurements to liquid metals.

The method is based on rapid self-heating of the specimen from room temperature to any desired high temperature (up to its melting point) in less than one second by the passage of electric currents through it; and on measuring, with millisecond resolution, experimental quantities such as current through the specimen, potential drop across the specimen, and specimen temperature.

A functional diagram of the system is shown in *Figure 8*. It consists of an

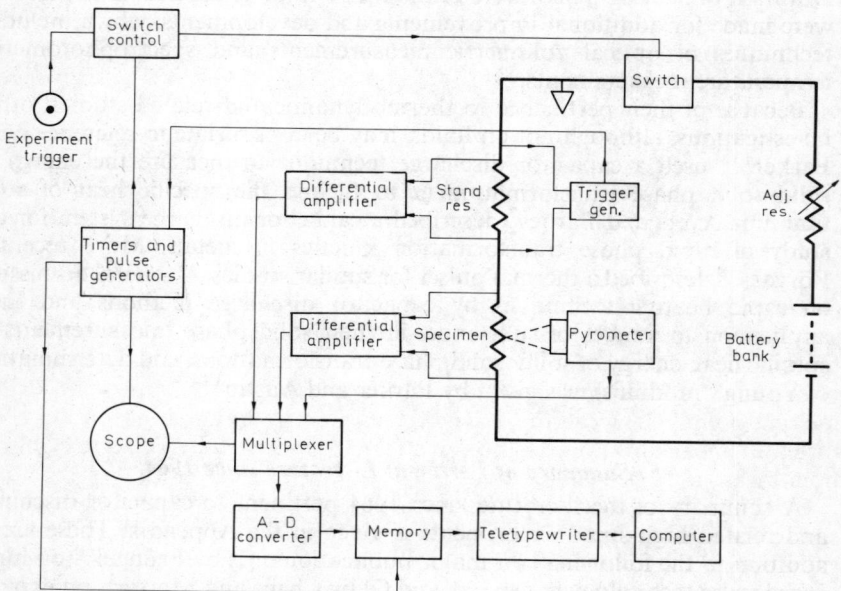

Figure 8. Functional diagram of a millisecond-resolution thermodynamic measurement system.

electric power-pulsing circuit and associated control, measuring, and calibration circuits. The pulsing circuit includes the specimen in series with a battery bank (28 V), a standard resistance (0.001 Ω), an adjustable resistance, and a switch. The specimen (for measurements in the solid phase) is a tube of the following nominal dimensions: length, 100 mm; outside diameter, 6 mm; wall thickness. 0.5 mm. A small rectangular hole (1 × 0.5 mm) in the wall at the middle of the specimen provides an approximation to blackbody conditions for optical temperature measurements. Depending on the investigations, specimens having other geometrical forms (rods, strips, wires) may be used. The test chamber is designed for conducting experiments with the specimen in a vacuum or in a controlled atmosphere. Timing of various events, such as closing and opening of the switch and triggering of electronic equipment, is achieved automatically by utilizing logic circuits and a series of time-delay units.

Current through the specimen is determined from the measurement of the potential difference across the standard resistance placed in series with the specimen. Potential difference across the middle two thirds of the specimen is measured using spring-loaded, knife-edge probes. A high-speed photoelectric pyrometer[57], constructed specifically for this system, permits 1 200 evaluations of the specimen temperature per second.

Experimental quantities are recorded with a high-speed digital data acquisition system consisting of a multiplexer, an analog-to-digital converter (13 bits), a core memory (16 bits per word) and various control and interfacing equipment. The system is capable of recording sets of data corresponding to temperature, voltage, and current approximately every 0.4 ms with a full-scale signal resolution of one part in 8 192 ($8 192 = 2^{13}$). In a subsecond duration experiment, up to 2 048 data points can be digitized and stored in the memory. At the end of the experiment, data are retrieved via a teletypewriter and processed immediately using a time-shared computer.

Details of the design and operational characteristics of the entire high-speed measurement system are given in an earlier publication[27].

The above system (with a tubular specimen) was used to measure specific heat, electrical resistivity, hemispherical total emittance, normal spectral emittance of selected refractory metals (niobium[28], molybdenum[34], tantalum[36], and tungsten[37]) in the range from approximately 2 000 K to near their respective melting points.

If the current during the pulse heating is not interrupted, the specimen (in tubular form) reaches its melting point and then it collapses under gravitational forces, thereby opening the electrical circuit. However, it has been possible to observe a plateau in temperature corresponding to the region of solid and liquid equilibria. The results of such experiments yielded the melting points of niobium[30], molybdenum[35], and tungsten[31]. Duration of the experiments (heating the specimen from room temperature to its melting point) was, in all cases, less than 1 s.

By utilizing a specimen in the form of a strip, it is possible to measure surface radiance temperature at the melting point in a subsecond duration experiment. This, in addition to a separate measurement of the melting point, yields normal spectral emittance at the melting point. Results for tantalum[26] showed that emittance decreased almost linearly as melting progressed.

However, similar experiments on niobium[30] did not show any variation in normal spectral emittance during melting.

Estimated inaccuracies in measured properties in a typical experiment are: 3 per cent in specific heat, 0.5 per cent in electrical resistivity, 3 per cent in thermal radiation properties, and 10 K in melting point (at 3000 K).

2. *An Example of Thermodynamic Measurements at and above Melting Points*

In the following paragraphs a preliminary experiment is described which shows the possibility of applying the above-described millisecond-resolution system for thermodynamic investigations at and above the melting point of metals[33].

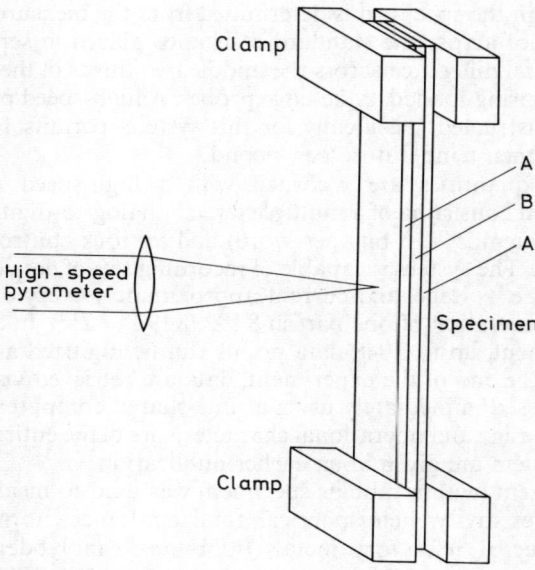

Figure 9. Schematic diagram showing the arrangement of the composite specimen, clamps, and surface radiance temperature measurement system (dimensions are not to scale). Strip of material 'B' is sandwiched between two strips of material 'A'. The melting point of 'B' is lower than that of 'A'.

In this experiment, a composite specimen consisting of three strips in parallel was used (*Figure 9*). The two outer strips were made of tantalum, and the inner strip was made of niobium. Other elements could also have been used as long as the melting point of the inner strip was lower than that of the outer strips. The nominal dimensions of the strips were: length, 100 mm; width, 6 mm; thickness, 0.25 mm. The function of the two outer strips was to contain the niobium during its melting and postmelting periods.

The specimen was pulse heated from room temperature to approximately 3000 K in 0.8 s. During this period, the surface radiance temperature of tantalum was measured with the high-speed pyrometer. The circular area viewed by the pyrometer was 0.2 mm in diameter. The results when plotted as surface radiance temperature versus time (*Figure 10*) show a plateau that corresponds to the melting of the niobium.

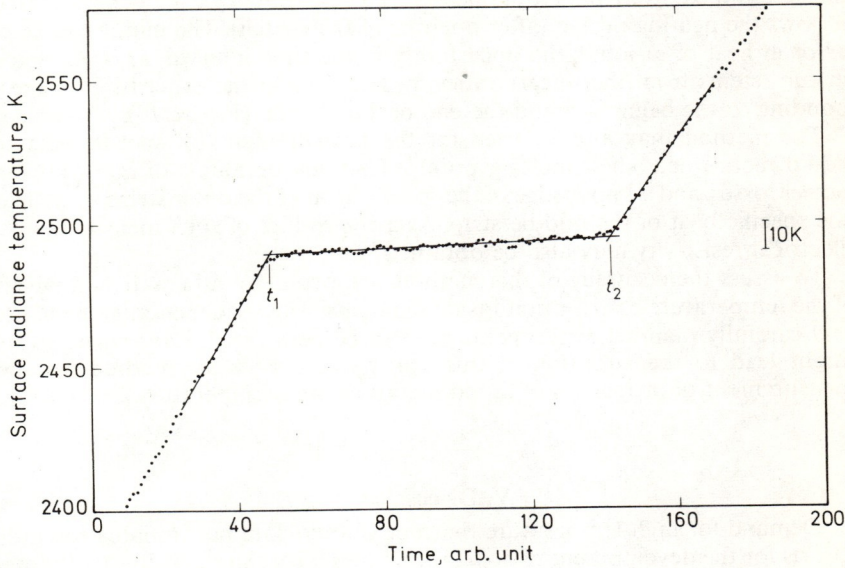

Figure 10. Variation of surface radiance temperature of a composite specimen, similar to that shown in *Figure 9*, as a function of time during its heating period. The inner strip is niobium and the two outer strips are tantalum. The plateau corresponds to melting of niobium. One time unit is 0.833 ms.

Using the least squares method, the temperature data in the three regions (premelting, during melting, and postmelting of niobium) were separately fitted to linear functions of time. The corresponding standard deviations (individual point) were 0.7 K, 0.8 K, and 0.6 K, respectively. A difference (increase) of approximately 6 K in temperature was observed between the beginning and the end of the plateau. This may be partly due to temperature non-uniformities in the composite specimen.

Data from the above experiment may yield the following properties for the substance forming the middle strip: melting point, heat of fusion, specific heat (solid and liquid phases), and electrical resistivity (solid and liquid phases).

The melting point may be obtained from measurements of the surface radiance temperature (at the plateau) of the outer strip and from a knowledge of the normal spectral emittance (which may be obtained from separate experiments) for the outer strip. However, this may not be an accurate method of measuring the melting point for most substances.

The technique shows considerable promise for measuring heats of fusion of substances having high melting points. Heat of fusion may be obtained by measuring power input to the specimen (from measurements of current and voltage) and integrating it over the duration of the plateau, that is over the period $t_2 - t_1$ as shown in *Figure 10*. A correction for power losses, which in this case may be mostly due to thermal radiation, should be made. This may be determined from data obtained during the initial cooling period that follows the heating period (after opening of the switch). The major source of error in heat of fusion is the uncertainty in the time interval, $t_2 - t_1$, which in turn stems from the uncertainties in determining the exact times corresponding to the beginning and the end of the plateau (*Figure 10*).

The method may also be used for the measurements of specific heat of liquid metals near their melting points. The measurements of input power, power losses, and a knowledge of the specific heat of the outer strips may yield the specific heat of the middle strip. As a byproduct of such measurements, electrical resistivity may also be obtained.

To assess the accuracy of this method, a thorough mathematical analysis of the temperature distribution inside the composite specimen must be made and carefully planned experiments have to be performed. Favorable results might lead to the adoption of this scheme as a possible method for the measurement of properties of liquid metals at high temperatures.

V. Discussion

Demand for high temperature thermodynamic data has initiated research efforts for the development of novel measurement techniques. In this chapter discharge techniques were discussed and their potentials as tools for thermodynamic and related research were pointed out. In addition to the transient techniques, other methods (steady-state and quasi steady-state) have also been explored by various investigators. Among these the following may be included: (1) levitation calorimetry for measuring heat of fusion and enthalpy of solids and liquids[14, 41, 42, 138], and (2) ohmic heating of metals as rotating liquid pipes[64]. However, both of the above have temperature limitations, since they require that the specimen be exposed to high temperatures for relatively long periods (minutes or more). External heating may also be considered in specific applications. A discussion on rapid heating by pulsed laser techniques is given in the literature[129].

Although some of these techniques may be used in high temperature research, it is likely that the ultimate development of truly high temperature thermodynamic measurement facilities will involve capacitor discharge techniques. At the present time no systematic effort has been made in this direction. However, the results of a few preliminary investigations have been encouraging and have shown the feasibility of thermodynamic measurements in experiments having microsecond resolution. Integration of existing capacitor discharge and related techniques with some modifications and further advances in the areas of high-speed measurement of temperature and electrical quantities may result in the development of accurate microsecond- and even submicrosecond-resolution thermodynamic measurement systems.

ELECTRICAL DISCHARGE TECHNIQUES

VI. Appendix

Summary of literature since 1964 pertinent to capacitor discharge and related high-speed experiments.

1. Capacitor discharge systems (Components and their characteristics)

Investigator	Year	Ref.	Description
Broadbent	1964	20	Design of a spark gap, which is composed of two main electrodes and a heated filament, is given. By adjusting temperature of the filament, triggering of the gap over a wide range of voltages is achieved. Results of tests up to 250 kV are reported.
Harraway	1964	68	A spark gap, triggered with an exploding wire and operating in the range 0.5 to 6 kV, is described.
Kassirov and Koval'chuk	1964	80	Time lag of the discharge in the electrical breakdown of vacuum gaps (0.1 to 1 mm) with different degrees of overvoltage on the gap is investigated.
Nesterikhin et al.	1964	104	Pulsed breakdown of small gaps in the nanosecond range are investigated.
Schrank and Henry	1964	125	Construction and operation of a trigger system designed to activate a spark gap (30 kV, 5000 A) are described.
Issinskii and Myznikov	1965	76	A spark gap, capable of switching 1 to 60 kA in 40 to 150 kV circuits, is described. Switching time is reported to be reproducible within 5 ns.
James et al.	1965	77	Development of solid dielectric switches for 100 kV, 1 MA, capacitor discharge experiments is described.
Medley et al.	1965	100	A system employing a pulsed ultra-violet light source for triggering a spark gap is described.
Oertel and Williams	1965	109	An optical method is described for monitoring the breakdown of a series of spark gaps with a time resolution of 1 ns.
Ornstein et al.	1965	111	A vacuum spark gap is described, which is triggered by plasmoids and can be operated in the range 50 V to 18 kV.
Pendleton	1965	117	Investigations of a laser-triggered spark gap are reported. Delay between the laser pulse and current flow across the gap is reported to be less than 10 ns.
Konotop et al.	1965	81	Design of a low inductance capacitor and spark gap combination is described.
Bishop and Edmonds	1969	17	An air spark gap, capable of operating to 40 kV with peak current of 250 kA, is described.
Alcock et al.	1970	1	A laser-triggered spark gap, having subnanosecond switching time, is described.
Emel'yanov et al.	1970	54	Volt/coulomb characteristics of a spark gap are obtained.
Kuswa and Stallings	1970	86	An air spark gap for switching voltages up to 30 kV at currents to 500 kA is described.
Ware et al.	1971	141	Design and operational characteristics of a vacuum spark gap, which is capable of operating in the range 0.1 to 50 kV for currents up to 400 kA, is described.
Dewhurst et al.	1972	46	Characteristics of laser-triggered spark gaps are investigated.
Post and Chen	1972	119	A fast closing switch capable of handling 36 kJ at 85 kV is described.

2. Electrical Measurements

Investigator	Year	Ref.	Description
Belkin and Lukashev	1964	9	Pulse voltage measurements with a capacitive voltage divider are described for single voltage pulses having a duration of approximately 1 µs and amplitudes of up to 1500 kV.
Leavitt et al.	1965	88	A fast high voltage probe is described which utilizes a capacitive divider. Operates up to 600 kV.
Cassidy et al.	1966	22	A voltage measurement technique in capacitor discharge experiments is described. The scheme utilizes two dividers and an adjustable coil to compensate for the induced voltages in the measuring circuit. Time resolved measurements of current, power, and energy are also included.
Markiewicz and Emmett	1966	99	A technique for the measurement of pulse currents is described, which is based on measuring the time derivative of current (using a coil) and obtaining the current by electronic integration of this signal.
Dunn and Maitland	1967	53	An inductively loaded probe for monitoring current pulses (0.1 to 10 µs duration) is described. The device can discriminate between current and rate of change of current.
Malewski	1968	97	A method of improving the response time of a coaxial shunt is described. The faster response time (9 ns) is achieved by including within the measuring tube a compensating pickup loop.
Takeshita and Sasano	1968	133	A technique for the measurement of pulse currents based on the Faraday rotation of a laser beam is described.
Tawara	1968	134	A method for measuring pulsed current and magnetic field utilizing the Faraday effect is described.
Andreev and Liukonen	1970	5	The design of a shunt, in which the active element is a thin layer of Aquadag, is described. The reported response time is 10^{-11} s.
Cassidy et al.	1970	24	Development and evaluation of an electro-optical high-voltage pulse measurement technique are presented.
Thomas	1970	135	A review of various techniques used for the measurement of pulse currents and voltages is given.
Thornton	1970	136	The design of a metal-foil current shunt for the measurement of pulse currents of the order of 100 kA and 100 ns is presented.
Pellinen and Heurlin	1971	115	Pulse voltage measurements using a resistive voltage divider are described (nanosecond rise time for voltages up to 1 MV).
Pellinen and Spence	1971	116	The use of an integrating Rogowski coil for measuring mega-ampere current pulses is presented. The reported response time is about 2 ns.
Schwab	1971	126	The design of low-resistance shunts for pulse current measurements is presented.
Creed et al.	1972	43	Evaluation of impulse measurements is discussed.
Harada et al.	1972	67	A voltage divider using optoelectronics for impulse voltage measurements is described (response time is less than 5 ns).
Simon and Leroy	1972	128	A discussion of impulse voltage measuring systems is given.

3. Temperature measurements

Investigator	Year	Ref.	Description
Cassidy et al.	1968	23	A technique based on measurements at four wavelengths using a spectrograph is described. Performance is demonstrated in exploding wire experiments.
Korneff	1971	82	A three color optical pyrometer is described which has 1 µs time resolution. Performance is demonstrated in exploding wire experiments.
Polyakov and Kotel'nikova	1971	118	A procedure is described for measuring the temperature of a surface in contact with a hot radiating gas. The performance is tested in a shock tube during reflection of a shock wave in air.
Lapworth et al.	1972	87	An emission–absorption technique is described, which operates at 4.7 µm. Performance is demonstrated in shock tube experiments. In the range 600 to 4000 K, reported inaccuracy in temperature measurements is 1.5 to 2 per cent.
Urtiew and Grover	1972	139	A technique based on total radiation detection utilizing a solid state detector is described. Performance is demonstrated in shock compression experiments on metals in the range 3.5×10^5 to 5.5×10^5 bar and approximately 1600 to 3000 K.

4. Capacitor discharge studies (general)

Investigator	Year	Ref.	Description
Loginov	1964	94	An apparatus is described for the study of electronic absorption spectra utilizing the exploding wire technique.
Gol'ts and Sadkovich	1965	61	Description of a technique to explode a mercury jet is given. The scheme provides repeated electrical explosions.
Moesta and Breuer	1965	102	Description of a method for the explosion of liquid metals is given. Explosion can be repeated every 6 s.
Oktay	1965	110	Optimum discharge in an exploding wire circuit is investigated in terms of wire cross-sectional area.
Bennett	1966	11	Non-linear, ordinary differential equations applicable to damped oscillatory circuit with exploding wires are developed. Approximate solutions to these equations are obtained.
Cassidy and Abramowitz	1967	25	A spectroscopic technique for continuous and time-resolved study of the atomic and molecular species produced by electrically exploded wires is described. Results are presented from experiments with aluminum wires exploded in various controlled atmospheres.
Chace and Fish	1967	38	Change in the shape of a non-cylindrical conductor during explosion is investigated using a pulsed x-ray method. The change is attributed mainly to unbalanced magnetic pressures.
Vlastos	1967	140	Discharge characteristics of exploding wires with particular reference to 'dwell time', are investigated.
Cassidy et al.	1968	23	Various investigations on exploding wires are described. These include: time-resolved measurements of electrical energy, power, voltage, and current

4. Capacitor discharge studies (general)—*continued*			
Investigator	Year	Ref.	Description
Lebedev	1968	89	during discharges; time-resolved photography; time-resolved measurement of intensity and spectral distribution of emitted radiation; and time-resolved absorption spectra from the products of the discharge. Possibility of using the exploding wire technique for investigations of metals at high temperatures is discussed. Preliminary experiments to measure the jump in electrical resistivity of metals during melting are described.
Ben-Yosef and Rubin	1969	10	Measurement of electrical resistivity of liquid copper as a function of particle density in an exploding wire experiment is described.
Calker and Erb	1969	21	Characteristics of exploding wires with particular emphasis on vaporization and the causes of re-ignition of the metal vapor are studied.
Erb	1969	55	Expansion of exploding wires is investigated using flash x-ray and high-speed photographic techniques.
Lebedev et al.	1969	90	Particle size during explosion of tungsten wires is studied.
Neumann et al.	1969	105	Exploding wire experiments are described which can be performed under constant current conditions over the entire period of the event. An inductive storage system is used.
Stenerhag et al.	1969	131	Experiments on exploding tungsten wires as a flash x-ray source for taking radiographs of high-speed events are described.
Weber and Shear	1969	142	Studies on measuring particle size in exploding wire experiments utilizing a light scattering technique are described.
Ya'akobi	1969	145	Expansion measurements on exploding wires at atmospheric pressure using a high-speed camera are reported.
Graham et al.	1970	62	Shock wave measurements resulting from exploding wire experiments are described.
Lundquist and Vlastos	1970	96	Behavior of the temperature, the electron density, and the thermal conductivity of the restrike channel of an exploding wire is studied.
Salge et al.	1970	123	Exploding wire experiments in high inductance capacitor discharge circuits are described.
Bennett	1971	12	Non-uniformity of current density during typical wire explosions is estimated.
Bhat and Jordan	1971	15	Experiments on explosion of bare and insulated copper wires are described.
Erb and Calker	1971	56	Explosion characteristics of nickel, molybdenum and tungsten are investigated.
Hoffman	1971	71	An apparatus for conducting investigations of fast chemical kinetics in solution from the microsecond to the nanosecond time range is described. A sample solution can be resistively heated a maximum of 10 K within 50 ns using the discharge of a coaxial cable capacitor.
Korneff and Chace	1971	83	A method utilizing a twin tube flash x-ray unit is described for determining density/radius relations during wire explosions.

ELECTRICAL DISCHARGE TECHNIQUES

4. Capacitor discharge studies (general)—*continued*

Investigator	Year	Ref.	Description
Malkin and Pyshnov	1971	98	Dependence of the parameters of a high-current pulsed discharge on the current and on the initial pressure is investigated.
Stenerhag *et al.*	1971	132	Effect of wire parameters on the emission of hard x-rays from exploding wires is studied.
Bless	1972	18	Production of pressures up to 3×10^5 bar in solid rods using capacitor discharge-powered magnetic pinch is discussed.

5. Capacitor discharge studies (thermodynamics related)

Investigator	Year	Ref.	Description
Kroepelin and Neumann	1968	84	An apparatus for performing exploding wire experiments (to generate strong cylindrical shock waves) in liquids is described. The system is developed for possible experimental determination of the equation of state of liquids at high pressures (up to 2×10^5 bar).
Neumann *et al.*	1970	106	Extension of work described in the above item. System development and operational characteristics are described.
Dikhter and Lebedev	1970	49	An application of exploding wire techniques to the measurement of specific heat (solid and liquid states) and heat of fusion of tungsten is described.
Dikhter and Lebedev	1971	50	Results of measurements of specific heat and electrical resistivity of molybdenum (solid and liquid phases) are reported.
Lebedev *et al.*	1971	91	Results of measurements of heat of fusion and resistivity ratio (liquid to solid) at the melting point for selected refractory metals are given.
Henry	1971	70	A capacitor discharge technique is described for the measurement of pressure, volume, enthalpy, and resistivity of equilibrium thermodynamic states of liquid metals at high temperatures and high pressures.

6. High-speed photography

Investigator	Year	Ref.	Description
Zavoisky and Fanchenko	1965	146	A review of image converter high-speed photography is given.
Hecht *et al.*	1966	69	A technique for high-speed stroboscopic photography utilizing a Kerr cell modulated laser source is described.
Holland *et al.*	1966	73	A review of high-speed photographic instruments is given.
Hyzer	1966	75	High-speed photography is reviewed.
Nilsson and Högberg	1968	107	Various developments related to high-speed photography are presented (proceedings of a conference on high-speed photography).

6. High-speed photography—*continued*

Investigator	Year	Ref.	Description
Bird *et al.*	1969	16	Photography (general). The state-of-the-art of the efficiency of radiation detection by photographic films is presented.
Kryder and Humphrey	1969	85	Description of a Kerr magneto-optic camera is given.
Davies	1970	44	A technique which utilizes an optically modulated laser carrier beam for calibrating streak cameras is described.
Price	1970	120	A framing camera is described which is used to observe the events during the breaking of electrical contacts.
Duguay and Mattick	1971	52	An ultrafast camera with 10 ps framing time is described.
Lieber and Sutphin	1971	93	A framing camera, with an effective rate of 10^8 frames per second, is described.

7. High-speed radiation modulation

Investigator	Year	Ref.	Description
Jones	1964	79	A review of methods of modulating light at extreme frequencies is given. Mechanical shutters as well as electro-optical modulators are discussed.
Brixner	1965	19	Performance characteristics of rotating steel mirrors are given.
Dike and Kemp	1965	48	A closing (less than 10 µs) mechanical light shutter is described.
Roberts	1965	122	A pulsed image intensifying tube is described for use as a shutter.
Weiss	1965	143	An opening (less than 0.5 µs) shutter is described. Its operation is based on exploding a thin film (originally deposited on a glass) to clear the optical path.
Schenck	1966	124	Development and application of optical shutters based on the Kerr effect are discussed.
Avellone *et al.*	1967	6	A mechanical sliding shutter operated by electromagnetic induction is described (80 µs is required for 1 mm opening).
Steele	1968	130	Theory of electro-optical modulators is presented.
Michon *et al.*	1969	101	An electro-optical shutter, with a switching time less than 1 ns, is described.
Fralick and Zipf	1970	59	An electro-optical shutter, having a cut-off time of 25 ns, is described.
Wood and Miller	1970	144	A mechanical shutter (speed 70 m s^{-1}), activated by an exploding wire, is described.
Hoffman and Jovin	1971	72	A mechanical chopper, used for laser light and having nanosecond risetime, is described.

8. Short light pulses

Investigator	Year	Ref.	Description
Andreev and Vanyukov	1965	4	Production of very short (of the order of nanoseconds) light pulses utilizing an exploding wire circuit is described.
De Maria et al.	1967	45	Generation of ultrashort (down to 10^{-12} s duration) light pulses using lasers is discussed.
Rentzepis and Mitschele	1970	121	Generation of picosecond duration light pulses and their application in spectroscopy are discussed.
Morgan and Peacock	1971	103	An electro-optic switching circuit is described which can produce a pulse of duration 1 ns or less from a laser.

VII. References

[1] Alcock, A. J., M. C. Richardson and K. Leopold, *Rev. Sci. Instrum.* **41**, 1028 (1970).
[2] Amsel, G. and C. Zajde, *Rev. Sci. Instrum.* **35**, 1538 (1964).
[3] Anderson, L. K. and B. J. McMurtry, *Applied Optics*, **5**, 1573 (1966).
[4] Andreev, S. I. and M. P. Vanyukov, *Soviet Phys-Tech. Phys.* **9**, 1443 (1965).
[5] Andreev, S. I. and R. A. Liukonen, *Instruments and Experimental Techniques (USSR)*, No. 3, 868 (1970).
[6] Avellone, J. C., P. L. Byard and F. L. Damm. *Rev. Sci. Instrum.* **38**, 1808 (1967).
[7] Ayers, J. D. and W. C. Barnes, *Rev. Sci. Instrum.* **42**, 302 (1971).
[8] Beckett, C. W. and A. Cezairliyan, 'High-speed thermodynamic measurements and related techniques', in *Experimental Thermodynamics*, Vol. I, *Calorimetry of Non-Reacting Systems*, p 551. J. P. McCullough and D. W. Scott, eds. Butterworths: London (1968).
[9] Belkin, V. M. and A. A. Lukashev, *Instruments and Experimental Techniques (USSR)*, No. 6, 1231 (1964).
[10] Ben-Yosef, N. and A. G. Rubin, *Phys. Rev. Letters*, **23**, 289 (1969).
[11] Bennett, F. D., *Physics of Fluids*, **9**, 471 (1966).
[12] Bennett, F. D., *J. Appl. Phys.* **42**, 2835 (1971).
[13] Bennett, F. K. and C. G. Klotz, 'A high-speed data acquisition system', in *Proceedings of the Symposium on Engineering Problems of Controlled Thermonuclear Research*, p 91. Lawrence Radiation Laboratory, Livermore, California (1965).
[14] Berezin, B. Ya., V. Ya. Chekhovskoi and A. E. Sheindlin, *High Temp. High Press.* **3**, 287 (1971).
[15] Bhat, B. K. and I. B. Jordan, *J. Appl. Phys.* **42**, 809 (1971).
[16] Bird, G. R., R. C. Jones and A. E. Ames, *Applied Optics*, **8**, 2389 (1969).
[17] Bishop, A. E. and G. D. Edmonds, *J. Sci. Instrum. (J. Phys. E)*, **2**, 414 (1969).
[18] Bless, S. J., *J. Appl. Phys.* **43**, 1580 (1972).
[19] Brixner, B., *Rev. Sci. Instrum.* **36**, 1297 (1965).
[20] Broadbent, T. E., *Brit. J. Appl. Phys.* **15**, 97 (1964).
[21] Calker, J. van and W. Erb, *Z. Angew. Phys.* **26**, 291 (1969).
[22] Cassidy, E. C., S. W. Zimmerman and K. K. Neumann, *Rev. Sci. Instrum.* **37**, 210 (1966).
[23] Cassidy, E. C., S. Abramowitz and C. W. Beckett, *US Nat. Bur. Stand. Monogr. No. 109* (1968).
[24] Cassidy, E. C., H. N. Cones and S. R. Booker, *IEEE Trans. on Instrum. and Meas.* **19**, 395 (1970).
[25] Cassidy, E. C. and S. Abramowitz, *Appl. Spectrosc.* **21**, 360 (1967).
[26] Cezairliyan, A., *High Temp. High Press.* **2**, 501 (1970).
[27] Cezairliyan, A., *J. Res. Nat. Bur. Stand.* **75C** (Eng. and Instr.), 7 (1971).
[28] Cezairliyan, A., *J. Res. Nat. Bur. Stand.* **75A** (Phys. and Chem.), 565 (1971).
[29] Cezairliyan, A., *Applied Optics*, **10**, 1178 (1971).
[30] Cezairliyan, A., *High Temp. High Press.* **4**, 453 (1972).
[31] Cezairliyan, A., *High Temperature Science*, **4**, 248 (1972).

[32] Cezairliyan, A., 'Measuring transient high temperatures by optical pyrometry', in *Temperature*, Vol. IV, p 657. H. H. Plumb, ed. Instrument Society of America: Pittsburgh (1972).
[33] Cezairliyan, A., in preparation.
[34] Cezairliyan, A., M. S. Morse, H. A. Berman and C. W. Beckett, *J. Res. Nat. Bur. Stand.* **74A** (Phys. and Chem.), 65 (1970).
[35] Cezairliyan, A., M. S. Morse and C. W. Beckett, *Rev. Int. Hautes Tempér. et Réfract.* **7**, 382 (1970).
[36] Cezairliyan, A., J. L. McClure and C. W. Beckett, *J. Res. Nat. Rur. Stand.* **75A** (Phys. and Chem.), 1 (1971).
[37] Cezairliyan, A. and J. L. McClure, *J. Res. Nat. Bur. Stand.* **75A** (Phys. and Chem.), 283 (1971).
[38] Chace, W. G. and C. V. Fish, *J. Appl. Phys.* **38**, 3986 (1967).
[39] Chace, W. G. and H. K. Moore, eds. *Exploding Wires*, Plenum: New York, Vol. III (1964), Vol. IV (1968).
[40] Chaney, N. K., V. C. Hamister and S. W. Glass, *Trans. Electrochem. Soc.* **67**, 107 (1935).
[41] Chaudhuri, A. K., D. W. Bonnell, L. A. Ford and J. L. Margrave, *High Temperature Science*, **2**, 203 (1970).
[42] Chekhovskoi, V. Ya., A. E. Sheindlin and B. Ya. Berezin, *High Temp. High Press.* **2**, 301 (1970).
[43] Creed, F. C., M. M. C. Collins, Aa. Pedersen and P. Lausen, *IEEE Trans. on Power Apparatus and Systems*, **91**, 485 (1972).
[44] Davies, T. J., *Rev. Sci. Instrum.* **41**, 920 (1970).
[45] De Maria, A. J., D. A. Stetser and W. H. Glenn Jr, *Science*, **156**, 1557 (1967).
[46] Dewhurst, R. J., G. J. Pert and S. A. Ramsden, *J. Phys. D: Appl. Phys.* **5**, 97 (1972).
[47] DiDomenico, M., W. M. Sharpless and J. J. McNicol, *Applied Optics*, **4**, 677 (1965).
[48] Dike, R. S. and E. L. Kemp, *Rev. Sci. Instrum.* **36**, 1256 (1965).
[49] Dikhter, I. Ya. and S. V. Lebedev, *High Temp. High Press.* **2**, 55 (1970).
[50] Dikhter, I. Ya. and S. V. Lebedev, *High Temperature*, **9**, 845 (1971).
[51] Drawin, H. W., *High Temp. High Press.* **2**, 359 (1970).
[52] Duguay, M. A. and A. T. Mattick, *Applied Optics*, **10**, 2162 (1971).
[53] Dunn, M. H. and A. Maitland, *J. Sci. Instrum.* **44**, 555 (1967).
[54] Emel'yanov, Yu. M., V. G. Babayan and Z. I. Ashurly, *Soviet Phys.-Tech. Phys.* **14**, 933 (1970).
[55] Erb, W., *Z. Angew. Phys.* **26**, 295 (1969).
[56] Erb, W. and J. van Calker, *Z. Angew. Phys.* **31**, 71 (1971).
[57] Foley, G. M. *Rev. Sci. Instrum.* **41**, 827 (1970).
[58] Forgacs, R. L. *Rev. Sci. Instrum.* **43**, 302 (1972).
[59] Fralick, R. D. and E. C. Zipf. *Rev. Sci. Instrum.* **41**, 47 (1970).
[60] Früngel, F. *High-Speed Pulse Technology*, Volumes I and II. Academic Press: New York (1965).
[61] Gol'ts, E. Ya. and N. P. Sadkovich, *JETP (Letters)*, **2**, 288 (1965).
[62] Graham, M. E., R. E. Wengler and D. V. Keller, *Instrum. Soc. Amer. Trans.* **9**, 133 (1970).
[63] Griem, H. R. *Plasma Spectroscopy*, McGraw-Hill: New York (1964).
[64] Grosse, A. V., J. A. Cahill, W. L. Liddell, W. J. Murphy and C. S. Stokes. *Science*, **160**, 528 (1968).
[65] Grover, F. W. *Inductance Calculations*, Van Nostrand: New York (1946).
[66] Hanson, R. K. *Rev. Sci. Instrum.* **43**, 394 (1972).
[67] Harada, T., T. Kawamura, K. Kishi, Y. Aoshima, N. Ohira, K. Takigami and Y. Horiko, *IEEE Trans. on Power Apparatus and Systems*, **91**, 494 (1972).
[68] Harraway, R. A. *J. Sci. Instrum.* **41**, 399 (1964).
[69] Hecht, G. J., G. B. Steel and A. K. Oppenheim, *Instrum. Soc. Amer. Trans.* **5**, 133 (1966).
[70] Henry, K. W. *A Technique for Measuring the Pressure, Volume, Enthalpy, and Resistance of Equilibrium Thermodynamic States of Liquid Metals at High Temperatures and Pressures*, University of California, Lawrence Radiation Laboratory, *UCRL-51035* (1971).
[71] Hoffmann, G. W. *Rev. Sci. Instrum.* **42**, 1643 (1971).
[72] Hoffmann, G. W. and T. M. Jovin. *Applied Optics*, **10**, 218 (1971).
[73] Holland, T. E., T. J. Healey and C. H. Bagley, *Instrum. Soc. Amer. Trans.* **5**, 5 (1966).
[74] Hoyaux, M. F. *High Temp. High Press.* **2**, 17 (1970).
[75] Hyzer, W. G. *Instrum. Soc. Amer. Trans.* **5**, 1 (1966).
[76] Issinskii, I. B. and K. P. Myznikov, *Instruments and Experimental Techniques (USSR)*, No. 3, 605 (1965).

77 James, T. E., K. Harries and R. D. Medford, 'Development of fast 100-kV, 1-MA, solid dielectric switches and associated triggering studies', *Proceedings of the Symposium on Engineering Problems of Controlled Thermonuclear Research*, p 77. Lawrence Radiation Laboratory (1965).
78 Jobes, F. C. *Rev. Sci. Instrum.* **39**, 1429 (1968).
79 Jones, O. C. *J. Sci. Instrum.* **41**, 653 (1964).
80 Kassirov, G. M. and B. M. Koval'chuk, *Soviet. Phys.-Tech. Phys.* **9**, 377 (1964).
81 Konotop, V. V., V. Ya. Linetskii and S. M. Fertik, *Instruments and Experimental Techniques (USSR)*, No. 6, 1430 (1965).
82 Korneff, T. *Rev. Sci. Instrum.* **42**, 1561 (1971).
83 Korneff, T. and W. Chace. *Rev. Sci. Instrum.* **42**, 1184 (1971).
84 Kroepelin, H. and K. K. Neumann, 'An exploding-wire-driven, shock-wave generator for thermodynamic measurements in liquids at very high pressures', in *Proceedings of the Fourth Symposium on Thermophysical Properties*, p 446. J. R. Moszynski, ed., American Society of Mechanical Engineers: New York (1968).
85 Kryder, M. H. and F. B. Humphrey. *Rev. Sci. Instrum.* **40**, 829 (1969).
86 Kuswa, G. and C. Stallings, *Rev. Sci. Instrum.* **41**, 1429 (1970).
87 Lapworth, K. C., L. A. Allnutt and J. R. Pendlebury. 'Short duration temperature measurements by infra-red emission–absorption', in *Temperature*, Vol. IV, p 665. H. H. Plumb, ed. Instrument Society of America: Pittsburgh (1972).
88 Leavitt, G. E., J. D. Shipman Jr and I. M. Vitkovitsky. *Rev. Sci. Instrum.* **36**, 1371 (1965).
89 Lebedev, S. V. *High Temperature*, **6**, 150 (1968).
90 Lebedev, S. V., B. V. Lukin, A. E. Rautbort and A. I. Savvatimskii. *High Temperature*, **7**, 951 (1969).
91 Lebedev, S. V., A. I. Savvatimskii and Yu. B. Smirnov. *High Temperature*, **9**, 578 (1971).
92 Lee, R. D. and E. Lewis. *Applied Optics*, **5**, 1858 (1966).
93 Lieber, A. J. and H. D. Sutphin. *Rev. Sci. Instrum.* **42**, 1663 (1971).
94 Loginov, V. A. *Instruments and Experimental Techniques (USSR)*, No. 1, 181 (1964).
95 Lucovsky, G. and R. B. Emmons. *Applied Optics*, **4**, 697 (1965).
96 Lundquist, S. and A. E. Vlastos. *J. Appl. Phys.* **41**, 4830 (1970).
97 Malewski, R. *Rev. Sci. Instrum.* **39**, 90 (1968).
98 Malkin. O. A. and A. V. Pyshnov. *High Temperature*, **9**, 802 (1971).
99 Markiewicz, J. P. and J. L. Emmett. *Applied Optics*, **5**, 1687 (1966).
100 Medley, S. S., F. L. Curzon and C. C. Daughney. *Rev. Sci. Instrum.* **36**, 713 (1965).
101 Michon, M., H. Guillet, D. Le Goff and S. Raynaud. *Rev. Sci. Instrum.* **40**, 263 (1969).
102 Moesta, H. and D. Breuer. *Rev. Sci. Instrum.* **36**, 1372 (1965).
103 Morgan, P. D. and N. J. Peacock. *J. Phys. (E)–Scientific Instruments*, **4**, 677 (1971).
104 Nesterikhin, Yu. E., V. S. Komel'kov and E. Z. Meilikhov. *Soviet Phys.–Tech. Phys.* **9**, 29 (1964).
105 Neumann, K. K., J. Salge, R. Brilka and T. Redeker, *Z. Angew. Phys.* **28**, 65 (1969).
106 Neumann, K. K., J. Salge, H. Kroepelin and R. Willms, 'Exploding-wire experiments in high-inductive discharge circuits for thermodynamic measurements', in *Proceedings of the Fifth Symposium on Thermophysical Properties*, p 209. C. F. Bonilla, ed., American Society of Mechanical Engineers: New York (1970).
107 Nilsson, N. R. and L. Högberg, eds. *Proceedings of the Eighth International Congress on High-Speed Photography*, Wiley: New York (1968).
108 Null, M. R. and W. W. Lozier. *J. Opt. Soc. Amer.* **52**, 1156 (1962).
109 Oertel, G. K. and M. D. Williams. *Rev. Sci. Instrum.* **36**, 672 (1965).
110 Oktay, E. *Rev. Sci. Instrum.* **36**, 1327 (1965).
111 Ornstein, L. Th. M., C. A. J. Hugenholtz and H. A. van der Laan. *J. Sci. Instrum.* **42**, 659 (1965).
112 Park, J. H. *J. Res. Nat. Bur. Stand.* **39**, 191 (1947).
113 Parker, R. *Trans. Met. Soc. A⋯IE*, **233**, 1545 (1965).
114 Parker, R. and A. L. Austin. 'Rapid-heating techniques by capacitor-discharge methods and their applications', in *Techniques of Material Preparation and Handling*, p 307. R. F. Bunshah, ed., Interscience: New York (1968).
115 Pellinen, D. G. and S. Heurlin. *Rev. Sci. Instrum.* **42**, 824 (1971).
116 Pellinen, D. G. and P. W. Spence. *Rev. Sci. Instrum.* **42**, 1699 (1971).
117 Pendleton, W. K. *Rev. Sci. Instrum.* **36**, 1546 (1965).
118 Polyakov, Yu. A. and N. V. Kotel'nikova. *High Temperature*, **9**, 737 (1971).
119 Post, R. S. and Y. G. Chen. *Rev. Sci. Instrum.* **43**, 622 (1972).

[120] Price, M. J. *J. Phys. (E)–Sci. Instr.* **3**, 521 (1970).
[121] Rentzepis, P. M. and C. J. Mitschele. *Analyt. Chem.* **42**, 20 (1970).
[122] Roberts, J. R. *Applied Optics*, **4**, 1179 (1965).
[123] Salge, J., N. Pauls and K. K. Neumann, *Z. Angew. Phys.* **29**, 339 (1970).
[124] Schenck, W. J. *Instrum. Soc. Amer. Trans.* **5**, 14 (1966).
[125] Schrank, G. and G. Henry. *Rev. Sci. Instrum.* **35**, 1326 (1964).
[126] Schwab, A. J. *IEEE Trans. on Power Apparatus and Systems*, **90**, 2251 (1971).
[127] Shumaker, J. B. and C. H. Popenoe. *J. Res. Nat. Bur. Stand.* **76A** (Phys. and Chem.), 71 (1972).
[128] Simon, M. F. and G. L. Leroy. *IEEE Trans. on Power Apparatus and Systems*, **91**, 478 (1972).
[129] Speich, G. R. and A. Szirmae. 'Rapid heating by laser techniques', in *Techniques of Materials Preparation and Handling*, p 335. R. F. Bunshah, ed., Interscience: New York (1968).
[130] Steele, E. L. *Optical Lasers in Electronics*. Wiley: New York (1968).
[131] Stenerhag, B., S. K. Händel and B. Göhle. *Rev. Sci. Instrum.* **40**, 563 (1969).
[132] Stenerhag, B., S. K. Händel and B. Göhle *J. Appl. Phys.* **42**, 1876 (1971).
[133] Takeshita, S. and T. Sasano. *Proc. IEEE (Letters)*, 1404 (1968).
[134] Tawara, H. *Jap. J. Appl. Phys.* **7**, 1254 (1968).
[135] Thomas, R. J. *IEEE Trans. on Instrum. and Meas.* **19**, 102 (1970).
[136] Thornton, E. *J. Phys. E., Scientific Instruments*, **3**, 862 (1970).
[137] Topp, M. R., P. M. Rentzepis and R. P. Jones. *J. Appl. Phys.* **42**, 3415 (1971).
[138] Treverton, J. A. and J. L. Margrave. 'Thermodynamic properties of liquid molybdenum by levitation calorimetry', in *Proceedings of the Fifth Symposium on Thermophysical Properties*, p 489. C. F. Bonilla, ed., American Society of Mechanical Engineers: New York (1970).
[139] Urtiew, P. A. and R. Grover, 'Radiation temperature in solids under shock loading', in *Temperature*, Vol. IV, p 677. H. H. Plumb, ed. Instrument Society of America: Pittsburgh (1972).
[140] Vlastos, A. E. *J. Appl. Phys.* **38**, 4993 (1967).
[141] Ware, K. D., J. W. Mather, A. H. Williams, P. J. Bottoms and J. P. Carpenter. *Rev. Sci. Instrum.* **42**, 512 (1971).
[142] Weber, F. N. and D. D. Shear. *J. Appl. Phys.* **40**, 3854 (1969).
[143] Weiss, H. H. *Applied Optics*, **4**, 935 (1965).
[144] Wood, S. M. and M. H. Miller. *Rev. Sci. Instrum.* **41**, 1196 (1970).
[145] Ya'akobi, B. *J. Appl. Phys.* **40**, 4205 (1969).
[146] Zavoisky, E. K. and S. D. Fanchenko. *Applied Optics*, **4**, 1155 (1965).

CHAPTER 25

The Ballistic Compression and High Temperature Properties of Dense Gases

G. T. Lalos and G. L. Hammond

*Naval Ordnance Laboratory,
White Oak, Silver Spring, Maryland 20910, USA*

Contents

I.	Introduction	1193
II.	The Ballistic Piston Compressor	1195
	1. General Description	1195
	2. Operation	1198
	3. Instrumentation	1202
	A. Pressure Measurements	1202
	B. Volume Measurements	1203
	C. Temperature Measurements	1204
	D. Density Measurements	1207
	E. Spectroscopic Measurements	1209
III.	Physical Properties Studies	1210
	1. The Equation of State	1210
	2. Optical Studies	1212
	A. Line Broadening Experiments	1212
	B. Opacity Measurements	1213
	C. Fission Studies	1214
	D. Laser Pumping	1214
	3. Other Studies	1215
	A. Chemical Reactions	1215
	B. Dissociation Kinetics	1215
	C. Ignition Studies	1216
IV.	Summary	1216
V.	References	1217

I. Introduction

Gases at simultaneously high pressures and temperatures are finding increasing use in modern day technology. It is important, therefore, to develop new, improved techniques for producing and investigating the properties of hot, highly compressed gases. Historically, static gas compression with external heating was the first method employed. This technique, still in widespread use today, has the capability of generating gas pressures up to 15 000 atm, but is limited in temperature to the melting point of structural materials.

To circumvent this difficulty, it is necessary to employ transient heating of the gas under study. Two methods are available, i.e. the shock-tube, and the ballistic piston (free-piston) compressor. Both techniques rely on rapid gas compression of the test gas itself. In the shock-tube, a thin diaphragm which separates a high pressure (driver) gas from a low pressure (test) gas is ruptured and causes a shock wave to travel through the low pressure gas heating it to very high temperatures. This method is attractive because of its simplicity (no moving parts) and its ability to produce very high gas temperatures. It cannot, however, generate high pressures without recourse to explosive driving. The ballistic piston compressor, on the other hand, is a device that produces simultaneously high gas pressures and temperatures by firing a close-fitting piston into a closed-end tube containing the test gas under study. Two modes of operation are possible. In the normally employed uniform compression mode (heavy piston) proper choice of piston mass, tube length, and sound speed of the test gas results in a gas compression rate

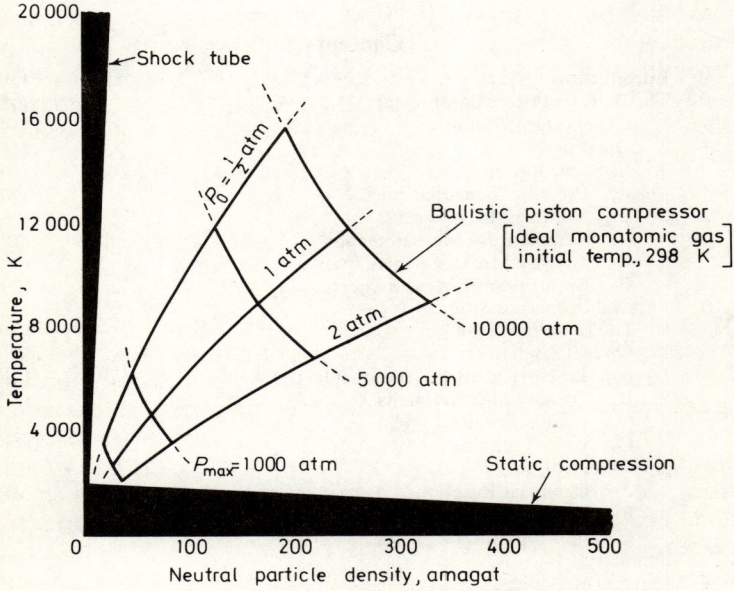

Figure 1. Modes of production of hot, dense gases.

rapid enough to make heat transfer to the tube walls small, but not so rapid as to produce shock waves in the test gas. A gas compressed in this manner remains essentially homogeneous throughout its volume, and its pressure, volume and temperature behavior can be approximated by that of an isentropically compressed gas. This mode of operation has been investigated extensively and will be described in some detail. The shock compression mode (light piston) is characterized by a plane hydrodynamic shock wave that is produced by the rapidly accelerating piston repeatedly traversing the test gas as it reflects back and forth between the end-plug and piston face during the compression–expansion cycle. Only a brief qualitative study of the shock compression mode has been made to date.

A comparison of the ranges of temperature and density that can be produced by these three techniques is shown in *Figure 1*. It is seen that static gas compression is capable of achieving a large density range, but that the temperature range is limited to about 2000 K. Internal heating by a heating element located within the container can extend the temperature range to about 3000 K. The shock tube régime, shown on the left, is characterized by very high temperatures and consequently high densities of charged particles. It is not capable, however, of producing high neutral particle densities. The ballistic piston compressor régime shown in the center for monatomic gases includes temperatures up to 10000 K and higher, with densities ranging up to 500 amagats. Polyatomic gases give lower temperatures but higher densities. Isentropes corresponding to initial gas pressures of 0.5, 1, and 2 atm and an initial temperature of 298 K are shown. Low initial pressures produce conditions approaching those generated by the shock-tube, whereas high initial pressures result in conditions approaching those generated by static compression. Intermediate values of the initial gas pressure will produce high values of neutral and charged particle density simultaneously. Calculations show that at these extreme conditions electron densities may approach 10^{19} cm^{-3}, while neutral particle densities can exceed 10^{21} cm^{-3}.

II. The Ballistic Piston Compressor

1. General Description

The Naval Ordnance Laboratory ballistic piston compressor[39, 20] is shown schematically in *Figure 2*. It consists of a gas reservoir, a piston release section, a piston, a 5 cm bore × 8 m long tube, and a high-pressure section with a design pressure limit of 10000 atm. The entire assembly is mounted on roller bearings located beneath the reservoir and high pressure test section to allow for recoil during firing. The reservoir gas is sealed off from the test gas when the plunger (located in piston release section) is in the forward position. This makes it possible to disconnect the tube from the piston release section and facilitates cleaning and making adjustments or changes to the piston, tube, and high-pressure section while retaining the driver gas

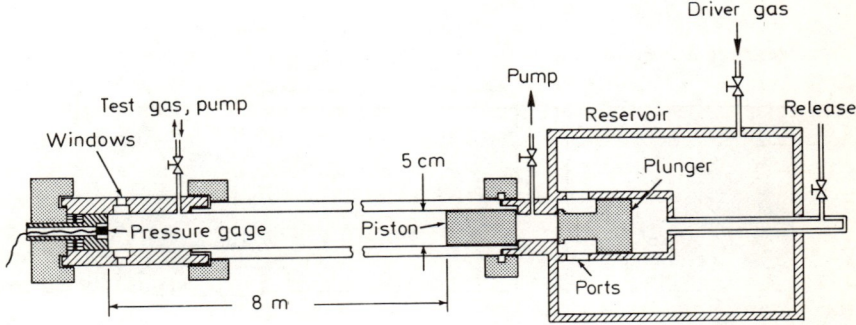

Figure 2. Schematic diagram of ballistic piston compressor.

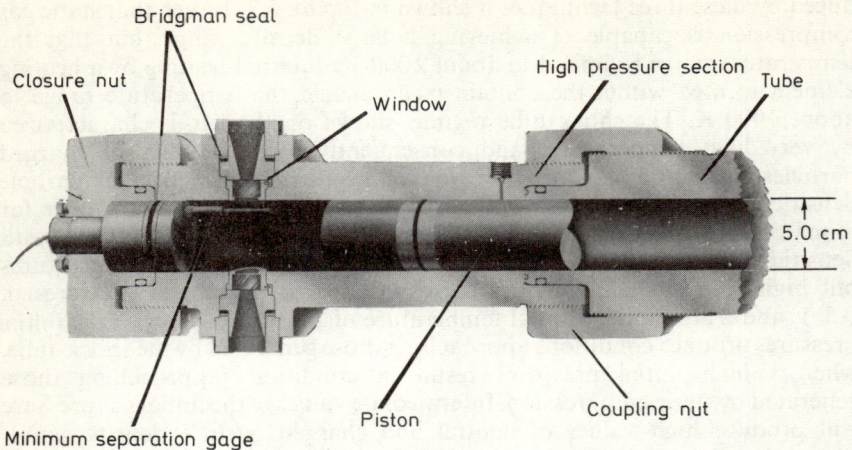

Figure 3. High pressure test section.

in the reservoir. The high-pressure section was made separate from the tube because the test gas pressure remains relatively low until after the piston enters the high-pressure section. The tube, therefore, is designed for relatively low pressures, whereas the high-pressure section must be designed for the maximum test gas pressure.

Figure 3 is a cross-sectional view of the high-pressure section instrumented with four diametrically opposite high pressure windows for simultaneous absorption and emission measurements, a high speed piezoelectric pressure gage, and a minimum end-plug/piston separation gage. Leakage of the high pressure test gas out of the high-pressure section is reduced to a minimum amount by use of O-rings, Poulter seals (windows), Bridgman seals (end-plug and window assemblies), and flexible cup-seals on the piston. The inner walls of the high-pressure section are chromium plated to suppress vaporization of the steel walls by the hot test gas. Synthetic sapphire windows are normally

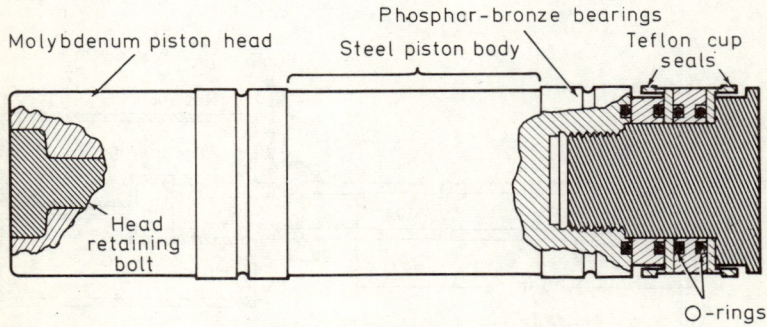

Figure 4. Piston with seals and high temperature resistant head.

employed, although glass and other transparent materials can also be used. Typical values of minimum separation between the end-plug and the piston face are 1 to 10 cm.

The piston, which consists of a steel body, a refractory metal head, and a sealing assembly is shown in *Figure 4*. The ends of the steel body are machined to accommodate two bearings (piston rings) made of phosphor-bronze, and contain a small circumferential groove that induces turbulence in the leaking gas and catches small particles that might otherwise cause galling. The radial clearance between the bearing surface and tube bore is 50 microns. The piston head, piston body, and sealing assembly are all undercut 150 microns from the tube bore. A molybdenum head is attached to the front of the piston with a retaining screw. This refractory head is used in order to prevent ablation of the front surface of the piston when it comes into contact with the hot test gas at the peak of the compression cycle. During this time period high heat transfer to the edge of the piston head occurs as a small quantity of the hot, highly compressed test gas flows at high velocity through the thin annular region between the tube walls and the piston head. Under these conditions materials of lower melting point exhibit severe ablation, whereas molybdenum shows only slight discoloration even at test gas temperatures as high as 6000 K. The rear of the piston is fitted with self-actuating double-cup seals. Leakage of reservoir gas across the piston (toward high pressure section) during the acceleration part of the compression stroke, and leakage of test gas across the piston (toward reservoir) during the deceleration part of the compression stroke is undesirable because it results in inability to predict accurately the ballistic action of the compressor, inaccuracies in physical property measurements due to poorly known compositions and quantity of test gas, and ablation of the piston and high heat transfer. Gas leakage can be reduced to a small value by reducing the clearance between the bearing surfaces of the piston and the bore of the compressor tube. Phosphor-bronze, a soft metal possessing good wear qualities, is used as the bearing material. The bearings slide on the compressor bore which is made of highly polished, hardened steel. Radial clearances smaller than approximately 25 microns result in excessive wear of the bearing surfaces and occasionally cause piston jamming. Further improvement in leakage reduction is made through the use of double-cup, flexible, tight-fitting seals. The diameter of the bearing surfaces of the seals is made approximately equal to the bore diameter of the tube. Assembly is made with the open-end of each seal facing the gas that is being sealed. A serrated ring and bolt head provides support for the cup-seal lips but also allows gas passage into the cups. The double seals are assembled with the cups back-to-back. As the compressor is fired the reservoir gas impinges on the rear of the piston and the rear cup is forced to expand radially reducing the clearance with the tube bore essentially to zero and thereby producing an effective gas seal. Near the end of the compression stroke, when the pressure differential across the piston reverses, the forward seal is actuated in a similar manner. Static tests with and without the cup-seals indicate a 100-fold reduction in leakage rate when the seals are used. The cup-seals are made of Rulon LD, a filled-Teflon material with substantially superior wear properties over regular Teflon, in order to increase the life of the cup-seals. The piston is long enough that the rear seals do not cross the junction between the tube and test section. The seals have been used up to a maximum dynamic pressure of 2 500 atm without failure.

2. *Operation*

In operation the chamber behind the plunger of the piston release section is pressurized so that the plunger moves to the forward position (left) and seals the reservoir from the tube. The tube is then filled with the test gas or gas mixture and its pressure and temperature adjusted. Next, the reservoir is filled with driver gas to the pressure necessary to produce the desired maximum pressure in the test gas. The compressor is fired by dumping the pressure behind the plunger in the piston release section with a solenoid actuated quick-opening valve. The reservoir pressure acting on the front of the plunger moves it to its rear position (right) removing the seal between the reservoir and the tube. Reservoir gas rushes through ports in the piston release section and impinges on the back of the piston driving it swiftly down

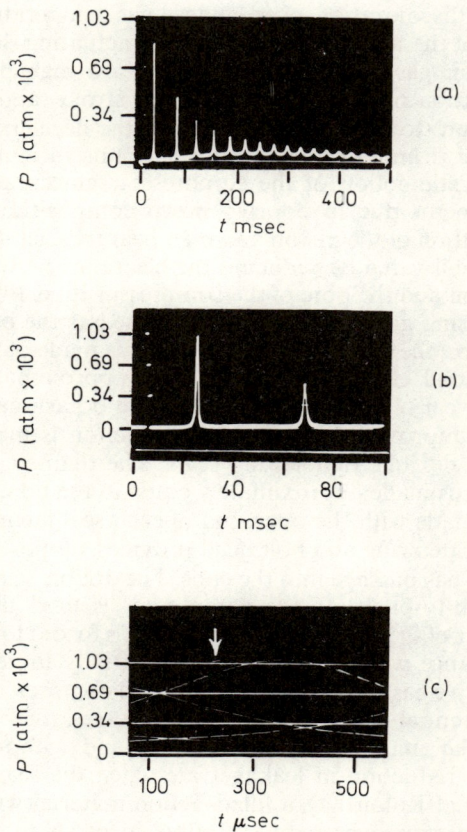

Figure 5. Oscillograms of test gas pressure versus time.

the tube. The seals on the rear of the piston prevent the reservoir gas from leaking into the tube and mixing with the test gas during a shot. During the peak pressure part of the compression cycle, a small amount of test gas leaks across the piston and mixes with the reservoir gas, but it does not have any adverse effects on the experiment. After the first compression cycle the piston oscillates back and forth until friction brings it to rest with equal gas pressure on its front and rear sides. The desired measurements are made during the peak pressure of the first compression stroke by an appropriately instrumented high-pressure section.

The transient nature of the gas compression produced by the free piston operating in the uniform compression (heavy piston) mode is shown in *Figure 5*. *Figure 5*(a) is an oscilloscope record of the pressure of the test gas versus time with an oscilloscope sweep speed sufficiently slow to show the complete history from the time of piston release to the time the piston comes to its final rest position. In *Figure 5*(b) the scope sweep speed was increased to show the pressure profile of the first two compression strokes in more detail. Pressure pulse half widths are typically a few milliseconds and can range from 0.5 to 10 msec by proper choice of piston mass. A further increase in scope sweep speed was made in *Figure 5*(c). The multiple traces on this record show the pressure/time history of the first compression cycle only. A pulse indicating the opening of a spectrograph slit by a high-speed electromechanical shutter can also be seen. The opening and closing times of this shutter are normally adjusted to allow gas emission to enter the spectrograph only during the time interval of the top ten per cent of the pressure pulse in order to assign a specific value of pressure to the recorded emission spectra. As mentioned earlier, physical property measurements are normally made only during the peak pressure of the first compression cycle.

When the ballistic piston compressor is operated in the uniform compression mode (heavy piston) with near ideal gases it is possible to approximate the variation of test gas density and temperature with test gas pressure from the ideal gas polytropic relations:

$$(p/p_0) = (n/n_0)^\gamma = (T/T_0)^{\gamma/(\gamma-1)} \quad (1)$$

where γ is the polytropic exponent, the ratio of specific heat capacities. Equating the work done on the piston by the expanding reservoir gas to the work done by the piston on the test gas being compressed, assuming that the reservoir pressure, p_r, is constant, and substituting equation 1 gives the following equation which shows that p_{max} does not depend on the length of the tube or piston diameter, but is a function of only p_0, p_r, and γ

$$\left(\frac{p_r}{p_0}\right)(\gamma - 1) = \frac{(p_{max}/p_0) - (p_{max}/p_0)^{1/\gamma}}{(p_{max}/p_0)^{1/\gamma} - 1} \quad (2)$$

For $(p_{max}/p_0) \gg 1$ this equation can be simplified to

$$(p_{max}/p_0) = [1 + (\gamma - 1)(p_r/p_0)]^{\gamma/\gamma - 1} \quad (3)$$

which illustrates that the peak pressure generated in the test gas has a power law increase with reservoir pressure, i.e. a relatively low reservoir pressure generates a high test gas pressure, and that low γ gases produce the highest peak test gas pressures for a given reservoir pressure. Typically, a reservoir pressure of 30 atm will generate a test gas pressure of 1 000 atm in a monatomic gas. The constant reservoir pressure assumption is quite valid for $V_{res} \gg V_{tube}$ and makes possible use of an inexpensive, inert gas such as helium as the reservoir gas regardless of the nature of the test gas under study.

Operation in the shock compression mode (light piston) is obtained by simply replacing the heavy piston with a light piston and employing test gases of low sound speed. This results in a compression–expansion cycle similar to that obtained in the uniform compression case on to which are superimposed pressure, density and temperature perturbations. These perturbations are produced by a hydrodynamic shock wave that emanates from the piston face as the piston accelerates into the test gas when it is first released at the beginning of the compression cycle. This shock wave traverses the test gas sample, reflects at the end-plug wall, retraverses the test gas, reflects at the piston face, and repeats this process during the entire course of the compression–expansion cycle.

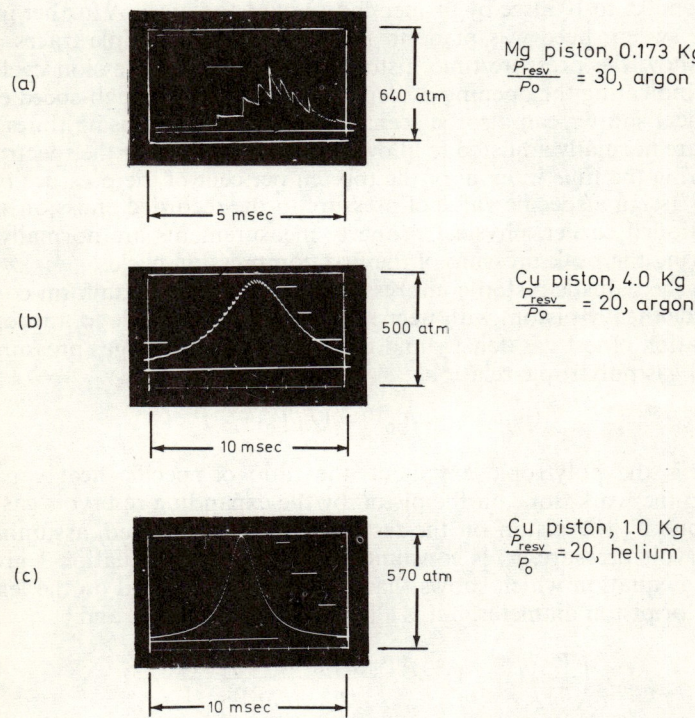

Figure 6. Pressure/time records showing shock structure.

Figure 6 shows three pressure/time records illustrating different strength shocks. In Figure 6(a) a light 0.173 kg piston was used to compress the heavy test gas argon resulting in very strong shock structure. Figure 6(b) shows that the shock strength in argon can be drastically reduced by employing a heavier 4.0 kg piston. *Figure 6*(c) shows that negligible shock structure is produced when the high sound speed test gas helium is compressed with a 1.0 kg piston. *Figure 6*(a) is an example of shock compression (light piston) operation, whereas *Figure 6*(b) and *6*(c) are examples of uniform compression (heavy piston) operation.

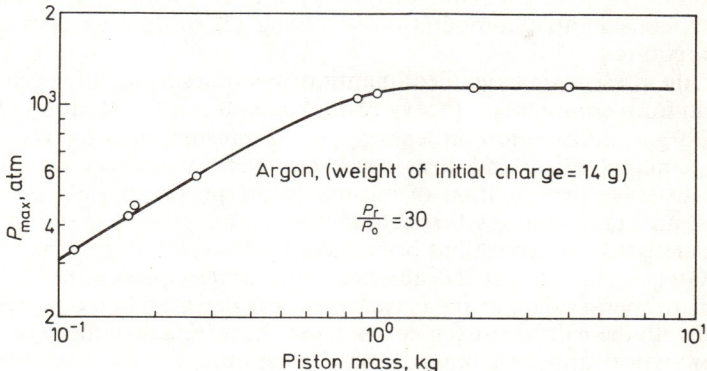

Figure 7. Maximum test gas pressure versus piston mass.

In the course of the shock compression (light piston) investigation it was found that the shock structure was accompanied by a reduction of p_{max}. Results of a study of the effects of piston mass on p_{max} for identical conditions of p_0 (1.0 atm) and p_r (30.0 atm) in argon are shown in *Figure 7*. Measurements of p_{max} were made at points of maximum pressure. This study showed that there exists a minimum value of the piston mass above which p_{max} is independent of piston mass, and shock strengths are negligibly small.

Some accounts of shock formation accompanying rapid gas compression may be found in the literature. Evans and Evans[15] treat the case of a piston moving into a closed cylinder with constant speed after being suddenly accelerated from rest. The resulting shock is assumed to divide the test gas into two uniform regions, shocked and unshocked. The entropy of the gas increases as each shock passes, and the product pV^γ no longer remains constant but becomes a function of entropy. For isentropic compression, a plot of log p versus log V results in a straight line with a slope of $-\gamma$. Evans and Evans[15] have shown that if shocks are also present, after a sufficient number of reflections have occurred the possible values of p and V lie along another straight line parallel to the isentropic one but displaced upward from it such that compression to a given volume V results in a pressure p greater than the corresponding isentropic pressure. They further found that the greatest relative increase in the entropy function occurs in the passage of the first shock wave.

Further work on the problem of shocks accompanying isentropic compression has been done more recently by Minardi and Schwartz[49]. They have shown that there exists a critical value of a piston mass parameter,

$\gamma W_p/W$, where γ is the ratio of specific heats of the test gas, W is the weight of the charge of test gas ahead of the piston, and W_p is the weight of the piston. For values of $\gamma W_p/W$ greater than the critical value, the peak pressure produced by a given reservoir driving pressure is not a function of $\gamma W_p/W$, but for values less than the critical value, peak pressures produced by a given driving pressure fall off sharply with decreasing piston mass. The results of *Figure 7* confirm the above relationship, but in order to calculate the critical value of the piston mass parameter using the treatment of Minardi and Schwartz, considerable computation involving the evaluation of frictional forces is required.

Up to the present time detailed quantitative studies have only been made for the uniform compression (heavy piston) case. If a valid equation of state of the test gas, information on leakage past the piston, data for viscosities, thermal conductivities, and heat capacities were available, it would be possible to solve the equation of motion of the piston to yield pressures, densities, and temperatures throughout the compression–expansion cycle. Such an analytical approach has been taken by Takeo *et al.*[62] at the University of Oregon. In view of the absence of accurate equation of state and transport properties data at the extreme test gas densities and temperatures achieved with the ballistic piston compressor, the accuracy of the predictions of the analytical approach must remain uncertain. We have adopted the alternative approach of developing experimental diagnostic techniques for direct measurement of the state variables. Details of these experimental techniques are given in the following section.

3. *Instrumentation*

A. Pressure Measurements

A commercial (Kistler) pressure transducer is mounted centrally in the end-wall (see *Figure 3*). This device is a quartz crystal piezoelectric transducer with an upper limit of 5 000 atm pressure. Its rise time of 1.5 μs allows it to follow accurately the test gas pressure pulse which typically has a half-width duration of 2 ms. The calibration supplied by the manufacturer is occasionally checked by measuring the response to a step function generated by a quick release of pressure accurately measured with a deadweight balance. These checks generally agree with the manufacturer's calibration to within two per cent.

The signal from the transducer is amplified by a charge amplifier, displayed on oscilloscope screens, and photographed with Polaroid cameras. A calibration pulse is simultaneously displayed and photographed to provide a reference signal, thus eliminating the need for precise oscilloscope calibration. However, non-linearities in the display introduce a maximum additional error of one per cent.

The pressure is considered to be uniform throughout the test gas volume in experiments with light gases because transit times for sound waves across the volume are short compared to the characteristic times in the compression cycle. Corrections due to the momentum flux arising from heat conduction to the walls will be negligibly small for thin boundary layers (Griem[18], see

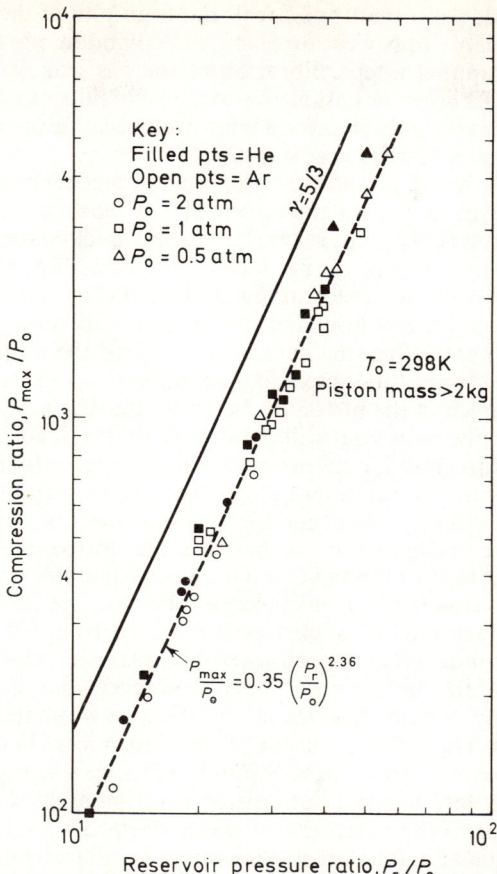

Figure 8. Maximum pressure versus reservoir pressure: Ar and He.

Chapter 9). Therefore, the accuracy of routine pressure measurements is ±3 per cent. This accuracy can readily be increased to better than ±1 per cent by taking extra care with transducer calibration and record reading.

Figure 8 illustrates the relationships between initial test gas pressure, p_0, reservoir pressure, p_r, and maximum pressure, p_{max}, for the monatomic gases argon and helium. The solid line is a plot of equation 3 with $\gamma = 1.667$, and the experimental points cluster tightly along a similar line. However, an increase of approximately 30 per cent in driving pressure over that predicted by equation 3 is needed to achieve a given compression ratio p_{max}/p_0. This degree of departure from ideal, isentropic compression emphasizes the need for accurate diagnostic schemes to measure directly the physical state variables in the hot, compressed gas.

B. Volume Measurements

In early work with the NOL compressor[53], the test gas volume was computed as a function of time from streak photography of the leading edge of the piston as it came into view through a side-window port. The molecular volume was computed after calibration of the gas leakage past the piston with a test gas of known equation of state. The results of those experiments suffered from uncertainties in gas temperature, because the temperature was inferred from the ideal, adiabatic gas law.

More recently, the minimum test gas volume has been monitored with a telescopic probe mounted on the end-wall[20]. It consists of two concentric interference-fit tubes. The outer tube is a solid thick-walled tube threaded on one end for attachment to the face of the end-plug. The inner tube is thin-walled and contains a slot along its length. One end is chamfered to facilitate insertion. Partial insertion of the inner tube into the outer tube is made prior to a shot. When the compressor is fired the motion of the piston forces the slotted inner tube into the outer tube to an extent determined by the closest approach of the piston to the end-plug. The minimum separation gage installed in the end-plug is illustrated in *Figure 3*. Its use neither interferes with pressure measurements nor with the recording of emission or absorption spectra. The minimum piston/end-plug separation is determined by removing the end-plug from the high pressure section and measuring the distance from the end-plug face to the tip of the slotted inner part.

The most recent development in volume measurements consists of the installation of a magnetic transducer in the test section side wall. This transducer is a permanent magnet pole-piece surrounded by a multi-turn coil. A voltage is induced in the coil when ferromagnetic material approaches or recedes from the pole-piece. The piston body has been modified by machining evenly spaced grooves in the steel between the two phosphor-bronze bearings. The voltage output of the transducer, as the piston passes the transducer location, is a series of pulses which can be correlated with the piston groove pattern to yield both piston position and velocity throughout the major portion of the compression cycle. The groove spacing is such that the piston position at maximum compression can be determined to ± 0.5 mm. This error corresponds to approximately one per cent uncertainty in the volume at a maximum pressure of 1 000 atm with monatomic gases.

C. Temperature Measurements

The spectroscopic 'brightness–emissivity' method is employed for gas temperature measurements. This scheme previously has been applied to flames in the infra-red spectral region[60], to internal combustion engines in the ultra-violet region[61], and to a variety of plasma diagnostic experiments in shock-tubes[31,17]. The widely used line-reversal technique is a special case of this more general method.

The scheme is based on the solution of the radiative transfer equation for a homogeneous gas sample in local thermodynamic equilibrium (LTE)

$$I_\lambda = B(\lambda, T_g)[1 - e^{-\tau_\lambda}] \qquad (4)$$

where I_λ is the intensity at some wavelength λ, $B(\lambda, T_g)$ is the Planck function for the gas temperature T_g, and τ_λ is the optical depth at that wavelength. To apply equation 4, I_λ and τ_λ are simultaneously measured and then 4 can be solved for T_g. The optical depth is measured by passing a chopped light beam through the hot gas and into a polychromator equipped with photomultiplier detectors. Exit slits in the polychromator are positioned to allow opacity measurements at several wavelengths simultaneously. The optical depth is obtained directly from attenuation measurements via the Lambert–Beer law. During the intervals when the chopper is closed, the emission signal I_λ from the hot, dense gas is detected in each wavelength channel of the polychromators. The emission intensity is put on an absolute basis by passing a light beam from a radiation standard through the same optics and into the polychromator, and taking care to fill identically all apertures by the beams from both the hot gas and the standard source. A schematic diagram of the temperature optics is shown in *Figure 9*.

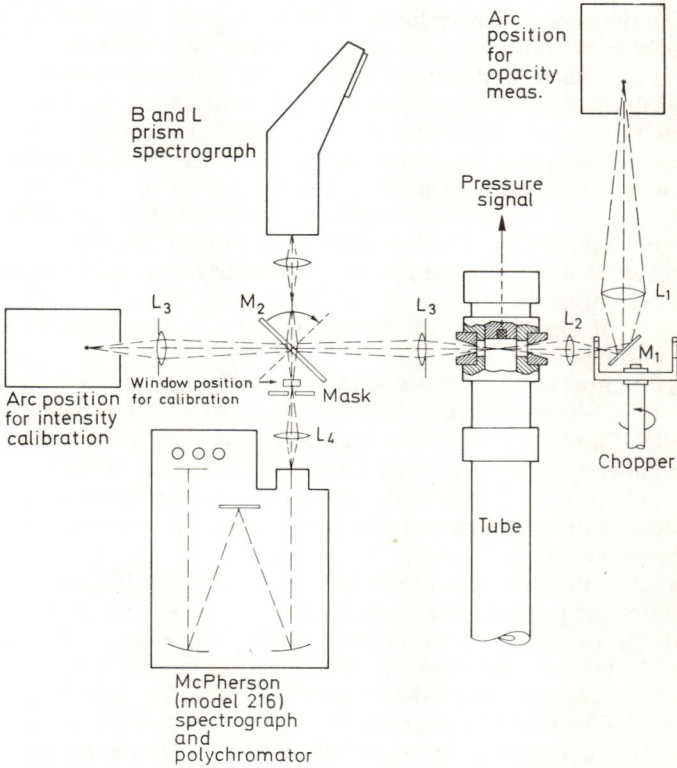

Figure 9. Optics for temperature measurements.

The standard source used in this work is the hot anode crater of a 12 A d.c. graphite arc manufactured by Mole-Richardson Company. The radiance temperature, spectral radiance and short-time fluctuations of the Mole-Richardson arc have been determined at NBS[41, 24, 7], and the spectral emissivity of the hot anode crater has most recently been measured by Schurer[59]. Those studies have shown that when the arc is operating within 0.5 A below the overload or 'hissing' point, the radiance temperature is 3800 K $\pm$ 10 K, this temperature is constant with wavelength over the range 3000–8500 Å except in the vicinity of the CN molecular bands originating in the plasma column, and the emissivity over this range is greater than 0.98 and equals 0.99 from 4000–6000 Å. Thus the anode crater is very nearly a blackbody radiator in this spectral region; its true temperature is only about 10 K greater than its brightness temperature. Short-term temperature fluctuations are approximately ± 5 K on a millisecond time scale.

Temperatures have been measured with the polychromator slots positioned at the peaks of several of the strong, severely broadened spectral lines. The bandwidth of 3 Å for each channel is much narrower than the linewidths, and thus the opacity and intensity over the bandwidth is nearly constant. Measurements have been made at the sodium 'D' lines, the calcium resonance line at 4226 Å, and the neutral copper line at 5105 Å which has its lower state 1.4 eV above the ground state. At temperatures greater than 4000 K, continuum emission is detectable and polychromator channels have been installed at line-free positions adjacent to the lines being measured. These channels allow corrections to be made for continua 'under' the lines so that the excitation temperature of the species radiating the line can be determined. Temperature measurements have also been made at the band head of the C_2 (o, o) Swan band at 5165 Å. This species is formed in the hot core of the gas sample from the decomposition of CH_4 which is introduced in trace amounts for thermometric studies. Nearly all the temperature measurements made to date have been with helium as the test gas.

The results of the temperature studies in helium are summarized by the following findings:

(1) When temperatures of more than one species are measured in a single experiment, the disagreement is less than four per cent between resonance lines (Na 'D', Ca 4226) throughout the compression cycle.

(2) For shots with temperatures greater than 4500 K, the C_2 band and non-resonant Cu(I) line yield higher temperatures than the resonance lines, but they agree with each other within five per cent. This result is attributable to boundary layer effects.

(3) The maximum temperatures occur approximately 100 μsec after p_{max}; at these times the pressure has decreased less than two per cent from p_{max}. This delay can be attributed to the mass leakage out of the test section past the piston during the high pressure portion of the compression cycle.

(4) A strong correlation exists between p_{max} and T_{max} over the pressure range 400–2500 atm and temperature range 2600–5200 K. *Figure 10* illustrates this correlation for a plot of log (T_{max}/T_0) versus log (p_{max}/p_0), where $T_0 = 298$ K. A linear least squares fit to these data yields a correlation coefficient of 0.985, and an r.m.s. vertical deviation from the best fit line corresponding to seven per cent of T_{max}. The equation for the best fit line is

$$\log(T/T_0) = 0.360 \log(p/p_0) + 0.015 \qquad (5)$$

Referring to equation 1 for comparison with an ideal isentropic process, the coefficient 0.360 in equation 5 should be compared with 0.400 corresponding to $\gamma = 1.667$ for the ideal isentropic process. The small temperature-axis intercept of 0.015 is only 10 degrees above the n.t.p. initial conditions. There-

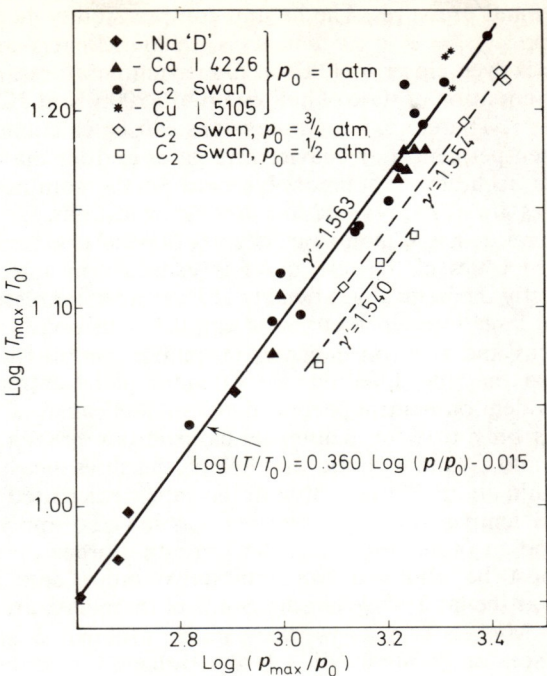

Figure 10. T_{max}/p_{max} correlation.

fore, we interpret equation 5 as one of polytropic form with an exponent $\gamma' = 1.563$. The data used to determine equation 5 were those for $p_0 = 1$ atm. Some temperature measurements on shots with $p_0 = 0.75$ and 0.5 atm have been made (see *Figure 10*), and the results indicate a weak dependence of γ' on p_0. Since T_{max} occurs well within the interval centered at p_{max} during which spectroscopic observations are made, we conclude that equation 5 provides an adequate temperature calibration for helium over these pressure and temperature ranges. We have used it for all spectroscopic data discussed here because it is not feasible to make both spectra and temperature measurements simultaneously with the same spectrograph.

The assumptions of homogeneity and thermodynamic equilibrium that are employed here to derive gas temperatures are discussed in a later section of this chapter.

D. Density Measurements

There is presently no density diagnostic scheme in operation on the NOL Compressor, and recourse must be made to the equation of state to calculate the density from pressure and temperature measurements. For experiments with helium, this procedure introduces a negligible error compared to the three per cent pressure error and seven per cent temperature error. The ideal gas equation of state, $p = NkT$, is appropriate for helium at compressor temperatures and pressures. The helium compressibility factor, $Z = p/NkT$, at 400 K and $N_{He} = 3 \times 10^{21}$ cm^{-3} is 1.006 (Michels and Wouters[46]), and it becomes even closer to 1.000 at higher temperatures. Also, ionization effects on the equation of state of helium up to 15000 K at 1000 atm pressure are negligible[43]. Therefore, helium densities can be calculated here with an accuracy of ten per cent. For heavier rare gases and for diatomic test gases, however, this accuracy is not possible because the compressibility factors for those gases are not known at compressor conditions.

Conventional density diagnostic schemes depend upon knowledge of the molecular properties of the test gas. For example, in a recent attempt to measure directly the neutral gas density in ballistic compressors[6], two-color, Fabry–Perot, laser interferometry was employed to obtain simultaneously the neutral gas and electron densities in helium compression experiments. They assumed that the difference in the index of refraction from unity is linearly dependent on neutral particle density (the Gladstone–Dale relation). Again, this is only true for helium at ballistic compressor conditions. In fact, Burgess and Grindlay found that helium densities measured in this way agreed to within about 20 per cent with the values calculated from measured pressures and temperatures via the ideal gas law. To apply this interferometric method to other test gases, the Lorentz–Lorenz function, $(n^2 - 1)/(n^2 + 1)\rho$, must be obtained from refractive index data at compressor densities and at the laser wavelengths. Some of these data are available from the work by Michels and co-workers at the van der Waals Laboratory, Amsterdam (see, for example, Michels and Botzen[47] for data on argon).

Burgess and Grindlay[6] also found that the electron density in their helium experiments was less than their detectability limit of 10^{16} cm^{-3}. This result agrees with our calculations with the Saha equilibrium equation for mixtures of rare gases with traces of metal contaminant vapors. We calculate electron densities between 10^{13} and 10^{14} cm^{-3} for typical compressor conditions. Thus, the ratio of electrons to neutral atoms is of the order of 10^{-8}.

A few other methods are currently under consideration here for density measurements. The Rayleigh scattering cross sections for He, H and H_2 are well known, and thus a laser scattering experiment, preferably at ultraviolet wavelengths, may be feasible for those species. An electron beam scattering experiment presents some difficulties in getting the beam into and out of the thick-walled test section, but a γ-ray absorption measurement appears feasible according to the recent results of Kislykh et al.[33]. Finally, the addition of a radioactive tracer element to the test gas would allow a density determination if a rugged particle detector can be located at the inner wall of the test section. This scheme has been applied to electric arcs by Ecker[14].

It must be assumed that no gas leaks past the piston in order to determine the gas density from measurements of the volumetric compression ratio. This approach has been taken by Lewis et al.[42] with the ballistic compressor at the von Kármán Institute for Fluid Dynamics. We have found, however, that this is a dubious assumption even with tight-fitting piston seals that do not leak under static pressure tests. We find that with helium at a maximum pressure of 1000 atm, the density obtained from the measured temperature and ideal gas law is 30 per cent higher than the density calculated by assuming an isentropic compression, and the density obtained from measurement of the volumetric compression ratio with the magnetic pickup transducer is 50 per cent higher than that obtained from the measured temperature and ideal gas law.

E. Spectroscopic Measurements

The compressor emission spectra are recorded photographically with a fast, moderate dispersion spectrograph. This instrument is a McPherson Model 216, 1 m focal length spectrograph with a 600 groove/mm plane grating in a Czerny–Turner mount. The nominal reciprocal dispersion is 16 Å/mm, and the effective aperture ratio is $f/9$. The optical components between the spectrograph and compressor are fused quartz lenses and an aluminized front-surface mirror, as shown in *Figure 9* except that lens L_4 is moved slightly to focus the compressor exit window on the entrance slit jaws.

In order to admit light into the spectrograph only during moments of maximum compression, a high-speed, electromechanical shutter system has been designed and mounted at the spectrograph entrance slit. This shutter[32] consists of a slotted plate driven horizontally past the entrance slit. The driving action results from the motion of a ring in contact with the slotted plate and positioned coaxially in front of a rigid stationary coil of flattened, enameled copper wire. A current pulse through the coil is initiated by trigger circuits monitoring the pressure transducer signal, and the induced current in the ring generates a magnetic field opposing that in the coil and the resulting force on the ring moves it and the shutter plate quickly away from the fixed coil. The open time during which light can pass through the slot and spectrograph slit is inversely proportional to the square of the voltage on the storage capacitor which is discharged through the fixed coil. The trigger point and open time are selected to admit light emitted during the interval centered at and within 95 per cent of p_{max}. The gas density and temperature vary only three per cent and two per cent, respectively, over this interval. The open times used in most experiments range from 150 to 500 μsec.

Absorption spectroscopy is accomplished with background light sources of high brightness temperature. Absorption coefficient measurements with the graphite d.c. arc have already been discussed in the temperature measurement section. For observations in the ultra-violet regions of the spectrum, the brightness temperature of the lamp must be extremely high to overcome the rapid decrease in transmission of optical components and air with decreasing wavelength. Conventional ultra-violet sources, such as high pressure hydrogen or deuterium lamps, are not bright enough for wavelengths less than 2000 Å, and high energy, transient discharges must be used for these vacuum-ultra-violet wavelengths. We use a device that satisfies all the requirements of high brightness temperature, line-free continuum spectra, reproducibility, ease of operation and compactness: the Garton–Charatis flashlamp[57, 8]. This lamp emits an intense light pulse of approximately 10 μsec duration which is synchronized with the peak of the compressor pressure pulse with the aid of electronic circuitry that monitors the pressure transducer signal.

The lamp consists of a 6.3 cm long capillary of 5 mm bore in a phenolic cylinder. The tungsten electrodes at each end of the capillary also have 5 mm diameter holes through which the light generated by the sliding spark discharge in the capillary passes to the ultra-violet transmitting Suprasil II windows at each end. The lamp is filled with air at approximately 5 mm Hg pressure. The application of 5000 V across the electrodes is sufficient to cause electrical breakdown in the capillary and the resulting discharge is characterized by a peak current density of approximately 1.4×10^3 A/cm^2, and a brightness temperature of 30000 K in the visible portion of the spectrum. The plasma generated in the capillary consists of a mixture of air and high density ionized organic material ablated from the capillary walls.

The most recent application of this lamp to compressor experiments has been the experimental search for the quasi-molecular hydrogen absorption in the vacuum ultra-violet that has been predicted by Doyle[13] and by Sando et al.[58]. The lamp is sufficiently intense, however, that a single flash will produce heavy exposures on sensitive spectroscopic emulsions in the visible region of the spectrum. Consequently, conventional absorption spectroscopy with this lamp can be done routinely on the compressor.

III. Physical Properties Studies

We consider here the measurement of equilibrium properties of gases at typical compressor conditions. Such conditions in monatomic gases range from 3000 to 6000 K and densities from 10 to 150 amagats, and in diatomic gases from 2000 to 4000 K and from 50 to 400 amagats. Measurements of transport properties over these ranges would also be of considerable interest, but they present severe practical problems that have not yet been seriously attacked by ballistic compressor workers. In this regard, however, we should mention the early work by Ryabinin[56] and co-workers on the electrical conductivity of hot, compressed gases.

1. The Equation of State

Although considerable theoretical progress on the equation of state (EOS) for simultaneously hot, high pressure gases has been made (see, for example, Rowlinson[55], Thiele[63], Klein[34]), we are not aware of any successful experimental EOS work at these extreme conditions. Here we regard hot as the temperatures exceeding those feasible in static, high pressure cells (generally, greater than 3000 K), and we regard as high pressures those of the order of several thousand atmospheres. There are, however, several areas of practical interest in EOS data for these conditions including detonation phenomena, certain aerodynamic problems, rocket motor performance, and some astrophysical problems. It is useful, therefore, to consider the prospects of EOS work with ballistic compressors.

By taking extra care in pressure and temperature measurements, errors of the order of one per cent in pressure and five per cent in temperature are feasible with the NOL compressor diagnostic schemes. With the development of a density diagnostic scheme capable of, say, three per cent accuracy, we can then detect departures from the ideal gas EOS in species with compressibilities differing from unity by more than about ten per cent. From inspection of tables of calculated compressibilities[26, 43, 1] we estimate that possible candidate species include most of the common diatomic gases and the heavy rare gases, but for argon the prospects may be marginal without further improvements in diagnostic accuracies.

An important consideration for these p–V–T measurements in ballistic compressors is the question of the existence of thermodynamic equilibrium. As a measure of the time scale in the compression cycle, we estimate 100 μs as an interval over which the physical conditions in the test gas change by a significant amount. At compressor densities and temperatures, gas kinetic collision frequencies are of the order of 10^{10} sec^{-1}. If we consider dissociation of diatomic molecules via collisions, say: $N_2 + M \rightarrow 2N + M$, where M is any third body, we see that with typical diatomic collisional dissociation cross sections of about 10^{-2} times gas kinetic values, dissociative equilibrium is very rapidly achieved compared with the compression cycle time scale. Vibrational and rotational equilibrium can also be expected in most common diatomic species, but little is known about electronic excitation rates via neutral particle collisions.

Collisional ionization occurs in most laboratory plasmas via excitation into intermediate states. The ionization equilibration time will then be controlled by the slowest step in the process, namely, the collisional excitation from the ground state. Since collisional excitation by electrons in partially ionized gases is far more efficient than by neutrals, we see that at low electron densities, ionization equilibration can proceed quite slowly. Using the approximate relations derived by Griem[18], we find for argon, with $N_e \approx 10^{13}$, the equilibration time is approximately 80 μs. On the other hand, ionization equilibrium of metallic vapors mixed with the test gases proceeds on sub-microsecond time scales. We must be cautious, therefore, if we assume complete ionization equilibrium in compressor experiments.

Some tests of excitation equilibrium have been conducted with the NOL compressor in connection with the spectroscopic temperature measurements. The existence of local thermodynamic equilibrium (LTE) in the compressor hinges on excitation rates for atom–atom collisions because the electron densities are so low. As a 'rule of thumb', the collisional excitation rate should exceed by a factor of ten the radiative decay rate for the energy level under consideration[18]. For, say, helium densities of 10^{21} cm^{-3}, relative velocities of 10^6 cm/s, and transition probabilities of 10^8 sec^{-1} appropriate for resonance lines of Ca(I) and Na(I), the excitation cross section for helium collisions should be greater than 10^{-18} cm^2 for LTE. From the meager experimental data available for alkali-rare gas collisional excitation (Faizullov and Sobolev[16]; Dodd et al.[10], who also reference earlier work), we can infer that these cross sections certainly are not much larger than 10^{-18} cm^2. The observed agreement in temperatures obtained with species of differing excitation cross sections is taken to be the best available evidence for the existence of LTE in the compressor.

Very little experimental evidence for chemical and thermal homogeneity exists. Boundary layer effects have already been mentioned. The existence of a uniform, hot gas core interior to a thin (approximately 1 mm) boundary layer is postulated upon consideration of helium thermal diffusivity under compressor conditions. The C_2 and non-resonant Cu(I) temperatures are expected to be representative of the hot core temperature, and the C_2 molecules are expected to be uniformly distributed in the hot core.

Although the feasibility of useful EOS work with ballistic compressors seems well established, insufficient attention has been devoted to date to measurements of transport properties at compressor conditions. P, V, T diagnostics are still required for such measurements, but it should be recalled that transport properties are generally more sensitive to the nature of the intermolecular potential than compressibility data or second virial coefficients[22, 23, 35]. The specification of the intermolecular potential is often the long range goal of thermodynamic studies, and it is quite attractive to set such a goal for compressor research. This is especially attractive when we realize that compressors operate in the temperature range where thermodynamic properties are sensitive to the form of the intermolecular potential.

2. Optical Studies

It is quite straightforward to record emission spectra from ballistic compressors with fast spectrographs and sensitive emulsions. At temperatures below 6000 K with monatomic test gases, the spectra consist of numerous metallic lines, and occasionally some molecular bands, emitted by contaminant species and by vapors in trace amounts formed by vaporization and ablation of compressor parts exposed to hot gas. Some continuum radiation is also detectable. At higher temperatures, the radiation from the carrier gas itself has been detected with xenon[12]. We have detected, with some difficulty, weak emission from pure N_2 at 5000 atm in the near infra-red, and Ryabinin[56] reported no difficulty in detecting emission spectra from air in his early compressor experiments. Some details of these and other optical experiments are discussed in the following sections.

A. Line Broadening Experiments

It is well-known that collision effects in neutral, dense gases result in perturbations to the electronic structure of atoms and molecules that reveal themselves as a broadening of the emitted spectral lines and a shift of the peak of the line profile away from its normal, low pressure wavelength. At extremely high densities, the line profiles also deviate from symmetrical, Lorentzian shape and exhibit extended wings on either the long or short wavelength side. Also, intensity fluctuations in the wings of some lines, called 'satellite bands', have been observed. Reviews of these observations and of some of the theoretical explanations have been written by Ch'en and Takeo[9], Traving[64, 65], Breene[4], Van Regemorter[54], and Hindmarsh[27, 28].

All of these effects have been observed in ballistic compressor spectra, and were first reported by Lalos and Hammond[37] with the NOL compressor. Shifts and widths of ionized barium lines broadened by argon and helium collisions have been measured by Hammond[19], and data for several lines of many metallic species perturbed by argon and helium have been reported by Holmes et al.[29] with the University of Oregon compressor. Widths of one of the sodium 'D' lines broadened by helium were measured by Burgess and Grindlay[6] with the Harvard College Observatory compressor. Recently,

Hammond[21] has completed a study of the widths and shifts of ionized calcium lines broadened by helium in the NOL compresor.

Many of these observations have direct application to astrophysical problems in stellar atmospheres. Compressor conditions simulate the pressures, temperatures and even chemical compositions of certain white dwarf stars[21]. Some of the low density compressor data[6] may be extrapolatable to solar photospheric conditions.

A fundamental goal of line broadening spectroscopists has been an understanding of the line broadening data in terms of the intermolecular potential between emitter (or absorber) and perturber. Since the form of this potential is only partially known for neutral perturbers, the hope has been to probe this potential with a variety of line broadening experiments. The procedure has been to insert a model potential into the line broadening theory and if the predictions of that theory then match the observations, that model potential then becomes a candidate for the 'real' potential function. Unfortunately, most experiments are sensitive only to small portions of the full potential function, and no clear picture of the 'real' potential for any interacting pair has yet emerged. One of the many problems encountered is the fact that line broadening data reflect the properties of the difference potential, the difference between the interaction potentials of the upper and lower quantum states of the electronic transition. These potentials must often be assumed to have the same functional form for the problem to be tractable, but there is evidence to the contrary[2].

Ballistic compressor experiments contribute to these investigations principally in two ways. First, they provide line broadening data over a temperature range far wider than is possible with absorption spectroscopy in high pressure cells, and the velocity dependence of shift and width can be as sensitive to the potential as the shift and width themselves[3, 30]. Secondly, they provide experimental access to electronic transitions between excited states that may be more hydrogenic in character and thus simpler to analyse theoretically. They also permit study of the spectra of ionized species, and thus provide an investigation of the influence of the charge state of the emitter on line broadening problems.

B. Opacity Measurements

Temperature measurements using the brightness–emissivity method routinely employ opacity measurements made at selected wavelengths. Absorption measurements over broader wavelength regions can be used to search for new photoabsorption processes. An experimental investigation with the NOL compressor[39, 40] has recently been made to measure the absorption coefficients of hot, highly compressed hydrogen in the ultra-violet. This study had as its goal the experimental confirmation of the theoretically predicted quasi-molecular association process[13]. In this process (not to be confused with singlet ground state absorption of molecular hydrogen below 1750 Å) the collision of two hydrogen atoms with parallel electron spins forms the repulsive ground state of the triplet system. Absorption by a free–bound process is expected to occur to the stable triplet sigma state. The absorption coefficient for this quasi-molecular association process is continuous and is predicted to occur in the wavelength region 1750–2500 Å. This photoabsorption process is predicted to dominate all other continuous absorption processes in this spectral region and may account for the anomalous absorption that occurs in the sun from about 4800 Å to shorter wavelengths.

C. Fission Studies

The rapid gas compression technique has been employed recently in a research program whose principal aim is the laboratory generation of a fissioning uranium plasma, and the measurement of its spectral absorption coefficients[48]. Aside from the general importance that is associated with the ability to generate a fissioning plasma in the laboratory, the information obtained from the investigation of its optical properties is expected to find immediate application in the radiative heat transfer calculations that are required in the design of gaseous core nuclear rocket engines. The initial experiments are designed to record the emission and absorption spectrum of non-fissioning ^{238}U using a ballistic piston compressor. These experiments employ a mixture composed of 100 torr UF_6 (room temperature vapor pressure) and sufficient He to give a gas temperature at maximum compression high enough to produce the desired spectra. At the completion of these initial studies analogous measurements will be made with fissionable ^{235}U in place of the ^{238}U and with the gas sample (compressor high pressure section) located in the thermal column of the University of Florida 100 kW nuclear reactor. It is expected that the neutron irradiation of the gaseous ^{235}U at the peak pressure of the compression stroke will produce a fissioning uranium plasma for times of the order of 1 msec, and make possible a detailed investigation of its spectral properties.

D. Laser Pumping

The versatility of the ballistic piston compressor was demonstrated recently by its being used as a unique high-intensity light source. Dowling et al.[11, 12] compressively heated room-temperature xenon initially at 40 torr pressure to peak pressures of from 800 to 2400 atm, with accompanying temperatures of 8000 to 15 000 K. Brightness measurements of the light pulses produced indicated blackbody radiation at these temperatures. The radiation was coupled into a neodymium-doped glass laser rod with resulting laser action.

Their compressor is in the form of a cylinder 204 cm long and of 38 cm i.d. The energy imparted to the 525 g piston is typically 250 to 1000 J. A 2.5 cm diameter quartz window is normally located in the high pressure end-wall of the cylinder. For laser pumping, the end cap was modified to accommodate a 6.35 mm diameter by 25.4 mm long laser rod by drilling a 6.35 mm diameter hole through the window proper and end cap. The outside of the window was aluminized and a neodymium laser rod was placed within the window. The rod was Schott LG55 glass doped with five per cent neodymium. The rod ends were dielectrically coated to 98 per cent and 100 per cent reflectivity at 1.06 μ. A silicon photodiode detector was used to monitor the laser output.

Conversion of piston kinetic energy to laser pumping energy with efficiencies as high as 55 per cent has been demonstrated. This system, therefore, affords a means of direct transfer of mechanical energy into radiant energy without the necessity of intermediate energy storage devices.

3. Other Studies

A. Chemical Reactions

Although this volume is restricted to non-reacting fluids it was thought worthwhile to include a brief description of some compressor studies employing reacting fluids in order to emphasize the versatility of the rapid gas compression technique for conducting basic research. The extremely high gas temperatures produced by rapid gas compression makes possible the study of fast, high-temperature chemical reactions. The very rapid cooling of the gas that occurs as the piston rebounds after having generated the maximum pressure and temperature can quench the reaction and freeze out its products. Rapid gas compression (and expansion) thus allows one to investigate chemical reactions at temperatures substantially higher than those employed in conventional high-temperature bomb techniques.

Longwell et al.[44] at the California Institute of Technology have made a number of studies of chemical reactions using a ballistic piston compressor. In an investigation of the formation and decomposition of nitric oxide at temperatures between 2200 and 4400 K (Olin and Sage[50]) they show that a cooling rate of at least 10^8 K/s is required to obtain more than 0.04 mole fraction nitric oxide. Pistons of different mass were used to vary the piston velocity and hence the gas expansion and cooling rate over a range of values. Other studies[45, 51, 52] included the thermal decomposition of n-hexane at high pressures; thermal rearrangement of methane, n-butane, n-hexane, and n-octane in the presence of hydrogen; and the thermal decomposition and partial oxidation of ammonia.

Tsiklis and Borodina[66] in the USSR, also using a ballistic piston compressor, have investigated the formation of acetylene on adiabatic compression of methane. They used mixtures of methane with nitrogen, neon, argon, xenon, krypton, and hydrogen and found that the greatest yield, approximately 30 to 35 per cent, was obtained for a mixture of 5 per cent methane in 95 per cent argon compressed to 5000 atm. In another study[67] they compressed a gas mixture consisting of 8.8 per cent methane, 2.5 per cent ammonia, and 88.7 per cent argon to 3000 atm and obtained a 20 per cent yield of hydrocyanic acid.

B. Dissociation Kinetics

A variation of the basic rapid compression technique employs a locked-stroke compressor, that is, one whose piston is mechanically arrested in its position of maximum travel. This compressor is also characterized by a rapid rise in gas pressure and temperature to their peak values but differs from the ballistic piston compressor in that the peak conditions are maintained for periods of time as long as 10 μs. The duration of this period is determined by the rate of heat transfer to the compressor walls and by the rate of gas leakage across the piston. The maximum temperature that can be attained without damage to the apparatus is somewhat less for the locked-stroke compressor than for the ballistic piston compressor. Henderson et al.[25] attempted to study the homogeneous dissociation kinetics of the fluorine oxidizers N_2F_2 and OF_2 by means of a locked-stroke compressor attached to the fast reaction cell of a Bendix time-of-flight mass spectrometer. They were interested in reactions in the temperature range of 500 to 1 200 K, with possible reaction times of up to 10 ms. These requirements, incidentally, ruled out use of the shock tube but were within the capabilities of the locked-stroke compressor. The procedure was to leak the hot gas into the mass spectrometer for continuous analysis of the dissociating species. In a typical experiment a few per cent of the gas to be dissociated, OF_2 or N_2F_2 for example, is mixed with helium containing one per cent argon. The argon serves as a non-dissociating quantitative reference for the mass spectra.

C. Ignition Studies

Another example of the use of a locked-stroke compressor for making physical measurements under conditions not practical with static compression techniques may be found in the work of Bryan and Noonan[5]. These researchers determined the minimum energy, delivered in a 3 ms interval, just sufficient to ignite a 1 cm^2 surface of seven different solid explosives. The experimentally determined values were found to vary from greater than 0.4 cal/cm^2 for TNT to less than 0.1 cal/cm^2 for lead styphnate.

IV. Summary

The ballistic piston compression method produces a unique set of physical conditions in monatomic and diatomic gases that have hitherto been unattainable in laboratory devices. These conditions of simultaneous high temperature, pressure and density are measurable to a degree of accuracy that permits detailed study of the thermodynamic properties of the gas sample under equilibrium conditions and free from electric and magnetic field perturbations. The extreme physical conditions persist for intervals sufficiently long that highly sophisticated diagnostic instrumentation is not required.

One of the most prominent limitations of the present apparatus is the damage suffered by compressor components exposed to hot gas. Some further engineering development is required, probably involving refractory metal coatings or liners, to permit routine operation at temperatures exceeding 5 000 K. Some further development is also necessary to improve the chemical purity of the gas sample. It is somewhat challenging to design an apparatus capable of containing several thousand atmospheres pressure and also capable of being evacuated to a few microns Hg. We are aware of no existing compressor that can be baked to remove adsorbed gases. Such a capability would also permit an increased range of test gas initial conditions and thus would yield an even wider range of final maximum conditions than those presently attainable.

Finally, it is clear that the study of thermodynamic properties by means of ballistic piston compression is still in relative infancy. In view of the numerous practical applications of such data, we expect to see a proliferation of compressors and an increasing variety of experimental techniques in the decade ahead.

V. References

[1] Baker, J. and H. Swift, *J. Appl. Phys.* **43**, 950 (1972).
[2] Baylis, W., *J. Chem. Phys.* **51**, 2665 (1969).
[3] Baylis, W., M. Pillon, G. Reck and R. Hood. *Bull. Amer. Phys. Soc.* **16**, 1349 (1971).
[4] Breene, R., *The Shift and Shape of Spectral Lines*. Pergamon: New York (1961).
[5] Bryan, G. J. and E. C. Noonan. *Proc. Roy. Soc.* **246**, 167 (1958).
[6] Burgess, D. and J. Grindlay, *Astrophys. J.* **161**, 343 (1970).
[7] Cezairliyan, A., *Applied Optics*, **10**, 1178 (1971).
[8] Charatis, G. and T. Hershey. University of Maryland IFDAM *Tech. Note BN-361* (1964).
[9] Ch'en, S. Y. and M. Takeo, *Rev. Mod. Phys.* **29**, 20 (1957).
[10] Dodd, J., E. Enemark and A. Gallagher. *J. Chem. Phys.* **50**, 4838 (1969).
[11] Dowling, J. A., J. Shumsky, J. Eckerman, R. E. Schlier and P. Kisatsky. *Appl. Phys. Letters*, **12**, 184 (1968).
[12] Dowling, J., J. Davis, J. Eckerman, R. E. Schlier, J. Shumsky and P. Kisatsky. *Applied Optics*, **8**, 1867 (1969).
[13] Doyle, R., *Astrophys. J.* **153**, 987 (1968).
[14] Ecker, G., *Z. Phys.* **130**, 585 (1951).
[15] Evans, C. and F. Evans. *J. Fluid Mech.* **1**, 399 (1956).
[16] Faizullov, F. and N. Sobolev. *Opt. Spectrosc.* **11**, 310 (1961).
[17] Grasdalen, G., M. Huber and W. Parkinson. *Astrophys. J.* **156**, 1153 (1969).
[18] Griem, H. R., *Plasma Spectroscopy*. McGraw-Hill: New York (1964).
[19] Hammond, G. L., *Astrophys. J.* **136**, 431 (1962).
[20] Hammond, G. L. and G. T. Lalos. *NOLTR 71-228* (1971).
[21] Hammond, G. L., University of Maryland Ph.D. Dissertation (unpublished) (1973).
[22] Hanley, H. and M. Klein. *Tech. Note US Nat. Bur. Stand No. 360* (1967).
[23] Hanley, H. and M. Klein. *J. Chem. Phys.* **50**, 4765 (1969).
[24] Hattenburg, A., *Applied Optics*, **6**, 95 (1967).
[25] Henderson, U. V., H. A. Rhodes and V. M. Barnes. *Rev. Sci. Instrum.* **37**, 294 (1966).
[26] Hilsenrath, J., C. Beckett, W. Benedict, L. Fano, H. Hoge, J. Masi, R. Nuttall, Y. Touloukian and H. Woolley. *Circ. US Nat. Bur. Stand No. 564*, Washington, D.C. (1955).
[27] Hindmarsh, W., A. Petford and G. Smith. *Proc. Roy. Soc. A*, **207**, 296 (1967).
[28] Hindmarsh, W., *Atomic Physics*, Vol. II, P. Sanders, Ed., Plenum: New York (1971).
[29] Holmes Q., S. Y. Ch'en and M. Takeo. *J. Quant. Spectrosc. Radiat. Transfer*, **9**, 749, 761, 769 (1969).
[30] Hood, R. and G. Reck, *J. Chem. Phys.* **56**, 4053 (1972).
[31] Huber, M. and F. Tobey. *Astrophys. J.* **152**, 609 (1968).
[32] Kendall, P., *Appl. Spectrosc.* **22**, 274 (1968).
[33] Kislykh, V., V. Vasil'ev and E. Verem'ev. *High Temperature* (Moscow), **9**, 836 (1972).
[34] Klein, M., *AEDC-TR-67-67*, Arnold Engineering Development Center, Tullahoma, Tennessee (1967).
[35] Klein, M. and H. Hanley. *Trans. Faraday Soc.* **64**, 2927 (1968).
[36] Lalos, G. T., *Rev. Sci. Instrum.* **33**, 214 (1962).
[37] Lalos, G. T. and G. L. Hammond. *Astrophys. J.* **135**, 616 (1962).
[38] Lalos, G. T. and G. L. Hammond. *Rev. Sci. Instrum.* **36**, 550 (1965).
[39] Lalos, G. T. and G. L. Hammond. *NASA CR-72116* (*NOLTR 66-202*) (1966).
[40] Lalos, G. T. and G. L. Hammond. *NASA CR-72589* (*NOLTR 70-15*) (1969).
[41] Lee, R. and E. Lewis. *Applied Optics*, **5**, 1858 (1966).
[42] Lewis, M., B. Roman and G. Rouel. *VKI Tech. Memo No. 23*, Rhode-Saint-Genese: Belgium (1971).
[43] Lick, W. and H. Emmons. *Thermodynamic Properties of Helium to 50000 K*. Harvard University Press: Cambridge, Mass. (1962).

[44] Longwell, P. A., H. H. Reamer, N. P. Wilburn and B. H. Sage. *Industr. Engng Chem.* **50**, 603 (1958).
[45] Longwell, P. A. and B. H. Sage. *J. Chem. Engng Data*, **5**, 322 (1960).
[46] Michels, A. and H. Wouters. *Physica*, **8**, 923 (1941).
[47] Michels, A. and A. Botzen. *Physica*, **15**, 769 (1949).
[48] Miller, B. E., R. T. Schneider, K. Thom and G. T. Lalos. Second Symposium on Uranium Plasmas: Research and Applications, Atlanta, Georgia (1971).
[49] Minardi, J. and R. Schwartz. Aerospace Research Laboratories. *Rep. No. ARL 63-167*, Wright-Patterson AFB, Ohio (1963).
[50] Olin, J. B. and B. H. Sage. *J. Chem. Engng Data*, **5**, 16 (1960).
[51] Olin, J. B. and B. H. Sage. *J. Chem. Engng Data*, **6**, 384 (1961).
[52] Olin, J. B., H. C. Wiese and B. H. Sage. *J. Chem. Engng Data*, **6**, 372 (1961).
[53] Price, D. and G. T. Lalos. *Industr. Engng Chem.* **49**, 1987 (1957).
[54] van Regemorter, H., *Ann. Rev. Astron. Astrophys.* **3**, 71 (1965).
[55] Rowlinson, J., *Molec. Phys.* **7**, 349 (1963).
[56] Ryabinin, Yu. N., *Gases at High Densities and Temperatures*, Pergamon: New York (1961).
[57] Samson, J., *Techniques of Vacuum Ultraviolet Spectroscopy*, Wiley: New York (1967).
[58] Sando, K., R. Doyle and A. Dalgarno. *Astrophys. J.* **157**, L143 (1969).
[59] Schurer, K., *Applied Optics*, **7**, 461 (1968).
[60] Silverman, S., *J. Opt. Soc. Amer.* **39**, 275 (1949).
[61] Smith, D. and E. Starkman. *Rev. Sci. Instrum.* **40**, 1541 (1969).
[62] Takeo, M., Q. Holmes and S. Y. Ch'en. *J. Appl. Phys.* **38**, 3544 (1967).
[63] Theile, E., *J. Chem. Phys.* **39**, 474 (1963).
[64] Traving, G., *Uber die Theorie der Druckverbreiterung von Spektrallinien*. Braun: Karlsruhe (1960).
[65] Traving, G., *Plasma Diagnostics*, Chap. 2. W. Lochte-Holtgreven, Ed., North-Holland: Amsterdam (1968).
[66] Tsiklis, D. S. and M. D. Borodina. *Dokl. Akad. Nauk SSSR*, **140**, 1376 (1961).
[67] Tsiklis, D. S. and M. D. Borodina. *Dokl. Akad. Nauk SSSR*, **147**, 860 (1962).

CHAPTER 26

Thermodynamic Properties of Fluids below 20 K*

WILLIAM E. KELLER

*Los Alamos Scientific Laboratory, University of California,
Los Alamos, New Mexico 87544, USA*

Contents

I.	Introduction	1219
II.	Temperature Scales and Thermometry below 20 K	1220
	1. The Basis for Thermodynamic Thermometry below 20 K	1221
	2. Primary Thermometry from 1 to 20 K	1221
	3. Primary Thermometry below 1 K	1222
	A. Mössbauer Effect Thermometry	1223
	B. Nuclear Orientation Thermometry	1225
	C. Nuclear Spin Thermometry	1226
	D. Noise Thermometry	1229
	4. Temperature Scales in Use below 20 K	1230
	A. The Scale IPTS-68 (T_{68}) from 13.81 to 20 K	1231
	B. 1958 ^{4}He Scale (T_{58}) and 1962 ^{3}He (T_{62}) Scale between 0.5 and 5.2 K	1232
	C. Scales Spanning the Range 1 to 20 K	1233
	5. Secondary Thermometry below 20 K	1234
	A. Capacitance Thermometry	1235
	B. ^{3}He Melting Curve Thermometry	1236
	C. Superconducting Transition Temperatures as Fixed Points	1239
	D. Superconducting Thin Film Thermometers	1240
III.	Refrigeration Techniques below 0.3 K	1240
	1. The ^{3}He–^{4}He Dilution Refrigerator	1241
	2. Nuclear Cooling	1244
	3. Cooling by Adiabatic Freezing of Liquid ^{3}He	1246
	4. Heat Transfer and Isolation at ULT	1248
IV.	Preparation of Helium and Hydrogen Samples	1251
V.	Calorimetry of Fluids below 20 K	1253
	1. Liquid ^{4}He near the Melting Curve	1253
	2. Liquid ^{4}He near the λ-Point	1254
	3. Liquid ^{3}He at ULT and at Pressures up to the Melting Curve	1256
	4. Latent Heats of Vaporization of Hydrogen and ^{4}He	1256
	5. Liquid ^{3}He–^{4}He Mixtures	1257
VI.	pVT Measurements below 20 K	1257
	1. Molar Volume	1258
	2. pVT Properties at Melting	1260
	3. Osmotic Pressure of ^{3}He–^{4}He Solutions	1262
VII.	References	1263

I. Introduction

The title of this chapter is in one sense very restrictive and in another

---

* Work done under the auspices of the US Atomic Energy Commission.

entirely open-ended. Below 20 K the only chemical species exhibiting fluid behavior are the isotopes of hydrogen and helium, often referred to as quantum fluids since quantum effects are responsible for many startling differences in behavior of these substances as compared with ordinary fluids. On the other hand, because both ^{3}He and ^{4}He are expected in the absence of externally applied pressure to remain as liquids even at the absolute zero of temperature, the temperature range permitted for discussion may extend downward without limit. In a third sense there are of course many additional degrees of freedom implicit in the words 'thermodynamic properties'. These circumstances combine to set this chapter apart organizationally from other chapters in this volume; others tend to describe measurement methods for a single variable, or class of related variables, as may be applied to a variety of substances, whereas the present chapter is to deal with techniques for measuring several different variables—over perhaps wide ranges—as applied to a few special substances.

Boundary conditions are clearly required here and the following have been established to limit the scope of this chapter: (1) the temperature range covered is from 1 mK to 20 K which, being more than four decades of the temperature scale, is nevertheless a very large span; (2) methods for investigating the thermodynamic properties of H_2, ^{4}He, ^{3}He and ^{3}He–^{4}He mixtures will be considered, with emphasis placed on those used for the pure helium isotopes, especially at temperatures below 1 K—in some instances, e.g. in the cases of thermometry and refrigeration methods, the techniques are applicable to low temperature systems other than those considered here; (3) standard experimental methods which merely have the added complication of being carried out at low temperatures will be omitted; and (4) an attempt will be made not to duplicate descriptions of techniques which may be found either in Volume I* of this series or elsewhere in this volume, although appropriate references may be provided; however, where some amplification of these descriptions seems desirable it will be included. Even within these limitations the diversity of topics to be considered remains staggering, rendering completeness impossible and brevity imperative. Hence the spirit of this chapter is to provide an indication of some of the more recent developments in laboratory methods which aid in furthering our understanding of the thermodynamic properties of quantum fluids (for a review of the properties of ^{3}He and ^{4}He the reader is referred to the recent monographs by Wilks[118] and Keller[58]; similarly, for H_2 a compendium edited by Scott[91] is useful; for more details on general cryogenic procedure, the books by Bailey[8], Croft[24], Rose-Innes[85] and Hoare et al.[46], for example, may be consulted).

II. Temperature Scales and Thermometry below 20 K

Of utmost importance to the determination of any thermodynamic property X is the proper assignment of a temperature T to an experimental observation of X. In practice one attempts to accomplish this through the use of a thermometer bearing a specific known relationship to a temperature

---

* Throughout this text, Volume I refers to *Experimental Thermodynamics* Vol. I (J. P. McCullough and D. W. Scott, Eds.), Butterworths: London (1968).

scale which, in turn, represents as closely as possible a scale defined according to any equation expressing the second law of thermodynamics. At best the steps in this process are taken with difficulty; but as the temperature is lowered, the obstacles become very much more severe. This section reviews first several methods of thermodynamic (primary) thermometry applicable below 20 K, then the temperature scales presently used for this region, and finally some practical (secondary) thermometric techniques.

1. The Basis for Thermodynamic Thermometry below 20 K

The bases for thermodynamic thermometry and the Kelvin temperature scale have been presented in Volume I (pp 16 ff), indicating the arbitrariness residing in numerical values of the scale with respect to the choice of the scale interval (the degree) and a standard reference temperature T_0 (the triple point of water = 273.16 K). It is also worthy of mention that every formula connecting experiments with attempts to realize the thermodynamic scale contains the natural proportionality constant k_B, or $N_A k_B = R$, which arises in statistical thermodynamic descriptions of the second law (k_B is the Boltzmann constant and N_A Avogadro's number). Swenson[103] has pointed out that $R = 8.31434 \pm 0.00035$ J/mol-K is likely the least accurately known fundamental constant. However, this is as yet not a serious concern for the scale below 20 K, where uncertainties as large as 10 mK (or >1/2000) persist. The latter is indeed serious as current techniques of secondary thermometry are capable of providing temperature measurements reproducible to 1 mK or better over much of the range considered here.

2. Primary Thermometry from 1 to 20 K

The main primary thermometric methods applicable to realizing the Kelvin scale in the range 1 to 20 K have been mentioned in Volume I (pp 18–23). These include four types of gas thermometry, ultrasonic thermometry using an acoustic interferometer, and magnetic thermometry from susceptibility measurements on paramagnetic salts. Swenson[103] has reviewed and assessed the current (1970) status of experimental results in these areas with respect to the scale and found that the situation remains largely unsatisfactory. Results for a given method but from different researchers generally lack consistency; and a similar deficiency occurs in the comparison of the results from different methods. Discrepancies of the order of 5 mK are routinely found and amount to as much as 10 mK at such key temperatures as the normal equilibrium hydrogen boiling point (near 20 K) and the normal helium boiling point (near 4 K). These difficulties are especially noticeable in those instances where data obtained over one part of the temperature range from one experiment join with those over an adjacent range from a different experiment. It is likely that a major source of the uncertainty of the results lies in the accounting for departures from ideality in the systems studied, i.e. from the ideal gas law (involving second and higher virial coefficients) or from Curie's law requiring inclusion of Weiss constant and shape factors). However, difficulties in making some of the associated measurements, such as of pressure and of gas absorption corrections, are not insignificant. In any event, improvement of the situation is urgent. Fortunately, as of this writing, workers at such institutions as Iowa State University, the National Bureau of Standards (NBS–USA), the National Physical Laboratory (NPL–UK), the Kamerlingh Onnes Laboratory (KOL—the Netherlands), the National Research Laboratory of Metrology (NRLM—Japan), and the National Standards Laboratory (NSL—Australia) are deeply involved in this probelm.

3. Primary Thermometry below 1 K

Within the last decade new refrigeration techniques (see section III) have made steady-state temperatures available down to the millikelvin range, well below temperatures achievable (~ 0.25 K) using pumped cryogenic liquids. This coupled with the intense interest in the ultralow temperature* properties of the quantum fluids ^{3}He and ^{4}He (and of many other systems) has produced demands both for a reliable representation of the thermodynamic scale in this region as well as for convenient secondary thermometry. Clearly because of the vanishingly small vapor pressures of both ^{4}He and ^{3}He at these temperatures, acoustic and gas thermometry are ruled out. On the other hand, work with $Ce_2Mg_3(NO_3)_{12} \cdot 24H_2O$ (CMN) in magnetic thermometry has been extensive; a single crystal of CMN has been found to be very nearly an ideal paramagnet, obeying Curie's law from 1 K down to ~ 6 mK[48], and is therefore a good candidate for use as an absolute thermometer. Although CMN does in fact presently provide the standard for thermometry below 1 K, because its use still requires calibration it is not yet a truly primary thermometer. For many applications it is precise, convenient and reliable, but as T decreases the problem of thermal contact with the test body becomes more difficult. Moreover, and perhaps most important, uncertainty concerning the proper shape correction which must be added to the Weiss constant has, at least for the present, severely diminished confidence in CMN as a thermometer below 6 mK and in addition has undesirable consequences on the scale up to 100 mK[3]. Another drawback of CMN is its large specific heat maximum near 3.5 mK, which seriously limits its use at lower temperatures. (See also Volume I, pp 18–19, 55–56, 284–287).

Accordingly there has been a search for other primary methods, useful well below 1 K, which not only avoid the limitations of CMN but fulfil in addition to the thermodynamic requirements already mentioned certain criteria which are generally valid but assume greater importance in ULT work[3]. To meet these criteria the thermometer should: (1) be capable of being placed in sufficiently intimate thermal contact with the test body so as to reflect properly the body's temperature; (2) possess a short internal relaxation time, for similar reasons; (3) not require for a measurement an amount of energy which would unduly perturb the system of interest; (4) have a small heat capacity; (5) be capable of at least one per cent accuracy; (6) be sufficiently sensitive to resolve $0.001\ T$; and (7) not be seriously affected by stray magnetic fields. In addition it would be highly desirable if most ULT experiments could conveniently carry their own primary thermometers serving simul-

---

* Any temperature below 1 K is arbitrarily here designated as 'ultralow', hereafter abbreviated as ULT.

taneously as interpolation devices. This appears to be quite plausible and would be an advantage not generally shared with experiments at higher temperatures, where most primary methods are exceedingly cumbersome. Although thermometry satisfying the above criteria for the range 1 mK to 1 K is far from fully developed, it should be useful to mention the principles of several techniques currently receiving serious consideration.

A. Mössbauer Effect Thermometry

The splitting energy Δ between adjacent nuclear spin sublevels is generally equivalent to temperatures much smaller than 1 K. For the ith nuclear state with spin I_i

$$\Delta_i = \mu H / I_i \tag{1}$$

where μ is the magnetic moment of the nucleus and H is the magnetic (hyperfine) field at the nucleus. Whereas for $\Delta \ll k_B T$ the levels (with magnetic quantum numbers $m = -I, \ldots, +I$) are equally populated, when $\Delta \gtrsim k_B T$ the equilibrium populations are proportional to the Boltzmann factor $\exp(-\Delta/k_B T)$; specifically, the population of the mth level is

$$a_m = \exp(-m\Delta/k_B T) / \sum_{m'=-I}^{I} \exp(-m'\Delta/k_B T) \tag{2}$$

It can be easily demonstrated that a determination of the population ratio for two such levels can provide a direct measure of the thermodynamic temperature. A particularly convenient method for observing this ratio

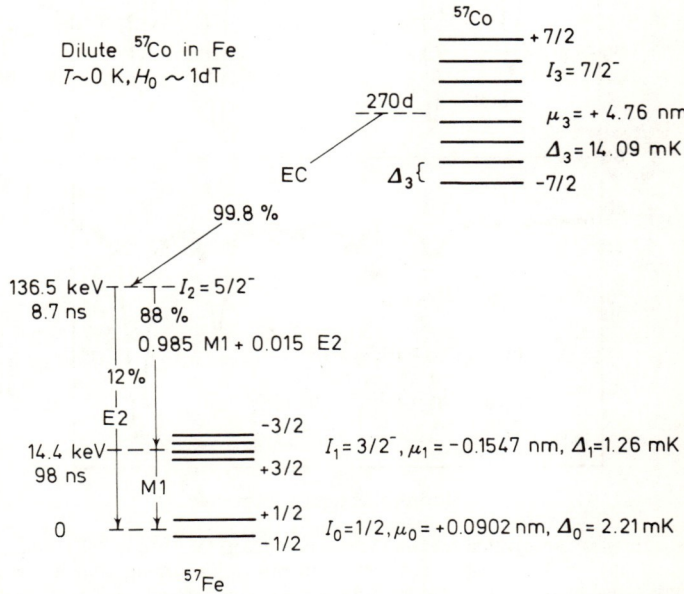

Figure 1. Nuclear decay scheme of ^{57}Co as a dilute impurity in Fe (from Taylor[106]).

involves the recoil-free emission and absorption of γ-rays (Mössbauer effect, or ME) for suitable radionuclei, such as ^{57}Fe, ^{197}Au, ^{119}Sn, and ^{193}In[54].

As an example of this technique let us consider nuclear resonance absorption (ME) spectra obtained with ^{57}Co nuclei (thermally diffused in a thin foil of iron enriched in ^{56}Fe) acting as a source and maintained at the low temperature of interest[106]. ^{57}Co decays according to the scheme shown in *Figure 1*. The transition from the $I_1 = 3/2$ level to the ground state proceeds via a Mössbauer 14.4 keV γ-ray. Because the selection rules for transitions between the hyperfine levels of two different states are $\Delta m = 0, \pm 1$, six components are allowed for the ME γ-ray. Now, at equilibrium the relative populations of the eight hyperfine levels of the parent ^{57}Co nucleus reflect the temperature of the surroundings; and because of the short lifetimes of the excited states of ^{57}Fe it is believed that this information is transmitted through the two nuclear transitions in a completely predictable, albeit complicated, manner[27]. Thus the relative intensities of the six ME spectral lines reflect the temperature of the ^{57}Co environment.

Figure 2(a) illustrates a typical ME resonance spectrum obtained by Taylor[106] for a ^{57}Co-in-Fe source maintained at 0.9 K in a 0.12 T transverse

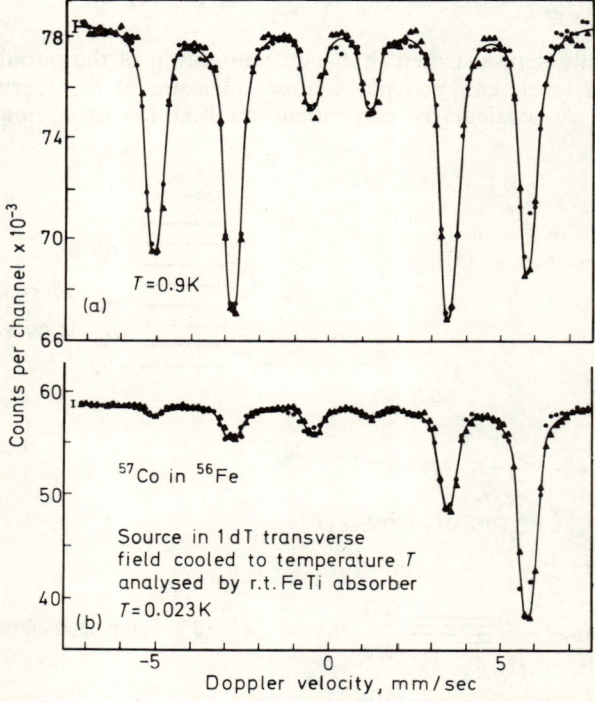

Figure 2. Resonant γ-ray absorption spectra (Mössbauer effect) for ^{57}Co in ^{56}Fe source at low temperature and an ^{56}FeTi absorber at room temperature illustrating changes of relative intensities of hyperfine lines as the source temperature is lowered from (a) 0.9 K to (b) 0.023 K (from Taylor[106]).

magnetic field and analysed with a FeTi absorber at room temperature*
Energy displacements are given in terms of Doppler velocity of the absorber
with respect to the source, and the intensities at various velocities are obtained
by well-established γ-ray counting techniques [use of highly efficient NaI(Tl)
scintillation detectors has been discussed by e.g. Neiler and Bell[78]; information on newer, high resolution Ge(Li) detectors can be obtained from the
manufacturers, e.g. Ortec, Inc., Oak Ridge, Tenn., USA; and a review article
on solid-state radiation detectors by Dearnaley[28] is of interest]. Absorption
occurs when the Doppler shifted energy of the unsplit absorber line overlaps
the various hyperfine split lines of the emitter, with the ME γ-ray producing
a transition in an absorber nucleus from the ground state to the 14.4 keV
state. Already at 0.9 K there is some asymmetry in the pattern with respect
to the outer pair of hyperfine lines. When the source is cooled to 0.023 K
[*Figure 2*(b)] this is greatly enhanced, with other pairs also showing large
asymmetries. In principle the accuracy of absolute temperature determination
is limited by the accuracy with which the intensity ratio of pairs of lines can
be measured and by how well the appropriate splitting energy (here Δ_0
of ^{57}Co in Fe) is known from other measurements (through H, Δ is sensitive
to the environment of the radionucleus and must be determined for each
system, usually by nuclear magnetic resonance methods—e.g. see Baker
et al.[9]—which can yield values of Δ accurate to about one part in 10^4).
With the system discussed here using current state-of-the-art techniques,
practical thermometry is achievable over the range 5 to 500 mK.

The theory and application of ME thermometry now rests on solid ground.
With further refinements in technique and variations in experimental methods
(source and absorber preparation and composition, counting times, source
strength, etc.) a number of ME systems should be able to provide suitable
thermometry satisfying most of the criteria mentioned earlier for work at
ULT. One of the principal limitations of the method lies in the self-heating
of the source. For a useful 1-millicurie source of ^{57}Co, for example, this
energy is about 0.9 μJ/s, which is large for ULT work, although cryostat
design can be arranged so that no more than about 25 per cent of this energy
is dissipated at low temperatures. However, it is possible to interchange
absorber and source[27]; and with the absorber cooled this heating can be
reduced by another factor of five.

B. Nuclear Orientation Thermometry

Atomic nuclei with spin $I \neq 0$ when cooled in a non-symmetrical force
field (crystal field, magnetic field, etc.) tend to orient in a preferred direction
with respect to that field. When such nuclei are radioactive the spatial distribution of decay product emission is similarly anisotropic. The degree of this
anisotropy is temperature dependent and through the fundamental theory
of nuclear angular momentum is directly related to the absolute temperature.

To date the directional distribution, $W(\theta)$, of γ-rays from oriented radio-

---

* In a ^{57}Co-in-Fe source the magnetic field at the Co nucleus due to Fe ions produces the hyperfine splitting. The external field is applied to the source in order to orient the magnetic domains. Fe dissolved in non-magnetic Ti provides an unsplit or monochromatic absorber.

nuclei has received the most attention with regard to thermometry. With $W(\theta)$ taken as the γ-ray counting rate at an angle θ with respect to the alignment axis in the case of axially symmetric orientation, the theory indicates[30]

$$W(\theta) = 1 + \sum_i A_i f_i P_i (\cos \theta) \tag{3}$$

The coefficients A_i are characteristic of the particular nuclear decay in question and are often completely defined by the level spin assignments; P_i is the ith Legendre polynomial; and f_i,[7] related to the relative populations of the magnetic sublevels, contains the temperature dependence, again through the Boltzmann factor. For example, in a magnetic field

$$f_1 = \Sigma m a_m / I \tag{4}$$

$$f_2 = [\Sigma m^2 a_m - \tfrac{1}{3} I(I + 1)]/I^2 \tag{5}$$

etc.

with a_m given by equation 2.

In practice the radionucleus—^{54}Mn, ^{60}Co, ^{57}Co, and ^{125}Sb are suitable choices since the required nuclear data are well known— is dissolved as an impurity in a magnetic host such as an iron foil. A field of the order of 0.1 to 0.2 T is applied to the foil in order to provide an orientation direction, and the γ-ray counting is performed at $\theta = 0°$ and $90°$ with respect to this direction. The magnetization of the Fe foil is approximately 98 per cent saturated at ~ 0.2 T, which is satisfactory for most purposes. However, when the thermometer is used in experiments requiring an additional field, $W(\theta)$ will be altered and corresponding corrections should be applied (above 98 per cent, saturation is approached as $1/H$). As in ME thermometry Δ must be determined for each system.

Sites et al.[95] have investigated the relative merits of ^{54}Mn, ^{60}Co, and ^{125}Sb as primary thermometers and concluded that the ^{54}Mn system furnishes the most reliable results in the range 2 to 40 mK and is accurate to five per cent down to 4 mK with a sensitivity of ~ 0.1 mK. They have also discussed some of the technical problems associated with achieving accurate data, especially with regard to choice of radionuclei with short intermediate state lifetimes, care in metallurgical preparation of foils, and proper geometric corrections which must be applied to counting rates to account for the finite size of the detectors. It is of interest to note that Taylor[106] has compared ^{57}Co-in-Fe ME thermometry with γ-ray anisotropy in ^{57}Co, ^{57}Mn, and ^{60}Co (two γ-rays) at 24 mK. The latter method gave consistently higher results than the former by a few mK; however, it is suspected that these deviations are not real, but rather are traceable to violation of the strict condition for anisotropy thermometry that all domains point in the $\theta = 0$ direction.

C. Nuclear Spin Thermometry

Because the magnetic moments of nuclei are small and obey Curie's law to temperatures several orders of magnitude lower than electron paramagnets, nuclear susceptibility measurements offer promise of considerable usefulness in ULT primary thermometry. Similar to electron paramagnetism, the susceptibility for a nuclear spin system is given by Curie's law

$$\chi_n = M/H_0 = C_n/T_s \qquad (6)$$

where M is the bulk magnetization of the sample, i.e. the macroscopic nuclear magnetic moment; H_0 is the steady applied magnetic field; T_s is the temperature of the spin system; and C_n is Curie's constant for nuclei

$$C_n = N\gamma^2\hbar^2 I(I+1)/3k_B \qquad (7)$$

in terms of number density of nuclear spins N and the average nuclear gyromagnetic ratio γ of the sample (isotopes of the same element generally differ in γ); $\hbar$ is Planck's constant divided by 2π. Equation 7 contains an approximation for the Brillouin expression for the magnetization valid when $\hbar\gamma H_0 \ll k_B T_s$; this condition is met primarily by diamagnetic or weakly paramagnetic materials and probably holds for nuclear magnetism to temperatures as low as 10^{-5} or 10^{-6} K.

The usefulness of equation 6 for absolute thermometry depends upon the extent to which $T_s = T$, the thermodynamic temperature, at the time of measurement. In other words, the spin system must be in thermal equilibrium with its surroundings. An indication of how easily and quickly this equilibrium can be attained is given by τ_1, the so-called spin-lattice relaxation time. For metals τ_1 increases with decreasing temperature according to the Korringa relation

$$\tau_1 T = A, \qquad (T = T_e) \qquad (8)$$

where T is equivalent to the temperature of the conduction electrons T_e; τ_1 ranges from about 10^{-3} to 10^2 s at high magnetic fields (at low fields τ decreases). For dielectric solids T would correspond to that of the phonon field but τ_1 becomes much too long for these materials to be of practical use. Several pertinent properties of metals which have been considered for nuclear spin thermometry are listed in *Table 1* (see also section III.2).

Table 1. Some properties of selected metals relevant to nuclear thermometry and nuclear cooling[a]

Metal	Nuclear spin I	Nuclear moment $\mu = \gamma\hbar I$ (nuclear magnetons)[b]	External field H (Tesla)	Local field h_d (Tesla)	h_q	Korringa constant A (K·s)	Cooling power[c] $\dot{Q}$ (µW/kg·atom)
^{63}Cu	3/2	2.2	0	3×10^{-4}		0.4	6.4×10^{-5}
^{63}Cu	3/2	2.2	6×10^{-2}	3×10^{-4}		1.1	1.0
^{115}In	9/2	5.5	0		2.54×10^{-1}	0.024	3.0×10^3
^{69}Ga	3/2	2.0	0		1.9×10^{-1}	1.0	4×10
^{195}Pt(33%)	1/2	0.6	6×10^{-2}			0.030	1.5×10^{-1}
^{119}Sn(17%)	1/2	−1.0	6×10^{-2}			0.034	1.7
^{205}Tl	1/2	1.6	6×10^{-2}			0.006	1.7×10^2

[a] Values of atomic weight, I, and μ are for most abundant isotope with $I \neq 0$ (e.g. see Kittel[64]); other entries taken from Table I of Symko[105].
[b] One nuclear magneton = 2.5427×10^{-2} m^{-1}·T^{-1}.
[c] Calculated for $T_e = 1.1\, T_s$.

Three different techniques for measuring nuclear susceptibilities have been employed. One of these is a magnetometer method which measures the temperature dependence of magnetic flux changes in a static magnetic field. The other two are resonance methods—steady state and pulsed nuclear magnetic resonance (n.m.r.). Both static methods respond to the total paramagnetic content of the bulk specimen and are therefore at a disadvantage with respect to the pulsed method, which ignores signals from spurious paramagnetic effects and picks out only those from the nuclei of interest. A particularly important feature of the pulsed n.m.r. is that the results represent the condition of the system prior to the pulse rather than one perturbed by the probe. In addition, the pulsed n.m.r. deposits far less r.f. heating in the sample than does steady-state n.m.r. An extensive treatment of the pulsed n.m.r. techniques applied to Cu and Na ($A \sim 5$ K·s) has been given by Walstedt et al.[113]. In these studies the authors chose to calibrate their thermometers against the ^{4}He vapor pressure at ~ 1 K, although in principle the method can be made independent of any calibration. Their results confirmed suggestions of earlier work[65] that the spin system could be cooled to $T_s \sim 10^{-5}$ K (measured with ± 10 per cent accuracy) while T_e remained at 20 mK and higher. Through this improvement in the understanding of nuclear spin systems a more confident approach may be taken toward nuclear spin thermometry at ULT.

The magnetometer method also has many appealing aspects. By the use of 'superconducting electronics' the sensitivity and precision can be made very large and the heat dissipation during a measurement very small (the fractional temperature rise, $\Delta T/T$, of ~ 1 per cent at 1 mK has been estimated for one experimental arrangement[44]). The method makes use of a superconducting quantum interference device (SQUID)—based on the Josephson effect[53] across a weak superconducting link—as the detector of magnetic flux, sensitive to about 10^{-4} of the magnetic flux quantum ϕ_0

$$\phi_0 = h/2e = 2.0679 \times 10^{-15} \, \text{T·m}^2 \tag{9}$$

(e is the electron charge). Descriptions of SQUIDs and the associated circuitry for this purpose have been given by Zimmerman et al.[122] (point contact Josephson junction devices) and by Goodkind and Stolfa[37] (thin film bridge Josephson junctions). Several configurations have been used in SQUID devices, one of which is shown in Figure 3(b) designed for another purpose. Use of a point contact device as a magnetometer exploits the fact that the total current I_j across such junctions is exceedingly sensitive to a magnetic field H applied perpendicular to I_j, with I_j varying sinusoidally in the flux ϕ_j produced by H (the interference effects between I_j and ϕ_j are formally described by the Fraunhofer diffraction formula with the argument ϕ_j modulated by $2\pi e/h$). A useful magnetometer configuration employs a point contact embedded between two straight cylindrical holes (symmetric structure[122]). The total flux trapped in the two holes is constant, and with magnetic material in one of these, changes in H shift flux from one hole to another. On the other hand the nuclear susceptibility measurements on pure copper by Hirschkoff et al.[44] employed a low fixed external field which would produce no shift of flux. Instead, the magnetic changes in the sample (maintained at ULT) were detected by the SQUID (at 4 K) via a superconducting flux transformer (more details of such a device capable of isolating the magnetic material of interest have been given by Stolfa and Goodkind[101]). The nuclear susceptibility data obtained in this way were compared against simultaneously measured CMN temperatures in the range 20 to 100 mK. The authors concluded that the nuclear system indeed would provide an excellent thermometer were it not for a background magnetic effect which was later[45] found to be due to an enamel coating of the wires.

D. Noise Thermometry

Part of the input energy to an electrical circuit always appears as an electrical fluctuation or noise, the magnitude of which is temperature dependent. Nyquist[79] first calculated the mean square voltage $\overline{V_n^2}$ of the these fluctuations, lying in the frequency range f to $f + \Delta f$ and produced in a resistance R at a given temperature, to be

$$\overline{V_n^2} = 4Rhf[\exp(hf/k_B T) - 1]^{-1}\Delta f \tag{10}$$

or more simply approximated, when $k_B T \gg hf$, as

$$\overline{V_n^2} = (4k_B TR)\Delta f \tag{11}$$

The Nyquist relation has served as a basis for noise thermometry in, for example, the range 4 to 300 K[111]; but because of experimental difficulties—e.g. with amplifier and multiplier characteristics, and with microphonics—it has generally been necessary to calibrate the thermometers. However, Kamper, and co-workers[55-57] have through the use of the Josephson effect been able to eliminate these uncertainties and to develop a method of ULT absolute thermometry. In this work, a small voltage V_j of direct current I across a point contact Josephson junction (SQUID) establishes an a.c. oscillation according to the Josephson relation

$$f = 2eV_j/h = V_j/\phi_0 \tag{12}$$

Because V_j may be considered as

$$V_j = IR + \text{noise} \tag{13}$$

fluctuations in V_j due to noise appear, through equation 12, as fluctuations in f from its mean value $\bar{f}$. Experimentally, the signal from the Josephson junction is amplified to drive a frequency counter with a set gate time τ. Fluctuations in f and thence T are then obtained through the variance σ of the fluctuations

$$\sigma^2 = \langle(f - \bar{f})^2\rangle = 2k_B TR/\tau\phi_0^2 \tag{14}$$

Details of SQUID construction for this application as well as associated circuitry for noise thermometry are depicted in *Figure 3*. Kamper and Zimmerman[57] have discussed the theoretical and experimental limitations of the method applied to absolute ULT thermometry, indicating that the most serious difficulties arise from extraneous non-thermal noise superimposed on the weak signal from the Josephson junction. Otherwise the method should be useful down to the submillikelvin range.

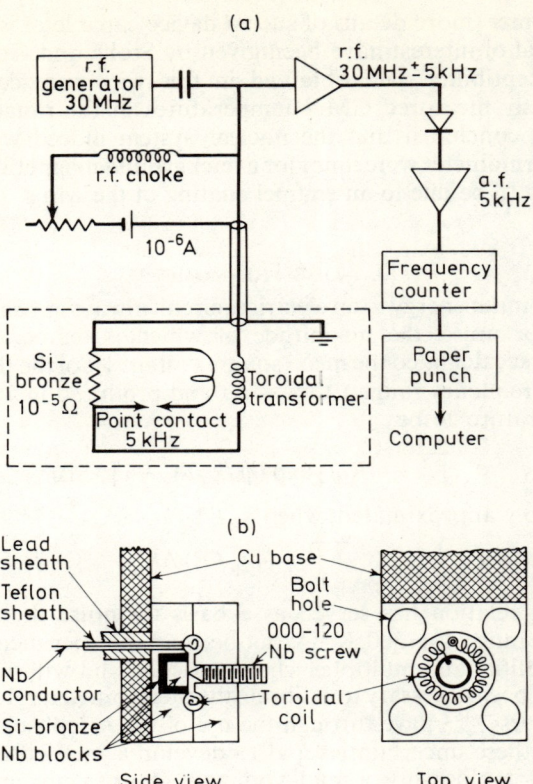

Figure 3. (a) Schematic circuit diagram for use of Josephson junction point contact device (SQUID) in noise thermometry; (b) construction details of section enclosed by dashed lines in (a) (drawn after Kamper et al.[56]).

As of this writing the most successful thermometry[56] using this method has employed a system with $V_j = 10^{-11}$ volt corresponding to $f = 5\,\text{kHz}$. In the neighborhood of 20 mK temperatures were accurate (compared with CMN) to several millikelvins. Clearly further work is required; this is difficult work, but it could be very rewarding.

4. Temperature Scales in Use below 20 K

There is a hierarchy of respectability among practical temperature scales. This ranges, beginning at the bottom, from a scale used only by an individual worker, to a laboratory scale, to a national scale, and finally to an international scale sanctioned by the Comité International des Poids et Mesures. Until October 1968 no scale of the highest rank existed for measurements below 90.188 K. However, at that time the Comité adopted the International Practical Temperature Scale of 1968 (IPTS-68) which extended the range of international agreement to 13.81 K. Previously, the Comité had recommended (an action conferring somewhat less prestige than adoption) the 1958 ^{4}He Scale and the 1962 ^{3}He Scale for use over ranges limited at the upper ends by the respective critical points of ^{4}He and ^{3}He and at the lower ends 'by the vapor pressures becoming too low for practical measurements'. These recommendations are included as Appendix II to the text of IPTS-68[49]. Thus not officially accounted for at present are the regions between 5.2 and 13.81 K and the infinite continuum below about 0.5 K. It should be emphasized that while the ultimate goal is to achieve the best possible representation of the true Kelvin scale, adoption of a scale by the Comité does not indicate this, but only that through compromise international agreement has been reached. This does not seriously impair the usefulness of an adopted scale, because if all measurements of T are referred to it, when at some future time the scale is refined (again by international agreement) it is a relatively easy matter to adjust collected data to the new scale. Nevertheless, of prime concern is the improvement—and extension—of IPTS-68, a concern which provides the impetus for the discussions of this section, where scales on the lower rungs of the hierarchy are compared with IPTS-68.

A. The Scale IPTS-68 (T_{68}) from 13.81 to 20 K

In the interval 13.81 to 90.188 K IPTS-68 was derived as a compromise among four national scales [NBS-55 (USA), NPL-61 (UK), PRMI-54 (USSR) and PSU-54 (USA)] in terms of a reference polynomial interpolation function[12] (called CCT-68) between argued-upon fixed points. CCT-68 relates T_{68} to the resistance ratio of a standard platinum resistance thermometer,

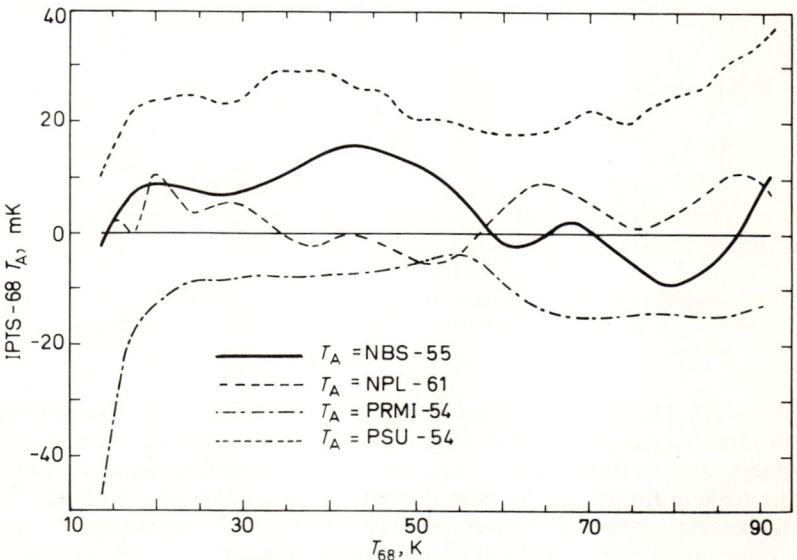

Figure 4. Deviations of IPTS-68 from the national temperature scales T_A used in its determination (drawn after Bedford et al.[11]).

from which any other such thermometer may be calibrated. A tabular comparison of the fixed points according to IPTS-68 and to the generating national scales appears in a paper by Bedford et al.[11]. It is sufficient to note here than an uncertainty of ± 0.01 K is ascribed to each of the three lowest fixed points, i.e. the triple point of equilibrium H_2 (eH_2): 13.81 K; equilibrium between the liquid and vapor phases of eH_2 at 25/76 standard atmospheres: 17.042 K; and the normal boiling point of eH_2: 20.28 K. *Figure 4*, adapted from ref. 11, indicates the deviations of the national scales from IPTS-68 which appear over the whole range of CCT-68. Additional discussions of IPTS-68 have been given by Swenson[103] and Durieux[31].

B. 1958 ^{4}He Scale (T_{58}) and 1962 ^{3}He Scale (T_{62}) between 0.5 and 5.2 K

The scales for T_{58} and T_{62} can conveniently be considered together because in the region of their overlap (0.9 to 3.3 K) they are believed to agree to within 0.3 mK or better[31]. Descriptions of these scales are given in Volume

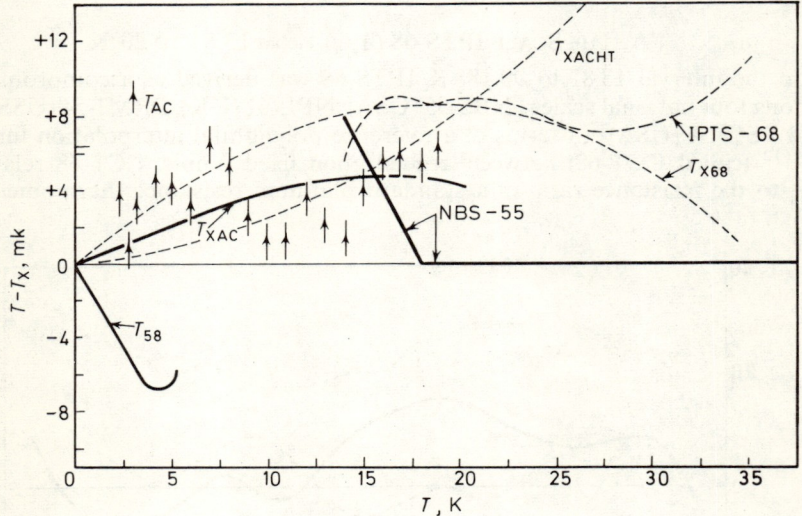

Figure 5. Deviations of several temperature scales from the magnetic scale T_x of Cetas and Swenson[21]—see text for meaning of symbols (drawn after Cetas and Swenson[21]).

I (pp 29–30). However, since the introduction of these scales, various determinations by gas thermometry and acoustic thermometry, as well as by magnetic and resistance secondary methods, have suggested serious errors in the scale (although some measurements have confirmed the scale). These difficulties have been reviewed by Keller[58] and more thoroughly and recently by Swenson[103]. The preponderance of data suggests that the corrections should have the shape and magnitude shown in *Figure 5*, although considerably more work needs to be done, especially in tying the scale to a correct reference value near 20 K (see discussion below), before a scale more nearly approaching the true thermodynamic scale can be achieved.

C. Scales Spanning the Range 1 to 20 K

Various proposals from national laboratories and others (e.g. Berry[16] and Barber[10] at NPL and Mitsui et al.[75] at NRLM, Japan—gas thermometry; Tiggleman and Durieux[109] at KOL—platinum thermometry; Cataland et al.[20] at NBS and Colclough[23] at NPL—acoustic thermometry; van Rijn and Durieux[110] at KOL—magnetic thermometry) have been advanced for improving and extending IPTS-68. Even a partial discussion of these is beyond the scope of this work. However, it should be instructive to examine the results of one such investigation—without prejudice to the others—to indicate the extent of the confusion and difficulties attendant to this matter. Cetas and Swenson[21] at Iowa State University have developed a magnetic scale, T_x, between 0.9 and 18 K using CMA [$Cr(CH_3NH_3)(SO_4)_2 \cdot 12H_2O$] and MAS [$Mn(NH_4)_2(SO_4)_2 \cdot 6H_2O$]—CMN, which incidentally is unsuitable for thermometry *above* 2.6 K because of Curie law deviations, was also used for $0.9 < T < 2.6$ K. The temperature may be related to the observed susceptibility χ through the relation

$$\chi = A + \frac{B}{T + \Delta + \delta/T} = A + \frac{B}{T} + \frac{C}{T^2} + \frac{D}{T^3} + \ldots \quad (15)$$

so that for each salt the evaluation of a set of four (non-physical) parameters is required to define a temperature scale; furthermore, the values of these parameters—and thus the resulting scale—depend upon what calibration points are taken. In the work of Cetas and Swenson both CMA and MAS were calibrated in the range 18 to 34 K against a platinum resistance thermometer reproducing the NBS-55 scale (see Volume I, p 28), and the resulting magnetic scale T_x is defined in terms of eight parameters, four each for CMA and MAS. The relationship between T_x and χ is given in terms of four germanium resistance thermometers which were thermally tied to the salt samples: two of these were calibrated by earlier constant volume gas thermometer methods[84] and another by NBS acoustic thermometry[81]. The results of this work are summarized in *Figure 5*, where the symbols have the following significance: T_x, the reference scale, is arbitrarily and for the sake of convenience standardized to NBS-55 for $T > 18$ K, although standardization to IPTS-68 would have been equally valid; the smoothed magnetic scale, had it been based on IPTS-68, is consistent with the latter up to ~ 27 K as shown by the comparison (T_{X68}, IPTS-68); T_{AC} are the individual points determining the NBS acoustic scale[81]; T_{XAC} is a magnetic interpolation curve through these points when T_x is read based on the germanium thermometer common to T_x and T_{AC} measurements; T_{XACHT} is derived similarly to T_{XAC} but is based in addition upon new high-temperature acoustic thermometry with ^{4}He and Ne[82]. Cetas and Swenson suggest that T_{XACHT} is the closest approximation to the true thermodynamic temperature, which would indicate that the boiling point of eH_2 given by IPTS-68 is within ~ 1 mK of being correct. Values of T_x are considered precise to ± 0.01 per cent of T_x; but because of uncertainty in the normalization the accuracy to which T_x represents the true scale is considerably less than this. Thus, for example, the true value of the correction to T_{58} near 4 K via T_x, as may now be appreciated, depends considerably upon what scale for normalization is selected—i.e. NBS-55 or IPTS-68 at 18 K, or T_{XACHT} or T_{XAC} at low temperature—and is accordingly uncertain by more than 3 mK.

5. Secondary Thermometry below $20°K$

By the term 'secondary thermometry' we signify here any means of temperature measurement which does not involve a thermodynamic definition of temperature but can be conveniently related to such a definition. However, this would include a potential primary thermometer—CMN for example—when its use includes calibration against some other temperature scale ultimately related to an accepted scale—such as T_{62}. In most experiments, as indicated earlier, a secondary rather than a primary thermometer is used because of undue complications associated with the latter; and generally the choice of a particular secondary thermometer is made on the basis of the specific demands of the experiment under consideration. Sometimes, for example, the needs are for a smooth interpolation device which can be fulfilled by vapor pressure, resistance, magnetic, or still other types of thermometry, depending largely on convenience and desired precision. Sometimes a highly sensitive differential thermometer is required. And at still other times it is desirable to have fixed or thermodynamic equilibrium points for reference—ultimately these may be related with sufficient reliability to the thermodynamic scale to warrant use as definitions for practical scales. To be most useful in any of these applications the temperature-sensitive property associated with the device should be stable and reproducible over long periods of time and with temperature cycling; and in some instances these requirements must be fulfilled under such adverse working conditions as high pressures and high magnetic fields. Of course, the criteria 1 through 4 mentioned in section 3 above in connection with primary thermometry, also apply to secondary thermometry.

Methods and techniques for vapor pressure, resistance, and magnetic thermometry have already been mentioned here and in Volume I. However, particular attention should be called to the germanium resistance thermometer (GRT) (Volume I, pp 282–284; see also e.g. Swenson[103] and Halverson and Johns[42] for recent reviews of the method) which during the past decade has reached a pre-eminent position as a versatile thermometer in the range 0.1 to 20 K. While any given GRT generally retains its calibration to within ± 1 mK over long periods of time and with thermal cycling, has resonable response times (can be ~ 0.01 s/K), small self-heating ($\sim 10^{-8}$ W) and high sensitivity (ΔT as low as $\sim 10^{-7}$ K), GRTs also possess several drawbacks, such as non-analytic resistance/temperature relations, magnetoresistance effects requiring recalibration for work in magnetic fields—important with respect to the thermodynamic properties of He at ULT—and very large resistances at ULT which can be measured only with relatively diminished precision. Accordingly the search continues for devices with the virtues of GRTs and none of the attendant deficiencies.

A. Capacitance Thermometry

One promising candidate for supplementing the uses of the GRT is the glass–ceramic capacitance thermometer (GCCT) recently developed and tested by Lawless and co-workers[67, 68, 88]. This device consists of a rod-like unit about $5 \times 2 \times 1$ mm in which thin (~ 0.025 mm) layers of dielectric—Al_2O_3–SiO_2 glass in which $SrTiO_3$ has been crystallized at temperatures of 1100° to 1200°C—alternate with Au–Pt alloy electrode layers to which the lead wires are attached. The process of forming the $SrTiO_3$ crystallites leaves them in the perovskite phase such that at low temperatures (below 60 K) the composite exhibits a large positive temperature coefficient of capacitance, dC/dT. *Figure 6* shows the C versus T relation over the range up to 20 K for

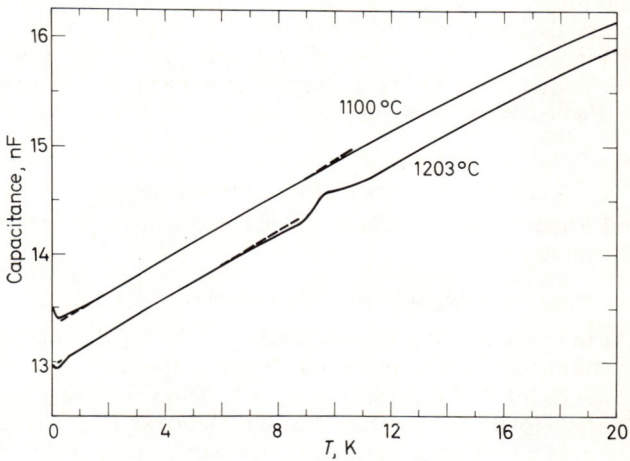

Figure 6. Capacitance versus temperature relations for two glass–ceramic capacitance thermometers, one crystallized at 1100°C, the other at 1203°C. Dashed lines indicate deviations from constant slope established by the linear portions from 1 to 5 K (drawn from normalized data of Lawless *et al.*[67, 68]).

two test GCCTs crystallized at 1100 C and at 1203 C. Three features of this plot are worthy of mention: (1) near 10 K a small but very undesirable hump appears for the 1203°C crystallized thermometer but not for the other (the reasons for this are still being investigated); (2) in the important region between about 1 and 5 K C versus T for both thermometers is linear, with dC/dT very nearly the same for both; and (3) at the lowest T—25 to 100 mK—dC/dT becomes negative; for the 1100°C crystallized thermometer C goes nearly as $1/T$ and is still useful for thermometry.

Characteristics which should make the GCCT valuable for thermometry below 20 K are: (1) the C versus T relation can be smooth and expressible by simple mathematical functions; (2) the sensitivity in C is of the order of 250 picofarads/kelvin; with the use of a three-terminal capacitance bridge method (as for example, the General Radio 1615-A transformer ratio-arm bridge) 0.05 pF is easily resolvable, which translates to ~ 0.2 mK sensitivity in T; (3) the self-heating can be extremely low, less than 0.1 nW at 4 K, and decreases as T decreases; (4) the response time is of the order 0.1 s/K; and, of special significance, (5) magnetic fields of up to 150 kOe do not disturb the thermometer calibration within the measurement uncertainty (<1 mK)[88].

On the other hand the prototype thermometers did possess two undesirable features: (1) on being subjected to a large change in temperature the thermometer may first display a drift (as large as 30 mK equivalent) for ~ 30 minutes and then stabilize; and (2) stabilized measurements are repeatable to only about ± 2 mK after thermal cycling between 1.7 and 40 K. These difficulties may possibly be due to He entering into the matrix of glass–ceramic, as the test units were in direct contact with He gas used for heat exchange purposes; if so, encapsulation of the units should remedy the problems.

Although further experiments with preparative methods and temperature responses are desirable, GCCTs developed at Corning Glass Works are now being marketed by Lake Shore Cryotronics, Inc. (9631 Sandrock Road, Eden, New York 14057). Other materials for capacitance thermometers are being investigated by other workers.

B. ^{3}He Melting Curve Thermometry

The equilibrium between solid and liquid is expressed through the Clausius–Clapeyron equation

$$(dp_m/dT) = -(S_s - S_l)/(V_l - V_s) \tag{16}$$

where p_m is the pressure of melting (freezing) at T, S is the molar entropy, V the molar volume and subscripts s and l refer to the solid and liquid phases. For ^{3}He $S_l > S_s$ for $T > 0.32$ K; but $S_s > S_l$ below 0.32 K. Over this range $(V_l - V_s)$ varies slowly and remains positive. Hence at $T_{min} = 0.32$ K $dp_m/dT = 0$ and there is a minimum in the melting curve pressure, p_{min}. With decreasing temperature p_m increases ($dp_m/dT < 0$) from $p_{min} = 28.94$ atm* at 0.32 K to 33.2 atm at 0.020 K, while dp_m/dT goes from zero to -37.85 atm/K[40]. According to the third law of thermodynamics $dp_m/dT \equiv 0$ at 0 K, so p_m cannot increase indefinitely; and in fact, evidence of nuclear spin ordering (antiferromagnetic with a Néel temperature of about 2 mK) in the solid begins to appear near 20 mK through a decrease in S_s, which in turn produces a flattening in the melting curve. These thermodynamic relations along the melting curve are illustrated in *Figure 7*.

There are two ways in which the ^{3}He melting curve may serve as a thermometer in the ULT region. First, the melting curve pressure may be measured and compared with $p_m(T)$ data in the literature. However, for a determination of T with one per cent this requires absolute pressure measurements to about one part in 3 000, nearly an order of magnitude greater than the accuracy with which $p_m(T)$ is presently known. The second method involves measurement of the melting curve slope dp_m/dT and a comparison with quantities entering into equation 16 made thermodynamically consistent. This procedure has the advantages over the former in that differential rather than absolute pressures are to be measured and little loss in sensitivity is suffered in the region of the

---

* 1 atm = 1.01325×10^5 N/m^2.

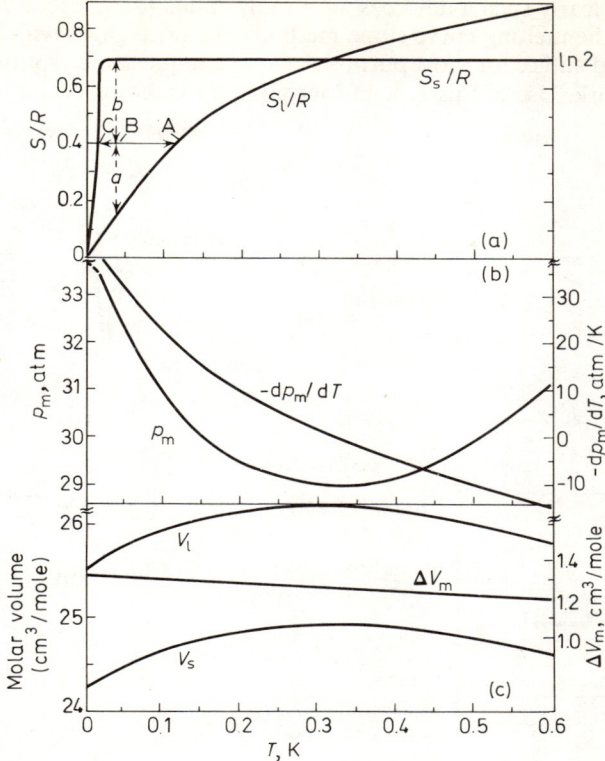

Figure 7. Thermal properties of ^{3}He at melting as a function of temperature. (a) Entropy of the liquid and solid; (b) melting pressure and slope of the melting curve; (c) molar volume of the liquid and solid and change of molar volume on melting.

minimum. Furthermore, through the use of iterative processes, beginning with the measured values of the entropy and volume changes on melting and of $p_m(T)$, a self-consistent table of the thermodynamic melting properties can be generated from which determinations of T to ± 1 per cent should be possible. The most recent such calculation has been made by Grilly[40] for $20 < T < 320$ mK. Alternatively, it is possible to achieve useful thermometry by calibrating dp_m/dT against CMN. The ^{3}He melting curve methods show promise of ultimately providing primary thermometry.

Essential to the melting curve technique is a precise pressure sensor. For this Scribner et al.[92] have developed a capacitive pressure gage, details of which are shown in *Figure 8(a)*. Briefly, changes of pressure in the sample region distort the diaphragm, physically attached to (with epoxy) but electrically insulated from the upper capacitor plate. Thus motion of the diaphragm alters the spacing between the upper and lower (fixed) capacitor plates. By measuring the resultant changes in capacitance with a three-terminal bridge (in this case the General Radio 1620 system) a pressure sensitivity is attained of about 3×10^{-6} atm, which corresponds to a temperature resolution of

10^{-7} K. Clearly such a device is admirably suited for temperature measurements via the melting curve slope method; but since gages with lower sensitivity would suffice for most purposes, the technique can be routinely applied in ULT work. Use of this type of thermometry is, however, limited to those

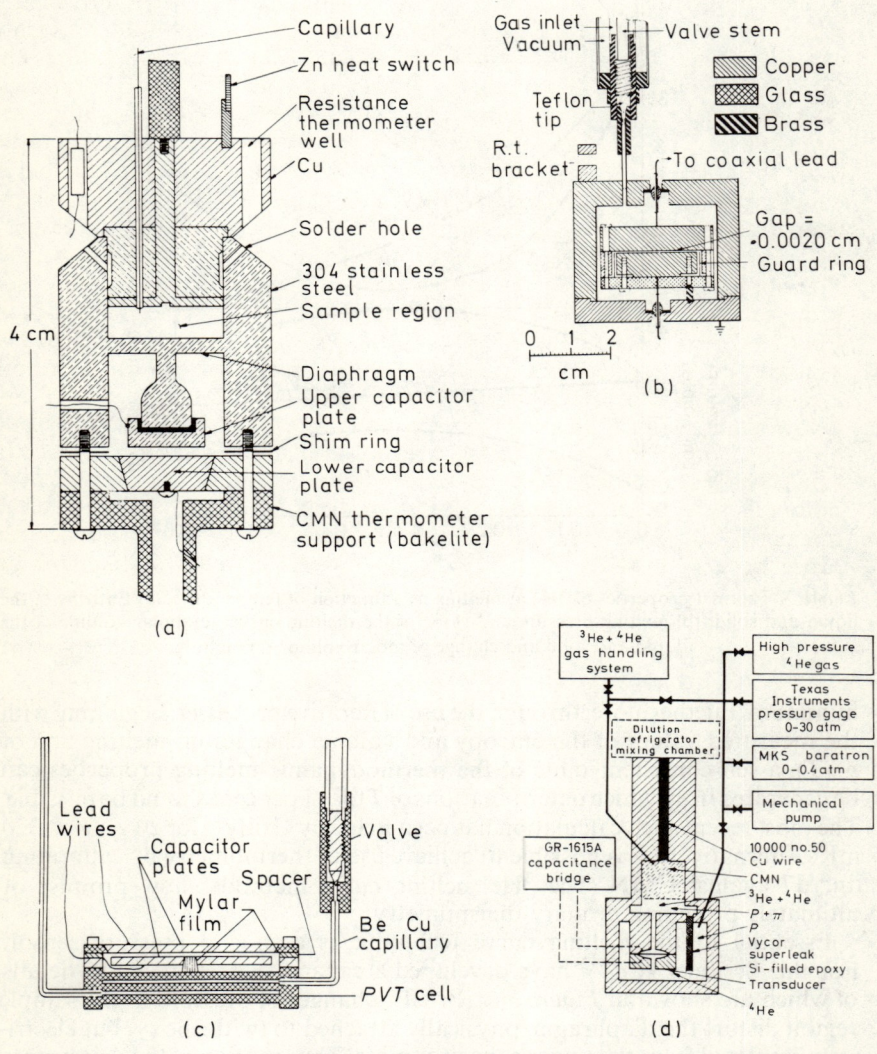

Figure 8. Four different types of capacitance gages for pVT work. (a) Schematic of capacitance strain gage for measuring (p_m, T_m) and ΔV_m of ^{3}He (drawn after Scribner *et al.*[92]); (b) cell for measuring fluid densities at low pressures (from Kerr and Sherman[61]); (c) schematic of pVT cell for moderate pressures and melting curve properties (from Grilly[40]); (d) schematic of apparatus for determining osmotic pressures of ^{3}He–^{4}He solutions (drawn after Landau *et al.*[66]).

experiments in which the sample of interest can be brought into intimate thermal contact with the working ^{3}He; on the other hand, it is a natural adjunct to the adiabatic compression method of cooling (section III.3).

C. Superconducting Transition Temperatures as Fixed Points

The usefulness of fixed-point temperatures has already been demonstrated in the above discussions of IPTS-68. In regions beyond the range of this scale (i.e. $T < 13.81$ K) the establishment of such points would provide a valuable means of calibrating to a single set of reference temperatures the various interpolation thermometry devices used in different laboratories; ultimately these may serve as bases for extending IPTS-68. Generally, fixed points are associated with thermodynamic phase transitions, which in order to be suitable for defining temperature scales must: (1) be reproducible in temperature for various samples of a given substance, (2) occur over a narrow temperature range, and (3) be practically and conveniently realizable in a laboratory experiment in conjunction with the instrument to be calibrated.

Schooley and Soulen[90] have provided supporting evidence for their proposal to use the superconducting transition temperatures T_c of Pb($T_c = 7.201$ K), In (3.417 K), Al (1.176 K), Zn (0.844 K), and Cd (0.515 K) as fixed points. The electrical resistivity of certain metals and alloys exhibits a sudden drop to effectively zero when these materials are cooled to T_c and below. Although the intrinsic temperature width of the transition is of the order 10^{-6} K, both this width as well as T_c itself are seriously influenced by small changes in material composition, metallurgical imperfections, and other inhomogeneities. Furthermore T_c is reduced by the presence of an external magnetic field (for $H = 0$ the transition is second order, but becomes first order for $H \neq 0$; the earth's field shifts T_c generally by a few millikelvins, but otherwise has an insignificant effect upon the transition). Specimens of the metals listed above have been prepared with sufficient purity (generally of six-nines-grade) and metallurgical quality (by annealing) to yield T_cs reproducible to $\sim \pm 2$ mK and with transition widths of 1 mK or less.

Among various methods for detecting the abrupt loss of electrical resistance, the authors considered an inductive technique as most advantageous and experimentally convenient, requiring no leads to be attached to the sample—which can be a small, self-supporting wire—and allowing continuous monitoring of the transition without disturbing the experimental environment. The sample is simply enclosed in a cylindrical mutual induction coil pair as part of a mutual induction bridge circuit. An essential property of a superconductor vis-à-vis an ordinary conductor is the exclusion of magnetic flux from the interior of a specimen in the superconducting state. Hence as T is lowered through T_c magnetic flux is expelled from the specimen causing an imbalance in the mutual inductance bridge. The midpoint of this imbalance is taken as the midpoint of the transition. Two pairs of Helmholtz coils can be used to nullify the earth's magnetic field.

Should acceptance of T_c fixed point thermometry be sufficiently widespread, the Office of Standard Reference Materials (NBS) is willing to serve as a supplier of sets of samples of very pure Pb, Al, Zn and Cd, with corresponding members of sets prepared from a single batch to insure uniformity.

D. Superconducting Thin Film Thermometers

Often it is useful to have extremely sensitive thermometers for measuring small temperature changes over a limited range. Resolution in T of the order 10^{-7} K is obtainable with GRTs and carbon resistance thermometers (see also Volume I, pp 35–36, 282–284); but these are not entirely suitable for applications requiring unobtrusively small physical size and heat capacity combined with extremely rapid response and small power dissipation—such applications as demanded by ^{4}He film experiments and thermal pulse detection in liquid ^{3}He and ^{4}He down to ULT. In response to those and similar needs superconducting thin film thermometers have been developed.

In comparison with bulk materials the superconducting transition for thin films both broadens (with respect to T) and shifts in temperature. Hence in the transition region the resistance R is a strong function of T and the film thickness d and is variable almost at the experimenter's will. Naturally, the R/T relation for each film must be calibrated against some temperature scale.

Early experiments by Golovashkin and Motulevich[36] with thin film thermometry were carried out with copper–lead alloys vaccum deposited on substrates in thicknesses of 0.04 to 0.17 μm. Although these films were useful over the range 1.5 to 4 K, depending on alloy composition, they nevertheless suffered from instability and non-reproducibility due to oxidation of the film. However, Liebenberg and Allen[70] have more recently found that pure aluminum films are stabilized by surface oxide formation and are entirely suitable as thermometers, reproducible to within better than ± 3 mK (absolute) after repeated temperature cyclings and exposures to air, and over periods longer than several months. One prescription for thermometer preparation involves sputtering gold leads ~ 2 cm apart on the surface of a dielectric material, such as clean glass, and then depositing by evaporation techniques a pure aluminum strip ~ 0.1-cm wide ($d \sim 0.1$ μm) across these leads. Wires for four terminal resistance measurements are attached to the gold by indium solder. Variations of this process include codeposition of aluminum and silicon oxide to produce a granular film with a larger normal state resistance R_N than the pure metal and deposition in 'staircase' and wedge configurations (with respect to d) to take advantage of the fact that for these films $T_c \sim d^{-1}$. In this way films have been obtained with dR/dT ranging from 10^2 to 10^5 Ω/K, transition widths (defined as between 0.25 R_N and 0.75 R_N) as large as 0.3 K, resolution in T as small as 5 μK, and thermometric usefulness from 1.2 to 2.0 K.

III. Refrigeration Techniques below 0.3 K

Various types of cryostats and refrigeration techniques for use below 20 K have been fully discussed in Volume I (pp 263–281). However, the ^{3}He–^{4}He dilution refrigerator was commented upon there only briefly. Because the dilution refrigerator has assumed such an important role in the exploration of the ULT properties of the He isotopes—as well as of other systems—more about this technique will be described here, including its use as an ULT stage for further cooling by nuclear demagnetization techniques and by adiabatic solidification of ^{3}He. In addition, some of the special problems of thermal isolation and contact at ULT will be briefly mentioned.

1. The ^{3}He–^{4}He Dilution Refrigerator

The principle of the ^{3}He–^{4}He dilution refrigerator and the ultimate low temperatures attainable in its operation depend upon the facts that first, below 0.83 K ^{3}He and ^{4}He are not miscible in all proportions but, depending upon the relative concentrations, separate into two phases; and second, for ^{3}He mole fractions $x \gtrsim 0.064$ phase separation does not occur at any temperature. The T/x phase diagram for these mixtures is shown in *Figure 9*. From

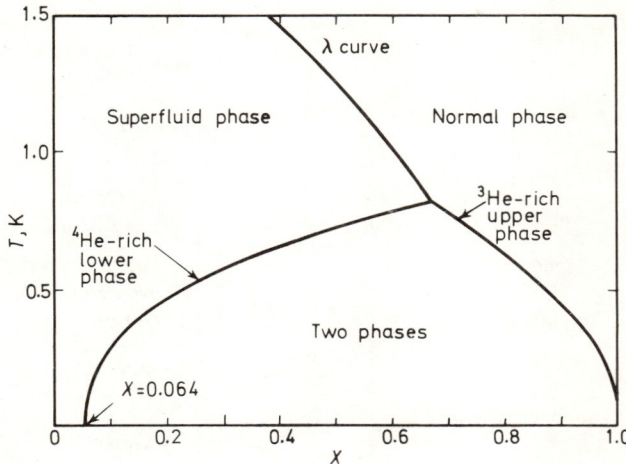

Figure 9. Temperature/composition phase diagram for ^{3}He–^{4}He mixtures at saturated vapor pressures; $x =$ mole fraction of ^{3}He.

thermodynamics the phase separation implies a heat and entropy of mixing; and experimentally it has been shown[97] that the mixing heat is positive, i.e. when the process is carried out adiabatically heat is absorbed and cooling results. For the refrigerator it is practical to use a mixture with $x \gtrsim 0.064$, so that when this is cooled (see *Figure 9*) to below about 0.1 K, the separation produces an ^{3}He-rich phase with $x \approx 1$ and an ^{4}He-rich phase with $x \approx 0.064$. The ^{3}He-rich phase, being less dense, floats on top of the ^{4}He-rich phase. Now, as pure ^{3}He at temperature T of the mixing chamber is forced to flow at a rate $\dot{n}_3$ through the concentrated upper phase and into the dilute lower phase, the heat of mixing ΔH is given by

$$\Delta H = H_3(x_1, T) - H_3^\circ(T) = T[S_3(x_1, T) - S_3^\circ(T)] \quad (17)$$

where $H_3(x_1, T)$ and $S_3(x_1, T)$ [$H_3^\circ(T)$ and $S_3^\circ(T)$] are the molar enthalpy and entropy of ^{3}He in the lower [upper] phase. Then the heat absorption rate in the mixing chamber is

$$\dot{Q} = \dot{n}_3 \Delta H \quad (18)$$

The maximum in $\dot{Q}/\dot{n}_3$ occurs for $x = 0.064$ and is roughly equal to 82 T^2 J/mol ^{3}He (if $x = 0.068$, for example, $\dot{Q}/\dot{n}_3 = 78\ T^2$ J/mol ^{3}He). In practice the ^{3}He enters the mixing chamber at $T_i > T$; then equation 18 becomes

$$\dot{Q}(T, T_i) = \dot{n}_3(94\,T^2 - 12\,T_i^2) \qquad (19)$$

In any case the cooling power of the continuous refrigerator is, in the zeroth approximation, proportional to the rate at which ^{3}He can be circulated.

Figure 10 schematically illustrates the flow path and the workings of the con-

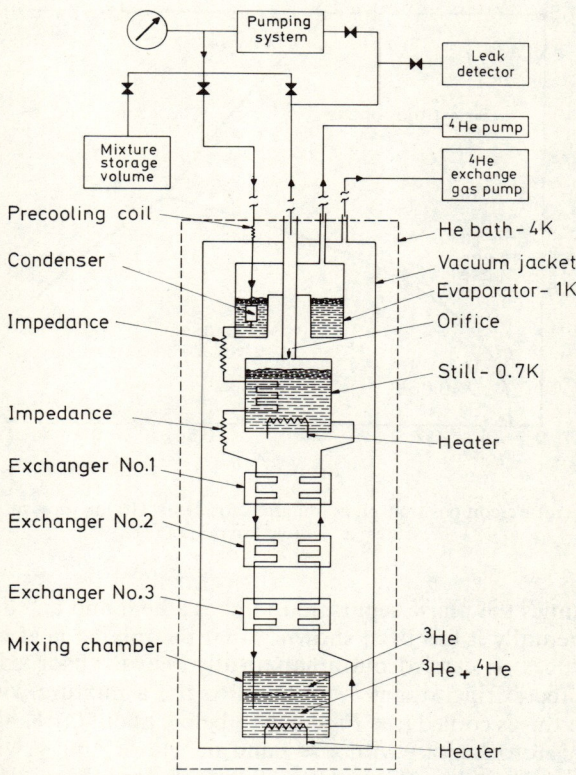

Figure 10. Schematic drawing of ^{3}He–^{4}He continuous dilution refrigerator with three lumped heat exchangers (drawn after Wheatley et al.[116]).

tinuous dilution refrigerator. Several features deserve additional comment. (For extensive details on design and operation of dilution refrigerators, the interested reader may consult two excellent papers by Wheatley and his collaborators[115, 116]).

Still. An osmotic pressure due to the difference in x between the still and mixing chamber produces a net flow of ^{3}He from the latter to the still and thus provides the driving force for the dilution process in the mixing chamber. A one per cent ^{3}He in ^{4}He solution is maintained at 0.7 K by heating the liquid and pumping off the vapor, which at this temperature is nearly pure ^{3}He. The orifice at the top of the still is required to reduce the heat leak via the ^{4}He superfluid film.

Condenser–capillary (impedance). These must be carefully designed so that the condensing rate and flow impedance of the capillary are matched. Otherwise vapor may pass through the capillary and place an unacceptable heat load on the still and the heat exchangers.

Heat exchangers. Probably more effort has been focused on the design and construction of heat exchangers between the counter streams of pure ^{3}He and the dilute mixture than on any other single component of the refrigerator. If the lowest temperature desired is ~ 0.04 K, then a simple coaxial, counter-flow, continuous heat exchanger such as reported by Anderson[6] (100-cm lengths of concentric 0.04-cm and 0.12-cm diameter tubing) will suffice. A wide variety of experiments with this type of exchanger has been reported by Wheatley et al.[115]. However, to achieve maximum cooling for a given $\dot{n}_3$, exchangers presenting much larger surface areas, and hence greater efficiencies, are required; invariably such exchangers are of the lumped type in several stages, as shown in *Figure 10*; a compromise must be reached between the number of stages and the area of each. Use of sintered copper (sponge)[116], stacked copper foils[77, 87], and spiral-wound corrugated copper sheet[115] represent some of the successful attempts at achieving increased surface area.

Although the continuous maintenance of temperatures down to ~ 15 mK using dilution refrigerators is now considered practically routine, stringent natural limitations of the method prevent its easy extension to lower temperatures. The two most important of these difficulties are the thermal boundary resistance and viscous heating of the circulating ^{3}He, both effects becoming more serious with decreasing temperature. The boundary (or Kapitza) resistance R_K (see also section III.4 below) is responsible for a temperature difference ΔT_K whenever an amount of heat $\dot{Q}_K$ flows across a liquid/metal boundary. ΔT_K is believed to arise from an impedance mismatch between the thermal phonons (acoustic modes) of the two substances and is proportional to $\dot{Q}_K/A$, where A is the surface area of the boundary. Hence increasing A should increase the heat exchanger efficiency. Another promising but difficult technique for minimizing ΔT_K has been introduced by Roubeau et al.[86] who have separated the two streams with a pure copper foil sufficiently thin (~ 20 μm) to allow direct passage of the heat-carrying phonons from one wall to the other (i.e. phonon–electron interactions are minimized). As yet no satisfactory means of avoiding viscous heating, $\dot{Q}_\eta$, have been found. The viscosity η of ^{3}He and of dilute solutions rises with falling T; also $\dot{Q}_\eta$ rises with flow rate $\dot{n}_3$. Wheatley et al.[116] have indicated that $\dot{Q}_\eta$ as measured by $\Delta T/T$ is about eight times greater for the dilute than for the concentrated phase under the same conditions and that $\Delta T/T$ increases as T^{-4}. For a single-cycle refrigerator (^{3}He is pumped from the still but not recirculated) $\dot{Q}_\eta$ does not depend upon $\dot{n}_3$; but, assuming laminar flow, the lowest T obtainable in the mixing chamber is given by $4d^{-\frac{1}{3}}$ mK (d is the tube diameter, in mm, of the lead from the mixing chamber), which in practice means ~ 4 mK.

H. London[71] is credited with suggesting in 1951 the principle of the dilution refrigerator; but it was not until 1965 that the first reports[26] of an operational refrigerator, achieving a lowest temperature of 0.22 K, were revealed. Since then numerous machines of varying characteristics—cool-down rate, lowest temperature attainable, refrigeration capacity, etc.—have been been built in many laboratories; and several models are commercially available from US and UK concerns (for example, Superconductivity Helium Electronics Manufacturing Corporation, San Diego, Helsinki, London; Oxford Instrument Company, Ltd, Oxford). To present some idea of the state of the art, it might be useful to give some records with respect to dilution refrigeration as uncovered by a literature search: Ehnholm and Gylling[34] report a refrigerator with the largest circulation rate, $\dot{n}_3 = 3 \times 10^{-4}$ mol/s; with $\dot{n}_3 = 1.1 \times 10^{-4}$ mol/s, but with extremely efficient heat exchangers, a machine built by Neganov et al.[77] has achieved the largest refrigeration power at 20 mK of 3.1 μW; as reported by Zinov'eva[123], Neganov probably holds the continuous low temperature record of 5.5 mK.

2. Nuclear Cooling

Many of the physical principles associated with nuclear thermometry (section II.3.B) are also involved in the method of cooling through the adiabatic demagnetization of a nuclear spin system (here we consider only metallic systems, although this type of cooling using dielectric crystals has been reported[22]). In the first approximation starting from an initial temperature T_i, the final temperature T_f reached after demagnetizing from field H_i to H_f is

$$T_f = H_f(T_i/H_i) \qquad (20)$$

As the ultimate hope for this method is to provide refrigeration to the submillikelvin region, it is first of all necessary to cool the metal specimen to $T_i \sim 10$ mK by other means, e.g. a dilution refrigerator, and to magnetize the specimen isothermally at T_i so as to start with a large value of (H_i/T_i). The cooling power of the nuclear stage, on the other hand, is limited (and defined) by $\dot{Q}_n$ the rate at which energy can be transferred to the nuclear spins at temperature T_s from the conduction electrons at T_e. With the help of the Korringa relation [equation 7] $\dot{Q}_n$ can be derived from the energy expression of the nuclear spins as

$$\dot{Q}_n = (C_n/A)(H^2 + h^2)(T_e - T_s)/T_s \qquad (21)$$

where h is the internal (local) magnetic field arising from either dipole–dipole interactions ($h = h_d$) or quadrupole interactions ($h = h_q$). Values of $\dot{Q}_n$ as well as h_d and h_q for several metals are shown in *Table 1*, where $\dot{Q}_n$ is calculated for $T_e = 1.1\ T_s$. For fixed $\dot{Q}_n$ and (H_i/T_i) when the demagnetization is to a finite field (demagnetization to $H_f = 0$ results in impractically low $\dot{Q}_n$—cf. data for copper in *Table 1*) it can be shown that the minimum T_e will be reached when

$$H_f = (A\dot{Q}_n/C_n)^{\frac{1}{2}} \qquad (22)$$

with

$$T_e = 2H_f(T_i/H_i) \qquad (23)$$

Successful experiments for reducing T_s to the microkelvin range were carried out first in 1956 by Kurti et al.[65] on copper; however, T_e probably remained close to $T_i = 0.012$ K and T_s warmed up to this temperature with an effective $\tau_1 = 40$ s. Although relatively few further attempts at nuclear cooling have been made[105, 113], one of the more recent has achieved notable progress as indicated in a preliminary report by Berglund et al.[13]. Using a dilution refrigerator these authors have cooled 1.5 kg of copper to a steady $T_i = 16$ mK magnetized in a field $H_i = 4.6$ T. The copper was in the form of thin (50 μm) long wires bonded together with Araldite, with one end of the bundle attached to the refrigerator mixing chamber—the experimental arrangement of the nuclear stage with respect to the mixing chamber is illustrated in *Figure 11*. Following demagnetization to $H_f = 7.27 \times 10^{-2}$ T the spin and conduction electron systems of the copper equilibrated at 0.62 mK and remained below 1 mK for more than four hours (temperatures were measured using the pulsed n.m.r. technique). The total heat leak per mole of nuclear cooling material was an amazingly small 5.7×10^{-10} J/s.

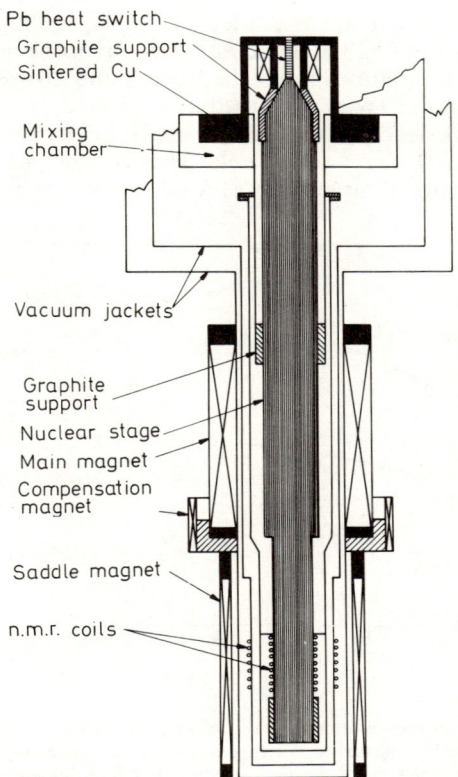

Figure 11. Schematic drawing of nuclear cooling apparatus using a bundle of fine copper wires in the nuclear stage starting from the mixing chamber of an ³He–⁴He dilution refrigerator (drawn after Berglund et al.[13]).

Nevertheless, several remediable circumstances prevented complete utilization of this degree of isolation, so that even better cooling characteristics may be expected from this apparatus. The authors emphasize the need for the large refrigeration capacity of the 16 mK stage satisfied at present only by the dilution refrigerator: during the isothermal magnetization step, up to 0.1 J must be transferred from the nuclear material in a convenient period of time. Whereas copper was chosen for the nuclear stage in these experiments because it is readily obtainable in pure fine wires, *Table 1* indicates that other metals, especially thallium and indium, should be superior nuclear cooling agents (however, the toxicity of thallium is a serious deterrent to its use).

3. Cooling by Adiabatic Freezing of Liquid ^{3}He

The melting curve properties of ^{3}He described by *Figure 7* imply that for $T < 0.32$ the amount of heat absorbed ΔQ_c in the isothermal conversion of n moles of liquid to the solid at a temperature T on the melting curve is

$$\Delta Q_c = nT(S_s - S_l) \qquad (24)$$

As pointed out by Johnson and Wheatley[52] (JW) the cooling power obtained by compressionally freezing the liquid is indeed favorable in relation to that of other refrigeration methods in use between 2 and 20 mK. A comparison of these methods due to JW is presented in *Figure 12*. However, in practice the conversion is not accomplished isothermally, but as nearly isentropically as possible so that ΔQ_c cools the sample. Thus one begins the compression at a point on the melting curve, such as A in *Figure 7*(a), with all liquid (i.e.,

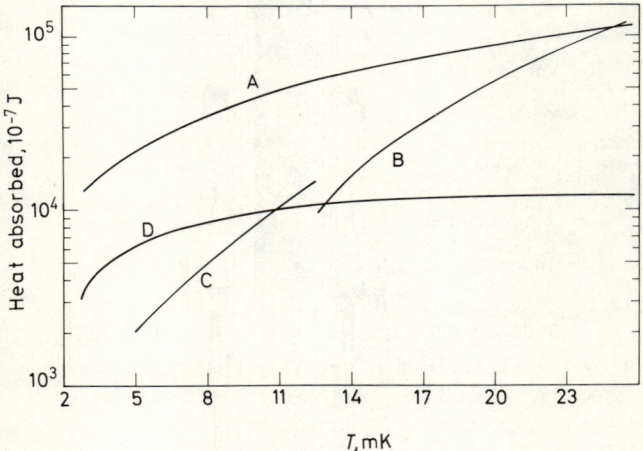

Figure 12. Comparison of refrigeration capacity of various methods in use for the low millikelvin range, given as the heat absorbed at temperature T under the conditions that: A, 0.1 mol liquid ^{3}He is converted to the solid at the melting curve; B, the refrigerator of Wheatley *et al.*[116] operates 10^4 s with $\dot{n}_3 = 3.2 \times 10^{-5}$ mol/s; C, 0.1 mol ^{3}He is removed from a typical single cycle dilution refrigerator; and D, 0.1 mol of CMN in zero magnetic field warms from 2 mK to T (note: the volumes occupied by 0.1 mol CMN and 0.1 mol liquid ^{3}He are respectively 38 cm^3 and 2.5 cm^3). (Drawn after Johnson and Wheatley[52].)

$y = 0$, where y is the molar fraction of ^{3}He in the solid form) in the refrigeration chamber. As the compression proceeds reaching lower temperatures along the isentrope ABC, y is given by the ratio a/b at each T. JW cite as a numerical example of this ideal process that if point A is at 24 mK and point B is at 3 mK, $y(B) = 0.183$ leaving more than 80 per cent of the original charge of liquid ^{3}He for further cooling capacity. However, because pdV work $[\Delta W = p(V_1 - V_s)]$ is done upon the ^{3}He during compression the process is not reversible; and this combined with non-elastic deformations of the cell can produce frictional heating which below ~ 10 mK can be very serious. JW have suggested that these effects can be minimized by the selection of cell construction materials which allow a five to six per cent elastic change in cell volume for a 20 atm variation in differential pressure between inside and outside the cell. Accordingly proper refrigeration chamber design becomes crucial to the success of the method.

Cooling by adiabatic compression of solid ^{3}He is also called the Pomeranchuk effect after the Russian scientist who in 1950 qualitatively predicted[83] the relationship between S_1 and S_s for ^{3}He at ULT shown in *Figure 7*(a). It was, however, not until 1965 that the effect was employed in a cooling experiment, when Anufriev[7] starting at 50 mK reached a temperature somewhat less than 20 mK by this method. Subsequently Goldstein[35] analysed melting curve data, together with their extension to lower T obtained using a molecular field model of the condensed ^{3}He magnetic system, to predict a lower limit in T attainable by adiabatic freezing. This turns out to be 0.524 mK, where once again $S_1 = S_s$. Two important additional experimental papers on this method of cooling have appeared: one by JW, referred to above, which presents a thorough analysis of the techniques, limitations and advantages associated with the method (application to ^{3}He melting curve determinations is given by Johnson *et al.*[51]); the other by Sites *et al.*[94] describing the method as used in an experimental determination of the magnetic susceptibility of solid ^{3}He. In both of these investigations a lower limit of about 2 mK was reached.

The apparatus for compressional cooling is complicated by the melting curve minimum: the ^{3}He cell cannot be directly pressurized by fluid in a capillary extending to room temperature because when an attempt is made to go to $T < T_{min}$ a solid plug will form in the transmitting capillary at $p = p_{min}$. Fortunately ^{4}He (which also exhibits a melting curve minimum) remains fluid at all T for $p < 24.9$ atm and therefore may serve as an intermediate pressurizing medium (of course ^{3}He could serve the same purpose, but ^{4}He is more convenient and economical). The ^{3}He must then be contained in a flexible vessel. *Figure 13* illustrates the system used by Sites *et al.*[94] to accomplish this. The regions of interest here are the three chambers, labelled I, II and III, separated by two concentric bellows. Chamber I is filled with ^{3}He and chambers II and III with ^{4}He, all at pressures near the respective melting curves when precooled to ~ 25 mK by a dilution refrigerator. To compress the ^{3}He contents of I, ^{4}He is slowly (over periods ranging from one to five hours) let out of II resulting in a downward motion of the top plate against the cover of chamber I. A combination of the directly read pressures of ^{4}He in II and III and the displacement of the top plate as monitored by the displacement capacitor allows the ^{3}He pressure in I to be determined.

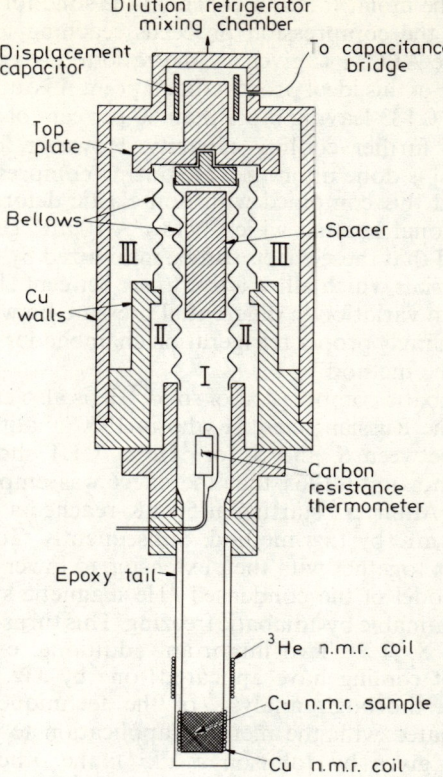

Figure 13. Schematic drawing of an arrangement for refrigeration by adiabatic solidification of liquid ^{3}He (drawn after Sites et al.[94]).

The design of the flexible ^{3}He chamber in the apparatus of JW is also ingenious. Here a flattened copper–nickel tube is used, spring-loaded by other axially slit tubes to prevent irreversible bulging of the ^{3}He tube. The cross section thus looks like this:

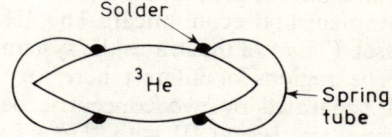

4. Heat Transfer and Isolation at ULT

The above discussions of refrigeration techniques at ULT have generally deferred the problem common to all such methods, namely, the control of the flow of heat. On the one hand there is the demand to deliver the cooling power of a refrigerator to the sample of interest (here fluid helium) while on the other hand the sample, as well as the cold stages of the refrigerator, must be protected from heat influx. Various methods of thermal contact and isolation have been well covered in Volume I (pp 272–281) mainly with respect to experiments carried out at $T > 0.3$ K. Whereas the same principles are valid below 0.3 K, all of the problems become more acute, especially in multistage cooling systems. Sarwinski[89] has reviewed these rather thoroughly. In the following, major sources of difficulty will be recapitulated along with brief mention of preferred corrective (or preventive) measures. It should be noted that with respect to ULT experimental cells it is often desirable to reduce the total heat leak to less than 10^{-9} J/s.

Heat leak due to vibration. A cryostat should be mounted on a heavy frame supported by vibration isolators with the minimum natural frequency ω_n possible—the ratio of the transmitted force to the disturbing force (of frequency ω), i.e. the transmissivity of the system, decreases with increasing ω/ω_n. Pumps or other sources of vibration should be removed from the frame and pumping lines fitted with vibration damping sections (rubber tubing, bellows, etc.). Tubing supporting inner parts of the cryostat must be rigid.

Heat leak due to stray r.f. fields. Electrical equipment inside a cryostat may pick up r.f. signals from nearby radio or television stations. About the only sure, albeit costly, method of avoiding the possibly severe heat generation from this source is to instal the cryostat in an electrically shielded room.

Heat leak due to thermal radiation. When it is considered that a surface at 4 K radiates $\sim 10^{-9}$/J·cm² to a surface at ULT, the need for thermal shielding tied to colder parts of the apparatus becomes apparent (these are of course in addition to other radiation shields and traps protecting the cold portions of the apparatus from room T radiation). In designing and fabricating these shields (usually from stainless steel or thin copper foil) care must be taken lest these interfere with the purposes of the experiments—e.g. a heavy copper shield is not compatible with measurements, such as of temperature, made using magnetic coils external to the shield.

Heat leak down supporting structures and electrical leads. For supporting structures the problem is to effect a compromise between the requirements for rigidity and small total cross-sectional area. For electrical leads the compromise is between resistive heating and small cross-sectional area of the wires. Of course materials with low thermal conductivity k must be used for both of these elements. In this regard use of plastics such as nylon, Teflon and castable epoxy resins are of considerable value in cryostat construction. One particularly interesting material is AGOT graphite which in addition to being easily machinable has a high k—of the same order as for copper—down to ~ 70 K but becomes an exceptionally good insulator at ULT[33, 96]. One useful application for this material is in the mounting and separation of the heat exchanger stages of a dilution refrigerator, where the graphite aids in cool-down of the apparatus but thermally isolates the stages under working conditions.

Heat leak down fill lines. When liquid ^{4}He below 2.2 K is contained in sample fill lines, heat leak due to the large effective thermal conductivity of the superfluid bulk liquid and/or of the mobile film must be minimized. This can be accomplished by using long narrow lines, by placing an orifice in a wide tube (see *Figure 10*), or by valving arrangements. Similar considerations apply to the superfluid phase of ^{3}He–^{4}He mixtures and to pure ^{3}He at low T (k for ^{3}He below 0.4 K increases with decreasing T and below ~ 0.1 K increases as T^{-1}—but no superfluid film is associated with pure ^{3}He).

Heat leak across vacuum spaces. Often ^{4}He is used as an exchange gas to effect heat transfer from a ^{4}He bath to the inner parts of a cryostat. However, difficulty can be experienced in pumping away the gas to achieve sufficiently low pressures for proper isolation at ULT—this arises chiefly from adsorption of gas on the cold surfaces. The best means of preventing these problems is to design the apparatus so as to avoid the use of exchange gas.

Thermal switches. Of the types of thermal switches discussed in Volume I (pp 273–279) superconducting switches are most suitable at ULT. Operation of the others below 0.3 K generally introduces excessive amounts of heat into the system.

Thermal contact at ULT. The concerns here are to insure first that the cooling power of a refrigeration device is delivered to the helium sample and second that thermometers in contact with the helium read true temperatures of the sample. Sometimes the first of these can be satisfied by having the working materials of the refrigerator double as the sample. Thus experiments on dilute ^{3}He–^{4}He mixtures and on pure ^{3}He might be carried out on the appropriate separated phases in the mixing chamber of a dilution refrigerator; or melting curve properties of ^{3}He can be studied using the ^{3}He charge of an adiabatic freezing refrigerator—which may function simultaneously as a thermometer[51]. More often, however, such fortunate coincidences do not occur and it is necessary to attach a sample cell to the refrigerator. Here there are two alternatives: either the cell is fabricated from the same piece of material as the lowest temperature stage of the refrigerator (e.g. the mixing chamber), or the cell is made separate and attached. In the former only the boundary resistance R_K between helium and the cell walls or the refrigeration material itself is involved, whereas in the latter an additional solid/solid interface must be considered. An example of each case is given respectively in *Figures 14* and *11*, which incidentally also illustrate several other ULT techniques mentioned above. *Figure 14* illustrates parts of a cryostat for cooling liquid ^{3}He to ULT, with the liquid in contact with powdered CMN acting both as a thermometer and the last cooling stage. Whereas for boundaries between liquid helium and most solids $R_K \sim T^{-b}$ with $b \sim 3$, quite interestingly and fortunately Abel et al.[1] have found that $R_K \sim T$ and is small for ^{3}He–CMN below about 10 mK. The drawing in *Figure 11* illustrates a way of staging a nuclear demagnetization from a dilution refrigerator via a lead heat switch and employing a graphite support member. Steyert[99] has devised a novel method of transferring the CMN demagnetization cooling to a sample by embedding a large number of pure copper wires into a block of highly

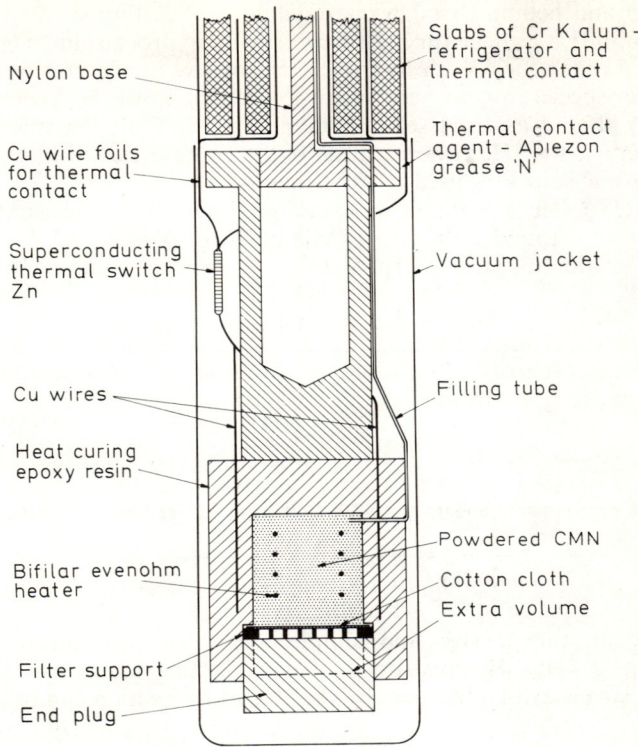

Figure 14. Schematic drawing of an apparatus for measuring the low temperature specific heat of liquid ³He in which CMN acts both as refrigerant and thermometer [drawn after Wheatley, J. C., in *Quantum Fluids*, p 183 (Brewer, D. F., Ed.), North-Holland: Amsterdam (1966)].

compressed, intimately mixed, fine particles of CMN and copper to provide a large contact area (a later version of this device[100] substitutes gold for copper because the surfaces of the copper were found to be corroded by the CMN). In the systems discussed here, solid–solid thermal contact is attained by close physical contact, via a layer of grease (glycerol can also be useful), or by solder[102]. For the last method special solders can be used, such as eutectic Cd–Bi[98]; but in any event the thickness of the solder layer should be kept small—of the order of tens of a micrometer.

IV. Preparation of Helium and Hydrogen Samples

Because helium and hydrogen boil at very low temperatures there are only a limited number of contaminants that need be considered in obtaining pure samples of these liquids. Gas separation using distillation and rectification techniques[91] serve to concentrate the species of interest and can be accomplished on an industrial scale. The experimenter is usually concerned with eliminating small quantities of foreign substances such as nitrogen, oxygen, methane, etc. For most such purposes passage of commercially obtainable hydrogen and helium through charcoal at ~ 77 K (liquid nitrogen normal boiling point) is satisfactory. However, with hydrogen and ^{3}He there are additional difficulties. With hydrogen the relative concentrations of the two spin-state species, ortho- and para-hydrogen, must be specified and if necessary altered, as discussed in Volume I (p 55). With ^{3}He it is often necessary to reduce the ^{4}He concentration of the commercially obtained material or of isotopic solutions used in experiments.

In the USA ^{3}He is available in several grades from the Monsanto Research Corporation, Mound Laboratory, Miamisburg, Ohio 45342. Current prices for these grades are given in *Table 2*.

Table 2. Costs for available grades of ^{3}He

Grade	Description*	Cost s.t.p. liter ($)
99.5%	$\geqslant 99.5\%$ ^{3}He in total He $\geqslant 99.5\%$ total He	195.00
99.98%	$\geqslant 99.98\%$ ^{3}He in total He $\geqslant 99.5\%$ total He	255.00
Vapor pressure standard	$\geqslant 99.98\%$ ^{3}He in total He $\geqslant 99.9\%$ total He	440.00

* All grades have $\leqslant 1 \times 10^{-11}\%$ tritium content.

Clearly an inexpensive and simple means of concentrating the ^{3}He fraction of a ^{3}He–^{4}He mixture is of considerable usefulness. Fortunately such a means is available, whereby the ^{4}He mole fraction can be readily and

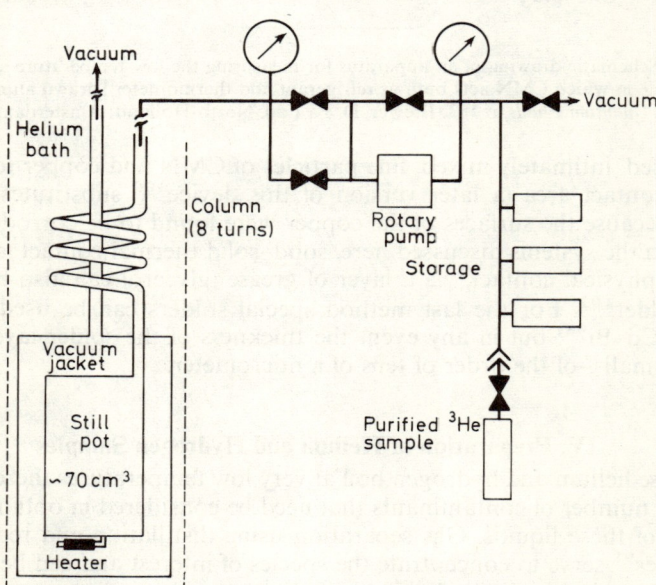

Figure 15. Schematic drawing of a distillation system for efficient enhancement of the ^{3}He concentration of ^{3}He–^{4}He mixtures (drawn after Sherman[93]).

efficiently reduced to a few parts per million. The method, due to Sherman[93], uses a distillation system shown schematically in *Figure 15*. In operation, the entire batch to be purified is condensed from the storage bank into the still, which is maintained at a temperature of 2.2 K, just above the ^{4}He λ-point, in order to eliminate problems associated with superfluidity. Heating the still produces vapor at the top of the column considerably richer in ^{3}He than the liquid charge. As examples of the performance, vapor samples with ^{3}He mole fractions x_g of 0.9996 and 0.99997 can be drawn from liquid batches starting with x_l of 0.827 and 0.966 respectively. The rate of withdrawal is of the order 5 s.t.p. liters/hour. With slight modifications the process can be made continuous rather than batch.

V. Calorimetry of Fluids below 20 K

Discussed below are several calorimetric techniques developed to meet specific problems related to the thermal properties of quantum fluids.

1. Liquid ^{4}He near the Melting Curve

The thermodynamics of ^{4}He near the melting curve at temperatures

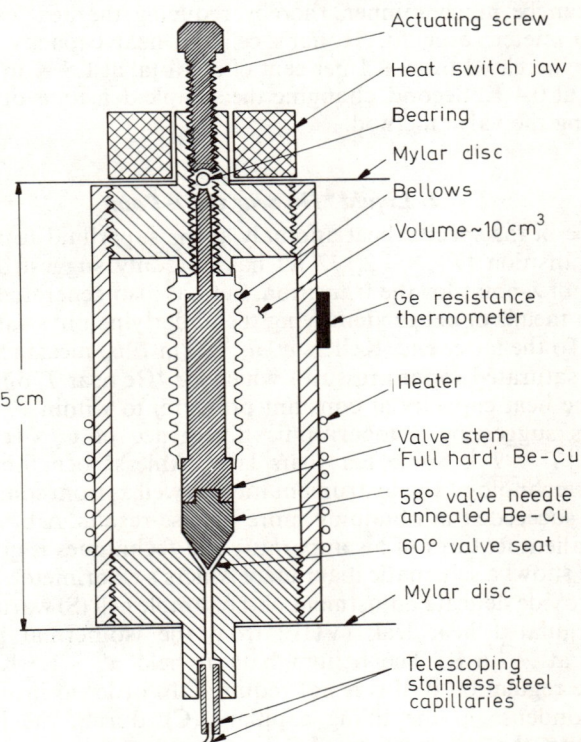

Figure 16. Schematic drawing of cell for measurements of constant volume heat capacity of ^{4}He under pressure (drawn after Hoffer[47]).

between 0.3 and 1.8 K are of interest with respect to the very shallow minimum of the melting curve ($p_{min} = 25$ atm, $T_{min} = 77$ K) and three solid–liquid triple points that occur within this region; in addition, the thermal properties of compressed fluid ^{4}He bear significantly upon the parameters of the liquid excitation spectrum. Specific heat measurements recently reported by Wiebes[117] and by Hoffer[47] provide the first precise and systematic studies of this region. In particular Hoffer has directly measured the specific heat at constant volume C_v using a novel sample cell, a sketch of which appears in *Figure 16*. In operation after the cell has been filled at low T through the stainless steel capillary, the valve is closed (tight to superfluid ^{4}He) by applying a removable wrench to the actuating screw. Liquid ^{4}He remaining in the capillary is pumped away to aid in thermally isolating the cell. Additional isolation is achieved by supporting the cell from five thin Mylar discs in a vacuum can and by using a mechanical heat switch to disconnect the cooling link between the cell and a pumped ^{3}He post (in closed position the jaws of the switch also prevent motion of the calorimeter against the torque generated by the valve-closing operation). The cell described here has at least two major advantages over more conventional constant volume calorimeters in which the cell is filled to high pressure and sealed at room temperature. First, because maximum cell pressures are considerably lower in the present design, the cell walls can be much thinner, thereby reducing the heat capacity of the empty calorimeter; even for Hoffer's cell the heat capacity of the empty calorimeter increased from ~ 1 per cent of the total at 1.5 K to ~ 25 per cent of the total at 0.4 K. Second, changing the sample density is obviously much simpler using the valve method.

2. Liquid ^{4}He near the λ-Point

The shape of the specific heat anomaly at the superfluid-to-normal (He II to He I) transition ($T_{\lambda 58} = 2.172$ K) has not only suggested the popular appellation of λ-point for the transition, but has also generated considerable interest as a means of better identifying the underlying physical nature of the transition. To the latter end, Kellers et al.[19, 60] in 1960 measured C_s (the heat capacity at saturated vapor pressure), which for ^{4}He near T_λ differs negligibly from C_p (the heat capacity at constant pressure) to within $\sim 10^{-6}$ K of T_λ with results suggesting a logarithmic divergence of C_s over about four decades of $|T - T_\lambda|$. Some ten years later, Ahlers[5] benefiting both from advances in technology and instrumentation as well as from some innovations of his own, succeeded in obtaining more precise results. A brief description of several salient features of his apparatus and techniques is given below.

Figure 17 shows a schematic diagram of Ahlers's calorimeter. The principle here is to provide delicate adjustment of the sample cell (S) warming rate by a carefully regulated heat leak (VHL) from the isothermal platform (IP) maintained at ~ 1.6 K. The helium bath is held at ~ 1.3 K, but careful temperature regulation of this is not required. In order to insure that liquid does not condense in the filling capillary (C), during the heat capacity measurements, the temperature of the point of emergence of C from the capillary vacuum (CV) into the main vacuum (MV) is controlled by a system of heaters (H) and heat leaks (HL). Conversely by pumping on the sample through C and monitoring the pressure, the vapor volume in S can be minimized and kept extremely small. The λ-transition point is determined with a precision of 10^{-7} K as follows. Earlier work by Ahlers[4] had shown that due to gravitational effects as a liquid ^{4}He sample is slowly warmed

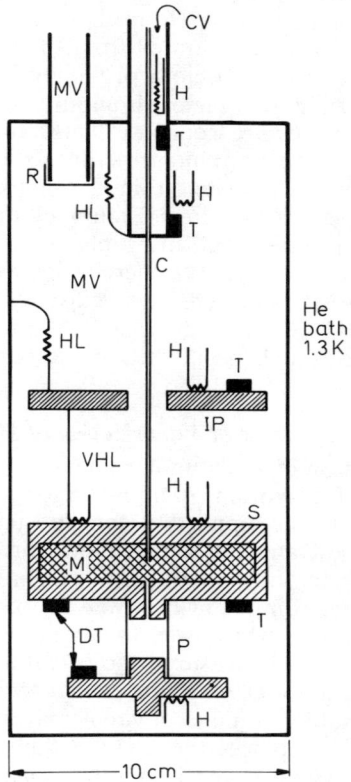

Figure 17. Schematic drawing of low temperature part of apparatus for high resolution heat capacity measurements of liquid ^{4}He near the λ transition: H, heater; T, resistance thermometer; DT, differential thermometer; HL, heat leak; VHL, variable heat leak; MV, main vacuum; CV, capillary vacuum; R, radiation trap; IP, isothermal platform; P, thermal conductivity probe; S, sample cell; M, copper mesh; C, filling capillary (drawn after Ahlers[5]).

through the transition, He I first forms at the bottom. Because of the great difference in thermal conductivity between He I and He II, as soon as some He I forms in the probe (P) of *Figure 17*, a thermal gradient appears (detected by DT, the differential thermometer) signalling the transition. This allows the transition temperature at the bottom and top of C to be calculated—the flat geometry of C is dictated to minimize this gravitational effect. The interested reader may find considerably more detail on the calorimeter components and experimental procedure in Ahlers's paper[5].

3. Liquid ^{3}He at ULT and at Pressures up to the Melting Curve

The heat capacity of liquid ^{3}He has been measured by Abel et al.[2] and by Mota et al.[76] in a cell as shown in *Figure 14* where the sample is located in the interstices of packed, finely powdered CMN. These authors devised a clever method of solving the commonly met problem in calorimetry of making corrections for the non-^{3}He background specific heat. The bottom plug of the sample cell was separated from the CMN by a thin perforated epoxy disc backed with cotton cloth providing holes too small to pass the CMN. The heat capacity of a metered quantity, n_1 moles of liquid ^{3}He was measured at 0.28 atm pressure from 6 to 50 mK; then the cell was pressurized to 27 atm and the same measurements performed on an increased number of moles n_2 of liquid; n_2 is calculated from n_1 using known values of the molar volumes at the respective pressures. Following this the bottom plug of the cell was carefully machined out and replaced with another plug having a cavity (dashed lines in *Figure 14*), thereby increasing the cell volume but otherwise leaving it undisturbed. Heat capacity measurements were then repeated at 0.28 and 27 atm with n'_1 and n'_2 moles respectively. Thus by differences the background specific heat could be eliminated completely.

4. Latent Heats of Vaporization of Hydrogen and ^{4}He

A direct observation of L, the molar latent heat of vaporization, involves the determination of the volume of liquid vaporized by the introduction of a measured amount of heat supplied at constant temperature for a measured length of time. While Dana and Onnes[25] measured L for ^{4}He in this way in 1926—and they are the only investigators to have done so—the accuracy of their experiment was limited by how well they could determine the volume change of their sample, which was to about ± 0.5 per cent. Subsequent investigators have instead measured the so-called 'apparent latent heat' L_a by determining the amount of vapor evolved when n moles of liquid are evaporated. The resulting n moles of vapor may be considered in two parts: (1) that vapor which replaces the liquid volume evaporated at low temperature, or nV_1/V_g, where V_1 and V_g are the molar volumes of the liquid and vapor; and (2) the remainder, $n(1 - V_1/V_g)$, which is collected and measured at room temperature in a volume-manometer (here the dead volume in the lines is neglected; the correction for this represents one of the largest sources of error in the experiments). Hence L is obtained at $L = L_a(1 - V_1/V_g)$. L_a, incidentally, is the quantity required for corrections to constant volume liquid calorimetry when the vapor phase is present. Berman et al.[14, 15] and Ter Harmsel[107] have measured L_a for ^{4}He between 2.2 and 5.1 K; similarly, Ter Harmsel et al.[107, 108] have obtained L_a for 20.4 K eH_2 from 13.9 to 23.5 K. In this work of Ter Harmsel the systematic errors have been reduced to below 0.1 per cent. In the conversion of L_a to L, V_g represents the largest uncertainty, which nevertheless still results in about a fivefold increase in accuracy over the direct method of determining L. On the other hand, through the Clausius–Clapeyron equation

$$\frac{dp}{dT} = \frac{L}{T(V_g - V_1)} = \frac{L_a}{TV_g} \tag{25}$$

V_g may be determined from L_a measurements if dp/dT is well-known from a vapor pressure temperature scale such as T_{58}.

In the vicinity of the ^{4}He λ-point and at lower T, the L_a method is difficult because of the heat leak arising from the superfluid film flow. In this region the only results for L are those of Dana and Onnes, but these are not well defined near the λ-point. However, it is possible to use the thermodynamic relation

$$\frac{dL}{dT} = \frac{L}{T} - C_{sl} + C_{sg} \approx \tfrac{5}{2}R - C_{sl} \tag{26}$$

together with heat capacity data to determine the behavior of L here (C_{sg} and C_{sl} are the specific heats of the saturated vapor and liquid respectively).

5. Liquid ^{3}He–^{4}He Mixtures

DeBruyn Ouboter et al.[29] have measured the specific heat of ^{3}He–^{4}He mixtures over the entire concentration range $0 \leqslant x \leqslant 1$ and at temperatures between 0.4 and 2 K. For the most part the calorimeter and measurement techniques used were conventional. However, the copper calorimeter itself was divided into two compartments, one containing the sample and the other pure ^{3}He as refrigerant. Pumping the ^{3}He brought the system to the lowest T, after which pumping was continued until no ^{3}He remained. The heat capacity of the sample could then be determined, without the complication of the refrigerant, by the warm-up characteristics in response to the introduction of metered amounts of heat.

The results of these heat capacity measurements are rich in information about the thermodynamic properties of the mixtures. This is of some importance because ^{3}He–^{4}He solutions obey no known model for non-ideal solutions; hence all thermodynamic properties must be obtained empirically and are generally expressed as 'excess' quantities representing the deviations from ideal solution behavior. From the heat capacity data in combination with vapor pressure data[104] it is possible to obtain for the entire concentration range values of the heat of mixing (excess enthalpy), excess Gibbs function, and the excess entropy. In addition, discontinuities and spikes in the observed specific heat curves as a function of T for various values of x indicate the stratification temperatures and the λ temperatures of the mixtures—i.e. they may be used to define the phase diagram of *Figure 9*. (It should be noted, however, that capacitance methods—similar to those discussed in section VI.1—which sense the different densities of the two mixture phases, provide more sensitive means of determining stratification temperatures[32, 39].)

VI. pVT Measurements below 20 K

pVT measurement methods at higher temperatures, discussed thoroughly elsewhere in this volume, are generally easily adaptable to fluid helium and hydrogen below 20 K. On the other hand, pVT properties along a melting curve have not been treated previously but will be considered here. In addition, it appears useful to include methods of molar volume (or density) determinations which, although adaptable to a wide range of pressure and temperature conditions, have been highly developed for precision low temperature work. Finally, included here is some mention of ^{3}He–^{4}He osmotic pressure measurements even though these are not strictly pVT properties. A secondary theme of this section dwells on the wide usefulness and variety of capacitance techniques for obtaining pVT data.

1. Molar Volume

Conventional procedures for fluid density determinations at low temperature, including pyknometer techniques (e.g. for liquid ^{4}He[63] and ^{3}He[62], the latter at T as low as 0.16 K) and fixed-volume cell methods (e.g. for gaseous ^{3}He and ^{4}He[59], liquid hydrogen[38, 50]) are laborious and time-consuming; all require elaborate room temperature volume-manometry and accurate cell volume calibration; and some require, in addition, complicated dead-volume corrections to be made. However, the availability of high precision capacitance bridges has provided a means of eliminating nearly all of these objectionable features while at the same time allowing considerable improvement in the results. Furthermore, the same method can be used for vapor or liquid phase samples over a wide range of pressure.

The Clausius–Mosotti equation relates the density ρ and the dielectric constant ε of a sample

$$\alpha\rho = (3M/4\pi)(\varepsilon - 1)/(\varepsilon + 2) \qquad (27)$$

where M is the molecular weight and α the molar polarizability of the material. For many purposes α can be considered a constant for a given substance; however, α is generally density-dependent, which should be taken into account for the most accurate work. For ^{3}He, over the range $0.004314 < \rho < 0.09285$ (both liquid and gas), Kerr and Sherman[61] have found; α (mol/cm^3) = $0.123413 - 0.002376\,\rho$ (g/cm^3). Because ε is defined as C/C_{vac} (the ratio of the capacitance of the sample in a given cell to that of the evacuated cell alone; for precise work a correction is required for deformation of the cell with pressure) the density can be directly determined by capacitance measurements independent of the cell geometry and without recourse to additional gas handling.

Figure 8(b) illustrates a pVT capacitance cell[61] designed to be structurally stable (and therefore capable of providing reproducible results with temperature cycling) and to minimize the effect of pressure on ε. These aims are accomplished by supporting both upper and lower plates from the same non-conducting (glass) platform, by proper electrical shielding so that the only capacitance measured is that between the gap formed by the two plates, and by arranging to have pressures exerted isotropically on all sensitive elements.

General Radio transformer ratio-arm capacitance bridges (Model 1615A used with a lock-in amplifier as a null detector; Model 1620, a measurement system which also includes a signal generator and a null detector; or the newer Model 1616—ten times more sensitive than the 1615—and its corresponding packaged system, Model 1621) are well suited for making the three-terminal measurements which eliminate all stray capacitances. The cell in *Figure 8(b)* was designed for such measurements. Even though the value of α for helium is small, the method is capable of providing density data precise to better than 0.1 per cent even at low density. Of comparable sensitivity is a method which monitors the resonant frequency f of an LC tank circuit, one element of which is the capacitor filled with the material of interest. The basis for this method, first applied to liquid helium by Lee et al.[69], is the fact that f^2 is proportional to C^{-1}. A typical circuit diagram for this purpose is shown schematically in *Figure 18*. Both the bridge and

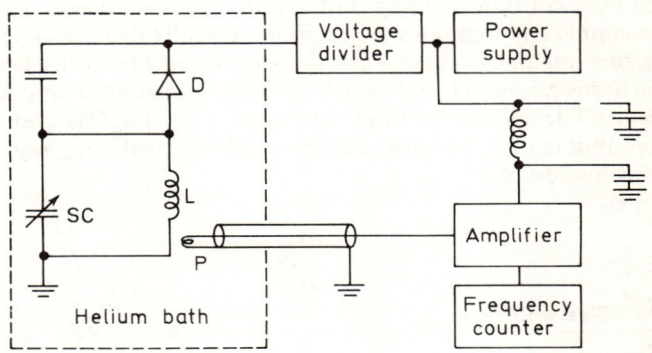

Figure 18. Schematic diagram of tank circuit for measuring the density of fluid ^{3}He and ^{4}He by the dielectric constant method: D, tunnel or backward diode; L, inductance; P, pickup coil; SC, sample cell capacitor.

tank circuit methods require careful temperature control to achieve their inherent maximum precision—the former at room temperatures (capacitive elements of the bridge are slightly temperature-sensitive) and the latter at liquid helium temperatures. Maintenance of the complete tank circuit at liquid helium temperatures serves to sharpen the circuit resonance in two ways: the resistance of the coil L and of the leads is reduced effectively to zero by plating the copper wires with tin–lead solder, which becomes superconducting; the diode (either a tunnel or backward diode) is operated in the amplifier mode which is enhanced as T decreases. Placement of the pickup coil P at the end of the coaxial cable is critical: it should be far enough from L to avoid adding a significant load to the tank circuit, yet just close enough to produce a signal which when amplified can be handled by the frequency counter. Stability of such a circuit can be as high as ± 0.5 Hz/min with f of the order 20 MHz/s. Measurements using the tank circuit method with fixed-spacing capacitor plates have yielded, for example, precision results on the density, the isobaric thermal expansion coefficient α_p, and entropy of compression of liquid ^{3}He[18] and ^{4}He[17], and on the equation of state of fluid ^{3}He in the vicinity of the critical point[112]. The entropy of compression may be obtained through the relation

$$S(p, T) - S(0, T) = - M \int_{p=0}^{p} \alpha_p \rho^{-1} \, dp \qquad (28)$$

2. pVT Properties at Melting

For determining the (p_m, T_m) relation of a melting curve, except in regions below the temperature of the minima in the cases of ^{3}He and ^{4}He, the blocked capillary technique, introduced in 1926 by Onnes and van Gulik[80], is probably the most satisfactory. Here the fluid sample is contained in a narrow capillary tube which extends from a pressurizing source at the room temperature environment down into the cryostat and then out again to a pressure gage. At the lowest point the tube is sharply bent, and this region is maintained at the constant temperature of interest. As p is increased past p_m the capillary becomes blocked by frozen material, presumably at the bend, as evidenced by a constant reading on the outlet pressure gage; decrease of p melts the sample and unblocks the capillary. Usually there is some hysteresis in the freezing–melting processes and the sample must be cycled several times before the hysteresis loop is sufficiently reduced to insure the desired accuracy of the results. Clearly this method is unsatisfactory for ^{3}He and ^{4}He at T_m below the minima of p_m because a block will be formed at p_m preventing the direct transmission of $p > p_m$.

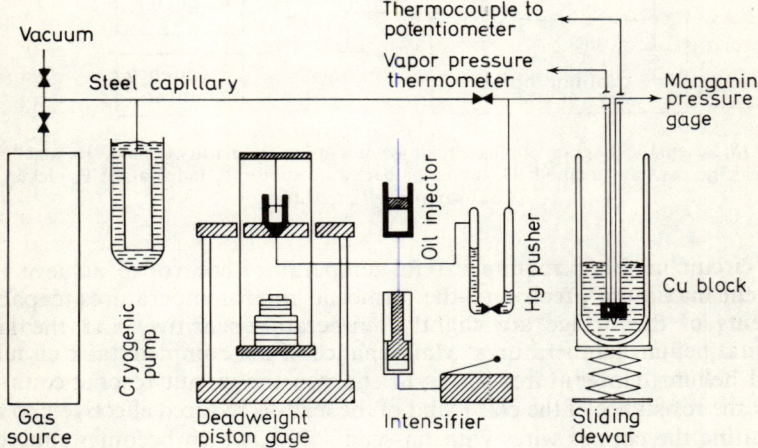

Figure 19. Schematic diagram of apparatus for melting curve measurements by the blocked capillary technique (drawn after Mills and Grilly[73]).

Mills and Grilly[73] have used a blocked capillary apparatus (shown schematically in *Figure 19*) to measure (p_m, T_m) for ^{3}He, ^{4}He, H$_2$, D$_2$, T$_2$, Ne, H$_2$, and O$_2$ at pressures up to 3.5×10^8 N/m^2 and covering a temperature range of 1.9 to 121 K. In their system the sample line was filled to a p just less than p_m via the cryogenic pump; then the sample-line valve was closed leaving the sample gas separated from the pressurizing fluid (oil) by a U-tube (non-magnetic stainless steel) partially filled with mercury. Weights were then added to the calibrated deadweight piston gage and the oil injector adjusted to maintain the piston location constant, thereby displacing the mercury levels and increasing p in the sample. Mercury level heights were read by magnetic search coils which sensed ferritic floats on the mercury surfaces. This formed the primary pressure measurement system. The monitor pressure gage on the outlet was either a Bourdon helix or a manganin resistance gage, the latter with a pressure resolution of $\pm 3 \times 10^4$ N/m^2. The sharp bend in the capillary forming the freezing and melting chamber was mounted in a small copper block with the tip of the bend extending below. Also in this block were embedded a thermocouple and a vapor pressure bulb for temperature measurements. In order to insure that the region of the copper block was the coldest part of the capillary, the dewar system was supported on an adjustable jack permitting the liquid level height of the refrigerant bath to be slowly and easily positioned with respect to the fixed block.

Using the cell shown in *Figure 8*(a), Scribner et al.[92] have also measured (p_m, T_m) of ^{3}He extending to $T = 0.017$ K, well below the minimum. For this purpose the sample space was initially charged with sufficient ^{3}He at 0.7 K to maintain a mixture of liquid and solid at all $T < 0.7$ K with at least one third of the sample as liquid at (p_{min}, T_{min}). This amount of ^{3}He was trapped in the sample chamber by the formation of a solid plug in the filling capillary when the dilution refrigerator to which the capillary was thermally linked (not shown in the figure) was cooled rapidly. In this way the equilibrium (p_m, T_m) could be determined directly by measuring p and T as T was varied.

Determinations of the volume change on melting, $\Delta V_m = V_l - V_s$, can be made[41] with only slight modifications of the apparatus shown in *Figure 19*, primarily by replacing the capillary bend with a calibrated volume. The cell is first charged with fluid which is then carefully solidified to insure that solid completely fills the cell. With the refrigerant bath just covering the cell its temperature is raised a few tenths of a kelvin above the equilibrium melting point, so that the melting proceeds from T_a(solid) to T_b(liquid) with $T_a < T_m < T_b$. A correction for the thermal expansion of the solid from T_a to T_m and of the liquid from T_m to T_b is required.

In order to extend ΔV_m experiments with ^{3}He from 1.2 K down to 0.3 K, Mills et al.[74] devised a double-chambered cell, where changes in the pressure of the liquid ^{3}He contained in the outer chamber were transmitted to the liquid ^{3}He contents of the inner chamber via a spring-loaded bellows—an arrangement similar in concept to that shown in *Figure 13*. Values of ΔV_m were obtained by measuring amounts of fluid withdrawn as required to maintain p_m constant during the melting process, starting with the inner chamber filled with solid ^{3}He. Absolute values of V_l were determined separately but again by a withdrawal procedure. These were used as calibration points by Scribner et al.[92] to carry the ΔV_m measurements down to 0.020 K. Their apparatus [again referring to *Figure 8*(a)] required separate determinations of V_l and V_s to obtain ΔV_m, which entailed different filling procedures. In the case of V_l the sample space was filled with liquid at $T > T_{min}$, $p > p_{min}$, followed by the formation of a solid plug in the capillary and thus fixing the liquid volume at V_i. On cooling along the fluid isochore corresponding to V_i the melting curve was intersected on the high side of the minimum and the data of Mills et al.[74] used to establish the value of V_i. On further cooling the sample partially freezes and ultimately completely melts again on the low side of the melting curve, with V_l at this point given by V_i. One serious problem with this method involved possible slippage of the plug in the capillary, particularly as (p_{min}, T_{min}) was passed in the cooling process. Several remedial steps were taken to minimize this slippage.

More recently Grilly[40] has constructed a pVT cell [*Figure 8(c)*] which simultaneously eliminates the problems of the blocked capillary for obtaining ΔV_m and (p_m, T_m) below the minimum and also avoids the necessity of thermal expansion corrections in determining ΔV_m. Furthermore, the cell is adaptable for pVT measurements over a wide range of conditions for both fluid and solid phases; and although designed for measurements on ^{3}He it can be used as well for other materials. It is constructed from three beryllium–copper discs welded together with spacers at the circumferences, thus forming two chambers. The upper chamber contains the sample; the lower, a pressurizing fluid. Deflections of the topmost diaphragm caused by volume changes of the sample move the lower capacitor plate relative to the fixed upper plate. Hence the thicknesses of the middle and upper discs must be designed to deform compatibly with the pressure range of interest and the desired sensitivity of the experiment. Again the General Radio capacitance bridges are appropriate for the capacitance measurements. The volume of the upper chamber V_U at pressure p_U can thus be changed by temperature variations at constant pressure of the lower chamber p_L (to obtain the isobaric expansivity α_p) or by variations of p_L at constant T (to obtain the isothermal compressibility K_T). Under any conditions V_U is given by

$$V_U = V_{UO} + (S_U + S_M) p_U - S_M p_L \qquad (29)$$

where V_{UO} is the calibrated volume of the upper cell with the diaphragms undistorted, and S_U and S_M represent the sensitivities of the upper and middle diaphragms as the volume change per unit pressure change. Hence

$$\alpha_P = \frac{1}{V_U}\frac{dV_U}{dT} = \frac{S_M}{V_U}\frac{dp_L}{dT} \qquad (30)$$

and

$$K_T = \frac{1}{V_U}\frac{dV_U}{dp_U} = \frac{S_M}{V_U}\left(\frac{dp_L}{dp_U} - \frac{S_U + S_M}{S_M}\right) \qquad (31)$$

Similarly, by maintaining $p_U = p_m$ at constant T_m and starting with liquid in the upper chamber, an increase in p_L will cause this liquid to freeze; ΔV_m may then be determined by

$$\Delta V_m = (V_l/V_U) S_M \Delta p_L \qquad (32)$$

3. Osmotic Pressure of ^{3}He–^{4}He Solutions

The properties of ^{3}He–^{4}He solutions in the temperature region below ~ 0.5 K, where both constituents exhibit quantum degeneracy, is of considerable theoretical interest. Critiques for models predicting the behavior of these solutions must rely on thermodynamic measurements, and one of the most useful of these is the osmotic pressure π of solutions with ^{3}He mole fraction $x_3 \gtrsim 10$ per cent. An Ohio State University group[66,119,121] has made such measurements extending over temperatures between 0.027 and

1.16 K and at pressures up to 20 atm, using most recently the apparatus shown in *Figure 8*(d)[66]. Two samples of liquid, one of pure ^{4}He and the other a dilute ^{3}He mixture, are condensed in two chambers separated by a Vycor glass superleak[120] acting as a semipermeable membrane—the structure of Vycor glass is exceedingly open, presenting a maze of channels ~ 40 to 60 Å in diameter through which superfluid ^{4}He readily flows but ^{3}He, restricted by its relatively large viscosity, does not. When the thermodynamic pressures and temperatures are equal in both chambers, the condition for thermodynamic equilibrium between the chambers can be expressed through the equality of the ^{4}He chemical potential μ_4 on the two sides of the superleak:

$$\mu_4(p, T, x) = \mu_4(p - \pi, T, 0) \tag{33}$$

From this a 'working definition' of π may be deduced as

$$\pi(p, T, x) V_4(p, T) = \mu_4(p, T, 0) - \mu_4(p, T, x) \tag{34}$$

where V_4 is the molar volume of pure ^{4}He.

Values of $\pi(p, T, x)$ are obtained by determining the pressure difference between the two chambers under various conditions of p, T, and x. Previous workers[72, 114] had measured π by balancing this pressure with the thermomechanical pressure, or fountain pressure, $p_f = \int_{T_1}^{T_2} S \, dT$, produced by maintaining the superfluid ^{4}He at different temperatures, T_1 and T_2, in the two chambers. A more practical method consists of direct measurement of π, such as reported by Wilson and Tough[121], using a commercial strain gage, the voltage output of which is proportional to the pressure (this same technique had been used earlier by Hammel and Keller[43] to measure directly the thermomechanical effect in pure ^{4}He). The apparatus in *Figure 8*(d) illustrates a considerable advance in sensitivity of the π measurement, again involving capacitance measurements between a fixed and a variable electrode. The latter consists of a 25 μm-thick gold-plated (one side) Mylar sheet stretched between and attached to two blocks of epoxy resin forming the sample chambers (the epoxy resin is mixed with silica as a filler so as to match the thermal contraction of the Mylar). As shown in the drawing the epoxy blocks are domed about the diaphragm; affixed to the upper dome is another section of gold-plated Mylar sheet, also 25 μm thick, which serves as the fixed electrode. Capacitance observations on this system resulted in differential pressure measurements having a sensitivity of 10^{-1} N/m^2. However, this instrument requires calibration against a known pressure for which Landau *et al.*[66] chose the absolute π measurements made by Wilson *et al.*[121] at 0.65 K using the strain gage mentioned above. Further corrections were required to compensate for the change in ε with x and for changes in geometry due to pressure and temperature variations.

VII. References

[1] Abel, W. R., A. C. Anderson, W. C. Black and J. C. Wheatley. *Phys. Rev. Letters*, **16**, 273 (1966).
[2] Abel, W. R., A. C. Anderson, W. C. Black and J. C. Wheatley. *Phys. Rev.* **147**, 111 (1966).
[3] Abraham, B. M., UTS*, p 37.
[4] Ahlers, G., *Phys. Rev.* **171**, 275 (1968).
[5] Ahlers, G., *Phys. Rev.* **A3**, 696 (1971).

6. Anderson, A. C., UTS*, p 82.
7. Anufriev, Yu. D., *Zh. Eksp. Teor. Fiz. Pis'ma*, **1**, 155 (1965); English translation, *JETP Letters*, **1**, 155 (1965).
8. Bailey, A. C. (Ed.), *Advanced Cryogenics*, Plenum: New York (1970).
9. Baker, J. M., B. Bleaney, P. M. Llewellyn and P. F. D. Shaw. *Proc. Phys. Soc.* (*London*), **69A**, 353 (1956).
10. Barber, C. R., *Fifth Symposium on Temperature†*, p 99.
11. Bedford, R. E., M. Durieux, R. Muijlwijk and C. R. Barber. *Metrologia*, **5**, 47 (1969).
12. Bedford, R. E., H. Preston-Thomas, M. Durieux and R. Muijlwijk. *Metrologia*, **5**, 45 (1969).
13. Berglund, P. M., G. J. Ehnholm, R. G. Gylling, O. V. Lounasmaa and R. P. Søvik. UTS*, p 113; also in *Proceedings of the Third International Cryogenic Engineering Conference* (*Berlin*), p 473. Iliffe: Guildford, UK (1970).
14. Berman, R. and C. F. Mate. *Phil. Mag., Ser. III*, **29**, 461 (1958).
15. Berman, R. and J. Poulter. *Phil. Mag.* **43**, 1047 (1952).
16. Berry, K. H., *Fifth Symposium on Temperature†*, p 323.
17. Boghosian, C. and H. Meyer. *Phys. Rev.* **152**, 200 (1966).
18. Boghosian, C., H. Meyer and J. E. Rives. *Phys. Rev.* **146**, 110 (1966).
19. Buckingham, M. J. and W. M. Fairbank, in *Progress in Low Temperature Physics*, Vol. III (Gorter, C. J., Ed.), p 80. North-Holland: Amsterdam (1961).
20. Cataland, G. and H. H. Plumb. *Fifth Symposium on Temperature†*, p. 183.
21. Cetas, T. C. and C. A. Swenson. *Phys. Rev. Letters*, **25**, 338 (1970); *Fifth Symposium on Temperature†*, p 57.
22. Chapellier, M., M. Goldman, V. H. Chau and A. Abragam. *CR Acad. Sci., Paris*, **268**, 1530 (1969).
23. Colclough, A. R., *Fifth Symposium on Temperature†*, p 365.
24. Croft, A. J., *Cryogenic Laboratory Equipment*, Plenum: New York (1969).
25. Dana, L. I. and H. K. Onnes. *Leiden Commun.* **179c** (1926).
26. Das, P., R. De Bruyn Ouboter and K. W. Taconis. *Low Temperature Physics LT9*, Part B, p 1253. Plenum: New York (1965).
27. Dash, J. G., R. D. Taylor, D. E. Nagle, P. P. Craig and W. M. Visscher. *Phys. Rev.* **122**, 1116 (1961).
28. Dearnaley, G., *Contemp. Phys.* **8**, 607 (1967).
29. De Bruyn Ouboter, R., K. W. Taconis, C. LePair and J. J. M. Beenakker. *Physica*, **26**, 853 (1960).
30. deGroot, S. R., H. A. Tolhoek and W. J. Huiskamp, in *Alpha-, Beta-, and Gamma-Ray Spectroscopy*, Vol. II, p 1199. (Siegbahn, K., Ed.), North-Holland: Amsterdam (1965).
31. Durieux, M., in *Progress in Low Temperature Physics*, Vol. VI, p 405. (Gorter, C. J., Ed.), North-Holland: Amsterdam (1970).
32. Edwards, D. O., E. M. Ifft and R. E. Sarwinski. *Phys. Rev.* **177**, 380 (1969).
33. Edwards, D. O., R. E. Sarwinski, P. Seligman and J. T. Tough. *Cryogenics*, **8**, 392 (1968).
34. Ehnholm, G. J. and R. G. Gylling. *Cryogenics*, **11**, 39 (1971).
35. Goldstein, L., *Phys. Rev.* **159**, 120 (1967); **188**, 349 (1969).
36. Golovashkin, A. N. and G. P. Motulevich. *Cryogenics*, **3**, 167 (1963).
37. Goodkind, J. M. and D. L. Stolfa, *Rev. Sci. Instrum.* **41**, 799 (1970).
38. Goodwin, R. D., *J. Res. Nat. Bur. Stand.* **65C**, 231 (1961).
39. Graf, E. H., D. M. Lee and J. D. Reppy. *Phys. Rev. Letters*, **19**, 417 (1967).
40. Grilly, E. R., *J. Low Temp. Phys.* **4**, 615 (1971).
41. Grilly, E. R. and R. L. Mills. *Phys. Rev.* **105**, 1140 (1957); *Ann. Phys.* (*NY*), **8**, 1 (1959).
42. Halverson, G. and D. A. Johns. *Fifth Symposium on Temperature†*, p 803.
43. Hammel, E. F. and W. E. Keller. *Phys. Rev.* **124**, 1641 (1961).
44. Hirschkoff, E. C., O. G. Symko and J. C. Wheatley. *J. Low Temp. Phys.* **2**, 653 (1970).
45. Hirschkoff, E. C., O. G. Symko and J. C. Wheatley. *J. Low Temp. Phys.* **4**, 111 (1971).
46. Hoare, F. E., L. C. Jackson and N. Kurti. *Experimental Cryophysics*, Butterworths: London (1961).
47. Hoffer, J. K., *Thesis* (UCRL-18319), University of California, Berkeley (1968).
48. Hudson, R. P. and R. S. Kaeser. *Physics*, **3**, 95 (1967).
49. International Practical Temperature Scale of 1968, *Metrologia*, **5**, 35 (1969).
50. Johnston, H. L., W. E. Keller and A. S. Friedman. *J. Amer. Chem. Soc.* **76**, 1482 (1954).
51. Johnson, R. T., O. V. Lounasmaa, R. Rosenbaum, O. G. Symko and J. C. Wheatley. *J. Low Temp. Phys.* **2**, 403 (1970).

[52] Johnson, R. T. and J. C. Wheatley. *J. Low Temp. Phys.* **2**, 423 (1970).
[53] Josephson, B. D., *Phys. Letters*, **1**, 251 (1962).
[54] Kalvins, G. M., T. E. Katila and O. V. Lounasmaa, in *Mössbauer Effect Methodology*, Vol. 5, pp 231–261. (Gruverman, I. J., Ed.), Plenum: New York (1970).
[55] Kamper, R. A., *Fifth Symposium on Temperature*†, p 349.
[56] Kamper, R. A., J. D. Siegwarth, R. Radebaugh and J. E. Zimmerman. *Proc. I.E.E.E.*, **59**, 1368 (1971).
[57] Kamper, R. A. and J. E. Zimmerman. *J. Appl. Phys.* **42**, 132 (1971).
[58] Keller, W. E., *Helium-3 and Helium-4*, Plenum: New York (1969).
[59] Keller, W. E., *Phys. Rev.* **97**, 1 (1955); **98**, 1571 (1955); **100**, 1790 (1955).
[60] Kellers, C. F., *Thesis*, Duke University (1960).
[61] Kerr, E. C. and R. H. Sherman. *J. Low Temp. Phys.* **3**, 451 (1970).
[62] Kerr, E. C. and R. D. Taylor. *Ann. Phys. (N.Y.)*, **20**, 450 (1962).
[63] Kerr, E. C. and R. D. Taylor. *Ann. Phys. (N.Y.)*, **26**, 292 (1964).
[64] Kittel, C., *Introduction to Solid State Physics*, 4th ed., p 579. Wiley: New York (1971).
[65] Kurti, N., F. N. H. Robinson, F. E. Simon and D. A. Spohr. *Nature (London)*, **178**, 450 (1956).
[66] Landau, J., J. T. Tough, N. R. Brubaker and D. O. Edwards. *Phys. Rev.* **2A**, 2472 (1970); *Rev Sci. Instrum.* **41**, 444 (1970).
[67] Lawless, W. N., *Rev. Sci. Instrum.* **42**, 561 (1971); *Fifth Symposium on Temperature*†, p 1143.
[68] Lawless, W. N., R. Radebaugh and R. J. Soulen. *Rev. Sci. Instrum.* **42**, 567 (1971).
[69] Lee, D. M., H. A. Fairbank and E. J. Walker. *Phys. Rev.* **121**, 1258 (1961).
[70] Liebenberg, D. H. and L. D. F. Allen. *J. Appl. Phys.* **41**, 4050 (1970); *Fifth Symposium on Temperature*†, p 875.
[71] London, H., *Proceedings of the International Conference on Low Temperature Physics*, p 157. Oxford, August 1951.
[72] London, H., G. R. Clarke and E. Mendoza. *Phys. Rev.* **128**, 1992 (1962).
[73] Mills, R. L. and E. R. Grilly. *Phys. Rev.* **99**, 480 (1955); **101**, 1246 (1956).
[74] Mills, R. L., E. R. Grilly and S. G. Sydoriak. *Ann. Phys. (N.Y.)*, **12**, 41 (1961).
[75] Mitsui, K., H. Sakurai and T. Mochizuki. *Fifth Symposium on Temperature*†, p 333.
[76] Mota, A. C., R. P. Platzeck, R. Rapp and J. C. Wheatley. *Phys. Rev.* **177**, 266 (1969).
[77] Neganov, B. S., N. Borisov and M. J. Liburg. *Zh. Eksp. Teor. Fiz.* **50**, 1445 (1966); English translation: *Soviet Phys. JETP*, **23**, 959 (1966). See also, Neganov, B. S., Institute for Nuclear Problems, Dubna, *Report P13*, 4014 (1968) (in Russian).
[78] Neiler, J. H. and P. R. Bell, in *Alpha-, Beta-, and Gamma-Ray Spectroscopy*, Vol. I, p 245. (Siegbahn, K., Ed.), North-Holland: Amsterdam (1965).
[79] Nyquist, H., *Phys. Rev.* **32**, 110 (1928).
[80] Onnes, H. K. and W. van Gulik. *Leiden Commun.* **184A** (1926).
[81] Plumb, H. H. and G. Cataland, *Metrologia*, **2**, 127 (1966).
[82] Plumb, H. H. and G. Cataland, to be published.
[83] Pomeranchuk, I., *Zh. Eksp. Teor. Fiz.* **20**, 919 (1950).
[84] Rogers, J. S., R. J. Tainsh, M. S. Anderson and C. A. Swenson. *Metrologia*, **4**, 47 (1968).
[85] Rose-Innes, A. C., *Low Temperature Techniques*, English Universities Press: London (1964).
[86] Roubeau, P., D. Le Fur and E. J. A. Varoquaux, in *Proceedings of the Third International Cryogenic Engineering Conference*, Berlin, p 315. Iliffe: Guildford, UK (1970).
[87] Roubeau, P. and E. J. A. Varoquaux. *Cryogenics*, **10**, 255 (1970).
[88] Rubin, L. G. and W. N. Lawless. *Rev. Sci. Instrum.* **42**, 571 (1971).
[89] Sarwinski, R. E., UTS*, p 48.
[90] Schooley, J. F. and R. J. Soulen. *Fifth Symposium on Temperature*†, p 169.
[91] Scott, R. B. (Ed.), *Technology and Uses of Liquid Hydrogen*, Macmillan: New York (1964).
[92] Scribner, R. A., M. F. Panczyk and E. D. Adams. *J. Low Temp. Phys.* **1**, 313 (1969).
[93] Sherman, R. H., *Proceedings of the Tenth International Conference on Low Temperature Physics*, Vol. I, p 188. (Malkov, M. P., Ed.), VENETE: Moscow (1967).
[94] Sites, J. R., D. D. Osheroff, R. C. Richardson and D. M. Lee. *Phys. Rev. Letters*, **23**, 836 (1969).
[95] Sites, J. R., H. A. Smith and W. A. Steyert. *J. Low Temp. Phys.* **4**, 605 (1971).
[96] Shore, F. J., V. L. Sailor, H. Marshak and C. A. Reynolds. *Rev. Sci. Instrum.* **31**, 970 (1960).
[97] Sommers, H. S., W. E. Keller and J. G. Dash. *Phys. Rev.* **92**, 1345 (1953).
[98] Steyert, W. A., *Rev. Sci. Instrum.* **38**, 964 (1967).
[99] Steyert, W. A., UTS*, p 90.
[100] Steyert, W. A., private communication.

101. Stolfa, D. L. and J. M. Goodkind. UTS*, p 120.
102. Suomi, M., A. C. Anderson and B. Holmstrom. *Physica*, **38**, 67 (1968).
103. Swenson, C. A., *CRC Crit. Rev. Solid State Sciences*, **1**, 99 (1970).
104. Sydoriak, S. G. and T. R. Roberts, *Phys. Rev.* **118**, 901 (1960).
105. Symko, O. G., *J. Low Temp. Phys.* **1**, 451 (1969).
106. Taylor, R. D., *Fifth Symposium on Temperature*, p. 1259.
107. Ter Harmsel, H., *Thesis*, Leiden (1966).
108. Ter Harmsel, H., H. Van Dijk and M. Durieux. *Physica*, **33**, 503 (1967).
109. Tiggelman, J. L. and M. Durieux. *Fifth Symposium on Temperature*†, p 849.
110. van Rijn, C. and M. Durieux. *Fifth Symposium on Temperature*†, p 73.
111. Wagner, R. R. and B. Bertman. Paper T-2 presented at Commission I, International Institute of Refrigeration, Tokyo (1970).
112. Wallace, B. and H. Meyer. *Phys. Rev.* **A2**, 1563 (1970).
113. Walstedt, R. E., E. L. Hahn, C. Froidevaux and E. Geissler. *Proc. Roy. Soc.* A, **284**, 499 (1965).
114. Wansink, D. H. N. and K. E. Taconis. *Physica*, **23**, 125 (1957).
115. Wheatley, J. C., R. E. Rapp and R. T. Johnson. *J. Low Temp. Phys.* **4**, 1 (1971).
116. Wheatley, J. C., D. E. Vilches and W. R. Abel. *Physics*, **4**, 1 (1968).
117. Wiebes, J., *Thesis*, Leiden (1969).
118. Wilks, J., *Liquid and Solid Helium*, Clarendon Press: Oxford (1967).
119. Wilson, M. F., D. O. Edwards and J. T. Tough. *Phys. Rev. Letters*, **19**, 1368 (1967).
120. Wilson, M. F., D. O. Edwards and J. T. Tough. *Rev. Sci. Instrum.*, **39**, 134 (1968).
121. Wilson, M. F. and J. T. Tough. *Phys. Rev.* **1A**, 914 (1970).
122. Zimmerman, J. E., P. Thiene and J. T. Harding. *J. Appl. Phys.* **41**, 1572 (1970).
123. Zinov'eva, K. N., as reported at the Grenoble Helium Conference (March 1970).

* *Proceedings* of the 1970 Ultralow Temperature Symposium, Naval Research Laboratory, Washington, (23–24 April 1970), *NRL Report 7133*.
† *Fifth Symposium on Temperature*, Washington, (21–24 June, 1971), published as *Temperature, Its Measurement and Control in Science and Industry*, Vol. IV (H. H. Plumb, ed.). Instrument Society of America: Pittsburgh (1972).

AUTHOR INDEX

Abbiss, C. P., 605, 684, 722
Abbott, L. H., 140, 145, 201, 464, 467, 474, 498, 832
Abbott, M. M., 343, 346
Abegg, R., 847, 898, 899
Abel, W. R., 1242, 1243, 1246, 1256, 1263, 1266
Abragam, A., 1264
Abraham, B. M., 574, 1263
Abramowitz, S., 1185, 1189
Abrams, S. T., 832
Achundov, T. S., 693, 723
Ackermann, Th., 429, 496
Adair, T. W., 576
Adam, N. K., 448, 449, 494, 652
Adamek, J., 652
Adams, E. D., 1237, 1238, 1261, 1265
Adams, J. C., 270, 997, 998, 1020
Adams, L. H., 438, 443, 447, 472, 494, 499, 654, 878, 898
Adams, W. A., 461, 462, 480, 494, 495
Adler, P. N., 142, 144
Admiraal, L. J., 832
Affsprung, H. E., 782, 784
Ahlers, G., 675, 695, 721, 1254, 1255, 1263
Aimé, G., 446, 447, 495
Ainsworth, J. E., 270
Akaishi, K., 297
Akhmatova, I. A., 973
Akimoto, S., 830
Akiyama, Y., 247, 248, 270
Aklerlof, G., 746
Alabyshev, A. F., 903, 950
Alani, G. H., 70, 623, 654
Alcock, A. J., 1183, 1189
Alcock, C. B., 654, 748
Alder, B. J., 747, 1141, 1152, 1154, 1155–1159
Alekhin, A. D., 715, 721
Alekseev, V. A., 984, 985, 990
Alexandrov, A. A., 395–397, 418
Alexejev, W., 746
Alexeyev, K. A., 211
Alieva, P. Z., 830
Allen, G., 720, 769, 770, 784
Allen, L. D. F., 1240, 1265
Allen, P. W., 652
Alley, C. O., 721
Allnutt, L. A., 1185, 1191
Almer, H. E., 168, 200
Alonso, J. I. F., 319
Alpert, D., 279, 283, 290, 297
Altman, A., 372, 380
Altman, R. L., 625, 638, 653

Al'tshuler, L. V., 1150, 1151, 1157
Altunin, V. V., 484, 500
Alwani, Z., 474, 495, 793, 795, 798, 800
Amadori, M., 867
Amagat, E. H., 150, 200, 358, 369, 371–373, 377, 451, 455–457, 495, 821, 830
Amakasu, F., 1067
Ambrose, D., 607, 623, 631, 634, 637, 652
Ames, A. E., 1188, 1189
Amey, R. L., 597, 600, 605
Amirkhanov, Kh. I., 695, 721
Ampola, G., 867
Amsel, G., 1189
Anderson, A., 577
Anderson, A. C., 1243, 1256, 1263, 1264, 1266
Anderson, H. U., 297
Anderson, J. R., 260, 270
Anderson, L. K., 1189
Anderson, M. S., 1265
Anderson, R. L., 130, 201, 270, 346
Andon, R. J. L., 345
Andreev, S. I., 1184, 1189
Andrews, L. E., 211
Andrews, T., 618, 652
Angus, S., 417
Anhorn, V. J., 1066
Anufriev, Yu. D., 1247, 1264
Aoshima, Y., 1184, 1190
Apgar, E., 297
Appelt, G., 297
Ardon, M., 870
Arens, P. L., 830
Armarego, W. L. F., 748
Armstrong, R. A., 299, 417, 1037, 1038, 1066
Arnaldi, F., 867
Aronson, M. H., 200
Arpshofen, T., 816, 832
Arrhenius, S., 839, 840, 898
Artyukhovskaya, L. M., 688, 721
Asamoto, R. R., 111
Ascah, R. G., 326, 346
Ashby, H. F., 112
Ashton, J. T., 746
Ashurly, Z. I., 1183, 1190
Ashworth, A. J., 778, 779, 784
Asinovskii, E. I., 494, 497
Assadulaieva, N. N., 693, 723
Aston, J. G., 326, 346, 655, 1024, 1066
Astrov, D. N., 345
Atanov, Yu. A., 203, 211, 397, 416
Aubry, B., 241, 243, 270
Austin, A. L., 1177, 1191

1267

AUTHOR INDEX

Austin, W. E., 261, 270, 654
Auwers, K., 867
Avellone, J. C., 1188, 1189
Ayers, J. D., 1177, 1189
Aylmore, D. W., 1065
Aziz, R. A., 551, 574, 575

Babayan, V. G., 1183, 1190
Babb, S. E., 144, 357, 369, 370, 372, 373, 377, 380, 391–393, 397, 399, 411, 412, 416, 418, 472, 482, 495, 823, 830
Babini, V., 867
Bacarella, A. L., 950
Back, R. A., 640, 652
Baconin, J., 1157, 1158
Baehr, H. D., 721
Baer, S., 746
Baes, C. F., 950, 951
Bagatskii, M. I., 695, 721
Bagley, C. H., 1187, 1190
Bailey, A. C., 1220, 1264
Bakanova, A. A., 1150, 1151, 1157
Baker, E. H., 438, 495
Baker, J. M., 1217, 1225, 1264
Bakhvalova, V. V., 211
Bakker, G., 994, 1020
Bakr, A. M., 785
Balchan, A. S., 144, 830
Baldini, G., 605
Baldwin, R. R., 746
Bale, H. D., 721
Baliga, B. T., 450, 495
Ball, A. F., 130, 831, 987, 988, 990
Balson, E. W., 652
Baly, E. C. C., 605
Balzarini, D., 689, 724
Bamberger, C. E., 950
Bannard, J. E., 438, 495
Banzhof, M., 767, 784
Barber, C. R., 111, 345, 652, 830, 1231–1233, 1264
Barieau, R. E., 371, 378, 721
Barker, J. A., 376, 378, 413, 416, 417, 568, 574, 773, 774, 784, 1157
Barmatz, M., 566, 574, 696, 718, 721, 722
Barnes, G., 297
Barnes, V. M., 1216, 1217
Barnes, W. C., 1177, 1189
Barnett, J. D., 138–145, 830
Baronetsky, E., 297
Barr, W. E., 1066
Barrall, E. M., 830
Barrell, H., 314, 319, 320
Barrer, R. M., 416, 1054, 1065
Barrett, C. S., 435, 495
Barriault, R. J., 357, 371, 377, 416
Bartholomew, N., 877

Bartholomew, R. F., 950
Bartlett, E. P., 357, 369–372, 377, 416
Bartocci, V., 952
Barton, A. F. M., 438, 442, 491, 495
Barton, J. L., 631, 652
Barton, R. S., 270
Bartsch, W., 867
Barus, C., 451, 455, 490, 495, 821, 830
Barz, A., 297
Bashforth, F., 270, 997, 998, 1020
Basin, A. C., 973
Basset, J. (2), 371, 372, 377, 398, 399, 416, 473, 495
Bassett, W. A., 140–142, 144, 145, 830
Bastide, J. P., 143, 144
Bates, R. G., 924, 950
Battino, R., 70, 652, 740, 746, 755, 767, 784
Batuecas, T., 319, 320
Bauer, H. J., 536, 574
Bauer, N., 422, 484, 495
Baughan, E. C., 867
Baxendale, J. H., 768–770, 784
Baxter, G. P., 647, 652
Bayard, R. T., 279, 283, 290, 297
Baylis, W., 1217
Beaglehole, D., 603, 605
Beams, J. W., 260, 270, 440, 485, 487, 496
Beattie, J. A., 70, 116, 130, 345, 357, 369–371, 373, 377, 378, 416, 418, 477, 478, 495, 503, 525, 620, 622, 623, 630, 652
Beaumont, D. W., 879, 880, 882, 899
Bechtler, A., 1107
Bechtold, M. F., 652
Beck, A. H., 297
Becker, B., 867
Becker, E. J., 297
Becker, J. A., 297
Becker, M., 1108
Becker, R., 1157
Beckett, C. W., 1161, 1185, 1189, 1190, 1217
Beckmann, E., 867
Bedford, R. E., 90, 111, 1231, 1232, 1264
Bedford, R. G., 784
Bedfort, J. G., 840, 859, 867, 898
Beebe, C. W., 747
Beebe, R. A., 1062, 1063, 1065
Beenakker, J. J. M., 338, 344–346, 1264
Behl, W. K., 950
Beitel, G. A., 297, 299
Beitz, J. V., 1107
Bekkedahl, N., 833
Belkin, V. M., 1184, 1189
Bell, J. M., 746
Bell, P. M., 93, 94, 112, 388, 417
Bell, P. R., 1225, 1265
Bell, T. N., 652, 767, 784
Bell, W. E., 655
Belova, M. M., 974

AUTHOR INDEX

Belton, J. W., 1020
Benaglia, C., 877
Bender, E., 517, 525
Bender, J., 976, 990
Bender, J. A., 748
Bendiner, W., 690, 721
Benedek, G. B., 70, 715, 717, 718, 721–723
Benedict, M., 130, 345, 363, 372, 378, 394, 417, 478, 484, 495, 503, 504, 525, 630, 652
Benedict, R. P., 112
Benedict, W., 1217
Benedictus, J., 867
Benjamin, L., 746
Ben-Naim, A., 746
Bennett, C. O., 201, 371, 378
Bennett, F. D., 1185, 1186, 1189
Bennett, G., 867
Bennett, G. M., 746
Bennett, M. J., 653
Bennewitz, H. G., 297
Benoît, J. R., 304, 320
Benson, B. B., 746, 747
Benson, G. C., 612, 655
Benson, H. K., 841, 899
Ben-Yosef, N., 1186, 1189
Berezin, B. Ya., 973, 1189
Berg, W. T., 1067
Berga, J., 859, 867, 899
Bergé, P., 717, 721
Berger, E. V., 746
Berger, J., 1148, 1157
Berglund, P. M., 1245, 1264
Bergmann, K., 1107
Bergmann, L., 549, 574
Bergeon, R., 406, 417
Bering, B. P., 1045, 1046, 1066
Berkeley, Earl of, 445, 495
Berkowitz-Mattuck, Joan B., 651, 653
Berman, H. A., 1190
Berman, R., 1256, 1264
Berry, K. H., 1233, 1264
Berry, R. J., 345
Berstein, J., 867
Bertman, B., 1266
Bethe, H. A., 1120, 1122, 1158
Bett, K. E., 70, 130, 144, 201, 434, 495
Betts, J., 640, 652
Beyer, R. T., 549, 574
Beynon, J. D. E., 651, 653
Bhat, B. K., 1186, 1189
Bhatia, A. B., 539, 541, 574
Bhatt, L. A., 370, 381
Biddiscombe, D. P., 612, 654
Bienaimé, R., 112
Bigg, P. H., 130, 324, 345
Biles, M. B., 211
Billett, F., 735, 736, 748
Bills, G., 301

Birch, F., 92, 112, 417, 830, 976, 990
Bird, G. R., 1188, 1189
Bird, R. B., 70, 270, 417, 560, 574, 575, 1158
Birdsall, C. M., 654
Bishop, A. E., 1183, 1189
Bitchkou, V. P., 1021
Björck, Äke., 525
Black, C., 746
Black, W. C., 1256, 1263
Blackburn, J. H., 1156, 1158
Blackburn, D. A., 90, 112, 417
Blackburn, G. F., 112
Blagoy, Yu. P., 466, 469, 495, 568, 570, 571, 575, 576, 688, 721
Blaisdell, B. E., 70, 116, 130, 236, 243, 270, 345, 630, 652
Blaisse, B., 369, 379, 482, 485, 498, 562, 576, 691, 692, 723, 832
Blakey, J. P., 832
Blancett, A. L., 374, 378
Blandamer, M. J., 575
Blander, M., 898, 950
Blank, G., 70
Blanke, M., 870
Blankenship, F. F., 950, 952
Blankenstein, F., 304, 320
Bleaney, B., 1225, 1264
Blears, J., 277, 297
Blend, H., 653
Bless, S. J., 1187, 1189
Blevin, W. R., 89, 112
Blitz, J., 549, 575
Bloch, D., 94, 112
Bloom, G. H., 1150, 1158
Bloom, H., 631, 652
Blosser, L. G., 112
Blumendal, H. B., 492, 498
Bobrowsky, A., 499
Bockris, J. O'M., 500, 654, 748, 833, 950–952
Boelhouwer, J. W. M., 466, 467, 495
Boeseken, J., 867
Bogan, M., 767, 784
Bogatyrenko, B. B., 998–1000, 1020
Bogdanov, V. S., 140, 144, 211
Boghosian, C., 1264
Bogomolov, V. M., 1151, 1159
Bogue, R. H., 830
Bohdansky, J., 973
Bohon, R. L., 746
Boiko, N. V., 485, 495
Boks, J. D. A., 357, 370, 378, 379
Boksha, S. S., 830
Bombi, G. G., 950, 952
Bonhoure, J., 130, 345
Bonilla, C. F., 70, 381, 973, 985, 986, 989, 990, 1191
Bonnell, D. W., 1190
Bonner, O. D., 867

Bonse, U., 305, 320
Booker, S. R., 1184, 1189
Boon, J. P., 535, 575
Booth, H. S., 653
Boren, M. D., 144, 397, 417
Borisoglebski, V. P., 493, 499
Borisov, N., 1244, 1265
Borodina, M. D., 1215, 1218
Borovick-Romanov, A. C., 345
Boschi, L. A., 1065
Bosco, C. D., 139, 144, 146
Bostanjoglo, O., 605
Böttcher, C. J. F., 582, 583, 588, 605
Bottomley, G. A., 324, 338, 340–342, 345, 784
Bottoms, P. J., 1183, 1192
Botzen, A., 582, 585, 600, 605, 1208, 1218
Bourassa, R. R., 90, 112, 417
Bourdon, E., 149, 201
Bourlange, C., 877
Boutaric, A., 868, 898
Boutry, G. A., 260, 262–264, 270
Bowen, N. L., 830
Bowman, D. H., 551, 574, 575
Bowman, A. H., 154, 162, 201
Bowman, H. A., 170, 201
Boyd, F. R., 93, 94, 112, 141, 142, 145, 800, 830
Boyd, M. E., 575
Boyd, R. J., 388, 417
Braak, C., 357, 371, 379
Bradfield, G., 435, 495
Bradley, D., 440, 466, 469, 500
Bradley, R. S., 418, 830, 1065, 1158
Bradshaw, A., 495
Brandes, R. G., 297
Brandt, L. W., 380
Brandt, N. B., 145
Braunstein, Helen, 901, 950
Braunstein, J., 883, 898, 901, 950–952
Bray, U. B., 748
Brayford, J. R., 867
Brazhnik, M. J., 1150, 1157
Brechlin, A., 870
Bredig, M. A., 831, 877, 878, 899, 950
Breene, R., 1212, 1217
Breitschwerdt, K. G., 1107
Breuer, D., 1185, 1191
Breusov, O. N., 1158
Brewer, D. F., 1251
Brewer, J., 344, 345
Brickwedde, F. G., 84, 831–833
Bridgeman, O. C., 503, 525
Bridgman, P. A., 130
Bridgman, P. W., 91, 95, 112, 135, 136, 140–143, 145, 152, 159, 161, 166, 201, 206, 211, 361, 369, 371, 372, 378, 434, 437, 441, 442, 447, 449, 456, 463–466, 470–476, 495, 496, 523, 525, 821, 822, 830, 976, 990
Brielles, J., 94, 96, 112, 113, 347, 388, 417
Brierley, J. S., 357, 371, 377, 416
Briggs, L. J., 488, 496
Briggs, T. C., 371, 378
Brilka, R., 1186, 1191
Brindley, S. R., 748
Brisbane, A. D., 297
Brixner, B., 1188, 1189
Broadbent, T. E., 1183, 1189
Broderick, B. E., 623, 652
Brombacher, W. G., 130, 274, 297
Brønsted, I. J. N., 770, 784
Bronstein, H. R., 950, 951
Brose, M., 492, 496
Brown, G. G., 373, 378
Brown, G. R., 695, 696, 698, 721
Brown, I., 765, 774, 784
Brown, J. R. C., 1039, 1067
Brown, P. G., 880, 882, 898
Brown, R. C., 1003, 1020
Brown, W. F., 585, 587, 605
Brubaker, N. R., 1238, 1263, 1265
Bruch, L. W., 558, 575
Bruneaux, A. M., 950, 951
Bruni, G., 867
Brunner, E., 71
Bruwell, R. L., 653
Bryan, G. J., 1216, 1217
Bryce, H. G., 867
Buback, M., 793, 796, 798, 800
Büchler, A., 651, 653
Buchmann, E., 496
Buck, U., 973
Buckingham, A. D., 585, 586, 605
Buckingham, M. J., 683, 695–698, 721–723, 1254, 1264
Buimova, I. P., 141, 143, 146, 167, 201
Buisson, J. M., 304, 320
Bunch, M. D., 112
Bundy, F. P., 93, 97, 104–107, 110, 112, 143–145, 394, 417, 497, 830–833
Bunge, O., 271
Bunger, W. B., 748
Bunn, C. W., 474, 498
Bunsen, R., 821, 830
Bunzl, K. W., 867
Burden, R. S., 270
Burdina, K. P., 141, 143, 146, 167, 201
Burdon, R. S., 999, 1020
Bureau, A. J., 297
Burgess, D., 1208, 1212, 1217
Burgess, G. K., 830
Burgess, H. A., 630, 655
Buritz, R. S., 297
Burkat, R., 488, 498
Burks, H., 372, 379
Burlew, J. S., 428, 496
Burnett, E. S., 335, 345, 367, 369, 378

AUTHOR INDEX

Burnham, C. W., 392–394, 399, 409, 417, 437, 438, 477, 482, 496
Burns, G. W., 111, 112, 831
Burova, L. L., 211
Burrows, G., 746
Burton, C. V., 445, 495
Bury, G. R., 867
Bury, P., 387, 417, 418
Busey, R. H., 130
Bushuk, W., 1054, 1065
Butcher, E. G., 370, 374, 378
Butko, A. E., 570, 571, 575, 576
Butuzov, V. P., 830
Byard, D. L., 1188, 1189
Bye, J., 894, 895, 898
Byl, A., 563, 576
Byrne, M. A., 338, 345

Cady, H. P., 746
Cahill, A., 951
Cahn, L., 1059, 1065
Cailletet, L., 437, 446, 496, 676, 721, 987, 988, 990
Cairns, R. B., 653
Caldin, E. F., 1107
Caldwell, F. R., 112
Calingaert, G., 653
Calmettes, P., 717, 721
Calzolari, F., 867
Campbell, A. N., 867
Canfield, F. B., 366, 371, 373, 374, 378
Canjar, L. N., 371, 378
Cannell, D. S., 715, 718, 721, 723
Canton, J., 430, 450, 496
Cantor, M., 1006, 1020
Cantor, S., 877
Carasso, J. I., 655
Carley Jr., C. T., 271, 300
Carlson, K. D., 653
Carlton, T. S., 877
Carmichael, J. H., 297
Carnaham, N. F., 375, 378
Carnavale, E. H., 569, 575
Carnazzi, P., 456, 496
Carome, E. F., 718, 722
Carpenter, H. C. H., 817, 830
Carpenter, J. P., 1183, 1192
Carr, H. Y., 703, 721
Carter, G., 297, 299
Carter, W. J., 832
Casado, F. L., 319, 345
Cassidy, E. C., 1184, 1185, 1189
Castellan, G. W., 1108
Cataland, G., 552, 557, 575, 576, 1233, 1264, 1265
Cater, E. D., 653
Caton Jr., R. D., 951
Catto, K. A., 831

Caubergh, M., 723
Cavallaro, L., 861, 898
Caw, W. A., 201
Caywood, L. P., 112
Cavero, B., 452, 496
Cedrone, N. P., 575
Cescon, P., 952
Céspiro, Z., 258, 270
Cessac, J.-J., 898
Cetas, T. C., 1232, 1233, 1264
Cezairliyan, Ared, 1161, 1189, 1190, 1217
Chace, W. G., 1177, 1178, 1185, 1186, 1190, 1191
Chaisse, F., 94, 112
Chalin, R., 898
Chalyi, A. V., 715, 721
Chance, B., 1108
Chandler, T. R. D., 111, 113
Chaney, N. K., 1190
Chaney, P. E., 823, 830
Chang, R. F., 721
Changeux, P., 1099, 1108
Chao, K. C., 778, 786
Chapellier, M., 1264
Chapman, J. A., 589, 591, 596, 605
Chapman, S., 575
Chappuis, P., 55, 70, 304, 319
Charatis, G., 1217
Charlton, T., 722
Chase, C. E., 566, 577, 690, 721, 724
Chashkin, Yu. R., 695, 724
Chatel, S., 998, 1020
Chattle, M. V., 130
Chau, V. H., 1264
Chaudhuri, A. K., 1190
Check, C. H., 748
Chekhovskoi, V. Ya., 973, 1189
Chen, S. H., 574, 575
Ch'en, S. Y., 1202, 1212, 1218
Chen, T., 211
Chen, W. T., 367, 381, 656, 722
Chen, Y. G., 1183, 1191
Chnébault, P., 1040, 1065
Cheng, P. Y., 488, 496
Cherney, B. J., 373, 378
Chester, P. F., 830
Chiang, Y., 833
Chick, B., 549, 577
Chihara, H., 1024, 1065
Chikalla, T. D., 656
Chiotti, P., 950
Cho, S. J., 488, 489, 500
Choumoff, P. S., 270, 299
Chris, M. D., 84
Christian, R. G., 260, 261, 270, 297
Christian, S. D., 782, 784
Chu, B., 715, 717, 721
Chubb, J. N., 270
Chun, K. W., 653

Churagulov, B. R., 830
Chynoweth, A. G., 539, 575, 718, 721
Claesson, S., 1107
Clark, A. L., 641, 653
Clark, J. B., 136, 145, 826, 827, 832
Clark, J. M., 952
Clark, J. S., 131
Clark, J. T., 1066
Clark, R. J., 270
Clark Jr., S. P., 823, 830
Clarke, A. M., 487, 495, 625, 653
Clarke, E. C. W., 612, 653
Clarke, G. R., 1265
Clarke, H. T., 952
Clarke, S. P., 417
Clausius, P., 643, 653
Claussen, W. F., 746, 747
Clay, F. P., 299
Clayton, G. T., 424, 497
Cleaver, J. S., 299
Clement, J. R., 84, 748
Clever, H. L., 70, 740, 746, 757, 784, 785
Clopper, P. R., 625, 638, 653
Cobic, B., 299
Coburn, J. F., 653
Cochran, D. R. F., 747
Cochrane, H., 1066
Cockburn, A., 748
Cohen, E., 456, 496, 747
Cohen, L. H., 826, 830
Cohen-Adad, R., 869, 870, 898
Cola, M., 868
Colarusso, V. G., 84
Colclough, A. R., 1233, 1264
Cole, G. H. A., 417
Cole, R. H., 466, 468, 469, 497, 597–600, 605
Coleman, T. C., 525
Colgate, S. O., 256, 257, 270
Collardeau, E., 987, 988, 990
Collings, A. F., 1155, 1158
Collins, M. M. C., 1184, 1190
Combs, R. L., 647, 655
Comings, E. W., 378
Compton, J. P., 345
Compy, E. M., 145
Conder, J. R., 650, 653, 778, 784
Cones, H. N., 1184, 1189
Connolly, J. F., 417, 618, 653
Constabaris, G., 345, 1043, 1046
Constantinov, V. A., 1021
Contré, M., 139–141, 143–145
Convers, L., 1020
Cook, A. H., 70, 116, 130, 303, 319, 320
Cook, D., 345, 575, 721
Cook, F. J., 676, 703, 720
Cook, G. A., 70
Cook, M. A., 1149, 1158
Cook, M. W., 747

Cook, S. S., 470, 472, 498
Coolidge, A. S., 1044, 1066
Coolidge, W. D., 437, 477, 496, 748
Coons, C. E., 631, 640, 655
Cooper, M. J., 676, 722
Cooper, R., 625, 626, 641, 653
Cope, J. O., 345
Copenhafer, D. T., 867
Cops, A., 569, 577
Corbett, J., 951
Coriell, A. S., 201
Corll, J. A., 112
Cornec, E., 894, 898
Corneliussen, R., 782, 784
Cornfeld, A. B., 703, 721
Cornish, R. H., 112
Cornish, R. M., 459, 496
Corrin, M. L., 1035, 1066
Corruccini, R. J., 831
Cosgarea, A., 656, 973, 974
Cottrell, F. G., 627, 653, 764, 784
Cottrell, T. L., 538, 575
Coumou, D. S., 784
Counsell, J. F., 653
Cowling, T. G., 575
Cowper, A. D., 474, 476, 496
Cowperthwaite, M., 1151, 1156, 1158
Cox, J. D., 345, 747
Craig, P. P., 1264
Craig, R. A., 867
Cramer, J. D., 374, 378
Crawford, R. K., 363, 364, 369, 378, 433, 485, 496
Creed, F. C., 1184, 1190
Croft, A. J., 1220, 1264
Crommelin, C. A., 357, 369–372, 379
Crooks, J. E., 1107
Cross, J. L., 130, 154, 199, 201, 270, 271
Crouthamel, C. E., 877, 951
Crowe, M. L., 458, 498
Crowell, A. D., 1024, 1067
Cruickshank, A. J. B., 429, 455, 490, 496, 621, 653, 747, 777, 784
Cubicciotti, D., 618, 655, 801
Cukor, P. M., 747
Cummins, H. Z., 717, 718, 721–723
Cupples, H. L., 357, 371, 372, 377, 416
Curl, R. F., 576
Curran, D. R., 575
Curry, R. W., 440, 452, 453, 498
Curtis, I., 320
Curtiss, C. F., 70, 270, 417, 575, 1158
Curzon, F. L., 1183, 1191
Cusack, N. E., 438, 487, 498, 978–980, 984, 990
Cussler, E. L., 456, 465, 499, 652, 767, 784, 833
Cutkosky, R. D., 721

Cutler, A. J. B., 621, 653
Cutler, W. G., 466, 496
Cutting, P. A., 1060, 1066
Czanderna, A. W., 1057, 1066
Czaplinski, A., 1066
Czerlinsky, G., 1075, 1107

Daane, A. H., 974
Dadson, R. S., 140, 145, 154, 155, 164, 184, 187, 201, 256, 257, 268, 270, 299, 368, 370, 374, 378, 1066
Dahl, D., 683, 721
Dainton, F. S., 626, 653
Dalgarno, A., 1210, 1218
Dalla Valle, J. M., 1067
Dalton, B. J., 371, 378
Damien, B. C., 821, 831
Damköhler, G., 880, 899
Damm, F. L., 1188, 1189
Dammers-DeKlerk, A., 952
Dana, L. I., 1256, 1257, 1264
Daniel, A. C., 553, 576
Daniel, S. G., 746
Daniels, W. B., 363, 364, 369, 378, 433, 485, 496
Danneil, A., 791, 801
Dannell, H., 370, 378
Dapoigny, J., 1155, 1158
Darken, L. S., 831
Darling, H. E., 211
Darmois, E., 841, 885, 898
D'Arrigo, G., 574, 575
Das, P., 1264
Dash, J. G., 1264, 1265
Dastur, S. P., 775, 784
Daughney, C. C., 1183, 1191
Dave, R. A., 746
Davies, C. W., 1097, 1107
Davies, G., 106, 112
Davies, L. A., 112
Davies, T. J., 1188, 1190
Davis, B., 831
Davis, C. M., 576
Davis, F. N., 252, 255, 271
Davis, F. W., 1031, 1067
Davis, J., 1214, 1217
Davis, L. A., 424, 496
Davis, N. F., 392–394, 399, 409, 417, 438, 482, 496
Davis, W. D., 299
Davison, R. R., 653
Dawson, J., 370, 379
Dawson, J. P., 620, 653
Dawson, L. R., 867
Dayantis, J., 488, 496
Dayton, B. B., 301
Dearnaley, G., 1264
Deans, H. A., 778, 785

de Boelpaep, J., 693, 722, 724
de Boer, J., 358, 372, 380, 575, 585, 605, 1043, 1044, 1067
De Bruyn Ouboter, R., 1264
Debye, P., 1085, 1096, 1107
Decker, D. L., 133, 140–142, 145
Dedit, A., 347
Defay, R., 839, 899
Deffet, L., 372, 378, 822, 831
Defretière, Mlle J., 898
Degiorgio, V., 721
de Graaff, Q. W., 369, 380
de Graaff, W., 70, 369–371, 373, 374, 379, 723
de Groot, S. R., 588, 606, 1264
de Grueter, J., 370, 379
de Haas, W. J., 357, 371, 379
Deitz, V. R., 1049, 1066
De Laet, W., 557, 577
Delahay, P., 951, 952
Delbart, R., 241, 243, 270
Delbove, F., 877
de Lépinay, J. M., 304, 320
Del Grosso, U. A., 552, 575
Delimarskii, Y. K., 903, 951
De Maeyer, L., 1075, 1095, 1107
De Maria, A. J., 1189, 1190
Deming, W. E., 721
Denbigh, K. G., 652
Denison, D. R., 301
Dennis, N. T. M., 270
Deschner, W. W., 373, 378
Desgoffe, C., 150, 201
Desmyter, A., 767, 785
Desre, R., 843, 899
de Swaan Arons, J., 793, 797, 801
de Vries, A. E., 271, 299
Devyatykh, G. G., 748
Dewar, J., 821, 831
Dewhurst, R. J., 1183, 1190
Dexter, D. L., 603, 605
Diaz Peña, M., 452, 496
Dick, R. D., 1149, 1150, 1156, 1158
Dickinson, W., 653
Dickson, A., 112
Di Domenico, M., 1190
Diepen, G. A. M., 653, 793, 797, 801
Dietz, V., 653
Dijkhuis, C. G. M., 951
Dike, R. S., 1188, 1190
Dikhter, I. Ya., 973, 1176, 1187, 1190
Diller, D. E., 373, 378, 684, 721, 722
Dillon, I. G., 70, 973, 977, 980, 985, 989, 990
Dimitrov, C., 655
Din, F., 70, 416, 417, 625, 653
Dittmar, P., 373, 378
Dixon, A. L., 631, 637, 655
Dixon, D. T., 781, 782, 784

Dixon, J. A., 466, 498
Di Zio, S. F., 343, 346
Dobbs, B. C., 721
Dodd, J., 1211, 1217
Dodge, B. F., 371, 374, 378, 379, 791, 800
Dohmann, H. D., 297
Dolly, L. G. F., 369, 370, 373, 379
Donaldson, E. E., 301
Doniach, S., 587, 605
Doolittle, A. K., 459, 496
Doran, D. G., 1158
Dornte, R. W., 832
Dorpema, B., 974
Dorrepaal, J. A., 373, 378
Dorsey, N. E., 997, 1020
Dorsman, C., 357, 373, 379
Doucet, J., 867, 898
Doucet, Y., 835, 870, 877, 898
Douglas Jr., D. H., 599, 606, 674, 690, 723
Douglas, E., 747
Douglas, T. B., 130, 831, 878, 898, 987, 988, 990
Douslin, D. R., 325, 346, 372, 373, 378, 416, 417, 618, 620, 625, 640, 643, 646, 653, 655
Dow, R. B., 823, 831
Dowling, J., 1214, 1217
Downer, L., 450, 496
Downs, J. L., 112
Doyle, R., 1210, 1217
Drain, L. E., 1066
Drawin, H. W., 1173, 1190
Dreisbach, R. R., 85
Dremin, A. N., 1144, 1158
Dressler, K., 605
Drew, C. M., 1052, 1066
Drickamer, H. G., 139, 143–145, 830, 832
Drossbach, P., 937, 951
Drost-Hansen, W., 440, 452, 453, 498
Dubinin, M. M., 1045, 1046, 1066
Dudley, J. D., 831
Duedall, I. W., 469, 496
Duff, R. E., 141, 145
Duffey, A. H., 1158
Dugdale, J. S., 112, 493, 496, 830
Duguay, M. A., 1188, 1190
Duk, Y., 112
Dumoulin, E., 795, 798–800
Dundon, M. L., 747
Dunlap, R. D., 767, 784
Dunlop, P. J., 652, 654, 767, 784
Dunlop, R. D., 748
Dunn, A. F., 605
Dunn, M. H., 1184, 1190
Dunn, R. A., 140, 145, 832
Dunoyer, L., 270
Dupinay, R., 371, 372, 377
Dupont, G., 867
Durand, J., 867

Durieux, M., 84, 373, 378, 1231–1233, 1256, 1264, 1266
Dushman, S., 270, 625, 653
Duty, R. C., 650, 653
Duvall, G. E., 145, 1125, 1155, 1156, 1158, 1159
Dworkin, A. S., 831, 877, 878, 899
Dymond, J. H., 417, 739, 747

Earley, J. E., 466, 468, 469, 497
Earnshaw, J., 299
Eatwell, A. J., 605
Ebert, H., 161, 201
Ecker, G., 1208, 1217
Eckerman, J., 1214, 1217
Eckert, C. A., 469, 499
Eckstein, Y., 574
Eden, D., 539, 540, 575, 718, 719, 722
Edeskuty, F. J., 462, 483, 499, 723
Edgecombe, F. H. C., 1066
Edmison, M. T., 831
Edmonds, G. D., 1183, 1189
Edmonds, T., 299, 1030, 1066
Edsinger, R. E., 130, 201, 270, 346
Eduljee, H. E., 456, 496
Edwards, C., 683, 695–698, 721, 723
Edwards, D. G., 417, 620, 652
Edwards, D. O., 1263–1266
Edwards, J. G., 642, 653
Edwards, J. L., 636, 653
Edwards, M. H., 592, 605, 685, 721
Edwards, T., 270
Edwards, W., 833
Egan, J. J., 907, 950, 951
Eggers, F., 551, 575, 1107
Eggerstein, F. T., 648, 653
Ehnholm, G. J., 1244, 1245, 1264
Ehrlich, G., 299
Ehrlich, P., 793, 800
Eichelberger, W. C., 867, 881, 899
Eigen, M., 1075, 1094, 1096, 1107
Einstein, A., 709
Eisenhauer, F. S., 748
Eisenstein, A. S., 715, 721, 722
Eisinger, J., 299
Elbaum, C., 549, 577
Eleffson, E. T., 747
El Hadi, Z. E. M. A., 373, 378
Ellington, R. T., 381
Elliot, J. H., 84
Elliott, K. W. T., 130, 256, 257, 268, 270, 299, 1066
Elliott, R. J., 605
Ellis, S. R. M., 764, 765, 775, 785
Elsey, H. M., 746
Elsey, P. G., 747
Elterman, P. B., 70
Elving, P. J., 747

Elwell, D., 690, 721
Embree, N. D., 654
Emel'yanov, Yu. M., 1183, 1190
Emerson, W. H., 747
Emmet, R. T., 440, 485, 487, 498
Emmett, J. L., 1184, 1191
Emmons, H., 1217
Emmons, R. B., 1191
Enagonio, D. P., 85
Endo, H., 976, 990
Enemark, E., 1211, 1217
England, J. L., 93, 94, 112, 141, 142, 145, 388, 417, 800, 830, 831
Englander-Golden, P., 300
Enüstün, B. V., 768–770, 784
Epain, R., 143, 144, 833
Epp, C. D., 391, 399, 416, 482, 495
Erb, W., 1186, 1189, 1190
Eremenko, V. N., 998–1000, 1020
Eriksen, D. P., 300
Ernsberger, F. M., 117, 130, 233, 256, 639, 640, 653, 1032, 1052, 1066
Errington, R. F., 248, 270
Ertl, G., 1096, 1107
Esin, O. A., 999, 1021
Essex, H., 456, 496
Estler, W. T., 722, 724
Eucken, A., 563, 575
Evans, C., 1201, 1217
Evans, E. J., 368, 378
Evans, E. L., 654
Evans, F., 1201, 1217
Evans, F. D., 652, 767, 784
Evans, J. P., 831
Evans, P. L., 1062, 1063, 1065
Evans, T., 106, 112
Evans, T. W., 747
Evans, W. H., 877
Even, U., 990
Everett, D. H., 652, 769, 770, 778, 779, 784, 785, 1024, 1066
Eversole, W. G., 359, 360, 369, 371, 374, 379
Evrard, R., 260, 262–264, 270
Ewans, M. G., 1020
Ewbank, E. K., 748
Ewing, C. T., 211, 402, 418, 973, 974, 980, 986, 988, 990
Ewing, M. B., 454, 496
Ewins, A. G., 747
Eykman, J. F., 867
Eyring, H., 417, 898, 899, 1158
Eyring, L., 656
Ezell, R., 952
Ezheleva, A. E., 748

Faeth, P. A., 1066
Fahey, P. F., 440, 485, 487, 496
Fairbank, W. M., 695, 698, 722, 1254, 1259, 1264, 1265

Fairchild, C. O., 831
Faizullou, F., 1211, 1217
Falciola, P., 867
Fanchenko, S. D., 1187, 1192
Fano, L., 1217
Farrington, P. S., 373, 378
Fateeva, N. S., 108–110, 112
Faust, O., 867
Feber, R., 973
Feke, G. T., 70, 718, 722
Feklichev, E. M., 830
Feldman, R. F., 1067
Felsing, W. A., 490, 497
Fender, B. E. F., 342, 346
Ferratini, A., 867
Fertik, S. M., 1183, 1191
Feynman, R. P., 1158
Fickett, W., 1155, 1158
Ficks, F., 372, 378
Field, A. L., 746
Filmer, D., 1099, 1108
Findlay, A., 867
Finel, A., 853, 899
Finiels, G., 867, 898
Finken, H., 867
Finnemore, P. C., 589, 596, 605
Fiorani, M., 950, 952
Firth, C. A., 1107
Fischer, A. K., 653
Fish, C. V., 1185, 1190
Fisher, M. E., 605, 722
Fitch, R. K., 299
Fixman, M., 539, 575, 719, 722
Flanick, A. P., 270
Flaschner, C., 747
Flengas, S. N., 951, 952
Fletcher, B., 299
Fleury, P. A., 535, 575
Flint, E. P., 618, 653
Flood, E. A., 1066, 1067
Flood, H., 877, 881, 898, 899
Flügge, S., 576, 831, 833
Flynn, G. W., 1107
Foex, M., 811, 815, 831
Fogler, M. F., 631, 653
Foley, G. M., 1190
Foote, F. D., 831
Ford, L. A., 1190
Ford, N. C., 715, 723
Forgacs, R. L., 575, 1177, 1190
Förland, T., 843, 881, 897–899, 911, 913, 950–952
Forsgren, K., 832
Forster, H. J., 1107
Foster, M. S., 951
Fougère, P. F., 552, 575
Fowler, L., 612, 653
Fowler, R. H., 747
Fowler, R. J., 764, 785

Fowles, G. R., 145, 1125, 1156, 1158
Fox, J. J., 747
Fralick, R. D., 1188, 1190
Franck, A., 648, 653
Franck, E. U., 70, 400, 417, 418, 438, 474, 477, 481, 492, 498, 791–793, 797, 798, 800, 801, 973, 975–977, 979, 981, 983–985, 987–990
Frantz, H. W., 490, 491, 500
Franzosini, P., 877
Fraser, M. V., 656
Fray, D. J., 438, 442, 491, 495
Frear, G. L., 747
Freeguard, G. F., 785
Freeman, R. D., 642, 653
Freeth, F., 372, 378
Freud, P. J., 92, 94, 95, 98, 99, 103, 105, 112
Freyland, W. F., 70, 800, 973, 982, 988–990
Fried, V., 651, 654, 787, 800
Friedman, A. S., 371, 378, 483, 497, 1264
Friedman, H. C., 1066, 1067
Friedman, H. L., 747
Friedrich, K., 653
Friess, S. L., 1107, 1108
Frisch, H. L., 413, 417
Fritsch, K., 718, 722
Fritsche, L., 551, 575
Fritz, J. N., 832
Frivoli, O. E., 867
Froidevaux, C., 1228, 1266
Frøyland, K., 952
Früngel, F., 1177, 1190
Fujisaki, M., 1157, 1158
Fujishiro, I., 101, 102, 112
Funtikov, A. I., 1150, 1157
Fuoss, R. M., 456, 465, 499, 833
Furukawa, G. J., 831
Furrow, S. D., 784
Fykse, O., 877

Gaddy, V. L., 371, 372, 381, 419, 483, 500, 748
Gaede, W., 254, 270
Gaines, J., 297
Gaines, Jr., J. M., 130, 345
Gainey, B. W., 747
Gal, I. J., 952
Galaktionov, V. A., 464, 499, 833
Gallagher, A., 1211, 1217
Gallagher, J. S., 70, 112, 418
Gallego, E., 1100, 1108
Gallon, T. E., 1066
Galy-Cazalat, M., 149, 201
Gamet, M. B., 483, 497
Garbe, S., 299
Garbuzov, A. I., 832
Garcia, E. A., 1065
Garden, L. A., 1066

Gardiner, K. E. S., 450, 496
Gardner, H. J., 877
Gardner, P. J., 853, 899
Garelli, F., 867
Garland, C. W., 539, 540, 567, 568, 573, 575, 718, 719, 722
Garn, P., 833
Garrabos, Y., 87
Garrett, A. B., 867
Garrison, J. B., 112
Garside, D. H., 589, 590, 596, 605
Garvin, D., 70
Gasparini, F., 695, 722
Gaune, M., 877
Gaune-Escard, Mme., 843, 898, 899
Gaur, H. C., 951
Geissler, E., 1228, 1266
Geld, P. V., 999, 1021
Geldermans, M., 370, 379
Genre, P. A., 256, 257, 270
Gerischer, H., 1096, 1107
Gerrard, W., 653
Gerry, H. T., 70, 116, 130, 373, 379, 653
Getman, F. H., 867
Getting, J. C., 94, 97, 99, 101, 103–107, 112, 143, 145, 388, 417
Giardini, A. A., 143, 145, 831–833
Giauque, W. F., 130
Gibbons, R. M., 372, 378
Gibbs, D. F., 95, 112
Gibby, C. W., 371, 378
Gibson, Q. H., 1107
Gibson, R., 358, 379
Gibson, R. E., 447, 448, 496, 654, 747, 823, 831
Giddings, J. C., 650, 654
Gielen, H., 693, 722, 724
Gielessen, J., 211
Giglio, M., 715, 717, 722, 723
Giguère, P. A., 429, 496
Gilchrist, A., 466, 468, 469, 497
Gildseth, W., 55, 70
Gilles, P. W., 608, 654
Gillespie, D. T. C., 765, 785
Gillespie, L. J., 653
Gilman, H. W., 763, 764, 785
Gingrich, N. S., 715, 722
Ginnings, D. C., 70, 637, 650, 654, 655, 831, 833
Ginnings, J., 987, 988, 990
Ginsburg, N. I., 145
Gitlesen, G., 831
Gjaldbaek, J. C., 747, 748
Glagoleva, A. A., 867
Glasgow, A. R., 849, 899
Glasgow, Jr., A. R., 831
Glass, S. W., 1190
Glassford, A. M. P., 371, 378
Glenn, Jr., W. H., 1189, 1190

Glew, D. H., 747
Glew, D. N., 612, 653
Gleysteen, L. F., 1049, 1066
Glick, J., 488, 498
Gobv, B., 867
Goddman, J. B., 748
Göhle, B., 1186, 1187, 1192
Göhre, H., 1067
Goig-Botella, S., 370, 378
Gold, V., 746
Goldfarb, I. J., 655
Goldhammer, D. A., 722
Goldman, K., 486, 497, 706, 722
Goldman, M., 1264
Goldsack, D. E., 1107
Goldstein, L., 1247, 1264
Golovashkin, A. N., 1240, 1264
Gol'ts, E. Ya., 1185, 1190
Golubev, I. F., 485, 486, 497
Gomer, R., 299, 1060, 1066
Gondberg, H., 867
Goncharov, A., K., 617, 654
Gonikberg, M. G., 466, 499, 830, 831
Goodall, D., 1100, 1108
Goodkind, J. M., 877, 1228, 1229, 1264, 1266
Goodwin, H. M., 877, 878, 899
Goodwin, R. D., 70, 357, 373, 378, 493, 497, 505, 526, 654, 722, 1264
Googin, J. M., 647, 655
Gopalaraman, C. P., 299
Göranson, R. W., 831
Gorbunova, V. G., 695, 724
Gordon, R. B., 112, 424, 496
Gordon, R. G., 538, 575
Goree, W. S., 112
Gorni, F., 867
Gorodetskii, J. G., 747
Gorter, C. J., 1264
Gosselin, C. M., 297, 299
Gosting, L. J., 688, 722
Goudeket, M., 370, 371, 379
Gould, F. A., 270, 346
Gourjault, J. J., 270
Gouy, M., 678, 688, 720, 722
Gowenlock, B. G., 1027, 1067
Graf, E. H., 1264
Graham, M. E., 1186, 1190
Graham, R. A., 1143, 1158
Grange, J., 602, 605
Grasdalen, G., 1217
Graves, A. D., 951
Graves, N. R., 373, 381
Gray, D., 951
Green, H. S., 831
Green, J. H. S., 653
Green, M. S., 70, 576, 577, 676, 720–724
Greenhorn, R. A., 409, 418
Greenspan, L., 170, 201

Greenspan, M., 535, 536, 575
Greenwood, D. A., 976, 983, 989, 990
Greenwood, H. J., 390, 391, 417, 476, 497
Greenwood, N. N., 654
Greer, W. L., 722
Greig, R. G. P., 140, 145, 154, 155, 164, 187, 201
Griem, H. R., 1173, 1190, 1202, 1211, 1217
Griffiths, E. J., 867
Griffiths, T., 301
Griggs, D., 831
Grigor'ev, A. M., 299
Grilly, E. R., 494, 497, 832, 1238, 1260–1262, 1264, 1265
Grimes, W. R., 952
Grimshaw, R. W., 831
Grimsrud, D. T., 558, 575
Grindlay, J., 1208, 1212, 1217
Grindley, T., 460, 497
Grisard, J. W., 655
Grjotheim, K., 872, 877, 888, 898, 899, 952
Groot, C., 747
Grosse, A. V., 976, 988–990
Groszkowski, J., 299
Grove, R. N., 576
Grover, F. W., 1190
Grover, R., 1185, 1192
Gruverman, I. J., 1265
Gugan, D., 112
Guggenheim, E. A., 426, 427, 497, 575, 747, 757, 785, 951, 1066
Guildner, L. A., 115, 130, 201, 233, 270, 346
Guillet, H., 1188, 1191
Gulbransen, E. A., 1054, 1055, 1066
Guptill, E. W., 551, 575
Gurtman, G. A., 1156, 1158
Gurry, R. W., 831
Gusak, V. G., 695, 721
Güsewell, D., 570, 575
Gutfreund, H., 1107
Gutt, W., 808, 831, 833
Guye, P. A., 320
Gylling, R. G., 1244, 1245, 1264
Gyorog, D. A., 560, 575

Haar, L., 406, 414, 415, 417, 623, 654
Haase, R., 841, 845, 846, 899
Habenschuss, A., 55, 70
Haber, F., 883, 899
Habgood, H. W., 680, 691, 692, 722
Hackham, R., 654
Hackspill, M. L., 615, 654
Hadlock, C., 357, 370, 373, 377
Haebel, E. U., 537, 575
Haefer, R., 299
Hahn, E. L., 1228, 1266
Hála, E., 651, 654, 775, 786, 787, 800
Halbedel, H. S., 653

Hales, J. L., 653
Hall, H. T., 138–143, 145, 830, 831
Hall, K. R., 366, 374, 378, 381
Hall, W. J., 113
Halperin, B. I., 722
Hals, L., 867
Halsey, G. D., 342, 345, 346
Halsey Jr., G. D., 1043, 1066
Halverson, G., 1234, 1264
Hamann, S. D., 1158
Hamer, W. J., 655, 951
Hamers, J., 684, 723
Hamister, V. C., 1190
Hammel, E. F., 1263, 1264
Hammes, G. G., 1107, 1108
Hammond, G. L., 1193, 1212, 1213, 1217
Hammons, B. E., 96, 112
Händel, S. K., 1186, 1187, 1192
Handley, R., 652
Hannay, J. B., 618, 654
Hanneman, R. E., 91, 101–107, 112, 388, 417, 831
Hansen, J. P., 376, 378
Hansen, N., 1066
Hanson, D. N., 747
Harada, T., 1184, 1190
Harbourn, C. L. A., 654
Harding, J. T., 1228, 1266
Harkins, W. D., 1002, 1020
Harraway, R. A., 1183, 1190
Harries, K., 1183, 1190
Harris, F. K., 951
Harris, K. R., 652, 654, 767, 784
Harris, R. W., 424, 497
Harrison, D., 490, 497
Harrison, J. P., 899
Harrison, R. H., 372, 373, 378, 417
Harrison, T. R., 831
Hart, K. R., 531, 577
Hart, M., 305, 320
Hartshorne, L., 951
Harvey Jr., A. E., 831
Hashimoto, H., 247, 248, 270
Hastings, J. R., 111, 113, 211, 367–370, 381, 472, 499
Hattenburg, A., 1217
Hauck, F., 563, 575
Hawkes, S. J., 654
Hawkinson, A., 880, 899
Haworth, W. S., 342–344, 346
Hayes, P. F., 70, 144, 201
Haygarth, J. D., 143, 145
Hayward, A. T. J., 435, 440, 453, 454, 456, 473, 488, 492, 497
Hayward, W. H., 299
Head, A. J., 747
Healey, T. J., 1187, 1190
Heath, J. B. R., 480, 497
Heaton, E., 831

Hecht, G. J., 1187, 1190
Hecker, J. C., 654
Heins, C., 371, 381, 419
Heins Jr., C., 748
Helisch, J., 1108
Hellemans, R., 574
Helmer, J. C., 299
Henderson, D., 376, 378, 413, 417
Henderson, U. V., 1216, 1217
Henderson, W. E., 747
Hengevoss, J., 299
Henius, A., 833
Henry, D. L., 717, 718, 722
Henry, G., 1183, 1192
Henry, K. W., 1177, 1187, 1190
Henry, W. F., 631, 640, 655
Hensel, F., 70, 800, 973, 975–977, 979, 981–985, 987–990
Herb, R. G., 300
Herd, A. C., 1065
Heric, H. L., 747
Herington, E. F. G., 345, 652, 760, 785
Hermsen, R. W., 774, 785
Herrick, C. C., 973
Herring, C., 1066
Herrington, T. M., 346
Hershey, T., 1217
Herzfeld, C. M., 88, 112, 381, 951
Herzfeld, K. F., 539, 541, 575
Hess, D., 1107
Hession, F. P., 301
Hetherington, H. C., 357, 369, 370, 377, 416
Heurlin, S., 1184, 1191
Heuse, E. O., 656
Hevesy, G., 747
Heydemann, P. L. M., 112, 116, 131, 145, 147, 165, 167, 201, 474, 497, 831
Heydweller, A., 996, 1020
Hibbard, W. R., 497, 832, 833
Hibsham, H. B., 823, 831
Hickman, K. C. D., 634, 654
Hickmott, T. W., 299
Hicks, C. P., 747
Higginbotham, I. G., 1066
Hildebrand, F. B., 508, 526
Hildebrand, J. H., 490, 491, 500, 575, 739, 747, 748, 758, 785
Hill, A. E., 747
Hill, A. V., 652, 654
Hill, D. L., 951
Hill, J. D., 154, 162, 201
Hill, J. E., 973
Hill, J. L. E., 791, 801
Hill, R. W., 493, 497
Hill, T. L., 785, 1024, 1066
Hillecke, D., 1057, 1066
Hills, G. J., 438, 491, 495, 950, 951
Hilsenrath, J., 346, 416, 417, 575, 1217
Hindmarsh, W., 1212, 1217

Hirai, N., 899
Hirschfelder, J. O., 70, 270, 417, 560, 574, 575, 1158
Hirschkoff, E. C., 1228, 1264
Hitch, B. F., 951
Hladik, J., 950, 951
Hnizda, V., 653
Ho, C. Y., 418
Ho, J. T., 675, 723
Hoare, F. E., 1220, 1264
Hobson, J. P., 270, 299, 1030, 1037, 1038, 1066, 1067
Hochman, J. M., 973, 986, 990
Hock, L., 867
Hocken, R., 684, 722
Hoerr, C. W., 747
Hoffer, J. K., 1253, 1254, 1264
Hoffmann, F., 831
Hoffmann, G. W., 1108, 1186, 1188, 1190
Hoffmann, H., 1108
Hogarth, J., 618, 654
Hogben, D., 201
Högberg, L., 1187, 1191
Hoge, H. J., 70, 654, 1217
Hohenberg, P. C., 696, 718, 721, 722
Holborn, L., 201, 357, 369, 371–373, 378, 417, 483, 497
Holbrook, R. D., 575
Holder, G. A., 440, 445, 456, 458, 461, 496, 497
Holkeboer, D. H., 299
Holland, T. E., 1187, 1190
Holloway, J. R., 392–394, 399, 409, 417, 438, 482, 496
Holm-Jensen, I., 654
Holmes, J. M., 1024, 1066
Holmes. Q., 1202, 1212, 1218
Holmstrom, B., 1266
Holser, W. T., 391, 417, 477, 497
Holst, G., 357, 373, 379
Honig, J. M., 1057, 1066
Hood, R., 1217
Hoohschagen, J., 832
Hooley, J. G., 1066
Hoover, W. G., 375, 380
Hopke, E. R., 614, 654, 655
Hopmann, R., 1108
Horie, Y., 831, 1158
Horiko, Y., 1184, 1190
Horiuti, J., 747
Horsford, A., 830
Houck, J., 142, 145, 167, 201, 474, 497
Hougen, O. A., 409, 418
Hovorka, F., 886, 899
Howello, W., 867
Hoyaux, M. F., 1173, 1190
Hoyt, C. K., 551, 575
Hryckowian, E., 139, 146
Hsu. C. C., 372. 378

Hsu, L., 831
Hu, J. H., 747
Hubbard, J. C., 550, 575
Huber, M., 1217
Huber, W. K., 299
Hudson, R. P., 721, 1264
Huff, J. A., 417
Hugenholtz, C. A. J., 1183, 1191
Huggins, C. M., 576
Huggins, M. L., 747
Huggins, R., 587, 605
Huiskamp, W. J., 1264
Hulbert, J. A., 112
Hulett, A., 747
Humenik, M., 999, 1020
Humphery, E. C., 1002, 1020
Humphrey, F. B., 1188, 1191
Huntress, A. H., 270
Hurst, R. E., 1107
Hust, J. G., 113, 508, 526
Hutson, A. R., 201
Hutton, U. O., 156, 157, 201, 832
Hyde, J. H., 463, 497
Hyne, J. B., 490, 491, 498
Hynne, F., 724
Hyzer, W. G., 1187, 1190

Iapteff, B., 299
Ibraguimov, Kh. I., 1020
Ifft, E. M., 1264
Ilgenfritz, G., 1108
Iliin, B., 371, 380
Il'ina, M. A., 145
Ingersoll, L. R., 602, 605
Ingraham, T. R., 951
Inman, D., 924, 950, 951
Irvine, Jr., J. W., 952
Isaac, P. J., 1066
Isaac, R., 371, 378
Isbell, J., 877
Isbell, W. M., 1158
Ishii, H., 254, 256, 257, 270, 299, 1031, 1066
Ishikawa, K., 300
Ishimura, H., 1053, 1056, 1066
Issinskii, I. B., 1183, 1190
Itakevich, E. S., 145
Ivanov, A. A., 1152, 1159
Ivanov, A. G., 1158
Ivanov, S. M., 1021
Ivanov, V. A., 473, 498
Ivanova, E. M., 211
Ivashchenko, Ju. P., 998–1000, 1020
Ives, D. J. G., 951
Izrailov, K. S., 125, 127, 129, 131

Jackson, L. C., 1220, 1264
Jacobus, D. D., 130, 345

Jaeckel, R., 300
Jaeger, F. M., 1009, 1020
Jaffe, I., 877
Jaffray, J., 831
Jahr, K. F., 868, 870, 894, 899
James, A. T., 1066
James, D. W., 877
James, L. H., 300
James, T. E., 1183, 1190
Jamieson, J. C., 1150, 1156, 1159
Jansen, C. G., 270
Jansen, L., 587, 588, 605
Jansoone, V., 693, 722, 724
Janz, G. J., 872, 873, 877, 899, 951
Jardine, D. A., 1066
Jarman, M., 95, 112
Jayaraman, A., 142, 143, 145, 201, 826, 827, 831, 832
Jeans, J. J., 575
Jeffrey, R. N., 94, 112, 138, 139, 141–143, 145, 388, 417
Jeffes, J. H. E., 952
Jen, J., 1108
Jenkin, C. F., 562, 563, 575
Jenkins, A. C., 654
Jenkins, H. W., 951
Jenner, G., 467, 498
Jepsen, R. L., 299
Jepson, W. B., 475, 497, 654, 1065
Jessup, R. S., 201, 369, 379, 442, 443, 451, 497, 831
Jimeno, E., 748
Jindal, H. L., 951
Jobes, F. C., 1191
Jogansen, A. V., 747
Johannin, P., 387, 417, 418
Johns, D. A., 1234, 1264
Johnson, C. A., 70, 116, 130
Johnson, C. E., 877
Johnson, C. S., 1108
Johnson, D. P., 130, 140, 145, 153, 154, 162, 165, 201, 241, 243, 244, 270, 271, 636, 653, 831, 832
Johnson, J. F., 830
Johnson, J. R., 782, 784
Johnson, J. W., 618, 655, 801
Johnson, R. T., 1243, 1246, 1247, 1264–1266
Johnson, V. J., 577
Johnston, D. R., 597, 605
Johnston, H. L., 346, 371, 378, 483, 497, 1264
Johnston, J., 472, 494, 747, 748
Joigneau, S., 1148, 1157
Jolley, E. L., 453, 497
Jona, T., 867
Jonassen, H. B., 495, 653
Jones, A. H., 1158
Jones, C. H., 791, 801

Jones, E. D., 831
Jones, G., 449, 498, 804, 831
Jones, G. O., 600, 605
Jones, M. R., 338, 345
Jones, O., 1188, 1191
Jones, R. C., 1188, 1189
Jones, W. M., 1066
Jordan, C. F., 867
Jordan, I. B., 1186, 1189
Joris, G. G., 746, 747
Jortner, J., 990
Joseph, R. I., 724
Josephson, B. D., 1228, 1264
Jost, A., 1108
Jost, W., 722
Joubert, D., 371, 379
Joule, J. P., 428, 497
Jovin, T. M., 1108, 1188, 1190
Joy, A. S., 1066
Joyner, L. G., 1042, 1043, 1066
Judd, N. F., 785
Jura, G., 143–145
Juza, J., 401, 402, 417

Kablunov, I. A., 747
Kadanoff, L. P., 669, 722
Kaeser, R. S., 1264
Kagan, D. N., 974
Kaganer, M. G., 369, 380, 483, 499
Kahn, J., 748
Kalarickal, S., 1108
Kalashnikov, Ya. A., 394, 395, 418, 433, 484, 499, 830
Kalfoglu, N. K., 374, 379
Kalidas, C., 1108
Kalmus, H. T., 877, 878, 899
Kalvins, G. M., 1264
Kamerlingh Onnes, H., 357, 358, 369–372, 377–380, 1256, 1257, 1260, 1264, 1265
Kamper, R. A., 1229, 1230, 1265
Kandalic, G. A., 618, 653
Kaneda, R., 131
Kang, T. L., 381
Kanomata, I., 1026, 1027, 1067
Karagunis, G., 880, 899
Karapet'yants, M. Kh., 617, 654
Kardos, G., 211
Kassirov, G. M., 1183, 1191
Kasterin, N. P., 996, 1020
Kastler, A., 721
Katila, T. E., 1264
Kats, G., 831
Katz, D. L., 781, 785
Katz, J. L., 413, 417, 898, 899
Kaufman, M. H., 640, 654
Kawamura, T., 1184, 1190

AUTHOR INDEX

Kawasaki, K., 539, 576, 683, 719, 722
Kay, N. B., 781, 785
Kay, W. B., 618, 622, 623, 654, 793, 800
Kay, W. C., 418
Kaye, J., 70, 116, 130, 345
Kazarnovskii, Ya. S., 417
Kazavchinskii, Y. A., 416, 418
Kedrinskii, V. K., 1144, 1158
Keeler, R. N., 1150, 1115, 1158
Keeling, B. F. E., 830
Keenan, A. G., 877
Kees, J., 297
Keesom, A. P., 492, 497
Keesom, W. H., 371, 373, 380, 492, 497, 557, 576, 577
Keevil, N. G., 248, 270
Kegeles, G., 688, 722
Keil, T. H., 588, 605
Kell, G. S., 57, 70, 417, 424, 434, 440, 442, 445, 451, 452, 469, 476–481, 487, 490
Keller, D. V., 1186, 1190
Keller, J. M., 297
Keller, R. A., 654
Keller, W. E., 371, 378, 379, 483, 497, 1219, 1220, 1232, 1263–1265
Kellers, C. F., 695, 698, 722, 1254, 1265
Kelley, K. K., 877, 878, 899
Kelly, F. J., 877
Kemme, H. R., 648, 654
Kemp, E. L., 1188, 1190
Kendall, B. J., 370, 379
Kendall, P., 1217
Kennard, E. H., 270
Kennedy, G. C., 94, 97, 99, 104–107, 112, 141–143, 145, 166, 201, 370, 379, 388, 390, 391, 398, 417, 474, 477, 497, 791, 801, 826, 827, 831–833
Kentamaa, J., 870, 882, 899
Kerimov, A. M., 695, 721
Kernaghan, M., 998, 1021
Kerr, E. C., 722, 1238, 1258, 1265
Kerr, S. L., 483, 497
Kershaw, R. W., 783, 785
Kertes, A. S., 725
Kerwin, L., 300
Kessler, L. H., 483, 497
Kestin, J., 407, 417
Ketelaar, J. A. A., 951, 952
Ketterson, J. B., 574
Keyes, D. B., 747, 748
Keyes, F. G., 369, 371–373, 379, 475–478, 490, 497
Keyes, P. H., 721, 723
Keyes, R. T., 1158
Khanina, I. F., 300
Khitarov, N. T., 832
Khodeeva, S. M., 795, 800
Kieffer, J., 1110, 1155, 1158, 1159
Kienitz, H., 71

Kierstead, H. A., 691, 693, 722
Kikoin, I. K., 973, 983–985, 989, 990
Kim, J. H., 656, 973, 974
Kim, M., 301
Kimberley, H. M., 626, 653
King, J. H., 368, 378
Kingery, W. D., 951, 999, 1020
Kington, G. L., 1024, 1066
Kini, K. A., 1066
Kippert, F., 491, 492, 499
Kirby, C. G. M., 328, 330, 331, 346
Kirenkov, I. I., 125, 127, 129, 131
Kirillov, G. A., 1150, 1151, 1158
Kirkwood, J. G., 584, 605
Kirschner, K., 1100, 1108
Kisatsky, P., 1214, 1217
Kishi, K., 1184, 1190
Kislykh, V., 1208, 1217
Kistemaker, J., 233, 270, 346
Kister, A. T., 757, 785
Kittel, C., 1265
Kiyama, R., 369, 379
Klein, F. K., 974
Klein, M., 4, 70, 418, 721, 1210, 1217
Klement, W., 832
Klement Jr., W., 142, 143, 145, 826, 830, 833
Klemperer, W., 538, 575
Klèppa, O. J., 898, 899
Klier, K., 1028, 1066
Kling, R., 490
Kljachko, Ju. A., 1020
Klopfer, A., 297
Klopfer, K., 299
Klots, C. E., 747
Klotz, C. G., 1189
Klumb, H., 264, 270, 625, 645, 654
Klundert, L., 370, 380
Kmonicek, V., 401, 402, 417
Knaap, H. F. P., 344, 346
Kneser, H. O., 539, 576
Knight, R. J., 772, 775, 786
Knight, W. L., 477, 497
Knobler, C. M., 344, 346, 605, 684, 722
Knoche, W., 1069, 1107, 1108
Knowles Middleton, W. E., 201
Knox, J. H., 440, 485, 487, 498
Knox, R. S., 603, 605
Knudsen, M., 270, 625, 642, 644, 645, 654
Kobatake, Y., 747
Kobayashi, R., 371–374, 378, 778, 785
Kobe, K. A., 371, 373, 381, 654, 748
Koch, P. P., 369, 379
Kocian, P., 297
Koefoed, J., 770, 784
Koffer, H., 1108
Kohlmuller, R., 877
Kohnstamm, Ph., 793, 800
Kolenkow, R. J., 1055, 1066
Kolesov, V. P., 951

Kolthoff, I. M., 747
Komada, E., 830
Komarek, K. L., 641, 643, 654
Komel'kov, V. S., 1183, 1191
Konotop, V. V., 1183, 1191
Konyaev, Yu. S., 166, 201, 833
Kormer, S. B., 1150, 1151, 1157, 1158
Korneff, T., 1185, 1186, 1191
Kornelsen, E. V., 300, 1067
Korshunov, Yu. S., 494, 497
Kortüm, G., 886, 900
Koshevnik, Ju. U., 997, 1020
Koshland, D. E., 1099, 1108
Kotel'nikova, N. V., 1185, 1191
Koval'chuk, B. M., 1183, 1191
Kozakevitch, P., 998, 999, 1020
Kozlov, A. D., 395–397, 418
Kozlowski, T. R., 877
Kraček, F. C., 831, 832
Kraftmakher, Y. A., 973
Kramer, G. M., 374, 379
Krase, N. W., 748
Kraus, C. A., 867, 881, 899
Kreglewski, A., 654
Kreis, R. W., 900
Kremnevskaia, E. A., 693, 723
Kreps, S. I., 648, 654
Kretschmer, C. B., 767, 785
Krichevskii, I. R., 362, 379, 789–791, 796, 800
Krishkevich, G. V., 1151, 1158
Kroepelin, H., 1176, 1187, 1191
Kruse, W., 1107
Kryder, M. H., 1188, 1191
Ku, H. H., 70
Ku, P. S., 374, 379
Kubaschewski, O., 654
Kubens, R., 868, 870, 894, 899
Kuchnir, M., 574
Kudchadker, A. P., 70, 623, 654
Kudrjawzew, B. B., 549, 576
Kuehn, C., 1108
Kuhl, W., 576
Kuhn, W. E., 1066
Kuivila, H. G., 867
Kulikova, A. I., 373, 381
Kumagai, A., 446, 497
Kummers, J. T., 1066, 1067
Kunavin, A. T., 494, 497
Kun Li, 371, 378
Kupke, D. W., 440, 485, 487, 496
Kurata, F., 748, 781, 785
Kuriakose, A. K., 474, 497, 832
Kurkchi, G. A., 747
Kurpen, J. J., 793, 800
Kurti, N., 1220, 1245, 1264, 1265
Kusakov, M. M., 997, 1020
Kushiro, I., 830, 832
Kustin, K., 1107

Kuswa, G., 1183, 1191
Kuzin, N. N., 146, 833
Kuypers, H. A., 373, 379
Kvalnes, H. M., 357, 369, 370, 372, 377, 416
Kvish, J. W., 1156, 1158

Labrie, R. J., 951
Lacam, A., 112, 576
Lacey, W. N., 369, 370, 372–374, 380, 442, 480, 482, 499
Lachman, J. C., 112, 951
Lachowicz, S. K., 747
Lacy, L. L., 553, 576
Lafferty, J. M., 270, 300
Lahr, P. H., 359, 360, 369, 371, 374, 379
Laidler, K. J., 480, 494
Laing, W., 1066
Laitinen, H. A., 951, 952
Laity, R. W., 903, 950, 951
Laj, C., 717, 721
Lallemand, M., 94, 112, 113, 347, 388, 417
Lalos, G. T., 1193, 1212, 1217, 1218
Lamb, A. B., 487, 497
Lamb, J., 541, 553, 576, 577
Lamb, V. A., 951
Lambert, B., 346, 1056, 1066
Lambert, R. H., 1020
La Mer, V. K., 880, 885, 886, 899
Lamers, K. W., 1028, 1066
Lamm, O., 1108
La Mori, P. N., 92, 94, 95, 98, 99, 105, 112, 141–143, 145, 166, 201, 474, 497, 826, 827, 832
Lampard, D. G., 606
Landau, J., 1238, 1263, 1265
Landau, L. D., 709, 716, 722, 1120, 1158
Landolt, H., 832
Lane, R. D. G., 853, 899
Lange, J., 854, 859, 861, 867, 880, 882, 899
Lange, W. J., 300
Langenberg, D. N., 28, 70, 346
Langer, S. H., 785
Langley, K. H., 70
Langmuir, I., 625, 641, 642, 654
Lannung, A., 747
Lansing, J. E., 647, 652
Lantraton, M. F., 903, 950
Laplace, P., 232, 233, 270, 271
Laplace, P. S., 992, 1020
Lapworth, K. C., 1185, 1191
Larsen, S. Y., 575, 606, 684, 722
Larson, D. B., 141, 145
Larson, E. U., 576
Lascar, A., 1109, 1157, 1158
Laskin, S., 1039, 1066
Laslett, L. S., 297
Lassiter, J. W., 70

AUTHOR INDEX

Lassiter, W. S., 300
Lastovka, J. B., 70, 721
Latter, A., 1156, 1157, 1158
Latter, R., 1156, 1157, 1158
Lausen, P., 1184, 1190
Lauterbach, K. E., 1039, 1066
Lavrov, I. A., 466–468, 499
Lawless, W. N., 1235, 1265
Lawrence, C. K., 357, 369, 377, 378, 416
Lawson, A. W., 112, 113
Lawson, R. W., 300
Lazarev, V. V., 1021
Lazarre, F., 112
Lazarus, D., 90, 94, 112, 388, 417
Leach, L., 1039, 1066
Leavitt, G. E., 1184, 1191
Lebedev, S. V., 973, 1176, 1186, 1187, 1190, 1191
Lebowitz, J. L., 413, 417
Leck, J. H., 261, 270, 271, 297, 299–301, 1067
Leclercq, R., 213
Lecocq, A., 404, 417
Lee, D. M., 1247, 1248, 1259, 1264, 1265
Lee, K. P., 551, 577
Lee, R., 1217
Lee, R. E., 487, 497
Lee, T. D., 658, 722
Lees, J., 828, 832
Lefèvre, J., 270
Le Fur, D., 1243, 1265
Leger, J. M., 832
Le Goff, D., 1188, 1191
Leighton, D. A., 747
Leitner, N., 877
Leland, T. W., 371–374, 378
Lemaire, M., 869–871, 886, 899
Le Neindre, B., 1, 87, 347, 387, 417, 418
Lentz, H., 474, 498, 793, 795, 800
Leonard, P. J., 416
Leopold, K., 1183, 1189
LePair, C., 1264
Lepple, F. K., 440, 498
Leroux, Y., 347
Leroy, G. L., 1184, 1192
Lespiau, R., 867
Lestz, S. S., 576
Letcher, S. V., 549, 574
Leussing, D. L., 748
Levchenko, V. G., 833
Levelt Sengers, J. M. H., 4, 70, 347, 369–371, 379, 380, 406, 417, 418, 424, 483, 499, 565, 576, 605, 623, 654, 657, 676, 722, 724
Levesque, D., 413, 418
Levich, V. G., 444, 498
Levin, E. M., 832
Levin, S. B., 831
Levin, Yu. L., 140, 144

Levine, S., 877
Levine, S. W., 416
Levinson, L. S., 973, 974
Levy, O., 725
Lewin, S. Z., 422, 484, 495
Lewis, A., 346
Lewis, E., 1217
Lewis, E. S., 1107, 1108
Lewis, G. N., 748
Lewis, L. C., 371, 379
Lewis, M., 1208, 1217
Leycuras, A., 385, 418
Leycuras, Y., 385, 418
Liang, S. Chu, 271, 624, 654, 1028, 1067
Liburg, M. J., 1244, 1265
Lichtenthaler, R. N., 340, 346
Lick, W., 1217
Liebenberg, D. H., 602, 605, 1240, 1265
Lieber, A. J., 1188, 1191
Lifshitz, E. M., 709, 722, 1120, 1158
Lim, C. C., 551, 574, 575
Lin, D. C. K., 370, 379
Lin, J. S., 715, 721
Lind, J. E., 460, 497
Linde, R. K., 1158
Lindgren, R. M., 950
Lindroos, A. E., 791, 800
Lindsley, D., 832
Linensberg, A., 870
Linetskii, V. Ya., 1183, 1191
Linford, R. G., 748
Linnenbom, V. J., 748
Linsen, B. G., 1043, 1044, 1067
Lipa, J. A., 683, 695–698, 721, 723
Lipmann, D. Z., 576
Lippens, B. C., 1043, 1044, 1067
Lippman, H., 211
Lisell, E., 211
Litovitz, T. A., 539, 541, 569, 575, 576
Litster, J. D., 675, 723
Little, J. W., 241, 243, 244, 271
Little, W. A., 695, 723
Liu, C. S., 974
Liukonen, R. A., 1184, 1189
Livingston, J., 841, 868, 899
Ljunggren, S., 1108
Llewellyn, P. M., 1225, 1264
Lloyd, E., 832
Lloyd, E. C., 144–146, 832
Lo, H. Y., 493, 498
Lochte-Holtgreven, W., 1218
Logan, J. K., 84
Loginov, V. A., 1185, 1191
Lohnstein, Th., 996, 1002, 1020
Lonberger, S. T., 85
London, H., 1265
Longuet-Higgins, H. C., 898, 899
Longwell, P. A., 1215, 1218
Loomis, E. H., 847, 899

1283

Loowerse, P., 370–374, 379
Loprest, F. J., 748
Lorentz, H. A., 580, 605
Lorentzen, H. L., 605, 686, 723
Lorenz, P. K., 951
Loria, G., 201
Lossing, F. P., 651, 656
Louginine, W., 867
Lounasmaa, O. V., 493, 497, 1245, 1247, 1264, 1265
Love, J., 1107
Lovejoy, D. R., 346, 748, 832
Lowe, F., 883, 899
Löwenhertz, R., 899
Lowenthal, G. C., 974
Lowry, T. M., 748
Lozier, W. W., 1191
Lubman, N. M., 997, 1020
Lückert, J., 645, 654
Lucovsky, G., 1191
Luedemann, H. D., 145
Luft, L., 399, 400, 418
Lukashev, A. A., 1184, 1189
Lukin, B. V., 1186, 1191
Lumsden, J., 898, 899, 950, 952
Lunacek, J. H., 715, 723
Lunbeck, R. J., 372, 373, 379
Lundberg, N. H., 952
Lundquist, S., 1186, 1191
Lupien, Y., 381, 419
Lupton, J. M., 370, 379
Luth, W. C., 800
Lydersen, A. L., 409, 418
Lynn, R. E., 654

Ma, C. K., 111
Maass, G., 1107
Maass, O., 654
McAuliffe, C., 746
McBain, J. W., 778, 779, 785, 1050, 1067
Macca, C., 952
Maccabee, B. S., 715, 723
McCarty, R. D., 501, 505, 508, 526
McClean, R. T. B., 649, 654
McClure, J. L., 1190
McCormack, K. E., 418
McCormick, H., 1003, 1020
McCoskey, R. E., 831
McCoubry, J. C., 538, 575
McCullough, J. M., 417
McCullough, J. P., 372, 373, 378, 748, 832, 951, 1189, 1220
McDaniel, A. S., 748
McDaniel, C. L., 85, 833
McDaniel, M. E., 823, 832
McDermott, C., 775, 785
MacDonald, D. B., 490, 491, 498

MacDonald, J. R., 489, 498
McDowell, B., 112
McEven, B., 747
McFadden, P. W., 242, 243, 271
McFee, J. H., 201
McGlashan, M. L., 70, 340, 346, 452, 496, 652, 748, 755, 766, 767, 770, 771, 781, 782, 784, 785, 951, 952
McGowan, W., 300
McGurty, J. A., 112, 951
Machin, W. D., 654, 1027, 1067
McIntyre, D., 715, 723, 724
Macintyre, J. D. E., 877
Mack, E., 821, 832
Mack, Jr., E., 747
Mackay, C. M., 654
Mackenzie, J. D., 500, 654, 748, 833, 952
McKetta, J. J., 371–373, 378, 381
McKinnon, J. R., 772, 775, 786
Mackle, H., 649, 654
Mackor, E. L., 784
McLachlan, D., 1155, 1159
McLaren, E. H., 85, 814, 832
MacLaurin, G. E., 417, 480, 497
McLeod, H., 245, 271
McMickle, R. H., 466, 496
McMurdie, H. F., 832
McMurty, B. J., 1189
McNicol, J. J., 1190
McNish, A. G., 951
McQueen, R. G., 832, 1158
McSkimin, H. J., 554, 576
MacWood, G. E., 747
Mader, W. J., 748
Madey, T. E., 300
Madigosky, W. M., 467, 498
Magnien, C., 112
Magnusson, N., 617, 655
Maier, S., 418, 477, 498
Maiden, C. J., 1158
Mailey, R. D., 437, 477, 496
Maines, R. G., 201
Mair, B. J., 655, 849, 899
Maitland, A., 1184, 1190
Major, A., 832
Makarova, O. P., 1021
Malbrunot, P., 94, 96, 112, 113, 211, 383, 385, 386, 388, 404, 406, 417, 418, 462, 493, 498
Malcolm, G. N., 490, 498, 783, 785
Malewski, R., 1184, 1191
Malik, K. L., 650, 654
Malischeva, V. T., 747
Malisoff, W. M., 747
Malkin, O. A., 1187, 1191
Malkov, M. P., 1265
Malmberg, M. S., 951
Mal'tsev, B. K., 429, 499
Malvorsen, T., 877

Mamantov, G., 950–952
Mameli, E., 867
Mandel, M., 586, 588, 605
Mangold, K., 438, 492, 498
Manning, D. L., 951
Manowitz, B., 748
Manuelli, C., 867
Mao, H. K., 144, 145
Marassi, R., 952
Marchman, H., 373, 378
Marcus, Y., 748
Marek, W. J., 131, 320
Marenko, I. N., 473, 498
Margolin, H., 142, 144
Margrave, J. L., 625, 638, 641, 651, 653–655, 748, 1190, 1191
Margules, M., 785
Markham, A. E., 748
Markiewicz, J. P., 1184, 1191
Markov, B. F., 903, 951
Markovits, G. Y., 725
Markus, W., 189, 201
Marple, S., 417
Marriner, R., 319
Marsh, K. N., 454, 496, 772, 775, 785
Marsh, S. P., 832
Marshak, H., 1265
Marshall, A. L., 656
Martel, C. S., 655
Martel Jr., C. S., 1067
Martin, A. J. P., 776, 785, 1066
Martin, J. F., 345, 612, 653, 654
Martinez, J. P., 372, 378
Martings, C. R., 1156, 1158
Martire, D. E., 785
Martire, J. F., 650, 654
Masalov, Ya. F., 418
Mascarelli, L., 867
Masi, J., 1217
Masi, J. F., 496, 499
Maslach, G. J., 211
Maslennikova, V. Ya., 791, 796, 801
Mason, E. A., 372, 380, 413, 418, 576
Mason, W. P., 549, 550, 575, 576
Mason, W. R., 574
Massalski, T. B., 435, 495
Massen, C. H., 1067
Massie, D. S., 345
Masson, I., 369–371, 373, 374, 378, 379
Mastrangelo, S. V. R., 832, 867
Mate, C. F., 1256, 1264
Mather, J. W., 1183, 1192
Matheson, A. J., 538, 541, 549, 576
Mathews, J., 867
Mathews, J. F., 374, 379
Mathias, E. C., 676, 721
Mathieu, J. C., 843, 899
Mathot, V., 785
Mathur, G. P., 462, 499

Mattick, A. T., 1188, 1190
Matukuma, A., 650, 654
Maxted, E. M., 735, 748
Maxwell, J. C., 254, 270, 271
May, A. D., 456, 499
Mayberry, W. R., 650, 653
Mayer, H., 1057, 1066, 1067
Mayer, J. E., 724
Mayer, L., 867
Mayers, C. H., 369, 379
Mazur, P., 586–588, 605
Mazzocchin, G. A., 950
Mears, W. H., 211
Meatyard, R., 634, 656
Medana, R., 1067
Medford, R. D., 1183, 1190
Medley, S. S., 1183, 1191
Meeks, A. C., 655
Meilikhov, E. Z., 1183, 1191
Meinke, C., 257, 258, 271, 300
Melfi, L. T., 299, 300
Mellors, G. W., 952
Melville, S. H., 1027, 1067
Mendelssohn, K., 785
Mendez, J., 952
Mendoza, E., 1265
Menon, P. G., 1046, 1067
Menzies, A. W. C., 616, 648, 655, 748
Mermin, N. D., 676, 723
Merrill, L., 140–142, 145
Merriman, R. W., 748
Merten, U., 655
Metlay, M., 653
Metropolis, N., 1158
Metzger, J., 860, 867, 899
Meunier, P., 385, 386, 418
Meunier, P. A., 211
Meyer, D. E., 1049, 1067
Meyer, E., 576
Meyer, E. A., 300
Meyer, H., 70, 112, 690, 692, 695, 696, 698, 721, 724, 1264, 1266
Meyer, J., 488, 498
Meyer, R., 860, 867, 899
Meyer, S. W., 877
Meyers, C. H., 201, 655, 833
Michel, A., 1046, 1067
Michels, A., 70, 160, 201, 349, 350, 358, 359, 363, 368–374, 379, 380, 437, 458, 482, 485, 493, 498, 562–565, 576, 582, 585, 600, 605, 655, 684, 691, 692, 723, 795, 798–800, 832, 1208, 1218
Michels, C., 369, 379, 458, 482, 485, 498, 562–564, 576, 691, 692, 723
Michelsen, A. A., 304
Michon, M., 1188, 1191
Mii, H., 101, 102, 112
Mikhailenko, S. A., 568, 570, 571, 575, 576
Miles, M. H., 1125, 1158

Miller, B. E., 1218
Miller, D. G., 952
Miller, E. E., 456, 498
Miller, G. A., 625, 655
Miller, J. E., 380
Miller, J. G., 374, 379, 419
Miller, J. R., 300
Miller, K. W., 746, 748
Miller, M. H., 1188, 1192
Miller, R. R., 211, 402, 418, 748, 973, 974, 980, 986, 988, 990
Miller, W. T., 867
Millero, F. J., 440, 452, 453, 485, 487, 498
Millet, M., 467, 498
Mills, R. L., 832, 1260, 1261, 1264, 1265
Milne, A. B., 320
Minardi, J., 1201, 1202, 1218
Minnich, B. H., 748
Minshall, F. S., 141, 145
Misdolea, C., 898, 900
Mishin, V. P., 832
Mistura, L., 539, 568, 573–576, 718, 719, 722, 723
Mitchell, A. C., 1150, 1158
Mitschele, C. J., 1189, 1192
Mitsui, K., 131, 1233, 1265
Mochel, J. M., 784, 785
Mochizuki, T., 131, 1233, 1265
Moelwyn-Hughes, E. A., 490, 497
Moessen, G. W., 326, 346
Moesta, H., 300, 1185, 1191
Mohri, M., 1064, 1067
Moiseev, B. N., 1150, 1158
Moldover, M. R., 683, 695, 696, 698, 720–723
Mcles, E., 748
Mølgard, H. V., 589, 590, 596, 601, 605, 606
Molody, T. K., 1003, 1020
Monchick, L., 413, 418
Monod, J., 1099, 1108
Montemartin, C., 867
Montgomery, H. A. C., 748
Montgomery, P. W., 498
Moon, C. H., 735, 748
Moore, G. E., 300
Moore, H. K., 1177, 1178, 1190
Moore, J. W., 260, 270
Moore, R. T., 372, 373, 378, 417, 620, 653
Mopsik, F. I., 466, 469, 498
Morachevskii, A. G., 903, 950
Morand, G., 950, 951
Morcom, K. W., 748
Moreau, C., 1054, 1067
Morey, D. C. D., 374, 379
Morgan, I., 320, 841, 868, 899
Morgan, P. D., 1189, 1191
Morgan, R., 868
Morris, E., 1149, 1150, 1158

Morrison, J. A., 1024, 1049, 1063, 1065–1067
Morrison, J. T., 735, 736, 748
Morse, M. S., 1190
Moser, H., 300, 346, 857, 899
Mota, A. C., 1256, 1265
Motard, R. L., 373, 379
Motulevich, G. P., 1240, 1264
Motylewski, S., 748
Motzfeldt, K., 645, 655, 831, 926, 952
Moulin, M., 848, 899
Mountain, R. D., 575, 606, 684, 709, 716, 720, 722, 723
Muan, A., 832
Mueller, E. P., 630, 655
Mueller, W., 372, 374, 380
Muendel, C. F., 271
Muijlwijk, R., 1231, 1232, 1264
Müller, H. J., 850, 870, 871, 894, 899
Munro, D., 830
Munster, V., 370, 378
Murphy, K. P., 211
Murray, J., 499
Musatty, I., 867
Mustajoki, A., 877
Myers, D. B., 757, 784, 785
Myznikov, K. P., 1183, 1190

Nagle, D. E., 1264
Najdich, Ju. V., 1020
Nakayama, K., 247, 248, 254, 256, 257, 270, 271, 299, 1031, 1066
Nakeo, I. F., 300
Nakowska, J., 767, 785
Natrella, M. G., 70
Naugle, D. G., 576
Naumenko, Zh. P., 688, 723
Nebuloni, M., 85
Nederbragt, G., 372, 374, 379
Neganov, B. S., 1244, 1265
Neil, D. E., 952
Neiler, J. H., 1225, 1265
Nejmark, B. E., 396, 418
Nelson, L. C., 409, 418
Nelson, P. A., 70, 973, 977, 980, 985, 989, 990
Nelson, R. R., 466, 498
Nemethy, G., 1099, 1108
Nernst, W., 847, 899
Nesland, A., 881, 899
Nesmeyanov, A. N., 655
Nesterikhin, Yu. E., 1183, 1191
Neuenschwender, E., 877
Neumann, C., 254, 270, 271
Neumann, K., 643, 655

AUTHOR INDEX

Neumann, K. K., 1176, 1184, 1186, 1187, 1189, 1191, 1192
Newhall, D. H., 140, 145, 152, 153, 162, 163, 201, 211, 464, 467, 474, 498, 832
Newitt, D. M., 70, 130, 144, 201, 424, 434, 456, 496, 498
Newman, L. T., 248, 270
Newman, M., 507, 526
Newman, M. S., 867
Newton, R. C., 142, 145, 826, 827, 831
Newton, R. F., 652, 950, 952
Nichiporovich, G. A., 300
Nicholson, G. A., 367, 380, 418
Niedermayer, R., 1067
Niesen, F., 370, 379
Nienhuis, K., 300
Nietz, A. H., 1020
Nihoul, J., 556, 577
Nijenko, V. I., 1020
Nijhoff, G. P., 371, 373, 380
Nikanarov, E. V., 974, 987, 988, 990
Nikitine, Ju. P., 1021
Nilsson, N. R., 1187, 1191
Nishibata, K., 131
Nogatkin, A. G., 211
Noonan, E. C., 1216, 1217
Norman, J. V., 655
Nottebohm, C. I., 867
Nottingham, W. B., 271, 279, 300
Novak, P. E., 111
Novikov, S. A., 1158
Noyes, A. A., 748
Noyes, M., 1108
Nozdrev, V. F., 549, 568, 576
Null, M. R., 1191
Nuttall, R., 1217
Nyquist, H., 1229, 1265
Nys, F., 574, 577

Obert, E. F., 409, 418, 560, 575
Oddo, G., 867
Oeder, D., 793, 794, 797, 800
Oersted, M., 450, 498
Oertel, G. K., 1183, 1191
Oguri, T., 1026, 1027, 1067
Ohira, N., 1184, 1190
Ohtsuki, T., 1067
Oishi, J., 369, 380
Oktay, E., 1185, 1191
Okada, S., 301
Olds, R. H., 369, 370, 372, 374, 380
Olin, J. B., 1215, 1218
Oliver, G. D., 655
Oliver, J. P., 1039, 1067
Onsager, L., 582, 606, 658, 688, 694, 722, 723, 1080, 1108
Opekunov, A., 831
Oppenheim, A. K., 1187, 1190

Orcutt, R. H., 597, 599, 605
Orlova, M. P., 345
Ornstein, L. S., 709, 712, 723
Ornstein, L. Th. M., 1183, 1191
Orr, C. T., 1067
Orye, R. V., 785
Osaka, H., 867
Osborn, A., 325, 346, 640, 653, 655
Osborne, N. S., 637, 655
Osheroff, D. D., 1226, 1247, 1248, 1265
Oster, G. F., 70, 985, 986, 989, 990
Osteryoung, R. A., 951
Ostromisslensky, P., 867
Østvold, T., 907, 911, 913, 951, 952
Otto, J., 346, 357, 369, 371–373, 378, 417, 483, 497
Oudemans, G. J., 597, 605
Ovcharenko, V. G., 984, 985, 990
Overberger, J. E., 655
Overdijk, S. D. J., 605
Owen, B. B., 55, 70, 748
Owen, P. T., 868, 899
Owens, B. B., 823, 830
Owens, C. L., 300
Øye, H., 952
Ozawa, S., 1067

Pace, E. L., 1067
Padoa, M., 867
Page, C., 70
Pak, J. C., 793, 800
Palmer, H. B., 688, 723
Pancyk, M. F., 1237, 1238, 1261, 1265
Paneth, F. A., 747
Panish, M. B., 952
Pankey, J. W., 951
Papadakis, E. P., 554, 576
Papadopoulos, D., 1108
Papetti, R. A., 1157, 1158
Parham, D. N., 490, 499
Park, J. H., 1170, 1191
Parker, P. D. M., 746
Parker, R., 1177, 1191
Parker, W. H., 28, 70, 346
Parkinson, W., 1217
Parlin, R. B., 1158
Parpia, D. Y., 589, 595–597, 606
Parsons, C. A., 470, 472, 498
Partington, J. R., 832
Parvatikar, K. G., 997, 1021
Pask, J. A., 832
Pastine, D. J., 145
Paterno, E., 867
Paul, W., 832
Paule, R. C., 655
Paulowich, S., 469, 496
Pauls, N., 1186, 1192
Pauly, H., 973

1287

Pavlichenko, L. A., 688, 721
Pavlov, P. P., 1003, 1020
Pavlovich, N. V., 372, 380, 485, 498
Payne, R. T., 112
Payne, W. H., 831
Peacock, N. J., 1189, 1191
Pearce, J. N., 647, 655
Pearson, E. G., 85
Peavey, S. T., 201
Pecceu, W., 576
Pecsok, R. L., 785
Pedersen, Aa., 1184, 1190
Pella, E., 85
Pellinen, D. G., 1184, 1191
Pelupessey, B., 1067
Pendlebury, J. R., 1185, 1191
Pendleton, W. K., 1183, 1191
Penney, M. F., 652, 769, 770, 784
Penning, F. M., 300, 369, 370, 380
Penther, C. J., 832
Pepela, C. N., 652, 767, 784
Perano, J. L., 877, 951
Perez-Albuerne, E., 832
Perkins, J., 430, 445, 463, 498
Perkins, P. G., 654
Perrin, D., 748
Pert, G. J., 1183, 1190
Peter, S., 791, 800
Peters, E. T., 107, 112
Petersen, D. E., 576
Petford, A., 1212, 1217
Petit, G., 877, 888, 889, 899
Petrov, V. R., 371, 381
Petrucci, S., 950
Petsev, N., 655
Pettit, L. D., 867
Pezzati, E., 952
Pfennig, H. W., 373, 380
Pfohl, F. W., 653
Philip, W. G., 746
Phillips, C. S. G., 346, 1056, 1066
Phillips, D., 1066
Pick, J., 651, 654, 787, 800
Pickford, P., 748
Pierret, E., 868
Pierotti, R. A., 1058, 1067
Piethorn, W. S., 1066
Pignocco, J. M., 656
Pikal, M. J., 952
Pilchikov, N. D., 994, 1020
Pillon, M., 1217
Pine, A. S., 576
Pings, C. J., 590–595, 600, 605, 606, 684, 685, 722, 723, 1155, 1158
Pistorius, M. C., 832
Pistorius, C. W. F. T., 136, 145, 474, 498, 803, 826, 827, 832
Pitman, H. W., 117, 130, 233, 256, 270, 639, 640, 653, 1032, 1066

Pitzer, K. S., 569, 576
Placzek, G., 715, 722
Plambeck, J. A., 952
Platzeck, R. P., 1256, 1265
Pleuger, F., 867
Plumb, H. H., 552, 557, 575, 576, 1190–1192, 1233, 1264–1266
Podgurski, H. H., 252, 255, 271, 1031, 1067
Podurets, M. A., 1150, 1158
Poffenberger, N., 357, 370, 373, 377
Pohl, F. M., 1108
Poindexter, F. E., 998, 1021
Pointud, Y., 950, 951
Pokrovskij, N. L., 998, 1020
Polglase, M. F., 747
Pollara, L. Z., 650, 654
Pollard, L. J., 458, 498
Pollet, A. P., 898
Polonsky, N., 574, 575
Poltz, H., 300
Polyakov, E. V., 369, 380, 397, 418
Polyakov, Yu. A., 1185, 1191
Pomeranchuk, I., 1247, 1265
Pompe, A., 416
Ponec, V., 652
Ponsot, A., 851, 878, 899
Ponyatovskii, E. G., 145, 830
Pool, R. A. H., 346
Popel, S. I., 999, 1021
Popenoe, C. H., 1173, 1192
Pople, J. A., 585, 605
Popov, L. V., 1150, 1158
Popov, V. A., 695, 724
Popov, V. H., 371, 381
Poppe, G., 800
Porta, H., 867
Porta, G., 898
Porter, A. W., 655, 996, 1021
Porter, G., 1107, 1108
Porter, G. B., 867
Porter, R. S., 830
Posey, C. D., 747
Post, R. S., 1183, 1191
Postill, D. R., 438, 487, 498, 978–980, 984, 990
Potier, A., 896, 899
Potter, D. J. B., 340, 346
Poulis, J. A., 1067
Poulter, J., 1256, 1264
Poulter, K. F., 300
Powell, R. E., 1158
Powell, R. F., 112
Powell, R. L., 112, 113
Power, B. D., 270
Poynting, J. H., 622, 655
Praesgaard, E., 413, 417
Prausnitz, J. M., 413, 418, 747, 758, 774, 775, 777, 785, 796, 800
Pray, H. A., 748

Predel, B., 816, 832
Preece, F. H., 746
Prengle, H., 373, 379
Prentiss, S., 880, 899
Presnall, D. C., 398, 418, 438, 483, 498
Preston-Thomas, H., 328, 330, 331, 346, 1264
Pribadi, K. S., 462, 498
Price, D., 1218
Price, M. J., 1188, 1191
Prigogine, I., 767, 785, 839, 899
Prins, C., 370, 380
Proskurin, V. B., 974
Provine, J. A., 374, 380
Prue, J. E., 766, 767, 785, 880, 882, 898
Prutton, M., 1066
Prydz, R., 70, 350, 351, 370, 380, 381, 493, 499, 505, 525, 655, 723
Puddington, I. E., 783, 785
Pugachevich, P. P., 991, 998, 1020
Pugh, H. L., 1159
Pugh, H. L. D., 416
Puglielli, V. G., 715, 723
Purnell, J. H., 653, 778, 784, 785
Pushin, N. A., 867
Pushkinskii, M. D., 466–468, 499
Pyshnov, A. V., 1187, 1191

Quincke, G., 996, 1001, 1003, 1021
Quinn, T. R., 111, 113

Rabinovich, V. A., 416, 418
Rabl, C.-R., 1108
Radebaugh, R., 1229, 1230, 1235, 1265
Raizer, Y. P., 1125, 1127, 1159
Ralston, A. W., 747
Rambeau, G., 256, 271
Ramsay, W., 369, 371–373, 380, 426, 498, 618, 634, 655, 867
Ramsden, S. A., 1183, 1190
Ramsteadt, F. E., 748
Ranbosch, G. M., 654, 781, 785
Rand, M. J., 1051, 1052, 1067
Randall, M., 880, 882, 899
Rands, Jr, R. D., 833
Rankin, G. A., 809, 832, 833
Ransom, L. D., 877
Rao, M. R., 569, 576
Raoult, F. M., 837, 839, 847, 848, 878, 887, 899
Rapoport, E., 136, 145, 826, 832
Rapp, D., 300
Rapp, R., 1243, 1256, 1265, 1266
Rapp, R. A., 653, 655
Rauch, A., 1109
Rautbort, A. E., 1186, 1191
Rayleigh, Lord, 993, 1021

Raymond, C. L., 763, 764, 785
Raynaud, S., 1188, 1191
Razumikhin, V. N., 485, 486, 495
Read, J. W., 656
Reamer, H. H., 369, 370, 372–374, 380, 1215, 1218
Recapet, J. M., 877
Reck, G., 1217
Redeker, T., 1186, 1191
Redhead, P. A., 273, 299, 300, 1067
Redlich, O., 757, 785
Ree, F. H., 375, 380
Ree, T., 899
Reed, T. M., 417
Rees, J. A., 271
Reeves, C. G., 338, 340–342, 345, 784
Rehbinder, P. A., 1007, 1021
Rehr, J. J., 723
Reich, G., 257, 258, 271, 300
Reid, R. C., 411, 418
Reinhartz, K., 791, 800
Reiss, H., 898, 899
Renkert, H., 987, 989, 990
Renn, R., 300
Renon, H., 759, 785
Rentzepis, P. M., 1189, 1192
Reppy, J. D., 1264
Rettinghaus, G., 299
Reutler, M., 867
Reynolds, C. A., 1265
Reynolds, O., 254, 270, 271
Rhodes, F. H., 748
Rhodes, H. A., 1216, 1217
Rhodin, T. N., 1054, 1067
Ricca, F., 1067
Riccardi, R., 877
Rice, M. H., 1140, 1149–1151, 1156, 1157, 1158, 1159
Rice, O. K., 705, 724
Rice, S. A., 782, 784
Richard, A. J., 488, 498
Richard, M., 211
Richards, L. W., 1062, 1063, 1065
Richards, T. W., 436, 449, 450, 498
Richardson, A., 372, 380
Richardson, E. G., 549, 576
Richardson, F. D., 748, 877, 952
Richardson, G. M., 867
Richardson, M. C., 1183, 1189
Richardson, M. J., 654, 655
Richardson, R. C., 1226, 1247, 1248, 1265
Riddick, J. A., 748
Rideal, E., 951
Rigby, H. A., 474, 498
Rigler, R., 1108
Riley, B., 811, 832
Rimatori, C., 867
Ringwood, A., 832
Ritchie, G. L. D., 490, 498

AUTHOR INDEX

Rives, J. E., 1264
Rivière, C. A., 987, 988, 990
Rivkin, S. L., 370, 380, 693, 723
Roach, P., 574
Roach, P. R., 599, 606, 674, 690, 723
Roark, R. J., 366, 380
Robb, J., 625, 653
Robens, E., 1060, 1066, 1067
Roberts, A. L., 831
Roberts, H. M., 867
Roberts, J. R., 1188, 1192
Roberts, T. R., 85, 1266
Roberts-Austen, W. C., 817, 832
Robbins, C. R., 832
Robertson, A. R., 90, 113
Robertson, C., 880, 885, 886, 899
Robertson, E. C., 830
Robertson, P. W., 867
Robertson, S. L., 357, 369, 370, 372, 373, 377, 380, 391, 392, 399, 411, 416, 418, 482, 495
Robin, S., 655
Robinson, D. K., 551, 575
Robinson, F. N. H., 1245, 1265
Robinson, R. A., 655, 891, 899, 1099, 1108
Robinson, R. H., 785
Robinson, R. L., 778, 786
Röck, H., 615, 655
Rodebush, H. J., 886, 899
Rodebush, W. H., 631, 637, 638, 640, 653, 655
Roder, H. M., 373, 378, 505, 526, 722
Rodgers, R. I., 370, 381
Rodionov, K. P., 486, 500
Roe, D. K., 951
Roebuck, J. R., 456, 498
Roehrig, J. R., 300
Roesner, W. F., 85
Rogers, B. L., 775, 777, 785
Rogers, J. S., 1265
Rogers, K. S., 488, 498
Rogers, L. A., 1149, 1158
Rogers, L. B., 830
Rogovaya, I. A., 369, 380, 483, 499
Rohr, W. G., 974
Rol, P. K., 271, 299
Rolin, M., 877
Rolla, M., 868, 877
Rollet, A. P., 870
Roloff, M., 851, 878, 899
Rolt, F. H., 314, 320
Roman, B., 1208, 1217
Romberger, K. A., 883, 898, 952
Rondeau, R. E., 655
Rony, P. R., 1028, 1066
Roof, J. G., 456, 500
Rose-Innes, A. C., 1220, 1265
Rosen, F. D., 626, 655
Rosen, J. S., 591, 606

Rosenbaum, E., 867
Rosenbaum, R., 1247, 1264
Rosenberg, A. J., 655, 1040, 1041, 1067
Rosenberg, P., 247, 270, 271
Rosenblum, C., 747
Rosenstreich, J., 952
Ross, M., 1152, 1154, 1158
Ross, R. G., 438, 487, 498, 976, 978–980, 983, 984, 989, 990
Ross, S., 1039, 1067
Rossini, F. D., 10, 70, 629, 655, 831, 833, 849, 850, 877, 899
Rossiter, B. W., 656
Rossmann, M. G., 655
Rossotti, H., 903, 952
Roth, W., 867
Rothe, E. W., 257, 271, 300
Rothe, R., 831
Rotondi, G., 867
Rott, L. A., 801
Roubeau, P., 1243, 1265
Rouel, G., 1208, 1217
Rouge, E., 867
Roughton, G. W., 1108
Rouquerol, J., 951
Rowlinson, J. S., 346, 414, 418, 475, 497, 575, 576, 621, 623, 654, 655, 676, 723, 724, 783, 785, 800, 974, 1210, 1218
Royce, E. B., 1158
Rubin, A. G., 1186, 1189
Rubin, B., 951
Rubin, L. C., 503, 504, 525, 1265
Ruchholz, E., 150, 201
Rüdorff, F., 847, 899
Rugta, K. K., 462, 499
Ruhemann, M., 833
Rühenbeck, A., 492, 499
Ruoff, A. L., 95, 112, 113, 973
Rusch, M., 271
Rushbrooke, G. S., 723
Ruthberg, S., 229, 271, 1035, 1066
Ryabchikov, I., 831
Ryabinin, Yu. N., 1210, 1212, 1218
Ryan, J. J., 107, 112
Ryzhkov, Yu. F., 984, 985, 990

Sabbah, R., 882, 899
Sacchetto, G. A., 952
Sadkov, V., 833
Sadkov, Yu. A., 146
Sadkovich, N. P., 1185, 1190
Sage, B. H., 369, 370, 372–374, 379, 380, 442, 480, 482, 499, 1215, 1218
Sage, M., 998, 1020
Sailor, V. L., 1265

Sainsbury, J. E. J., 766, 767, 785
Saito, T., 1062, 1067
Sakaida, S., 112
Sakurai, H., 131, 1233, 1265
Saladin, E., 833
Salge, J., 1186, 1187, 1191, 1192
Salkind, A., 497
Saltzman, P. K., 1155, 1158
Samara, G. A., 143, 145, 833
Sameshima, J., 369, 380, 764, 785
Samson, J., 1218
Samsonov, G. V., 70
Sanders, P., 1217
Sando, K., 1210, 1218
Sandstede, G., 1060, 1066, 1067
Sarwinski, R. E., 1264, 1265
Sasano, T., 1184, 1192
Sass, A., 374, 380
Satkiewicz, F. G., 773, 785
Sauerwald, F., 833, 974
Saurel, J. R., 112, 347, 380, 389, 403, 404, 418, 438, 442, 462, 493, 499
Saville, G., 321, 346, 367, 369, 381
Savvatimskii, A. I., 1186, 1187, 1191
Saylor, C. P., 85
Scatchard, G., 757, 758, 763, 773, 784, 785, 879, 880, 882, 899, 952
Scatchard, O., 655
Scatena, G. M., 211
Schaaffs, W., 549, 555, 561, 570, 576
Schaafsma, J. G., 373, 380
Schachman, H. K., 488, 496, 499
Schadow, E., 793, 801
Schäfer, K., 340, 346
Schairer, J. F., 810, 830, 833
Schamp, H., 372, 373, 380, 472, 499
Schanfoldt, D. W., 113
Shanks, H. R., 418
Scheel, K., 304, 320
Schenk, R., 867
Schenk, W. J., 1188, 1192
Scherick, R., 867
Scheuer, O., 867
Schiessler, R. W., 466, 496
Schinke, H., 833
Schins, H. E. J., 973, 974
Schinz, G., 149, 201
Schirber, J. E., 113
Schissel, P. O., 300
Schleicher, K. E., 495
Schlier, R. E., 1214, 1217
Schmeissner, F., 570, 575
Schmid, J., 570, 575
Schmid, L., 867
Schmidt, E. H. W., 686, 705, 723
Schmidt, G., 271
Schmidt, H. H., 695, 697, 723
Schmidt, L., 605
Schmidt, P. W., 715, 720, 721, 723

Schmidt, W., 299
Schmutzler, R. W., 990
Schneider, G. M., 474, 495, 785, 787, 793–798, 800, 801
Schneider, R. T., 1218
Schneider, S. J., 85, 807, 833
Schneider, W. G., 367, 380, 381, 418, 419, 438, 494, 500, 539, 575, 591, 606, 680, 691, 692, 701, 718, 721, 722, 724
Schnepp, O., 605
Schodder, G. R., 576
Schoen, J., 1107
Schofield, P., 673, 675, 723
Scholander, P. F., 748
Scholten, A., 990
Schomburg, G., 655
Schooley, J. F., 1239, 1265
Schoonover, R. M., 170, 201
Schornack, L. G., 469, 499
Schouten, J. A., 791, 801
Schramm, B., 340, 346
Schrank, G., 1183, 1192
Schreiner, E., 867
Schröder, F. K., 576
Schroen, W., 1067
Schuemann, W. C., 300
Schuhmann, S., 271, 300
Schultze, H., 201, 369, 371, 378, 1059, 1065
Schulz, F., 373, 378
Schulz, G. J., 300
Schulze, A., 211
Schumacher, E. E., 994, 1021
Schürenkämper, A., 1040, 1065
Schurer, K., 1205, 1218
Schuster, I., 1100, 1108
Schwab, A. J., 1184, 1192
Schwab, F. W., 857, 899
Schwartz, R., 1201, 1202, 1218
Schwarz, G., 1108
Schwarz, H., 264, 270, 654
Schweickert, C. E., 748
Schwenker, R., 833
Scott, D. W., 748, 832, 951, 1189, 1220
Scott, G. A., 357, 370, 381
Scott, G. J., 144, 369, 380, 391, 399, 411, 416, 418, 482, 495, 823, 832
Scott, G. N., 880, 899
Scott, R. B., 833, 1220, 1265
Scott, R. L., 575, 747, 757, 785
Scott, T. A., 112
Scrase, N. G., 706, 722
Scribner, R. A., 1237, 1238, 1261, 1265
Scroger, M., 111
Sears, F. W., 1085, 1107
Sears, G. W., 614, 654, 655
Sears, J. E., 131
Sechenkov, A. R., 494, 497, 973, 983–985, 989, 990
Seeder, W. A., 467, 499

Seeling, J., 1108
Seemann, H. J., 974
Seibert, E. E., 648, 653
Seifert, F., 801
Seitz, F., 1158
Seitz, Q. W. L., 1150, 1156, 1159
Seki, S., 951
Sekoyan, S. S., 140, 144
Seligman, P., 1264
Sell, H., 551, 576
Semerchan, A. A., 146, 833
Semon, H. A., 84
Senderoff, S., 952
Sengers, A. L., 655
Sengers, J. V., 655, 715, 720–723
Senoo, M., 101, 102, 112
Sensui, Y., 256, 271, 1028, 1029, 1032, 1067
Sereda, P. J., 1067
Serpan, C. Z., 974
Serpinsky, V. V., 1045, 1046, 1066
Sessions, R. F., 1050, 1067
Sessler, G., 576
Sette, D., 539, 541, 568, 573–577
Seward, R. P., 877
Seybert, R. A., 831
Seyss, L., 150, 201
Shakhovskoi, G. P., 466–468, 499, 830
Shamir, J., 748
Shapiro, H., 653
Sharp, W. E., 833
Sharpless, W. M., 1190
Shaw, J. A., 647, 655
Shaw, M. L., 300
Shaw, P. F. D., 1225, 1264
Shaw, R., 1151, 1156, 1158
Shchekochikhina, V. V., 695, 724
Shear, D. D., 1186, 1192
Sheffer, F. E. C., 653
Sheindlin, A. E., 973, 1189
Shenker, H. S., 414, 415, 417
Shepherd, E. S., 809, 833
Shepherd, M., 655, 748
Sherman, R. H., 85, 462, 483, 499, 722, 723, 1238, 1252, 1258, 1265
Sherwood, A. E., 413, 418
Sherwood, T. K., 411, 418
Shields, B. D. C., 346
Shields, F. D., 551, 577
Shimada, M., 833
Shimanskaya, E. T., 688, 721, 723
Shimanskii, Yu. I., 688, 721, 723
Shipley, R., 450, 498
Shipman Jr., J. D., 1184, 1191
Shmin, Yu. I., 140, 144
Shore, F. J., 1265
Shpilrain, E. E., 953, 974, 987, 988, 990
Shredov, K. K., 1144, 1158
Shumaker, J. B., 1173, 1192
Shumsky, J., 1217

Sidgwick, M. V., 748
Sidgwick, W. V., 748
Siebert, A. R., 1067
Siedentopf, H., 996, 1021
Siegbahn, K., 1264, 1265
Siegwarth, J. D., 1229, 1230, 1265
Sifner, O., 401, 402, 417
Silberberg, I. H., 371, 381
Silva, J., 801
Silva, W. J., 618, 655
Silverman, M. D., 950
Silverman, S., 1218
Simakov, G. V., 1150, 1158
Simard, G. L., 357, 370, 377, 378
Simkin, V. G., 695, 724
Simon, F., 491, 492, 499, 833
Simon, F. E., 493, 496, 1245, 1265
Simon, I., 113, 459, 496
Simon, M. F., 1184, 1192
Simon, N., 1006, 1021
Simonds, J. C., 300
Simons, J. H., 748
Sinclair, D. A., 655, 785
Singh, Jaswant, 612, 655
Singleton, J. H., 300, 345, 1043, 1066
Sinistri, C., 877, 952
Sinitsyn, M. V., 1150, 1151, 1158
Sinka, J. V., 211
Sinnige, L. R., 747
Sinnock, A. C., 589, 600, 606
Sinor, J. E., 748
Sirianni, A. F., 783, 785
Sirota, A. M., 429, 499
Sites, J. R., 1226, 1247, 1248, 1265
Siu, M. C. I., 271
Siwoboloff, A., 617, 655
Sjölander, A., 499
Skelton, G. F., 795, 798–800
Skidmore, I. C., 1149, 1150, 1158
Skinner, J. F., 456, 465, 499, 833
Skljarenko, L. I., 1020
Skurat, V. E., 747
Slesarev, V., 833
Slutskii, A. B., 832, 833
Slutsky, L. T., 1057, 1067
Smetana, F. O., 271, 300
Smethurst, B., 853, 899
Smirnov, S. P., 830
Smirnov, Yu. B., 1187, 1191
Smith, A., 297, 616, 655, 748
Smith, A. H., 113
Smith, A. J., 371, 378
Smith, A. L., 270
Smith, A. W., 451, 499
Smith, B. L., 579, 589–591, 595–597, 600, 605
Smith, C. S., 1066
Smith, D., 1218
Smith, E. B., 417, 746

Smith, G., 1212, 1217
Smith, G. P., 950, 952
Smith, H., 634, 656
Smith, H. A., 647, 655, 1265
Smith, H. M., 867
Smith, I. W., 715, 717, 723
Smith, J. S., 55, 70
Smith, L. B., 371–373, 379, 381, 418, 477, 499
Smith, R. A., 748
Smith, Jr., W. H., 653
Smod, E. I., 357, 371, 379
Smoluchowski, M. V., 709, 723
Smoluchowsky, R., 724
Smothers, W. J., 833
Smura, E. J., 552, 575
Smyth, F. H., 438, 499
Smyth, H. T., 819, 833
Snigirev, V. G., 695, 724
Snow, R. D., 647, 655
Snyder, P. S., 466, 499
Sobelev, N., 1211, 1217
Sobotka, H., 748
Sokhan, V. I., 688, 721
Solbrig, C. W., 381
Solem, J., 588, 605
Solomons, C., 872, 873, 877, 899
Soloukhin, R. J., 1144, 1158
Solov'ev, A. N., 973
Soloviev, A. N., 1021
Sommers, H. S., 1265
Sonntag, R. E., 372, 381
Sood, D. D., 952
Soriano, J., 748
Sorokin, V. A., 466, 469, 495, 570, 571, 576
Soroos, H., 653
Souchay, P., 895, 899
Soulen, R. J., 1235, 1239, 1265
Sourirajan, S., 801
Southard, J. C., 833
Søvik, R. P., 1245, 1264
Spann, J. R., 211, 402, 418, 973, 974, 980, 986, 988, 990
Sparks, L. L., 113
Spauschus, H. O., 624, 655
Spedding, F. H., 55, 69, 655, 974
Speich, G. R., 1191
Spencer, C. M., 301
Speranskaya, M. P., 1150, 1157
Spindler, H., 833
Spiridonov, G. A., 418
Spohr, D. A., 1245, 1265
Spotz, E. L., 560, 574
Sprake, C. H. S., 631, 634, 637, 652
Spurling, T. H., 324, 345, 576
Sridhar, R., 952
Stager, R. A., 462, 499
Staheli, D., 301
Stalkup, F. I., 778, 785

Stallings, C., 1183, 1191
Stammbach, K., 653
Stanley, H. E., 723
Stark, W., 143–145
Starkman, E., 1218
Starling, K. E., 574
Starling, M. E., 375, 378
Statton, W. O., 833
Staveley, L. A. K., 338, 345, 346, 483, 490, 499, 531, 577, 654
Stebnovskii, S. V., 1114, 1158
Steckel, F., 55, 70
Steckelmacher, W., 301
Steel, G. B., 1187, 1190
Steele, E. L., 1188, 1192
Steele, W. A., 655
Steinberg, M., 748
Steinberger, I. T., 605
Steiner, R., 793, 801
Steinfeld, J. I., 538, 575
Steinküller, E. W., 211, 402, 418, 973, 974, 988, 990
Stenerhag, B., 1186, 1187, 1192
Stepanov, V. A., 113, 385, 386, 418
Stephens, D. R., 144, 145
Sterett, K. F., 833
Stern, J., 768–770, 784
Stetser, D. A., 1189, 1190
Stevels, J. M., 833
Stevens, B., 538, 577
Stevenson, W. H., 242, 243, 271
Stewart, E. S., 577
Stewart, J. C., 951
Stewart, J. L., 552, 577
Stewart, J. W., 472, 499
Stewart, R. B., 526, 577
Steyert, W. A., 1250, 1265
Stickings, B. J., 746
Stickney, W. W., 301
Stiel, L. I., 493, 498
Stillman, M. H., 271
Stimson, H. F., 130, 201, 270, 346, 637, 655, 748
Stishov, S. M., 145, 473, 498
Stock, R., 785
Stockbridge, C. D., 1057, 1067
Stockmayer, W. H., 417
Stoddart, C. T. H., 785
Stokes, R. H., 454, 496, 785, 891, 899, 1099, 1108
Stolfa, D. L., 1228, 1229, 1264, 1266
Stolzenberg, H., 370, 378
Stone, J. P., 211, 402, 418, 973, 974, 980, 986, 988, 990
Stone, N. W. B., 70, 116, 130, 320
Stook, P., 830
Stortenbecker, W., 841, 900
Stranks, D. R., 625, 626, 641, 653

Straty. G. C., 70, 350, 351, 370, 380, 381, 493, 499, 505, 526, 655, 723
Straub, J., 70, 605, 686, 705, 722, 723
Strauss, W., 458, 498
Street, W. B., 372, 381, 483, 499, 791, 801
Strehlow, H., 1069, 1108
Streiff, A. J., 655, 831
Strelkov, P. G., 345
Strese, G., 373, 378
Strobridge, T. R., 504, 526
Stromberg, A. G., 748
Stromberg, H. D., 144, 145
Strong, H. M., 91, 101–107, 112, 388, 417, 497, 831–833
Stross, F. H., 648, 653, 832
Stroud, L., 380
Stryland, J. C., 456, 485, 499
Stubley, D., 655
Stückrath, P., 152
Stuehr, J., 1108
Stull, D. R., 951
Stull, W. N., 436, 450, 498
Stulla-Gotz, J., 131
Stünkel, D., 1067
Style, D. W. G., 346
Su, G. J., 357, 370, 377, 378
Suchodski, W. A., 450, 451, 499
Sudo, S., 131
Sugden, S., 1007–1009, 1021
Sullivan, J. A., 371, 372, 381
Sundheim, A. R., 952
Sundheim, B., 951
Sunner, S., 617, 655
Suomi, M., 1266
Susse, C., 143, 144, 832, 833
Sutin, N., 1107
Sutphin, H. D., 1188, 1191
Sutton, L. E., 342–344, 346
Swallow, J. C., 371, 378
Swanson, B. S., 70, 973, 977, 980, 985, 989, 990
Swenson, C. A., 1221, 1232–1234, 1264–1266
Swietoslawski, W., 627–629, 652, 655
Swift, H., 1217
Swinehart, C. F., 653
Swinehart, J. H., 1108
Swinkels, D., 952
Swinnerton, J. W., 748
Swinney, H. L., 717, 718, 721–723
Swofford, H. S., 952
Sydoriak, S. G., 85, 1261, 1265, 1266
Symko, O. G., 1227, 1228, 1247, 1264, 1266
Syono, Y., 832
Szapiro, S., 55, 70
Szirmae, A., 1191
Szita, C., 656

Taconis, K. W., 1264, 1266
Tajnsh, R. J., 1265
Tait, P. G., 446, 450, 456, 457, 499
Takács, J., 656
Takahashi, T., 144, 145, 830
Takaishi, T., 255, 256, 271, 1023, 1028, 1029, 1031, 1032, 1064, 1067
Takenouchi, S., 791, 801
Takeo, M., 1202, 1212, 1218
Takeshita, S., 1184, 1192
Takigami, K., 1184, 1190
Talbott, A. C., 450, 499
Tallmadge, J. A., 444, 499
Tal'rose, V. L., 747
Tamm, K., 1094, 1107
Tammann, G., 475, 476, 492, 496, 499, 821, 833
Tanaka, S., 833
Tanishita, I., 482, 499
Tanner, C. C., 369, 371, 374, 378, 379
Tarján, G., 656
Tartaglia, P., 574, 575
Tate, D. R., 169, 201
Tawara, H., 1184, 1192
Tawde, N. R., 997, 1021
Taylor, B. N., 28, 70, 346
Taylor, H. S., 746, 747
Taylor, I. K., 55, 70
Taylor, J. B., 783, 785
Taylor, L. S., 606
Taylor, N. W., 748
Taylor, R. D., 1223, 1224, 1226, 1264–1266
Taylor, R. E., 418
Taylor, R. S., 372, 381
Taylor, W. J., 656, 833, 850, 900
Teague, R. K., 590, 592–595, 600, 605, 606, 723
Telegin, G. S., 1150, 1159
Teller, E., 1158
Teloy, E., 300
Temkin, M., 842, 900
Temperley, H. N. V., 723
ten Seldam, C. A., 585, 588, 605, 1046, 1067
Ter Harmsel, H., 1256, 1266
Terrien, J., 115, 130, 131, 233, 345
Terry, R. E., 95, 113
Tettamanzi, A., 867
Thatcher, W. J., 299
Thiele, E., 1210, 1218
Thiene, P., 1228, 1266
Thiesen, M., 320
Thodos, G., 775, 784
Thoen, J., 540, 564, 566, 567, 569, 574, 575, 577, 718, 722, 723
Thom, K., 1218
Thom, N. S., 748
Thoma, R. E., 952
Thomaes, G., 346
Thomas, A. M., 241, 243, 244, 271

Thomas, D. B., 113
Thomas, J. E., 715, 723
Thomas, J. M., 1067
Thomas, L. H., 634, 656
Thomas, R. D., 654
Thomas, R. J., 1184, 1192
Thomas, W., 346, 370, 381, 493, 499
Thompson, A. M., 606
Thompson, D. R., 705, 724
Thompson, W. T., 952
Thomsen, E. S., 748
Thomsen, J. S., 1159
Thomson, C. W., 499
Thomson, G. W., 618, 625, 643, 646, 656
Thomson, W., 821, 833
Thornton, E., 1184, 1192
Thulin, L. U., 911, 913, 951, 952
Tichanovsky, J. J., 1005, 1021
Tichnor, L. B., 784, 785
Tickner, A. W., 651, 656
Tien, J. K., 112
Tiggelman, J. L., 1233, 1266
Tikhomirova, N. A., 145
Tilton, L. W., 55, 70
Timberg, G., 1021
Timmermans, J., 793, 800
Timrot, D. L., 372, 380, 485, 493, 498, 499
Tischer, R. P., 951
Tisza, L., 658, 694, 724
Tobey, F., 1217
Tobias, C. W., 951
Todd, J., 507, 526
Tödheide, K., 400, 417, 492, 496, 791, 792, 797, 801
Toenshoff, D. A., 113
Tokutaka, H., 1066
Tolhoek, H. A., 1264
Tomlinson, J. W., 438, 442, 491, 495
Tompkins, F. C., 653
Tonkov, E. Yu., 145
Topol, L. E., 877
Toriumi, T., 446, 497
Torney, Jr., F. L., 271
Torricelli, E., 201
Tough, J. T., 1263–1266
Touloukian, Y. S., 577, 1217
Tournarie, M., 717, 721
Townend, D. T. A., 370, 381
Townsend, R., 623, 652
Tracht, J., H., 370, 381
Trakhtengerts, M. S., 418
Trappeniers, N. J., 371, 381
Travers, M., 369, 371–373, 380
Traverse, J., 811, 815, 831
Traving, G., 1212, 1218
Treiber, G., 801
Treiman, L. H., 833
Trelin, J. S., 974
Tremearne, T. H., 357, 369–372, 377, 416

Trendelenburg, E. A., 297
Treverton, J. A., 1191
Troianovskaya, G. V., 693, 723
Trovanelli, A., 867
Truell, R., 549, 577
Truesdell, C., 533, 577
Trump, W. N., 612, 653
Trunin, R. F., 1150, 1151, 1157, 1158
Tsai, D. H., 496, 499
Tsederberg, H. B., 371, 381
Tsiklis, D. S., 113, 347, 362, 369, 372, 373, 379–381, 397, 418, 424, 499, 789–791, 796, 800, 801, 831, 833, 1215, 1218
Tsyganova, T. A., 974
Tufeu, R., 387, 417, 418
Tung, C. T., 1125, 1155, 1158
Tunnicliffe, R. J., 271
Tupman, W. I., 531, 577
Turcotte, R. P., 656
Turnbull, D., 1158
Turner, D. H., 1107
Tuttle, O. F., 800
Tuzi, Y., 301, 1061, 1067
Twentyman, M. E., 654
Tydings, J. E., 831
Tyrer, D., 489, 499

Ulmer, G. C., 498
Ulrich, D. V., 487, 499
Unterwald, F. C., 300
Urbain, G., 998, 999, 1020
Urbanek, K., 301
Urlin, Y. D., 1150–1152, 1158, 1159
Urnes, S., 877, 952
Ursenbach, W. O., 1158
Urtiew, P. A., 1185, 1192
Utterback, N. G., 301
Ury, J. F., 346

Vaisnays, J. R., 112
Vallet, C. E., 877, 896, 900, 950, 951
Van Agt, F. P., 370, 377
van Artsdalen, E. R., 898, 900
van Calker, J., 1186, 1189, 1190
Van Dael, W., 527, 544, 564, 566, 567, 569–574, 576, 577, 718, 723
van den Bosch, A., 1065
van der Gulik, P. S., 723
van der Laan, H. A., 1183, 1191
van der Maesen, F., 585, 605
van der Waals, J. D., 503, 526, 658, 661, 724, 779, 785
Vanderzee, C. E., 951
van Dijk, H., 84, 1256, 1266
van Dijk, J. J. Th., 800
Van Dugen, M. S., 655

Vanfleet, H. B., 138, 139, 141–143, 145, 146
Vangeel, E., 564, 566, 567, 570–574, 577, 718, 723
van Gulik, W., 1260, 1265
Van Itterbeek, A., 379, 424, 433, 483, 493, 499, 544, 556, 557, 569, 576, 577
van Laar, J. J., 722, 785
van Lieshout, A. K. W. A., 456, 496
Van Ness, H. C., 343, 346, 577
Van Oostrom, A., 301
van Regemorter, H., 1212, 1218
van Rijn, C., 1266
van Santen, J. J. M., 379
Vanselow, A. P., 882, 899
van Seventer, W., 369, 380
van Staaten, W., 370, 379
van Steenwinkel, R., 346
van Tets, A., 809, 883
van Thiel, M., 1141, 1154, 1155, 1158, 1159
Van't Hoff, J. H., 837, 900
van Urk, A. Th., 372, 379
van Valkenburg, A., 833
van Vleck, J. H., 586, 605, 606
van Wijk, R. W. M., 974
van Witzenburg, W., 485, 500
Vanyukov, M. P., 1189
Varekamp, F. H., 338, 346
Varner, R. N., 201
Varoquaux, E. J. A., 1243, 1265
Vasilesco, V., 381
Vasil'ev, B. N., 1045, 1046, 1066
Vasil'ev, V., 1208, 1217
Vasilyev, I. N., 974
Vassermann, A. A. A., 416, 418
Vassura, G., 201
Vaughan, W. E., 373, 381
Vaughn, G. W., 344, 345
Vecchiotti, L., 867
Venema, A., 270
Verbeke, O., 424, 433, 483, 493, 499, 693, 722, 724
Verem'ev, E., 1208, 1217
Vereshchagin, L. F., 108–110, 112, 141–143, 146, 167, 201, 385, 386, 394, 395, 418, 433, 464, 484, 499, 830, 833
Verhoek, F. H., 656
Verlet, L., 376, 381, 413, 418
Vermandé, M. J., 271
Vermesse, J., 347
Verreault, R., 270
Verschaffelt, J. E., 659, 724, 993, 1021
Verschoyle, T. T. H., 371, 372, 381
Vicentini-Missoni, M., 70, 722, 724
Vicker, R. K., 900
Vickers, T., 270, 346
Vidal, D., 94, 96, 112, 113, 347, 385, 386, 388, 417, 418
Vigoreux, P., 70
Vilbrandt, F. E., 748

Vilches, D. E., 1242, 1243, 1246, 1266
Vilcu, R., 898, 900
Vilevich, A. V., 394, 395, 418, 433, 484, 499
Vilim, O., 651, 654, 787, 800
Villey, J., 490, 500
Vingee, R. A., 881, 899
Vinogradov, Y. K., 974
Visscher, W. M., 1264
Visser, A., 372, 379
Vitkovitsky, I. M., 1184, 1191
Vladimirov, L. A., 1150, 1157
Vlastos, A. E., 1185, 1186, 1191, 1192
Vodar, B., 112, 201, 387, 404, 417, 418, 438, 442, 462, 493, 499, 655, 832, 833, 1110, 1155, 1158, 1159
Vogl, W. F., 374, 381
Vogler, C. E., 612, 653
Vohra, S. P., 371, 373, 381
Voityuk, B. V., 485, 495
Vold, M. J., 748
Vold, R. D., 748
Völker, E., 643, 655
Volmer, M., 656
Volochine, B., 717, 721
Volyak, L. D., 974
von Antropoff, A., 746
von Dorhn, M., 867
von Halban, H., 886, 900
von Köster, H., 481, 497, 973, 976, 977, 990
Vonnegut, B., 879, 880, 882, 899
von Nieuwenburg, C. J., 492, 498
Voronel, A. V., 659, 695, 721, 724
Voskoboinikov, I. M., 1151, 1159
Vukalovich, M. P., 395–397, 418, 484, 500

Wackerle, J., 1150, 1156, 1159
Wada, Y., 569, 577
Waddington, D., 575
Wade, W. H., 1057, 1067
Wagman, D. D., 877
Wagner, C., 903, 909, 951, 952
Wagner, H. J., 951
Wagner, R. R., 1266
Wakefield, G. F., 974
Walden, P., 867, 893, 900
Walker, A. C., 748
Walker, E. J., 1259, 1265
Walker, P. L., 1066
Wall, V. J., 437, 477, 482, 496
Wallace, B., 70, 690, 692, 724, 1266
Wallace, D., 626, 655
Walsh, J. M., 1140, 1149–1151, 1156–1159
Walstedt, R. E., 1228, 1266
Walter, G., 1060, 1066, 1067
Walters, R. I., 370, 381
Wansink, D. H. N., 1266

Ward, J. W., 656
Ware, K. D., 1183, 1192
Waring, J. L., 833
Warner, A. W., 1050, 1057, 1067
Warner, M. F., 832
Warschauer, D. M., 832
Washburn, E. W., 656
Wasilewski, S., 1054, 1065
Wassenaar, T., 349, 350, 363, 369–374, 380, 433, 493, 498, 500, 655
Watson, I. D., 772, 775, 786
Watson, W., 438, 482, 500
Watts, H. G., 370, 378
Watts, J. F., 299
Waugh, J., 1108
Waxman, M. M., 111, 113, 211, 367–370, 381, 656
Weale, K. E., 130, 434, 456, 495, 496, 796, 801
Weaver, J. S., 144
Webb, G. B., 503, 504, 525
Webb, W., 466, 496, 498
Weber, F. N., 1186, 1192
Weber, L. A., 53, 70, 373, 378, 381, 493, 497, 505, 526, 690, 705, 722, 724
Weber, R. L., 813, 833
Weber, S., 271
Webster, D. C., 370, 380
Webster, T. J., 656
Weddington, G., 620, 653
Weibe, R., 767, 785
Weier, R. D., 346, 367, 369, 381
Weimer, H. R., 346
Weinberger, E. B., 370, 381
Weinberger, M. A., 494, 500, 591, 606, 701, 724
Weinstein, A., 1050, 1067
Weir, C. E., 472, 474, 500
Weiss, H. H., 1188, 1192
Weiss, J. D., 94, 112, 388, 417
Weiss, J. J., 376, 381
Weissberger, A., 495, 653, 656, 748, 1107, 1108
Weissman, S., 472, 499
Welch, B. E., 147
Welch, J. H., 808, 833
Weller, A., 1108
Wells, F. W., 456, 500
Wells, J. E., 1049, 1067
Wendt, H., 1108
Wengler, R. E., 1186, 1190
Wentorf Jr., R. H., 100–102, 105, 113, 388, 419, 499, 1159
Werner, J. G., 301
Werntz, J. H., 558, 575
West, E. D., 833
West, J. M., 952
Westrum, E. F., 951
Westwater, W., 490, 491, 500

Wetterer, G., 271
Wexler, A., 170, 201
Weyl, W. A., 724
Whalley, E., 57, 70, 381, 417, 419, 421, 434, 439, 440, 442, 445, 451, 456, 458, 461, 469, 474, 476–481, 487, 500, 832
Wheatley, J. C., 1228, 1242, 1243, 1246, 1247, 1251, 1256, 1263–1266
Wheeler, J. C., 573, 575, 676, 722
White, A. D., 371, 378
White, D. W., 346, 999, 1021
White, J. A., 715, 723
White, J. L., 422, 484, 495, 500, 654, 748, 833, 952
White, J. R., 55, 70
White, W. P., 849, 858, 900
Whitfield, M., 830
Whitman, C. I., 656
Whitney, W. M., 577
Whittaker, A. G., 640, 654
Whitton, W. I., 785
Whytlaw-Gray, R., 338, 340–342, 345
Wichers, E., 857, 899
Wichterle, I., 775, 786
Widom, B., 659, 669, 676, 724
Wiebe, H. F., 202
Wiebe, R., 371, 381, 419, 483, 500, 748
Wiebes, J., 1254, 1266
Wiedemann, H. G., 645, 656, 809, 833
Wiener, O., 687, 724
Wiese, H. C., 1218
Wijker, H., 369, 380
Wilburn, N. P., 1215, 1218
Wilcox, L. R., 684, 689, 722, 724
Wiley, W. J., 551, 577
Wilhelm, E., 746, 767, 784
Wilkins, J. B., 747
Wilks, J., 1220, 1266
Williams, A. H., 1183, 1192
Williams, D. D., 402, 418, 748, 974, 988, 990
Williams, D. W., 419
Williams, J., 553, 577
Williams, M. D., 1183, 1191
Williamson, A. G., 749, 770–772, 775, 785–787, 801
Williamson, B. H. J., 828, 832
Williamson, E. D., 472, 494
Williamson, R. C., 566, 577
Willingham, C. B., 656, 1066
Willis, G. H., 867
Willms, R., 1187, 1191
Wilsdon, B. H., 748
Wilson, A. E., 656, 974
Wilson, D. C., 130
Wilson, G. M., 758, 773, 785, 786
Wilson, M. F., 1263, 1266
Wilson, W., 440, 466, 469, 500, 833
Wims, A. M., 724
Winchell, P., 655

AUTHOR INDEX

Windsor, M. L., 777, 784
Winkler, C. A., 1054, 1065
Winkler, R., 1094, 1107, 1108
Winnick, J., 466, 488, 499, 500
Winterbottom, W. L., 641, 656
Winters, H. F., 301
Wiswall, R. H., 951, 952
Witkowski, A. W., 369, 381
Witonsky, R. J., 419
Wittenberg, L. J., 974
Wleisteuber, T. C., 1062, 1063, 1065
Wolfe, C. R., 951
Wolkers, G. J., 370–374, 380, 565, 576, 605
Wolsky, S. P., 1056, 1067
Wong, L. Y., 577
Wood, R. H., 900
Wood, S. E., 655, 784, 785, 877, 951
Wood, S. M., 1188, 1192
Wood, W. W., 1155, 1158
Woodbrey, J. C., 784
Woodbury, W. C., 592, 605, 685, 721
Woodman, D. M., 256, 257, 268, 270, 299, 1066
Woolley, H., 1217
Woolsey, G. B., 867
Wooten, L. A., 1039, 1067
Wouters, H., 358, 369, 371, 372, 380, 458, 498, 1208, 1218
Wrentig, P., 867
Wright, F. E., 809, 832, 833
Wright, R., 748
Wu, Y., 271
Wurzbacher, G., 1060, 1067
Wyatt, P. A., 867
Wylie, P., 163, 202
Wyman, J., 1099, 1108
Wynn Jones, I., 346, 367, 369, 381

Ya'akobi, B. J., 1186, 1192
Yadorich, A., 778, 786
Yakimovich, K. A., 974
Yakuba, V. V., 570, 571, 575
Yamakawa, H., 782, 784
Yang, C. N., 658, 699, 722, 724
Yang, C. P., 666, 699, 724
Yao, Y. F. Y., 1067
Yarwood, J., 655
Yasunami, K., 140, 146
Yates, J. T., 300
Yeager, E., 497, 1108
Yekhlakov, A. D., 486, 500
Yntema, J. L., 419, 438, 500
Yoder Jr., H. S., 801, 833
York, R., 373, 378

York, S. G., 786
Yosim, I. J., 898
Young, C. L., 650, 656, 747, 777, 784
Young, D. M., 1024, 1049, 1050, 1063, 1067, 1068
Young, H. S., 112
Young, J. D., 1066
Young, J. L., 260, 270
Young, J. P., 950
Young, J. R., 301
Young, L., 950, 951
Young, S., 371, 381, 426, 498, 618, 623, 634, 655, 656, 703, 724
Young, T., 271
Yusa, A., 1067
Yushko, K. B., 1151, 1158
Yvanova, E. M., 397, 416
Yvon, J., 584, 606, 678, 724

Zajde, C., 1189
Zakrzewski, W. H., 299
Zandbergen, P., 345, 346
Zander, M., 370, 381
Zarzycki, J., 842, 877, 888, 889, 898, 900
Zavoisky, E. K., 1187, 1192
Zawidski, J., 893, 900
Zdanuk, E. J., 1056, 1067
Zel'dovich, Y. B., 1125, 1127, 1159
Zelhoffer, G. F., 748
Zelyanskaya, A. I., 748
Zernike, F., 709, 712, 723
Zeto, R. J., 138, 139, 141, 143, 146
Zhokhovskii, M. K., 140, 146, 202, 211, 441, 442, 500, 833
Zibello, D., 343, 346
Zielinski, E., 1066
Zimmerman, G. O., 690, 721, 724
Zimmerman Jr., H. K., 748
Zimmerman, J. E., 1228–1230, 1265, 1266
Zimmerman, S. W., 1184, 1189
Zink, H., 574, 577
Zinov'eva, K. N., 1244, 1266
Zipf, E. C., 1188, 1190
Zitzewitz, P. W., 1055, 1066
Zmaczynski, A., 627–629, 656
Zorin, A. D., 748
Zubarev, V. N., 395–397, 418, 1150, 1159
Zubova, E. V., 141, 143, 146, 167, 201
Zvereva (Belova), A. M., 974
Zwanzig, R. W., 376, 381, 413, 419, 586, 606, 684, 722
Zwietering, Th, N., 349, 350, 363, 380, 493, 498
Zwolinski, B. J., 70, 623, 654
Zysk, E. D., 90, 113

Subject Index

Bold numerals indicate principal entries.

A, Helmholtz energy, 15, 34
Absorption spectroscopy, 1209
Absorption vessel, 731, 732
Accuracy, 407
— and precision, 63
— gas–liquid solubility, 735–737
Accurate measurements of density under pressure, table of, 440
Acetanilide, melting point, 84
Acetone, 1156
Acoustical methods for measuring chemical relaxation, 1081
Acoustical thermometer, 557
Activation of metals by irradiation with thermal neutrons, 977, 979
Activity, 906, 917, 921, 939, 945
— absolute, 57, 62
— relative, 15, 61, 62
— of substance B, 15
Activity coefficient, 753, 754, 758, 783, 906, 916, 939, 941, 945, 949
— concentration basis, 19, 61
— mole fraction basis, 16, 61
Activity coefficients, practical, definition of, 891
— interconversion, 729
— rational, definition of, 891
— scales, 727
Adiabatic compressibility, 22, 35
— in density measurement, 963
Adiabatic compression over pressure range of one bar, 489
Adiabatic equipment for equilibrium method, 878, 881
Adiabatic process, 24, 29, 31, 33, 35
Adiabatic properties, 530
Adiabatic solidification of ^{3}He, 1240
Adipic acid, triple point, 84
Adsorbed gas, definitions for amount of, 1044
Adsorption, cleaning adsorbent surfaces, 1033
— equilibration times in, 1034
— volumetric methods of, 1036
— — at low pressures, 1038
— — at pressures below 10^{-6} Torr, 1038
— — at pressures between 10^{-3} and 10 Torr, 1040
Adsorption cells, 1048
Adsorption cryostats, 1063
Adsorption of gas, 334, 337, 341, 343
Adsorption isotherms, 1033–1063
— experimental methods for measuring, 1023–1068
— — beam type balance, 1054

Adsorption isotherms, experimental methods for measuring, determination of dead-space volume, 1046
— — flow methods, 1060
— — — very low pressure region, 1060
— — gas chromatographic method, 1062
— — gas dosing systems, 1049
— — gravimetric methods, 1050
— — helical spring balance, 1051
— — in high pressure region, 1044
— — in subatmospheric pressure region, 1042
— — sources of error and their correction, 1057
— — using resonating quartz crystals, 1057
Adsorption measurements, flow methods in, 1060
— gas chromatographic methods in, 1062
— gravimetric method in, 1050
— ultrahigh vacuum system for, 1037
Aimé's method of measuring compression of liquids, 446
Air, agitation, solubility effect, 880
— buoyancy correction, 1058
— critical constants, 38
— density, 170
— literature, 369, 408
Airspeed indicator, 224, 225, 227
Alcohol, 1151
— (ethyl), 1156
Altimeter, 223, 225
Alumina insulators, 981
Aluminum, freezing point, 77, 82
Aluminum–copper eutectic, freezing point, 77
Aluminum oxide, melting point, 82
Amagat density, 53
Ammonia, critical constants, 38
— literature, 369, 408
Amount of substance, 6, 10
Amplitudes of relaxation effects, 1091
Analytical method for determining phase equilibria, 789, 796, 797
Aneroid capsule, 214
Angle-dependent scattering, 710
Aniline–cyclohexane liquid mixture, 572
Anisic acid, triple point, 84
Anomalies, critical, 660
Anthraquinone, triple point, 84
Antimony, freezing point, 77
— influence of pressure, 78
Antisymmetry, μ–ρ, **665–667**, 668, 669, 672
Anvil system, 136, 139

SUBJECT INDEX

Apparatus of Alwani and Schneider, optical cell, 798
Apparatus of Buback and Franck, optical cell, 798
Apparatus for calibration of gas solubilities, 731
— degassing liquids, 730
— high pressure solubilities, 737
— manometric measurements, 729, 735
— microgasometry, 741
— mixing, 734
— percolation of gases, 743
— saturation of liquids, 743
— volume measurements, 746
Apparatus of de Swaan Arons and Diepen, optical cell, 797
Apparatus of Krichevskii and Tsiklis, 796
Apparatus of Michels, Skelton and Dumoulin, for investigating gas–liquid equilibria, 798
Apparatus of Oeder and Schneider, optical cell, 797
Apparatus, that requires a fixed point calibration, 133
Apparatus of Tödheide and Franck, 797
Apparatus of Tsiklis and Maslennikova, 796
Approach to equilibrium in v.p. measurement, 613
Aqueous eutectic solvents table, 870, 871
Aqueous solution equipment for adiabatic cryoscopy, 878
Area, effective, 154, 171, 192
Argon, 555, 564, 565, 1155
— as pressure-transmitting medium, 978
— coexistence curve, 593
— critical constants, 38
— dielectric constant, 599–600
— equation of state, 412
— in pressure measurement, 1203
— literature, 369, 408
— magneto-optical effect, 602
— refractive index, 592, 594
— second virial coefficient, 40
— solubility of, 737
Asbestos diaphragm, 934
Assay of sample in gas-saturation method, 648
Asymmetry potential, 921
Atmosphere, standard, 223
Atomic weights of elements, 36, 37
Attenuation, sound, 718, 719
Autoclaves, 981
— externally heated, 976
— internally heated, 978, 979
Avogadro's number, 26, 305
— use of single crystals of pure silicon in measurement of, 318
Azeotrope, 608, 614, 631

Babb's equation of state, 411

Balances, beam-type vacuum, 1054
— helical spring, 1051
— McBain, 778
— resonating quartz crystal, 1057
Ball gages, 162
Ballistic piston compressor, 1195
Ba I–II transition, 136, 139, 140, 142
Ba II–III transition, 144
Barotropic systems, 790, 792
Bayard–Alpert gages, 275
Beam-type vacuum balances, 1054
Beattie–Bridgman equation of state, 410
Beattie's apparatus for determination of pVT properties of water, 620
Beckmann thermometer, 854
Behavior, change from metallic to non-metallic, 983, 986, 987, 989, 990
Benedict–Webb–Rubin equation of state, 410, 504, 544
Bent-beam gage, 294
Benzanilide, melting point, 84
Benzene, 1149, 1151, 1155
— equation of state, 1156
Benzoic acid, freezing point, 77, 82
Benzophenone, freezing point, 82
Berthelot's equation, 543
Beryllium oxide, 980
— container, 980
BET areas, 1042
Bias, 63
Binary liquid mixtures, 705, 710, 714, 717
Binary systems, 905, 910, 916, 920, 941
Biphenyl, freezing point, 82
BIPM manometer, 118
Bi I–II transition, 136–141
Bi III–V transition, 136, 137, 139, 140, 143
Bi II–III transition, 142
Bismuth, freezing point, 77
— influence of pressure, 78
Bismuth alloy electrode, 940
Blears effect, 277
Blackbody, radiance of, 76, 80
Blocked capillary technique, 1260
Boiling fluid cryostat, 349
Boiling temperature in ebulliometric measurements, 626
Boltzmann constant, 17
Bourdon (gage) tube manometer (quartz), 204, 325, 330, 339, 980, 1261
Boyle temperature, 52
Brillouin effect, stimulated, 1150
Brillouin lines, 716, 718
Bubble-point, 607
Bubble-point measurements, 780–782
Buckingham potential, 1153
Bulk modulus, 326
Bunsen coefficient, 729
Buoyancy corrections, 1058
Buoyancy method, 978

SUBJECT INDEX

Burnett expansion method, 326, 335, 341, 356, 365

Cadmium, freezing point, 77, 82
— influence of pressure, 78
Calibration, controlled clearance gage, 168
— cross-float, 188
— of gages, 286
— of manometers, 731
— similarity method, 184
Calorimetric methods of measuring compression of liquids, 493
Calorimetric methods for melting point determination, 811
Calorimetry, 681–683, **693–696**, 1253
— constant volume, 1254
Capacitance bridge, 3-terminal, 1235, 1237
Capacitance cell, pVT, 1258
Capacitance diaphragm gage, 268
Capacitance manometer, 121
Capacitance pressure gage, 1237
Capacitance thermometry, 1235
Capacitor discharge systems, 1164, 1183
Capacitors, 219, 1164
Capillary depression, 117, 992, 994
Capsule, of niobium–zirconium alloy, 980
Carbazole, triple point, 84
Carbon dioxide, 562
— compressibility, 562
— critical constants, 38
— literature, 369, 408
— second virial coefficient, 41, 42
— solubility of, 737, 739
— sublimation point, 77
— vapor pressure, 636
Carbon disulfide, 1149
— equation of state, 1156
— melting point, 84
— solubility of, 737
Carbon monoxide, solubility of, 737, 739
Carbon tetrachloride, 1149
— equation of state, 1156
— temperature determination in, 1151
Carbon tetrafluoride, solubility of, 737
Carbon thermometer, 88
Cathetometer, 117
Cavity field in dielectric, 583
Cell, adsorption, 1048
— electrical resistance for mercury, 981
— junction, 907
— resistance, 924
— Weston, 924
Cell materials, all metal, 929–931
— alumina, 931
— boron nitride, 925, 929, 931
— fused silica, 929, 933
— graphite, 929
— Pyrex, 928, 929, 933, 946

Cell materials, silica, 931
Cell models, 1153
Cells, classification of, 903
— concentration, 904, 906, 907, 939, 941, 945
— formation, 903, 906, 907, 921, 933
— Jacob–Daniell, 908, 914
— membrane, 906, 907, 918, 920
— mixing, 903, 931, 932
Cesium, compressibility, 976, 985
— conductivity at sub- and supercritical temperatures, 986
— critical region, 38, 987
— density, 977, 980, 983, 985, 986
— literature, 408
— thermal ionization of vapor, 986
— thermoelectric power, 990
— vapor pressure, 980, 981, 986
Chapman–Jouguet pressure, 1134
Chemical methods in gas analysis, 729, 739
Chemical potential of substance B, 21, 31, 57, 60, 62, 664, 669, 678, 752, 755, 756
Chemical relaxation, 1069
Chemical relaxation techniques, 1071, 1076
— acoustical methods, 1081
— dielectric methods, 1086
— field-jump methods, 1080
— pressure-jump methods, 1078
— temperature-jump methods, 1075
2-Chloroanthraquinone, triple point, 84
Chlorobenzene, melting point, 84
Circulating percolator, 743
Classical behavior, 662
Classification of vapor pressure methods, 609
Clausius–Clapeyron relationships, 517, 522, 1236, 1256
Clausius–Mosotti relation, **689**, 690, 1258
Cleaning, 194
Cobalt, freezing point, 77
Coefficient of heat transfer, 16
Coexistence curve, 661, 662, 666, 670, 676, **700–709**
— of argon, 593
Coexisting liquid phase, 562
Cold-cathode ionization gages, 291
Comparative methods of v.p. measurement, 617, 624, 627
Complexes, of metal ions, formation of, 1096
Composition analysis, 556
Compressibility, 116, 659, 661, 670, 671, 677, 678, 691, 710, 976
Compressibility diagrams, generalized, 409
Compressibility factor Z, 348
Compressibility of liquid carbon dioxide, 562
Compression, heat of, 436
— isentropic, 1201, 1203
— relative expansion and, 431
Compression of liquids, 421–500

SUBJECT INDEX

Compression wave, 1112
— stability of, 1117
Compressor, ballistic piston, 1195
Computable cross capacitor, 598
Computational methods, 501ff
Concentration gradients, 678, 680, 913
Concentration measurement, chemical analysis method, 886
— conductivity method, 882
— interferometric method, 883
— photometric method, 885
— polarimetric method, 884
— potentiometric method, 882
Concentration scales, 727
— conversion of, 728
Condensation temperature in ebulliometric measurements, 626
Condensed gas mixtures, investigation of, 799
Condenser–capillary (impedance), 1243
Conductivity, electrical, 981, 986, 987, 990
Confidence intervals, 515
Connections, 197, 198
Constant pressure measurements, 405, 406
Constant pressure methods, 399
Constant temperature methods, 389, 405
— displacement method, 397, 405
— injection, 389
— metallic bellows, 391
— weighing technique and constant volume piezometer, 394, 406
Constant volume apparatus, 405
Constant volume calorimeters, 1254
Constant volume methods, 400
Constant-volume vessel for measuring compressibility of amorphous materials, 474
— bellows volumometer, 482
— direct weighing of fluid removed, 482
— displacement of a solid piston, 476
— displacement of a liquid piston, 480
— volumetric measurement of gas at low pressure, 483
— volumetric measurement of liquid at low pressure, 483
Connodes, 793
Container materials, compatible with potassium or sodium, 982
— tungsten–rhenium, 982
Controlled clearance, 152, 161, 165
Convection, 978
Convection of gas, as source of error in microbalance system, 1057
Conversion factors, tables of, 66–69
Cooling, nuclear, 1244
— by adiabatic freezing of liquid ^{3}He, 1246
Copper, freezing point, 77, 82
Copper–aluminum eutectic, freezing point, 77
Copper–silver eutectic, freezing point, 83
Correlation length, 660, 684, **714**

Corresponding states, law of, 409
— treatment for sound velocity in gases, 559, 568
Corrosion, of cell materials, 928
— of electrodes, 928, 936
Corrosive liquids, 436
Cottrell pump, 627, 766, 767
Covers, 200
Critical constants, 38
Critical curves, 788
— gas–liquid, 788, 789
Critical data of metals, 976, 981, 985–987
Critical dispersion, 539, 567
Critical endpoint, 789
Critical exponents, 567, 662, 663, 669, 670, 671
Critical isochore, 661, 677
Critical isotherm, 661, 669, 676
Critical opalescence, 685, **714, 715**
Critical phenomena, of fluid binary mixtures, 788, 800
— gas–liquid systems, 789
— liquid–liquid systems, 789
Critical point, classical, 621, 662
— non-classical, 676
Critical pressure, 788
Critical region, 565, 711, 715
Critical solution temperature, 573, 759
Cross-float, 173, 186, 188, 189
Cryoscopic common ion rule, 841, 861
Cryoscopic constant in molal scale, formula 9, p838
Cryoscopic Debye equation, formula 20, p841
Cryoscopic equation of Van Laar—Schröder—Le Chatelier, formula 4, p838
— formula 10, p839
Cryoscopic limiting law for electrolyte solutions, formulas 17 and 18, p840
Cryoscopic Raoult law, formula 8, p838
— formula 11, p839
Cryoscopic titrations, 894
Cryostat, see chapter 7
Cube of fused silica, volume of, 315
Cube of tungsten carbide, volume of, 311
Cubic expansion coefficient, 19
Curie constant, 216
Cylinder, volume of, 305
— elastic distortion, 153

Dead-space volume, determination of, 1046
Debye–Sears effect, 1085
Definitions for amount of adsorbed gas, 1044
Degassing, 613, 767
— of liquids, 730
Demagnetization, nuclear, 1240
Dense gases, 1194

SUBJECT INDEX

Density, co-existing, 665, **700–709**
— conversion factors for, 53, 66
— determination of, 303
— — using x-rays or neutrons, 961
— local, 680–684, 686, 690
— of mercury, 54, 116
— — in critical point region, 962
— of water, 55, 304, 318
Density fluctuations, 660, 709–712
Density gradient, **678, 679**, 687, 694
Density measurements, 957, 1207
— of liquid metals, 961, 963, 976–979, 985, 986
— — extension to critical and supercritical region, 979, 983
— — of superheated vapors of Cs, Na, K, 980
— using uranium in zirconia pyknometer, 959
— using x-rays, 961
Density profile, 680, 685, 689, 690, 701, 702
Derivatives of pressure, volume and temperature, 425
Detector, position-sensitive, 217
Determination of dead-space volume, 1046
Determination of heat capacity of liquid metals using drop-calorimeter, 954
Determination of intermolecular potential constants, 560
Determination of molecular weights, 652
Determination of pVT properties of water, 620
Determination of specific heat of liquid metals, by direct heating, 953
— by exploding wire method, 956
Deuterium, critical constants, 38
— literature, 370
— second virial coefficient, 43
Deviation function, 72, 75
Dew-point, 607
— measurements, 780–782
Diameter, rectilinear, 661, 668, 676, **708**
Diaphragm manometer, 330, 339
Dielectric constant, 588, 689, 705
— transformer bridge method for determining, 597
Dielectric methods for measuring chemical relaxation, 1086
Dielectric virial coefficients, 598
— first and second, of argon, 599, 600
— — of helium, 599, 600
— — of krypton, 599, 600
— — of neon, 599, 600
Differential coil transformer, 220
Differential manometers, 203, 210, 783
Differential method for properties of dilute solutions, 783
Differential pressure gages, 768
Differential pressure transducer, 189
Differential thermal analysis, 648

Differential thermal analysis, at normal pressure, 816
Differential transformer, 220
Differential voltmeter, 925
Diffusion controlled reaction, 1096
Diffusion gradient, 909, 916, 917
Diffusivity, thermal, 15, 659, 677, 678, 681, 716, 717
Digital data acquisition, high-speed, 1175, 1179
Dilatation of the vessel, 432
Dilatometer method for measuring liquid metal densities, 963
Dilute solutions, 916, 939, 945
Dimethyl terephthalate, freezing point, 82
— melting point, 84
Diphenyl ether, triple point, 77
Displacement measurements, differential coil transformer, 220
— direct reading, 220
— potentiometer, 218
— pressure effect, 218
— strain gages, 222
— synchro-detection, 218
— variable capacitors, 219
— variation of self-inductance, 221
Dissociation on vaporization, 608
Distribution of pore sizes, 1043
Divided piezometer method, 357
Drop calorimeter, 954
Dynamic measurements without a buffer gas, 635
Dynamic scaling, 717
Dynamic shock, 137

Ebulliometric measurements at pressures below 2 kPa, 633
Ebulliometric method of v.p. measurement, 626
Effect of presence of mercury, 621
Effect of pressure on thermocouples, 387
Effect of temperature on relative volatility, 614
Elastic constants, 174
Elastic distortion of pVT vessel, 326
Elastic gage, 611
Elastic shock wave, 1125
Elevation of vapor pressure under applied pressure, 622
Electrical contact piezometer, 449
Electrical discharge techniques for measuring thermodynamic properties of fluids at high temperatures, 1161, 1183
Electrical feedthrough techniques, 95
Electrical measurements, transient, 1170, 1184
Electrical resistivity, of metals, by pulse technique, 1179
Electrical switching, high-speed, 1165, 1183

SUBJECT INDEX

Electrodes, alloy, 904, 939, 940
— dissolution of, 915
— — gas, 921, 930, 931, 934
— halogen, 904, 931–936
— metal, 902, 904, 907, 921, 932, 937, 939, 947, 948
— nitrate, 928
— polarizable, 924
— reference, 917
— reversible, 924, 927, 939, 947
— second kind, 946
— silver, 934, 937, 938, 941
— silver halogenide, 938, 947, 948
— standard hydrogen, 909
— third kind, 927, 948
Electrode potential, mixed, 949
— single, 903, 916
Electrode processes, 909, 915, 917, 922, 946
Electrolytes, aqueous, 909
— solubility of, 728, 743
Electrolyte solution properties, determination by isopiestic method, 782
Electromagnetic absorption, 1133
Electromagnetic levitation, 955
Electromotive force, sign conventions for, 908, 909
— measurements, 902, 903, 950
— — enthalpy and entropy from, 902
— — errors in, 920
— — instruments for, 924, 925
— — phase diagrams and, 941
— — supercooling and, 942, 943
— series, 950
— thermal, 922
Electronic desorption, 280
Electro-optical and magneto-optical effects, 600
Emission spectrum, 1133, 1209
Emittance, normal spectral, 1179
— total, hemispherical, 1179
Empirical equations, 409
End melting point method, 850
Energy, conversion factors for, 69
— free, 34
— internal, U, 15, 18, 31
— specific, conversion factors for, 67
Enthalpy, 16, 32, 520, 521, 955
— of formation, 54
— of fusion, measurement, 888
— of mixing, 755
— specific, conversion factors for, 67
Entropy, 18, 32, 520, 521
— conversion factors for, 67
— of fusion, measurement, 888
— of liquid metals, measurement by drop method, 955
— standard, 54
Equations deduced from Percus–Yevick theory, 414

Equations of the perturbation type, 413
Equations of state, 29, 136, 348, 409, 501, 502, 512, 541, 1110, 1202, 1207, 1208, 1210
— by Babb, 411
— by Beattie–Bridgman, 410
— by Benedict–Webb–Rubin, 410, 504, 544
— estimation of parameters for, 505
— — least squares, 506
— — — with constraints, 508
— — statistical aspects, 511
— — non-linear, 509
— for gases, 1156
— for liquids at very high pressures, 1146
— for Ga, K, Na, 976
— for Hg, 980
— by Haar and Shenker, 414
— Mie–Grüneisen, 1155
— by Redlich–Kwong, 410
— by Rowlinson, 414
— theoretical models, 1152
— by van der Waals, 503
— virial, 510
— by Zwanzig, 413
Equilibration, **681–683**, 692, 693
— times in adsorption, 1034
Equilibrium, apparatus, 767
— approach to, 932, 934
— association, 944, 946, 948, 949
— dissolution, mixing and, 931, 932
— gas flow and, 932
— isobaric, 754, 755, 758
— isothermal, 754
— liquid–vapor, 753, 755
— method, 851, 878
— of mixtures, 751
— thermodynamic, 1204, 1207, 1211
Equilibrium still, 651, 763–765
Error assignment, 672, 707
Errors, sources of, in physicochemical measurement, 439
Estimate of standard error, 64
Ethane, critical constants, 38
— literature, 370
— solubility of, 737
Ether, 1156
Ethyl alcohol, 1156
Ethyl ether, 1149
Ethylene, solubility of, 737
Eucken factor, 534
Eutectic cryoscopic apparatus, 868
Eutectic solvents tables, 870, 871
Evaluation, 191
Evaporation methods for low pressures, 641
Excess chemical potential, 754, 755, 758
— calculation, 896
Excess functions, 548, 756
Excess Gibbs free energy, 755–757
Exchange current, 922, 938, 939

SUBJECT INDEX

Expansion, compression, expansivity and compressibility, 431
— relative, and compression, 431
Experimental data on high temperature–high pressure gases, 407
Experimental methods, for determining liquid–vapor equilibrium data, 761–775
— — dynamic, 763, 767
— — multicomponent, 775
— — static, 767, 774, 775
— — ternary, 775
— for measuring adsorption isotherms, 1023–1068
— — pressure measurements, 1025
— — vacuum systems, 1024
Exploding wires, discharge characteristics of, with particular reference to 'dwell time', 1177, 1185
— method, 956
Exponents, critical, 662, 663, 669–671
External heating, 384
External and internal heating of pressure vessels, 437
Extractor gage, 280

Fall rate, measurement, 178, 180
Faraday effect, 601
Fe α–ϵ transition, 143
Field-jump methods for measuring chemical relaxation, 1080
First law of thermodynamics, 31
Fittings, 197, 198
Fixed path interferometer, 550
Fixed points, 71, 134, 135, 138, 140, 143, 208
— calibration, 133
— choice of, 134
— measurement of pressure at, 135
— present set for pressure calibration, 140
— — above 100 kbar, 143
— obtainment, 856
— primary, 71
— superconducting transition temperatures as, 1239
— thermometric, 71, 81
— use of, in high pressure calibration, 138
Flight control instruments, 223
— airspeed indicator, 224, 227
— altimeter, 223, 225
— Machmeter, 224, 227
Floating index method of measuring density of liquid ammonia, 446
Floats, 705
Flow methods for measuring adsorption isotherms, 1060
Fluctuations, 678, 709–712
Fluid metals, properties, at medium and low pressures, 953

Fluids, 166, 193, 198, 199
— drying agents, 197
— filters, 197
— surface tension, 171
— viscous drag, 154, 185
Force, conversion factors for, 66
— exerted by vapor, 638
— measurements. Pressure effect, 217
— of molecular beam for v.p. measurement, 645
Förland theory, 843
Formation of metal ion complexes, 1096
'Forward' x-ray effect, 280
Free capsule, 214
Free energy, 34
— and e.m.f. measurement, 924
— of formation, 902–906, 914, 920, 921
— Gibbs, 34, 57
— Helmholtz, 34
— of mixing, 902, 903
— of transfer, 904, 907
Free piston gage, 136, 354
Freezing curve, 360
Freezing point, determination, 78
— influence of pressure on, 78
Friction, 164, 185
Fugacity, 16, 18, 57, 60
Fugacity coefficient, 60, 61
Fundamental constants, 25
Furnaces, 926
— baffles for, 927
— configuration, 926
— cover gas, 927, 930, 931
— insulation in, 926, 927
— materials for, 926, 928
— temperature control, 927
— windings, 926
Fused salts, 929, 933
— preparation and purification, 932–934
— reactivity and decomposition, 928
— water detection and removal from, 933
Fused silica, 929, 933
— volume of cube of, 315

G–M counter, 1028
Gages, Bayard–Alpert, 275
— bent-beam, 294
— Bourdon, 204
— calibration, 286
— extractor, 280
— hot-cathode magnetron, 296
— inverted magnetron, 293
— McLeod, 286
— magnetron, 292
— manganin, 206–210
— Penning, 291
— Pirani, 1026
— shielded collector, 294

Gages—*contd.*
— spoon, 1027
— thermistor, 1042
— trigger discharge, 292
Gage sensitivity, relative, for different gases, 284
Gallium, equation of state, 976
Gamma irradiation of metals, activation, 977, 979
Gamma ray emission, from mercury, 980
Gas, electrodes, 921
— inert cover, 930, 931
— standard state, 904
Gas adsorption, 334, 337, 341, 343
Gas chromatographic measurements, 778, 779
Gas chromatographic methods for measuring adsorption isotherms, 1062
Gas chromatography, 649, 739, 743
Gas constant R, 26, 322, 323
Gas density, microbalance, 323, 342
Gas dosing systems, 1049
Gas expansion method, 356
Gas–gas equilibria, 788, 789, 794, 796
— first kind, 789, 795
— second kind, 789
Gas head, 678, 690, 692
Gas heat capacity, 521
Gas–liquid equilibria, 787, 789, 795, 796, 798
Gas–liquid chromatographic analysis, 775, 776
Gas–liquid critical curves, 788
Gas/oil separator, 363
Gas-operated pressure balance, 324
Gas-saturation method for v.p. measurement, 645
Gas-saturator, 647
Gas thermometer, 9, 321–323, 326, 330, 332, 333, 341
Generalized compressibility diagrams, 409
Germanium resistance thermometer, 88, 1234
Gibbs–Duhem relation, 755, 759–761, 906, 913, 915, 917
Gibbs energy relationship, 517
Gibbs free energy, 16, 34, 751, 752, 754, 757, 758, 782, 784
Gladstone–Dale relation, 1208
Glass membranes, 906, 920, 921, 938
Glycerin, 1151, 1156
Gold, freezing point, 73, 76, 78, 80, 82
Graphite arc, 1174
Gravimetric method of adsorption measurement, 1050
Gravitational effects on liquid ^{4}He, 1255
Gravity, 169, 678–680, 691, 694, 696, 697, 700, 702, 718
Grüneisen constant, 1155, 1156
Guns, 1136

Haar and Shenker's equation of state, 414
Haase R. theory, 845
Hall coefficient, 990
Hard sphere system, 376
Heat, 18
Heat capacity, 15, 34, 520
— of liquid ^{3}He, 1256
— of ^{4}He at saturated vapor pressure, 1254
— specific, conversion factors for, 67
Heat exchangers, 1243
Heat of compression, 436
Heat of fusion, of metals, by pulse technique, 1187
— of tungsten, 1176
Heat leak, vibration, 1249
— stray r.f. fields, 1249
— thermal radiation, 1249
— down supporting structures, 1249
— down electrical leads, 1249
— down fill lines, 1250
— across vacuum spaces, 1250
Heat pipe problems at high temperatures, 973
Heat transfer, isolation, 1248
Heavy water, critical constants, 38
— second virial coefficient, 43
Helical spring balance, 1051
Helium, critical constants, 38
— dielectric constant, 599
— in temperature measurement, 1206
— literature, 370, 408
— samples, preparation of, 1251
— solubility of, 737–739
— transition, II–I, specific heat anomaly, 1254
— 1962 ^{3}He scale (T_{62}), 80, 1232
— ^{3}He melting curve thermometry, 1236
— ^{3}He, liquid, specific heat of, 1251
— — — heat capacity of, 1256
— ^{3}He–^{4}He dilution refrigerator, 1240, 1241
— ^{3}He–^{4}He mixtures, specific heat of, 1257
— — heat of mixing of, 1257
— — solutions, osmotic pressure of, 1262
— 1958 ^{4}He scale (T_{58}), 80, 1232
— ^{4}He, gravitational effects on liquid, 1255
— — heat capacity at saturated vapor pressure, 1254
— — latent heat of vaporization of, 1256
— — liquid, near lambda point, 1254
— — thermodynamics of, 1253
Helmholtz energy, A, 15, 34
Henry's law, 730
n-Heptane, melting point, 84
Hexachlorobenzene, melting point, 84
High intensity lamp, 1209, 1214
High-pressure piezometer, 796
High pressure vessels, 354, 737
High pressure windows, 1196
High pressures, measurement of thermodynamic properties of liquids at, 1176

SUBJECT INDEX

High-speed digital data acquisition, 1179
High-speed photoelectric pyrometer, 1179
High-speed photography, for diagnosis and measurement, 1174
— — literature summary, 1187
High-speed radiation modulation, 1188
High-speed recording, by oscilloscopic techniques, 1175
High temperatures, 1161ff, 1204
— thermodynamic properties of fluids at, measurement by electrical discharge techniques, 1161
High voltage, 1161ff
Hittorf internal frame of reference, 920
Hittorf transference numbers, 912, 945
Homogeneity, **669,** 672
Hot-cathode ionization gages, 290
— bent-beam, 294
— Orbitron, 283
— shielded collector, 294
Hot-cathode magnetron gage, 296
Hot gases, 1204, 1206, 1210
Hugoniot curve, 137, 138, 1110, 1116, 1118, 1120, 1121, 1138, 1146
— equation of, 1115
Hydrocarbon–methane system, behavior, 798
Hydrocarbon–water mixtures, 798
Hydrogen, 1155
— boiling point, 72, 73, 77, 78
— critical constants, 38
— latent heat of vaporization of, 1256
— literature, 371, 408
— preparation of samples of, 1251
— second virial coefficient, 45
— triple point, 72, 73, 77, 78
— vapor pressure, 79
Hydrogen cyanide, solubility of, 739
Hydrogen fluoride, solubility of, 737, 742
Hydrolysis constant calculation, 893
Hypersonic sound velocity, 535
Hypersound, **718**

Ideal activity of ionic component, from Haase, 845
— from Temkin, 842
— from Förland, 843
Ideal gases, 541, 1199
Ideal mixing, 545, 752, 753
Imidazole, melting point, 84
Impedance match method, 1131, 1138
Imprecision, 64, 65
Incongruent melting, 805
Indium, freezing point, 77
— influence of pressure, 78
Infinitely dilute solution, 61
Insulating low density vapor phase, 985
Insulators, electrical, of alumina, 981
Intensity of x-radiation, 962

Intercepts, isothermal, **703, 704**
— isochoric, **703, 704**
Interference method for refractive index, 592
Interferometer, 119
— methods, 549
— Michelson, 119
— ultrasonic, 125
Intermolecular potential, 558, 1212, 1213
Internal energy, 15, 18, 31
Internal heating, 384, 385
Internal heating devices, temperature gradients in, 387
International Committee on Weights and Measures, 76, 80
International Practical Temperature Scale (IPTS-68), 71, 77, 352
Interpolation instrument, 72
Interphase surface tension, 991
— method of measurement, by capillary rise or depression, 992, 994
— — by floating sheet, 992
— — by maximal pressure in a drop or bubble, 992
— — by oscillating jets, 992
— — by oscillating drops, 992
— — by ring or plate rupture, 992, 1003
— — by shape of hanging drop or bubble, 992
— — by shape of sessile drop or bubble, 992, 995
— — by weight or volume of a drop, 992, 1001
Inverted magnetron gage, 293
p-Iodobenzoic acid, melting point, 84
Ionic dissociation calculation, 891
Ionic membranes, La F_3, 919
Ionic mobility, 907, 916
Ionic strength, 16
Ionic transport, 909, 910
Ionization, 1208, 1211
Ionization gages, calibration of, 286
— cold cathode, Penning, 291
— — magnetron, 292
— — inverted magnetron, 293
— hot cathode, 290
— — bent-beam, 294
— — shielded collector, 294
— residual currents in, 278
— — due to electronic desorption, 280
— — due to soft x-ray photoemission, 279
— — measurement of, 283
— — x-ray limit, 283
IPTS-68, 71, 77, 80, 1230
[International Practical Temperature Scale of 1968]
Iridium, freezing point, 77
Isentrope, touching Hugoniot curve, 1118
Isentropic compression, 1201, 1203
Ising model, 658, 665, 694

SUBJECT INDEX

Isobar, 790
Isobaric data, 760
Isochore, 979
Isochore method, 356, 362
— for measuring density of xenon, 595
Isochoric thermal pressure coefficient, 490
— by constant-volume vessel with direct pressure measurement, 491
— by constant-volume vessel with indirect pressure measurement, 492
— by electrical contact method, 490
— by piezometric methods, 490
Isolation, heat transfer, 1248
Iso-optic systems, 792
Isopiestic method for comparing vapor pressures of solutions, 652
Isopiestic method of determining electrolyte solution properties, 782
Isopleth, 791, 792
Isoteniscope, 616
Isotherms, 790
— adsorption, 1033
Isothermal compressibility, 20, 35
Isothermal data, 760
Isothermal method, 356
Isotonic method for comparing vapor pressures of solutions, 652

Jacket pressure, 162
— coefficients, 175, 177
— zero clearance, 178, 182
Jacobi–Daniell cells, 908, 914
Josephson effect, 1228
Joule–Thomson coefficient, 21, 35
Junction materials, asbestos plug or fiber, 947
— alumina spiral, 908, 936
— glass frit, 936
— porous frit, 907, 931
— Pyrex membrane, 909
— Vycor membrane, 936

Kay's apparatus for static v.p. measurement at elevated temperatures and pressures, 618
Kinetic methods, 847
Knudsen's method for v.p. measurement, 642, 966
— radiometer, 263
Korringa relation, 1227
Krypton, critical constants, 38
— dielectric constant, 599–600
— literature, 371, 408
— magneto-optical effect, 602
— second virial coefficient, 44

Lambda transitions, 694

Langmuir's method for v.p. measurement, 641
Latent heat of vaporization of hydrogen and ^{4}He, 1256
Lattice gas, 665, 666
Law of corresponding states, 409
Laws of thermodynamics, 28
— zeroth, 29
— first, 31
— second, 32
— third, 33
Lead, freezing point, 77, 82
— influence of pressure, 78
Least squares (linear) parameter estimation, 506, 510–512
— with constraints, 508
Length, conversion factors for, 66
Lennard–Jones potential, 1153
Levitation, electromagnetic, 955
Ligand binding on an allosteric enzyme, 1099
Light modulation, high-speed, 1188
Light pulses, 1189
Light scattering, 678, **709–717**
— measurements, 783, 784
Line broadening, 1212
Linear model, **675**, 696
Liquefaction, in shock waves, 1123
Liquefied gases, 1150
— nitrogen, 1150
Liquid junction, 907, 910, 918, 921, 945
— e.m.f., 911, 913, 916–923, 945
— expressions for, 910
— formation of, 909
— potential, 911
Liquid metals, 1180, 1187
— density measurements in, 961, 963, 976–979, 985, 986
— measurement of volume, enthalpy and resistance of, at high temperatures, 1177
— resistance, at high temperatures, 1177
Liquid mixtures, 753
Liquids, compression of, 421–500
Liquid–liquid phase separation, 759
Liquid–liquid equilibria, 788, 789, 794
Liquid–vapor equilibrium data, 755
— experimental methods, 761
— multicomponent, 775
— phase diagram, 781
Local density, 680–684, 686, 690
Lorentz model, 581
Lorentz–Lorenz function, **684**, 690, 1208
Lower limit of pressure for ebulliometric method, 634

McBain balance, 778
Machmeter, Mach number, 224, 225, 227
McLeod gage, 245, 286, 612
Magnesium oxide window, 980
Magnetic forces, 1169

SUBJECT INDEX

Magnetic thermometry, 1222, 1233
Magnetometer, 1228
Magneto-optical effects in argon, krypton, xenon, 602
Magneto-optical and electro-optical effects, 600
Manganin resistance gage, 137, 206–210, 1261
Manganin wire transducer, 1142
Manometer, BIPM, 118
— Bourdon tube, 325, 330, 339
— capacitance, 121
— diaphragm, 330, 339
— differential, 203, 210
— mechanical, 1027
— mercury absolute, 115, 135, 323, 328, 335
— NBS, 121
— NRLM, 120
— null, 329, 335
— VNIIM, 126
Manometric method of analysis, 730
— accuracy of, 735
Mass, apparent to true conversion, 196, 197
— conversion factors for, 66
Mass spectrometry, 651
— in gas analysis, 740
Massieu function, 16
Measurement of density of liquid cesium, 957
Measurement of density of liquid metals, direct method, 957
— by pyknometer, 958
— by indirect methods, 963
Measurement of gas flows, 647
Measurement of pressure, 611
Measurement of sound velocity in liquid, aluminum, 964
— bismuth, 965
— cadmium, 965
— indium, 965
— lead, 965
— mercury, 964, 965
— sodium, potassium and their alloys, 966
— tin, 965
Measurement of ultrasonic absorption, 1085
Measurement of vapor pressure, Knudsen method, 966
— of alkali metals, 968
Measurement of very low vapor pressures, 625
Mechanical manometer, 1027
Melting, in shock waves, 1123
Melting curve, of mercury, 136, 140
Melting curve pressure, minimum, 1236, 1246, 1253
Melting point, 803, 1179
— of metals, by pulse technique, 1179
Membrane systems, 910, 936, 938
Meniscus disappearance, **702, 703**
Mercury, 304, 1155, 1156

Mercury, absolute manometers, 115ff
— boiling point, 77
— compressibility, 116, 984, 985
— critical region, 38, 983, 987
— density, 116, 962, 963, 978, 983, 985
— equation of state, 54, 980
— freezing point, 77
— gamma ray emission from, 980
— influence of pressure, 78
— liquid, 980, 983
— as pressure-transmitting medium, 329, 337, 354, 403, 404
— measurement of density, 979, 983
— melting curve, 136, 140,
— meniscus, ripples, 122
— sound velocity, 964–965
— temperature in VNIIM manometer, 128
— thermal expansivity, 116
— thermoelectric power, 990
— use as a confining fluid, 337, 341
— vapor pressure, 117, 981
Mercury capillary depression, 329
Mercury gages, 354
Mercury manometer, 323, 328, 335
Mercury meniscus co-volume, 329
Mercury (separator), 406
Mercury vapor drag effect, 1031
Metal diaphragm, 980
Metastable melting, 807
Methane, critical constants, 38
— equation of state, 412
— literature, 371, 372
— second virial coefficient, 45
— solubility of, 737, 739
Methanol, 1156
Method of capillary rise or depression, 992–994
Method of floating sheet, 992
Method of maximal pressure in drop or bubble, 992, 1006–1009
Method of oscillating drops, 992
Method of oscillating jets, 992
Method of ring or disc rupture, 992, 1003–1005
Method of shape of hanging drop or bubble, 992
Method of shape of sessile drop or bubble, 992, 995–1001
Method of weight or volume of a bubble, 992, 1001–1003
Methods of cloud point determination, 742, 744
— instrumental analysis, 746
— liquid-film saturation, 736
— manometric measurements, 729, 736
— saturation, 743
— volume measurements, 741, 746
Methods for determining volume changes on melting, 829

SUBJECT INDEX

Methods of measuring composition, by density, 762
— by differential method, 783
— by ebulliometry, 763
— by gas–liquid chromatography, 762, 775
— by isopiestic method, 782
— by refractive index, 762
Methods of measuring compressibility, based on radioactivity, 494
— using variable-volume vessel, 494
Methods of measuring melting points at high pressures in apparatus using hydrostatic pressure media, 821
— plugged-capillary method, 823
Methods of measuring melting points at high pressures in apparatus using quasi-hydrostatic pressure media, 824
Methods of measuring melting points at high pressures in multiple-anvil apparatus, 828
Methods of measuring melting points at high pressures in opposed-anvil apparatus, 827
Methods of measuring surface tension, 991
— capillary rise or depression, 992–994
— floating sheet, 992
— shape of hanging drop or bubble, 992
— shape of sessile drop or bubble, 992, 995–1001
— ring or disc rupture, 992, 1003–1005
— weight or volume of a bubble, 992, 1001–1003
— maximal pressure in drop or bubble, 992, 1006–1009
— oscillating jets, 992
— oscillating drops, 992
Methods used for melting point determination at high pressure, 820
1-Methylnaphthalene, melting point, 84
Micro-balance, gas density, 323
Micro-cryoscopic apparatus, 860
Micromanometry, 229–271
Microscopical methods for melting point determination, 808
Mie–Grüneisen equation of state, 1155
Mixed crystals study, 888
Mixing in gas–liquid systems, 734
Mixtures, 60
— equilibrium of, 751
— of gases, 753
— of liquids, 753
— multicomponent, 759
— of small and highly symmetrical molecules, theoretical treatment for, 799
— showing phase separation, 572
— thermodynamic relations for, 751
Mode–Mode coupling, 539
Modulation methods (of pressure measurement), 282
Molality, 61, 62
— definition of, 727

Molar change in electric moment, 1087
Molar masses measurement, 887
Molar volume, 1258
Molarity, definition of, 727
Mole, 6
Mole fraction, definition of, 727
Molecular polarizability, 587
— variation of, with density, 587
Molecular sound velocity, 569
Molecular weight in v.p. measurement, 608
Molten salt cryoscopic, table, 875
— apparatus, 871
— procedure, 874
Molybdenum, cell, 980, 981
— electrical resistivity of, 1179
— hemispherical total emittance of, 1179
— melting point, 1179
— normal spectral emittance of, 1179
— solid, 979
— specific heat of, 1179
Mössbauer effect thermometry, 1223, 1224
Moulin correction for supercooling error, 848
Multicomponent liquid–vapor equilibrium data, 775
Multicomponent systems, 759, 902, 910, 916

NBS correction for supercooling error, 849
NBS equation, **673**
NBS manometer, 121
NRLM manometer, 120
Naphthalene, freezing point, 82
— triple point, 84
Negative pressures of compressed liquids, 488
Neon, boiling point, 72, 73, 77, 79
— critical constants, 38
— dielectric constant, 599–600
— literature, 372, 408
— second virial coefficient, 47
— triple point, 77
— vapor pressure, 79
Neopentane, transition point, 82
— triple point, 82
Nernst law, 922, 940, 947, 949
Neutralization reaction, 1095
Nickel, freezing point, 77
Niobium, specific heat of, 1179
— electrical resistivity of, 1179
— hemispherical total emittance of, 1179
— normal spectral emittance of, 1179
— melting point of, 1179
Niobium–zirconium alloy, for capsule, 980
Nitrates, decomposition of, 928
— electrodes for, 928
— association equilibria in, 945
Nitrogen, 1155
— boiling point, 77
— critical constants, 39

— equation of state, 412, 1156
— liquefied, 1150
— literature, 372, 408
— second virial coefficient, 48
— solubility of, 737–740
— triple point, 77
Nitrogen trifluoride, solubility of, 737
Nitromethane, equation of state, 1156
p-Nitrotoluene, triple point, 84
Noise thermometry, 1229
Non-aqueous solvents equipment for adiabatic cryoscopy, 881
Non-classical critical point, 662
Non-electrolytes, mutual solubility of, 727
Non-ideal mixtures, 548
Non-linear parameter estimation, 509
Non-uniform environments, pressure measurement in, 277
Normal density, 53
Normal volume, 53
Noxious volumes, 678, 693
Nuclear cooling, 1244
Nuclear demagnetization, 1240
Nuclear magnetic resonance, pulsed, 1228
Nuclear orientation thermometry, 1225
Nuclear spin thermometry, 1226
— susceptibility, 1227
Null manometer, 329, 335

O–Z plot, 713, 715
Onsager–Böttcher theory, 582
Opalescence, critical, 685, **714, 715**
Optical absorption, 990
Optical pyrometer, 10, 89, 108
Optical spectra, 1212
Orbitron gage, 283
Organic and inorganic solvents, table, 862–867
Oscillator, 220
Osmotic coefficient, 62, 889
Osmotic pressure, 21
Osmotic pressure of ^{3}He–^{4}He solutions, 1262
Ostwald coefficient, 729
Oxygen, boiling point, 73, 75, 77, 79
— critical constants, 39
— literature, 373
— purity of, 79
— second virial coefficient, 49
— solubility of, 737–740
— triple point, 72, 73, 77, 79
— vapor pressure, 79

Palladium, freezing point, 77
Palladium electrode, 948
Parallelepiped, volume of, 305
Parametric form, 673, 675, 696
Partial pressure, 60

Partition function, 30, 32–35
Pascal, 115
Penning gages, 291
— trigger discharge, 292
Percolation in gas–liquid systems, 743
Perfect gas equation, 751–753
Perturbation method, 376
Perturbation-type equations, 413
Petroleum industry, low temperature measurements for industrial applications, 800
— high temperature, high pressure measurements for industrial applications, 800
Phase behavior from pVT data, 794
— dynamic method, 795, 798
Phase change, 981
Phase changes in shock waves, 1123, 1148, 1149
Phase diagrams for fluid binary mixtures, 788
Phase rule, 750
Phase transitions, 134, 136–144
— Ba I–II, 142
— Bi I–II, 141
— Bi III–V, 143
— Tl II–III, 142
Phenanthrene, freezing point, 83
Phenomenological shell model, 586
Phlegmatic liquids, 634
Photography, high-speed, 1187
Phthalic anhydride, melting point, 84
Piezometer, 389
— seasoning of, 434
— with contacts, 358
Piezometric methods of measuring p–V–T properties of liquids—liquid piston, 441
— single-point measurements, 445
— — electrical contact method, 449
— — floating index method, 446
— multiple-point methods, 450
— — capacitance method, 460
— — differential transformer method, 458
— — electrical contact methods, 456
— — magnetic float method, 462
— — radiated signal method, 462
— — visual method, 450
Piezometric methods of measuring p–V–T properties of liquids—bellows, 465
— solid piston, 463
Piezo-transistors, 222
Pirani gages, 1026
Piston, ball, 162
— ballistic, compressor, 1195
— effective area, 171
— — temperature correction of, 173
— — change with jacket pressure, 175
— — change with zero clearance jacket pressure, 178
— elastic distortion under load, 153, 154, 185
— grooved, 163

SUBJECT INDEX

Piston—*contd.*
— rotation, 172
Piston and cylinder apparatus, 136, 138
Piston displacement method, 359
Piston gage, applications, 193
— ball, 162, 163
— basic equations, 152
— calibration of primary standards, 168
— calibration, of cross-float, 188
— — evaluation, 191
— — reference levels, 191
— choice of weights, 196
— cleaning, 194
— connections, 197
— controlled clearance, 151, 161, 165, 168
— — calibration of, 168
— — total uncertainty of pressure, 183
— designs, 155
— differential, 159
— fluids for, 198
— high pressure, 165ff
— historical review, 149
— mechanical support and covers, 200
— re-entrant, 156, 159
— rotation of piston or cylinder, 199
— temperature control and measurement, 195
— simple, 155
— tilting, 156
— use of, for pressure measurement, 193
— — procedures and methods, 194
— vacuum-backed, 157
— very high pressure, 165
Planck function, 19, 1204
Planck's radiation law, 89, 109
Plane wave generator, 1133
Plastic shock wave, 1125
Platinum, freezing point, 77, 83
Platinum resistance thermometer, 9, 72, 75–78, 88, 89, 353, 854, 858
Platinum–10% rhodium/platinum thermocouple, 72, 874
— electromotive force, 76
Plumbing, 197, 198
Poisson's ratio, 326
Polarization of electrodes, 926
Polymers, 804
Porcelain membranes, 906
Pore sizes, distribution of, 1043
Potassium, chloride, temperature determination in, 1150
— compressibility, 976
— critical region, 987
— density, 977, 980
— equation of state, 976
— literature, 408
— resistivity, 982
— thermoelectric power, 990
— thermoelectric voltage, 982

— vapor pressure, 980–982, 988
— vaporization entropy, 988
Potentiometer, 218, 924–926
Power, conversion factors for, 67
— thermoelectric, 990
Power laws, 566, 659–662, 669, 670, 675, 706
Power measurements, 1170
Poynting equation, 622
Practical (secondary) thermometry, 1221
Practical temperature scale, 80
Precise temperature measurements, 387, 407
Precision, 407
— accuracy and, 63
Precongelation method for adiabatic cryoscopy, 886
Preparation of helium samples, 1251
Preparation of hydrogen samples, 1251
Pressure, conversion factors for, 67
— derivative of, 425
— generated by piston gage, 153, 168, 188
— in absolute units, 304
— measurement of, at fixed points, 135
— units, 7, 12, 67, 115
Pressure balance, 611, 638
— gas-operated, 324
Pressure coefficient, 20
— of effective area, 153–156, 173, 174, 192
Pressure-jump methods for measuring chemical relaxation, 1078
Pressure maximum in a gas bubble, 963
Pressure measurements, chapter 4, 193, 1025, 1202
— below 10^{-10} Torr, 294
— mechanical manometer for, 1027
— sources of error and their correction, 1028
— with bent-beam gages, 294
— with hot-cathode magnetron gage, 296
— with Pirani gages, 1026
— with shielded-collector gages, 294
Pressure sensitive elements for instruments, aneroid, 214
— bellows and springs, 215
— Bourdon tube, 215
— flattened tube, 215
— free capsule, 214
— gimlet shaped tube, 216
— restrained capsule, 214
— piezoelectric quartz, 216
Pressure uncertainty, 183, 187, 193
Pressure vessels, external and internal heating of, 437
— seasoning of, 434
Primary fixed points, 71, 72
— temperature of, 73
Primary piston gages, calibration of, 168
— — similarity method, 184
Principle of corresponding states, 989

SUBJECT INDEX

Pseudo-equilibrium method for measuring concentration, 886
Pulse currents, literature summary, 1184
Pulse heating, 1161ff
Pulse interferometer, 965
Pulsed nuclear magnetic resonance, 1228
Pulse methods, for sound velocity measurement, 552, 964, 965
Pulse voltage measurements, literature summary, 1184
Pumping and re-emission in gages, 274–276
Purity of sample in v.p. measurement, 608, 610, 614, 631
pVT, capacitance cell, 1258
— measurements, chapters 6, 7, 8 and p 1257
— properties at melting, 1260
— values, 347
Pyknometer, 958, 959, 976
— overflowing, 958, 959
Pyknometer technique, 958, 976
Pyrex, 928, 929, 933
— cells, 928, 929, 946
— junction membranes, 909
Pyrometer, 72, 76, 80
— high-speed, 1172, 1179, 1185
— optical, 10, 89, 108
Pyrometric methods for melting point determination, for use above 2000°C, 810
Pyrophyllite, 978

Quartz Bourdon gage, 268
Quartz crystal, 550
Quartz thermometer, 89, 858
Quartz transducer, 1143
Quasi-static method of v.p. measurement, 637
Quenching methods for melting point determination, 809

Radiation detection, photoelectric, 1172
— photographic, 1173
Radiation detectors, solid state, 1172
Radiation modulation, high-speed, 1188
Radiation shields in ebulliometric measurements, 630
Radioactive source for measurement of density of liquid alkali metals, 961
Radioactive tracers in v.p. measurement, 625
Ramsey and Young's method of v.p. measurement, 634
Raoult correction for supercooling error, 847
Raoult–van't Hoff equation, 806
Rarefaction wave, 1112, 1113, 1117, 1125, 1126, 1128, 1132
Rayleigh line, 716, 717, 1123
Rayleigh ratio, 710–712
Rayleigh scattering, 784, 1208
Rayleigh's law, 225

Reaction field in dielectric, 583
Real mixture, 753
Recirculating still, 764, 766, 767
Recording, high-speed, 1175
Rectilinear diameter, 661, 668, 676, **708**
— law, 983, 985, 988
Redlich–Kwong equation of state, 410
Reference function, 72
Reference level, 191
Reference material, 71
Reflected shock waves, 1130
Refractive index, 684–686, 688, 705
Refractive index results for argon, 594
Refrigerator, ^{3}He–^{4}He dilution, 1240, 1241
Regular solutions, theory of, 726, 758
Relative expansion and compression, 431
Relative volatility, 767
— in polymer solutions, 767
Relaxation, 683, 695, 696
Relaxation time, 536, 1073
— spin-lattice, 1227
Relaxation titration, 1094
Residual currents, 278, 283
Resistance, electrical, measurement of, 982
Resistance gage, manganin, 1261
Resistance, of liquid metals, at high temperatures, 1177
Resistance ratio, 72
— values of, 74, 75
Resistance thermometer, 72, 75–78
— Ge, 1234
Resistivity, electrical, 982
Resonating quartz crystal balance, 1057
Restrained capsule, 214
'Reverse' x-ray effect, 280
Reversible and irreversible processes, 29
Reversibility, criteria for, 924, 927, 939, 947
— polarizability, 924
Rhodium, freezing point, 77
Riemann integral, 1129
Roberteau thermometer, 854
Rotating mirror camera, 1140
Rotation of piston or cylinder, 162, 199
Rotational equilibrium, 1087
Rotational relaxation, 536
Rowlinson's equation of state, 414
Rubidium, compressibility, 976
— density, 977, 986
— saturated liquid and vapor, 986
Rüddorff method of measuring starting temperature of freezing, 847

Saha equation, 1208
Saint Venant's law, 224, 225
Salt hydrate and aqueous eutectic cryoscopy, 861
Salt hydrate solvents, table, 868
Salt media, 841, 861

SUBJECT INDEX

Samples of helium and hydrogen, preparation of, 1251
Saturation curve, 986
Saturation in gas–liquid systems, 743
Scaling, 661, **669,** 672, 675, 699
— dynamic, 717
— extended, 670, **676,** 699
Scattering, angle dependence, 711
— light, 678, **711–717**
— x-ray, **714, 715**
Schlieren techniques, 685, **687, 688**
Seals, 1197
Seasoning of pressure vessels and piezometers, 434
Second law of thermodynamics, 32
Second shock wave, 1127
Secondary reference points, 76
— temperatures of, 77
Seebeck coefficient, 90, 91
Self-inductance, 221
Semiconductor pressure transducer, 1144
Sensor (of pressure), 213ff
Shear viscosity, 534
Shielded collector gage, 294
Shock accumulation, 1135
Shock impedance, 1131
Shock wave, 1113, 1200
— compression, 1112, 1117
— — phase changes in, 1123, 1148, 1149
— — — liquefaction, 1123
— — — melting, 1123
— — plastic, 1125
— elastic, 1125
— equations, 1114, 1115
— generators, 1133
— rarefaction, 1112, 1113, 1117, 1125, 1126, 1128, 1132
— reflected, 1130
— second, 1127
— stability of, 1111, 1113, 1116, 1117
— thickness, 1113, 1125
— very intense, 1119
— weak, 1116
Short light pulses, 1189
Significance tests, 514
Silicon, single crystals of pure, 318
Silicone, liquid, equation of state, 1156
Silver, freezing point, 73, 76, 78, 80, 83
Silver–copper eutectic, freezing point, 83
Simple mixture, strictly regular, 757
Simple piston–cylinder method of measuring compressibility of liquids, 469
Single crystal membrane, 919
Single crystals of pure silicon, used for measuring Avogadro's number, 318
Single ion activities, 903
Siwoloboff's method for v.p. measurement (small samples), 617
Skin effect, 1168

Sn I–II transition, 140, 143, 144
Sodium, compressibility, 976
— critical constants, 39
— density, 977, 980
— equation of state, 976
— freezing point, 83
— literature, 408
— vapor pressure, 980
Sodium chloride, temperature determination in, 1150
Soft x-ray photoemission, 279
Solidification, adiabatic, of ^{3}He, 1240
Solid sphere, volume of, 305
Solubility of gases in liquids, 632
Solubility at high pressure, 737
— at high temperature, 738
— definition of, 725
— of electrolytes, 727, 743
— of gases in liquids, 729
— of liquids in liquids, 741
— of non-electrolytes, 727
— of solids in liquids, 743
Solubility, of metals in molten salts, 921, 923
— of silver halogenides, 938, 947, 948
Solubility parameter, 727
Solute activity coefficient calculation, 890
Solute particles numeration, 889
Solutions, 61
— infinitely dilute, 61
Solvation calculation, 891
Solvent activity coefficient calculation, 895
Solvent particles numeration, 888
Solvents, degassing of, 730
— purification of, 726
Sound absorption, 533
Sound dispersion, 532, 533, 719
Sound propagation in water, 1150
Sound velocity, 532, 718, 964
— in density measurements, 963
— in gases, 555
— — corresponding states treatment, 559
— general behavior, 555
— in homogeneous mixtures of liquids, 570
— in liquid mercury, 965
— in liquid potassium or sodium, 966
— in mixtures of gases or liquids, 544
— in mixtures showing phase separation, 572
— in pure liquids, 561
— thermodynamic relations, 532
Sources of error in physicochemical measurements, 439
Spark gaps, 1165, 1183
Specific energy, conversion factors for, 67
Specific enthalpy, conversion factors for, 67
Specific entropy, conversion factors for, 67
Specific heat, 536, 563, 564, 658, 659, 662, 670, 671, 675, 680, **693–699,** 953
— of helium, anomaly, 1254
— of ^{3}He–^{4}He mixtures, 1257

SUBJECT INDEX

— of liquid ^{3}He, 1251
— of metals, high temperature measurement of, 1176
— — by pulse techniques, 1179, 1187
— of niobium, molybdenum, tantalum, and tungsten 1179
Specific heat capacity, 15
— conversion factors for, 67
— at constant pressure, 519
— at constant saturation, 519
— at constant volume, 519
Specific volumes of water, 58, 59
Spectral line broadening, 1212
Spectrometric method of minimum deviation for refractive index, 589
Spectroscopy, 1209
Spectrum, scattered light, 709, 715–717
Speed, conversion factors for, 68
Speed of sound, 35, 1111, 1122, 1123
— see also: Sound velocity
Spin-lattice relaxation time, 1227
Spoon gage, 1027
SQUID, 1228
Stability of shock waves, 1111, 1113, 1116, 1117
Standard atmosphere, 223
Standard error, estimate of, 64
Standard reference function, 72
Standard reference samples, 81
Standard resistance for pulse currents, 1170
Standard state, definition of, 726, 904
Standards of length, 8
Standards of mass, 8
Static expansion, 264
Static measurements of vapor pressure, 610
— at elevated temperatures and pressures, 618
Static vapor pressure apparatus, 767–769, 772, 775
Statistical aspects of least squares estimation, 511
Still, 1242
Stimulated Brillouin effect, 1150
Stirring liquid cryostat, 349
Stortenbecker theorem, 841
Strain gages, 222, 354, 1263
Stratification temperatures, 1257
Stress relief of pVT vessel, 333
Structure factor, 712
Substance, amount of, 6, 10
Sulfur, boiling point, 83
Sulfur tetrafluoride, solubility of, 737
Superconducting quantum interference device, 1228
Superconducting thin film thermometers, 1240
Superconducting transition temperatures as fixed points, 1239

Supercooling, 942, 943
Supercooling error, 848
Supercritical temperatures, 984, 986
Support, 200
Surface finish, 1033, 1034
Surface tension, 991
Susceptibility for a nuclear spin system, 1227
Switching, high currents, 1165, 1183
Symmetry, gas–liquid, 663, **664**, 665
Synchro-detection, 218
Synthetic method for determining phase equilibria, 791, 797, 798
Systematic error, 63–65

Tank circuit, LC, 1259
Tantalum, electrical resistivity of, 1179
— hemispherical total emittance of, 1179
— normal spectral emittance of, 1179
— specific heat of, 1179
Tare error, 171
Temkin theory, 842
Temperature, conversion factors for, 68
— derivative of, 425
Temperature control, 195, 331
— of furnaces, 927
Temperature-jump methods for measuring chemical relaxation, 1075
Temperature measurements, 195, 726, 1204
— by accurate microsecond resolution techniques, 1172
— high-speed, 1172, 1185
— literature summary, 1185
Temperature modulation, 954
Temperature scale of 1958, 558
Temperature scales below 20 K, 1220
Temperature, thermodynamic, 5, 9, 18
— absolute, 18
— Celsius, 18
Tensometers, 1003, 1005
Ternary aqueous eutectic solvents, 871
Ternary liquid–vapor equilibrium measurements, 775
1,2,4,5-Tetrachlorobenzene, freezing point, 83
Thermal arrest methods for melting point determination, 813
Thermal boundary resistance, 1243
Thermal conductivity, 17, 717
— conversion factors for, 68
Thermal contact, 1250
Thermal diffusivity, 15, 659, 677, 678, 716, 717
Thermal expansion, 153, 173, 174, 1169
Thermal expansion coefficients, 563
Thermal expansivity of mercury, 116
Thermal ionization of Cs vapor, 986
Thermal noise method, 100

SUBJECT INDEX

Thermal switches, 1250
Thermal transpiration, 254, 278, 624, 1028
Thermionic emission, 1169
Thermistor gage, 1042
Thermistor thermometer, 859
Thermocouple, 88, 89, 91, 96, 98, 104, 106, 107
— Au–Co/Cu [or Ag], 88
— chromel–alumel, 94, 96, 102, 104, 105, 107
— copper–constantan, 94, 102, 105
— Ni/Ni–18% Mo, 105
— Pt–10% Rh/Pt, 9, 89, 93–96, 100–107
Thermocouples, 72, 76, 77, 80, 982
— for adiabatic cryoscopy (*Figure 19*), 879
— Au/Pt–40% Rh, 106
— construction of, calculations for, 855
— — insulation of, 854
— — metal wires for, 854
— — use for measurement, 856
— effect of pressure on, 387
— e.m.f., 76
— Fe–V, Fe–Ni, Fe–constantan, 106
— W/Ir, 106
Thermodynamic consistency, 759, 760
Thermodynamic equilibrium, 29, 1204, 1207, 1211
— in arcs, 1174
Thermodynamic laws, 28
Thermodynamic measurements, in liquids, at high pressures, 1176
Thermodynamic properties of fluids at high temperatures, by electrical discharge techniques, measurement of, 1161
Thermodynamic property equations, 518
Thermodynamic relations for mixtures, 751
Thermodynamic system, 28
— closed, 28, 32, 33
Thermodynamic temperature scale, 352, 557
Thermodynamic (primary) thermometry, 1221
Thermodynamics of ^{4}He, 1253
Thermodynamics of volume changes, 425
Thermoelectric effect (thermal e.m.f.), 922, 923, 936, 1169
Thermoelectric power of Cs. K. Hg. 990
Thermoelectric voltage of K, 982
Thermometer, carbon, 88
— gas, 9
— germanium, 88
— platinum resistance, 9, 88, 89
— quartz, 89
— thermistor, 859
Thermometric probe calibration, by comparison KCl, 861
— by cryoscopy, 861
Thermometry, below 20 K, 1220
— capacitance, 1235
— ^{3}He melting curve, 1236

Thermometry—*contd.*
— magnetic, 1222, 1233
— Mössbauer effect, 1223
— noise, 1229
— nuclear orientation, 1225
— nuclear spin, 1226
— primary, 1221
— secondary, 1221
— superconducting thin film, 1240
Thermomolecular force, 1057
Thermophysical properties of solid electrical conductors above 1500 K, 1178
Thermostat, ice bath, 332
— stirred liquid, 332
— vacuum isolated, 332
Thickness of shock wave, 1113, 1125
Third law of thermodynamics, 33
Thomas–Fermi–Dirac model, 1152
Time, conversion factors for, 68
Tin, freezing point, 73, 75, 77, 80, 83
— influence of pressure, 78
Tl II–III transition, 140, 142
Torsion–effusion method for v.p. measurement, 643
Total vapor pressure measurements, 783
Transducer, 204, 214
Transformer bridge method for dielectric constant, 597
Transient current measurement, 1170, 1184
Transient temperature measurement, 1172, 1185
Transient voltage measurement, 1170, 1184
Translational relaxation, 534
— time, 534
Transmitter (of pressure), 214
Transpiration method for v.p. measurement, 646
Transport method for v.p. measurement, 646
Transport processes, 909, 911, 915, 917, 946
— external transference number, 912, 916, 941
— frame of reference, 912, 920
— internal transference number, 912, 916, 941
— ionic mobility, 907, 916
— reference constituent, 912, 917
Trigger discharge gage, 292
Triphenylene, melting point, 84
Tritium, solubility of, 742
Tungsten, electrical resistivity of, 1179
— carbide, volume of cube of, 311
— freezing point, 83
— heat of fusion, 1176
— hemispherical total emittance of, 1179
— index points, 328, 337, 340
— melting point, 77, 1179
— normal spectral emittance of, 1179
— specific heat of, 1179
Tungsten–rhenium container, 982

SUBJECT INDEX

Turbidity, 715
— by visual observation, 792

Ultracentrifuge method of measuring compression of a liquid, 488
Ultrahigh vacuum system for adsorption measurements, 1037
Ultra low pressure measurement (below 10^{-6} Torr), 273–301
— in non-uniform environments, 277–278
Ultrasonic absorption, measurement of, 1085
— in water–dioxane mixtures, 1103
Ultrasonic interferometer, 125
Ultrasound, 718
Uncertainty, of final results, 64
Units, SI system, 4, 5, 11
— derived, 6
— outside SI system, 11
— supplementary, 7
Universal cryoscopic apparatus (*Figure 9*), 853
Universality, 660, 662, 719
Upper critical solution temperature (UCST), 789
Upper critical temperature solutions, 797
Upper limit of temperature for application of ebulliometric method, 631, 632
Uranium plasma, 1214

Valves, 191, 197, 198
Van der Waals equation of state, 543
Van Laar–Schröder equation, 838
Van't Hoff coefficient, formula 19, p841
Vapor density, 976
Vapor pressure, 691, **693**, 753, 754, 772, 966, 976, 1214
— of alkali metals, 615, 968, 980, 981
— of carbon dioxide, 636
— curves, 986, 987
— of mercury, 117, 972
— of potassium, sodium, 968
— subcritical liquid and vapor phase at equilibrium, 985
Vapor pressure measurement, approach to equilibrium in, 613
— in calorimetric experiments, 624, 637
— comparative methods for, 617, 624, 627
— ebulliometric method for, 626, 971
— force of molecular beam for, 645
— gas-saturation method for, 645
— Kay's apparatus for, 618
— Knudsen's method for, 642, 966, 967
— Langmuir's method for, 641
— Langmuir–Taylor method, 967
— molecular weight in, 608
— quasi-static method for, 637
— radioactive tracers in, 625

— Ramsay and Young's method for, 634
— Siwoloboff's method for, 617
— static methods, 967, 968
— torsion–effusion method for, 643
— transpiration method for, 646
— transport method for, 646
Vapor pressures, 607
— of liquefied gases, 623
— of mixtures, 608, 651
— of solids, 609, 613
— of solutions, 652
Vapor recirculation, 766
Vapor stream pumping, 254
Vaporization criterion, 1127
Variable capacitor, 219
Variable path interferometer, 552
Variation of Lorentz–Lorenz functions with temperature, 596
Verdet constant, 601
Vibrational relaxation, 536
Virial coefficients, 30, 32, 33, 35, 39, 322, 502, 503, 558
— first, dielectric, 599
— second, 323, 333, 336, 340, 344
— — dielectric, 585, 599
— in density series, 366
Virial equation of state, 510, 541
Virial expansion, 412, 503
Viscosity, dynamic, conversion factors for, 68
— gage, 258
— kinematic, conversion factors for, 68
Visual methods for melting point determination, 807
Voltage, thermoelectric, 982
Voltage divider, 1170
Volume, absolute measurement of, 303, 1203
— conversion factors for, 68
— of cube of fused silica, 315
— of cube of tungsten carbide, 311
— derivative of, 425
Volume calibration, 327
Volume change on mixing two gases, 344
Volume fraction, definition of, 727
— measurements, 741, 746
— percentage, definition of, 727
— reading, 727
Volume measurements, 1203
Volume viscosity, 536
Volumes of vessels, 304
— solid spheres, 305
— cylinders, 305
— parallelepipeds, 305
Volumetric methods of adsorption, 1036

Water, boiling point, 73, 75, 77, 79
— critical constants, 39
— density of, 55, 304, 318
— freezing point, 77

SUBJECT INDEX

Water—*contd.*
— isotherm equation, 1156
— literature, 408
— Rice and Walsh's equation of state, 1156
— second virial coefficient of vapor, 52
— to ice VII, 1151
— triple point, 73, 75, 857
— vapor pressure, 79
Water–triethylamine liquid mixture, 573
Weak shock waves, 1116, 1148
Weight fraction, definition of, 727
— percentage, definition of, 727
Weight method, 356, 363
Weight methods for determining density of a liquid, 484
— direct weighing methods, 484
— hydrostatic weighing, 485
Weights, 169, 196
— hydraulic loading, 162
— loading, 197
Windows, high pressure, 1196
Work, 19
— conversion factors for, 69

Xenon, 540
— critical constants, 39
— isochoric method for refractive index of, 595
— literature, 373, 408
— magneto-optical effect, 602
— second virial coefficient, 50–53
X-Ray effect (forward, reverse), 280
X-Ray interference measurements, 305
X-Ray limit, 283
X-Ray scattering, **714, 715**

Yang–Yang relation. 666. 698. 699
Young's modulus, 326, 327, 356

Zero concentration extrapolation, 839, 887
Zeroth law of thermodynamics, 29
Zinc, freezing point, 73, 75, 78, 80, 83
— influence of pressure, 78
Zwanzig's equation of state, 413